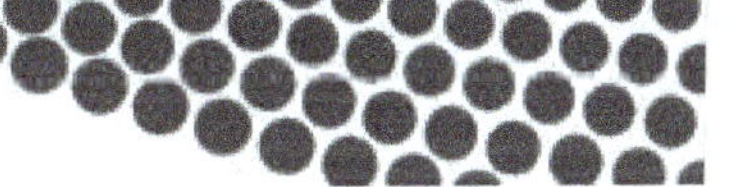

Blackboard | WileyPLUS

# Welcome to your *WileyPLUS* and Blackboard Course

## tudent Registration

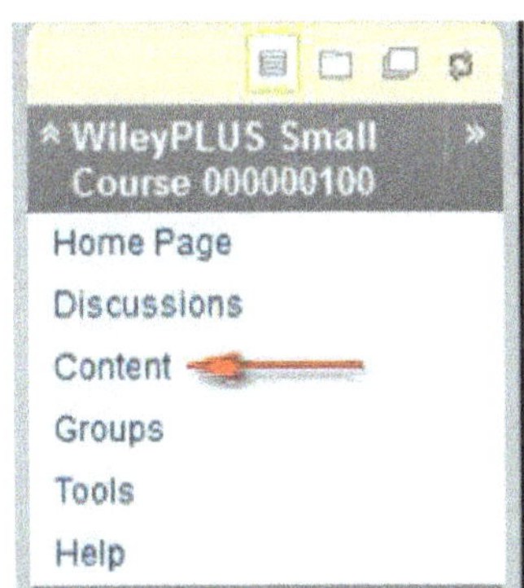


1. First, log in to Blackboard, locate your course and click the Content link in the Course menu to access course content.
2. Click on any link to *WileyPLUS* material (indicated by the *WileyPLUS* icon).
3. If your instructor has not created any *WileyPLUS* assignments yet, you can find *WileyPLUS* under the TOOLS section in your navigation panel.
4. Follow the directions and enter your registration code when prompted. If you prefer to buy direct access to *WileyPLUS*, you can do so online by simply following the directions.

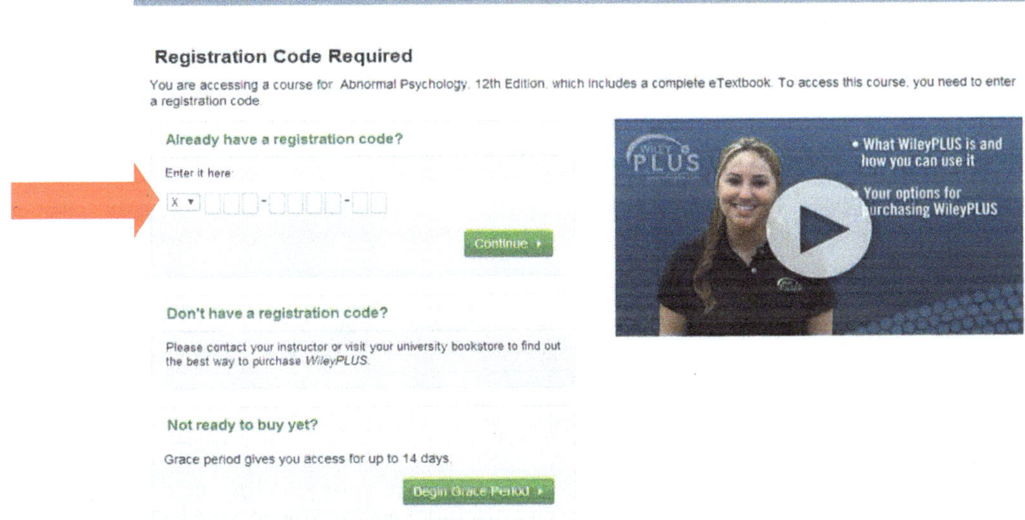


5. You will now be directed to *WileyPLUS*. To navigate back, click on the 'Return to Blackboard' tab and from there you should be all set!

Return to Blackboard

## Need Help?

**LIVE CHAT!** Technical Support: **www.wileyplus.com/support**

Need help registering? www.wileyplus.com/go/bb/register

# Organic Chemistry

## Second Edition

David Klein

Iowa State University

Wiley Custom Learning Solutions

Cover Image © isak55 / Shutterstock

Printing identification and country of origin will either be included on this page and/or the end of the book. In addition, if the ISBN on this page and the back cover do not match, the ISBN on the back cover should be considered the correct ISBN.



To order books or for customer service, please call 1(800)-CALL-WILEY (225-5945).

Printed in the United States of America.

ISBN 978-1-119-30331-2

Printed and bound by Strategic Content Imaging

10 9 8 7 6 5 4 3

# Contents

# Approximate p$K_a$ Values for Commonly Encountered Structural Types

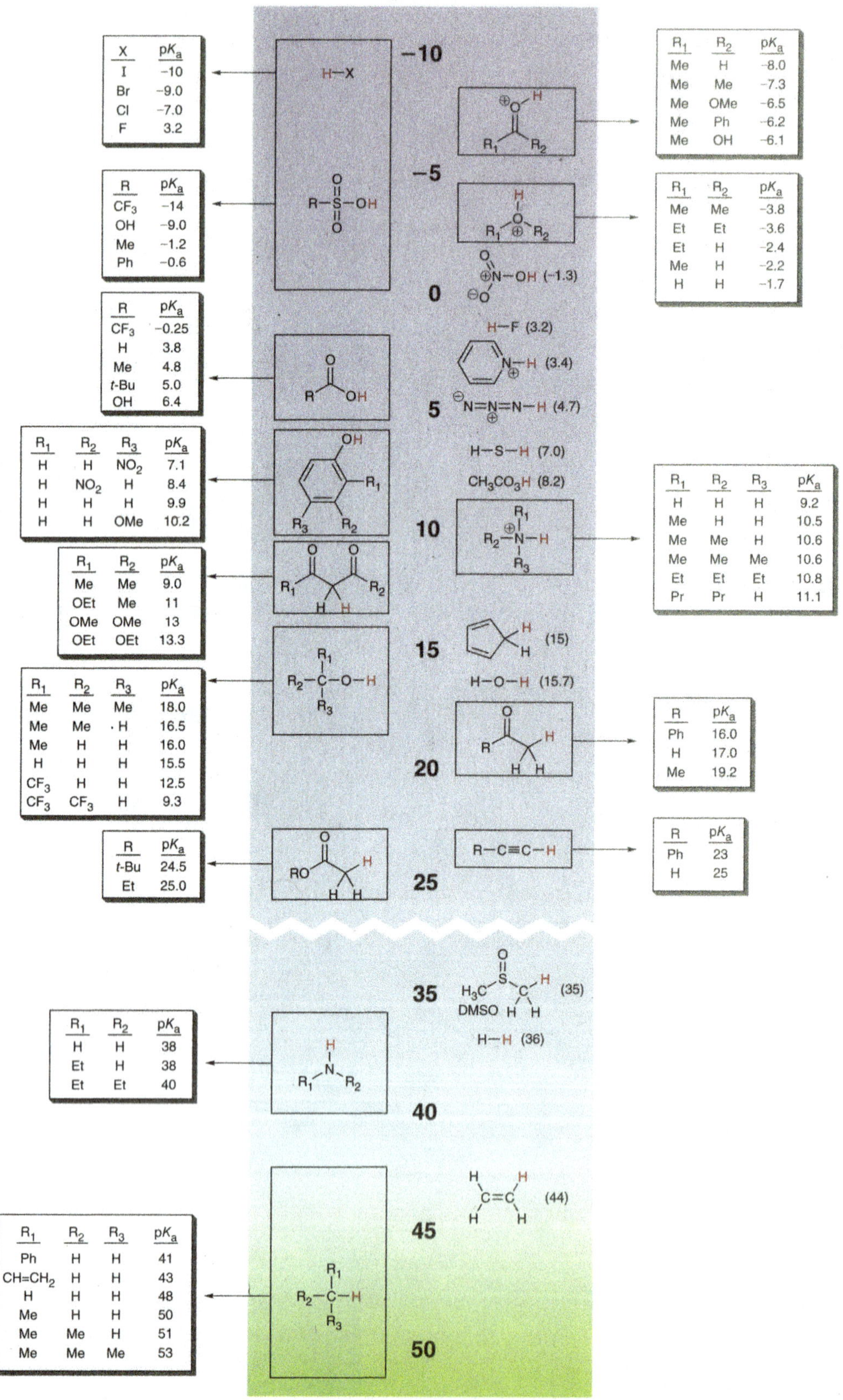

## EXAMPLES OF COMMON FUNCTIONAL GROUPS

| FUNCTIONAL GROUP* | CLASSIFICATION | EXAMPLE | CHAPTER |
|---|---|---|---|
| R—X: (X = Cl, Br or I) | Alkyl halide | *n*-Propyl chloride | 7 |
| $R_2C=CR_2$ | Alkene | 1-Butene | 8, 9 |
| R—C≡C—R | Alkyne | 1-Butyne | 10 |
| R—OH | Alcohol | 1-Butanol | 13 |
| R—O—R | Ether | Diethyl ether | 14 |
| R—SH | Thiol | 1-Butanethiol | 14 |
| R—S—R | Sulfide | Diethyl sulfide | 14 |
| (benzene ring) | Aromatic (or arene) | Methylbenzene | 18, 19 |
| RC(=O)R | Ketone | 2-Butanone | 20 |
| RC(=O)H | Aldehyde | Butanal | 20 |
| RC(=O)OH | Carboxylic acid | Pentanoic acid | 21 |
| RC(=O)X | Acyl halide | Acetyl chloride | 21 |
| RC(=O)OC(=O)R | Anhydride | Acetic anhydride | 21 |
| RC(=O)OR | Ester | Ethyl acetate | 21 |
| RC(=O)N(R)H | Amide | Butanamide ($NH_2$) | 21 |
| $R_3N$ | Amine | Diethylamine | 23 |

* The "R" refers to the remainder of the compound, usually carbon and hydrogen atoms.

# Organic Chemistry

SECOND EDITION

DAVID KLEIN
Johns Hopkins University

WILEY

VICE PRESIDENT, PUBLISHER Petra Recter
SPONSORING EDITOR Joan Kalkut
SENIOR MARKETING MANAGER Kristine Ruff
SENIOR PROJECT EDITOR Jennifer Yee
SENIOR PRODUCT DESIGNER Geraldine Osnato
SENIOR ASSOCIATE EDITOR Veronica Armour
DESIGN DIRECTOR Harold Nolan
SENIOR DESIGNER Maureen Eide/DeMarinis Design LLC
COVER DESIGN Thomas Nery
SENIOR PHOTO EDITOR Lisa Gee
EDITORIAL ASSISTANTS Mallory Fryc, Susan Tsui
SENIOR CONTENT MANAGER Kevin Holm
SENIOR MEDIA SPECIALIST Daniela DiMaggio
SENIOR PRODUCTION EDITOR Elizabeth Swain

Cover/preface photo credits: flask 1-(yellow and red pill capsule) rya sick/iStockphoto, (tiny pills) coulee/iStockphoto; flask 2-(syringe) Stockcam/Getty Images, (liquid) mkurthas/iStockphoto; flask 3-(key) Gary S. Chapman/Photographer's Choice/Getty Images, (key pile) Olexiy Bayev/Shutterstock; flask 4-(poppy) Margaret Rowe/Garden Picture Library/Getty Images, (poppies) Kuttelvaserova Stuchelova/ Shutterstock; flask 5-(tomato evolution)Alena Brozova/Shutterstock, (cherry tomatoes) Natalie Erhova (summerky)/Shutterstock, (cherry tomatoes pile) © Jessica Peterson/Tetra Images/Corbis; flask 6-(ribbon) macia/Getty Images (ribbon symbol) JamesBrey/iStockphoto; flask 7-(scoop) Fuse/Getty Images (coffee beans) Vasin Lee/Shutterstock, (espresso) Rob Stark/Shutterstock; flask 8-(smoke) stavklem/ Shutterstock, (chili peppers) Ursula Alter/Getty Images.

This book was set in 10/12 Garamond by Prepare and manufactured in the United States by RR Donnelley.

Founded in 1807, John Wiley & Sons, Inc. has been a valued source of knowledge and understanding for more than 200 years, helping people around the world meet their needs and fulfill their aspirations. Our company is built on a foundation of principles that include responsibility to the communities we serve and where we live and work. In 2008, we launched a Corporate Citizenship Initiative, a global effort to address the environmental, social, economic, and ethical challenges we face in our business. Among the issues we are addressing are carbon impact, paper specifications and procurement, ethical conduct within our business and among our vendors, and community and charitable support. For more information , please visit our website: www.wiley.com/go/citizenship.

The paper in this book was manufactured by a mill whose forest management programs include sustained yield-harvesting of its timberlands. Sustained yield harvesting principles ensure that the number of trees cut each year does not exceed the amount of new growth.



ISBN
Main edition: 978-1-118-45228-8
Binder version: 978-1-118-45431-2

Printed in the United States of America
10 9 8 7 6 5 4

# Dedication

*To Larry,*

*By inspiring me to pursue a career in organic chemistry instruction, you served as the spark for the creation of this book. You showed me that any subject can be fascinating (even organic chemistry!) when presented by a masterful teacher. Your mentorship and friendship have profoundly shaped the course of my life, and I hope that this book will always serve as a source of pride and as a reminder of the impact you've had on your students.*

*To my wife, Vered,*

*This book would not have been possible without your partnership. As I worked for years in my office, you shouldered all of our life responsibilities, including taking care of all of the needs of our five amazing children. This book is our collective accomplishment and will forever serve as a testament of your constant support that I have come to depend on for everything in life. You are my rock, my partner, and my best friend. I love you.*

# Brief Contents

Scan this code to view Chapter 27, *Synthetic Polymers* *Also accessible at www.wiley.com/college/klein*

# Contents

## 1 A Review of General Chemistry: Electrons, Bonds, and Molecular Properties 1



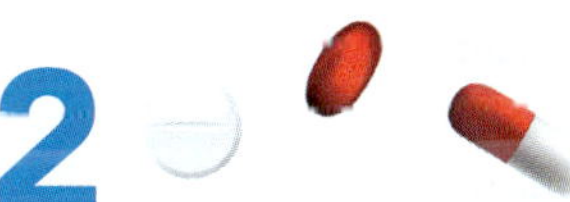

## 2 Molecular Representations 51

## 3 Acids and Bases 96

## 4 Alkanes and Cycloalkanes 139

## 5 Stereoisomerism 192

## 6 Chemical Reactivity and Mechanisms 236



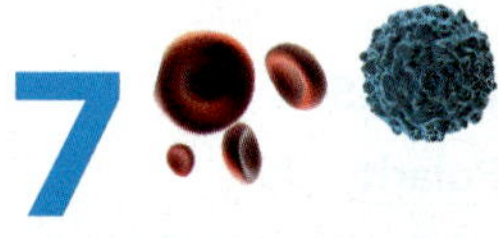

## 7 Substitution Reactions 287

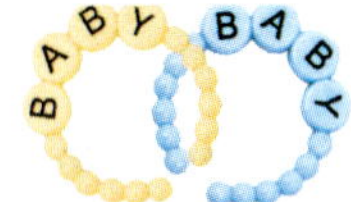
BABY
BABY

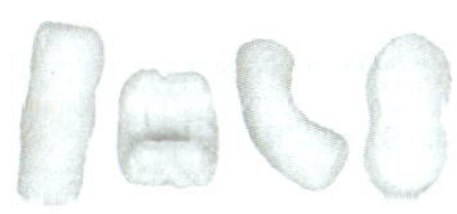

# 15

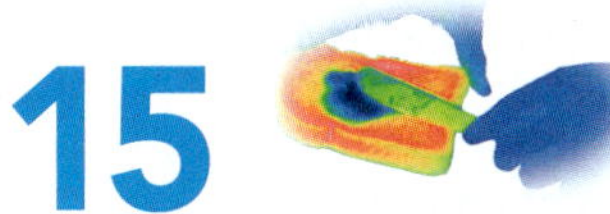

## Infrared Spectroscopy and Mass Spectrometry 683



# 16

## Nuclear Magnetic Resonance Spectroscopy 731

# 17

## Conjugated Pi Systems and Pericyclic Reactions 782

E
C B
D L F
P T E O

## 21 Carboxylic Acids and Their Derivatives 984

## 22 Alpha Carbon Chemistry: Enols and Enolates 1044

## 23 Amines 1102

## 24 Carbohydrates 1151

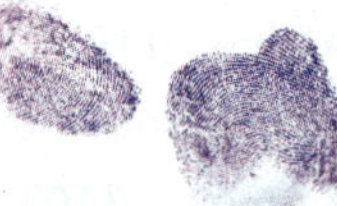



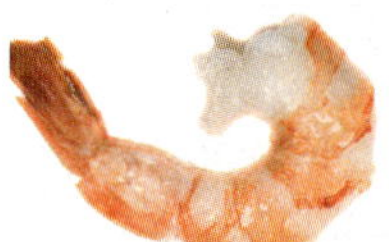

Scan this code to view Chapter 27, *Synthetic Polymers*
*Also accessible at www.wiley.com/college/klein*

# Preface

## WHY I WROTE THIS BOOK

Students who perform poorly on organic chemistry exams often report having invested countless hours studying. Why do many students have difficulty preparing themselves for organic chemistry exams? Certainly, there are several contributing factors, including inefficient study habits, but perhaps the most dominant factor is a fundamental *disconnect* between what students learn in the lecture hall and the tasks expected of them during an exam. To illustrate the disconnect, consider the following analogy.

Imagine that a prestigious university offers a course entitled "Bike-Riding 101." Throughout the course, physics and engineering professors explain many concepts and principles (for example, how bicycles have been engineered to minimize air resistance). Students invest significant time studying the information that was presented, and on the last day of the course, the final exam consists of riding a bike for a distance of 100 feet. A few students may have innate talents and can accomplish the task without falling. But most students will fall several times, slowly making it to the finish line, bruised and hurt; and many students will not be able to ride for even one second without falling. Why? Because there is a *disconnect* between what the students learned and what they were expected to do for their exam.

Many years ago, I noticed that a similar disconnect exists in traditional organic chemistry instruction. That is, learning organic chemistry is much like bicycle riding; just as the students in the bike-riding analogy were expected to ride a bike after attending lectures, it is often expected that organic chemistry students will independently develop the necessary skills for solving problems. While a few students have innate talents and are able to develop the necessary skills independently, most students require guidance. This guidance was not consistently integrated within existing textbooks, prompting me to write the first edition of my textbook, *Organic Chemistry, 1e*. The main goal of my text was to employ a skills-based approach to bridge the gap between theory (concepts) and practice (problem-solving skills). The phenomenal success of the first edition has been extremely gratifying because it provides strong evidence that my skills-based approach is indeed effective at bridging the gap described above.

I firmly believe that the scientific discipline of organic chemistry is NOT merely a compilation of principles, but rather, it is a disciplined method of thought and analysis. Students must certainly understand the concepts and principles, but more importantly, students must learn to think like organic chemists . . . that is, they must learn to become proficient at approaching new situations methodically, based on a repertoire of skills. That is the true essence of organic chemistry.

## A SKILLS-BASED APPROACH

To address the disconnect in organic chemistry instruction, I have developed a *skills-based approach* to instruction. The textbook includes all of the concepts typically covered in an organic chemistry textbook, complete with *conceptual checkpoints* that promote mastery of the concepts, but special emphasis is placed on skills development through SkillBuilders to support these concepts. Each SkillBuilder contains 3 parts:

**Learn the Skill:** contains a solved problems that demonstrates a particular skill.

**Practice the Skill:** includes numerous problems (similar to the solved problem in *Learn the Skill*) that give students valuable opportunities to practice and master the skill.

**Apply the Skill:** contains one or two more challenging problems in which the student must apply the skill in a slightly different environment. These problems include conceptual, cumulative, and applied problems that encourage students to think outside of the box. Sometimes problems that foreshadow concepts introduced in later chapters are also included.

At the end of each SkillBuilder, a *Need More Practice?* reference suggests end-of-chapter problems that students can work to practice the skill.

This emphasis upon skills development will provide students with a greater opportunity to develop proficiency in the key skills necessary to succeed in organic chemistry. Certainly, not all necessary skills can be covered in a textbook. However, there are certain skills that are fundamental to all other skills.

As an example, resonance structures are used repeatedly throughout the course, and students must become masters of resonance structures early in the course. Therefore a significant portion of Chapter 2 is devoted to pattern-recognition for drawing resonance structures. Rather than just providing a list of rules and then a few follow-up problems, the skills-based approach provides students with a series of skills, each of which must be mastered in sequence. Each skill is reinforced with numerous practice problems. The sequence of skills is designed to foster and develop proficiency in drawing resonance structures.

As another example of the skills-based approach, Chapter 7, Substitution Reactions, places special emphasis on the skills necessary for drawing all of the mechanistic steps for $S_N2$ and $S_N1$ processes. Students are often confused when they see an $S_N1$ process whose mechanism is comprised of four or five mechanistic steps (proton transfers, carbocation rearrangements, etc.). This chapter contains a novel approach that trains students to identify the number of mechanistic steps required in a substitution process. Students are provided with numerous examples and are given ample opportunity to practice drawing mechanisms.

The skills-based approach to organic chemistry instruction is a unique approach. Certainly, other textbooks contain tips for problem solving, but no other textbook consistently presents skills development as the primary vehicle for instruction.

## WHAT'S NEW IN THIS EDITION

Peer review played a very strong role in the development of the first edition of *Organic Chemistry*. Specifically, the first edition manuscript was reviewed by nearly 500 professors and over 5,000 students. In preparing the second edition, peer review has played an equally prominent role. We have received a tremendous amount of input from the market, including surveys, class tests, diary reviews, and phone interviews. All of this input has been carefully culled and has been instrumental in identifying the focus of the second edition.

### Literature-based Challenge Problems

The first edition of my textbook, *Organic Chemistry 1e*, was written to address a gap between theory (concepts) and practice (problem-solving skills). In *Organic Chemistry 2e*, I have endeavored to bridge yet another gap between theory and practice. Specifically, students who have studied organic chemistry for an entire year are often left profoundly disconnected from the dynamic and exciting world of research in the field of organic chemistry. That is, students are not exposed to actual research performed by practicing organic chemists around the world. To bridge this gap and to address market feedback suggesting that the text would benefit from a larger number of challenge problems, I've created literature-based Challenge Problems for this edition. These problems will expose students to the fact that organic chemistry is an evolving, active branch of science, central to addressing global challenges.

The literature-based Challenge Problems are more challenging than the problems presented in the text's SkillBuilders because they require the students to think "outside the box" and to predict or explain an unexpected observation. Over 225 new literature-based Challenge Problems have been added in *Organic Chemistry, 2e*. All of these problems are based on the chemical literature and include references. The problems are all designed to be thought-provoking puzzles that are challenging, but possible to solve with the principles and skills developed in the textbook. The inclusion of literature-based problems will expose students to exciting real-world examples of chemical research being conducted in real laboratories. Students will see that organic chemistry is a vibrant field of study, with endless possibilities for exploration and research that can benefit the world in very concrete ways. Most chapters of *Organic Chemistry, 2e* will have 8-10 literature-based Challenge Problems. These problems are all coded for assigning and grading in

*WileyPLUS*. In addition, within the *WileyPLUS* course for *Organic Chemistry, 2e*, I've created problem solving videos that provide key strategies for solving a subset of these problems.

### Rewriting for Clarity

In response to market feedback a few sections in the textbook have been rewritten for clarity:

*Chapter 7: Substitution Reactions/Section 7.5: The $S_N1$ Mechanism*

- The discussion of the rate-determining step has been revised to focus on the highest energy transition state. A more detailed discussion of the thermodynamic principles involved is now included.

*Chapter 20: Aldehydes and Ketones/Section 20.7: Mechanism Strategies*

- The section on hydrolysis, as well as the corresponding SkillBuilder, have been rewritten for clarity.

*Chapter 20: Aldehydes and Ketones/Section 20.10: Carbon Nucleophiles*

- The discussion of the Wittig reaction mechanism has been revised to better reflect the observations and insights discussed in the literature.

### Applications and Chapter Openers

Much like the literature-based Challenge Problems underscore the relevance of organic chemistry to current research in the field, the Medically Speaking and Practically Speaking applications demonstrate how the first principles of organic chemistry are relevant to practicing physicians and have everyday commercial applications. We have received very positive feedback from the market regarding these applications. In recognition of the fact that some applications generate more interest than others, we've replaced approximately 10% of the applications, to make them even more relevant and exciting. Since these applications are often foreshadowed in the Chapter Openers, many Chapter Openers have been revised as well.

### Reference Materials

An appendix containing rules for naming polyfunctional compounds as well as a reference table of $pK_a$ values are now included.

In addition, all known errors, inaccuracies, or ambiguities have been corrected in the second edition.

## TEXT ORGANIZATION

The sequence of chapters and topics in *Organic Chemistry, 2e* do not differ markedly from that of other organic chemistry textbooks. Indeed, the topics are presented in the traditional order, based on functional groups (alkenes, alkynes, alcohols, ethers, aldehydes and ketones, carboxylic acid derivatives, etc.). Despite this traditional order, a strong emphasis is placed on mechanisms, with a focus on pattern recognition to illustrate the similarities between reactions that would otherwise appear unrelated (for example, acetal formation and enamine formation, which are mechanistically quite similar). No shortcuts were taken in any of the mechanisms, and all steps are clearly illustrated, including all proton transfer steps.

Two chapters (6 and 12) are devoted almost entirely to skill development and are generally not found in other textbooks. Chapter 6, *Chemical Reactivity and Mechanisms*, emphasizes skills that are necessary for drawing mechanisms, while Chapter 12, *Synthesis*, prepares the students for proposing syntheses. These two chapters are strategically positioned within the traditional order described above and can be assigned to the students for independent study. That is, these two chapters do not need to be covered during precious lecture hours, but can be, if so desired.

The traditional order allows instructors to adopt the skills-based approach without having to change their lecture notes or methods. For this reason, the spectroscopy chapters (Chapters 15 and 16) were written to be stand-alone and portable, so that instructors can cover these chapters in any order desired. In fact, five of the chapters (Chapters 2, 3, 7, 13, and 14) that precede the spectroscopy chapters include end-of-chapter spectroscopy problems, for those students who covered spectroscopy earlier. Spectroscopy coverage also appears in subsequent functional group chapters, specifically Chapter 18 (Aromatic Compounds), Chapter 20 (Aldehydes and Ketones), Chapter 21 (Carboxylic Acids and Their Derivatives), Chapter 23 (Amines), Chapter 24 (Carbohydrates), and Chapter 25 (Amino Acids, Peptides, and Proteins).

## THE *WileyPLUS* ADVANTAGE

*WileyPLUS* is a research-based online environment for effective teaching and learning. *WileyPLUS* is packed with interactive study tools and resources, including the complete online textbook.

### New to *WileyPLUS* for *Organic Chemistry, 2e*

*WileyPLUS* for *Organic Chemistry, 2e* highlights David Klein's innovative pedagogy and teaching style:

- *NEW* Solved Problem Videos by David Klein, extends the SkillBuilding pedagogy by walking students through problem solving strategies for the new end of chapter literature-based Challenge Problems.
- *NEW* Guided Online (GO) Tutorials walk students step-by-step through solving the problem using David Klein's problem-solving pedagogy.
- *NEW* Do you Remember? Practice Quizzes help students prepare for chapter course materials by evaluating students' foundational knowledge.

*WileyPLUS* for *Organic Chemistry, 2e* is now supported by an adaptive learning module called **ORION**. Based on cognitive science, ORION provides students with a personal, adaptive learning experience so they can build proficiency in concepts and use their study time effectively. *WileyPLUS* with ORION helps students learn by learning about them.

***WileyPLUS* with ORION** is great as:

- An adaptive **pre-lecture tool** that assesses your students' conceptual knowledge so they come to class better prepared.
- A **personalized study guide** that helps students understand both strengths and areas where they need to invest more time, especially in preparation for quizzes and exams.

Unique to ORION, students **begin** by taking a quick **diagnostic** for any chapter. This will determine each student's baseline proficiency on each topic in the chapter. Students see their individual diagnostic report to help them decide what to do next with the help of ORION's recommendations.

For each topic, students can either Study, or Practice. **Study** directs the students to the specific topic they choose in *WileyPLUS*, where they can read from the e-textbook, or use the variety of relevant resources available there. Students can also **practice**, using questions and feedback powered by ORION's adaptive learning engine. Based on the results of their diagnostic and ongoing practice, ORION will present students with questions appropriate for their current level of understanding, and will continuously adapt to each student, helping them build their proficiency.

ORION includes a number of reports and ongoing recommendations for students to help them maintain their proficiency over time for each topic. Students can easily access ORION from multiple places within *WileyPLUS*. It does not require any additional registration, and there will not be any additional charge for students using this adaptive learning system.

## Hallmark Features of *WileyPLUS* for *Organic Chemistry, 2e*

Breadth and Depth of Assessment: Four unique silos of assessment are available to Instructors for creating online homework and quizzes.

WILEYPLUS ASSESSMENT — FOR ORGANIC CHEMISTRY

| WileyPLUS Assessment | For Organic Chemistry |
|---|---|
| REACTION EXPLORER | MEANINGFUL PRACTICE OF MECHANISMS AND SYNTHESIS PROBLEMS (A DATABASE OF OVER 100,000 QUESTIONS) |
| IN CHAPTER/EOC ASSESSMENT | 100% OF REVIEW PROBLEMS AND END OF CHAPTER QUESTIONS ARE CODED FOR ON LINE ASSESSMENT |
| CONCEPT MASTERY | PRE-BUILD CONCEPT MASTERY ASSIGNMENTS ( FROM DATABASE OF OVER 12,500 QUESTIONS) |
| TEST BANK | RICH TESTBANK CONSISTING OF OVER 3,000 QUESTIONS |

# ADDITIONAL INSTRUCTOR RESOURCES

**Testbank** Authored by Kevin Minbiole, *Villanova University* and Vidyullata Waghulde, *St. Louis Community College, Meramec.*

**PowerPoint Lecture Slides with Answer Slides** Authored by James Beil, *Lorain County Community College.*

**PowerPoint Art Slides** Images selected by Christine Hermann, *Radford University.*

**Personal Response System ("Clicker") Questions** Authored by Cynthia Lamberty, *Cloud County Community College, Geary County Campus*, Neal Tonks, *College of Charleston*, Christine Whitlock, *Georgia Southern University.*

# STUDENT RESOURCES

**Student Study Guide and Solutions Manual** (ISBN 9781118700815) Authored by David Klein. The second edition of the *Student Study Guide and Solutions Manual* to accompany *Organic Chemistry, 2e* contains:

- More detailed explanations within the solutions for every problem.
- Concept Review Exercises
- SkillBuilder Review Exercises
- Reaction Review Exercises
- A list of new reagents for each chapter, with a description of their function.
- A list of "Common Mistakes to Avoid" in every chapter.

Scan this code to view a sample Chapter of the Student Study/ Solutions Manual for *Organic Chemistry, 2e*

**Molecular Visions™ Model Kit** To support the learning of organic chemistry concepts and allow students the tactile experience of manipulating physical models, we offer a molecular modeling kit from the Darling Company. The model kit can be bundled with the textbook or purchased stand alone.

# CUSTOMIZATION AND FLEXBILE OPTIONS TO MEET YOUR NEEDS

## All Access Packs

**The All Access Pack** for *Organic Chemistry, 2e* by David Klein gives today's students everything they need for their course anytime, anywhere, and on any device. The All Access Pack includes:

- *WileyPLUS*
- Wiley eText powered by VitalSource
- Student Solutions Manual in a binder-ready looseleaf format

### Wiley Custom Select

**Wiley Custom Select** allows you to create a textbook with precisely the content you want, in a simple, three-step online process that brings your students a cost-efficient alternative to a traditional textbook. Select from an extensive collection of content at http://customselect.wiley.com, upload your own materials as well, and select from multiple delivery formats—full color or black and white print with a variety of binding options, or eBook. Preview the full text online, get an instant price quote, and submit your order; we'll take it from there.

### WileyFlex

**WileyFlex** offers content in flexible and cost-saving options to students. Our goal is to deliver our learning materials to our customers in the formats that work best for them, whether it's traditional text, eTextbook, *WileyPLUS*, loose-leaf binder editions, or customized content through Wiley Custom Select.

## CONTRIBUTORS TO *ORGANIC CHEMISTRY, 2E*

I owe special thanks to my contributors for their collaboration, hard work and creativity. Many of the new, literature-based, challenge problems were written by Kevin Caran, *James Madison University*; Danielle Jacobs, *Rider University*; William Maio, *New Mexico State University, Las Cruces*; Kensaku Nakayama, *California State University, Long Beach*; and Justin Wyatt, *College of Charleston*. Many of the new Medically Speaking and Practically Speaking applications throughout the text were written by Susan Lever, *University of Missouri, Columbia*; Glenroy Martin, *University of Tampa*; John Sorensen, *University of Manitoba*; and Ron Swisher, *Oregon Institute of Technology*.

## ACKNOWLEDGEMENTS

The feedback received from both faculty and students supported the creation, development, and execution of both the first and second editions of *Organic Chemistry*. I wish to extend sincere thanks to my colleagues (and their students) who have graciously devoted their time to offer valuable comments that helped shape this textbook.

## SECOND EDITION

### Reviewers

ALABAMA
Marco Bonizzoni, The University of Alabama
Richard Rogers, University of South Alabama
Kevin Shaughnessy, The University of Alabama
Timothy Snowden, The University of Alabama

ARIZONA
Satinder Bains, Paradise Valley Community College
Cindy Browder, Northern Arizona University
John Pollard, University of Arizona

CALIFORNIA
Dianne A. Bennett, Sacramento City College
Megan Bolitho, University of San Francisco
Elaine Carter, Los Angeles City College
Carl Hoeger, University of California, San Diego
Ling Huang, Sacramento City College
Marlon Jones, Long Beach City College
Jens Kuhn, Santa Barbara City College
Barbara Mayer, California State University, Fresno
Hasan Palandoken, California Polytechnic State University
Teresa Speakman, Golden West College
Linda Waldman, Cerritos College

COLORADO
Kenneth Miller, Fort Lewis College

FLORIDA
Eric Ballard, University of Tampa
Mapi Cuevas, Santa Fe College
Donovan Dixon, University of Central Florida
Andrew Frazer, University of Central Florida
Randy Goff, University of West Florida
Harpreet Malhotra, Florida State College, Kent Campus
Glenroy Martin, University of Tampa
Tchao Podona, Miami Dade College
Bobby Roberson, Pensacola State College

GEORGIA
Vivian Mativo, Georgia Perimeter College
Michele Smith, Georgia Southwestern State University

INDIANA
Hal Pinnick, Purdue University Calumet

KANSAS
Cynthia Lamberty, Cloud County Community College

KENTUCKY
Lili Ma, Northern Kentucky University
Tanea Reed, Eastern Kentucky University
Chad Snyder, Western Kentucky University

LOUISIANA
Kathleen Morgan, Xavier University of Louisiana
Sarah Weaver, Xavier University of Louisiana

MAINE
Amy Keirstead, University of New England

MARYLAND
Jesse More, Loyola University Maryland
Benjamin Norris, Frostburg State University

MASSACHUSETTS
Rich Gurney, Simmons College

MICHIGAN
Dalia Kovacs, Grand Valley State University

MISSOURI
Eike Bauer, University of Missouri, St. Louis
Alexei Demchenko, University of Missouri, St. Louis
Donna Friedman, St. Louis Community College at Florissant Valley
Jack Lee Hayes, State Fair Community College
Vidyullata Waghulde, St. Louis Community College, Meramec

NEBRASKA
James Fletcher, Creighton University

NEVADA
Pradip Bhowmik, University of Nevada, Las Vegas

NEW JERSEY
Thomas Berke, Brookdale Community College
Danielle Jacobs, Rider University

NEW YORK
Michael Aldersley, Rensselaer Polytechnic Institute
Brahmadeo Dewprashad, Borough of Manhattan Community College
Eric Helms, SUNY Geneseo
Ruben Savizky, Cooper Union

NORTH CAROLINA
Deborah Pritchard, Forsyth Technical Community College

OHIO
James Beil, Lorain County Community College
Adam Keller, Columbus State Community College
Mike Rennekamp, Columbus State Community College

OKLAHOMA
Steven Meier, University of Central Oklahoma

OREGON
Gary Spessard, University of Oregon

PENNSYLVANIA
Rodrigo Andrade, Temple University
Geneive Henry, Susquehanna University
Michael Leonard, Washington & Jefferson College
William Loffredo, East Stroudsburg University
Gloria Silva, Carnegie Mellon University
Marcus Thomsen, Franklin & Marshall College
Eric Tillman, Bucknell University
William Wuest, Temple University

SOUTH CAROLINA
Rick Heldrich, College of Charleston

SOUTH DAKOTA
Grigoriy Sereda, University of South Dakota

TENNESSEE
Phillip Cook, East Tennessee State University

TEXAS
Frank Foss, University of Texas at Arlington
Scott Handy, Middle Tennessee State University
Carl Lovely, University of Texas at Arlington
Javier Macossay, The University of Texas-Pan American
Patricio Santander, Texas A&M University
Claudia Taenzler, University of Texas, Dallas

VIRGINIA
Joyce Easter, Virginia Wesleyan College
Christine Hermann, Radford University

CANADA
Ashley Causton, University of Calgary
Michael Chong, University of Waterloo
Isabelle Dionne, Dawson College
Paul Harrison, McMaster University
Edward Lee-Ruff, York University
R. Scott Murphy, University of Regina
John Sorensen, University of Manitoba
Jackie Stewart, The University of British Columbia

## Class Testers

Steve Gentemann, Southwestern Illinois College
Laurel Habgood, Rollins College
Shane Lamos, St. Michael's College
Brian Love, East Carolina University
James Mackay, Elizabethtown College
Tom Russo, Florida State College, Kent Campus
Ethan Tsui, Metropolitan State University of Denver

## Focus Group Participants

Beverly Clement, Blinn College
Greg Crouch, Washington State University
Ishan Erden, San Francisco State University
Henry Forman, University of California, Merced
Chammi Gamage-Miller, Blinn College
Randy Goff, University of West Florida
Jonathan Gough, Long Island University
Thomas Hughes, Siena College
Willian Jenks, Iowa State University
Paul Jones, Wake Forest University
Phillip Lukeman, St. John's University
Andrew Morehead, East Carolina University
Joan Muyanyatta-Comar, Georgia State University
Christine Pruis, Arizona State University
Laurie Starkey, California Polytechnic University at Pomona
Don Warner, Boise State University

## Accuracy Checkers

Eric Ballard, University of Tampa
Kevin Caran, James Madison University
James Fletcher, Creighton University
Michael Leonard, Washington and Jefferson College
Kevin Minbiole, Villanova University
John Sorenson, University of Manitoba

## FIRST EDITION REVIEWERS: CLASS TEST PARTICIPANTS, FOCUS GROUP PARTICIPANTS, AND ACCURACY CHECKERS

Philip Albiniak, Ball State University
Thomas Albright, University of Houston
Michael Aldersley, Rensselaer Polytechnic Institute
David Anderson, University of Colorado, Colorado Springs
Merritt Andrus, Brigham Young University
Laura Anna, Millersville University
Ivan Aprahamian, Dartmouth College
Yiyan Bai, Houston Community College
Satinder Bains, Paradise Valley Community College
C. Eric Ballard, University of Tampa
Edie Banner, Richmond University
James Beil, Lorain County Community College
Peter Bell, Tarleton State University
Dianne Bennet, Sacramento City College
Thomas Berke, Brookdale Community College
Daniel Bernier, Riverside Community College
Narayan Bhat, University of Texas Pan American
Gautam Bhattacharyya, Clemson University
Silas Blackstock, University of Alabama
Lea Blau, Yeshiva University
Megan Bolitho, University of San Francisco
Matthias Brewer, The University of Vermont
David Brook, San Jose State University
Cindy Browder, Northern Arizona University
Pradip Browmik, University of Nevada, Las Vegas
Banita Brown, University of North Carolina Charlotte
Kathleen Brunke, Christopher Newport University
Timothy Brunker, Towson University
Jared Butcher, Ohio University
Arthur Cammers, University of Kentucky, Lexington
Kevin Cannon, Penn State University, Abington
Kevin Caran, James Madison University
Jeffrey Carney, Christopher Newport University
David Cartrette, South Dakota State University
Steven Castle, Brigham Young University
Brad Chamberlain, Luther College
Paul Chamberlain, George Fox University
Seveda Chamras, Glendale Community College
Tom Chang, Utah State University
Dana Chatellier, University of Delaware
Sarah Chavez, Washington University
Emma Chow, Palm Beach Community College
Jason Chruma, University of Virginia
Phillip Chung, Montefiore Medical Center
Steven Chung, Bowling Green State University
Nagash Clarke, Washtenaw Community College
Adiel Coca, Southern Connecticut State University
Jeremy Cody, Rochester Institute of Technology
Phillip Cook, East Tennessee State University
Jeff Corkill, Eastern Washington University
Sergio Cortes, University of Texas at Dallas
Philip J. Costanzo, California Polytechnic State University, San Luis Obispo
Wyatt Cotton, Cincinnati State College
Marilyn Cox, Louisiana Tech University
David Crich, University of Illinois at Chicago
Mapi Cuevas, Sante Fe Community College
Scott Davis, Mercer University, Macon
Frank Day, North Shore Community College
Peter de Lijser, California State University, Fullerton
Roman Dembinski, Oakland University
Brahmadeo Dewprashad, Borough of Manhattan Community College
Preeti Dhar, SUNY New Paltz
Bonnie Dixon, University of Maryland, College Park
Theodore Dolter, Southwestern Illinois College
Norma Dunlap, Middle Tennessee State University
Joyce Easter, Virginia Wesleyan College
Jeffrey Elbert, University of Northern Iowa
J. Derek Elgin, Coastal Carolina University
Derek Elgin, Coastal Carolina University
Cory Emal, Eastern Michigan University
Susan Ensel, Hood College
David Flanigan, Hillsborough Community College
James T. Fletcher, Creighton University
Francis Flores, California Polytechnic State University, Pomona
John Flygare, Stanford University
Frantz Folmer-Andersen, SUNY New Paltz
Raymond Fong, City College of San Francisco
Mark Forman, Saint Joseph's University
Frank Foss, University of Texas at Arlington
Annaliese Franz, University of California, Davis
Andrew Frazer, University of Central Florida
Lee Friedman, University of Maryland, College Park
Steve Gentemann, Southwestern Illinois College
Tiffany Gierasch, University of Maryland, Baltimore County
Scott Grayson, Tulane University
Thomas Green, University of Alaska, Fairbanks
Kimberly Greve, Kalamazoo Valley Community College
Gordon Gribble, Dartmouth College
Ray A. Gross, Jr., Prince George's Community College
Nathaniel Grove, University of North Carolina, Wilmington
Yi Guo, Montefiore Medical Center
Sapna Gupta, Palm Beach State College
Kevin Gwaltney, Kennesaw State University
Asif Habib, University of Wisconsin, Waukesha
Donovan Haines, Sam Houston State University
Robert Hammer, Louisiana State University
Scott Handy, Middle Tennessee State University
Christopher Hansen, Midwestern State University
Kenn Harding, Texas A&M University
Matthew Hart, Grand Valley State University
Jack Hayes, State Fair Community College
Eric Helms, SUNY Geneseo
Maged Henary, Georgia State University, Langate

Amanda Henry, Fresno City College
Christine Hermann, Radford University
Patricia Hill, Millersville University
Ling Huang, Sacramento City College
John Hubbard, Marshall University
Roxanne Hulet, Skagit Valley College
Christopher Hyland, California State University, Fullerton
Danielle Jacobs, Rider University
Christopher S. Jeffrey, University of Nevada, Reno
Dell Jensen, Augustana College
Yu Lin Jiang, East Tennessee State University
Richard Johnson, University of New Hampshire
Marlon Jones, Long Beach City College
Reni Joseph, St. Louis Community College, Meramec Campus
Cynthia Judd, Palm Beach State College
Eric Kantorowski, California Polytechnic State University, San Luis Obispo
Andrew Karatjas, Southern Connecticut State University
Adam Keller, Columbus State Community College
Mushtaq Khan, Union County College
James Kiddle, Western Michigan University
Kevin Kittredge, Siena College
Silvia Kolchens, Pima Community College
Dalila Kovacs, Grand Valley State University
Jennifer Koviach-Côté, Bates College
Paul J. Kropp, University of North Carolina, Chapel Hill
Jens-Uwe Kuhn, Santa Barbara City College
Silvia Kölchens, Pima County Community College
Massimiliano Lamberto, Monmouth University
Cindy Lamberty, Cloud County Community College, Geary County Campus
Kathleen Laurenzo, Florida State College
William Lavell, Camden County College
Iyun Lazik, San Jose City College
Michael Leonard, Washington & Jefferson College
Sam Leung, Washburn University
Michael Lewis, Saint Louis University
Scott Lewis, James Madison University
Deborah Lieberman, University of Cincinnati
Harriet Lindsay, Eastern Michigan University
Jason Locklin, University of Georgia
William Loffredo, East Stroudsburg University
Robert Long, Eastern New Mexico University
Rena Lou, Cerritos College
Brian Love, East Carolina University
Douglas Loy, University of Arizona
Frederick A. Luzzio, University of Louisville
Lili Ma, Northern Kentucky University
Javier Macossay-Torres, University of Texas Pan American
Kirk Manfredi, University of Northern Iowa
Ned Martin, University of North Carolina, Wilmington
Vivian Mativo, Georgia Perimeter College, Clarkston
Barbara Mayer, California State University, Fresno
Dominic McGrath, University of Arizona
Steven Meier, University of Central Oklahoma
Dina Merrer, Barnard College
Stephen Milczanowski, Florida State College
Nancy Mills, Trinity University
Kevin Minbiole, James Madison University
Thomas Minehan, California State University, Northridge
James Miranda, California State University, Sacramento
Shizue Mito, University of Texas at El Paso
David Modarelli, University of Akron
Jesse More, Loyola College
Andrew Morehead, East Carolina University
Sarah Mounter, Columbia College of Missouri
Barbara Murray, University of Redlands
Kensaku Nakayama, California State University, Long Beach
Thomas Nalli, Winona State University
Richard Narske, Augustana College
Donna Nelson, University of Oklahoma
Nasri Nesnas, Florida Institute of Technology
William Nguyen, Santa Ana College
James Nowick, University of California, Irvine
Edmond J. O'Connell, Fairfield University
Asmik Oganesyan, Glendale Community College
Kyungsoo Oh, Indiana University, Purdue University Indianapolis
Greg O'Neil, Western Washington University
Edith Onyeozili, Florida Agricultural & Mechanical University
Catherine Owens Welder, Dartmouth College
Anne B. Padias, University of Arizona
Hasan Palandoken, California Polytechnic State University, San Luis Obispo
Chandrakant Panse, Massachusetts Bay Community College
Sapan Parikh, Manhattanville College
James Parise Jr., Duke University
Edward Parish, Auburn University
Keith O. Pascoe, Georgia State University
Michael Pelter, Purdue University, Calumet
Libbie Pelter, Purdue University, Calumet
H. Mark Perks, University of Maryland, Baltimore County
John Picione, Daytona State College
Chris Pigge, University of Iowa
Harold Pinnick, Purdue University, Calumet
Tchao Podona, Miami Dade College
John Pollard, University of Arizona
Owen Priest, Northwestern University, Evanston
Paul Primrose, Baylor University
Christine Pruis, Arizona State University
Martin Pulver, Bronx Community College
Shanthi Rajaraman, Richard Stockton College of New Jersey
Sivappa Rasapalli, University of Massachusetts, Dartmouth
Cathrine Reck, Indiana University, Bloomington
Ron Reese, Victoria College
Mike Rennekamp, Columbus State Community College
Olga Rinco, Luther College
Melinda Ripper, Butler County Community College
Harold Rogers, California State University, Fullerton
Mary Roslonowski, Brevard Community College
Robert D. Rossi, Gloucester County College
Eriks Rozners, Northeastern University
Gillian Rudd, Northwestern State University
Thomas Russo, Florida State College—Kent Campus
Lev Ryzhkov, Towson University
Preet-Pal S. Saluja, Triton College
Steve Samuel, SUNY Old Westbury
Patricio Santander, Texas A&M University
Gita Sathianathan, California State University, Fullerton
Sergey Savinov, Purdue University, West Lafayette
Amber Schaefer, Texas A&M University
Kirk Schanze, University of Florida
Paul Schueler, Raritan Valley Community College
Alan Schwabacher, University of Wisconsin, Milwaukee
Pamela Seaton, University of North Carolina, Wilmington
Jason Serin, Glendale Community College
Gary Shankweiler, California State University, Long Beach
Kevin Shaughnessy, The University of Alabama
Emery Shier, Amarillo College
Richard Shreve, Palm Beach State College
John Shugart, Coastal Carolina University
Edward Skibo, Arizona State University
Douglas Smith, California State University, San Bernadino
Michelle Smith, Georgia Southwestern State University
Rhett Smith, Clemson University

Irina Smoliakova, University of North Dakota
Timothy Snowden, University of Alabama
Chad Snyder, Western Kentucky University
Scott Snyder, Columbia University
Vadim Soloshonok, University of Oklahoma
John Sowa, Seton Hall University
Laurie Starkey, California Polytechnic State University, Pomona
Mackay Steffensen, Southern Utah University
Mackay Steffensen, Southern Utah University
Richard Steiner, University of Utah
Corey Stephenson, Boston University
Nhu Y Stessman, California State University, Stanislaus
Erland Stevens, Davidson College
James Stickler, Allegany College of Maryland
Robert Stockland, Bucknell University
Jennifer Swift, Georgetown University
Ron Swisher, Oregon Institute of Technology
Carole Szpunar, Loyola University Chicago
Claudia Taenzler, University of Texas at Dallas
John Taylor, Rutgers University, New Brunswick
Richard Taylor, Miami University
Cynthia Tidwell, University of Montevallo
Eric Tillman, Bucknell University
Bruce Toder, University of Rochester
Ana Tontcheva, El Camino College
Jennifer Tripp, San Francisco State University
Adam Urbach, Trinity University
Melissa Van Alstine, Adelphi University
Christopher Vanderwal, University of California, Irvine
Aleskey Vasiliev, East Tennessee State University
Heidi Vollmer-Snarr, Brigham Young University
Edmir Wade, University of Southern Indiana
Vidyullata Waghulde, St. Louis Community College
Linda Waldman, Cerritos College
Kenneth Walsh, University of Southern Indiana
Reuben Walter, Tarleton State University
Matthew Weinschenk, Emory University
Andrew Wells, Chabot College
Peter Wepplo, Monmouth University`
Lisa Whalen, University of New Mexico
Ronald Wikholm, University of Connecticut, Storrs
Anne Wilson, Butler University
Michael Wilson, Temple University
Leyte Winfield, Spelman College
Angela Winstead, Morgan State University
Penny Workman, University of Wisconsin, Marathon County
Stephen Woski, University of Alabama
Stephen Wuerz, Highland Community College
Linfeng Xie, University of Wisconsin, Oshkosh
Hanying Xu, Kingsborough Community College of CUNY
Jinsong Zhang, California State University, Chico
Regina Zibuck, Wayne State University

CANADA

Ashley Causton, University of Calgary
Michael Chong, University of Waterloo
Andrew Dicks, University of Toronto
Torsten Hegmann, University of Manitoba
Ian Hunt, University of Calgary
Norman Hunter, University of Manitoba
Michael Pollard, York University
Stanislaw Skonieczny, University of Toronto
Jackie Stewart, University of British Columbia
Shirley Wacowich-Sgarbi, Langara College

This book could not have been created without the incredible efforts of the following people at John Wiley & Sons, Inc.

Photo Editor Lisa Gee helped identify exciting photos. Maureen Eide and freelance designer Anne DeMarinis conceived of a visually refreshing and compelling interior design and cover. Senior Production Editor Elizabeth Swain kept this book on schedule and was vital to ensuring such a high-quality product. Joan Kalkut, Sponsoring Editor was invaluable in the creation of both editions of this book. Her tireless efforts, together with her day-to-day guidance and insight, made this project possible. Senior Product Designer Geraldine Osnato and Senior Associate Editor Veronica Armour conceived of and built a compelling ***WileyPLUS*** course. Senior Marketing Manager Kristine Ruff enthusiastically created an exciting message for this book. Editorial assistants Mallory Fryc and Susan Tsui helped to manage many facets of the review and supplements process. Senior Project Editor Jennifer Yee managed the creation of the second edition of the solutions manual. Publisher Petra Recter provided strong vision and guidance in bringing this book to market.

Scan this code to learn more about David Klein and *Organic Chemistry, 2e*

Despite my best efforts, as well as the best efforts of the reviewers, accuracy checkers, and class testers, errors may still exist. I take full responsibility for any such errors and would encourage those using my textbook to contact me with any errors that you may find.

**David R. Klein, Ph.D.**
*Johns Hopkins University*
klein@jhu.edu

# A Review of General Chemistry

1

*ELECTRONS, BONDS, AND MOLECULAR PROPERTIES*



## DID YOU EVER WONDER...
### what causes lightning?

Believe it or not, the answer to this question is still the subject of debate (that's right... scientists have not yet figured out everything, contrary to popular belief). There are various theories that attempt to explain what causes the buildup of electric charge in clouds. One thing is clear, though—lightning involves a flow of electrons. By studying the nature of electrons and how electrons flow, it is possible to control where lightning will strike. A tall building can be protected by installing a lightning rod (a tall metal column at the top of the building) that attracts any nearby lightning bolt, thereby preventing a direct strike on the building itself. The lightning rod on the top of the Empire State Building is struck over a hundred times each year.

Just as scientists have discovered how to direct electrons in a bolt of lightning, chemists have also discovered how to direct electrons in chemical reactions. We will soon see that although organic chemistry is literally defined as the study of compounds containing carbon atoms, its true essence is actually the study of electrons, not atoms. Rather than thinking of reactions in terms of the motion of atoms, we

continued >

must recognize that *reactions occur as a result of the motion of electrons*. For example, in the following reaction the curved arrows represent the motion, or flow, of electrons. This flow of electrons causes the chemical change shown:

$$:\ddot{\underset{..}{Cl}}:^{\ominus} + CH_3{-}\ddot{\underset{..}{Br}}: \longrightarrow :\ddot{\underset{..}{Cl}}{-}CH_3 + :\ddot{\underset{..}{Br}}:^{\ominus}$$

Throughout this course, we will learn how, when, and why electrons flow during reactions. We will learn about the barriers that prevent electrons from flowing, and we will learn how to overcome those barriers. In short, we will study the behavioral patterns of electrons, enabling us to predict, and even control, the outcomes of chemical reactions.

This chapter reviews some relevant concepts from your general chemistry course that should be familiar to you. Specifically, we will focus on the central role of electrons in forming bonds and influencing molecular properties.

## 1.1 Introduction to Organic Chemistry

In the early nineteenth century, scientists classified all known compounds into two categories: *Organic compounds* were derived from living organisms (plants and animals), while *inorganic compounds* were derived from nonliving sources (minerals and gases). This distinction was fueled by the observation that organic compounds seemed to possess different properties than inorganic compounds. Organic compounds were often difficult to isolate and purify, and upon heating, they decomposed more readily than inorganic compounds. To explain these curious observations, many scientists subscribed to a belief that compounds obtained from living sources possessed a special "vital force" that inorganic compounds lacked. This notion, called vitalism, stipulated that it should be impossible to convert inorganic compounds into organic compounds without the introduction of an outside vital force. Vitalism was dealt a serious blow in 1828 when German chemist Friedrich Wöhler demonstrated the conversion of ammonium cyanate (a known inorganic salt) into urea, a known organic compound found in urine:

$$NH_4OCN \xrightarrow{\text{Heat}} H_2N{-}C({=}O){-}NH_2$$

**Ammonium cyanate** (Inorganic) **Urea** (Organic)

**BY THE WAY**
There are some carbon-containing compounds that are traditionally excluded from organic classification. For example, ammonium cyanate (seen on this page) is still classified as inorganic, despite the presence of a carbon atom. Other exceptions include sodium carbonate ($Na_2CO_3$) and potassium cyanide (KCN), both of which are also considered to be inorganic compounds. We will not encounter many more exceptions.

Over the decades that followed, other examples were found, and the concept of vitalism was gradually rejected. The downfall of vitalism shattered the original distinction between organic and inorganic compounds, and a new definition emerged. Specifically, organic compounds became defined as those compounds containing carbon atoms, while inorganic compounds generally were defined as those compounds lacking carbon atoms.

Organic chemistry occupies a central role in the world around us, as we are surrounded by organic compounds. The food that we eat and the clothes that we wear are comprised of organic compounds. Our ability to smell odors or see colors results from the behavior of organic compounds. Pharmaceuticals, pesticides, paints, adhesives, and plastics are all made from organic compounds. In fact, our bodies are constructed mostly from organic compounds (DNA, RNA, proteins, etc.) whose behavior and function are determined by the guiding principles of organic chemistry. The responses of our bodies to pharmaceuticals are the results of reactions guided by the principles of organic chemistry. A deep understanding of those principles enables the design of new drugs that fight disease and improve the overall quality of life and longevity. Accordingly, it is not surprising that organic chemistry is required knowledge for anyone entering the health professions.

## 1.2 The Structural Theory of Matter

In the mid-nineteenth century three individuals, working independently, laid the conceptual foundations for the structural theory of matter. August Kekulé, Archibald Scott Couper, and Alexander M. Butlerov each suggested that substances are defined by a specific arrangement of atoms. As an example, consider the following two compounds:

```
    H       H              H   H
    |       |              |   |
H — C — O — C — H      H — C — C — O — H
    |       |              |   |
    H       H              H   H
```

**Dimethyl ether** — Boiling point = −23°C

**Ethanol** — Boiling point = 78.4°C

These compounds have the same molecular formula ($C_2H_6O$), yet they differ from each other in the way the atoms are connected—that is, they differ in their constitution. As a result, they are called **constitutional isomers**. Constitutional isomers have different physical properties and different names. The first compound is a colorless gas used as an aerosol spray propellant, while the second compound is a clear liquid, commonly referred to as "alcohol," found in alcoholic beverages.

According to the structural theory of matter, each element will generally form a predictable number of bonds. For example, carbon generally forms four bonds and is therefore said to be **tetravalent**. Nitrogen generally forms three bonds and is therefore **trivalent**. Oxygen forms two bonds and is **divalent**, while hydrogen and the halogens form one bond and are **monovalent** (Figure 1.1).

| *Tetravalent* | *Trivalent* | *Divalent* | *Monovalent* |
|---|---|---|---|
| —C— (four bonds) | —N— (three bonds) | —O— | H— X— (where X = F, Cl, Br, or I) |
| Carbon generally forms ***four*** bonds. | Nitrogen generally forms ***three*** bonds. | Oxygen generally forms ***two*** bonds. | Hydrogen and halogens generally form ***one*** bond. |

**FIGURE 1.1** Valencies of some common elements encountered in organic chemistry.

# SKILLBUILDER

### 1.1 DETERMINING THE CONSTITUTION OF SMALL MOLECULES

**LEARN the skill**

There is only one compound that has molecular formula $C_2H_5Cl$. Determine the constitution of this compound.

**SOLUTION**

The molecular formula indicates which atoms are present in the compound. In this example, the compound contains two carbon atoms, five hydrogen atoms, and one chlorine atom. Begin by determining the valency of each atom that is present in the compound. Each carbon atom is expected to be tetravalent, while the chlorine and hydrogen atoms are all expected to be monovalent:

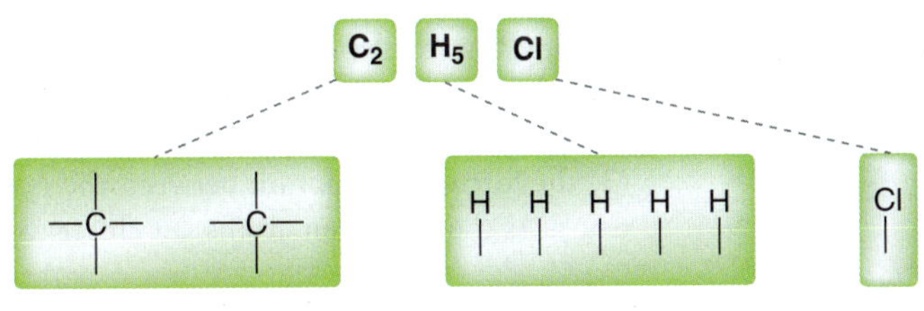


**STEP 1**
Determine the valency of each atom in the compound.

**STEP 2**
Determine how the atoms are connected—atoms with the highest valency should be placed at the center and monovalent atoms should be placed at the periphery.

Now we must determine how these atoms are connected. The atoms with the most bonds (the carbon atoms) are likely to be in the center of the compound. In contrast, the chlorine atom and hydrogen atoms can each form only one bond, so those atoms must be placed at the periphery. In this example, it does not matter where the chlorine atom is placed. All six possible positions are equivalent.

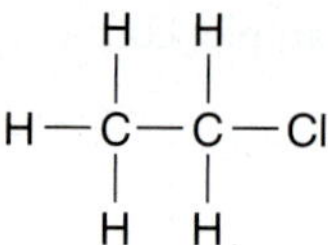

**PRACTICE the skill** **1.1** Determine the constitution of the compounds with the following molecular formulas:

**(a)** $CH_4O$ **(b)** $CH_3Cl$ **(c)** $C_2H_6$ **(d)** $CH_5N$

**(e)** $C_2F_6$ **(f)** $C_2H_5Br$ **(g)** $C_3H_8$

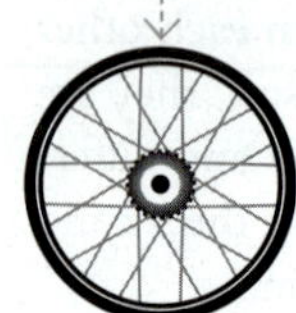

**APPLY the skill**

**1.2** There are two constitutional isomers with molecular formula $C_3H_7Cl$, because there are two possibilities for the location of the chlorine atom. It can be connected either to the central carbon atom or to one of the other two (equivalent) carbon atoms. Draw both isomers.

**1.3** Draw three constitutional isomers that have molecular formula $C_3H_8O$.

**1.4** Draw all constitutional isomers that have molecular formula $C_4H_{10}O$.

need more **PRACTICE?** **Try Problems 1.34, 1.46, 1.47, 1.54**

## 1.3 Electrons, Bonds, and Lewis Structures

### What Are Bonds?

As mentioned, atoms are connected to each other by bonds. That is, bonds are the "glue" that hold atoms together. But what is this mysterious glue and how does it work? In order to answer this question, we must focus our attention on electrons.

The existence of the electron was first proposed in 1874 by George Johnstone Stoney (National University of Ireland), who attempted to explain electrochemistry by suggesting the existence of a particle bearing a unit of charge. Stoney coined the term *electron* to describe this particle. In 1897, J. J. Thomson (Cambridge University) demonstrated evidence supporting the existence of Stoney's mysterious electron and is credited with discovering the electron. In 1916, Gilbert Lewis (University of California, Berkeley) defined a **covalent bond** as the result of *two atoms sharing a pair of electrons*. As a simple example, consider the formation of a bond between two hydrogen atoms:

Each hydrogen atom has one electron. When these electrons are shared to form a bond, there is a decrease in energy, indicated by the negative value of $\Delta H$. The energy diagram in Figure 1.2 plots the energy of the two hydrogen atoms as a function of the distance between them. Focus on the right side of the diagram, which represents the hydrogen atoms separated by a large distance. Moving toward the left on the diagram, the hydrogen atoms approach each other, and there are several forces that must be taken into account: (1) the force of repulsion between the two negatively charged electrons, (2) the force of repulsion between the two positively charged nuclei, and (3) the forces of attraction between the positively charged nuclei and the negatively charged electrons. As the hydrogen atoms get closer to each other, all of these forces get stronger. Under these

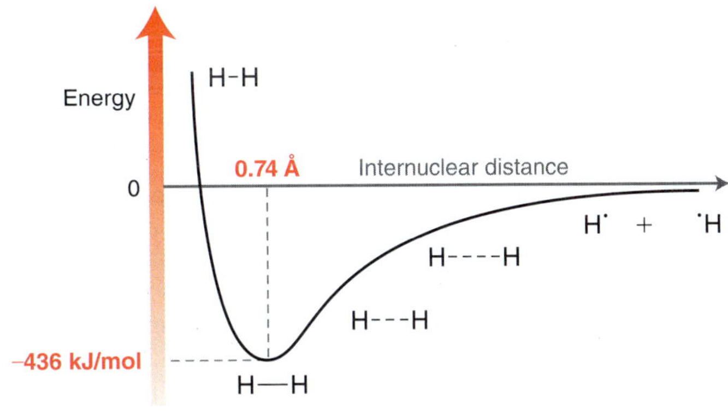


**FIGURE 1.2**
An energy diagram showing the energy as a function of the internuclear distance between two hydrogen atoms.

circumstances, the electrons are capable of moving in such a way so as to minimize the repulsive forces between them while maximizing their attractive forces with the nuclei. This provides for a net force of attraction, which lowers the energy of the system. As the hydrogen atoms move still closer together, the energy continues to be lowered until the nuclei achieve a separation (internuclear distance) of 0.74 angstroms (Å). At that point, the force of repulsion between the nuclei begins to overwhelm the forces of attraction, causing the energy of the system to increase. The lowest point on the curve represents the lowest energy (most stable) state. This state determines both the bond length (0.74 Å) and the bond strength (436 kJ/mol).

**BY THE WAY**
1 Å = $10^{-10}$ meters.

## Drawing the Lewis Structure of an Atom

Armed with the idea that a bond represents a pair of shared electrons, Lewis then devised a method for drawing structures. In his drawings, called **Lewis structures**, the electrons take center stage. We will begin by drawing individual atoms, and then we will draw Lewis structures for small molecules. First, we must review a few simple features of atomic structure:

- The nucleus of an atom is comprised of protons and neutrons. Each proton has a charge of +1, and each neutron is electrically neutral.
- For a neutral atom, the number of protons is balanced by an equal number of electrons, which have a charge of −1 and exist in shells. The first shell, which is closest to the nucleus, can contain two electrons, and the second shell can contain up to eight electrons.
- The electrons in the outermost shell of an atom are called the valence electrons. The number of valence electrons in an atom is identified by its group number in the periodic table (Figure 1.3).

The Lewis dot structure of an individual atom indicates the number of valence electrons, which are placed as dots around the periodic symbol of the atom (C for carbon, O for oxygen, etc.). The placement of these dots is illustrated in the following SkillBuilder.

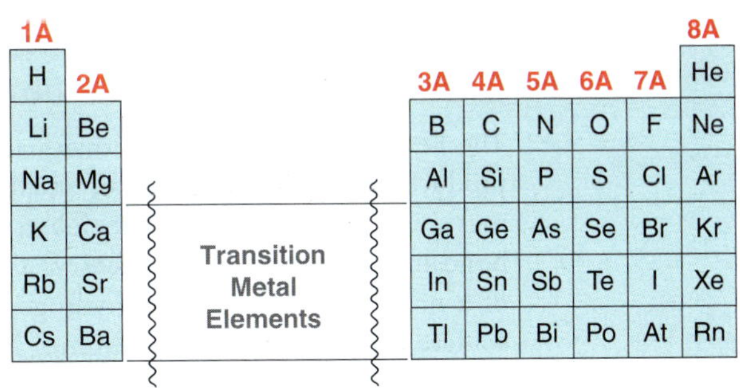


**FIGURE 1.3**
A periodic table showing group numbers.

# SKILLBUILDER

## 1.2 DRAWING THE LEWIS DOT STRUCTURE OF AN ATOM

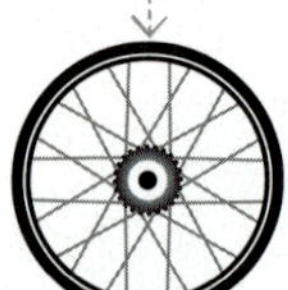

**LEARN the skill** Draw the Lewis dot structure of (a) a boron atom and (b) a nitrogen atom.

**SOLUTION**

**STEP 1**
Determine the number of valence electrons.

**(a)** In a Lewis dot structure, only valence electrons are drawn, so we must first determine the number of valence electrons. Boron belongs to group 3A on the periodic table, and it therefore has three valence electrons. The periodic symbol for boron (B) is drawn, and each electron is placed by itself (unpaired) around the B, like this:

•B•
 •

**STEP 2**
Place one valence electron by itself on each side of the atom.

**(b)** Nitrogen belongs to group 5A on the periodic table, and it therefore has five valence electrons. The periodic symbol for nitrogen (N) is drawn, and each electron is placed by itself (unpaired) on a side of the N until all four sides are occupied:

 •
•N•
 •

**STEP 3**
If the atom has more than four valence electrons, the remaining electrons are paired with the electrons already drawn.

Any remaining electrons must be paired up with the electrons already drawn. In the case of nitrogen, there is only one more electron to place, so we pair it up with one of the four unpaired electrons (it doesn't matter which one we choose):

**PRACTICE the skill** **1.5** Draw a Lewis dot structure for each of the following atoms:

**(a)** Carbon **(b)** Oxygen **(c)** Fluorine **(d)** Hydrogen
**(e)** Bromine **(f)** Sulfur **(g)** Chlorine **(h)** Iodine

**APPLY the skill**

**1.6** Compare the Lewis dot structure of nitrogen and phosphorus and explain why you might expect these two atoms to exhibit similar bonding properties.

**1.7** Name one element that you would expect to exhibit bonding properties similar to boron. Explain.

**1.8** Draw a Lewis structure of a carbon atom that is missing one valence electron (and therefore bears a positive charge). Which second-row element does this carbon atom resemble in terms of the number of valence electrons?

**1.9** Draw a Lewis structure of a carbon atom that has one extra valence electron (and therefore bears a negative charge). Which second-row element does this carbon atom resemble in terms of the number of valence electrons?

## Drawing the Lewis Structure of a Small Molecule

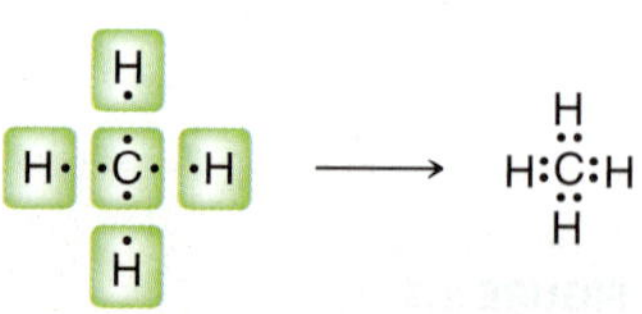

The Lewis dot structures of individual atoms are combined to produce Lewis dot structures of small molecules. These drawings are constructed based on the observation that atoms tend to bond in such a way so as to achieve the electron configuration of a noble gas. For example, hydrogen will form one bond to achieve the electron configuration of helium (two valence electrons), while second-row elements (C, N, O, and F) will form the necessary number of bonds so as to achieve the electron configuration of neon (eight valence electrons).

This observation, called the **octet rule**, explains why carbon is tetravalent. As just shown, it can achieve an octet of electrons by using each of its four valence electrons to form a bond. The octet rule also explains why nitrogen is trivalent. Specifically, it has five valence electrons and requires three bonds in order to achieve an octet of electrons. Notice that the nitrogen atom contains one pair of unshared, or nonbonding, electrons, called a **lone pair**.

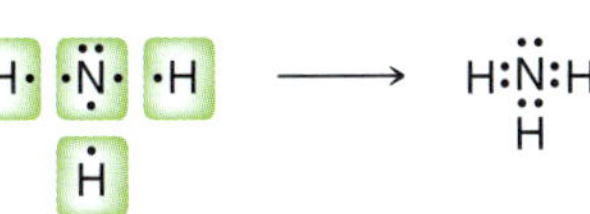


In the next chapter, we will discuss the octet rule in more detail; in particular, we will explore when it can be violated and when it cannot be violated. For now, let's practice drawing Lewis structures.

# SKILLBUILDER

## 1.3 DRAWING THE LEWIS STRUCTURE OF A SMALL MOLECULE

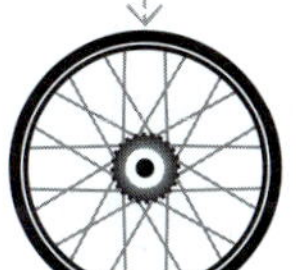

**LEARN the skill** Draw the Lewis structure of $CH_2O$.

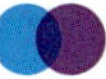

**SOLUTION**

**STEP 1** Draw all individual atoms.

There are four discrete steps when drawing a Lewis structure: First determine the number of valence electrons for each atom:

**STEP 2** Connect atoms that form more than one bond.

Then, connect any atoms that form more than one bond. Hydrogen atoms only form one bond each, so we will save those for last. In this case, we connect the C and the O:

·Ċ:Ö:

**STEP 3** Connect the hydrogen atoms.

Next, connect all hydrogen atoms. We place the hydrogen atoms next to carbon, because carbon has more unpaired electrons than oxygen:

H:Ċ:Ö: (with H below C)

**STEP 4** Pair any unpaired electrons so that each atom achieves an octet.

Finally, check to see if each atom (except hydrogen) has an octet. In fact, neither the carbon nor the oxygen has an octet, so in a situation like this, the unpaired electrons are shared as a double bond between carbon and oxygen:

H:Ċ:Ö: ⟶ H:C::O: (with H below C)

Now all atoms have achieved an octet. When drawing Lewis structures, remember that you cannot simply add more electrons to the drawing. For each atom to achieve an octet the existing electrons must be shared. The total number of valence electrons should be correct when you are finished. In this example, there was one carbon atom, two hydrogen atoms, and one oxygen atom, giving a total of 12 valence electrons (4 + 2 + 6). The drawing above MUST have 12 valence electrons, no more and no less.

**PRACTICE the skill** **1.10** Draw a Lewis structure for each of the following compounds:

**(a)** $C_2H_6$ **(b)** $C_2H_4$ **(c)** $C_2H_2$

**(d)** $C_3H_8$ **(e)** $C_3H_6$ **(f)** $CH_3OH$

**APPLY the skill**

**1.11** Borane ($BH_3$) is very unstable and quite reactive. Draw a Lewis structure of borane and explain the source of the instability.

**1.12** There are four constitutional isomers with molecular formula $C_3H_9N$. Draw a Lewis structure for each isomer and determine the number of lone pairs on the nitrogen atom in each case.

need more **PRACTICE?** **Try Problems 1.35, 1.38, 1.42**

## 1.4 Identifying Formal Charges

A **formal charge** is associated with any atom that does not exhibit the appropriate number of valence electrons. When such an atom is present in a Lewis structure, the formal charge must be drawn. Identifying a formal charge requires two discrete tasks:

1. Determine the appropriate number of valence electrons for an atom.
2. Determine whether the atom exhibits the appropriate number of electrons.

The first task can be accomplished by inspecting the periodic table. As mentioned earlier, the group number indicates the appropriate number of valence electrons for each atom. For example, carbon is in group 4A and therefore has four valence electrons. Oxygen is in group 6A and has six valence electrons.

After identifying the appropriate number of electrons for each atom in a Lewis structure, the next task is to determine if any of the atoms exhibit an unexpected number of electrons. For example, consider the following structure:

:Ö:
H—C—H
H

Each line represents two shared electrons (a bond). For our purposes, we must split each bond apart equally, and then count the number of electrons on each atom:

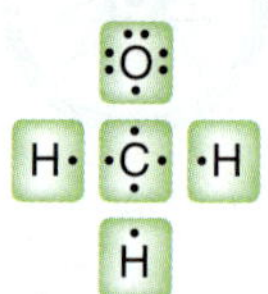

Each hydrogen atom has one valence electron, as expected. The carbon atom also has the appropriate number of valence electrons (four), but the oxygen atom does not. The oxygen atom in this structure exhibits seven valence electrons, but it should only have six. In this case, the oxygen atom has one extra electron, and it must therefore bear a negative formal charge, which is indicated like this:

:Ö:⊖
H—C—H
H

# SKILLBUILDER

## 1.4 CALCULATING FORMAL CHARGE

**LEARN the skill** Consider the nitrogen atom in the structure below and determine if it has a formal charge:

H
H—N—H
H

**SOLUTION**

**STEP 1** Determine the appropriate number of valence electrons.

We begin by determining the appropriate number of valence electrons for a nitrogen atom. Nitrogen is in group 5A of the periodic table, and it should therefore have five valence electrons.

**STEP 2** Determine the actual number of valence electrons in this case.

Next, we count how many valence electrons are exhibited by the nitrogen atom in this particular example:

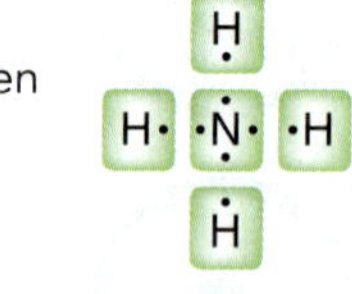

In this case, the nitrogen atom exhibits only four valence electrons. It is missing one electron, so it must bear a positive charge, which is shown like this:

**STEP 3** Assign a formal charge.

H
H—N⊕—H
H

**PRACTICE the skill** **1.13** Identify any formal charges in the structures below:

(a) 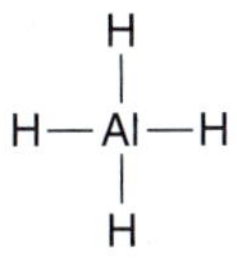 (b) 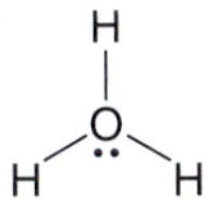 (c) 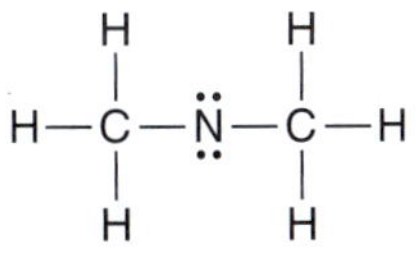 (d) 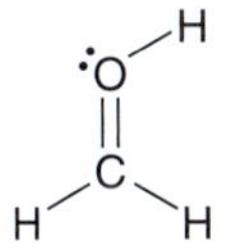 (e)

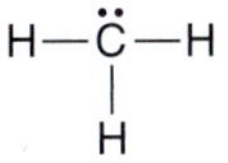

(f) 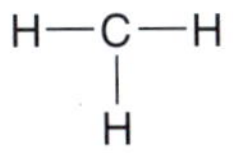 (g) 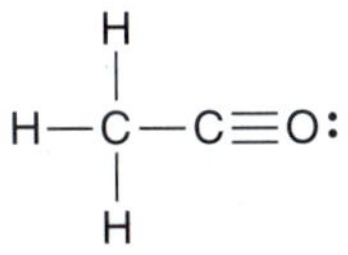 (h) 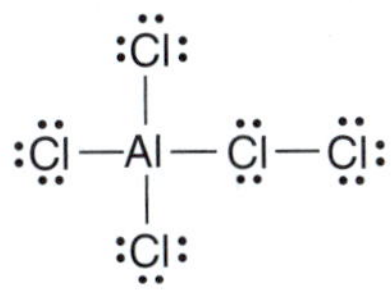 (i)

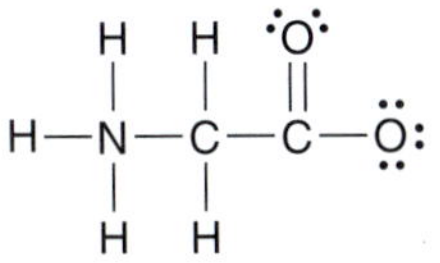

**APPLY the skill** **1.14** Draw a structure for each of the following ions; in each case, indicate which atom possesses the formal charge:

**(a)** $BH_4^-$ **(b)** $NH_2^-$ **(c)** $C_2H_5^+$

need more **PRACTICE?** **Try Problem 1.41**

## 1.5 Induction and Polar Covalent Bonds

Chemists classify bonds into three categories: (1) covalent, (2) polar covalent, and (3) ionic. These categories emerge from the electronegativity values of the atoms sharing a bond. Electronegativity is a measure of the ability of an atom to attract electrons. Table 1.1 gives the electronegativity values for elements commonly encountered in organic chemistry.

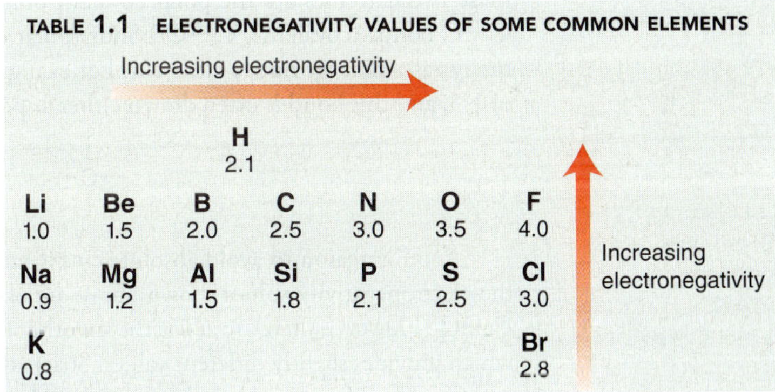

**TABLE 1.1 ELECTRONEGATIVITY VALUES OF SOME COMMON ELEMENTS**

Increasing electronegativity (→ across a row; ↑ up a column)

| | | | | | | |
|---|---|---|---|---|---|---|
| | | H 2.1 | | | | |
| Li 1.0 | Be 1.5 | B 2.0 | C 2.5 | N 3.0 | O 3.5 | F 4.0 |
| Na 0.9 | Mg 1.2 | Al 1.5 | Si 1.8 | P 2.1 | S 2.5 | Cl 3.0 |
| K 0.8 | | | | | | Br 2.8 |

When two atoms form a bond, one critical consideration allows us to classify the bond: What is the difference in the electronegativity values of the two atoms? Below are some rough guidelines:

***If the difference in electronegativity is less than 0.5***, the electrons are considered to be equally shared between the two atoms, resulting in a covalent bond. Examples include C—C and C—H:

The C—C bond is clearly covalent, because there is no difference in electronegativity between the two atoms forming the bond. Even a C—H bond is considered to be covalent, because the difference in electronegativity between C and H is less than 0.5.

***If the difference in electronegativity is between 0.5 and 1.7***, the electrons are not shared equally between the atoms, resulting in a **polar covalent bond**. For example, consider a bond between carbon and oxygen (C—O). Oxygen is significantly more electronegative (3.5) than carbon (2.5), and therefore oxygen will more strongly attract the electrons of the bond.

The withdrawal of electrons toward oxygen is called **induction**, which is often indicated with an arrow like this:

$$\overset{\longmapsto}{\mathrm{C{-}O}}$$

Induction causes the formation of partial positive and partial negative charges, symbolized by the Greek symbol delta (δ). The partial charges that result from induction will be very important in upcoming chapters:

$$\overset{\delta+}{\mathrm{C}}-\overset{\delta-}{\mathrm{O}}$$

***If the difference in electronegativity is greater than 1.7***, the electrons are not shared at all. For example, consider the bond between sodium and oxygen in sodium hydroxide (NaOH):

$$\overset{\oplus}{\mathrm{Na}} \quad \overset{\ominus}{:\ddot{\mathrm{O}}\mathrm{H}}$$

The difference in electronegativity between O and Na is so great that both electrons of the bond are possessed solely by the oxygen atom, rendering the oxygen negatively charged and the sodium positively charged. The bond between oxygen and sodium, called an **ionic bond**, is the result of the force of attraction between the two oppositely charged ions.

The cutoff numbers (0.5 and 1.7) should be thought of as rough guidelines. Rather than viewing them as absolute, we must view the various types of bonds as belonging to a spectrum without clear cutoffs (Figure 1.4).

| Covalent | | | Polar covalent | | | Ionic |
|---|---|---|---|---|---|---|
| C—C | C—H | N—H | C—O | Li—C | Li—N | NaCl |
| Small difference in electronegativity | | | | | | Large difference in electronegativity |

**FIGURE 1.4** The nature of various bonds commonly encountered in organic chemistry.

This spectrum has two extremes: covalent bonds on the left and ionic bonds on the right. Between these two extremes are the polar covalent bonds. Some bonds fit clearly into one category, such as C—C bonds (covalent), C—O bonds (polar covalent), or NaCl bonds (ionic). However, there are many cases that are not so clear-cut. For example, a C—Li bond has a difference in electronegativity of 1.5, and this bond is often drawn either as polar covalent or as ionic. Both drawings are acceptable:

$$-\overset{|}{\underset{|}{\mathrm{C}}}-\mathrm{Li} \quad \textit{or} \quad -\overset{|}{\underset{|}{\mathrm{C}}}:^{\ominus}\ {}^{\oplus}\mathrm{Li}$$

Another reason to avoid absolute cutoff numbers when comparing electronegativity values is that the electronegativity values shown above are obtained via one particular method developed by Linus Pauling. However, there are at least seven other methods for calculating electronegativity values, each of which provides slightly different values. Strict adherence to the Pauling scale would suggest that C—Br and C—I bonds are covalent, but these bonds will be treated as polar covalent throughout this course.

# SKILLBUILDER

## 1.5 LOCATING PARTIAL CHARGES RESULTING FROM INDUCTION

**LEARN** the skill

Consider the structure of methanol. Identify all polar covalent bonds and show any partial charges that result from inductive effects:

$$\mathrm{H}-\overset{\mathrm{H}}{\underset{\mathrm{H}}{\overset{|}{\underset{|}{\mathrm{C}}}}}-\ddot{\underset{\cdot\cdot}{\mathrm{O}}}-\mathrm{H}$$

**Methanol**

## SOLUTION

First identify all polar covalent bonds. The C—H bonds are considered to be covalent because the electronegativity values for C and H are fairly close. It is true that carbon is more electronegative than hydrogen, and therefore, there is a small inductive effect for each C—H bond. However, we will generally consider this effect to be negligible for C—H bonds.

The C—O bond and the O—H bond are both polar covalent bonds:

**STEP 1**
Identify all polar covalent bonds.

Polar covalent

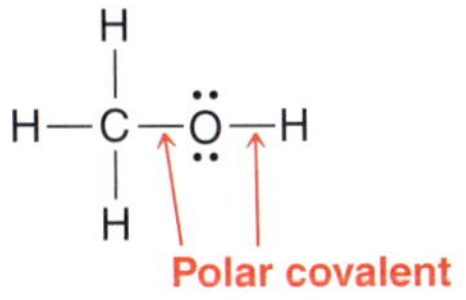

Now determine the direction of the inductive effects. Oxygen is more electronegative than C or H, so the inductive effects are shown like this:

**STEP 2**
Determine the direction of each dipole.

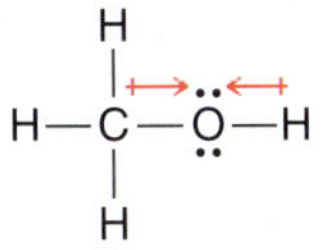

These inductive effects dictate the locations of the partial charges:

**STEP 3**
Indicate the location of partial charges.

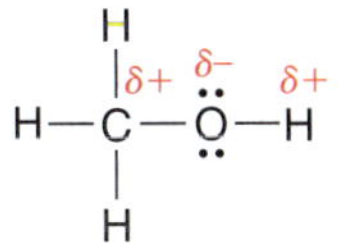

$\delta+$ $\delta-$ $\delta+$

**PRACTICE the skill**

**1.15** For each of the following compounds, identify any polar covalent bonds by drawing $\delta+$ and $\delta-$ symbols in the appropriate locations:

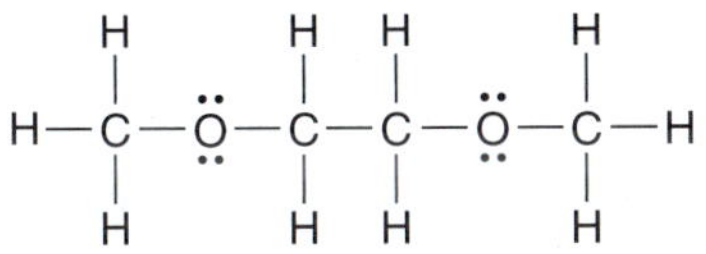

(a) 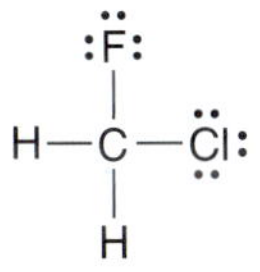 (b)

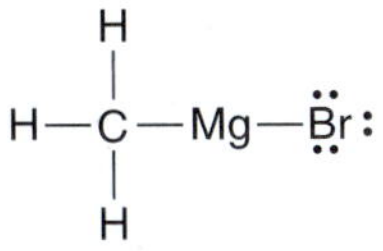

(c) 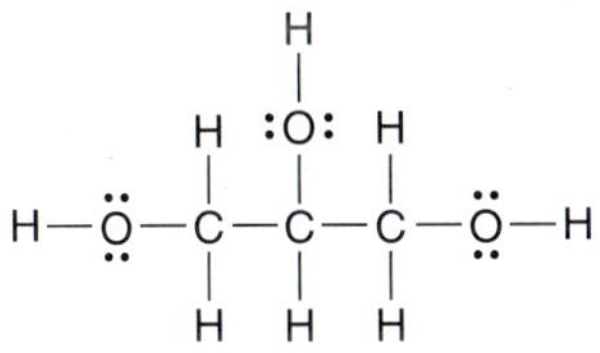 (d)

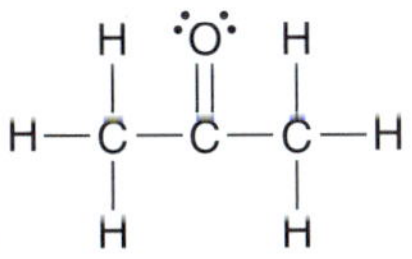

(e) 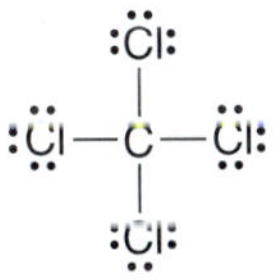 (f)

**APPLY the skill**

**1.16** The regions of $\delta+$ in a compound are the regions most likely to be attacked by an anion, such as hydroxide ($HO^-$). In the compound below, identify the two carbon atoms that are most likely to be attacked by a hydroxide ion:

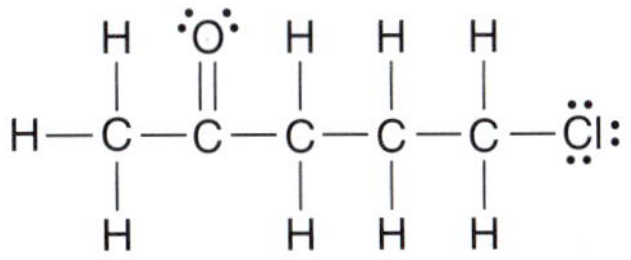

need more **PRACTICE?** **Try Problems 1.36, 1.37, 1.48, 1.57**

## Electrostatic Potential Maps

Partial charges can be visualized with three-dimensional, rainbow-like images called **electrostatic potential maps**. As an example, consider the electrostatic potential map of chloromethane:

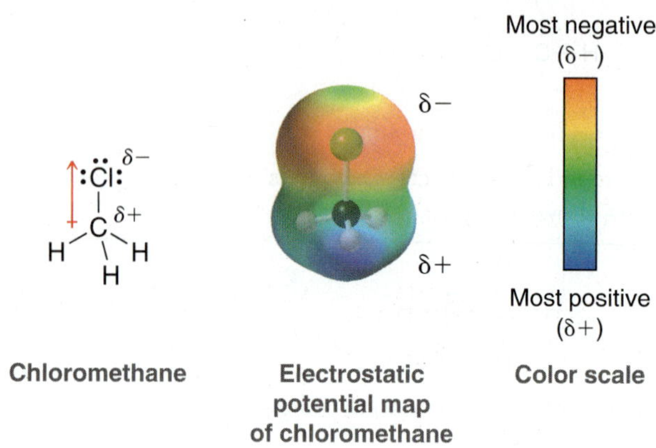


In the image, a color scale is used to represent areas of δ− and δ+. As indicated, red represents a region that is δ−, while blue represents a region that is δ+. In reality, electrostatic potential maps are rarely used by practicing organic chemists when they communicate with each other; however, these illustrations can often be helpful to students who are learning organic chemistry. Electrostatic potential maps are generated by performing a series of calculations. Specifically, an imaginary point positive charge is positioned at various locations, and for each location, we calculate the potential energy associated with the attraction between the point positive charge and the surrounding electrons. A large attraction indicates a position of δ−, while a small attraction indicates a position of δ+. The results are then illustrated using colors, as shown.

A comparison of any two electrostatic potential maps is only valid if both maps were prepared using the same color scale. Throughout this book, care has been taken to use the same color scale whenever two maps are directly compared to each other. However, it will not be useful to compare two maps from different pages of this book (or any other book), as the exact color scales are likely to be different.

## 1.6 Atomic Orbitals

### Quantum Mechanics

By the 1920s, vitalism had been discarded. Chemists were aware of constitutional isomerism and had developed the structural theory of matter. The electron had been discovered and identified as the source of bonding, and Lewis structures were used to keep track of shared and unshared electrons. But the understanding of electrons was about to change dramatically.

In 1924, French physicist Louis de Broglie suggested that electrons, heretofore considered as particles, also exhibited wavelike properties. Based on this assertion, a new theory of matter was born. In 1926, Erwin Schrödinger, Werner Heisenberg, and Paul Dirac independently proposed a mathematical description of the electron that incorporated its wavelike properties. This new theory, called *wave mechanics*, or **quantum mechanics**, radically changed the way we viewed the nature of matter and laid the foundation for our current understanding of electrons and bonds.

Quantum mechanics is deeply rooted in mathematics and represents an entire subject by itself. The mathematics involved is beyond the scope of our course, and we will not discuss it here. However, in order to understand the nature of electrons, it is critical to understand a few simple highlights from quantum mechanics:

- An equation is constructed to describe the total energy of a hydrogen atom (i.e., one proton plus one electron). This equation, called the wave equation, takes into account the wavelike behavior of an electron that is in the electric field of a proton.
- The wave equation is then solved to give a series of solutions called wavefunctions. The Greek symbol psi ($\psi$) is used to denote each wavefunction ($\psi_1$, $\psi_2$, $\psi_3$, etc.). Each of these wavefunctions corresponds to an allowed energy level for the electron. This result is incredibly important because it suggests that an electron, when contained in an atom, can only exist at discrete energy levels ($\psi_1$, $\psi_2$, $\psi_3$, etc.). In other words, the energy of the electron is *quantized*.
- Each wavefunction is a function of spatial location. It provides information that allows us to assign a numerical value for each location in three-dimensional space relative to the nucleus.

The square of that value ($\psi^2$ for any particular location) has a special meaning. It indicates the probability of finding the electron in that location. Therefore, a three-dimensional plot of $\psi^2$ will generate an image of an atomic orbital (Figure 1.5).

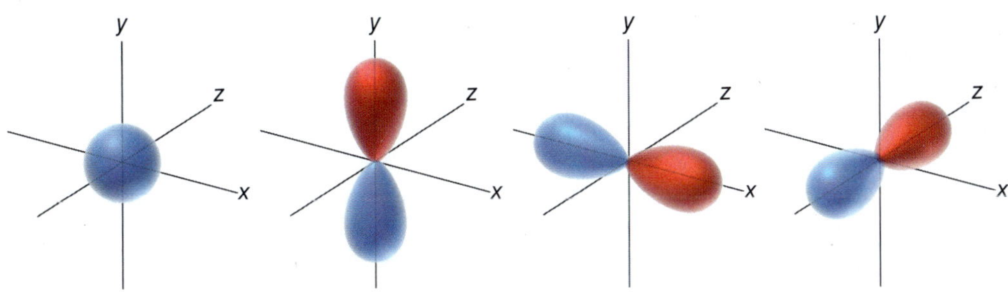


**FIGURE 1.5** Illustrations of an *s* orbital and three *p* orbitals.

## Electron Density and Atomic Orbitals

An *orbital* is a region of space that can be occupied by an electron. But care must be taken when trying to visualize this. There is a statement from the previous section that must be clarified because it is potentially misleading: "*$\psi^2$ represents the probability of finding an electron in a particular location.*" This statement seems to treat an electron as if it were a particle flying around within a specific region of space. But remember that an electron is not purely a particle—it has wavelike properties as well. Therefore, we must construct a mental image that captures both of these properties. That is not easy to do, but the following analogy might help. We will treat an occupied orbital as if it is a cloud—similar to a cloud in the sky. No analogy is perfect, and there are certainly features of clouds that are very different from orbitals. However, focusing on some of these differences between electron clouds (occupied orbitals) and real clouds makes it possible to construct a better mental model of an electron in an orbital:

- Clouds in the sky can come in any shape or size. However, electron clouds only come in a small number of shapes and sizes (as defined by the orbitals).
- A cloud in the sky is comprised of billions of individual water molecules. An electron cloud is not comprised of billions of particles. We must think of an electron cloud as a single entity, even though it can be thicker in some places and thinner in other places. This concept is critical and will be used extensively throughout the course in explaining reactions.
- A cloud in the sky has edges, and it is possible to define a region of space that contains 100% of the cloud. In contrast, an electron cloud does not have defined edges. We frequently use the term **electron density**, which is associated with the probability of finding an electron in a particular region of space. The "shape" of an orbital refers to a region of space that contains 90–95% of the electron density. Beyond this region, the remaining 5–10% of the electron density tapers off but never ends. In fact, if we want to consider the region of space that contains 100% of the electron density, we must consider the entire universe.

In summary, we must think of an orbital as a region of space that can be occupied by electron density. An occupied orbital must be treated as a *cloud of electron density*. This region of space is called an **atomic orbital** (AO), because it is a region of space defined with respect to the nucleus of a single atom. Examples of atomic orbitals are the *s*, *p*, *d*, and *f* orbitals that were discussed in your general chemistry textbook.

## Phases of Atomic Orbitals

Our discussion of electrons and orbitals has been based on the premise that electrons have wavelike properties. As a result, it will be necessary to explore some of the characteristics of simple waves in order to understand some of the characteristics of orbitals.

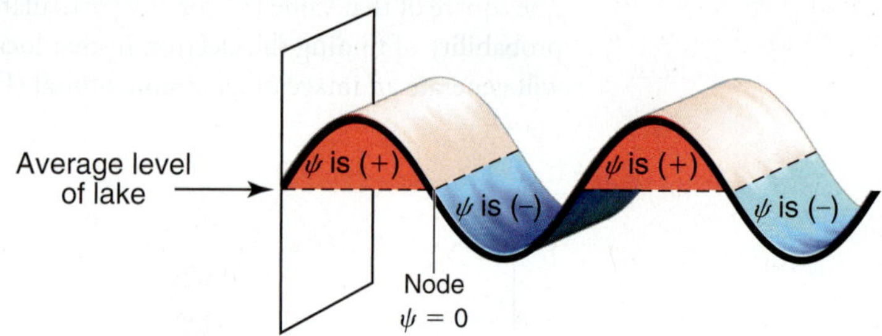


**FIGURE 1.6**
Phases of a wave moving across the surface of a lake.

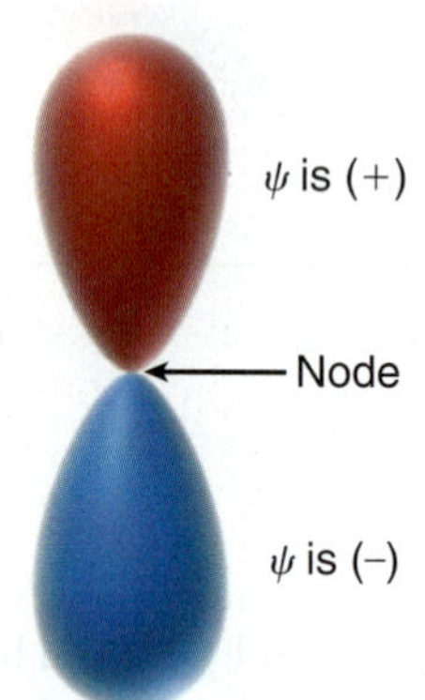


**FIGURE 1.7**
The phases of a *p* orbital.

Consider a wave that moves across the surface of a lake (Figure 1.6). The wavefunction ($\psi$) mathematically describes the wave, and the value of the wavefunction is dependent on location. Locations above the average level of the lake have a positive value for $\psi$ (indicated in red), and locations below the average level of the lake have a negative value for $\psi$ (indicated in blue). Locations where the value of $\psi$ is zero are called **nodes**.

Similarly, orbitals can have regions where the value of $\psi$ is positive, negative, or zero. For example, consider a *p* orbital (Figure 1.7). Notice that the *p* orbital has two lobes: The top lobe is a region of space where the values of $\psi$ are positive, while the bottom lobe is a region where the values of $\psi$ are negative. Between the two lobes is a location where $\psi = 0$. This location represents a node.

Be careful not to confuse the sign of $\psi$ (+ or −) with electrical charge. A positive value for $\psi$ does not imply a positive charge. The value of $\psi$ (+ or −) is a mathematical convention that refers to the *phase* of the wave (just like in the lake). Although $\psi$ can have positive or negative values, nevertheless $\psi^2$ (which describes the electron density as a function of location) will always be a positive number. At a node, where $\psi = 0$, the electron density ($\psi^2$) will also be zero. This means that there is no electron density located at a node.

From this point forward, we will draw the lobes of an orbital with colors (red and blue) to indicate the phase of $\psi$ for each region of space.

## Filling Atomic Orbitals with Electrons

The energy of an electron depends on the type of orbital that it occupies. Most of the organic compounds that we will encounter will be composed of first- and second-row elements (H, C, N, and O). These elements utilize the 1*s* orbital, the 2*s* orbital, and the three 2*p* orbitals. Our discussions will therefore focus primarily on these orbitals (Figure 1.8). Electrons are lowest in energy when they occupy a 1*s* orbital, because the 1*s* orbital is closest to the nucleus and it has no nodes (the more nodes that an orbital has, the greater its energy). The 2*s* orbital has one node and is farther away form the nucleus; it is therefore higher in energy than the 1*s* orbital. After the 2*s* orbital, there are three 2*p* orbitals that are all equivalent in energy to one another. Orbitals with the same energy level are called **degenerate orbitals.**

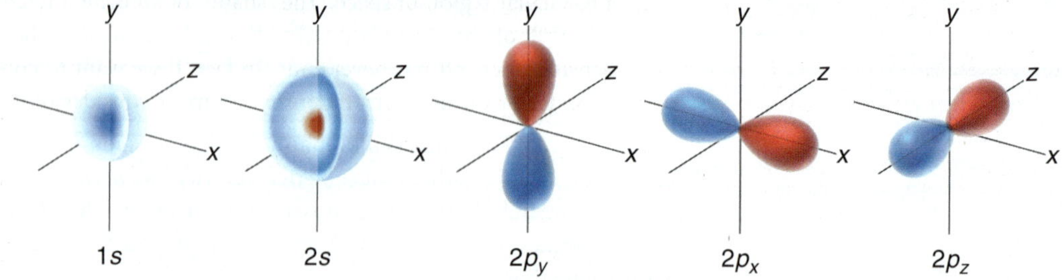


**FIGURE 1.8**
Illustrations of *s* orbitals and three *p* orbitals.

As we move across the periodic table, starting with hydrogen, each element has one more electron than the element before it (Figure 1.9). The order in which the orbitals are filled by electrons is determined by just three simple principles:

1. The **Aufbau principle.** The lowest energy orbital is filled first.
2. The **Pauli exclusion principle.** Each orbital can accommodate a maximum of two electrons that have opposite spin. To understand what "spin" means, we can imagine an electron spinning

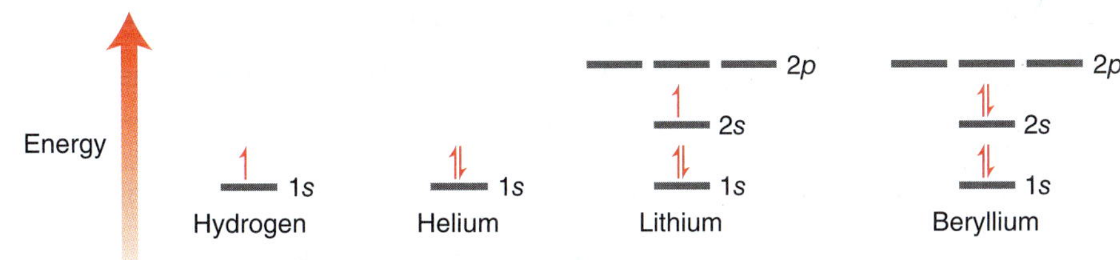


**FIGURE 1.9** Energy diagrams showing the electron configurations for H, He, Li, and Be.

in space (although this is an oversimplified explanation of the term "spin"). For reasons that are beyond the scope of this course, electrons only have two possible spin states (designated by ⇃ or ↾). In order for the orbital to accommodate two electrons, the electrons must have opposite spin states.

3. **Hund's rule.** When dealing with degenerate orbitals, such as *p* orbitals, one electron is placed in each degenerate orbital first, before electrons are paired up.

The application of the first two principles can be seen in the electron configurations shown in Figure 1.9 (H, He, Li, and Be). The application of the third principle can be seen in the electron configurations for the remaining second-row elements (Figure 1.10).

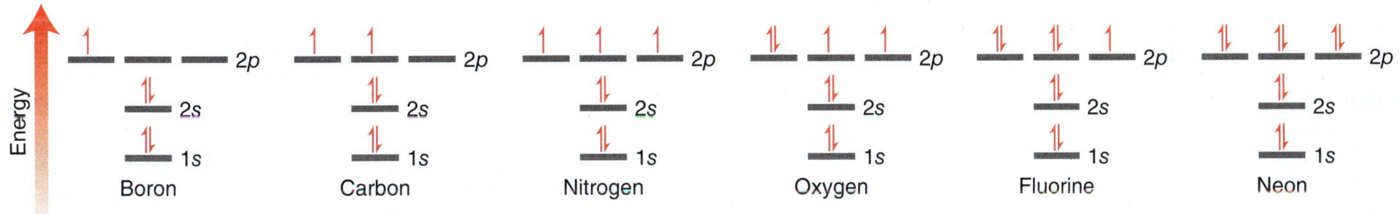


**FIGURE 1.10**
Energy diagrams showing the electron configurations for B, C, N, O, F, and Ne.

## SKILLBUILDER

### 1.6 IDENTIFYING ELECTRON CONFIGURATIONS

**LEARN** the skill

Identify the electron configuration of a nitrogen atom.

**SOLUTION**

**STEP 1**
Place the valence electrons in atomic orbitals using the Aufbau principle, the Pauli exclusion principle, and Hund's rule.

The electron configuration indicates which atomic orbitals are occupied by electrons. Nitrogen has a total of seven electrons. These electrons occupy atomic orbitals of increasing energy, with a maximum of two electrons in each orbital:

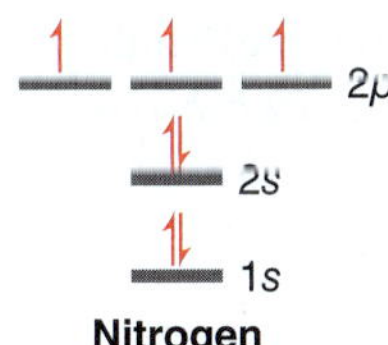


**STEP 2**
Identify the number of valence electrons in each atomic orbital.

Two electrons occupy the 1*s* orbital, two electrons occupy the 2*s* orbital, and three electrons occupy the 2*p* orbitals. This is summarized using the following notation:

$$1s^2 2s^2 2p^3$$

**PRACTICE** the skill

**1.17** Determine the electron configuration for each of the following atoms:

**(a)** Carbon **(b)** Oxygen **(c)** Boron **(d)** Fluorine **(e)** Sodium **(f)** Aluminum

**APPLY the skill**

**1.18** Determine the electron configuration for each of the following ions:

**(a)** A carbon atom with a negative charge

**(b)** A carbon atom with a positive charge

**(c)** A nitrogen atom with a positive charge

**(d)** An oxygen atom with a negative charge

need more **PRACTICE?** **Try Problem 1.44**

## 1.7 Valence Bond Theory

With the understanding that electrons occupy regions of space called orbitals, we can now turn our attention to a deeper understanding of covalent bonds. Specifically, a covalent bond is formed from the overlap of atomic orbitals. There are two commonly used theories for describing the nature of atomic orbital overlap: valence bond theory and molecular orbital (MO) theory. The valence bond approach is more simplistic in its treatment of bonds, and therefore we will begin our discussion with valence bond theory.

If we are going to treat electrons as waves, then we must quickly review what happens when two waves interact with each other. Two waves that approach each other can interfere in one of two possible ways—constructively or destructively. Similarly, when atomic orbitals overlap, they can interfere either constructively (Figure 1.11) or destructively (Figure 1.12).

An electron is like a wave — An electron is like a wave — Internuclear distance — *Bring these waves closer together...* — Internuclear distance — *...and the waves reinforce each other* — **Constructive interference**

**FIGURE 1.11** Constructive interference resulting from the interaction of two electrons.

**Constructive interference** produces a wave with larger amplitude. In contrast, destructive interference results in waves canceling each other, which produces a node (Figure 1.12).

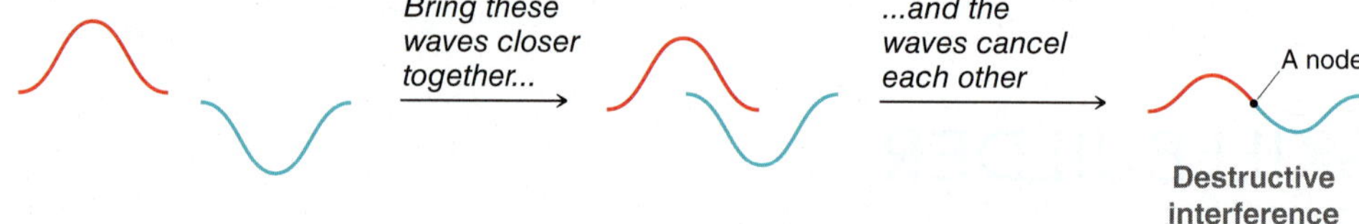


**FIGURE 1.12** Destructive interference resulting from the interaction of two electrons.

According to **valence bond theory**, a bond is simply the sharing of electron density between two atoms as a result of the constructive interference of their atomic orbitals. Consider, for example, the bond that is formed between the two hydrogen atoms in molecular hydrogen ($H_2$). This bond is formed from the overlap of the 1*s* orbitals of each hydrogen atom (Figure 1.13).

The electron density of this bond is primarily located on the bond axis (the line that can be drawn between the two hydrogen atoms). This type of bond is called a **sigma (σ) bond** and is characterized by circular symmetry with respect to the bond axis. To visualize what this means, imagine a plane that is drawn perpendicular to the bond axis. This plane will carve out a circle (Figure 1.14). This is the defining feature of σ bonds and will be true of all purely single bonds. Therefore, *all single bonds are σ bonds.*

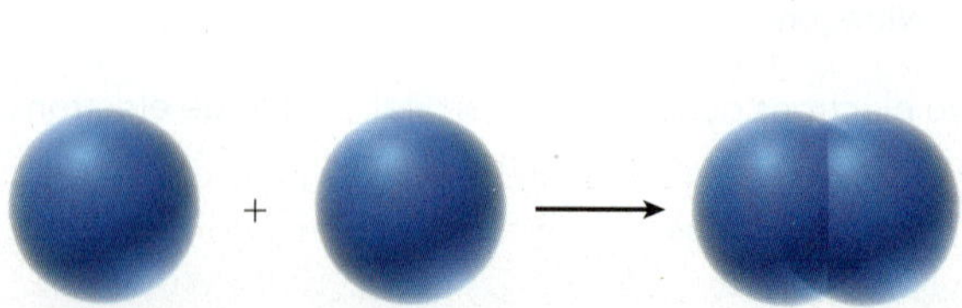

**FIGURE 1.13** The overlap of the 1s atomic orbitals of two hydrogen atoms, forming molecular hydrogen ($H_2$).

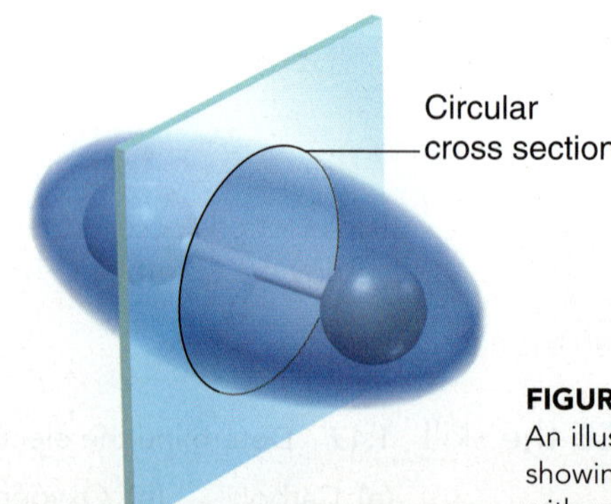


**FIGURE 1.14** An illustration of a sigma bond, showing the circular symmetry with respect to the bond axis.

## 1.8 Molecular Orbital Theory

In most situations, valence bond theory will be sufficient for our purposes. However, there will be cases in the upcoming chapters where valence bond theory will be inadequate to describe the observations. In such cases, we will utilize molecular orbital theory, a more sophisticated approach to viewing the nature of bonds.

Much like valence bond theory, **molecular orbital (MO) theory** also describes a bond in terms of the constructive interference between two overlapping atomic orbitals. However, MO theory goes one step further and uses mathematics as a tool to explore the consequences of atomic orbital overlap. The mathematical method is called the linear combination of atomic orbitals (LCAO). According to this theory, atomic orbitals are mathematically combined to produce new orbitals, called **molecular orbitals**.

It is important to understand the distinction between atomic orbitals and molecular orbitals. Both types of orbitals are used to accommodate electrons, but an atomic orbital is a region of space associated with an individual atom, while a molecular orbital is associated with an entire molecule. That is, the molecule is considered to be a single entity held together by many electron clouds, some of which can actually span the entire length of the molecule. These molecular orbitals are filled with electrons in a particular order in much the same way that atomic orbitals are filled. Specifically, electrons first occupy the lowest energy orbitals, with a maximum of two electrons per orbital. In order to visualize what it means for an orbital to be associated with an entire molecule, we will explore two molecules: molecular hydrogen ($H_2$) and bromomethane ($CH_3Br$).

Consider the bond formed between the two hydrogen atoms in molecular hydrogen. This bond is the result of the overlap of two atomic orbitals (*s* orbitals), each of which is occupied by one electron. According to MO theory, when two atomic orbitals overlap, they cease to exist. Instead, they are replaced by two molecular orbitals, each of which is associated with the entire molecule (Figure 1.15).

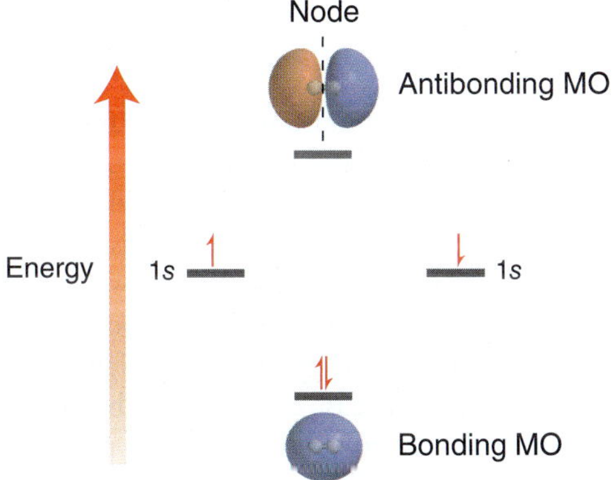


**FIGURE 1.15** An energy diagram showing the relative energy levels of bonding and antibonding molecular orbitals.

In the energy diagram shown in Figure 1.15, the individual atomic orbitals are represented on the right and left, with each atomic orbital having one electron. These atomic orbitals are combined mathematically (using the LCAO method) to produce two molecular orbitals. The lower energy molecular orbital, or **bonding MO**, is the result of constructive interference of the original two atomic orbitals. The higher energy molecular orbital, or **antibonding MO**, is the result of destructive interference. Notice that the antibonding MO has one node, which explains why it is higher in energy. Both electrons occupy the bonding MO in order to achieve a lower energy state. This lowering in energy is the essence of the bond. For an H—H bond, the lowering in energy is equivalent to 436 kJ/mol. This energy corresponds with the bond strength of an H—H bond (as shown in Figure 1.2).

Now let's consider a molecule such as $CH_3Br$, which contains more than just one bond. Valence bond theory continues to view each bond separately, with each bond being formed from two overlapping atomic orbitals. In contrast, MO theory treats the bonding electrons as being associated with the entire molecule. The molecule has many molecular orbitals, each of which can be occupied by two electrons. Figure 1.16 illustrates one of the many molecular orbitals of $CH_3Br$. This molecular orbital is capable of accommodating up to two electrons. Red and blue regions indicate the different phases, as described in Section 1.6. As we saw with molecular hydrogen, not all molecular orbitals

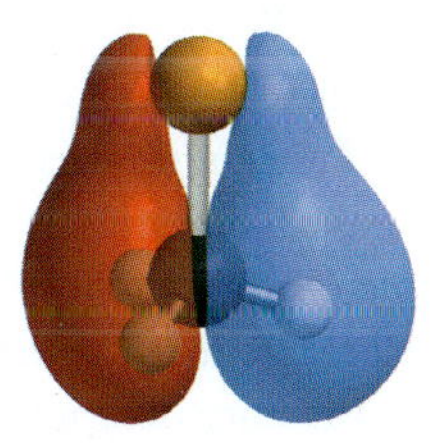

**FIGURE 1.16** A low-energy molecular orbital of $CH_3Br$. Red and blue regions indicate the different phases, as described in Section 1.6. Notice that this molecular orbital is associated with the entire molecule, rather than being associated with two specific atoms.

will be occupied. The bonding electrons will occupy the lower energy molecular orbitals (such as the one shown in Figure 1.16), while the higher energy molecular orbitals remain unoccupied. For every molecule, two of its molecular orbitals will be of particular interest: (1) the highest energy orbital from among the occupied orbitals is called the **highest occupied molecular orbital**, or **HOMO**, and (2) the lowest energy orbital from among the unoccupied orbitals is called the **lowest unoccupied molecular orbital**, or **LUMO**. For example, in Chapter 7, we will explore a reaction in which $CH_3Br$ is attacked by a hydroxide ion ($HO^-$). In order for this process to occur, the hydroxide ion must transfer its electron density into the lowest energy, empty molecular orbital, or LUMO, of $CH_3Br$ (Figure 1.17). The nature of the LUMO (i.e., number of nodes, location of nodes, etc.) will be useful in explaining the preferred direction from which the hydroxide ion will attack.

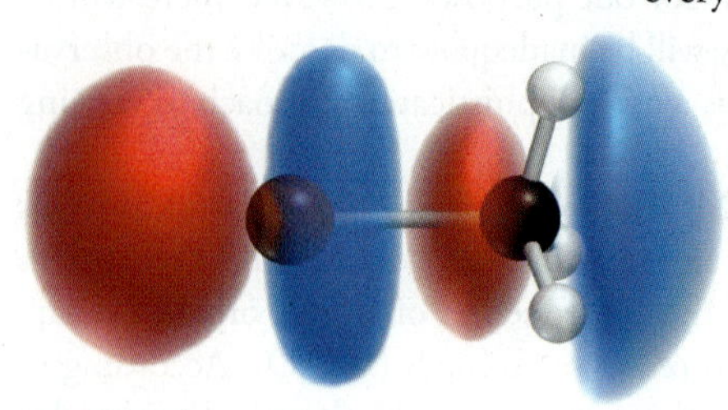

**FIGURE 1.17**
The LUMO of $CH_3Br$.

We will use MO theory several times in the chapters that follow. Most notably, in Chapter 17, we will investigate the structure of compounds containing several double bonds. For those compounds, valence bond theory will be inadequate, and MO theory will provide a more meaningful understanding of the bonding structure. Throughout this textbook, we will continue to develop both valence bond theory and MO theory.

## 1.9 Hybridized Atomic Orbitals

### Methane and $sp^3$ Hybridization

Let us now apply valence bond theory to the bonds in methane:

```
    H
    |
H—C—H
    |
    H
```

**Methane**

Recall the electron configuration of carbon (Figure 1.18). This electron configuration cannot satisfactorily describe the bonding structure of methane ($CH_4$), in which the carbon atom has four separate C—H bonds, because the electron configuration shows only two atomic orbitals capable of forming bonds (each of these orbitals has one unpaired electron). This would imply that the carbon atom will form only two bonds, but we know that it forms four bonds. We can solve this problem by imagining an excited state of carbon (Figure 1.19): a state in which a 2*s* electron has been promoted to a higher energy 2*p* orbital. Now the carbon atom has four atomic orbitals capable of forming bonds, but there is yet another problem here. The geometry of the 2*s* and three 2*p* orbitals does not satisfactorily explain the observed three-dimensional geometry of methane (Figure 1.20). All bond angles are 109.5°, and the four bonds point away from each other in a perfect tetrahedron. This geometry cannot be explained by an excited state of carbon because the *s* orbital and the three *p* orbitals do not occupy a tetrahedral geometry. The *p* orbitals are separated from each other by only 90° (as seen in Figure 1.5) rather than 109.5°.

Energy 2p 2s 1s

**FIGURE 1.18**
An energy diagram showing the electron configuration of carbon.

This problem was solved in 1931 by Linus Pauling, who suggested that the electronic configuration of the carbon atom in methane does not necessarily have to be the same as the electronic configuration of a free carbon atom. Specifically, Pauling mathematically averaged, or *hybridized*, the 2*s* orbital and the three 2*p* orbitals, giving four degenerate hybridized atomic orbitals (Figure 1.21). The hybridization process in Figure 1.21 does not represent a real physical process that the orbitals undergo. Rather, it is a mathematical procedure that is used to arrive at a satisfactory description of

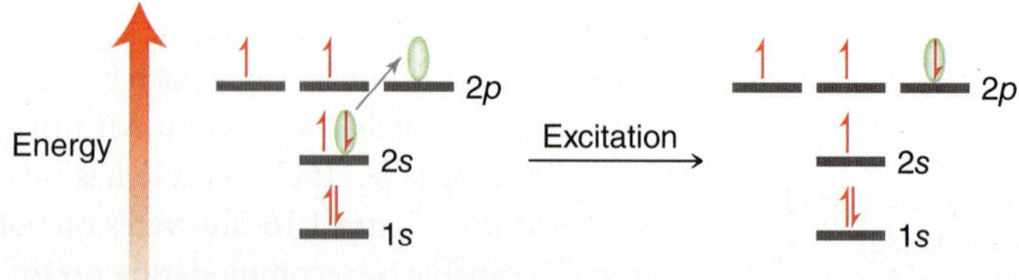


**FIGURE 1.19**
An energy diagram showing the electronic excitation of an electron in a carbon atom.

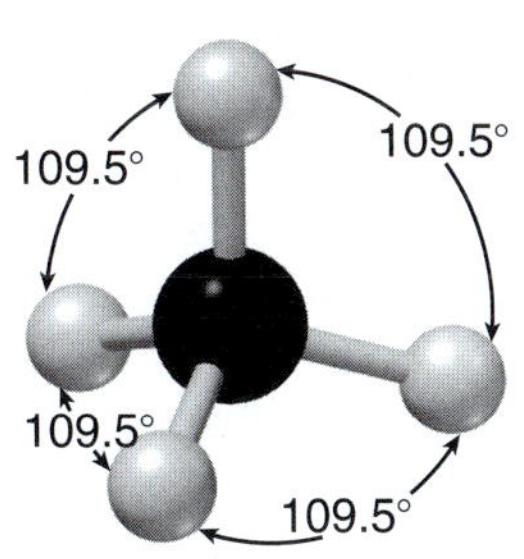


**FIGURE 1.20**
The tetrahedral geometry of methane. All bond angles are 109.5°.

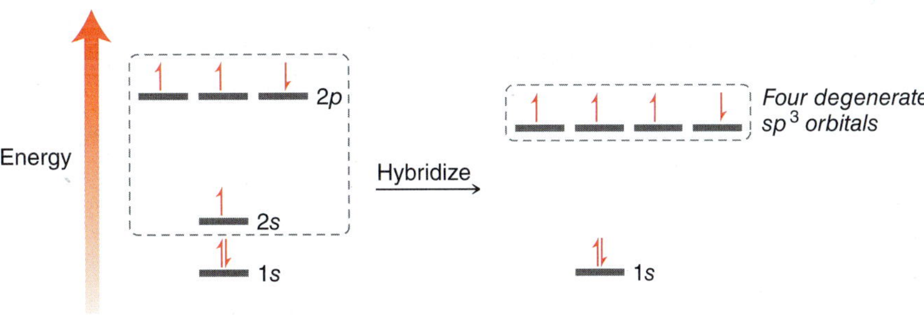


**FIGURE 1.21**
An energy diagram showing four degenerate hybridized atomic orbitals.

the observed bonding. This procedure gives us four orbitals that were produced by averaging one *s* orbital and three *p* orbitals, and therefore we refer to these atomic orbitals as ***sp*$^3$-hybridized orbitals**. Figure 1.22 shows an *sp*$^3$-hybridized orbital. If we use these hybridized atomic orbitals to describe the bonding of methane, we can successfully explain the observed geometry of the bonds. The four *sp*$^3$-hybridized orbitals are equivalent in energy (degenerate) and will therefore position themselves as far apart from each other as possible, achieving a tetrahedral geometry. Also notice that hybridized atomic orbitals are unsymmetrical. That is, hybridized atomic orbitals have a larger front lobe (shown in red in Figure 1.22) and a smaller back lobe (shown in blue). The larger front lobe enables hybridized atomic orbitals to be more efficient than *p* orbitals in their ability to form bonds.

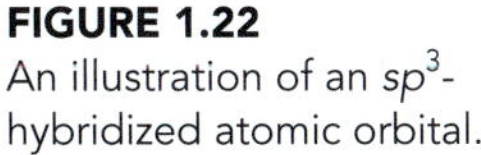

**FIGURE 1.22**
An illustration of an *sp*$^3$-hybridized atomic orbital.

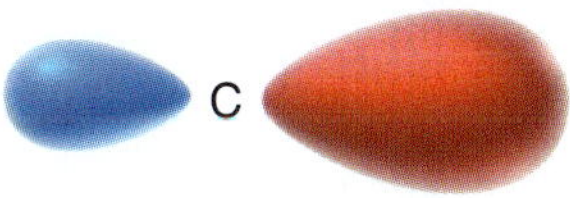


Using valence bond theory, each of the four bonds in methane is represented by the overlap between an *sp*$^3$-hybridized atomic orbital from the carbon atom and an *s* orbital from a hydrogen atom (Figure 1.23). For purposes of clarity the back lobes (blue) have been omitted from the images in Figure 1.23.

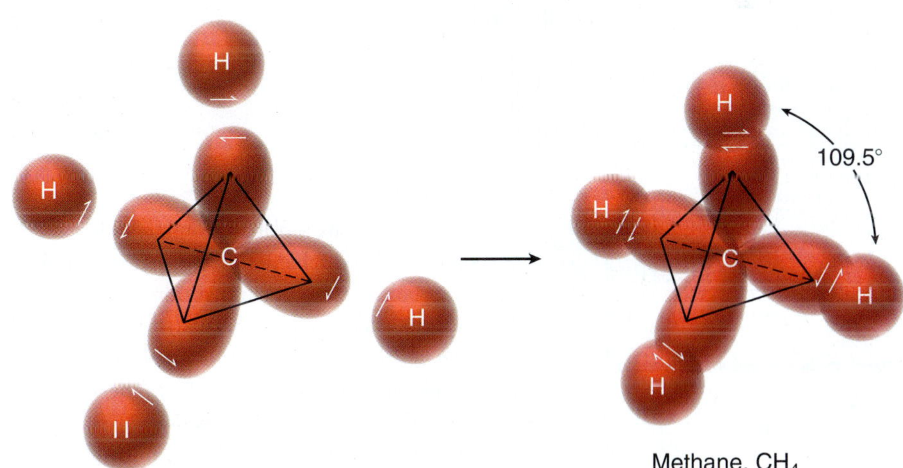


**FIGURE 1.23**
A tetrahedral carbon atom using each of its four *sp*$^3$-hybridized orbitals to form a bond.

The bonding in ethane is treated in much the same way:

```
    H   H
    |   |
H — C — C — H
    |   |
    H   H
```

**Ethane**

All bonds in this compound are single bonds, and therefore they are all σ bonds. Using the valence bond approach, each of the bonds in ethane can be treated individually and is represented by the overlap of atomic orbitals (Figure 1.24).

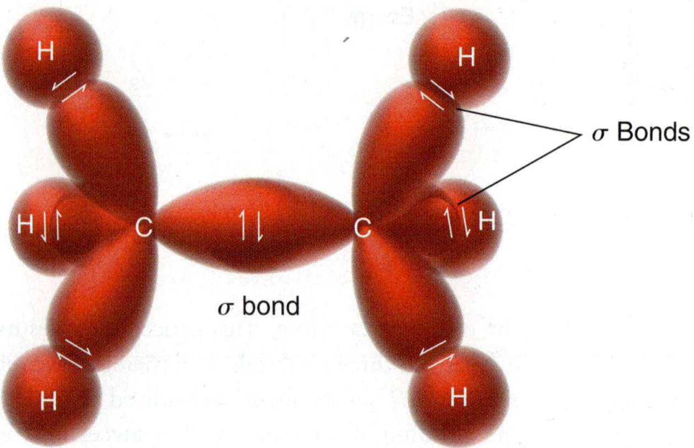


**FIGURE 1.24**
A valence bond picture of the bonding in ethane.

## CONCEPTUAL CHECKPOINT

**1.19** Cyclopropane is a compound in which the carbon atoms form a three-membered ring:

**Cyclopropane**

Each of the carbon atoms in cyclopropane is $sp^3$ hybridized. Cyclopropane is more reactive than other cyclic compounds (four-membered rings, five-membered rings, etc.). Analyze the bond angles in cyclopropane and explain why cyclopropane is so reactive.

## Double Bonds and $sp^2$ Hybridization

Now let's consider the structure of a compound bearing a double bond. The simplest example is ethylene:

**Ethylene**

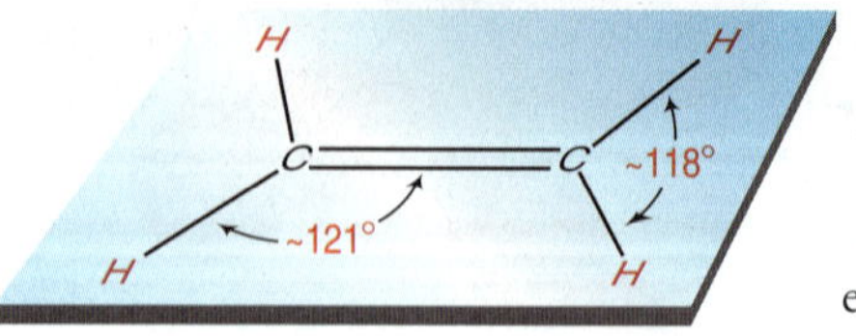


**FIGURE 1.25**
All six atoms of ethylene are in one plane.

Ethylene exhibits a planar geometry (Figure 1.25). A satisfactory model for explaining this geometry can be achieved by the mathematical maneuver of hybridizing the *s* and *p* orbitals of the carbon atom to obtain hybridized atomic orbitals. When we did this procedure earlier to explain the bonding in methane, we hybridized the *s* orbital and all three *p* orbitals to produce four equivalent $sp^3$-hybridized orbitals. However, in the case of ethylene, each carbon atom only needs to form bonds with three atoms, not four. Therefore, each carbon atom only needs three hybridized orbitals. So in this case we will mathematically average the *s* orbital with only two of the three *p* orbitals (Figure 1.26). The remaining *p* orbital will remain unaffected by our mathematical procedure.

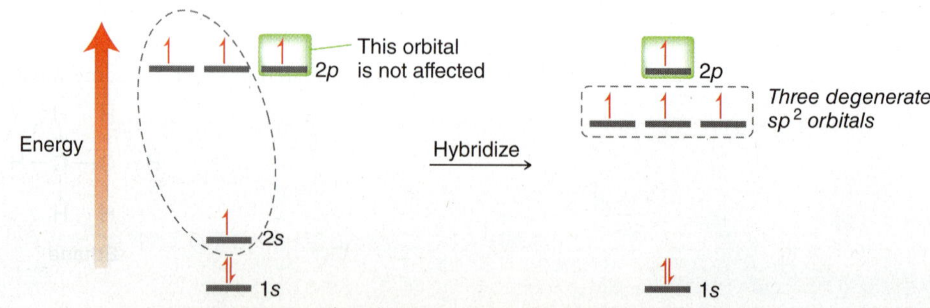


**FIGURE 1.26**
An energy diagram showing three degenerate $sp^2$-hybridized atomic orbitals.

The result of this mathematical operation is a carbon atom with one *p* orbital and three ***sp*²-hybridized orbitals** (Figure 1.27). In Figure 1.27 the *p* orbital is shown in red and blue, and the hybridized orbitals are shown in gray (for clarity, only the front lobe of each hybridized orbital is shown). They are called *sp*²-hybridized orbitals to indicate that they were obtained by averaging one *s* orbital and two *p* orbitals. As shown in Figure 1.27, each of the carbon atoms in ethylene is *sp*² hybridized, and we can use this hybridization state to explain the bonding structure of ethylene.

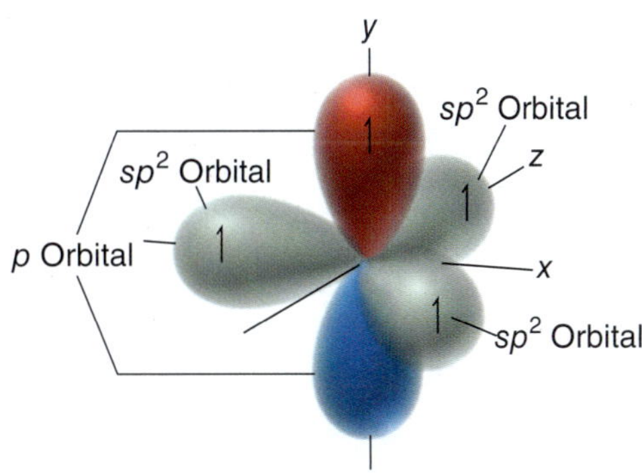


**FIGURE 1.27**
An illustration of an *sp*²-hybridized carbon atom.

Each carbon atom in ethylene has three *sp*²-hybridized orbitals available to form σ bonds (Figure 1.28). One σ bond forms between the two carbon atoms, and then each carbon atom also forms a σ bond with each of its neighboring hydrogen atoms.

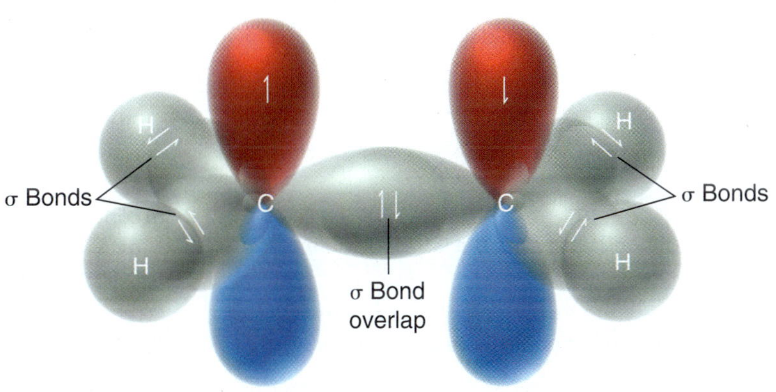


**FIGURE 1.28**
An illustration of the σ bonds in ethylene.

In addition, each carbon atom has one *p* orbital (shown in Figure 1.28 with blue and red lobes). These *p* orbitals actually overlap with each other as well, which is a separate bonding interaction called a **pi (π) bond** (Figure 1.29). Do not be confused by the nature of this type of bond. It is true that the π overlap occurs in two places above the plane of the molecule (in red) and below the plane (in blue). Nevertheless, these two regions of overlap represent only one interaction called a π bond.

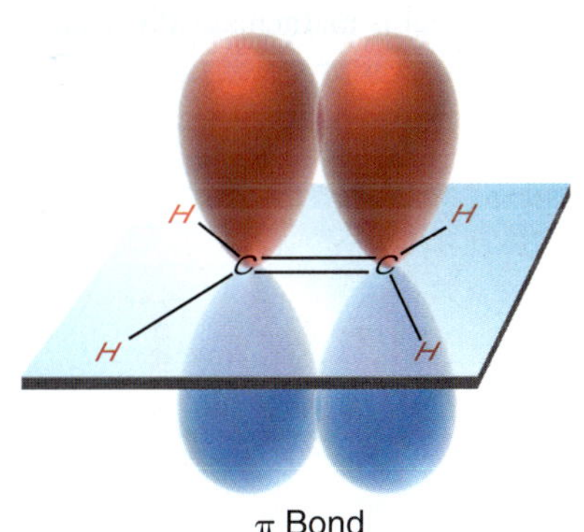


**FIGURE 1.29**
An illustration of the π bond in ethylene.

The picture of the π bond in Figure 1.29 is based on the valence bond approach (the *p* orbitals are simply drawn overlapping each other). Molecular orbital theory provides a fairly similar image of a π bond. Compare Figure 1.29 with the bonding MO in Figure 1.30.

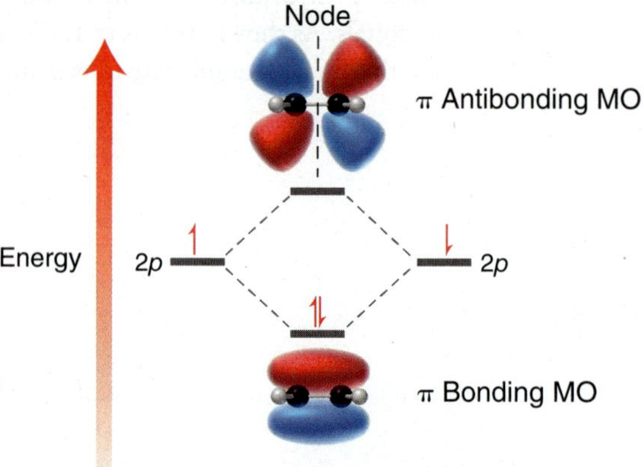


**FIGURE 1.30** An energy diagram showing images of bonding and antibonding MOs in ethylene.

To summarize, we have seen that the carbon atoms of ethylene are connected via a σ bond and a π bond. The σ bond results from the overlap of $sp^2$-hybridized atomic orbitals, while the π bond results from the overlap of *p* orbitals. These two separate bonding interactions (σ and π) comprise the double bond of ethylene.

## CONCEPTUAL CHECKPOINT

**1.20** Consider the structure of formaldehyde:

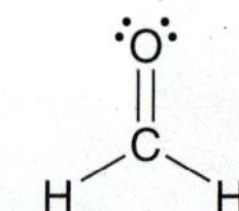


**Formaldehyde**

(a) Identify the type of bonds that form the C═O double bond.

(b) Identify the atomic orbitals that form each C—H bond.

(c) What type of atomic orbitals do the lone pairs occupy?

**1.21** Sigma bonds experience free rotation at room temperature:

In contrast, π bonds do not experience free rotation. Explain. (**Hint:** Compare Figures 1.24 and 1.29, focusing on the orbitals used in forming a σ bond and the orbitals used in forming a π bond. In each case, what happens to the orbital overlap during bond rotation?)

## Triple Bonds and *sp* Hybridization

Now let's consider the bonding structure of a compound bearing a triple bond, such as acetylene:

H—C≡C—H

**Acetylene**

A triple bond is formed by ***sp*-hybridized** carbon atoms. To achieve *sp* hybridization, one *s* orbital is mathematically averaged with only one *p* orbital (Figure 1.31). This leaves two *p* orbitals

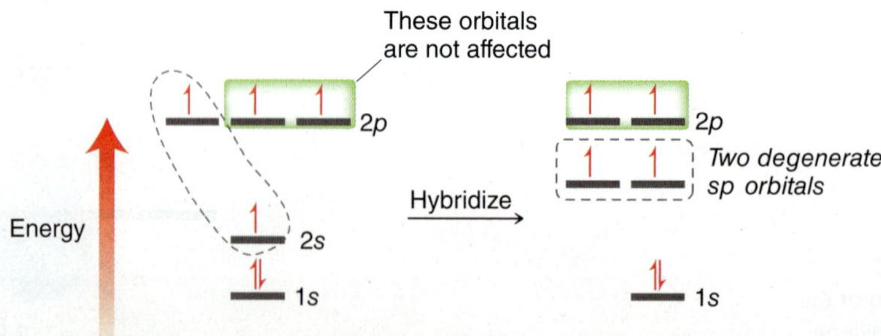


**FIGURE 1.31** An energy diagram showing two degenerate *sp*-hybridized atomic orbitals.

unaffected by the mathematical operation. As a result, an *sp*-hybridized carbon atom has two *sp* orbitals and two *p* orbitals (Figure 1.32).

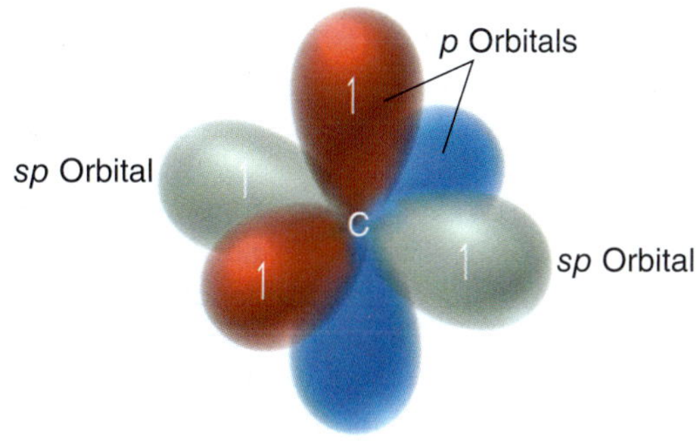


**FIGURE 1.32**
An illustration of an *sp*-hybridized carbon atom. The *sp*-hybridized orbitals are shown in gray.

The two *sp*-hybridized orbitals are available to form σ bonds (one on either side), and the two *p* orbitals are available to form π bonds, giving the bonding structure for acetylene shown in Figure 1.33. A triple bond between two carbon atoms is therefore the result of three separate bonding interactions: one σ bond and two π bonds. The σ bond results from the overlap of *sp* orbitals, while each of the two π bonds result from overlapping *p* orbitals. As shown in Figure 1.33, the geometry of the triple bond is linear.

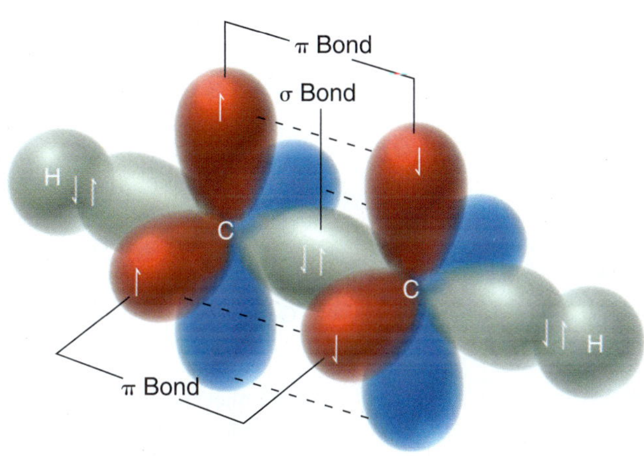


**FIGURE 1.33**
An illustration of the σ bonds and π bonds in acetylene.

## SKILLBUILDER

### 1.7 IDENTIFYING HYBRIDIZATION STATES

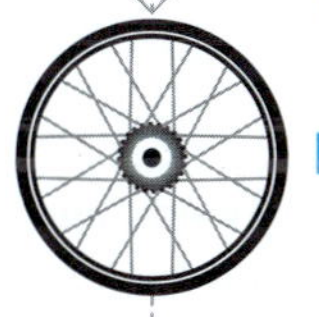

**LEARN the skill** Identify the hybridization state of each carbon atom in the following compound:

**SOLUTION**

To determine the hybridization state of an uncharged carbon atom, simply count the number of σ bonds and π bonds:

*sp*³ *sp*² *sp*

A carbon atom with four single bonds (four σ bonds) will be *sp*³ hybridized. A carbon atom with three σ bonds and one π bond will be *sp*² hybridized. A carbon atom with two σ bonds and two π bonds will be *sp* hybridized. Carbon atoms bearing a positive or negative charge will be discussed in more detail in the upcoming chapter.

Using the simple scheme above, the hybridization state of most carbon atoms can be determined instantly:

**PRACTICE the skill** **1.22** Below are the structures of two common over-the-counter pain relievers. Determine the hybridization state of each carbon atom in these compounds:

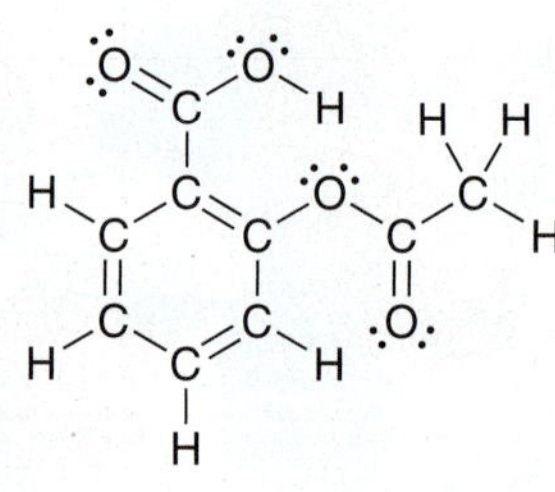

(a) **Acetylsalicylic acid** (aspirin)

(b) **Ibuprofen** (Advil or Motrin)

**APPLY the skill** **1.23** Determine the hybridization state of each carbon atom in the following compounds:

(a) (b)

need more **PRACTICE?** **Try Problems 1.55, 1.56, 1.58**

## Bond Strength and Bond Length

The information we have seen in this section allows us to compare single bonds, double bonds, and triple bonds. A single bond has only one bonding interaction (a σ bond), a double bond has two bonding interactions (one σ bond and one π bond), and a triple bond has three bonding interactions (one σ bond and two π bonds). Therefore, it is not surprising that a triple bond is stronger than a double bond, which in turn is stronger than a single bond. Compare the strengths and lengths of the C—C bonds in ethane, ethylene, and acetylene (Table 1.2).

**TABLE 1.2 COMPARISON OF BOND LENGTHS AND BOND ENERGIES FOR ETHANE, ETHYLENE, AND ACETYLENE**

| | ETHANE | ETHYLENE | ACETYLENE |
|---|---|---|---|
| Structure | 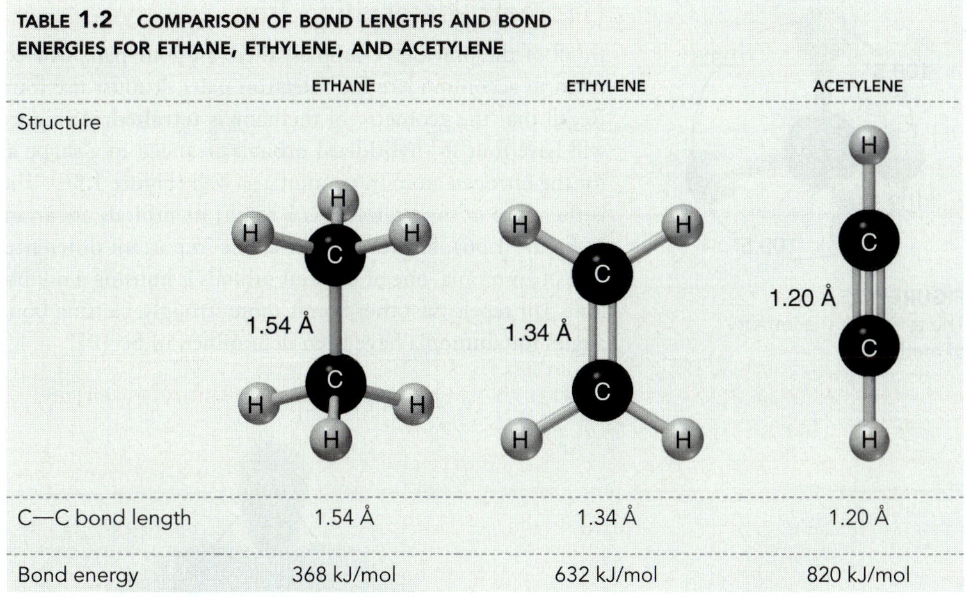 | | |
| C—C bond length | 1.54 Å | 1.34 Å | 1.20 Å |
| Bond energy | 368 kJ/mol | 632 kJ/mol | 820 kJ/mol |

## CONCEPTUAL CHECKPOINT

**1.24** Rank the indicated bonds in terms of increasing bond length:

## 1.10 VSEPR Theory: Predicting Geometry

In order to predict the geometry of a small compound, we focus on the central atom and count the number of σ bonds and lone pairs. The total (σ bonds plus lone pairs) is called the **steric number**. Figure 1.34 gives several examples in which the steric number is 4 in each case.

| Methane | Ammonia | Water |
|---|---|---|
| # of σ bonds = 4 | # of σ bonds = 3 | # of σ bonds = 2 |
| # of lone pairs = 0 | # of lone pairs = 1 | # of lone pairs = 2 |
| Steric number = 4 | Steric number = 4 | Steric number = 4 |

**FIGURE 1.34** Calculation of the steric number for methane, ammonia, and water.

The steric number indicates the number of electron pairs (bonding and nonbonding) that are repelling each other. The repulsion causes the electron pairs to arrange themselves in three-dimensional space so as to achieve maximal distance from each other. As a result, the geometry of the central atom will be determined by the steric number. This principle is called the ***v**alence **s**hell **e**lectron **p**air **r**epulsion* (**VSEPR**) theory. Let's take a closer look at the geometry of each of the compounds above.

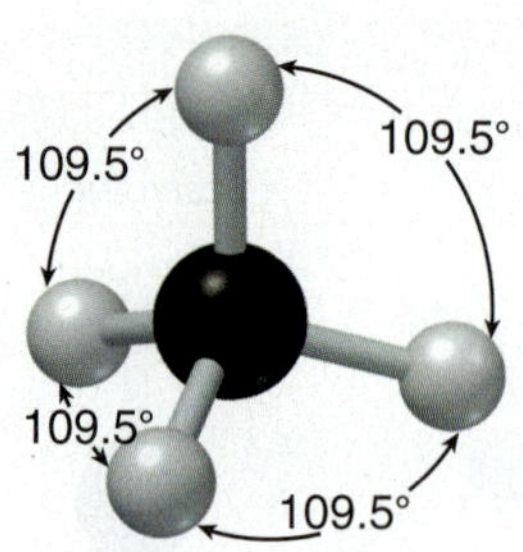


**FIGURE 1.35**
The tetrahedral geometry of methane.

## Geometries Resulting from $sp^3$ Hybridization

In all of the previous examples, there are four pairs of electrons (steric number 4). In order for an atom to accommodate four electron pairs, it must use four orbitals and is therefore $sp^3$ hybridized. Recall that the geometry of methane is **tetrahedral** (Figure 1.35). In fact, any $sp^3$-hybridized atom will have four $sp^3$-hybridized orbitals arranged in a shape approximating a tetrahedron. This is true for the nitrogen atom in ammonia as well (Figure 1.36). The nitrogen atom is using four orbitals and is therefore $sp^3$ hybridized. As a result, its orbitals are arranged in a tetrahedron (shown on the left in Figure 1.36). However, there is one important difference between ammonia and methane. In the case of ammonia, one of the four orbitals is housing a nonbonding pair of electrons (a lone pair). This lone pair repels the other bonds more strongly, causing bond angles to be smaller than 109.5°. Bond angles for ammonia have been determined to be 107°.

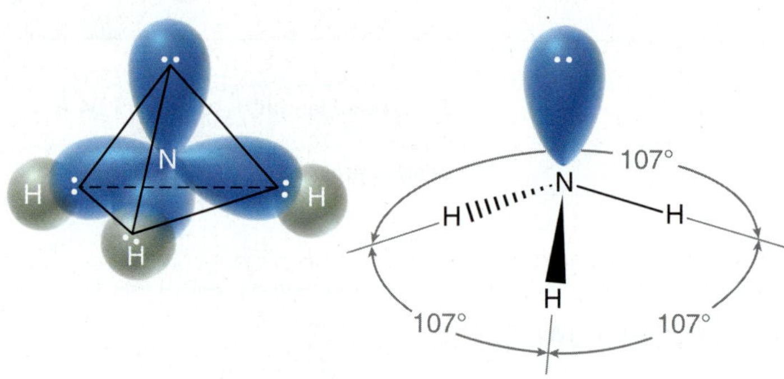


**FIGURE 1.36**
The orbitals of ammonia are arranged in a tetrahedral geometry.

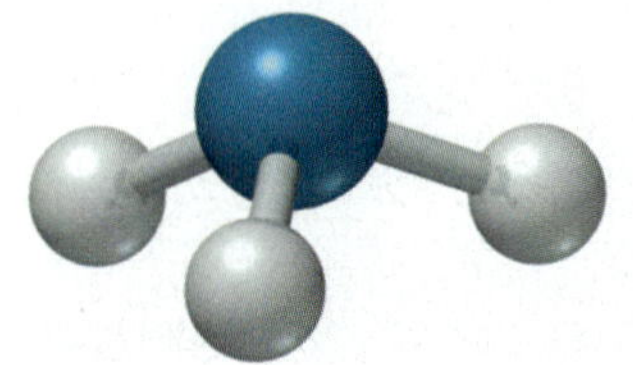

**FIGURE 1.37**
The geometry of ammonia is trigonal pyramidal.

The term "geometry" does not refer to the arrangement of electron pairs. Rather, it refers to the arrangement of atoms. When we show only the positions of the atoms (ignoring the lone pairs), ammonia appears as in Figure 1.37. This geometry is called **trigonal pyramidal** (Figure 1.37). "Trigonal" indicates the nitrogen atom is connected to three other atoms, and "pyramidal" indicates the compound is shaped like a pyramid, with the nitrogen atom sitting at the top of the pyramid.

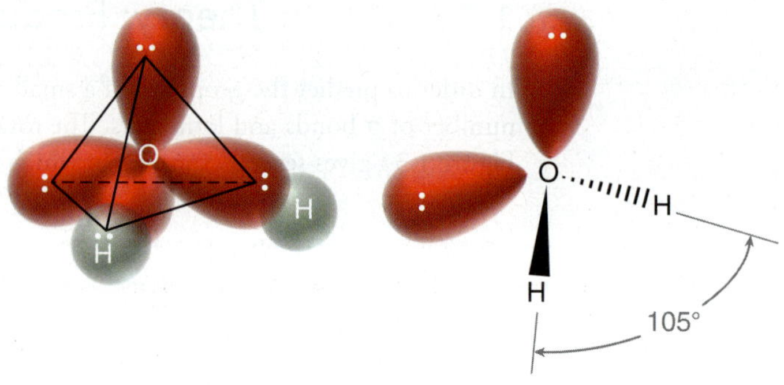


**FIGURE 1.38**
The orbitals of $H_2O$ are arranged in a tetrahedral geometry.

Another example of $sp^3$ hybridization is water ($H_2O$). The oxygen atom has a steric number of 4, and it therefore requires the use of four orbitals. As a result, it must be $sp^3$ hybridized, with its four orbitals in a tetrahedral arrangement (Figure 1.38). Once again, lone pairs repel each other more strongly than bonds, causing the bond angle between the two O—H bonds to be even smaller than the bond angles in ammonia. The bond angle of water has been determined to be 105°. In order to describe the geometry, we ignore the lone pairs and focus only on the arrangement of atoms, which gives a **bent** geometry in this case (Figure 1.39). In summary, there are only three different types of

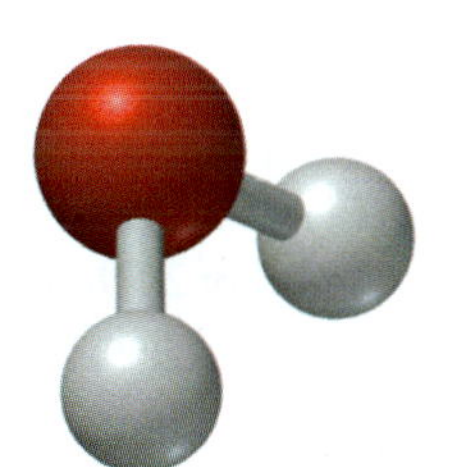

**FIGURE 1.39**
The geometry of water is bent.

geometry arising from $sp^3$ hybridization: tetrahedral, trigonal pyramidal, and bent. In all cases, the electrons were arranged in a tetrahedron, but the lone pairs were ignored when describing geometry. Table 1.3 summarizes this information.

**TABLE 1.3 GEOMETRIES RESULTING FROM $sp^3$ HYBRIDIZATION**

| EXAMPLE | STERIC NUMBER | HYBRIDIZATION | ARRANGEMENT OF ELECTRON PAIRS | ARRANGEMENT OF ATOMS (GEOMETRY) |
|---|---|---|---|---|
| $CH_4$ | 4 | $sp^3$ | Tetrahedral | Tetrahedral |
| $NH_3$ | 4 | $sp^3$ | Tetrahedral | Trigonal pyramidal |
| $H_2O$ | 4 | $sp^3$ | Tetrahedral | Bent |

## Geometries Resulting from $sp^2$ Hybridization

When the central atom of a small compound has a steric number of 3, it will be $sp^2$ hybridized. As an example, consider the structure of $BF_3$. Boron has three valence electrons, each of which is used to form a bond. The result is three bonds and no lone pairs, giving a steric number of 3. The central boron atom therefore requires three orbitals, rather than four, and must be $sp^2$ hybridized. Recall that $sp^2$-hybridized orbitals achieve maximal separation in a **trigonal planar** arrangement (Figure 1.40): "trigonal" because the boron is connected to three other atoms and "planar" because all atoms are found in the same plane (as opposed to trigonal pyramidal).

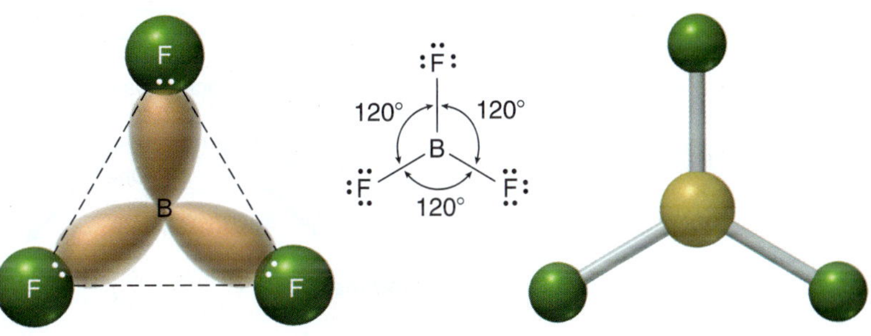


**FIGURE 1.40**
The geometry of $BF_3$ is trigonal planar.

As another example, consider the nitrogen atom of an imine:

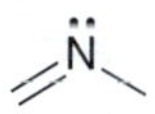

**An imine**

**LOOKING AHEAD**
We will explore imines in more detail in Chapter 20.

To determine the geometry of the nitrogen atom, we first consider the steric number, which is not affected by the presence of the $\pi$ bond. Why not? Recall that a $\pi$ bond results from the overlap of $p$ orbitals. The steric number of an atom is meant to indicate how many hybridized orbitals are necessary ($p$ orbitals are not included in this count). The steric number in this case is 3 (Figure 1.41). As a result, the nitrogen atom must be $sp^2$ hybridized. The $sp^2$ hybridization state is always characterized by a trigonal planar arrangement of electron pairs, but when describing geometry, we focus only on the atoms (ignoring any lone pairs). The geometry of this nitrogen atom is therefore bent:

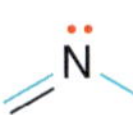

# of $\sigma$ bonds = 2
# of lone pairs = 1
Steric number = 3

**FIGURE 1.41**
The steric number of the nitrogen atom of an imine.

## Geometry Resulting from *sp* Hybridization

When the central atom of a small compound has a steric number of 2, the central atom will be *sp* hybridized. As an example, consider the structure of $BeH_2$. Beryllium has two valence electrons, each of which is used to form a bond. The result is two bonds and no lone pairs, giving a steric number of 2. The central beryllium atom therefore requires only two orbitals and must be *sp* hybridized. Recall that *sp*-hybridized orbitals achieve maximal separation when they are linear (Figure 1.42).

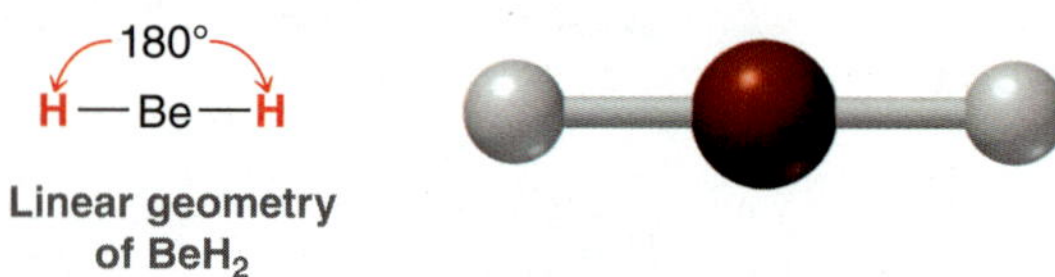


**FIGURE 1.42**
The geometry of $BeH_2$ is linear.

As another example of *sp* hybridization, consider the structure of $CO_2$:

$$:\ddot{O}=C=\ddot{O}:$$

Once again, the π bonds do *not* impact the calculation of the steric number, so the steric number is 2. The carbon atom must be *sp* hybridized and is therefore linear.

As summarized in Figure 1.43, the three hybridization states give rise to five common geometries.

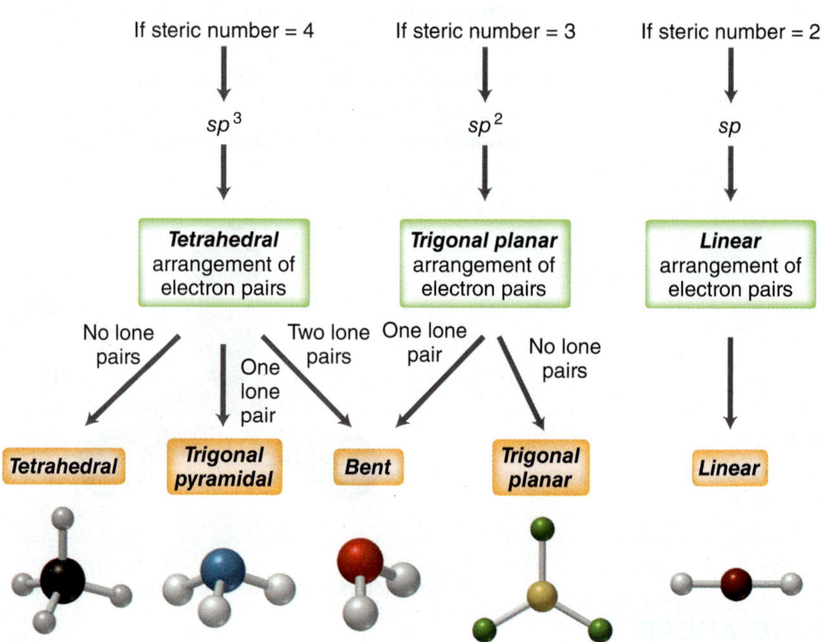


**FIGURE 1.43**
A decision tree for determining geometry.

# SKILLBUILDER

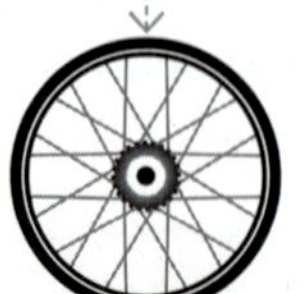

### 1.8 PREDICTING GEOMETRY

**LEARN** the skill

Predict the geometry for all atoms (except hydrogen) in the compound below:

## SOLUTION

For each atom, the following three steps are followed:

**STEP 1**
Determine the steric number.

1. Determine the steric number by counting the number of lone pairs and σ bonds.

**STEP 2**
Determine the hybridization state and electronic arrangement.

2. Use the steric number to determine the hybridization state and electronic arrangement:
   - If the steric number is 4, then the atom will be $sp^3$ hybridized, and the electronic arrangement will be tetrahedral.
   - If the steric number is 3, then the atom will be $sp^2$ hybridized, and the electronic arrangement will be trigonal planar.
   - If the steric number is 2, then the atom will be *sp* hybridized, and the electronic arrangement will be linear.

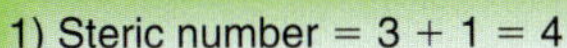

**STEP 3**
Ignore lone pairs and describe the resulting geometry.

3. Ignore any lone pairs and describe the geometry only in terms of the arrangement of atoms:

1) Steric number = 3 + 1 = 4
2) 4 = $sp^3$ = electronically tetrahedral
3) Arrangement of atoms = trigonal pyramidal

1) Steric number = 4 + 0 = 4
2) 4 = $sp^3$ = electronically tetrahedral
3) Arrangement of atoms = tetrahedral

1) Steric number = 2 + 2 = 4
2) 4 = $sp^3$ = electronically tetrahedral
3) Arrangement of atoms = bent

1) Steric number = 3 + 0 = 3
2) 3 = $sp^2$ = electronically trigonal planar
3) Arrangement of atoms = trigonal planar

1) Steric number = 4 + 0 = 4
2) 4 = $sp^3$ = electronically tetrahedral
3) Arrangement of atoms = tetrahedral

1) Steric number = 3 + 0 = 3
2) 3 = $sp^2$ = electronically trigonal planar
3) Arrangement of atoms = trigonal planar

It is not necessary to describe the geometry of hydrogen atoms. Each hydrogen atom is monovalent, so the geometry is irrelevant. Geometry is only relevant when an atom is connected to at least two other atoms. For our purposes, we can also disregard the geometry of the oxygen atom in a C═O double bond because it is connected to only one atom:

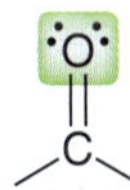

Can disregard the geometry of this oxygen

**PRACTICE the skill** **1.25** Predict the geometry for the central atom in each of the compounds below:

**(a)** $NH_3$ **(b)** $H_3O^+$ **(c)** $BH_4^-$ **(d)** $BCl_3$ **(e)** $BCl_4^-$ **(f)** $CCl_4$ **(g)** $CHCl_3$ **(h)** $CH_2Cl_2$

**1.26** Predict the geometry for all atoms except hydrogen in the compounds below:

(a) (b) (c)

**APPLY the skill** **1.27** Compare the structures of a carbocation and a carbanion:

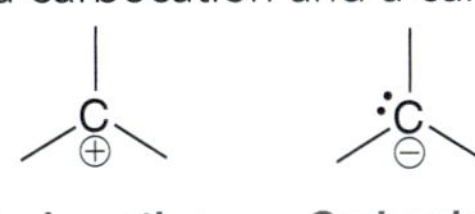

In one of these ions, the central carbon atom is trigonal planar; in the other it is trigonal pyramidal. Assign the correct geometry to each ion.

**1.28** Identify the hybridization state and geometry of each carbon atom in benzene. Use that information to determine the geometry of the entire molecule:

**Benzene**

need more **PRACTICE?** **Try Problems 1.39–1.41, 1.50, 1.55, 1.56, 1.58**

## 1.11 Dipole Moments and Molecular Polarity

Recall that induction is caused by the presence of an electronegative atom, as we saw earlier in the case of chloromethane. In Figure 1.44a the arrow shows the inductive effect of the chlorine atom. Figure 1.44b is a map of the electron density, revealing that the molecule is polarized. Chloromethane is said to exhibit a dipole moment, because *the center of negative charge and the center of positive charge are separated from one another by a certain distance*. The **dipole moment** ($\mu$) is used as an indicator of polarity, where $\mu$ is defined as the amount of partial charge ($\delta$) on either end of the dipole multiplied by the distance of separation ($d$):

$$\mu = \delta \times d$$

Partial charges ($\delta+$ and $\delta-$) are generally on the order of $10^{-10}$ esu (electrostatic units) and the distances are generally on the order of $10^{-8}$ cm. Therefore, for a polar compound, the dipole moment ($\mu$) will generally have an order of magnitude of around $10^{-18}$ esu · cm. The dipole moment of chloromethane, for example, is $1.87 \times 10^{-18}$ esu · cm. Since most compounds will have a dipole moment on this order of magnitude ($10^{-18}$), it is more convenient to report dipole moments with a new unit, called a **debye (D)**, where

$$1 \text{ debye} = 10^{-18} \text{ esu} \cdot \text{cm}$$

Using these units, the dipole moment of chloromethane is reported as 1.87 D. The debye unit is named after Dutch scientist Peter Debye, whose contributions to the fields of chemistry and physics earned him a Nobel Prize in 1936.

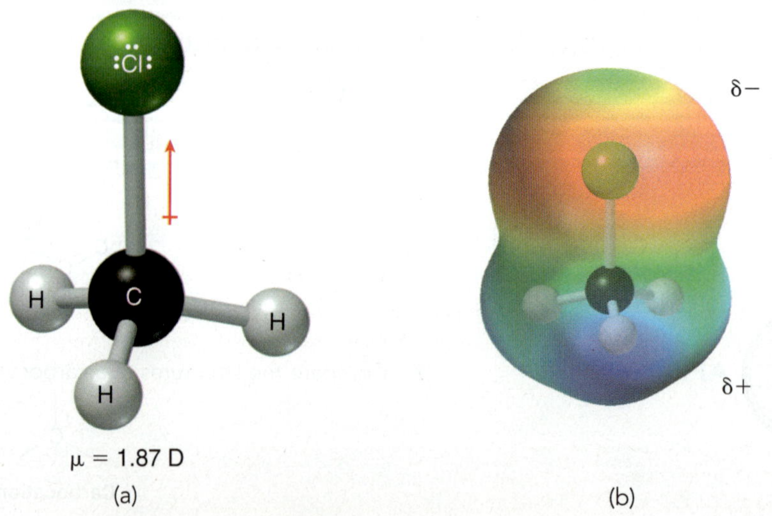


**FIGURE 1.44**
(a) Ball-and-stick model of chloromethane showing the dipole moment.
(b) An electrostatic potential map of chloromethane.

Measuring the dipole moment of a particular bond allows us to calculate the percent ionic character of that bond. As an example, let's analyze a C—Cl bond. This bond has a bond length of $1.772 \times 10^{-8}$ cm, and an electron has a charge of $4.80 \times 10^{-10}$ esu. If the bond were 100% ionic, then the dipole moment would be

$$\begin{aligned} \mu &= e \times d \\ &= (4.80 \times 10^{-10}\ \text{esu}) \times (1.772 \times 10^{-8}\ \text{cm}) \\ &= 8.51 \times 10^{-18}\ \text{esu} \cdot \text{cm} \end{aligned}$$

or 8.51 D. In reality, the bond is not 100% ionic. The experimentally observed dipole moment is measured at 1.87 D, and we can use this value to calculate the percent ionic character of a C—Cl bond:

$$\frac{1.87\ \text{D}}{8.51\ \text{D}} \times 100\% = \boxed{22\%}$$

Table 1.4 shows the percent ionic character for a few of the bonds that we will frequently encounter in this text. Take special notice of the C=O bond. It has considerable ionic character, rendering it extremely reactive. Chapters 20–22 are devoted exclusively to the reactivity of compounds containing C=O bonds.

**TABLE 1.4 PERCENT IONIC CHARACTER FOR SEVERAL BONDS**

| BOND | BOND LENGTH ($\times 10^{-8}$ CM) | OBSERVED $\mu$ (D) | PERCENT IONIC CHARACTER |
|---|---|---|---|
| C—O | 1.41 | 0.7 D | $\frac{(0.7 \times 10^{-18}\ \text{esu} \cdot \text{cm})}{(4.80 \times 10^{-10}\ \text{esu})(1.41 \times 10^{-8}\ \text{cm})} \times 100\% = \boxed{10\%}$ |
| O—H | 0.96 | 1.5 D | $\frac{(1.5 \times 10^{-18}\ \text{esu} \cdot \text{cm})}{(4.80 \times 10^{-10}\ \text{esu})(0.96 \times 10^{-8}\ \text{cm})} \times 100\% = \boxed{33\%}$ |
| C=O | 1.227 | 2.4 D | $\frac{(2.4 \times 10^{-18}\ \text{esu} \cdot \text{cm})}{(4.80 \times 10^{-10}\ \text{esu})(1.23 \times 10^{-8}\ \text{cm})} \times 100\% = \boxed{41\%}$ |

Chloromethane was a simple example, because it has only one polar bond. When dealing with a compound that has more than one polar bond, it is necessary to take the vector sum of the individual dipole moments. The vector sum is called the **molecular dipole moment**, and it takes into account both the magnitude and the direction of each individual dipole moment. For example, consider the structure of dichloromethane (Figure 1.45). The individual dipole moments partially cancel, but not completely. The vector sum produces a dipole moment of 1.14 D, which is significantly smaller than the dipole moment of chloromethane because the two dipole moments here partially cancel each other.

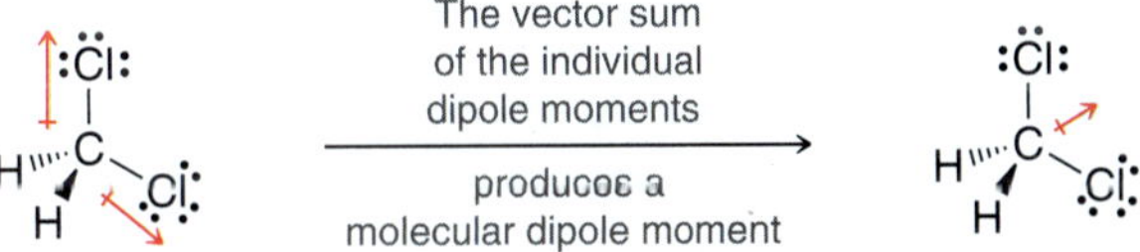


**FIGURE 1.45**
The molecular dipole moment of dichloromethane is the net sum of all dipole moments in the compound.

The presence of a lone pair has a significant effect on the molecular dipole moment. The two electrons of a lone pair are balanced by two positive charges in the nucleus, but the lone pair is separated from the nucleus by some distance. There is, therefore, a dipole moment associated with every lone pair. Common examples are ammonia and water (Figure 1.46).

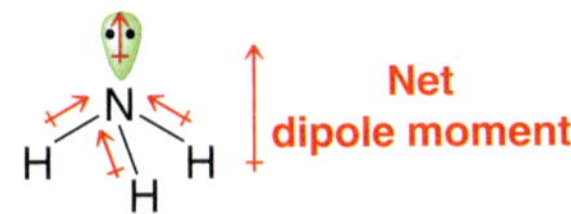


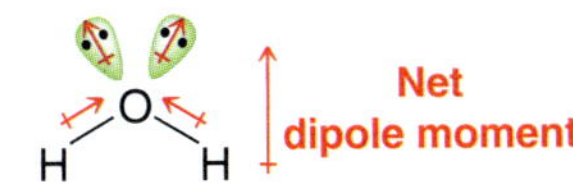


**FIGURE 1.46**
The net dipole moments of ammonia and water.

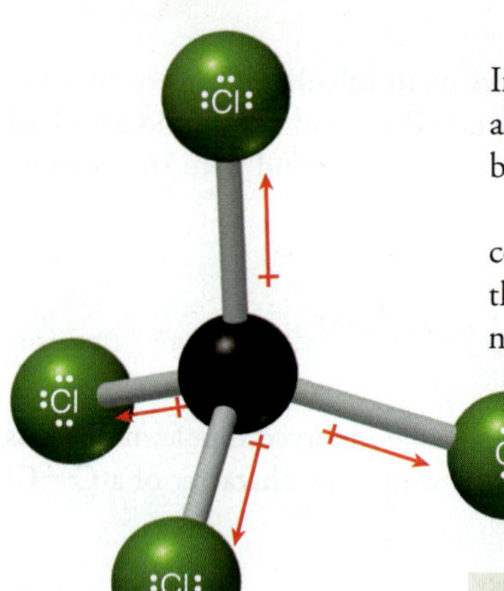


**FIGURE 1.47**
A ball-and-stick model of carbon tetrachloride. The individual dipole moments cancel to give a zero net dipole moment.

In this way, the lone pairs contribute significantly to the magnitude of the molecular dipole moment, although they do not alter its direction. That is, the direction of the molecular dipole moment would be the same with or without the contribution of the lone pairs.

Table 1.5 shows experimentally observed molecular dipole moments (at 20°C) for several common solvents. Notice that carbon tetrachloride ($CCl_4$) has no molecular dipole moment. In this case, the individual dipole moments cancel each other completely to give the molecule a zero net dipole moment ($\mu = 0$). This example (Figure 1.47) demonstrates that we must take geometry into account when assessing molecular dipole moments.

**TABLE 1.5 DIPOLE MOMENTS FOR SOME COMMON SOLVENTS (AT 20°C)**

| COMPOUND | STRUCTURE | DIPOLE MOMENT | COMPOUND | STRUCTURE | DIPOLE MOMENT |
|---|---|---|---|---|---|
| Acetone | $H_3C-C(=\ddot{O}:)-CH_3$ | 2.69 D | Ammonia | $:NH_3$ | 1.47 D |
| Chloromethane | $CH_3Cl$ | 1.87 D | Diethyl ether | $H-CH_2-CH_2-\ddot{O}-CH_2-CH_2-H$ | 1.15 D |
| Water | $H_2O$ | 1.85 D | Methylene chloride | $CH_2Cl_2$ | 1.14 D |
| Methanol | $CH_3OH$ | 1.69 D | Pentane | $H-CH_2-CH_2-CH_2-CH_2-CH_2-H$ | 0 D |
| Ethanol | $H-CH_2-CH_2-\ddot{O}-H$ | 1.66 D | Carbon tetrachloride | $CCl_4$ | 0 D |

## SKILLBUILDER

### 1.9 IDENTIFYING THE PRESENCE OF MOLECULAR DIPOLE MOMENTS

**LEARN the skill**

Identify whether each of the following compounds exhibits a molecular dipole moment. If so, indicate the direction of the net molecular dipole moment:

**(a)** $CH_3CH_2OCH_2CH_3$ **(b)** $CO_2$

**SOLUTION**

**(a)** In order to determine whether the individual dipole moments cancel each other completely, we must first predict the molecular geometry. Specifically, we need to know if the geometry around the oxygen atom is linear or bent:

**STEP 1**
Predict the molecular geometry.

$H_3C-CH_2-\ddot{O}-CH_2-CH_3$

**Linear**

$H_3C-CH_2-\ddot{O}-CH_2-CH_3$ (bent at O)

**Bent**

To make this determination, we use the three-step method from the previous section:

1. The steric number of the oxygen atom is 4.
2. Therefore, the hybridization state must be $sp^3$, and the arrangement of electron pairs must be tetrahedral.
3. Ignore the lone pairs, and the oxygen atom has a bent geometry.

After determining the molecular geometry, now draw all dipole moments and determine whether they cancel each other. In this case, they do not fully cancel each other:

**STEP 2** Identify the direction of all dipole moments.

**STEP 3** Draw the net dipole moment.

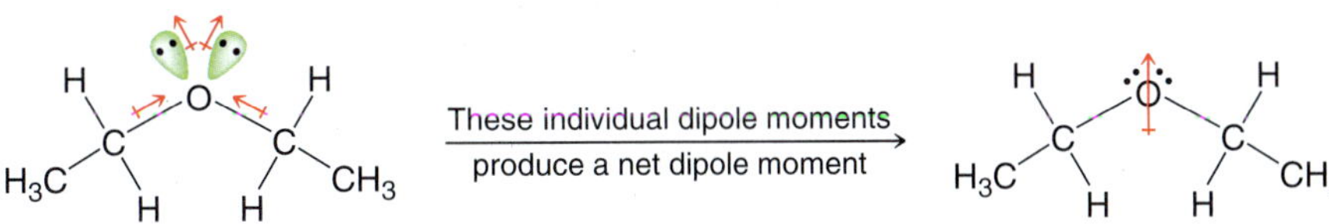


This compound does in fact have a net molecular dipole moment, and the direction of the moment is shown above.

**(b)** Carbon dioxide ($CO_2$) has two C═O bonds, each of which exhibits a dipole moment. In order to determine whether the individual dipole moments cancel each other completely, we must first predict the molecular geometry. We apply our three-step method: The steric number is 2, the hybridization state is *sp*, and the compound has a linear geometry. As a result, we expect the dipole moments to fully cancel each other:

:O═C═O:

In a similar way, the dipole moments associated with the lone pairs also cancel each other, and therefore $CO_2$ does not have a net molecular dipole moment.

**PRACTICE the skill** **1.29** Identify whether each of the following compounds exhibits a molecular dipole moment. For compounds that do, indicate the direction of the net molecular dipole moment:

**(a)** $CHCl_3$ **(b)** $CH_3OCH_3$ **(c)** $NH_3$ **(d)** $CCl_2Br_2$

**(e)** **(f)** **(g)** **(h)**

**(i)** **(j)** **(k)** **(l)**

**APPLY the skill** **1.30** Which of the following compounds has the larger dipole moment? Explain your choice:

$CHCl_3$ or $CBrCl_3$

**1.31** Bonds between carbon and oxygen (C—O) are more polar than bonds between sulfur and oxygen (S—O). Nevertheless, sulfur dioxide ($SO_2$) exhibits a dipole moment while carbon dioxide ($CO_2$) does not. Explain this apparent anomaly.

need more **PRACTICE?** **Try Problems 1.37, 1.40, 1.43, 1.61, 1.62**

## 1.12 Intermolecular Forces and Physical Properties

The physical properties of a compound are determined by the attractive forces between the individual molecules, called **intermolecular forces**. It is often difficult to use the molecular structure alone to predict a precise melting point or boiling point for a compound. However, a few simple trends will allow us to compare compounds to each other in a relative way, for example, to predict which compound will boil at a higher temperature.

All intermolecular forces are *electrostatic*—that is, these forces occur as a result of the attraction between opposite charges. The electrostatic interactions for neutral molecules (with no formal charges) are often classified as (1) **dipole-dipole interactions**, (2) hydrogen bonding, and (3) fleeting dipole-dipole interactions.

### Dipole-Dipole Interactions

Compounds with net dipole moments can either attract each other or repel each other, depending on how they approach each other in space. In the solid phase, the molecules align so as to attract each other (Figure 1.48).

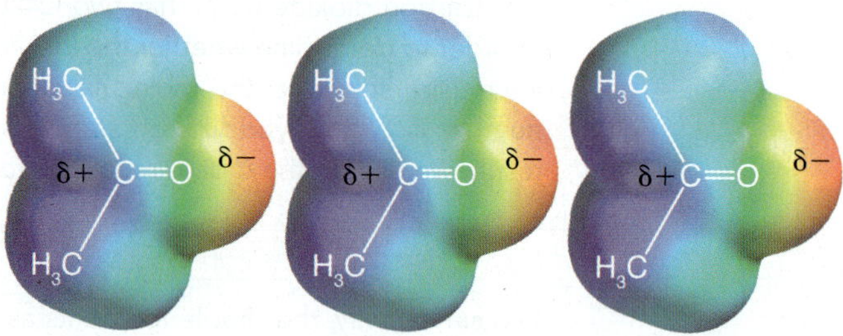


**FIGURE 1.48** In solids, molecules align themselves so that their dipole moments experience attractive forces.

In the liquid phase, the molecules are free to tumble in space, but they do tend to move in such a way so as to attract each other more often than they repel each other. The resulting net attraction between the molecules results in an elevated melting point and boiling point. To illustrate this, compare the physical properties of isobutylene and acetone:

**Isobutylene**
Melting point = −140.3°C
Boiling point = −6.9°C

**Acetone**
Melting point = −94.9°C
Boiling point = 56.3°C

Isobutylene lacks a significant dipole moment, but acetone does have a net dipole moment. Therefore, acetone molecules will experience greater attractive interactions than isobutylene molecules. As a result, acetone has a higher melting point and higher boiling point than isobutylene.

### Hydrogen Bonding

The term **hydrogen bonding** is misleading. A hydrogen bond is not actually a "bond" but is just a specific type of dipole-dipole interaction. When a hydrogen atom is connected to an electronegative atom, the hydrogen atom will bear a partial positive charge (δ+) as a result of induction. This δ+ can then interact with a lone pair from an electronegative atom of another molecule. This can be illustrated with water or ammonia (Figure 1.49). This attractive interaction can occur with any

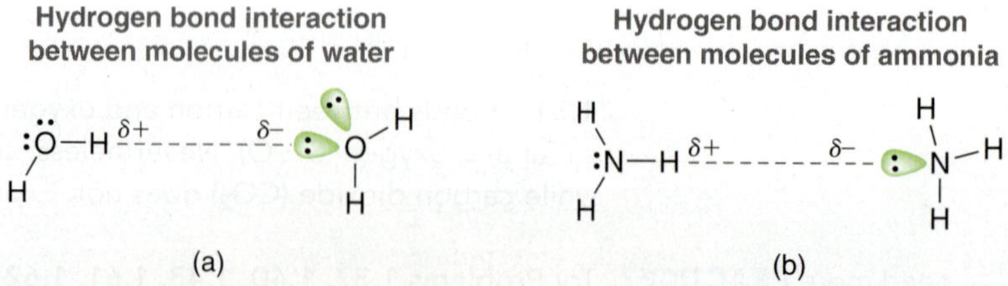

Hydrogen bond interaction between molecules of water

(a)

Hydrogen bond interaction between molecules of ammonia

(b)

**FIGURE 1.49** (a) Hydrogen bonding between molecules of water. (b) Hydrogen bonding between molecules of ammonia.

protic compound, that is, any compound that has a proton connected to an electronegative atom. Ethanol, for example, exhibits the same kind of attractive interaction (Figure 1.50).

**FIGURE 1.50** Hydrogen bonding between molecules of ethanol.

This type of interaction is quite strong because hydrogen is a relatively small atom, and as a result, the partial charges can get very close to each other. In fact, the effect of hydrogen bonding on physical properties is quite dramatic. At the beginning of this chapter, we briefly mentioned the difference in properties between the following two constitutional isomers:

**Ethanol**
Boiling point = 78.4°C

**Methoxymethane**
Boiling point = −23°C

These compounds have the same molecular formula, but they have very different boiling points. Ethanol experiences intermolecular hydrogen bonding, giving rise to a very high boiling point. Methoxymethane does not experience intermolecular hydrogen bonding, giving rise to a relatively lower boiling point. A similar trend can be seen in a comparison of the following amines:

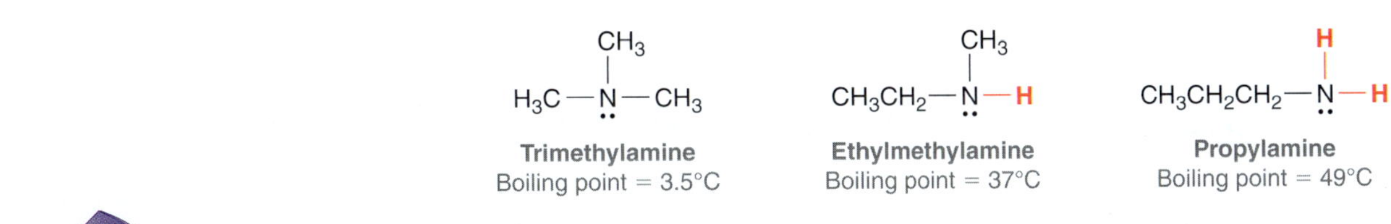


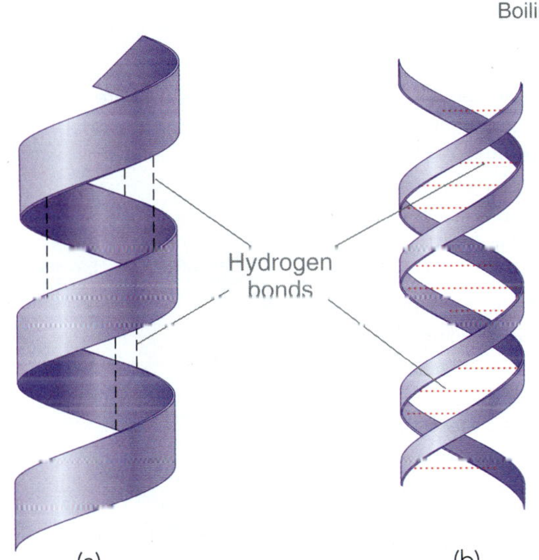


**FIGURE 1.51**
(a) An alpha helix of a protein.
(b) The double helix in DNA.

Once again, all three compounds have the same molecular formula ($C_3H_9N$), but they have very different properties as a result of the extent of hydrogen bonding. Trimethylamine does not exhibit any hydrogen bonding and has a relatively low boiling point. Ethylmethylamine does exhibit hydrogen bonding and therefore has a higher boiling point. Finally, propylamine, which has the highest boiling point of the three compounds, has two N—H bonds and therefore exhibits even more hydrogen-bonding interactions.

Hydrogen bonding is incredibly important in determining the shapes and interactions of biologically important compounds. Chapter 25 will focus on proteins, which are long molecules that coil up into specific shapes under the influence of hydrogen bonding (Figure 1.51a). These shapes ultimately determine their biological function. Similarly, hydrogen bonds hold together individual strands of DNA to form the familiar double-helix structure.

As mentioned earlier, hydrogen "bonds" are not really bonds. To illustrate this, compare the energy of a real bond with the energy of a hydrogen-bonding interaction. A typical single bond (C—H, N—H, O—H) has a bond strength of approximately 400 kJ/mol. In contrast, a hydrogen-bonding interaction has an average strength of approximately 20 kJ/mol. This leaves us with the obvious question: Why do we call them hydrogen *bonds* instead of just hydrogen *interactions*? To answer this question, consider the double-helix structure of DNA (Figure 1.51b). The two strands are joined by hydrogen-bonding interactions that function like rungs of a very long, twisted ladder. The net sum of these interactions is a significant factor that contributes to the structure of the double helix, in which the hydrogen-bonding interactions appear *as if* they were actually bonds. Nevertheless, it is relatively easy to "unzip" the double helix and retrieve the individual strands.

**LOOKING AHEAD**
The structure of DNA is explored in more detail in Section 24.9.

## Fleeting Dipole-Dipole Interactions

Some compounds have no permanent dipole moments, and yet analysis of boiling points indicates that they must have fairly strong intermolecular attractions. To illustrate this point, consider the following compounds:

**Butane**
($C_4H_{10}$)
Boiling point = 0°C

**Pentane**
($C_5H_{12}$)
Boiling point = 36°C

**Hexane**
($C_6H_{14}$)
Boiling point = 69°C

**LOOKING AHEAD**
Hydrocarbons will be discussed in more detail in Chapters 4, 17, and 18.

These three compounds are hydrocarbons, compounds that contain only carbon and hydrogen atoms. If we compare the properties of the hydrocarbons above, an important trend becomes apparent. Specifically, the boiling point appears to increase with increasing molecular weight. This trend can be justified by considering the fleeting, or transient, dipole moments that are more prevalent in larger hydrocarbons. To understand the source of these temporary dipole moments, we consider the electrons to be in constant motion, and therefore, the center of negative charge is also constantly moving around within the molecule. On average, the center of negative charge coincides with the center of positive charge, resulting in a zero dipole moment. However, at any given instant, the center of negative charge and the center of positive charge might not coincide. The resulting transient dipole moment can then induce a separate transient dipole moment in a neighboring molecule, initiating a fleeting attraction between the two molecules (Figure 1.52). These attractive forces are called **London dispersion forces**, named after German-American physicist Fritz London. Large hydrocarbons have

## practically speaking Biomimicry and Gecko Feet

The term biomimicry describes the notion that scientists often draw creative inspiration from studying nature. By investigating some of nature's processes, it is possible to mimic those processes and to develop new technology. One such example is based on the way that geckos can scurry up walls and along ceilings. Until recently, scientists were baffled by the curious ability of geckos to walk upside down, even on very smooth surfaces such as polished glass.

As it turns out, geckos do not use any chemical adhesives, nor do they use suction. Instead, their abilities arise from the intermolecular forces of attraction between the molecules in their feet and the molecules in the surface on which they are walking. When you place your hand on a surface, there are certainly intermolecular forces of attraction between the molecules of your hand and the surface, but the microscopic topography of your hand is quite bumpy. As a result, your hand only makes contact with the surface at perhaps a few thousand points. In contrast, the foot of a gecko has approximately half a million microscopic flexible hairs, called *setae*, each of which has even smaller hairs.

When a gecko places its foot on a surface, the flexible hairs allow the gecko to make extraordinary contact with the surface, and the resulting London dispersion forces are collectively strong enough to support the gecko.

In the last decade, many research teams have drawn inspiration from geckos and have created materials with densely packed microscopic hairs. For example, some scientists are developing adhesive bandages that could be used in the healing of surgical wounds, while other scientists are developing special gloves and boots that would enable people to climb up walls (and perhaps walk upside down on ceilings). Imagine the possibility of one day being able to walk on walls and ceilings like Spiderman.

There are still many challenges that we must overcome before these materials will show their true potential. It is a technical challenge to design microscopic hairs that are strong enough to prevent the hairs from becoming tangled but flexible enough to allow the hairs to stick to any surface. Many researchers believe that these challenges can be overcome, and if they are right, we might have the opportunity to see the world turned literally upside down within the next decade.

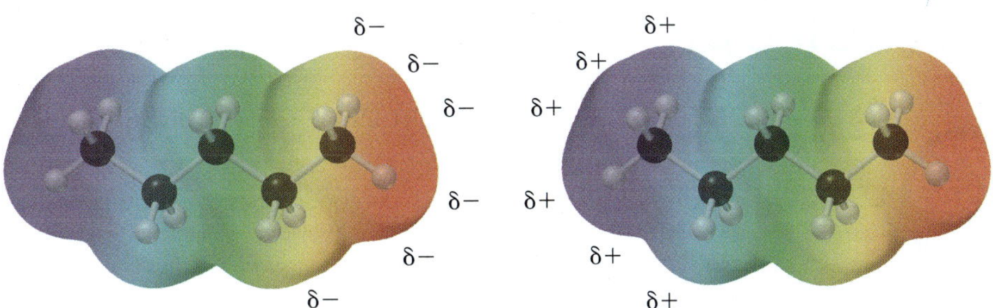


**FIGURE 1.52** The fleeting attractive forces between two molecules of pentane.

more surface area than smaller hydrocarbons and therefore experience these attractive forces to a larger extent.

London dispersion forces are stronger for higher molecular weight hydrocarbons because these compounds have larger surface areas that can accommodate more interactions. As a result, compounds of higher molecular weight will generally boil at higher temperatures. Table 1.6 illustrates this trend.

A branched hydrocarbon generally has a smaller surface area than its corresponding straight-chain isomer, and therefore, branching causes a decrease in boiling point. This trend can be seen by comparing the following constitutional isomers of $C_5H_{12}$:

```
  H   H   H   H   H
  |   |   |   |   |
H—C — C — C — C — C—H
  |   |   |   |   |
  H   H   H   H   H
```

**Pentane**
Boiling point = 36°C

```
               H
             H  \ / H
  H   H          C
  |   |         /
H—C — C — C—H
  |   |         \
  H   H          C
             H  / \ H
               H
```

**2-Methylbutane**
Boiling point = 28°C

```
    H    H
  H \|   |/ H
  H—C    C—H
     \  /
      C
     /  \
  H—C    C—H
  H /|   |\ H
    H    H
```

**2,2-Dimethylpropane**
Boiling point = 10°C

**TABLE 1.6 BOILING POINTS FOR HYDROCARBONS OF INCREASING MOLECULAR WEIGHT**

| STRUCTURE | BOILING POINT (°C) | STRUCTURE | BOILING POINT (°C) |
|---|---|---|---|
| H<br>H—C—H<br>H | −164 | H H H H H H<br>H—C—C—C—C—C—C—H<br>H H H H H H | 69 |
| H H<br>H—C—C—H<br>H H | −89 | H H H H H H H<br>H—C—C—C—C—C—C—C—H<br>H H H H H H H | 98 |
| H H H<br>H—C—C—C—H<br>H H H | −42 | H H H H H H H H<br>H—C—C—C—C—C—C—C—C—H<br>H H H H H H H H | 126 |
| H H H H<br>H—C—C—C—C—H<br>H H H H | 0 | H H H H H H H H H<br>H—C—C—C—C—C—C—C—C—C—H<br>H H H H H H H H H | 151 |
| H H H H H<br>H—C—C—C—C—C—H<br>H H H H H | 36 | H H H H H H H H H H<br>H—C—C—C—C—C—C—C—C—C—C—H<br>H H H H H H H H H H | 174 |

# SKILLBUILDER

## 1.10 PREDICTING PHYSICAL PROPERTIES OF COMPOUNDS BASED ON THEIR MOLECULAR STRUCTURE

**LEARN the skill** Determine which compound has the higher boiling point, neopentane or 3-hexanol:

**Neopentane** **3-Hexanol**

### SOLUTION

When comparing boiling points of compounds, we look for the following factors:

**STEP 1** Identify all dipole-dipole interactions in both compounds.

**STEP 2** Identify all H-bonding interactions in both compounds.

**STEP 3** Identify the number of carbon atoms and extent of branching in both compounds.

**1.** Are there any dipole-dipole interactions in either compound?

**2.** Will either compound form hydrogen bonds?

**3a.** How many carbon atoms are in each compound?

**3b.** How much branching is in each compound?

The second compound above (3-hexanol) is the winner in all of these categories. It has a dipole moment, while neopentane does not. It will experience hydrogen bonding, while neopentane will not. It has six carbon atoms, while neopentane only has five. And, finally, it has a straight chain, while neopentane is highly branched. Each of these factors alone would suggest that 3-hexanol should have a higher boiling point. When we consider all of these factors together, we expect that the boiling point of 3-hexanol will be significantly higher than neopentane.

When comparing two compounds, it is important to consider all four factors. However, it is not always possible to make a clear prediction because in some cases there may be competing factors. For example, compare ethanol and heptane:

**Ethanol** **Heptane**

Ethanol will exhibit hydrogen bonding, but heptane has many more carbon atoms. Which factor dominates? It is not easy to predict. In this case, heptane has the higher boiling point, which is perhaps not what we would have guessed. In order to use the trends to make a prediction, there must be a clear winner.

**PRACTICE the skill** **1.32** For each of the following pairs of compounds, identify the higher boiling compound and justify your choice:

(a)

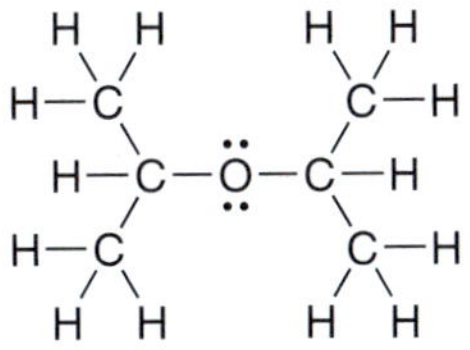

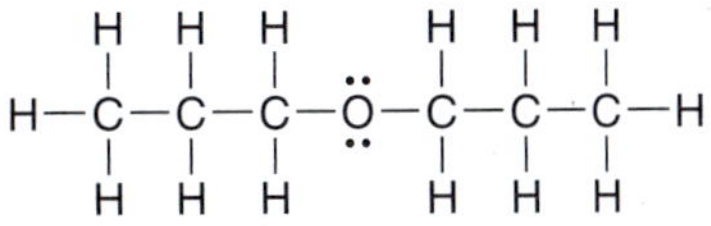

(b)

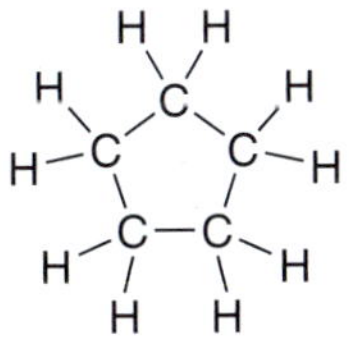

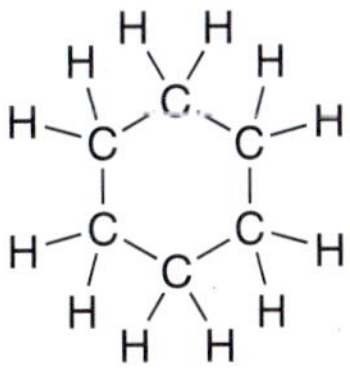

(c)

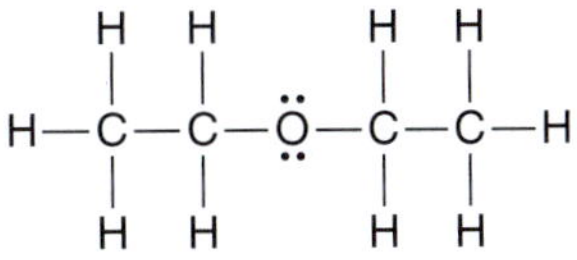

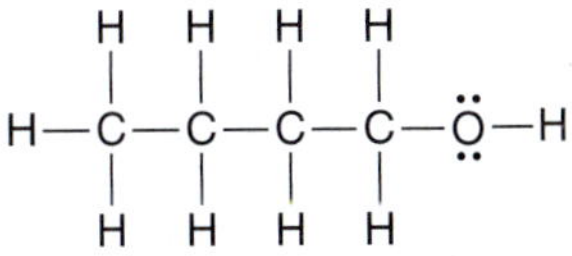

(d)

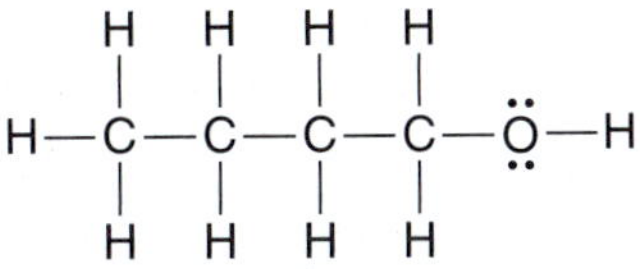

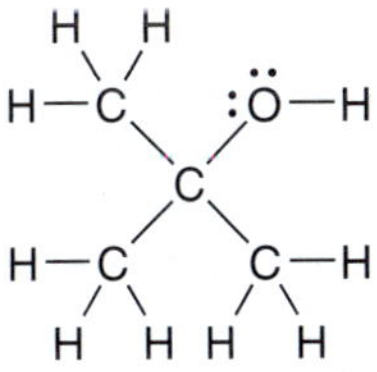

**APPLY the skill** **1.33** Arrange the following compounds in order of increasing boiling point:

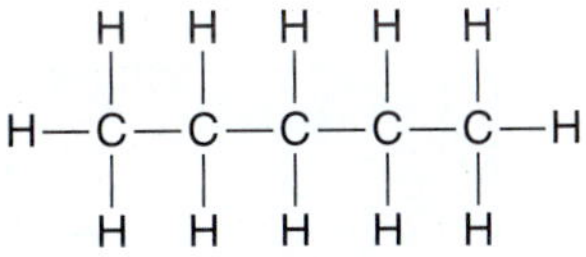

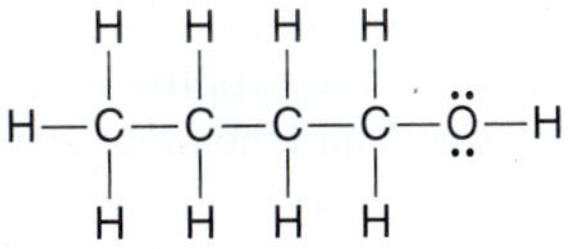

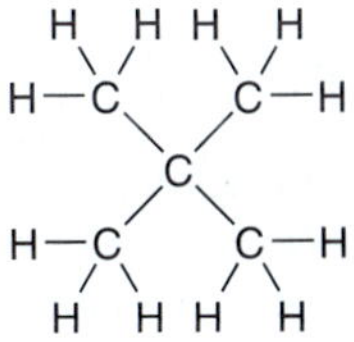

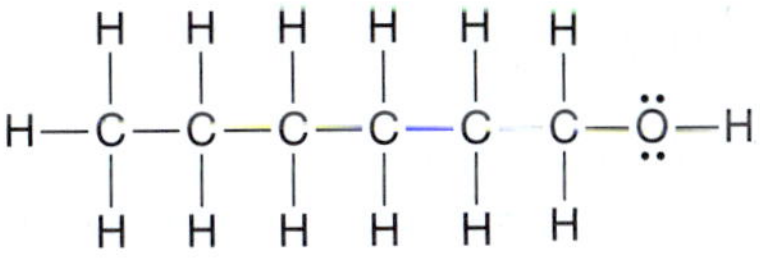

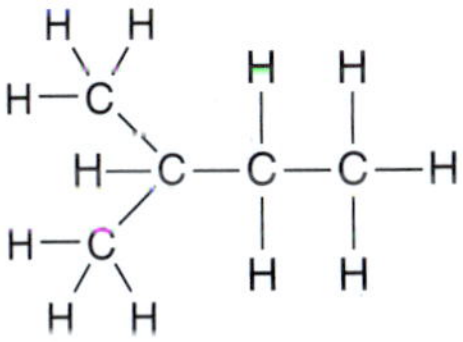

need more **PRACTICE?** **Try Problems 1.52, 1.53, 1.60**

## Drug-Receptor Interactions

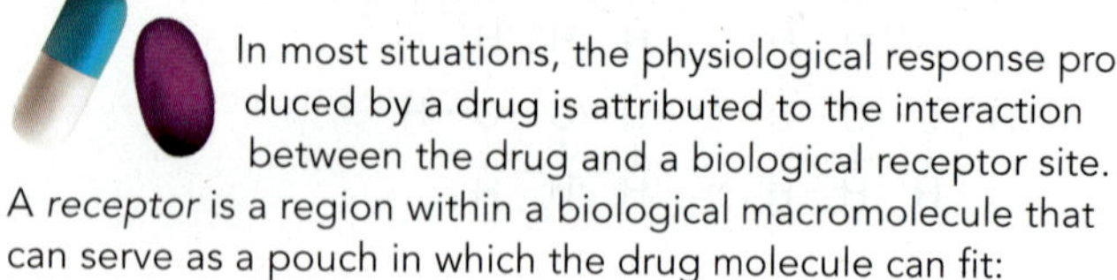

In most situations, the physiological response produced by a drug is attributed to the interaction between the drug and a biological receptor site. A *receptor* is a region within a biological macromolecule that can serve as a pouch in which the drug molecule can fit:

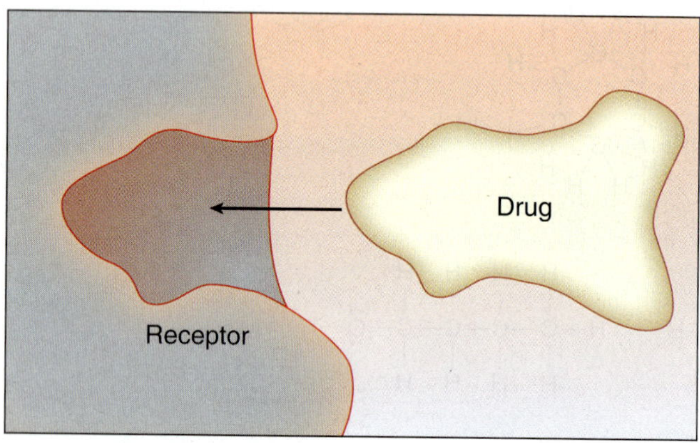


Initially, this mechanism was considered to work much like a lock and key. That is, a drug molecule would function as a key, either fitting or not fitting into a particular receptor. Extensive research on drug-receptor interactions has forced us to modify this simple lock-and-key model. It is now understood that both the drug and the receptor are flexible, constantly changing their shapes. As such, drugs can bind to receptors with various levels of efficiency, with some drugs binding more strongly and other drugs binding more weakly.

How does a drug bind to a receptor? In some cases, the drug molecule forms covalent bonds with the receptor. In such cases, the binding is indeed very strong (approximately 400 kJ/mol for each covalent bond) and therefore irreversible. We will see an example of irreversible binding when we explore a class of anticancer agents called nitrogen mustards (Chapter 7). For most drugs, however, the desired physiological response is meant to be temporary, which can only be accomplished if a drug can bind *reversibly* with its target receptor. This requires a weaker interaction between the drug and the receptor (at least weaker than a covalent bond). Examples of weak interactions include hydrogen-bonding interactions (20 kJ/mol) and London dispersion forces (approximately 4 kJ/mol for each carbon atom participating in the interaction). As an example, consider the structure of a benzene ring, which is incorporated as a structural subunit in many drugs:

**Benzene**

In the benzene ring, each carbon is $sp^2$ hybridized and therefore trigonal planar. As a result, a benzene ring represents a flat surface:

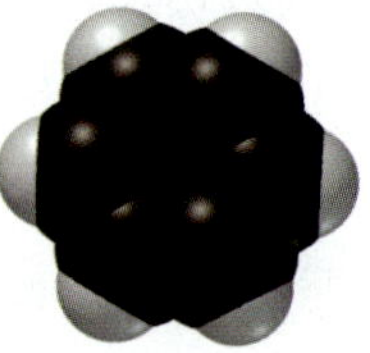

If the receptor also has a flat surface, the resulting London dispersion forces can contribute to the reversible binding of the drug to the receptor site:

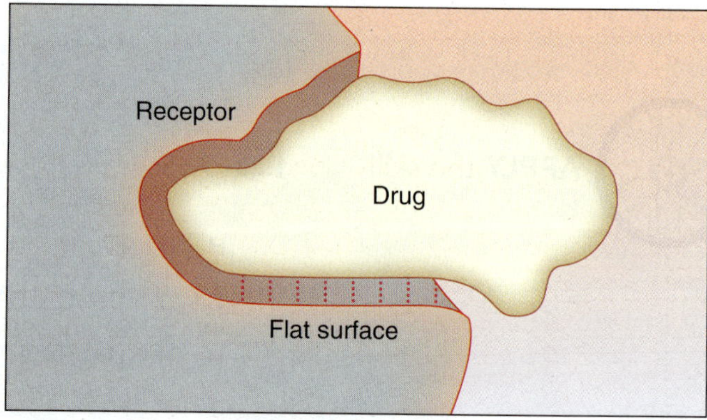


This interaction is roughly equivalent to the strength of a single hydrogen-bonding interaction. The binding of a drug to a receptor is the result of the sum of the intermolecular forces of attraction between a portion of the drug molecule and the receptor site. We will have more to say about drugs and receptors in the upcoming chapters. In particular, we will see how drugs make their journey to the receptor, and we will explore how drugs flex and bend when interacting with a receptor site.

## 1.13 Solubility

Solubility is based on the principle that "like dissolves like." In other words, polar compounds are soluble in polar solvents, while nonpolar compounds are soluble in nonpolar solvents. Why is this so? A polar compound experiences dipole-dipole interactions with the molecules of a polar solvent, allowing the compound to dissolve in the solvent. Similarly, a nonpolar compound experiences London dispersion forces with the molecules of a nonpolar solvent. Therefore, if an article of clothing is stained with a polar compound, the stain can generally be washed away with water (like dissolves like). However, water will be insufficient for cleaning clothing stained with nonpolar compounds, such as oil or grease. In a situation like this, the clothes can be cleaned with soap or by dry cleaning.

### Soap

Soaps are compounds that have a polar group on one end of the molecule and a nonpolar group on the other end (Figure 1.53).

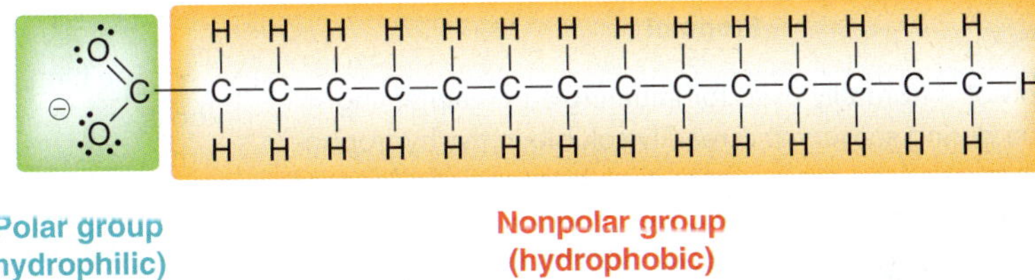


**FIGURE 1.53** The hydrophilic and hydrophobic ends of a soap molecule

The polar group represents the **hydrophilic** region of the molecule (literally, "loves water"), while the nonpolar group represents the **hydrophobic** region of the molecule (literally, "afraid of water"). Oil molecules are surrounded by the hydrophobic tails of the soap molecules, forming a **micelle** (Figure 1.54).

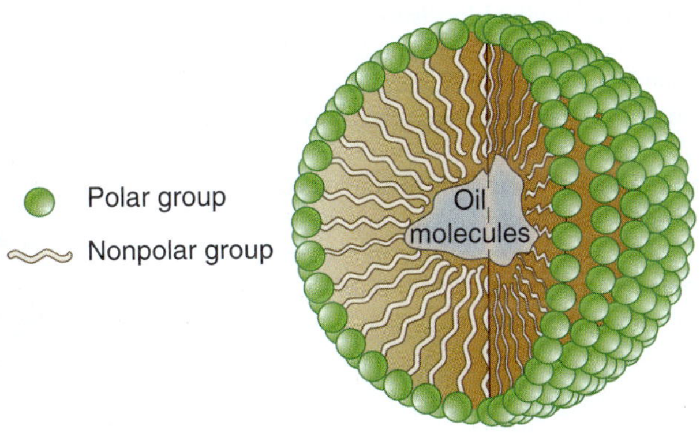


**FIGURE 1.54** A micelle is formed when the hydrophobic tails of soap molecules surround the nonpolar oil molecules.

The surface of the micelle is comprised of all of the polar groups, rendering the micelle water soluble. This is a clever way to dissolve the oil in water, but this technique only works for clothing that can be subjected to water and soap. Some clothes will be damaged in soapy water, and in those situations, dry cleaning is the preferred method.

### Dry Cleaning

Rather than surrounding the nonpolar compound with a micelle so that it will be water soluble, it is actually conceptually simpler to use a nonpolar solvent. This is just another application of the principle of "like dissolves like." Dry cleaning utilizes a nonpolar solvent, such as tetrachloroethylene, to dissolve the nonpolar compounds. This compound is nonflammable, making it an ideal choice as a solvent. Dry cleaning allows clothes to be cleaned without coming into contact with water or soap.

Tetrachloroethylene

## medically speaking — Propofol: The Importance of Drug Solubility

The drug propofol received a lot of publicity in 2009 as one of the drugs implicated in the death of Michael Jackson:

Propofol

Propofol is normally used for initiating and maintaining anesthesia during surgery. It is readily soluble in the hydrophobic membranes of the brain, where it inhibits the firing (or excitation) of brain neurons. To be effective, propofol must be administered through intravenous injection, yet this poses a solubility problem. Specifically, the hydrophobic region of the drug is much larger than the hydrophilic region, and consequently, the drug does not readily dissolve in water (or in blood). Propofol *does* dissolve very well in soybean oil (a complex mixture of hydrophobic compounds, discussed in Section 26.3), but injecting a dose of soybean oil into the bloodstream would result in an oil globule, which could be fatal. To overcome this problem, a group of compounds called lecithins (discussed in Section 26.5) are added to the mixture. Lecithins are compounds that exhibit hydrophobic regions as well as a hydrophilic region. As such, lecithins form micelles, analogous to soap (as described in Section 1.13), which encapsulate the mixture of propofol and soybean oil. This solution of micelles can then be injected into the bloodstream. The propofol readily passes out of the micelles, crosses the hydrophobic membranes of the brain, and reaches the target neurons.

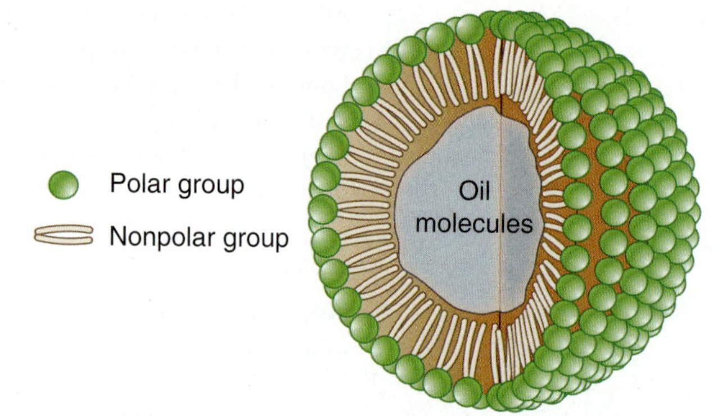


The high concentration of micelles results in a solution that looks very much like milk, and ampules of propofol are sometimes referred to as "milk of amnesia."

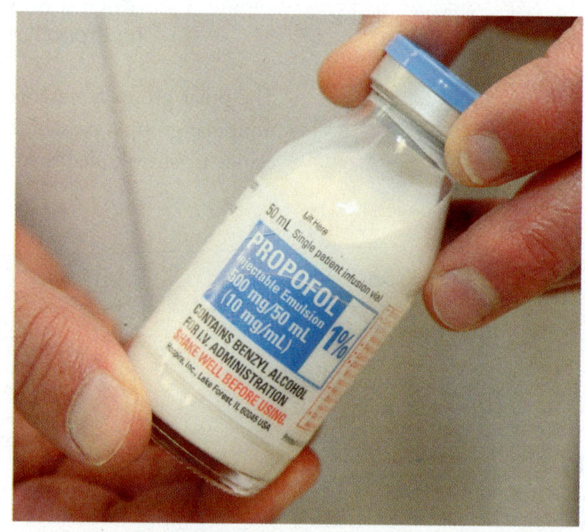


# REVIEW OF CONCEPTS AND VOCABULARY

### SECTION 1.1

- Organic compounds contain carbon atoms.

### SECTION 1.2

- **Constitutional isomers** share the same molecular formula but have different connectivity of atoms and different physical properties.
- Each element will generally form a predictable number of bonds. Carbon is generally **tetravalent**, nitrogen **trivalent**, oxygen **divalent**, and hydrogen and the halogens **monovalent**.

### SECTION 1.3

- A **covalent bond** results when two atoms share a pair of electrons.
- Covalent bonds are illustrated using **Lewis structures**, in which electrons are represented by dots.
- Second-row elements generally obey the **octet rule**, bonding to achieve noble gas electron configuration.
- A pair of unshared electrons is called a **lone pair**.

SECTION 1.4

- A **formal charge** occurs when atoms do not exhibit the appropriate number of valence electrons; formal charges must be drawn in Lewis structures.

SECTION 1.5

- Bonds are classified as (1) **covalent**, (2) **polar covalent**, or (3) **ionic.**
- Polar covalent bonds exhibit **induction**, causing the formation of **partial positive charges** ($\delta+$) and **partial negative charges** ($\delta-$). **Electrostatic potential maps** present a visual illustration of partial charges.

SECTION 1.6

- **Quantum mechanics** describes electrons in terms of their wavelike properties.
- A **wave equation** describes the total energy of an electron when in the vicinity of a proton. Solutions to wave equations are called **wavefunctions** ($\psi$), where $\psi^2$ represents the probability of finding an electron in a particular location.
- **Atomic orbitals** are represented visually by generating three-dimensional plots of $\psi^2$; nodes indicate that the value of $\psi$ is zero.
- An occupied orbital can be thought of as a cloud of **electron density**.
- Electrons fill orbitals following three principles: (1) the **Aufbau principle**, (2) the **Pauli exclusion principle**, and (3) **Hund's rule**. Orbitals with the same energy level are called degenerate orbitals.

SECTION 1.7

- **Valence bond theory** treats every bond as the sharing of electron density between two atoms as a result of the constructive interference of their atomic orbitals. **Sigma** ($\sigma$) **bonds** are formed when the electron density is located primarily on the bond axis.

SECTION 1.8

- **Molecular orbital theory** uses a mathematical method called the **linear combination of atomic orbitals** (LCAO) to form molecular orbitals. Each molecular orbital is associated with the entire molecule, rather than just two atoms.
- The bonding MO of molecular hydrogen results from constructive interference between its two atomic orbitals. The antibonding MO results from destructive interference.
- An **atomic orbital** is a region of space associated with an individual atom, while a molecular orbital is associated with an entire molecule.
- Two molecular orbitals are the most important to consider: (1) the **highest occupied molecular orbital**, or HOMO, and (2) the **lowest unoccupied molecular orbital**, or LUMO.

SECTION 1.9

- Methane's tetrahedral geometry can be explained using four degenerate **$sp^3$-hybridized orbitals** to achieve its four single bonds.
- Ethylene's planar geometry can be explained using three degenerate **$sp^2$-hybridized orbitals**. The remaining *p* orbitals overlap to form a separate bonding interaction, called a pi ($\pi$) bond. The carbon atoms of ethylene are connected via a $\sigma$ bond, resulting from the overlap of $sp^2$-hybridized atomic orbitals, and via a $\pi$ bond, resulting from the overlap of *p* orbitals, both of which comprise the double bond of ethylene.
- Acetylene's linear geometry is achieved via ***sp*-hybridized** carbon atoms in which a triple bond is created from the bonding interactions of one $\sigma$ bond, resulting from overlapping *sp* orbitals, and two $\pi$ bonds, resulting from overlapping *p* orbitals.
- Triple bonds are stronger and shorter than double bonds, which are stronger and shorter than single bonds.

SECTION 1.10

- The geometry of small compounds can be predicted using valence shell electron pair repulsion (**VSEPR**) theory, which focuses on the number of $\sigma$ bonds and lone pairs exhibited by each atom. The total, called the steric number, indicates the number of electron pairs that repel each other.
- A tetrahedral arrangement of orbitals indicates **$sp^3$ hybridization** (steric number 4). A compound's geometry depends on the number of lone pairs and can be tetrahedral, trigonal pyramidal, or bent.
- A trigonal planar arrangement of orbitals indicates **$sp^2$ hybridization** (steric number 3); however, the geometry may be bent, depending on the number of lone pairs.
- Linear geometry indicates ***sp* hybridization** (steric number 2).

SECTION 1.11

- **Dipole moments** ($\mu$) occur when the center of negative charge and the center of positive charge are separated from one another by a certain distance; the dipole moment is used as an indicator of polarity (measured in debyes).
- The percent ionic character of a bond is determined by measuring its dipole moment. The vector sum of individual dipole moments in a compound determines the **molecular dipole moment**.

SECTION 1.12

- The physical properties of compounds are determined by intermolecular forces, the attractive forces between molecules.
- **Dipole-dipole interactions** occur between two molecules that possess permanent dipole moments. Hydrogen bonding, a special type of dipole-dipole interaction, occurs when the lone pairs of an electronegative atom interact with an electron-poor hydrogen atom. Compounds that exhibit hydrogen bonding have higher boiling points than similar compounds that lack hydrogen bonding.
- **London dispersion forces** result from the interaction between transient dipole moments and are stronger for larger alkanes due to their larger surface area and ability to accommodate more interactions.

SECTION 1.13

- Polar compounds are soluble in polar solvents; nonpolar compounds are soluble in nonpolar solvents.
- Soaps are compounds that contain both **hydrophilic** and **hydrophobic** regions. The hydrophobic tails surround nonpolar compounds, forming a water-soluble micelle.

# SKILLBUILDER REVIEW

## 1.1 DETERMINING THE CONSTITUTION OF SMALL MOLECULES

**STEP 1** Determine the valency of each atom in the compound.

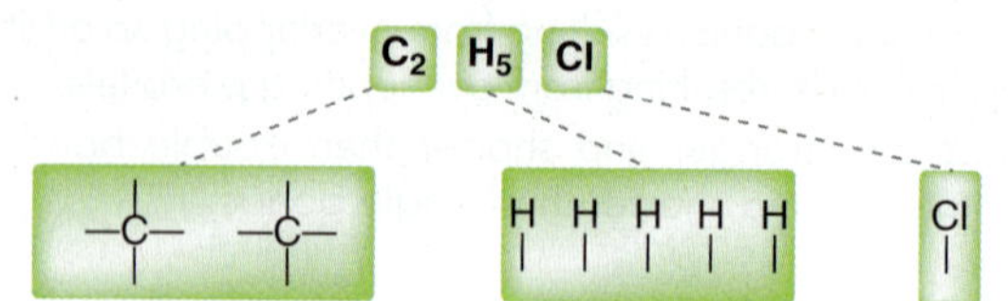


**STEP 2** Determine how the atoms are connected. Atoms with highest valency should be placed at the center, and monovalent atoms should be placed at the periphery.

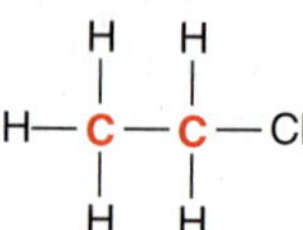

Try Problems 1.1–1.4, 1.34, 1.46, 1.47, 1.54

## 1.2 DRAWING THE LEWIS DOT STRUCTURE OF AN ATOM

**STEP 1** Determine the number of valence electrons.

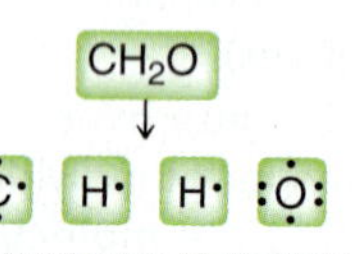


**STEP 2** Place one electron by itself on each side of the atom.

**STEP 3** If the atom has more than four valence electrons, the remaining electrons are paired with the electrons already drawn.

Try Problems 1.5–1.9

## 1.3 DRAWING THE LEWIS STRUCTURE OF A SMALL MOLECULE

**STEP 1** Draw all individual atoms.

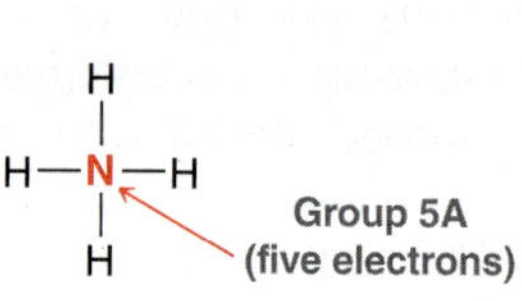


**STEP 2** Connect atoms that form more than one bond.

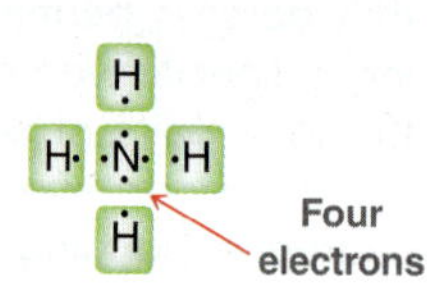

**STEP 3** Connect hydrogen atoms.

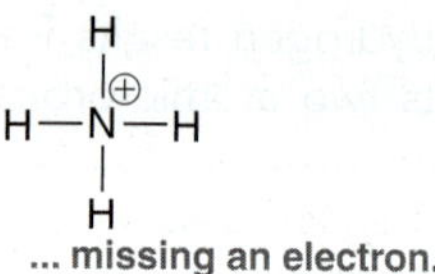

**STEP 4** Pair any unpaired electrons, so that each atom achieves an octet.

Try Problems 1.10–1.12, 1.35, 1.38, 1.42

## 1.4 CALCULATING FORMAL CHARGE

**STEP 1** Determine appropriate number of valence electrons.

Group 5A (five electrons)

**STEP 2** Determine the number of valence electrons in this case.

Four electrons

**STEP 3** Assign a formal charge.

... missing an electron.

Try Problems 1.13, 1.14, 1.41

## 1.5 LOCATING PARTIAL CHARGES

**STEP 1** Identify all polar covalent bonds.

Polar covalent

**STEP 2** Determine the direction of each dipole.

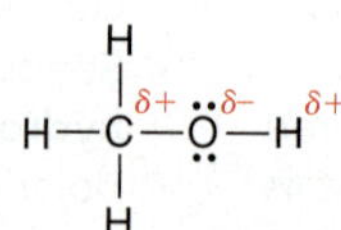

**STEP 3** Indicate location of partial charges.

Try Problems 1.15, 1.16, 1.36, 1.37, 1.48, 1.57

### 1.6 IDENTIFYING ELECTRON CONFIGURATIONS

**STEP 1** Fill orbitals using the Aufbau principle, the Pauli exclusion principle, and Hund's rule.

Nitrogen ⟶ 2p, 2s, 1s

**STEP 2** Summarize using the following notation:

$1s^2 2s^2 2p^3$

Try Problems **1.17, 1.18, 1.44**

### 1.7 IDENTIFYING HYBRIDIZATION STATES

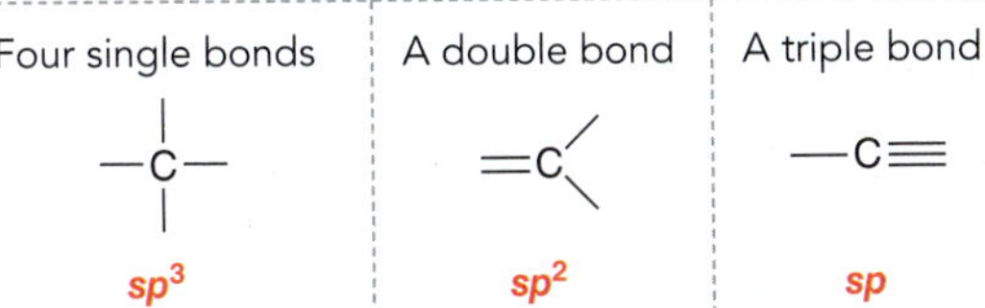


Try Problems **1.22, 1.23, 1.55, 1.56, 1.58**

### 1.8 PREDICTING GEOMETRY

**STEP 1** Determine the steric number by adding the number of $\sigma$ bonds and lone pairs.

H—N—H, H

\# of $\sigma$ bonds = 3

**# of lone pairs = 1**

***Steric number* = 4**

**STEP 2** Use steric number to determine hybridization state and electronic geometry.

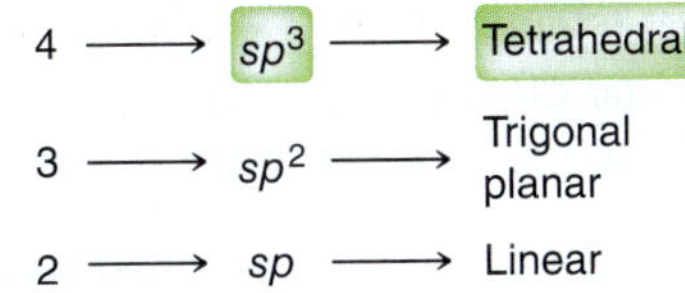


**STEP 3** Ignore lone pairs and describe geometry.

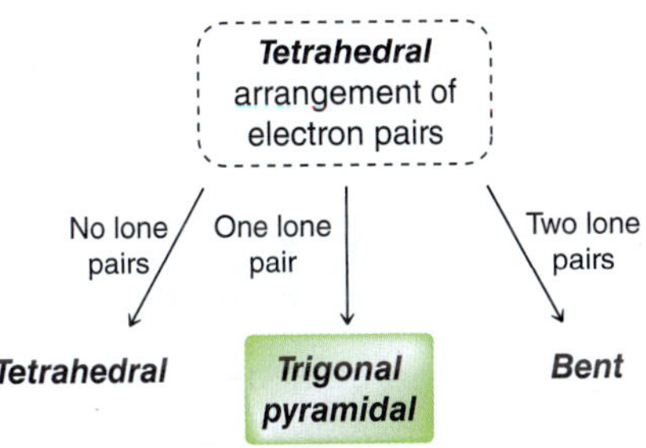


Try Problems **1.25–1.28, 1.39–1.41, 1.50, 1.55, 1.56, 1.58**

### 1.9 IDENTIFYING MOLECULAR DIPOLE MOMENTS

**STEP 1** Predict geometry.

**Bent**

**STEP 2** Identify direction of all dipole moments.

**STEP 3** Draw net dipole moment.

Try Problems **1.29–1.31, 1.37, 1.40, 1.43, 1.61, 1.62**

### 1.10 PREDICTING PHYSICAL PROPERTIES

**STEP 1** Identify dipole-dipole interactions.

$H_2C=C(CH_3)_2$ vs. $(CH_3)_2C=O$

**Higher boiling point**

**STEP 2** Identify H-bonding interactions.

$CH_3OCH_3$ vs. $CH_3CH_2OH$

**Higher boiling point**

**STEP 3** Identify number of carbon atoms and extent of branching.

$CH_3CH_2CH_3$ vs. $CH_3CH_2CH_2CH_2CH_3$

**Higher boiling point**

Try Problems **1.32, 1.33, 1.52, 1.53, 1.60**

## PRACTICE PROBLEMS

**Note: Most of the Problems are available within WileyPLUS, an online teaching and learning solution.**

**1.34** Draw structures for all constitutional isomers with the following molecular formulas:

(a) $C_4H_{10}$ (b) $C_5H_{12}$ (c) $C_6H_{14}$
(d) $C_2H_5Cl$ (e) $C_2H_4Cl_2$ (f) $C_2H_3Cl_3$

**1.35** Draw structures for all constitutional isomers with the molecular formula $C_4H_8$ that have:

(a) Only single bonds (b) One double bond

**1.36** For each compound below, identify any polar covalent bonds and indicate the direction of the dipole moment using the symbols δ+ and δ−:

(a) HBr (b) HCl (c) $H_2O$ (d) $CH_4O$

**1.37** For each pair of compounds below, identify the one that would be expected to have more ionic character. Explain your choice.

(a) NaBr or HBr (b) BrCl or FCl

**1.38** Draw a Lewis dot structure for each of the following compounds:

(a) $CH_3CH_2OH$ (b) $CH_3CN$

**1.39** Predict the geometry of each atom except hydrogen in the compounds below:

```
      H  H  H                       H
  ⊖   |  |  |               ⊕      |
H—C̈—C—C—C—H            H—Ö—C—C=C—H
   |  |  |  |                |  |  |  |
(a) H  H  H  H          (b)  H  H  H  H

    H  H  H                 H  H  H
  ⊕ |  |  |                 |  |  |    ⊖
H—N—C—C—Ö—H             H—C—C—C—Ö:
   |  |  |                  |  |  |
(c) H  H  H             (d) H  H  H
```

**1.40** Draw a Lewis structure for a compound with molecular formula $C_4H_{11}N$ in which three of the carbon atoms are bonded to the nitrogen atom. What is the geometry of the nitrogen atom in this compound? Does this compound exhibit a molecular dipole moment? If so, indicate the direction of the dipole moment.

**1.41** Draw a Lewis structure of the anion $AlBr_4^-$ and determine its geometry.

**1.42** Draw the structure for the only constitutional isomer of cyclopropane:

```
    H H
     C
 H  / \  H
   C———C
  H     H
```

**Cyclopropane**

**1.43** Determine whether each compound below exhibits a molecular dipole moment:

(a) $CH_4$ (b) $NH_3$ (c) $H_2O$
(d) $CO_2$ (e) $CCl_4$ (f) $CH_2Br_2$

**1.44** Identify the neutral element that corresponds with each of the following electron configurations:

(a) $1s^22s^22p^4$ (b) $1s^22s^22p^5$ (c) $1s^22s^22p^2$
(d) $1s^22s^22p^3$ (e) $1s^22s^22p^63s^23p^5$

**1.45** In the compounds below, classify each bond as covalent, polar covalent, or ionic:

(a) NaBr (b) NaOH (c) $NaOCH_3$
(d) $CH_3OH$ (e) $CH_2O$

**1.46** Draw structures for all constitutional isomers with the following molecular formulas:

(a) $C_2H_6O$ (b) $C_2H_6O_2$ (c) $C_2H_4Br_2$

**1.47** Draw structures for any five constitutional isomers with molecular formula $C_2H_6O_3$.

**1.48** For each type of bond below, determine the direction of the expected dipole moment:

(a) C—O (b) C—Mg (c) C—N (d) C—Li
(e) C—Cl (f) C—H (g) O—H (h) N—H

**1.49** Predict the bond angles for all bonds in the following compounds:

(a) $CH_3CH_2OH$ (b) $CH_2O$ (c) $C_2H_4$ (d) $C_2H_2$
(e) $CH_3OCH_3$ (f) $CH_3NH_2$ (g) $C_3H_8$ (h) $CH_3CN$

**1.50** Identify the expected hybridization state and geometry for the central atom in each of the following compounds:

```
      H            H             H
      |            |             |
(a) H⁄N̈⁀H    (b) H⁄B⁀H    (c) H⁄C⊕⁀H

      H            H
      | ⊖          | ⊕
(d) H⁄C̈⁀H    (e) H⁄Ö⁀H
```

**1.51** Count the total number of σ bonds and π bonds in the compound below:

```
          H                  H
          |                  |
H—Ö—C—C=C—C≡C—C—N̈—H
          |  |  |            |  |
          H  H  H            H  H
```

**1.52** For each pair of compounds below, predict which compound will have the higher boiling point and explain your choice:

(a) $CH_3CH_2CH_2OCH_3$ or $CH_3CH_2CH_2CH_2OH$
(b) $CH_3CH_2CH_2CH_3$ or $CH_3CH_2CH_2CH_2CH_3$

```
     H  :O:  H            H  H  H
     |  ||   |            |  |  |
  H—C—C—C—H    or   H—C—C—C—H
     |       |            |  |  |
(c)  H       H            H  H  H
```

**1.53** Which of the following pure compounds will exhibit hydrogen bonding?

(a) $CH_3CH_2OH$ (b) $CH_2O$ (c) $C_2H_4$ (d) $C_2H_2$
(e) $CH_3OCH_3$ (f) $CH_3NH_2$ (g) $C_3H_8$ (h) $NH_3$

**1.54** For each case below, identify the most likely value for x:

(a) $BH_x$ (b) $CH_x$ (c) $NH_x$ (d) $CH_2Cl_x$

**1.55** Identify the hybridization state and geometry of each carbon atom in the following compounds:

(a) (b) (c)

**1.56** Ambien™ is a sedative used in the treatment of insomnia. It was discovered in 1982 and brought to market in 1992 (it takes a long time for new drugs to undergo the extensive testing required to receive approval from the Food and Drug Administration). Identify the hybridization state and geometry of each carbon atom in the structure of this compound:

Zolpidem (Ambien™)

**1.57** Identify the most electronegative element in each of the following compounds:

(a) $CH_3OCH_2CH_2NH_2$ (b) $CH_2ClCH_2F$ (c) $CH_3Li$

**1.58** Nicotine is an addictive substance found in tobacco. Identify the hybridization state and geometry of each of the nitrogen atoms in nicotine:

Nicotine

**1.59** Below is the structure of caffeine, but its lone pairs are not shown. Identify the location of all lone pairs in this compound:

Caffeine

**1.60** There are two different compounds with molecular formula $C_2H_6O$. One of these isomers has a much higher boiling point than the other. Explain why.

**1.61** Identify which compounds below possess a molecular dipole moment and indicate the direction of that dipole moment:

(a) (b) (c) (d)

**1.62** Methylene chloride ($CH_2Cl_2$) has fewer chlorine atoms than chloroform ($CHCl_3$). Nevertheless, methylene chloride has a larger molecular dipole moment than chloroform. Explain.

# INTEGRATED PROBLEMS

**1.63** Consider the three compounds shown below and then answer the questions that follow:

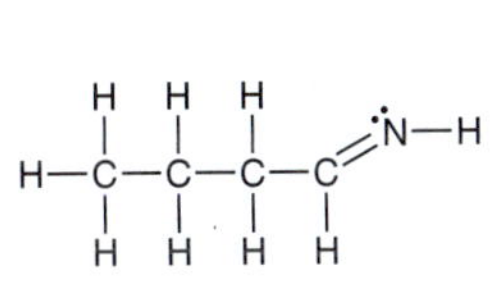

Compound A

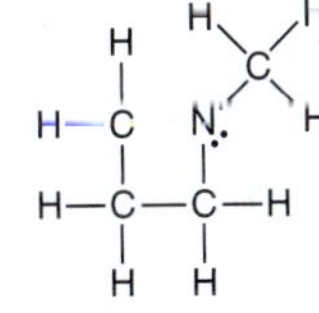

Compound B

Compound C

(a) Which two compounds are constitutional isomers?

(b) Which compound contains a nitrogen atom with trigonal pyramidal geometry?

(c) Identify the compound with the greatest number of σ bonds.

(d) Identify the compound with the fewest number of σ bonds.

(e) Which compound contains more than one π bond?

(f) Which compound contains an $sp^2$-hybridized carbon atom?

(g) Which compound contains only $sp^3$-hybridized atoms (in addition to hydrogen atoms)?

(h) Which compound do you predict will have the highest boiling point? Explain.

**1.64** Propose at least two different structures for a compound with six carbon atoms that exhibits the following features:

(a) All six carbon atoms are $sp^2$ hybridized.

(b) Only one carbon atom is *sp* hybridized, and the remaining five carbon atoms are all $sp^3$ hybridized (remember that your compound can have elements other than carbon and hydrogen).

(c) There is a ring, and all of the carbon atoms are $sp^3$ hybridized.

(d) All six carbon atoms are *sp* hybridized, and the compound contains no hydrogen atoms (remember that a triple bond is linear and therefore cannot be incorporated into a ring of six carbon atoms).

**1.65** Draw all constitutional isomers with molecular formula $C_5H_{10}$ that possess one π bond.

**1.66** With current spectroscopic techniques (discussed in Chapters 15–17), chemists are generally able to determine the structure of an unknown organic compound in just one day. These techniques have only been available for the last several decades. In the first half of the twentieth century, structure determination was a very slow and painful process in which the compound under investigation would be subjected to a variety of chemical reactions. The results of those reactions would provide chemists with clues about the structure of the compound. With enough clues, it was sometimes (but not always) possible to determine the structure. As an example, try to determine the structure of an unknown compound, using the following clues:

- The molecular formula is $C_4H_{10}N_2$.
- There are no π bonds in the structure.
- The compound has no net dipole moment.
- The compound exhibits very strong hydrogen bonding.

You should find that there are at least two constitutional isomers that are consistent with the information above. (**Hint:** Consider incorporating a ring in your structure.)

**1.67** A compound with molecular formula $C_5H_{11}N$ has no π bonds. Every carbon atom is connected to exactly two hydrogen atoms. Determine the structure of the compound.

**1.68** Isonitriles (**A**) are an important class of compounds because of the versatile reactivity of the functional group, enabling the preparation of numerous new compounds and natural products. Isonitriles can be converted to isonitrile dihalides (**B**), which represents a useful procedure for temporarily hiding the reactivity of an isonitrile (*Tetrahedron Lett.* **2012**, *53*, 4536–4537).

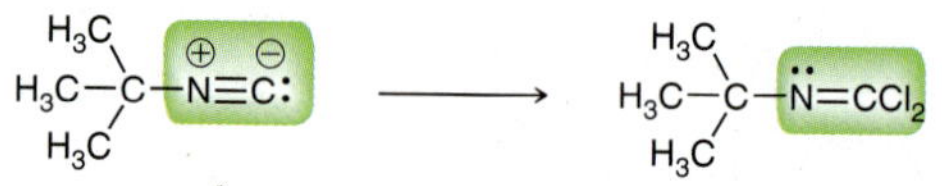

A: $(H_3C)_3C{-}\overset{\oplus}{N}{\equiv}\overset{\ominus}{C}{:}$ → B: $(H_3C)_3C{-}\ddot{N}{=}CCl_2$

(a) Identify the hybridization state for each highlighted atom in **A**.

(b) One of the carbon atoms in **A** exhibits a lone pair. In what type of atomic orbital does this lone pair reside?

(c) Predict the C—N—C bond angle in compound **A**.

(d) Identify the hybridization state for each highlighted atom in **B**.

(e) The nitrogen atom in **B** exhibits a lone pair. In what type of atomic orbital does this lone pair reside?

(f) Predict the C—N—C bond angle in compound **B**.

**1.69** In Section 1.12 we discussed the effect that branching can have on the boiling point of a compound. In certain instances, branching may also affect how a molecule can react with different molecules. We use the term "steric hindrance" to describe how branching can influence reactivity. For example, greater "steric crowding" may decrease the reactivity of a C=C π bond (*Org. Lett.* **1999**, *1*, 1123–1125). In the following molecule, identify each π bond and determine which has a greater degree of steric crowding. (**Note:** This is not the same as "steric number".)

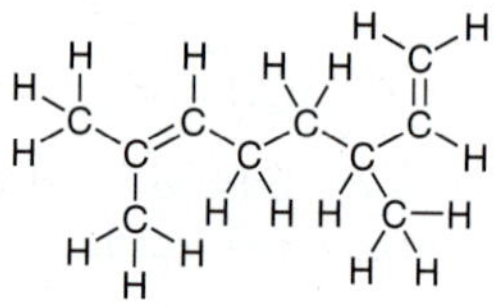

# CHALLENGE PROBLEMS

**1.70** Epichlorohydrin (**1**) is an epoxide used in the production of plastic, epoxy glues, and resins (reactions of epoxides will be discussed in Chapter 14). When epichlorohydrin is treated with phenol (**2**), two products are formed (**3** and **4**). At room temperature, these liquid products are difficult to separate from each other, but upon heating, these compounds are easily separated from one another. Explain these observations (*Tetrahedron* **2006**, *62*, 10968–10979).

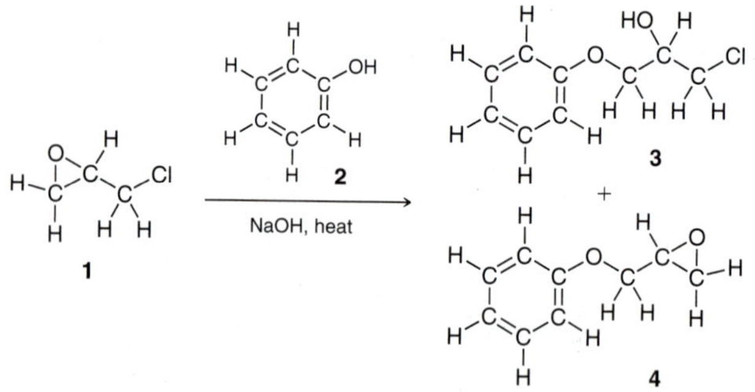

1 + 2 → (NaOH, heat) 3 + 4

**1.71** Consider the following table that provides bond lengths for a variety of C—X bonds (measured in Å).

| | X = F | X = Cl | X = Br | X = I |
|---|---|---|---|---|
| 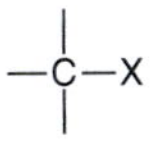 —C—X | 1.40 | 1.79 | 1.97 | 2.16 |
| 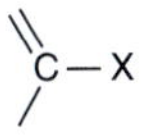 =C—X | 1.34 | | 1.88 | 2.10 |
| ≡C—X | 1.27 | 1.63 | 1.79 | |

(a) Compare the data for bonds of the type $C_{sp^3}$—X. Describe the trend that you observe and offer an explanation for this trend.

(b) Compare the data for bonds of the following types: $C_{sp^3}$—F, and $C_{sp^2}$—F, and $C_{sp}$—F. Describe the trend that you observe and offer an explanation.

(c) When comparing the bond lengths for $C_{sp^2}$—Cl and $C_{sp}$—I, the trends established in the first two parts of this problem appear to be in conflict with each other. Describe how the two trends oppose each other, and then use the data in the table to determine which bond you predict to be longer. Explain your choice.

**1.72** Positron emission tomography (PET) is a medical imaging technique that produces a three-dimensional picture of functional processes in the body, such as the brain uptake of glucose. PET imaging requires the introduction of [$^{18}$F]-fluorine (a radioactive isotope of fluorine) into molecules and can be achieved by several routes, such as the following (*Tetrahedron Lett.* **2003**, *44*, 9165–9167):

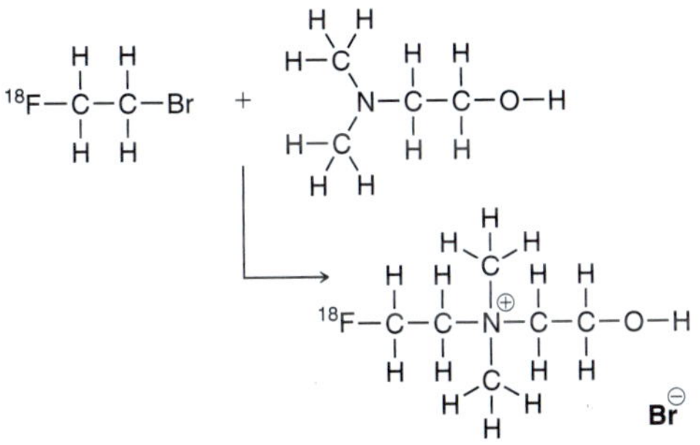

(a) Identify the hybridization for every heteroatom (atom other than C or H) in the three compounds. Note: Throughout this chapter, all lone pairs were drawn. We will soon see (Chapter 2) that lone pairs are often omitted from structural drawings, because they can be inferred by the presence or absence of formal charges. In this problem, none of the lone pairs have been drawn.

(b) Predict the bond angle of each C—N—C bond in the product.

(c) Identify the type of atomic orbital(s) that contain the lone pairs on the oxygen atom in the product.

**1.73** Phenalamide $A_2$ belongs to a class of natural products that are of interest because of their antibiotic, antifungal, and antiviral activity. In the first total synthesis of this compound, the following boronate ester was utilized (*Org. Lett.* **1999**, *1*, 1713–1715):

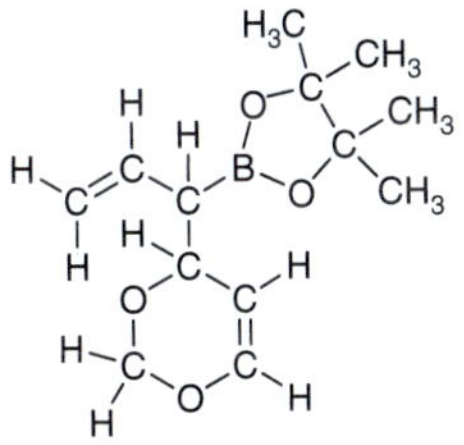

(a) Determine the hybridization state of the boron atom.

(b) Predict the O—B—O bond angle and then suggest a reason why the actual bond angle might deviate from the predicted value in this case.

(c) The lone pairs have not been drawn. Draw all of them. (**Hint:** Note that the structure has no formal charges.)

**1.74** The formation of a variety of compounds called oxazolidinones is important for the synthesis of many different natural products and other compounds that have potential use as future medicines. One method for preparing oxazolidinones involves the conversion of a hydroximoyl chloride, such as compound **1**, into a nitrile oxide, such as compound **2** (*J. Org. Chem.* **2009**, *74*, 1099–1113):

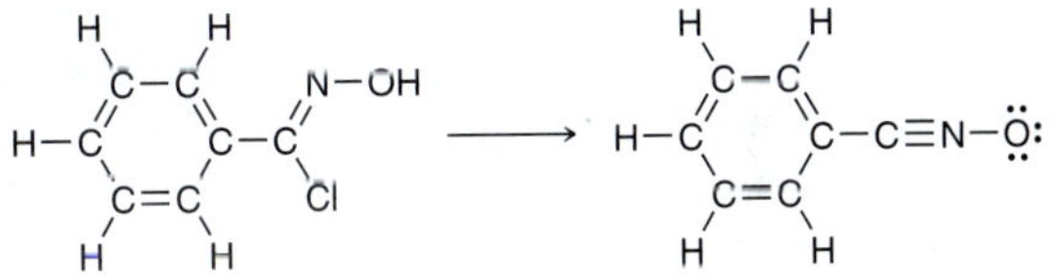

(a) Identify any formal charges that are missing from the structures of **1** and **2**.

(b) Determine which compound is expected to be more soluble in a polar solvent, and justify your choice.

(c) Determine the amount by which the C—C—N bond angle increases as a result of the conversion from **1** to **2**.

**1.75** The following compound belongs to a class of compounds, called estradiol derivatives, which show promise in the treatment of breast cancer (*Tetrahedron Lett.* **2001**, *42*, 8579–8582):

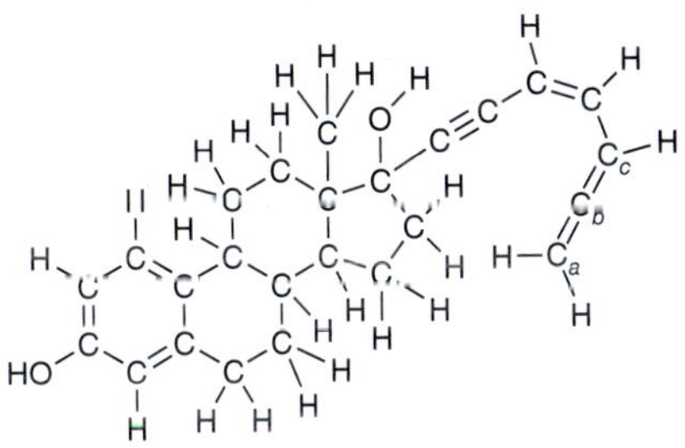

(a) Determine the hybridization state of $C_a$, $C_b$, and $C_c$.

(b) Determine the H—$C_a$—$C_b$ bond angle.

(c) Determine the $C_a$—$C_b$—$C_c$ bond angle.

(d) $C_b$ exhibits two $\pi$ bonds: one to $C_a$, and the other to $C_c$. Draw a picture of $C_a$, $C_b$ and $C_c$ that shows the relative orientation of the two different *p* orbitals that $C_b$ is utilizing to form its two $\pi$ bonds. Also show the *p* orbitals on $C_a$ and $C_c$ that are being used by each of those atoms. Describe the relative orientation of the *p* orbitals on $C_a$ and $C_c$.

**1.76** The study of analogues of small peptide chains (Chapter 25), called peptidomimetics, provides models to understand the observed stabilizing effects in larger peptides, including enzymes (nature's catalysts). The following structure shows promise for studying how enzymes coil up into very discrete shapes that endow them with catalytic function (*Org. Lett.* **2001,** *3,* **3843–3846**):

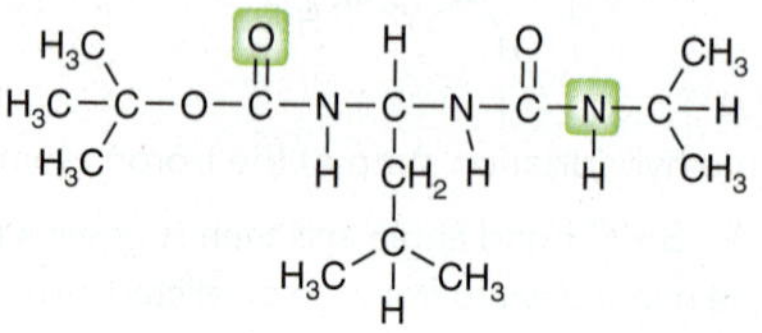

(a) This compound has two N—C—N units, with differing bond angles. Predict the difference in bond angles between these two units and explain the source of the difference.

(b) When this compound was prepared and investigated, it demonstrated a preference for adopting a three-dimensional shape in which the two highlighted regions were in close proximity. Describe the interaction that occurs and create a drawing that illustrates this interaction.

**1.77** Organic polymers (molecules with very high molecular weight) can be designed with pores (cavities) that are specifically tailored to fit a particular guest molecule due to complementary size, shape, and functional group interactions (intermolecular forces). Such systems are often inspired by binding sites in natural enzymes, which display a high degree of specificity for a particular guest. Several porous polymers were designed to bind bisphenol A (compound **1**), a compound that is of environmental concern due to its ability to mimic natural hormones such as estrogen (*J. Am. Chem. Soc.* **2009,** *131,* **8833–8838**). Schematics of the binding sites of two of these polymers are shown below. Polymer **A** was found to have only limited selectivity in that it binds **1** or **2** (a control guest molecule) to a similar degree. Polymer **B** is more selective in that it binds strongly to **1** and weakly to **2**. Draw pictures showing how each guest (**1**, **2**) may bind in the pores of each polymer (**A**, **B**), and propose an explanation for the differences in binding specificity described above.

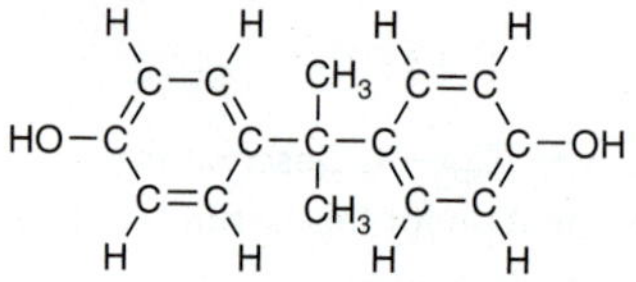

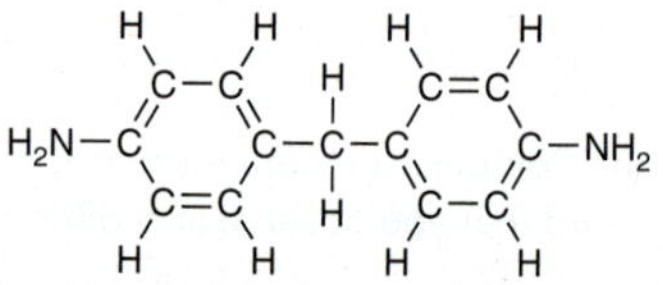

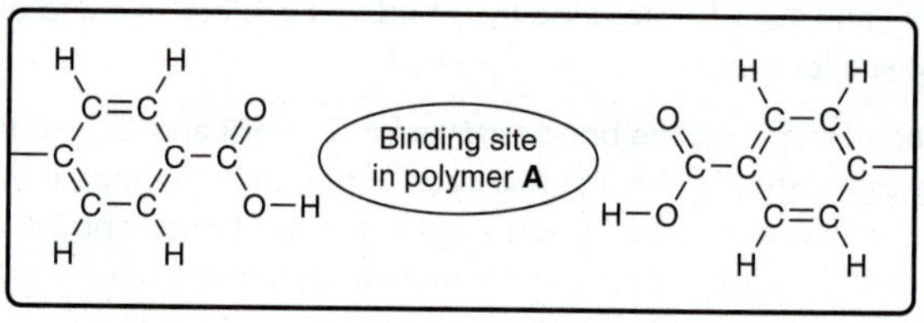

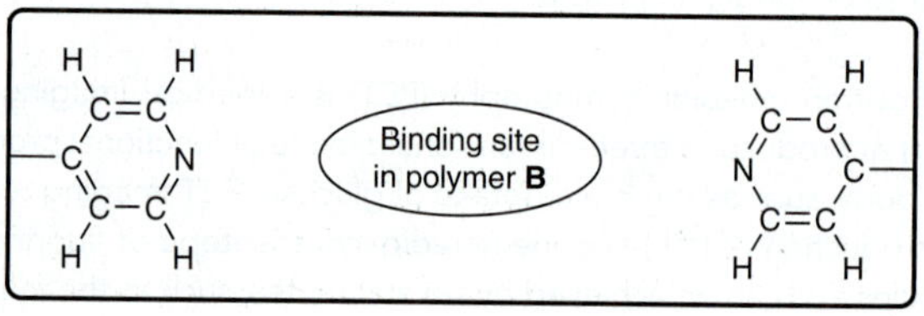

2

# Molecular Representations



## DID YOU EVER WONDER...

**how new drugs are designed?**

Scientists employ many techniques in the design of new drugs. One such technique, called lead modification, enables scientists to identify the portion of a compound responsible for its medicinal properties and then to design similar compounds with better properties. We will see an example of this technique, specifically, where the discovery of morphine led to the development of a whole family of potent analgesics (codeine, heroin, methadone, and many others).

In order to compare the structures of the compounds being discussed, we will need a more efficient way to draw the structures of organic compounds. Lewis structures are only efficient for small molecules, such as those we considered in the previous chapter. The goal of this chapter is to master the skills necessary to use and interpret the drawing method most often utilized by organic chemists and biochemists. These drawings, called bond-line structures, are fast to draw and easy to read, and they focus our attention on the reactive centers in a compound. In the second half of this chapter, we will see that bond-line structures are inadequate in some circumstances, and we will explore the technique that chemists employ to deal with the inadequacy of bond-line structures.

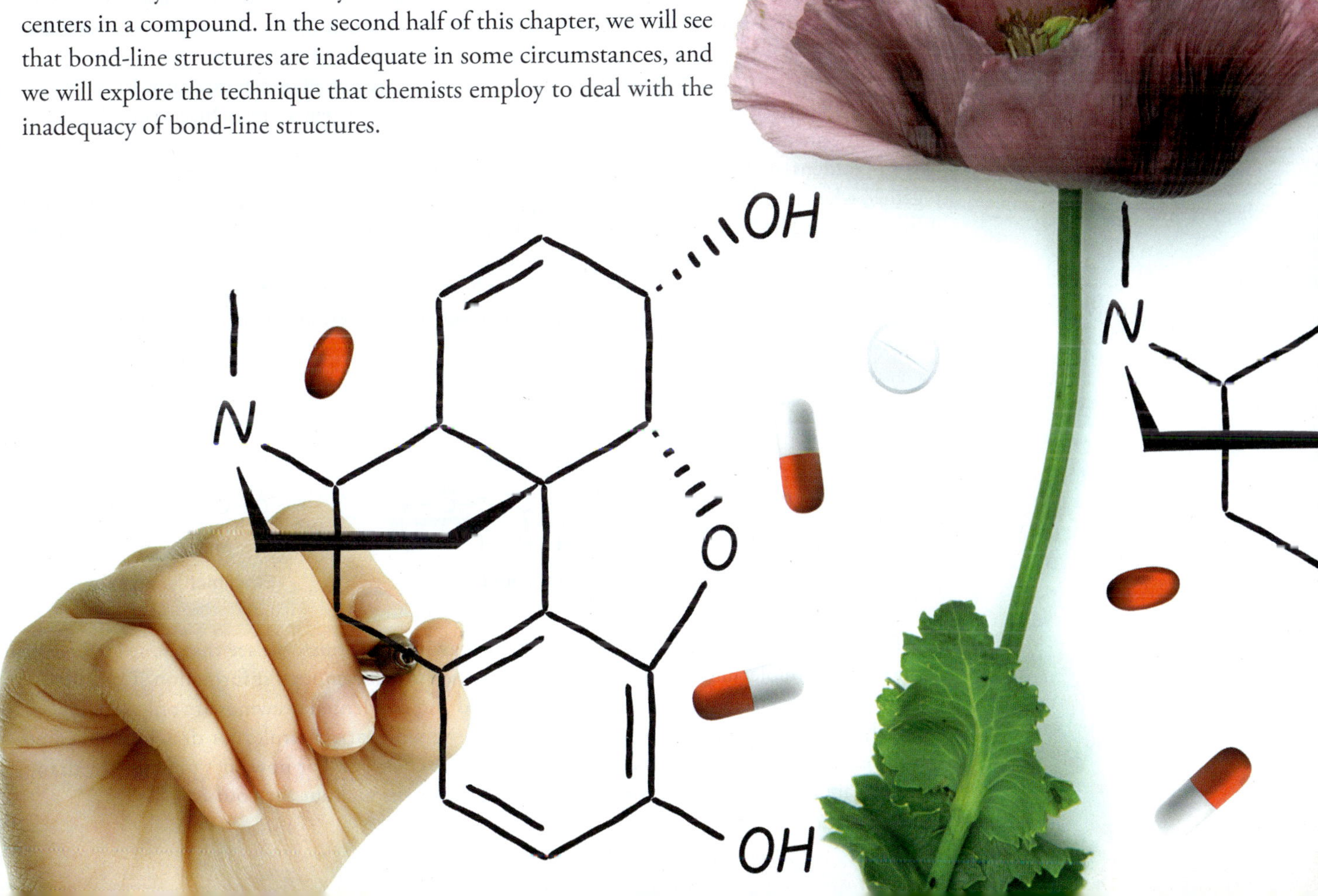

## DO YOU REMEMBER?

*Before you go on, be sure you understand the following topics.*
*If necessary, review the suggested sections to prepare for this chapter:*

- Electrons, Bonds, and Lewis Structures (Section 1.3)
- Identifying Formal Charges (Section 1.4)
- Molecular Orbital Theory (Section 1.8)

Take the DO YOU REMEMBER? QUIZ in **WileyPLUS** to check your understanding.

## 2.1 Molecular Representations

Chemists use many different styles to draw molecules. Let's consider the structure of isopropanol, also called isopropyl rubbing alcohol, which is used as a disinfectant in sterilizing pads. The structure of this compound is shown below in a variety of drawing styles:

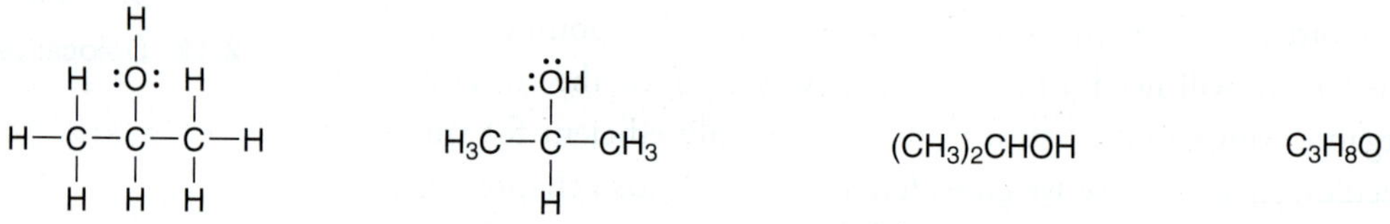

$H_3C-CH(OH)-CH_3$ $(CH_3)_2CHOH$ $C_3H_8O$

**Lewis structure** **Partially condensed structure** **Condensed structure** **Molecular formula**

Lewis structures were discussed in the previous chapter. The advantage of Lewis structures is that all atoms and bonds are explicitly drawn. However, Lewis structures are only practical for very small molecules. For larger molecules, it becomes extremely burdensome to draw out every bond and every atom.

In **partially condensed structures**, the C—H bonds are not all drawn explicitly. In the example above, $CH_3$ refers to a carbon atom with bonds to three hydrogen atoms. Once again, this drawing style is only practical for small molecules.

In **condensed structures**, single bonds are not drawn. Instead, groups of atoms are clustered together, when possible. For example, isopropanol has two $CH_3$ groups, both of which are connected to the central carbon atom, shown like this: $(CH_3)_2CHOH$. Once again, this drawing style is only practical for small molecules with simple structures.

The molecular formula of a compound simply shows the number of each type of atom in the compound ($C_3H_8O$). No structural information is provided. There are actually three constitutional isomers with molecular formula $C_3H_8O$:

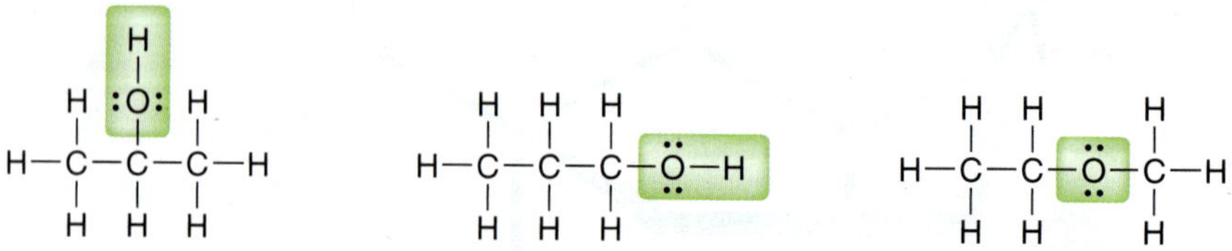

**Isopropanol** **Propanol** **Ethyl methyl ether**

In reviewing some of the different styles for drawing molecules, we see that none are convenient for larger molecules. Molecular formulas do not provide enough information, Lewis structures take too long to draw, and partially condensed and condensed drawings are only suitable for relatively simple molecules. In upcoming sections, we will learn the rules for drawing bond-line structures, which are most commonly used by organic chemists. For now, let's practice the drawing styles above, which will be used for small molecules throughout the course.

# SKILLBUILDER

## 2.1 CONVERTING BETWEEN DIFFERENT DRAWING STYLES

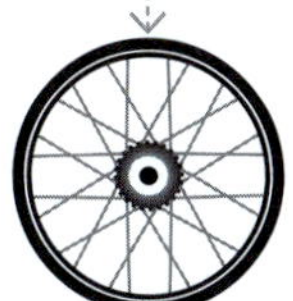

**LEARN the skill**

Draw a Lewis structure for the following compound:

$$(CH_3)_2CH\ddot{\underset{\cdot\cdot}{O}}CH_2CH_3$$

### SOLUTION

This compound is shown in condensed format. In order to draw a Lewis structure, begin by drawing out each group separately, showing a partially condensed structure:

**STEP 1**
Draw each group separately.

$(CH_3)_2CH\ddot{O}CH_2CH_3$ - - - - - -> $H_3C-\underset{\underset{CH_3}{|}}{\overset{H}{C}}-\ddot{O}-CH_2CH_3$

**Condensed structure** **Partially condensed structure**

Then draw all C—H bonds:

**STEP 2**
Draw all C—H bonds.

$H_3C-\underset{\underset{CH_3}{|}}{\overset{H}{C}}-\ddot{O}-CH_2CH_3$ - - - - - -> Lewis structure

**Partially condensed structure** **Lewis structure**

**PRACTICE the skill**

**2.1** Draw a Lewis structure for each of the compounds below:

**(a)** $CH_2{=}CHOCH_2CH(CH_3)_2$

**(b)** $(CH_3CH_2)_2CHCH_2CH_2OH$

**(c)** $(CH_3CH_2)_3COH$

**(d)** $(CH_3)_2C{=}CHCH_2CH_3$

**(e)** $CH_2{=}CHCH_2OCH_2CH(CH_3)_2$

**(f)** $(CH_3CH_2)_2C{=}CH_2$

**(g)** $(CH_3)_3CCH_2CH_2OH$

**(h)** $CH_3CH_2CH_2CH_2CH_2CH_3$

**(i)** $CH_3CH_2CH_2OCH_3$

**(j)** $(CH_3CH_2CH_2)_2CHOH$

**(k)** $(CH_3CH_2)_2CHCH_2OCH_3$

**(l)** $(CH_3)_2CHCH_2OH$

**APPLY the skill**

**2.2** Identify which two compounds below are constitutional isomers:

$(CH_3)_3C\ddot{O}CH_3$ $(CH_3)_2CH\ddot{O}CH_3$ $(CH_3)_2CH\ddot{O}CH_2CH_3$

**2.3** Identify the number of $sp^3$-hybridized carbon atoms in the following compound:

$$(CH_3)_2C{=}CHC(CH_3)_3$$

**2.4** Cyclopropane (shown below) has only one constitutional isomer. Draw a condensed structure of that isomer.

$H_2C$, $CH_2$, $H_2C$ (three-membered ring)

need more **PRACTICE?** **Try Problems 2.49, 2.50**

## 2.2 Bond-Line Structures

It is not practical to draw Lewis structures for all compounds, especially large ones. As an example, consider the structure of amoxicillin, one of the most commonly used antibiotics in the penicillin family:

**Amoxicillin**

Previously fatal infections have been rendered harmless by antibiotics such as the one above. Amoxicillin is not a large compound, yet drawing this compound is time consuming. To deal with this problem, organic chemists have developed an efficient drawing style that can be used to draw molecules very quickly. **Bond-line structures** not only simplify the drawing process but also are easier to read. The following is a bond-line structure of amoxicillin.

$NH_2$ H N S O O N HO O HO

Most of the atoms are not drawn, but with practice, these drawings will become very user-friendly. Throughout the rest of this textbook, most compounds will be drawn in bond-line format, and therefore, it is absolutely critical to master this drawing technique. The following sections are designed to develop this mastery.

**BY THE WAY**

You may find it worthwhile to purchase or borrow a molecular model set. There are several different kinds of molecular model sets on the market, and most of them are comprised of plastic pieces that can be connected to generate models of small molecules. Any one of these model sets will help you to visualize the relationship between molecular structures and the drawings used to represent them.

### How to Read Bond-Line Structures

Bond-line structures are drawn in a zigzag format (), where each corner or endpoint represents a carbon atom. For example, each of the following compounds has six carbon atoms (count them!):

Double bonds are shown with two lines, and triple bonds are shown with three lines:

Notice that triple bonds are drawn in a linear fashion rather than in a zigzag format, because triple bonds involve *sp*-hybridized carbon atoms, which have linear geometry (Section 1.9). The two carbon atoms of a triple bond and the two carbon atoms connected to them are drawn in a straight line. All other bonds are drawn in a zigzag format; for example, the following compound has eight carbon atoms:

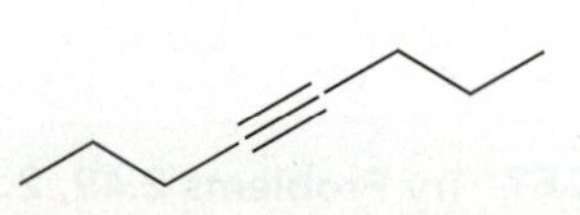

Hydrogen atoms bonded to carbon are also not shown in bond-line structures, because it is assumed that each carbon atom will possess enough hydrogen atoms so as to achieve a total of four bonds. For example, the highlighted carbon atom below appears to have only two bonds:

The drawing indicates only two bonds connected to this carbon atom

Therefore, we can infer that there must be two more bonds to hydrogen atoms that have not been drawn (to give a total of four bonds). In this way, all hydrogen atoms are inferred by the drawing:

With a bit of practice, it will no longer be necessary to count bonds. Familiarity with bond-line structures will allow you to "see" all of the hydrogen atoms even though they are not drawn. This level of familiarity is absolutely essential, so let's get some practice.

# SKILLBUILDER

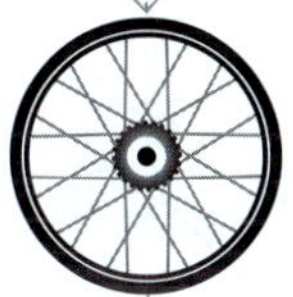

## 2.2 READING BOND-LINE STRUCTURES

### LEARN the skill

Consider the structure of diazepam, first marketed by the Hoffmann-La Roche Company under the trade name Valium:

**Diazepam (Valium)**

Valium is a sedative and muscle relaxant used in the treatment of anxiety, insomnia, and seizures. Identify the number of carbon atoms in diazepam, then fill in all the missing hydrogen atoms that are inferred by the drawing.

### SOLUTION

Remember that each corner and each endpoint represents a carbon atom. This compound therefore has 16 carbon atoms, highlighted below:

**STEP 1**
Count the carbon atoms, which are represented by corners or endpoints.

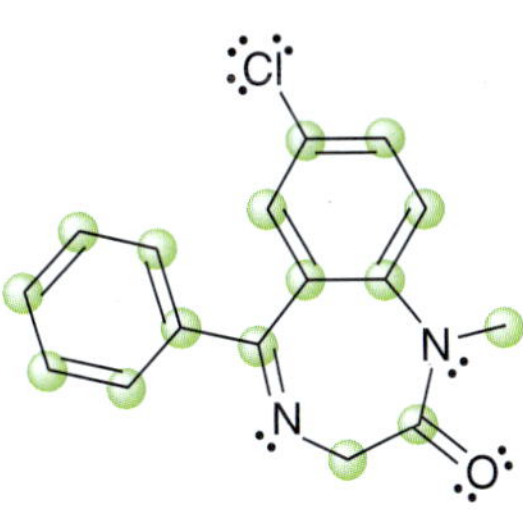

Each carbon atom should have four bonds. We therefore draw enough hydrogen atoms in order to give each carbon atom a total of four bonds. Any carbon atoms that already have four bonds will not have any hydrogen atoms:

**STEP 2**
Count the hydrogen atoms. Each carbon atom will have enough hydrogen atoms to have exactly four bonds.

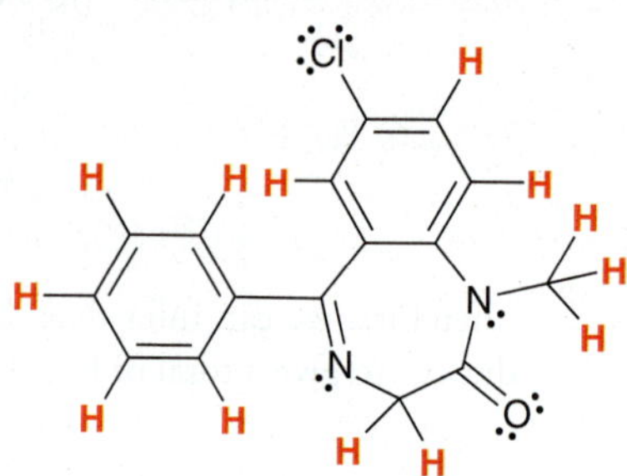

**PRACTICE the skill** **2.5** For each of the following molecules, determine the number of carbon atoms present and then determine the number of hydrogen atoms connected to each carbon atom:

(a) 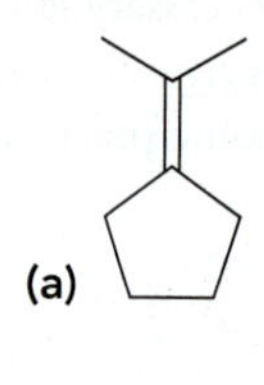

(b) 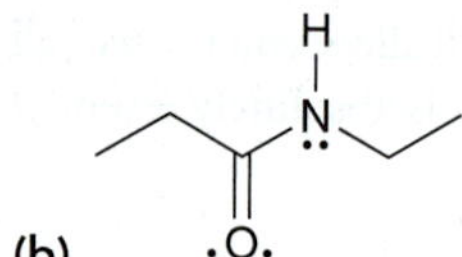


(c)

(d) 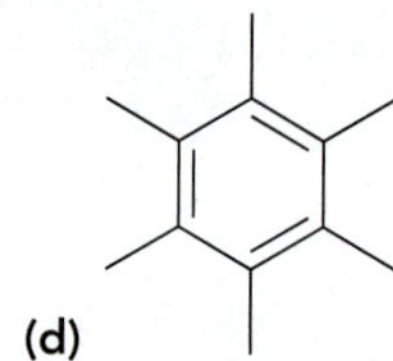

(e) (f) 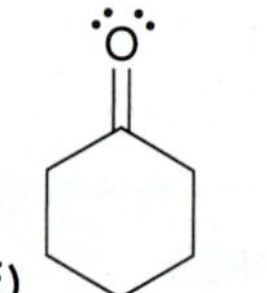


(g) (h) (i) (j) (k) (l)

**APPLY the skill**

**2.6** Each transformation below shows a starting material being converted into a product (the reagents necessary to achieve the transformation have not been shown). For each transformation, determine whether the product has more carbon atoms, fewer carbon atoms, or the same number of carbon atoms as the starting material. In other words, determine whether each transformation involves an increase, decrease, or no change in the number of carbon atoms.

(a) (b) (c) (d)

**2.7** Identify whether each transformation below involves an increase, a decrease, or no change in the number of hydrogen atoms:

(a) (b)

need more **PRACTICE?** **Try Problems 2.39, 2.49, 2.52**

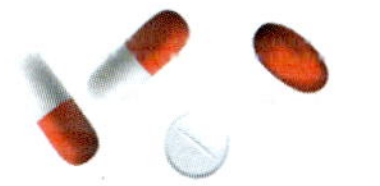

## How to Draw Bond-Line Structures

It is certainly important to be able to read bond-line structures fluently, but it is equally important to be able to draw them proficiently. When drawing bond-line structures, the following rules should be observed:

**WATCH OUT**

Notice that the first structure includes atom labels for each carbon (C) and hydrogen (H), while the second structure does not show any atom labels for carbon and hydrogen. Both structures are valid drawings, but it is incorrect to label the carbon atoms without also labeling the hydrogen atoms (for example: C—C—C—C). Either draw every C and every H, as in the first structure or don't draw any of those labels, as in the second structure.

1. Carbon atoms in a straight chain should be drawn in a zigzag format:

   is drawn like this:

2. When drawing double bonds, draw all bonds as far apart as possible:

   is much better than

   Bad

3. When drawing single bonds, the direction in which the bonds are drawn is irrelevant:

   is the same as

   These two drawings do not represent constitutional isomers—they are just two drawings of the same compound. Both are perfectly acceptable.

4. All *heteroatoms* (atoms other than carbon and hydrogen) must be drawn, and any hydrogen atoms attached to a heteroatom must also be drawn. For example:

5. Never draw a carbon atom with more than four bonds. Carbon only has four orbitals in its valence shell, and therefore carbon atoms can only form four bonds.

# SKILLBUILDER

## 2.3 DRAWING BOND-LINE STRUCTURES

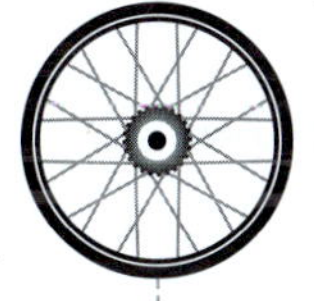

**LEARN the skill**

Draw a bond-line structure for the following compound:

## SOLUTION

Drawing a bond-line structure requires just a few conceptual steps. First, delete all hydrogen atoms except for those connected to heteroatoms:

**STEP 1**
Delete hydrogen atoms, except for those connected to heteroatoms.

Then, place the carbon skeleton in a zigzag arrangement, making sure that any triple bonds are drawn as linear:

**STEP 2**
Draw in zigzag format, keeping triple bonds linear.

Finally, delete all carbon atoms:

**STEP 3**
Delete carbon atoms.

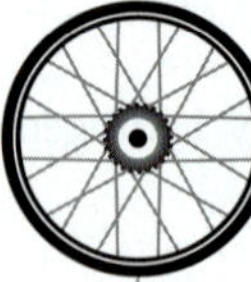

**PRACTICE** the skill

**2.8** Draw a bond-line structure for each of the following compounds:

(a) (b) (c)

**(d)** $(CH_3)_3C{-}C(CH_3)_3$ **(e)** $CH_3CH_2CH(CH_3)_2$ **(f)** $(CH_3CH_2)_3COH$

**(g)** $(CH_3)_2CHCH_2OH$ **(h)** $CH_3CH_2CH_2OCH_3$ **(i)** $(CH_3CH_2)_2C{=}CH_2$

**(j)** $CH_2{=}CHOCH_2CH(CH_3)_2$ **(k)** $(CH_3CH_2)_2CHCH_2CH_2NH_2$

**(l)** $CH_2{=}CHCH_2OCH_2CH(CH_3)_2$ **(m)** $CH_3CH_2CH_2CH_2CH_2CH_3$

**(n)** $(CH_3CH_2CH_2)_2CHCl$ **(o)** $(CH_3)_2C{=}CHCH_2CH_3$

**(p)** $(CH_3CH_2)_2CHCH_2OCH_3$ **(q)** $(CH_3)_3CCH_2CH_2OH$

**(r)** $(CH_3CH_2CH_2)_3COCH_2CH_2CH{=}CHCH_2CH_2OC(CH_2CH_3)_3$

**APPLY** the skill

**2.9** Draw bond-line structures for all constitutional isomers of the following compound:

$$CH_3CH_2CH(CH_3)_2$$

**2.10** In each of the following compounds, identify all carbon atoms that you expect will be deficient in electron density (δ+). If you need help, refer to Section 1.5.

(a)

(b)

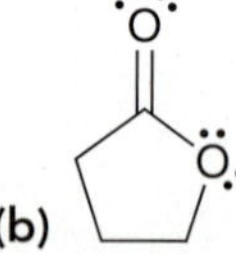

(c)

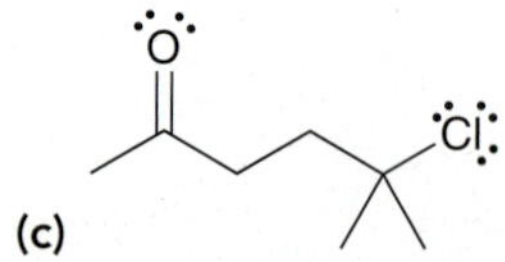

need more **PRACTICE?** **Try Problems 2.40, 2.41, 2.54, 2.58**

## 2.3 Identifying Functional Groups

Bond-line drawings are the preferred drawing style used by practicing organic chemists. In addition to being more efficient, bond-line drawings are also easier to read. As an example, consider the following reaction:

$$(CH_3)_2CHCH{=}C(CH_3)_2 \xrightarrow[Pt]{H_2} (CH_3)_2CHCH_2CH(CH_3)_2$$

When the reaction is presented in this way, it is somewhat difficult to see what is happening. It takes time to digest the information being presented. However, when we redraw the same reaction using bond-line structures, it becomes very easy to identify the transformation taking place:

$$\xrightarrow[Pt]{H_2}$$

It is immediately apparent that a double bond is being converted into a single bond. With bond-line drawings, it is easier to identify the functional group and its location. A **functional group** is a characteristic group of atoms/bonds that possess a predictable chemical behavior. In each of the reactions below, the starting material has a carbon-carbon double bond, which is a functional group. Compounds with carbon-carbon double bonds typically react with molecular hydrogen ($H_2$) in the presence of a catalyst (such as Pt). Both of the starting materials below have a carbon-carbon double bond, and consequently, they exhibit similar chemical behavior.

$$\xrightarrow[Pt]{H_2}$$

$$\xrightarrow[Pt]{H_2}$$

The chemistry of every organic compound is determined by the functional groups present in the compound. Therefore, the classification of organic compounds is based on their functional groups. For example, compounds with carbon-carbon double bonds are classified as *alkenes*, while compounds possessing an OH group are classified as *alcohols*. Many of the chapters in this book are organized by functional group. Table 2.1 provides a list of common functional groups and the corresponding chapters in which they appear.

### CONCEPTUAL CHECKPOINT

**2.11** Atenolol and enalapril are drugs used in the treatment of heart disease. Both of these drugs lower blood pressure (albeit in different ways) and reduce the risk of heart attack. Using Table 2.1, identify and label all functional groups in these two compounds:

$NH_2$ O N H OH O

**Atenolol**

O O N N H HO O O

**Enalapril**

**TABLE 2.1 EXAMPLES OF COMMON FUNCTIONAL GROUPS**

| FUNCTIONAL GROUP* | CLASSIFICATION | EXAMPLE | CHAPTER |
|---|---|---|---|
| R—X: (X=Cl, Br, or I) | Alkyl halide |  *n*-Propyl chloride | 7 |
| 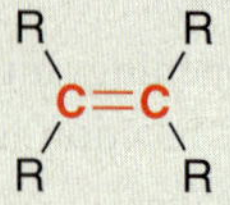 $R_2C{=}CR_2$ | Alkene | 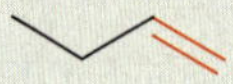 1-Butene | 8, 9 |
| R—C≡C—R | Alkyne | 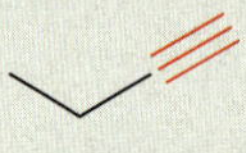 1-Butyne | 10 |
| R—OH | Alcohol | 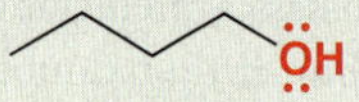 1-Butanol | 13 |
| R—O—R | Ether |  Diethyl ether | 14 |
| R—SH | Thiol | 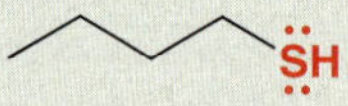 1-Butanethiol | 14 |
| R—S—R | Sulfide |  Diethyl sulfide | 14 |
|  (benzene ring) | Aromatic (or arene) | 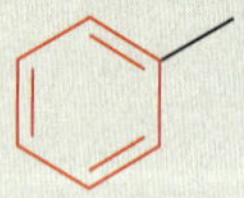 Methylbenzene | 18, 19 |
| 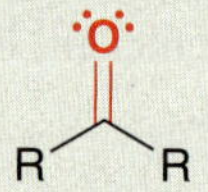 R(C=O)R | Ketone | 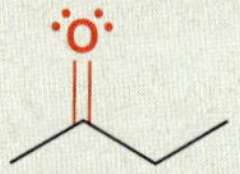 2-Butanone | 20 |
| 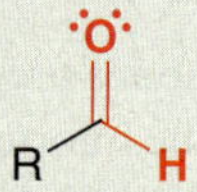 R(C=O)H | Aldehyde | 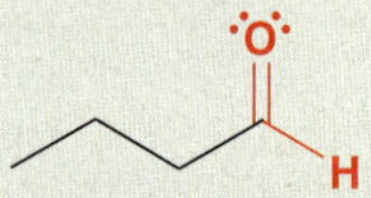 Butanal | 20 |
| 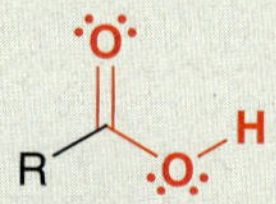 R(C=O)OH | Carboxylic acid | 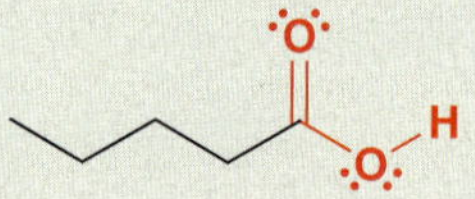 Pentanoic acid | 21 |
| 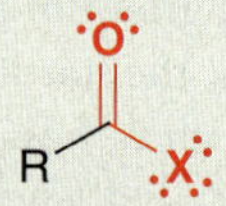 R(C=O)X | Acyl halide | 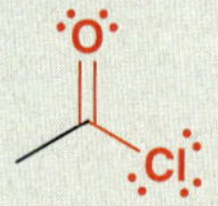 Acetyl chloride | 21 |
| 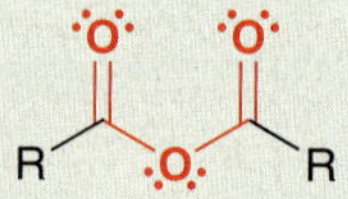 R(C=O)O(C=O)R | Anhydride | 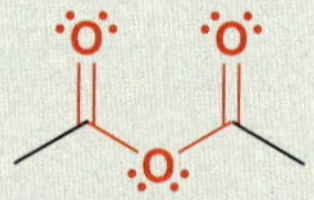 Acetic anhydride | 21 |
| 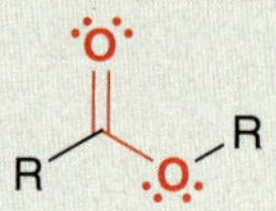 R(C=O)OR | Ester | 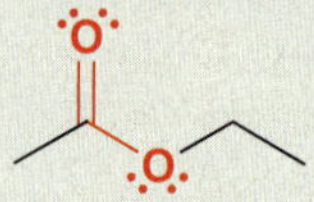 Ethyl acetate | 21 |
| 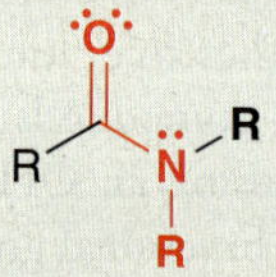 R(C=O)NR₂ | Amide | 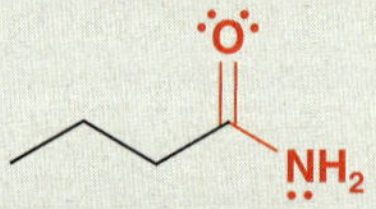 Butanamide | 21 |
| 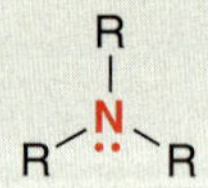 $R_3N$ | Amine | 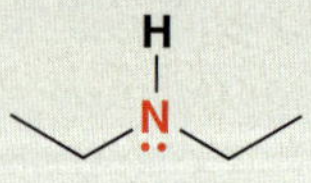 Diethylamine | 23 |

* The "R" refers to the remainder of the compound, usually carbon and hydrogen atoms.

medically speaking

## Marine Natural Products

The field of marine natural products (MNP) continues to expand rapidly as researchers explore the ocean's rich biodiversity in search of new pharmaceuticals. It is a subdiscipline of natural product chemistry and has recently seen some success stories of new drugs. In the early days before the dawn of SCUBA, marine drug discovery efforts were aimed primarily at easily accessible marine life like red algae, sponges, and soft corals that were not far from the ocean's shoreline.

Since then more emphasis was placed on the previously overlooked deep sea organisms as well as marine microbes associated with ocean sediments and macroorganisms. These sources have yielded both structurally diverse and bioactive compounds. Moreover, the marine microbes are taught to be the real producers of MNP that were previously isolated from their macroorganism hosts like mollusks, sponges, and tunicates.

The early stage of MNP drug discovery was met with supply issues due to limited marine samples that yielded small quantities of the bioactive natural products. This demand was addressed with innovative methods such as aquaculture (farming of aquatic organisms), total synthesis, and biosynthesis. The use of genomics (DNA sequencing of organisms) and proteomics (study of protein structure and function), for example, has provided new insights into the production and distribution of these natural products.

Of all the MNP and MNP-derived drug candidates identified to date, 7 have been approved by the Food and Drug Administration (FDA) for public use, with at least 13 others at various stages of clinical trials. Examples of FDA approved drugs are: eribulin mesylate (E7389), omega-3-acid ethyl ester, and trabectedin (ET-743). See structures below.

Eribulin mesylate is a synthetic analogue of the marine natural product halichondrin B which was isolated from the black sponge *Halichondria okadai*. It is a polyether analogue that is used in the treatment of cancer which spreads from the breast to other body organs. It acts by blocking cellular growth, which results in the death of the cancer cells.

The second drug, omega-3-acid ethyl ester, contains derivatives of long-chain, unsaturated carboxylic acids (omega-3 fatty acids) that were isolated from fish oil. They consist primarily of ethyl esters of eicosapentanoic acid (EPA) and docosahexanoic acid (DHA). The drug is used to treat persons with high levels of fat (lipid) in their bloodstream as this condition can lead to heart disease and stroke.

The third therapeutic agent, trabectedin (ET-743), was isolated from the marine tunicate *Ecteinascidia turbinata*. It is used to treat patients with soft tissue sarcoma (STS) and those with recurring ovarian cancer. Trabectedin interferes with the DNA minor groove and causes cell death.

The future of MNP is indeed bright. With the increasing technological advances in high-throughput screening (HTS), compound libraries, mass spectrometry (MS), nuclear magnetic resonance (NMR) spectroscopy, genomics, compound biosynthesis, and more, scientists will continue to push the envelope in the hopes of finding those important marine natural products that will serve as future drug candidates.

**Halichondrin B**

**Eribulin mesylate**

**EPA ethyl ester**

**DHA ethyl ester**

**Trabectedin (ET-743)**

## 2.4 Carbon Atoms with Formal Charges

In Section 1.4 we saw that a formal charge is associated with any atom that does not exhibit the appropriate number of valence electrons. Formal charges are extremely important, and they must be shown in bond-line structures. A missing formal charge renders a bond-line structure incorrect and therefore useless. Accordingly, let's quickly practice identifying formal charges in bond-line structures.

### CONCEPTUAL CHECKPOINT

**2.12** For each of the following structures determine whether any of the nitrogen atoms bear a formal charge:

(a) 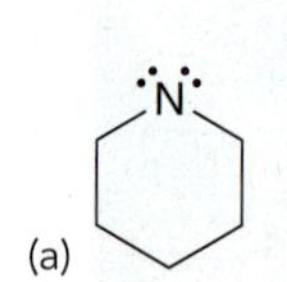


(b) 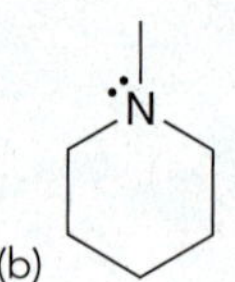


(c) 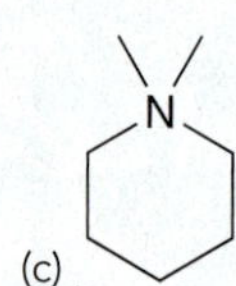


(d) 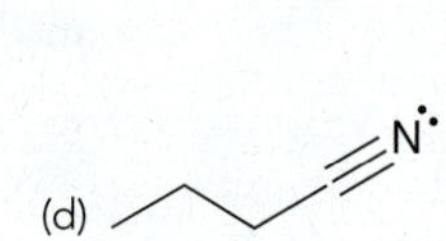


**2.13** For each of the following structures determine whether any of the oxygen atoms bear a formal charge:

(a) 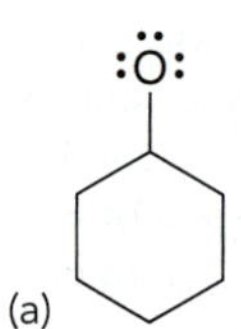


(b) 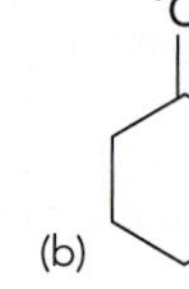


(c) 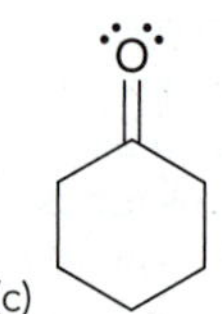


(d) 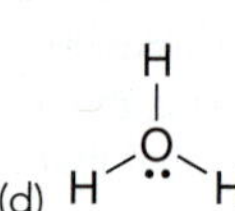


Now let's consider formal charges on carbon atoms. We have seen that carbon generally has four bonds, which allows us to "see" all of the hydrogen atoms even though they are not explicitly shown in bond-line structures. Now we must modify that rule: *A carbon atom will generally have four bonds only when it does not have a formal charge.* When a carbon atom bears a formal charge, either positive or negative, it will have three bonds rather than four. To understand why, let's first consider $C^+$, and then we will consider $C^-$.

Recall that the appropriate number of valence electrons for a carbon atom is four. In order to have a positive formal charge, a carbon atom must be missing an electron. In other words, it must have only three valence electrons. Such a carbon atom can only form three bonds. This must be taken into account when counting hydrogen atoms:

**No hydrogen atoms on this $C^+$**

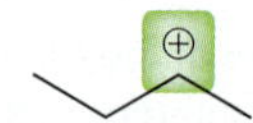

**One hydrogen atom on this $C^+$**

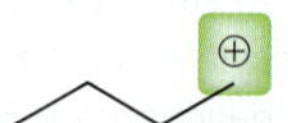

**Two hydrogen atoms on this $C^+$**

Now let's focus on negatively charged carbon atoms. In order to have a negative formal charge, a carbon atom must have one extra electron. In other words, it must have five valence electrons. Two of those electrons will form a lone pair, and the other three electrons will be used to form bonds:

$$H-\overset{\displaystyle H}{\underset{\displaystyle H}{\overset{|}{\underset{|}{C}}}}:^{\ominus}$$

In summary, both $C^+$ and $C^-$ will have only three bonds. The difference between them is the nature of the fourth orbital. In the case of $C^+$, the fourth orbital is empty. In the case of $C^-$, the fourth orbital holds a lone pair of electrons.

## 2.5 Identifying Lone Pairs

In order to determine the formal charge on an atom, we must know how many lone pairs it has. On the flip side, we must know the formal charge in order to determine the number of lone pairs on an atom. To understand this, let's examine a case where neither the lone pairs nor the formal charges are drawn:

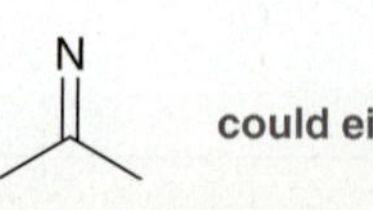


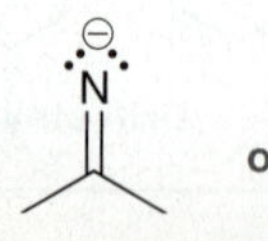


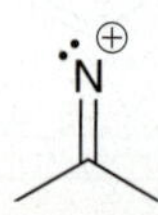

If the lone pairs were shown, then we could determine the charge (two lone pairs would mean a negative charge, and one lone pair would mean a positive charge). Alternatively, if the formal charge were shown, then we could determine the number of lone pairs (a negative charge would mean two lone pairs, and a positive charge would mean one lone pair).

Therefore, a bond-line structure will only be clear if it contains either all of the lone pairs or all of the formal charges. Since there are typically many more lone pairs than formal charges in any one particular structure, chemists have adopted the convention of always drawing formal charges, which allows us to leave out the lone pairs.

**WATCH OUT**

Formal charges must always be drawn and can never be omitted, unlike lone pairs, which may be omitted from a bond-line structure.

Now let's get some practice identifying lone pairs when they are not drawn. The following example will demonstrate the thought process:

In order to determine the number of lone pairs on the oxygen atom, we simply use the same two-step process described in Section 1.4 for calculating formal charges:

1. *Determine the appropriate number of valence electrons for the atom.* Oxygen is in group 6A of the periodic table, and therefore, it should have six valence electrons.
2. *Determine if the atom actually exhibits the appropriate number of electrons.* This oxygen atom has a negative formal charge, which means it must have one extra electron. Therefore, this oxygen atom must have $6 + 1 = 7$ valence electrons. One of those electrons is being used to form the C—O bond, which leaves six electrons to be housed as lone pairs. This oxygen atom must therefore have three lone pairs:

is the same as

The process above represents an important skill; however, it is even more important to become familiar enough with atoms that the process becomes unnecessary. There are just a handful of patterns to recognize. Let's go through them methodically, starting with oxygen. Table 2.2 summarizes the important patterns that you will encounter for oxygen atoms.

- A negative charge corresponds with one bond and three lone pairs.
- The absence of charge corresponds with two bonds and two lone pairs.
- A positive charge corresponds with three bonds and one lone pair.

TABLE **2.2** FORMAL CHARGE ON AN OXYGEN ATOM ASSOCIATED WITH A PARTICULAR NUMBER OF BONDS AND LONE PAIRS

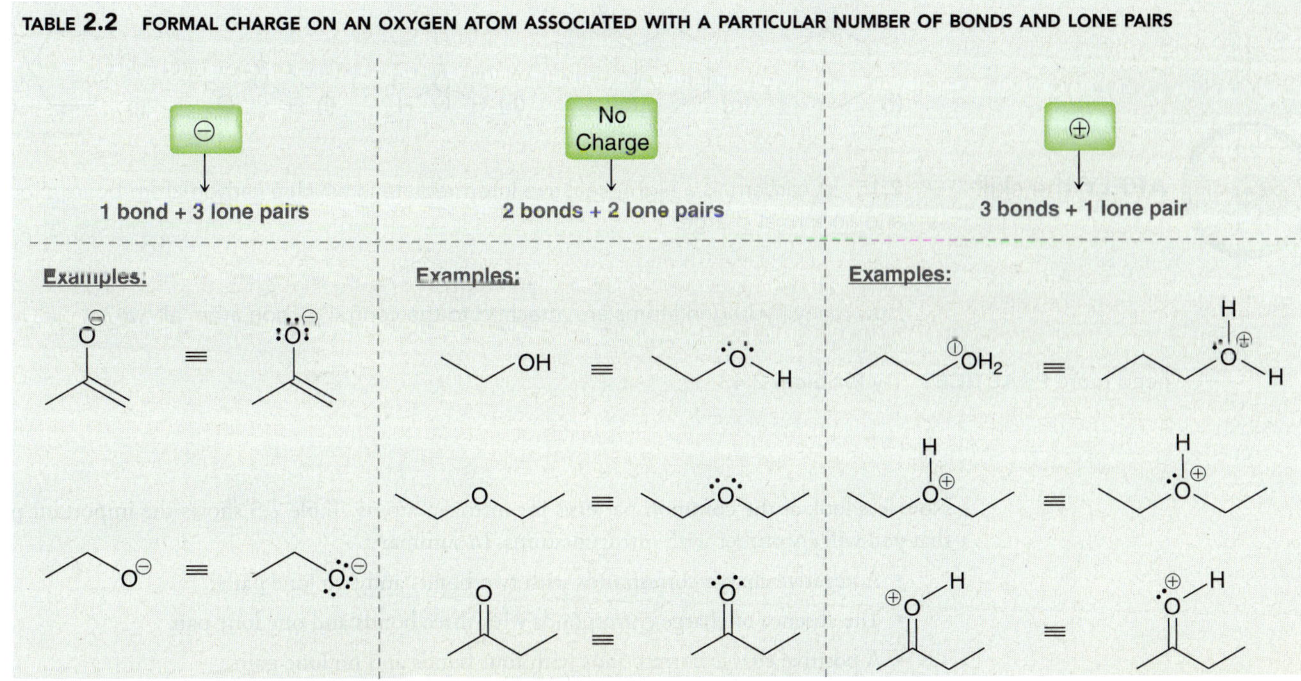

# SKILLBUILDER

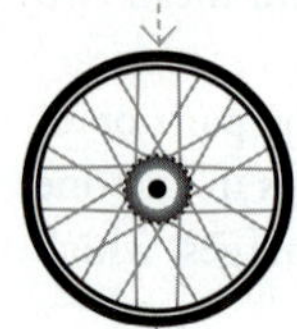

## 2.4 IDENTIFYING LONE PAIRS ON OXYGEN ATOMS

**LEARN the skill**

Draw all of the lone pairs in the following structure:

### SOLUTION

**STEP 1**
Determine the appropriate number of valence electrons.

**STEP 2**
Analyze the formal charge and determine the actual number of valence electrons.

**STEP 3**
Count the number of bonds and determine how many of the actual valence electrons must be lone pairs.

The oxygen atom above has a positive formal charge and three bonds. It is preferable to recognize the pattern—that a positive charge and three bonds must mean that the oxygen has just one lone pair:

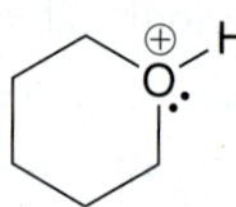

Alternatively, and less preferably, it is possible to calculate the number of lone pairs using the following two steps. First, determine the appropriate number of valence electrons for the atom. Oxygen is in group 6A of the periodic table, and therefore, it should have six valence electrons. Then, determine if the atom actually exhibits the appropriate number of electrons. This oxygen atom has a positive charge, which means it is missing an electron: $6 - 1 = 5$ valence electrons. Three of these five electrons are being used to form bonds, which leaves just two electrons for a lone pair. This oxygen atom has only one lone pair.

**PRACTICE the skill**

**2.14** Draw all lone pairs on each of the oxygen atoms in the following structures. Before doing this, review Table 2.2 and then come back to these problems. Try to identify all lone pairs without having to count. Then, count to see if you were correct.

(a) (b) (c) (d) (e)

(f) (g) (h) (i) (j)

**APPLY the skill**

**2.15** A carbene is a highly reactive intermediate in which a carbon atom bears a lone pair and no formal charge:

How many hydrogen atoms are attached to the central carbon atom above?

need more **PRACTICE?** **Try Problem 2.43**

Now let's look at the common patterns for nitrogen atoms. Table 2.3 shows the important patterns that you will encounter with nitrogen atoms. In summary:

- A negative charge corresponds with two bonds and two lone pairs.
- The absence of charge corresponds with three bonds and one lone pair.
- A positive charge corresponds with four bonds and no lone pairs.

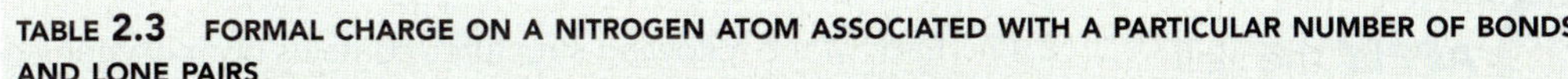


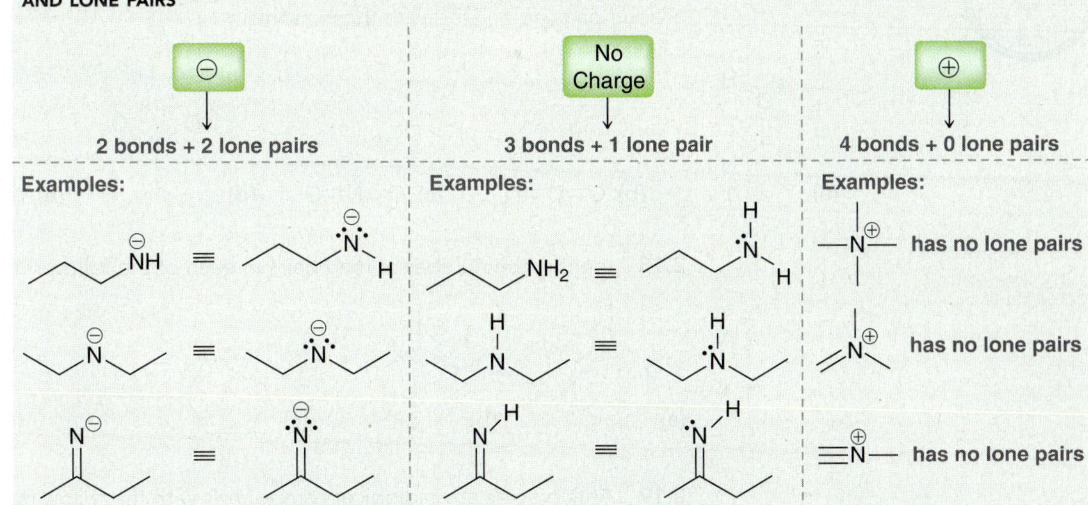


# SKILLBUILDER

## 2.5 IDENTIFYING LONE PAIRS ON NITROGEN ATOMS

**LEARN the skill**

Draw any lone pairs associated with the nitrogen atoms in the following structure:

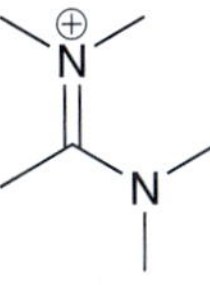

### SOLUTION

**STEP 1**
Determine the appropriate number of valence electrons.

**STEP 2**
Analyze the formal charge and determine the actual number of valence electrons.

**STEP 3**
Count the number of bonds and determine how many of the actual valence electrons must be lone pairs.

The top nitrogen atom has a positive formal charge and four bonds. The bottom nitrogen has three bonds and no formal charge. It is preferable to simply recognize that the top nitrogen atom must have no lone pairs and the bottom nitrogen atom must have one lone pair:

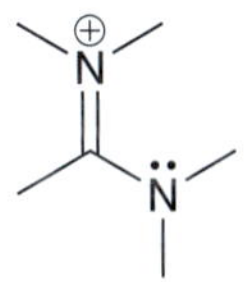


Alternatively, and less preferably, it is possible to calculate the number of lone pairs using the following two steps. First, determine the appropriate number of valence electrons for the atom. Each nitrogen atom should have five valence electrons. Next, determine if each atom actually exhibits the appropriate number of electrons. The top nitrogen atom has a positive charge, which means it is missing an electron. This nitrogen atom actually has only four valence electrons. Since the nitrogen atom has four bonds, it is using each of its four electrons to form a bond. This nitrogen atom does not possess a lone pair. The bottom nitrogen atom has no formal charge, so this nitrogen atom must be using five valence electrons. It has three bonds, which means that there are two electrons left over, forming one lone pair.

**PRACTICE the skill**

**2.16** Draw all lone pairs on each of the nitrogen atoms in the following structures. First, review Table 2.3 and then come back to these problems. Try to identify all lone pairs *without* having to count. Then, count to see if you were correct.

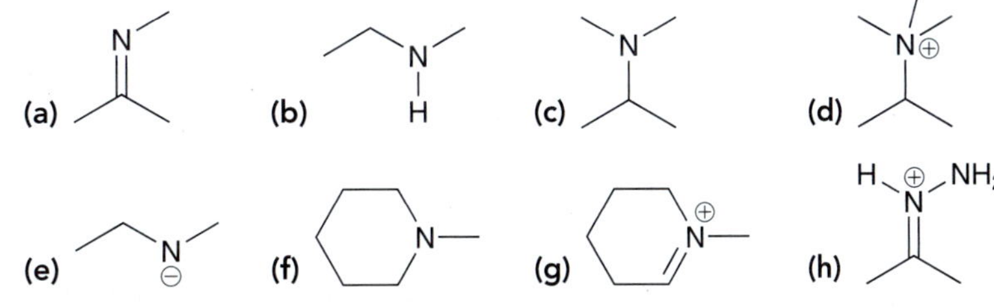

**APPLY the skill**

**2.17** Each of the following structures contains both oxygen and nitrogen atoms. Identify all lone pairs associated with those atoms.

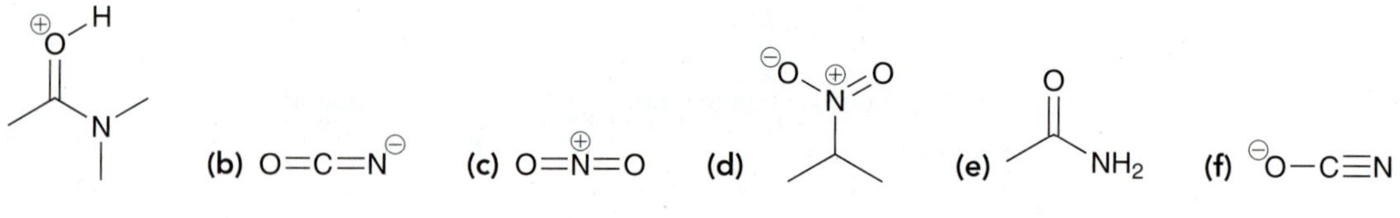

**2.18** Identify the number of lone pairs in each of the following structures:

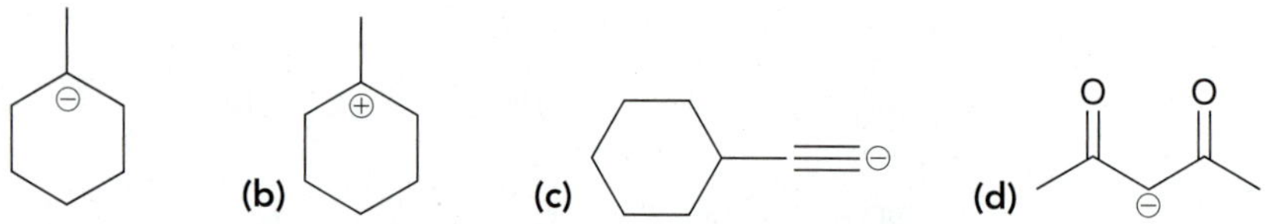

**2.19** Amino acids are biological compounds with the following structure, where the R group can vary. The structure and biological function of amino acids will be discussed in Chapter 25. Identify the total number of lone pairs present in an amino acid, assuming that the R group does not contain any atoms with lone pairs.

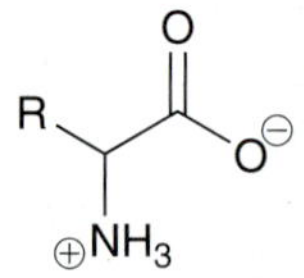

**An amino acid**

need more **PRACTICE?** **Try Problem 2.39**

## 2.6 Three-Dimensional Bond-Line Structures

Throughout this book, we will use many different kinds of drawings to represent the three-dimensional geometry of atoms. The most common method is a bond-line structure that includes **wedges** and **dashes** to indicate three dimensionality. These structures are used for all types of compounds, including acyclic, cyclic, and bicyclic compounds (Figure 2.1). In the drawings in Figure 2.1, a wedge represents a group coming out of the page, and a dash represents a group going behind the page. We will use wedges and dashes extensively in Chapter 5 and thereafter.

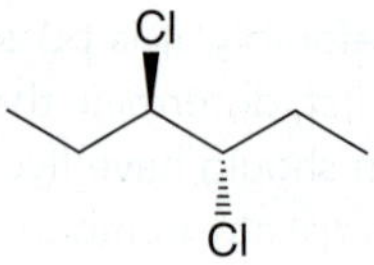

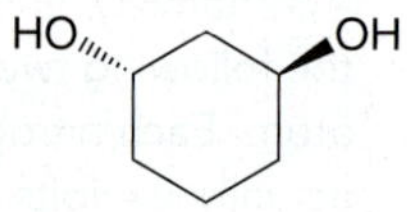

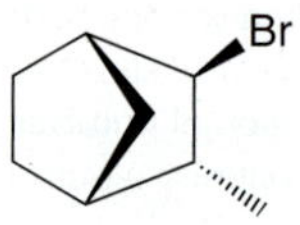

**FIGURE 2.1** Bond-line structures with wedges and dashes to indicate three dimensionality.

In certain circumstances, there are other types of drawings that can be used, all of which also indicate three-dimensional geometry (Figure 2.2).

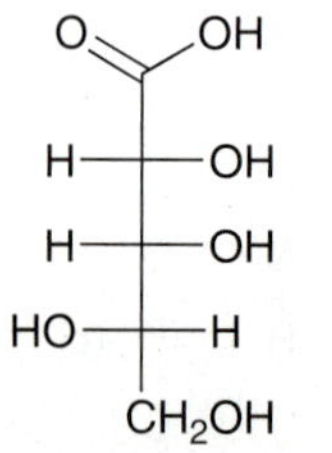

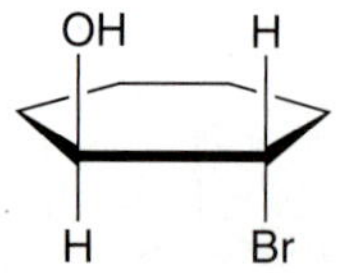

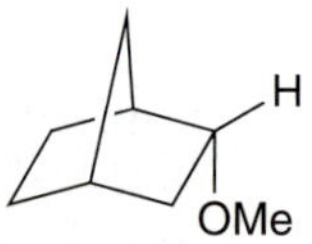

**FIGURE 2.2** Common drawing styles that show three dimensionality for acylic, cyclic, and bicyclic compounds.

**Fischer projections** are used for acyclic compounds while **Haworth projections** are used exclusively for cyclic compounds. Each of these drawing styles will be used several times throughout this book, particularly in Chapters 5, 9, and 24.

## medically speaking | Identifying the Pharmacophore

As mentioned in the chapter opener, there are many techniques that scientists employ in the design of new drugs. One such technique is called *lead modification*, which involves modifying the structure of a compound known to exhibit desirable medicinal properties. The known compound "leads" the way to the development of other similar compounds and is therefore called the *lead compound*. The story of morphine provides a good example of this process.

Morphine is a very potent analgesic (pain reliever) that is known to act on the central nervous system as a depressant (causing sedation and slower respiratory function) and as a stimulant (relieving symptoms of anxiety and causing an overall state of euphoria). Because morphine is addictive, it is primarily used for the short-term treatment of acute pain and for terminally ill patients suffering from extreme pain. The analgesic properties of morphine have been exploited for over a millennium. It is the major component of opium, obtained from the unripe seed pods of the poppy plant, *Papaver somniferum*. Morphine was first isolated from opium in 1803, and by the mid-1800s, it was used heavily to control pain during and after surgical procedures. By the end of the 1800s, the addictive properties of morphine became apparent, which fueled the search for nonaddictive analgesics.

In 1925, the structure of morphine was correctly determined. This structure functioned as a lead compound and was modified to produce other compounds with analgesic properties. Early modifications focused on replacing the hydroxyl (OH) groups with other functional groups. Examples include heroin and codeine. Heroin exhibits stronger activity than morphine and is extremely addictive. Codeine shows less activity than morphine and is less addictive. Codeine is currently used as an analgesic and cough suppressant.

Morphine

Heroin

Codeine

In 1938, the analgesic properties of meperidine, also known as Demerol, were fortuitously discovered. As the story goes, meperidine was originally prepared to function as an antispasmodic agent (to suppress muscle spasms). When administered to mice, it curiously caused the tails of the mice to become erect. It was already known that morphine and related compounds produced a similar effect in mice, so meperidine was further tested and found to exhibit analgesic properties. This discovery generated much interest by providing new insights in the search for other analgesics. By comparing the structures of morphine, meperidine, and their derivatives, scientists were able to determine which structural features are essential for analgesic activity, shown in red:

Morphine

Meperidine

When morphine is drawn in this way, its structural similarity to meperidine becomes more apparent. Specifically, the bonds indicated in red represent the portion of each compound responsible for the analgesic activity. This part of the compound is called the *pharmacophore*. If any part of the pharmacophore is removed or changed, the resulting compound will not be capable of binding effectively to the appropriate biological receptor, and the compound will not exhibit analgesic properties. The term *auxophore* refers to the rest of the compound (the bonds shown in black). Removing any of these bonds may or may not affect the strength with which the pharmacophore binds to the receptor, thereby affecting the compound's analgesic potency. When modifying a lead compound, the auxophoric regions are the portions targeted for modification. For example, the auxophoric regions of morphine were modified to develop methadone and etorphine.

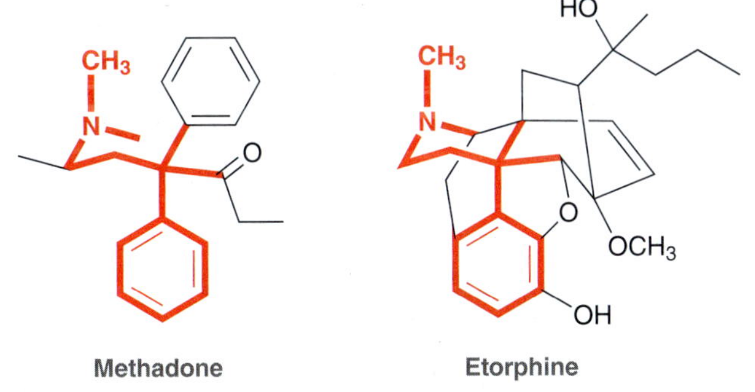


Methadone

Etorphine

Methadone, developed in Germany during World War II, is used to treat heroin addicts suffering from withdrawal symptoms. Methadone binds to the same receptor as heroin, but it has a longer retention time in the body, thereby enabling the body to cope with the decreasing levels of drug that normally cause withdrawal symptoms. Etorphine is over 3000 times more potent than morphine and is used exclusively in veterinary medicine to immobilize elephants and other large mammals.

Scientists are constantly searching for new lead compounds. In 1992, researchers at NIH (National Institutes of Health) in Bethesda, Maryland, isolated epibatidine from the skin of the Ecuadorian frog, *Epipedobates tricolor*. Epibatidine was found to be an analgesic that is 200 times more potent than morphine. Further studies indicated that epibatidine and morphine bind to different receptors. This discovery is very exciting, because it means that epibatidine can serve as a new lead compound. Although this compound is too toxic for clinical use, a significant number of researchers are currently working to identify the pharmacophore of epibatidine and to develop nontoxic derivatives. This area of research indeed looks promising.

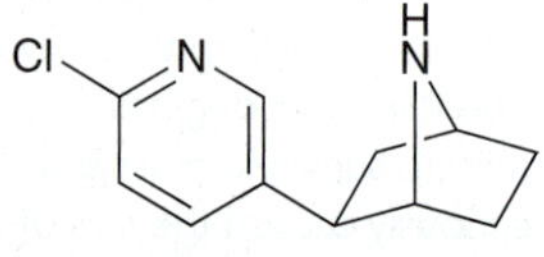

Epibatidine

CONCEPTUAL CHECKPOINT

**2.20** Troglitazone, rosiglitazone, and pioglitazone, all antidiabetic drugs introduced to the market in the late 1990s, are believed to act on the same receptor:

Troglitazone

Rosiglitazone

Pioglitazone

(a) Based on these structures, try to identify the likely pharmacophore that is responsible for the antidiabetic activity of these drugs.

(b) Consider the structure of rivoglitazone (below). This compound is currently being studied for potential antidiabetic activity. Based on your analysis of the likely pharmacophore, do you believe that rivoglitazone will exhibit antidiabetic properties?

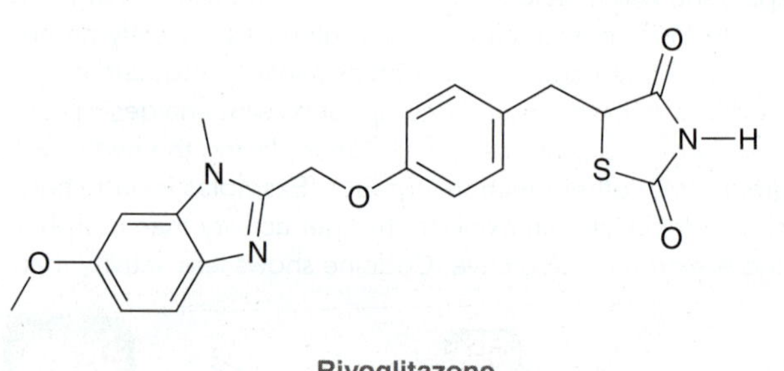

Rivoglitazone

## 2.7 Introduction to Resonance

### The Inadequacy of Bond-Line Structures

We have seen that bond-line structures are generally the most efficient and preferred way to draw the structure of an organic compound. Nevertheless, bond-line structures suffer from one major defect. Specifically, a pair of bonding electrons is always represented as a line that is drawn between two atoms, which implies that the bonding electrons are confined to a region of space directly in between two atoms. In some cases, this assertion is acceptable, as in the following structure:

In this case, the π electrons are in fact located where they are drawn, in between the two central carbon atoms. But in other cases, the electron density is spread out over a larger region of the molecule. For example, consider the following ion, called an *allyl carbocation*:

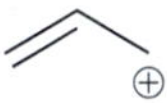

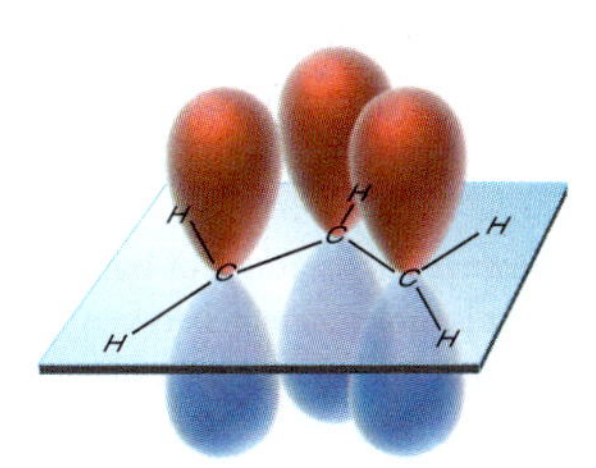

**FIGURE 2.3**
The overlapping *p* orbitals of an allyl carbocation.

It might seem from the drawing above that there are two π electrons on the left side and a positive charge on the right side. But this is not the entire picture, and the drawing above is inadequate. Let's take a closer look and first analyze the hybridization states. Each of the three carbon atoms above is $sp^2$ hybridized. Why? The two carbon atoms on the left side are each $sp^2$ hybridized because each of those carbon atoms is utilizing a *p* orbital to form the π bond (Section 1.9). The third carbon atom, bearing the positive charge, is also $sp^2$ hybridized because it has an empty *p* orbital. Figure 2.3 shows the three *p* orbitals associated with an allyl carbocation.

This image focuses our attention on the continuous system of *p* orbitals, which functions as a "conduit," allowing the two π electrons to be associated with all three carbon atoms. Valence bond theory is inadequate for analysis of this system because it treats the electrons as if they were confined between only two atoms. A more appropriate analysis of the allyl cation requires the use of molecular orbital (MO) theory (Section 1.8), in which electrons are associated with the molecule as a whole, rather than individual atoms. Specifically, in MO theory, the entire molecule is treated as one entity, and all of the electrons in the entire molecule occupy regions of space called molecular orbitals. Two electrons are placed in each orbital, starting with the lowest energy orbital, until all electrons occupy orbitals.

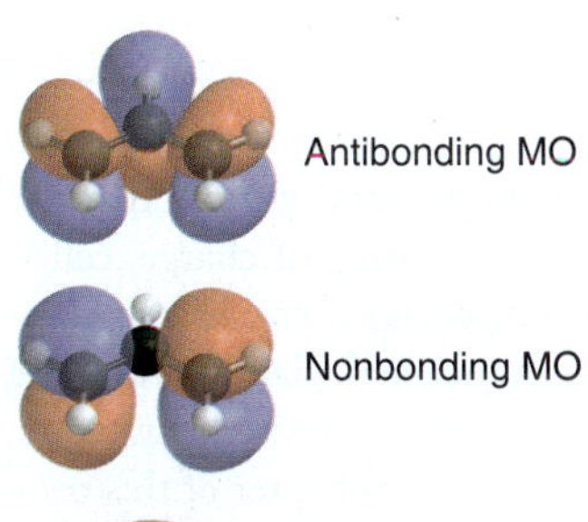


**FIGURE 2.4**
The molecular orbitals associated with the π electrons of an allylic system.

According to MO theory, the three *p* orbitals shown in Figure 2.3 no longer exist. Instead, they have been replaced by three MOs, illustrated in Figure 2.4 in order of increasing energy. Notice that the lowest energy MO, called the *bonding molecular orbital*, has no vertical nodes. The next higher energy MO, called the *nonbonding molecular orbital*, has one vertical node. The highest energy MO, called the *antibonding molecular orbital*, has two vertical nodes. The π electrons of the allyl system will fill these MOs, starting with the lowest energy MO. How many π electrons will occupy these MOs? The allyl carbocation has only two π electrons, rather than three, because one of the carbon atoms bears a positive formal charge indicating that one electron is missing. The two π electrons of the allyl system will occupy the lowest energy MO (the bonding MO). If the missing electron were to return, it would occupy the next higher energy MO, which is the nonbonding MO. Focus your attention on the nonbonding MO.

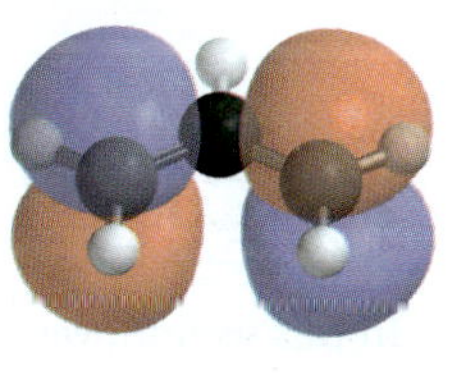

The nonbonding molecular orbital (from Figure 2.4) associated with the π electrons of an allylic system.

There should be an electron occupying this nonbonding MO, but the electron is missing. Therefore, the colored lobes are empty and represent regions of space that are electron deficient. In conclusion, MO theory suggests that the positive charge of the allyl carbocation is associated with the two ends of the system, rather than just one end.

In a situation like this, any single bond-line structure that we draw will be inadequate. How can we draw a positive charge that is spread out over two locations, and how can we draw two π electrons that are associated with three carbon atoms?

## Resonance

The approach that chemists use to deal with the inadequacy of bond-line structures is called **resonance**. According to this approach, we draw more than one bond-line structure and then mentally meld them together:

These drawings are called **resonance structures**, and they show that the positive charge is spread over two locations. Notice that we separate resonance structures with a straight, two-headed arrow, and we place brackets around the structures. The arrow and brackets indicate that the drawings are resonance structures *of one entity*. This one entity, called a **resonance hybrid**, is *not* flipping back and forth between the different resonance structures. To better understand this, consider the following analogy: A person who has never before seen a nectarine asks a farmer to describe a nectarine. The farmer answers:

> Picture a *peach* in your mind, and now picture a *plum* in your mind. Well, a *nectarine* has features of both fruits: the inside tastes like a peach, the outside is smooth like a plum, and the color is somewhere in between the color of a peach and the color of a plum. So take your image of a peach together with your image of a plum and *meld them together* in your mind into one image. That's a nectarine.

Here is the important feature of the analogy: The nectarine does not vibrate back and forth every second between being a peach and being a plum. A nectarine is a nectarine all of the time. The image of a peach by itself is not adequate to describe a nectarine. Neither is the image of a plum. But by combining certain characteristics of a peach with certain characteristics of a plum, it is possible to imagine the features of a nectarine. Similarly, with resonance structures, no single drawing adequately describes the nature of the electron density spread out over the molecule. To deal with this problem, we draw several drawings and then meld them together in our minds to obtain one image, or hybrid, just as we did to obtain an image for a nectarine.

Don't be confused by this important point: The term "resonance" does not describe something that is happening. Rather, it is a term that describes the way we deal with the inadequacy of our bond-line drawings.

### Resonance Stabilization

We developed the concept of resonance using the allyl cation as an example, and we saw that the positive charge of an allyl cation is spread out over two locations. This spreading of charge, called **delocalization**, is a stabilizing factor. That is, *the delocalization of either a positive charge or a negative charge stabilizes a molecule.* This stabilization is often referred to as **resonance stabilization**, and the allyl cation is said to be *resonance stabilized*. Resonance stabilization plays a major role in the outcome of many reactions, and we will invoke the concept of resonance in almost every chapter of this textbook. The study of organic chemistry therefore requires a thorough mastery of drawing resonance structures, and the following sections are designed to foster the necessary skills.

## 2.8 Curved Arrows

In this section, we will focus on **curved arrows**, which are the tools necessary to draw resonance structures properly. Every curved arrow has a *tail* and *head:*

Tail ⌒ Head

Curved arrows used for drawing resonance structures do not represent the motion of electrons—they are simply tools that allow us to draw resonance structures with ease. These tools treat the electrons *as if* they were moving, even though the electrons are actually not moving at all. In Chapter 3, we will encounter curved arrows that actually do represent the flow of electrons. For now, keep in mind that all curved arrows in this chapter are just tools and do not represent a flow of electrons.

It is essential that the tail and head of every arrow be drawn in precisely the proper location. The tail shows where the electrons are coming from, and the head shows where the electrons are going (remember, the electrons aren't really going anywhere, but we treat them as if they were for the purpose of drawing the resonance structures). We will soon learn patterns for drawing proper curved arrows. But, first, we must learn where not to draw curved arrows. There are two rules that must be followed when drawing curved arrows for resonance structures:

1. *Avoid breaking a single bond.*
2. *Never exceed an octet for second-row elements.*

Let's explore each of these rules:

1. *Avoid breaking a single bond when drawing resonance structures.* By definition, resonance structures must have all the same atoms connected in the same order. Breaking a single bond would change this—hence the first rule:

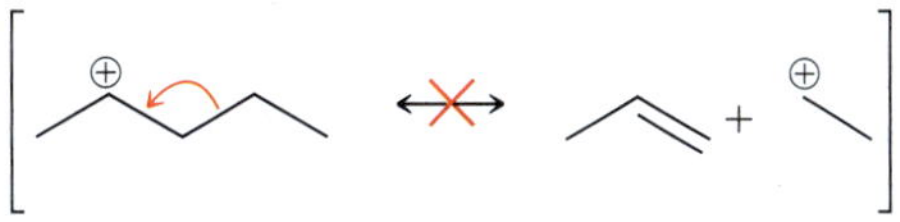

**Don't break a single bond**

There are very few exceptions to this rule, and we will only violate it two times in this textbook (both in Chapter 9). Each time, we will explain why it is permissible in that case. In all other cases, the tail of an arrow should never be placed on a single bond.

2. *Never exceed an octet for second-row elements.* Elements in the second row (C, N, O, F) have only four orbitals in their valence shell. Each orbital can either form a bond or hold a lone pair. Therefore, for second-row elements the total of the number of bonds plus the number of lone pairs can never be more than four. They can never have five or six bonds; the most is four. Similarly, they can never have four bonds and a lone pair, because this would also require five orbitals. For the same reason, they can never have three bonds and two lone pairs. Let's see some examples of curved arrows that violate this second rule. In each of these drawings, the central atom cannot form another bond because it does not have a fifth orbital that can be used.

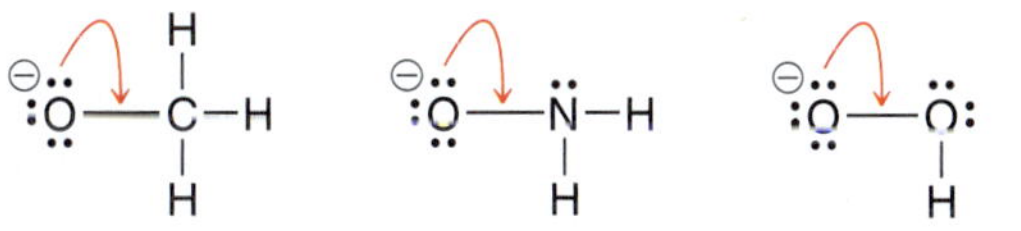

**Bad** arrow **Bad** arrow **Bad** arrow

The violation in each example above is clear, but with bond-line structures, it can be more difficult to see the violation because the hydrogen atoms are not drawn (and, very often, neither are the lone pairs). Care must be taken to "see" the hydrogen atoms even when they are not drawn:

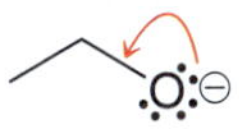

**Bad** arrow is the same as **Bad** arrow

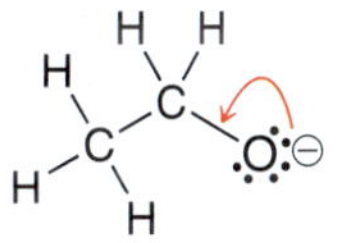

At first it is difficult to see that the curved arrow on the left structure is violating the second rule. But when we count the hydrogen atoms, it becomes clear that the curved arrow above would create a carbon atom with five bonds.

From now on, we will refer to the second rule as the *octet rule*. But be careful—for purposes of drawing resonance structures, it is only considered a violation if a second-row element has *more* than an octet of electrons. However, it is not a violation if a second-row element has *less* than an octet of electrons. For example:

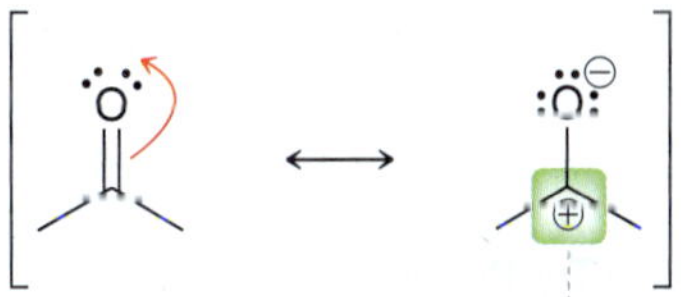

**This carbon atom does not have an octet**

This second drawing above is perfectly acceptable, even though the central carbon atom has only six electrons surrounding it. For our purposes, we will only consider the octet rule to be violated if we exceed an octet.

Our two rules (avoid breaking a single bond and never exceed an octet for a second-row element) reflect the two features of a curved arrow: the tail and the head. A poorly placed arrow tail violates the first rule, and a poorly directed arrow head violates the second rule.

# SKILLBUILDER

## 2.6 IDENTIFYING VALID RESONANCE ARROWS

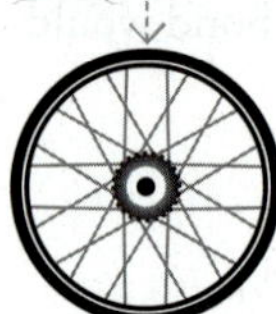

**LEARN the skill**

Inspect the arrow drawn on the following structure and determine whether it violates either of the two rules for drawing curved arrows:

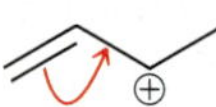

**SOLUTION**

In order to determine if either rule has been broken, we must look carefully at the tail and the head of the curved arrow. The tail is placed on a double bond, and therefore, this curved arrow does not break a single bond. So the first rule is not violated.

Next, we look at the head of the arrow: Has the octet rule been violated? Is there a fifth bond being formed here? Remember that a carbocation ($C^+$) only has three bonds, not four. Two of the bonds are shown, which means that the $C^+$ has only one bond to a hydrogen atom:

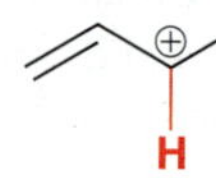


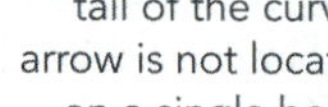

**STEP 1**
Make sure that the tail of the curved arrow is not located on a single bond.

**STEP 2**
Make sure that the head of the curved arrow does not violate the octet rule.

Therefore, the curved arrow will give the carbon atom a fourth bond, which does not violate the octet rule.

The curved arrow is valid, because the two rules were not violated. Both the tail and head of the arrow are acceptable.

**PRACTICE the skill**

**2.21** For each of the problems below, determine whether each curved arrow violates either of the two rules and describe the violation, if any. (Don't forget to count all hydrogen atoms and all lone pairs.)

(a) 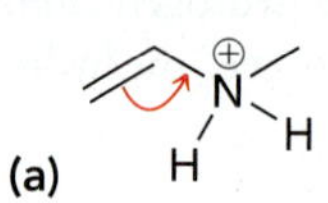


(b) 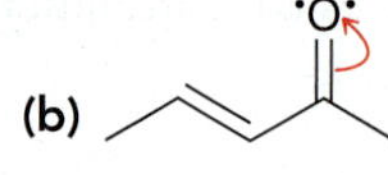


(c) 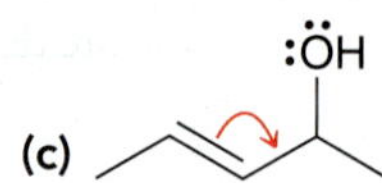


(d) 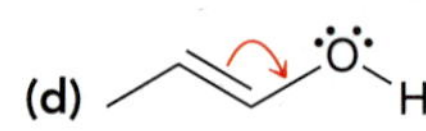


(e) 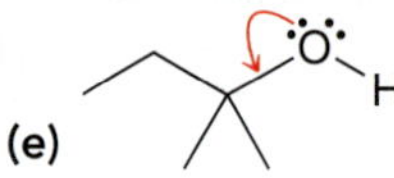


(f)

(g) 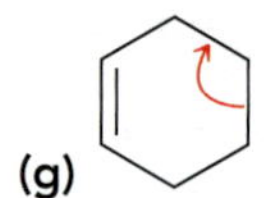

(h) 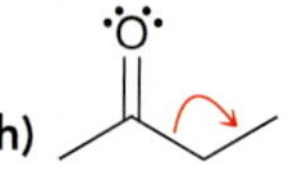


(i) C≡N:

(j) $H_3C—\overset{\oplus}{N}≡N:$

(k) O R

(l) N

**APPLY the skill**

**2.22** Drawing the resonance structure of the following compound requires one curved arrow. The head of this curved arrow is placed on the oxygen atom, and the tail of the curved arrow can only be placed in one location without violating the rules for drawing curved arrows. Draw this curved arrow.

need more **PRACTICE?** **Try Problem 2.51**

Whenever more than one curved arrow is used, all curved arrows must be taken into account in order to determine if any of the rules have been violated. For example, the following arrow violates the octet rule:

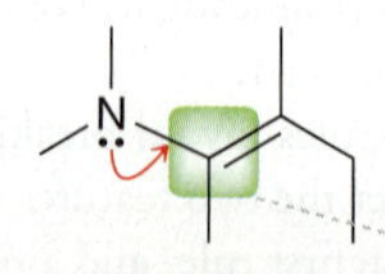

However, by adding another curved arrow, we remove the violation:

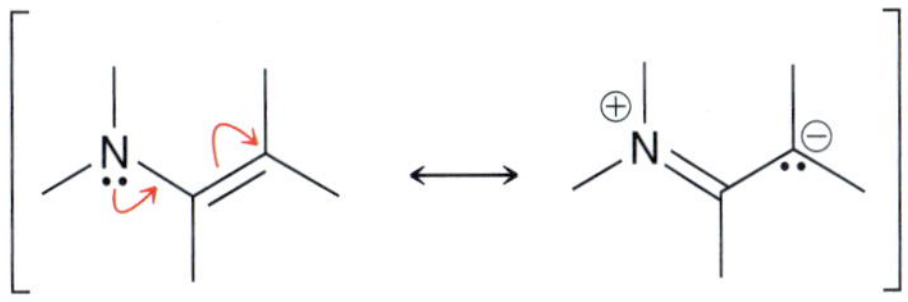

The second curved arrow removes the violation of the first curved arrow. In this example, both arrows are acceptable, because taken together, they do not violate our rules.

Arrow pushing is much like bike riding. The skill of bike riding cannot be learned by watching someone else ride. Learning to ride a bike requires practice. Falling occasionally is a necessary part of the learning process. The same is true with arrow pushing. The only way to learn is with practice. This chapter is designed to provide ample opportunity for practicing and mastering resonance structures.

## 2.9 Formal Charges in Resonance Structures

In Section 1.4, we learned how to calculate formal charges. Resonance structures very often contain formal charges, and it is absolutely critical to draw them properly. Consider the following example:

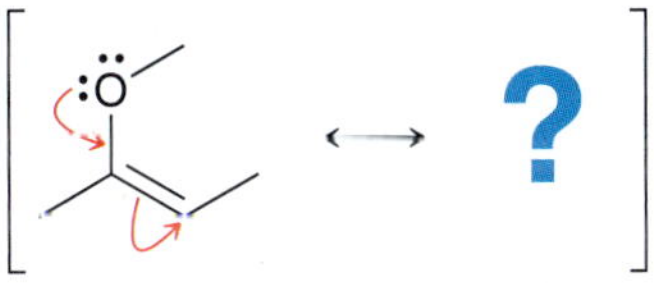

**WATCH OUT**
The electrons are not really moving. We are just treating them as if they were.

In this example, there are two curved arrows. The first arrow pushes one of the lone pairs to form a bond, and the second arrow pushes the π bond to form a lone pair on a carbon atom. When both arrows are pushed at the same time, neither of the rules is violated. So, let's focus on how to draw the resonance structure by following the instructions provided by the curved arrows. We delete one lone pair from oxygen and place a π bond between carbon and oxygen. Then we must delete the C—C π bond and place a lone pair on carbon:

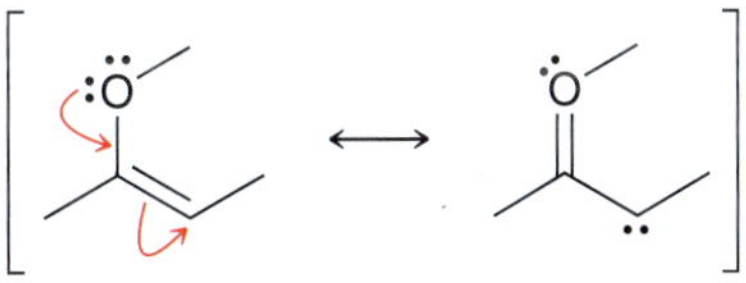

The arrows are really a language, and they tell us what to do. However, the structure is not complete without drawing formal charges. If we apply the rules of assigning formal charges, oxygen acquires a positive charge and carbon acquires a negative charge:

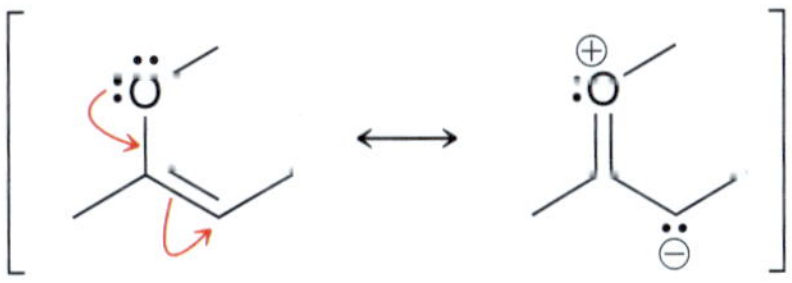

Another way to assign formal charges is to think about what the arrows are indicating. In this case, the curved arrows indicate that the oxygen atom is losing a lone pair and gaining a bond. In other words, it is losing two electrons and only gaining one back. The net result is the loss of one electron, indicating that oxygen must incur a positive charge in the resonance structure. A similar analysis for the carbon atom on the bottom right shows that it must incur a negative charge. Notice that the overall net charge is the same in each resonance structure. Let's practice assigning formal charges in resonance structures.

# SKILLBUILDER

## 2.7 ASSIGNING FORMAL CHARGES IN RESONANCE STRUCTURES

**LEARN the skill**

Draw the resonance structure below. Be sure to include formal charges.

### SOLUTION

The arrows indicate that one of the lone pairs on oxygen is coming down to form a bond, and the C═C double bond is being pushed to form a lone pair on a carbon atom. This is very similar to the previous example. The arrows indicate that we must delete one lone pair on oxygen, place a double bond between carbon and oxygen, delete the carbon-carbon double bond, and place a lone pair on carbon:

**STEP 1**
Carefully read what the curved arrows indicate.

Finally, we must assign formal charges. In this case, oxygen started with a negative charge, and this charge has now been pushed down (as the arrows indicate) onto a carbon atom. Therefore, the carbon atom must now bear the negative charge:

**STEP 2**
Assign formal charges.

Earlier in this chapter, we said that it is not necessary to draw lone pairs, because they are implied by bond-line structures. In the example above, the lone pairs are shown for clarity. This raises an obvious question. Look at the first curved arrow above: The tail is drawn on a lone pair. If the lone pairs had not been drawn, how would the curved arrow be drawn? In situations like this, organic chemists will sometimes draw the curved arrow coming from the negative charge:

is the same as

Nevertheless, you should avoid this practice, because it can easily lead to mistakes in certain situations. It is highly preferable to draw the lone pairs and then place the tail of the curved arrow on a lone pair, rather than placing it on a negative charge.

After drawing a resonance structure and assigning formal charges, it is always a good idea to count the total charge on the resonance structure. This total charge MUST be the same as on the original structure (conservation of charge). If the first structure had a negative charge, then the resonance structure must also have a net negative charge. If it doesn't, then the resonance structure cannot possibly be correct. The total charge must be the same for all resonance structures, and there are no exceptions to this rule.

**PRACTICE the skill** **2.23** For each of the structures below, draw the resonance structure that is indicated by the curved arrows. Be sure to include formal charges.

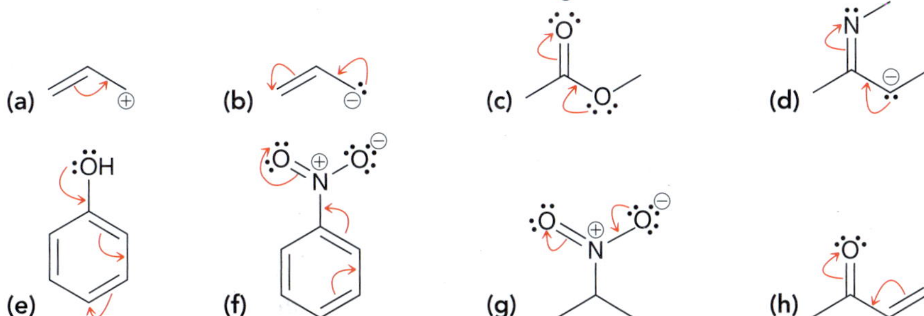

**APPLY the skill** **2.24** In each case below, draw the curved arrow(s) required in order to convert the first resonance structure into the second resonance structure. In each case, begin by drawing all lone pairs and then use the formal charges to guide you.

(a) (b)

(c) (d)

need more **PRACTICE?** **Try Problems 2.44, 2.53**

## 2.10 Drawing Resonance Structures via Pattern Recognition

In order to become truly proficient at drawing resonance structures, you must learn to recognize the following five patterns: (1) an allylic lone pair, (2) an allylic positive charge, (3) a lone pair adjacent to a positive charge, (4) a π bond between two atoms of differing electronegativity, and (5) conjugated π bonds in a ring.

We will now explore each of these five patterns, with examples and practice problems.

1. *An allylic lone pair.* Let's begin with some important terminology that we will use frequently throughout the remainder of the text. When a compound contains a carbon-carbon double bond, the two carbon atoms bearing the double bond are called **vinylic** positions, while the atoms connected directly to the vinylic positions are called **allylic** positions:

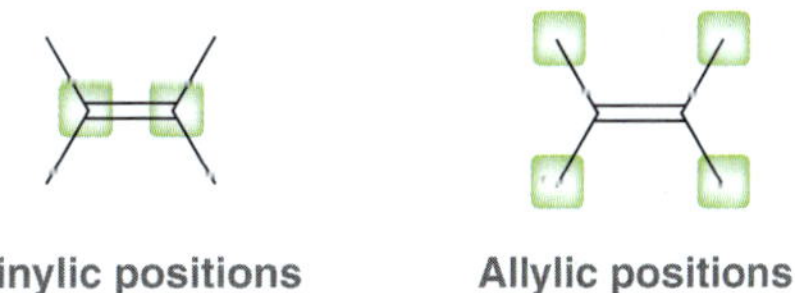


We are specifically looking for lone pairs in an allylic position. As an example, consider the following compound, which has two lone pairs:

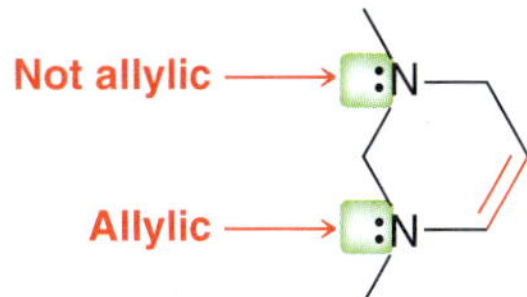

We must learn to identify lone pairs in allylic positions. Here are several examples:

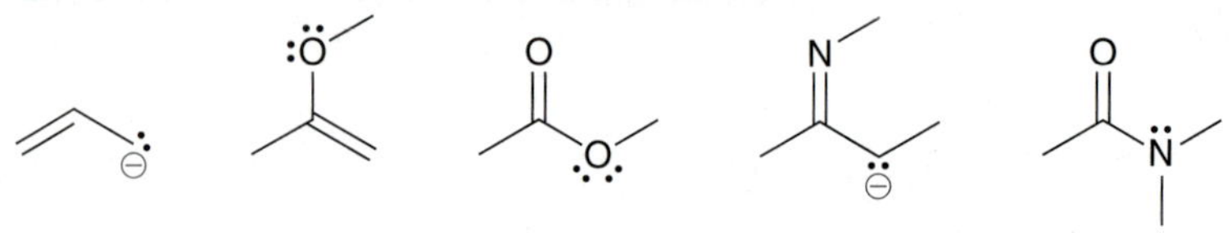

In the last three cases above, the lone pairs are not next to a *carbon-carbon* double bond and are technically not allylic lone pairs (an allylic position is the position next to a carbon-carbon double bond and not any other type of double bond). Nevertheless, for purposes of drawing resonance structures, we will treat these lone pairs in the same way that we treat allylic lone pairs. Specifically, all of the examples above exhibit at least one lone pair next to a $\pi$ bond.

For each of the examples above, there will be a resonance structure that can be obtained by drawing exactly two curved arrows. The first curved arrow goes from the lone pair to form a $\pi$ bond, while the second curved arrow goes from the $\pi$ bond to form a lone pair:

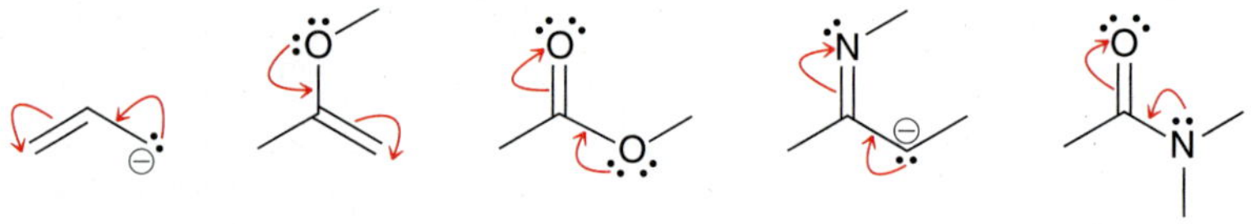

Let's carefully consider the formal charges produced in each of the cases above. When the atom with the lone pair has a negative charge, then it transfers its negative charge to the atom that ultimately receives a lone pair:

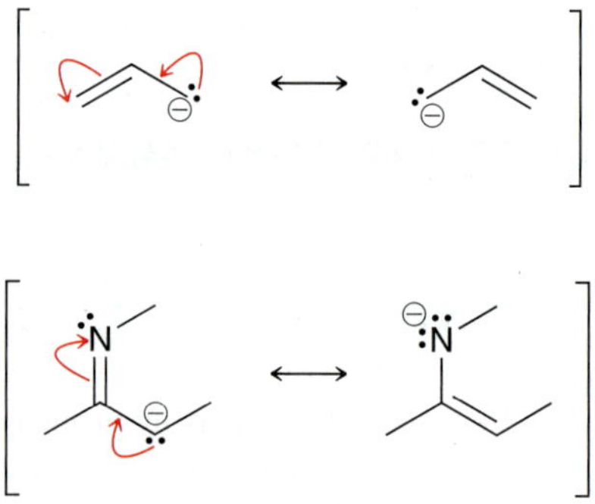

When the atom with the lone pair does not have a negative charge, then it will incur a positive charge, while the atom receiving the lone pair will incur a negative charge:

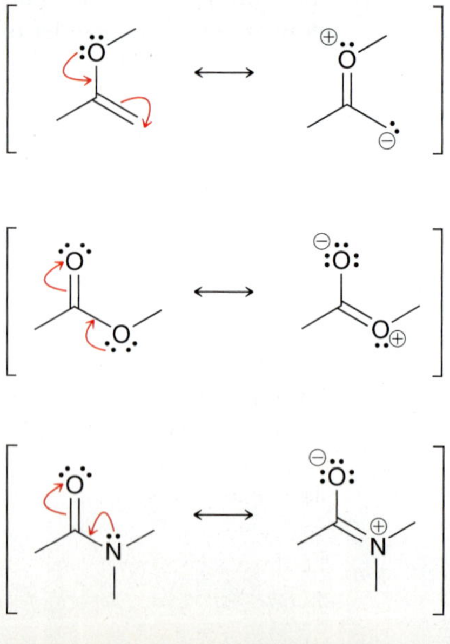

Recognizing this pattern (a lone pair next to a π bond) will save time in calculating formal charges and determining if the octet rule is being violated.

## CONCEPTUAL CHECKPOINT

**2.25** For each of the compounds below, locate the pattern we just learned (lone pair next to a π bond) and draw the appropriate resonance structure:

(a) (b) (c) (d) (e)

(f) (g) (h)

**Acetylcholine (a neurotransmitter)**

**5-Amino-4-oxopentanoic acid (used in therapy and diagnosis of hepatic tumors)**

**2.** *An allylic positive charge.* Again we are focusing on allylic positions, but this time, we are looking for a positive charge located in an allylic position:

**Allylic positive charge**

When there is an allylic positive charge, only one curved arrow will be required; this arrow goes from the π bond to form a new π bond:

Notice what happens to the formal charge in the process. The positive charge is moved to the other end of the system.

In the previous example, the positive charge was next to one π bond. The following example contains two π bonds, which are said to be **conjugated**, because they are separated from each other by exactly one σ bond (we will explore conjugated π systems in more detail in Chapter 17).

In this situation, we push each of the double bonds over, one at a time:

It is not necessary to waste time recalculating formal charges for each resonance structure, because the arrows indicate what is happening. Think of a positive charge as a hole of electron density—a place that is missing an electron. When we push π electrons to plug up the hole, a new hole is created nearby. In this way, the hole is simply moved from one location to another. Notice that in the above structures the tails of the curved arrows are placed on the π bonds, not on the positive charge. *Never place the tail of a curved arrow on a positive charge* (that is a common mistake).

## CONCEPTUAL CHECKPOINT

**2.26** Draw the resonance structure(s) for each of the compounds below:

(a) (b) (c) (d)

3. *A lone pair adjacent to a positive charge.* Consider the following example:

The oxygen atom exhibits three lone pairs, all of which are adjacent to the positive charge. This pattern requires only one curved arrow. The tail of the curved arrow is placed on a lone pair, and the head of the arrow is placed to form a π bond between the lone pair and the positive charge:

Notice what happens with the formal charges above. The atom with the lone pair has a negative charge in this case, and therefore the charges end up canceling each other. Let's consider what happens with formal charges when the atom with the lone pair does not bear a negative charge. For example, consider the following:

Once again, there is a lone pair adjacent to a positive charge. Therefore, we draw only one curved arrow: The tail goes on the lone pair, and the head is placed to form a π bond. In this case, the oxygen atom did not start out with a negative charge. Therefore, it will incur a positive charge in the resonance structure (remember conservation of charge).

## CONCEPTUAL CHECKPOINT

**2.27** For each of the compounds below, locate the lone pair adjacent to a positive charge and draw the resonance structure:

(a) (b) (c)

In one of the previous problems, a negative charge and a positive charge are seen canceling each other to become a double bond. However, there is one situation where it is not possible to combine charges to form a double bond—this occurs with the nitro group. The structure of the nitro group looks like this:

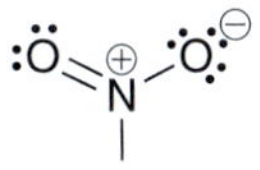

In this case, there is a lone pair adjacent to a positive charge, yet we cannot draw a single curved arrow to cancel out the charges:

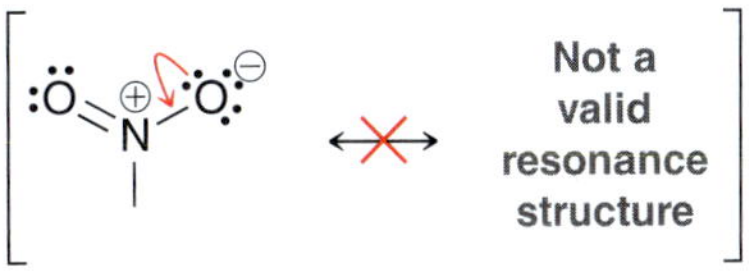

Not a valid resonance structure

Why not? The curved arrow shown above violates the octet rule, because it would give the nitrogen atom five bonds. Remember that second-row elements can never have more than four bonds. There is only one way to draw the curved arrow above without violating the octet rule—we must draw a second curved arrow, like this:

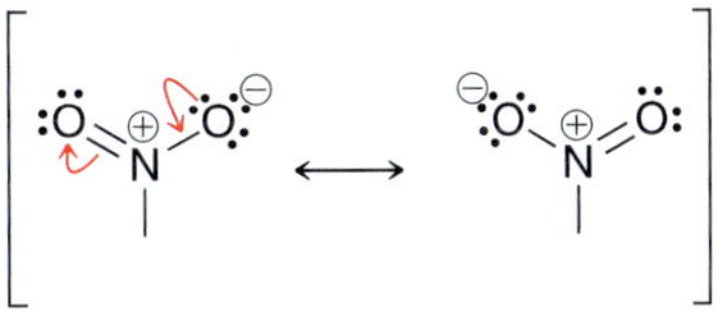

Look carefully. These two curved arrows are simply our first pattern (a lone pair next to a π bond). Notice that the charges have not been canceled. Rather, the location of the negative charge has moved from one oxygen atom to the other. The two resonance structures above are the only two valid resonance structures for a nitro group. In other words, the nitro group must be drawn with charge separation, even though the nitro group is overall neutral. The structure of the nitro group cannot be drawn without the charges.

4. *A π bond between two atoms of differing electronegativity.* Recall that electronegativity measures the ability of an atom to attract electrons. A chart of electronegativity values can be found in Section 1.11. For purposes of recognizing this pattern, we will focus on C=O and C=N double bonds.

In these situations, we move the π bond up onto the electronegative atom to become a lone pair.

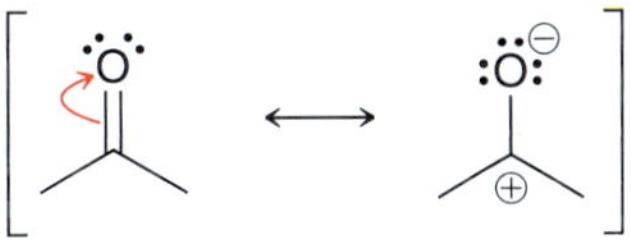

Notice what happens with the formal charges. A double bond is being separated into a positive and negative charge (this is the opposite of our third pattern, where the charges came together to form a double bond).

## CONCEPTUAL CHECKPOINT

**2.28** Draw a resonance structure for each of the compounds below.

(a) (b) (c)

**2.29** Draw a resonance structure of the compound below, which was isolated from the fruits of *Ocotea corymbosa*, a native plant of the Brazilian Cerrado.

**2.30** Draw a resonance structure of the compound shown below, called 2-heptanone, which is found in some kinds of cheese.

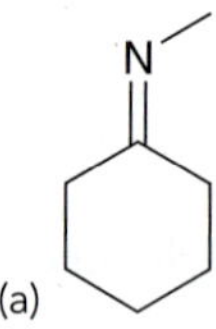

**5.** *Conjugated π bonds enclosed in a ring.* In one of the previous patterns, we referred to π bonds as being *conjugated* when they are separated from each other by one σ bond (i.e., C=C—C=C).

When conjugated π bonds are enclosed in a ring of alternating double and single bonds, we push all of the π bonds over by one position:

**LOOKING AHEAD**

In this molecule, called benzene, the electrons are delocalized. As a result, benzene exhibits significant resonance stabilization. We will explore the pronounced stability of benzene in Chapter 18.

When drawing the resonance structure above, all of the π bonds can be pushed clockwise or they can all be pushed counterclockwise. Either way achieves the same result.

## CONCEPTUAL CHECKPOINT

**2.31** Fingolimod is a novel drug that has recently been developed for the treatment of multiple sclerosis. In April of 2008, researchers reported the results of phase III clinical trials of fingolimod, in which 70% of patients who took the drug daily for three years were relapse free. This is a tremendous improvement over previous drugs that only prevented relapse in 30% of patients. Draw a resonance structure of fingolimod:



**Fingolimod**

Figure 2.5 summarizes the five patterns for drawing resonance structures. Take special notice of the number of curved arrows used for each pattern. When drawing resonance structures, always begin by looking for the patterns that utilize only one curved arrow. Otherwise, it is possible to miss a resonance structure. For example, consider the resonance structures of the following compound:

Notice that each pattern used in this example involves only one curved arrow. If we had started by recognizing a lone pair next to a π bond (which utilizes two curved arrows), then we might have missed the middle resonance structure above:

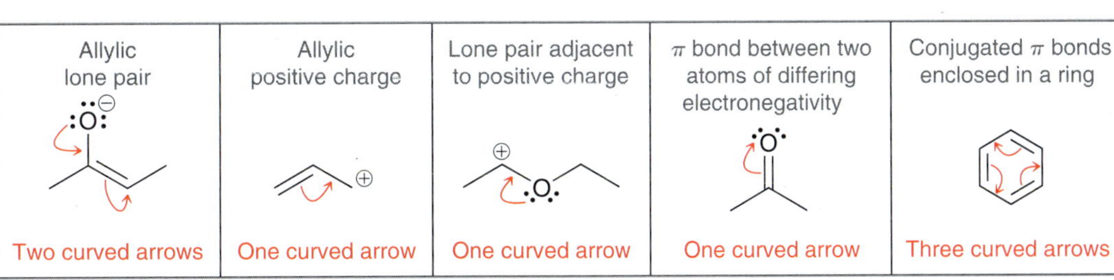


**FIGURE 2.5** A summary of the five patterns for drawing resonance structures.

## CONCEPTUAL CHECKPOINT

**2.32** Draw resonance structures for each of the following compounds:

## 2.11 Assessing Relative Importance of Resonance Structures

Not all resonance structures are equally significant. A compound might have many valid resonance structures (structures that do not violate the two rules), but it is possible that one or more of the structures is insignificant. To understand what we mean by "insignificant," let's revisit the analogy that we used earlier in the chapter to explain the concept of resonance.

Recall the analogy in which we merged the image of a peach with the image of a plum to obtain an image of a nectarine (Section 2.7). Now let's modify that analogy slightly. Imagine that we create a new type of fruit that is a hybrid between *three* fruits: a peach, a plum, and a kiwi. Suppose that the hypothetical hybrid fruit has the following character: 65% peach character, 34% plum character, and 1% kiwi character. This hybrid fruit is expected to look almost exactly like a nectarine, because the amount of kiwi character is too small to affect the nature of the resulting hybrid. Even though the new fruit is actually a hybrid of all three fruits, it will look like a hybrid of only two fruits—because the kiwi character is *insignificant*.

A similar concept exists when comparing resonance structures. For example, a compound could have three resonance structures, but the three structures might not contribute equally to the overall resonance hybrid. One resonance structure might be the major contributor (like the peach), while another might be insignificant (like the kiwi). In order to understand the true nature of the compound, we must be able to compare resonance structures and determine which structures are major contributors and which structures are not significant.

Three rules will guide us in determining the significance of resonance structures:

1. *Minimize charges.* The best kind of structure is one without any charges. It is acceptable to have one or two charges, but structures with more than two charges should be avoided, if possible. Compare the following two cases:

Both compounds have a lone pair next to a C=O double bond. So we might expect these compounds to have the same number of significant resonance structures. But they do not. Let's see why. Consider the resonance structures of the first compound:

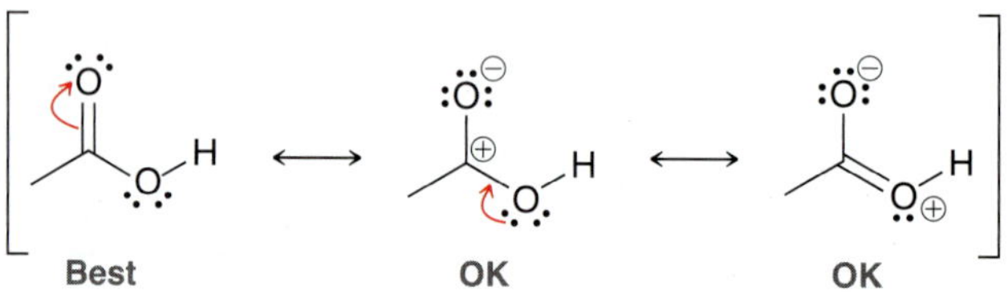

Best OK OK

The first resonance structure is the major contributor to the overall resonance hybrid, because it has no charge separation. The other two drawings have charge separation, but there are only two charges in each drawing, so they are both significant resonance structures. They might not contribute as much character as the first resonance structure does, but they are still significant. Therefore, this compound has three significant resonance structures.

Now, let's try the same approach for the other compound:

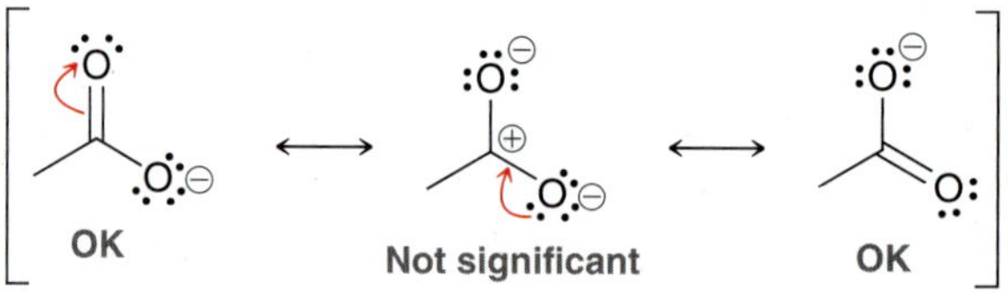

OK Not significant OK

The first and last structures are acceptable (each has only one charge), but the middle resonance structure has too many charges. This resonance structure is not significant, and therefore, it will not contribute much character to the overall resonance hybrid. It is like the kiwi in our analogy above. This compound has only two significant resonance structures.

One notable exception to this rule involves compounds containing the nitro group (—$NO_2$), which have resonance structures with more than two charges. Why? We saw earlier that the structure of the nitro group must be drawn with charge separation in order to avoid violating the octet rule:

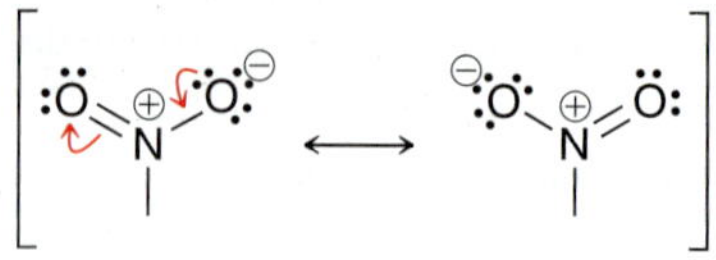

Therefore, the two charges of a nitro group don't really count when we are counting charges. Consider the following case as an example:

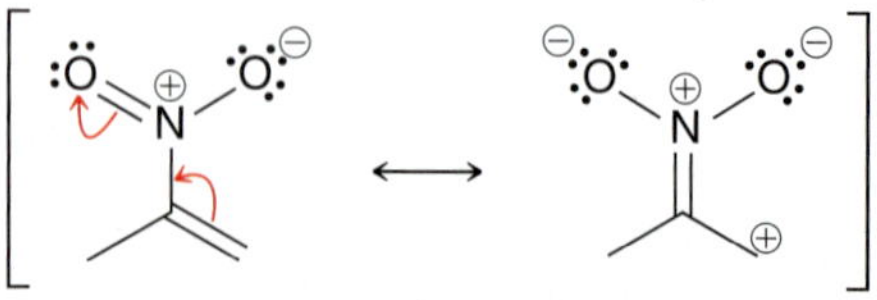

If we apply the rule about limiting charge separation to no more than two charges, then we might say that the second resonance structure above appears to have too many charges to be significant. But it actually is significant, because the two charges associated with the nitro group are not included in the count. We would consider the resonance structure above as if it only had two charges, and therefore it is significant.

2. *Electronegative atoms, such as N, O, and Cl, can bear a positive charge, but only if they possess an octet of electrons.* Consider the following as an example:

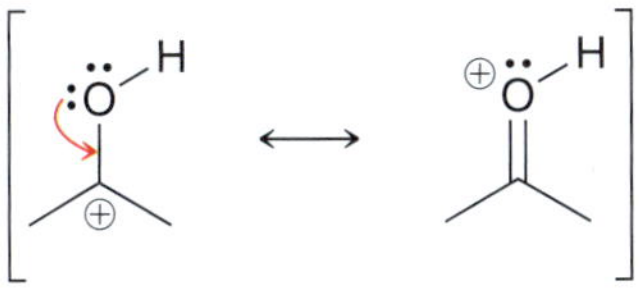

The second resonance structure is significant, even though it has a positive charge on oxygen. Why? Because the positively charged oxygen has an octet of electrons (three bonds plus one lone pair = 6 + 2 = 8 electrons). In fact, the second resonance structure is even more significant than the first resonance structure. We might have thought otherwise, because the first resonance structure has a positive charge on carbon, which is generally much better than having a positive charge on an electronegative atom. Nevertheless, the second resonance structure is more significant because all of its atoms achieve an octet. In the first structure, oxygen has its octet, but carbon only has six electrons. In the second resonance structure, both oxygen and carbon have an octet, which makes that structure more significant, even though the positive charge is on oxygen.

Here is another example, this time with the positive charge on nitrogen:

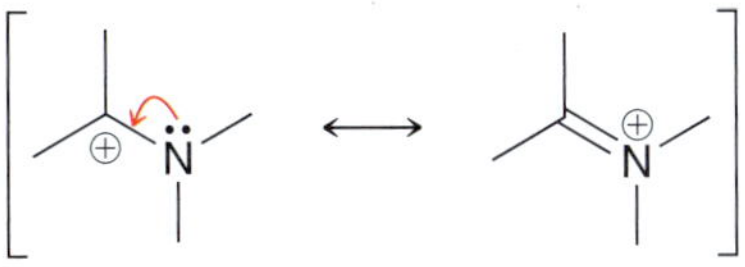

Once again, the second structure is significant, in fact, even more significant than the first. *In summary, the most significant resonance structures are generally those in which all atoms have an octet.*

3. *Avoid drawing a resonance structure in which two carbon atoms bear opposite charges.* Such resonance structures are generally insignificant, for example:

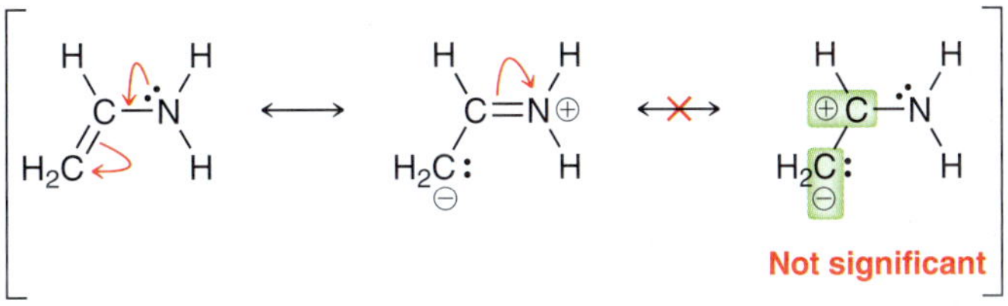

Not significant

In this case, the third resonance structure is insignificant because it has both a $C^+$ and a $C^-$. The presence of carbon atoms with opposite charges, whether close to each other (as in the example above) or far apart, renders the structure insignificant. Throughout this text, we will see only one exception to this rule (in Problem 18.54).

# SKILLBUILDER

## 2.8 DRAWING SIGNIFICANT RESONANCE STRUCTURES

**LEARN** the skill

Draw all significant resonance structures of the following compound:

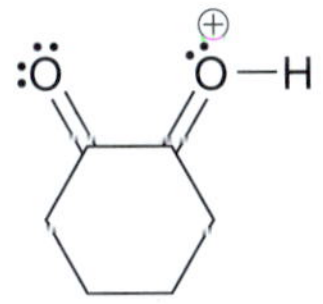

### SOLUTION

**STEP 1**
Using the five patterns, identify a resonance structure.

We begin by looking for any of the five patterns. This compound contains a C=O bond (a π bond between two atoms of differing electronegativity), and we can therefore draw the following resonance structure:

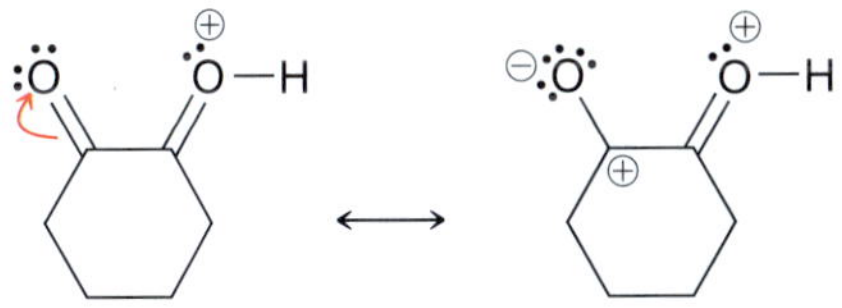

This resonance structure is valid, because it was generated using one of the five patterns. However, it has too many charges, and it is therefore not significant. In general, try to avoid drawing resonance structures with three or more charges.

**STEP 2**
Identify if the resonance structure is significant by inspecting the number of charges and the number of electrons on heteroatoms.

Next, we look at the other C=O bond, and we try the same pattern:

To determine if this resonance structure is significant, we ask three questions:

1. *Does this structure have an acceptable number of charges?* Yes, it has only one charge (on the carbon atom), which is perfectly acceptable.
2. *Do all electronegative atoms have an octet?* Yes, both oxygen atoms have an octet of electrons.
3. *Does the structure avoid having carbon atoms with opposite charges?* Yes.

This resonance structure passes the test, and therefore it is a significant resonance structure.

Now that we have found a significant resonance structure, we analyze it to see if any of the five patterns will allow us to draw another resonance structure. In this case, there is a positive charge next to a $\pi$ bond. So we draw one curved arrow, generating the following resonance structure:

**STEP 3**
Repeat steps 1 and 2 to identify any other significant resonance structures.

To determine whether the structure is significant, we first check to see whether it has an acceptable number of charges. It has only one charge, which is perfectly acceptable. Next we check whether all electronegative atoms have an octet. The oxygen atom bearing the positive charge does not have an octet of electrons, which is not acceptable and means that this resonance structure is not significant. In summary, this compound has the following significant resonance structure:

**PRACTICE the skill** **2.33** Draw all significant resonance structures for each of the following compounds:

(a) (b) (c) (d) (e) (f)

(g) (h) (i) (j) (k) (l)

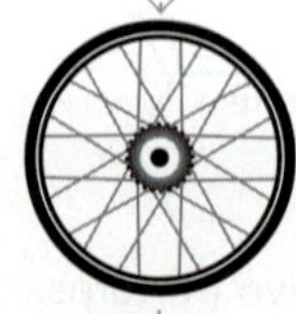

**APPLY the skill** **2.34** Use resonance structures to help you identify all sites of low electron density (δ+) in the following compound:

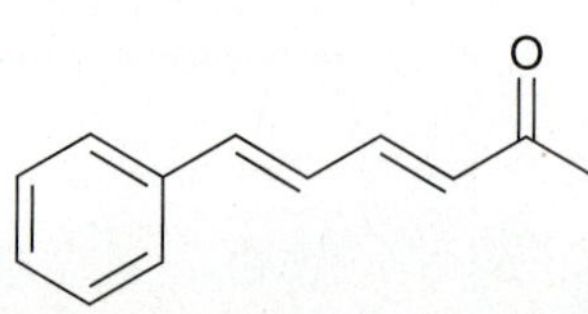

**2.35** Use resonance structures to help you identify all sites of high electron density (δ−) in the following compound:

OH

need more **PRACTICE?** **Try Problems 2.45, 2.48, 2.59, 2.60, 2.62, 2.65, 2.66**

## 2.12 Delocalized and Localized Lone Pairs

In this section, we will explore some important differences between lone pairs that participate in resonance and lone pairs that do not participate in resonance.

### Delocalized Lone Pairs

Recall that one of our five patterns was a lone pair that is allylic to a π bond. Such a lone pair will participate in resonance and is said to be **delocalized**. When an atom possesses a delocalized lone pair, the geometry of that atom is affected by the presence of the lone pair. As an example, consider the structure of an amide:

**An amide**

The rules we learned in Section 1.10 would suggest that the nitrogen atom should be $sp^3$ hybridized and trigonal pyramidal, but this is not correct. Instead, the nitrogen atom is actually $sp^2$ hybridized and trigonal planar. Why? The lone pair is participating in resonance and is therefore delocalized:

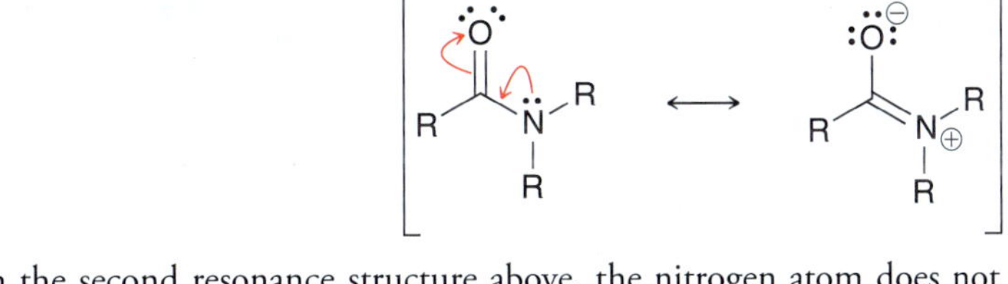

In the second resonance structure above, the nitrogen atom does not bear a lone pair. Rather, the nitrogen atom bears a *p* orbital being used to form a π bond. In that resonance structure, the nitrogen atom is clearly $sp^2$ hybridized. This creates a conflict: How can the nitrogen atom be $sp^3$ hybridized in one resonance structure and $sp^2$ hybridized in the other structure? That would imply that the geometry of the nitrogen atom is flipping back and forth between trigonal pyramidal and trigonal planar. This cannot be the case, because resonance is not a physical process. The nitrogen atom is actually $sp^2$ hybridized and trigonal planar in both resonance structures. How? The nitrogen atom has a delocalized lone pair, and it therefore occupies a *p* orbital (rather than a hybridized orbital), so that it can overlap with the *p* orbitals of the π bond (Figure 2.6).

**FIGURE 2.6** An illustration of the overlapping atomic *p* orbitals of an amide.

Whenever a lone pair participates in resonance, it will occupy a *p* orbital rather than a hybridized orbital, and this must be taken into account when predicting geometry. This will be extremely important in Chapter 25 when we discuss the three-dimensional shape of proteins.

### Localized Lone Pairs

A **localized lone pair**, by definition, is a lone pair that does not participate in resonance. In other words, the lone pair is not allylic to a π bond:

**Localized** **Delocalized**

In some cases, a lone pair might appear to be delocalized even though it is actually localized. For example, consider the structure of pyridine:

Lone pair →

π bond →

**Pyridine**

The lone pair in pyridine appears to be allylic to a π bond, and it is tempting to use our pattern to draw the following resonance structure:

**Not a valid resonance structure**

However, this resonance structure is not valid. Why not? In this case, the lone pair on the nitrogen atom is actually not participating in resonance, even though it is next to a π bond. Recall that in order for a lone pair to participate in resonance, it must occupy a *p* orbital that can overlap with the neighboring *p* orbitals, forming a "conduit" (Figure 2.7).

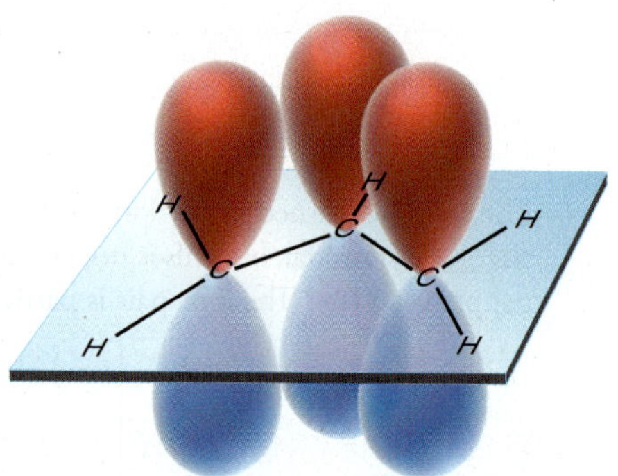

**FIGURE 2.7** Resonance applies to systems that involve overlapping *p* orbitals that form a "conduit."

In the case of pyridine, the nitrogen atom is already using a *p* orbital for the π bond (Figure 2.8). The nitrogen atom can only use one *p* orbital to join in the conduit shown in Figure 2.8, and that *p* orbital is already being utilized by the π bond. As a result, the lone pair cannot join in the conduit, and therefore it cannot participate in resonance. In this case, the lone pair occupies an $sp^2$-hybridized orbital, which is in the plane of the ring.

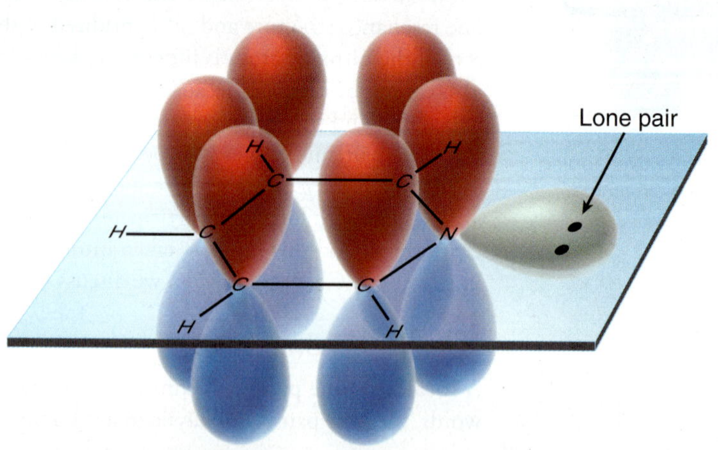


**FIGURE 2.8** The overlapping *p* orbitals of pyridine.

Here is the bottom line: *Whenever an atom possesses both a π bond and a lone pair, they will not both participate in resonance.* In general, only the π bond will participate in resonance, and the lone pair will not.

Let's get some practice identifying localized and delocalized lone pairs and using that information to determine geometry.

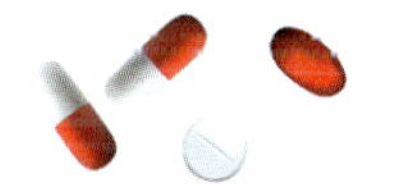

# SKILLBUILDER

## 2.9 IDENTIFYING LOCALIZED AND DELOCALIZED LONE PAIRS

### LEARN the skill

Histamine is a compound that plays a key role in many biological functions. Most notably, it is involved in immune responses, where it triggers the symptoms of allergic reactions:

**Histamine**

Each nitrogen atom exhibits a lone pair. In each case, identify whether the lone pair is localized or delocalized and then use that information to determine the hybridization state and geometry for each nitrogen atom in histamine.

### SOLUTION

Let's begin with the nitrogen on the right side of the compound. This lone pair is localized, and therefore we can use the method outlined in Section 1.10 to determine the hybridization state and geometry:

There are 3 bonds and 1 lone pair, and therefore:

1) **Steric number = 3 + 1 = 4**
2) **4 = $sp^3$ = electronically tetrahedral**
3) **Arrangement of atoms = <u>trigonal pyramidal</u>**

This lone pair is not participating in resonance, so our method accurately predicts the geometry to be trigonal pyramidal.

Now, let's consider the nitrogen atom on the left side of the compound. The lone pair on that nitrogen atom is delocalized by resonance:

Therefore, this lone pair is actually occupying a *p* orbital, rendering the nitrogen atom $sp^2$ hybridized, rather than $sp^3$ hybridized. As a result, the geometry is trigonal planar.

Now let's consider the remaining nitrogen atom:

This nitrogen atom already has a π bond participating in resonance. Therefore, the lone pair cannot also participate in resonance. In this case, the lone pair must be localized. The nitrogen atom is in fact $sp^2$ hybridized and exhibits bent geometry.

To summarize, each of the nitrogen atoms in histamine has a different geometry:

Trigonal pyramidal

Trigonal planar

Bent

**PRACTICE the skill** **2.36** For each compound below, identify all lone pairs and indicate whether each lone pair is localized or delocalized. Then, use that information to determine the hybridization state and geometry for each atom that exhibits a lone pair.

(a) (b) (c) (d) (e) (f)

**APPLY the skill** **2.37** Nicotine is a toxic substance present in tobacco leaves.

**Nicotine**

There are two lone pairs in the structure of nicotine. In general, localized lone pairs are much more reactive than delocalized lone pairs. With this information in mind, do you expect both lone pairs in nicotine to be reactive? Justify your answer.

**2.38** Isoniazid is used in the treatment of tuberculosis and multiple sclerosis. Identify each lone pair as either localized or delocalized. Justify your answer in each case.

**Isoniazid**

need more **PRACTICE?** **Try Problems 2.47, 2.61**

## REVIEW OF CONCEPTS AND VOCABULARY

### SECTION 2.1

- Chemists use many different drawing styles to communicate structural information, including Lewis structures, **partially condensed structures**, and **condensed structures**.
- The molecular formula does not provide structural information.

### SECTION 2.2

- In **bond-line structures**, carbon atoms and most hydrogen atoms are not drawn.
- Bond-line structures are faster to draw and easier to interpret than other drawing styles.

### SECTION 2.3

- A **functional group** is a characteristic group of atoms/bonds that show a predictable chemical behavior.
- The chemistry of every organic compound is determined by the functional groups present in the compound.

### SECTION 2.4

- A formal charge is associated with any atom that does not exhibit the appropriate number of valence electrons.
- When a carbon atom bears either a positive or negative charge, it will have only three, rather than four, bonds.

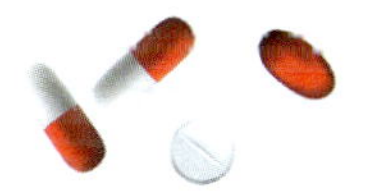

### SECTION 2.5

- Lone pairs are often not drawn in bond-line structures. It is important to recognize that these lone pairs are present.

### SECTION 2.6

- In bond-line structures, a **wedge** represents a group coming out of the page, and a **dash** represents a group behind the page.
- Other drawings used to show three dimensionality include **Fischer projections** and **Haworth projections**.

### SECTION 2.7

- Bond-line structures are inadequate in some situations, and an approach called **resonance** is required.
- **Resonance structures** are separated by double-headed arrows and surrounded by brackets:

- **Resonance stabilization** refers to the **delocalization** of either a positive charge or a negative charge via resonance.

### SECTION 2.8

- **Curved arrows** are tools for drawing resonance structures.
- When drawing curved arrows for resonance structures, avoid breaking a single bond and never exceed an octet for second-row elements.

### SECTION 2.9

- All formal charges must be shown when drawing resonance structures.

### SECTION 2.10

- Resonance structures are most easily drawn by looking for the following five patterns:
  1. An **allylic** lone pair
  2. An allylic positive charge
  3. A lone pair adjacent to a positive charge
  4. A $\pi$ bond between two atoms of differing electronegativity
  5. **Conjugated $\pi$ bonds** enclosed in a ring
- When drawing resonance structures, always begin by looking for the patterns that utilize only one curved arrow.

### SECTION 2.11

- There are three rules for identifying significant **resonance structures**:
  1. Minimize charge.
  2. Electronegative atoms (N, O, Cl, etc.) can bear a positive charge, but only if they possess an octet of electrons.
  3. Avoid drawing a resonance structure in which two carbon atoms bear opposite charges.

### SECTION 2.12

- A **delocalized lone pair** participates in resonance and occupies a *p* orbital.
- A **localized lone pair** does not participate in resonance.
- Whenever an atom possesses both a $\pi$ bond and a lone pair, they will not both participate in resonance.

# SKILLBUILDER REVIEW

## 2.1 CONVERTING BETWEEN DIFFERENT DRAWING STYLES

Draw the Lewis structure of this compound.

$(CH_3)_3C\ddot{O}CH_3$

**STEP 1** Draw each group separately.

**STEP 2** Draw all C—H bonds.

Try Problems 2.1–2.4, 2.49, 2.50

## 2.2 READING BOND-LINE STRUCTURES

Identify all carbon atoms and hydrogen atoms.

**STEP 1** The end of every line represents a carbon atom.

**STEP 2** Each carbon atom will possess enough hydrogen atoms in order to achieve four bonds.

Try Problems 2.5–2.7, 2.39, 2.49, 2.52

### 2.3 DRAWING BOND-LINE STRUCTURES

| Draw a bond-line drawing of this compound. | **STEP 1** Delete all hydrogen atoms except for those connected to heteroatoms. | **STEP 2** Draw in zig-zag format, keeping triple bonds linear. | **STEP 3** Delete all carbon atoms. |
|---|---|---|---|

**Try Problems** 2.8–2.10, 2.40, 2.41, 2.54, 2.58

### 2.4 IDENTIFYING LONE PAIRS ON OXYGEN ATOMS

Oxygen with a negative charge...

...has three lone pairs.

A neutral oxygen atom...

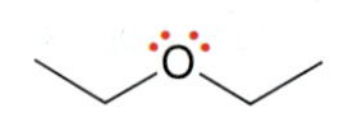

...has two lone pairs.

Oxygen with a positive charge...

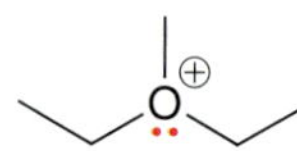

...has one lone pair.

**Try Problems** 2.14, 2.15, 2.43

### 2.5 IDENTIFYING LONE PAIRS ON NITROGEN ATOMS

Nitrogen with a negative charge...

...has two lone pairs.

A neutral nitrogen atom...

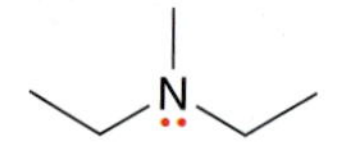

...has one lone pair.

Nitrogen with a positive charge...

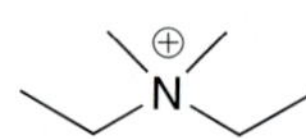

...has no lone pairs.

**Try Problems** 2.16–2.19, 2.39

### 2.6 IDENTIFYING VALID RESONANCE ARROWS

**RULE 1:** The tail of a curved arrow should not be placed on a single bond.

**RULE 2:** The head of a curved arrow cannot result in a bond causing a second-row element to exceed an octet.

**Try Problems** 2.21, 2.22, 2.51

### 2.7 ASSIGNING FORMAL CHARGES IN RESONANCE STRUCTURES

Read the curved arrows.

This negative charge...

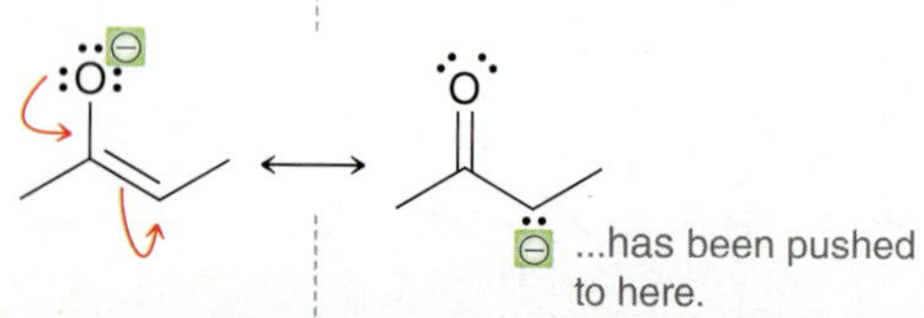


**Try Problems** 2.23, 2.24, 2.44, 2.53

### 2.8 DRAWING SIGNIFICANT RESONANCE STRUCTURES

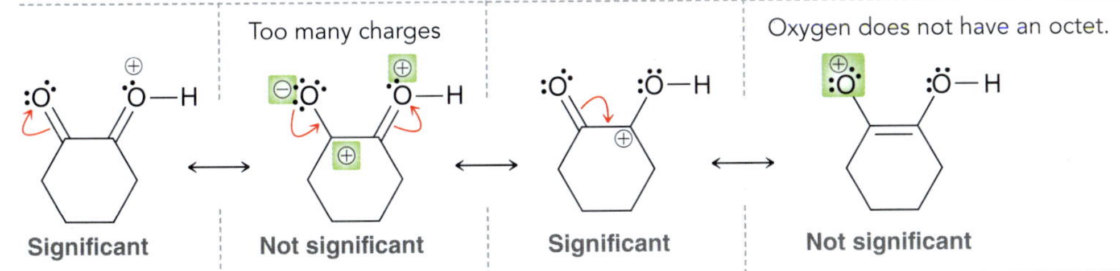


Try Problems 2.33–2.35, 2.45, 2.48, 2.59, 2.60, 2.62, 2.65, 2.66

### 2.9 IDENTIFYING LOCALIZED AND DELOCALIZED LONE PAIRS

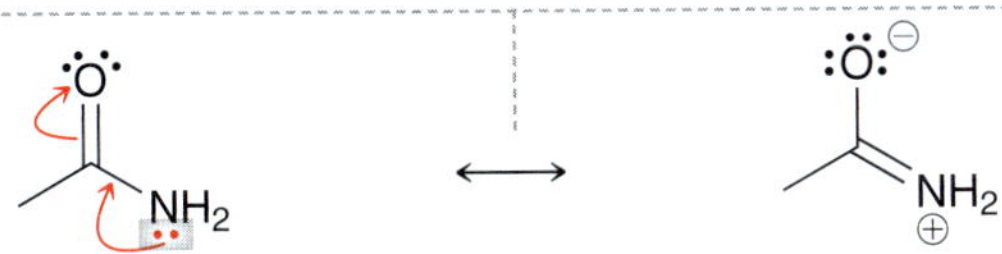

The lone pair on this nitrogen atom is delocalized by resonance, and it therefore occupies a *p* orbital.

As a result, the nitrogen atom is $sp^2$ hybridized and is therefore trigonal planar.

Try Problems 2.36–2.38, 2.47, 2.61

## PRACTICE PROBLEMS

**Note: Most of the Problems are available within WileyPLUS, an online teaching and learning solution.**

**2.39** Draw all carbon atoms, hydrogen atoms, and lone pairs for the following compounds:

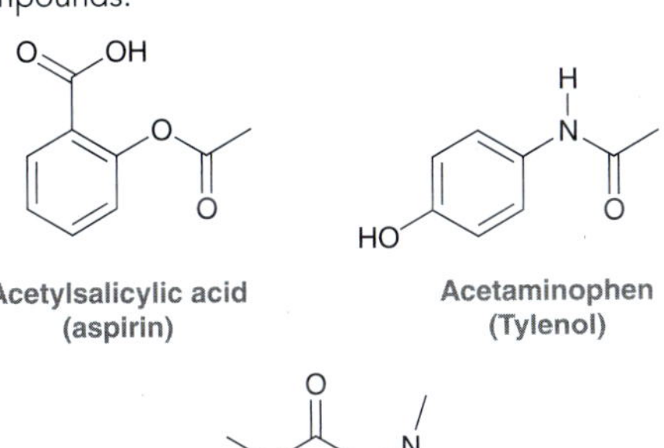

**Acetylsalicylic acid (aspirin)**

**Acetaminophen (Tylenol)**

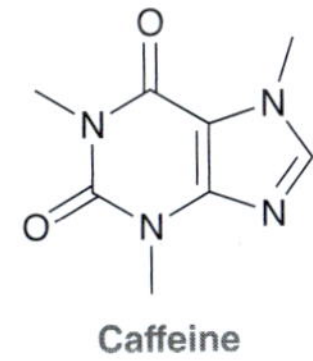

**Caffeine**

**2.40** Draw bond-line structures for all constitutional isomers of $C_4H_{10}$.

**2.41** Draw bond-line structures for all constitutional isomers of $C_5H_{12}$.

**2.42** Draw bond-line structures for vitamin A and vitamin C:

**Vitamin A**

**Vitamin C**

**2.43** How many lone pairs are found in the structure of vitamin C?

**2.44** Identify the formal charge in each case below:

**2.45** Draw significant resonance structures for the following compound:

**2.46** *Learning to extract structural information from molecular formulas:*

(a) Write out the molecular formula for each of the following compounds:

Compare the molecular formulas for the above compounds and fill in the blanks in the following sentence: The number of hydrogen atoms is equal to ________ times the number of carbon atoms plus ________.

(b) Now write out the molecular formula for each of these compounds:

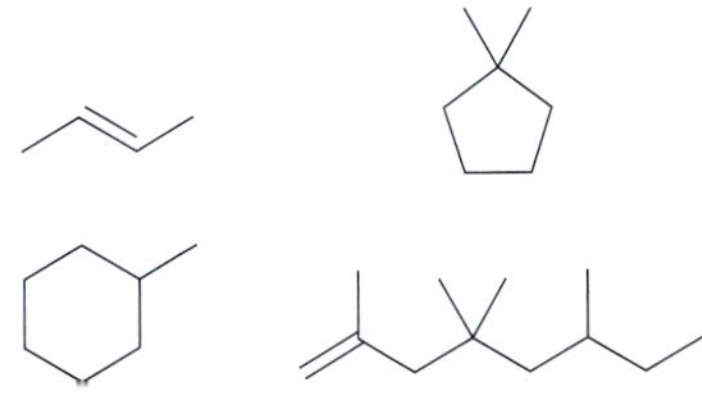

Each of the compounds above has *either* a double bond *or* a ring. Compare the molecular formulas for each of these compounds. In each case, the number of hydrogen atoms is __________ times the number of carbon atoms. Fill in the blank.

(c) Now write out the molecular formula for each of these compounds:

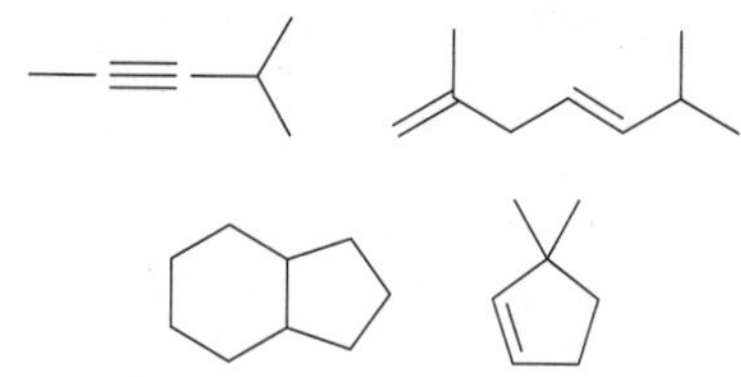

Each of the compounds above has *either* a triple bond *or* two double bonds *or* two rings *or* a ring and a double bond. Compare the molecular formulas for each of these compounds. In each case, the number of hydrogen atoms is __________ times the number of carbon atoms minus __________. Fill in the blanks.

(d) Based on the trends above, answer the following questions about the structure of a compound with molecular formula $C_{24}H_{48}$. Is it possible for this compound to have a triple bond? Is it possible for this compound to have a double bond?

(e) Draw all constitutional isomers that have the molecular formula $C_4H_8$.

**2.47** Each compound below exhibits one lone pair. In each case, identify the type of atomic orbital in which the lone pair is contained.

(a) (b) (c)

**2.48** Draw all significant resonance structures for each of the following compounds:

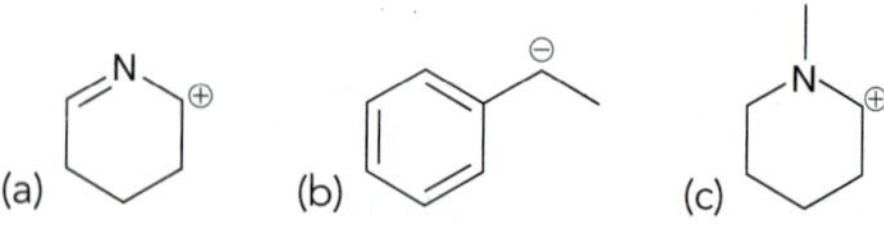

(a) (b) (c)

**2.49** Write a condensed structural formula for each of the following compounds:

(a) (b) (c)

**2.50** What is the molecular formula for each compound in the previous problem?

**2.51** Which of the following drawings is not a resonance structure for 1-nitrocyclohexene? Explain why it cannot be a valid resonance structure.

(a) (b) (c) (d)

**2.52** Identify the number of carbon atoms and hydrogen atoms in the compound below:

**2.53** Identify any formal charges in the following structures:

(a) (b) (c) (d)

**2.54** Draw bond-line structures for all constitutional isomers with molecular formula $C_4H_9Cl$.

**2.55** Draw resonance structures for each of the following:

(a) (b) (c) (d)

(e) (f) (g) (h)

(i) (j)

**2.56** Determine the relationship between the two structures below. Are they resonance structures or are they constitutional isomers?

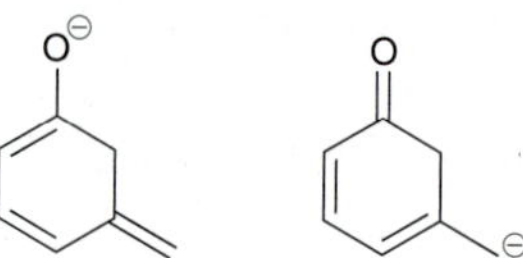

**2.57** Consider each pair of compounds below and determine whether the pair represent the same compound, constitutional isomers, or different compounds that are not isomeric at all:

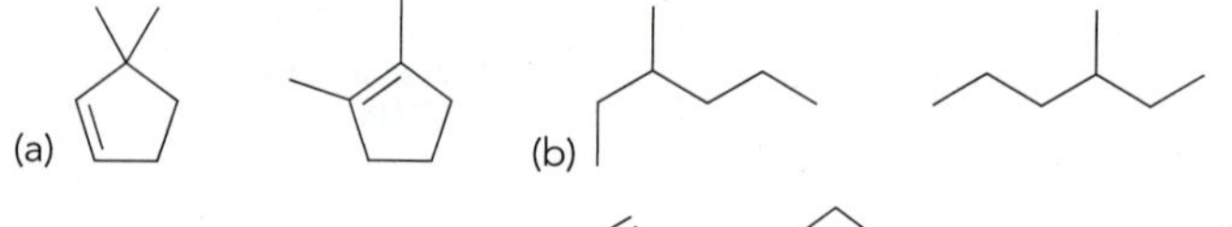

(a) (b)

(c) (d)

**2.58** Draw a bond-line structure for each of the following compounds:

(a) $CH_2{=}CHCH_2C(CH_3)_3$ (b) $(CH_3CH_2)_2CHCH_2CH_2OH$

(c) $CH{\equiv}COCH_2CH(CH_3)_2$ (d) $CH_3CH_2OCH_2CH_2OCH_2CH_3$

(e) $(CH_3CH_2)_3CBr$ (f) $(CH_3)_2C{=}CHCH_3$

**2.59** A mixture of sulfuric acid and nitric acid will produce small quantities of the nitronium ion ($NO_2^+$):

Does the nitronium ion have any significant resonance structures? Why or why not?

**2.60** Consider the structure of ozone:

Ozone is formed in the upper atmosphere, where it absorbs short-wavelength UV radiation emitted by the sun, thereby protecting us from harmful radiation. Draw all significant resonance structures for ozone. (**Hint:** Begin by drawing all lone pairs.)

**2.61** Melatonin is an animal hormone believed to have a role in regulating the sleep cycle:

The structure of melatonin incorporates two nitrogen atoms. What are the hybridization state and geometry of each nitrogen atom? Explain your answer.

**Melatonin**

**2.62** Draw all significant resonance structures for each of the following compounds:

**Estradiol** (Female sex hormone)

**Testosterone** (Male sex hormone)

# INTEGRATED PROBLEMS

**2.63** Cycloserine is an antibiotic isolated from the microbe *Streptomyces orchidaceous*. It is used in conjunction with other drugs for the treatment of tuberculosis.

**Cycloserine**

(a) What is the molecular formula of this compound?

(b) How many $sp^3$-hybridized carbon atoms are present in this structure?

(c) How many $sp^2$-hybridized carbon atoms are present in this structure?

(d) How many *sp*-hybridized carbon atoms are present in this structure?

(e) How many lone pairs are present in this structure?

(f) Identify each lone pair as localized or delocalized.

(g) Identify the geometry of each atom (except for hydrogen atoms).

(h) Draw all significant resonance structures of cycloserine.

**2.64** Ramelteon is a hypnotic agent used in the treatment of insomnia:

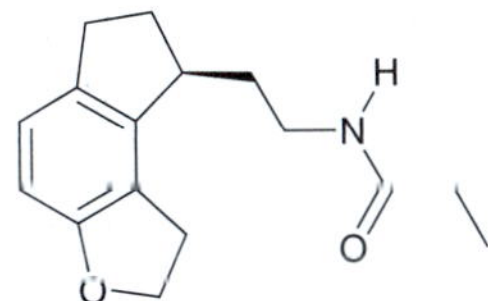

**Ramelteon**

(a) What is the molecular formula of this compound?

(b) How many $sp^3$-hybridized carbon atoms are present in this structure?

(c) How many $sp^2$-hybridized carbon atoms are present in this structure?

(d) How many *sp*-hybridized carbon atoms are present in this structure?

(e) How many lone pairs are present in this structure?

(f) Identify each lone pair as localized or delocalized.

(g) Identify the geometry of each atom (except for hydrogen atoms).

**2.65** In the compound below, identify all carbon atoms that are electron deficient (δ+) and all carbon atoms that are electron rich (δ−). Justify your answer with resonance structures.

**2.66** Consider the following two compounds:

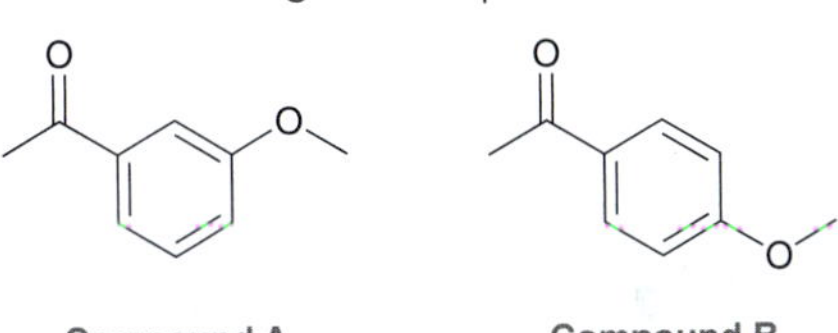

**Compound A** **Compound B**

(a) Identify which of these two compounds has greater resonance stabilization.

(b) Would you expect compound C (below) to have a resonance stabilization that is more similar to compound A or to compound B?

**Compound C**

**2.67** Single bonds generally experience free rotation at room temperature (as will be discussed in more detail in Chapter 4):

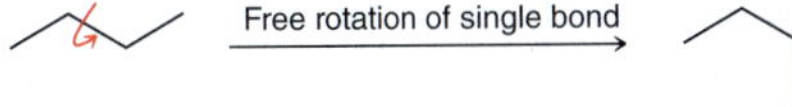


Nevertheless, the "single bond" shown below exhibits a large barrier to rotation. In other words, the energy of the system is greatly increased if that bond is rotated. Explain the source of this energy barrier. (**Hint:** Think about the atomic orbitals being used to form the "conduit.")

**Problems 2.68–2.69 are intended for students who have already covered IR spectroscopy (Chapter 15).**

**2.68** Polymers are very large compounds, assembled from smaller units, called monomers. For example, polymer **2**, called poly(vinyl acetate), is made from vinyl acetate (**1**). Polymer **2** can be converted to polymer **4**, called poly(vinyl alcohol), or PVA, via a process called hydrolysis (explored further in Chapter 21). Complete hydrolysis of **2** leads to **4**, while incomplete hydrolysis of **2** leads to **3** in which the polymer chain still contains some residual acetate groups. Polymer **2** is one of the main ingredients in white glue, while polymer **3** is present in aqueous PVA glues.

Describe how IR spectroscopy can be used to monitor the conversion of **3** to **4**. Specifically, describe what you would look for to confirm complete hydrolysis of the acetate groups (*J. Chem. Ed.* **2006,** *83,* **1534–1536**).

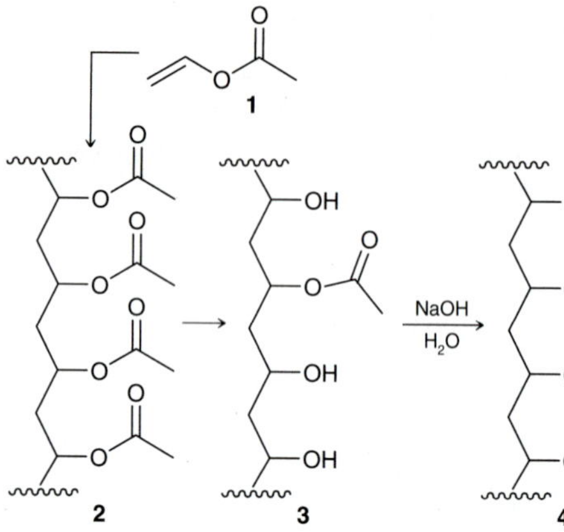

**2.69** Coumarin and its derivatives exhibit a broad array of industrial applications, including, but not limited to, cosmetics, food preservatives, and fluorescent laser dyes. Some derivatives of coumarin, such as warfarin, exhibit antithrombic activity and are currently used as blood thinners (to prevent the formation of potentially fatal blood clots). Identify how you might use IR spectroscopy to monitor the following reaction, in which compound **1** is converted into a coumarin derivative (compound **2**). Describe at least three different signals that you could analyze to confirm the transformation of **1** to **2** (*J. Chem. Ed.* **2006,** *83,* **287–289**).

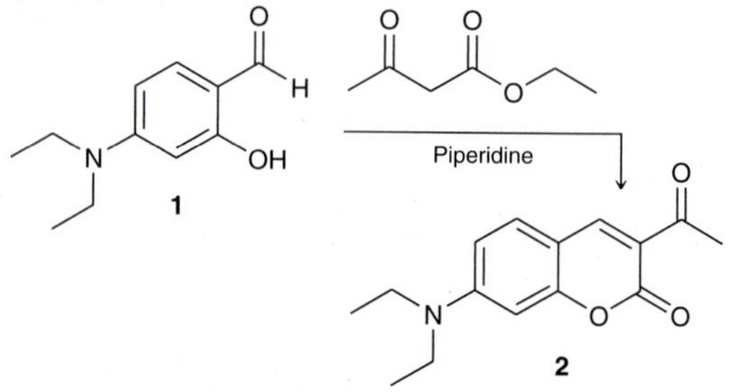

# CHALLENGE PROBLEMS

**2.70** CL-20 and HMX are both powerful explosives. CL-20 produces a more powerful blast but is generally considered too shock-sensitive for practical use. HMX is significantly less sensitive and is used as a standard military explosive. When a 2:1 mixture of the two compounds is cocrystallized, the resulting explosive is expected to be more powerful than HMX alone, but with a sensitivity similar to HMX (*Cryst. Growth Des.* **2012,** *12,* **4311–4314**).

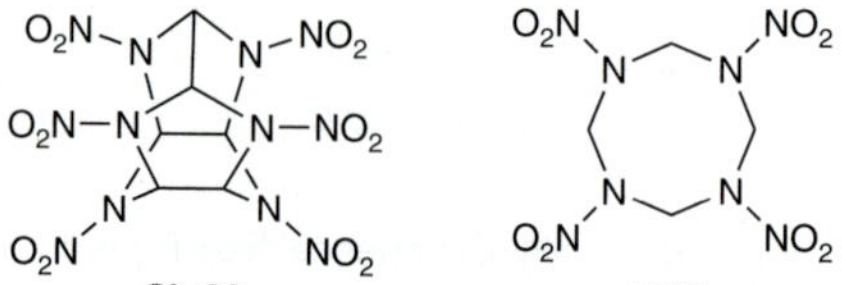

(a) What are the molecular formulas for CL-20 and HMX?

(b) Consider the lone pair of electrons on one of the nitrogen atoms within the ring(s) for either molecule. Is the lone pair localized or delocalized?

**2.71** Progesterone is a female hormone that plays a critical role in the menstrual cycle by preparing the lining of the uterus for implantation of an egg. During W. S. Johnson's biomimetic synthesis of progesterone (a synthesis that draws inspiration from and mimics naturally occurring, biosynthetic pathways), one of the final reactions in the synthesis was believed to have proceeded via the following intermediate (*J. Am. Chem. Soc.* **1971,** *93,* **4332–4334**). Use resonance structures to explain why this intermediate is particularly stable.

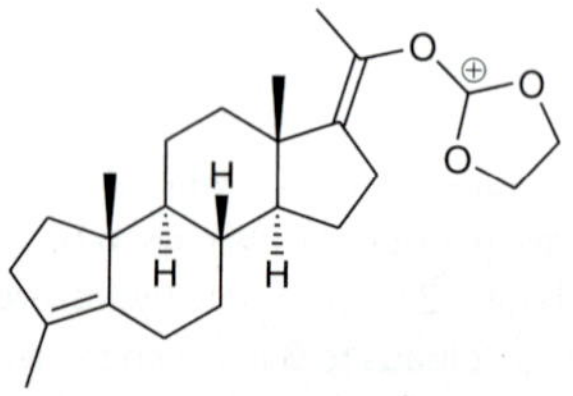

**2.72** Many compounds with desirable medicinal properties are isolated from natural sources and are thus referred to as natural products. However, a compound's medicinal properties are often known before the structure of the compound has been determined. Below are examples of compounds where the first proposed structure was incorrect (*Angew. Chem. Int. Ed.* **2005,** *44,* **1012–1044**). In each case, the corresponding correct structure is also shown. Identify all functional groups in each pair of compounds and then compare the similarities and differences between their molecular structures.

(a)

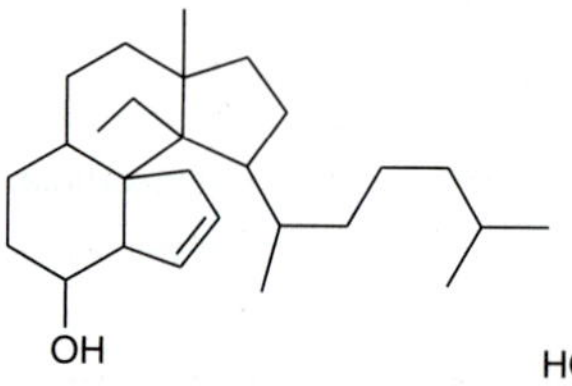

Proposed structure of cholesterol

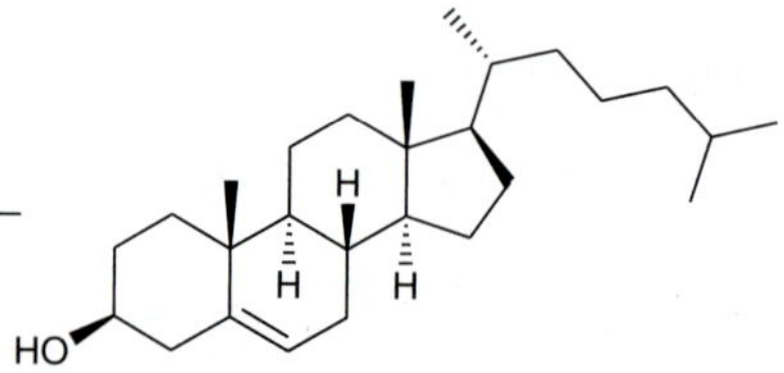

Correct structure of cholesterol

(b)

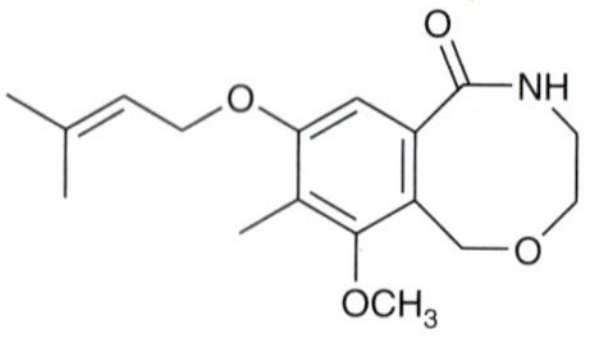

Proposed structure of porritoxin

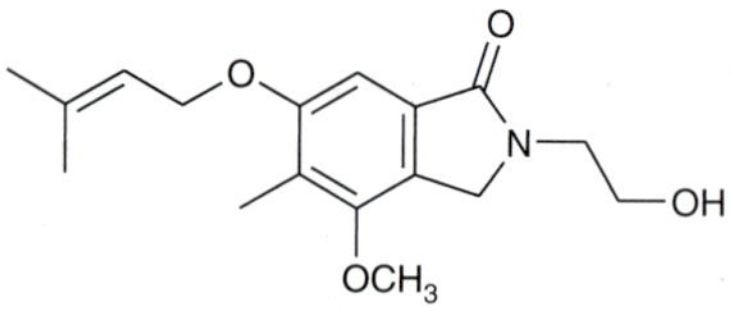

Correct structure of porritoxin

**2.73** The following compound is an intermediate in the synthesis of a *gelator*, which is a compound capable of self-assembling to form a gel in an organic liquid (*Soft Matter* **2012,** *8,* **5486–5492**).

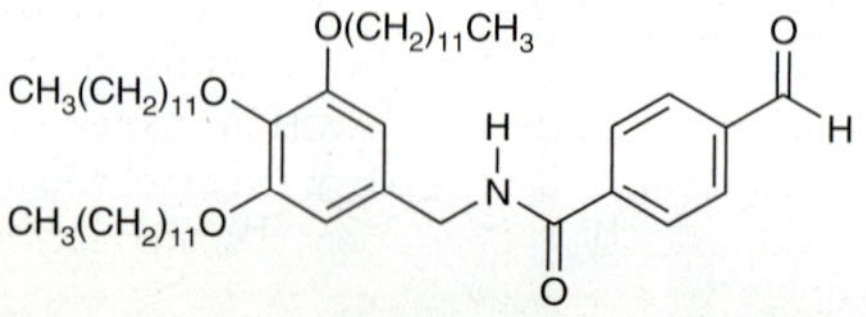

(a) Identify each functional group in the molecule.

(b) In Chapter 4, we will learn that molecules generally rotate freely around single bonds, while rotation around double bonds is significantly restricted. Consider the four single bonds connecting the two rings in this structure. Two of them can rotate freely, while the other two cannot. Identify the two with restricted rotation and justify your answer by drawing resonance structures.

**2.74** The reaction between an aldehyde and an amine results in the formation of a functional group called an *imine*, as shown in the following general reaction:

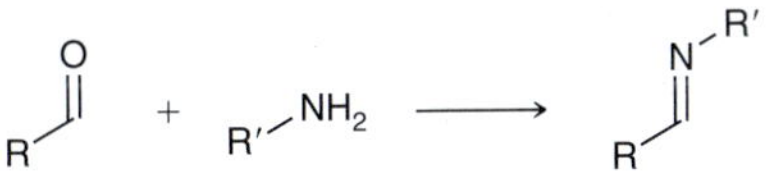

**An imine**

Consider the reactants **A–D**, which have been used in the synthesis of novel self-assembling molecules with unique and dynamic architectures (*Langmuir* **2012**, *28*, 14567–14572).

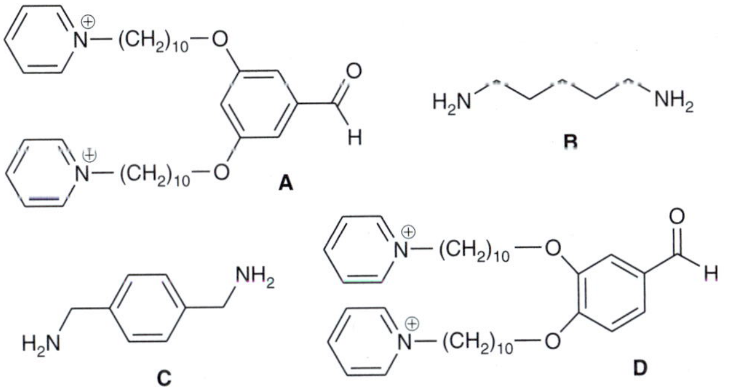

**A** **B** **C** **D**

(a) Draw the product of the reaction between:

i. **A** and **C** (2:1 molar ratio)

ii. **C** and **D** (1:2 molar ratio)

iii. **A** and **B** (2:1 molar ratio)

iv. **B** and **D** (1:2 molar ratio)

(b) What is the relationship between the products of reactions iii and iv?

**2.75** The following compound is an amino acid derivative (Chapter 25). In solution, molecules of this compound show a tendency to "stick" together, or *self-assemble*, via a series of intermolecular hydrogen bonds. During self-assembly, the growing aggregate can entrap molecules of liquid, thereby forming a gel (*J. Am. Chem. Soc.* **2009**, *131*, 11478–11484).

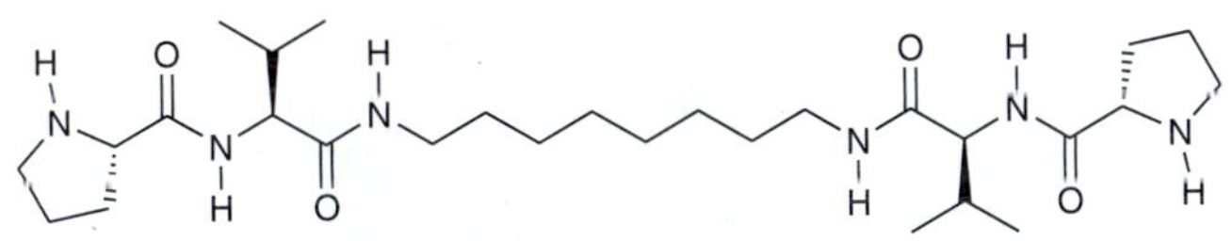

(a) Each of the six hydrogen atoms bound directly to nitrogen can form a hydrogen bond, but H's that are on the amides can form stronger hydrogen bonds than H's on the amines. Explain why this is so using resonance.

(b) Draw the intermolecular hydrogen bonds that form during self-assembly. Do this by drawing three molecules stacked directly on top of each other, each in the orientation shown. Then, draw hydrogen bonds from the central molecule to each of the other two molecules, showing eight intermolecular H bonds between amide H's on one molecule to amide O's on adjacent molecules.

**2.76** In Chapter 3, we will explore the factors that render compounds acidic or basic. Tropolone **(1)** is a compound that is both fairly acidic and fairly basic (*J. Org. Chem.* **1997**, *62*, 3200–3207). It is acidic because it is capable of losing a proton ($H^+$) to form a relatively stable anion **(2)**, while it is basic because of its ability to receive a proton to form cation **3**:

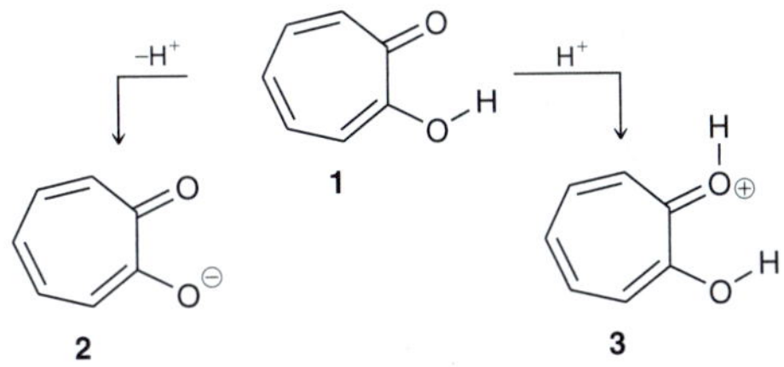

**2** **1** **3**

(a) Draw all significant resonance structures of anion **2** and of cation **3** and explain why each of these ions is stabilized.

(b) In compound **1**, the bond lengths for the C—C bonds vary greatly and alternate in length (long, short, long, etc.). But in ions **2** and **3**, the lengths of the bonds are similar. Explain.

(c) The intramolecular hydrogen bonding interaction in **3** is significantly diminished in comparison to **1**. Explain.

**2.77** Triphenylmethane dyes are among the first synthetic dyes developed for commercial use. A comparison of the structures of these compounds reveals that even small differences in structure can lead to large differences in color (*Chem. Rev.* **1993**, *93*, 381–433). The structures of three such dyes are shown below.

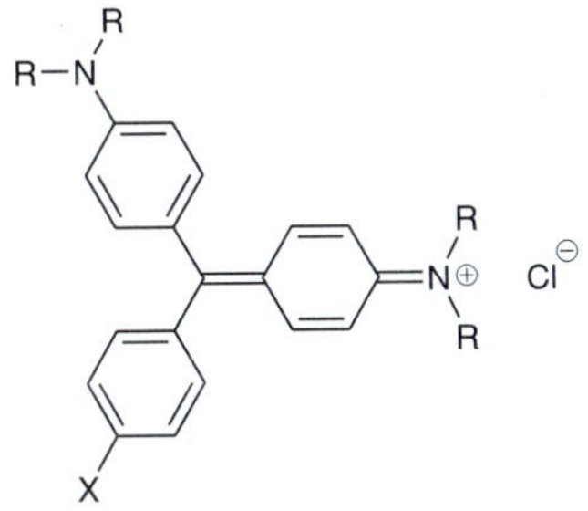

| | X | R |
|---|---|---|
| Basic green 4 | H | $CH_3$ |
| Basic red 9 | $NH_2$ | H |
| Basic violet 4 | $N(C_2H_5)_2$ | $C_2H_5$ |

(a) Draw resonance structures for basic green 4 that illustrate the delocalization of the positive charge.

(b) Determine whether basic green 4 or basic violet 4 is expected to have greater resonance stabilization. Justify your choice.

**2.78** Basic red 1 is a tetracyclic compound (it has four rings) that shares many structural similarities with the dyes in the previous problem (*Chem. Rev.* **1993**, *93*, 381–433). This compound has many significant resonance structures, and the positive charge is highly delocalized. While resonance structures can be drawn in which the positive charge is spread throughout all four rings, nonetheless, one of the rings likely bears very little of the charge relative to the other rings. Identify the ring that is not participating as effectively in resonance and suggest an explanation.

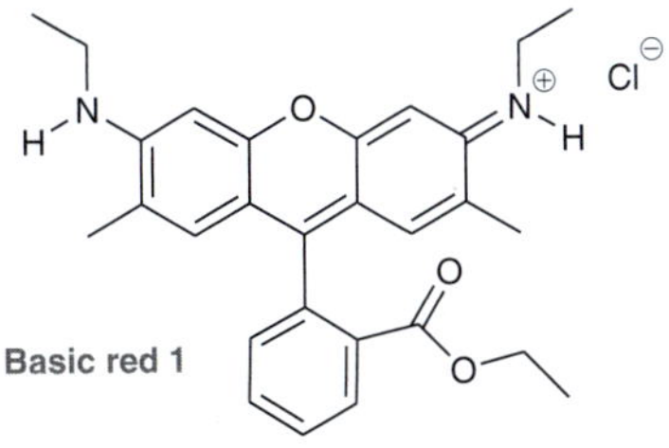

**Basic red 1**

# 3

# Acids and Bases



## DID YOU EVER WONDER...

### how dough rises to produce fluffy (leavened) rolls and bread?

Dough rises fairly quickly in the presence of a leavening agent, such as yeast, baking powder, or baking soda. All of these leavening agents work by producing bubbles of carbon dioxide gas that get trapped in the dough, causing it to rise. Then, upon heating in an oven, these gas bubbles expand, creating holes in the dough. Although leavening agents work in similar ways, they differ in how they produce the $CO_2$. Yeast produces $CO_2$ as a by-product of metabolic processes, while baking soda and baking powder produce $CO_2$ as a by-product of acid-base reactions. Later in this chapter, we will take a closer look at the acid-base reactions involved, and we will discuss the difference between baking soda and baking powder. An understanding of the relevant reactions will lead to a greater appreciation of food chemistry.

In this chapter, our study of acids and bases will serve as an introduction to the role of electrons in ionic reactions. An *ionic reaction* is a reaction in which ions participate as reactants, intermediates, or products. These reactions represent 95% of the reactions covered in this textbook. In order to prepare ourselves for the study of ionic reactions, it is critical to be able to identify acids and bases. We will learn how to draw acid-base reactions and to compare the acidity or basicity of compounds. These tools will enable us to predict when acid-base reactions are likely to occur and to choose the appropriate reagent to carry out any specific acid-base reaction.

## DO YOU REMEMBER?

*Before you go on, be sure you understand the following topics.*
*If necessary, review the suggested sections to prepare for this chapter.*

- Drawing and Interpreting Bond-Line Structures (Section 2.2)
- Identifying Formal Charges (Sections 1.4 and 2.4)
- Identifying Lone Pairs (Section 2.5)
- Drawing Resonance Structures (Section 2.10)

Take the DO YOU REMEMBER? QUIZ in **WileyPLUS** to check your understanding.

## 3.1 Introduction to Brønsted-Lowry Acids and Bases

This chapter will focus primarily on Brønsted-Lowry acids and bases. There is also a brief section dealing with Lewis acids and bases, a topic that is revisited in Chapter 6 and subsequent chapters.

The definition of **Brønsted-Lowry acids and bases** is based on the transfer of a proton ($H^+$). An *acid* is defined as a *proton donor*, while a *base* is defined as a *proton acceptor*. As an example, consider the following acid-base reaction:

H—Cl + $H_2O$ ⇌ $Cl^-$ + $H_3O^+$

Acid (proton donor) | Base (proton acceptor)

In the reaction above, HCl functions as an acid because it donates a proton to $H_2O$, while $H_2O$ functions as a base because it accepts the proton from HCl. The products of a proton transfer reaction are called the **conjugate base** and the **conjugate acid**:

| HCl | + | $H_2O$ | ⇌ | $Cl^-$ | + | $H_3O^+$ |
|---|---|---|---|---|---|---|
| Acid | | Base | | Conjugate base | | Conjugate acid |

In this reaction, $Cl^-$ is the conjugate base of HCl. In other words, the conjugate base is what remains of the acid after it has been deprotonated. Similarly, in the reaction above, $H_3O^+$ is the conjugate acid of $H_2O$. We will use this terminology throughout the rest of this chapter, so it is important to know these terms well.

In the example above, $H_2O$ served as a base by accepting a proton, but in other situations, it can serve as an acid by donating a proton. For example:

| Base | + | Acid | ⇌ | Conjugate acid | + | Conjugate base |
|---|---|---|---|---|---|---|
| $(CH_3)_3CO^-$ | | $H_2O$ | | $(CH_3)_3COH$ | | $HO^-$ |

In this case, water functions as an acid rather than a base. Throughout this course, we will see countless examples of water functioning either as a base or as an acid, so it is important to understand that both are possible and very common. When water functions as an acid, as in the reaction above, the conjugate base is $HO^-$.

## 3.2 Flow of Electron Density: Curved-Arrow Notation

All reactions are accomplished via a flow of electron density (the motion of electrons). Acid-base reactions are no exception. The flow of electron density is illustrated with curved arrows:

$$B:^- + H—A \rightleftharpoons B—H + :A^-$$

Although these curved arrows look exactly like the curved arrows used for drawing resonance structures, there is an important difference. When drawing resonance structures, curved arrows are used simply as tools and do not represent any real physical process. But in the reaction above, the curved arrows do represent an actual physical process. There is a flow of electron density that causes a proton to be transferred from one reagent to another; the curved arrows illustrate this flow. The arrows show the **reaction mechanism**, that is, they show how the reaction occurs in terms of the motion of electrons.

Notice that the mechanism of a proton transfer reaction involves electrons from a base deprotonating an acid. This is an important point, because acids do not lose protons without the help of a base. It is necessary for a base to abstract the proton. Here is a specific example:

Base Acid Conjugate acid Conjugate base

In this example, hydroxide ($HO^-$) functions as a base to abstract a proton from the acid. Notice that there are exactly two curved arrows. The mechanism of a proton transfer always involves at least two curved arrows.

In Chapter 6, reaction mechanisms will be introduced and explored in more detail. Mechanisms represent the core of organic chemistry, and by proposing and comparing mechanisms, we will discover trends and patterns that define the behavior of electrons. These trends and patterns will enable us to predict how electron density flows and to explain new reactions. For almost every reaction throughout this book, we will propose a mechanism and then analyze it in detail.

Most mechanisms involve one or more proton transfer steps. For example, one of the first reactions to be covered (Chapter 8) is called an elimination reaction, and it is believed to occur via the following mechanism:

Proton transfer — $-H_2O$ — Proton transfer

This mechanism has many steps, each of which is shown with curved arrows. Notice that the first and last steps are simply protons transfers. In the first step, $H_3O^+$ functions as an acid, and in the last step, water is acting as a base.

Clearly, proton transfers play an integral role in reaction mechanisms. Therefore, in order to become proficient in drawing mechanisms, it is essential to master proton transfers. Important skills to be mastered include drawing curved arrows properly, being able to predict when a proton transfer is likely or unlikely, and being able to determine which acid or base is appropriate for a specific situation. Let's get some practice drawing the mechanism of a proton transfer.

## SKILLBUILDER

### 3.1 DRAWING THE MECHANISM OF A PROTON TRANSFER

**LEARN the skill**

Draw a mechanism for the following acid-base reaction. Label the acid, base, conjugate acid, and conjugate base:

$$H_2O + CH_3O^- \rightleftharpoons HO^- + CH_3OH$$

Water Methoxide Hydroxide Methanol

## SOLUTION

**STEP 1** Identify the acid and the base.

We begin by identifying the acid and the base. Water is losing a proton to form hydroxide. Therefore, water is functioning as a proton donor, rendering it an acid. Methoxide ($CH_3O^-$) is accepting the proton to form methanol ($CH_3OH$). Therefore, methoxide is the base.

To draw the mechanism properly, remember that there must be two curved arrows. The tail of the first curved arrow is placed on a lone pair of the base and the head is placed on the proton of the acid. This first curved arrow shows the base abstracting the proton. The next curved arrow always comes from the X—H bond being broken and goes to the atom connected to the proton:

**STEP 2** Draw the first curved arrow.

**STEP 3** Draw the second curved arrow.

$H_2O$ (Acid) + $CH_3O^-$ (Base) ⇌ $HO^-$ + $CH_3OH$

Make sure that the head and tail of each arrow are positioned in exactly the right place or the mechanism will be incorrect.

When water loses a proton, hydroxide is generated. Therefore, hydroxide is the conjugate base of water. When methoxide receives the proton, methanol is generated. Methanol is therefore the conjugate acid of methoxide:

$H_2O$ (Acid) + $CH_3O^-$ (Base) ⇌ $HO^-$ (Conjugate base) + $CH_3OH$ (Conjugate acid)

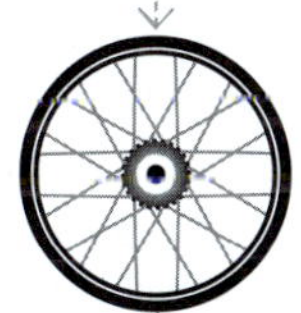

**PRACTICE the skill** **3.1** All of the following acid-base reactions are reactions that we will study in greater detail in the chapters to follow. For each one, draw a mechanism and then clearly label the acid, base, conjugate acid, and conjugate base:

(a) phenol + $^{\ominus}OH$ ⟶ phenoxide + $H_2O$

(b) acetone + $H_3O^+$ ⇌ protonated acetone + $H_2O$

(c) ketone + diisopropylamide ⟶ enolate + diisopropylamine

(d) benzoic acid + $^{\ominus}OH$ ⟶ benzoate + $H_2O$

**APPLY the skill** **3.2** Each of the following mechanisms contains one or more errors—that is, the curved arrows may or may not be correct. In each case, identify the errors and then describe what modification would be necessary in order to make the curved arrows correct. Explain your suggested modification in each case (the following examples represent common student errors, so it is in your best interests to identify these errors, recognize them, and then avoid these mistakes):

(a) $R_3N$ + H—Cl ⟶ $R_3N^+$—H + $Cl^-$

(b) 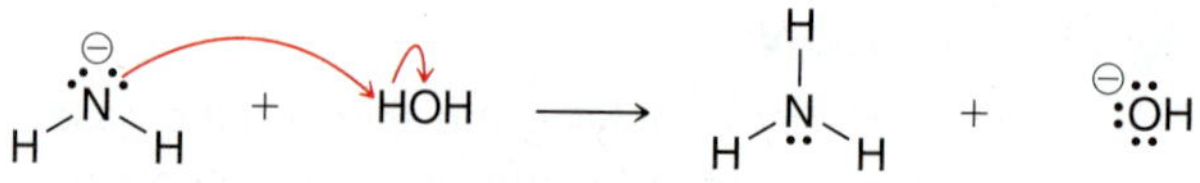 $H_2N^-$ + HOH ⟶ $NH_3$ + $^-OH$

(c) 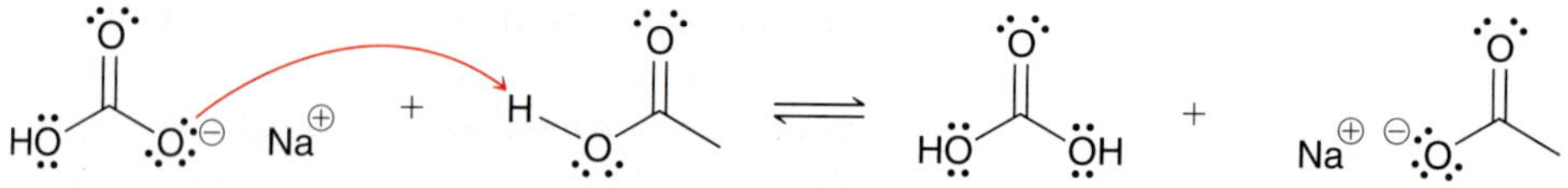 $HOCO_2^-$ $Na^+$ + $CH_3COOH$ ⇌ $H_2CO_3$ + $Na^+$ $^-OOCCH_3$

**3.3** In an *intramolecular proton transfer reaction*, the acidic site and the basic site are tethered together in the same structure, and a proton is passed from the acidic region of the structure to the basic region of the structure, as shown below:

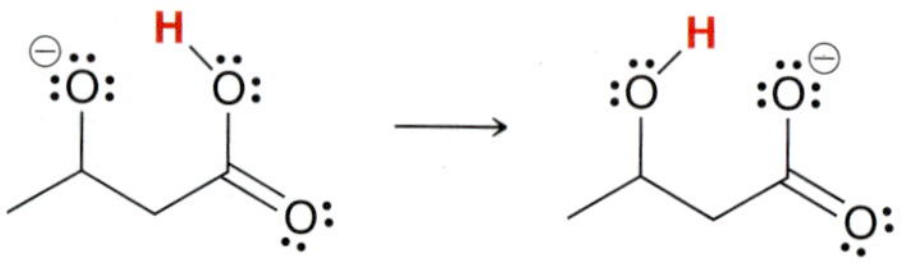

Draw a mechanism for this process.

need more **PRACTICE?** **Try Problem 3.44**

## medically speaking

## Antacids and Heartburn

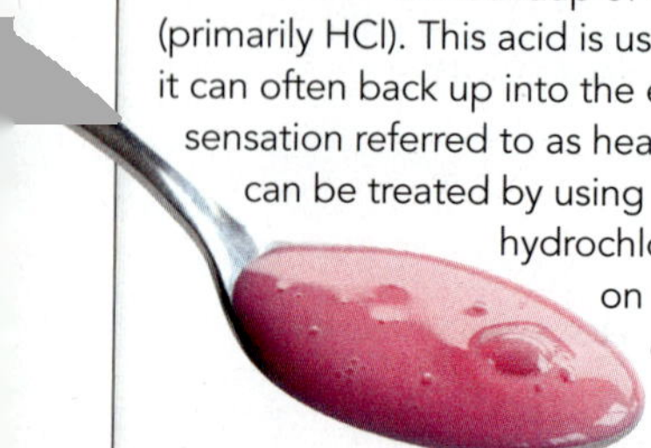

Most of us have experienced occasional heartburn, especially after eating pizza. Heartburn is caused by the buildup of excessive amounts of stomach acid (primarily HCl). This acid is used to digest the food we eat, but it can often back up into the esophagus, causing the burning sensation referred to as heartburn. The symptoms of heartburn can be treated by using a mild base to neutralize the excess hydrochloric acid. Many different antacids are on the market and can be purchased over the counter. Here are just a few that you will probably recognize:

**Alka Seltzer**

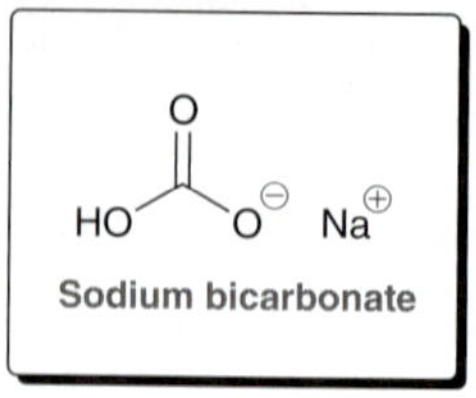

**Sodium bicarbonate**

**Tums or Rolaids**

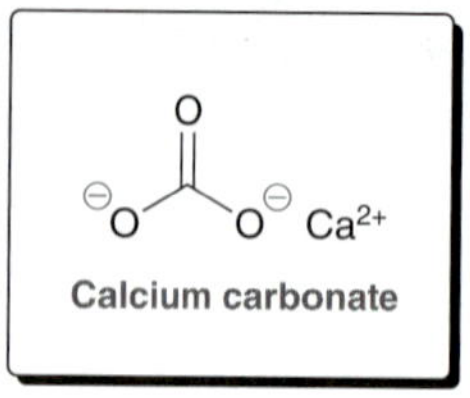

**Calcium carbonate**

**Pepto Bismol**

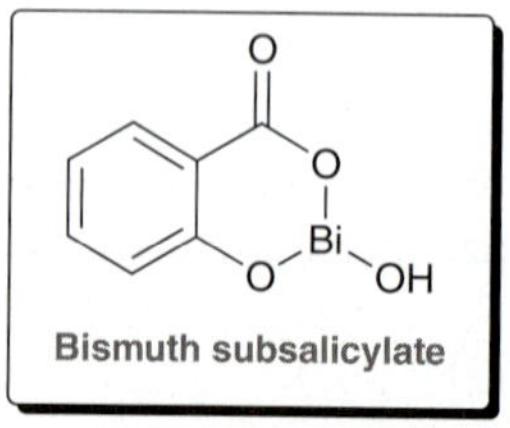

**Bismuth subsalicylate**

**Maalox or Mylanta**

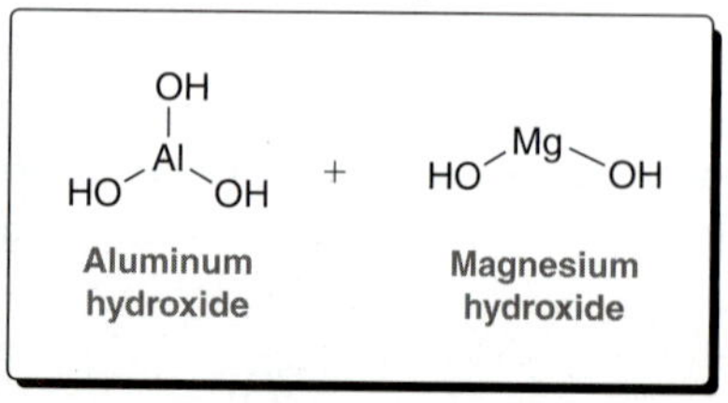

**Aluminum hydroxide** + **Magnesium hydroxide**

All of these work in similar ways. They are all mild bases that can neutralize HCl in a proton transfer reaction. For example, sodium bicarbonate deprotonates HCl to form carbonic acid:

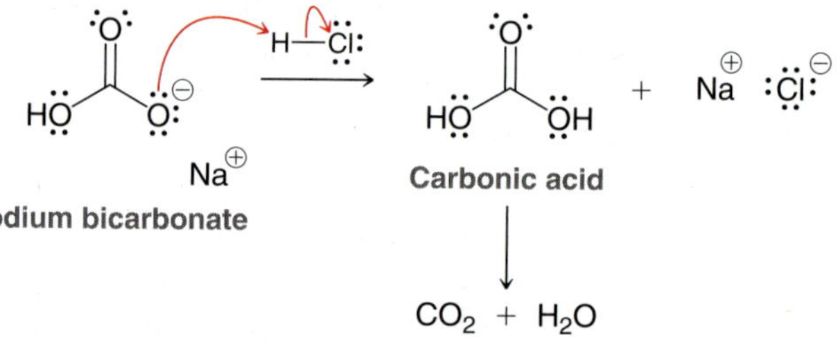

**Sodium bicarbonate** + H–Cl ⟶ **Carbonic acid** + $Na^+$ $Cl^-$

Carbonic acid ⟶ $CO_2$ + $H_2O$

Carbonic acid then quickly degrades into carbon dioxide and water (a fact that we will discuss again later in this chapter).

If you ever find yourself in a situation where you have heartburn and no access to any of the antacids above, a substitute can be found in your kitchen. Baking soda is just sodium bicarbonate (the same compound found in Alka Seltzer). Take a teaspoon of baking soda, dissolve it in a glass of water by stirring with a spoon, and then drink it down. The solution will taste salty, but it will alleviate the burning sensation of heartburn. Once you start burping, you know it is working; you are releasing the carbon dioxide gas that is produced as a by-product of the acid-base reaction shown above.

## 3.3 Brønsted-Lowry Acidity: Quantitative Perspective

There are two ways to predict when a proton transfer reaction will occur: (1) via a quantitative approach (comparing p$K_a$ values) or (2) via a qualitative approach (analyzing the structures of the acids). It is essential to master both methods. In this section, we focus on the first method, and in the upcoming sections we will focus on the second method.

### Using p$K_a$ Values to Compare Acidity

The terms $K_a$ and p$K_a$ were defined in your general chemistry textbook, but it is worthwhile to quickly review their definitions. Consider the following general acid-base reaction between HA (an acid) and $H_2O$ (functioning as a base in this case):

$$\mathrm{HA} + \mathrm{H_2O} \rightleftharpoons \mathrm{A^-} + \mathrm{H_3O^+}$$

The reaction is said to have reached **equilibrium** when there is no longer an observable change in the concentrations of reactants and products. At equilibrium, the rate of the forward reaction is exactly equivalent to the rate of the reverse reaction, which is indicated with two arrows pointing in opposite directions, as shown above. The position of equilibrium is described by the term $\boldsymbol{K_{eq}}$, which is defined in the following way:

$$K_{eq} = \frac{[\mathrm{H_3O^+}]\,[\mathrm{A^-}]}{[\mathrm{HA}]\,[\mathrm{H_2O}]}$$

It is the product of the equilibrium concentrations of the products divided by the product of the equilibrium concentrations of the reactants. When an acid-base reaction is carried out in dilute aqueous solution, the concentration of water is fairly constant (55.5M) and can therefore be removed from the expression. This gives us a new term, $\boldsymbol{K_a}$:

$$K_a = K_{eq}\,[\mathrm{H_2O}] = \frac{[\mathrm{H_3O^+}]\,[\mathrm{A^-}]}{[\mathrm{HA}]}$$

The value of $K_a$ measures the strength of the acid. Very strong acids can have a $K_a$ on the order of $10^{10}$ (or 10,000,000,000), while very weak acids can have a $K_a$ on the order of $10^{-50}$ (or 0.00000000000000000000000000000000000000000000000001). The values of $K_a$ are often very small or very large numbers. To deal with this, chemists often express p$K_a$ values, rather than $K_a$ values, where p$K_a$ is defined as

$$\mathrm{p}K_a = -\log K_a$$

When p$K_a$ is used as the measure of acidity, the values will generally range from −10 to 50. We will deal with p$K_a$ values extensively throughout this chapter, and there are two things to keep in mind: (1) A strong acid will have a low p$K_a$ value, while a weak acid will have a high p$K_a$ value. For example, an acid with a p$K_a$ of 10 is more acidic than an acid with a p$K_a$ of 16. (2) Each unit represents an order of magnitude. An acid with a p$K_a$ of 10 is six orders of magnitude (one million times) more acidic than an acid with a p$K_a$ of 16. Table 3.1 provides p$K_a$ values for many of the compounds commonly encountered in this course. A p$K_a$ table with more data can be found on the inside cover of this book.

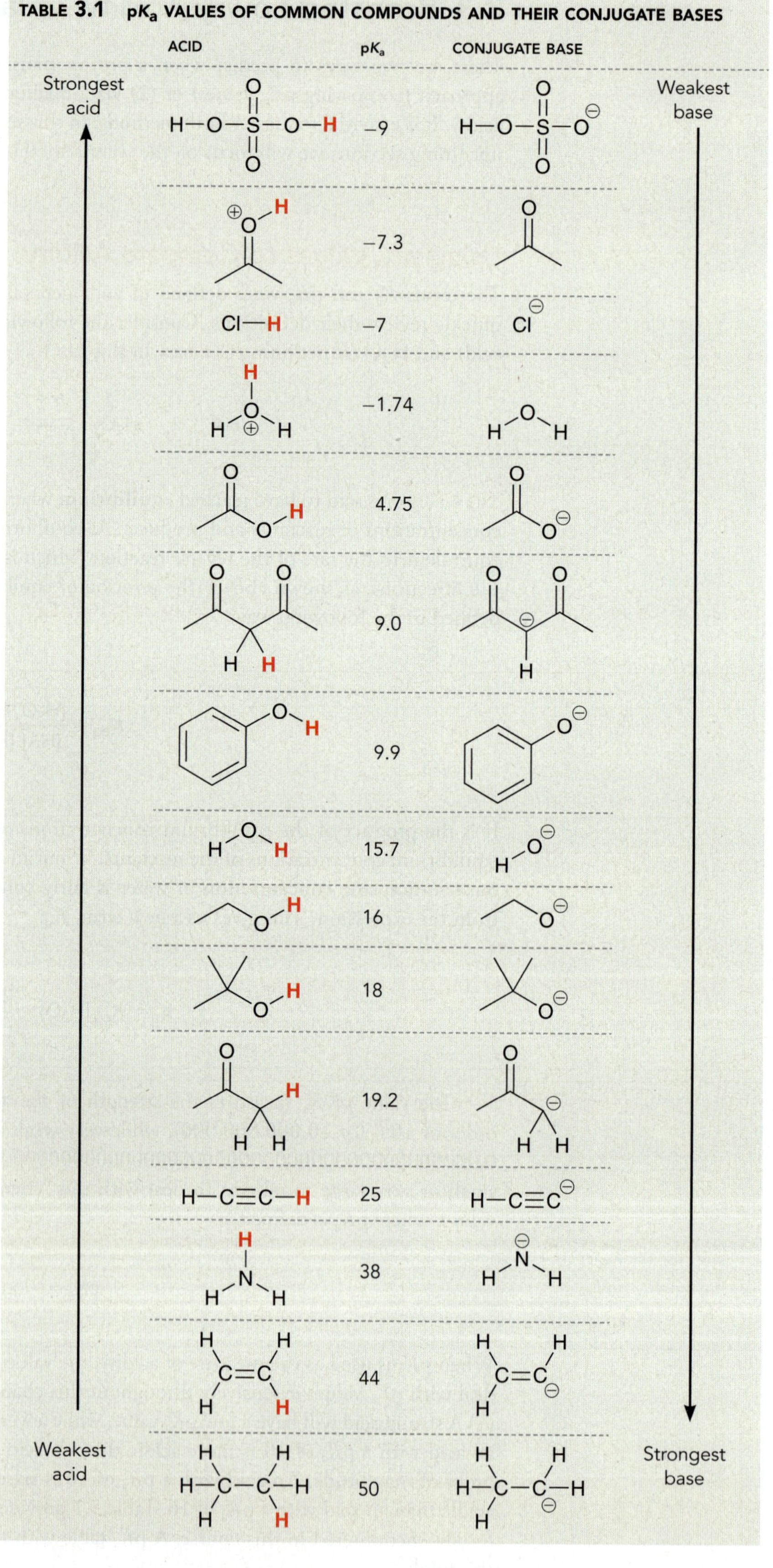

**TABLE 3.1** p$K_a$ **VALUES OF COMMON COMPOUNDS AND THEIR CONJUGATE BASES**

Strongest acid ↑ ... Weakest acid; Weakest base ↓ ... Strongest base

| ACID | p$K_a$ | CONJUGATE BASE |
|---|---|---|
| H—O—S(=O)(=O)—O—H | −9 | H—O—S(=O)(=O)—O⊖ |
| | −7.3 | |
| Cl—H | −7 | Cl⊖ |
| | −1.74 | |
| | 4.75 | |
| | 9.0 | |
| | 9.9 | |
| | 15.7 | |
| | 16 | |
| | 18 | |
| | 19.2 | |
| H—C≡C—H | 25 | H—C≡C⊖ |
| | 38 | |
| | 44 | |
| | 50 | |

# SKILLBUILDER

## 3.2 USING p$K_a$ VALUES TO COMPARE ACIDS

**LEARN the skill**

Acetic acid is the main constituent in vinegar solutions and acetone is a solvent often used in nail polish remover:

**Acetic acid** **Acetone**

Using the p$K_a$ values in Table 3.1, identify which of these two compounds is more acidic.

**SOLUTION**

Acetic acid has a p$K_a$ of 4.75, while acetone has a p$K_a$ of 19.2. The compound with the lower p$K_a$ is more acidic, and therefore, acetic acid is more acidic. In fact, when we compare the p$K_a$ values, we see that acetic acid is approximately 14 orders of magnitude ($10^{14}$) more acidic than acetone (or approximately 100,000,000,000,000 times more acidic). We will discuss the reason for this in the upcoming sections of this chapter.

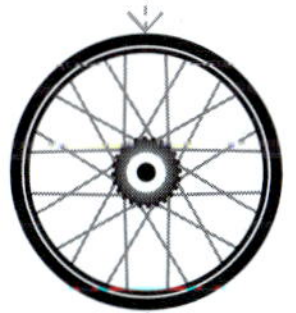

**PRACTICE the skill**

**3.4** For each pair of compounds below, identify the more acidic compound:

(a) (b)

(c) H—C≡C—H (d) Cl—H

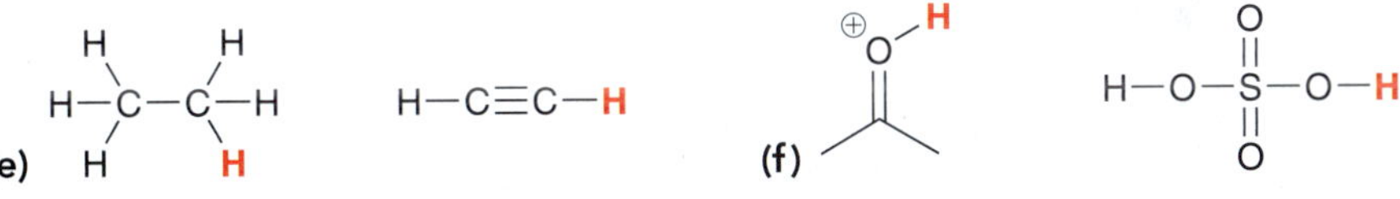

(e) H—C≡C—H (f)

**APPLY the skill**

**3.5** Propranolol is an antihypertensive agent (used to treat high blood pressure). Using Table 3.1, identify the two most acidic protons in the compound and indicate the approximate expected p$K_a$ for each proton.

**Propranolol**

**3.6** L-dopa is used in the treatment of Parkinson's disease. Using Table 3.1, identify the four most acidic protons in the compound and then arrange them in order of increasing acidity (two of the protons will be very similar in acidity and difficult to distinguish at this point in time):

**L-dopa**

need more **PRACTICE?** **Try Problem 3.38**

## Using p$K_a$ Values to Compare Basicity

We have seen how to use p$K_a$ values to compare acids, but it is also possible to use p$K_a$ values to compare bases to one another. It is not necessary to use a separate chart of p$K_b$ values. The following example will demonstrate how to use p$K_a$ values to compare basicity.

# SKILLBUILDER

### 3.3 USING p$K_a$ VALUES TO COMPARE BASICITY

**LEARN the skill** Using the p$K_a$ values in Table 3.1, identify which of the following anions is a stronger base:

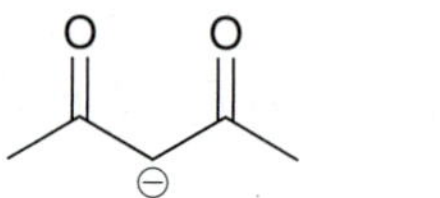

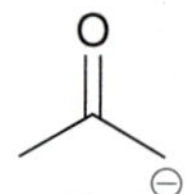

**SOLUTION**

Each of these bases can be thought of as the conjugate base of some acid. We just need to compare the p$K_a$ values of those acids. To do this, we imagine protonating each of the bases above, producing the following conjugate acids:

**STEP 1** Draw the conjugate acid of each base.

We then look up the p$K_a$ values of these acids and compare them:

**STEP 2** Compare p$K_a$ values.

p$K_a$ = 9.0 p$K_a$ = 19.2

**STEP 3** Identify the stronger base.

The first compound has a lower p$K_a$ value and is therefore a stronger acid than the second compound. Remember that the stronger acid always generates the weaker base. As a result, the conjugate base of the first compound will be a weaker base than the conjugate base of the second compound:

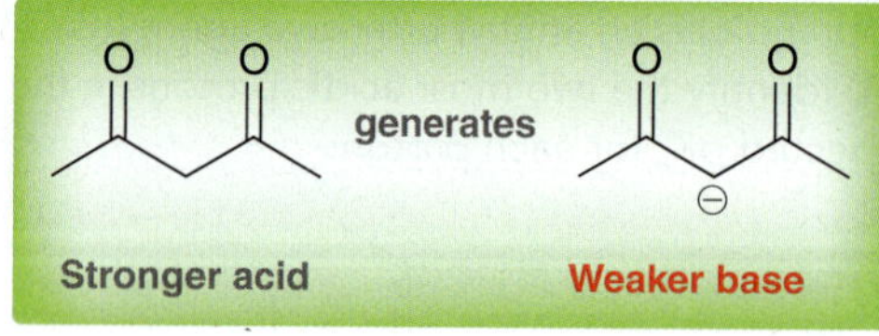


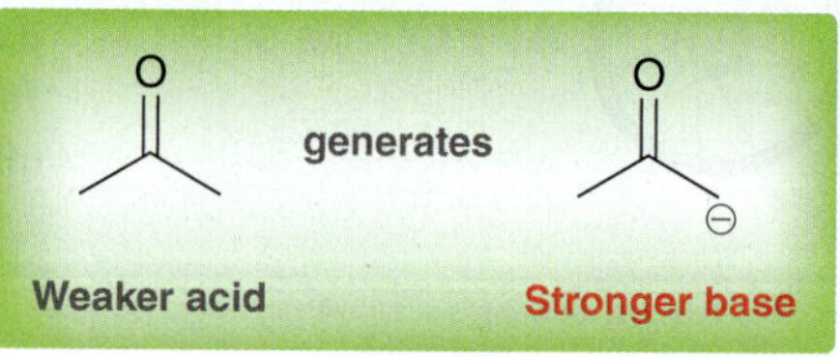

**PRACTICE the skill** **3.7** Identify the stronger base in each of the following cases:

(a) $H-C\equiv C^{\ominus}$ $H-\overset{\ominus}{N}-H$ (b)

(c) $H-O-H$ (d) $^{\ominus}OH$

(e) $H_3C-CH_2^{\ominus}$ $H-C\equiv C^{\ominus}$ (f) $Cl^{\ominus}$ $^{\ominus}OH$

**3.8** The following compound has three nitrogen atoms:

Each of the nitrogen atoms exhibits a lone pair that can function as a base (to abstract a proton from an acid). Rank these three nitrogen atoms in terms of increasing base strength using the following information:

$pK_a = 3.4$ $pK_a = 3.8$ $pK_a = 10.5$

**3.9** Consider the following $pK_a$ values and then answer the following questions:

$pK_a = -2.2$ $pK_a = 10.5$

**(a)** For the following compound, will the lone pair on the nitrogen atom be more or less basic than the lone pair on the oxygen atom?

**(b)** Fill in the blanks: The lone pair on the __________ atom is __________ orders of magnitude more basic than the lone pair on the __________ atom.

need more **PRACTICE?** **Try Problem 3.50**

## Using $pK_a$ Values to Predict the Position of Equilibrium

Using the chart of $pK_a$ values, we can also predict the position of equilibrium for any acid-base reaction. The equilibrium will always favor formation of the weaker acid (higher $pK_a$ value). For example, consider the following acid-base reaction:

**Base** + **Acid** ⇌ **Conjugate acid** + **Conjugate base**

Acid: $pK_a = 15.7$   Conjugate acid: $pK_a = 18$

The equilibrium for this reaction will lean to the right side, favoring formation of the weaker acid.

For some reactions, the $pK_a$ values are so vastly different that for practical purposes the reaction is treated not as an equilibrium process but rather as one that goes to completion. For example, consider the following reaction:

Base + Acid ⟶ Conjugate acid + Conjugate base

$pK_a = 15.7$ (acid, $H_2O$) $pK_a = 50$ (conjugate acid)

The reverse process is negligible, and for such reactions, organic chemists often draw an irreversible arrow, rather than the traditional equilibrium arrows. Technically, it is true that all proton transfers are equilibrium processes, but in the case above, the $pK_a$ values are so vastly different (34 orders of magnitude) that we can essentially ignore the reverse reaction.

## SKILLBUILDER

### 3.4 USING $pK_a$ VALUES TO PREDICT THE POSITION OF EQUILIBRIUM

**LEARN the skill**

Using $pK_a$ values from Table 3.1, determine the position of equilibrium for each of the following two proton transfer reactions:

(a) acetone + $H_3O^+$ ⇌ protonated acetone + $H_2O$

(b) 2,4-pentanedione + $^-OH$ ⇌ enolate anion + $H_2O$

**SOLUTION**

(a) We begin by identifying the acid on either side of the reaction, and then we compare their $pK_a$ values:

**STEP 1** Identify the acid on each side of the equilibrium.

$pK_a = -1.74$ ($H_3O^+$) $pK_a = -7.3$ (protonated acetone)

**STEP 2** Compare $pK_a$ values.

In this reaction, the C=O bond receives a proton (protonation of C=O bonds will be discussed in more detail in Chapter 20). The equilibrium always favors the weaker acid (the acid with the higher $pK_a$ value). The $pK_a$ values shown above are both negative numbers, so it can be confusing as to which is the higher $pK_a$ value: −1.74 is a larger number than −7.3. Therefore, the equilibrium will favor the left side of the reaction, which is drawn like this:

acetone + $H_3O^+$ ⇌ protonated acetone + $H_2O$ (equilibrium favoring the left)

The difference in $pK_a$ values represents a difference in acidity of nearly six orders of magnitude. This means that at any given moment in time, approximately one out of every million C=O bonds will bear a proton. When we study this reaction later on, we will see that protonation of a C=O bond can serve as a way to catalyze a number of reactions. For catalytic purposes, it is sufficient to have only a small percentage of the C=O bonds protonated.

**(b)** We must first identify the acid on either side of the reaction and then compare their p$K_a$ values:

**STEP 1**
Identify the acid on each side of the equilibrium.

**p$K_a$ = 9.0** **p$K_a$ = 15.7**

**STEP 2**
Compare p$K_a$ values.

This reaction shows the deprotonation of a β-diketone (a compound with two C═O bonds separated from each other by one carbon atom). Once again, this is a proton transfer that we will study in more depth later in the course. The equilibrium will favor the side of the weaker acid (the side with the higher p$K_a$ value). Therefore, the equilibrium will favor the right side of the reaction:

The difference in p$K_a$ values represents a difference in acidity of six orders of magnitude. In other words, when hydroxide is used to deprotonate a β-diketone, the vast majority of the diketone molecules are deprotonated (only one out of every million is not deprotonated). We can conclude from this analysis that hydroxide is a suitable base to accomplish this deprotonation.

**PRACTICE** the skill **3.10** Determine the position of equilibrium for each acid-base reaction below:

(a)

(b) + $H_2O$

(c)

(d) H—C≡C—H + $^{\ominus}NH_2$ ⇌ H—C≡C:$^{\ominus}$ + $NH_3$

**APPLY** the skill **3.11** Hydroxide is not a suitable base for deprotonating acetylene:

H—C≡C—H + $^{\ominus}$:OH ⇌ H—C≡C:$^{\ominus}$ + H—Ö—H

**Acetylene** **Hydroxide**

Explain why not. Can you propose a base that would be suitable?

need more **PRACTICE?** **Try Problem 3.48**

## Drug Distribution and $pK_a$

Most drugs will endure a very long journey before reaching their site of action. This journey involves several transitions between polar environments and nonpolar environments. In order for a drug to reach its intended target, it must be capable of being distributed in both types of environments along the journey. A drug's ability to transition between environments is, in most cases, a direct result of the drug's acid-base properties. In fact, most drugs today are acids or bases and, as such, are in equilibrium between charged and uncharged forms. As an example, consider the structure of aspirin and its conjugate base:

Aspirin — Uncharged form ⇌ Conjugate base — Charged form + $H^+$

In the equilibrium above, the left side represents the uncharged form of aspirin, while the right side represents the charged form (the conjugate base). The position of this equilibrium, or *percent ionization*, will depend on the pH of the solution. The $pK_a$ of aspirin is approximately 3.0. At a pH of 3.0 (when pH = $pK_a$), aspirin and its conjugate base will be present in equal amounts. That is, 50% ionization occurs. At a pH below 3, the uncharged form will predominate. At a pH above 3, the charged form will predominate.

With this in mind, consider the journey that aspirin takes after you ingest it. This journey begins in your stomach, where the pH can be as low as 2. Under these very acidic conditions, aspirin is mostly in its uncharged form. That is, there is very little of the conjugate base present. The uncharged form of aspirin is absorbed by the nonpolar environment of the gastric mucosa in your stomach and the intestinal mucosa in the intestinal tract. After passing through these nonpolar environments, the molecules of aspirin enter the blood, which is a polar (aqueous) environment with a pH of approximately 7.4. At that pH, aspirin exists mainly in the charged form (the conjugate base), and it is distributed throughout the circulatory system in this form. Then, in order to pass the blood-brain barrier, or a cell membrane, the molecules must be converted once again into the uncharged form so that they can pass through the necessary nonpolar environments. The drug is capable of successfully reaching the target because of its ability to exist in two different forms (charged and uncharged). This ability allows it to pass through polar environments as well as nonpolar environments.

The case above (aspirin) was an example of an acid that achieves biodistribution as a result of its ability to lose a proton. In contrast, some drugs are bases, and they achieve biodistribution as a result of their ability to gain a proton. For example, codeine (discussed in the previous chapter) can function as a base and accept a proton:

Codeine — Uncharged form + H–Cl → Codeine — Charged form ($Cl^-$)

Once again, the drug exists in two forms: charged and uncharged. But in this case a low pH favors the charged form rather than the uncharged form. Consider the journey that codeine takes after you ingest it. The drug first encounters the acidic environment of the stomach, where it is protonated and exists mostly in its charged form:

$pK_a = 8.2$

With a $pK_a$ of 8.2, the charged form predominates at low pH. It cannot pass through nonpolar environments and is therefore not absorbed by the gastric mucosa in the stomach. When the drug reaches the basic conditions of the intestines, it is deprotonated, and the uncharged form predominates. Only then can it transition at an appreciable rate into a nonpolar environment.

Accordingly, the efficacy of any drug is highly dependent on its acid-base properties. This must be taken into account in the design of new drugs. It is certainly important that a drug can bind with its designated receptor, but it is equally important that its acid-base properties allow it to reach the receptor efficiently.

## CONCEPTUAL CHECKPOINT

**3.12** Amino acids, such as glycine, are the key building blocks of proteins and will be discussed in greater detail in Chapter 25. At the pH of the stomach, glycine exists predominantly in a protonated form in which there are two acidic protons of interest. The p$K_a$ values for these protons are shown. Using this information, draw the form of glycine that will predominate at physiological pH of 7.4.

p$K_a$ = 9.87 p$K_a$ = 2.35

Glycine

# 3.4 Brønsted-Lowry Acidity: Qualitative Perspective

In the previous section, we learned how to compare acids or bases by comparing p$K_a$ values. In this section, we will now learn how to make such comparisons by analyzing and comparing their structures and without the use of p$K_a$ values.

## Conjugate Base Stability

In order to compare acids without the use of p$K_a$ values, we must look at the conjugate base of each acid:

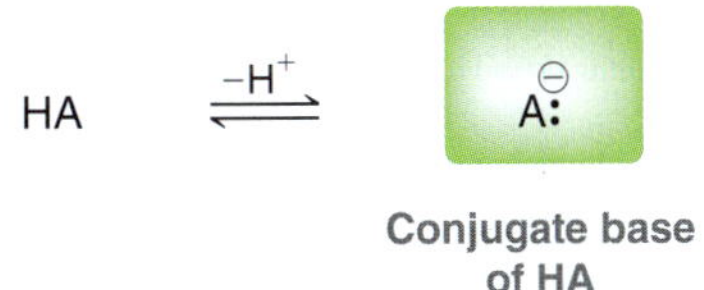


Conjugate base of HA

If $A^-$ is very stable (weak base), then HA must be a strong acid. If, on the other hand, $A^-$ is very unstable (strong base), then HA must be a weak acid. As an illustration of this point let's consider the deprotonation of HCl:

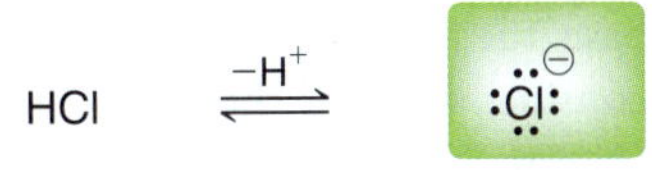


Conjugate base

Chlorine is an electronegative atom, and it can therefore stabilize a negative charge. The chloride ion ($Cl^-$) is in fact very stable, and therefore, HCl is a strong acid. HCl can serve as a proton donor because the conjugate base left behind is stabilized.

Let's look at one more example. Consider the structure of butane:

Conjugate base

When butane is deprotonated, a negative charge is generated on a carbon atom. Carbon is not a very electronegative element and is generally not capable of stabilizing a negative charge. Since this $C^-$ is very unstable, we can conclude that butane is not very acidic.

**LOOKING AHEAD**

In Section 22.2, we will see an exceptional case of a stabilized negative charge on a carbon atom. Until then, most instances of $C^-$ are considered to be very unstable.

This approach can be used to compare the acidity of two compounds, HA and HB. We simply look at their conjugate bases, $A^-$ and $B^-$, and compare them to each other:

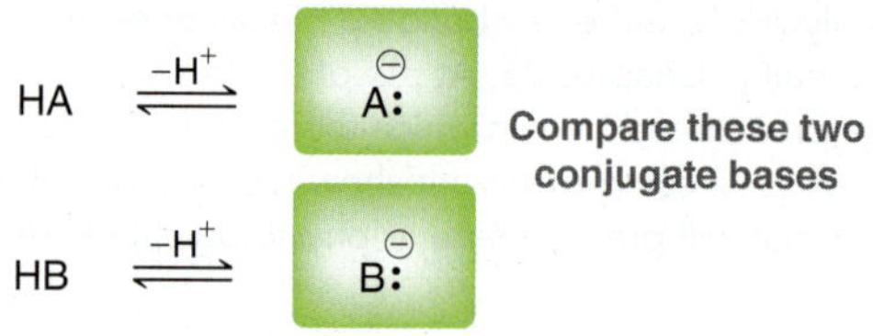

HA $\xrightleftharpoons{-H^+}$ A:⊖

HB $\xrightleftharpoons{-H^+}$ B:⊖

Compare these two conjugate bases

By determining the more stable conjugate base, we can identify the stronger acid. For example, if we determine that $A^-$ is more stable than $B^-$, then HA must be a stronger acid than HB. This approach does not allow us to predict exact p$K_a$ values, but it does allow us to compare the relative acidity of two compounds quickly, without the need for a chart of p$K_a$ values.

## Factors Affecting the Stability of Negative Charges

A qualitative comparison of acidity requires a comparison of the stability of negative charges. The following discussion will develop a methodical approach for comparing negative charge stability. Specifically, we will consider four factors: (1) the atom bearing the charge, (2) resonance, (3) induction, and (4) orbitals.

1. ***Which atom bears the charge?*** The first factor involves comparing the atoms bearing the negative charge in each conjugate base. For example, consider the structures of butane and propanol:

Butane　　Propanol

In order to assess the relative acidity of these two compounds, we must first deprotonate each of these compounds and draw the conjugate bases:

Now we compare these conjugate bases by looking at where the negative charge is located in each case. In the first conjugate base, the negative charge is on a carbon atom. In the second conjugate base, the negative charge is on an oxygen atom. To determine which of these is more stable, we must consider whether these elements are in the same row or in the same column of the periodic table (Figure 3.1).

**FIGURE 3.1** Examples of elements in the same row or in the same column of the periodic table.

Increasing electronegativity

C N O F

P S Cl

Br

I

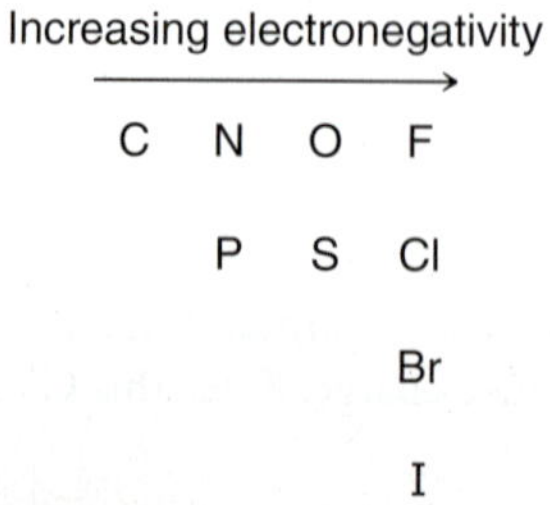

**FIGURE 3.2** Electronegativity trends in the periodic table.

For example, $C^-$ and $O^-$ appear in the same row of the periodic table. When two atoms are in the same row, electronegativity is the dominant effect. Recall that electronegativity measures an atom's affinity for electrons (how willing the atom is to accept a new electron), and electronegativity increases across a row (Figure 3.2). Oxygen is more electronegative than carbon, so oxygen is more capable of stabilizing the negative charge. Therefore, a proton on oxygen is more acidic than a proton on carbon:

More acidic

The story is different when comparing two atoms in the same column of the periodic table. For example, let's compare the acidity of water and hydrogen sulfide:

$H_2O$ $H_2S$

In order to assess the relative acidity of these two compounds, we deprotonate each of them and compare their conjugate bases:

$^{\ominus}$OH $^{\ominus}$SH

In this example, we are comparing $O^-$ and $S^-$, which appear in the same column of the periodic table. In such a case, electronegativity is not the dominant effect. Instead, the dominant effect is size (Figure 3.3). Sulfur is larger than oxygen and can therefore better stabilize a negative charge by spreading the charge over a larger volume of space. As such, $HS^-$ is more stable than $HO^-$, and therefore, $H_2S$ is a stronger acid than $H_2O$. We can verify this prediction by looking at p$K_a$ values (the p$K_a$ of $H_2S$ is 7.0, while the p$K_a$ of $H_2O$ is 15.7).

C N O F
P S Cl
Br
I
Increasing electronegativity (↑)

C N O F
P S Cl
Br
I
Increasing size (↓)

**FIGURE 3.3** Competing trends in the periodic table: size vs. electronegativity.

To summarize, there are two important trends: electronegativity (for comparing atoms in the same row) and size (for comparing atoms in the same column).

# SKILLBUILDER

## 3.5 ASSESSING RELATIVE STABILITY: FACTOR 1—ATOM

**LEARN the skill**

Compare the two protons that are shown in the following compound. Which one is more acidic?

### SOLUTION

The first step is to deprotonate each location and draw the two possible conjugate bases:

**STEP 1** Draw the conjugate bases.

**STEP 2** Compare the location of the charge in each case.

We now compare these two conjugate bases and determine which one is more stable. The first conjugate base has a negative charge on nitrogen, while the second conjugate base has a negative charge on oxygen. Nitrogen and oxygen are in the same row of the periodic table, so electronegativity is the determining factor. Oxygen is more electronegative than

**STEP 3**
The more stable conjugate base corresponds with the more acidic proton.

nitrogen and can better stabilize the negative charge. Therefore, the proton on the oxygen can be removed with greater ease than the proton on the nitrogen:

**More acidic**

**PRACTICE the skill** **3.13** In each compound below, two protons are clearly identified. Determine which of the two protons is more acidic:

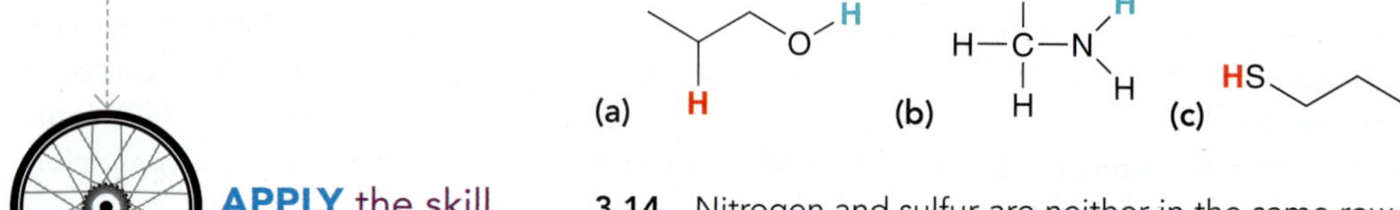


**APPLY the skill**

**3.14** Nitrogen and sulfur are neither in the same row nor in the same column of the periodic table. Nevertheless, you should be able to identify which proton below is more acidic. Explain your choice:

need more **PRACTICE?** **Try Problems 3.45b, 3.47h, 3.51b,h, 3.53**

2. ***Resonance.*** The second factor for comparing conjugate base stability is resonance. To illustrate the role of resonance in charge stability, let's consider the structures of ethanol and acetic acid:

**Ethanol** **Acetic acid**

In order to compare the acidity of these two compounds, we must deprotonate each of them and draw the conjugate bases:

In both cases, the negative charge is on oxygen. Therefore, factor 1 does not indicate which proton is more acidic. But there is a critical difference between these two negative charges. The first conjugate base has no resonance structures, while the second conjugate base does:

In this case, the charge is delocalized over both oxygen atoms. Such a negative charge will be more stable than a negative charge localized on one oxygen atom:

**Charge is localized (less stable)** **Charge is delocalized (more stable)**

For this reason, compounds containing a C=O bond directly next to an OH are generally mildly acidic, because their conjugate bases are resonance stabilized:

A carboxylic acid

Resonance-stabilized conjugate base

These compounds are called *carboxylic acids*. The R group above simply refers to the rest of the molecule that has not been drawn. Carboxylic acids are actually not very acidic at all compared with inorganic acids such as $H_2SO_4$ or HCl. Carboxylic acids are only considered to be acidic when compared with other organic compounds. The "acidity" of carboxylic acids highlights the fact that *acidity is relative.*

## SKILLBUILDER

### 3.6 ASSESSING RELATIVE STABILITY: FACTOR 2—RESONANCE

**LEARN the skill** Compare the two protons shown in the following compound. Which one is more acidic?

**SOLUTION**

Begin by drawing the respective conjugate bases:

**STEP 1** Draw the conjugate bases.

In the first conjugate base, the charge is localized on nitrogen:

**STEP 2** Look for resonance stabilization.

Charge is localized
NOT resonance stabilized

In the second conjugate base, the charge is delocalized by resonance:

Charge is resonance stabilized

**STEP 3**
The more stable conjugate base corresponds with the more acidic proton.

The charge is distributed over two atoms, N and O. The delocalization of the charge makes it more stable, and therefore, this proton is more acidic:

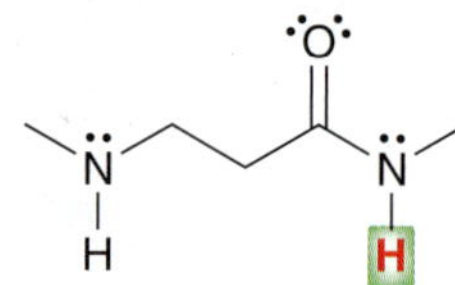

**PRACTICE the skill** **3.15** In each compound below, two protons are clearly identified. Determine which of the two protons is more acidic:

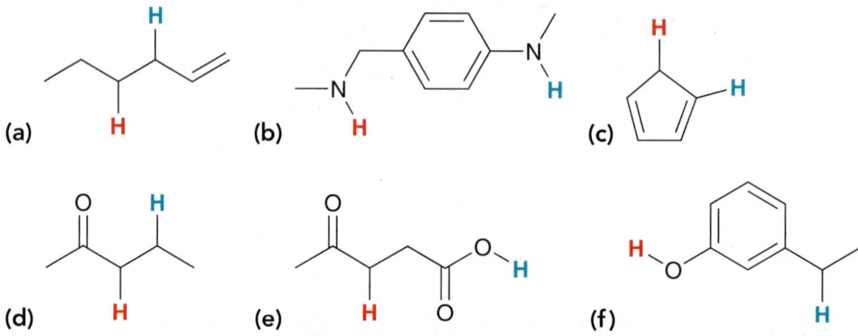

**APPLY the skill**

**3.16** Ascorbic acid (vitamin C) does not contain a traditional carboxylic acid group, but it is, nevertheless, still fairly acidic ($pK_a = 4.2$). Identify the acidic proton and explain your choice using resonance structures, if necessary:

**Ascorbic acid**
(Vitamin C)

**3.17** In the following compound two protons are clearly identified. Determine which of the two is more acidic. After comparing the conjugate bases, you should get stuck on the following question: Is it more stabilizing for a negative charge to be spread out over one oxygen atom and three carbon atoms or to be spread out over two oxygen atoms? Draw all of the resonance structures of each conjugate base and then take a look at the $pK_a$ values listed in Table 3.1.

need more **PRACTICE?** **Try Problems 3.45a, 3.46a, 3.47b,e–g, 3.51c–f**

3. ***Induction.*** The two factors we have examined so far do not explain the difference in acidity between acetic acid and trichloroacetic acid:

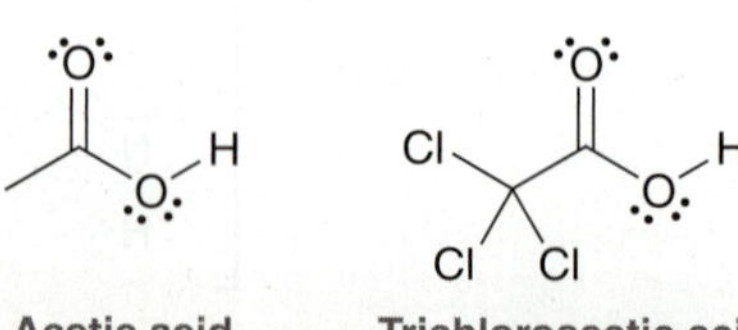

Which compound is more acidic? In order to answer this question without help from a chart of p$K_a$ values, we must draw the conjugate bases of the two compounds and then compare them:

Factor 1 does not answer the question because the negative charge is on oxygen in both cases. Factor 2 also does not answer the question because there are resonance structures that delocalize the charge over two oxygen atoms in both cases. The difference between these conjugate bases is clearly the chlorine atoms. Recall that each chlorine atom withdraws electron density via induction:

The net effect of the chlorine atoms is to withdraw electron density away from the negatively charged region of the structure, thereby stabilizing the negative charge. Therefore the conjugate base of trichloroacetic acid is more stable than the conjugate base of acetic acid:

More stable

From this, we can conclude that trichloroacetic acid is more acidic:

More acidic

We can verify this prediction by looking up p$K_a$ values. In fact, we can use p$K_a$ values to verify the individual effect of each Cl:

p$K_a$ = 4.75 p$K_a$ = 2.87 p$K_a$ = 1.25 p$K_a$ = 0.70

Notice the trend. With each additional Cl, the compound becomes more acidic.

# SKILLBUILDER

## 3.7 ASSESSING RELATIVE STABILITY: FACTOR 3—INDUCTION

**LEARN the skill** Identify which of the protons shown below is more acidic:

## SOLUTION

Begin by drawing the respective conjugate bases:

**STEP 1**
Draw the conjugate bases.

**STEP 2**
Look for inductive effects.

**STEP 3**
The more stable conjugate base corresponds with the more acidic proton.

In the conjugate base on the left, the charge is stabilized by the inductive effects of the nearby fluorine atoms. In contrast, the conjugate base on the right lacks this stabilization. Therefore, we predict that the conjugate base on the left is more stable. And as a result, we conclude that the proton near the fluorine atoms will be more acidic:

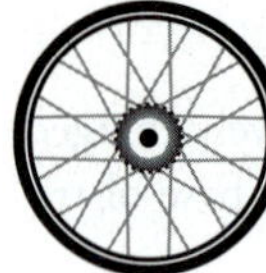

**PRACTICE the skill** **3.18** Identify the most acidic proton in each of the following compounds and explain your choice:

(a) (b)

**3.19** For each pair of compounds below, identify which compound is more acidic and explain your choice:

(a) (b)

**APPLY the skill** **3.20** Consider the structure of 2,3-dichloropropanoic acid:

This compound has many constitutional isomers.

**(a)** Draw a constitutional isomer that is slightly more acidic and explain your choice.

**(b)** Draw a constitutional isomer that is slightly less acidic and explain your choice.

**(c)** Draw a constitutional isomer that is significantly (at least 10 orders of magnitude) less acidic and explain your choice.

**3.21** Consider the two protons highlighted in the following compound:

Do you expect these protons to be equivalent or is one proton more acidic than the other? Explain your choice. (*Hint:* Think carefully about the geometry at the central carbon atom.)

need more **PRACTICE?** **Try Problems 3.46b, 3.47c, 3.51g**

4. ***Orbitals.*** The three factors we have examined so far will not explain the difference in acidity between the two identified protons in the following compound:

Draw the conjugate bases to compare them:

In both cases, the negative charge is on a carbon atom, so factor 1 does not help. In both cases, the charge is not stabilized by resonance, so factor 2 does not help. In both cases, there are no inductive effects to consider, so factor 3 does not help. The answer here comes from looking at the hybridization states of the orbitals that accommodate the charges. Recall from Chapter 1 that a carbon with a triple bond is *sp* hybridized, a carbon with a double bond is $sp^2$ hybridized, and a carbon with all single bonds is $sp^3$ hybridized. The first conjugate base (above left) has a negative charge on an $sp^2$-hybridized carbon atom, while the second conjugate base (above right) has a negative charge on an *sp*-hybridized carbon atom. What difference does this make? Let's quickly review the shapes of hybridized orbitals (Figure 3.4).

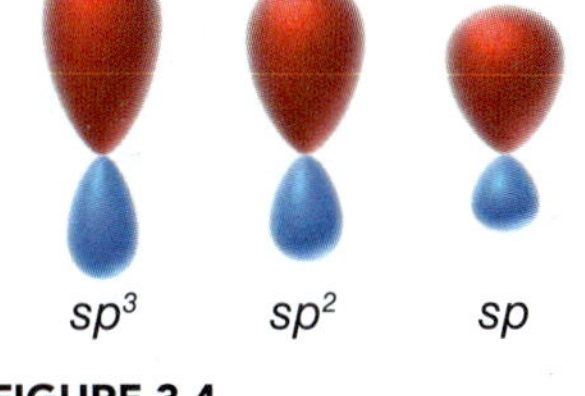


**FIGURE 3.4** Relative shapes of hybridized orbitals.

A pair of electrons in an *sp*-hybridized orbital is held closer to the nucleus than a pair of electrons in an $sp^2$- or $sp^3$-hybridized orbital. As a result, electrons residing in an *sp* orbital are stabilized by being close to the nucleus. Therefore, a negative charge on an *sp*-hybridized carbon is more stable than a negative charge on an $sp^2$-hybridized carbon:

More stable

We conclude that a proton on a triple bond will be more acidic than a proton on a double bond, which in turn will be more acidic than a proton on a carbon with all single bonds. We can verify this trend by looking at the p$K_a$ values in Figure 3.5. These p$K_a$ values suggest that this effect is very significant; acetylene is 19 orders of magnitude more acidic than ethylene.

Ethane p$K_a$ = 50 | Ethylene p$K_a$ = 44 | Acetylene p$K_a$ = 25

**FIGURE 3.5** p$K_a$ values for ethane, ethylene, and acetylene.

## SKILLBUILDER

### 3.8 ASSESSING RELATIVE STABILITY: FACTOR 4—ORBITALS

**LEARN the skill** Determine which of the protons identified below is more acidic:

1-Pentene

## SOLUTION

Begin by drawing the respective conjugate bases:

**STEP 1**
Draw the conjugate bases.

In both cases, the negative charge is on a carbon atom, so factor 1 does not help. In both cases, the charge is not stabilized by resonance, so factor 2 does not help. In both cases, there are no inductive effects to consider, so factor 3 does not help. The answer here comes from looking at the hybridization states of the orbitals that accommodate the charges. The first conjugate base has the negative charge in an $sp^3$-hybridized orbital, while the second conjugate base has the negative charge in an $sp^2$-hybridized orbital. An $sp^2$-hybridized orbital is closer to the nucleus than an $sp^3$-hybridized orbital and therefore better stabilizes a negative charge. So we conclude that the vinylic proton is more acidic:

**STEP 2**
Analyze the orbital that accommodates the charge in each case.

**STEP 3**
The more stable conjugate base corresponds with the more acidic proton.

**PRACTICE the skill** **3.22** Identify which of the following compounds is more acidic. Explain your choice:

**3.23** Identify the most acidic proton in each of the following compounds:

**APPLY the skill** **3.24** Amines contain C—N single bonds, while imines contain C—N double bonds:

Amine Imine

Most simple amines have a p$K_a$ somewhere in the range between 35 and 45. Based on this information, predict which statement is most likely to be true and explain the reasoning behind your selection:

**(a)** Most imines will have a p$K_a$ below 35.

**(b)** Most imines will have a p$K_a$ above 45.

**(c)** Most imines will have a p$K_a$ in the range between 35 and 45.

need more **PRACTICE?** **Try Problem 3.45c**

## Ranking the Factors That Affect the Stability of Negative Charges

We have thus far examined four factors that affect the stability of negative charges. We must now consider their order of priority—in other words, which factor takes precedence when two or more factors are present?

Generally speaking, the order of priority is the order in which the factors were presented:

1. *Atom.* Which atom bears the charge? (How do the atoms compare in terms of electronegativity and size? Remember the difference between comparing atoms in the same row vs. atoms in the same column.)
2. *Resonance.* Are there any resonance effects that make one conjugate base more stable than the other?
3. *Induction.* Are there any inductive effects that stabilize one of the conjugate bases?
4. *Orbital.* In what orbital do we find the negative charge for each conjugate base?

A helpful way to remember the order of these four factors is to take the first letter of each factor, giving the following mnemonic device: *ARIO.*

As an example, let's compare the protons shown in the following two compounds:

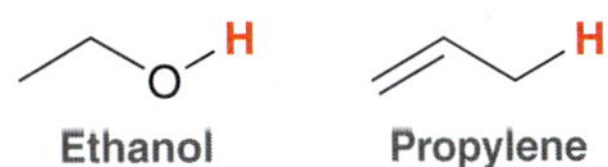


We compare these compounds by drawing their conjugate bases:

Factor 1 suggests that the first conjugate base is more stable ($O^-$ better than $C^-$). However, factor 2 suggests that the second conjugate base is more stable (resonance that delocalizes the charge). This leaves us with an important question: Is a negative charge more stable when it is localized on one oxygen atom or is a negative charge more stable when it is delocalized over two carbon atoms? The answer is: In general, factor 1 beats factor 2. A negative charge is more stable on one oxygen than on two carbon atoms. We can verify this assertion by comparing $pK_a$ values (Figure 3.6).

**FIGURE 3.6**
$pK_a$ values for ethanol and ethylene.

In fact, the $pK_a$ values indicate that a negative charge on one oxygen atom is 27 orders of magnitude (a billion billion billion times) more stable than a negative charge on two carbon atoms.

This prioritization scheme (ARIO) will often be helpful, but strict adherence to it can sometimes produce the wrong prediction. In other words, there are many exceptions. As an example, compare the structures of acetylene and ammonia:

H—C≡C—H **Acetylene** $:NH_3$ **Ammonia**

To determine which compound is more acidic, we draw the conjugate bases:

$H{-}C{\equiv}C{:}^{\ominus}$ $\ominus{:}NH_2$

When comparing these two negative charges, there are two competing factors. Factor 1 suggests that the second conjugate base is more stable ($N^-$ is more stable than $C^-$), but factor 4 suggests that the first conjugate base is more stable (an *sp*-hybridized orbital can stabilize a negative charge better than an $sp^3$-hybridized orbital). In general, factor 1 wins over the others. But this case is an exception, and factor 4 (orbitals) actually predominates here. In this case, the negative charge is more stable on the carbon atom, even though nitrogen is more electronegative than carbon:

$H{-}C{\equiv}C{:}^{\ominus}$ $\ominus{:}NH_2$

**More stable**

In fact, for this reason, $H_2N^-$ is often used as a base to deprotonate a triple bond:

$$\text{H}-\text{C}\equiv\text{C}-\text{H} \; + \; {}^{\ominus}\!:\!\ddot{\text{N}}\text{H}_2 \longrightarrow \text{H}-\text{C}\equiv\text{C}:^{\ominus} \; + \; :\text{NH}_3$$

$pK_a = 25$ (acetylene) $\qquad pK_a = 38$ (ammonia)

We see from the $pK_a$ values that acetylene is 13 orders of magnitude more acidic than ammonia. This explains why $H_2N^-$ is a suitable base for deprotonating acetylene.

There are, of course, other exceptions to the ARIO prioritization scheme, but the exception shown above is the most common. In the vast majority of cases, it would be a safe bet to apply the four factors in the order ARIO to provide a qualitative assessment of acidity. However, to be certain, it is always best to look up $pK_a$ values and verify your prediction.

## SKILLBUILDER

### 3.9 ASSESSING RELATIVE STABILITY: ALL FOUR FACTORS

**LEARN the skill**

Determine which of the two protons identified below is more acidic and explain why:

**SOLUTION**

We always begin by drawing the respective conjugate bases:

**STEP 1**
Draw the conjugate bases.

**STEP 2**
Analyze all four factors.

Now we consider all four factors (ARIO) in comparing the stability of these negative charges:

1. *Atom.* In both cases, the charge is on an oxygen atom, so this factor doesn't help.
2. *Resonance.* The conjugate base on the left is resonance stabilized while the conjugate base on the right is not. Based on this factor alone, we would say the conjugate base on the left is more stable.
3. *Induction.* The conjugate base on the right has an inductive effect that stabilizes the charge while the conjugate base on the left does not. Based on this factor alone, we would say the conjugate base on the right is more stable.
4. *Orbital.* This factor does not help.

Our analysis reveals a competition between two factors. In general, resonance will beat induction. Based on this, we predict that the conjugate base on the left is more stable. Therefore, we conclude that the following proton is more acidic:

**STEP 3**
The more stable conjugate base corresponds with the more acidic proton.

**PRACTICE the skill** **3.25** In each compound below, two protons are clearly identified. Determine which of the two protons is more acidic:

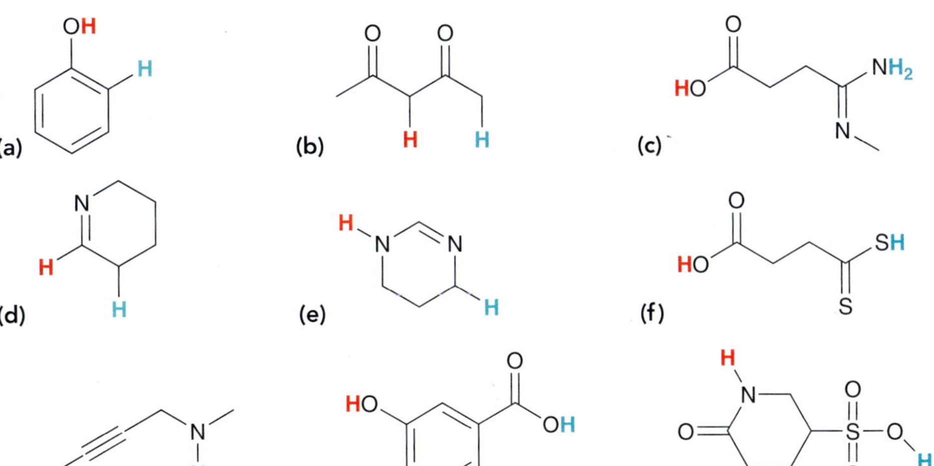


**3.26** For each pair of compounds below, predict which will be more acidic:

**(a)** HCl HBr **(b)** $H_2O$ $H_2S$ **(c)** $NH_3$ $CH_4$

**(d)** H—≡—H $H_2C{=}CH_2$ **(e)** $Cl_3C$–CH(OH)–$CCl_3$

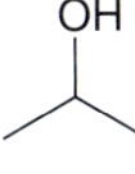

**APPLY the skill** **3.27** The following compound is one of the strongest known acids:

**(a)** Explain why it is such a strong acid.

**(b)** Suggest a modification to the structure that would render the compound even more acidic.

**3.28** Amphotericin B is a powerful antifungal agent used for intravenous treatment of severe fungal infections. Identify the most acidic proton in this compound:

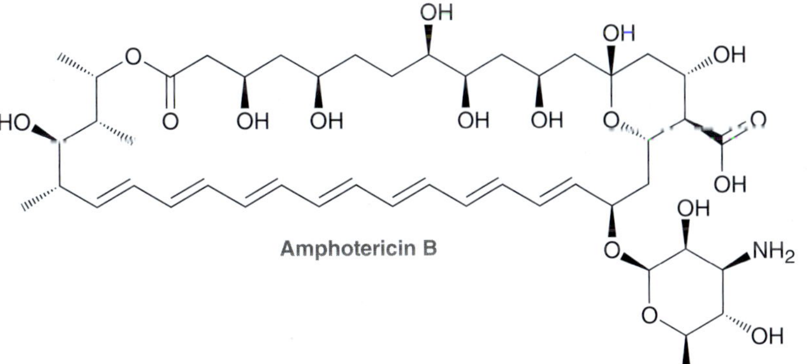


need more **PRACTICE?** **Try Problems 3.47d, 3.51a, 3.57–3.61**

## 3.5 Position of Equilibrium and Choice of Reagents

Earlier in this chapter, we learned how to use p$K_a$ values to determine the position of equilibrium. In this section, we will learn to predict the position of equilibrium just by comparing conjugate bases without using p$K_a$ values. To see how this works, let's examine a generic acid-base reaction:

$$\text{H—A} + \text{B}^{\ominus} \rightleftharpoons \text{A}^{\ominus} + \text{HB}$$

This equilibrium represents the competition between two bases ($A^-$ and $B^-$) for $H^+$. The question is whether $A^-$ or $B^-$ is more capable of stabilizing the negative charge. The equilibrium will always favor the more stabilized negative charge. If $A^-$ is more stable, then the equilibrium will favor formation of $A^-$. If $B^-$ is more stable, then the equilibrium will favor formation of $B^-$. Therefore, the position of equilibrium can be predicted by comparing the stability of $A^-$ and $B^-$. Let's see an example of this.

## SKILLBUILDER

### 3.10 PREDICTING THE POSITION OF EQUILIBRIUM WITHOUT USING p$K_a$ VALUES

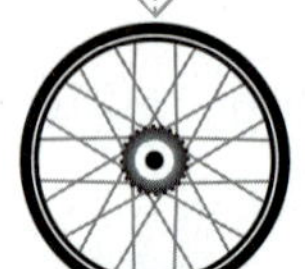

**LEARN** the skill

Predict the position of equilibrium for the following reaction:

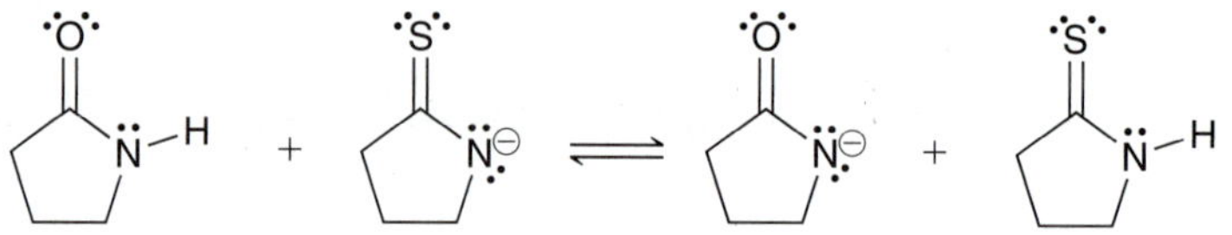

### SOLUTION

We look at both sides of the equilibrium and compare the stability of the base on each side:

**STEP 1**
Identify the base on either side of the equilibrium.

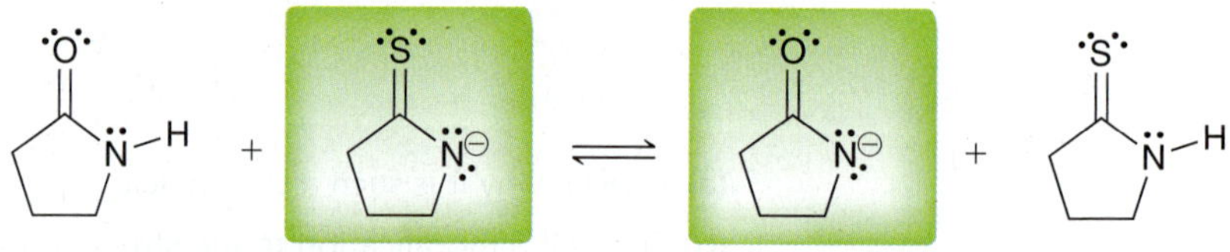

To determine which of these bases is more stable, we use the four factors (ARIO):

1. *Atom.* In both cases, the negative charge appears to be on a nitrogen atom, so this factor is not relevant.

2. *Resonance.* Both of these bases are resonance stabilized, but we do expect one of them to be more stable:

**STEP 2**
Compare the stability of these two bases using all four factors.

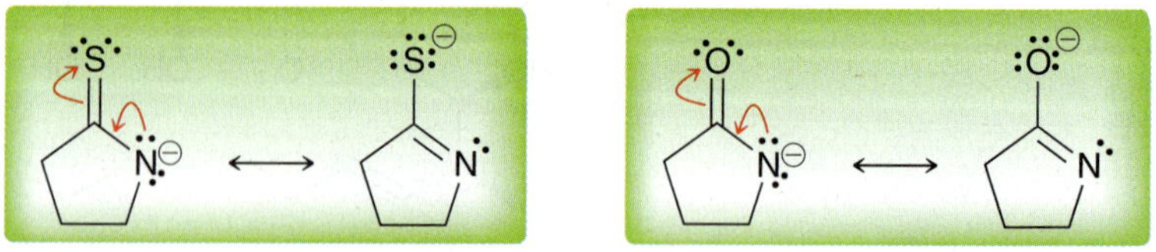

The first base has the negative charge delocalized over N and S. The second base has the negative charge delocalized over N and O. Because of its size, sulfur is more efficient than oxygen at stabilizing a negative charge, so we expect the first base to be more stable.

3. *Induction.* Neither of these bases is stabilized by inductive effects.

4. *Orbital.* Not a relevant factor in this case.

Based on factor 2, we conclude that the base on the left side is more stable, and therefore the equilibrium favors the left side of the reaction. Our prediction can be verified if we look up p$K_a$ values for the acid on either side of the reaction:

**STEP 3** The equilibrium will favor the more stable base.

p$K_a$ = 15 (Weaker acid)

p$K_a$ = 13 (Stronger acid)

The equilibrium favors production of the weaker acid (higher p$K_a$), so the left side is favored, just as we predicted.

**PRACTICE the skill** **3.29** Predict the position of equilibrium for each of the following reactions:

(a)

(b)

(c)

**APPLY the skill** **3.30** As we will learn in Chapter 21, treating a lactone (a cyclic ester) with sodium hydroxide will initially produce an anion:

NaOH

Initially formed

This anion rapidly undergoes an intramolecular proton transfer (see Problem 3.3), in which the negatively charged oxygen atom abstracts the nearby acidic proton. Draw the product of this intramolecular acid-base process and then identify which side of the equilibrium is favored. Explain your answer.

need more **PRACTICE?** **Try Problems 3.49, 3.52**

The process described in the previous SkillBuilder can also be used to determine whether a specific reagent is suitable for accomplishing a particular proton transfer, as shown in SkillBuilder 3.11.

## SKILLBUILDER

### 3.11 CHOOSING THE APPROPRIATE REAGENT FOR A PROTON TRANSFER REACTION

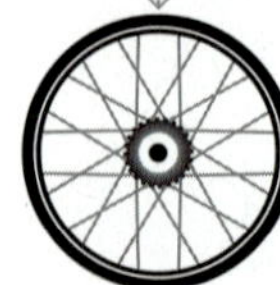

**LEARN the skill**

Determine whether $H_2O$ would be a suitable reagent for protonating the acetate ion:

**SOLUTION**

We begin by drawing the acid-base reaction that occurs when the base is protonated by water:

**STEP 1**
Draw the equilibrium.

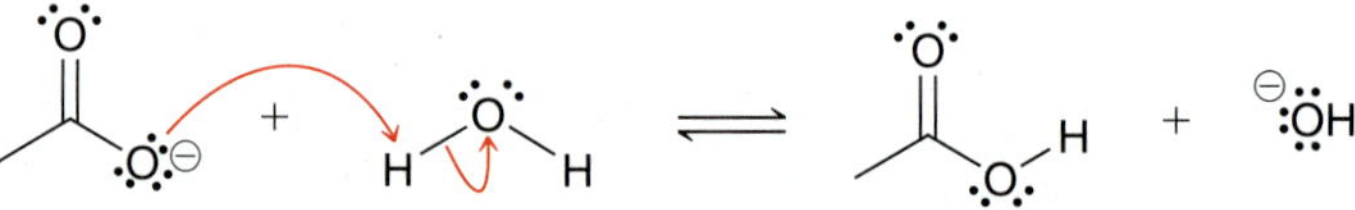

**STEP 2**
Compare the stability of the bases on either side of the equilibrium using all four factors.

Now compare the bases on either side of the reaction and ask which is more stable:

**STEP 3**
The reagent is only suitable if the equilibrium favors the desired products.

Applying the four factors (ARIO), we see that the base on the left side is more stable because of resonance. Therefore, the equilibrium will favor the left side of the reaction. This means that $H_2O$ is not a suitable proton source in this situation. In order to protonate this base, an acid stronger than water is required. A suitable acid would be $H_3O^+$.

**PRACTICE the skill**

**3.31** In each of the following cases, identify whether the reagent shown is suitable to accomplish the task described. Explain why or why not:

**(a)** To protonate using $H_2O$

**(b)** To protonate using

**(c)** To deprotonate using $^{\ominus}NH_2$

**(d)** To protonate using $H_2O$

**(e)** To protonate [structure] using $H_2O$

**(f)** To deprotonate H—C≡C—H using $^{\ominus}NH_2$

**APPLY** the skill

**3.32** We will learn the following reactions in upcoming chapters. For each of these reactions, notice that the product is an anion (ignore the positively charged ion in each case). In order to obtain a neutral product, this anion must be treated with a proton source in a process called "working up the reaction." For each of the following reactions, identify whether water will be a suitable proton source for working up the reaction:

(a) 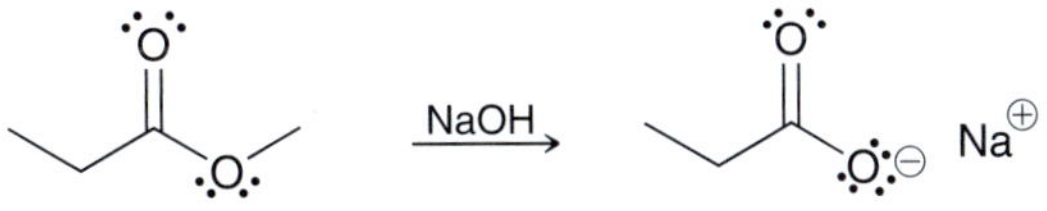 NaOH → Na⊕ Chapter 21

(b) 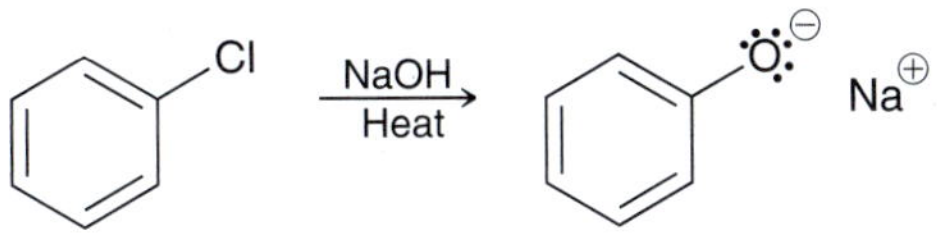 Cl; NaOH, Heat → Na⊕ Chapter 19

need more **PRACTICE?** **Try Problems 3.40, 3.42**

## 3.6 Leveling Effect

Bases stronger than hydroxide cannot be used when the solvent is water. To illustrate why, consider what happens if we mix the amide ion ($H_2N^-$) and water:

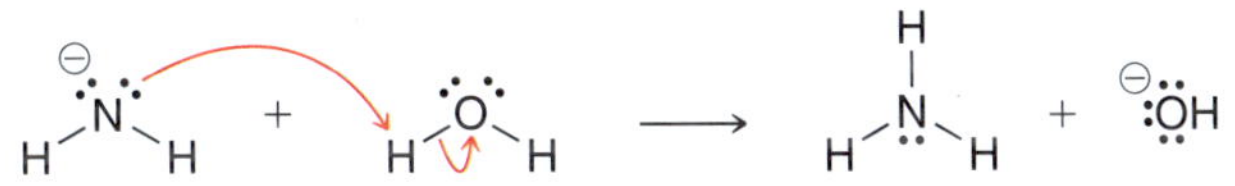

The amide ion is a strong enough base to deprotonate water, forming a hydroxide ion ($HO^-$). A hydroxide ion is more stable than an amide ion, so the equilibrium will favor formation of hydroxide. In other words, the amide ion is destroyed by the solvent and replaced with a hydroxide ion. In fact, this is true of any base stronger than $HO^-$. If a base stronger than $HO^-$ is dissolved in water, the base reacts with water to produce hydroxide. This is called the **leveling effect**.

In order to work with bases that are stronger than hydroxide, a solvent other than water must be employed. For example, in order to work with an amide ion as a base, we use liquid ammonia ($NH_3$) as a solvent. If a specific situation requires a base even stronger than an amide ion, then liquid ammonia cannot be used as the solvent. Just as before, if a base stronger than $H_2N^-$ is dissolved in liquid ammonia, the base will be destroyed and converted into $H_2N^-$. Once again, the leveling effect prevents us from having a base stronger than an amide ion in liquid ammonia. In order to use a base that is even stronger than $H_2N^-$, we must use a solvent that cannot be readily deprotonated. There are a number of solvents with high p$K_a$ values, such as hexane and THF, that can be used to dissolve very strong bases:

**Hexane**  **Tetrahydrofuran (THF)**

Throughout the course, we will see other examples of solvents suitable for working with very strong bases.

The leveling effect is also observed in acidic solutions. For example, consider the following equilibrium that is established in an aqueous solution of $H_2SO_4$:

$$H_2SO_4 \; (pK_a = -9) + H_2O \rightleftharpoons HSO_4^- + H_3O^+ \; (pK_a = -1.7)$$

Although this is indeed an equilibrium process, consider the difference in p$K_a$ values (approximately seven p$K_a$ units). This indicates that $H_2SO_4$ is $10^7$ (or 10 million) times more acidic than a hydronium ion ($H_3O^+$). As such, there is very little $H_2SO_4$ that is actually present in an aqueous solution of $H_2SO_4$. Specifically, there will be one molecule of $H_2SO_4$ for every 10 million hydronium ions. That is, 99.99999% of all sulfuric acid molecules will have transferred their protons to water molecules. A similar situation occurs with aqueous HCl or any other strong acid that is dissolved in water. In other words, an aqueous solution of either $H_2SO_4$ or HCl can simply be viewed as an aqueous solution of $H_3O^+$. The main difference between concentrated $H_2SO_4$ and dilute $H_2SO_4$ is the concentration of $H_3O^+$.

## 3.7 Solvating Effects

In some cases, solvent effects are invoked to explain small differences in p$K_a$ values. For example, compare the acidity of *tert*-butanol and ethanol:

***tert*-Butanol**
**p$K_a$ = 18**

**Ethanol**
**p$K_a$ = 16**

The p$K_a$ values indicate that *tert*-butanol is less acidic than ethanol by two orders of magnitude. In other words, the conjugate base of *tert*-butanol is less stable than the conjugate base of ethanol. This difference in stability is best explained by considering the interactions between each conjugate base and the surrounding solvent molecules (Figure 3.7). Compare the way in which each conjugate base interacts with solvent molecules. The *tert*-butoxide ion is very bulky, or **sterically hindered**, and is less capable of interacting with the solvent. The ethoxide ion is not as sterically hindered so it can accommodate more solvent interactions. As a result, ethoxide is better solvated and is therefore more stable than *tert*-butoxide (Figure 3.7). This type of solvent effect is generally weaker than the other effects we have encountered in this chapter (ARIO).

### CONCEPTUAL CHECKPOINT

**3.33** Predict which of the following compounds is more acidic:

After making your prediction, use the p$K_a$ values from Table 3.1 to determine whether your prediction was correct.

**Ethanol** **Water**

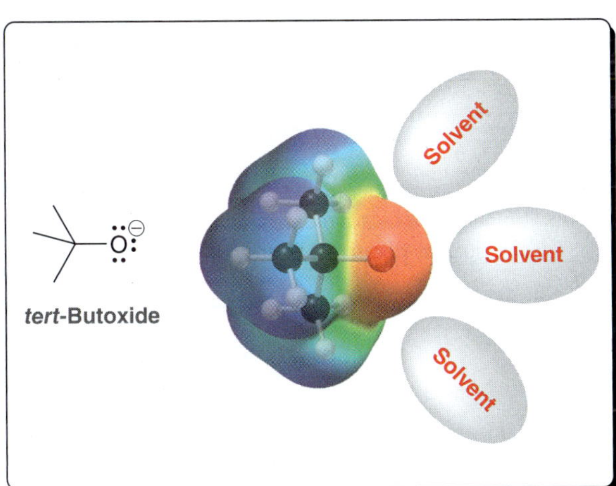


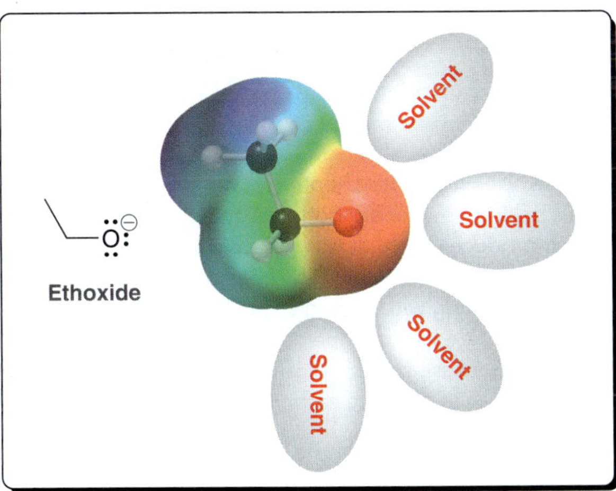


**FIGURE 3.7**
Electrostatic potential maps of *tert*-butoxide and ethoxide.

## 3.8 Counterions

**LOOKING BACK**
Your general chemistry textbook likely used the term *spectator ion* to refer to a counterion.

Negatively charged bases are always accompanied by positively charged species, called **cations** (pronounced CAT-EYE-ONZ). For example, $HO^-$ must be accompanied by a counterion, such as $Li^+$, $Na^+$, or $K^+$. We will often see the following reagents: LiOH, or NaOH, or KOH. Don't be alarmed. All of these reagents are simply $HO^-$ with the counterion indicated. Sometimes it is shown; sometimes not. Even when the counterion is not shown, it is still there. It is just not indicated because it is largely irrelevant. Up to this point in this chapter, counterions have not been shown, but from here on they will be. For example, consider the following reaction:

$$H_2N^- + H_2O \longrightarrow NH_3 + HO^-$$

This reaction might be shown like this:

$$NaNH_2 + H_2O \longrightarrow NH_3 + NaOH$$

It is important to become accustomed to ignoring the cations when they are indicated and to focus on the real players—the bases. Although counterions generally do not play a significant role in reactions, they can, under some circumstances, influence the course of a reaction. We will only see one or two such examples throughout this course. The overwhelming majority of reactions that we encounter are not significantly affected by the choice of counterion.

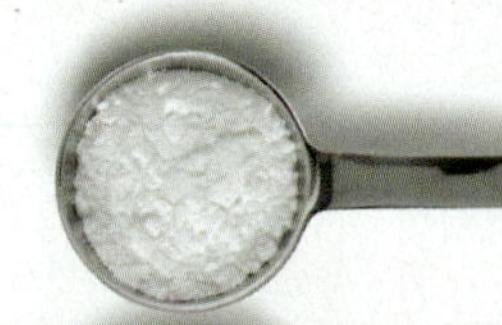

## practically speaking: Baking Soda versus Baking Powder

In the opening of this chapter, we mentioned that baking soda and baking powder are both leavening agents. That is, they both produce $CO_2$ that will make a dough or batter fluffy. We will now explore how each of these compounds accomplishes its task, beginning with baking soda. As mentioned earlier in this chapter, baking soda is the household name for sodium bicarbonate. Because sodium bicarbonate is mildly basic, it will react with an acid to produce carbonic acid, which in turn degrades into $CO_2$ and water:

**Sodium bicarbonate (baking soda)** + H–A ⇌ **Carbonic acid** + $Na^{\oplus}$ $:A^{\ominus}$ → $CO_2$ + $H_2O$

The mechanism for the conversion of carbonic acid into $CO_2$ and water will be discussed in Chapter 21. Looking at the reaction above, it is clear that an acid must be present in order for baking soda to do its job. Many breads and pastries include ingredients that naturally contain acids. For example, buttermilk, honey, and citrus fruits (such as lemons) all contain naturally occurring organic acids:

**Lactic acid** (found in buttermilk)

**Gluconic acid** (found in honey)

**Citric acid** (found in citrus fruits)

When an acidic compound is present in the dough or batter, baking soda can be protonated, causing liberation of $CO_2$. However, when acidic ingredients are absent, the baking soda cannot be protonated, and $CO_2$ is not produced. In such a situation, we must add both the base (baking soda) and some acid. Baking powder does exactly that. It is a powder mixture that contains both sodium bicarbonate and an acid salt, such as potassium bitartrate:

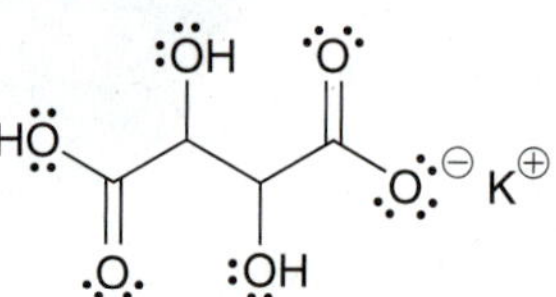

**Potassium bitartrate**

Baking powder also contains some starch to keep the mixture dry, which prevents the acid and base from reacting with each other. When mixed with water, the acid and the base can react with each other, ultimately producing $CO_2$:

**Sodium bicarbonate** + **Potassium bitartrate** ⇌ **Carbonic acid** + $Na^{\oplus}$ (bitartrate dianion) $K^{\oplus}$ → $CO_2$ + $H_2O$

Baking powder is often used when making pancakes, muffins, and waffles. It is an essential ingredient in the recipe if you want your pancakes to be fluffy. In any recipe, the exact ratio of acid and base is important. Excess base (sodium bicarbonate) will impart a bitter taste, while excess acid will impart a sour taste. In order to get the ratio just right, a recipe will often call for some specific amount of baking soda and some specific amount of baking powder. The recipe is taking into account the amount of acidic compounds present in the other ingredients, so that the final product will not be unnecessarily bitter or sour. Baking is truly a science!

## 3.9 Lewis Acids and Bases

The Lewis definition of acids and bases is broader than the Brønsted-Lowry definition. According to the Lewis definition, acidity and basicity are described in terms of electrons, rather than protons. A **Lewis acid** is defined as an *electron acceptor*, while a **Lewis base** is defined as an *electron donor*. As an illustration, consider the following Brønsted-Lowry acid-base reaction:

Base (electron donor) Acid (electron acceptor)

HCl is an acid according to either definition. It is a Lewis acid because it serves as an electron acceptor, and it is a Brønsted-Lowry acid because it serves as a proton donor. But the Lewis definition is an expanded definition of acids and bases, because it includes reagents that would otherwise not be classified as acids or bases. For example, consider the following reaction:

Base (electron donor) Acid (electron acceptor)

According to the Brønsted-Lowry definition, $BF_3$ is not considered an acid because it is has no protons and cannot serve as a proton donor. However, according to the Lewis definition, $BF_3$ can serve as an electron acceptor, and it is therefore a Lewis acid. In the reaction above, $H_2O$ is a Lewis base because it serves as an electron donor.

Take special notice of the curved-arrow notation. There is only one curved arrow in the reaction above, not two.

Chapter 6 will introduce the skills necessary to analyze reactions, and in Section 6.7 we will revisit the topic of Lewis acids and bases. In fact, we will see that most of the reactions in this textbook occur as the result of the reaction between a Lewis acid and a Lewis base. For now, let's get some practice identifying Lewis acids and Lewis bases.

## SKILLBUILDER

### 3.12 IDENTIFYING LEWIS ACIDS AND LEWIS BASES

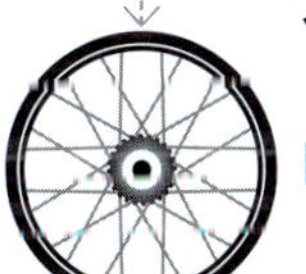

**LEARN the skill**

Identify the Lewis acid and the Lewis base in the reaction between $BH_3$ and THF:

THF

**SOLUTION**

We must first decide the direction of electron flow. Which reagent is serving as the electron donor and which reagent is serving as the electron acceptor? To answer this question, we analyze each reagent and look for a lone pair of electrons. Boron is in the third column of the periodic table and only has three valence electrons. It is using all three valence electrons to form bonds, which means that it does not have a lone pair of electrons. Rather, it has an empty *p* orbital (for a review of the structure of $BH_3$, see Section 1.10). Oxygen does have a lone pair. So, we conclude that oxygen attacks boron:

**STEP 1** Identify the direction of the flow of electrons.

**STEP 2** Identify the electron acceptor as the Lewis acid and the electron donor as the Lewis base.

$BH_3$ is the electron acceptor (Lewis acid), and THF is the electron donor (Lewis base).

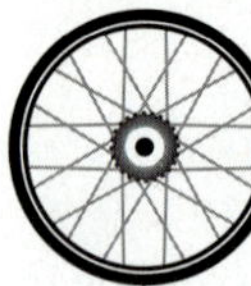

**PRACTICE the skill** **3.34** In each case below, identify the Lewis acid and the Lewis base:

(a)

(b)

(c)

(d)

(e)

**APPLY the skill** **3.35** Identify the compounds below that can function as Lewis bases:

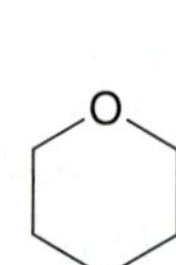


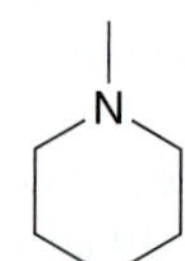


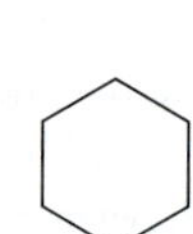

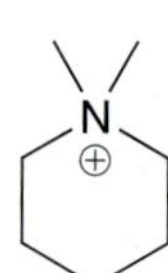


need more **PRACTICE?** **Try Problem 3.39**

## REVIEW OF CONCEPTS AND VOCABULARY

**SECTION 3.1**

- A **Brønsted-Lowry acid** is a proton donor, while a **Brønsted-Lowry base** is a proton acceptor.
- A Brønsted-Lowry acid-base reaction produces a **conjugate acid** and a **conjugate base**.

**SECTION 3.2**

- Curved arrows show the **reaction mechanism**.
- The mechanism of proton transfer always involves at least two curved arrows.

**SECTION 3.3**

- For an acid-base reaction occurring in water, the position of equilibrium is described using $K_a$ rather than $K_{eq}$.
- Typical p$K_a$ values range from −10 to 50.
- A strong acid has a low p$K_a$, while a weak acid has a high p$K_a$.
- **Equilibrium** always favors formation of the weaker acid (higher p$K_a$).

**SECTION 3.4**

- Relative acidity can be predicted (qualitatively) by analyzing the structure of the conjugate base. If $A^-$ is very stable, then HA must be a strong acid. If $A^-$ is very unstable, then HA must be a weak acid.
- To compare the acidity of two compounds, HA and HB, simply compare the stability of their conjugate bases.
- There are four factors to consider when comparing the stability of conjugate bases:
  1. Which atom bears the charge? For elements in the same row of the periodic table, electronegativity is the dominant effect. For elements in the same column, size is the dominant effect.

2. Resonance—a negative charge is stabilized by resonance.
3. Induction—electron-withdrawing groups, such as halogens, stabilize a nearby negative charge via induction.
4. Orbital—a negative charge in an *sp*-hybridized orbital will be closer to the nucleus and more stable than a negative charge in an $sp^3$-hybridized orbital.

- When multiple factors compete, ARIO (atom, resonance, induction, orbital) is generally the order of priority, but there are exceptions.

### SECTION 3.5

- The equilibrium of an acid-base reaction always favors the more stabilized negative charge.

### SECTION 3.6

- A base stronger than hydroxide cannot be used when the solvent is water because of the **leveling effect**. If a stronger base is desired, a solvent other than water must be employed.

### SECTION 3.7

- In some cases, solvent effects explain small differences in p$K_a$'s. For example, bases that are bulky, or **sterically hindered**, are generally less efficient at forming stabilizing solvent interactions.

### SECTION 3.8

- Negatively charged bases are always accompanied by positively charged species called **cations**.
- The choice of the counterion does not affect most reactions encountered in this book.

### SECTION 3.9

- A **Lewis acid** is an electron acceptor, while a **Lewis base** is an electron donor.

# SKILLBUILDER REVIEW

## 3.1 DRAWING THE MECHANISM OF A PROTON TRANSFER

**STEP 1** Identify the acid and the base.

$CH_3O^-$ Base   $H_2O$ Acid

**STEP 2** Draw the first curved arrow....
(a) Place tail on lone pair (of base).
(b) Place head on proton (from acid).

Base   Acid

**STEP 3** Draw the second curved arrow....
(a) Place tail on O—H bond.
(b) Place head on O.

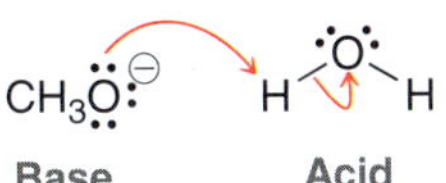


Base   Acid

Try Problems 3.1–3.3, 3.44

## 3.2 USING p$K_a$ VALUES TO COMPARE ACIDS

The compound with the lower p$K_a$ is more acidic.

p$K_a$ = 19.2   More acidic p$K_a$ = 4.75

Try Problems 3.4–3.6, 3.38

## 3.3 USING p$K_a$ VALUES TO COMPARE BASICITY

**EXAMPLE** Compare the basicity of these two anions.

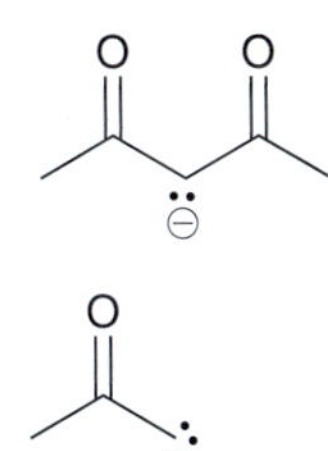

**STEP 1** Draw the conjugate acid of each.

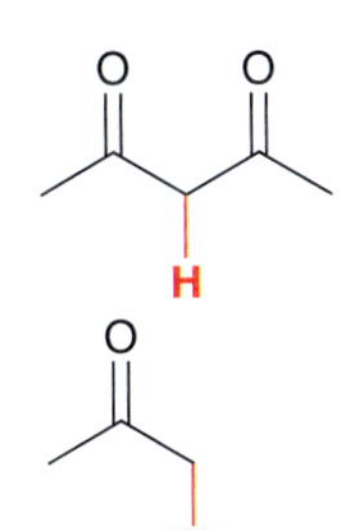


**STEP 2** Compare p$K_a$ values.

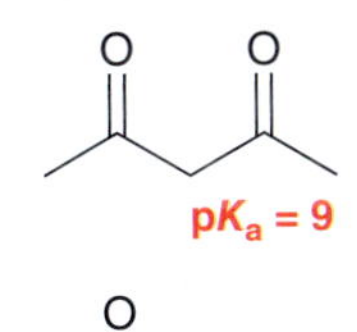


p$K_a$ = 19

**STEP 3** Identify the stronger base.

Weaker acid   Stronger base

Try Problems 3.7–3.9, 3.50

### 3.4 USING p$K_a$ VALUES TO PREDICT THE POSITION OF EQUILIBRIUM

**STEP 1** Identify the acid on each side of the equilibrium.

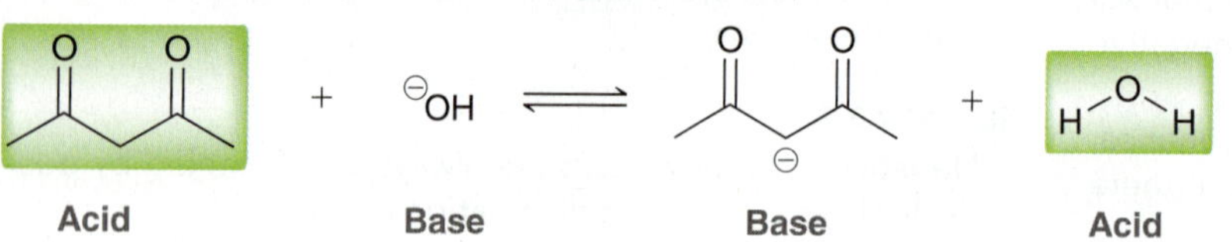


**Acid** **Base** **Base** **Acid**

**STEP 2** Compare p$K_a$ values.

H O H

p$K_a$ = 9.0 p$K_a$ = 15.7

**The equilibrium will favor the weaker acid.**

Try Problems 3.10–3.11, 3.48

### 3.5 ASSESSING RELATIVE STABILITY: FACTOR 1—ATOM

**STEP 1** Draw the conjugate bases...

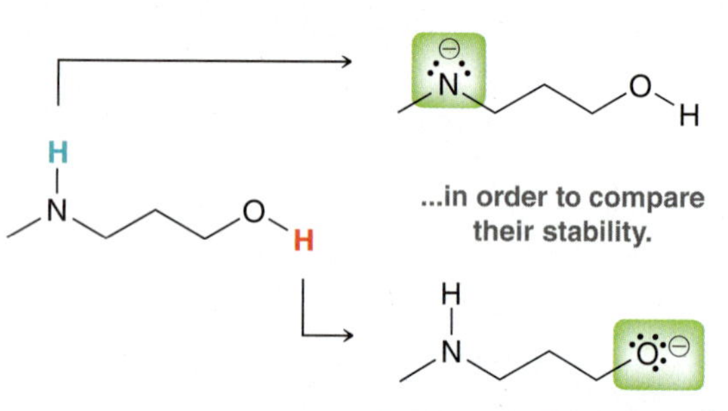


**...in order to compare their stability.**

**STEP 2** Compare location of charge taking into account two trends.

**Electronegativity**

C N O F

P S Cl

Br

I

**Size**

**STEP 3** The more stable conjugate base...

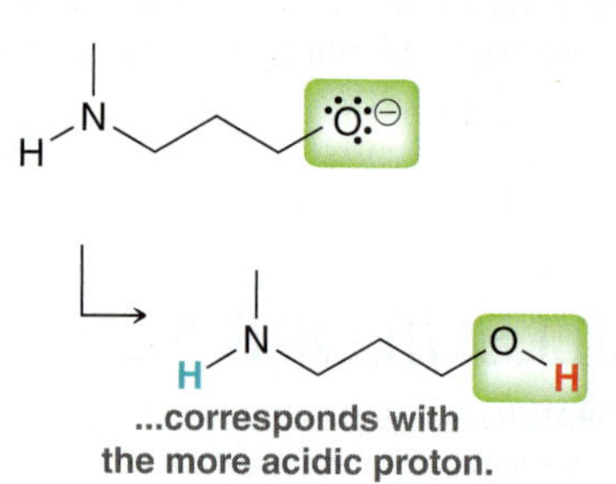


**...corresponds with the more acidic proton.**

Try Problems 3.13, 3.14, 3.45b, 3.47h, 3.51b, h, 3.53

### 3.6 ASSESSING RELATIVE STABILITY: FACTOR 2—RESONANCE

**STEP 1** Draw the conjugate bases...

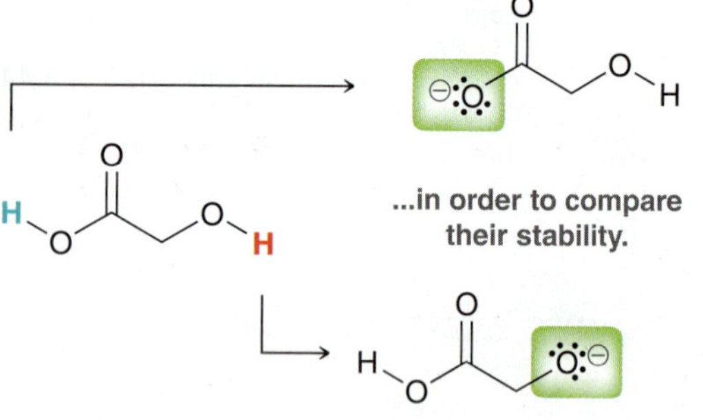


**...in order to compare their stability.**

**STEP 2** Look for resonance stabilization.

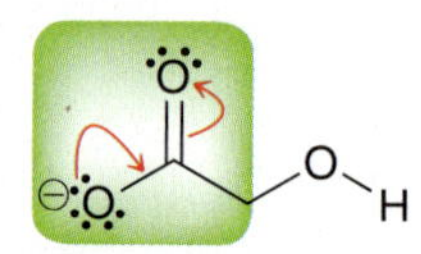


**Resonance stabilized**

H O

**Not resonance stabilized**

**STEP 3** The more stable conjugate base...

O H

H O O H

**...corresponds with the more acidic proton.**

Try Problems 3.15–3.17, 3.45a, 3.46a, 3.47b,e–g, 3.51c–f

### 3.7 ASSESSING RELATIVE STABILITY: FACTOR 3—INDUCTION

**STEP 1** Draw the conjugate bases...

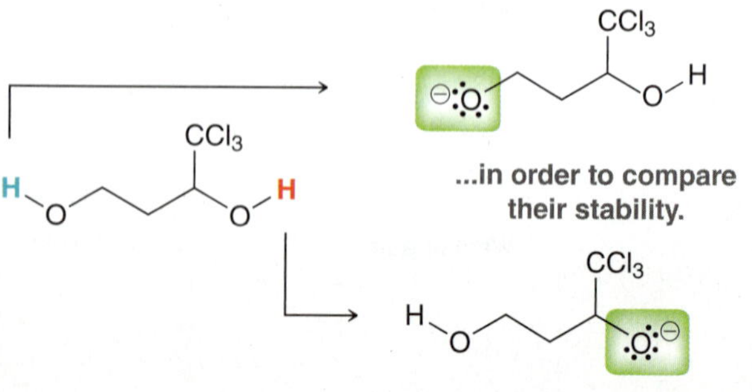


**...in order to compare their stability.**

**STEP 2** Look for inductive effects.

$CCl_3$

H O

**More stable**

**STEP 3** The more stable conjugate base...

$CCl_3$

H O

$CCl_3$

H O O H

**...corresponds with the more acidic proton.**

Try Problems 3.18–3.21, 3.46b, 3.47c, 3.51g

### 3.8 ASSESSING RELATIVE STABILITY: FACTOR 4—ORBITALS

**STEP 1** Draw the conjugate bases...

...in order to compare their stability.

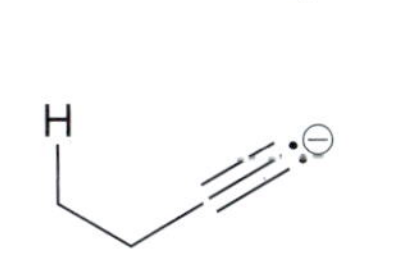


**STEP 2** Analyze orbitals.

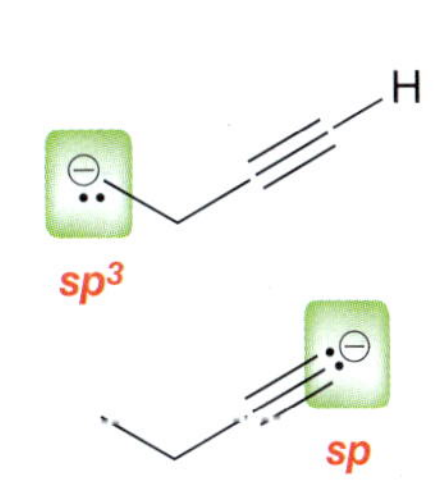


**STEP 3** The more stable conjugate base...

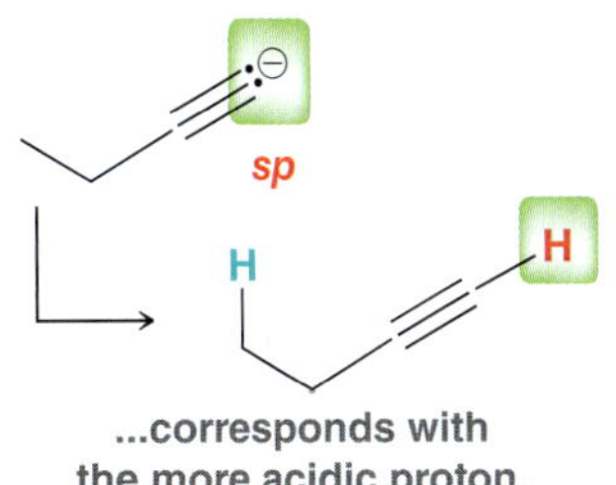


...corresponds with the more acidic proton.

Try Problems **3.22–3.24, 3.45c**

### 3.9 ASSESSING RELATIVE STABILITY: ALL FOUR FACTORS

**STEP 1** Draw the conjugate bases...

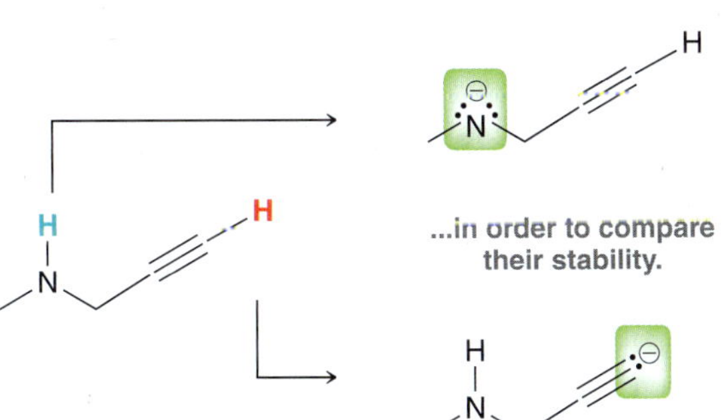


...in order to compare their stability.

**STEP 2** Analyze all four factors in this order:

Atom

Resonance

Induction

Orbital

**Identify all factors that apply.**

**STEP 3** Take into account exceptions to the order of priority (ARIO) and determine the more stable base...

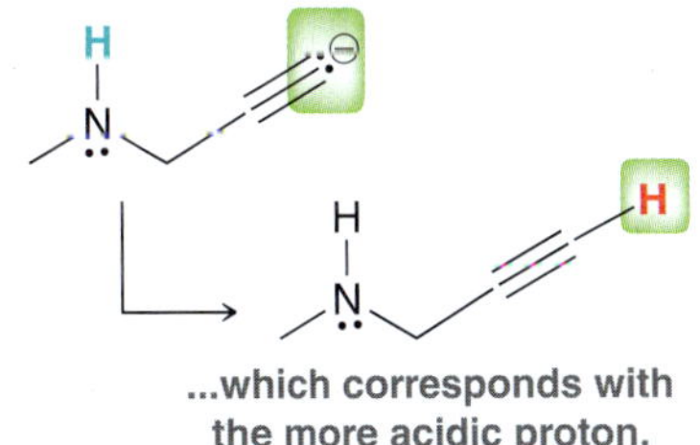


...which corresponds with the more acidic proton.

Try Problems **3.25–3.28, 3.47d, 3.51a, 3.58–3.62**

### 3.10 PREDICTING THE POSITION OF EQUILIBRIUM WITHOUT USING p$K_a$ VALUES

**STEP 1** Identify the base on either side of the equilibrium.

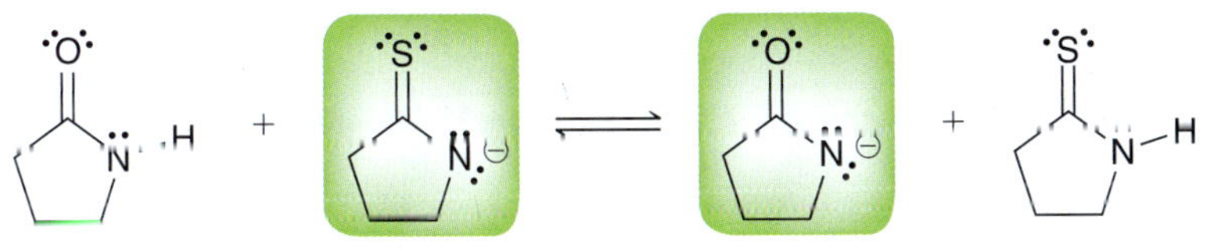


**STEP 2** Compare the stability of these conjugate bases using all four factors, in this order:

Atom

Resonance

Induction

Orbital

**STEP 3** Equilibrium will favor the more stable base.

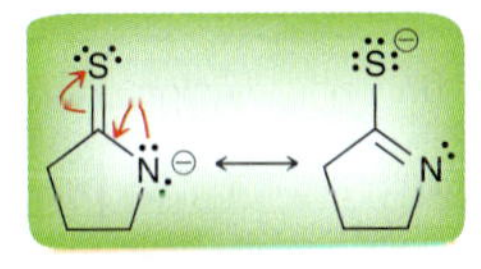


Try Problems **3.29, 3.30, 3.49, 3.52**

### 3.11 CHOOSING THE APPROPRIATE REAGENT FOR A PROTON TRANSFER REACTION

**STEP 1** Draw the equilibrium and identify the base on either side.

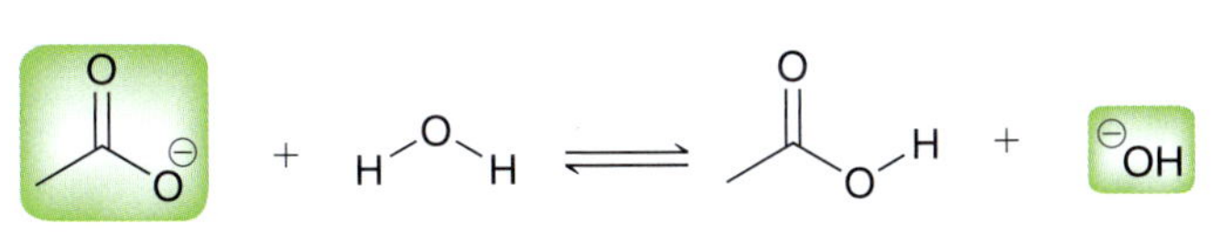


**STEP 2** Compare the stability of these conjugate bases using all four factors, in this order:

Atom

Resonance →

Induction

Orbital

More stable

**STEP 3** Equilibrium will favor the more stable base. If products are favored, the reaction is useful. If starting materials are favored, the reaction is not useful.

**In this case, water is not a suitable proton source.**

Try Problems **3.31, 3.32, 3.40, 3.42**

### 3.12 IDENTIFYING LEWIS ACIDS AND LEWIS BASES

**STEP 1** Identify the direction of the flow of electrons.

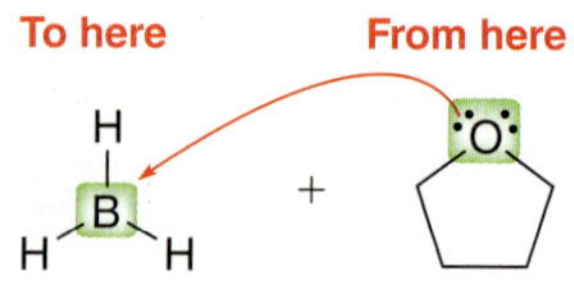


**STEP 2** Identify the electron acceptor as the Lewis acid and the electron donor as the Lewis base.

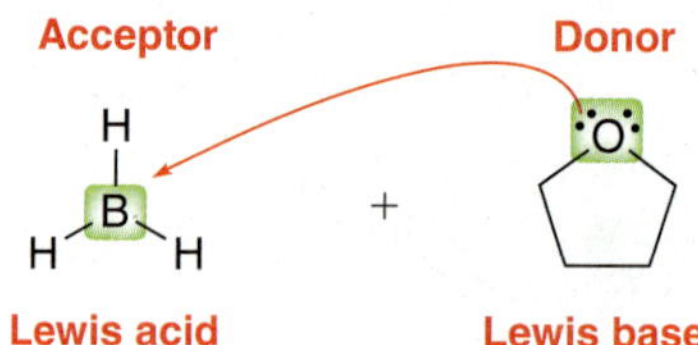


Try Problems 3.34, 3.35, 3.39

## PRACTICE PROBLEMS

Note: Most of the Problems are available within **WileyPLUS**, an online teaching and learning solution.

**3.36** Draw the conjugate base for each of the following acids:

(a) (b) (c) $NH_3$ (d) $H_3O^+$

(e) (f) (g) $NH_4^+$

**3.37** Draw the conjugate acid for each of the following bases:

(a) (b) (c) $NaNH_2$ (d) $H_2O$

(e) (f) (g) (h) NaOH

**3.38** Compound A has a p$K_a$ of 7 and compound B has a p$K_a$ of 10. Compound A is how many times more acidic than compound B?

(a) 3 (b) 3000 (c) 1000

**3.39** In each reaction, identify the Lewis acid and the Lewis base:

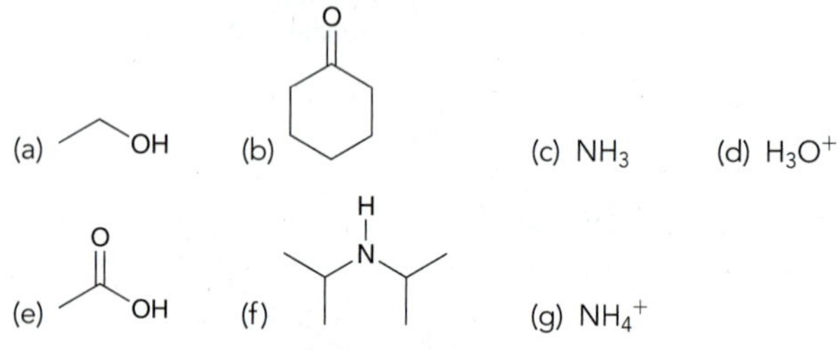

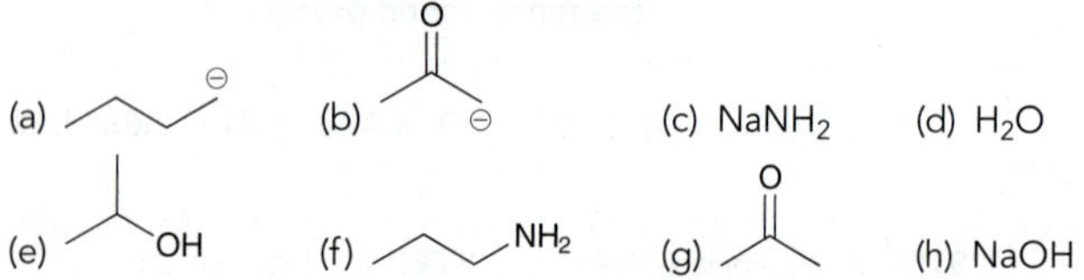

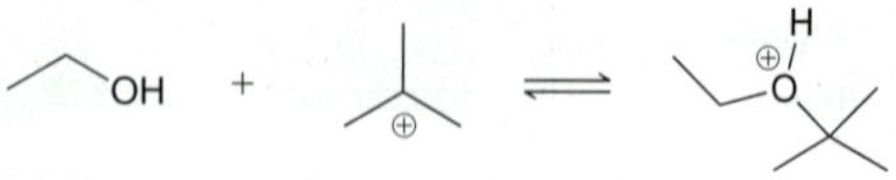

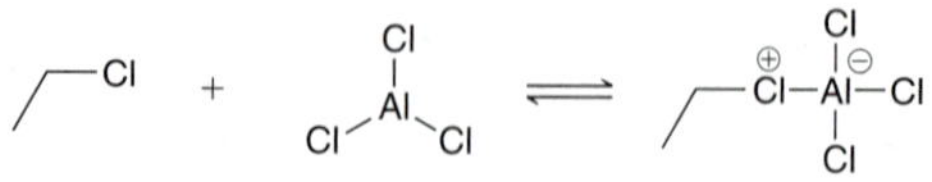

**3.40** What reaction will take place if $H_2O$ is added to a mixture of $NaNH_2/NH_3$?

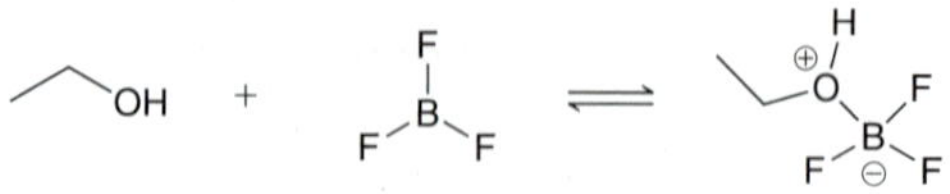

**3.41** Would ethanol ($CH_3CH_2OH$) be a suitable solvent in which to perform the following proton transfer? Explain your answer:

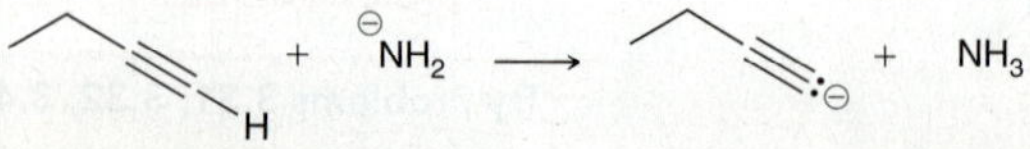

**3.42** Would water be a suitable proton source to protonate the following compound?

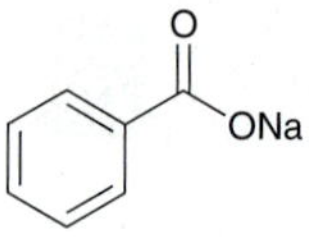

**3.43** Write an equation for the proton transfer reaction that occurs when each of the following acids reacts with water. In each case, draw curved arrows that show the mechanism of the proton transfer:

(a) HBr (b) 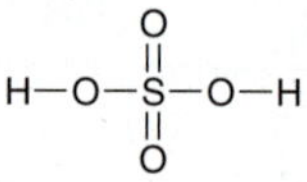 (c)

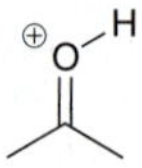

**3.44** Write an equation for the proton transfer reaction that occurs when each of the following bases reacts with water. In each case, draw curved arrows that show the mechanism of the proton transfer:

(a)  (b) 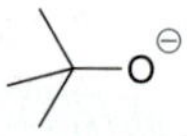 (c) 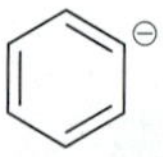 (d)

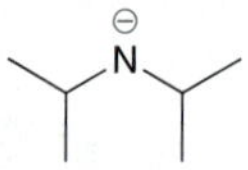

**3.45** In each case, identify the more stable anion. Explain why it is more stable:

(a)  vs.

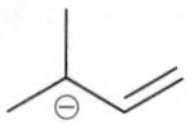

(b)  vs.

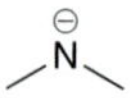

(c)  vs.

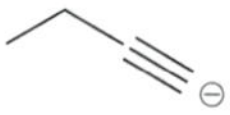

**3.46** In each group of compounds below, select the most acidic compound:

(a)

(b)

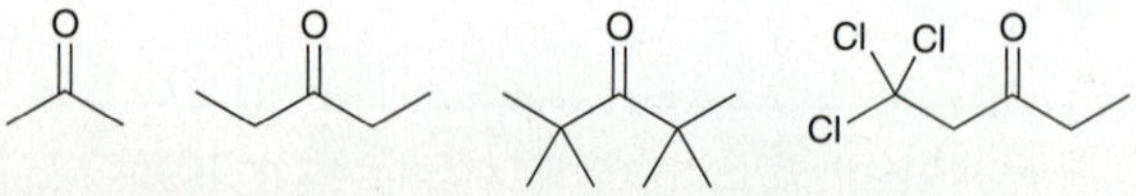

**3.47** For each pair of compounds below, identify the more acidic compound:

(a) 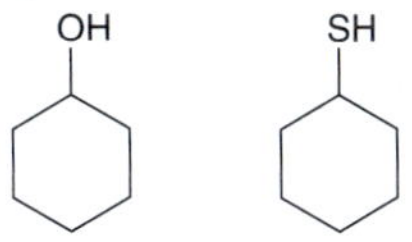 (b) 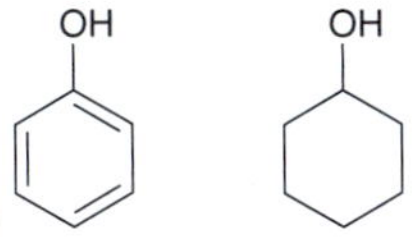 (c) 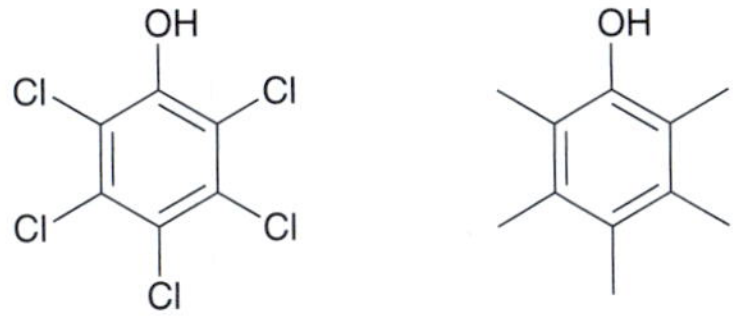 (d) 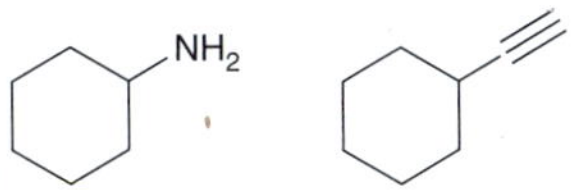 (e) 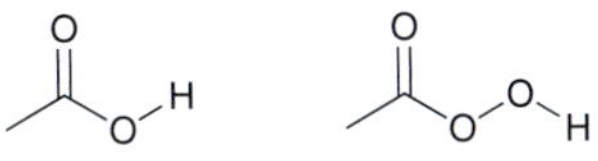 (f)  (g)  (h)

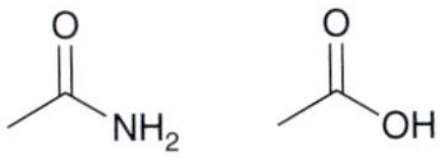

**3.48** HA has a p$K_a$ of 15, while HB has a p$K_a$ of 5. Draw the equilibrium that would result upon mixing HB with NaA. Does the equilibrium favor formation of HA or of HB?

**3.49** For each reaction below, draw the mechanism (curved arrows) and then predict which side of the reaction is favored under equilibrium conditions:

(a) 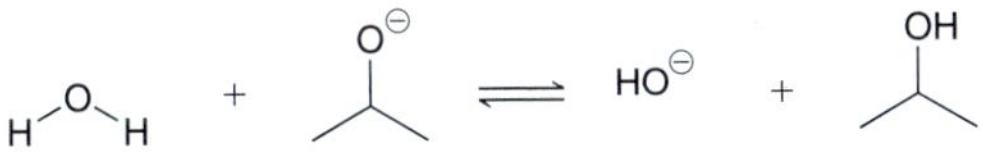 (b) 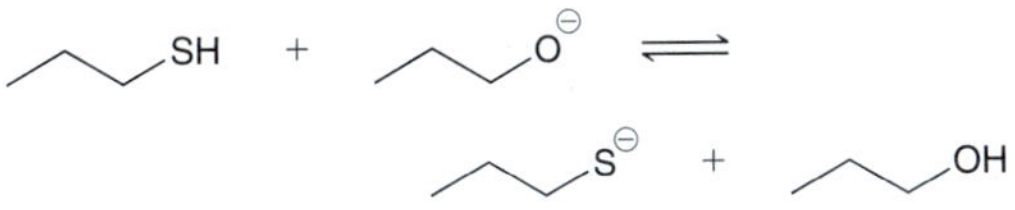 (c)  (d)

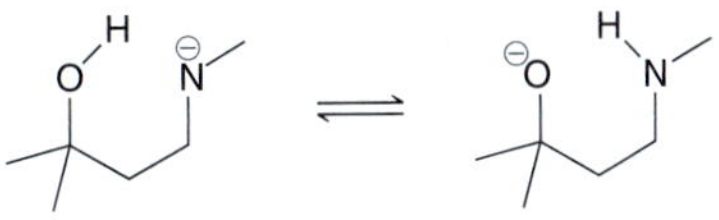

**3.50** Rank the following anions in terms of increasing basicity:

$HO^{\ominus}$

**3.51** For each compound below, identify the most acidic proton in the compound:

(a) 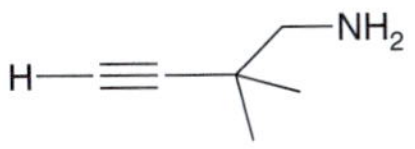 (b) 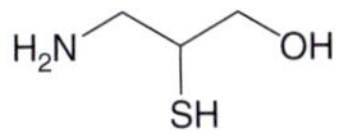 (c) 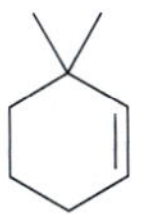 (d) 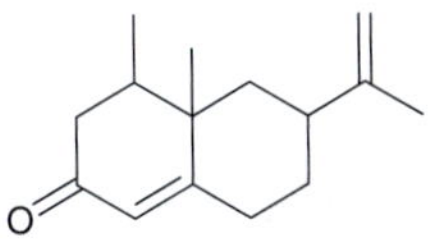 (e) 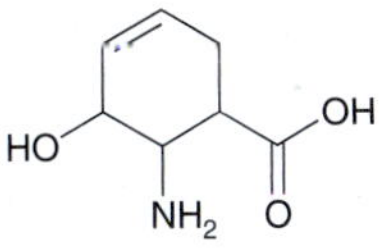 (f) 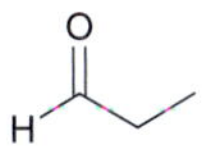 (g) 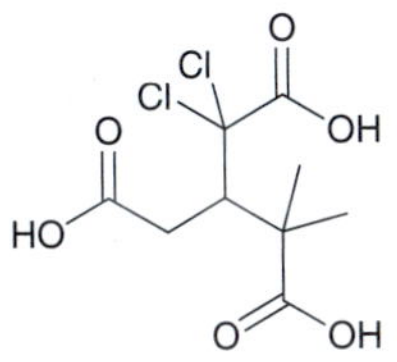 (h)

# INTEGRATED PROBLEMS

**3.52** In each case below, identify the acid and the base. Then draw the curved arrows showing a proton transfer reaction. Draw the products of that proton transfer and then predict the position of equilibrium:

(a) 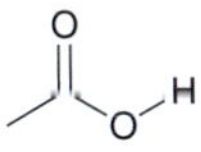 + LiOH

(b)  + $H_2O$

(c) 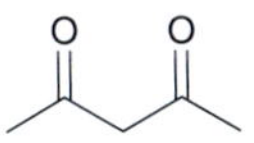 + NaOH

**3.53** Draw all constitutional isomers with molecular formula $C_2H_6S$ and rank them in terms of increasing acidity.

**3.54** Draw all constitutional isomers with molecular formula $C_3H_8O$ and rank them in terms of increasing acidity.

**3.55** Consider the structure of cyclopentadiene and then answer the following questions:

**Cyclopentadiene**

(a) How many $sp^3$-hybridized carbon atoms are present in the structure of cyclopentadiene?

(b) Identify the most acidic proton in cyclopentadiene. Justify your choice.

(c) Draw all resonance structures of the conjugate base of cyclopentadiene.

(d) How many $sp^3$-hybridized carbon atoms are present in the conjugate base?

(e) What is the geometry of the conjugate base?

(f) How many hydrogen atoms are present in the conjugate base?

(g) How many lone pairs are present in the conjugate base?

**3.56** In Section 3.4, we learned four factors (ARIO) for comparing the relative acidity of compounds. When two of these factors are in competition, the order of priority is the order in which these factors were covered ("atom" being the most important factor and "orbital" being the least important). However, we also mentioned that there are exceptions to this order of priority. One such exception was covered at the end of Section 3.4. Compare the two compounds below and determine if they constitute another exception. Justify your choice:

$pK_a = 4.75$ $pK_a = 10.6$

**3.57** Consider the $pK_a$ values of the following constitutional isomers:

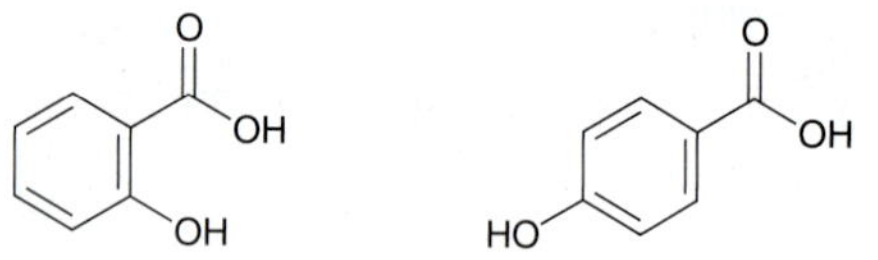

**Salicylic acid** $pK_a = 3.0$

***para*-Hydroxybenzoic acid** $pK_a = 4.6$

Using the rules that we developed in this chapter (ARIO), we might have expected these two compounds to have the same $pK_a$. Nevertheless, they are different. Salicylic acid is apparently more acidic than its constitutional isomer. Can you offer an explanation for this observation?

**3.58** Consider the following compound with molecular formula $C_4H_8O_2$:

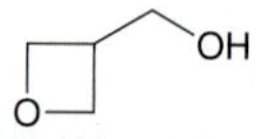

$C_4H_8O_2$

(a) Draw a constitutional isomer that you expect will be approximately one trillion ($10^{12}$) times more acidic than the compound above.

(b) Draw a constitutional isomer that you expect will be less acidic than the compound above.

(c) Draw a constitutional isomer that you expect will have approximately the same $pK_a$ as the compound above.

**3.59** There are only four constitutional isomers with molecular formula $C_4H_9NO_2$ that contain a nitro group ($—NO_2$). Three of these isomers have similar $pK_a$ values, while the fourth isomer has a much higher $pK_a$ value. Draw all four isomers and identify which one has the higher $pK_a$. Explain your choice.

**3.60** Predict which of the following compounds is more acidic and explain your choice:

**3.61** Below is the structure of rilpivirine, a promising new anti-HIV drug that combats resistant strains of HIV. Its ability to side-step resistance will be discussed in the upcoming chapter.

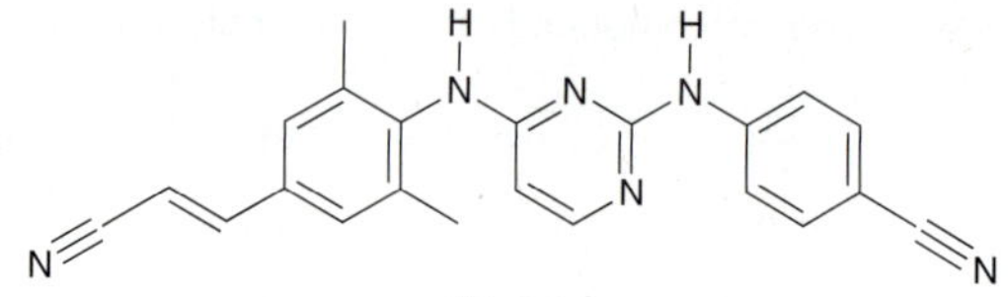

**Rilpivirine**

(a) Identify the two most acidic protons in rilpivirine.

(b) Identify which of these two protons is more acidic. Explain your choice.

**3.62** Most common amines ($RNH_2$) exhibit $pK_a$ values between 35 and 45. R represents the rest of the compound (generally carbon and hydrogen atoms). However, when R is a cyano group, the $pK_a$ is found to be drastically lower:

$pK_a = 35–45$ $pK_a = 17$

(a) Explain why the presence of the cyano group so drastically impacts the $pK_a$.

(b) Can you suggest a different replacement for R that would lead to an even stronger acid ($pK_a$ lower than 17)?

**3.63** In one step of a recent total synthesis of (−)-seimatopolide A, a potential antidiabetic drug, the following two structures reacted with each other in an acid-base reaction (*Tetrahedron Lett.* **2012**, *53*, 5749–5752):

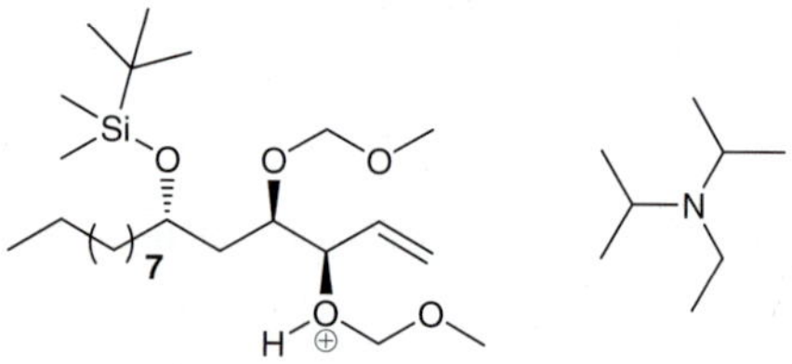

(a) Identify the acid and the base, draw the products of the reaction, and show a mechanism for their formation.

(b) Using the $pK_a$ values provided in Problem 3.9 as a rough guideline, predict the position of equilibrium for this acid-base reaction.

**Problems 3.64–3.65 are intended for students who have already covered IR spectroscopy (Chapter 15).**

**3.64** Deuterium (D) is an isotope of hydrogen, in which the nucleus has one proton and one neutron. This nucleus, called a deuteron, behaves very much like a proton, although there are observed differences in the rates of reactions involving either protons or deuterons (an effect called the kinetic isotope effect). Deuterium can be introduced into a compound via the process below:

(a) The C—Mg bond in compound **3** can be drawn as ionic. Redraw **3** as an ionic species, with $BrMg^+$ as a counterion, and then draw the mechanism for the conversion of **3** to **4**.

(b) The IR spectrum of compound **4** exhibits a group of signals between 1250 and 1500 $cm^{-1}$, a signal at 2180 $cm^{-1}$, and another group of signals between 2800 and 3000 $cm^{-1}$. Identify the location of the C-D signal in the spectrum and explain your reasoning (*J. Chem. Ed.* **1981**, *58*, 79–80).

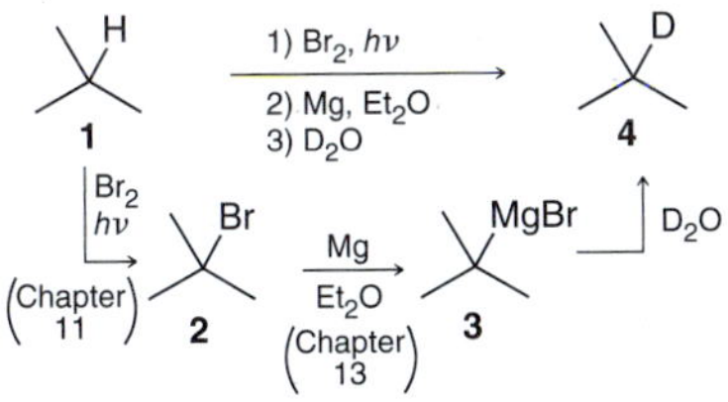

(a) Draw the structure of **5** and show a mechanism for its formation from **3**.

(b) Use a quantitative argument ($pK_a$ values) to verify that **4** is an appropriate base for this transformation.

(c) Explain how you would use IR spectroscopy to verify the conversion of **1** to **5**.

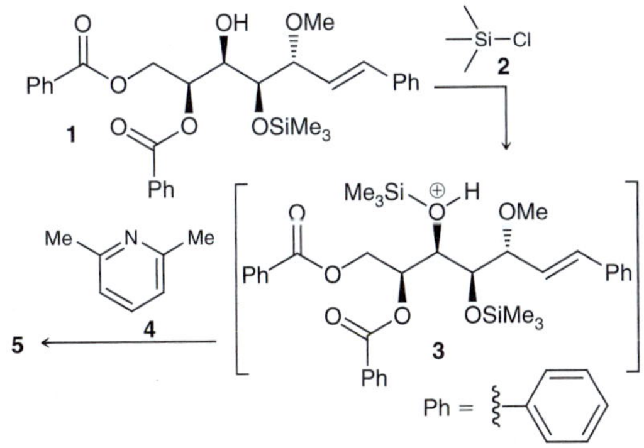

**3.65** The bengamides are a series of natural products that have shown inhibitory effects on the enzyme methionine aminopeptidase, which plays a key role in the growth of new blood vessels, a necessary process for the progression of diseases such as solid tumor cancers and rheumatoid arthritis. During the synthesis of bengamides, it is often required to convert OH groups into other, less reactive groups, called protecting groups, which can be converted back into OH groups when desired. For example, compound **1** is protected upon treatment with compound **2** in the presence of compound **4** (*Tetrahedron Lett.* **2007**, *48*, 8787–8789). First, **1** reacts with **2** to give intermediate **3**, which is then deprotonated by **4** to give **5**:

# CHALLENGE PROBLEMS

**3.66** Asteltoxin, isolated from the cultures of *Aspergillus stellatus*, exhibits a potent inhibitory effect on the activity of *E. coli* $BF_1$-ATPase. During S. L. Schreiber's synthesis of asteltoxin, compound **1** was treated with a strong base to form anion **2** (*J. Am. Chem. Soc.* **1984**, *106*, 4186–4188):

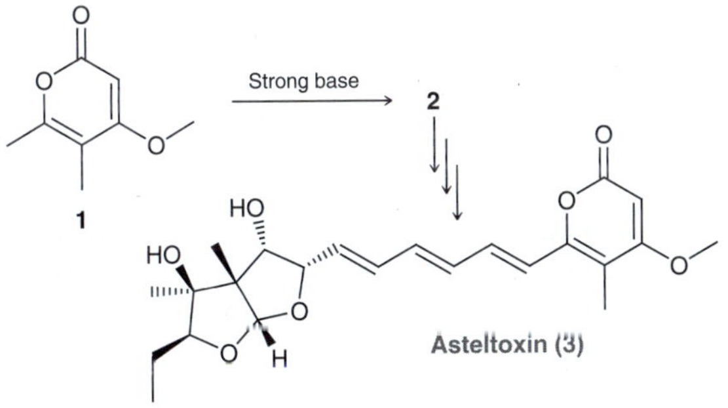

(a) Identify the most acidic proton in **1** and justify your choice using any necessary drawings.

(b) Determine whether each of the following bases would be suitable for deprotonating compound **1** and explain your decision in each case: (i) NaOH, (ii) $NaNH_2$, or (iii) $CH_3CO_2Na$

**3.67** In a key step during a recent synthesis of 7-desmethoxyfusarentin, a compound that is known to display cytotoxicity against breast cancer cells (MCF-7), an acid-base reaction was employed, in which the following two structures reacted with each other (*Tetrahedron Lett.* **2012**, *53*, 4051–4053). Predict the products of this reaction, and propose a mechanism for their formation:

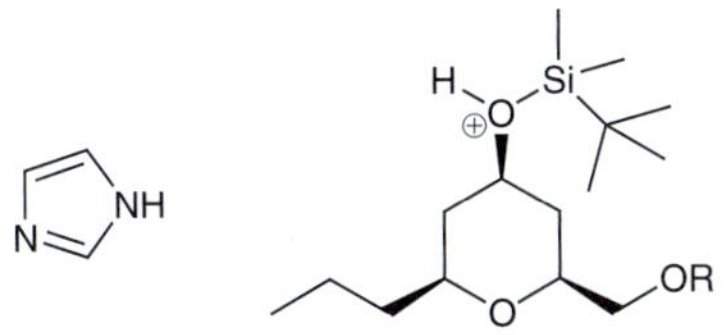

**3.68** During a synthesis of (+)-coronafacic acid, a key component in the plant toxin coronatine, the following reaction was performed, in which a ketone was converted into an acetal (the acetal functional group will be covered in Chapter 20). In this case, *p*-toluenesulfonic acid (*p*-TsOH) functions as an acid catalyst (*J. Org. Chem.* **2009**, *74*, 2433–2437):

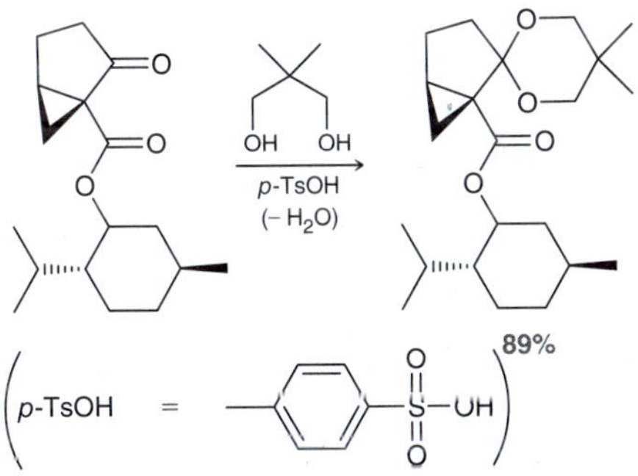

(a) The mechanism for this process begins with a proton transfer step in which the ketone is protonated. Draw this mechanistic step using curved arrows.

(b) For this protonation step, predict whether the equilibrium favors the protonated ketone. Justify your prediction with estimated $pK_a$ values.

**3.69** Phakellin (**3**), a natural product isolated from marine organisms, has been studied for its potential use as an antibiotic agent. During studies aimed at developing a strategy for the synthesis of phakellin and its derivatives, compound **1** was investigated as a potential precursor (*Org. Lett.* **2002**, *4*, 2645–2648):

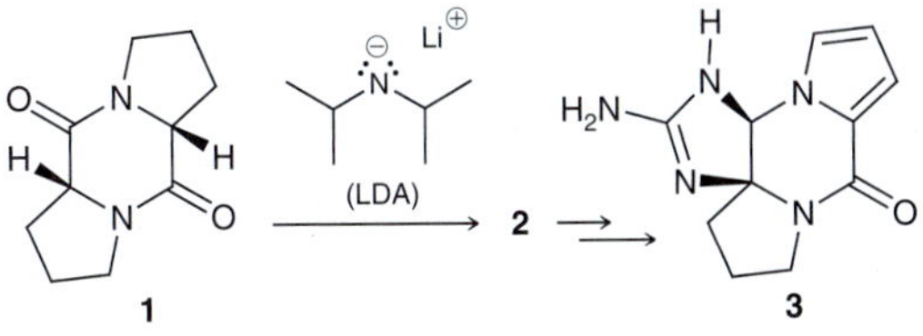

(a) Identify the most acidic proton in compound **1**, draw the corresponding conjugate base, **2**, and justify your choice.

(b) Using a quantitative argument based on p$K_a$ values as well as a qualitative argument based on structural comparisons, justify why lithium diisopropyl amide (LDA) is a suitable base to deprotonate **1**.

(c) Draw a mechanism for the conversion of **1** to **2**.

**3.70** The Nazarov cyclization is a versatile method for making five-membered rings, a common feature in many natural products. This process has been used successfully in the preparation of many complex structures with a wide variety of biological activities, including antibiotic and anticancer properties (*Org. Lett.* **2003**, *5*, **4931–4934**). The key step of the Nazarov cyclization involves a Lewis acid, such as $AlCl_3$. In the first step of the mechanism, compound 1 interacts with $AlCl_3$ to form a Lewis acid-Lewis base complex. Determine the site within compound 1 that interacts most strongly with the Lewis acid and justify your choice by exploring the resonance structures of the resulting complex:

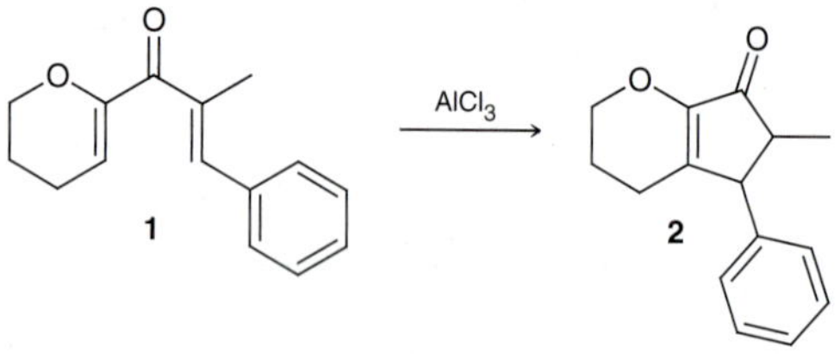

$AlCl_3$

1 2

**3.71** Crude extracts from the ginkgo tree, *Ginkgo biloba*, have been used for centuries to alleviate symptoms associated with asthma. There are four principal components of *Ginkgo* extracts, called ginkgolide A, B, D, and M. During E. J. Corey's classic synthesis of ginkgolide B, compound **1** was converted into compound **5** (*J. Am. Chem. Soc.* **1988**, *110*, **649–651**):

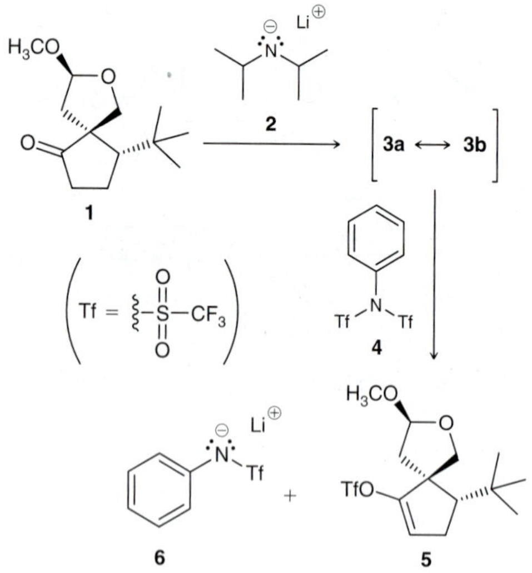

$H_3CO$ Li⊕ 2 [3a ⟷ 3b] 1 Tf = $-SO_2-CF_3$ Tf–N(Ph)–Tf 4 6 5 TfO

(a) Draw the resonance-stabilized anion (**3**) that is expected when compound **1** is treated with LDA (compound **2**), which was introduced in Problem 3.69.

(b) Draw a mechanism for the conversion of **1** to **3** and explain why this step is irreversible.

(c) Upon treatment with **4**, anion **3** is converted into compound **5**, and byproduct **6** is formed. Describe the factors that render **6** a particularly stable anion.

**3.72** The p$K_a$ of the most acidic $CH_2$ group in each of the following compounds was measured in DMSO as solvent (*J. Org. Chem.* **1981**, *46*, **4327–4331**).

| Acid | p$K_a$ |
|---|---|
| $CH_3COCH_2CO_2CH_2CH_3$, **1** | 14.1 |
| $PhCH_2CN$, **2** | 21.9 |
| $PhCH_2CO_2CH_2CH_3$, **3** | 22.7 |
| $PhCH_2CON(CH_3)_2$, **4** | 26.6 |

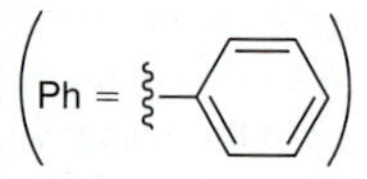

(Ph = phenyl)

Given the data above, determine which of the following two compounds (**5** or **6**) is more acidic by comparing the stability of the corresponding conjugate bases. Include all necessary resonance contributors in your discussion. Consider the given p$K_a$ values in determining which resonance contributors are more effective in stabilizing the conjugate bases.

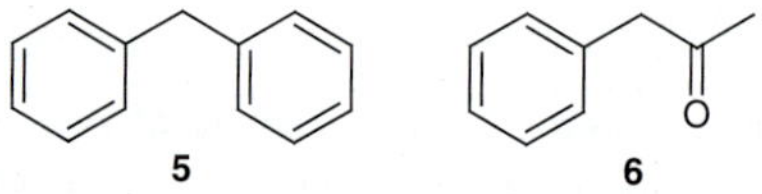

5 6

**3.73** The following compound has been designed to allow for labeling of a specific site of a protein (*J. Am. Chem. Soc.* **2009**, *131*, **8720–8721**):

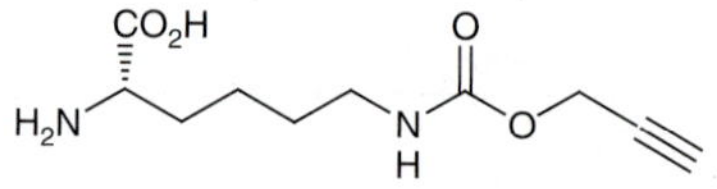

$CO_2H$ $H_2N$ N H O O

Consider each of the possible positions on this molecule that can be deprotonated. Predict the likely product of an acid/base reaction of one mole of this molecule with each of the following:

(a) One mole of EtNa
(b) Two moles of EtNa
(c) Three moles of EtNa
(d) Four moles of EtNa
(e) One mole of EtONa
(f) Two moles of EtONa
(g) Three moles of EtONa
(h) Four moles of EtONa

# Alkanes and Cycloalkanes

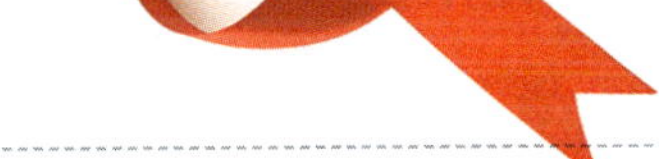

4



## DID YOU EVER WONDER...

**why scientists have not yet developed a cure for AIDS?**

As you probably know, acquired immunodeficiency syndrome (AIDS) is caused by the human immunodeficiency virus (HIV). Although scientists have not yet developed a way to destroy HIV in infected people, they have developed drugs that significantly slow the progress of the virus and the disease. These drugs interfere with the various processes by which the virus replicates itself. Yet, anti-HIV drugs are not 100% effective, primarily because HIV has the ability to mutate into forms that are drug resistant. Recently, scientists have developed a class of drugs that show great promise in the treatment of patients infected with HIV. These drugs are designed to be *flexible*, which apparently enables them to evade the problem of drug resistance.

A flexible molecule is one that can adopt many different shapes, or conformations. The study of the three-dimensional shapes of molecules is called conformational analysis. This chapter will introduce only the most basic principles of conformational analysis, which we will use to analyze the flexibility of molecules. To simplify our discussion, we will explore compounds that lack a functional group, called alkanes and cycloalkanes. Analysis of these compounds will enable us to understand how molecules achieve flexibility. Specifically, we will explore how alkanes and cycloalkanes change their three-dimensional shape as a result of the rotation of C—C single bonds.

Our discussion of conformational analysis will involve the comparison of many different compounds and will be more efficient if we can refer to compounds by name. A system of rules for naming alkanes and cycloalkanes will be developed prior to our discussion of molecular flexibility.

## DO YOU REMEMBER?

*Before you go on, be sure you understand the following topics.*
*If necessary, review the suggested sections to prepare for this chapter.*

- Molecular Orbital Theory (Section 1.8)
- Predicting Geometry (Section 1.10)
- Bond-Line Structures (Section 2.2)
- Three-Dimensional Bond-Line Structures (Section 2.6)

Take the DO YOU REMEMBER? QUIZ in **WileyPLUS** to check your understanding.

## 4.1 Introduction to Alkanes

Recall that hydrocarbons are compounds comprised of just C and H; for example:

**Ethane** $C_2H_6$ | **Ethylene** $C_2H_4$ | **Acetylene** $C_2H_2$ | **Benzene** $C_6H_6$

Ethane is unlike the other examples in that it has no π bonds. Hydrocarbons that lack π bonds are called **saturated hydrocarbons**, or **alkanes**. The names of these compounds usually end with the suffix "-ane," as seen in the following examples:

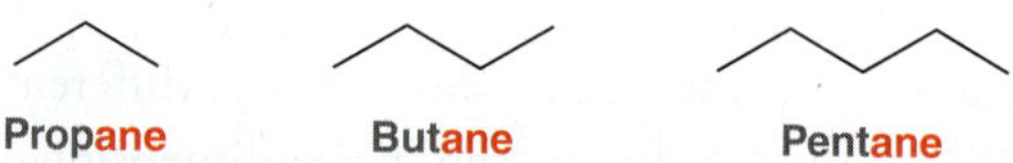


This chapter will focus on alkanes, beginning with a procedure for naming them. The system of naming chemical compounds, or **nomenclature**, will be developed and refined throughout the remaining chapters of this book.

## 4.2 Nomenclature of Alkanes

### An Introduction to IUPAC Nomenclature

In the early nineteenth century, organic compounds were often named at the whim of their discoverers. Here are just a few examples:

**Formic acid**
Isolated from ants and named after the Latin word for ant, *formica*

**Urea**
Isolated from urine

**Morphine**
A painkiller named after the Greek God of dreams, Morpheus

**Barbituric acid**
Adolf von Baeyer named this compound in honor of a woman named Barbara

A large number of compounds were given names that became part of the common language shared by chemists. Many of these common names are still in use today.

As the number of known compounds grew, a pressing need arose for a systematic method for naming compounds. In 1892, a group of 34 European chemists met in Switzerland and developed a system of organic nomenclature called the Geneva rules. The group ultimately became known as the International Union of Pure and Applied Chemistry, or **IUPAC** (pronounced "I–YOU–PACK"). The original Geneva rules have been regularly revised and updated and are now called IUPAC nomenclature.

Names produced by IUPAC rules are called **systematic names**. There are many rules, and we cannot possibly study all of them. The upcoming sections are meant to serve as an introduction to IUPAC nomenclature.

## Selecting the Parent Chain

The first step in naming an alkane is to identify the longest chain, called the parent chain:

Choose longest chain

Parent has 9 carbon atoms

In this example, the parent chain has nine carbon atoms. When naming the parent chain of a compound, the names in Table 4.1 are used. These names will be used very often in this course. Parent chains of more than 10 carbon atoms will be less common, so it is essential to commit to memory at least the first 10 parents on the list in Table 4.1.

If there is a competition between two chains of equal length, then choose the chain with the greater number of substituents. **Substituents** are the groups connected to the parent chain:

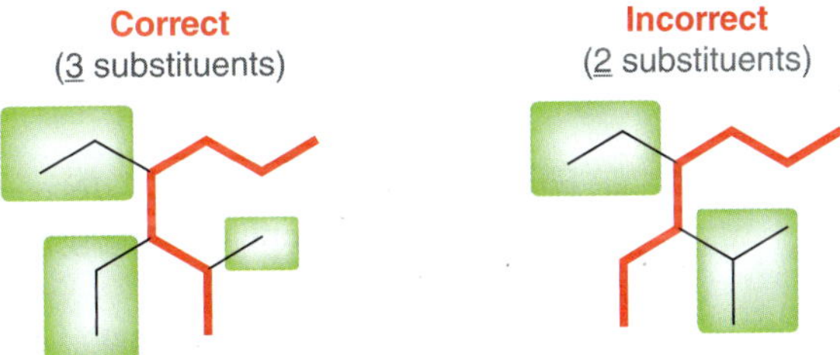


Parent has 7 carbon atoms    Parent has 7 carbon atoms

**TABLE 4.1 PARENT NAMES FOR ALKANES**

| NUMBER OF CARBON ATOMS | PARENT | NAME OF ALKANE | NUMBER OF CARBON ATOMS | PARENT | NAME OF ALKANE |
|---|---|---|---|---|---|
| 1 | *meth* | methane | 11 | *undec* | undecane |
| 2 | *eth* | ethane | 12 | *dodec* | dodecane |
| 3 | *prop* | propane | 13 | *tridec* | tridecane |
| 4 | *but* | butane | 14 | *tetradec* | tetradecane |
| 5 | *pent* | pentane | 15 | *pentadec* | pentadecane |
| 6 | *hex* | hexane | 20 | *eicos* | eicosane |
| 7 | *hept* | heptane | 30 | *triacont* | triacontane |
| 8 | *oct* | octane | 40 | *tetracont* | tetracontane |
| 9 | *non* | nonane | 50 | *pentacont* | pentacontane |
| 10 | *dec* | decane | 100 | *hect* | hectane |

The term "cyclo" is used to indicate the presence of a ring in the structure of an alkane. For example, these compounds are called **cycloalkanes**:

Cyclopropane Cyclobutane Cyclopentane

## SKILLBUILDER

### 4.1 IDENTIFYING THE PARENT

**LEARN the skill**

Identify and provide a name for the parent in the following compound:

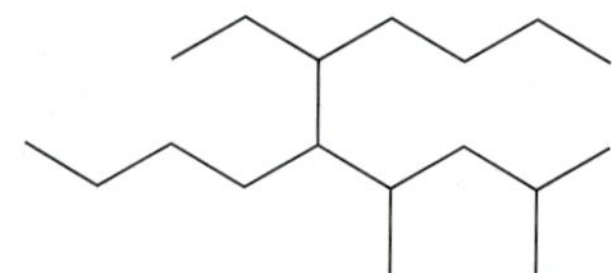

**SOLUTION**

Locate the longest chain. In this case, there are two choices that have 10 carbon atoms:

**STEP 1**
Choose the longest chain.

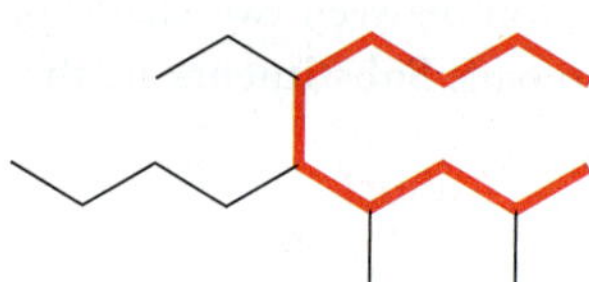

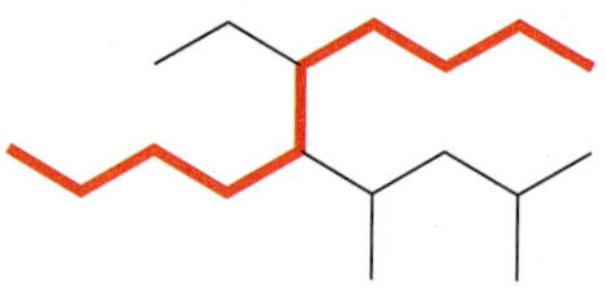

Either way, the parent will be *decane*, but we must choose the correct parent chain. The correct parent is the one with the greatest number of substituents:

**STEP 2**
When two chains compete, choose the chain with more substituents.

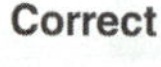

Correct

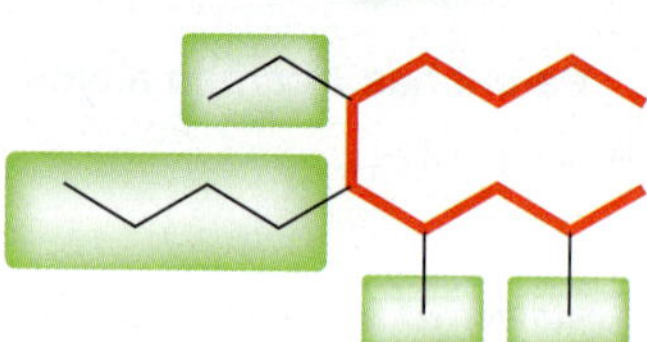

4 Substituents

Incorrect

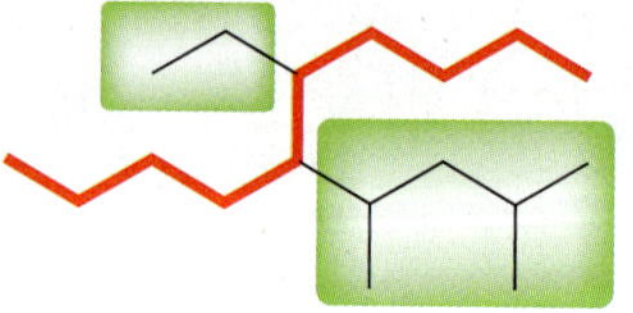

2 Substituents

**PRACTICE the skill** **4.1** Identify and name the parent in each of the following compounds:

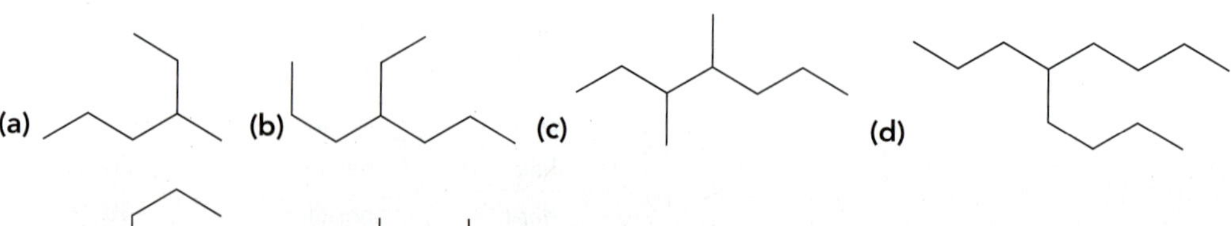

**APPLY** the skill

**4.2** Identify the two compounds below that have the same parent:

**4.3** There are five constitutional isomers with molecular formula $C_6H_{14}$. Draw a bond-line structure for each isomer and identify the parent in each case.

**4.4** There are 18 constitutional isomers with molecular formula $C_8H_{18}$. *Without drawing all 18 isomers*, determine how many of the isomers will have a parent name of heptane.

need more **PRACTICE?** **Try Problem 4.39**

## Naming Substituents

Once the parent has been identified, the next step is to list all of the substituents:

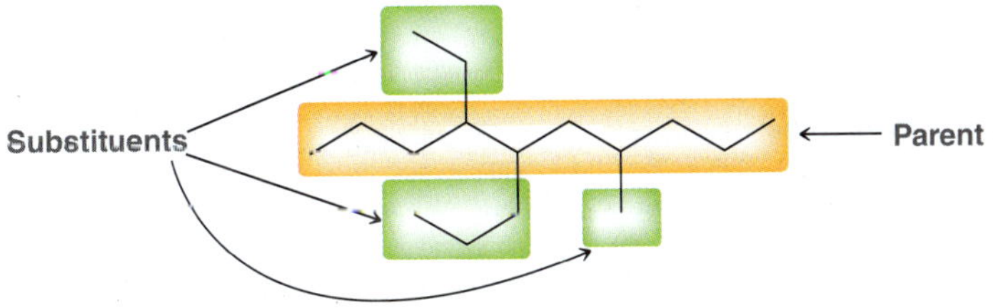


**TABLE 4.2 NAMES OF ALKYL GROUPS**

| NUMBER OF CARBON ATOMS IN SUBSTITUENT | TERMINOLOGY |
|---|---|
| 1 | *Methyl* |
| 2 | *Ethyl* |
| 3 | *Propyl* |
| 4 | *Butyl* |
| 5 | *Pentyl* |
| 6 | *Hexyl* |
| 7 | *Heptyl* |
| 8 | *Octyl* |
| 9 | *Nonyl* |
| 10 | *Decyl* |

Substituents are named with the same terminology used for naming parents, only we add the letters "yl." For example, a substituent with one carbon atom (a $CH_3$ group) is called a methyl group. A substituent with two carbon atoms is called an ethyl group. These groups are generically called **alkyl groups**. A list of alkyl groups is given in Table 4.2. In the example shown earlier, the substituents would have the following names:

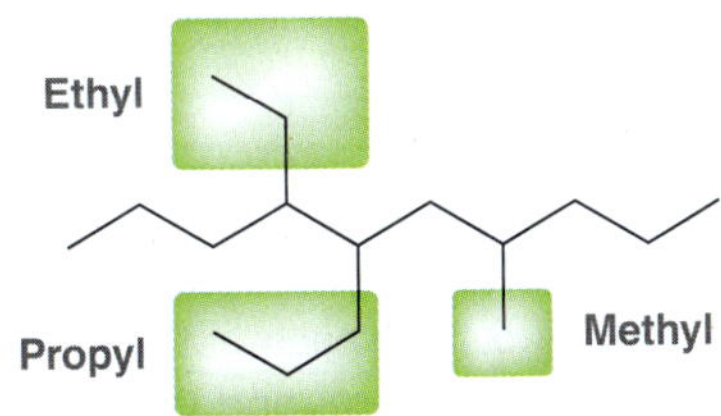


When an alkyl group is connected to a ring, the ring is generally treated as the parent:

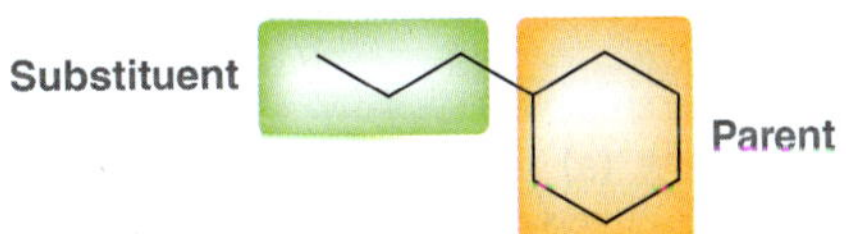


**Propylcyclohexane**

However, this is only true when the ring is comprised of more carbon atoms than the alkyl group. In the example above, the ring is comprised of six carbon atoms, while the alkyl group has only three carbon atoms. In contrast, consider the following example in which the alkyl group has more carbon atoms than the ring. As a result, the ring is named as a substituent and is called a cyclopropyl group:

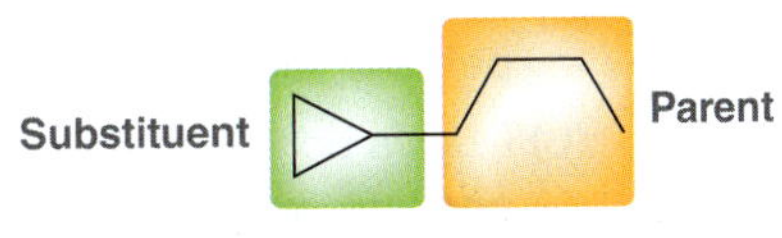


**1-Cyclopropylbutane**

# SKILLBUILDER

## 4.2 IDENTIFYING AND NAMING SUBSTITUENTS

**LEARN the skill** Identify and name all substituents in the following compound:

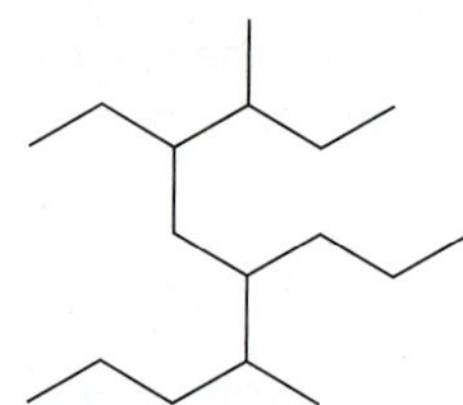

**SOLUTION**

First identify the parent by looking for the longest chain:

**STEP 1** Identify the parent.

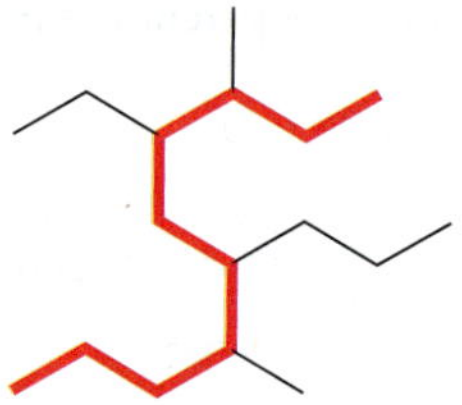

In this case, the parent has 10 carbon atoms (decane). Everything connected to the chain is a substituent, and we use the names from Table 4.2 to name each substituent:

**STEP 2** Identify all alkyl substituents connected to the parent.

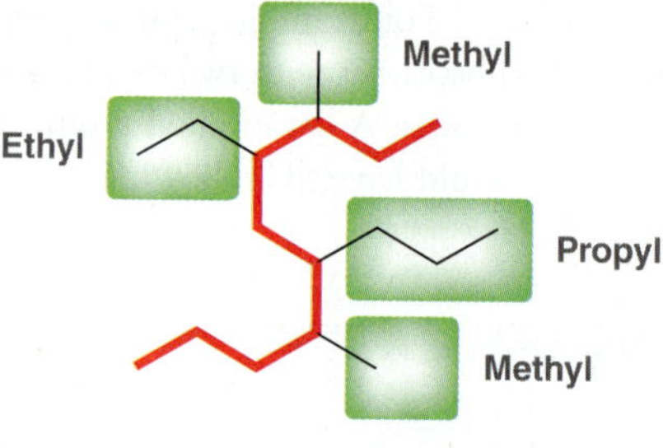

**PRACTICE the skill** **4.5** For each of the following compounds, identify all groups that would be considered substituents and then indicate how you would name each substituent:

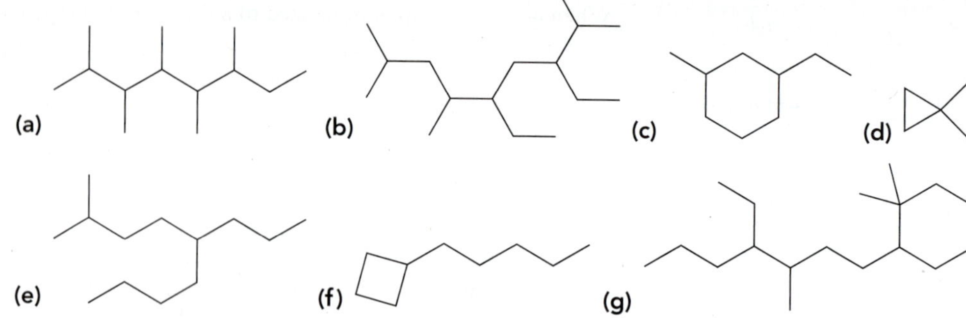

**APPLY the skill** **4.6** There are nine constitutional isomers with molecular formula $C_7H_{16}$.

**(a)** Draw a bond-line structure for all isomers that fit the following criteria: the parent = pentane, and there are two methyl groups connected to the parent.

**(b)** Draw a bond-line structure for the isomer that fits the following criteria: the parent = pentane, and there is one ethyl group connected to the parent.

need more **PRACTICE?** **Try Problems 4.40a, 4.40c**

## practically speaking | Pheromones: Chemical Messengers

Many animals use chemicals, called *pheromones*, to communicate with each other. In fact, insects use pheromones as their primary means of communication. An insect secretes the pheromone, which then binds to a receptor in another insect, triggering a biological response. Some insects use compounds that warn of danger (alarm pheromones), while other insects use compounds that promote aggregation among members of the same species (aggregation pheromones). Pheromones are also used by many insects to attract members of the opposite sex for mating purposes (sex pheromones).

For example, 2-methylheptadecane is a sex pheromone used by some moths, and undecane is used as an aggregation pheromone by some cockroaches.

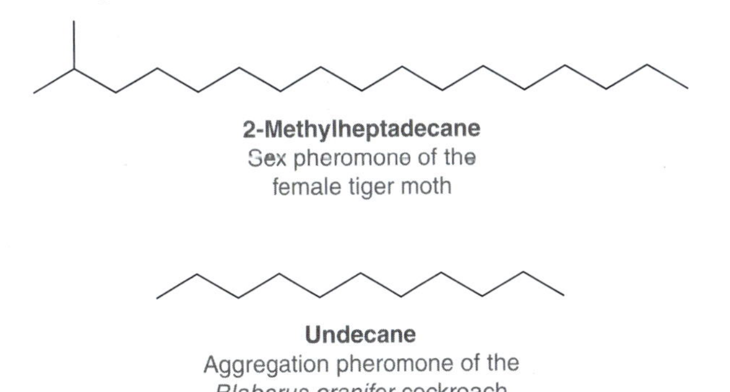

**2-Methylheptadecane**
Sex pheromone of the female tiger moth

**Undecane**
Aggregation pheromone of the *Blaberus cranifer* cockroach

Both of these examples above are alkanes, but many pheromones exhibit one or more functional groups. Examples of such pheromones will appear throughout this book.

## Naming Complex Substituents

Naming branched alkyl substituents is more complex than naming straight-chain substituents. For example, consider the following substituent:

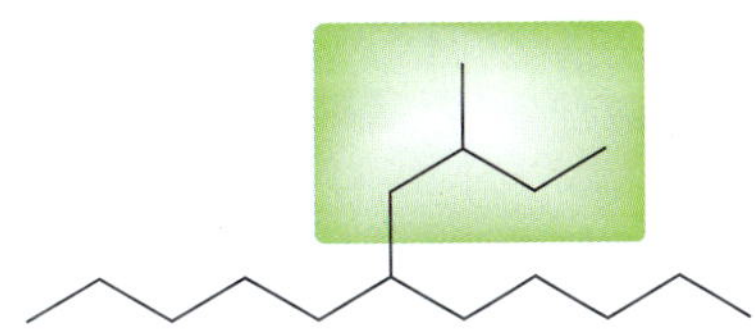

How do we name this substituent? It has five carbon atoms, but it cannot simply be called a pentyl group, because it is not a straight-chain alkyl group. In situations like this, the following method is employed: Begin by placing numbers on the substituent, going *away* from the parent chain:

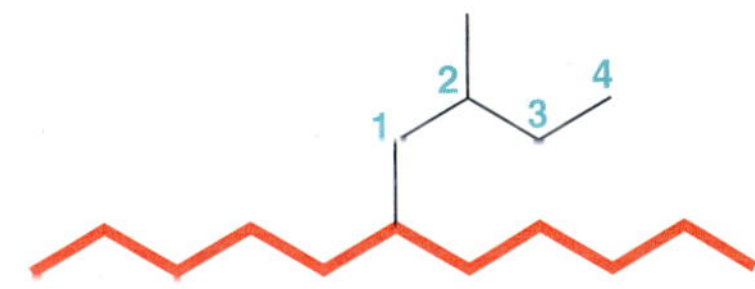


Place the numbers on the longest straight chain present in the substituent, and in this case, there are four carbon atoms. This group is therefore considered to be a butyl group that has one methyl group attached to it at the 2 position. Accordingly, this group is called a (2-methylbutyl) group. In essence, treat the complex substituent as if it is a miniparent with its own substituents. When naming a complex substituent, place parentheses around the name of the substituent. This will avoid confusion, as we will soon place numbers on the parent chain and we don't want to confuse those numbers with the numbers on the substituent chain.

Some complex substituents have common names. These common names are so well entrenched that IUPAC allows them. It would be wise to commit the following common names to memory, as they will be used frequently throughout the course.

An alkyl group bearing three carbon atoms can only be branched in one way, and it is called an *isopropyl group*:

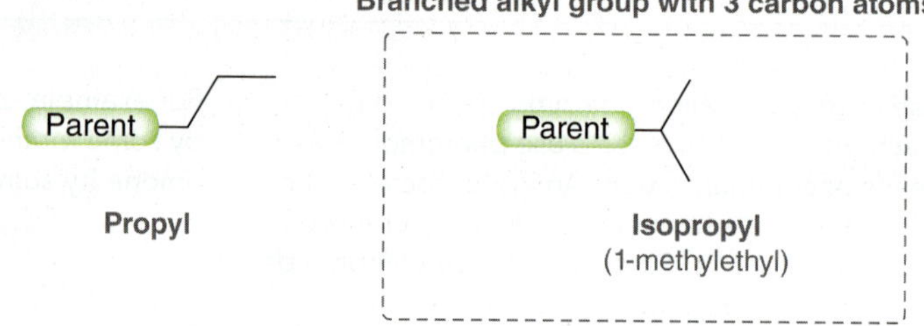


Alkyl groups bearing four carbon atoms can be branched in three different ways:

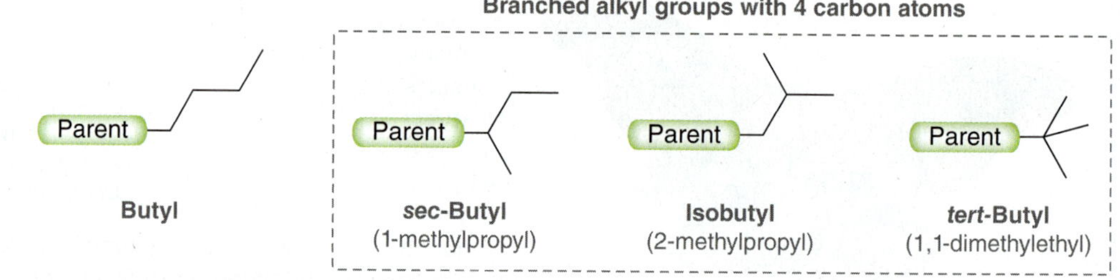


Alkyl groups bearing five carbon atoms can be branched in many more ways. Here are two common ways:

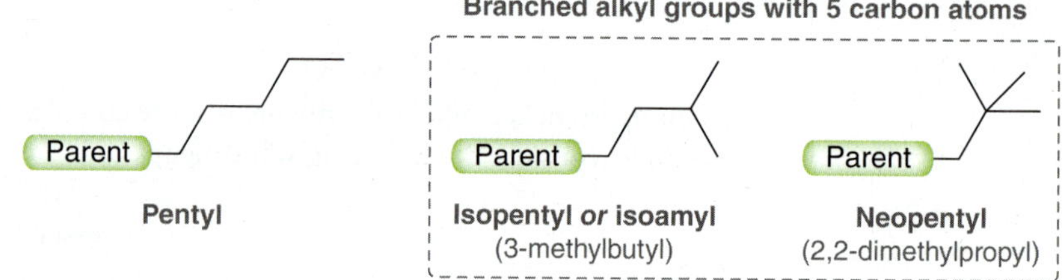


## SKILLBUILDER

### 4.3 IDENTIFYING AND NAMING COMPLEX SUBSTITUENTS

**LEARN the skill**

In the following compound, identify all groups that would be considered substituents and then indicate the systematic name as well as the common name for each substituent:

**SOLUTION**

First identify the parent:

**STEP 1**
Identify the parent.

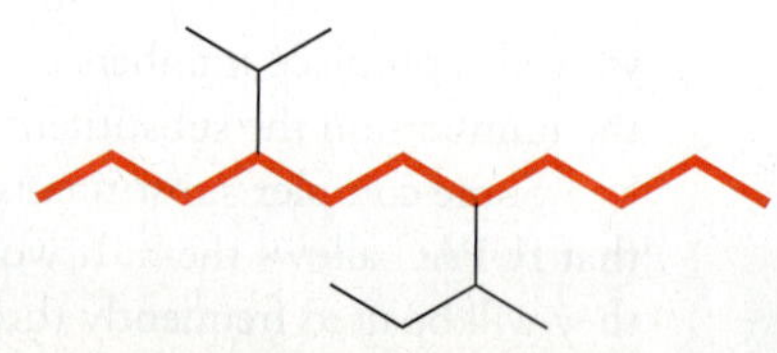

**STEP 2**
Place numbers on the complex substituent.

In this example, there are two complex substituents. To name them, we place numbers on each substituent, going away from the parent in each case. Then, we consider each complex substituent to be composed of "a substituent on a substituent," and we use the numbers for that purpose:

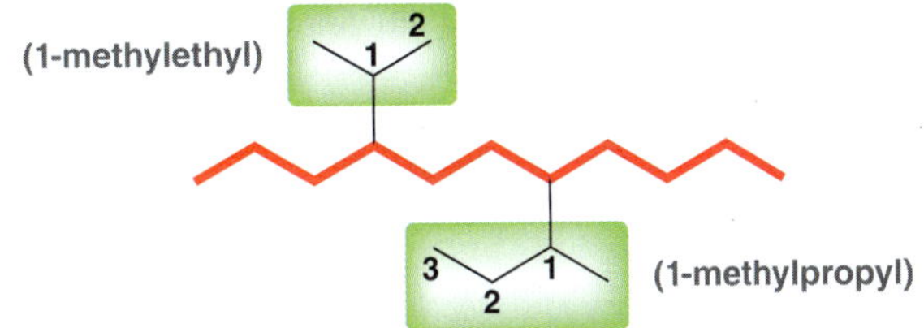


**STEP 3**
Name the complex substituent.

Alternatively, IUPAC rules allow us to use the following common names:

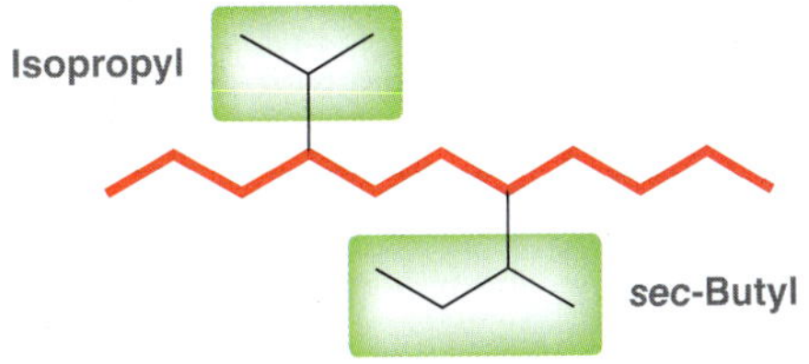

**PRACTICE the skill** **4.7** For each of the following compounds, identify all groups that would be considered substituents and then indicate the systematic name as well as the common name for each substituent:

(a) (b) (c) (d) (e)

**APPLY the skill** **4.8** The following substituent is called a phenyl group:

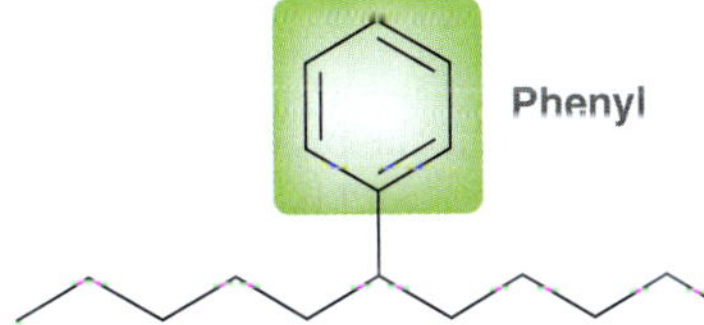


With this in mind, identify the systematic name for each substituent below:

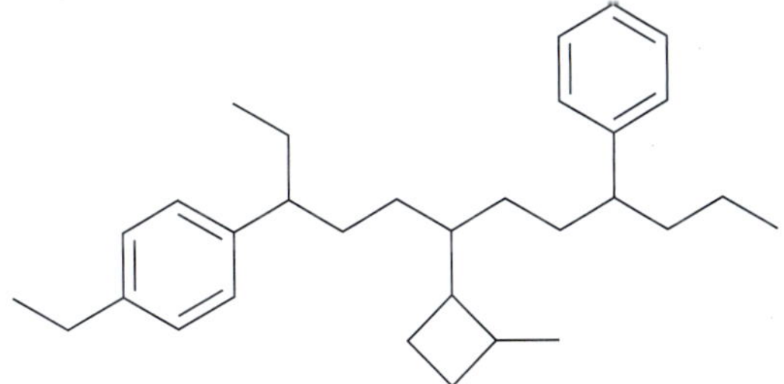

**4.9** Draw all possible alkyl substituents that possess exactly five carbon atoms and no ring. Provide a systematic name for each of these substituents.

need more **PRACTICE?** **Try Problems 4.40b, 4.40d**

## Assembling the Systematic Name of an Alkane

In order to assemble the systematic name of an alkane, the carbon atoms of the parent chain are numbered, and these numbers are used to identify the location of each substituent. As an example, consider the following two compounds:

**2-Methylpentane** **3-Methylpentane**

In each case, the location of the methyl group is clearly identified with a number, called a **locant**. In order to assign the correct locant, we must number the parent chain properly, which can be done by following just a few rules:

- If one substituent is present, it should be assigned the lower possible number. In this example, we place the numbers so that the methyl group is at C2 rather than C6:

- When multiple substituents are present, assign numbers so that the first substituent receives the lower number. In the following case, we number the parent chain so that the substituents are 2,5,5 rather than 3,3,6 because we want the first locant to be as low as possible:

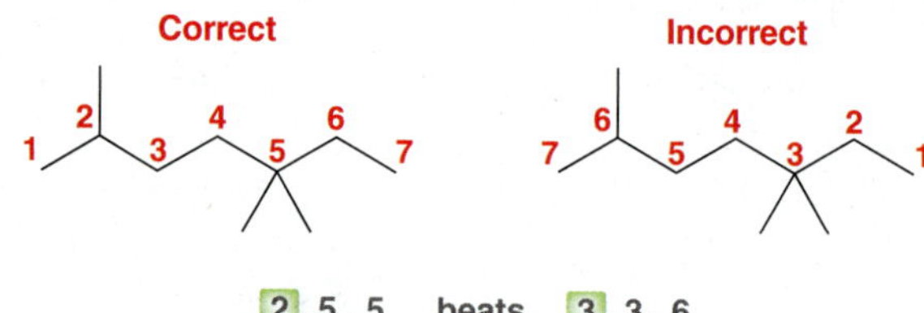


- If there is a tie, then the second locant should be as low as possible:

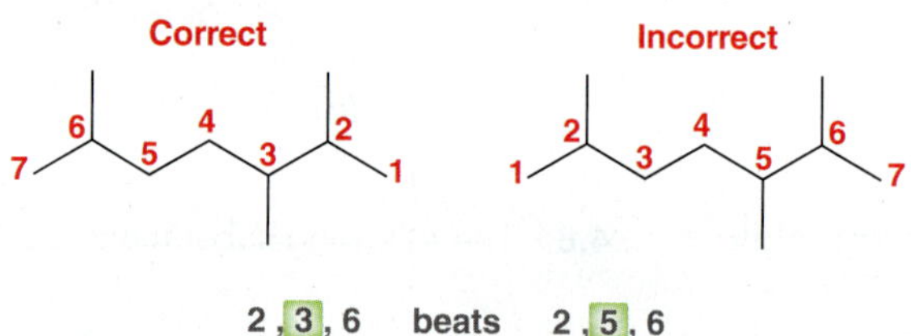


- If the previous rule does not break the tie, such as in the case below, then the lowest number should be assigned alphabetically:

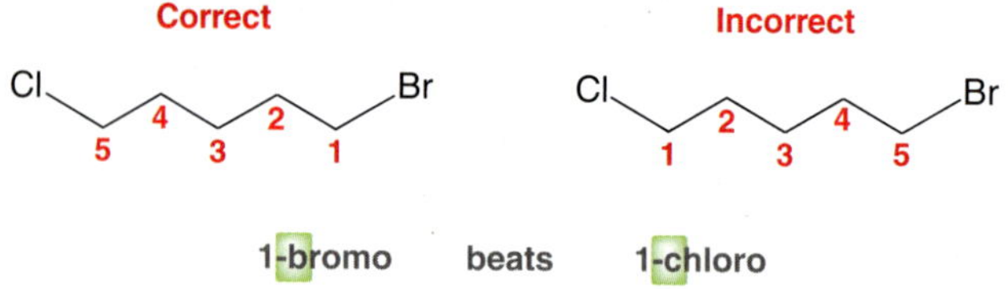


- When dealing with cycloalkanes, all of the same rules apply: for example:

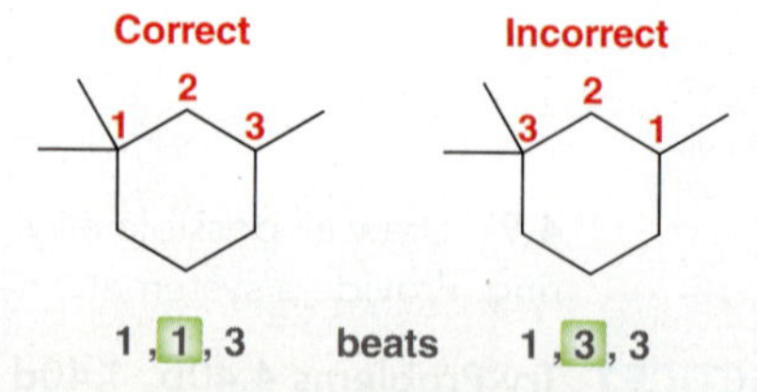

- When a substituent appears more than once in a compound, a prefix is used to identify how many times that substituent appears in the compound (di = 2, tri = 3, tetra = 4, penta = 5, and hexa = 6). For example, the previous compound would be called *1,1,3-trimethyl*cyclohexane. Notice that a hyphen is used to separate numbers from letters, while commas are used to separate two numbers from each other.
- Once all substituents have been identified and assigned the proper locants, they are placed in alphabetical order. Prefixes (di, tri, tetra, penta, and hexa) are not included as part of the alphabetization scheme. In other words, "dimethyl" is alphabetized as if it started with the letter "m" rather than "d." Similarly, *sec* and *tert* are also ignored for alphabetization purposes, however, iso is not ignored. In other words, *sec*-butyl is alphabetized as a "b" while isobutyl is alphabetized as an "i."

In summary, four discrete steps are required when assigning the name of an alkane:

1. ***Identify the parent chain***: Choose the longest chain. For two chains of equal length, the parent chain should be the chain with the greater number of substituents.
2. ***Identify and name the substituents.***
3. ***Number the parent chain and assign a locant to each substituent***: Give the first substituent the lower possible number. If there is a tie, choose the chain in which the second substituent has the lower number.
4. ***Arrange the substituents alphabetically***. Place locants in front of each substituent. For identical substituents, use di, tri, or tetra, which are ignored in alphabetizing.

## SKILLBUILDER

### 4.4 ASSEMBLING THE SYSTEMATIC NAME OF AN ALKANE

**LEARN** the skill

Provide a systematic name for the following compound:

**SOLUTION**

Assembling a systematic name requires four discrete steps. The first two steps are to identify the parent and the substituents:

**STEP 1** Identify the parent.

**STEP 2** Identify and name substituents.

Methyl

Methyl

Ethyl

Then, assign a locant to each substituent and arrange the substituents alphabetically:

**STEP 3** Assign locants.

**STEP 4** Arrange the substituents alphabetically.

4-Ethyl-2,3-dimethyloctane

Notice that **e**thyl is placed before di**m**ethyl in the name. Also, make sure that hyphens separate letters from numbers, while commas separate numbers from each other.

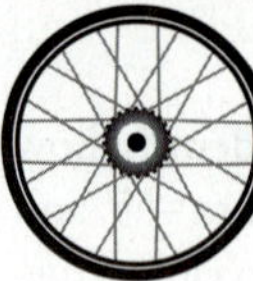

**PRACTICE** the skill **4.10** Provide a systematic name for each of the following compounds:

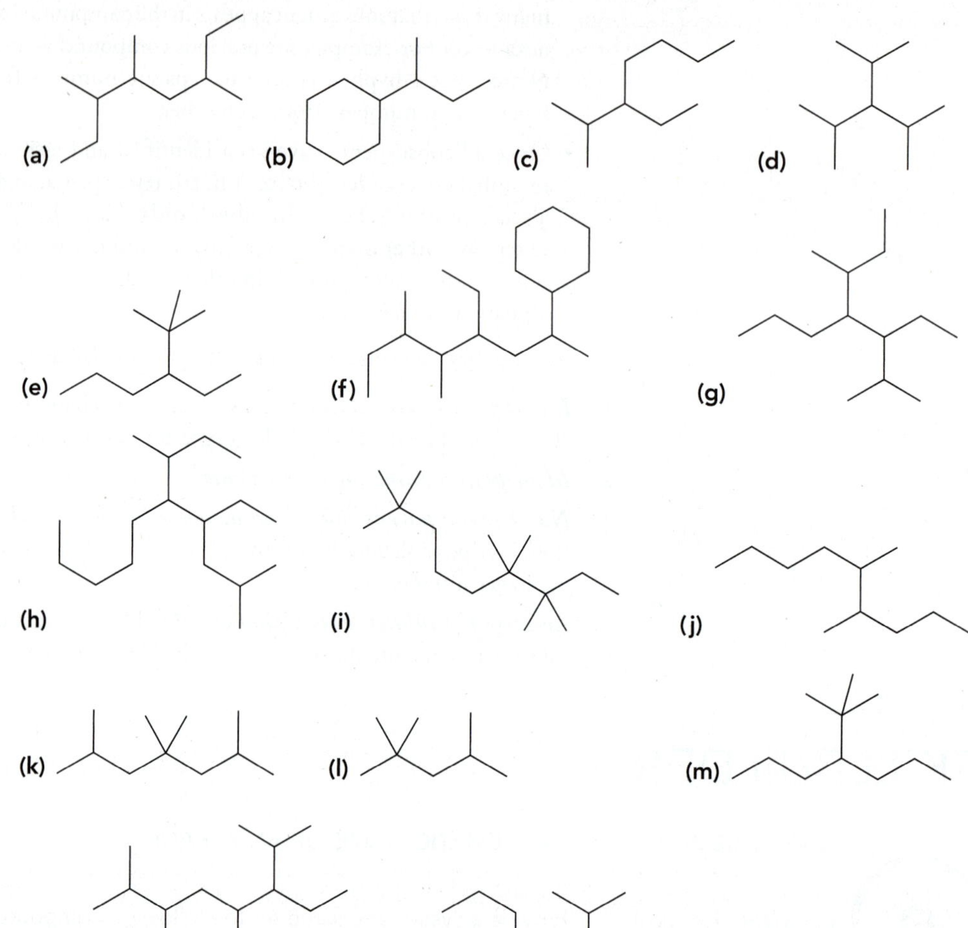

**APPLY** the skill

**4.11** Draw a bond-line drawing for each of the following compounds:

**(a)** 3-Isopropyl-2,4-dimethylpentane **(b)** 4-Ethyl-2-methylhexane

**(c)** 1,1,2,2-Tetramethylcyclopropane

need more **PRACTICE?** **Try Problems 4.41a, 4.41b, 4.45a, 4.45b**

## Naming Bicyclic Compounds

Compounds that contain two fused rings are called **bicyclic** compounds, and they can be drawn in different ways:

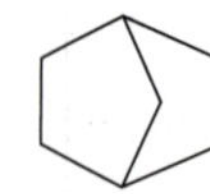 **is the same as** 

The second drawing style implies the three dimensionality of the molecule, a topic that will be covered in more detail in the upcoming chapter. For now, we will focus on naming bicyclic systems, which is very similar to naming alkanes and cycloalkanes. We follow the same four-step procedure outlined in the previous section, but there are differences in naming and numbering the parent. Let's start with naming the parent.

For bicyclic systems, the term "bicyclo" is introduced in the name of the parent; for example, the compound above is comprised of seven carbon atoms and is therefore called bicycloheptane (where the parent is *bicyclohept*). The problem is that this parent is not specific enough. To illustrate this, consider the following two compounds, both of which are called bicycloheptane:

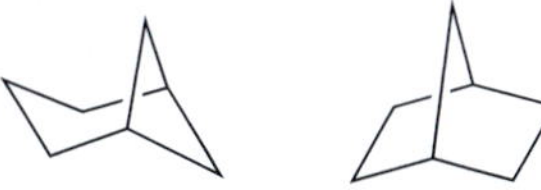

Both compounds consist of two fused rings and seven carbon atoms. Yet, the compounds are clearly different, which means that the name of the parent needs to contain more information. Specifically, it must indicate the way in which the rings are constructed (the constitution of the compound). In order to do this, we must identify the two **bridgeheads**. These are the two carbon atoms where the rings are fused together:

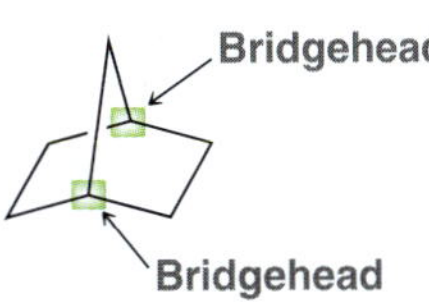


There are three different paths connecting these two bridgeheads. For each path, count the number of carbon atoms, excluding the bridgeheads themselves. In the compound above, one path has two carbon atoms, another path has two carbon atoms, and the third (shortest path) has only one carbon atom. These three numbers, ordered from largest to smallest, [2.2.1], are then placed in the middle of the parent surrounded by brackets:

Bicyclo[2.2.1]heptane

These numbers provide the necessary specificity to differentiate the compounds shown earlier:

**Bicyclo[3.1.1]heptane** **Bicyclo[2.2.1]heptane**

If a substituent is present, the parent must be numbered properly in order to assign the correct locant to the substituent. To number the parent, start at one of the bridgeheads and begin numbering along the longest path, then go to the second longest path, and finally go along the shortest path. For example, consider the following bicyclic system:

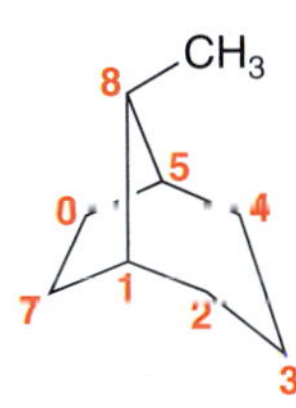


In this example, the methyl substituent did not get a low number. In fact, it got the highest number possible because of its location. Specifically, it is on the shortest path connecting the bridgeheads. Regardless of the position of substituents, the parent must be numbered beginning with the longest path first. The only choice is which bridgehead will be counted as C1; for example:

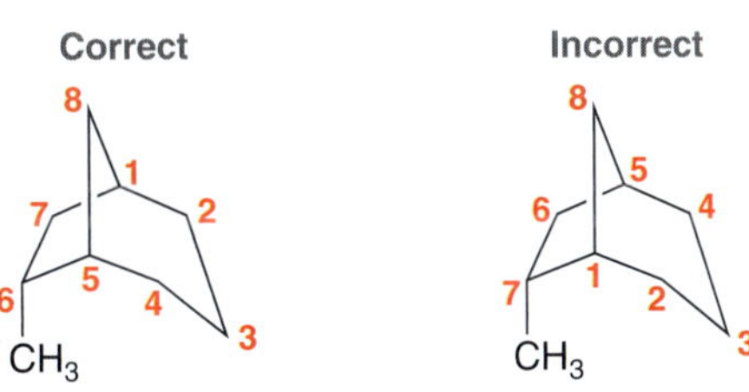


Either way, the numbers begin along the longest path. However, we must start numbering at the bridgehead that gives the substituent the lowest possible number. In the example above, the correct path places the substituent at C6 rather than at C7. The name of this compound is 6-methylbicyclo[3.2.1]octane.

# SKILLBUILDER

## 4.5 ASSEMBLING THE NAME OF A BICYCLIC COMPOUND

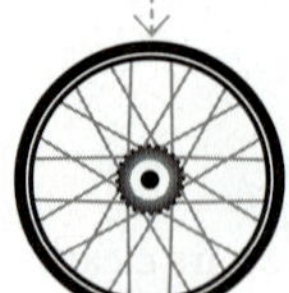

**LEARN the skill**

Assign a name for the following compound:

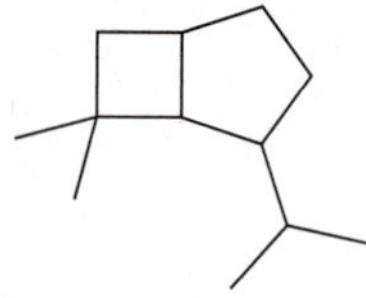

### SOLUTION

**STEP 1** Identify the parent.

Once again, we use our four-step procedure. First identify the parent. In this case, we are dealing with a bicyclic system, so we count the total number of carbon atoms comprising both ring systems. There are seven carbon atoms in both rings combined, so the parent must be "bicyclo-hept." The two bridgeheads are highlighted. Now count the number of carbon atoms along each of the three possible paths that connect the bridgeheads. The longest path (on the right side) has three carbon atoms in between the bridgeheads. The second longest path (on the left side) has two carbon atoms in between the bridgeheads. The shortest path has no carbon atoms in between the bridgeheads; that is, the bridgeheads are connected directly to each other. Therefore the parent is "bicyclo[3.2.0]hept." Next, identify and name the substituents:

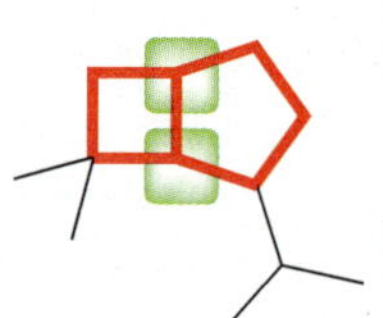

**STEP 2** Identify and name substituents.

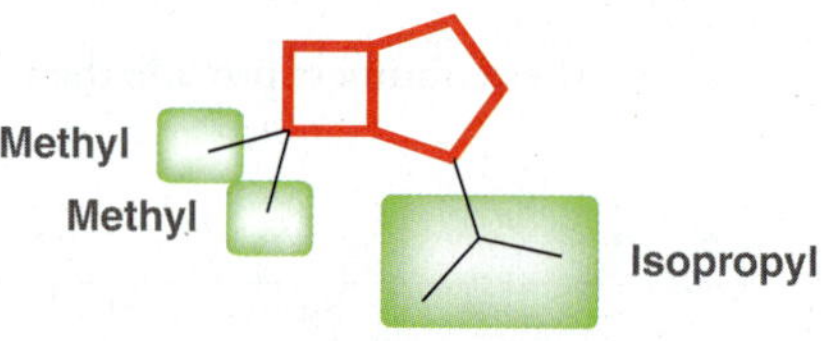


**STEP 3** Assign locants.

Then number the parent, and assign a locant to each substituent. Start at one of the bridgeheads and continue numbering along the longest path that connects the bridgeheads. In this case, we start at the lower bridgehead, so as to give the isopropyl group the lower number:

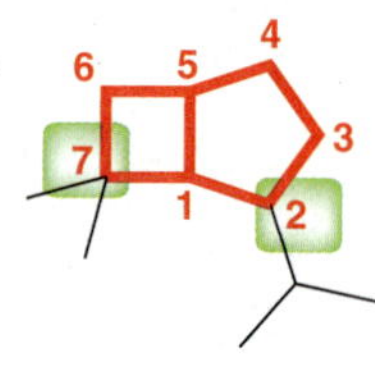


**STEP 4** Arrange the substituents alphabetically.

Finally, arrange the substituents alphabetically:

2-Isopropyl-7,7-dimethylbicyclo[3.2.0]heptane

**PRACTICE the skill**

**4.12** Name each of the following compounds:

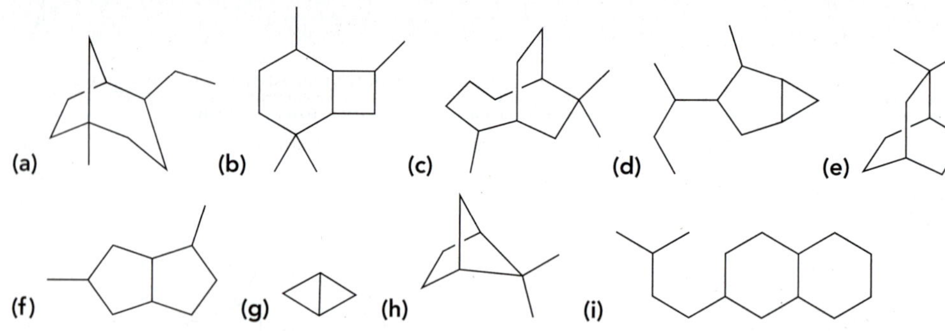


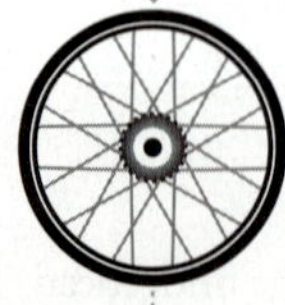

**APPLY the skill**

**4.13** Draw a bond-line structure for each of the following compounds:

**(a)** 2,2,3,3-Tetramethylbicyclo[2.2.1]heptane **(b)** 8,8-Diethylbicyclo[3.2.1]octane

**(c)** 3-Isopropylbicyclo[3.2.0]heptane

need more **PRACTICE?** **Try Problems 4.41c, 4.41d, 4.45c**

## Naming Drugs

Pharmaceuticals often have cumbersome IUPAC names and are therefore given shorter names, called *generic names*. For example, consider the following compound:

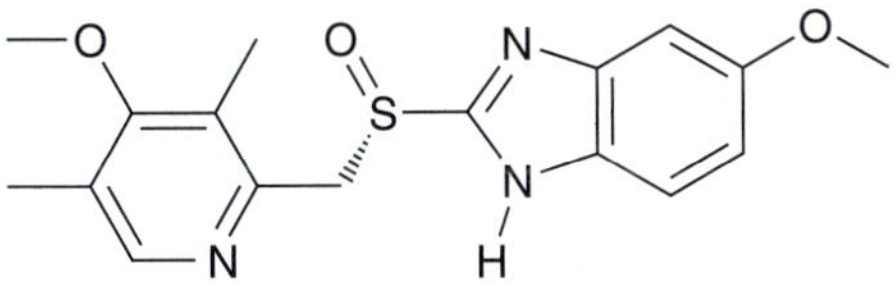

**(*S*)-5-Methoxy-2-[(4-methoxy-3,5-dimethyl pyridin-2-yl)methylsulfinyl]-3*H*-benzoimidazole**

The IUPAC name for this compound is quite a mouthful, so a generic name, *esomeprazole*, has been assigned and accepted by the international community. For marketing purposes, drug companies will also select a catchy name, called a *trade name*. The trade name of esomeprazole is Nexium. This compound is a proton-pump inhibitor used in the treatment of reflux disease.

In summary, pharmaceuticals have three important names: (1) trade names, (2) generic names, and (3) systematic IUPAC names. Table 4.3 lists several common drugs whose trade names are likely to sound familiar.

**TABLE 4.3 NAMES OF COMMON PHARMACEUTICALS**

| TRADE NAME | GENERIC NAME | STRUCTURE AND IUPAC NAME | USES |
|---|---|---|---|
| Aspirin | Acetylsalicylic acid | 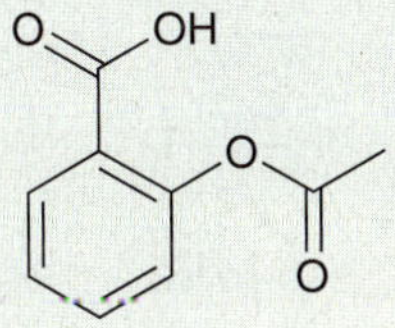 **2-Acetoxybenzoic acid** | Analgesic, antipyretic (reduces fever), anti-inflammatory |
| Advil or Motrin | Ibuprofen | 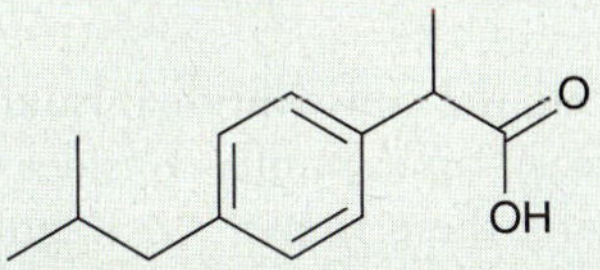 **2-[4-(2-Methylpropyl)phenyl]propanoic acid** | Analgesic, antipyretic, anti-inflammatory |
| Demerol | Pethidine | 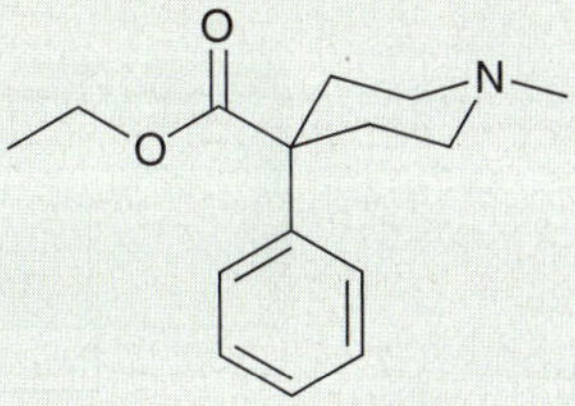 **Ethyl 1-methyl-4-phenylpiperidine-4-carboxylate** | Analgesic |
| Dramamine | Meclizine | 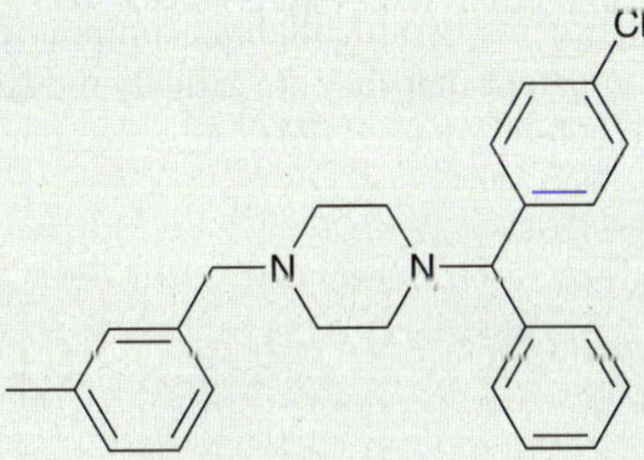 **1-[(4-Chlorophenyl)-phenyl-methyl]-4-[(3-methylphenyl)methyl]piperazine** | Antiemetic (inhibits nausea and vomiting) |
| Tylenol | Acetaminophen | 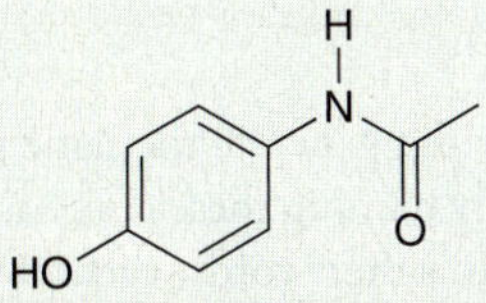 ***N*-(4-Hydroxyphenyl)ethanamide** | Analgesic, antipyretic |

## 4.3 Constitutional Isomers of Alkanes

For an alkane, the number of possible constitutional isomers increases with increasing molecular size. Table 4.4 illustrates this trend.

**TABLE 4.4 NUMBER OF CONSTITUTIONAL ISOMERS FOR VARIOUS ALKANES**

| MOLECULAR FORMULA | NUMBER OF CONSTITUTIONAL ISOMERS |
|---|---|
| $C_3H_8$ | 1 |
| $C_4H_{10}$ | 2 |
| $C_5H_{12}$ | 3 |
| $C_6H_{14}$ | 5 |
| $C_7H_{16}$ | 9 |
| $C_8H_{18}$ | 18 |
| $C_9H_{20}$ | 35 |
| $C_{10}H_{22}$ | 75 |
| $C_{15}H_{32}$ | 4,347 |
| $C_{20}H_{42}$ | 366,319 |
| $C_{30}H_{62}$ | 4,111,846,763 |
| $C_{40}H_{82}$ | 62,481,801,147,341 |

When drawing the constitutional isomers of an alkane, make sure to avoid drawing the same isomer twice. As an example, consider $C_6H_{14}$, for which there are five constitutional isomers. When drawing these isomers, it might be tempting to draw more than five structures. For example:

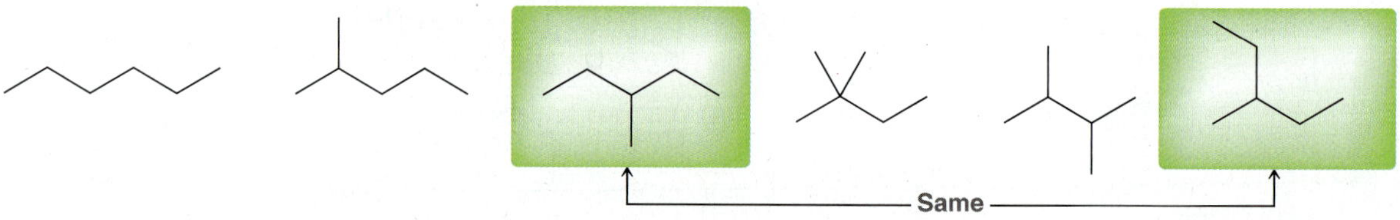

Same

At first glance, the two highlighted compounds seem to be different. But upon further inspection, it becomes apparent that they are actually the same compound. To avoid drawing the same compound twice, it is helpful to use IUPAC rules to name each compound. If there are duplicates, it will become apparent:

3-Methylpentane    3-Methylpentane

These two drawings generate the same name, and therefore, they must be the same compound. Even without formally naming these compounds, it is helpful to simply look at molecules from an IUPAC point of view. In other words, each of these compounds should be viewed as a parent chain of five carbon atoms, with a methyl group at C3. Viewing molecules in this way (from an IUPAC point of view) will prove to be helpful in some cases.

# SKILLBUILDER

## 4.6 IDENTIFYING CONSTITUTIONAL ISOMERS

**LEARN the skill** Identify whether the following two compounds are constitutional isomers or whether they are simply different drawings of the same compound:

### SOLUTION

Use the rules of nomenclature to name each compound. In each case, identify the parent, locate the substituents, number the parent, and assemble a name:

**STEP 1** Name each compound.

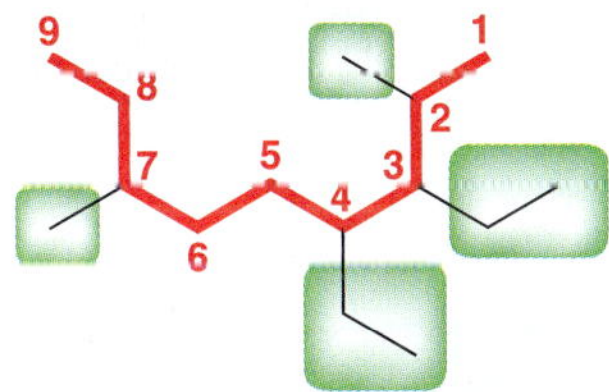

3,4-Diethyl-2,7-dimethylnonane

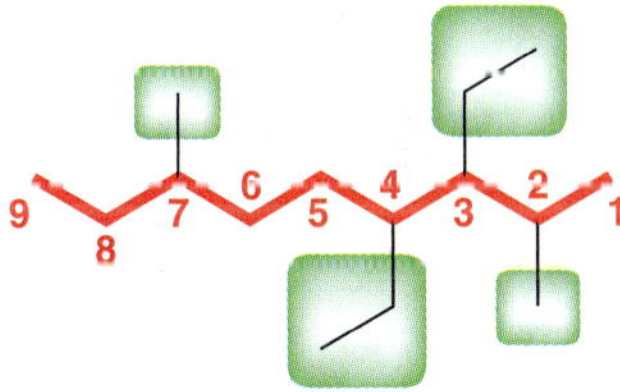


3,4-Diethyl-2,7-dimethylnonane

**STEP 2** Compare the names.

These compounds have the same name, and therefore, they are not constitutional isomers. They are actually two representations of the same compound.

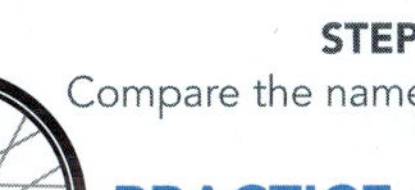

**PRACTICE the skill** **4.14** For each pair of compounds, identify whether they are constitutional isomers or two representations of the same compound:

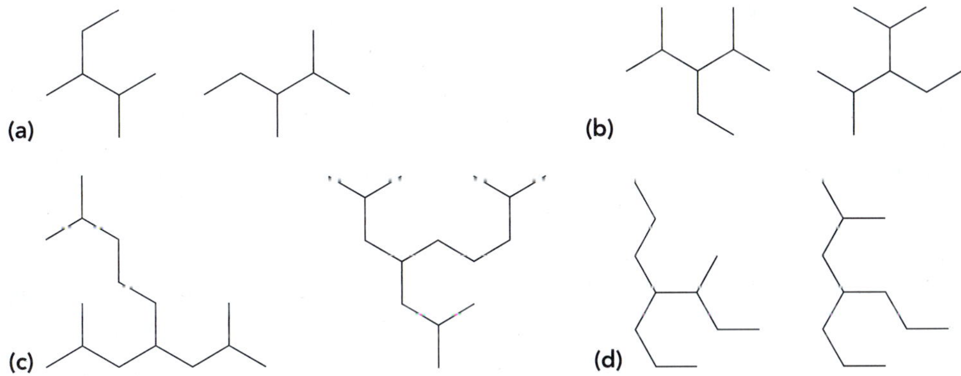

**APPLY the skill** **4.15** Table 4.4 indicates the number of constitutional isomers with molecular formula $C_7H_{16}$. Draw each of the isomers, making sure not to draw the same compound twice.

need more **PRACTICE?** **Try Problems 4.42, 4.66b,d,k,l**

## 4.4 Relative Stability of Isomeric Alkanes

In order to compare the stability of constitutional isomers, we look at the heat liberated when they each undergo combustion. For an alkane, combustion describes a reaction in which the alkane reacts with oxygen to produce $CO_2$ and water. Consider the following example:

$$+ \ 8\,O_2 \longrightarrow 5\,CO_2 + 6\,H_2O \qquad \Delta H° = -3509 \text{ kJ/mol}$$

In this reaction, an alkane (pentane) is ignited in the presence of oxygen, and the resulting reaction is called combustion. The value shown above, $\Delta H°$ for this reaction, is the *change in enthalpy* associated with the complete combustion of 1 mol of pentane in the presence of oxygen. We will revisit the concept of enthalpy in more detail in Chapter 6, but for now, we will simply think of it as the heat given off during the reaction. For a combustion process, $-\Delta H°$ is called the **heat of combustion**.

Combustion can be conducted under experimental conditions using a device called a calorimeter, which can measure heats of combustion accurately. Careful measurements reveal that the heats of combustion for two isomeric alkanes are different, even though the products of the reactions are identical:

$$+ \quad 12\tfrac{1}{2}\, O_2 \longrightarrow 8\, CO_2 + 9\, H_2O \qquad -\Delta H° = 5470 \text{ kJ/mol}$$

$$+ \quad 12\tfrac{1}{2}\, O_2 \longrightarrow 8\, CO_2 + 9\, H_2O \qquad -\Delta H° = 5452 \text{ kJ/mol}$$

Notice that although the two reactions shown above yield the same number of moles of $CO_2$ and water, the heats of combustion for the two reactions are different. We can use this difference to compare the stability of isomeric alkanes (Figure 4.1). By comparing the amount of heat given off by each combustion process, we can compare the potential energy that each isomer had before combustion. This analysis leads to the conclusion that branched alkanes are lower in energy (more stable) than straight-chain alkanes.

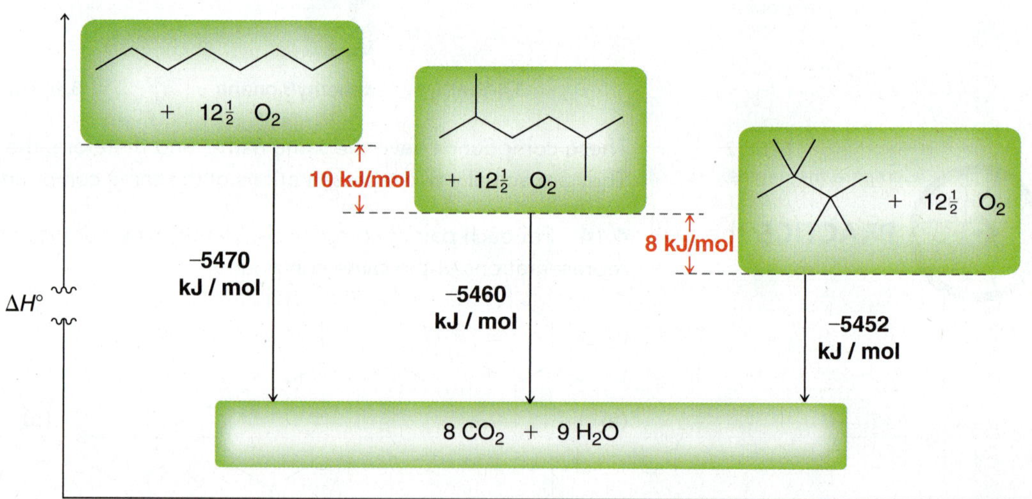


**FIGURE 4.1** An energy diagram comparing the heats of combustion for three constitutional isomers of octane.

Heats of combustion are an important way to determine the relative stability of compounds. We will use this technique several times throughout this book to compare the stability of compounds.

## 4.5 Sources and Uses of Alkanes

The main source of alkanes is petroleum, which comes from the Latin words *petro* ("rock") and *oleum* ("oil"). Petroleum is a complex mixture of hundreds of hydrocarbons, most of which are alkanes (ranging in size and constitution). It is believed that Earth's petroleum deposits were formed slowly, over millions of years, by the decay of prehistoric bioorganic material (such as plants and forests).

The first oil well was drilled in Pennsylvania in 1859. The petroleum obtained was separated into its various components by distillation, the process by which the components of a mixture are separated from each other based on differences in their boiling points. At the time, kerosene, one of the high-boiling fractions, was considered to be the most important product obtained from petroleum. Automobiles based on the internal combustion engine were still waiting to be invented (some 50 years

**TABLE 4.5 INDUSTRIAL USES OF PETROLEUM FRACTIONS**

| BOILING RANGE OF FRACTION (° C) | NUMBER OF CARBON ATOMS IN MOLECULES | USE |
|---|---|---|
| Below 20 | $C_1$–$C_4$ | Natural gas, petrochemicals, plastics |
| 20–100 | $C_5$–$C_7$ | Solvents |
| 20–200 | $C_5$–$C_{12}$ | Gasoline |
| 200–300 | $C_{12}$–$C_{18}$ | Kerosene, jet fuel |
| 200–400 | $C_{12}$ and higher | Heating oil, diesel |
| Nonvolatile liquids | $C_{20}$ and higher | Lubricating oil, grease |
| Nonvolatile solids | $C_{20}$ and higher | Wax, asphalt, tar |

later), so gasoline was not yet a coveted petroleum product. There was a large market opportunity for kerosene, as kerosene lamps produced better night light than standard candles. Over time, other uses were found for the other fractions of petroleum. Today, every precious drop of petroleum is put to use in one way or another (Table 4.5). The process of separating crude oil (petroleum) into commercially available products is called *refining*. A typical refinery can process 100,000 barrels of crude oil a day (1 barrel = 42 gallons). The most important product is currently the gasoline fraction ($C_5$–$C_{12}$), yet this fraction only represents approximately 19% of the crude oil. This amount does not satisfy the current demand for gasoline, and therefore, two processes are employed that increase the yield of gasoline from every barrel of crude oil.

1. *Cracking* is a process by which C—C bonds of larger alkanes are broken, producing smaller alkanes suitable for gasoline. This process effectively converts more of the crude oil into compounds suitable for use as gasoline. Cracking can be achieved at high temperature (*thermal cracking*) or with the aid of catalysts (*catalytic cracking*). Cracking generally yields straight-chain alkanes. Although suitable for gasoline, these alkanes tend to give rise to preignition, or *knocking*, in automobile engines.

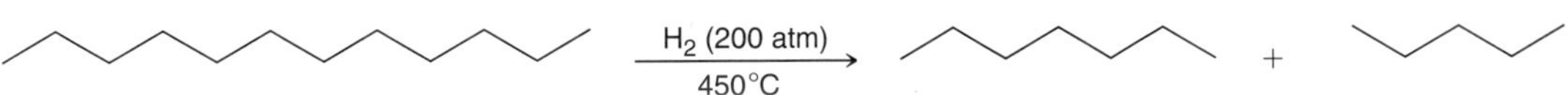


2. *Reforming* is a process involving many different types of reactions (such as dehydrogenation and isomerization reactions) with the goal of converting straight-chain alkanes into branched hydrocarbons and aromatic compounds (discussed in Chapter 18):

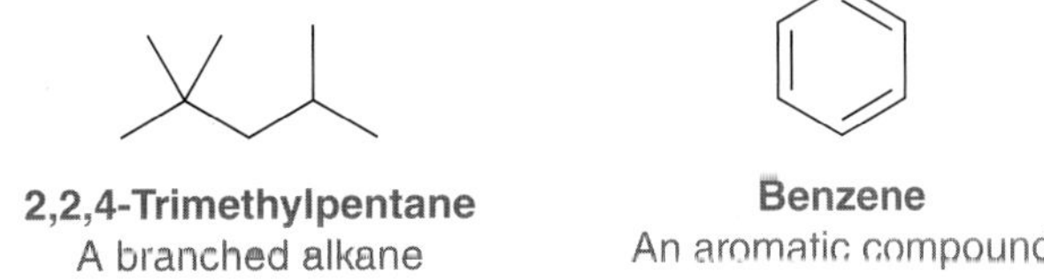


Branched hydrocarbons and aromatic hydrocarbons show less of a tendency for knocking. It is therefore desirable to convert some petroleum into branched alkanes and aromatic compounds and then blend them with straight-chain alkanes. The combination of cracking and reforming effectively increases the gasoline yield from 19 to 47% for every barrel of crude oil. Gasoline is therefore a sophisticated blend of straight-chain alkanes, branched alkanes, and aromatic hydrocarbons. The precise blend is dependent on a number of conditions. In colder climates, for example, the blend must be appropriate for temperatures below zero. Therefore gasoline used in Chicago is not the same as the gasoline used in Houston.

Petroleum is not a renewable energy source. At our current rate of consumption it is estimated that Earth's supply of petroleum will be exhausted by 2060. We may find more petroleum deposits, but that will just delay the inevitable. Petroleum is also the primary source of a wide variety of organic compounds used for making plastics, pharmaceuticals, and numerous other products. It is vital that Earth's supply of petroleum not be completely exhausted, although such a dire picture is unlikely. As the supply of petroleum dwindles and the demand increases, the price of crude oil will rise (and so will the price of gasoline at the pump). Eventually, the price of petroleum will surpass the price of alternative energy sources, at which point a major shift will occur. What do you think will replace petroleum as the next global energy source?

## practically speaking — An Introduction to Polymers

Table 4.5 reveals that low-molecular-weight alkanes (such as methane or ethane) are gases at room temperature, alkanes of slightly higher molecular weight (such as hexane and octane) are liquids at room temperature, and alkanes of very high molecular weight (such as hectane, with 100 carbon atoms) are solids at room temperature. This trend is explained by the increased London dispersion forces experienced by higher molecular weight alkanes (as described in Section 1.12). With this in mind, consider an alkane composed of approximately 100,000 carbon atoms. It should not be surprising that such an alkane should be a solid at room temperature. This material, called polyethylene, is used for a variety of purposes, including garbage containers, plastic bottles, packaging material, bulletproof vests, and toys. As its name implies, polyethylene is produced from polymerization of ethylene:

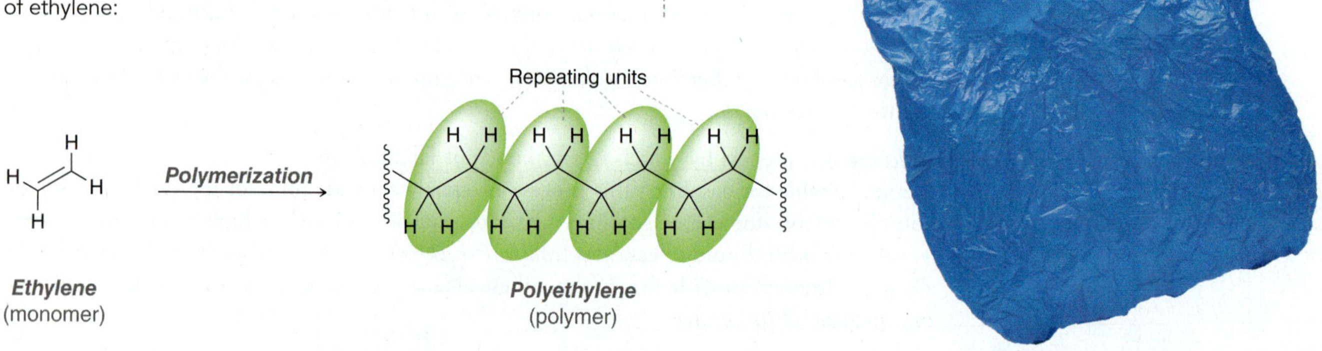


Polyethylene is an example of a *polymer*, because it is created by the joining of small molecules called *monomers*. Over 100 billion pounds of polyethylene are produced worldwide each year.

In our everyday lives, we are surrounded by a variety of polymers. From carpet fibers to plumbing pipes, our society has clearly become dependent on polymers. Polymers are discussed in more detail in later chapters.

## 4.6 Drawing Newman Projections

We will now turn our attention to the way in which molecules change their shape with time. Rotation about C—C single bonds allows a compound to adopt a variety of possible three-dimensional shapes, called **conformations**. Some conformations are higher in energy, while others are lower in energy. In order to draw and compare conformations, we will need to use a new kind of drawing—one specially designed for showing the conformation of a molecule. This type of drawing is called a **Newman projection** (Figure 4.2). To understand what a Newman projection represents, consider the wedge and dash drawing of ethane in Figure 4.2. Begin rotating it about the vertical axis drawn in gray so that all of the red H's come out in front of the page and all of the blue H's go back behind the page.

**FIGURE 4.2** Three drawings of ethane: (a) wedge and dash, (b) sawhorse, and (c) a Newman projection.

Wedge and dash | Sawhorse | Newman projection

Back carbon | Front carbon

**FIGURE 4.3** A Newman projection of ethane, showing the front carbon and the back carbon.

The second drawing (the sawhorse) represents a snapshot after 45° of rotation, while the Newman projection represents a snapshot after 90° of rotation. One carbon is directly in front of the other, and each carbon atom has three H's attached to it (Figure 4.3). The point at the center of the drawing in Figure 4.3 represents the front carbon atom, while the circle represents the back carbon. We will use Newman projections extensively throughout the rest of this chapter, so it is important to master both drawing and reading them.

# SKILLBUILDER

## 4.7 DRAWING NEWMAN PROJECTIONS

**LEARN the skill** Draw a Newman projection of the following compound, as viewed from the angle indicated:

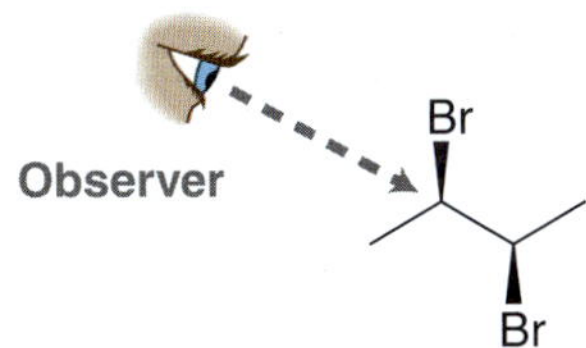

### SOLUTION

Identify the front and back carbon atoms. From the angle of the observer, the front and back carbon atoms are:

Front
Back
Observer
Br
Br

Now we must ask: From the perspective of the observer, what is connected to the front carbon? The observer will see a methyl group, a bromine, and a hydrogen atom. Remember that a wedge is coming out of the page, and a dash is going back behind the page. So, from the perspective of the observer, the front carbon atom looks like this:

**STEP 1** Identify the three groups connected to the front carbon atom.

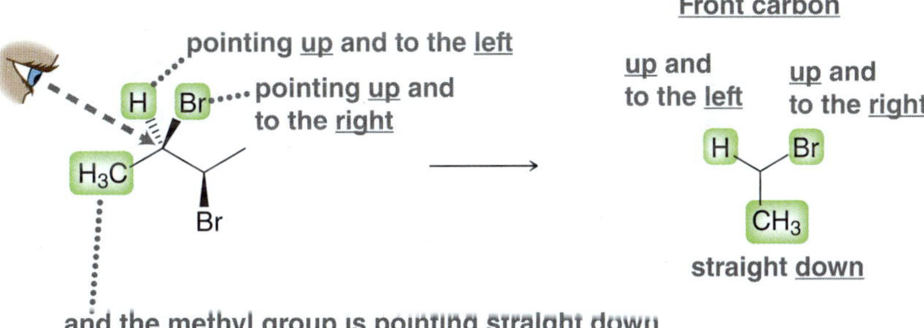


Now let's focus our attention on the back carbon atom. The back carbon has one $CH_3$ group, one Br, and one H. From the perspective of the observer, it looks like this:

**STEP 2** Identify the three groups connected to the back carbon atom.

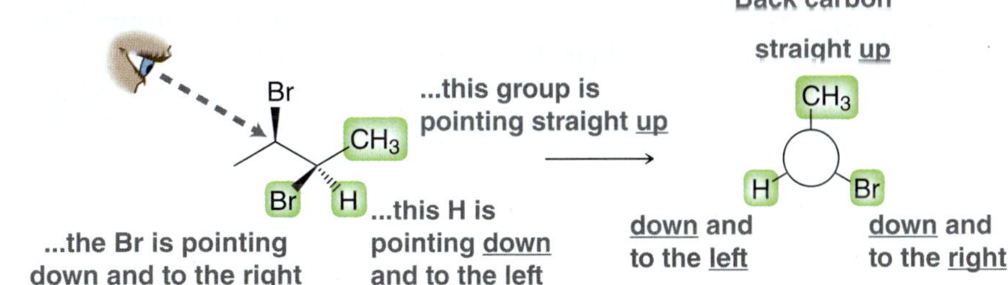


**STEP 3** Draw the Newman projection.

Now we put both pieces of our drawing together:

CH3
H Br
H Br
CH3

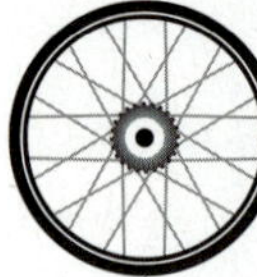

**PRACTICE the skill** **4.16** In each case below, draw a Newman projection as viewed from the angle indicated:

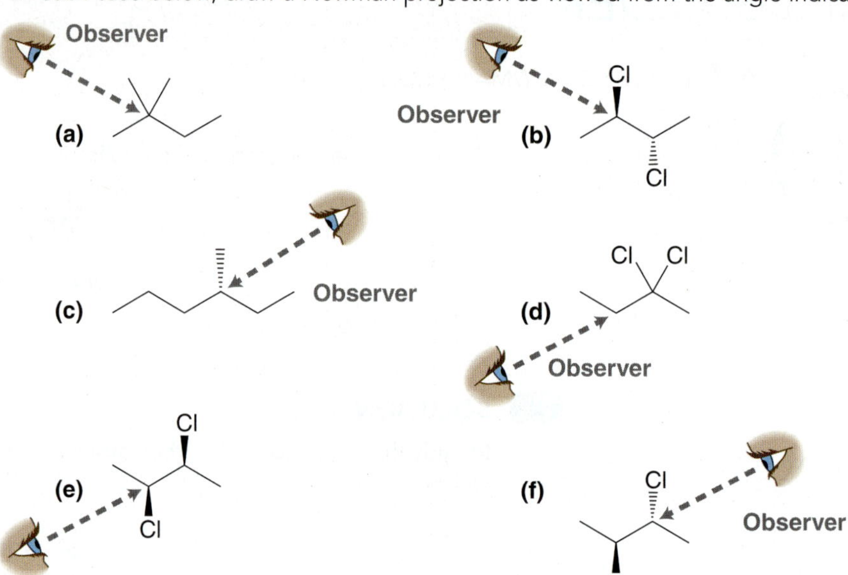

**APPLY the skill** **4.17** Draw a bond-line structure for each of the following compounds:

(a) (b) (c)

**4.18** Determine whether the following compounds are constitutional isomers:

need more **PRACTICE?** **Try Problem 4.56**

## 4.7 Conformational Analysis of Ethane and Propane

Dihedral angle = 60°

**FIGURE 4.4** The dihedral angle between two hydrogen atoms in a Newman projection of ethane.

Consider the two hydrogen atoms shown in red in the Newman projection of ethane (Figure 4.4). These two hydrogen atoms appear to be separated by an angle of 60°. This angle is called the **dihedral angle** or **torsional angle**. This dihedral angle changes as the C—C bond rotates—for example, if the front carbon rotates clockwise while the back carbon is held stationary. The value for the dihedral angle between two groups can be any value between 0° and 180°. Therefore, there are an infinite number of possible conformations. Nevertheless, there are two conformations that require our special attention: the lowest energy conformation and the highest energy conformation (Figure 4.5). The **staggered conformation** is the lowest in energy, while the **eclipsed conformation** is the highest in energy.

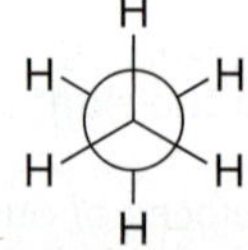

**Staggered conformation**
**Lowest in energy**

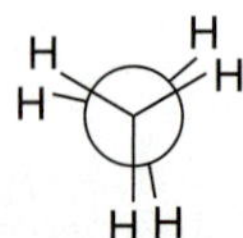

**Eclipsed conformation**
**Highest in energy**

**FIGURE 4.5** Staggered and eclipsed conformations of ethane.

The difference in energy between staggered and eclipsed conformations of ethane is 12 kJ/mol, as shown in the energy diagram in Figure 4.6. Notice that all staggered conformations of ethane are **degenerate**; that is, all of the staggered conformations have the same amount of energy. Similarly, all eclipsed conformations of ethane are degenerate.

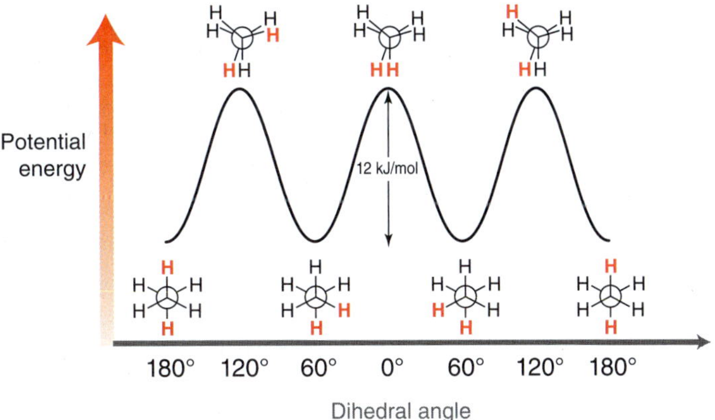


**FIGURE 4.6** An energy diagram showing the conformational analysis of ethane.

**LOOKING BACK**
For a review of bonding and antibonding molecular orbitals, see Section 1.8.

The difference in energy between staggered and eclipsed conformations of ethane is referred to as **torsional strain**, and its cause has been somewhat debated over the years. Based on recent quantum mechanical calculations, it is now believed that the staggered conformation possesses a favorable interaction between an occupied, bonding MO and an unoccupied, antibonding MO (Figure 4.7).

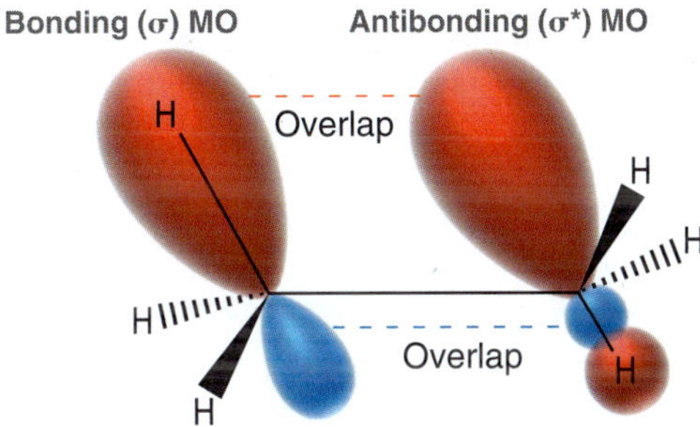


**FIGURE 4.7** In the staggered conformation, favorable overlap occurs between a bonding MO and an antibonding MO.

This interaction lowers the energy of the staggered conformation. This favorable interaction is only present in the staggered conformation. When the C—C bond is rotated (going from a staggered to an eclipsed conformation), the favorable overlap above is temporarily disrupted, causing an increase in energy. In ethane, this increase amounts to 12 kJ/mol. Since there are three separate eclipsing interactions, it is reasonable to assign 4 kJ/mol to each pair of eclipsing H's (Figure 4.8).

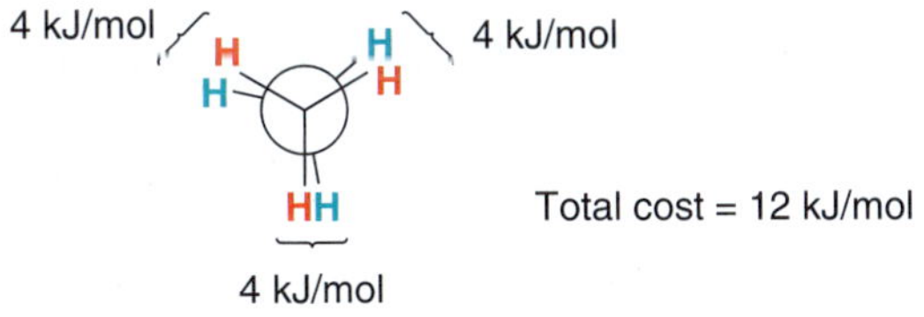

4 kJ/mol 4 kJ/mol

Total cost = 12 kJ/mol

4 kJ/mol

**FIGURE 4.8** The total energy cost associated with the eclipsed conformation of ethane (relative to the staggered conformation) amounts to 12 kJ/mol.

This energy difference is significant. At room temperature, a sample of ethane gas will have approximately 99% of its molecules in staggered conformations at any given instant.

The energy diagram of propane (Figure 4.9) is very similar to that of ethane, except that the torsional strain is 14 kJ/mol rather than 12 kJ/mol. Once again, notice that all staggered conformations are degenerate, as are all eclipsed conformations.

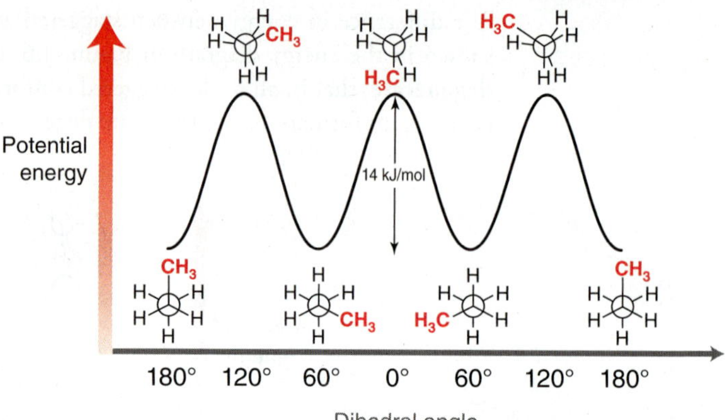


**FIGURE 4.9** An energy diagram showing the conformational analysis of propane.

We already assigned 4 kJ/mol to each pair of eclipsing H's. If we know that the torsional strain of propane is 14 kJ/mol, then it is reasonable to assign 6 kJ/mol to the eclipsing of an H and a methyl group. This calculation is illustrated in Figure 4.10.

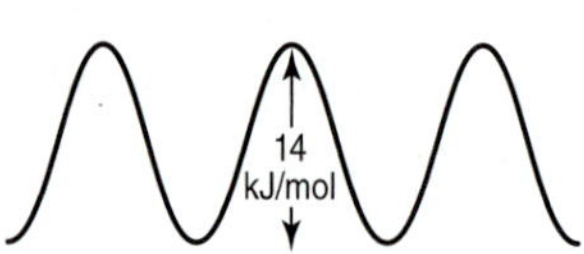


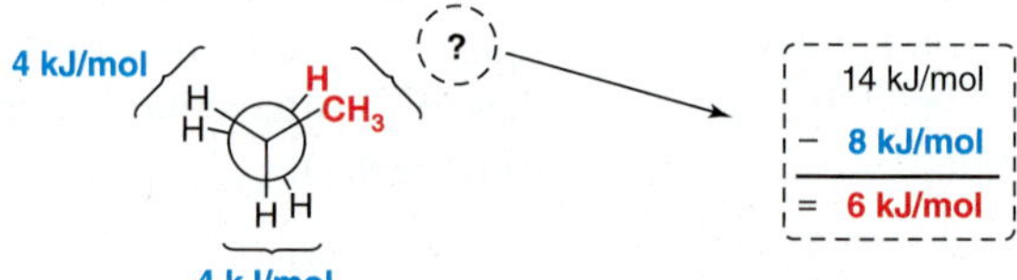


If the total energy cost is 14 kJ/mol...

...and we already know that each pair of eclipsing H's has an energy cost of 4 kJ/mol...

...then we can conclude that the energy cost of an H eclipsing a $CH_3$ group must be 6 kJ/mol

**FIGURE 4.10** The energy cost associated with a methyl group eclipsing a hydrogen atom amounts to 6 kJ/mol.

## CONCEPTUAL CHECKPOINT

**4.19** For each of the following compounds, predict the energy barrier to rotation (looking down any one of the C—C bonds). Draw a Newman projection and then compare the staggered and eclipsed conformations. Remember that we assigned 4 kJ/mol to each pair of eclipsing H's and 6 kJ/mol to an H eclipsing a methyl group:

(a) 2,2-Dimethylpropane (b) 2-Methylpropane

## 4.8 Conformational Analysis of Butane

Conformational analysis of butane is a bit more complex than the conformational analysis of either ethane or propane. Look carefully at the shape of the energy diagram for butane (Figure 4.11), and then we will analyze it step by step.

The three highest energy conformations are the eclipsed conformations, while the three lowest energy conformations are the staggered conformations. In this way, the energy diagram above is similar to the energy diagrams of ethane and propane. But in the case of butane, notice that one eclipsed conformation (where dihedral angle = 0) is higher in energy than the other two eclipsed conformations. In other words, the three eclipsed conformations are not degenerate. Similarly, one staggered conformation (where dihedral angle = 180°) is lower in energy than the other two staggered conformations. Clearly, we need to compare the staggered conformations to each other, and we need to compare the eclipsed conformations to each other.

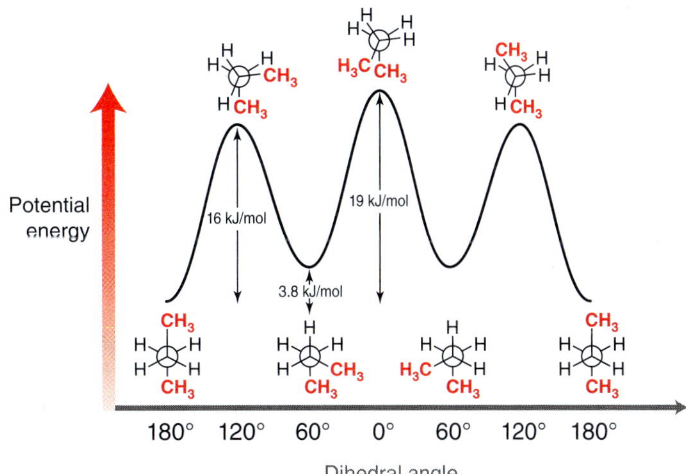


**FIGURE 4.11** An energy diagram showing the conformational analysis of butane.

Let's begin with the three staggered conformations. The conformation with a dihedral angle of 180° is called the ***anti*** **conformation**, and it represents the lowest energy conformation of butane. The other two staggered conformations are 3.8 kJ/mol higher in energy than the *anti* conformation. Why? We can more easily see the answer to this question by drawing Newman projections of all three staggered conformations (Figure 4.12).

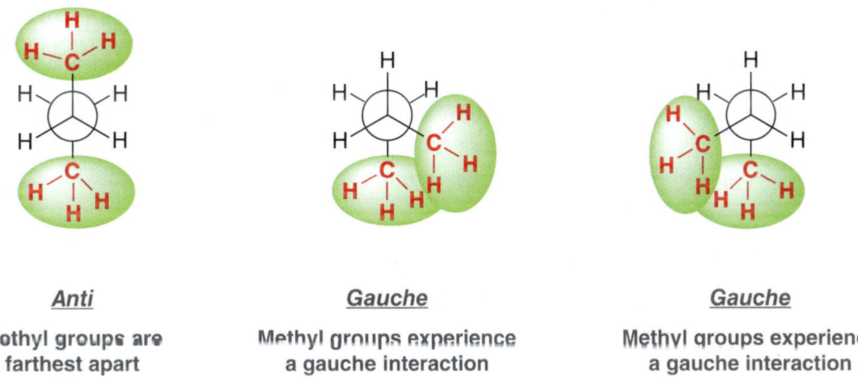


**FIGURE 4.12** Two of the three staggered conformations of butane exhibit *gauche* interactions.

In the *anti* conformation, the methyl groups achieve maximum separation from each other. In the other two conformations, the methyl groups are closer to each other. Their electron clouds are repelling each other (trying to occupy the same region of space), causing an increase in energy of 3.8 kJ/mol. This unfavorable interaction, called a ***gauche*** **interaction**, is a type of *steric* interaction, and it is different from the concept of torsional strain. The two conformations above that exhibit this interaction are called ***gauche*** **conformations**, and they are degenerate (Figure 4.13).

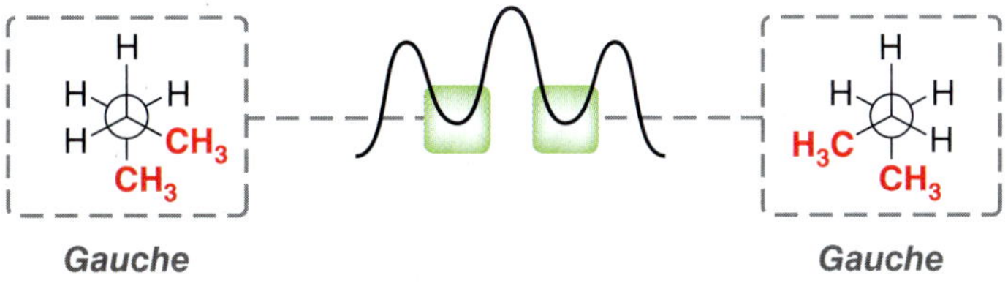

**FIGURE 4.13** The two staggered conformations that exhibit *gauche* interactions are degenerate.

Now let's turn our attention to the three eclipsed conformations. One eclipsed conformation is higher in energy than the other two. Why? In the highest energy conformation, the methyl groups are eclipsing each other. Experiments suggest that this conformation has a total energy cost of 19 kJ/mol. Since we already assigned 4 kJ/mol to each H—H eclipsing interaction, it is reasonable to assign 11 kJ/mol to the eclipsing interaction of two methyl groups. This calculation is illustrated in Figure 4.14. The conformation with the two methyl groups eclipsing each other is the highest energy conformation. The other two eclipsed conformations are degenerate (Figure 4.15).

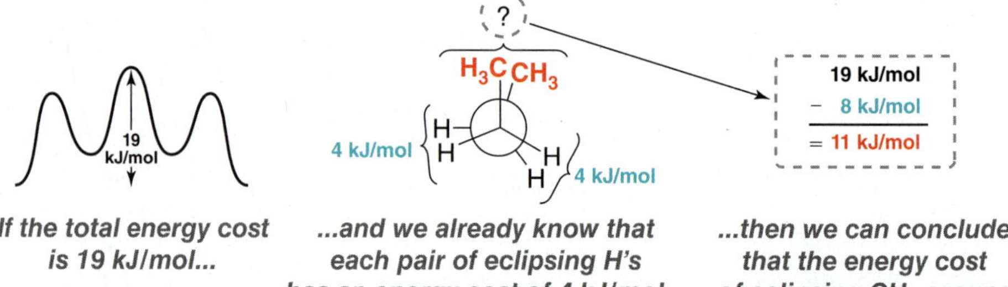


**FIGURE 4.14** The energy cost associated with two methyl groups eclipsing each other amounts to 11 kJ/mol.

**FIGURE 4.15** Two of the eclipsed conformations of butane are degenerate.

4 kJ/mol
6 kJ/mol
6 kJ/mol
Total cost = 16 kJ/mol

**FIGURE 4.16** The total energy cost associated with the degenerate eclipsed conformations of butane amounts to 16 kJ/mol.

In each case, there is one pair of eclipsing H's and two pairs of eclipsing $H/CH_3$. We have all the information necessary to calculate the energy of these conformations. We know that eclipsing H's are 4 kJ/mol, and each set of eclipsing $H/CH_3$ is 6 kJ/mol. Therefore, we calculate a total energy cost of 16 kJ/mol (Figure 4.16).

To summarize, we have seen just a few numbers that can be helpful in analyzing energy costs. With these numbers, it is possible to analyze an eclipsed conformation or a staggered conformation and determine the energy cost associated with each conformation. Table 4.6 summarizes these numbers.

**TABLE 4.6 ENERGY COSTS FOR COMPARING THE RELATIVE ENERGY OF CONFORMATIONS**

| INTERACTION | TYPE OF STRAIN | ENERGY COST (KJ/MOL) |
|---|---|---|
| H/H Eclipsed | Torsional strain | 4 |
| $CH_3$/H Eclipsed | Torsional strain | 6 |
| $CH_3/CH_3$ Eclipsed | Torsional strain + steric interaction | 11 |
| $CH_3/CH_3$ *Gauche* | Steric interaction | 3.8 |

# SKILLBUILDER

## 4.8 IDENTIFYING RELATIVE ENERGY OF CONFORMATIONS

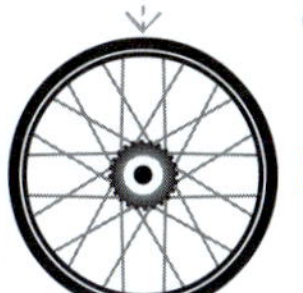

**LEARN the skill**

Consider the following compound:

**(a)** Rotating only the C3—C4 bond, identify the lowest energy conformation.

**(b)** Rotating only the C3—C4 bond, identify the highest energy conformation.

**SOLUTION**

**STEP 1** Draw a Newman projection.

**(a)** Begin by drawing a Newman projection, looking along the C3—C4 bond:

Observer → $CH_2CH_3$, H, H, H, $CH_3$, $CH_2CH_3$ ≡ Et, H, H, H, Me, Et

**BY THE WAY** The symbol "Et" is commonly used for an ethyl group and the symbol "Me" is used for a methyl group.

To determine the lowest energy conformation, compare all three staggered conformations. To draw them, we can either rotate the groups on the back carbon or rotate the groups on the front carbon. It will be easier to rotate the back carbon since the back carbon has only one group. It is easier to keep track of only one group. Notice that the difference between these three conformations is the position of the ethyl group on the back carbon:

**STEP 2** Compare all three staggered conformations. Look for the fewest or least severe *gauche* interactions.

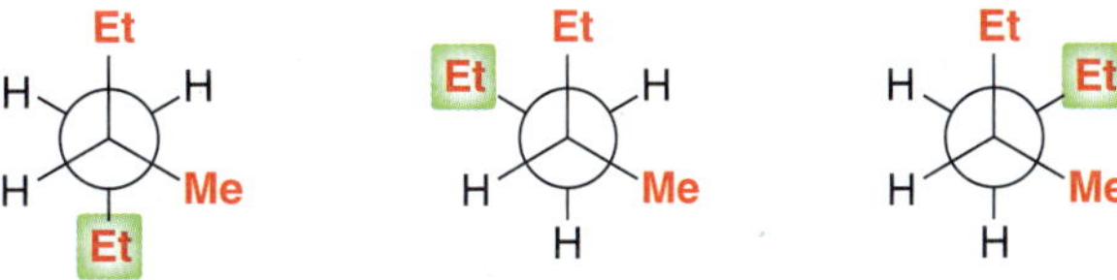


Now compare these three conformations by looking for *gauche* interactions. The first conformation has one Et/Me *gauche* interaction. The second conformation has one Et/Et *gauche* interaction. The third conformation has two *gauche* interactions: one Et/Et interaction and one Et/Me interaction. Choose the one with the fewest and least severe *gauche* interactions. The first conformation (with one Et/Me interaction) will be the lowest in energy.

**(b)** To determine the highest energy conformation, we will need to compare all three *eclipsed* conformations. To draw them, simply take the three staggered conformations and turn each of them into an eclipsed conformation by rotating the back carbon 60°. Notice once again that the difference between these three conformations is the position of the Et group on the back carbon:

**STEP 3** Compare all three eclipsed conformations. Look for the highest energy interactions.

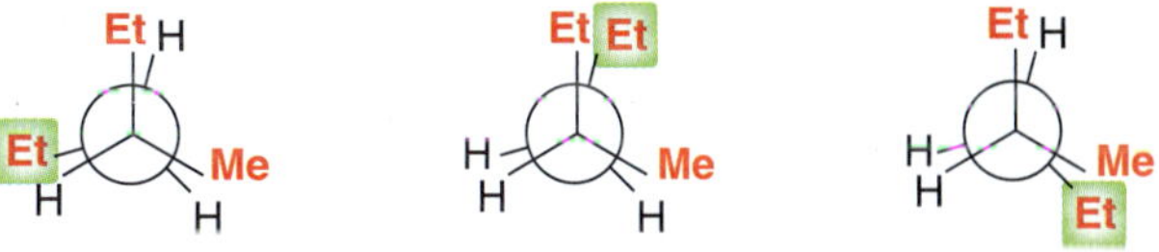


Now compare these three conformations by looking for eclipsing interactions. In the first conformation, none of the alkyl groups are eclipsing each other (they are all eclipsed by H's). In the second conformation, the ethyl groups are eclipsing each other. In the third conformation, a methyl and ethyl are eclipsing each other. Of the three possibilities, the highest energy conformation will be the one in which the two ethyl groups are eclipsing each other (the second conformation).

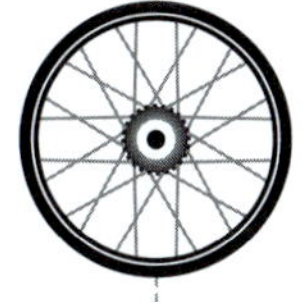

**PRACTICE the skill**

**4.20** In each case below, identify the highest and lowest energy conformations. In cases where two or three conformations are degenerate, draw only one as your answer.

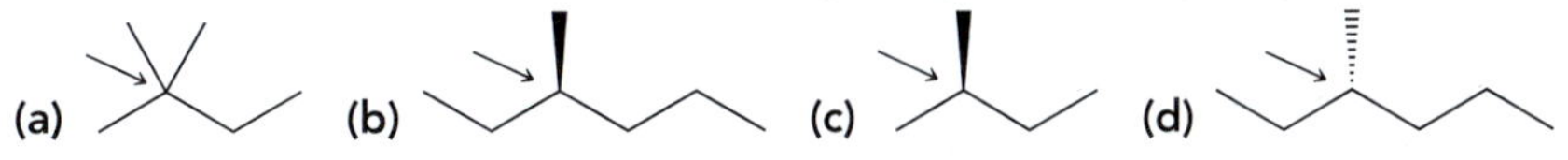

**APPLY the skill**

**4.21** Compare the three staggered conformations of ethylene glycol. The *anti* conformation of ethylene glycol is not the lowest energy conformation. The other two staggered conformations are actually lower in energy than the *anti* conformation. Suggest an explanation.

HO–CH2–CH2–OH
**Ethylene glycol**

need more **PRACTICE?** Try Problems 4.47, 4.59

## medically speaking — Drugs and Their Conformations

Recall from Chapter 2 that a drug will bind with a biological receptor if the drug possesses a specific three-dimensional arrangement of functional groups, called a pharmacophore. For example, the pharmacophore of morphine is shown in red:

**Morphine**

Morphine is a very rigid molecule, because it has very few bonds that undergo free rotation. As a result, the pharmacophore is locked in place. In contrast, flexible molecules are capable of adopting a variety of conformations, and only some of those conformations can bind to the receptor. For example, methadone has many single bonds, each of which undergoes free rotation:

**Methadone**

Methadone is used to treat heroin addicts suffering from withdrawal symptoms. Methadone binds to the same receptor as heroin, and it is widely believed that the active conformation is the one in which the position of the functional groups matches the pharmacophore of heroin (and morphine):

**Methadone** **Heroin**

Other, more open conformations of methadone are probably incapable of binding to the receptor.

This explains how it is possible for one drug to produce several physiological effects. In many cases, one conformation binds to one receptor, while another conformation binds to an entirely different receptor. Conformational flexibility is therefore an important consideration in the study of how drugs behave in our bodies.

As mentioned in the chapter opener, conformational flexibility has recently received much attention in the design of novel compounds to treat viral infections. To treat the symptoms of a virus, we must study the structure and behavior of that particular virus and then design drugs that interfere with the key steps in the replication process for that virus.

The vast majority of antiviral research has focused on designing drugs to treat those viral infections that are life threatening, such as HIV. Many anti-HIV drugs have been developed over the last few decades. Nevertheless, these drugs are not 100% effective, because HIV can undergo genetic mutations that effectively change the geometry of the cavity where the drugs are supposed to bind. The new strain of the virus is then drug resistant, because the drugs cannot bind with their intended receptors.

A new class of compounds, exhibiting conformational flexibility, appears to evade the problem of drug resistance. One such example, called rilpivirine, exhibits five single bonds whose rotation would lead to a conformational change:

**Rilpivirine**

Bonds shown in red can undergo rotation without a significant energy cost, rendering the compound very flexible. The flexibility of rilpivirine enables it to bind to the desired receptor and tolerate changes to the geometry of the cavity resulting from virus mutation. In this way, rilpivirine makes it more difficult for the virus to develop resistance to it.

Rilpivirine was approved by the FDA (Food and Drug Administration) in 2011. Compounds like rilpivirine have changed the way scientists approach drug design of antiviral agents. It is now clear that conformational flexibility plays an important role in the design of effective drugs.

## 4.9 Cycloalkanes

In the nineteenth century, chemists were aware of many compounds containing five-membered rings and six-membered rings, but no compounds with smaller rings were known. Many unfruitful attempts to synthesize smaller or larger rings fueled speculation regarding the feasibility of ever creating such compounds. Toward the end of the nineteenth century Adolph von Baeyer proposed a theory describing cycloalkanes in terms of **angle strain**, the increase in energy associated with a bond angle that has deviated from the preferred angle of 109.5°. Baeyer's theory was based on the angles found in geometric shapes (Figure 4.17). Baeyer reasoned that five-membered rings should contain almost no angle strain, while other rings would be strained (both smaller rings and larger rings). He also reasoned that very large cycloalkanes cannot exist, because the angle strain associated with such large bond angles would be prohibitive.

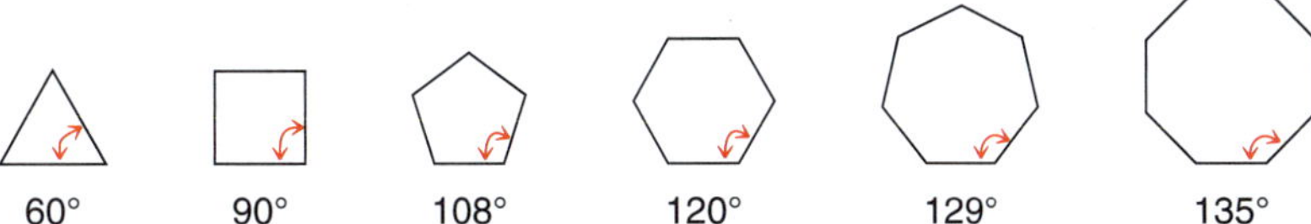


**FIGURE 4.17**
Bond angles found in geometric shapes.

Evidence refuting Baeyer's conclusions came from thermodynamic experiments. Recall from earlier in this chapter that heats of combustion can be used to compare isomeric compounds in terms of their total energy. It is not fair to compare the heats of combustion for rings of different sizes since heats of combustion are expected to increase with each additional $CH_2$ group. We can more accurately compare rings of different sizes by dividing the heat of combustion by the number of $CH_2$ groups in the compound, giving a heat of combustion per $CH_2$ group. Table 4.7 shows heats of combustion per $CH_2$ group for various ring sizes. The conclusions from these data are more easily seen when plotted (Figure 4.18). Notice that a six-membered ring is lower in energy than a five-membered ring, in contrast with Baeyer's theory. In addition, the relative energy level does not increase with increasing ring size, as Baeyer predicted. A 12-membered ring is in fact much lower in energy than an 11-membered ring.

**TABLE 4.7 HEATS OF COMBUSTION PER $CH_2$ GROUP FOR CYCLOALKANES**

| CYCLOALKANE | NUMBER OF $CH_2$ GROUPS | HEAT OF COMBUSTION (KJ / MOL) | HEAT OF COMBUSTION PER $CH_2$ GROUP (KJ / MOL) |
|---|---|---|---|
| Cyclopropane | 3 | 2091 | 697 |
| Cyclobutane | 4 | 2721 | 680 |
| Cyclopentane | 5 | 3291 | 658 |
| Cyclohexane | 6 | 3920 | 653 |
| Cycloheptane | 7 | 4599 | 657 |
| Cyclooctane | 8 | 5267 | 658 |
| Cyclononane | 9 | 5933 | 659 |
| Cyclodecane | 10 | 6587 | 659 |
| Cycloundecane | 11 | 7273 | 661 |
| Cyclododecane | 12 | 7845 | 654 |

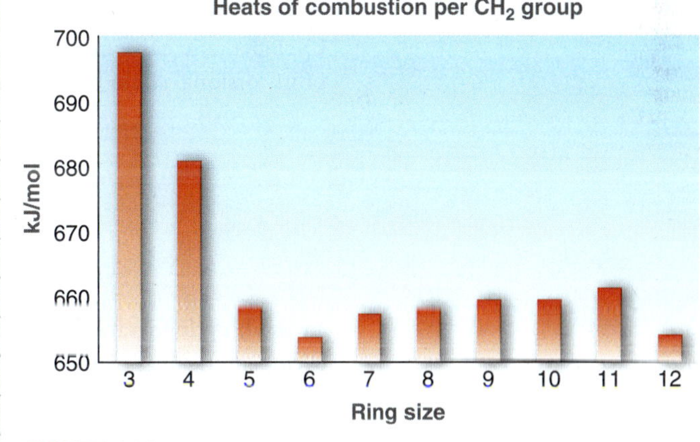


**FIGURE 4.18**
Heats of combustion per $CH_2$ group for cycloalkanes.

Baeyer's conclusions did not hold because they were based on the incorrect assumption that cycloalkanes are planar, like the geometric shapes shown earlier. In reality, the bonds of a larger cycloalkane can position themselves three dimensionally so as to achieve a conformation that minimizes the total energy of the compound. We will soon see that angle strain is only one factor that contributes to the energy of a cycloalkane. We will now explore the main factors contributing to the energy of various ring sizes, starting with cyclopropane.

### Cyclopropane

The angle strain in cyclopropane is severe. Some of this strain can be alleviated if the orbitals making up the bonds bend outward, as in Figure 4.19. Not all of the angle strain is removed, however, because there is an increase in energy associated with inefficient overlap of the orbitals. Although some of the angle strain is reduced, cyclopropane still has significant angle strain.

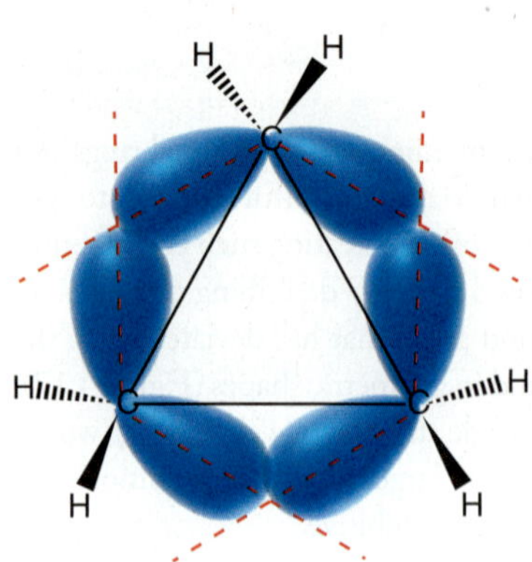

**FIGURE 4.19** The C—C bonds of cyclopropane bend outward (on the dotted red lines) to alleviate some of the angle strain.

In addition, cyclopropane also exhibits significant torsional strain, which can best be seen in a Newman projection:

Notice that the ring is locked in an eclipsed conformation, with no possible way of achieving a staggered conformation.

In summary, cyclopropane has two main factors contributing to its high energy: angle strain (from small bond angles) and torsional strain (from eclipsing H's). This large amount of strain makes three-membered rings highly reactive and very susceptible to ring-opening reactions. In Chapter 14, we will explore many ring-opening reactions of a special class of three-membered rings called epoxides:

**An epoxide**

## Cyclobutane

Cyclobutane has less angle strain than cyclopropane. However, it has more torsional strain, because there are four sets of eclipsing H's rather than just three. To alleviate some of this additional torsional strain, cyclobutane can adopt a slightly puckered conformation without gaining too much angle strain:

88°

## Cyclopentane

Cyclopentane has much less angle strain than cyclobutane or cyclopropane. It can also reduce much of its torsional strain by adopting the following conformation:

In total, cyclopentane has much less total strain than cyclopropane or cyclobutane. Nevertheless, cyclopentane does exhibit some strain. This is in contrast with cyclohexane, which can adopt a conformation that is nearly strain free. We will spend the remainder of the chapter discussing conformations of cyclohexane.

**medically speaking** | **Cyclopropane as an Inhalation Anesthetic**

In the mid-1840s, surgeons began utilizing the anesthetic properties of diethyl ether and chloroform to anesthetize patients during surgery:

**Diethyl ether** **Chloroform**

Despite the known risks and side effects associated with each of these anesthetic agents, they were commonly used for nearly 100 years, because a suitable alternative was not yet available. Diethyl ether is highly flammable, it is irritating to the respiratory tract, and it causes nausea and vomiting (which can lead to serious lung damage in an unconscious patient). In addition, the onset and recovery from anesthesia are slow. Chloroform is nonflammable, which was greatly appreciated by anesthesiologists and surgeons, but it caused a substantial incidence of heart arrhythmias, sometimes fatal. It also caused dangerous drops in blood pressure and liver damage with prolonged exposure.

With time, other anesthetic agents were discovered. One such example is cyclopropane, which became commercially available as an inhalation anesthetic in the mid-1930s:

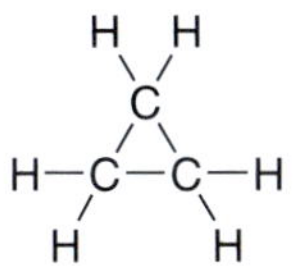

It did not induce vomiting, did not cause liver damage, provided rapid onset and recovery from anesthesia, and maintained normal blood pressure. Its major shortcoming was its instability due to the ring strain associated with the three-membered ring. Even a static spark could trigger an explosion, with disastrous consequences for the patient. Anesthesiologists took meticulous measures to prevent explosions, and accidents were very rare, but the use of cyclopropane was not for the faint-hearted. It was largely replaced in the 1960s when other inhalation anesthetics were discovered. The inhalation anesthetics most commonly used today are discussed in a medically speaking application in Section 14.3.

## 4.10 Conformations of Cyclohexane

Cyclohexane can adopt many conformations, as we will soon see. For now, we will explore two conformations: the **chair conformation** and the **boat conformation** (Figure 4.20). In both conformations, the bond angles are fairly close to 109.5°, and therefore, both conformations possess very little angle strain. The significant difference between them can be seen when comparing torsional strain. The chair conformation has no torsional strain. This can best be seen with a Newman projection (Figure 4.21). Notice that all H's are staggered. None are eclipsed. This is not the case in a boat conformation, which has two sources of torsional strain (Figure 4.22). Many of the H's are eclipsed (Figure 4.22a), and the H's on either side of the ring experience steric interactions called **flagpole interactions,** as shown in Figure 4.22b. The boat can alleviate some of this torsional strain by twisting (very much the way cyclobutane puckers to alleviate some of its torsional strain), giving a conformation called a **twist boat** (Figure 4.23).

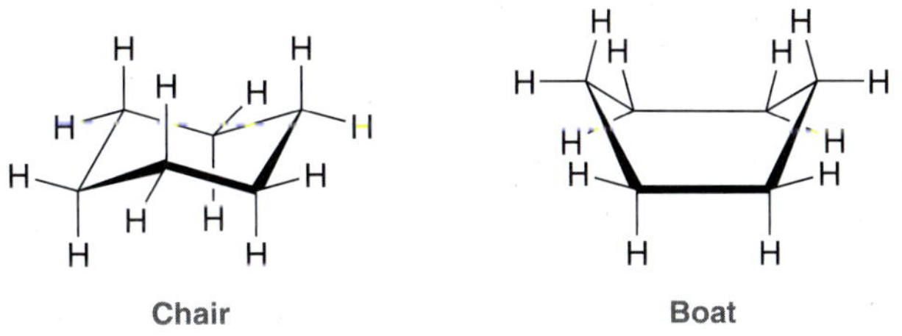

**FIGURE 4.20** The chair and boat conformations of cyclohexane.

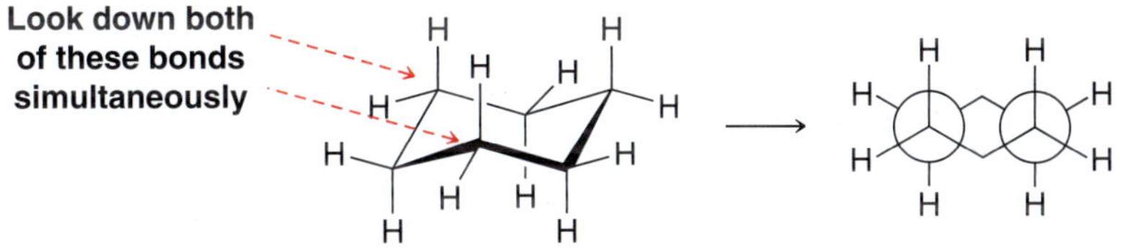

**FIGURE 4.21** A Newman projection of cyclohexane in a chair conformation.

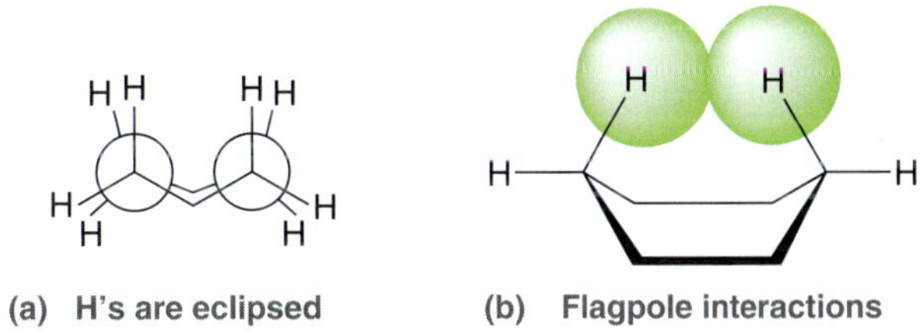

**FIGURE 4.22** (a) A Newman projection of cyclohexane in a boat conformation. (b) Flagpole interactions in the boat conformation.

**FIGURE 4.23** The twist boat conformation of cyclohexane.

In fact, cyclohexane can adopt many different conformations, but the most important is the chair conformation. There are actually two different chair conformations that rapidly interchange via a pathway that passes through many different conformations, including a high-energy half-chair conformation, as well as twist boat and boat conformations. This is illustrated in Figure 4.24, which is an energy diagram summarizing the relative energy levels of the various conformations of cyclohexane.

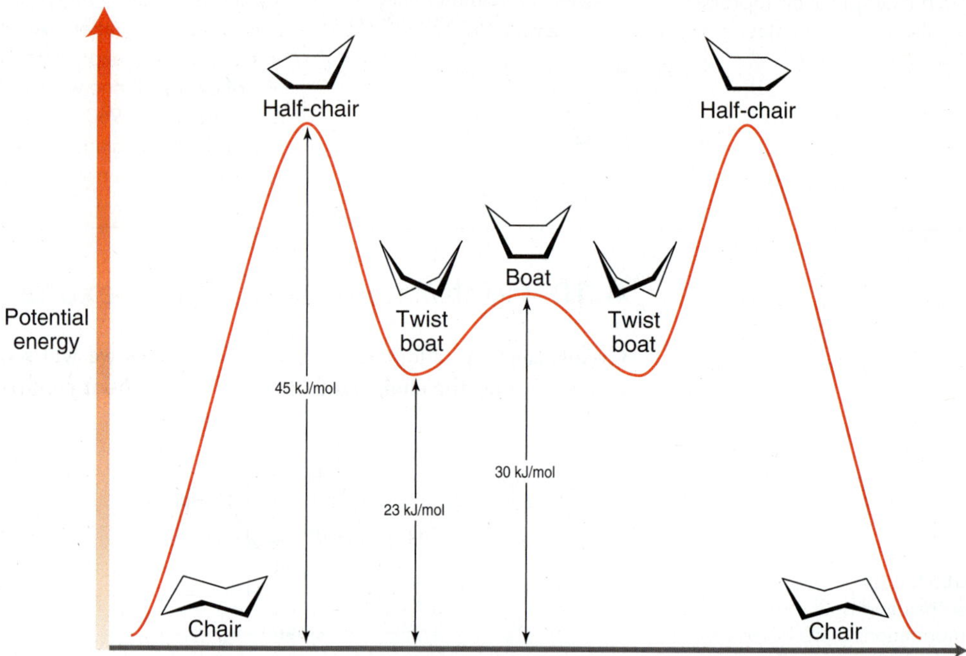


**FIGURE 4.24** An energy diagram showing the conformational analysis of cyclohexane.

The lowest energy conformations are the two chair conformations, and therefore, cyclohexane will spend the majority of its time in a chair conformation. Accordingly, the remainder of our treatment of cyclohexane will focus on chair conformations. Our first step is to master drawing them.

## 4.11 Drawing Chair Conformations

When drawing a chair conformation, it is important to draw it precisely. Make sure that you avoid drawing sloppy chairs, because it will be difficult to draw the substituents correctly if the skeleton is not precise.

### Drawing the Skeleton of a Chair Conformation

Let's get some practice drawing chairs.

## SKILLBUILDER

### 4.9 DRAWING A CHAIR CONFORMATION

**LEARN the skill**

Draw a chair conformation of cyclohexane:

**SOLUTION**

The following procedure outlines a step-by-step method for drawing the skeleton of a chair conformation precisely:

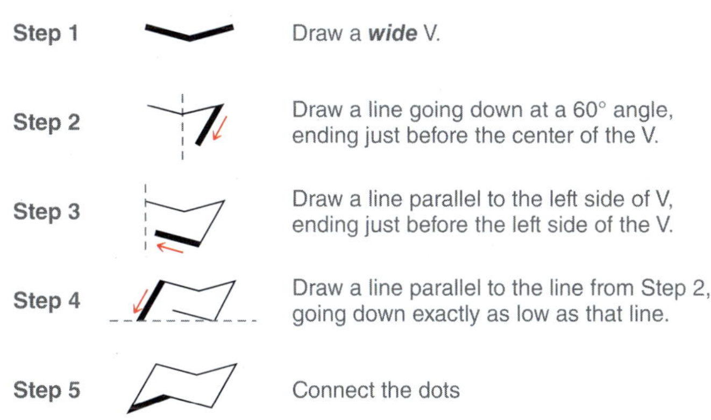


When you are finished drawing a chair, it should contain three sets of parallel lines. If your chair does not contain three sets of parallel lines, then it has been drawn incorrectly.

**PRACTICE the skill** **4.22** Practice drawing a chair several times using a blank piece of paper. Repeat the procedure until you can do it without looking at the instructions above. For each of your chairs, make sure that it contains three sets of parallel lines.

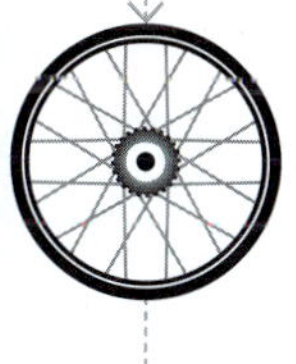

**APPLY the skill** **4.23** Draw a chair conformation for each of the following compounds:

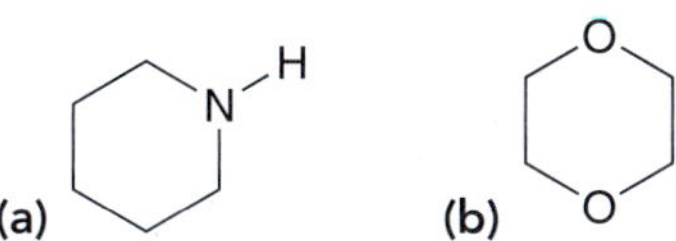


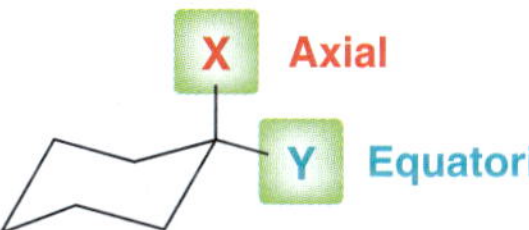


**FIGURE 4.25** Axial and equatorial positions in a chair conformation.

## Drawing Axial and Equatorial Substituents

Each carbon atom in a cyclohexane ring can bear two substituents (Figure 4.25). One group is said to occupy an **axial position**, which is parallel to a vertical axis passing through the center of the ring. The other group is said to occupy an **equatorial position**, which is positioned approximately along the equator of the ring (Figure 4.25). In order to draw a substituted cyclohexane, we must first practice drawing all axial and equatorial positions properly.

# SKILLBUILDER

## 4.10 DRAWING AXIAL AND EQUATORIAL POSITIONS

**LEARN the skill** Draw all axial and all equatorial positions on a chair conformation of cyclohexane.

### SOLUTION

Let's begin with the axial positions, as they are easier to draw. Begin at the right side of the V and draw a vertical line pointing up. Then, go around the ring, drawing vertical lines, alternating in direction (up, down, up, etc.):

**STEP 1**
Draw all axial positions as vertical lines alternating in direction.

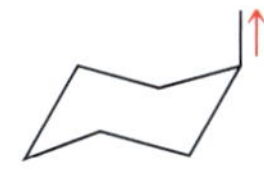

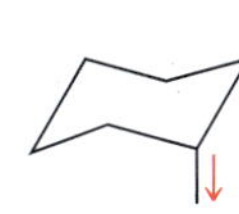

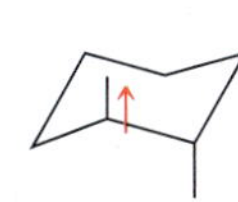

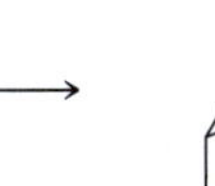

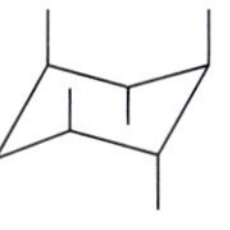

These are the six axial positions. All six lines are vertical.

Now let's draw the six equatorial positions. The equatorial positions are more difficult to draw properly, but mistakes can be avoided in the following way. We saw earlier that a properly drawn chair skeleton is composed of three pairs of parallel lines:

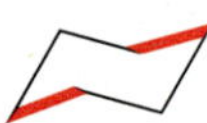

Now we will use these pairs of parallel lines to draw the equatorial positions. In between each pair of red lines above, draw two equatorial groups that are parallel to (but not directly touching) the red lines:

**STEP 2**
Draw all equatorial positions as pairs of parallel lines.

Notice that all equatorial positions are drawn going to the outside of, or away from, the ring, not going into the ring. Now let's summarize by drawing all six axial positions and all six equatorial positions:

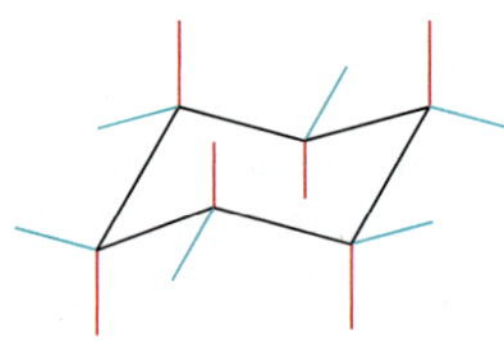

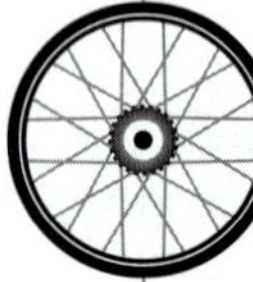

**PRACTICE the skill** **4.24** Practice drawing a chair conformation with all six axial positions. Repeat until you can draw all six positions without looking at the instructions above.

**4.25** Practice drawing a chair conformation with all six equatorial positions. Repeat until you can draw all six positions without looking at the instructions above.

**4.26** Practice drawing a chair conformation with all 12 positions (6 axial and 6 equatorial). Practice several times on a blank piece of paper. Repeat until you can draw all 12 positions without looking at the instructions above.

**APPLY the skill** **4.27** In the following compound, identify the number of hydrogen atoms that occupy axial positions as well as the number of hydrogen atoms that occupy equatorial positions:

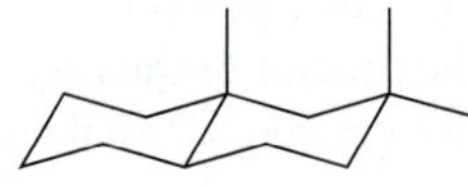

need more **PRACTICE?** **Try Problem 4.62**

## 4.12 Monosubstituted Cyclohexane

### Drawing Both Chair Conformations

Consider a ring containing only one substituent. Two possible chair conformations can be drawn: The substituent can be in an axial position or in an equatorial position. These two possibilities represent two different conformations that are in equilibrium with each other:

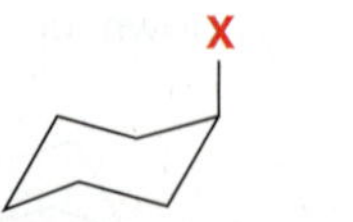


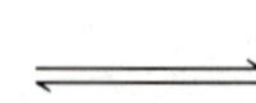

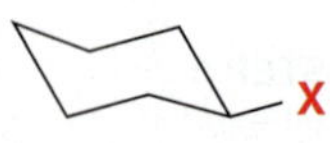


The term "ring flip" is used to describe the conversion of one chair conformation into the other. This process is not accomplished by simply flipping the molecule like a pancake. Rather, a **ring flip** is a conformational change that is accomplished only through a rotation of all C—C single bonds. This can be seen with a Newman projection (Figure 4.26). Let's get some practice drawing ring flips.

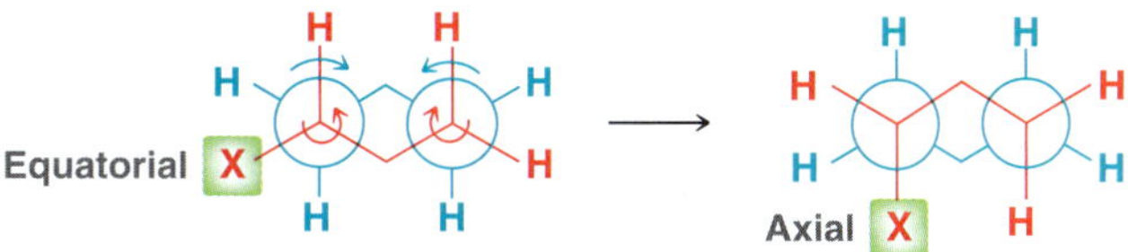


**FIGURE 4.26** A ring flip drawn with Newman projections.

## SKILLBUILDER

### 4.11 DRAWING BOTH CHAIR CONFORMATIONS OF A MONOSUBSTITUTED CYCLOHEXANE

**LEARN** the skill

Draw both chair conformations of bromocyclohexane:

Br

**SOLUTION**

**STEP 1** Draw a chair conformation.

**STEP 2** Place the substituent.

Begin by drawing the first chair conformation. Then, place the bromine in any position:

In this chair conformation, the bromine occupies an axial position. In order to draw the other chair conformation (the result of a ring flip), redraw the skeleton of the chair. Only this time, we will draw the skeleton differently. The first chair was drawn by following these steps:

The second chair must now be drawn by *mirroring* these steps:

**STEP 3** Draw a ring flip, and the axial group should become equatorial.

In the second chair, the bromine will occupy an equatorial position:

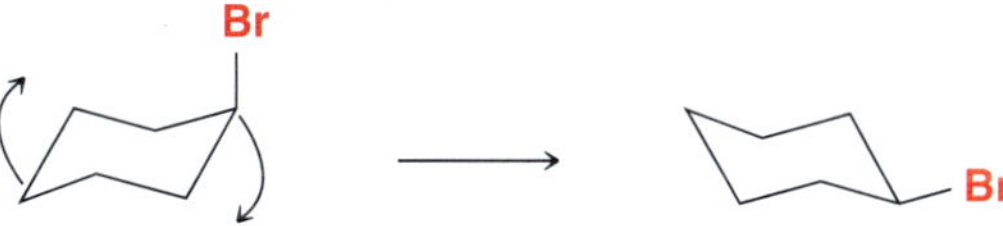


**PRACTICE** the skill

**4.28** Draw both chair conformations for each of the following compounds:

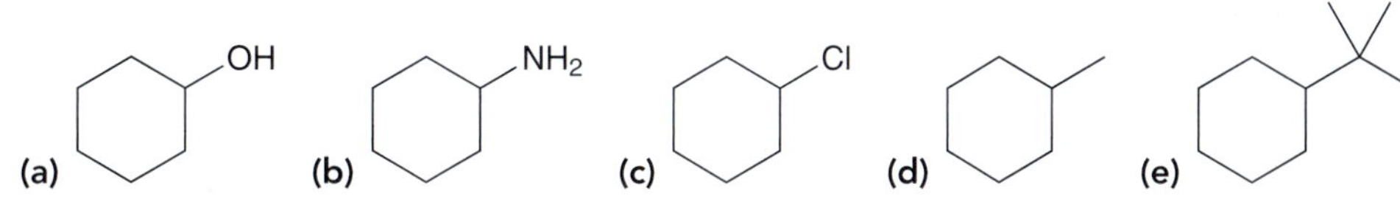

**APPLY the skill**

**4.29** Consider the following chair conformation of bromocyclohexane:

**(a)** Identify whether the bromine atom occupies an axial position or an equatorial position in the conformation above.

**(b)** Draw a bond-line drawing of this chair conformation (without Newman projections).

**(c)** Draw a bond-line drawing of the other chair conformation (after a ring flip).

need more **PRACTICE?** **Try Problem 4.54a**

## Comparing the Stability of Both Chair Conformations

When two chair conformations are in equilibrium, the lower energy conformation will be favored. For example, consider the two chair conformations of methylcyclohexane:

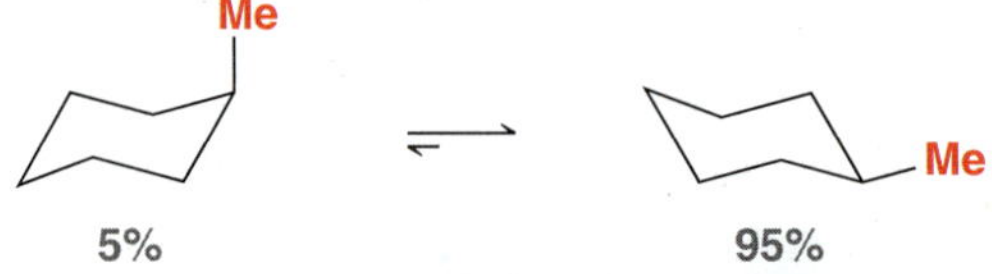


At room temperature, 95% of the molecules will be in the chair conformation that has the methyl group in an equatorial position. This must therefore be the lower energy conformation, but why? When the substituent is in an axial position, there are steric interactions with the other axial H's on the same side of the ring (Figure 4.27).

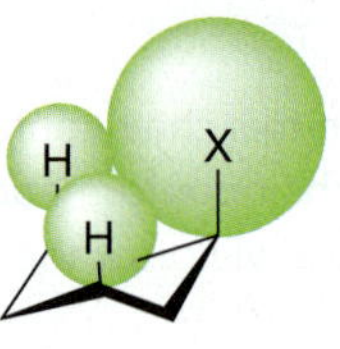


**FIGURE 4.27** Steric interactions that occur when a substituent occupies an axial position.

The substituent's electron cloud is trying to occupy the same region of space as the H's that are highlighted, causing steric interactions. These interactions are called **1,3-diaxial interactions**, where the numbers "1,3" describe the distance between the substituent and each of the H's. When the chair conformation is drawn in a Newman projection, it becomes clear that most 1,3-diaxial interactions are nothing more than *gauche* interactions. Compare the *gauche* interaction in butane with one of the 1,3-diaxial interactions in methylcyclohexane (Figure 4.28).

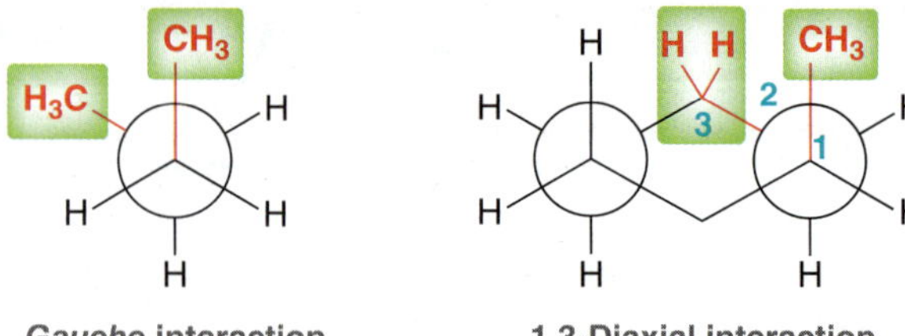


**FIGURE 4.28** An illustration showing that 1,3-diaxial interactions are really just *gauche* interactions.

The presence of 1,3-diaxial interactions causes the chair conformation to be higher in energy when the substituent is in an axial position. In contrast, when the substituent is in an equatorial position, these 1,3-diaxial (*gauche*) interactions are absent (Figure 4.29).

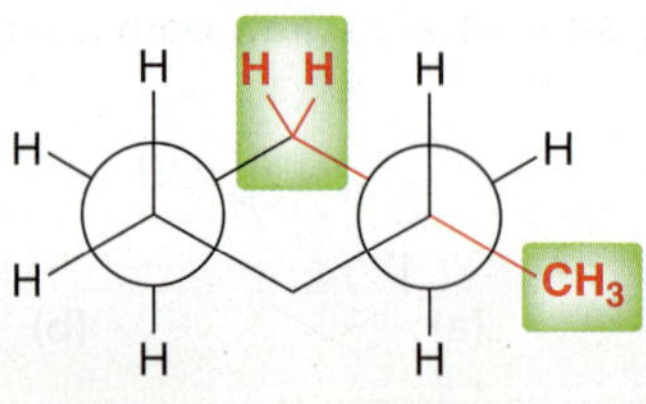


**FIGURE 4.29** When a substituent is in an equatorial position, it experiences no *gauche* interactions.

For this reason, the equilibrium between the two chair conformations will generally favor the conformation with the equatorial substituent. The exact equilibrium concentrations of the two chair conformations will depend on the size of the substituent. Larger groups will experience greater steric interactions, and the equilibrium will more strongly favor the equatorial substituent. For example, the equilibrium of *tert*-butylcyclohexane almost completely favors the chair conformation with an equatorial *tert*-butyl group:

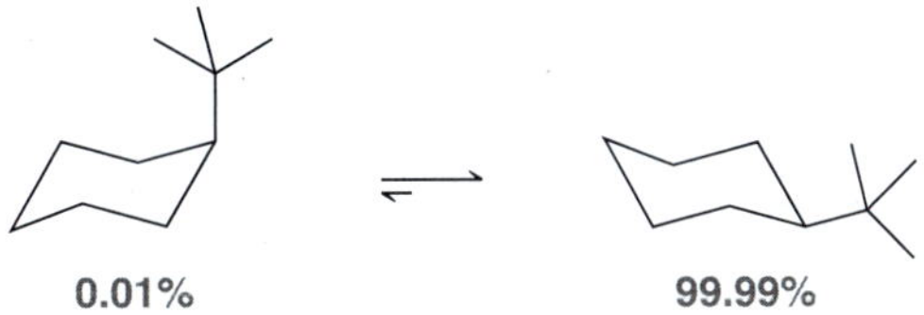


Table 4.8 shows the steric interactions associated with various groups as well as the equilibrium concentrations that are achieved.

**TABLE 4.8 1,3-DIAXIAL INTERACTIONS FOR SEVERAL COMMON SUBSTITUENTS**

| SUBSTITUENT | 1,3-DIAXIAL INTERACTIONS (KJ/MOL) | EQUATORIAL-AXIAL RATIO (AT EQUILIBRIUM) |
|---|---|---|
| —Cl | 2.0 | 70 : 30 |
| —OH | 4.2 | 83 : 17 |
| $—CH_3$ | 7.6 | 95 : 5 |
| $—CH_2CH_3$ | 8.0 | 96 : 4 |
| $—CH(CH_3)_2$ | 9.2 | 97 : 3 |
| $—C(CH_3)_3$ | 22.8 | 9999 : 1 |

## CONCEPTUAL CHECKPOINT

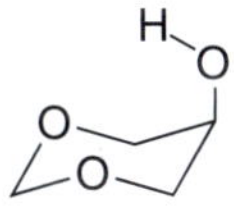


**4.30** The most stable conformation of 5-hydroxy-1,3-dioxane has the OH group in an axial position, rather than an equatorial position. Provide an explanation for this observation.

# 4.13 Disubstituted Cyclohexane

## Drawing Both Chair Conformations

When drawing chair conformations of a compound that has two or more substituents, there is an additional consideration. Specifically, we must also consider the three-dimensional orientation, or *configuration*, of each substituent. To illustrate this point, consider the following compound:

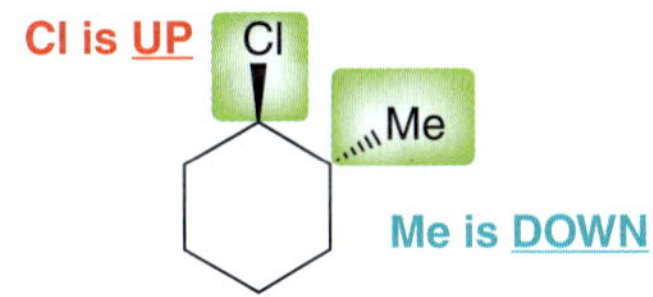


Notice that the chlorine atom is on a wedge, which means that it is coming out of the page: it is UP. The methyl group is on a dash, which means that it is below the ring, or DOWN. The two chair conformations for this compound are as follows:

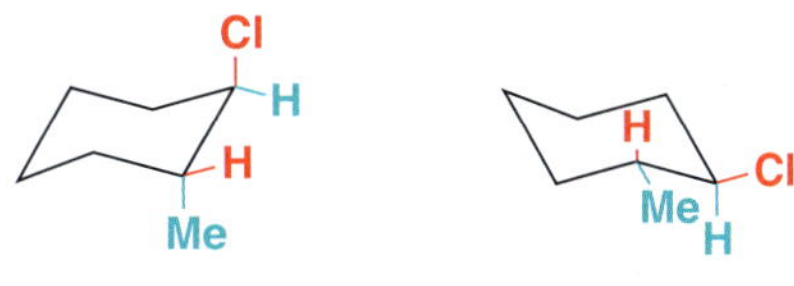

Notice that the chlorine atom is above the ring (UP) in both chair conformations, and the methyl group is below the ring (DOWN) in both chair conformations. The configuration (i.e., UP or DOWN) does not change during a ring flip. It is true that the chlorine is axial in one conformation and equatorial in the other conformation, but a ring flip does not change configuration. The chlorine atom must be UP in both chair conformations. Similarly, the methyl group must be DOWN in both chair conformations. Let's get some practice using these new descriptors (UP and DOWN) when drawing chair conformations.

## SKILLBUILDER

### 4.12 DRAWING BOTH CHAIR CONFORMATIONS OF DISUBSTITUTED CYCLOHEXANES

**LEARN** the skill

Draw both chair conformations of the following compound:

#### SOLUTION

Begin by numbering the ring and identifying the location and three-dimensional orientation of each substituent:

**STEP 1**
Determine the location and configuration of each substituent.

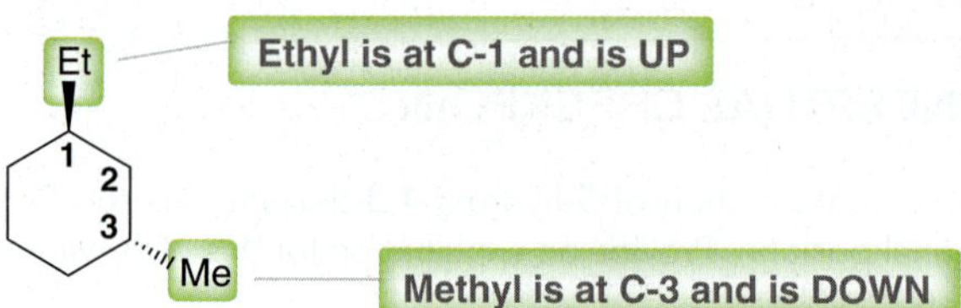


This numbering system does not need to be in accordance with IUPAC rules. It does not matter where the numbers are placed; these numbers are just tools used to compare positions in the original drawing and in the chair conformation to ensure that all substituents are placed correctly. The numbers can be placed either clockwise or counterclockwise, but they must be consistent. If the numbers are placed clockwise in the compound, then they must be placed clockwise as well when drawing the chair:

Once the numbers have been assigned, place the substituents in the correct locations and with the correct configuration. Ethyl is at C-1 and must be UP, while the methyl group is at C-3 and must be DOWN:

**STEP 2**
Place the substituents on the first chair using the information from step 1.

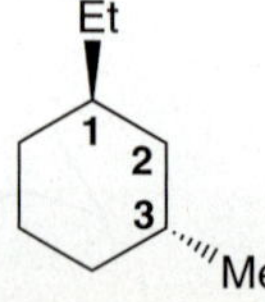


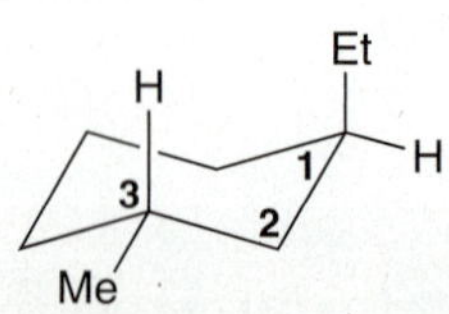

Notice that it is not possible to draw substituents properly without being able to draw all 12 positions on the cyclohexane ring. If you do not feel comfortable drawing all 12 positions, it would be a worthwhile investment of time to go back to that section of the chapter and practice. The drawing above represents the first chair conformation. In order to draw the second chair conformation, begin by drawing the other skeleton and numbering it:

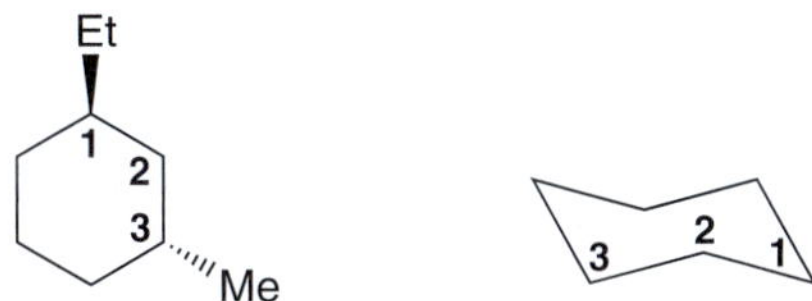


Then, once again, place the substituents so that the ethyl is at C-1 and is UP, while the methyl is at C-3 and is DOWN:

**STEP 3**
Place the substituents on the second chair using the information from step 1.

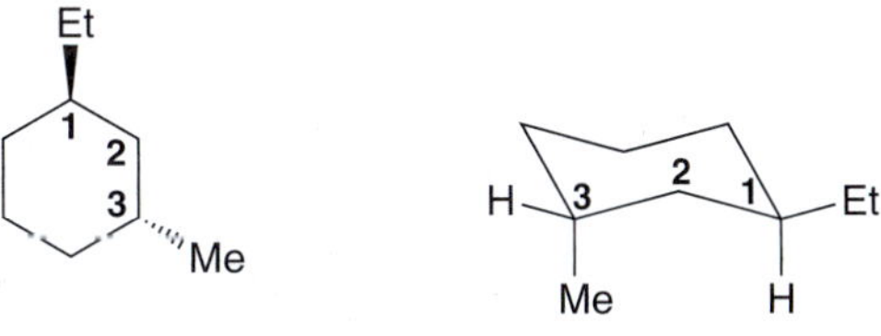


Therefore, the two chair conformations of this compound are:

Et
Me
Et
Me

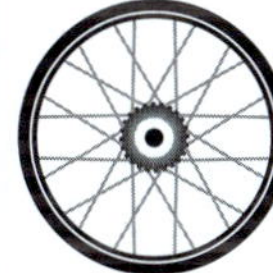

**PRACTICE the skill** **4.31** Draw both conformations for each of the following compounds:

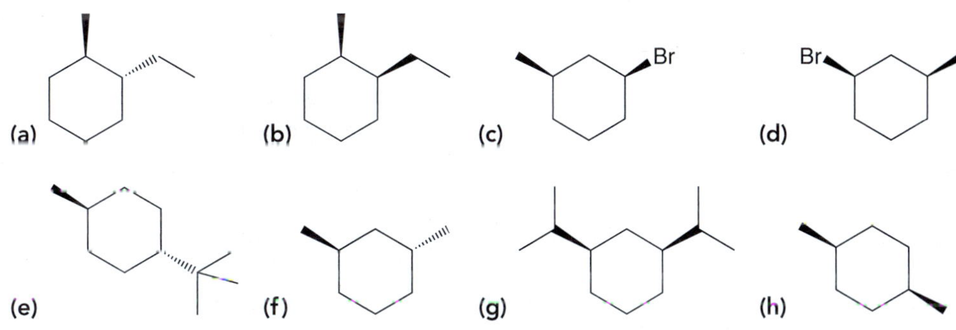

**APPLY the skill** **4.32** Lindane (hexachlorocyclohexane) is an agricultural insecticide that can also be used in the treatment of head lice. Draw both chair conformations of lindane.

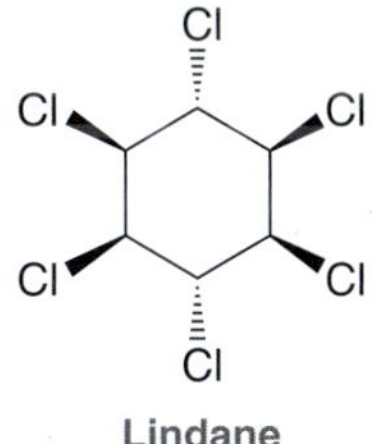

**Lindane**

need more **PRACTICE?** **Try Problems 4.54b–d, 4.66g**

## Comparing the Stability of Chair Conformations

Let's compare the stability of chair conformations once again, this time for compounds that bear more than one substituent. Consider the following example:

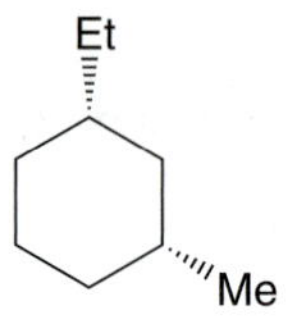

The two chair conformations of this compound are:

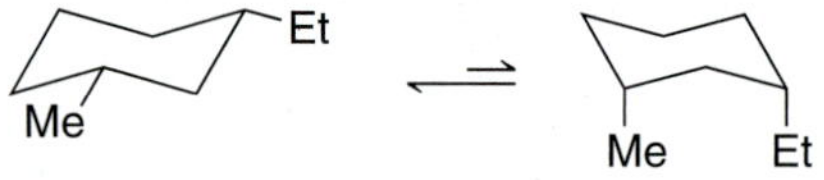

In the first conformation, both groups are equatorial. In the second conformation, both groups are axial. In the previous section, we saw that chair conformations will be lower in energy when substituents are in equatorial positions (avoiding 1,3-diaxial interactions). Therefore, the first chair will certainly be more stable.

In some cases, two groups might be in competition with each other. For example, consider the following compound:

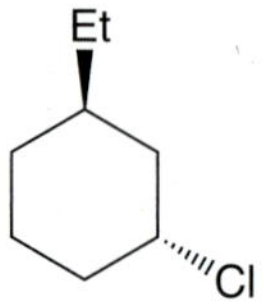

The two chair conformations of this compound are:

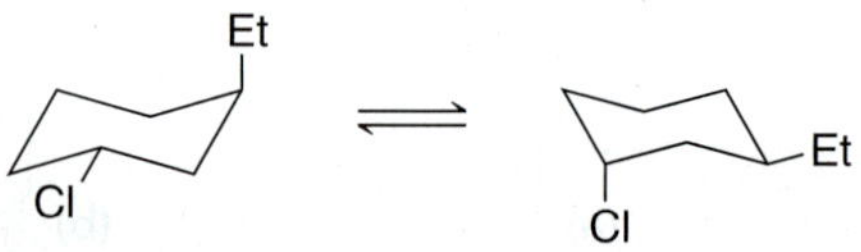

In this example, neither conformation has two equatorial substituents. In the first conformation, the chlorine is equatorial, but the ethyl group is axial. In the second conformation, the ethyl group is equatorial, but the chlorine is axial. In a situation like this, we must decide which group exhibits a greater preference for being equatorial: the chlorine atom or the ethyl group. To do this, we use the numbers from Table 4.8:

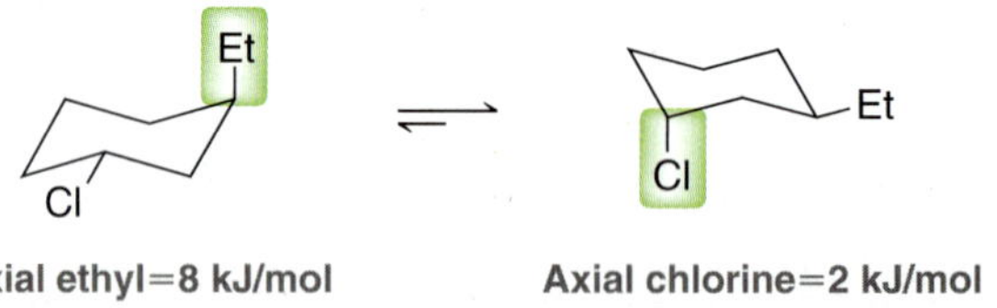

Axial ethyl=8 kJ/mol     Axial chlorine=2 kJ/mol

Both conformations will exhibit 1,3-diaxial interactions, but these interactions are less pronounced in the second conformation. The energy cost of having a chlorine atom in an axial position is lower than the energy cost of having an ethyl group in an axial position. Therefore, the second conformation is lower in energy. Let's get some practice with this.

# SKILLBUILDER

## 4.13 DRAWING THE MORE STABLE CHAIR CONFORMATION OF POLYSUBSTITUTED CYCLOHEXANES

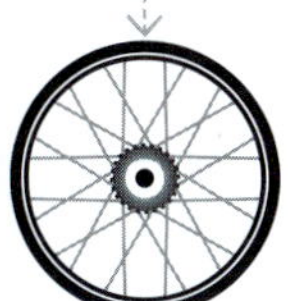

**LEARN the skill** Draw the more stable chair conformation of the following compound:

### SOLUTION

Begin by drawing both chair conformations using the method from the previous section. Place numbers on the ring, and for each substituent, identify its location and configuration:

**STEP 1**
Determine the location and configuration of each substituent.

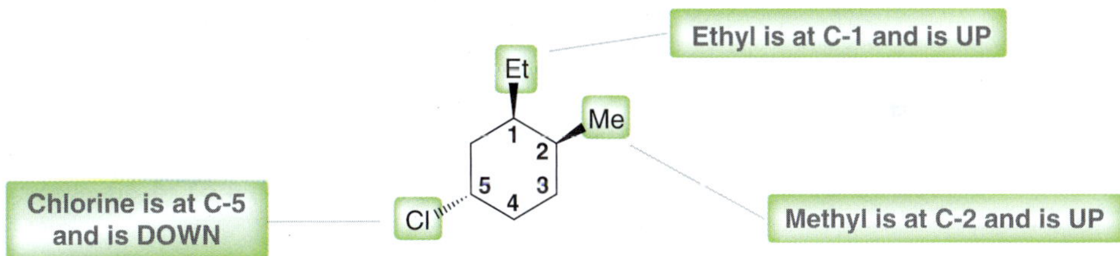


Now draw the skeleton of the first chair conformation, placing the substituents in the correct locations and with the correct configuration:

Then draw the skeleton of the second chair conformation, number it, and once again, place the substituents in the correct locations and with the correct configuration:

**STEP 2**
Draw both chair conformations.

Therefore, the two chair conformations of this compound are:

**STEP 3**
Assess the energy cost of each axial group.

Now we can compare the relative energy of these two chair conformations. In the first conformation, there is one ethyl group in an axial position. According to Table 4.8, the energy cost associated with an axial ethyl group is 8.0 kJ/mol. In the second conformation, two groups are in axial positions: a methyl group and a chlorine. According to Table 4.8, the total energy cost is 7.6 kJ/mol + 2.0 kJ/mol = 9.6 kJ/mol. According to this calculation, the energy cost is lower for the first conformation (with an axial ethyl group). The first conformation is therefore lower in energy (more stable).

**PRACTICE the skill** **4.33** Draw the lowest energy conformation for each of the following compounds:

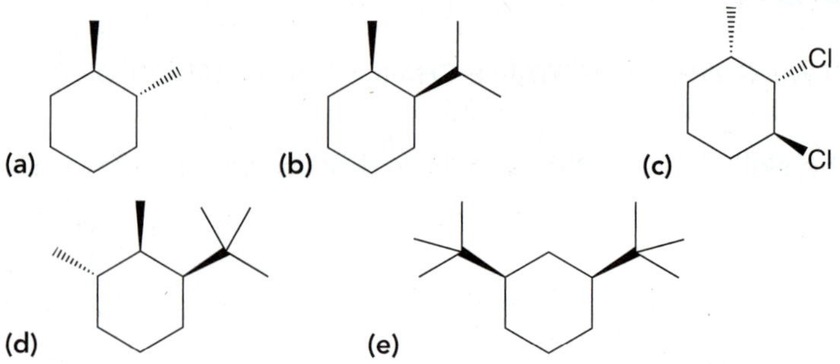

**APPLY the skill**

**4.34** In Problem 4.32, you drew the two chair conformations of lindane. Carefully inspect them and predict the difference in energy between them, if any.

**4.35** Compound A exists predominantly in a chair conformation, while compound B exists predominantly in a twist boat conformation. Explain.

**Compound A** **Compound B**

need more **PRACTICE?** **Try Problems 4.53, 4.55, 4.57, 4.61, 4.68**

## 4.14 *cis-trans* Stereoisomerism

When dealing with cycloalkanes, the terms *cis* and *trans* are used to signify the relative spatial relationship of similar substituents:

$CH_3$ $CH_3$ / H H — ***cis*-1,2-Dimethylcyclohexane**

$CH_3$ H / H $CH_3$ — ***trans*-1,2-Dimethylcyclohexane**

The term *cis* is used to signify that the two groups are on the same side of the ring, while the term *trans* signifies that the two groups are on opposite sides of the ring. The drawings above are *Haworth projections* (as seen in Section 2.6) and are used to clearly identify which groups are above the ring and which groups are below the ring. These drawings are planar representations and do not represent conformations. Each compound above is better represented as an equilibrium between two chair conformations (Figure 4.30). *cis*-1,2-Dimethylcyclohexane and *trans*-1,2-dimethylcyclohexane are **stereoisomers** (as we will see in the next chapter). They are different compounds with different physical properties, and they cannot be interconverted via a conformational change. *trans*-1,2-Dimethylcyclohexane is more stable, because it can adopt a chair conformation in which both methyl groups are in equatorial positions.

$CH_3$ $CH_3$ / H H
$CH_3$ / $CH_3$ ⇌ $CH_3$ / $CH_3$

H $CH_3$ / $CH_3$ H
$CH_3$ / $CH_3$ ⇌ $CH_3$ / $CH_3$

**FIGURE 4.30** Each stereoisomer of 1,2-dimethylcyclohexane has two chair conformations.

## CONCEPTUAL CHECKPOINT

**4.36** Draw Haworth projections for *cis*-1,3-dimethylcyclohexane and *trans*-1,3-dimethylcyclohexane. Then, for each compound, draw the two chair conformations. Use these conformations to determine whether the *cis* isomer or the *trans* isomer is more stable.

**4.37** Draw Haworth projections for *cis*-1,4-dimethylcyclohexane and *trans*-1,4-dimethylcyclohexane. Then for each compound, draw the two chair conformations. Use these conformations to determine whether the *cis* isomer or the *trans* isomer is more stable.

**4.38** Draw Haworth projections for *cis*-1,3-di-*tert*-butylcyclohexane and *trans*-1,3-di-*tert*-butylcyclohexane. One of these compounds exists in a chair conformation, while the other exists primarily in a twist boat conformation. Offer an explanation.

## 4.15 Polycyclic Systems

Decalin is a bicyclic system composed of two fused six-membered rings. The structures of *cis*-decalin and *trans*-decalin are as follows:

***cis*-Decalin** ***trans*-Decalin**

The relationship between these compounds is stereoisomeric (as in the previous section). These two compounds are not interconvertible by ring flipping. They are two different compounds with different physical properties. Many naturally occurring compounds, such as steroids, incorporate decalin systems into their structures. Steroids are a class of compounds comprised of four fused rings (three six-membered rings and one five-membered ring). Below are two examples of steroids:

**Testosterone** **Estradiol**

Testosterone is an androgenic hormone (male sex hormone) produced in the testes, and estradiol is an estrogenic hormone (female sex hormone) produced from testosterone in the ovaries. Both compounds play a number of biological roles, ranging from the development of secondary sex characteristics to the promotion of tissue and muscle growth.

Another common polycyclic system is **norbornane**. Norbornane is the common name for bicyclo[2.2.1]heptane. We can think of this compound as a six-membered ring locked into a boat conformation by a $CH_2$ group that serves as a bridge. Many naturally occurring compounds are substituted norbornanes, such as camphor and camphene:

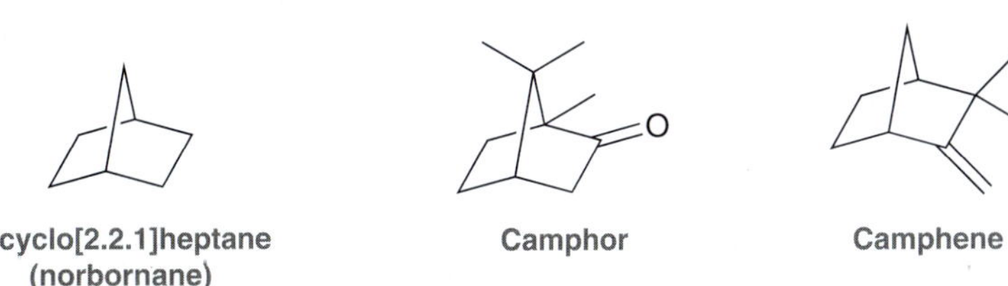


**Bicyclo[2.2.1]heptane (norbornane)** **Camphor** **Camphene**

Camphor is a strongly scented solid that is isolated from evergeen trees in Asia. It is used as a spice as well as for medicinal purposes. Camphene is a minor constituent in many natural oils, such as pine oil and ginger oil. It is used in the preparation of fragrances.

Polycyclic systems based on six-membered rings are also found in nonbiological materials. Most notably, the structure of diamond is based on fused six-membered rings locked in chair conformations. The drawing in Figure 4.31 represents a portion of the diamond structure. Every carbon atom is bonded to four other carbon atoms forming a three-dimensional lattice of chair conformations. In this way, a diamond is one large molecule. Diamonds are one of the hardest known substances, because cutting a diamond requires the breaking of billions of C—C single bonds.

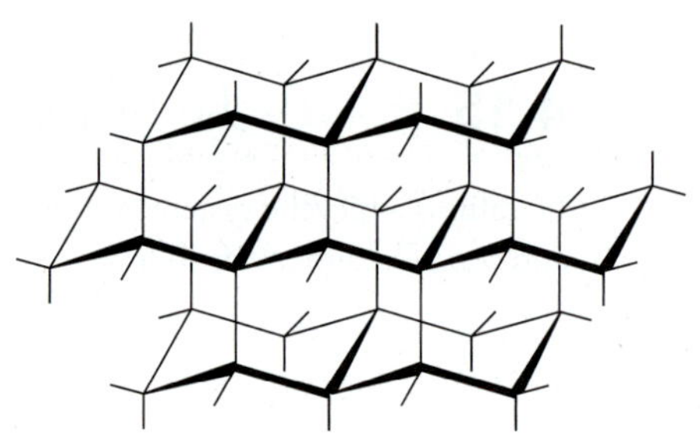

**FIGURE 4.31**
The structure of diamonds.

## REVIEW OF CONCEPTS AND VOCABULARY

### SECTION 4.1

- Hydrocarbons that lack $\pi$ bonds are called **saturated hydrocarbons** or **alkanes**.
- The system of rules for naming compounds is called **nomenclature**.

### SECTION 4.2

- Although **IUPAC** rules provide a systematic way for naming compounds, many common names are still in use. Assigning a **systematic name** involves four discrete steps:
  1. Identify the parent compound. Alkanes containing a ring are called **cycloalkanes**.
  2. Name the **substituents**, which can be either simple **alkyl groups** or branched alkyl groups, called complex substituents. Many common names for complex substituents are allowed according to IUPAC rules.
  3. Number the carbon atoms of the parent and assign a **locant** to each substituent.
  4. Assemble the substituents alphabetically, placing locants in front of each substituent. For identical substituents, use di, tri, tetra, penta, or hexa, which are ignored when alphabetizing.
- **Bicyclic** compounds are named just like alkanes and cycloalkanes, with just two subtle differences:
  1. The term "bicyclo" is used, and bracketed numbers indicate how the **bridgeheads** are connected.
  2. To number the parent, travel first along the longest path connecting the bridgeheads.

### SECTION 4.3

- The number of possible constitutional isomers for an alkane increases with increasing molecular size.
- When drawing constitutional isomers, use IUPAC rules to avoid drawing the same compound twice.

### SECTION 4.4

- For an alkane, the **heat of combustion** is the negative of the *change in enthalpy* ($-\Delta H°$) associated with the complete combustion of 1 mol of alkane in the presence of oxygen.
- Heats of combustion can be measured experimentally and used to compare the stability of isomeric alkanes.

### SECTION 4.5

- Petroleum is a complex mixture of hydrocarbons, most of which are alkanes (ranging in size and constitution).
- These compounds are separated into fractions via distillation (separation based on differences in boiling points). The refining of crude oil separates it into many commercial products.
- The yield of useful gasoline can be improved in two ways:
  1. Cracking is a process by which C—C bonds of larger alkanes are broken, producing smaller alkanes suitable for gasoline.
  2. Reforming is a process in which straight-chain alkanes are converted into branched hydrocarbons and aromatic compounds, which exhibit less *knocking* during combustion.

### SECTION 4.6

- Rotation about C—C single bonds allows a compound to adopt a variety of **conformations**.
- **Newman projections** are often used to draw the various conformations of a compound.

### SECTION 4.7

- In a Newman projection, the **dihedral angle**, or **torsional angle,** describes the relative positions of one group on the back carbon and one group on the front carbon.
- **Staggered conformations** are lower in energy, while **eclipsed conformations** are higher in energy.
- In the case of ethane, all staggered conformations are **degenerate** (equivalent in energy), and all eclipsed conformations are degenerate.
- The difference in energy between staggered and eclipsed conformations of ethane is referred to as **torsional strain.** The torsional strain for propane is larger than that of ethane.

### SECTION 4.8

- For butane, one of the eclipsed conformations is higher in energy than the other two.
- One staggered conformation (the ***anti*** **conformation**) is lower in energy than the other two staggered conformations, because they possess ***gauche*** **interactions**.

### SECTION 4.9

- **Angle strain** occurs in cycloalkanes when bond angles are less than the preferred 109.5°.
- Angle strain and torsional strain are components of the total energy of a cycloalkane, which can be assessed by measuring heats of combustion per $CH_2$ group.

### SECTION 4.10

- The **chair conformation** of cyclohexane has no torsional strain and very little angle strain.
- The **boat conformation** of cyclohexane has significant torsional strain (from eclipsing H's as well as **flagpole interactions**). The boat can alleviate some of its torsional strain by twisting, giving a conformation called a **twist boat**.
- Cyclohexane is found in a chair conformation most of the time.

### SECTION 4.11

- Each carbon atom in a cyclohexane ring can bear two substituents.
- One substituent is said to occupy an **axial position**, while the other substituent is said to occupy an **equatorial position**.

### SECTION 4.12

- When a ring has one substituent, the substituent can occupy either an axial or an equatorial position. These two possibilities represent two different conformations that are in equilibrium with each other.
- The term **ring flip** is used to describe the conversion of one chair conformation into the other.
- The equilibrium will favor the chair conformation with the substituent in the equatorial position, because an axial substituent generates **1,3-diaxial interactions**.

### SECTION 4.13

- To draw the two chair conformations of a disubstituted cyclohexane, each substituent must be identified as being either UP or DOWN. The three-dimensional orientation of the substituents (UP or DOWN) does not change during a ring flip.
- After drawing both chair conformations, the relative energy levels can be determined by comparing the energy cost associated with all axial groups.

### SECTION 4.14

- The terms *cis* and *trans* signify the relative spatial relationship of similar substituents, as can be seen clearly in Haworth projections.
- *cis*-1,2-Dimethylcyclohexane and *trans*-1,2-dimethylcyclohexane are **stereoisomers**. They are different compounds with different physical properties, and they cannot be interconverted via a conformational change.

### SECTION 4.15

- The relationship between *cis*-decalin and *trans*-decalin is stereoisomeric. These two compounds are not interconvertable by ring flipping.
- **Norbornane** is the common name for bicyclo[2.2.1]heptane and is a commonly encountered bicyclic system.
- Polycyclic systems based on six-membered rings are also found in the structure of diamonds.

## SKILLBUILDER REVIEW

### 4.1 IDENTIFYING THE PARENT

**STEP 1** Choose the longest chain.

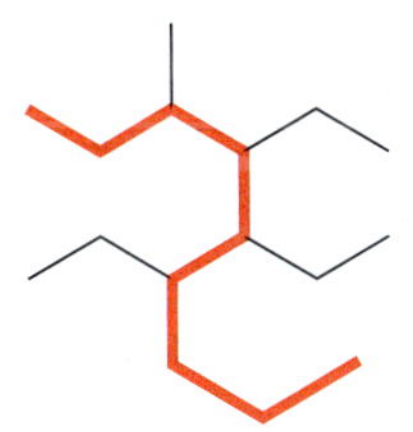

**STEP 2** In a case where two chains compete, choose the chain with more substituents.

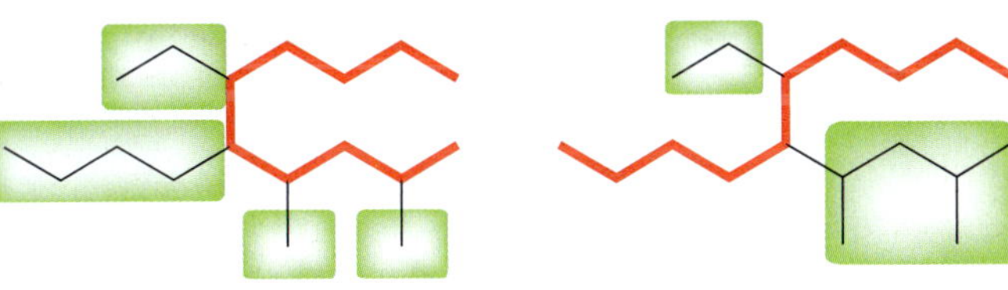


Try Problems **4.1–4.4, 4.39**

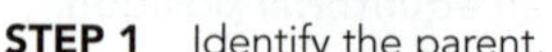

### 4.2 IDENTIFYING AND NAMING SUBSTITUENTS

**STEP 1** Identify the parent.

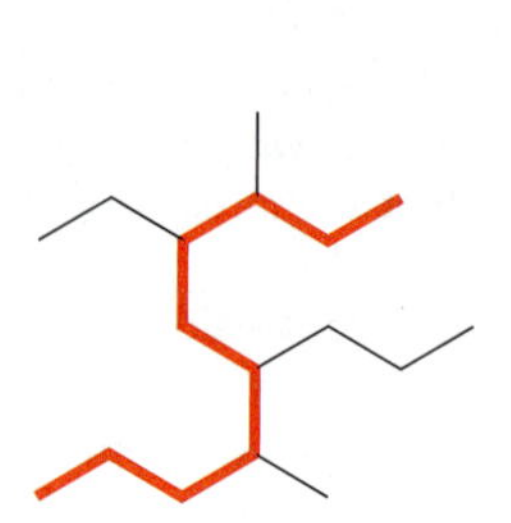

**STEP 2** Identify all alkyl substituents connected to the parent.

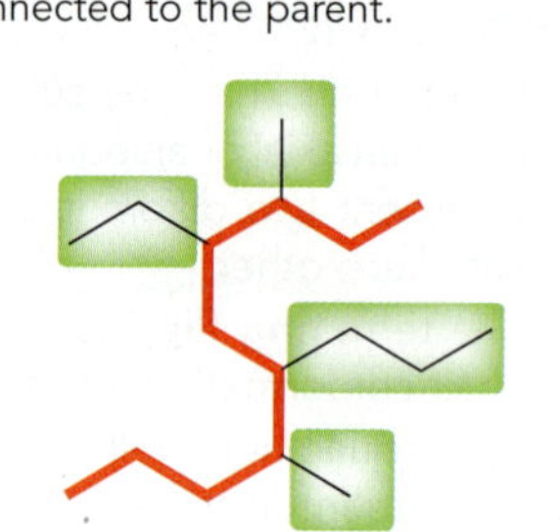

**STEP 3** Name each substituent using the names in Table 4.2.

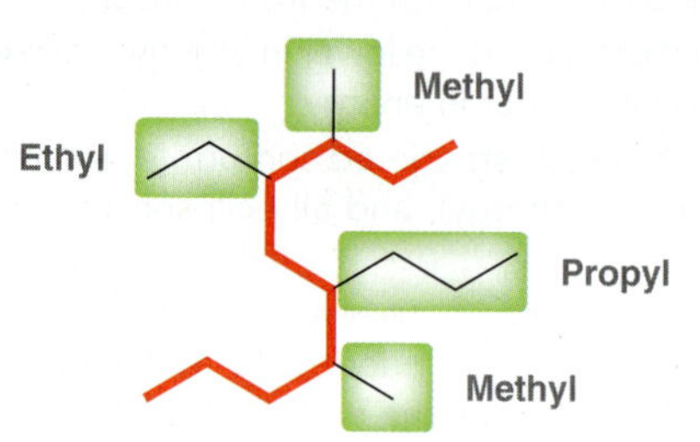


Try Problems **4.2, 4.3, 4.40a, 4.40c**

### 4.3 IDENTIFYING AND NAMING COMPLEX SUBSTITUENTS

**STEP 1** Identify the parent.

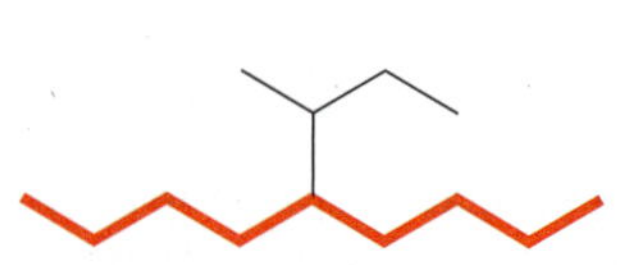

**STEP 2** Place numbers on the complex substituent, going away from the parent chain.

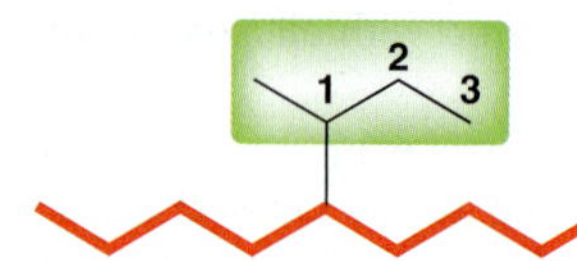


**STEP 3** Treat the entire group as a substituent on a substituent.

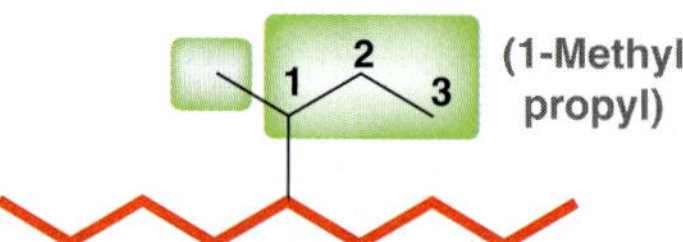


Try Problems **4.7–4.9, 4.40b, 4.40d**

### 4.4 ASSEMBLING THE SYSTEMATIC NAME OF AN ALKANE

**STEP 1** Identify the parent.

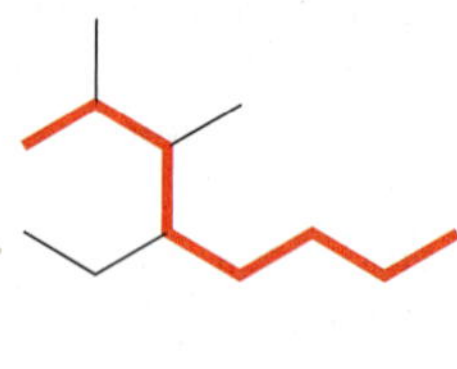

**STEP 2** Identify and name substituents.

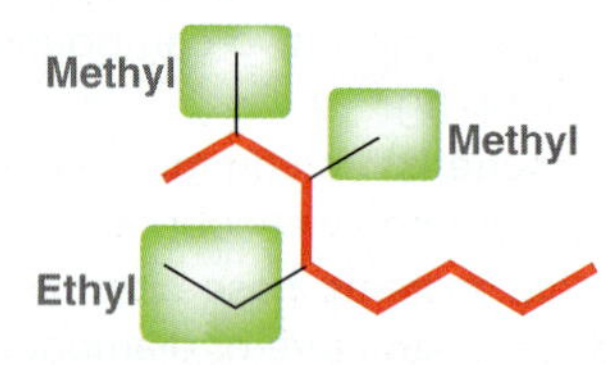


**STEP 3** Number the parent chain and assign a locant to each substituent.

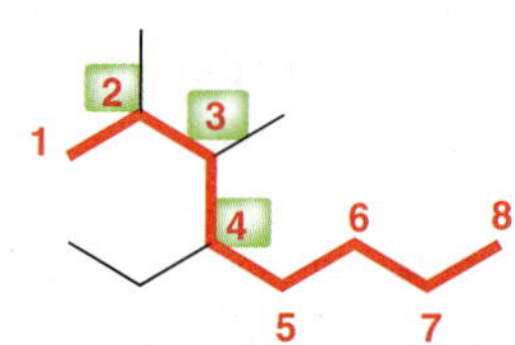


**STEP 4** Arrange the substituents alphabetically.

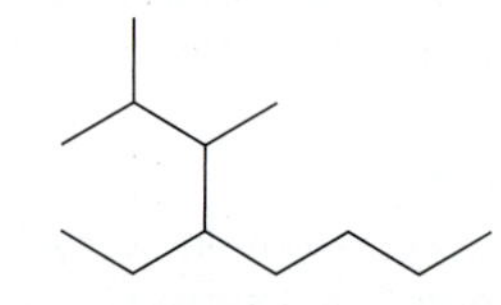

**4-Ethyl-2,3-dimethyloctane**

Try Problems **4.10, 4.11, 4.41a, 4.41b, 4.45a, 4.45b**

### 4.5 ASSEMBLING THE NAME OF A BICYCLIC COMPOUND

**STEP 1** Identify the bicyclo parent and then indicate how the bridgeheads are connected.

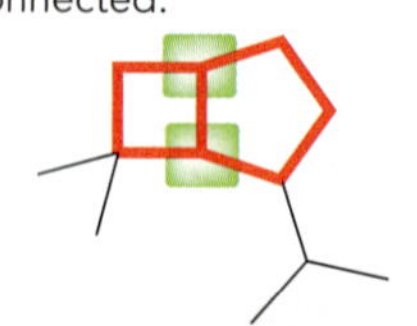

**Bicyclo[3.2.0]heptane**

**STEP 2** Identify and name substituents.

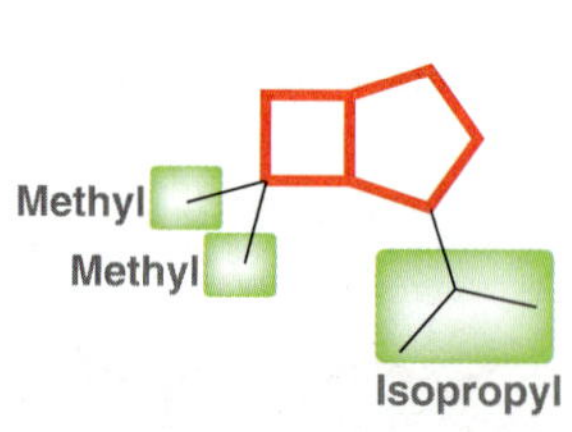


**STEP 3** Number the parent chain (start with longest bridge) and assign a locant to each substituent.

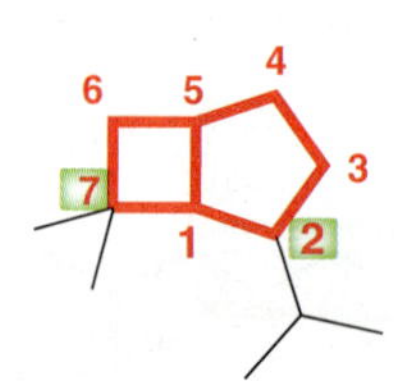


**STEP 4** Assemble the substituents alphabetically.

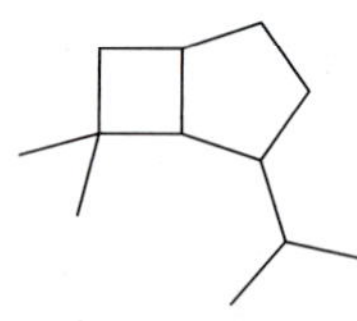

**2-Isopropyl-7,7-dimethyl bicyclo[3.2.0]heptane**

Try Problems **4.12, 4.13, 4.41c, 4.41d, 4.45c**

## 4.6 IDENTIFYING CONSTITUTIONAL ISOMERS

Constitutional isomers have different names. If two compounds have the same name, then they are the same compound.

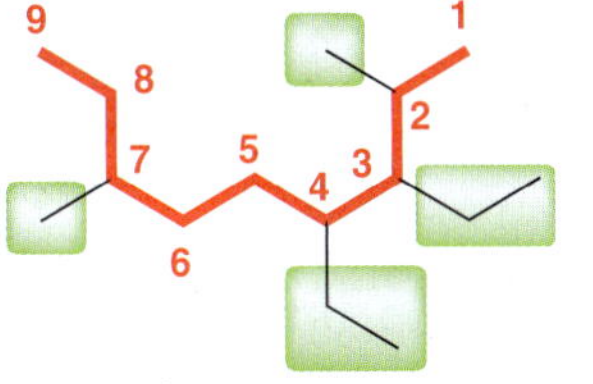


3,4-Diethyl-2,7-dimethylnonane

3,4-Diethyl-2,7-dimethylnonane

Try Problems 4.14, 4.15, 4.42, 4.66b,d,k,l

## 4.7 DRAWING NEWMAN PROJECTIONS

**STEP 1** Identify the three groups connected to the front carbon atom.

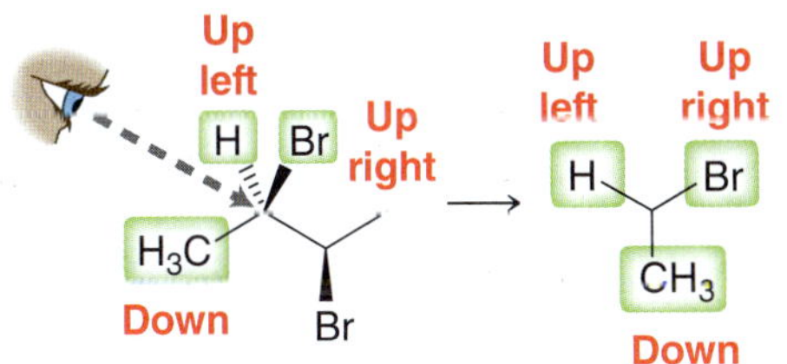


**STEP 2** Identify the three groups connected to the back carbon atom.

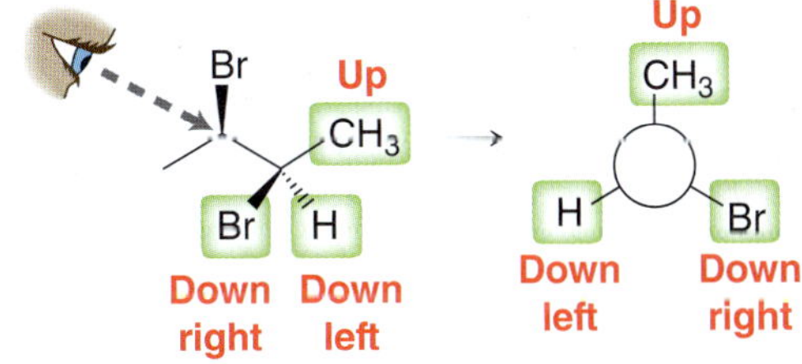


**STEP 3** Assemble the Newman projection from the two pieces obtained in the previous steps.

$CH_3$ H Br H Br $CH_3$

Try Problems 4.16–4.18, 4.56

## 4.8 IDENTIFYING RELATIVE ENERGY OF CONFORMATIONS

**STEP 1** Draw a Newman projection.

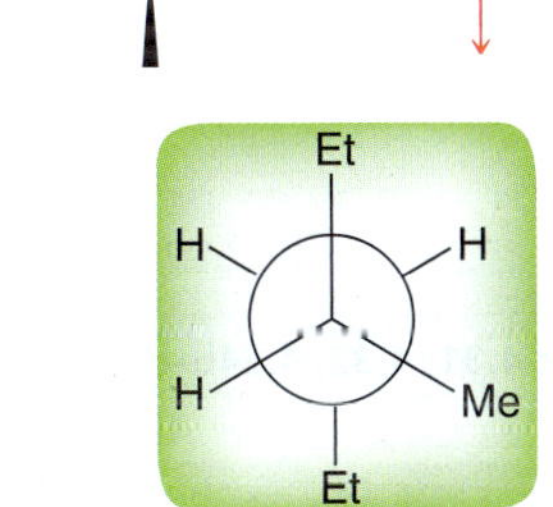


**STEP 2** Draw all three staggered conformations and determine which one has the fewest or least severe *gauche* interactions.

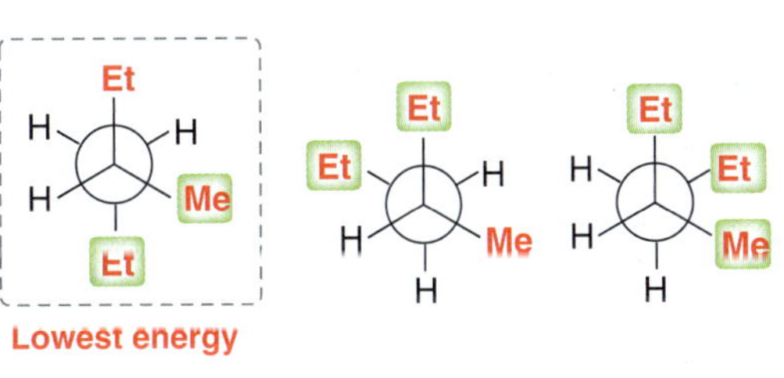


**STEP 3** Draw all three eclipsed conformations and determine which one has the highest energy interactions.

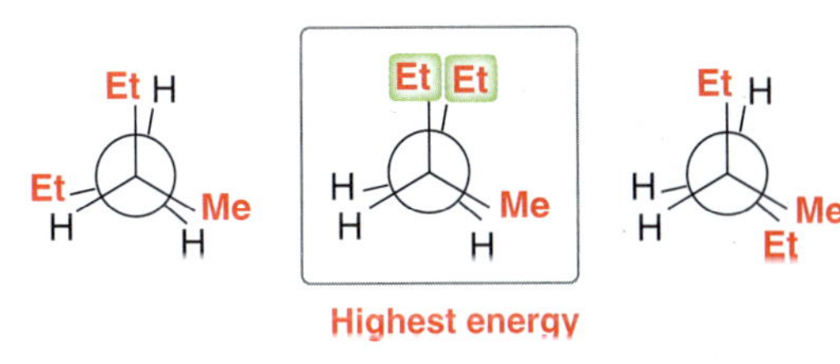


Try Problems 4.20, 4.21, 4.47, 4.59

## 4.9 DRAWING A CHAIR CONFORMATION

**STEP 1** Draw a wide V.

**STEP 2** Draw a line going down at a 60° angle, ending just before the center of the V.

**STEP 3** Draw a line parallel to the left side of the V ending just before the left side of the V.

**STEP 4** Draw a line parallel to the line from step 2, going down exactly as low as that line.

**STEP 5** Connect the dots.

Try Problems 4.22, 4.23

### 4.10 DRAWING AXIAL AND EQUATORIAL POSITIONS

**STEP 1** Draw all axial positions as parallel lines, alternating in directions.

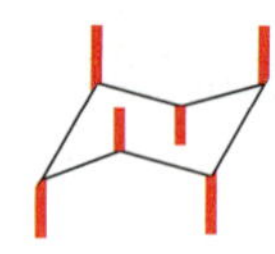

**STEP 2** Draw all equatorial positions as pairs of parallel lines.

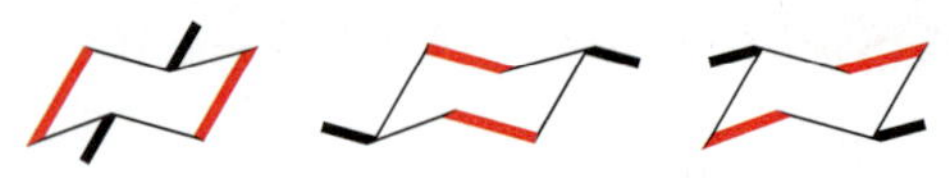

**SUMMARY** All substituents are drawn like this:

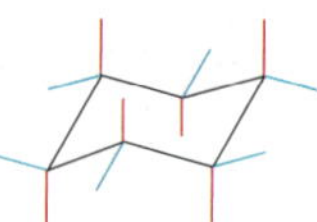

Try Problems 4.24–4.27, 4.62

### 4.11 DRAWING BOTH CHAIR CONFORMATIONS OF A MONOSUBSTITUTED CYCLOHEXANE

**STEP 1** Draw a chair conformation.

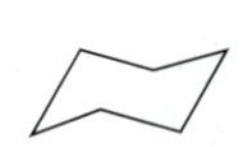

**STEP 2** Place the substituent in an axial position.

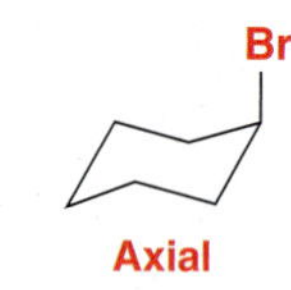


**STEP 3** Draw the ring flip and the axial group becomes equatorial.

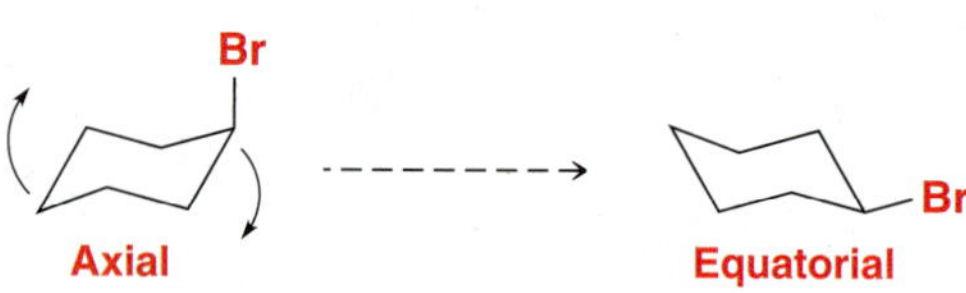


Try Problems 4.28, 4.29, 4.54a

### 4.12 DRAWING BOTH CHAIR CONFORMATIONS OF DISUBSTITUTED CYCLOHEXANES

**STEP 1** Using a numbering system, determine the location and configuration of each substituent.

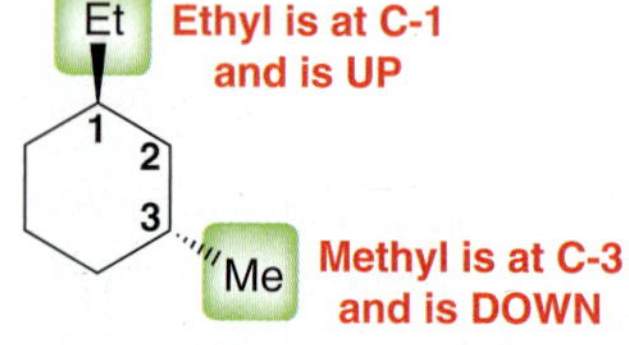


**STEP 2** Place the substituents on the first chair using the information from step 1.

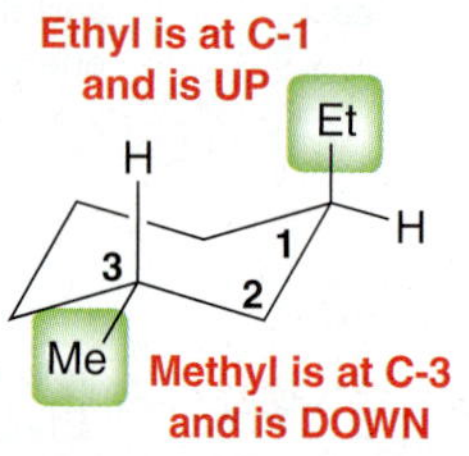


**STEP 3** Draw the second chair skeleton, and place substituents using the information from step 1.

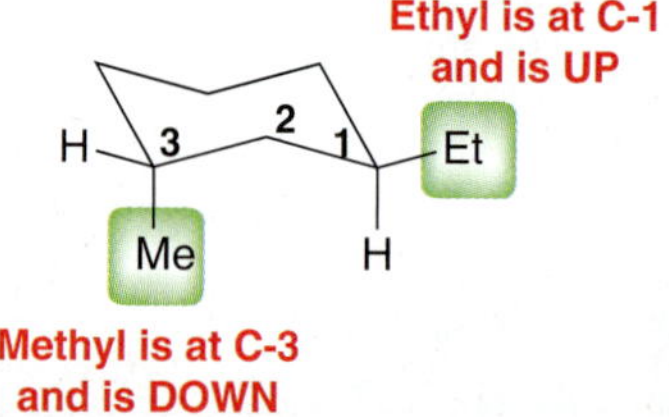


Try Problems 4.31, 4.32, 4.54b–d, 4.66g

### 4.13 DRAWING THE MORE STABLE CHAIR CONFORMATION OF POLYSUBSTITUTED CYCLOHEXANES

**STEP 1** Using a numbering system, determine the location and configuration of each substituent.

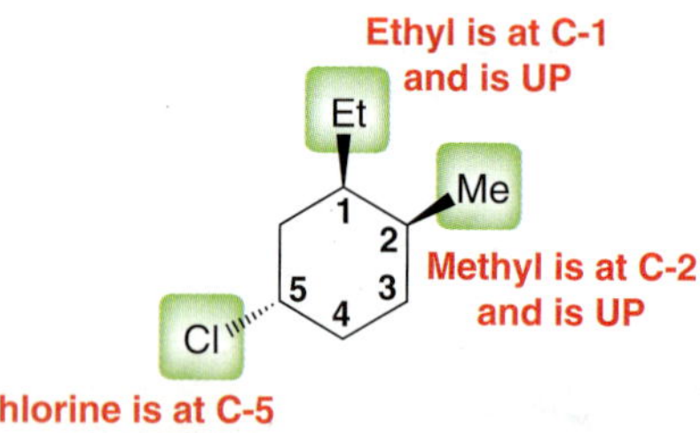


**STEP 2** Using the information from step 1, draw both chair conformations.

Cl Et Me

Me Et Cl

**STEP 3** Assess the energy cost of each axial group.

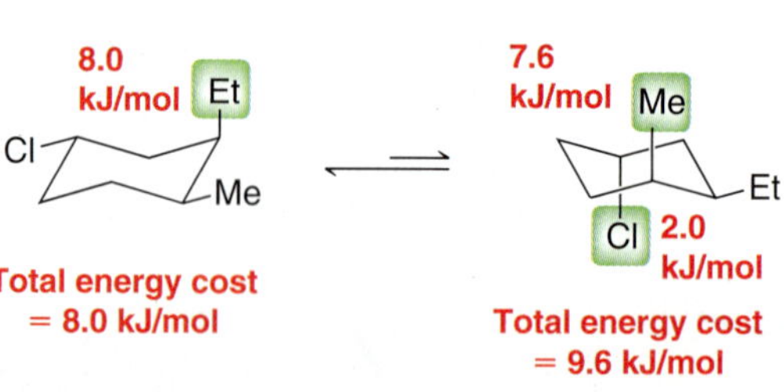


Try Problems 4.33–4.35, 4.53, 4.55, 4.57, 4.61, 4.68

# PRACTICE PROBLEMS

**Note: Most of the Problems are available within WileyPLUS, an online teaching and learning solution.**

**4.39** Identify the name of the parent for each of the following compounds:

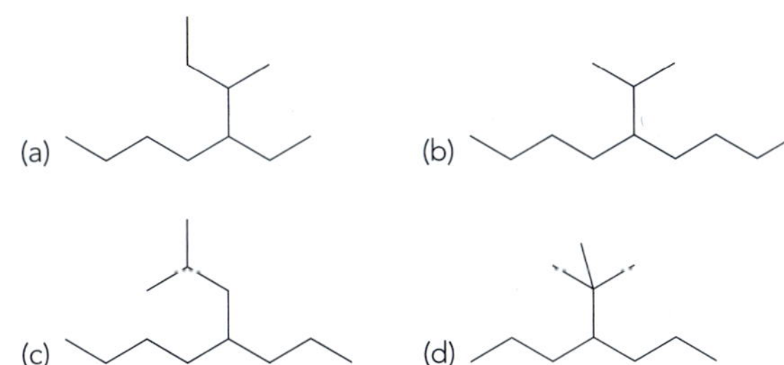

**4.40** Each of the structures in the previous problem has one or more substituents connected to the parent.

(a) Identify the name of each substituent in 4.39a.

(b) Identify the common name and the IUPAC name of the complex substituent in 4.39b.

(c) Identify the name of each substituent in 4.39c.

(d) Identify the common name and the IUPAC name of the complex substituent in 4.39d.

**4.41** What is the systematic name for each of the following compounds:

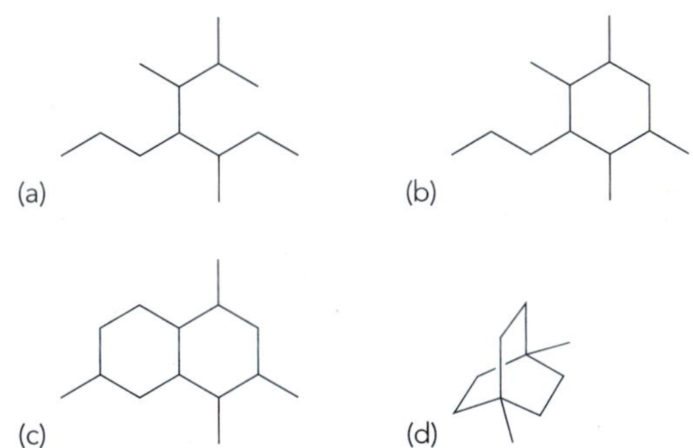

**4.42** For each of the following pairs of compounds, identify whether the compounds are constitutional isomers or different representations of the same compound:

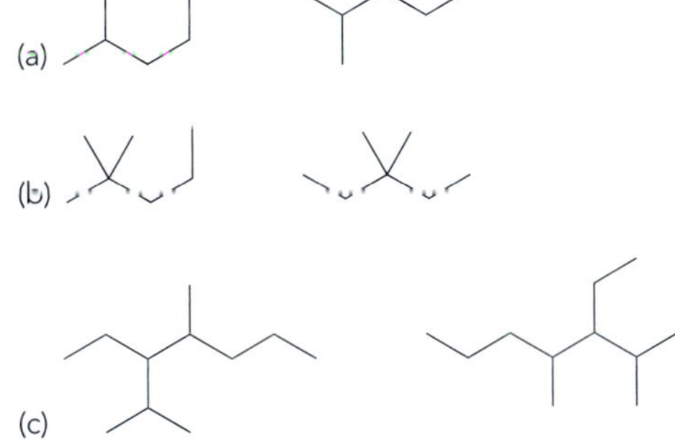

**4.43** Use a Newman projection to draw the most stable conformation of 3-methylpentane, looking down the C2—C3 bond.

**4.44** Identify which of the following compounds is expected to have the larger heat of combustion:

**4.45** Draw each of the following compounds:

(a) 2,2,4-Trimethylpentane

(b) 1,2,3,4-Tetramethylcycloheptane

(c) 2,2,4,4-Tetraethylbicyclo[1.1.0]butane

**4.46** Sketch an energy diagram that shows a conformational analysis of 2,2-dimethylpropane. Does the shape of this energy diagram more closely resemble the shape of the energy diagram for ethane or for butane?

**4.47** What are the relative energy levels of the three staggered conformations of 2,3-dimethylbutane when looking down the C2—C3 bond?

**4.48** Draw the ring flip for each of the following compounds:

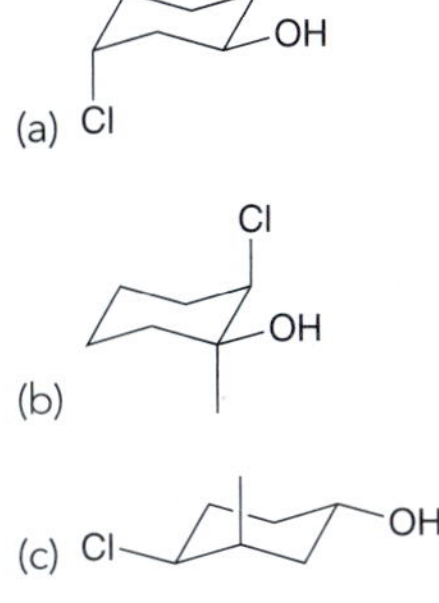


**4.49** For each of the following pairs of compounds, identify the compound that would have the higher heat of combustion:

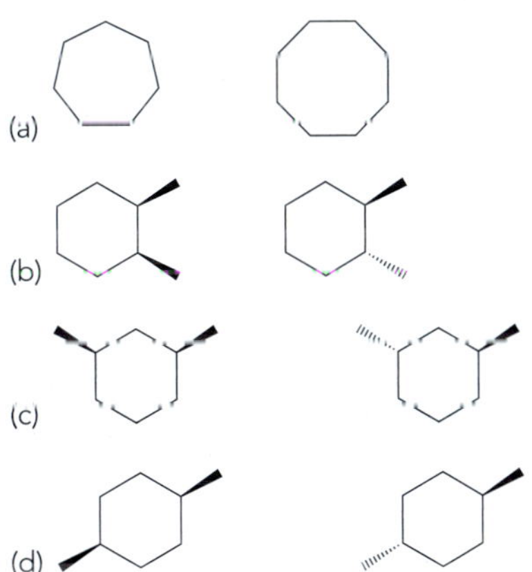

**4.50** Draw a relative energy diagram showing a conformational analysis of 1,2-dichloroethane. Clearly label all staggered conformations and all eclipsed conformations with the corresponding Newman projections.

**4.51** Assign IUPAC names for each of the following compounds:

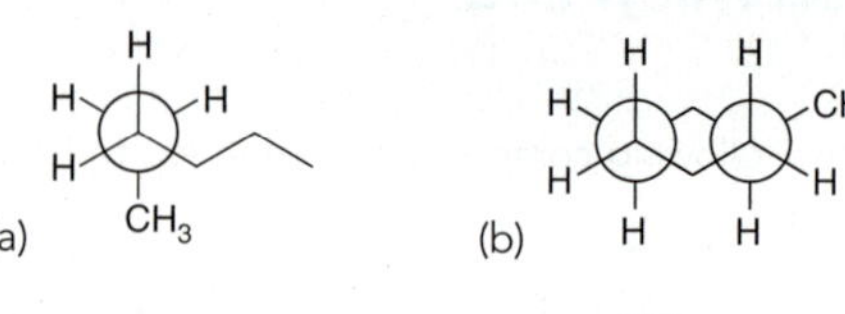


(c) (d)

**4.52** The barrier to rotation of bromoethane is 15 kJ/mol. Based on this information, determine the energy cost associated with the eclipsing interaction between a bromine atom and a hydrogen atom.

**4.53** Menthol, isolated from various mint oils, is used in the treatment of minor throat irritation. Draw both chair conformations of menthol and indicate which conformation is lower in energy.

**Menthol**

**4.54** Draw both chair conformations for each of the following compounds. In each case, identify the more stable chair conformation:

(a) Methylcyclohexane

(b) *trans*-1,2-Diisopropylcyclohexane

(c) *cis*-1,3-Diisopropylcyclohexane

(d) *trans*-1,4-Diisopropylcyclohexane

**4.55** For each of the following pairs of compounds, determine which compound is more stable (you may find it helpful to draw out the chair conformations):

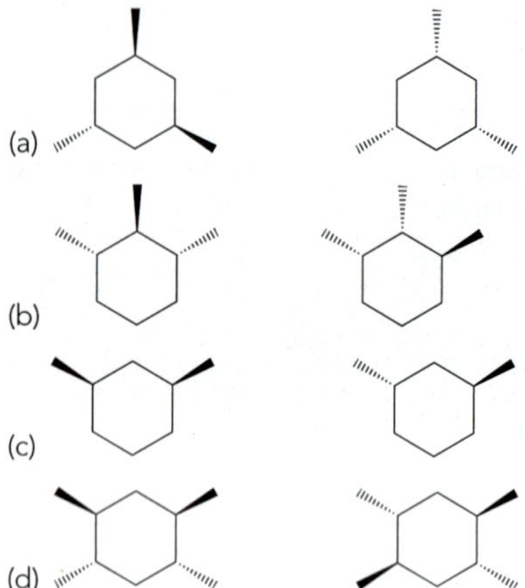


**4.56** Draw a Newman projection of the following compound as viewed from the angle indicated:

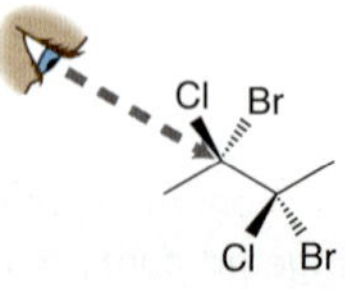


**4.57** Glucose (a sugar) is produced by photosynthesis and is used by cells to store energy. Draw the most stable conformation of glucose:

**Glucose**

**4.58** Sketch an energy diagram showing a conformational analysis of 2,2,3,3-tetramethylbutane. Use Table 4.6 to determine the energy difference between staggered and eclipsed conformations of this compound.

**4.59** Rank the following conformations in order of increasing energy:

**4.60** Consider the following two conformations of 2,3-dimethylbutane. For each of these conformations, use Table 4.6 to determine the total energy cost associated with all torsional strain and steric strain.

(a) (b)

**4.61** *myo*-Inositol is a polyol (a compound containing many OH groups) that serves as the structural basis for a number of secondary messengers in eukaryotic cells. Draw the more stable chair conformation of *myo*-inositol.

**4.62** Below is the numbered skeleton of *trans*-decalin:

Identify whether each of the following substituents would be in an equatorial position or an axial position:

(a) A group at the C-2 position, pointing UP

(b) A group at the C-3 position, pointing DOWN

(c) A group at the C-4 position, pointing DOWN

(d) A group at the C-7 position, pointing DOWN

(e) A group at the C-8 position, pointing UP

(f) A group at the C-9 position, pointing UP

**4.63** Propylene is produced by cracking petroleum and is a very useful precursor in the production of many useful polymers. Propylene has one constitutional isomer. Draw that isomer and identify its systematic name.

**Propylene**

# INTEGRATED PROBLEMS

**4.64** *trans*-1,3-Dichlorocyclobutane has a measurable dipole moment. Explain why the individual dipole moments of the C—Cl bonds do not cancel each other to produce a zero net dipole moment.

***trans*-1,3-Dichlorocyclobutane**

**4.65** Do you expect cyclohexene to adopt a chair conformation? Why or why not? Explain.

**Cyclohexene**

**4.66** For each pair of compounds below, determine whether they are identical compounds, constitutional isomers, stereoisomers, or different conformations of the same compound:

(a)

(b)

(c)

(d)

(e)

(f)

(g)

(h)

(i)

(j)

(k)

(l)

**4.67** Consider the structures of *cis*-1,2-dimethylcyclopropane and *trans*-1,2-dimethylcyclopropane:

(a) Which compound would you expect to be more stable? Explain your choice.

(b) Predict the difference in energy between these two compounds.

**4.68** Consider the following tetra-substituted cyclohexane:

(a) Draw both chair conformations of this compound.

(b) Determine which conformation is more stable.

(c) At equilibrium, would you expect the compound to spend more than 95% of its time in the more stable chair conformation?

**4.69** Consider the structures of *cis*-decalin and *trans*-decalin:

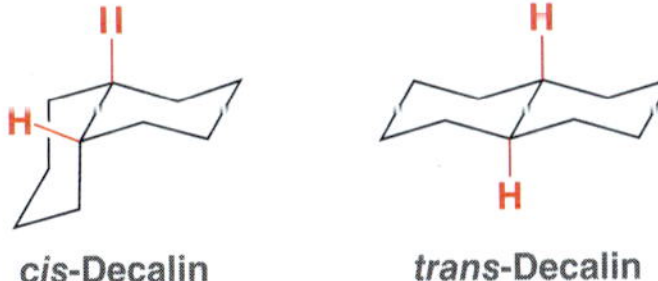


***cis*-Decalin** ***trans*-Decalin**

(a) Which of these compounds would you expect to be more stable?

(b) One of these two compounds is incapable of ring flipping. Identify it and explain your choice.

# CHALLENGE PROBLEMS

**4.70** A biosynthetic pathway was recently proposed for the polycyclic, cytotoxic compound aspernomine, isolated from the fungus *Aspergillus nomius* (*J. Am. Chem. Soc.* **2012,** *134,* **8078–8081**):

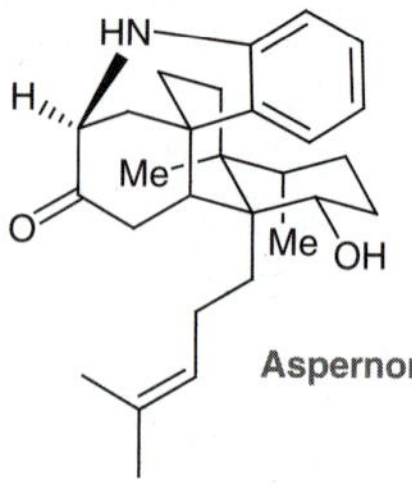

Aspernomine

(a) Two of the six-membered rings are represented in chair conformations. Is the connection between these two rings analogous to that in *cis*- or *trans*-decalin?

(b) What is the approximate dihedral angle between the two methyl groups directly bound to the chairs?

(c) Is the aromatic ring in an axial or equatorial position?

(d) Consider the six-carbon acyclic substituent bound to one of the bridgehead carbons. What is its relationship (axial or equatorial) to each of the chairs?

**4.71** Draw all three staggered conformations for the following compound, viewed along the $C_2$—$C_3$ bond. Determine which conformation is the most stable, taking into account *gauche* interactions and hydrogen-bonding interactions (*J. Phys. Chem. A* **2001,** *105,* **6991–6997**). Provide a reason for your choice by identifying all of the interactions that led to your decision.

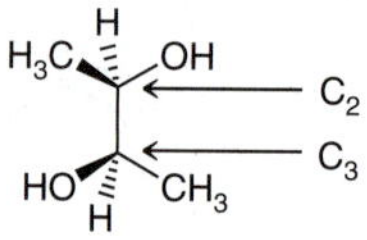

**4.72** The conformations of methanol, viewed along the C—O bond, include both the staggered and the eclipsed forms, just as discussed in the case for ethane. The barrier for interconversion between these two conformations has been determined to be approximately 4.2 kJ/mol (or 1 kcal/mol) (*J. Phys. Chem. A* **2002,** *106,* **1642–1646**).

(a) Draw a Newman projection for each conformation.

(b) Determine the energy cost (amount of torsional strain) associated with the eclipsing interaction between a lone pair and a C—H bond.

**4.73** The following compound was designed as a molecular switch, in that it can be induced to access any of the four orientations shown (**1–4**) using pH as a switch (*Org. Lett.* **2011,** *13,* **30–33**). Describe the changes in (i) bonding and (ii) conformation and (iii) intramolecular hydrogen bonding shown as this compound is converted from **1** → **2** → **3** → **4**:

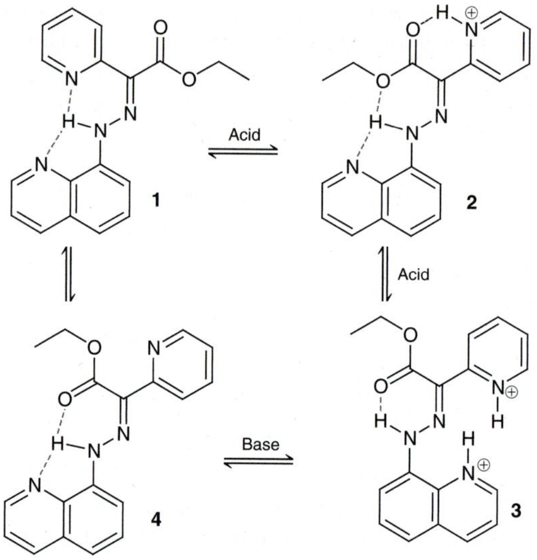

**4.74** Organic compounds with extended regions of consecutive $sp^2$-hybridized atoms will sometimes exhibit an ability to absorb UV light and subsequently emit visible light, a property called fluorescence. We will explore a related phenomenon (absorbance in UV/visible wavelengths) in Chapter 17. The compound below (R = polyether group) is known to be fluorescent (*J. Org. Chem.* **2012,** *77,* **7479–7486**):

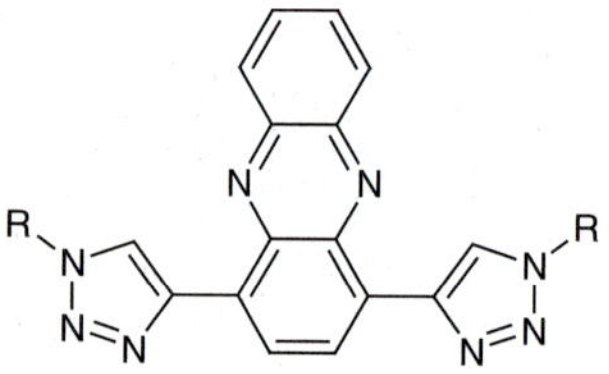

**Planar conformation A** 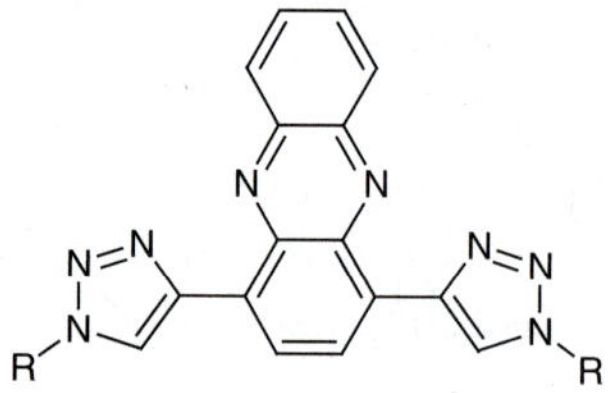 **Planar conformation B**

(a) Identify the σ bonds that must undergo free rotation in order for conformation A to be converted into conformation B and describe what happens to the geometry of the molecule during this conformational change.

(b) Conformation A is more stable than conformation B by 8.96 kcal/mol. Propose an explanation for this observation.

**4.75** The conformations of (+)-epichlorohydrin **(1)**, viewed along the $C_a$—$C_b$ bond, can be analyzed in exactly the same manner as the acyclic alkanes discussed in Chapter 4 (*J. Phys. Chem. A* **2000,** *104,* **6189–6196**).

(a) Draw all staggered conformations for **1** viewed along this bond.

(b) Identify the least stable staggered conformation for **1**.

(c) The most stable conformation of 1,2-dichloropropane exhibits a methyl group that is *anti* to a chlorine atom. Using this information, identify the most stable conformation of **1** and justify your choice.

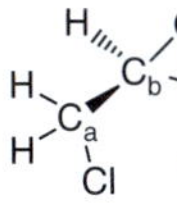

**4.76** The all-*trans*-1,2,3,4,5,6-hexaethylcyclohexane **(1)** prefers the all-equatorial conformation while the all-*trans*-1,2,3,4,5,6-hexaisopropylcyclohexane **(2)** possesses a severely destabilized all-equatorial conformation (*J. Am. Chem. Soc.* **1990,** *112,* **893–894**):

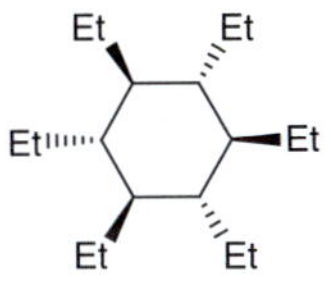

(a) By examining a molecular model of cyclohexane with several all-trans-equatorial isopropyl groups and another model with several all-*trans*-equatorial ethyl groups, determine why adjacent equatorially oriented isopropyl groups experience severe steric interactions which are lacking in the ethyl case. Draw a chair conformation of the former case which illustrates these severe steric interactions. Also draw a Newman projection looking down one of the C—C bonds connecting the cyclohexyl ring to an equatorial isopropyl group and illustrate a conformation with severe steric strain.

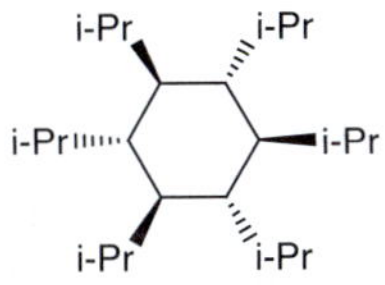

(b) The all-axial conformation of **2** possesses a fairly stable conformation where the steric repulsion among the axial isopropyl groups can be minimized. Again, use models to examine this chair conformation and draw it. Also, draw a Newman projection looking down one of the C—C bonds connecting the cyclohexyl ring to an axial isopropyl group and illustrate why this conformation is lower than the all-equatorial chair conformation in steric destabilization.

**4.77** Alkyne **1** can adopt the coplanar conformation **1a** or the perpendicular conformation **1b** (*Chem. Rev.* **2010,** *110,* **5398–5424**). Predict which form is more favorable for **1** and provide a complete explanation for your choice.

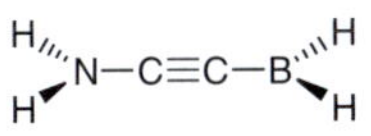

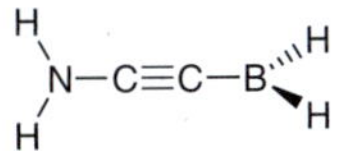

**4.78** The structural unit below has been incorporated into a synthetic polymer designed to mimic the skeletal muscle protein *titin* (*J. Am. Chem. Soc.* **2009,** *131,* **8766–8768**). In its most stable conformation (not the one shown), it forms one intramolecular hydrogen bond and four intermolecular hydrogen bonds to an identical unit in the same conformation. Draw two equivalents of this molecule and clearly show all of the interactions described above.

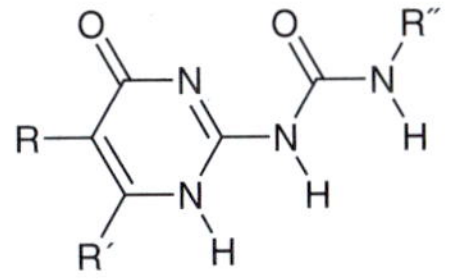

**4.79** Compounds **1** and **2** were prepared, and the difference in their respective heats of combustion was found to be 17.2 kJ/mol (*J. Am. Chem. Soc.* **1961,** *83,* **606–614**):

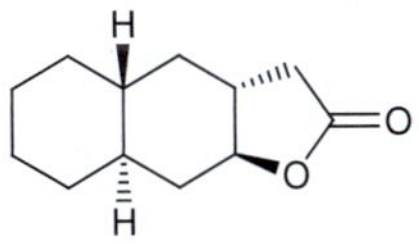

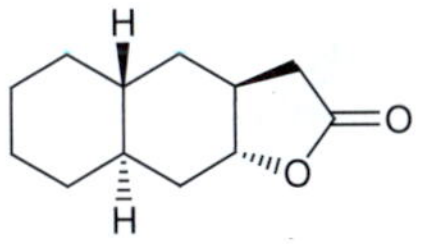

(a) Redraw compounds **1** and **2** showing chair conformations for the six-membered rings. In each case, draw the lowest energy conformation for the compound.

(b) Identify which compound has the larger heat of combustion. Explain your reasoning.

# 5 Stereoisomerism

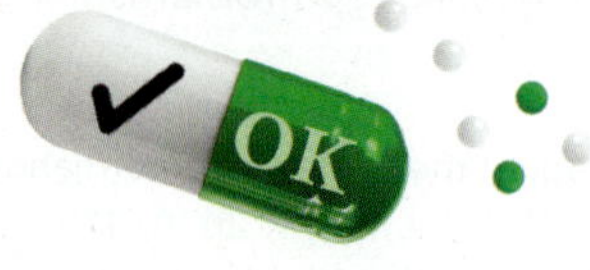




## DID YOU EVER WONDER...

### whether pharmaceuticals are really safe?

As mentioned in the previous chapter, most drugs cause multiple physiological responses. In general, one response is desirable, while the rest are undesirable. The undesirable responses can range in severity and can even result in death. This might sound scary, but the FDA (Food and Drug Administration) has strict guidelines that must be followed before a drug is approved for sale to the public. Every potential drug must first undergo animal testing followed by three separate clinical trials involving human patients (phases I, II, and III). Phase I involves a small number of patients (20–80), phase II can involve several hundred patients, and phase III can involve several thousand patients. The FDA will only approve drugs that have passed all three phases of clinical trials. Even after receiving approval to be sold, the effects of a drug are still monitored for potential long-term adverse effects. In some cases, a drug must be pulled off the market after adverse side effects are observed in a small percentage of the population. One such example is Vioxx, an anti-inflammatory drug that was heavily used in the treatment of osteoarthritis and acute pain. Vioxx was approved by the FDA in 1999 and instantly became very popular, providing its manufacturer (Merck & Co.) with over two billion dollars in annual sales. In 2004, Merck had to remove Vioxx from the market because of concerns that long-term use increased the risk of heart attacks and strokes. This example highlights the fact that pharmaceutical safety cannot be absolutely guaranteed. Nevertheless, the development of pharmaceuticals has vastly improved our quality of life and longevity, and the positive effects significantly outweigh the rare examples of negative effects.

The effects of any particular drug are determined by a host of factors. We have already seen many of these factors, including the role of the pharmacophore, conformational flexibility, and acid-base properties. In this chapter, we will take a closer look at the three-dimensional structure of compounds, and we will see that this characteristic is arguably one of the most important factors to be considered when designing drugs and assessing

continued >

their safety. In particular, we will explore compounds that differ from each other only in the three-dimensional, spatial arrangement of their atoms, but not in the connectivity of their atoms. Such compounds are called *stereoisomers*, and we will explore the connection between stereoisomerism and drug action.

This chapter will focus on the different kinds of stereoisomers. We will learn to identify stereoisomers, and we will learn several drawing styles that will allow us to compare stereoisomers. The upcoming chapters will focus on reactions that produce stereoisomers.

## DO YOU REMEMBER?

*Before you go on, be sure you understand the following topics.*
*If necessary, review the suggested sections to prepare for this chapter.*

- Constitutional Isomerism (Section 1.2)
- Tetrahedral Geometry (Section 1.10)
- Three-Dimensional Representations (Section 2.1)
- Drawing and Interpreting Bond-Line Structures (Section 2.2)

Take the DO YOU REMEMBER? QUIZ in **WileyPLUS** to check your understanding.

## 5.1 Overview of Isomerism

The term *isomers* comes from the Greek words *isos* and *meros*, meaning "made of the same parts." That is, isomers are compounds that are constructed from the same atoms (same molecular formula) but that still differ from each other. We have already seen two kinds of isomers: constitutional isomers (Section 4.3) and stereoisomers (Section 4.14), as illustrated in Figure 5.1.

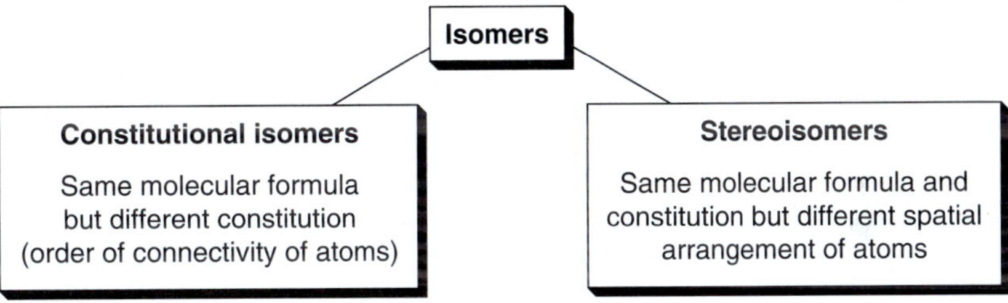


**FIGURE 5.1** The main categories of isomers.

Constitutional isomers differ in the connectivity of their atoms; for example:

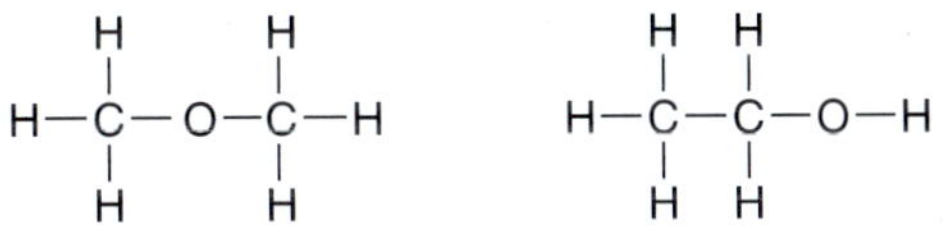

**Methoxymethane**
Boiling point = −23°C

**Ethanol**
Boiling point = 78.4°C

The two compounds above have the same molecular formula, but they differ in their constitution. As a result, they are different compounds with different physical properties.

Stereoisomers are compounds that have the same constitution but differ in the spatial arrangement of their atoms. In the previous chapters, we discussed one example of *cis-trans* stereoisomerism among disubstituted cycloalkanes:

***cis*-1,2-Dimethylcyclohexane**

***trans*-1,2-Dimethylcyclohexane**

The *cis* stereoisomer exhibits groups on the same side of the ring, while the *trans* stereoisomer exhibits groups on opposite sides of the ring. In addition to the examples above, the terms *cis* and *trans* are also used to describe stereoisomerism among double bonds:

$H_3C$ $CH_3$ H H
**cis-2-Butene**
Boiling point = 4°C

$H_3C$ H H $CH_3$
***trans*-2-Butene**
Boiling point = 1°C

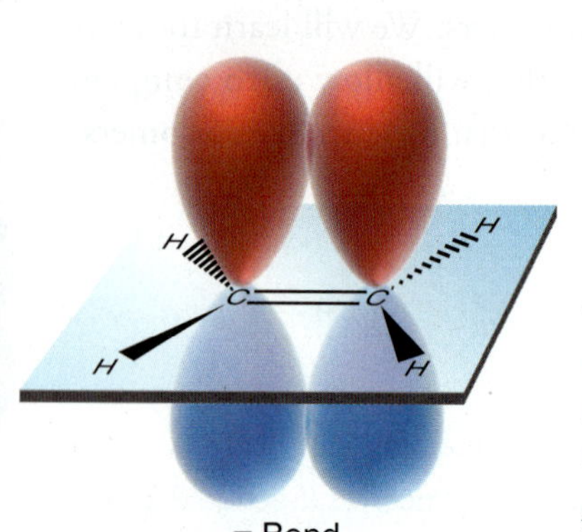


**FIGURE 5.2**
An illustration of *p* orbitals overlapping to form a π bond.

The *cis* stereoisomer exhibits groups on the same side of the double bond, while the *trans* stereoisomer exhibits groups on opposite sides of the double bond. The two drawings above represent different compounds with different physical properties, because the double bond does not experience free rotation as single bonds do. Why not? Recall that a π bond is formed from the overlap of two *p* orbitals (Figure 5.2). Rotation about the C—C double bond would effectively destroy the overlap between the *p* orbitals. Therefore, the C—C double bond does not experience free rotation at room temperature.

In order to use the *cis-trans* terminology to differentiate stereoisomers, there must be two identical groups to compare:

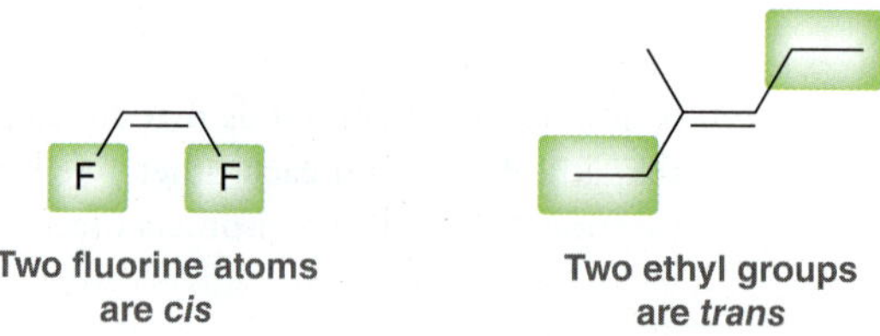


Hydrogen atoms can also be used to assign *cis-trans* terminology; for example:

$H_3C$ F **is *trans* because of the H's:** $H_3C$ H F H

When two identical groups are connected to the same position, there cannot be *cis-trans* isomerism. For example, consider the following compound:

Cl Cl **is the same as** Cl Cl

These two drawings represent the same compound. This compound has two chlorine atoms connected to the same position, and therefore, the compound does not exhibit stereoisomerism. This is true any time there are two identical groups connected to the same position. Here is one more example:

Do not try to identify this compound as *cis* or *trans*. Two identical groups (methyl groups) are connected to the same position, and therefore, this compound is neither *cis* nor *trans*.

# SKILLBUILDER

## 5.1 IDENTIFYING *CIS-TRANS* STEREOISOMERISM

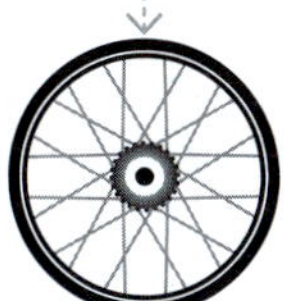

**LEARN the skill**

Determine whether the stereoisomer shown below exhibits a *cis* configuration or a *trans* configuration:

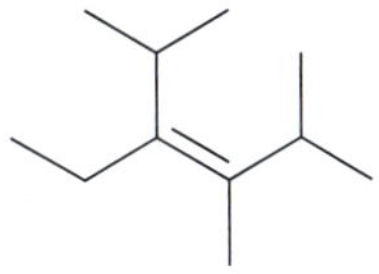

**SOLUTION**

Begin by circling the four groups attached to the double bond and try to name them:

**STEP 1**
Identify and name all four groups attached to the double bond.

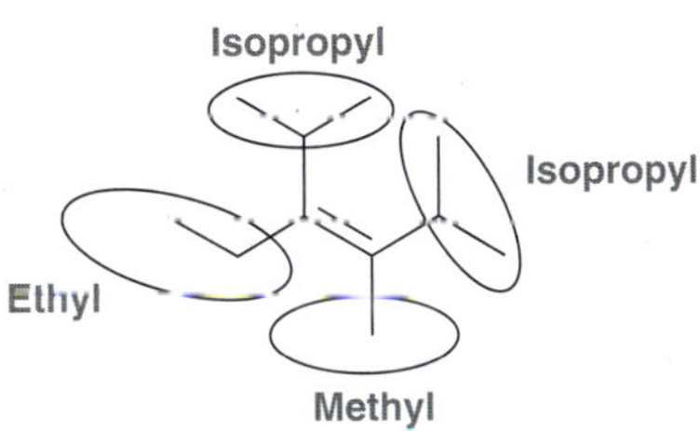


**STEP 2**
Look for two identical groups on different vinylic positions and assign the configuration as *cis* or *trans*.

Naming the four groups helps to identify identical groups. There are always four groups (even if some of those groups are just hydrogen atoms). In this case, naming the groups makes it evident that there are two isopropyl groups that are *cis* to each other.

**PRACTICE the skill**

**5.1** For each of the following compounds determine whether it exhibits a *cis* configuration or a *trans* configuration or whether it is simply not stereoisomeric:

(a)  (b) 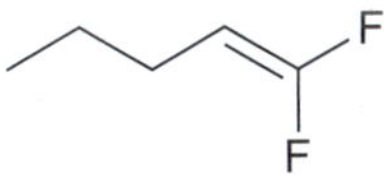 F F (c)

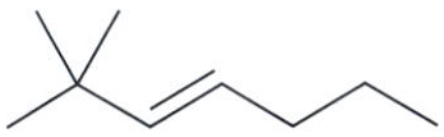

(d) 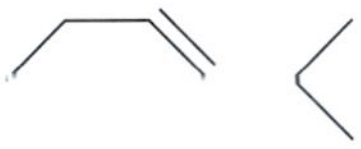 (e)

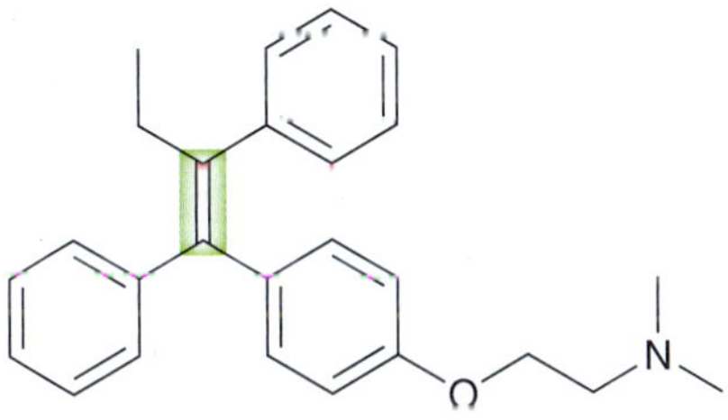

O N

**Tamoxifen**
Used in treatment of breast cancer

(f) 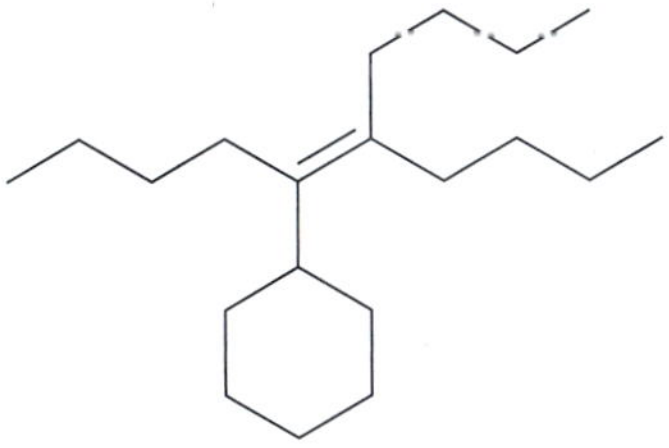 (g)

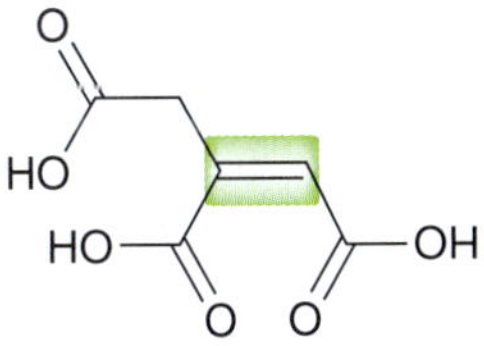

O HO HO OH O O

**Aconitic acid**
Involved in metabolism

**5.2** Identify the number of stereoisomers that are possible for a compound with the following constitution: $H_2C{=}CHCH_2CH_2CH_2CH{=}CH_2$.

**5.3** Compound X and compound Y are constitutional isomers with molecular formula $C_5H_{10}$. Compound X possesses a carbon-carbon double bond in the *trans* configuration, while compound Y possesses a carbon-carbon double bond that is not stereoisomeric:

**(a)** Identify the structure of compound X.

**(b)** Identify four possible structures for compound Y.

need more **PRACTICE?** **Try Problem 5.35**

## 5.2 Introduction to Stereoisomerism

In the previous section, we reviewed *cis-trans* stereoisomerism, but there are many other kinds of stereoisomers. We begin our exploration of the various kinds of stereoisomers by investigating the relationship between an object and its mirror image.

### Chirality

Any object can be viewed in a mirror, revealing its mirror image. Take, for example, a pair of sunglasses (Figure 5.3). For many objects, like the sunglasses in Figure 5.3, the mirror image is identical to the actual object. The object and its mirror image are said to be **superimposable**. This is not the case if we remove one of the lenses (Figure 5.4). The object and its mirror image are now different. One pair is missing the right lens, while the other pair is missing the left lens. In this case, the object and its mirror image are *nonsuperimposable*. Many familiar objects, such as hands, are nonsuperimposable on their mirror images. A right hand and a left hand are mirror images of one another, but they are not identical; they are not superimposable on one another. A left hand will not fit into a right-handed glove, and a right hand will not fit into a left-handed glove.

Objects, like hands, that are not superimposable on their mirror images are called **chiral** objects, from the Greek word *cheir* (meaning "hand"). All three-dimensional objects can be classified as either chiral or achiral. Molecules are three-dimensional objects and can therefore also be classified into one of these two categories. Chiral molecules are like hands: They are nonsuperimposable on their mirror images. Achiral molecules are not like hands: They are superimposable on their mirror images. What makes a molecule chiral?

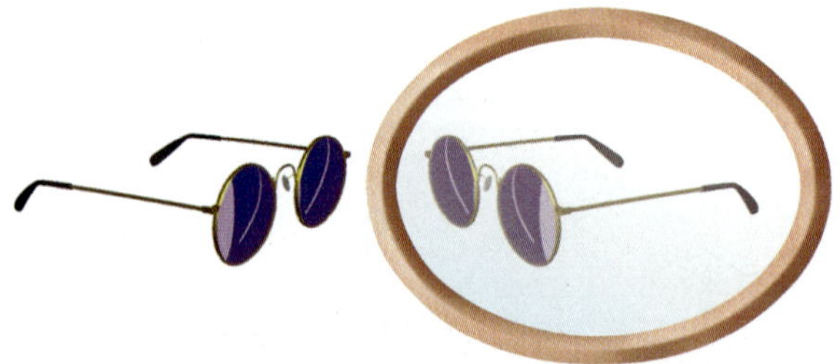

**FIGURE 5.3**
An object that is superimposable on its mirror image.

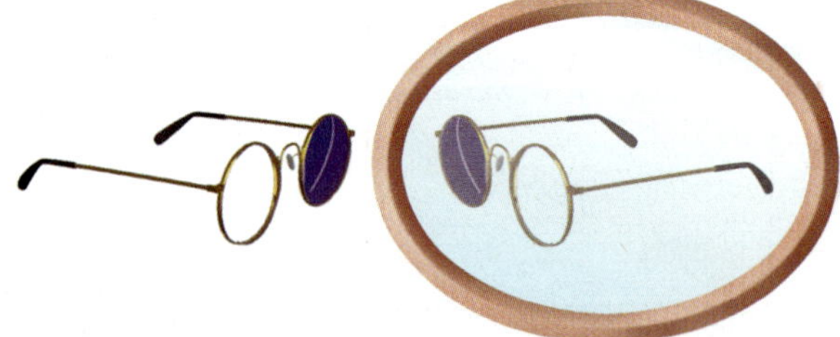

**FIGURE 5.4**
An object that is not superimposable on its mirror image.

**LOOKING AHEAD**
For other sources of molecular chirality, see Problems 5.58–5.60 at the end of this chapter.

## Chirality Centers

The most common source of molecular chirality is the presence of a carbon atom bearing four different groups. There are two different ways to arrange four groups around a central carbon atom (Figure 5.5). These two arrangements are nonsuperimposable mirror images.

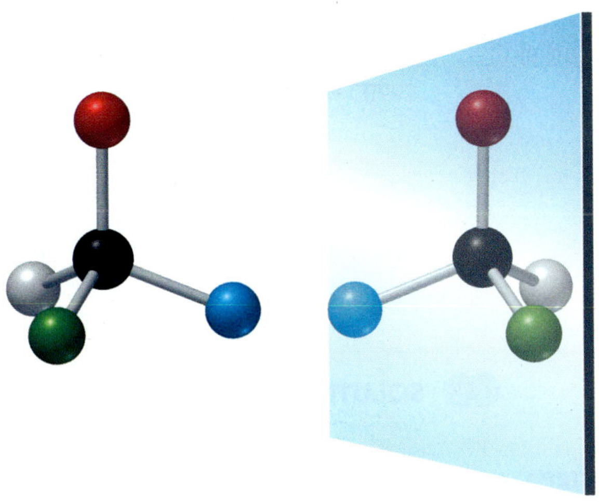

**FIGURE 5.5** There are two ways to arrange four different groups around a carbon atom.

**BY THE WAY**
To see that these two compounds are nonsuperimposable, build a molecular model using any one of the commercially available molecular model kits.

Consider, for example, the structure of 2-butanol, which can be arranged in two ways in three-dimensional space:

OH Et Me H | Mirror | OH H Et Me

These two compounds are nonsuperimposable mirror images, and they represent two different compounds. These compounds differ from each other only in the spatial arrangement of their atoms, and therefore, they are stereoisomers.

In 1996, the IUPAC recommended that a tetrahedral carbon bearing four different groups be called a **chirality center**. Many other names in common use include *chiral center*, *stereocenter*, *stereogenic center*, and *asymmetric center*. For the remainder of our discussion, we will use the IUPAC-recommended term. Below are several examples of chirality centers:

Cl   OH   Br

**BY THE WAY**
Despite the IUPAC recommendation to use the term "chirality center," the terms stereocenter and stereogenic center are more commonly used. However, these terms have a broader definition: A stereocenter, or stereogenic center, is defined as a location at which the interchange of two substituents will generate a stereoisomer. This definition includes chirality centers, but it also includes *cis* and *trans* double bonds, which are not chirality centers.

Each of the highlighted carbon atoms bears four different groups. In the last compound, the carbon atom is a chirality center because one path around the ring is different than the other path (one path encounters the double bond sooner). In contrast, the following compound does not have a chirality center. In this case, the clockwise path and the counterclockwise path are identical.

# SKILLBUILDER

## 5.2 LOCATING CHIRALITY CENTERS

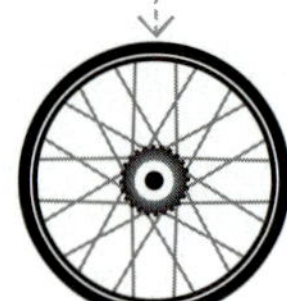

**LEARN the skill**

Propoxyphene, sold under the trade name Darvon, is an analgesic (painkiller) and antitussive (cough suppressant). Identify all chirality centers in propoxyphene:

### SOLUTION

**STEP 1**
Ignore $sp^2$- and $sp$-hybridized carbon atoms.

We are looking for a tetrahedral carbon atom that bears four different groups. Each of the $sp^2$-hybridized carbon atoms are connected to only three groups, rather than four. So, none of these carbon atoms can be chirality centers:

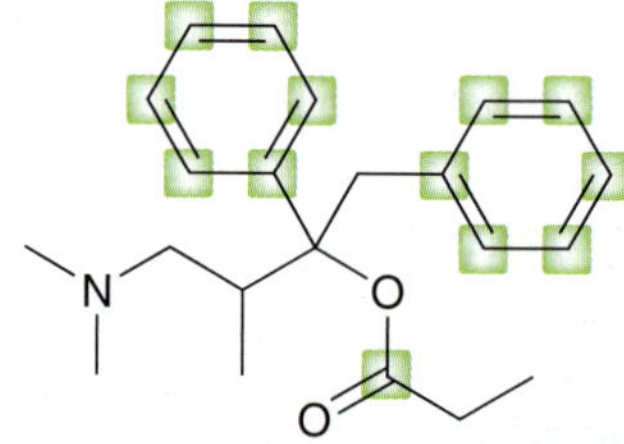


**These carbon atoms cannot be chirality centers**

We can also rule out any $CH_2$ groups or $CH_3$ groups, as these carbon atoms do not bear four *different* groups:

**STEP 2**
Ignore $CH_2$ and $CH_3$ groups.

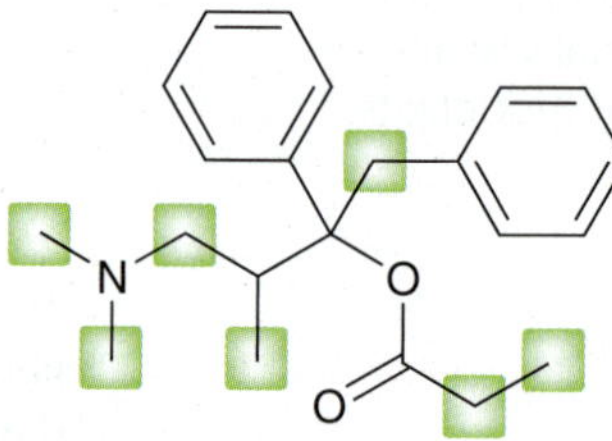


**These carbon atoms cannot be chirality centers**

It is helpful to become proficient at identifying carbon atoms that cannot be chirality centers. This will make it easier to spot the chirality centers more quickly. In this example, there are only two carbon atoms worth considering:

**STEP 3**
Identify any carbon atoms bearing four different groups.

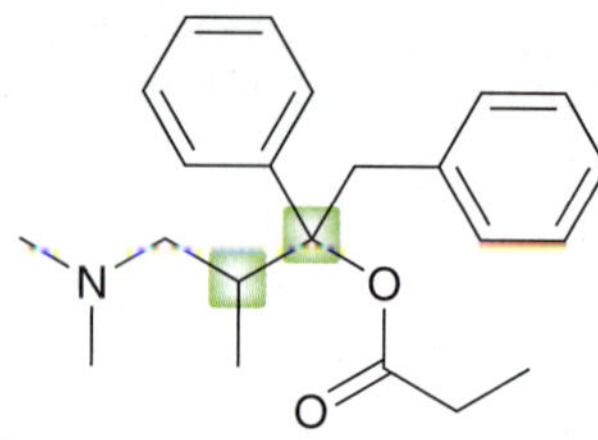


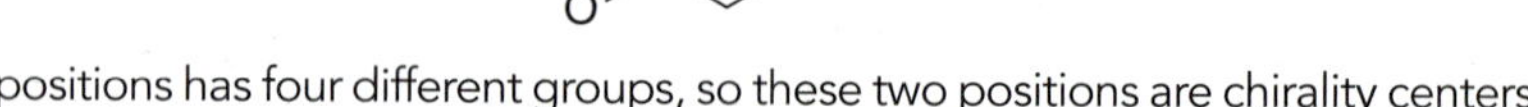

Each of these positions has four different groups, so these two positions are chirality centers.

**PRACTICE the skill** **5.4** Identify all chirality centers in each of the following compounds:

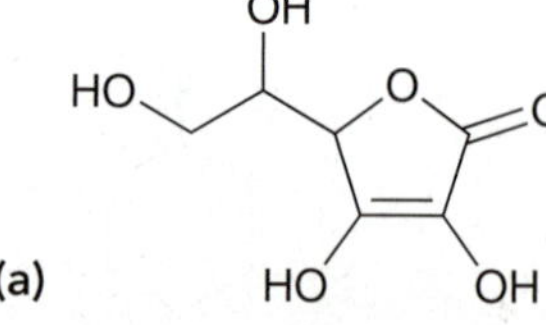


(a) **Ascorbic acid (vitamin C)**

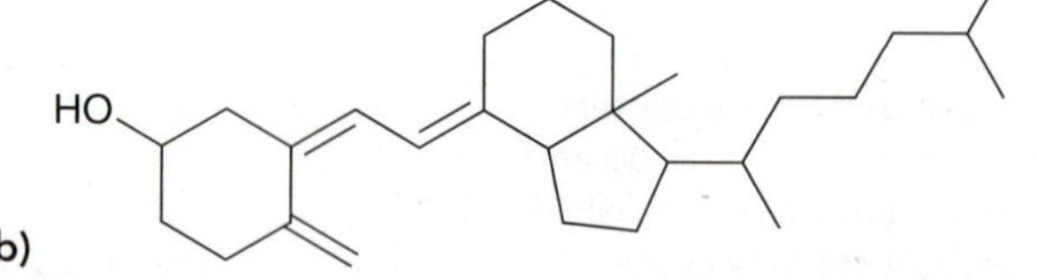


(b) **Vitamin $D_3$**

(c)

OH

O

**Mestranol**
An oral contraceptive

HO

OH

N

O

OH

(d)

**Fexofenadine**
Nonsedating antihistamine

**5.5** Draw all constitutional isomers of $C_4H_9Br$ and identify the isomer(s) that possess chirality centers.

**5.6** Do you expect the following compound to be chiral? Explain your answer (consider whether this compound is superimposable on its mirror image).

P

need more **PRACTICE?** **Try Problems 5.34b, 5.48**

## Enantiomers

When a compound is chiral, it will have one nonsuperimposable mirror image, called its **enantiomer** (from the Greek word meaning "opposite"). The compound and its mirror image are said to be a pair of enantiomers. The word "enantiomer" is used in speech in the same way that the word "twin" is used in speech. When two children are *a pair of twins*, each one is said to be *the twin of the other*. Similarly, when two compounds are *a pair of enantiomers*, each compound is said to be *the enantiomer of the other*. A chiral compound will have exactly one enantiomer, never more and never less. Let's practice drawing the enantiomer of chiral compounds.

# SKILLBUILDER

## 5.3 DRAWING AN ENANTIOMER

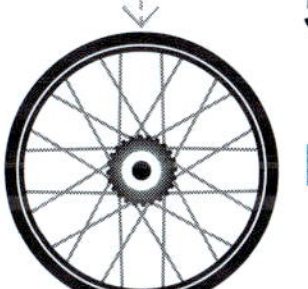


Amphetamine is a prescription stimulant used in the treatment of ADHD (attention-deficit hyperactivity disorder) and chronic fatigue syndrome. During World War II it was used heavily by soldiers to reduce fatigue and increase alertness. Draw the enantiomer of amphetamine.

$NH_2$

**Amphetamine**

### SOLUTION

This problem is asking us to draw the mirror image of the compound above. There are three ways to do this, because there are three places where we can imagine placing a mirror: (1) behind, (2) next to, or (3) beneath the molecule:

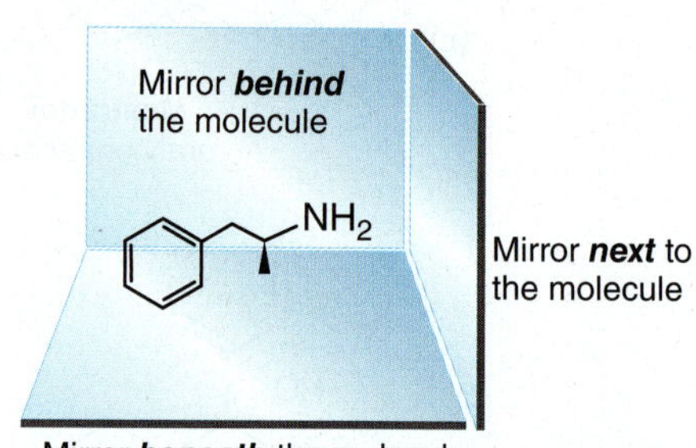


**BY THE WAY**
In most cases, it is easiest to draw a mirror image by placing the mirror behind the molecule.

It is easiest to place the mirror behind the molecule, because the skeleton of the molecule is drawn in exactly the same way except that *all dashes become wedges and all wedges become dashes.* Every other aspect of the molecule is drawn in exactly the same way:

Mirror **behind** the molecule

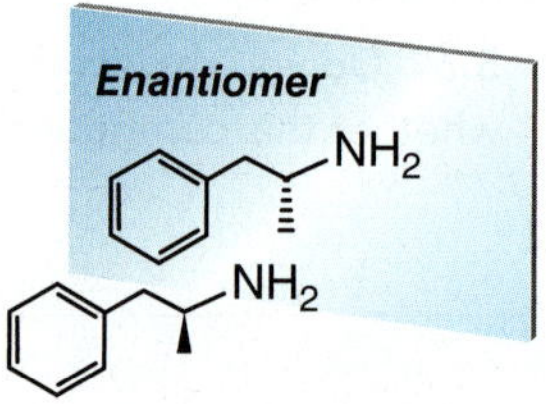


The second way to draw an enantiomer is to place the mirror on the side of the molecule. When doing so, draw the mirror image of the skeleton, but all dashes remain dashes and all wedges remain wedges:

Mirror **next** to the molecule

$NH_2$ $H_2N$

**Enantiomer**

Finally, we can place the mirror under the molecule. When doing so, draw the mirror image of the skeleton; once again, all dashes remain dashes and all wedges remain wedges:

$NH_2$

Mirror **beneath** the molecule

$NH_2$

**Enantiomer**

Of the three ways to draw an enantiomer, only the first method involves interchanging wedges and dashes. Even so, make sure to understand that all three of these methods produce the same answer:

$NH_2$ is the same as $H_2N$ is the same as $NH_2$

All three drawings represent the same compound. They might look different, but rotation of any one drawing will generate the other drawings. Remember that a molecule can only have one enantiomer, so all three methods must produce the same answer.

In general, it is easiest to use the first method. Simply redraw the compound, replacing all dashes with wedges and all wedges with dashes. However, this method will not work in all situations. In some molecular drawings, wedges and dashes are not drawn, because the three-dimensional geometry is implied by the drawing. This is the case for bicyclic compounds. When dealing with bicyclic compounds, it will be easier to use one of the other two methods (placing the mirror either on the side of the molecule or below the molecule). For example:

Mirror

H Cl — H Cl

***Compound A*** ***Compound B***

In this case, it is easiest to place the mirror on the side of compound A, which allows us to draw compound B (the enantiomer of compound A).

**PRACTICE the skill** **5.7** Draw the enantiomer of each of the following compounds:

(a)

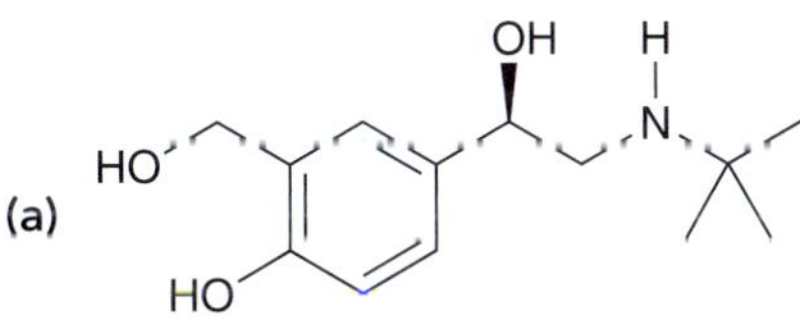


**Albuterol**
**(sold under the trade name Ventolin)**
A bronchodilator used
in the treatment of asthma

(b) O OH N H

**Propranolol**
A beta blocker used in the
treatment of hypertension

(c)

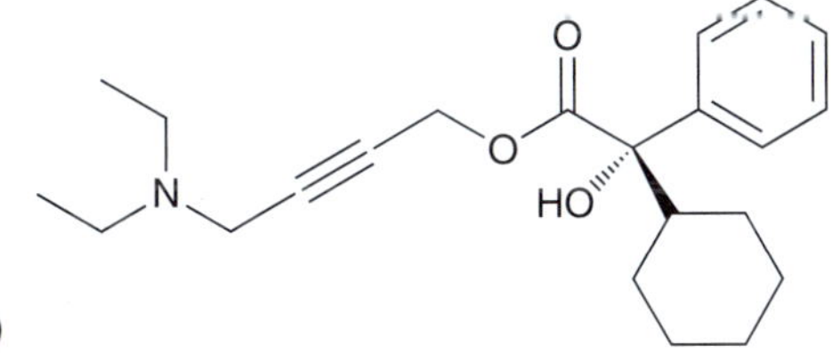


**Oxybutynin**
Used to treat urinary and bladder disorders

(d) HO HO OH H N $CH_3$

**Adrenaline**
A hormone that acts
as a bronchodilator

(e) Ketamine
An anesthetic

(f) 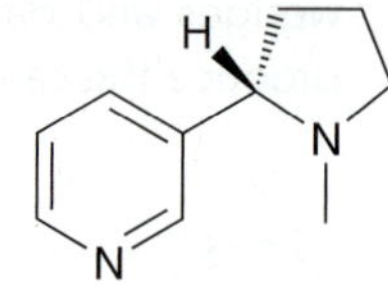

Nicotine
An addictive substance present in tobacco

(g) 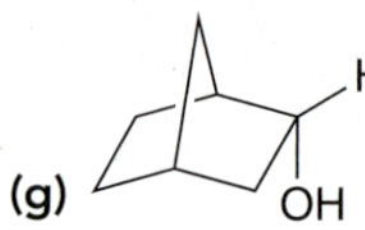

**5.8** Ixabepilone is a cytotoxic compound approved by the FDA in 2007 for the treatment of advanced breast cancer. Bristol-Myers Squibb is marketing this drug under the trade name Ixempra. Draw the enantiomer of this compound:

Ixabepilone

need more **PRACTICE?** **Try Problems 5.33, 5.34a, 5.38a–f, i–l**

## practically speaking — The Sense of Smell

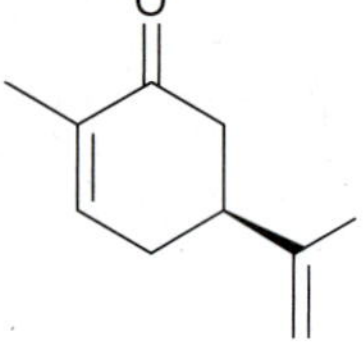

In order for an object to have an odor, it must release organic compounds into the air. Most plastic and metal items do not release molecules into the air at room temperature and are therefore odorless. Spices, on the other hand, have very strong odors, because they release many organic compounds. These compounds enter your nose when you inhale, where they encounter receptors that detect their presence. The compounds bind to the receptors, causing the transmission of nerve signals that the brain interprets as an odor.

A specific compound can bind to a number of different receptors, creating a pattern that the brain identifies as a particular odor. Different compounds generate different patterns, enabling us to distinguish over 10,000 odors from one another. This mechanism has many fascinating features. In particular, "mirror-image" compounds (enantiomers) will often bind to different receptors, thereby creating different patterns which are interpreted as unique odors. To understand the reason for this, consider the following analogy. A right hand will fit into a brown paper bag just as easily as a left hand will, because the bag is not chiral. However, a right hand will not fit into a left-handed glove just as easily as a left hand will, because the glove itself is chiral. Similarly, if a receptor is chiral, as is usually the case, then we might expect that only one enantiomer of a pair will effectively bind to it.

As an example, carvone has one chirality center and therefore has two enantiomeric forms:

**(*R*)-Carvone**
(odor of spearmint)

**(*S*)-Carvone**
(odor of caraway seeds)

One enantiomer is responsible for the odor of spearmint, while the other is responsible for the odor of caraway seeds (the seeds found in seeded rye bread). In order for us to detect different odors, our receptors must themselves be chiral. This illustrates the chiral environment of the human body (a sea of chiral molecules interacting with each other). We will soon explore how drug action is also determined by the chiral nature of most biological receptors.

## 5.3 Designating Configuration Using the Cahn-Ingold-Prelog System

In the previous section, we used bond-line structures to illustrate the difference between a pair of enantiomers:

A pair of enantiomers

**LOOKING AHEAD**
A summary of the Cahn-Ingold-Prelog system appears immediately prior to SkillBuilder 5.4.

In order to communicate more effectively, we also need a system of nomenclature for identifying each enantiomer individually. This system is named after the chemists who devised it: Cahn, Ingold, and Prelog. The first step of this system involves assigning priorities to each of the four groups attached to the chirality center, based on atomic number. The atom with the highest atomic number is assigned the highest priority (1), while the atom with the lowest atomic number is assigned the lowest priority (4). As an example, consider one of the enantiomers from above:

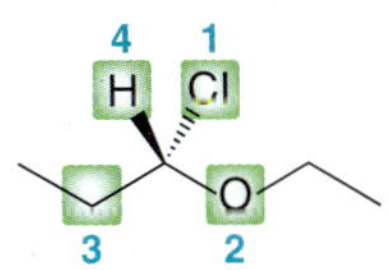


Of the four atoms attached to the chirality center, chlorine has the highest atomic number so it is assigned priority 1. Oxygen has the second highest atomic number, so it is assigned priority 2. Carbon is assigned priority 3, and finally, hydrogen is priority 4 because it has the lowest atomic number.

After assigning priorities to all four groups, we then rotate the molecule so that the fourth priority is directed behind the page (on a dash):

**BY THE WAY**
To visualize this rotation, you may find it helpful to build a molecular model using any one of the commercially available molecular model kits.

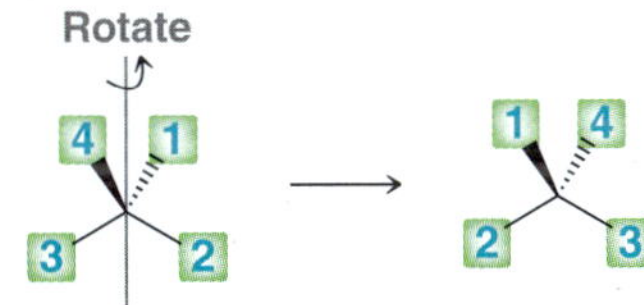


Then, we look to see if the sequence 1-2-3 is clockwise or counterclockwise. In this enantiomer, the sequence is counterclockwise:

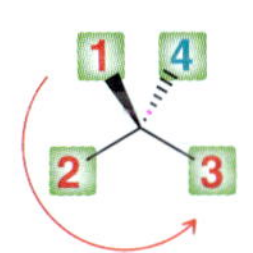


Counterclockwise = ***S***

A counterclockwise sequence is designated as ***S*** (from the Latin word *sinister*, meaning "left"). The enantiomer of this compound will have a clockwise sequence and is designated as ***R*** (from the Latin word *rectus*, meaning "right"):

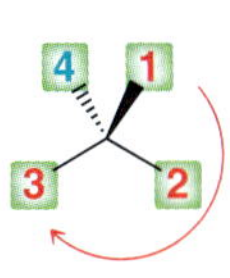


Clockwise = ***R***

The descriptors *R* and *S* are used to describe the **configuration** of a chirality center. Assigning the configuration of a chirality center therefore requires three distinct steps. The following sections will deal with the subtle nuances of these steps.

1. Prioritize all four groups connected to the chirality center.
2. If necessary, rotate the molecule so that the fourth priority group is on a dash.
3. Determine whether the sequence 1-2-3 is clockwise or counterclockwise.

## Assigning Priorities to All Four Groups

The previous example was fairly straightforward because all four atoms attached to the chirality center were different (Cl, O, C, and H). It is more common to encounter two or more atoms with the same atomic number. For example:

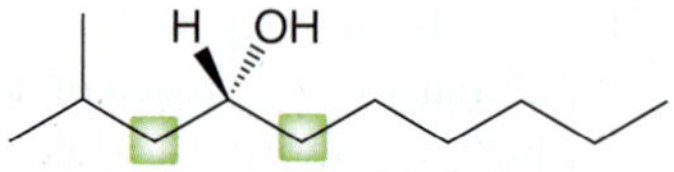

In this case, two carbon atoms are directly connected to the chirality center. Certainly, the oxygen is assigned priority 1 and the hydrogen is assigned priority 4. But which carbon atom is assigned priority 2? In order to make this determination, notice that each carbon atom is connected to the chirality center *and to three other atoms*. Make a list of the three atoms on either side in decreasing order of atomic number and compare:

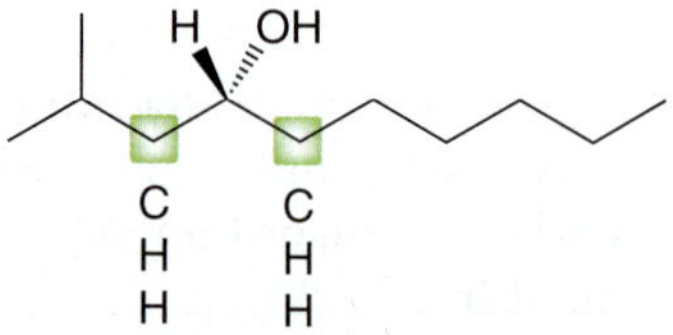

These lists are identical, so we move farther away from the chirality center and repeat the process:

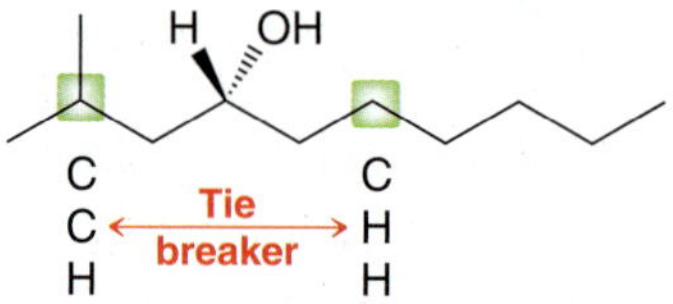

We are specifically looking for the first point of difference. In this case, the left side will be assigned the higher priority, because C has a higher atomic number than H.

Here is another example that illustrates an important point:

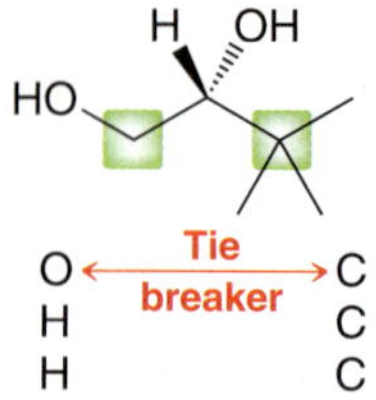

In this case, the left side will be assigned the higher priority, because oxygen has a higher atomic number than carbon. We do not compare the sum total of each list. It might be true that three carbon atoms have a greater sum total of atomic numbers (6+6+6) than one oxygen atom and two hydrogen atoms (8+1+1). However, the deciding factor is the first point of difference, and in this case, oxygen beats carbon.

When assigning priorities, a double bond is considered as two separate single bonds. The following carbon atom (highlighted) is treated as if it is connected to two oxygen atoms:

The list for this carbon atom is

O
O
H

The oxygen atom counts twice

The same rule is applied for any type of multiple bond, as illustrated in the following example:

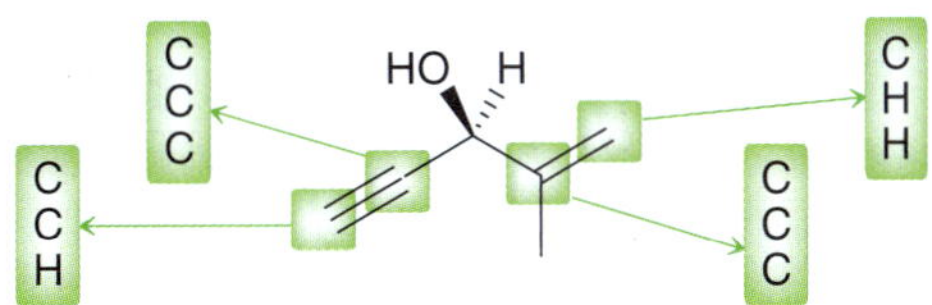


## Rotating the molecule

Some students have trouble rotating a molecule and redrawing it in the proper perspective (with the fourth priority on a dash). If you are having trouble, there is a helpful technique based on the following principle: Switching any two groups on a chirality center will invert the configuration:

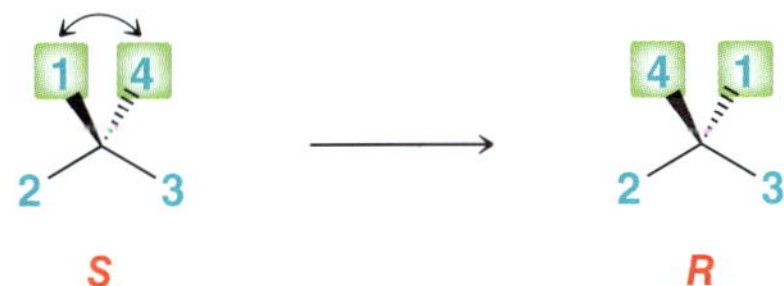


It doesn't matter which two groups are switched. Switching any two groups will change the configuration. Then, switching any two groups a second time will switch the configuration back. Based on this idea, we can devise a simple procedure that will allow us to rotate a molecule without actually having to visualize the rotation. Consider the following example:

To assign the configuration of this chirality center, first find the fourth priority and switch it with the group that is on the dash. Then, switch the other two groups:

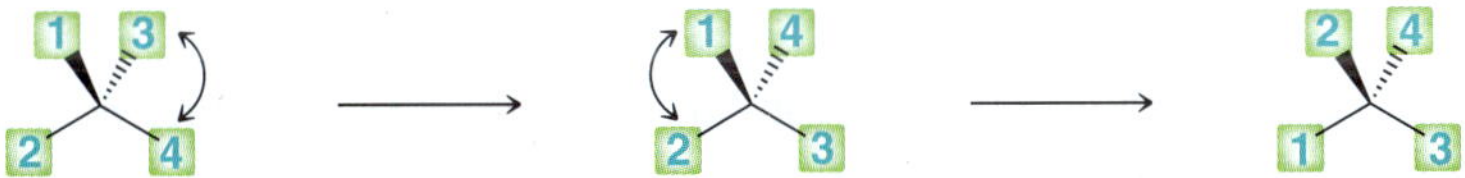


Two switches have been performed, so the final drawing must have the same configuration as the original drawing. But now, the molecule has been redrawn in a perspective that exhibits the fourth priority on a dash. This technique provides a method for redrawing the molecule in the appropriate perspective (as if the molecule had been rotated). In this example, the sequence 1-2-3 is clockwise, so the configuration is *R*. A summary of the procedure for assigning configuration follows.

**A REVIEW OF CAHN-INGOLD-PRELOG RULES: ASSIGNING THE CONFIGURATION OF A CHIRALITY CENTER**

| STEP 1 | STEP 2 | STEP 3 | STEP 4 | STEP 5 |
|---|---|---|---|---|
| Identify the four atoms directly attached to the chirality center. | Assign a priority to each atom based on its atomic number. The highest atomic number receives priority 1, and the lowest atomic number (often a hydrogen atom) receives priority 4. | If two atoms have the same atomic number, move away from the chirality center looking for the first point of difference. When constructing lists to compare, remember that a double bond is treated as two separate single bonds. | Rotate the molecule so that the fourth priority is on a dash (going behind the plane of the page). | Determine whether the sequence 1-2-3 follows a clockwise order (*R*) or a counterclockwise order (*S*). |

# SKILLBUILDER

## 5.4 ASSIGNING THE CONFIGURATION OF A CHIRALITY CENTER

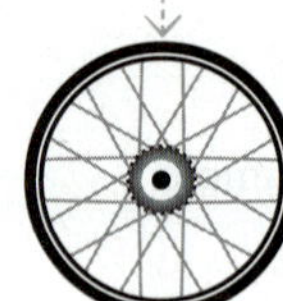

**LEARN the skill**

The following bond-line structure represents one enantiomer of 2-amino-3-(3,4-dihydroxyphenyl)propanoic acid, used in the treatment of Parkinson's disease. Assign the configuration of the chirality center in this compound:

**SOLUTION**

Begin by identifying the four atoms directly attached to the chirality center:

**STEP 1**
Identify the four atoms attached to the chirality center and prioritize by atomic number.

The four atoms are N, C, C, and H. Nitrogen has the highest atomic number, so it is assigned priority 1. Hydrogen is assigned priority 4. We must now decide which carbon atom is assigned priority 2. Make a list in each case and look for the first point of difference:

**STEP 2**
If two or more atoms are carbon, look for the first point of difference.

**Tie breaker**: O, O, O vs. C, H, H

**STEP 3**
Redraw the chirality center showing only priorities.

Oxygen has a higher atomic number than carbon, so the left side is assigned the higher priority. The priorities are therefore arranged in the following way:

In this case, the fourth priority is not on a dash, so the molecule must be rotated into the appropriate perspective:

**STEP 4**
Rotate the molecule so that the fourth priority is on a dash.

Rotate

The sequence of 1-2-3 is counterclockwise, so the configuration is *S*. If the rotation step is difficult to see, the technique described earlier will help. Switch the 4 and the 1, so that the 4 is on a dash. Then switch the 2 and the 3. After two successive switches, the configuration remains the same:

**STEP 5**
Assign the configuration based on the order of the sequence 1-2-3.

Now the 4 is on a dash, where it needs to be. The sequence of 1-2-3 is counterclockwise, so the configuration is *S*.

**PRACTICE the skill** **5.9** Each of the following compounds possesses carbon atoms that are chirality centers. Locate each of these chirality centers and identify the configuration of each one:

(a)

**Ephedrine**
A bronchodilator and decongestant obtained from the Chinese plant *Ephedra sinica*

(b)

**Halomon**
An antitumor agent isolated from marine organisms

(c)

**Streptimidone**
An antibiotic

(d)

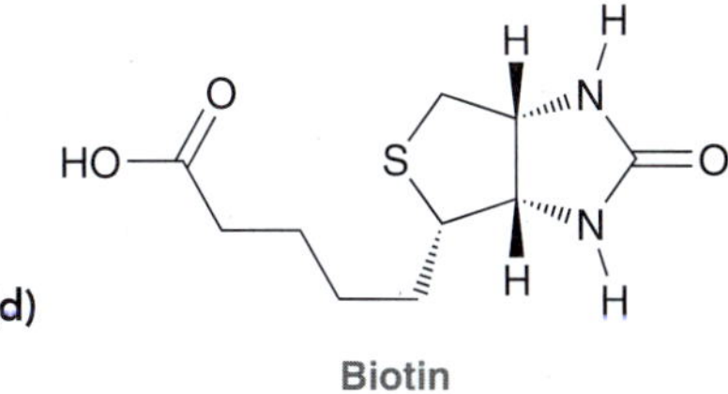


**Biotin**
(vitamin $B_7$)

(e)

**Kumepaloxane**
A signal agent produced by *Haminoea cymbalum*, a snail indigenous to Guam

(f)

**Chloramphenicol**
An antibiotic agent isolated from the *Streptomyces venezuelae* bacterium

**APPLY the skill**

**5.10** Assign the configuration of the chirality center in the following compound:

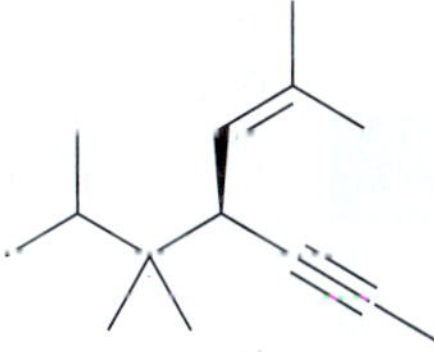

**5.11** Carbon is not the only element that can function as a chirality center. In Problem 5.6 we saw an example in which a phosphorus atom is a chirality center. In such a case, the lone pair is always assigned the fourth priority. Using this information, assign the configuration of the chirality center in this compound:

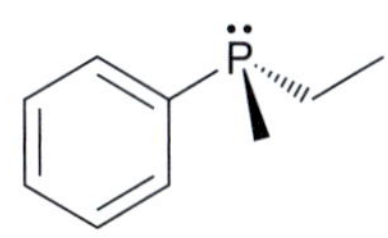


need more **PRACTICE?** **Try Problems 5.32, 5.39a–g,i, 5.45**

## Designating Configuration in IUPAC Nomenclature

When naming a chiral compound, the configuration of the chirality center is indicated at the beginning of the name, italicized, and surrounded by parentheses:

(*R*)-2-Butanol (*S*)-2-Butanol

When multiple chirality centers are present, each configuration must be preceded by a locant (a number) to indicate its location on the parent chain:

(2*R*,3*S*)-3-Methyl-2-pentanol

## medically speaking | Chiral Drugs

Thousands of drugs are marketed throughout the world. The origins of these drugs fall into three categories:

1. natural products—compounds isolated from natural sources, such as plants or bacteria,
2. natural products that have been chemically modified in the laboratory, or
3. synthetic compounds (made entirely in the laboratory).

Most drugs obtained from natural sources consist of a single enantiomer. It is important to realize that a pair of enantiomers will rarely exhibit the same potency. We have seen in previous chapters that drug action is usually the result of drug-receptor binding. If the drug binds to the receptor in at least three places (called three-point binding), then one enantiomer of the drug may be more capable of binding with the receptor:

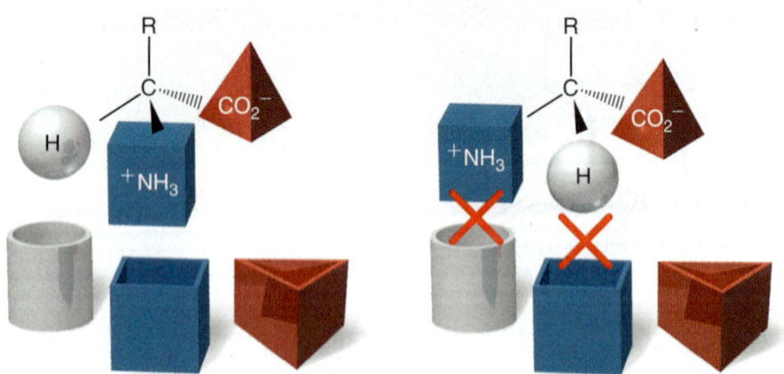


The first compound (left) can bind with the receptor, while its enantiomer (right) cannot bind with the receptor. For this reason, enantiomers will rarely produce the same biological response. As an example, consider the enantiomers of ibuprofen:

(*S*)-Ibuprofen (*R*)-Ibuprofen

Ibuprofen is an analgesic (painkiller) with anti-inflammatory properties. The *S* enantiomer is the active agent, while the *R* enantiomer is inactive. Nevertheless, ibuprofen is sold as a mixture of both enantiomers (under the trade names Advil and Motrin), because the benefit of separating the enantiomers is not clear. In fact, there is evidence that the human body is capable of slowly converting the *R* enantiomer into the desired *S* enantiomer. Many synthetic drugs are sold as a mixture of enantiomers, because of the high cost and difficulty associated with separating enantiomers.

In many cases, enantiomers can trigger entirely different physiological responses. For example, consider the enantiomers of Timolol:

(*S*)-Timolol (*R*)-Timolol

The *S* enantiomer treats angina and high blood pressure, while the *R* enantiomer is useful in treating glaucoma. In this example, both enantiomers produce desirable, albeit different, results. In other cases, one enantiomer can produce an undesirable response. For example, consider the enantiomers of penicillamine:

(*R*)-Penicillamine (*S*)-Penicillamine

The *S* enantiomer was used to treat chronic arthritis (until other, more effective drugs were developed), while the *R* enantiomer is highly toxic. In such a case, the drug cannot be sold as a mixture of enantiomers.

Another example is naproxen:

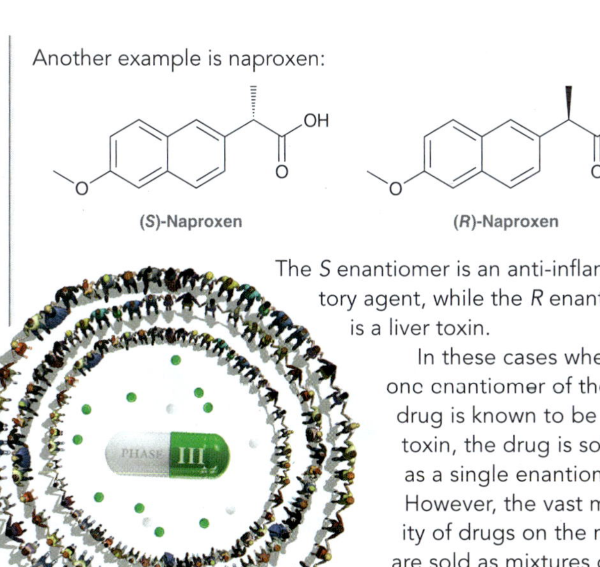


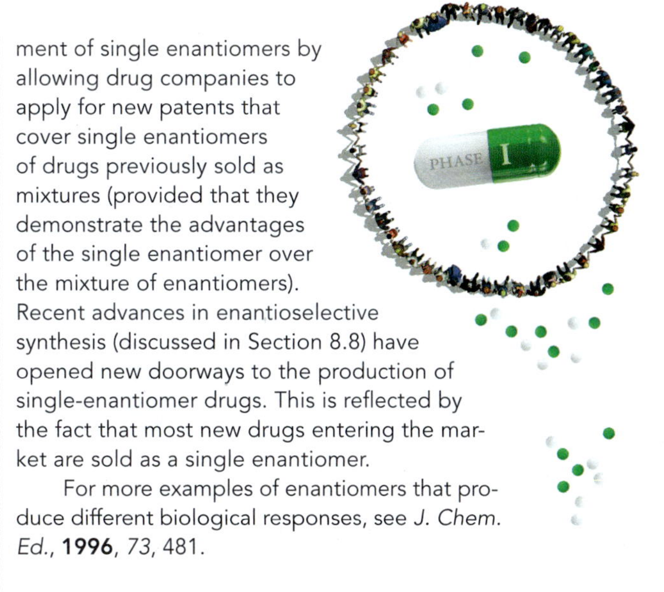


The *S* enantiomer is an anti-inflammatory agent, while the *R* enantiomer is a liver toxin.

In these cases where one enantiomer of the drug is known to be a toxin, the drug is sold as a single enantiomer. However, the vast majority of drugs on the market are sold as mixtures of enantiomers. The FDA has recently encouraged development of single enantiomers by allowing drug companies to apply for new patents that cover single enantiomers of drugs previously sold as mixtures (provided that they demonstrate the advantages of the single enantiomer over the mixture of enantiomers). Recent advances in enantioselective synthesis (discussed in Section 8.8) have opened new doorways to the production of single-enantiomer drugs. This is reflected by the fact that most new drugs entering the market are sold as a single enantiomer.

For more examples of enantiomers that produce different biological responses, see *J. Chem. Ed.*, **1996**, *73*, 481.

## 5.4 Optical Activity

Enantiomers exhibit identical physical properties. For example compare the melting and boiling points for the enantiomers of carvone:

| (*R*)-Carvone | (*S*)-Carvone |
| --- | --- |
| Melting point = 25°C | Melting point = 25°C |
| Boiling point = 231°C | Boiling point = 231°C |

This should make sense, because physical properties are determined by intermolecular interactions, and the intermolecular interactions of one enantiomer are just the mirror image of the intermolecular interactions of the other enantiomer. Nevertheless, enantiomers do exhibit different behavior when exposed to plane-polarized light. To explore this difference, let's first quickly review the nature of light.

### Plane-Polarized Light

Electromagnetic radiation (light) is comprised of oscillating electric and magnetic fields propagating through space (Figure 5.6). Notice that each oscillating field is located in a plane, and these planes

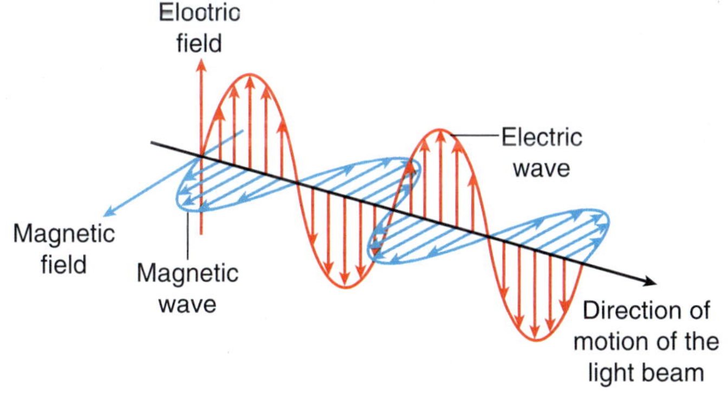


**FIGURE 5.6** Light waves consist of perpendicular, oscillating electric and magnetic fields.

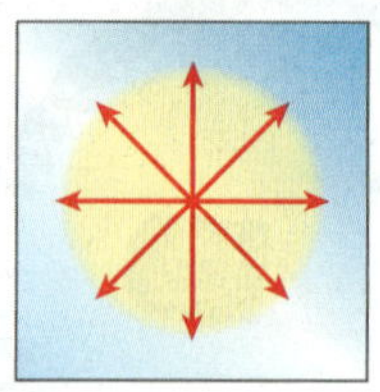

**FIGURE 5.7**
Unpolarized light.

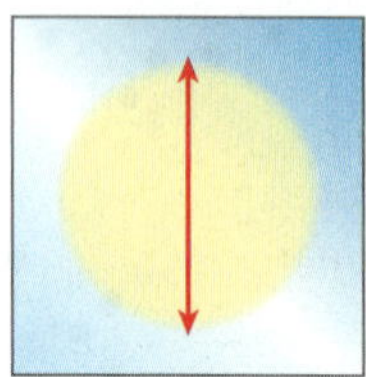

**FIGURE 5.8**
Plane-polarized light.

are perpendicular to each other. The orientation of the electric field (shown in red) is called the **polarization** of the light wave. When many waves of light are traveling in the same direction, they each have a different polarization, randomly oriented with respect to one another (Figure 5.7). When light passes through a polarizing filter, only photons of a particular polarization are allowed to pass through the filter, giving rise to **plane-polarized light** (Figure 5.8). When plane-polarized light is passed through a second polarizing filter, the orientation of the filter will determine whether light passes through or is blocked (Figure 5.9).

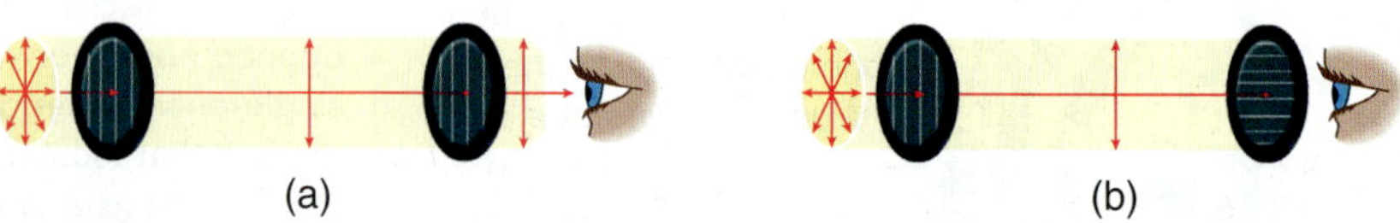


**FIGURE 5.9**
Plane-polarized light passing through (a) two parallel polarizing filters or (b) two perpendicular polarizing filters.

## Polarimetry

In 1815, French scientist Jean Baptiste Biot was exploring the nature of light by passing plane-polarized light through various solutions of organic compounds. In so doing, he discovered that solutions of certain organic compounds (such as sugar) rotate the plane of plane-polarized light. These compounds were therefore said to be **optically active**. He also noted that only some organic compounds possess this quality. Compounds lacking this ability were said to be **optically inactive**.

The rotation of plane-polarized light caused by optically active compounds can be measured using a device called a **polarimeter** (Figure 5.10). The light source is generally a sodium lamp, which emits light at a fixed wavelength of 589 nm, called the D line of sodium. This light then passes through a polarizing filter, and the resulting plane-polarized light continues through a tube containing a solution of an optically active compound, which causes the plane to rotate. The polarization of the emerging light can then be determined by rotating the second filter and observing the orientation that allows the light to pass.

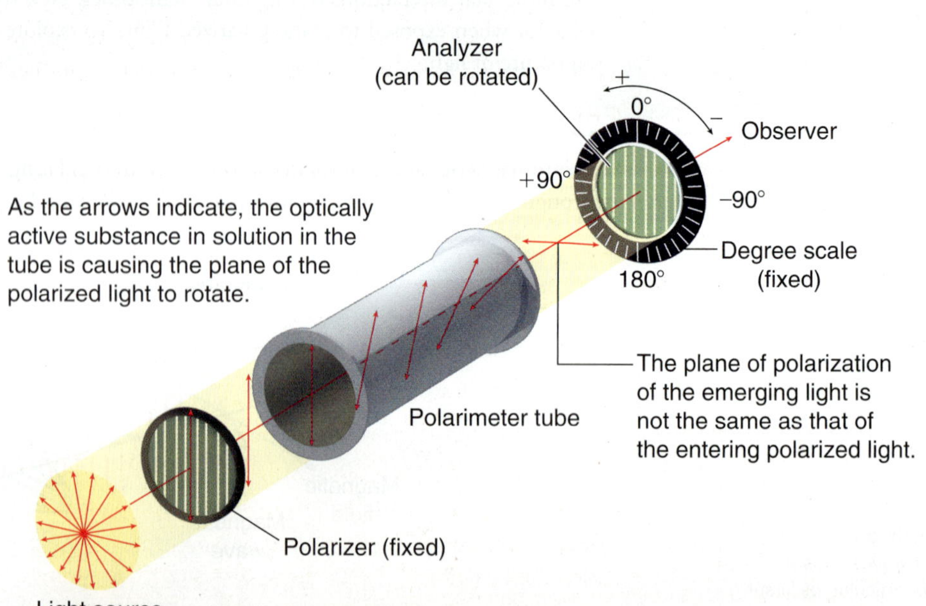


**FIGURE 5.10**
The components of a polarimeter.

## Source of Optical Activity

In 1847, an explanation for the source of optical activity was proposed by French scientist Louis Pasteur. Pasteur's investigation of tartrate salts (discussed later in this chapter) led him to the conclusion that optical activity is a direct consequence of chirality. That is, chiral compounds are optically active, while achiral compounds are not. Moreover, Pasteur noted that enantiomers (nonsuperimposable mirror images) will rotate the plane of plane-polarized light in equal amounts but in opposite directions. We will now explore this idea in more detail.

## Specific Rotation

When a solution of a chiral compound is placed in a polarimeter, the **observed rotation** (symbolized by the Greek letter alpha, $\alpha$) will be dependent on the number of molecules that the light encounters as it travels through the solution. If the concentration of the solution is doubled, the observed rotation will be doubled. The same is true for the distance that the light travels through the solution (the pathlength). If the pathlength is doubled, the observed rotation will be doubled. In order to compare the rotations for various compounds, scientists had to choose a set of standard conditions. By using a standard concentration (1 g/mL) and a standard pathlength (1 dm) for measuring rotations, it is possible to make meaningful comparisons between compounds. The **specific rotation** for a compound is defined as the observed rotation under these standard conditions.

It is not always practical to use a concentration of 1 g/mL when measuring the optical activity of a compound, because chemists often work with very small quantities of a compound (milligrams, rather than grams). Very often, optical activity must be measured using a very dilute solution. To account for the use of nonstandard conditions, the specific rotation can be calculated in the following way:

$$\text{Specific rotation} = [\alpha] = \frac{\alpha}{c \times l}$$

where $[\alpha]$ is the specific rotation, $\alpha$ is the observed rotation, $c$ is the concentration (measured in grams per milliliter), and $l$ is the pathlength (measured in decimeters, where 1 dm = 10 cm). In this way, the specific rotation for a compound can be calculated for nonstandard concentrations or pathlengths. The specific rotation of a compound is a physical constant, much like its melting point or boiling point.

The specific rotation of a compound is also sensitive to temperature and wavelength, but these factors cannot be incorporated into the equation, because the relationship between these factors and the specific rotation is not a linear relationship. In other words, doubling the temperature does not necessarily double the observed rotation; in fact, it can sometimes cause a decrease in the observed rotation. The wavelength chosen can also affect the observed rotation in a nonlinear fashion. Therefore, these two factors are simply reported in the following format:

$$[\alpha]_{\lambda}^{T}$$

where $T$ is the temperature (in degrees Celsius) and $\lambda$ is the wavelength of light used. Some examples are as follows:

Br Br

(*R*)-2-Bromobutane (*S*)-2-Bromobutane

$[\alpha]_{D}^{20} = -23.1$ $[\alpha]_{D}^{20} = +23.1$

where D stands for the D line of sodium (589 nm). Notice that the specific rotations for enantiomers are equal in magnitude but opposite in direction. A compound exhibiting a positive rotation (+) is called **dextrorotatory,** or ***d***, while a compound exhibiting a negative rotation (−) is called **levorotatory,** or ***l***. In the example above, the first enantiomer is levorotatory and is therefore called (−)-2-bromobutane, while the second enantiomer is dextrorotatory and is therefore called (+)-2-bromobutane. There is no direct relationship between the *R/S* system of nomenclature and the sign of the specific rotation (+ or −). The descriptors (*R*) and (*S*) refer to the configuration (the three-dimensional arrangement) of a chirality center, which is independent of conditions. In contrast, the descriptors (+) and (−) refer to the direction in which plane-polarized light is rotated, which is dependent on conditions. As an example, consider the structure of (*S*)-aspartic acid:

The configuration of the chirality center above is *S*, regardless of temperature. But the optical activity of this compound is very sensitive to temperature: It is levorotatory at 20°C, but it is dextrorotatory at 100°C. This example illustrates that we cannot use the configuration (*R* or *S*) to predict with certainty whether a compound will be (+) or (−). The magnitude and direction of optical activity can only be determined experimentally.

# SKILLBUILDER

## 5.5 CALCULATING SPECIFIC ROTATION

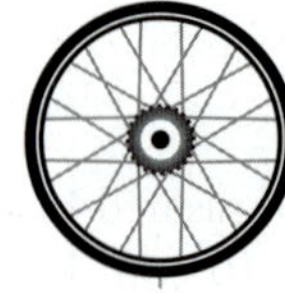

**LEARN** the skill

When 0.300 g of sucrose is dissolved in 10.0 mL of water and placed in a sample cell 10.0 cm in length, the observed rotation is +1.99° (using the D line of sodium at 20°C). Calculate the specific rotation of sucrose.

### SOLUTION

**STEP 1** Use the following equation.

The following equation can be used to calculate the specific rotation:

$$\text{Specific rotation} = [\alpha] = \frac{\alpha}{c \times l}$$

and the problem provides all of the necessary values to plug into this equation. We just need to make sure that the units are correct: *Concentration* (c) must be in units of grams per milliliter. The problem states that 0.300 g is dissolved in 10.0 mL. Therefore, the concentration is 0.300 g/10.0 mL = 0.03 g/mL. *Pathlength* (*l*) must be reported in decimeters (where 1 dm = 10 cm). The problem states that the pathlength is 10.0 cm, which is equivalent to 1.00 dm. Now, simply plug these values into the equation:

**STEP 2** Plug in the given values.

$$\text{Specific rotation} = [\alpha] = \frac{\alpha}{c \times l} = \frac{+1.99°}{0.03\ \text{g/mL} \times 1.00\ \text{dm}} = +66.3$$

The temperature (20°C) and wavelength (D line of sodium) are reported as a superscript and subscript, respectively. Notice that the specific rotation is generally reported as if it were a unitless number:

$$[\alpha]_D^{20} = +66.3$$

**PRACTICE** the skill

**5.12** When 0.575 g of monosodium glutamate (MSG) is dissolved in 10.0 mL of water and placed in a sample cell 10.0 cm in length, the observed rotation at 20°C (using the D line of sodium) is +1.47°. Calculate the specific rotation of MSG.

**5.13** When 0.095 g of cholesterol is dissolved in 1.00 mL of ether and placed in a sample cell 10.0 cm in length, the observed rotation at 20°C (using the D line of sodium) is −2.99°. Calculate the specific rotation of cholesterol.

**5.14** When 1.30 g of menthol is dissolved in 5.00 mL of ether and placed in a sample cell 10.0 cm in length, the observed rotation at 20°C (using the D line of sodium) is +0.57°. Calculate the specific rotation of menthol.

**5.15** Predict the value for the specific rotation of the following compound. Explain your answer.

**5.16** The specific rotation of (*S*)-2-butanol is +13.5. If 1.00 g of its enantiomer is dissolved in 10.0 mL of ethanol and placed in a sample cell with a length of 1.00 dm, what observed rotation do you expect?

need more **PRACTICE?** **Try Problems 5.44, 5.50c, 5.52**

## Enantiomeric Excess

A solution containing a single enantiomer is said to be **optically pure**, or **enantiomerically pure**. That is, the other enantiomer is entirely absent.

A solution containing equal amounts of both enantiomers is called a **racemic mixture** and will be optically inactive. When plane-polarized light passes through such a mixture, the light encounters one molecule at a time, rotating slightly with each interaction. Since there are equal amounts of both enantiomers, the net rotation will be zero degrees. Although the individual compounds are optically active, the mixture is not optically active.

A solution containing both enantiomers in unequal amounts will be optically active. For example, imagine a solution of 2-butanol, containing 70% (*R*)-2-butanol and 30% (*S*)-2-butanol. In such a case, there is a 40% excess of the *R* stereoisomer (70% − 30%). The remainder of the solution is a racemic mixture of both enantiomers. In this case, the **enantiomeric excess** (*ee*) is 40%.

The % *ee* of a compound is experimentally measured in the following way:

$$\% \, ee = \frac{|\text{observed } \alpha|}{|\alpha \text{ of pure enantiomer}|} \times 100\%$$

Let's get some practice using this expression to calculate % *ee*.

# SKILLBUILDER

## 5.6 CALCULATING % *ee*

**LEARN the skill**

The specific rotation of optically pure adrenaline in water (at 25°C) is −53. A chemist devised a synthetic route to prepare optically pure adrenaline, but it was suspected that the product was contaminated with a small amount of the undesirable enantiomer. The observed rotation was found to be −45°. Calculate the % *ee* of the product.

**SOLUTION**

We use the following equation to calculate the % *ee*:

**STEP 1**
Use the following equation.

$$\% \, ee = \frac{|\text{observed } \alpha|}{|\alpha \text{ of pure enantiomer}|} \times 100\%$$

and then plug in the given values:

**STEP 2**
Plug in the given values.

$$\% \text{ ee} = \frac{45}{53} \times 100\% = 85\%$$

An 85% *ee* indicates that 85% of the product is adrenaline, while the remaining 15% is a racemic mixture of adrenaline and its enantiomer (7.5% adrenaline and 7.5% of the enantiomer). An 85% *ee* therefore indicates that the product is comprised of 92.5% adrenaline and 7.5% of the undesired enantiomer.

**PRACTICE the skill** **5.17** The specific rotation of L-dopa in water (at 15°C) is −39.5. A chemist prepared a mixture of L-dopa and its enantiomer, and this mixture had a specific rotation of −37. Calculate the % *ee* of this mixture.

**5.18** The specific rotation of ephedrine in ethanol (at 20°C) is −6.3. A chemist prepared a mixture of ephedrine and its enantiomer, and this mixture had a specific rotation of −6.0. Calculate the % *ee* of this mixture.

**5.19** The specific rotation of vitamin $B_7$ in water (at 22°C) is +92. A chemist prepared a mixture of vitamin $B_7$ and its enantiomer, and this mixture had a specific rotation of +85. Calculate the % *ee* of this mixture.

**APPLY the skill** **5.20** The specific rotation of L-alanine in water (at 25°C) is +2.8. A chemist prepared a mixture of L-alanine and its enantiomer, and 3.50 g of the mixture was dissolved in 10.0 mL of water. This solution was then placed in a sample cell with a pathlength of 10.0 cm and the observed rotation was +0.78. Calculate the % *ee* of the mixture.

need more **PRACTICE?** **Try Problems 5.40, 5.50a,b**

## 5.5 Stereoisomeric Relationships: Enantiomers and Diastereomers

In Section 5.1, we saw that stereoisomers have the same constitution (connectivity of atoms) but are nonsuperimposable (they differ in their spatial arrangement of atoms). Stereoisomers can be subdivided into two categories, as shown in Figure 5.11.

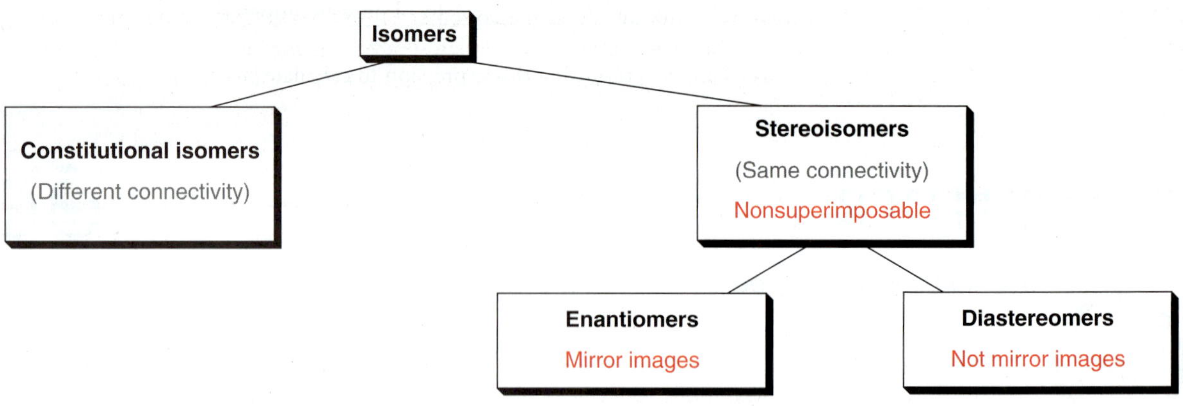


**FIGURE 5.11**
The main categories of stereoisomers.

Enantiomers are stereoisomers that are mirror images of one another, while **diastereomers** are stereoisomers that are not mirror images of one another. According to these definitions, we can understand why *cis-trans* isomers (discussed at the beginning of this chapter) are said to be diastereomers, rather than enantiomers.

Consider, once again, the structures of *cis*-2-butene and *trans*-2-butene:

*cis*-2-Butene    *trans*-2-Butene

They are stereoisomers, but they are not mirror images of one another and are therefore diastereomers. An important difference between enantiomers and diastereomers is that enantiomers have the same physical properties (as seen in Section 5.4), while diastereomers have different physical properties (as seen in Section 5.1).

The difference between enantiomers and diastereomers becomes especially relevant when we consider compounds containing more than one chirality center. As an example, consider the following structure:

OH 1 2 Me

This compound has two chirality centers. Each one can have either the *R* configuration or the *S* configuration, giving rise to four possible stereoisomers (two pairs of enantiomers):

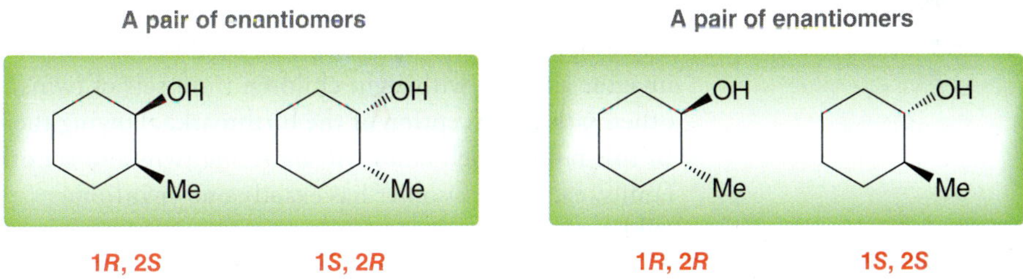


To describe the relationship between these four stereoisomers, we will look at one of them and describe its relationship to the other three stereoisomers. The first stereoisomer listed above has the configuration (1*R*, 2*S*). This stereoisomer has only one mirror image, or enantiomer, which has the configuration (1*S*, 2*R*). The third stereoisomer is not a mirror image of the first stereoisomer and is therefore its diastereomer. In a similar way, there is a diastereomeric relationship between the first stereoisomer and the fourth stereoisomer above because they are not mirror images of one another.

To help better visualize the relationship between the four compounds above we will use an analogy. Imagine a family with four children (two sets of twins). The first pair of twins are identical to each other in almost every way, except for the placement of one birthmark. One child has the birthmark on the right cheek, while the other child has the birthmark on the left cheek. These twins can be distinguished from each other based on the position of the birthmark. They are nonsuperimposable mirror images of each other. The second pair of twins look very different from the first pair. But the second pair of twins are once again identical to each other in every way, except the position of the birthmark on the cheek. They are nonsuperimposable mirror images of each other.

In this family of four children, each child has one twin and two other siblings. The same relationship exists for the four stereoisomers shown above. In this molecular family, each stereoisomer has exactly one enantiomer (mirror-image twin) and two diastereomers (siblings).

Now consider a case with three chirality centers:

OH 1 2 Me 3 Cl

Once again, each chirality center can have either the *R* configuration or the *S* configuration, giving rise to a family of eight possible stereoisomers:

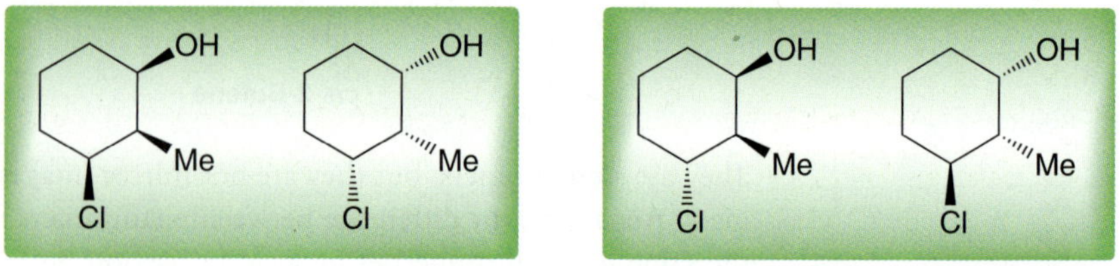

**1*R*, 2*R*, 3*S*** **1*S*, 2*S*, 3*R*** **1*R*, 2*R*, 3*R*** **1*S*, 2*S*, 3*S***

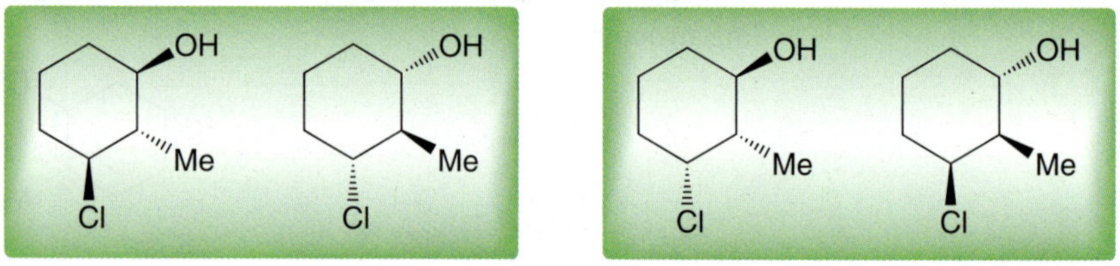

**1*R*, 2*S*, 3*S*** **1*S*, 2*R*, 3*R*** **1*R*, 2*S*, 3*R*** **1*S*, 2*R*, 3*S***

These eight stereoisomers are arranged above as four pairs of enantiomers. To help visualize this, imagine a family with eight children (four sets of twins). Each pair of twins are identical to each other with the exception of the birthmark, allowing them to be distinguished from one another. In this family, each child will have one twin and six other siblings. Similarly, in the molecular family, each stereoisomer has exactly one enantiomer (mirror-image twin) and six diastereomers (siblings).

Note that the presence of three chirality centers produces a family of four pairs of enantiomers. A compound with four chirality centers will generate a family of eight pairs of enantiomers. This begs the obvious question: What is the relationship between the number of chirality centers and the number of stereoisomers in the family? This relationship can be summarized as follows:

$$\text{Maximum number of stereoisomers} = 2^n$$

where $n$ refers to the number of chirality centers. A compound with four chirality centers can have a maximum of $2^4$ stereoisomers = 16 stereoisomers, or eight pairs of enantiomers. As another example, consider the structure of cholesterol:

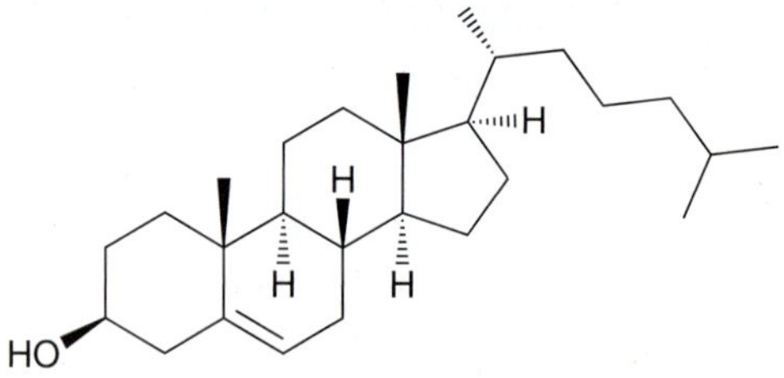

**Cholesterol**

Cholesterol has eight chirality centers, giving rise to a family of $2^8$ stereoisomers (256 stereoisomers). The specific stereoisomer shown has only one enantiomer and 254 diastereomers, although the structure shown is the only stereoisomer produced by nature.

# SKILLBUILDER

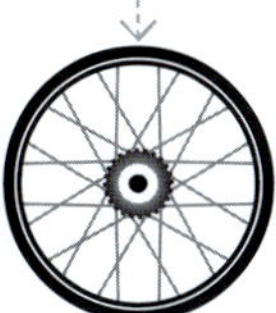

## 5.7 DETERMINING THE STEREOISOMERIC RELATIONSHIP BETWEEN TWO COMPOUNDS

**LEARN the skill**

Identify whether each pair of compounds are enantiomers or diastereomers:

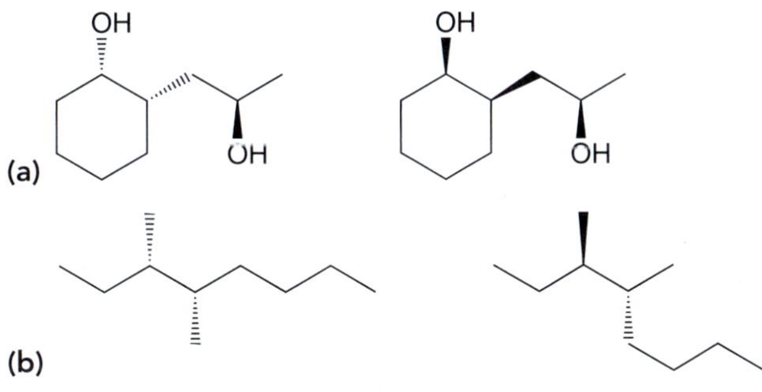


### SOLUTION

**(a)** In SkillBuilder 5.2, we saw how to draw the enantiomer of a compound. In particular, we saw that the easiest method is to redraw the compound and invert all chirality centers. That is, all dashes are redrawn as wedges, and all wedges are redrawn as dashes. Remember that two compounds can only be enantiomers if all of their chirality centers have opposite configuration. Let's carefully analyze all three chirality centers in each compound. Two of the chirality centers do have opposite configurations:

**STEP 1**
Compare the configuration of each chirality center.

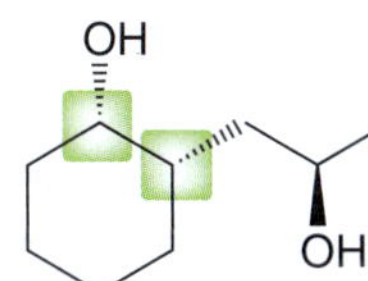


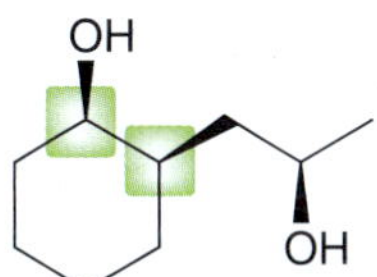


But there remains one chirality center that has the same configuration in both stereoisomers:

**STEP 2**
If only some of the chirality centers have opposite configuration, then the compounds are diastereomers.

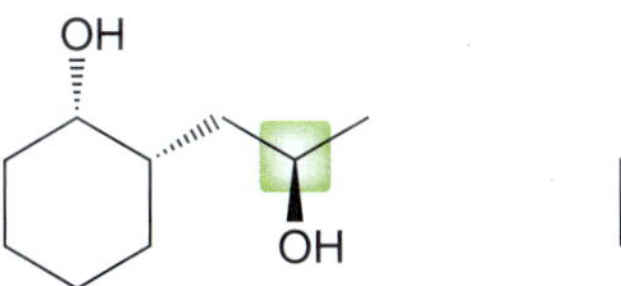


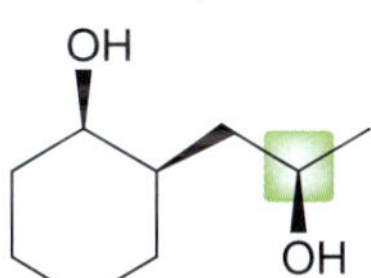


Therefore, these stereoisomers are not mirror images of each other; they must be diastereomers.

**(b)** Once again, compare the configurations of the chirality centers in both compounds. The first chirality center is easily identified as having a different configuration in each compound:

**STEP 1**
Compare the configuration of each chirality center.

The second chirality center is harder to compare with a brief inspection, because the skeleton is drawn in a different conformation (remember that single bonds are always free to rotate). When in doubt, you can always just assign a configuration (*R* or *S*) to each chirality center and compare them that way:

**STEP 2**
If all chirality centers have opposite configuration, then the compounds are enantiomers.

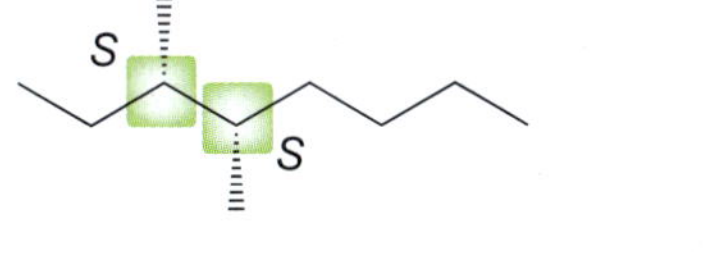


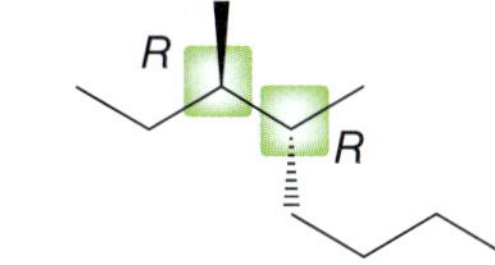

When comparing these stereoisomers, both chirality centers have opposite configurations, so these compounds are enantiomers.

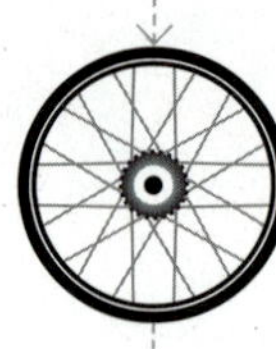

**PRACTICE** the skill **5.21** Identify whether each of the following pairs of compounds are enantiomers or diastereomers:

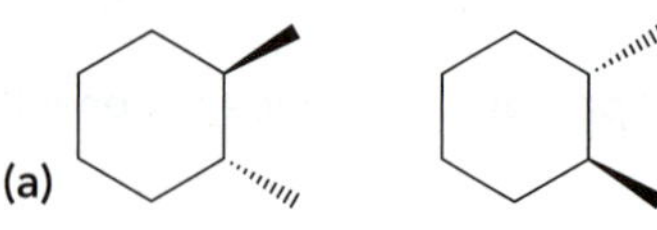

(a) (b)

(c) OH OH Cl Cl (d) OH OH

(e)

**APPLY** the skill **5.22** Identify the relationship between the following two compounds:

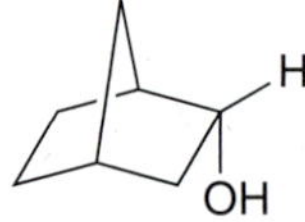


**Hint:** Take careful notice of the number of chirality centers.

need more **PRACTICE?** **Try Problems 5.36a–j,l, 5.41c, 5.49a,c,e–j, 5.57**

## 5.6 Symmetry and Chirality

### Rotational Symmetry Versus Reflectional Symmetry

In this section we will learn how to determine whether a compound is chiral or achiral. It is true that any compound with a single chirality center must be chiral. But the same statement is not necessarily true for compounds with two chirality centers. Consider the *cis* and *trans* isomers of 1,2-dimethylcyclohexane:

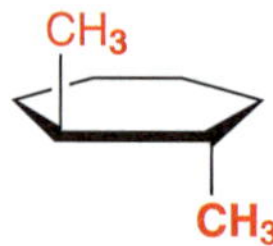


*trans*-
1,2-Dimethylcyclohexane

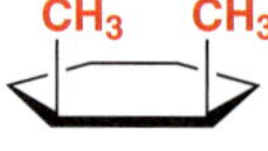


*cis*-
1,2-Dimethylcyclohexane

Each of these compounds has two chirality centers, but the *trans* isomer is chiral, while the *cis* isomer is not chiral. To understand why, we must explore the relationship between symmetry and chirality. Let's quickly review the different types of symmetry.

Broadly speaking, there are only two types of symmetry: *rotational symmetry* and *reflectional symmetry*. The first compound above (the *trans* isomer) exhibits rotational symmetry. To visualize this, imagine that the cyclohexane ring is pierced with an imaginary stick. Then, while your eyes are closed, the stick is rotated:

CH3 CH3 CH3 CH3

Rotate 180° about this axis

This operation generated the same exact image

If, after opening your eyes, it is impossible to determine whether or not the rotation occurred, then the molecule possesses rotational symmetry. The imaginary stick is called an **axis of symmetry**.

Now let's consider the structure of the *cis* isomer:

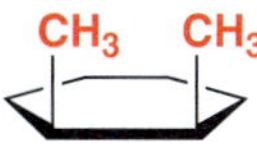


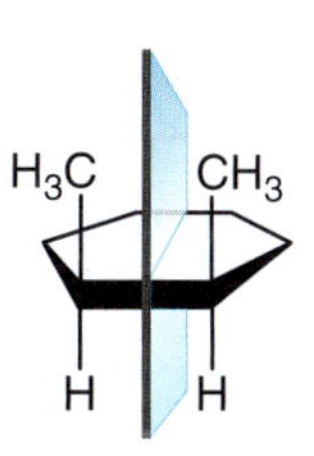


**FIGURE 5.12**
An illustration showing the plane of symmetry that is present in *cis*-1,2-dimethylcyclohexane.

This compound does not have the same axis of symmetry that the *trans* isomer exhibits. If we pierce the molecule with an imaginary stick, we would have to rotate 360° in order to regenerate the exact same image. The same is true if we pierce the molecule from any angle. Therefore, this compound does not exhibit rotational symmetry. However, the compound does exhibit reflectional symmetry. Imagine closing your eyes while the molecule is reflected about the plane shown in Figure 5.12. That is, everything on the right side is reflected to the left side, and everything on the left side is reflected to the right side. When you open your eyes, you will not be able to determine whether or not the reflection took place. This molecule possesses a **plane of symmetry**.

To summarize, a molecule with an axis of symmetry possesses rotational symmetry, and a molecule with a plane of symmetry possesses reflectional symmetry. With an understanding of these two types of symmetry, we can now explore the relationship between symmetry and chirality. In particular, *chirality is not dependent in any way on rotational symmetry*. That is, the presence or absence of an axis of symmetry is completely irrelevant when determining whether a compound is chiral or achiral. We saw that *trans*-1,2-dimethylcyclohexane possesses rotational symmetry; nevertheless, the compound is still chiral and exists as a pair of enantiomers:

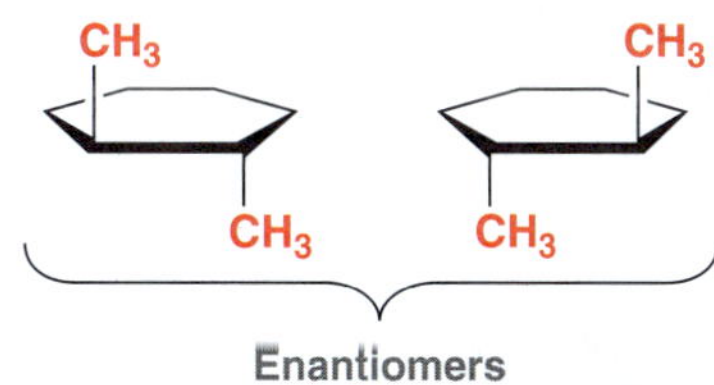


Even the presence of several different axes of symmetry does not indicate whether the compound is chiral or achiral. Chirality is only dependent on the presence or absence of reflectional symmetry. *Any compound that possesses a plane of symmetry in any conformation will be achiral.* We saw that the *cis* isomer of 1,2-dimethylcyclohexane exhibits a plane of symmetry, and therefore, the compound is achiral. It is identical to its mirror image. It does not have an enantiomer.

Although the presence of a plane of symmetry renders a compound achiral, the converse is not always true—that is, the absence of a plane of symmetry does not necessarily mean that the compound is chiral. Why? Because a plane of symmetry is only one type of reflectional symmetry. There are other types of reflectional symmetry, and the presence of any kind of reflectional symmetry will render a compound achiral. As an example, consider the following compound:

This compound does not have a plane of symmetry, but it does have another kind of reflectional symmetry. Instead of reflecting about a plane, imagine reflecting about a point at the center of the compound. Reflection about a point (rather than a plane) is called *inversion*. During the inversion process, the methyl group at the top of the drawing (on a wedge) is reflected to the bottom of the drawing (on a dash). Similarly, the methyl group at the bottom of the drawing (on a dash) is reflected to the top of the drawing (on a wedge). All other groups are also reflected about the center of the compound. If your eyes are closed when the inversion takes place, you would not be able to determine whether anything had happened. The compound is said to exhibit a center of inversion, which endows it with reflectional symmetry. As a result, the compound is achiral even though it lacks a plane of symmetry.

There are also other types of reflectional symmetry, but they are beyond the scope of our discussion. For our purposes, it will be sufficient to look for a plane of symmetry. We can summarize the relationship between symmetry and chirality with the following three statements:

- The presence or absence of rotational symmetry is irrelevant to chirality.
- A compound that has a plane of symmetry will be achiral.
- A compound that lacks a plane of symmetry will most likely be chiral (although there are rare exceptions, which can mostly be ignored for our purposes).

We will now practice identifying planes of symmetry.

## CONCEPTUAL CHECKPOINT

**5.23** For each of the following objects determine whether or not it possesses a plane of symmetry:

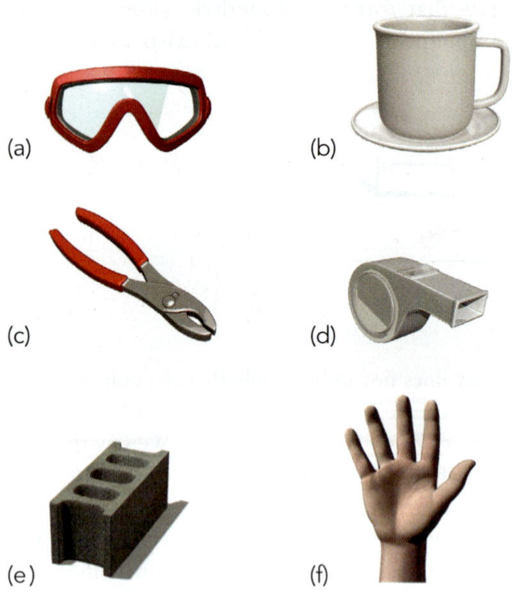


**5.24** One object above has three planes of symmetry. Identify that object.

**5.25** Each of the following molecules has one plane of symmetry. Find the plane of symmetry in each case: (*Hint*: A plane of symmetry can slice atoms in half.)

(a) Me, Me, Cl, Br (b) (c) Me, Me (d) HO, OH (e) (f) Cl, Cl

## *Meso* Compounds

We have seen that the presence of chirality centers does not necessarily render a compound chiral. Specifically, a compound that exhibits reflectional symmetry will be achiral even though it has chirality centers. Such compounds are called ***meso* compounds**. A family of stereoisomers containing a *meso* compound will have fewer than $2^n$ stereoisomers. As an example, consider the following structure:

This structure has two chirality centers and therefore we would expect $2^2$ (=4) stereoisomers. In other words, we expect two pairs of enantiomers:

A pair of enantiomers

But these two drawings represent the same compound

The first pair of enantiomers meets our expectations. But the second pair is actually just one compound. It has reflectional symmetry and is therefore a *meso* compound:

Plane of symmetry

The compound is not chiral, and it does not have an enantiomer. To help visualize this, imagine a family with three children: a pair of twins and a single child. The pair of twins are identical to each other in almost every way, except for the placement of one birthmark. One child has the birthmark on the right cheek, while the other child has the birthmark on the left cheek. They are nonsuperimposable mirror images of each other. In contrast, the single child has two birthmarks: one on each cheek. This child is superimposable on his own mirror image; he does not have a twin. This family of three children is similar to the molecular family of three stereoisomers shown above.

## SKILLBUILDER

### 5.8 IDENTIFYING *MESO* COMPOUNDS

**LEARN the skill** Draw all possible stereoisomers of 1,3-dimethylcyclopentane:

**SOLUTION**

This compound has two chirality centers, so we expect four possible stereoisomers (two pairs of enantiomers):

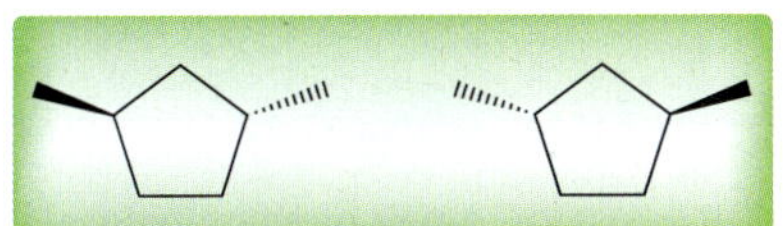

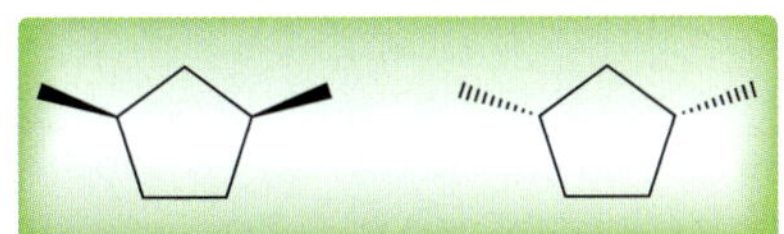

Before concluding that there are four stereoisomers here, we must look to see if either pair of enantiomers is actually just a *meso* compound. The first pair does in fact represent a pair of enantiomers.

Students often have trouble seeing that these two compounds are nonsuperimposable. It is a common mistake to believe that rotating the first compound will generate the second compound. This is not the case! Recall that a methyl group on a dash is pointing away from us (behind the page) while a methyl group on a wedge is pointing toward us (in front of the page). When the first compound above is rotated about a vertical axis, the methyl group on the left side ends up on the right side of the compound, but it does not remain as a wedge. Why? Before the rotation, it was pointing toward us, but after the rotation, it is pointing away from us (a molecular model will help you see this). As a result, that methyl group ends up as a dash on the right side of the compound. The fate of the other methyl group is similar. Before the rotation, it is on a dash on the right side of the compound, but after the rotation, it ends up as a wedge on the left side of the compound. As a result, rotating this compound will only regenerate the same drawing again:

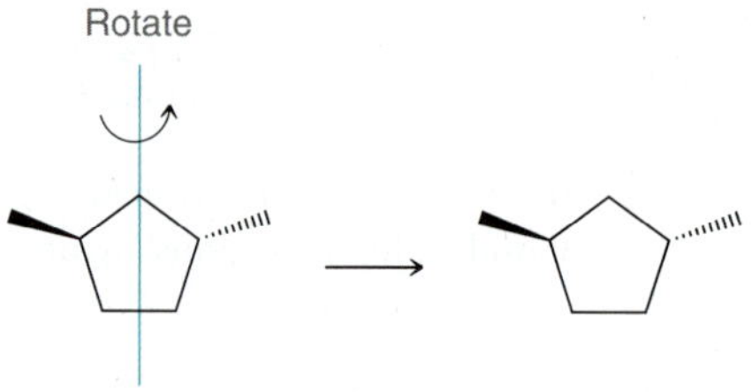

Rotate

Rotating this compound will never give its enantiomer:

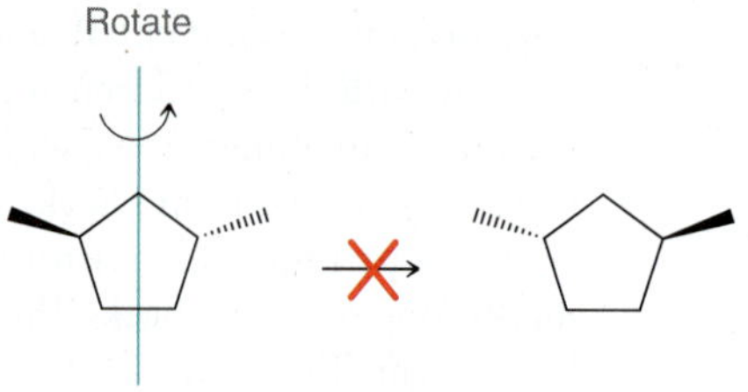

Rotate

These compounds are enantiomers because they lack reflectional symmetry. They only possess rotational symmetry, which is completely irrelevant.

In contrast, the second pair of compounds do not represent a pair of enantiomers; instead, they represent two drawings of the same compound (a *meso* compound):

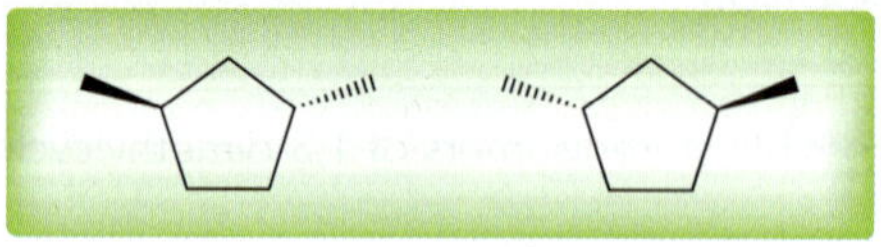

A pair of enantiomers

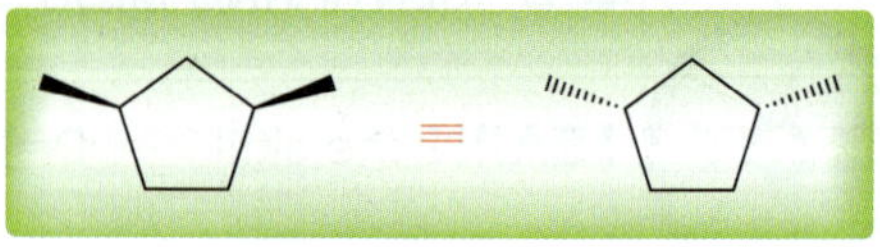

These two drawings represent the same compound

This molecular family is therefore comprised of only three stereoisomers—a pair of enantiomers and one *meso* compound:

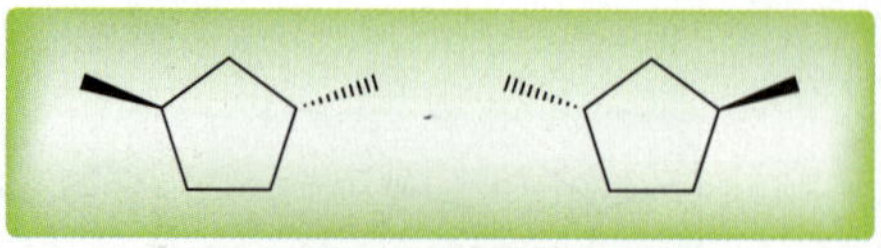

A pair of enantiomers

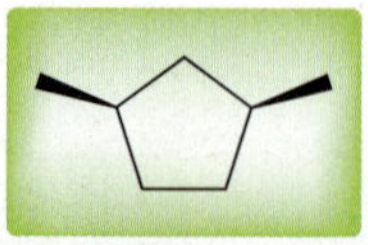

A *meso* compound

**PRACTICE the skill** **5.26** Draw all possible stereoisomers for each of the following compounds. Each possible stereoisomer should be drawn only once:

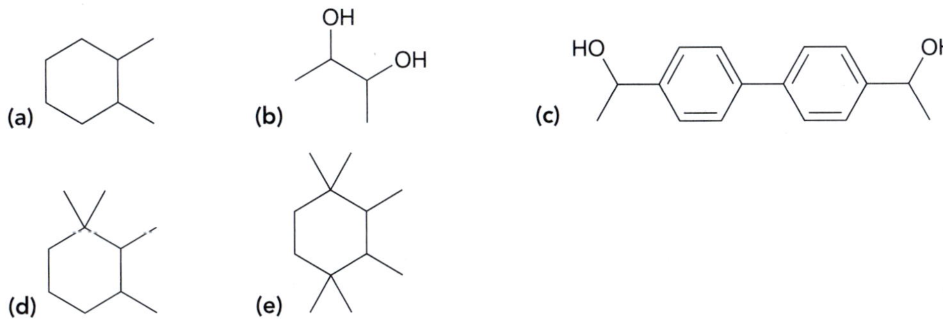

**APPLY the skill**

**5.27** There are only two stereoisomers of 1,4-dimethylcyclohexane. Draw them and explain why only two stereoisomers are observed.

**5.28** How many stereoisomers do you expect for the following compound? Draw all of the stereoisomers.

need more **PRACTICE?** **Try Problems 5.37, 5.46, 5.47, 5.51, 5.53a,b,d,f–h**

## 5.7 Fischer Projections

There is another drawing style that is often used when dealing with compounds bearing multiple chirality centers. These drawings, called **Fischer projections**, were devised by the German chemist Emil Fischer in 1891. Fischer was investigating sugars, which have multiple chirality centers. In order to quickly draw these compounds, he developed a faster method for drawing chirality centers:

Two chirality centers | Three chirality centers | Four chirality centers

For each chirality center in a Fischer projection, the horizontal lines are considered to be coming out of the page, and the vertical lines are considered to be going behind the page:

Fischer projections are primarily used for analyzing sugars (Chapter 24). In addition, Fischer projections are also helpful for quickly comparing the relationship between stereoisomers:

Enantiomers

Diastereomers

The first pair of compounds are enantiomers because all chirality centers have opposite configuration. The second pair of compounds are diastereomers. Recall that two compounds will be enantiomers only if all of their chirality centers have opposite configurations. Students often have trouble assigning configurations to chirality centers in a Fischer projection, so let's get some practice with that skill.

## SKILLBUILDER

### 5.9 ASSIGNING CONFIGURATION FROM A FISCHER PROJECTION

**LEARN the skill**

Assign the configuration of the following chirality center:

**SOLUTION**

Remember what a Fischer projection represents. All horizontal lines are wedges and all vertical lines are dashes:

is the same as

Until now, chirality centers have not been shown in this way. We are more accustomed to assigning configurations when a chirality center is shown with only one dash and one wedge:

**STEP 1**
Draw one horizontal line as a wedge and draw one vertical line as a dash.

The chirality center of a Fischer projection can easily be converted into this more familiar type of drawing. Simply choose one horizontal line in the Fischer projection and redraw it as a wedge; then choose one vertical line and draw it as a dash. The answer will be the same, regardless of which two lines are chosen:

**STEP 2**
Assign the configuration using the steps in SkillBuilder 5.4.

The configuration of this chirality center will be *R* in all of the drawings. To see why this works, build a molecular model and rotate it around to convince yourself.

When a Fischer projection has multiple chirality centers, this process is simply repeated for each chirality center.

**PRACTICE** the skill **5.29** Identify the configuration of the chirality centers shown below:

(a) O, OH; H—NH$_2$; $CH_2OH$ (b) O, H; HO—H; $CH_3$ (c) $CH_2OH$; HO—H; $CH_2CH_3$ (d) $CH_2OH$; Br—H; $CH_3$

**APPLY** the skill **5.30** Determine the configuration for every chirality center in each of the following compounds:

(a) O, OH; H—OH; HO—H; H—OH; $CH_2OH$ (b) O, OH; HO—H; HO—H; HO—H; $CH_2OH$ (c) O, OH; HO—H; H—OH; HO—H; $CH_2OH$

**5.31** Draw the enantiomer of each compound in the previous problem.

need more **PRACTICE?** **Try Problems 5.39h, 5.41a,b, 5.43c, 5.53c,e, 5.54–5.56**

## 5.8 Conformationally Mobile Systems

In the previous chapter, we learned how to draw Newman projections, and we used them to compare the various conformations of butane. Recall that butane can adopt two staggered conformations that exhibit *gauche* interactions:

H, H, H, H, $CH_3$, $CH_3$ ⇌ H, H, H, H, $CH_3$, $CH_3$

These two conformations are nonsuperimposable mirror images of each other, and their relationship is therefore enantiomeric. Nevertheless, butane is not a chiral compound. It is optically inactive, because these two conformations are constantly interconverting via single-bond rotation (which occurs with a very low energy barrier). The temperature would have to be extremely low to prevent interconversion between these two conformations. In contrast, a chirality center cannot invert its configuration via single-bond rotations. (*R*)-2-Butanol cannot be converted into (*S*)-2-butanol via a conformational change.

In the previous chapter, we also saw that substituted cyclohexanes adopt various conformations. Consider, for example, (*cis*)-1,2-dimethylcyclohexane. Below are both chair conformations:

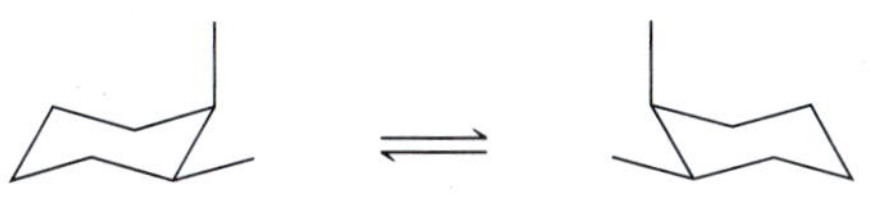

The second chair has been rotated in space to illustrate the relationship between these two chair conformations. These conformations are mirror images, yet they are not superimposable on one another. To see this more clearly, compare them in the following way: Traveling *clockwise* around the ring, the first chair has an axial methyl group first, followed by an equatorial methyl group. The second chair, going *clockwise* again, has an equatorial methyl group first, followed by an axial methyl. The order has been reversed. These two conformations are distinguishable from one another; that is, they are not superimposable. The relationship between these two chair conformations is therefore enantiomeric; however, this compound is optically inactive, because the conformations rapidly interconvert at room temperature.

The following procedure will be helpful for determining whether or not a cyclic compound is optically active: (1) Either draw a Haworth projection of the compound or simply draw a ring with dashes and wedges for all substituents and then (2) look for a plane of symmetry in either one of these drawings. For example, (*cis*)-1,2-dimethylcyclohexane can be drawn in either of the following ways:

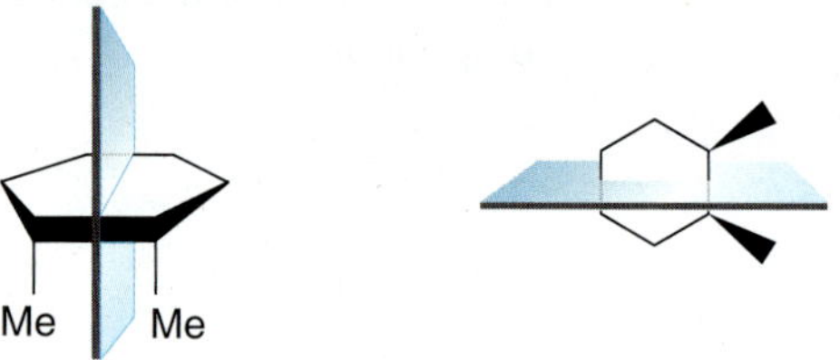


**Internal plane of symmetry** **Internal plane of symmetry**

In each drawing, a plane of symmetry is apparent, and the presence of that symmetry plane indicates that the compound is not optically active.

## 5.9 Resolution of Enantiomers

As mentioned earlier, enantiomers have the same physical properties (boiling point, melting point, solubility, etc.). Since traditional separation techniques generally rely on differences in physical properties, they cannot be used to separate enantiomers from each other. The **resolution** (separation) of enantiomers can be achieved in a variety of other ways.

### Resolution via Crystallization

The first resolution of enantiomers occurred in 1847, when Pasteur successfully separated enantiomeric tartrate salts from each other. Tartaric acid is a naturally occurring, optically active compound found in grapes and easily obtained during the wine-making process:

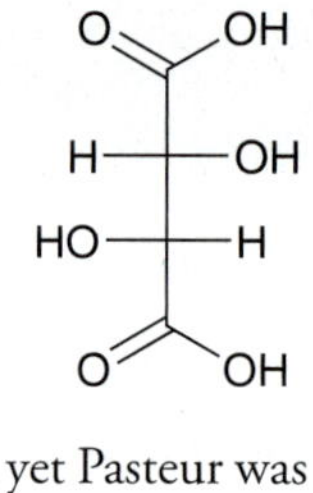

Only this stereoisomer is found in nature, yet Pasteur was able to obtain a racemic mixture of tartrate salts from the owner of a chemical plant:

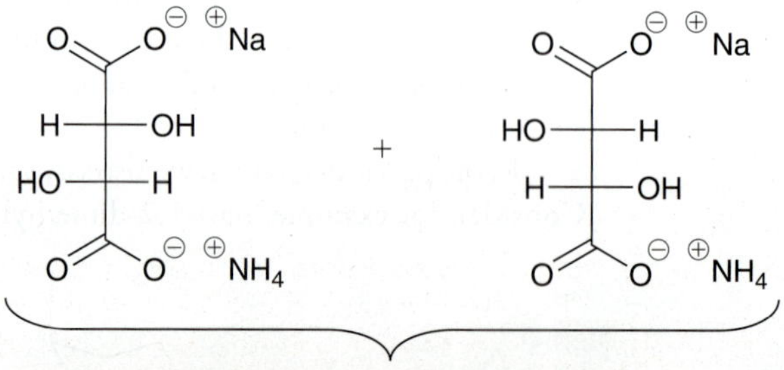

**Racemic mixture of tartrate salts**

The tartrate salts were then allowed to crystallize, and Pasteur noticed that the crystals had two distinct shapes that were nonsuperimposable mirror images of each other. Using only a pair of tweezers, he then physically separated the crystals into two piles. He dissolved each pile in water and placed each solution in a polarimeter to discover that their specific rotations were equal in amount but opposite in sign. Pasteur correctly concluded that the molecules themselves must be nonsuperimposable mirror images of each other. He was the first to describe molecules as having this property and is therefore credited with discovering the relationship between enantiomers.

Most racemic mixtures are not easily resolved into mirror-image crystals when allowed to crystallize, so other methods of resolution are required. Two common ways will now be discussed.

## Chiral Resolving Agents

When a racemic mixture is treated with a single enantiomer of another compound, the resulting reaction produces a pair of diastereomers (rather than enantiomers):

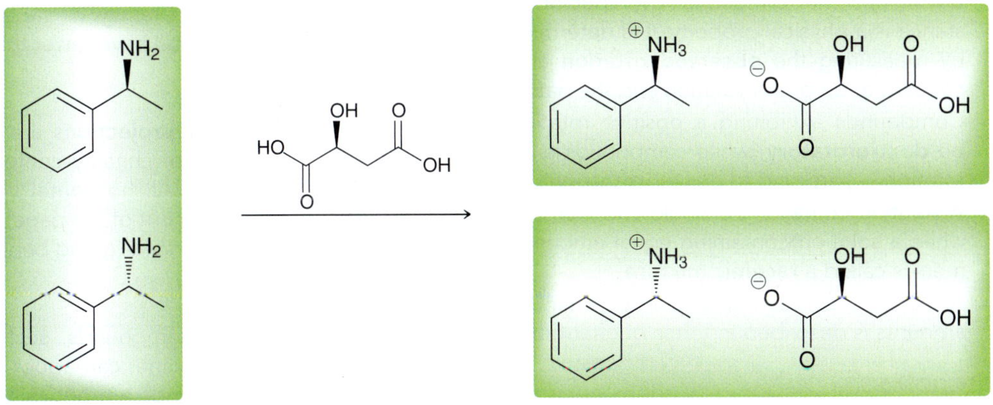

A pair of enantiomers

Diastereomeric salts

On the left are the enantiomers of 1-phenylethylamine. When a racemic mixture of these enantiomers is treated with (*S*)-malic acid, a proton transfer reaction produces diastereomeric salts. Diastereomers have different physical properties and can therefore be separated by conventional means (such as crystallization). Once separated, the diastereomeric salts can then be converted back into the original enantiomers by treatment with a base. Thus (*S*)-malic acid is said to be a **resolving agent** in that it makes it possible to resolve the enantiomers of 1-phenylethylamine.

## Chiral Column Chromatography

Resolution of enantiomers can also be accomplished with **column chromatography**. In column chromatography, compounds are separated from each other based on a difference in the way they interact with the medium (the adsorbent) through which they are passed. Some compounds interact strongly with the adsorbent and move very slowly through the column; other compounds interact weakly and travel at a faster rate through the column. When enantiomers are passed through a traditional column, they travel at the same rate because their properties are identical. However, if a chiral adsorbent is used, the enantiomers interact with the adsorbent differently, causing them to travel through the column at different rates. Enantiomers are often separated in this way.

# REVIEW OF CONCEPTS AND VOCABULARY

### SECTION 5.1

- Constitutional isomers have the same molecular formula but differ in their connectivity of atoms.
- Stereoisomers have the same connectivity of atoms but differ in their spatial arrangement. The terms *cis* and *trans* are used to differentiate stereoisomeric alkenes as well as disubstituted cycloalkanes.

### SECTION 5.2

- **Chiral** objects are not **superimposable** on their mirror images.
- The most common source of molecular chirality is the presence of a **chirality center**, a carbon atom bearing four different groups.
- A compound with one chirality center will have one nonsuperimposable mirror image, called its **enantiomer**.

SECTION 5.3

- The Cahn-Ingold-Prelog system is used to assign the **configuration** of a chirality center. The four groups are each assigned a priority, based on atomic number, and the molecule is then rotated into a perspective in which the fourth priority is on a dash. A clockwise sequence of 1-2-3 is designated as ***R***, while a counterclockwise sequence is designated as ***S***.

SECTION 5.4

- A **polarimeter** is a device used to measure the ability of chiral organic compounds to rotate the plane of **plane-polarized light**. Such compounds are said to be **optically active**. Compounds that do not rotate plane-polarized light are said to be **optically inactive**.
- Enantiomers rotate plane-polarized light in equal amounts but in opposite directions. The **specific rotation** of a substance is a physical property. It is determined experimentally by measuring the **observed rotation** and dividing by the concentration of the solution and the pathlength.
- Compounds exhibiting a positive rotation (+) are said to be **dextrorotatory**, while compounds exhibiting a negative rotation (−) are said to be **levorotatory**.
- A solution containing a single enantiomer is **optically pure**, while a solution containing equal amounts of both enantiomers is called a **racemic mixture**.
- A solution containing a pair of enantiomers in unequal amounts is described in terms of **enantiomeric excess**.

SECTION 5.5

- For a compound with multiple chirality centers, a family of stereoisomers exists. Each stereoisomer will have at most one enantiomer, with the remaining members of the family being **diastereomers**.
- The number of stereoisomers of a compound can be no larger than $2^n$, where $n$ = the number of chirality centers.
- Enantiomers are mirror images. Diastereomers are not mirror images.

SECTION 5.6

- There are two kinds of symmetry: *rotational symmetry* and *reflectional symmetry*.
- The presence or absence of an **axis of symmetry** (rotational symmetry) is irrelevant.
- A compound that possesses a **plane of symmetry** will be achiral.
- A compound that lacks a plane of symmetry will most likely be chiral (although there are rare exceptions, which can mostly be ignored for our purposes).
- A ***meso* compound** contains multiple chirality centers but is nevertheless achiral, because it possesses reflectional symmetry. A family of stereoisomers containing a *meso* compound will have fewer than $2^n$ stereoisomers.

SECTION 5.7

- **Fischer projections** are drawings that convey the configurations of chirality centers, without the use of wedges and dashes. All horizontal lines are understood to be wedges (coming out of the page) and all vertical lines are understood to be dashes (going behind the page).

SECTION 5.8

- Some compounds, such as butane and (*cis*)-1,2-dimethylcyclohexane, can adopt enantiomeric conformations. These compounds are optically inactive because the enantiomeric conformations are in equilibrium.

SECTION 5.9

- **Resolution** (separation) of enantiomers can be accomplished in a number of ways, including the use of chiral **resolving agents** and chiral **column chromatography**.

## SKILLBUILDER REVIEW

### 5.1 IDENTIFYING *CIS-TRANS* STEREOISOMERISM

**STEP 1** Identify and name all four groups attached to the π bond.

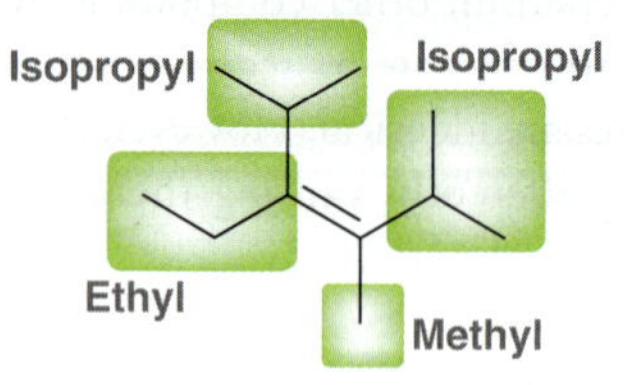


**STEP 2** Look for two identical groups on different vinylic positions and assign the configuration as *cis* or *trans*.

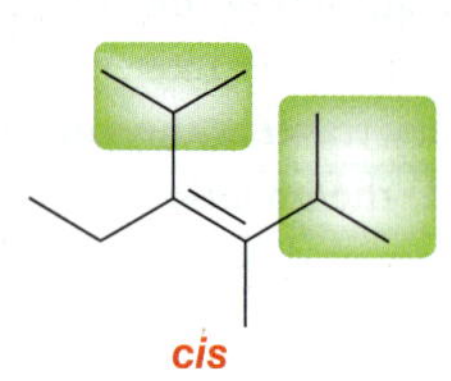


Try Problems 5.1–5.3, 5.35

### 5.2 LOCATING CHIRALITY CENTERS

**STEP 1** Ignore $sp^2$- and $sp$-hybridized centers.

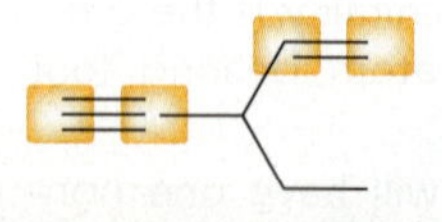

**STEP 2** Ignore $CH_2$ and $CH_3$ groups.

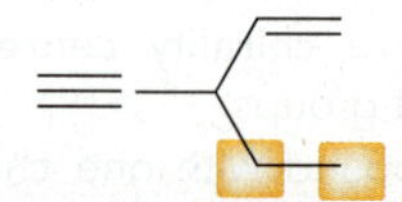

**STEP 3** Identify any carbon atoms bearing four different groups.

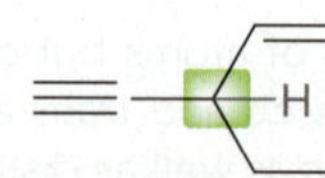


Try Problems 5.4–5.6, 5.34b, 5.48

## 5.3 DRAWING AN ENANTIOMER

| Either place the mirror behind the compound... | or place the mirror on the side of the compound... | or place the mirror below the compound. |
|---|---|---|
| 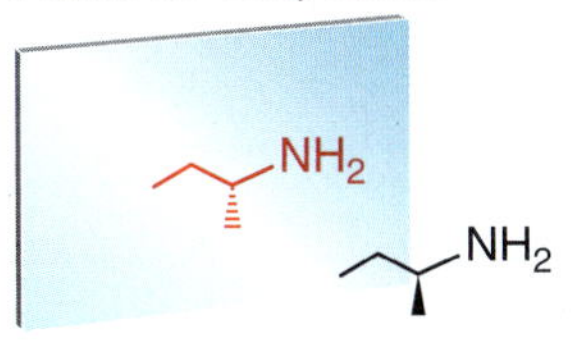  | 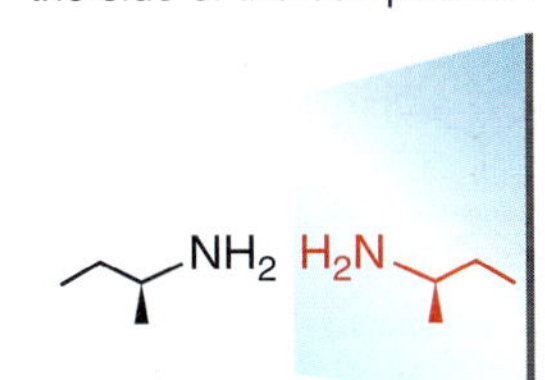  | 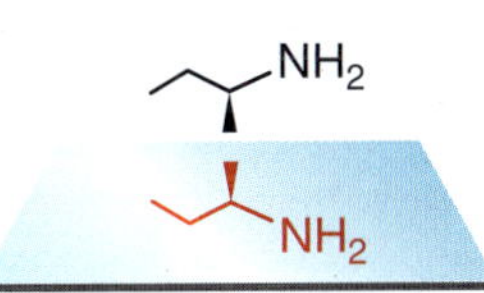  |

Try Problems **5.7, 5.8, 5.33, 5.34a, 5.38a–f,i–l**

## 5.4 ASSIGNING CONFIGURATION

**STEP 1** Identify the four atoms attached to the chirality center and prioritize by atomic number.

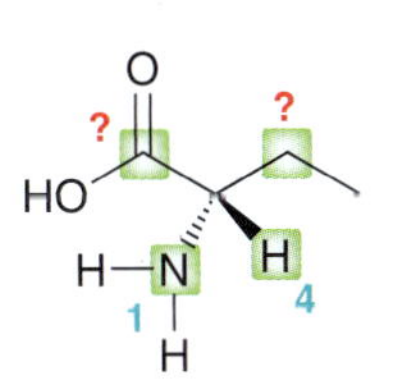


**STEP 2** If two (or more) atoms are identical, make a list of substituents and look for the first point of difference.

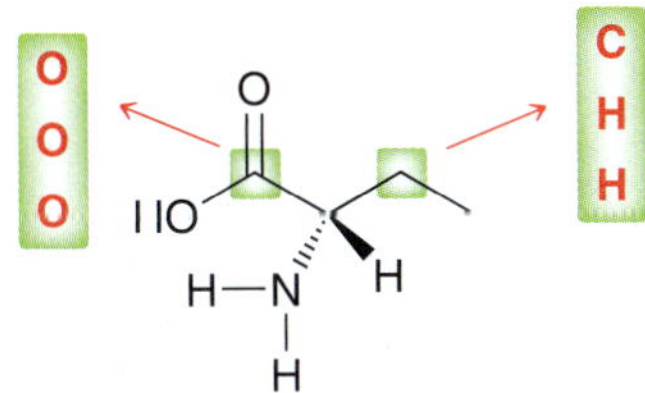


**STEP 3** Redraw the chirality center, showing only the priorities.

**STEP 4** Rotate molecule so that the fourth priority is on a dash.

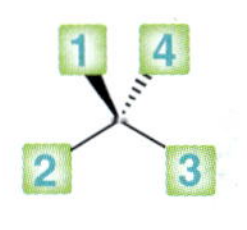


**STEP 5** Identify direction of 1-2-3 sequence: clockwise is *R*, and counterclockwise is *S*.

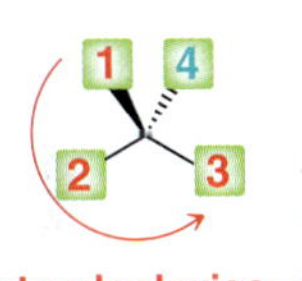


**Counterclockwise = S**

Try Problems **5.8–5.10, 5.32, 5.39a–g,i, 5.45**

## 5.5 CALCULATING SPECIFIC ROTATION

**EXAMPLE** Calculate the specific rotation given the following information:

- 0.300 g sucrose dissolved in 10.0 mL of water
- Sample cell = 10.0 cm
- Observed rotation = +1.99°

**STEP 1** Use the following equation:

$$\text{Specific rotation} = [\alpha] = \frac{\alpha}{c \times l}$$

**STEP 2** Plug in the given values:

$$= \frac{+1.99°}{0.03 \text{ g/mL} \times 1.00 \text{ dm}}$$
$$= +66.3$$

Try Problems **5.12–5.16, 5.44, 5.50c, 5.52**

## 5.6 CALCULATING % *ee*

**EXAMPLE** Given the following information, calculate the enantiomeric excess:

- The specific rotation of optically pure adrenaline is −53. A mixture of (*R*)- and (*S*)-adrenaline was found to have a specific rotation of −45. Calculate the % *ee* of the mixture.

**STEP 1** Use the following equation:

$$\% \, ee = \frac{|\text{observed } \alpha|}{|\alpha \text{ of pure enantiomer}|} \times 100\%$$

**STEP 2** Plug in the given values:

$$= \frac{45}{53} \times 100\% = 85\%$$

Try Problems **5.17–5.20, 5.40, 5.50a,b**

## 5.7 DETERMINING THE STEREOISOMERIC RELATIONSHIP BETWEEN TWO COMPOUNDS

**STEP 1** Compare the configuration of each chirality center.

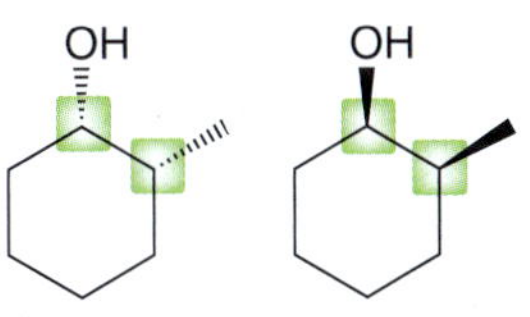


**STEP 2** If all chirality centers have opposite configuration, the compounds are enantiomers. If only some of the chirality centers have opposite configuration, then the compounds are diastereomers.

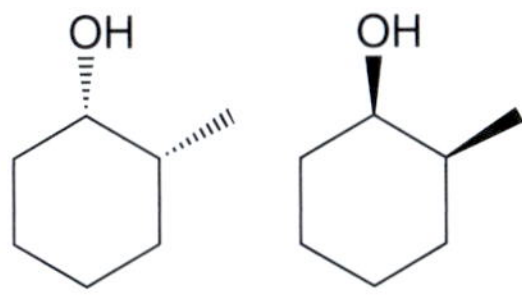


**Enantiomers**

Try Problems **5.21, 5.22, 5.36a–j,l, 5.41c, 5.49a,c,e–j**

### 5.8 IDENTIFYING *MESO* COMPOUNDS

**Example**

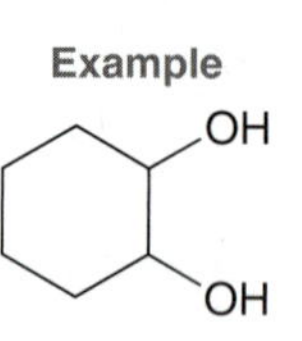


Draw all possible stereoisomers and then look for a plane of symmetry in any of the drawings. The presence of a plane of symmetry indicates a *meso* compound.

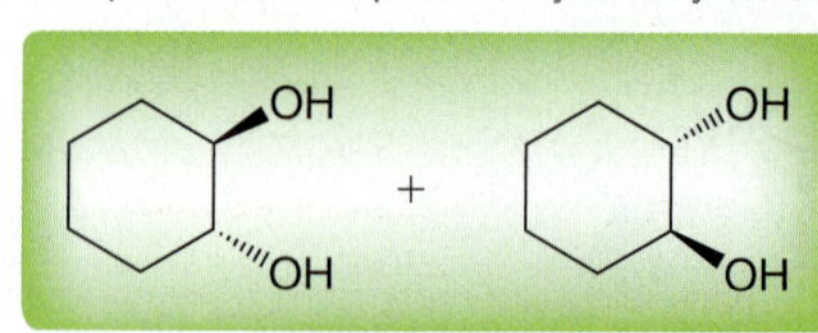


**Enantiomers**

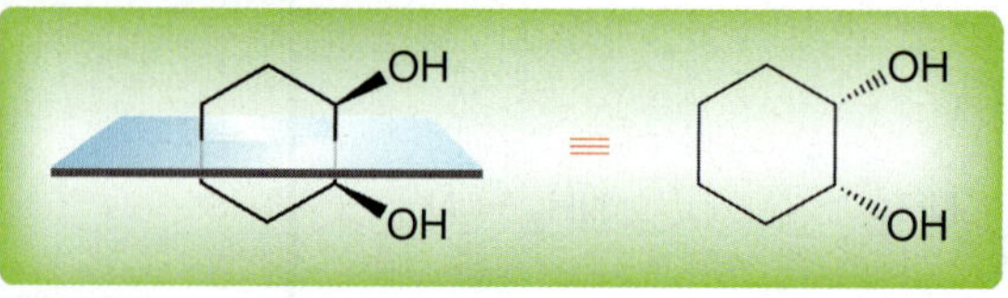


***Meso***

Try Problems 5.26–5.28, 5.37, 5.46, 5.47, 5.51, 5.53a,b,d,f–h

### 5.9 ASSIGNING CONFIGURATION FROM A FISCHER PROJECTION

**EXAMPLE** Assign the configuration of this chirality center.

O OH
H — OH
$CH_2OH$

**STEP 1** Choose one horizontal line and draw it as a wedge. Choose one vertical line and draw it as a dash.

O OH
H — OH
$CH_2OH$

**STEP 2** Prioritize.

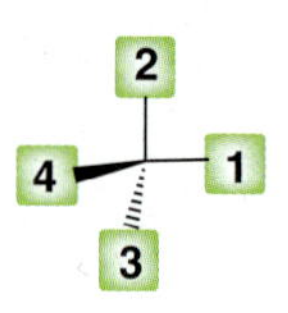


**STEP 3** Rotate so that the fourth priority is on a dash.

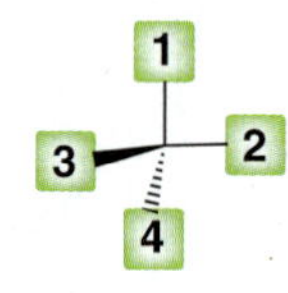


**STEP 4** Assign configuration.

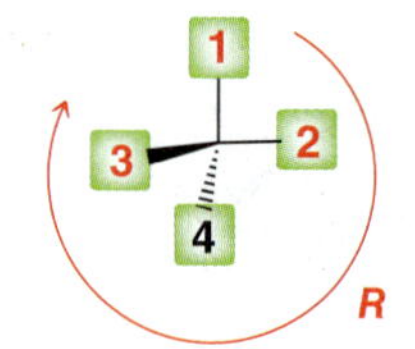


Try Problems 5.29–5.31, 5.39h, 5.41a,b, 5.43c, 5.53c,e, 5.54–5.56

## PRACTICE PROBLEMS

Note: Most of the Problems are available within **WileyPLUS**, an online teaching and learning solution.

**5.32** Atorvastatin is sold under the trade name Lipitor and is used for lowering cholesterol. Annual global sales of this compound exceed $13 billion. Assign a configuration to each chirality center in atorvastatin:

**5.33** Atropine, extracted from the plant *Atropa belladonna*, has been used in the treatment of bradycardia (low heart rate) and cardiac arrest. Draw the enantiomer of atropine:

**5.34** Paclitaxel (marketed under the trade name Taxol) is found in the bark of the Pacific yew tree, *Taxus brevifolia*, and is used in the treatment of cancer:

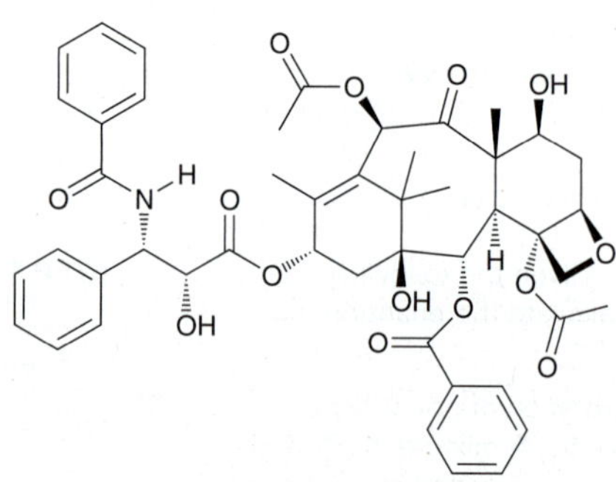


(a) Draw the enantiomer of paclitaxel. (b) How many chirality centers does this compound possess?

**5.35** Classify each of the following compounds below as *cis*, *trans*, or not stereoisomeric:

HO HO OH

**5.36** For each of the following pairs of compounds, determine the relationship between the two compounds:

(a)

(b)

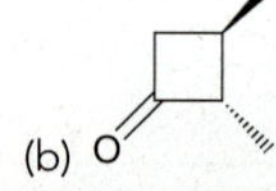


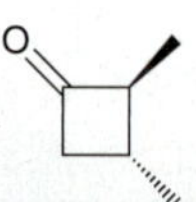

(c) Br Br

(d) (e)

(f)

(g) (h)

(i) OH OH

(j) OH OH

(k) Cl Cl Cl Cl

(l)

**5.37** Identify the number of stereoisomers expected for each of the following:

(a) OH Cl OH (b) OH OH (c) Me Me Cl

(d) (e) HO OH (f)

**5.38** Draw the enantiomer for each of the following compounds:

(a) Cl OH (b) (c)

(d) Et OH (e) Cl F OH Me (f) Cl F

(g) O H, H–OH, H–OH, HO–H, $CH_3$ (h) O OH, HO–H, H–OH, HO–H, $CH_2OH$

(i) OH OH Cl Cl OH O Me Me

(j) OH (k) OH Cl (l) OH

**5.39** Identify the configuration of each chirality center in the following compounds:

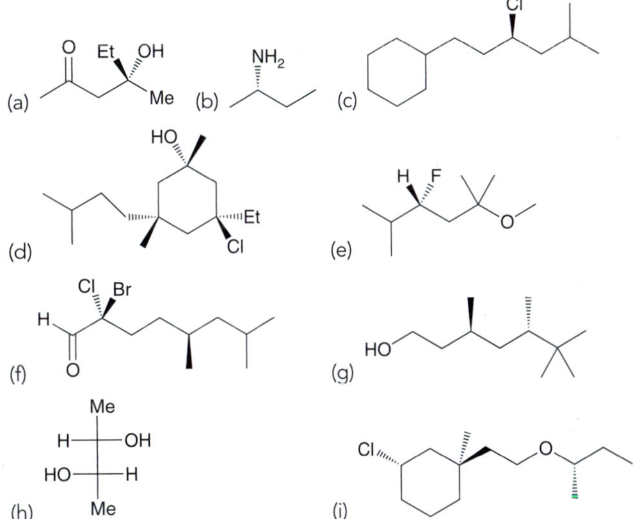


(f) Cl Br H O (g) HO

(h) Me, H–OH, HO–H, Me (i) Cl O

**5.40** You are given a solution containing a pair of enantiomers (A and B). Careful measurements show that the solution contains 98% A and 2% B. What is the *ee* of this solution?

**5.41** For each of the following pairs of compounds, determine the relationship between the two compounds:

(a) O H, HO–H, H–OH, $CH_2OH$ / O H, HO–H, HO–H, $CH_2OH$ (b) O OH, H–OH, H–OH, $CH_2OH$ / O OH, H–OH, HO–H, $CH_2OH$

(c) Cl Cl / Cl Cl (d)

(e)

**5.42** For each of the following pairs of compounds, determine the relationship between the two compounds:

(a) Cl Cl / Cl Cl

(b)

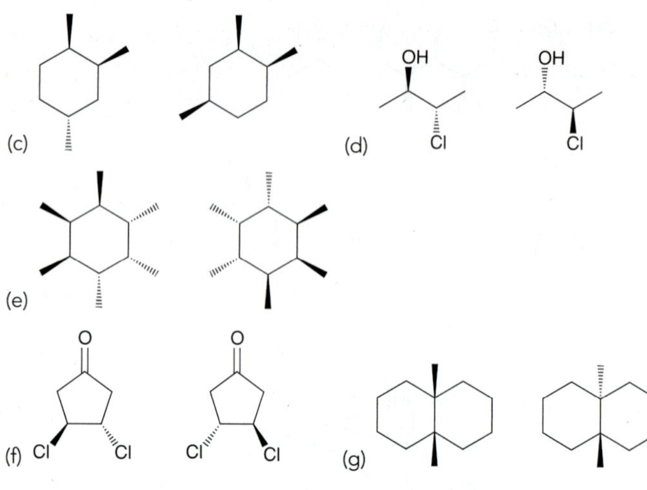


**5.43** Determine whether each statement is true or false:

(a) A racemic mixture of enantiomers is optically inactive.

(b) A *meso* compound will have exactly one nonsuperimposable mirror image.

(c) Rotating the Fischer projection of a molecule with a single chirality center by 90° will generate the enantiomer of the original Fischer projection:

A, D, B, C ⟶ D, C, A, B

**5.44** When 0.075 g of penicillamine is dissolved in 10.0 mL of pyridine and placed in a sample cell 10.0 cm in length, the observed rotation at 20°C (using the D line of sodium) is −0.47°. Calculate the specific rotation of penicillamine.

**5.45** (*R*)-Limonene is found in many citrus fruits, including oranges and lemons:

For each of the following compounds identify whether it is (*R*)-limonene or its enantiomer, (*S*)-limonene:

(a) (b) (c) (d)

**5.46** Each of the following compounds possesses a plane of symmetry. Find the plane of symmetry in each compound. In some cases, you will need to rotate a single bond to place the molecule into a conformation where you can more readily see the plane of symmetry.

(a) (b) HO, OH

(c) $H_3C$, Br, H, Br, H, $CH_3$ (d) Cl, H, Me, Me, H, Cl

**5.47** *cis*-1,3-Dimethylcyclobutane has two planes of symmetry. Draw the compound and identify both planes of symmetry.

**5.48** Consider the following two compounds. These compounds are stereoisomers of 1,2,3-trimethylcyclohexane. One of these compounds has three chirality centers, while the other compound has only two chirality centers. Identify which compound has only two chirality centers and explain why.

**5.49** For each of the following pairs of compounds, determine the relationship between the two compounds:

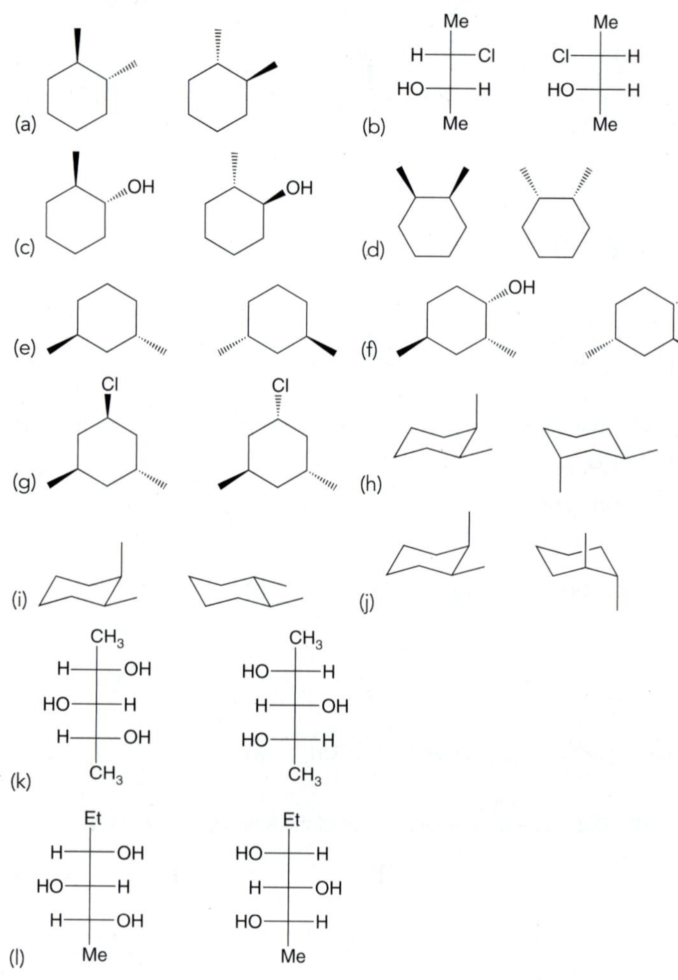


**5.50** The specific rotation of (*S*)-carvone (at 20°C) is +61. A chemist prepared a mixture of (*R*)-carvone and its enantiomer, and this mixture had an observed rotation of −55°.

(a) What is the specific rotation of (*R*)-carvone at 20°C?

(b) Calculate the % *ee* of this mixture.

(c) What percentage of the mixture is (*S*)-carvone?

**5.51** Identify whether each of the following compounds is chiral or achiral:

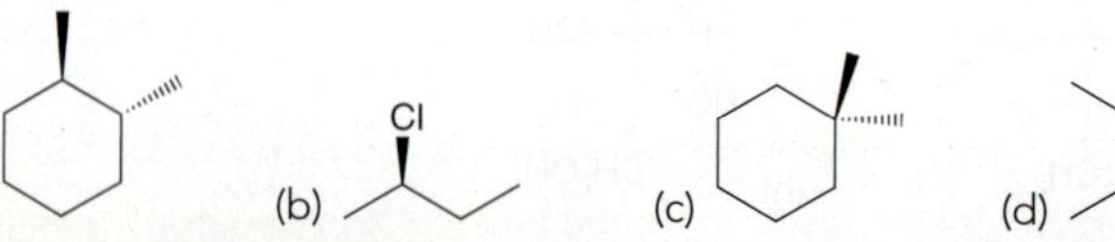

(e) 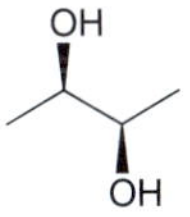 (f) 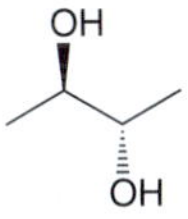 (g) 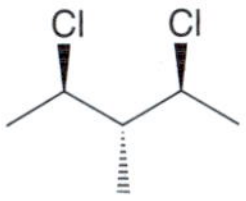 (h)

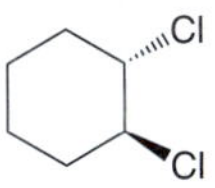

(i) 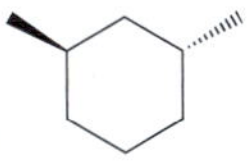 (j) 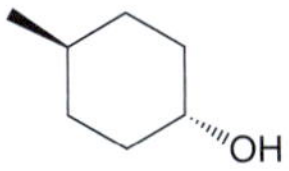 (k) 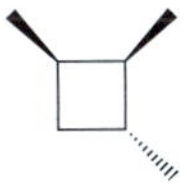 (l)

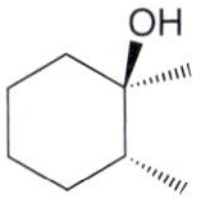

(m) 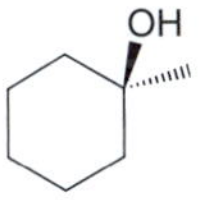 (n)  (o) 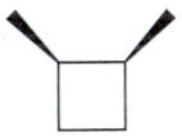 (p)

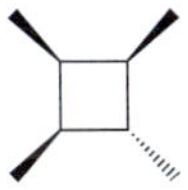

**5.52** The specific rotation of vitamin C (using the D line of sodium, at 20°C) is +24. Predict what the observed rotation would be for a solution containing 0.100 g of vitamin C dissolved in 10.0 mL of ethanol and placed in a sample cell with a length of 1.00 dm.

# INTEGRATED PROBLEMS

**5.53** Determine whether each of the following compounds is optically active or optically inactive:

(a) 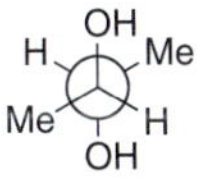 (b) 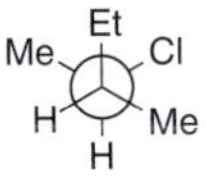 (c) 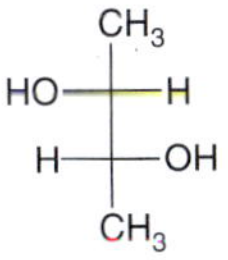 (d)  (e) 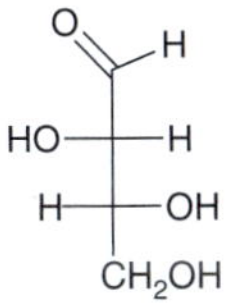 (f) 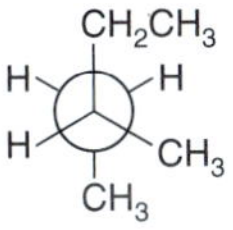 (g)  (h)

**5.54** Draw bond-line structures using wedges and dashes for the following compounds:

(a) 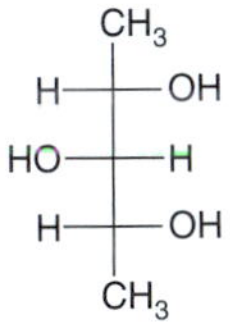 (b) 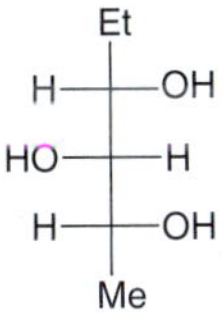 (c) 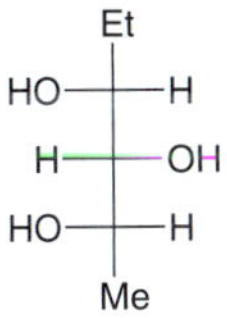 (d) 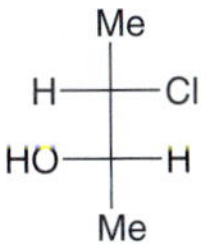 (e)

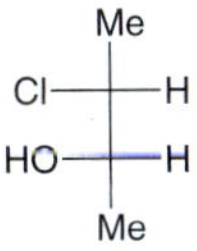

**5.55** The following questions apply to the five compounds in Problem 5.54.

(a) Which compound is *meso*?

(b) Would an equal mixture of compounds b and c be optically active?

(c) Would an equal mixture of compounds d and e be optically active?

**5.56** Draw a Fischer projection for each of the following compounds, placing the $—CO_2H$ group at the top:

(a) 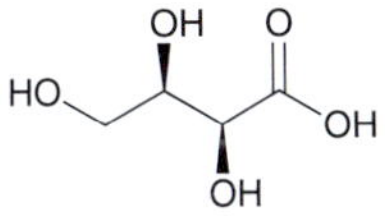 (b) 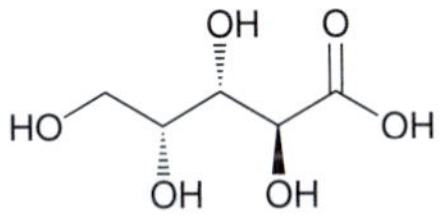 (c)

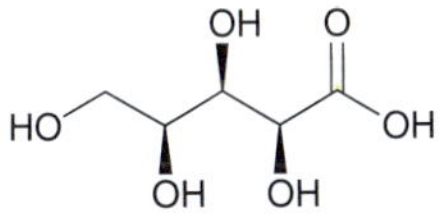

**5.57** For each of the following pairs of compounds, determine the relationship between the two compounds:

(a) 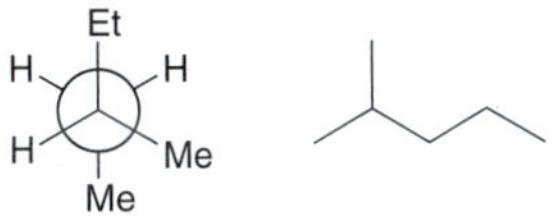 (b)

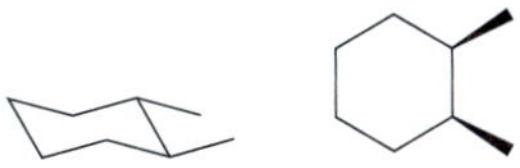

**5.58** It is possible for a compound to be chiral even though it lacks a carbon atom with four different groups. For example, consider the structure of the following compound which belongs to a class of compounds called allenes. This allene is chiral. Draw its enantiomer and explain why this compound is chiral.

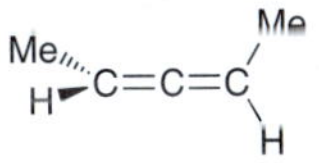

**5.59** Based on your analysis in the previous problem, determine whether the following allene is expected to be chiral:

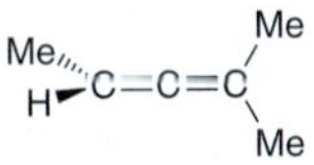

**5.60** As mentioned in Problem 5.58, some molecules are chiral even though they lack a chirality center. For example, consider the following two compounds and explain the source of chirality in each case:

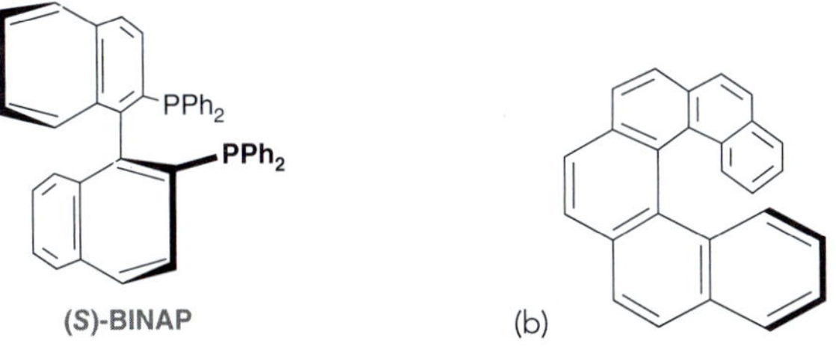

**5.61** The following compound is known to be chiral. Draw its enantiomer and explain the source of chirality.

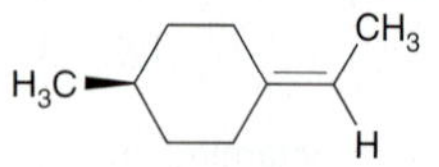

**5.62** The following compound is optically inactive. Explain why.

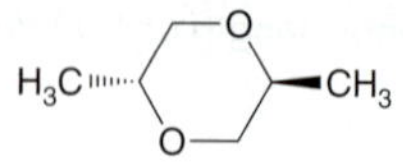

**5.63** The following compound, whose central ring is referred to as a 1,2,4-trioxane, is an anticancer agent that demonstrates activity against canine osteosarcoma (*J. Am. Chem. Soc.* **2012,** *134,* **13554–13557**). Assign the configuration of each chirality center and then draw all possible stereoisomers of this compound, showing the specific stereochemical relationship between each of your drawings.

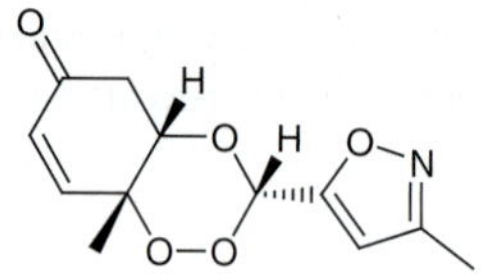

**5.64** Coibacin B is a natural product that exhibits potent anti-inflammatory activity and potential activity in the treatment of leishmaniasis, a disease caused by certain parasites (*Org. Lett.* **2012,** *14,* **3878–3881**):

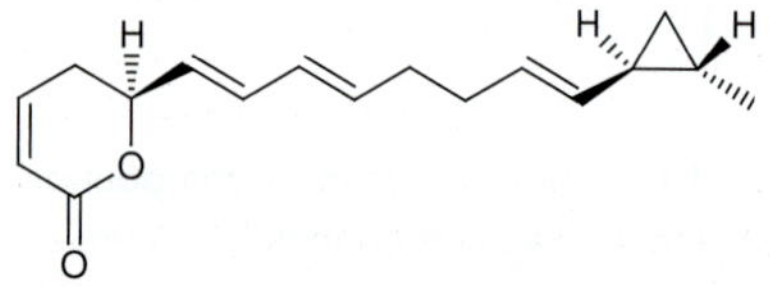

(a) Assign the configuration of each chirality center in coibacin B.

(b) Identify the number of possible stereoisomers for this compound, assuming that the geometry of the alkenes are fixed.

**5.65** Chiral catalysts can be designed to favor the formation of a single enantiomer in reactions where a new chirality center is formed. Recently, a novel type of chiral copper catalyst was developed that is "redox reconfigurable" in that the *S* enantiomer of the product predominates when the $Cu^{I}$ form of the catalyst is used, and the *R* enantiomer is the main product when the $Cu^{II}$ form is used (see below). This is not a general phenomenon, and only works for a limited number of reactions (*J. Am. Chem. Soc.* **2012,** *134,* **8054–8057**).

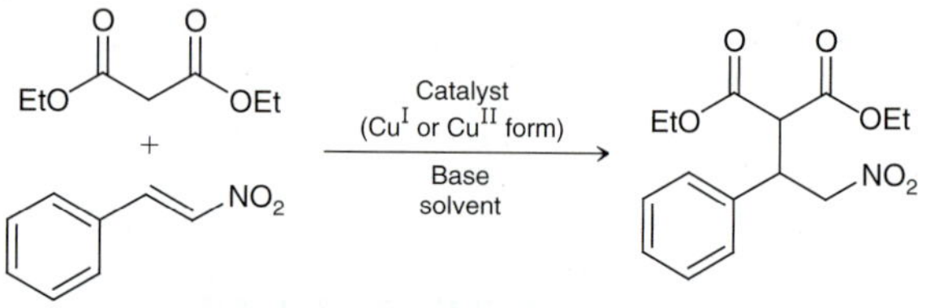

Catalyst ($Cu^{I}$ or $Cu^{II}$ form), Base, solvent

(a) Draw the major products that form when the $Cu^{I}$ or the $Cu^{II}$ catalyst is used, specifying the stereochemistry and the catalyst required to make each product.

(b) Consider the chart below, which summarizes the results of the reaction using different solvents. For each solvent, calculate the %*S* and %*R*. Which solvent produces the optimal results?

| Solvent | % *ee* of (*S*)-product | % yield of product |
|---|---|---|
| Toluene | 24 | 55 |
| Tetrahydrofuran | 48 | 33 |
| $CH_3CN$ | 72 | 55 |
| $CHCl_3$ | 30 | 40 |
| $CH_2Cl_2$ | 46 | 44 |
| Hexane | 51 | 30 |

## CHALLENGE PROBLEMS

**5.66** In 2010 the structure of the compound (+)-trigonoliimine A (isolated from the leaves of a plant in the Yunnan province of China) was reported as shown below. In 2011, a synthesis of the reported structure was accomplished, and the configuration of a chirality center in the compound was found to have been incorrectly assigned in 2010 (*J. Am. Chem. Soc.* **2011,** *133,* **10768–10771**):

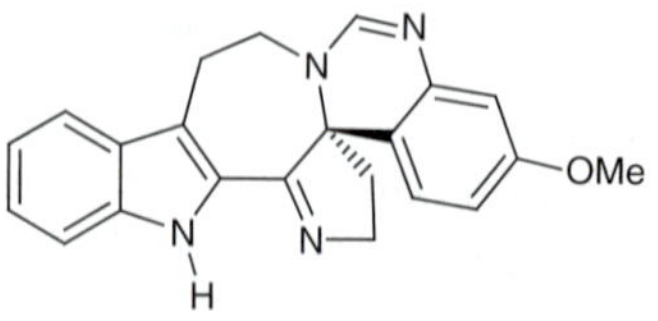

**Reported structure of (+)-trigonoliimine A**

(a) Draw the *correct* structure of (+)-trigonoliimine A and provide its absolute stereochemical configuration.

(b) The 2011 synthesis enabled the investigators to determine that the original assignment of configuration (of the natural product) was incorrect. Explain how this determination could have been made.

**5.67** Consider the reaction below, which involves both a dehydrogenation (removal of two neighboring hydrogen atoms) and a Diels-Alder reaction (which we will learn about in Chapter 17). Utilizing a specially designed catalyst, the achiral starting materials are converted to a total of four stereoisomeric products—two major and two minor (*J. Am. Chem. Soc.* **2011,** *133,* **14892–14895**). One of the major products is shown:

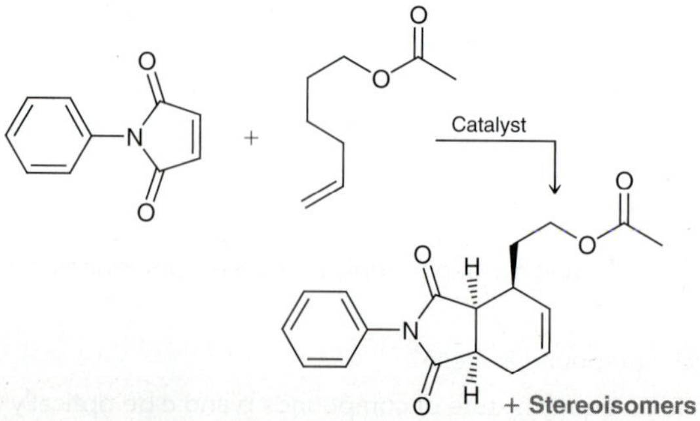

Catalyst

+ Stereoisomers

(a) Draw the other major product, which is the enantiomer of the product shown.

(b) The two minor products retain the *cis* connectivity at the bridgehead carbons but differ from the major products in terms of the relative stereochemistry of the third chirality center. Draw the two minor products.

(c) What is the relationship between the two minor products?

(d) What is the relationship between the major products and the minor products?

**5.68** When a toluene solution of the sugar-derived compound below is cooled, the molecules self-assemble into fibrous aggregates which work in concert with the surface tension of the solvent to form a stable gel (*Langmuir*, **2012**, *28*, 14039–14044). Redraw the structure showing both nonaromatic rings in chair conformations with the bridgehead hydrogen atoms in axial positions. Be sure to conserve the correct absolute stereochemistry in your answer.

**5.69** A key step of the Wittig reaction (Chapter 20) involves the decomposition of an oxaphosphetane, such as **A**, into two products (*J. Org. Chem.* **1986**, *51*, 3302–3308). If both chirality centers in **A** have the *R* configuration, determine whether the C=C π bond in **B** will have a *cis* or *trans* configuration.

**5.70** Consider the structure of the following ketone (*J. Org. Chem.* **2001**, *66*, 2072–2077):

(a) Does this compound exhibit rotational symmetry?

(b) Does this compound exhibit reflectional symmetry?

(c) Is the compound chiral? If so, draw its enantiomer.

**5.71** The natural product meloscine can be prepared via a 19-step synthesis, featuring the following allene (see Problems 5.58 and 5.59) as a key intermediate (*Org. Lett.* **2012**, *14*, 934–937):

(a) Draw a Newman projection of the allene when viewed from the left side of the C=C=C unit. Note that in this case the "front" and "back" carbons of the Newman projection are not directly attached to each other but instead are separated by an intervening *sp*-hybridized carbon atom.

(b) Draw the enantiomer of this compound in bond-line format (using wedges/dashes to show stereochemistry where appropriate) and as a Newman projection.

(c) Draw two diastereomers of this compound in bond-line and Newman formats.

**5.72** Each of 10 stereoisomeric sugar derivatives can be prepared via a multiple-step synthesis starting from either glucuronolactone **1D** or its enantiomer **1L**. Depending on the specific series of reactions used, the configuration at carbons 2, 3, and 5 can be selectively retained or inverted over the course of the synthesis. This protocol does not allow for inversion at C4, but selection of **1D** or **1L** allows access to products with either configuration at this chirality center (*J. Org. Chem.* **2012**, *77*, 7777–7792).

**1D (or enantiomer)** **(10 possible stereoisomers)**

(a) Draw the structure of **1L**.

(b) How many stereoisomers of **1D** are possible?

(c) Draw the structures of the eight chiral products (as pairs of enantiomers) and the two achiral products.

(d) Only four of the products (including the two achiral ones) can theoretically be prepared from either reactant, **1D** or **1L**. Identify these four compounds and explain your answer.

# 6



# Chemical Reactivity and Mechanisms

## DID YOU EVER WONDER...

**how Alfred Nobel made the fortune that he used to fund all of the Nobel Prizes (each of which includes a cash prize of over one million dollars)?**

As we will see later in this chapter, Alfred Nobel earned his vast fortune by developing and marketing dynamite. Many of the principles that inform the design of explosives are the same principles that govern all chemical reactions. In this chapter, we will explore some of the key features of chemical reactions (including explosions). We will explore the factors that cause reactions to occur as well as the factors that speed up reactions. We will practice drawing and analyzing energy diagrams, which will be used heavily throughout this book to compare reactions. Most importantly, we will focus on the skills necessary to draw and interpret the individual steps of a reaction. This chapter will develop the core concepts and critical skills necessary for understanding chemical reactivity.

## DO YOU REMEMBER?

*Before you go on, be sure you understand the following topics. If necessary, review the suggested sections to prepare for this chapter.*

- Identifying Lone Pairs (Section 2.5)
- Curved Arrows in Drawing Resonance Structures (Section 2.8)
- Drawing Resonance Structures via Pattern Recognition (Section 2.10)
- The Flow of Electron Density: Curved-Arrow Notation (Section 3.2)

Take the DO YOU REMEMBER? QUIZ in **WileyPLUS** to check your understanding.

## 6.1 Enthalpy

In Chapter 1, we discussed the nature of electrons and their ability to form bonds. In particular, we saw that electrons achieve a lower energy state when they occupy a bonding molecular orbital (Figure 6.1). It stands to reason, then, that breaking a bond requires an input of energy.

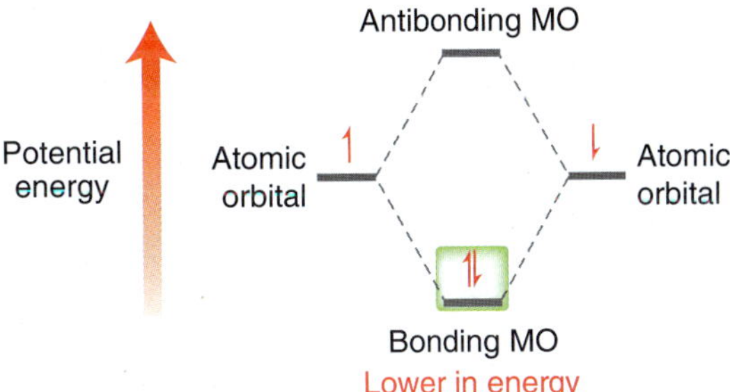


**FIGURE 6.1** An energy diagram showing that bonding electrons occupy a bonding MO.

For a bond to break, the electrons in the bonding MO must receive energy from their surroundings. Specifically, the surrounding molecules must transfer some of their kinetic energy to the system (the bond being broken). The term **enthalpy** is used to measure this exchange of energy:

$$\Delta H = q \quad \text{(at constant pressure)}$$

The change in enthalpy ($\Delta H$) for any process is defined as the exchange of kinetic energy, also called heat ($q$), between a system and its surroundings under conditions of constant pressure. For a bond-breaking reaction, $\Delta H$ is primarily determined by the amount of energy necessary to break the bond *homolytically*. **Homolytic bond cleavage** generates two uncharged species, called **radicals**, each of which bears an unpaired electron (Figure 6.2). Notice the use of single headed curved arrows, often called fish-hooks. Radicals and fish-hook arrows will be discussed in more detail in Chapter 11. In contrast, **heterolytic bond cleavage** is illustrated with a two-headed curved arrow, generating charged species, called ions (Figure 6.3).

**FIGURE 6.2** Homolytic bond cleavage produces two radicals.

**FIGURE 6.3** Heterolytic bond cleavage produces ions.

The energy required to break a covalent bond via homolytic bond cleavage is called the **bond dissociation energy**. The term $\Delta H°$ (with the "naught" symbol, or small circle, next to the *H*) refers to the bond dissociation energy when measured under standard conditions (i.e., where the pressure is 1 atm and the compound is in its standard state: a gas, a pure liquid, or a solid). Table 6.1 gives $\Delta H°$ values for a variety of bonds commonly encountered in this text.

**TABLE 6.1 BOND DISSOCIATION ENERGIES ($\Delta H°$) OF COMMON BONDS**

| Bond | KJ/MOL | KCAL/MOL |
|---|---|---|
| **Bonds to H** | | |
| H—H | 435 | 104 |
| H—$CH_3$ | 435 | 104 |
| H—$CH_2CH_3$ | 410 | 98 |
| H—$CH(CH_3)_2$ | 397 | 95 |
| H—$C(CH_3)_3$ | 381 | 91 |
| H—$C_6H_5$ (phenyl) | 473 | 113 |
| H—$CH_2C_6H_5$ (benzyl) | 356 | 85 |
| H—$CH{=}CH_2$ (vinyl) | 464 | 111 |
| H—$CH_2CH{=}CH_2$ (allyl) | 364 | 87 |
| H—F | 569 | 136 |
| H—Cl | 431 | 103 |
| H—Br | 368 | 88 |
| H—I | 297 | 71 |
| H—OH | 498 | 119 |
| H—$OCH_2CH_3$ | 435 | 104 |
| **C—C bonds** | | |
| $CH_3$—$CH_3$ | 368 | 88 |
| $CH_3CH_2$—$CH_3$ | 356 | 85 |
| $(CH_3)_2CH$—$CH_3$ | 351 | 84 |
| $H_2C{=}CH$—$CH_3$ | 385 | 92 |
| $HC{\equiv}C$—$CH_3$ | 489 | 117 |
| **Bonds to methyl** | | |
| $CH_3$—H | 435 | 104 |
| $CH_3$—F | 456 | 109 |
| $CH_3$—Cl | 351 | 84 |
| $CH_3$—Br | 293 | 70 |
| $CH_3$—I | 234 | 56 |
| $CH_3$—OH | 381 | 91 |
| **$H_3C$—C(H)(H)—X** | | |
| $CH_3CH_2$—H | 410 | 98 |
| $CH_3CH_2$—F | 448 | 107 |
| $CH_3CH_2$—Cl | 339 | 81 |
| $CH_3CH_2$—Br | 285 | 68 |
| $CH_3CH_2$—I | 222 | 53 |
| $CH_3CH_2$—OH | 381 | 91 |
| **$H_3C$—C($CH_3$)(H)—X** | | |
| $(CH_3)_2CH$—H | 397 | 95 |
| $(CH_3)_2CH$—F | 444 | 106 |
| $(CH_3)_2CH$—Cl | 335 | 80 |
| $(CH_3)_2CH$—Br | 285 | 68 |
| $(CH_3)_2CH$—I | 222 | 53 |
| $(CH_3)_2CH$—OH | 381 | 91 |
| **$H_3C$—C($CH_3$)($CH_3$)—X** | | |
| $(CH_3)_3C$—H | 381 | 91 |
| $(CH_3)_3C$—F | 444 | 106 |
| $(CH_3)_3C$—Cl | 331 | 79 |
| $(CH_3)_3C$—Br | 272 | 65 |
| $(CH_3)_3C$—I | 209 | 50 |
| $(CH_3)_3C$—OH | 381 | 91 |
| **X—X bonds** | | |
| F—F | 159 | 38 |
| Cl—Cl | 243 | 58 |
| Br—Br | 193 | 46 |
| I—I | 151 | 36 |
| HO—OH | 213 | 51 |

Most reactions involve the breaking and forming of several bonds. In such cases, we must take into account each bond being broken or formed. The total change in enthalpy ($\Delta H°$) for the reaction is referred to as the **heat of reaction**. The sign of $\Delta H°$ for a reaction (whether it is positive or negative) indicates the direction in which the energy is exchanged and is determined from the perspective of the system. A positive $\Delta H°$ indicates that the system increased in energy (it received energy from the surroundings), while a negative $\Delta H°$ indicates that the system decreased in energy (it gave energy to the surroundings). The direction of energy exchange is described by the terms *endothermic* and *exothermic*. In an **exothermic** process, the system gives energy to the surroundings ($\Delta H°$ is negative). In an **endothermic** process, the system receives energy from the surroundings ($\Delta H°$ is positive).

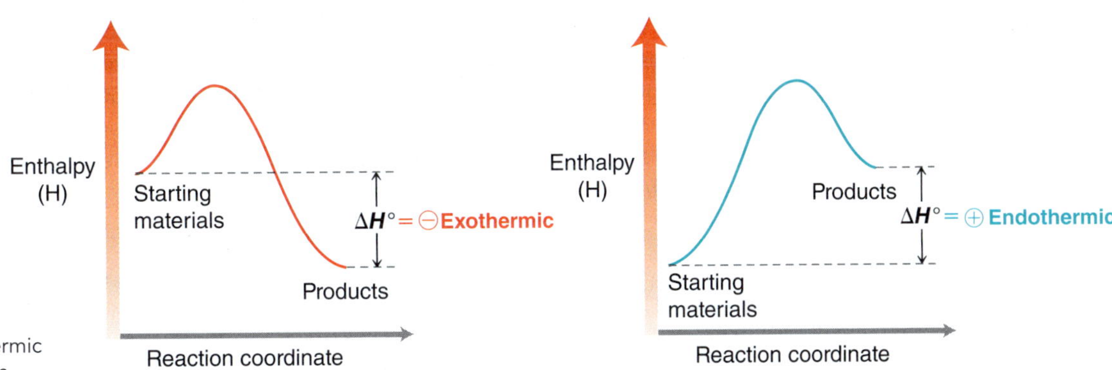


**FIGURE 6.4**
Energy diagrams of exothermic and endothermic processes.

This is best illustrated with energy diagrams (Figure 6.4) in which the curve represents the change in energy of the system as the reaction proceeds. In these diagrams, the progress of the reaction (the *x* axis of the diagram) is referred to as the *reaction coordinate*. Students are often confused by the sign of $\Delta H°$, and there is a valid reason for this confusion. In physics, the signs are the reverse of those shown here. Physicists think of $\Delta H°$ in terms of the surroundings rather than the system. They care about how devices have an impact on the environment—how much work the device can perform. Chemists, on the other hand, think in terms of the system. When chemists run a reaction, they care about the reactants and the products; they don't care about how the reaction ever so slightly changes the temperature in the laboratory. If you are currently enrolled in a physics course and find yourself confused about the sign of $\Delta H°$, just remember that chemists think of $\Delta H°$ from the perspective of the reaction (the system). For a chemist, an exothermic process involves the system losing energy to the surroundings, so $\Delta H°$ is negative.

Let's get some practice predicting the sign and value of $\Delta H°$ for a reaction.

## SKILLBUILDER

### 6.1 PREDICTING $\Delta H°$ OF A REACTION

**LEARN** the skill

Predict the sign and magnitude of $\Delta H°$ for the following reaction. Give your answer in units of kilojoules per mole, and identify whether the reaction is expected to be endothermic or exothermic.

$+ \; Cl_2 \longrightarrow$ Cl $+ \; HCl$

**SOLUTION**

Identify all bonds that are either broken or formed:

**Bonds broken** — **Bonds formed**

$H_3C-C(CH_3)_2-H \; + \; Cl-Cl \longrightarrow H_3C-C(CH_3)_2-Cl \; + \; H-Cl$

**STEP 1**
Identify the bond dissociation energy of each bond that is broken or formed.

Then use Table 6.1 to find the bond dissociation energies for each of these bonds:

| Bond | kJ/mol |
|---|---|
| $H-C(CH_3)_3$ | 381 |
| $Cl-Cl$ | 243 |
| $(CH_3)_3C-Cl$ | 331 |
| $H-Cl$ | 431 |

**STEP 2**
Determine the appropriate sign for each value from step 1.

Now we must decide what sign (+ or −) to place in front of each value. Remember that $\Delta H°$ is defined with respect to the system. For each bond broken, the system must receive energy in order for the bond to break, so $\Delta H°$ must be positive. For each bond formed, the electrons are going to a lower energy state, and the system releases energy to the surroundings, so $\Delta H°$ must be negative. Therefore, $\Delta H°$ for this reaction will be the sum total of the following numbers:

**STEP 3**
Add the values for all bonds broken and formed.

| Bonds broken | kJ/mol | Bonds formed | kJ/mol |
|---|---|---|---|
| $H—C(CH_3)_3$ | +381 | $(CH_3)_3C—Cl$ | −331 |
| Cl—Cl | +243 | H—Cl | −431 |

That is, $\Delta H° = -138$ kJ/mol. For this reaction $\Delta H°$ is negative, which means that the system is losing energy. It is giving off energy to the environment, so the reaction is exothermic.

**PRACTICE the skill** **6.1** Using the data in Table 6.1, predict the sign and magnitude of $\Delta H°$ for each of the following reactions. In each case, identify whether the reaction is expected to be endothermic or exothermic:

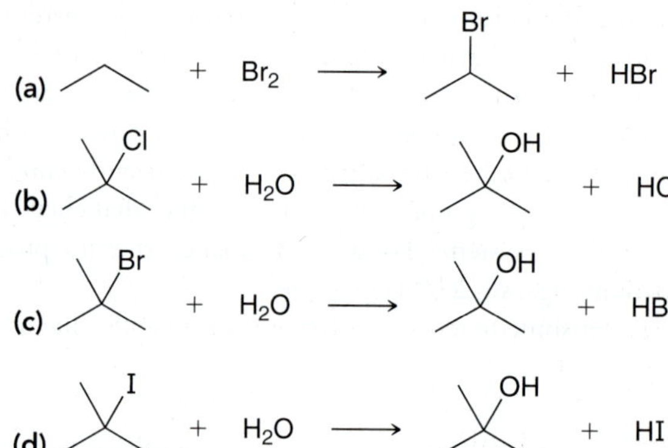

**APPLY the skill** **6.2** Recall that a C=C bond is comprised of a σ bond and a π bond. These two bonds together have a combined BDE (bond dissociation energy) of 632 kJ/mol. Use this information to predict whether the following reaction is exothermic or endothermic:

$H_2C=CH_2 + H_2O \longrightarrow CH_3CH_2OH$

need more **PRACTICE?** **Try Problem 6.21a**

## 6.2 Entropy

The sign of $\Delta H°$ is not the ultimate measure of whether or not a reaction can or will occur. Although exothermic reactions are more common, there are still plenty of examples of endothermic reactions that readily occur. This begs the question: What is the ultimate measure for determining whether or not a reaction can occur? The answer to this question is *entropy*, which is the underlying principle guiding all physical, chemical, and biological processes. **Entropy** is informally defined as the measure of disorder associated with a system, although this definition is overly simplistic. Entropy is more accurately described in terms of probabilities. To understand this, consider the following analogy. Imagine four coins, lined up in a row. If we toss all four coins, the chances that exactly half of the coins will land on heads are much greater than the chances that all four coins will land on heads. Why? Compare the number of possible ways to achieve each result (Figure 6.5). There is only one state in which all four coins land on heads, yet there are six different states in which exactly two coins land on heads. The probability of exactly two coins landing on heads is six times greater than the probability of all four coins landing on heads.

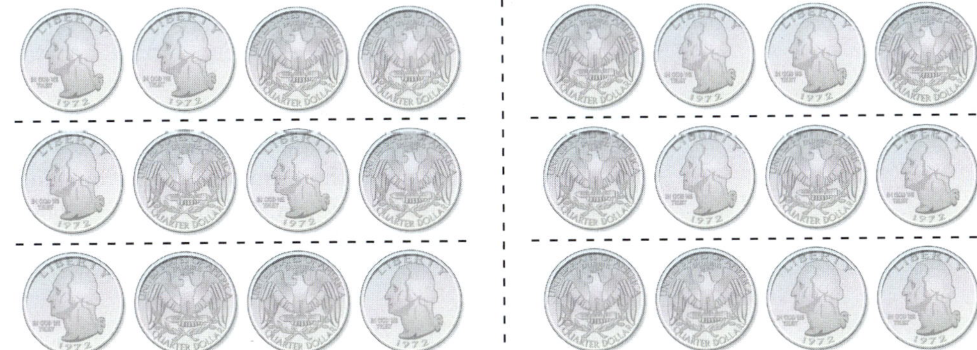


**FIGURE 6.5**
Comparing the number of ways to achieve various outcomes in a coin-tossing experiment.

Expanding the analogy to six coins, we find that the probability of exactly half of the coins landing on heads is 20 times greater than the probability of all six coins landing on heads. Expanding the analogy further to eight coins, we find that the probability of exactly half of the coins landing on heads is 70 times greater than the probability of all eight coins landing on heads. A trend is apparent: As the number of coins being tossed increases, the chances of all coins landing on heads becomes less likely. Imagine a floor covered with a billion coins. What are the chances that flipping all of them will result in all heads? The chances are miniscule (you have better chances of winning the lottery a hundred times in a row). It is much more likely that approximately half of the coins will land on heads because there are so many more ways of accomplishing that.

Now apply the same principle to describe the behavior of gas molecules in a two-chamber system, as shown in Figure 6.6. In the initial condition, one of the chambers is empty, and a divider prevents the gas molecules from entering that chamber. When the divider between the chambers is removed, the gas molecules undergo a free expansion. Free expansion occurs readily, but the reverse process is never observed. Once spread out into the two chambers, the gas molecules will not suddenly collect into the first chamber, leaving the second chamber empty. This scenario is very similar to the coin-tossing analogy. At any moment in time, each molecule can be in either chamber 1 or chamber 2 (just like each coin can be either heads or tails). As we increase the number of molecules, the chances become less likely that all of the molecules will be found in one chamber. When dealing with a mole ($6 \times 10^{23}$ molecules), the chances are practically negligible, and we do not observe the molecules suddenly collecting into one chamber (at least not in our lifetimes).

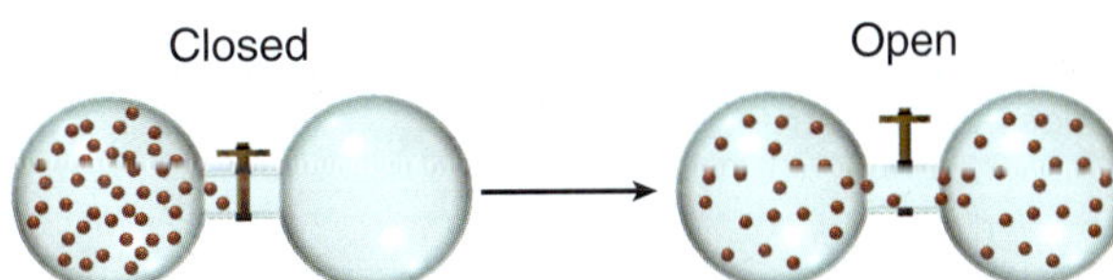


**FIGURE 6.6**
Free expansion of a gas.

Free expansion is a classic example of entropy. When molecules occupy both chambers, the system is said to be at a higher state of entropy, because the number of states in which the molecules are spread between both chambers is so much greater than the number of states in which the molecules are all found in one chamber. Entropy is really nothing more than an issue of likelihood and probability.

A process that involves an increase in entropy is said to be **spontaneous**. That is, the process can and will occur, given enough time. Chemical reactions are no exception, although the considerations are slightly more complex than in a simple free expansion. In the case of free expansion, we only had to consider the change in entropy of the system (of the gas particles). The surroundings were unaffected by the free expansion. However, in a chemical reaction, the surroundings are affected. We must take into account not only the change in entropy of the system but also the change in entropy of the surroundings:

$$\Delta S_{tot} = \Delta S_{sys} + \Delta S_{surr}$$

where $\Delta S_{tot}$ is the total change in entropy associated with the reaction. In order for a process to be spontaneous, the total entropy must increase. The entropy of the system (the reaction) can actually decrease, as long as the entropy of the surroundings increases by an amount that offsets the decreased entropy of the system. As long as $\Delta S_{tot}$ is positive, the reaction will be spontaneous. Therefore, if we wish to assess whether a particular reaction will be spontaneous, we must assess the values of $\Delta S_{sys}$ and $\Delta S_{surr}$. For now, we will focus our attention on $\Delta S_{sys}$; we will discuss $\Delta S_{surr}$ in the next section.

The value of $\Delta S_{sys}$ is affected by a number of factors. The two most dominant factors are shown in Figure 6.7. In the first example in Figure 6.7, one mole of reactant produces two moles of product. This represents an increase in entropy, because the number of possible ways to arrange the molecules increases when there are more molecules (as we saw when we expanded our coin-tossing analogy by increasing the number of coins). In the second example, a cyclic compound is being converted into an acyclic compound. Such a process also represents an increase in entropy, because acyclic compounds have more freedom of motion than cyclic compounds. An acyclic compound can adopt a larger number of conformations than a cyclic compound, and once again, the larger number of possible states corresponds with a larger entropy.

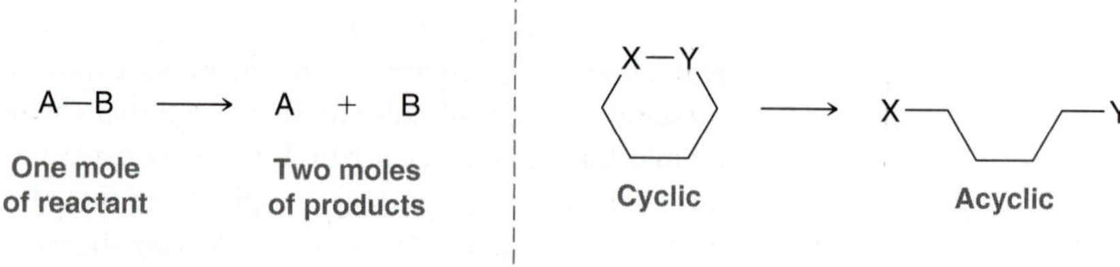


**FIGURE 6.7**
Two ways in which the entropy of a chemical system can increase.

## CONCEPTUAL CHECKPOINT

**6.3** For each of the following processes predict the sign of $\Delta S$ for the reaction. In other words, will $\Delta S_{sys}$ be positive (an increase in entropy) or negative (a decrease in entropy)?

(a) (b) (c)

(d) Br ⟶ ⊕ + $Br^{\ominus}$ (e) ⊕ H–O–H ⟶ H–O⊕–H (f) H–O⊕ ⟶ ⊕ OH

## 6.3 Gibbs Free Energy

In the previous section, we saw that entropy is the one and only criterion that determines whether or not a chemical reaction will be spontaneous. But it is not enough to consider $\Delta S$ of the system alone; we must also take into account $\Delta S$ of the surroundings:

$$\Delta S_{tot} = \Delta S_{sys} + \Delta S_{surr}$$

The total change in entropy (system plus surroundings) must be positive in order for the process to be spontaneous. It is fairly straightforward to assess $\Delta S_{sys}$ using tables of standard entropy values (a skill that you likely learned in your general chemistry course). However, the assessment of $\Delta S_{surr}$ presents more of a challenge. It is certainly not possible to observe the entire universe, so how can we possibly measure $\Delta S_{surr}$? Fortunately, there is a clever solution to this problem.

Under conditions of constant pressure and temperature, it can be shown that:

$$\Delta S_{surr} = -\frac{\Delta H_{sys}}{T}$$

**BY THE WAY**
Organic chemists generally carry out reactions under constant pressure (atmospheric pressure) and at a constant temperature. For this reason, the conditions most relevant to practicing organic chemists are conditions of constant pressure and temperature.

Notice that $\Delta S_{surr}$ is now defined *in terms of the system*. Both $\Delta H_{sys}$ and $T$ (temperature in Kelvin) are easily measured, which means that $\Delta S_{surr}$ can in fact be measured. Plugging this expression for $\Delta S_{surr}$ into the equation for $\Delta S_{tot}$, we arrive at a new equation for $\Delta S_{tot}$ for which all terms are measurable:

$$\Delta S_{tot} = \left(-\frac{\Delta H_{sys}}{T}\right) + \Delta S_{sys}$$

This expression is still the ultimate criterion for spontaneity ($\Delta S_{tot}$ must be positive). As a final step, we multiply the entire equation by $-T$, which gives us a new term, called **Gibbs free energy**:

$$-T\Delta S_{tot} = \Delta H_{sys} - T\Delta S_{sys}$$

$$\downarrow$$

$$\Delta G$$

In other words, $\Delta G$ is nothing more than a repackaged way of expressing total entropy:

$$\Delta G = \Delta H - T\Delta S$$

**Associated with the change in entropy of the surroundings** ($\Delta H$)

**Associated with the change in entropy of the system** ($T\Delta S$)

In this equation, the first term ($\Delta H$) is associated with the change in entropy of the surroundings (a transfer of energy to the surroundings increases the entropy of the surroundings). The second term ($T\Delta S$) is associated with the change in entropy of the system. The first term ($\Delta H$) is often much larger than the second term ($T\Delta S$), so for most processes, $\Delta H$ will determine the sign of $\Delta G$. Remember that $\Delta G$ is just $\Delta S_{tot}$ multiplied by *negative T*. So if $\Delta S_{tot}$ must be positive in order for a process to be spontaneous, then $\Delta G$ must be negative. *In order for a process to be spontaneous, $\Delta G$ for that process must be negative.* This requirement is called the second law of thermodynamics.

**BY THE WAY**
You may recall from your general chemistry course that the first law of thermodynamics is the requirement for conservation of energy during any process.

In order to compare the two terms contributing to $\Delta G$, we will sometimes present the equation in a nonstandard way, with a plus sign between the two terms:

$$\Delta G = \Delta H + (-T\Delta S)$$

## Explosives

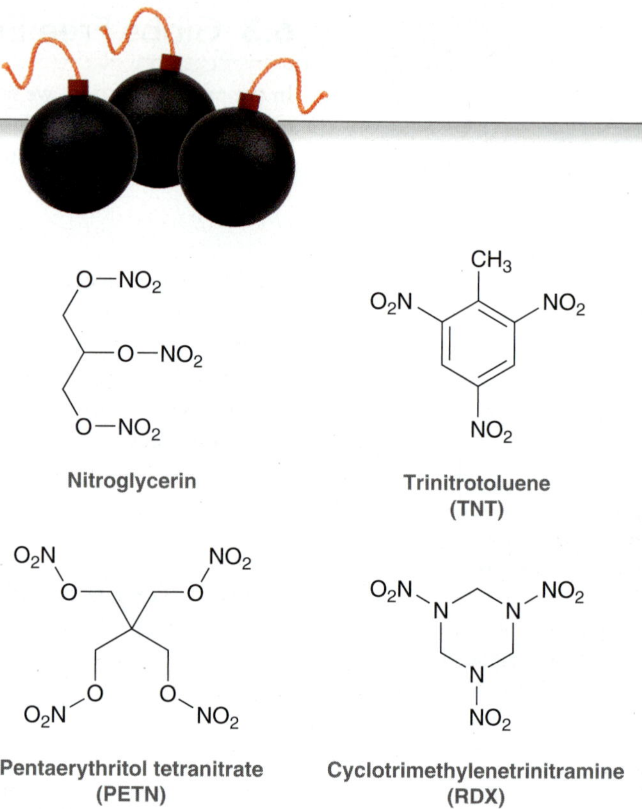


Explosives are compounds that can be detonated, generating sudden large pressures and liberating excessive heat. Explosives have several defining characteristics:

1. In the presence of oxygen, an explosive will produce a reaction for which both expressions contribute to a very favorable $\Delta G$:

$$\Delta G = \Delta H - T\Delta S$$

   That is, the changes in entropy of both the surroundings and the system are extremely favorable. As a result, both terms contribute to a very large and negative value of $\Delta G$. This makes the reaction extremely favorable.

2. Explosives must generate large quantities of gas very quickly in order to produce the sudden increase in pressure observed during an explosion. Compounds containing multiple nitro groups are capable of liberating gaseous $NO_2$. Several examples are shown.

3. When detonated, the reaction must occur very rapidly. This aspect of explosives will be discussed later in this chapter.

In some cases, this nonstandard presentation will allow for a more efficient analysis of the competition between the two terms. As an example, consider the following reaction:

[Reaction: 1,3-butadiene + ethylene → cyclohexene]

In this reaction, two molecules are converted into one molecule, so $\Delta S_{sys}$ is not favorable. However, $\Delta S_{surr}$ is favorable. Why? $\Delta S_{surr}$ is determined by $\Delta H$ of the reaction. In the first section of this chapter, we learned how to estimate $\Delta H$ for a reaction by looking at the bonds being formed and broken. In the reaction above, three $\pi$ bonds are broken while one $\pi$ bond and two $\sigma$ bonds are formed. In other words, two $\pi$ bonds have been transformed into two $\sigma$ bonds. Sigma bonds are stronger (lower in energy) than $\pi$ bonds, and therefore, $\Delta H$ for this reaction must be negative (exothermic). Energy is transferred to the surroundings, and this increases the entropy of the surroundings. To recap, $\Delta S_{sys}$ is not favorable, but $\Delta S_{surr}$ is favorable. Therefore, this process exhibits a competition between the change in entropy of the system and the change in entropy of the surroundings:

$$\Delta G = \underset{\ominus}{\Delta H} + \underset{\oplus}{(-T\Delta S)}$$

Remember that $\Delta G$ must be negative in order for a process to be spontaneous. The first term is negative, which is favorable, but the second term is positive, which is unfavorable. The term that is larger will determine the sign of $\Delta G$. If the first term is larger, then $\Delta G$ will be negative, and the reaction will be spontaneous. If the second term is larger, then $\Delta G$ will be positive, and the reaction will not be spontaneous. Which term dominates? The value of the second term is dependent on $T$, and therefore, the outcome of this reaction will be very sensitive to temperature. Below a certain $T$, the process will be spontaneous. Above a certain $T$, the reverse process will be favored.

Any process with a negative $\Delta G$ will be spontaneous. Such processes are called **exergonic**. Any process with a positive $\Delta G$ will not be spontaneous. Such processes are called **endergonic** (Figure 6.8).

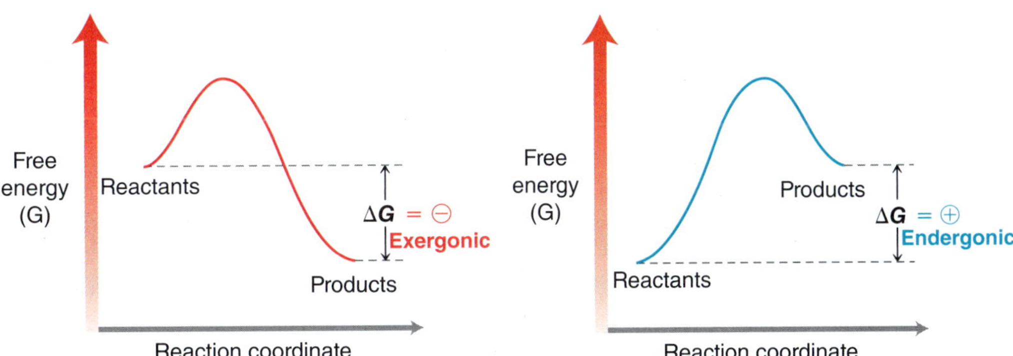


**FIGURE 6.8** Energy diagrams of an exergonic process and an endergonic process.

These energy diagrams are incredibly useful when analyzing reactions, and we will use these diagrams frequently throughout the course. The next few sections of this chapter deal with some of the more specific details of these diagrams.

## practically speaking — Do Living Organisms Violate the Second Law of Thermodynamics?

Students often wonder whether life is a violation of the second law of thermodynamics. It is certainly a legitimate question. Our bodies are highly ordered, and we are capable of imposing order on our surroundings. We are able to take raw materials and build tall skyscrapers. How can that be? In addition, many of the reactions necessary to synthesize the macromolecules necessary for life, such as DNA and proteins, are endergonic processes ($\Delta G$ is positive). How can those reactions occur?

Earlier in this chapter, we saw that it is possible for a reaction to exhibit a decrease in entropy and still be spontaneous if and only if the entropy of the surroundings increases such that $\Delta S_{tot}$ for the process is positive. In other words, we have to take the whole picture into account when determining if a process will be spontaneous. Yes, it is true that many of the reactions employed by life are not spontaneous *by themselves*. But when they are coupled with other highly favorable reactions, such as the metabolism of food, the total entropy (system plus surroundings) does actually increase. As an example, consider the metabolism of glucose:

$$\text{Glucose} + 6\,O_2 \longrightarrow 6\,CO_2 + 6\,H_2O \qquad \Delta G^\circ = -2880 \text{ kJ/mol}$$

This transformation has a large and negative $\Delta G$. Processes like this one are ultimately responsible for driving otherwise nonspontaneous reactions that are necessary for life.

Living organisms do not violate the second law of thermodynamics—quite the opposite, in fact. Living organisms are prime examples of entropy at its finest. Yes, we do impose order on our environment, but that is only permitted because we are *entropy machines*. We give off heat to our surroundings and we consume highly ordered molecules (food) and break them down into smaller, more stable compounds with much lower free energy. These features allow us to impose order on our surroundings, because our net effect is to increase the entropy of the universe.

## CONCEPTUAL CHECKPOINT

**6.4** For each of the following reactions predict the sign of $\Delta G$. If a prediction is not possible because the sign of $\Delta G$ will be temperature dependent, describe how $\Delta G$ will be affected by raising the temperature.

(a) An endothermic reaction for which the system exhibits an increase in entropy

(b) An exothermic reaction for which the system exhibits an increase in entropy

(c) An endothermic reaction for which the system exhibits a decrease in entropy

(d) An exothermic reaction for which the system exhibits a decrease in entropy

**6.5** At room temperature, molecules spend most of their time in lower energy conformations. In fact, there is a general tendency for any system to move toward lower energy. As another example, electrons form bonds because they "prefer" to achieve a lower energy state. We can now appreciate that the reason for this preference is based on entropy. When a compound assumes a lower energy conformation or when electrons assume a lower energy state, $\Delta S_{tot}$ increases. Explain.

## 6.4 Equilibria

Consider the energy diagram in Figure 6.9, showing a reaction in which the reactants, A and B, are converted into products, C and D. The reaction exhibits a negative $\Delta G$ and therefore will be spontaneous. Accordingly, we might expect a mixture of A and B to be converted completely into C and D. But this is not the case. Rather, an equilibrium is established in which all four compounds are present. Why should this be the case? If C and D are truly lower in free energy than A and B, then why is there any amount of A and B present when the reaction is complete?

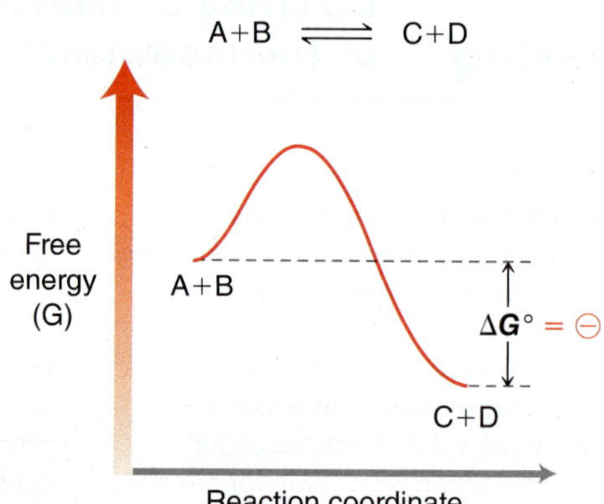


**FIGURE 6.9** The energy diagram of an exergonic reaction in which reactants (A and B) are converted into products (C and D).

To answer this question, we must consider the effect of having a large number of molecules. The energy diagram in Figure 6.9 describes the reactions between one molecule of A and one molecule of B. However, when dealing with moles of A and B, the changing concentrations have an effect on the value of $\Delta G$. When the reaction begins, only A and B are present. As the reaction proceeds, the concentrations of A and B decrease, and the concentrations of C and D increase. This has an effect on $\Delta G$, which can be illustrated with the diagram in Figure 6.10. As the reaction proceeds, the free energy decreases until it reaches a minimum value at very particular concentrations of reactants and products. If the reaction were to proceed further in either direction, the result would be an increase in free energy, which is not spontaneous. At this point, no further change is observed, and the system is said to have reached equilibrium. The exact position of equilibrium for any reaction is described by the equilibrium constant, $K_{eq}$,

$$K_{eq} = \frac{[\text{products}]}{[\text{reactants}]} = \frac{[C][D]}{[A][B]}$$

where $K_{eq}$ is defined as the equilibrium concentrations of products divided by the equilibrium concentrations of reactants.

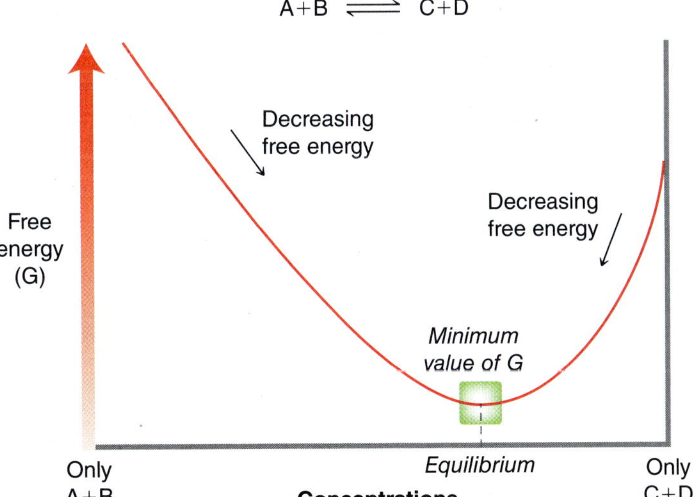


**FIGURE 6.10** An energy diagram illustrating the relationship between $\Delta G$ and concentration. A minimum in free energy is achieved at particular concentrations (equilibrium).

If the concentration of products is greater than the concentration of reactants, then $K_{eq}$ will be greater than 1. On the other hand, if the concentration of products is less than the concentration of reactants, then $K_{eq}$ will be less than 1. The term $K_{eq}$ indicates the exact position of the equilibrium, and it is related to $\Delta G$ in the following way, where $R$ is the gas constant (8.314 J/mol · K) and $T$ is the temperature measured in Kelvin:

$$\Delta G = -RT \ln K_{eq}$$

For any process, whether it is a reaction or a conformational change, the relationship between $K_{eq}$ and $\Delta G$ is defined by the equation above. Table 6.2 gives a few examples and illustrates that $\Delta G$ determines the maximum yield of products. If $\Delta G$ is negative, the products will be favored ($K_{eq} > 1$). If $\Delta G$ is positive, then the reactants will be favored ($K_{eq} < 1$). In order for a reaction to be useful (in order for products to dominate over reactants), $\Delta G$ must be negative; that is, $K_{eq}$ must be greater than 1. The values in Table 6.2 indicate that a small difference in free energy can have a significant impact on the ratio of reactants to products.

**TABLE 6.2 SAMPLE VALUES OF $\Delta G$ AND CORRESPONDING $K_{eq}$**

| $\Delta G°$ (KJ/MOL) | $K_{eq}$ | % PRODUCTS AT EQUILIBRIUM |
|---|---|---|
| −17 | $10^3$ | 99.9% |
| −11 | $10^2$ | 99% |
| −6 | $10^1$ | 90% |
| 0 | 1 | 50% |
| +6 | $10^{-1}$ | 10% |
| +11 | $10^{-2}$ | 1% |
| +17 | $10^{-3}$ | 0.1% |

In this section, we have investigated the relationship between $\Delta G$ and equilibrium—a topic that falls within the realm of **thermodynamics**. Thermodynamics is the study of how energy is distributed under the influence of entropy. For chemists, the thermodynamics of a reaction specifically refers to the study of the relative energy levels of reactants and products (Figure 6.11). This difference in free energy ($\Delta G$) ultimately determines the yield of products that can be expected for any reaction.

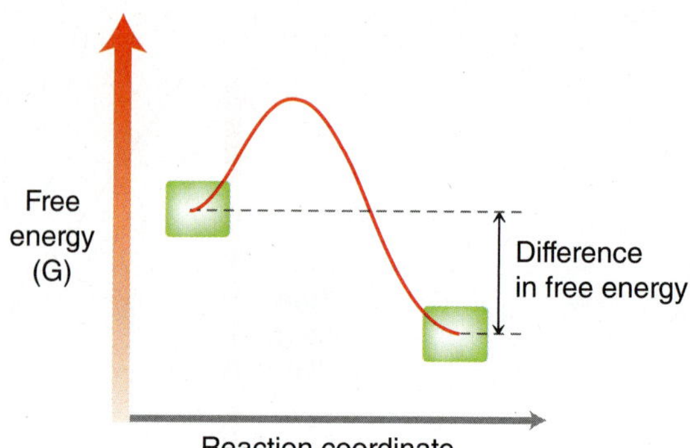


**FIGURE 6.11** Thermodynamics of a reaction is based on the difference in energy between starting materials and products.

## CONCEPTUAL CHECKPOINT

**6.6** In each of the following cases, use the data given to determine whether the reaction favors reactants or products:

(a) A reaction for which $\Delta G = +1.52$ kJ/mol

(b) A reaction for which $K_{eq} = 0.5$

(c) A reaction carried out at 298 K, for which $\Delta H = +33$ kJ/mol and $\Delta S = +150$ J/mol · K

(d) An exothermic reaction with a positive value for $\Delta S_{sys}$

(e) An endothermic reaction with a negative value for $\Delta S_{sys}$

## 6.5 Kinetics

In the previous sections, we saw that a reaction will be spontaneous if $\Delta G$ for the reaction is negative. The term *spontaneous* does not mean that the reaction will occur suddenly. Rather, it means that the reaction is thermodynamically favorable; that is, the reaction favors formation of products. Spontaneity has nothing to do with the speed of the reaction. For example, the conversion of diamonds into graphite is a spontaneous process at standard pressure and temperature. In other words, all diamonds are turning into graphite at this very moment. But even though this process is spontaneous, it is nevertheless very, very slow. It will take millions of years, but ultimately, all diamonds will eventually turn into graphite!

Why is it that some spontaneous processes are fast, like explosions, while others are slow, like diamonds turning into graphite? The study of reaction rates is called **kinetics**. In this section, we will explore issues related to reaction rates.

### Rate Equations

The rate of any reaction is described by a **rate equation,** which has the following general form:

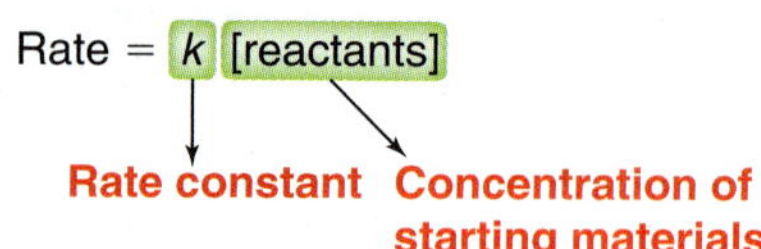


The general equation above indicates that the rate of a reaction is dependent on the rate constant ($k$) and on the concentration of the reactants. The rate constant ($k$) is a value that is specific to each reaction and is dependent on a number of factors. We will explore those factors in the next section. For now, we will focus on the effect of concentrations on the rate.

A reaction is the result of a collision between reactants, and therefore, it makes sense that increasing the concentrations of reactants should increase the frequency of collisions that lead to a reaction, thereby increasing the rate of the reaction. However, the precise effect of concentration on rate must be determined experimentally:

$$\text{Rate} = k\,[A]^x\,[B]^y$$

In this rate equation notice that the concentrations of A and B are shown with exponents ($x$ and $y$). These exponents must be experimentally determined, by exploring how the rate is affected when the concentrations of A and B are each separately doubled. Here are just a few of the possibilities that we might discover for any particular reaction (Figure 6.12).

**FIGURE 6.12** Rate equations for first-order, second-order, and third-order reactions.

| Rate $= k[A]$ | Rate $= k[A]\,[B]$ | Rate $= k[A]^2\,[B]$ |
|---|---|---|
| **First order** | **Second order** | **Third order** |

**LOOKING AHEAD**

If you are having trouble understanding how the rate would not be affected by the concentration of a reactant, don't worry. This will be explained in Chapter 7.

The first possibility has [B] absent from the rate equation. This is a situation where doubling the concentration of A has the effect of doubling the rate of the reaction, but doubling the concentration of B has no effect at all. In such a case, the sum of the exponents is 1, and the reaction is said to be **first order**.

The second possibility above represents a situation where doubling [A] has the effect of doubling the rate, while doubling [B] also has the effect of doubling the rate. In such a case, both [A] and [B] are present in the rate equation, and the exponent of each is 1. The sum of the exponents in this case is 2, and the reaction is said to be **second order**. The third possibility above represents a situation where doubling [A] has the effect of *quadrupling* the rate, while doubling [B] has the effect of doubling the rate. In such a case, the exponent of [A] is 2 and the exponent of [B] is 1. The sum of the exponents in this case is 3, and the reaction is said to be **third order**.

## Factors Affecting the Rate Constant

As seen in the previous section, the rate of a reaction is dependent on the rate constant ($k$):

$$\text{Rate} = k\,[A]^x\,[B]^y$$

A relatively fast reaction is associated with a large rate constant, while a relatively slow reaction is associated with a small rate constant. The value of the rate constant is dependent on three factors: the energy of activation, temperature, and steric considerations.

1. **Energy of Activation.** The energy barrier (the hump) between the reactants and the products is called the **energy of activation**, or $E_a$ (Figure 6.13). This energy barrier represents the minimum amount of energy required for a reaction to occur between two reactants that collide. If a collision between the reactants does not involve this much energy, they will not react with each other to form products. The number of successful collisions is therefore dependent on the number of molecules that have a certain threshold kinetic energy.

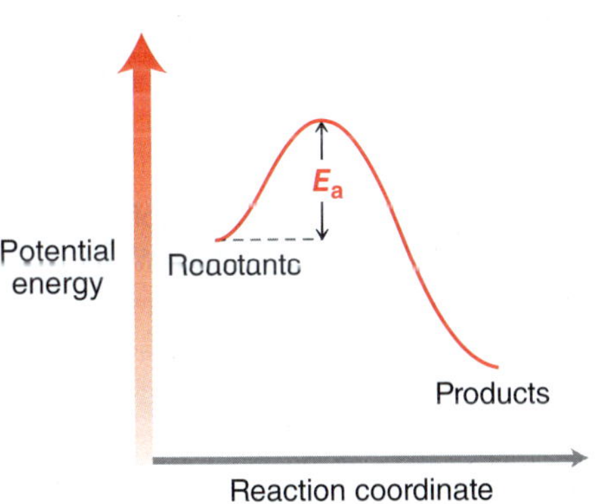


**FIGURE 6.13** An energy diagram showing the energy of activation ($E_a$) of a reaction.

At any specific temperature, the reactants will have a specific average kinetic energy, but not all molecules will possess this average energy. In fact, most molecules have either less than the

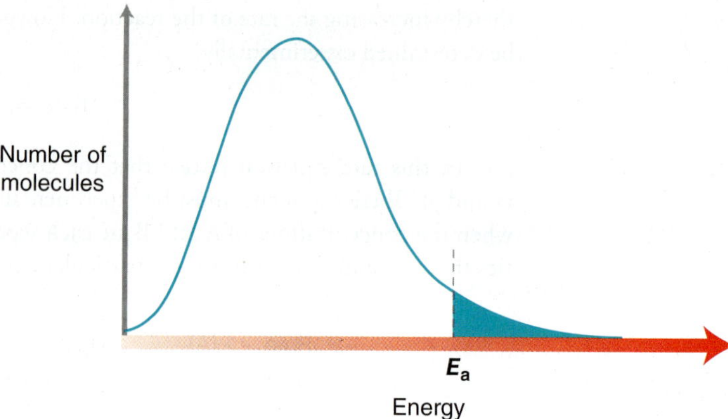


**FIGURE 6.14** Distribution of kinetic energy. Shown in blue is the fraction of molecules that possess enough energy to produce a reaction.

average or more than the average, giving rise to a distribution as shown in Figure 6.14. Notice that only a certain number of the molecules will have the minimum energy necessary to produce a reaction. The number of molecules with this energy will be dependent on the value of $E_a$. If $E_a$ is small, then a large percentage of the molecules will have the threshold energy necessary to produce a reaction. Therefore, a low $E_a$ will lead to a faster reaction (Figure 6.15).

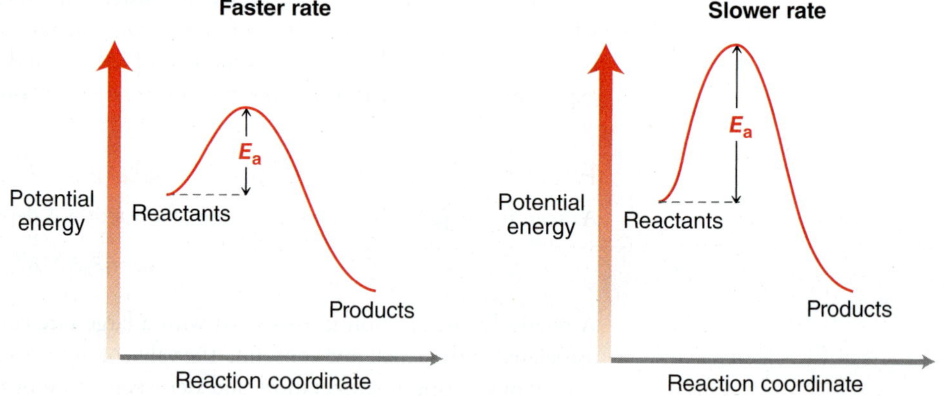


**FIGURE 6.15** The rate of a reaction is dependent on the size of $E_a$.

2. **Temperature.** The rate of a reaction is also very sensitive to temperature (Figure 6.16). Raising the temperature of a reaction will cause the rate to increase, because the molecules will have more kinetic energy at a higher temperature. At higher temperature, a larger number of molecules will have the kinetic energy sufficient to produce a reaction. As a rule of thumb, raising the temperature by 10 °C causes the rate to double.

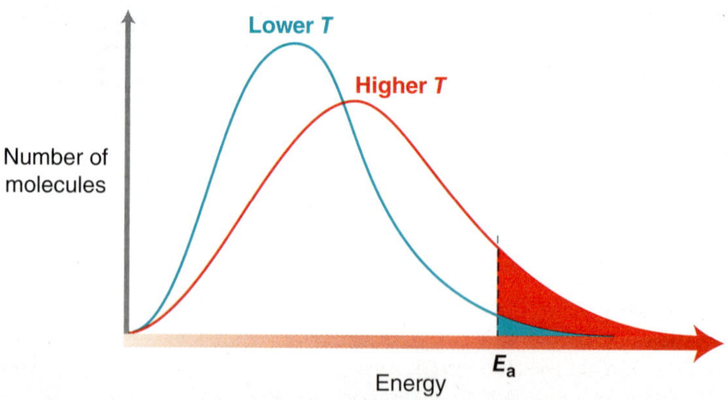


**FIGURE 6.16** Increasing the temperature increases the number of molecules with sufficient energy to produce a reaction.

3. **Steric Considerations.** The geometry of the reactants and the orientation of their collision can also have an impact on the frequency of collisions that lead to a reaction. This factor will be explored in greater detail in Section 7.4.

## medically speaking | Nitroglycerin: An Explosive with Medicinal Properties

Explosives can be divided into two categories: primary explosives and secondary explosives. Primary explosives are very sensitive to shock or heat, and they detonate very easily. That is, the energy of activation for the explosion is very small and easily overcome. In contrast, secondary explosives have a larger activation energy and are therefore more stable. Secondary explosives are often detonated with a very small amount of primary explosive.

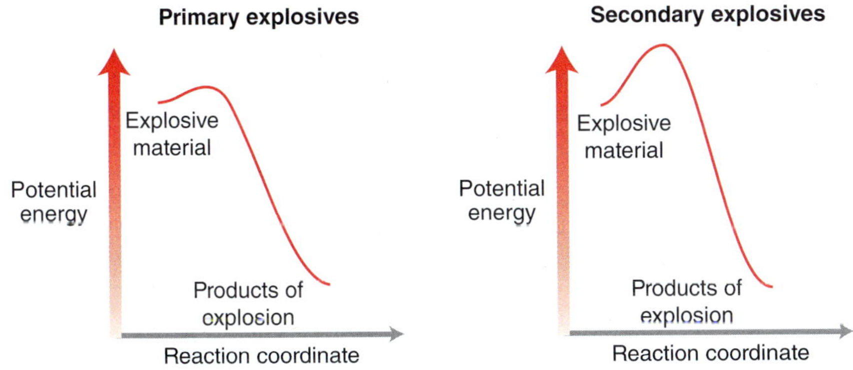


The first commercial secondary explosive was produced by Alfred Nobel in the mid-1800s. At the time, there were no strong explosives safe enough to handle. Nobel focused his efforts on finding a way to stabilize nitroglycerin:

**Nitroglycerin**

Nitroglycerin is very shock sensitive and unsafe to handle. In an effort to find a formulation of nitroglycerin that would have a larger energy of activation, Nobel conducted many experiments. Many of those experiments caused explosions in the Nobel factory, one of which killed his younger brother, Emil, along with several co-workers. Ultimately, Nobel was successful in stabilizing nitroglycerin by mixing it with diatomaceous earth (a type of sedimentary rock that easily crumbles into a fine powder). The resulting mixture, which he called dynamite, became the first commercial secondary explosive, and numerous factories across Europe were soon built in order to produce dynamite in large quantities. As mentioned in the chapter opener, Alfred Nobel became fabulously wealthy from his invention and used some of the proceeds to fund the Nobel prizes, with which we are all so familiar.

Over time, it was discovered that workers in the dynamite production factories experienced several physiological responses associated with prolonged exposure to nitroglycerin. Most importantly, workers who suffered from heart problems found that their conditions greatly improved. For many decades, the reason for this response was not clear, but a clear pattern had been established. As a result, doctors began to treat patients experiencing heart problems by giving them small quantities of nitroglycerin to ingest. Ultimately, Nobel himself suffered from heart problems, and his doctor suggested that he eat nitroglycerin. Nobel refused to ingest what he considered to be an explosive, and he ultimately died of heart complications.

With the passing decades, it became clear that nitroglycerin serves as a vasodilator (dilates blood vessels) and therefore reduces the chances of blockage that leads to a heart attack. However, it was not known how nitroglycerin functions as a vasodilator. The study of drug action belongs to a broader field of study called pharmacology. A scientist at UCLA by the name of Louis Ignarro was interested in this pharmacological question and heavily investigated the action of nitroglycerin in the body. He discovered that metabolism of nitroglycerin produces nitric oxide (NO) and that this small compound is ultimately responsible for a large number of physiological processes. At first, his discovery was met with skepticism by the scientific community since nitric oxide was known to be an atmospheric contaminant (present in smog), and it was difficult to believe that the very same compound could be responsible for the medicinal value of nitroglycerin. Ultimately, his ideas were verified, and he was credited with discovering the mechanism for the physiological effects of nitroglycerin. Ignarro's discovery ultimately led to the development of many new commercial drugs, including Viagra. Viagra is a vasodilator that treats impotency, a condition that afflicts 9% of all adult males in the United States.

Perhaps if Alfred Nobel had been privy to Ignarro's research, he might have followed his doctor's instructions after all and eaten nitroglycerin for its medicinal value. It is therefore quite interesting that Ignarro was awarded the 1998 Nobel Prize in Physiology or Medicine, a prize that was funded by Nobel with the fortune that he accrued for his discovery of stabilized nitroglycerin. It appears that history does in fact have a sense of irony.

## Catalysts and Enzymes

A **catalyst** is a compound that can speed up the rate of a reaction without itself being consumed by the reaction. A catalyst works by providing an alternate pathway with a smaller activation energy (Figure 6.17). Notice that a catalyst does not change the energy of reactants or products, and therefore, the position of equilibrium is not affected by the presence of a catalyst. Only the rate of reaction is affected by the catalyst. We will see many examples of catalysts in the coming chapters.

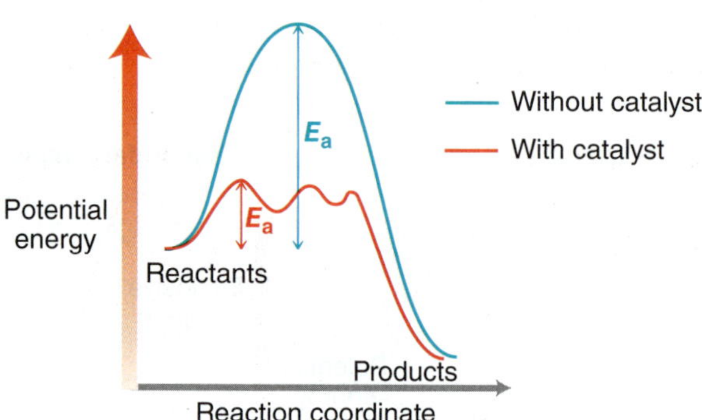


**FIGURE 6.17** An energy diagram showing an uncatalyzed pathway and a catalyzed pathway.

Nature employs catalysis in many biological functions as well. Enzymes are naturally occurring compounds that catalyze very specific biologically important reactions. Enzymes will be discussed in greater detail in Chapter 25.

**practically speaking** | **Beer Making**

Beer making relies on the fermentation of sugars (produced from grains such as wheat or barley). The fermentation process produces ethanol and is thermodynamically favorable ($\Delta G$ is negative). The direct conversion of sugar into ethanol has a very large activation energy and therefore does not occur by itself at an appreciable rate:

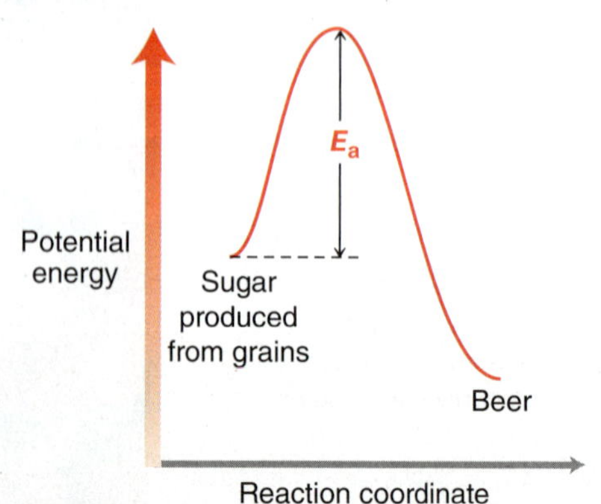


When yeast is added to the mixture, the energy of activation is lowered, and the process takes place at an observable rate. Yeast is a microorganism that utilizes many enzymes that catalyze many different reactions. Some of these enzymes enable the yeast to metabolize sugars, producing ethanol and $CO_2$ as waste products. The ethanol generated in the process is toxic to the yeast, so as the concentration of ethanol rises, the fermentation slows. As a result, it is difficult to achieve an alcohol concentration greater than 12% when using standard brewing yeast.

Without the catalytic action of yeast, beer might never have been discovered.

## 6.6 Reading Energy Diagrams

As we begin to study reactions in the next chapter, we will rely heavily on the use of energy diagrams. Let's quickly review some of the features of energy diagrams.

### Kinetics Versus Thermodynamics

Don't confuse kinetics and thermodynamics—they are two entirely separate concepts (Figure 6.18).

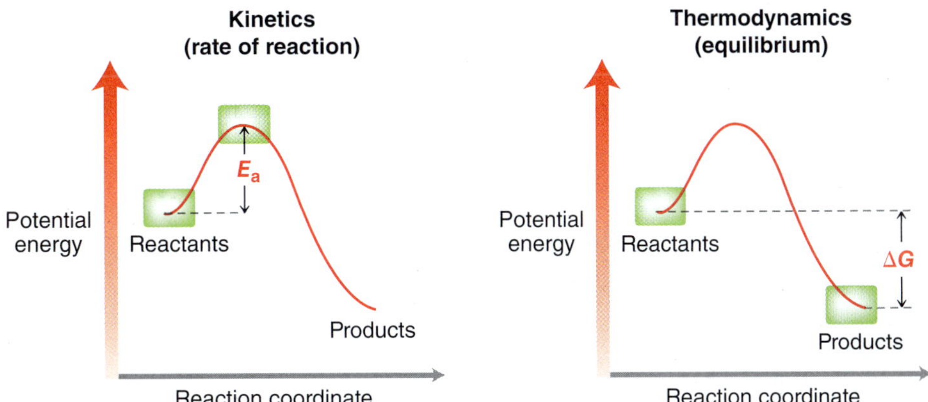


**FIGURE 6.18** Energy diagrams showing the difference between kinetics and thermodynamics.

Kinetics refers to the rate of a reaction, while thermodynamics refers to the equilibrium concentrations of reactants and products. Suppose that two compounds, A and B, can react with each other in one of two possible pathways (Figure 6.19). The pathway determines the products. Notice that products C and D are thermodynamically favored over products E and F, because C and D are lower in energy. In addition, C and D are also kinetically favored over E and F because formation of C and D involves a smaller activation energy. In summary, C and D are favored by thermodynamics as well as kinetics. When reactants can react with each other in two possible ways, it is often the case that one reaction pathway is both thermodynamically and kinetically favored, although there are many cases in which thermodynamics and kinetics oppose each other. Consider the energy diagram in Figure 6.20 showing two possible reactions between A and B. In this case, products C and D are favored by thermodynamics because they are lower in energy. However, products E and F are favored by kinetics because formation of E and F involves a lower energy of activation. In such a case, temperature will play a pivotal role. At low temperature, the reaction that forms E and F will be more rapid, even though this reaction does not produce the most stable products. At high temperature,

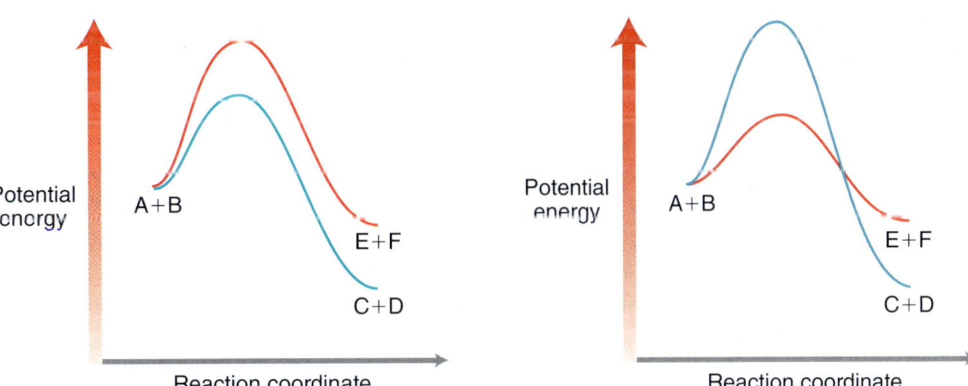


**FIGURE 6.19** An energy diagram showing two possible pathways for the reaction between A and B.

**FIGURE 6.20** An energy diagram showing two possible pathways for the reaction between A and B. In this case, C and D are the thermodynamic products, while E and F are the kinetic products.

equilibrium concentrations will be quickly achieved, favoring formation of C and D. We will see several examples of kinetics versus thermodynamics throughout this course.

## Transition States Versus Intermediates

Reactions often involve multiple steps. In the energy diagram of a multistep process, all local minima (valleys) represent *intermediates*, while all local maxima (peaks) represent *transition states* (Figure 6.21). It is important to understand the difference between transition states and intermediates.

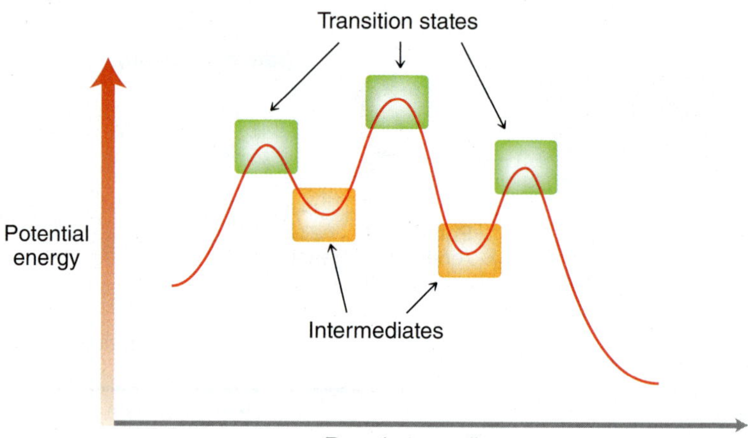


**FIGURE 6.21** In an energy diagram, all peaks are transition states, and all valleys are intermediates.

A **transition state**, as the name implies, is a state through which the reaction passes. Transition states cannot be isolated. In this high-energy state, bonds are in the process of being broken and/or formed simultaneously, as shown in Figure 6.22.

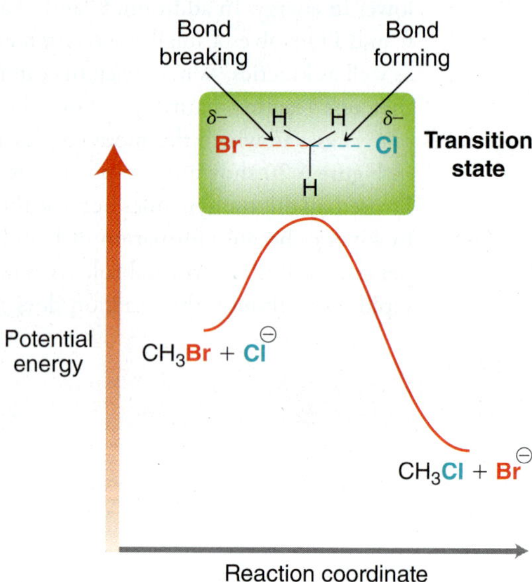


**FIGURE 6.22** In a transition state, bonds are being broken and/or formed.

As a crude analogy for the difference between a transition state and an intermediate, consider jumping in the air as high as you can while your friend takes a photograph of the highest point of your trajectory. The picture shows the height that you achieved, although it would not be fair to say that you were able to spend any considerable amount of time at that height (to hover in midair). It was a state through which you passed. In contrast, imagine jumping onto a desk and then jumping back down. A photograph of you standing on the desk will be very different from the previous picture. It is actually possible to stand on a desk for a period of time, but it is not possible to remain hovering in the air for any reasonable period of time. The picture of you standing on the desk is similar to the picture of a reaction intermediate. **Intermediates** have a certain, albeit short, lifetime.

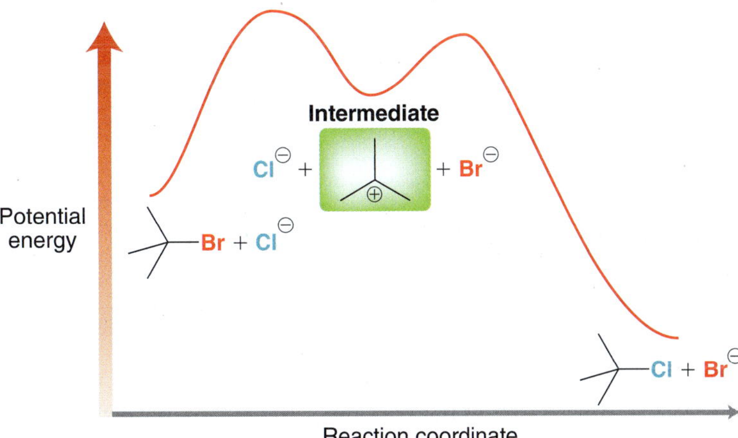


**FIGURE 6.23**
An energy diagram of a reaction that exhibits an intermediate characterized by a valley in the energy diagram.

An intermediate is not in the process of forming or breaking bonds, for example (Figure 6.23). Intermediates, such as the one shown in Figure 6.23, are very common, and we will encounter them hundreds of times throughout this course.

## The Hammond Postulate

Consider the two points on the energy diagram in Figure 6.24. Because these two points are close to each other on the curve, they are close in energy and are therefore structurally similar.

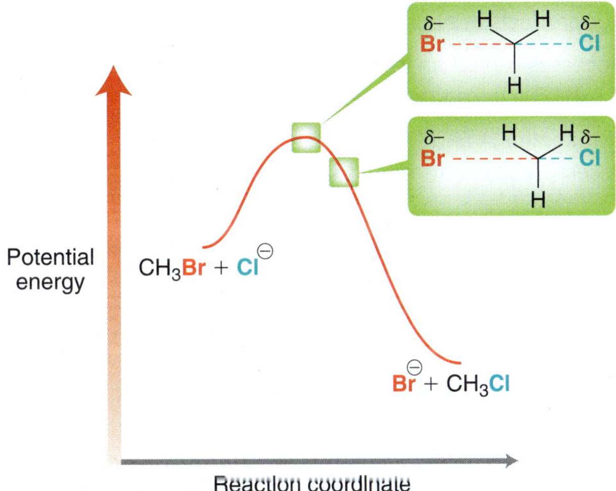


**FIGURE 6.24**
On an energy diagram, two points that are close together will represent structurally similar states.

Using this principle, we can make a generalization about the structure of a transition state in any exothermic or endothermic process (Figure 6.25). In an exothermic process the transition state is closer in energy to the reactants than to the products, and therefore, the structure of the transition

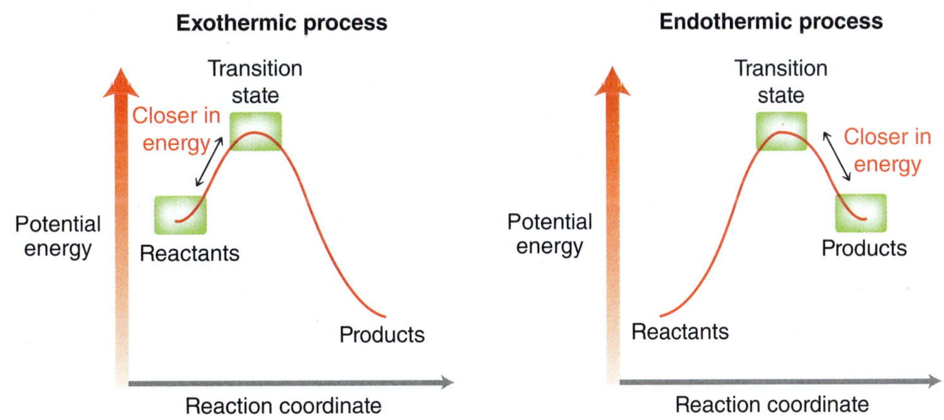


**FIGURE 6.25**
A transition state will be closer in energy to the starting materials in an exothermic process, but it will be closer in energy to the products in an endothermic process.

state more closely resembles the reactants. In contrast, the transition state in an endothermic process is closer in energy to the products, and therefore, the transition state more closely resembles the products. This principle is called the **Hammond postulate**. We will use this principle many times in our discussions of transition states in the upcoming chapters.

## CONCEPTUAL CHECKPOINT

**6.7** Consider the relative energy diagrams for four different processes:

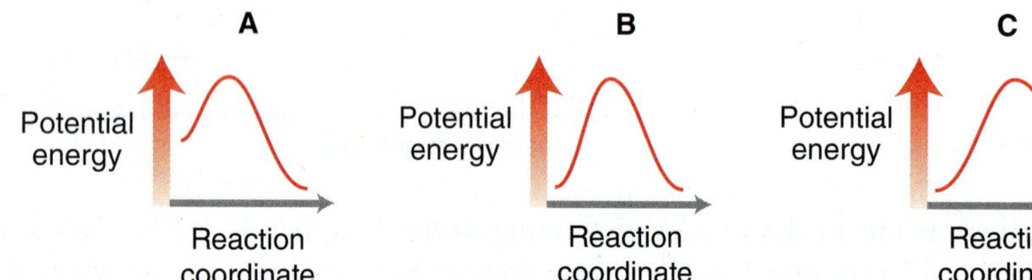


(a) Compare energy diagrams A and D. Assuming all other factors (such as concentrations and temperature) are identical for the two processes, identify which process will occur more rapidly. Explain.

(b) Compare energy diagrams A and B. Which process will more greatly favor products at equilibrium? Explain.

(c) Do any of the processes exhibit an intermediate? Do any of the processes exhibit a transition state? Explain.

(d) Compare energy diagrams A and C. In which case will the transition state resemble reactants more than products? Explain.

(e) Compare energy diagrams A and B. Assuming all other factors (such as concentrations and temperature) are identical for the two processes, identify which process will occur more rapidly. Explain.

(f) Compare energy diagrams B and D. Which process will more greatly favor products at equilibrium? Explain.

(g) Compare energy diagrams C and D. In which case will the transition state resemble products more than reactants? Explain.

## 6.7 Nucleophiles and Electrophiles

**Ionic reactions**, also called **polar reactions**, involve the participation of ions as reactants, intermediates, or products. In most cases, the ions are present as intermediates. These reactions represent most (approximately 95%) of the reactions that we will encounter in this course. The other two major categories, radical reactions and pericyclic reactions, occupy a much smaller focus in the typical undergraduate organic chemistry course but will be discussed in upcoming chapters. The remainder of this chapter will focus on ionic reactions.

Ionic reactions occur when one reactant has a site of high electron density and the other reactant has a site of low electron density. For example, consider the electrostatic potential maps of methyl chloride and methyllithium (Figure 6.26). Each compound exhibits an inductive effect, but in opposite directions.

**Methyl chloride**

Cl, C, H, H, H, δ+

The carbon atom is electron deficient

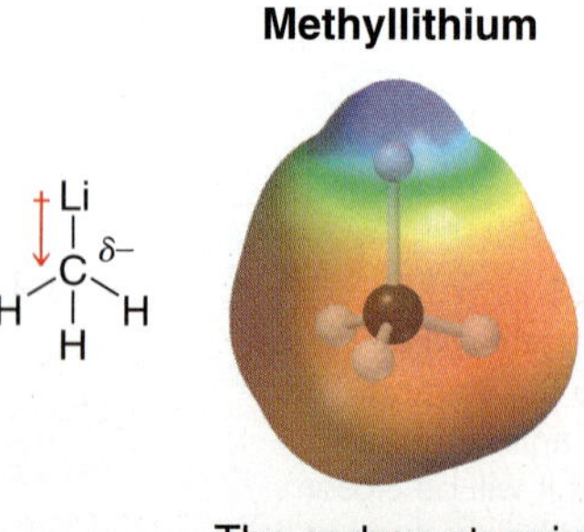


**FIGURE 6.26** Electrostatic potential maps of methyl chloride and methyllithium, clearly indicating the inductive effects.

**LOOKING BACK**
For a review of inductive effects, see Section 1.11.

The carbon atom in methyl chloride represents a site of low electron density, while the carbon atom in methyllithium represents a site of high electron density. Since opposite charges attract, these two compounds will react with each other. The type of reaction that occurs between methyl chloride and methyllithium will be explored in the upcoming chapter. For now, we will focus on the nature of each reactant.

An electron-rich center, such as the carbon atom in methyllithium, is called a **nucleophile**, which comes from Greek meaning "nucleus lover." That is, a nucleophilic center is characterized by its ability to react with a positive charge or partial positive charge. In contrast, an electron-deficient center, such as the carbon atom in methyl chloride, is called an **electrophile**, which comes from Greek meaning "electron lover." That is, an electrophilic center is characterized by its ability to react with a negative charge or partial negative charge.

Throughout this text, we will focus extensively on the behavior of nucleophiles and electrophiles. Ultimately, there are just a handful of principles that will enable us to explain, and even predict, most reactions. However, in order to learn and use these principles, it will first be necessary to become proficient in identifying the nucleophilic and electrophilic centers in any compound. This skill is arguably one of the most important in organic chemistry. We have said before that the essence of organic chemistry is to study and predict how electron density flows during a reaction. It will not be possible to make any intelligent predictions without knowing where the electron density can be found and where it is likely to go. The following sections will explore the nature of nucleophiles and electrophiles in more depth.

## Nucleophiles

**LOOKING BACK**
For a review of Lewis bases, see Section 3.9.

A nucleophilic center is an electron-rich atom that is capable of donating a pair of electrons. Notice that this definition is very similar to the definition of a Lewis base. In fact, the terms "nucleophile" and "Lewis base" are synonymous.

Two examples of nucleophiles are as follows:

Ethoxide Ethanol

Each of these examples has lone pairs on an oxygen atom. Ethoxide bears a negative charge and is therefore more nucleophilic than ethanol. Nevertheless, ethanol can still function as a nucleophile (albeit weak), because the lone pairs in ethanol represent regions of high electron density. Any atom that possesses a localized lone pair can be nucleophilic.

We will see in Chapter 9 that π bonds can also function as nucleophiles, because a π bond is a region in space of high electron density (Figure 6.27).

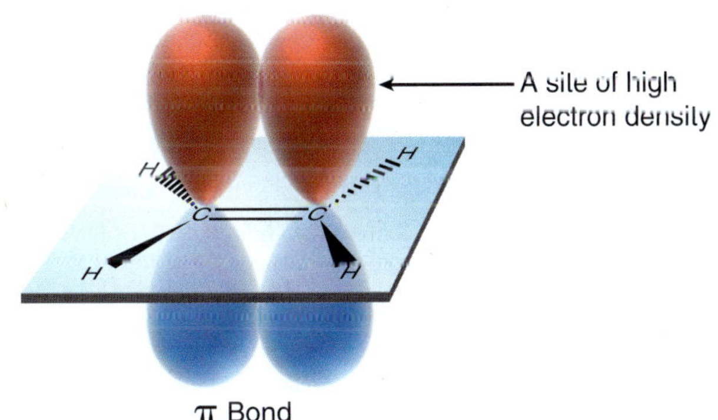


**FIGURE 6.27**
A π bond is a region in space of high electron density.

The strength of a nucleophile is affected by many factors, including polarizability. **Polarizability**, loosely defined, describes the ability of an atom to distribute its electron density unevenly in response to external influences. Polarizability is directly related to the size of the atom (and more specifically, the number of electrons that are distant from the nucleus). For example, sulfur is very large and has many electrons that are distant from the nucleus, and its electron density can be unevenly distributed when it comes near an electrophile. Iodine shares the same feature. As a result, $I^-$ and $HS^-$ are particularly strong nucleophiles. This fact will be revisited in the upcoming chapter.

## Electrophiles

An electrophilic center is an electron-deficient atom that is capable of accepting a pair of electrons. Notice that this definition is very similar to the definition for a Lewis acid. In fact, the terms "electrophile" and "Lewis acid" are synonymous.

Two examples of electrophiles are as follows:

δ+ Cl δ−

The first compound exhibits an electrophilic carbon atom as a result of the inductive effects of the chlorine atom. The second example exhibits a positively charged carbon atom and is called a **carbocation**. A carbocation has an empty *p* orbital (Figure 6.28). The empty *p* orbital functions as a site that can accept a pair of electrons, rendering the compound electrophilic. We will discuss carbocations in more depth later in this chapter.

**LOOKING BACK**
For a review of Lewis acids, see Section 3.9.

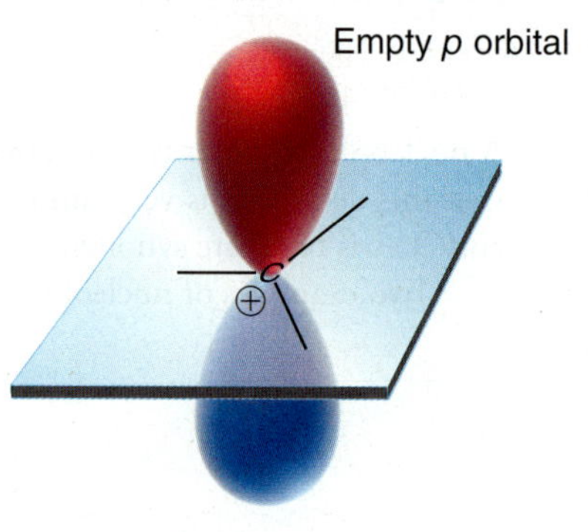


**FIGURE 6.28**
The empty *p* orbital of a carbocation.

Table 6.3 provides a summary of some common features that render a compound nucleophilic or electrophilic.

**TABLE 6.3 A SUMMARY OF SOME COMMON NUCLEOPHILIC CENTERS AND ELECTROPHILIC CENTERS**

| NUCLEOPHILES | | ELECTROPHILES | |
|---|---|---|---|
| FEATURE | EXAMPLE | FEATURE | EXAMPLE |
| Inductive effects | H–C(H)(H)–Li (δ−) | Inductive effects | H–C(H)(H)–Cl (δ+) |
| Lone pair | H–Ö–H | Empty *p* orbital | ⊕ |
| π Bond | | | |

# SKILLBUILDER

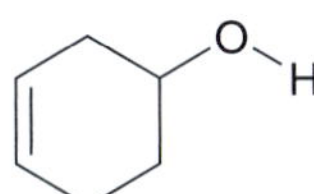

## 6.2 IDENTIFYING NUCLEOPHILIC AND ELECTROPHILIC CENTERS

### LEARN the skill

Identify all nucleophilic centers and all electrophilic centers in the following compound:

**STEP 1** Identify nucleophilic centers by looking for inductive effects, lone pairs, or π bonds.

**SOLUTION**

Let's first look for nucleophilic centers. Specifically, we are looking for inductive effects, lone pairs, or π bonds. In this compound, there are two nucleophilic centers:

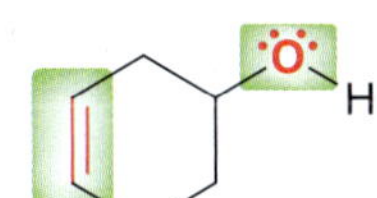

**STEP 2** Identify electrophilic centers by looking for inductive effects or empty *p* orbitals.

Next, we look for electrophilic centers. Specifically, we are looking for inductive effects or an empty *p* orbital. None of the atoms in this compound exhibit an empty *p* orbital, but there are inductive effects:

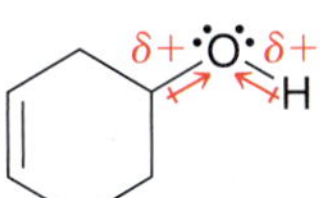


The oxygen atom is electronegative, causing the adjacent atoms to be electron deficient. In this case, there are two adjacent atoms that incur a partial positive charge (δ+): the adjacent carbon atom and the adjacent hydrogen atom. In general, a hydrogen atom with a partial positive charge is described as acidic, rather than electrophilic. Therefore, this compound is considered to have only one electrophilic center.

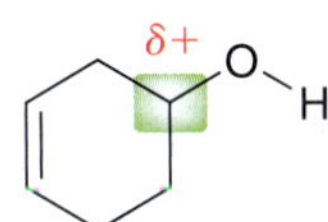

### PRACTICE the skill

**6.8** Identify all of the nucleophilic centers in each of the following compounds:

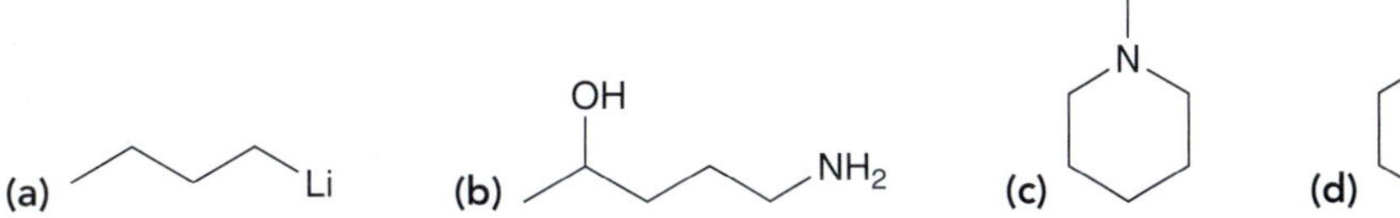


**6.9** Identify all of the electrophilic centers in each of the following compounds:

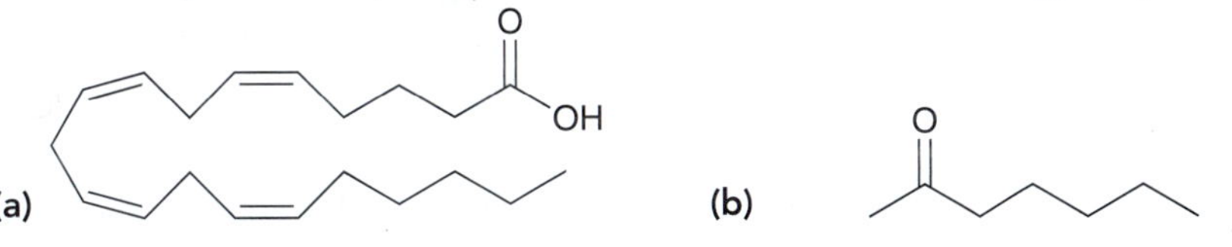


**Arachidonic acid**
A precursor in the biosynthesis of many hormones

**2-Heptanone**
Used to control the population of *Varroa* mites in honey bee colonies

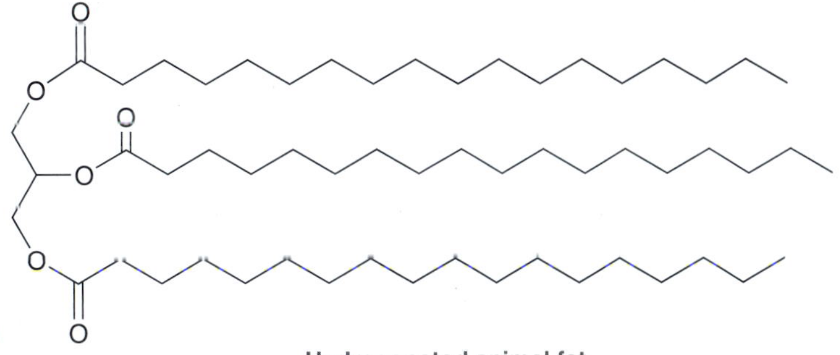


**Hydrogenated animal fat**

### APPLY the skill

**6.10** The following hypothetical compound cannot be prepared or isolated, because it has a very reactive nucleophilic center and a very reactive electrophilic center, and the two sites would react with each other rapidly. Identify the nucleophilic center and electrophilic center in this hypothetical compound:

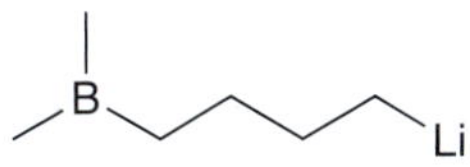

**6.11** Each of the following compounds exhibits two electrophilic centers. Identify both centers in each compound. (**Hint:** You will need to draw resonance structures in each case.)

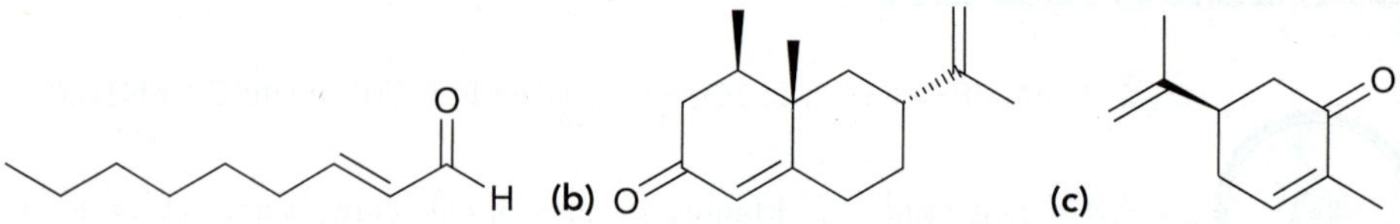

(a) A cockroach repellant found in cucumbers

(b) Nootkatone
Found in grapefruits

(c) (*R*)-Carvone
Responsible for the odor of spearmint

need more **PRACTICE?** Try Problem 6.53

## 6.8 Mechanisms and Arrow Pushing

Recall from Chapter 3 that a mechanism shows how a reaction takes place using curved arrows to illustrate the flow of electrons. The tail of every curved arrow shows where the electrons are coming from, and the head of every curved arrow shows where the electrons are going:

$$B:^{\ominus} + H{-}A \rightleftharpoons B{-}H + A:^{\ominus}$$

**LOOKING BACK**
Recall that we also used a handful of patterns for arrow pushing to master resonance structures in Chapter 2.

In order to master ionic mechanisms, it will be helpful to become familiar with characteristic patterns for arrow pushing. We will now learn patterns of electron flow, and these patterns will empower us to understand reaction mechanisms and even propose new mechanisms. There are only four characteristic patterns, and all ionic mechanisms are simply combinations of these four steps. Let's go through them one by one.

### Nucleophilic Attack

The first pattern is **nucleophilic attack**, characterized by a nucleophile attacking an electrophile; for example:

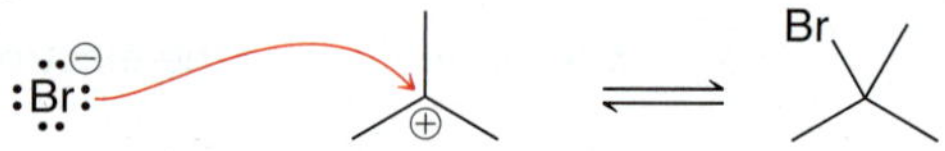

Nucleophile Electrophile

Bromide is a nucleophile because it possesses a lone pair, and the carbocation is an electrophile because of its empty *p* orbital. In this example, the attack of the nucleophile on the electrophile requires just one curved arrow. The tail of this curved arrow is placed on the nucleophilic center and the head is placed on the electrophilic center. It is also common to see a nucleophilic attack that utilizes more than one curved arrow; for example:

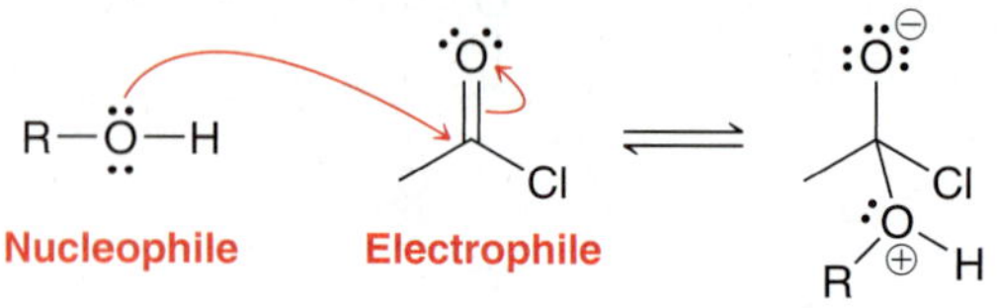

Nucleophile Electrophile

In this case, there are two curved arrows. The first shows the nucleophile attacking the electrophile, but what is the function of the second curved arrow? There are a couple of ways to view this second curved arrow. We can simply think of it as a resonance arrow: We can imagine first drawing the resonance structure of the electrophile and then having the nucleophile attack:

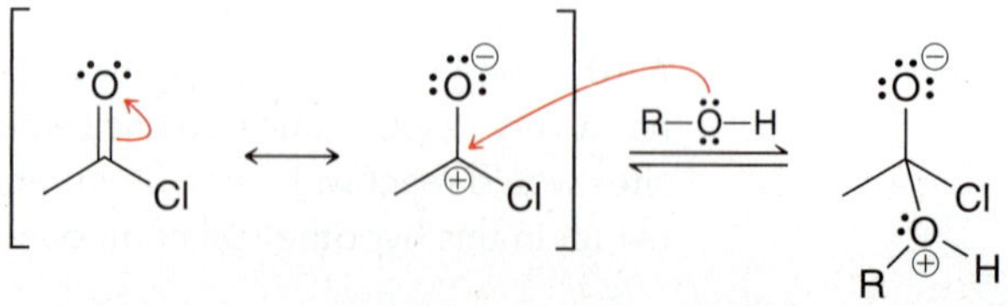

From this perspective, it seems that only one of the curved arrows is actually showing the nucleophilic attack. The other curved arrow can be thought of as a resonance curved arrow.

Alternatively, we can think of the second curved arrow as an actual flow of electron density that goes up onto the oxygen atom when the nucleophile attacks:

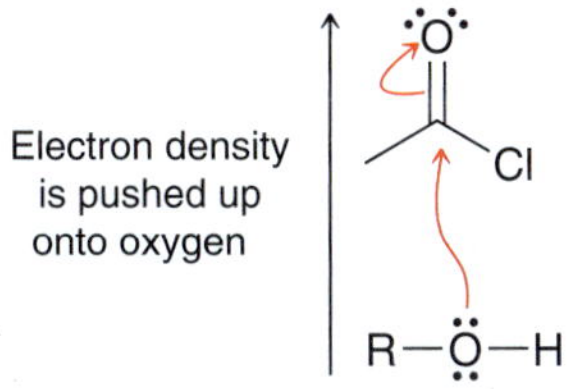

Electron density is pushed up onto oxygen

This perspective is perhaps more accurate, but we must keep in mind that both curved arrows are showing just one arrow-pushing pattern: *nucleophilic attack.*

π Bonds can also serve as nucleophiles; for example:

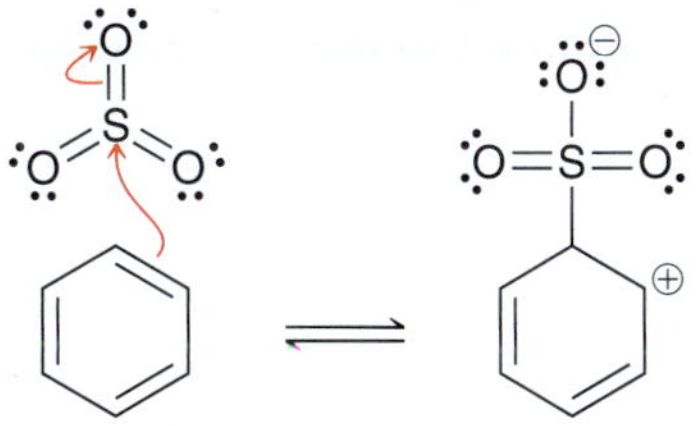

## Loss of a Leaving Group

The second pattern for arrow pushing is characterized by the loss of a leaving group; for example:

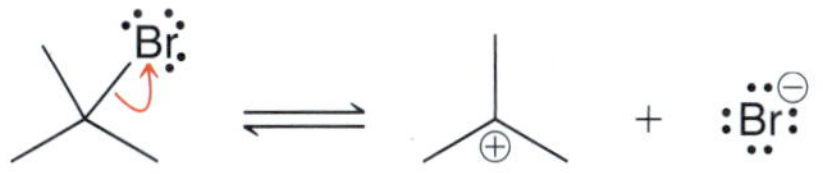

This step requires one curved arrow, although it is common to see more than one curved arrow being used to show the loss of a leaving group. For example:

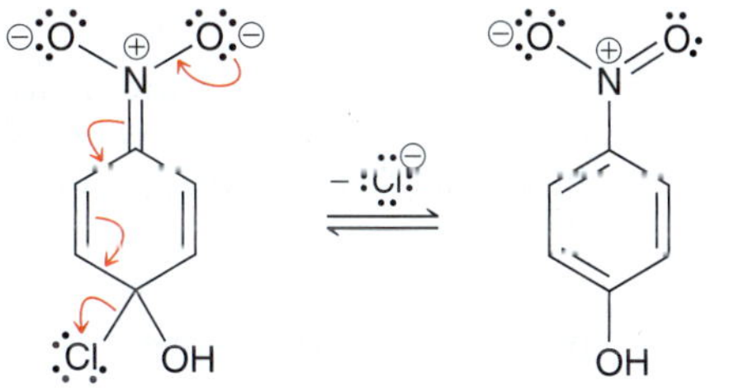

Only one of the curved arrows above actually shows the chloride leaving (the curved arrow at the very bottom). The remaining curved arrows can be viewed in two different ways (just as in the previous section). We can view the other curved arrows as resonance arrows, drawn after the leaving group leaves:

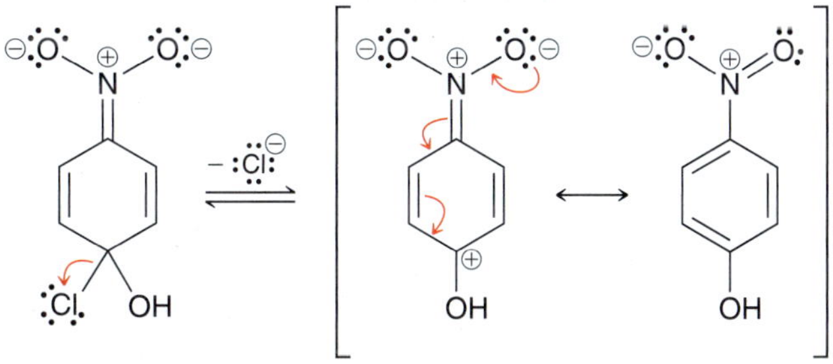

or we can view the process as a flow of electron density that pushes out the leaving group:

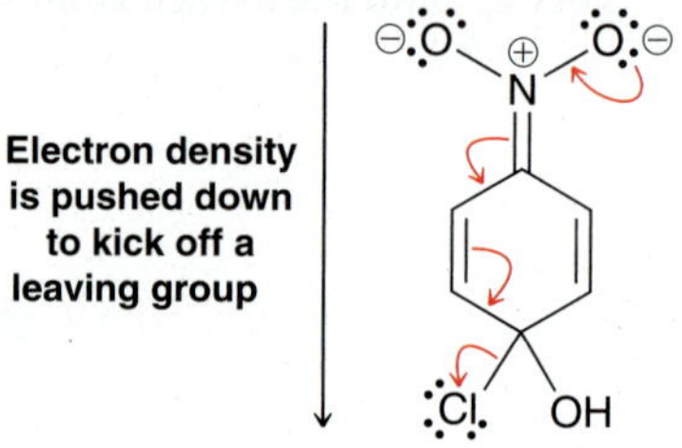

**Electron density is pushed down to kick off a leaving group**

Regardless of how we view the process, it is important to recognize that all of these curved arrows together show only one arrow-pushing pattern: **loss of a leaving group**.

## Proton Transfers

The third pattern for arrow pushing has already been discussed in detail in Chapter 3. Recall that a **proton transfer** is characterized by two curved arrows:

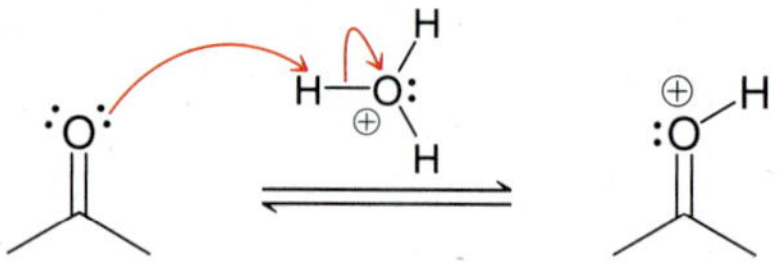

In this example, a ketone is being protonated, which is shown with two curved arrows. The first is drawn from the ketone to the proton. The second curved arrow shows what happens to the electrons that were previously holding the proton.

A proton transfer step is illustrated with two curved arrows, whether the compound is protonated (as above) or deprotonated, as shown below:

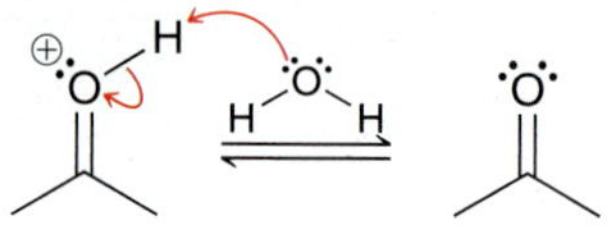

Sometimes, proton transfers are shown with only one curved arrow:

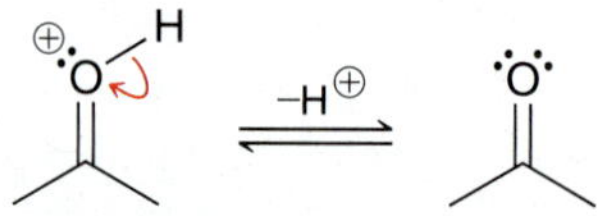

In this case, the base that removes the proton has not been indicated. Chemists will sometimes use this approach for clarity of presentation, even though a proton doesn't fall off into space by itself. In order for a compound to lose a proton, a base must be involved in order to deprotonate the compound. In general, it is preferable to show the base involved and to use *two* curved arrows.

At times, *more* than two curved arrows are used for a proton transfer step. For example, the following case has three curved arrows:

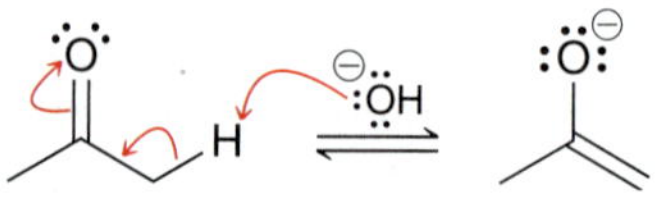

Once again, this step can be viewed in one of two ways. We can view this as a simple proton transfer followed by drawing a resonance structure:

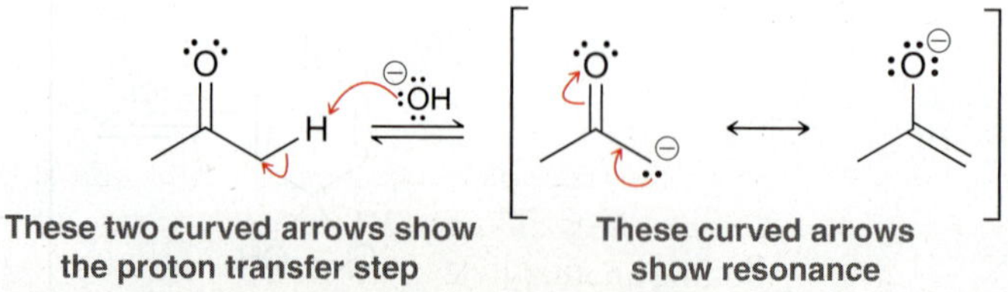

**These two curved arrows show the proton transfer step**

**These curved arrows show resonance**

or we can argue that all of the curved arrows together show the flow of electron density that takes place during the proton transfer:

Electron density flows up onto the oxygen of the ketone

For proton transfers, both perspectives are equally valid.

## Rearrangements

The fourth, and final, pattern is characterized by a **rearrangement**. There are several kinds of rearrangements, but at this point, we will focus exclusively on *carbocation rearrangements*. In order to discuss carbocation rearrangements, we must first explore one feature of carbocations. Specifically, carbocations are stabilized by neighboring alkyl groups (Figure 6.29).

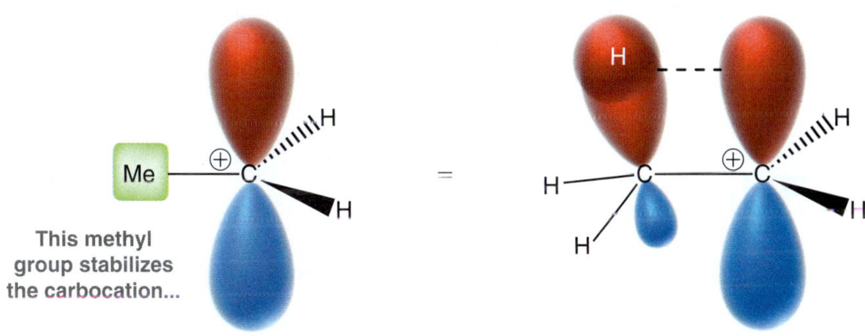

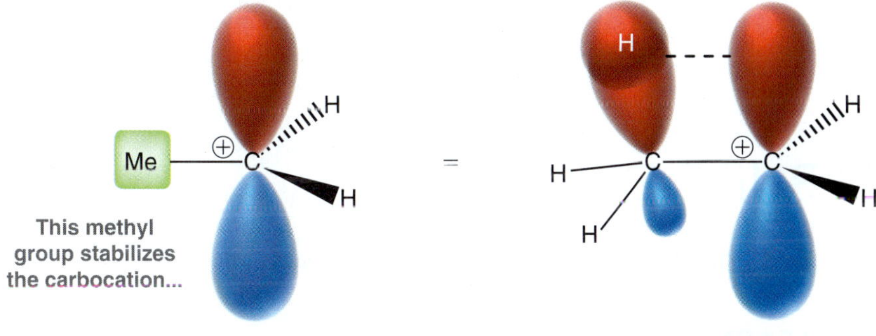


**FIGURE 6.29** Neighboring alkyl groups will stabilize a carbocation through hyperconjugation.

The bonding MO associated with a neighboring C—H bond slightly overlaps with the empty *p* orbital, placing some of its electron density in the empty *p* orbital. This effect, called **hyperconjugation**, stabilizes the empty *p* orbital. This explains the observed trend in Figure 6.30. The terms **primary**, **secondary**, and **tertiary** refer to the number of alkyl groups attached directly to the positively charged carbon atom. Tertiary carbocations are more stable (lower in energy) than secondary carbocations, which are more stable than primary carbocations.

Increasing stability

Methyl | Primary | Secondary | Tertiary

**FIGURE 6.30** Carbocations with more alkyl groups will be more stable than carbocations with fewer alkyl groups.

At the end of this chapter, we will learn to predict when rearrangements are likely to occur. For now, we will just focus on recognizing carbocation rearrangements. There are two common ways in which carbocation rearrangements are accomplished: via either a **hydride shift** or a **methyl shift**.

A hydride shift involves the migration of $H^-$:

**WATCH OUT**
Hydride is $H:^-$. It is a hydrogen atom with an extra electron (two electrons total). Don't confuse the hydride ion with a proton ($H^+$), which is a hydrogen atom missing its electron.

In this example, a secondary carbocation is transformed into a more stable, tertiary carbocation. How does this occur? As an analogy, imagine that there is a hole in the ground. Now imagine digging another hole nearby and using the dirt to fill up the first hole. The first hole can be filled, but in its place, a new hole has been created nearby. A hydride shift is a similar concept. The carbocation is a hole (a place where electron density is missing). The neighboring hydrogen atom takes both of its electrons and migrates over to fill the empty *p* orbital. This process generates a new, more stable, empty *p* orbital nearby.

A carbocation rearrangement can also be accomplished via a methyl shift:

$H_3C$ ⟶ $CH_3$

Once again, a secondary carbocation is being converted into a tertiary carbocation. But this time, it is a methyl group (rather than a hydride) that migrates with its two electrons to plug up the hole. In order for a methyl shift to occur, the methyl group must be attached to the carbon atom that is adjacent to the carbocation.

The two examples above (hydride shift and methyl shift) are the most common types of carbocation rearrangements.

In summary, we have seen only four characteristic patterns for arrow pushing in ionic reactions: (1) nucleophilic attack, (2) loss of a leaving group, (3) proton transfer, and (4) rearrangement. Let's get some practice identifying them.

## SKILLBUILDER

### 6.3 IDENTIFYING AN ARROW-PUSHING PATTERN

**LEARN the skill**

Consider the following step:

Identify which arrow-pushing pattern is utilized in this case.

**SOLUTION**

Read the curved arrows. The first three curved arrows show double bonds moving over, and the last curved arrow shows chloride being ejected from the compound:

In other words, chloride is functioning as a leaving group. The other curved arrows can be viewed as resonance arrows:

**Loss of a leaving group** **Resonance**

Alternatively, we can view the π bonds as "pushing out" the chloride ion:

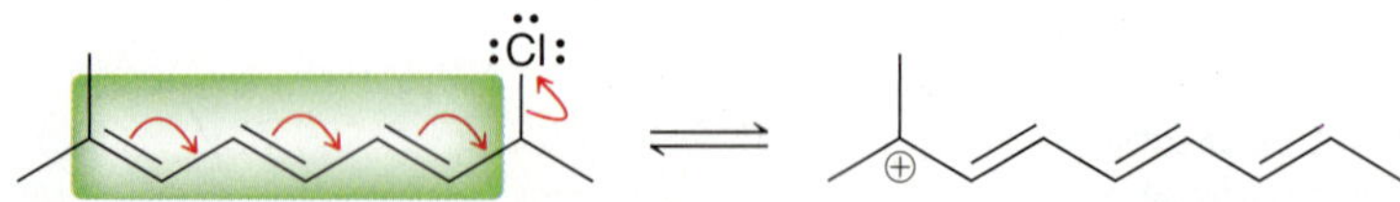


**π Bonds pushing out the chloride ion**

Either way, there is only one arrow-pushing pattern being utilized here: *loss of a leaving group*.

**PRACTICE the skill** **6.12** For each of the following cases, read the curved arrows and identify which arrow-pushing pattern is utilized:

(a)

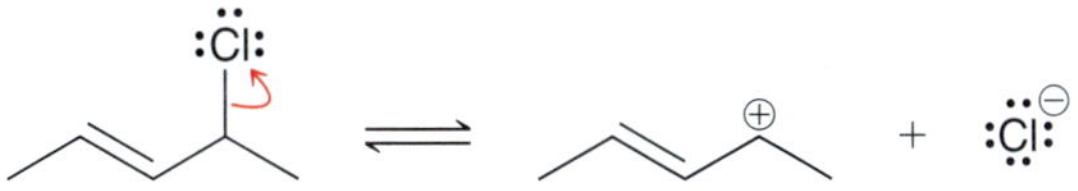

(b)

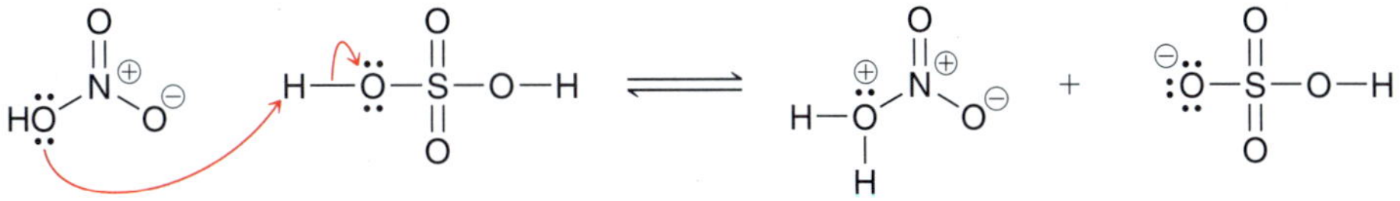

(c)

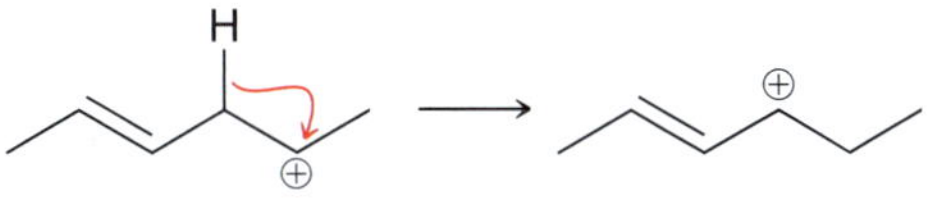

(d)

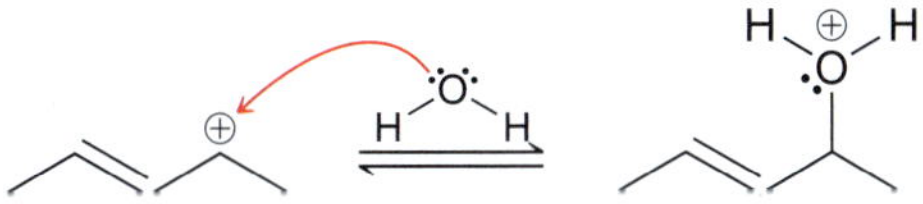

(e)

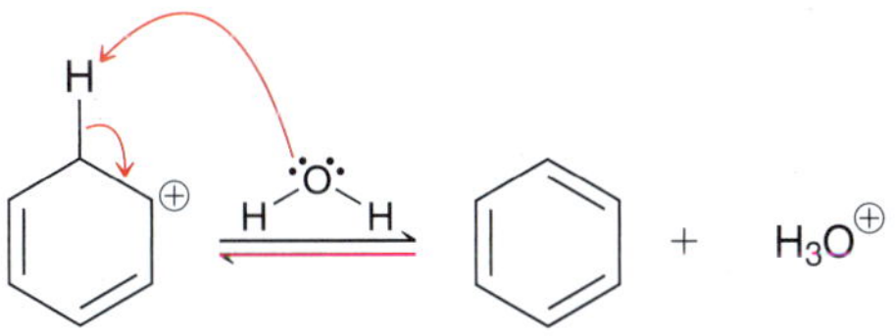

(f)

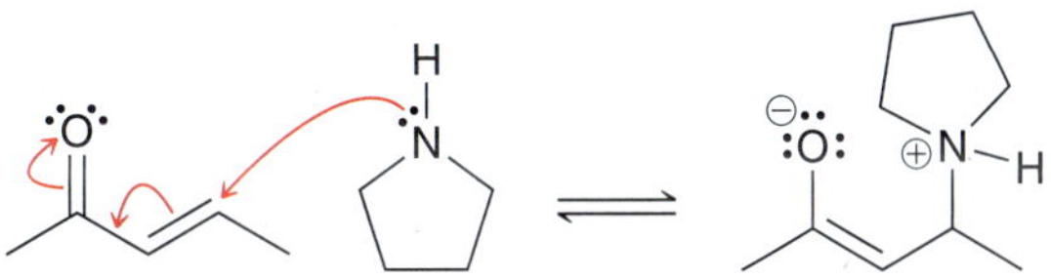

(g)

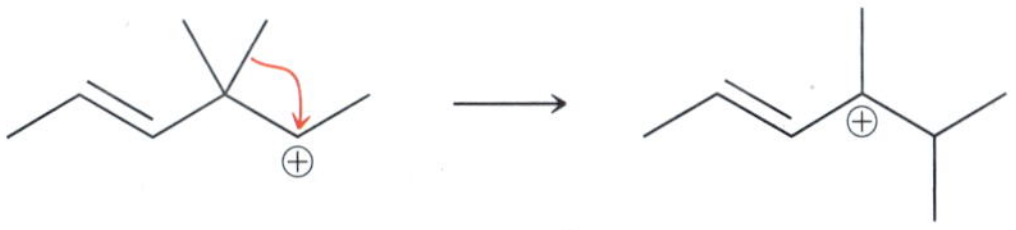

(h)

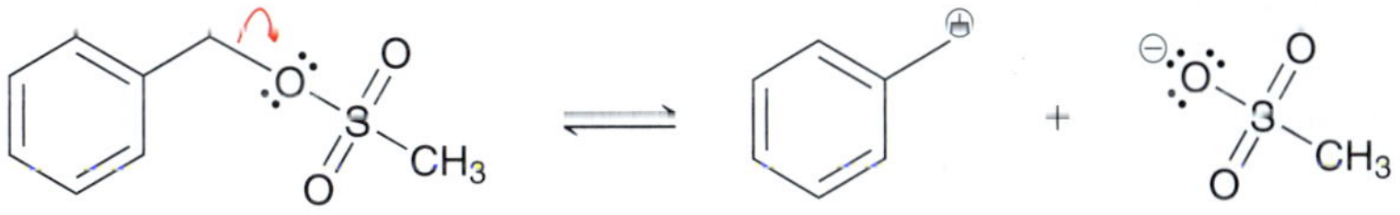

(i)

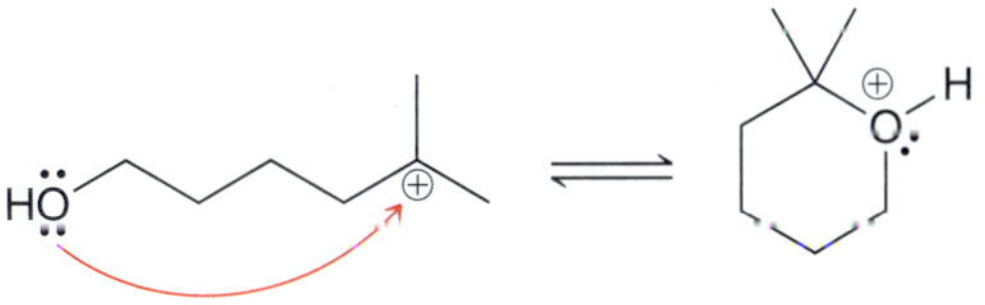

**APPLY the skill** **6.13** Identify which arrow-pushing pattern is utilized in the following step:

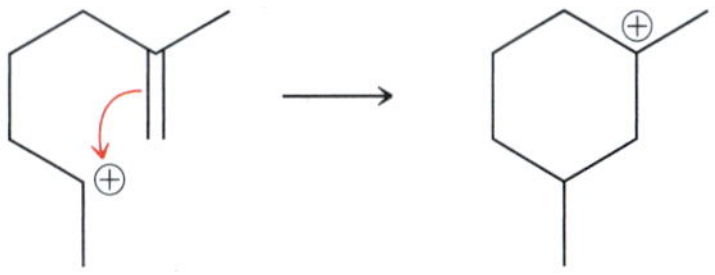

need more **PRACTICE?** **Try Problem 6.30**

## 6.9 Combining the Patterns of Arrow Pushing

All ionic mechanisms, regardless of how complex, are just different combinations of the four characteristic patterns seen in the previous section. As an example, consider the reaction in Figure 6.31, which we will explore in greater detail in Chapter 7.

Proton transfer — Loss of a leaving group — Carbocation rearrangement — Nucleophilic attack

**FIGURE 6.31**
A reaction containing all four characteristic patterns for arrow pushing.

The mechanism in Figure 6.31 has four steps, each of which is one of the characteristic patterns from the previous section. Sometimes, a single mechanistic step will involve two simultaneous patterns; for example:

There are two curved arrows shown here. One curved arrow, coming from the chloride ion, shows a nucleophilic attack. The second curved arrow shows loss of a leaving group. In this case, we are utilizing two arrow-pushing patterns simultaneously. This is called a concerted process. In the next chapter, we will explore some of the important differences between concerted and stepwise mechanisms. For now, we will simply focus on recognizing the various patterns that can arise as a result of combining the various arrow-pushing patterns. By doing so, similarities will emerge between apparently different mechanisms.

## SKILLBUILDER

### 6.4 IDENTIFYING A SEQUENCE OF ARROW-PUSHING PATTERNS

**LEARN the skill** Identify the sequence of arrow-pushing patterns in the following reaction:

**SOLUTION**

In the first step, hydroxide is attacking C=O in a nucleophilic attack:

In the next step, an alkoxide ion (RO$^-$) is being ejected as a leaving group:

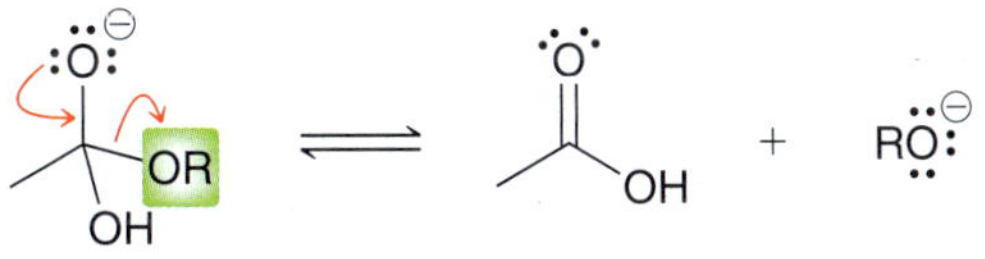

Finally, the last step is a proton transfer, which always involves two curved arrows:

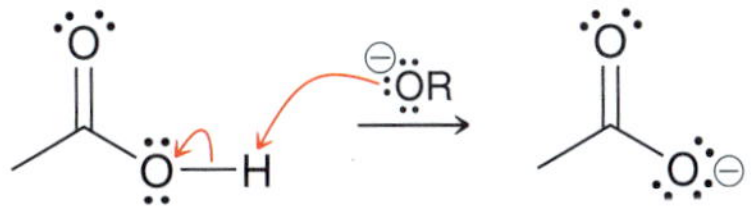

The sequence of this reaction is therefore (1) nucleophilic attack, (2) loss of a leaving group, and (3) proton transfer. Throughout this book, and especially in Chapter 21, we will see dozens of reactions that follow this three-step sequence. By viewing all of these reactions in this format (as a sequence of characteristic patterns), it will be easier to see the similarities between apparently different reactions.

**PRACTICE the skill** **6.14** For each of the following multistep reactions, read the curved arrows and identify the sequence of arrow–pushing patterns:

(a)

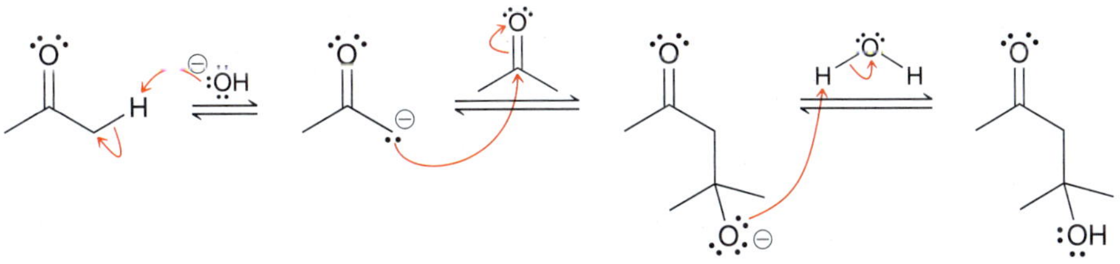

(b)

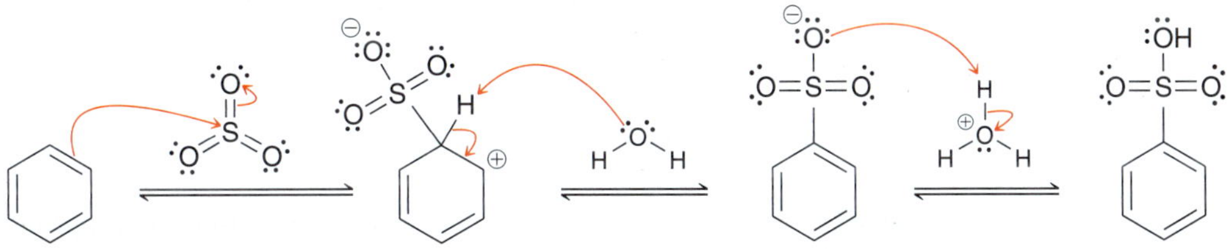

(c)

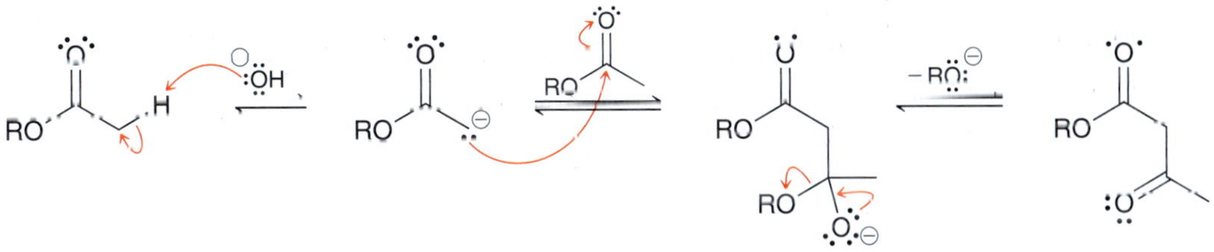

(d)

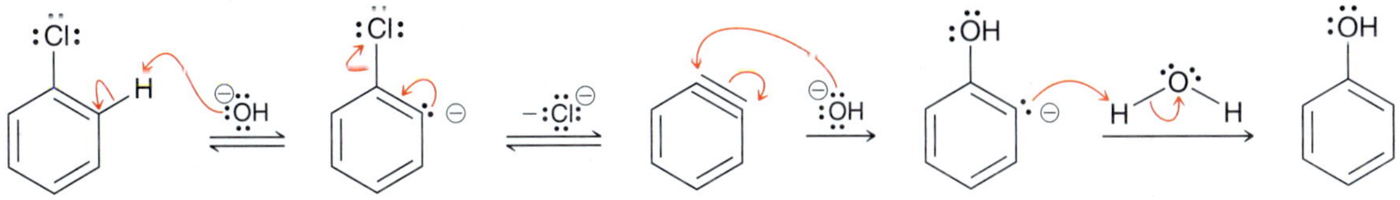

(e)

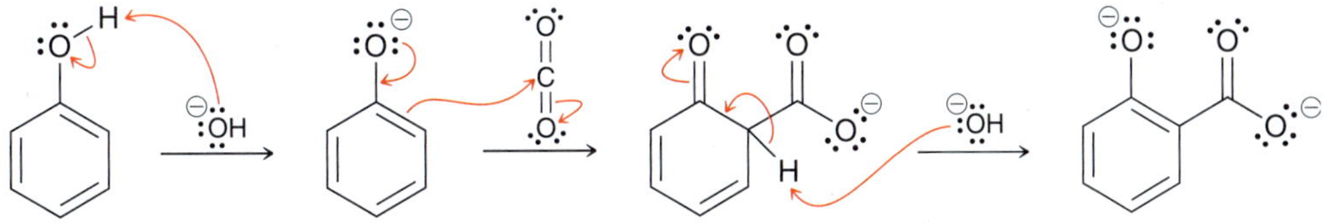

**APPLY the skill**

**6.15** The following two reactions will be explored in different chapters. Yet, they are very similar. Identify and compare the sequences of arrow-pushing patterns for the two reactions.

**Reaction 1 (Chapter 20)**

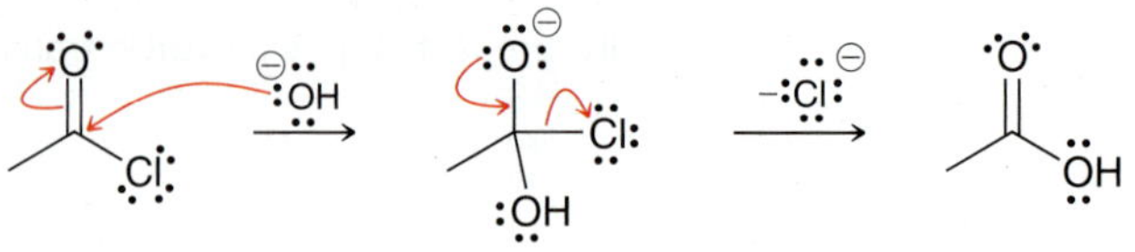

**Reaction 2 (Chapter 18)**

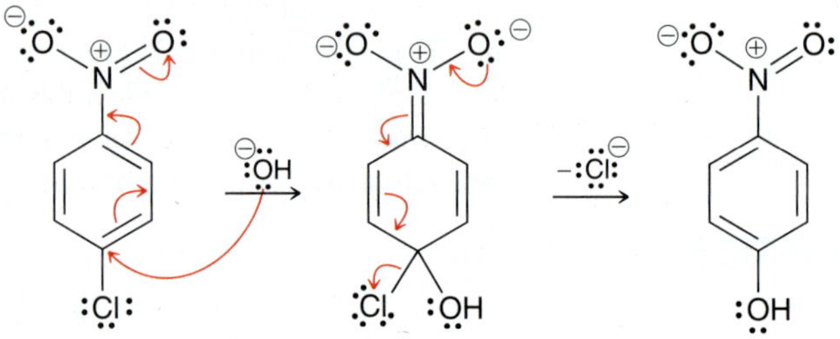

need more **PRACTICE?** **Try Problems 6.32–6.41**

There may be more than 100 mechanisms in a full organic chemistry course, but there are fewer than a dozen different sequences of arrow-pushing patterns in these mechanisms. As we proceed through the chapters, we will learn rules for determining when each pattern can and cannot be utilized. These rules will empower us to propose mechanisms for new reactions.

## 6.10 Drawing Curved Arrows

Curved arrows have a very precise meaning, and they must be drawn precisely. Avoid sloppy arrows. Make sure to be deliberate when drawing the tail and the head of every arrow. The tail must be placed on either a bond or a lone pair. For example, the following reaction employs two curved arrows. One of the curved arrows has its tail placed on a lone pair, and the other has its tail placed on a bond:

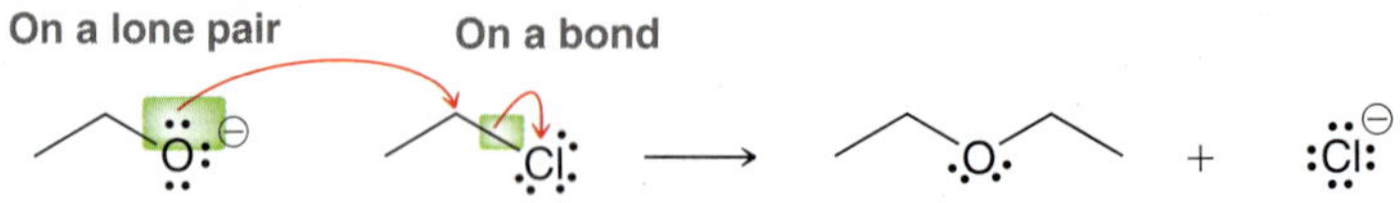

On a lone pair　　On a bond

These are the only two possible locations where the tail of a curved arrow can be placed. The tail shows where electrons are coming from, and electrons can only be found in lone pairs or bonds. *Never place the tail of a curved arrow on a positive charge.*

The head of a curved arrow must be placed so that it shows either the formation of a bond or the formation of a lone pair:

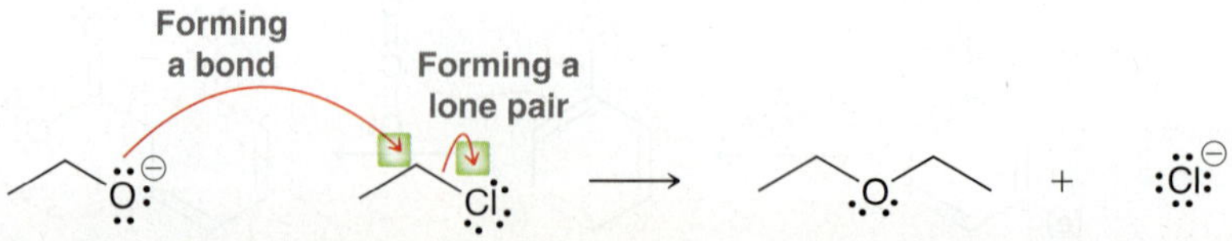

Forming a bond　　Forming a lone pair

**WATCH OUT**
C, N, and O are the second-row elements that require special attention. Never give one of these elements more than four orbitals.

When drawing the head of a curved arrow, make sure to avoid drawing an arrow that violates the octet rule. Specifically, never draw an arrow that gives more than four orbitals to a second-row element:

Violates octet rule | Does not violate octet rule

In the first example the head of the curved arrow is giving a fifth bond to the carbon atom. This violates the octet rule. The second example does not violate the octet rule, because the carbon atom is gaining one bond but losing another. In the end, the carbon atom never has more than four bonds.

Make sure that all the curved arrows you draw accomplish one of the four characteristic arrow-pushing patterns.

Nucleophilic attack

Loss of a leaving group

Avoid drawing arrows like the following. This arrow violates the octet rule, and it does not accomplish any one of the four characteristic arrow-pushing patterns.

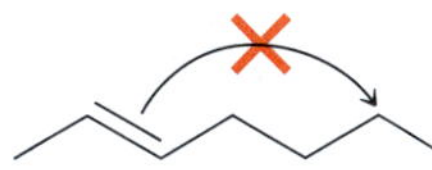

# SKILLBUILDER

## 6.5 DRAWING CURVED ARROWS

**LEARN the skill** Draw the curved arrows that accomplish the following transformation:

### SOLUTION

**STEP 1**
Identify which of the four arrow-pushing patterns to use.

We begin by identifying which characteristic arrow-pushing pattern to use in this case. Look carefully, and notice that $H_3O^+$ is losing a proton. That proton has been transferred to a carbon atom. Therefore, this is a proton transfer step, where the π bond is functioning as the base to deprotonate $H_3O^+$. A proton transfer step requires two curved arrows. Make sure to properly place the head and the tail of each curved arrow. The first curved arrow must originate on the π bond and end on a proton:

**STEP 2**
Draw the curved arrows focusing on the proper placement of the tail and head of each curved arrow.

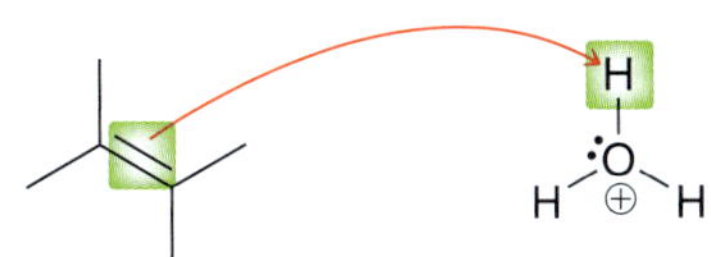

Don't forget to draw the second curved arrow (students will often leave out the second arrow, producing an incomplete mechanism). The second curved arrow shows what happens to the electrons that were previously holding the proton. Specifically, the tail is placed on the O—H bond, and the head is placed on the oxygen atom:

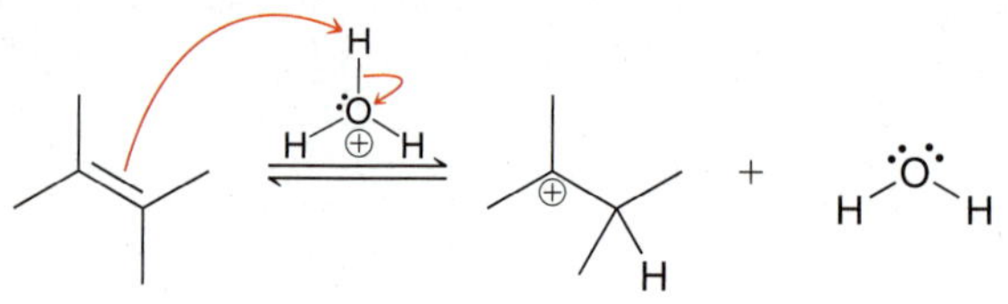

**PRACTICE the skill** **6.16** Draw the curved arrows that accomplish each of the following transformations:

(a)

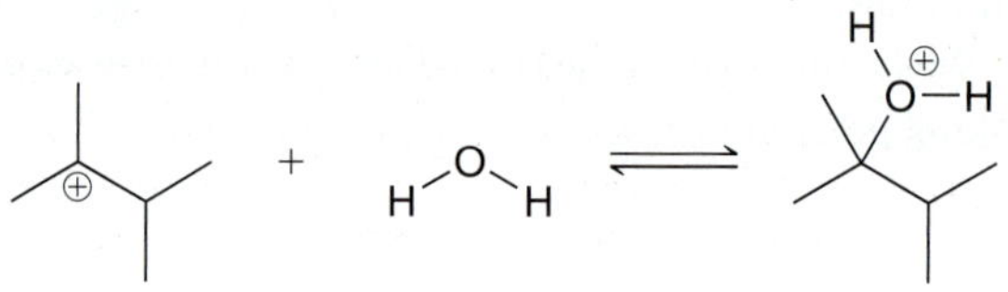

(b)

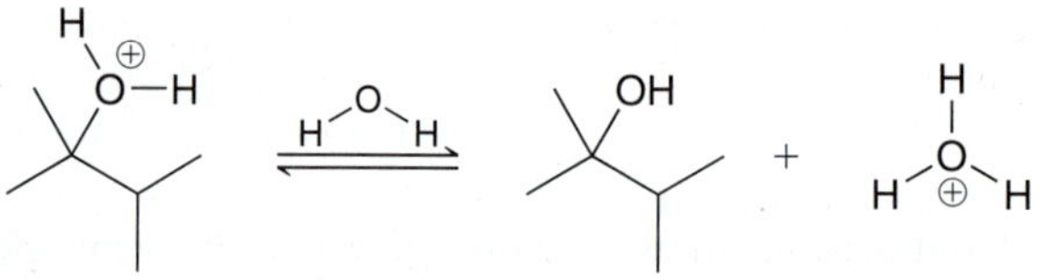

(c)

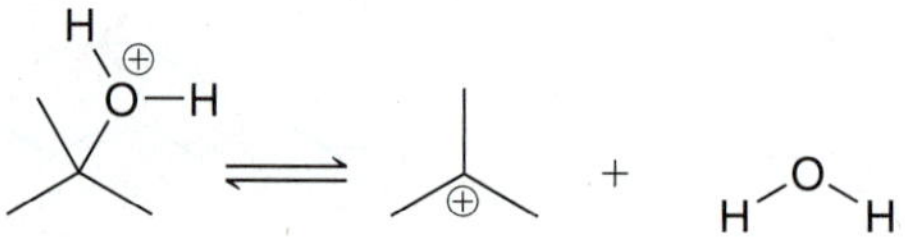

**APPLY the skill**

**6.17** The following four reactions will be the focus of the upcoming chapters (substitution and elimination reactions). Draw the curved arrows that accomplish each of the transformations shown:

(a)

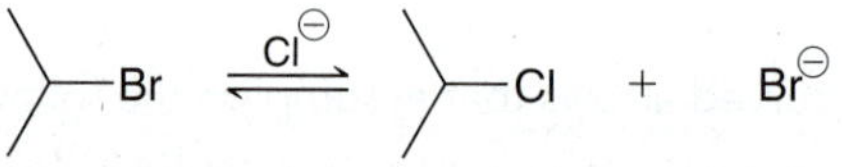

(b)

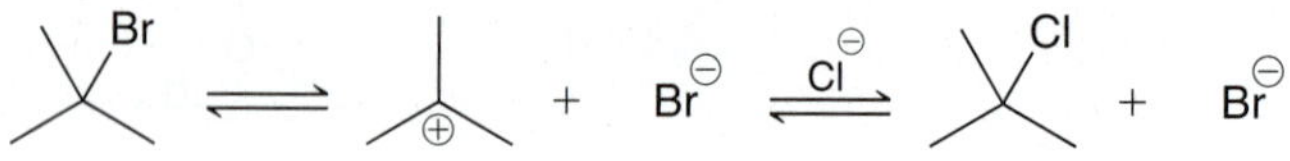

(c)

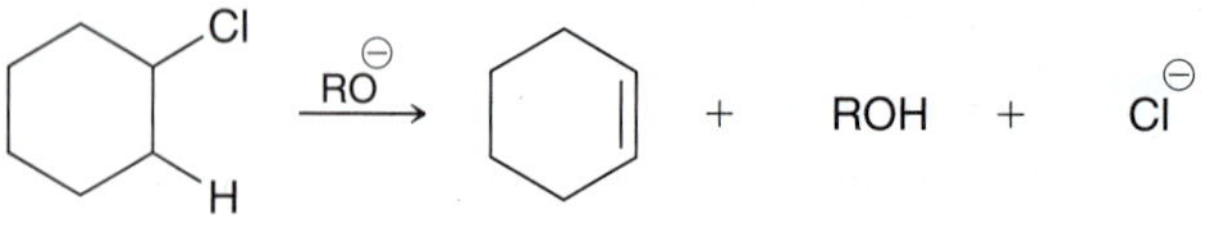

(d)

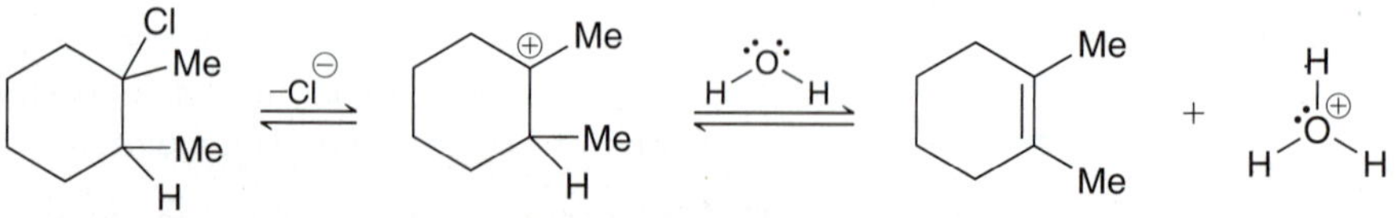

need more **PRACTICE?** **Try Problems 6.42–6.47, 6.54**

## 6.11 Carbocation Rearrangements

Throughout this course, we will encounter many examples of carbocation rearrangements, so we must be able to predict when these rearrangements will occur. Recall that the two common types of carbocation rearrangement are hydride shifts and methyl shifts (Figure 6.32).

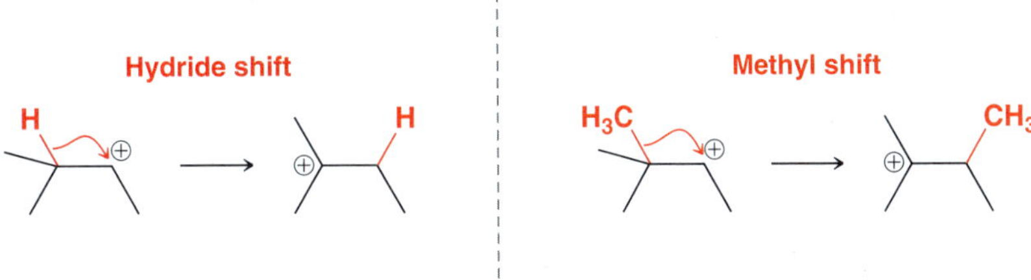


**FIGURE 6.32** Hydride shifts and methyl shifts are the two most common types of carbocation rearrangements.

In both cases, a secondary carbocation is converted into a more stable, tertiary carbocation. Stability is the key. In order to predict when a carbocation rearrangement will occur, we must determine whether the carbocation can become more stable via a rearrangement. For example, consider the following carbocation:

In order to determine if this carbocation can undergo a rearrangement, we must identify any hydrogen atoms or methyl groups attached directly to the neighboring carbon atoms:

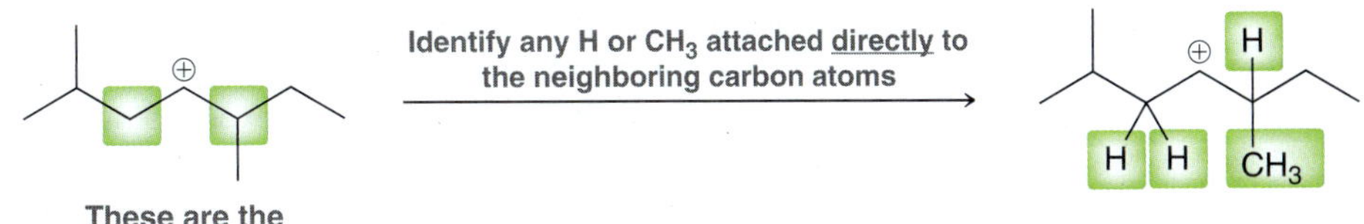


There are four candidates. Now imagine each one of these groups shifting over to plug the carbocation, generating a new carbocation. Would the new carbocation be more stable? In this example only one group can migrate to produce a more stable carbocation:

Hydride shift

If this hydride migrates, it will generate a new, tertiary carbocation. Therefore, we do expect a hydride shift in this case.

Carbocation rearrangements generally do not occur when the carbocation is already tertiary unless a rearrangement will produce a resonance stabilized carbocation; for example:

This tertiary carbocation will undergo rearrangement... ...because this carbocation is resonance stabilized

In this case, the original carbocation is tertiary. However, the newly formed carbocation is tertiary and it is stabilized by resonance. Such a carbocation is called an **allylic carbocation** because the positive charge is located at an allylic position.

**LOOKING BACK**

Recall that the term allylic describes the positions connected directly to the π bond:

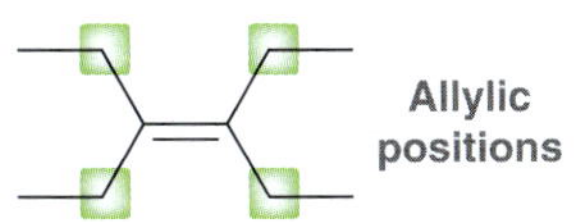

# SKILLBUILDER

## 6.6 PREDICTING CARBOCATION REARRANGEMENTS

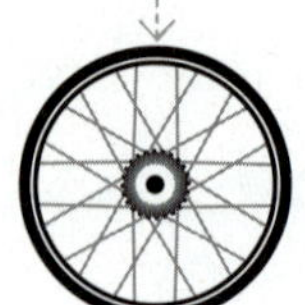

**LEARN** the skill

Predict whether the following carbocation will rearrange, and if so, draw the curved arrow showing the carbocation rearrangement:

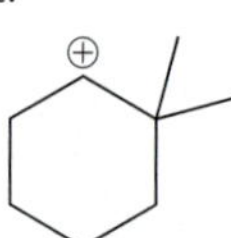

### SOLUTION

**STEP 1**
Identify neighboring carbon atoms.

This carbocation is secondary, so it certainly has the potential to rearrange. We must look to see if a carbocation rearrangement can produce a more stable, tertiary carbocation. Begin by identifying the carbon atoms neighboring the carbocation:

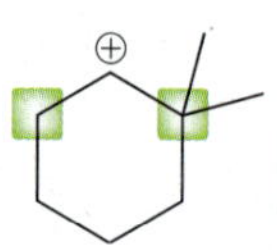

**STEP 2**
Identify any H or $CH_3$ groups attached directly to the neighboring carbon atoms.

Look at all hydrogen atoms or methyl groups connected to these carbon atoms:

H H ⊕

**STEP 3**
Determine which of these groups can migrate to generate a more stable carbocation.

Consider if migration of any of these groups will generate a more stable, tertiary carbocation. Migration of either of the neighboring hydride groups will only generate another secondary carbocation. So, we do not expect a hydride shift to occur.

Secondary C+ ⟶ (✗) Secondary C+

**STEP 4**
Draw a curved arrow showing the carbocation rearrangement and draw the new carbocation.

However, if either of the methyl groups migrates, a tertiary carbocation is generated. Therefore, we expect a methyl shift to take place, generating a more stable tertiary carbocation:

Secondary C+ ⟶ Tertiary C+

**PRACTICE** the skill

**6.18** For each of the following carbocations determine if it will rearrange, and if so, draw the carbocation rearrangement with a curved arrow:

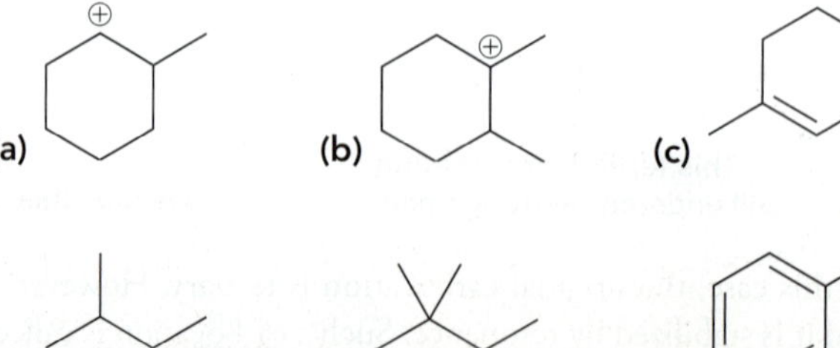

(a) (b) (c) (d)

(e) (f) (g) (h)

**6.19** Occasionally, carbocation rearrangements can be accomplished via the migration of a carbon atom other than a methyl group. Such an example follows. Identify the group that is migrating and draw the curved arrow that shows the migration:

need more **PRACTICE?** **Try Problem 6.48**

## 6.12 Reversible and Irreversible Reaction Arrows

As we explore ionic reaction mechanisms in the upcoming chapters, you might notice that some steps will be drawn with reversible reaction arrows (equilibrium arrows), while other steps will be drawn with irreversible reaction arrows; for example:

Reversible

Irreversible

With each new mechanism that we explore, you might wonder how to determine whether it is more appropriate to use a reversible or an irreversible reaction arrow. In this section, we will explore some general rules for each of the four patterns (nucleophilic attack, loss of a leaving group, proton transfer, and carbocation rearrangement).

### Nucleophilic Attack

For a step in which a nucleophile attacks an electrophile, a reversible reaction arrow is generally used if the nucleophile is capable of functioning as a good leaving group after the attack has occurred, while an irreversible reaction arrow is used if the nucleophile is a poor leaving group. To illustrate this concept, consider the following two examples:

Reversible

Nucleophile

Good leaving group

Irreversible

Nucleophile

Bad leaving group

In the first case, the nucleophile (water) can function as a good leaving group after it has attacked. Therefore, a reversible reaction arrow was used to indicate that the following reverse process occurs at an appreciable rate:

In contrast, the second example employs a nucleophile that is a poor leaving group. As such, an irreversible reaction arrow was used to indicate that the following reverse process does not occur:

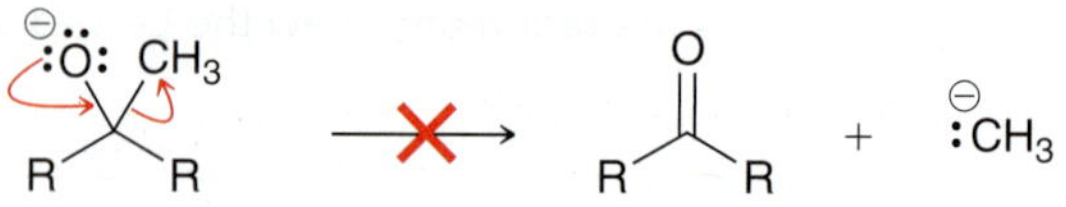

In Section 7.8, we will learn to distinguish between good leaving groups and poor leaving groups. Specifically, we will see that weak bases (such as $H_2O$) are good leaving groups, while strong bases (such as $H_3C^-$) are poor leaving groups.

## Loss of a Leaving Group

For a step involving the loss of a leaving group, a reversible reaction arrow is generally used if the leaving group is capable of functioning as a good nucleophile; for example:

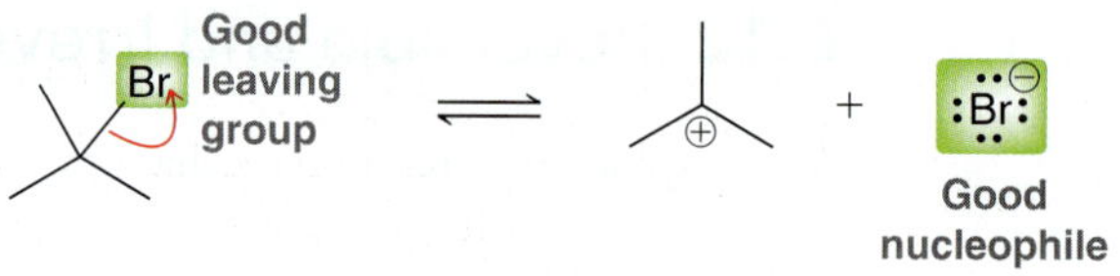

Good leaving group

Good nucleophile

Most leaving groups that we will encounter in this course can also function as nucleophiles, so loss of a leaving group is rarely drawn with an irreversible reaction arrow.

## Proton Transfer

It should be noted that all proton transfer steps are technically reversible. However, there are many cases that can be effectively treated as irreversible. For example, consider the following example, in which water is deprotonated by a very strong base:

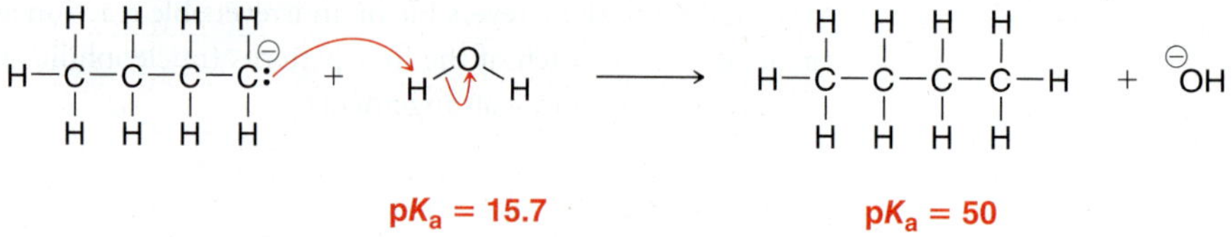

p$K_a$ = 15.7 p$K_a$ = 50

In this case, the difference between the p$K_a$ values of the two acids (water and butane) is approximately 34. This indicates that there should be $10^{34}$ molecules of butane for every one molecule of water. In a solution containing one mole ($\sim 10^{23}$ molecules), the chances are exceedingly small that even a single molecule of butane will be deprotonated by a hydroxide ion at any given moment in time. As such, the process shown above is effectively irreversible.

Now consider a proton transfer step in which the p$K_a$ values are more similar; for example:

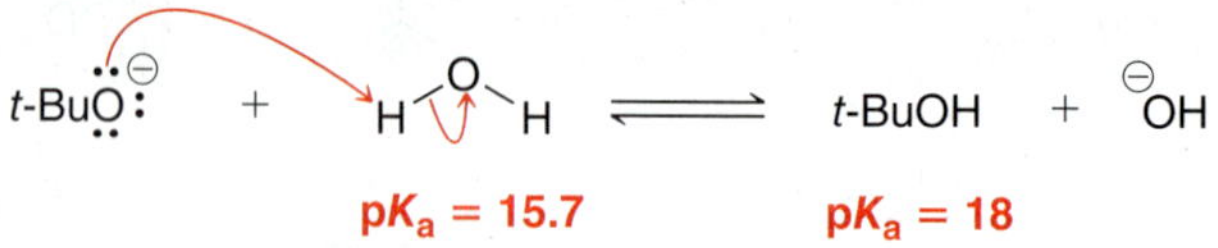

p$K_a$ = 15.7 p$K_a$ = 18

In such a case, a reversible reaction arrow is more appropriate, because an equilibrium is established in which both acids are present (the compound with the higher p$K_a$ is favored, but both are present in substantial quantities). This begs the question: What is the cutoff? That is, how large must the difference in p$K_a$ values be in order to justify the use of an irreversible reaction arrow. Unfortunately, there is no clear answer for this question, as it often depends on the context of the discussion. Generally speaking, irreversible reaction arrows are used for reactions in which the acids differ in strength by more than 10 p$K_a$ units, as seen in the following example:

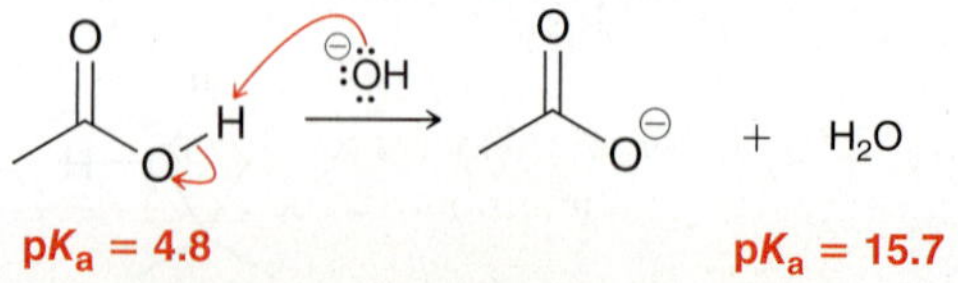

p$K_a$ = 4.8 p$K_a$ = 15.7

When the difference in $pK_a$ values is between 5 and 10 $pK_a$ units, either reversible or irreversible reaction arrows might be used, depending on the context of the discussion.

## Carbocation Rearrangement

Technically speaking, whenever a carbocation rearrangement can occur, an equilibrium will be established in which all of the possible carbocations are present, with the most stable carbocation dominating the equilibrium. The difference in energy between the possible carbocations (for example, secondary vs. tertiary) is often significant, so carbocation rearrangements will generally be drawn as irreversible processes:

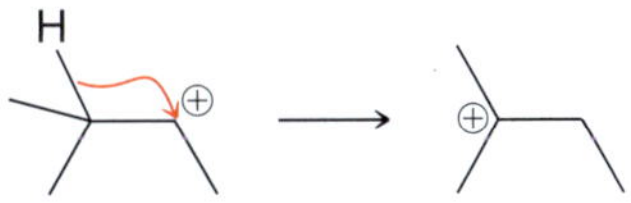

The reverse process (the transformation of a tertiary carbocation to a secondary carbocation) is effectively negligible.

In this section, we have seen many rules for determining when it is appropriate to draw a reversible reaction arrow and when it is appropriate to draw an irreversible reaction arrow. These rules are meant to serve as general guidelines, and there will certainly be exceptions. Keep in mind that there are other relevant factors that were not presented in this section, so strict adherence to the rules in this section is not advisable. For example, when a step involves the liberation of a gas, which is free to exit the reaction vessel, the reaction is expected to proceed to completion (Le Châtelier's principle). To illustrate this concept, consider the following step, which we will encounter in Chapter 20 (Mechanism 20.8):

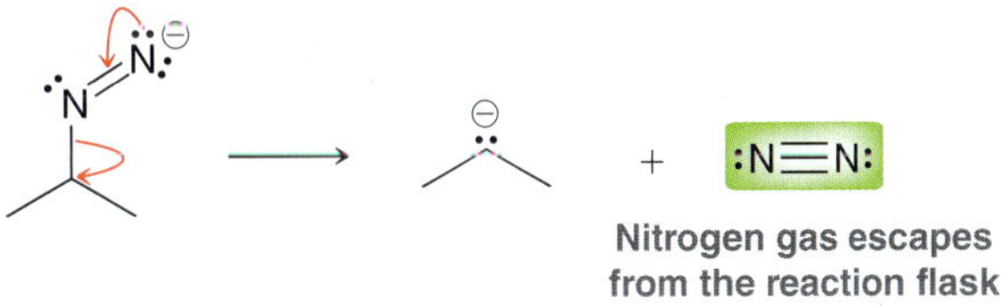

Nitrogen gas escapes from the reaction flask

This step proceeds to completion, despite the formation of a high-energy species, because the evolution of nitrogen gas serves as a driving force for the reaction.

Most of the steps in most of the mechanisms that we will explore in this textbook will conform to the rules described in this section. You may find it helpful to revisit this section periodically, as you continue to progress through the subsequent chapters.

# REVIEW OF CONCEPTS AND VOCABULARY

### SECTION 6.1

- The total change in **enthalpy** for any reaction ($\Delta H$), also called the **heat of reaction**, is a measure of the energy exchanged between the system and its surroundings.
- Each type of bond has a unique **bond dissociation energy**, which is the amount of energy necessary to accomplish **homolytic bond cleavage**, producing **radicals**.
- **Exothermic** reactions involve a transfer of energy from the system to the surroundings, while **endothermic** reactions involve a transfer of energy from the surroundings to the system.

### SECTION 6.2

- **Entropy** is loosely defined as the disorder of a system and is the ultimate criterion for spontaneity. In order for a reaction to be **spontaneous**, the total change in entropy ($\Delta S_{sys} + \Delta S_{surr}$) must be positive.
- Reactions with a positive $\Delta S_{sys}$ involve an increase in the number of molecules, or an increase in the amount of conformational freedom.

### SECTION 6.3

- In order for a process to be spontaneous, the change in **Gibbs free energy** ($\Delta G$) must be negative.
- A process with a negative $\Delta G$ is called **exergonic**, while a process with a positive $\Delta G$ is called **endergonic**.

### SECTION 6.4

- Equilibrium concentrations of a reaction represent the point of lowest free energy available to the system.
- The exact position of equilibrium is described by the equilibrium constant, $K_{eq}$, and is a function of $\Delta G$.
- If $\Delta G$ is negative, the reaction will favor products over reactants, and $K_{eq}$ will be greater than 1. If $\Delta G$ is positive, the reaction will favor reactants over products, and $K_{eq}$ will be less than 1.
- The study of relative energy levels ($\Delta G$) and equilibrium concentrations ($K_{eq}$) is called **thermodynamics**.

### SECTION 6.5

- **Kinetics** is the study of reaction rates. The rate of any reaction is described by a **rate equation**. A reaction can be **first order, second order**, or **third order**, depending on whether the sum of the exponents in the rate equation is 1, 2, or 3, respectively.
- A low **energy of activation,** $E_a$, corresponds with a fast rate.
- Increasing the temperature will increase the number of molecules that possess the minimum necessary kinetic energy for a reaction, thereby increasing the rate of reaction.
- **Catalysts** speed up the rate of a reaction by providing an alternate pathway with a lower $E_a$.

### SECTION 6.6

- Kinetics refers to the rate of a reaction and is dependent on the relative energy levels of the reactants and the transition state.
- Thermodynamics refers to the equilibrium concentrations and is dependent on the relative energy levels of the reactants and products.
- On an energy diagram, each peak represents a **transition state** while each valley represents an **intermediate**. Transition states cannot be isolated; intermediates have a finite lifetime.
- The **Hammond postulate** states that a transition state will resemble the reactants for an exergonic process but will resemble the products for an endergonic process.

### SECTION 6.7

- **Ionic reactions**, also called **polar reactions**, involve the participation of ions as reactants, intermediates, or products—in most cases, as intermediates.
- A **nucleophile** has an electron-rich atom that is capable of donating a pair of electrons. Nucleophilic centers include atoms with lone pairs, π bonds, or atoms that are electron rich due to inductive effects.
- An **electrophile** has an electron-deficient atom that is capable of accepting a pair of electrons. Electrophilic centers include atoms that are electron deficient due to inductive effects as well as **carbocations**, which have an empty *p* orbital.
- **Polarizability** describes the ability of an atom to distribute its electron density unevenly as a result of external influences.

### SECTION 6.8

- In drawing a mechanism, the tail of a curved arrow shows where the electrons are coming from, and the head of the arrow shows where the electrons are going.
- There are four characteristic arrow-pushing patterns: (1) **nucleophilic attack**, (2) **loss of a leaving group**, (3) **proton transfer**, and (4) **rearrangement**.
- The most common type of rearrangement is a carbocation rearrangement, in which a carbocation undergoes either a **hydride shift** or a **methyl shift** to produce a more stable carbocation. As a result of **hyperconjugation**, **tertiary** carbocations are more stable than **secondary** carbocations, which are more stable than **primary** carbocations.

### SECTION 6.9

- All ionic mechanisms, regardless of complexity, are different combinations of the four characteristic arrow-pushing patterns.
- When a single mechanistic step involves two simultaneous arrow-pushing patterns, it is called a concerted process.

### SECTION 6.10

- The tail of a curved arrow must be placed either on a bond or a lone pair, while the head of a curved arrow must be placed so that it shows either formation of a bond or of a lone pair.
- Never draw an arrow that gives a fifth orbital to a second-row element (C, N, O, F).

### SECTION 6.11

- A carbocation rearrangement will occur if it leads to a more stable carbocation.
- Tertiary carbocations generally do not rearrange, unless a rearrangement will produce a resonance-stabilized carbocation, such as an **allylic carbocation.**

### SECTION 6.12

- Irreversible reaction arrows are generally used in the following circumstances: 1) for a nucleophilic attack involving a strong nucleophile, or 2) for a proton transfer step in which there is a vast difference in $pK_a$ values between the acids on either side of the equilibrium, or 3) for a carbocation rearrangement.
- Reversible reaction arrows are used in most other instances.

## SKILLBUILDER REVIEW

### 6.1 PREDICTING $\Delta H°$ OF A REACTION

**EXAMPLE** Calculate $\Delta H°$ for this reaction.

$Cl_2$ → Cl + H—Cl

**STEPS 1 AND 2** Identify the bond dissociation energy of each bond broken and formed and then determine the appropriate sign for each value.

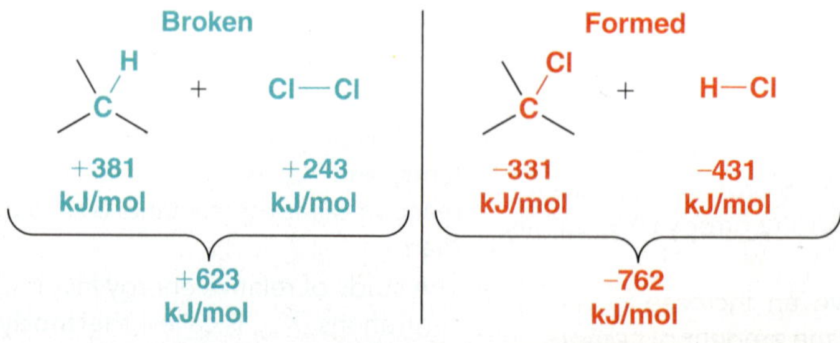


**STEP 3** Take the sum of steps 1 and 2.

(+624 kJ/mol) + (−762 kJ/mol)

−138 kJ/mol Exothermic

Try Problems **6.1, 6.2, 6.21a**

## 6.2 IDENTIFYING NUCLEOPHILIC AND ELECTROPHILIC CENTERS

**STEP 1** Identify nucleophilic centers by looking for inductive effects, lone pairs, or π bonds:

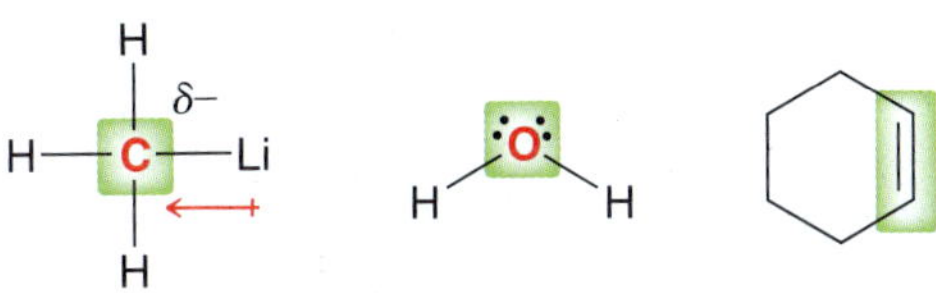


Inductive effect

Lone pair

π bond

**STEP 2** Identify electrophilic centers by looking for inductive effects or empty *p* orbitals:

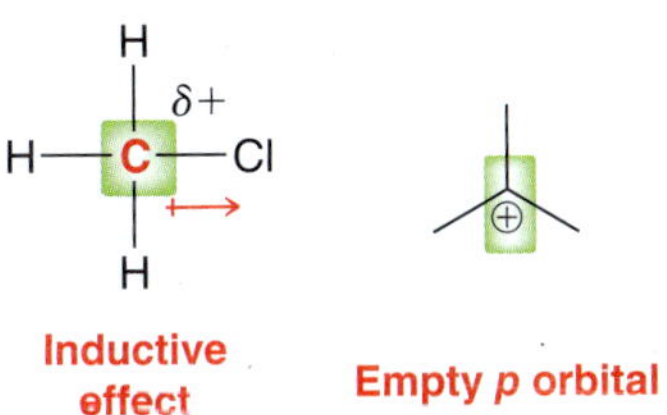


Inductive effect

Empty *p* orbital

Try Problems 6.8–6.11, 6.53

## 6.3 IDENTIFYING AN ARROW-PUSHING PATTERN

**Nucleophilic attack**

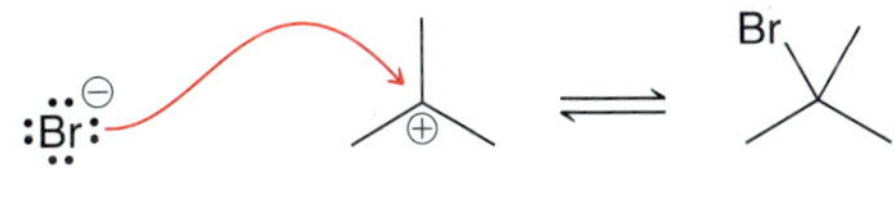


**Loss of a leaving group**

**Proton transfer**

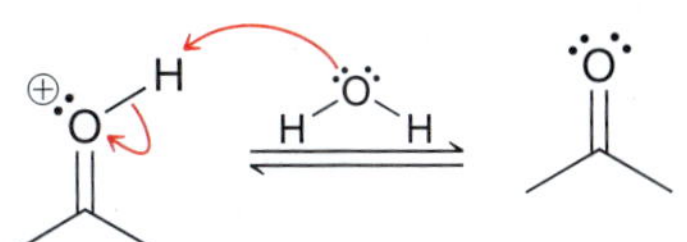


**C+ rearrangement**

$H_3C$ … $CH_3$

Try Problems 6.12, 6.13, 6.30

## 6.4 IDENTIFYING A SEQUENCE OF ARROW-PUSHING PATTERNS

Identify each characteristic pattern of arrow pushing.

Proton transfer

H–Br

$-H_2O$

Loss of a leaving group

C+ rearrangement

Nucleophilic attack

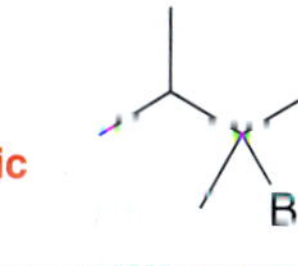


Try Problems 6.14, 6.15, 6.32–6.41

## 6.5 DRAWING CURVED ARROWS

**STEP 1** Identify which of the four arrow-pushing patterns to use.

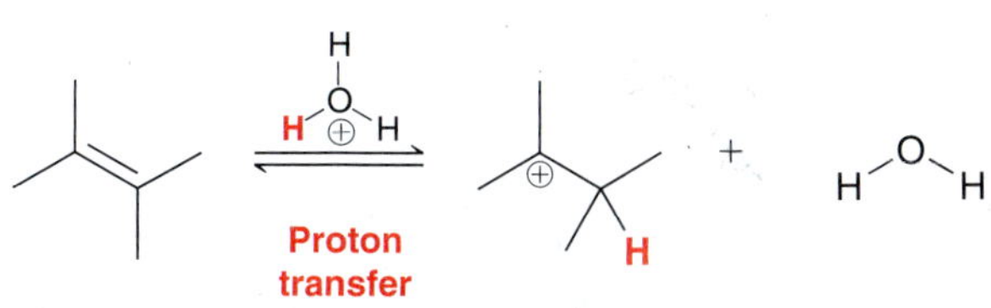


**STEP 2** Draw the curved arrow, focusing on the proper placement of the tail and head of each curved arrow.

Try Problems 6.16, 6.17, 6.42–6.47, 6.54

### 6.6 PREDICTING CARBOCATION REARRANGEMENTS

**STEP 1** Identify neighboring carbon atoms.

**STEP 2** Identify any H or $CH_3$ attached directly to the neighboring carbon atoms.

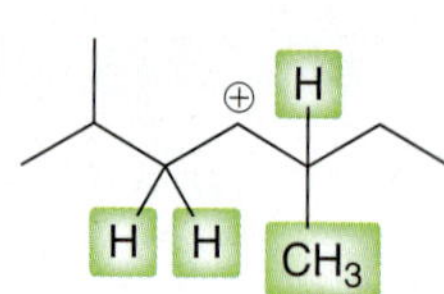

**STEP 3** Find any groups that can migrate to generate a more stable C+.

**STEP 4** Draw a curved arrow showing the C+ rearrangement and then draw the new carbocation.

Try Problems 6.18, 6.19, 6.48

## PRACTICE PROBLEMS

Note: Most of the Problems are available within **WileyPLUS**, an online teaching and learning solution.

**6.20** In each of the following cases compare the bonds identified in red, and determine which bond you would expect to have the largest bond dissociation energy:

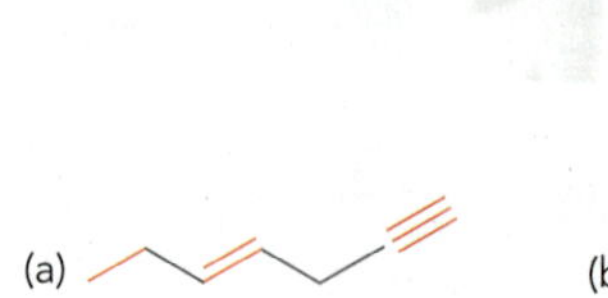

(a) (b)

**6.21** Consider the following reaction:

(a) Use Table 6.1 to estimate $\Delta H$ for this reaction.

(b) $\Delta S$ of this reaction is positive. Explain.

(c) Determine the sign of $\Delta G$.

(d) Is the sign of $\Delta G$ dependent on temperature?

(e) Is the magnitude of $\Delta G$ dependent on temperature?

**6.22** For each of the following cases use the information given to determine whether or not the equilibrium will favor products over reactants:

(a) A reaction with $K_{eq} = 1.2$

(b) A reaction with $K_{eq} = 0.2$

(c) A reaction with a positive $\Delta G$

(d) An exothermic reaction with a positive $\Delta S$

(e) An endothermic reaction with a negative $\Delta S$

**6.23** Which value of $\Delta G$ corresponds with $K_{eq} = 1$?

(a) +1 kJ/mol (b) 0 kJ/mol (c) −1 kJ/mol

**6.24** Which value of $\Delta G$ corresponds with $K_{eq} < 1$?

(a) +1 kJ/mol (b) 0 kJ/mol (c) −1 kJ/mol

**6.25** For each of the following reactions determine whether $\Delta S$ for the reaction ($\Delta S_{sys}$) will be positive, negative, or approximately zero:

(a) + HBr ⟶

(b) ⟶ + HCl

(c) + $H_2O$ ⟶ + HCl

(d) ⟶

(e) ⟶

**6.26** Draw an energy diagram of a reaction with the following characteristics:

(a) A one-step reaction with a negative $\Delta G$

(b) A one-step reaction with a positive $\Delta G$

(c) A two-step reaction with an overall negative $\Delta G$, where the intermediate is higher in energy than the reactants and the first transition state is higher in energy than the second transition state

**6.27** Consider the following four energy diagrams:

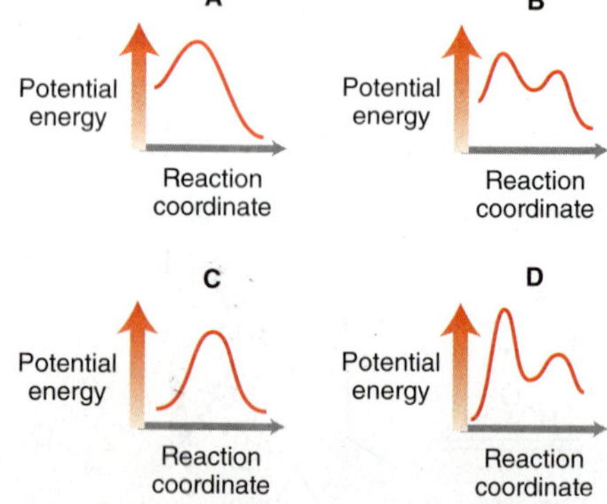


(a) Which diagrams correspond with a two-step mechanism?

(b) Which diagrams correspond with a one-step mechanism?

(c) Compare energy diagrams A and C. Which has a relatively larger $E_a$?

(d) Compare diagrams A and C. Which has a negative $\Delta G$?

(e) Compare diagrams A and D. Which has a positive $\Delta G$?

(f) Compare all four energy diagrams. Which one exhibits the largest $E_a$?

(g) Which processes will have a value of $K_{eq}$ that is greater than 1?

(h) Which process will have a value of $K_{eq}$ that is roughly equal to 1?

**6.28** Identify all transition states and intermediates on the following energy diagram:

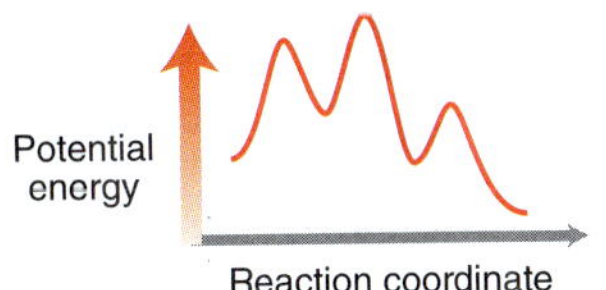


**6.29** Consider the following reaction:

This reaction has been determined to be second order.

(a) What is the rate equation for this reaction?

(b) How will the rate be affected if the concentration of hydroxide is tripled?

(c) How will the rate be affected if the concentration of chloroethane is tripled?

(d) How will the rate be affected if the temperature is raised by 40°C?

**6.30** For each of the following reactions identify the arrow-pushing pattern that is being utilized:

(a)

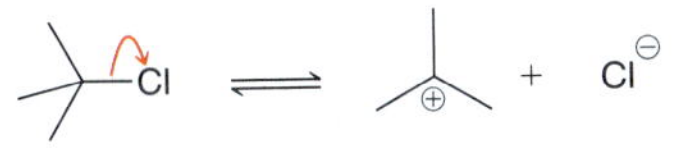

(b)

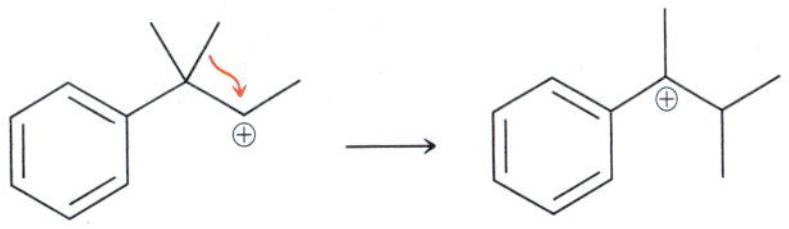

(c)

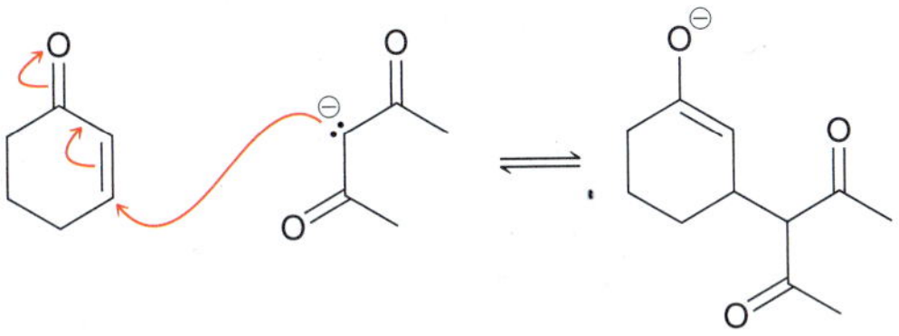

(d)

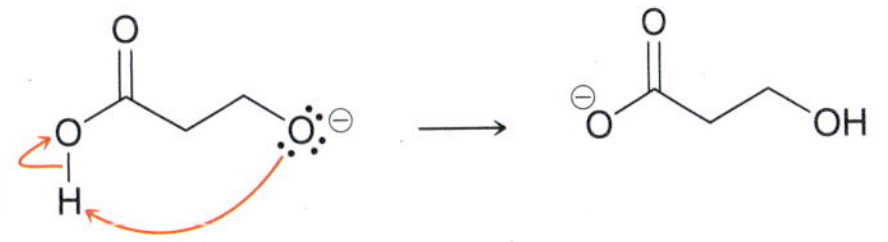

**6.31** Rank the three carbocations shown in terms of increasing stability:

(a)

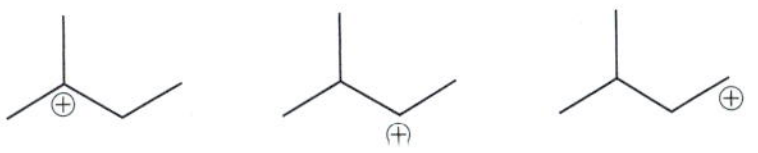

(b)

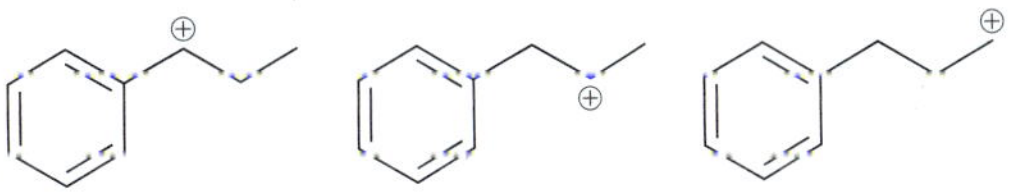

**6.32** For the following mechanism, identify the sequence of arrow-pushing patterns:

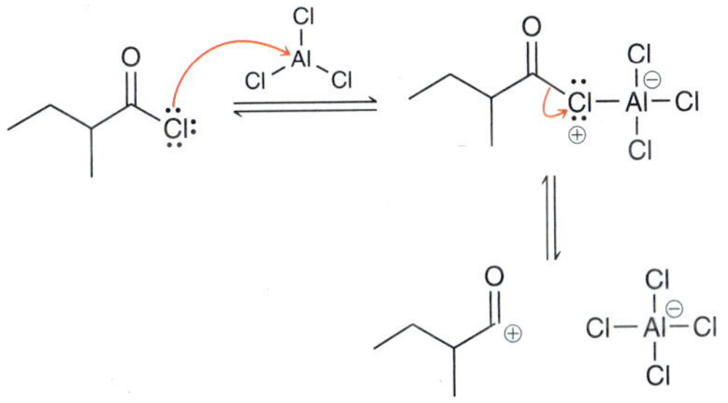

**6.33** For the following mechanism, identify the sequence of arrow pushing patterns:

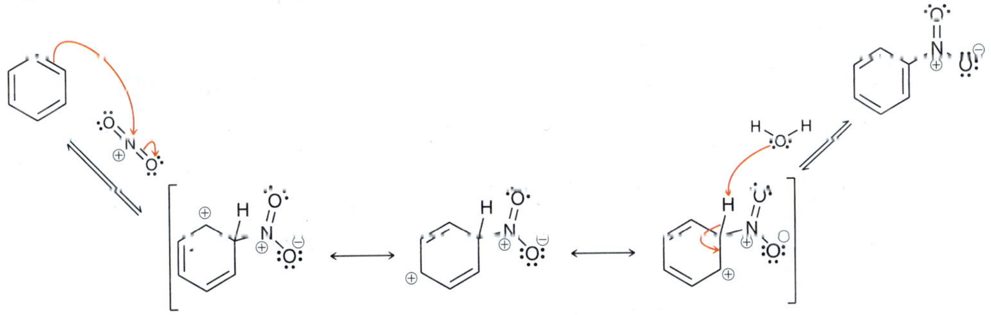

**6.34** For the following mechanism, identify the sequence of arrow-pushing patterns:

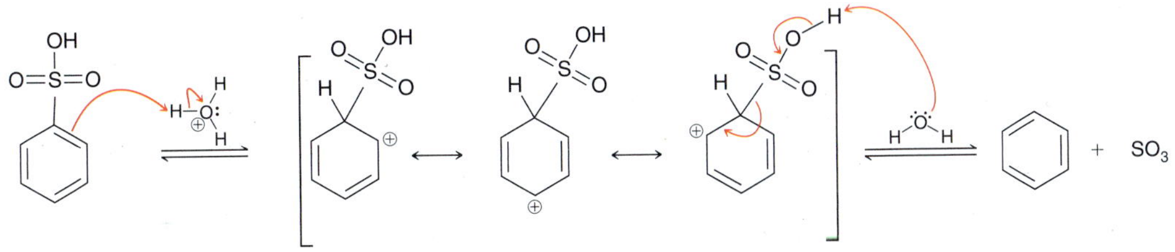

**6.35** For the following mechanism, identify the sequence of arrow-pushing patterns:

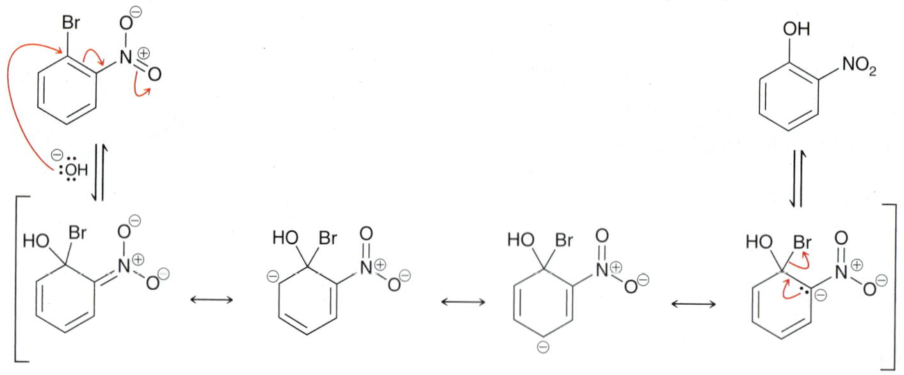

**6.36** For the following mechanism, identify the sequence of arrow-pushing patterns:

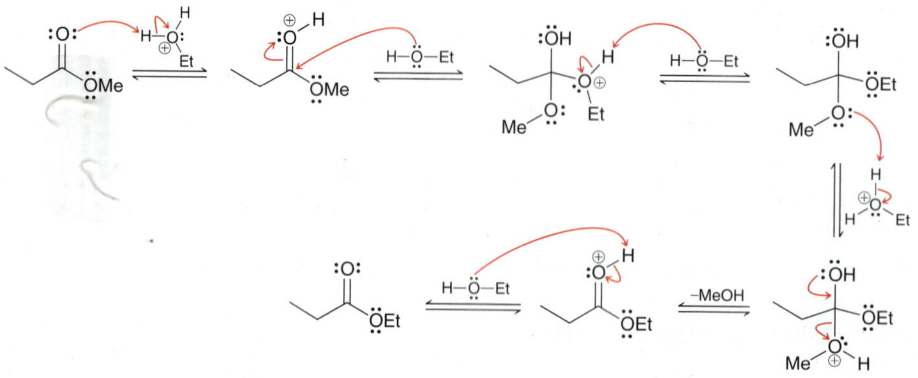

**6.37** For the following mechanism, identify the sequence of arrow-pushing patterns:

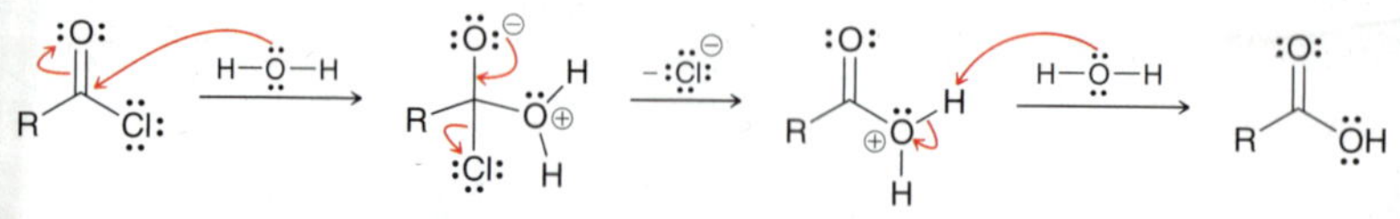

**6.38** For the following mechanism, identify the sequence of arrow-pushing patterns:

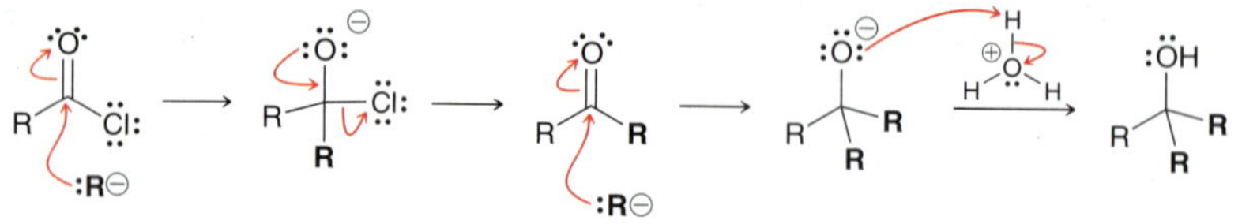

**6.39** For the following mechanism, identify the sequence of arrow-pushing patterns:

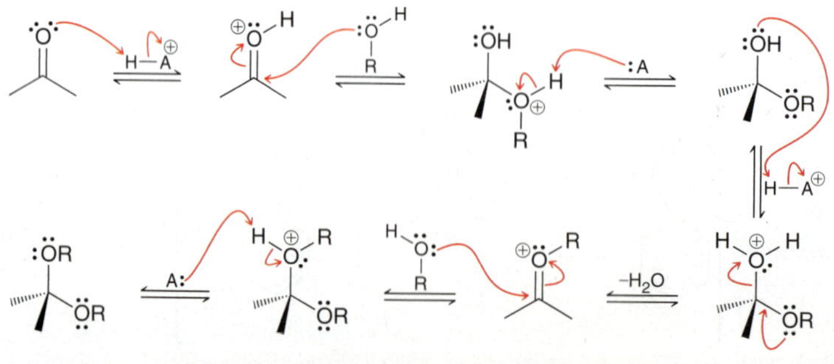

**6.40** For the following mechanism, identify the sequence of arrow-pushing patterns:

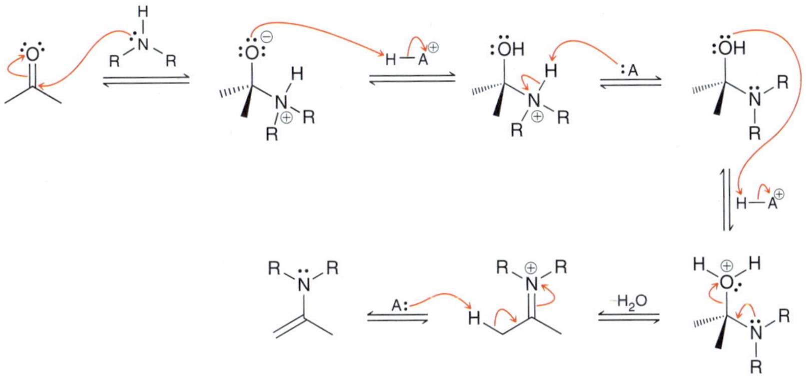

**6.41** For the following mechanism, identify the sequence of arrow-pushing patterns:

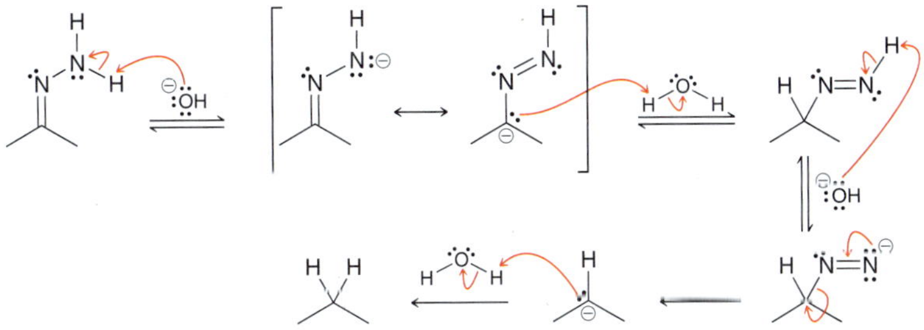

**6.42** Draw curved arrows for each step of the following mechanism:

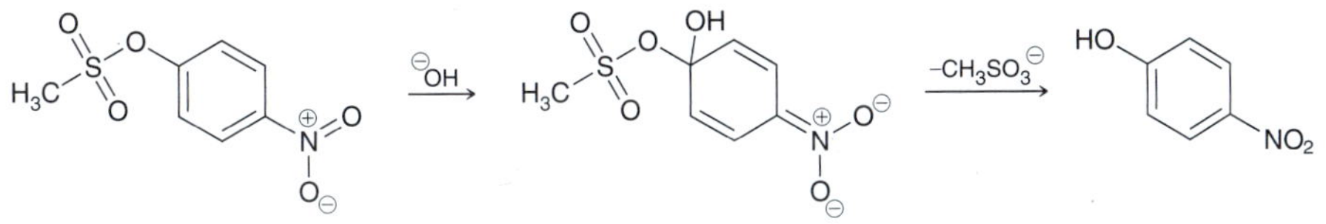

**6.43** Draw curved arrows for each step of the following mechanism:

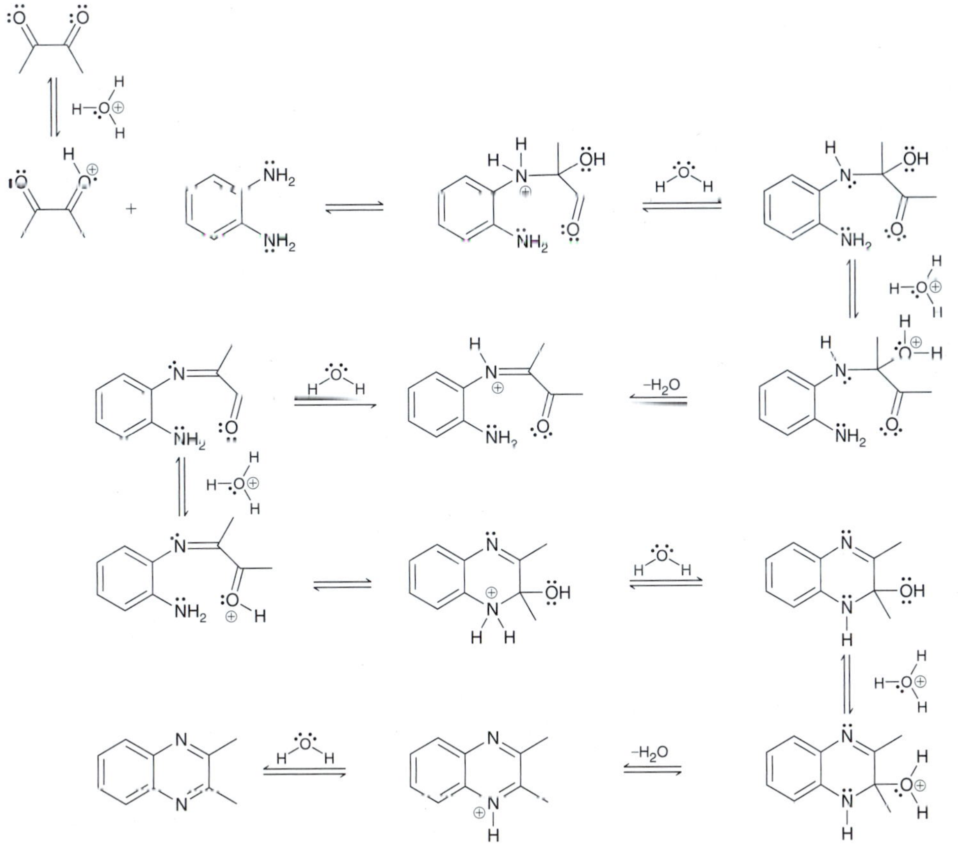

**6.44** Draw curved arrows for each step of the following mechanism:

**6.45** Draw curved arrows for each step of the following mechanism:

**6.46** Draw curved arrows for each step of the following mechanism:

**6.47** Draw curved arrows for each step of the following mechanism:

$-H_2O$

**6.48** Predict whether each of the following carbocations will rearrange. If so, draw the expected rearrangement using curved arrows.

(a) (b) (c) (d) (e) (f) (g)

# INTEGRATED PROBLEMS

**6.49** Consider the following reaction:

$$HO^{-} + CH_3Br \longrightarrow HOCH_3 + Br^{-}$$

The following rate equation has been experimentally established for this process:

$$\text{Rate} = k[HO^-][CH_3Br]$$

The energy diagram for this process is shown below:

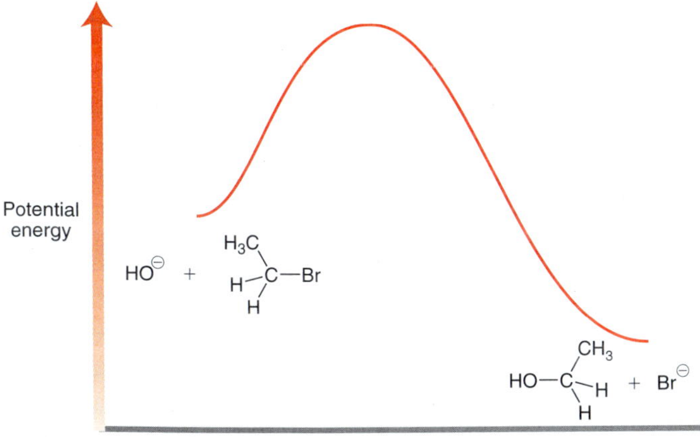


(a) Draw the curved arrows showing a mechanism for this process.

(b) Identify the two characteristic arrow-pushing patterns that are required for this mechanism.

(c) Would you expect this process to be exothermic or endothermic? Explain.

(d) Would you expect $\Delta S_{sys}$ for this process to be positive, negative, or approximately zero?

(e) Is $\Delta G$ for this process positive or negative?

(f) Would you expect an increase in temperature to have a significant impact on the position of equilibrium (equilibrium concentrations)? Explain.

(g) Draw the transition state of this process and identify its location on the energy diagram.

(h) Is the transition state closer in structure to the reactants or products? Explain.

(i) Is the reaction first order or second order?

(j) How will the rate be affected if the concentration of hydroxide is doubled?

(k) Will the rate be affected by an increase in temperature?

**6.50** Identify whether each of the following factors will affect the rate of a reaction:

(a) $K_{eq}$ (b) $\Delta G$ (c) Temperature

(d) $\Delta H$ (e) $E_a$ (f) $\Delta S$

**6.51** In the presence of a special type of catalyst, hydrogen gas will add across a triple bond to produce a double bond:

+ $H_2$ —Catalyst→

The process is exothermic. Do you expect a high temperature to favor products or reactants?

**6.52** Consider the following reaction. Predict whether an increase in temperature will favor reactants or products. Justify your prediction.

**6.53** When an amine is protonated, the resulting ammonium ion is not electrophilic:

An amine — An ammonium ion

However, when an imine is protonated, the resulting iminium ion is highly electrophilic:

An imine — An iminium ion

Explain this difference in reactivity between an ammonium ion and an iminium ion.

**6.54** Draw the curved arrows that accomplish the following transformation:

$\ominus$OH → → + N≡N + Cl$^{\ominus}$

# CHALLENGE PROBLEMS

**6.55** Under basic conditions (catalytic $MeO^-$ in MeOH), compound **1** rearranges through the mechanistic sequence shown below to a propellane-type isomer **5** (*J. Am. Chem. Soc.* **2012,** *134,* **17877–17880**). Note that $MeO^-$ is catalytic in that it is consumed in the first step of the mechanism and regenerated in the final step. Provide curved arrows consistent with each mechanistic step and the resonance structure of **2** that is the greatest contributor to the resonance hybrid.

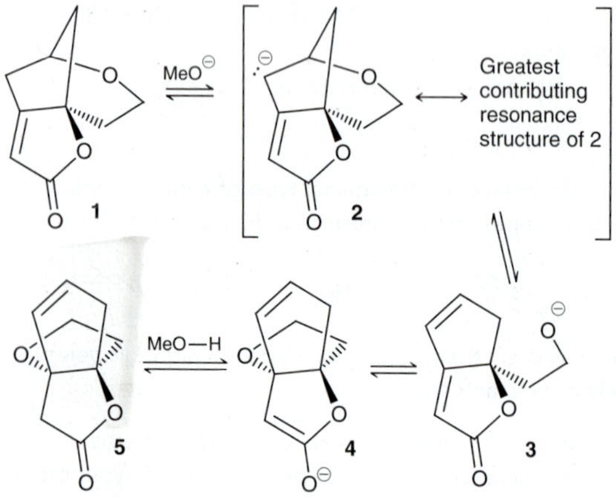

Greatest contributing resonance structure of 2

MeO⊖ MeO—H

**6.56** Compound **1** has been prepared and studied to investigate a novel type of intramolecular elimination mechanism (*J. Org. Chem.* **2007,** *72,* **793–798**). The proposed mechanistic pathway for this transformation is presented below. Complete the mechanism by drawing curved arrows consistent with the change in bonding in each step.

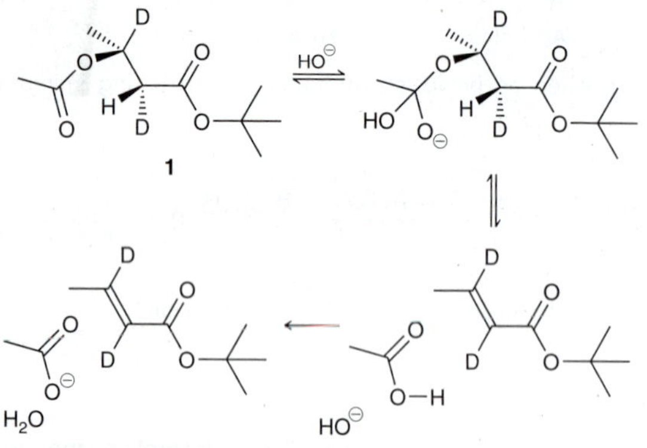

HO⊖ $H_2O$

**6.57** Indolizomycin is a potent antibiotic agent produced by a microorganism called SK2-52. Of particular interest is the fact that SK2-52 is a mutant microorganism that was created in a laboratory from two different strains, *Streptomyces teryimanensis* and *Streptomyces grisline*, neither of which is antibiotic producing. This raises the possibility of using mutant microorganisms as factories for producing novel structures with a variety of medicinal properties. During S. J. Danishefsky's synthesis of racemic indolizomycin, compound **1** was treated with triphenylphosphine ($PPh_3$) to afford compound **2** (*J. Am. Chem. Soc.* **1990,** *112,* **2003–2005**). The reaction between **2** and **3** then gave compound **4**. These processes are believed to proceed via the following intermediates. Complete the mechanism by drawing all curved arrows and identify the arrow-pushing pattern employed in each step.

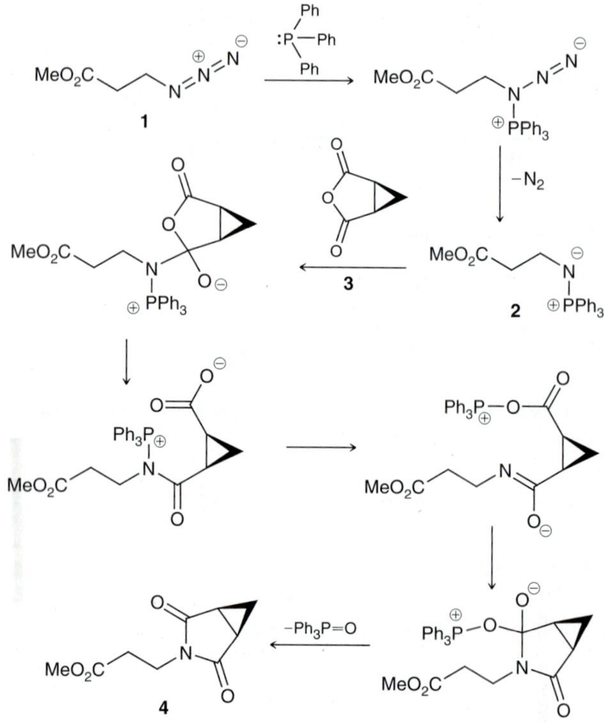

$-N_2$ $-Ph_3P{=}O$

**6.58** The following sequence was utilized in a biosynthetically inspired synthesis of preussomerin, an antifungal agent isolated from a coprophilous (dung-loving) fungus (*Org. Lett.* **1999,** *1,* **3–5**):

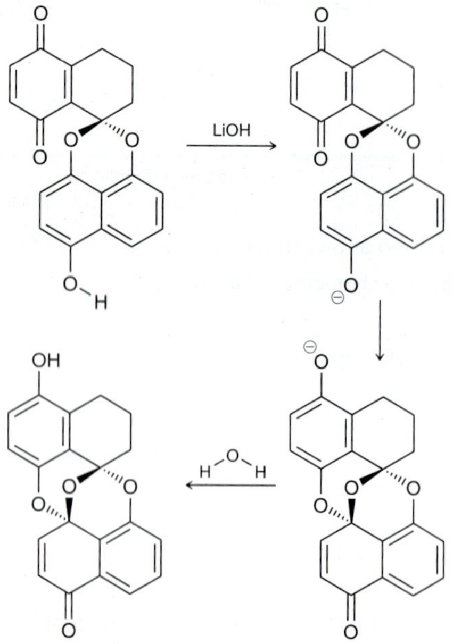

LiOH

(a) Draw curved arrows for each step of the mechanism.
(b) Include an explanation for the observed stereochemistry.

**6.59** This chapter focused on the mechanisms of ionic reactions, which represent the vast majority of the reactions throughout this textbook. In Chapter 17, we will explore another class of reactions, called *pericyclic* reactions, for which the changes in bonding can be described by electrons moving in a concerted fashion within a closed loop. For example, in the presence of an enzyme, chorismate (**1**) rearranges via a pericyclic reaction to produce compound **2**; this is an important step in the biosynthesis of several naturally occurring amino acids. In the absence of the enzyme, a different conformation of chorismate (**1a**) can undergo a pericyclic reaction to yield **3** and **4** (*J. Am. Chem. Soc.* **2005**, *127*, 12957–12964). The mechanism for each of these pericyclic reactions involves three curved arrows, representing the movement of six electrons, in a loop.

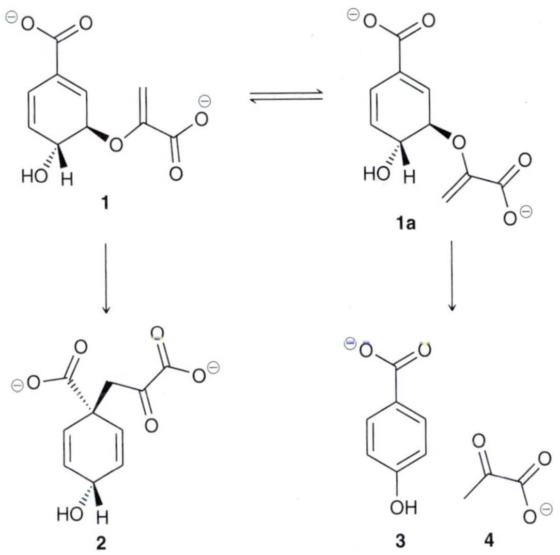

(a) Draw a concerted mechanism for the conversion of **1** → **2**.

(b) Describe the conformational change from **1** → **1a**.

(c) Draw a concerted mechanism for the conversion of **1a** → **3** + **4**.

**6.60** (+)-Aureol is a natural product that shows selective anticancer activity against certain strains of lung cancer and colon cancer. A key step in the biosynthesis of (+)-aureol (how nature makes the molecule) is believed to involve the conversion of carbocation **A** to carbocation **B** (*Org. Lett.* **2012**, *14*, 4710–4713). Propose a possible mechanism for this transformation and explain the observed stereochemical outcome.

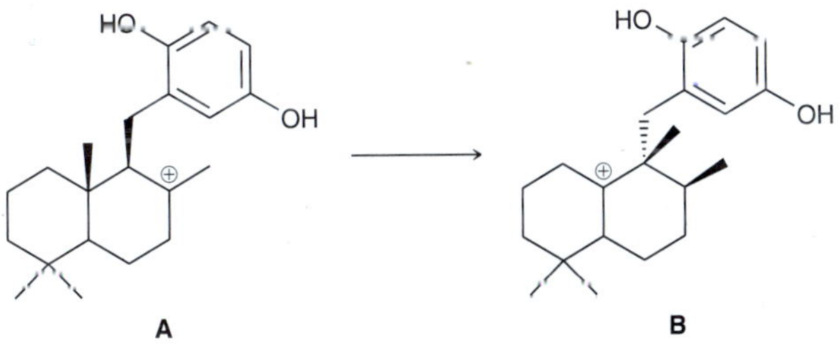

**6.61** Strychnine (**6**), a notorious poison isolated from the *strychnos* genus, is commonly used as a pesticide in the treatment of rodent infestations. The sophisticated structure of strychnine, first elucidated in 1946, served as a tantalizing goal for synthetic organic chemists. The first synthesis of strychnine was achieved by R. B. Woodward (Harvard Univ.) in 1954 (*J. Am. Chem. Soc.* **1954**, *76*, 4749–4751). In the last step of Woodward's synthesis, **1** was converted into **6** via intermediates **2–5**, as shown:

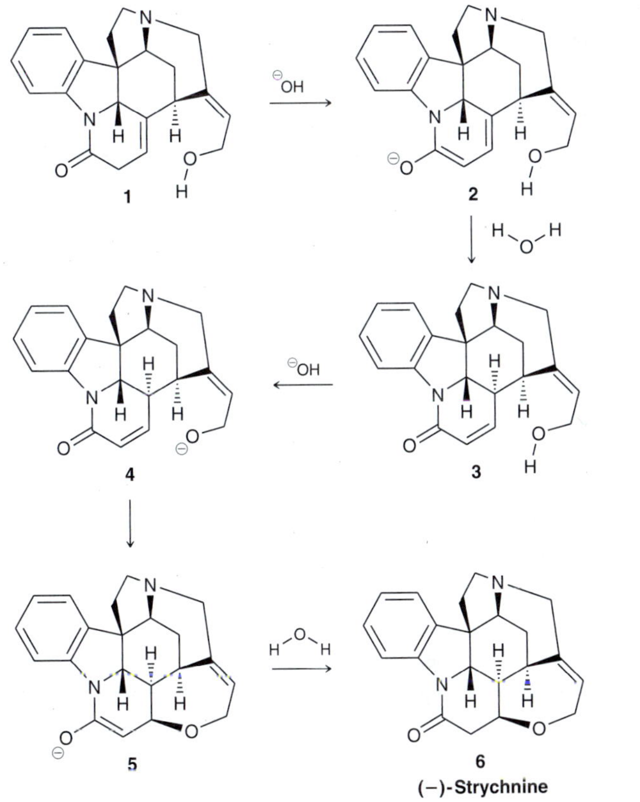

**6**
**(−)-Strychnine**

(a) Draw all curved arrows that show the entire transformation of **1** into **6** and identify the entire sequence of arrow-pushing patterns.

(b) For the step in which **4** is transformed into **5**, identify the nucleophilic center and the electrophilic center and use resonance structures to justify why the latter is electron poor.

(c) The conversion of **4** into **5** involves the creation of a new chirality center. Determine its configuration and explain why the other possible configuration is not observed.

**6.62** There are many examples of carbocation rearrangements that cannot be classified as a methyl shift or hydride shift. For example, consider the following rearrangement that was employed in the synthesis of isocomene, a tricyclic compound isolated from the perennial plant *Isocoma wrightii* (*J. Am. Chem. Soc.* **1979**, *101*, 7130–7131):

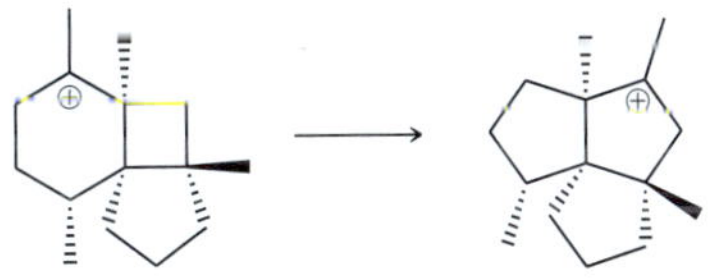

(a) Identify the carbon atom that is migrating and draw a mechanism for this carbocation rearrangement.

(b) This process generates a new chirality center, yet only one configuration is observed. That is, the process is said to be *diastereoselective*, because only one of the two possible diastereomeric carbocations is formed. Draw the diastereomeric carbocation that is not formed, and provide a reason for its lack of formation (**Hint:** You may find it helpful to construct a molecular model).

(c) In this case, a tertiary carbocation is rearranging to give a different tertiary carbocation. Suggest a driving force for this process (explain why this process is favorable).

**6.63** Compound **1** undergoes a thermal elimination of nitrogen at 250 °C to form nitrile **4** (*Org. Lett.* **1999,** *1,* **537–539**). One proposed (and subsequently refuted) mechanism for this transformation involves intermediates **2** and **3**:

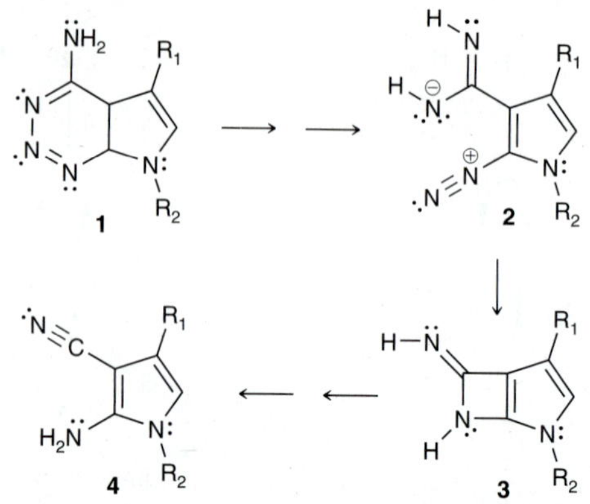

(a) Draw curved arrows that show the conversion of **2** to **3**.

(b) To determine whether or not the proposed mechanism operates, one of the nitrogen atoms was isotopically labeled as $^{15}N$, and its location in the product was determined. The absence of **4b** in the product mixture demonstrates that the proposed mechanism is not operating. Using resonance structures, explain why the proposed mechanism predicts that **4b** should be formed.

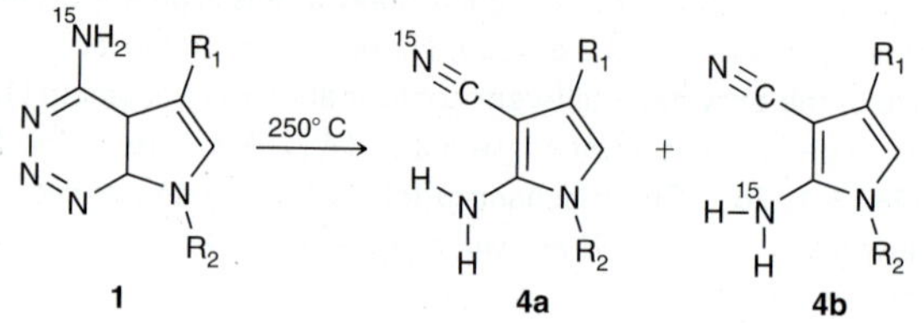

**6.64** In the presence of a Lewis acid, compound **1** rearranges, via intermediate **2,** to afford compound **3** (*Org. Lett.* **2005,** *7,* **515–517**):

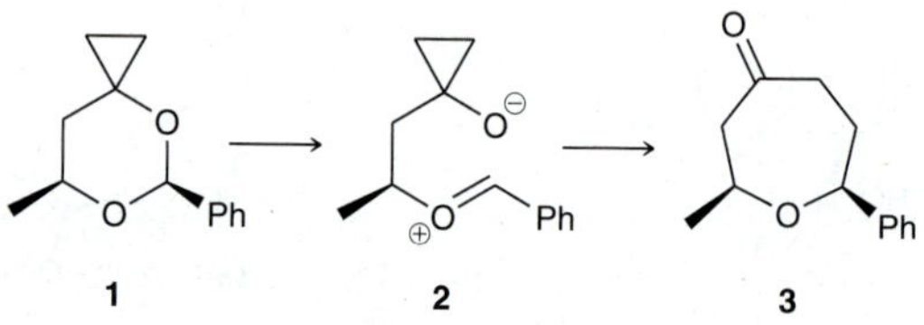

(a) Draw curved arrows showing how **1** is transformed into **2**. Note that the Lewis acid has been left out for simplicity.

(b) Draw curved arrows showing how **2** is transformed into **3**. (**Hint:** It may be helpful to redraw **2** in a different conformation.)

# 7 Substitution Reactions

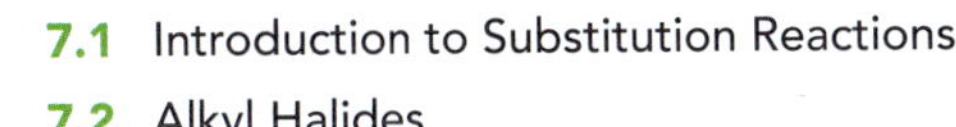



## DID YOU EVER WONDER...

### what chemotherapy is?

As its name implies, chemotherapy is the use of chemical agents in the treatment of cancer. Dozens of chemotherapy drugs are currently in clinical use, and researchers around the world are currently working on the design and development of new drugs for treating cancer. The primary goal of most chemotherapeutic agents is to cause irreparable damage to cancer cells while causing only minimal damage to normal, healthy cells. Since cancer cells grow much faster than most other cells, many anticancer drugs have been designed to interrupt the growth cycle of fast-growing cells. Unfortunately, some healthy cells are also fast growing, such as hair follicles and skin cells. For this reason, chemotherapy patients often experience a host of side effects, including hair loss and rashes.

The field of chemotherapy began in the mid-1930s, when scientists realized that a chemical warfare agent (sulfur mustard) could be modified and used to attack tumors. The action of sulfur mustard (and its derivatives) was thoroughly investigated and was found to involve a series of reactions called substitution reactions. Throughout this chapter, we will explore many important features of substitution reactions. At the end of the chapter, we will revisit the topic of chemotherapy by exploring the rational design of the first chemotherapeutic agents.

## DO YOU REMEMBER?

*Before you go on, be sure you understand the following topics.*
*If necessary, review the suggested sections to prepare for this chapter.*

- The Cahn-Ingold-Prelog System (Section 5.3)
- Kinetics and Energy Diagrams (Sections 6.5, 6.6)
- Nucleophiles and Electrophiles (Section 6.7)
- Arrow Pushing and Carbocation Rearrangements (Sections 6.8–6.11)

Take the DO YOU REMEMBER? QUIZ in **WileyPLUS** to check your understanding.

## 7.1 Introduction to Substitution Reactions

This chapter introduces a class of reactions, called **substitution reactions**, in which one group is exchanged for another, while Chapter 8 introduces **elimination** reactions, characterized by the formation of a $\pi$ bond:

X — Substitution → Y **(Chapter 7)**
X — Elimination → alkene **(Chapter 8)**

Substitution and elimination reactions often compete with each other; in fact, the last three sections of Chapter 8 will explore the main factors that govern this competition. In this chapter, we will focus our attention exclusively on substitution reactions, and then we will broaden our discussion in Chapter 8 to include elimination reactions.

A substitution reaction can occur when a suitable electrophile is treated with a nucleophile, as in the following example:

**LOOKING BACK**
For a review of nucleophiles and electrophiles see Section 6.7.

Cl + $^{\ominus}$SH → SH + $^{\ominus}$Cl

**Electrophile** (substrate) **Nucleophile**

Organic chemists often use the term **substrate** when referring to the electrophile in a substitution reaction. In order for an electrophile to function as a substrate in a substitution reaction, it must contain a **leaving group**, which is a group capable of separating from the substrate. In the example above, chloride functions as the leaving group. A leaving group serves two critical functions:

1. The leaving group withdraws electron density via induction, rendering the adjacent carbon atom electrophilic. This can be visualized with electrostatic potential maps of various methyl halides (Figure 7.1). In each image, the blue color indicates a region of low electron density.

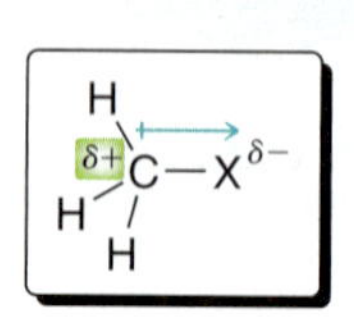


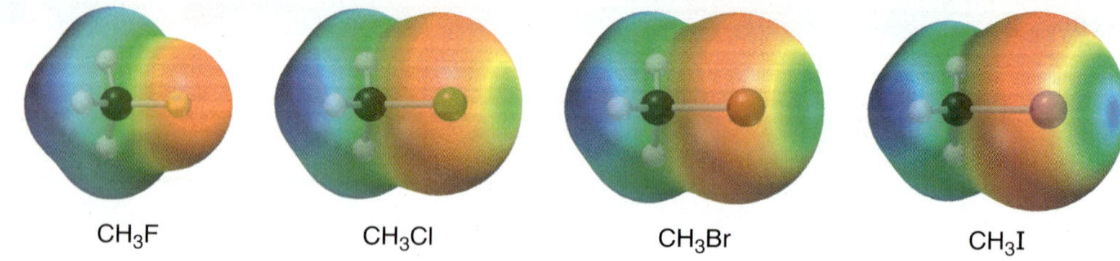


**FIGURE 7.1** Electrostatic potential maps of methyl halides.

2. The leaving group can stabilize any negative charge that may develop as the result of the leaving group separating from the substrate:

:Cl: ⇌ ⊕ + :Cl:$^{\ominus}$ **Stabilized charge**

Halogens (Cl, Br, and I) are very common leaving groups.

## 7.2 Alkyl Halides

Halogenated organic compounds are commonly used as electrophiles in substitution reactions. Although other compounds can also serve as electrophiles, we will focus our attention for now on compounds containing halogens.

### Naming Halogenated Organic Compounds

Recall from Section 4.2 that systematic (IUPAC) names of alkanes are assigned using four discrete steps:

1. Identify and name the parent.
2. Identify and name the substituents.
3. Number the parent chain and assign a locant to each substituent.
4. Assemble the substituents alphabetically.

The same exact four-step procedure is used to name compounds that contain halogens, and all of the rules discussed in Chapter 4 apply here as well. Halogens are simply treated as substituents and receive the following names: fluoro-, chloro-, bromo-, and iodo-. Below are two examples:

As we saw in Chapter 4, the parent is the longest chain, and it should be numbered so that the first substituent receives the lower number:

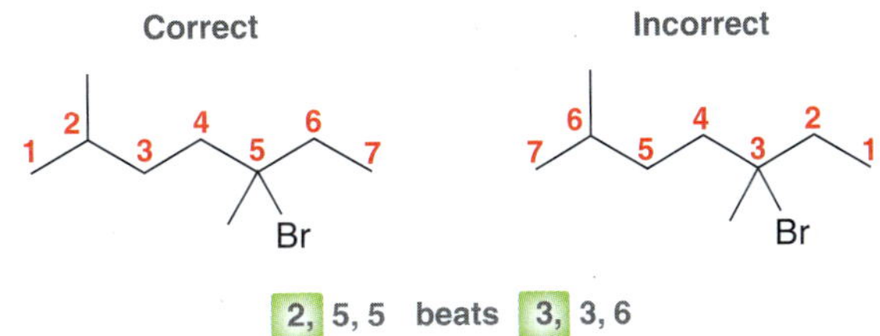

CONCEPTUAL CHECKPOINT

7.1 Assign a systematic name for each of the following compounds.

(a) Cl (b) Br (c) Br Br Cl (d) F

When a chirality center is present in the compound, the configuration must be indicated at the beginning of the name:

1 2 3 4 5 6 7 Br

(*R*)-5-Bromo-2,3,3-trimethylheptane

In addition to systematic names, IUPAC nomenclature also recognizes common names for many halogenated organic compounds.

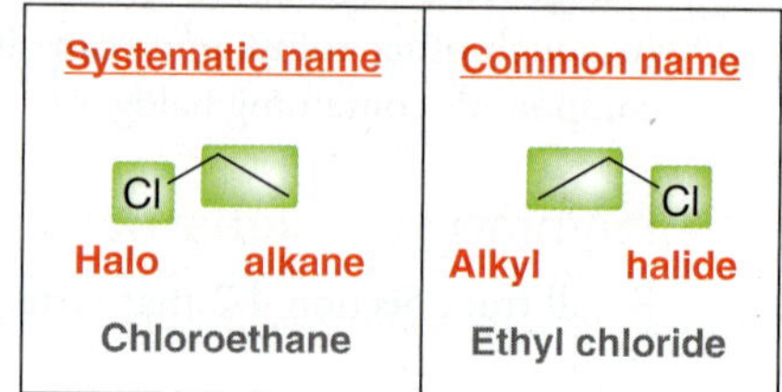


The systematic name treats a halogen as a substituent, calling the compound a **haloalkane**. The common name treats the compound as an alkyl substituent connected to a halide, and the compound is called an **alkyl halide** or an **organohalide**.

## Structure of Alkyl Halides

Each carbon atom is described in terms of its proximity to the halogen using letters of the Greek alphabet. The **alpha (α) position** is the carbon atom connected directly to the halogen, while the **beta (β) positions** are the carbon atoms connected to the alpha position:

An alkyl halide will have only one α position, but there can be as many as three β positions. This chapter focuses on reactions that occur at the α position, and the next chapter will focus on reactions involving the β position.

Alkyl halides are classified as **primary** (1°), **secondary** (2°), or **tertiary** (3°) based on the number of alkyl groups connected to the α position (Figure 7.2).

**FIGURE 7.2** Classification of alkyl halides as primary, secondary, or tertiary.

## Uses of Organohalides

Many organohalides are toxic and have been used as insecticides:

DDT (dichlorodiphenyltrichloroethane) was developed in the late 1930s and became one of the first insecticides to be used around the globe. It was found to exhibit strong toxicity for insects but rather low toxicity for mammals. DDT was used as an insecticide for many decades and has been credited with saving more than half a billion lives by killing mosquitos that carry deadly diseases. Unfortunately, it was found that DDT does not degrade quickly and persists in the environment.

Rising concentrations of DDT in wildlife began to threaten the survival of many species. In response, the Environmental Protection Agency (EPA) banned the use of DDT in 1972, and it was replaced with other, environmentally safer, insecticides.

Lindane is used in shampoos designed to treat head lice, while chlordane and methyl bromide have been used to prevent and treat termite infestations. The use of methyl bromide has recently been regulated due to its role in the destruction of the ozone layer (for more on the hole in the ozone layer, see Section 11.8).

Organohalides are particularly stable compounds, and many of them, like DDT, persist and accumulate in the environment. PCBs (polychlorinated biphenyls) represent another well-known example. Biphenyl is a compound that can have up to 10 substituents:

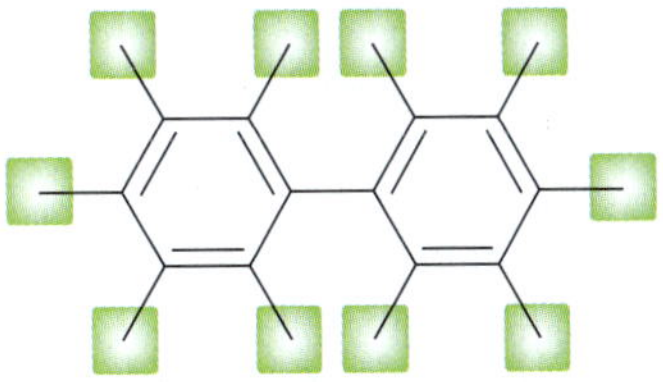

**Biphenyl**

PCBs are compounds in which many of these positions contain chlorine atoms. PCBs were originally produced as coolants and insulating fluids for industrial transformers and capacitors. They were also used as hydraulic fluids and as flame retardants. But their accumulation in the environment began to threaten wildlife, and their use was banned.

The above examples have contributed to the bad reputation of organohalides. As a result, organohalides are often viewed as man-made poisons. However, research over the last 20 years has indicated that organohalides are actually more common in nature than had previously been thought. For example, methyl chloride is the most abundant organohalide in the atmosphere. It is produced in large quantities by evergreen trees and marine organisms, and it is consumed by many bacteria, such as *Hyphominocrobium* and *Methylobacterium*, that convert methyl chloride into $CO_2$ and $Cl^-$.

Many organohalides are also produced by marine organisms. Over 5000 such compounds have already been identified, and several hundred new compounds are discovered each year. Here are two examples:

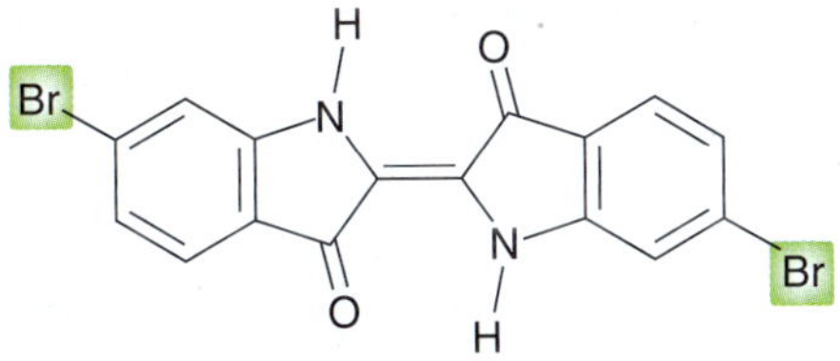

**Tyrian purple**

Isolated from the sea snail *Hexaplex trunculus*, this compound is one of the oldest known dyes and was used to make royal clothing thousands of years ago.

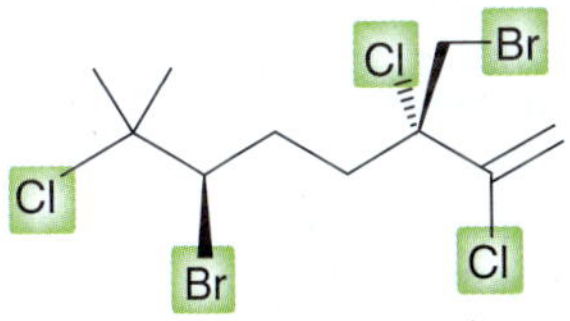

**Halomon**

Isolated from the red algae *Portieria hornemannii*, this compound is currently in clinical trials for use as an antitumor agent.

Organohalides serve a variety of functions in living organisms. In sponges, corals, snails, and seaweeds organohalides are used as a defense mechanism against predators (a form of chemical warfare). Here are two such examples:

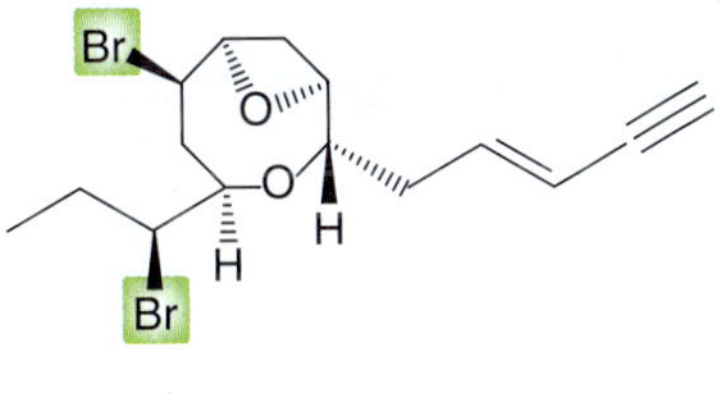

**(3*E*)-Laureatin**

Used by the red algae *Laurencia nipponina*

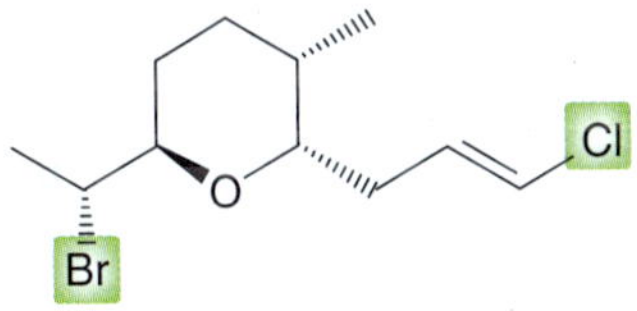

**Kumepaloxane**

Used by the snail *Haminoea cymbalum*

Both of these compounds are used to ward off predators. In many kinds of organisms organohalides act as hormones (chemical messengers that act only on specific target cells). Examples include the following:

**2,6-Dichlorophenol**
Used as a sex hormone by the lone star tick *Amblyomma americanum*

**2,6-Dibromophenol**
Isolated from the acorn worm *Balanoglossus biminiensis*, likely used as a hormone

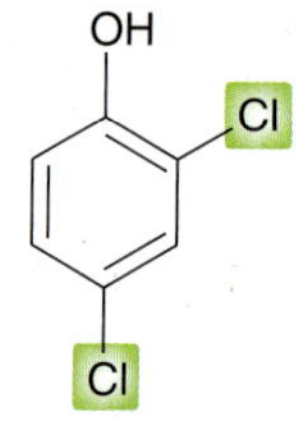


**2,4-Dichlorophenol**
Used as a growth hormone by *Penicillium* molds

Not all halogenated compounds are toxic. In fact, many organohalides have clinical applications. For example, the following compounds are widely used and have contributed much to the improvement of physical and psychological health:

**Bronopol**
**(2-Bromo-2-nitropropane-1,3-diol)**
A powerful antimicrobial compound safe enough to use in baby-wipes

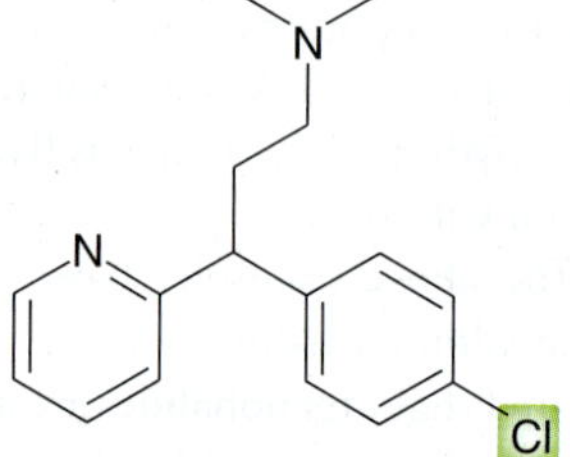


**Chlorpheniramine**
An antihistamine, sold under the trade name Chlor-Trimeton

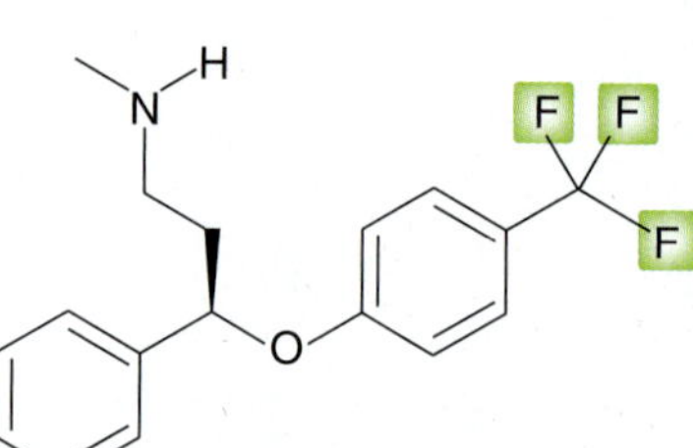


**(*R*)-Fluoxetine**
An antidepressant, sold under the trade name Prozac

**BY THE WAY**

Sucralose was discovered by accident in 1976 when a British company (Tate and Lyle) was conducting research on potential uses of chlorinated sugars. A foreign graduate student participating in the research misunderstood a request to "test" one of the compounds and instead thought he was being asked to "taste" the compound. The graduate student reported an intensely sweet taste, and it was later found to be safe to consume.

Some organohalides have even been used in the food industry. Consider, for example, the structure of sucralose, shown here. Sucralose contains three chlorine atoms, but it is known not to be toxic. It is several hundred times sweeter than sugar and is sold as an artificial, low-calorie sweetener under the trade name Splenda.

**Sucralose**
An artificial sweetener, sold under the trade name Splenda

## 7.3 Possible Mechanisms for Substitution Reactions

Recall from Chapter 6 that ionic mechanisms are comprised of only four types of arrow-pushing patterns (Figure 7.3). All four of these steps will be used in this chapter, so it might be wise to review Sections 6.7–6.10.

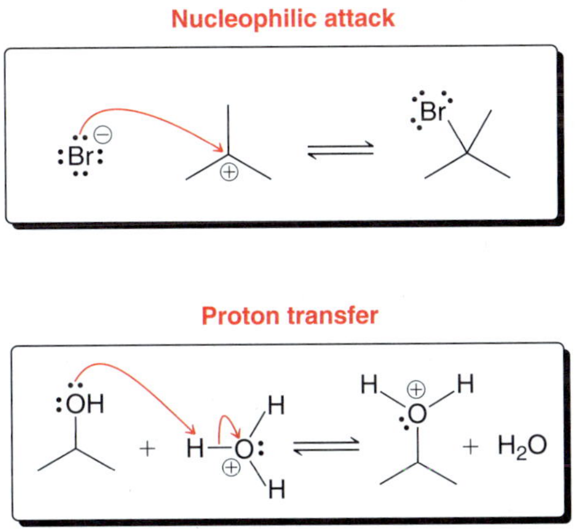

Nucleophilic attack

Loss of a leaving group

Proton transfer

Rearrangement

**FIGURE 7.3**
The four arrow-pushing patterns for ionic processes.

Every substitution reaction exhibits at least two of the four patterns—nucleophilic attack and loss of a leaving group:

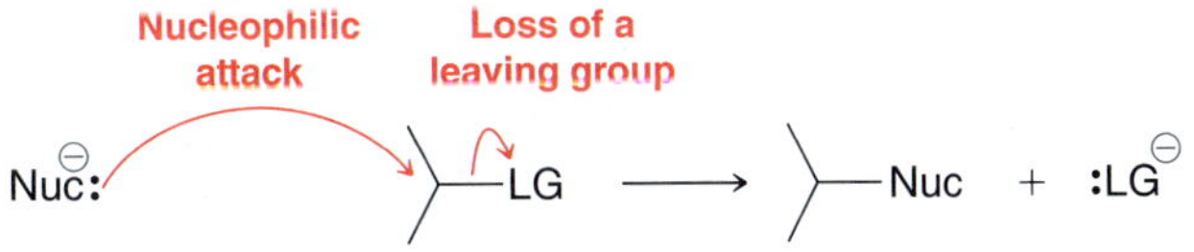

Nucleophilic attack — Loss of a leaving group

But consider the order of these events. Do they occur simultaneously (in a concerted fashion), as shown above, or do they occur in a stepwise fashion, as shown below?

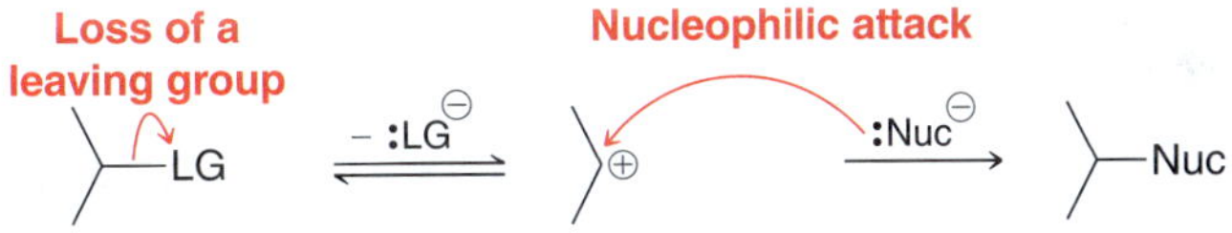

Loss of a leaving group — Nucleophilic attack

In the stepwise mechanism, the leaving group leaves, generating an intermediate carbocation, which is then attacked by the nucleophile. The nucleophile cannot attack before the leaving group leaves, because that would violate the octet rule:

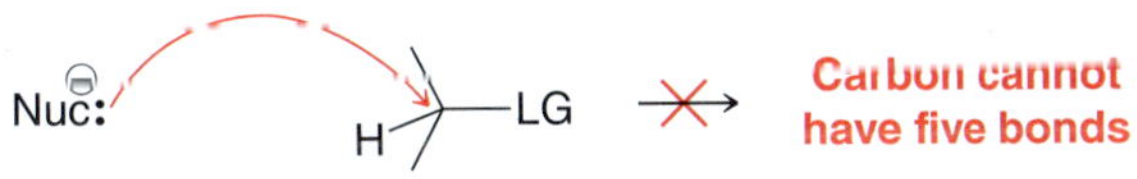

Carbon cannot have five bonds

Therefore, there are two possible mechanisms for a substitution reaction:

- In a *concerted process*, nucleophilic attack and loss of the leaving group occur simultaneously.
- In a *stepwise process*, loss of the leaving group occurs first followed by nucleophilic attack.

We will see that both of these mechanisms do occur, but under different conditions. We will explore each mechanism in the next section, but first let's practice drawing the curved arrows for the two mechanisms.

# SKILLBUILDER

## 7.1 DRAWING THE CURVED ARROWS OF A SUBSTITUTION REACTION

**LEARN the skill**

Below are two substitution reactions. Experimental evidence suggests that the first reaction proceeds via a concerted process, while the second reaction proceeds via a stepwise process. Draw a mechanism for each reaction:

(a) butyl bromide (Br) + NaOH ⟶ butanol (OH) + NaBr

(b) *tert*-butyl bromide (Br) + NaCl ⟶ *tert*-butyl chloride (Cl) + NaBr

### SOLUTION

**(a)** First identify the substrate, the leaving group, and the nucleophile. Here, the substrate is butyl bromide, the leaving group is bromide, and the nucleophile is a hydroxide ion:

**STEP 1** Identify the substrate, leaving group, and nucleophile.

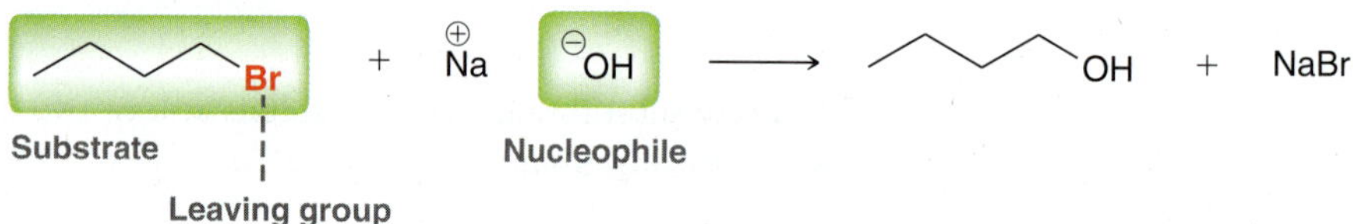


When you see NaOH, remember that the reagent is a hydroxide ion ($HO^{\ominus}$). $Na^{+}$ is the counter-ion, and its role in the reaction does not concern us in most cases. In a concerted process, nucleophilic attack and loss of a leaving group occur simultaneously. This process requires two curved arrows—one to show the nucleophilic attack and one to show the loss of the leaving group. When drawing the first curved arrow, place the tail on a lone pair of the nucleophile and place the head on the carbon atom bearing the leaving group:

**STEP 2** Draw two curved arrows, showing nucleophilic attack and loss of the leaving group.

⊖:OH (Nucleophile) + butyl bromide :Br: (Substrate) ⟶ butanol OH + :Br:⊖

**WATCH OUT** Be very precise in placing the head and tail of every curved arrow.

**(b)** A stepwise process involves two separate mechanistic steps: (1) loss of a leaving group to form a carbocation intermediate followed by (2) nucleophilic attack. To draw these steps, we must identify the substrate, leaving group, and nucleophile. Here, the substrate is *tert*-butyl bromide, the leaving group is a bromide ion, and the nucleophile is a chloride ion:

**STEP 1** Identify the substrate, leaving group, and nucleophile.

*tert*-butyl bromide, Br (Substrate; Leaving group) + ⊕Na ⊖Cl (Nucleophile) ⟶ *tert*-butyl chloride (Cl) + NaBr

The first step of the mechanism requires one curved arrow showing the loss of the leaving group. The tail of this curved arrow is placed on the bond that is broken (the C—Br bond); the head of the arrow is placed on the bromine atom.

**STEP 2** Draw a curved arrow showing loss of the leaving group and then draw the resulting carbocation.

**Loss of a leaving group**

*tert*-butyl bromide :Br: ⇌ ($-Br^{\ominus}$) *tert*-butyl carbocation ⊕

**STEP 3**
Draw a curved arrow showing a nucleophilic attack.

The second step of the mechanism requires one curved arrow showing the nucleophilic attack in which the carbocation intermediate is captured by the nucleophile (chloride):

Nucleophilic attack

The complete mechanism can therefore be drawn like this:

Loss of a leaving group — Nucleophilic attack

**BY THE WAY**
Draw all three groups of a tertiary carbocation as far apart from each other as possible:

not

**PRACTICE the skill**

**7.2** For each of the following reactions, assume a concerted process is taking place and draw the mechanism:

(a) + NaSH ⟶ + NaBr

(b) + NaOMe ⟶ + NaI

**7.3** For each of the following reactions assume a stepwise process is taking place and draw the mechanism:

(a) + ⟶ + $Br^{\ominus}$

(b) + NaCl ⟶ + NaI

**APPLY the skill**

**7.4** When a nucleophile and electrophile are tethered to each other (that is, both present in the same compound), an *intramolecular substitution reaction* can occur, as shown. Assume that this reaction occurs via a concerted process and draw the mechanism.

⟶ + $Br^{\ominus}$

**7.5** For the substitution reaction shown below, assume a stepwise process is taking place and draw the mechanism. (**Hint:** Review the rules for drawing resonance structures, Section 2.10.)

+ NaCl ⟶ + NaBr

need more **PRACTICE?** **Try Problem 7.64a**

## 7.4 The $S_N2$ Mechanism

During the 1930s, Sir Christopher Ingold and Edward D. Hughes (University College, London) investigated substitution reactions in an effort to elucidate their mechanism. Based on kinetic and stereochemical observations, Ingold and Hughes proposed a concerted mechanism for many of the substitution reactions that they investigated. We will now explore the observations that led them to propose a concerted mechanism.

### Kinetics

For most of the reactions that they investigated, Ingold and Hughes found the rate of reaction to be dependent on the concentrations of both the substrate and the nucleophile. This observation is summarized in the following rate equation:

$$\text{Rate} = k\,[\text{substrate}]\,[\text{nucleophile}]$$

Specifically, they found that doubling the concentration of the nucleophile caused the reaction rate to double. Similarly, doubling the concentration of the substrate also caused the rate to double. The rate equation above is described as **second order**, because the rate is linearly dependent on the concentrations of two different compounds. Based on their observations, Ingold and Hughes concluded that the mechanism must exhibit a step in which the substrate and the nucleophile collide with each other. Because that step involves two chemical entities, it is said to be **bimolecular**. Ingold and Hughes coined the term $\mathbf{S_N2}$ to refer to bimolecular substitution reactions:

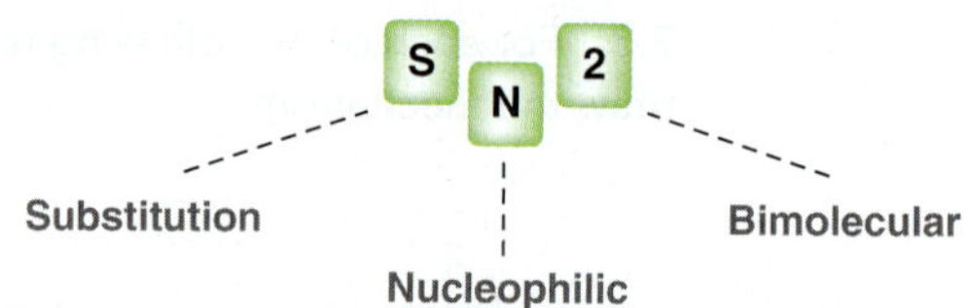


The experimental observations for $S_N2$ reactions are consistent with a concerted mechanism, because a concerted mechanism exhibits only one mechanistic step, involving both the nucleophile and the substrate:

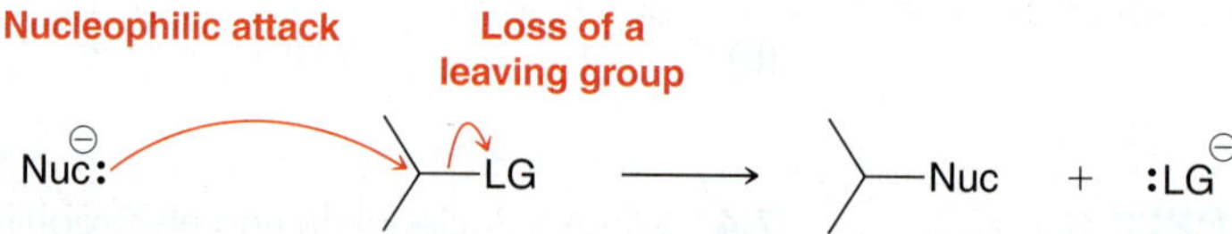


It makes sense that the rate should be dependent on the concentrations of both the nucleophile and the substrate.

### CONCEPTUAL CHECKPOINT

**7.6** The reaction below exhibits a second-order rate equation:

(1-iodopropane) I + NaOH ⟶ (1-propanol) OH + NaI

(a) What happens to the rate if the concentration of 1-iodopropane is tripled and the concentration of sodium hydroxide remains the same?

(b) What happens to the rate if the concentration of 1-iodopropane remains the same and the concentration of sodium hydroxide is doubled?

(c) What happens to the rate if the concentration of 1-iodopropane is doubled and the concentration of sodium hydroxide is tripled?

## Stereospecificity of $S_N2$ Reactions

There is another crucial piece of evidence that led Ingold and Hughes to propose the concerted mechanism. When the α position is a chirality center, a change in configuration is generally observed, as illustrated in the following example:

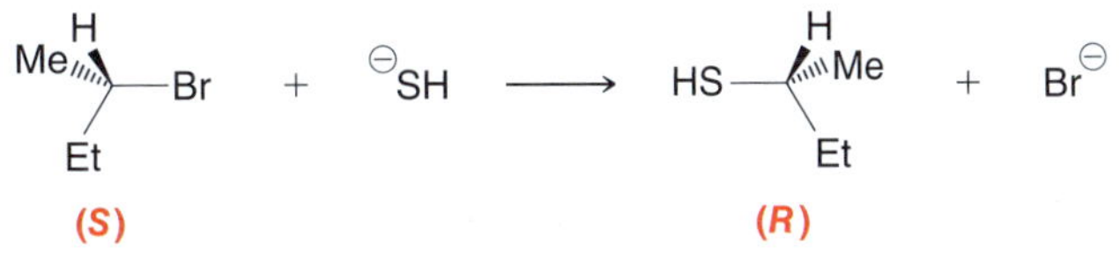

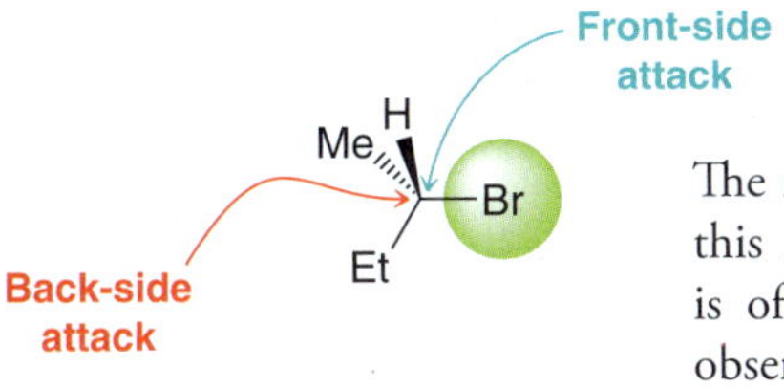

(S) Me, H, Et–C–Br + ⊖SH ⟶ HS–C(H, Me, Et) (R) + Br⊖

The reactant exhibits the *S* configuration, while the product exhibits the *R* configuration. That is, this reaction is said to proceed with **inversion of configuration**. This stereochemical outcome is often called a Walden inversion, named after Paul Walden, the German chemist who first observed it.

Front-side attack / Back-side attack

**FIGURE 7.4** The front and back sides of a substrate.

The requirement for inversion of configuration means that the nucleophile can only attack from the back side (the side opposite the leaving group) and never from the front side (Figure 7.4). There are two ways to explain why the reaction proceeds through **back-side attack**:

1. The lone pairs of the leaving group create regions of high electron density that effectively block the front side of the substrate, so the nucleophile can only approach from the back side.
2. Molecular orbital (MO) theory provides a more sophisticated answer. Recall that molecular orbitals are associated with the entire molecule (as opposed to *atomic* orbitals, which are associated with individual atoms). According to MO theory, the electron density flows from the HOMO of the nucleophile into the LUMO of the electrophile. As an example let's focus our attention on the LUMO of methyl bromide (Figure 7.5). If a nucleophile attacks methyl bromide from the front side, the nucleophile will encounter a node, and as a result, no net bonding will result from the overlap between the HOMO of the nucleophile and the LUMO of the electrophile. In contrast, nucleophilic attack from the back side allows for efficient overlap between the HOMO of the nucleophile and the LUMO of the electrophile.

**LOOKING BACK**
For a review of molecular orbital theory and the terms HOMO and LUMO, see Section 1.8.

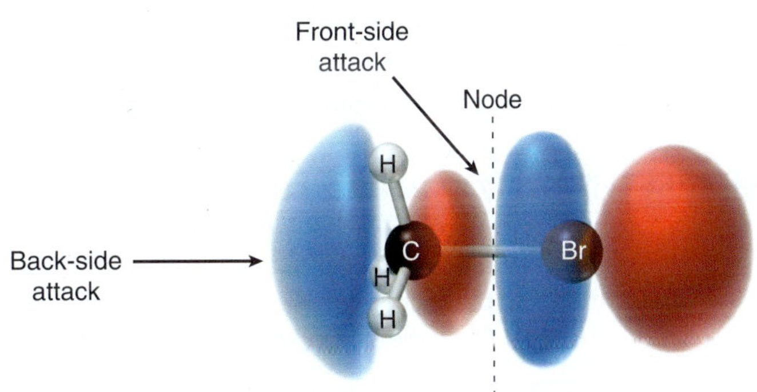


**FIGURE 7.5** The lowest unoccupied molecular orbital (LUMO) of methyl bromide.

The observed stereochemical outcome for an $S_N2$ process (inversion of configuration) is consistent with a concerted mechanism. The nucleophile attacks with simultaneous loss of the leaving group. This causes the chirality center to behave like an umbrella flipping in the wind:

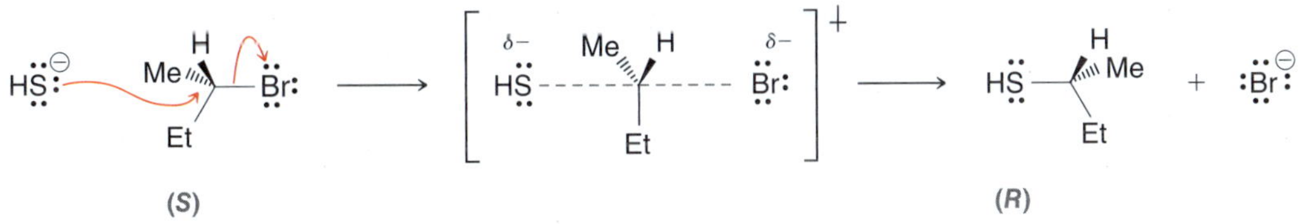

HS:⊖ + (S) Me, H, Et–C–Br: ⟶ [HS(δ−)----C(Me, H, Et)----Br:(δ−)]‡ ⟶ HS–C(H, Me, Et) (R) + :Br:⊖

The transition state (drawn in brackets) will be discussed in more detail in the coming section. This reaction is said to be **stereospecific**, because the configuration of the product is dependent on the configuration of the starting material.

# SKILLBUILDER

## 7.2 DRAWING THE PRODUCT OF AN $S_N2$ PROCESS

**LEARN** the skill

When (*R*)-2-bromobutane is treated with a hydroxide ion, a mixture of products is obtained. An $S_N2$ process is responsible for generating one of the minor products, while the major product is generated via an elimination process, as will be discussed in the next chapter. Draw the $S_N2$ product that is obtained when (*R*)-2-bromobutane reacts with a hydroxide ion.

**SOLUTION**

First draw the reagents described in the problem:

Br + $^{\ominus}$OH ⟶ ?

**(*R*)-2-Bromobutane** **Hydroxide**

Now identify the nucleophile and the substrate. Bromobutane is the substrate and hydroxide is the nucleophile. When hydroxide attacks, it will eject the bromide ion as a leaving group. The net result is that the Br will be replaced with an OH group:

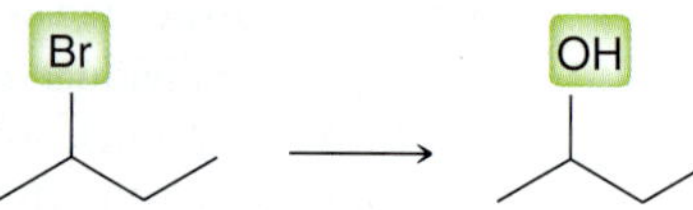


**WATCH OUT**

In an $S_N2$ reaction, if the $\alpha$ position is a chirality center, make sure to draw an inversion of configuration in the product.

In this case, the $\alpha$ position is a chirality center, so we expect inversion:

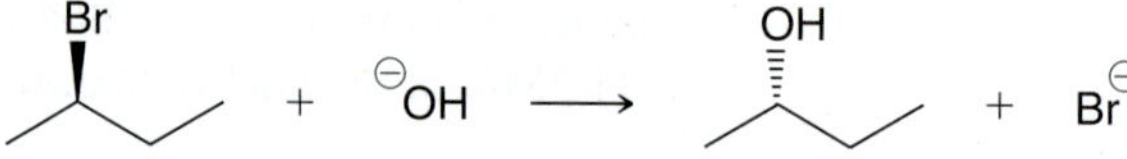

**PRACTICE** the skill

**7.7** Draw the product for each of the following $S_N2$ reactions:

**(a)** (*S*)-2-Chloropentane and NaSH **(b)** (*R*)-3-Iodohexane and NaCl

**(c)** (*R*)-2-Bromohexane and sodium hydroxide

**APPLY** the skill

**7.8** When (*S*)-1-bromo-1-fluoroethane reacts with sodium methoxide, an $S_N2$ reaction takes place in which the bromine atom is replaced by a methoxy group (OMe). The product of this reaction is (*S*)-1-fluoro-1-methoxyethane. How can it be that the starting material and the product both have the *S* configuration? Shouldn't $S_N2$ involve a change in the configuration? Draw the starting material and the product of inversion, and then explain the anomaly.

need more **PRACTICE?** **Try Problems 7.45, 7.56, 7.61**

### Structure of the Substrate

For $S_N2$ reactions, Ingold and Hughes also found the rate to be sensitive to the nature of the starting alkyl halide. In particular, methyl halides and primary alkyl halides react most quickly with nucleophiles. Secondary alkyl halides react more slowly, and tertiary alkyl halides are essentially unreactive toward $S_N2$ (Figure 7.6). This trend is consistent with a concerted process in which the nucleophile is expected to encounter steric hindrance as it approaches the substrate.

To understand the nature of the steric effects that govern $S_N2$ reactions, we must explore the transition state for a typical $S_N2$ reaction, shown in general form in Figure 7.7. Recall that a transition state is represented by a peak in an energy diagram. Consider, for example, an energy diagram showing the reaction between a cyanide ion and methyl bromide (Figure 7.8). The highest point on the curve represents the transition state. The superscript symbol that looks like a telephone pole outside the brackets indicates that the drawing shows a transition state rather than an intermediate. The relative energy of this transition state determines the rate of the reaction. If the transition state is high in energy, then $E_a$

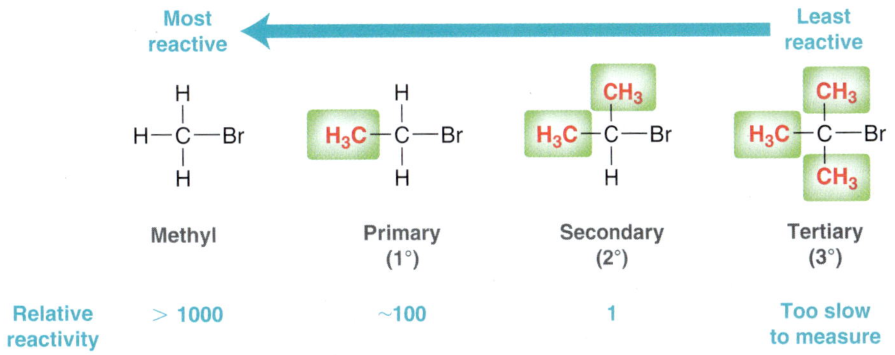

Most reactive ← Least reactive

| | Methyl | Primary (1°) | Secondary (2°) | Tertiary (3°) |
|---|---|---|---|---|
| Substrate | $CH_3Br$ | $H_3C$–$CH_2$–Br | $(H_3C)_2CH$–Br | $(H_3C)_3C$–Br |
| Relative reactivity | > 1000 | ~100 | 1 | Too slow to measure |

**FIGURE 7.6** The relative reactivity of various substrates toward $S_N2$.

will be large, and the rate will be slow. If the transition state is low in energy, then $E_a$ will be small, and the rate will be fast. With this in mind, we can now explore the effects of steric hindrance in slowing down the reaction rate and explain why tertiary substrates are unreactive.

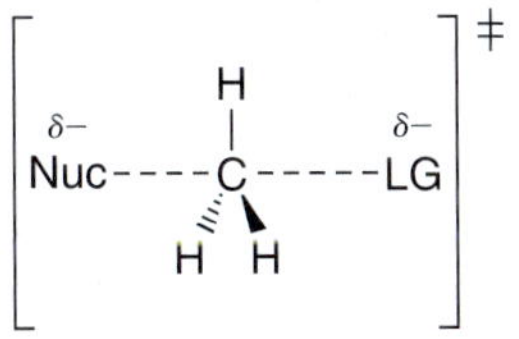

[ $\delta-$ Nuc----C(H)(H)(H)----LG $\delta-$ ]$^\ddagger$

**FIGURE 7.7** The generic form of a transition state in an $S_N2$ process.

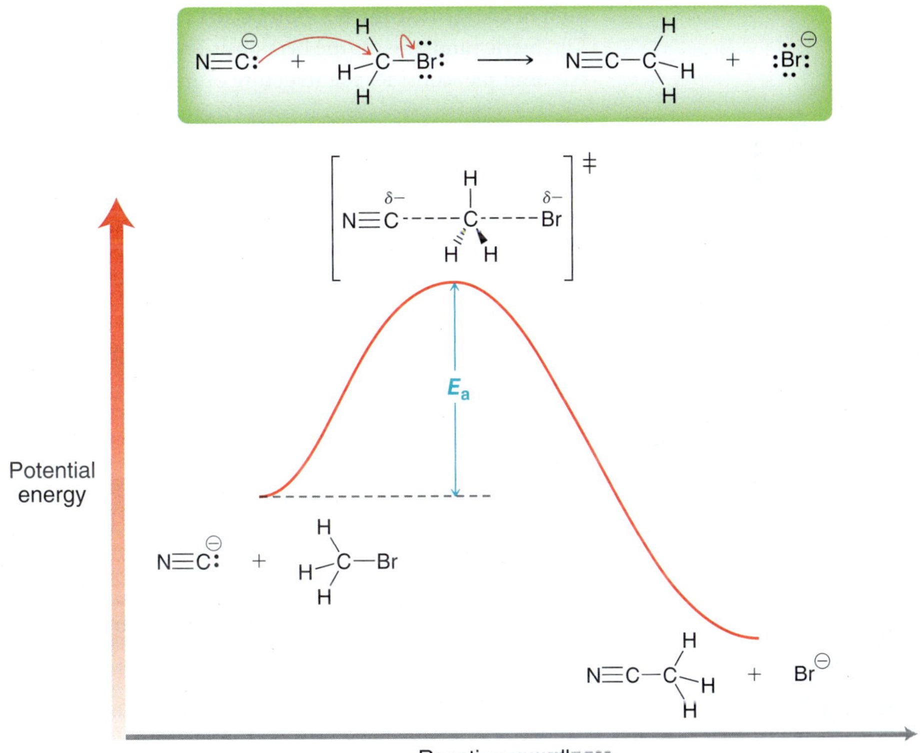


**FIGURE 7.8** An energy diagram of the $S_N2$ reaction that occurs between methyl bromide and a cyanide ion.

Take a close look at the transition state. The nucleophile is in the process of forming a bond with the substrate, and the leaving group is in the process of breaking its bond with the substrate. Notice that there is a partial negative charge on either side of the transition state. This can be seen more clearly in an electrostatic potential map of the transition state (Figure. 7.9). If the hydrogen atoms

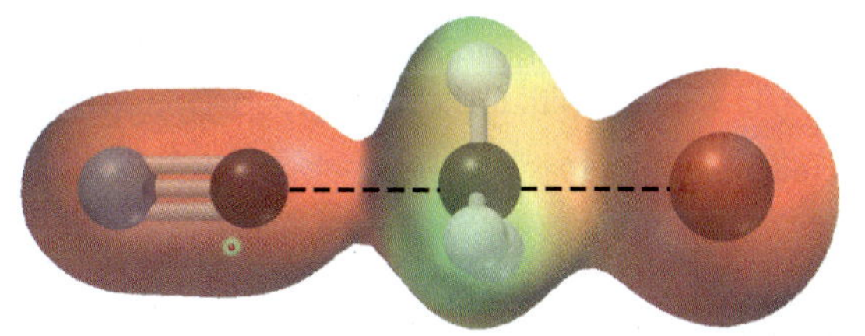


**FIGURE 7.9** An electrostatic potential map of the transition state from Figure 7.7. The red areas represent regions of high electron density.

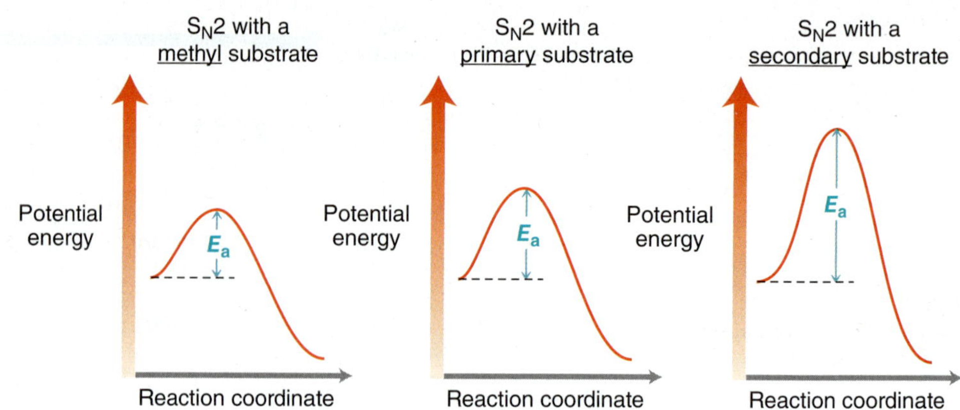


**FIGURE 7.10** Energy diagrams comparing $S_N2$ processes for methyl, primary, and secondary substrates.

in Figure 7.9 are replaced with alkyl groups, steric interactions cause the transition state to be higher in energy, raising $E_a$ for the reaction. Compare the energy diagrams for reactions involving methyl, primary, and secondary substrates (Figure 7.10). With a tertiary substrate, the transition state is so high in energy that the reaction occurs too slowly to observe.

Steric hindrance at the beta position can also decrease the rate of reaction. For example, consider the structure of neopentyl bromide:

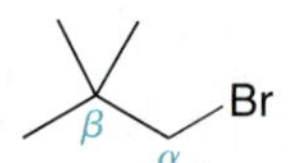


Neopentyl bromide

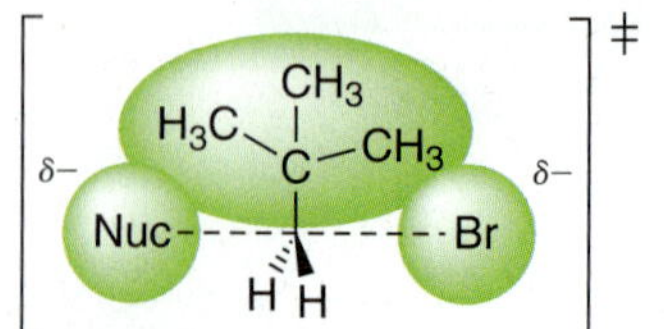


**FIGURE 7.11** The transition state for an $S_N2$ process involving a neopentyl substrate.

This compound is a primary alkyl halide, but it has three methyl groups attached to the beta position. These methyl groups provide steric hindrance that causes the energy of the transition state to be very high (Figure 7.11). Once again, the rate is very slow. In fact, the rate of a neopentyl substrate is similar to the rate of a tertiary substrate in $S_N2$ reactions. This is an interesting example, because the substrate is a primary alkyl halide that essentially does not undergo an $S_N2$ reaction. This example illustrates why it is best to understand concepts in organic chemistry rather than memorize rules without knowing what they mean.

## SKILLBUILDER

### 7.3 DRAWING THE TRANSITION STATE OF AN $S_N2$ PROCESS

**LEARN the skill** Draw the transition state of the following reaction:

**SOLUTION**

**STEP 1** Identify the nucleophile and the leaving group.

First identify the nucleophile and the leaving group. These are the two groups that will be on either side of the transition state:

Nucleophile Leaving group

**STEP 2** Draw a carbon atom connected by dotted lines to the nucleophile and the leaving group.

The transition state will need to show a bond forming with the nucleophile and a bond breaking with the leaving group. Dotted lines are used to show the bonds that are breaking and forming:

The δ− symbol is placed on both the incoming nucleophile and the outgoing leaving group to indicate that the negative charge is spread out over both locations.

Now we must draw all of the alkyl groups connected to the α position. In our example, the α position has one $CH_3$ group and two H's:

**STEP 3**
Draw the three groups attached to the carbon atom.

Cl = $H_3C$ (α, H, H) Cl

So we draw these groups in the transition state connected to the α position. One group is placed on a straight line, and the other two groups are placed on a wedge and on a dash:

**STEP 4**
Place brackets as well as the symbol that indicates a transition state.

[ δ− HS - - - C($CH_3$)(H)(H) - - - Cl δ− ]‡

It does not matter whether the $CH_3$ group is placed on the line, wedge, or dash. But don't forget to indicate that the drawing is a transition state by surrounding it with brackets and using the symbol that indicates a transition state.

**PRACTICE the skill** **7.9** Draw the transition state for each of the following $S_N2$ reactions:

(b) I + O, O⊖ ⟶ O, O + I⊖

(c) Cl + NaOH ⟶ OH + NaCl

(d) Br + NaSH ⟶ SH + NaBr

**7.10** In Problem 7.4, we saw that an *intramolecular* substitution reaction can occur when the nucleophilic center and electrophilic center are present in the same compound. Draw the transition state of the reaction in Problem 7.4.

**7.11** Treatment of 5-hexen-1-ol with bromine affords a cyclic product:

HO ⟶ ($Br_2$) Br, O + NaBr

The mechanism of this reaction involves several steps, one of which is an intramolecular $S_N2$-like process:

HÖ Br⊕ ⟶ O⊕ H Br:

In this step, a bond is in the process of breaking, while another bond is in the process of forming. Draw the transition state of this $S_N2$-like process, and identify which bond is being broken and which bond is being formed. Can you offer an explanation as to why this step is favorable?

need more **PRACTICE?** **Try Problems 7.46, 7.64e**

## $S_N2$ Reactions in Biological Systems—Methylation

In the laboratory, the transfer of a methyl group is accomplished via an $S_N2$ process using methyl iodide:

$$Nuc:^{\ominus} + H_3C{-}I \longrightarrow Nuc{-}CH_3 + :I:^{\ominus}$$

This process is called *alkylation*, because an alkyl group has been transferred to the nucleophile. It is an $S_N2$ process, which means there are limitations on the type of alkyl group that can be used. Tertiary alkyl groups cannot be transferred. Secondary alkyl groups can be transferred, but slowly. Primary alkyl groups and methyl groups are transferred most readily. The alkylation process shown above is the transfer of a methyl group and is therefore called *methylation*. Methyl iodide is ideally suited for this task, because iodide is an excellent leaving group and because methyl iodide is a liquid at room temperature. This makes it easier to work with than methyl chloride or methyl bromide, which are gases at room temperature.

Methylation reactions also occur in biological systems, but instead of $CH_3I$, the methylating agent is a compound called SAM (*S*-adenosylmethionine). Your body produces SAM via an $S_N2$ reaction between ATP and the amino acid methionine:

Leaving group · Nucleophile · Adenosine triphosphate (ATP) · Methionine · *S*-Adenosylmethionine (SAM)

In this reaction, methionine acts as a nucleophile and attacks adenosine triphosphate (ATP), kicking off a triphosphate leaving group. The resulting product, called SAM, is able to function as a methylating agent, very much like $CH_3I$. Both $CH_3I$ and SAM exhibit a methyl group attached to an excellent leaving group.

| Methyl iodide | *S*-Adenosylmethionine (SAM) |
| --- | --- |
| $H_3C{-}I$ <br> Iodide is a relatively simple leaving group. | This leaving group is more complex. |

SAM is the biological equivalent of $CH_3I$. The leaving group is much larger, but SAM functions in the same way as $CH_3I$. When SAM is attacked by a nucleophile, an excellent leaving group is expelled:

$$Nuc:^{\ominus} + H_3C{-}S^{\oplus}R_2 \longrightarrow Nuc{-}CH_3 + SR_2$$

SAM plays a role in the biosynthesis of many compounds, such as adrenaline, which is released into the bloodstream in response to danger or excitement. Adrenaline is produced via a methylation reaction that takes place between noradrenaline and SAM in the adrenal gland:

Noradrenaline · (SAM) · $-H^+$ · Adrenaline (after loss of a proton)

After being released into the bloodstream, adrenaline increases heart rate, elevates sugar levels to provide a boost of energy, and increases levels of oxygen reaching the brain. These physiological responses prepare the body for "fight or flight."

## CONCEPTUAL CHECKPOINT

**7.12** Nicotine is an addictive compound found in tobacco, and choline is a compound involved in neurotransmission. The biosynthesis of each of these compounds involves the transfer of a methyl group from SAM. Draw a mechanism for both of these transformations:

(a) [structure] $\xrightarrow{\text{SAM}}$ Nicotine

(b) [structure] $\xrightarrow{\text{SAM}}$ Choline

# 7.5 The $S_N1$ Mechanism

The second possible mechanism for a substitution reaction is a stepwise process in which there is (1) loss of the leaving group to form a carbocation intermediate followed by (2) nucleophilic attack on the carbocation intermediate:

Loss of a leaving group (− :LG⁻) → Intermediate carbocation → Nucleophilic attack (:Nuc⁻) → product with Nuc

An energy diagram for this type of process (Figure 7.12) is expected to exhibit two humps, one for each step. Notice that the first hump is taller than the second hump, indicating that the transition state for the first step is higher in energy than the transition state for the second step. This is extremely important, because for any process, the step with the highest energy transition state determines the rate of the overall process. That step is therefore called the **rate-determining step** (RDS). In Figure 7.12, we can see that the rate-determining step is loss of the leaving group.

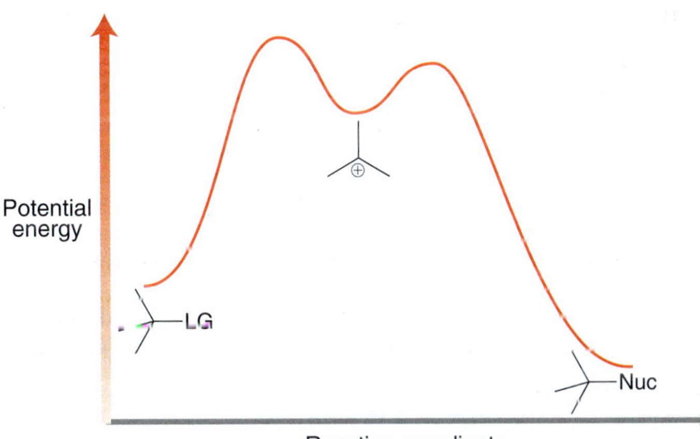


**FIGURE 7.12** An energy diagram of an $S_N1$ process.

Many substitution reactions appear to follow this stepwise mechanism. There are several pieces of evidence that support this stepwise mechanism in those cases. This evidence will now be explored.

## Kinetics

Many substitution reactions do not exhibit second-order kinetics. Consider the following example:

tert-butyl iodide + NaBr ⟶ tert-butyl bromide + NaI

In the reaction above, the rate is dependent only on the concentration of the substrate. The rate equation has the following form:

$$\text{Rate} = k\,[\text{substrate}]$$

Increasing or decreasing the concentration of the nucleophile has no measurable effect on the rate. The rate equation is said to be **first order**, because the rate is linearly dependent on the concentration of only one compound. In such cases, the mechanism must exhibit a rate-determining step in which the nucleophile does not participate. Because that step involves only one chemical entity, it is said to be **unimolecular**. Ingold and Hughes coined the term $\mathbf{S_N1}$ to refer to unimolecular substitution reactions:

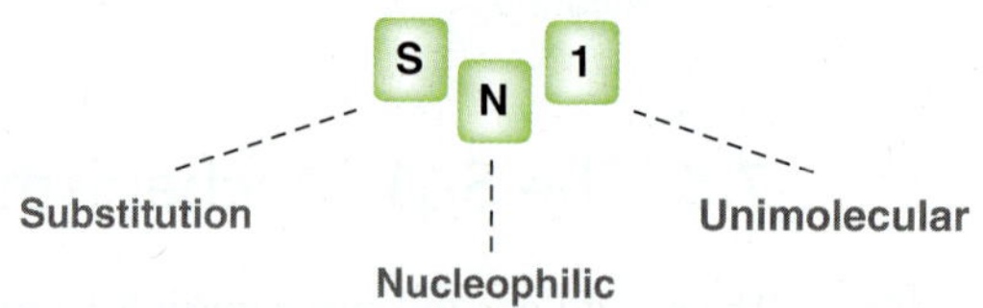


When we use the term unimolecular, we don't mean that the nucleophile is completely irrelevant. Clearly, the nucleophile is necessary, or there won't be a reaction. The term *unimolecular* simply describes the fact that only one chemical entity participates in the rate-determining step of the reaction, and as a result, the rate of the reaction is not affected by how much nucleophile is present. That is, the rate of an $S_N1$ process is dependent only on the rate at which the leaving group leaves. As a result, the rate of an $S_N1$ process will only be affected by factors that affect the rate of that step. Increasing the concentration of the nucleophile has no impact on the rate at which the leaving group leaves. It is true that the nucleophile must be present in order to obtain the product, but an excess of nucleophile will not speed up the reaction. A unimolecular substitution reaction is therefore consistent with a stepwise mechanism in which the first step is the rate-determining step.

## CONCEPTUAL CHECKPOINT

**7.13** The following reaction occurs via an $S_N1$ mechanistic pathway:

(tert-butyl iodide) —NaCl→ (tert-butyl chloride) + NaI

(a) What happens to the rate if the concentration of *tert*-butyl iodide is doubled and the concentration of sodium chloride is tripled?

(b) What happens to the rate if the concentration of *tert*-butyl iodide remains the same and the concentration of sodium chloride is doubled?

## Structure of Substrate

The rate of an $S_N1$ reaction is highly dependent on the nature of the substrate, but the trend is the reverse of the trend we saw for $S_N2$ reactions. With $S_N1$ reactions, tertiary substrates react most quickly, while methyl and primary substrates are mostly unreactive (Figure 7.13). This observation supports a stepwise mechanism ($S_N1$). Why? With $S_N2$ reactions, steric hindrance was the issue

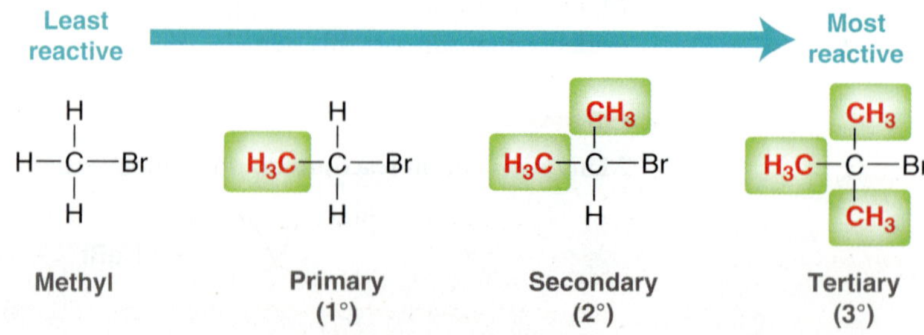


**FIGURE 7.13** The relative reactivity of various substrates toward $S_N1$.

because the nucleophile was directly attacking the substrate. In contrast, in $S_N1$ reactions the nucleophile does not attack the substrate directly. Instead, the leaving group leaves first, resulting in the formation of a carbocation, and that step is the rate-determining step. Once the carbocation forms, the nucleophile captures it very quickly. The rate is only dependent on how quickly the leaving group leaves to form a carbocation. Steric hindrance is not at play, because the rate-determining step does not involve nucleophilic attack. The dominant factor now becomes carbocation stability.

Recall that carbocations are stabilized by neighboring alkyl groups (Figure 7.14).

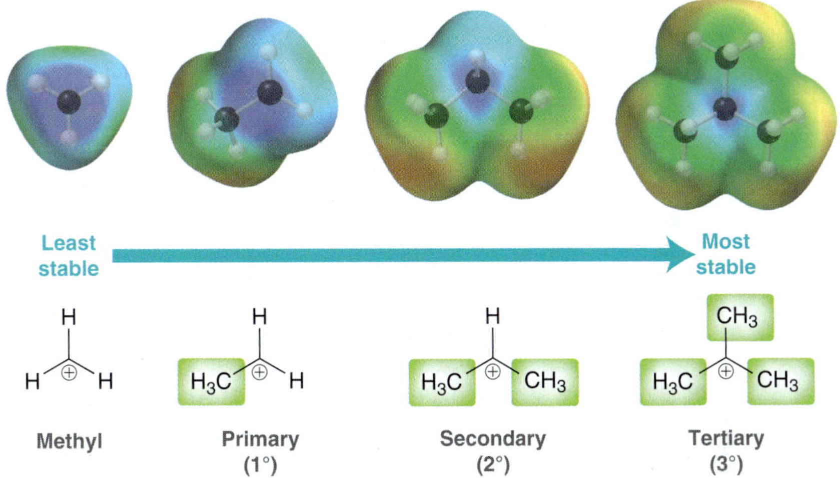


**FIGURE 7.14**
Electrostatic potential maps of various carbocations. The alkyl groups help spread the positive charge, thereby stabilizing the charge.

Tertiary carbocations are more stable than secondary carbocations, which are more stable than primary carbocations. Therefore, formation of a tertiary carbocation will have a smaller $E_a$ than formation of a secondary carbocation (Figure 7.15). The larger $E_a$ associated with formation of a secondary carbocation can be explained by the Hammond postulate (Section 6.6). Specifically, the transition state for formation of a tertiary carbocation will be close in energy to a tertiary carbocation, while the transition state for formation of a secondary carbocation will be close in energy to a secondary carbocation. Therefore, formation of a tertiary carbocation will involve a smaller $E_a$.

The bottom line is that tertiary substrates generally undergo substitution via an $S_N1$ process, while primary substrates generally undergo substitution via an $S_N2$ process. Secondary substrates can proceed via either pathway ($S_N1$ *or* $S_N2$) depending on other factors, which are discussed later in this chapter.

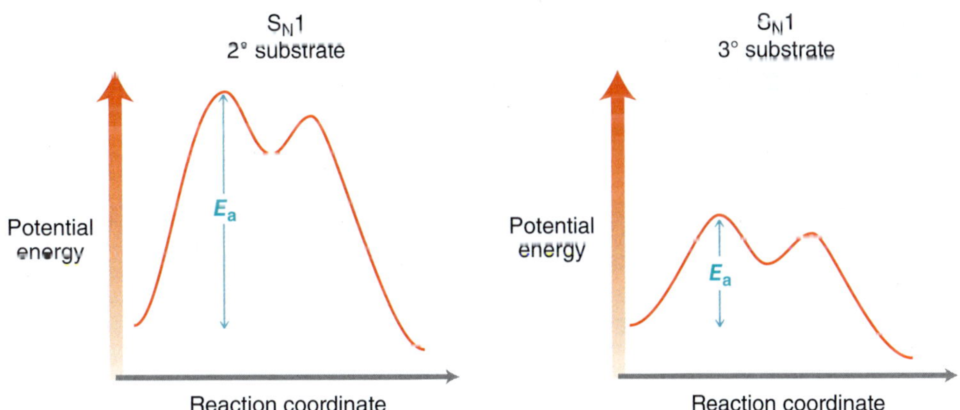


**FIGURE 7.15**
Energy diagrams comparing $S_N1$ processes for secondary and tertiary substrates.

# SKILLBUILDER

## 7.4 DRAWING THE CARBOCATION INTERMEDIATE OF AN $S_N1$ PROCESS

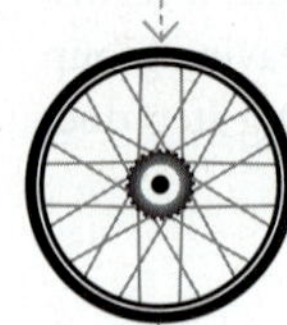

**LEARN** the skill

Draw the carbocation intermediate of the following $S_N1$ reaction:

**SOLUTION**

First identify the leaving group:

**STEP 1**
Identify the leaving group.

Loss of the leaving group will produce a carbocation and a chloride ion. To keep track of the electrons, it is helpful to draw the curved arrow that shows the flow of electrons:

When drawing the carbocation intermediate, make sure that all three groups on the carbocation are drawn as far apart as possible. Remember that a carbocation has trigonal planar geometry, and the drawing should reflect that:

**STEP 2**
Draw all three groups pointing away from each other.

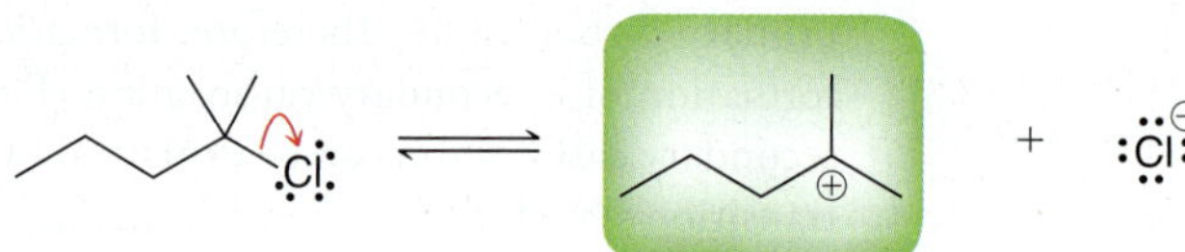

**PRACTICE** the skill

**7.14** Draw the carbocation intermediate generated by each of the following substrates in an $S_N1$ reaction:

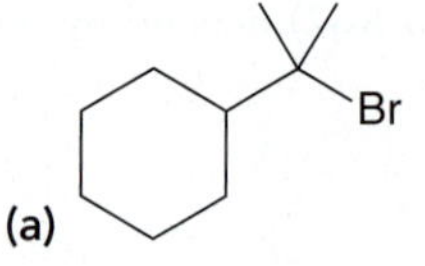


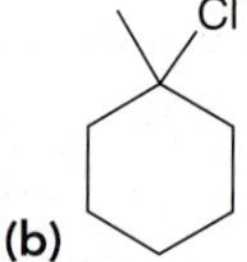

(d) Br

**APPLY** the skill

**7.15** Identify which of the following substrates will undergo an $S_N1$ reaction more rapidly. Explain your choice.

Br or Br

need more **PRACTICE?** Try Problems 7.50, 7.51

## Stereochemistry of $S_N1$ Reactions

Recall that $S_N2$ reactions proceed via an inversion of configuration:

(*S*)-2-Bromobutane + NaCl ⟶ (*R*)-2-Chlorobutane + NaBr

In contrast, $S_N1$ reactions involve formation of an intermediate carbocation, which can then be attacked from either side (Figure 7.16), leading to both inversion of configuration and **retention of configuration**.

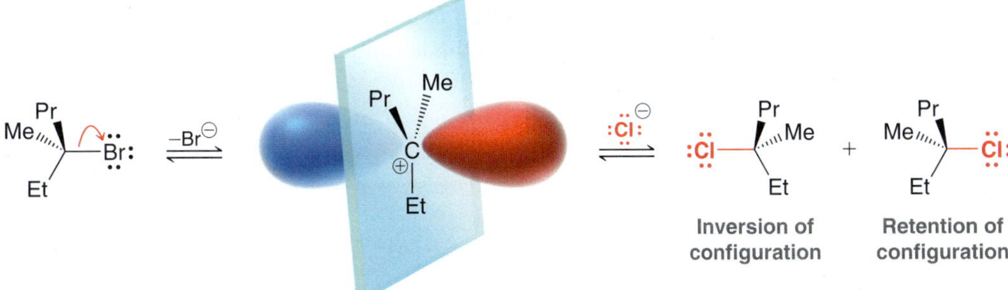


A carbocation is planar, and either side of the plane can be attacked by the nucleophile with equal likelihood.

**FIGURE 7.16** The intermediate of an $S_N1$ process is a planar carbocation.

Since the carbocation can be attacked on either side with equal likelihood, we should expect $S_N1$ reactions to produce a racemic mixture (equal mixture of inversion and retention). In practice, though, $S_N1$ reactions rarely produce exactly equal amounts of inversion and retention products. There is usually a slight preference for the inversion product. The accepted explanation involves the formation of ion pairs. When the leaving group first leaves, it is initially very close to the intermediate carbocation, forming an intimate ion pair (Figure 7.17). If the nucleophile attacks the carbocation while it is still participating in an ion pair, then the leaving group effectively blocks one face of the carbocation. The other side of the carbocation can experience unhindered attack by a nucleophile. As a result, the nucleophile will attack more often on the side opposite the leaving group, leading to a slight preference for inversion over retention.

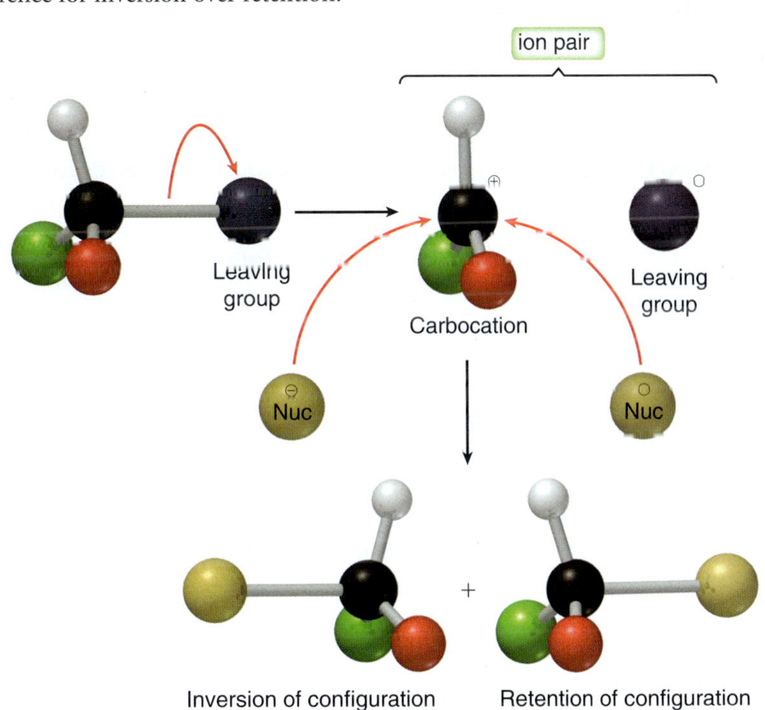


**FIGURE 7.17** Loss of a leaving group initially forms an ion pair, which hinders attack on one face of the carbocation.

# SKILLBUILDER

## 7.5 DRAWING THE PRODUCTS OF AN $S_N1$ PROCESS

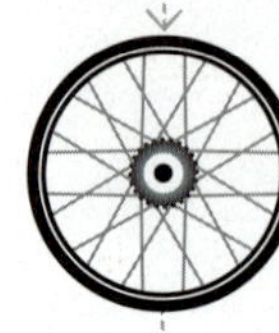

**LEARN the skill**

Draw the products of the following $S_N1$ reaction:

I NaBr ?

### SOLUTION

First identify the leaving group and the nucleophile that will attack once the leaving group has left:

**STEP 1** Identify the nucleophile and the leaving group.

Leaving group

Nucleophile

In an $S_N1$ process, the leaving group leaves first, generating a carbocation that is then attacked by the nucleophile:

**STEP 2** Replace the leaving group with the nucleophile.

$-I^{\ominus}$ :Br: Br

In this example, substitution is taking place at a chirality center, so we must consider the stereochemical outcome. In an $S_N1$ process, both enantiomers are expected as products, with a slight preference for the enantiomer resulting from inversion of configuration:

**STEP 3** If the reaction takes place at a chirality center, draw both possible enantiomers.

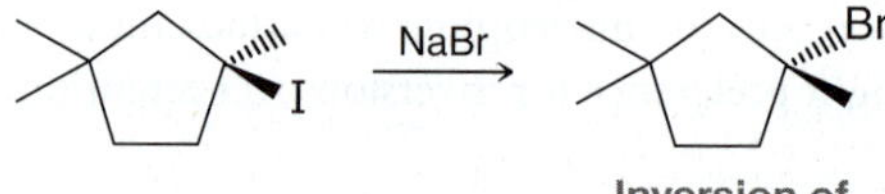


Inversion of configuration > 50%

Retention of configuration < 50%

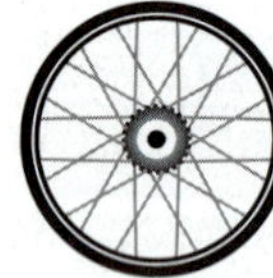

**PRACTICE the skill** **7.16** Draw the products that you expect in each of the following $S_N1$ reactions:

(a)

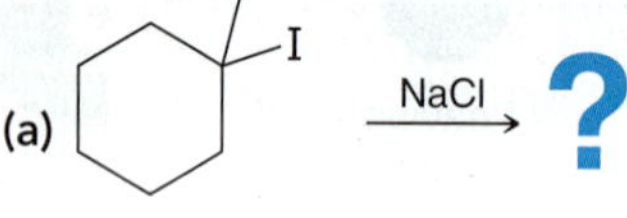


(b) Br ⊖SH ?

(c)

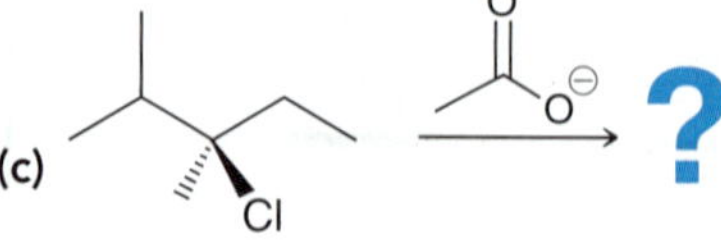

**APPLY the skill** **7.17** Draw the two products that you expect in the following $S_N1$ reaction and describe their stereoisomeric relationship:

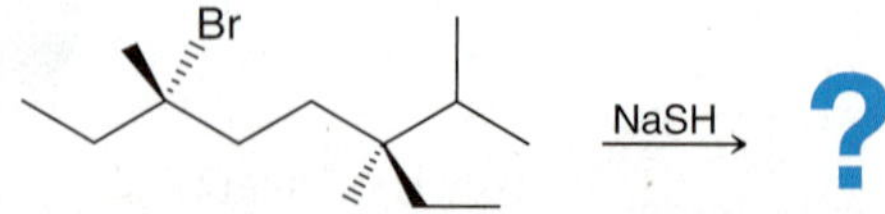


need more **PRACTICE?** Try Problem 7.54b

Let's now summarize the differences that we have seen between $S_N2$ and $S_N1$ processes (Table 7.1).

**TABLE 7.1 A COMPARISON OF $S_N2$ AND $S_N1$ PROCESSES**

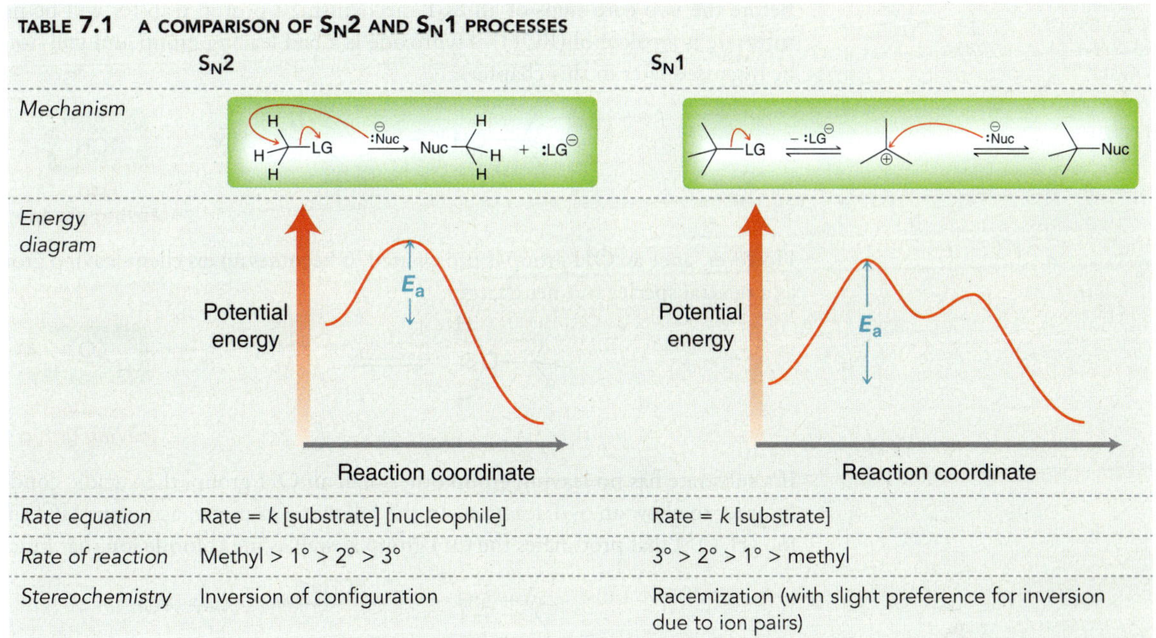


| | $S_N2$ | $S_N1$ |
|---|---|---|
| *Mechanism* | | |
| *Energy diagram* | | |
| *Rate equation* | Rate = $k$ [substrate] [nucleophile] | Rate = $k$ [substrate] |
| *Rate of reaction* | Methyl > 1° > 2° > 3° | 3° > 2° > 1° > methyl |
| *Stereochemistry* | Inversion of configuration | Racemization (with slight preference for inversion due to ion pairs) |

## 7.6 Drawing the Complete Mechanism of an $S_N1$ Reaction

We have now seen that substitution reactions can occur through either a concerted mechanism ($S_N2$) or a stepwise mechanism ($S_N1$) (Figure 7.18). When drawing the mechanism of an $S_N2$ or $S_N1$ process, additional mechanistic steps will sometimes be required. In this section, we will focus on the additional steps that can accompany an $S_N1$ process. Recall from Chapter 6 that ionic mechanisms are constructed using only four different types of arrow-pushing patterns. This will now be important, as all four patterns can play a role in $S_N1$ processes.

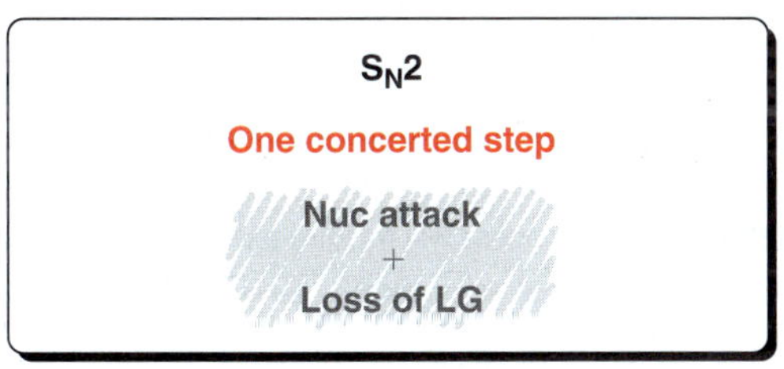


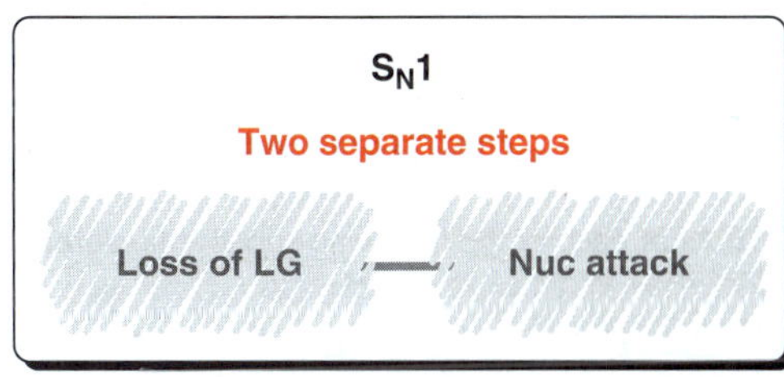


**FIGURE 7.18** The mechanistic steps in $S_N2$ and $S_N1$ processes.

As seen in Figure 7.18, every $S_N1$ mechanism exhibits two separate steps: (1) loss of a leaving group and (2) nucleophilic attack. In addition to these two core steps, some $S_N1$ processes are also accompanied by additional steps (highlighted in blue in Figure 7.19), which can occur before, between, or after the two core steps:

1. *Before the two core steps*—a proton transfer step is possible.
2. *Between the two core steps*—a carbocation rearrangement is possible.
3. *After the two core steps*—a proton transfer step is possible.

Proton transfer — Loss of LG — Carbocation rearrangement — Nuc attack — Proton transfer

**FIGURE 7.19** The two core steps (gray) and the three possible additional steps (blue) that can accompany an $S_N1$ process.

We will now explore each of these three possibilities, and we will learn how to determine whether any of the three additional steps should be included when proposing the mechanism for a transformation that occurs via an $S_N1$ process.

## Proton Transfer at the Beginning of an $S_N1$ Process

+H⁺ — −LG — Nuc attack

Before the two core steps of an $S_N1$ mechanism, a proton transfer will be necessary whenever the substrate is an alcohol (ROH). Hydroxide is a bad leaving group and will not leave by itself (as will be discussed later in this chapter):

OH ⟶ + :ÖH⊖

Bad leaving group

However, once an OH group is protonated, it becomes an excellent leaving group because it can leave as a neutral species (no net charge):

Excellent leaving group

If a substrate has no leaving group other than an OH group, then acidic conditions will be required in order to allow an $S_N1$ reaction. In the following example, acqueous HCl supplies the hydronium ion ($H_3O^+$) that protonates the OH group as well as the chloride ion that functions as a nucleophile:

HCl (aq)

Proton transfer

Loss of leaving group

$-H_2O$

Nucleophilic attack

Notice that the two core steps of this $S_N1$ process are preceded by a proton transfer, giving a total of three mechanistic steps:

+ H⁺ — −LG — Nuc attack

### CONCEPTUAL CHECKPOINT

**7.18** For each of the following substrates, determine whether an $S_N1$ process will require a proton transfer at the beginning of the mechanism:

(a) I (b) OH (c) Br (d) OH (e) OH (f) Cl

## Proton Transfer at the End of an $S_N1$ Process

−LG — Nuc attack — −H⁺

After the two core steps of an $S_N1$ mechanism, a proton transfer will be necessary whenever the nucleophile is neutral (not negatively charged). For example:

Loss of leaving group

$-Cl^{\ominus}$

Nucleophilic attack

Proton transfer

In this case, the nucleophile is water ($H_2O$), which does not possess a negative charge. In such a case, nucleophilic attack of the carbocation will produce a positively charged species. Removal of the positive charge requires a proton transfer. Notice that the mechanism above has three steps:

–LG — Nuc attack — $-H^+$

Any time the attacking nucleophile is neutral, a proton transfer is necessary at the end of the mechanism. Below is one more example. Reactions like this, in which the solvent functions as the nucleophile, are called **solvolysis** reactions.

Loss of leaving group | Nucleophilic attack | Proton transfer

## CONCEPTUAL CHECKPOINT

**7.19** Will an $S_N1$ process involving each of the following nucleophiles require a proton transfer at the end of the mechanism?

(a) NaSH (b) $H_2S$ (c) $H_2O$ (d) EtOH (e) NaCN (f) NaCl
(g) $NaNH_2$ (h) $NH_3$ (i) NaOMe (j) NaOEt (k) MeOH (l) KBr

## Carbocation Rearrangements during an $S_N1$ Process

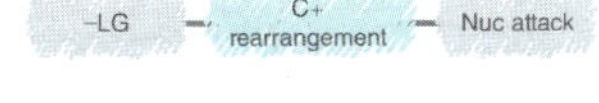


The first core step of an $S_N1$ process is loss of a leaving group to generate a carbocation. Recall from Chapter 6 that carbocations are susceptible to rearrangement via either a hydride shift or a methyl shift. Here is an example of an $S_N1$ mechanism with a carbocation rearrangement:

NaCl | Loss of leaving group | Carbocation rearrangement | Nucleophilic attack | 2° carbocation | 3° carbocation

Notice that the carbocation rearrangement occurs between the two core steps of the $S_N1$ process:

–LG — C+ rearrangement — Nuc attack

In reactions where a carbocation rearrangement is possible, a mixture of products is generally obtained. The following products are obtained from the above reaction:

Br → NaCl → Cl + Cl

**This is the product if the carbocation is captured by the nucleophile *before* rearrangement.**

**This is the product if the carbocation is captured by the nucleophile *after* rearrangement.**

The product distribution (ratio of products) depends on how fast the rearrangement takes place and how fast the nucleophile attacks the carbocation. If the rearrangement occurs faster than attack by the nucleophile, then the rearranged product will predominate. However, if the nucleophile attacks the carbocation faster than rearrangement (if it attacks before rearrangement occurs), then the unrearranged product will predominate. In most cases, the rearranged product predominates. Why? A carbocation rearrangement is an intramolecular process, while nucleophilic attack is an intermolecular process. In general, intramolecular processes occur more rapidly than intermolecular processes.

## CONCEPTUAL CHECKPOINT

**7.20** For each of the following substrates, determine whether an $S_N1$ process is likely to involve a carbocation rearrangement or not:

(a) I (b) OH (c) OH (d) OH (e) Br (f) Cl

## Summary of the $S_N1$ Process and Its Energy Diagram

+H⁺ — −LG — C+ rearrangement — Nuc attack — −H⁺

We have seen that an $S_N1$ process has two core steps and can be accompanied by three additional steps, as summarized in Mechanism 7.1.

## MECHANISM 7.1 THE $S_N1$ PROCESS

**Two core steps**

Loss of LG — Nuc attack

LG ⇌ (−:LG⁻) carbocation ⇌ (:Nuc⁻) Nuc

**In the first core step of an $S_N1$ process, the leaving group leaves to form a carbocation.**

**In the second core step of an $S_N1$ process, a nucleophile attacks the carbocation.**

***Possible additional steps***

Proton transfer — Loss of LG — Carbocation rearrangement — Nuc attack — Proton transfer

**If the substrate is an alcohol, then the OH group must be protonated before it can leave.**

**If the carbocation initially formed can rearrange to generate a more stable carbocation, then a rearrangement will occur.**

**If the nucleophile is neutral, a proton transfer is required to remove the positive charge that is generated.**

Here is an example of an $S_N1$ process that is accompanied by all three additional steps:

$H_3O^+$/MeOH

Proton transfer

Loss of leaving group $-H_2O$

Carbocation rearrangement

Nucleophilic attack

Proton transfer

2° carbocation

3° carbocation

Since this mechanism has five steps, we expect the energy diagram for this reaction to exhibit five humps (Figure 7.20). The number of humps in the energy diagram of an $S_N1$ process will always be equal to the number of steps in the mechanism. Since the number of steps can range anywhere from two to five, the energy diagram of an $S_N1$ process can have anywhere from two to five humps. The $S_N1$ processes encountered most frequently will have two or three steps.

A few aspects of the energy diagram in Figure 7.21 are worth special mention:

- The tertiary carbocation is lower in energy than the secondary carbocation.
- The $E_a$ for the carbocation rearrangement is shown to be very small because a carbocation rearrangement is generally a very fast process.
- Oxonium ions (intermediates with a positively charged oxygen atom) are generally lower in energy than carbocations (because the oxygen atom of an oxonium ion has an octet of electrons, while the carbon atom of a carbocation does not have an octet).

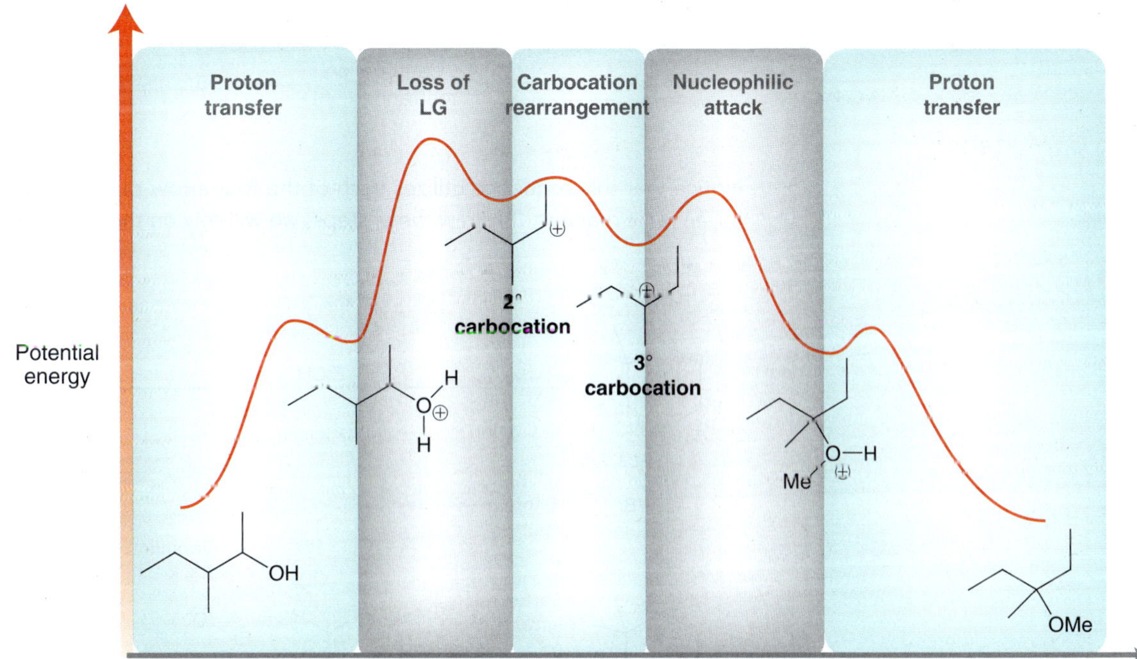


**FIGURE 7.20**
An energy diagram of an $S_N1$ process that is accompanied by three additional steps. In total, there are five steps, giving an energy diagram with five humps.

# SKILLBUILDER

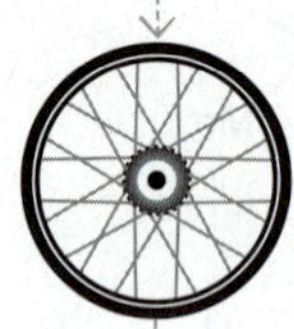

## 7.6 DRAWING THE COMPLETE MECHANISM OF AN $S_N1$ PROCESS

**LEARN** the skill

Draw the mechanism of the following $S_N1$ process:

Br → (EtOH) → OEt

### SOLUTION

An $S_N1$ process must always exhibit two core steps: loss of a leaving group and nucleophilic attack. But we must consider whether any of the other three possible steps will occur:

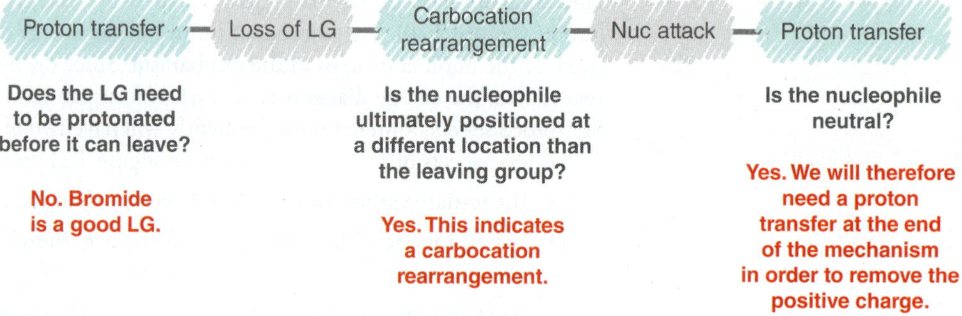


The mechanism will not begin with a proton transfer, but there will be a carbocation rearrangement, and there will be a proton transfer at the end of the mechanism. Therefore, the mechanism will have four steps:

Loss of LG — Carbocation rearrangement — Nuc attack — Proton transfer

Notice that this sequence utilizes each of the four arrow-pushing patterns that are possible for an ionic reaction. To draw these steps, we will rely on the skills we learned in Chapter 6:

:Br: → (EtOH) → ÖEt

$-Br^{\ominus}$ **Loss of leaving group**

**Carbocation rearrangement**

**Nucleophilic attack** Et–Ö–H

**Proton transfer** Et–Ö–H

2° carbocation

3° carbocation

**PRACTICE the skill**

**7.21** Draw the mechanism for each of the following $S_N1$ processes:

(a) OH → HBr → Br

(b) OH → HBr → Br

(c) Br → $H_2O$ → OH

(d) I → EtOH → OEt

(e) OH → $H_2SO_4$, MeOH → OMe

(f) OH → $H_2SO_4$, MeOH → MeO

(g) Br → NaSH → SH

(h) I → EtOH → OEt

**APPLY the skill**

**7.22** Identify the number of steps (patterns) for the mechanisms in Problems 7.21a–h. For example, the patterns for the first two are:

7.21a: $+H^+$ — LG — Nuc attack This mechanism exhibits a proton transfer before the two core steps.

7.21b: $+H^+$ — -LG — C+ rearrangement — Nuc attack This mechanism exhibits a proton transfer before the two core steps as well as a carbocation rearrangement in between the two core steps.

These patterns are not identical. Draw patterns for the other six problems. Then compare the patterns. There is only one pattern that is repeated. Identify the two problems that exhibit the same pattern and then describe in words why those two reactions are so similar.

**7.23** Treatment of (2*R*,3*R*)-3-methyl-2-pentanol with $H_3O^+$ affords a compound with no chirality centers. Predict the product of this reaction and draw the mechanism of its formation. Use your mechanism to explain how both chirality centers are destroyed.

HO

**(2*R*,3*R*)-3-methyl-2-pentanol**

need more **PRACTICE?** **Try Problems 7.48, 7.49, 7.52, 7.54, 7.65**

In some rare cases, loss of the leaving group and carbocation rearrangement can occur in a concerted fashion (Figure 7.21). For example, neopentyl bromide cannot directly lose its leaving group, as that would generate a primary carbocation, which is too high in energy to form:

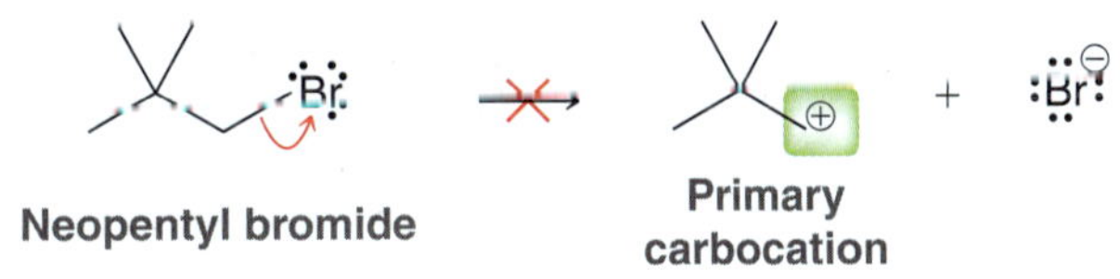


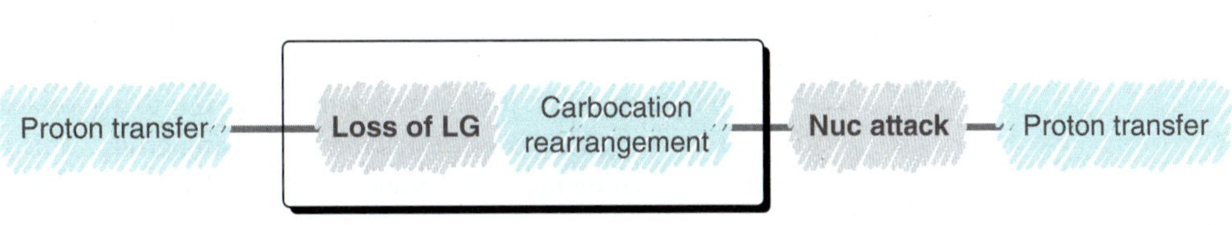


**FIGURE 7.21** In an $S_N1$ process, loss of the leaving group and carbocation rearrangement can occur in a concerted fashion.

But it is possible for the leaving group to leave as a result of a methyl shift:

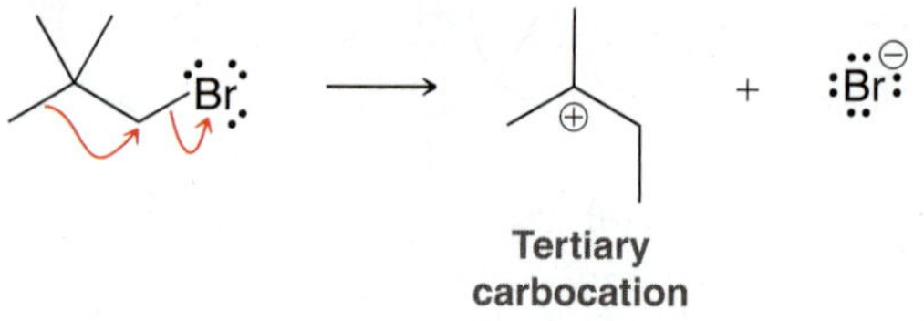

Tertiary carbocation

This is essentially a concerted process in which loss of the leaving group occurs simultaneously with a carbocation rearrangement. Examples like this are less common. In the vast majority of cases, each step of an $S_N1$ process occurs separately.

## 7.7 Drawing the Complete Mechanism of an $S_N2$ Reaction

Nuc attack + loss of LG

**FIGURE 7.22** The one concerted step of an $S_N2$ process.

In the previous section, we analyzed the additional steps that can accompany an $S_N1$ process. In this section, we analyze the additional steps that can accompany an $S_N2$ process. Recall that an $S_N2$ reaction is a concerted process in which nucleophilic attack and loss of the leaving group occur simultaneously (Figure 7.22). No carbocation is formed, so there can be no carbocation rearrangement. In an $S_N2$ process, the only two possible additional steps are proton transfers (Figure 7.23). There can be a proton transfer before and/or after the concerted step. Proton transfers will accompany $S_N2$ processes for the same reasons that they accompany $S_N1$ processes.

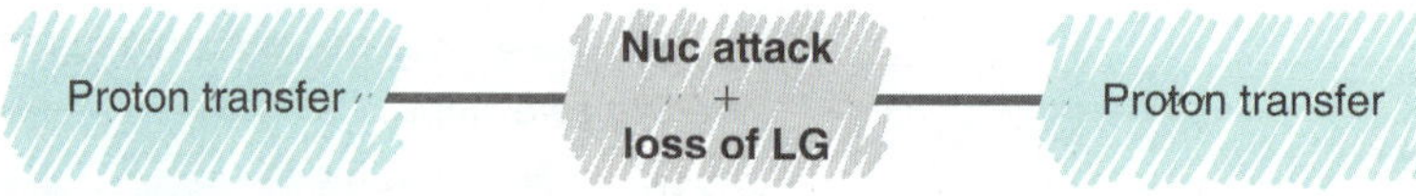


**FIGURE 7.23** The concerted step and the two possible additional steps of an $S_N2$ process.

Specifically, a proton transfer is required at the beginning of a mechanism if the substrate is an alcohol, and a proton transfer is required at the end of the mechanism if the nucleophile is neutral. Let's see examples of each.

### Proton Transfer at the Beginning of an $S_N2$ Process

+H⁺ — $S_N2$

A proton transfer is necessary at the beginning of an $S_N2$ process if the substrate is an alcohol, although this type of process will rarely be encountered throughout the remainder of this textbook. An example of this reaction is the conversion of methanol to methyl chloride, a reaction first performed by the French chemists Jean-Baptiste Dumas and Eugene Peligot in 1835. This transformation was achieved by boiling a mixture of methanol, sulfuric acid, and sodium chloride.

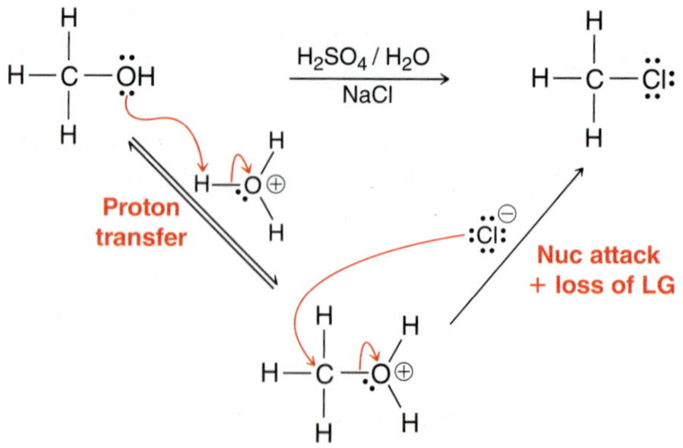

$H_2SO_4 / H_2O$, NaCl

Proton transfer

Nuc attack + loss of LG

The OH group is first protonated, converting it into a good leaving group, and then the chloride ion attacks in an $S_N2$ process, displacing the leaving group. This type of process, in which an alcohol is used as a substrate in an $S_N2$ reaction, has very limited utility and is generally only utilized in certain industrial applications. Methyl chloride is prepared commercially by such a process (using aqueous HCl as the source of $H_3O^+$ and $Cl^-$):

$$CH_3OH \xrightarrow{HCl} CH_3Cl$$

### Proton Transfer at the End of an $S_N2$ Process

A proton transfer will occur at the end of an $S_N2$ process if the nucleophile is neutral. For example, consider the following solvolysis reaction:

Nuc attack + loss of LG

Proton transfer

The substrate is primary, and therefore, the reaction must proceed via an $S_N2$ process. In this case, the solvent (ethanol) is functioning as the nucleophile, so this is a solvolysis reaction. Since the nucleophile is neutral, a proton transfer is required at the end of the mechanism in order to remove the positive charge from the compound.

### Proton Transfer Before and After an $S_N2$ Process

Throughout this course, we will see other examples of $S_N2$ processes that are accompanied by proton transfers. For example, the following reaction will be explored in Sections 9.16 and 14.10:

$H_3O^+$

Proton transfer

$S_N2$ attack

Proton transfer

This reaction involves two proton transfer steps—one before and one after the $S_N2$ attack—as seen in Mechanism 7.2.

## MECHANISM 7.2 THE $S_N2$ PROCESS

**One concerted step**

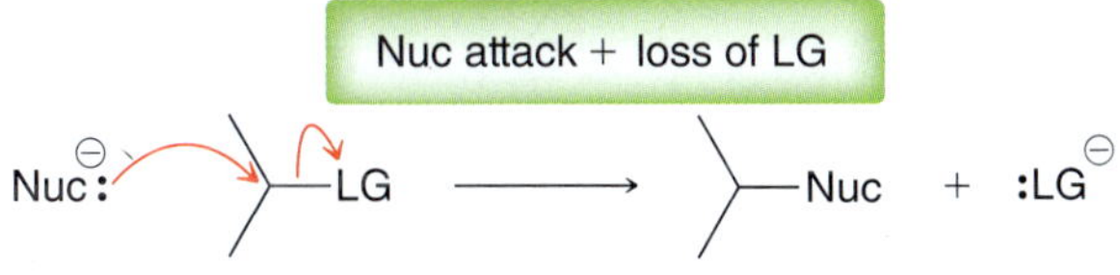


An $S_N2$ process is comprised of just one concerted step in which the nucleophile attacks with simultaneous loss of the leaving group.

***Possible additional steps***

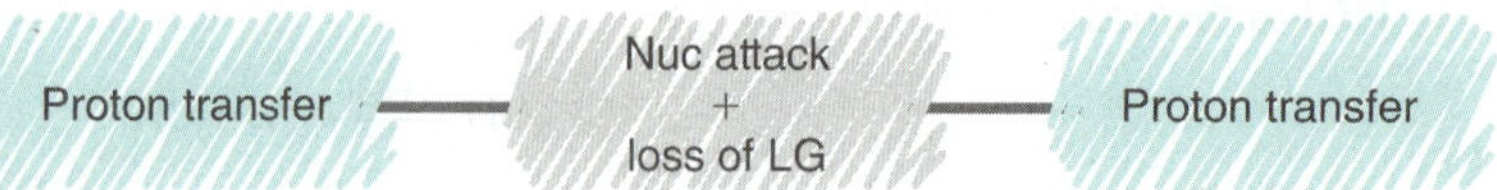


**If the substrate is an alcohol, then the OH group must be protonated before it can leave.**

**If the nucleophile is neutral, a proton transfer is required to remove the positive charge that is generated.**

# SKILLBUILDER

## 7.7 DRAWING THE COMPLETE MECHANISM OF AN $S_N2$ PROCESS

**LEARN** the skill

Ethyl bromide was dissolved in water and heated, and the following solvolysis reaction was observed to occur slowly, over a long period of time. Propose a mechanism for this reaction.

Br $\xrightarrow{H_2O}$ OH

**SOLUTION**

The substrate is primary, so the reaction must proceed via an $S_N2$ process, rather than $S_N1$. An $S_N1$ mechanism cannot be invoked in this case, because a primary carbocation would be too unstable to form.

In an $S_N2$ process, there is one concerted step, and we must determine whether a proton transfer will be necessary either before or after the concerted step:

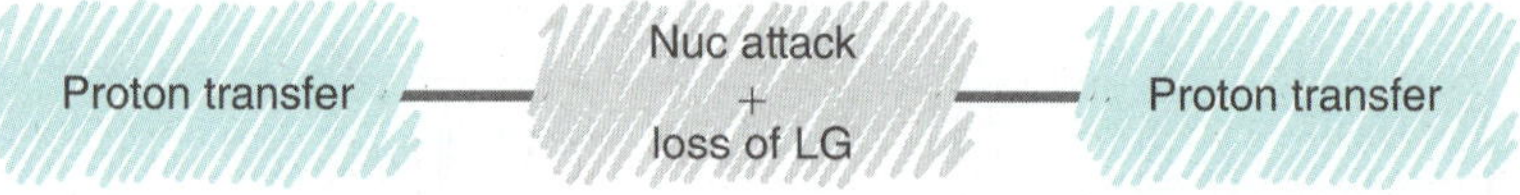


**Does the LG need to be protonated first?**

**No. Bromide is a good LG.**

**Is the nucleophile neutral?**

**Yes. We will therefore need a proton transfer at the end of the mechanism in order to remove the positive charge.**

The mechanism does not require a proton transfer before the concerted step, and even if it did, the reagents are not acidic and could not donate a proton anyway. In order to have a proton transfer at the beginning of a mechanism, an acid is required to serve as a proton source. At the end of the mechanism, there will be a proton transfer because the attacking nucleophile ($H_2O$) is neutral. Therefore, the mechanism will have two steps:

Nuc attack + loss of LG — Proton transfer

The first step involves a simultaneous nucleophilic attack and loss of leaving group, and the second step is a proton transfer:

Br $\xrightarrow[\text{(solvolysis)}]{H_2O}$ OH

**Nuc attack + loss of LG**

**Proton transfer**

**PRACTICE the skill** **7.24** Draw the mechanism for each of the following solvolysis reactions:

(a) MeOH (solvolysis) → OMe

(b) Br → EtOH (solvolysis) → O

(c) I → $H_2O$ (solvolysis) → OH

(d) Cl → OH (solvolysis) → O

**APPLY the skill** **7.25** In Chapter 23, we will learn that treatment of ammonia with excess methyl iodide produces a quaternary ammonium salt. This transformation is the result of four sequential $S_N2$ reactions. Use the tools we have learned in this chapter to draw the mechanism of this transformation. Your mechanism should have seven steps.

$NH_3$ → Excess MeI → $Me_4N^{\oplus}$ $I^{\ominus}$

**Quaternary ammonium salt**

need more **PRACTICE?** **Try Problems 7.53, 7.64, 7.66**

## medically speaking — Radiolabeled Compounds in Diagnostic Medicine

Recall from your general chemistry course that isotopes are atoms of the same element that differ from each other only in the number of neutrons. For example, carbon has three isotopes that are found in nature: $^{12}C$ (called carbon 12), $^{13}C$ (called carbon 13), and $^{14}C$ (called carbon 14). Each of these isotopes has six protons and six electrons, but they differ in their number of neutrons. They have six, seven, and eight neutrons, respectively. Of these isotopes, $^{12}C$ is the most abundant, constituting 98.9% of all carbon atoms found on Earth. The second most abundant isotope of carbon is $^{13}C$, constituting approximately 1.1% of all carbon atoms. The amount of $^{14}C$ found in nature is very small (0.0000000001%).

The element fluorine (F) has many isotopes, although only one is found to be stable ($^{19}F$). Other isotopes of fluorine can be produced, but they are unstable and will undergo radioactive decay. One such example is $^{18}F$, which has a half-life ($t_{1/2}$) of approximately 110 minutes. If a compound possesses $^{18}F$, then the location of that compound can be tracked by observing the decay process. For this reason, *radiolabeled compounds* (compounds containing an unstable isotope, such as $^{18}F$) have found great utility in the field of medicine. One such application will now be described.

Consider the structure of glucose as well as the radiolabeled glucose derivative, called $^{18}F$-2-fluorodeoxyglucose (FDG):

Glucose

$^{18}F$-2-Fluorodeoxyglucose (FDG)

Glucose is an important compound that represents the primary source of energy for our bodies. Our bodies are capable of metabolizing glucose molecules in a series of enzymatic steps called *glycolysis*, a process in which the high-energy C—C and C—H bonds in glucose are broken and converted into lower energy C—O and C=O bonds. The difference in energy is significant, and our bodies are capable of capturing that energy and storing it for later use:

$$\text{Glucose} + 6\,O_2 \longrightarrow 6\,CO_2 + 6\,H_2O \qquad \Delta G° = -2880\text{ kJ}$$

FDG is a radiolabeled compound that is a derivative of glucose (an OH group has been replaced with $^{18}F$). In our bodies, the first step of glycolysis occurs with FDG just as it does with glucose. But in the case of FDG, the second step of glycolysis does not proceed at an appreciable rate. Therefore, *areas of the body that utilize glucose will also accumulate FDG*. This accumulation can be monitored, because $^{18}F$ decays via a process that ultimately releases high-energy photons (gamma rays) that can be detected. The details of the decay process are beyond the scope of our discussion, but for our purposes, it can be summarized in the following way: $^{18}F$ decays via a process called *positron emission* ($\beta^+$ decay), in which an antiparticle called a *positron* is created, travels a short distance, and then encounters an electron and annihilates it, causing two coincident gamma ($\gamma$) rays to be emitted from the body. These $\gamma$ rays are then detected using special instrumentation, resulting in an image. This imaging technique is called positron emission tomography (PET).

As an example, the brain metabolizes glucose, so administering FDG to a patient will cause an accumulation of FDG in specific locations of the brain corresponding to the level of metabolic activity. Consider the following PET/$^{18}$FDG scans of a healthy volunteer and a cocaine abuser:

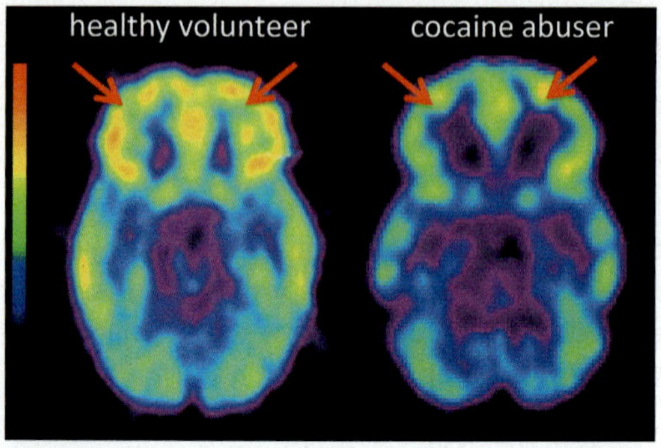


Red/orange indicates regions of high metabolism, while purple/blue indicates regions of low metabolism.

Notice that the cocaine abuser shows less accumulation of FDG in the orbitofrontal cortex (indicated with red arrows). This indicates a lower metabolism in that region, which is known to play a large role in cognitive function and decision making. This example demonstrates how neurologists can use radiolabeled compounds, such as FDG, to explore the brain and learn more about its function in normal and disease states. This technique has found great utility in diagnostic medicine, because cancerous tissues exhibit enhanced metabolic rates of glucose relative to surrounding normal tissues and therefore will also accumulate more FDG. As a result, cancerous tissues can be visualized and monitored with a PET/$^{18}$FDG scan.

FDG has been approved by the U.S. Food and Drug Administration (FDA) to assist oncologists in the diagnosis and staging of cancer, in addition to monitoring the response to therapy. The increasing role of FDG in the field of medicine has fueled the demand for the daily production of FDG. Because $^{18}$F is a short-lived radionuclide, a fresh batch of FDG must be made on a daily basis. Fortunately, the rate of decay is slow enough to permit synthesis at a regional radiopharmacy followed by distribution to regional hospitals for imaging studies.

The synthesis of FDG utilizes many of the principles covered in this chapter, because one important step in the process is an $S_N2$ reaction:

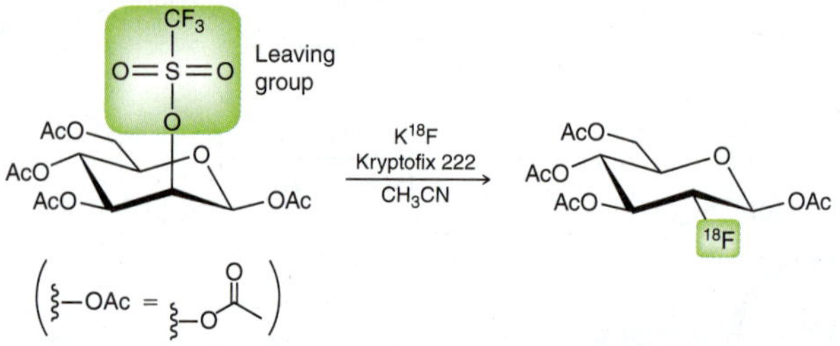


This reaction has a few important features worthy of our attention:

- KF is the source of fluoride ions, which function as the nucleophilic agent in this reaction. Generally, fluoride ions are not good nucleophiles, but Kryptofix interacts with the $K^+$ ions, thereby freeing the fluoride ions to function as nucleophiles. The ability of Kryptofix to enhance the nucleophilicity of fluoride ions is consistent with the action of crown ethers (a topic that will be explored in more detail in Section 14.4).
- This reaction proceeds with inversion of configuration, as expected for an $S_N2$ process.
- The solvent used for this reaction, acetonitrile ($CH_3CN$), is a polar aprotic solvent, and it is used to speed up the rate of the process (as described in Section 7.8).
- Notice that the OH groups (normally found in glucose) have been converted into acetate groups (OAc). This was done in order to minimize side reactions. These acetate groups can be removed easily upon treatment with aqueous acid (we will explore this process, called hydrolysis, in more detail in Section 21.11).

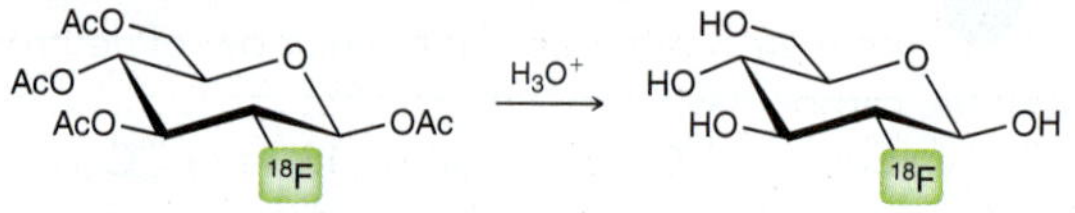


The effective application of FDG in PET scans certainly requires the contribution from many different disciplines, but organic chemistry has played the most critical role: The synthesis of FDG is achieved via an $S_N2$ process!

## 7.8 Determining Which Mechanism Predominates

In order to draw the products of a specific substitution reaction, we must first identify the reaction mechanism as either $S_N2$ or $S_N1$. This information is important in the following two ways:

- If substitution is taking place at a chirality center, then we must know whether to expect inversion of configuration ($S_N2$) or racemization ($S_N1$).
- If the substrate is susceptible to carbocation rearrangement, then we must know whether to expect rearrangement ($S_N1$) or whether rearrangement is not possible ($S_N2$).

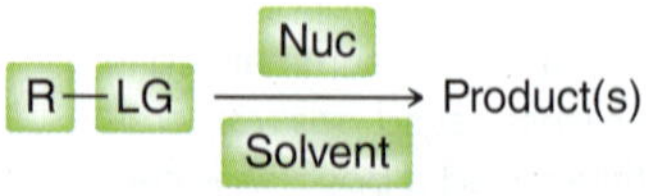


**FIGURE 7.24** The four factors that determine which mechanism predominates.

Four factors have an impact on whether a particular reaction will occur via an $S_N2$ or an $S_N1$ mechanism: (1) the substrate, (2) the leaving group, (3) the nucleophile, and (4) the solvent (Figure 7.24). We must learn to look at all four factors, one by one, and to determine whether the factors favor $S_N1$ or $S_N2$.

## The Substrate

The identity of the substrate is the most important factor in distinguishing between $S_N2$ and $S_N1$. Earlier in the chapter, we saw different trends for $S_N2$ and $S_N1$ reactions. These trends are compared in the charts in Figure 7.25.

The trend in $S_N2$ reactions is due to issues of steric hindrance in the transition state, while the trend in $S_N1$ reactions is due to carbocation stability. The bottom line is that methyl and primary substrates favor $S_N2$, while tertiary substrates favor $S_N1$. Secondary substrates can proceed via either mechanism, so a secondary substrate does not indicate which mechanism will predominate. In such a case, you must move on to the next factor, the nucleophile (covered in the next section).

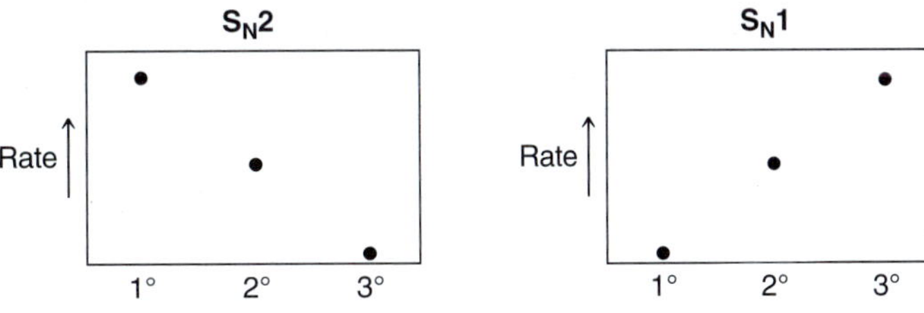


**FIGURE 7.25** Substrate effects on the rates of $S_N2$ and $S_N1$ processes.

*Allylic halides* and *benzylic halides* can react either via $S_N2$ or via $S_N1$ processes:

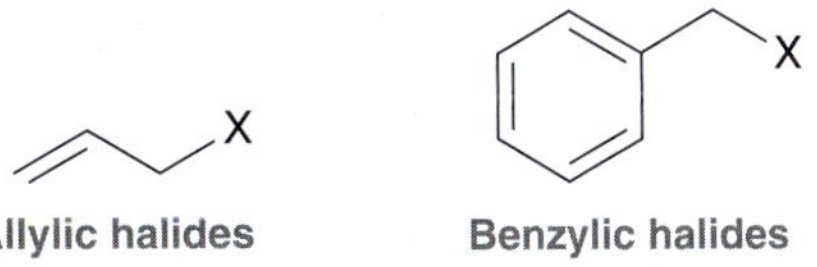

Allylic halides    Benzylic halides

These substrates can react via an $S_N2$ mechanism because they are relatively unhindered, and they can react via an $S_N1$ mechanism because loss of a leaving group generates a resonance-stabilized carbocation:

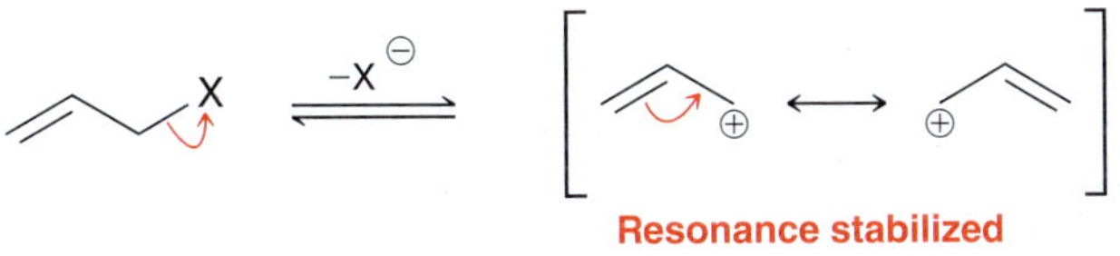

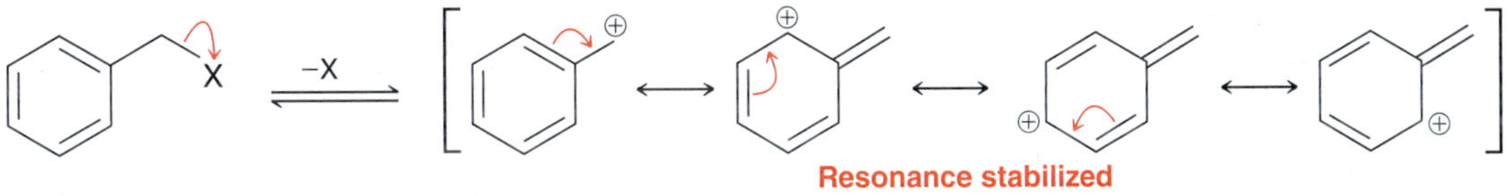

Resonance stabilized

Resonance stabilized

In contrast, *vinyl halides* and *aryl halides* are unreactive in substitution reactions:

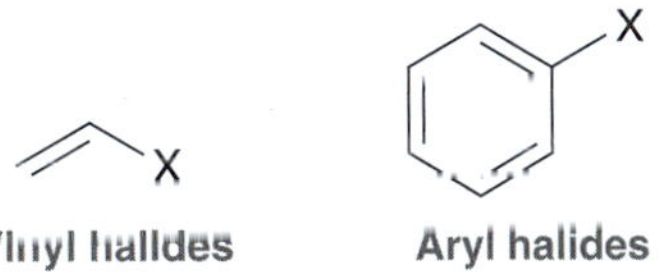

Vinyl halides    Aryl halides

$S_N2$ reactions are generally not observed at $sp^2$-hybridized centers, because back-side attack is sterically encumbered. In addition, vinyl halides and aryl halides are also unreactive toward $S_N1$, because loss of a leaving group would generate an unstable carbocation:

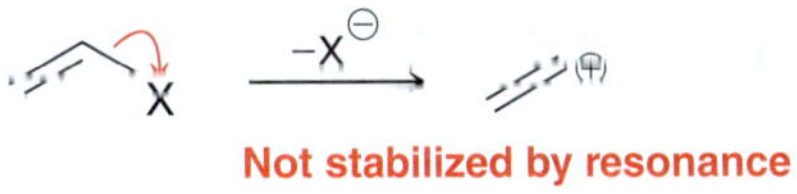

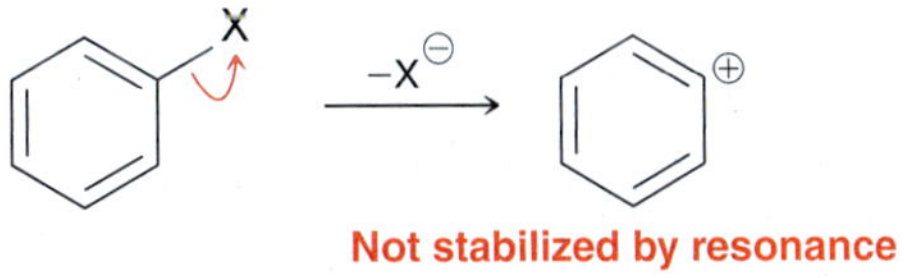

Not stabilized by resonance

Not stabilized by resonance

To summarize:

- Methyl and primary substrates favor $S_N2$.
- Tertiary substrates favor $S_N1$.
- Secondary substrates and allylic and benzylic substrates can often react via either mechanism.
- Vinyl and aryl substrates do not react via either mechanism.

## CONCEPTUAL CHECKPOINT

**7.26** Identify whether each of the following substrates favors $S_N2$, $S_N1$, both, or neither:

(a) (b) (c) (d) (e) (f) (g)

## The Nucleophile

Recall that the rate of an $S_N2$ process is dependent on the concentration of the nucleophile. For the same reason, $S_N2$ processes are also dependent on the *strength* of the nucleophile. A strong nucleophile will speed up the rate of an $S_N2$ reaction, while a weak nucleophile will slow down the rate of an $S_N2$ reaction. In contrast, an $S_N1$ process is not affected by the concentration or strength of the nucleophile because the nucleophile does not participate in the rate-determining step. In summary, the nucleophile has the following effect on the competition between $S_N2$ and $S_N1$:

- A strong nucleophile favors $S_N2$.
- A weak nucleophile disfavors $S_N2$ (and thereby allows $S_N1$ to compete successfully).

We must therefore learn to identify nucleophiles as strong or weak. The strength of a nucleophile is determined by many factors that were first discussed in Section 6.7. Figure 7.26 shows some strong and weak nucleophiles that we will encounter.

**Common nucleophiles**

| Strong | | | Weak |
|---|---|---|---|
| $I^{\ominus}$ | $HS^{\ominus}$ | $HO^{\ominus}$ | $F^{\ominus}$ |
| $Br^{\ominus}$ | $H_2S$ | $RO^{\ominus}$ | $H_2O$ |
| $Cl^{\ominus}$ | RSH | $N{\equiv}C^{\ominus}$ | ROH |

**FIGURE 7.26** Some common nucleophiles grouped according to strength. Note: The designation of fluoride as a weak nucleophile is only accurate in protic solvents. In polar aprotic solvents, fluoride is in fact a strong nucleophile. This issue will be discussed in more detail later in this section, just before Table 7.3.

## CONCEPTUAL CHECKPOINT

**7.27** Does each of the following nucleophiles favor $S_N2$ or $S_N1$?

(a) OH (b) SH (c) $O^{\ominus}$ (d) NaOH (e) NaCN

## The Leaving Group

Both $S_N1$ and $S_N2$ mechanisms are sensitive to the identity of the leaving group. If the leaving group is bad, then neither mechanism can operate, but $S_N1$ reactions are generally more sensitive to the leaving group than $S_N2$ reactions. Why? Recall that the rate-determining step of an $S_N1$ process is loss of a leaving group to form a carbocation and a leaving group:

LG $\xrightarrow{\text{RDS}}$ $\oplus$ + $:LG^{\ominus}$

We have already seen that the rate of this step is very sensitive to the stability of the carbocation, but it is also sensitive to the stability of the leaving group. The leaving group must be highly stabilized in order for an $S_N1$ process to be effective.

What determines the stability of a leaving group? As a general rule, good leaving groups are the conjugate bases of strong acids. For example, iodide ($I^-$) is the conjugate base of a very strong acid (HI):

$$H{-}I + H_2O \rightleftharpoons I^- + H_3O^+$$

Strong acid | Conjugate base (weak)

Iodide is a very weak base because it is highly stabilized. As a result, iodide can function as a good leaving group. In fact, iodide is one of the best leaving groups. Figure 7.27 shows a list of good leaving groups, all of which are the conjugate bases of strong acids. In contrast, hydroxide is a bad leaving group, because it is not a stabilized base. In fact, hydroxide is a relatively strong base, and therefore, it rarely functions as a leaving group. It is a bad leaving group.

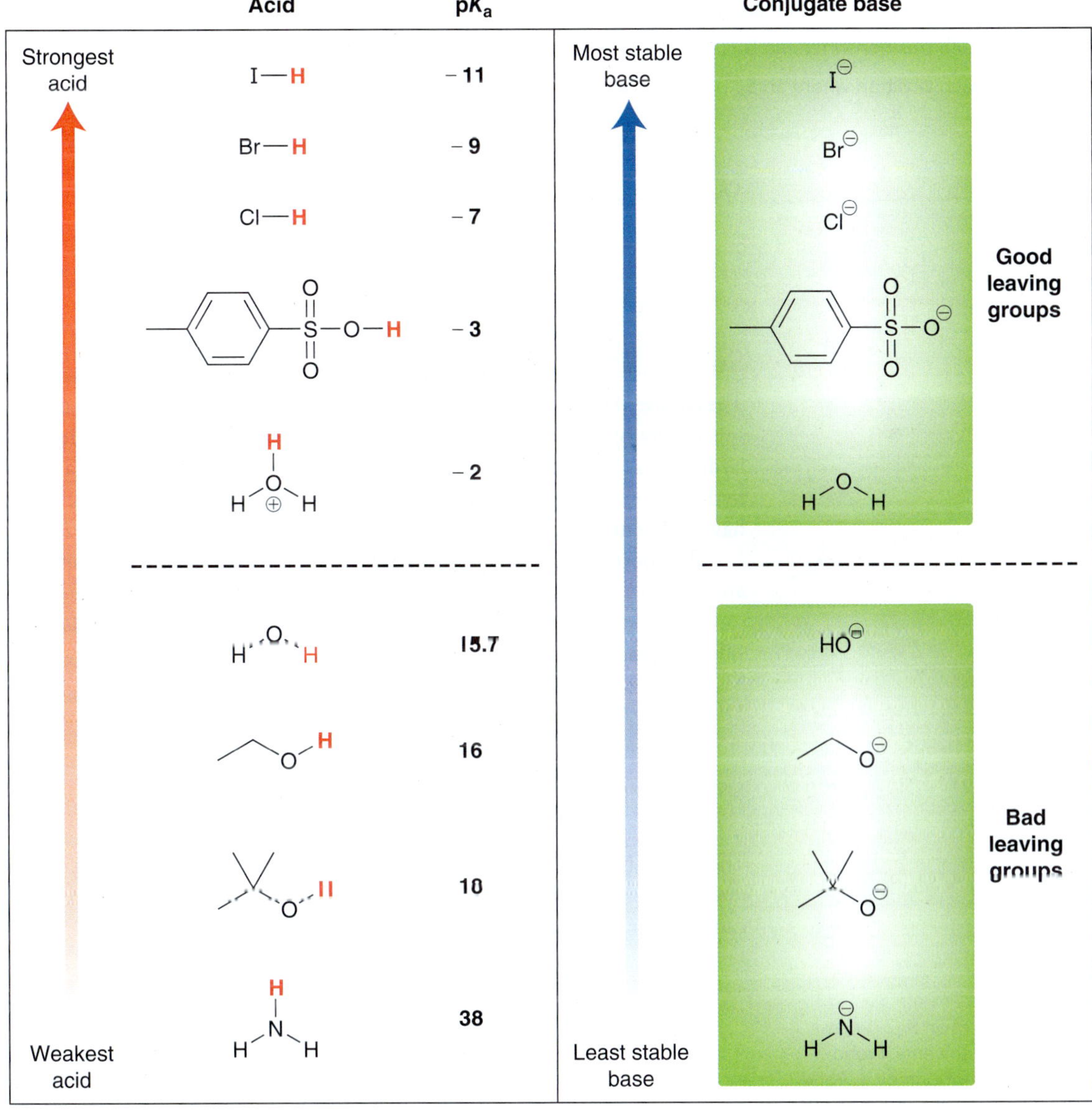


**FIGURE 7.27**
The conjugate base of a strong acid will generally be a good leaving group. The conjugate base of a weak acid will not be a good leaving group.

The most commonly used leaving groups are halides and **sulfonate ions** (Figure 7.28). Among the halides, iodide is the best leaving group because it is a weaker base (more stable) than bromide or chloride. Among the sulfonate ions, the best leaving group is the triflate group, but the most commonly used is the **tosylate** group. It is abbreviated as OTs. When you see OTs connected to a compound, you should recognize the presence of a good leaving group.

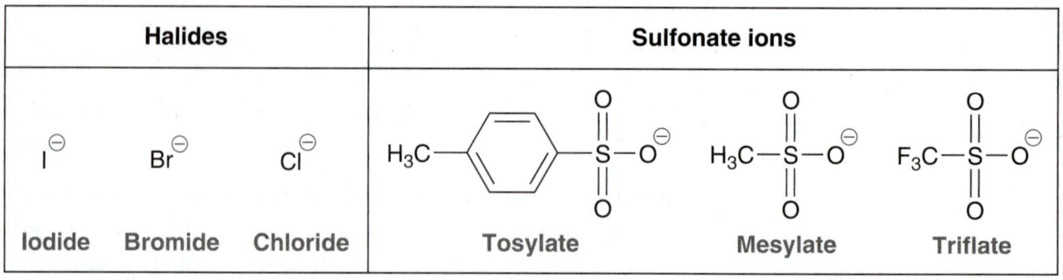

| Halides | | | Sulfonate ions | | |
|---|---|---|---|---|---|
| $I^{\ominus}$ | $Br^{\ominus}$ | $Cl^{\ominus}$ | $H_3C{-}C_6H_4{-}SO_2{-}O^{\ominus}$ | $H_3C{-}SO_2{-}O^{\ominus}$ | $F_3C{-}SO_2{-}O^{\ominus}$ |
| **Iodide** | **Bromide** | **Chloride** | **Tosylate** | **Mesylate** | **Triflate** |

**FIGURE 7.28**
Common leaving groups.

## CONCEPTUAL CHECKPOINT

**7.28** Consider the structure of the compound below.

(a) Identify each position where an $S_N2$ reaction is likely to occur if the compound were treated with hydroxide.

(b) Identify each position where an $S_N1$ reaction is likely to occur if the compound were treated with water.

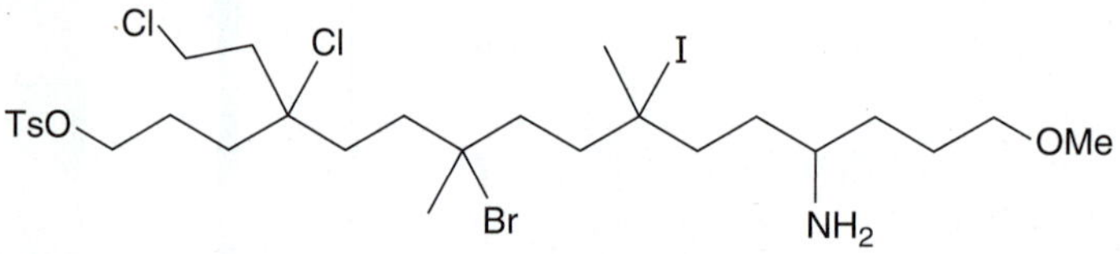

TsO, Cl, Cl, I, Br, $NH_2$, OMe

## Solvent Effects

The choice of solvent can have a profound effect on the rates of $S_N1$ and $S_N2$ reactions. We will focus specifically on the effects of protic and polar aprotic solvents. **Protic solvents** contain at least one hydrogen atom connected directly to an electronegative atom. **Polar aprotic solvents** contain no hydrogen atoms connected directly to an electronegative atom. These two different kinds of solvents have different effects on the rates of $S_N1$ and $S_N2$ processes. Table 7.2 summarizes these effects.

The bottom line is that protic solvents are used for $S_N1$ reactions, while polar aprotic solvents are used to favor $S_N2$ reactions.

The effect of polar aprotic solvents on the rate of $S_N2$ reactions is significant. For example, consider the reaction between bromobutane and an azide ion:

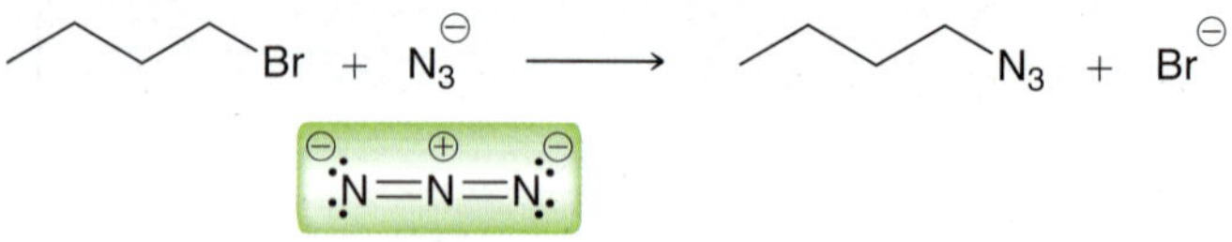

$$\text{Br} + N_3^{\ominus} \longrightarrow \text{N}_3 + Br^{\ominus}$$

$$^{\ominus}\ddot{N}{=}\overset{\oplus}{N}{=}\ddot{N}^{\ominus}$$

Azide

The rate of this reaction is highly dependent on the choice of solvent. Figure 7.29 shows the relative rates of this $S_N2$ reaction in various solvents. From these data, we see that $S_N2$ reactions are significantly faster in polar aprotic solvents than in protic solvents.

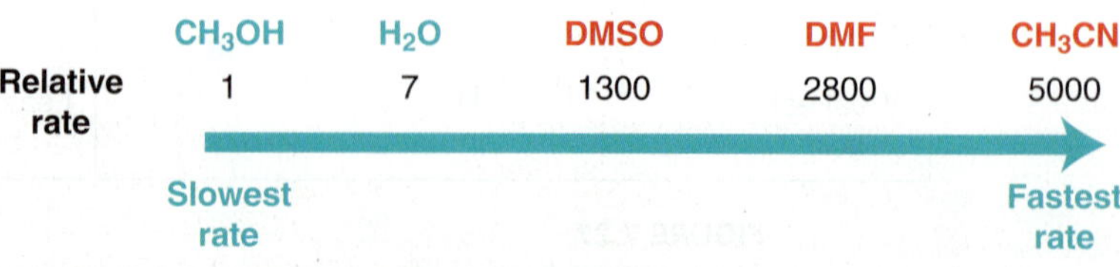


**FIGURE 7.29**
Relative rates of an $S_N2$ process in a variety of solvents. Protic solvents are shown in blue. Polar aprotic solvents are shown in red.

**TABLE 7.2 THE EFFECTS OF PROTIC SOLVENTS AND POLAR APROTIC SOLVENTS**

| | PROTIC | POLAR APROTIC |
|---|---|---|
| Definition | Protic solvents contain at least one hydrogen atom connected directly to an electronegative atom. | Polar aprotic solvents contain no hydrogen atoms connected directly to an electronegative atom. |
| Examples | **Water**, **Methanol**, **Ethanol**, **Acetic acid**, **Ammonia** | **Dimethylsulfoxide (DMSO)**, **Acetonitrile**, **Dimethylformamide (DMF)**, **Hexamethylphosphoramide (HMPA)** |
| Function | Protic solvents stabilize cations and anions. Cations are stabilized by lone pairs from the solvent, while anions are stabilized by H-bonding interactions with the solvent: **The lone pairs on the oxygen atoms of $H_2O$ stabilize the cation.** **Hydrogen-bonding interactions stabilize the anion.** As a result, anions and cations are both solvated and surrounded by a solvent shell. | Polar aprotic solvents stabilize cations, but not anions. Cations are stabilized by lone pairs from the solvent, while anions are not stabilized by the solvent: **The lone pairs on the oxygen atoms of DMSO stabilize the cation.** **The anion is not stabilized by the solvent.** Cations are solvated and surrounded by a solvent shell, but anions are not. As a result, nucleophiles are higher in energy when placed in a polar aprotic solvent. |
| Effects | Favors $S_N1$. Protic solvents favor $S_N1$ by stabilizing polar intermediates and transition states. 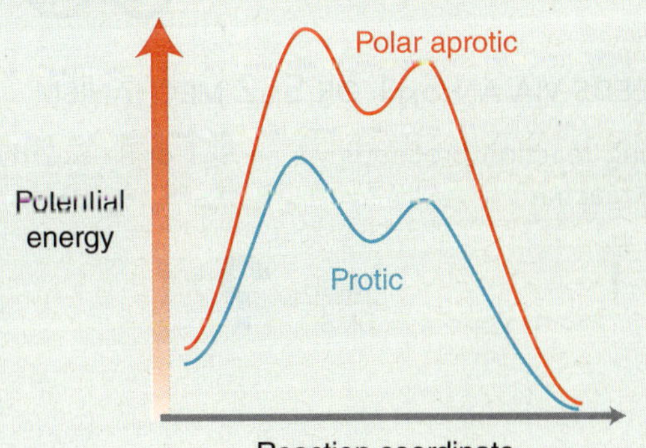  | Favors $S_N2$. Polar aprotic solvents favor $S_N2$ by raising the energy of the nucleophile, giving a smaller $E_a$: 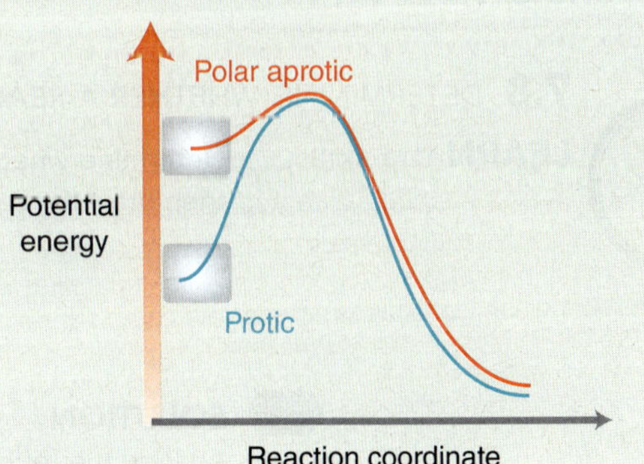  |

## CONCEPTUAL CHECKPOINT

**7.29** Does each of the following solvents favor an $S_N2$ reaction or an $S_N1$ reaction? (See Table 7.2.)

(a) ethanol (OH) (b) DMSO (S=O) (c) acetic acid (O, OH) (d) DMF (O, H, N)

(e) MeOH (f) $CH_3CN$ (g) HMPA (h) $NH_3$

**7.30** When used as a solvent, will acetone favor an $S_N2$ or an $S_N1$ mechanism? Explain.

Acetone

The choice of solvent can also have an impact on the order of reactivity of the halides. If we compare the nucleophilicity of the halides, we find that it is dependent on the solvent. In protic solvents, the following order is observed:

$$I^{\ominus} > Br^{\ominus} > Cl^{\ominus} > F^{\ominus}$$

Iodide is the strongest nucleophile, and fluoride is the weakest. However, in polar aprotic solvents, the order is reversed:

$$F^{\ominus} > Cl^{\ominus} > Br^{\ominus} > I^{\ominus}$$

Why the reversal of order? Fluoride is the strongest because it is the least stable anion. In protic solvents, fluoride is the most tightly bound to its solvent shell and is the least available to function as a nucleophile (it would have to shed part of its solvent shell, which it does not do often). In such an environment, it is a weak nucleophile. However, when a polar aprotic solvent is used, there is no solvent shell, and fluoride is free to function as a strong nucleophile.

### Summary of Factors Affecting $S_N2$ and $S_N1$ Mechanisms

Table 7.3 summarizes what we have learned in this section about the four factors that affect $S_N2$ and $S_N1$ processes. Now let's get some practice analyzing all four factors:

**TABLE 7.3 FACTORS THAT FAVOR $S_N2$ AND $S_N1$ PROCESSES**

| FACTOR | FAVORS $S_N2$ | FAVORS $S_N1$ |
|---|---|---|
| Substrate | Methyl or primary | Tertiary |
| Nucleophile | Strong nucleophile | Weak nucleophile |
| Leaving group | Good leaving group | Excellent leaving group |
| Solvent | Polar aprotic | Protic |

## SKILLBUILDER

### 7.8 DETERMINING WHETHER A REACTION PROCEEDS VIA AN $S_N1$ OR $S_N2$ MECHANISM

**LEARN the skill**

Determine whether the following reaction proceeds via an $S_N1$ or an $S_N2$ mechanism and then draw the product(s) of the reaction:

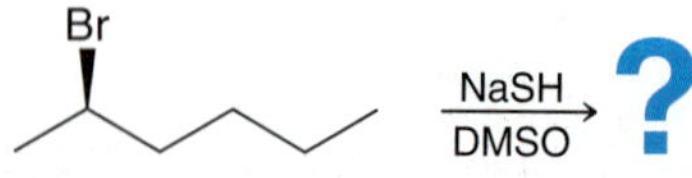

**SOLUTION**

Analyze the four factors one by one:

**(a)** *Substrate*. The substrate is secondary. If it were primary, we would predict $S_N2$, and if it were tertiary, we would predict $S_N1$. But with a secondary substrate, it could be either, so we move on to the next factor.

(b) *Nucleophile*. NaSH indicates that the nucleophile is $HS^-$ (remember that $Na^+$ is just the counter ion). $HS^-$ is a strong nucleophile, which favors $S_N2$.

(c) *Leaving Group*. $Br^-$ is a good leaving group. This factor alone does not indicate a preference for either $S_N1$ or $S_N2$.

(d) *Solvent*. DMSO is a polar aprotic solvent, which favors $S_N2$.

Weighing all four factors, there is a preference for $S_N2$ because both the nucleophile and the solvent favor $S_N2$. Therefore, we expect inversion of configuration:

Br → (NaSH, DMSO) → SH

**PRACTICE the skill** **7.31** Determine whether each of the following reactions proceeds via an $S_N1$ or $S_N2$ mechanism and then draw the product(s) of the reaction:

(a) I, MeOH → ?

(b) Br, $Cl^{\ominus}$ / HMPA → ?

(c) O—H, H—Br → ?

(d) OTs, NaCN / DMF → ?

(e) I, $H_2O$ → ?

(f) Br, NaCN / DMSO → ?

**APPLY the skill** **7.32** In Chapter 23, we will learn several methods for making primary amines ($RNH_2$). Each of these methods utilizes a different approach for forming the C—N bond. One of these methods, called the Gabriel synthesis, forms the C—N bond by treating potassium phthalimide with an alkyl halide:

O, N$^{\ominus}$, O, K$^{\oplus}$ → (R—X) → N—R → → H, N—R, H

The first step of this process occurs via an $S_N2$ mechanism. Using this information, determine whether the Gabriel synthesis can be used to prepare the following amine. Explain your answer.

$NH_2$

need more **PRACTICE?** **Try Problems 7.37, 7.38, 7.40, 7.41, 7.44, 7.55, 7.57, 7.58**

## 7.9 Selecting Reagents to Accomplish Functional Group Transformation

As mentioned at the beginning of the chapter, substitution reactions can be utilized to accomplish functional group transformation:

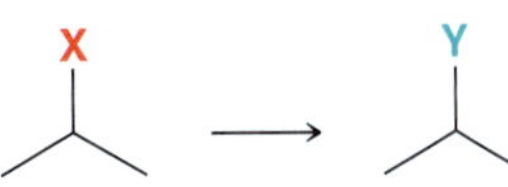


A wide range of nucleophiles can be used, providing a great deal of versatility in the type of products that can be formed with substitution reactions. Figure 7.30 shows some of the types of compounds

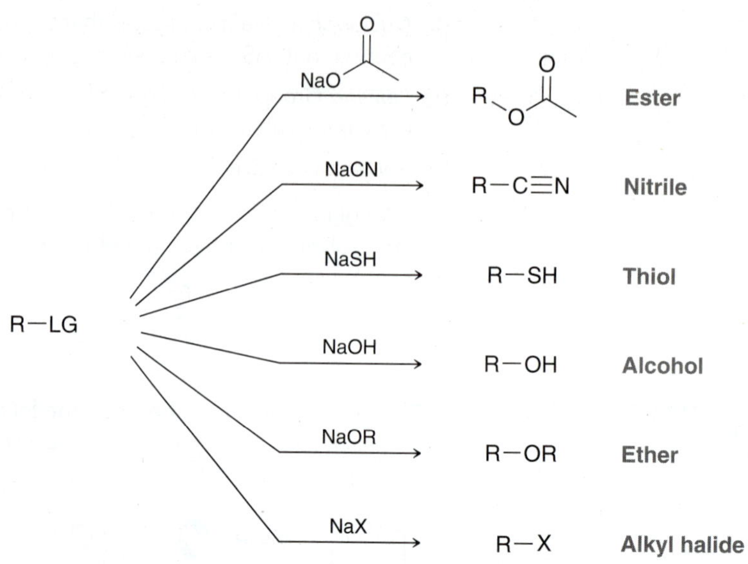


**FIGURE 7.30** Various products that can be obtained via substitution reactions.

that can be prepared using substitution reactions. When selecting reagents for a substitution reaction, remember the following tips:

- *Substrate.* The identity of the substrate indicates which process to use. If the substrate is methyl or primary, an $S_N2$ process must be used. If the substrate is tertiary, an $S_N1$ process must be used. If the substrate is secondary, generally try to use an $S_N2$ process because it avoids the issue of carbocation rearrangement and provides greater control over the stereochemical outcome.
- *Nucleophile and Solvent.* Once you have decided whether you want to use an $S_N1$ or an $S_N2$ process (based on the substrate), make sure to choose a nucleophile and solvent that are consistent with that mechanism. For an $S_N1$ reaction, use a weak nucleophile in a protic solvent. For an $S_N2$ reaction, use a strong nucleophile in a polar aprotic solvent.
- *Leaving Group.* Remember that OH is a bad leaving group and will not leave as is. It must first be converted into a good leaving group. In an $S_N1$ process, use an acid to protonate the OH group, converting it into an excellent leaving group. In an $S_N2$ reaction, the OH group is generally converted into a tosylate, an excellent leaving group, rather than protonating the OH group. This transformation is accomplished with tosyl chloride and pyridine (and is discussed in more detail in Chapter 13):

R⁀OH —[$H_3C$–C₆H₄–$SO_2$–Cl, Pyridine]→ R⁀OTs

# SKILLBUILDER

## 7.9 IDENTIFYING REAGENTS NECESSARY FOR A SUBSTITUTION REACTION

**LEARN** the skill

Identify the reagents you would use to accomplish the following transformation:

(OH-substituted secondary carbon) ⟶ (CN-substituted secondary carbon)

**SOLUTION**

First determine which mechanism to use by looking at the substrate:

**STEP 1**
Analyze the substrate and the stereochemical outcome.

**(a)** *Substrate*. The substrate is secondary, so it could go either way. In general, choose $S_N2$, because it gives greater control. In this particular case, an $S_N2$ pathway must be used because the product is formed only through inversion of configuration. Now choose reagents that favor $S_N2$.

**STEP 2**
Analyze the leaving group.

**(b)** *Leaving Group*. The OH is a bad leaving group and must be converted into a better leaving group. When performing an $S_N2$ reaction, the OH should be converted into a tosylate by using TsCl and pyridine:

OH —TsCl, pyridine→ OTs

Notice that the configuration does not change when the OH group is converted into a tosylate group. If the OH is on a wedge, then the tosylate group will also be on a wedge.

**STEP 3**
Choose conditions that favor the required mechanism.

**(c)** *Nucleophile*. In order to accomplish the desired transformation, the nucleophile needs to be cyanide ($CN^-$). Cyanide is a strong nucleophile, which supports an $S_N2$ process.

**(d)** *Solvent*. In order to favor an $S_N2$ process, a polar aprotic solvent such as DMSO should be used:

OH —1) TsCl, pyridine; 2) NaCN / DMSO→ CN

Notice that the conversion of the OH into a tosylate and the $S_N2$ reaction are two separate synthetic steps, so we place the numbers 1 and 2 before the sets of reagents to indicate that they are separate reactions.

**PRACTICE the skill** **7.33** Identify the reagents you would use to accomplish each of the following transformations:

(a) I → OH
(b) OH → Br
(c) OH → I
(d) Br → SH
(e) Br → O(C=O)
(f) OH → Br
(g) I → O
(h) Br → OH

**APPLY the skill** **7.34** What reagents would you use to accomplish a substitution with retention of configuration; for example:

OH → SH

(*R*)-2-Butanol (*R*)-2-Butanethiol

need more **PRACTICE?** **Try Problems 7.59, 7.60, 7.63**

## medically speaking — Pharmacology and Drug Design

*Pharmacology* is the study of how drugs interact with biological systems, including the mechanisms that explain drug action. Pharmacology is a very important field of study because it serves as the basis for the design of new drugs. In this section, we will explore one specific example, the design and development of chlorambucil, an antitumor agent:

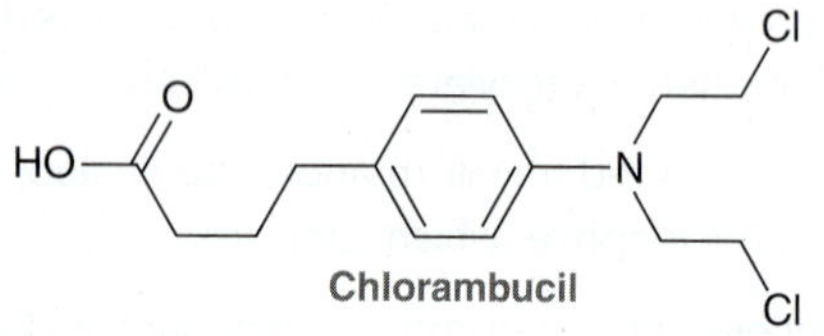

Chlorambucil

Chlorambucil was designed by chemists using principles that we have learned in this and previous chapters. The story of chlorambucil begins with a toxic compound called sulfur mustard.

This compound was first used as a chemical weapon in World War I. It was sprayed as an aerosol mixture with other chemicals and exhibited a characteristic odor similar to that of mustard plants, thus the name *mustard gas*. Sulfur mustard is a powerful alkylating agent. The mechanism of alkylation involves a sequence of two substitution reactions:

Sulfur mustard

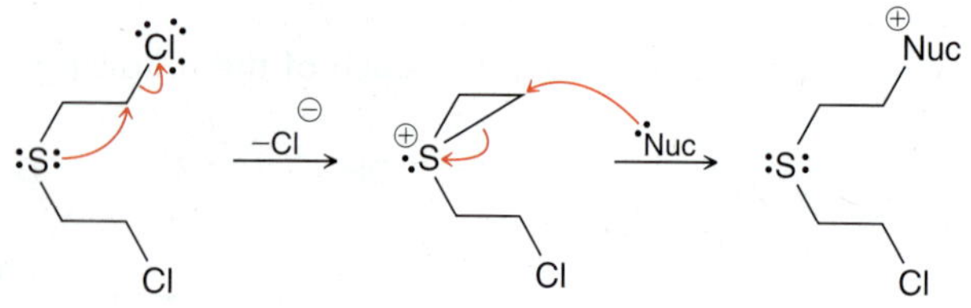

The first substitution reaction is an intramolecular $S_N2$-type process in which a lone pair on the sulfur serves as a nucleophile, expelling chloride as a leaving group. The second reaction is another $S_N2$ process involving attack of an external nucleophile. The net result is the same as if the nucleophile had attacked directly:

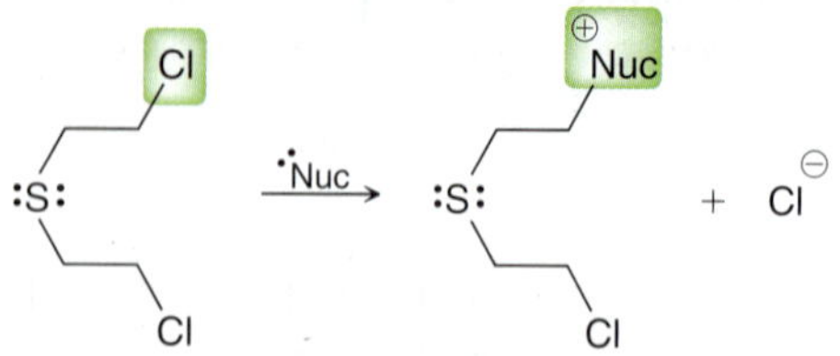

The reaction occurs much more rapidly than a regular $S_N2$ process on a primary alkyl chloride because the sulfur atom assists in ejecting chloride as a leaving group. The effect that sulfur has on the rate of reaction is called *anchimeric assistance*.

Each molecule of sulfur mustard has two chloride ions and is, therefore, capable of alkylating DNA two times. This causes individual strands of DNA to cross-link (Figure 7.31). Cross-linking of DNA prevents the DNA from replicating and ultimately leads to cell death. The profound impact of sulfur mustard on cell function inspired research on the use of this compound as

DNA DNA Cross-linked

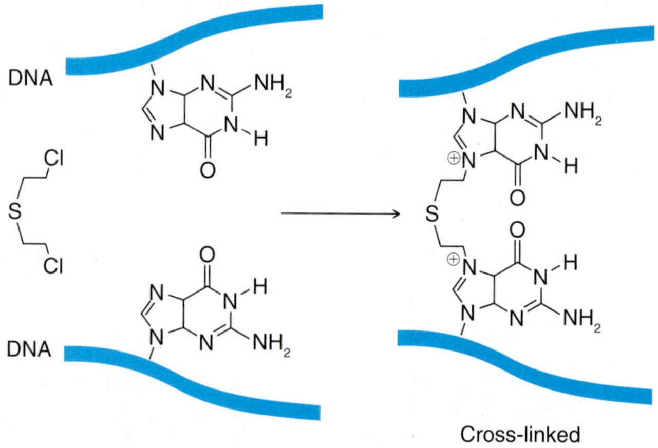

**FIGURE 7.31**
Sulfur mustard can alkylate two different strands of DNA, causing cross-linking.

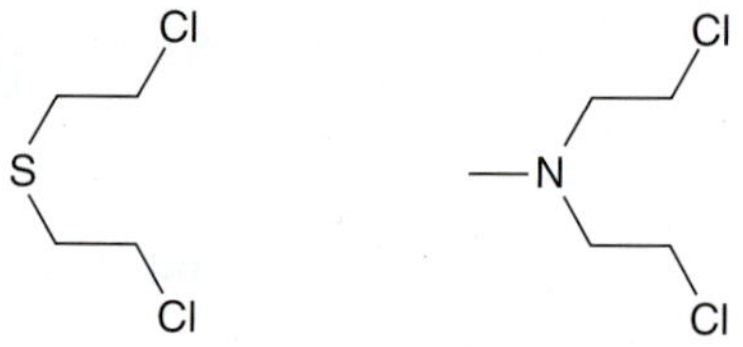

an antitumor agent. In 1931, sulfur mustard was injected directly into tumors with the intention of stopping tumor growth by interrupting the rapid division of the cancerous cells. Ultimately, sulfur mustard was found to be too toxic for clinical use, and the search began for a similar, less toxic, compound. The first such compound to be produced was a nitrogen analogue called mechlorethamine:

Sulfur mustard — Nitrogen mustard (mechlorethamine)

Mechlorethamine is a "nitrogen mustard" that reacts with nucleophiles in the same way as sulfur mustard, via two successive substitution reactions. The first reaction is an intramolecular $S_N2$-type process; the second reaction is another $S_N2$ process involving attack of an external nucleophile:

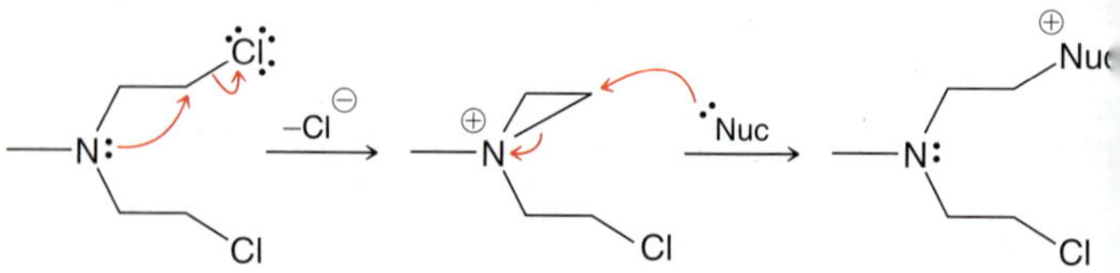

This nitrogen mustard is also capable of alkylating DNA, causing cell death, but is less toxic than sulfur mustard. The discovery of mechlorethamine launched the field of chemotherapy, the use of chemical agents to treat cancer.

Mechlorethamine is still in use today, in combination with other agents, for the treatment of advanced Hodgkin's lymphoma and chronic lymphocytic leukemia (CLL). The use of mechlorethamine is limited, though, by its high rate of reactivity with water. This limitation led to a search for other analogues. Specifically, it was found that replacing the methyl group with an aryl group had the effect of delocalizing the lone pair through resonance, rendering the lone pair less nucleophilic:

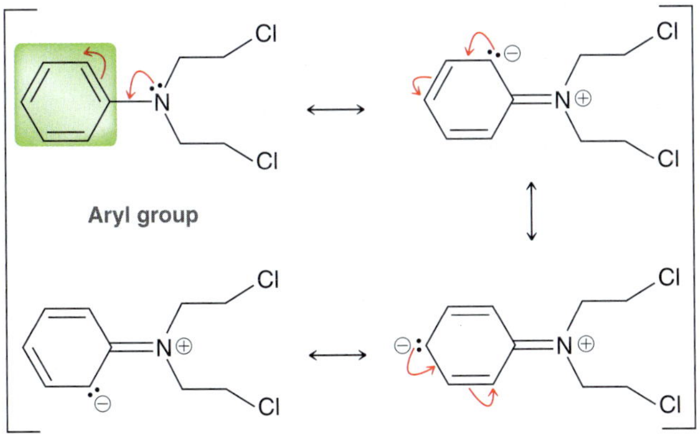

Aryl group

The resonance structures above all exhibit a negative charge on a carbon atom, and therefore, these resonance structures do not contribute very much to the overall resonance hybrid. Nevertheless, they are valid resonance structures, and they do contribute some character. As a result, the lone pair on the nitrogen atom is delocalized (it is spread out over the aryl group) and less nucleophilic. This decreased nucleophilicity is manifested in a decrease of anchimeric assistance from the nitrogen atom. The compound can still function as an antitumor agent, but its rate of reactivity with water is reduced.

Introduction of the aryl group (in place of the methyl group) might have solved one problem, but it created another problem. Specifically, this new compound was not water soluble, which prevented intravenous administration. This problem was solved by introducing a carboxylate group, which rendered the compound water soluble:

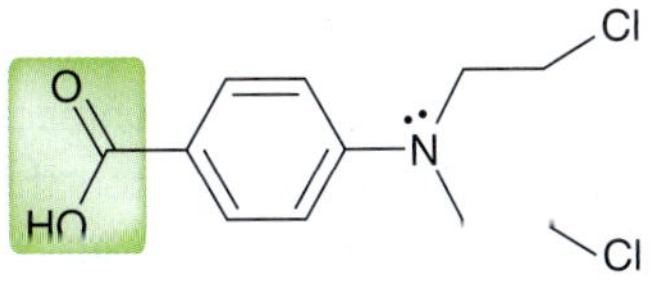

**This group renders the compound water soluble.**

But, once again, solving one problem created another. Now, the lone pair on the nitrogen atom was too delocalized, because of the following resonance structure:

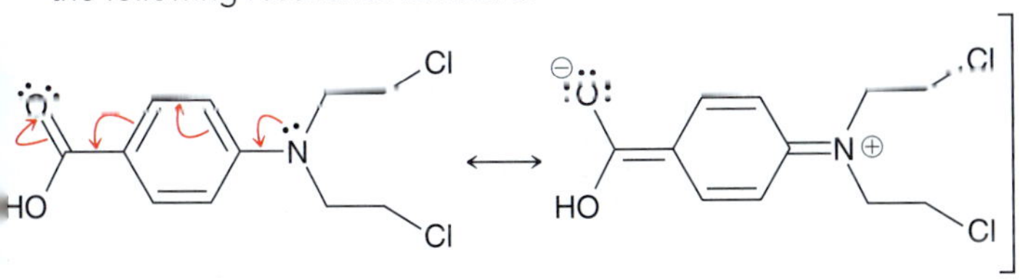

The lone pair was delocalized onto an oxygen atom, and a negative charge on an oxygen atom is much more stable than a negative charge on a carbon atom. The delocalization effect was so pronounced that the reagent no longer functioned as an antitumor agent. The lone pair on the nitrogen atom was not sufficiently nucleophilic to provide anchimeric assistance. Solving all these problems required a way to maintain water solubility without overly stabilizing the lone pair on the nitrogen atom. This was achieved by placing methylene groups ($CH_2$ groups) between the carboxylate group and the aryl group:

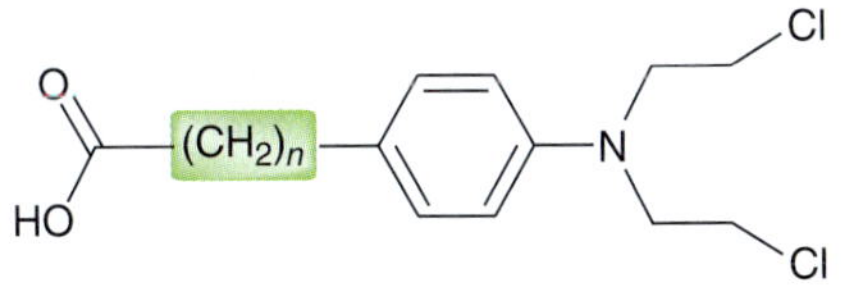

This way, the nitrogen lone pair is no longer participating in resonance with the carboxylate group, but the presence of the carboxylate group is still able to render the compound water soluble. This final change solves all of the problems. In theory, only one methylene group is needed to ensure that the nitrogen lone pair is not overly delocalized by resonance. But in practice, research with various compounds indicated that optimal reactivity was achieved when three methylene groups were placed between the carboxylate group and the aryl group:

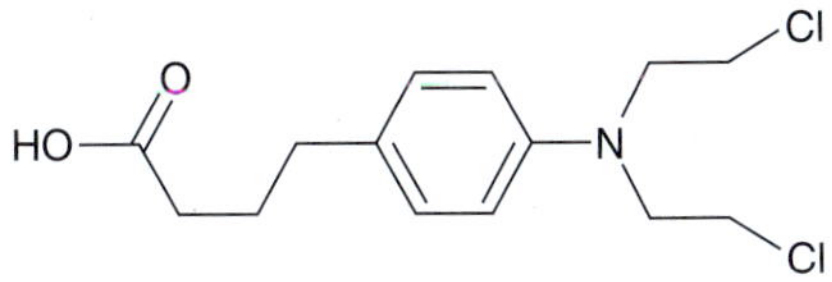

**Chlorambucil**

The resulting compound, called chlorambucil, was marketed under the trade name Lukeran™ by GlaxoSmithKline. It was mainly used for treatment of CLL, until other, more powerful agents were discovered.

The design and development of chlorambucil is just one example of drug design, but it demonstrates how an understanding of pharmacology, coupled with an understanding of the first principles of organic chemistry, enables chemists to design and create new drugs. Each year, organic chemists and biochemists make enormous strides in the exciting fields of pharmacology and drug design.

## CONCEPTUAL CHECKPOINT

**7.35** Melphalan is a chemotherapy drug used in the treatment of multiple myeloma and ovarian cancer. Melphalan is an alkylating agent belonging to the nitrogen mustard family. Draw a likely mechanism for the alkylation process that occurs when a nucleophile reacts with melphalan:

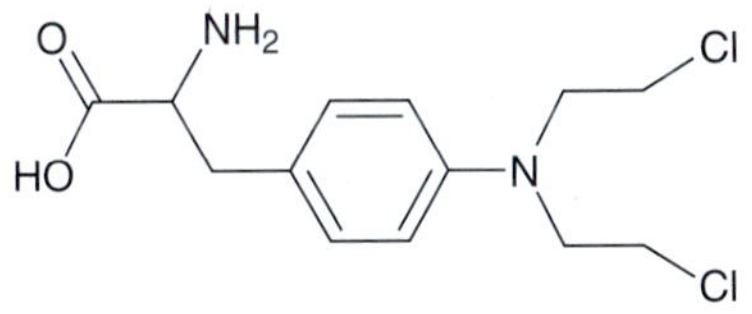

**Melphalan**

# REVIEW OF REACTIONS

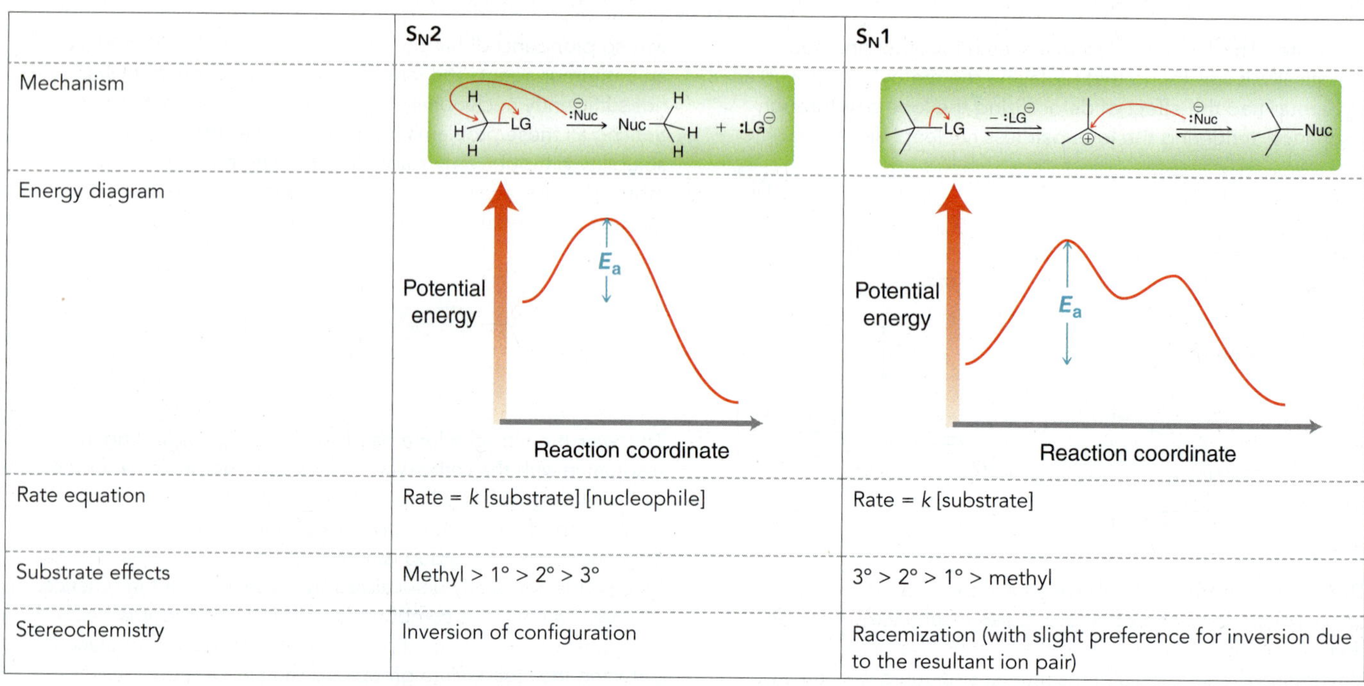


| | $S_N2$ | $S_N1$ |
|---|---|---|
| Mechanism | | |
| Energy diagram | | |
| Rate equation | Rate = $k$ [substrate] [nucleophile] | Rate = $k$ [substrate] |
| Substrate effects | Methyl > 1° > 2° > 3° | 3° > 2° > 1° > methyl |
| Stereochemistry | Inversion of configuration | Racemization (with slight preference for inversion due to the resultant ion pair) |

# REVIEW OF CONCEPTS AND VOCABULARY

### SECTION 7.1

- **Substitution reactions** exchange one functional group for another.
- The electrophile is called the **substrate**, and it must contain a **leaving group**.

### SECTION 7.2

- There are two ways to name halogenated organic compounds. The systematic name treats the compound as a **haloalkane**, while the common name treats the compound as an **alkyl halide**.
- The **alpha (α) position** is the carbon atom connected directly to the halogen, while the **beta (β) positions** are the carbon atoms connected to the α position.
- Alkyl halides are classified as **primary**, **secondary**, and **tertiary** according to the number of alkyl groups connected to the α position.

### SECTION 7.3

- In a ***concerted process***, nucleophilic attack and loss of the leaving group occur simultaneously.
- In a ***stepwise process***, first the leaving group leaves, and then the nucleophile attacks.

### SECTION 7.4

- Evidence for the concerted mechanism, called $\mathbf{S_N2}$, includes the observation of a **second-order** rate equation. The $S_N2$ process is said to be **bimolecular**.
- Methyl halides and primary alkyl halides react most quickly, while tertiary alkyl halides are essentially unreactive toward $S_N2$.
- When the α position is a chirality center, the reaction proceeds with **inversion of configuration**. The preference for **back-side attack** stems from the need for constructive overlap of orbitals (according to MO theory). $S_N2$ reactions are said to be **stereospecific** because the configuration of the product is determined by the configuration of the substrate.

### SECTION 7.5

- Evidence for the stepwise mechanism, called $\mathbf{S_N1}$, includes the observation of a **first-order** rate equation. These reactions are said to be **unimolecular**.
- The first step of an $S_N1$ process (loss of a leaving group) is the **rate-determining step**.
- The carbocation intermediate can be attacked from either side, leading to **inversion of configuration** and **retention of configuration**. There is usually a slight preference for the inversion product due to the formation of ion pairs.

### SECTION 7.6

- A proton transfer is necessary at the beginning of an $S_N1$ mechanism if the substrate is an alcohol.
- A carbocation rearrangement can take place if it will lead to a more stable carbocation intermediate.
- A proton transfer is necessary at the end of an $S_N1$ mechanism if the nucleophile is neutral (not negatively charged).
- When the solvent functions as a nucleophile, the reaction is called a **solvolysis** reaction.

### SECTION 7.7

- A proton transfer is necessary at the beginning of an $S_N2$ mechanism if the leaving group is an OH group.
- A proton transfer must occur at the end of an $S_N2$ mechanism if the nucleophile is neutral.

### SECTION 7.8

- There are four factors that impact the competition between the $S_N2$ mechanism and $S_N1$: (1) the substrate, (2) the nucleophile, (3) the leaving group, and (4) the solvent.
- The most common leaving groups are halides and **sulfonate ions**. Of the sulfonate ions, the most common is the **tosylate** group.
- **Protic** solvents favor $S_N1$, while **polar aprotic** solvents favor $S_N2$.

### SECTION 7.9

- Depending on the type of nucleophile used, substitution reactions can be used to produce a wide range of different compounds.

# SKILLBUILDER REVIEW

## 7.1 DRAWING THE CURVED ARROWS OF A SUBSTITUTION REACTION

**CONCERTED MECHANISM** Two curved arrows drawn in one step. Nucleophilic attack is accompanied by simultaneous loss of a leaving group.

Nuc attack

Loss of a leaving group

Nuc:⁻ + LG → (−:LG⁻) → Nuc

**STEPWISE MECHANISM** Two curved arrows drawn in two separate steps. Leaving group leaves to form a carbocation intermediate, followed by a nucleophilic attack.

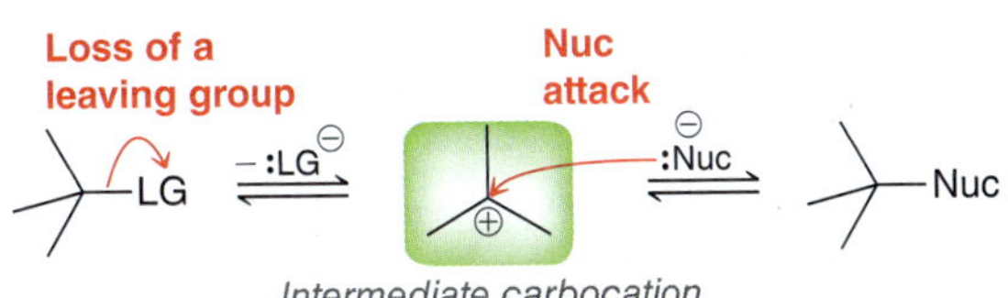


*Intermediate carbocation*

Try Problems 7.2–7.5, 7.64a

## 7.2 DRAWING THE PRODUCT OF AN $S_N2$ PROCESS

Replace the LG with the Nuc, and draw inversion of configuration.

Br + ⁻OH ⟶ OH + Br⁻

Try Problems 7.7, 7.8, 7.45, 7.56, 7.61

## 7.3 DRAWING THE TRANSITION STATE OF AN $S_N2$ PROCESS

**EXAMPLE** Draw the transition state.

Cl —NaSH→ SH

**STEP 1** Identify the nucleophile and the leaving group.

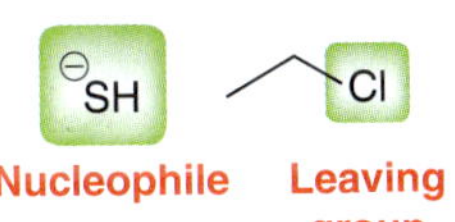


**STEP 2** Draw the carbon atom with the Nuc and LG on either side.

$\delta-$ HS - - - C - - - Cl $\delta-$

Bond forming    Bond breaking

**STEP 3** Draw the three groups attached to the carbon atom. Place brackets and the symbol indicating a transition state.

$\left[ \overset{\delta-}{HS} \cdots C(CH_3)(H)(H) \cdots \overset{\delta-}{Cl} \right]^{\ddagger}$

Try Problems 7.9–7.11, 7.46, 7.64e

## 7.4 DRAWING THE CARBOCATION INTERMEDIATE OF AN $S_N1$ PROCESS

**STEP 1** Identify the leaving group.

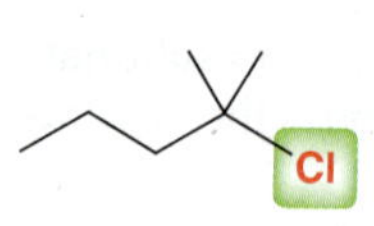


**STEP 2** Draw all three groups pointing away from each other.

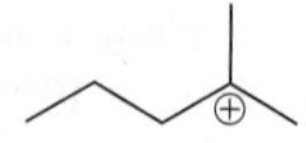

Try Problems 7.14, 7.15, 7.50, 7.51

## 7.5 DRAWING THE PRODUCTS OF AN $S_N1$ PROCESS

**EXAMPLE** Predict the products.

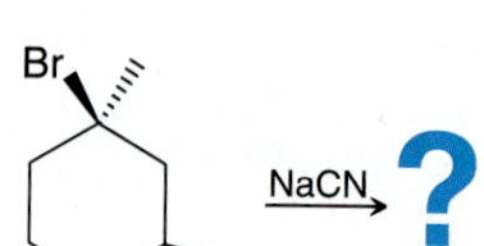


**STEP 1** Identify the nucleophile and the leaving group.

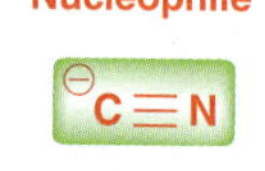

**STEP 2** Replace LG with Nuc.

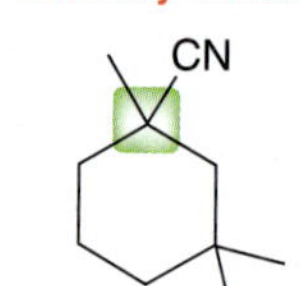


**STEP 3** If chirality center, then $S_N1$ will produce pair of enantiomers.

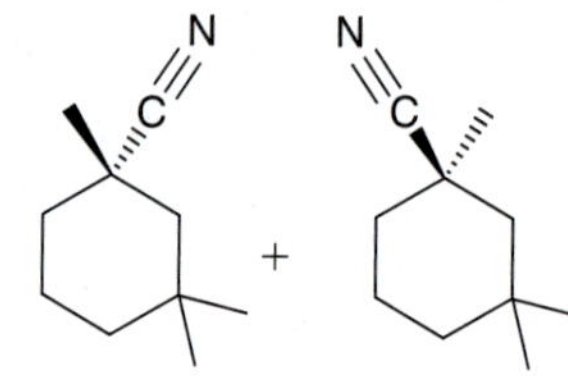


Try Problems 7.16, 7.17, 7.54b

## 7.6 DRAWING THE COMPLETE MECHANISM OF AN $S_N1$ PROCESS

There are two core steps (gray) and three possible additional steps (blue).

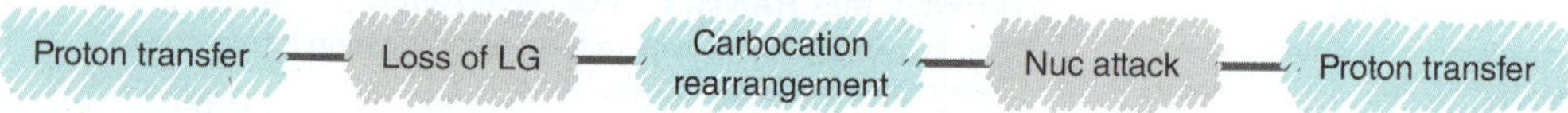


**Does the LG need to be protonated before it can leave?** If OH group, then yes.

**Is the nucleophile ultimately positioned at a different location than the leaving group?** If yes, then there is a carbocation rearrangement.

**Is the nucleophile neutral?** If yes, then there will be a proton transfer at the end of the mechanism in order to remove the positive charge.

Try Problems 7.21, 7.22, 7.23, 7.48, 7.49, 7.52, 7.54, 7.65

## 7.7 DRAWING THE COMPLETE MECHANISM OF AN $S_N2$ PROCESS

There is one core step (concerted) and two possible additional steps.

**Does the LG need to be protonated first?** If OH group, then yes.

**Is the nucleophile neutral?** If yes, then there will be a proton transfer at the end of the mechanism in order to remove the positive charge.

Try Problems 7.24, 7.25, 7.53, 7.64, 7.66

## 7.8 DETERMINING WHETHER A REACTION PROCEEDS VIA AN $S_N1$ MECHANISM OR AN $S_N2$ MECHANISM

**RELEVANT FACTORS**

| | $S_N2$ | $S_N1$ |
|---|---|---|
| *Substrate* | Methyl or primary | Tertiary |
| *Nuc* | Strong Nuc | Weak Nuc |
| *LG* | Good LG | Excellent LG |
| *Solvent* | Polar aprotic | Protic |

**EXAMPLE**

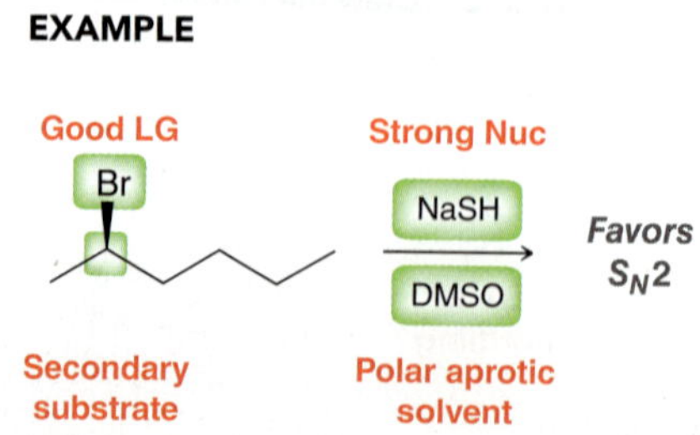


***Favors $S_N2$***

Try Problems 7.31, 7.32, 7.37, 7.38, 7.40, 7.41, 7.44, 7.55, 7.57, 7.58

## 7.9 IDENTIFYING THE REAGENTS NECESSARY FOR A SUBSTITUTION REACTION

**EXAMPLE**

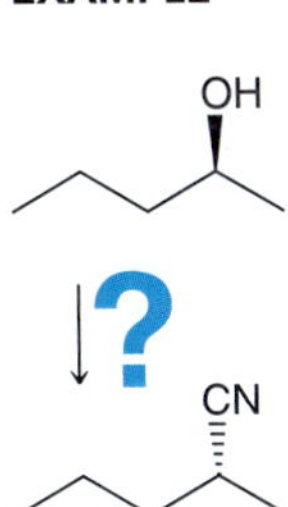


**STEP 1** Analyze the substrate and the stereochemistry.

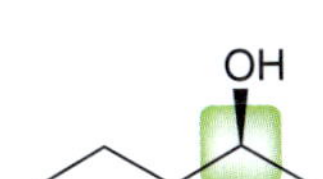


Secondary substrate. Inversion of configuration. Must be $S_N2$.

**STEP 2** Analyze the LG.

Bad LG. Must convert to tosylate using TsCl and pyridine.

**STEP 3** Use conditions that favor $S_N2$: strong Nuc (NaCN) and a polar aprotic solvent (DMSO).

Reagents:
1. TsCl, pyridine
2. NaCN, DMSO

Try Problems 7.33, 7.34, 7.59, 7.60, 7.63

# PRACTICE PROBLEMS

Note: Most of the Problems are available within **WileyPLUS**, an online teaching and learning solution.

**7.36** List the systematic name and common name for each of the following compounds:

(a) 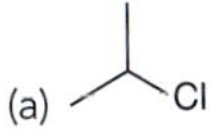

(b) 

(c) 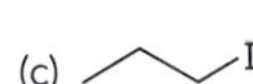

(d) 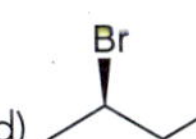

(e) Cl

**7.37** Draw all isomers of $C_4H_9I$ and then arrange them in order of increasing reactivity toward an $S_N2$ reaction.

**7.38** For each of the following pairs of compounds, identify which compound would react more rapidly in an $S_N2$ reaction. Explain your choice in each case.

(a) Cl, Cl (b) Br, Br (c) Cl, Cl (d) Br, I

**7.39** In Chapter 10, we will see that an acetylide ion (formed by treatment of acetylene with a strong base) can serve as a nucleophile in an $S_N2$ reaction:

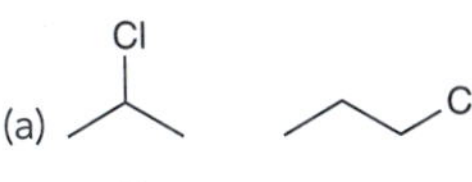

This reaction provides a useful method for making a variety of substituted alkynes. Determine whether this process can be used to make the following alkyne. Explain your answer.

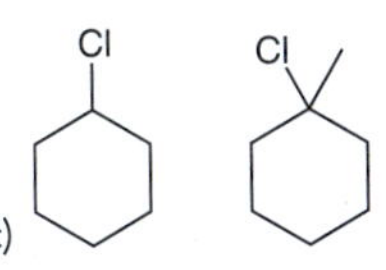


**7.40** Identify the stronger nucleophile:

(a) NaSH vs. $H_2S$

(b) Sodium hydroxide vs. water

(c) Methoxide dissolved in methanol vs. methoxide dissolved in DMSO

**7.41** For each pair of the following compounds, identify which compound would react more rapidly in an $S_N1$ reaction. Explain your choice in each case.

(a) Cl, Cl (b) Br, Br (c) Cl, Cl (d) Cl, OTs

**7.42** Consider the following reaction:

Br —NaCN, DMSO→ CN + NaBr

(a) How would the rate be affected if the concentration of the alkyl halide is doubled?

(b) How would the rate be affected if the concentration of sodium cyanide is doubled?

**7.43** Consider the following reaction:

OH —HBr→ Br + $H_2O$

(a) How would the rate be affected if the concentration of the alcohol is doubled?

(b) How would the rate be affected if the concentration of HBr is doubled?

**7.44** Classify each of the following solvents as protic or aprotic:

(a) DMF
(b) Ethanol
(c) DMSO
(d) Water
(e) Ammonia

**7.45** Consider the following $S_N2$ reaction:

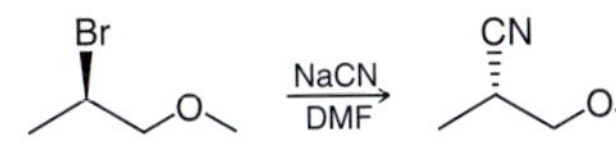

(a) Assign the configuration of the chirality center in the substrate.

(b) Assign the configuration of the chirality center in the product.

(c) Does this $S_N2$ process proceed with inversion of configuration? Explain.

**7.46** Draw the transition state for the reaction between ethyl iodide and sodium acetate ($CH_3CO_2Na$).

**7.47** (*S*)-2-Iodopentane undergoes racemization in a solution of sodium iodide in DMSO. Explain.

**7.48** When the following optically active alcohol is treated with HBr, a racemic mixture of alkyl bromides is obtained:

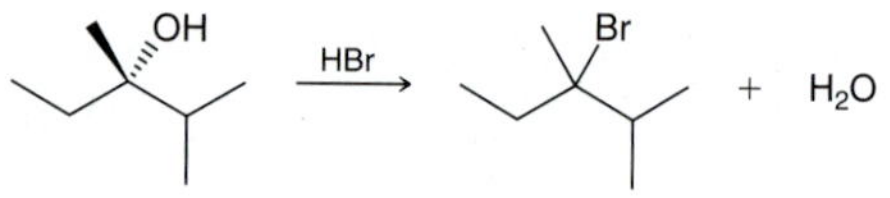

OH HBr Br + $H_2O$

Racemic mixture

Draw the mechanism of the reaction, and explain the stereochemical outcome.

**7.49** (*R*)-2-Pentanol racemizes when placed in dilute sulfuric acid. Draw a mechanism that explains this stereochemical outcome, and draw an energy diagram of the process.

**7.50** List the following carbocations in order of increasing stability:

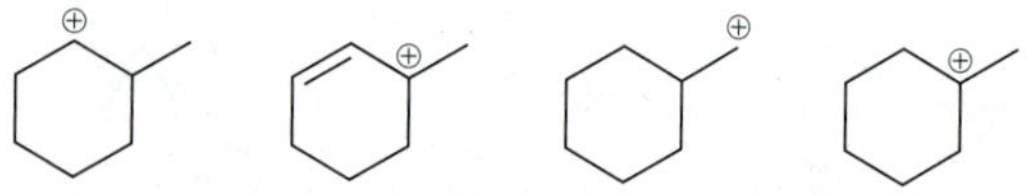

**7.51** Draw the carbocation intermediate that would be formed if each of the following substrates would participate in an $S_N1$ reaction. In each case, identify the carbocation as being primary, secondary, or tertiary.

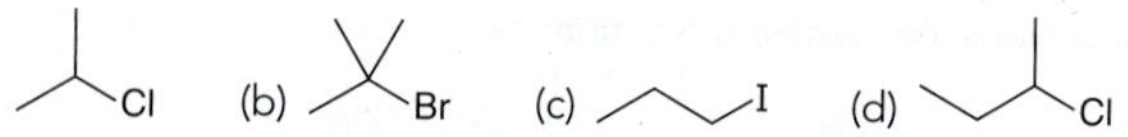

(a) Cl (b) Br (c) I (d) Cl

**7.52** Propose a mechanism for the following transformation:

OH HCl Cl + $H_2O$

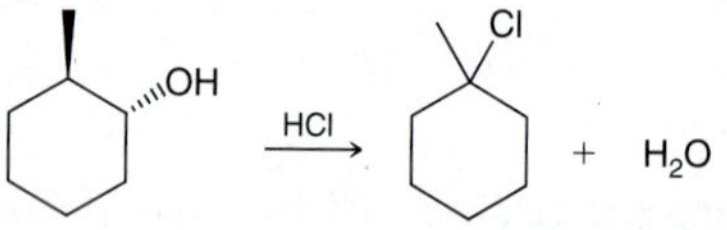

**7.53** Draw the mechanism of the following reaction:

Br O ONa O O + NaBr

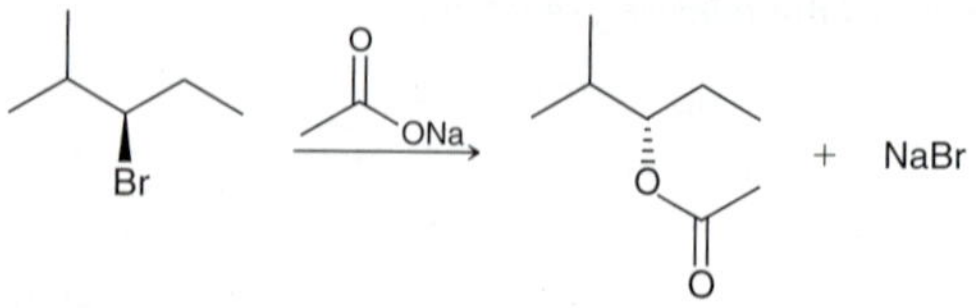

**7.54** Each of the following reactions proceeds via an $S_N1$ mechanism and will have anywhere from two to five steps, as discussed in Section 7.6. Determine the number of steps for each reaction and then draw the mechanism in each case:

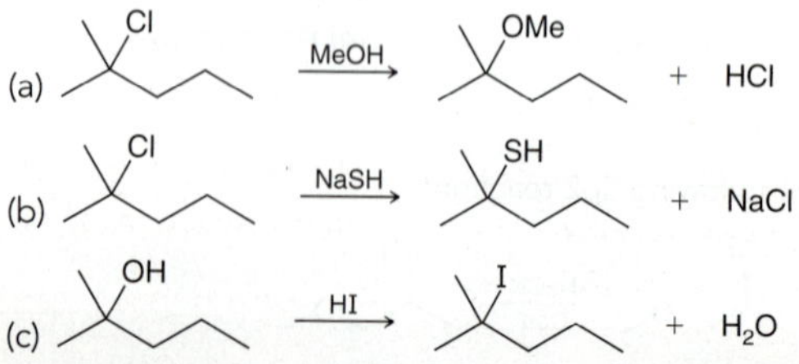

(a) Cl MeOH OMe + HCl

(b) Cl NaSH SH + NaCl

(c) OH HI I + $H_2O$

(d) OTs EtOH OEt + TsOH

**7.55** Identify the product(s) in each of the following reactions:

(a) Br EtOH ? (b) OTs NaBr ?

(c) OH HCl ? (d) I NaCN DMSO ?

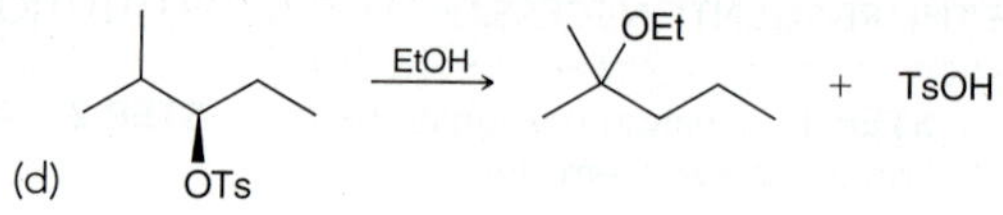

**7.56** Identify the product of the following reaction:

NaO ONa Br Br $C_4H_8O_2$ + 2 NaBr

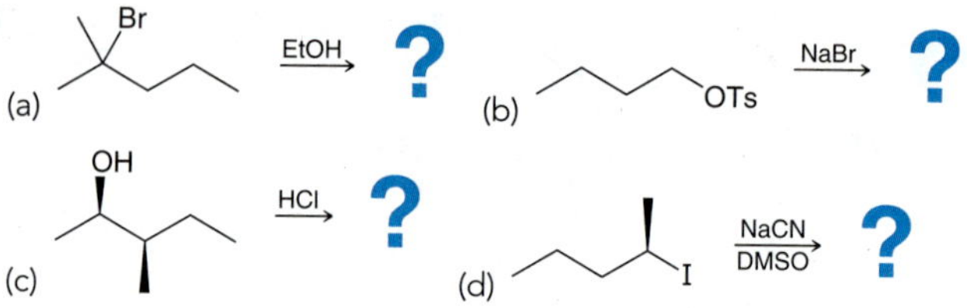

**7.57** The following reaction is very slow. Identify the mechanism and explain why the reaction is so slow.

Br NaOH, $H_2O$ OH

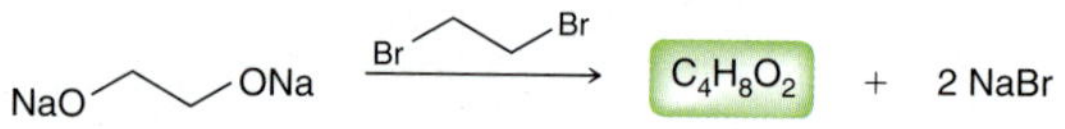

**7.58** The following reaction is very slow:

Br $H_2O$ OH + HBr

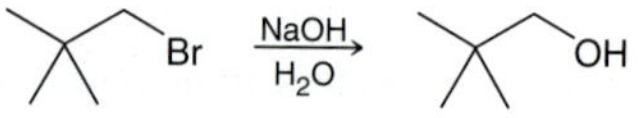

(a) Identify the mechanism.

(b) Explain why the reaction is so slow.

(c) When hydroxide is used instead of water, the reaction is very rapid. Draw the mechanism of this reaction and explain why it is so fast.

**7.59** Identify the reagents you would use to achieve each of the following transformations:

(a) OTs → OH

(b) OH → CN

(c) OH → Br

(d) Cl → SH

(e) Br → O O

**7.60** Each of the following compounds can be prepared with an alkyl iodide and a suitable nucleophile. In each case, identify the alkyl iodide and the nucleophile that you would use.

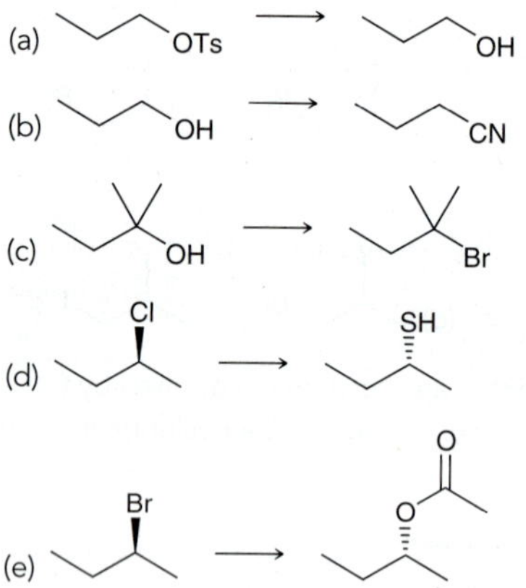

(a) OH (b) O O

(c) CN (d) SH

(e) OH (f) SH

**7.61** What products would you expect from the reaction between (*S*)-2-iodobutane and each of the following nucleophiles?

(a) NaSH (b) NaSEt (c) NaCN

**7.62** Below are two potential methods for preparing the same ether, but only one of them is successful. Identify the successful approach and explain your choice.

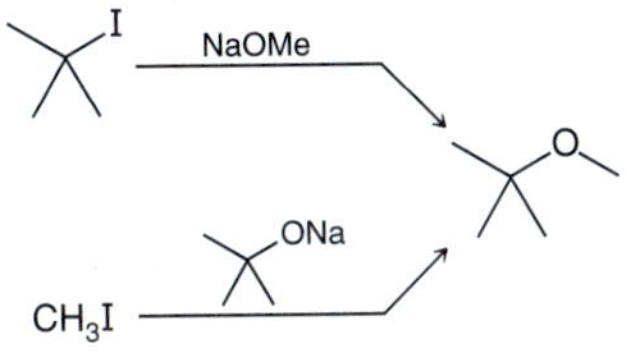

$CH_3I$ + *tert*-butoxide (ONa) ; *tert*-butyl iodide + NaOMe

**7.63** Identify the reagent you would use to accomplish each of the following transformations:

(a) 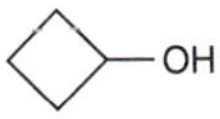 cyclobutanol (—OH) ⟶ bromocyclobutane

(b) $(CH_3)_3COH$ ⟶ *tert*-butyl chloride

(c) $CH_3CH_2Cl$ ⟶ $CH_3CH_2OH$

## INTEGRATED PROBLEMS

**7.64** Consider the following $S_N2$ reaction:

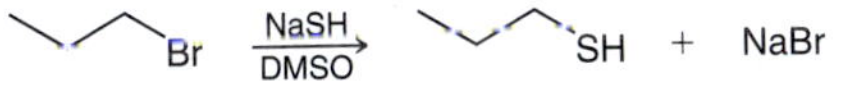

Br —(NaSH, DMSO)→ SH + NaBr

(a) Draw the mechanism of this reaction.

(b) What is the rate equation of this reaction?

(c) What would happen to the rate if the solvent is changed from DMSO to ethanol?

(d) Draw an energy diagram of the reaction above.

(e) Draw the transition state of this reaction.

**7.65** Consider the following substitution reaction:

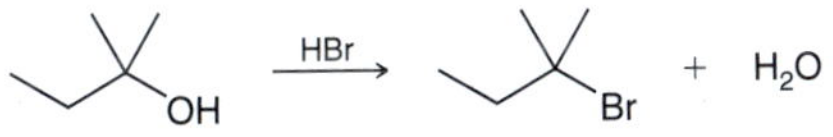

OH —(HBr)→ Br + $H_2O$

(a) Determine whether this reaction proceeds via an $S_N1$ or $S_N2$ process.

(b) Draw the mechanism of this reaction.

(c) What is the rate equation of this reaction?

(d) Would the reaction occur at a faster rate if sodium bromide were added to the reaction mixture?

(e) Draw an energy diagram of this reaction.

**7.66** Consider the following substitution reaction:

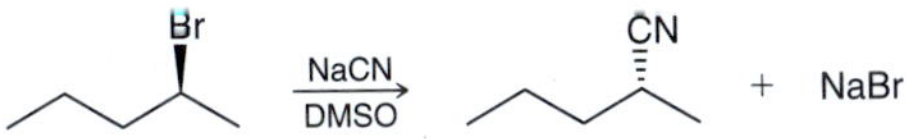

Br —(NaCN, DMSO)→ CN + NaBr

(a) Determine whether this reaction proceeds via an $S_N1$ or $S_N2$ process.

(b) Draw the mechanism of this reaction.

(c) What is the rate equation of this reaction?

(d) Would the reaction occur at a faster rate if the concentration of cyanide were doubled?

(e) Draw an energy diagram of the reaction above.

**7.67** Propose a mechanism for the following transformation:

I —($H_2O$)→ OH

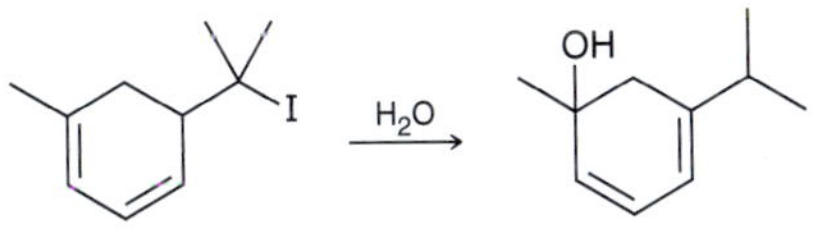

**7.68** When the following ester is treated with lithium iodide in DMF, a carboxylate ion is obtained:

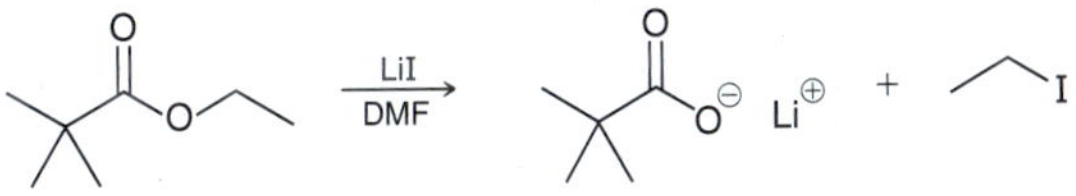

O —(LiI, DMF)→ O⊖ Li⊕ + I

(a) Draw the mechanism of this reaction.

(b) When the methyl ester is used as the substrate, the reaction is 10 times faster:

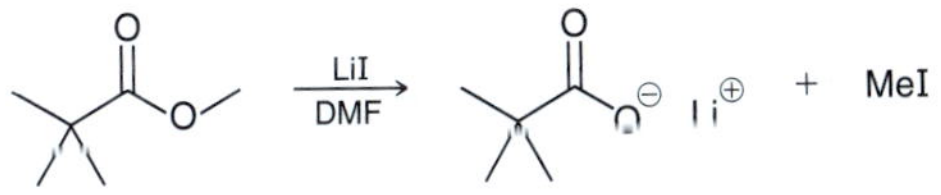

O —(LiI, DMF)→ O⊖ Li⊕ + MeI

Explain the increase in rate.

**7.69** When (1*R*,2*R*)-2-bromocyclohexanol is treated with a strong base, an epoxide (cyclic ether) is formed. Suggest a mechanism for formation of the epoxide:

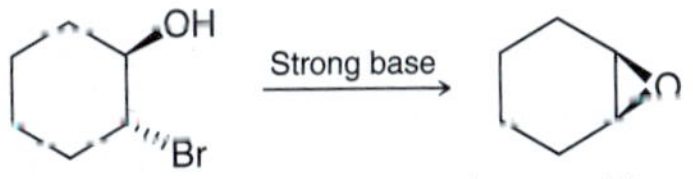

OH, Br —(Strong base)→ O

**An epoxide**

**7.70** When butyl bromide is treated with sodium iodide in ethanol, the concentration of iodide quickly decreases but then slowly returns to its original concentration. Identify the major product of the reaction.

**7.71** The following compound can react rapidly via an $S_N1$ process. Explain why this primary substrate will undergo an $S_N1$ reaction so rapidly.

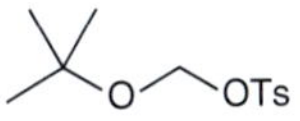

O OTs

**7.72** Consider the reaction below. The rate of this reaction is markedly increased if a small amount of sodium iodide is added to the reaction mixture. The sodium iodide is not consumed by the reaction and is therefore considered to be a catalyst. Explain how the presence of iodide can speed up the rate of the reaction.

NaCN, DMSO

**7.73** Propose a mechanism for the following transformation:

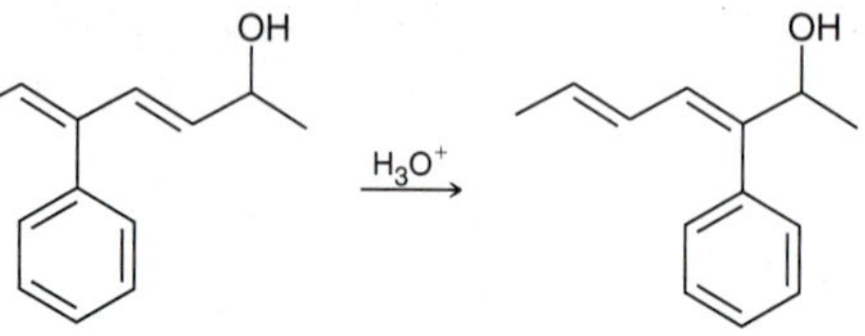

$H_3O^+$

**7.74** The following reaction sequence was part of a stereocontrolled synthesis of cyoctol, used in the treatment of male pattern baldness (*Tetrahedron* **2004,** *60,* **9599–9614**). The third step in this process employs an uncharged nucleophile, affording an ion pair as the product (anion and cation).

(a) Draw the product of the reaction sequence, and describe the factors that make the third step favorable.

(b) Suggest a reason for the function of the second step of this process.

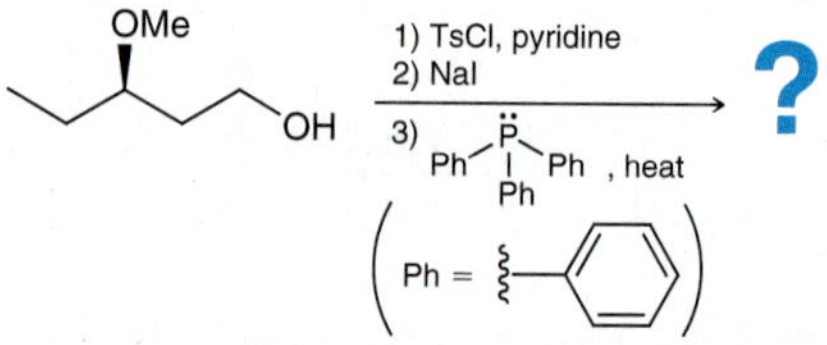

1) TsCl, pyridine
2) NaI
3) $PPh_3$, heat

**Problems 7.75–7.76 are intended for students who have already covered IR spectroscopy (Chapter 15).**

**7.75** Upon treatment with a strong base (such as NaOH), 2-naphthol (compound **1**) is converted to its conjugate base (anion **2**), which is then further converted into compound **3** upon treatment with butyl iodide (*J. Chem. Ed.* **2009,** *86,* **850–852**).

(a) Draw the structure of **2**, including a mechanism for its formation, and justify why hydroxide is a sufficiently strong base to achieve the conversion of **1** to **2**.

(b) Draw the structure of **3**, including a mechanism of its formation.

(c) Compound **3** is most easily purified via recrystallization in water. As such, the product would have to be thoroughly dried in order to utilize IR spectral analysis to verify the conversion of **1** to **3**. Explain why a wet sample would interfere with the utility of IR spectral analysis in this case and explain how you would use IR spectral analysis to verify complete conversion of **1** to **3**, assuming the product was properly dried.

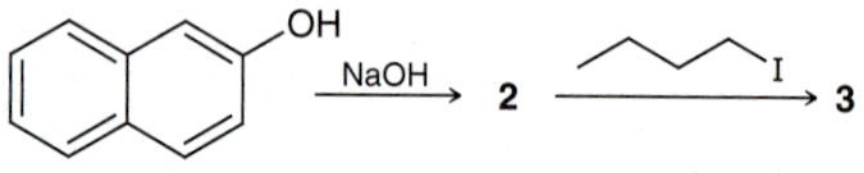

**7.76** Bromotriphenylmethane (compound **1**) can be converted to **2a** or **2b** or **2c** upon treatment with the appropriate nucleophile (*J. Chem. Ed.* **2009,** *86,* **853–855**).

(a) Draw a mechanism for the conversion of **1** to **2b**.

(b) In all three cases, conversion of **1** to **2** is observed to be nearly instantaneous (the reaction occurs extremely rapidly). Justify this observation with any drawings that you feel are necessary.

(c) IR spectroscopy is an ideal tool for monitoring the conversion of **1** to **2a**, while other forms of spectroscopy will be better suited for monitoring the conversion of **1** → **2b** or **1** → **2c**. Explain.

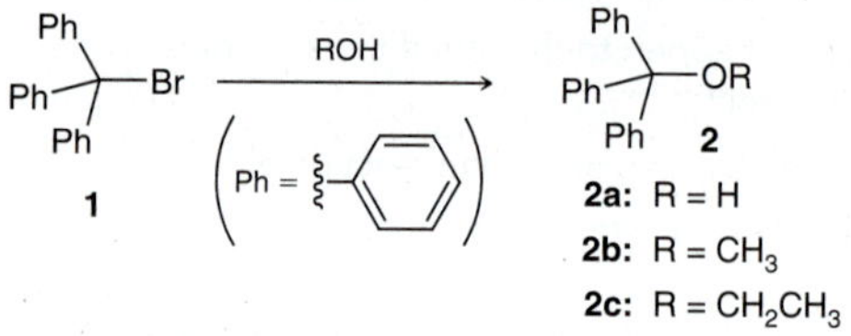

ROH

**2a:** R = H
**2b:** R = $CH_3$
**2c:** R = $CH_2CH_3$

# CHALLENGE PROBLEMS

**7.77** A common method for confirming the proposed structure and stereochemistry of a natural product is to achieve a total synthesis of the proposed structure and then compare its spectroscopic properties (Chapters 15 and 16) with those of the natural product. This technique was used to verify the structure of (−)-cameroonanol, a compound isolated from the essential oil of the flowering plant *Echniops giganteus* (*Org. Lett.* **2000,** *2,* **2717–2719**). During the synthesis of (±)-cameroonanol, the following reaction was employed. Draw a plausible mechanism for this transformation.

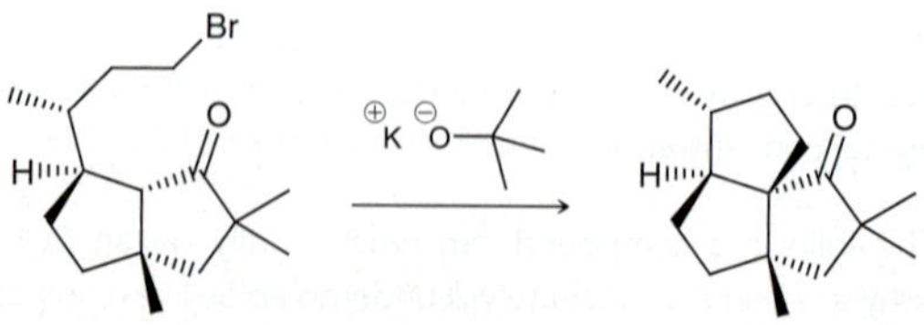

**7.78** Thienamycin is a potent antibacterial agent isolated from the fermentation broth of the soil microorganism *Streptomyces cattleya*. The following $S_N2$ process was utilized in a synthesis of thienamycin (*J. Am. Chem. Soc.* **1980,** *102,* **6161–6163**).

(a) Draw the product of this process (compound **3**).

(b) The nucleophile exhibits a six-membered ring with two sulfur atoms, called a dithiane ring. Explain why the dithiane ring in compound **3** exists primarily in two conformations, while the dithiane ring in compound **2** exists primarily in one conformation.

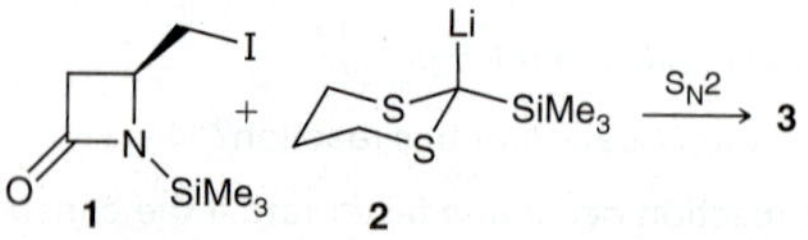

**7.79** Optically pure 2-octyl sulfonate was treated with varying mixtures of water and dioxane, and the optical purity of the resulting product (2-octanol) was found to vary with the ratio of water to dioxane, as shown in the following table (*J. Am. Chem. Soc.* **1965,** *87,* **287–291**). Given that dioxane possesses fairly nucleophilic oxygen atoms, provide a complete mechanism that explains the variation in the product's optical purity due to changes in solvent composition.

Excellent leaving group

(dioxane)

$H_2O$

**(R)-2-Octyl sulfonate (optically pure)**

**(S)-2-Octanol**

| Solvent ratio (water : dioxane) | Optical purity of (S)-2-octanol |
|---|---|
| 25 : 75 | 77% |
| 50 : 50 | 88% |
| 75 : 25 | 95% |
| 100 : 0 | 100% |

**7.80** Cyclopropyl chloride (**1**) cannot generally be converted into cyclopropanol (**4**) through a direct substitution reaction, because undesired, ring-opening reactions occur. The following represents an alternative method for preparing cyclopropanol (*Tetrahedron Lett.* **1967,** *8,* **4941–4944**).

NaOH

Mg

$H_3O^+$

MgCl

(a) Compound **2** is a powerful nucleophile, and for our purposes, we will treat $MgCl^+$ as a counterion. The transformation of **2** into **3** is accomplished via an $S_N2$-type process. Draw a mechanism for this process and identify the leaving group.

(b) Explain why the conversion of **2** to **3** is an irreversible process.

(c) Under aqueous acidic conditions, **3** can be converted into **4** either via an $S_N1$ process or via an $S_N2$ process. Draw a complete mechanism for each of these pathways.

(d) During conversion of **3** to **4**, another alcohol (ROH) is formed as a byproduct. Draw the structure of this alcohol.

**7.81** Compound **1** was prepared during a recent synthesis of 1-deoxynojirimycin, a compound with application to HIV chemotherapy (*Org. Lett.* **2010,** *12,* **136–139**). Upon formation, compound **1** rapidly undergoes ring contraction in the presence of chloride ion to form compound **2**. Propose a plausible mechanism that includes a justification for the stereochemical outcome.

**7.82** Halogenated derivatives of toluene will undergo hydrolysis via an $S_N1$ process:

The rate of hydrolysis is dependent on two main factors: (1) the stability of the leaving group and (2) the stability of the intermediate carbocation. The following are rates of hydrolysis ($\times 10^4$/min) for halogenated derivatives of toluene at 30 °C in 50% aqueous acetone (*J. Am. Chem. Soc.* **1951,** *73,* **22–23**):

| | Z = H | Z = Cl | Z = Br |
|---|---|---|---|
| X = H, Y = Cl | 0.22 | 2.21 | 31.1 |
| X = Cl, Y = Cl | 2.21 | 110.5 | 2122 |
| X = Br, Y = Br | 6.85 | 1803 | 1131 |

Using these data, answer the following questions:

(a) Using Figure 7.28, determine whether chloride or bromide is the better leaving group and explain your choice. Then, determine whether the hydrolysis data support your choice. Explain.

(b) Determine whether a carbocation is stabilized by an adjacent chloro group (i.e., a chlorine atom attached directly to $C^+$). Justify your choice by drawing resonance structures for the carbocation.

(c) Determine whether a carbocation is stabilized by an adjacent bromo group (i.e., a bromine atom attached directly to $C^+$). Justify your choice by drawing resonance structures for the carbocation.

(d) Determine whether a carbocation is more greatly stabilized by an adjacent chloro group or an adjacent bromo group.

(e) For these hydrolysis reactions, determine which factor is more important in determining the rate of hydrolysis: the stability of the leaving group or the stability of the carbocation. Explain your choice.

**7.83** Bimolecular substitution reactions commonly occur at $sp^3$-hybridized centers but generally do not occur at $sp^2$-hybridized centers. This selectivity is clearly observed in the conversion of **2** to **3** in the reaction sequence shown below. However, the conversion of **3** to **4** appears to be a rare example of an $S_N2$-type process occurring at an $sp^2$ hybridized center (*J. Am. Chem. Soc.* **2004,** *126,* **6868–6869**).

(a) Draw the structures of **2** and **3**.

(b) In the conversion of **3** to **4**, sodium amide ($NaNH_2$) functions as a base, and ammonia is the solvent. Draw a mechanism for the conversion of **3** to **4** and make sure to show the transition state for this process.

(c) Explain how the stereochemical outcome of this transformation is consistent with an $S_N2$-type process.

TsCl, pyridine; $NaNH_2$, $NH_3$, −78 °C

**7.84** When the following alkyl bromide is treated with sodium acetate in $CH_3CN$, two products are formed. The minor product retains the three-membered ring of the starting material, whereas the major product features a four-membered ring (*J. Org. Chem.* **2012,** *77,* **3181–3190**). Provide a plausible mechanism that explains the formation of these two products. (**Hint:** You might find inspiration for your answer in the medically speaking application at the end of the chapter.)

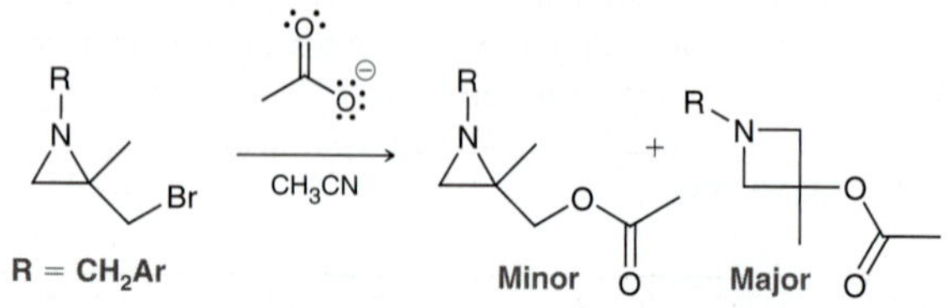

R = $CH_2Ar$ $\xrightarrow{CH_3CN}$ Minor + Major

**7.85** Biotin (compound **4**) is an essential vitamin that plays a vital role in several important physiological processes. A total synthesis of biotin, developed by scientists at Hoffmann-La Roche, involved the preparation of compound **1** (*J. Am. Chem. Soc.* **1982,** *104,* **6460–6462**). Conversion of **1** to **4** required removal of the OH group, which was achieved in several steps. First, the OH group was replaced with Cl by treating **1** with $SOCl_2$ to give **3**. The mechanism for the conversion of **1** to **3** proceeds via intermediate **2**, which has an excellent leaving group ($SO_2Cl$). Ejection of this leaving group causes the liberation of $SO_2$ gas and a chloride ion, which can occur if **2** is attacked by a chloride ion in an $S_N2$ reaction. As such, we expect the transformation of **2** to **3** to proceed via inversion of configuration. However, X-ray crystallographic analysis of compound **3** revealed that the transformation occurred with a net retention of configuration, as shown. Propose a mechanism that explains this curious result. (**Hint:** You might find it helpful to reference the medically speaking application in Section 7.9.)

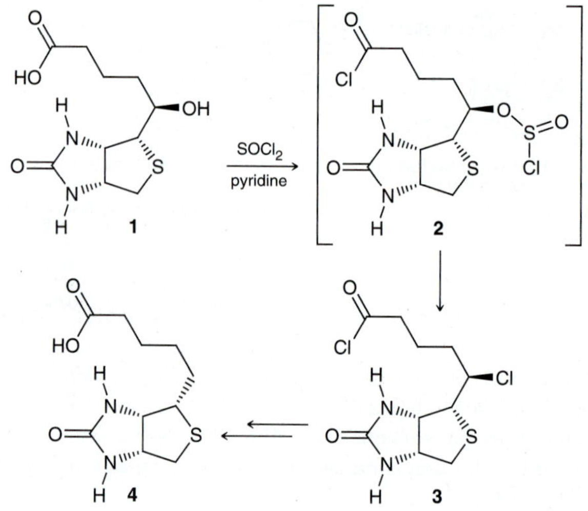

**1** $\xrightarrow[\text{pyridine}]{SOCl_2}$ [**2**] → **3** → **4**

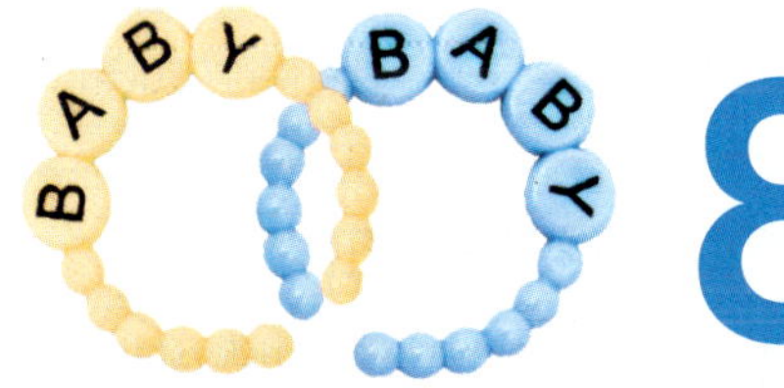


# 8

# Alkenes: Structure and Preparation via Elimination Reactions



## DID YOU EVER WONDER...

**what causes the skin of some newborn babies to turn yellow-orange (a condition called neonatal jaundice) and why exposure to blue light is the most effective remedy?**

The distinct color associated with neonatal jaundice is caused by an accumulation of a yellow-orange compound called bilirubin. If left untreated, the increasing levels of bilirubin in the body can interfere with the development of the baby's brain cells, causing permanent retardation. As we will see later in this chapter (in a Medically Speaking application), exposure to blue light causes a change in the configuration of a carbon-carbon double bond in bilirubin, which renders the compound more water soluble. As a result, the excess bilirubin is more readily excreted in the urine and feces, causing a decrease in bilirubin levels.

Like bilirubin, many naturally occurring compounds contain a carbon-carbon double bond. This chapter will focus on the properties and synthesis of compounds that possess carbon-carbon double bonds.

## DO YOU REMEMBER?

*Before you go on, be sure you understand the following topics.*
*If necessary, review the suggested sections to prepare for this chapter.*

- Kinetics and Energy Diagrams (Sections 6.5, 6.6)
- Arrow Pushing and Carbocation Rearrangements (Sections 6.8–6.11)
- Substitution Reactions (Sections 7.3–7.8)
- Introduction to Stereoisomerism (Section 5.2)
- Chair Conformations of Cyclohexane (Sections 4.11–4.13)

Take the DO YOU REMEMBER? QUIZ in **WileyPLUS** to check your understanding.

## 8.1 Introduction to Elimination Reactions

In the previous chapter, we saw that a substitution reaction can occur when a compound possesses a leaving group. In this chapter, we will explore another type of reaction, called *elimination*, commonly observed for compounds with leaving groups. Although elimination reactions occur in a variety of different contexts, for now we will examine them as a method for forming alkenes. Consider the difference between substitution and elimination reactions (Figure 8.1).

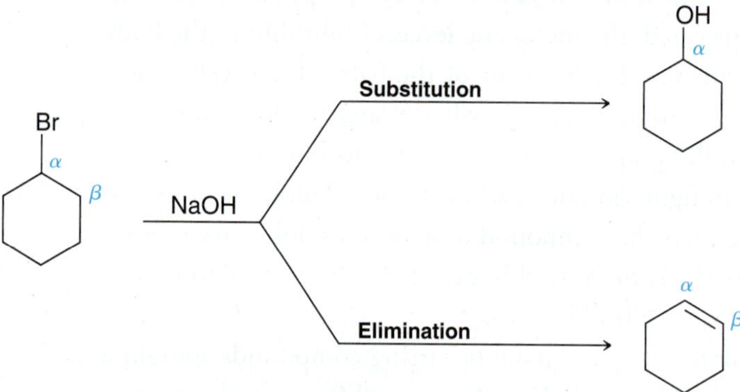


**FIGURE 8.1** The products of substitution and elimination reactions.

A substitution reaction occurs when the leaving group is replaced with a nucleophile. But in an elimination reaction, a proton from the beta (β) position is removed together with the leaving group, forming a double bond. This type of reaction is called a **beta elimination**, or **1,2-elimination**, and can be accomplished with any good leaving group. Specific classes of beta eliminations were characterized before chemists understood they had a common set of mechanisms. Thus, some types of beta elimination reactions are named on the basis of the leaving group. For example, when the leaving group is specifically a halide, the reaction is also called a **dehydrohalogenation**; when the leaving group is water, the reaction is also called **dehydration**. Regardless of the name, the key is to remember that all of these reactions have a common set of mechanisms that can be used to describe their underlying processes.

In the context of this discussion, the product of an elimination reaction possesses a C=C double bond and is called an **alkene**. The chapter will focus on the structure of alkenes and their preparation via beta elimination reactions.

## 8.2 Alkenes in Nature and in Industry

Alkenes are abundant in nature. Here are just a few examples, all acylic compounds (compounds that do not contain a ring):

O S S OH

**Allicin**
Responsible for the odor of garlic

**Geraniol**
Isolated from roses and used in perfumes

**α-Farnesene**
Found in the natural waxy coating on apple skins

Nature also produces many cyclic, bicyclic, and polycyclic alkenes:

**Limonene**
Responsible for the strong smell of oranges

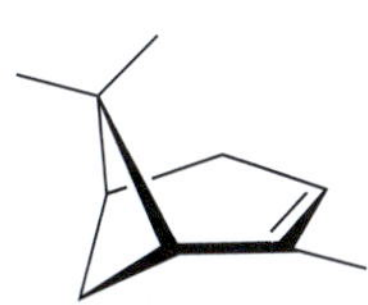

**α-Pinene**
Isolated from pine resin; a primary constituent of turpentine (paint thinner)

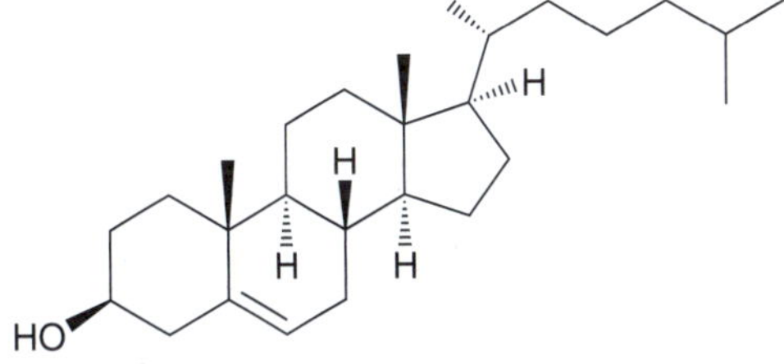


**Cholesterol**
Produced by all animals; this compound plays a pivotal role in many biological processes

Double bonds are also often found in the structures of pheromones. Recall that pheromones are chemicals used by living organisms to trigger specific behavioral responses in other members of the same species. For example, alarm pheromones are used to signal danger, while sex pheromones are used to attract the opposite sex for mating.

## practically speaking | Pheromones to Control Insect Populations

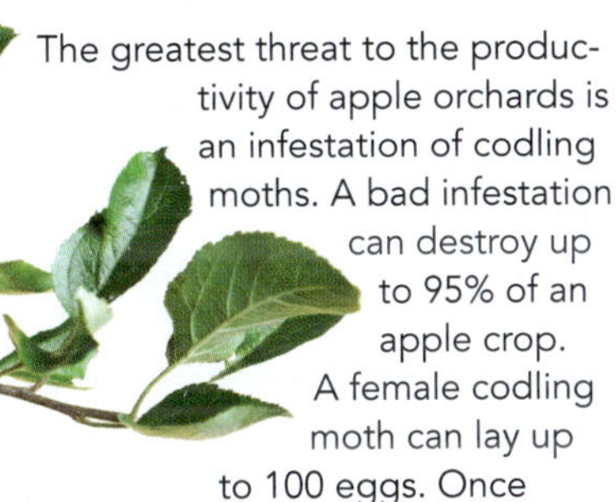

The greatest threat to the productivity of apple orchards is an infestation of codling moths. A bad infestation can destroy up to 95% of an apple crop. A female codling moth can lay up to 100 eggs. Once hatched, the larvae dig into the apples, where they are shielded from insecticides. The so-called "worm" in an apple is generally the larva of a codling moth. The main tool for dealing with these pests involves using one of the sex pheromones of the female to disrupt mating:

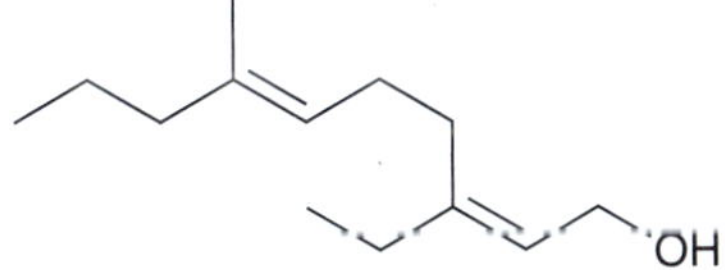


**(2Z,6E)-3-Ethyl-7-methyldeca-2,6-dien-1-ol**
A sex pheromone of the codling moth

This compound is easily produced in a laboratory. It can then be sprayed in an apple orchard where its presence interferes with the ability of the male to find the female. The sex pheromone is also used to lure the males into traps, enabling farmers to monitor populations and time the use of insecticides to coincide with the time period during which females are laying eggs, thereby increasing the efficacy of insecticides. Current research focuses on new compounds that attract both males and females. One such example is ethyl (2*E*,4*Z*)-2,4-decadienoate, known as the pear ester:

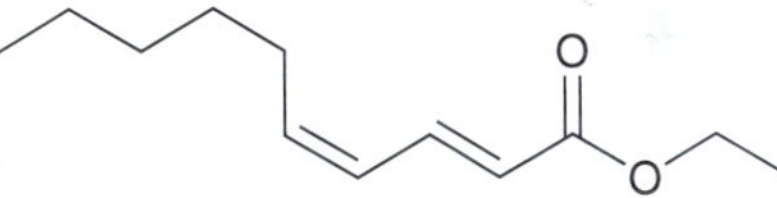


**Ethyl (2*E*,4*Z*)-2,4-decadienoate**
Pear ester

Researchers at the USDA (U.S. Department of Agriculture) have discovered that the pear ester can be potentially more effective in controlling codling moth populations than other compounds. This compound will potentially allow farmers to target the females and their eggs with greater precision. One of the major advantages of using pheromones as insecticides is that they tend to be less toxic to humans and less harmful to the environment.

Below are several examples of pheromones that contain double bonds:

**Muscalure**
Sex pheromone of the common housefly

**Ectocarpene**
A pheromone released by the eggs of the seaweed *Ectocarpus siliculosus* to attract sperm cells

**β-Farnesene**
An aphid alarm pheromone

Alkenes are also important precursors in the chemical industry. The two most important industrial alkenes, ethylene and propylene, are formed from cracking petroleum and are used as starting materials for preparing a wide variety of compounds (Figure 8.2).

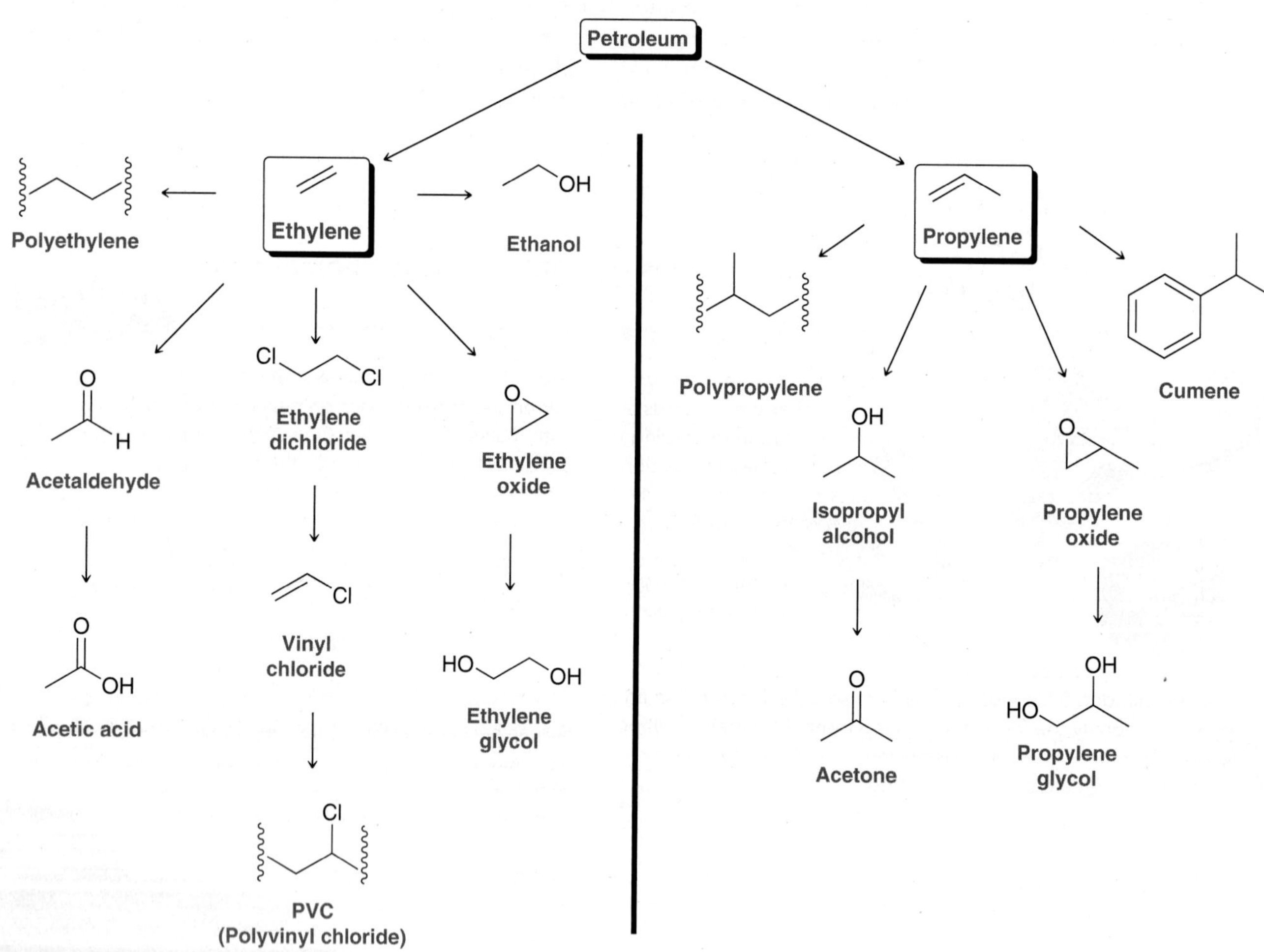


**FIGURE 8.2**
Industrially important compounds produced from ethylene and propylene. The squiggly lines (shown in the structures of polyethylene, polypropylene, and PVC) are used to indicate the repeating unit of a polymer. For example, PVC is comprised of repeating units, which can also be drawn in the following way: —$(CH_2CHCl)_n$—, where *n* is a very large number.

Each year, over 200 billion pounds of ethylene and 70 billion pounds of propylene are produced globally and used to make many compounds, including those shown in Figure 8.2.

## 8.3 Nomenclature of Alkenes

Recall from Chapter 4 that naming alk*anes* requires four discrete steps (Section 4.2):

1. Identify the parent.
2. Identify the substituents.
3. Assign a locant to each substituent.
4. Arrange the substituents alphabetically.

Alk*enes* are named using the same four steps, with the following additional guidelines:

When naming the parent, replace the suffix "ane" with "ene" to indicate the presence of a C=C double bond:

When we learned to name alkanes, we chose the parent by identifying the longest chain. When choosing the parent of an alkene, use the longest chain that includes the π bond:

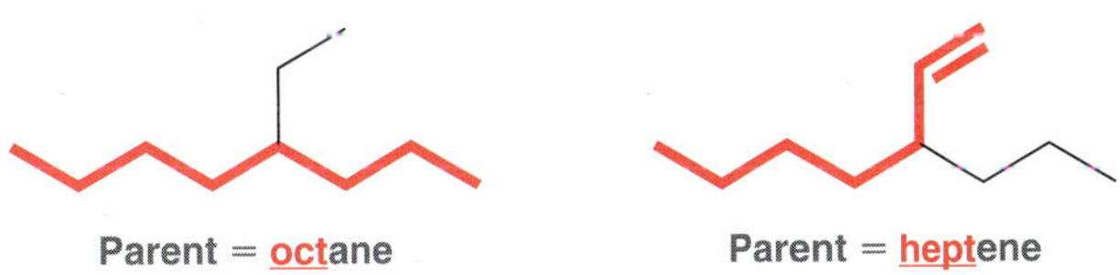


When numbering the parent chain of an alkene, the π bond should receive the lowest number possible despite the presence of alkyl substituents:

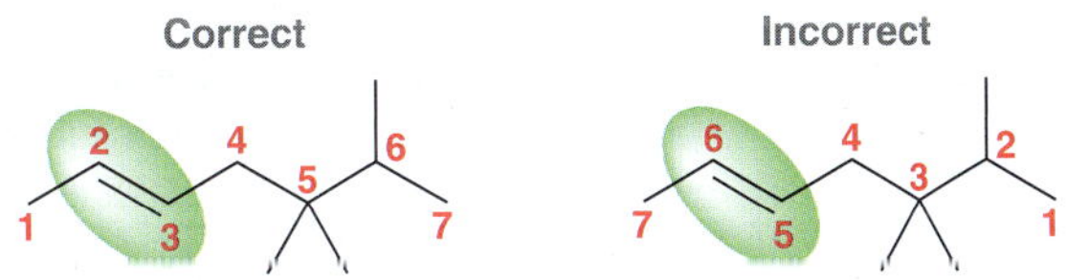


The position of the double bond is indicated using a single locant, rather than two locants. Consider the previous example where the double bond is between C2 and C3 on the parent chain. In this case, the position of the double bond is indicated with number 2. The IUPAC rules published in 1979 dictate that this locant be placed immediately before the parent, while the IUPAC recommendations released in 1993 and 2004 allow for the locant to be placed before the suffix "ene." Both names are acceptable:

5,5,6-Trimethyl-2-heptene
*or*
5,5,6-Trimethylhept-2-ene

# SKILLBUILDER

## 8.1 ASSEMBLING THE SYSTEMATIC NAME OF AN ALKENE

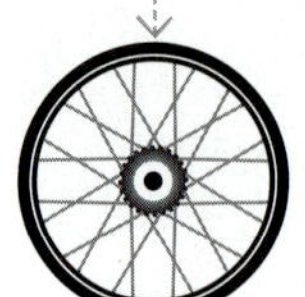

**LEARN the skill**

Provide a systematic name for the following compound:

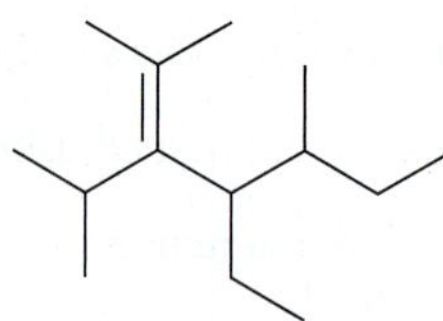

**SOLUTION**

Assembling a systematic name requires four discrete steps. First, identify the parent. Choose the longest chain that includes both carbon atoms of the $\pi$ bond:

**STEP 1**
Identify the parent.

Heptene

**STEP 2**
Identify and name the substituents.

Next, identify and name the substituents connected to the parent:

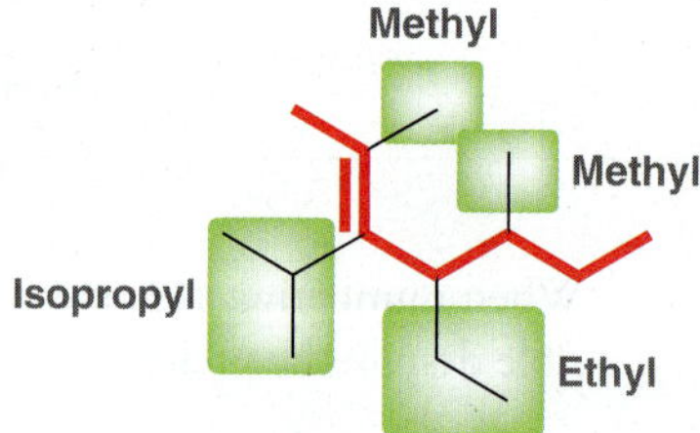


**STEP 3**
Number the parent chain and assign a locant to each substituent.

**STEP 4**
Arrange the substituents alphabetically.

**LOOKING BACK**
Recall that the prefix "iso" is counted as part of the alphabetization scheme, while the prefix "di" is ignored. For a review, see Section 4.2.

Then number the parent chain and assign a locant to each substituent. The parent is numbered starting from the side that is closest to the $\pi$ bond. This numbering scheme is used to determine the locant for each substituent:

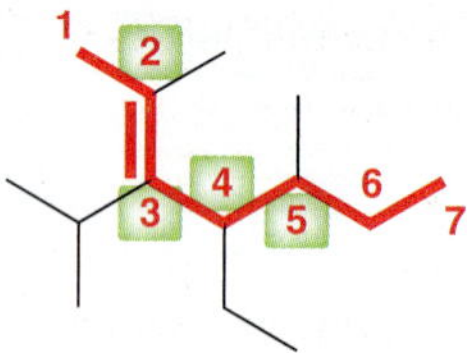


Finally, arrange the substituents alphabetically, making sure to include a locant that identifies the position of the double bond:

4-Ethyl-3-isopropyl-2,5-dimethyl-2-heptene

Notice that ***e***thyl appears first, then ***i***sopropyl, and finally di***m***ethyl. Also, make sure that hyphens separate letters from numbers, while commas separate numbers from each other.

**PRACTICE the skill** **8.1** Provide a systematic name for each of the following compounds:

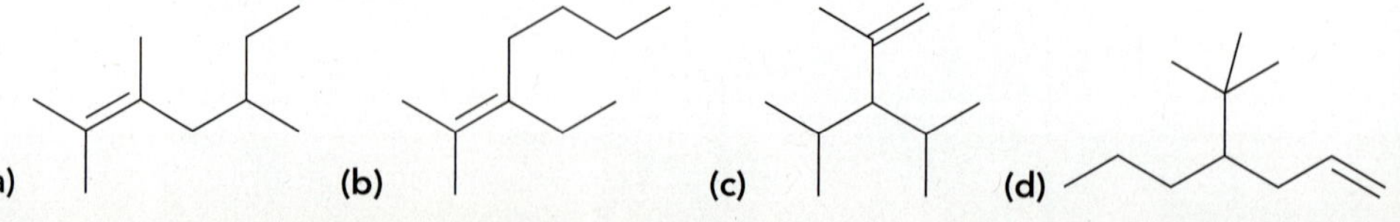

**APPLY the skill**

**8.2** Draw a bond-line structure for each of the following compounds:

**(a)** 3-Isopropyl-2,4-dimethyl-2-pentene

**(b)** 4-Ethyl-2-methyl-2-hexene

**(c)** 1,2-Dimethylcyclobutene (The name of a cycloalkene will often not include a locant to specify the position of the π bond, because, by definition, it is assumed to be between C1 and C2.)

**8.3** In Section 4.2, we learned how to name bicyclic compounds. Using those rules, together with the rules discussed in this section, provide a systematic name for the following bicyclic compound. In a case like this, the lowest number is assigned to a bridgehead (not to the π bond).

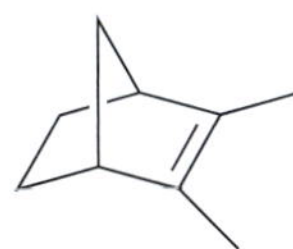

need more **PRACTICE?** **Try Problem 8.50b,c**

IUPAC nomenclature also recognizes common names for many simple alkenes. Here are three such examples:

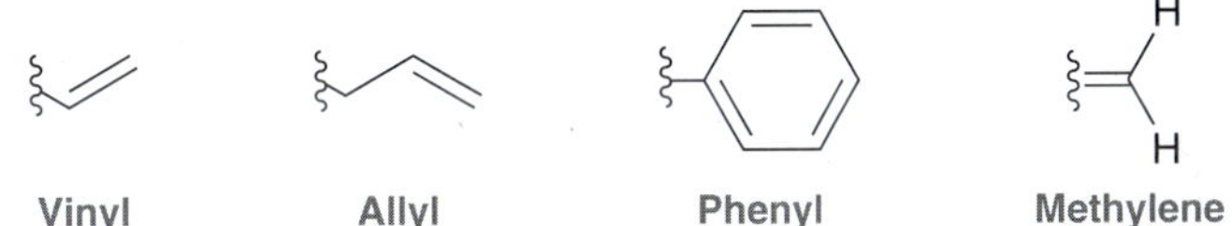


IUPAC nomenclature recognizes common names for the following groups when they appear as substituents in a compound:

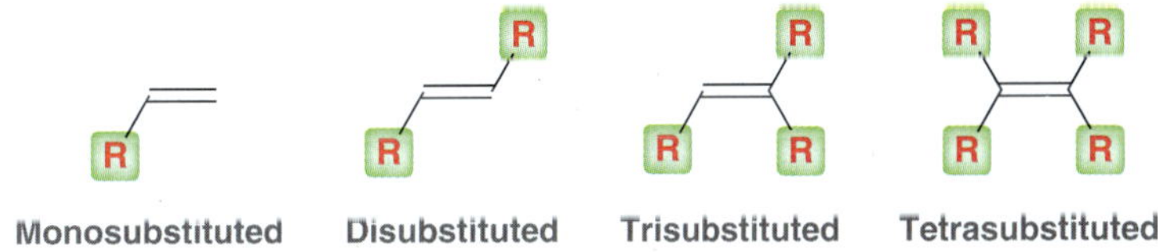


In addition to their systematic and common names, alkenes are also classified by their **degree of substitution** (Figure 8.3). Do not confuse the word *substitution* with the type of reaction discussed in the previous chapter. The same word has two different meanings. In the previous chapter, the word "substitution" referred to a reaction involving replacement of a leaving group with a nucleophile. In the present context, the word "substitution" refers to the number of alkyl groups connected to a double bond.

**FIGURE 8.3** The degree of substitution indicates the number of alkyl groups connected to the double bond.

R Monosubstituted
R R Disubstituted
R R R Trisubstituted
R R R R Tetrasubstituted

CONCEPTUAL CHECKPOINT

**8.4** Classify each of the following alkenes as monosubstituted, disubstituted, trisubstituted, or tetrasubstituted:

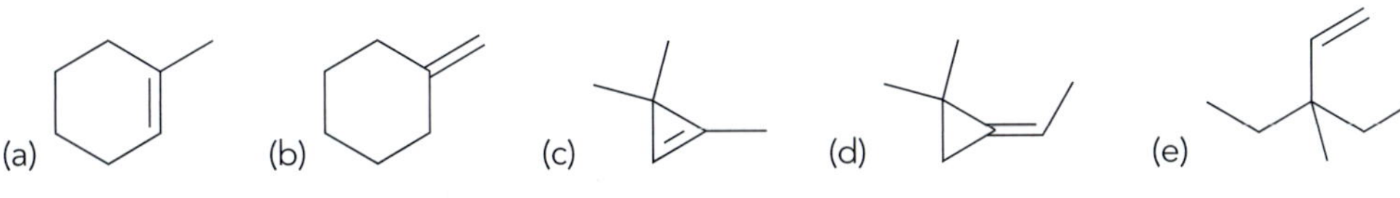

## 8.4 Stereoisomerism in Alkenes

### Using *cis* and *trans* Designations

Recall that a double bond is comprised of a σ bond and a π bond (Figure 8.4).

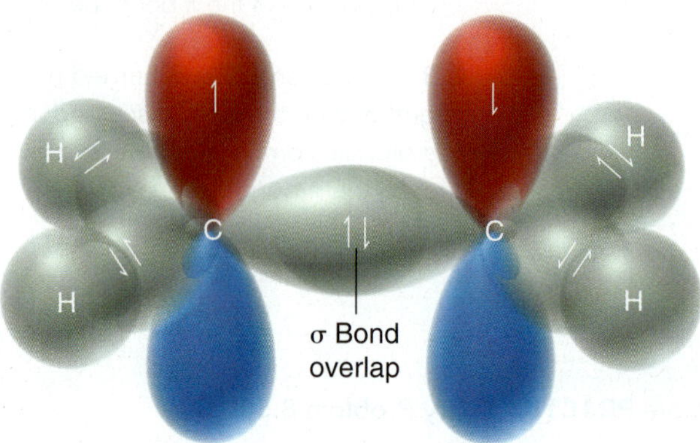


**FIGURE 8.4**
The σ and π bonds of a C=C double bond.

The σ bond is the result of overlapping $sp^2$-hybridized orbitals, while the π bond is the result of overlapping *p* orbitals. We have seen that double bonds do not exhibit free rotation at room temperature, giving rise to stereoisomerism:

Cycloalkenes comprised of fewer than seven carbon atoms cannot accommodate a *trans* π bond. These rings can only accommodate a π bond in a *cis* configuration:

In these examples, there is no need to identify the stereoisomerism of the double bond when naming each of these compounds, because the stereoisomerism is inferred. For example, the last compound is called cyclohexene, rather than *cis*-cyclohexene. A seven-membered ring containing a *trans* double bond has been prepared, but this compound (*trans*-cycloheptene) is not stable at room temperature. An eight-membered ring is the smallest ring that can accommodate a *trans* double bond and be stable at room temperature:

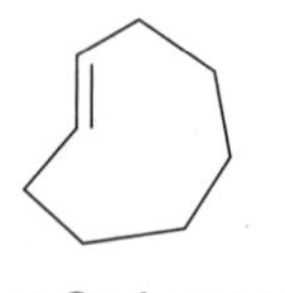


This rule also applies to bridged bicyclic compounds and is called **Bredt's rule**. Specifically, Bredt's rule states that it is not possible for a bridgehead carbon of a bicyclic system to possess a C=C double bond if it involves a *trans* π bond being incorporated in a small ring. For example, the following compound is too unstable to form:

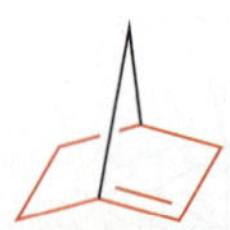

This compound would require a *trans* double bond in a six-membered ring, highlighted in red. The compound is not stable because the geometry of the bridgehead prevents it from maintaining the parallel overlap of *p* orbitals necessary to keep the π bonding intact. As a result, this type of compound is extremely high in energy, and its existence is fleeting.

Bridged bicyclic compounds can only exhibit a double bond at a bridgehead position if one of the rings has at least eight carbon atoms. For example, the following compound can maintain parallel overlap of the *p* orbitals, and as a result, it is stable at room temperature:

**This compound is stable**

## Using *E* and *Z* Designations

The stereodescriptors *cis* and *trans* can only be used to indicate the relative arrangement of similar groups. When an alkene possesses nonsimilar groups, usage of *cis-trans* terminology would be ambiguous. Consider, for example, the following two compounds:

These two compounds are not the same; they are stereoisomers. But which compound should be called *cis* and which should be called *trans*? In situations like this, IUPAC rules provide us with a method for assigning different, unambiguous stereodescriptors. Specifically, we look at the two groups on each vinylic position and choose which of the two groups is assigned the higher priority:

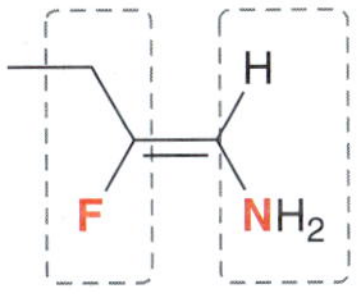

**F gets priority over C** **N gets priority over H**

In each case, a priority is assigned by the same method used in Section 5.3. Specifically, priority is given to the element with the higher atomic number. In this case, F has priority over C, while N has priority over H. We then compare the position of the higher priority groups. If they are on the same side (as shown above), the configuration is designated with the letter ***Z*** (for the German word *zussamen*, meaning "together"); if they are on opposite sides, the configuration is designated with the letter ***E*** (for the German word *entgegen*, meaning "opposite"):

*Z* *E*

These examples are fairly straightforward, because the atoms connected directly to the vinylic positions all have different atomic numbers. Other examples may require the comparison of two carbon atoms. In those cases, we will use the same tie-breaking rules that we used when assigning the configurations of chirality centers in Chapter 6. The following SkillBuilder will serve as a reminder of those rules.

## SKILLBUILDER

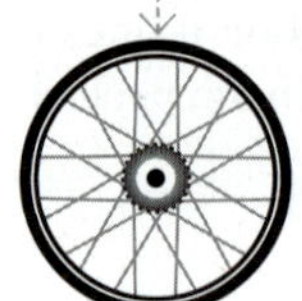

### 8.2 ASSIGNING THE CONFIGURATION OF A DOUBLE BOND

**LEARN the skill**

Identify the configuration of the following alkene:

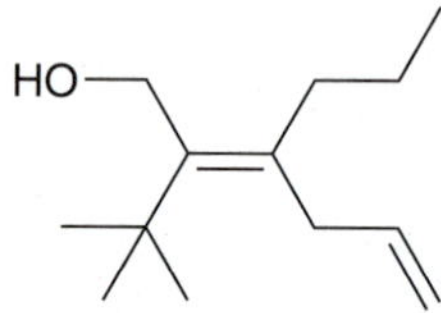

**SOLUTION**

This compound has two π bonds, but only one of them is stereoisomeric. The π bond shown on the bottom right is not stereoisomeric because it has two hydrogen atoms connected to the same vinylic position.

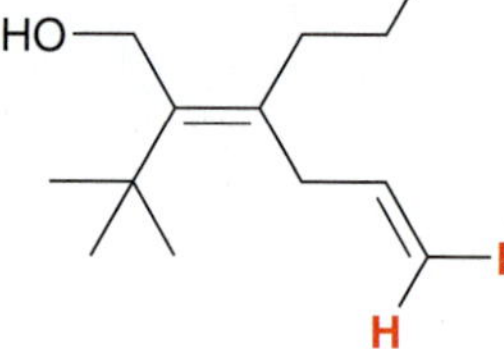


**STEP 1**
Identify the two groups connected to one vinylic position and then determine which group has priority.

Let's focus on the other double bond and try to assign its configuration. Consider each vinylic position separately. Let's begin with the vinylic position on the left side. Compare the two groups connected to it and identify which group should be assigned the higher priority.

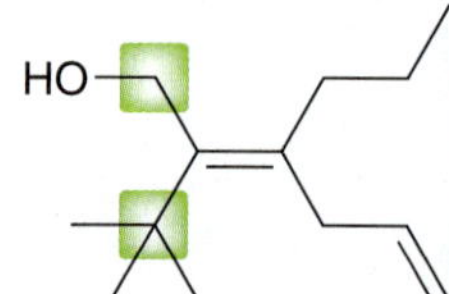


In this case, we are comparing two carbon atoms (which have the same atomic number), so we construct a list for each carbon atom (just as we did in Chapter 6) and look for the first point of difference:

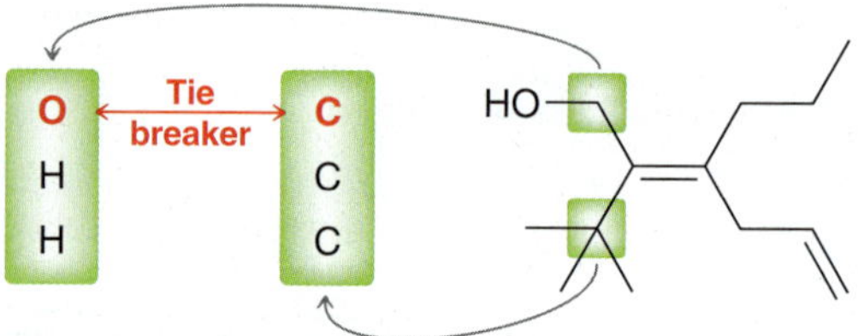


Remember that we do not add atomic numbers of the atoms in each list, but rather, we look for the first point of difference. O has a higher atomic number than C, so the group in the upper left corner is assigned priority over the *tert*-butyl group.

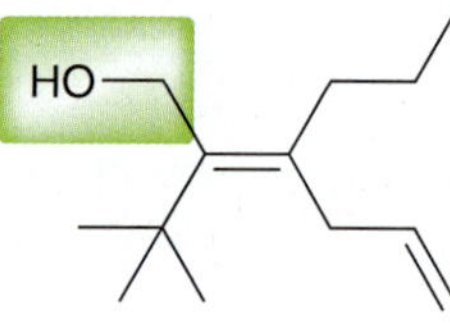


**STEP 2**
Identify the two groups connected to the other vinylic position and then determine which group has priority.

Now we turn our attention to the right side of the double bond. Using the same rules, we try to identify which group gets the priority.

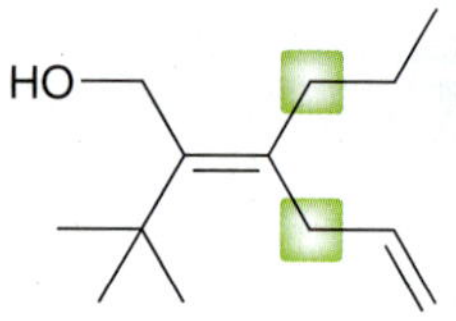


Once again, we are comparing two carbon atoms, but in this case, constructing lists does not provide us with a point of difference. Both lists are identical:

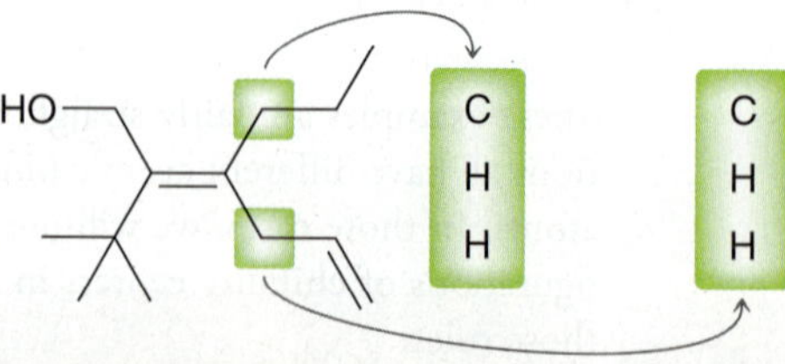

So, we move farther away from the stereoisomeric double bond and again construct lists and look for the first point of difference. Recall that a double bond is treated as two separate single bonds:

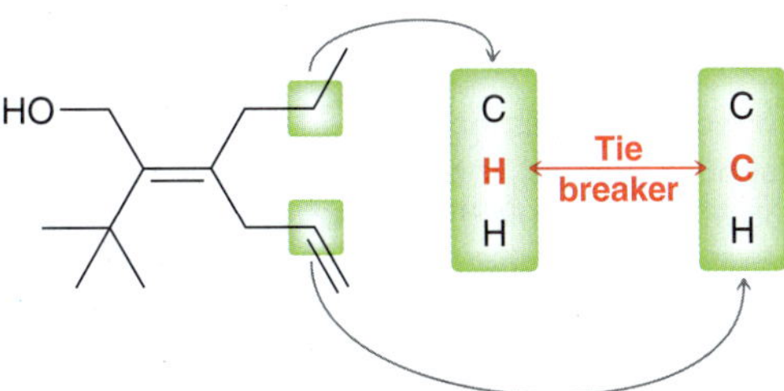


C has a higher atomic number than H, so the group in the bottom right corner is assigned priority over the propyl group.

Finally, compare the relative positions of the priority groups. These groups are on opposite sides of the double bond, so the configuration is *E*:

**STEP 3**
Determine whether the priorities are on the same side (*Z*) or opposite sides (*E*) of the double bond.

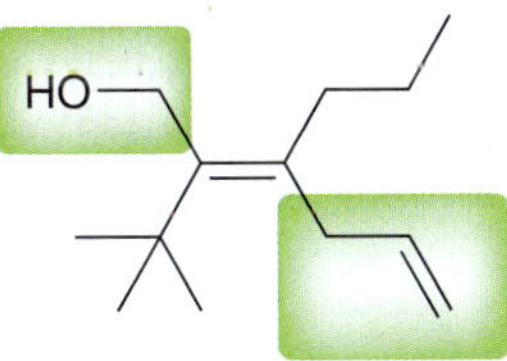

**PRACTICE the skill**

**8.5** For each of the following alkenes, assign the configuration of the double bond as either *E* or *Z*:

(a) (b)

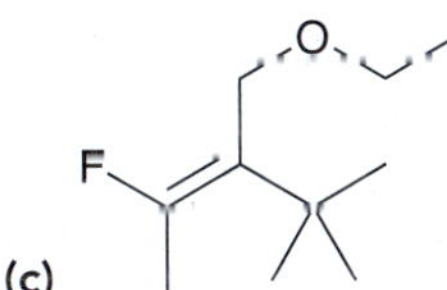


(c) (d)

**APPLY the skill**

**8.6** When a double bond has two identical groups (one on each vinylic position), we can use either *cis-trans* terminology or *E-Z* terminology, as seen in the following example:

*trans* or *E*

The configuration of this double bond can be designated as either *trans* or *E*. Both designations are acceptable. In most cases, *cis* = *Z* and *trans* = *E*. However, there are exceptions. For example:

This compound is *trans*, but *Z*. Explain this exception and then provide two other examples of exceptions.

need more **PRACTICE?** **Try Problem 8.51**

## medically speaking | Phototherapy Treatment for Neonatal Jaundice

When red blood cells reach the end of their lifetime (about 3 months), they release red-colored *hemoglobin*, the molecule that is responsible for transporting oxygen in the blood. The heme portion of hemoglobin is metabolized into a yellow-orange-colored compound called *bilirubin*:

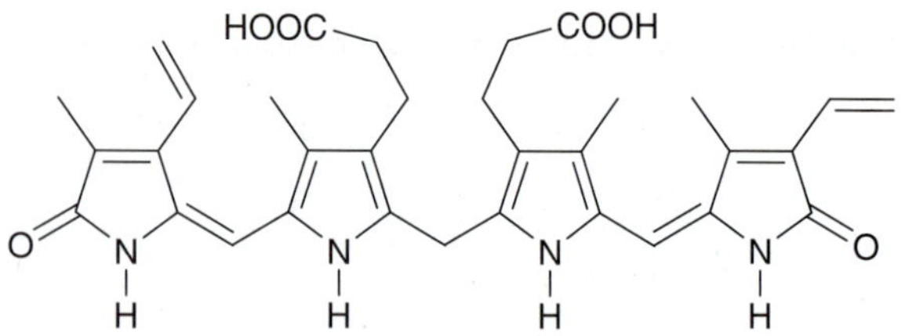

HOOC COOH

**Bilirubin**

Bilirubin contains many polar functional groups, and we might therefore expect it to be highly soluble in water. But surprisingly, the solubility of bilirubin in water (and hence urine and feces) is very small. This is attributed to the ability of bilirubin to adopt a low-energy conformation in which the carboxylic acid groups and the amide groups form internal hydrogen bonds:

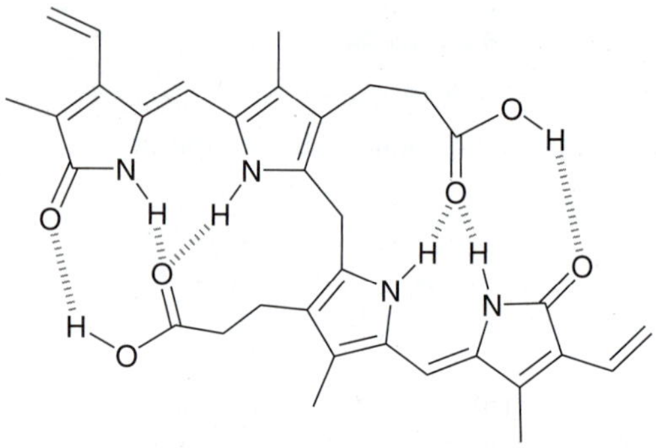

With all of the polar groups pointing toward the interior of the molecule, the exterior surface of the molecule is primarily hydrophobic. Therefore, bilirubin behaves much like a nonpolar compound and is only sparingly soluble in water. As a result, high concentrations of bilirubin build up in the fatty tissues and membranes in the body (which are nonpolar environments). Since bilirubin is a yellow-orange-colored compound, this produces *jaundice*, a condition characterized by yellowing of the skin.

The major route for excretion of bilirubin out of the body is in feces (stools). But since stools are mostly water, bilirubin must be converted into a more water-soluble compound in order to be excreted. This is achieved when the liver enzyme *glucuronyl transferase* is used to covalently attach two very polar molecules of *glucuronic acid* (or *glucuronate*) onto the nonpolar bilirubin molecule to make *bilirubin glucuronide*, more commonly called *conjugated bilirubin*. (Note that this use of the term "conjugated" means that the bilirubin molecule has been covalently bonded to another molecule, in this case glucuronic acid.)

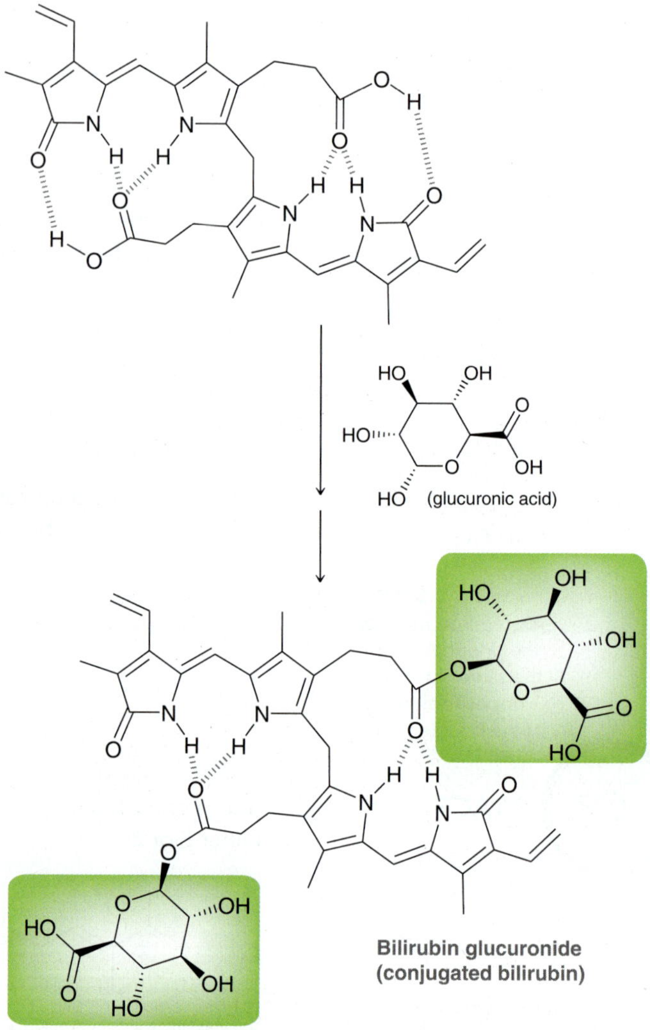

(glucuronic acid)

**Bilirubin glucuronide (conjugated bilirubin)**

Bilirubin glucuronide (conjugated bilirubin) is much more water soluble and is excreted from the liver in bile and from there into the small intestine. In the large intestine bacteria metabolize it in a series of steps into *stercobilin* (from the Greek word *sterco* for poop), which has a brown color. However, if liver function is inadequate, there may be insufficient levels of the glucuronyl transferase enzyme to carry out the conjugation of glucuronic acid molecules to bilirubin. And as a result, the bilirubin cannot be excreted in the stools. The accumulation of bilirubin in the body is the cause of jaundice. There are three common situations in which liver function is impaired sufficiently to produce jaundice:

1. Hepatitis
2. Cirrhosis of the liver due to alcoholism
3. Immature liver function in babies, especially common in premature babies, resulting in *neonatal jaundice*. The following picture shows a baby with yellow skin associated with neonatal jaundice:

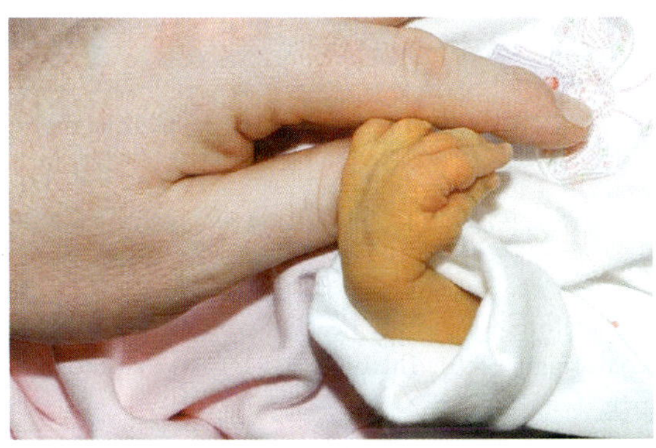

Neonatal jaundice has serious long-term consequences for the baby's mental growth because high levels of bilirubin in the brain can inhibit development of the baby's brain cells and cause permanent retardation. Consequently, if blood bilirubin levels get too high (more than about 15 mg/dL for substantial lengths of time), action must be taken to reduce blood bilirubin levels. In the past, this involved expensive blood transfusions. A major advance took place when it was noticed that babies with neonatal jaundice would demonstrate increased urinary excretion rates of bilirubin after being exposed to natural sunlight. Further studies revealed that exposure to blue light causes the increased bilirubin excretion in the urine. Neonatal jaundice is now treated with a *bili-light* that exposes the baby's skin to high-intensity blue light.

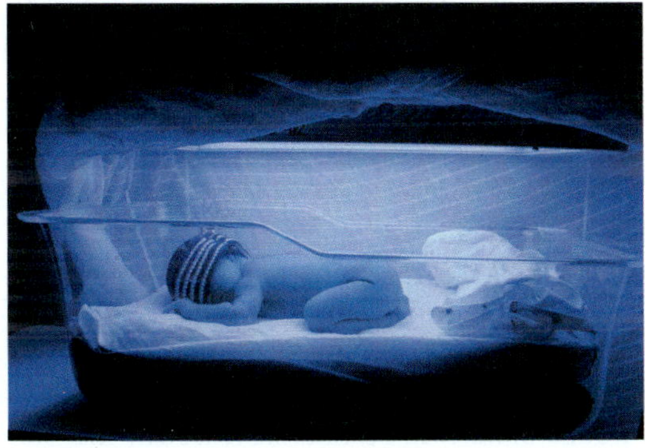

How does the bili-light work? Upon exposure to blue light, one of the C=C double bonds (adjacent to the five-membered ring in bilirubin) undergoes *photoisomerization*, and the configuration changes from *Z* to *E*):

*Z*

Blue light

*E*

In the process, more polar groups are exposed to the environment. This renders the bilirubin considerably more water soluble, even though the composition of the molecule has not changed (only the configuration of a C=C $\pi$ bond has changed). Bilirubin excretion in the urine increases substantially and neonatal jaundice decreases!

## 8.5 Alkene Stability

In general, a *cis* alkene will be less stable than its stereoisomeric *trans* alkene. The source of this instability is attributed to steric strain exhibited by the *cis* isomer. This steric strain can be visualized by comparing space-filling models of *cis*-2-butene and *trans*-2-butene (Figure 8.5). The methyl groups are only able to avoid a steric interaction when they occupy a *trans* configuration.

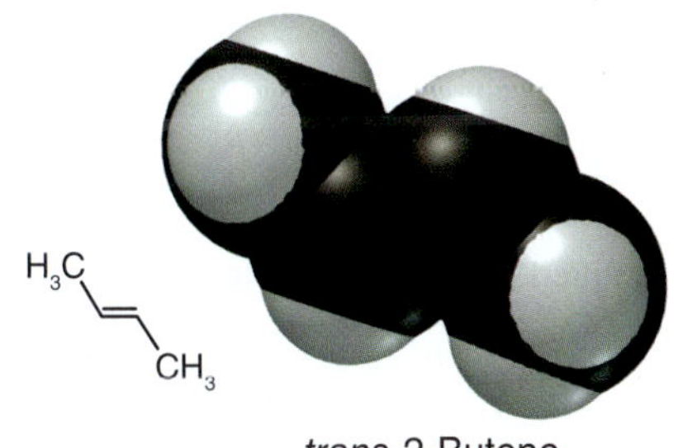


*trans*-2-Butene

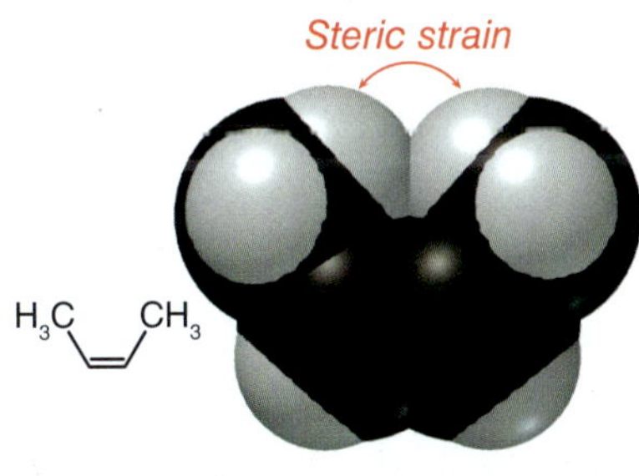


*cis*-2-Butene

**FIGURE 8.5**
Space-filling models of the stereoisomers of 2-butene.

**LOOKING BACK**
For a review of heats of combustion, see Section 4.4.

The difference in energy between stereoisomeric alkenes can be quantified by comparing their heats of combustion:

$$\text{(trans-2-butene)} + 6\,O_2 \longrightarrow 4\,CO_2 + 4\,H_2O \qquad \Delta H° = -2682\ \text{kJ/mol}$$

$$\text{(cis-2-butene)} + 6\,O_2 \longrightarrow 4\,CO_2 + 4\,H_2O \qquad \Delta H° = -2686\ \text{kJ/mol}$$

Both reactions yield the same products. Therefore, we can use heats of combustion to compare the relative energy levels of the starting materials (Figure 8.6). This analysis suggests that the *trans* isomer is 4 kJ/mol more stable than the *cis* isomer.

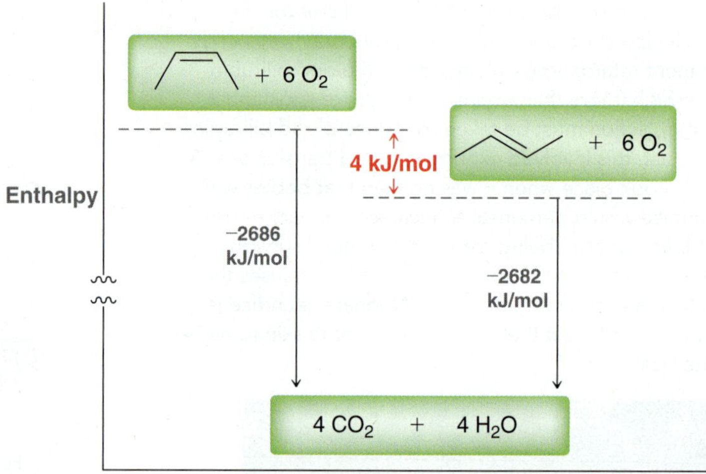


**FIGURE 8.6** Heats of combustion of the stereoisomers of 2-butene.

When comparing the stability of alkenes, another factor must be taken into account, in addition to steric effects. We must also consider the degree of substitution. By comparing heats of combustion for isomeric alkenes (all with the same molecular formula, $C_6H_{12}$), the trend in Figure 8.7 emerges. Alkenes are more stable when they are highly substituted. Tetrasubstituted alkenes are more stable than trisubstituted alkenes. The reason for this trend is not a steric effect (through space), but rather an electronic effect (through bonds). In Section 6.11, we saw that alkyl groups can stabilize a carbocation by donating electron density to the neighboring $sp^2$-hybridized carbon atom ($C^+$). This effect, called hyperconjugation, is a stabilizing effect because it enables the delocalization of electron density. In a similar way, alkyl groups can also stabilize the neighboring $sp^2$-hybridized carbon atoms of a π bond. Once again, the resulting delocalization of electron density is a stabilizing effect.

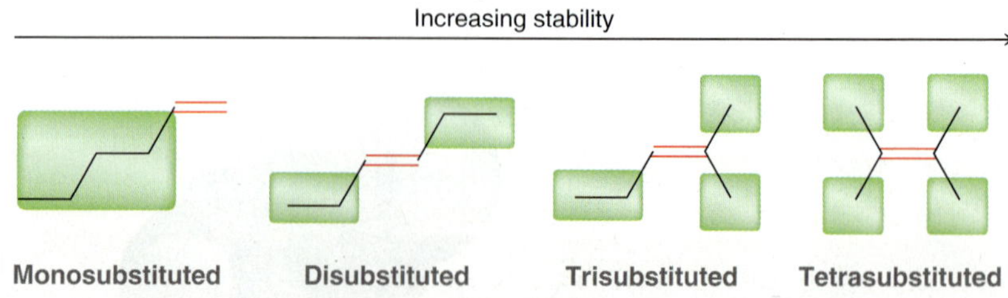


**FIGURE 8.7** The relative stability of isomeric alkenes with varying degrees of substitution.

# SKILLBUILDER

## 8.3 COMPARING THE STABILITY OF ISOMERIC ALKENES

**LEARN** the skill

Arrange the following isomeric alkenes in order of stability:

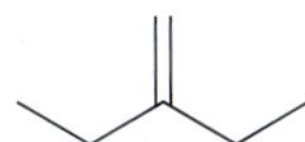 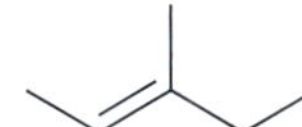 

**SOLUTION**

First identify the degree of substitution for each alkene:

**STEP 1**
Identify the degree of substitution for each alkene.

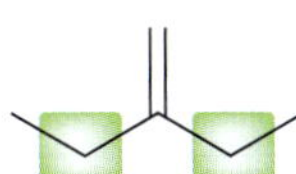


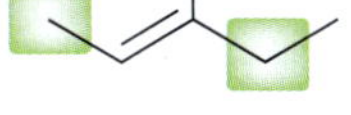


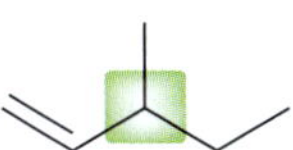


The most highly substituted alkene will be the most stable, and therefore, the following order of stability is expected:

**STEP 2**
Select the alkene that is more substituted.

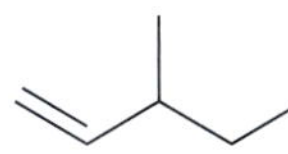

**PRACTICE** the skill

**8.7** Arrange each set of isomeric alkenes in order of stability:

(a) 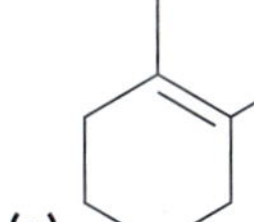 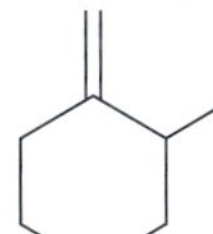 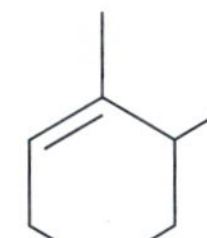

(b) 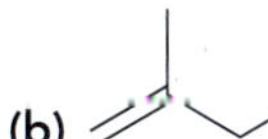 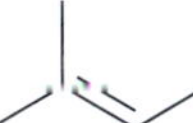

**APPLY** the skill

**8.8** Consider the following two isomeric alkenes. The first isomer is a monosubstituted alkene, while the second isomer is a disubstituted alkene. We might expect the second isomer to be more stable, yet heats of combustion for these two compounds indicate that the first isomer is more stable. Offer an explanation.

need more **PRACTICE?** **Try Problem 8.53**

## 8.6 Possible Mechanisms for Elimination

As mentioned at the beginning of the chapter, alkenes can be prepared via elimination reactions, in which a proton and a leaving group are removed to form a π bond:

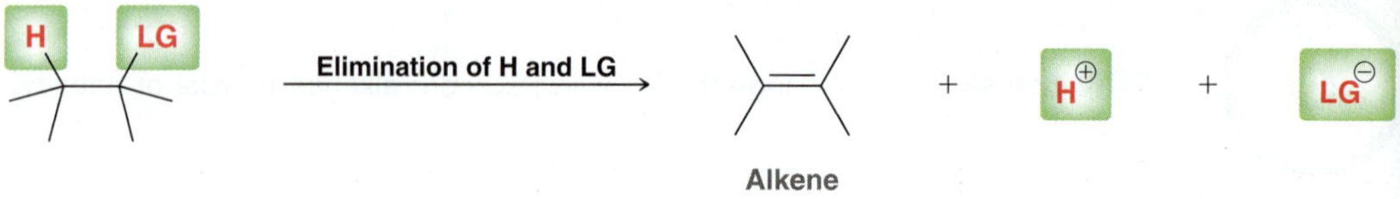

H LG Elimination of H and LG → Alkene + $H^{\oplus}$ + $LG^{\ominus}$

As we begin to explore the possible mechanisms for elimination reactions, recall from Chapter 6 that ionic mechanisms are comprised of only four types of fundamental arrow-pushing patterns (Figure 8.8). All four of these steps will appear in this chapter, so it might be wise to review SkillBuilders 6.3 and 6.4.

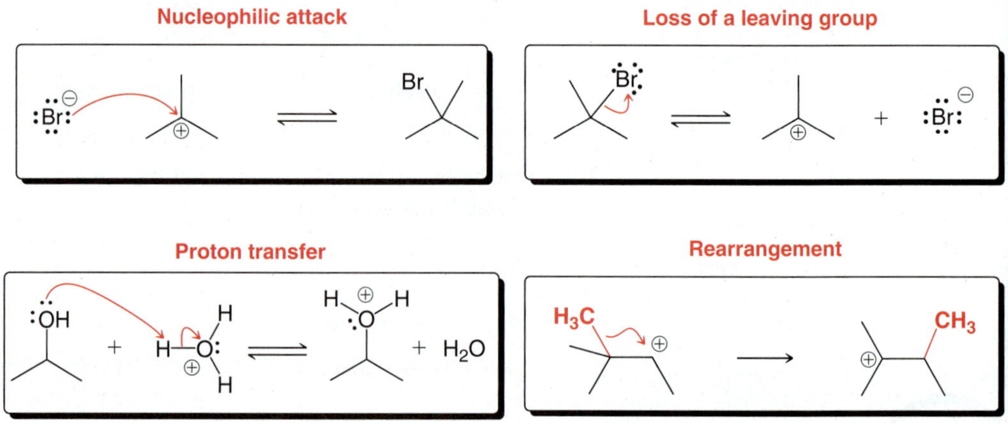

Nucleophilic attack

Loss of a leaving group

Proton transfer

Rearrangement

**FIGURE 8.8**
The four fundamental electron transfer steps for ionic processes.

All elimination reactions exhibit at least two of the four patterns: (1) proton transfer and (2) loss of a leaving group:

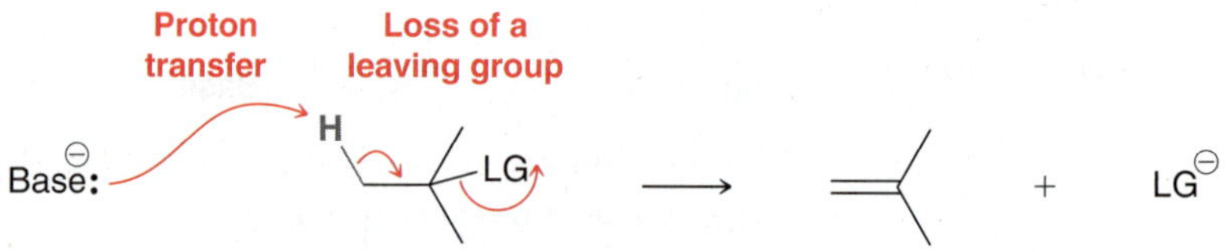

Proton transfer   Loss of a leaving group

Every elimination reaction features a proton transfer as well as loss of a leaving group. But consider the order of events of these two steps. In the figure above, they occur simultaneously (in a concerted fashion). Alternatively, we can imagine them occurring separately, in a stepwise fashion:

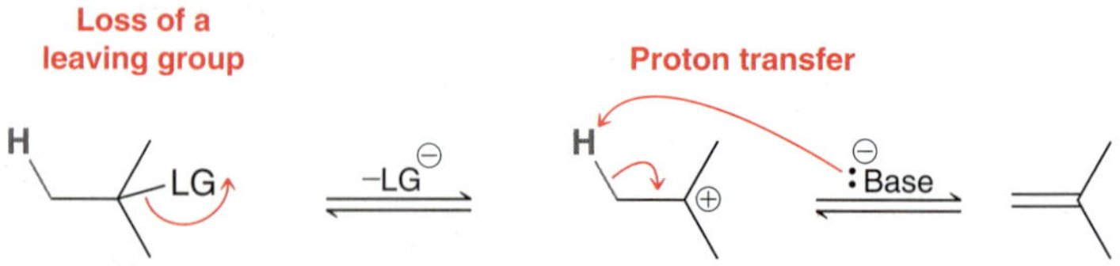

Loss of a leaving group   Proton transfer

In this stepwise mechanism the leaving group leaves, generating an intermediate carbocation (much like an $S_N1$ reaction), which is then deprotonated by a base to produce an alkene.

The key difference between these two mechanisms is summarized below:

- In a *concerted process*, a base abstracts a proton and the leaving group leaves simultaneously.
- In a *stepwise process*, first the leaving group leaves, and then the base abstracts a proton.

**LOOKING AHEAD**
There is another possible mechanism for elimination, in which deprotonation occurs first, followed by loss of the leaving group. That type of elimination process will be discussed in Chapter 22 (Mechanism 22.6).

In this chapter, we will explore these two mechanisms in more detail, and we will see that each mechanism predominates under a different set of conditions. In preparation, we must first practice drawing curved arrows.

# SKILLBUILDER

## 8.4 DRAWING THE CURVED ARROWS OF AN ELIMINATION REACTION

**LEARN the skill**

Assume that the following elimination reaction proceeds via a concerted process and draw the mechanism:

Br + NaOMe ⟶ + MeOH + NaBr

**SOLUTION**

First identify the base and the substrate. The base in this case is methoxide (remember that $Na^+$ is the counterion, and its role in the reaction does not concern us in most cases):

**WATCH OUT**

When drawing mechanisms, be very precise about the placement of the head and tail of each curved arrow. Students often draw this arrow in the wrong direction.

In a concerted mechanism, the base abstracts a proton and the leaving group leaves simultaneously. This requires a total of three curved arrows. When drawing the first curved arrow, make sure to place the tail on a lone pair of the base, and place the head on the proton that is being removed:

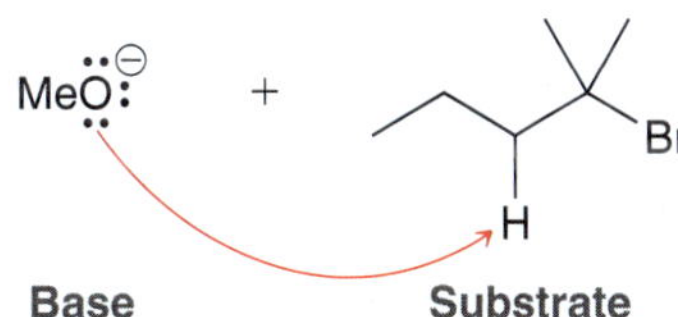


For the other two curved arrows, one shows the formation of the double bond, and the other shows loss of the leaving group:

MeO: + H ... Br ⟶ + :Br: + MeOH

**PRACTICE the skill**

**8.9** For each of the following elimination reactions, assume a concerted process is taking place and draw the mechanism:

(a) Cl —NaOH→

(b) OTs —NaOEt→

(c) Br —NaOMe→

**8.10** For each of the following elimination reactions, assume a stepwise process is taking place and draw the mechanism (first loss of the leaving group and then proton transfer):

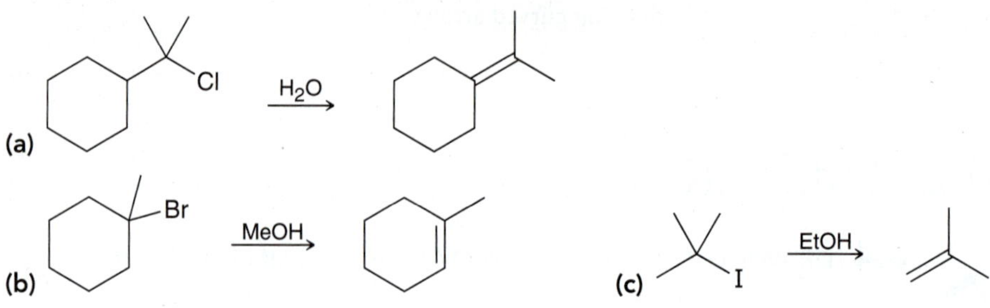

**8.11** Carefully read the following curved arrows shown and draw the expected alkene that is produced by this elimination reaction. Is this mechanism concerted or stepwise?

EtÖ: OTs H → ?

**8.12** Draw the intermediate of the following elimination reaction, and then draw the curved arrows for the second step (the reaction between the intermediate and ethanol):

Br ⇌ ? EtÖH →

need more **PRACTICE?** **Try Problem 8.69**

## 8.7 The E2 Mechanism

In this section we explore the concerted mechanism, or E2, pathway:

Proton transfer — Loss of a leaving group

Base: H LG → + $LG^{\ominus}$

### Kinetic Evidence for a Concerted Mechanism

Kinetic studies show that many elimination reactions exhibit second-order kinetics with a rate equation that has the following form:

$$\text{Rate} = k\,[\text{substrate}]\,[\text{base}]$$

Much like $S_N2$ reactions, the rate is linearly dependent on the concentrations of two different compounds (the substrate and the base). This observation suggests that the mechanism must exhibit a step in which the substrate and the nucleophile collide with each other. This is consistent with a concerted mechanism in which there is only one mechanistic step involving both the substrate and the base. Because that step involves two chemical entities, it is said to be bimolecular. Bimolecular elimination reactions are called **E2** reactions:

## CONCEPTUAL CHECKPOINT

**8.13** The following reaction exhibits a second-order rate equation:

Chlorocyclopentane $\xrightarrow{\text{NaOH}}$ cyclopentene

(a) What happens to the rate if the concentration of chlorocyclopentane is tripled and the concentration of sodium hydroxide remains the same?

(b) What happens to the rate if the concentration of chlorocyclopentane remains the same and the concentration of sodium hydroxide is doubled?

(c) What happens to the rate if the concentration of chlorocyclopentane is doubled and the concentration of sodium hydroxide is tripled?

## Effect of Substrate

In Chapter 7 we saw that the rate of an $S_N2$ process with a tertiary substrate is generally so slow that it can be assumed that the substrate is inert under such reaction conditions. It might therefore come as a surprise that tertiary substrates undergo E2 reactions quite rapidly. To explain why tertiary substrates will undergo E2 but not $S_N2$ reactions, we must recognize that the key difference between substitution and elimination is the role played by the reagent. A substitution reaction occurs when the reagent functions as a nucleophile and attacks an electrophilic position, while an elimination reaction occurs when the reagent functions as a base and abstracts a proton. With a tertiary substrate, steric hindrance prevents the reagent from functioning as a nucleophile at an appreciable rate, but the reagent can still function as a base without encountering much steric hindrance (Figure 8.9).

**Substitution**

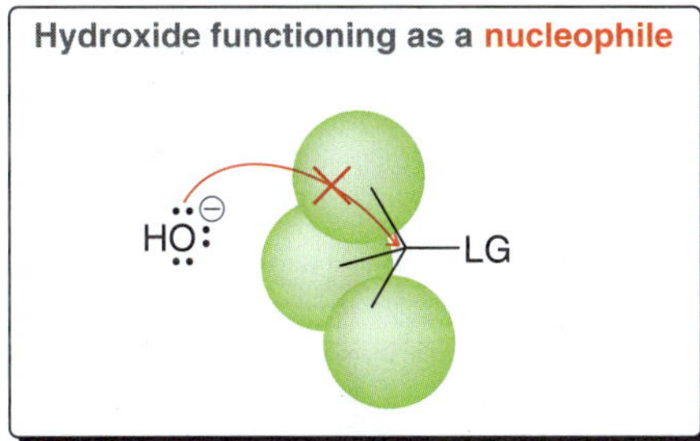


A tertiary substrate is too sterically hindered. The nucleophile cannot penetrate and attack.

**Elimination**

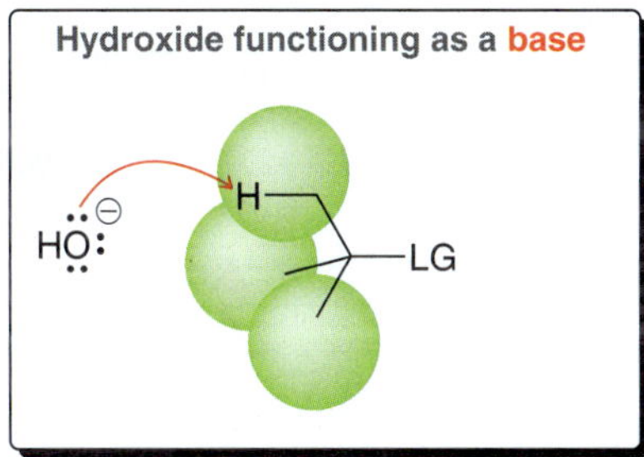


A $\beta$ proton can easily be abstracted even though the substrate is tertiary.

**FIGURE 8.9** Steric effects in $S_N2$ and E2 reactions.

Tertiary substrates react readily in E2 reactions; in fact, they react even more rapidly than primary substrates (Figure 8.10).

Rate of reactivity →

| Primary (1°) | Secondary (2°) | Tertiary (3°) |
| --- | --- | --- |
| H, H, $H_3C$, Br | $H_3C$, H, $H_3C$, Br | $H_3C$, $CH_3$, $H_3C$, Br |

**FIGURE 8.10** Relative rates of reactivity for substrates in an E2 reaction.

To understand the reason for this trend, let's look at an energy diagram of an E2 process (Figure 8.11). We focus our attention on the transition state (Figure 8.12). For a tertiary substrate, both R groups (shown in red in Figure 8.12) are alkyl groups. For a primary substrate, both of these R groups are just hydrogen atoms. In the transition state, a C=C double bond is forming. For a tertiary substrate, the transition state exhibits a partial double bond that is more highly substituted, and therefore, the transition state will be lower in energy. Compare the energy diagrams for E2 reactions involving primary, secondary, and

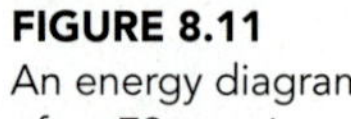

**FIGURE 8.11** An energy diagram of an E2 reaction.

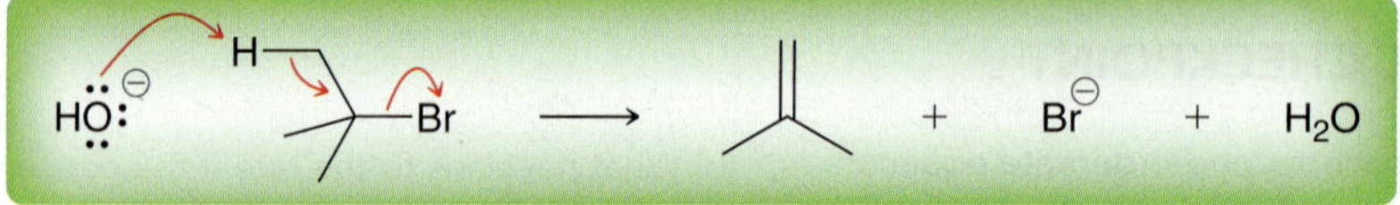


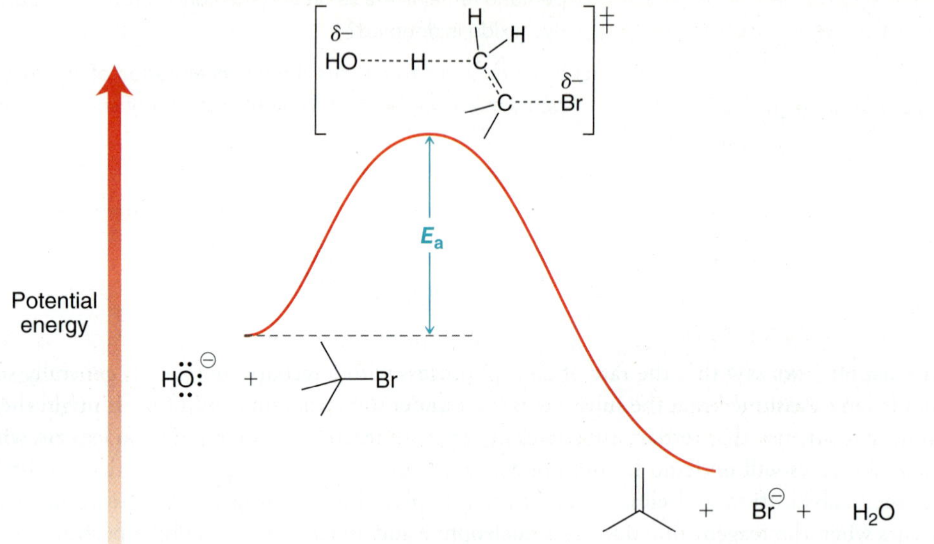


δ− HO ------ H ------ C (H, H) ⋯ C (R, R) ------ Br δ− ‡

**FIGURE 8.12** The transition state of an E2 reaction.

tertiary substrates (Figure 8.13). The transition state is lowest in energy when a tertiary substrate is used, and therefore, the energy of activation is lowest for tertiary substrates. This explains the observation that tertiary substrates react most rapidly in E2 reactions. That doesn't mean that primary substrates are slow to react in E2 reactions. In fact, primary substrates readily undergo E2 reactions. Tertiary substrates simply react more rapidly under the same conditions.

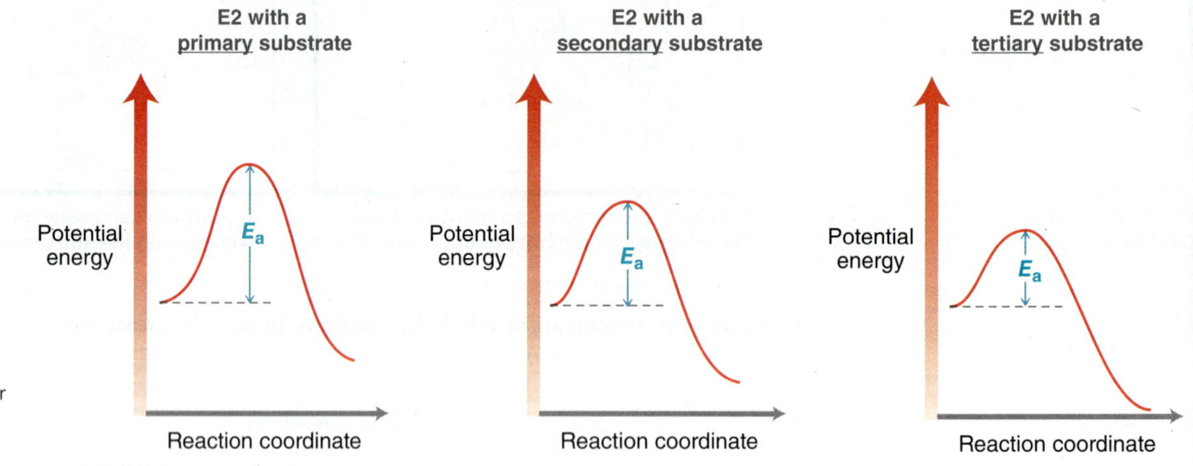


**FIGURE 8.13** Energy diagrams for E2 reactions with various substrates.

## CONCEPTUAL CHECKPOINT

**8.14** Arrange each set of compounds in order of reactivity toward an E2 process:

(a) Br Br Br

(b) Cl Cl Cl

## Regioselectivity of E2 Reactions

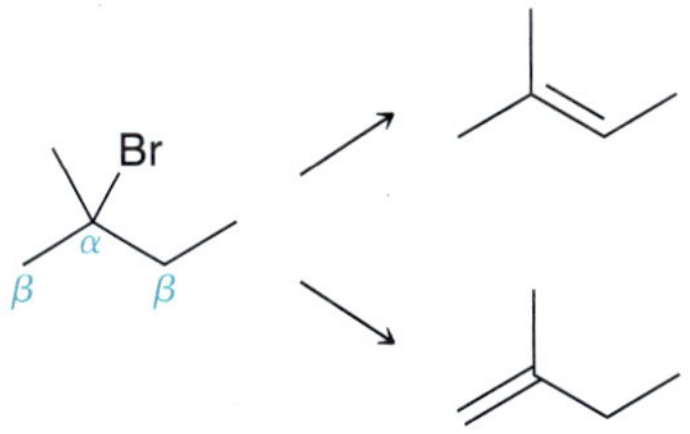

In many cases, an elimination reaction can produce more than one possible product. In this example, the β positions are not identical, so the double bond can form in two different regions of the molecule. This consideration is an example of **regiochemistry**, and the reaction is said to produce two different regiochemical outcomes. Both products are formed, but the more substituted alkene is generally observed to be the major product. For example:

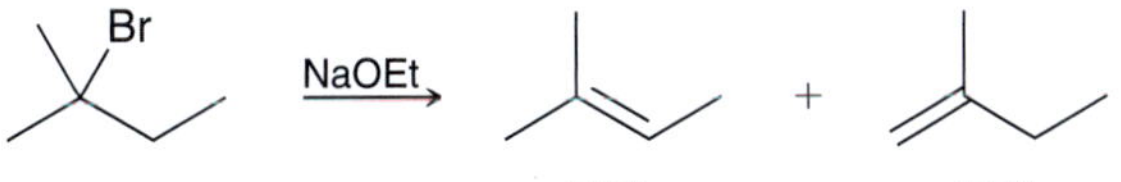

Br — α — β — β

Br — NaOEt → 71% + 29%

The reaction is said to be **regioselective**. This trend was first observed in 1875 by Russian chemist Alexander M. Zaitsev (University of Kazan), and as a result, the more substituted alkene is called the **Zaitsev product**. However, many exceptions have been observed in which the Zaitsev product is the minor product. For example, when both the substrate and the base are sterically hindered, the less substituted alkene is the major product:

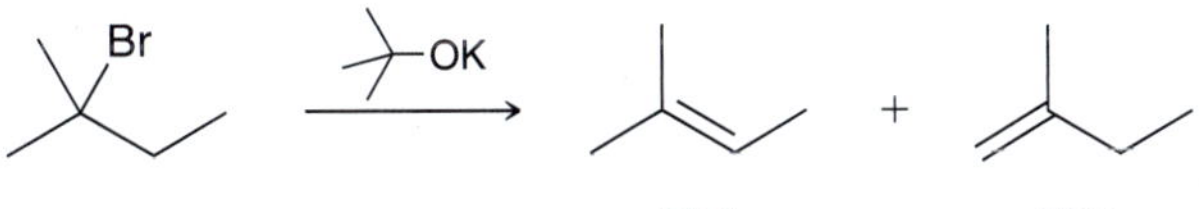

Br — OK → 28% + 72%

The less substituted alkene is often called the **Hofmann product**. The product distribution (relative ratio of Zaitsev and Hofmann products) is dependent on a number of factors and is often difficult to predict. The choice of base (how sterically hindered the base is) certainly plays an important role. For example, the product distribution for the reaction above is highly dependent on the choice of base (Table 8.1).

**TABLE 8.1 PRODUCT DISTRIBUTION OF AN E2 REACTION AS A FUNCTION OF BASE**

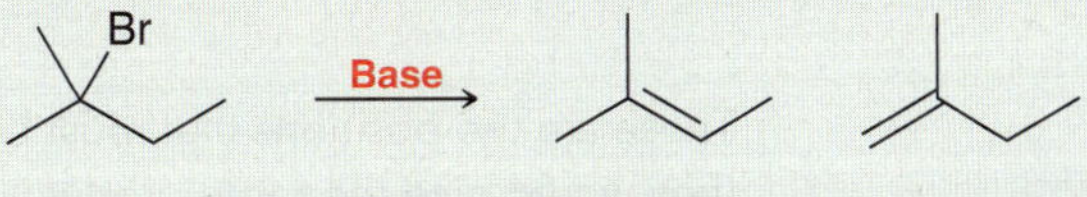

Br — Base →

| | ZAITSEV | HOFMANN |
|---|---|---|
| $EtO^{\ominus}$ | 71% | 29% |
| 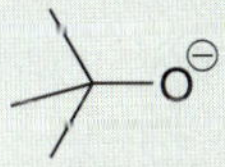 $O^{\ominus}$ (tert-butoxide) | 28% | 72% |
| 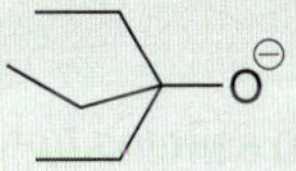 $O^{\ominus}$ | 8% | 92% |

When ethoxide is used, the Zaitsev product is the major product. But when sterically hindered bases are used, the Hofmann product becomes the major product. This case illustrates a critical concept: *The regiochemical outcome of an E2 reaction can often be controlled by carefully choosing the base.* Sterically hindered bases are employed in a variety of reactions, not just elimination, so it is useful to recognize a few sterically hindered bases (Figure 8.14).

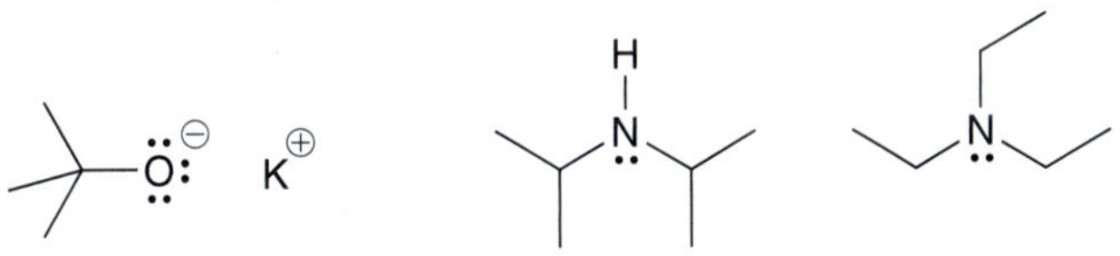

$\ddot{O}:^{\ominus}$ $K^{\oplus}$ | H–N | N

Potassium *tert*-butoxide (*t*-BuOK) | Diisopropylamine | Triethylamine

**FIGURE 8.14** Commonly used sterically hindered bases.

## SKILLBUILDER

### 8.5 PREDICTING THE REGIOCHEMICAL OUTCOME OF AN E2 REACTION

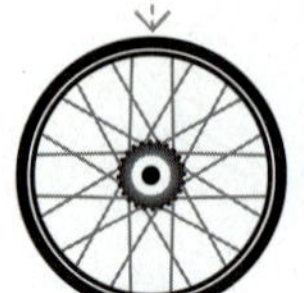

**LEARN** the skill

Identify the major and minor products of the following E2 reaction:

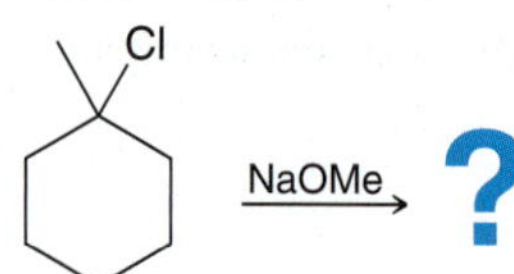

**SOLUTION**

First identify the α position. This is the position bearing the leaving group:

**STEP 1**
Identify the α position.

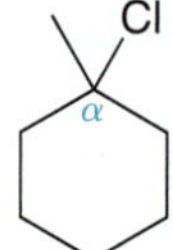


Next, identify all β positions that have protons:

**STEP 2**
Identify all β positions with protons.

These are the positions that must be explored. Two of these positions are identical, because they would give the same product:

**STEP 3**
Take notice of β positions that lead to the same product.

**Abstracting a proton from either of these two positions affords the same product**

**These two drawings represent the same product**

Abstracting a proton from the other β position will produce the following alkene:

**Abstracting a proton from this position...**

**...affords this product**

Therefore, there are two possible products. In order to determine which product predominates, analyze each of the products and determine the degree of substitution:

**STEP 4**
Remove a proton from each β position and compare the degree of substitution of the resulting alkenes.

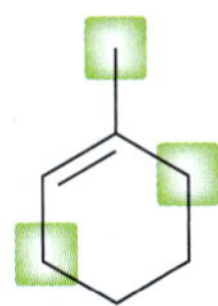

**Trisubstituted (Zaitsev)**

**Disubstituted (Hofmann)**

The more substituted alkene is the Zaitsev product, and the less substituted alkene is the Hofmann product. The Zaitsev product is generally the major product, unless a sterically hindered base is used. This example does not employ a sterically hindered base, so we expect that the Zaitsev product will be the major product:

**STEP 5**
Analyze the base to determine which product predominates.

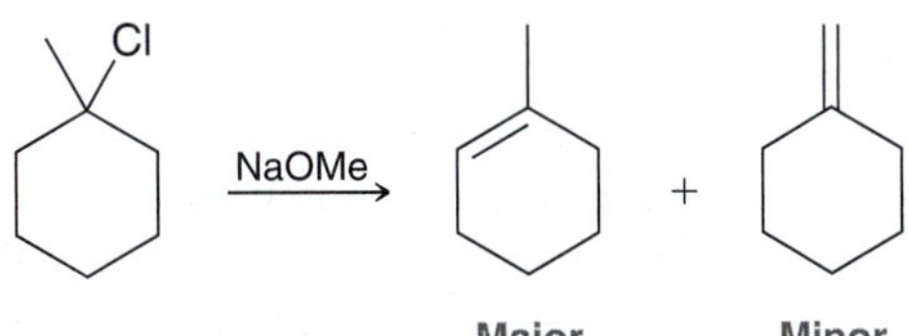


**Major** **Minor**

**PRACTICE** the skill

**8.15** Identify the major and minor products for each of the following E2 reactions:

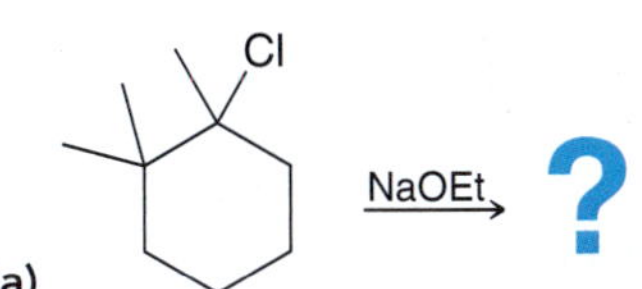


(a)

(b) Cl, *t*-BuOK, ?

(c) I, NaOH, ?

(d) I, *t*-BuOK, ?

(e) Br, NaOH, ?

(f) Br, *t*-BuOK, ?

**APPLY** the skill

**8.16** For each of the following reactions, identify whether you would use hydroxide or *tert*-butoxide to accomplish the desired transformation:

(a) Cl →

(b) Br →

**8.17** Show two different methods for preparing each of the following alkenes using a sterically hindered base:

(a) (b)

need more **PRACTICE?** **Try Problems 8.60, 8.61b–d, 8.64, 8.66a,d, 8.74**

## Stereoselectivity of E2 Reactions

The examples in the previous sections focused on regiochemistry. We will now focus our attention on stereochemistry. For example, consider performing an E2 reaction with 3-bromopentane as the substrate:

Br

This compound has two identical β positions so regiochemistry is not an issue in this case. Deprotonation of either β position produces the same result. But in this case, stereochemistry is relevant, because two possible stereoisomeric alkenes can be obtained:

Br —NaOEt→ Major + Minor

Both stereoisomers (*cis* and *trans*) are produced, but the *trans* product predominates. Consider an energy diagram showing formation of the *cis* and *trans* products (Figure 8.15). Using the Hammond postulate (Section 6.6), we can show that the transition state for formation of the *trans* alkene is more stable than the transition state for formation of the *cis* alkene. This reaction is said to be **stereoselective** because the substrate produces two stereoisomers in unequal amounts.

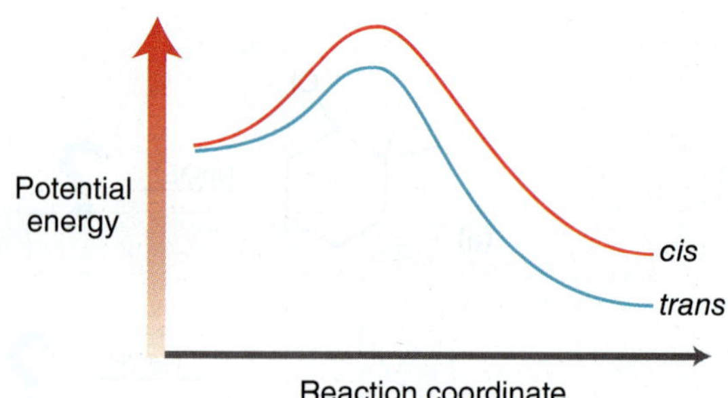


**FIGURE 8.15** An energy diagram showing formation of *cis* and *trans* products in an E2 reaction.

## Stereospecificity of E2 Reactions

In the previous example, the β position had two different protons:

H H
Br

In such a case, both the *cis* and the *trans* isomers were produced, with the *trans* isomer being favored. Now let's consider a case where the β position contains only one proton. For example, consider performing an elimination with the following substrate:

Br
Me
β
H
H
β

In this example, there are two β positions. One of these positions has no protons at all, and the other position has only one proton. In such a case, a mixture of stereoisomers is not obtained. In this case, there will only be one stereoisomeric product:

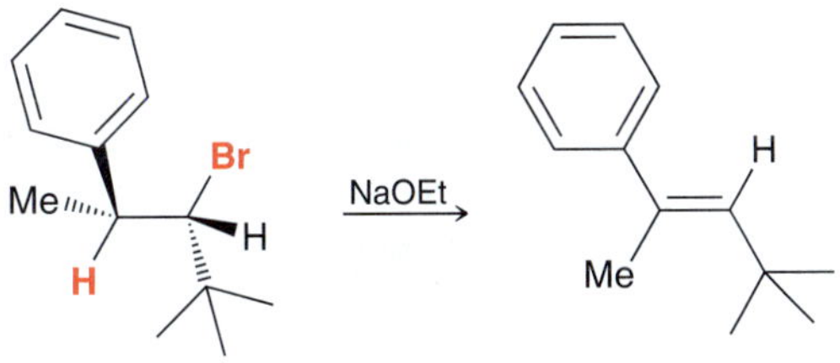

Me, H, Br, H, NaOEt, H, Me

Why is the other possible stereoisomer not obtained? To understand the answer to this question, we must explore the alignment of orbitals in the transition state. In the transition state, a π bond is forming. Recall that a π bond is comprised of overlapping *p* orbitals. Therefore, the transition state must involve the formation of *p* orbitals that are positioned such that they can overlap with each other as they are forming. In order to achieve this kind of orbital overlap, the following four atoms must all lie in the same plane: the proton at the β position, the leaving group, and the two carbon atoms that will ultimately bear the π bond. These four atoms must all be **coplanar**:

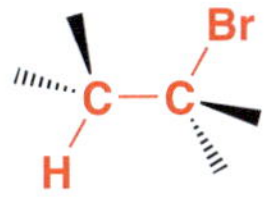

Br, C—C, H

**These four atoms (shown in red) must all lie in the same plane**

Recall that the C—C single bond is free to rotate before the reaction occurs. If we imagine rotating this bond, we see that there are two ways to achieve a coplanar arrangement:

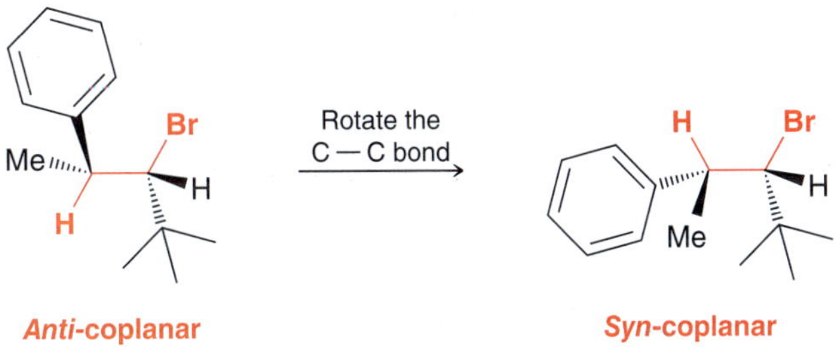

Me, Br, H, H — Rotate the C—C bond → H, Br, H, Me

***Anti*-coplanar** ***Syn*-coplanar**

The first conformation is called ***anti*-coplanar**, while the second conformation is called ***syn*-coplanar**. In this context, the terms *anti* and *syn* refer to the relative positions of the proton and the leaving group, which can be seen more clearly with Newman projections (Figure 8.16).

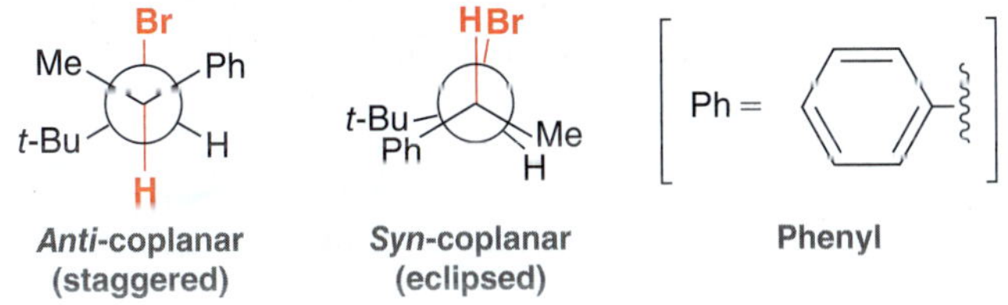

Br, Me, Ph, t-Bu, H, H — ***Anti*-coplanar (staggered)**

HBr, t-Bu, Ph, Me, H — ***Syn*-coplanar (eclipsed)**

Ph = — **Phenyl**

**FIGURE 8.16** Newman projections of *anti* coplanar and *syn*-coplanar conformations.

**LOOKING BACK**
For practice drawing Newman projections, see Section 4.7.

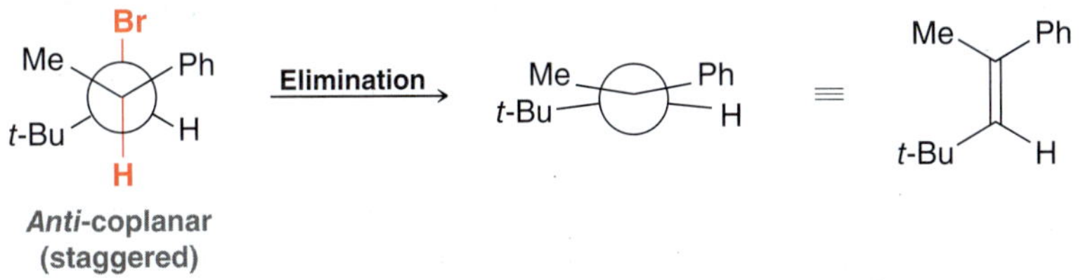

When viewed in this way, we can see that the *anti*-coplanar conformation is staggered, while the *syn*-coplanar conformation is eclipsed. Elimination via the *syn*-coplanar conformation would involve a transition state of higher energy as a result of the eclipsed geometry. Therefore, elimination occurs more rapidly via the *anti*-coplanar conformation. In fact, in most cases, elimination is observed to occur exclusively via the *anti*-coplanar conformation, which leads to one specific stereoisomeric product:

Br, Me, Ph, t-Bu, H, H — Elimination → Me, Ph, t-Bu, H ≡ Me, Ph, t-Bu, H

***Anti*-coplanar (staggered)**

The requirement for coplanarity is not entirely absolute. That is, small deviations from coplanarity can be tolerated. If the dihedral angle between the proton and the leaving group is not exactly 180°, the reaction can still proceed as long as the dihedral angle is close to 180°. The term **periplanar** (rather than coplanar) is used to describe a situation in which the proton and leaving group are *nearly* coplanar (for example, a dihedral angle of 178° or 179°). In such a conformation, the orbital overlap is significant enough for an E2 reaction to occur. Therefore, it is not absolutely necessary for the proton and the leaving group to be *anti*-coplanar. Rather, it is sufficient for the proton and the leaving group to be ***anti*-periplanar**. From now on, we will use the term *anti*-periplanar when referring to the stereochemical requirement for an E2 process.

The requirement for an *anti*-periplanar arrangement will determine the stereoisomerism of the product. In other words, *the stereoisomeric product of an E2 process depends on the configuration of the starting alkyl halide*:

It would be absolutely wrong to say that the product will always be the *trans* isomer. The product obtained depends on the configuration of the starting alkyl halide. The only way to predict the configuration of the resulting alkene is to analyze the substrate carefully, draw a Newman projection, and then determine which stereoisomeric product is obtained. The E2 reaction is said to be *stereospecific*, because the stereoisomerism of the product is dependent on the stereoisomerism of the substrate. The stereospecificity of an E2 reaction is only relevant when the β position has only one proton:

In such a case, the β proton must be arranged *anti*-periplanar to the leaving group in order for the reaction to occur, and that requirement will determine the stereoisomeric product obtained. If, however, the β position has two protons, then either of these two protons can be arranged so that it is *anti*-periplanar to the leaving group. As a result, both stereoisomeric products will be obtained:

In such a case, the more stable isomeric alkene will predominate. This is an example of stereoselectivity, rather than stereospecificity. The difference between these two terms is often misunderstood, so let's spend a moment on it. The key is to focus on the nature of the substrate (Figure 8.17):

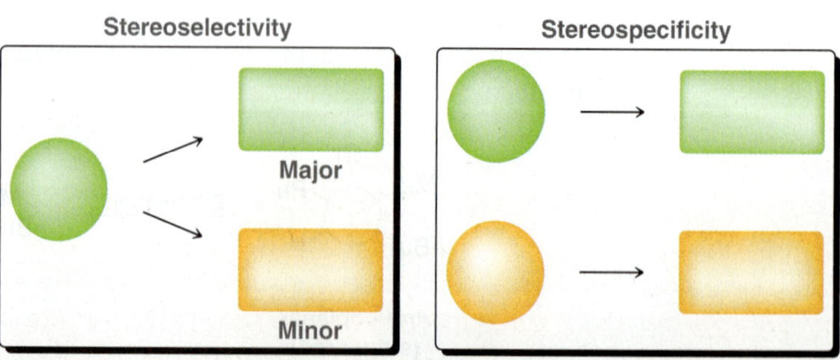


**FIGURE 8.17**
An illustration of the difference between stereoselectivity and stereospecificity in an E2 reaction.

- In a *stereoselective* E2 reaction: The substrate itself is not necessarily stereoisomeric; nevertheless, this substrate can produce two stereoisomeric products, and it is found that one stereoisomeric product is formed in higher yield.
- In a *stereospecific* E2 reaction: The substrate is stereoisomeric, and the stereochemical outcome is dependent on which stereoisomeric substrate is used.

# SKILLBUILDER

## 8.6 PREDICTING THE STEREOCHEMICAL OUTCOME OF AN E2 REACTION

**LEARN the skill**

Identify the major and minor products for the E2 reaction that occurs when each of the following substrates is treated with a strong base:

(a) Cl (b) Cl

**SOLUTION**

**(a)** In this case, regiochemistry is irrelevant because there is only one β position that bears protons:

**STEP 1**
Identify all β positions with protons.

H H β β Cl

In order to determine if the reaction is stereoselective or stereospecific, count the number of protons at the β position. In this case, the β position has two protons. Therefore, we expect the reaction to be stereoselective. That is, we expect both *cis* and *trans* isomers, with a preference for *trans*:

**STEP 2**
If the β position has two protons, expect *cis* and *trans* isomers.

Cl ⟶ Major + Minor

**(b)** In this case, the regiochemistry is also irrelevant, because there is once again only one β position that has protons:

**STEP 1**
Identify all β positions with protons.

H β Cl

In order to determine if the reaction is stereoselective or stereospecific, count the number of protons at the β position. In this case, the β position has only one proton. Therefore, we expect the reaction to be stereospecific. That is, we expect only one particular stereoisomeric product, not both. To determine which product to expect, begin by drawing the Newman projection:

**STEP 2**
If the β position has only one proton, draw a Newman projection.

H Cl Observer = Me H H Cl Et

In this conformation, the proton and the leaving group (Cl) are not *anti*-periplanar. In order to achieve the appropriate conformation, rotate the C—C single bond so that the proton and the chlorine atom are *anti*-periplanar:

**STEP 3**
Rotate the C—C single bond to achieve an *anti*-periplanar conformation.

Now use this Newman projection to draw the product:

**STEP 4**
Use the Newman projection as a guide for drawing the product.

In this case, the product is the *Z* isomer. The *E* isomer is not obtained, because that would require elimination from a *syn*-periplanar conformation, which is eclipsed and too high in energy.

**PRACTICE the skill** **8.18** Identify the major and minor products for the E2 reaction that occurs when each of the following substrates is treated with a strong base:

(a) (b) (c)

(d) (e) (f)

(g) (h)

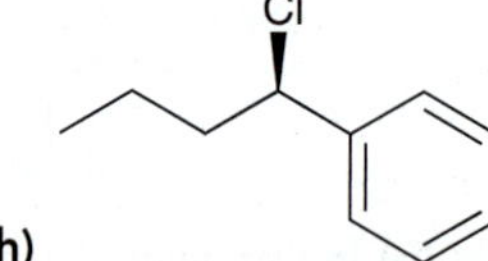

**APPLY the skill** **8.19** Identify an alkyl halide that could be used to make the following alkene:

need more **PRACTICE?** **Try Problems 8.56, 8.63, 8.82**

## Stereospecificity of E2 Reactions on Substituted Cyclohexanes

In the previous section, we explored the requirement that an E2 reaction proceed via an *anti*-periplanar conformation. That requirement has special significance when dealing with substituted cyclohexanes. Recall that a substituted cyclohexane ring can adopt two different chair conformations:

**LOOKING BACK**
For practice drawing chair conformations, see Section 4.12.

In one chair conformation, the leaving group occupies an axial position. In the other chair conformation, the leaving group occupies an equatorial position. The requirement for an *anti*-periplanar conformation demands that an E2 reaction can only occur from the chair conformation in which the leaving group occupies an axial position. To see this more clearly, consider the Newman projections for each chair conformation:

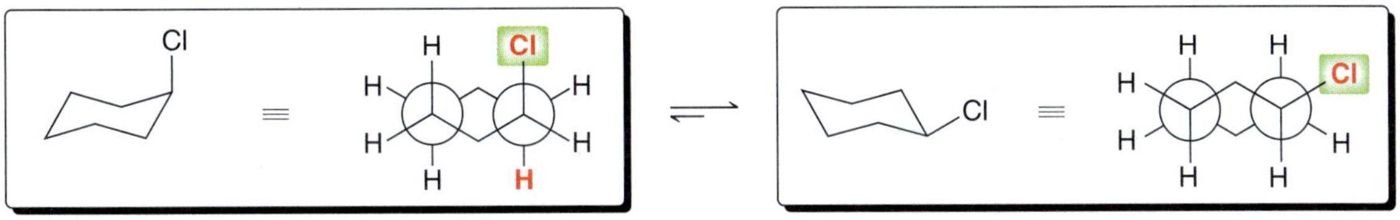

When Cl is axial, it can be *anti*-periplanar with a neighboring hydrogen atom

When Cl is equatorial, it cannot be *anti*-periplanar with any of its neighboring hydrogen atoms

When the leaving group occupies an axial position, it is *anti*-periplanar with a neighboring proton. Yet, when the leaving group occupies an equatorial position, it is not *anti*-periplanar with any neighboring protons. Therefore, on a cyclohexane ring, an E2 reaction only occurs from the chair conformation in which the leaving group is axial. From this, it follows that an E2 reaction can only take place when the leaving group and the proton are on opposite sides of the ring (one on a wedge and the other on a dash):

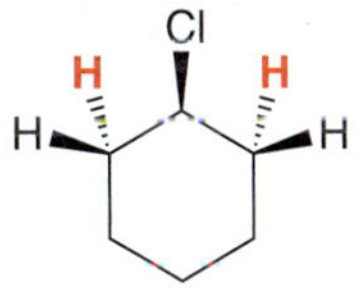

Only the two hydrogen atoms shown in red can participate in an E2 reaction

If there are no such protons, then an E2 reaction cannot occur at an appreciable rate. As an example, consider the following compound:

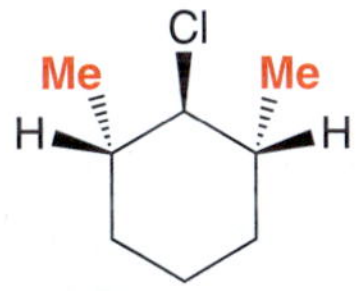

This compound will not undergo an E2 reaction

This compound has two β positions, and each β position bears a proton. But, neither of these protons can be *anti*-periplanar with the leaving group. Since the leaving group is on a wedge, an E2 reaction will only occur if there is a neighboring proton on a dash.

Consider another example:

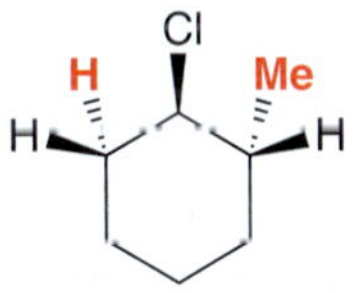

In this case, only one proton can be *anti*-periplanar with the leaving group, so in this case, only one product is obtained:

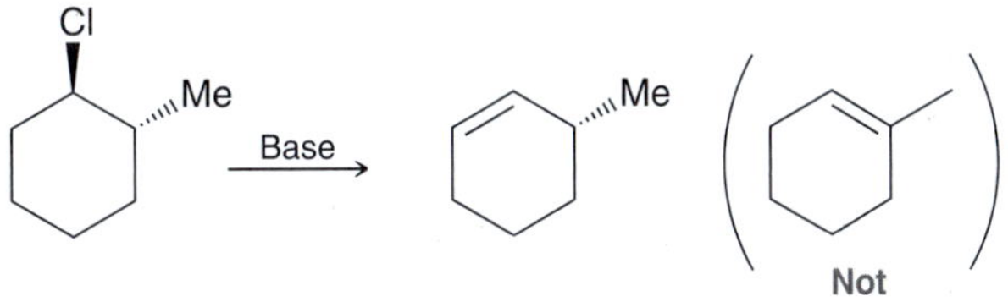

Base

Not observed

The other product (the Zaitsev product) is not formed in this case, even though it is the more substituted alkene.

For substituted cyclohexanes, the rate of an E2 reaction is greatly affected by the amount of time that the leaving group spends in an axial position. As an example, compare the following two compounds:

Neomenthyl chloride Menthyl chloride

Neomenthyl chloride is 200 times more reactive toward an E2 process. Why? Draw both chair conformations of neomenthyl chloride:

More stable

The more stable chair conformation has the larger isopropyl group occupying an equatorial position. In this chair conformation, the chlorine occupies an axial position, which is set up nicely for an E2 elimination. In other words, neomenthyl chloride spends most of its time in the conformation necessary for an E2 process to occur. In contrast, menthyl chloride spends most of its time in the wrong conformation:

More stable

The large isopropyl group occupies an equatorial position in the more stable chair conformation. In this conformation, the leaving group occupies an equatorial position as well, which means that the leaving group spends most of its time in an equatorial position. In this case, an E2 process can only occur from the higher energy chair conformation, whose concentration is small at equilibrium. As a result, menthyl chloride will undergo an E2 reaction more slowly.

## CONCEPTUAL CHECKPOINT

**8.20** When menthyl chloride is treated with a strong base, only one elimination product is observed. Yet, when neomenthyl chloride is treated with a strong base, two elimination products are observed. Draw the products and explain.

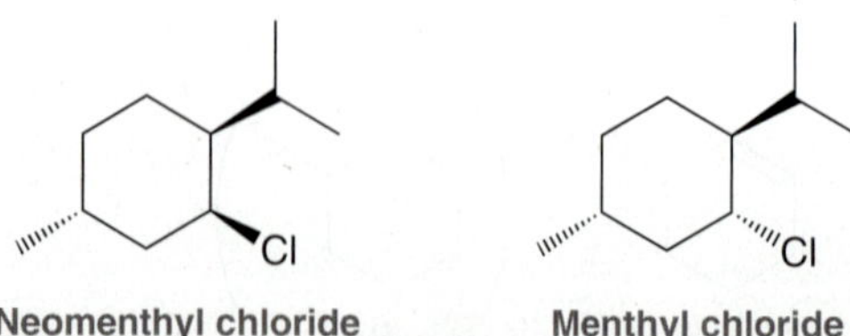


**8.21** Predict which of the following two compounds will undergo an E2 reaction more rapidly:

## 8.8 Drawing the Products of an E2 Reaction

As we have seen, predicting the products of an E2 reaction is often more complex than predicting the products of an $S_N2$ reaction. With an E2 reaction, two major issues must be considered before drawing the products: regiochemistry (Section 8.4) and stereochemistry (Section 8.5). Both of these issues are illustrated in the following example.

# SKILLBUILDER

### 8.7 DRAWING THE PRODUCTS OF AN E2 REACTION

**LEARN** the skill

Predict the product(s) of the following E2 reaction:

Br

NaOEt → ?

**SOLUTION**

We must consider both the regiochemistry and the stereochemistry of this reaction. We will start with the regiochemistry. First identify all of the β positions that possess protons:

**STEP 1**
Assess the regiochemical outcome.

Br
β β

There are two β positions. The base (ethoxide) is not sterically hindered, so we expect the Zaitsev product (the more substituted alkene) as the major product. The less substituted alkene will be the minor product. Next, we must identify the stereochemistry of formation of each of the products. Let's begin with the minor product (the less substituted alkene), because its double bond does not exhibit stereoisomerism:

**STEP 2**
Assess the stereochemical outcome.

This alkene is neither *E* nor *Z*, so we don't need to worry about stereoisomerism of the double bond in this product. But with the major product we must predict which stereoisomer will be obtained. To do this, draw a Newman projection:

Observer

Br ≡ Et, H, Br, Me, H, Me

Then rotate the C—C single bond so that the proton and the leaving group are *anti*-coplanar, and draw the product:

Et, H, Br, Me, H, Me → Me, H, Br, H, Et, Me (*Anti*-coplanar) —Base→ Me, H, Me, Et ≡ Me, H, Et, Me

To summarize, we expect the following products:

Br — NaOEt → **Major** + **Minor**

**PRACTICE the skill** **8.22** Predict the major and minor products for each of the following E2 reactions:

(a) Cl — NaOEt → ?

(b) Br — NaOEt → ?

(c) Br — NaOEt → ?

(d) I — NaOEt → ?

(e) Br — *t*-BuOK → ?

(f) Br — *t*-BuOK → ?

**APPLY the skill** **8.23** Draw an alkyl halide that will produce only one alkene upon treatment with a strong base.

**8.24** Draw an alkyl halide that will produce exactly two stereoisomeric alkenes upon treatment with a strong base.

**8.25** Draw an alkyl halide that will produce two constitutional isomers (but no stereoisomers) upon treatment with a strong base.

need more **PRACTICE?** **Try Problems 8.59–8.61, 8.64, 8.66**

## 8.9 The E1 Mechanism

In this section we explore the stepwise mechanism, or E1, pathway:

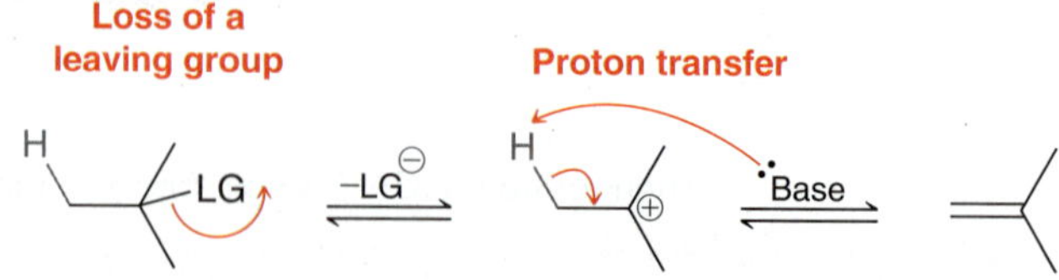


### Kinetic Evidence for a Stepwise Mechanism

Kinetic studies show that many elimination reactions exhibit first-order kinetics, with a rate equation that has the following form:

$$\text{Rate} = k\,[\text{substrate}]$$

Much like $S_N1$ reactions, the rate is linearly dependent on the concentration of only one compound (the substrate). This observation is consistent with a stepwise mechanism, in which the rate-determining step does not involve the base. The rate-determining step is the first step in the mechanism (loss of the leaving group), just as we saw in $S_N1$ reactions. The base does not participate in this step, and therefore, the concentration of the base does not affect the rate. Because this step involves only one chemical entity, it is said to be unimolecular. Unimolecular elimination reactions are called **E1** reactions:

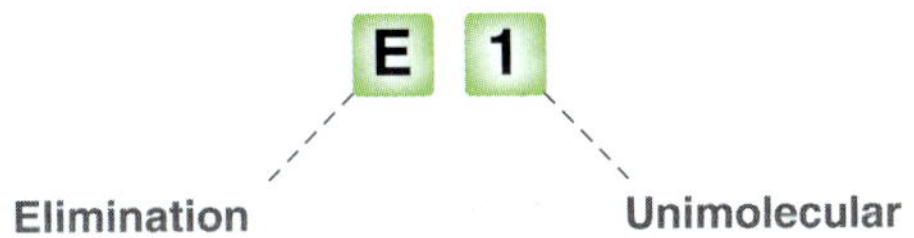

## CONCEPTUAL CHECKPOINT

**8.26** The following reaction occurs via an E1 mechanistic pathway:

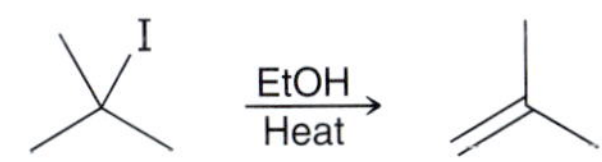


(a) What happens to the rate if the concentration of *tert*-butyl iodide is doubled and the concentration of ethanol is tripled?

(b) What happens to the rate if the concentration of *tert*-butyl iodide remains the same and the concentration of ethanol is doubled?

## Effect of Substrate

For E1 reactions, the rate is found to be very sensitive to the nature of the starting alkyl halide, with tertiary halides reacting most readily (Figure 8.18). This trend is identical to the trend we saw for $S_N1$ reactions, and the reason for the trend is the same as well. A stepwise mechanism involves formation of a carbocation intermediate, and the rate of reaction will be dependent on the stability of the carbocation.

Least reactive → Most reactive

Primary (1°) Secondary (2°) Tertiary (3°)

**FIGURE 8.18** The rate of reactivity of various substrates in an E1 reaction.

**LOOKING BACK**
For a review of carbocation stability and hyperconjugation, see Section 6.11.

Recall that tertiary carbocations are more stable than secondary carbocations (Figure 8.19), as a result of hyperconjugation. Compare the energy diagrams for E1 reactions involving secondary and tertiary substrates (Figure 8.20). Tertiary substrates exhibit a lower energy of activation during an E1 process and therefore react more rapidly. Primary substrates are generally unreactive toward an E1 mechanism, because a primary carbocation is too unstable to form.

Least stable → Most stable

Primary (1°) Secondary (2°) Tertiary (3°)

**FIGURE 8.19** Relative stability of primary, secondary, and tertiary carbocations.

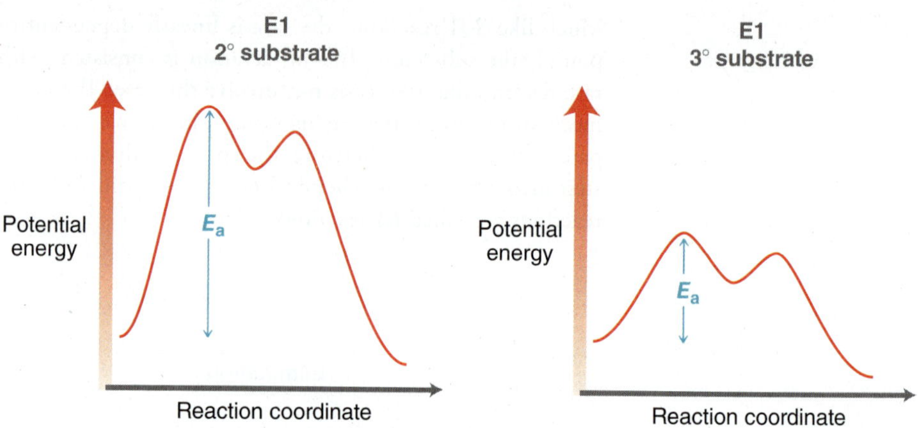


**FIGURE 8.20** A comparison of energy diagrams for the E1 reactions of secondary and tertiary substrates.

The first step of an E1 process is identical to the first step of an $S_N1$ process. In each process, the first step involves loss of the leaving group to form a carbocation intermediate:

Loss of a leaving group — $-LG^{\ominus}$ — Carbocation intermediate — Nucleophile → Substitution; Base → Elimination

An E1 reaction is generally accompanied by a competing $S_N1$ reaction, and a mixture of products is generally obtained.

In the previous chapter, we saw that an OH group is a terrible leaving group and that an $S_N1$ reaction can only occur if the OH group is first protonated to give a better leaving group:

Bad leaving group — H—X — Excellent leaving group

The same is true with an E1 process. If the substrate is an alcohol, a strong acid will be required in order to protonate the OH group. Concentrated aqueous sulfuric acid is commonly used for this purpose:

$$\text{R-OH} \xrightarrow[\text{Heat}]{\text{conc. } H_2SO_4} \text{alkene} + H_2O$$

The reaction above involves elimination of water to form an alkene and is therefore called a dehydration reaction.

## CONCEPTUAL CHECKPOINT

**8.27** Draw the carbocation intermediate generated by each of the following substrates in an E1 reaction:

(a) Br (b) Cl (c) I (d) Br

**8.28** Draw the carbocation intermediate generated when each of the following substrates is treated with sulfuric acid:

(a) (b) (c) (d)

## Regioselectivity of E1 Reactions

E1 processes exhibit a regiochemical preference for the Zaitsev product, as was observed for E2 reactions. For example:

conc. $H_2SO_4$, 80°C → 90% + 10%

The more substituted alkene (Zaitsev product) is the major product. However, there is one critical difference between the regiochemical outcomes of E1 and E2 reactions. Specifically, we have seen that the regiochemical outcome of an E2 reaction can often be controlled by carefully choosing the base (sterically hindered or not sterically hindered). In contrast, the regiochemical outcome of an E1 process cannot be controlled. The Zaitsev product will generally be obtained.

# SKILLBUILDER

## 8.8 PREDICTING THE REGIOCHEMICAL OUTCOME OF AN E1 REACTION

**LEARN the skill** Identify the major and minor products of the following E1 reaction:

conc. $H_2SO_4$, Heat → ?

**SOLUTION**

First identify all β positions that have protons.

**STEP 1** Identify all β positions that have protons.

These are the positions that must be explored. Two of these positions are identical, because they would give the same product:

**STEP 2** Take notice of the β positions that lead to the same product.

Abstracting a proton from either of these two positions yields the same product

These two drawings represent the same product

The other β position will produce the following alkene:

OH

**Abstracting a proton from this position...** **...yields this product**

Therefore, there are two possible products, and we expect to obtain a mixture of both products. The major product will be the more substituted alkene:

**STEP 3**
Draw all products and label the more substituted alkene as the major product.

OH

conc. $H_2SO_4$ / Heat → **Major** + **Minor**

**PRACTICE the skill** **8.29** Identify the major and minor products for each of the following E1 reactions:

(a) OH, conc. $H_2SO_4$ / Heat → ?

(b) Br, EtOH / Heat → ?

(c) OH, conc. $H_2SO_4$ / Heat → ?

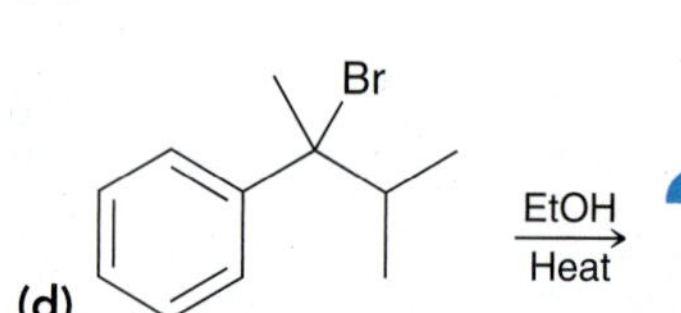

(d) Br, EtOH / Heat → ?

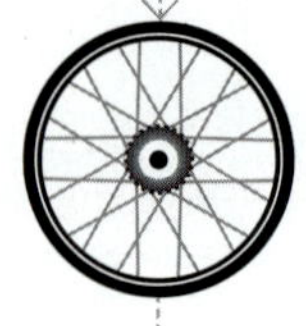

**APPLY the skill** **8.30** Identify two different starting alcohols that could be used to make 1-methylcyclohexene. Then determine which alcohol would be expected to react more rapidly under acidic conditions. Explain your choice.

need more **PRACTICE?** **Try Problem 8.62**

## Stereoselectivity of E1 Reactions

E1 reactions are not stereospecific—that is, they do not require *anti*-periplanarity in order for the reaction to occur. Nevertheless, E1 reactions are stereoselective. When *cis* and *trans* products are possible, we generally observe a preference for formation of the *trans* stereoisomer:

OH

conc. $H_2SO_4$ / Heat → 75% + 25%

CONCEPTUAL CHECKPOINT

**8.31** Draw only the major product for each of the following E1 reactions:

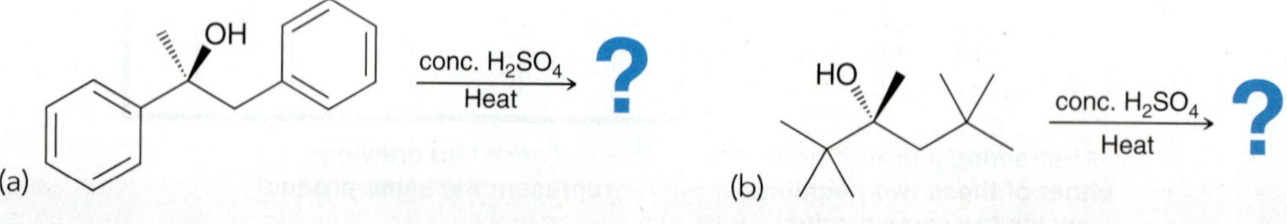

## 8.10 Drawing the Complete Mechanism of an E1 Process

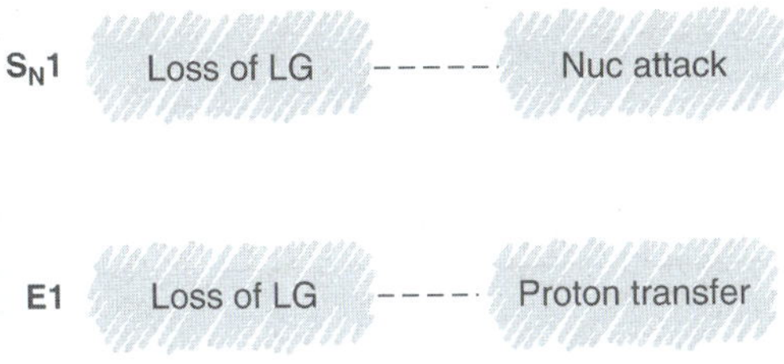

$S_N1$ Loss of LG ----- Nuc attack

E1 Loss of LG ----- Proton transfer

**FIGURE 8.21**
A comparison of the core steps of $S_N1$ and E1 processes.

Compare the two core steps of an E1 mechanism with the two core steps of an $S_N1$ mechanism (Figure 8.21). In Section 7.6, we saw that an $S_N1$ process can have up to three additional steps (Figure 8.22). Similarly, we might expect that an E1 mechanism could also have up to three additional steps. However, in practice, nothing ever happens after the second core step. An E1 process can only have up to two additional steps (Figure 8.23):

- Proton transfer before the first core step
- Carbocation rearrangement in between the two core steps

Can you explain why a proton transfer is never required after the second core step of an E1 process? Think about what circumstances require proton transfer at the end of an $S_N1$ mechanism and then consider why that doesn't happen in an E1 process.

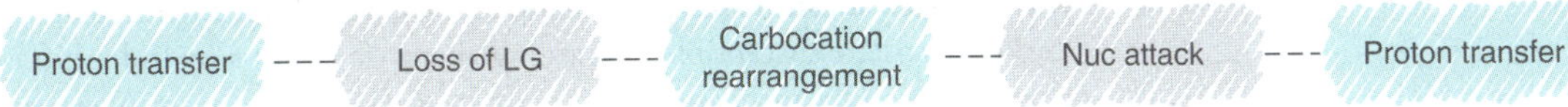


**FIGURE 8.22**
The core steps (gray) and additional steps (blue) of an $S_N1$ process.

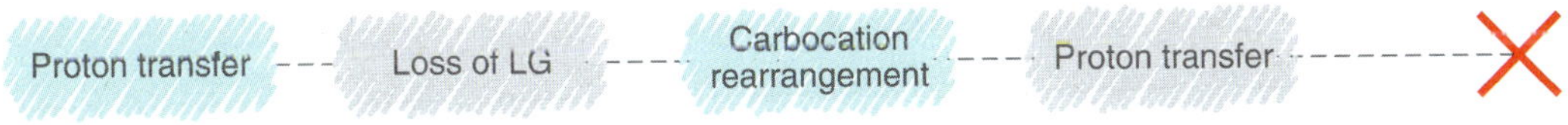

Proton transfer --- Loss of LG ---- Carbocation rearrangement ---- Proton transfer ---------- ✗

**FIGURE 8.23**
The core steps (gray) and additional steps (blue) of an E1 process.

### Proton Transfer at the Beginning of an E1 Mechanism

$+H^+$ --- $-LG$ --- $-H^+$

A proton transfer is required at the beginning of an E1 process for the same reason that it was required at the beginning of an $S_N1$ process. Specifically, a proton transfer is necessary whenever the leaving group is an OH group. Hydroxide is a bad leaving group and will not leave by itself. Therefore, an alcohol will only participate in an E1 reaction if the OH group is first protonated, which means that an acid is needed. Here is an example:

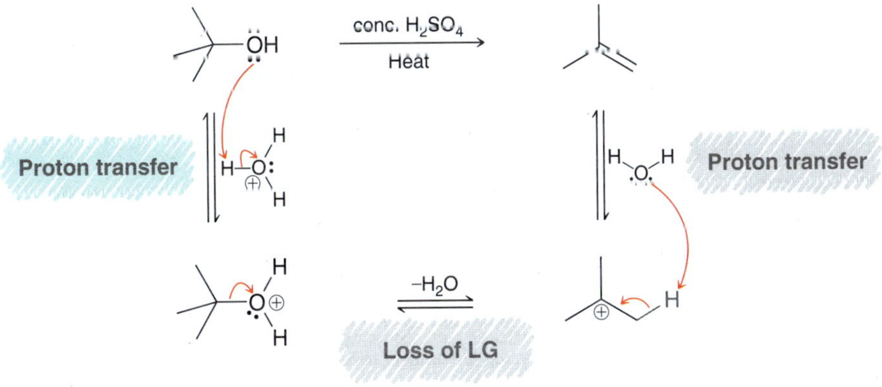

conc. $H_2SO_4$
Heat

Proton transfer

Proton transfer

$-H_2O$

Loss of LG

Notice that this mechanism has three steps:

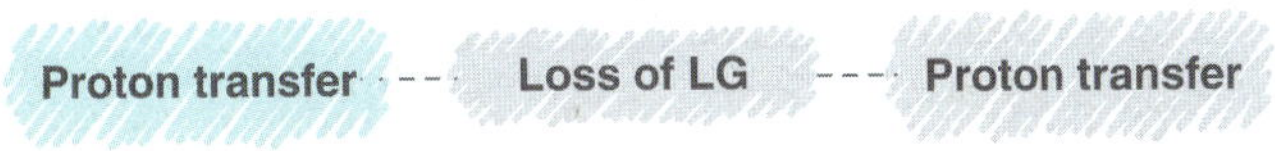

Proton transfer --- Loss of LG --- Proton transfer

CONCEPTUAL CHECKPOINT

**8.32** For each of the following substrates, determine whether an E1 process will require the use of an acid:

(a) (b) OH (c) OTs (d) OH (e) Br (f) OTs

## Carbocation Rearrangement During an E1 Mechanism

The E1 mechanism involves formation of a carbocation intermediate. Recall from Chapter 6 that carbocations are susceptible to rearrangement via either a hydride shift or a methyl shift. Here is an example of an E1 mechanism that contains a carbocation rearrangement:

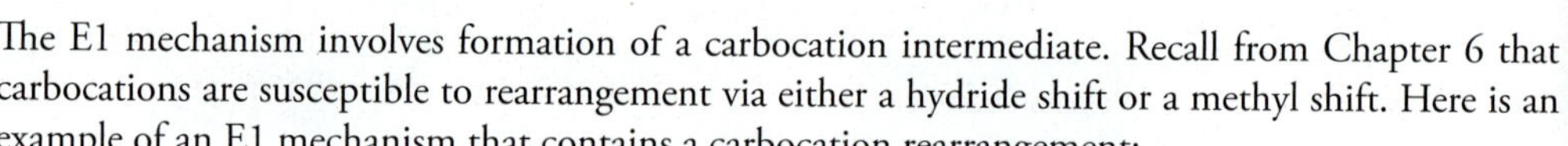

$H_2O$ Heat

Loss of LG

Proton transfer

2° Carbocation

C+ Rearrangement

3° Carbocation

**LOOKING BACK**
For practice with hydride shifts and methyl shifts, see Section 6.11.

Notice that the carbocation rearrangement occurs between the two core steps of an E1 mechanism:

Loss of LG ---- C+ Rearrangement ---- Proton transfer

CONCEPTUAL CHECKPOINT

**8.33** For each of the following substrates, determine whether an E1 process might involve a carbocation rearrangement or not:

(a) I (b) OH (c) OH (d) OH (e) Br (f) Cl

## An E1 Mechanism Can Have Many Steps

As we have seen, an E1 mechanism must have at least two core steps (loss of leaving group and proton transfer) and can also have up to two additional steps:

1. *Before the two core steps*—It is possible to have a proton transfer.
2. *In between the two core steps*—It is possible to have a carbocation rearrangement.

## MECHANISM 8.1 THE E1 MECHANISM

### Two core steps

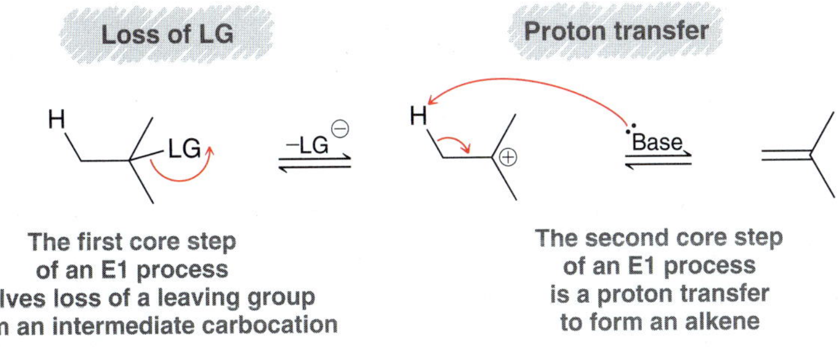


### *Possible additional steps*

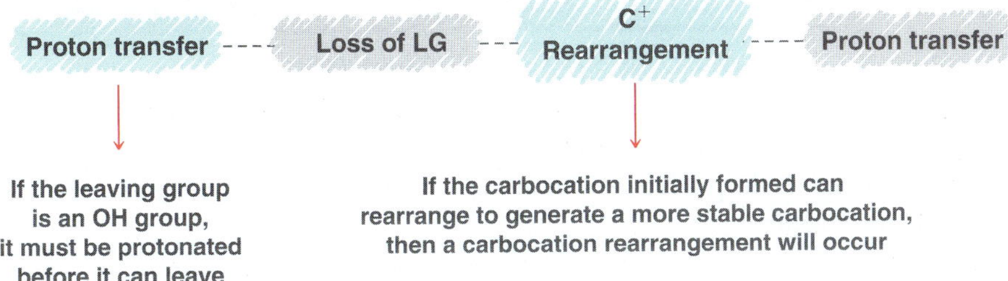


An E1 process can have one or both of the additional steps shown in Mechanism 8.1. Here is an example of an E1 process with four steps:

conc. $H_2SO_4$

Heat

Proton transfer

$-H_2O$

Loss of LG

C+

Rearrangement

2° Carbocation

3° Carbocation

Proton transfer

This mechanism has four steps, and therefore, we expect the energy diagram of this reaction to exhibit four humps (Figure 8.24). The number of humps in the energy diagram of an E1 process will always be equal to the number of steps in the mechanism. As we have seen, the number of steps can range from two to four, which means that the energy diagram of an E1 process can have from two to four humps.

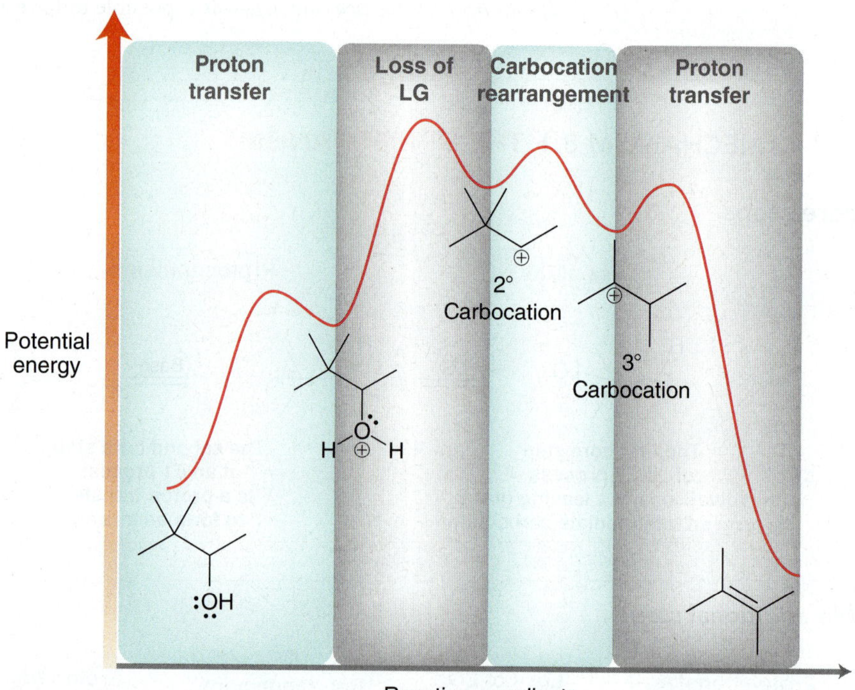


**FIGURE 8.24** The energy diagram of an E1 process with four steps.

In cases where a carbocation rearrangement is possible, we expect to obtain the product(s) from rearrangement as well as the product(s) without rearrangement. In the previous example, the following products are obtained:

OH —conc. $H_2SO_4$, Heat→ [64% + 33%] + [3%]

**Products from rearrangement** **Product from no rearrangement**

# SKILLBUILDER

## 8.9 DRAWING THE COMPLETE MECHANISM OF AN E1 REACTION

**LEARN the skill** Draw a mechanism for the following E1 process:

OH —conc. $H_2SO_4$, Heat→

## SOLUTION

An E1 process must always involve at least two core steps: loss of a leaving group and proton transfer. But we must consider whether the other two possible steps will occur:

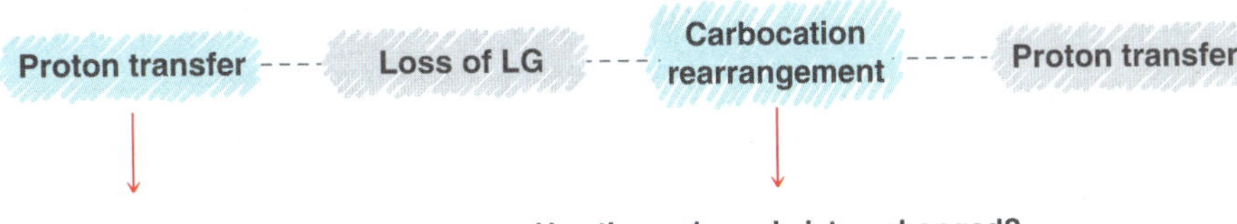


Does the LG need to be protonated before it can leave? —Yes. Hydroxide is a bad LG and must be protonated.

Has the carbon skeleton changed? —Yes. This indicates a carbocation rearrangement.

**LOOKING BACK**
For practice with arrow pushing, see Sections 6.8–6.10).

In this case, the mechanism should exhibit all four steps (the two core steps plus the two additional steps). To draw these steps, we will rely on the skills we learned in Chapter 6:

conc. $H_2SO_4$ / Heat

Proton transfer

Proton transfer

$-H_2O$

Loss of LG

2° Carbocation

$C^+$ Rearrangement

3° Carbocation

**PRACTICE the skill** **8.34** Draw a mechanism for each of the following E1 processes:

(a) conc. $H_2SO_4$ / Heat

(b) EtOH / Heat

(c) EtOH / Heat

(d) conc. $H_2SO_4$ / Heat

**APPLY the skill** **8.35** Identify the pattern for each mechanism in Problem 8.34. For example, the pattern for Problem 8.34a is:

This mechanism is comprised of a proton transfer followed by the two core steps of an E1 process (loss of leaving group and then proton transfer).

Draw patterns for the other three problems (8.34b–8.34d). Then compare the patterns. Identify which two reactions exhibit the same exact pattern and then describe in words why those two reactions are so similar.

**8.36** Identify which of the following methods is more efficient for producing 3,3-dimethylcyclohexene. Explain your choice.

Br

NaOEt
EtOH

OH

conc. $H_2SO_4$
Heat

need more **PRACTICE?** **Try Problems 8.68a, 8.76a–d, 8.81, 8.83, 8.84**

## 8.11 Drawing the Complete Mechanism of an E2 Process

In the previous section, we saw that additional steps can accompany an E1 process. In contrast, an E2 process consists of one concerted step and is rarely accompanied by any other steps. A carbocation is never formed, and therefore, there is no possibility for a carbocation rearrangement. In addition, E2 conditions generally require the use of a strong base, and an OH group cannot be protonated under such conditions. It is therefore not common to see an E2 process with a proton transfer at the beginning of the mechanism. It is much more common to see a proton transfer at the beginning of an E1 process (a dehydration reaction). All of the E2 processes that we will encounter in this textbook will be comprised of just one concerted step, as seen in Mechanism 8.2.

### MECHANISM 8.2 THE E2 MECHANISM

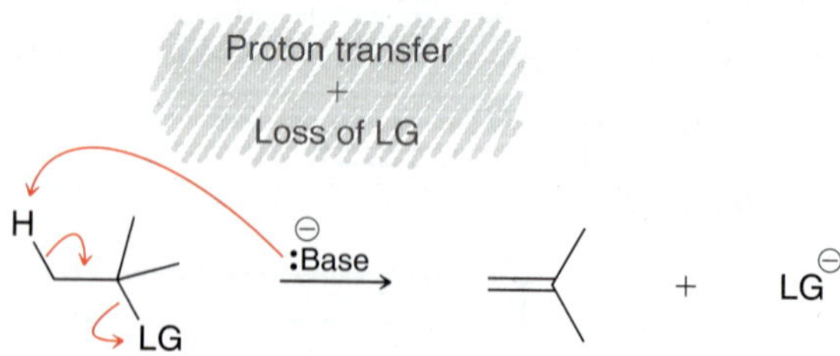


**An E2 process is comprised of just one concerted step in which the base abstracts a proton and the leaving group leaves at the same time**

The E2 mechanism requires only three curved arrows, all of which must be carefully placed. When drawing the first curved arrow, the tail is placed on a lone pair of the base and the head is drawn pointing to a proton at the β position. When drawing the second curved arrow, the tail is placed on the C—H bond that is breaking, and the head is placed on the bond between the α and β positions. Finally, the third curved arrow is drawn to show the loss of the leaving group.

## CONCEPTUAL CHECKPOINT

**8.37** Draw a mechanism for each of the following E2 reactions:

(a) Cl; NaOMe →

(b) Br; *t*-BuOK →

(c) Br; NaOH →

**8.38** In the next chapter, we will learn a method for preparing alkynes (compounds containing C≡C triple bonds). In the following reaction, a dihalide (a compound with two halogen atoms) is treated with an excess of strong base (sodium amide), resulting in two successive E2 reactions. Draw a mechanism for this transformation.

Cl Cl R R H H; Excess $NaNH_2$ → R—≡—R

## 8.12 Substitution vs. Elimination: Identifying the Reagent

Substitution and elimination reactions are almost always in competition with each other. In order to predict the products of a reaction, it is necessary to determine which mechanisms are likely to occur. In some cases, only one mechanism will predominate:

I; NaOMe / MeOH →

**From E2** (only product)

In other cases, two or more mechanisms will compete; for example:

I; $H_2O$ / Heat → + OH

**From E1** **From $S_N1$**

Don't fall into the trap of thinking that there must always be one clear winner. Sometimes there is, but sometimes there are multiple products. The goal is to predict all of the products and to predict which products are major and which are minor. To accomplish this goal, three steps are required:

1. Determine the function of the reagent.
2. Analyze the substrate and determine the expected mechanism(s).
3. Consider any relevant regiochemical and stereochemical requirements.

Each of these three steps will be explored, in detail, in the remainder of this chapter. This section will focus on the skills necessary to achieve the first step, while the second and third steps will be discussed in the following sections. The ultimate goal is proficiency in predicting products, and the skills in this section represent the first step in achieving that goal.

We have seen earlier in this chapter that the main difference between substitution and elimination is the function of the reagent. A substitution reaction occurs when the reagent functions as a nucleophile, while an elimination reaction occurs when the reagent functions as a base. Accordingly, the first step in any specific case is to determine whether the reagent is a strong nucleophile or a weak nucleophile and whether it is a strong base or a weak base. Nucleophilicity and basicity are not the same concepts, as first described in Section 6.7. Nucleophilicity is a kinetic phenomenon and refers to the rate of reaction, while basicity is a thermodynamic phenomenon and refers to the position of equilibrium. We will soon see that it is possible for a reagent to be a weak nucleophile and a strong base. Similarly, it is possible for a reagent to be a strong nucleophile and a weak base. That is, basicity and nucleophilicity do not always parallel each other. We will now explore nucleophilicity and basicity in more detail in order to develop the skills necessary to identify the expected function of a reagent.

## Nucleophilicity

Nucleophilicity refers to the rate at which a particular nucleophile will attack an electrophile. There are many factors that contribute to nucleophilicity, as first described in Section 6.7. One such factor is charge, which can be illustrated by comparing a hydroxide ion and water. Hydroxide is a strong nucleophile, while water is a weak nucleophile, as seen in Section 7.8.

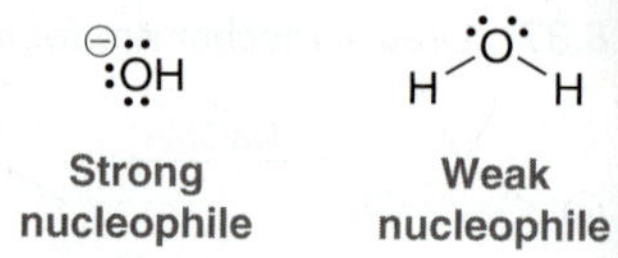

Strong nucleophile | Weak nucleophile

Another factor that impacts nucleophilicity is *polarizability*, which is often even more important than charge. Recall that polarizability describes the ability of an atom to distribute its electron density unevenly as a result of external influences (Section 6.7). Polarizability is directly related to the size of the atom and, more specifically, to the number of electrons that are distant from the nucleus. A sulfur atom is very large and has many electrons that are distant from the nucleus, and it is therefore highly polarizable. Most halogens share this same feature. For this reason, $H_2S$ is a much stronger nucleophile than $H_2O$:

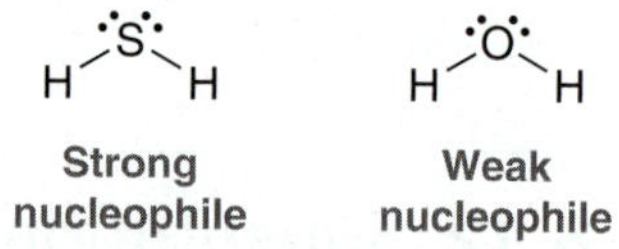

Strong nucleophile | Weak nucleophile

Notice that $H_2S$ lacks a negative charge but is nevertheless a strong nucleophile because it is highly polarizable. The relationship between polarizability and nucleophilicity also explains why the hydride ion ($H^-$) of NaH does not function as a nucleophile, even though it has a negative charge. The hydride ion is extremely small (hydrogen is the smallest atom), and it is not sufficiently polarizable to function as a nucleophile. We will see, however, that the hydride ion does function as a very strong base. Let's now review the factors that affect basicity.

## Basicity

Unlike nucleophilicity, basicity is not a kinetic phenomenon and does not refer to the rate of a process. Rather, basicity is a thermodynamic phenomenon and refers to the position of equilibrium:

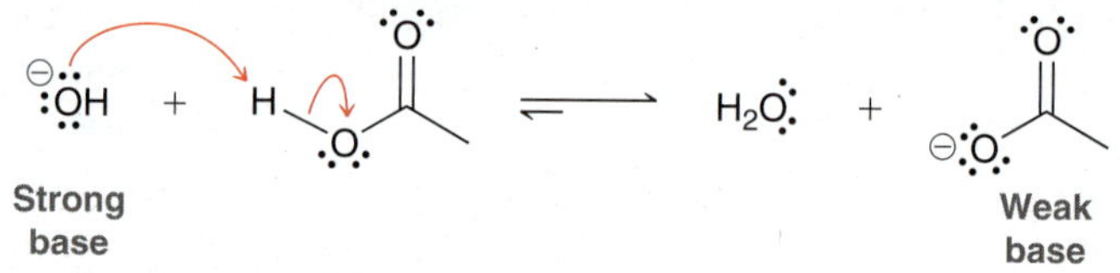

Strong base | Weak base

In a proton transfer process, the equilibrium will favor the weaker base.

There are several methods for determining whether a base is strong or weak. Chapter 3 discussed two such methods: a quantitative approach and a qualitative approach. The former requires access to $pK_a$ values. Specifically, the strength of a base is determined by assessing the $pK_a$ of its conjugate acid. For example, a chloride ion is an extremely weak base, because its conjugate acid (HCl) is strongly acidic ($pK_a = -7$). The other approach, the qualitative method, was described in Section 3.4 and involves the use of four factors for determining the relative stability of a base containing a negative charge. For example, a chloride ion has a negative charge on a large, electronegative atom and is therefore highly stabilized (a weak base). Notice that both the quantitative approach and qualitative approach provide the same prediction: that a chloride ion is a weak base.

As another example, consider the bisulfate ion ($HSO_4^-$). The quantitative approach identifies this ion as a weak base, because its conjugate acid is strongly acidic ($pK_a = -9$). The qualitative approach gives a similar prediction, because the bisulfate ion is highly resonance stabilized:

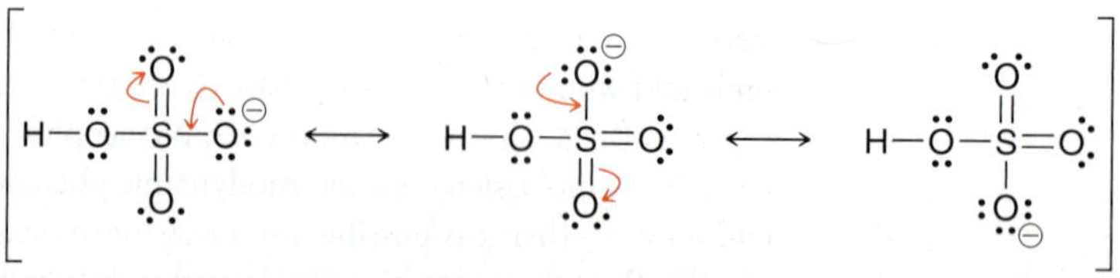

Notice once again that the quantitative approach and qualitative approach are consistent. Either approach can be used. The specific case will usually dictate which method is easier to use.

## Nucleophilicity vs. Basicity

After reviewing the main factors contributing to nucleophilicity and basicity, we can now categorize all reagents into four groups (Figure 8.25). Let's quickly review each of these four categories. The first category contains reagents that function only as nucleophiles. That is, they are strong nucleophiles because they are highly polarizable, but they are weak bases because their conjugate acids are fairly acidic. The use of a reagent from this category signifies that a substitution reaction is occurring (not elimination).

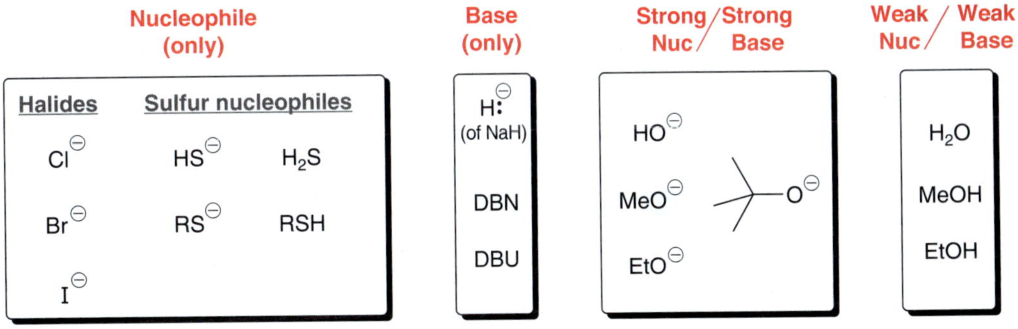

| Nucleophile (only) | | | Base (only) | Strong Nuc / Strong Base | | Weak Nuc / Weak Base |
|---|---|---|---|---|---|---|
| **Halides** | **Sulfur nucleophiles** | | | | | |
| $Cl^{\ominus}$ | $HS^{\ominus}$ | $H_2S$ | $H:^{\ominus}$ (of NaH) | $HO^{\ominus}$ | $(CH_3)_3CO^{\ominus}$ | $H_2O$ |
| $Br^{\ominus}$ | $RS^{\ominus}$ | RSH | DBN | $MeO^{\ominus}$ | | MeOH |
| $I^{\ominus}$ | | | DBU | $EtO^{\ominus}$ | | EtOH |

**FIGURE 8.25** Classification of common reagents used for substitution/elimination.

**LOOKING AHEAD**
The hydride ion will be used as a strong base in other applications throughout this book.

The second category contains reagents that function only as bases, not as nucleophiles. The first reagent on this list is the hydride ion, usually shown as NaH, where $Na^+$ is the counterion. As mentioned earlier, the hydride ion is not a nucleophile because it is not sufficiently polarizable; nevertheless, it is a very strong base because its conjugate acid (H—H) is an extremely weak acid. The use of a hydride ion as the reagent indicates that elimination will occur rather than substitution. There are indeed many reagents, other than hydride, that also function exclusively as bases (and not as nucleophiles). Two commonly used examples are DBN and DBU:

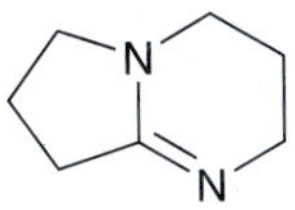

1,5-Diazabicyclo[4.3.0]non-5-ene (DBN)

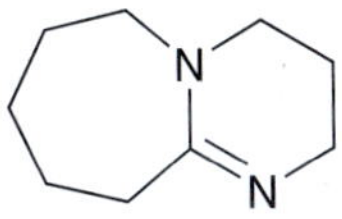

1,8-Diazabicyclo[5.4.0]undec-7-ene (DBU)

These two compounds are very similar in structure. When either of these compounds is protonated, the resulting positive charge is resonance stabilized:

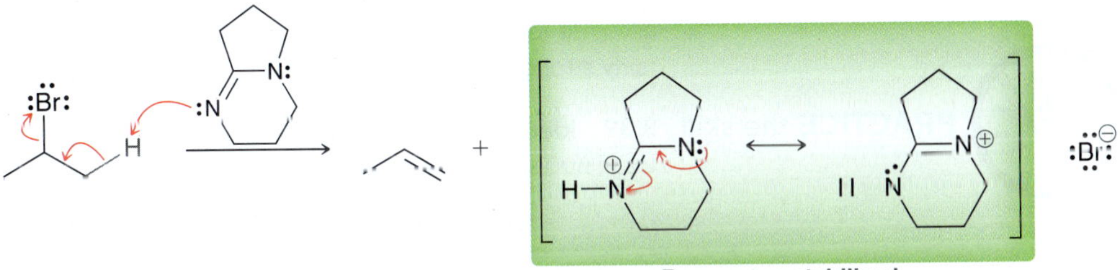

Resonance stabilized

The positive charge is spread over two nitrogen atoms, rather than one. The conjugate acid is stabilized and is therefore not very acidic. As a result, both DBN and DBU are strongly basic. These examples demonstrate that it is possible for a neutral compound to be a strong base.

The third category in Figure 8.25 contains reagents that are both strong nucleophiles and strong bases. These reagents include hydroxide ($HO^-$) and alkoxide ions ($RO^-$). Such reagents are generally used for bimolecular processes ($S_N2$ and E2).

The fourth and final category contains reagents that are weak nucleophiles and weak bases. These reagents include water ($H_2O$) and alcohols (ROH). Such reagents are generally used for unimolecular processes ($S_N1$ and E1).

In order to predict the products of a reaction, the first step is determining the identity and nature of the reagent. That is, you must analyze the reagent and determine the category to which it belongs. Let's get some practice with this critical skill.

# SKILLBUILDER

## 8.10 DETERMINING THE FUNCTION OF A REAGENT

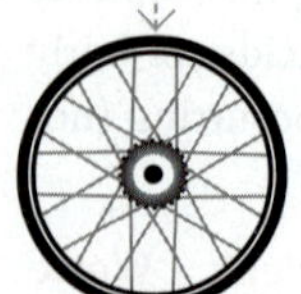

**LEARN the skill**

Consider the phenolate ion. We will explore the reactivity of this ion in Chapter 19. From the following list, identify the category to which the phenolate ion belongs:

**(a)** Strong nucleophile and weak base

**(b)** Weak nucleophile and strong base

**(c)** Strong nucleophile and strong base

**(d)** Weak nucleophile and weak base

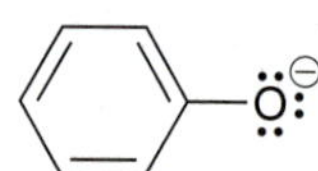

**Phenolate**

### SOLUTION

**STEP 1**
Determine whether or not the reagent is a strong nucleophile by looking for charge and/or polarizability.

Let's begin by exploring its nucleophilicity. We look for charge and polarizability. The reagent certainly has a negative charge, which should render it a strong nucleophile. The oxygen atom is not highly polarizable, but it is also not small enough of an atom (like hydrogen) to render it non-nucleophilic. Accordingly, we predict that the phenolate ion should be a fairly strong nucleophile. This feature will be an important aspect of its reactivity in Chapter 19.

**STEP 2**
Determine whether or not the reagent is a strong base using either the quantitative method or the qualitative method.

Next, we explore the basicity of the phenolate ion. There are two methods that can be used, and their results should agree. Using the quantitative approach, we look up the $pK_a$ value of the conjugate acid (called phenol):

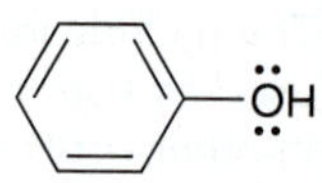

**Phenol**

Table 3.1 indicates that the $pK_a$ of phenol is 9.9, which is much more acidic than a typical alcohol (ROH usually has a $pK_a$ in the range of 16–18). Since phenol is fairly acidic, we expect the phenolate ion to be a fairly weak base. Alternatively, we could draw the same conclusion by using the qualitative approach. Specifically, we note that the phenolate ion is resonance stabilized:

As a result, the phenolate ion is expected to be a weaker base (more stable) than a regular alkoxide ion ($RO^-$). In conclusion, our analysis predicts that the phenolate ion should be a fairly strong nucleophile and a fairly weak base.

**PRACTICE the skill**

**8.39** Identify whether each of the following reagents would be a strong nucleophile or a weak nucleophile and also indicate whether it would be a strong base or a weak base:

**(a)**  **(b)** SH **(c)** $O^{\ominus}$ **(d)** $Br^{\ominus}$

**(e)** LiOH **(f)** MeOH **(g)** NaOMe **(h)** DBN

**APPLY the skill**

**8.40** We have seen that NaH is a strong base but a weak nucleophile. In contrast, lithium aluminum hydride (LAH) is a reagent that can serve as a source of nucleophilic hydride ion:

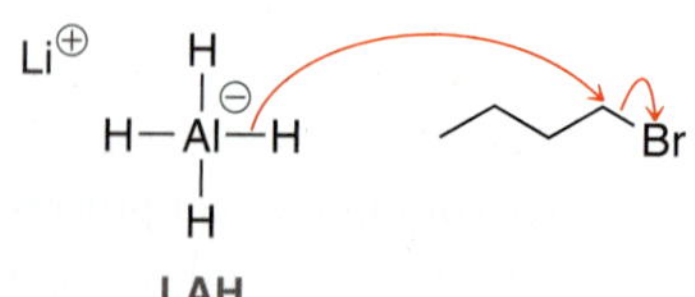


In this case, LAH functions as a delivery agent of a nucleophilic hydride ion. We will see this reagent in many upcoming chapters. Explain why LAH is capable of functioning as a strong nucleophile while NaH is not.

need more **PRACTICE?** **Try Problem 8.55**

## 8.13 Substitution vs. Elimination: Identifying the Mechanism(s)

We mentioned that there are three main steps for predicting the products of substitution and elimination reactions. In the previous section, we explored the first step (identifying the function of the reagent). In this section, we now explore the second step in which we analyze the substrate and identify which mechanism(s) operate.

As described in the previous section, there are four categories of reagents. For each category, we must explore the expected outcome with a primary, secondary, or tertiary substrate. Then, we will summarize all of the information in one master flow chart. Let's begin with reagents that function only as nucleophiles.

### Possible Outcomes for Reagents That Function Only as Nucleophiles

When the reagent functions exclusively as a nucleophile (and not as a base), only substitution reactions will occur (not elimination), as illustrated in Figure 8.26. The substrate will determine which mechanism operates. $S_N2$ will predominate for primary substrates, and $S_N1$ will predominate for tertiary substrates. For secondary substrates, both $S_N2$ and $S_N1$ pathways are viable, although $S_N2$ is generally favored. The rate of an $S_N2$ process can be further enhanced by using a polar aprotic solvent, as described in Section 7.8.

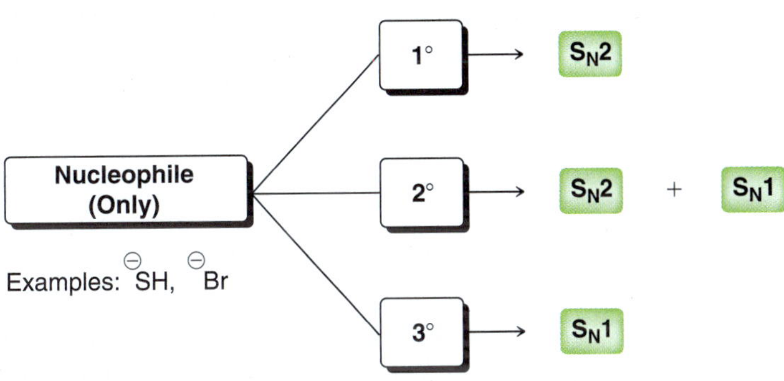


**FIGURE 8.26** The outcomes to be expected when a primary, secondary, or tertiary substrate is treated with a reagent that functions only as a nucleophile.

### Possible Outcomes for Reagents That Function Only as Bases

When the reagent functions exclusively as a base (and not as a nucleophile), only elimination reactions will occur (not substitution), as illustrated in Figure 8.27. Such reagents are generally strong bases, resulting in an E2 process. As seen in Section 8.7, an E2 process can occur for primary, secondary, or tertiary substrates.

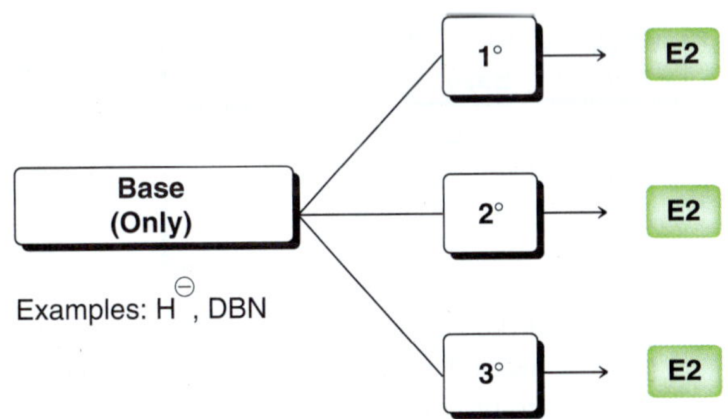


**FIGURE 8.27** An E2 process is the expected outcome when a primary, secondary, or tertiary substrate is treated with a reagent that functions only as a base.

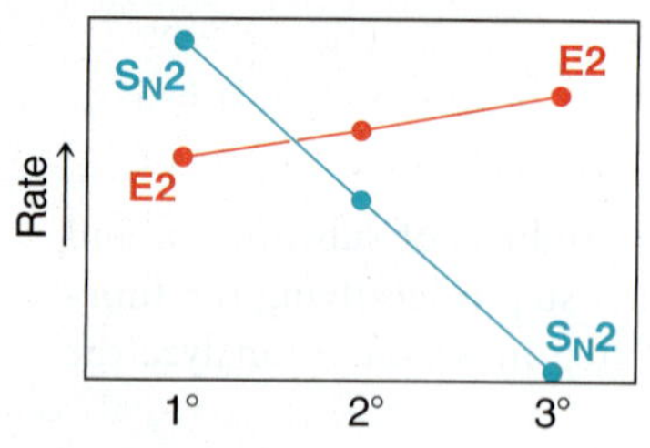


**FIGURE 8.28** The effects of substrate on the competition between $S_N2$ and E2 processes.

## Possible Outcomes for Reagents That Are Strong Bases and Strong Nucleophiles

When the reagent is both a strong nucleophile and a strong base, bimolecular mechanisms will dominate ($S_N2$ and E2), and the substrate plays a critical role in the competition between $S_N2$ and E2. The nature of this competition is illustrated in Figure 8.28.

The rates of $S_N2$ and E2 processes are affected differently by the substrate. Notice that the $S_N2$ process predominates when the substrate is primary, while the E2 process predominates when the substrate is secondary. For tertiary substrates, only E2 is observed, because the $S_N2$ pathway is too sterically hindered to occur. Recall that the E2 pathway is not sensitive to steric hindrance. This competition between $S_N2$ and E2 is summarized in Figure 8.29.

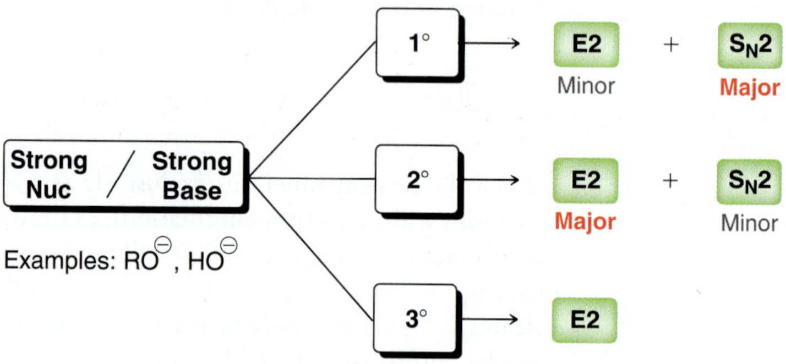


**FIGURE 8.29** The outcomes to be expected when a primary, secondary, or tertiary substrate is treated with a reagent that functions both as a strong base and as a strong nucleophile.

Notice that when a primary substrate is treated with an alkoxide ion ($RO^-$), the $S_N2$ pathway predominates over the E2 pathway. There is one notable exception to this general rule. Specifically, *tert*-butoxide is a sterically hindered alkoxide, and it favors E2 over $S_N2$, even when the substrate is primary. This exception is useful because it enables the conversion of a primary alkyl halide into an alkene:

X $\xrightarrow{t\text{-BuOK}}$

## Possible Outcomes for Reagents That Are Weak Bases and Weak Nucleophiles

Now let's consider the possible outcomes when the reagent is both a weak nucleophile and a weak base, as illustrated in Figure 8.30. These conditions are not practical for primary and secondary substrates. Primary substrates generally react slowly with weak nucleophiles and weak bases, and secondary substrates produce a mixture of too many products. However, when the substrate is tertiary, it can be effective to use a reagent that is both a weak base and a weak nucleophile. In such a case, unimolecular pathways predominate ($S_N1$ and E1). $S_N1$ products are generally favored, although the product distribution is sensitive to reaction conditions, with temperature often being the most significant factor. An increase in temperature will enhance the rates of both $S_N1$ and E1, but E1 is generally more greatly enhanced. As a result, the E1 pathway will sometimes predominate over the $S_N1$ pathway at elevated temperatures.

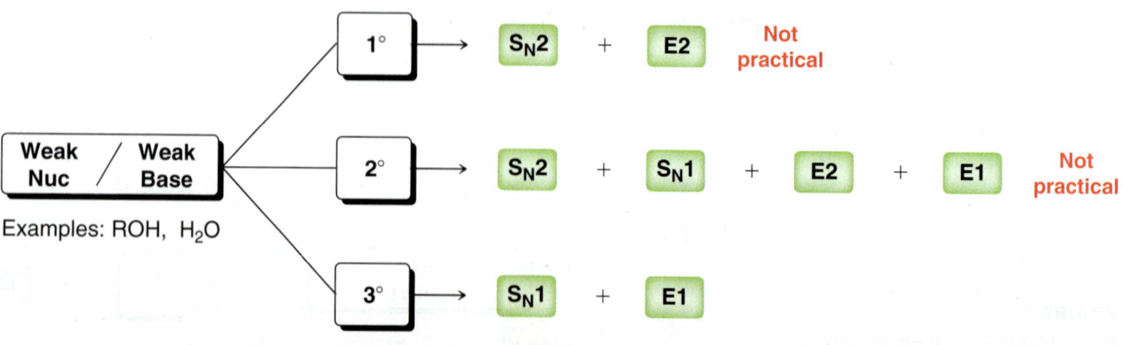


**FIGURE 8.30** The outcomes to be expected when a primary, secondary, or tertiary substrate is treated with a reagent that functions both as a weak base and as a weak nucleophile.

If the substrate is a secondary or tertiary alcohol, then treatment with concentrated sulfuric acid and heat will produce an E1 reaction (Section 8.9). In such a case, sulfuric acid serves as a proton source (to protonate the OH group), and water serves as a weak base that completes the E1 process.

All of the outcomes previously described are now summarized in one master flow chart (Figure 8.31). It is important to know this flow chart extremely well, but be careful not to memorize it. It is more important to "understand" the reasons for all of these outcomes. A proper understanding will prove to be far more useful on an exam than simply memorizing a set of rules. Figure 8.31 can be used to determine which mechanism(s) operate for any specific case.

**Reagent** | **Substrate** | **Mechanism** | **Other factors / comments**

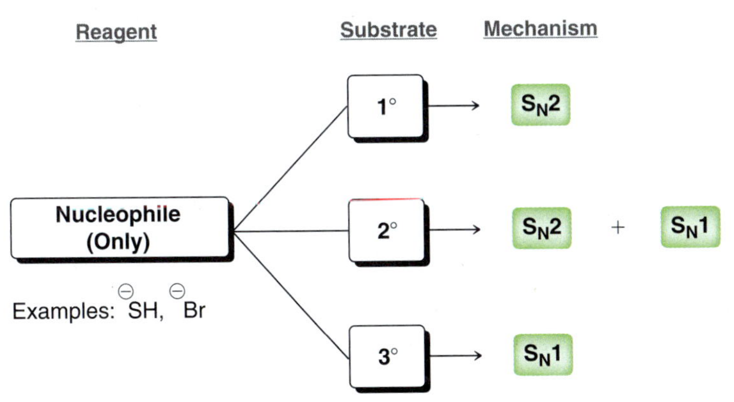


When the reagent functions exclusively as a nucleophile (and not as a base), only substitution reactions occur (not elimination). The substrate determines which mechanism operates. $S_N2$ predominates for primary substrates, and $S_N1$ predominates for tertiary substrates. For secondary substrates, both $S_N2$ and $S_N1$ can occur, although $S_N2$ is generally favored (especially when a polar aprotic solvent is used).

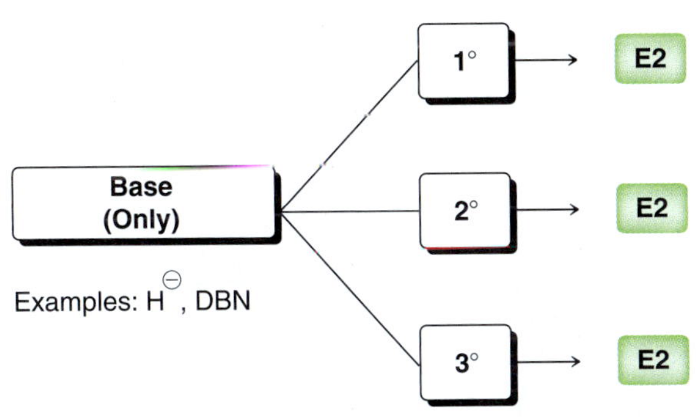


When the reagent functions exclusively as a base (and not as a nucleophile), only elimination reactions occur. Such reagents are generally strong bases, resulting in an E2 process. This mechanism can occur for primary, secondary, and tertiary substrates.

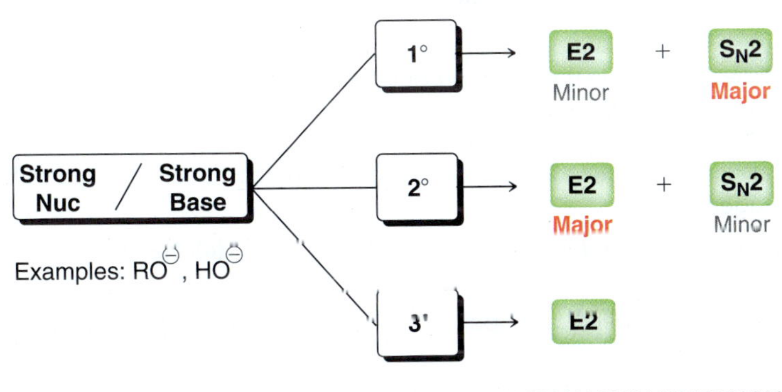


When the reagent is both a strong nucleophile and a strong base, bimolecular reactions ($S_N2$ and E2) are favored. For *primary* substrates, $S_N2$ predominates over E2, unless *t*-BuOK is used as the reagent, in which case E2 predominates. For *secondary* substrates, E2 predominates. For *tertiary* substrates, only E2 is observed, because the $S_N2$ pathway is too sterically hindered to occur.

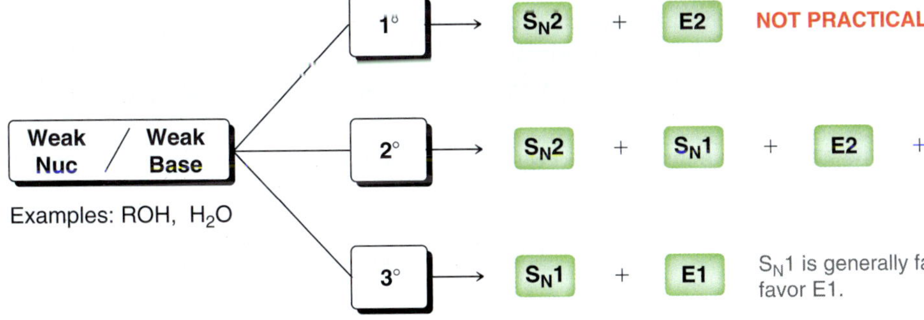


**NOT PRACTICAL**, because the reactions are too slow.

**NOT PRACTICAL**, because the reactions are too slow, and too many products are formed. A secondary alcohol will undergo an E1 reaction when treated with concentrated sulfuric acid and heat (Section 8.9).

$S_N1$ is generally favored, although elevated temperatures might favor E1.

**FIGURE 8.31**
A flow chart for determining the outcome of substitution / elimination reaction.

# SKILLBUILDER

## 8.11 IDENTIFYING THE EXPECTED MECHANISM(S)

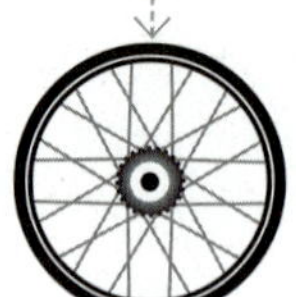

**LEARN the skill**

Identify the mechanism(s) expected to occur when bromocyclohexane is treated with sodium ethoxide.

**SOLUTION**

**STEP 1** Identify the function of the reagent.

First identify the function of the reagent. Using the skills developed in the previous section, we can determine that sodium ethoxide is both a strong nucleophile and a strong base:

Next, we identify the substrate. Bromocyclohexane is a secondary substrate, and therefore, we expect E2 and $S_N2$ mechanisms to operate:

**STEP 2** Identify the substrate and determine the mechanisms that operate.

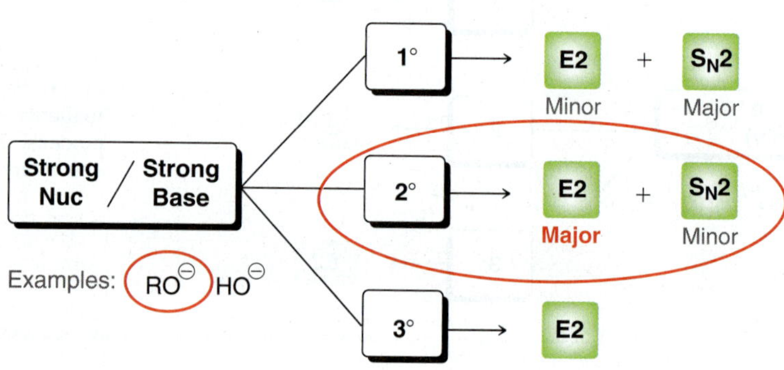


The E2 pathway is expected to provide the major product because the $S_N2$ pathway is more sensitive to steric hindrance provided by secondary substrates.

**PRACTICE the skill**

**8.41** Identify the mechanism(s) expected to operate when 1-bromobutane is treated with each of the following reagents:

**(a)** NaOH **(b)** NaSH **(c)** *t*-BuOK **(d)** DBN **(e)** NaOMe

**8.42** Identify the mechanism(s) expected to operate when 2-bromopentane is treated with each of the following reagents:

**(a)** NaOEt **(b)** NaI/DMSO **(c)** DBU **(d)** NaOH **(e)** *t*-BuOK

**8.43** Identify the mechanism expected to operate when 2-bromo-2-methylpentane is treated with each of the following reagents:

**(a)** EtOH **(b)** *t*-BuOK **(c)** NaI **(d)** NaOEt **(e)** NaOH

**APPLY the skill**

**8.44** When 1-chlorobutane is treated with ethanol, neither elimination process (E1 or E2) is observed at an appreciable rate:

**(a)** Explain why an E2 reaction does not occur.

**(b)** Explain why an E1 reaction does not occur.

**(c)** Modifying the reactants can have a profound effect on the rate of elimination. What modification would you suggest to enhance the rate of an E2 process?

**(d)** What modification would you suggest to enhance the rate of an E1 process?

**8.45** When 3-chloro-2,2,4,4-tetramethylpentane is treated with sodium hydroxide, neither E2 nor $S_N2$ products are formed. Explain. (**Hint:** See Figure 7.11 and the associated discussion.)

need more **PRACTICE?** **Try Problems 8.68, 8.69, 8.78a**

## 8.14 Substitution vs. Elimination: Predicting the Products

We mentioned that predicting the products of substitution and elimination reactions requires three discrete steps:

1. Determine the function of the reagent.
2. Analyze the substrate and determine the expected mechanism(s).
3. Consider any relevant regiochemical and stereochemical requirements.

In the previous two sections, we explored the first two steps of this process. In this last section of the chapter, we explore the third and final step. After determining which mechanism(s) operate, the final step is to consider the regiochemical and stereochemical outcomes for each of the expected mechanisms. Table 8.2 provides a summary of guidelines that must be followed when drawing products. Table 8.2 does not contain any new information. All of the information can be found in this chapter and the previous chapter. The table is meant only as a summary of all of the relevant information, so that it is easily accessible in one location. The following SkillBuilder provides the opportunity to practice applying these guidelines.

**TABLE 8.2 GUIDELINES FOR DETERMINING THE REGIOCHEMICAL AND STEREOCHEMICAL OUTCOME OF SUBSTITUTION AND ELIMINATION REACTIONS**

| | REGIOCHEMICAL OUTCOME | STEREOCHEMICAL OUTCOME |
|---|---|---|
| $S_N2$ | The nucleophile attacks the α position, where the leaving group is connected. | The nucleophile replaces the leaving group with inversion of configuration. |
| $S_N1$ | The nucleophile attacks the carbocation, which is where the leaving group was originally connected, unless a carbocation rearrangement takes place. | The nucleophile replaces the leaving group with racemization. |
| E2 | The Zaitsev product is generally favored over the Hofmann product, unless a sterically hindered base is used, in which case the Hofmann product will be favored. | This process is both stereoselective and stereospecific. When applicable, a *trans* disubstituted alkene will be favored over a *cis* disubstituted alkene. When the β position of the substrate has only one proton, the stereoisomeric alkene resulting from *anti*-periplanar elimination will be obtained (exclusively, in most cases). |
| E1 | The Zaitsev product is always favored over the Hofmann product. | The process is stereoselective. When applicable, a *trans* disubstituted alkene will be favored over a *cis* disubstituted alkene. |

# SKILLBUILDER

## 8.12 PREDICTING THE PRODUCTS OF SUBSTITUTION AND ELIMINATION REACTIONS

**LEARN the skill**

Predict the product(s) of the following reaction and identify the major and minor products:

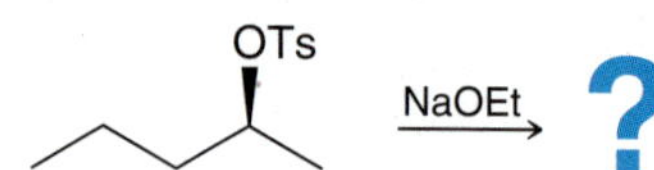


### SOLUTION

In order to draw the products, the following three steps must be followed:

1. Determine the function of the reagent.
2. Analyze the substrate and determine the expected mechanism(s).
3. Consider any relevant regiochemical and stereochemical requirements.

**STEP 1** Determine the function of the reagent.

In the first step, we analyze the reagent. The ethoxide ion is both a strong base and a strong nucleophile. Next, we analyze the substrate. In this case, the substrate is secondary, so we would expect E2 and $S_N2$ pathways to compete with each other:

**STEP 2** Analyze the substrate and determine the expected mechanisms.

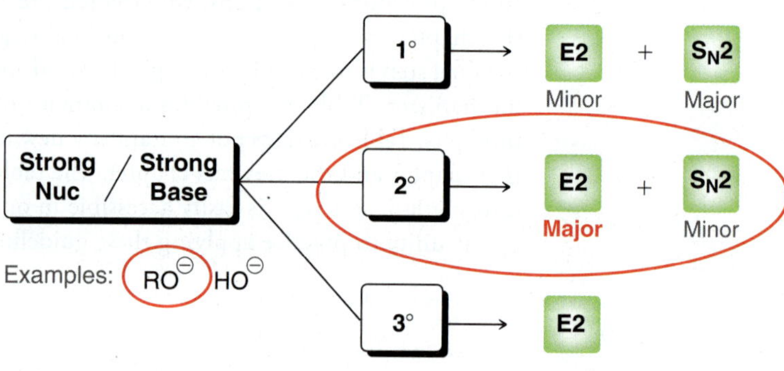


The E2 pathway would be expected to predominate, because it is less sensitive to steric hindrance than the $S_N2$ pathway. Therefore, we would expect the major product(s) to be generated via an E2 process and the minor product(s) to be generated via an $S_N2$ process. In order to draw the products, we must complete the final step. That is, we must consider the regiochemical and stereochemical outcomes for both the E2 and $S_N2$ processes. Let's begin with the E2 process.

**STEP 3** Consider regiochemical and stereochemical requirements.

For the regiochemical outcome, we expect the Zaitsev product to be the major product, because the reaction does not utilize a sterically hindered base. For the stereochemical outcome, we know that the E2 process is stereoselective, so we expect *cis* and *trans* isomers, with a predominance of the *trans* isomer:

OTs NaOEt → Major + Minor + Minor

The E2 process is also stereospecific, but in this case, the β position has more than one proton, so the stereospecificity of this reaction is not relevant.

Now consider the $S_N2$ product. This case involves a chirality center, so we expect inversion of configuration:

OTs NaOEt → OEt

In summary, we expect the following products:

OTs NaOEt → Major + Minor + Minor + Minor (OEt)

**PRACTICE the skill** **8.46** Identify the major and minor product(s) that are expected for each of the following reactions:

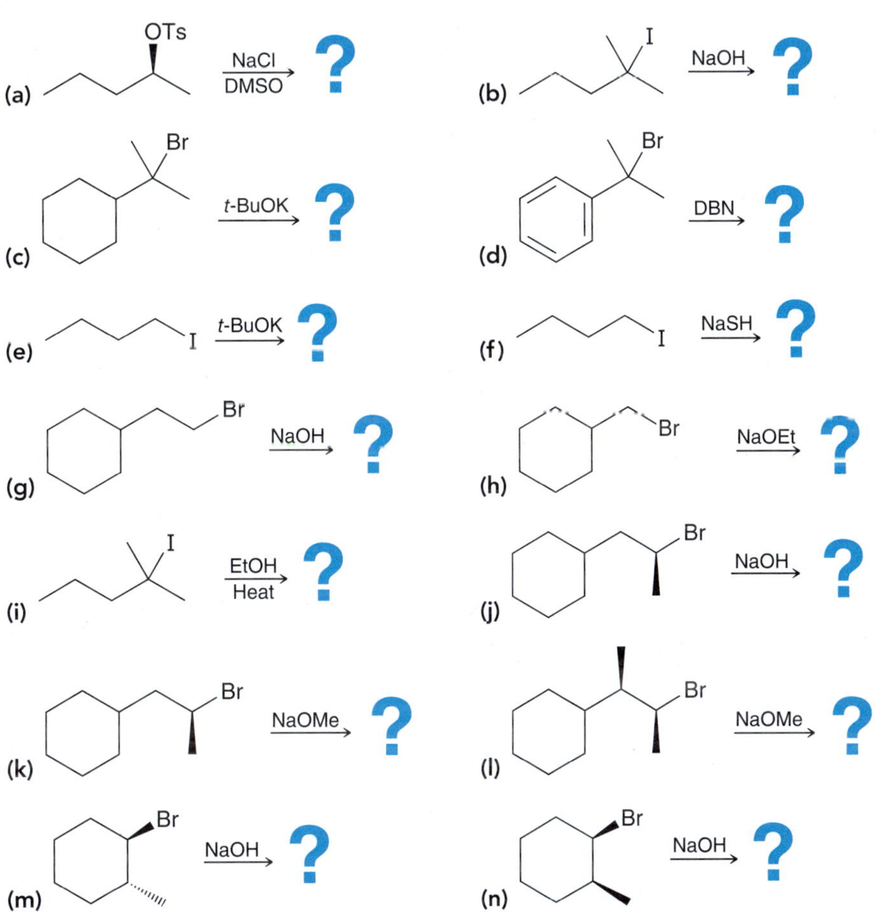

**APPLY the skill** **8.47** Compound A and compound B are constitutional isomers with molecular formula $C_3H_7Cl$. When compound A is treated with sodium methoxide, a substitution reaction predominates. When compound B is treated with sodium methoxide, an elimination reaction predominates. Propose structures for compounds A and B.

**8.48** Compound A and compound B are constitutional isomers with molecular formula $C_4H_9Cl$. Treatment of compound A with sodium methoxide gives *trans*-2-butene as the major product, while treatment of compound B with sodium methoxide gives a different disubstituted alkene as the major product.

**(a)** Draw the structure of compound A.

**(b)** Draw the structure of compound B.

**8.49** An unknown compound with molecular formula $C_6H_{13}Cl$ is treated with sodium ethoxide to produce 2,3-dimethyl-2-butene as the major product. Identify the structure of the unknown compound.

need more **PRACTICE?** **Try Problems 8.77, 8.79**

# REVIEW OF REACTIONS

## SYNTHETICALLY USEFUL ELIMINATION REACTIONS

### Dehydration of an Alcohol via an E1 Process

OH

conc. $H_2SO_4$
Heat

### Dehydrohalogenation via an E2 Process—Formation of the Zaitsev Product

Br

NaOEt

### Dehydrohalogenation via an E2 Process—Formation of the Hofmann Product

Br

*t*-BuOK

# REVIEW OF CONCEPTS AND VOCABULARY

### SECTION 8.1

- **Alkenes** are compounds that possess a C=C double bond and can be prepared via **beta elimination**, or **1,2-elimination**.
- When the leaving group is specifically a halide, the reaction is also called a **dehydrohalogenation**.

### SECTION 8.2

- Alkenes are abundant in nature.
- Ethylene and propylene, both formed from cracking petroleum, are used as starting materials for a wide variety of compounds.

### SECTION 8.3

- Alkenes are named much like alkanes, with the following additional rules:
  - The suffix "ane" is replaced with "ene."
  - The parent is the longest chain that includes the $\pi$ bond.
  - The $\pi$ bond should receive the lowest number possible.
  - The position of the double bond is indicated with a single locant placed either before the parent or before the suffix "ene."
- **Degree of substitution** refers to the number of alkyl groups connected to the double bond.

### SECTION 8.4

- Double bonds do not exhibit free rotation at room temperature, giving rise to stereoisomerism.
- An eight-membered ring is the smallest size ring that can accommodate a *trans* double bond. This rule also applies to bridged bicyclic compounds and is called **Bredt's rule**.
- The stereodescriptors *cis* and *trans* can only be used to indicate the relative arrangement of similar groups.
- When an alkene possesses nonsimilar groups, the stereodescriptors *E* and *Z* must be used. ***Z*** indicates priority groups on the same side, while ***E*** indicates priority groups on opposite sides.

### SECTION 8.5

- A *trans* alkene is generally more stable than its stereoisomeric *cis* alkene.
- Alkene stability increases with increasing degree of substitution.

### SECTION 8.6

- In a *concerted process*, a base abstracts a proton and the leaving group leaves simultaneously.
- In a *stepwise process*, first the leaving group leaves, and then the base abstracts a proton.

### SECTION 8.7

- Evidence for the concerted **E2** mechanism includes the observation of a second-order rate equation.
- Tertiary halides react most readily toward E2.
- **Regiochemistry** is an issue in E2 reactions, which are said to be **regioselective**, because the more substituted alkene, called the **Zaitsev product**, is generally the major product.
- When both the substrate and the base are sterically hindered, the less substituted alkene, called the **Hofmann product**, is the major product.
- E2 reactions are *stereospecific*, because they generally occur via the ***anti*-coplanar** conformation, rather than the ***syn*-coplanar** conformation. Small deviations from coplanarity can be tolerated, and it is sufficient for the proton and the leaving group to be ***anti*-periplanar**.

- Substituted cyclohexanes only undergo E2 reactions from the chair conformation in which the leaving group and the proton both occupy axial positions.
- E2 reactions are also **stereoselective**, favoring *trans* disubstituted alkenes over *cis* disubstituted alkenes.

**SECTION 8.8**

- Regiochemistry and stereochemistry must be considered when predicting the products of an E2 reaction.

**SECTION 8.9**

- Evidence for the stepwise **E1** mechanism includes the observation of a first-order rate equation.
- Tertiary halides react most readily via an E1 process while primary halides are unreactive toward an E1 process.
- An E1 reaction is generally accompanied by a competing $S_N1$ reaction, and a mixture of products is generally obtained.
- Elimination of water to form an alkene is called a **dehydration** reaction.
- E1 reactions exhibit a regiochemical preference for the Zaitsev product.
- E1 reactions are not stereospecific, but they are stereoselective.

**SECTION 8.10**

- If the substrate is an alcohol, a strong acid is required to protonate the OH group.
- During an E1 process, a carbocation rearrangement occurs if it will produce a more stable carbocation intermediate.

**SECTION 8.11**

- An E2 process is comprised of just one concerted step.

**SECTION 8.12**

- Strong nucleophiles are compounds that contain a negative charge and/or are polarizable.
- Strong bases are compounds whose conjugate acids are not very acidic.
- All reagents can be classified into four categories: (1) only nucleophiles, (2) only bases, (3) strong nucleophiles and strong bases, and (4) weak nucleophiles and weak bases.

**SECTION 8.13**

- The reagent and the substrate determine the expected mechanism(s).

**SECTION 8.14**

- Predicting the products of substitution and elimination reactions requires three discrete steps:
  1. Determine the function of the reagent.
  2. Analyze the substrate and determine the expected mechanism(s).
  3. Consider any relevant regiochemical and stereochemical requirements.

# SKILLBUILDER REVIEW

## 8.1 ASSEMBLING THE SYSTEMATIC NAME OF AN ALKENE

**STEP 1** Identify the parent: choose the longest chain that includes the $\pi$ bond.

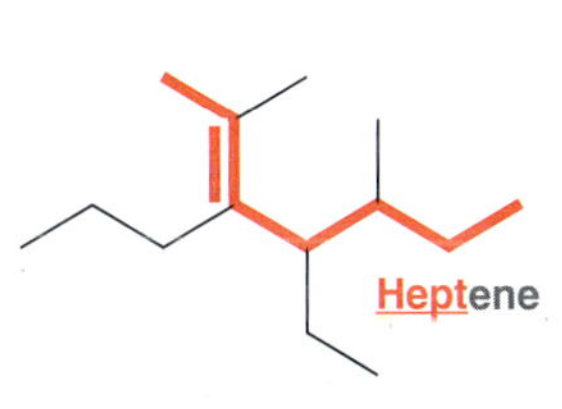


**STEP 2** Identify and name the substituents.

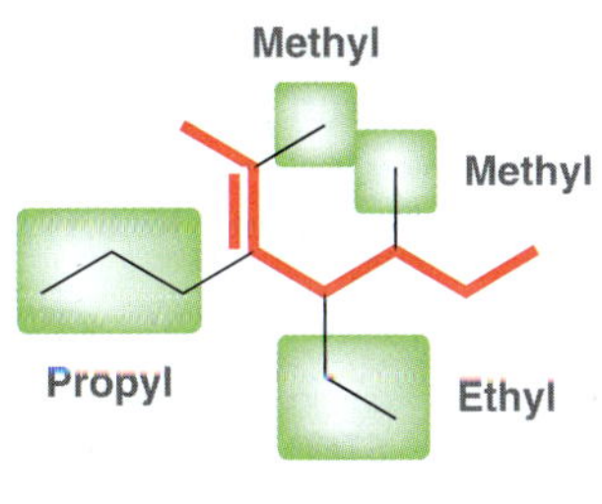


**STEP 3** Number the parent chain and assign a locant to each substituent.

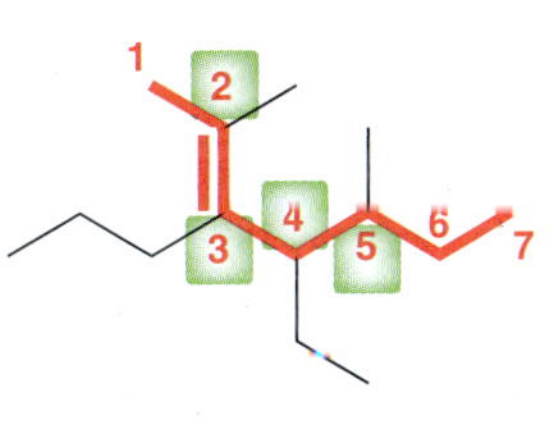


**STEP 4** Arrange the substituents alphabetically.

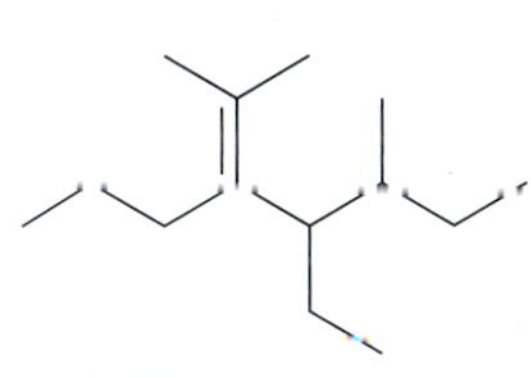

4-Ethyl-2,5-dimethyl-3-propyl-2-heptene

Try Problems 8.1–8.3, 8.50b,c

## 8.2 ASSIGNING THE CONFIGURATION OF A DOUBLE BOND

**STEP 1** Identify the two groups connected to one vinylic position and then determine which group has priority.

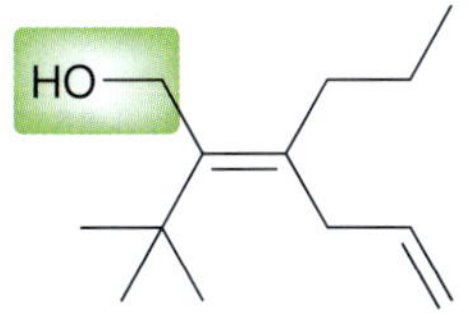


**STEP 2** Repeat step 1 for the other vinylic position, moving away from the double bond and looking for the first point of difference.

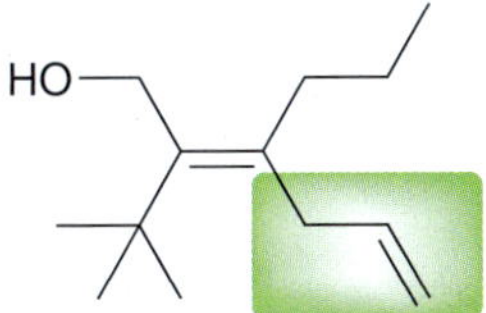


**STEP 3** Determine whether the priorities are on the same side (*Z*) or opposite sides (*E*) of the double bond.

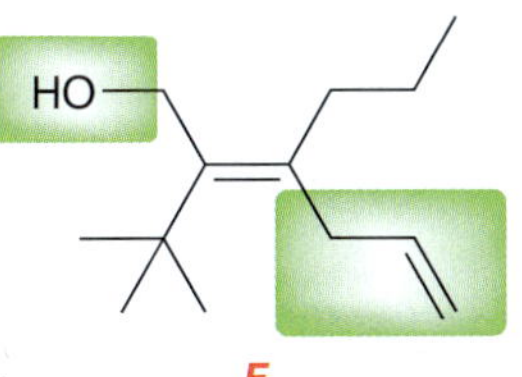


*E*

Try Problems 8.5, 8.6, 8.51

### 8.3 COMPARING THE STABILITY OF ISOMERIC ALKENES

**STEP 1** Identify the degree of substitution for each alkene.

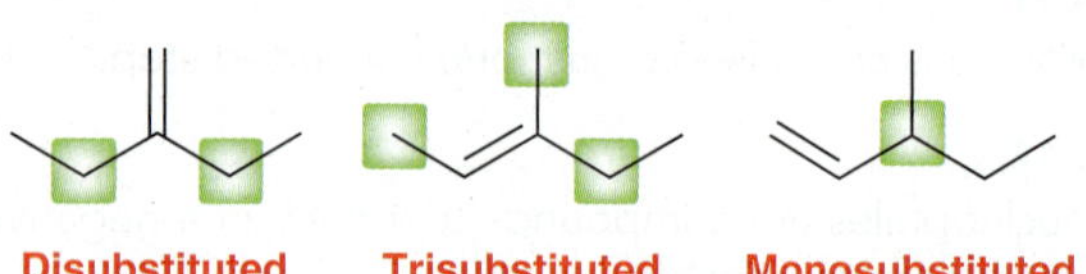


**STEP 2** Select the alkene that is most substituted.

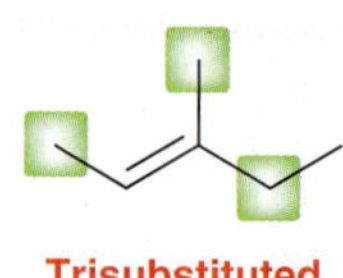


Try Problems 8.7, 8.8, 8.53

### 8.4 DRAWING THE CURVED ARROWS OF AN ELIMINATION REACTION

**CONCERTED MECHANISM** Three curved arrows, all drawn in one step. Proton transfer is accompanied by simultaneous loss of a leaving group.

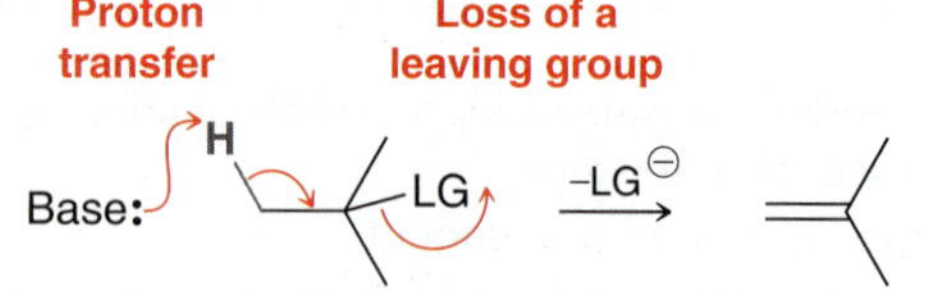


**STEPWISE MECHANISM** Three curved arrows drawn in two separate steps. Leaving group leaves to form a carbocation intermediate, followed by a proton transfer.

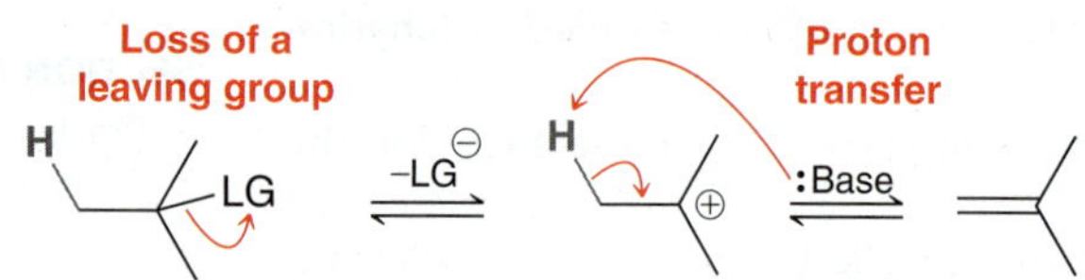


Try Problems 8.9–8.12, 8.69

### 8.5 PREDICTING THE REGIOCHEMICAL OUTCOME OF AN E2 REACTION

**STEP 1** Identify the $\alpha$ position.

**STEP 2** Identify all $\beta$ positions with protons.

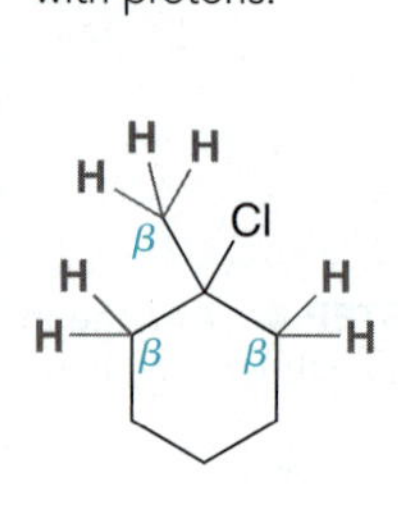


**STEP 3** Take notice of $\beta$ positions that lead to same product.

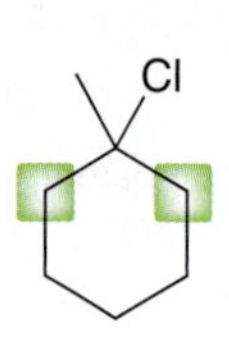


**STEP 4** Remove a proton from each $\beta$ position to draw each product. Then, identify the degree of substitution.

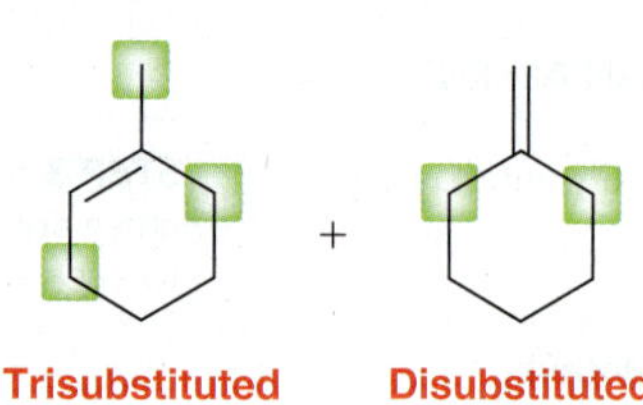


**STEP 5** Analyze the base to determine which product predominates.

| | Zaitsev | Hofmann |
|---|---|---|
| Not sterically hindered | Major | Minor |
| Sterically hindered | Minor | Major |

Try Problems 8.15–8.17, 8.60, 8.61b–d, 8.64, 8.66a,d, 8.74

### 8.6 PREDICTING THE STEREOCHEMICAL OUTCOME OF AN E2 REACTION

**STEP 1** Identify all $\beta$ positions with protons.

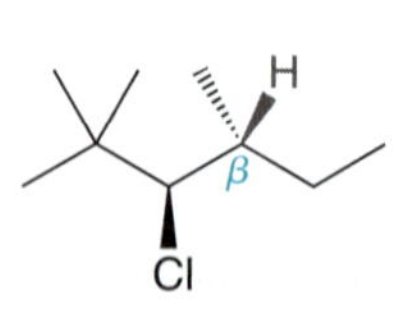


**STEP 2** If $\beta$ position has two protons, expect *cis* and *trans* isomers. If $\beta$ position has only one proton, then draw the Newman projection.

**STEP 3** Rotate to achieve *anti*-periplanar conformation.

**STEP 4** Use Newman projection to draw product.

Try Problems 8.18, 8.19, 8.56, 8.63, 8.82

### 8.7 DRAWING THE PRODUCTS OF AN E2 REACTION

**EXAMPLE**

Br

NaOEt → ?

**STEP 1** Assess regiochemical outcomes using the method from SkillBuilder 8.5.

Br

$\beta$ $\beta$

**STEP 2** Assess stereochemical outcomes using the method from SkillBuilder 8.6.

\+

**Major** **Minor**

Try Problems **8.22–8.25, 8.59–8.61, 8.64, 8.66**

### 8.8 PREDICTING THE REGIOCHEMICAL OUTCOME OF AN E1 REACTION

**EXAMPLE** Predict the regiochemical outcome.

OH

conc. $H_2SO_4$, Heat → ?

**STEP 1** Identify all $\beta$ positions that have protons.

$\beta$ OH

$\beta$ $\beta$

**STEP 2** Take notice of the $\beta$ positions that lead to the same product.

OH

**STEP 3** Draw all products and label the more substituted alkene as the major product.

\+

**Major** **Minor**

Try Problems **8.29, 8.30, 8.62**

### 8.9 DRAWING THE COMPLETE MECHANISM OF AN E1 REACTION

There are two core steps (shown in gray) and two possible additional steps (shown in blue).

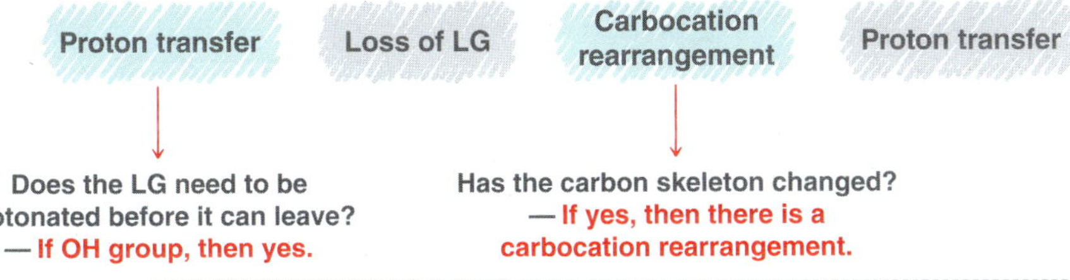


Try Problems **8.34–8.36, 8.60a, 8.76a–d, 8.81, 8.83, 8.84**

### 8.10 DETERMINING THE FUNCTION OF A REAGENT

**Nucleophile (only)**

| Halides | Sulfur nucleophiles | |
|---|---|---|
| $Cl^{\ominus}$ | $HS^{\ominus}$ | $H_2S$ |
| $Br^{\ominus}$ | $RS^{\ominus}$ | RSH |
| $I^{\ominus}$ | | |

**Base (only)**

- $H:^{\ominus}$
- DBN
- DBU

**Strong Nuc / Strong Base**

- $HO^{\ominus}$
- $MeO^{\ominus}$
- $EtO^{\ominus}$
- $(CH_3)_3CO^{\ominus}$

**Weak Nuc / Weak Base**

- $H_2O$
- MeOH
- EtOH

Try Problems **8.39, 8.40, 8.55**

**8.11** IDENTIFYING THE EXPECTED MECHANISM(S)

| **EXAMPLE** Identify the mechanism(s) expected to occur. | **STEP 1** Identify the function of the reagent. | **STEP 2** Identify the substrate and determine the mechanisms that operate. |
|---|---|---|
| 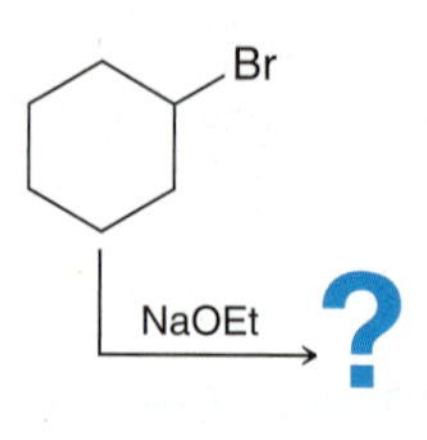 |  | 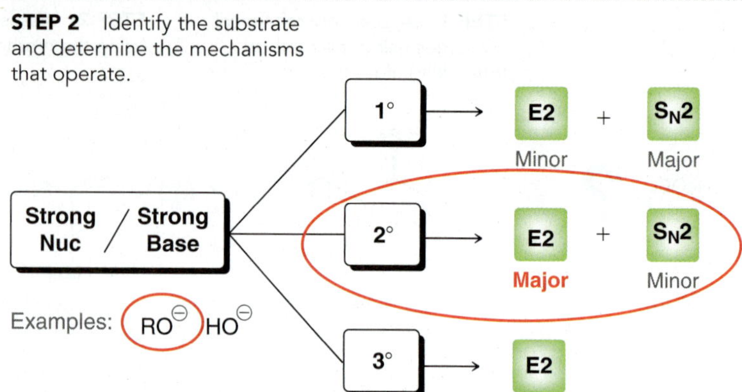 |





Strong Nuc
Strong base



Try Problems 8.41–8.45, 8.68, 8.69, 8.78a

**8.12** PREDICTING THE PRODUCTS OF SUBSTITUTION AND ELIMINATION REACTIONS

| **STEP 1** Determine the function of the reagent (Section 8.12). | **STEP 2** Analyze the substrate and determine the expected mechanism(s) (Section 8.13). | **STEP 3** Consider any relevant regiochemical and stereochemical requirements (Section 8.14). |
|---|---|---|

Try Problems 8.46–8.49, 8.77, 8.79

## PRACTICE PROBLEMS

Note: Most of the Problems are available within **WileyPLUS**, an online teaching and learning solution.

**8.50** Assign a systematic (IUPAC) name for each of the following compounds:

(a) 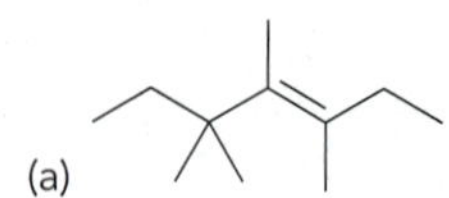 (b) 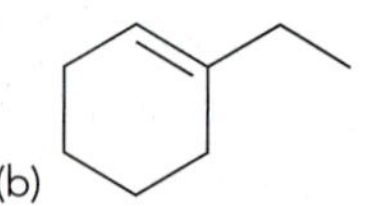 (c) 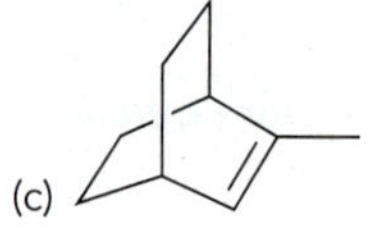

**8.51** Using *E-Z* designators, identify the configuration of each C=C double bond in the following compound:



**Dactylyne**
A natural product
isolated from marine sources

**8.52** There are two stereoisomers of 1-*tert*-butyl-4-chloro-cyclohexane. One of these isomers reacts with sodium ethoxide in an E2 reaction that is 500 times faster than the reaction of the other isomer. Identify the isomer that reacts faster and explain the difference in rate for these two isomers.

**8.53** Arrange the following alkenes in order of increasing stability:

 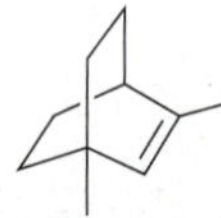 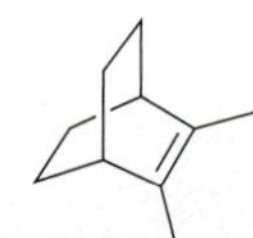 

**8.54** For each pair of compounds identify which compound would react more rapidly in an E1 reaction:

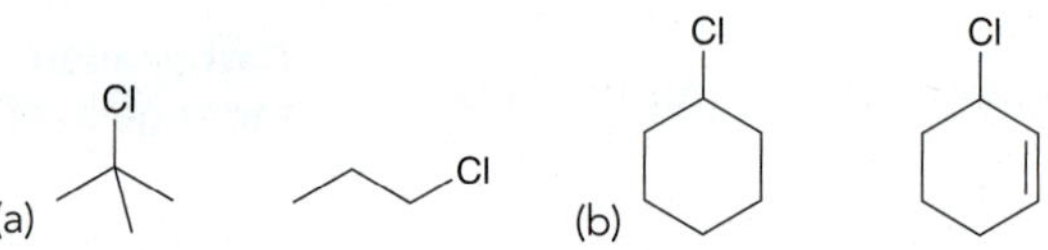



**8.55** Identify the stronger base:

(a) NaOH vs. $H_2O$

(b) Sodium ethoxide vs. ethanol

**8.56** *(2S,3S)*-2-Bromo-3-phenylbutane undergoes an E2 reaction when treated with a strong base to produce *(E)*-2-phenyl-2-butene. Use Newman projections to explain the stereochemical outcome of this reaction.

**8.57** Consider the following reaction:

Br —NaOEt, EtOH→

(a) How would the rate be affected if the concentration of *tert*-butyl bromide is doubled?

(b) How would the rate be affected if the concentration of sodium ethoxide is doubled?

**8.58** Consider the following reaction:

Br —EtOH→

(a) How would the rate be affected if the concentration of *tert*-butyl bromide is doubled?

(b) How would the rate be affected if the concentration of ethanol is doubled?

**8.59** When (*R*)-3-bromo-2,3-dimethylpentane is treated with sodium hydroxide, four different alkenes are formed. Draw all four products and rank them in terms of stability. Which product do you expect to be the major product?

**8.60** When 3-bromo-2,4-dimethylpentane is treated with sodium hydroxide, only one alkene is formed. Draw the product and explain why this reaction has only one regiochemical outcome.

**8.61** Predict the major product for each of the following E2 reactions:

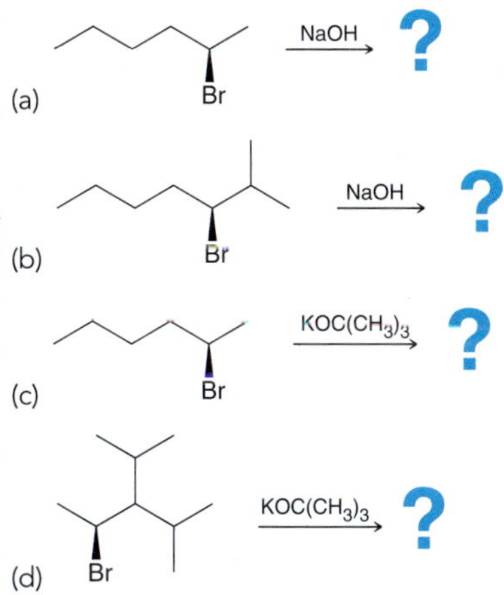


**8.62** Draw the most stable E1 product that can be produced in each of the following reactions:

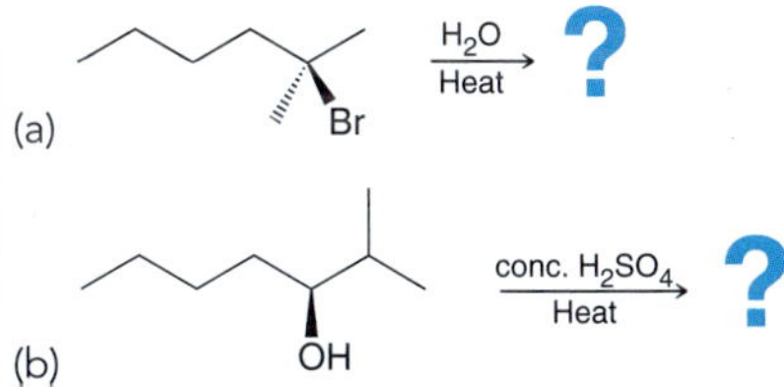


**8.63** Predict the stereochemical outcome for each of the following E2 reactions. In each case, draw only the major product of the reaction.

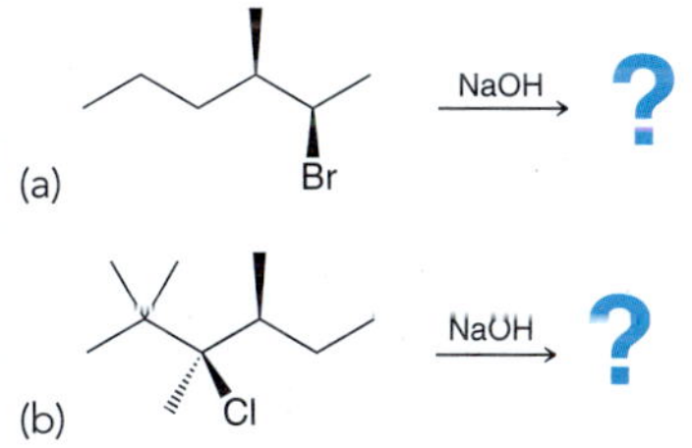


**8.64** Identify the sole product of the following reaction:

Br — NaOEt → $C_{10}H_{20}$

**8.65** For each of the following descriptions draw the structure of a compound that fits the description. (*Note*: There are many correct answers for each of these problems.)

(a) An alkyl halide that produces four different alkenes when treated with a strong base

(b) An alkyl halide that produces three different alkenes when treated with a strong base

(c) An alkyl halide that produces two different alkenes when treated with a strong base

(d) An alkyl halide that produces only one alkene when treated with a strong base

**8.66** How many different alkenes will be produced when each of the following substrates is treated with a strong base?

(a) 1-Chloropentane

(b) 2-Chloropentane

(c) 3-Chloropentane

(d) 2-Chloro-2-methylpentane

(e) 3-Chloro-3-methylhexane

**8.67** In each of the following cases draw the structure of an alkyl halide that will undergo an E2 elimination to yield *only* the indicated alkene:

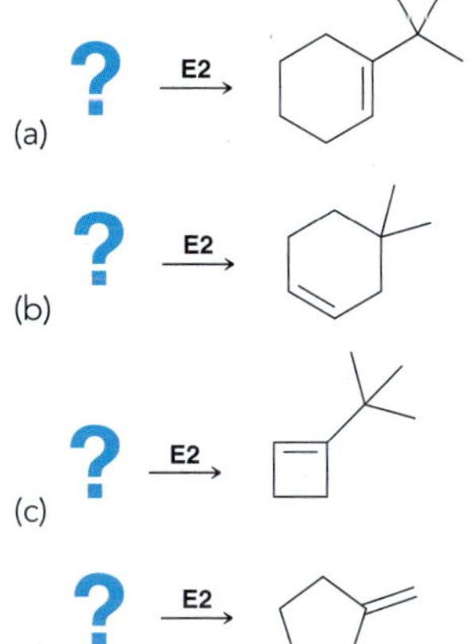


**8.68** Consider the following reaction:

OH — conc. $H_2SO_4$, Heat →

(a) Draw a mechanism for this reaction.

(b) What is the rate equation of this reaction?

(c) Draw an energy diagram for the reaction.

**8.69** Draw the carbocation intermediate that would be formed if each of the following substrates participated in a stepwise elimination process (E1). In each case, identify the intermediate carbocation as being primary, secondary, or tertiary. One of the substrates does not undergo an E1 reaction. Identify which one and explain why.

(a) Cl (b) Br (c) I (d) Cl

**8.70** Draw the transition state for the reaction between *tert*-butyl chloride and sodium hydroxide.

**8.71** The following three reactions are similar, differing only in the configuration of the substrate. One of these reactions is very fast, one is very slow, and the other does not occur at all. Identify each reaction and explain your choice.

Br NaOH

Br NaOH

Br NaOH

**8.72** 1-Bromobicyclo[2.2.2]octane does not undergo an E2 reaction when treated with a strong base. Explain why not.

**8.73** For each pair of the following compounds identify which compound would react more rapidly in an E2 reaction:

(a) Cl Cl (b) Br Br

**8.74** Indicate whether you would use sodium ethoxide or potassium *tert*-butoxide to achieve each of the following transformations:

(a) Br →

(b) Br →

(c) Br → (d) Br →

**8.75** Explain why the following reaction yields the Hofmann product exclusively (no Zaitsev product at all) even though the base is not sterically hindered:

Br NaOEt / EtOH

**8.76** Propose a mechanism for each of the following transformations:

(a) OH conc. $H_2SO_4$ / Heat

(b) OH conc. $H_2SO_4$ / Heat

(c) OH conc. $H_2SO_4$ / Heat

(d) I EtOH / Heat

(e) I NaOEt / EtOH

## INTEGRATED PROBLEMS

**8.77** *Substitution vs. Elimination*: Identify the major and minor product(s) for each of the following reactions:

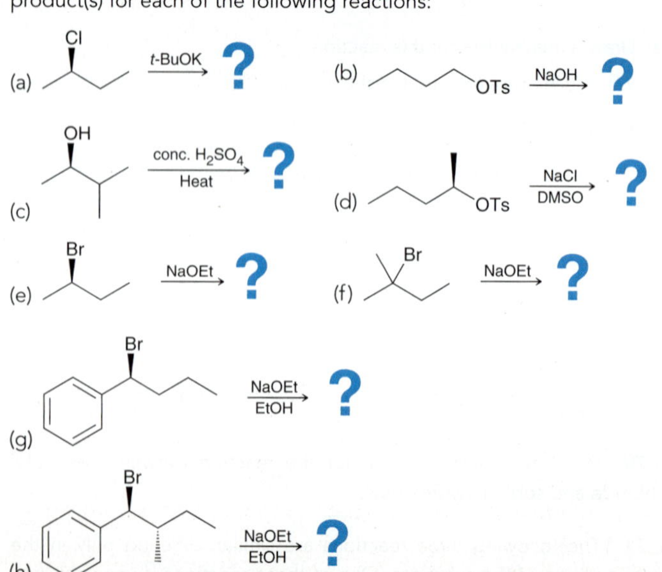


(i) Br NaOEt / EtOH ?

(j) Br $KOC(CH_3)_3$ ?

(k) Br NaOMe ? (l) Cl NaOH ?

**8.78** When 2-bromo-2-methylhexane is treated with sodium ethoxide in ethanol, the major product is 2-methyl-2-hexene.

(a) Draw a mechanism for this reaction.

(b) What is the rate equation of this reaction?

(c) What would happen to the rate if the concentration of base is doubled?

(d) Draw an energy diagram for this reaction.

(e) Draw the transition state of this reaction.

**8.79** Predict the major product for each of the following reactions:

(a) 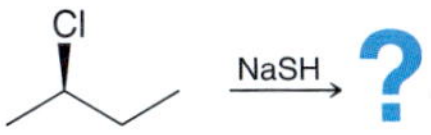 NaSH ?

(b)  OTs DBN ?

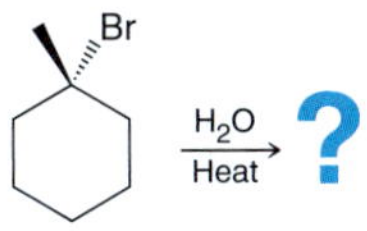

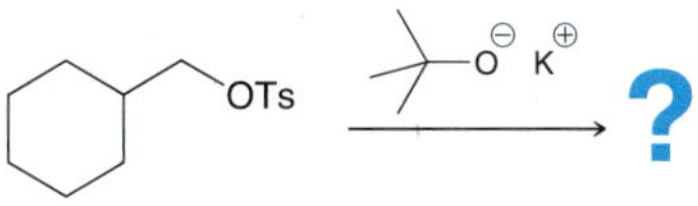

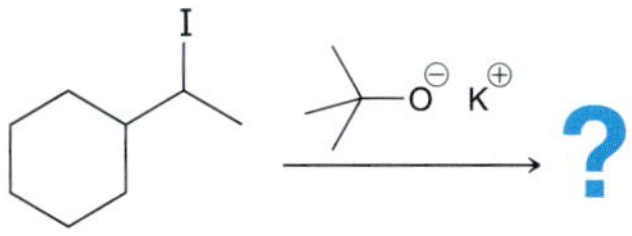

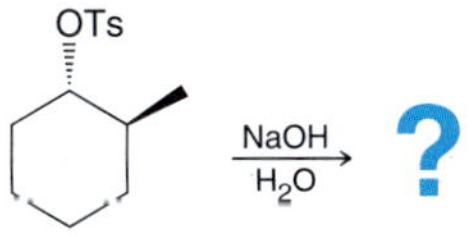

(c) 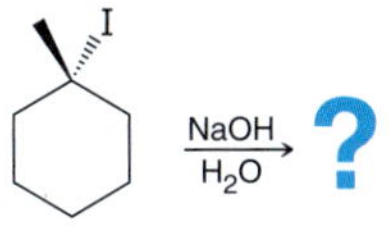 I NaOH, $H_2O$ ?

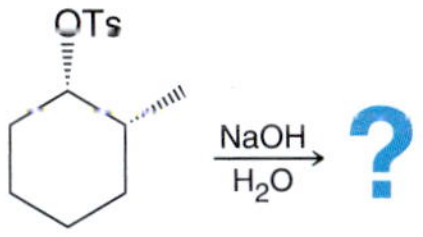

(d) Br $H_2O$, Heat ?

(e) OTs ?

(f) I ?

(g) OTs NaOH, $H_2O$ ?

(h) OTs NaOH, $H_2O$ ?

(i) Br NaOMe, MeOH ?

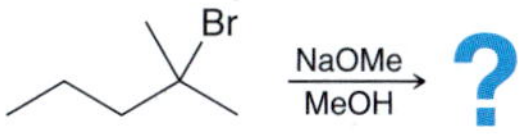

(j) 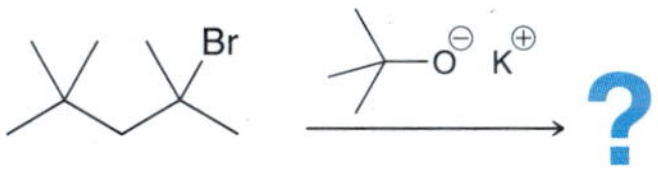 Br ?

**8.80** Draw all constitutional isomers of $C_4H_9Br$ and then arrange them in order of increasing reactivity toward an E2 reaction.

**8.81** Propose a mechanism that explains formation of the following product:

OH conc. $H_2SO_4$, Heat

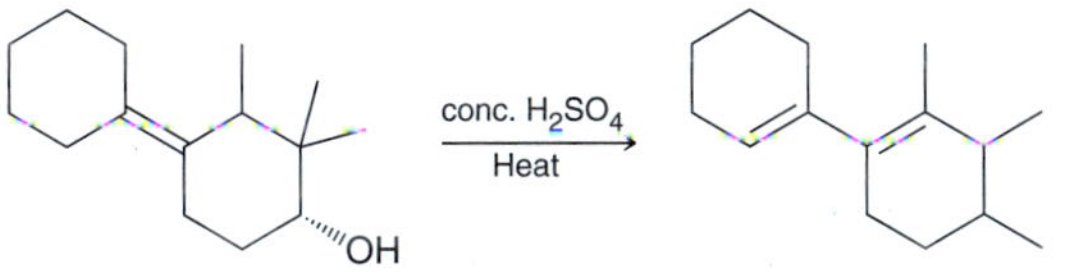

**8.82** (*S*)-1-Bromo-1,2-diphenylethane reacts with a strong base to produce *cis*-stilbene and *trans*-stilbene:

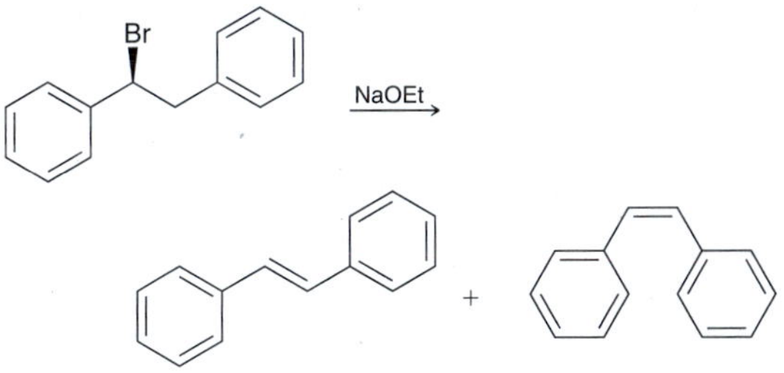

Br NaOEt

***trans*-Stilbene (major product)** + ***cis*-Stilbene**

(a) This reaction is stereoselective, and the major product is *trans*-stilbene. Explain why the *trans* isomer is the predominant product. To do so, draw the Newman projections that lead to formation of each product and compare their stability.

(b) When (*R*)-1-bromo-1,2-diphenylethane is used as the starting substrate, the stereochemical outcome does not change. That is, *trans*-stilbene is still the major product. Explain.

**8.83** Propose a mechanism for the following transformation:

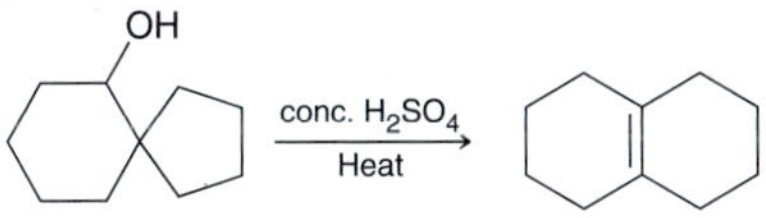

OH conc. $H_2SO_4$, Heat

**8.84** Propose a mechanism of formation for each of the following products:

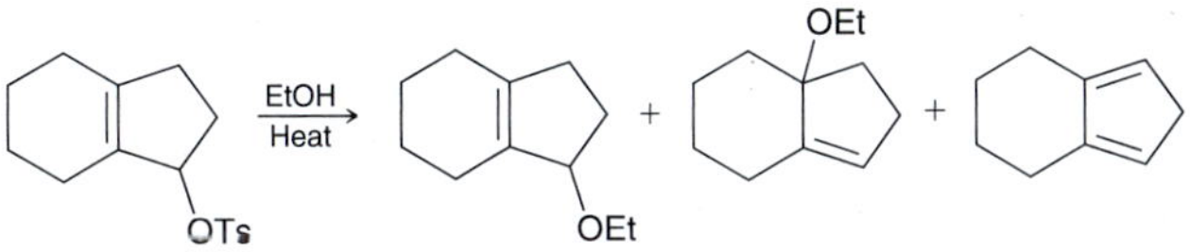

OTs EtOH, Heat → OEt + OEt +

**8.85** There are many stereoisomers of 1,2,3,4,5,6-hexachlorocyclohexane. One of those stereoisomers undergoes E2 elimination thousands of times more slowly than the other stereoisomers. Identify which stereoisomer and explain why it is so slow toward E2.

**8.86** Predict which of the following substrates will undergo an E1 reaction more quickly. Explain your choice.

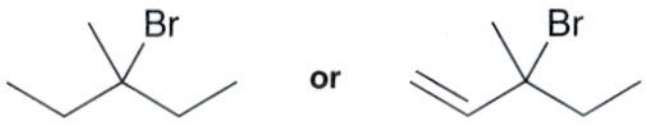

Br or Br

**8.87** Pladienolide B is a macrocyclic (large-ring) natural product isolated from an engineered strain of the bacterium *Streptomyces platensis*. An enantioselective, 31-step synthesis of it was reported in 2012—an improvement over a previously reported 59-step approach (*Org. Lett.* **2012**, *14*, **4730–4733**):

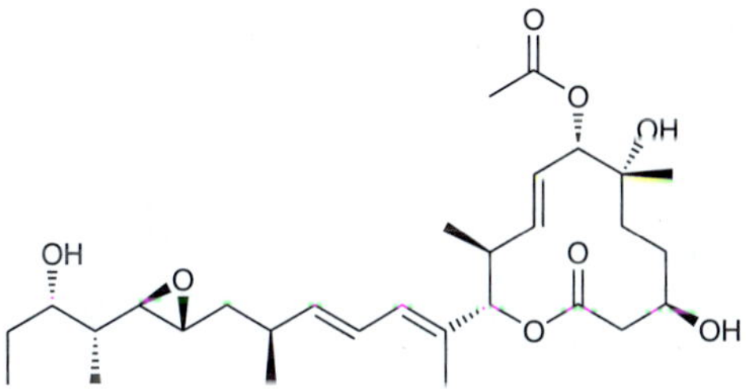

O O OH OH O O O OH

(a) Identify the trisubstituted π bond and determine whether it has the *E* or *Z* configuration.

(b) How many total stereoisomers are possible for pladienolide B?

(c) Draw the enantiomer of pladienolide B.

(d) Draw the diastereomer that is identical to pladienolide B except for the configuration of each of the chirality centers directly connected to an ester oxygen.

(e) Draw the diastereomer that is identical to pladienolide B except for the configuration of the disubstituted π bond that is outside of the ring.

# CHALLENGE PROBLEMS

**8.88** Steroids (covered in Chapter 26) and their derivatives are among the most widely used therapeutic agents. They are used in birth control, hormone replacement therapy, and the treatment of inflammatory conditions and cancer. New steroid derivatives are discovered regularly by systematically modifying the structure of known steroids and testing the resulting derivatives for therapeutic properties. As part of a synthetic strategy for preparing a class of promising steroid derivatives, compound **1a** was treated with TsCl and pyridine followed by sodium acetate ($CH_3CO_2Na$) to give compound **2a** (*Tetrahedron Lett.* **2010,** *51,* **6948–6950**).

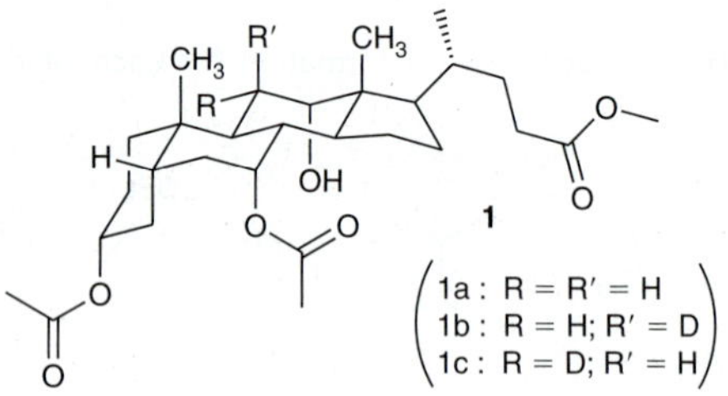

1a : R = R′ = H
1b : R = H; R′ = D
1c : R = D; R′ = H

(a) Sodium acetate functions as a base in this instance. Draw the structure of **2a**.

(b) Deuterium (D) is an isotope of hydrogen, and deuterons will typically behave very much like protons (although small differences in reaction rates are typically observed). If **1b** or **1c** were treated with TsCl and pyridine, followed by sodium acetate, **2b** or **2c** would be produced, respectively. Identify which product (**2b** or **2c**) is expected to contain deuterium and justify your choice.

**8.89** Compounds **1** and **2** exhibit an oxetane ring (a four-membered ring containing an oxygen atom), which is a very important structural feature in a number of natural products. For example, an oxetane ring is present in the structure of taxol, used in the treatment of breast cancer. When either **1** or **2** is treated with TsCl and pyridine, followed by *t*-BuOK, the Zaitsev elimination product dominates, despite the use of a sterically hindered base (*Tetrahedron Lett.* **2007,** *48,* **8353–8355**). Determine the stereochemical outcome for each of these processes and draw the expected product in each case. Use Newman projections to justify your answers.

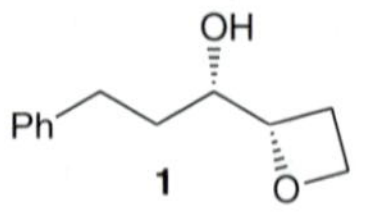

**8.90** Compound **1** is observed to undergo *debromination* upon treatment with DMF to afford an alkene. The *E* isomer (compound **2**) is obtained exclusively (none of the *Z* isomer is observed). A concerted mechanism (shown below) has been refuted by the observation that diastereomers of **1** also afford the *E* isomer exclusively when treated with DMF (*J. Org. Chem.* **1991,** *56,* **2582–2584**):

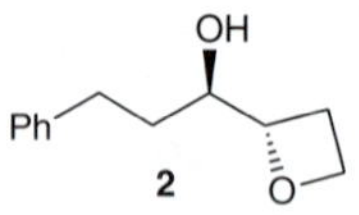

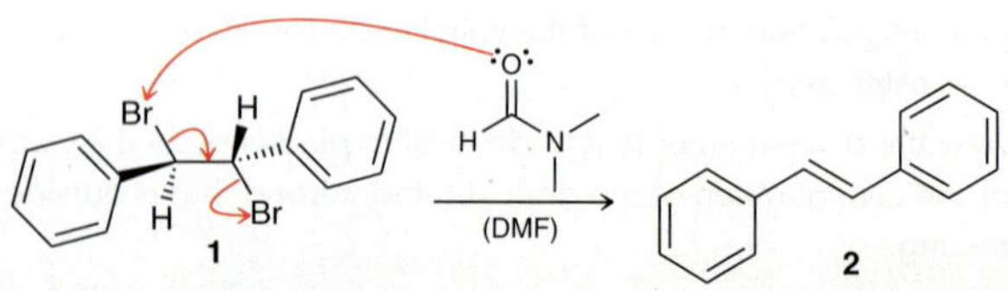

(a) Explain why the mechanism shown is inconsistent with this observation.

(b) Classify this reaction as stereoselective or stereospecific and explain your choice.

**8.91** The following sequence of reactions was performed during a synthesis of (+)-coronafacic acid, a key component in the plant toxin coronatine (*J. Org. Chem.* **2009,** *74,* **2433–2437**). Predict the product of this reaction sequence and justify the regiochemical outcome of the second reaction.

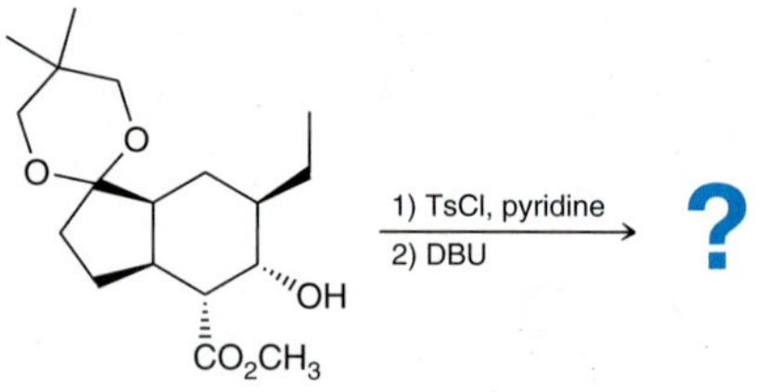

1) TsCl, pyridine
2) DBU
?

**8.92** We have seen that an E2 reaction generally proceeds from a conformation in which the β proton is *anti*-periplanar to the leaving group. A similar geometric requirement is also observed for other types of reactions such as the one below (*J. Am. Chem. Soc.* **2010,** *132,* **2530–2531**). Under acidic conditions, compound **1** rearranges to form mechanistic intermediate **2,** which then undergoes three additional mechanistic steps (along with the loss of an $N_2$ leaving group) to give product **3**. The conversion of **1** → **2** is beyond the scope of our current discussion. Draw a mechanism for the final three steps, as described: (x) The alkyl group *anti*-periplanar to the $N_2^+$ group undergoes a 1,2-shift to allow a backside displacement of $N_2$. (y) The alcohol oxygen attacks the resulting carbocation. (z) The resulting oxonium ion is deprotonated.

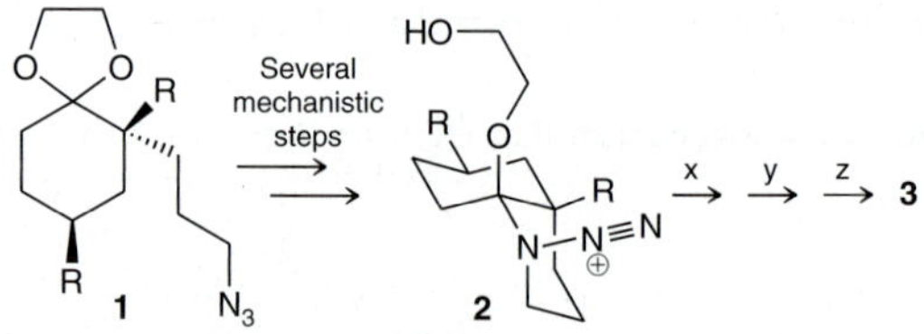

Several mechanistic steps

**8.93** When 2-iodobutane is treated with a variety of oxyanion bases in DMSO at 50 °C, the percentage of 1-butene formed among total butenes is found to be dependent on the choice of base, as seen in the chart below (*J. Am. Chem. Soc.* **1973,** *95,* **3405–3407**):

| NAME OF BASE | STRUCTURE OF BASE | % 1-BUTENE (IN TOTAL BUTENES) |
|---|---|---|
| Potassium benzoate | 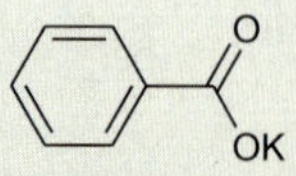 $C_6H_5CO_2K$ | 7.2 |
| Potassium phenoxide | 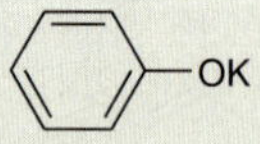 $C_6H_5OK$ | 11.4 |
| Sodium 2,2,2-trifluoroethoxide | $NaOCH_2CF_3$ | 14.3 |
| Sodium ethoxide | NaOEt | 17.1 |

(a) Identify the trend observed by comparing the basicity of the oxyanion bases and then describe the correlation between reactivity (base strength) and selectivity (specifically regioselectivity).

(b) The p$K_a$ of 4-nitrophenol is 7.1. Based on the trend in the previous part of this problem, provide an estimate for the percentage of 1-butene that is expected if 2-iodobutane is treated with the conjugate base of 4-nitrophenol in DMSO at 50 °C.

**8.94** During a recent total synthesis of (+)-aureol, a potential anticancer agent isolated from a marine sponge, the following Lewis acid catalyzed rearrangement was employed (*Org. Lett.* **2012,** *14,* 4710–4713). Provide a plausible mechanism for this biosynthetically inspired process (**Hint:** review Section 3.9) and make sure to include a rationale for the observed stereochemistry of the product.

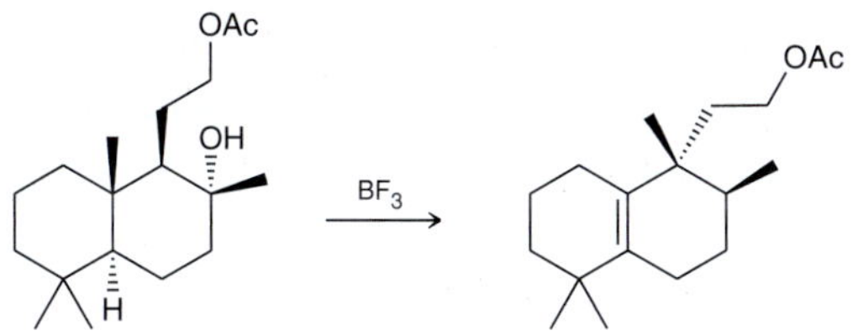

OAc OH H $\xrightarrow{BF_3}$ OAc

**8.95** When the given 2-butyl compounds are treated with *t*-BuOK in *t*-BuOH at 50 °C, the major product is generally 1-butene (as expected with a sterically hindered base), while the minor products are *cis*-2-butene and *trans*-2-butene. In cases where the leaving group is a halogen (I, Br, or Cl), the *trans* alkene is favored over the *cis* alkene. However, the *cis* isomer becomes the favored 2-alkene when the leaving group is a tosylate group (*J. Am. Chem. Soc.* **1965,** *87,* 5517–5518). Provide a transition state argument, including relevant Newman projections, that explains this difference.

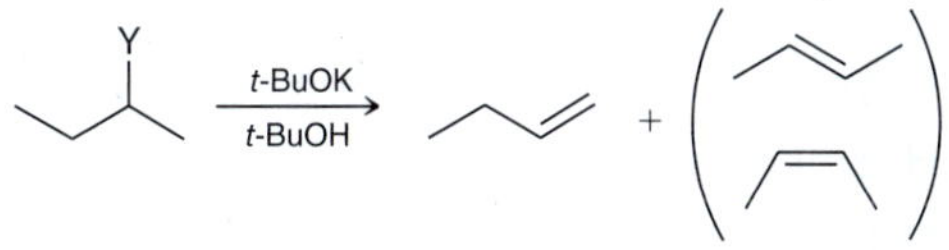

Y $\xrightarrow[t\text{-BuOH}]{t\text{-BuOK}}$ +

| Y | *trans:cis* |
|---|---|
| I | 2.2 |
| Br | 1.4 |
| Cl | 1.3 |
| OTs | 0.6 |

# 9



# Addition Reactions of Alkenes

## DID YOU EVER WONDER...
**what Styrofoam is and how it is made?**

Packing peanuts, coffee cups, and coolers are made from a material that most people erroneously refer to as Styrofoam. Styrofoam, a trademark of the Dow Chemical Company, is a blue product that is used primarily in the production of building materials for insulating walls, roofs, and foundations. No coffee cups, packing peanuts, or coolers are made from Styrofoam. Instead, they are made from a similar material called foamed polystyrene. Polystyrene is a polymer that can be prepared either in a rigid form or as a foam. The rigid form is used in the production of computer housings, CD and DVD cases, and disposable cutlery, while foamed polystyrene is used in the production of foam coffee cups and packaging materials. Polystyrene is made by polymerizing styrene via a type of reaction called an addition reaction. This chapter will explore many different types of addition reactions. In the course of our discussion, we will revisit the structure of Styrofoam and foamed polystyrene.

## DO YOU REMEMBER?

*Before you go on, be sure you understand the following topics.*
*If necessary, review the suggested sections to prepare for this chapter.*

- Energy Diagrams (Sections 6.5, 6.6)
- Nucleophiles and Electrophiles (Section 6.7)
- Arrow Pushing and Carbocation Rearrangements (Sections 6.8–6.11)

Take the DO YOU REMEMBER? QUIZ in **WileyPLUS** to check your understanding.

## 9.1 Introduction to Addition Reactions

In the previous chapter, we learned how to prepare alkenes via elimination reactions. In this chapter, we will explore **addition reactions**, common reactions of alkenes, characterized by the addition of two groups across a double bond. In the process, the pi ($\pi$) bond is broken:

Addition of X and Y

Some addition reactions have special names that indicate the identity of the two groups that were added. Several examples are listed in Table 9.1.

**TABLE 9.1 SOME COMMON TYPES OF ADDITION REACTIONS**

| TYPE OF ADDITION REACTION | NAME | SECTION |
|---|---|---|
| Addition of H and X | Hydrohalogenation (X=Cl, Br, or I) | 9.3 |
| Addition of H and OH | Hydration | 9.6 |
| Addition of H and H | Hydrogenation | 9.7 |
| Addition of X and X | Halogenation (X=Cl or Br) | 9.8 |
| Addition of OH and X | Halohydrin formation (X=Cl, Br, or I) | 9.8 |
| Addition of OH and OH | Dihydroxylation | 9.9, 9.10 |

Many different addition reactions are observed for alkenes, enabling them to serve as synthetic precursors for a wide variety of functional groups. The versatility of alkenes can be directly attributed to the reactivity of $\pi$ bonds, which can function either as weak bases or as weak nucleophiles:

As a base

As a nucleophile

**LOOKING BACK**
For a review of the subtle differences between basicity and nucleophilicity, see Section 8.12.

The first process illustrates that π bonds can be readily protonated, while the second process illustrates that π bonds can attack electrophilic centers. Both processes will appear many times throughout this chapter.

## 9.2 Addition vs. Elimination: A Thermodynamic Perspective

In many cases, an addition reaction is simply the reverse of an elimination reaction:

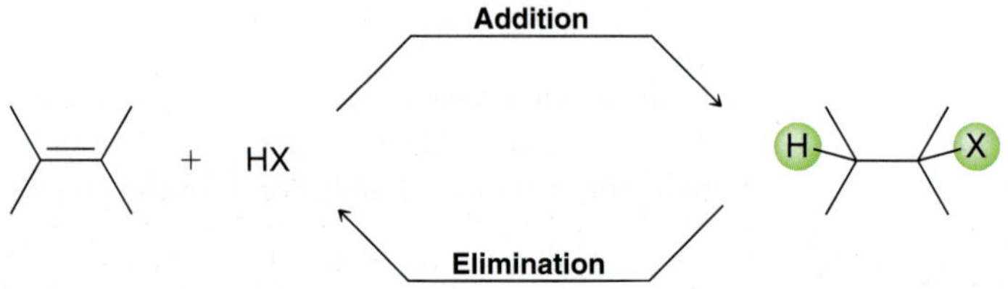

Addition

+ HX

H X

Elimination

These two reactions represent an equilibrium that is temperature dependent. Addition is favored at low temperature, while elimination is favored at high temperature. To understand the reason for this temperature dependence, recall that the sign of $\Delta G$ determines whether the equilibrium favors reactants or products (Section 6.3). $\Delta G$ must be negative for the equilibrium to favor products. The sign of $\Delta G$ depends on two terms:

$$\Delta G = \underbrace{(\Delta H)}_{\text{Enthalpy term}} + \underbrace{(-T\Delta S)}_{\text{Entropy term}}$$

Let's consider these terms individually, beginning with the enthalpy term ($\Delta H$). Many factors contribute to the sign and magnitude of $\Delta H$, but the dominant factor is generally bond strength. Compare the bonds broken and formed in an addition reaction:

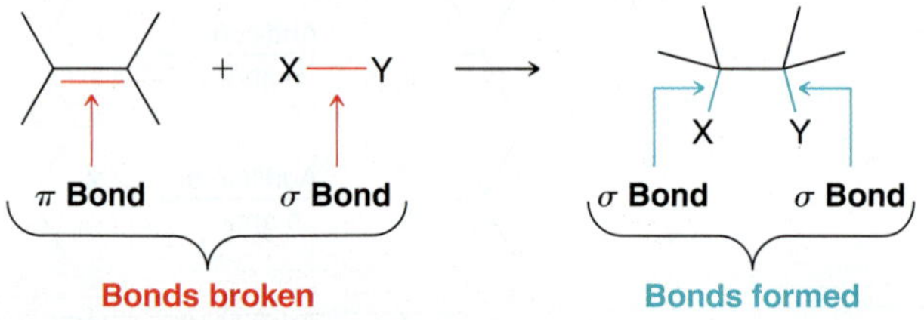

+ X—Y ⟶ X Y

π Bond σ Bond σ Bond σ Bond

Bonds broken Bonds formed

Notice that one π bond and one σ bond are broken, while two σ bonds are formed. Recall from Section 1.9 that σ bonds are stronger than π bonds, and therefore, the bonds being formed are stronger than the bonds being broken. Consider the following example:

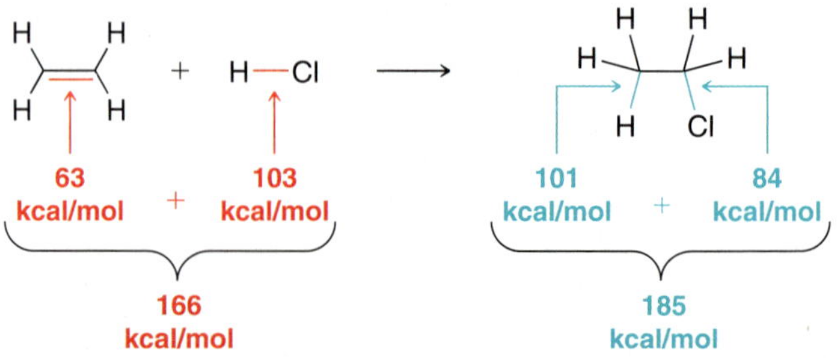

H H H H
+ H—Cl ⟶ H H
H H H Cl

63 kcal/mol + 103 kcal/mol 101 kcal/mol + 84 kcal/mol

166 kcal/mol 185 kcal/mol

$$\text{Bonds broken} - \text{bonds formed} = 166\ \text{kcal/mol} - 185\ \text{kcal/mol} = -19\ \text{kcal/mol}$$

The important feature here is that $\Delta H$ has a negative value. In other words, this reaction is exothermic, which is generally the case for addition reactions.

**BY THE WAY**

The actual $\Delta H$ for this reaction, in the gas phase, has been measured to be −17 kcal/mol, confirming that bond strengths are, in fact, the dominant factor in determining the sign and magnitude of $\Delta H$.

Now let's consider the entropy term $-T\,\Delta S$. This term will always be positive for an addition reaction. Why? In an addition reaction, two molecules are joining together to produce one molecule of product. As described in Section 6.2, this situation represents a decrease in entropy, and $\Delta S$ will have a negative value. The temperature component, $T$ (measured in Kelvin), is always positive. As a result, $-T\,\Delta S$ will be positive for addition reactions.

Now let's combine the enthalpy and entropy terms. The enthalpy term is negative and the entropy term is positive, so the sign of $\Delta G$ for an addition reaction will be determined by the competition between these two terms:

$$\Delta G = \underbrace{(\Delta H)}_{\text{Enthalpy term } \ominus} + \underbrace{(-T\,\Delta S)}_{\text{Entropy term } \oplus}$$

In order for $\Delta G$ to be negative, the enthalpy term must be larger than the entropy term. This competition between the enthalpy term and the entropy term is temperature dependent. At low temperature, the entropy term is small, and the enthalpy term dominates. As a result, $\Delta G$ will be negative, which means that products are favored over reactants (the equilibrium constant $K$ will be greater than 1). In other words, addition reactions are thermodynamically favorable at low temperature.

However, at high temperature, the entropy term will be large and will dominate the enthalpy term. As a result, $\Delta G$ will be positive, which means that reactants will be favored over products (the equilibrium constant $K$ will be less than 1). In other words, the reverse process (elimination) will be thermodynamically favored at high temperature:

Low T
+ H—X
H X
High T

**BY THE WAY**

Not all addition reactions are reversible at high temperature, because in many cases high temperature will cause the reactants and/or products to undergo thermal degradation.

For this reason, the addition reactions discussed in this chapter are generally performed below room temperature.

## 9.3 Hydrohalogenation

### Regioselectivity of Hydrohalogenation

The treatment of alkenes with HX (where X = Cl, Br, or I) results in an addition reaction called **hydrohalogenation**, in which H and X are added across the π bond:

+ H—Br
−30°C
Br
(76%)

In this example the alkene is symmetrical. However, in cases where the alkene is unsymmetrical, the ultimate placement of H and X must be considered. In the following example, there are two possible vinylic positions where X can be placed:

H—X
?
?
Vinylic positions
Where does X go?

This *regiochemical* issue was investigated over a century ago.

In 1869, Vladimir Markovnikov, a Russian chemist, investigated the addition of HBr across many different alkenes, and he noticed that the *H is generally placed at the vinylic position already bearing the larger number of hydrogen atoms*. For example:

This vinylic position currently bears **one** H
This vinylic position currently bears **no** H's
H—Br
H is placed here

Markovnikov described this regiochemical preference in terms of where the hydrogen atom is ultimately positioned. However, Markovnikov's observation can alternatively be described in terms of where the halogen (X) is ultimately positioned. Specifically, the *halogen is generally placed at the more substituted position*:

Less substituted
More substituted
H—Br
Br is placed here

The vinylic position bearing more alkyl groups is more substituted, and that is where the Br is ultimately placed. This regiochemical preference, called **Markovnikov addition,** is also observed for addition reactions involving HCl and HI. Reactions that proceed with a regiochemical preference are said to be *regioselective.*

Interestingly, attempts to repeat Markovnikov's observations occasionally met with failure. In many cases involving the addition of HBr, the observed regioselectivity was, in fact, the opposite of what was expected—that is, the bromine atom would be installed at the less substituted carbon, which came to be known as ***anti*-Markovnikov addition.** These curious observations fueled much speculation over the underlying cause, with some researchers even suggesting that the phases of the moon had an impact on the course of the reaction. Over time, it was realized that purity of reagents was the critical feature. Specifically, Markovnikov addition was observed whenever purified reagents were used, while the use of impure reagents sometimes led to *anti*-Markovnikov addition. Further investigation revealed the identity of the impurity that most greatly affected the regioselectivity of the reaction. It was found that peroxides (ROOR), even in trace amounts, would cause HBr to add across an alkene in an *anti*-Markovnikov fashion:

H—Br
ROOR
(95%)
*Anti*-Markovnikov addition

In the following section we will explore Markovnikov addition in more detail and propose a mechanism that involves ionic intermediates. In contrast, *anti*-Markovnikov addition of HBr is known to proceed via an entirely different mechanism, one involving radical intermediates. This radical pathway (*anti*-Markovnikov addition) is efficient for the addition of HBr but not HCl or HI, and the details of that process will be discussed in Section 11.10. For now, we will simply note that the regiochemical outcome of HBr addition can be controlled by choosing whether or not to use peroxides:

HBr
Markovnikov addition
HBr
ROOR
*Anti*-Markovnikov addition

## CONCEPTUAL CHECKPOINT

**9.1** Draw the expected major product for each of the following reactions:

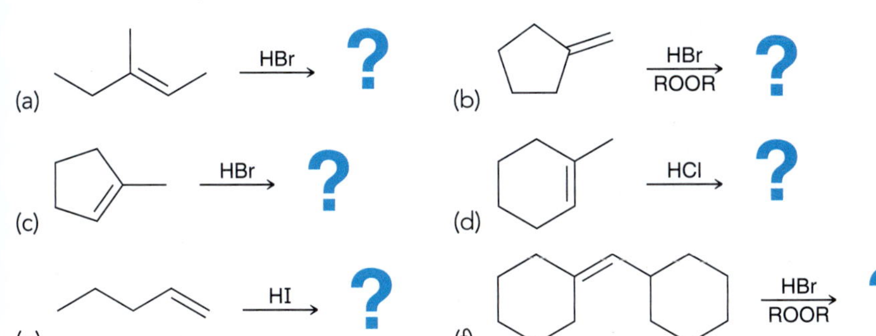


**9.2** Identify the reagents that you would use to achieve each of the following transformations:

(a) Br

(b) Br

## A Mechanism for Hydrohalogenation

Mechanism 9.1 accounts for the Markovnikov addition of HX to alkenes.

## MECHANISM 9.1 HYDROHALOGENATION

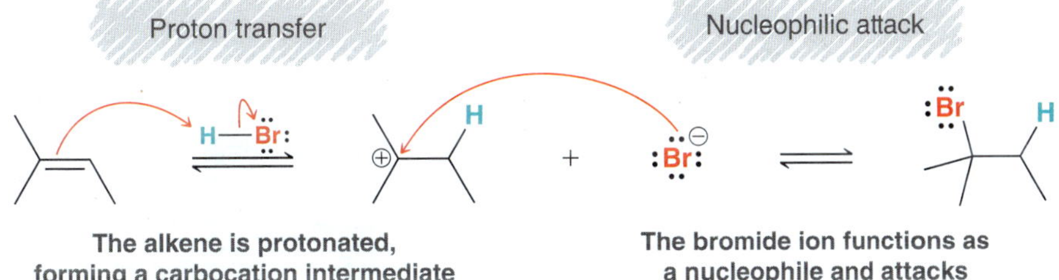


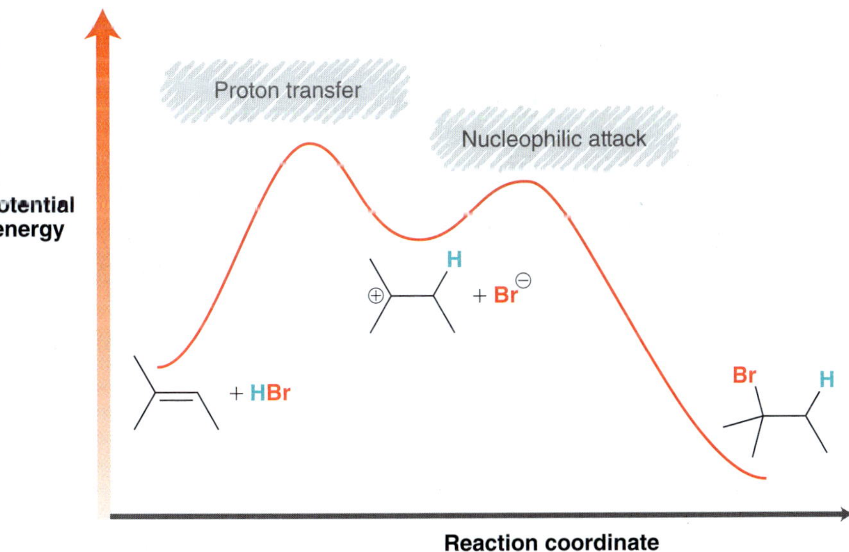


In the first step, the π bond of the alkene is protonated, generating a carbocation intermediate. In the second step, this intermediate is attacked by a bromide ion. Figure 9.1 shows an energy diagram for this two-step process. The observed regioselectivity for this process can be attributed to the first step of the mechanism (proton transfer), which is the rate-determining step because it exhibits a higher transition state energy than the second step of the mechanism.

**FIGURE 9.1** An energy diagram for the two steps involved in the addition of HBr across an alkene.

In theory, protonation can occur with either of two regiochemical possibilities. It can either occur to form the less substituted, secondary carbocation,

Secondary carbocation

or occur to form the more substituted, tertiary carbocation,

Tertiary carbocation

Recall that tertiary carbocations are more stable than secondary carbocations, due to hyperconjugation (Section 6.11). Figure 9.2 shows an energy diagram comparing the two possible pathways for the first step of HX addition. The blue curve represents HX addition proceeding via a secondary carbocation, while the red curve represents HX addition proceeding via a tertiary carbocation. Look closely at the transition states through which the carbocation intermediates are formed. Recall that the Hammond postulate (Section 6.6) suggests that each of these transition states has significant carbocationic character. Therefore, the transition state for formation

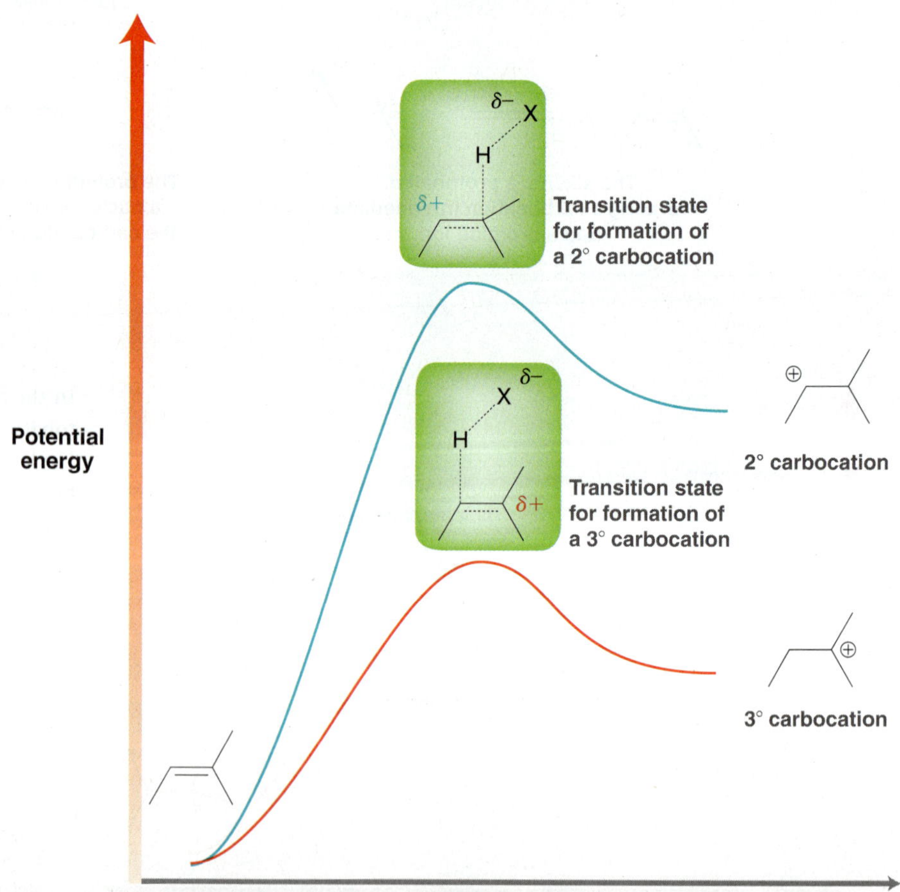


**FIGURE 9.2** An energy diagram illustrating the two possible pathways for the first step of hydrohalogenation. One pathway generates a secondary carbocation, while the other pathway generates a tertiary carbocation.

of the tertiary carbocation will be significantly lower in energy than the transition state for formation of the secondary carbocation. The energy barrier for formation of the tertiary carbocation will be smaller than the energy barrier for formation of the secondary carbocation, and as a result, the reaction will proceed more rapidly via the more stable carbocation intermediate. The proposed mechanism therefore provides a theoretical explanation (based on first principles) for the regioselectivity that Markovnikov observed. Specifically, the *regioselectivity of an ionic addition reaction is determined by the preference for the reaction to proceed through the more stable carbocation intermediate.*

# SKILLBUILDER

## 9.1 DRAWING A MECHANISM FOR HYDROHALOGENATION

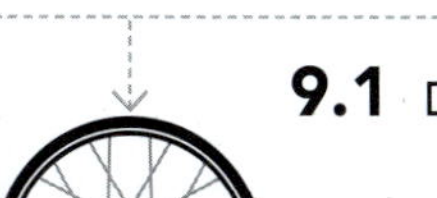

**LEARN the skill** Draw a mechanism for the following transformation:

HCl

Cl

### SOLUTION

**STEP 1**
Using two curved arrows, protonate the alkene to form the more stable carbocation.

In this reaction, a hydrogen atom and a halogen are added across an alkene. The halogen (Cl) is ultimately positioned at the more substituted carbon, which verifies that this reaction takes place via an ionic mechanism (Markovnikov addition). The ionic mechanism for hydrohalogenation has two steps: (1) *protonation* of the alkene to form the more stable carbocation and (2) *nucleophilic attack*. Each step must be drawn precisely.

When drawing the first step of the mechanism (protonation), make sure to use two curved arrows:

H—Cl:  ⇌  +  :Cl:⊖

**WATCH OUT**
Take special notice of the arrow with its head placed on the proton; it is a common mistake to draw this arrow in the wrong direction. Remember that curved arrows show the flow of electrons, not atoms.

One curved arrow is drawn with its tail on the $\pi$ bond and its head on the proton. The second curved arrow is drawn with its tail on the H—Cl bond and its head on the Cl.

When drawing the second step of the mechanism (nucleophilic attack of chloride), only one curved arrow is needed. The chloride ion, formed in the previous step, functions as a nucleophile and attacks the carbocation:

**STEP 2**
Using one curved arrow, draw the halide ion attacking the carbocation.

:Cl:⊖  ⇌  :Cl:

**PRACTICE the skill** **9.3** Draw a mechanism for each of the following transformations:

(a) HBr → Br

(b) HCl → Cl

(c) HCl → Cl

**APPLY the skill**

**9.4** Draw the intermediate carbocation that is formed when each of the following compounds is treated with HBr:

**(a)** **(b)** **(c)** **(d)**

**9.5** When 1-methoxy-2-methylpropene is treated with HCl, the major product is 1-chloro-1-methoxy-2-methylpropane. Although this reaction proceeds via an ionic mechanism, the Cl is ultimately positioned at the less substituted carbon. Draw a mechanism that is consistent with this outcome and then explain why the less substituted carbocation intermediate is more stable in this case.

HCl

**1-Methoxy-2-methylpropene** **1-Chloro-1-methoxy-2-methylpropane**

need more **PRACTICE?** **Try Problems 9.51c, 9.64b, 9.69, 9.72**

## Stereochemistry of Hydrohalogenation

In many cases, hydrohalogenation involves the formation of a chirality center; for example:

HCl → Cl — **Chirality center**

In this reaction, one new chirality center is formed. Therefore, two possible products are expected, representing a pair of enantiomers:

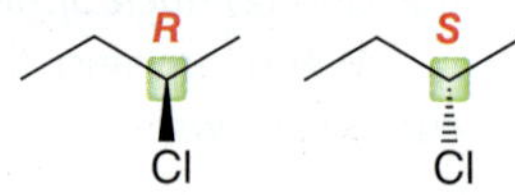


The two enantiomers are produced in equal amounts (a racemic mixture). The stereochemical outcome for this reaction can also be explained by the proposed mechanism, which identifies the key intermediate as a carbocation. Recall that a carbocation is trigonal planar, with an empty *p* orbital orthogonal to the plane. This empty *p* orbital is subject to attack by a nucleophile from either side (the lobe above the plane or the lobe below the plane), as seen in Figure 9.3. Both faces of the plane can be attacked with equal likelihood, and therefore, both enantiomers are produced in equal amounts.

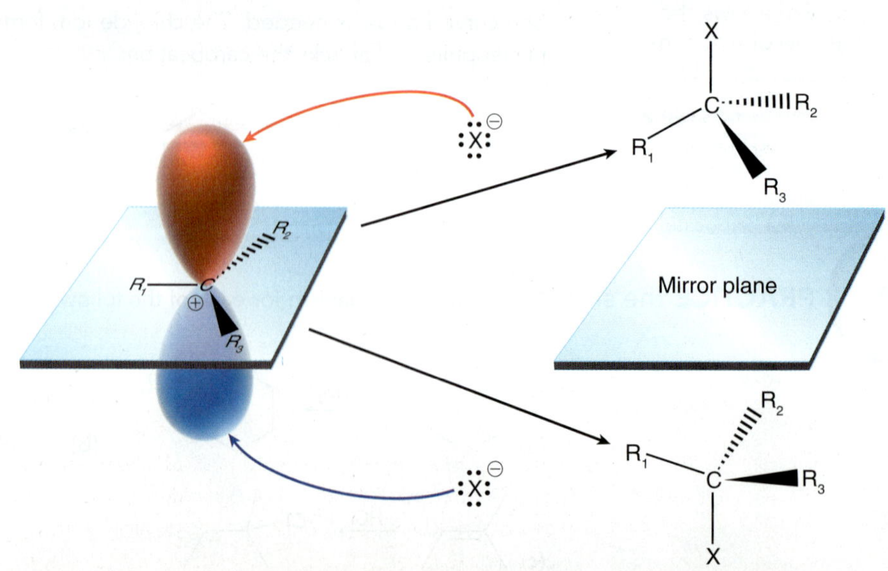


**FIGURE 9.3** In the second step of hydrohalogenation, the carbocation intermediate is planar and can be attacked from either face, leading to a pair of mirror-image products (enantiomers).

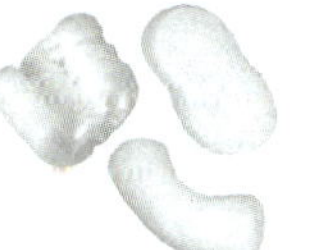

## CONCEPTUAL CHECKPOINT

**9.6** Predict the products for each of the following reactions. In some cases, the reaction produces a new chirality center, while in other cases, no new chirality center is formed. Consider that when drawing the product(s).

(a) HBr ? (b) HCl ? (c) HBr ?

(d) HI ? (e) HCl ? (f) HCl ?

## Hydrohalogenation with Carbocation Rearrangements

In Section 6.11, we discussed the stability of carbocations and their ability to rearrange via either a methyl shift or a hydride shift. The problems in that section focused on predicting when and how carbocations rearrange. That skill will be essential now, because the mechanism for HX addition involves formation of an intermediate carbocation. Therefore, HX additions are subject to carbocation rearrangements. Consider the following example, in which the π bond is protonated to generate the more stable, secondary carbocation, rather than the less stable, primary carbocation:

As we might expect, this carbocation can be captured by a chloride ion. However, it is also possible for this secondary carbocation to rearrange before encountering a chloride ion. Specifically, a hydride shift will produce a more stable, tertiary carbocation, which can then be captured by a chloride ion:

Secondary Tertiary

In cases where rearrangements are possible, HX additions produce a mixture of products, including those resulting from a carbocation rearrangement as well as those formed without rearrangement (i.e., if the nucleophile captures the carbocation before rearrangement has a chance to occur):

H—Cl → Cl 40% + Rearrangement product Cl 60%

The product ratio shown above (40 : 60) can be moderately affected by altering the concentration of HCl, but a mixture of products is unavoidable. The bottom line is: *When carbocation rearrangements can occur, they do occur.* As a result, HX addition is only synthetically useful in situations where carbocation rearrangements are not possible.

# SKILLBUILDER

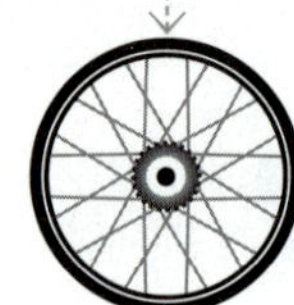

## 9.2 DRAWING A MECHANISM FOR A HYDROHALOGENATION PROCESS WITH A CARBOCATION REARRANGEMENT

### LEARN the skill

Draw a mechanism for the following transformation:

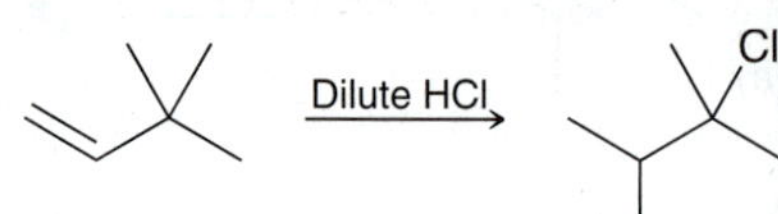

**SOLUTION**

The reaction is a hydrohalogenation. However, this product is not the expected product from a simple addition. Rather, the carbon skeleton of the product is different from the carbon skeleton of the reactant, and therefore, a carbocation rearrangement is likely.

In the first step, the alkene is protonated to form the more stable carbocation (secondary rather than primary):

**STEP 1**
Using two curved arrows, protonate the alkene to form the more stable carbocation.

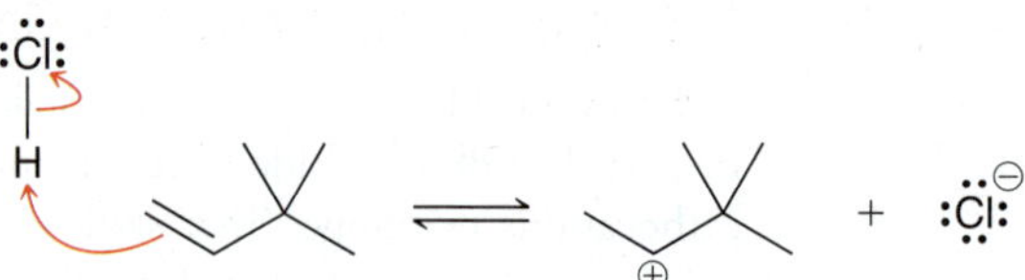


**STEP 2**
Using one curved arrow, draw the carbocation rearrangement.

Now look to see if a carbocation rearrangement is possible. In this case, a methyl shift can produce a more stable, tertiary carbocation:

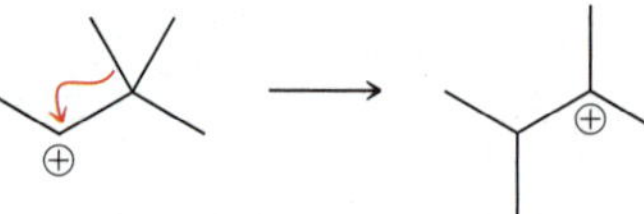

Secondary Tertiary

**STEP 3**
Using one curved arrow, draw the halide ion attacking the carbocation.

Finally, the chloride ion attacks the tertiary carbocation to give the product:

:Cl: :Cl:

### PRACTICE the skill

**9.7** Draw a mechanism for each of the following transformations:

(a) Dilute HBr → Br

(b) Dilute HBr → Br

(c) Dilute HCl →

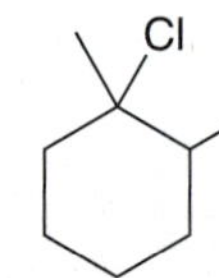

### APPLY the skill

**9.8** The mechanism of the following transformation involves a carbocation intermediate that rearranges in a way that we have not yet seen. Rather than occurring via a methyl shift or a hydride shift, a carbon atom of the ring migrates, thereby converting a secondary carbocation into a more stable, tertiary carbocation. Using this information, draw the mechanism of the following transformation:

**9.9** The following transformation proceeds through two successive carbocation rearrangements. Draw the mechanism and explain why each of the carbocation rearrangements is favorable:

Dilute HBr → Br

need more PRACTICE? Try Problems 9.51d, 9.76

## Cationic Polymerization and Polystyrene

Polymers were first introduced in Chapter 4, and discussions of polymers appear throughout the book. We saw that polymers are large molecules formed by the joining of monomers. More than 200 billion pounds of synthetic organic polymers are produced in the United States every year. They are used in a variety of applications, including automobile tires, carpet fibers, clothing, plumbing pipes, plastic squeeze bottles, cups, plates, cutlery, computers, pens, televisions, radios, CDs and DVDs, paint, and toys. Our society has clearly become dependent on polymers.

Synthetic organic polymers are commonly formed via one of three possible mechanistic pathways: *radical polymerization*, *anionic polymerization*, or *cationic polymerization*. Radical polymerization will be discussed in Chapter 11 and anionic polymerization will be discussed in Chapter 27. In this section, we will briefly introduce cationic polymerization.

The first step of cationic polymerization is similar to the first step of hydrohalogenation. Specifically, an alkene is protonated in the presence of an acid to produce a carbocation intermediate. The most commonly used acid catalyst for cationic polymerization is formed by treating $BF_3$ with water:

Acid catalyst

In the absence of halide ions, hydrohalogenation cannot occur. Instead, the carbocation intermediate must react with a nucleophile other than a halide ion. Under these conditions, the only nucleophile present in abundance is the starting alkene, which can attack the carbocation to form a new carbocation. This process repeats itself, adding one monomer at a time; for example:

Polymer

For monomers that undergo cationic polymerization, the process is terminated when the carbocation intermediate is deprotonated by a base or attacked by a nucleophile other than an alkene (such as water).

Many alkenes polymerize readily via cationic polymerization, making possible the production of a wide range of polymers with varying properties and applications. For example, polystyrene (mentioned in the chapter opener) can be prepared via cationic polymerization:

Styrene

Polystyrene

The resulting polymer is a rigid plastic that can be heated, molded into a desired shape, and then cooled again. This feature makes polystyrene ideal in the production of a variety of plastic items, including plastic cutlery, toy pieces, CD jewel cases, computer housings, radios, and televisions. In the early 1940s, an accidental discovery led to the development of Styrofoam, which is also made from polystyrene. Ray McIntire, working for the Dow Chemical Company, was searching for a polymeric material that could serve as a flexible electrical insulator. His research efforts led to the discovery of a type of foamed polystyrene which is comprised of approximately 95% air and is therefore much lighter (per unit volume) than regular polystyrene. This new material was found to have many useful physical properties. In particular, foamed polystyrene was found to be moisture resistant, unsinkable, and a poor conductor of heat (and therefore an excellent insulator). Dow called its material Styrofoam and has continuously improved the method of its production over the last half of a century. Coffee cups, coolers, and packing materials are also made from foamed polystyrene, but the method of preparation is not the same as the process used by Dow, and as a result, the properties are not exactly the same as the properties of Styrofoam.

Foamed polystyrene is generally prepared by heating polystyrene and then using hot gases, called blowing agents, to expand the polystyrene into a foam. In the past, CFCs (chlorofluorocarbons) were primarily used as blowing agents, but they are no longer in use because of their suspected role in destroying the ozone layer. The alternative blowing agents currently in use are safer for the environment.

## 9.4 Acid-Catalyzed Hydration

The next few sections discuss three methods for adding water (H and OH) across a double bond, a process called **hydration**. The first two methods proceed through Markovnikov additions, while the third method proceeds through an *anti*-Markovnikov addition. In this section, we will explore the first of these three reactions.

### Experimental Observations

Addition of water across a double bond in the presence of an acid is called **acid-catalyzed hydration**. For most simple alkenes, this reaction proceeds via a Markovnikov addition, as shown here. The net result is an addition of H and OH across the $\pi$ bond, with the OH group positioned at the more substituted carbon.

$H_3O^+$ OH (90%)

The reagent, $H_3O^+$, represents the presence of both water ($H_2O$) and an acid source, such as sulfuric acid. These conditions can be shown in the following way:

$+ \ H_2O \xrightarrow{[H_2SO_4]}$ OH

where the brackets indicate that the proton source is not consumed in the reaction. It is a catalyst, and, therefore, this reaction is said to be an *acid-catalyzed hydration.*

The rate of an acid-catalyzed hydration is very much dependent on the structure of the starting alkene. Compare the relative rates of the following three reactions and analyze the effects of an alkyl substituent on the relative rate of each reaction. With each additional alkyl group, the reaction rate increases by many orders of magnitude. Also take special notice of where the OH is placed in each of the cases above: *at the more substituted position.*

|  | Relative rate |
|---|---|
| $\xrightarrow{H_3O^+}$ HO | 1 |
| $\xrightarrow{H_3O^+}$ OH | $10^6$ |
| $\xrightarrow{H_3O^+}$ OH | $10^{11}$ |

### Mechanism and Source of Regioselectivity

Mechanism 9.2 is consistent with the observed regiochemical preference for Markovnikov addition as well as the effect of alkene structure on the rate of the reaction.

#### MECHANISM 9.2 ACID-CATALYZED HYDRATION

Proton transfer — H–O⊕(H)H — The alkene is protonated, forming a carbocation intermediate.

**Carbocation** — Nucleophilic attack — H–O–H — Water functions as a nucleophile and attacks the carbocation intermediate.

**Oxonium ion** — Proton transfer — H–O–H — Water functions as a base and deprotonates the oxonium ion, yielding the product.

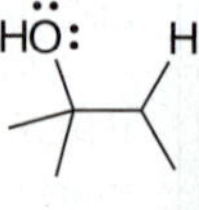

**LOOKING AHEAD**
A proposed mechanism must always be consistent with the stated conditions, as we see here with the use of $H_2O$ in acidic conditions rather than hydroxide for the last step of the mechanism. This principle will appear many more times throughout this text.

The first two steps of this mechanism are virtually identical to the mechanism we proposed for hydrohalogenation. Specifically, the alkene is first protonated to generate a carbocation intermediate, which is then attacked by a nucleophile. However, in this case, the attacking nucleophile is neutral ($H_2O$) rather than an anion ($X^-$), and therefore, a charged intermediate is generated as a result of nucleophilic attack. This intermediate is called an **oxonium ion**, because it exhibits an oxygen atom with a positive charge. In order to remove this charge and form an electrically neutral product, the mechanism must conclude with a proton transfer. Notice that the base used to deprotonate the oxonium ion is $H_2O$ rather than a hydroxide ion ($HO^-$). Why? In acidic conditions, the concentration of hydroxide ions is extremely low, but the concentration of $H_2O$ is quite large.

The proposed mechanism is consistent with the experimental observations discussed earlier in this section. The reaction proceeds via a Markovnikov addition, just as we saw for hydrohalogenation, because there is a strong preference for the reaction to proceed via the more stable carbocation intermediate. Similarly, reaction rates for substituted alkenes can be justified by comparing the carbocation intermediates in each case. Reactions that proceed via tertiary carbocations will generally occur more rapidly than reactions that proceed via secondary carbocations.

## CONCEPTUAL CHECKPOINT

**9.10** In each of the following cases identify the alkene that is expected to be more reactive toward acid-catalyzed hydration.

(a) or

(b) 2-Methyl-2-butene or 3-methyl-1-butene

## Controlling the Position of Equilibrium

**LOOKING BACK**
Acid-catalyzed dehydration was first discussed in Section 8.9.

Carefully examine the proposed mechanism for acid-catalyzed hydration. Notice the equilibrium arrows ($\rightleftharpoons$ rather than $\rightarrow$). These arrows indicate that the reaction actually goes in both directions. Consider the reverse path (starting from the alcohol and ending with the alkene). That process is an elimination reaction in which an alcohol is converted into an alkene. More specifically, it is an E1 process called acid-catalyzed dehydration. The truth is that most reactions represent equilibrium processes; however, organic chemists generally draw equilibrium arrows only in situations where the equilibrium can be easily manipulated (allowing for control over the product distribution). Acid-catalyzed dehydration is an excellent example of such a reaction.

Earlier in this chapter, thermodynamic arguments were presented to explain why low temperature favors addition, while high temperature favors elimination. But in acid-catalyzed dehydration there is yet another way to easily control the equilibrium. The equilibrium is sensitive not only to temperature but also to the concentration of water that is present. By controlling the amount of water present (using either concentrated acid or dilute acid), one side of the equilibrium can be favored over the other:

+ $H_2O$ — **Dilute** $H_2SO_4$ (more $H_2O$) → HO ; ← **Conc.** $H_2SO_4$ (less $H_2O$)

Control over this equilibrium derives from an understanding of Le Châtelier's principle, which states that a system at equilibrium will adjust in order to minimize any stress placed on the system. To understand how this principle applies, consider the above process after equilibrium has been established. Notice that water is on the left side of the reaction. How would the introduction of more water affect the system? The concentrations would no longer be at equilibrium, and the system would have to adjust to reestablish new equilibrium concentrations. The introduction of more water will cause the position of equilibrium to move in such a way that more of the alcohol is produced. Therefore, dilute acid (which is mostly water) is used to convert an alkene into an alcohol. On the flip side, removing water from the system would cause the equilibrium to favor the alkene. Therefore, concentrated acid (very little water) is used to favor formation of the alkene. Alternatively, water can also be removed from the system via a distillation process, which would also favor the alkene.

In summary, the outcome of a reaction can be greatly affected by carefully choosing the reaction conditions and concentrations of reagents.

## CONCEPTUAL CHECKPOINT

**9.11** Identify whether you would use dilute sulfuric acid or concentrated sulfuric acid to achieve each of the following transformations. In each case, explain your choice.

(a) + $H_2O$ $\xrightarrow{[H_2SO_4]}$ OH

(b) HO $\xrightarrow{[H_2SO_4]}$ + $H_2O$

## Stereochemistry of Acid-Catalyzed Hydration

The stereochemical outcome of acid-catalyzed hydration is similar to the stereochemical outcome of hydrohalogenation. Once again, the intermediate carbocation can be attacked from either side with equal likelihood (Figure 9.4). Therefore, when a new chirality center is generated, a racemic mixture of enantiomers is expected:

$\xrightarrow{H_3O^+}$ OH + HO

50 / 50

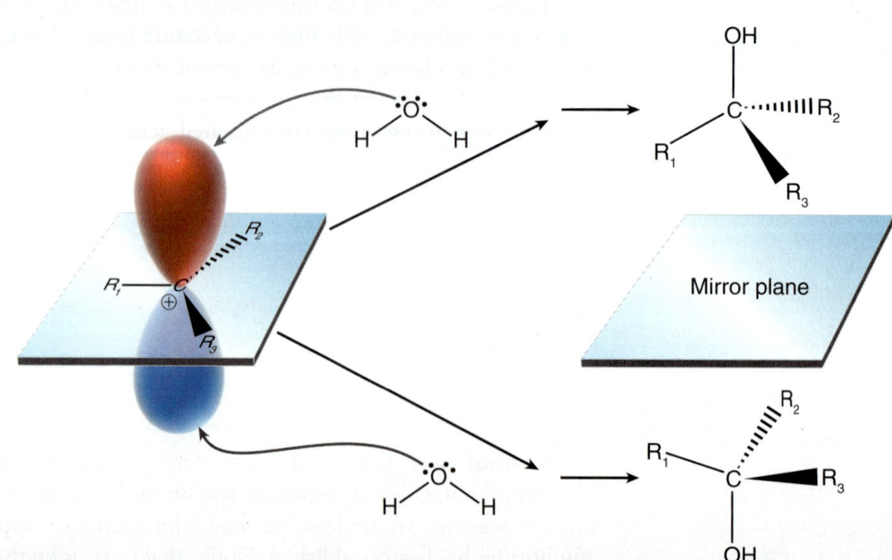


**FIGURE 9.4**
In the second step of acid-catalyzed hydration, the carbocation intermediate is planar and can be attacked from either face, leading to a pair of mirror-image products (enantiomers).

# SKILLBUILDER

## 9.3 DRAWING A MECHANISM FOR AN ACID-CATALYZED HYDRATION

**LEARN the skill** Draw a mechanism for the following transformation:

$\xrightarrow{H_3O^+}$ OH

### SOLUTION

In this reaction, water is added across an alkene in a Markovnikov fashion under acid-catalyzed conditions. As a result, the OH is positioned at the more substituted carbon. To draw a mechanism for this process, recall that the proposed mechanism for acid-catalyzed hydration has three steps: (1) protonation to give a carbocation, (2) nucleophilic attack of water to give an oxonium ion, and (3) deprotonation to generate a neutral product. When drawing the first step of the mechanism (protonation), make sure to use two curved arrows and make sure to form the more stable carbocation:

**STEP 1**
Using two curved arrows, protonate the alkene to form the more stable carbocation.

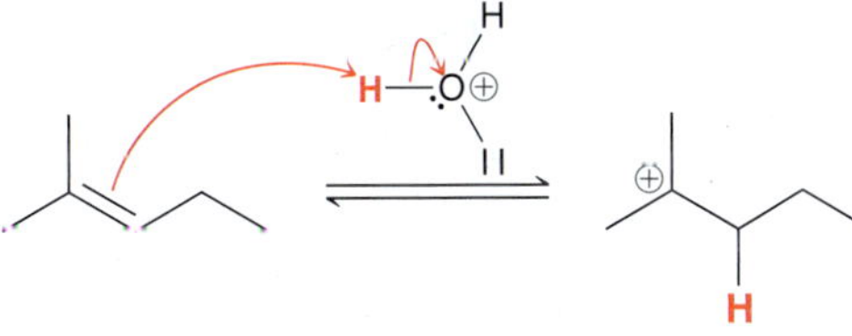

**WATCH OUT**
Make sure to draw both curved arrows and be very precise when placing the head and tail of each curved arrow.

One curved arrow is drawn with its tail on the π bond and its head on the proton, while the second curved arrow is drawn with its tail on an O—H bond and its head on the oxygen atom of water. When drawing the second step of the mechanism (nucleophilic attack of water), only one curved arrow is required. The tail should be placed on a lone pair of water, and the head should be placed on the carbocation:

**STEP 2**
Using one curved arrow, draw a water molecule attacking the carbocation.

In the final step of the mechanism, water (not hydroxide) functions as a base and abstracts a proton from the oxonium ion. Like all proton transfer steps, this process requires two curved arrows. One curved arrow is drawn with its tail on a lone pair of water and its head on the proton, while the second curved arrow is drawn with its tail on the O—H bond and its head on the oxygen atom:

**STEP 3**
With two curved arrows, deprotonate the oxonium ion using water as a base.

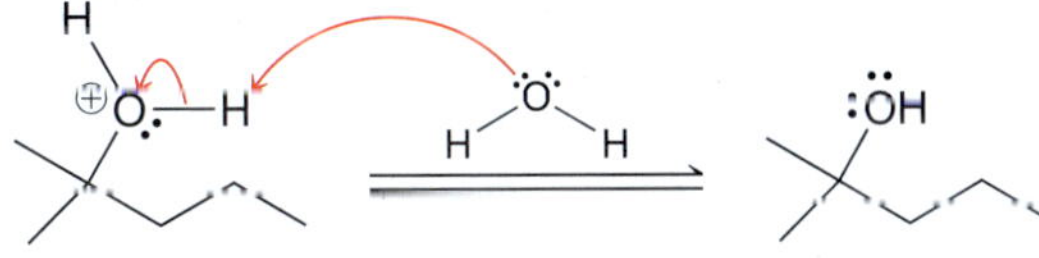

**PRACTICE the skill** **9.12** Draw a mechanism for each of the following transformations:

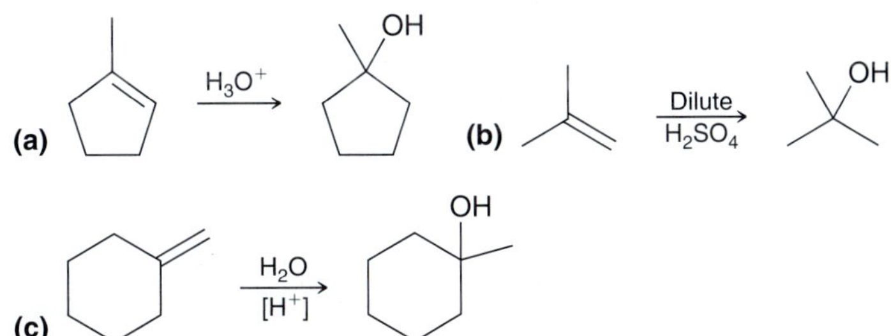

**9.13** If an alkene is protonated and the solvent is an alcohol rather than water, a reaction takes place that is very similar to acid-catalyzed hydration, but in the second step of the mechanism the alcohol functions as a nucleophile instead of water. Draw a plausible mechanism for the following process:

MeOH, $[H_2SO_4]$ → OMe

**9.14** Using the reaction in the previous problem as a reference, propose a plausible mechanism for the following intramolecular reaction:

OH → $[H_2SO_4]$ → O

need more **PRACTICE?** **Try Problems 9.51a,b, 9.64a**

**Industrial Production of Ethanol**

Ethanol found in alcoholic beverages is generally produced via the fermentation of sugars by yeast. However, there are many other important industrial uses for ethanol (as a gasoline additive, as a solvent, for perfumes and paints, etc.), and therefore, an alternative, more efficient process for producing pure ethanol is required. This can be accomplished by the acid-catalyzed hydration of ethylene obtained from petroleum:

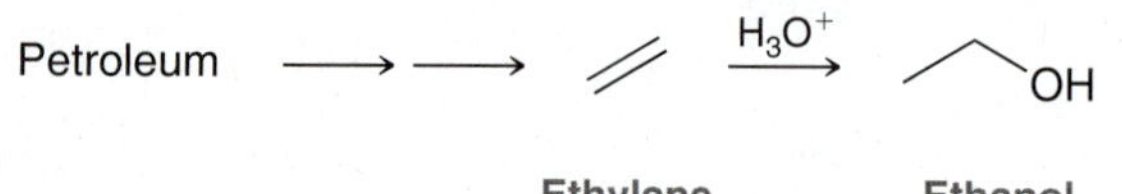


Phosphoric acid is generally employed as the source of protons. The United States produces more than five billion gallons of ethanol each year via this process.

## 9.5 Oxymercuration-Demercuration

The previous section explored how acid-catalyzed hydration can be used to achieve a Markovnikov addition of water across an alkene. The utility of that process is somewhat diminished by the fact that carbocation rearrangements can produce a mixture of products:

$H_3O^+$ → OH (No rearrangement) + OH (Rearrangement)

In cases where protonation of the alkene ultimately leads to carbocation rearrangements, acid-catalyzed hydration is an inefficient method for adding water across the alkene. Many other methods can achieve a Markovnikov addition of water across an alkene without carbocation rearrangements. One of the oldest and perhaps best known methods is called **oxymercuration-demercuration**:

Oxymercuration → OH, HgOAc → Demercuration → OH

To understand this process, we must explore the reagents employed. The process begins when mercuric acetate, $Hg(OAc)_2$, dissociates to form a mercuric cation:

Mercuric acetate

Mercuric cation

This mercuric cation is a powerful electrophile and is subject to attack by a nucleophile, such as the π bond of an alkene. When a π bond attacks a mercuric cation, the nature of the resulting intermediate is quite different from the nature of the intermediate formed when a π bond is simply protonated. Let's compare:

**When a π bond is protonated:**

**A carbocation**

**When a π bond attacks a mercuric cation:**

**A mercurinium ion**

When a π bond is protonated, the intermediate formed is simply a carbocation, as we have seen many times in this chapter. In contrast, when a π bond attacks a mercuric cation, the resulting intermediate cannot be considered as a carbocation, because the mercury atom has electrons that can interact with the nearby positive charge to form a bridge. This intermediate, called a **mercurinium ion**, is more adequately described as a hybrid of two resonance structures. A mercurinium ion has some of the character of a carbocation, but it also has some of the character of a bridged, three-membered ring. This dual character can be illustrated with the following drawing:

**BY THE WAY**
As mentioned in Chapter 2, we generally avoid breaking single bonds when drawing resonance structures. However, this is one of the rare exceptions. We will see one other such exception in the next section of this chapter.

**Mercurinium ion**

The more substituted carbon atom bears a partial positive charge (δ+), rather than a full positive charge. As a result, this intermediate will not readily undergo carbocation rearrangements, but it is still subject to attack by a nucleophile:

Notice that the attack takes place at the more substituted position, ultimately leading to Markovnikov addition. After attack of the nucleophile, the mercury can be removed through a process called *demercuration*, which can be accomplished with sodium borohydride. There is much evidence that

**LOOKING AHEAD**
Radical processes are discussed in more detail in Chapter 11.

demercuration occurs via a radical process. The net result is the addition of H and a nucleophile across an alkene:

1) $Hg(OAc)_2$, Nuc-H
2) $NaBH_4$

Many nucleophiles can be used, including water:

1) $Hg(OAc)_2$, $H_2O$
2) $NaBH_4$

This reaction sequence provides for a two-step process that enables the hydration of an alkene without carbocation rearrangements:

1) $Hg(OAc)_2$, $H_2O$
2) $NaBH_4$

OH
(94%)
**No rearrangement**

## CONCEPTUAL CHECKPOINT

**9.15** Predict the product for each reaction and predict the products if an acid-catalyzed hydration had been performed rather than an oxymercuration-demercuration:

(a) 1) $Hg(OAc)_2$, $H_2O$ 2) $NaBH_4$ ?

(b) 1) $Hg(OAc)_2$, $H_2O$ 2) $NaBH_4$ ?

(c) 1) $Hg(OAc)_2$, $H_2O$ 2) $NaBH_4$ ?

**9.16** In the first step of the process (oxymercuration), nucleophiles other than water may be used. Predict the product when each of the following nucleophiles is used instead of water:

(a) 1) $Hg(OAc)_2$, EtOH 2) $NaBH_4$ ?

(b) 1) $Hg(OAc)_2$, $EtNH_2$ 2) $NaBH_4$ ?

# 9.6 Hydroboration-Oxidation

## An Introduction to Hydroboration-Oxidation

The previous sections covered two different methods for achieving a Markovnikov addition of water across a π bond: (1) acid-catalyzed hydration and (2) oxymercuration-demercuration. In this section, we will explore a method for achieving an *anti*-Markovnikov addition of water. This process, called **hydroboration-oxidation**, places the OH group at the less substituted carbon:

1) $BH_3 \cdot THF$
2) $H_2O_2$, NaOH

OH
(90%)

**Less substituted vinylic position**

***Anti*-Markovnikov addition**

The stereochemical outcome of this reaction is also of particular interest. Specifically, when two new chirality centers are formed, the addition of water (H and OH) is observed to occur in a way that places the H and OH on the same face of the π bond:

1) $BH_3 \cdot THF$
2) $H_2O_2$, NaOH

Enantiomers

This mode of addition is called a ***syn* addition**. The reaction is said to be stereospecific because only two of the four possible stereoisomers are formed. That is, the reaction does not produce the two stereoisomers that would result from adding H and OH to opposite sides of the π bond:

These stereoisomers are not formed

Any mechanism that we propose for hydroboration-oxidation must explain both the regioselectivity (*anti*-Markovnikov addition) as well as the stereospecificity (*syn* addition). We will soon propose a mechanism that explains both observations. But first, we must explore the nature of the reagents used for hydroboration-oxidation.

## Reagents for Hydroboration-Oxidation

The structure of borane ($BH_3$) is similar to that of a carbocation, but without the charge:

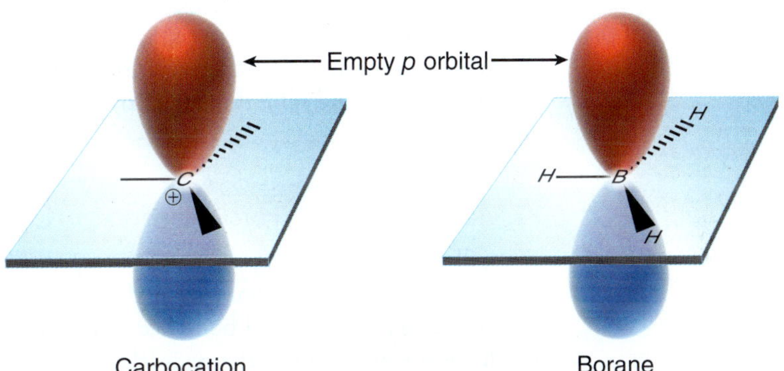


The boron atom lacks an octet of electrons and is therefore very reactive. In fact, one borane molecule will even react with another borane molecule to form a dimeric structure called **diborane**. This dimer is believed to possess a special type of bonding that is unlike anything we have seen thus far. It can be more easily understood by drawing the following resonance structures:

As with the mercurinium ion (Section 9.5), this is another of those rare cases where we break a single bond when drawing the resonance structures. Careful examination of these resonance structures shows that each of the hydrogen atoms, colored in red and blue above, is partially bonded to two boron atoms using a total of two electrons:

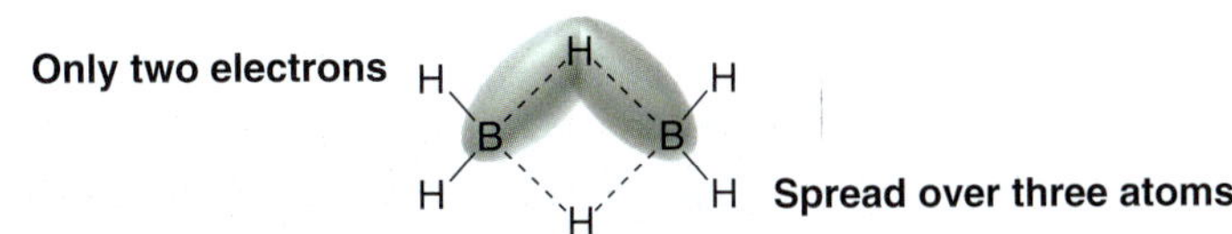

Such bonds are called **three-center, two-electron bonds**. In the world of organic chemistry, there are many other examples of three-center, two-electron bonds; however, we will not encounter many other examples in this text.

Borane and diborane coexist in the following equilibrium:

$$BH_3 + BH_3 \rightleftharpoons B_2H_6$$

Borane Diborane

This equilibrium lies very much to the side of diborane ($B_2H_6$), leaving very little borane ($BH_3$) present at equilibrium. It is possible to stabilize $BH_3$, thereby increasing its concentration at equilibrium, by using a solvent such as THF (tetrahydrofuran), which can donate electron density into the empty *p* orbital of boron:

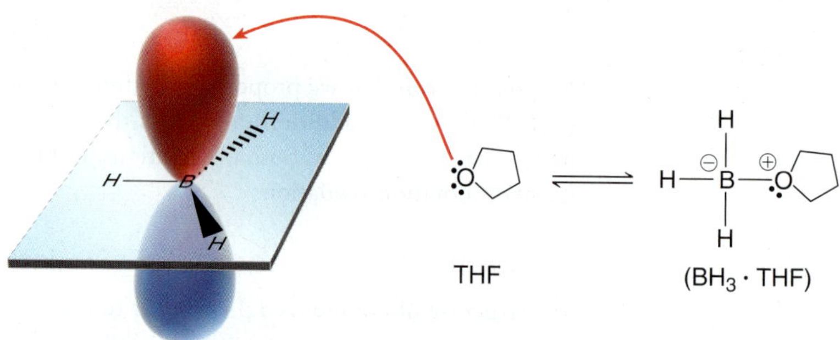


Although the boron atom does receive some electron density from the solvent, it is nevertheless still very electrophilic and subject to attack by the π bond of an alkene. A mechanism for hydroboration-oxidation follows.

## A Mechanism for Hydroboration-Oxidation

We will begin by focusing on the first step of hydroboration, in which borane is attacked by a π bond, triggering a simultaneous hydride shift. In other words, formation of the C—$BH_2$ bond and formation of the C—H bond occur together in a concerted process. This step of the proposed mechanism explains both the regioselectivity (*anti*-Markovnikov addition) as well as the stereospecificity (*syn* addition) for this process. Each of these features will now be discussed in more detail.

## Regioselectivity of Hydroboration-Oxidation

As seen in Mechanism 9.3, a $BH_2$ group is installed at the less substituted carbon and is ultimately replaced with an OH group, leading to the observed regioselectivity. The preference for $BH_2$ to be positioned at the less substituted carbon can be explained in terms of electronic considerations or steric considerations. Both explanations are presented below:

1. ***Electronic considerations***: In the first step of the proposed mechanism, attack of the π bond triggers a simultaneous hydride shift. However, this process does not have to be perfectly simultaneous. As the π bond attacks the empty *p* orbital of boron, one of the vinylic positions can begin to develop a partial positive charge (δ+). This developing δ+ then triggers a hydride shift:

δ− δ+ ‡ $BH_2$

## MECHANISM 9.3 HYDROBORATION-OXIDATION

### Hydroboration

The first step is repeated for each B—H bond

A trialkylborane

### Oxidation

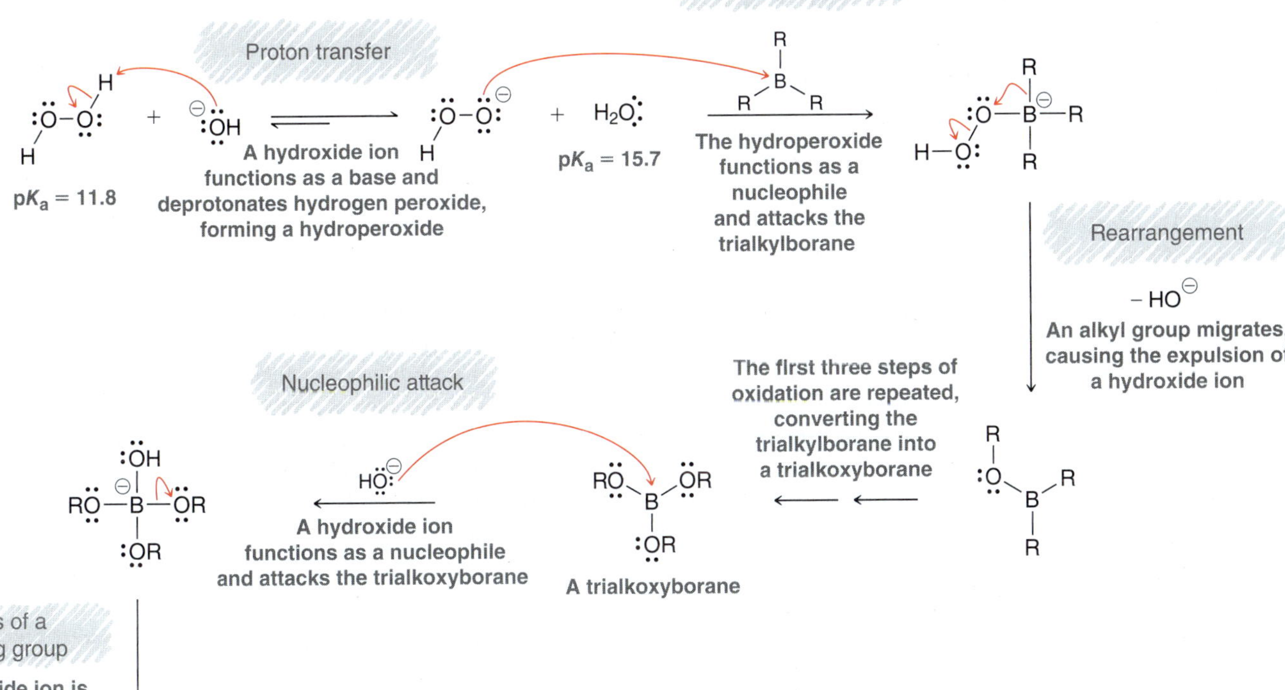


Proton transfer

The alkoxide ion is protonated

Transition state *anti*-Markovnikov addition

Transition state Markovnikov addition

More crowded

**FIGURE 9.5** A comparison of the transition states for hydroboration via Markovnikov addition or *anti*-Markovnikov addition. The latter will be lower in energy because of decreased steric crowding.

In this way, one of the vinylic carbon atoms develops a partial positive charge when the alkene begins to interact with borane. There will be a preference (as we have seen previously in this chapter) for any positive character to develop at the more substituted carbon. In order to accomplish this, the $BH_2$ group must be positioned at the less substituted carbon atom.

2. ***Steric considerations***: In the first step of the proposed mechanism, both H and $BH_2$ are adding across the double bond simultaneously. Since $BH_2$ is bigger than H, the transition state will be less crowded and lower in energy if the $BH_2$ group is positioned at the less sterically hindered position (Figure 9.5). It is likely that both electronic and steric factors contribute to the observed regioselectivity for hydroboration-oxidation.

## CONCEPTUAL CHECKPOINT

**9.17** Below are several examples of hydroboration-oxidation. In each case, consider the expected regioselectivity and then draw the product:

(a) 1) $BH_3 \cdot THF$ 2) $H_2O_2$, NaOH → ?

(b) 1) $BH_3 \cdot THF$ 2) $H_2O_2$, NaOH → ?

(c) 1) $BH_3 \cdot THF$ 2) $H_2O_2$, NaOH → ?

**9.18** Compound A has molecular formula $C_5H_{10}$. Hydroboration-oxidation of compound A produces 2-methylbutan-1-ol. Draw the structure of compound A:

**Compound A ($C_5H_{10}$)** 1) $BH_3 \cdot THF$ 2) $H_2O_2$, NaOH → OH

## Stereospecificity of Hydroboration-Oxidation

The observed stereospecificity for hydroboration-oxidation is consistent with the first step of the proposed mechanism, in which H and $BH_2$ are simultaneously added across the π bond of the alkene. The concerted nature of this step requires that both groups add across the same face of the alkene, giving a *syn* addition. In this way, the proposed mechanism explains not only the regiochemistry but also the stereochemistry.

When drawing the products of hydroboration-oxidation, it is essential to consider the number of chirality centers that are created during the process. If no chirality centers are formed, then the stereospecificity of the reaction is not relevant:

1) $BH_3 \cdot THF$ 2) $H_2O_2$, NaOH → OH **No chirality centers**

In this case, only one product is formed, rather than a pair of enantiomers, and the *syn* requirement is irrelevant.

Now consider the stereochemical outcome in a case where one chirality center is formed:

1) $BH_3 \cdot THF$ 2) $H_2O_2$, NaOH → OH **One chirality center**

In this case, both enantiomers are obtained, because *syn* addition can take place from either face of the alkene with equal likelihood:

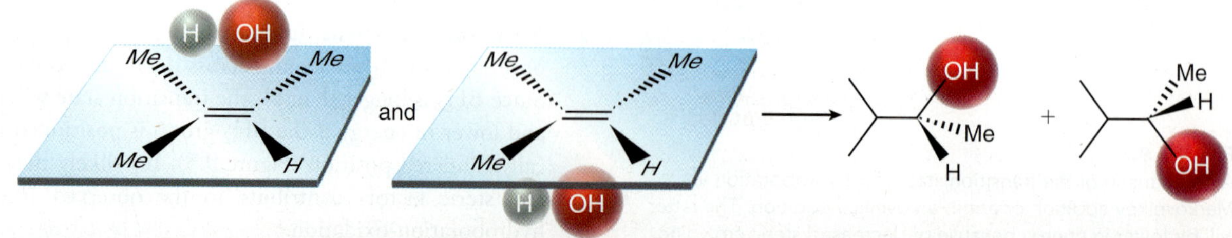

Now consider a case in which two chirality centers are formed:

1) $BH_3 \cdot THF$
2) $H_2O_2$, NaOH

OH

Two chirality centers

In such a case, the requirement for *syn* addition determines which pair of enantiomers is obtained:

1) $BH_3 \cdot THF$
2) $H_2O_2$, NaOH

H, H, OH + H, H, OH

A pair of enantiomers

The other two possible stereoisomers are not obtained.

Under special circumstances, it is possible to achieve *enantioselective addition reactions*, that is, reactions where one enantiomer predominates over the other (and there is an observed % *ee*, or enantiomeric excess). However, we have not yet seen any methods for accomplishing this. We will see one such example later in this chapter, but as of yet, none of the reactions in this chapter have been enantioselective. As a result, all of the reactions presented in this chapter will produce a product mixture that is optically inactive.

# SKILLBUILDER

## 9.4 PREDICTING THE PRODUCTS OF HYDROBORATION-OXIDATION

**LEARN** the skill

Predict the product(s) for the following transformation:

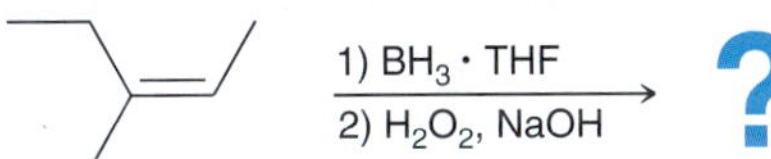

### SOLUTION

This process is a hydroboration-oxidation, which will add water across the double bond. The first step is to determine the regiochemical outcome. That is, we must determine where the OH is positioned. Recall that hydroboration oxidation produces an *anti*-Markovnikov addition, which means that the OH is positioned at the less substituted carbon:

**STEP 1**
Determine the regiochemical outcome based on the requirement for *anti*-Markovnikov addition.

1) $BH_3 \cdot THF$
2) $H_2O_2$, NaOH

OH

Less substituted

Now we know where to place the OH group, but the drawing is still not complete, because the stereochemical outcome must be indicated. Specifically, we must ask if two new chirality centers are formed in this reaction. In this example, there are, in fact, two new chirality centers being formed. So, the stereospecificity of this process (*syn* addition) is relevant and must be taken into account when drawing the products. The reaction will produce only the pair of enantiomers that results from a *syn* addition:

**STEP 2**
Determine the stereochemical outcome based on the requirement for *syn* addition.

H

1) $BH_3 \cdot THF$
2) $H_2O_2$, NaOH

H, H, OH + Enantiomer

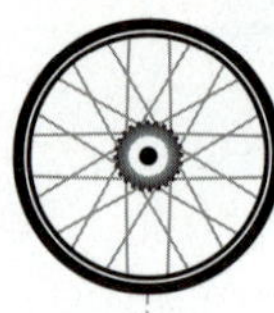

**PRACTICE the skill** **9.19** Predict the product(s) for each of the following transformations:

(a) 1) $BH_3 \cdot THF$ 2) $H_2O_2$, NaOH ?

(b) 1) $BH_3 \cdot THF$ 2) $H_2O_2$, NaOH ?

(c) 1) $BH_3 \cdot THF$ 2) $H_2O_2$, NaOH ?

(d) 1) $BH_3 \cdot THF$ 2) $H_2O_2$, NaOH ?

(e) 1) $BH_3 \cdot THF$ 2) $H_2O_2$, NaOH ?

(f) 1) $BH_3 \cdot THF$ 2) $H_2O_2$, NaOH ?

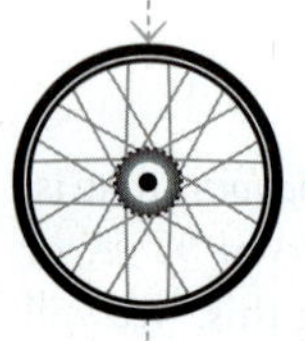

**APPLY the skill** **9.20** Draw the product(s) obtained from hydroboration-oxidation of (*E*)-3-methyl-3-hexene.

**9.21** The products obtained from hydroboration-oxidation of *cis*-2-butene are identical to the products obtained from hydroboration-oxidation of *trans*-2-butene. Draw the products and explain why the configuration of the starting alkene is not relevant in this case.

**9.22** Compound A has molecular formula $C_5H_{10}$. Hydroboration-oxidation of compound A produces an alcohol with no chirality centers. Draw two possible structures for compound A.

need more **PRACTICE?** **Try Problem 9.66**

## 9.7 Catalytic Hydrogenation

**LOOKING BACK**
The hydrogen atoms installed during this process are not explicitly drawn in the product, because the presence of hydrogen atoms can be inferred from bond-line drawings. For more practice "seeing" the hydrogen atoms that are not explicitly drawn, review the exercises in Section 2.2.

**Catalytic hydrogenation** involves the addition of molecular hydrogen ($H_2$) across a double bond in the presence of a metal catalyst; for example:

$H_2$ / Pt

**(100%)**

The net result of this process is to reduce an alkene to an alkane. Paul Sabatier was the first to demonstrate that catalytic hydrogenation could serve as a general procedure for reducing alkenes, and for his pioneering work, he was a corecipient of the 1912 Nobel Prize in Chemistry.

### Stereospecificity of Catalytic Hydrogenation

In the previous reaction there are no chirality centers in the product, so the stereospecificity of the process is irrelevant. In order to explore whether a reaction displays stereospecificity, we must examine a case where two new chirality centers are being formed. For example, consider the following case:

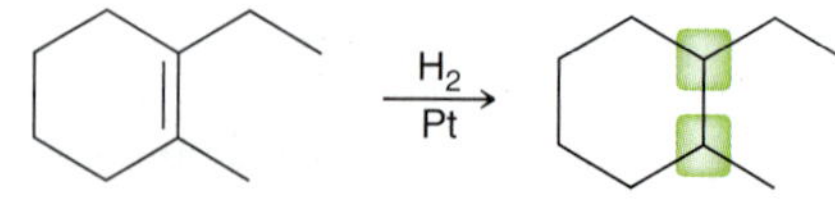


With two chirality centers, there are four possible stereoisomeric products (two pairs of enantiomers):

Pair of enantiomers (observed)

Pair of enantiomers (not observed)

However, the reaction does not produce all four products. Only one pair of enantiomers is observed, the pair that results from a *syn* addition. To understand the reason for the observed stereospecificity, we must take a close look at the reagents and their proposed interactions.

## The Role of the Catalyst in Catalytic Hydrogenation

Catalytic hydrogenation is accomplished by treating an alkene with $H_2$ gas and a metal catalyst, often under conditions of high pressure. The role of the catalyst is illustrated by the energy diagram in Figure 9.6. The pathway without the metal catalyst (blue) has a very large energy of activation ($E_a$), rendering the reaction too slow to be of practical use. The presence of a catalyst provides a pathway (red) with a lower energy of activation, thereby allowing the reaction to occur more rapidly.

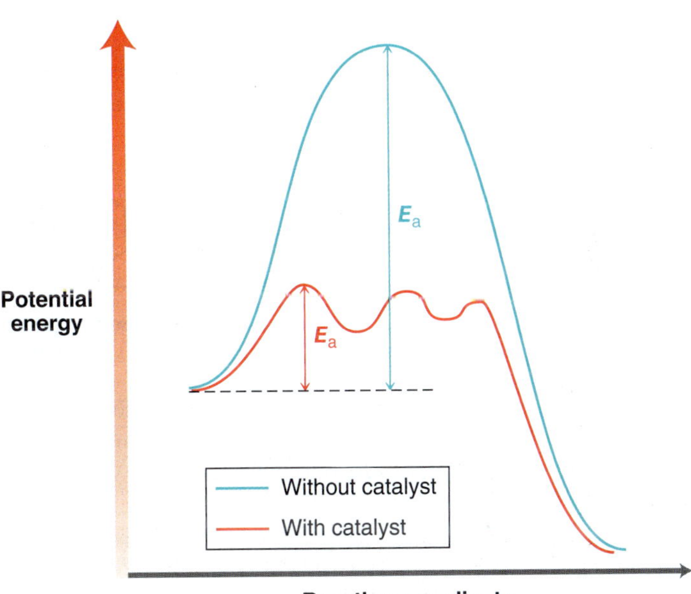


**FIGURE 9.6**
An energy diagram showing hydrogenation with a catalyst (red) and without a catalyst (blue). The latter has a lower energy of activation and therefore occurs more rapidly.

A variety of metal catalysts can be used, such as Pt, Pd, or Ni. The process is believed to begin when molecular hydrogen ($H_2$) interacts with the surface of the metal catalyst, effectively breaking the H—H bonds and forming individual hydrogen atoms adsorbed to the surface of the metal. The alkene coordinates with the metal surface, and surface chemistry allows for the reaction between the π bond and two hydrogen atoms, effectively adding H and H across the alkene (Figure 9.7). In this process, both hydrogen atoms add to the same face of the alkene, explaining the observed stereospecificity (*syn* addition).

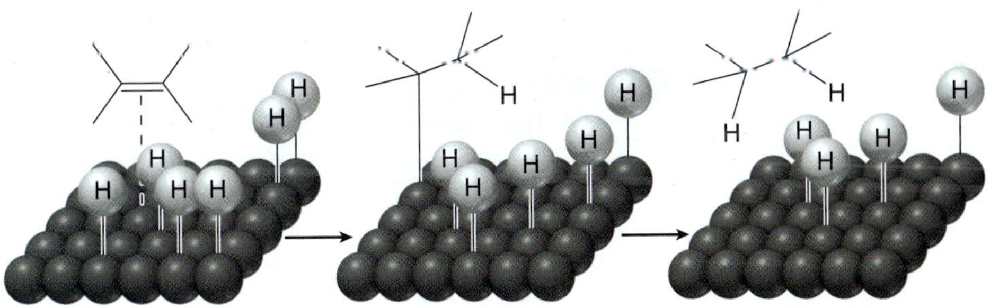

**FIGURE 9.7**
The addition process takes place on the surface of a metal catalyst.

**TABLE 9.2 SUMMARY OF RELATIONSHIP BETWEEN NUMBER OF CHIRALITY CENTERS FORMED AND STEREOCHEMICAL OUTCOME OF CATALYTIC HYDROGENATION**

| | |
|---|---|
| Zero chirality centers | *Syn* requirement is not relevant. Only one product formed. |
| One chirality center | Both possible enantiomers are formed. |
| Two chirality centers | The requirement for *syn* addition determines which pair of enantiomers is obtained. |

In any given case, *the stereochemical outcome is dependent on the number of chirality centers formed in the process*, as summarized in Table 9.2. Care must be taken when applying the simple paradigm in Table 9.2, because symmetrical alkenes will produce a *meso* compound rather than a pair of enantiomers. Consider the following case:

$\xrightarrow[Pt]{H_2}$ ?

In this example, two new chirality centers are formed, and therefore, we might expect that *syn* addition will produce a pair of enantiomers. However, in this case, there is only one product from *syn* addition, not a pair of enantiomers:

is the same as

***Syn* addition produces only one product: a *meso* compound**

A *meso* compound, by definition, does not have an enantiomer. A *syn* addition on one face of the alkene generates exactly the same compound as a *syn* addition on the other face of the alkene. And, therefore, care must be taken not to write "+ Enantiomer" or "+ En" for short:

$\xrightarrow[Pt]{H_2}$ + En

# SKILLBUILDER

## 9.5 PREDICTING THE PRODUCTS OF CATALYTIC HYDROGENATION

**LEARN** the skill

Predict the products of each of the following reactions:

(a) $\xrightarrow[Pt]{H_2}$ ?

(b) $\xrightarrow[Pt]{H_2}$ ?

### SOLUTION

(a) This reaction is a catalytic hydrogenation process, in which H and H are added across the alkene. Since both groups are identical (H and H), it is not necessary to consider regiochemical issues. However, stereochemistry is a factor that must be considered. In order to properly draw the products, it is first necessary to determine how many chirality centers are formed as a result of this reaction:

**STEP 1**
Determine the number of chirality centers formed.

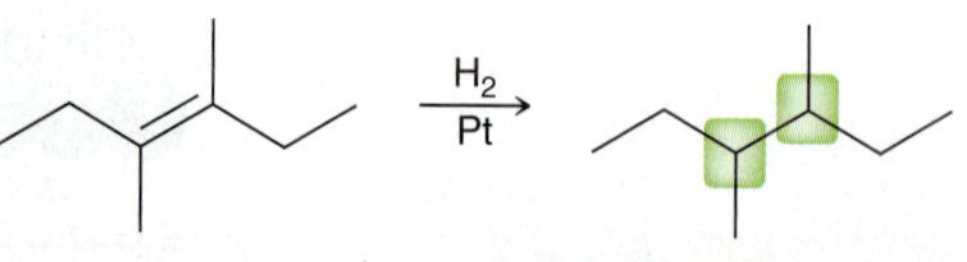

In this case, two chirality centers are formed. Therefore, we expect only the pair of enantiomers that would result from a *syn* addition:

**STEP 2**
Determine the stereochemical outcome based on the requirement for *syn* addition.

As a final check, just make sure that the products do not represent a single *meso* compound. The compounds above lack an internal plane of symmetry and do not represent a single *meso* compound.

**STEP 3**
Verify that the products do not represent a single *meso* compound.

**LOOKING BACK**
For practice identifying *meso* compounds, review the exercises in Section 5.6.

**(b)** Like the previous example, this reaction is a catalytic hydrogenation process, in which H and H are added across the alkene. In order to properly draw the products, it is first necessary to determine how many chirality centers are formed as a result of this reaction:

**One chirality center**

In this case, only one chirality center is formed. Therefore, we expect both possible enantiomers, because *syn* addition can occur from either face of the π bond with equal likelihood:

It is impossible for a compound containing exactly one chirality center to be a *meso* compound, so the compounds above are a pair of enantiomers.

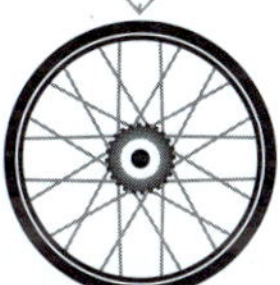

**PRACTICE the skill** **9.23** Predict the product(s) for each of the following reactions:

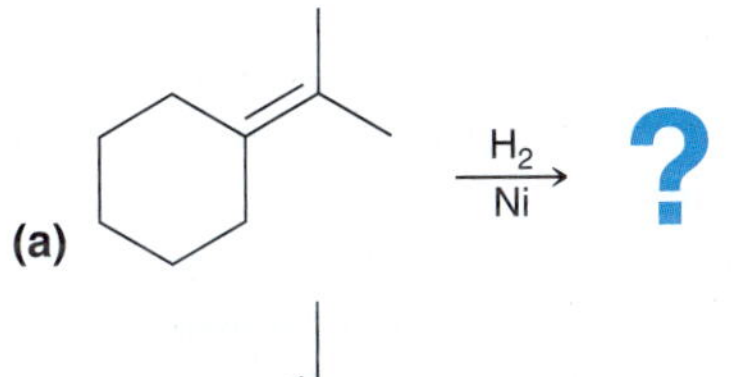


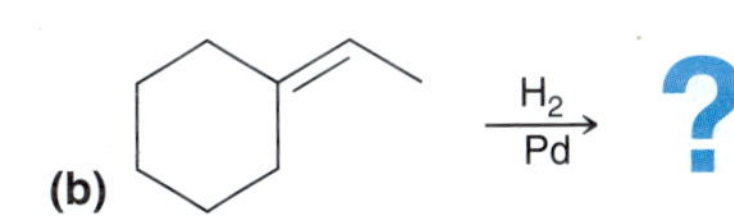


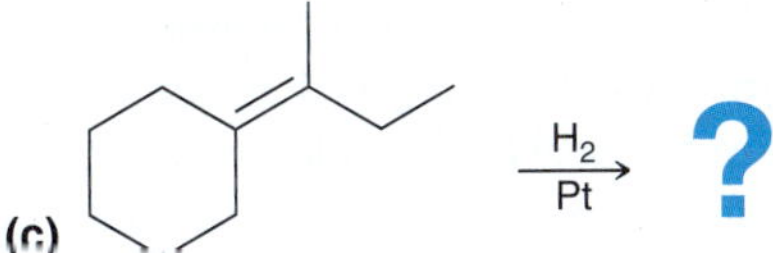


(d) $\xrightarrow[Ni]{H_2}$ ?

(f) $\xrightarrow[Pd]{H_2}$ ?

**APPLY the skill** **9.24** In much the same way that they react with $H_2$, alkenes also react with $D_2$ (deuterium is an isotope of hydrogen). Use this information to predict the product(s) of the following reaction:

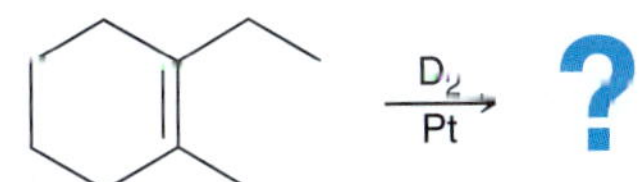


**9.25** Compound X has molecular formula $C_5H_{10}$. In the presence of a metal catalyst, compound X reacts with one equivalent of molecular hydrogen to yield 2-methylbutane.

**(a)** Suggest three possible structures for compound X.

**(b)** Hydroboration-oxidation of compound X yields a product with no chirality centers. Identify the structure of compound X.

need more **PRACTICE?** **Try Problems 9.55, 9.75**

## Homogeneous Catalysts

The catalysts described so far (Pt, Pd, Ni) are all called **heterogeneous catalysts** because they do not dissolve in the reaction medium. In contrast, **homogeneous catalysts** are soluble in the reaction medium. The most common homogeneous catalyst for hydrogenation is called Wilkinson's catalyst:

$(Ph_3P)_3RhCl$

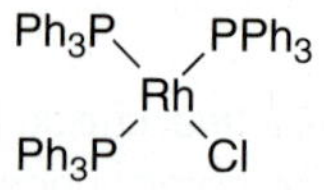

**Wilkinson's catalyst**

With homogeneous catalysts, a *syn* addition is also observed:

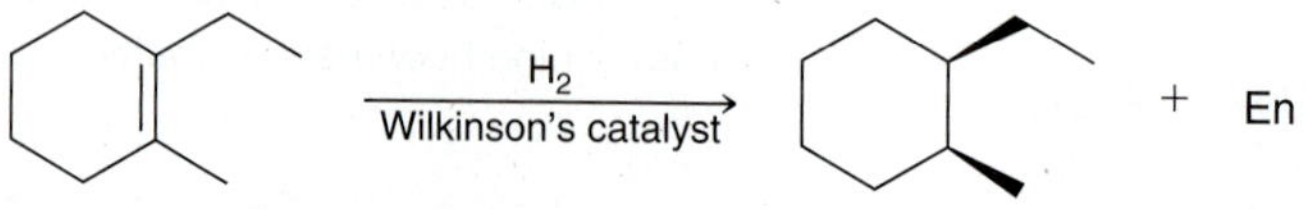

$\xrightarrow[\text{Wilkinson's catalyst}]{H_2}$ + En

## Asymmetric Catalytic Hydrogenation

As seen earlier in this section, when hydrogenation involves the creation of either one or two chirality centers, a pair of enantiomers is expected:

**Creating one chirality center:**

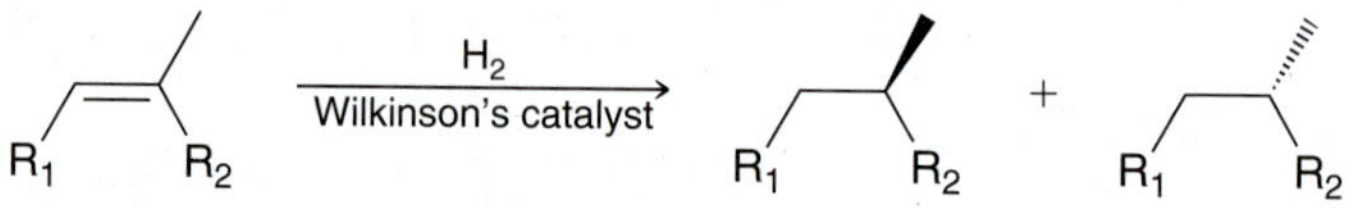

$R_1$ $R_2$ $\xrightarrow[\text{Wilkinson's catalyst}]{H_2}$ $R_1$ $R_2$ + $R_1$ $R_2$

**Creating two chirality centers:**

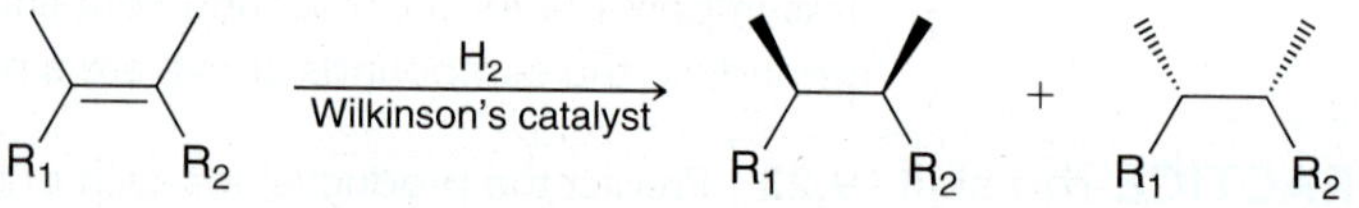

$R_1$ $R_2$ $\xrightarrow[\text{Wilkinson's catalyst}]{H_2}$ $R_1$ $R_2$ + $R_1$ $R_2$

In both reactions a racemic mixture is formed. This begs the obvious question: Is it possible to create only one enantiomer rather than a pair of enantiomers? In other words, is it possible to perform an **asymmetric hydrogenation**?

Before the 1960s, asymmetric catalytic hydrogenation had not been achieved. However, a major breakthrough came in 1968 when William S. Knowles, working for the Monsanto Company, developed a method for asymmetric catalytic hydrogenation. Knowles realized that asymmetric induction might be possible through the use of a chiral catalyst. He reasoned that a chiral catalyst should be capable of lowering the energy of activation for formation of one enantiomer more dramatically than the other enantiomer (Figure 9.8). In this way, a chiral catalyst could theoretically favor the production of one enantiomer over another, leading to an observed enantiomeric excess (*ee*).

Knowles succeeded in developing a chiral catalyst by preparing a cleverly modified version of Wilkinson's catalyst. Recall that Wilkinson's catalyst has three triphenylphosphine ligands:

Wilkinson's catalyst: $Ph_3P$, $PPh_3$, Rh, $Ph_3P$, Cl — Three triphenylphosphine ligands

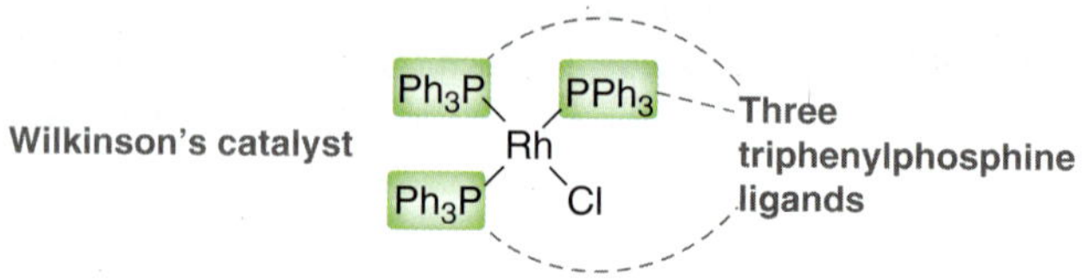

Knowles' idea was to use chiral phosphine ligands, rather than symmetrical phosphine ligands:

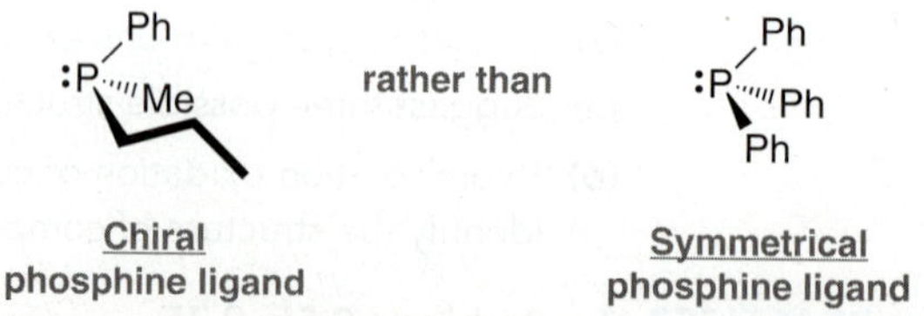

:P(Ph)(Me)(propyl) — **Chiral phosphine ligand** — rather than — :P(Ph)(Ph)(Ph) — **Symmetrical phosphine ligand**

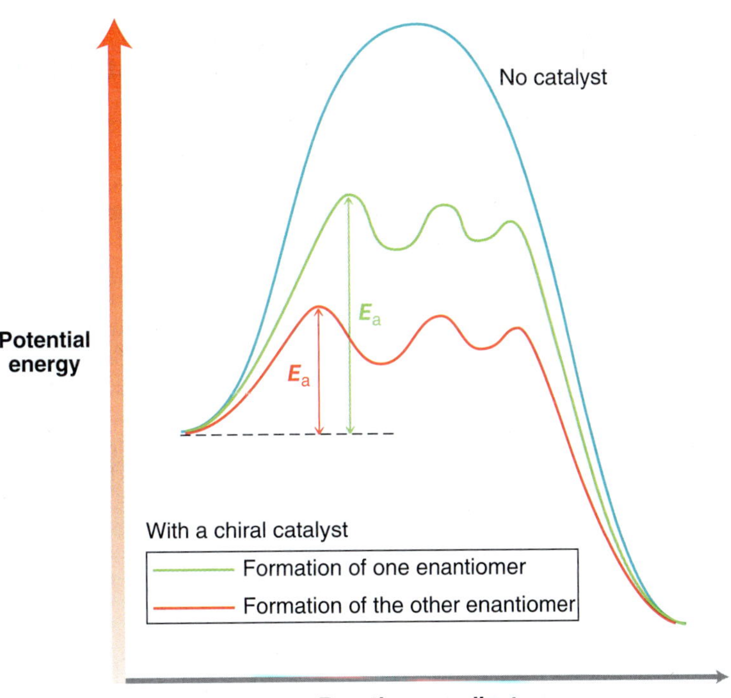


**FIGURE 9.8** An energy diagram showing the ability of a chiral catalyst to favor formation of one enantiomer (red pathway) over the other enantiomer (green pathway)

Using chiral phosphine ligands, Knowles prepared a chiral version of Wilkinson's catalyst. He used his chiral catalyst in a hydrogenation reaction and demonstrated a modest enantiomeric excess. He didn't get exclusively one enantiomer, but he got enough of an enantiomeric excess to prove the point—that asymmetric catalytic hydrogenation was in fact possible.

Knowles developed other catalysts capable of much higher enantiomeric excess, and he then set out to use asymmetric catalytic hydrogenation to develop an industrial synthesis of the amino acid L-dopa:

Asymmetric hydrogenation of this $\pi$ bond

COOH HN O O O O —— $H_2$ / Chiral catalyst ——> (S) COOH HN O O O O —— Several additional steps ——> (S) COOH $NH_2$ HO OH

(S)-3,4-Dihydroxyphenylalanine
(L-dopa)

**LOOKING AHEAD**
The L designation describes the configuration of the chirality center present in the compound. This notation will be described in greater detail in Section 25.2.

L-Dopa had been shown to be effective in treating dopamine deficiencies associated with Parkinson's disease (a discovery that earned Arvid Carlsson the 2000 Nobel Prize in Physiology or Medicine). Dopamine is an important neurotransmitter in the brain. A dopamine deficiency cannot be treated by administering dopamine to the patient, because dopamine cannot cross the blood-brain barrier. However, L-dopa can cross this barrier, and it is subsequently converted into dopamine in the central nervous system. This provides a way to increase the levels of dopamine in the brains of patients with Parkinson's disease, thereby providing temporary relief from some of the symptoms associated with the disease. But there is one catch. The enantiomer of L-dopa is believed to be toxic, and therefore, an enantioselective synthesis of L-dopa is required.

$NH_2$ HO OH

Dopamine

## practically speaking Partially Hydrogenated Fats and Oils

Earlier in this chapter, we focused on the role of hydrogenation in the pharmaceutical industry. However, the best known role of hydrogenation is in the food industry, where hydrogenation is used to prepare partially hydrogenated fats and oils.

Naturally occurring fats and oils, such as vegetable oil, are generally mixtures of compounds called *triglycerides*, which contain three long alkyl chains:

**A triglyceride**

These important compounds are discussed in more detail in Section 26.3. For now, we focus our attention on the π bonds present in the alkyl chains. Hydrogenation of some of these π bonds alters the physical properties of the oil. For example, cottonseed oil is a liquid at room temperature; however, *partially hydrogenated* cottonseed oil (Crisco) is a solid at room temperature. This is advantageous because it gives the oil a longer shelf-life. Margarine is prepared in a similar way from a variety of animal and vegetable oils.

Partially hydrogenated oils do have issues, though. There is much evidence that the catalysts present during the hydrogenation process can often isomerize some of the double bonds, producing *trans* double bonds:

$H_2$ catalyst

*trans*

These so-called *trans* fats are believed to cause an increase in LDL (low-density lipoprotein) cholesterol levels, which leads to an increased rate of cardiovascular disease. In response, the food industry is now making an effort to minimize or completely remove the *trans* fats in food products. Current food labels often boast: "0 grams of *trans* fats."

### LOOKING AHEAD

For a discussion of the connection between cholesterol and cardiovascular disease, see the Medically Speaking box in Section 26.5.

### BY THE WAY

The other half of the 2001 Nobel Prize in Chemistry was awarded to K. Barry Sharpless, for his work on enantioselective synthesis, described in Section 14.9.

Asymmetric catalytic hydrogenation turned out to be an efficient way to prepare L-dopa with high optical purity. For his work, Knowles shared half of the 2001 Nobel Prize in Chemistry, together with Ryoji Noyori (Nagoya University, Japan), who was independently investigating a wide variety of chiral catalysts that could produce an asymmetric catalytic hydrogenation.

Noyori varied both the metal and the ligands attached to the metal, and he was able to create chiral catalysts that achieved enantioselectivity close to 100% *ee*. One example, commonly used in synthesis these days, is based on the chiral ligand called BINAP:

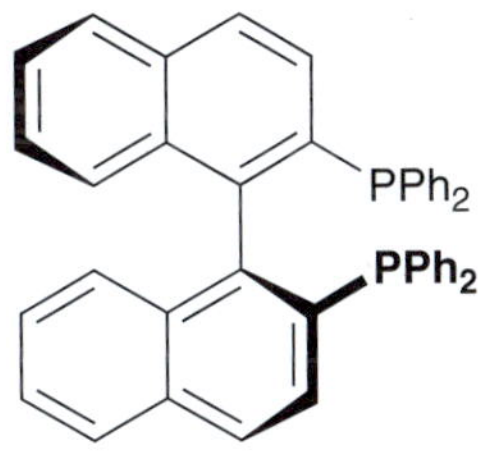

(S)-(−)-BINAP
(S)-2,2′-Bis(diphenylphosphino)-1,1′-binaphthyl

BINAP does not have a chirality center, but nevertheless, it is a chiral compound because the single bond joining the two ring systems does not experience free rotation (as a result of steric hindrance). BINAP can be used as a chiral ligand to form a complex with ruthenium, producing a chiral catalyst capable of achieving very pronounced enantioselectivity:

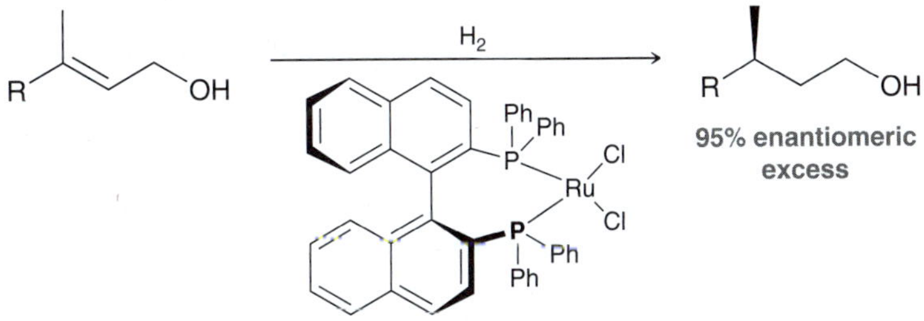

$H_2$ / Ru(BINAP)$Cl_2$ — 95% enantiomeric excess

## 9.8 Halogenation and Halohydrin Formation

### Experimental Observations

**Halogenation** involves the addition of $X_2$ (either $Br_2$ or $Cl_2$) across an alkene. As an example, consider the chlorination of ethylene to produce dichloroethane:

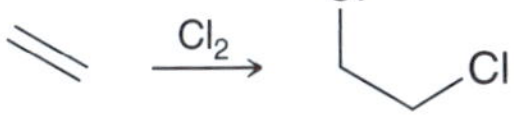

$Cl_2$ (97%)

This reaction is a key step in the industrial preparation of polyvinylchloride (PVC):

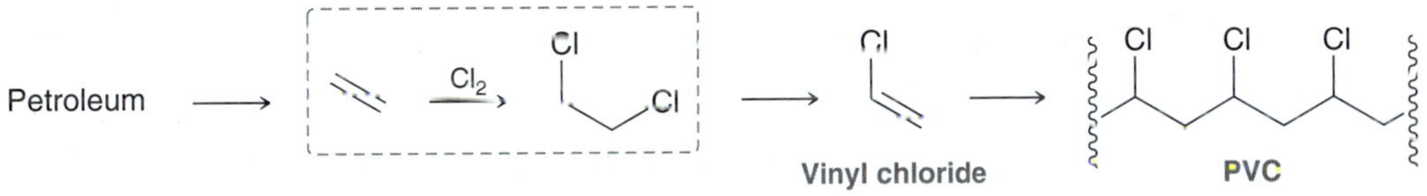

Petroleum → $Cl_2$ → Vinyl chloride → PVC

Halogenation of alkenes is only practical for the addition of chlorine or bromine. The reaction with fluorine is too violent, and the reaction with iodine often produces very low yields.

The stereospecificity of halogenation reactions can be explored in a case where two new chirality centers are formed. For example, consider the products that are formed when cyclopentene is treated with molecular bromine ($Br_2$):

**WATCH OUT**
The products of this reaction are a pair of enantiomers. They are not the same compound. Students are often confused with this particular example. For a review of enantiomers, go to Section 5.5.

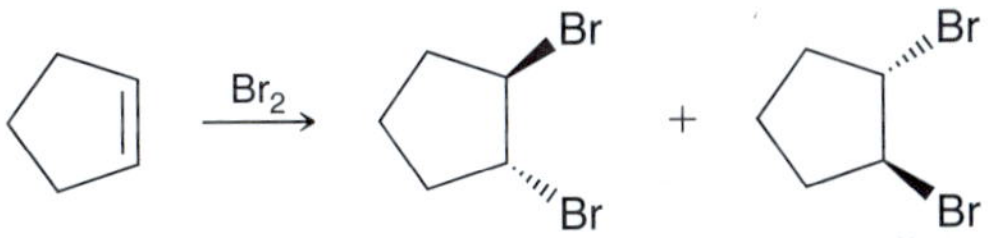

$Br_2$ → + (Br, Br enantiomers)

Notice that addition occurs in a way that places the two halogen atoms on opposite sides of the π bond. This mode of addition is called an ***anti* addition**. *For most simple alkenes, halogenation appears to proceed primarily via an anti addition.* Any proposed mechanism must be consistent with this observation.

## A Mechanism for Halogenation

Molecular bromine is a nonpolar compound, because the Br—Br bond is covalent. Nevertheless, the molecule is polarizable, and the proximity of a nucleophile can cause a temporary, induced dipole moment (Figure 9.9).

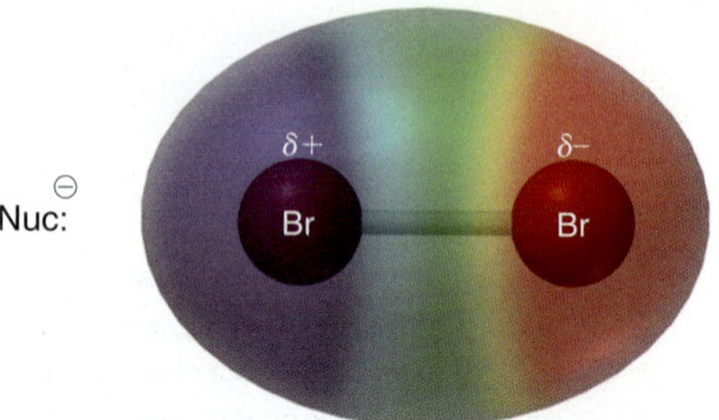


**FIGURE 9.9** An electrostatic potential map showing the temporary induced dipole moment of molecular bromine when it is in the vicinity of a nucleophile.

This effect places a partial positive charge on one of the bromine atoms, rendering the molecule electrophilic. Many nucleophiles are known to react with bromine:

$$\text{Nuc}:^{\ominus} \; + \; \overset{\delta+}{:\text{Br}}-\overset{\delta-}{\text{Br}:}$$

We have seen that π bonds are nucleophilic, and therefore, it is reasonable to expect an alkene to attack bromine as well:

$$\text{alkene} \; + \; \overset{\delta+}{:\text{Br}}-\overset{\delta-}{\text{Br}:} \longrightarrow \text{carbocation (with Br)} \; + \; :\text{Br}:^{\ominus}$$

Although this step seems plausible, there is a fatal flaw in this proposal. Specifically, the production of a free carbocation is inconsistent with the observed *anti* stereospecificity of halogenation. If a free carbocation were produced in the process, then both *syn* and *anti* addition would be expected to occur, because the carbocation could be attacked from either side:

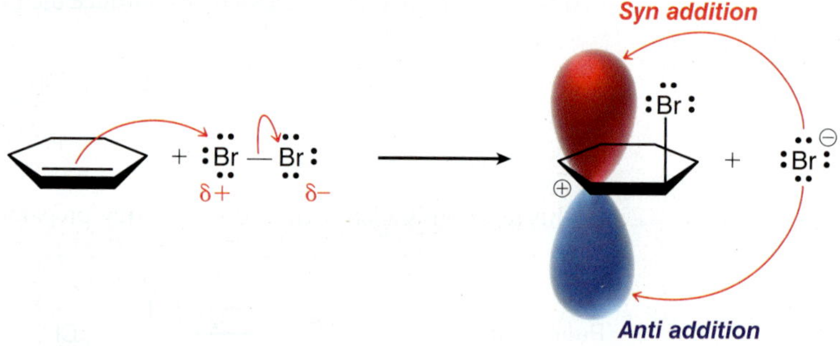


This mechanism does not account for the observed *anti* stereospecificity of halogenation. The modified mechanism shown in Mechanism 9.4 is consistent with an *anti* addition:

### MECHANISM 9.4 HALOGENATION

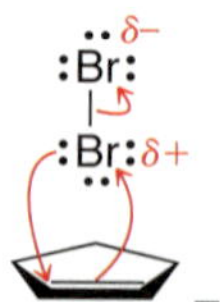


Nucleophilic attack + Loss of a leaving group

The alkene functions as a nucleophile and attacks molecular bromine, expelling bromide as a leaving group and forming a bridged intermediate, called a bromonium ion

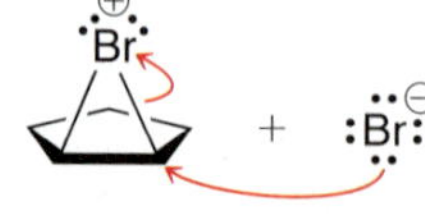


Bromonium ion

Nucleophilic attack

Bromide functions as a nucleophile and attacks the bromonium ion in an $S_N2$ process

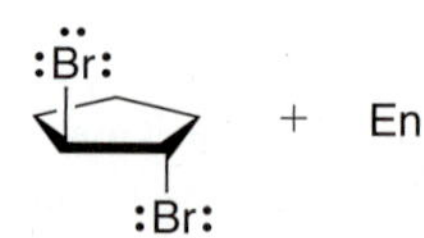


+ En

In this mechanism, an additional curved arrow has been introduced, forming a bridged intermediate rather than a free carbocation. This bridged intermediate, called a **bromonium ion**, is similar in structure and reactivity to the mercurinium ion discussed in Section 9.5. Compare their structures:

**Bromonium ion** **Mercurinium ion**

In the second step of the proposed mechanism, the bromonium ion is attacked by the bromide ion that was produced in the first step. This step is an $S_N2$ process and must therefore proceed via a back-side attack (as seen in Section 7.4). The requirement for back-side attack explains the observed stereochemical requirement for *anti* addition.

The stereochemical outcome for halogenation reactions is dependent on the configuration of the starting alkene. For example, *cis*-2-butene will yield different products than *trans*-2-butene:

$Br_2$ → Br Br + Br Br

***cis*-2-Butene**

$Br_2$ → Br Br *Meso* (= Br Br)

***trans*-2-Butene**

*Anti* addition across *cis*-2-butene leads to a pair of enantiomers, while *anti* addition across *trans*-2-butene leads to a *meso* compound. These examples illustrate that *the configuration of the starting alkene determines the configuration of the product for halogenation reactions.*

## CONCEPTUAL CHECKPOINT

**9.26** Predict the major product(s) for each of the following reactions:

(a) $Br_2$ → ? (b) $Br_2$ → ? (c) $Br_2$ → ? (d) $Br_2$ → ?

### Halohydrin Formation

When bromination occurs in a non-nucleophilic solvent, such as $CHCl_3$, the result is the addition of $Br_2$ across the π bond (as seen in the previous sections). However, when the reaction is performed in the presence of water, the bromonium ion that is initially formed can be captured by a water molecule, rather than bromide:

**Bromonium ion** H–O–H → :Br: + En

The intermediate bromonium ion is a high-energy intermediate and will react with any nucleophile that it encounters. When water is the solvent, it is more likely that the bromonium ion will be captured by a water molecule before having a chance to react with a bromide ion (although some dibromide product is likely to be formed as well). The resulting oxonium ion is then deprotonated to give the product (Mechanism 9.5).

## MECHANISM 9.5 HALOHYDRIN FORMATION

Nucleophilic attack + loss of a leaving group

The alkene attacks $Br_2$, expelling bromide as a leaving group and forming a bromonium ion

Bromonium ion

Nucleophilic attack

Water functions as a nucleophile and attacks the bromonium ion in an $S_N2$ process

Proton transfer

Water serves as a base and deprotonates the oxonium ion

The net result is the addition of Br and OH across the alkene. The product is called a **bromohydrin**. When chlorine is used in the presence of water, the product is called a **chlorohydrin**:

A chlorohydrin

These reactions are generally referred to as **halohydrin formation**.

## Regiochemistry of Halohydrin Formation

In most cases, halohydrin formation is observed to be a regioselective process. Specifically, the OH is generally positioned at the more substituted position:

The proposed mechanism for halohydrin formation can justify the observed regioselectivity. Recall that in the second step of the mechanism the bromonium ion is captured by a water molecule:

Bromonium ion is captured by water

Bromonium ion

Focus carefully on the position of the positive charge throughout the reaction. Think of the positive charge as a hole (or more accurately, a site of electron deficiency) that is passed from one place to

another. It begins on the bromine atom and ends up on the oxygen atom. In order to do so, the positive charge must pass through a carbon atom in the transition state:

In other words, the transition state for this step will bear partial carbocationic character. This explains why the water molecule is observed to attack the more substituted carbon. The more substituted carbon is more capable of stabilizing the partial positive charge in the transition state. As a result, the transition state will be lower in energy when the attack takes place at the more substituted carbon atom. The proposed mechanism is therefore consistent with the observed regioselectivity of halohydrin formation.

# SKILLBUILDER

## 9.6 PREDICTING THE PRODUCTS OF HALOHYDRIN FORMATION

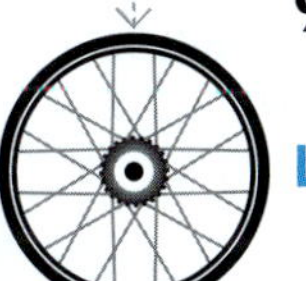

**LEARN** the skill

Predict the major product(s) for the following reaction:

**SOLUTION**

The presence of water indicates halohydrin formation (addition of Br and OH). The first step is to identify the regiochemical outcome. Recall that the OH group is expected to be positioned at the more substituted carbon:

**STEP 1**
Determine the regiochemical outcome. The OH group should be placed at the more substituted carbon.

OH group is placed here at the more substituted position

The next step is to identify the stereochemical outcome. In this case, two new chirality centers are formed, so we expect only the pair of enantiomers that would result from *anti* addition. That is, the OH and Br will be installed on opposite sides of the $\pi$ bond:

**STEP 2**
Determine the stereochemical outcome based on the requirement for *anti* addition.

When drawing the products of halohydrin formation, make sure to consider both the regiochemical outcome and the stereochemical outcome. It is not possible to draw the products correctly without considering both of these issues.

**PRACTICE the skill** **9.27** Predict the major product(s) that are expected when each of the following alkenes is treated with $Br_2/H_2O$:

(a) 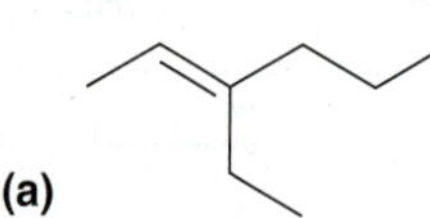

(b) 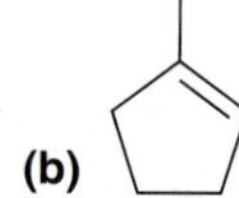

(c) 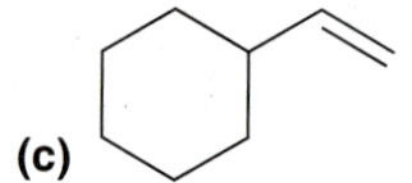

(d) 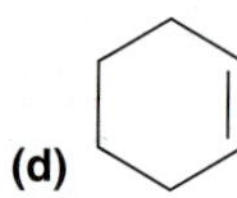

**APPLY the skill** **9.28** Bromonium ions can be captured by nucleophiles other than water. Predict the products of each of the following reactions:

(a) $Br_2$, OH → ?  (b) $Br_2$, $EtNH_2$ → ?

**9.29** When *trans*-1-phenylpropene is treated with bromine, some *syn* addition is observed. Explain why the presence of a phenyl group causes a loss of stereospecificity.

$Br_2$ Br Br + En + Br Br + En

***trans*-1-Phenylpropene**

***Anti* addition products (83%)**

***Syn* addition products (17%)**

need more **PRACTICE?** Try Problem 9.73

## 9.9 *Anti* Dihydroxylation

As mentioned in the introductory section of this chapter, **dihydroxylation** reactions are characterized by the addition of OH and OH across an alkene. As an example, consider the dihydroxylation of ethylene to produce ethylene glycol:

**Dihydroxylation** HO OH

**Ethylene** **Ethylene glycol**

There are a number of reagents well suited to carry out this transformation. Some reagents provide for an *anti* dihydroxylation, while others provide for a *syn* dihydroxylation. In this section, we will explore a two-step procedure for achieving *anti* dihydroxylation:

$RCO_3H$ O $H_3O^+$ OH OH + En

**An epoxide** **A *trans* diol**

The first step of the process involves conversion of the alkene into an epoxide, and the second step involves opening the epoxide to form a *trans* diol (Mechanism 9.6). An **epoxide** is a three-membered, cyclic ether.

## MECHANISM 9.6 *ANTI* DIHYDROXYLATION

### Formation of an epoxide

Transition state

An epoxide

### Acid-catalyzed opening of an epoxide

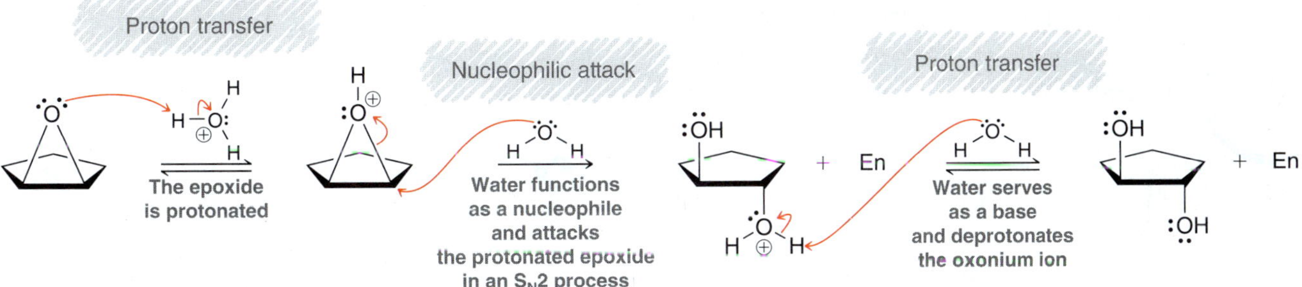


In the first part of the process, a peroxy acid ($RCO_3H$) reacts with the alkene to form an epoxide. Peroxy acids resemble carboxylic acids in structure, possessing just one additional oxygen atom. Two common peroxy acids are shown below:

Peroxyacetic acid

*meta*-Chloroperoxybenzoic acid (MCPBA)

Peroxy acids are strong oxidizing agents and are capable of delivering an oxygen atom to an alkene in a single step. The product is an epoxide.

Once the epoxide has been formed, it can then be opened with water under either acid-catalyzed or base-catalyzed conditions. Both sets of conditions are explored and compared in more detail in Section 14.10. For now, we will explore only the acid catalyzed opening of epoxides, as seen in Mechanism 9.6. Under these conditions, the epoxide is first protonated to produce an intermediate that is very similar to a bromonium or mercurinium ion. All three cases involve a three-membered ring bearing a positive charge:

A protonated epoxide    A bromonium ion    A mercurinium ion

**LOOKING AHEAD**

You might be wondering why a weak nucleophile can participate in an $S_N2$ process at a secondary substrate. In fact, this reaction can occur even if the center being attacked is tertiary. This will be explained in Section 14.10.

We have seen that bromonium and mercurinium ions can be attacked by water from the back side. In much the same way, a protonated epoxide can also be attacked by water from the back side, as seen in Mechanism 9.6. The necessity for back-side attack ($S_N2$) explains the observed stereochemical preference for *anti* addition.

In the final step of the mechanism, the oxonium ion is deprotonated to yield a *trans* diol. Once again, notice that water is used to deprotonate (rather than hydroxide) in order to stay consistent with the conditions. In acidic conditions, hydroxide ions are not present in sufficient quantity to participate in the reaction, and therefore, they cannot be used when drawing the mechanism.

## SKILLBUILDER

### 9.7 DRAWING THE PRODUCTS OF *ANTI* DIHYDROXYLATION

**LEARN the skill**

Predict the major product(s) for each of the following reactions:

(a) 1) MCPBA 2) $H_3O^+$ ?  (b) 1) MCPBA 2) $H_3O^+$ ?

**SOLUTION**

**STEP 1** Determine the number of chirality centers formed.

**STEP 2** Determine the stereochemical outcome based on the requirement for *anti* addition.

**(a)** These reagents achieve dihydroxylation, which means that OH and OH will add across the alkene. Since the two groups are identical (OH and OH), we don't need to consider the regiochemistry of this process. However, the stereochemistry must be considered. Begin by determining the number of chirality centers formed. In this case, two new chirality centers are formed, so we expect only the pair of enantiomers that result from an *anti* addition. That is, the OH groups will be added to opposite sides of the $\pi$ bond:

H — 1) MCPBA 2) $H_3O^+$ → HO OH H + HO OH H

**(b)** In this example, only one new chirality center is formed. It is true that the reaction proceeds through an *anti* addition of OH and OH. However, with only one chirality center in the product, the preference for *anti* addition becomes irrelevant. Both possible enantiomers are formed:

1) MCPBA 2) $H_3O^+$ → OH OH + OH OH

**PRACTICE the skill** **9.30** Predict the products that are expected when each of the following alkenes is treated with a peroxy acid (such as MCPBA) followed by aqueous acid:

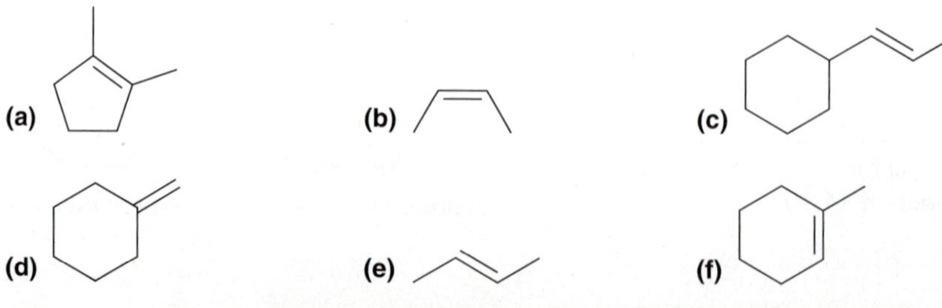

**APPLY** the skill

**9.31** Under acid-catalyzed conditions, epoxides can be opened by a variety of nucleophiles other than water, such as alcohols. In such a case, the nucleophile will generally attack at the more substituted position. Using this information, predict the products for each of the following reactions:

(a) 1) MCPBA; 2) $[H_2SO_4]$, ethanol (OH) → ?

(b) epoxide, phenol (OH), $[H_2SO_4]$ → ?

**9.32** Compound A and compound B both have molecular formula $C_6H_{12}$. Both compounds produce epoxides when treated with MCPBA.

**(a)** The epoxide resulting from compound A was treated with aqueous acid ($H_3O^+$) and the resulting diol had no chirality centers. Propose two possible structures for compound A.

**(b)** The epoxide resulting from compound B was treated with $H_3O^+$ and the resulting diol was a *meso* compound. Draw the structure of compound B.

need more **PRACTICE?** **Try Problem 9.65d**

## 9.10 *Syn* Dihydroxylation

The previous section described a method for *anti* dihydroxylation of alkenes. In this section, we will explore two different sets of reagents that can accomplish *syn* dihydroxylation of alkenes. When an alkene is treated with osmium tetroxide ($OsO_4$), a cyclic osmate ester is produced:

**A cyclic osmate ester**

Osmium tetroxide adds across the alkene in a concerted process. In other words, both oxygen atoms attach to the alkene simultaneously. This effectively adds two groups across the same face of the alkene; hence, the *syn* addition. The cyclic osmate ester that results can be isolated and then treated with either aqueous sodium sulfite ($Na_2SO_3$) or sodium bisulfite ($NaHSO_3$) to produce a diol:

$Na_2SO_3/H_2O$ or $NaHSO_3/H_2O$

This method can be used to convert alkenes into diols with fairly high yields, but there are several disadvantages. In particular, $OsO_4$ is expensive and toxic. To deal with these issues, several methods have been developed that use a co-oxidant that serves to regenerate $OsO_4$ as it is consumed during the reaction. In this way, $OsO_4$ functions as a catalyst so that even small quantities can produce large quantities of the diol. Typical co-oxidants include *N*-methylmorpholine *N*-oxide (NMO) and *tert*-butyl hydroperoxide:

$OsO_4$, (NMO) or $OsO_4$, tert-butyl hydroperoxide (OOH) → cis-diol **(60–90%)**

A different, although mechanistically similar, method for achieving *syn* dihydroxylation involves treatment of alkenes with cold potassium permanganate under basic conditions:

Not isolated

NaOH

(85%)

Once again, a concerted process adds both oxygen atoms simultaneously across the double bond. Notice the similarity between the mechanisms of these two methods ($OsO_4$ vs. $KMnO_4$).

Potassium permanganate is fairly inexpensive; however, it is a very strong oxidizing agent, and it often causes further oxidation of the diol. Therefore, synthetic organic chemists often choose to use $OsO_4$ together with a co-oxidant to achieve *syn* dihydroxylation.

## CONCEPTUAL CHECKPOINT

**9.33** Predict the product(s) for each of the following reactions. In each case, make sure to consider the number of chirality centers being formed.

(a) $OsO_4$ (catalytic) / NMO → ?

(b) 1) $OsO_4$ 2) $NaHSO_3$ / $H_2O$ → ?

(c) $KMnO_4$, NaOH / Cold → ?

(d) $KMnO_4$, NaOH / Cold → ?

(e) $OsO_4$ (catalytic) / *tert*-BuOOH → ?

(f) $OsO_4$ (catalytic) / NMO

## 9.11 Oxidative Cleavage

There are many reagents that will add across an alkene and completely cleave the C=C bond. In this section, we will explore one such reaction, called **ozonolysis**. Consider the following example:

1) $O_3$ 2) DMS

Notice that the C=C bond is completely split apart to form two C=O double bonds. Therefore, issues of stereochemistry and regiochemistry become irrelevant. In order to understand how this reaction occurs, we must first explore the structure of ozone.

Ozone is a compound with the following resonance structures:

Ozone is formed primarily in the upper atmosphere where oxygen gas ($O_2$) is bombarded with ultraviolet light. The ozone layer in our atmosphere serves to protect us from harmful UV radiation from the sun. Ozone can also be prepared in the laboratory, where it can serve a useful purpose. Ozone will react with an alkene to produce an initial, primary ozonide (or molozonide), which undergoes rearrangement to produce a more stable ozonide:

**LOOKING AHEAD**
The role of ozone in atmospheric chemistry is discussed in Section 11.8.

Initial ozonide (molozonide)

Ozonide

When treated with a mild reducing agent, the ozonide is converted into products:

Mild reducing agent

Ozonide

Common examples of reducing agents include dimethyl sulfide (DMS) or $Zn/H_2O$:

1) $O_3$
2) DMS
*or* $Zn/H_2O$

The following worked example illustrates a simple method for drawing the products of ozonolysis.

# SKILLBUILDER

## 9.8 PREDICTING THE PRODUCTS OF OZONOLYSIS

**LEARN the skill**

Predict the products of the following reaction:

1) $O_3$
2) DMS

?

### SOLUTION

**STEP 1**
Redraw the C=C double bonds longer than normal.

There are two C=C double bonds in this compound. Begin by redrawing the compound with the C=C double bonds longer than normal:

**STEP 2**
Erase the center of each C=C double bond and place two oxygen atoms in the space.

Then, erase the center of each C=C double bond and place two oxygen atoms in the space:

This simple procedure can be used for quickly drawing the products of any ozonolysis reaction.

**PRACTICE the skill** **9.34** Predict the products that are expected when each of the following alkenes is treated with ozone followed by DMS:

(a) 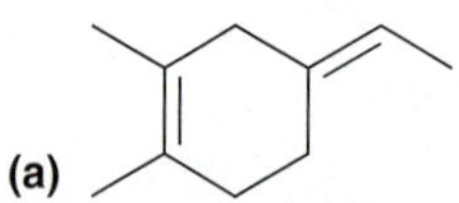 (b) 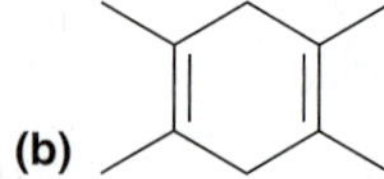 (c)

(d) 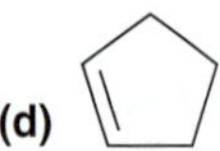 (e)  (f)

**APPLY the skill** **9.35** Identify the structure of the starting alkene in each of the following cases:

(a) $C_8H_{14}$ $\xrightarrow[\text{2) DMS}]{\text{1) } O_3}$

(b) $C_{10}H_{16}$ $\xrightarrow[\text{2) DMS}]{\text{1) } O_3}$

(c) $C_{10}H_{16}$ $\xrightarrow[\text{2) DMS}]{\text{1) } O_3}$

need more **PRACTICE?** **Try Problems 9.68, 9.77**

## 9.12 Predicting the Products of an Addition Reaction

Many addition reactions have been covered in this chapter, and in each case, there are several factors to take into account when predicting the products of a reaction. Let's now summarize the factors that are common to all addition reactions: In order to predict products properly, the following three questions must be considered:

1. What are the identities of the groups being added across the double bond?
2. What is the expected regioselectivity (Markovnikov or *anti*-Markovnikov addition)?
3. What is the expected stereospecificity (*syn* or *anti* addition)?

Answering all three questions requires a careful analysis of both the starting alkene and the reagents employed. It is absolutely essential to recognize reagents, and while that might sound like it involves a lot of memorization, it actually does not. By understanding the proposed mechanism for each reaction, you will intuitively understand all three pieces of information for each reaction. Remember that a proposed mechanism must explain the experimental observations. Therefore, the mechanism for each reaction can serve as the key to remembering the three pieces of information listed above.

## SKILLBUILDER

### 9.9 PREDICTING THE PRODUCTS OF AN ADDITION REACTION

**LEARN the skill** Predict the products of the following reaction:

$\xrightarrow[\text{2) } H_2O_2\text{, NaOH}]{\text{1) } BH_3 \cdot \text{THF}}$ ?

## SOLUTION

In order to predict the products, the following three questions must be answered:

**STEP 1**
Identify the two groups being added across the π bond.

1. ***Which two groups are being added across the double bond?*** To answer this question, it is necessary to recognize that these reagents achieve hydroboration-oxidation, which adds H and OH across a π bond. It would be impossible to solve this problem without being able to recognize the reagents. But even with that recognition, there are still two more questions that must be answered in order to arrive at the correct answer.

2. ***What is the expected regioselectivity (Markovnikov or anti-Markovnikov)?*** We have seen that hydroboration-oxidation is an *anti*-Markovnikov process. Think about the mechanism of this process and recall that there were two different explanations for the observed regioselectivity, one based on a steric argument and the other based on an electronic argument. An *anti*-Markovnikov addition means that the OH group is placed at the less substituted carbon:

**STEP 2**
Identify the expected regioselectivity.

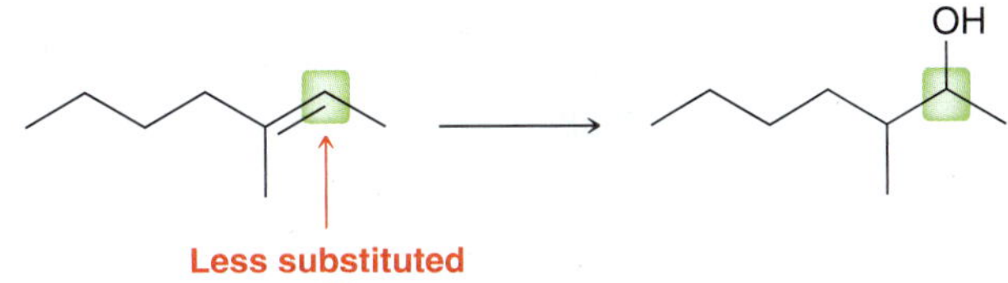


**STEP 3**
Identify the expected stereospecificity.

3. ***What is the expected stereospecificity (syn or anti)?*** We have seen that hydroboration-oxidation produces a *syn* addition. Think about the mechanism of this process and recall the reason that was given for *syn* addition. In the first step, $BH_2$ and H are added to the same face of the alkene in a concerted process. To determine if this *syn* requirement is even relevant in this case, we must analyze how many chirality centers are being formed in the process:

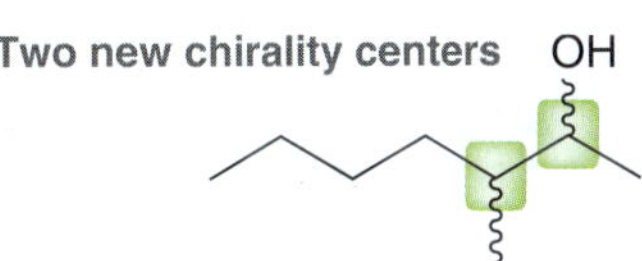


Two new chirality centers are formed, and therefore, the requirement for *syn* addition is relevant. The reaction is expected to produce only the pair of enantiomers that would result from a *syn* addition:

1) $BH_3 \cdot THF$
2) $H_2O_2$, NaOH
H, OH, Me, H + En

**PRACTICE the skill** **9.36** Predict the products of each of the following reactions:

(a) 1) $BH_3 \cdot THF$ 2) $H_2O_2$, NaOH → ?

(b) $H_2$, Pt → ?

(c) 1) $CH_3CO_3H$ 2) $H_3O^+$ → ?

(d) 1) $OsO_4$ 2) $NaHSO_3/H_2O$ → ?

(e) $H_3O^+$ → ?

(f) HBr → ?

(g) 1) MCPBA 2) $H_3O^+$ ?

(h) 1) $BH_3 \cdot THF$ 2) $H_2O_2$, NaOH ?

(i) $OsO_4$ (catalytic) NMO ?

**APPLY the skill**

**9.37** *Syn* dihydroxylation of the compound below yields two products. Draw both products and describe their stereoisomeric relationship (i.e., are they enantiomers or diastereomers?):

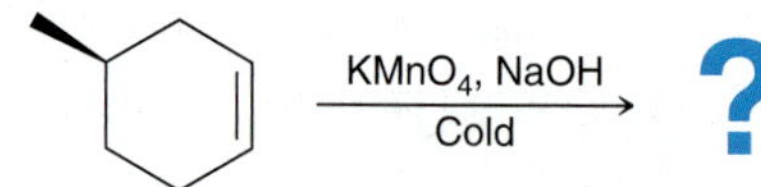


**9.38** Determine whether *syn* dihydroxylation of *trans*-2-butene will yield the same products as *anti* dihydroxylation of *cis*-2-butene. Draw the products in each case and compare them.

**9.39** Compound A has molecular formula $C_5H_{10}$. Hydroboration-oxidation of compound A produces a pair of enantiomers, compounds B and C. When treated with HBr, compound A is converted into compound D, which is a tertiary alkyl bromide. When treated with $O_3$ followed by DMS, compound A is converted into compounds E and F. Compound E has three carbon atoms, while compound F has only two carbon atoms. Identify the structures of compounds A, B, C, D, E, and F.

need more **PRACTICE?** **Try Problems 9.49, 9.50, 9.65, 9.73**

## 9.13 Synthesis Strategies

In order to begin practicing synthesis problems, it is essential to master all of the individual reactions covered thus far. We will begin with one-step synthesis problems and then progress to cover multistep problems.

### One-Step Syntheses

Until now, we have covered substitution reactions ($S_N1$ and $S_N2$), elimination reactions (E1 and E2), and addition reactions of alkenes. Let's quickly review what these reactions can accomplish. *Substitution reactions* convert one group into another:

X → Y

*Elimination reactions* can be used to convert alkyl halides or alcohols into alkenes:

X →

*Addition reactions* are characterized by two groups adding across a double bond:

→ X Y

It is essential to familiarize yourself with the reagents employed for each type of addition reaction covered in this chapter.

# SKILLBUILDER

## 9.10 PROPOSING A ONE-STEP SYNTHESIS

**LEARN** the skill

Identify the reagents that you would use to accomplish the following transformation:

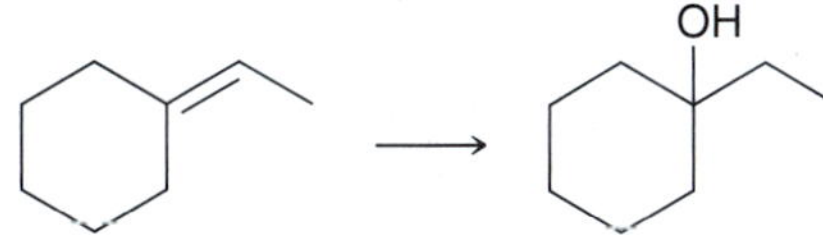


**STEP 1**
Identify the two groups being added across the π bond.

**STEP 2**
Identify the regioselectivity.

**STEP 3**
Identify the stereospecificity.

### SOLUTION

We approach this problem using the same three questions developed in the previous section:

1. Which two groups are being added across the double bond?—H and OH.
2. What is the regioselectivity?—Markovnikov addition.
3. What is the stereospecificity?—Not relevant (no chirality centers formed).

**STEP 4**
Identify reagents that will achieve the details described in the first three steps.

This transformation requires reagents that will give a Markovnikov addition of H and OH. This can be accomplished in either of two ways: acid-catalyzed hydration or oxymercuration demercuration. Acid-catalyzed hydration might be more simple in this case, because there is no chance of a carbocation rearrangement:

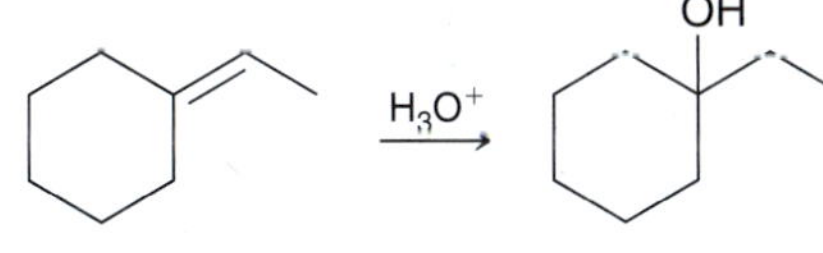


If rearrangement were possible, then oxymercuration-demercuration would have been the preferred route.

**PRACTICE** the skill

**9.40** Identify the reagents that you would use to accomplish each of the following transformations:

(a) + En (b) (c) (d) (e) (f) + En (g) (h)

**APPLY** the skill

**9.41** Identify what reagents you would use to achieve each transformation:

**(a)** Conversion of 2-methyl-2-butene into a secondary alkyl halide

**(b)** Conversion of 2-methyl-2-butene into a tertiary alkyl halide

**(c)** Conversion of *cis*-2-butene into a *meso* diol

**(d)** Conversion of *cis*-2-butene into enantiomeric diols

need more **PRACTICE?** **Try Problems 9.60, 9.62, 9.70, 9.71**

## Changing the Position of a Leaving Group

Now let's get some practice combining the reactions covered thus far. As an example, consider the following transformation:

Br → Br

The net result is the change in position of the Br atom. How can this type of transformation be achieved? This chapter did not present a one-step method for moving the location of a bromine atom. However, this transformation can be accomplished in two steps: an elimination reaction followed by an addition reaction:

Br —Elimination→ —Addition→ Br

When performing this two-step sequence, there are a few important issues to keep in mind. In the first step (elimination), the product can be the more substituted alkene (Zaitsev product) or the less substituted alkene (Hofmann product):

NaOMe → **Zaitsev**

Br

*t*-BuOK → **Hofmann**

Note that a careful choice of reagents makes it possible to control the regiochemical outcome, as we first saw in Section 8.7. With a strong base, such as sodium methoxide (NaOMe) or sodium ethoxide (NaOEt), the product is the more substituted alkene. With a strong, sterically hindered base, such as potassium *tert*-butoxide (*t*-BuOK), the product is the less substituted alkene. After forming the double bond, the regiochemical outcome of the next step (addition of HBr) can also be controlled by the reagents used. HBr produces a Markovnikov addition while HBr/ROOR produces an *anti*-Markovnikov addition.

When changing the location of a leaving group, make sure to remember that OH is a terrible leaving group. Let's see what to do when dealing with an OH group. As an example, consider how the following transformation might be achieved:

OH → OH

Our strategy would suggest the following steps:

OH —Elimination→ —Addition of H and OH→ OH

However, this sequence presents a serious obstacle in that the first step is an elimination reaction in which OH would be the leaving group. It is possible to protonate an OH group using concentrated acid, which converts a bad leaving group into an excellent leaving group (see Section 7.6). However, that reaction cannot be used here because it is an E1 process, which always yields the

Zaitsev product, not the Hofmann product. The regiochemical outcome of an E1 process cannot be controlled. The regiochemical outcome of an E2 process can be controlled, but an E2 process cannot be used in this example, because OH is a bad leaving group. It is not possible to protonate the OH group with a strong acid and then use a strong base to achieve an E2 reaction, because when mixed together, a strong base and a strong acid will simply neutralize each other. The question remains: How can an E2 process be performed when the desired leaving group is an OH group?

In Chapter 7, we explored a method that would allow an E2 process in this example, which maintains control over the regiochemical outcome. The OH group can first be converted into a tosylate, which is a much better leaving group than OH (for a review of tosylates, see Section 7.8). After the OH is converted into a tosylate, the strategy outlined above can be followed. Specifically, a strong, sterically hindered base is used for the elimination reaction, followed by *anti*-Markovnikov addition of H—OH:

Conversion of OH into a good leaving group — TsCl, pyridine (OH → OTs); Elimination — *t*-BuOK; Addition — 1) $BH_3 \cdot THF$ 2) $H_2O_2$, NaOH (→ OH)

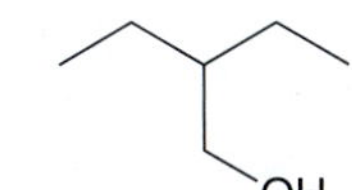

# SKILLBUILDER

## 9.11 CHANGING THE POSITION OF A LEAVING GROUP

**LEARN** the skill

Identify the reagents you would use to accomplish the following transformation:

Br → Br

**SOLUTION**

This problem shows the Br changing its position. This can be accomplished with a two-step process: (1) elimination to form a double bond followed by (2) addition across that double bond:

Br —Elimination→ —Addition→ Br

**STEP 1**
Control the regiochemical outcome of elimination by choosing the appropriate base.

**STEP 2**
Control the regiochemical outcome of addition by choosing the appropriate reagents.

Care must be taken to control the regiochemical outcome in each of these steps. In the elimination step, the product with the less substituted double bond (i.e., the Hofmann product) is desired, and therefore, a sterically hindered base must be used. In the addition step, the Br must be positioned at the less substituted carbon (*anti*-Markovnikov addition), and therefore, we must use HBr with peroxides. This gives the following overall synthesis:

Br —1) *t*-BuOK 2) HBr, ROOR→ Br

**PRACTICE the skill** **9.42** Identify the reagents you would use to accomplish each of the following transformations:

**APPLY the skill** **9.43** Identify the reagents you would use to achieve each of the following transformations:

**(a)** Convert *tert*-butyl bromide into a primary alkyl halide

**(b)** Convert 2-bromopropane into 1-bromopropane

**9.44** Identify the reagents you would use to accomplish the following transformation:



need more **PRACTICE?** **Try Problems 9.57, 9.58b**

## Changing the Position of a π Bond

In the previous section, we combined two reactions in one synthetic strategy—*eliminate and then add*—which enabled us to change the location of a halogen or hydroxyl group. Now, let's focus on another type of strategy—*add and then eliminate*:



This two-step sequence makes it possible to change the position of a π bond. When using this strategy, the regioselectivity of each step can be carefully controlled. In the first step (addition), Markovnikov addition is achieved by using HBr, while *anti*-Markovnikov addition is achieved by using HBr with peroxides. In the second step (elimination), the Zaitsev product can be obtained by using a strong base, while the Hofmann product can be obtained by using a strong, sterically hindered base.

# SKILLBUILDER

## 9.12 CHANGING THE POSITION OF A π BOND

**LEARN the skill** Identify the reagents you would use to accomplish the following transformation:

### SOLUTION

This example involves moving the position of a π bond. We have not seen a way to accomplish this transformation in one step. However, it can be accomplished in two steps—addition followed by elimination:

**STEP 1** Control the regiochemical outcome of addition by choosing the appropriate reagents.

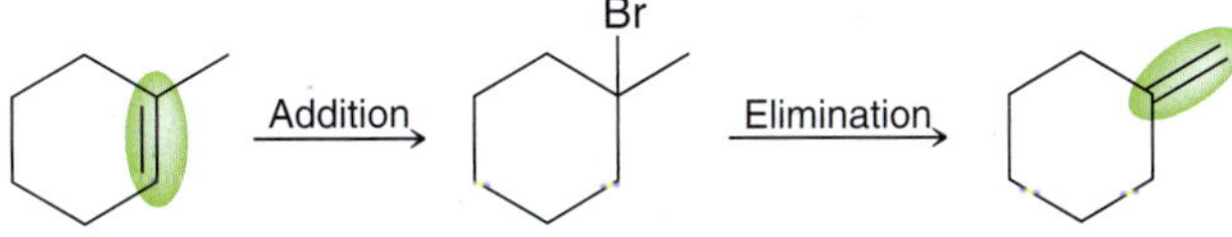


**STEP 2** Control the regiochemical outcome of elimination by choosing the appropriate base.

In the first step, a Markovnikov addition is required (Br must be placed at the more substituted carbon), which can be accomplished by using HBr. The second step requires an elimination to give the Hofmann product, which can be accomplished with a sterically hindered base, such as *tert*-butoxide. Therefore, the overall synthesis is:

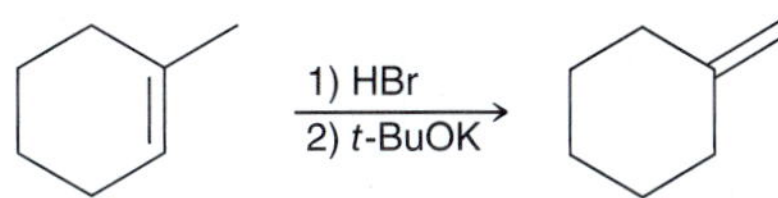

**PRACTICE the skill** **9.45** Identify the reagents you would use to accomplish each of the following transformations:

**(a)** 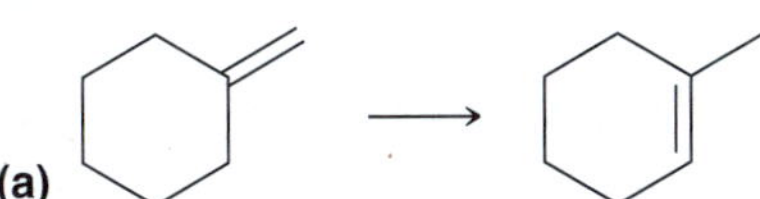

**(b)** 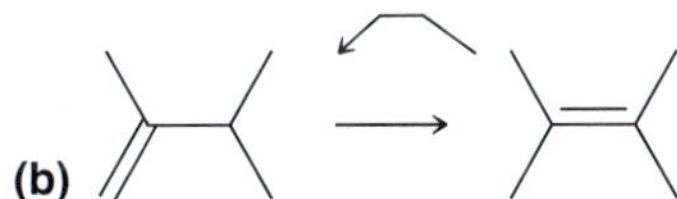

**APPLY the skill** **9.46** Identify the reagents you would use to accomplish each of the following transformations:

**(a)** Convert 2-methyl-2-butene into a monosubstituted alkene

**(b)** Convert 2,3-dimethyl-1-hexene into a tetrasubstituted alkene

**9.47** Identify the reagents you would use to accomplish each of the following transformations:

**(a)** 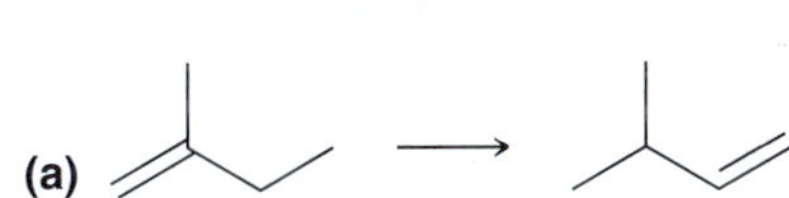

**(b)** 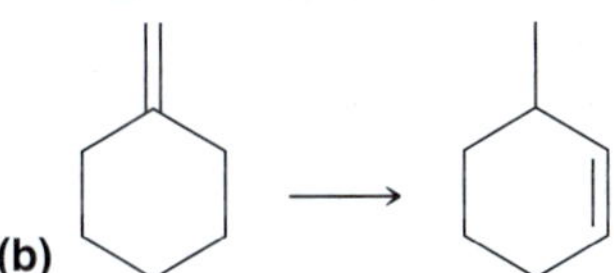

need more **PRACTICE?** **Try Problems 9.53, 9.58a**

## REVIEW OF REACTIONS

**1.** Hydrohalogenation (Markovnikov)

**2.** Hydrohalogenation (*anti*-Markovnikov)

**3.** Acid-catalyzed hydration and oxymercuration-demercuration

**4.** Hydroboration-oxidation

**5.** Hydrogenation

**6.** Bromination

**7.** Halohydrin formation

**8.** *Anti* dihydroxylation

**9.** *Syn* dihydroxylation

**10.** Ozonolysis

## REVIEW OF CONCEPTS AND VOCABULARY

### SECTION 9.1

- **Addition reactions** are characterized by the addition of two groups across a double bond.

### SECTION 9.2

- Addition reactions are thermodynamically favorable at low temperature and disfavored at high temperature.

### SECTION 9.3

- **Hydrohalogenation** reactions are characterized by the addition of H and X across a $\pi$ bond.
- For unsymmetrical alkenes, the placement of the halogen represents an issue of regiochemistry. Hydrohalogenation reactions are *regioselective*, because the halogen is generally placed at the more substituted position, called **Markovnikov addition**.
- In the presence of peroxides, addition of HBr proceeds via an ***anti*-Markovnikov addition**.
- The regioselectivity of an ionic addition reaction is determined by the preference for the reaction to proceed through the more stable carbocation intermediate.
- When one new chirality center is formed, a racemic mixture of enantiomers is obtained.
- Hydrohalogenation reactions are only efficient when carbocation rearrangements are not possible.

SECTION 9.4

- Addition of water (H and OH) across a double bond is called **hydration**.
- Addition of water in the presence of an acid is called **acid-catalyzed hydration**, which generally proceeds via a Markovnikov addition.
- Acid-catalyzed hydration proceeds via a carbocation intermediate, which is attacked by water to produce an **oxonium ion**, followed by deprotonation.
- Acid-catalyzed hydration is inefficient when carbocation rearrangements are possible.
- Dilute acid favors formation of the alcohol, while concentrated acid favors the alkene.
- When generating a new chirality center, a racemic mixture of enantiomers is expected.

SECTION 9.5

- **Oxymercuration-demercuration** achieves hydration of an alkene without carbocation rearrangements.
- The reaction is believed to proceed via a bridged intermediate called a **mercurinium ion**.

SECTION 9.6

- **Hydroboration-oxidation** can be used to achieve an *anti*-Markovnikov addition of water across an alkene. The reaction is stereospecific and proceeds via a ***syn* addition**.
- Borane ($BH_3$) exists in equilibrium with its dimer, **diborane**, which exhibits **three-center, two-electron bonds**.
- Hydroboration proceeds via a concerted process, in which borane is attacked by a π bond, triggering a *simultaneous* hydride shift.

SECTION 9.7

- **Catalytic hydrogenation** involves the addition of $H_2$ across an alkene in the presence of a metal catalyst.
- The reaction proceeds via a *syn* addition.
- **Heterogeneous catalysts** are not soluble in the reaction medium, while **homogeneous catalysts** are.
- **Asymmetric hydrogenation** can be achieved with a chiral catalyst.

SECTION 9.8

- **Halogenation** involves the addition of $X_2$ (either $Br_2$ or $Cl_2$) across an alkene.
- Bromination proceeds via a bridged intermediate, called a **bromonium ion**, which is opened by an $S_N2$ process that produces an ***anti* addition**.
- In the presence of water, the product is a **bromohydrin** or a **chlorohydrin**, and the reaction is called **halohydrin formation**.

SECTION 9.9

- **Dihydroxylation** reactions are characterized by the addition of OH and OH across an alkene.
- A two-step procedure for *anti* dihydroxylation involves conversion of an alkene to an **epoxide**, followed by acid-catalyzed ring opening.

SECTION 9.10

- *Syn* dihydroxylation can be achieved with osmium tetroxide or potassium permanganate.

SECTION 9.11

- **Ozonolysis** can be used to cleave a double bond and produce two carbonyl groups.

SECTION 9.12

- In order to predict products properly, the following three questions must be considered:
  - What is the identity of the groups being added across the double bond?
  - What is the expected regioselectivity (Markovnikov or *anti*-Markovnikov addition)?
  - What is the expected stereospecificity (*syn* or *anti* addition)?

SECTION 9.13

- The position of a leaving group can be changed via elimination followed by addition.
- The position of a π bond can be changed via addition followed by elimination.

# SKILLBUILDER REVIEW

## 9.1 DRAWING A MECHANISM FOR HYDROHALOGENATION

**STEP 1** Using two curved arrows, protonate the alkene to form the more stable carbocation.

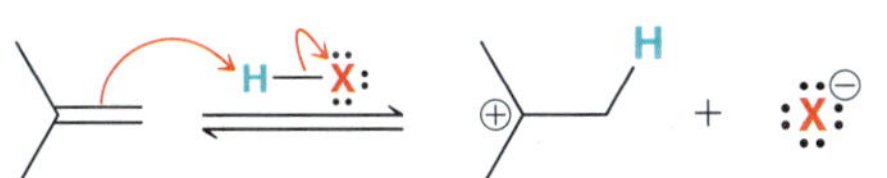


**STEP 2** Using one curved arrow, draw the halide ion attacking the carbocation.

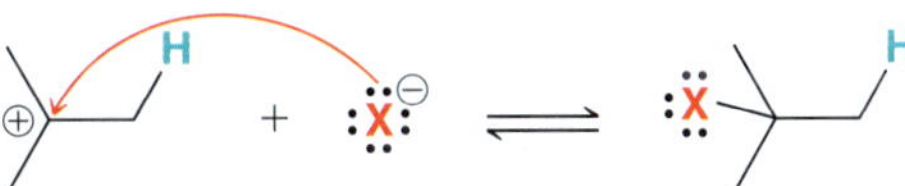


Try Problems 9.3–9.5, 9.51c, 9.64b, 9.69, 9.72

## 9.2 DRAWING A MECHANISM FOR HYDROHALOGENATION WITH A CARBOCATION REARRANGEMENT

**STEP 1** Using two curved arrows, protonate the alkene to form the more stable carbocation.

**STEP 2** Using one curved arrow, draw a carbocation rearrangement that forms a more stable carbocation, via either a hydride shift or a methyl shift.

**STEP 3** Using one curved arrow, draw the halide ion attacking the carbocation.

H—Cl: + :Cl:⊖

Secondary → Tertiary

Try Problems 9.7–9.9, 9.51d, 9.76

## 9.3 DRAWING A MECHANISM FOR AN ACID-CATALYZED HYDRATION

**STEP 1** Using two curved arrows, protonate the alkene to form the more stable carbocation.

**STEP 2** Using one curved arrow, draw water attacking the carbocation.

**STEP 3** Using two curved arrows, deprotonate the oxonium ion using water as a base.

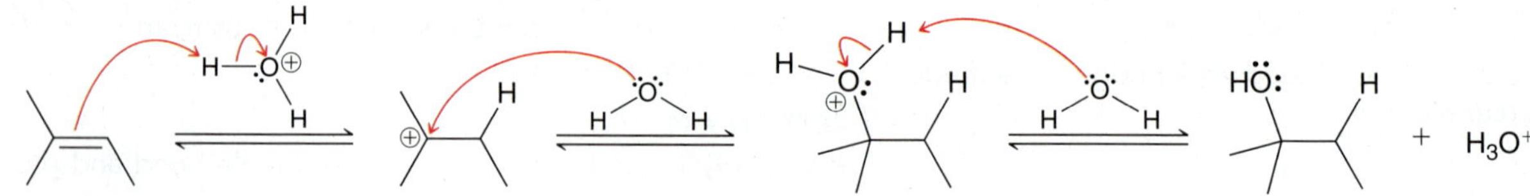


Try Problems 9.12–9.14, 9.51a,b, 9.64a

## 9.4 PREDICTING THE PRODUCTS OF HYDROBORATION-OXIDATION

**STEP 1** Determine the regiochemical outcome based on the requirement for *anti*-Markovnikov addition.

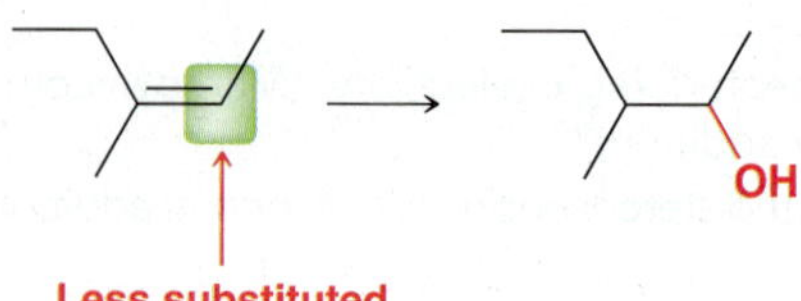


**STEP 2** Determine the stereochemical outcome based on the requirement for *syn* addition.

1) $BH_3 \cdot THF$
2) $H_2O_2$, NaOH

Enantiomers

Try Problems 9.19–9.22, 9.66

## 9.5 PREDICTING THE PRODUCTS OF CATALYTIC HYDROGENATION

**STEP 1** Determine the number of chirality centers formed in the process.

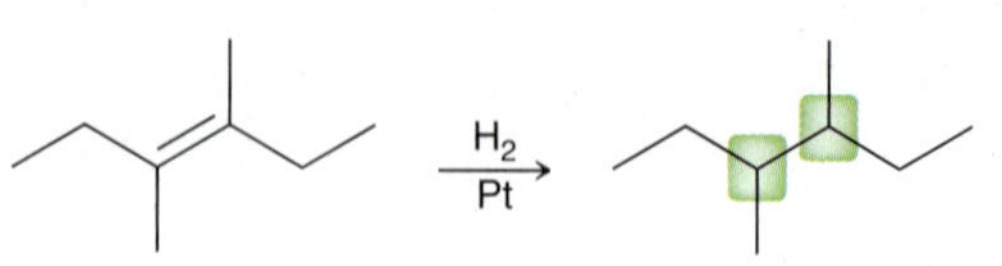


**STEP 2** Determine the stereochemical outcome based on the requirement for *syn* addition.

Enantiomers

**STEP 3** Verify that the products do not represent a single *meso* compound.

Try Problems 9.23–9.25, 9.55, 9.75

## 9.6 PREDICTING THE PRODUCTS OF HALOHYDRIN FORMATION

**STEP 1** Determine the regiochemical outcome. The OH group should be placed at the more substituted position.

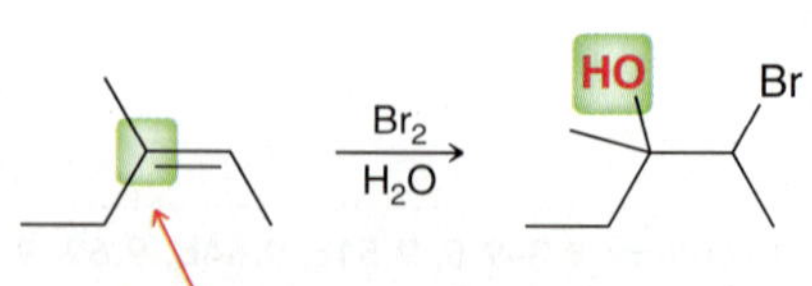


**STEP 2** Determine the stereochemical outcome based on the requirement for *anti* addition.

$Br_2$
$H_2O$

Enantiomers

Try Problems 9.27–9.29, 9.73

### 9.7 DRAWING THE PRODUCTS OF *ANTI* DIHYDROXYLATION

**STEP 1** Determine the number of chirality centers formed in the process.

1) MCPBA
2) $H_3O^+$

HO OH H

**Two stereocenters**

**STEP 2** Determine the stereochemical outcome based on the requirement for *anti* addition.

HO OH H + HO OH H

**Enantiomers**

Try Problems 9.30–9.32, 9.65d

### 9.8 PREDICTING THE PRODUCTS OF OZONOLYSIS

**STEP 1** Redraw the compound with the C═C double bonds longer than normal.

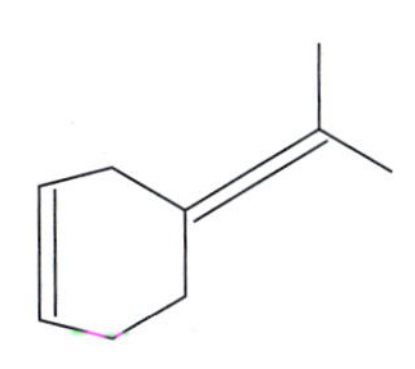

**STEP 2** Erase the center of each C═C double bond and place two oxygen atoms in the space.

Try Problems 9.34, 9.35, 9.68, 9.77

### 9.9 PREDICTING THE PRODUCTS OF AN ADDITION REACTION

**STEP 1** Identify the two groups being added across the π bond.

1) $BH_3 \cdot THF$
2) $H_2O_2$, NaOH

**Adds H and OH**

**STEP 2** Identify the expected regioselectivity.

**OH is placed here at the less substituted position**

**STEP 3** Identify the expected stereospecificity.

H OH Me H + En

Try Problems 9.3–9.39, 9.49, 9.50, 9.65, 9.73

### 9.10 PROPOSING A ONE-STEP SYNTHESIS

**STEP 1** Identify which two groups were added across the π bond.

OH H

**STEP 2** Identify the regioselectivity.

OH H

**Markovnikov addition**

**STEP 3** Identify the stereospecificity.

OH H

**No chirality centers Not relevant**

**STEP 4** Identify reagents that will achieve the details specified in the first three steps.

$H_3O^+$ OH

Try Problems 9.40, 9.41, 9.60, 9.62, 9.70, 9.71

### 9.11 CHANGING THE POSITION OF A LEAVING GROUP

**STEP 1** Control the regiochemical outcome of the first step by choosing the appropriate base.

**STEP 2** Control the regiochemical outcome of the second step by choosing the reagents to achieve either Markovnikov or *anti*-Markovnikov addition.

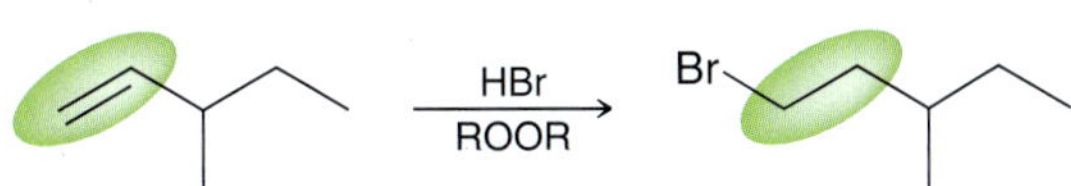


Try Problems 9.42–9.44, 9.57, 9.58b

**9.12 CHANGING THE POSITION OF A π BOND**

**STEP 1** Control the regiochemical outcome of the first step by choosing the reagents to achieve either Markovnikov or *anti*-Markovnikov addition.

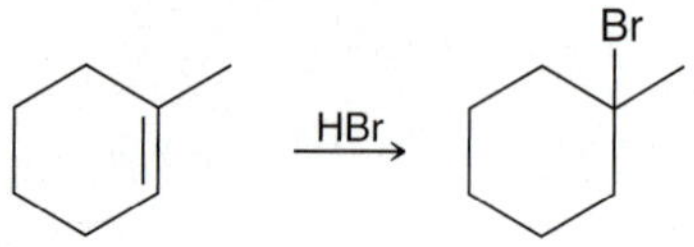

HBr

Br

**STEP 2** Control the regiochemical outcome of the second step by choosing the appropriate base.

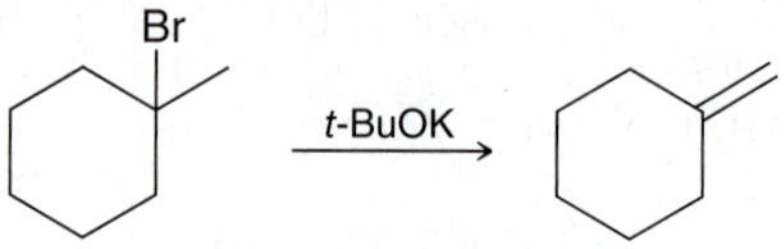

Br

*t*-BuOK

Try Problems 9.45–9.47, 9.53, 9.58

# PRACTICE PROBLEMS

**Note: Most of the Problems are available within WileyPLUS, an online teaching and learning solution.**

**9.48** At high temperatures, alkanes can undergo *dehydrogenation* to produce alkenes. For example:

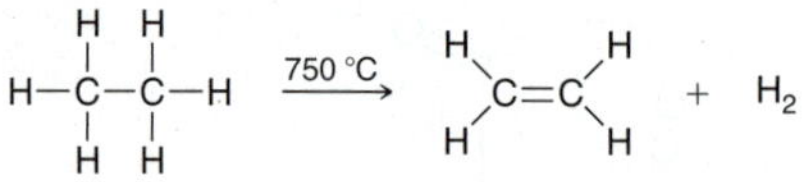

$$\text{H}_3\text{C–CH}_3 \xrightarrow{750\ °\text{C}} \text{H}_2\text{C=CH}_2 + \text{H}_2$$

**Ethane** **Ethylene** **Hydrogen gas**

This reaction is used industrially to prepare ethylene while simultaneously serving as a source of hydrogen gas. Explain why dehydrogenation only works at high temperatures.

**9.49** Predict the major product(s) for each of the following reactions:

$KMnO_4$, NaOH, cold → ?

1) $Hg(OAc)_2$, $H_2O$ 2) $NaBH_4$ → ?

HCl → ?

$H_2$, Pt → ?

$Br_2$, $H_2O$ → ?

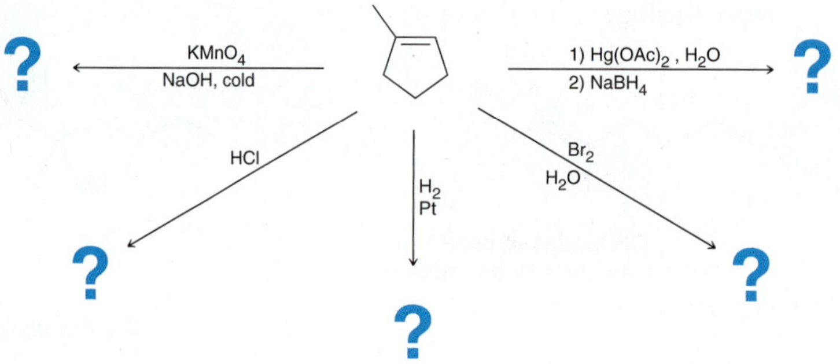

**9.50** Predict the major product(s) for each of the following reactions:

1) MCPBA 2) $H_3O^+$ → ?

1) $BH_3 \cdot THF$ 2) $H_2O_2$, NaOH → ?

HBr → ?

$H_2$, Pt → ?

$Br_2$ → ?

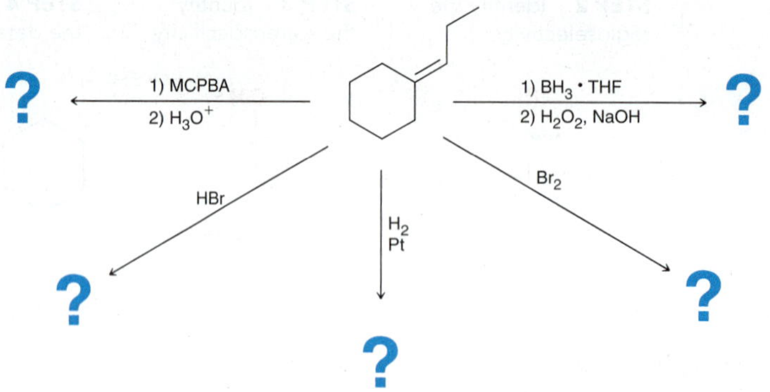

**9.51** Propose a mechanism for each of the following reactions:

(a) 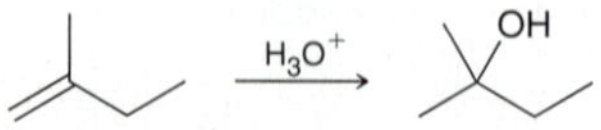 $\xrightarrow{H_3O^+}$ OH

(b) 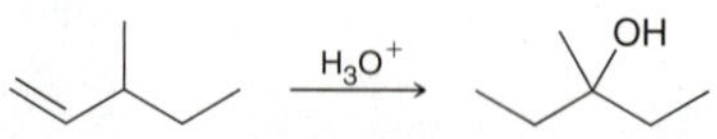 $\xrightarrow{H_3O^+}$ OH

(c) 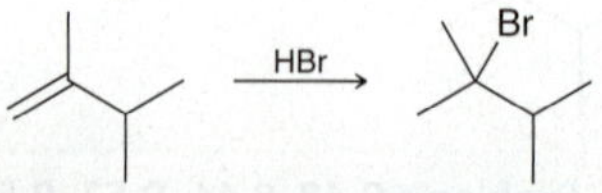 $\xrightarrow{HBr}$ Br

(d) 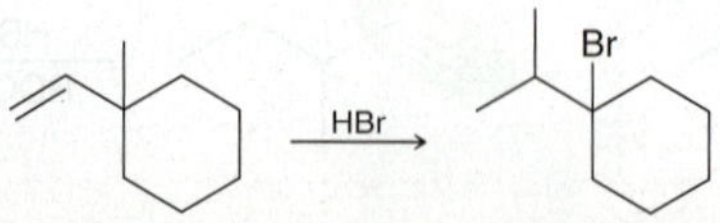 $\xrightarrow{HBr}$ Br

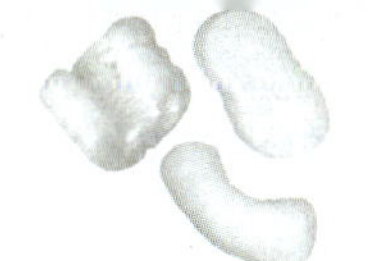

**9.52** Compound A reacts with one equivalent of $H_2$ in the presence of a catalyst to give methylcyclohexane. Compound A can be formed upon treatment of 1-bromo-1-methylcyclohexane with sodium methoxide. What is the structure of compound A?

**9.53** Suggest an efficient synthesis for each of the following transformations:

(a)

(b)

**9.54** Suggest an efficient synthesis for the following transformation:

OH

**9.55** How many different alkenes will produce 2,4-dimethylpentane upon hydrogenation? Draw them.

**9.56** Compound A is an alkene that was treated with ozone (followed by DMS) to yield only $(CH_3CH_2CH_2)_2C{=}O$. Identify the major product that is expected when compound A is treated with MCPBA followed by aqueous acid ($H_3O^+$).

**9.57** Suggest an efficient synthesis for each of the following transformations:

(a) OH OH

(b) Br Br

(c) Cl

(d) OH OH OH

**9.58** Suggest an efficient synthesis for each of the following transformations:

(a)

(b) OH OH

**9.59** Compound A has molecular formula $C_7H_{15}Br$. Treatment of compound A with sodium ethoxide yields only one elimination product (compound B) and no substitution products. When compound B is treated with dilute sulfuric acid, compound C is obtained, which has molecular formula $C_7H_{16}O$. Draw the structures of compounds A, B, and C.

**9.60** Suggest suitable reagents to perform each of the following transformations:

Br + En

Br

HO + En

OH

**9.61** (*R*)-Limonene is found in many citrus fruits, including oranges and lemons:

Draw the structures and identify the relationship of the two products obtained when (*R*)-limonene is treated with excess hydrogen in the presence of a catalyst.

**9.62** Suggest suitable reagents to perform the following transformation:

OH

Racemic

**9.63** Propose a mechanism for the following transformation:

O $\xrightarrow[\text{MeOH}]{[H_2SO_4]}$ MeO OH

**9.64** Propose a mechanism for each of the following transformations:

(a) $\xrightarrow{H_3O^+}$ OH

(b) $\xrightarrow{\text{HBr}}$ Br

**9.65** Predict the major product(s) for each of the following reactions:

(a) $\xrightarrow[(PPh_3)_3RhCl]{H_2}$ ?

(b) $\xrightarrow{H_3O^+}$ ?

(c) $\xrightarrow[\text{2) } H_2O_2\text{, NaOH}]{\text{1) } BH_3 \cdot \text{THF}}$ ?

(d) $\xrightarrow[\text{2) } H_3O^+]{\text{1) MCPBA}}$ ?

**9.66** Explain why each of the following alcohols cannot be prepared via hydroboration-oxidation:

(a) OH (b) OH (c) OH

**9.67** Compound X is treated with $Br_2$ to yield *meso*-2,3-dibromobutane. What is the structure of compound X?

**9.68** Identify the alkene that would yield the following products via ozonolysis:

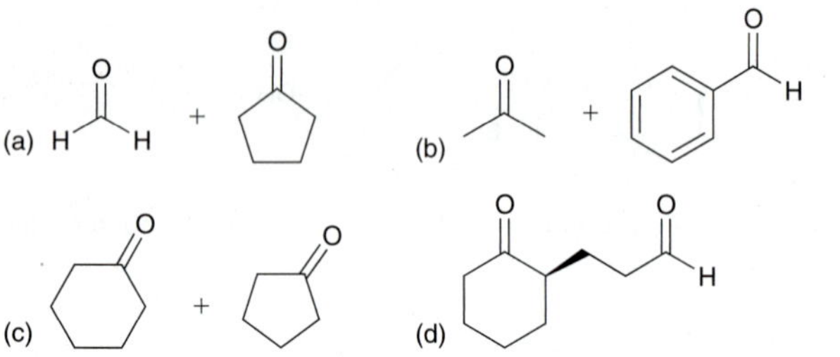

(a) (b) (c) (d)

**9.69** The following reaction is observed to be regioselective. Draw a mechanism for the reaction and explain the source of regioselectivity in this case:

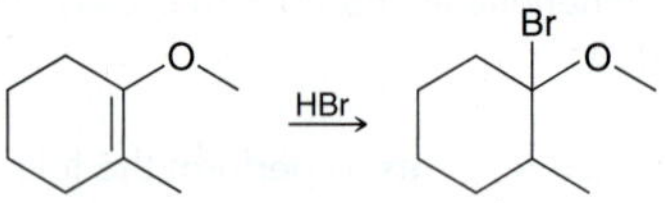

HBr

**9.70** Identify the reagents you would use to accomplish each of the following transformations:

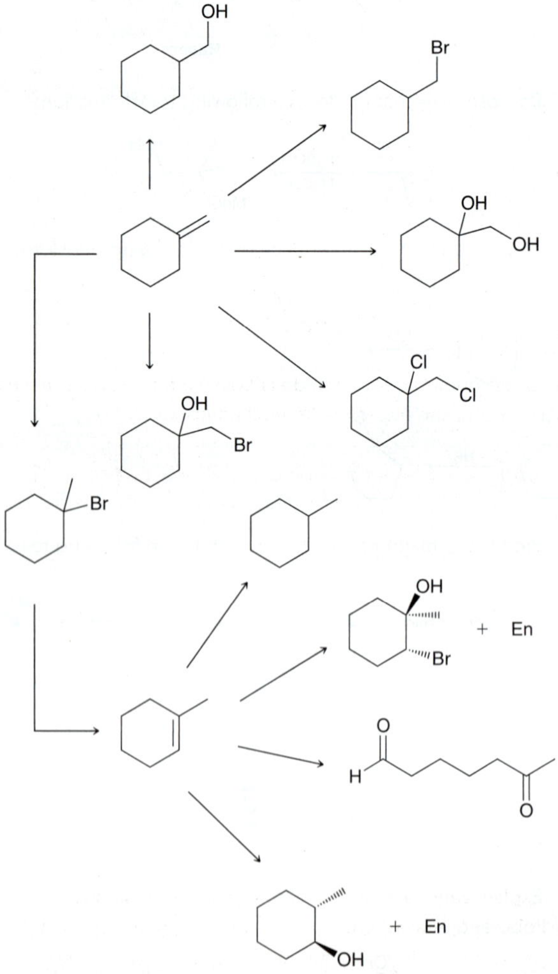

**9.71** Identify the reagents you would use to accomplish each of the following transformations:

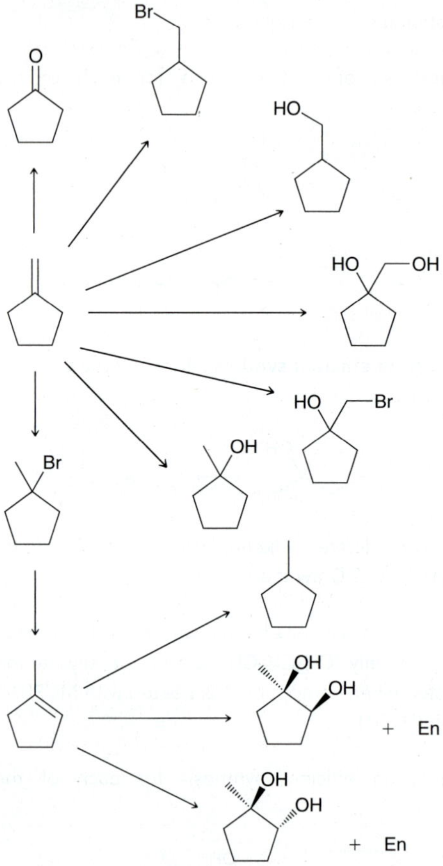

**9.72** Identify which of the following two reactions you would expect to occur more rapidly: (1) addition of HBr to 2-methyl-2-pentene or (2) addition of HBr to 4-methyl-1-pentene. Explain your choice.

**9.73** Predict the major product(s) of the following reaction:

$Br_2$, $H_2S$ ?

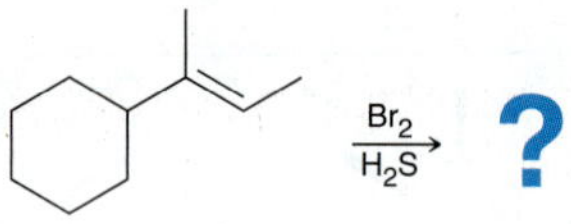

# INTEGRATED PROBLEMS

**9.74** Suggest an efficient synthesis for the following transformation:

OH

Br

\+ En

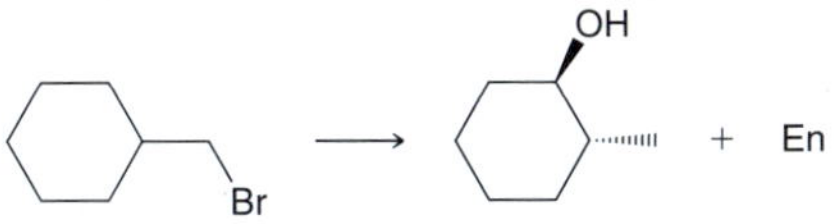

**9.75** Compound X has molecular formula $C_7H_{14}$. Hydrogenation of compound X produces 2,4-dimethylpentane. Hydroboration-oxidation of compound X produces a racemic mixture of 2,4-dimethyl-1-pentanol (shown below). Predict the major product(s) obtained when compound X is treated with aqueous acid ($H_3O^+$).

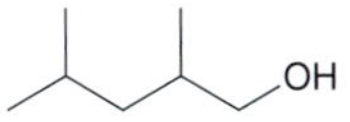

OH

**2,4-Dimethyl-1-pentanol**

**9.76** When (*R*)-2-chloro-3-methylbutane is treated with potassium *tert*-butoxide, a monosubstituted alkene is obtained. When this alkene is treated with HBr, a mixture of products is obtained. Draw all of the expected products.

**9.77** Compound Y has molecular formula $C_7H_{12}$. Hydrogenation of compound Y produces methylcyclohexane. Treatment of compound Y with HBr in the presence of peroxides produces the following compound:

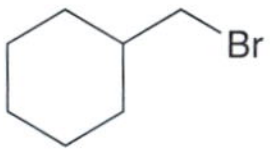

Br

Predict the products when compound Y undergoes ozonolysis.

**9.78** Muscalure is the sex pheromone of the common housefly and has the molecular formula $C_{23}H_{46}$. When treated with $O_3$ followed by DMS, the following two compounds are produced. Draw two possible structures for muscalure.

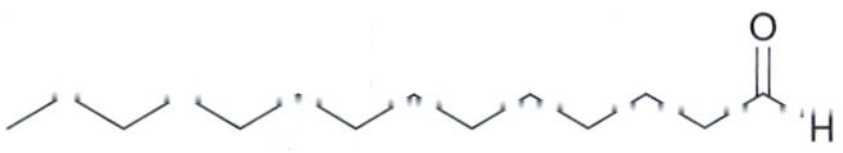

O

H

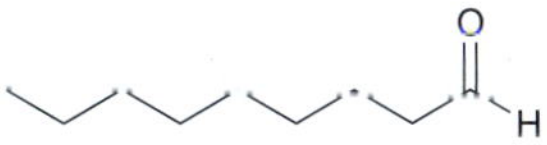

O

H

**9.79** Propose a plausible mechanism for each of the following reactions:

(a) OH O $\xrightarrow{[H_2SO_4]}$ HO O

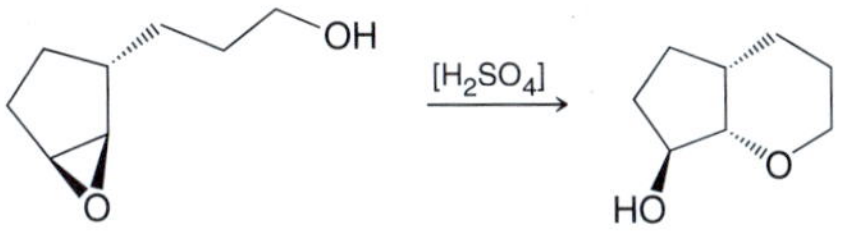

(b) O $\xrightarrow{\text{Conc. } H_2SO_4}$ HO

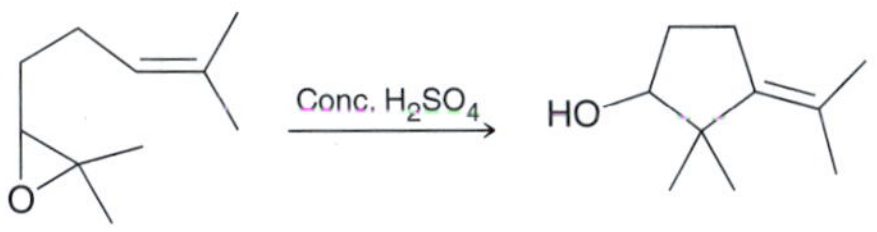

**9.80** Suggest an efficient synthesis for the following transformation:

Cl → H O O H

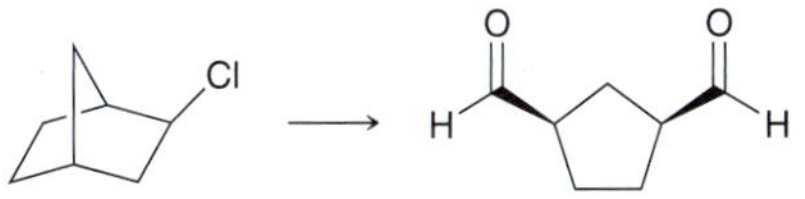

**9.81** Propose a plausible mechanism for the following reaction:

OH $\xrightarrow{Br_2}$ Br O

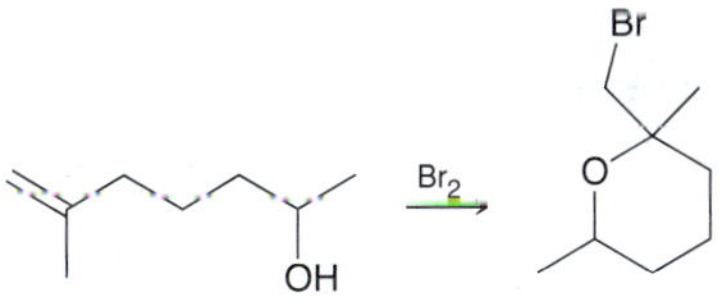

**9.82** Propose a plausible mechanism for the following process, called iodolactonization:

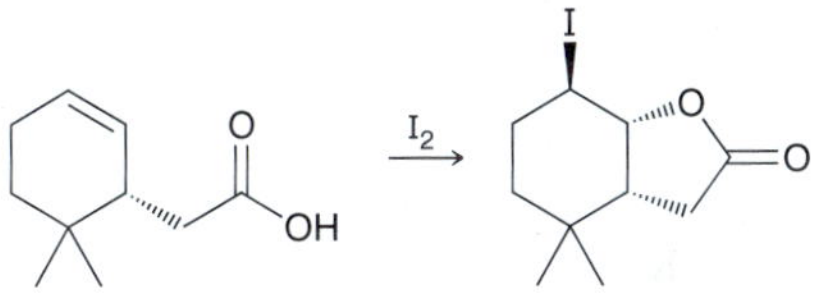

O OH $\xrightarrow{I_2}$ I O O

**9.83** When 3-bromocyclopentene is treated with HBr, the observed product is a racemic mixture of *trans*-1,2-dibromocyclopentane. None of the corresponding *cis*-dibromide is observed. Propose a mechanism that accounts for the observed stereochemical outcome:

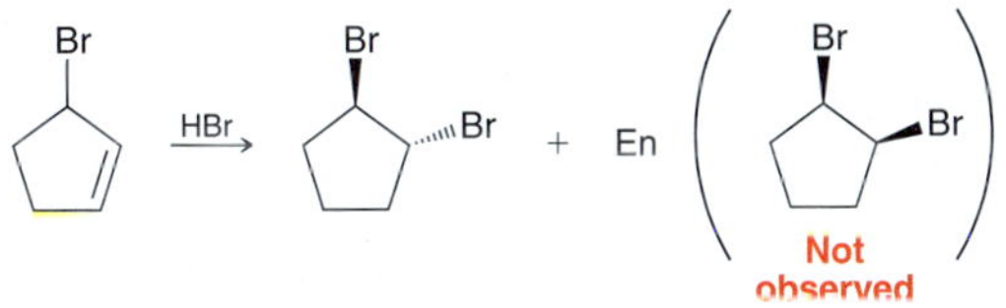

Br $\xrightarrow{HBr}$ Br Br + En ( Br Br Not observed )

# CHALLENGE PROBLEMS

**9.84** Taxol (compound **3**) can be isolated from the bark of the Pacific yew tree, *Taxus brevifolia*, and is currently used in the treatment of several kinds of cancer, including breast cancer. Each yew tree contains a very small quantity of the precious compound (approximately 300 mg), enough for just one dose for one person. This fueled great interest in a synthetic route to taxol, and indeed, several total syntheses have been reported. During K. C. Nicolaou's classic synthesis of taxol, compound **1** was treated with $BH_3$ in THF, followed by oxidative workup, to afford alcohol **2**. The formation of **2** is observed to occur regioselectively as well as stereoselectively (*Nature* **1994**, *367*, 630–634):

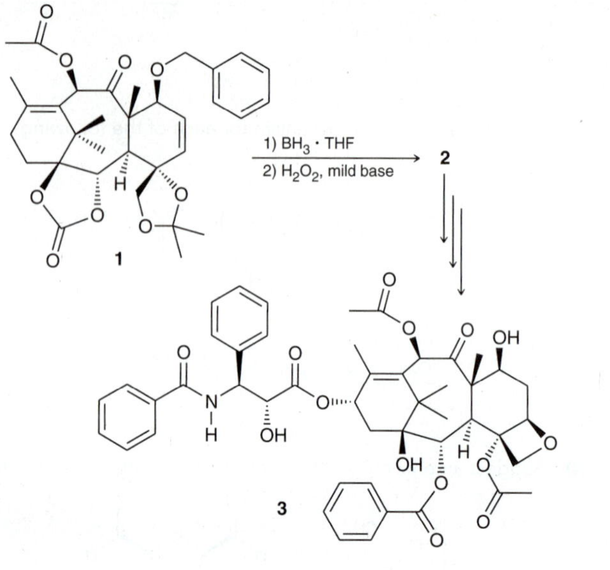

1) $BH_3 \cdot THF$
2) $H_2O_2$, mild base

(a) Identify the functional group in **1** that will react with $BH_3$ and justify your choice.

(b) Predict the regiochemical and stereochemical outcome of the process and draw the structure of compound **2**.

**9.85** When adamantylideneadamantane, shown below, is treated with bromine, the corresponding bromonium ion forms in high yield. This intermediate is stable and does not undergo further nucleophilic attack by bromide ion to give the dibromide (*J. Am. Chem. Soc.* **1985,** *107,* **4504–4508**). Draw the structure of the bromonium ion and explain why it is so inert toward further nucleophilic attack. Draw a reaction coordinate diagram that is consistent with your explanation and compare and contrast it to a reaction coordinate diagram for a more typical dibromination, such as that of propylene.

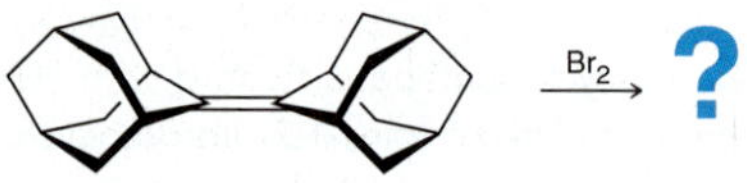

**9.86** The following table provides relative rates of oxymercuration for a variety of alkenes with mercuric acetate (*J. Am. Chem. Soc.* **1989,** *111,* **1414–1418**). Provide structural explanations for the trend observed in the relative rates of reactivity.

| Alkenes | Relative Reactivity |
|---|---|
| $CH_2{=}C(CH_3)(CH_2CH_2CH_3)$, **1** | 1000 |
| $CH_2{=}CHCH_2CH_2CH_2CH_3$, **2** | 100 |
| $CH_2{=}CHCH_2OMe$, **3** | 31.8 |
| $CH(CH_3){=}C(CH_3)_2$, **4** | 25.8 |
| $CH_2{=}CHCH_2Cl$, **5** | 2.4 |

**9.87** Diisopinocampheylborane ($Ipc_2BH$) is a chiral organoborane, readily employed for the production of many asymmetric products used in total synthesis. It is a crystalline material that can be prepared as a single enantiomer via the hydroboration of two equivalents of α-pinene with borane (*J. Org. Chem.* **1984,** *49,* **945–947**). Explain why only one enantiomer of $Ipc_2BH$ is formed.

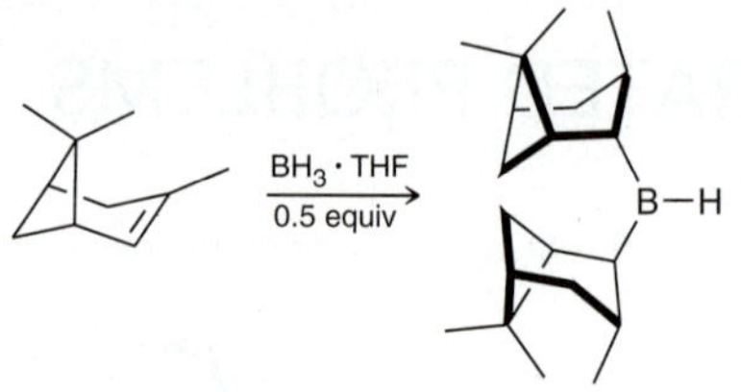

$BH_3 \cdot THF$
0.5 equiv

**9.88** 9-Borabicyclo[3.3.1]nonane (9-BBN) is a reagent commonly used in the hydroboration of alkynes (Section 10.7), but it can also be employed in reactions with alkenes. The following table provides the relative rates of hydroboration (using 9-BBN) for a variety of alkenes (*J. Am. Chem. Soc.* **1989,** *111,* **1414–1418**):

| Alkenes | Relative Reactivity |
|---|---|
| $CH_2{=}CHOBu$, **1** | 1615 |
| $CH_2{=}CHBu$, **2** | 100 |
| $CH_2{=}CHCH_2OMe$, **3** | 32.5 |
| $CH_2{=}CHOAc$, **4** | 22.8 |
| $CH_2{=}CHCH_2OAc$, **5** | 21.9 |
| $CH_2{=}CHCH_2CN$, **6** | 5.9 |
| $CH_2{=}CHCH_2Cl$, **7** | 4.0 |
| *cis*-2-Butene, **8** | 0.95 |
| *trans*-$CH_3CH_2CH{=}CHCl$, **9** | 0.003 |

(a) Provide a structural explanation for the relative rates of reactivity for compounds **1–5**. Be sure to explore resonance effects as well as inductive effects.

(b) Provide a structural explanation for the relative rates of reactivity for compounds **5, 6**, and **7**.

(c) Provide a structural explanation for the relative rates of reactivity for compounds **2, 8**, and **9**.

**9.89** Compound **1** undergoes *syn*-dihydroxylation to give diastereomers **2** and **3**:

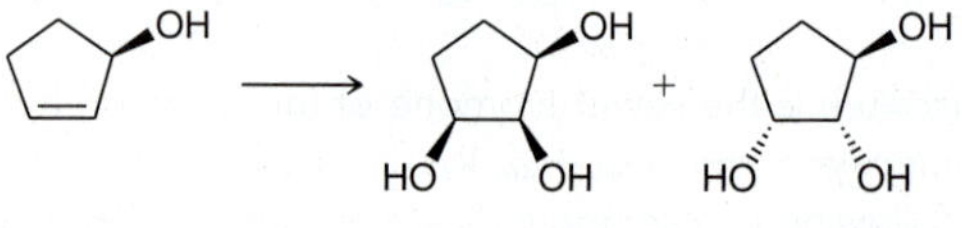

The ratio of products was found to be highly sensitive to the conditions employed (*Tetrahedron Lett.* **1997,** *38,* **5027–5030**). Under typical catalytic conditions (NMO and catalytic $OsO_4$), the ratio of compounds **2** and **3** was found to be 1:4. In contrast, the ratio was observed to be 7:1 under conditions in which stoichiometric amounts of both $OsO_4$ and tetramethylethylene diamine (TMEDA) were used. The investigators noted that TMEDA, a diamine, likely forms a bidentate complex with $OsO_4$ in the manner shown below. Provide an explanation for the observed diastereoselectivity under each set of conditions.

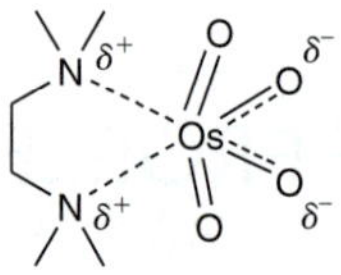

**9.90** Propose a plausible mechanism for the following transformation, which was used as the key step in the formation of the natural product heliol (*Org. Lett.* **2012,** *14,* **2929–2931**):

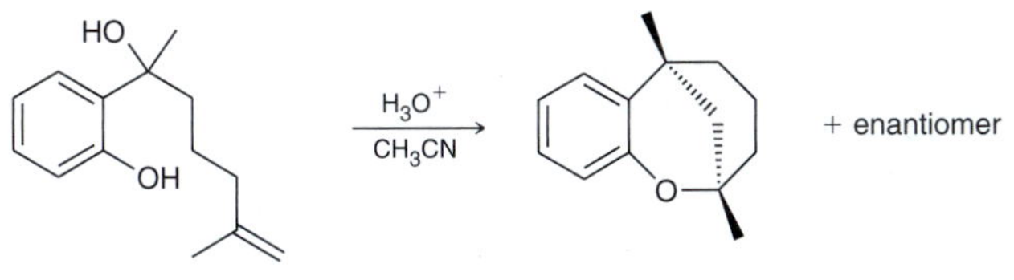

**9.91** Zaragozic acids are a group of structurally related, natural products that were first isolated from fungal cultures in 1992. These compounds have been shown to reduce cholesterol levels in primates, an observation that has fueled interest in the synthesis of zaragozic acids and their derivatives. During a synthesis of zaragozic acid A, compound **1** was treated with catalytic $OsO_4$ and NMO to afford intermediate **2**, which quickly rearranged to give compound **3** (*Angew. Chem. Int. Ed.* **1994,** *33,* **2187–2190**). The conversion of **1** to **2** was observed to proceed in a diastereoselective fashion. That is, *syn* addition only occurs on the top face of the π bond, not on the bottom face. Provide a justification for this facial selectivity. You might find it helpful to build a molecular model.

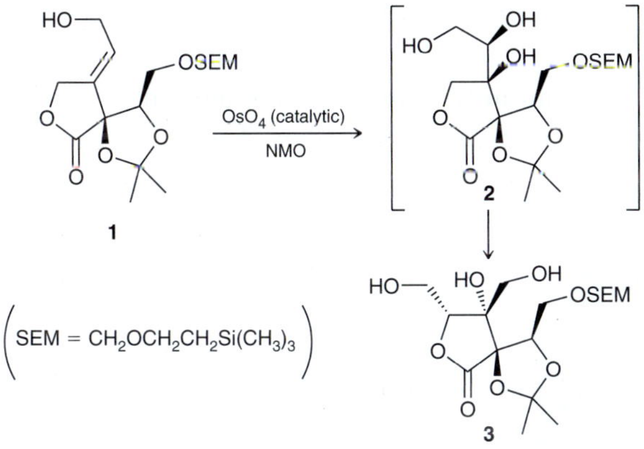

($SEM = CH_2OCH_2CH_2Si(CH_3)_3$)

**9.92** When alcohol **1** is treated with propionaldehyde in the presence of mercuric acetate, two initial intermediates are formed (**A** and **B**). But from these, only one product results: compound **A** forms a cyclic product, while **B** does not (*Org. Lett.* **2000,** *2,* **403–405**):

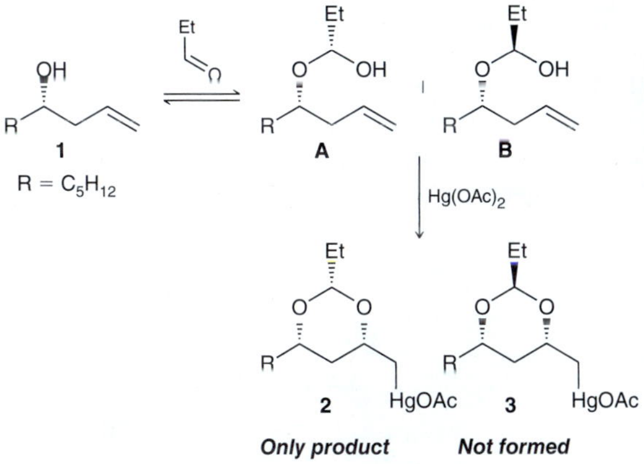

$R = C_5H_{12}$

**2** *Only product* **3** *Not formed*

(a) Propose a plausible mechanism to account for the conversion of intermediate **A** into compound **2**.

(b) Assuming that the reaction proceeds via a transition state that resembles a chair conformation, explain why only intermediate **A** is able to form the cyclic product.

**9.93** Monensin is a potent antibiotic compound isolated from *Streptomyces cinnamonensis*. The following reaction was employed during W. C. Still's synthesis of monensin (*J. Am. Chem. Soc.* **1980,** *102,* **2118–2120**). Sodium bicarbonate ($NaHCO_3$) functions as a mild base to deprotonate the carboxylic acid group. The transformation is said to be diastereoselective, because only one diastereomeric product (5*S*,6*S*) is obtained. The other expected diastereomer (5*R*,6*R*) is not observed.

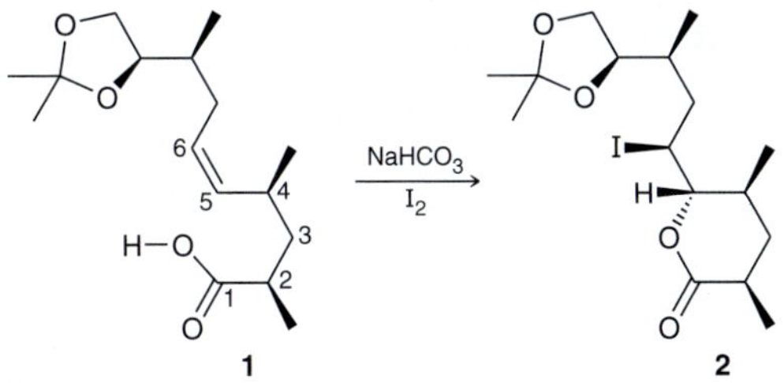

(a) Draw a mechanism for this transformation.

(b) The stereochemical outcome can be rationalized with a conformational analysis of compound **1**. Draw a Newman projection looking down the C4–C5 bond and identify the lowest energy conformation. Use your findings to rationalize the observed diastereoselectivity.

**9.94** Reserpine can be isolated from the extracts of the Indian snakeroot *Rauwolfia serpentina* and has been used in the effective treatment of mental disorders. R. B. Woodward employed the following reaction sequence in his classic 1958 synthesis of reserpine (*Tetrahedron* **1958,** *2,* **1–57**). After compound **1** was prepared, it was treated with molecular bromine, followed by sodium methoxide, to afford compound **2**, which rapidly undergoes a Michael reaction to give compound **3** (Michael reactions will be covered in Chapter 22). Propose a plausible mechanism for the conversion of **1** to the short-lived intermediate **2**.

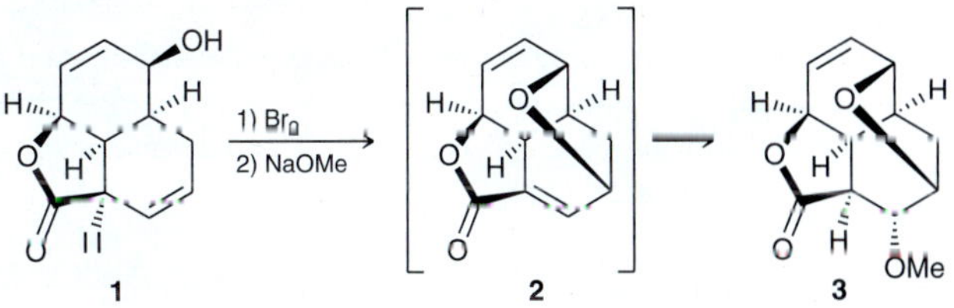

1) $Br_2$ 2) NaOMe

# 10 Alkynes



## DID YOU EVER WONDER . . .
### what causes Parkinson's disease and how it is treated?

Parkinson's disease is a motor system disorder that affects an estimated 3% of the U.S population over the age of 60. The main symptoms of Parkinson's disease include trembling and stiffness of the limbs, slowness of movement, and impaired balance. Parkinson's disease is a neurodegenerative disease, because the symptoms are caused by the degeneration of neurons (brain cells). The neurons most affected by the disease are located in a region of the brain called the *substantia nigra*. When these neurons die, they cease to produce dopamine, the neurotransmitter used by the brain to regulate voluntary movement. The symptoms described above begin to appear when 50–80% of dopamine-producing neurons have died. There is no known cure for the disease, which is progressive. However, the symptoms can be treated through a variety of methods. One such method utilizes a drug called selegiline, whose molecular structure contains a C≡C triple bond. We will see that the presence of the triple bond plays an important role in the action of this drug. This chapter explores the properties and reactivity of compounds with C≡C triple bonds called *alkynes*.

## DO YOU REMEMBER?

*Before you go on, be sure you understand the following topics.*
*If necessary, review the suggested sections to prepare for this chapter.*

- Brønsted-Lowry Acidity (Sections 3.4, 3.5)
- Nucleophiles and Electrophiles (Section 6.7)
- Energy Diagrams (Sections 6.5, 6.6)
- Arrow Pushing (Sections 6.8–6.10)

Take the DO YOU REMEMBER? QUIZ in **WileyPLUS** to check your understanding.

## 10.1 Introduction to Alkynes

### Structure and Geometry of Alkynes

The previous chapter explored the reactivity of alkenes. In this chapter, we expand that discussion to include the reactivity of **alkynes**, compounds containing a C≡C triple bond. Recall from Section 1.9 that a triple bond is comprised of three separate bonds: a σ bond and two π bonds. The σ bond results from the overlap of *sp*-hybridized orbitals, while each of the π bonds results from overlapping *p* orbitals (Figure 10.1). Each carbon atom is *sp* hybridized and exhibits linear geometry. An electrostatic potential map of acetylene (H—C≡C—H) reveals a cylindrical region of high electron density encircling the triple bond (Figure 10.2). This region (shown in red) explains why alkynes react with compounds that bear a region of low electron density. As a result, alkynes are similar to alkenes in their ability to function either as bases or as nucleophiles. We will see examples of both behaviors in this chapter.

**FIGURE 10.1**
The atomic orbitals used to form a triple bond. The σ bond is formed from the overlap of two hybridized orbitals, while each of the two π bonds is formed from overlapping *p* orbitals.

**FIGURE 10.2**
An electrostatic potential map of acetylene, indicating a cylindrical region of high electron density (red).

### Alkynes in Industry and Nature

The simplest alkyne, acetylene (H—C≡C—H), is a colorless gas that undergoes combustion to produce a high-temperature flame (2800°C) and is used as a fuel for welding torches. Acetylene is also used as a precursor for the preparation of higher alkynes, as we will see in Section 10.10.

Alkynes are less common in nature than alkenes, although more than 1000 different alkynes have been isolated from natural sources. One such example is histrionicotoxin, which is one of the many toxins secreted by the South American frog, *Dendrobates histrionicus*, as a defense against predators. For many centuries, South American tribes have extracted the mixture of toxins from the frog's skin and placed it on the tips of arrows to produce poison arrows.

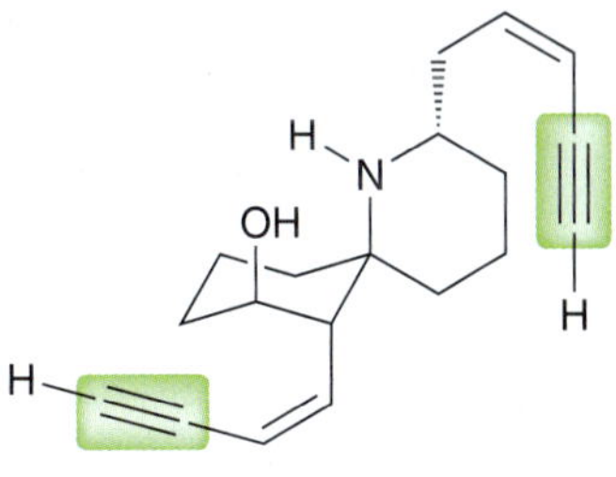


**Histrionicotoxin**

## medically speaking — The Role of Molecular Rigidity

As mentioned in the chapter opening, Parkinson's disease is a neurodegenerative disorder marked by a decreased production of dopamine in the brain. Since dopamine regulates motor function, the decrease in dopamine leads to impaired motor control. Although there is no cure for Parkinson's disease, the symptoms can be treated by a variety of methods. The most effective course of treatment is to administer a drug called L-dopa, which is converted into dopamine in the brain:

(*S*)-3,4-Dihydroxyphenylalanine (L-dopa) → → Dopamine

This method was described previously in Section 9.7 and is effective because it replenishes the supply of dopamine in the brain. Another method for increasing dopamine levels is to slow the rate at which dopamine is removed from the brain. Dopamine is primarily metabolized under the influence of an enzyme (a biological catalyst), called monoamine oxidase B (MAO B). Any drug that inactivates this enzyme will effectively slow the rate at which dopamine is metabolized, thereby slowing the rate at which dopamine levels decrease in the brain. Unfortunately, a closely related enzyme, called MAO A, is used for the metabolism of other compounds, and any drug that inactivates MAO A leads to significant cardiovascular side effects. Therefore, the selective inactivation of MAO B (but not MAO A) is required. The first selective MAO B inactivator, called selegiline, was approved by the FDA in 1989 for the treaent of Parkinson's disease:

Selegiline

Selegiline, sold under the trade name Eldepryl, is often prescribed in combination with L-dopa. The combination of these two drugs offers a more effective method for combating the diminishing supply of dopamine in the brain.

Notice that the structure of selegiline exhibits a C≡C triple bond, which serves an important function. Specifically, its linear geometry imparts structural rigidity to the compound. The aromatic ring on the other side of the compound also imparts structural rigidity, and both of these structural subunits enable the compound to selectively bind to MAO B, thereby causing its inactivation. Triple bonds appear in several other FDA-approved drugs, where they often serve a similar function. In order for a drug to bind to its target receptor effectively, it must have the appropriate balance of structural rigidity and flexibility. In the design of new drugs, triple bonds are sometimes used to achieve that balance.

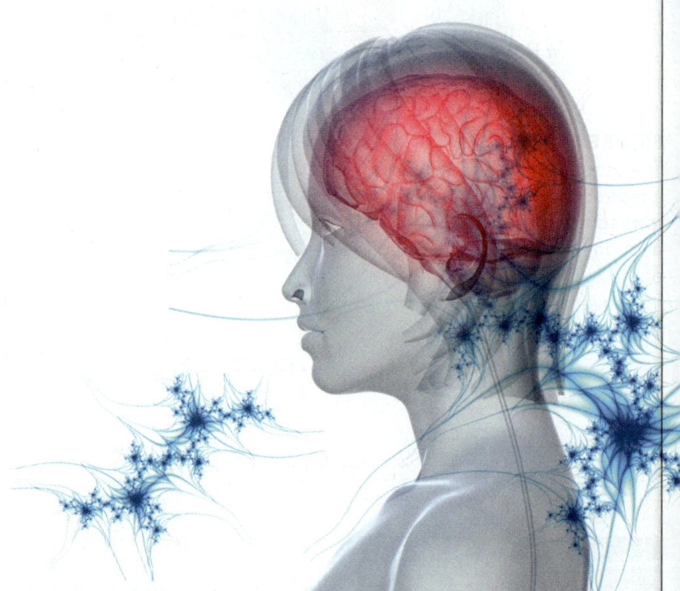

In addition to the alkynes found in nature, many synthetic alkynes (prepared in the laboratory) are of particular interest. One such example is ethynylestradiol, found in many birth control formulations. This synthetic oral contraceptive elevates hormone levels in women and prevents ovulation. The presence of the triple bond renders this compound a more potent contraceptive than its natural analogue, which lacks a triple bond. The effect of the triple bond is attributed to the additional structural rigidity that it imparts to the compound (as described in the box on molecular rigidity).

Ethynylestradiol

## practically speaking Conducting Organic Polymers

Polyacetylene, which can be prepared via the polymerization of gaseous acetylene, was the first known example of an organic polymer capable of conducting electricity.

Acetylene —Polymerization→ Polyacetylene

In practice, polyacetylene only poorly conducts electricity. However, when the polymer is prepared as a cation or anion (a process called *doping*), it can conduct electricity almost as well as a copper wire. Both the cationic polymer and the anionic polymer are resonance stabilized, and they each conduct electricity very efficiently. This discovery effectively opened the doorway to the exciting field of conducting organic polymers, and the 2000 Nobel Prize in Chemistry was awarded to its discoverers: Alan Heeger, Alan MacDiarmid, and Hideki Shirakawa. Polyacetylene is used in the packaging materials for computer parts, due to its ability to dissipate static charges that could damage sensitive circuitry.

Polyacetylene itself has limited application because of its sensitivity to air and moisture, but many other conducting polymers have been developed. Consider the structure of poly(*p*-phenylene vinylene), or PPV, which is a conducting polymer:

Poly(*p*-phenylene vinylene)

Conducting polymers such as PPV are sometimes used in LED (light emitting diode) displays. When subjected to an electric field, LEDs emit light, a phenomenon called electroluminescence. Organic LED systems, or OLEDs, are generally less effective than inorganic LED systems, although many useful OLEDs have been developed over the past several decades. They are used in cell phone displays, digital cameras, and the displays of MP3 players. LEDs are used in a wide range of applications, including traffic lights, flat-panel TV displays, and alarm clock displays.

Polyacetylene —Remove one electron→ Resonance stabilized

Polyacetylene —Add one electron→ Resonance stabilized

## 10.2 Nomenclature of Alkynes

Recall from Sections 4.2 and 8.3 that naming alkanes and alkenes requires four discrete steps:

1. Identify and name the parent.
2. Identify and name the substituents.
3. Number the parent chain and assign a locant to each substituent.
4. Assemble the substituents alphabetically.

Alkynes are named using the same four steps, with the following additional rules:
When naming the parent, the suffix "yne" is used to indicate the presence of a C≡C triple bond:

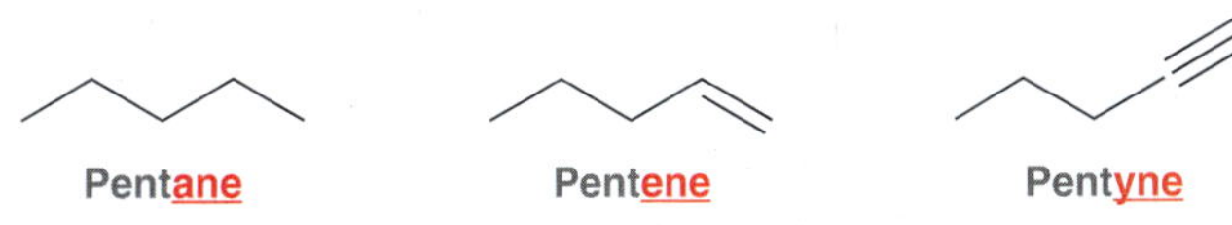

The parent of an alkyne is the longest chain *that includes the C≡C triple bond:*

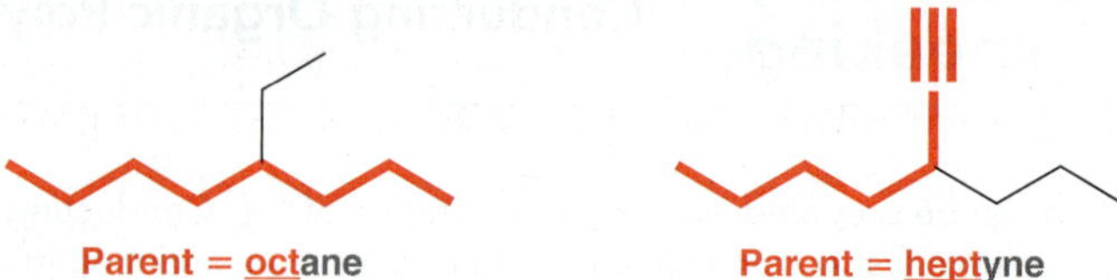


When numbering the parent chain of an alkyne, the triple bond should receive the lowest number possible, despite the presence of alkyl substituents:

The position of the triple bond is indicated using a single locant, not two. In the previous example, the triple bond is between C2 and C3 on the parent chain. In this case, the position of the triple bond is indicated with the number 2. The IUPAC rules published in 1979 dictate that this locant be placed immediately before the parent, while the IUPAC recommendations released in 1993 and 2004 allow for the locant to be placed before the suffix "yne." Both names are acceptable IUPAC names:

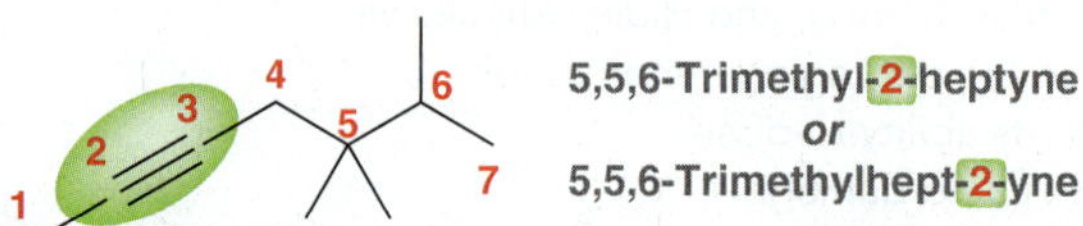


In addition to IUPAC nomenclature, chemists use common names for many alkynes. Ethyne (H—C≡C—H) is called acetylene, while larger alkynes have common names that identify the alkyl groups attached to the parent acetylene:

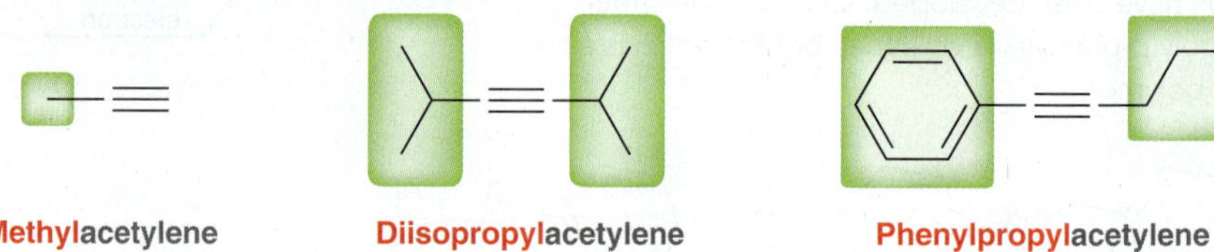


Notice that the first example is monosubstituted; it possesses only one alkyl group. Monosubstituted acetylenes are called **terminal alkynes**, while disubstituted acetylenes are called **internal alkynes**. This distinction will be important in the upcoming sections of this chapter.

## SKILLBUILDER

### 10.1 ASSEMBLING THE SYSTEMATIC NAME OF AN ALKYNE

**LEARN the skill** Provide a systematic name for the following compound:

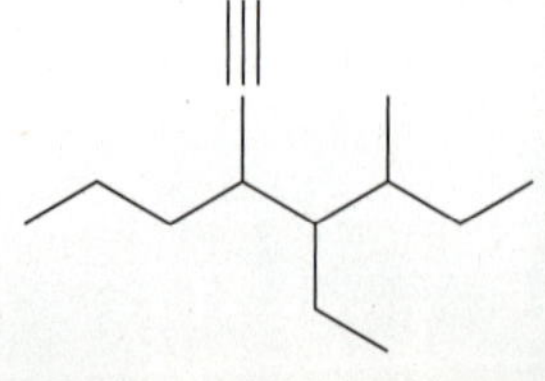

## SOLUTION

**STEP 1**
Identify the parent.

Assembling a systematic name requires four discrete steps: Begin by identifying the parent. Choose the longest chain that includes the triple bond:

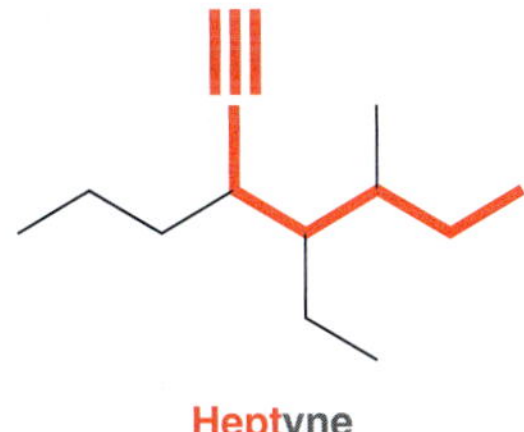


**STEP 2**
Identify and name the substituents.

Then, identify and name the substituents:

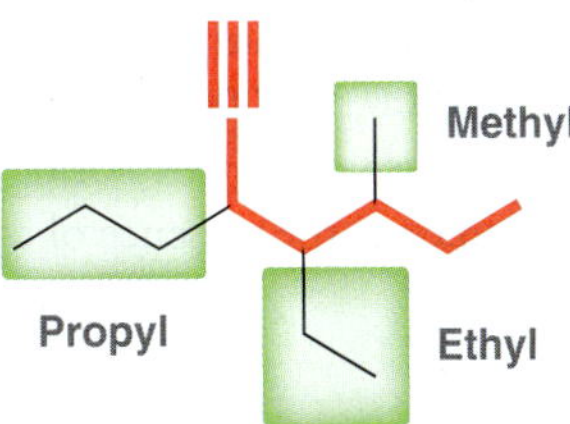


**STEP 3**
Number the parent chain and assign a locant to each substituent.

Next, number the parent chain and assign a locant to each substituent:

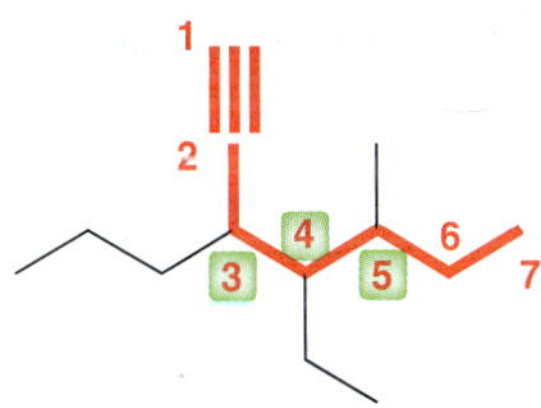


**STEP 4**
Assemble the substituents alphabetically.

Finally, assemble the substituents alphabetically, making sure to include a locant that identifies the position of the triple bond:

4-Ethyl-5-methyl-3-propyl-1-heptyne

Make sure that hyphens separate letters from numbers, while commas separate numbers from each other.

**PRACTICE the skill**

**10.1** Provide a systematic name for each of the following compounds:

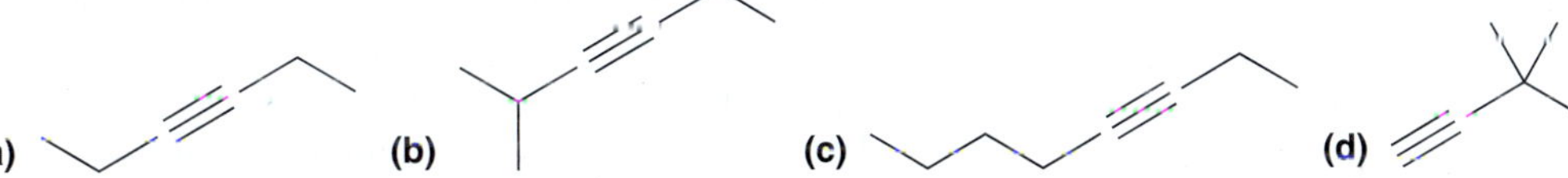

**APPLY the skill**

**10.2** Draw a bond-line structure for each of the following compounds:

**(a)** 4,4-Dimethyl-2-pentyne **(b)** 5-Ethyl-2,5-dimethyl-3-heptyne

**10.3** Although triple bonds have linear geometry, they do have some flexibility and can be incorporated in large rings. Cycloalkynes containing more than eight carbon atoms have been isolated and are stable at room temperature. When naming cycloalkynes that lack any other functional groups, the triple bond does not require a locant, because it is assumed to be between C1 and C2. Draw the structure of (*R*)-3-methylcyclononyne.

**10.4** Draw and name four terminal alkynes with molecular formula $C_6H_{10}$.

need more **PRACTICE?** **Try Problems 10.35, 10.36**

## 10.3 Acidity of Acetylene and Terminal Alkynes

Compare the p$K_a$ values for ethane, ethylene, and acetylene:

| Ethane | Ethylene | Acetylene |
|---|---|---|
| p$K_a$ = 50 | p$K_a$ = 44 | p$K_a$ = 25 |

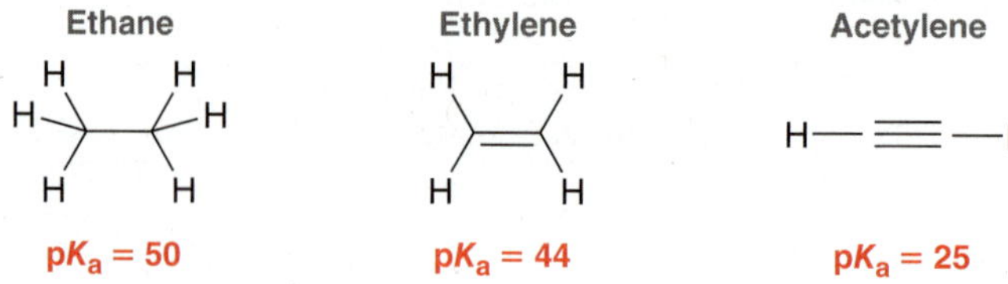

Recall that a lower p$K_a$ corresponds to a greater acidity. Therefore, acetylene (p$K_a$ = 25) is significantly more acidic than ethane or ethylene. To be precise, acetylene is 19 orders of magnitude (10,000,000,000,000,000,000 times) more acidic than ethylene. The relative acidity of acetylene can be explained by exploring the stability of its conjugate base, called an **acetylide ion** (the suffix "ide" indicates the presence of a negative charge):

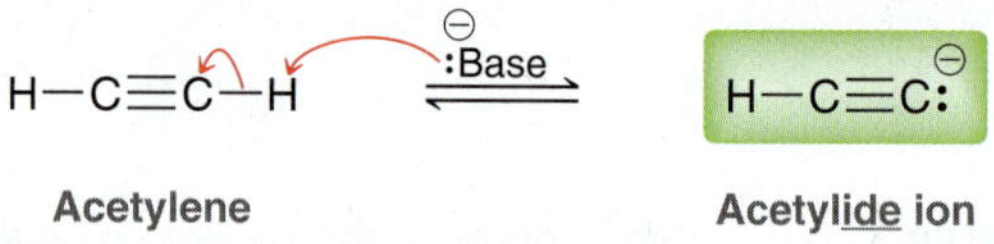

H—C≡C—H + :Base⁻ ⇌ H—C≡C:⁻

Acetylene — Acetylide ion

LOOKING BACK
This effect was first discussed in Section 3.4.

The stability of an acetylide ion can be rationalized using hybridization theory, in which the negative charge is considered to be associated with a lone pair that occupies an *sp*-hybridized orbital. Compare the conjugate bases for ethane, ethylene, and acetylene (Figure 10.3).

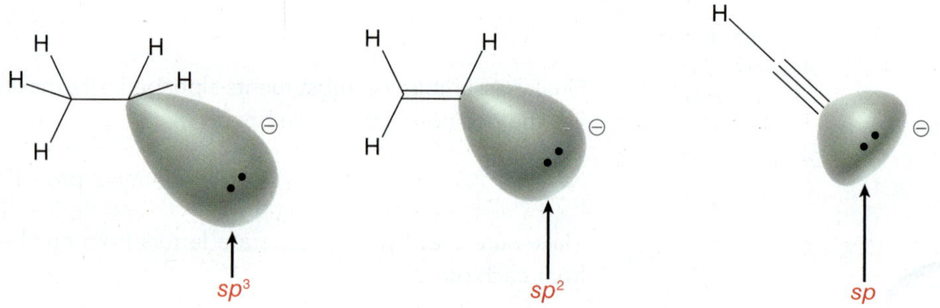

**FIGURE 10.3** The conjugate base of ethane exhibits a lone pair in an $sp^3$-hybridized orbital. The conjugate base of ethylene has the lone pair in an $sp^2$-hybridized orbital, and the conjugate base of acetylene has the lone pair in an *sp*-hybridized orbital.

In an *sp*-hybridized orbital, the electron density is closer to the positively charged nucleus and is therefore more stable.

Now let's consider the equilibrium that is established when a strong base is used to deprotonate acetylene. Recall that the equilibrium of an acid-base reaction will always favor formation of the weaker acid and weaker base. For example, consider the equilibrium established when an amide ion ($H_2N^-$) is used as a base to deprotonate acetylene:

$H_2N^-$ $Na^+$ + H—C≡C—H ⇌ $H_2N$—H + $Na^+$ ⁻:C≡C—H

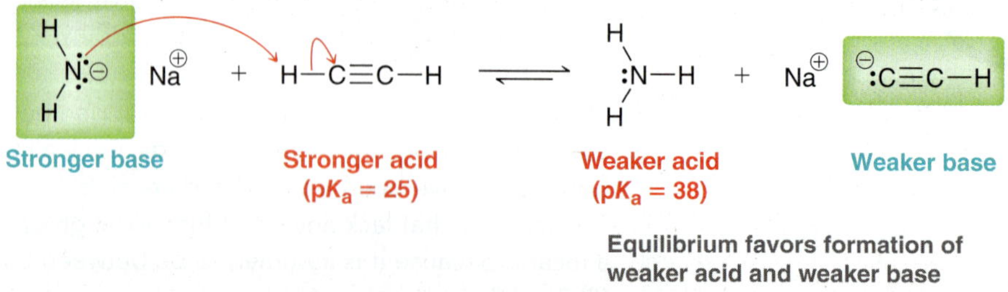

In this case, the equilibrium favors formation of the acetylide ion, because it is more stable (a weaker base) than the amide ion. In contrast, consider what happens when a hydroxide ion is used as the base:

$$\text{HO}^{\ominus}\ \text{Na}^{\oplus} + \text{H—C}\equiv\text{C—H} \rightleftharpoons \text{H}_2\text{O} + \text{Na}^{\oplus}\ {}^{\ominus}\text{:C}\equiv\text{C—H}$$

Weaker base | Weaker acid ($pK_a$ = 25) | Stronger acid ($pK_a$ = 15.7) | Stronger base

**Equilibrium favors the weaker acid and weaker base**

In this case, the equilibrium does not favor formation of the acetylide ion, because the acetylide ion is less stable (a stronger base) than the hydroxide ion. Therefore, hydroxide is not sufficiently basic to produce a significant amount of the acetylide ion. That is, hydroxide cannot be used to deprotonate acetylene.

Much like acetylene, terminal alkynes are also acidic and can be deprotonated with a suitable base:

$$\text{R—C}\equiv\text{C—H} \xrightleftharpoons{:\text{Base}^{\ominus}} \text{R—C}\equiv\text{C:}^{\ominus}$$

An alkyne | An alkynide ion

The conjugate base of a terminal alkyne, called an **alkynide ion**, can only be formed with a sufficiently strong base. Sodium hydroxide (NaOH) is not a suitable base for this purpose, but sodium amide ($NaNH_2$) can be used. There are several bases that can be used to deprotonate acetylene or terminal alkynes, as seen in Table 10.1. The three bases shown on the top left of the chart are all strong enough to deprotonate a terminal alkyne, and all three are commonly used to do so. Notice the position of the negative charge in each of these cases ($N^-$, $H^-$, or $C^-$). In contrast, the three bases on the bottom left of the chart all have the negative charge on an oxygen atom, which is not strong enough to deprotonate a terminal alkyne.

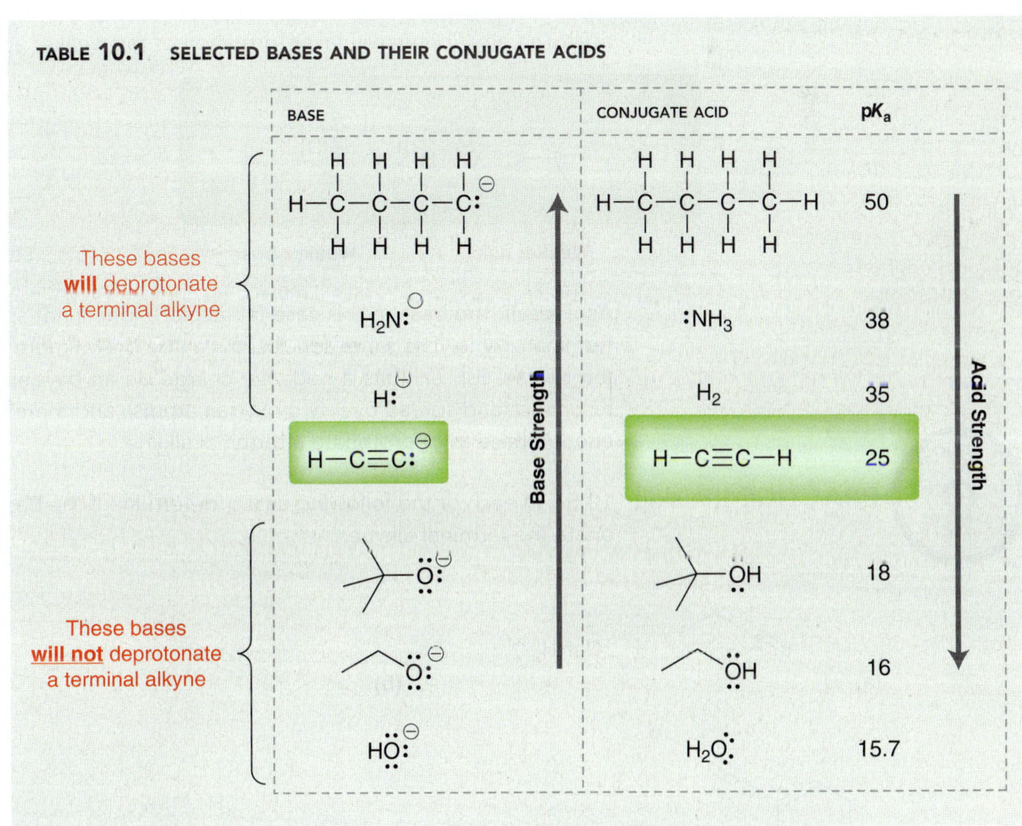

**TABLE 10.1 SELECTED BASES AND THEIR CONJUGATE ACIDS**

| | BASE | CONJUGATE ACID | $pK_a$ |
|---|---|---|---|
| These bases **will** deprotonate a terminal alkyne | $CH_3CH_2CH_2CH_2^{\ominus}$ | $CH_3CH_2CH_2CH_3$ | 50 |
| | $H_2N:^{\ominus}$ | $:NH_3$ | 38 |
| | $H:^{\ominus}$ | $H_2$ | 35 |
| | $\text{H—C}\equiv\text{C:}^{\ominus}$ | $\text{H—C}\equiv\text{C—H}$ | 25 |
| These bases **will not** deprotonate a terminal alkyne | $(CH_3)_3CO^{\ominus}$ | $(CH_3)_3COH$ | 18 |
| | $CH_3CH_2O^{\ominus}$ | $CH_3CH_2OH$ | 16 |
| | $HO^{\ominus}$ | $H_2O$ | 15.7 |

# SKILLBUILDER

## 10.2 PREDICTING THE POSITION OF EQUILIBRIUM FOR THE DEPROTONATION OF A TERMINAL ALKYNE

**LEARN the skill**

Using the information in Table 3.1, determine whether sodium acetate ($CH_3CO_2Na$) is a strong enough base to deprotonate a terminal alkyne. That is, determine whether the following equilibrium would favor formation of the alkynide ion:

$$\text{C}_6\text{H}_{11}\text{–C}\equiv\text{C–H} + \text{CH}_3\text{CO}_2^{\ominus}\ \text{Na}^{\oplus} \rightleftharpoons \text{C}_6\text{H}_{11}\text{–C}\equiv\text{C:}^{\ominus}\ \text{Na}^{\oplus} + \text{CH}_3\text{CO}_2\text{H}$$

### SOLUTION

Begin by identifying the acid and base on each side of the equilibrium:

**STEP 1** Identify the acid and base on each side of the equilibrium.

$$\text{C}_6\text{H}_{11}\text{–C}\equiv\text{C–H} + \text{CH}_3\text{CO}_2^{\ominus} \rightleftharpoons \text{C}_6\text{H}_{11}\text{–C}\equiv\text{C:}^{\ominus} + \text{CH}_3\text{CO}_2\text{H}$$

Acid — Base — Base — Acid

**BY THE WAY**
$Na^+$ is simply the counterion for each base and can be ignored in most cases.

**STEP 2** Determine which acid is weaker.

Next, compare the $pK_a$ values of the two acids in order to determine which one is the weaker acid. These values can be found in Table 3.1:

$\text{C}_6\text{H}_{11}\text{–C}\equiv\text{C–H}$ ($pK_a \sim 25$) $\quad$ $\text{CH}_3\text{CO}_2\text{H}$ ($pK_a = 4.75$)

**STEP 3** Identify the position of equilibrium.

Recall from Section 3.3 that a higher $pK_a$ indicates a weaker acid, so the alkyne is the weaker acid. The equilibrium will favor formation of the weaker acid and weaker base:

$$\text{C}_6\text{H}_{11}\text{–C}\equiv\text{C–H} + \text{CH}_3\text{CO}_2^{\ominus} \rightleftharpoons \text{C}_6\text{H}_{11}\text{–C}\equiv\text{C:}^{\ominus} + \text{CH}_3\text{CO}_2\text{H}$$

Weaker acid — Weaker base — Stronger base — Stronger acid

As a result, the base in this case (an acetate ion) is not sufficiently strong to deprotonate a terminal alkyne. The same conclusion can be drawn more quickly by simply recognizing that the acetate ion exhibits a negative charge on an oxygen atom (actually, it is stabilized by resonance and spread over two oxygen atoms), and therefore, it cannot possibly be a strong enough base to deprotonate a terminal alkyne.

**PRACTICE the skill**

**10.5** In each of the following cases, determine if the base is sufficiently strong to deprotonate the terminal alkyne:

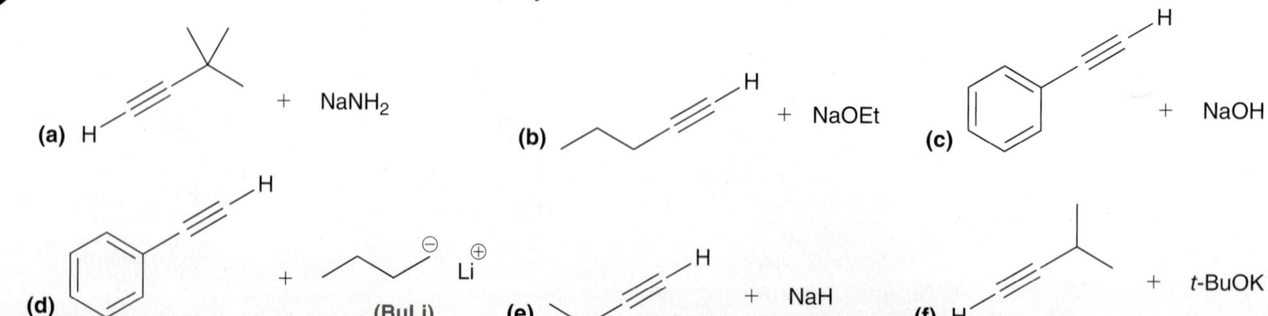

**APPLY the skill**

**10.6** The p$K_a$ of $CH_3NH_2$ is 40, while the p$K_a$ of HCN is 9.

**(a)** Explain this difference in acidity.

**(b)** Can the cyanide anion (the conjugate base of HCN) be used as a base to deprotonate a terminal alkyne? Explain.

need more **PRACTICE?** **Try Problems 10.39, 10.42**

## 10.4 Preparation of Alkynes

Just as alkenes can be prepared from alkyl halides, alkynes can be prepared from alkyl dihalides:

R–C(Br)(R)–C(R)(H)–R —Strong base→ $R_2C{=}CR_2$ (Section 8.7)

**An alkyl halide**

R–C(Br)(Br)–C(H)(H)–R —Strong base→ R–C≡C–R

**An alkyl dihalide**

An alkyl dihalide has two leaving groups, and the transformation is accomplished via two successive elimination (E2) reactions:

R–C(:Br:)(:Br:)–C(H)(H)–R —:Base⁻, E2→ RC(Br)=C(H)R —:Base⁻, E2→ R–C≡C–R

In this example, the dihalide used is a **geminal** dihalide, which means that both halogens are connected to the same carbon atom. Alternatively, alkynes can also be prepared from **vicinal** dihalides, in which the two halogens are connected to adjacent carbon atoms:

R–C(Br)(H)–C(Br)(H)–R —Strong base→ R–C≡C–R

Whether the starting dihalide is geminal or vicinal, the alkyne is obtained as the result of two successive elimination reactions. The first elimination can be readily accomplished using many different bases, but the second elimination requires a very strong base. Sodium amide ($NaNH_2$), dissolved in liquid ammonia ($NH_3$), is a suitable base for achieving two successive elimination reactions in a single reaction vessel. This method is used most frequently for the preparation of terminal alkynes, because the strongly basic conditions favor production of an alkynide ion, which serves as a driving force for the overall process:

R–C(H)(H)–C(Br)(Br)–H —2 $NaNH_2$ / $NH_3$→ R–C≡C–H —$NaNH_2$ / $NH_3$→ R–C≡C:⁻ $Na^+$

**Alkynide ion**

In total, three equivalents of the amide ion are required: two equivalents for the two E2 reactions and one equivalent to deprotonate the terminal alkyne and form the alkynide ion. After the alkynide ion has formed and the reaction is complete, a proton source can be introduced into the reaction vessel, thereby protonating the alkynide ion to regenerate the terminal alkyne:

$$R{-}C{\equiv}C:^{\ominus}\ Na^{\oplus} + H{-}\ddot{O}{-}H \rightleftharpoons R{-}C{\equiv}C{-}H + Na^{\oplus}\ {:}\ddot{O}H^{\ominus}$$

Stronger base — Stronger acid ($pK_a = 15.7$) — Weaker acid ($pK_a = 25$) — Weaker base

**Equilibrium favors formation of weaker acid and weaker base**

Comparison of the $pK_a$ values indicates that water is a sufficient proton source.

In summary, a terminal alkyne can be prepared by treating a dihalide with excess (xs) sodium amide followed by water:

1) xs $NaNH_2/NH_3$ 2) $H_2O$ → **(60%)**

## CONCEPTUAL CHECKPOINT

**10.7** For each of the following transformations, predict the major product and draw a mechanism for its formation:

(a) 1) xs $NaNH_2/NH_3$ 2) $H_2O$ → ?

(b) 1) xs $NaNH_2/NH_3$ 2) $H_2O$ → ?

**10.8** When 3,3-dichloropentane is treated with excess sodium amide in liquid ammonia, the initial product is 2-pentyne:

$$CH_3CH_2CCl_2CH_2CH_3 \xrightarrow{\text{xs } NaNH_2} CH_3CH_2C{\equiv}CCH_3$$

**2-Pentyne**

However, under these conditions, this internal alkyne quickly isomerizes to form a terminal alkyne that is subsequently deprotonated to form an alkynide ion:

$$CH_3CH_2C{\equiv}CCH_3 \underset{}{\overset{\text{xs } NaNH_2}{\rightleftharpoons}} CH_3CH_2CH_2C{\equiv}C{-}H \overset{\text{xs } NaNH_2}{\rightleftharpoons} CH_3CH_2CH_2C{\equiv}C:^{\ominus}$$

**2-Pentyne** — **1-Pentyne** — **Alkynide ion**

The isomerization process is believed to occur via a mechanism with the following four steps: (1) deprotonate, (2) protonate, (3) deprotonate, and (4) protonate. Using these four steps as a guide, try to draw the mechanism for isomerization using resonance structures whenever possible. Explain why the equilibrium favors formation of the terminal alkyne.

## 10.5 Reduction of Alkynes

### Catalytic Hydrogenation

Alkynes undergo many of the same addition reactions as alkenes. For example, alkynes will undergo catalytic hydrogenation just as alkenes do:

(100%)

(100%)

In the process, the alkyne consumes two equivalents of molecular hydrogen:

Under these conditions, the *cis* alkene is difficult to isolate because it is even more reactive toward further hydrogenation than the starting alkyne. The obvious question then is whether it is possible to add just one equivalent of hydrogen to form the alkene. With the catalysts we have seen thus far (Pt, Pd, or Ni), this is difficult to achieve. However, with a partially deactivated catalyst, called a *poisoned catalyst*, it is possible to convert an alkyne into a *cis* alkene (without further reduction):

There are many poisoned catalysts. One common example is called *Lindlar's catalyst:*

Lindlar's catalyst = Quinoline, Pd / $BaSO_4$, $CH_3OH$

Another common example is a nickel-boron complex ($Ni_2B$), which is often called the P-2 catalyst. A poisoned catalyst will catalyze the conversion of an alkyne into a *cis* alkene, but it will not catalyze the subsequent reduction to form the alkane (Figure 10.4). Therefore, a poisoned catalyst can be used to convert an alkyne into a *cis* alkene.

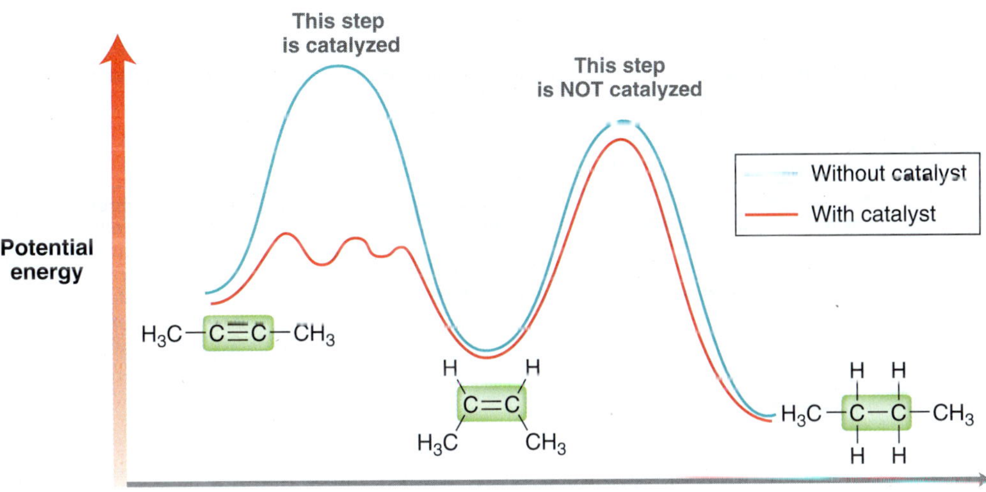


**FIGURE 10.4** An energy diagram showing the effect of a poisoned catalyst. Hydrogenation of the alkyne is catalyzed, but subsequent hydrogenation of the alkene is not catalyzed.

This process does not produce any *trans* alkene. The stereochemical outcome of alkyne hydrogenation can be rationalized in the same way that we rationalized the outcome of alkene hydrogenation (Section 9.7). Both hydrogen atoms are added to the same face of the alkene (*syn* addition) to give the *cis* alkene as the major product.

## CONCEPTUAL CHECKPOINT

**10.9** Draw the major product expected from each of the following reactions:

(a) $H_2$, Lindlar's catalyst → ?; $H_2$, Pt → ?

(b) $H_2$, $Ni_2B$ → ?; $H_2$, Ni → ?

## Dissolving Metal Reduction

### WATCH OUT

These reagents should not be confused with sodium amide ($NaNH_2$), which we saw earlier in this chapter. $NaNH_2$ is a source of $NH_2^-$, a very strong base. In contrast, the reagents employed here (Na, $NH_3$) represent a source of electrons.

In the previous section, we explored the conditions that enable the reduction of an alkyne to a *cis* alkene. Alkynes can also be reduced to *trans* alkenes via an entirely different reaction called **dissolving metal reduction**:

Na, $NH_3$ (*l*) → **(80%)**

The reagents employed are sodium metal (Na) in liquid ammonia ($NH_3$). Ammonia has a very low boiling point (−33 °C), so use of these reagents requires low temperature. When dissolved in liquid ammonia, sodium atoms serve as a source of electrons:

$$Na^{\cdot} \longrightarrow Na^{+} + e^{\ominus}$$

Under these conditions, reduction of alkynes is believed to proceed via Mechanism 10.1.

## MECHANISM 10.1 DISSOLVING METAL REDUCTIONS



In the first step of the mechanism, a single electron is transferred to the alkyne, generating an intermediate that is called a **radical anion**. It is an anion because of the charge associated with the lone pair, and it is a radical because of the unpaired electron:

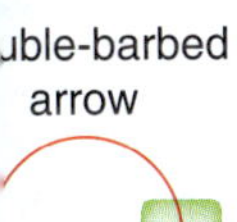


ws the motion
two electrons

Shows the motion
of one electron

**FIGURE 10.5**
Arrows used in ionic mechanisms (double barbed) and radical mechanisms (single barbed).

Radicals and their chemistry are discussed in more detail in Chapter 11. For now, we will just point out that mechanistic steps involving radicals utilize single-barbed curved arrows, often called **fishhook arrows**, rather than double-barbed curved arrows (Figure 10.5). Single-barbed curved arrows indicate the movement of one electron, while double-barbed arrows indicate the movement of two electrons. Notice the use of single-barbed curved arrows in the first step of the mechanism to form the intermediate radical anion. The nature of this intermediate explains the stereochemical preference for formation of a *trans* alkene. Specifically, the intermediate achieves a lower energy state when the paired and unpaired electrons are positioned as far apart as possible, minimizing their repulsion.

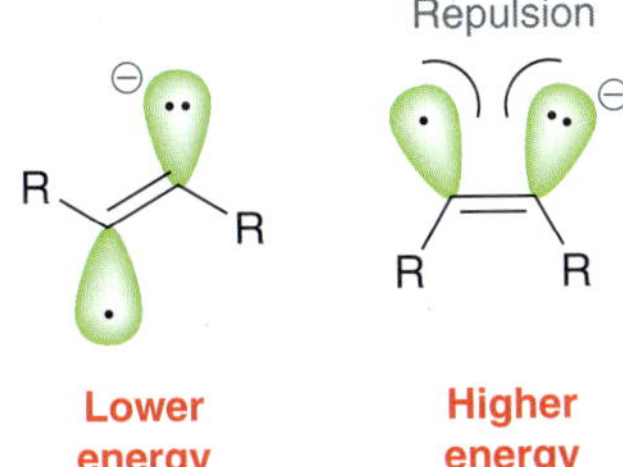


The reaction proceeds more rapidly through the lower energy intermediate, which is then protonated by ammonia under these conditions. In the remaining two steps of the mechanism, another electron is transferred, followed by one more proton transfer. The mechanism therefore is comprised of the following four steps: (1) electron transfer, (2) proton transfer, (3) electron transfer, and (4) proton transfer. That is, the net addition of molecular hydrogen ($H_2$) is achieved via the installation of two electrons and two protons in the following order: $e^-$, $H^+$, $e^-$, $H^+$. The net result is an *anti* addition of two hydrogen atoms across the alkyne:

Na
$NH_3$ (*l*)
H
H

## CONCEPTUAL CHECKPOINT

**10.10** Draw the major product expected when each of the following alkynes is treated with sodium in liquid ammonia:

(a)
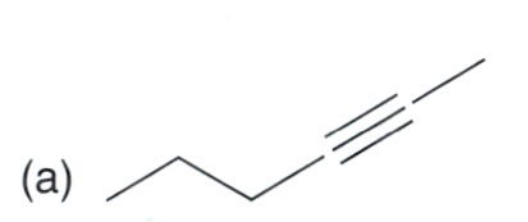

(b)
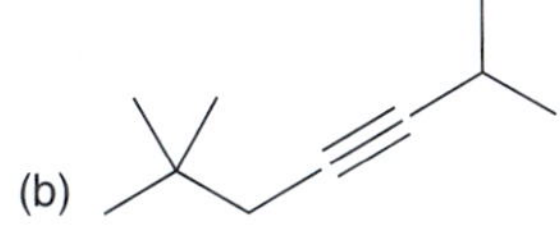

(c)

(d)

## Catalytic Hydrogenation vs. Dissolving Metal Reduction

Figure 10.6 summarizes the various methods we have seen for reducing an alkyne. This diagram illustrates how the outcome of alkyne reduction can be controlled by a careful choice of reagents:

- To produce an *alkane*, an alkyne can be treated with $H_2$ in the presence of a metal catalyst, such as Pt, Pd, or Ni.
- To produce a *cis alkene*, an alkyne can be treated with $H_2$ in the presence of a poisoned catalyst, such as Lindlar's catalyst or $Ni_2B$.
- To produce a *trans alkene*, an alkyne can be treated with sodium in liquid ammonia.

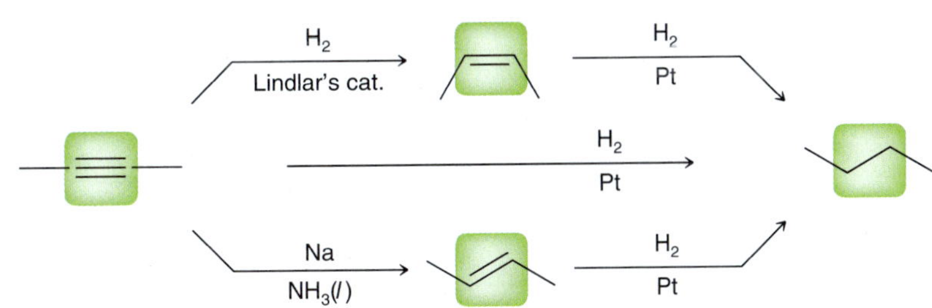


**FIGURE 10.6**
A summary of the various reagents that can be used to reduce an alkyne.

## CONCEPTUAL CHECKPOINT

**10.11** Identify the reagents you would use to achieve each of the following transformations:

(a)

(b)

**10.12** An alkyne with molecular formula $C_5H_8$ was treated with sodium in liquid ammonia to give a disubstituted alkene with molecular formula $C_5H_{10}$. Draw the structure of the alkene.

# 10.6 Hydrohalogenation of Alkynes

## Experimental Observations

In the previous chapter, we saw that alkenes will react with HX via a Markovnikov addition, thereby installing a halogen at the more substituted position:

HX

X

A similar Markovnikov addition is observed when alkynes are treated with HX:

HX

X

**(60–80%)**

Once again, the halogen is installed at the more substituted position.

When the starting alkyne is treated with excess HX, two successive addition reactions occur, producing a geminal dihalide:

Excess HX

X X

## A Mechanism for Hydrohalogenation

In Section 9.3, we proposed the following two-step mechanism for HX addition to alkenes: (1) protonation of the alkene to form the more stable carbocation intermediate followed by (2) nucleophilic attack:

Proton transfer

Nucleophilic attack

**Intermediate carbocation**

A similar mechanism can be proposed for addition of HX to a triple bond: (1) protonation to form a carbocation followed by (2) nucleophilic attack:

Proton transfer

Nucleophilic attack

Vinylic carbocation

This proposed mechanism invokes an intermediate **vinylic carbocation** (*vinyl* = a carbon atom bearing a double bond) and can successfully explain the observed regioselectivity. Specifically, the reaction is expected to proceed via the more stable, secondary vinylic carbocation, rather than via the less stable, primary vinylic carbocation:

more stable than

Secondary vinylic

Primary vinylic

Unfortunately, this proposed mechanism is not consistent with all of the experimental observations. Most notably, studies in the gas phase indicate that vinylic carbocations are not particularly stable. Secondary vinylic carbocations are believed to be similar in energy to regular primary carbocations. Accordingly, we would expect HX addition of alkynes to be significantly slower than HX addition of alkenes. A difference in rate is, in fact, observed, but this difference is not as large as expected—HX addition to alkynes is only slightly slower than HX addition to alkenes. Accordingly, other mechanisms have been proposed that avoid formation of a vinylic carbocation. For example, it is possible that the alkyne interacts with two molecules of HX simultaneously. This process is said to be **termolecular** (involving three molecules) and proceeds through the following transition state:

Transition state

This one-step mechanism avoids the formation of a vinylic carbocation but still invokes a transition state that exhibits some partial carbocationic character (notice the δ+ shown in the transition state). The development of a partial positive charge can effectively explain the observed regioselectivity, because the transition state will be lower in energy when this partial positive charge forms at the more substituted position. This more complex mechanism is supported in many cases by kinetic studies in which the rate expression is found to be overall third order:

$$\text{Rate} = k\,[\text{alkyne}]\,[\text{HX}]^2$$

This rate expression is consistent with the termolecular process proposed above.

It is believed that in most cases the addition of HX to alkynes probably occurs through a variety of mechanistic pathways, all occurring at the same time and competing with each other. The vinylic carbocation probably does play some limited role, but it cannot, by itself, explain all of the observations. In several of the mechanisms throughout this chapter, we may invoke a vinylic carbocation as an intermediate, although it should be understood that other, more complex mechanisms are likely operating simultaneously.

**LOOKING BACK**

Recall from Section 6.5 that "third order" means that the sum of the exponents is 3 for the rate expression.

## Radical Addition of HBr

Recall that in the presence of peroxides, HBr undergoes an *anti*-Markovnikov addition across an alkene:

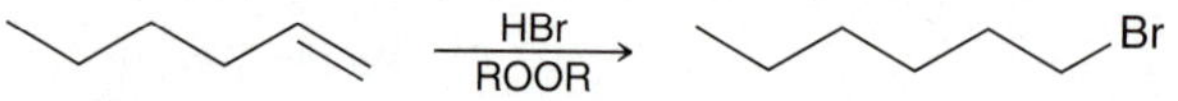


The Br is installed at the less substituted carbon, and the reaction is believed to proceed via a radical mechanism (explored in more detail in Section 11.10). A similar reaction is observed for alkynes. When a terminal alkyne is treated with HBr in the presence of peroxides, an *anti*-Markovnikov addition is observed. The Br is installed at the terminal position, producing a mixture of *E* and *Z* isomers:

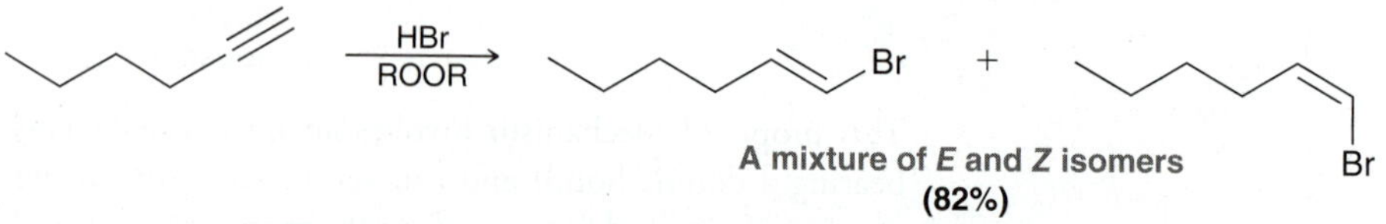


Radical addition only occurs with HBr (not with HCl or HI), as will be explained in Section 11.10.

## Interconversion of Dihalides and Alkynes

The reactions discussed thus far enable the interconversion between dihalides and terminal alkynes:

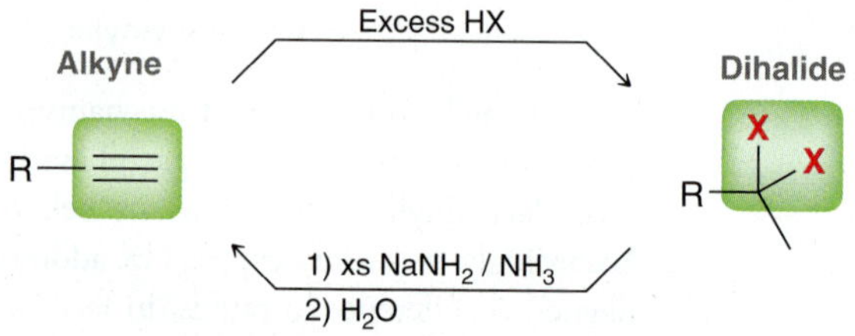


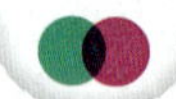

### CONCEPTUAL CHECKPOINT

**10.13** Predict the major product expected for each of the following reactions:

(a) 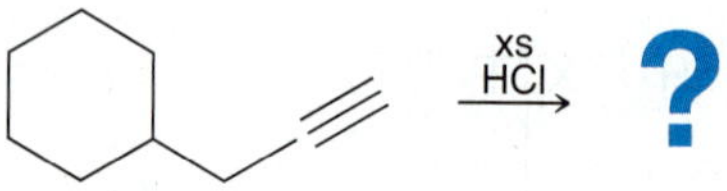


(b) 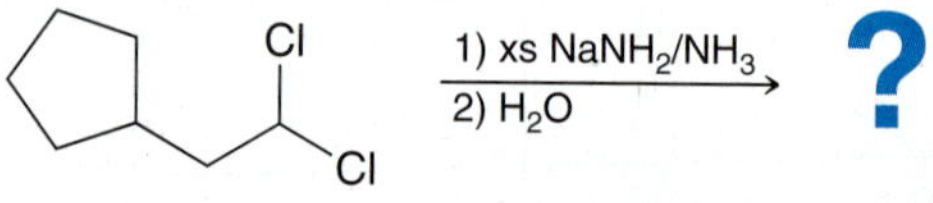


(c) 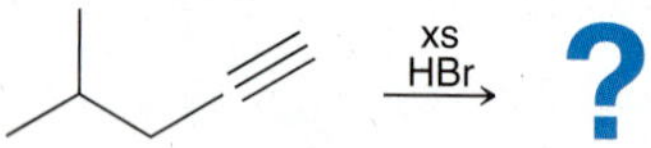


(d) 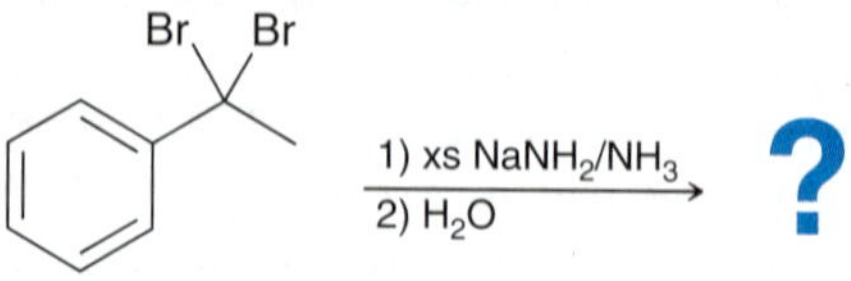


(e) 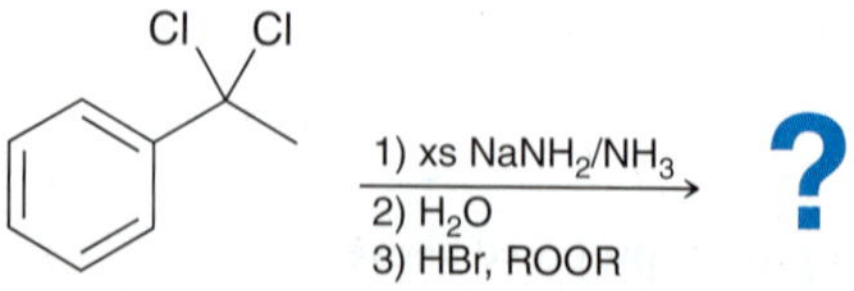


(f) 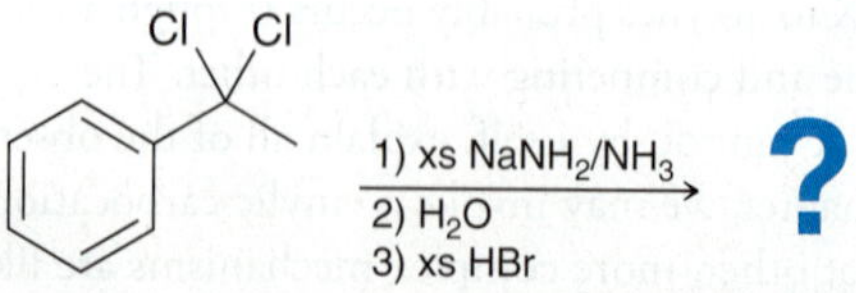


**10.14** Suggest reagents that would achieve the following transformation:

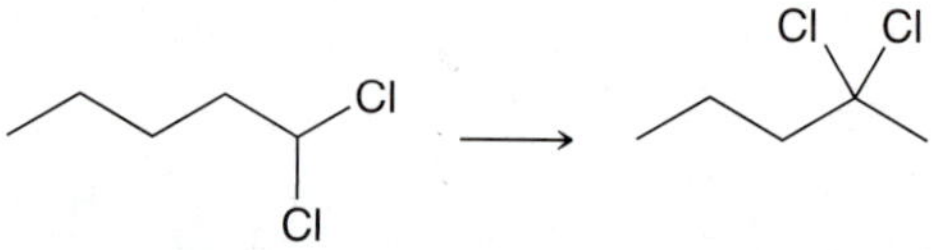


**10.15** An alkyne with molecular formula $C_5H_8$ is treated with excess HBr, and two different products are obtained, each of which has molecular formula $C_5H_{10}Br_2$.

(a) Identify the starting alkyne.

(b) Identify the two products.

## 10.7 Hydration of Alkynes

### Acid-Catalyzed Hydration of Alkynes

In the previous chapter, we saw that alkenes will undergo acid-catalyzed hydration when treated with aqueous acid ($H_3O^+$). The reaction proceeds via a Markovnikov addition, thereby installing a hydroxyl group at the more substituted position:

R R $\xrightarrow{H_3O^+}$ HO R R

Alkynes are also observed to undergo acid-catalyzed hydration, but the reaction is slower than the corresponding reaction with alkenes. As noted earlier in this chapter, the difference in rate is attributed to the high-energy, vinylic carbocation intermediate that is formed when an alkyne is protonated. The rate of alkyne hydration is markedly enhanced in the presence of mercuric sulfate ($HgSO_4$), which catalyzes the reaction:

R $\xrightarrow[HgSO_4]{H_2SO_4,\ H_2O}$ [OH R] → O R

Enol (Not isolated) — Ketone

The initial product of this reaction has a double bond (***en***) and an OH group (***ol***) and is therefore called an **enol**. But the enol cannot be isolated because it is rapidly converted into a ketone. The conversion of an enol into a ketone will appear again in many subsequent chapters and therefore warrants further discussion. Acid-catalyzed conversion of an enol to a ketone occurs via two steps (Mechanism 10.2).

### MECHANISM 10.2 ACID-CATALYZED TAUTOMERIZATION

Proton transfer — Proton transfer

Enol — The π bond of the enol is protonated, generating a resonance-stabilized intermediate — Resonance-stabilized intermediate — The resonance-stabilized intermediate is deprotonated to give a ketone — Ketone

The π bond of the enol is first protonated, generating a resonance-stabilized intermediate, which is then deprotonated to give the ketone. Notice that both steps of this mechanism are proton transfers. The result of this process is the migration of a proton from one location to another, accompanied by a change in location of the π bond:

Enol — Ketone

The enol and ketone are said to be **tautomers**, which are constitutional isomers that rapidly interconvert via the migration of a proton. The interconversion between an enol and a ketone is called **keto-enol tautomerization**. Tautomerization is an equilibrium process, which means that the equilibrium will establish specific concentrations for both the enol and the ketone. Generally, the ketone is highly favored, and the concentration of enol will be quite small. Be very careful not to confuse tautomers with resonance structures. Tautomers are constitutional isomers that exist in equilibrium with one another. Once the equilibrium has been reached, the concentrations of ketone and enol can be measured. In contrast, resonance structures are not different compounds and they are not in equilibrium with one another. Resonance structures simply represent different drawings of one compound.

**LOOKING BACK**
For a review of resonance structures, go to Section 2.7.

Keto-enol tautomerization is an equilibrium process that is catalyzed by even trace amounts of acid (or base). Glassware that is scrupulously cleaned will still have trace amounts of acid or base adsorbed to its surface. As a result, it is extremely difficult to prevent a keto-enol tautomerization from reaching equilibrium.

## SKILLBUILDER

### 10.3 DRAWING THE MECHANISM OF ACID-CATALYZED KETO-ENOL TAUTOMERIZATION

**LEARN the skill**

Under normal conditions, 1-cyclohexenol cannot be isolated or stored in a bottle, because it undergoes rapid tautomerization to yield cyclohexanone. Draw a mechanism for this tautomerization:

**SOLUTION**

The mechanism of acid-catalyzed keto-enol tautomerization has two steps: (1) protonate and then (2) deprotonate. To draw this mechanism, it is essential to remember where to protonate in the first step. There are two places where protonation could possibly occur: the OH group or the double bond. In fact, under these conditions, both sites are reversibly protonated (in acidic conditions, protons are transferred back and forth, wherever possible). In order to choose where to protonate, let's carefully consider the positions of the protons involved in this transformation:

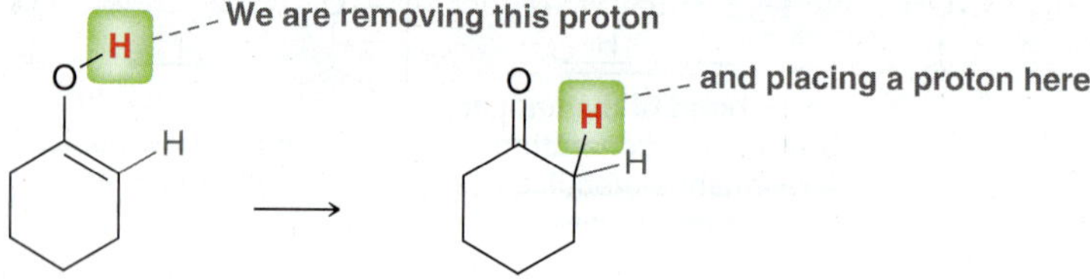


**WATCH OUT**
Do not give the OH another proton, as that pathway will not quickly lead to the ketone. Students who attempt to protonate the OH group invariably continue by losing water as a leaving group to form a vinylic carbocation. That pathway is too high in energy and simply does not lead to the product.

In acidic conditions, we must protonate first and only then deprotonate. In order to accomplish the transformation above, the double bond must be protonated, rather than the OH group. It might be tempting to protonate the OH group first (and this is a very common mistake), but look carefully at the OH group in the enol. The OH must lose its proton in the course of the reaction (to form the ketone). Make sure to begin by protonating the π bond rather than the OH group:

**STEP 1**
Protonate the π bond of the enol.

As is the case for all proton transfer steps, two curved arrows are required. Be sure to draw both of them.

Next, draw the resonance structure of the intermediate formed when the enol is protonated:

**STEP 2**
Draw the resonance structures of the intermediate.

Finally, remove a proton to form the ketone. Once again, two curved arrows are required:

**STEP 3**
Remove a proton to form the ketone.

**PRACTICE the skill** **10.16** The following enols cannot be isolated. They rapidly tautomerize to produce ketones. In each case, draw the expected ketone and show a mechanism for its formation under acid-catalyzed conditions ($H_3O^+$).

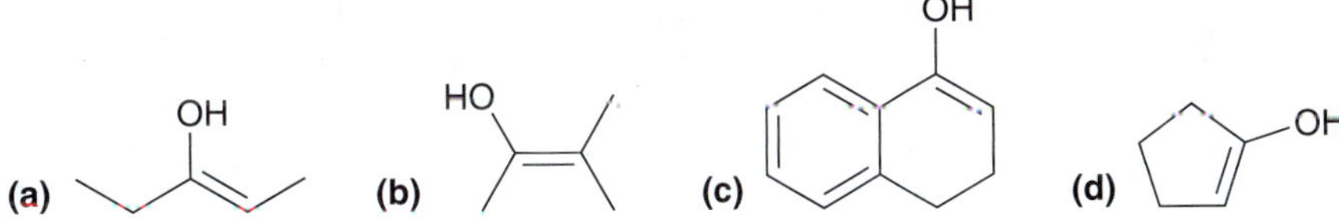


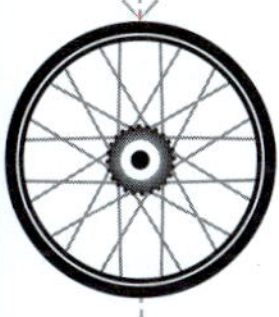

**APPLY the skill** **10.17** Consider the following equilibrium:

$H_3O^+$

These constitutional isomers rapidly interconvert in the presence of even trace amounts of acid and are therefore said to be tautomers of each other. Draw a mechanism for this transformation.

need more **PRACTICE?** **Try Problems 10.49b, 10.63, 10.64, 10.67**

We have seen that even trace amounts of acid will catalyze a keto-enol tautomerization. Therefore, the enol initially produced by hydration of an alkyne will immediately tautomerize to form a ketone. When asked to predict the product of alkyne hydration, do not make the mistake of drawing an enol. Simply draw the ketone.

Acid-catalyzed hydration of unsymmetrical internal alkynes yields a mixture of ketones:

$$R_1-\equiv-R_2 \xrightarrow[HgSO_4]{H_2SO_4,\ H_2O} R_1C(=O)CH_2R_2 + R_1CH_2C(=O)R_2$$

The lack of regiochemical control renders this process significantly less useful. It is most often used for the hydration of a terminal alkyne, which generates a methyl ketone as the product:

$$R-\equiv-H \xrightarrow[HgSO_4]{H_2SO_4,\ H_2O} RC(=O)CH_3$$

## CONCEPTUAL CHECKPOINT

**10.18** Draw the major product(s) expected when each of the following alkynes is treated with aqueous acid in the presence of mercuric sulfate ($HgSO_4$):

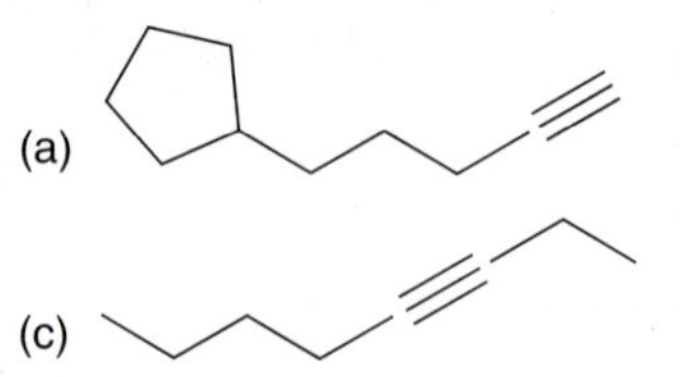

(a) (b) (c) (d) (e)

**10.19** Identify the alkyne you would use to prepare each of the following ketones via acid-catalyzed hydration:

(a) (b) (c)

## Hydroboration-Oxidation of Alkynes

In the previous chapter, we saw that alkenes will undergo hydroboration-oxidation (Section 9.6). The reaction proceeds via an *anti*-Markovnikov addition, thereby installing a hydroxyl group at the less substituted position:

1) $BH_3 \cdot THF$
2) $H_2O_2$, NaOH

Alkynes are also observed to undergo a similar process:

1) $BH_3 \cdot THF$
2) $H_2O_2$, NaOH

The initial product of this reaction is an enol that cannot be isolated because it is rapidly converted into an aldehyde via tautomerization. As we saw in the previous section, tautomerization cannot be prevented, and it is catalyzed by either acid or base. In this case, basic conditions are employed, so the tautomerization process occurs via a base-catalyzed mechanism (Mechanism 10.3).

## MECHANISM 10.3 BASE-CATALYZED TAUTOMERIZATION

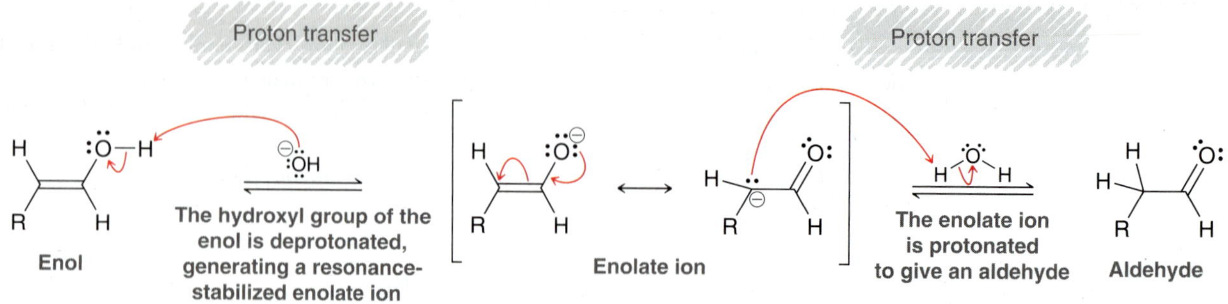


**LOOKING AHEAD**

Enolates are very useful intermediates, and we will explore their chemistry in Chapter 22.

Notice that the order of events under base-catalyzed conditions is the reverse of the order under acid-catalyzed conditions. That is, the enol is first deprotonated, and only then is it protonated (under acid-catalyzed conditions, the first step is to protonate the enol). In basic conditions, deprotonation of the enol leads to a resonance-stabilized anion called an *enolate ion*, which is then protonated to generate the aldehyde.

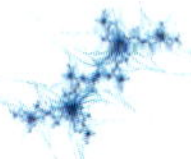



Hydroboration-oxidation of alkynes is believed to proceed via a mechanism that is similar to the mechanism invoked for hydroboration-oxidation of alkenes (Mechanism 9.3). Specifically, borane adds to the alkyne in a concerted process that gives an *anti*-Markovnikov addition. There is, however, one critical difference. Unlike an alkene, which only possesses one π bond, an alkyne possesses two π bonds. As a result, two molecules of $BH_3$ can add across the alkyne. To prevent the second addition, a dialkyl borane ($R_2BH$) is employed instead of $BH_3$. The two alkyl groups provide steric hindrance that prevents the second addition. Two commonly used dialkyl boranes are disiamylborane and 9-BBN:

H—B

**Disiamylborane**

**9-BBN**
**(9-Borabicyclo[3.3.1]nonane)**

With these modified borane reagents, hydroboration-oxidation is an efficient method for converting a terminal alkyne into an aldehyde:

1) Disiamylborane
2) $H_2O_2$, NaOH

Notice that the ultimate product of this reaction sequence is an aldehyde rather than a ketone.

## CONCEPTUAL CHECKPOINT

**10.20** Draw the major product for each of the following reactions:

(a) 1) 9-BBN 2) $H_2O_2$, NaOH ?

(b) 1) Disiamylborane 2) $H_2O_2$, NaOH ?

(c) 1) 9-BBN 2) $H_2O_2$, NaOH ?

**10.21** Identify the alkyne you would use to prepare each of the following compounds via hydroboration-oxidation:

(a)

(b)

(c)

## Controlling the Regiochemistry of Alkyne Hydration

In the previous sections, we explored two methods for the hydration of a terminal alkyne:

R—≡

$H_2SO_4$, $H_2O$, $HgSO_4$ → **A methyl ketone**

1) $R_2BH$ 2) $H_2O_2$, NaOH → **An aldehyde**

Acid-catalyzed hydration of a terminal alkyne produces a methyl ketone, while hydroboration-oxidation produces an aldehyde. In other words, the regiochemical outcome of alkyne hydration can be controlled by the choice of reagents. Let's get some practice determining which reagents to use.

## SKILLBUILDER

### 10.4 CHOOSING THE APPROPRIATE REAGENTS FOR THE HYDRATION OF AN ALKYNE

**LEARN the skill** Identify the reagents you would use to achieve the following transformation:

**SOLUTION**

The starting material is a terminal alkyne and the product has a C=O bond. This reaction can therefore be accomplished via hydration of the starting alkyne. To determine what reagents to use, carefully inspect the regiochemical outcome of the process. Specifically, the oxygen atom is placed at the more substituted position to produce a methyl ketone:

**STEP 1** Identify the regiochemical outcome.

More substituted position

A methyl ketone

This transformation requires a Markovnikov addition, which can be accomplished via an acid-catalyzed hydration:

**STEP 2** Identify the reagents that achieve the desired regiochemical outcome.

$H_2SO_4$, $H_2O$ / $HgSO_4$

**PRACTICE the skill** **10.22** Identify the reagents you would use to achieve each of the following transformations:

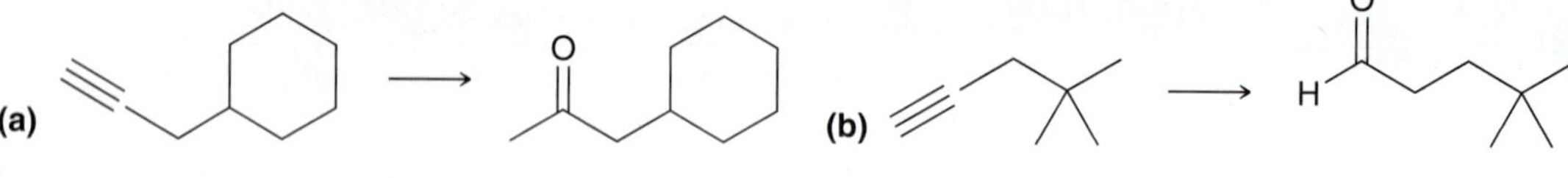


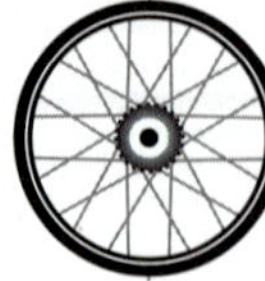

**APPLY the skill** **10.23** Identify the reagents you would use to achieve each of the following transformations:

(a) (b)

**10.24** In the upcoming chapters, we will learn a two-step method for achieving the following transformation. In the meantime, use reactions we have already learned to achieve this transformation:

need more **PRACTICE?** **Try Problem 10.57b**

## 10.8 Halogenation of Alkynes

In the previous chapter, we saw that alkenes will react with $Br_2$ or $Cl_2$ to produce a dihalide. In much the same way, alkynes are also observed to undergo halogenation. The one major difference is that alkynes have two π bonds rather than one and can, therefore, add two equivalents of the halogen to form a tetrahalide:

$$R-C\equiv C-R \xrightarrow[CCl_4]{\text{excess } X_2} R-CX_2-CX_2-R \quad (60\text{–}70\%)$$

(X = Cl or Br)

In some cases, it is possible to add just one equivalent of halogen to produce a dihalide. Such a reaction generally proceeds via an *anti* addition (just as we saw with alkenes), producing the *E* isomer as the major product:

$$R-C\equiv C-R \xrightarrow[CCl_4]{X_2 \text{ (one equivalent)}} (E)\text{-}RXC{=}CXR \text{ (Major)} + (Z)\text{-}RXC{=}CXR \text{ (Minor)}$$

The mechanism of alkyne halogenation is not entirely understood.

## 10.9 Ozonolysis of Alkynes

When treated with ozone followed by water, alkynes undergo oxidative cleavage to produce carboxylic acids:

$$R_1-C\equiv C-R_2 \xrightarrow[2)\ H_2O]{1)\ O_3} R_1-COOH + HO-C(=O)-R_2$$

When a terminal alkyne undergoes oxidative cleavage, the terminal side is converted into carbon dioxide:

$$R-C\equiv C-H \xrightarrow[2)\ H_2O]{1)\ O_3} R-COOH + O=C=O$$

Decades ago, chemists used oxidative cleavage to help with structural determinations. An unknown alkyne would be treated with ozone followed by water, and the resulting carboxylic acids would be identified. This technique allowed chemists to identify the location of a triple bond in an unknown alkyne. However, the advent of spectroscopic methods (Chapters 15 and 16) has rendered this process obsolete as a tool for structural determination.

### CONCEPTUAL CHECKPOINT

**10.25** Draw the major products that are expected when each of the following alkynes is treated with $O_3$ followed by $H_2O$:

(a)

(b)

(c)

(d)

**10.26** An alkyne with molecular formula $C_6H_{10}$ was treated with ozone followed by water to produce only one type of carboxylic acid. Draw the structure of the starting alkyne and the product of ozonolysis.

**10.27** An alkyne with molecular formula $C_4H_6$ was treated with ozone followed by water to produce a carboxylic acid and carbon dioxide. Draw the expected product when the alkyne is treated with aqueous acid in the presence of mercuric sulfate.

## 10.10 Alkylation of Terminal Alkynes

In Section 10.3, we saw that a terminal alkyne can be deprotonated in the presence of a sufficiently strong base, such as sodium amide ($NaNH_2$):

$$\text{R}-\text{C}\equiv\text{C}-\text{H} \xrightarrow{:\overset{\ominus}{\text{N}}\text{H}_2} \text{R}-\text{C}\equiv\text{C}:^{\ominus}$$

**Alkynide ion**

This reaction has powerful synthetic utility, because the resulting alkynide ion can function as a nucleophile when treated with an alkyl halide:

$$\text{R}-\text{C}\equiv\text{C}:^{\ominus} \xrightarrow{\text{R}-\text{X}} \text{R}-\text{C}\equiv\text{C}-\text{R}$$

This transformation proceeds via an $S_N2$ reaction and provides a method to install an alkyl group on a terminal alkyne. This process is called **alkylation**, and it is achieved in just two steps; for example:

$$\text{(1) } NaNH_2 \quad \text{(2) } \text{CH}_3\text{CH}_2\text{CH}_2\text{I}$$

*This process is only efficient with methyl or primary alkyl halides.* When secondary or tertiary alkyl halides are used, the alkynide ion functions primarily as a base and elimination products are obtained. This observation is consistent with the pattern we saw in Section 8.14 (substitution vs. elimination).

Acetylene possesses two terminal protons (one on either side) and can therefore undergo alkylation twice:

$$\text{H}-\text{C}\equiv\text{C}-\text{H} \xrightarrow[\text{2) RX}]{\text{1) NaNH}_2} \text{R}-\text{C}\equiv\text{C}-\text{H} \xrightarrow[\text{2) RX}]{\text{1) NaNH}_2} \text{R}-\text{C}\equiv\text{C}-\text{R}$$

Notice that two separate alkylations are required. One side of acetylene is first alkylated, and then, in a separate process, the other side is alkylated. This repetition is required because $NaNH_2$ and RX cannot be placed into the reaction flask at the same time. Doing so would produce unwanted substitution and elimination products resulting from the reactions between $NaNH_2$ and RX. It might seem burdensome to require two separate alkylation processes, but this requirement does provide additional synthetic utility. Specifically, it enables the installation of two different alkyl groups; for example:

$$\text{H}-\text{C}\equiv\text{C}-\text{H} \xrightarrow[\text{2) EtI}]{\text{1) NaNH}_2} \text{Et}-\text{C}\equiv\text{C}-\text{H} \xrightarrow[\text{2) MeI}]{\text{1) NaNH}_2} \text{Et}-\text{C}\equiv\text{C}-\text{Me}$$

## SKILLBUILDER

### 10.5 ALKYLATING TERMINAL ALKYNES

**LEARN the skill** Identify the reagents necessary to convert acetylene into 7-methyl-3-octyne.

**SOLUTION**

Whenever confronted with a problem that is written only in words, with no corresponding structures, the first step will always be to draw the structures described in the problem. In this case, the task is to identify the reagents necessary to accomplish the following transformation:

**STEP 1** Draw the structures described in the problem.

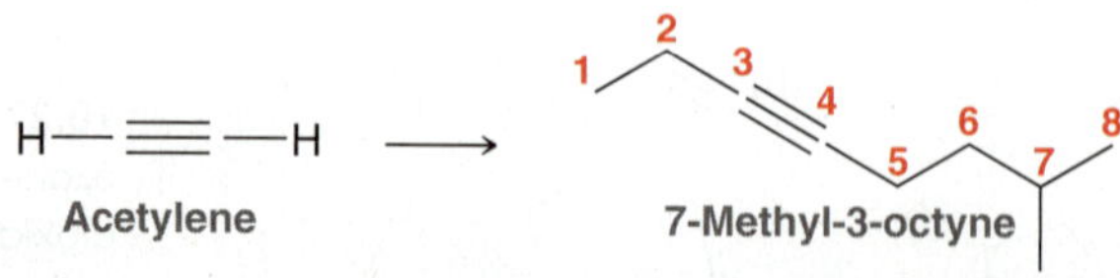


When drawn out, it becomes clear that two alkylations are required in this case. The order of events (which alkyl group is installed first) is not important. Just make sure to install each alkyl group separately. The first alkylation is accomplished by treating acetylene with $NaNH_2$ followed by an alkyl halide:

**STEP 2**
Install the first alkyl group.

H—≡—H 1) $NaNH_2$ 2) EtI →

The terminal alkyne is then alkylated once again using $NaNH_2$ followed by the appropriate alkyl halide:

**STEP 3**
Install the second alkyl group.

1) $NaNH_2$ 2) I →

The overall synthesis therefore has four steps:

H—≡—H 1) $NaNH_2$ 2) EtI 3) $NaNH_2$ 4) I →

Both alkyl halides are primary, so the process would be expected to be efficient.

**PRACTICE the skill** **10.28** Starting with acetylene, show the reagents you would use to prepare the following compounds:

**(a)** 1-Butyne **(b)** 2-Butyne **(c)** 3-Hexyne **(d)** 2-Hexyne **(e)** 1-Hexyne
**(f)** 2-Heptyne **(g)** 3-Heptyne **(h)** 2-Octyne **(i)** 2-Pentyne

**(j)** 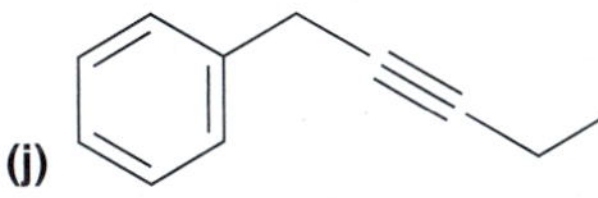 **(k)** 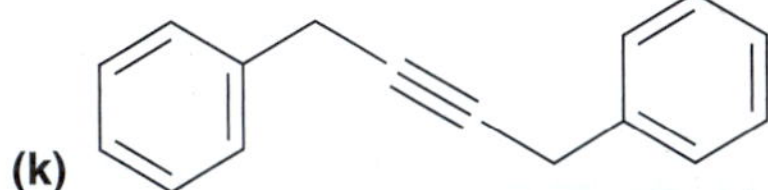

**APPLY the skill** **10.29** Preparation of 2,2-dimethyl-3-octyne cannot be achieved via alkylation of acetylene. Explain.

**10.30** A terminal alkyne was treated with $NaNH_2$ followed by propyl iodide. The resulting internal alkyne was treated with ozone followed by water, giving only one type of carboxylic acid. Provide a systematic, IUPAC name for the internal alkyne.

**10.31** Using acetylene as your only source of carbon atoms, outline a synthesis for 3-hexyne.

need more **PRACTICE?** **Try Problems 10.40f, 10.46, 10.50, 10.59**

## 10.11 Synthesis Strategies

Earlier in this chapter, we saw how to control the reduction of an alkyne. In particular, a triple bond can be converted into a double bond (either *cis* or *trans*) or into a single bond:

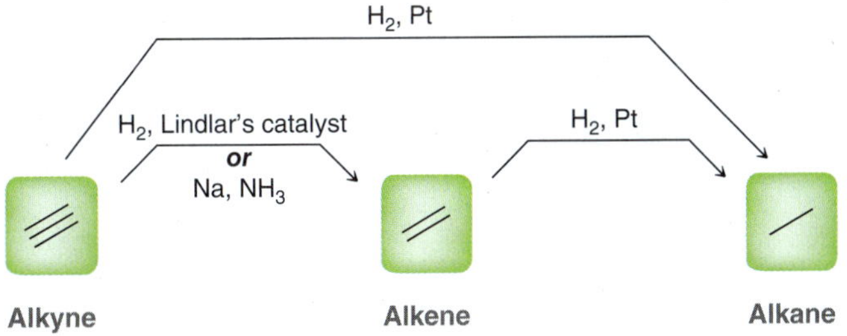

Now let's consider how to go in the other direction, that is, to convert a single bond or double bond into a triple bond:

Thus far, we have not learned the reactions necessary to convert an alkane into an alkene, although we will explore a method in Chapter 11. We did, however, learn how to convert an alkene into an alkyne. Specifically, an alkene can be converted into an alkyne via bromination followed by elimination:

$$\text{alkene} \xrightarrow[CCl_4]{Br_2} \text{Br–CH}_2\text{CH}_2\text{–Br} \xrightarrow[2)\ H_2O]{1)\ xs\ NaNH_2} \text{alkyne}$$

This now provides us with the flexibility to interconvert single, double, and triple bonds (Figure 10.7).

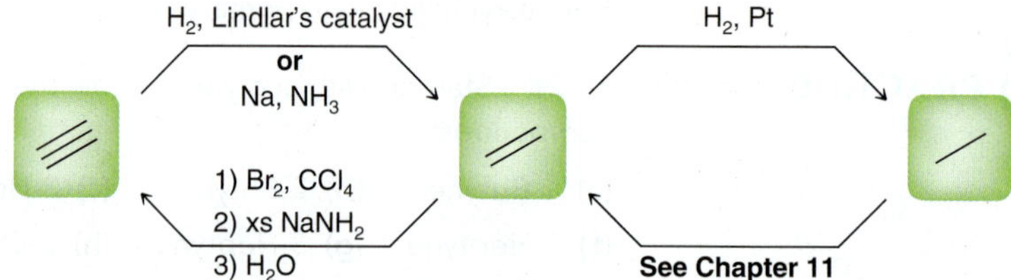


**FIGURE 10.7** A summary of the reagents that can be used to interconvert alkanes, alkenes, and alkynes.

## SKILLBUILDER

### 10.6 INTERCONVERTING ALKANES, ALKENES, AND ALKYNES

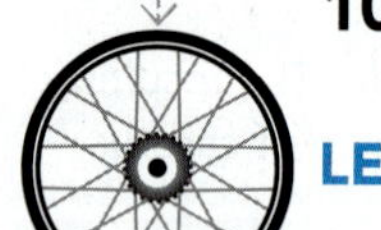

**LEARN the skill** Propose an efficient synthesis for the following transformation:

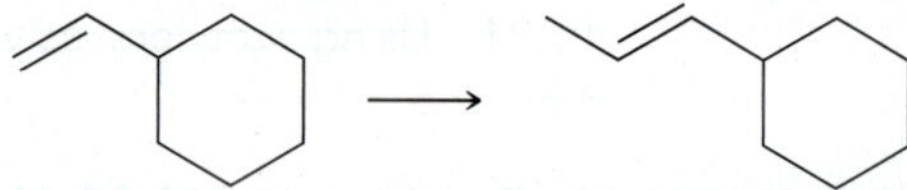

**SOLUTION**

This problem requires the installation of a methyl group at a vinylic position (i.e., alkylation of an alkene). We have not yet seen a direct way to alkylate an alkene. However, we did learn how to alkylate an alkyne. The ability to interconvert alkenes and alkynes therefore enables us to achieve the desired transformation. Specifically, the alkene can first be converted into an alkyne, which can be readily alkylated. Then, after the alkylation, the alkyne can be converted back into an alkene:

The first part of this strategy (conversion of the alkene into an alkyne) can be accomplished via bromination of the alkene to give a dibromide followed by elimination with excess $NaNH_2$. The resulting alkyne can be purified and isolated and then alkylated upon treatment with $NaNH_2$ followed by MeI. Finally, a dissolving metal reduction will convert the alkyne into a *trans* alkene:

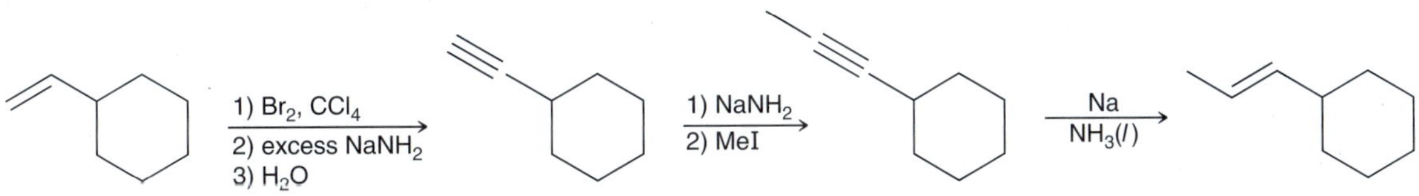

1) $Br_2$, $CCl_4$ 2) excess $NaNH_2$ 3) $H_2O$ → ; 1) $NaNH_2$ 2) MeI → ; Na, $NH_3(l)$ →

**PRACTICE the skill** **10.32** Propose an efficient synthesis for each of the following transformations:

(a) 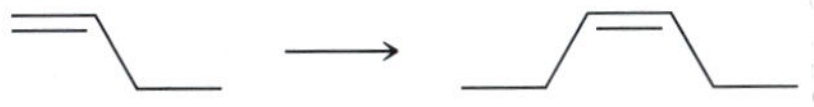 →

(b) 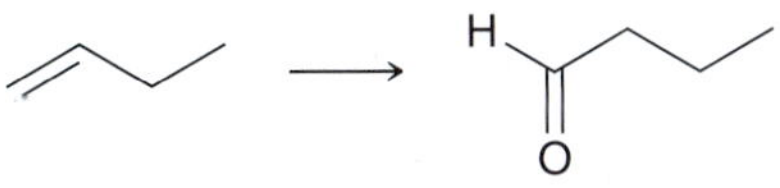 → H, O

(c) 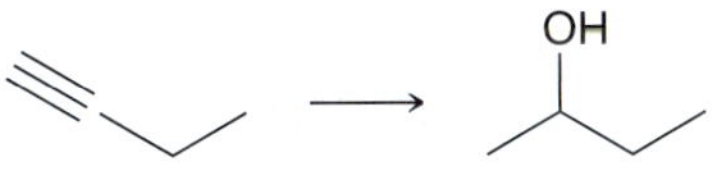 → OH

(d) 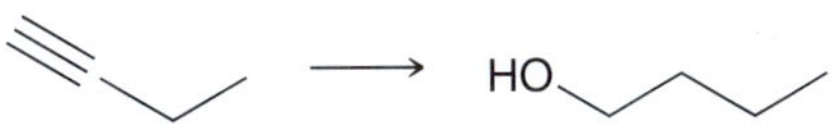 → HO

(e) 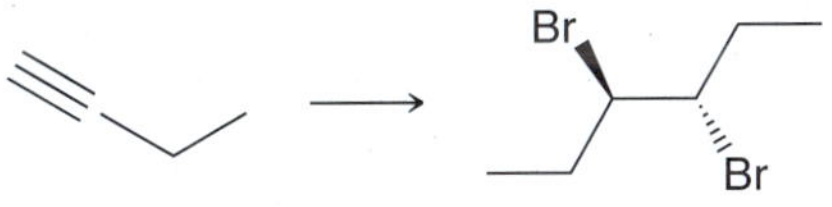 → Br, Br

(f) 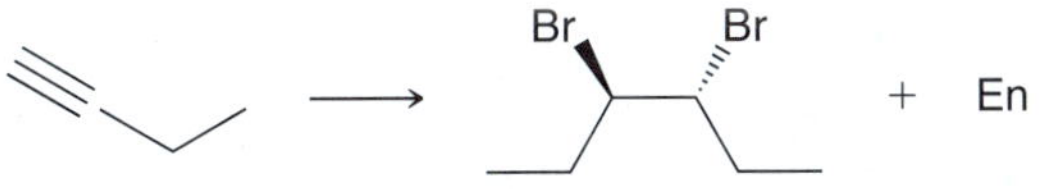 → Br, Br + En

**APPLY the skill** **10.33** Identify the reagents you would use to achieve each of the following transformations:

**(a)** Conversion of all of the carbon atoms in bromoethane into $CO_2$ gas

**(b)** Conversion of all of the carbon atoms in 2-bromopropane into acetic acid ($CH_3COOH$)

**10.34** Using ethylene ($H_2C{=}CH_2$) as your only source of carbon atoms, outline a synthesis for 3-hexanone ($CH_3CH_2COCH_2CH_2CH_3$).

need more **PRACTICE?** **Try Problems 10.57, 10.60**

## REVIEW OF REACTIONS

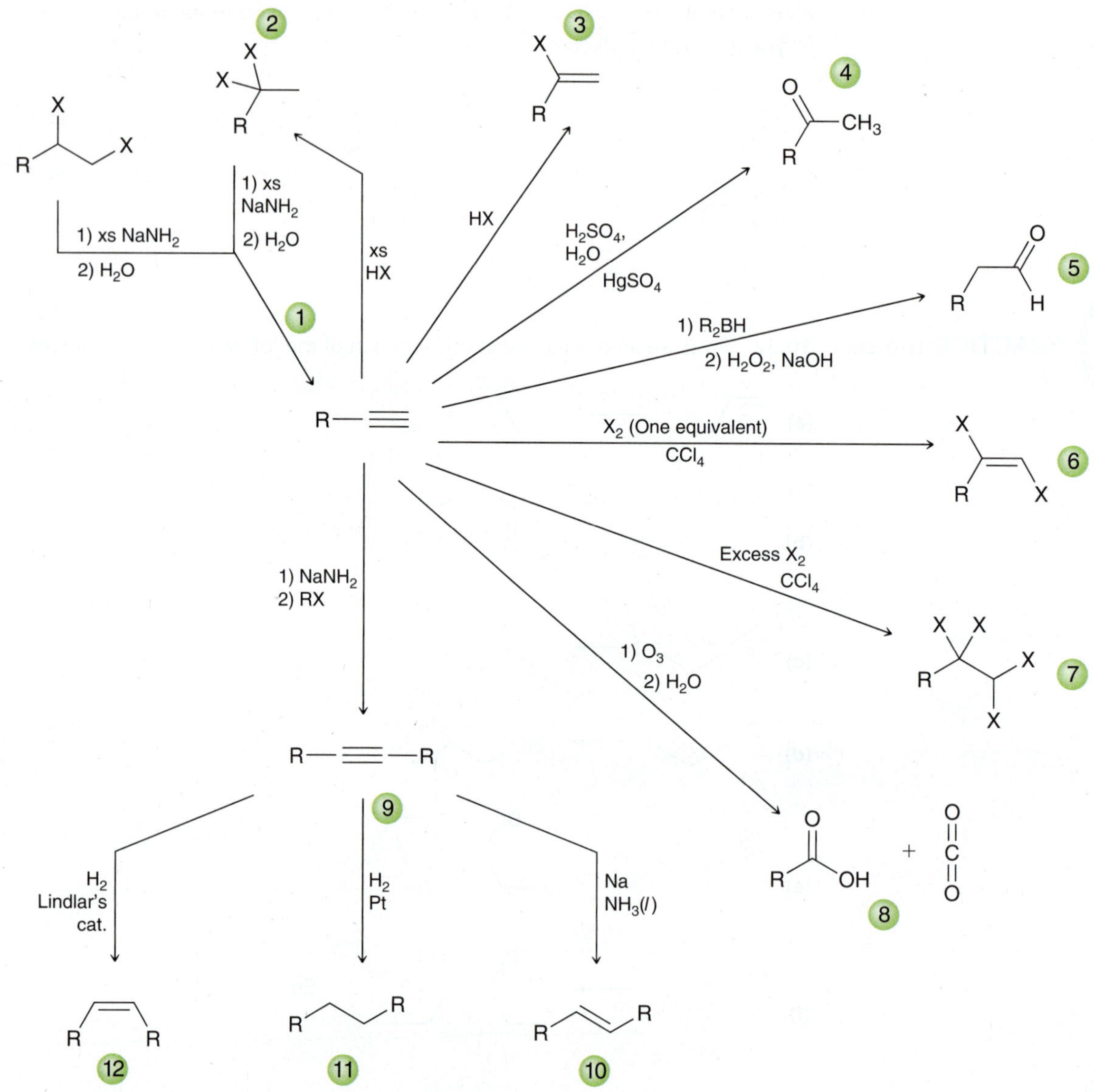


**1.** Elimination
**2.** Hydrohalogenation (two equivalents)
**3.** Hydrohalogenation (one equivalent)
**4.** Acid-catalyzed hydration
**5.** Hydroboration-oxidation
**6.** Halogenation (one equivalent)
**7.** Halogenation (two equivalents)
**8.** Ozonolysis
**9.** Alkylation
**10.** Dissolving metal reduction
**11.** Hydrogenation
**12.** Hydrogenation with a poisoned catalyst

## REVIEW OF CONCEPTS AND VOCABULARY

### SECTION 10.1

- A triple bond is comprised of three separate bonds: one σ bond and two π bonds.
- Alkynes exhibit linear geometry and can function either as bases or as nucleophiles.

### SECTION 10.2

- Alkynes are named much like alkanes, with the following additional rules:
- The suffix "ane" is replaced with "yne."
- The parent is the longest chain that includes the C≡C triple bond.

- The triple bond should receive the lowest number possible.
- The position of the triple bond is indicated with a single locant placed either before the parent or the suffix.
- Monosubstituted acetylenes are **terminal alkynes**, while disubstituted acetylenes are **internal alkynes**.

### SECTION 10.3

- The conjugate base of acetylene, called an **acetylide ion**, is relatively stabilized because the lone pair occupies an *sp*-hybridized orbital.
- The conjugate base of a terminal alkyne is called an **alkynide ion**, which can only be formed with a sufficiently strong base, such as $NaNH_2$.

### SECTION 10.4

- Alkynes can be prepared from either **geminal** or **vicinal** dihalides via two successive E2 reactions.

### SECTION 10.5

- Catalytic hydrogenation of an alkyne yields an alkane.
- Catalytic hydrogenation in the presence of a poisoned catalyst (Lindlar's catalyst or $Ni_2B$) yields a *cis* alkene.
- A **dissolving metal reduction** will convert an alkyne into a *trans* alkene. The reaction involves an intermediate **radical anion** and employs **fishhook arrows**, which indicate the movement of only one electron.

### SECTION 10.6

- Alkynes react with HX via a Markovnikov addition.
- One possible mechanism for the hydrohalogenation of alkynes involves a **vinylic carbocation**, while another possible mechanism is **termolecular**.
- Addition of HX to alkynes probably occurs through a variety of mechanistic pathways all of which are occurring at the same time and competing with each other.

- Treatment of a terminal alkyne with HBr and peroxides gives an *anti*-Markovnikov addition of HBr.

### SECTION 10.7

- Acid-catalyzed hydration of alkynes is catalyzed by mercuric sulfate ($HgSO_4$) to produce an **enol** that cannot be isolated because it is rapidly converted into a ketone.
- Enols and ketones are **tautomers**, which are constitutional isomers that rapidly interconvert via the migration of a proton.
- The interconversion between an enol and a ketone is called **keto-enol tautomerization** and is catalyzed by trace amounts of acid or base.
- Hydroboration-oxidation of a terminal alkyne proceeds via an *anti*-Markovnikov addition to produce an enol that is rapidly converted into an aldehyde via tautomerization.
- In basic conditions, tautomerization proceeds via a resonance-stabilized anion called an *enolate ion*.

### SECTION 10.8

- Alkynes can undergo halogenation to form a tetrahalide.

### SECTION 10.9

- When treated with ozone followed by water, internal alkynes undergo oxidative cleavage to produce carboxylic acids.
- When a terminal alkyne undergoes oxidative cleavage, the terminal side is converted into carbon dioxide.

### SECTION 10.10

- Alkynide ions undergo **alkylation** when treated with an alkyl halide (methyl or primary).
- Acetylene possesses two terminal protons and can undergo two separate alkylations.

### SECTION 10.11

- An alkene can be converted into an alkyne via bromination followed by elimination with excess $NaNH_2$.

## SKILLBUILDER REVIEW

### 10.1 ASSEMBLING THE SYSTEMATIC NAME OF AN ALKYNE

**STEP 1** Identify the parent: choose the longest chain that includes the triple bond.

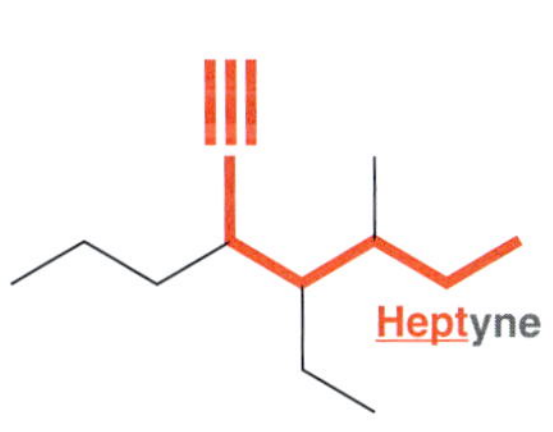


**STEP 2** Identify and name substituents.

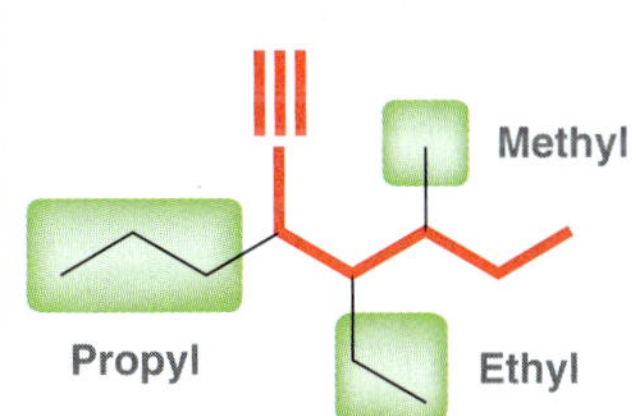


**STEP 3** Number the parent chain and assign a locant to each substituent.

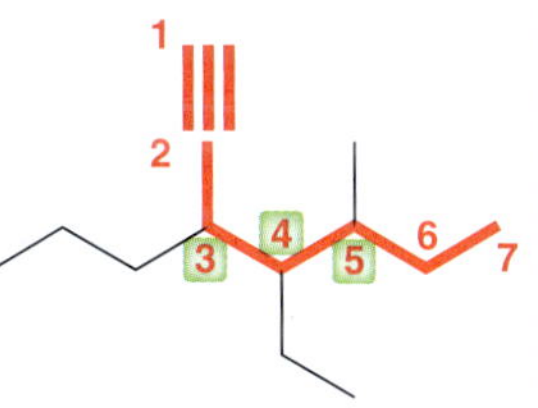


**STEP 4** Assemble the substituents alphabetically.

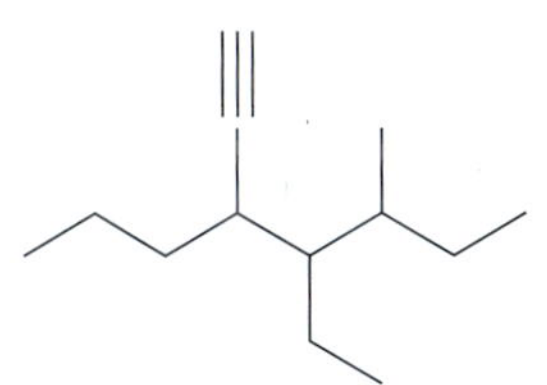


Try Problems 10.1–10.4, 10.35, 10.36

### 10.2 PREDICTING THE POSITION OF EQUILIBRIUM FOR THE DEPROTONATION OF A TERMINAL ALKYNE

**STEP 1** Identify the acid and base on each side of the equilibrium.

R—C≡C—H (Acid) + $^{\ominus}$:ÖH (Base) ⇌ R—C≡C:$^{\ominus}$ (Base) + $H_2O$ (Acid)

**STEP 2** Compare the two acids and determine which is the weaker acid (higher p$K_a$).

R—C≡C—H (p$K_a$ ~ 25) $H_2O$ (p$K_a$ = 15.7)

**STEP 3** Equilibrium favors weaker acid and weaker base. In this case, the equilibrium favors the left side.

Try Problems 10.5, 10.6, 10.39, 10.42

### 10.3 DRAWING THE MECHANISM OF ACID-CATALYZED KETO-ENOL TAUTOMERIZATION

**STEP 1** Protonate the π bond of the enol (do not protonate the hydroxyl group).

**STEP 2** Draw the resonance structure of the intermediate.

**STEP 3** Remove a proton to form the ketone.

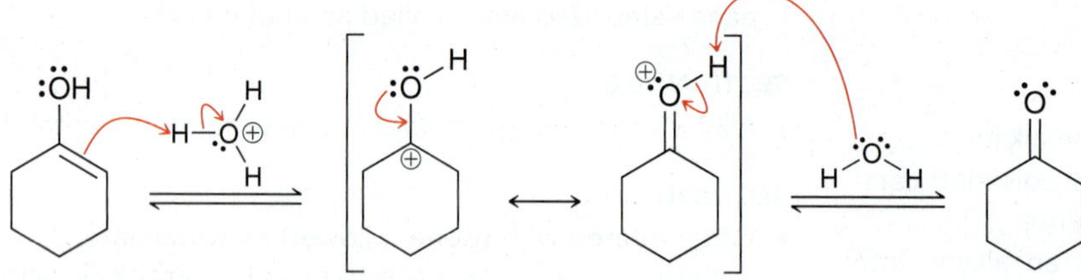

Try Problems 10.16, 10.17, 10.49b, 10.63, 10.64, 10.67

### 10.4 CHOOSING THE APPROPRIATE REAGENTS FOR THE HYDRATION OF AN ALKYNE

**STEP 1** Identify the regiochemical outcome.

**STEP 2** Choose the reagents that achieve that outcome.

R—≡ with $H_2SO_4$, $H_2O$, $HgSO_4$ → A methyl ketone

R—≡ with 1) $R_2BH$ 2) $H_2O_2$, NaOH → An aldehyde

Try Problems 10.22–10.24, 10.57b

### 10.5 ALKYLATING TERMINAL ALKYNES

**STEP 1** Draw the structures described in the problem.

H—≡—H → 7-Methyl-3-octyne

**STEP 2** Install the first alkyl group using sodium amide, followed by the appropriate alkyl halide.

H—≡—H → 1) $NaNH_2$ 2) EtI

**STEP 3** Install the second alkyl group (if necessary).

1) $NaNH_2$ 2) I-(3-methylbutyl)

Try Problems 10.28–10.31, 10.40f, 10.46, 10.50, 10.59

### 10.6 INTERCONVERTING ALKANES, ALKENES, AND ALKYNES

Reagents for interconversion:

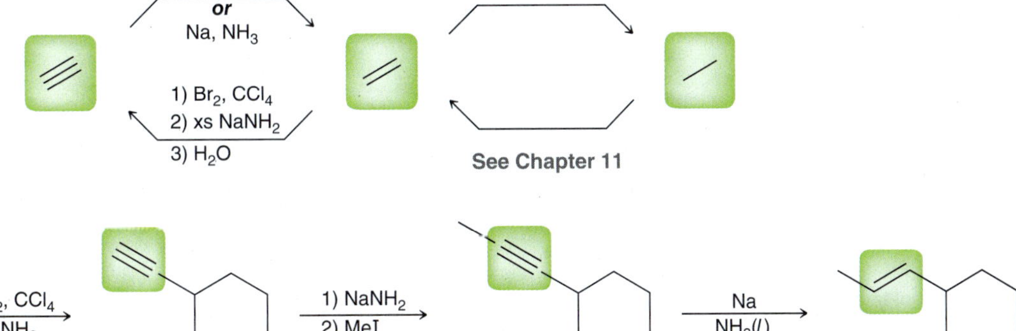


Example:

1) $Br_2$, $CCl_4$ 2) $NaNH_2$ 3) $H_2O$ → 1) $NaNH_2$ 2) MeI → Na, $NH_3(l)$ →

Try Problems **10.32–10.34, 10.57, 10.60**

## PRACTICE PROBLEMS

Note: Most of the Problems are available within **WileyPLUS**, an online teaching and learning solution.

**10.35** Provide a systematic name for each of the following compounds:

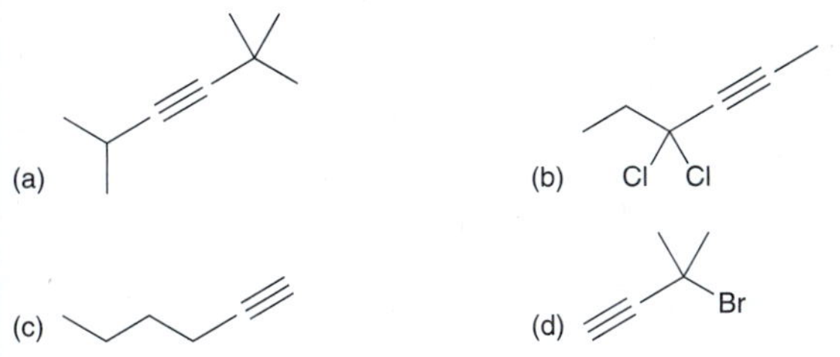

(a) (b) Cl Cl (c) (d) Br

**10.36** Draw a bond-line structure for each of the following compounds:

(a) 2-Heptyne

(b) 2,2-Dimethyl-4-octyne

(c) 3,3-Diethylcyclodecyne

**10.37** Predict the product for each of the following reactions:

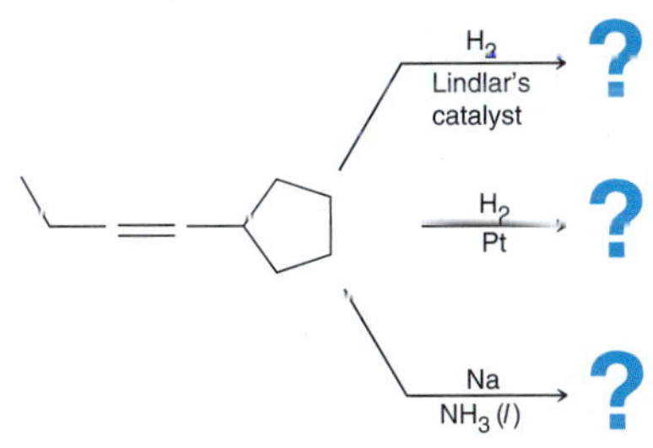


**10.38** Predict the product for each of the following reactions:

$H_2$, Lindlar's catalyst → ?

$H_2$, Pt → ?

**10.39** Draw the products of each of the following acid-base reactions and then predict the position of equilibrium in each case:

(a) H–C≡C–H + $CH_3CH_2CH_2^{\ominus}$ $Li^{\oplus}$ (b) (alkyne)–H + NaH

**10.40** Predict the products obtained when 1-pentyne reacts with each of the following reagents:

(a) $H_2SO_4$, $H_2O$, $HgSO_4$

(b) 9-BBN followed by $H_2O_2$, NaOH

(c) Two equivalents of HBr

(d) One equivalent of HCl

(e) Two equivalents of $Br_2$ in $CCl_4$

(f) $NaNH_2$ in $NH_3$ followed by MeI

(g) $H_2$, Pt

**10.41** Identify the reagents you would use to achieve each of the following transformations:

Cl, Br, Br, O, OH, O, H, O, $CH_3$

**10.42** Identify which of the following bases can be used to deprotonate a terminal alkyne:

(a) $NaOCH_3$ (b) NaH (c) BuLi (d) NaOH (e) $NaNH_2$

**10.43** Identify which of the following compounds represent a pair of keto-enol tautomers:

(a) (b) (c) (d)

**10.44** Draw the enol tautomer of each of the following ketones:

(a) (b) (c)

**10.45** Oleic acid and elaidic acid are isomeric alkenes:

Oleic acid

Elaidic acid

Oleic acid, a major component of butter fat, is a colorless liquid at room temperature. Elaidic acid, a major component of partially hydrogenated vegetable oils, is a white solid at room temperature. Oleic acid and elaidic acid can both be prepared in the laboratory by reduction of an alkyne called stearolic acid. Draw the structure of stearolic acid and identify the reagents you would use to convert stearolic acid into oleic acid or elaidic acid.

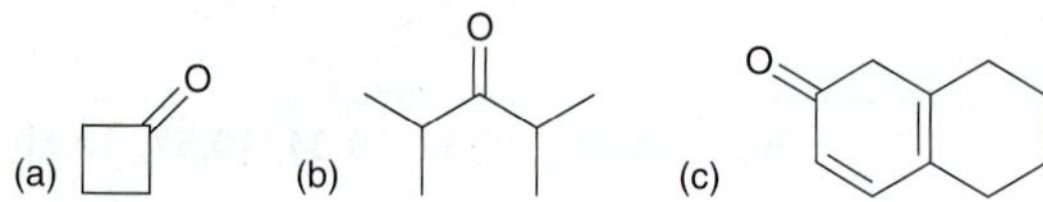

**10.46** Predict the final product(s) for each sequence of reactions:

(a) 1) Excess $NaNH_2$ 2) EtCl 3) $H_2$, Lindlar's catalyst → ?

(b) H—C≡C—H 1) $NaNH_2$ 2) MeI 3) 9-BBN 4) $H_2O_2$, NaOH → ?

(c) H—C≡C—H 1) $NaNH_2$ 2) EtI 3) $HgSO_4$, $H_2SO_4$, $H_2O$ → ?

(d) H—C≡C—H 1) $NaNH_2$ 2) MeI 3) $NaNH_2$ 4) EtI 5) Na, $NH_3$ (*l*) → ?

**10.47** When (*R*)-4-bromohept-2-yne is treated with $H_2$ in the presence of Pt, the product is optically inactive. Yet, when (*R*)-4-bromohex-2-yne is treated with the same conditions, the product is optically active. Explain.

**10.48** Draw the structure of an alkyne that can be converted into 3-ethylpentane upon hydrogenation. Provide a systematic name for the alkyne.

**10.49** Propose a mechanism for each of the following transformations:

(a) Na, $NH_3$ (*l*)

(b) $H_3O^+$

**10.50** Draw the expected product of each of the following reactions, showing stereochemistry where appropriate:

$NaNH_2$, $NH_3$ → ? → ?

**10.51** Compound A is an alkyne that reacts with two equivalents of $H_2$ in the presence of Pd to give 2,4,6-trimethyloctane.

(a) Draw the structure of compound A.

(b) How many chirality centers are present in compound A?

(c) Identify the locants for the methyl groups in compound A. Explain why the locants are not 2, 4, and 6 as seen in the product of hydrogenation.

**10.52** Compound A has molecular formula $C_7H_{12}$. Hydrogenation of compound A produces 2-methylhexane. Hydroboration-oxidation of compound A produces an aldehyde. Draw the structure of compound A and draw the structure of the aldehyde produced upon hydroboration-oxidation of compound A.

**10.53** Propose a plausible synthesis for each of the following transformations:

(a) (b) (c) (d) (e) (f)

**10.54** 1,2-Dichloropentane reacts with excess sodium amide in liquid ammonia (followed by water workup) to produce compound X. Compound X undergoes acid-catalyzed hydration to produce a ketone. Draw the structure of the ketone produced upon hydration of compound X.

**10.55** An unknown alkyne is treated with ozone (followed by hydrolysis) to yield acetic acid and carbon dioxide. What is the structure of the alkyne?

**10.56** Compound A is an alkyne with molecular formula $C_5H_8$. When treated with aqueous sulfuric acid and mercuric sulfate, two different products with molecular formula $C_5H_{10}O$ are obtained in equal amounts. Draw the structure of compound A, and draw the two products obtained.

**10.57** Propose an efficient synthesis for each of the following transformations:

(a)

(b)

(c)

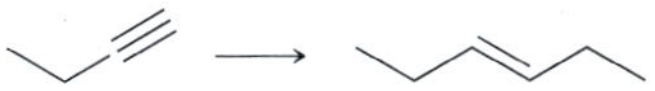

(d)

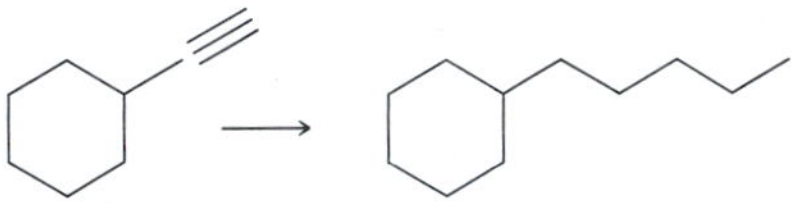

**10.58** Draw the structure of each possible dichloride that can be used to prepare the following alkyne via elimination:

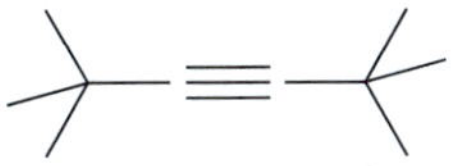

**10.59** Draw the structures of compounds **A** to **D**:

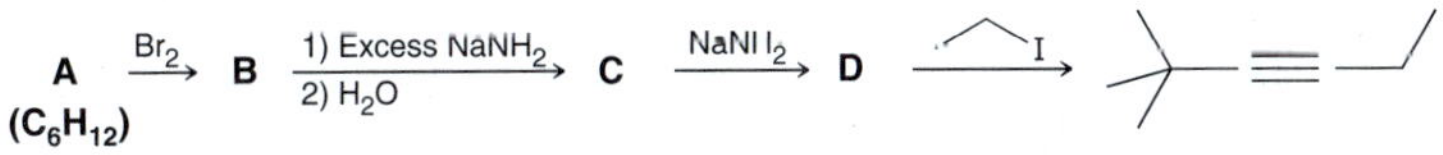

**A** ($C_6H_{12}$) —$Br_2$→ **B** —1) Excess $NaNH_2$ 2) $H_2O$→ **C** —$NaNH_2$→ **D** —I→

## INTEGRATED PROBLEMS

**10.60** Identify the reagents necessary to achieve each transformation below. In each case, you will need to use at least one reaction from this chapter and at least one reaction from the previous chapter. The essence of each problem is to choose reagents that will achieve the desired stereochemical outcome:

(a) 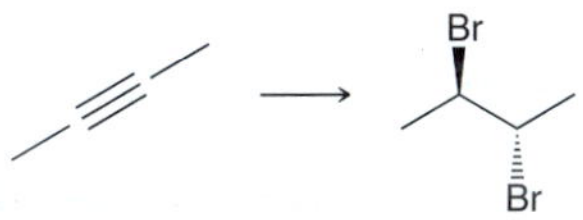 Br Br

(b) 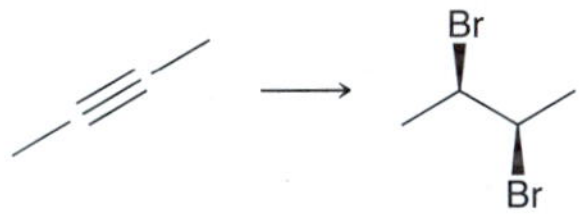 Br Br + En

(c) 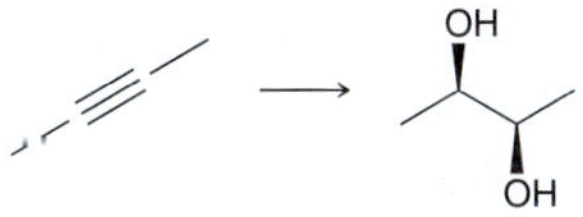 OH OH + En

(d) 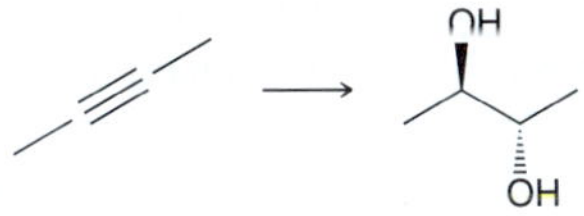 OH OH

(e) 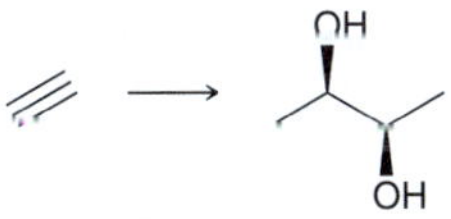 OH OH + En (f) 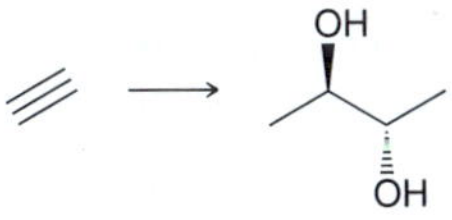 OH OH

**10.61** Identify the reagents necessary to achieve each transformation below:

(a) 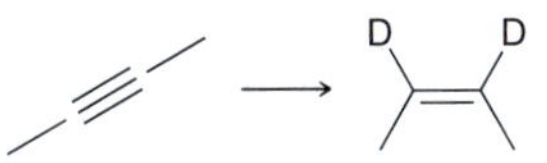 D D (b) 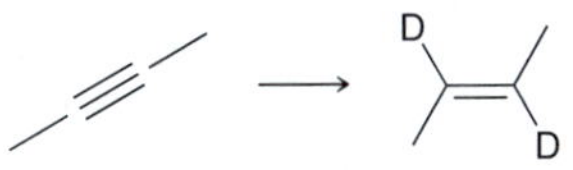 D D

(c) 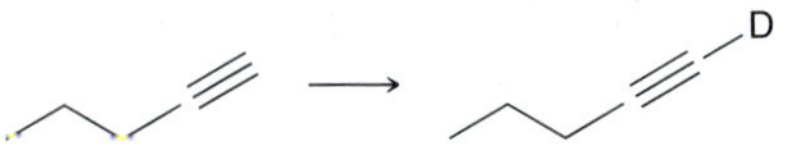 D

**10.62** The following reaction does not produce the desired product but does produce a product that is a constitutional isomer of the desired product. Draw the product that is obtained and propose a mechanism for its formation:

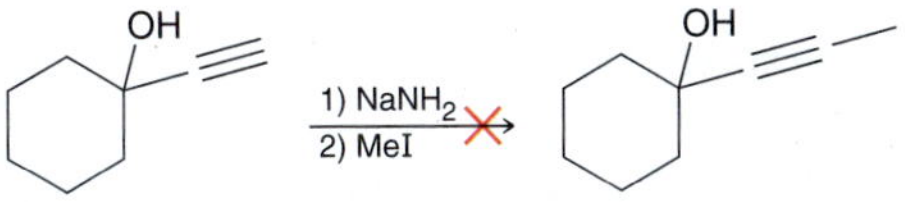

OH —1) $NaNH_2$ 2) MeI ✗→ OH

**10.63** Propose a plausible mechanism for the following transformation:

O OH —$H_3O^+$→ O OH

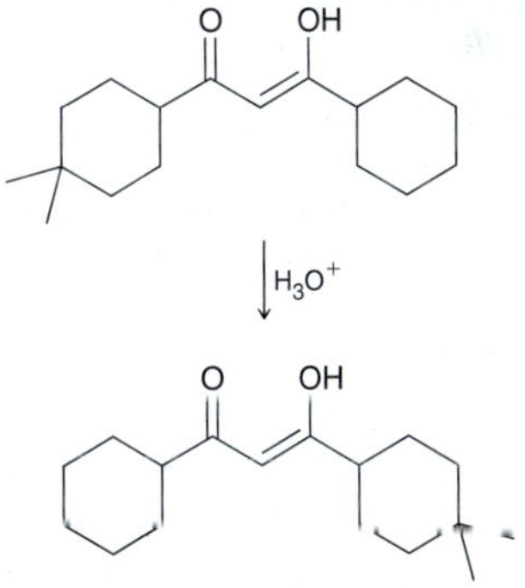

**10.64** Propose a plausible mechanism for the following tautomerization process:

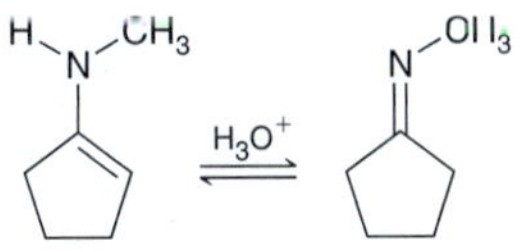

H–N–$CH_3$ ⇌($H_3O^+$) N–$CH_3$

**10.65** Using acetylene and methyl bromide as your only sources of carbon atoms, propose a synthesis for each of the following compounds:

(a) 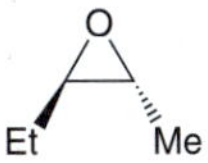 Et O Me + En (b) 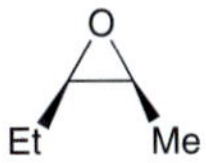 Et O Me + En

**10.66** Propose a plausible mechanism for the following transformation:

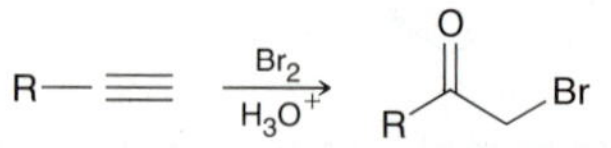

**10.67** Propose a plausible mechanism for the following transformation:

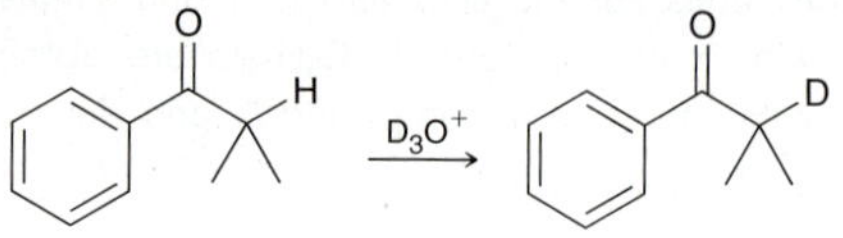

# CHALLENGE PROBLEMS

**10.68** Roquefortine C belongs to a class of natural products, called roquefortines, first isolated from cultures of the fungus *Penicillium roqueforti*. Roquefortine C, which is also present in blue cheese, exhibits bacteriostatic activity (it prevents bacteria from reproducing), and it is known to exist as a mixture of two tautomers, as shown below (*J. Am. Chem. Soc.* **2008,** *130,* **6281–6287**):

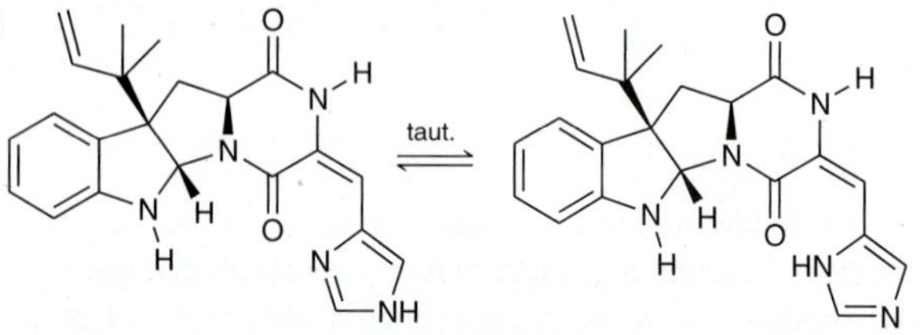

Roquefortine C

(a) Draw a base-catalyzed mechanism for the tautomerization of roquefortine C.

(b) Draw an acid-catalyzed mechanism for the tautomerization of roquefortine C.

(c) Predict which tautomer is more stable and explain your rationale.

**10.69** A small class of natural products, called α,α-disubstituted α-amino acids (Chapter 25), have been the targets of several synthetic techniques, because these compounds are structurally complex (representing a challenge for synthetic organic chemists), and they exhibit notable effects on biological activity. The following transformation was part of one such synthetic technique (*Org. Lett.* **2003,** *5,* **4017–4020**): (a) Identify reagents that can be used to achieve the following transformation and (b) assign the configuration of the chirality center in the starting material and in the final product.

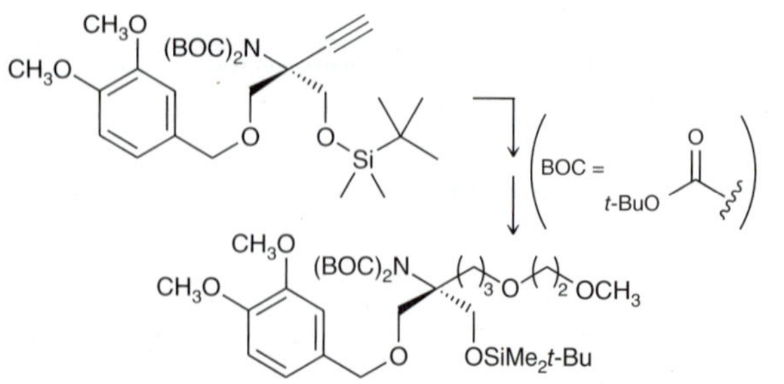

**10.70** Natural products that contain the *N*-1,1-dimethyl-2-propenyl group (called an *N*-reverse prenyl group) often exhibit antitumor or antifungal activity. The synthesis of a particular *N*-reverse prenylated indole antifungal compound begins with the two steps shown below (*Tetrahedron Lett.* **2001,** *42,* **7277–7280**):

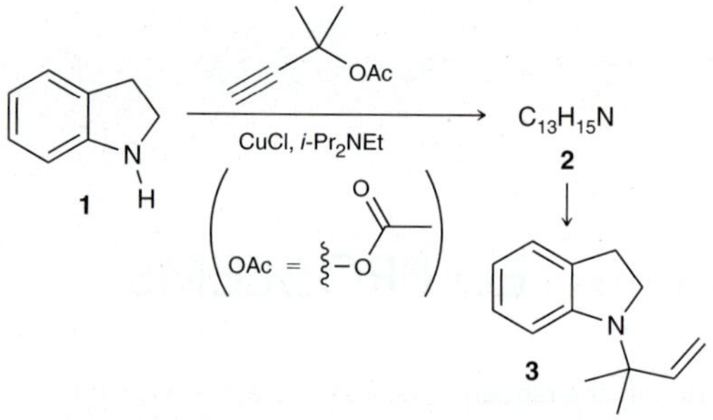

(a) Draw the structure of compound **2** and show a reasonable mechanism for its formation. Note that *i*-$Pr_2NEt$ is a base, and CuCl is used as a catalyst. The latter can be ignored for our purposes.

(b) Identify the reagents necessary for the conversion of **2** to **3**.

**10.71** Treatment of one mole of dimethyl sulfate ($CH_3OSO_3CH_3$) with two moles of sodium acetylide results in the formation of two moles of propyne as the major product (*J. Org. Chem.* **1959,** *24,* **840–842**):

(a) Draw the inorganic, ionic species that is generated as a byproduct of this reaction and show a mechanism for its formation.

(b) 2-Butyne is observed as a minor product of this reaction. Draw a mechanism accounting for the formation of this minor product and explain how your proposed mechanism is consistent with the observation that acetylene is present among the reaction products.

(c) Predict the major and minor products that are expected if diethyl sulfate is used in place of dimethyl sulfate.

**10.72** Salvinorin A, isolated from the Mexican plant *Salvia divinorum*, is known to bind with opioid receptors, thereby generating a powerful hallucinogenic effect. It has been suggested that salvinorin A may be useful in the treatment of drug addiction. Terminal alkyne **2**, shown below, was used in a total synthesis of salvinorin A (*J. Am. Chem. Soc.* **2007,** *129,* **8968–8969**). Propose a plausible mechanism for the formation of **2** from alkyne **1**. (*Hint:* refer to Problem 10.8).

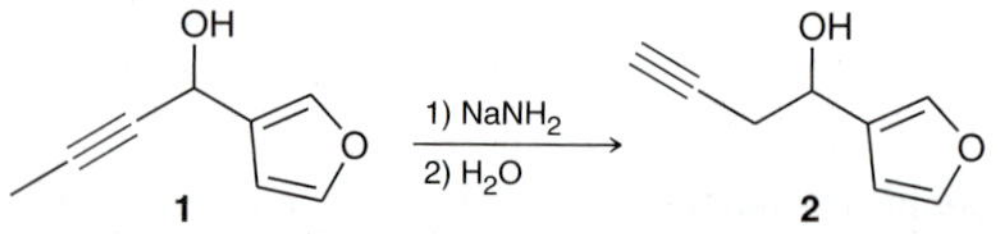

**10.73** Halogenation of alkynes with $Cl_2$ or $Br_2$ can generally be achieved with high yields, while halogenation of alkynes with $I_2$ typically gives low yields. However, the following reaction is successfully completed with $I_2$ in high yields (94%) to afford a potentially useful functionalized and conformationally constrained diiododiene synthetic intermediate

(*Tetrahedron Lett.* **1996,** *37,* **1571–1574**). Propose a plausible mechanism for this transformation.

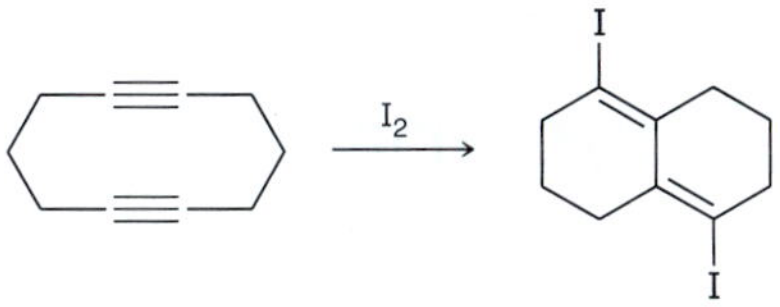

**10.74** The following compounds exhibit tautomerism, with a particularly high enol concentration. Compound **1** exhibits an enol concentration of 9.1%, as compared with the enol concentration of 0.014% for $(CH_3)_2CHCHO$. Compound **2** exhibits an enol concentration of 95% (*Acc. Chem. Res.* **1988,** *21,* **442–449**):

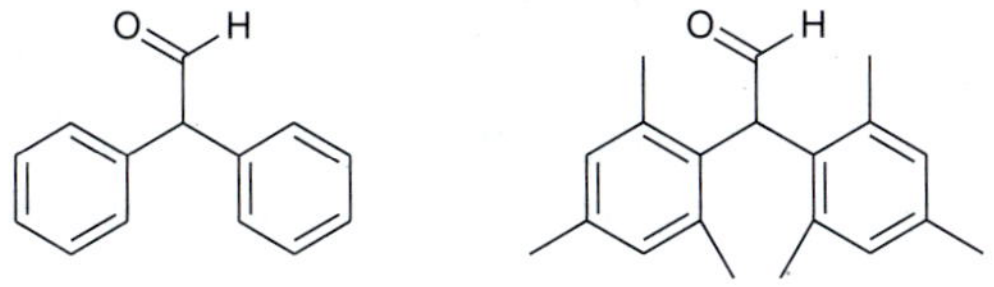

(a) Draw the enol of compound **1** and offer an explanation for the high enol concentration, relative to $(CH_3)_2CHCHO$.

(b) Draw the enol of compound **2**, and explain why the enol is favored over the aldehyde in this case.

**10.75** The two lowest energy conformations of pentane are the *anti-anti* and the *anti-gauche* forms, in terms of arrangements around the two central C—C bonds. A recent study analyzed the conformations of 3-heptyne as an "elongated" analogue of pentane, where a carbon-carbon triple bond is "inserted" between C2 and C3 of pentane (*J. Phys. Chem. A.* **2007,** *111,* **3513–3518**). Interestingly, the researchers found that in each of the two most stable conformations of 3-heptyne, C1 and C6 are nearly eclipsed (looking down the alkyne group). In one of these conformations, C4 and C7 are *anti* to each other (looking down the C5–C6 bond); and in the other conformation, C4 and C7 experience a *gauche* interaction. Draw the following:

(a) A wedge-and-dash structure for each of the two lowest energy conformations of pentane

(b) A wedge-and-dash structure of the conformer of 3-heptyne that is analogous to *anti-anti* pentane

(c) A Newman projection that illustrates the eclipsed nature of the low-energy conformations of 3-heptyne

(d) Newman projections that illustrate the difference between the two lowest energy conformations of 3-heptyne

**10.76** A variety of phenyl-substituted acetylenes (**1a–d**) were treated with HCl to give a mixture of *E* and *Z* isomers, as shown below (*J. Am. Chem. Soc.* **1976,** *98,* **3295–3300**):

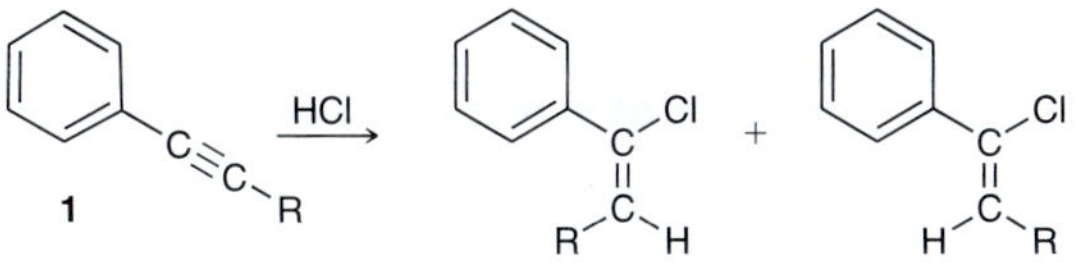

| | *E*:*Z* Ratio |
|---|---|
| **1a** (R = Me) | 70:30 |
| **1b** (R = Et) | 80:20 |
| **1c** (R = *i*-Pr) | 95:5 |
| **1d** (R = *t*-Bu) | 100:0 |

(a) Draw the two possible vinyl carbocations that can form when phenyl-substituted acetylenes are protonated by HCl and explain the regioselectivity observed in these reactions.

(b) By comparing transition state stabilities for the step in which the most stable vinyl carbocation is captured by a chloride ion, provide an explanation for the stereoselectivity (the *E*:*Z* ratio) observed in the reactions of **1a–d**.

**10.77** The compound 1,2-bis-(phenylethynyl)benzene (**1**) undergoes reactions with aqueous acid or bromine, as shown below (*J. Am. Chem. Soc.* **1966,** *88,* **4525–4526**). Propose a reasonable arrow-pushing mechanism for both of these reactions. Also provide an explanation for the configuration of the exocyclic alkene (the one that is external to the ring) in the product of the bromination reaction.

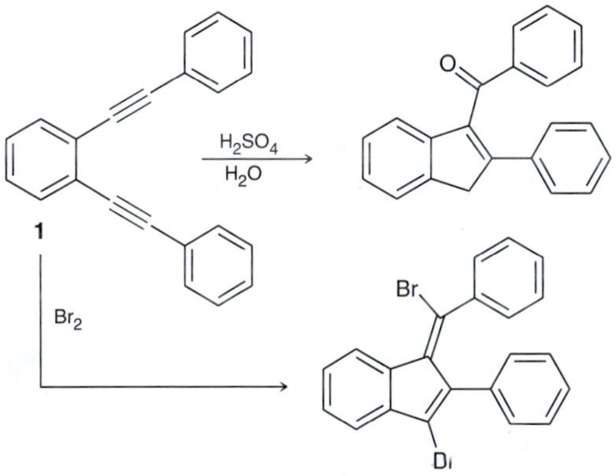

# 11

# Radical Reactions



## DID YOU EVER WONDER...

**how certain chemicals are able to extinguish fires more successfully than water?**

Fire is a chemical reaction called combustion in which organic compounds are converted into $CO_2$ and water together with the liberation of heat and light. The combustion process is a chain (self-perpetuating) process that is believed to occur via free radical intermediates. Understanding the nature of these free radicals is the key to understanding how fires can be extinguished most effectively.

This chapter will focus on radicals. We will learn about their structure and reactivity, and we will explore some of the important roles that radicals play in the food and chemical industries and in our overall health. We will also return to the topic of fire to explain how specially designed chemicals are able to destroy the radical intermediates in a fire, thereby stopping the combustion process and extinguishing the fire.

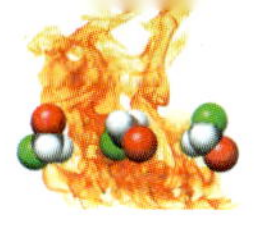

## DO YOU REMEMBER?

*Before you go on, be sure you understand the following topics.*
*If necessary, review the suggested sections to prepare for this chapter.*

- Enthalpy (Section 6.1)
- Entropy (Section 6.2)
- Gibbs Free Energy (Section 6.3)
- Reading Energy Diagrams (Section 6.6)

Take the DO YOU REMEMBER? QUIZ in **WileyPLUS** to check your understanding.

# 11.1 Radicals

## Introduction to Radicals

In Section 6.1, we mentioned that a bond can be broken in two different ways: heterolytic bond cleavage forms ions, while homolytic bond cleavage forms radicals (Figure 11.1). Until now, we have focused mostly on ionic reactions—that is, we have been exploring mechanisms that involve ions. This chapter will focus exclusively on radicals. Look carefully at the curved arrows used in the two

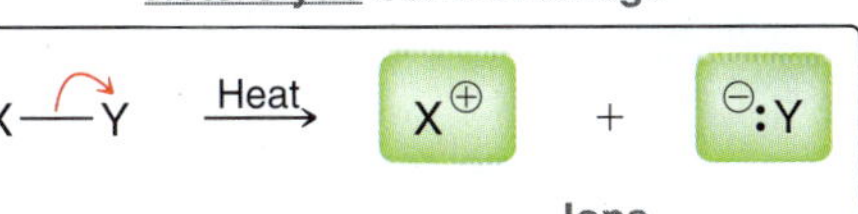


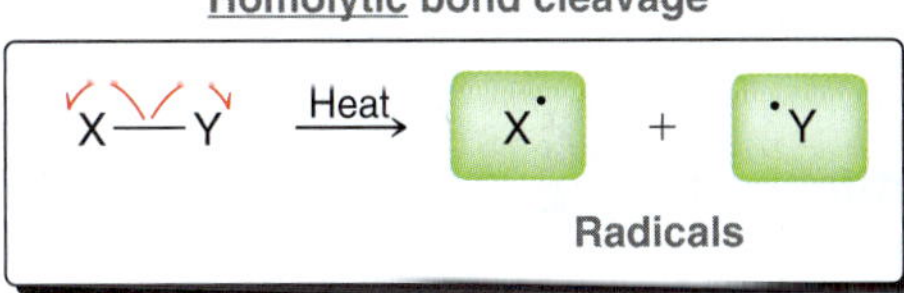


**FIGURE 11.1** An illustration of the difference between homolytic and heterolytic bond cleavage.

processes above. Specifically, an ionic process employs double-barbed curved arrows, while a radical process employs single-barbed arrows (Figure 11.2). A double-barbed curved arrow represents the motion of two electrons, while a single-barbed arrow represents the motion of only one electron. Single-barbed arrows are called *fishhook arrows* because of their appearance. This chapter will use fishhook arrows exclusively.

**FIGURE 11.2** Arrows used in ionic mechanisms are double-barbed, while arrows used in radical mechanisms are single-barbed.

## Structure and Geometry of Radicals

In order to understand the structure and geometry of a radical, we must quickly review the structures of carbocations and carbanions (Figure 11.3). A carbocation is $sp^2$ hybridized and trigonal planar,

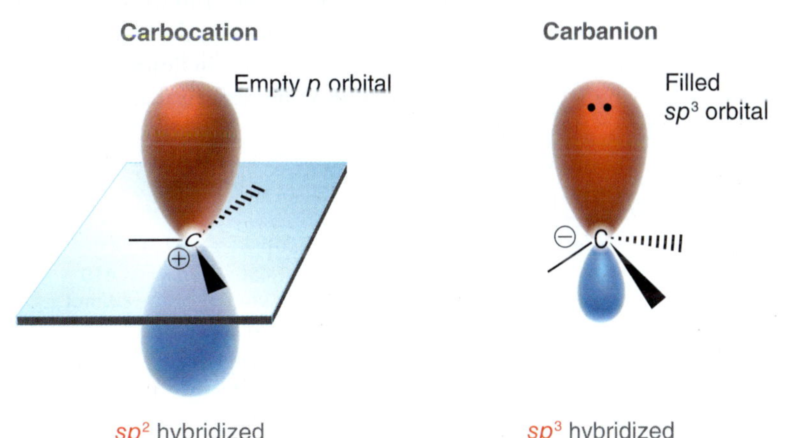


**FIGURE 11.3** A comparison of the geometries of a carbocation and a carbanion.

while a carbanion is $sp^3$ hybridized and trigonal pyramidal. The difference in geometry results from the difference in the number of nonbonding electrons. A carbocation has zero nonbonding electrons, while a carbanion has two nonbonding electrons. A carbon radical is between these two cases, because it has one nonbonding electron. It therefore might seem reasonable to expect that the geometry of a carbon radical would be somewhere in between trigonal planar and trigonal pyramidal. Experiments suggest that carbon radicals either are trigonal planar or exhibit a very shallow pyramid (nearly planar) with a very low barrier to inversion (Figure 11.4). Either way, carbon radicals can be treated as trigonal planar entities. This geometry will have significance later in the chapter when we deal with the stereochemical outcomes of radical reactions.

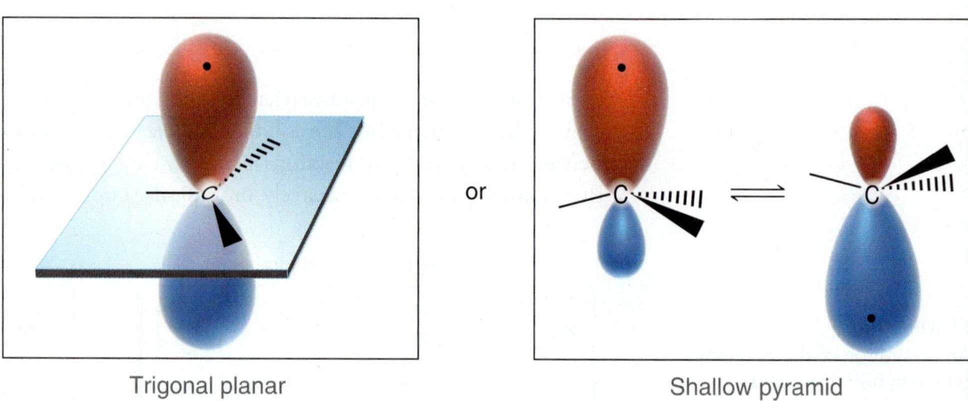


**FIGURE 11.4** The geometry of a carbon radical.

**LOOKING BACK**
For a review of carbocation stability and hyperconjugation, see Section 6.8.

The order of stability for radicals follows the same trend exhibited by carbocations in which tertiary radicals are more stable than secondary radicals, which in turn are more stable than primary radicals (Figure 11.5). The explanation for this trend is similar to the explanation for carbocation

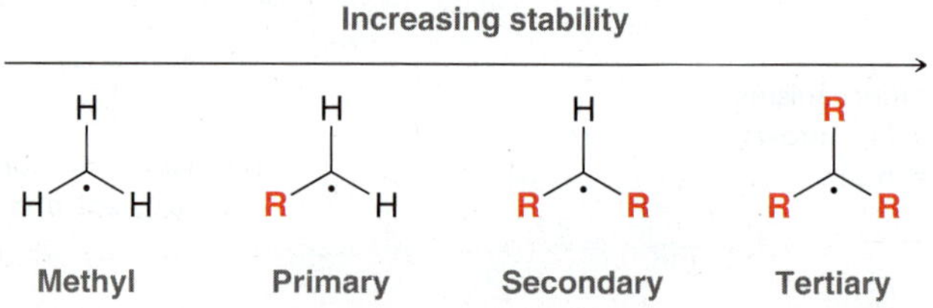

**Increasing stability**

Methyl — Primary — Secondary — Tertiary

**FIGURE 11.5** The order of stability for carbon radicals.

stability. Specifically, alkyl groups are capable of stabilizing the unpaired electron via a delocalization effect, called hyperconjugation. This trend in stability is supported by a comparison of bond dissociation energies (BDEs). Notice that the bond dissociation energy is lowest for a C—H bond involving a tertiary carbon (Figure 11.6), indicating that it is easiest to cleave this C—H bond homolytically. These observed BDE values suggest that a tertiary radical is about 16 kJ/mol more stable than a secondary radical.

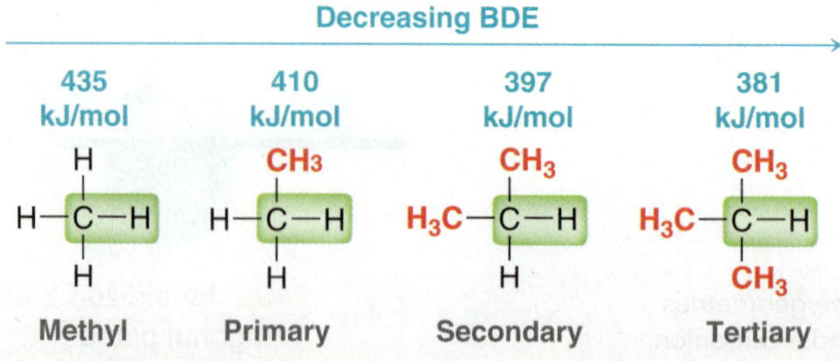


**FIGURE 11.6** Bond dissociation energies for various C—H bonds.

## CONCEPTUAL CHECKPOINT

**11.1** Rank the following radicals in order of stability:

(a)

(b)

### Resonance Structures of Radicals

In Chapter 2, we saw five patterns for drawing resonance structures. There are also several patterns for drawing resonance structures of radicals. However, the vast majority of situations require only one pattern, which is characterized by an unpaired electron next to a $\pi$ bond, in an allylic position:

Allylic position

In such a case, the unpaired electron is resonance stabilized, and three fishhook arrows are used to draw the resonance structure:

Benzylic positions also exhibit the same pattern, with several resonance structures contributing to the overall resonance hybrid:

Resonance-stabilized radicals are even more stable than tertiary radicals, as can be seen by comparing the BDE values in Figure 11.7. A C—H bond at a benzylic or allylic position is more easily cleaved than a C—H bond at a tertiary position. These observed BDE values suggest that resonance-stabilized radicals are about 70–80 kJ/mol more stable than tertiary radicals. This is a large difference when compared with the 16 kJ difference between secondary and tertiary radicals.

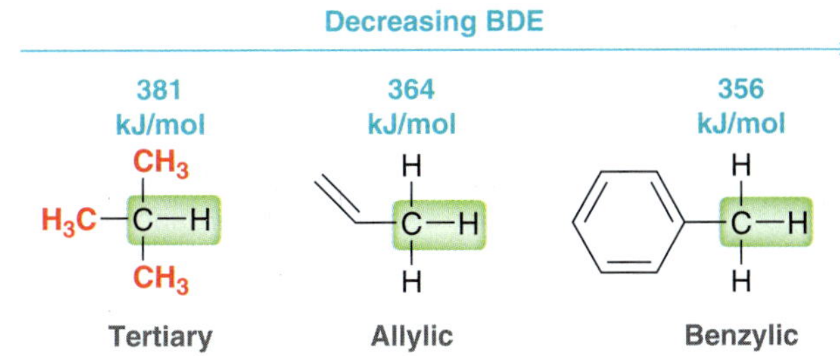


**FIGURE 11.7** Bond dissociation energies for various C—H bonds.

# SKILLBUILDER

## 11.1 DRAWING RESONANCE STRUCTURES OF RADICALS

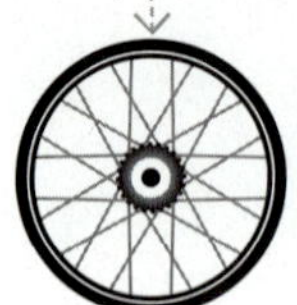

**LEARN** the skill

Draw all resonance structures of the following radical and make sure to show the necessary fishhook arrows:

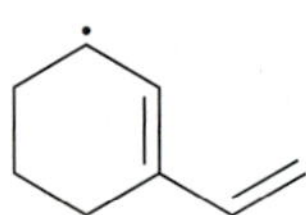

**SOLUTION**

**STEP 1**
Look for an unpaired electron next to a π bond.

**STEP 2**
Draw three fishhook arrows and then draw the corresponding resonance structure.

When drawing resonance structures of a radical, look for an unpaired electron next to a π bond. In this case, the unpaired electron is in fact located at an allylic position, so we draw the following three fishhook arrows:

The new resonance structure exhibits an unpaired electron that is allylic to another π bond, so again we draw three fishhook arrows to arrive at another resonance structure:

In total, there are three resonance structures:

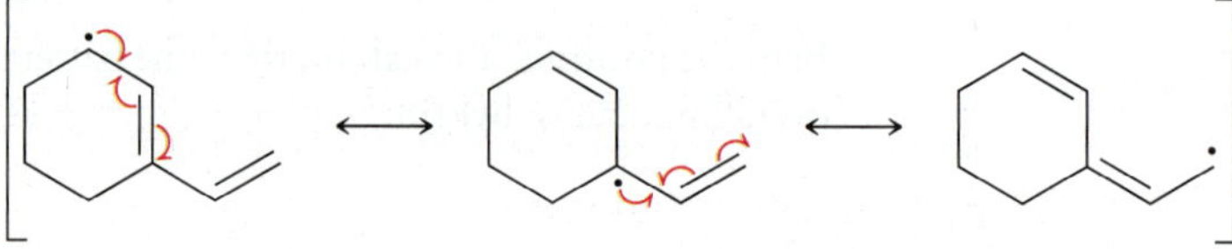

**PRACTICE** the skill

**11.2** Draw resonance structures for each of the following radicals:

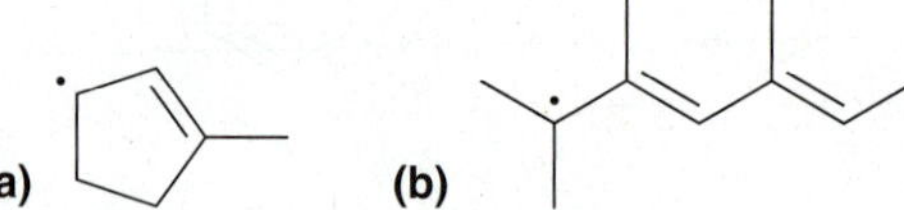

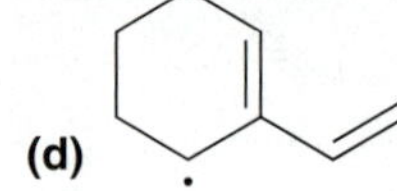

(a) (b) (c) (d)

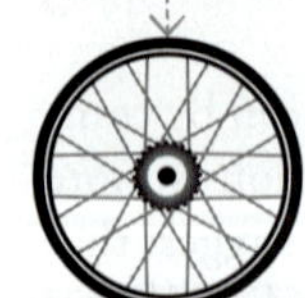

**APPLY** the skill

**11.3** The triphenylmethyl radical was the first radical to be observed. Draw all resonance structures of this radical and explain why this radical is unusually stable:

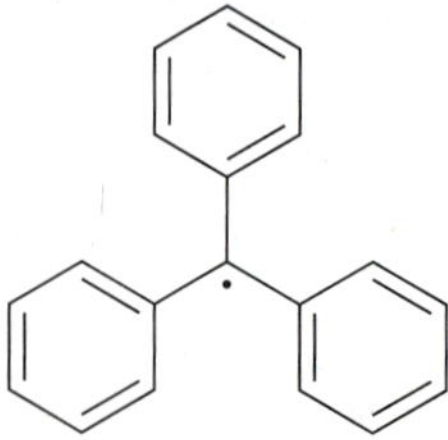

**11.4** 5-Methylcyclopentadiene undergoes homolytic bond cleavage of a C—H bond to form a radical that exhibits five resonance structures. Determine which hydrogen atom is abstracted and draw all five resonance structures of the resulting radical.

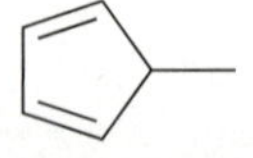

**5-Methylcyclopentadiene**

need more **PRACTICE?** **Try Problems 11.22, 11.25**

In this section, we have seen that allylic and benzylic radicals are stabilized by resonance. When analyzing the stability of a radical, make sure not to confuse allylic and vinylic positions:

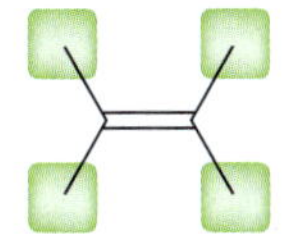

Allylic positions

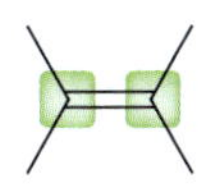

Vinylic positions

A vinylic radical is not resonance stabilized and does not have a resonance structure:

In fact, a vinylic radical is even less stable than a primary radical. This can be seen by comparing BDE values (Figure 11.8). A C—H bond at a vinylic position requires even more energy to cleave than a C—H bond at a primary position. These observed BDE values suggest that a vinylic radical is more than 50 kJ/mol higher in energy than a primary radical (a very large difference).

Decreasing BDE

464 kJ/mol — Vinylic
410 kJ/mol — Primary ($H_3C$—C—H)
364 kJ/mol — Allylic

**FIGURE 11.8** Bond dissociation energies for various C—H bonds.

## SKILLBUILDER

### 11.2 IDENTIFYING THE WEAKEST C—H BOND IN A COMPOUND

**LEARN the skill** Identify the weakest C—H bond in the following compound:

**SOLUTION**

Imagine homolytically breaking each different type of C—H bond in the compound. This would produce the following radicals:

**STEP 1**
Consider all possible radicals that would result from homolytic cleavage of a C—H bond.

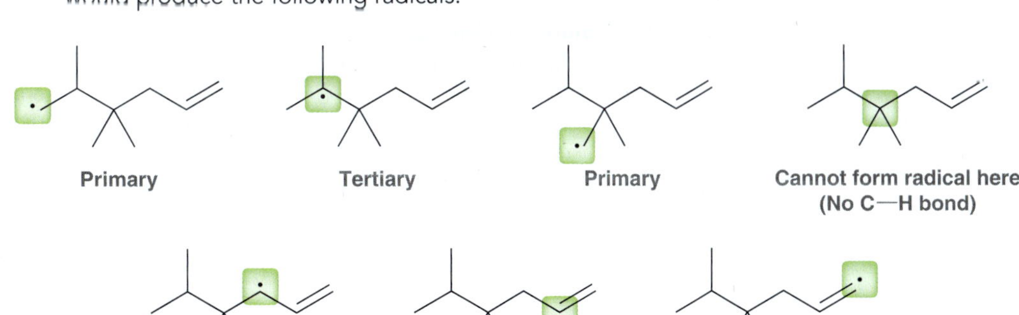

**STEP 2**
Identify the most stable radical.

Of all the possibilities, the allylic radical will be the most stable radical. The weakest C—H bond is the one that leads to the most stable radical. Therefore, the C—H bond at the allylic position is the weakest C—H bond. It is the easiest bond to break homolytically:

**STEP 3**
Determine which bond is the weakest.

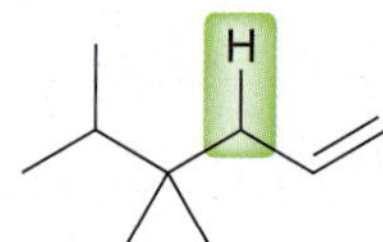

**PRACTICE the skill** **11.5** Identify the weakest C—H bond in each of the following compounds:

(a) 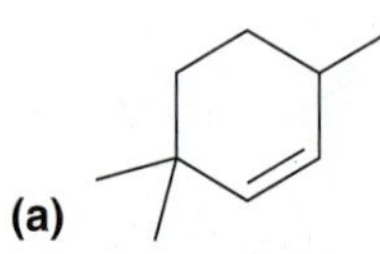 (b) 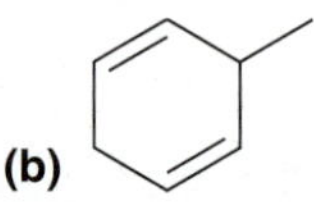 (c) 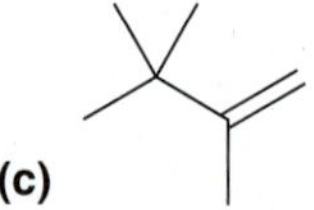 (d) 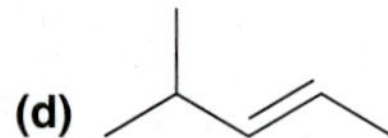

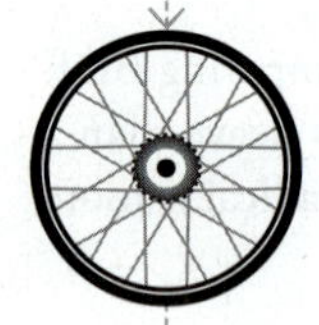

**APPLY the skill** **11.6** The C—H bonds shown in red exhibit very similar BDEs, because homolytic cleavage of either bond results in a resonance-stabilized radical. Nevertheless, one of these C—H bonds is weaker than the other. Identify the weaker bond and explain your choice:

$H_a$ $H_b$

need more **PRACTICE?** **Try Problem 11.23**

## 11.2 Common Patterns in Radical Mechanisms

In Chapter 6, we saw that ionic mechanisms are comprised of only four different kinds of arrow-pushing patterns (nucleophilic attack, loss of a leaving group, proton transfer, and rearrangement). Similarly, radical mechanisms are also comprised of only a few different kinds of arrow-pushing patterns, although these patterns are very different from the patterns in ionic mechanisms. For example, carbocations can undergo rearrangement (as seen in Section 6.11), but radicals are not observed to undergo rearrangement:

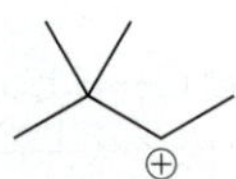

**This carbocation will rearrange to produce a more stable tertiary carbocation**

**This radical will not rearrange to produce a more stable tertiary radical**

There are six different kinds of arrow-pushing patterns that comprise radical reactions. We will now explore these patterns one by one:

1. **Homolytic cleavage**: Homolytic cleavage requires a large input of energy. This energy can be supplied in the form of heat (Δ) or light ($h\nu$).

2. **Addition to a π bond**: A radical adds to a π bond, thereby destroying the π bond and generating a new radical.

3. **Hydrogen abstraction**: A radical can abstract a hydrogen atom from a compound, generating a new radical. Don't confuse this step with a proton transfer, which is an ionic step. In a proton

transfer, only the nucleus of the hydrogen atom (a proton, $H^+$) is being transferred. Here, the entire hydrogen atom (proton and electron, $H^\bullet$ ) is being transferred from one location to another.

$$X^\bullet \; H{-}R \longrightarrow X{-}H \; ^\bullet R$$

4. **Halogen abstraction**: A radical can abstract a halogen atom, generating a new radical. This step is similar to hydrogen abstraction, only a halogen atom is being abstracted instead of a hydrogen atom.

$$R^\bullet \; X{-}X \longrightarrow R{-}X \; ^\bullet X$$

5. **Elimination**: The position bearing the unpaired electron is called the alpha position. In an elimination step, a double bond forms between the alpha ($\alpha$) and beta ($\beta$) positions. As a result, a single bond at the $\beta$ position is cleaved, causing the compound to fragment into two pieces.

6. **Coupling**: Two radicals join together and form a bond.

$$X^\bullet \; ^\bullet X \longrightarrow X{-}X$$

Figure 11.9 summarizes the six common steps in radical mechanisms. This might seem like a lot to remember. It is therefore helpful to group them. For example, notice that the first step (homolytic cleavage) and the sixth step (coupling) are just the reverse of each other. Homolytic cleavage creates radicals, while coupling destroys them:

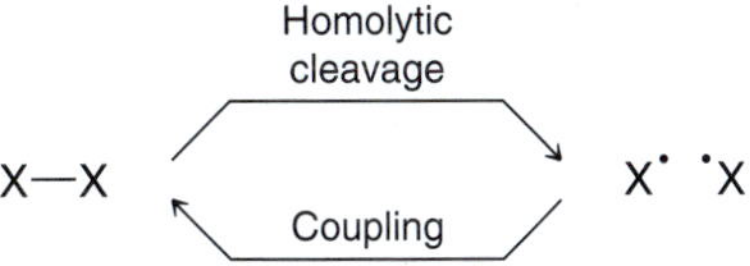


The second and fifth steps shown in Figure 11.9 are also the reverse of each other. Addition to a $\pi$ bond is just the reverse of elimination:

Addition to a $\pi$ bond

Elimination

That leaves only two more steps, both of which are called abstractions (hydrogen abstraction and halogen abstraction). These two steps are not the reverse of each other. The reverse of a hydrogen abstraction is simply a hydrogen abstraction:

Hydrogen abstraction

$$X^\bullet \; H{-}R \rightleftharpoons X{-}H \; ^\bullet R$$

Hydrogen abstraction

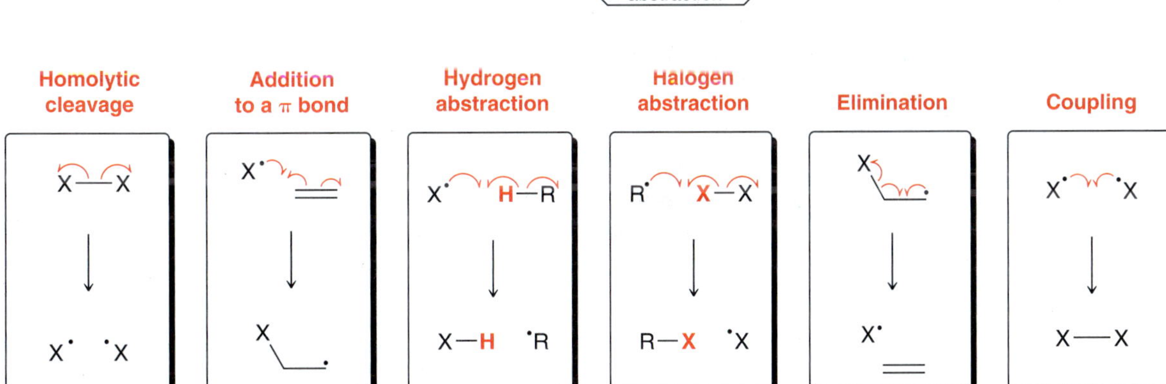


**FIGURE 11.9** Six steps common in radical mechanisms.

The same is true for halogen abstraction. That is, the reverse of a halogen abstraction is simply a halogen abstraction.

When drawing a radical mechanism, remember that every step will generally have either two or three fishhook arrows. Take another close look at the six steps shown in Figure 11.9. The first and last step (homolytic cleavage and coupling) each requires exactly two fishhook arrows, because two electrons are moving in each case. In contrast, all other steps have three electrons moving and therefore require three fishhook arrows. In any step, the number of fishhook arrows must correspond to the number of electrons moving. Let's get some practice drawing fishhook arrows.

## SKILLBUILDER

### 11.3 DRAWING FISHHOOK ARROWS FOR A RADICAL PROCESS

**LEARN the skill**

Draw the appropriate fishhook arrows for the following radical process:

$$\text{(1-methylcyclohexyl radical)} + Br_2 \longrightarrow \text{1-bromo-1-methylcyclohexane} + \cdot Br:$$

**SOLUTION**

The following procedure can be used for all six types of radical steps (not just the one in this problem). First identify the type of process taking place. A radical is reacting with $Br_2$, and the result is the transfer of a bromine atom:

**STEP 1** Identify the process.

Therefore, this step is a *halogen abstraction.*

**STEP 2** Determine the number of fishhook arrows required.

Determine the number of fishhook arrows that must be used (either two or three). Recall that there are six different kinds of steps. Cleavage and coupling require two fishhooks, while all other steps require three fishhook arrows. This process (halogenation) must have *three* fishhook arrows, which means that three electrons are moving. Each of the fishhooks will have its tail placed on one of the electrons that is moving.

Next, identify any bonds being broken or formed:

**STEP 3** Identify any bonds being broken or formed.

Bond broken

Bond formed

If a bond is being formed, draw the two fishhook arrows that show bond formation:

**STEP 4** For a bond being formed, draw two fishhook arrows.

Finally, if a bond is being broken, make sure to have two fishhook arrows, one for each electron of the bond broken. In this case, one fishhook arrow is already there, so we only need to place the remaining fishhook arrow:

**STEP 5** For a bond being broken, make sure that two fishhook arrows are shown.

**PRACTICE the skill** **11.7** Draw the appropriate fishhook arrows for each of the following radical processes.

(a) + HBr ⟶ + ·Br:

(b) + ·Br: ⟶ Br

(c) + ·$CH_3$ ⟶

(d) + ·Br: ⟶ + HBr

(c) ⟶ +

(f) R, O:, :O, R, O—O ⟶ 2 R, O, O:

**APPLY the skill** **11.8** Draw a mechanism for the following intramolecular process:

**11.9** The triphenylmethyl radical reacts with itself to form the following dimer:

2 ⟶

Identify the type of radical process taking place and draw the appropriate fishhook arrows.

**(Hint:** It will help to draw just a few resonance structures of the triphenylmethyl radical.)

need more **PRACTICE?** **Try Problem 11.48**

Each of the six patterns shown in Figure 11.9 can be placed in one of three categories based on the fate of the radicals (Figure 11.10). **Initiation** is when radicals are created, while **termination** is when two radicals annihilate each other by forming a bond. The other four steps are generally **propagation** steps, in which the location of the unpaired electron is moved from place to place. Later in this chapter, we will encounter situations that will require us to refine our definitions for initiation, propagation, and termination, but for now, these simplistic (and incomplete) definitions provided will be helpful to us as we begin exploring radical mechanisms in the upcoming sections.

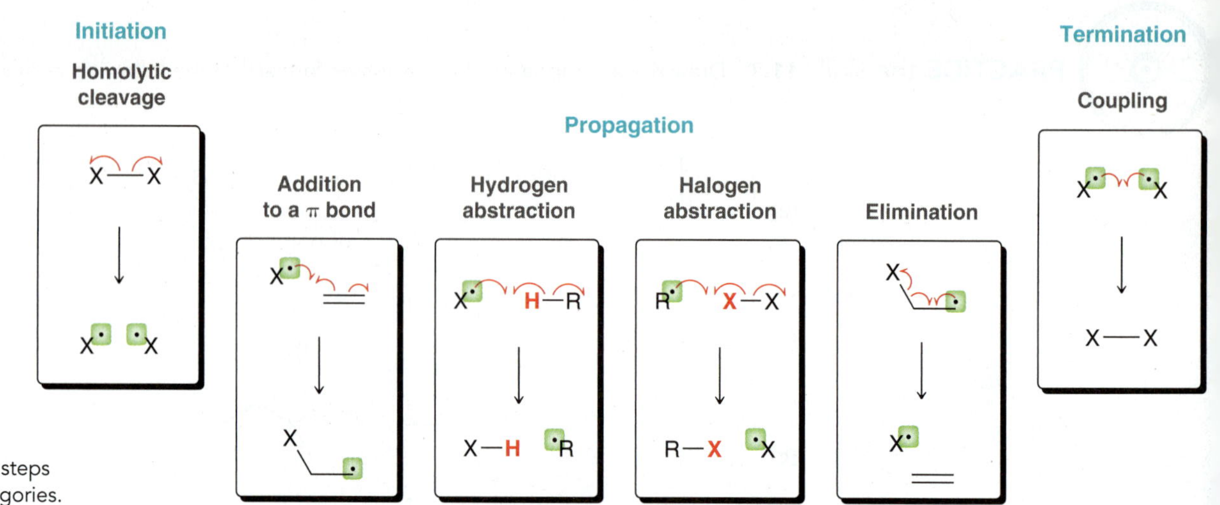


**FIGURE 11.10** The six common steps divided into categories.

## 11.3 Chlorination of Methane

### A Radical Mechanism for Chlorination of Methane

We will now use the skills developed in the previous section to explore the mechanisms of radical reactions. As a first example, consider the reaction between methane and chlorine to form methyl chloride:

$$\underset{\textbf{Methane}}{CH_4} \xrightarrow[h\nu]{Cl_2} \underset{\textbf{Methyl chloride}}{CH_3Cl} + HCl$$

Evidence suggests that this reaction proceeds via a radical mechanism (Mechanism 11.1).

### MECHANISM 11.1 RADICAL CHLORINATION

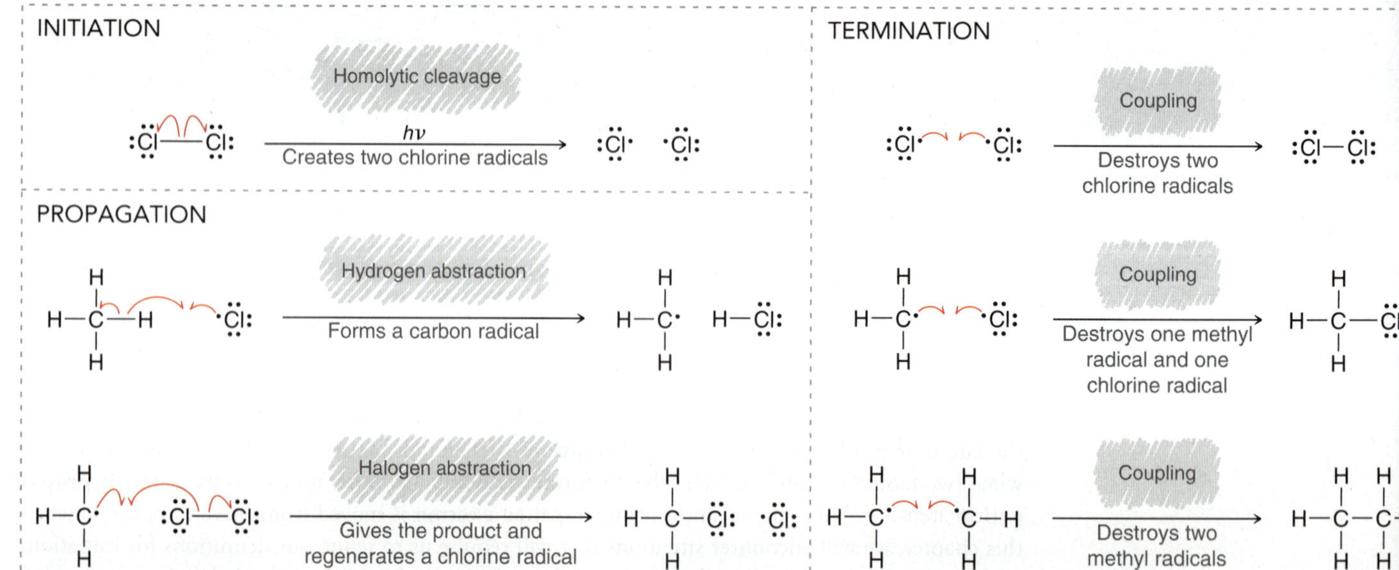

The reaction mechanism for radical chlorination is divided into three distinct stages. The initiation stage involves creation of radicals, while the termination stage involves destruction of radicals. The propagation stage is the most important, because the propagation steps are responsible for the observed reaction. The first propagation step is a hydrogen abstraction, and the second is a halogen abstraction. Notice that the sum of these propagation steps gives the net reaction:

Hydrogen abstraction $CH_4 + \cdot Cl \longrightarrow \cdot CH_3 + HCl$

Halogen abstraction $\cdot CH_3 + Cl_2 \longrightarrow CH_3Cl + \cdot Cl$

Net reaction $CH_4 + Cl_2 \longrightarrow CH_3Cl + HCl$

This now provides us with a new, refined definition for propagation steps. Specifically, the sum of the propagation steps gives the net chemical reaction. All other steps must be either initiation or termination, not propagation. We will revisit this refined definition later in the chapter.

There is one more critical aspect of the propagation steps shown above. Notice that the first propagation step *consumes* a chlorine radical, while the second propagation step *regenerates* a chlorine radical. In this way, one chlorine radical can ultimately cause thousands of molecules of methane to be converted into chloromethane (assuming enough $Cl_2$ is present). Therefore, the reaction is called a **chain reaction.**

When excess $Cl_2$ is present, polychlorination is observed:

$CH_4 \xrightarrow[h\nu]{Cl_2} CH_3Cl \xrightarrow[h\nu]{Cl_2} CH_2Cl_2 \xrightarrow[h\nu]{Cl_2} CHCl_3 \xrightarrow[h\nu]{Cl_2} CCl_4$

Methyl chloride, Methylene chloride, Chloroform, Carbon tetrachloride

The initial product, methyl chloride ($CH_3Cl$), is even more reactive toward radical halogenation than methane. As methyl chloride is formed, it reacts with chlorine to produce methylene chloride. The process continues until carbon tetrachloride is produced. In order to produce methyl chloride as the major product (monohalogenation), it is necessary to use an excess of methane and a small amount of $Cl_2$. Unless otherwise indicated, the conditions of a halogenation reaction are generally controlled so as to produce monohalogenation.

## SKILLBUILDER

### 11.4 DRAWING A MECHANISM FOR RADICAL HALOGENATION

**LEARN the skill**

Draw the mechanism of the radical chlorination of methyl chloride to produce methylene chloride:

$CH_3Cl \xrightarrow[h\nu]{Cl_2} CH_2Cl_2 + HCl$

Methyl chloride, Methylene chloride

**SOLUTION**

The mechanism will have three distinct stages. The first stage is initiation, which creates chlorine radicals. This initiation step involves homolytic bond cleavage and should employ only two fishhook arrows:

**STEP 1** Draw the initiation step(s).

$:\ddot{Cl}-\ddot{Cl}: \xrightarrow{h\nu} :\ddot{Cl}\cdot \quad \cdot\ddot{Cl}:$

The next stage involves the propagation steps. There are two propagation steps: hydrogen abstraction to remove the hydrogen atom followed by halogen abstraction to attach the chlorine atom. Each of these steps should have three fishhook arrows:

**STEP 2**
Draw the propagation steps.

**Hydrogen abstraction**

$$H_2CCl{-}H \ + \ \cdot Cl \longrightarrow H_2CCl\cdot \ + \ H{-}Cl$$

**Halogen abstraction**

$$H_2CCl\cdot \ + \ Cl{-}Cl \longrightarrow H_2CCl{-}Cl \ + \ \cdot Cl$$

These two abstractions together represent the core of the reaction. They show how the product is formed.

Termination ends the process. There are a number of possible termination steps. When drawing a mechanism for a radical reaction, it is generally not necessary to draw all possible termination steps unless specifically asked to do so. It is sufficient to draw the one termination step that also produces the desired product:

**STEP 3**
Draw at least one termination step.

$$H_2CCl\cdot \ + \ \cdot Cl \longrightarrow H_2CCl{-}Cl$$

**11.10** Draw a mechanism for each of the following processes:

**(a)** Chlorination of methylene chloride to produce chloroform

**(b)** Chlorination of chloroform to produce carbon tetrachloride

**(c)** Chlorination of ethane to produce ethyl chloride

**(d)** Chlorination of 1,1,1-trichloroethane to produce 1,1,1,2-tetrachloroethane

**(e)** Chlorination of 2,2-dichloropropane to produce 1,2,2-trichloropropane

**APPLY the skill**

**11.11** In practice, the chlorination of methane often produces many by-products. For example, ethyl chloride is obtained in small quantities. Can you suggest a mechanism for the formation of ethyl chloride?

## Radical Initiators

Energy is required to initiate a radical chain reaction:

$$:\ddot{Cl}{-}\ddot{Cl}: \xrightarrow[\text{or } h\nu]{\Delta} 2\ :\ddot{Cl}\cdot \qquad \Delta H° = 243 \text{ kJ/mol}$$

The change in enthalpy for homolytic bond cleavage of $Cl_2$ is 243 kJ/mol. In order to break the Cl—Cl bond homolytically, this energy must be supplied in the form of either heat or light. Photochemical initiation requires the use of UV light, while thermal initiation requires a very high temperature (several hundred degrees Celsius) in order to produce the requisite energy necessary to cause homolytic bond cleavage of a Cl—Cl bond. Achieving thermal initiation at more moderate temperatures requires the use of a **radical initiator,** a compound with a weak bond that undergoes homolytic bond cleavage with greater ease. For example, consider the homolytic bond cleavage of **peroxides**, which are compounds containing an O—O bond:

$$R\ddot{O}{-}\ddot{O}R \xrightarrow[\text{or } h\nu]{\Delta} 2\ R\ddot{O}\cdot \qquad \Delta H° = 159 \text{ kJ/mol}$$

**A peroxide**

Notice that this process requires less energy, because the O—O bond is weaker than the Cl—Cl bond. This process can be achieved at 80°C, and the radicals produced (RO•) can start the chain process.

**Acyl peroxides** are often used as radical initiators, because the O—O bond is especially weak:

$\Delta H° = 121$ kJ/mol

An acyl peroxide

The energy required to homolytically cleave this bond is only 121 kJ/mol. This bond has a particularly small BDE because the radical produced is resonance stabilized:

## Radical Inhibitors

While some compounds, like peroxides, help initiate radical reactions, compounds called radical inhibitors have the opposite effect. A **radical inhibitor** is a compound that prevents a chain process from either getting started or continuing. Radical inhibitors effectively destroy radicals and are therefore also called radical scavengers. For example, molecular oxygen ($O_2$) is a diradical:

Molecular oxygen is able to couple with other radicals, thereby destroying them. Each molecule of oxygen is capable of destroying two radicals. As a result, radical chain reactions generally won't occur rapidly until all of the available oxygen has been consumed.

Another example of a radical inhibitor is hydroquinone. When hydroquinone meets a radical, a hydrogen abstraction can take place, generating a resonance-stabilized radical that is less reactive than the original radical:

Hydrogen abstraction

Hydroquinone

Resonance stabilized

This resonance-stabilized radical can then destroy another radical via another hydrogen abstraction to form benzoquinone:

Hydrogen abstraction

Benzoquinone

In Section 11.9, we will discuss the role of radical inhibitors in biological processes.

# 11.4 Thermodynamic Considerations for Halogenation Reactions

In the previous section, we explored the proposed mechanism for the chlorination of methane. We will now explore whether this reaction can be accomplished with other halogens as well. Is it possible to achieve a radical fluorination, bromination, or iodination? To answer this question, we must explore some thermodynamic aspects of halogenation.

In Section 6.3, we saw that $\Delta G$ (the change in Gibbs free energy) determines whether or not a reaction is thermodynamically favorable. In order for a reaction to favor products over starting materials, the reaction must exhibit a negative $\Delta G$. If $\Delta G$ is positive, starting materials will be favored, and the reaction will not produce the desired products. We will now use this information to determine whether halogenation can be accomplished with halogens other than chlorine.

Recall that $\Delta G$ is comprised of two components—enthalpy and entropy:

$$\Delta G = \underbrace{(\Delta H)}_{\text{Enthalpy term}} + \underbrace{(-T\Delta S)}_{\text{Entropy term}}$$

For halogenation of an alkane, the entropy component is negligible because two molecules of starting material are converted into two molecules of products:

$$\underbrace{CH_4 + X_2}_{\text{Two molecules}} \xrightarrow{h\nu} \underbrace{CH_3X + HX}_{\text{Two molecules}}$$

Since the change in entropy of a halogenation reaction is negligible, we can assess the value of $\Delta G$ by analyzing the enthalpy term alone:

$$\Delta G \approx \Delta H$$

The enthalpy term is determined by a number of factors, but the most important factor is bond strength. We can estimate $\Delta H$ by comparing the energy of the bonds broken and the energy of bonds formed:

$$H_3C{-}H + X{-}X \xrightarrow{h\nu} H_3C{-}X + H{-}X$$

Bonds broken (C—H, X—X)     Bonds formed (C—X, H—X)

To make the comparison, we look at bond dissociation energies. The relevant values are shown in Table 11.1. Using these values, we can estimate the sign of $\Delta H$ (positive or negative) associated with each halogenation reaction. Remember that:

- Bonds broken (shown in blue) require an input of energy in order to be broken, contributing to a positive value of $\Delta H$ (the system increases in energy).
- Bonds formed (shown in red) release energy when they are formed, contributing to a negative value of $\Delta H$ (the system decreases in energy).

**TABLE 11.1 BOND DISSOCIATION ENERGIES OF RELEVANT BONDS (kJ/mol)**

| | $H_3C{-}H$ | X—X | $H_3C{-}X$ | H—X |
|---|---|---|---|---|
| F | 435 | 159 | 456 | 569 |
| Cl | 435 | 243 | 351 | 431 |
| Br | 435 | 193 | 293 | 368 |
| I | 435 | 151 | 234 | 297 |

The data in Table 11.1 can be used to make the following predictions:

$$CH_4 + F_2 \xrightarrow{h\nu} CH_3F + HF \qquad \Delta H^\circ = -431 \text{ kJ/mol}$$

$$CH_4 + Cl_2 \xrightarrow{h\nu} CH_3Cl + HCl \qquad \Delta H^\circ = -104 \text{ kJ/mol}$$

$$CH_4 + Br_2 \xrightarrow{h\nu} CH_3Br + HBr \qquad \Delta H^\circ = -33 \text{ kJ/mol}$$

$$CH_4 + I_2 \xrightarrow{h\nu} CH_3I + HI \qquad \Delta H^\circ = +55 \text{ kJ/mol}$$

All of the processes above have a negative value for $\Delta H$ and are exothermic except for iodination. Iodination of methane has a positive $\Delta H$, which means that $\Delta G$ will also be positive for that reaction. As a result, iodination is not thermodynamically favorable, and the reaction simply does not occur. The other halogenation reactions are all thermodynamically favorable, but fluorination is so exothermic that the reaction is too violent to be of practical use. Therefore, only chlorination and bromination are practical in the laboratory.

When this type of thermodynamic analysis is applied to alkanes other than methane, similar results are obtained. For example, ethane will undergo both radical chlorination and radical bromination:

$$H_3C{-}CH_3 + X_2 \xrightarrow[\text{or ROOR, } \Delta]{h\nu} H_3C{-}CH_2{-}X + HX \qquad (X = Br \text{ or } Cl)$$

Radical fluorination of ethane is too violent to be practical, and radical iodination of ethane does not occur.

A comparison of chlorination and bromination reveals that bromination is generally a much slower process than chlorination. To understand why, we must take a closer look at the individual propagation steps. We will compare $\Delta H$ of each propagation step for the chlorination and bromination of ethane:

**First propagation step: hydrogen abstraction**

$$H_3C{-}CH_3 + {}^{\bullet}\ddot{X}{:} \longrightarrow H_3C{-}\dot{C}H_2 + H{-}\ddot{X}{:}$$

**Second propagation step: halogen abstraction**

$$H_3C{-}\dot{C}H_2 + {:}\ddot{X}{-}\ddot{X}{:} \longrightarrow H_3C{-}CH_2{-}\ddot{X}{:} + {}^{\bullet}\ddot{X}{:}$$

| $\Delta H$ (kJ/mol) | X = Cl | X = Br |
|---|---|---|
| First propagation step | −21 | +42 |
| Second propagation step | −96 | −92 |
| Net reaction: | −117 kJ/mol | −50 kJ/mol |

In each case, the net reaction is exothermic. The estimated values for $\Delta H$ for chlorination and bromination are −117 and −50 kJ/mol, respectively. Therefore, both of these reactions are thermodynamically favorable. But notice that the first step of bromination is an endothermic process ($\Delta H$ has a positive value). This endothermic step does not prevent bromination from occurring, because the net reaction is still exothermic. However, the endothermic first step does greatly affect the rate of the reaction. Compare the energy diagrams for the chlorination and bromination of ethane (Figure 11.11). Each

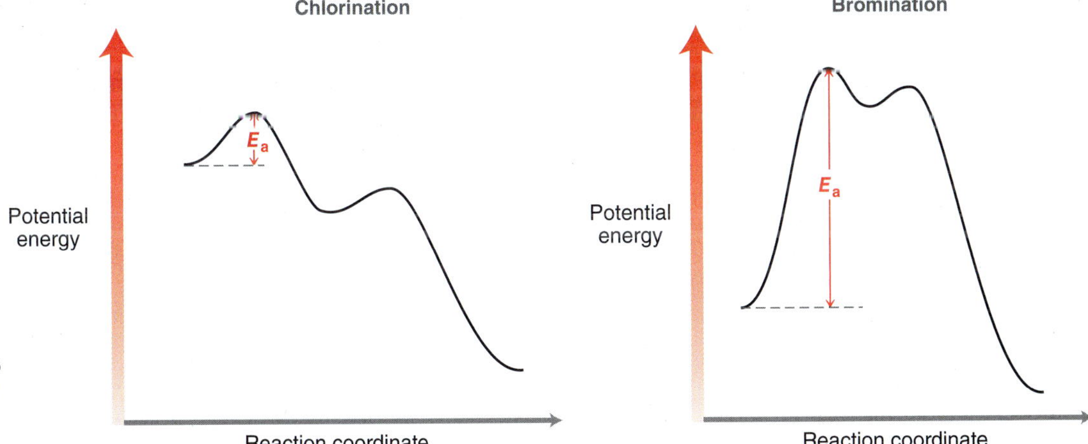


**FIGURE 11.11** Energy diagrams for the two propagation steps of radical chlorination and radical bromination of ethane.

energy diagram displays both propagation steps. In each case, the first propagation step (hydrogen abstraction) is the rate-determining step. For chlorination, the rate-determining step is exothermic, and the energy of activation ($E_a$) is relatively small. In contrast, the rate-determining step of the bromination process is endothermic, and the energy of activation ($E_a$) is relatively large. As a result, bromination occurs more slowly than chlorination.

According to this analysis, it might seem to be a disadvantage that the first step of bromination is endothermic. We will see in the next section that this endothermic step actually makes bromination more useful (albeit slower) than chlorination.

## 11.5 Selectivity of Halogenation

When propane undergoes radical halogenation, there are two possible products, and both are formed:

$$\text{propane} \xrightarrow[h\nu]{X_2} \text{2-halopropane} + \text{1-halopropane}$$

Statistically, we should be able to predict what the product distribution should be (how much of each product to expect), based on the number of each type of hydrogen atom:

Two hydrogen atoms → 2-halopropane

Six hydrogen atoms → 1-halopropane

Based on this analysis, we should expect halogenation to occur at the primary position three times as often as it does at the secondary position. However, these expectations are not supported by observations:

$$\text{propane} \xrightarrow[h\nu]{Cl_2} \text{2-chloropropane } (\sim 60\%) + \text{1-chloropropane } (\sim 40\%)$$

These results show that halogenation occurs at the secondary position more readily than statistics alone would suggest. Why? Recall that the rate-determining step is the first propagation step—hydrogen abstraction. We therefore focus our attention on that step. Specifically, compare the transition state for abstraction at the primary position with the transition state for abstraction at the secondary position (Figure 11.12). Compare the energy levels of the transition states highlighted in green in Figure 11.12. In each transition state, an unpaired electron (radical) is developing on a

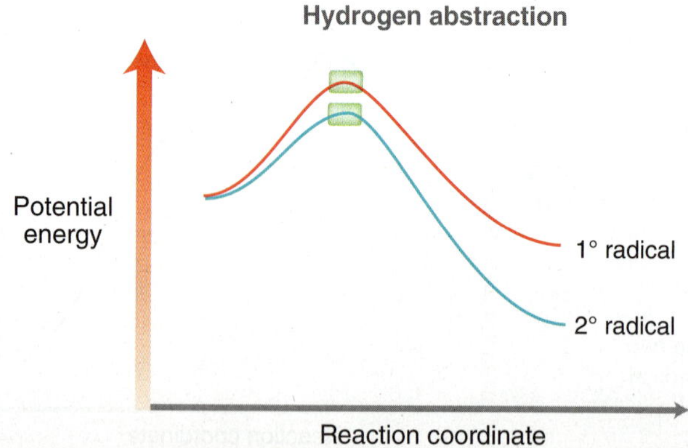


**FIGURE 11.12** An energy diagram showing hydrogen abstraction producing either a primary or a secondary radical.

carbon atom. The developing radical is more stable at the secondary position than at the primary position. As a result, $E_a$ is lower for hydrogen abstraction at the secondary position, and the reaction occurs more rapidly at that site. This explains the observation that chlorination at the secondary position occurs more readily than is predicted by statistics alone.

When propane undergoes radical bromination (rather than chlorination), the following product distribution is observed:

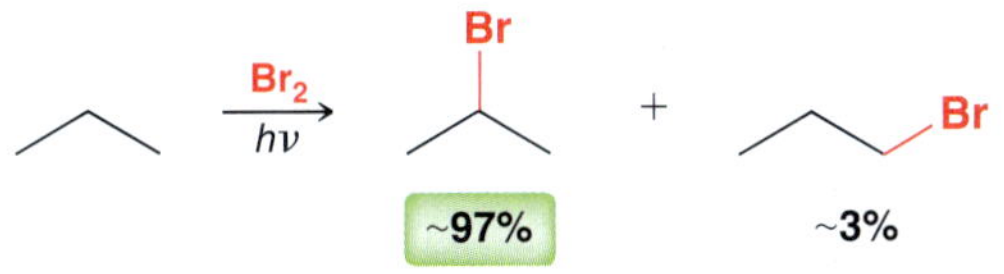

$Br_2$, $h\nu$ → Br (~97%) + Br (~3%)

These results indicate that the tendency for bromination at the secondary position is much more pronounced, with 97% of the product occurring from reaction at the secondary position. Bromination is said to be more *selective* than chlorination. To understand the reason for this selectivity, recall that the rate-determining step of chlorination is exothermic, while the rate-determining step of bromination is endothermic. In the previous section, we used that fact to explain why bromination is slower than chlorination. Now, we will use that fact to explain why bromination is more selective than chlorination.

Recall the Hammond postulate (Section 6.6), which describes the nature of the transition state in each case. For the chlorination process, the rate-determining step is exothermic, and therefore, the energy and structure of the transition state more closely resemble reactants than products (Figure 11.13).

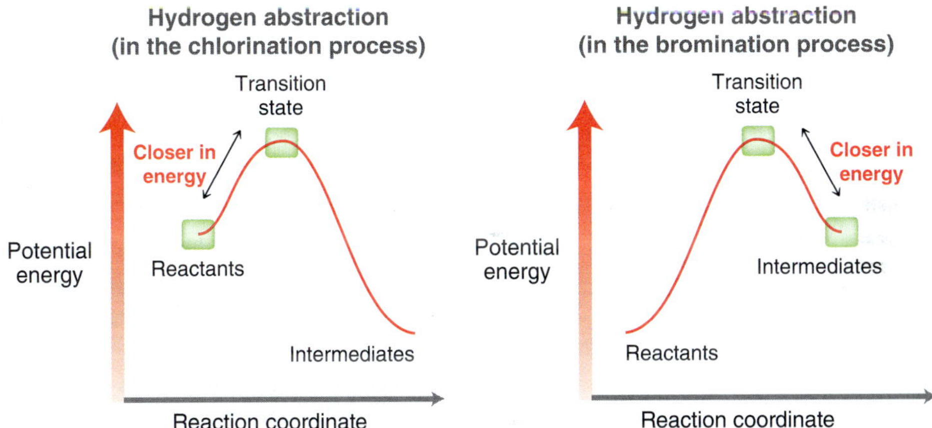


**FIGURE 11.13** Use of the Hammond postulate to compare the transition states in the first propagation step of radical chlorination and radical bromination.

That means that the C—H bond has only begun to break, and the carbon atom has very little radical character. In the case of bromination, the rate-determining step is endothermic, and therefore, the energy and structure of the transition state more closely resemble intermediates than reactants (Figure 11.13). As a result, the C—H bond is almost completely broken in the transition state, and the carbon atom has significant radical character (Figure 11.14). In both cases, the carbon atom has partial radical character (δ•), but during chlorination, the amount of radical character is small. During bromination, the amount of radical character is much larger. As a result, the transition state will be more sensitive to the nature of the substrate during bromination. In the energy diagram for

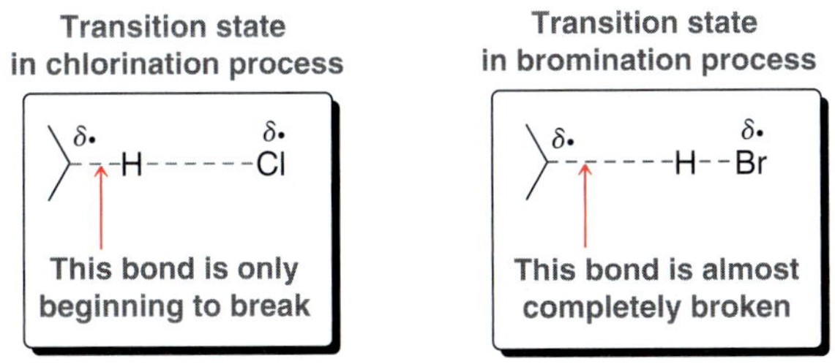

Transition state in chlorination process

δ• H------Cl δ•

This bond is only beginning to break

Transition state in bromination process

δ• ------H--Br δ•

This bond is almost completely broken

**FIGURE 11.14** A comparison of the transition states during the first propagation step of radical chlorination and radical bromination.

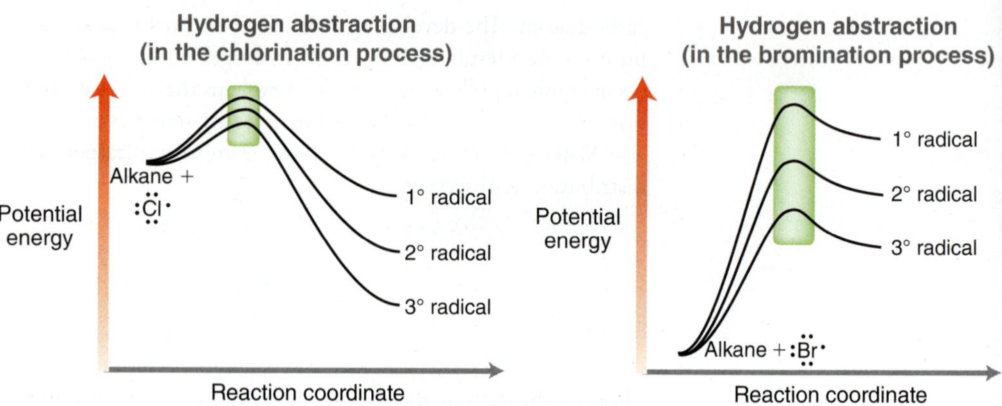


**FIGURE 11.15** Energy diagrams for the first propagation step of chlorination and bromination, comparing the difference in transition states leading to primary, secondary, or tertiary radicals.

the chlorination process, there is only a small difference in energy between the transition states leading to primary, secondary, or tertiary radicals (Figure 11.15, left). In contrast, the energy diagram for the bromination process shows a large difference in energy between the transition states leading to primary, secondary, or tertiary radicals (Figure 11.15, right). As a result, bromination shows greater selectivity than chlorination.

Here is another example that illustrates the difference in selectivity between chlorination and bromination:

$Cl_2$, $h\nu$: Cl (tertiary) ~35% + Cl (primary) ~65%

$Br_2$, $h\nu$: Br (tertiary) ~100% + Br (primary) ~0%

In this case, the selectivity of bromination is extremely pronounced because there are only two possible outcomes: halogenation at a primary position or halogenation at a tertiary position. For the bromination process, the difference in energy of the transition states will be very significant.

Table 11.2 summarizes the relative selectivity of fluorination, chlorination, and bromination. Fluorine only shows a very small selectivity for tertiary over primary halogenation. Fluorine is the most reactive of all the halogens, and it is therefore the least selective. In contrast, bromine shows a very large selectivity for halogenation at the tertiary position (1600 times greater than at the primary position). Bromine is less reactive than fluorine and chlorine, and it is therefore the most selective. The relationship between reactivity and selectivity is a general trend that we will observe many times throughout this course (not just with radical reactions). Specifically, there is an inverse relationship between reactivity and selectivity: *Reagents that are the least reactive will generally be the most selective.*

**TABLE 11.2 THE RELATIVE SELECTIVITY OF FLUORINATION, CHLORINATION, AND BROMINATION**

| | PRIMARY | SECONDARY | TERTIARY |
|---|---|---|---|
| F | 1 | 1.2 | 1.4 |
| Cl | 1 | 4.5 | 5.1 |
| Br | 1 | 82 | 1600 |

# SKILLBUILDER

## 11.5 PREDICTING THE SELECTIVITY OF RADICAL BROMINATION

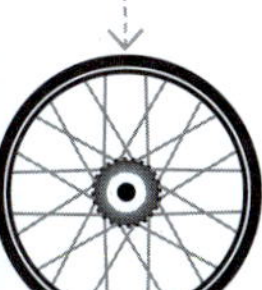

**LEARN the skill**

Predict the major product obtained upon radical bromination of 2,2,4-trimethylpentane:

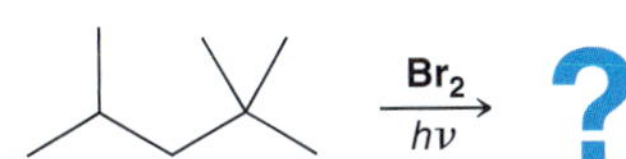

**SOLUTION**

Analyze all possible positions that can possibly undergo bromination and identify each position as primary, secondary, or tertiary:

**STEPS 1 AND 2**
Identify all possible positions and determine the most stable location for a radical.

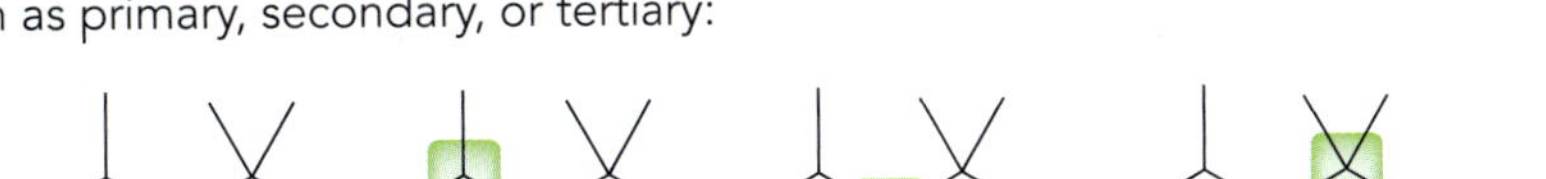


Make sure to avoid selecting a quaternary position, as quaternary positions do not possess a C—H bond. Bromination cannot occur at a quaternary site. The major product is expected to result from bromination at the tertiary position:

**STEP 3**
Draw the product of bromination at the location of the most stable radical.

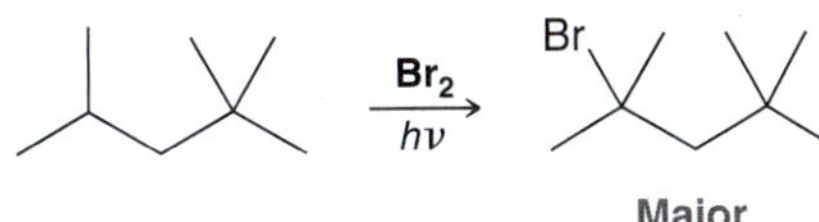

**PRACTICE the skill**

**11.12** Predict the major product obtained upon radical bromination of each of the following compounds:

(a) 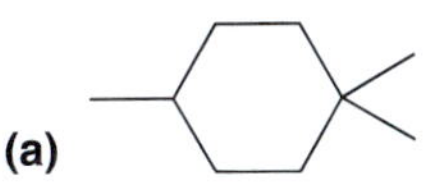 (b) 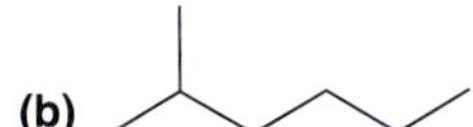 (c) 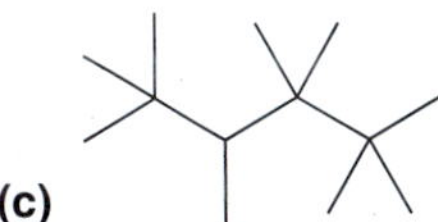

**APPLY the skill**

**11.13** Compound A has molecular formula $C_5H_{12}$ and undergoes monochlorination to produce four different constitutional isomers.

(a) Draw the structure of compound A.

(b) Draw all four monochlorination products.

(c) If compound A undergoes monobromination (instead of monochlorination), one product predominates. Draw that product.

need more **PRACTICE?** **Try Problems 11.27, 11.33a–c,f, 11.38, 11.40, 11.44**

## 11.6 Stereochemistry of Halogenation

We now focus our attention on the stereochemical outcome of radical halogenation. We will investigate two different situations: (1) halogenation that creates a new chirality center and (2) halogenation that occurs at an existing chirality center.

## Halogenation That Creates a New Chirality Center

When butane undergoes radical chlorination, two constitutional isomers are obtained:

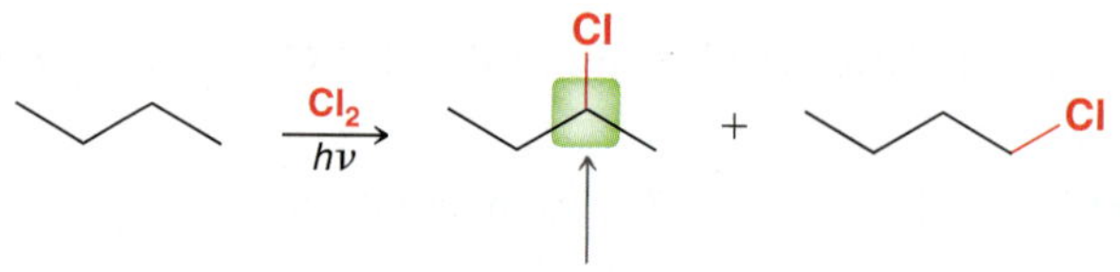


The products are 2-chlorobutane and 1-chlorobutane. The former has a new chirality center that was created by the reaction. In this case, a racemic mixture of 2-chlorobutane is obtained. Why? Consider the structure of the radical intermediate:

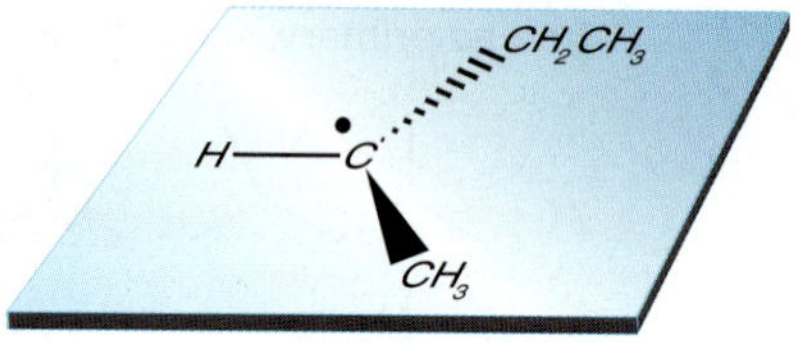


Trigonal planar

At the beginning of this chapter, we saw that a carbon radical either is trigonal planar or is a shallow pyramid that is rapidly inverting. Either way, it can be treated as trigonal planar. We therefore expect that halogen abstraction can occur on either face of the plane with equal likelihood, leading to a racemic mixture of 2-chlorobutane:

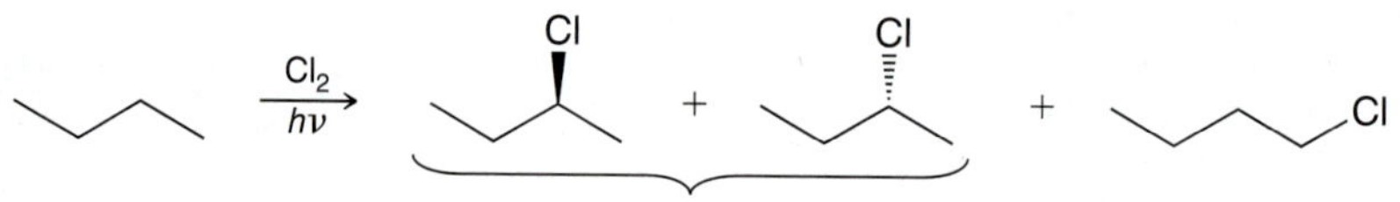


## Halogenation at an Existing Chirality Center

In some case, halogenation will occur at an existing chirality center. Consider the following example:

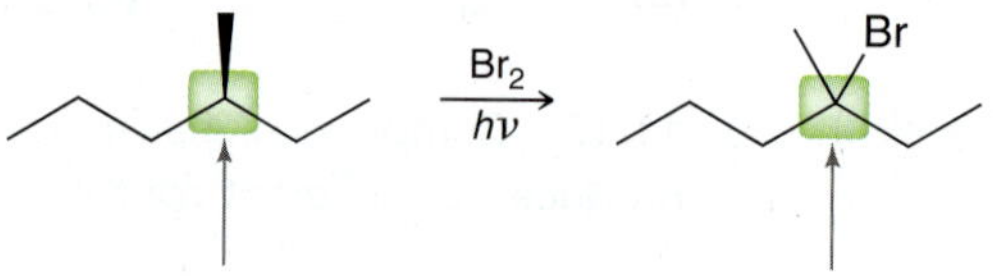


We expect bromination to occur at the tertiary position, and this position is already a chirality center before the reaction takes place. In such a case, what happens to the chirality center? Once again, we expect a racemic product, regardless of the configuration of the starting material. The first propagation step is removal of a hydrogen atom from the chirality center to form a radical intermediate that can be treated as planar:

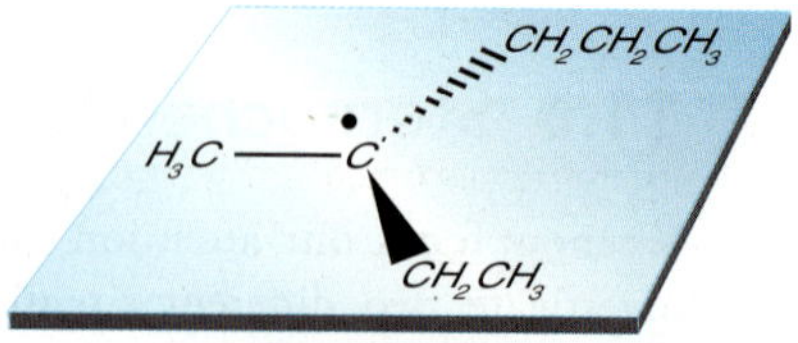


Trigonal planar

At this point, the configuration of the starting alkane has been lost. The second propagation step can now occur on either face of the plane with equal likelihood, leading to a racemic mixture:

$$\xrightarrow[h\nu]{Br_2}$$

Racemic mixture

# SKILLBUILDER

## 11.6 PREDICTING THE STEREOCHEMICAL OUTCOME OF RADICAL BROMINATION

**LEARN the skill**

Predict the stereochemical outcome of radical bromination of the following alkane:

**SOLUTION**

First identify the regiochemical outcome—that is, identify the position that will undergo bromination. In this example, there is only one tertiary position:

**STEP 1**
Identify the location where bromination will occur.

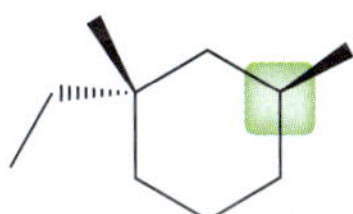

**STEP 2**
If the location of bromination is a chirality center or will become a chirality center, draw both possible stereoisomers.

This position is an existing chirality center, so we expect loss of configuration to produce both possible stereoisomers:

Notice that the other chirality center was unaffected by the reaction. The products of this reaction are not enantiomers and cannot be called a racemic mixture. In this case, the products are diastereomers.

**PRACTICE the skill** **11.14** Predict the stereochemical outcome of radical bromination of the following alkanes:

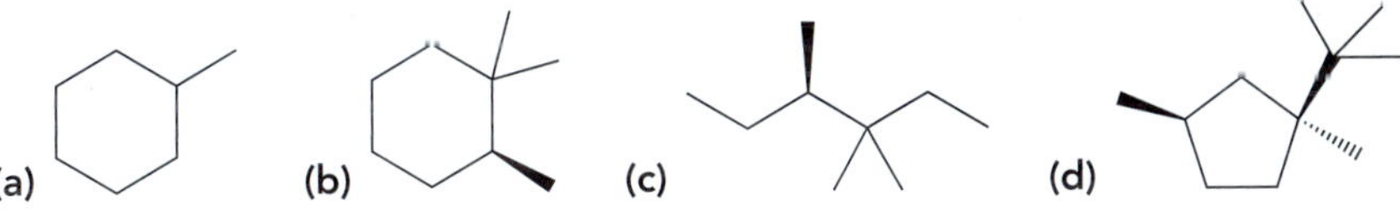

**APPLY the skill** **11.15** Compound A has molecular formula $C_5H_{11}Br$. When compound A is treated with bromine in the presence of UV light, the major product is 2,2-dibromopentane. Treatment of compound A with NaSH (a strong nucleophile) produces a compound with one chirality center having the *R* configuration. What is the structure of compound A?

need more **PRACTICE?** **Try Problems 11.35, 11.41**

## 11.7 Allylic Bromination

Until now, we have focused on reactions of alkanes. Now let's consider the radical halogenation of alkenes. For example, consider what outcome you might expect when cyclohexene undergoes radical bromination. Begin by comparing all C—H bonds to identify the bond that is most easily broken. Specifically, compare the BDE for each type of C—H bond in cyclohexene:

H 444 kJ/mol
H 364 kJ/mol
H 402 kJ/mol

Of the three different types of C—H bonds in cyclohexene, the C—H bond at an allylic site has the lowest BDE, because hydrogen abstraction at that site generates a resonance-stabilized allylic radical:

Hydrogen abstraction

Therefore, we expect bromination of cyclohexene to produce the allylic bromide:

$Br_2$, $h\nu$ — Br

This reaction is called an **allylic bromination**, and it suffers from one serious flaw. When the reaction is performed with $Br_2$ as the reagent, a competition occurs between allylic bromination and ionic addition of bromine across the $\pi$ bond (as we saw in Section 9.8):

$Br_2$ — Br, Br + **En**

To avoid this competitive reaction, the concentration of bromine ($Br_2$) must be kept as low as possible throughout the reaction. This can be accomplished by using ***N*-bromosuccinimide** (NBS) as a reagent instead of $Br_2$. NBS is an alternate source of the bromine radical:

$h\nu$

*N*-Bromosuccinimide (NBS)

Stabilized by resonance

The N—Br bond is weak and is easily cleaved to produce a bromine radical, which achieves the first propagation step:

Hydrogen abstraction + HBr

Resonance stabilized

The HBr produced in this step then reacts with NBS in an ionic reaction that produces $Br_2$. This $Br_2$ is then used in the second propagation step to form the product:

Halogen abstraction

Throughout the process, concentrations of HBr and $Br_2$ are kept at a minimum. Under these conditions, ionic addition of $Br_2$ does not successfully compete with radical bromination.

With many alkenes, a mixture of products is obtained, because the allylic radical initially formed is resonance stabilized and can undergo halogen abstraction at either site:

NBS
$h\nu$

# SKILLBUILDER

## 11.7 PREDICTING THE PRODUCTS OF ALLYLIC BROMINATION

**LEARN the skill**

Predict the products that are obtained when methylenecyclohexane is treated with NBS and irradiated with UV light:

### SOLUTION

First identify any allylic positions:

**STEP 1**
Identify the allylic position.

Allylic position

In this case, there is only one unique allylic position, because the allylic position on the other side of the compound is identical (it will lead to the same products).

Next, remove a hydrogen atom from the allylic position and draw the resonance structures of the resulting allylic radical:

**STEP 2**
Remove a hydrogen atom and draw resonance structures.

Finally, use these resonance structures to determine the products of the second propagation step (halogen abstraction). Simply place a bromine at the position of the unpaired electron in each resonance structure, giving the following products:

**STEP 3**
Place a bromine at each location that bears an unpaired electron.

**(Racemic mixture)**

The first product is expected to be obtained as a racemic mixture of enantiomers, as described in Section 11.6.

**PRACTICE the skill** **11.16** Predict the products when each of the following compounds is treated with NBS and irradiated with UV light:

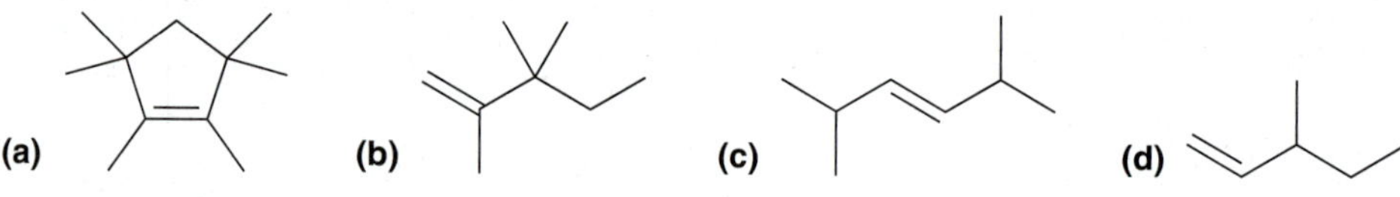

**APPLY the skill** **11.17** When 2-methyl-2-butene is treated with NBS and irradiated with UV light, five different monobromination products are obtained, one of which is a racemic mixture of enantiomers. Draw all five monobromination products and identify the product that is obtained as a racemic mixture.

need more **PRACTICE?** **Try Problems 11.26, 11.28, 11.33d,e, 11.36, 11.39**

## 11.8 Atmospheric Chemistry and the Ozone Layer

Ozone ($O_3$) is constantly produced and destroyed in the stratosphere, and its presence plays a vital role in shielding us from harmful UV radiation emitted by the sun. It is believed that life could not have flourished on land without this protective ozone layer, and instead, life would have been restricted to the depths of the ocean. The ability of ozone to protect us from harmful radiation is believed to result from the following mechanism:

$$O_3 \xrightarrow{h\nu} O_2 + \cdot\ddot{O}\cdot$$

$$O_2 + \cdot\ddot{O}\cdot \longrightarrow O_3 + \text{heat}$$

In the first step above, ozone absorbs UV light and splits into two pieces. In the second step, these two pieces recombine to release energy. There is no net chemical change, but there is one important consequence of this process: Harmful UV light is converted into another form of energy (Figure 11.16). This process exemplifies the role of entropy in nature. Light and heat are both forms of energy, but

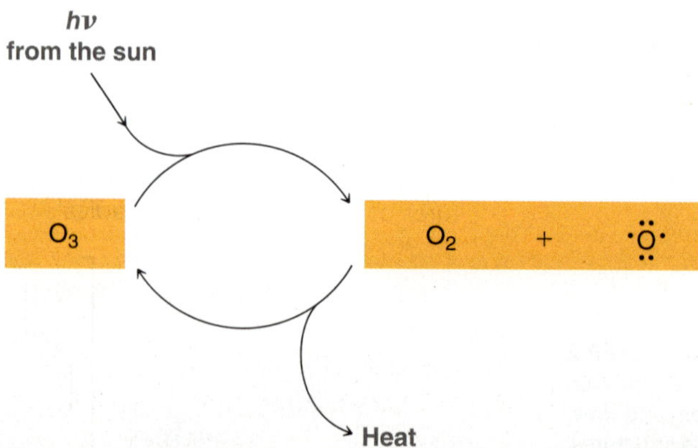


**FIGURE 11.16**
An illustration of the ability of stratospheric ozone to convert light into heat.

heat is a more disordered form of energy. The driving force for the conversion of light into heat is an increase in entropy. Ozone is simply the vehicle by which ordered energy (light) is converted into disordered energy (heat).

Measurements over the last several decades have indicated a rapid decrease of stratospheric ozone. This decrease has been most drastic over Antarctica, where the ozone layer is now almost completely absent (Figure 11.17). Stratospheric ozone over the rest of the planet has decreased at a rate of about 6% each year. Each year, we are exposed to increasing levels of harmful UV radiation that are linked with skin cancer and other health issues. While many factors contribute to ozone depletion, it is believed that the main culprits are compounds called **chlorofluorocarbons** (**CFCs**). CFCs are compounds containing only carbon, chlorine, and fluorine. In the past they were heavily used for a wide variety of commercial applications, including as refrigerants, as propellants, in the production of foam insulation, as fire-fighting materials, and for many other useful applications. They were sold under the trade name "**Freons**":

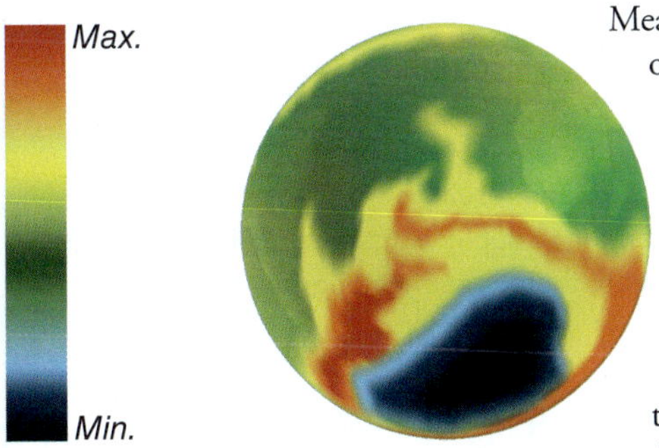

Max.

Min.

**FIGURE 11.17**
An illustration of the hole in the ozone layer over Antarctica.

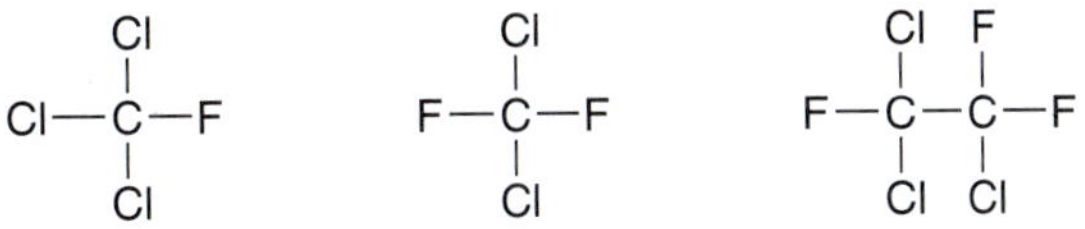

CFC-11 (Freon 11): $CCl_3F$

CFC-12 (Freon 12): $CCl_2F_2$

CFC-113 (Freon 113): $CCl_2F{-}CClF_2$

The harmful effects of CFCs on the ozone layer were elucidated in the late 1960s and early 1970s by Mario Molina, Frank Rowland, and Paul Crutzen, who shared the 1995 Nobel Prize in Chemistry for their work. CFCs are stable compounds that do not undergo chemical change until they reach the stratosphere. In the stratosphere, they interact with high-energy UV light and undergo homolytic cleavage, forming chlorine radicals. These radicals are then believed to destroy ozone by the following mechanism:

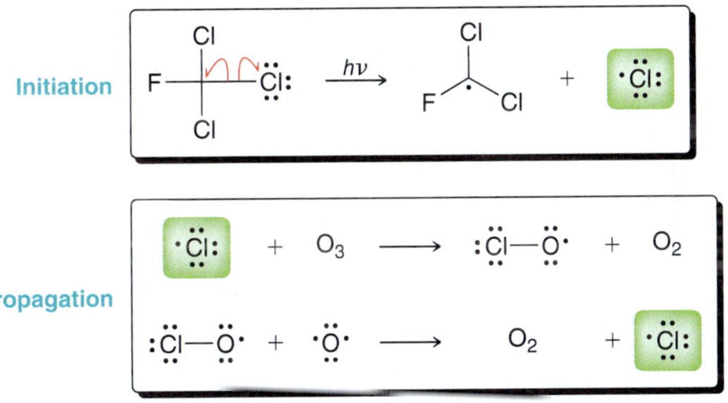

**Initiation**

$$CFCl_2{-}Cl \xrightarrow{h\nu} \cdot CFCl_2 + \cdot Cl$$

**Propagation**

$$\cdot Cl + O_3 \longrightarrow Cl{-}O\cdot + O_2$$

$$Cl{-}O\cdot + \cdot O\cdot \longrightarrow O_2 + \cdot Cl$$

The second propagation step regenerates a chlorine radical, which continues the chain reaction. In this way, each CFC molecule can destroy thousands of ozone molecules. Awareness of the effect of CFCs led to a ban on their production in most countries as of January 1, 1996, as part of a global treaty called the Montreal Protocol on Substances that Deplete the Ozone Layer. As a result, much research has been directed toward finding suitable substitutes, such as the following two categories of compounds.

**Hydrochlorofluorocarbons** (**HCFCs**) are compounds that possess at least one C—H bond. Below are a few examples:

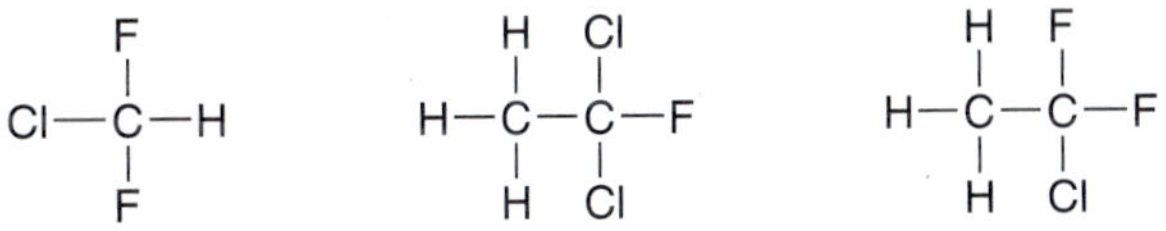

HCFC-22: $CHClF_2$

HCFC-141b: $CH_3{-}CCl_2F$

HCFC-142b: $CH_3{-}CClF_2$

These compounds are believed to be less destructive to the ozone layer, because the C—H bond allows them to decompose before they reach the stratosphere. HCFC-22 and HCFC-141b have replaced CFC-11 in the production of foam insulation.

## practically speaking — Fighting Fires with Chemicals

As mentioned in the chapter-opening paragraphs, the combustion process is believed to involve free radicals. When subjected to excessive heat, the single bonds in organic compounds undergo homolytic bond cleavage to produce radicals, which then react with molecular oxygen in a coupling reaction. A series of radical chain reactions continue until most of the C—C and C—H bonds have been broken to produce $CO_2$ and $H_2O$. In order for this chain process to continue, a fire needs three essential ingredients: fuel (such as wood, which is comprised of organic compounds), oxygen, and heat. In order to extinguish a fire, we must deprive the fire of at least one of these three ingredients. Alternatively, we can find a way to stop the radical chain reaction by destroying the radical intermediates.

Many reagents can be used to extinguish a fire. Common examples found in many small fire extinguishers are $CO_2$, water, and argon. A sudden discharge of $CO_2$ or argon gas will deprive a fire of oxygen, while water deprives a fire of heat (by absorbing the heat in order to evaporate or boil).

The most powerful reagents for extinguishing fires are called halons, because they are organic compounds containing halogen atoms. These compounds are generally either CFCs or BFCs (bromofluorocarbons). Here are just two examples of halons that have been extensively used as fire suppression agents:

$CF_2ClBr$ — **Halon 1211 (Freon 12B1)**

$CF_3Br$ — **Halon 1301 (Freon 13B1)**

Halons are extremely effective, because they fight fire in three different ways:

1. Halons are gases, so a sudden discharge of a halon gas will deprive the fire of oxygen.
2. Halons absorb heat to undergo homolytic bond cleavage:

$$CF_2Cl{-}Br \xrightarrow{\text{Heat}} CF_2Cl\cdot + \cdot Br$$

By absorbing heat, they deprive the fire of one of its essential ingredients.

3. Homolytic bond cleavage results in the formation of free radicals, which can then couple with the radicals participating in the chain reaction, thereby terminating the process:

$$CF_2Cl\cdot + \cdot R \longrightarrow CF_2Cl{-}R$$

Halons are therefore able to speed up the rate of termination steps so that they compete with propagation steps.

For all of these reasons halons are extremely effective as firefighting agents and were heavily used over the last three decades. Halons also have the added benefit of not leaving behind any residue after being used, rendering them particularly useful for fighting fires involving sensitive electronic equipment or documents. Unfortunately, halons have been shown to contribute to ozone depletion, and their production is now banned by the Montreal Protocol. The ban only applies to production, and it is still permitted to use existing stockpiles of halon gases. These stockpiles have been designated for use in special situations involving sensitive equipment, such as on planes or in control rooms. For all other situations, halon gases have been replaced with alternative gases that do not contribute to ozone depletion but are less effective as firefighting agents. One such example is called FM-200:

$CF_3{-}CHF{-}CF_3$

**FM-200**

**Hydrofluorocarbons (HFCs)** are compounds that contain only carbon, fluorine, and hydrogen (no chlorine). Below are a few examples:

HFC-32 HFC-125 HFC-134a

HFCs are not believed to contribute to ozone depletion, as they do not contain chlorine and therefore do not produce chlorine radicals. Nevertheless, they are believed to contribute to global warming. HFC-134a has replaced CFC-12 in refrigeration systems and medical aerosols.

## CONCEPTUAL CHECKPOINT

**11.18** Most supersonic planes produce exhaust of hot gases containing many compounds, including nitric oxide (NO). Nitric oxide is a radical that is believed to play a role in ozone depletion. Propose propagation steps that show how nitric oxide can destroy ozone in a chain process.

## 11.9 Autooxidation and Antioxidants

### Autooxidation

In the presence of atmospheric oxygen, organic compounds are known to undergo a slow oxidation process called **autooxidation**. For example, consider the reaction of cumene with oxygen to form a **hydroperoxide** (ROOH):

$O_2$

Cumene Cumene hydroperoxide

Autooxidation is believed to proceed via Mechanism 11.2.

## MECHANISM 11.2 AUTOOXIDATION

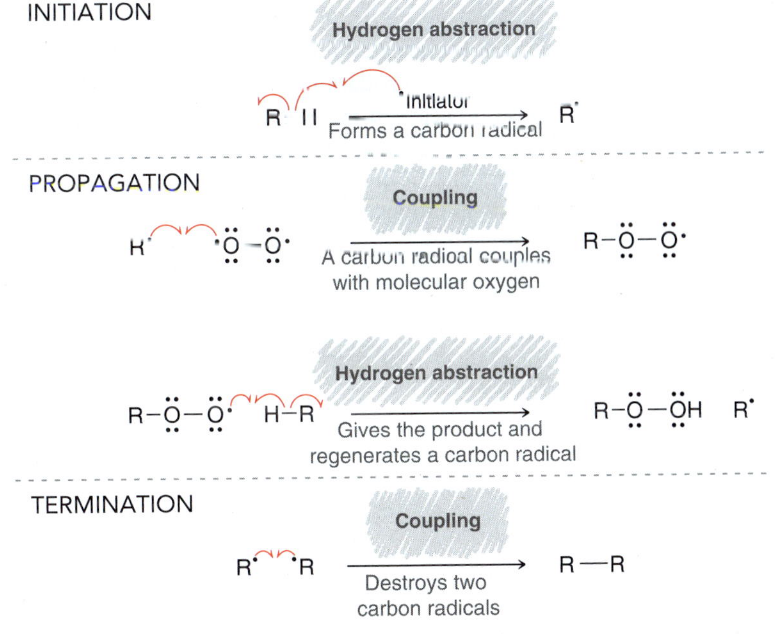

The initiation step might seem abnormal at first glance, because it is not a homolytic bond cleavage. The initiation step in this case is a hydrogen abstraction, which is a step that we normally associate with propagation. Moreover, look closely at the first propagation step. That step is a coupling reaction, which is a step that we have associated with termination until now. These peculiarities force us to refine our definitions of initiation, propagation, and termination. Propagation steps are defined as the steps that produce the net chemical reaction. In other words, the net reaction must be the sum of the propagation steps. In this case, the net reaction is the sum of the following two steps:

Coupling: R· + ·Ö—Ö· ⟶ R—Ö—Ö·

Hydrogen abstraction: R—Ö—Ö· + H—R ⟶ R—Ö—ÖH + R·

Net reaction: R—H + $O_2$ ⟶ R—O—O—H

These two steps are therefore the propagation steps, and any other steps must be classified as either initiation or termination steps.

Many organic compounds, such as ethers, are particularly susceptible to autooxidation. For example, consider the reaction between diethyl ether and oxygen to form a hydroperoxide:

Diethyl ether —$O_2$→ A hydroperoxide (OOH)

This reaction is slow, but old bottles of ether will invariably contain a small concentration of hydroperoxides, rendering the solvent very dangerous to use. Hydroperoxides are unstable and decompose violently when heated. Many laboratory explosions have been caused by distillation of ether that was contaminated with hydroperoxides. For this reason, ethers used in the laboratory must be dated and used in a timely fashion.

Most organic compounds are subject to autooxidation, which is a process initiated by light. In the absence of light, autooxidation occurs at a much slower rate. For this reason, organic chemicals are typically packaged in dark bottles. Vitamins are often sold in brown bottles for the same reason.

Compounds with benzylic or allylic hydrogen atoms are particularly sensitive to autooxidation, because the initiation step produces a resonance-stabilized radical:

H —Hydrogen abstraction→ Resonance stabilized ⟶ OOH

H —Hydrogen abstraction→ Resonance stabilized ⟶ OOH

## Antioxidants as Food Additives

Naturally occurring fats and oils, such as vegetable oil, are generally mixtures of compounds called triglycerides, which contain three long alkyl chains:

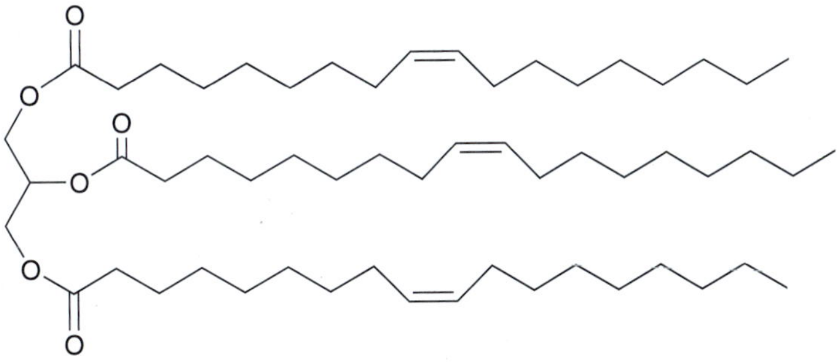

**A triglyceride**

The alkyl chains generally contain double bonds, which render them susceptible to autooxidation, specifically at the allylic positions where hydrogen abstraction can occur more readily:

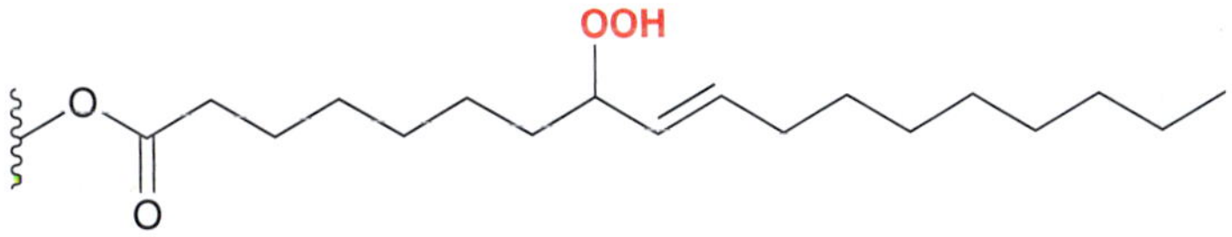

OOH

The resulting hydroperoxides contribute to the rancid smell that develops over time in foods containing unsaturated oils. Moreover, hydroperoxides are also toxic. Food products with unsaturated oils therefore have a short shelf life unless radical inhibitors are used to slow the autooxidation process. Many radical inhibitors are used as food preservatives, including BHT and BHA:

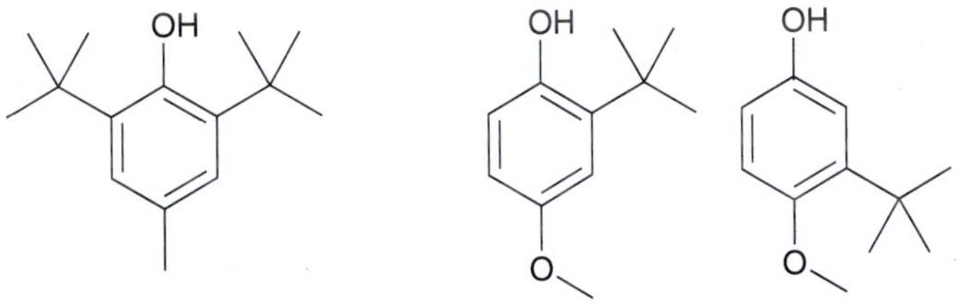

**Butylated hydroxytoluene (BHT)**

**Butylated hydroxyanisole (BHA)**

BHA is a mixture of constitutional isomers. These compounds function as radical inhibitors, because they react with radicals to generate resonance-stabilized radicals:

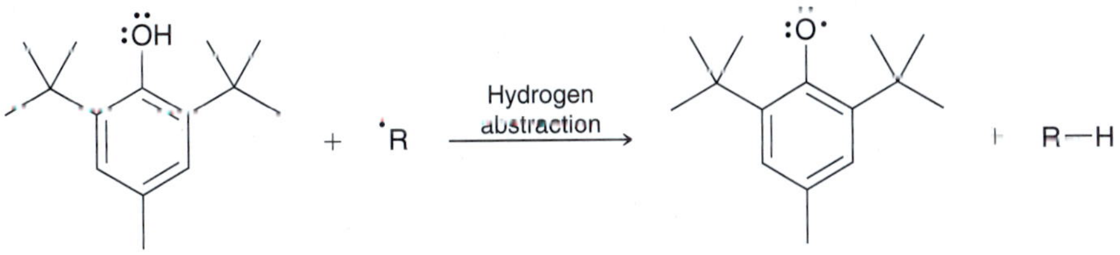

+ •R —Hydrogen abstraction→ + R—H

**Resonance stabilized**

The *tert*-butyl groups provide steric hindrance, which further reduces the reactivity of the stabilized radical. BHT and BHA effectively scavenge and destroy radicals. They are called **antioxidants** because one molecule of a radical scavenger can prevent the autooxidation of thousands of oil molecules by not allowing the chain process to begin.

## Why Is an Overdose of Acetaminophen Fatal?

Tylenol (acetaminophen) can be very helpful in alleviating pain, but it is well known that an overdose of Tylenol can be fatal. The source of this toxicity (in high doses) involves radical chemistry. Our bodies utilize many radical reactions for a wide variety of functions, but these reactions are controlled and localized. If uncontrolled, free radicals are very dangerous and are capable of damaging DNA and enzymes, ultimately leading to cell death. Free radicals are regularly produced as by-products of metabolism, and our bodies utilize a variety of compounds to destroy these radicals. One such example is glutathione, which bears a mercapto group (SH), highlighted in red:

**Glutathione**

Glutathione, often abbreviated GSH, is produced in the liver and plays many key roles, including its ability to function as a radical scavenger. The SH bond is particularly susceptible to hydrogen abstraction, which produces a stabilized radical on a sulfur atom:

$$GS{-}H + {}^{\bullet}R \longrightarrow GS^{\bullet} + H{-}R$$

**Glutathione** → **Glutathione radical**

The original radical is destroyed and replaced by a glutathione radical. Two glutathione radicals, formed in this way, can then couple with each other to produce a compound called glutathione disulfide, abbreviated GSSG, which is then ultimately converted back into glutathione:

$$GS^{\bullet} + {}^{\bullet}SG \longrightarrow GS{-}SG$$

**Glutathione disulfide**

↓ **Reduction**

$$2\ GS{-}H$$

**Glutathione**

The important function of glutathione as a radical scavenger is compromised by an overdose of acetaminophen (Tylenol).

Acetaminophen is metabolized in the liver via a process that consumes glutathione, thereby causing a temporary reduction in glutathione levels. For people with healthy livers, the body is able to replenish the supply of glutathione quickly by biosynthesizing more of this important compound. In this way, the concentration of glutathione never reaches dangerously low levels. However, an overdose of acetaminophen can cause a temporary depletion of glutathione. During that time, free radicals are uncontrolled and cause a host of problems that lead to irreversible liver damage. If untreated, an overdose of acetaminophen can lead to liver failure and death within a few days. Early intervention can prevent irreversible liver damage and consists of treatment with *N*-acetylcysteine (NAC). NAC acts as an antidote for an acetaminophen overdose by delivering to the liver a high concentration of cysteine:

***N*-acetylcysteine (NAC)** → **Cysteine**

Glutathione is biosynthesized in the liver from cysteine, glycine, and glutamic acid:

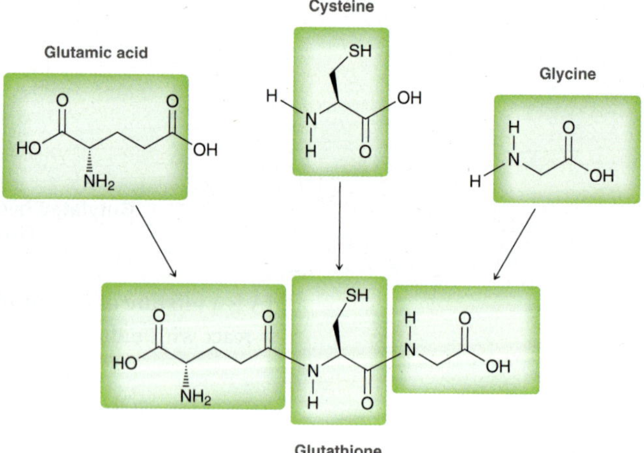


Glycine and glutamic acid are abundant, while cysteine is the limiting reagent. By supplying the liver with excess cysteine, the body is able to produce glutathione rapidly and bring the glutathione levels back up to healthy concentrations.

## Naturally Occurring Antioxidants

**LOOKING AHEAD**
The structure of cell membranes is discussed in more detail in Section 26.5.

Nature employs many of its own natural antioxidants to prevent the oxidation of cell membranes and protect a variety of biologically important compounds. Examples of natural antioxidants include vitamin E and vitamin C:

Vitamin E

Vitamin C

Vitamin E has a long alkyl chain and is therefore hydrophobic, making it capable of reaching the hydrophobic regions of cell membranes. Vitamin C is a small molecule with multiple OH groups, rendering it water soluble. It functions as an antioxidant in hydrophilic regions, such as in blood.

Each of these compounds can destroy a reactive radical by undergoing hydrogen abstraction, transferring a hydrogen atom to the radical, thereby generating a more stable, less reactive radical. A connection has been suggested between the aging process and the natural oxidation processes that occur in the body. This suggestion has led to the widespread use of antioxidant products that include compounds such as vitamin E, despite the lack of evidence for a measurable connection between these products and the rate of aging.

**CONCEPTUAL CHECKPOINT**

**11.19** Compare the structure of vitamin E with the structures of BHT and BHA and then determine which hydrogen atom is most easily abstracted from vitamin E.

# 11.10 Radical Addition of HBr: *Anti*-Markovnikov Addition

## Regiochemical Observations

In Chapter 9, we saw that an alkene will react with HBr via an ionic addition that installs a bromine atom at the more substituted position (Markovnikov addition):

H—Br → Markovnikov addition

Recall that purity of reagents was found to be a critical feature of HBr additions. With impure reagents the reaction proceeds via *anti-Markovnikov addition*, where the halogen appears in the product at the less substituted position:

\+ HBr → (Impure reagents)

Further investigation revealed that even trace amounts of peroxides (ROOR) can cause this regiochemical preference for *anti*-Markovnikov addition:

H—Br / ROOR → (95%) *Anti*-Markovnikov addition

## A Mechanism for *Anti*-Markovnikov Addition of HX

*Anti*-Markovnikov addition of HBr in the presence of peroxides can be explained by invoking a radical mechanism (Mechanism 11.3).

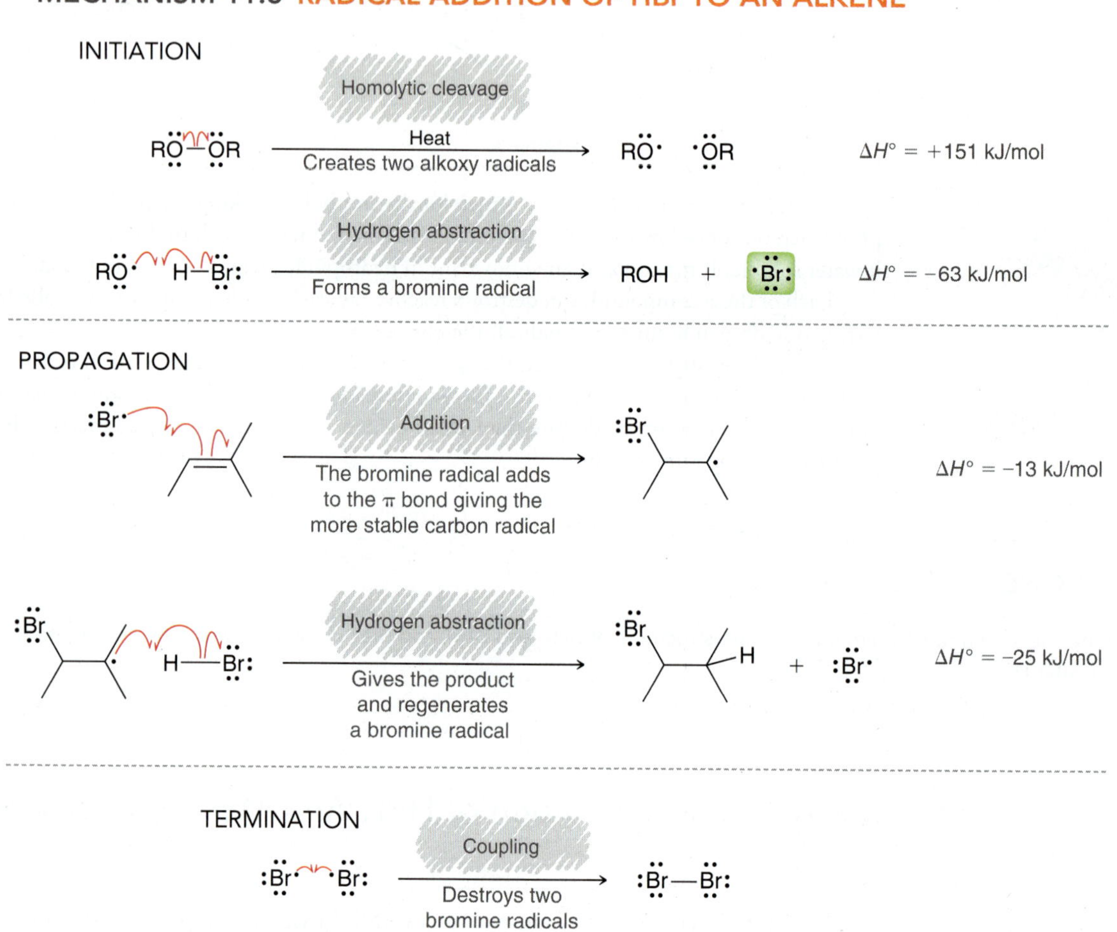


The two initiation steps generate a bromine radical. The two propagation steps are responsible for formation of the observed product, so we must focus on these two steps in order to understand the regiochemical preference for *anti*-Markovnikov addition. Notice that the intermediate formed in the first propagation step is a tertiary carbon radical, rather than a secondary radical:

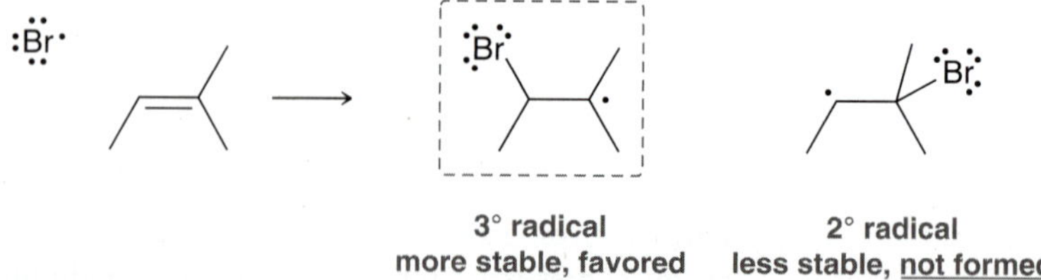


In Chapter 9, we saw that the regiochemistry of ionic addition of HBr is determined by the tendency to proceed via the more stable carbocation intermediate. Similarly, the regiochemistry of radical addition of HBr is also determined by the tendency to proceed via the more stable intermediate. But here, the intermediate is a radical, rather than a carbocation. To see this more

clearly, compare the intermediate involved in an ionic mechanism with the intermediate of a radical mechanism:

**Ionic mechanism** — H — Tertiary carbocation intermediate

**Radical mechanism** — Br — Tertiary radical intermediate

Each reaction proceeds via the lowest energy pathway available—via either a tertiary carbocation or a tertiary radical. But take special notice of the fundamental difference. In the ionic mechanism, the alkene reacts with the proton first, while in the radical mechanism, the alkene reacts with the bromine first. Therefore:

- An ionic mechanism results in a Markovnikov addition.
- A radical mechanism results in an *anti*-Markovnikov addition.

In each case, the regioselectivity is based on the tendency to proceed via the most stable possible intermediate.

## Thermodynamic Considerations

Markovnikov addition can be accomplished with HCl, HBr, or HI. In contrast, *anti*-Markovnikov addition can only be accomplished with HBr. The radical pathway is not thermodynamically favorable for addition of either HCl or HI. To understand the reason for this, we must explore thermodynamic considerations for each step of the radical mechanism.

In Section 6.3 (and earlier in this chapter), we saw that the sign of $\Delta G$ must be negative in order for a reaction to be spontaneous. Recall that the two components of $\Delta G$ are enthalpy and entropy:

$$\Delta G = \underbrace{(\Delta H)}_{\text{Enthalpy term}} + \underbrace{(-T\Delta S)}_{\text{Entropy term}}$$

In order to assess the sign of $\Delta G$ for any process, we must evaluate the signs of both the enthalpy and entropy terms. At the beginning of this chapter, we posed this type of thermodynamic argument to explore the halogenation of alkanes. In this section, we will explore each propagation step of the radical addition mechanism separately.

Let's begin with the first propagation step. In the following chart, the enthalpy and entropy terms are evaluated individually for the cases of HCl, HBr, and HI:

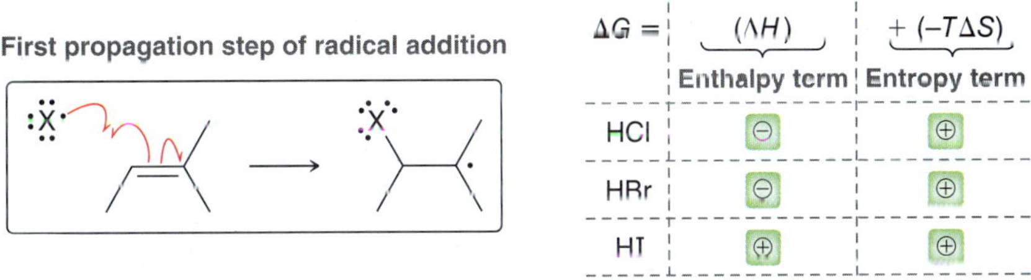


| $\Delta G =$ | $(\Delta H)$ Enthalpy term | $+ (-T\Delta S)$ Entropy term |
|---|---|---|
| HCl | ⊖ | ⊕ |
| HBr | ⊖ | ⊕ |
| HI | ⊕ | ⊕ |

For the cases of HCl and HBr, the sign of $\Delta G$ is determined by a competition between the enthalpy and entropy terms. At high temperature, the entropy term will dominate, and $\Delta G$ will be positive. At low temperature, the enthalpy term will dominate, and $\Delta G$ will be negative. Therefore, the process will be thermodynamically favorable at low temperatures.

The case of HI is fundamentally different. Look at the previous chart and notice that the enthalpy term is positive in the case of HI (the step is endothermic). The enthalpy and entropy terms are not in competition here. Regardless of temperature $\Delta G$ will be positive. Therefore, radical addition of HI (*anti*-Markovnikov addition) is not observed at all.

Now let's perform the same type of analysis for the second propagation step in the radical mechanism:

**Second propagation step of radical addition**

| | $\Delta G =$ $(\Delta H)$ Enthalpy term | $+ (-T\Delta S)$ Entropy term |
|---|---|---|
| HCl | ⊕ | ~0 |
| HBr | ⊖ | ~0 |
| HI | ⊖ | ~0 |

In each of the cases above, the second term ($-T\Delta S$) is insignificant. Why? In each case, *two* chemical entities are reacting with each other to produce *two* new chemical entities. Based on this, we might expect $\Delta S = 0$. However, there is still some small value of $\Delta S$ in each case, due to changes in vibrational degrees of freedom. This effect will be very small though, and as a result, $\Delta S$ will be close to zero. Therefore, the sign of $\Delta H$ will be the dominant factor in determining the sign of $\Delta G$ in each case. In the case of HCl, $\Delta H$ has a positive value. Therefore, it will be difficult to find a temperature at which the second term ($-T\Delta S$) can overcome the $\Delta H$ term. As a result, radical addition of HCl is not an effective process.

To summarize our analysis for radical addition of HX across a π bond: The first propagation step cannot occur in the case of HI, and the second propagation step is unlikely to occur in the case of HCl. Only in the case of HBr are both propagation steps thermodynamically favorable.

## Stereochemistry for Radical Addition of HBr

For some alkenes, radical addition of HBr results in the formation of one new chirality center:

In cases like this, there is no reason to expect one enantiomer to prevail over the other, as the first propagation step can occur from either face of the alkene. Therefore, the reaction will produce a racemic mixture of enantiomers:

50% 50%

# SKILLBUILDER

## 11.8 PREDICTING THE PRODUCTS FOR RADICAL ADDITION OF HBr

**LEARN the skill**

Predict the products for the following reaction:

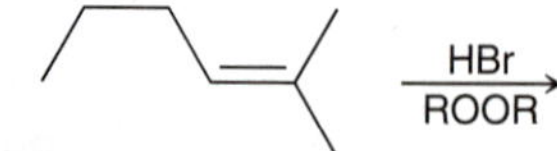


**SOLUTION**

**STEP 1**
Identify the two groups being added across the double bond.

In Section 9.12, we saw that predicting the products of an addition reaction requires the following three questions:

1. Which two groups are being added across the double bond? HBr indicates an addition of H and Br across the double bond.

**STEP 2**
Identify the expected regioselectivity.

2. What is the expected regioselectivity (Markovnikov or *anti*-Markovnikov)? The presence of peroxides indicates that the reaction proceeds via an *anti*-Markovnikov addition. That is, the Br is expected to be placed at the less substituted position:

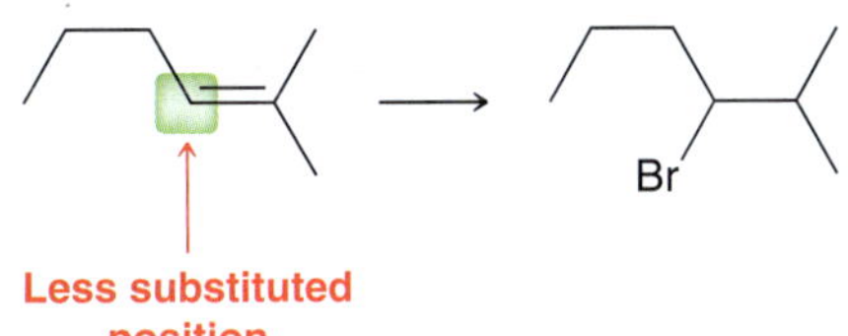


**STEP 3**
Identify the expected stereospecificity.

3. What is the expected stereospecificity? In this case, one new chirality center is created, which results in a racemic mixture of the two possible enantiomers:

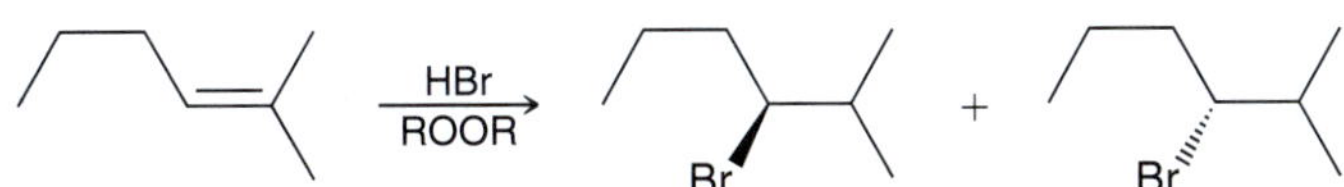

**PRACTICE the skill** **11.20** Predict the products for each reaction. In each case, be sure to consider whether a chirality center is being generated and then draw all expected stereoisomers.

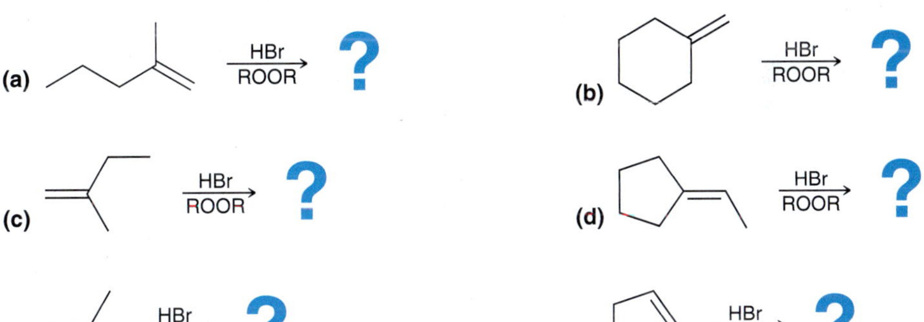


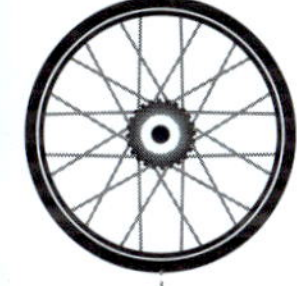

**APPLY the skill**

**11.21** The initiation step for radical addition of HBr is highly endothermic:

$$\text{RO–OR} \xrightarrow[\text{or heat}]{h\nu} \text{RO}\cdot \quad \cdot\text{OR} \qquad \Delta H^\circ = +151\ \text{kJ/mol}$$

**(a)** Explain how this step can be thermodynamically favorable at high temperature even though it is endothermic.

**(b)** Explain why this step is not thermodynamically favorable at low temperature.

## 11.11 Radical Polymerization

### Radical Polymerization of Ethylene

In Chapter 9 we explored polymerization that occurs via an ionic mechanism. In this section, we will see that polymerization can also occur via a radical process. Consider, for example, the radical polymerization of ethylene to form polyethylene:

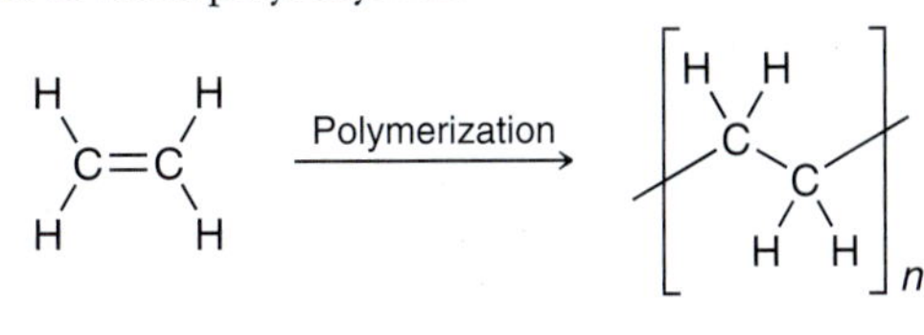

This polymerization can occur via a radical mechanism (Mechanism 11.4).

## MECHANISM 11.4 RADICAL POLYMERIZATION

INITIATION

Homolytic cleavage

RO–OR —Heat→ RO· ·OR

Creates two alkoxy radicals

Addition

RO· + ethylene → RO–CH₂CH₂·

Forms a carbon radical

PROPAGATION

Addition

The carbon radical adds to the π bond giving a new carbon radical, thereby adding one monomer at a time.

Addition

Addition

**Polymer** ← ← ←

TERMINATION

Coupling

Destroys two radicals

Coupling

Destroys two radicals

When ethylene is heated in the presence of peroxides, two initiation steps produce a carbon radical. This carbon radical then attacks another molecule of ethylene, generating a new carbon radical. These propagation steps continue, adding one monomer at a time, until a termination step occurs. If the conditions are carefully controlled to favor propagation over termination, the process can generate very large molecules of polyethylene with over 10,000 repeating units.

During the polymerization process, **chain branching** inevitably occurs. The following scheme shows a mechanism for the formation of chain branches, which begins with a hydrogen abstraction that occurs on an existing chain:

The extent of chain branching determines the physical properties of the resulting polymer. For example, a flexible plastic squeeze bottle and its relatively inflexible cap are both made from polyethylene, but the former has extensive branching while the latter has relatively little. These processes will be discussed in more detail in Chapter 27, and we will see that the extent of branching can be controlled with specialized catalysts, although a small amount of chain branching is generally unavoidable.

## Radical Polymerization of a Substituted Ethylene

A substituted ethylene (ethylene that bears a substituent) will generally undergo radical polymerization to produce a polymer with the following structure:

For example, vinyl chloride is polymerized to form **polyvinyl chloride (PVC)**, which is a very hard polymer used to make plumbing pipes. PVC can be softened if polymerization takes place in the presence of compounds called plasticizers, as we will explore in more detail in Chapter 27. PVC made with plasticizers is a bit more flexible (although durable and strong) and is used for a wide variety of purposes, such as garden hoses, plastic raincoats, and shower curtains. Table 11.3 shows a number of common polymers produced from substituted ethylenes.

**TABLE 11.3 COMMON POLYMERS FORMED FROM ETHYLENE AND SUBSTITUTED ETHYLENE**

| MONOMER | POLYMER | APPLICATION | MONOMER | POLYMER | APPLICATION |
|---|---|---|---|---|---|
| Ethylene | Polyethylene | Plastic bottles and packaging material, insulation, Tupperware | Vinyl chloride | Poly(vinyl chloride) | PVC piping for plumbing, CDs, garden hoses, raincoats, shower curtains |
| Propylene | Polypropylene | Carpet fibers, appliances, car tires | Tetrafluoroethylene | Teflon | Nonstick coating for frying pans |
| Styrene | Polystyrene | Televisions, radios, Styrofoam | | | |

## 11.12 Radical Processes in the Petrochemical Industry

Radical processes are used heavily in chemical industries, particularly in the petrochemical industry. One such example is the process called *cracking*. In Chapter 4, we mentioned that cracking of petroleum converts large alkanes into smaller alkanes that are more suitable for use as gasoline. Cracking is a radical process:

When cracking is performed in the presence of hydrogen gas, alkanes are produced and the process is called **hydrocracking**:

$H_2$

Heat

Reforming, also mentioned in Chapter 4, is a process that causes straight-chain alkanes to become more highly branched, and this process also involves radical intermediates:

Heat

## 11.13 Halogenation as a Synthetic Technique

In this chapter, we have seen that radical chlorination and radical bromination are both thermodynamically favorable processes. Bromination is slower but is more selective than chlorination. Both reactions are used in synthesis. When the starting compound has only one kind of hydrogen (all hydrogen atoms are equivalent), chlorination can be accomplished because selectivity is not necessary:

$Cl_2$

$h\nu$

Cl

When different types of hydrogen atoms are present in the compound, it is best to use bromination because it is more selective and it avoids a mixture of products:

$Br_2$

$h\nu$

Br

In truth, even radical bromination has very limited utility in synthesis. Its greatest utility is to serve as a method for introducing a functional group into an alkane. When the starting material is an alkane, there is very little that can be done other than radical halogenation. By introducing a functional group into the compound, the door is opened for a wide variety of reactions:

NBS

$h\nu$

Br

Substitution

X

Elimination

Chapter 12 is devoted entirely to synthesis techniques, and we will revisit the role of radical halogenation in designing syntheses.

# REVIEW OF REACTIONS

## SYNTHETICALLY USEFUL RADICAL REACTIONS

### Bromination of Alkanes

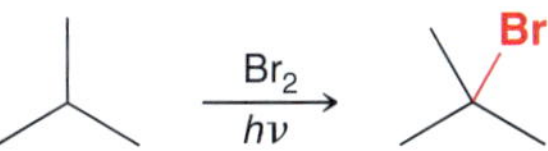


### *Anti*-Markovnikov Addition of HBr to Alkenes

HBr
ROOR
Br

### Allylic Bromination

NBS
$h\nu$
Br

# REVIEW OF CONCEPTS AND VOCABULARY

**SECTION 11.1**

- Radical mechanisms utilize *fishhook arrows*, each of which represents the flow of only one electron.
- The order of stability for radicals follows the same trend exhibited by carbocations.
- Allylic and benzylic radicals are resonance stabilized. Vinylic radicals are not.

**SECTION 11.2**

- Radical mechanisms are characterized by six different kinds of steps: (1) **homolytic cleavage**, (2) **addition to a π bond**, (3) **hydrogen abstraction**, (4) **halogen abstraction**, (5) **elimination**, and (6) **coupling**.
- Every step in a radical mechanism can be classified as **initiation**, **propagation**, or **termination**.

**SECTION 11.3**

- Methane reacts with chlorine via a radical mechanism.
- The sum of the two propagation steps gives the net chemical reaction. These steps together represent a **chain reaction**.
- A **radical initiator** is a compound with a weak bond that readily undergoes homolytic bond cleavage. Examples include **peroxides** and **acyl peroxides**.
- A **radical inhibitor**, also called a radical scavenger, is a compound that prevents a chain process from either getting started or continuing. Examples include molecular oxygen and hydroquinone.

**SECTION 11.4**

- Only radical chlorination and radical bromination have practical use in the laboratory.
- Bromination is generally a much slower process than chlorination.

**SECTION 11.5**

- Halogenation occurs more readily at substituted positions. Bromination is more selective than chlorination.
- In general, there is an inverse relationship between reactivity and selectivity.

**SECTION 11.6**

- When a new chirality center is created during a radical halogenation process, both possible stereoisomers are obtained.
- When a halogenation reaction takes place at a chirality center, a racemic mixture is obtained regardless of the configuration of the starting material.

**SECTION 11.7**

- Alkenes can undergo **allylic bromination**, in which bromination occurs at the allylic position.
- To avoid a competing ionic addition reaction, ***N*-bromosuccinimide** (NBS) can be used instead of $Br_2$.

**SECTION 11.8**

- Ozone is produced and destroyed by a radical process that shields the earth's surface from harmful UV radiation.
- A rapid decrease of stratospheric ozone is attributed to the use of **CFCs**, or **chlorofluorocarbons**, sold under the trade name **Freons.**
- A ban on CFCs prompted a search for viable substitutes, such as **hydrochlorofluorocarbons (HCFCs)** and **hydrofluorocarbons (HFCs)**.

**SECTION 11.9**

- Organic compounds undergo oxidation in the presence of atmospheric oxygen to produce **hydroperoxides**. This process, called **autooxidation**, is believed to proceed via a radical mechanism.
- **Antioxidants**, such as BHT and BHA, are used as food preservatives to prevent autooxidation of unsaturated oils.
- Natural antioxidants prevent the oxidation of cell membranes and protect a variety of biologically important compounds. Vitamins E and C are natural antioxidants.

**SECTION 11.10**

- Alkenes will react with HBr in the presence of peroxides to produce a radical addition reaction.

**SECTION 11.11**

- Polymerization of ethylene via a radical process generally involves **chain branching**.
- When vinyl chloride is polymerized, **polyvinyl chloride (PVC)** is obtained.

**SECTION 11.12**

- Radical processes are used heavily in the chemical industry, particularly in the petrochemical industry. Examples include cracking and reforming. When cracking is performed in the presence of hydrogen, it is called **hydrocracking**.

**SECTION 11.13**

- Radical halogenation provides a method for introducing functionality into an alkane.
- When the starting compound has only one kind of hydrogen atom, chlorination can be used.
- When different types of hydrogen atoms are present in the compound, it is best to use bromination in order to control the regiochemical outcome and avoid a mixture of products.

# SKILLBUILDER REVIEW

## 11.1 DRAWING RESONANCE STRUCTURES OF RADICALS

**STEP 1** Look for an unpaired electron next to a π bond.

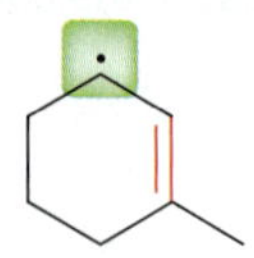

**STEP 2** Draw three fishhook arrows, and then draw the corresponding resonance structure.

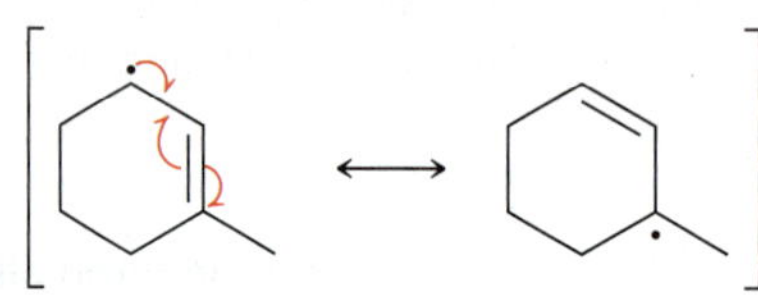

Try Problems 11.2–11.4, 11.22, 11.25

## 11.2 IDENTIFYING THE WEAKEST C—H BOND IN A COMPOUND

**STEP 1** Consider all possible radicals that would result from homolytic cleavage of a C—H bond.

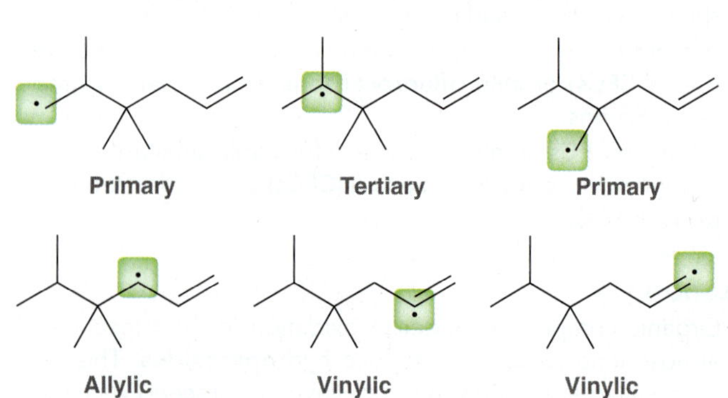


**STEP 2** Identify the most stable radical.

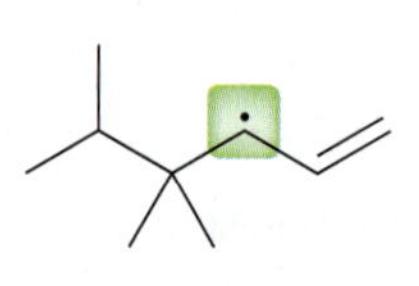


**STEP 3** The most stable radical indicates which bond is the weakest.

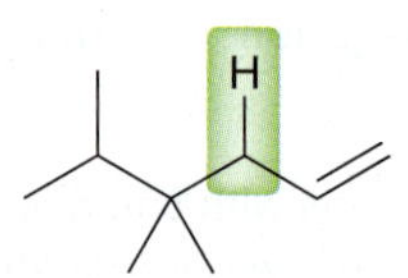


Try Problems 11.5, 11.6, 11.23

## 11.3 IDENTIFYING A RADICAL PATTERN AND DRAWING FISHHOOK ARROWS

There are six characteristic patterns to recognize.

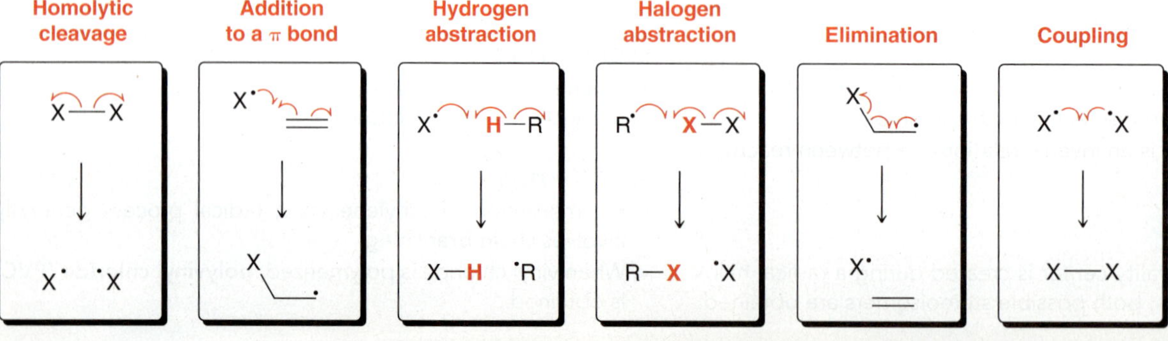

**STEPS 1–3** 1. Identify the type of process—halogen abstraction.
2. Determine the number of fishhook arrows required—three.
3. Identify any bonds being broken or formed.

Bond broken

Bond formed

**STEP 4** For a bond formed, draw two fishhook arrows.

**STEP 5** For a bond broken, make sure to have both fishhook arrows.

Try Problems 11.7–11.9, 11.48

## 11.4 DRAWING A MECHANISM FOR RADICAL HALOGENATION

**Initiation** Homolytic bond cleavage creates radicals, thereby initiating a radical process.

**Propagation**
- Hydrogen abstraction forms a radical on a carbon atom.
- Halogen abstraction forms the product and regenerates a halogen radical.

Hydrogen abstraction

Halogen abstraction

**Termination** Coupling can give the product, but this step destroys radicals.

Try Problems 11.10, 11.11

## 11.5 PREDICTING THE SELECTIVITY OF RADICAL BROMINATION

**STEP 1** Identify all possible positions.

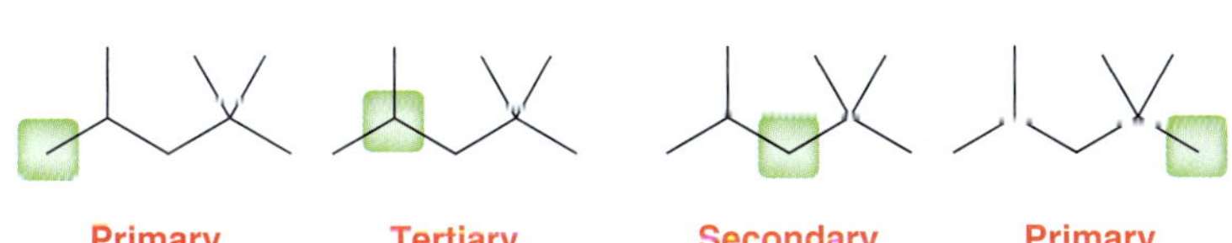


**STEP 2** Identify the most stable location for a radical.

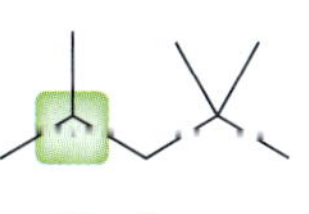


**STEP 3** Bromination occurs at the location of the most stable radical.

Br

Major product

Try Problems 11.12, 11.13, 11.27, 11.33a–c,f, 11.38, 11.40, 11.44

## 11.6 PREDICTING THE STEREOCHEMICAL OUTCOME OF RADICAL BROMINATION

**STEP 1** First identify the location where bromination will occur.

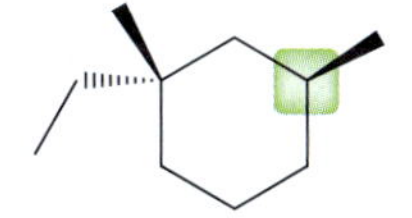

**STEP 2** If this position is a chirality center, or if it will become a chirality center, expect a loss of configuration to yield both possible stereoisomers.

Br + Br

Try Problems 11.14, 11.15, 11.35, 11.41

### 11.7 PREDICTING THE PRODUCTS OF ALLYLIC BROMINATION

**STEP 1** Identify the allylic position.

**STEP 2** Remove a hydrogen atom and draw resonance structures.

**STEP 3** Place a bromine at each location that bears an unpaired electron.

Br

Br

+

Racemic

Try Problems 11.16, 11.17, 11.26, 11.28, 11.33d,e, 11.36, 11.39

### 11.8 PREDICTING THE PRODUCTS FOR RADICAL ADDITION OF HBr

**STEP 1** Identify the two groups being added across the double bond.

$\xrightarrow[\text{ROOR}]{\text{HBr}}$ **Adds H and Br**

**STEP 2** Identify the expected regioselectivity.

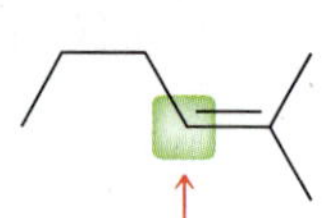

**Br is placed here at the less substituted position**

**STEP 3** Identify the expected stereospecificity.

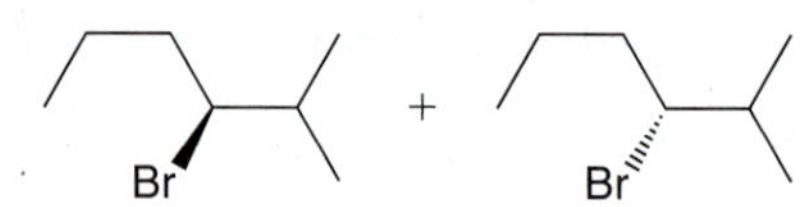


Try Problems 11.20, 11.21

## PRACTICE PROBLEMS

**Note: Most of the Problems are available within WileyPLUS, an online teaching and learning solution.**

**11.22** Draw all resonance structures for each of the following radicals:

(a) 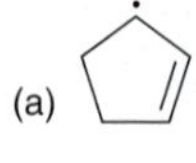 (b) 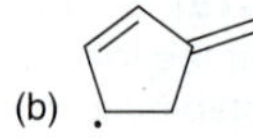 (c)

(d)  (e)

**11.23** Consider all of the different C—H bonds in cyclopentene and rank them in order of increasing bond strength:

H H H

**11.24** Rank each group of radicals in order of increasing stability:

(a)

(b)

**11.25** Draw all resonance structures of the radical produced when a hydrogen atom is abstracted from the OH group in BHT:

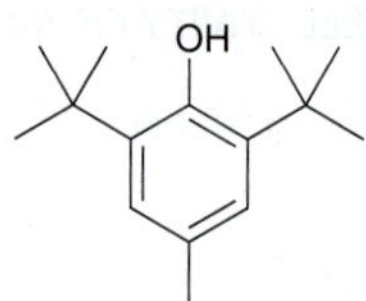


**Butylated hydroxytoluene (BHT)**

**11.26** When isopropylbenzene (cumene) is treated with NBS and irradiated with UV light, only one product is obtained. Propose a mechanism and explain why only one product is formed.

$\xrightarrow[h\nu]{\text{NBS}}$ Br

**11.27** There are three constitutional isomers with the molecular formula $C_5H_{12}$. Chlorination of one of these isomers yields only one product. Identify the isomer and draw the product of chlorination.

**11.28** When ethylbenzene is treated with NBS and irradiated with UV light, two stereoisomeric compounds are obtained in equal amounts. Draw the products and explain why they are obtained in equal amounts.

$\xrightarrow[h\nu]{\text{NBS}}$ **two products**

**11.29** AIBN is an azo compound (a compound with a N=N double bond) that is often used as a radical initiator. Upon heating, AIBN liberates nitrogen gas to produce two identical radicals:

**AIBN**

(a) Give two reasons why these radicals are so stable.

(b) Explain why the following azo compound is not useful as a radical initiator:

**11.30** Triphenylmethane readily undergoes autooxidation to produce a hydroperoxide:

**Triphenylmethane**

(a) Draw the expected hydroperoxide.

(b) Explain why triphenylmethane is so susceptible to autooxidation.

(c) In the presence of phenol ($C_6H_5OH$), triphenylmethane undergoes autooxidation at a much slower rate. Explain this observation.

**11.31** For each of the products shown in the following reaction, propose a mechanism that explains its formation:

**11.32** Abstraction of a hydrogen atom from 3-ethylpentane can yield three different radicals, depending on which hydrogen atom is abstracted. Draw all three radicals and rank them in order of increasing stability.

**11.33** Identify the major product(s) for each of the following reactions. If any of the reactions do not yield a product, indicate "no reaction."

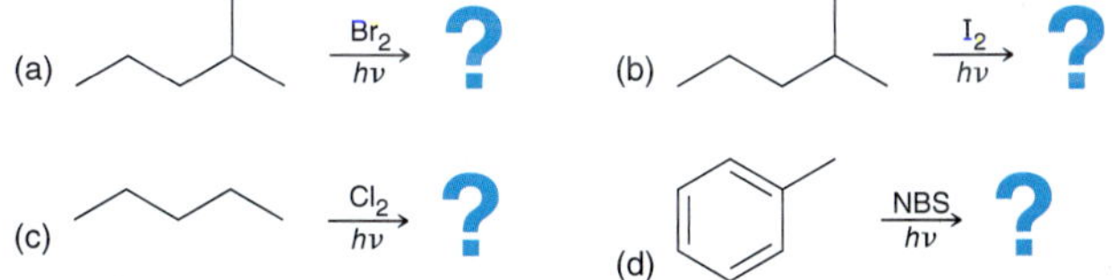

**11.34** Chlorination of (*S*)-2-chloropentane produces a mixture of isomers with molecular formula $C_5H_{10}Cl_2$. How many isomers are obtained (consider stereoisomers as separate products)? Draw all of the products.

**11.35** Predict the major product(s) obtained upon bromination of (*S*)-3-methylhexane.

**11.36** Identify all products expected for each of the following reactions. Take stereochemistry into account and draw expected stereoisomer(s), if any:

**11.37** Draw the propagation steps that achieve the autooxidation of diethyl ether to form a hydroperoxide:

**Diethyl ether** → **A hydroperoxide**

**11.38** Compound A has molecular formula $C_5H_{12}$, and monobromination of compound A produces only compound B. When compound B is treated with a strong base, a mixture is obtained containing compound C and compound D. Using this information, answer the following questions:

(a) Draw the structure of compound A.

(b) Draw the structure of compound B.

(c) Draw the structures of compounds C and D.

(d) When compound B is treated with potassium *tert*-butoxide, which product predominates: C or D? Explain your choice.

(e) When compound B is treated with sodium ethoxide, which product predominates: C or D? Explain your choice.

**11.39** Draw the products obtained when 3,3,6-trimethylcyclohexene is treated with NBS and irradiated with UV light.

**11.40** When 2-methylpropane is treated with bromine in the presence of UV light, one product predominates.

(a) Identify the structure of the product.

(b) Draw the structure of the expected minor product.

(c) Draw a mechanism for formation of the major product.

(d) Draw a mechanism for formation of the minor product.

(e) Using the mechanisms that you just drew, explain why we expect very little of the minor product.

**11.41** Consider the structure of the following compound:

(a) When this compound is treated with bromine under conditions that favor monobromination, two stereoisomeric products are obtained. Draw them and identify whether they are enantiomers or diastereomers.

(b) When this compound is treated with bromine under conditions that favor dibromination, three stereoisomeric products are obtained. Draw them and explain why there are only three products and not four.

# INTEGRATED PROBLEMS

**11.42** The rate at which two methyl radicals couple to form ethane is significantly faster than the rate at which two *tert*-butyl radicals couple. Offer two explanations for this observation.

**11.43** How many constitutional isomers are obtained when each of the following compounds undergoes monochlorination?

(a) 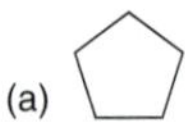 (b) 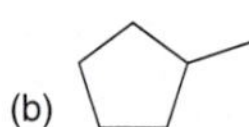 (c) 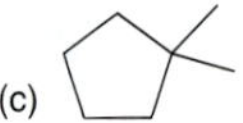

(d) 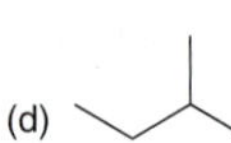 (e) 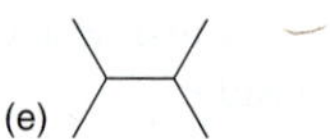 (f)

(g) (h) 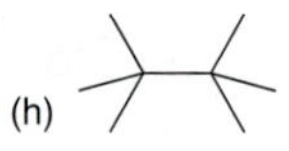 (i)

(j)

**11.44** Using acetylene and 2-methylpropane as your only sources of carbon atoms, propose a plausible synthesis for $CH_3COCH_2CH(CH_3)_2$. You will need to utilize many reactions from previous chapters.

**11.45** Propose an efficient synthesis for each of the following transformations. You might find it useful to review Section 11.13 before attempting to solve these problems.

(a) → Cl

(b) → I

(c) →

(d) → Br, Br + En

(e) →

**11.46** Consider the following two compounds. Monochlorination of one of these compounds produces twice as many stereoisomeric products as the other. Draw the products in each case and identify which compound yields more products upon chlorination.

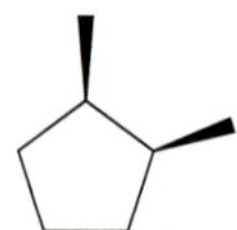

**11.47** When butane reacts with $Br_2$ in the presence of $Cl_2$, both brominated and chlorinated products are obtained. Under such conditions, the usual selectivity of bromination is not observed. In other words, the ratio of 2-bromobutane to 1-bromobutane is very similar to the ratio of 2-chlorobutane to 1-chlorobutane. Can you offer an explanation as to why we do not observe the normal selectivity expected for bromination?

**11.48** When an acyl peroxide undergoes homolytic bond cleavage, the radicals produced can liberate carbon dioxide to form alkyl radicals:

Δ or $h\nu$ → R˙ + $CO_2$

**An acyl peroxide**

Using this information, provide a mechanism of formation for each of the following products:

Δ

**11.49** Consider the following radical process, called a thiol-ene coupling reaction, in which an alkene is treated with a thiol (RSH) in the presence of a radical initiator:

R + H—S—R′ → (Radical initiator, Heat or light) R, S—R′, H, H, H, H

(a) Draw a plausible mechanism for this process, showing the initiation and propagation steps.

(b) Biochemists have become increasingly interested in using organic reactions to modify proteins (Chapter 26) and study their behavior. The thiol-ene coupling reaction has recently been used to couple proteins together in an effort to elucidate the function of specific enzymes (*J. Am. Chem. Soc.*, **2012**, *134*, 6916–6919). Predict the product that is expected when the following proteins undergo a thiol-ene coupling reaction:

PROTEIN—C(=O)—N(H)— … H—S—PROTEIN

## CHALLENGE PROBLEMS

**11.50** Commercial samples of pentacene (**A**) generally contain small amounts of impurities, including 6,13-dihydropentacene (**B**). Upon vacuum sublimation at high temperature (>300 °C), a series of radical reactions are initiated leading to the formation of peripentacene (**F**) (*J. Am. Chem. Soc.* **2007,** *129,* **6536–6546**). Several of the initial steps of a proposed mechanism of this conversion are shown below. Draw curved arrows consistent with the changes in bonding for each of the following steps: **A** + **B** → 2 **C**; **C** + **C** → **D**; **C** + **A** → **E**.

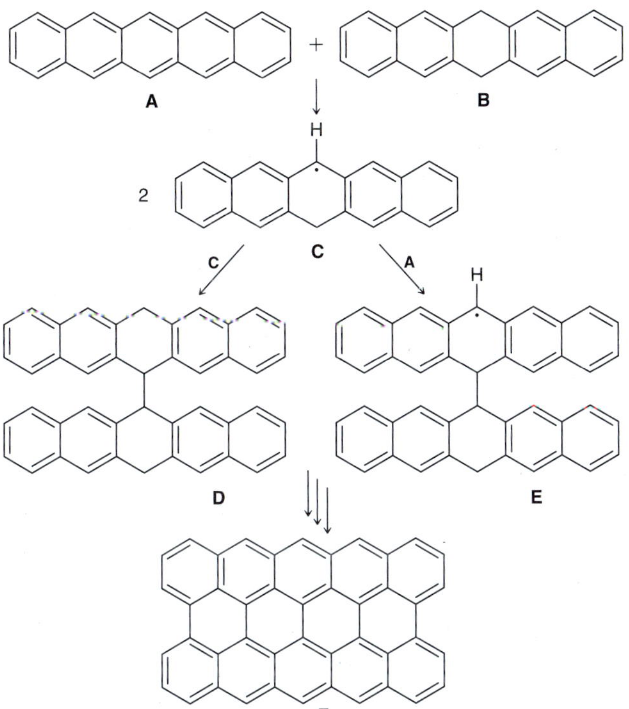

A B C D E F

**11.51** The alkyl halide 1-bromopropane is one of a number of compounds being considered as a replacement for chlorofluorocarbons as an industrial cleaning solvent. In a computational study of its atmospheric oxidation products, bromoacetone (structure below) was determined to be the major product (*J. Phys. Chem. A* **2008,** *112,* **7930–7938**). The proposed mechanism involves four steps: (1) hydrogen abstraction by an OH radical, (2) formation of a peroxy radical by coupling with $O_2$, (3) abstraction of an oxygen atom by NO, thus forming $NO_2$ and an alkoxy radical, and (4) abstraction of a hydrogen atom by $O_2$. Draw the mechanism that is consistent with this description.

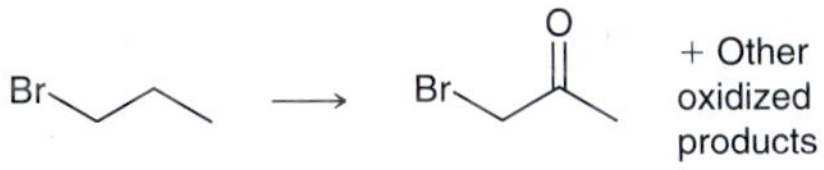

+ Other oxidized products

**11.52** One strategy for the synthesis of complicated polycyclic ring systems involves radical-mediated cyclization processes. For example, when the following radical is generated, it undergoes a spontaneous cyclization to produce two new radicals (*J. Am. Chem. Soc.* **1986,** *108,* **1708–1709**). Propose a plausible mechanism to account for the formation of each tricyclic ring system.

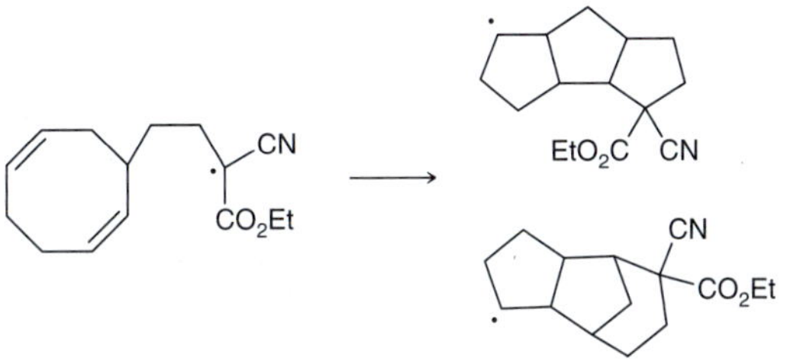

**11.53** Compounds with the structure ArN=NCON=NAr (where Ar represents an aromatic group) are used in liquid crystal formulations for LCDs (liquid crystal displays) as well as for optical switches and image storage because of their high resolution and sensitivity. However, it has been found recently that, upon irradiation with light at 350 nm, bisphenyl carbodiazone (PhN=NCON=NPh) undergoes homolytic decomposition to give radicals that can be trapped with a tetramethylpiperidinoxyl (TEMPO) radical (*Tetrahedron Lett.* **2006,** *47,* **2115–2118**). Propose a mechanism that explains the formation of all three products from this process, shown below.

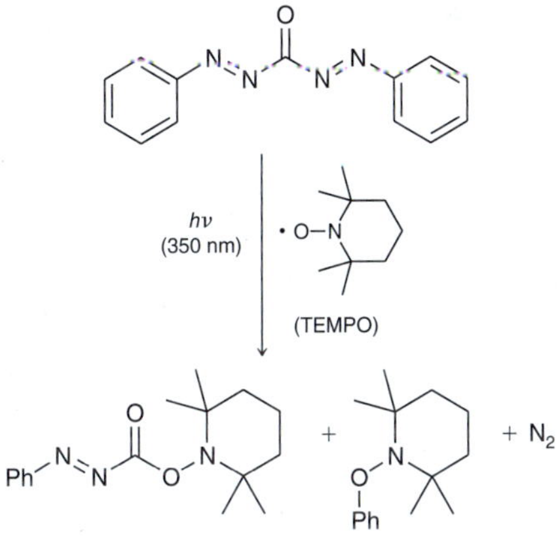

$h\nu$ (350 nm) (TEMPO) + $N_2$

**11.54** As seen in this chapter, hydrocarbons typically do not undergo radical iodination in the presence of $I_2$ (compound **2a**). Furthermore, radical halogenation (even chlorination) of strained hydrocarbons, like cubane (**1**), is problematic if the reaction produces a halogen radical (X•). An alternative method circumvents both of these issues by replacing the elemental halogen, $X_2$, with a tetrahalomethane, $CX_4$ (*J. Am. Chem. Soc.* **2001,** *123,* **1842–1847**). The reaction is initiated by formation of a trihalomethyl radical, •$CX_3$.

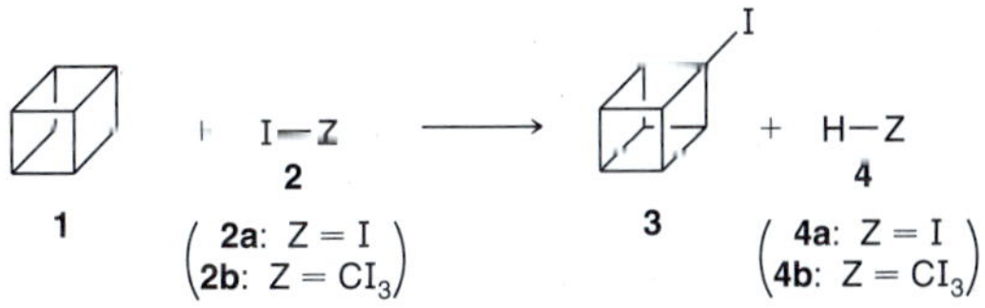

**1** + I–Z **2** (**2a**: Z = I; **2b**: Z = $CI_3$) → **3** + H–Z **4** (**4a**: Z = I; **4b**: Z = $CI_3$)

(a) Propose a propagation cycle for the production of iodocubane (**3**) from cubane (**1**) and tetraiodomethane (**2b**).

(b) Provide three possible termination steps.

(c) The bond dissociation energy (BDE) for I—$CI_3$ is 192 kJ/mol, while the BDE for H—$CI_3$ is 423 kJ/mol. Using this information, provide a thermodynamic argument that explains why **2b** is an effective iodination reagent while **2a** is not.

**11.55** Resveratrol is a naturally occurring antioxidant found in a broad range of plants including grapes, mulberries, and peanuts. It demonstrates useful activity against heart disease and cancer. In the presence of an alkoxy radical (RO•), a hydrogen atom is abstracted from resveratrol (*J. Org. Chem.* **2009,** *74,* **5025–5031**). Predict which hydrogen atom is preferentially abstracted and justify your choice.

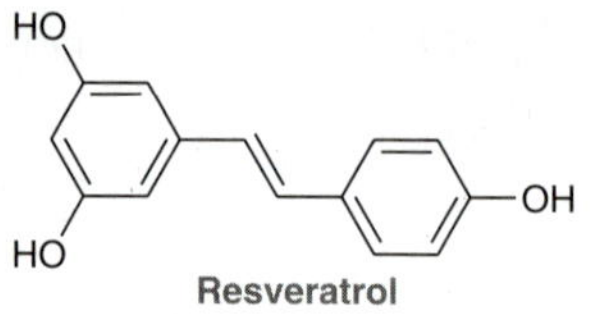

HO
OH
HO
Resveratrol

**11.56** In Chapter 2, we saw several patterns for drawing resonance structures of ions or uncharged compounds. There are also several patterns for drawing resonance structures of radicals, although we have only encountered one such pattern (allylic or benzylic radicals). One other pattern is characterized by a lone pair on an atom that is adjacent to an atom with an unpaired electron, as seen in the following example. The unpaired electron is delocalized by resonance, with radical character being spread over the nitrogen and phosphorus atoms. Note that phosphorus is a third-row element, so it can have more than eight electrons. We will make use of a related pair of resonance structures in this problem.

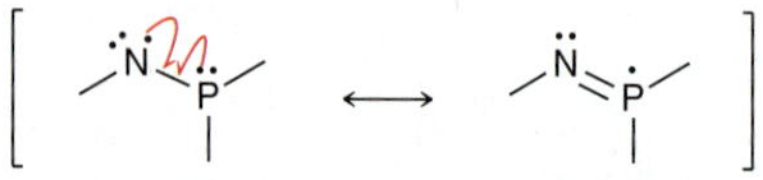

The rearrangement shown below (Ar = aromatic group) represents the conversion of a sulfenate to a sulfoxide, which is initiated by heat. The transformation is believed to occur via a two-step process in which the C—O bond undergoes homolytic cleavage followed by a recombination of the resulting radicals to form the product (*J. Am. Chem. Soc.* **2000,** *122,* **3367–3374**):

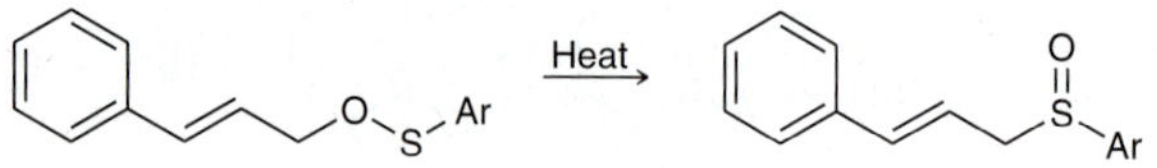

Heat

(a) Draw a mechanism of this rearrangement, consistent with the description provided.

(b) Provide a reasonable rationalization for the fact that the C—O bond is cleaved preferentially over the O—S bond in the first step of the mechanism.

**11.57** Compound **2** was used as a key intermediate in a synthesis of oseltamivir, an anti-influenza agent (*J. Am. Chem. Soc.* **2006;** *128,* **6310–6311**). Propose a plausible mechanism for the conversion of **1** to **2**, as shown, and explain the stereochemical outcome:

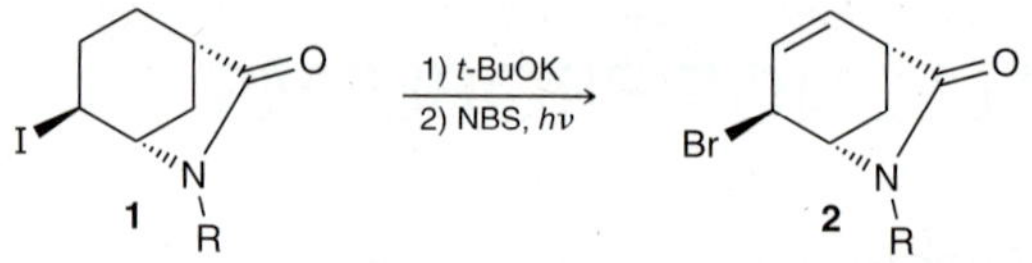

1) *t*-BuOK
2) NBS, *hν*

**11.58** The Kharasch reaction is a radical process in which carbon tetrachloride is added across an alkene:

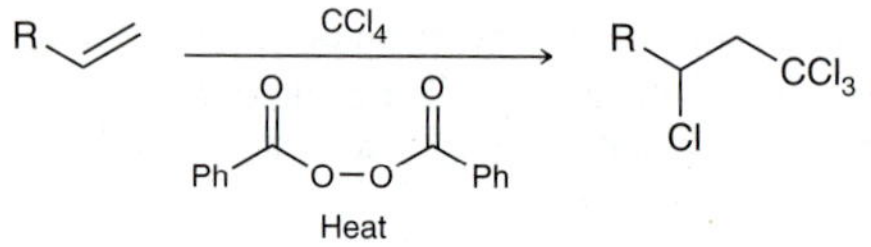

$CCl_4$
Heat

(a) Draw a mechanism showing initiation and propagation steps for this reaction and explain the observed regiochemical outcome.

(b) A variation of the Kharasch reaction was employed during a synthesis of a key piece of paclitaxel, an anticancer agent used in the treatment of breast cancer (*Tetrahedron Lett.* **2002,** *43,* **8587–8590**). Propose a plausible mechanism (initiation and propagation steps) for this ring-closure reaction, shown here:

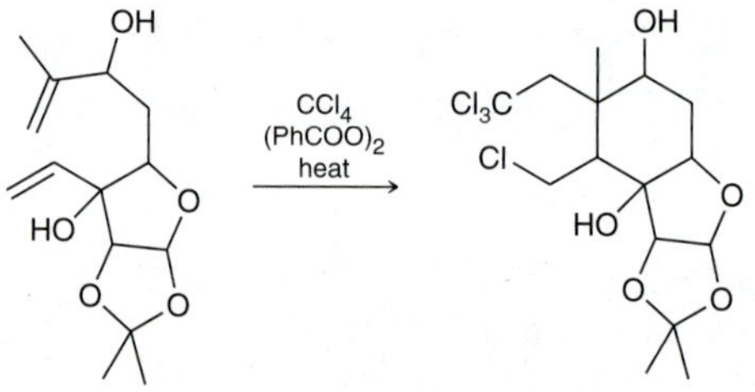

$CCl_4$
$(PhCOO)_2$
heat

**11.59** The ring-forming reaction below (initiated by heat) is believed to undergo a stepwise radical mechanism via the intermediate diradical shown (*J. Am. Chem. Soc.* **2009,** *131,* **653–661**). Propose a reasonable mechanism to account for the formation of the product from the intermediate diradical.

?

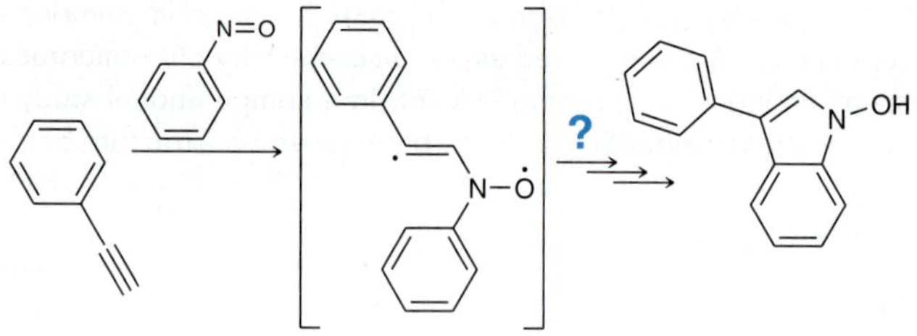

# Synthesis

# 12



## DID YOU EVER WONDER...

### what vitamins are and why we need them?

Vitamins are essential nutrients that our bodies require in order to function properly, and a deficiency of particular vitamins can lead to diseases, many of which can be fatal. Later in this chapter, we will learn more about the discovery of vitamins, and we will see that the laboratory synthesis of one particular vitamin represented a landmark event in the history of synthetic organic chemistry. This chapter serves as a brief introduction to organic synthesis.

Until this point in the text, we have only seen a limited number of reactions (a few dozen, at most). In this chapter, our modest repertoire of reactions will allow us to develop a methodical, step-by-step process for proposing syntheses. We will begin with one-step synthesis problems and then progress toward more challenging multistep problems. The goal of this chapter is to develop the fundamental skills required for proposing a multistep synthesis.

It should be noted that a carefully designed synthesis will not always work as planned. In fact, it is very common for a planned synthesis to fail, requiring the investigator to devise a strategy to circumvent the failed step. When designing a synthesis, many factors must be taken into account, including (but not limited to) the cost of reagents and the ease of purification of products. A high-yield synthesis can often be ruined by the formation of intractable side products. For our purposes in this chapter, we will adopt the inaccurate assumption that all reactions covered in previous chapters can be utilized reliably.

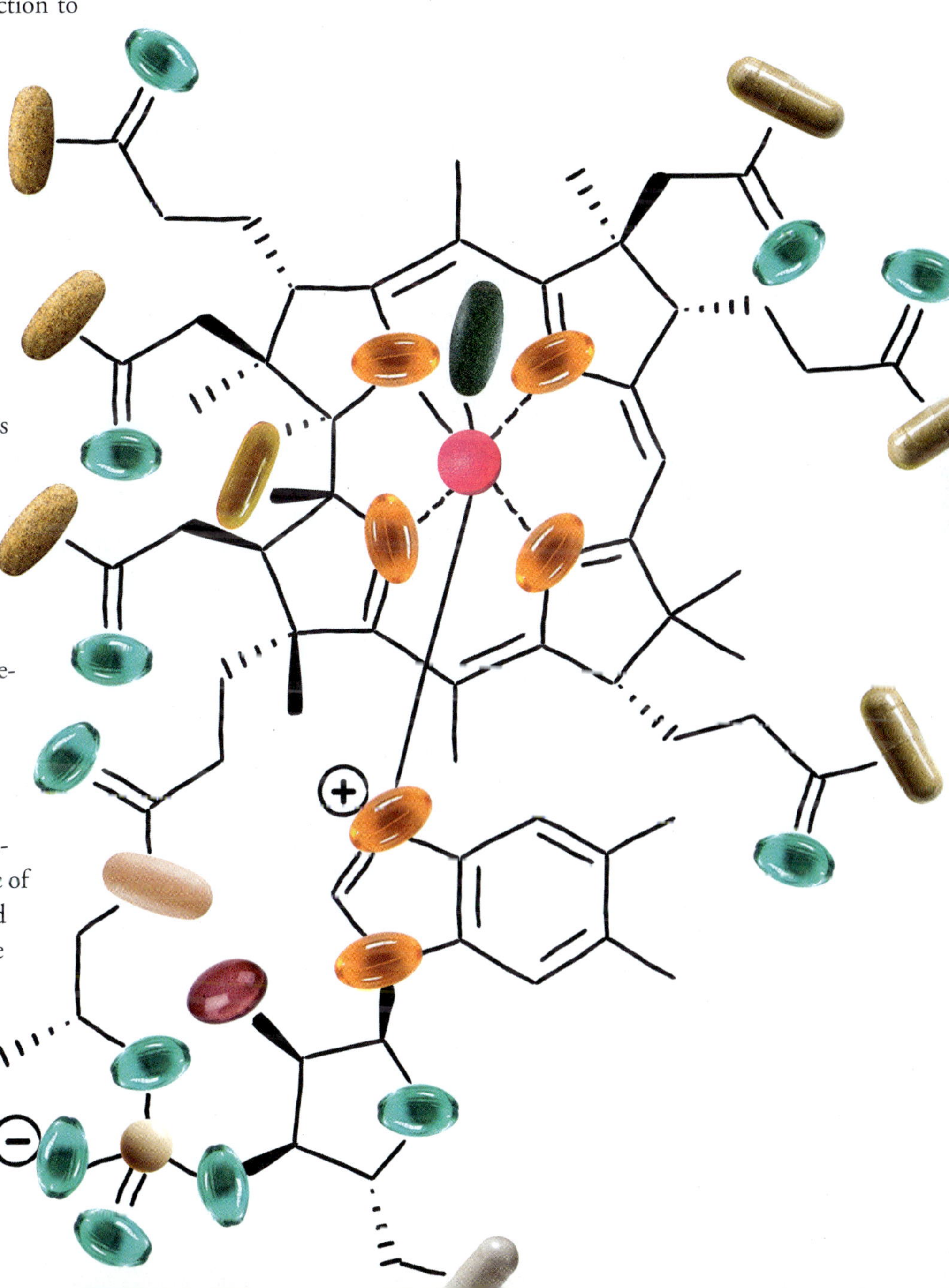

## DO YOU REMEMBER?

*Before you go on, be sure you understand the following topics.*
*If necessary, review the suggested sections to prepare for this chapter.*

- Selecting Reagents to Accomplish Functional Group Transformation (Section 7.9)
- Synthesis Strategies for Alkenes and Alkynes (Sections 9.13, 10.11)
- Substitution vs. Elimination (Section 8.13)

Take the DO YOU REMEMBER? QUIZ in **WileyPLUS** to check your understanding

## 12.1 One-Step Syntheses

The most straightforward synthesis problems are the ones that can be solved in just one step. For example, consider the following:

This transformation can be accomplished by treating the alkene with $Br_2$ in an inert solvent, such as $CCl_4$. Other synthesis problems might require more than a single step, and those problems will be more challenging. Before approaching multistep synthesis problems, it is absolutely essential to become comfortable with one-step syntheses. In other words, it is critical to achieve mastery over all reagents described in the previous chapters. If you can't identify the reagents necessary for a one-step synthesis problem, then certainly you will be unable to solve more complex problems. The following exercises represent a broad review of the reactions in previous chapters. These exercises are designed to help you identify which reagents are still not at the forefront of your consciousness.

### CONCEPTUAL CHECKPOINT

**12.1** Identify the reagents necessary to accomplish each of the transformations shown below. If you are having trouble, the reagents for these transformations appear in the review of reactions at the end of Chapter 9, but you should first try to identify the reagents yourself without help:

**12.2** Identify the reagents necessary to accomplish each of the following transformations. If you are having trouble, the reagents for these transformations appear in the review of reactions at the end of Chapter 10, but you should first try to identify the reagents yourself without help:



## 12.2 Functional Group Transformations

In the previous few chapters, we developed several synthesis strategies that enable us to move the location of a functional group or change its identity. Let's briefly review these techniques, as they will be extremely helpful when solving multistep synthesis problems.

In Chapter 9, we developed a technique for changing the position of a halogen by performing an elimination reaction followed by an addition reaction. For example:



In this two-step process, the halogen is removed and then reinstalled at a different location. The regiochemical outcome of each step must be carefully controlled. The choice of base in the elimination step determines whether the more substituted or the less substituted alkene is formed. In the addition step, the decision whether or not to use peroxides will determine whether a Markovnikov addition or an *anti*-Markovnikov addition occurs.

As we saw in Chapter 9, this technique must be slightly modified when the functional group is a hydroxyl group (OH). In such a case, the hydroxyl group must first be converted into a better leaving group, such as a tosylate, and only then can the technique be employed (elimination followed by addition):

OH — Convert OH into a better leaving group → OTs — Elimination → — Addition → HO

After converting the hydroxyl group into a tosylate, the regiochemical outcome for elimination and addition can be carefully controlled, as summarized below:

HO — 1) TsCl, pyridine 2) NaOEt → — $H_3O^+$ → OH

1) $BH_3 \cdot THF$ 2) $H_2O_2$, NaOH

1) $Hg(OAc)_2$, $H_2O$ 2) $NaBH_4$

1) TsCl, pyridine 2) *t*-BuOK → — 1) $BH_3 \cdot THF$ 2) $H_2O_2$, NaOH → HO

In Chapter 9, we also developed a two-step technique for moving the position of a double bond. For example:

Addition → Br — Elimination →

Once again, the regiochemical outcome of each step can be controlled by choice of reagents, as summarized below:

HBr → Br — *t*-BuOK →

NaOEt

NaOEt

HBr ROOR → Br — *t*-BuOK →

**BY THE WAY**
If your starting material is an alkane, the only useful reaction you should consider is a radical halogenation.

In Chapter 11, we developed one other important technique: installing functionality in a compound with no functional groups:

Radical bromination → Br ... Elimination →

This procedure, together with the other reactions covered in the previous chapters, enables the interconversion between single, double, and triple bonds:

$H_2$, Lindlar's catalyst *or* Na, $NH_3$ ; $H_2$, Pt ; 1) $Br_2/CCl_4$ 2) xs $NaNH_2$ 3) $H_2O$ ; 1) $Br_2$, $h\nu$ 2) NaOEt

We will soon learn a new way of approaching synthesis problems (rather than relying on a few, pre-canned techniques). For now, let's ensure mastery over the reactions and techniques that allow us to change the identity or position of a functional group.

# SKILLBUILDER

## 12.1 CHANGING THE IDENTITY OR POSITION OF A FUNCTIONAL GROUP

**LEARN the skill**

Propose an efficient synthesis for the following transformation:

→ Br

**SOLUTION**

We begin by analyzing the identity and location of the functional groups in both the starting material and the product. The identity of the functional group has certainly changed (an alkene has been converted into an alkyl halide), and the position of the functional group has also changed. This can be seen more easily by numbering the parent chain:

1 2 3 4 5 → Br 1 2 3 4 5

C-2 and C-3 are functionalized

C-1 is functionalized

The C-2 and C-3 positions are functionalized (have a functional group) in the starting material, but the C-1 position is functionalized in the product. Therefore, both the location and the identity of the functional group must be changed. That is, we must find a way to functionalize the C-1 position by moving the existing functional group. We have already seen a two-step method for changing the location of a $\pi$ bond:

→

This strategy would achieve the desired goal of functionalizing the C-1 position, and this type of transformation can be accomplished via addition followed by elimination. In order to move the position of the π bond, the reagents must be chosen carefully for each step of the process. Specifically, an *anti*-Markovnikov addition must be followed by a Hofmann elimination:

*Anti*-Markovnikov addition → X → Hofmann elimination →

For the first step of this process, we have learned only two reactions that proceed via *anti*-Markovnikov addition: either (a) addition of HBr with peroxides or (b) hydroboration-oxidation:

HBr, ROOR → Br (Racemic)

1) $BH_3 \cdot THF$ 2) $H_2O_2$, NaOH → OH (Racemic)

Both reactions will produce an *anti*-Markovnikov addition. However, when hydroboration oxidation is used, the resulting hydroxyl group must then be converted to a tosylate prior to the elimination process:

HBr, ROOR → Br → *t*-BuOK →

1) $BH_3 \cdot THF$ 2) $H_2O_2$, NaOH → OH → TsCl, Pyridine → OTs → *t*-BuOK →

Both routes will work, although the first route might be more efficient, because it requires fewer steps.

Now that the C-1 position has been functionalized, the last step is to install a bromine atom at the C-1 position. This can be accomplished with an *anti*-Markovnikov addition of HBr:

HBr, ROOR → Br

In summary, there are two plausible routes to achieve the desired synthetic transformation:

1) HBr, ROOR
2) *t*-BuOK
3) HBr, ROOR

Br

1) $BH_3 \cdot THF$
2) $H_2O_2$, NaOH
3) TsCl, pyridine
4) *t*-BuOK
5) HBr, ROOR

It is very common to find that multiple routes can be used to achieve a desired transformation. Don't fall into the trap of thinking that there is only one correct solution to a synthesis problem. There are almost always multiple pathways that are feasible.

**PRACTICE the skill** **12.3** Propose an efficient synthesis for each of the following transformations:

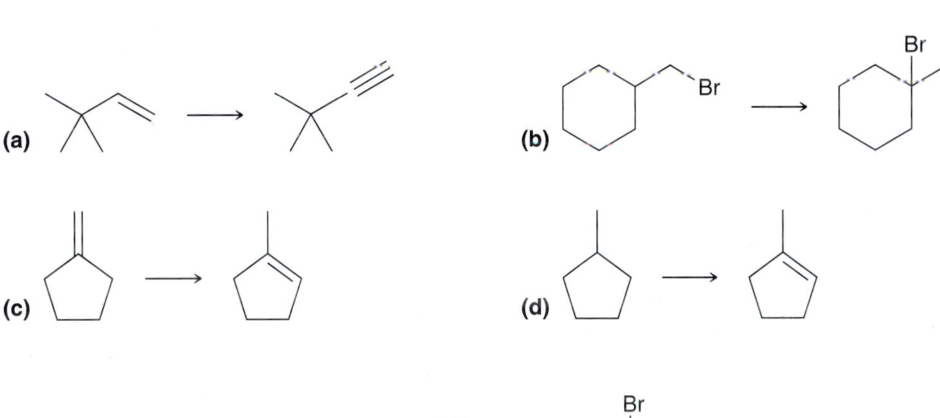


(e) OH → OH (f) Br → OH

(g) OH → Br (h) OH → OH

**APPLY the skill** **12.4** Identify the reagents you would use to convert 2-bromo-2-methylbutane into 3-methyl-1-butyne.

**12.5** Identify the reagents you would use to convert 1-pentene into a geminal dibromide ("geminal" indicates that both bromine atoms are connected to the same carbon atom).

**12.6** Identify the reagents you would use to convert methylcyclohexane into each of the following:

**(a)** A 3° alkyl halide **(b)** A trisubstituted alkene **(c)** 3-Methylcyclohexene

need more **PRACTICE?** **Try Problems 12.17, 12.21, 12.22**

## 12.3 Reactions That Change the Carbon Skeleton

In all of the problems in the previous section, the functional group changed its identity or location, but the carbon skeleton always remained the same. In this section, we will focus on examples in which the carbon skeleton changes. In some cases, the number of carbon atoms in the skeleton increases, and in other cases, the number of carbon atoms decreases.

If the size of the carbon skeleton increases, then a C—C bond-forming reaction is required. Thus far, we have only learned one reaction that can be used to introduce an alkyl group onto an existing carbon skeleton. Alkylation of a terminal alkyne (Section 10.10) will increase the size of a carbon skeleton:

$$CH_3CH_2C{\equiv}C:^{\ominus}\ Na^{\oplus} + X{-}CH_2CH_2CH_3 \longrightarrow CH_3CH_2C{\equiv}C{-}CH_2CH_2CH_3 + NaX$$

**Four** carbon atoms | **Three** carbon atoms | **Seven** carbon atoms

Over time, we will see many other C—C bond-forming reactions, but for now, we have only seen one such reaction. This should greatly simplify the problems in this section, enabling a smooth transition into the world of synthetic organic chemistry.

If the size of the carbon skeleton decreases, then a C—C bond-breaking reaction, called **bond cleavage**, is required. Once again, we have only seen one such reaction. Ozonolysis of an alkene (or alkyne) achieves bond cleavage at the location of the π bond:

$$CH_3CH_2CH_2CH{=}CH_2 \xrightarrow[\text{2) DMS}]{\text{1) } O_3} CH_3CH_2CH_2CHO + H_2C{=}O$$

**Five** carbon atoms | **Four** carbon atoms | **One** carbon atom

Over time, we will see other reactions that involve C—C bond cleavage. For now, the knowledge that we have only seen one such reaction should greatly simplify the problems in this section.

# SKILLBUILDER

## 12.2 CHANGING THE CARBON SKELETON

**LEARN the skill**

Identify reagents that can be used to achieve the following transformation:

(cyclohexyl)CH₂CH₂Br ⟶ (cyclohexyl)CH₂CH₂C≡CH

### SOLUTION

Count the carbon atoms in the starting material and in the desired product. There are seven carbon atoms in the starting material, and there are nine carbon atoms in the product. Therefore two carbon atoms must be installed. We have only learned one reaction capable of installing two carbon atoms on an existing carbon skeleton. This process requires the use of an alkynide ion and an alkyl halide:

$$R{-}C{\equiv}C:^{\ominus}\ Na^{\oplus} \quad R{-}X \longrightarrow R{-}C{\equiv}C{-}R + NaX$$

**Alkynide** | **Alkyl halide**

Until now, this reaction has always been viewed *from the perspective of the alkyne*. That is, the alkyne is the starting material, and the alkyl halide is used as a reagent in the second step of the process, thereby achieving the installation of an alkyl group (R):

$$\underset{\textbf{Starting material}}{\mathrm{H{-}C{\equiv}C{-}H}} \xrightarrow[\text{2) R}-\text{X}]{\text{1) NaNH}_2} \mathrm{H{-}C{\equiv}C{-}R}$$

Alternatively, this reaction can be viewed from the perspective of the alkyl halide. That is, the alkyl halide is the starting material, and an alkynide ion is used to achieve the installation of a triple bond onto an existing carbon skeleton:

$$\underset{\textbf{Starting material}}{\mathrm{R{-}X}} \xrightarrow{\overset{\oplus}{\text{Na}}\ \overset{\ominus}{:}\text{C}{\equiv}\text{C}{-}\text{H}} \mathrm{R{-}C{\equiv}C{-}H}$$

When viewed in this way, the alkylation process represents a technique for introducing an acetylenic group. This is exactly what is needed to solve this problem:

Br $\xrightarrow{\overset{\oplus}{\text{Na}}\ \overset{\ominus}{:}\text{C}{\equiv}\text{C}{-}\text{H}}$ C≡C–H

In fact, this one step provides the answer. This is just a one-step synthesis problem.

**PRACTICE the skill** **12.7** Identify reagents that can be used to achieve each of the following transformations:

(a) ⟶

(b) Br ⟶

(c) ⟶ O

**APPLY the skill** **12.8** Propose an efficient synthesis for each of the following transformations:

(a) Br ⟶

(b) Br ⟶ O H

(c) Br ⟶ Br Br

**12.9** When 3-bromo-3-ethylpentane is treated with sodium acetylide, the major products are 3-ethyl-2-pentene and acetylene. Explain why the carbon skeleton does not change in this case and justify the formation of the observed products.

need more **PRACTICE?** **Try Problems 12.18, 12.19, 12.20, 12.23, 12.26**

## medically speaking | Vitamins

Vitamins are compounds that our bodies require for normal functioning and must be obtained from food. An inadequate intake of certain vitamins causes specific diseases. This phenomenon had been observed long before the exact role of vitamins was understood. For example, sailors who remained at sea for extended periods would suffer from a disease called scurvy, characterized by the loss of teeth, swollen limbs, and bruising. If left untreated, the disease would be fatal. In 1747, a British naval physician named James Lind demonstrated that the effects of scurvy could be reversed by eating oranges and lemons. It was recognized that oranges and lemons must contain some "factor" that our bodies require, and lime juice became a normal part of a sailor's diet. For this reason, British sailors were called "Limeys."

Other studies revealed that a variety of foods contained mysterious "growth factors." For example, Gowland Hopkins (University of Cambridge) conducted a series of experiments in which he controlled the dietary intake of rats. He fed them carbohydrates, proteins, and fats (covered in Chapters 24, 25, and 26, respectively). This mixture was insufficient to sustain the rats, and it was necessary to add a few drops of milk to their dietary intake to achieve sustainable growth. These experiments demonstrated that milk contained some unknown ingredient, or factor, necessary for promoting proper growth.

Similar observations were made by a Dutch physician, Christiaan Eijkman, in the Dutch colonies in Indonesia. While investigating the cause for a massive outbreak of beriberi, a disease characterized by paralysis, he noted that the hens in the laboratory began to exhibit a form of paralysis as well. He discovered that they were being fed rice from which the fibrous husk had been removed (called polished rice). When he fed the hens raw whole rice, their condition improved drastically. It was therefore realized that the fibrous husk of rice contained some vital growth factor. In 1912, Polish biochemist Casimir Funk isolated the active compound from rice husks. Careful studies revealed that the structure contained an amino group and therefore belongs to a class of compounds called amines (Chapter 23):

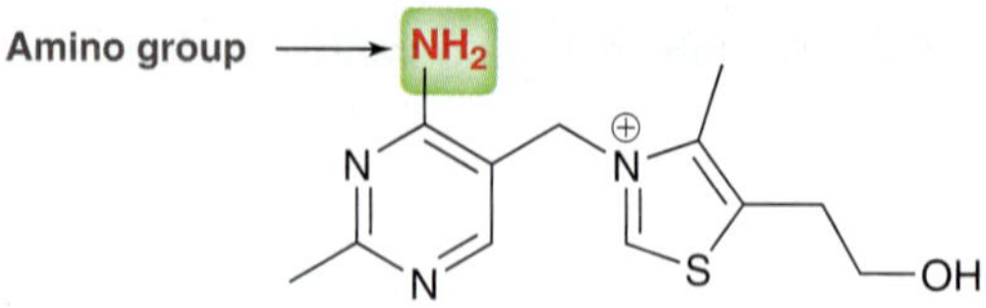

Amino group → $NH_2$

Thiamine (vitamin $B_1$)

It was originally believed that all vital growth factors were amines, so they were called vitamins (a combination of *vital* and *amine*). Further research demonstrated that most vitamins lack an amino group, yet the term "vitamins" persisted. For their discovery of the role vitamins play in nutrition, Eijkman and Hopkins were awarded the 1929 Nobel Prize in Physiology or Medicine.

Vitamins are grouped into families based on their behavior (rather than their structure); each family is designated with a letter. For example, vitamin B is a family of many compounds, each designated with a letter and a number ($B_1$, $B_2$, $B_3$, etc.). The table shows representative vitamins from several families:

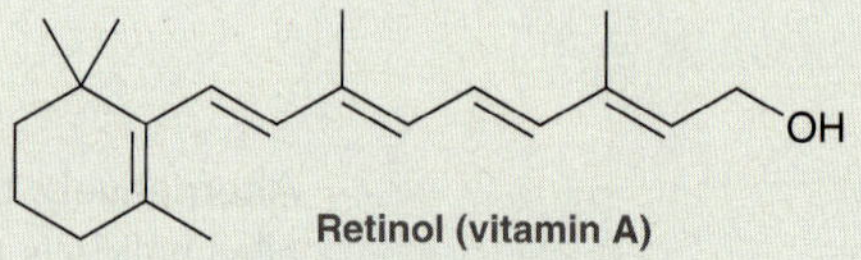

Retinol (vitamin A)

*Sources:* Milk, eggs, fruit, vegetables, and fish
*Deficiency disease:* Night-blindness (see Section 17.13)

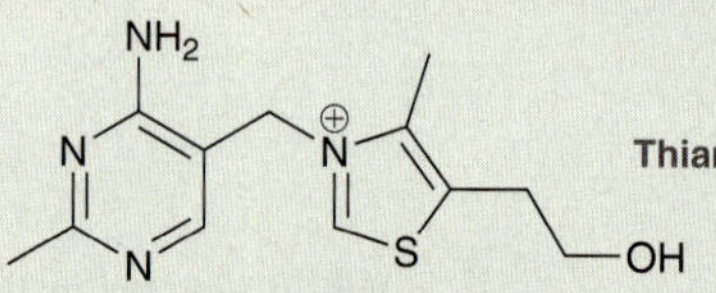

Thiamine (vitamin $B_1$)

*Sources:* Liver, potatoes, whole grains, and legumes
*Deficiency disease:* Beriberi

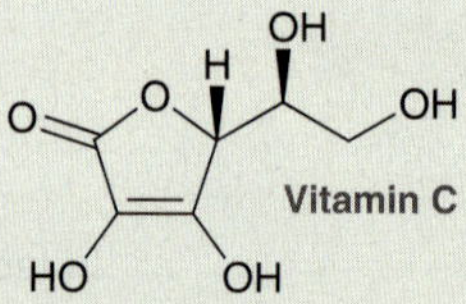

Vitamin C

*Sources:* Citrus fruits, bell peppers, tomatoes, and broccoli
*Deficiency disease:* Scurvy

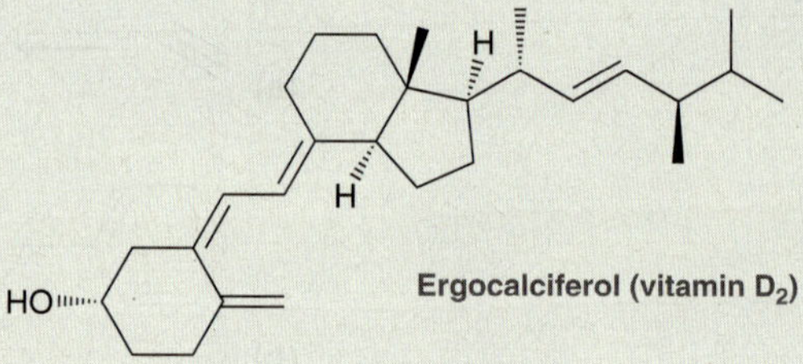

Ergocalciferol (vitamin $D_2$)

*Sources:* Fish, produced by the body when exposed to sunlight
*Deficiency disease:* Rickets (see Section 17.10)

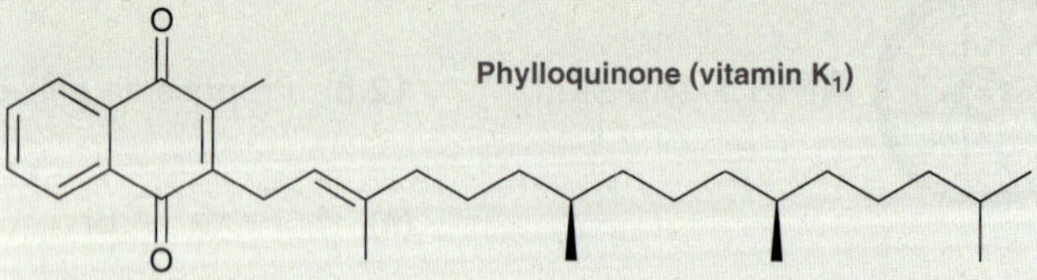

Phylloquinone (vitamin $K_1$)

*Sources:* Soybean oil, green vegetables, and lettuce
*Deficiency disease:* Hemorrhaging (internal bleeding)

Early studies revealed the structure of vitamin C, which enabled chemists to devise a synthetic strategy for preparing vitamin C in the laboratory. Success led to the preparation of vitamin C on an industrial scale. Such early successes encouraged the notion that chemists would soon elucidate the structures of all vitamins and devise methods for their preparation in the laboratory. The synthesis of vitamin $B_{12}$, however, would prove to be more complex, as we will see in the next Medically Speaking box.

## 12.4 How to Approach a Synthesis Problem

In the previous two sections, we covered two critical skills: (1) functional group transformations and (2) changing the carbon skeleton. In this section, we will explore synthesis problems that require both skills. From this point forward, every synthesis problem should be approached by asking the following two questions:

1. ***Is there a change in the carbon skeleton?*** Compare the starting material with the product to determine if the carbon skeleton is gaining or losing carbon atoms.
2. ***Is there a change in the identity or location of the functional group?*** Is one functional group converted into another, and does the position of functionality change?

The following example demonstrates how these two questions should be applied.

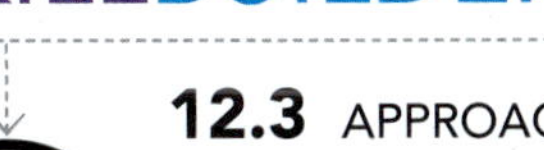

## SKILLBUILDER

### 12.3 APPROACHING A SYNTHESIS PROBLEM

**LEARN** the skill

Propose an efficient synthesis for the following transformation:

**SOLUTION**

Every synthesis problem should always be analyzed through the lens of the following two questions:

1. *Is there a change in the carbon skeleton?* The starting compound has five carbon atoms, and the product has seven carbon atoms. This transformation therefore requires the installation of two carbon atoms:

When numbering the carbon atoms, as shown above, it is not necessary to follow IUPAC rules for assigning locants. If we were naming the starting alkyne, we would be compelled to use proper locants (the triple bond would be between C-1 and C-2, rather than between C-4 and C-5). But the numbers here are only tools and can be used in whatever way is easiest for you to count.

2. *Is there a change in the identity or location of the functional group?* Certainly, the identity of the functional group has changed (a triple bond has been converted into a double bond), but consider the location of the functional group:

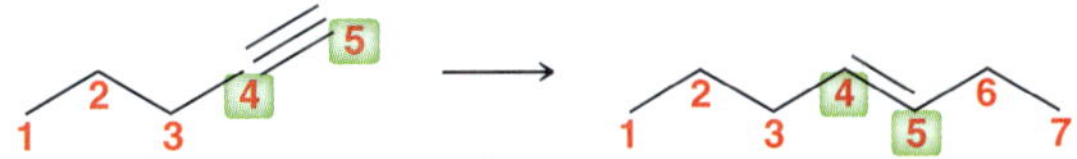


Once again, these numbers are not IUPAC numbers; rather, they are tools that help us identify that the double bond in the product occupies the same position as the triple bond in the starting material. In this case, strict adherence to the IUPAC numbering system would have been confusing and misleading.

By asking both questions, the following two tasks have been identified: (1) two carbon atoms must be installed and (2) the triple bond must be converted into a double bond in its current location. For each of these tasks, we must determine what reagents to use:

1. What reagents will add two carbon atoms to a skeleton?
2. What reagents will convert a triple bond into a *trans* double bond?

Two new carbon atoms can be introduced via alkylation of the starting alkyne:

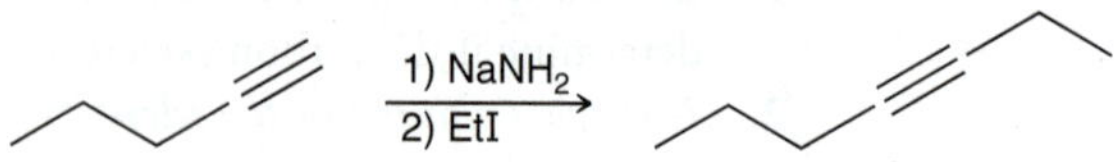

1) $NaNH_2$
2) EtI

Now that the correct carbon skeleton has been established, reduction of the triple bond can be accomplished via a dissolving metal reduction to afford the *trans* alkene:

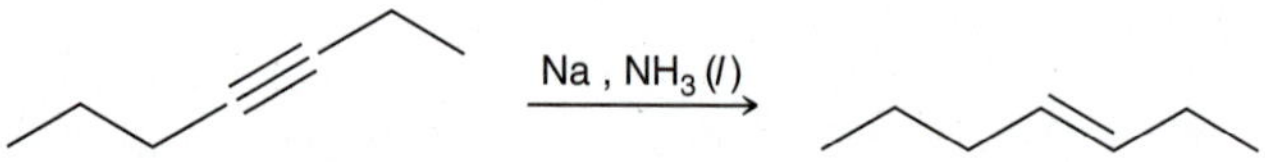

Na , $NH_3$ (*l*)

The solution to this problem requires two steps: (1) alkylation of the alkyne followed by (2) conversion of the triple bond into a double bond. Notice the order of events. If the triple bond had first been converted into a double bond, the alkylation process would not work. Only a terminal triple bond can be alkylated, not a terminal double bond.

**PRACTICE** the skill **12.10** Identify reagents that can be used to achieve each of the following transformations:

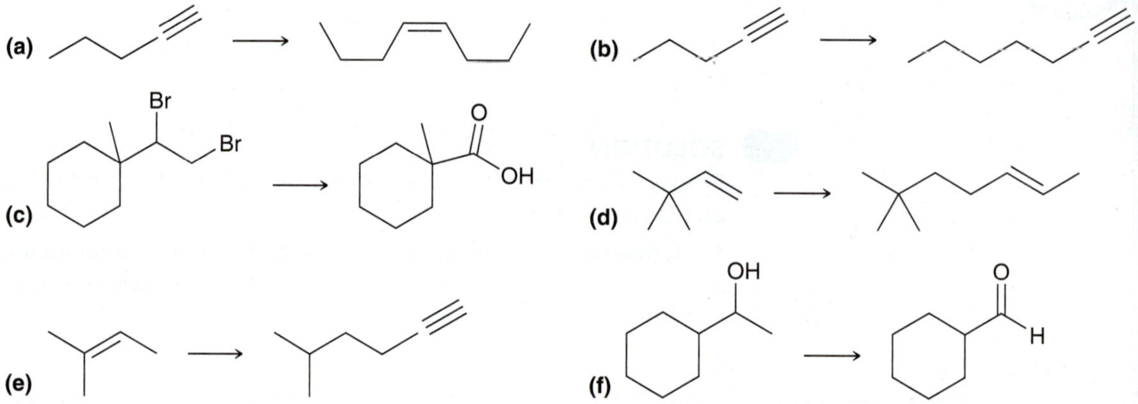

(a) (b)

Br Br O OH

(c) (d)

OH O H

(e) (f)

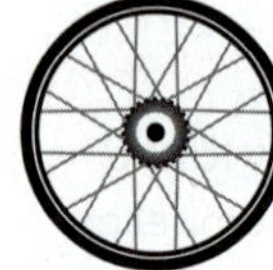

**APPLY** the skill **12.11** Propose an efficient synthesis for the following transformation (many steps are required):

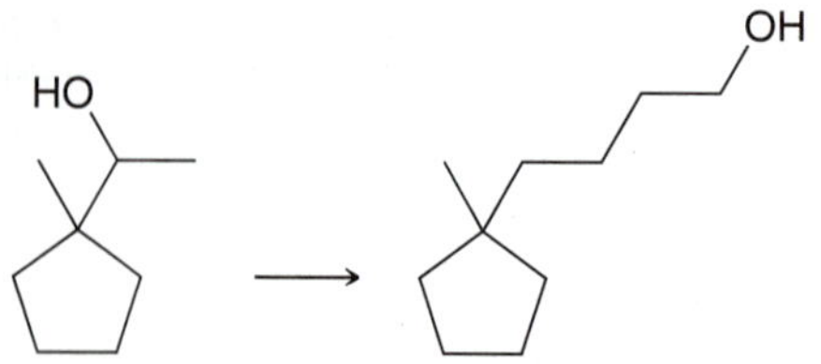

HO OH

**12.12** Propose an efficient synthesis for the following transformation, in which the carbon skeleton is increased by only one carbon atom:

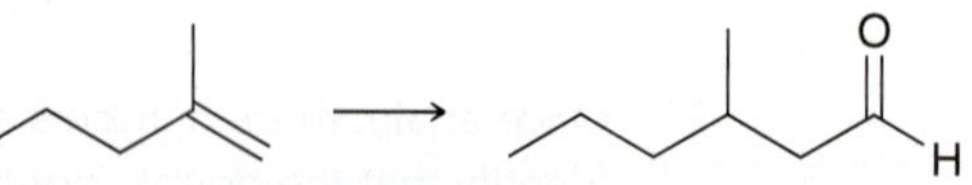

O H

need more **PRACTICE?** **Try Problems 12.19–12.26**

## medically speaking — The Total Synthesis of Vitamin $B_{12}$

The story of vitamin $B_{12}$ began when physicians recognized that anemia, a fatal disease caused by a low concentration of red blood cells, could often be treated effectively by feeding the patient liver. This finding sparked a race to extract the compounds in liver and isolate the vitamin capable of treating some forms of anemia. In 1947, vitamin $B_{12}$ was first isolated and purified as deep red crystals by Ed Rickes (a scientist working at the Merck Chemical Company). Efforts then focused on determining the structure. Some structural features were initially elucidated, but the complete structure was successfully determined by Dorothy Crowfoot Hodgkin (Oxford University) using X-ray crystallography. She found that the structure of vitamin $B_{12}$ is built upon a corrin ring system, which is similar to the porphyrin ring system present in chlorophyll, the green pigment that plants use for photosynthesis (see above).

Chlorophyll *a* — Porphyrin ring system

Vitamin $B_{12}$ — Corrin ring system

The corrin ring system of vitamin $B_{12}$ is also comprised of four heterocycles (rings containing a heteroatom, such as nitrogen) joined together in a macrocycle. However, the corrin ring system of vitamin $B_{12}$ is constructed around a central cobalt atom instead of the magnesium of chlorophyll, and vitamin $B_{12}$ contains many more chirality centers than chlorophyll. The determination of this complex structure pushed the boundaries of X-ray crystallography, which until then had never been used to elucidate such a complex structure. Hodgkin was a pioneer in the field of X-ray crystallography and identified the structures of many important biochemical compounds. For her efforts, she was awarded the 1964 Nobel Prize in Chemistry.

With the structure of vitamin $B_{12}$ elucidated, the stage was set for its total synthesis. At the time, the complexity of vitamin $B_{12}$ represented the greatest challenge to synthetic organic chemists, and a couple of talented synthetic organic chemists viewed the challenge as irresistible. Robert B. Woodward (Harvard University) had already established himself as a leading player in the field of organic chemistry with his successful total synthesis of many important natural products, including quinine (used in the treatment of malaria), cholesterol and cortisone (steroids that will be discussed in Section 26.6), strychnine (a poison), reserpine (a tranquilizer), and chlorophyll. With these impressive accomplishments under his belt, Woodward eagerly embraced the challenge of vitamin $B_{12}$. He began working on methods for constructing the corrin ring system as well as the stereochemically demanding side chain. Meanwhile, Albert Eschenmoser (at the ETH in Zurich, Switzerland) was also working on a synthesis of the vitamin. However, the two men were developing different strategies for constructing the corrin ring system. Woodward's A → B route involved forming the macrocycle between rings A and B, while Eschenmoser's A → D route involved forming the macrocycle between rings A and D:

A→B Approach: Macrocyclization occurs between ring A and ring B

A→D Approach: Macrocyclization occurs between ring A and ring D

*(continues on the next page)*

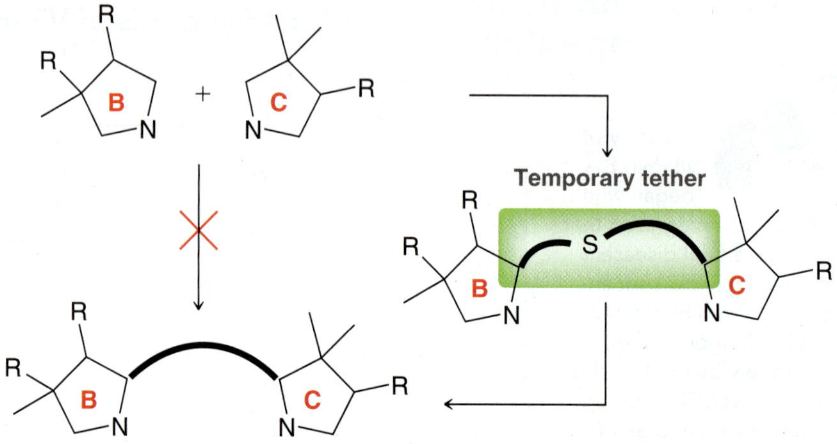

During the development of each pathway, unexpected obstacles emerged that required the development of new strategies and techniques. For example, Eschenmoser had successfully demonstrated a method for coupling the heterocycles, enabling construction of a simple corrin system (without the large, bulky side groups found in vitamin $B_{12}$); however, this approach failed to couple the heterocycles that needed to be joined to form the corrin ring system of vitamin $B_{12}$. This failure was attributed to the steric bulk of the substituents on each heterocycle. To circumvent the problem, Eschenmoser developed an ingenious method by which he tethered two rings together with a temporary sulfur bridge. By doing so, the coupling process became an *intra*molecular process rather than an *inter*molecular process (see above).

The sulfur bridge was readily expelled during the coupling process, and the overall process became known as "sulfide contraction." This is just one example of the creative solutions that synthetic chemists must develop when a planned synthetic route fails. A number of obstacles still faced both Woodward and Eschenmoser, and a partnership was forged in 1965 to tackle the problem together. In fact, that was the same year that Woodward was awarded the Nobel Prize in Chemistry for his contributions to the field of synthetic organic chemistry.

Woodward and Eschenmoser continued to work together for another seven years, often spending an entire year optimizing the conditions for an individual step. The intense effort, which involved nearly 100 graduate students working for a decade, would ultimately be rewarded. Woodward's team completed the assembly of the stereochemically demanding side chain, and the two groups combined some of the best methods and practices developed during studies aimed at construction of the corrin system. The pieces were finally joined together, and the synthesis was completed to produce vitamin $B_{12}$ in 1972. This landmark event represents one of the greatest achievements in the history of synthetic organic chemistry, and it demonstrated that organic chemists could prepare any compound, regardless of complexity, given enough time.

During his journey toward the total synthesis of vitamin $B_{12}$, Woodward encountered a class of reactions that were known to proceed with unexplained stereochemical outcomes. Together with his colleague Roald Hoffmann, he developed a theory and a set of rules that would successfully explain the stereochemical outcomes of an entire area of organic chemistry called pericyclic reactions. This class of reactions will be covered in Chapter 17 together with the so-called Woodward–Hoffmann rules used to describe and predict the stereochemical outcomes for these reactions. The development of these rules led to another Nobel Prize in 1981, for which Woodward would have been a co-recipient had he not died two years earlier.

The story of vitamin $B_{12}$ is a wonderful example of how organic chemistry progresses. During the total synthesis of a structurally complex compound, there is inevitably a point at which the planned route fails, requiring a creative method for circumventing the obstacle. In this way, new ideas and techniques are constantly developed. In the decades since the total synthesis of vitamin $B_{12}$, thousands of synthetic targets, most of them pharmaceuticals, have been constructed. New techniques, reagents, and principles constantly emerge from these endeavors. With time, synthetic targets are getting more and more complex, and the leaders in the field of organic chemistry are constantly pushing the boundaries of synthetic organic chemistry, which continues to evolve on a daily basis.

## 12.5 Retrosynthetic Analysis

As we progress through the course and increase our repertoire of reactions, synthesis problems will become increasingly more challenging. To meet this challenge, a modified approach will be necessary. The same two fundamental questions (as described in the previous section) will continue to serve as a starting point for analyzing all synthesis problems, but instead of trying to identify the first step of the synthesis, we will begin by trying to identify the last step of the synthesis. Analysis of the following synthesis will illustrate this process:

An alcohol

An alkyne

Rather than focusing on what can be done with an alcohol that will ultimately lead to an alkyne, we instead focus on reactions that can generate an alkyne:

Focus on this step

In this way, we work backward until arriving at the starting material. Chemists have intuitively used this approach for many years, but E. J. Corey (Harvard University) was the first to develop a systematic set of principles for application of this approach, which he called **retrosynthetic analysis**. Let's use a retrosynthetic analysis to solve the problem above.

We must always begin by determining whether there is a change in the carbon skeleton or in the identity or location of the functional group. In this case, both the starting material and the product contain six carbon atoms, and the carbon skeleton is not changing in this instance. However, there is a change in the functional group. Specifically, an alcohol is converted into an alkyne but remains in the same location in the skeleton. We have not learned a way to do this in just one step. In fact, using reactions covered so far, this transformation cannot be accomplished even in two steps. So we approach this problem *backward* and ask: "How are triple bonds made?" We have only covered one way to make a triple bond. Specifically, a dihalide undergoes two successive E2 eliminations in the presence of excess $NaNH_2$ (Section 10.4). Any one of the following three dihalides could be used to form the desired alkyne:

Geminal dibromide

Vicinal dibromide

Geminal dibromide

The geminal dibromides can be ruled out, because we only saw one way to make a geminal dihalide—and that was starting from an alkyne. We certainly do not want to start with an alkyne in order to produce the very same alkyne:

Therefore, the last step of our synthesis must be formation of the alkyne from a *vicinal* dihalide:

A special retrosynthetic arrow is used by chemists to indicate this type of "backward" thinking:

Don't be confused by this retrosynthetic arrow. It indicates a hypothetical synthetic pathway *thinking backward* from the product (alkyne). In other words, the previous figure should be read as: "In the last step of our synthesis, the alkyne *can be made from* a vicinal dibromide."

Now let's try to go backward one more step. We have learned only one way to make a vicinal dihalide, starting with an alkene:

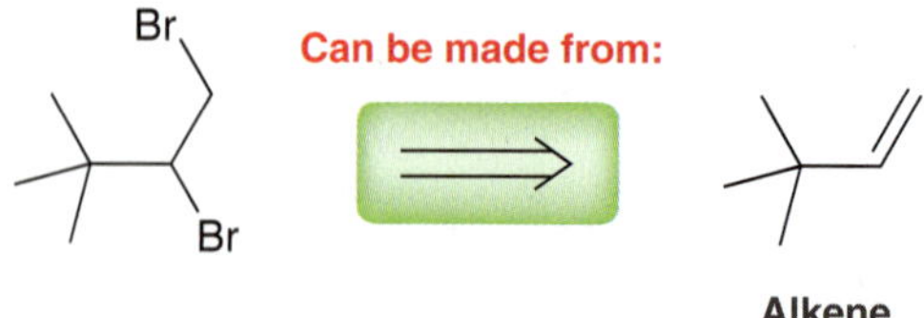


Notice again the retrosynthetic arrow. The figure indicates that the vicinal dibromide can be made from an alkene. In other words, the alkene can be used as a precursor to prepare the desired dibromide.

Therefore, our retrosynthetic analysis, so far, looks like this:

**Product**

Br

Br

**Alkene**

This scheme indicates that the product (alkyne) can be prepared from the alkene. This sequence of events represents one of the strategies discussed earlier in this chapter—converting a double bond into a triple bond:

$H_2$ , Lindlar's catalyst or Na, $NH_3$

$H_2$ , Pt

1) $Br_2/CCl_4$
2) xs $NaNH_2$
3) $H_2O$

1) $Br_2$, $h\nu$
2) NaOMe

In order to complete the synthesis, the starting material must be converted into the alkene. At this point, we can think forward, in an attempt to converge with the pathway revealed by the retrosynthetic analysis:

OH

?

1) xs $NaNH_2$
2) $H_2O$

Br

$Br_2$ / $CCl_4$

Br

This step can be accomplished with an E2 elimination. Just remember that the hydroxyl group must first be converted into a tosylate (a better leaving group). Then, an E2 elimination will create an alkene, which bridges the gap between the starting material and the product:

OH

TsCl
pyridine

1) xs $NaNH_2$
2) $H_2O$

Br

OTs

*t*-BuOK

$Br_2$ / $CCl_4$

Br

The synthesis seems complete. However, before recording the answer, it is always helpful to review all of the proposed steps and make sure that the regiochemistry and stereochemistry of each step will lead to the desired product *as a major product*. It would be inefficient to involve any steps that would rely on the formation of a minor product. We should only use steps that produce the desired product as the major product. After reviewing every step of the proposed synthesis, the answer can be recorded like this:

1) TsCl, py
2) *t*-BuOK
3) $Br_2/CCl_4$
4) xs $NaNH_2$
5) $H_2O$

# SKILLBUILDER

## 12.4 RETROSYNTHETIC ANALYSIS

**LEARN** the skill

Propose a plausible synthesis for each of the following transformations:

**(a)**

**(b)**

### SOLUTION

**(a)** First determine whether there is a change in the carbon skeleton. The starting compound has four carbon atoms while the product has six. This transformation therefore requires the installation of two carbon atoms:

As we have seen many times in this chapter, the only method we have learned for achieving this transformation is the alkylation of a terminal alkyne. Our synthesis must therefore involve an alkylation step. This should be taken into account when performing a retrosynthetic analysis.

Second, we need to determine whether there is a change in the identity or location of the functional group. In this case, the functional group has changed both its identity and its location. The product is an aldehyde. As seen in Section 10.7, an aldehyde can be made via hydroboration-oxidation of an alkyne. As shown with the retrosynthetic arrow, a terminal alkyne can be converted into an aldehyde.

**Can be made from:**

Continuing with a retrosynthetic analysis, we must consider the step that might be used to produce the alkyne. Recall that our synthesis must contain a step involving the alkylation of an alkyne. We therefore propose the following retrosynthetic step:

**Can be made from:**

X + $Na^{\oplus}$ $^{\ominus}$:C≡C—H

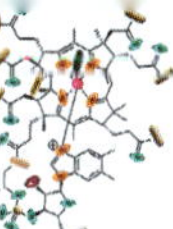



This step accomplishes the necessary installation of two carbon atoms.

To complete the synthesis, we just need to bridge the gap between the starting material and the alkyl halide:

?

$Na^{\oplus}$ $:C^{\ominus}\equiv C-H$

1) $R_2BH$
2) $H_2O_2$ , NaOH

Working forward now, it is apparent that the necessary transformation can be achieved with an *anti*-Markovnikov addition, which can be accomplished with HBr in the presence of peroxides. In summary, we propose the following synthesis.

1) HBr, ROOR
2) $H-C\equiv CNa$
3) $R_2BH$
4) $H_2O_2$ , NaOH

**(b)** Always begin by determining whether there is a change in the carbon skeleton. The starting compound has a two-carbon chain connected to the ring, while the product has a three-carbon chain connected to the ring:

This transformation therefore requires the installation of one carbon atom.

Next determine whether there is a change in the identity or location of the functional group. In this case the identity of the functional group does not change, but its location does change.

In order to achieve the necessary change in the carbon skeleton, our synthesis must involve an alkyne that will be alkylated and then converted into the product. Working backward, we focus on the alkyne:

The product can be made from this alkyne via a dissolving metal reduction. Recall the reason for going through the alkyne in the first place: to introduce a carbon atom via alkylation. The last two steps of our synthesis can therefore be written in the following way, using retrosynthetic arrows:

Now we must bridge the gap:

1) $NaNH_2$
2) MeI

Na, $NH_3$ (*l*)

To bridge the gap, a double bond must be converted into a triple bond and its position must be moved. We cannot convert it into a triple bond in its current location—that would give a pentavalent carbon, which is impossible:

**Never draw a carbon atom with five bonds**

**WATCH OUT**
Never draw a pentavalent carbon because that would violate the octet rule. Recall that carbon only has four orbitals with which to form bonds, and as a result, it can never form more than four bonds.

We must first move the position of the double bond and only then convert the double bond into a triple bond. The location of the double bond can be moved using the technique discussed earlier in the chapter—addition followed by elimination:

Addition → Br → Elimination

Take special note of the regiochemistry in each case. In the first step, an *anti*-Markovnikov addition of HBr is needed, so peroxides are required. Then, in the second step, the Hofmann product (the less substituted alkene) is needed, so a sterically hindered base is required:

HBr, ROOR → Br → *t*-BuOK

Let's summarize what we have so far, working forward:

HBr ROOR

Br

*t*-BuOK

1) $NaNH_2$
2) MeI

Na, $NH_3$ (*l*)

To bridge the gap, a double bond must be converted into a triple bond. Once again, this is one of the techniques reviewed earlier in this chapter. It is accomplished through bromination of the alkene followed by two successive elimination reactions to produce an alkyne:

1) $Br_2/CCl_4$
2) xs $NaNH_2$
3) $H_2O$

In summary, the desired transformation can be achieved with the following reagents:

1) HBr, ROOR
2) *t*-BuOK
3) $Br_2/CCl_4$
4) xs $NaNH_2$
5) $H_2O$
6) $NaNH_2$
7) MeI
8) Na, $NH_3$ (*l*)

**PRACTICE the skill** **12.13** Propose an efficient synthesis for each of the following transformations:

(a) Br OH

(b) Br O

(c) Br Br O

(d) O

(e) OH

(f) Br Br O OH

(g) Br OH + En

(h)

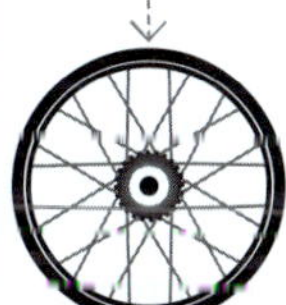

**APPLY the skill** **12.14** Using acetylene as your only source of carbon atoms, design a synthesis of *trans*-5-decene:

**12.15** Using acetylene as your only source of carbon atoms, design a synthesis of *cis*-3-decene:

**12.16** Using acetylene as your only source of carbon atoms, design a synthesis of pentanal (note: pentanal has an odd number of carbon atoms, while acetylene has an even number of carbon atoms):

O H

need more **PRACTICE?** **Try Problems 12.19–12.26**

## Retrosynthetic Analysis

Thus far, we have only explored a limited number of reactions, and the complexity of synthesis problems is therefore limited. At this stage, it might be difficult to see why retrosynthetic analysis is so critical. However, as we expand our repertoire of reactions, we will begin to explore the synthesis of more sophisticated structures, and the need for retrosynthetic analysis will become more apparent.

E. J. Corey first developed his retrosynthetic methodology as a faculty member at the University of Illinois. During his sabbatical year, which he spent at Harvard University, he shared his ideas with Robert Woodward (the very same Woodward who developed a total synthesis of vitamin $B_{12}$, described in the previous Medically Speaking box). Corey demonstrated his ideas by presenting a retrosynthetic analysis of longifolene, a natural compound that represented a challenging target because of its unique tricyclic structure. Corey identified the following strategic bond of longifolene for disconnection:

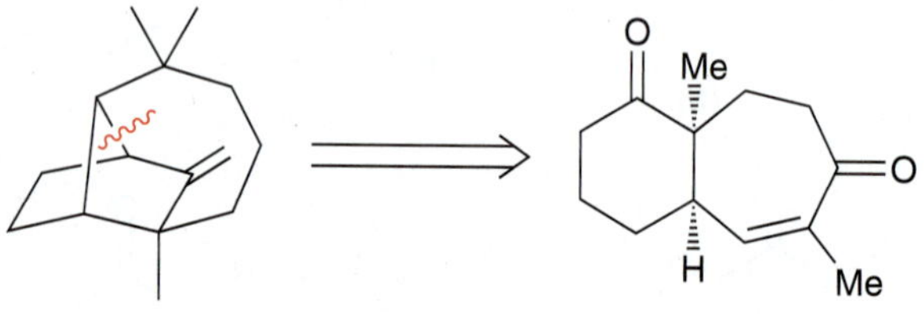

Longifolene

The functional groups in the precursor serve as handles enabling formation of the strategic bond (utilizing a reaction that we will explore in Chapter 22). Longifolene represented an excellent example of the power of Corey's retrosynthetic disconnection approach. For most of the bonds in longifolene, disconnection would produce a complex tricyclic precursor. That is, there are not many strategic bonds that can be disconnected to provide a simpler precursor. Woodward was impressed by Corey's insights and immediately recognized that Corey would soon be another leader in the field. This likely played a large role in the offer that Corey subsequently received to join the chemistry department at Harvard as a professor.

Corey developed his ideas further and established a set of rules and principles for proposing retrosynthetic analysis. He also spent much time creating computer programs to assist chemists in performing retrosynthetic analysis. It is clear that computers cannot replace the chemist altogether, as the chemist must provide the insight and creativity in determining which strategic bonds to disconnect. In addition, the chemist must exhibit creativity when a planned route fails. As discussed in the previous box, this is how new ideas, reagents, and techniques are developed. Corey used his approach for the total synthesis of dozens of compounds. In the process, he developed many new reagents, strategies, and reactions that now bear his name. Corey was an instrumental player in developing the field of synthetic organic chemistry. Retrosynthetic analysis is now taught to all students of organic chemistry. Similarly, nearly all reports of total synthesis in the chemical literature begin with a retrosynthetic analysis. For his contributions to the development of synthetic organic chemistry, Corey was awarded the Nobel Prize in Chemistry in 1990.

## 12.6 Practical Tips for Increasing Proficiency

### Organizing a Synthetic "Toolbox" of Reactions

All of the reactions in this course will collectively represent your "toolbox" for proposing syntheses. It will be very helpful to prepare two lists that parallel the two questions that must be considered in every synthesis problem (*change in the carbon skeleton and change in the functional group*). The first list should contain C—C bond-forming reactions and C—C bond-breaking reactions. At this point, this list is very small. We have only seen one of each: alkylation of alkynes for C—C bond forming and ozonolysis for C—C bond breaking. As the text progresses, more reactions will be added to the list, which will remain relatively small but very powerful. The second list should contain functional group transformations, and it will be longer.

As we move through the text, both lists will grow. For solving synthesis problems, it will be helpful to have the reactions categorized in this way in your mind.

### Creating Your Own Synthesis Problems

A helpful method for practicing synthesis strategies is to construct your own problems. The process of designing problems often uncovers patterns and new ways of thinking about the reactions. Begin by

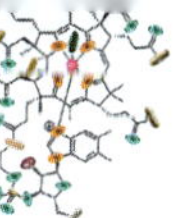



choosing a starting compound. To illustrate this, let's begin by choosing a simple starting compound, such as acetylene. Then choose a reaction expected for a triple bond, perhaps an alkylation:

1) $NaNH_2$
2) MeI

Next, choose another reaction, perhaps another alkylation:

1) $NaNH_2$
2) MeI

1) $NaNH_2$
2) MeI

Then, treat the alkyne with another reagent. Look at the list of reactions of alkynes and choose one, perhaps hydrogenation with a poisoned catalyst:

1) $NaNH_2$
2) MeI

1) $NaNH_2$
2) MeI

$H_2$
Lindlar's catalyst

Finally, simply erase everything except for the starting compound and the final product. The result is a synthesis problem:

Once you have created your own synthesis problem, it might be a really good problem, but it won't be helpful for you to solve it. You already know the answer! Nevertheless, the process of creating your own synthesis problems will be very helpful to you in sharpening your synthesis skills.

Once you have created several of your own problems, try to find a study partner who is also willing to create several problems. Each of you can swap problems, try to solve each other's problems, and then get back together to discuss the answers. You are likely to find that exercise to be very rewarding.

## Multiple Correct Answers

Remember that most synthesis problems will have numerous correct answers. As an example, *anti*-Markovnikov hydration of an alkene can be achieved through either of the following two possible routes:

1) $BH_3 \cdot THF$
2) $H_2O_2$, NaOH

OH

1) HBr, ROOR
2) NaOH

The first pathway represents hydroboration-oxidation of the alkene to achieve *anti*-Markovnikov addition of water. The second pathway represents a radical (*anti*-Markovnikov) addition of HBr followed by an $S_N2$ reaction to replace the halogen with a hydroxyl group. Each of these pathways represents a valid synthesis. As we learn more reactions, it will become more common to encounter synthesis problems with several perfectly correct answers. The goal should always be efficiency. A 3-step synthesis will generally be more efficient than a 10-step synthesis.

## medically speaking | Total Synthesis of Taxol

Taxol is a powerful anticancer agent that was first isolated from the bark of the Pacific yew tree (*Taxus brevifolia*) in 1967. The structure of Taxol, also known as paclitaxel, wasn't elucidated until 1971, and it wasn't until the late 1970s that the full anticancer potential was realized. Taxol prevents tumor growth by inhibiting the division of cells, or *mitosis*, in some types of cancer cells; and in other types of cancer cells, it induces cell death, or *apoptosis*. Taxol is now commonly prescribed as a treatment for breast and ovarian cancers.

**Taxol (paclitaxel)**

One of the major challenges associated with the use of Taxol as an anticancer drug is the small quantity that is produced by the rare Pacific yew tree. The average tree produces just over 0.5 g of Taxol, and at least 2 g is required to treat just one patient. In addition, it may take up to 100 years for the Pacific yew to mature and it is only found in small climate regions, such as the Pacific Northwest of the United States. For these reasons, the Pacific yew tree cannot serve as a practical source of Taxol for clinical use.

To deal with the lack of supply of Taxol, several efforts were undertaken to identify an alternate natural source of Taxol or a method of preparing it. The complicated structure of Taxol made it an especially challenging target for chemists who specialize in the synthesis of complex natural products. In early 1994, two research groups, one led by Robert Holton at Florida State University and the other led by Kyriacos Nicolaou at the Scripps Research Institute in California, reported (within days of each other) the synthesis of Taxol starting from simple precursors. Although both of these synthetic routes produced Taxol, each method involved several complicated steps that would be lengthy and expensive to use in the production of the large amounts of Taxol needed for anticancer treatment. For example, the Holton synthesis gave Taxol in a total yield of approximately 5%. However, these two syntheses served an essential purpose as they helped confirm that the structure proposed for the natural Taxol was indeed correct.

Despite the success of the synthetic routes to Taxol, chemists continued to look for other methods to produce enough of the drug for clinical applications. A search of other natural sources of Taxol (or related compounds) resulted in the identification of 10-deacetylbacatin III in the needles of the European yew *Taxus baccata*, an evergreen common to large parts of Europe:

**10-Deacetylbacatin III**

This compound is similar in structure to Taxol, and it can be readily converted into Taxol via a chemical transformation that was employed by the Holton group in the final steps of their Taxol synthesis. Since 10-deacetylbacatin III can be isolated from the needles, the plant survives the harvesting process and can continue to produce more needles, rendering it a sustainable source of 10-deacetylbacatin III. More importantly, the discovery of a source of 10-deacetylbacatin III facilitated the development of analogues of Taxol with different biological activity. One such example is Taxotere, also known as docetaxel, which can be easily prepared from 10-deacetylbacatin III. Taxotere is very effective in the treatment of certain types of breast cancer and lung cancer that Taxol is less effective in treating.

**Taxotere (docetaxel)**

The story of Taxol serves as a powerful example of how new drugs can be developed through a combination of natural product discovery and synthetic organic chemistry.

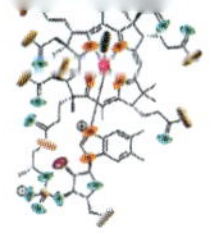


# REVIEW OF CONCEPTS AND VOCABULARY

### SECTION 12.1

- It is essential to achieve mastery over all reagents described in the previous chapters.

### SECTION 12.2

- The position of a halogen can be moved by performing elimination followed by addition.
- The position of a $\pi$ bond can be moved by performing addition followed by elimination.
- An alkane can be functionalized via radical bromination.

### SECTION 12.3

- If the size of the carbon skeleton increases, then a C—C bond-forming reaction is required.
- If the size of the carbon skeleton decreases, then a C—C bond-breaking reaction, called **bond cleavage**, is required.

### SECTION 12.4

- Every synthesis problem should be approached by asking the following two questions:
  - Is there any change in the carbon skeleton?
  - Is there any change in the identity or location of the functional group?

### SECTION 12.5

- In a **retrosynthetic analysis**, the last step of the synthetic route is first established, and the remaining steps are determined, working backward from the product.

### SECTION 12.6

- Most synthesis problems will have numerous correct answers, although efficiency should always be a top priority (fewer steps are preferable).

# SKILLBUILDER REVIEW

## 12.1 CHANGING THE IDENTITY OR POSITION OF A FUNCTIONAL GROUP

Change the position of a halogen.

Br
Elimination
Addition
Br

Change the position of a $\pi$ bond.

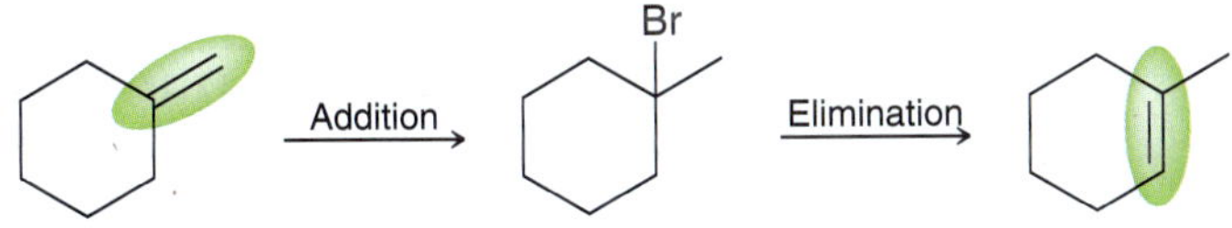


Install a functional group.

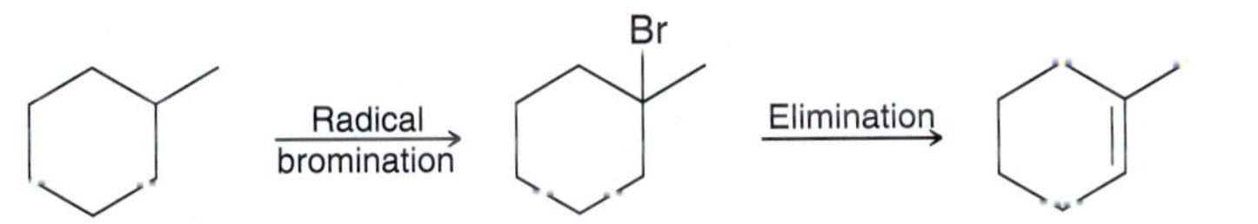


Interconvert between single, double, and triple bonds.

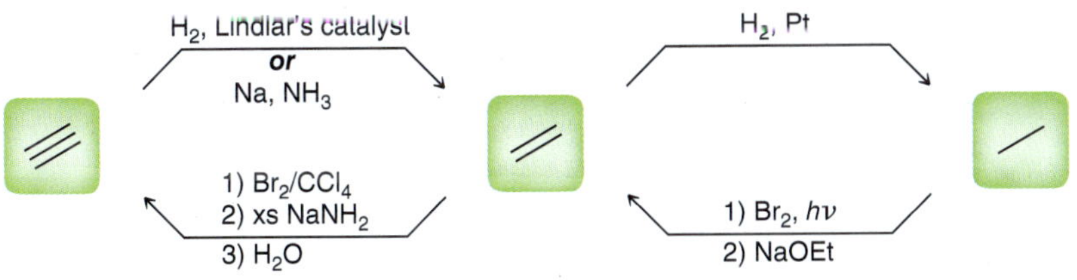


Try Problems 12.3–12.6, 12.17, 12.21, 12.22

### 12.2 CHANGING THE CARBON SKELETON

C—C bond formation.

**Four** carbon atoms + **Three** carbon atoms ⟶ **Seven** carbon atoms

C—C bond cleavage.

**Five** carbon atoms $\xrightarrow[\text{2) DMS}]{\text{1) } O_3}$ **Four** carbon atoms + **One** carbon atom

Try Problems **12.7–12.9, 12.18, 12.19, 12.20, 12.23, 12.26**

### 12.3 APPROACHING A SYNTHESIS PROBLEM BY ASKING TWO QUESTIONS

| 1. Is there any change in the carbon skeleton? | 2. Is there any change in the identity or location of the functional group? |
|---|---|
| Compare the starting material with the product to detemine if the carbon skeleton is gaining or losing any carbon atoms. | In other words, is one functional group converted into another, and does the position of functionality change? |

Try Problems **12.10–12.12, 12.19–12.26**

### 12.4 RETROSYNTHETIC ANALYSIS

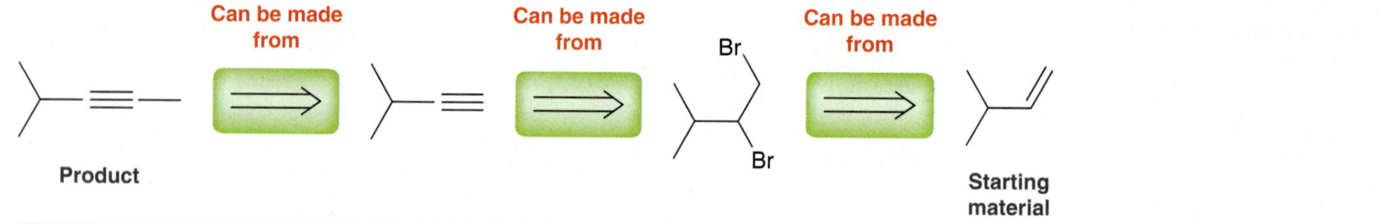


Try Problems **12.13–12.16, 12.19–12.26**

## PRACTICE PROBLEMS

**Note: Most of the Problems are available within WileyPLUS, an online teaching and learning solution.**

**12.17** Identify the reagents necessary to achieve each of the following transformations:

**12.18** Identify the reagents necessary to achieve each of the following transformations:

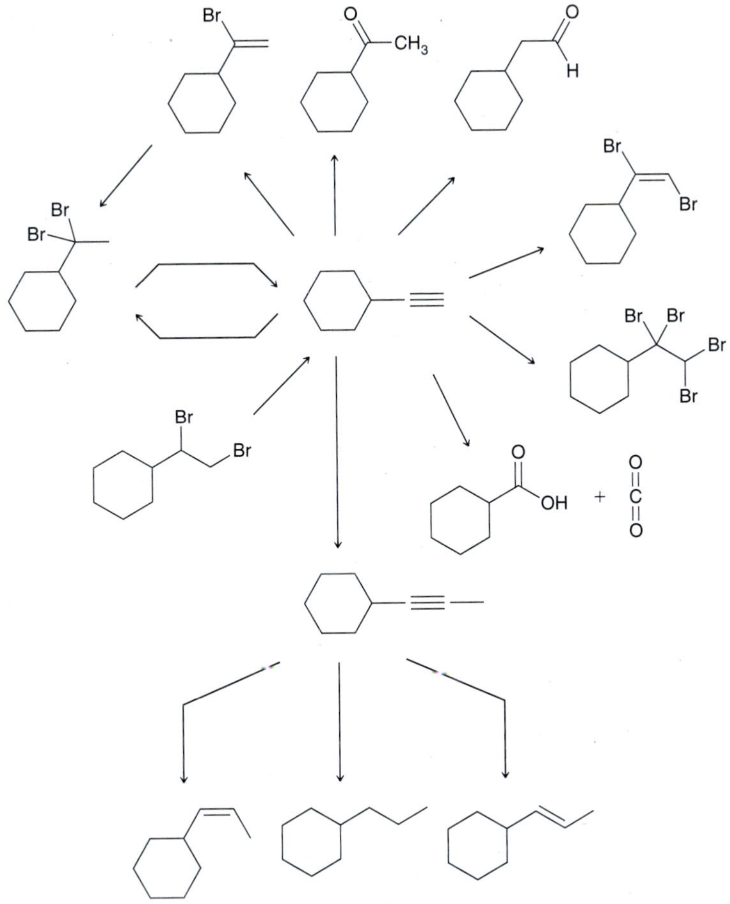

**12.19** Using acetylene as your only source of carbon atoms, identify a synthetic route for the production of 2-bromobutane.

**12.20** Using acetylene as your only source of carbon atoms, identify a synthetic route for the production of 1-bromobutane.

**12.21** Propose an efficient synthesis for each of the following transformations:

(a) 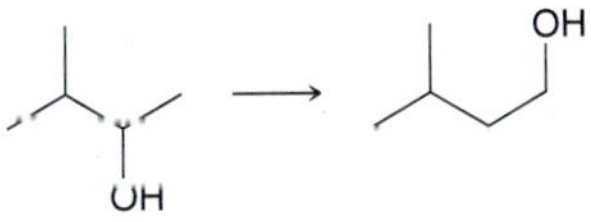 (b)

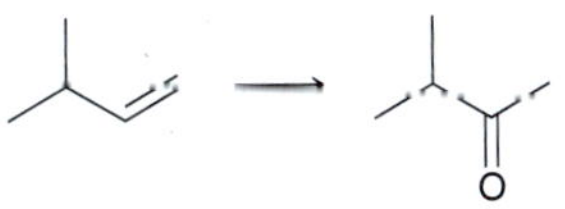

**12.22** Using any reagents you like, show a way to convert 1-methylcyclopentene into 3-methylcyclopentene.

**12.23** Propose an efficient synthesis for each of the following transformations:

(a)

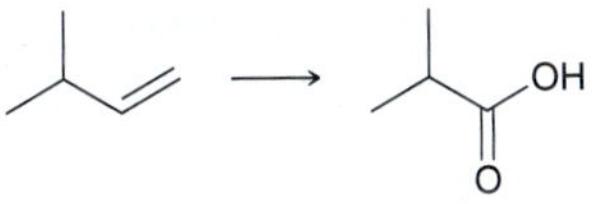

(b)

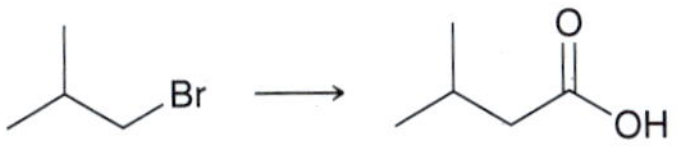

(c)

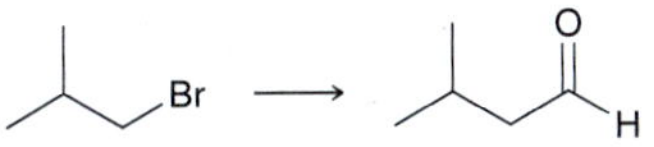

(d)

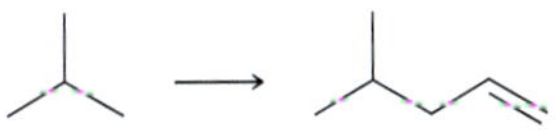

**12.24** Using any compounds that contain two carbon atoms or fewer, show a way to prepare a racemic mixture of (2*R*,3*R*)-2,3-dihydroxypentane and (2*S*,3*S*)-2,3-dihydroxypentane.

**12.25** Using any compounds that contain two carbon atoms or fewer, show a way to prepare a racemic mixture of (2*R*,3*S*)- and (2*S*,3*R*)-2,3-dihydroxypentane.

**12.26** Propose an efficient synthesis for each of the following transformations:

(a)

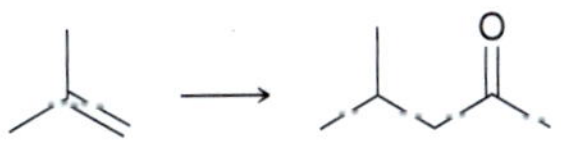

(b)

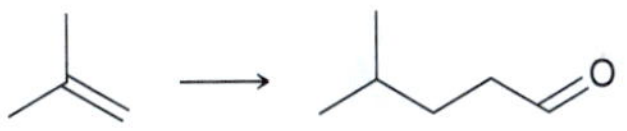

(c)

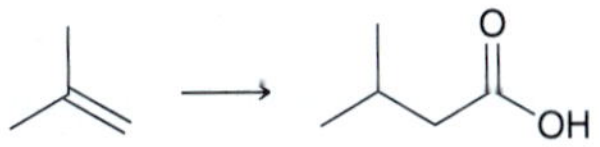

(d)

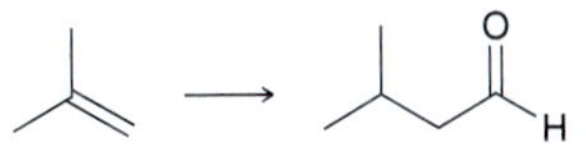

(e)

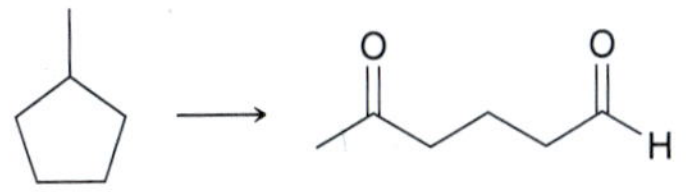

# INTEGRATED PROBLEMS

**12.27** In this chapter, we have seen that an acetylide ion can function as a nucleophile and attack an alkyl halide in an $S_N2$ process. More generally speaking, the acetylide ion can attack other electrophiles as well. For example, we will see in Chapter 14 that epoxides function as electrophiles and are subject to attack by a nucleophile. Consider the following reaction between an acetylide ion (the nucleophile) and an epoxide (the electrophile):

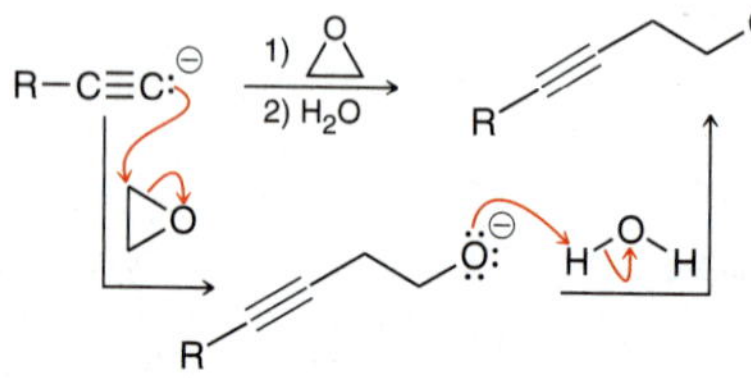

The acetylide ion attacks the epoxide, opening up the strained, three-membered ring and creating an alkoxide ion. After the reaction is complete, a proton source is used to protonate the alkoxide ion. In a synthesis, these two steps must be shown separately, because the acetylide ion will not survive in the presence of $H_2O$. Using this information, propose an efficient synthesis for the following compound using acetylene as your only source of carbon atoms:

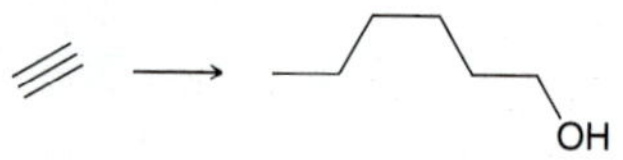

**12.28** In the previous problem, we saw that an acetylide ion can attack a variety of electrophiles. In Chapter 20, we will see that a C=O bond can also function as an electrophile. Consider the following reaction between an acetylide ion (the nucleophile) and a ketone (the electrophile):

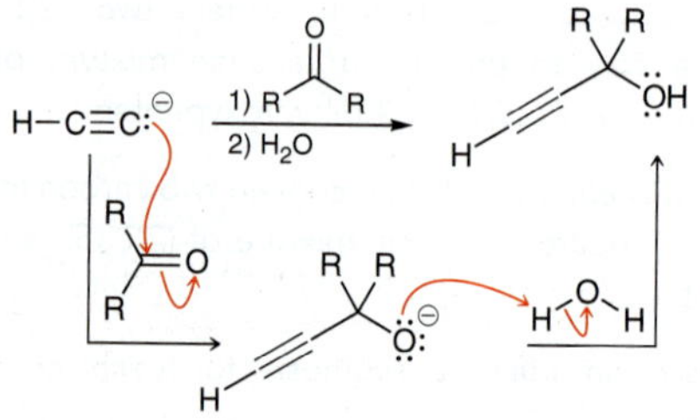

The acetylide ion attacks the ketone, generating an alkoxide ion. After the reaction is complete, a proton source is used to protonate the alkoxide ion. In a synthesis, these two steps must be shown separately, because the acetylide ion will not survive in the presence of $H_2O$. Using this information, propose a plausible synthesis for allyl alcohol, using acetylene as your only source of carbon atoms:

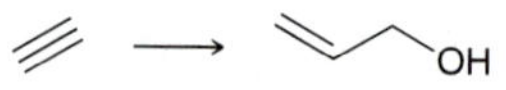

**12.29** Identify the reagents you might use to achieve the following transformation:

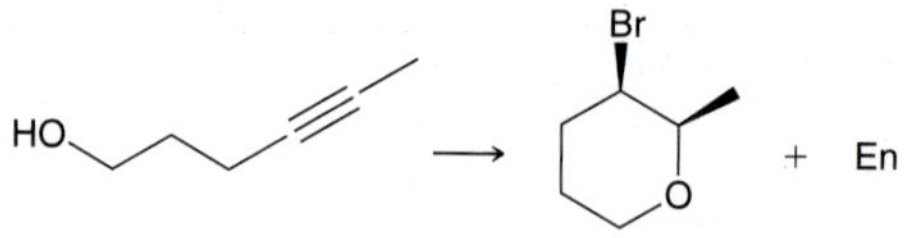

**12.30** Starting with acetylene as your only source of carbon atoms, identify how you would prepare each member of the following homologous series of aldehydes:

(a) $CH_3CHO$
(b) $CH_3CH_2CHO$
(c) $CH_3CH_2CH_2CHO$
(d) $CH_3CH_2CH_2CH_2CHO$

**12.31** Starting with acetylene as your only source of carbon atoms, propose a plausible synthesis for 1,4-dioxane:

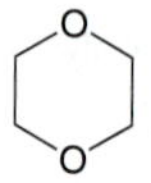

**1,4-Dioxane**

# CHALLENGE PROBLEMS

**12.32** In a study measuring the volatile components of 150 pounds of fried chicken, a total of 130 distinct compounds were identified, including the linear alkane tetradecane (*J. Agric. Food Chem.* **1983**, *31*, 1287–1292). Propose a synthesis of this compound using acetylene as your only source of carbon atoms.

**12.33** A study of compounds contained in the gas phase above a range of bacterial strains revealed the presence of 254 unique volatile organic compounds representing a broad range of structural classes. The two isomers below were among the compounds identified (*J. Nat. Prod.* **2012**, *75*, 1765–1776). Propose a synthesis of each of these compounds using alkenes with fewer than six carbon atoms as your only source of carbon.

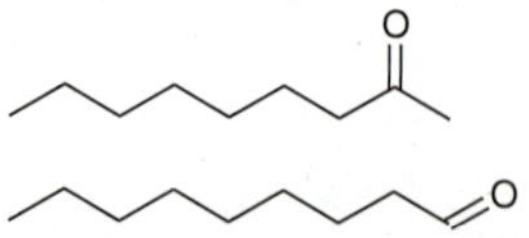

**12.34** In a study of volatile compounds extracted from cinnamon trees native to Sri Lanka, 27 compounds were identified, including 3-phenylpropyl acetate (*J. Agric. Food Chem.* **2003,** *51,* **4344–4348**). Propose a multiple-step synthesis of this compound using toluene and any two-carbon reactants as your only sources of carbon.

**Sources of carbon**

**3-Phenylpropyl acetate** ⟹ **Toluene** + Other 2C reactants

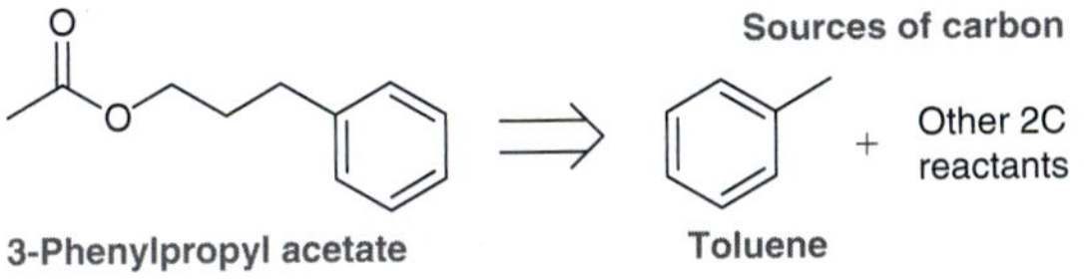

**12.35** The compound (*Z*)-hexenyl acetate is one of several volatile organic signaling compounds released from *Arabidopsis* leaves when they are damaged (*Nat. Prod. Rep.* **2012,** *29,* **1288–1303**). Propose a synthesis of this compound using ethylene ($H_2C{=}CH_2$) and acetic acid ($CH_3CO_2H$) as your only sources of carbon atoms. Use the reaction described in Problem 12.27 as one of the steps of the synthesis.

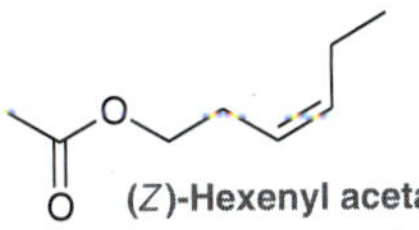

**(*Z*)-Hexenyl acetate**

**12.36** Kakkon is the root of an Asian plant that has traditionally been used in Chinese medicine. It smells fruity, winey, and medicinal. A study of the volatile components of kakkon (those responsible for the scent) resulted in the identification of 33 compounds, including 2,5-hexanedione (*Agric. Biol. Chem.* **1988,** *52,* **1053–1055**). Propose a synthesis of this diketone starting from 3,4-dimethylcyclobutene.

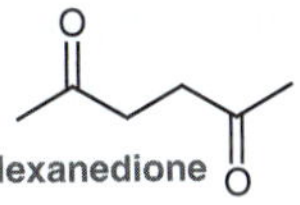

**2,5-Hexanedione**

**12.37** When a consumer purchases a tomato, smell is one of the factors affecting selection. Researchers in Japan have thus analyzed the volatile components of tomatoes using a novel method. Among the complex mixture of 367 compounds detected were 3-methylbutanal and hexanal (*J. Agric. Food Chem.* **2002,** *50,* **3401–3404**). Propose a single synthesis that produces an equimolar mixture of these two compounds (equal amounts of both products) starting with one 1°, one 2°, and one 3° alcohol, each with fewer than six carbons. (Note that the designations 1°, 2°, and 3° for alcohols are analogous to those of alkyl halides.)

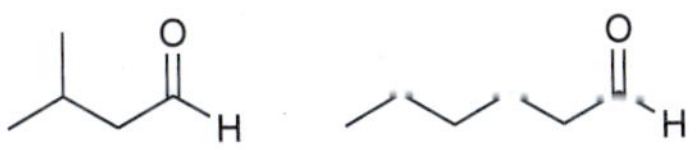

**3-Methylbutanal** **Hexanal**

**12.38** Methyl 4-methylpentanoate is one of 120 volatile compounds identified in a study of metabolites from *Streptomyces* bacteria (*Actinomycetologica* **2009,** *23,* **27–33**). Propose a synthesis of this ester starting from 2-methylbutane.

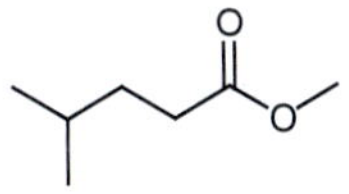

**Methyl 4-methylpentanoate**

**12.39** Odor is one of the principal criteria used by consumers when purchasing food products such as fish. Researchers thus compared relative amounts of various volatile substances present in samples of European freshwater catfish raised in an indoor concrete pool to those raised in an outdoor pond. The compound 1-penten-3-ol was amongst the 59 volatile molecules detected (*J. Agric. Food Chem.* **2005,** *53,* **7204–7211**). Propose a multiple-step synthesis of this compound using acetylene as your only source of carbons. Note that you will have to use a reaction similar to the one described in the introduction to Problem 12.28.

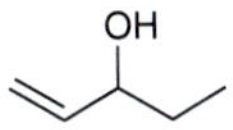

**1-Penten-3-ol**

**12.40** The aroma, taste, and general quality of wine are tied closely to the stage of development of the grapes from which it is made. In order to develop a molecular-level understanding of this phenomenon, the volatile components of cabernet sauvignon grapes were tested throughout their growth period. Among the many molecules detected, the concentration of *E*-2-hexenal was found to increase slowly for eight weeks from the flowering of the vine, then to spike for four weeks before decreasing again (*J. Agric. Food Chem.* **2009,** *57,* **3818–3830**). Using the reaction described in Problem 12.28, propose a synthesis of *E*-2-hexenal starting from 1,1-dibromopentane.

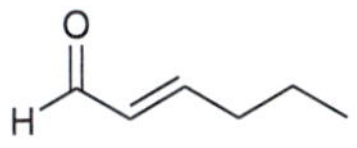

**(*E*)-2-Hexenal**

**12.41** The following sequence of reactions was employed during a synthesis of the sex pheromone of the dermestid beetle (*Tetrahedron* **1977,** *33,* **2447–2450**). The conversion of compound **5** to compound **6** involves the removal of the THP group (ROTHP → ROH), which is accomplished in acidic conditions (TsOH). Provide the structures of compounds **2, 4, 5, 6,** and **7**:

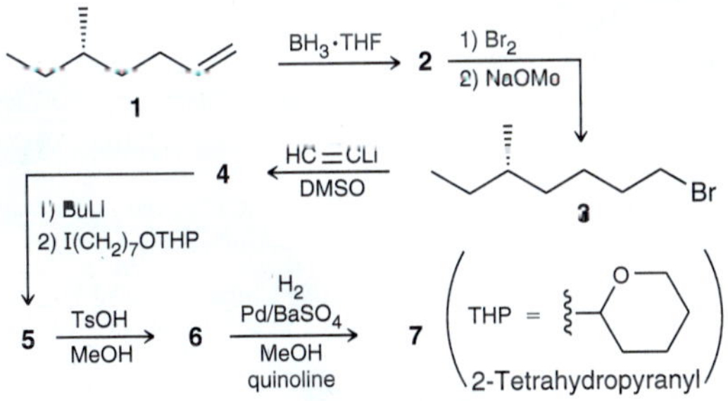

**1** $\xrightarrow{BH_3\cdot THF}$ **2** $\xrightarrow{\text{1) } Br_2 \text{ 2) NaOMe}}$ **3**

**3** $\xrightarrow{HC{\equiv}CLi,\ DMSO}$ **4** $\xrightarrow{\text{1) BuLi 2) } I(CH_2)_7OTHP}$ **5**

**5** $\xrightarrow{TsOH,\ MeOH}$ **6** $\xrightarrow{H_2,\ Pd/BaSO_4,\ MeOH,\ quinoline}$ **7**

(THP = 2-Tetrahydropyranyl)

# 13

# Alcohols and Phenols



## DID YOU EVER WONDER...

**what causes the hangover associated with drinking alcohol and whether anything can be done to prevent a hangover?**

*Veisalgia*, the medical term for a hangover, refers to the unpleasant physiological effects that result from drinking too much alcohol. These effects include headache, nausea, vomiting, fatigue, and a heightened sensitivity to light and noise. Hangovers are caused by a multitude of factors. These factors include (but are not limited to) dehydration caused by the stimulation of urine production, the loss of vitamin B, and the production of acetaldehyde in the body. Acetaldehyde is a product of the oxidation of ethanol. Oxidation is just one of the many reactions that alcohols undergo. In this chapter, we will learn about alcohols and their reactions. Then, we will revisit the topic of acetaldehyde production in the body, and we will see if anything can be done to prevent a hangover.

## DO YOU REMEMBER?

*Before you go on, be sure you understand the following topics.*
*If necessary, review the suggested sections to prepare for this chapter.*

- Brønsted-Lowry Acidity (Sections 3.3–3.4)
- Mechanisms and Curved Arrows (Sections 6.8–6.11)
- E2 and E1 Reactions (Sections 8.6–8.11)
- Designating the Configuration of a Chirality Center (Section 5.3)
- $S_N2$ and $S_N1$ Reactions (Sections 7.4–7.8)

Take the DO YOU REMEMBER? QUIZ in **WileyPLUS** to check your understanding.

## 13.1 Structure and Properties of Alcohols

**Alcohols** are compounds that possess a **hydroxyl group** (OH) and are characterized by names ending in "ol":

OH OH

**Ethanol** 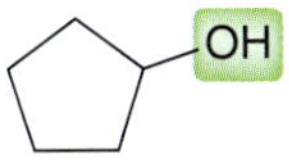 **Cyclopentanol**

A vast number of naturally occurring compounds contain the hydroxyl group. Here are just a few examples:

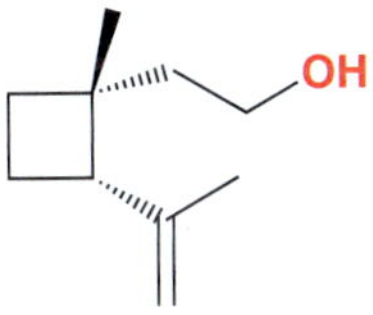

OH OH OH Cl $O_2N$ H N Cl O OH

**Grandisol**
The sex pheromone of the male boll weevil

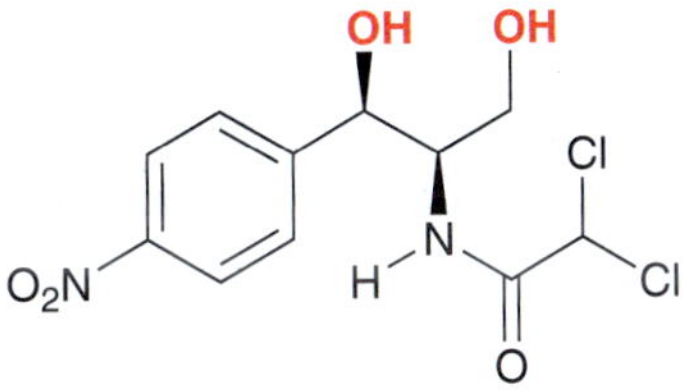

**Chloramphenicol**
An antibiotic isolated from the *Streptomyces venezuelae* bacterium. Potent against typhoid fever

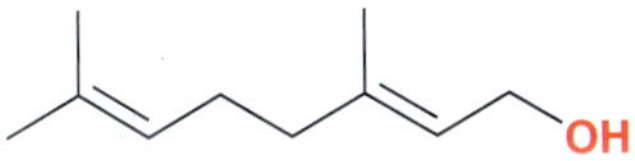

**Geraniol**
Isolated from roses and geraniums. Used in perfumes

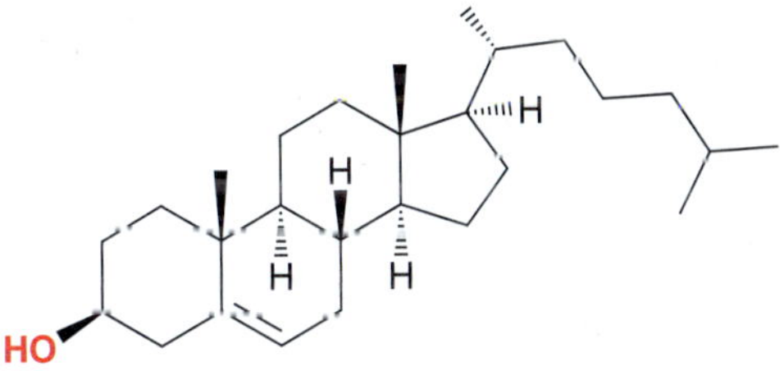

H H H H HO H H HO H

**Cholesterol**
Plays a vital role in the biosynthesis of many steroids

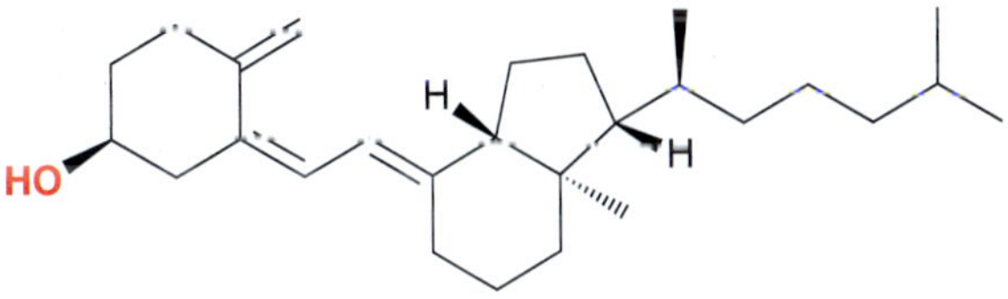

**Cholecalciferol (vitamin $D_3$)**
Regulates calcium levels and helps to form and maintain strong bones

Phenol is a special kind of alcohol. It is comprised of a hydroxyl group attached directly to a phenyl ring. Substituted phenols are extremely common in nature and exhibit a wide variety of properties and functions, as seen in the following examples:

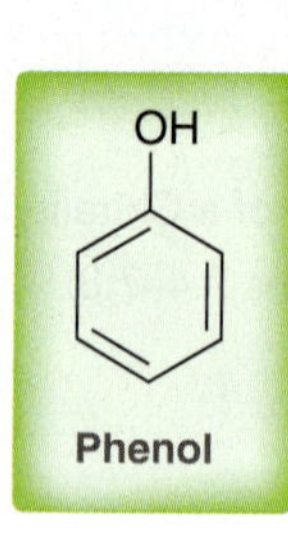


**Capsaicin**
The compound responsible for the spicy hot flavor of chili peppers

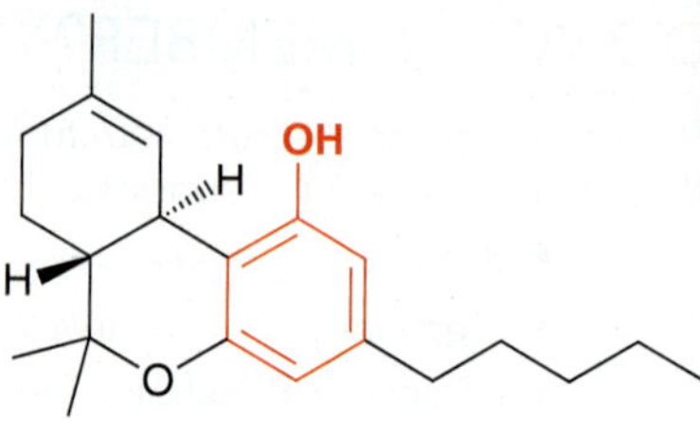


**Tetrahydrocannabinol (THC)**
The psychoactive drug found in marijuana (cannabis)

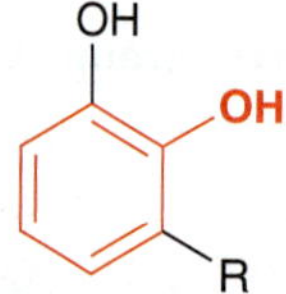


**Urushiols**
Present in the leaves of poison ivy and poison oak. Causes skin irritation

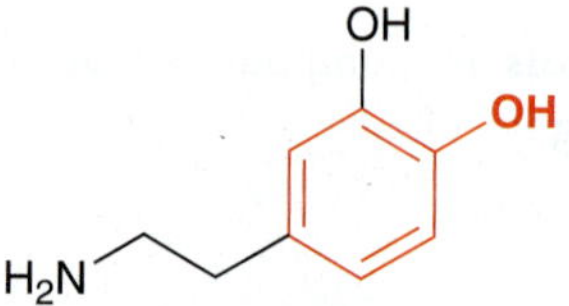


**Dopamine**
A neurotransmitter that is deficient in Parkinson's disease

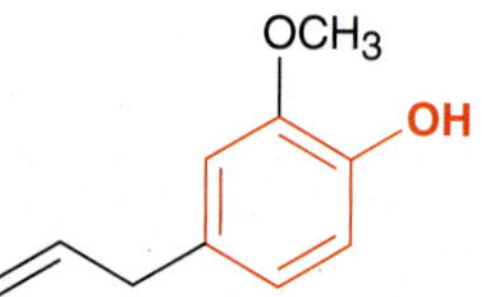


**Eugenol**
Isolated from cloves and used in perfumes and as a flavor additive

## Nomenclature

Recall that four discrete steps are required to name alkanes, alkenes, and alkynes:

1. Identify and name the parent.
2. Identify and name the substituents.
3. Assign a locant to each substituent.
4. Assemble the substituents alphabetically.

Alcohols are named using the same four steps and applying the following rules:

- When naming the parent, replace the suffix "e" with "ol" to indicate the presence of a hydroxyl group:

**Pentane** **Pentanol**

- When choosing the parent of an alcohol, identify the longest chain that includes the carbon atom connected to the hydroxyl group.

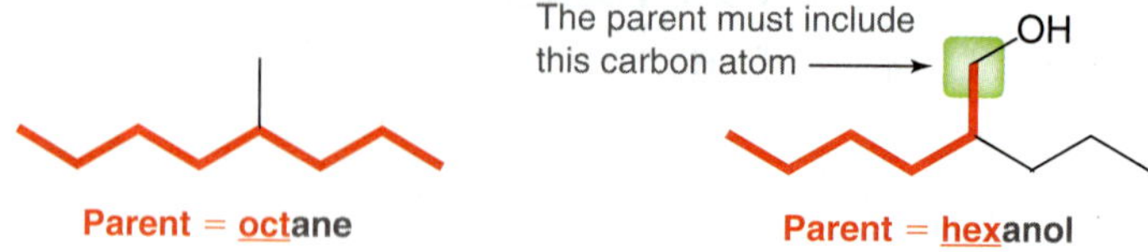


- When numbering the parent chain of an alcohol, the hydroxyl group should receive the lowest number possible, despite the presence of alkyl substituents or π bonds.

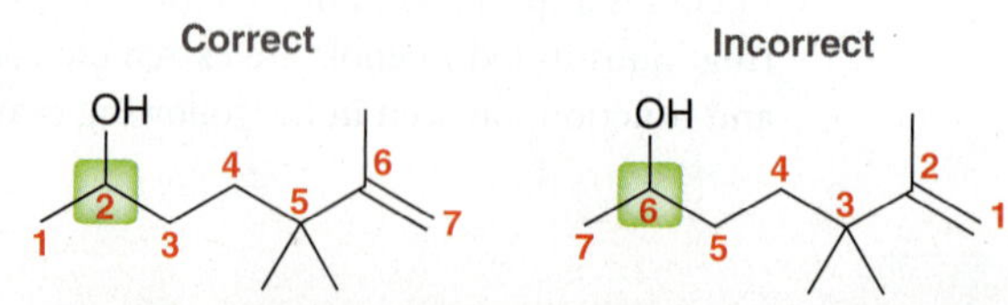

- The position of the hydroxyl group is indicated using a locant. The IUPAC rules published in 1979 dictate that this locant be placed immediately before the parent, while the IUPAC recommendations released in 1993 and 2004 allow for the locant to be placed before the suffix "ol." Both names are acceptable IUPAC names.

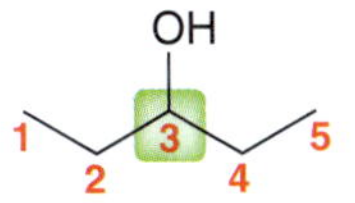

OH, 1, 2, 3, 4, 5

**3-Pentanol**
**or**
**Pentan-3-ol**

- When a chirality center is present, the configuration must be indicated at the beginning of the name; for example:

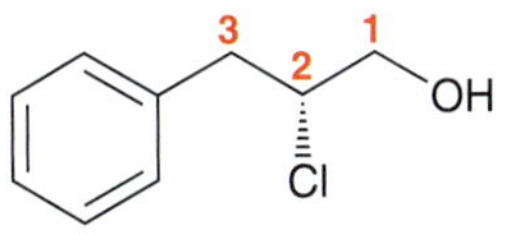

3, 2, 1, OH, Cl

**(*R*)-2-Chloro-3-phenyl-1-propanol**

- Cyclic alcohols are numbered starting at the position bearing the hydroxyl group, so there is no need to indicate the position of the hydroxyl group; it is understood to be at C–1.

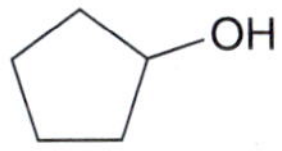

OH — **Cyclopentanol**

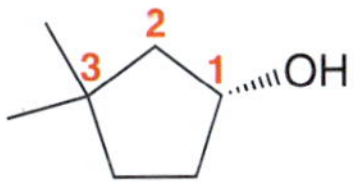

3, 2, 1, OH — **(*R*)-3,3-Dimethylcyclopentanol**

IUPAC nomenclature recognizes the common names of many alcohols, such as the following three examples:

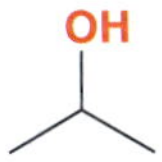

| OH | OH | OH |
|---|---|---|
| **Isopropyl alcohol (2-propanol)** | ***tert*-Butyl alcohol (2-methyl-2-propanol)** | **Benzyl alcohol (phenylmethanol)** |

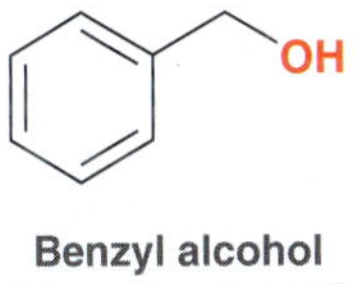

Alcohols are also designated as primary, secondary, or tertiary, depending on the number of alkyl groups attached directly to the alpha (α) position (the carbon atom bearing the hydroxyl group).

| 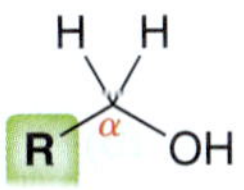 H H, R, α, OH | 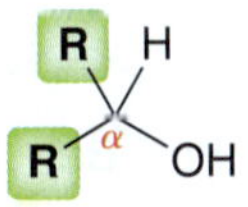 R H, R, α, OH | 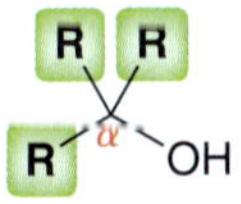 R R, R, α, OH |
|---|---|---|
| **Primary** | **Secondary** | **Tertiary** |

The word "phenol" is used to describe a specific compound (hydroxybenzene) but is also used as the parent name when substituents are attached.

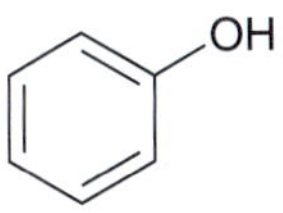

OH — **Phenol**

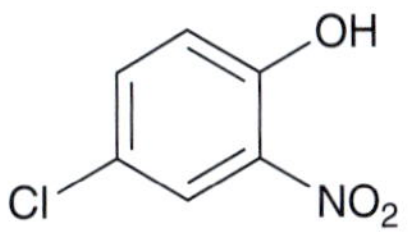

OH, Cl, $NO_2$ — **4-Chloro-2-nitrophenol**

# SKILLBUILDER

## 13.1 NAMING AN ALCOHOL

**LEARN the skill**

Assign an IUPAC name for the following alcohol:

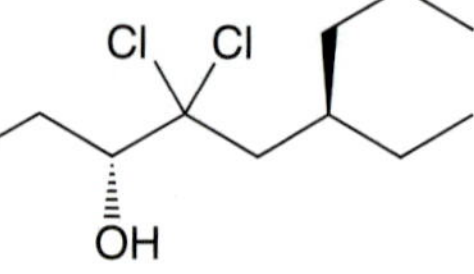


### SOLUTION

Begin by identifying and naming the parent. Choose the longest chain that includes the carbon atom connected to the hydroxyl group and then number the chain to give the hydroxyl group the lowest number possible.

**STEP 1**
Identify and name the parent.

**3-Nonanol**

Then identify the substituents and assign locants.

**STEPS 2 AND 3**
Identify the substituents and assign locants.

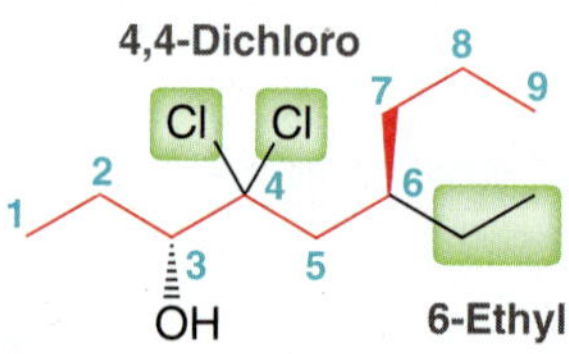


**STEP 4**
Assemble the substituents alphabetically.

Next, assemble the substituents alphabetically: 4,4-dichloro-6-ethyl-3-nonanol.

Before concluding, we must always check to see if there are any chirality centers This compound has two chirality centers. Using the skills from Section 5.3, we can assign the *R* configuration to both chirality centers.

**STEP 5**
Assign the configuration of any chirality center.

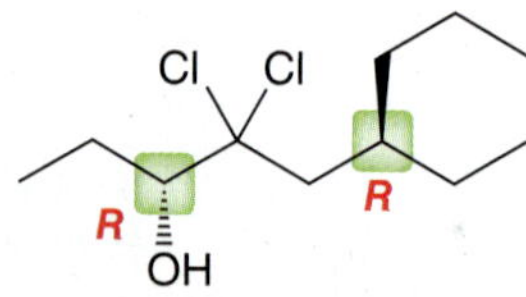


Therefore, the complete name is (3*R*,6*R*)-4,4-dichloro-6-ethyl-3-nonanol.

**PRACTICE the skill**

**13.1** Provide an IUPAC name for each of the following alcohols:

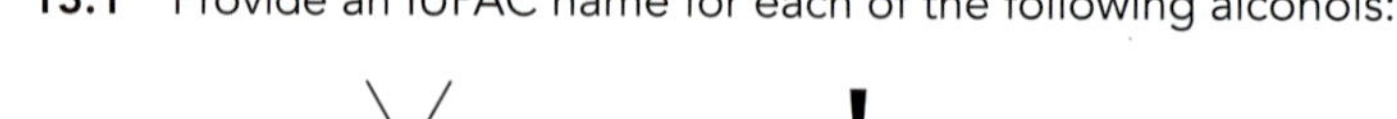

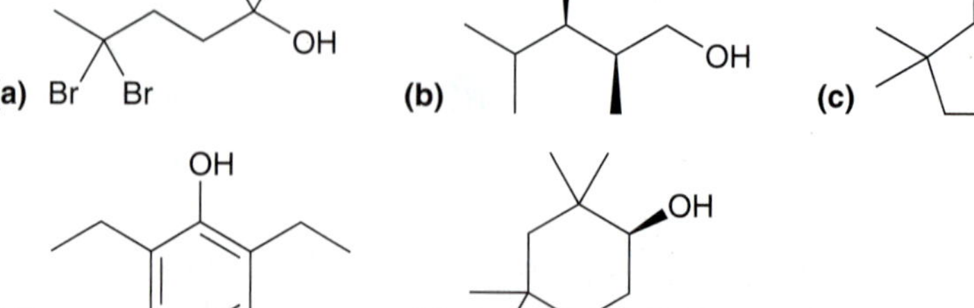

**APPLY the skill**

**13.2** Draw the structure of each of the following compounds:

**(a)** (*R*)-3,3-Dibromocyclohexanol **(b)** (*S*)-2,3-Dimethyl-3-pentanol

need more **PRACTICE?** **Try Problems 13.30–13.31a–d,f, 13.32**

## Commercially Important Alcohols

Methanol ($CH_3OH$) is the simplest alcohol. It is toxic, and ingestion can cause blindness and death, even in small quantities. Methanol can be obtained from heating wood in the absence of air and has therefore been called "wood alcohol." Industrially, methanol is prepared by the reaction between carbon dioxide ($CO_2$) and hydrogen gas ($H_2$) in the presence of suitable catalysts. Each year, the United States produces approximately 2 billion gallons of methanol, which is used as a solvent and as a precursor in the production of other commercially important compounds.

Methanol can also be used as a fuel to power combustion engines. In the second lap of the 1964 Indianapolis 500, a bad accident involving seven cars resulted in a large fire that claimed the lives of two drivers. That accident prompted the decision that all race cars switch from gasoline-powered engines to methanol-powered engines, because methanol fires produce no smoke and are more easily extinguished. In 2006, the Indianapolis 500 switched the choice of fuels for the race cars again, replacing methanol with ethanol.

Ethanol ($CH_3CH_2OH$), also called grain alcohol, is obtained from the fermentation of grains or fruits, a process that has been widely used for thousands of years. Industrially, ethanol is prepared via the acid-catalyzed hydration of ethylene. Each year, the United States produces approximately 5 billion gallons of ethanol, used as a solvent and as a precursor for the production of other commercially important compounds. Ethanol that is suitable for drinking is highly taxed by most governments. To avoid these taxes, industrial-grade ethanol is contaminated with small quantities of toxic compounds (such as methanol) that render the mixture unfit for human consumption. The resulting solution is called "denatured alcohol."

Isopropanol, also called rubbing alcohol, is prepared industrially via the acid-catalyzed hydration of propylene. Isopropanol has antibacterial properties and is used as a local antiseptic. Sterilizing pads typically contain a solution of isopropanol in water. Isopropanol is also used as an industrial solvent and as a gasoline additive.

## Physical Properties of Alcohols

The physical properties of alcohols are quite different from the physical properties of alkanes or alkyl halides. For example, compare the boiling points for ethane, chloroethane, and ethanol.

| Ethane | Chloroethane | Ethanol |
|---|---|---|
| bp = –89°C | bp = 12°C | bp = 78°C |

The boiling point of ethanol is much higher than the other two compounds as a result of the hydrogen-bonding interactions that occur between molecules of ethanol.

These interactions are fairly strong intermolecular forces, and they are also critical in understanding how alcohols interact with water. For example, methanol is **miscible** with water, which means that methanol can be mixed with water in any proportion (they will never separate into two layers like a mixture of water and oil). However, not all alcohols are miscible with water. To understand why, we must realize that every alcohol has two regions. The *hydrophobic* region does *not* interact well with water, while the *hydrophilic* region *does* interact with water via hydrogen bonding. Figure 13.1 shows the hydrophobic and hydrophilic regions of methanol and octanol. In the case of methanol, the hydrophobic end of the molecule is fairly small. This is true even of

**FIGURE 13.1** The hydrophobic and hydrophilic regions of methanol and octanol.

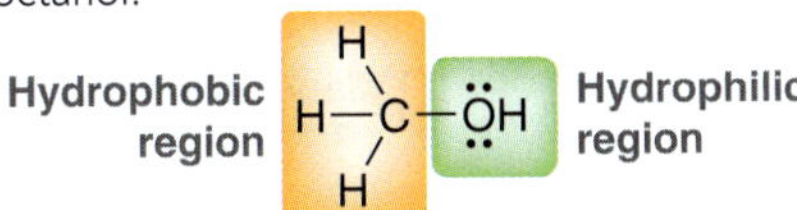


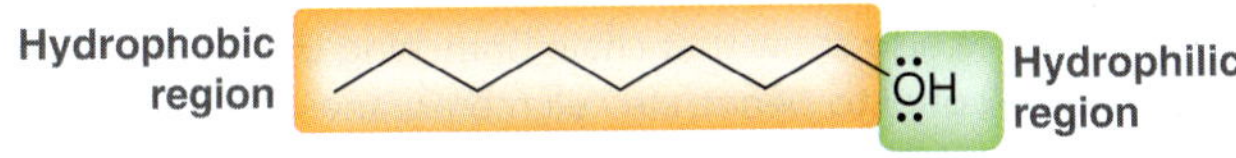

ethanol and propanol, but it is not true of butanol. The hydrophobic end of the butanol molecule is large enough to prevent miscibility. Water can still be mixed with butanol, but not in all proportions. In other words, butanol is considered to be soluble in water, rather than miscible. The term **soluble** means that only a certain volume of butanol will dissolve in a specified amount of water at room temperature.

As the size of the hydrophobic region increases, solubility in water decreases. For example, octanol exhibits extremely low solubility in water at room temperature. Alcohols with more than eight carbon atoms, such as nonanol and decanol, are considered to be insoluble in water.

## medically speaking — Chain Length as a Factor in Drug Design

Primary alcohols (methanol, ethanol, propanol, butanol, etc.) exhibit antibacterial properties. Research indicates that the antibacterial potency of primary alcohols increases with increasing molecular weight, and this trend continues up to an alkyl chain length of eight carbon atoms (octanol). Beyond eight carbon atoms, the potency decreases. That is, nonanol is less potent than octanol, and dodecanol (12 carbon atoms) has very little potency.

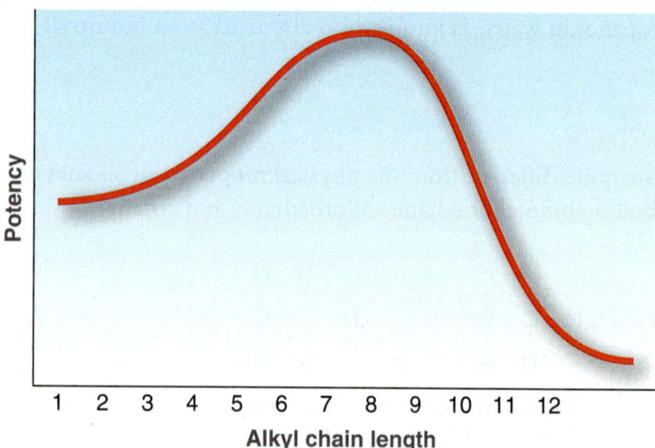


Two trends explain the observations:

- An alcohol with a larger alkyl chain (hydrophobic region) exhibits a greater ability to penetrate microbial membranes, which are composed of molecules with hydrophobic regions. According to this trend, potency should continue to increase with increased alkyl chain length, even beyond eight carbon atoms.
- A compound with a larger alkyl chain exhibits lower solubility in water, decreasing its ability to be transported through aqueous media. This trend explains why the potency of alcohols decreases steeply as the alkyl chain becomes larger than eight carbon atoms. A larger alcohol simply cannot reach its destination and therefore has low potency.

The balance between these two trends is achieved for octanol, which has the highest antibacterial potency of the primary alcohols.

Studies also show that chain branching decreases the ability of an alcohol to penetrate cell membranes. Accordingly, 2-propanol is actually less potent as an antibacterial agent than 1-propanol. Nevertheless, 2-propanol (rubbing alcohol) is used because it is less expensive to produce than 1-propanol, and the difference in antibacterial potency does not justify the added expense of production.

Many other antibacterial agents are specifically designed with alkyl chains that enable them to penetrate cell membranes. The design of these agents is optimized by carefully striking the right balance between the two trends previously discussed. Various chain lengths are tested to determine optimal potency. In most cases, the optimal chain length is found to be between five and nine carbon atoms. Consider, for example, the structure of resorcinol.

HO, OH — **Resorcinol** HO, OH — **Hexylresorcinol**

Resorcinol is a weak antiseptic (antimicrobial agent) used in the treatment of skin conditions such as eczema and psoriasis. Placing an alkyl chain on the ring increases its potency as an antiseptic. Studies indicate that the optimal potency is achieved with a six-carbon chain length. Hexylresorcinol exhibits bactericidal and fungicidal properties and is used in many throat lozenges.

### CONCEPTUAL CHECKPOINT

**13.3** Mandelate esters exhibit spasmolytic activity (they act as muscle relaxants). The nature of the alkyl group (R) greatly affects potency. Research indicates that the optimal potency is achieved when R is a nine-carbon chain (a nonyl group). Explain why nonyl mandelate is more potent than either octyl mandelate or decyl mandelate.

OH, O, R, O

**Mandelate esters (R = alkyl chain)**

## 13.2 Acidity of Alcohols and Phenols

### Acidity of the Hydroxyl Functional Group

> **LOOKING BACK**
> The factors affecting the stability of a negative charge were first discussed in Section 3.4.

> **LOOKING BACK**
> Recall that a strong acid will have a low p$K_a$ value. To review the relationship between p$K_a$ and acidity, see SkillBuilder 3.2.

As we learned in Chapter 3, the acidity of a compound can be qualitatively evaluated by analyzing the stability of its conjugate base:

R—Ö—H $\xrightarrow{-H^+}$ R—Ö:$^{\ominus}$

**To evaluate the acidity of this compound...** **...deprotonate...** **...and assess the stability of the conjugate base (an alkoxide ion)**

The conjugate base of an alcohol is called an **alkoxide** ion, and it exhibits a negative charge on an oxygen atom. A negative charge on an oxygen atom is more stable than a negative charge on a carbon or nitrogen atom but less stable than a negative charge on a halogen, X (Figure 13.2).

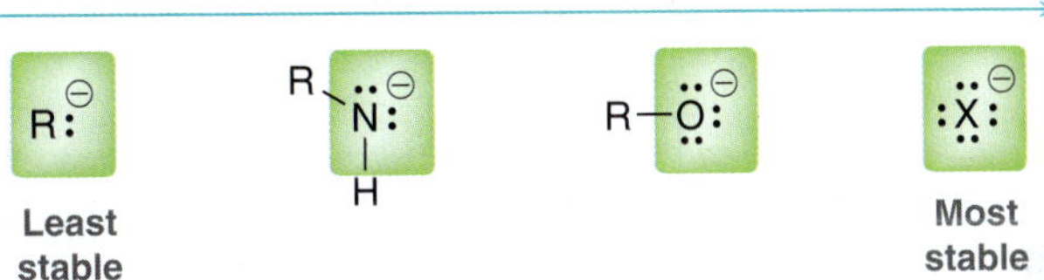


**FIGURE 13.2** The relative stability of various anions.

Therefore, alcohols are more acidic than amines and alkanes but less acidic than hydrogen halides (Figure 13.3). The p$K_a$ for most alcohols falls in the range of 15–18.

**Increasing acidity** →

| R—H | R—NH$_2$ | R—OH | :X—H |
|---|---|---|---|
| p$K_a$ between 45 and 50 | p$K_a$ between 35 and 40 | p$K_a$ between 15 and 18 | p$K_a$ between −10 and 3 |

**FIGURE 13.3** The relative acidity of alkanes, amines, alcohols, and hydrogen halides.

### Reagents for Deprotonating an Alcohol

There are two common ways to deprotonate an alcohol, forming an alkoxide ion.

1. A strong base can be used to deprotonate the alcohol. A commonly used base is sodium hydride (NaH), because hydride (H$^-$) deprotonates the alcohol to generate hydrogen gas, which bubbles out of solution:

   Ethanol + Sodium hydride ($Na^{\oplus}$ :H$^{\ominus}$) ⟶ Sodium ethoxide ($Na^{\oplus}$) + Hydrogen gas ($H_2$↑)

2. Alternatively, it is often more practical to use Li, Na, or K. These metals react with the alcohol to liberate hydrogen gas, producing the alkoxide ion.

   Ethanol $\xrightarrow{Na}$ ethoxide $Na^{\oplus}$ + $\frac{1}{2}$ $H_2$↑

**CONCEPTUAL CHECKPOINT**

**13.4** Draw the alkoxide formed in each of the following cases:

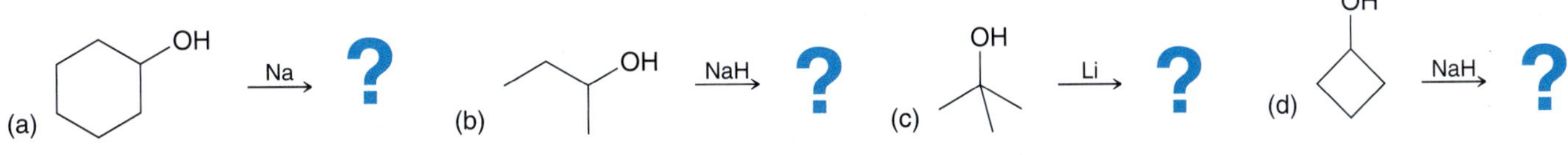

## Factors Affecting the Acidity of Alcohols and Phenols

How can we predict which, of a number of alcohols, is more acidic? In this section, we will explore three factors for comparing the acidity of alcohols.

1. *Resonance.* One of the most significant factors affecting the acidity of alcohols is resonance. As a striking example, compare the p$K_a$ values of cyclohexanol and phenol:

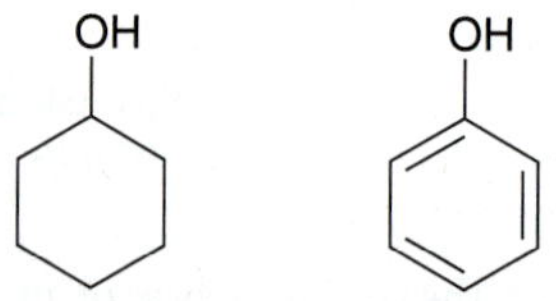

**Cyclohexanol** (**p$K_a$ = 18**)

**Phenol** (**p$K_a$ = 10**)

When phenol is deprotonated, the conjugate base is stabilized by resonance.

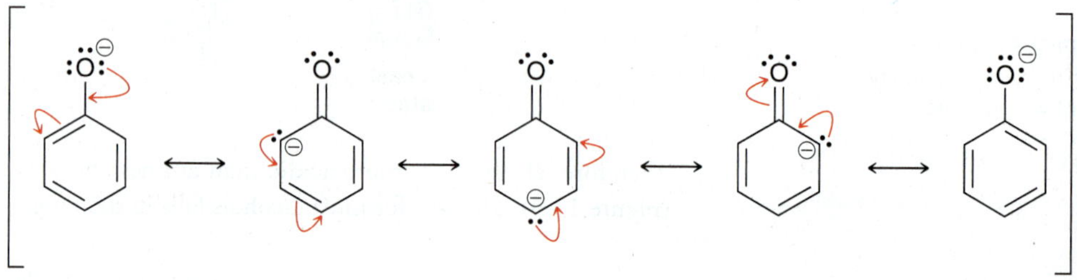

This resonance-stabilized anion is called a **phenolate**, or a **phenoxide** ion. Resonance stabilization of the phenoxide ion explains why phenol is eight orders of magnitude (100,000,000 times) more acidic than cyclohexanol. As a result, phenol does not need to be deprotonated with a very strong base like sodium hydride. Instead, it can be deprotonated by hydroxide.

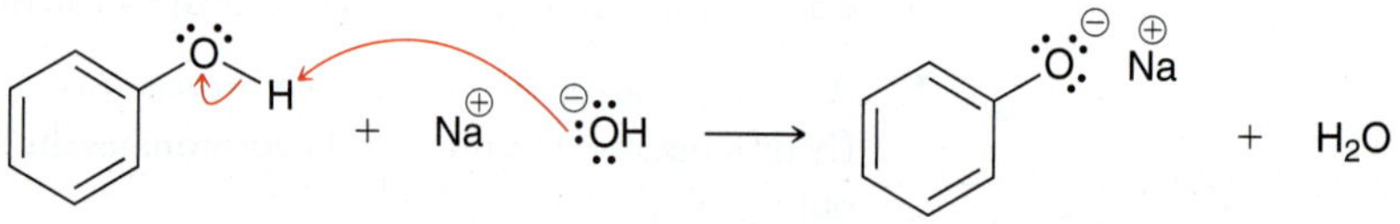

(**p$K_a$ = 10**) + $Na^+$ $^-OH$ ⟶ + $H_2O$ (**p$K_a$ = 15.7**)

The acidity of phenols is one of the reasons that phenols are a special category of alcohols. Later in this chapter and again in Chapter 19 we will see other reasons why phenols belong to a class of their own.

2. *Induction.* Another factor in comparing the acidity of alcohols is induction. As an example, compare the p$K_a$ values of ethanol and trichloroethanol.

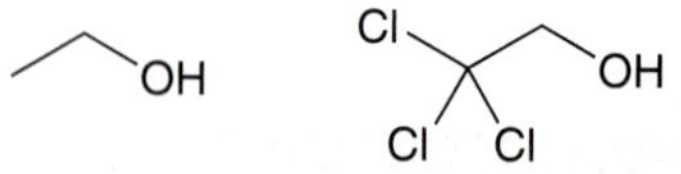

**Ethanol** (**p$K_a$ = 16**)

**Trichloroethanol** (**p$K_a$ = 12.2**)

Trichloroethanol is four orders of magnitude (10,000 times) more acidic than ethanol, because the conjugate base of trichloroethanol is stabilized by the electron-withdrawing effects of the nearby chlorine atoms.

LOOKING BACK
For a review of inductive effects, see Section 3.4.

3. *Solvation effects.* To explore the effect of alkyl branching, compare the acidity of ethanol and *tert*-butanol.

**Ethanol** ($pK_a = 16$)    ***tert*-Butanol** ($pK_a = 18$)

The $pK_a$ values indicate that *tert*-butanol is less acidic than ethanol, by two orders of magnitude. This difference in acidity is best explained by a steric effect. The ethoxide ion is not sterically hindered and is therefore easily solvated (stabilized) by the solvent, while *tert*-butoxide is sterically hindered and is less easily solvated (Figure 13.4). The conjugate base of *tert*-butanol is less stabilized than the conjugate base of ethanol, rendering *tert*-butanol less acidic.

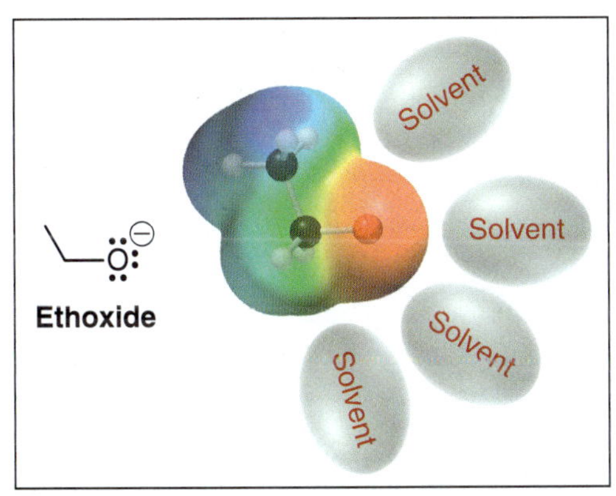


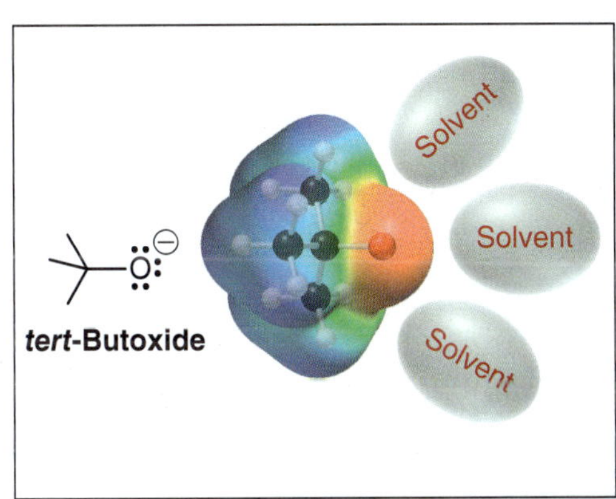


**FIGURE 13.4** An ethoxide ion is stabilized by the solvent to a greater extent than *tert*-butoxide is stabilized by the solvent.

# SKILLBUILDER

## 13.2 COMPARING THE ACIDITY OF ALCOHOLS

**LEARN the skill** Identify which of the following compounds is expected to be more acidic:

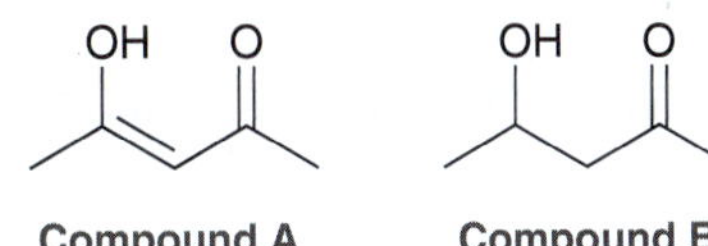

**Compound A**    **Compound B**

**SOLUTION**

Begin by drawing the conjugate base of each and then compare the stability of those conjugate bases.

**Conjugate base of compound A**    **Conjugate base of compound B**

The conjugate base of compound **B** is not resonance stabilized, but the conjugate base of compound **A** is resonance stabilized.

**Conjugate base of compound A**

The conjugate base of compound **A** will be more stable than the conjugate base of compound **B**. Therefore, compound **A** will be more acidic.

We expect compound **B** to have a p$K_a$ somewhere in the range of 15–18 (the range expected for alcohols). The p$K_a$ of compound **A** will be more difficult to predict. However, we can say with certainty that it will be lower (more acidic) than a regular alcohol. In other words, the p$K_a$ value will be lower than 15.

**PRACTICE** the skill **13.5** For each of the following pairs of alcohols, identify the one that is more acidic and explain your choice:

**(a)** OH F F OH **(b)** OH OH

Cl Cl OH Cl Cl Cl OH O

**(c)** **(d)** OH OH

OH OH

**(e)**

**APPLY** the skill **13.6** Consider the structures of 2-nitrophenol and 3-nitrophenol. These compounds have very different p$K_a$ values. Predict which one has the lower p$K_a$ and explain why. (**Hint:** In order to solve this problem, you must draw the structure of each nitro group.)

OH $NO_2$ OH $NO_2$

**2-Nitrophenol** **3-Nitrophenol**

need more **PRACTICE?** **Try Problems 13.33, 13.34**

## 13.3 Preparation of Alcohols via Substitution or Addition

### Substitution Reactions

As we saw in Chapter 7, alcohols can be prepared by substitution reactions in which a leaving group is replaced by a hydroxyl group.

$$R{-}X \longrightarrow R{-}OH$$

A primary substrate will require $S_N2$ conditions (a strong nucleophile), while a tertiary substrate will require $S_N1$ conditions (a weak nucleophile).

**Primary:**

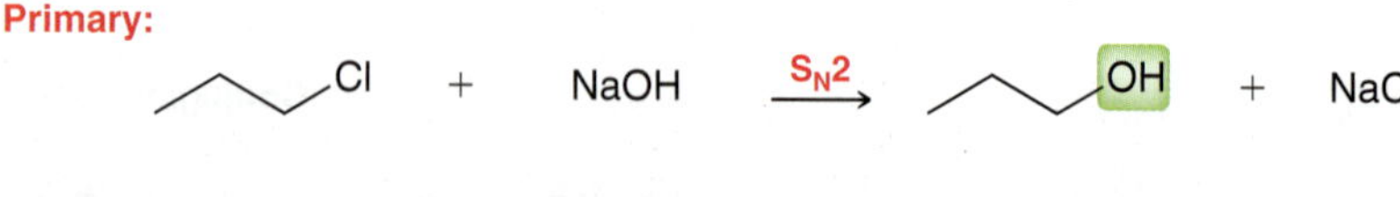


**Tertiary:**

Cl + H–O–H $\xrightarrow{S_N1}$ OH + HCl

**LOOKING BACK**
For a review of the factors that affect substitution vs. elimination, see Section 8.14.

With a secondary substrate, neither $S_N2$ nor $S_N1$ is particularly effective for preparing a secondary alcohol. Under $S_N1$ conditions, the reaction is generally too slow, while $S_N2$ conditions (use of hydroxide as the nucleophile) will generally favor elimination over substitution.

## Addition Reactions

In Chapter 9, we learned several addition reactions that produce alcohols.

Dilute $H_2SO_4$ → OH — **Acid-catalyzed hydration (Section 9.4)**

1) $Hg(OAc)_2$, $H_2O$ 2) $NaBH_4$ → OH — **Oxymercuration-demercuration (Section 9.5)**

1) $BH_3 \cdot THF$ 2) $H_2O_2$, NaOH → OH — **Hydroboration-oxidation (Section 9.6)**

Acid-catalyzed hydration proceeds with Markovnikov addition (Section 9.4). That is, the hydroxyl group is positioned at the more substituted carbon. It is a useful method if the substrate is not susceptible to carbocation rearrangements (Section 6.11). In a case where the substrate can possibly rearrange, oxymercuration-demercuration can be employed. This approach also proceeds via Markovnikov addition, but it does not involve carbocation rearrangements. Hydroboration-oxidation is used to achieve an *anti*-Markovnikov addition of water.

### CONCEPTUAL CHECKPOINT

**13.7** Identify the reagents that you would use to accomplish each of the following transformations:

(a) Br → OH

(b) Br → OH

(c) → OH

(d) → OH

(e) → OH

(f) → HO

**13.8** Identify reagents that can be used to achieve each of the following transformations:

(a) To convert 1-hexene into a primary alcohol

(b) To convert 3,3-dimethyl-1-hexene into a secondary alcohol

(c) To convert 2-methyl-1-hexene into a tertiary alcohol

## 13.4 Preparation of Alcohols via Reduction

In this section, we will explore a new method for preparing alcohols. This method involves a change in **oxidation state**, so let's spend a few moments to understand oxidation states and their relationship with formal charges.

## Oxidation States

An oxidation state refers to a method of electron bookkeeping. Chapter 1 developed a different method for electron bookkeeping called formal charge. To calculate the formal charge on an atom, we treat all bonds to that atom as if they were purely *covalent*, and we break them *homolytically*. For calculating oxidation states, we will take another extreme approach. We treat all bonds as if they were purely *ionic*, and we break them *heterolytically*, giving each pair of electrons to the more electronegative atom in each case. Formal charges and oxidation states therefore represent two extreme methods of electron bookkeeping. In Figure 13.5, the formal charge of the carbon atom is zero, because we count four electrons on the central carbon atom, which is equivalent to the number of valence electrons a carbon atom is supposed to have. In contrast, the same carbon atom has an oxidation state of −2, because we count six electrons on the carbon atom, which is two more electrons than it is supposed to have.

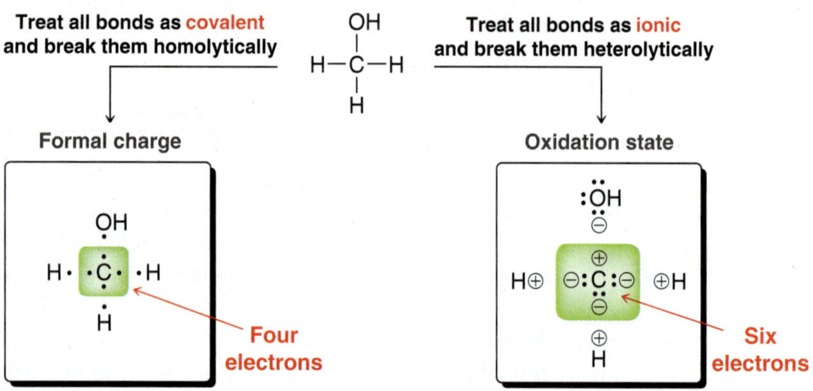

**FIGURE 13.5** Two different electron bookkeeping methods: formal charge vs. oxidation state.

A carbon atom with four bonds will always have no formal charge, but its oxidation state can range anywhere from −4 to +4.

| Methane | Methanol | Formaldehyde | Formic acid | Carbon dioxide |
|---|---|---|---|---|
| −4 | −2 | 0 | +2 | +4 |

A reaction involving an increase in oxidation state is called an oxidation. For example, when methanol is converted into formaldehyde, we say that methanol was oxidized. In contrast, a reaction involving a decrease in oxidation state is called a **reduction**. For example, when formaldehyde is converted into methanol, we say that formaldehyde was reduced. Let's get some practice identifying oxidations and reductions.

SKILLBUILDER

### 13.3 IDENTIFYING OXIDATION AND REDUCTION REACTIONS

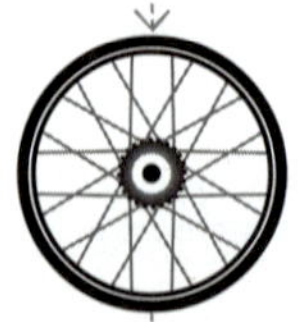

**LEARN** the skill

In the following transformation, identify whether the compound has been oxidized, reduced, or neither:

**SOLUTION**

Focus on the carbon atom where a change has occurred and determine whether the oxidation state has changed as a result of the transformation. Let's begin with the starting material:

**STEP 1**
Determine the oxidation state of the starting material.

$H_3C$—C(=O)—$CH_3$ → Break all bonds heterolytically, except for C—C bonds → $H_3C\cdot$ .C. $\cdot CH_3$

Two electrons

Each C—O bond is broken heterolytically, giving all four electrons of the C=O bond to the oxygen atom. Each C—C bond cannot be broken heterolytically, because C and C have the same electronegativity. For each C—C bond, just divide the electrons between the two carbon atoms, breaking the bond homolytically. This leaves a total of two electrons on the central carbon atom. Compare this number with the number of valence electrons that a carbon atom is supposed to have (four). The carbon atom in this example is missing two electrons. Therefore, it has an oxidation state of +2.

**STEP 2**
Determine the oxidation state of the product.

**STEP 3**
Determine if there has been a change in oxidation state.

Now analyze the carbon atom in the product. The same result is obtained: an oxidation state of +2. This should make sense, because the reaction has simply exchanged one C=O bond for two C—O single bonds. The oxidation state of the carbon atom has not changed during the reaction, and therefore the starting material has been neither oxidized nor reduced.

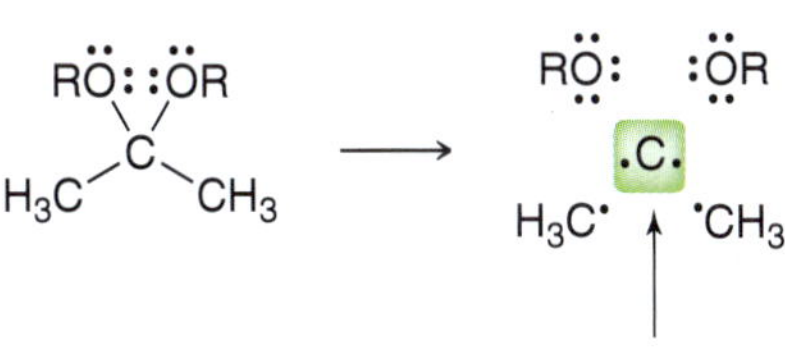


**PRACTICE** the skill **13.9** In each of the following transformations, identify whether the starting material has been oxidized, reduced, or neither. Try to determine the answer without calculating oxidation states and then use the calculations to see if your intuition was correct.

(a) (b) (c) (d) (e) (f)

**APPLY** the skill **13.10** In Chapter 9, we learned about addition of water across a π bond. Identify whether the alkene has been oxidized, reduced, or neither.

**(Hint:** First look at each carbon atom separately and then look at the net change for the alkene as a whole.)

$H_3O^+$

**13.11** In the following reaction, determine whether the alkyne has been oxidized, reduced, or neither. Using the answer from the previous problem, try to determine the answer without calculating oxidation states and then use the calculations to see if your intuition was correct.

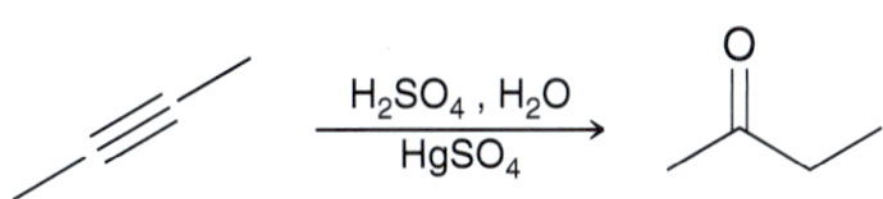


need more **PRACTICE?** **Try Problem 13.62**

## Reducing Agents

The conversion of a ketone (or aldehyde) to an alcohol is a reduction.

O — Reduction → OH

The reaction requires a **reducing agent,** which is itself oxidized as a result of the reaction. In this section, we will explore three reducing agents that can be used to convert a ketone or aldehyde to an alcohol:

1. In Chapter 9, we learned that an alkene can undergo hydrogenation in the presence of a metal catalyst such as platinum, palladium, or nickel. A similar reaction can occur for ketones or aldehydes, although more forcing conditions are generally required (higher temperature and pressure). This process works well for both ketones and aldehydes, often with fairly good yields.

O — $H_2$ / Pt, Pd, or Ni → OH

95%

2. Sodium borohydride ($NaBH_4$) is another common reducing agent that can be used to reduce ketones or aldehydes.

O — $NaBH_4$ / EtOH, MeOH, or $H_2O$ → OH

90%

**Sodium borohydride ($NaBH_4$)**

Sodium borohydride functions as a source of hydride ($H:^-$) and the solvent functions as the source of a proton ($H^+$). The solvent can be ethanol, methanol, or water. The precise mechanism of action has been heavily investigated and is somewhat complex. Nevertheless, Mechanism 13.1 presents a simplified version that will be sufficient for our purposes. The first step involves the transfer of hydride to the **carbonyl group** (the C=O bond), and the second step is a proton transfer.

### MECHANISM 13.1 REDUCTION OF A KETONE OR ALDEHYDE WITH $NaBH_4$

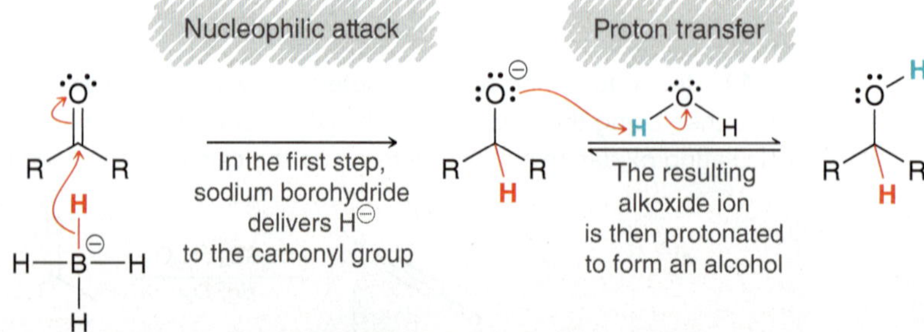

**LOOKING BACK**
For a review of the factors affecting nucleophilicity, see Section 6.7.

Hydride ($H:^-$), by itself, is not a good nucleophile because it is not polarizable. As a result, the reaction above cannot be achieved by using NaH (sodium hydride). NaH only functions as a base, not as a nucleophile. But $NaBH_4$ does function as a nucleophile. Specifically, $NaBH_4$ functions as a *delivery agent* of nucleophilic $H:^-$.

3. Lithium aluminum hydride (LAH) is another common reducing agent, and its structure is very similar to $NaBH_4$.

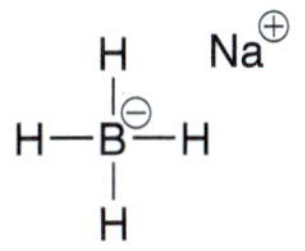

**Sodium borohydride ($NaBH_4$)** 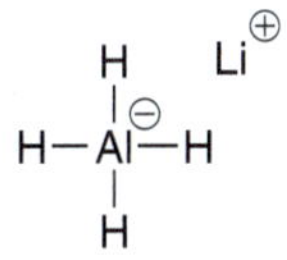 **Lithium aluminum hydride (LAH)**

Lithium aluminum hydride is commonly abbreviated either as LAH or as $LiAlH_4$. It is also a delivery agent of $H:^-$, but it is a much stronger reagent. It reacts violently with water, and therefore, a protic solvent cannot be present together with LAH in the reaction flask. First the ketone or aldehyde is treated with LAH, and then, in a separate step, the proton source is added to the reaction flask. Water ($H_2O$) can serve as a proton source, although $H_3O^+$ can also be used as a proton source:

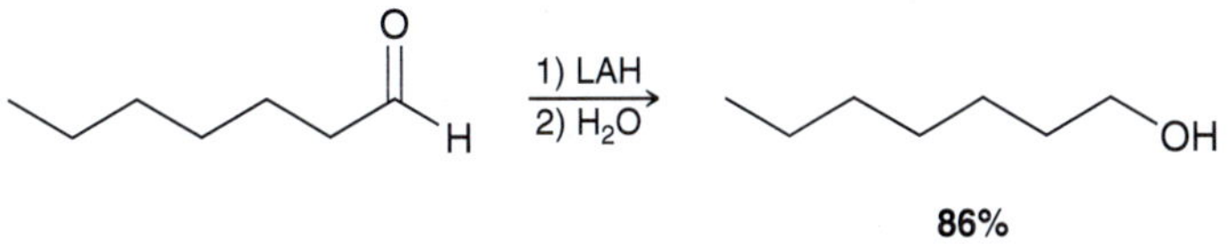

1) LAH
2) $H_2O$

**86%**

Notice that LAH and the proton source are listed as two separate steps. The mechanism of reduction with LAH (Mechanism 13.2) is similar to the mechanism of reduction with $NaBH_4$.

## MECHANISM 13.2 REDUCTION OF A KETONE OR ALDEHYDE WITH LAH

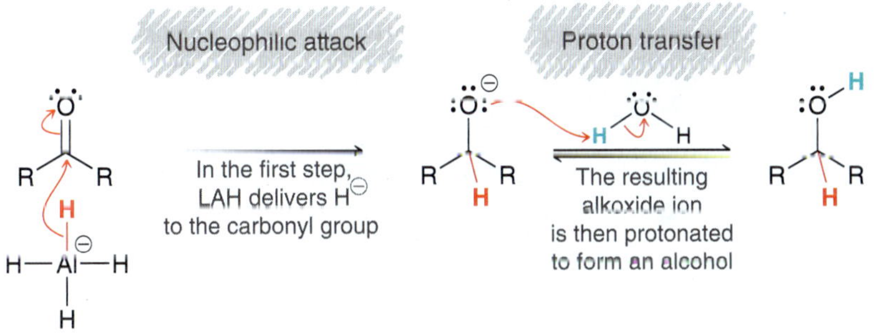

Nucleophilic attack

In the first step, LAH delivers $H^\ominus$ to the carbonyl group

Proton transfer

The resulting alkoxide ion is then protonated to form an alcohol

This simplified mechanism does not take into account many important observations, such as the role of the lithium cation ($Li^+$). However, a full treatment of the mechanism of hydride reducing agents is beyond the scope of this text, and this simplified version will suffice.

Both $NaBH_4$ and LAH will reduce ketones or aldehydes. These hydride delivery agents offer one significant advantage over catalytic hydrogenation in that they can selectively reduce a carbonyl group in the presence of a C=C bond. Consider the following example:

NaBH₄, MeOH → OH (90–95%)

1) LAH 2) $H_2O$ → 90–95%

$H_2$, Pt, Pd, or Ni → OH (90–95%)

When treated with a hydride reducing agent, only the carbonyl group is reduced. In contrast, catalytic hydrogenation will also reduce the C=C bond under the conditions required to reduce the carbonyl group (high temperature and pressure). For this reason, hydride reducing agents such as $NaBH_4$ and LAH are generally preferred over catalytic hydrogenation. Many hydride reducing reagents are commercially available. Some are even more reactive than LAH and others are even milder than $NaBH_4$. Many of these reagents are derivatives of $NaBH_4$ and LAH.

$Li^{\oplus}$ R–Al⁻(R)(R)–H    $Na^{\oplus}$ R–B⁻(R)(R)–H

Each R group can be an alkyl group, a cyano group, an alkoxy group, or any one of a number of groups. By carefully choosing the three R groups to be either electron donating or electron withdrawing, it is possible to modify the reactivity of the hydride reagent. Hundreds of different hydride reducing agents are available, and each has its own selectivity and advantages. For now, we will simply focus our attention on the differences between LAH and $NaBH_4$.

We mentioned that LAH is much more reactive than $NaBH_4$. As a result, LAH is less selective. LAH will react with a carboxylic acid or an ester to produce an alcohol, but $NaBH_4$ will not.

**BY THE WAY**
Although $NaBH_4$ does not reduce esters under mild conditions, it is observed that $NaBH_4$ will sometimes reduce esters when more forcing conditions are employed, such as high temperature.

RCOOH (**Carboxylic acid**) —1) Excess LAH, 2) $H_2O$→ $RCH_2OH$

RCOOMe (**Ester**) —1) Excess LAH, 2) $H_2O$→ $RCH_2OH$ + MeOH

The mechanism for reduction of the ester involves the transfer of hydride two times (Mechanism 13.3).

## MECHANISM 13.3 REDUCTION OF AN ESTER WITH LAH

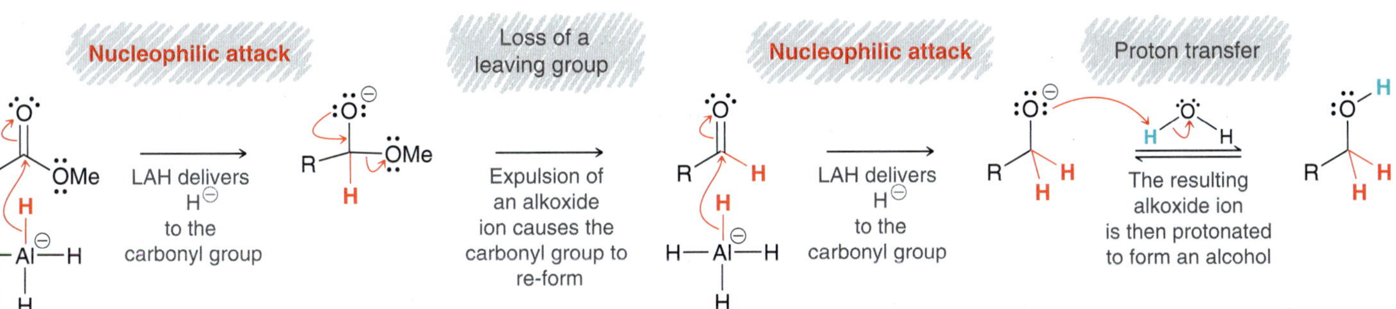


LAH delivers hydride to the carbonyl group, but then loss of a leaving group causes the carbonyl group to re-form. In the presence of LAH, the newly formed carbonyl group can be attacked again by hydride. The leaving group in the second step of the mechanism is a methoxide ion, which is generally not a good leaving group. For example, methoxide does not function as a leaving group in E2 or $S_N2$ reactions. The reason that it can function as a leaving group in this case stems from the nature of the intermediate after the first attack of hydride. That intermediate is high in energy and already exhibits a negatively charged oxygen atom. This intermediate is therefore capable of ejecting methoxide, because such a step is not uphill in energy.

**High-energy intermediate**

In Chapter 21, we will discuss this reaction in greater detail as well as the mechanism for the reaction of a carboxylic acid with LAH.

## SKILLBUILDER

### 13.4 DRAWING A MECHANISM AND PREDICTING THE PRODUCTS OF HYDRIDE REDUCTIONS

**LEARN the skill** Draw a mechanism and predict the product for the following reaction:

1) Excess LAH
2) $H_2O$
?

**SOLUTION**

LAH is a hydride reducing agent, and the starting material is a ketone. When hydride reducing agents react with a ketone or aldehyde, the mechanism consists of a nucleophilic attack followed by proton transfer. In the first step, draw the structure of LAH and show a hydride being delivered to the carbonyl group. Do not simply draw H:⁻ as the reagent, because H:⁻ is not nucleophilic by itself. It must be delivered by the delivery agent (LAH). Draw the complete structure of LAH (showing all bonds) and then draw a curved arrow that shows the electrons coming from one Al—H bond and attacking the carbonyl group.

**STEPS 1 AND 2**
Draw the two curved arrows that show the delivery of a hydride and then draw the resulting alkoxide ion.

LAH delivers a hydride to the carbonyl

In the second step of the mechanism, the alkoxide ion is protonated by a proton source, in this case, water.

**STEP 3**
Draw the two curved arrows showing the alkoxide ion being protonated by the proton source.

Proton transfer

The product is a secondary alcohol, which is what we expect from reduction of a ketone.

**PRACTICE the skill** **13.12** Draw a mechanism and predict the major product for each reaction:

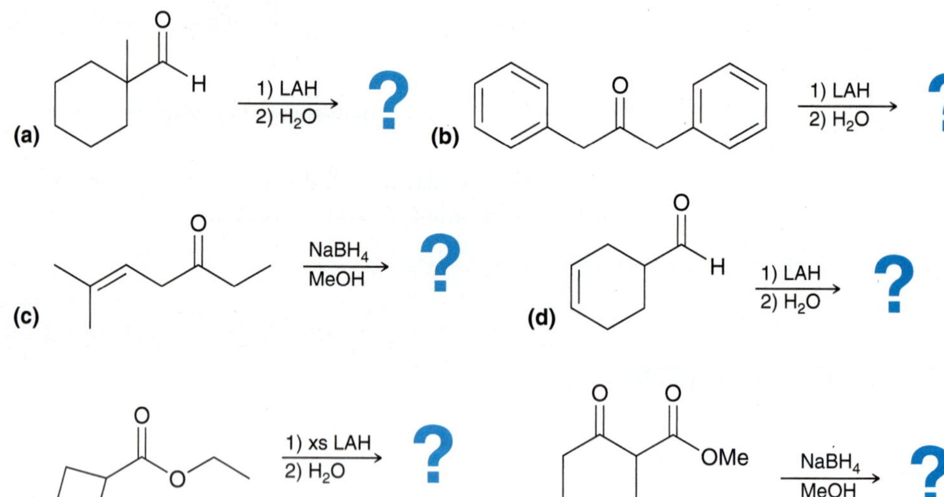


**APPLY the skill**

**13.13** Draw a mechanism and predict the major product of the following reaction:

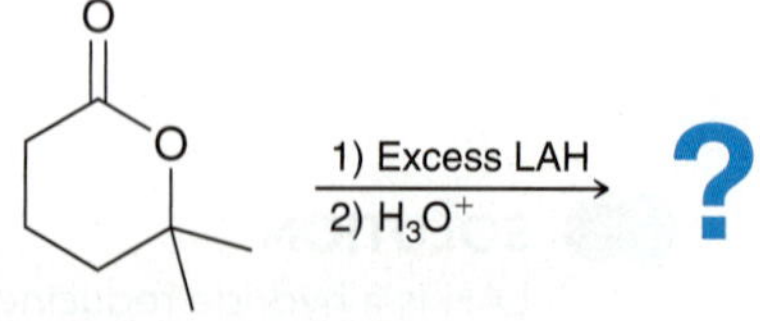


need more **PRACTICE?** **Try Problems 13.46, 13.47c, 13.48e,f, 13.60**

## 13.5 Preparation of Diols

**Diols** are compounds with two hydroxyl groups, and the following additional rules are used to name them:

1. The positions of both hydroxyl groups are identified with numbers placed before the parent.
2. The suffix "diol" is added to the end of the name:

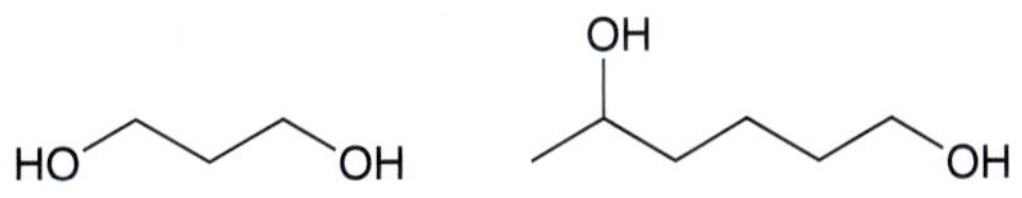

HO–CH₂CH₂CH₂–OH: **1,3-Propanediol**

**1,5-Hexanediol**

Notice that an "e" appears in between the parent and the suffix. In a regular alcohol, the "e" is dropped (i.e., propanol or hexanol). A few simple diols have common names that are accepted by IUPAC nomenclature.

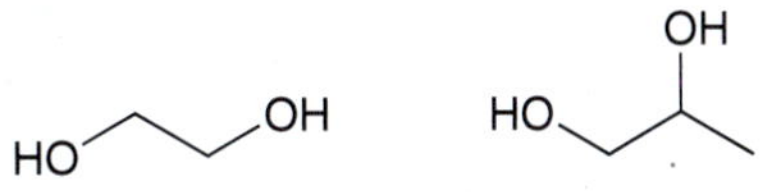

**Ethylene glycol** **Propylene glycol**

The term glycol indicates the presence of two hydroxyl groups. Diols can be prepared from diketones via reduction using any of the reducing agents that we have seen.

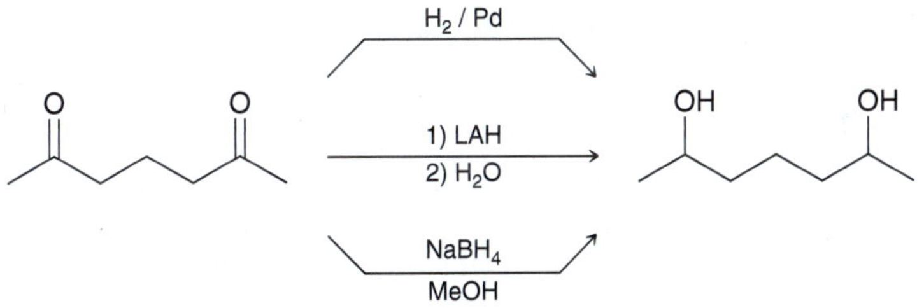

$H_2$ / Pd; 1) LAH 2) $H_2O$; $NaBH_4$ / MeOH

Alternatively, diols can be made via dihydroxylation of an alkene. In Chapter 9, we explored reagents for achieving either *syn* or *anti* dihydroxylation.

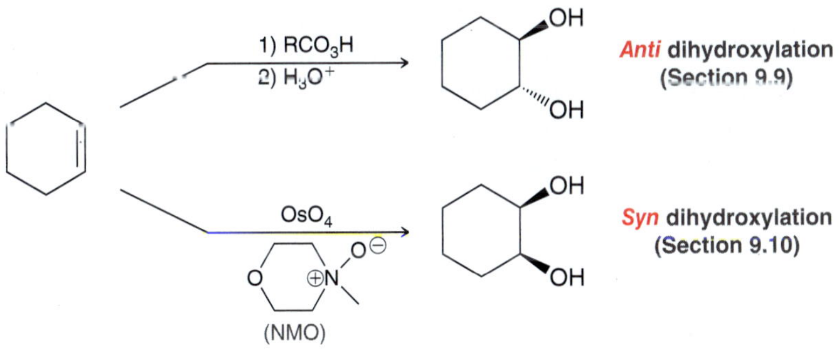

1) $RCO_3H$ 2) $H_3O^+$ — *Anti* **dihydroxylation (Section 9.9)**

$OsO_4$, (NMO) — *Syn* **dihydroxylation (Section 9.10)**

## practically speaking — Antifreeze

Automobiles are powered by internal combustion engines, and various parts of the engines can become very hot. To prevent damage caused by overheating, a liquid coolant is used to transfer some of the heat away from the sensitive engine parts. In most climates, pure water is unsuitable as a coolant because it can freeze if outdoor temperatures drop below 0°C. To prevent freezing, a coolant called antifreeze is used. Antifreeze is a solution of water and other compounds that significantly lower the freezing point of the mixture. Ethylene glycol and propylene glycol are the two most commonly used compounds for this purpose.

HO, OH — **Ethylene glycol**    HO, OH — **Propylene glycol**

## 13.6 Preparation of Alcohols via Grignard Reagents

In this section, we will discuss formation of alcohols using Grignard reagents. A **Grignard reagent** is formed by the reaction between an alkyl halide and magnesium.

$$R{-}X \xrightarrow{Mg} R{-}Mg{-}X$$

**Grignard reagent**

These reagents are named after the French chemist Victor Grignard, who demonstrated their utility in preparing alcohols. For his achievements, he was awarded the 1912 Nobel Prize in Chemistry. Below are a couple of specific examples of Grignard reagents.

Br → (Mg) → MgBr    Cl → (Mg) → MgCl

A Grignard reagent is characterized by the presence of a C—Mg bond. Carbon is more electronegative than magnesium, so the carbon atom withdraws electron density from magnesium via induction. This gives rise to a partial negative charge (δ−) on the carbon atom. In fact, the difference in electronegativity between C and Mg is so large that the bond can be treated as ionic.

δ− δ+ MgBr ≡ ⊖ ⊕MgBr

Grignard reagents are carbon nucleophiles capable of attacking a wide range of electrophiles, including the carbonyl group of ketones or aldehydes (Mechanism 13.4).

## MECHANISM 13.4 THE REACTION BETWEEN A GRIGNARD REAGENT AND A KETONE OR ALDEHYDE

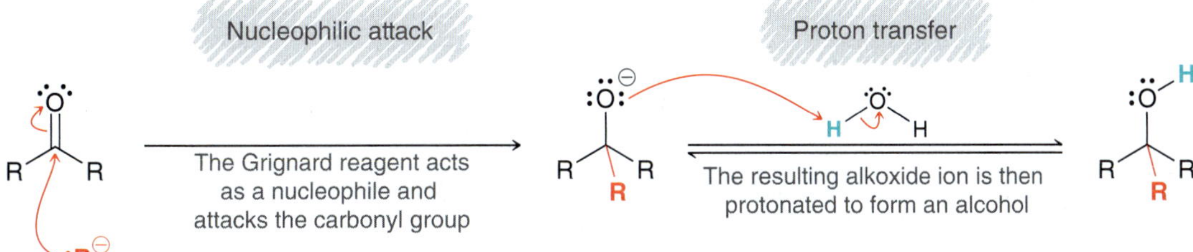


The product is an alcohol, and the mechanism presented here is similar to the mechanism that was presented for reduction via hydride reagents (LAH or $NaBH_4$). In fact, the reaction here is also a reduction, but it involves introduction of an R group.

O → 1) RMgX, 2) $H_2O$ → OH, R

Notice that water is added to the reaction flask in a separate step (similar to reduction with LAH). Water cannot be present together with the Grignard reagent, because the Grignard reagent is also a strong base and will deprotonate water.

$$\overset{\ominus}{R}:\ \overset{\oplus}{MgX} + H-\ddot{O}-H \longrightarrow R-H + HOMgX$$

(p$K_a$ = 15.7) (p$K_a$ ~ 50)

The difference in p$K_a$ values is so vast that the reaction is essentially irreversible. Every water molecule present in the reaction flask will destroy one molecule of Grignard reagent. Grignard reagents will even react with the moisture in the air, so care must be taken to use conditions that scrupulously avoid the presence of water. After the Grignard reagent has attacked the ketone, then water can be added to protonate the alkoxide.

A Grignard reagent will react with a ketone or an aldehyde to produce an alcohol.

O → 1) $CH_3MgBr$, 2) $H_2O$ → $H_3C$ OH

O, H → 1) $CH_3MgBr$, 2) $H_2O$ → OH, H, $CH_3$

Grignard reagents also react with esters to produce alcohols, with introduction of two R groups.

O, OMe → 1) Excess $CH_3MgBr$, 2) $H_2O$ → OH, $CH_3$, $CH_3$ + MeOH

The mechanism presented for this process (Mechanism 13.5) is similar to the mechanism presented for reduction of an ester with LAH (Mechanism 13.3).

## MECHANISM 13.5 THE REACTION BETWEEN A GRIGNARD REAGENT AND AN ESTER

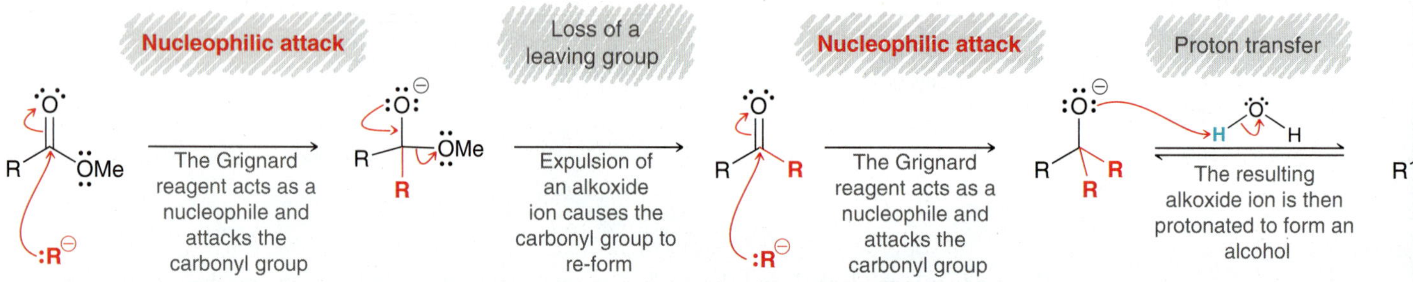


We will explore this reaction and others like it in much more detail in Chapter 21. A Grignard reagent will not attack the carbonyl group of a carboxylic acid. Instead, the Grignard reagent will simply function as a base to deprotonate the carboxylic acid.

$$R:^{\ominus}\ {}^{\oplus}MgX + H{-}O{-}C(=O){-}R \longrightarrow R{-}H + {}^{\ominus}:O{-}C(=O){-}R\ \ {}^{\oplus}MgX$$

In other words, a Grignard reagent is incompatible with a carboxylic acid. For similar reasons, it is not possible to form a Grignard reagent in the presence of even mildly acidic protons, such as the proton of a hydroxyl group.

Not compatible

OH ... Br $\xrightarrow[\times]{Mg}$ OH ... $\ominus$ $^{\oplus}$MgBr

**Cannot form this Grignard reagent**

This Grignard reagent cannot be formed, as it will simply attack itself to produce an alkoxide. In the next section, we will learn how to circumvent this problem. But first, let's get some practice preparing alcohols via Grignard reactions.

### 13.5 PREPARING AN ALCOHOL VIA A GRIGNARD REACTION

**LEARN the skill** Show how you would use a Grignard reaction to prepare the following compound:

OH

## SOLUTION

First identify the carbon atom attached directly to the hydroxyl group (the α position):

**STEP 1**
Identify the α position.

OH

**STEP 2**
Identify the three groups connected to the α position.

Next, identify all groups connected to this position. There are three: phenyl, methyl, and ethyl. We would need to start with a ketone in which two of the groups are already present, and we introduce the third group as a Grignard reagent. Here are all three possibilities:

**STEP 3**
Show how each group could have been installed via a Grignard reaction.

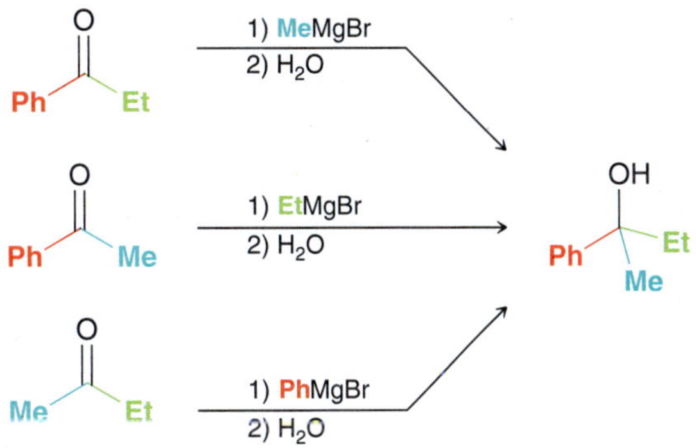


This problem illustrates an important point—that there are three perfectly correct answers to this problem. In fact, synthesis problems will rarely have only one solution. More often, synthesis problems will have multiple correct solutions.

**PRACTICE the skill** **13.14** Show how you would use a Grignard reaction to prepare each compound below:

(a) OH (b) OH (c) OH

(d) OH (e) OH (f) OH

**APPLY the skill**

**13.15** Two of the compounds from Problem 13.14 can be prepared from the reaction between a Grignard reagent and an ester. Identify those two compounds and explain why the other four compounds cannot be prepared from an ester.

**13.16** Three of the compounds from Problem 13.14 can be prepared from the reaction between a hydride reducing agent ($NaBH_4$ or LAH) and a ketone or aldehyde. Identify those three compounds and explain why the other three compounds cannot be prepared via a hydride reducing agent.

**13.17** Draw a mechanism and predict the product for the following reaction. In this case, $H_3O^+$ must be used as a proton source instead of water. Explain why.

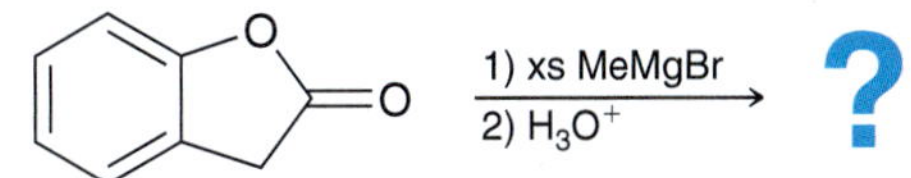


need more **PRACTICE?** **Try Problems 13.38, 13.40b, 13.52b–d,j,l–r, 13.58**

## 13.7 Protection of Alcohols

Consider the following transformation:

To achieve this transformation via a Grignard reaction, the following Grignard reagent would be required:

As we saw in the previous section, it is not possible to form this Grignard reagent, because of incompatibility with the hydroxyl group. To circumvent this problem, we employ a three-step process.

1. Protect the hydroxyl group by removing its proton and converting the hydroxyl group into a new group, called a **protecting group**, that is compatible with a Grignard reagent.
2. Form the Grignard reagent and perform the desired Grignard reaction.
3. Deprotect, by converting the protecting group back into a hydroxyl group.

Protect

Form Grignard reagent and perform Grignard reaction

Deprotect

Protecting group

The protecting group enables us to perform the desired Grignard reaction. One such example of a protecting group is the conversion of the hydroxyl group into a trimethylsilyl ether.

Protecting group

Trimethylsilyl ether

This protecting group is abbreviated as OTMS. The trimethylsilyl ether is formed via the reaction between an alcohol and trimethylsilyl chloride, abbreviated TMSCl.

(TMSCl)

$Et_3N$

This reaction is believed to proceed via an $S_N2$-like process (called $S_N2$-Si), in which the hydroxyl group functions as a nucleophile to attack the silicon atom and a chloride ion is expelled as a leaving group. A base, such as triethylamine, is then used to remove the proton connected to the oxygen atom. Notice that the first step involves an $S_N2$-like process occurring at a tertiary substrate. This should seem surprising because in Chapter 7 we learned that a nucleophile cannot effectively attack a sterically hindered substrate. This case is different because the electrophilic center is a silicon atom rather than a carbon atom. Bonds to silicon atoms are typically much longer than bonds to carbon atoms, and this longer bond length opens up the back side for attack. To visualize this, compare space-filling models of *tert*-butyl chloride and trimethylsilyl chloride (Figure 13.6).

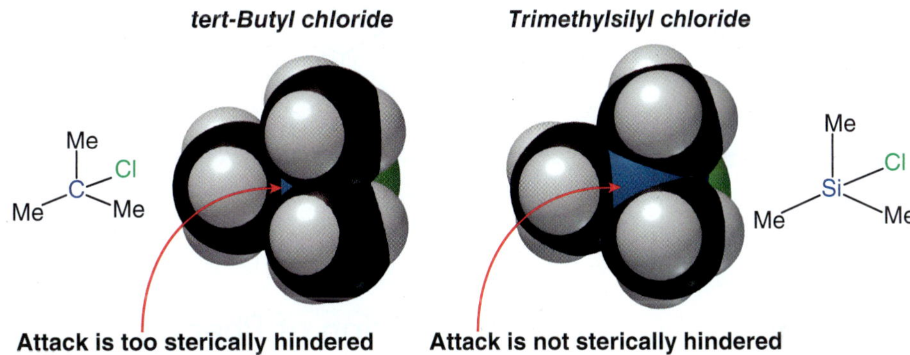


**FIGURE 13.6** Space-filling models of *tert*-butyl chloride and trimethylsilyl chloride. The latter is less sterically hindered and is susceptible to attack by a nucleophile.

After the desired Grignard reaction has been performed, the trimethylsilyl group can be removed easily with either $H_3O^+$ or fluoride ion.

$$\text{R—O—TMS} \xrightarrow[\text{or } F^{\ominus}]{H_3O^{\oplus}} \text{R—OH}$$

A commonly used source of fluoride ion is tetrabutylammonium fluoride (TBAF).

**Tetrabutylammonium fluoride (TBAF)**

The overall process is shown below.

Br—(CH₂)₄—OH → Protect: TMS—Cl, $Et_3N$ → Br—(CH₂)₄—OTMS

Form Grignard reagent and perform Grignard reaction: 1) Mg; 2) R(C=O)R; 3) $H_2O$ → HO–C(R)(R)–(CH₂)₄–OTMS

Deprotect: TBAF → HO–C(R)(R)–(CH₂)₄–OH

CONCEPTUAL CHECKPOINT

**13.18** Identify the reagents you would use to achieve each of the following transformations:

(a)

(b)

## 13.8 Preparation of Phenols

Phenol is prepared industrially via a multistep process involving the formation and oxidation of cumene.

Benzene → (propene, $H_3PO_4$) → Cumene → ($O_2$) → Cumene hydroperoxide → ($H_3O^+$) → Phenol + Acetone

A by-product of this process is acetone, which is also a commercially important compound. Over two million tons of phenol is produced each year in the United States. Phenol is used as a precursor in the synthesis of a wide variety of pharmaceuticals and other commercially useful compounds, including bakelite (a synthetic polymer made from phenol and formaldehyde), adhesives for plywood, and antioxidant food additives (BHT and BHA, discussed in Chapter 11).

Butylated hydroxytoluene (BHT)

Butylated hydroxyanisole (BHA)

**Phenols as Antifungal Agents**

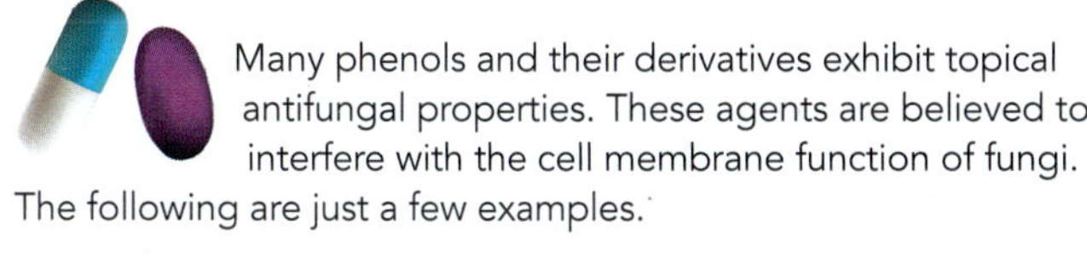

Many phenols and their derivatives exhibit topical antifungal properties. These agents are believed to interfere with the cell membrane function of fungi. The following are just a few examples.

These compounds, as well as many others, are used in the treatment of athlete's foot, jock itch, and ringworm. Tolnaftate is the active ingredient in Tinactin, Odor Eaters, and Desenex.

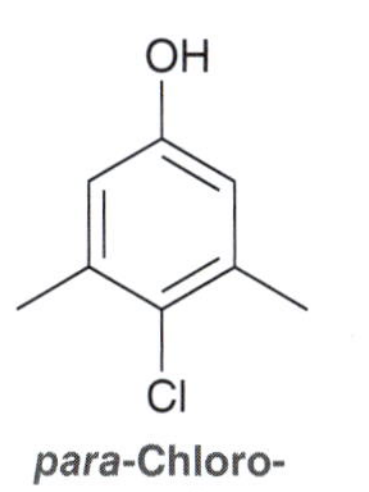


*para*-Chloro-*meta*-xylenol

Clioquinol

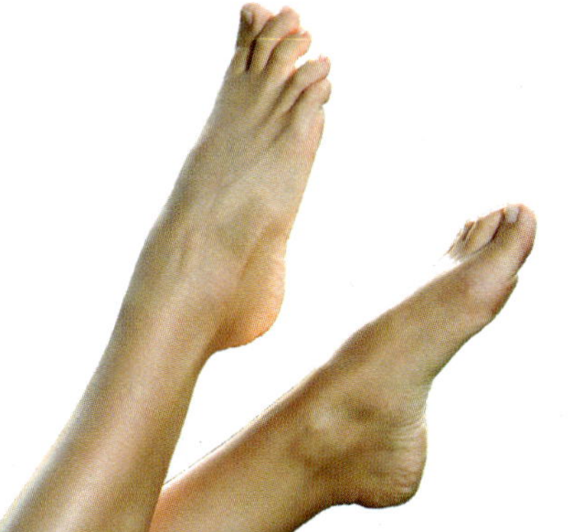


Tolnaftate

## 13.9 Reactions of Alcohols: Substitution and Elimination

### $S_N1$ Reactions with Alcohols

As seen in Section 7.5, tertiary alcohols will undergo a substitution reaction when treated with a hydrogen halide.

$$\text{(CH}_3)_3\text{C–OH} \xrightarrow{\text{HX}} \text{(CH}_3)_3\text{C–X} + H_2O$$

We saw that this reaction proceeds via an $S_N1$ mechanism.

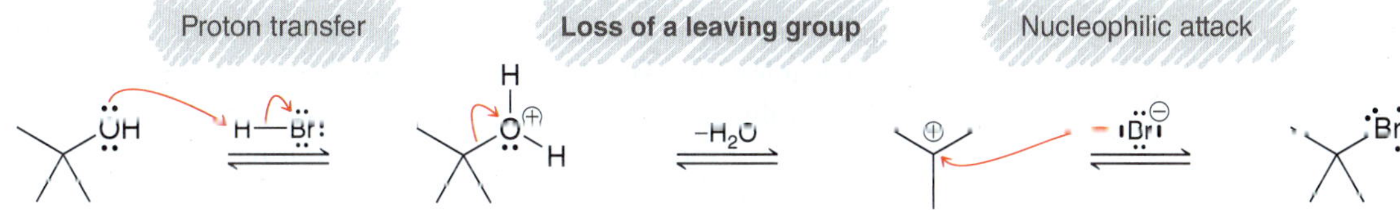


Recall that an $S_N1$ mechanism has two core steps (loss of leaving group and nucleophilic attack). When the starting material is an alcohol, we saw that an additional step is required in order to protonate the hydroxyl group first. This reaction proceeds via a carbocation intermediate and is therefore most appropriate for tertiary alcohols. Secondary alcohols undergo $S_N1$ more slowly, and primary alcohols will not undergo $S_N1$ at an appreciable rate. When dealing with a primary alcohol, an $S_N2$ pathway is required in order to convert an alcohol into an alkyl halide.

### $S_N2$ Reactions with Alcohols

Primary and secondary alcohols will undergo substitution reactions with a variety of reagents, all of which proceed via an $S_N2$ process. In this section, we will explore three such reactions that all employ an $S_N2$ process.

1. Primary alcohols will react with HBr via an $S_N2$ process.

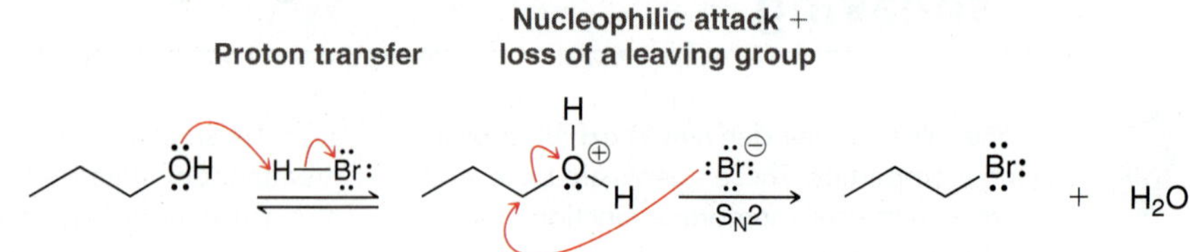


The hydroxyl group is first protonated, converting it into an excellent leaving group, followed by an $S_N2$ process. This reaction works well for HBr but does not work well for HCl. To replace the hydroxyl group with chloride, $ZnCl_2$ is used as a catalyst.

OH → (HCl, $ZnCl_2$) → Cl

The catalyst is a Lewis acid that converts the hydroxyl group into a better leaving group.

Nuc attack on a Lewis acid

Nuc attack + loss of LG

2. As seen in Section 7.8, an alcohol can be converted into a tosylate, followed by nucleophilic attack.

Bad leaving group

TsCl, py

py = pyridine

Good leaving group

$X^-$, $S_N2$

Using tosyl chloride and pyridine, the hydroxyl group is converted into a tosylate group (an excellent leaving group), which is susceptible to an $S_N2$ process. Notice the stereochemical outcome of the previous reaction. The configuration of the chirality center is not inverted during formation of the tosylate, but it is inverted during the $S_N2$ process. The net result is inversion of configuration.

*R* → 1) TsCl, py 2) NaBr → *S*

3. Primary and secondary alcohols react with $SOCl_2$ or $PBr_3$ via an $S_N2$ process.

OH → ($SOCl_2$, py) → Cl

OH → ($PBr_3$) → Br

The mechanisms for these two pathways are very similar. The first few steps convert a bad leaving group into a good leaving group, then the halide attacks in an $S_N2$ process (Mechanism 13.6).

## MECHANISM 13.6 THE REACTION BETWEEN $SOCl_2$ AND AN ALCOHOL



The reaction mechanism for $PBr_3$ has similar characteristics, including conversion of the hydroxyl group into a better leaving group followed by nucleophilic attack (Mechanism 13.7).

## MECHANISM 13.7 THE REACTION BETWEEN $PBr_3$ AND AN ALCOHOL



Notice the similarity among all of the $S_N2$ processes that we have seen in this section. All involve the conversion of the hydroxyl group into a better leaving group followed by nucleophilic attack. If any of these reactions occurs at a chirality center, inversion of configuration is to be expected.

# SKILLBUILDER

## 13.6 PROPOSING REAGENTS FOR THE CONVERSION OF AN ALCOHOL INTO AN ALKYL HALIDE

**LEARN** the skill

Identify the reagents you would use to accomplish the following transformation:

OH → Cl

### SOLUTION

**STEP 1**
Analyze the substrate.

**STEP 2**
Analyze the stereochemical outcome and determine whether $S_N2$ is required.

**STEP 3**
Identify reagents that achieve the process determined in step 2.

This transformation is a substitution process, so we must decide whether to use $S_N1$ conditions or $S_N2$ conditions. The key factor is the substrate. The substrate here is secondary, so either $S_N1$ or $S_N2$ could theoretically work. However, we don't have a choice in this case. The transformation requires inversion of configuration, which can only be accomplished via $S_N2$ (an $S_N1$ process would give racemization, as we saw in Section 7.5). In addition, the substrate would likely undergo a carbocation rearrangement if subjected to $S_N1$ conditions. For both of these reasons, an $S_N2$ process is required.

Either of the following two methods can be used to afford the desired product:

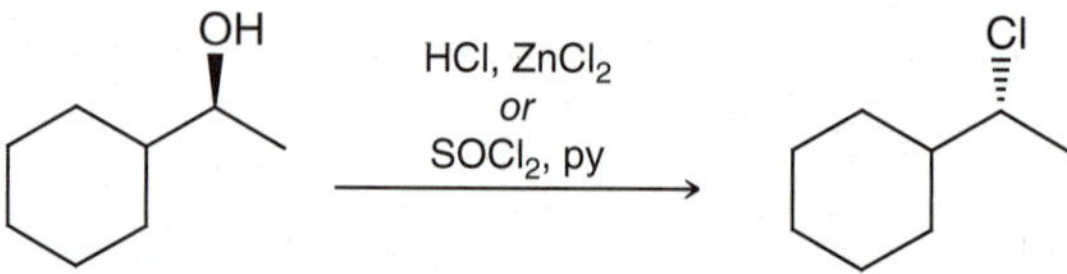

**PRACTICE** the skill

**13.19** Identify the reagents you would use to accomplish each of the following transformations:

(a) OH → Br (b) OH → Br

(c) OH → Cl (d) OH → Br

(e) OH → Cl (f) OH → Br

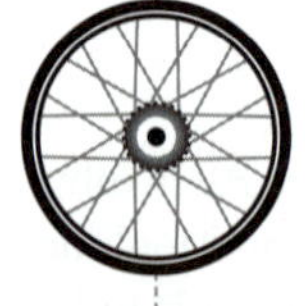

**APPLY** the skill

**13.20** Identify the reagents you would use to accomplish the following transformation:

OH → Cl

need more **PRACTICE?** **Try Problems 13.35a–c, 13.44f, 13.52r**

## practically speaking — Drug Metabolism

Drug metabolism refers to the set of reactions by which drugs are converted into other compounds that are either used by the body or excreted. Our bodies dispose of the medications we ingest by a variety of metabolic pathways.

One common metabolic pathway is called glucuronic acid conjugation, or simply glucuronidation. This process is very similar to all of the $S_N2$ processes that we investigated in the previous section. Specifically, a bad leaving group (hydroxyl) is first converted into a good leaving group, followed by nucleophilic attack. Glucuronidation exhibits the same two key steps.

1. Formation of UDPGA (uridine-5′-diphospho-α-D-glucuronic acid) from glucose involves conversion of a bad leaving group into a good leaving group.

D-Glucose — Bad LG (OH) → → UDPGA — Good LG (called UDP)

UDPGA is a compound with a very good leaving group. This large leaving group is called UDP. In this transformation, one of the hydroxyl groups in glucose has been converted into a good leaving group.

2. Next, an $S_N2$ process occurs in which the drug being metabolized (such as an alcohol) attacks UDPGA, expelling the good leaving group.

O–UDP —(RÖH, UDP-glucuronyl-transferase)→ —($-H^+$)→ β-Glucuronide

This $S_N2$ process requires an enzyme (a biological catalyst) called UDP-glucuronyl transferase. The reaction proceeds via inversion of configuration (as expected of an $S_N2$ process) to produce a β-glucuronide, which is highly water soluble and is readily excreted in the urine.

Many functional groups undergo glucuronidation, but alcohols and phenols are the most common classes of compounds that undergo this metabolic pathway. For example, morphine, acetaminophen, and chloramphenicol are all metabolized via glucuronidation:

**Morphine**
An opiate analgesic used to treat severe pain

**Acetaminophen**
An analgesic (pain-relieving) and antipyretic (fever-reducing) agent, sold under the trade name Tylenol

**Chloramphenicol**
An antibiotic used in eye drops to treat bacterial conjuctivitis

In each of these three compounds, the highlighted hydroxyl group attacks UDPGA. Glucuronidation is the main metabolic pathway by which these drugs are eliminated from the body.

## E1 and E2 Reactions with Alcohols

Recall from Section 8.9 that alcohols undergo elimination reactions in acidic conditions.

OH $\xrightarrow[\text{heat}]{\text{conc. } H_2SO_4}$ + $H_2O$

This transformation follows an E1 mechanism:

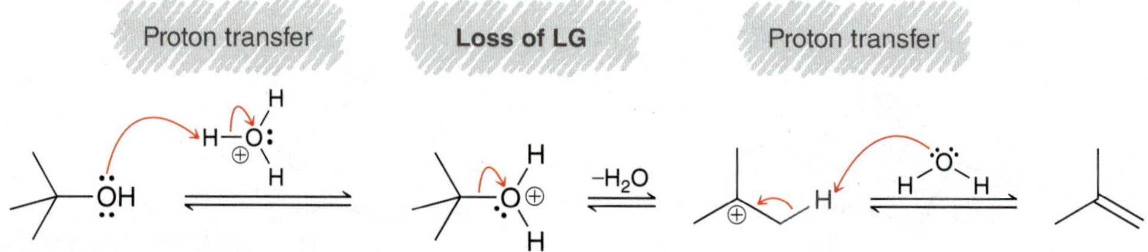


Recall that the two core steps of an E1 mechanism are loss of a leaving group followed by a proton transfer. However, when the starting material is an alcohol, an additional step is first required in order to protonate the hydroxyl group. This reaction proceeds via a carbocation intermediate and is therefore best for tertiary alcohols. Also recall that elimination generally favors the more substituted alkene.

OH $\xrightarrow[\text{heat}]{\text{conc. } H_2SO_4}$ + 

**Major** **Minor**

This transformation can also be accomplished via an E2 pathway if the hydroxyl group is first converted into a better leaving group, such as a tosylate. A strong base can then be employed to accomplish an E2 reaction.

OH $\xrightarrow[\text{py}]{\text{TsCl}}$ OTs $\xrightarrow[\text{E2}]{\text{NaOEt}}$

The E2 process also generally produces the more substituted alkene, and no carbocation rearrangements are observed in E2 processes.

### CONCEPTUAL CHECKPOINT

**13.21** Predict the products for each of the following transformations:

## 13.10 Reactions of Alcohols: Oxidation

In Section 13.4, we saw that alcohols can be formed via a reduction process. In this section, we will explore the reverse process, called **oxidation**, which involves an increase in oxidation state.

Oxidation / Reduction

The outcome of an oxidation process depends on whether the starting alcohol is primary, secondary, or tertiary. Let's first consider the oxidation of a primary alcohol.

[O] [O]

**Primary alcohol** **Aldehyde** **Carboxylic acid**

Notice that a primary alcohol has two protons at the α position (the carbon atom bearing the hydroxyl group). As a result, primary alcohols can be oxidized twice. The first oxidation produces an aldehyde, and then oxidation of the aldehyde produces a carboxylic acid.

Secondary alcohols only have one proton at the α position so they can only be oxidized once, forming a ketone.

[O]

**Secondary alcohol** **Ketone**

**LOOKING AHEAD**

For an exception to this general rule, see Section 20.11, where we will learn about a special oxidizing reagent that can oxidize a ketone to form an ester.

Generally speaking, the ketone is not further oxidized. Tertiary alcohols do not have any protons at the α position, and as a result, they generally do not undergo oxidation:

[O] **no reaction**

A large number of reagents are available for oxidizing primary and secondary alcohols. The most common oxidizing reagent is chromic acid ($H_2CrO_4$), which can be formed either from chromium trioxide ($CrO_3$) or from sodium dichromate ($Na_2Cr_2O_7$) in aqueous acidic solution.

$H_3O^+$, Acetone; $H_2SO_4$, $H_2O$

**Chromium trioxide** **Sodium dichromate** **Chromic acid**

The mechanism of oxidation with chromic acid has two main steps (Mechanism 13.8). The first step involves formation of a chromate ester, and the second step is an E2 process to form a carbon-oxygen π bond (rather than a carbon-carbon π bond).

## MECHANISM 13.8 OXIDATION OF AN ALCOHOL WITH CHROMIC ACID

**STEP 1**

$$R_2CH{-}OH + HO{-}Cr(=O)_2{-}OH \xrightleftharpoons{\text{Fast and reversible}} R_2CH{-}O{-}Cr(=O)_2{-}OH + H_2O$$

**Chromate ester**

**STEP 2**

$$R_2CH{-}O{-}Cr(=O)_2{-}OH \xrightarrow[\text{E2}]{H_2O} R_2C{=}O + H_3O^{\oplus} + HCrO_3^{\ominus}$$

**Chromate ester**

## practically speaking Breath Tests to Measure Blood Alcohol Level

Ethanol is a primary alcohol. Therefore, ethanol will react with potassium dichromate in acidic conditions to produce acetic acid.

$$CH_3CH_2OH + Cr_2O_7^{2-} \xrightarrow{H^+} CH_3COOH + Cr^{3+}$$

**Ethanol** — **Reddish orange** — **Acetic acid** — **Green**

In the process, ethanol is oxidized, and the chromium reagent is reduced. The new chromium compound is a different color, and therefore, the progress of the reaction can be monitored by the color change from reddish-orange to green. This reaction formed the basis for many of the early breath tests that assessed the level of blood alcohol. Breath tests that utilize this reaction are still commercially available. The test consists of a tube, where one end is fitted with a mouthpiece and the other end has a bag. The inside of the tube contains sodium dichromate that is adsorbed onto the surface of an inert solid, such as silica gel. As the user blows air through the tube and fills up the bag, the alcohol in the user's breath reacts with the oxidizing agent in the tube, and a color change occurs. The extent of the color change gives an indication of the blood alcohol content.

The Breathalyzer is based on the same idea, but it is more accurate. A measured volume of breath is bubbled through an aqueous acidic solution of potassium dichromate, and the change in color is measured with an ultraviolet-visible (UV-VIS) spectrophotometer (Section 17.11). Contrary to popular belief, it is not possible to beat the test by sucking a breath mint or rinsing with a breath freshener. These techniques might fool a person, but they won't fool the potassium dichromate.

When a primary alcohol is oxidized with chromic acid, a carboxylic acid is obtained. It is generally difficult to control the reaction to produce the aldehyde.

$$\text{Alcohol} \xrightarrow[H_2SO_4,\ H_2O]{Na_2Cr_2O_7} [\text{Aldehyde}] \longrightarrow \text{Carboxylic acid}$$

Alcohol — Aldehyde (Difficult to isolate) — Carboxylic acid

In order to produce the aldehyde as the final product, it is necessary to use a more selective oxidizing reagent, one that will react with the alcohol but will not react with the aldehyde. Many such reagents are available, including pyridinium chlorochromate (PCC). PCC is formed from the reaction between pyridine, chromium trioxide, and hydrochloric acid.

$$\text{Pyridine} + CrO_3 + HCl \longrightarrow \text{Pyridinium } (N^{+}\text{—H})\ \ \text{Chlorochromate } (CrO_3Cl^{-})$$

PCC

When PCC is used as the oxidizing agent, an aldehyde is produced as the major product. Under these conditions, the aldehyde is not further oxidized to the carboxylic acid.

$$\text{Primary alcohol} \xrightarrow[CH_2Cl_2]{PCC} \text{Aldehyde}$$

Methylene chloride ($CH_2Cl_2$) is typically the solvent used when PCC is employed.

Secondary alcohols are oxidized only once to form a ketone, which is stable under oxidizing conditions. Therefore, a secondary alcohol can be oxidized either with chromic acid or with PCC.

$$\text{Secondary alcohol} \xrightarrow[H_2SO_4,\ H_2O]{Na_2Cr_2O_7} \text{Ketone}$$

$$\text{Secondary alcohol} \xrightarrow[CH_2Cl_2]{PCC} \text{Ketone}$$

Sodium dichromate is less expensive, but PCC is more gentle and often preferred if other sensitive functional groups are present in the compound.

# SKILLBUILDER

## 13.7 PREDICTING THE PRODUCTS OF AN OXIDATION REACTION

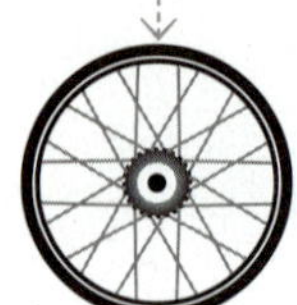

**LEARN the skill** Predict the major organic product of the following reaction:

$$\xrightarrow[H_3O^+,\ \text{acetone}]{CrO_3}\ ?$$

### SOLUTION

In order to predict the product, identify whether the alcohol is primary or secondary. In this case, the alcohol is primary, and therefore, we must decide whether the process will produce an aldehyde or a carboxylic acid:

**STEP 1** Identify whether the alcohol is primary or secondary.

**STEP 2** If the alcohol is primary, consider whether an aldehyde or a carboxylic acid is obtained.

OH $\xrightarrow{[O]}$ Aldehyde $\xrightarrow{[O]}$ Carboxylic acid

The product is determined by the choice of reagent. Chromium trioxide in acidic conditions forms chromic acid, which will oxidize the alcohol twice to give the carboxylic acid. In order to stop at the aldehyde, PCC would be used instead.

**STEP 3** Draw the product.

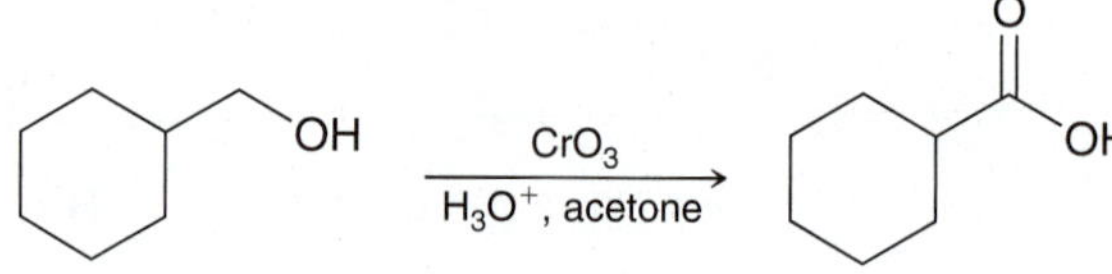

**PRACTICE the skill** **13.22** Predict the major organic product for each of the following reactions:

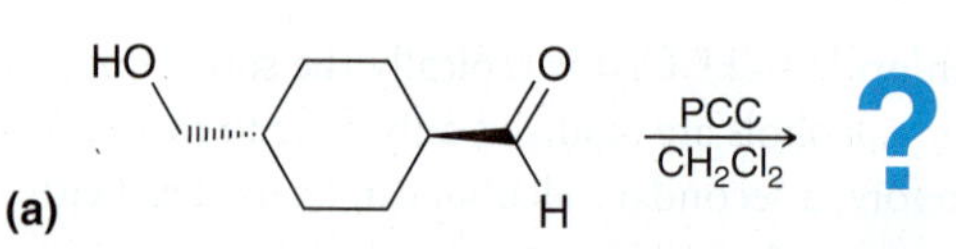


(a) PCC, $CH_2Cl_2$ ?

(b) $Na_2Cr_2O_7$, $H_2SO_4$, $H_2O$ ?

(c) xs $CrO_3$, $H_3O^+$, acetone ?

(d) PCC, $CH_2Cl_2$ ?

(e) PCC, $CH_2Cl_2$ ?

(f) $Na_2Cr_2O_7$, $H_2SO_4$, $H_2O$ ?

**APPLY the skill** **13.23** Propose an efficient synthesis for each of the following transformations:

(a) Br ⟶ O, H

(b) ⟶ O, H

(c) ⟶ O

(d) ⟶ O

need more **PRACTICE?** **Try Problems 13.35e–f, 13.37, 13.48**

## 13.11 Biological Redox Reactions

In this chapter, we have seen several reducing agents and several oxidizing reagents used by chemists in the laboratory. Nature employs its own reducing and oxidizing agents, although they are generally much more complex in structure and much more selective in their reactivity. One important biological reducing agent is NADH, which is a molecule comprised of many parts, shown in Figure 13.7.

**BY THE WAY**

NADH is much less reactive than $NaBH_4$ or LAH and requires a catalyst to do its job. Nature's catalysts are called enzymes (Section 25.8).

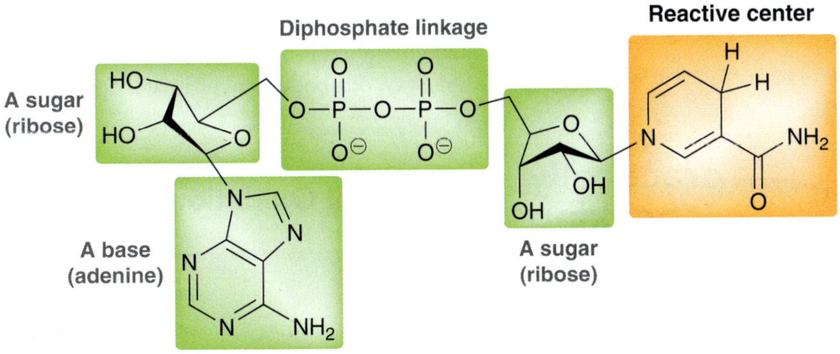

Diphosphate linkage
Reactive center
A sugar (ribose)
A base (adenine)
A sugar (ribose)

**FIGURE 13.7** The structure of NADH, a biological reducing agent.

The reactive center in NADH (highlighted in orange) functions as a hydride delivery agent (very much like $NaBH_4$ or LAH) and can reduce ketones or aldehydes to form alcohols. NADH acts as a reducing agent, and in the process, it is oxidized. The oxidized form is called $NAD^+$.

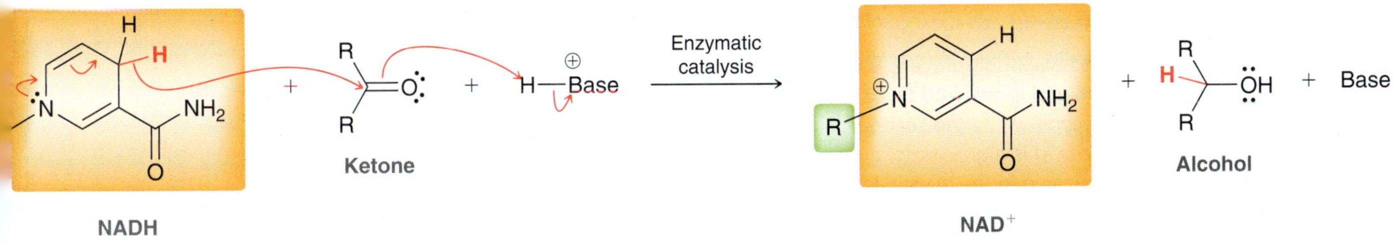

NADH + Ketone + H—Base → (Enzymatic catalysis) → $NAD^+$ + Alcohol + Base

The reverse process can also occur. That is, $NAD^+$ can act as an oxidizing agent by accepting a hydride from an alcohol, and in the process, $NAD^+$ is reduced to give NADH.

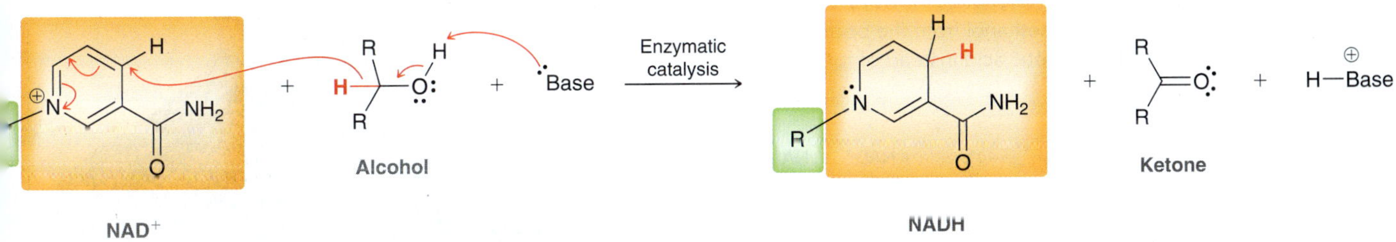

$NAD^+$ + Alcohol + Base → (Enzymatic catalysis) → NADH + Ketone + H—Base

**BY THE WAY**

To remember which is the reducing agent and which is the oxidizing agent, simply remember that NADH has an H at the end of its name, because it can function as a delivery agent of hydride.

$NAD^+$ and NADH are present in all living cells and function in a wide variety of redox reactions. NADH is a reducing agent, and $NAD^+$ is an oxidizing agent. Two important biological processes demonstrate the critical role of NADH and $NAD^+$ in biological systems (Figure 13.8). The *citric acid cycle* is part of the process by which food is metabolized. This process involves the conversion of $NAD^+$ to NADH. Another important biological process is the conversion of ADP (adenosine diphosphate) to ATP (adenosine triphosphate), called *ATP synthesis*, which involves the conversion of NADH to $NAD^+$. ATP is a molecule that is higher in energy than ADP, and the conversion of ADP to ATP is the way that our bodies store the energy that is obtained from metabolism of the food we eat. The energy stored in ATP can be released by conversion back into ADP, and that stored energy is used to power all of our life functions.

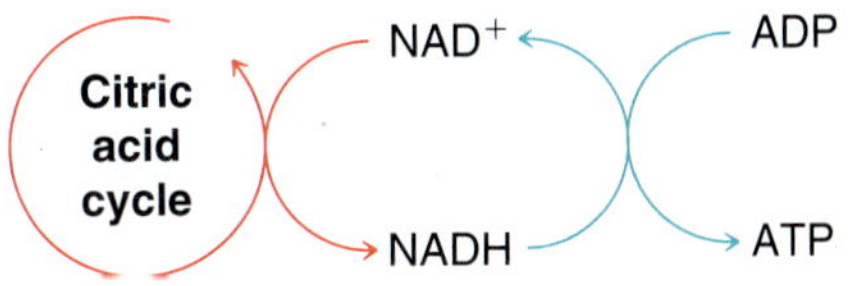

Citric acid cycle
$NAD^+$
NADH
ADP
ATP

**FIGURE 13.8** NADH and $NAD^+$ play an important role in the citric acid cycle as well as in ATP synthesis.

The citric acid cycle and ATP synthesis are linked into one story. That story can be summarized like this: Energy from the sun is absorbed by vegetation and is used to convert $CO_2$ molecules into larger organic compounds (a process called carbon dioxide fixing). These organic compounds have bonds (such as C—C and C—H) that are higher in energy than the C=O bonds in $CO_2$. Therefore, energy will be released if the organic compounds are converted back into $CO_2$ (this is precisely why the combustion of gasoline releases energy). Humans eat the vegetation (or we eat the animals that ate the vegetation), and our bodies utilize a series of chemical reactions that break down those large molecules into $CO_2$, thereby releasing energy. That energy is once again captured and stored in the form of high-energy ATP molecules, which are then used to provide energy for our biological processes. In this way, our bodies are ultimately powered by energy from the sun.

## practically speaking | Biological Oxidation of Methanol and Ethanol

Methanol is oxidized in our bodies by $NAD^+$.

$$\underset{\textbf{Methanol}}{CH_3OH} \xrightarrow[\text{Alcohol dehydrogenase}]{NAD^+ \rightarrow NADH} \underset{\textbf{Formaldehyde}}{HCHO} \xrightarrow{NAD^+ \rightarrow NADH} \underset{\textbf{Formic acid}}{HCOOH}$$

The enzyme that catalyzes this oxidation process is called alcohol dehydrogenase. Methanol is oxidized twice. The first oxidation produces formaldehyde, while the second oxidation produces formic acid. Formic acid is highly toxic, even in small quantities. A buildup of formic acid in the eyes leads to blindness, and a buildup of formic acid in other organs leads to organ failure and death.

A methanol overdose is typically treated by administering ethanol to the patient. Ethanol undergoes oxidation more rapidly than methanol, and the resulting decreased rate of methanol oxidation allows other metabolic pathways (such as glucuronidation) to remove the methanol from the body. The oxidation of ethanol produces acetic acid instead of formic acid, and acetic acid is not toxic.

$$\underset{\textbf{Ethanol}}{CH_3CH_2OH} \xrightarrow[\text{Alcohol dehydrogenase}]{NAD^+ \rightarrow NADH} \underset{\textbf{Acetaldehyde}}{CH_3CHO} \xrightarrow{NAD^+ \rightarrow NADH} \underset{\textbf{Acetic acid}}{CH_3COOH}$$

Ethanol is a primary alcohol and is oxidized twice. The first oxidation produces acetaldehyde, while the second oxidation produces acetic acid. Acetic acid can be used by the body for a variety of functions, but acetaldehyde is less useful. When a person drinks large quantities of ethanol (binge drinking), the concentration of acetaldehyde temporarily builds up. A high concentration of acetaldehyde causes nausea, vomiting, and other nasty symptoms.

We mentioned in the chapter opener that the effects of a hangover are caused by a variety of factors. The impact of some of those factors, such as dehydration, can be reduced by drinking a glass of water in between drinks. But other factors, such as the buildup of acetaldehyde, are the unavoidable consequences of binge drinking. The only way to avoid high concentrations of acetaldehyde is to drink small quantities of alcohol over a long period of time. Binge drinking will always produce the unpleasant effects of a hangover. There are many products on the market that claim to prevent hangovers, but there is little scientific evidence that any of these products are effective. The only way to prevent a hangover is to drink responsibly. That is, to drink small quantities of alcohol over a long period of time, together with plenty of water.

In addition to the unpleasant effects of a hangover, binge drinking can also potentially cause a host of more serious health issues, such as an irregular heart rhythm or acute pancreatitis (inflammation of the pancreas), both of which can be life threatening.

## 13.12 Oxidation of Phenol

In the previous two sections, we discussed reactions that achieve the oxidation of primary and secondary alcohols. In this section, we consider the oxidation of phenol. Based on our discussion thus far, we might expect that phenol would not readily undergo oxidation because it lacks a proton at the α position, much like a tertiary alcohol.

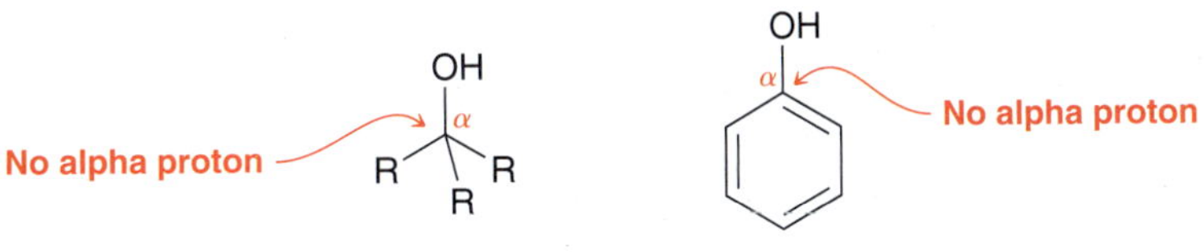

No alpha proton
OH
α
R R
R
3° alcohol

OH
α
No alpha proton
Phenol

Nevertheless, phenol is observed to undergo oxidation even more readily than primary and secondary alcohols. The product is benzoquinone.

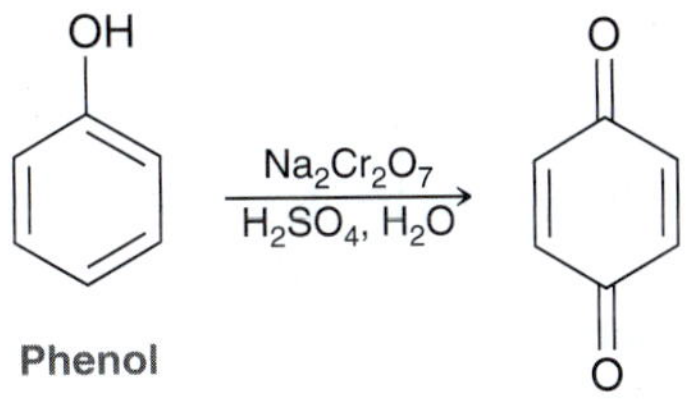

OH
Phenol
$Na_2Cr_2O_7$
$H_2SO_4$, $H_2O$
O
O
Benzoquinone

Quinones are important because they are readily converted to hydroquinones.

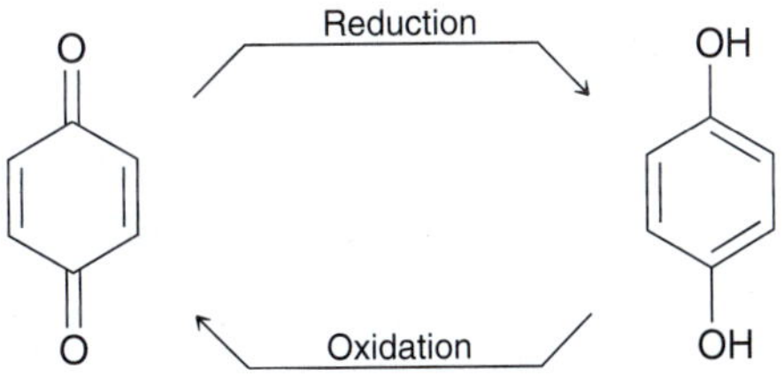

O
O
Benzoquinone
Reduction
Oxidation
OH
OH
Hydroquinone

The reversibility of this process is critical for **cellular respiration**, a process by which molecular oxygen is used to convert food into $CO_2$, water, and energy. Key players in this process are a group of quinones called ubiquinones.

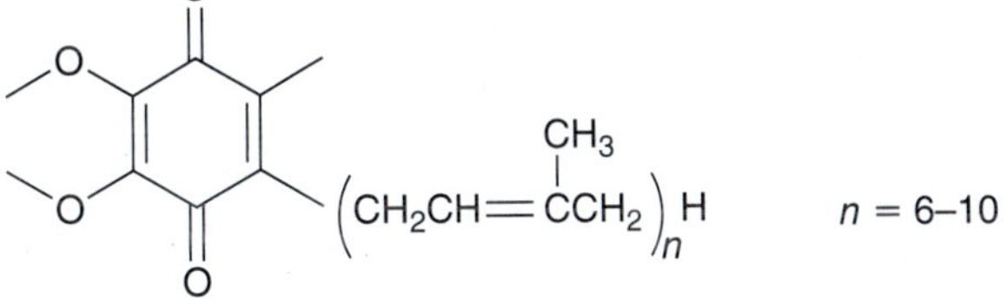

O
O
O
O
$CH_3$
$(CH_2CH{=}CCH_2)_nH$
$n = 6–10$
Ubiquinones

These compounds are called ubiquinones because they are ubiquitous in nature—that is, they are found in all cells. The redox properties of ubiquinones are utilized to convert molecular oxygen into water.

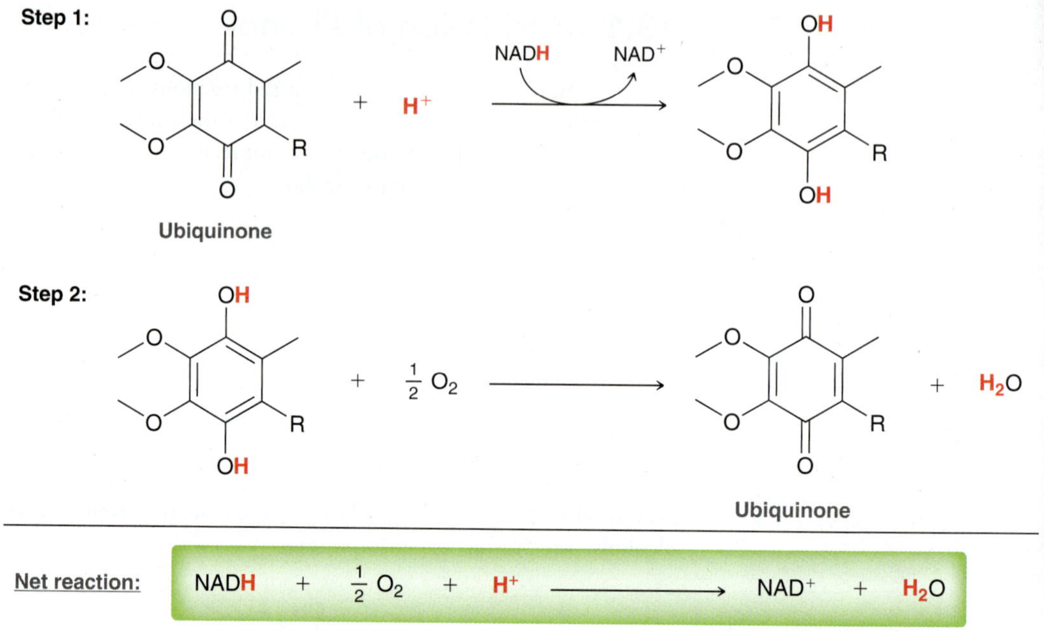

**Step 1:**

Ubiquinone + $H^+$ —(NADH → $NAD^+$)→ hydroquinone

**Step 2:**

hydroquinone + $\frac{1}{2}O_2$ → Ubiquinone + $H_2O$

**Net reaction:** $NADH + \frac{1}{2}O_2 + H^+ \longrightarrow NAD^+ + H_2O$

This process involves two steps. In the first step, a ubiquinone is reduced to a hydroquinone. In the second step the hydroquinone is oxidized to regenerate the ubiquinone. In this way, the ubiquinone is not consumed by the process. It is a catalyst for the conversion of molecular oxygen into water, a critical step in the breakdown of food molecules to release the energy stored in their chemical bonds. British biochemist Peter Mitchell was awarded the 1978 Nobel Prize in Chemistry for the discovery of the role that ubiquinones play in energy production (ATP synthesis).

## 13.13 Synthesis Strategies

Recall from Chapter 12 that there are two issues to consider when proposing a synthesis:

1. A change in the carbon skeleton
2. A change in the functional group

In this chapter, we have learned valuable skills for dealing with both of these issues. Let's focus on them one at a time.

### Functional Group Interconversion

In Chapter 12, we saw how to interconvert triple bonds, double bonds, and single bonds (Figure 13.9).

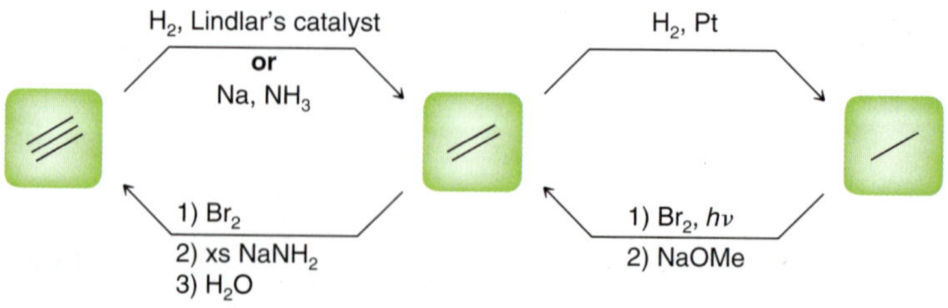

$H_2$, Lindlar's catalyst or Na, $NH_3$ (alkyne → alkene); $H_2$, Pt (alkene → alkane); 1) $Br_2$, $h\nu$ 2) NaOMe (alkane → alkene); 1) $Br_2$ 2) xs $NaNH_2$ 3) $H_2O$ (alkene → alkyne)

**FIGURE 13.9** A map showing the reactions that enable the interconversion between alkanes, alkenes, and alkynes.

In this chapter, we learned how to interconvert between ketones and secondary alcohols. This level of control will be very important later, as we will see many reactions involving compounds containing carbonyl groups.

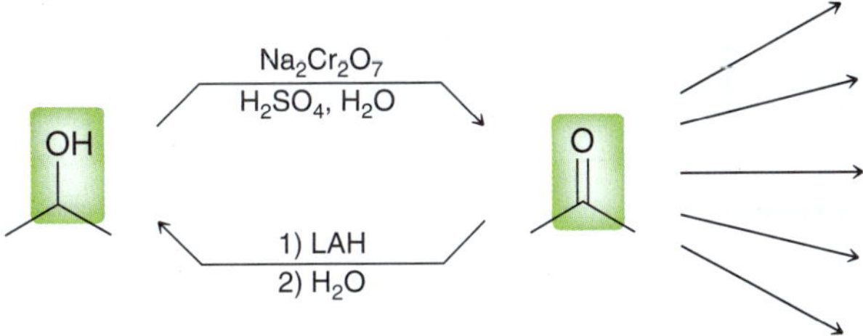


Primary alcohols and aldehydes can also be interconverted. PCC is used in place of chromic acid to convert an alcohol into an aldehyde.

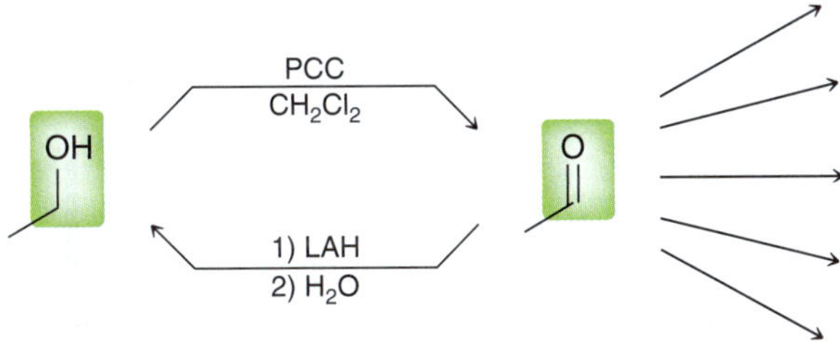


Let's now combine some important reactions we have seen in the previous few chapters into one map that will help us organize reactions that interconvert functional groups. We will focus on the six functional groups highlighted in Figure 13.10. These six functional groups have been separated into categories based on oxidation state. Conversion of a functional group from one category to another category is either a reduction or an oxidation. For example, converting an alkane into an alkyl halide is an oxidation reaction. In contrast, interconverting functional groups within a category is neither oxidation nor reduction. For example, conversion of an alcohol into an alkyl halide does not involve a change in oxidation state.

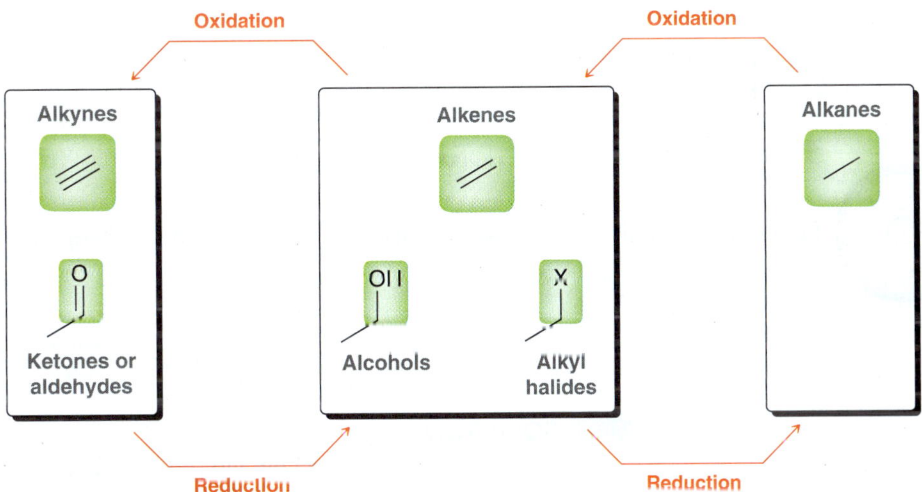


**FIGURE 13.10** A diagram showing six different functional groups and their relative oxidation states. The interconversion between alkenes, alcohols, and alkyl halides does not constitute oxidation or reduction.

We have seen many reagents that allow us to interconvert functional groups, both within a category and across categories (oxidation/reduction). Those reagents are shown in Figure 13.11. It is worth your time to study Figure 13.11 carefully, as it summarizes many of the key reactions that we have seen until now. Note that each functional group can be converted into any one of the other functional groups. Some conversions can be done in one step, such as the conversion of an alcohol into a ketone. Other conversions require multiple steps, such as the conversion of an alkane into a ketone. Having a clear mental picture of this chart will be extremely helpful when solving synthesis problems. Let's see an example.

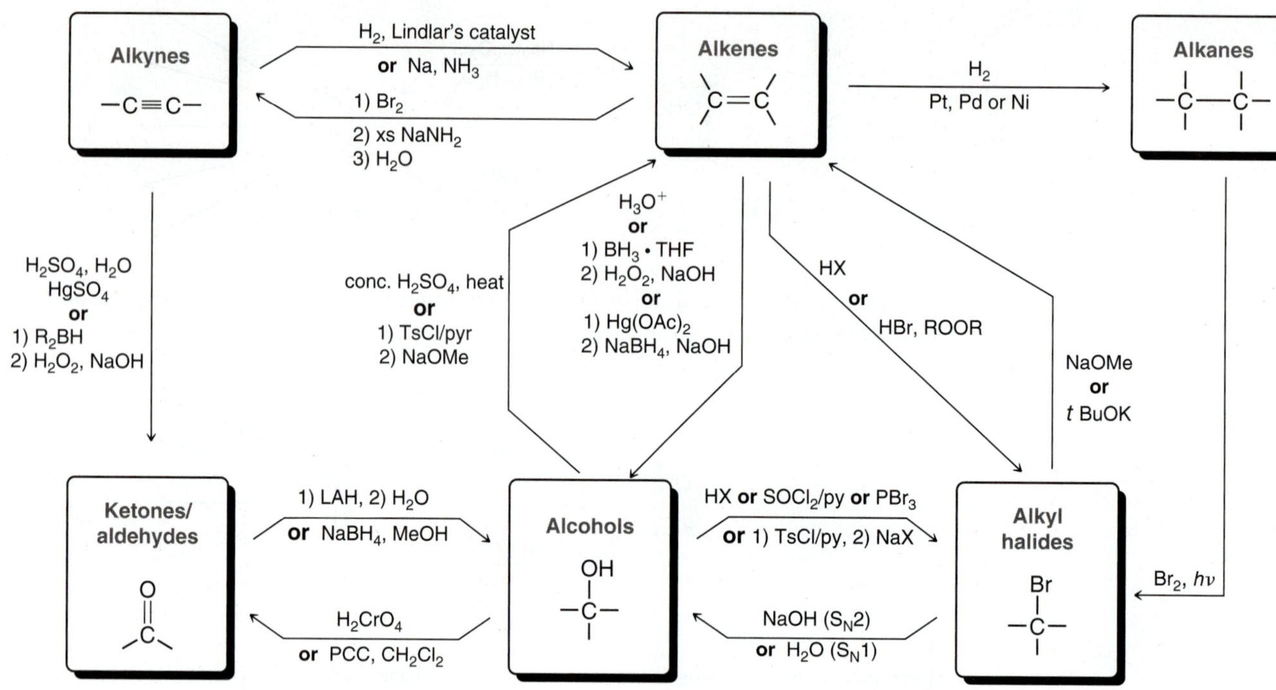


**FIGURE 13.11**
A map of reactions for interconversion between alkanes, alkenes, alkynes, alkyl halides, alcohols, and ketones/aldehydes.

# SKILLBUILDER

## 13.8 CONVERTING FUNCTIONAL GROUPS

**LEARN the skill** Propose an efficient synthesis for the following transformation:

**SOLUTION**

In this example, the reactant and product have the same carbon skeleton and differ only in the identity of the functional group. The reactant has a carbon-carbon double bond, while the product has a carbon-oxygen double bond (a carbonyl group). It is always preferable to achieve a transformation in the fewest number of steps possible, so we first consider whether there is a single step that will achieve this transformation. As seen in Figure 13.11, we have not yet discussed a one-step method for achieving the desired transformation. Therefore, we must consider whether the reactant can be converted into the product using a two-step synthesis. Figure 13.11 shows two possible routes.

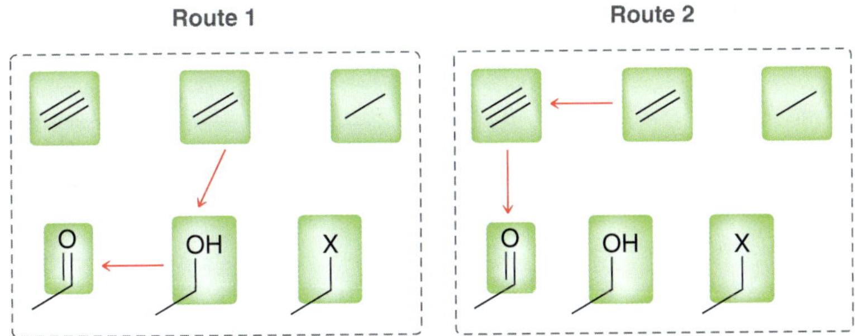


The first route passes through the alcohol, and the second route passes through an alkyne. Therefore, there are (at least) two perfectly acceptable answers to this problem.

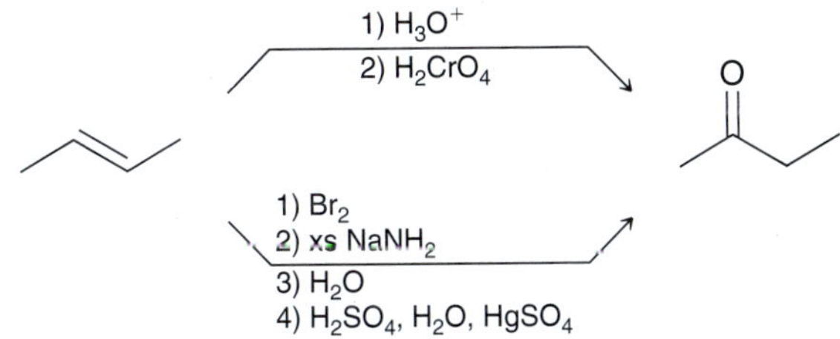

**PRACTICE the skill** **13.24** Propose an efficient synthesis for each of the following transformations:

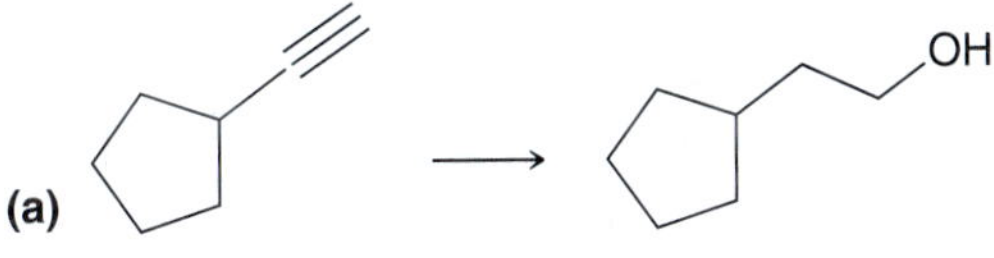


(b)  OH

(c) 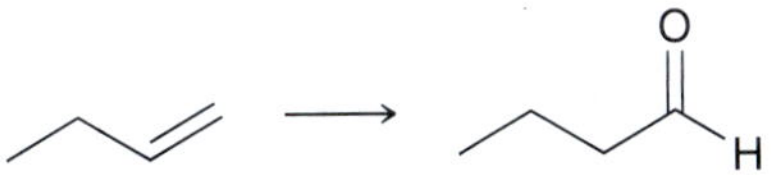 O H

(d)  OH

(e) 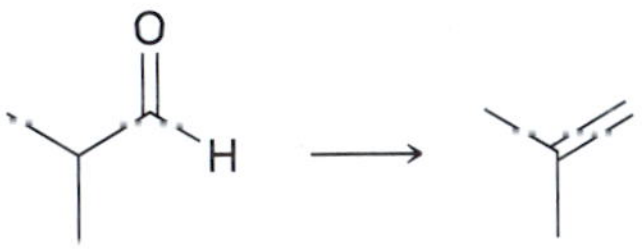 O H

(f) 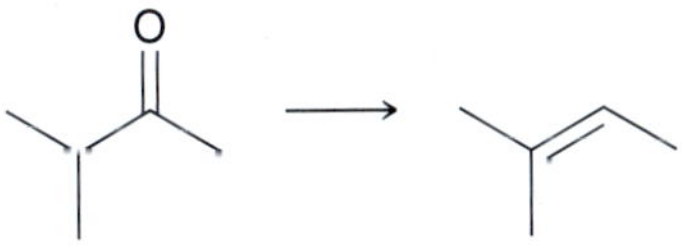 O

**APPLY the skill**

**13.25** Show at least two different methods for preparing 1-methylcyclohexene from 1-methylcyclohexanol.

**13.26** Using any reagents of your choosing, show how you would convert *tert*-butyl alcohol into 2-methyl-1-propanol.

need more **PRACTICE?** **Try Problems 13.35, 13.39, 13.48, 13.51**

## C—C Bond Formation

In this chapter, we learned a new way to form C—C bonds using the reaction between a Grignard reagent and a ketone or aldehyde.

1) $CH_3MgBr$
2) $H_2O$

1) $CH_3MgBr$
2) $H_2O$

We also saw that a Grignard reagent will attack an ester twice to form an alcohol. In this reaction, two new C—C bonds have been formed.

1) Excess $CH_3MgBr$
2) $H_2O$
+ MeOH

A Grignard reaction accomplishes more than just C—C bond formation; it also reduces the carbonyl group to an alcohol in the process. Now consider the following two steps: Suppose we perform a Grignard reaction with an aldehyde and then oxidize back up to a ketone.

Attack with Grignard → [O] →

The net result of these two steps is to convert an aldehyde into a ketone. This is very helpful, because we have not yet seen a direct way to convert aldehydes into ketones.

1) RMgBr
2) $H_2O$
3) $Na_2Cr_2O_7$, $H_2SO_4$, $H_2O$

CONCEPTUAL CHECKPOINT

**13.27** Identify the reagents that you would need to accomplish each of the following transformations:

(a)

(b)

## Functional Group Transformations and C—C Bond Formation

In this final section of the chapter, we will combine the skills from the previous two sections and work on synthesis problems that involve functional group interconversion and C—C bond formation. Recall from Chapter 12 that it is very helpful to approach a synthesis problem working forward as well as working backward (retrosynthetic analysis). We will use all of these skills in the following example.

# SKILLBUILDER

## 13.9 PROPOSING A SYNTHESIS

**LEARN** the skill

Propose an efficient synthesis for the following transformation:

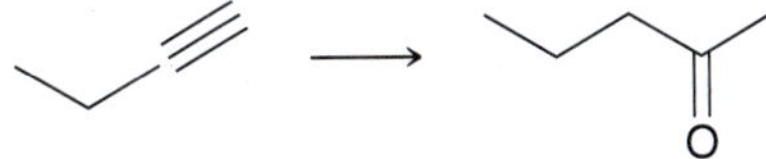

### SOLUTION

Always approach a synthesis problem by initially asking two questions.

**1.** Is there a change in the carbon skeleton? Yes, the carbon skeleton is increasing in size by one carbon atom.

**2.** Is there a change in the functional groups? Yes, the starting material has a triple bond, and the product has a carbonyl group.

To introduce one carbon atom, we will need to use a reaction that forms a C—C bond. So far, we have learned to form a C—C bond by alkylation of an alkyne (Section 10.10) and by using a Grignard reagent to attack a carbonyl group. Since our starting material is a terminal alkyne, let's first try alkylating the alkyne and see where that takes us.

Our first proposed strategy is to alkylate the alkyne and then hydrate the triple bond.

It is always critical to analyze any proposed strategy and make sure that the regiochemistry and stereochemistry of each step are correct. This particular strategy does not have any issues of stereochemistry (there are no chirality centers or stereoisomeric alkenes). However, the regiochemistry is problematic. There is a serious flaw with the second step (hydration of the alkyne to give a ketone). Specifically, it is not possible to control the regiochemical outcome of that reaction. That step will likely produce a mixture of two possible ketones (2-pentanone and 3-pentanone). This represents a critical flaw. We cannot propose a synthesis in which we have no control over the regiochemistry of one of the steps. All of our steps should lead to the desired product as the major product.

In the failed strategy above, we attempted to form the C—C bond via alkylation of an alkyne. So, let's now focus on our second method of forming a C—C bond, using a Grignard reagent to attack a carbonyl group. In other words, the last two steps of our synthesis could look like this:

$Na_2Cr_2O_7$, $H_2SO_4$, $H_2O$

H

1) $CH_3MgBr$
2) $H_2O$

O

OH

We are now using a retrosynthetic approach to this problem, after the forward approach proved to be unfruitful. Specifically, we have recognized that the final product can be made from an aldehyde via a two-step process. Now we just need to bridge the gap between the starting material and the aldehyde, which can be accomplished via hydroboration-oxidation.

1) $R_2BH$
2) $H_2O_2$, NaOH

$Na_2Cr_2O_7$, $H_2SO_4$, $H_2O$

1) $CH_3MgBr$
2) $H_2O$

This proposed strategy exhibits the correct regiochemistry for each step. Our answer is:

1) $R_2BH$
2) $H_2O_2$, NaOH
3) $CH_3MgBr$
4) $H_2O$
5) $Na_2Cr_2O_7$, $H_2SO_4$, $H_2O$

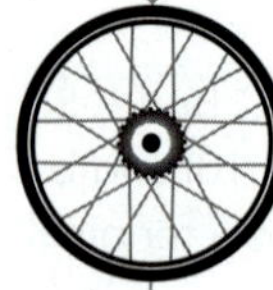

**PRACTICE the skill** **13.28** Propose an efficient synthesis for each of the following transformations:

(a) (b) (c) (d) (e) (f)

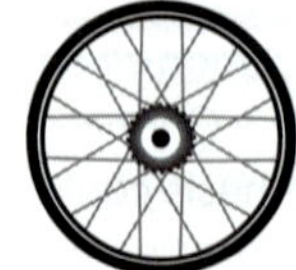

**APPLY the skill**

**13.29** Using any compounds that have no more than *two* carbon atoms, identify a method for preparing each of the following compounds:

(a) (b) (c) (d)

need more **PRACTICE?** **Try Problems 13.37, 13.38, 13.40, 13.45, 13.52, 13.59**

# REVIEW OF REACTIONS

## Preparation of Alkoxides

$$\text{ROH} \xrightarrow[\textit{or}\ \text{Na}]{\text{NaH}} \text{RO}^{\ominus}\ \text{Na}^{\oplus}$$

## Preparation of Alcohols via Reduction

$$\text{RCHO} \xrightarrow{\text{NaBH}_4\text{, MeOH} \ \textit{or}\ \text{H}_2\text{, Pt, Pd, or Ni} \ \textit{or}\ \text{1) LAH; 2) H}_2\text{O}} \text{RCH}_2\text{OH}$$

$$\text{RCOR} \xrightarrow{\text{NaBH}_4\text{, MeOH} \ \textit{or}\ \text{H}_2\text{, Pt, Pd, or Ni} \ \textit{or}\ \text{1) LAH; 2) H}_2\text{O}} \text{RCH(OH)R}$$

$$\text{RCOOH} \xrightarrow[\text{2) H}_2\text{O}]{\text{1) Excess LAH}} \text{RCH}_2\text{OH}$$

$$\text{RCOOMe} \xrightarrow[\text{2) H}_2\text{O}]{\text{1) Excess LAH}} \text{RCH}_2\text{OH} + \text{MeOH}$$

## Preparation of Alcohols via Grignard Reagents

$$(\text{CH}_3)_2\text{C=O} \xrightarrow[\text{2) H}_2\text{O}]{\text{1) RMgX}} (\text{CH}_3)_2\text{C(OH)R}$$

$$\text{RCOOMe} \xrightarrow[\text{2) H}_2\text{O}]{\text{1) Excess RMgX}} \text{RC(OH)R}_2 + \text{MeOH}$$

## Protection and Deprotection of Alcohols

$$\text{R—OH} \xrightarrow{\text{TMSCl, Et}_3\text{N}} \text{R—O—TMS}$$

$$\text{R—O—TMS} \xrightarrow{\text{TBAF}} \text{R—OH}$$

## $S_N1$ Reactions with Alcohols

$$\text{R}_3\text{C—OH} \xrightarrow{\text{HX}} \text{R}_3\text{C—X} + \text{H}_2\text{O}$$

## $S_N2$ Reactions with Alcohols

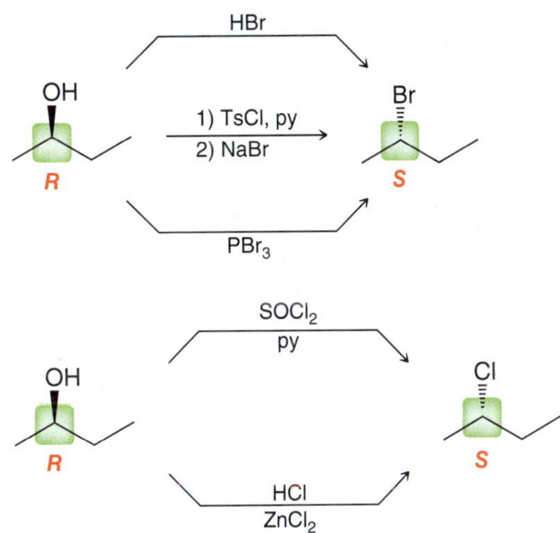


## E1 and E2 Reactions with Alcohols

$$(\text{CH}_3)_3\text{C—OH} \xrightarrow[\text{Heat}]{\text{conc. H}_2\text{SO}_4} (\text{CH}_3)_2\text{C=CH}_2 + \text{H}_2\text{O}$$

$$(\text{CH}_3)_3\text{C—OH} \xrightarrow[\text{py}]{\text{TsCl}} (\text{CH}_3)_3\text{C—OTs} \xrightarrow{\text{NaOEt}} (\text{CH}_3)_2\text{C=CH}_2$$

## Oxidation of Alcohols and Phenols

With Chromic Acid

$$\underset{\text{Secondary alcohol}}{\text{RCH(OH)R}} \xrightarrow[\text{H}_2\text{SO}_4\text{, H}_2\text{O}]{\text{Na}_2\text{Cr}_2\text{O}_7} \underset{\text{Ketone}}{\text{RCOR}}$$

$$\underset{\text{Primary alcohol}}{\text{RCH}_2\text{OH}} \xrightarrow[\text{H}_2\text{SO}_4\text{, H}_2\text{O}]{\text{Na}_2\text{Cr}_2\text{O}_7} \underset{\text{Carboxylic acid}}{\text{RCOOH}}$$

Phenol $\xrightarrow[H_2SO_4,\ H_2O]{Na_2Cr_2O_7}$ Benzoquinone

With PCC

Primary alcohol $\xrightarrow[CH_2Cl_2]{PCC}$ Aldehyde

## REVIEW OF CONCEPTS AND VOCABULARY

### SECTION 13.1

- Compounds that have a **hydroxyl group** (OH) are called **alcohols**.
- When naming an alcohol, the parent is the longest chain containing the hydroxyl group.
- All alcohols possess a *hydrophilic* region and a *hydrophobic* region. Small alcohols (methanol, ethanol, propanol) are **miscible** with water. A substance is said to be **soluble** in water when only a certain volume of the substance will dissolve in a specified amount of water at room temperature. Butanol is soluble in water.

### SECTION 13.2

- The conjugate base of an alcohol is called an **alkoxide** ion.
- The $pK_a$ for most alcohols falls in the range of 15–18.
- Alcohols are commonly deprotonated with either sodium hydride (NaH) or an alkali metal (Na, Li, or K).
- Several factors determine the relative acidity of alcohols, including resonance, induction, and solvating effects.
- The conjugate base of phenol is called a **phenolate**, or **phenoxide** ion.

### SECTION 13.3

- When preparing an alcohol via a substitution reaction, primary substrates will require $S_N2$ conditions, while tertiary substrates will require $S_N1$ conditions.
- Addition reactions that will produce alcohols include acid-catalyzed hydration, oxymercuration-demercuration, and hydroboration-oxidation.

### SECTION 13.4

- Alcohols can be formed by treating a **carbonyl group** (C=O bond) with a **reducing agent**. The resulting reaction involves a decrease in **oxidation state** and is called **reduction**.
- LAH is more reactive than $NaBH_4$. In fact, LAH will reduce carboxylic acids and esters, while $NaBH_4$ will not.

### SECTION 13.5

- **Diols** are compounds with two hydroxyl groups.
- Diols can be prepared from diketones via reduction using a reducing agent.
- Diols can also be made via *syn* dihydroxylation or *anti* dihydroxylation of an alkene.

### SECTION 13.6

- **Grignard reagents** are carbon nucleophiles that are capable of attacking a wide range of electrophiles, including the carbonyl group of ketones or aldehydes, to produce an alcohol.
- Grignard reagents also react with esters to produce alcohols with introduction of two R groups.

### SECTION 13.7

- **Protecting groups**, such as the trimethylsilyl group, can be used to circumvent the problem of Grignard incompatibility and can be easily removed after the desired Grignard reaction has been performed.

### SECTION 13.8

- Phenol, also called hydroxybenzene, is used as a precursor in the synthesis of a wide variety of pharmaceuticals and other commercially useful compounds.

### SECTION 13.9

- Tertiary alcohols will undergo an $S_N1$ reaction when treated with a hydrogen halide.
- Primary and secondary alcohols will undergo an $S_N2$ process when treated with HX, $SOCl_2$, or $PBr_3$ or when the hydroxyl group is converted into a tosylate group followed by nucleophilic attack.
- Tertiary alcohols undergo E1 elimination when treated with sulfuric acid.
- For an E2 process, the hydroxyl group must first be converted into a tosylate or an alkyl halide.

### SECTION 13.10

- Primary alcohols can undergo **oxidation** twice to give a carboxylic acid.
- Secondary alcohols are oxidized only once to give a ketone.
- Tertiary alcohols do not undergo oxidation.
- The most common oxidizing reagent is chromic acid ($H_2CrO_4$), which can be formed either from chromium trioxide ($CrO_3$) or from sodium dichromate ($Na_2Cr_2O_7$) in aqueous acidic solution.
- PCC is used to convert a primary alcohol into an aldehyde.

### SECTION 13.11

- NADH is a biological reducing agent that functions as a hydride delivery agent (very much like $NaBH_4$ or LAH), while $NAD^+$ is an oxidizing agent.

- NADH and $NAD^+$ play critical roles in biological systems. Examples include the **citric acid cycle** and **ATP** synthesis.

**SECTION 13.12**

- Phenols undergo oxidation to quinones. Quinones are biologically important because their redox properties play a significant role in **cellular respiration**.

**SECTION 13.13**

- There are two key issues to consider when proposing a synthesis:
  1. A change in the carbon skeleton
  2. A change in the functional group

# SKILLBUILDER REVIEW

## 13.1 NAMING AN ALCOHOL

**STEP 1** Choose the longest chain containing the OH group and number the chain starting from the end closest to the OH group.

3-Nonanol

**STEPS 2 AND 3** Identify the substituents and assign locants.

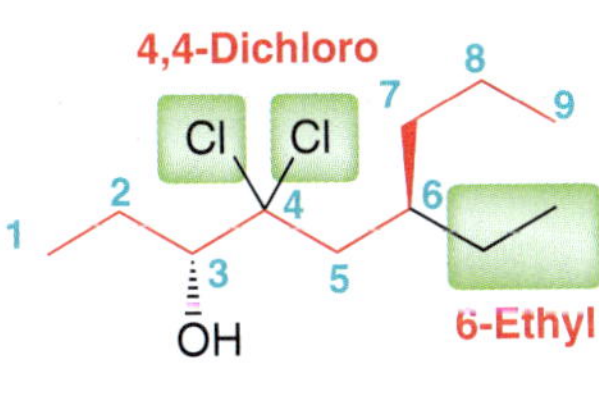


**STEP 4** Assemble the substituents alphabetically.

4,4-Dichloro-6-ethyl-3-nonanol

**STEP 5** Assign the configuration of any chirality center.

(3*R*,6*R*)-4,4-dichloro-6-ethyl-3-nonanol

Try Problems 13.1, 13.2, 13.30, 13.31a–d,f, 13.32

## 13.2 COMPARING THE ACIDITY OF ALCOHOLS

Look for **resonance effects**; for example:

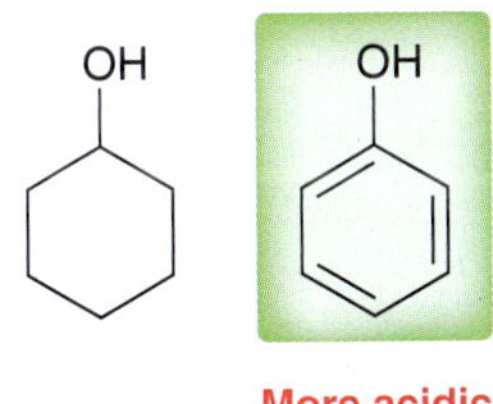


More acidic

Look for **inductive effects**; for example:

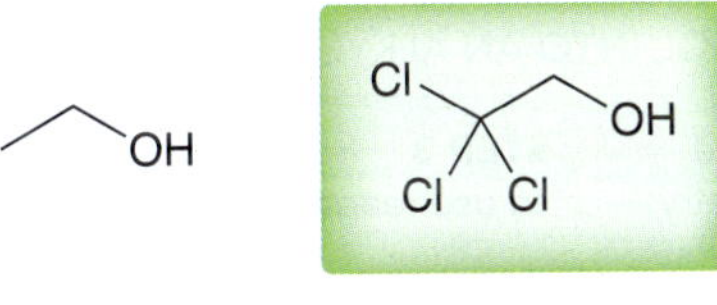


More acidic

Look for **solvating effects**; for example:

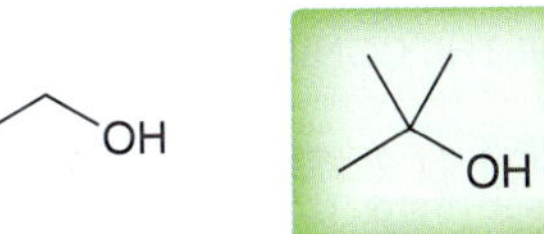


Less acidic

Try Problems 13.5, 13.6, 13.33, 13.34

## 13.3 IDENTIFYING OXIDATION AND REDUCTION REACTIONS

**EXAMPLE** Determine whether starting material has been oxidized, reduced, or neither

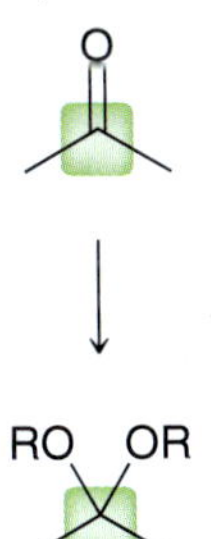


**STEP 1** Determine oxidation state of starting material. Break all bonds heterolytically, except for C—C bonds.

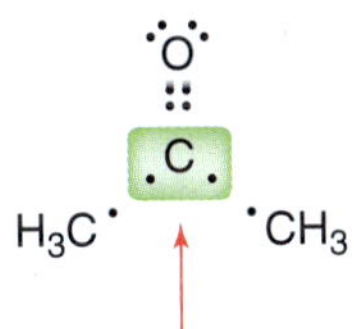


Two electrons, but carbon should have four. This carbon is missing two electrons

Oxidation state = +2

**STEP 2** Determine oxidation state of product. Break all bonds heterolytically, except for C—C bonds.

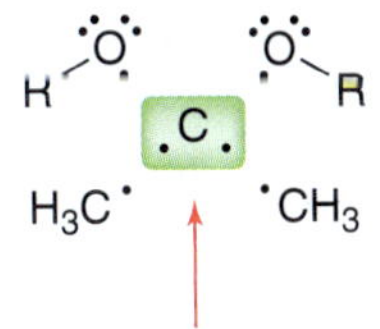


Two electrons, but carbon should have four. This carbon is missing two electrons

Oxidation state = +2

**STEP 3** Determine if there has been a change in oxidation state.

Increase = oxidation
Decrease = reduction
No change = neither

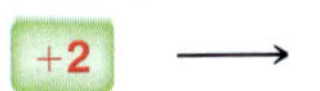


This example is neither an oxidation nor a reduction

Try Problems 13.9–13.11, 13.62

### 13.4 DRAWING A MECHANISM AND PREDICTING THE PRODUCTS OF HYDRIDE REDUCTIONS

**STEP 1** Draw the complete structure of LAH, and draw two curved arrows that show the delivery of hydride to the carbonyl group.

**STEP 2** Draw the alkoxide intermediate.

**STEP 3** Draw two curved arrows showing the alkoxide intermediate being protonated by the proton source.

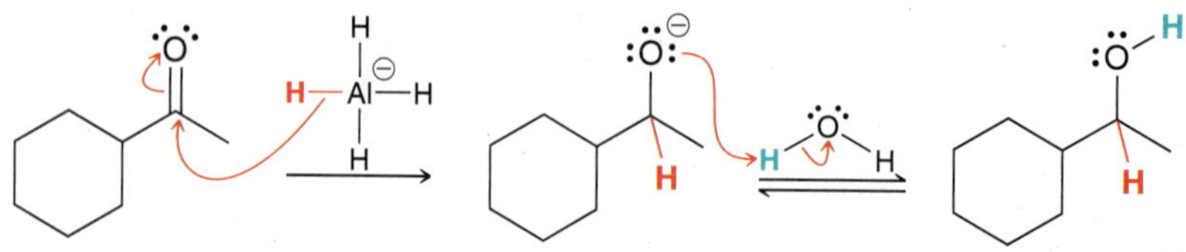

**Try Problems** 13.12, 13.13, 13.46, 13.47c, 13.48e,f, 13.60

### 13.5 PREPARING AN ALCOHOL VIA A GRIGNARD REACTION

**STEP 1** Identify the $\alpha$ position.

**STEP 2** Identify the three groups connected to the $\alpha$ position.

**STEP 3** Show how each group could have been installed via a Grignard reaction.

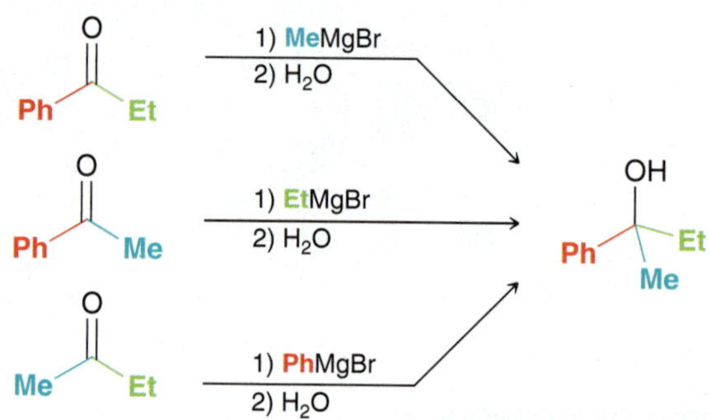


**Try Problems** 13.14–13.17, 13.38, 13.40b, 13.52b–d,j,l–r, 13.58

### 13.6 PROPOSING REAGENTS FOR THE CONVERSION OF AN ALCOHOL INTO AN ALKYL HALIDE

**EXAMPLE** Identify the necessary reagents.

**STEP 1** Analyze the substrate:

Primary = $S_N2$
Tertiary = $S_N1$

Substrate is secondary

**STEP 2** Analyze the stereochemistry: inversion = $S_N2$.

**STEP 3** Reaction must occur via $S_N2$ so use reagents that favor $S_N2$.

HCl, $ZnCl_2$

1) TsCl, py
2) NaCl

$SOCl_2$, py

**Try Problems** 13.19, 13.20, 13.35a–c, 13.44f, 13.52r

## 13.7 PREDICTING THE PRODUCTS OF AN OXIDATION REACTION

**EXAMPLE**

OH

$CrO_3$, $H_3O^+$, acetone → ?

**STEP 1** Identify whether the alcohol is primary or secondary.

H H OH

Primary

**STEP 2** A primary alcohol can be oxidized either to an aldehyde or to a carboxylic acid, depending on the reagents.

O H — Aldehyde

O OH — Carboxylic acid

**STEP 3** Analyze reagents. PCC is used to form the aldehyde. Chromic acid is used to form the carboxylic acid.

O OH

Try Problems 13.22, 13.23, 13.35e–f, 13.37, 13.48

## 13.8 CONVERTING FUNCTIONAL GROUPS

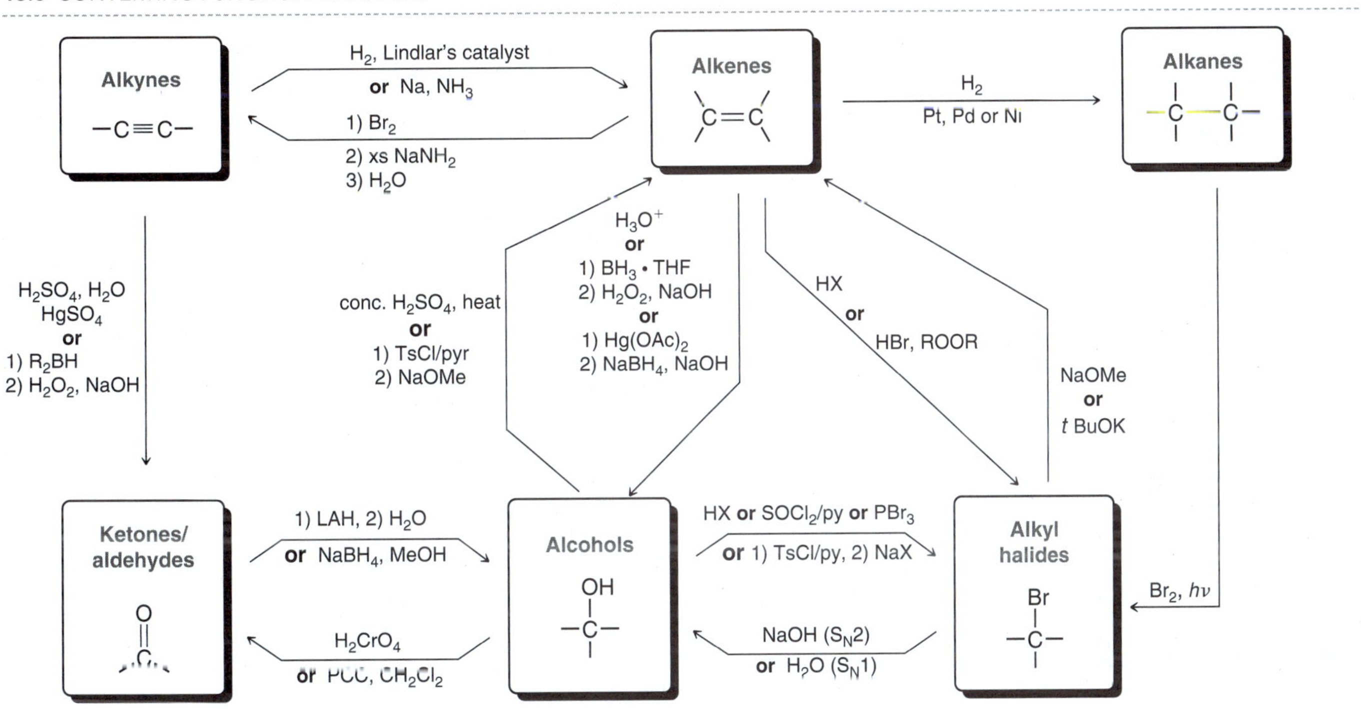


Try Problems 13.24–13.26, 13.39, 13.48, 13.51

## 13.9 PROPOSING A SYNTHESIS

**STEP 1** Is there a change in the carbon skeleton?

Keep track of all the C—C bond-forming reactions that you have learned until now.

**STEP 2** Is there a change in the functional groups?

The chart from the previous SkillBuilder summarizes many of the important functional group interconversions that we have seen.

**STEP 3** After proposing a synthesis, analyze your answer with the following two questions:

- Is the regiochemical outcome of each step correct?
- Is the stereochemical outcome of each step correct?

**MORE TIPS** Remember that the desired product should be the major product of your proposed synthesis.

Always think backwards (retrosynthetic analysis) as well as forwards and then try to bridge the gap.

Most synthesis problems will have multiple correct answers. Do not feel that you have to find the "one" correct answer.

Try Problems 13.28, 13.29, 13.37, 13.38, 13.40, 13.45, 13.52, 13.59

## PRACTICE PROBLEMS

**Note: Most of the Problems are available within WileyPLUS, an online teaching and learning solution.**

**13.30** Assign an IUPAC name for each of the following compounds:

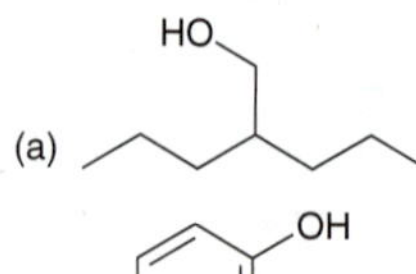


(b) (c) (d)

**13.31** Draw the structure of each compound:

(a) *cis*-1,2-Cyclohexanediol (b) Isobutanol

(c) 2,4,6-Trinitrophenol (d) (*R*)-2,2-Dimethyl-3-heptanol

(e) Ethylene glycol (f) (*S*)-2-Methyl-1-butanol

**13.32** Draw and name all constitutionally isomeric alcohols with molecular formula $C_4H_{10}O$.

**13.33** Rank each set of alcohols below in order of increasing acidity.

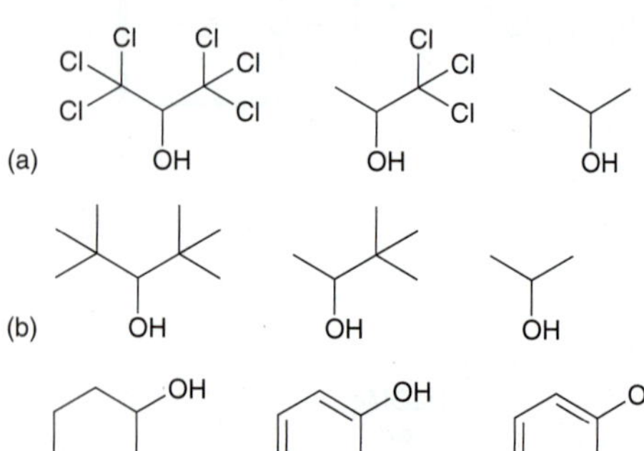


**13.34** Draw resonance structures for each of the following anions:

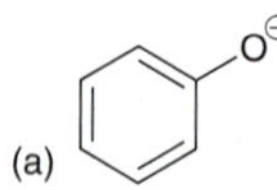


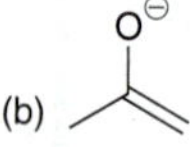


(c)

**13.35** Predict the major product of the reaction between 1-butanol and:

(a) $PBr_3$ (b) $SOCl_2$, py

(c) HCl, $ZnCl_2$ (d) Conc. $H_2SO_4$, heat

(e) PCC, $CH_2Cl_2$ (f) $Na_2Cr_2O_7$, $H_2SO_4$, $H_2O$

(g) Li (h) NaH

(i) TMSCl, $Et_3N$ (j) TsCl, pyridine

(k) Na (l) Potassium *tert*-butoxide

**13.36** Acid-catalyzed hydration of 3,3-dimethyl-1-butene produces 2,3-dimethyl-2-butanol. Show a mechanism for this reaction.

**13.37** Starting with 1-butanol, show the reagents you would use to prepare each of the following compounds:

(a) (b) (c) (d) (e)

**13.38** Using a Grignard reaction, show how you could prepare each of the following alcohols:

(a) (b) (c) (d)

**13.39** Each of the following alcohols can be prepared via reduction of a ketone or aldehyde. In each case, identify the aldehyde or ketone that would be required.

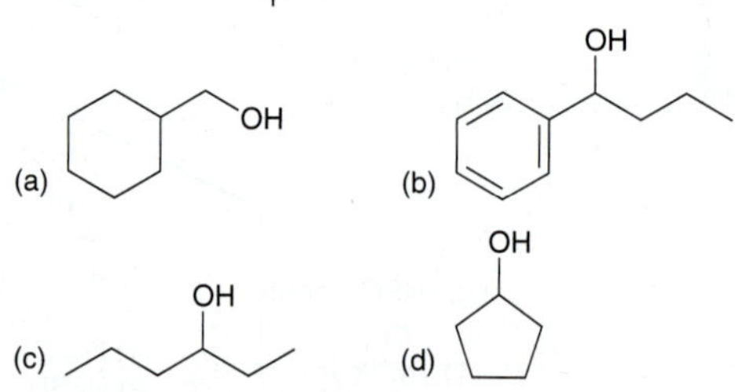


**13.40** What reagents would you use to perform each of the following transformations?

(a) (b)

**13.41** Propose a mechanism for the following reaction:

Br ~~~ OH —NaH→ (cyclic ether)

**13.42** Acid-catalyzed hydration of 1-methylcyclohexene yields two alcohols. The major product does not undergo oxidation, while the minor product will undergo oxidation. Explain.

**13.43** Consider the following sequence of reactions and identify the structures of compounds **A**, **B**, and **C**:

Compound **A** ($C_6H_{11}Br$) —Mg→ Compound **B** —1) acetone; 2) $H_2O$→ Compound **C** —conc. $H_2SO_4$, Heat→ (isopropylidenecyclohexane)

**13.44** Consider the following sequence of reactions and identify the reagents a–h:

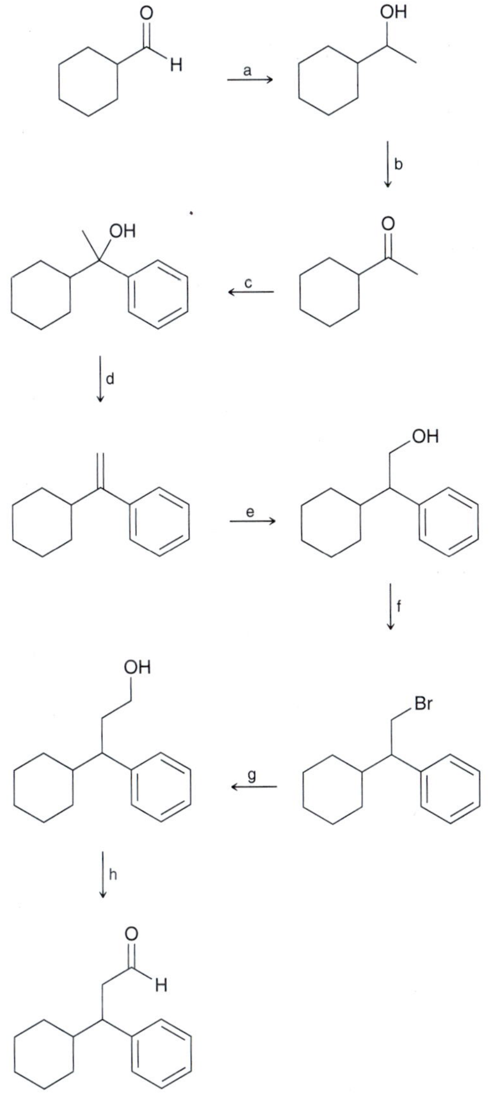

a, b, c, d, e, f, g, h

OH, O, H, OH, Br

**13.45** Using 2-propanol as your only source of carbon, show how you would prepare 2-methyl-2-pentanol.

**13.46** Predict the product and draw a mechanism for each of the following reactions:

(a) 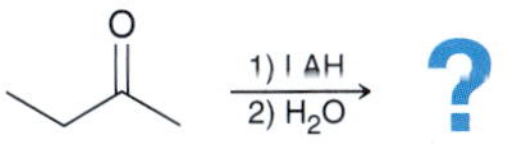 1) LAH 2) $H_2O$ ?

(b) 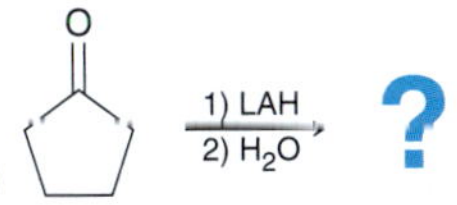 1) LAH 2) $H_2O$ ?

(c) 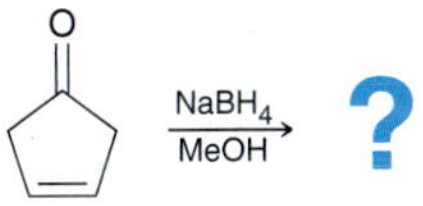 $NaBH_4$ MeOH ?

**13.47** Draw a mechanism for each of the following reactions:

(a)  OH → $SOCl_2$, py → Cl

(b) 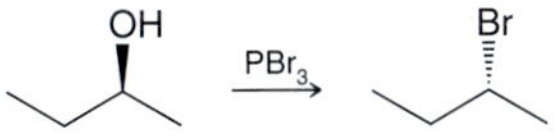 OH → $PBr_3$ → Br

(c) 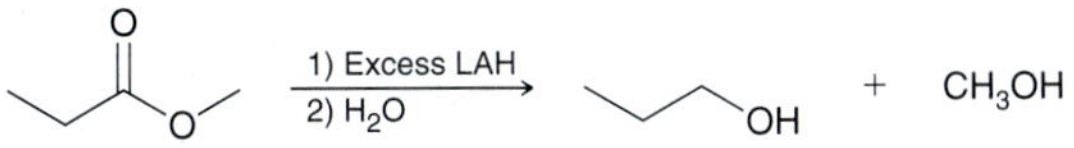 1) Excess LAH 2) $H_2O$ → OH + $CH_3OH$

**13.48** Identify the reagents you would use to accomplish each of the following transformations:

(a)  OH → O

(b) 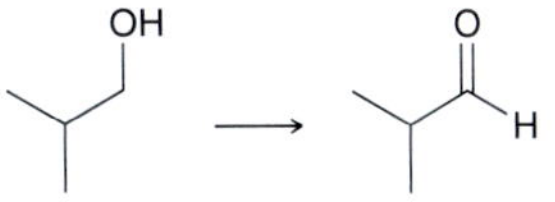 OH → O, H

(c) 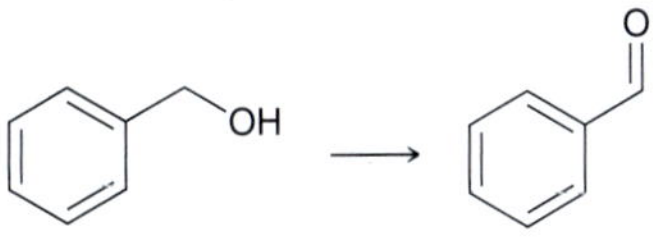 OH → O

(d) 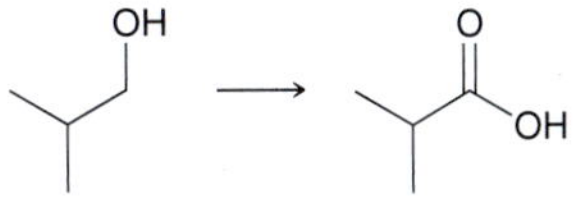 OH → O, OH

(e) 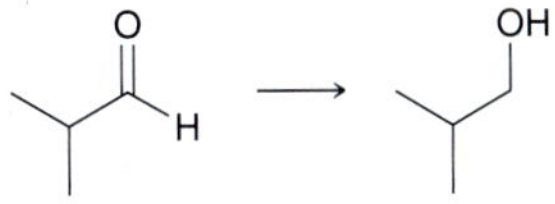 O, H → OH

(f) 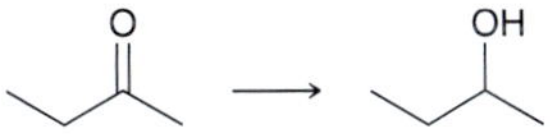 O → OH

**13.49** Predict the products for each of the following:

(a) 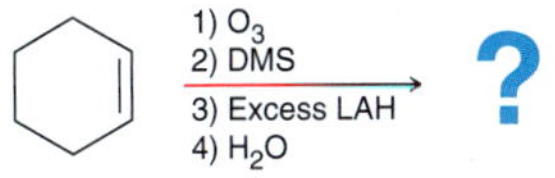 1) $O_3$ 2) DMS 3) Excess LAH 4) $H_2O$ ?

(b) 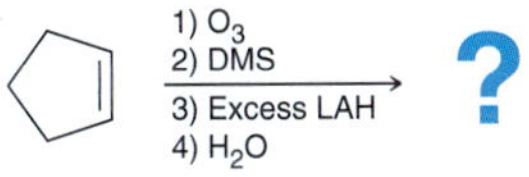 1) $O_3$ 2) DMS 3) Excess LAH 4) $H_2O$ ?

(c) 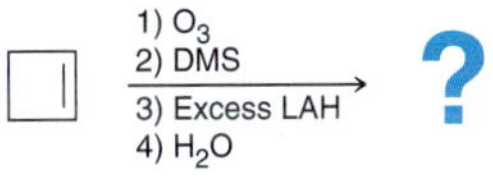 1) $O_3$ 2) DMS 3) Excess LAH 4) $H_2O$ ?

(d) 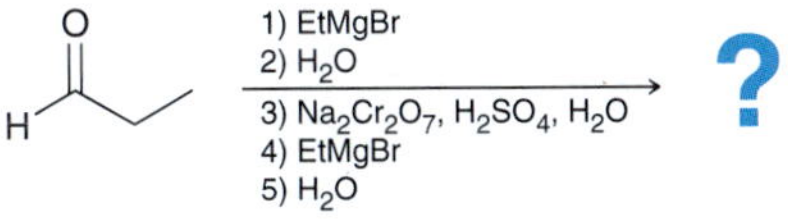 1) EtMgBr 2) $H_2O$ 3) $Na_2Cr_2O_7$, $H_2SO_4$, $H_2O$ 4) EtMgBr 5) $H_2O$ ?

(e) 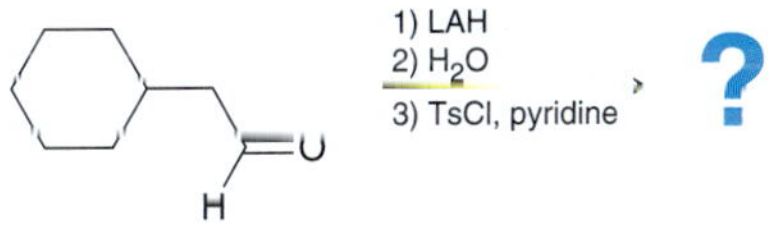 1) LAH 2) $H_2O$ 3) TsCl, pyridine ?

(f) 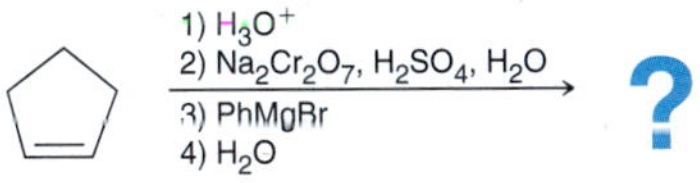 1) $H_3O^+$ 2) $Na_2Cr_2O_7$, $H_2SO_4$, $H_2O$ 3) PhMgBr 4) $H_2O$ ?

**13.50** Propose a plausible mechanism for each of the following transformations:

(a) 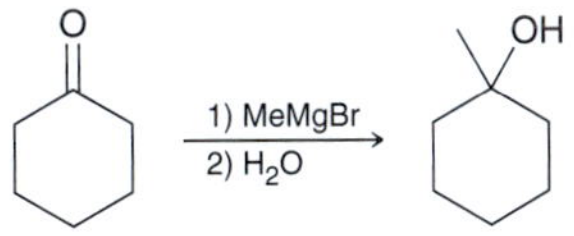 1) MeMgBr 2) $H_2O$ → OH

(b) 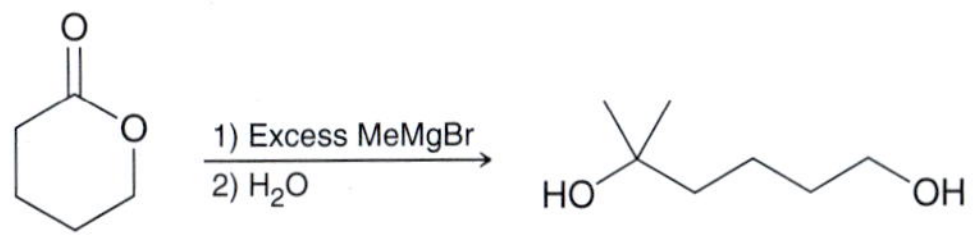 1) Excess MeMgBr 2) $H_2O$ → HO, OH

# INTEGRATED PROBLEMS

**13.51** Identify the reagents that would be necessary to accomplish each of the transformations shown here:

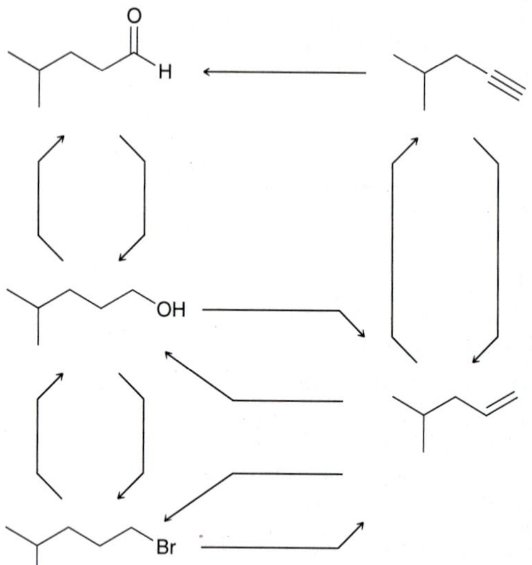

**13.52** Propose an efficient synthesis for each of the following transformations:

(a)

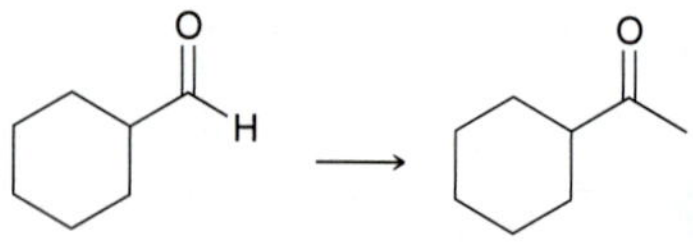

(b)

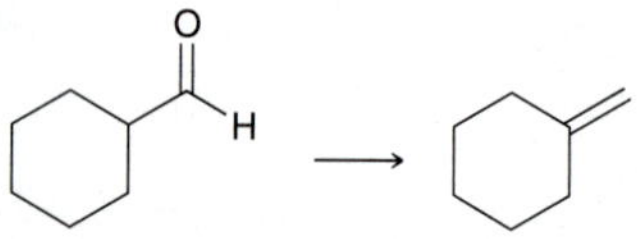

(c)

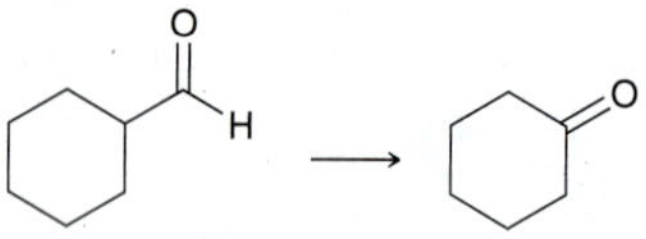

(d)

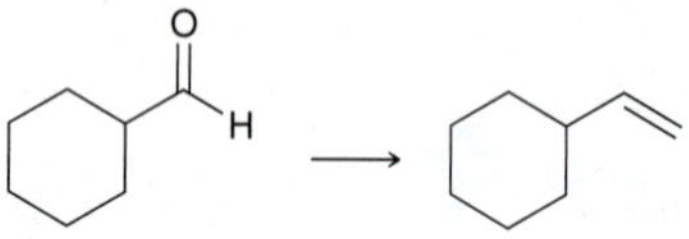

(e)

(f)

(g)

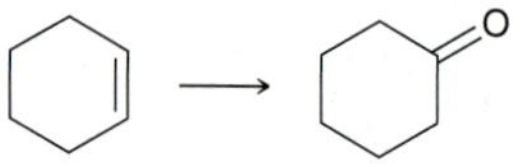

(h)

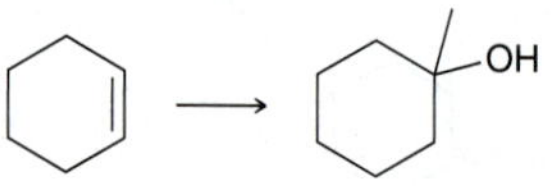

(i)

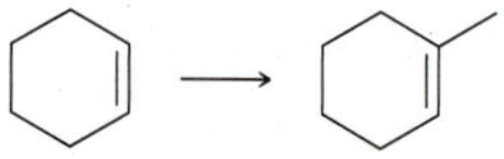

(j)

(k)

(l)

(m)

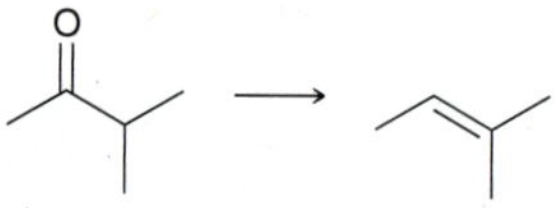

(n)

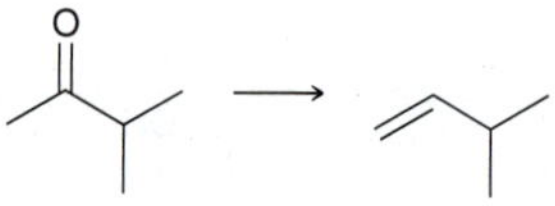

(o)

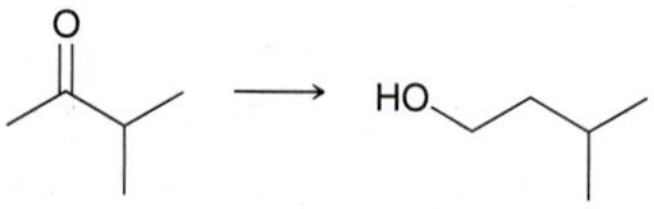

(p)

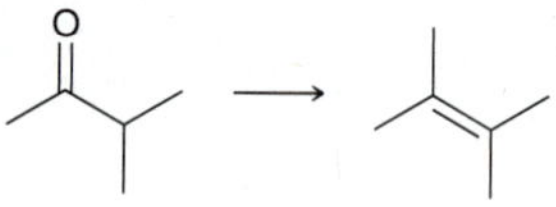

(q)

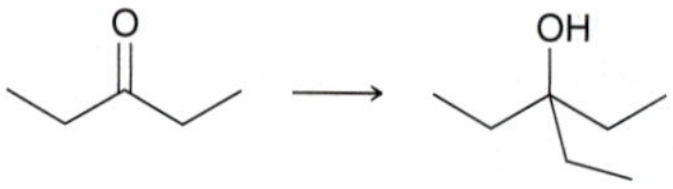

(r)

(s)

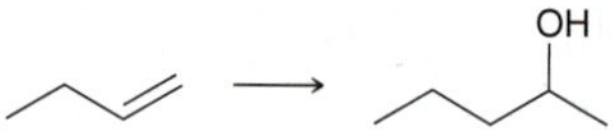

**Problems 13.53–13.56 are intended for students who have already covered spectroscopy (Chapters 15 and 16).**

**13.53** Propose a structure for a compound with molecular formula $C_{10}H_{14}O$ that exhibits the following $^1H$ NMR spectrum:

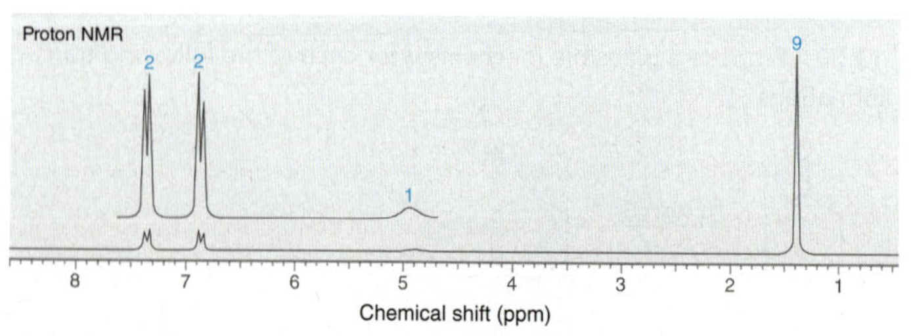


**13.54** Propose a structure for a compound with molecular formula $C_3H_8O$ that exhibits the following $^1H$ NMR and $^{13}C$ NMR spectra:

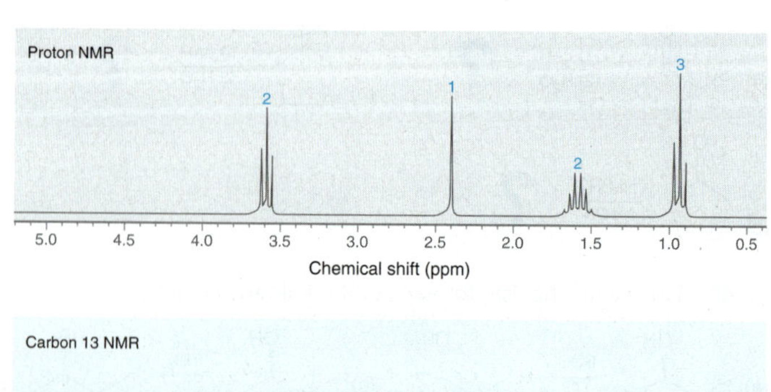


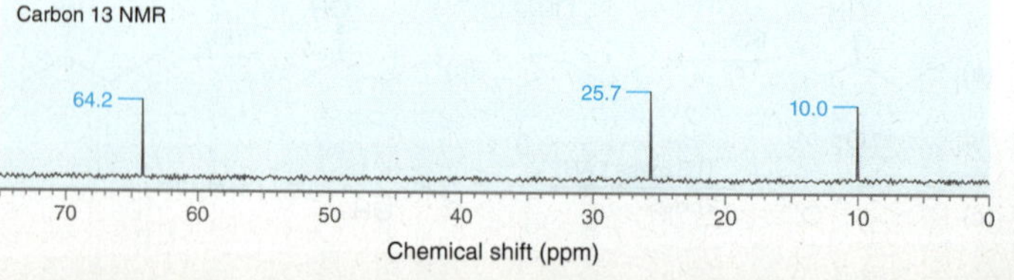

**13.55** Propose two possible structures for a compound with molecular formula $C_5H_{12}O$ that exhibits the following $^{13}C$ NMR and IR spectra:

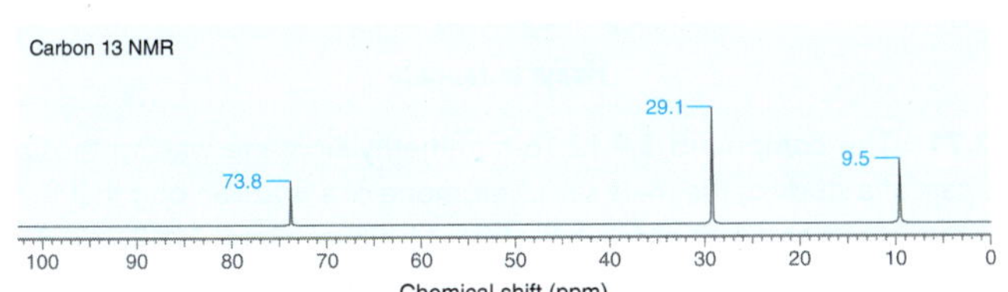


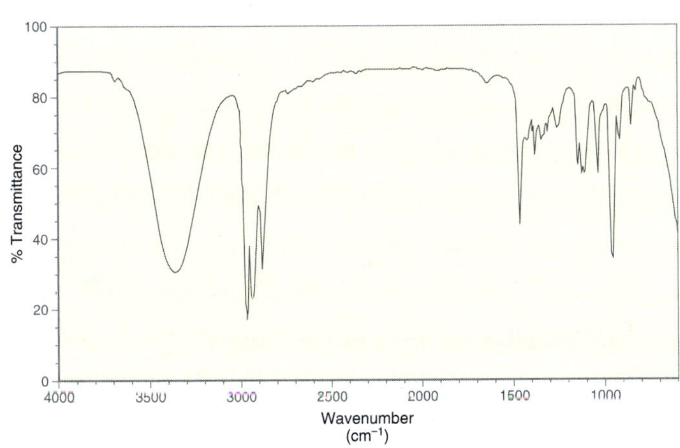


**13.56** Propose a structure for a compound with molecular formula $C_8H_{10}O$ that exhibits the following $^1H$ NMR spectrum:

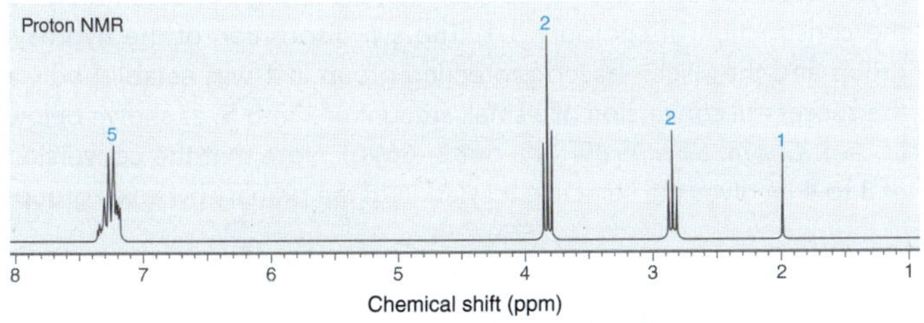


**13.57** Propose a mechanism for the following transformation:

1) Excess LAH
2) $H_2O$

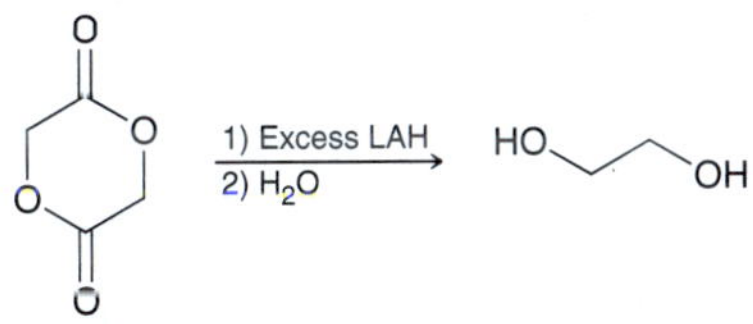

**13.58** Propose a mechanism for the following transformation:

1) Excess MeMgBr
2) $H_2O$

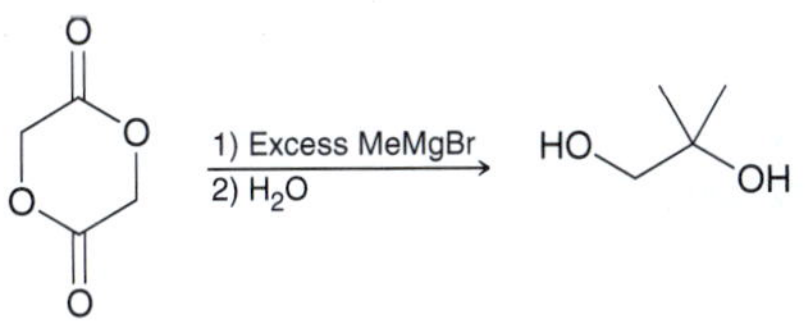

**13.59** Show the reagents you would use to achieve the following transformation:

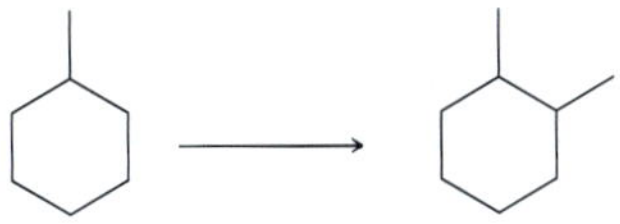

**13.60** Propose a mechanism for the following transformation:

1) Excess LAH
2) $H_2O$
$CH_3OH$

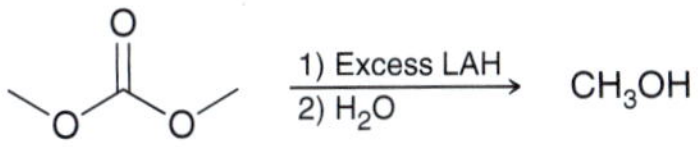

**13.61** A carbocation is resonance stabilized when it is adjacent to an oxygen atom:

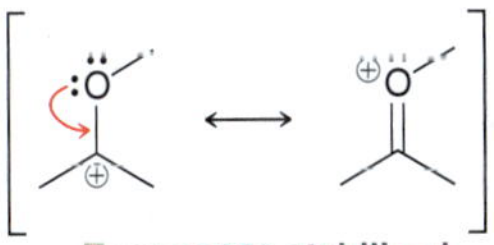

Resonance stabilized

Such a carbocation is even more stable than a tertiary carbocation. Using this information, propose a mechanism for the following transformation exhibited by a diol. This reaction is called a pinacol rearrangement:

conc. $H_2SO_4$
Heat

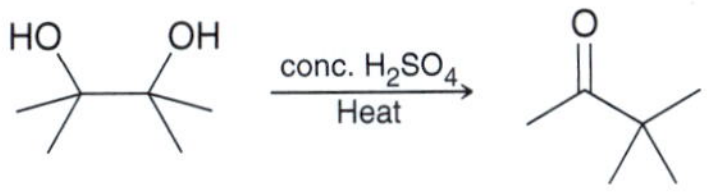

**13.62** Determine whether the pinacol rearrangement, shown in the previous problem, is a reduction, an oxidation, or neither.

**13.63** (*S*)-Gizzerosine is an amino acid that is believed to be responsible for a serious disease called "black vomit" in chickens. However, the same compound is a potential drug for the treatment of osteoporosis and acid buildup in the stomach. Provide a two-step synthesis for achieving the following transformation, which was employed during a recent synthesis of (*S*)-gizzerosine (*Tetrahedron Lett.* **2007**, *48*, 8479–8481).

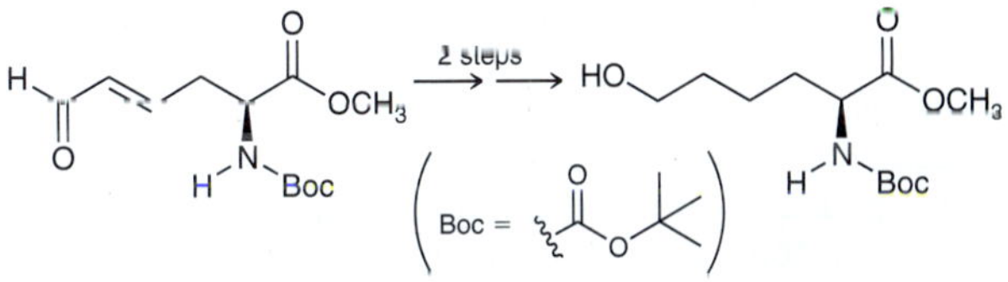

2 steps

(Boc = )

## CHALLENGE PROBLEMS

**13.64** Briarellin E (compound **3**) is produced by coral in the Caribbean and Mediterranean seas and belongs to a larger family of marine natural products that are currently being investigated as potential anticancer agents. During a recent synthesis of **3**, compound **1** was treated with TBAF to give compound **2** (*J. Org. Chem.* **2009**, *74*, 5458–5470). Draw the structure of compound **2**.

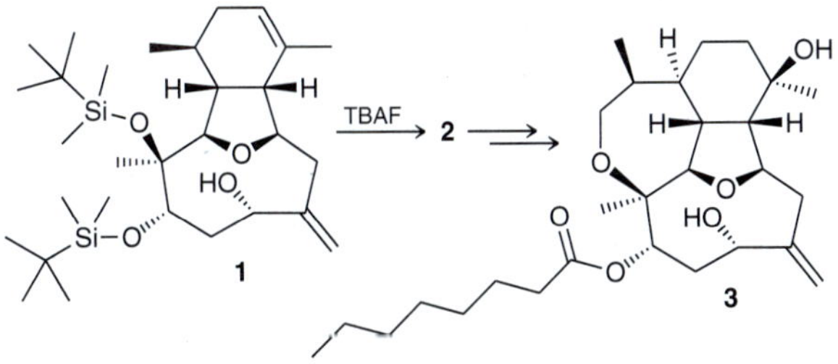

TBAF → **2** →

**13.65** Lepadiformine is a cytotoxic agent (toxic to cells) isolated from a marine tunicate. During a recent synthesis of lepadiformine, the investigators observed the formation of an interesting byproduct (**3**) while treating diol **1** with a reagent similar in function to $PBr_3$ (*J. Org. Chem.* **2012,** *77,* **3390–3400**):

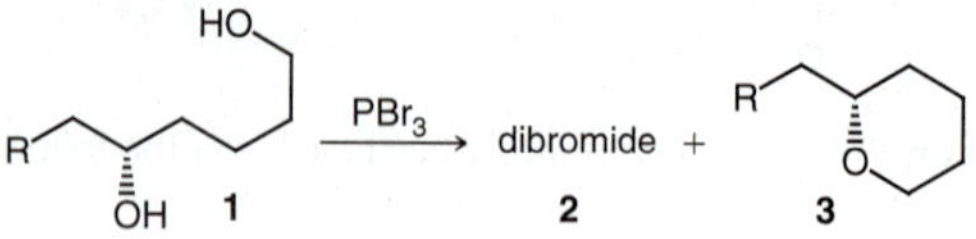

(a) Draw a structure for the desired product (**2**).

(b) Propose a plausible mechanism for formation of byproduct **3** using $PBr_3$ as the reagent. Make sure that your mechanism provides a justification for the stereochemical outcome.

**13.66** The compound duryne was one of several structurally related compounds isolated from a marine sponge (*J. Nat. Prod.* **2011,** *74,* **1262–1267**). Propose an efficient synthesis of duryne starting with any compounds containing eleven or fewer carbon atoms.

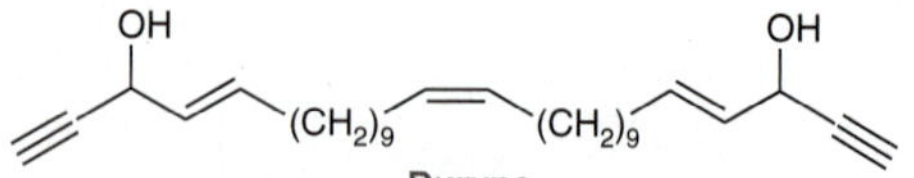

Duryne

**13.67** Estragole is an insect repellant that has been isolated from the leaves of the *Clausena anisata* tree (*J. Nat. Prod.* **1987,** *50,* **990–991**). Propose a synthesis of estragole starting from 4-methylphenol.

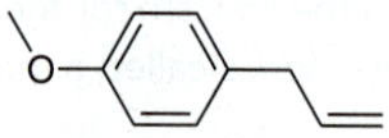

Estragole

**13.68** The following two isomeric ketones were among the 68 compounds isolated from the steam-distilled volatile oil of fresh and air-dried marijuana buds (*J. Nat. Prod.* **1996,** *59,* **49–51**). Propose a separate synthesis for each of these two compounds using only disubstituted alkenes containing four carbon atoms as starting materials.

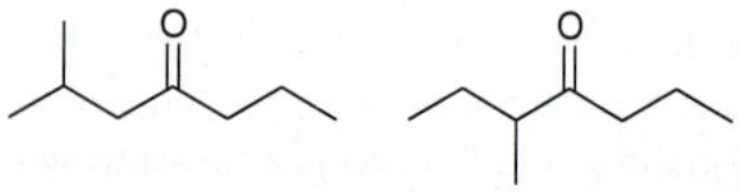

**13.69** Lithium borohydride ($LiBH_4$) is a useful reducing agent, as it is more selective than LAH but less selective than $NaBH_4$. Unlike the latter, $LiBH_4$ will reduce esters. However, it will not reduce amides (LAH reduces amides, as we will see in Chapter 21). This selectivity is frequently employed by synthetic organic chemists. For example, during a synthesis of (−)-croalbinecine, a natural product with cytotoxic properties, compound **1** was converted to compound **3**, as shown below (*J. Org. Chem.* **2000,** *65,* **9249–9251**). Draw the structures of **2** and **3**.

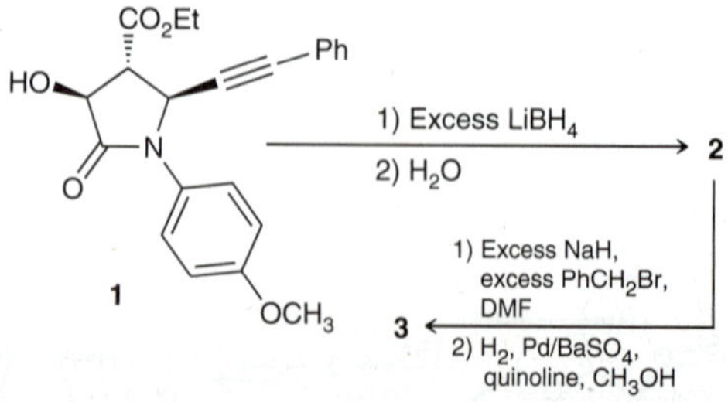

**13.70** In a study of the chemical components of salted and pickled prunes, hexyl butanoate was identified as one of 181 detected compounds (*J. Agric. Food Chem.* **1986,** *34,* **140–144**). Propose a synthesis of this compound using acetylene as your only source of carbon atoms.

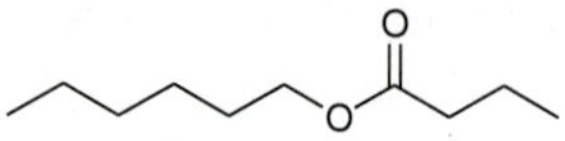

Hexyl butanoate

**13.71** The compound 5,9,12,16-tetramethyleicosane was synthesized as part of a study of the male sex pheromone of a Brazilian bug that feeds on the leaves and fruit of tomato plants (*J. Chem. Ecol.* **2012,** *38,* **814–824**). The synthesis of this alkane was reported using compounds **A–C**.

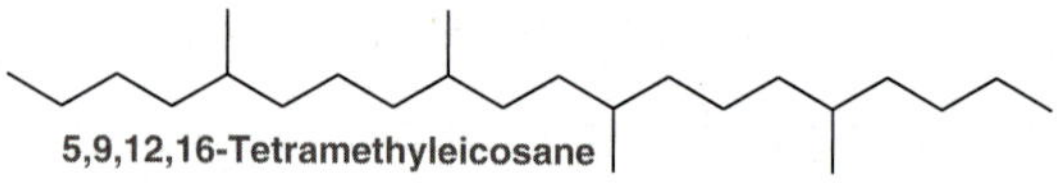

5,9,12,16-Tetramethyleicosane

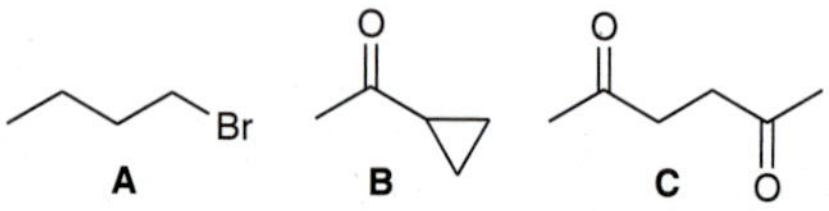

A key step in the synthesis is shown below, in which a mixture of *E* and *Z* isomers is obtained:

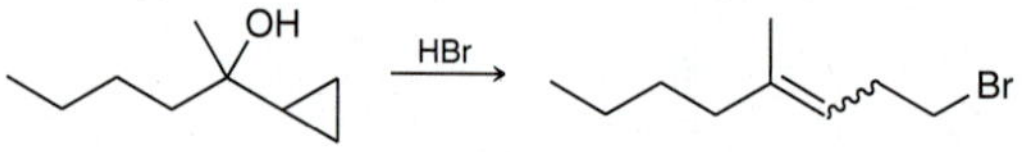

(a) Draw a plausible mechanism for this process.

(b) Propose an efficient synthesis of 5,9,12,16-tetramethyleicosane using **A**, **B**, and **C** as your only sources of carbon.

**13.72** Brevetoxin B (compound **2**) is produced by *Ptychodiscus brevis* Davis, a marine organism responsible for red tides. Brevetoxin B is a potent neurotoxin, as a result of its ability to bind to sodium channels and force them to remain open. Its biological action, together with its complex architecture, has made it an attractive synthetic target. During K. C. Nicolaou's synthesis of brevetoxin B, the K ring was assembled via the construction of compound **1**. The *syn* disposition of the hydroxyl group and the silicon-based protecting group in **1** was established via the successful conversion of a small amount of **1** into **5**, as shown below (*J. Am. Chem. Soc.* **1989,** *111,* **6682–6690**). Note that the conversion of **3** to **4** involves the selective tosylation of the primary hydroxyl group:

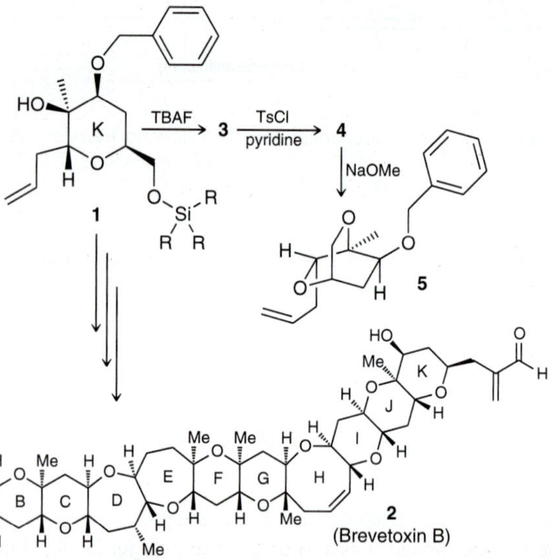

2 (Brevetoxin B)

(a) Draw the structures of **3** and **4**.

(b) Draw a mechanism for the conversion of **4** to **5** upon treatment with NaOMe.

(c) Explain why the successful conversion of **1** to **5** verifies the *syn* relationship between the hydroxyl group and the silyl protecting group in **1**.

# 14

# Ethers and Epoxides; Thiols and Sulfides



## DID YOU EVER WONDER...

### how cigarettes cause cancer?

Cigarette smoke contains many compounds that have been shown to cause cancer. In this chapter, we will explore one of those compounds and the sequence of chemical reactions that ultimately leads to the formation of cancer cells. These reactions involve the formation and reaction of high-energy compounds called epoxides.

Epoxides are a special category of ethers, and these compounds comprise the main topic of this chapter. We will learn about the properties and reactions of ethers, epoxides, and other related compounds. Then, we will return to the subject of this chapter opener to learn how the reactions presented in this chapter play a role in the development of cancer.

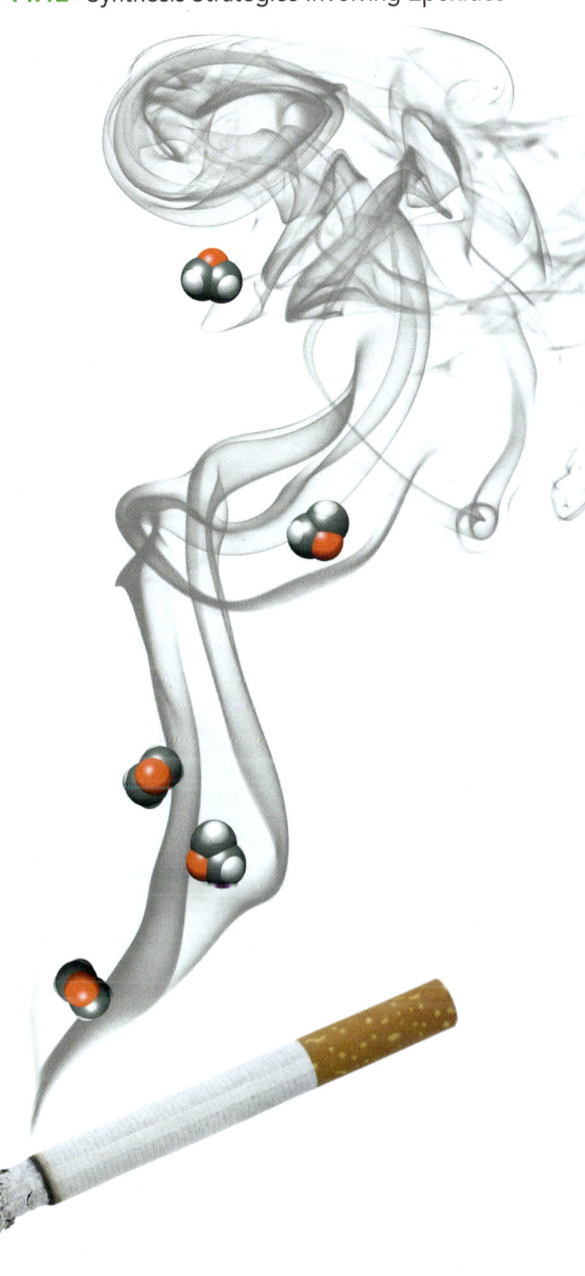

## DO YOU REMEMBER?

*Before you go on, be sure you understand the following topics.*
*If necessary, review the suggested sections to prepare for this chapter.*

- Reading Energy Diagrams (Section 6.6)
- $S_N2$ Reactions (Sections 7.4, 7.7)
- Halohydrin Formation (Section 9.8)
- Mechanisms and Curved Arrows (Sections 6.8–6.11)
- Oxymercuration-Demercuration (Section 9.5)

Take the DO YOU REMEMBER? QUIZ in **WileyPLUS** to check your understanding

## 14.1 Introduction to Ethers

**Ethers** are compounds that exhibit an oxygen atom bonded to two R groups, where each R group can be an alkyl, aryl, or vinyl group:

R–O–R

**An ether**

The ether group is a common structural feature of many natural compounds; for example:

**Melatonin**
A hormone that is believed to regulate the sleep cycle

**Morphine**
An opiate analgesic used to treat severe pain

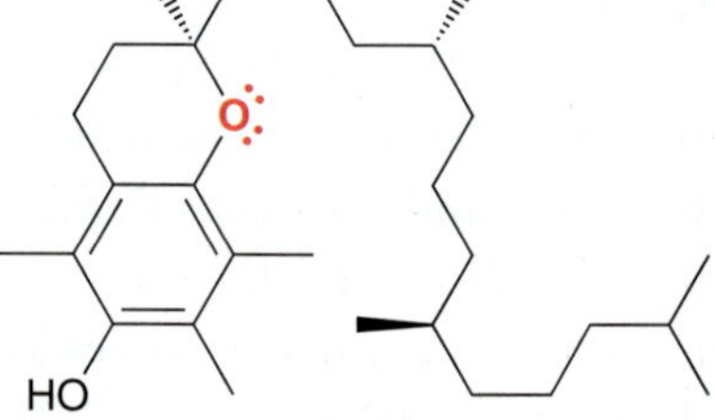


**Vitamin E**
An antioxidant

Many pharmaceuticals also exhibit an ether group; for example:

**(*R*)-Fluoxetine**
A powerful antidepressant sold under the trade name Prozac

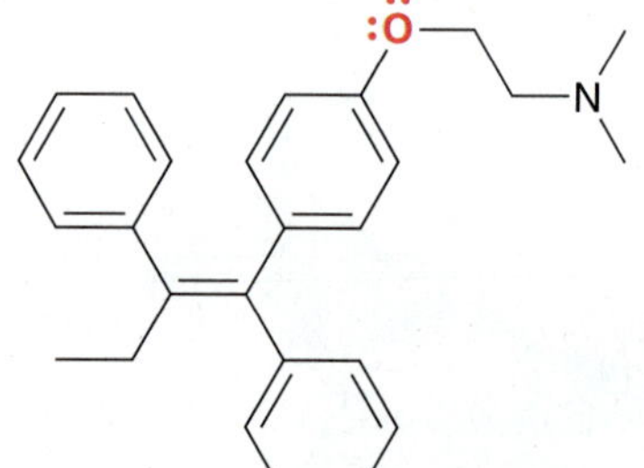


**Tamoxifen**
Inhibits the growth of some breast tumors

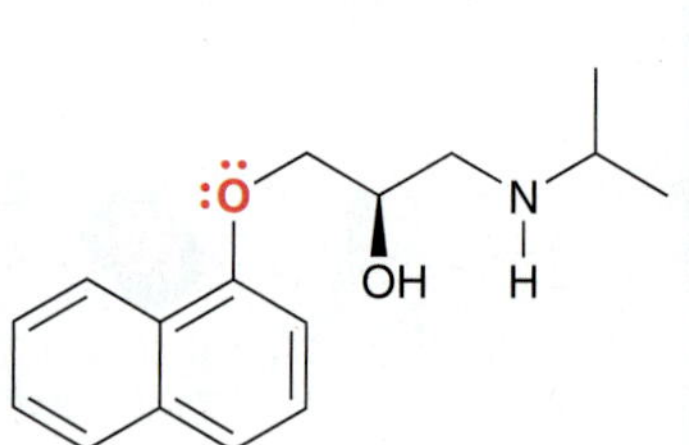


**Propanolol**
Used in the treatment of high blood pressure

## 14.2 Nomenclature of Ethers

IUPAC rules allow two different methods for naming ethers.

1. A common name is constructed by identifying each R group, arranging them in alphabetical order, and then adding the word "ether"; for example:

**Ethyl** **Methyl** **Ethyl methyl ether**

***tert*-Butyl** **Methyl** ***tert*-Butyl methyl ether**

In these examples, the oxygen atom is connected to two different alkyl groups. Such compounds are called **unsymmetrical ethers**. When the two alkyl groups are identical, the compound is called a **symmetrical ether** and is named as a *di*alkyl ether.

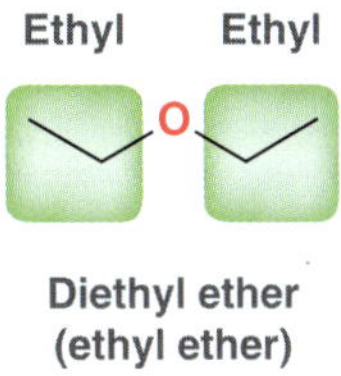


2. A systematic name is constructed by choosing the larger group to be the parent alkane and naming the smaller group as an **alkoxy substituent**.

Systematic names must be used for complex ethers that exhibit multiple substituents and/or chirality centers. Let's see some examples.

## SKILLBUILDER

### 14.1 NAMING AN ETHER

**LEARN the skill** Name the following compounds:

(a) Me–O–phenyl (b) 3,3-dichlorocyclopentyl ethyl ether structure

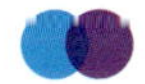

**SOLUTION**

**(a)** To assign a common name, identify each group on either side of the oxygen atom, arrange them in alphabetical order, and then add the word "ether."

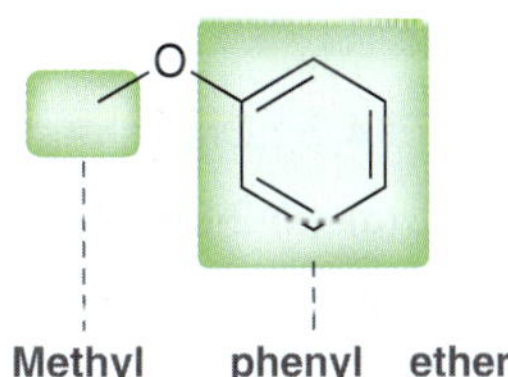


To assign a systematic name, choose the more complex (larger) group as the parent and name the smaller group as an alkoxy substituent.

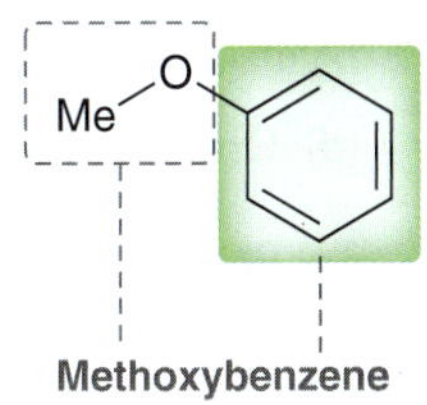

This compound therefore can be called methyl phenyl ether or methoxybenzene. Both names are accepted by IUPAC rules.

**(b)** The second compound is more complex. It has a chirality center and several substituents. Therefore, it will not have a common name. To assign a systematic name, begin by choosing the more complex group as the parent.

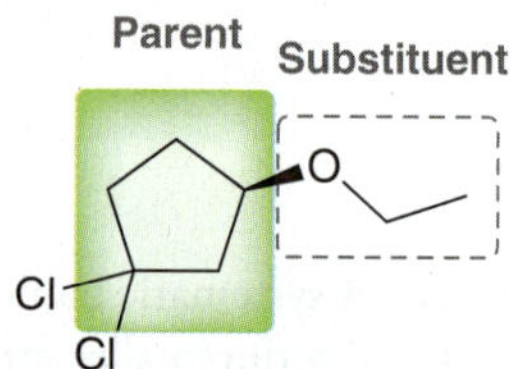


The cyclopentane ring becomes the parent, and the ethoxy group is listed as one of the three substituents on the cyclopentane ring. Locants are then assigned so as to give the lowest possible numbers to all three substituents (1,1,3 rather than 1,3,3):

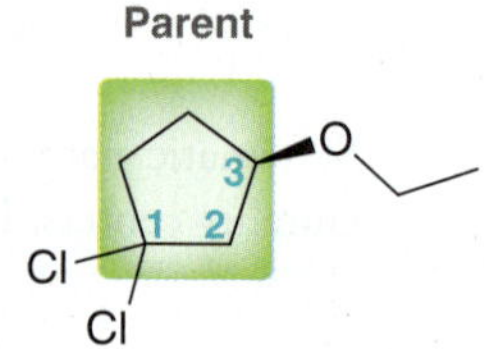


The configuration of the chirality center is identified at the beginning of the name:

**(*R*)-1,1-Dichloro-3-ethoxycyclopentane**

**PRACTICE** the skill **14.1** Provide an IUPAC name for each of the following compounds:

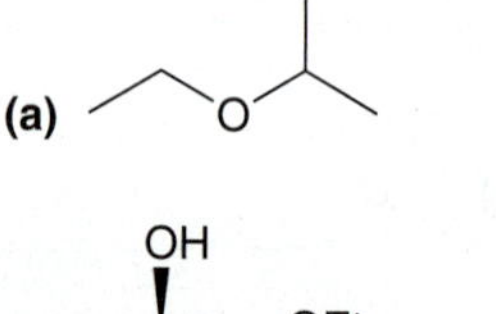

(a) O

(b) O Cl

(c) Cl Cl O

(d) OH OEt

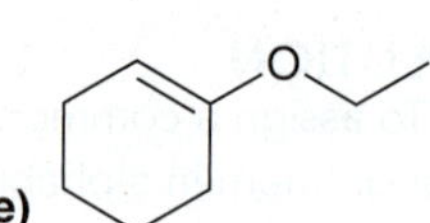

(e) O

**APPLY** the skill

**14.2** Draw the structure of each of the following compounds:

**(a)** (*R*)-2-Ethoxy-1,1-dimethylcyclobutane

**(b)** Cyclopropyl isopropyl ether

**14.3** There are six ethers with molecular formula $C_5H_{12}O$ that are constitutional isomers.

**(a)** Draw all six constitutional isomers.

**(b)** Provide a systematic name for each of the six compounds.

**(c)** Provide a common name for each of the six compounds.

**(d)** Only one of these compounds has a chirality center. Identify that compound.

need more **PRACTICE?** **Try Problems 14.30, 14.32**

## 14.3 Structure and Properties of Ethers

The geometry of an oxygen atom is similar for water, alcohols, and ethers. In all three cases, the oxygen atom is $sp^3$ hybridized, and the orbitals are arranged in a nearly tetrahedral shape. The exact bond angle depends on the groups attached to the oxygen atom, with ethers having the largest bond angles:

105° 109° 112°

**Water** **Methanol** **Dimethyl ether**

In the previous chapter, we saw that alcohols have relatively high boiling points due to the effects of intermolecular hydrogen bonding.

An ether can act as a hydrogen bond acceptor and can interact with the proton of an alcohol.

**An ether** (H bond acceptor) **An alcohol** (H bond donor)

However, ethers cannot function as hydrogen bond donors, and therefore, ethers cannot form hydrogen bonds with each other. As a result, the boiling points of ethers are significantly lower than their isomeric alcohols.

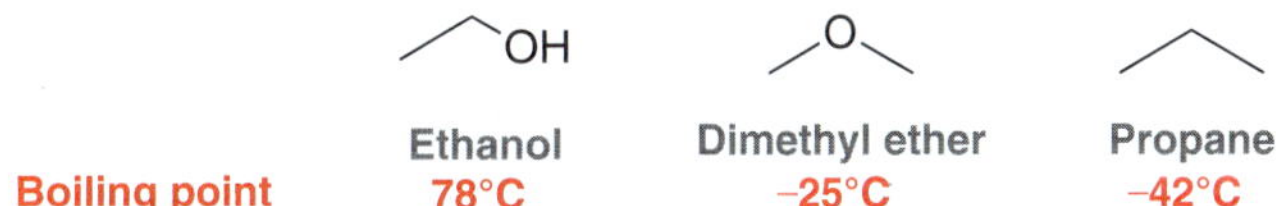


In fact, the boiling point of dimethyl ether is almost as low as the boiling point of propane. Both dimethyl ether and propane lack the ability to form hydrogen bonds. The slightly higher boiling point of dimethyl ether can be explained by considering the net dipole moment.

These individual dipole moments produce a net dipole moment

Ethers therefore exhibit dipole-dipole interactions, which slightly elevate the boiling point relative to propane. Ethers with larger alkyl groups have higher boiling points due to London dispersion forces between the alkyl groups on different molecules. This trend is significant.

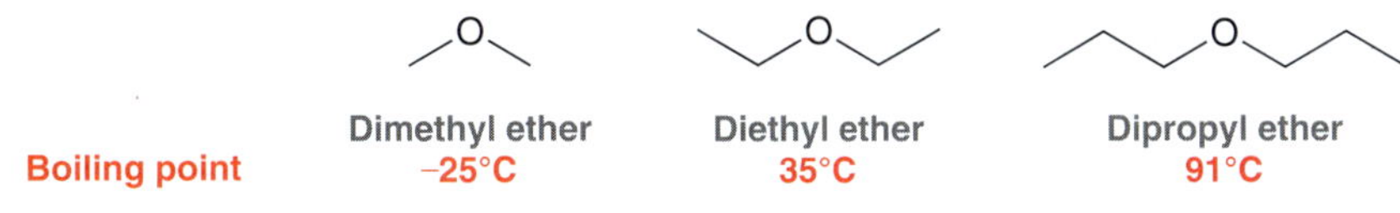

Ethers are often used as solvents for organic reactions, because they are fairly unreactive, they dissolve a wide variety of organic compounds, and their low boiling points allow them to be readily evaporated after a reaction is complete. Below are three common solvents.

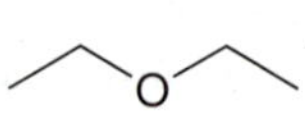

**Diethyl ether**  **Tetrahydrofuran** 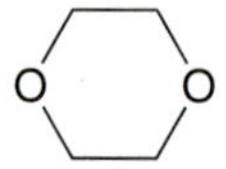 **1,4-Dioxane**

## medically speaking — Ethers as Inhalation Anesthetics

Diethyl ether was once used as an inhalation anesthetic, but the side effects were unpleasant, and the recovery was often accompanied by nausea and vomiting. Diethyl ether was eventually replaced by halogenated ethers, such as the ones shown below.

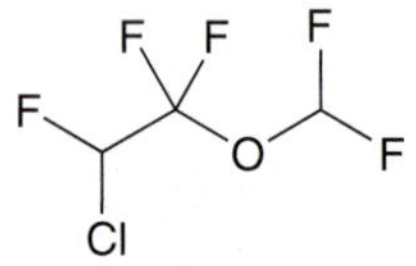

**Enflurane** 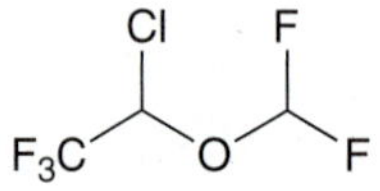 **Isoflurane**

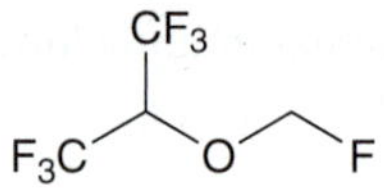

**Sevoflurane** 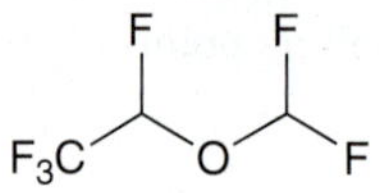 **Desflurane**

Enflurane was introduced in the mid-1970s and was eventually replaced by isoflurane. The use of isoflurane is now also declining as the newer generation ethers (sevoflurane and desflurane) are being more heavily used.

Inhalation anesthetics are introduced into the body via the lungs and distributed by the circulatory system. They specifically target the nerve endings in the brain. Nerve endings, which are separated by a synaptic gap, transmit signals across the gap by means of small organic compounds called neurotransmitters (shown as blue balls in the following figure).

A change in ionic conductance (electrical signal) causes the presynaptic cell to release neurotransmitters, which travel across the synaptic gap until they reach the receptors at the postsynaptic cell. When the neurotransmitters bind to the receptors, a change in conductance is triggered once again. In this way, a signal either can be relayed across the synaptic gap or can be stopped, depending on whether the neurotransmitters are allowed to do their job. Several factors are involved that either can inhibit or increase the function of the neurotransmitters. By controlling whether signals are sent or stopped at each synaptic gap, the nervous system is able to control the various systems in the body (similar to the way a computer uses zeros and ones to perform all of its functions).

Inhalation anesthetics disrupt the normal synaptic transmission process. Many mechanisms of action for anesthetics have been suggested, including the following:

1. Interfering with the release of neurotransmitters from the presynaptic nerve cell
2. Interfering with the binding of the neurotransmitters at the postsynaptic receptors
3. Affecting the ionic conductance (the electrical signal that causes neurotransmission)
4. Affecting reuptake of neurotransmitters into the presynaptic cell

The main mechanism of action is likely to be a combination of many of these factors.

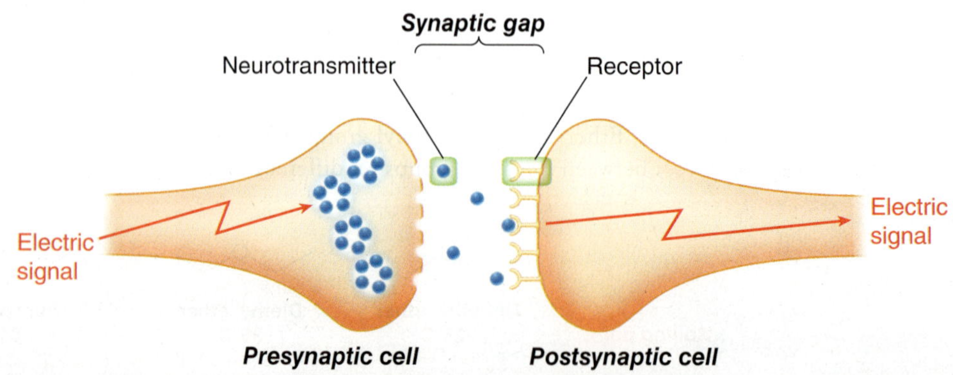

## 14.4 Crown Ethers

**BY THE WAY**

Although, like ethers, alcohols (ROH) also have an oxygen atom with lone pairs, they cannot be used to stabilize Grignard reagents, because alcohols possess acidic protons. As we saw in Section 13.6, Grignard reagents cannot be prepared in the presence of acidic protons.

Ethers can interact with metals that have either a full positive charge or a partial positive charge. For example, Grignard reagents are formed in the presence of an ether, such as diethyl ether. The lone pairs on the oxygen atom serve to stabilize the charge on the magnesium atom. The interaction is weak, but it is necessary in order to form a Grignard reagent.

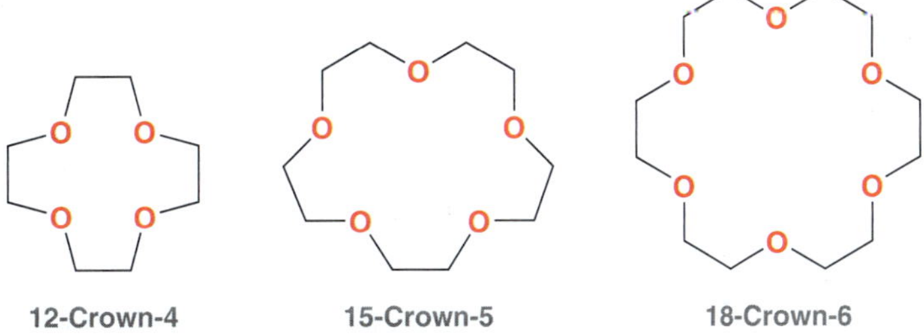


Charles J. Pedersen, working for Du Pont, discovered that the interaction between ethers and metal ions is significantly stronger for compounds with multiple ether groups. Such compounds are called **polyethers**. Pedersen prepared and investigated the properties of many cyclic polyethers, such as the following examples. Pedersen called them **crown ethers** because their molecular models resemble crowns.

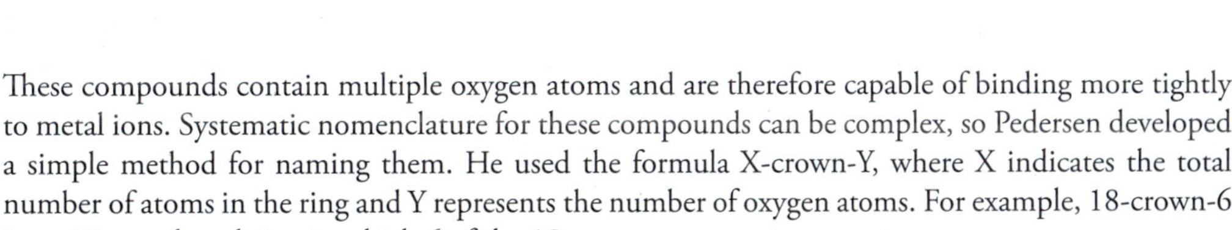


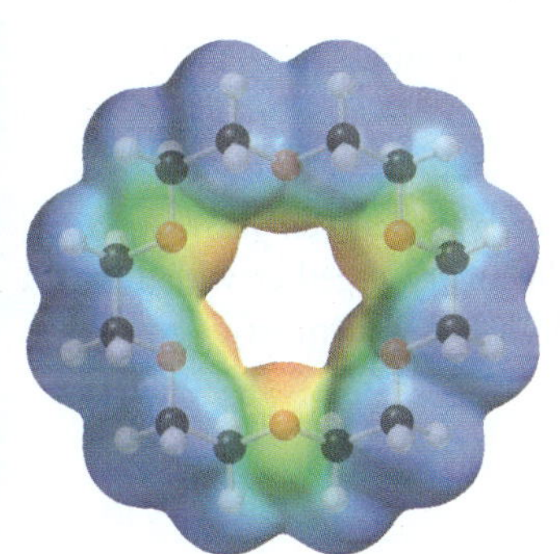

**FIGURE 14.1a**
An electrostatic potential map of 18-crown-6 shows the oxygen atoms facing the inside of the internal cavity.

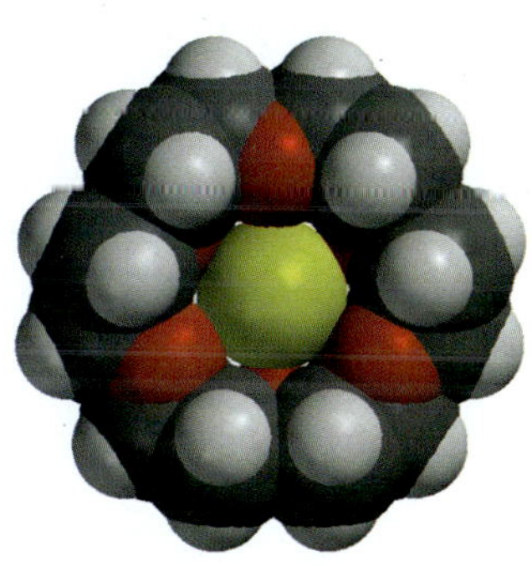

**FIGURE 14.1b**
A space-filling model of 18-crown-6 shows that a potassium cation can fit nicely inside the internal cavity.

These compounds contain multiple oxygen atoms and are therefore capable of binding more tightly to metal ions. Systematic nomenclature for these compounds can be complex, so Pedersen developed a simple method for naming them. He used the formula X-crown-Y, where X indicates the total number of atoms in the ring and Y represents the number of oxygen atoms. For example, 18-crown-6 is an 18-membered ring in which 6 of the 18 atoms are oxygen atoms.

The unique properties of these compounds derive from the size of their internal cavities. For example, the internal cavity of 18-crown-6 comfortably hosts a potassium cation ($K^+$). In the electrostatic potential map in Figure 14.1a, it is clear that the oxygen atoms all face toward the inside of the cavity, where they can bind to the metal cation. The space-filling model in Figure 14.1b shows how a potassium cation fits perfectly into the internal cavity. Once inside the cavity, the entire complex has an outer surface that resembles a hydrocarbon, rendering the complex soluble in organic solvents. In this way, 18-crown-6 is capable of solvating potassium ions in organic solvents. Normally, the metal cation by itself would not be soluble in a nonpolar solvent. The ability of crown ethers to solvate metal cations has enormous implications, in both the field of synthetic organic chemistry and the field of medicinal chemistry. As an example, consider what happens when KF and 18-crown-6 are mixed together in benzene (a common organic solvent).

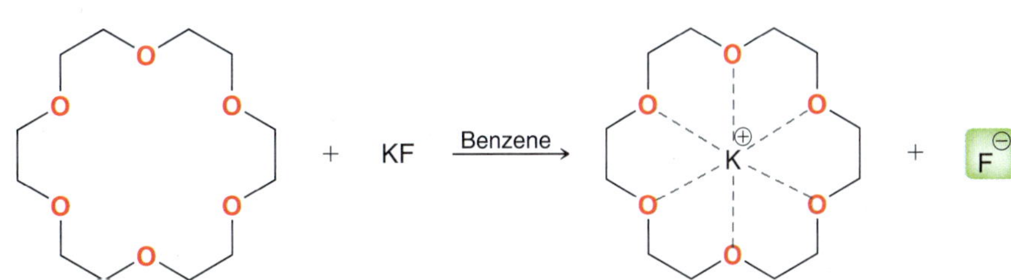

Without the crown ether, KF would simply not dissolve in benzene. The presence of 18-crown-6 generates a complex that dissolves in benzene. The result is a solution containing fluoride ions, which enables us to perform substitution reactions with $F^-$ as a nucleophile. Generally, it is too difficult to use $F^-$ as a nucleophile, because it will usually interact too strongly with the polar solvents in which it dissolves. The strong interaction between fluoride ions and polar solvents makes it difficult for $F^-$ to become "free" to serve as a nucleophile. However, the use of 18-crown-6 allows the creation of free fluoride ions in a nonpolar solvent, making substitution reactions possible. For example:

Br —(KF, benzene / 18-Crown-6)→ F
92%

Another example is the ability of 18-crown-6 to dissolve potassium permanganate ($KMnO_4$) in benzene. Such a solution is very useful for performing a wide variety of oxidation reactions.

Other metal cations can be solvated by other crown ethers. For example, a lithium ion is solvated by 12-crown-4, and a sodium ion is solvated by 15-crown-5.

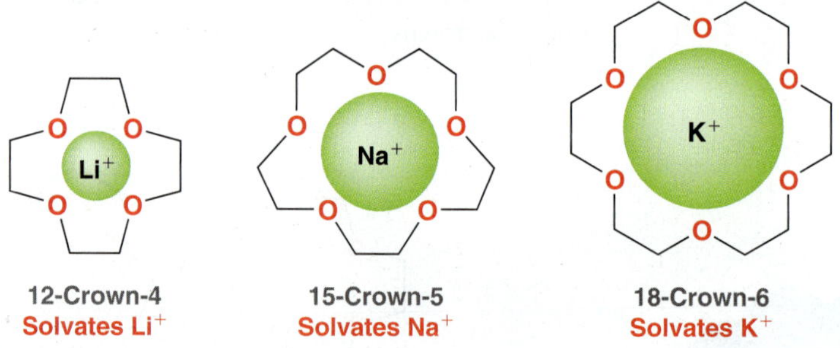


The discovery of these compounds led to a whole new field of chemistry, called *host-guest chemistry*. For his contribution, Pedersen shared the 1987 Nobel Prize in Chemistry together with Donald Cram and Jean-Marie Lehn, who were also pioneers in the field of host-guest chemistry.

## CONCEPTUAL CHECKPOINT

**14.4** Identify the missing reagent needed to achieve each of the following transformations:

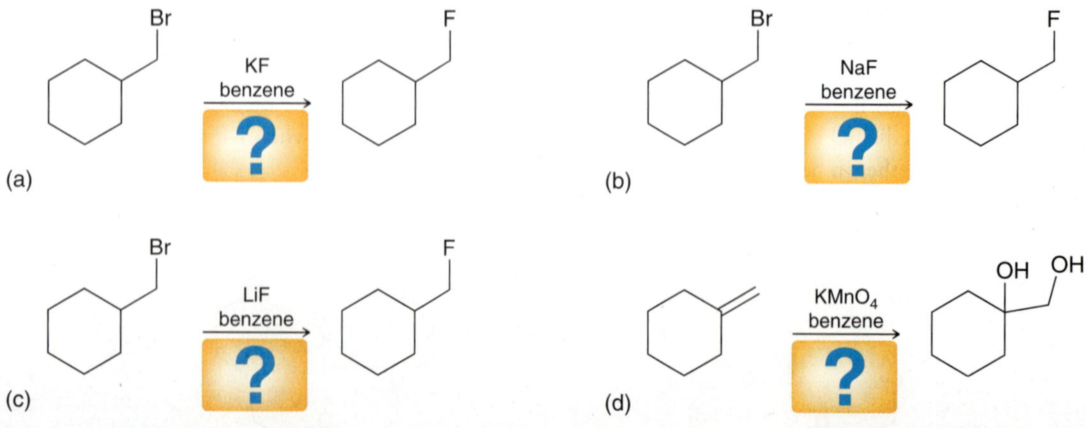

## medically speaking | Polyether Antibiotics

Some antibiotics function very much like crown ethers. For example, consider the structures of nonactin and monensin.

**Nonactin**

**Monensin**

These compounds are polyethers and therefore are capable of serving as hosts for metal cations, much like crown ethers. These polyethers are called *ionophores* because the internal cavity is capable of binding a metal ion. The outside surface of the ionophore is hydrocarbon-like (or *lipophilic*), allowing it to pass through cell membranes readily.

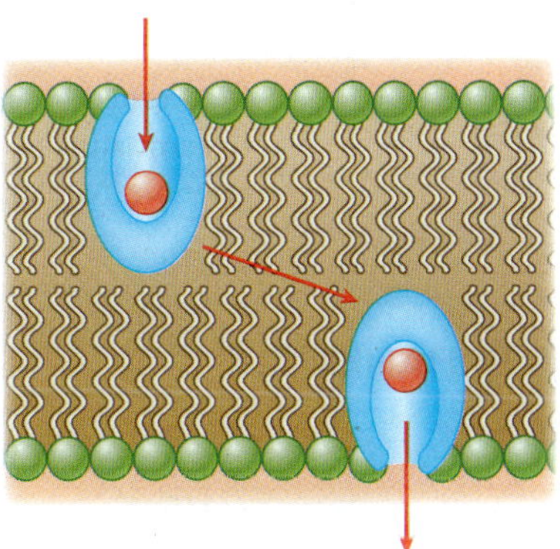

In order to function properly, cells must maintain a gradient between the concentration of sodium and potassium ions inside and outside the cell. That gradient is established because ions are not free to pass through the cell membrane, except through special ion channels where $K^+$ ions are pumped into the cell and $Na^+$ ions are pumped out of the cell. Ionophores effectively render the cell membrane permeable to these ions. The ionophores serve as hosts to the ions, carrying them across the cell membrane and destroying the necessary concentration gradient. In this way, ionophores interrupt cell function, thereby killing bacteria. Many new ionophores are currently under investigation as new potential antibiotics.

## 14.5 Preparation of Ethers

### Industrial Preparation of Diethyl Ether

Diethyl ether is prepared industrially via the acid-catalyzed dehydration of ethanol. The mechanism of this process is believed to involve an $S_N2$ process.

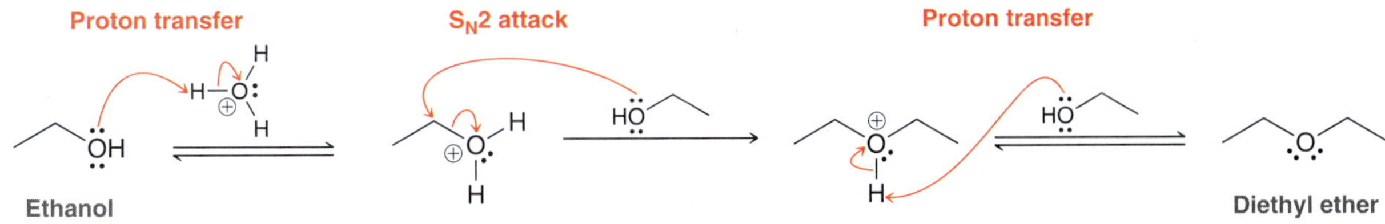


A molecule of ethanol is protonated and then attacked by another molecule of ethanol in an $S_N2$ process. As a final step, deprotonation generates the product. Notice that a proton is used in the first step of the mechanism, and then another proton is liberated in the last step of the mechanism. The acid is therefore a catalyst (not consumed by the reaction) that enables the $S_N2$ process to proceed.

This process has many limitations. For example, it only works well for primary alcohols (since it proceeds via an $S_N2$ pathway), and it produces symmetrical ethers. As a result, this process for preparing ethers is too limited to be of any practical value for organic synthesis.

## Williamson Ether Synthesis

Ethers can be readily prepared via a two-step process called a **Williamson ether synthesis.**

$$\mathrm{R{-}OH} \xrightarrow[\text{2) RX}]{\text{1) NaH}} \mathrm{R{-}O{-}R}$$

We learned both of these steps in the previous chapter. In the first step, the alcohol is deprotonated to form an alkoxide ion. In the second step, the alkoxide ion functions as a nucleophile in an $S_N2$ reaction (Mechanism 14.1).

### MECHANISM 14.1 THE WILLIAMSON ETHER SYNTHESIS

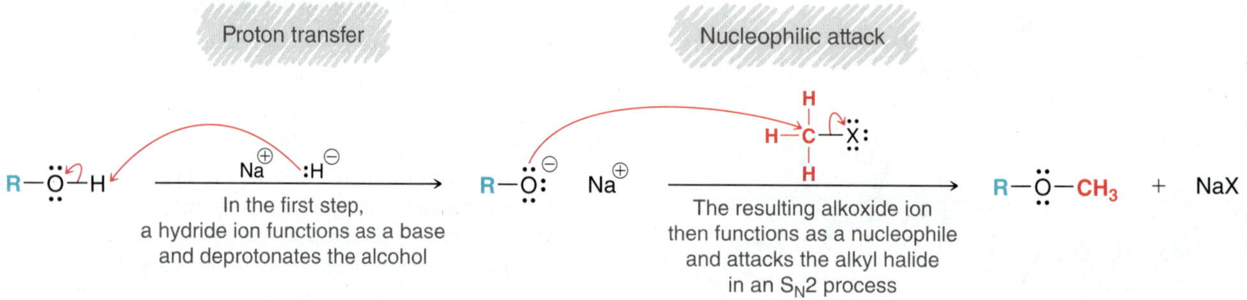


**BY THE WAY**
The *tert*-butyl group is alphabetized by the letter "b" rather than "t," and therefore, the *tert*-butyl group precedes the methyl group in the name. This compound is commonly called MTBE, which is an acronym of the incorrect name.

***tert*-Butyl methyl ether (MTBE)**

This process is named after Alexander Williamson, a British scientist who first demonstrated this method in 1850 as a way of preparing diethyl ether. Since the second step is an $S_N2$ process, steric effects must be considered. Specifically, the process works best when methyl or primary alkyl halides are used. Secondary alkyl halides are less efficient because elimination is favored over substitution, and tertiary alkyl halides cannot be used. This limitation must be taken into account when choosing which C—O bond to form. For example, consider the structure of *tert*-butyl methyl ether. MTBE was used heavily as a gasoline additive until concerns emerged that it might contribute to groundwater contamination. As a result, its use has declined in recent years. There are two possible routes to consider in the preparation of MTBE, but only one is efficient.

OH — 1) NaH, 2) $CH_3I$ → MTBE

$CH_3OH$ — 1) NaH, 2) I ✗ → MTBE

The first route is efficient because it employs a methyl halide, which is a suitable substrate for an $S_N2$ process. The second route does not work because it employs a tertiary alkyl halide, which will undergo elimination rather than substitution.

# SKILLBUILDER

## 14.2 PREPARING AN ETHER VIA A WILLIAMSON ETHER SYNTHESIS

**LEARN the skill**

Show the reagents you would use to prepare the following ether via a Williamson ether synthesis:

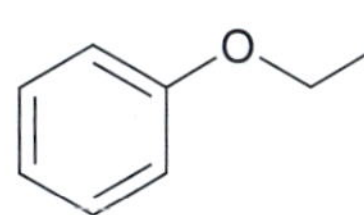

### SOLUTION

In order to prepare this ether via a Williamson ether synthesis, we must decide which starting alcohol and which starting alkyl halide to use. To make the proper choice, we analyze the positions on either side of the oxygen atom:

**STEP 1**
Classify the groups on either side of the oxygen.

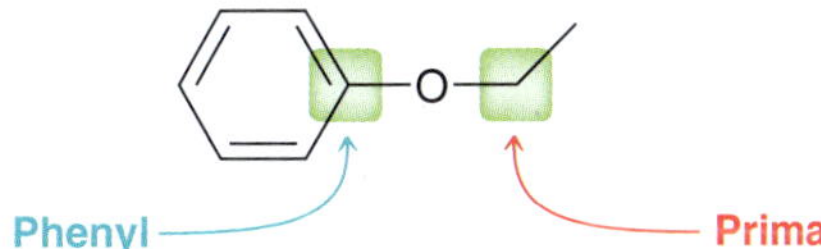


**STEP 2**
Determine which side is more capable of serving as a substrate in an $S_N2$ reaction.

The phenyl position is $sp^2$ hybridized, and $S_N2$ processes do not occur at $sp^2$-hybridized centers. The other position is a primary position, and $S_N2$ processes can occur readily at primary substrates. Therefore, we must start with phenol and an ethyl halide.

OH + X

**LOOKING BACK**
For a review of leaving groups, see Section 7.8.

The X can be any good leaving group, such as I, Br, Cl, or OTs. In general, iodide and tosylate are the best leaving groups. The alcohol in this case is phenol, which can be deprotonated with sodium hydroxide (as we saw in Section 13.2). We therefore propose the following synthesis:

**STEP 3**
Use a base to deprotonate the alcohol and then introduce the alkyl halide.

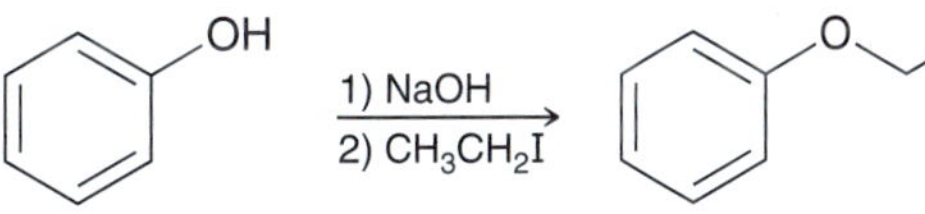

**PRACTICE the skill**

**14.5** Show what reagents you would use to prepare each of the following ethers via a Williamson ether synthesis and explain your reasoning:

(a) (b) (c)

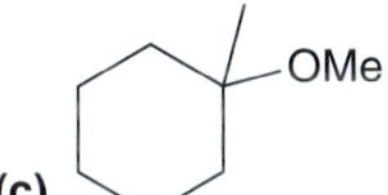


**APPLY the skill**

**14.6** The following cyclic ether can be prepared via an intramolecular Williamson ether synthesis. Show what reagents you would use to make this ether.

**14.7** Can the following compound be prepared via a Williamson ether synthesis? Explain your answer.

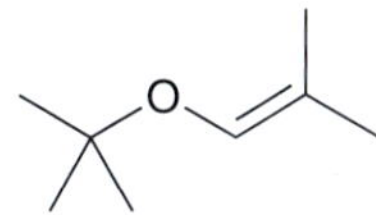


need more **PRACTICE?** **Try Problems 14.33a,c, 14.37, 14.40, 14.42d, 14.43b**

### Alkoxymercuration-Demercuration

Recall from Section 9.5 that alcohols can be prepared from alkenes via a process called oxymercuration-demercuration.

1) $Hg(OAc)_2$, $H_2O$
2) $NaBH_4$

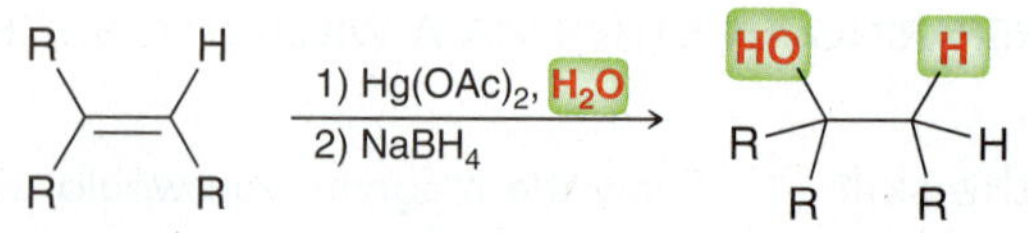

The net result is a Markovnikov addition of water (H and OH) across an alkene. That is, the hydroxyl group is ultimately placed at the more substituted position. The mechanism for this process was discussed in Section 9.3.

If an alcohol (ROH) is used in place of water, then the result is a Markovnikov addition of the alcohol (RO and H) across the alkene. This process is called **alkoxymercuration-demercuration**, and it can be used as another way of preparing ethers.

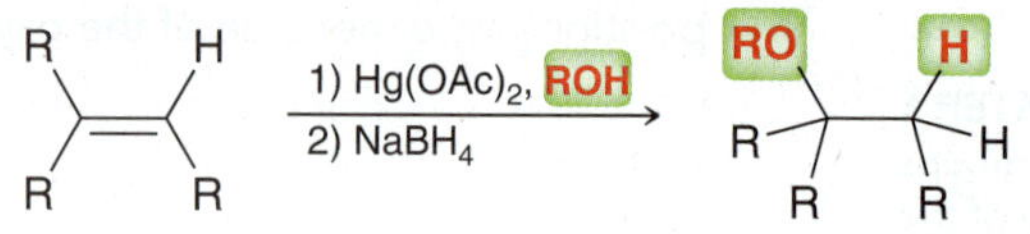

1) $Hg(OAc)_2$, ROH
2) $NaBH_4$

## CONCEPTUAL CHECKPOINT

**14.8** Show what reagents you would use to prepare each of the following ethers via an alkoxymercuration-demercuration:

(a) 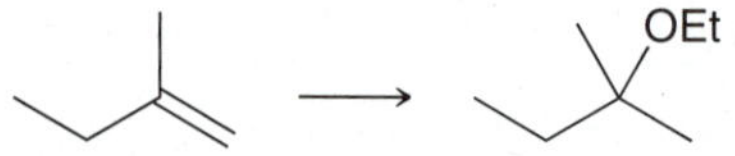 OEt

(b)

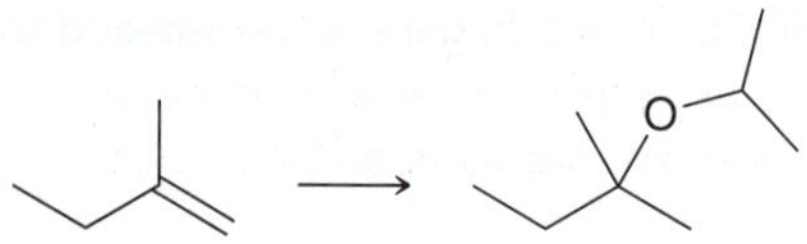

(c) 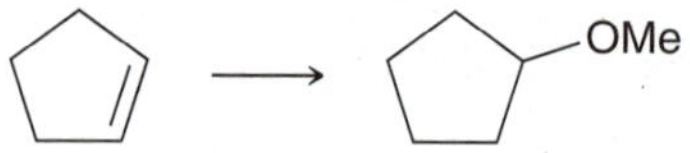 OMe

(d) 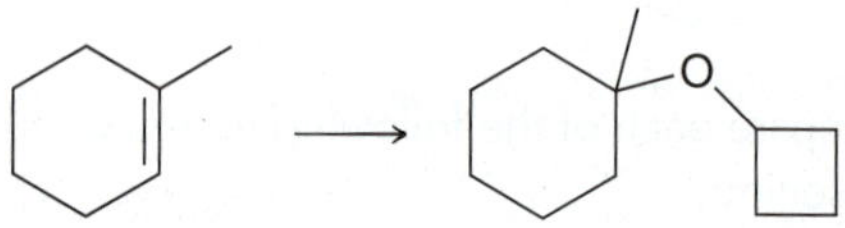 O

**14.9** How would you use an alkoxymercuration-demercuration to prepare dicyclopentyl ether using cyclopentene as your only source of carbon?

**14.10** Show how you would use an alkoxymercuration-demercuration to prepare isopropyl propyl ether using propene as your only source of carbon and any other reagents of your choosing.

## 14.6 Reactions of Ethers

Ethers are generally unreactive under basic or mildly acidic conditions. As a result, they are an ideal choice as solvents for many reactions. Nevertheless, ethers are not completely unreactive, and two reactions of ethers will be explored in this section.

### Acidic Cleavage

When heated with a concentrated solution of a strong acid, an ether will undergo **acidic cleavage**, in which the ether is converted into two alkyl halides:

$$R{-}O{-}R \xrightarrow[\text{Heat}]{\text{Excess HX}} R{-}X + R{-}X + H_2O$$

This process involves two substitution reactions (Mechanism 14.2).

## MECHANISM 14.2 ACIDIC CLEAVAGE OF AN ETHER

### FORMATION OF FIRST ALKYL HALIDE

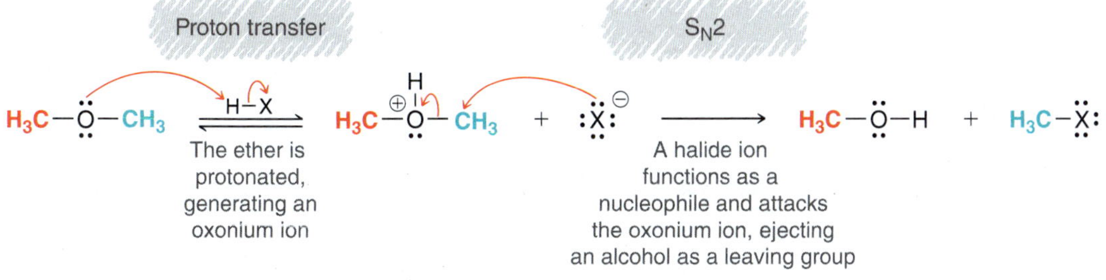


### FORMATION OF SECOND ALKYL HALIDE

Proton transfer

$S_N2$

$H_3C-\ddot{O}-H$ ⇌ (H–X) $H_3C-\overset{\oplus}{O}(H)-H$ + $:\ddot{X}:^{\ominus}$ ⟶ $H_3C-\ddot{X}:$ + $H_2O$

The alcohol is protonated, generating an oxonium ion

A halide ion functions as a nucleophile and attacks the oxonium ion, ejecting water as a leaving group

The formation of the first alkyl halide begins with protonation of the ether to form a good leaving group, followed by an $S_N2$ process in which a halide ion functions as a nucleophile and attacks the protonated ether. The second alkyl halide is then formed with the same two steps—protonation followed by an $S_N2$ attack. If either R group is tertiary, then substitution is more likely to proceed via an $S_N1$ process rather than $S_N2$.

When a phenyl ether is cleaved under acidic conditions, the products are phenol and an alkyl halide.

Excess HX, Heat ⟶ Phenol + RX

**Phenol**

The phenol is not further converted into a halide, because neither $S_N2$ nor $S_N1$ processes are efficient at $sp^2$-hybridized centers.

Both HI and HBr can be used to cleave ethers. HCl is less efficient, and HF does not cause acidic cleavage of ethers. This reactivity is a result of the relative nucleophilicity of the halide ions.

**LOOKING BACK**
For a review of the relative nucleophilicity of the halide ions, see Section 7.8.

## CONCEPTUAL CHECKPOINT

**14.11** Predict the products for each of the following reactions:

(a) Excess HBr, Heat ?

(b) Excess HI, Heat ?

(c) Excess HBr, Heat ?

(d) Excess HI, Heat ?

(e) Excess HI, Heat ?

(f) Excess HBr, Heat ?

## Autooxidation

Recall from Section 11.9 that ethers undergo autooxidation in the presence of atmospheric oxygen to form hydroperoxides:

$O_2$ (slow)

OOH

**A hydroperoxide**

This process occurs via a radical mechanism which is initiated by a hydrogen abstraction (Mechanism 14.3).

## MECHANISM 14.3 AUTOOXIDATION OF ETHERS

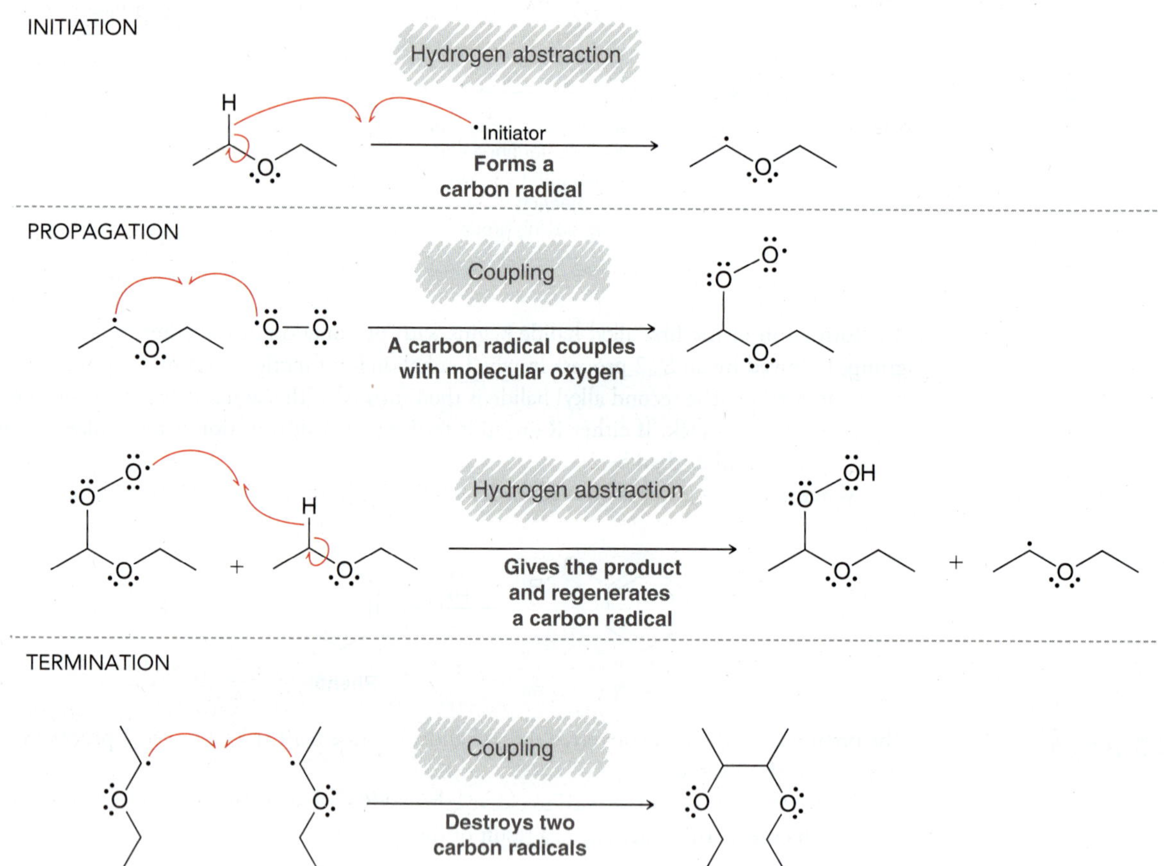


As with all radical mechanisms, the net reaction is the sum of the propagation steps:

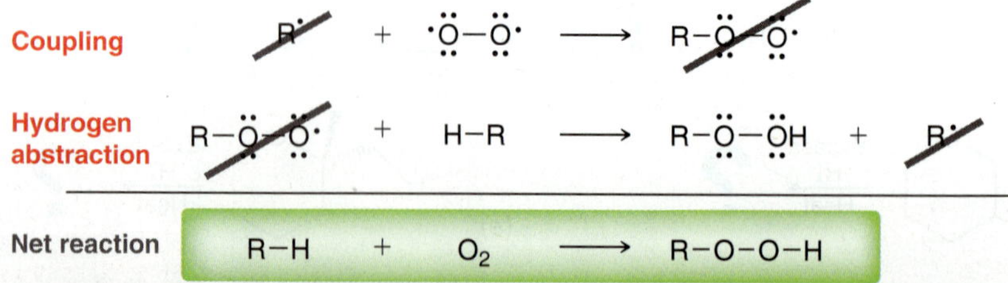

The reaction is slow, but old bottles of ether will invariably contain a small concentration of hydroperoxides, rendering the solvent very dangerous to use. Hydroperoxides are unstable and decompose violently when heated. Many laboratory explosions have been caused by distillation of ethers that were contaminated with hydroperoxides. Ethers used in the laboratory must be frequently tested for the presence of hydroperoxides and purified prior to use.

## 14.7 Nomenclature of Epoxides

Cyclic ethers are compounds that contain an oxygen atom incorporated in a ring. Special parent names are used to indicate the ring size:

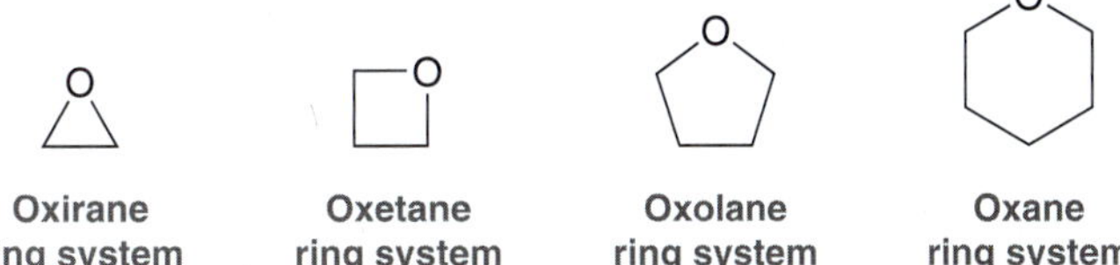


For our current discussion, we will focus on **oxiranes,** cyclic ethers containing a three-membered ring system. This ring system is more reactive than other ethers because it has significant ring strain. Substituted oxiranes, which are also called *epoxides*, can have up to four R groups. The simplest epoxide (no R groups) is often called by its common name, ethylene oxide:

**BY THE WAY**
The common name "ethylene oxide" denotes the fact that it is produced from ethylene.

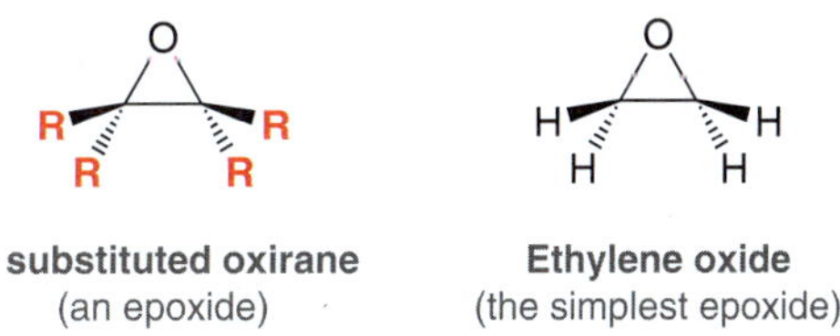


**A substituted oxirane**
(an epoxide)

**Ethylene oxide**
(the simplest epoxide)

Epoxides, although strained, are commonly found in nature. The following are two examples:

**Disparlure**
Sex pheromone of the female gypsy moth

**Periplanone B**
Sex pheromone of the female American cockroach

There are two methods for naming epoxides. In the first method, the oxygen atom is considered to be a substituent on the parent chain, and the exact location of the epoxide group is identified with two numbers followed by the term "epoxy."

**3-Ethyl-2-methyl-2,3-epoxypentane**

In the second method, the parent is considered to be the oxirane ring, and any groups connected to the oxirane ring are listed as substituents.

**2,2-Diethyl-3,3-dimethyloxirane**

## CONCEPTUAL CHECKPOINT

**14.12** Assign a name for each of the following compounds:

(a) (b) (c)

**14.13** Assign a name for each of the following compounds. Be sure to assign the configuration of each chirality center and indicate the configuration(s) at the beginning of the name.

(a) (b) (c)

## medically speaking — Epothilones as Novel Anticancer Agents

Epothilones are a class of novel compounds first isolated from the bacterium *Sorangium cellulosum* in Southern Africa.

**Epothilone A** (R=H)
**Epothilone B** (R=Me)

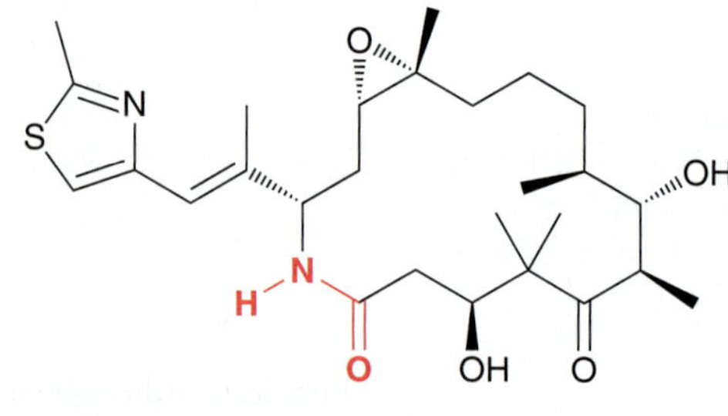


**Ixabepilone**

The discovery of the antitumor behavior of these naturally occurring epoxides led to a search for related compounds that might exhibit enhanced potency and selectivity. In October 2007, the FDA approved one such derivative, called ixabepilone, for treatment of advanced breast cancer. Ixabepilone is an analog of epothilone B, in which the ester linkage is replaced with an amide linkage, highlighted in red.

Ixabepilone is currently being marketed by Bristol-Myers Squibb under the trade name Ixempra. Several other analogs of the epothilones are currently undergoing clinical trials for treatment of many different forms of cancer. The next decade is likely to witness several epothilone analogs emerge as new anticancer agents.

## 14.8 Preparation of Epoxides

### Preparation with Peroxy Acids

Recall from Section 9.9 that alkenes can be converted into epoxides upon treatment with peroxy acids (see Mechanism 9.6).

Any peroxy acid can be used, although MCPBA and peroxyacetic acid are the most common:

*meta*-Chloroperoxybenzoic acid (MCPBA)

Peroxyacetic acid

The process is stereospecific. Specifically, substituents that are *cis* to each other in the starting alkene remain *cis* to each other in the epoxide, and substituents that are *trans* to each other in the starting alkene remain *trans* to each other in the epoxide:

*cis* → MCPBA → *cis*

*trans* → MCPBA → *trans*

## Preparation from Halohydrins

Recall from Section 9.8 that alkenes can be converted into halohydrins when treated with a halogen in the presence of water (see Mechanism 9.5).

$Br_2$ / $H_2O$ + Enantiomer

Halohydrins can be converted into epoxides upon treatment with a strong base:

NaOH

The process is achieved via an intramolecular Williamson ether synthesis. An alkoxide ion is formed, which then functions as a nucleophile in an intramolecular $S_N2$-like process (Mechanism 14.4).

### MECHANISM 14.4 EPOXIDE FORMATION FROM HALOHYDRINS

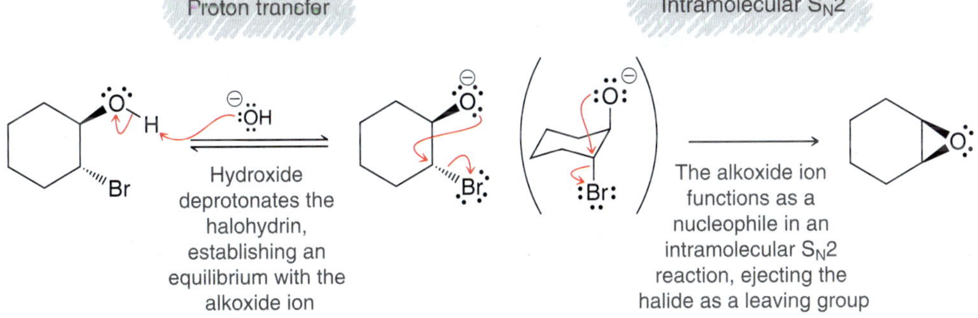

This provides us with another way of forming an epoxide from an alkene:

1) $Br_2$, $H_2O$
2) NaOH

The overall stereochemical outcome is the same as direct epoxidation with MCPBA. That is, substituents that are *cis* to each other in the starting alkene remain *cis* to each other in the epoxide, and substituents that are *trans* to each other in the starting alkene remain *trans* to each other in the epoxide:

MCPBA

1) $Br_2$, $H_2O$
2) NaOH

*cis* *cis*

# SKILLBUILDER

## 14.3 PREPARING EPOXIDES

**LEARN the skill** Show what reagents you would use to prepare the following epoxide:

Me Et Me + En

### SOLUTION

Begin by analyzing all four groups connected to the epoxide.

**STEP 1**
Identify the four groups attached to the epoxide.

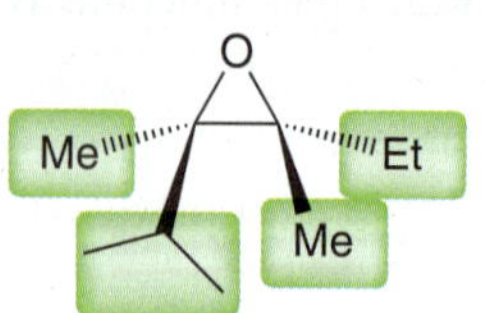


**STEP 2**
Identify the relative configuration of the four groups in the starting alkene.

The starting alkene must contain these four groups. Look carefully at the relative configuration of these groups. The methyl groups are *trans* to each other in the epoxide, which means that they must have been *trans* to each other in the starting alkene. To convert this alkene into the epoxide, we can use either of the following acceptable methods:

MCPBA

1) $Br_2$, $H_2O$
2) NaOH

*trans* *trans* + En

**PRACTICE the skill** **14.14** Identify the reactants you would use to form a racemic mixture of each of the following epoxides:

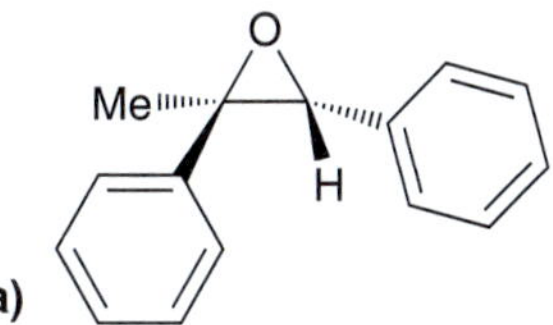


(a) (b) (c) (d)

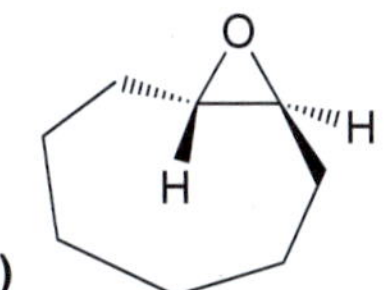

**APPLY the skill** **14.15** Consider the following two compounds. When treated with NaOH, one of these compounds forms an epoxide quite rapidly, while the other forms an epoxide very slowly. Identify which compound reacts more rapidly and explain the difference in rate between the two reactions. (**Hint:** You may find it helpful to review the conformations of substituted cyclohexanes in Section 4.13.)

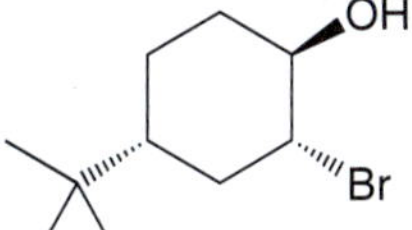


**Compound A** **Compound B**

need more **PRACTICE?** **Try Problems 14.39, 14.51t**

## medically speaking — Active Metabolites and Drug Interactions

Carbamazepine (Tegretol) is an anticonvulsant and mood-stabilizing drug used in the treatment of epilepsy and bipolar disorder. It is also used to treat ADD (attention-deficit disorder). It is metabolized in the liver to produce an epoxide.

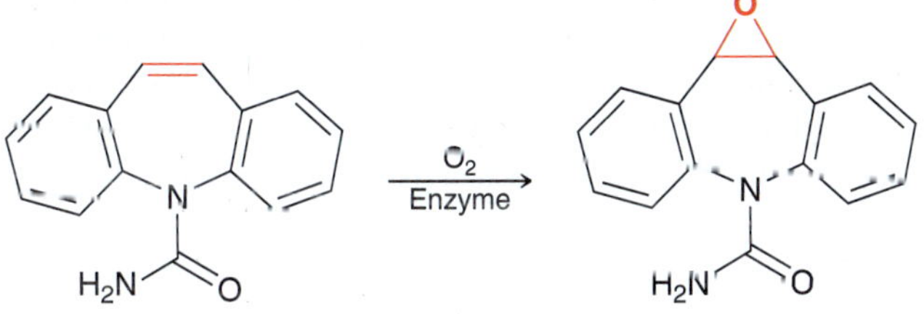


**Carbamazepine** **Carbamazepine-10,11-epoxide**

This epoxide is further metabolized by the enzyme epoxide hydroxylase to form a *trans* diol, which undergoes glucuronidation to produce a water-soluble adduct that can be excreted in the urine.

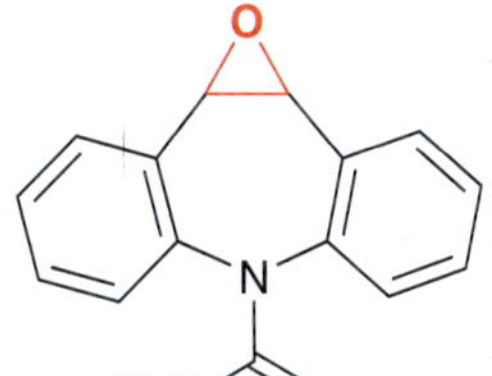


$H_2O$ / Epoxide hydroxylase → Glucuronidation → **Water-soluble adduct that is excreted in urine**

**Carbamazepine-10,11-epoxide** ***trans*-10,11-Dihydroxy-carbamazepine**

The epoxide metabolite is believed to exhibit activity similar to that of the parent compound and therefore contributes substantially to the overall therapeutic effects of carbamazepine. This fact must be taken into consideration when a patient is taking other medications. For instance, the antibiotic clarithromycin has been found to inhibit the action of the enzyme epoxide hydroxylase. This causes the concentration of the epoxide to be higher than normal, increasing the potency of carbamazepine. Before a physician prescribes carbamazepine for a patient, potential drug interactions must be taken into account. This is an example of one important factor that practicing physicians must consider—specifically, the effect that one drug can have on the potency of another drug.

## 14.9 Enantioselective Epoxidation

When forming an epoxide that is chiral, both of the previous methods will provide a racemic mixture:

MCPBA

1) $Br_2$, $H_2O$
2) NaOH

**Racemic mixture**

That is, the two enantiomers are formed in equal amounts, because the epoxide can be formed on either face of the alkene with equal likelihood.

Epoxide can form from above the plane

or epoxide can form from below the plane

If only one enantiomer is desired, then the methods we have learned for preparing epoxides will be inefficient, as half of the product is unusable and must be separated from the desired product. To favor formation of just one enantiomer, we must somehow favor epoxidation at one face of the alkene. K. Barry Sharpless, currently at the Scripps Research Institute, recognized that this could be accomplished with a chiral catalyst. He reasoned that a chiral catalyst could, in theory, create a chiral environment that would favor epoxidation at one face of the alkene. Specifically, a chiral catalyst can lower the energy of activation for formation of one enantiomer more dramatically than the other enantiomer (Figure 14.2). In this way, a chiral catalyst favors the production of one enantiomer over the other, leading to an observed enantiomeric excess (*ee*). Sharpless succeeded in developing such a catalyst for the *enantioselective epoxidation* of allylic alcohols. An allylic alcohol is an alkene in which a hydroxyl group is attached to an allylic position. Recall that the allylic position is the position next to a C=C bond.

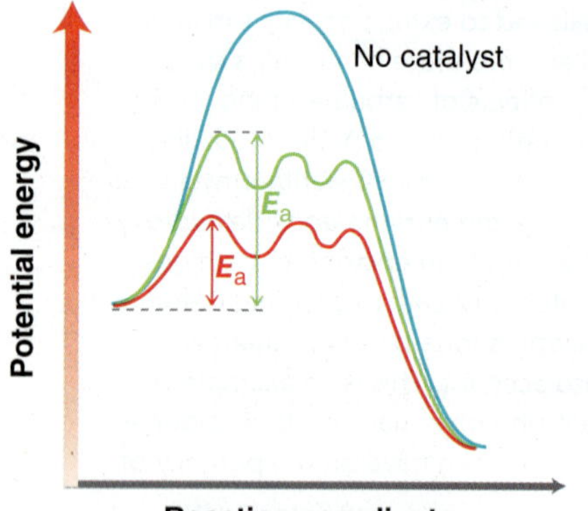


**FIGURE 14.2** An energy diagram that depicts the effect of a chiral catalyst. The formation of one enantiomer is more effectively catalyzed than the other.

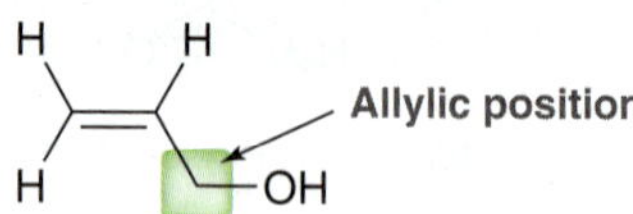


Sharpless' catalyst is comprised of titanium tetraisopropoxide and one enantiomer of diethyl tartrate (DET):

$Ti[OCH(CH_3)_2]_4$
**Titanium tetraisopropoxide**

**(+)-DET** or **(−)-DET**

Titanium tetraisopropoxide forms a chiral complex with either (+)-DET or (–)-DET, and this complex serves as the chiral catalyst. In the presence of such a catalyst, an oxidizing agent such as *tert*-butyl hydroperoxide (ROOH, where R = *tert*-butyl) can be employed to convert the alkene into an epoxide. The stereochemical outcome of the reaction depends on whether the chiral catalyst was formed with (+)-DET or (–)-DET. Both enantiomers of DET are readily available, and either one can be used. By choosing between (+)-DET or (–)-DET, it is possible to control which enantiomer is obtained:

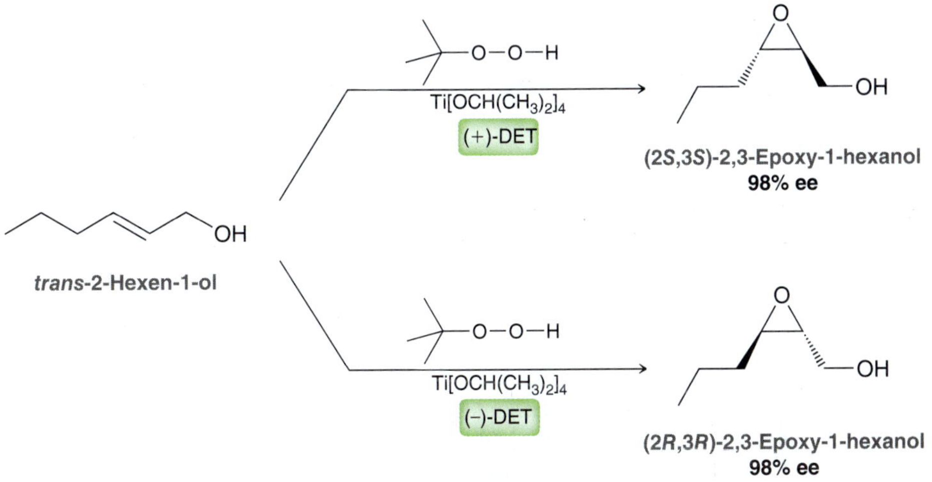

*trans*-2-Hexen-1-ol

$Ti[OCH(CH_3)_2]_4$, (+)-DET → **(2*S*,3*S*)-2,3-Epoxy-1-hexanol** **98% ee**

$Ti[OCH(CH_3)_2]_4$, (–)-DET → **(2*R*,3*R*)-2,3-Epoxy-1-hexanol** **98% ee**

This process is highly enantioselective and is extremely successful for a wide range of allylic alcohols. The double bond in the starting material can be mono-, di-, tri-, or tetrasubstituted. This process is extremely useful for the practicing synthetic organic chemist because it allows the introduction of a chirality center with enantioselectivity. Since Sharpless pioneered the field of enantioselective epoxidation, many more reagents have been developed for the asymmetric epoxidation of other alkenes that do not require the presence of an allylic hydroxyl group. Sharpless was instrumental in opening an important door and was a corecipient of the 2001 Nobel Prize in Chemistry (the other corecipients were Knowles and Noyori, who used similar reasoning to develop chiral catalysts for asymmetric hydrogenation reactions, as was discussed in Section 9.7).

To predict the product of a **Sharpless asymmetric epoxidation**, orient the molecule so that the allylic hydroxyl group appears in the upper right corner (Figure 14.3). When positioned in this way, (+)-DET gives epoxide formation above the plane, and (–)-DET gives epoxide formation below the plane.

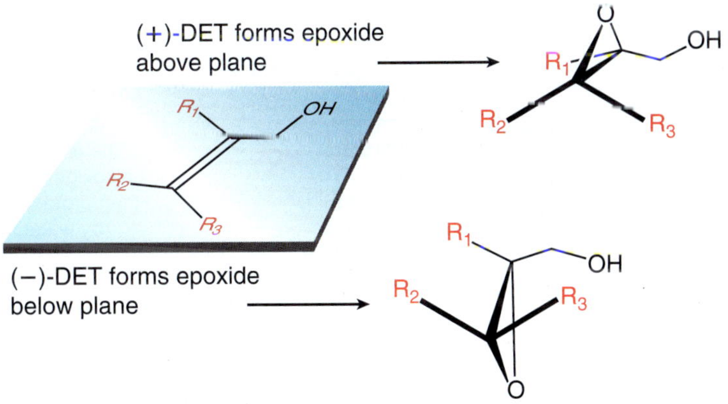

(+)-DET forms epoxide above plane

(−)-DET forms epoxide below plane

**FIGURE 14.3**
A method for predicting the product of a Sharpless epoxidation.

## CONCEPTUAL CHECKPOINT

**14.16** Predict the products for each of the following reactions:

(a) OH —[tert-butyl–O–O–H, Ti[OCH($CH_3$)$_2$]$_4$, (+)-DET]→ ?

(b) OH —[tert-butyl–O–O–H, Ti[OCH($CH_3$)$_2$]$_4$, (−)-DET]→ ?

(c) OH —[tert-butyl–O–O–H, Ti[OCH($CH_3$)$_2$]$_4$, (+)-DET]→ ?

(d) OH —[tert-butyl–O–O–H, Ti[OCH($CH_3$)$_2$]$_4$, (−)-DET]→ ?

## 14.10 Ring-Opening Reactions of Epoxides

Epoxides have significant ring strain, and as a result, they exhibit unique reactivity. Specifically, epoxides undergo reactions in which the ring is opened, which alleviates the strain. In this section, we will see that epoxides can be opened under conditions involving a strong nucleophile or under acid-catalyzed conditions.

### Reactions of Epoxides with Strong Nucleophiles

When an epoxide is subjected to attack by a strong nucleophile, a *ring-opening reaction* occurs. For example, consider the opening of ethylene oxide by a hydroxide ion.

O —[NaOH, $H_2O$]→ HO–CH₂CH₂–OH

The transformation involves two mechanistic steps (Mechanism 14.5).

## MECHANISM 14.5 EPOXIDE RING OPENING WITH A STRONG NUCLEOPHILE

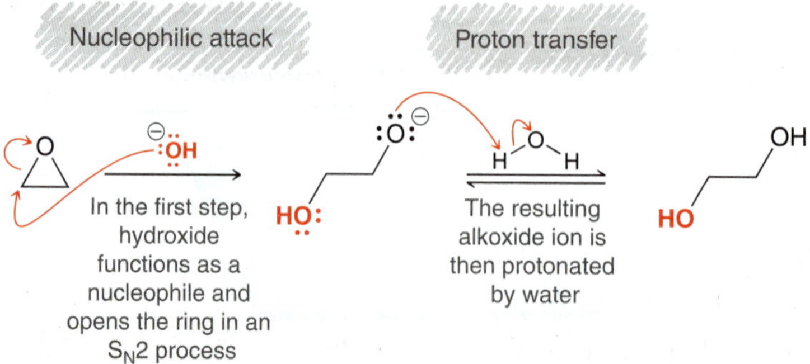


The first step of the mechanism is an $S_N2$ process, involving an alkoxide ion functioning as a leaving group. Although we learned in Chapter 7 that alkoxide ions do not function as leaving groups in $S_N2$ reactions, the exception here can be explained by focusing on the substrate. In this case,

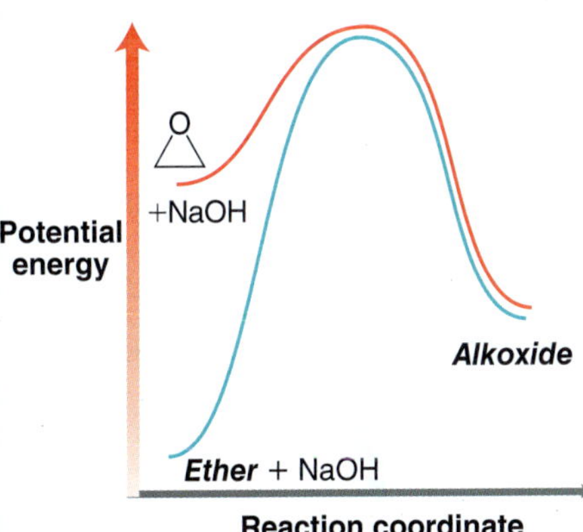


**FIGURE 14.4** An energy diagram showing the effect of using a high-energy substrate in an $S_N2$ reaction.

the substrate is an epoxide that exhibits significant ring strain and is therefore higher in energy than the substrates we encountered when we first learned about $S_N2$ reactions. The effects of a high-energy substrate are illustrated in the energy diagram in Figure 14.4. The blue curve represents a hypothetical $S_N2$ process in which the substrate is an ether and the leaving group is an alkoxide ion. The energy of activation for such a process is quite large, and more importantly, the products are higher in energy than the starting materials, so the equilibrium does not favor products. In contrast, when the starting substrate is an epoxide (red curve), the increased energy of the substrate has two pronounced effects: (1) the energy of activation is reduced, allowing the reaction to occur more rapidly, and (2) the products are now lower in energy than the starting materials, so the reaction is thermodynamically favorable. That is, the equilibrium will favor products over starting materials. For these reasons, an alkoxide ion can function as a leaving group in the ring-opening reactions of epoxides.

Many strong nucleophiles can be used to open an epoxide.

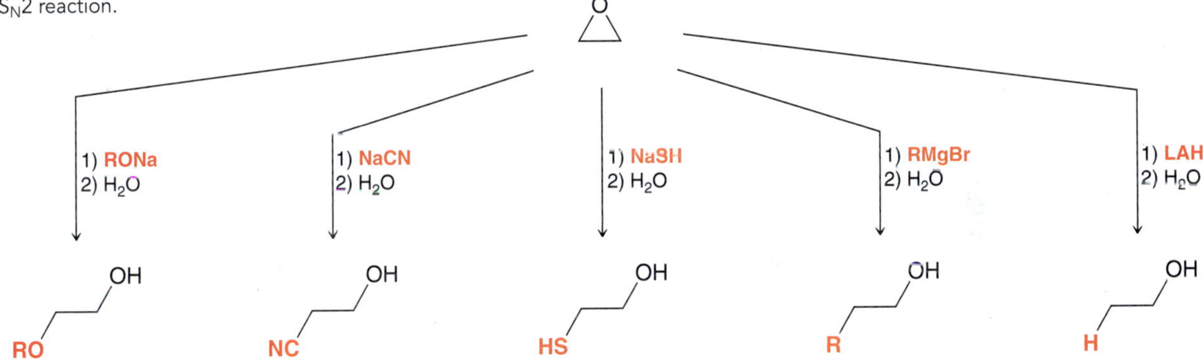


All of these nucleophiles are reagents that we have previously encountered, and they can all open epoxides. These reactions exhibit two important features that must be considered, regiochemistry and stereochemistry:

1. *Regiochemistry*. When the starting epoxide is unsymmetrical, the nucleophile attacks at the less substituted (less hindered) position.

O
1) Nuc⊖
2) $H_2O$
HO
Nuc

This position is less hindered, so the nucleophile attacks here

This steric effect is what we would expect from an $S_N2$ process.

2. *Stereochemistry*. When the attack takes place at a chirality center, inversion of configuration is observed.

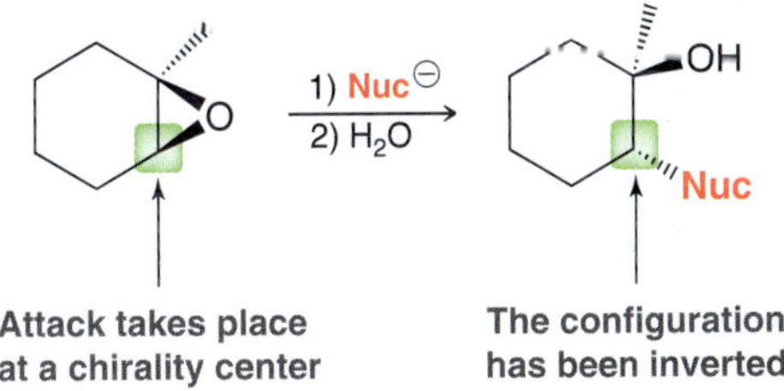


This result is also expected for an $S_N2$ process as a consequence of the requirement for back-side attack of the nucleophile. Notice that the configuration of the other chirality center is not affected by the process. Only the center being attacked undergoes an inversion of configuration.

# SKILLBUILDER

## 14.4 DRAWING THE MECHANISM AND PREDICTING THE PRODUCT OF THE REACTION BETWEEN A STRONG NUCLEOPHILE AND AN EPOXIDE

**LEARN the skill**

Predict the product of the following reaction and draw the likely mechanism for its formation:

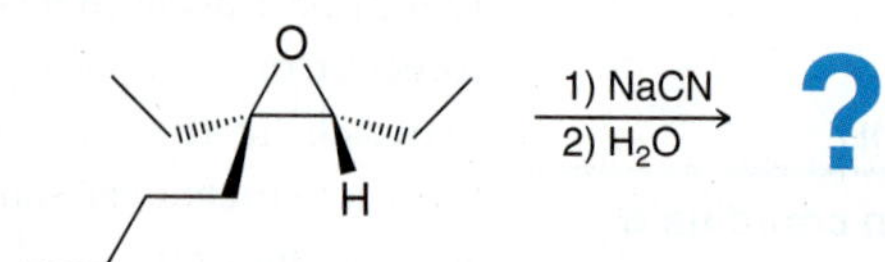


### SOLUTION

Cyanide is a strong nucleophile, so we expect a ring-opening reaction. In order to draw the product, we must consider the regiochemistry and stereochemistry of the reaction:

**STEP 1** Identify the regiochemistry by selecting the less hindered position as the site of nucleophilic attack.

1. To predict the regiochemistry, identify the less substituted (less hindered) position:

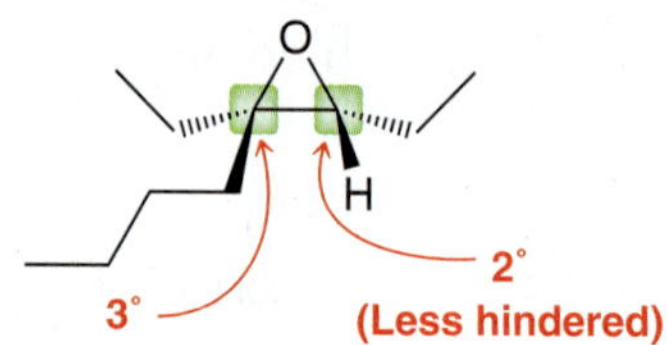


Recall that the reaction proceeds via an $S_N2$ process and is therefore highly sensitive to steric hindrance. As a result, we expect the nucleophile to attack the less hindered, secondary position, rather than the more hindered, tertiary position.

**STEP 2** Identify the stereochemistry by determining whether the nucleophile attacks a chirality center. If so, expect inversion of configuration.

2. Next identify the stereochemical outcome. Look at the center being attacked and determine if it is a chirality center. In this case, it is a chirality center. An $S_N2$ process is expected to proceed via back-side attack to give inversion of configuration at that center.

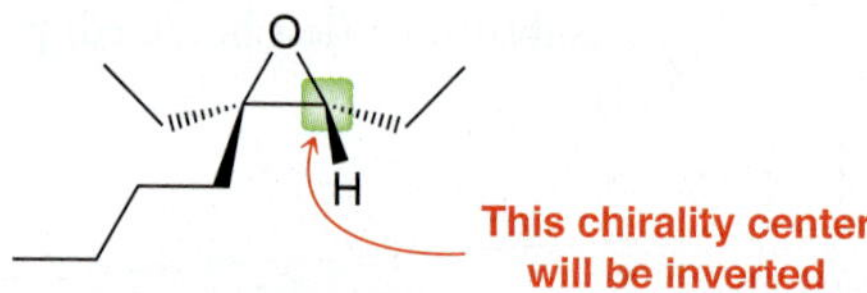


Now that we have predicted the regiochemical and stereochemical outcomes, we are ready to draw the mechanism. There are two separate steps: attack of the nucleophile to open the ring, followed by protonation of the alkoxide.

**STEP 3** Draw both steps of the mechanism.

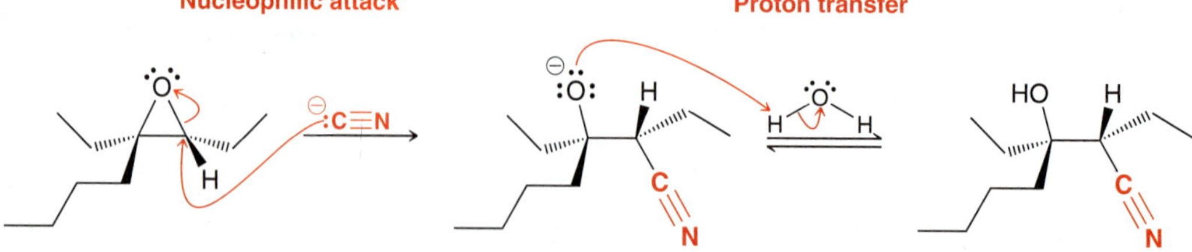

**PRACTICE the skill**

**14.17** For each of the following reactions, predict the product and draw a mechanism for its formation:

(c) 1) NaSH, 2) $H_2O$ → ?

(d) 1) LAH, 2) $H_2O$ → ?

(e) 1) NaSH, 2) $H_2O$ → ?

(f) 1) LAH, 2) $H_2O$ → ?

**APPLY** the skill

**14.18** When the following chiral epoxide is treated with aqueous sodium hydroxide, only one product is obtained, and that product is achiral. Draw the product and explain why only one product is formed.

NaOH, $H_2O$ → ?

**14.19** When *meso*-2,3-epoxybutane is treated with aqueous sodium hydroxide, two products are obtained. Draw both products and describe their relationship.

need more **PRACTICE?** **Try Problems 14.42a, 14.42c, 14.42e–h, 14.43a,c**

## practically speaking — Ethylene Oxide as a Sterilizing Agent for Sensitive Medical Equipment

Ethylene oxide is a colorless, flammable gas that is often used to sterilize temperature-sensitive medical equipment. The gas easily diffuses through porous materials and effectively kills all forms of microorganisms, even at room temperature. The mechanism of action likely involves a functional group in DNA attacking the ring and causing a ring opening of the epoxide, effectively alkylating that site.

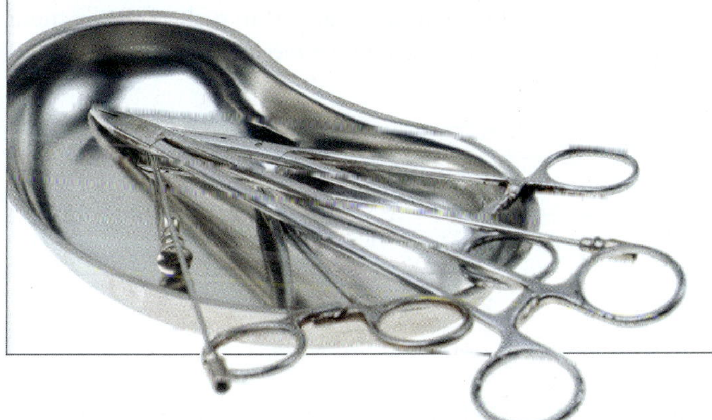

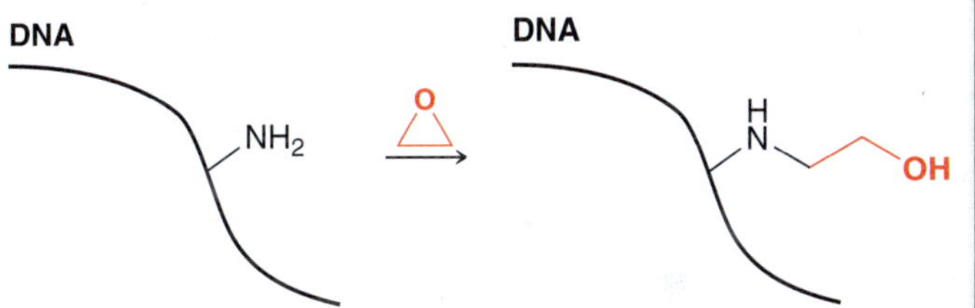


This alkylation process interferes with the normal function of DNA, thereby killing the microorganisms. The use of pure ethylene oxide presents a hazard, because it mixes with atmospheric oxygen and becomes susceptible to explosion. This problem is circumvented by using a mixture of ethylene oxide and carbon dioxide, which is no longer explosive. Such mixtures are sold commercially for the sterilization of medical equipment and agricultural grains. One such mixture is called Carboxide and is comprised of 10% ethylene oxide and 90% $CO_2$. Carboxide can be exposed to air without the danger of explosion.

## Acid-Catalyzed Ring Opening

In the previous section, we saw the reactions of epoxides with strong nucleophiles. The driving force for such reactions was the removal of ring strain associated with the three-membered ring of an epoxide. Ring-opening reactions can also occur under acidic conditions. As an example, consider the reaction between ethylene oxide and HX.

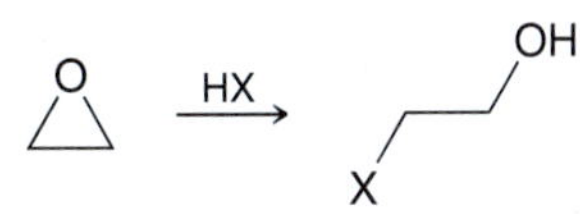

This transformation involves two mechanistic steps (Mechanism 14.6).

## MECHANISM 14.6 ACID-CATALYZED RING OPENING OF AN EPOXIDE

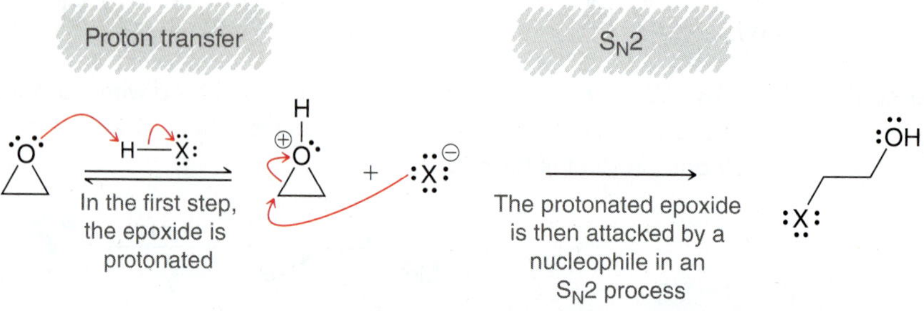


The first step is a proton transfer, and the second step is nucleophilic attack ($S_N2$) by a halide ion. This reaction can be accomplished with HCl, HBr, or HI. Other nucleophiles such as water or an alcohol can also open an epoxide ring under acidic conditions. A small amount of acid (often sulfuric acid) is used to catalyze the reaction.

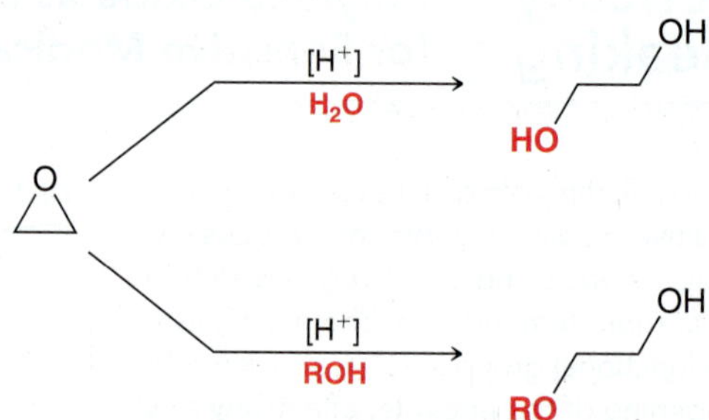


The brackets around the $H^+$ indicate that the acid functions as a catalyst. In each of the reactions above, the mechanism involves a proton transfer as the final step of the mechanism.

Proton transfer — $S_N2$ — Proton transfer

The first two steps are analogous to the two steps in Mechanism 14.6. The additional proton transfer step at the end of the mechanism is required to remove the charge formed after the attack of a neutral nucleophile. The process above is used for the mass production of ethylene glycol.

$[H_2SO_4]$, $H_2O$

**Ethylene oxide** → **Ethylene glycol**

Each year, over three million tons of ethylene glycol are produced in the United States via the acid-catalyzed ring opening of ethylene oxide. Most of the ethylene glycol is used as antifreeze.

As we saw in the previous section, there are two important features of ring-opening reactions: the regiochemical outcome and the stereochemical outcome. We'll begin with regiochemistry. When the

starting epoxide is unsymmetrical, the regiochemical outcome depends on the nature of the epoxide. If one side is primary and the other side is secondary, then attack takes place at the less hindered primary position, just as we would expect for an $S_N2$ process.

Nucleophile attacks here (less hindered)

However, when one side of the epoxide is a tertiary position, the reaction is observed to occur at the more substituted, tertiary site.

Nucleophile attacks here, even though it is more hindered

Why should this be the case? It is true that the primary site is less hindered, but there is a factor that is even more dominant than steric effects. That factor is an *electronic* effect. A protonated epoxide is positively charged, and the positively charged oxygen atom withdraws electron density from the two carbon atoms of the epoxide.

Each of the carbon atoms bears a partial positive charge (δ+). That is, they both have partial carbocationic character. Nevertheless, these two carbon atoms are not equivalent in their ability to support a partial positive charge. The tertiary position is significantly better at supporting a partial positive charge, so the tertiary position has significantly more partial carbocationic character than the primary position. The protonated epoxide is therefore more accurately drawn in the following way:

There are two important consequences of this analysis: (1) the more substituted carbon is a stronger electrophile and is therefore more susceptible to nucleophilic attack and (2) the more substituted carbon has significant carbocationic character, which means that its geometry is described as somewhere between tetrahedral and trigonal planar, allowing nucleophilic attack to occur at that position even though it is tertiary.

To summarize, the regiochemical outcome of acid-catalyzed ring opening depends on the nature of the epoxide.

Primary vs. secondary

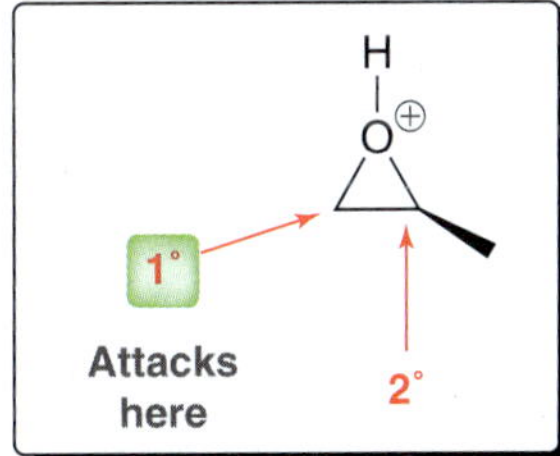


Dominant factor = steric effect

Primary vs. tertiary

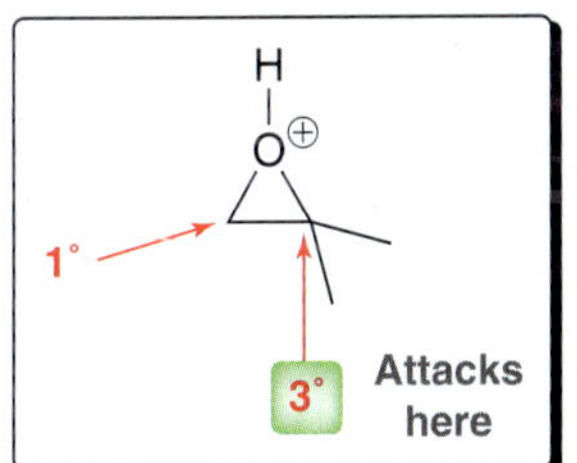


Dominant factor = electronic effect

There are two factors competing to control the regiochemistry: electronic effects vs. steric effects. The former favors attack at the more substituted position, while the latter favors attack at the less substituted position. To determine which factor is dominant, we must analyze the epoxide. When the epoxide possesses a tertiary position, the electronic effect will be dominant. When the epoxide possesses only primary and secondary positions, the steric effect will be dominant. The regiochemistry of acid-catalyzed ring opening is just one example where steric effects and electronic effects compete. As we progress through the course, we will see other examples of electronic vs. steric effects.

In the previous section (ring opening with strong nucleophiles), the regiochemistry was more straightforward, because electronics was not a factor at all. The epoxide was attacked by a nucleophile before being protonated, so the epoxide did not bear a positive charge when it was attacked. In such a case, steric hindrance was the only consideration.

Now let's turn our attention to the stereochemistry of ring-opening reactions under acid-catalyzed conditions. When the attack takes place at a chirality center, inversion of configuration is observed. This result is consistent with an $S_N2$-like process involving back-side attack of the nucleophile.

O, Me $\xrightarrow[ROH]{[H^+]}$ OH, Me, OR

**Attack takes place at a chirality center** — **The configuration has been inverted**

# SKILLBUILDER

## 14.5 DRAWING THE MECHANISM AND PREDICTING THE PRODUCT OF ACID-CATALYZED RING OPENING

**LEARN** the skill

Predict the product of the reaction below and draw the likely mechanism for its formation:

O, Et, Et, Me, H $\xrightarrow[EtOH]{[H_2SO_4]}$ ?

**SOLUTION**

The presence of sulfuric acid indicates that the epoxide is opened under acid-catalyzed conditions. In order to draw the product, we must consider the regiochemistry and stereochemistry of the reaction.

**STEP 1**
Identify the regiochemistry by determining whether steric or electronic effects will dominate.

1. To predict the regiochemical outcome, analyze the epoxide.

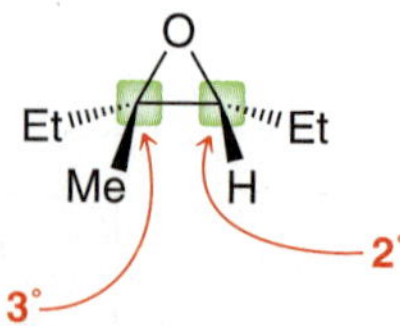


One side is tertiary, so we expect that electronic effects will control the regiochemistry, and we predict that nucleophilic attack will occur at the more substituted, tertiary position.

**STEP 2**
Identify the stereochemistry by determining if the nucleophile attacks a chirality center. If so, expect inversion.

2. Next identify the stereochemical outcome. Look at the center being attacked and determine if it is a chirality center. In this case, it is a chirality center, so we expect back-side attack of the nucleophile to invert the configuration of that center.

This chirality center will be inverted as a result of back-side attack

Now that we have predicted the regiochemical and stereochemical outcomes, we are ready to draw a mechanism. The epoxide is first protonated and then attacked by a nucleophile (EtOH).

**STEP 3**
Draw all three steps of the mechanism.

Proton transfer

$S_N2$

Since the attacking nucleophile was neutral (EtOH), an additional proton transfer step will be required to remove the charge and generate the final product.

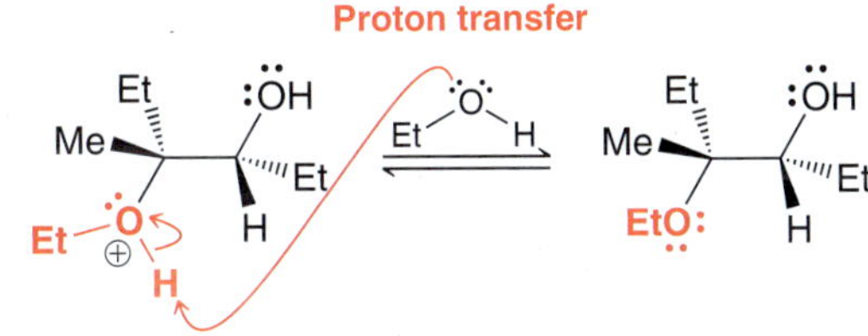

**PRACTICE the skill** 14.20 For each reaction, predict the product and draw a mechanism for its formation:

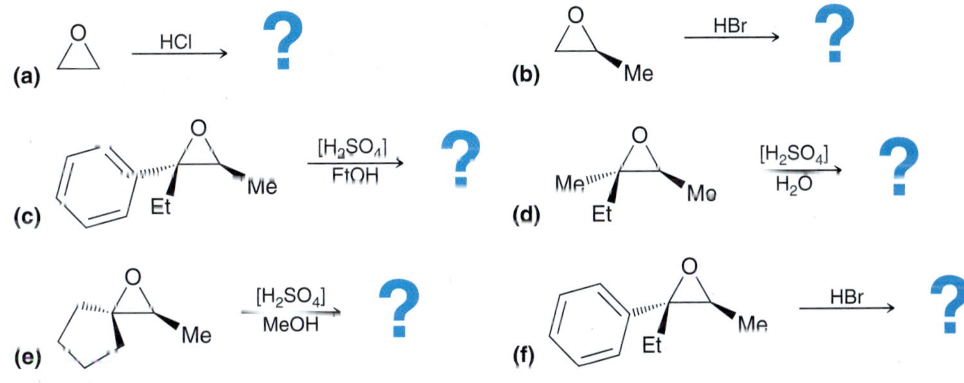

**APPLY the skill** 14.21 Propose a mechanism for the following transformation:

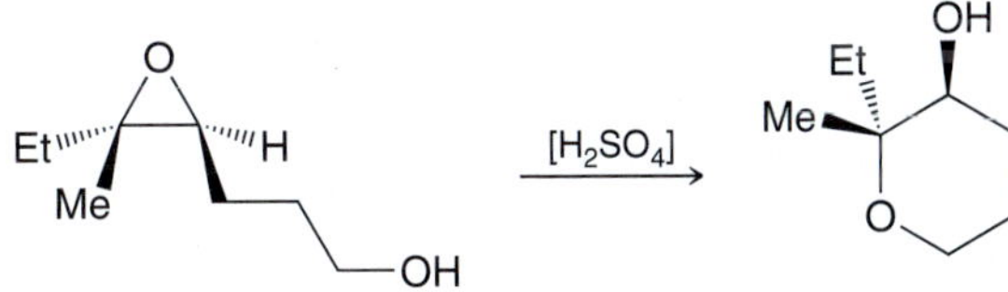


need more **PRACTICE?** **Try Problems 14.43d, 14.49, 14.50**

## Cigarette Smoke and Carcinogenic Epoxides

As we have seen many times in previous chapters, the combustion of organic materials should produce $CO_2$ and water. But combustion rarely produces only these two products. Usually, incomplete combustion produces organic compounds that account for the smoke that is observed to emanate from a fire. One of those compounds is called benzo[a]pyrene. This highly carcinogenic compound is produced from the burning of organic materials, such as gasoline, wood, and cigarettes. In recent years, extensive research has elucidated the likely mechanistic pathway for the carcinogenicity of this compound, and it has been shown that benzo[a]pyrene itself is not the compound that causes cancer. Rather, it is one of the metabolites (one of the compounds produced during the metabolism of benzo[a]pyrene) that is a highly reactive intermediate capable of alkylating DNA and thereby interfering with the normal function of DNA.

When benzo[a]pyrene is metabolized, an initial epoxide is formed (called an arene oxide) which is then opened by water to give a diol.

**Benzo[a]pyrene**

Cytochrome $P_{450}$ | $O_2$

O

**An arene oxide**

Epoxide hydrolase | $H_2O$

HO, OH

These two steps should be very familiar to you. The first step is epoxide formation, and the second step is a ring-opening reaction in the presence of a catalyst to form a diol. This diol can then undergo epoxidation another time to give the following diol epoxide:

HO, OH — $O_2$ / Cytochrome $P_{450}$ → O, HO, OH

**A diol epoxide**

This diol epoxide is carcinogenic because it is capable of alkylating DNA. Studies show that the amino group of deoxyguanosine (in DNA) attacks the epoxide. This reaction changes the structure of DNA and causes genetic code alterations that ultimately lead to the formation of cancer cells.

Benzo[a]pyrene is not the only carcinogenic compound in cigarette smoke. Throughout this book, we will see several other carcinogenic compounds that are also present in cigarette smoke.

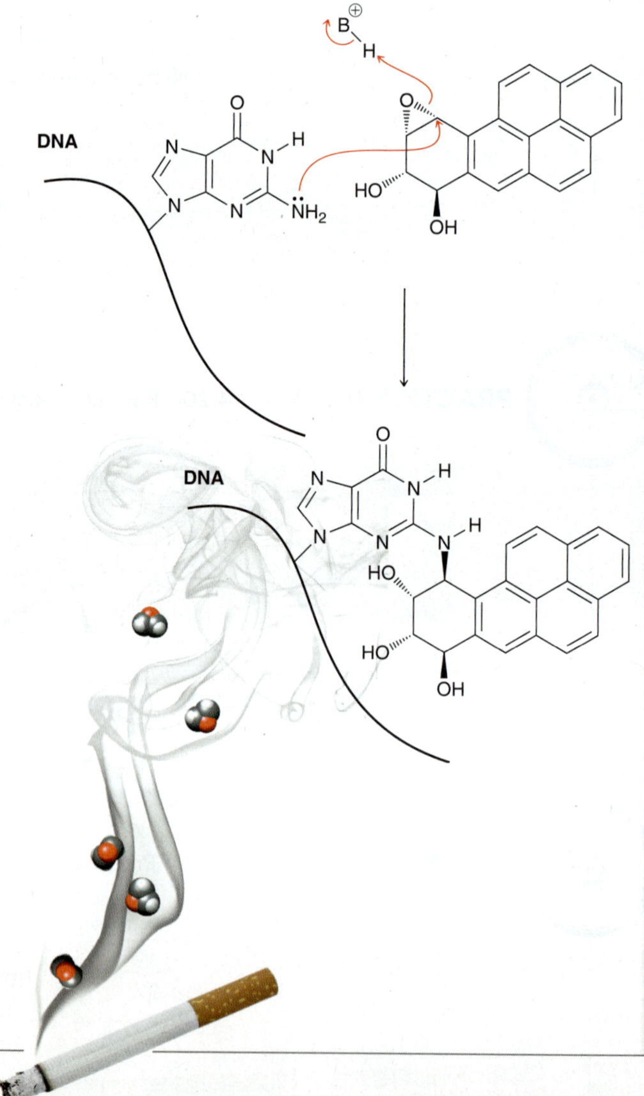

## 14.11 Thiols and Sulfides

### Thiols

Sulfur is directly below oxygen in the periodic table (in the same column), and therefore, many oxygen-containing compounds have sulfur analogs. Sulfur analogs of alcohols contain an SH group in place of an OH group and are called **thiols**. The nomenclature of thiols is similar to that of alcohols, but the suffix of the name is "thiol" instead of "ol":

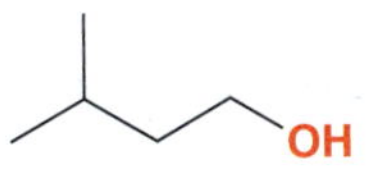

OH SH

3-Methyl-1-butanol 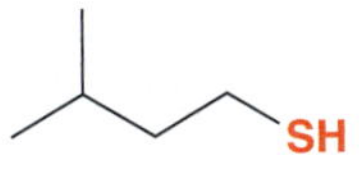 3-Methyl-1-butanethiol

Notice that the "e" is kept before the suffix "thiol." When another functional group is present in the compound, the SH group is named as a substituent and is called a **mercapto group**:

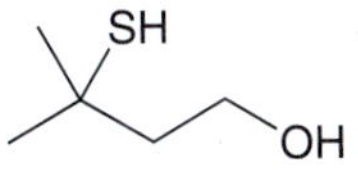

SH OH

3-Mercapto-3-methyl-1-butanol

The name "mercapto" is derived from the fact that thiols were once called mercaptans. This terminology was abandoned by IUPAC several decades ago, but old habits die hard, and many chemists still refer to thiols as mercaptans. The term is derived from the Latin *mercurium captans* (capturing mercury) and describes the ability of thiols to form complexes with mercury as well as other metals. This ability is put to good use by the drug called dimercaprol, which is used to treat mercury and lead poisoning.

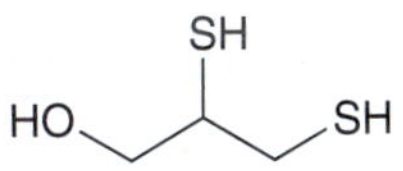

SH HO SH

Dimercaprol
(2,3-dimercapto-1-propanol)

Thiols are most notorious for their pungent, unpleasant odors. Skunks use thiols as a defense mechanism to ward off predators by spraying a mixture that delivers a mighty stench. Methanethiol is added to natural gas so that gas leaks can be easily detected. If you have ever smelled a gas leak, you were smelling the methanethiol ($CH_3SH$) in the natural gas, as natural gas is odorless. Surprisingly, scientists who have worked with thiols report that the nasty odor actually becomes pleasant after prolonged exposure. The author of this textbook can attest to this fact.

Thiols can be prepared via an $S_N2$ reaction between sodium hydrosulfide (NaSH) and a suitable alkyl halide; for example:

Br $\xrightarrow{\text{NaSH}}$ SH + NaBr

This reaction can occur even at secondary substrates without competing E2 reactions, because the hydrosulfide ion ($HS^-$) is an excellent nucleophile and a poor base. When this nucleophile attacks a chirality center, inversion of configuration is observed.

Br $R_1$ $R_2$ $\xrightarrow{\text{NaSH}}$ SH $R_1$ $R_2$

## CONCEPTUAL CHECKPOINT

**14.22** What reagents would you use to prepare each of the following thiols?

(a) (b) (c)

Thiols easily undergo oxidation to produce **disulfides**.

$$\text{EtSH} + \text{HSEt} \xrightarrow{\text{NaOH/H}_2\text{O, Br}_2} \text{EtS–SEt}$$

**A disulfide**

The conversion of thiols into disulfides requires an oxidizing reagent, such as bromine in aqueous hydroxide. The process begins with deprotonation of the thiol to generate a **thiolate** ion (Mechanism 14.7). Hydroxide is a strong enough base that the equilibrium favors formation of the thiolate ion. This thiolate ion is an excellent nucleophile and can attack molecular bromine in an $S_N2$ process. A second $S_N2$ process then produces the disulfide.

## MECHANISM 14.7 OXIDATION OF THIOLS

DEPROTONATION OF THE THIOL

Proton transfer

R–S–H + Na⊕ ⊖:OH ⇌ R–S:⊖ Na⊕ + $H_2O$

Thiol ($pK_a$=10.5) (Stronger acid) | Hydroxide | In the first step, the thiol is deprotonated to form a thiolate ion | Thiolate ion | Water ($pK_a$=15.7) (Weaker acid)

FORMATION OF THE DISULFIDE

$S_N2$ — R–S:⊖ Na⊕ + :Br–Br: → R–S–Br: + NaBr

A thiolate ion functions as a nucleophile and attacks molecular bromine

$S_N2$ — R–S–Br: + Na⊕ ⊖:S–R → R–S–S–R + NaBr

Another thiolate functions as a nucleophile in a second $S_N2$ process

There are many oxidizing agents that can be used to convert thiols into disulfides. In fact, the reaction is accomplished with so much ease that atmospheric oxygen can function as an oxidizing agent to produce disulfides. Catalysts can be used to speed up the process. Disulfides are also easily reduced back to thiols when treated with a reducing agent, such as HCl in the presence of zinc.

$$\text{EtS–SEt} \xrightarrow[\text{[Reduction]}]{\text{HCl, Zn}} \text{EtSH} + \text{HSEt}$$

The ease of interconversion between thiols and disulfides is attributed to the nature of the S—S bond. It has a bond strength of approximately 220 kJ/mol, which is about half the strength of many other covalent bonds. The interconversion between thiols and disulfides is extremely important in determining the shape of many biologically active compounds, as we will explore in Section 25.4.

## Sulfides

The sulfur analogs of ethers are called **sulfides,** or thioethers.

R–O–R — An ether

R–S–R — A sulfide (thioether)

Nomenclature of sulfides is similar to that of ethers. Common names are assigned using the suffix "sulfide" instead of "ether."

Diethyl ether — Diethyl sulfide

More complex sulfides are named systematically, much the way ethers are named, with the alkoxy group being replaced by an **alkylthio group**.

$OCH_3$ Methoxy group — 1,1-Dichloro-4-methoxycyclohexane

$SCH_3$ Methylthio group — 1,1-Dichloro-4-(methylthio)cyclohexane

Sulfides can be prepared from thiols in the following way:

$$\text{R—SH} \xrightarrow[\text{2) RX}]{\text{1) NaOH}} \text{R—S—R}$$

This process is essentially the sulfur analog of the Williamson ether synthesis (Mechanism 14.8).

### MECHANISM 14.8 PREPARATION OF SULFIDES FROM THIOLS

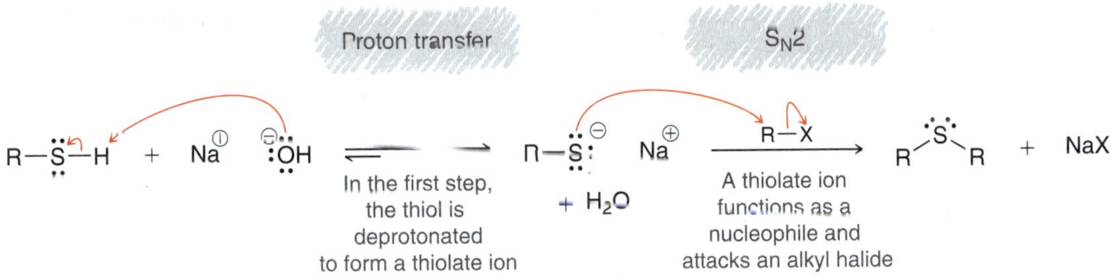


In the first step, hydroxide is used to deprotonate the thiol and produce a thiolate ion. The thiolate ion then functions as a nucleophile and attacks an alkyl halide, producing the sulfide. The process follows an $S_N2$ pathway, so the regular restrictions apply. The reaction works well with methyl and primary alkyl halides, can often be accomplished with secondary alkyl halides, and does not work for tertiary alkyl halides.

Since sulfides are structurally similar to ethers, we might expect sulfides to be as unreactive as ethers, but this is not the case. Sulfides undergo several important reactions.

1. Sulfides will attack alkyl halides in an $S_N2$ process.

The product of this step is a powerful alkylating agent, because it is capable of transferring a methyl group to a nucleophile.

**LOOKING BACK**
For a review of alkylation reactions with SAM, see Section 7.4.

We have already seen one such example of this process in Chapter 7. Recall that SAM is a biological methylating agent.

S-Adenosylmethionine (SAM)

2. Sulfides also undergo oxidation to give **sulfoxides** and then **sulfones**.

Sulfide Sulfoxide Sulfone

The initial product is a sulfoxide. If the oxidizing agent is strong enough and present in excess, then the sulfoxide is oxidized further to give the sulfone. For good yields of the sulfoxide without further oxidation to the sulfone, it is necessary to use an oxidizing reagent that will not oxidize the sulfoxide. Many such reagents are available, including sodium *meta*-periodate, $NaIO_4$.

Methyl phenyl sulfide Methyl phenyl sulfoxide

If the sulfone is the desired product, then two equivalents of hydrogen peroxide are used.

Methyl phenyl sulfide Methyl phenyl sulfone

The S=O bonds in sulfoxides and sulfones actually have little double-bond character. The $3p$ orbital of a sulfur atom is much larger than the $2p$ orbital of an oxygen, and therefore, the orbital overlap of the π bond between these two atoms is not effective. Consequently, sulfoxides and sulfones are often drawn with each S—O bond as a single bond.

Sulfoxides can be drawn as either one of these resonance structures

Sulfones can be drawn as either one of these resonance structures

The ease with which sulfides are oxidized renders them ideal reducing agents in a wide variety of applications. For example, recall that DMS (dimethyl sulfide) is used as a reducing agent in ozonolysis (Section 9.11). The by-product is dimethyl sulfoxide (DMSO).

## CONCEPTUAL CHECKPOINT

**14.23** Predict the products for each of the following reactions:

(a) 1) NaOH 2) Br ?

(b) SNa ?

(c) $NaIO_4$ ?

(d) $2\ H_2O_2$ ?

## 14.12 Synthesis Strategies Involving Epoxides

Recall from Chapter 12 that the two issues to consider when proposing a synthesis are whether there are changes in the carbon skeleton or in the functional group. In this chapter, we have learned valuable skills in both categories. Let's focus on them one at a time.

### Installing Two Adjacent Functional Groups

The most useful synthetic techniques that we learned in this chapter involve epoxides. We saw several ways to form epoxides and many reagents that open epoxides. Note that opening an epoxide provides two functional groups on adjacent carbon atoms.

Whenever you see two adjacent functional groups, you should think of epoxides. Let's see an example.

# SKILLBUILDER

## 14.6 INSTALLING TWO ADJACENT FUNCTIONAL GROUPS

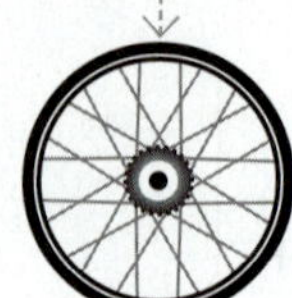

**LEARN** the skill

Propose a synthesis for the following transformation:

+ Enantiomer

### SOLUTION

Always approach a synthesis problem by initially asking two questions.

1. Is there a change in the carbon skeleton? No, the carbon skeleton is not changing.
2. Is there a change in the functional groups? Yes, the starting material has no functional groups, and the product has two adjacent functional groups.

The answers to these questions dictate what must be done. Specifically, we must install two adjacent functional groups. This suggests that we consider using a ring-opening reaction of an epoxide. Using a retrosynthetic analysis, we draw the epoxide that would be necessary.

The regiochemistry of this step requires that the methoxy group must be placed at the more substituted position. This dictates that the epoxide must be opened under acidic conditions to ensure that the nucleophile (MeOH) attacks at the more substituted position.

Our next step is to determine how to make the epoxide. We have seen a couple of ways to make epoxides, both of which start with an alkene.

At this point, we can start working forward, focusing on converting the starting material into the desired alkene. The starting material has no functional groups, and we have seen only one method for introducing a functional group into an alkane. Specifically, we must employ a radical bromination.

$Br_2$, $h\nu$

At this point, we just need to bridge the gap.

+ Enantiomer

$Br_2$, $h\nu$

MeOH, [ $H^+$ ]

?

MCPBA

A strong base, such as ethoxide, will produce an elimination reaction that will form the desired alkene. So, our proposed synthesis is:

1) $Br_2$, *hν*
2) NaOEt
3) MCPBA
4) $[H^+]$, MeOH

OMe, OH + Enantiomer

**PRACTICE the skill** **14.24** Propose an efficient synthesis for each of the following transformations:

(a) → HO, CN (b) → OH, OH + En

(c) → OH, SH (d) → SH, OH

(e) O →  O

**APPLY the skill** **14.25** In the previous chapter, we saw that ethylene glycol is one of the main components of automobile antifreeze. Using iodoethane as your starting material, show how you could prepare ethylene glycol.

need more **PRACTICE?** **Try Problem 14.51**

## Grignard Reagents: Controlling the Location of the Resulting Functional Group

In this chapter, we have seen a new way to make a C—C bond by using a Grignard reagent to open an epoxide. Until now, we have thought about this reaction from the point of view of the epoxide. That is, the epoxide is considered to be the starting material, and the Grignard reagent is used to modify the structure of the starting material. For example:

1) **R**MgBr, diethyl ether
2) $H_2O$

Starting material → Product (R, OH)

In this reaction, the epoxide is opened and an alkyl group (R) is introduced into the structure. Another way to think about this type of reaction is from the point of view of the alkyl halide. That is, the alkyl halide can be considered to be the starting material, and the epoxide is then used to introduce carbon atoms into the structure of the alkyl halide.

R Br (Starting material)

1) Mg, diethyl ether
2) epoxide
3) $H_2O$

R, OH (Product)

It is important to see this reaction from this point of view as well, as it highlights an important feature of the process. Specifically, this process can be used to introduce a chain of carbon atoms that possess a built-in functional group at the second carbon atom.

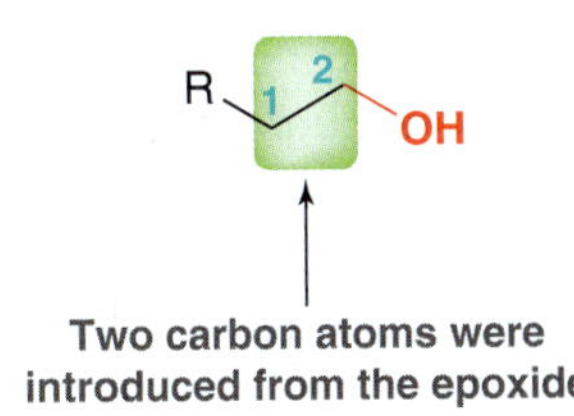

Notice the position of the functional group. It is on the second carbon atom of the newly installed chain. That is an extremely important feature, because we obtain a different outcome when a Grignard reagent attacks a ketone or aldehyde (Section 13.6). Specifically, the functional group appears on the first carbon atom of the new chain that was introduced.

R—Br → 1) Mg, diethyl ether; 2) acetaldehyde; 3) $H_2O$ →

Notice the difference. You must train your eyes to look at the precise location of a functional group when it appears on a newly introduced alkyl fragment. Let's see a specific example of this.

SKILLBUILDER

## 14.7 CHOOSING THE APPROPRIATE GRIGNARD REACTION

**LEARN the skill** Propose a synthesis for each of the following two transformations:

(a) (b)

### SOLUTION

**(a)** Always approach a synthesis problem by initially asking whether there are changes in the carbon skeleton or in the functional groups. In this case, there are changes both in the carbon skeleton and in the functional group. A three-carbon chain is introduced, and both the identity and location of the functional group have changed.

From the answers to these two questions we know that we must introduce three carbon atoms and a functional group (C═O). It would be inefficient to think of these two tasks as separate. If we first attach a three-carbon chain and only then think about how to introduce the C═O bond in exactly the correct position, we will find it very difficult to install the carbonyl group. It is more efficient to introduce the three carbon atoms in such a way that a functional group is already in the correct location. So let's analyze the precise location.

If we look at the three carbon atoms that are being introduced, the functional group is on the second carbon atom. That should signal an epoxide opening.

This one step introduces the three-carbon chain and simultaneously places a functional group in the proper location. Granted, the functional group is a hydroxyl group, rather than a carbonyl group. But once the hydroxyl group is in the correct location, it is easy enough to perform an oxidation and form the carbonyl group.

[O]

Always remember that functional groups can be easily interconverted. The important consideration is how to place a functional group in the desired location. In summary, our proposed synthesis is:

1) Mg, diethyl ether
2) (epoxide)
3) $H_2O$
4) $Na_2Cr_2O_7$, $H_2SO_4$, $H_2O$

**(b)** This problem is very similar to the previous problem. Once again, we are introducing three carbon atoms, but this time, the functional group is located at the first carbon atom on the chain that was introduced.

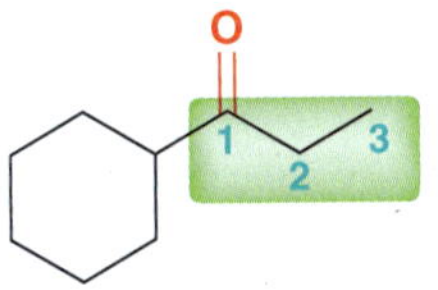


This means that the compound cannot be formed using an epoxide ring opening. In this case, the Grignard reagent must attack an aldehyde rather than an epoxide.

This sequence places the functional group in the desired location. It is then easy enough to oxidize the alcohol to form the desired ketone, as in the previous problem. In summary, our proposed synthesis is:

1) Mg, diethyl ether
2) (propanal)
3) $H_2O$
4) $Na_2Cr_2O_7$, $H_2SO_4$, $H_2O$

Notice the difference between the two syntheses above. Both involve introduction of carbon atoms via a Grignard reagent, but the precise location of the functional group dictates whether the Grignard reagent should attack an epoxide or an aldehyde.

**PRACTICE the skill** **14.26** Propose an efficient synthesis for each of the following transformations:

(a) (b) (c) (d) (e) (f) (g) (h) (i)

**APPLY the skill**

**14.27** Propose an efficient synthesis for 1,4-dioxane using acetylene as your only source of carbon atoms.

**1,4-Dioxane**

**14.28** Dimethoxyethane (DME) is a polar aprotic solvent often used for $S_N2$ reactions. Propose an efficient synthesis for DME using acetylene and methyl iodide as your only sources of carbon atoms.

**Dimethoxyethane**

**14.29** Using compounds that possess no more than two carbon atoms, propose an efficient synthesis for the following compound:

need more **PRACTICE?** **Try Problem 14.51**

# REVIEW OF REACTIONS

## Preparation of Ethers

**Williamson Ether Synthesis**

R—OH → (1) NaH, 2) RX) → R—O—R

**Alkoxymercuration-Demercuration**

1) $Hg(OAc)_2$, ROH
2) $NaBH_4$

## Reactions of Ethers

**Acidic Cleavage**

R—O—R → (Excess HX, Heat) → R—X + R—X + $H_2O$

Ph—O—R → (Excess HX, Heat) → Ph—OH + R—X

**Autooxidation**

$O_2$ (slow)

**A hydroperoxide**

## Preparation of Epoxides

MCPBA

1) $Br_2$, $H_2O$
2) NaOH

*cis* → *cis*

### Enantioselective Epoxidation

R–CH=CH–CH$_2$OH $\xrightarrow[\text{(+)-DET}]{(CH_3)_3COOH,\ Ti[OCH(CH_3)_2]_4}$ epoxide (R, OH)

R–CH=CH–CH$_2$OH $\xrightarrow[\text{(−)-DET}]{(CH_3)_3COOH,\ Ti[OCH(CH_3)_2]_4}$ epoxide (R, OH)

### Ring-Opening Reactions of Epoxides

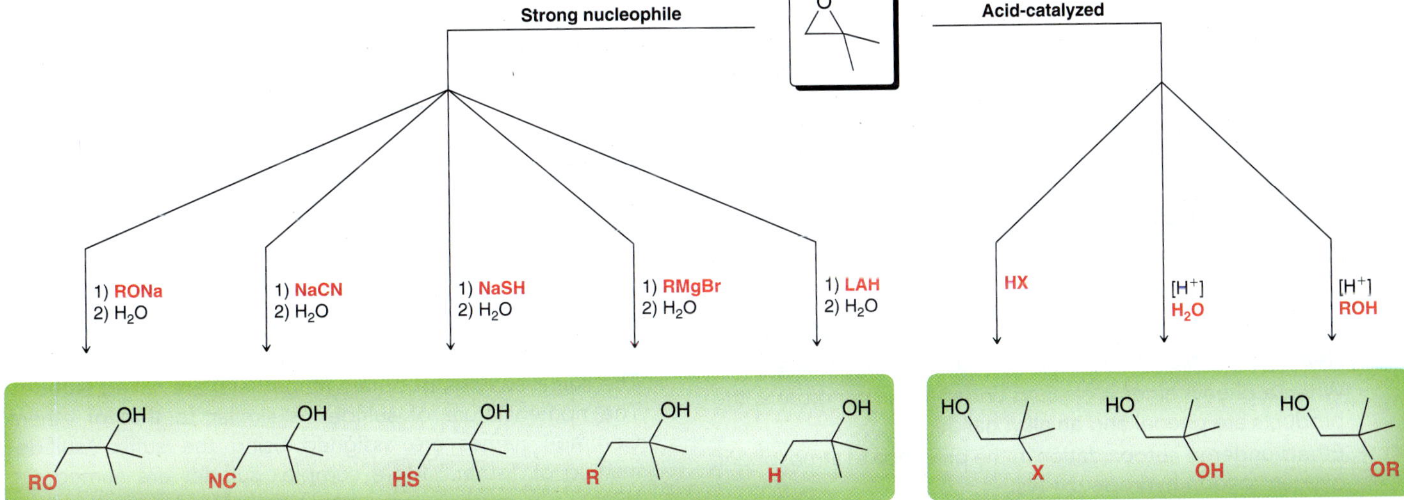


### Thiols and Sulfides

**Thiols**

$R_1$–CHBr–$R_2$ $\xrightarrow{\text{NaSH}}$ $R_1$–CHSH–$R_2$

RSH + RSH $\xrightarrow{\text{NaOH}/H_2O,\ Br_2}$ R–S–S–R (A disulfide); R–S–S–R $\xrightarrow{\text{HCl, Zn}}$ RSH + RSH

**Sulfides**

R–SH $\xrightarrow[\text{2) RX}]{\text{1) NaOH}}$ R–S–R

R–S–R $\xrightarrow{\text{MeX}}$ $R_2S^{\oplus}$–Me + $X^{\ominus}$

Sulfide $\xrightarrow{NaIO_4}$ Sulfoxide $\xrightarrow{H_2O_2}$ Sulfone; Sulfide $\xrightarrow{H_2O_2}$ Sulfone

## REVIEW OF CONCEPTS AND VOCABULARY

**SECTION 14.1**

- **Ethers** are compounds that have an oxygen atom bonded to two groups, which can be alkyl, aryl, or vinyl groups.
- The ether group is a common structural feature of many natural compounds and pharmaceuticals.

**SECTION 14.2**

- **Unsymmetrical ethers** have two different alkyl groups, while **symmetrical ethers** have two identical groups.
- The common name of an ether is constructed by assigning a name to each R group, arranging them in alphabetical order, and then adding the word "ether."

- The systematic name of an ether is constructed by choosing the larger group to be the parent alkane and naming the smaller group as an **alkoxy substituent**.

**SECTION 14.3**

- Ethers of low molecular weight have low boiling points, while ethers with larger alkyl groups have higher boiling points due to London dispersion forces between the alkyl groups.
- Ethers are often used as solvents for organic reactions.

**SECTION 14.4**

- The interaction between ethers and metal ions is very strong for **polyethers,** compounds with multiple ether groups.
- Cyclic polyethers, or **crown ethers**, are capable of solvating metal ions in organic (nonpolar) solvents.

**SECTION 14.5**

- Ethers can be readily prepared from the reaction between an alkoxide ion and an alkyl halide, a process called a **Williamson ether synthesis**. This process works best for methyl or primary alkyl halides. Secondary alkyl halides are significantly less efficient, and tertiary alkyl halides cannot be used.
- Ethers can be prepared from alkenes via **alkoxymercuration-demercuration**, which results in a Markovnikov addition of RO and H across an alkene.

**SECTION 14.6**

- When treated with a strong acid, an ether will undergo **acidic cleavage** in which it is converted into two alkyl halides.
- When a phenyl ether is cleaved under acidic conditions, the products are phenol and an alkyl halide.
- Ethers undergo autooxidation in the presence of atmospheric oxygen to form hydroperoxides.

**SECTION 14.7**

- A three-membered cyclic ether is called **oxirane**. It possesses significant ring strain and is therefore more reactive than other ethers.
- Substituted oxiranes are also called *epoxides*, which are named in either of two ways:
  - The oxygen atom is considered to be a substituent on the parent chain, and the exact location of the epoxide group is identified with two numbers followed by the term "epoxy."
  - The parent is considered to be the epoxide (parent = oxirane), and any groups connected to the epoxide are listed as substituents.

**SECTION 14.8**

- Alkenes can be converted into epoxides by treatment with peroxy acids or via halohydrin formation and subsequent epoxidation. Both procedures are stereospecific.
- Substituents that are *cis* to each other in the starting alkene remain *cis* to each other in the epoxide, and substituents that are *trans* to each other in the starting alkene remain *trans* to each other in the epoxide.

**SECTION 14.9**

- Chiral catalysts can be used to achieve the enantioselective epoxidation of allylic alcohols.
- In a **Sharpless asymmetric epoxidation**, the catalyst favors the production of one enantiomer over the other, leading to an observed enantiomeric excess.

**SECTION 14.10**

- Epoxides will undergo *ring-opening reactions* either (1) in conditions involving a strong nucleophile or (2) under acid-catalyzed conditions.
- When a strong nucleophile is used, the nucleophile attacks at the less substituted (less hindered) position.
- Under acid-catalyzed conditions, the regiochemical outcome is dependent on the nature of the epoxide and is explained in terms of a competition between *electronic* effects and *steric effects*.
- The stereochemical outcome involves inversion of configuration under all conditions.

**SECTION 14.11**

- Sulfur analogs of alcohols contain an SH group rather than an OH group and are called **thiols**.
- When another functional group is present in the compound, the SH group is named as a substituent and is called a **mercapto group**.
- Thiols can be prepared via an $S_N2$ reaction between sodium hydrosulfide (NaSH) and a suitable alkyl halide.
- Thiols easily undergo oxidation to produce **disulfides**, and disulfides are also easily reduced back to thiols when treated with a reducing agent.
- The sulfur analogs of ethers (thioethers) are called **sulfides**. The nomenclature of sulfides is similar to that of ethers. Common names are assigned using the suffix "sulfide" instead of "ether." More complex sulfides are named systematically, much the way ethers are named, with the alkoxy group being replaced by an **alkylthio group.**
- Sulfides can be prepared from thiols in a process that is essentially the sulfur analog of the Williamson ether synthesis, involving a **thiolate** ion, rather than an alkoxide.
- Sulfides will attack alkyl halides to produce alkylating agents, such as the biological alkylating agent SAM.
- Sulfides undergo oxidation to give **sulfoxides** and then **sulfones**.

**SECTION 14.12**

- Ring opening of an epoxide produces a compound with two functional groups on adjacent carbon atoms. Whenever you see two adjacent functional groups, you should think of epoxides.
- When a Grignard reagent reacts with an epoxide, a C—C bond is formed. This reaction can be used to introduce a chain of carbon atoms that possess a built-in functional group at the second carbon atom.
- In contrast, when a Grignard reagent attacks a ketone or aldehyde, the functional group appears on the first carbon atom that was introduced.
- You must train your eyes to look at the precise location of a functional group when it appears on a newly introduced alkyl fragment.

# SKILLBUILDER REVIEW

## 14.1 NAMING AN ETHER

**COMMON NAME**
Treat both sides as substituents and list them alphabetically.

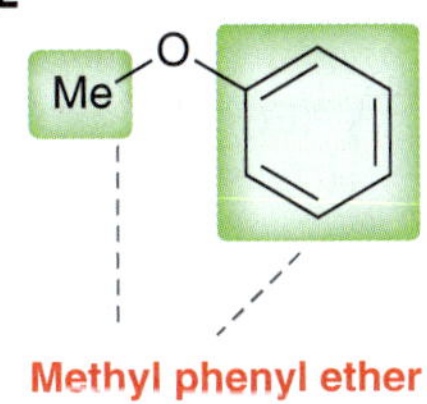


**SYSTEMATIC NAME**
1) Choose the parent (the more complex side).
2) Identify all substituents (including alkoxy group).
3) Assign locants.
4) Assemble substituents alphabetically with locants.
5) Assign the configuration of any chirality centers.

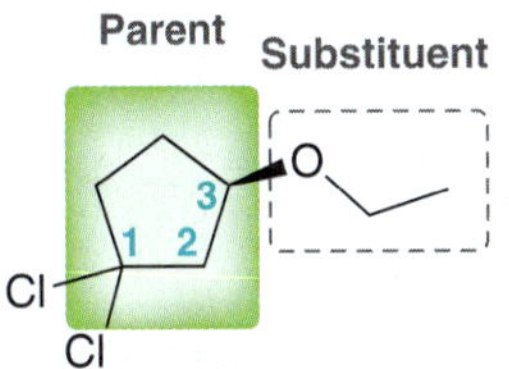


**(*R*)-1,1-Dichloro-3-ethoxycyclopentane**

Try Problems 14.1–14.3, 14.30, 14.32

## 14.2 PREPARING AN ETHER VIA A WILLIAMSON ETHER SYNTHESIS

**STEP 1** Identify the two groups on either side of the oxygen atom.

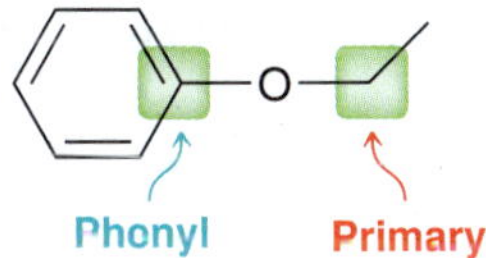


**STEP 2** Determine which side is more capable of serving as a substrate in an $S_N2$ reaction.

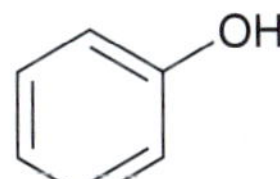


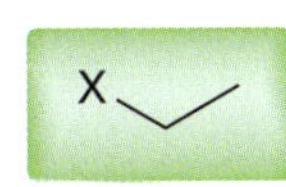


**STEP 3** Use a base to deprotonate the alcohol and then identify the alkyl halide.

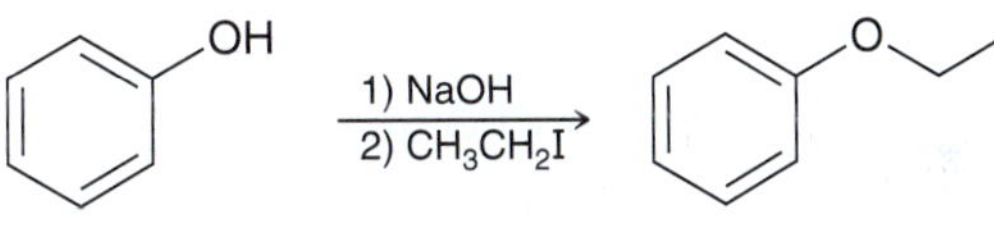


Try Problems 14.5–14.7, 14.33a,c, 14.37, 14.40, 14.42d, 14.43b

## 14.3 PREPARING EPOXIDES

**STEP 1** Identify the four groups attached to the epoxide.

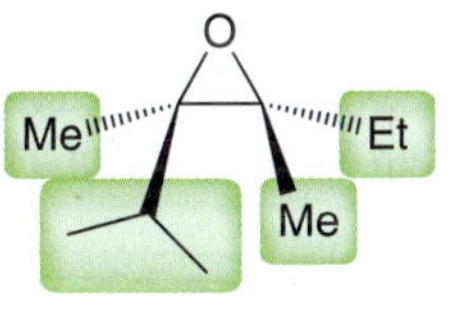


**STEP 2** Identify the relative configuration of the four groups and draw the starting alkene.

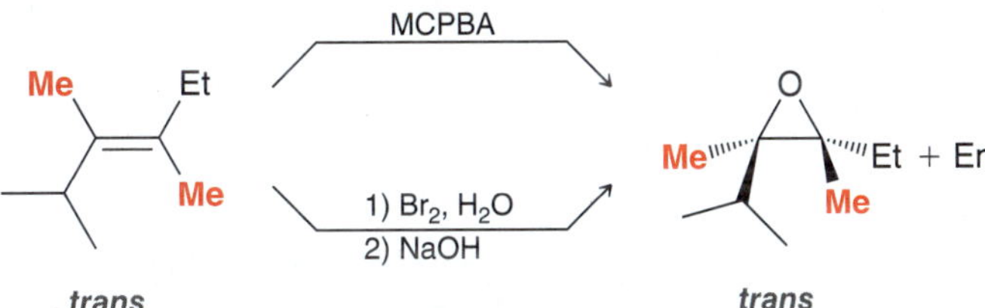


Try Problems 14.14, 14.15, 14.39, 14.51t

## 14.4 DRAWING A MECHANISM AND PREDICTING THE PRODUCT OF THE REACTION BETWEEN A STRONG NUCLEOPHILE AND AN EPOXIDE

**EXAMPLE** Predict the product and propose a mechanism for its formation.

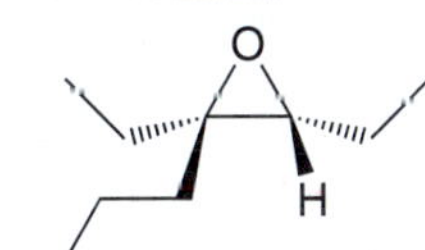


1) NaCN
2) $H_2O$

?

**STEP 1** Identify the regiochemistry by selecting the less-hindered position as the site of nucleophilic attack.

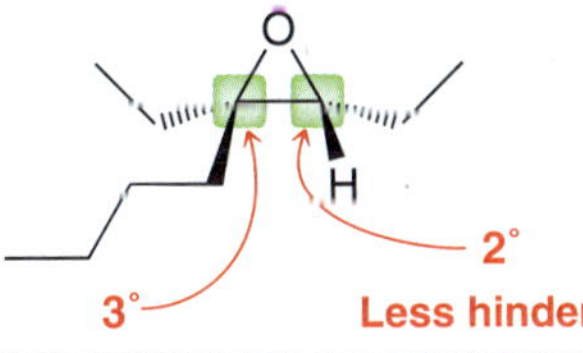


**STEP 2** Identify the stereochemistry by determining whether the nucleophile attacks a chirality center. If so, expect inversion of configuration.

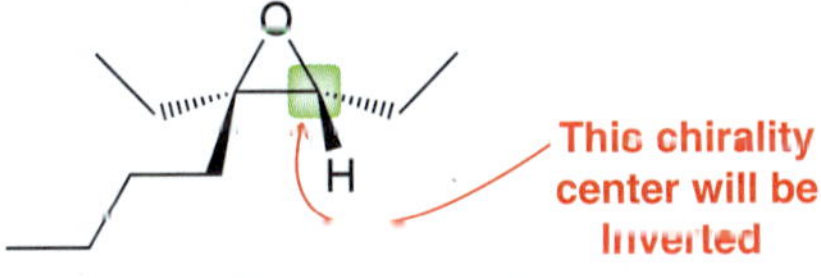


**STEP 3** Draw both steps of the mechanism:

Nucleophilic attack

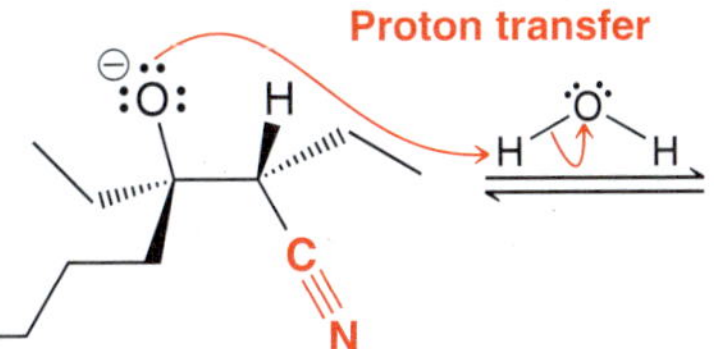


Proton transfer

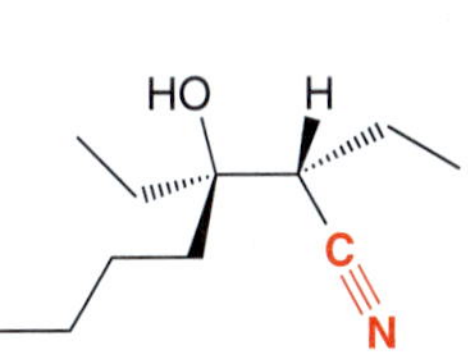


HO H C N

Try Problems 14.17–14.19, 14.42a,c,e–h, 14.43

### 14.5 DRAWING A MECHANISM AND PREDICTING THE PRODUCT OF ACID-CATALYZED RING OPENING

**EXAMPLE** Predict the product and propose a mechanism for its formation.

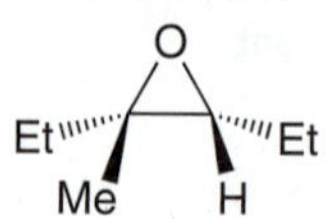


$[H_2SO_4]$
EtOH

?

**STEP 1** Identify the regiochemistry: determine whether steric or electronic effects will dominate.

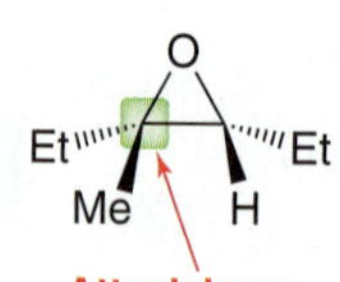


One side is tertiary, so electronic effects dominate

**STEP 2** Identify the stereochemistry: if nucleophile attacks a chirality center, expect inversion of configuration.

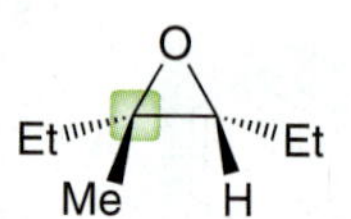


This chirality center will be inverted

**STEP 3** Draw all three steps of the mechanism:
(1) proton transfer, (2) nucleophilic attack, and (3) proton transfer.

Proton transfer

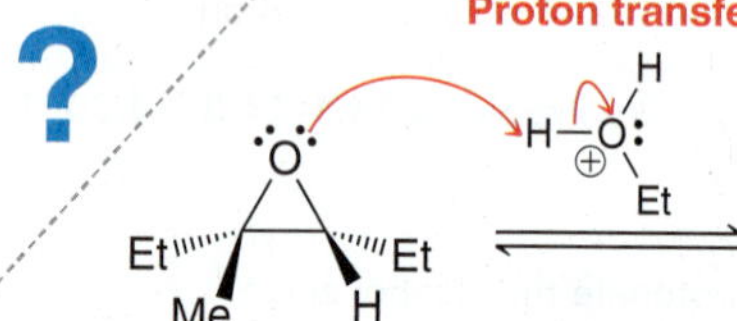


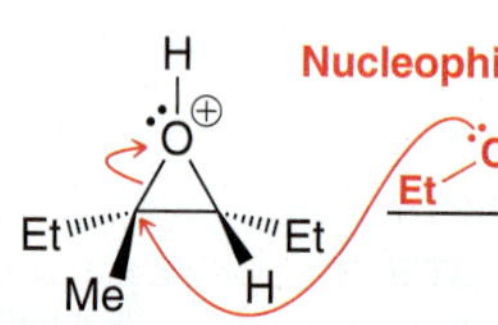


Nucleophilic attack



Proton transfer

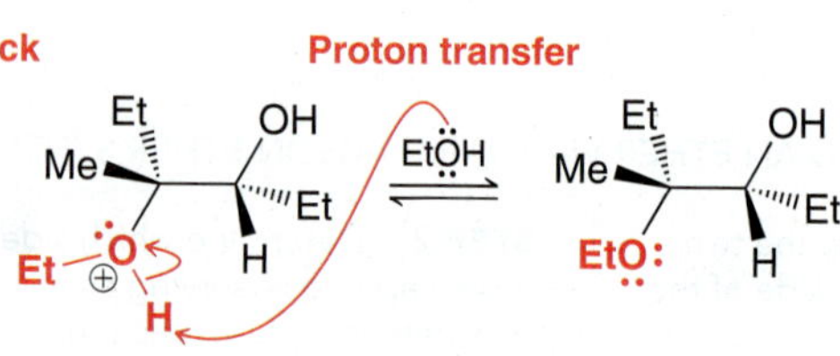


Try Problems **14.20, 14.21, 14.43d, 14.49, 14.50**

### 14.6 INSTALLING TWO ADJACENT FUNCTIONAL GROUPS

Convert an alkene into an epoxide. | Open the epoxide with regiochemical control.

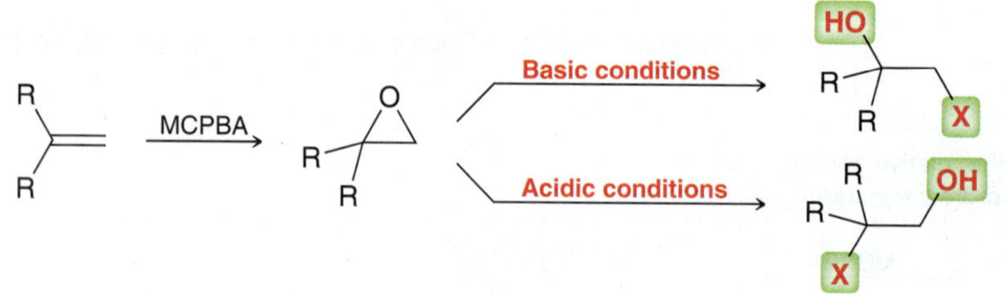


Try Problems **14.24, 14.25, 14.51**

### 14.7 CHOOSING THE APPROPRIATE GRIGNARD REACTION

Grignard reagent attacking an epoxide:

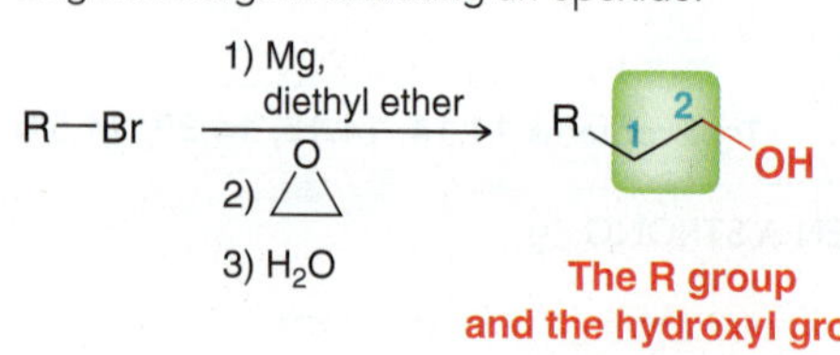


Grignard reagent attacking an aldehyde or ketone:

R—Br 1) Mg, diethyl ether; 2) O, H; 3) $H_2O$ → R, OH, 1, 2

The R group and the hydroxyl group are connected to the same carbon atom

Try Problems **14.26–14.29, 14.51**

## PRACTICE PROBLEMS

Note: Most of the Problems are available within **WileyPLUS**, an online teaching and learning solution.

**14.30** Assign an IUPAC name for each of the following compounds:

(a) (b) (c) SH (d) O S (e) O (f) OMe OMe (g) S

**14.31** Predict the products that are expected when each of the following compounds is heated with concentrated HBr:

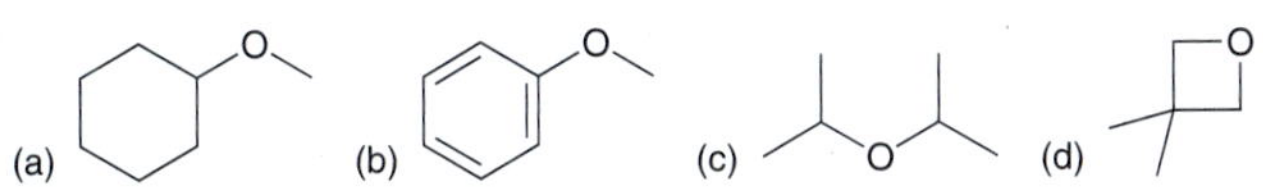


**14.32** Draw all constitutionally isomeric ethers with molecular formula $C_4H_{10}O$. Provide a common name and a systematic name for each isomer.

**14.33** Starting with cyclohexene and using any other reagents of your choice, show how you would prepare each of the following compounds:

(a) (b) OH, OMe (c)

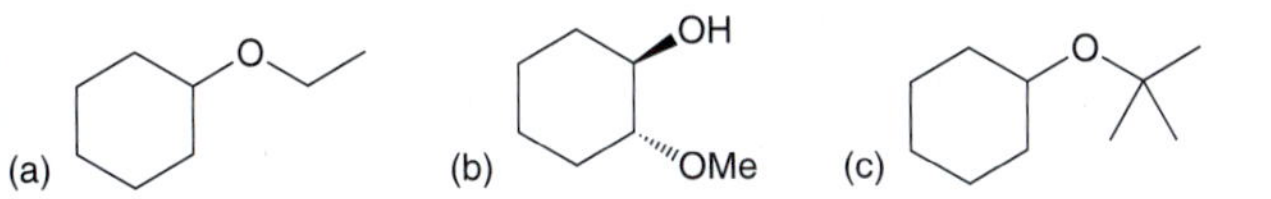

**14.34** When 1,4-dioxane is heated in the presence of HI, compound **A** is obtained:

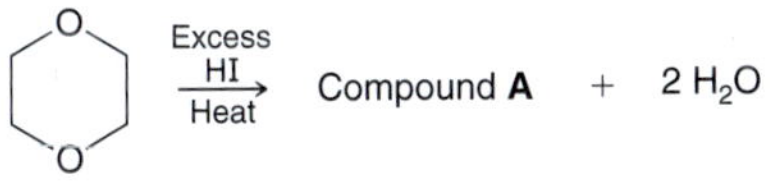

Excess HI, Heat → Compound **A** + 2 $H_2O$

(a) Draw the structure of compound **A**.

(b) If one mole of dioxane is used, how many moles of compound **A** are formed?

(c) Show a plausible mechanism for the conversion of dioxane into compound **A**.

**14.35** Tetrahydrofuran (THF) can be formed by treating 1,4-butanediol with sulfuric acid. Propose a mechanism for this transformation:

HO–(CH₂)₄–OH $\xrightarrow{H_2SO_4}$ THF

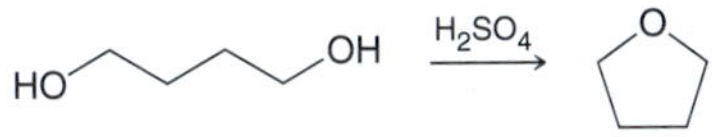

**1,4-Butanediol** **Tetrahydrofuran (THF)**

**14.36** When ethylene glycol is treated with sulfuric acid, 1,4-dioxane is obtained. Propose a mechanism for this transformation:

HO–CH₂CH₂–OH $\xrightarrow{H_2SO_4}$ 1,4-dioxane

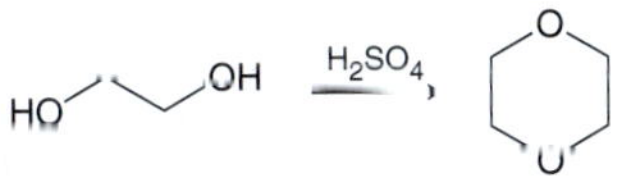

**Ethylene glycol** **1,4-Dioxane**

**14.37** The Williamson ether synthesis cannot be used to prepare *tert*-butyl phenyl ether.

(a) Explain why this method cannot be used in this case.

(b) Suggest an alternative method for preparing *tert*-butyl phenyl ether.

**14.38** Methylmagnesium bromide reacts rapidly with ethylene oxide, it reacts slowly with oxetane, and it does not react at all with tetrahydrofuran.

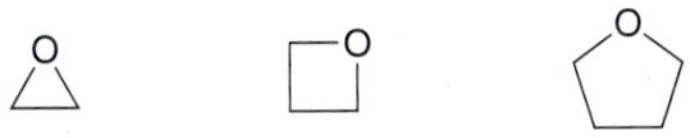

**Ethylene oxide** **Oxetane** **Tetrahydrofuran (THF)**

Explain this difference in reactivity.

**14.39** Identify the reagents necessary to accomplish each of the following transformations:

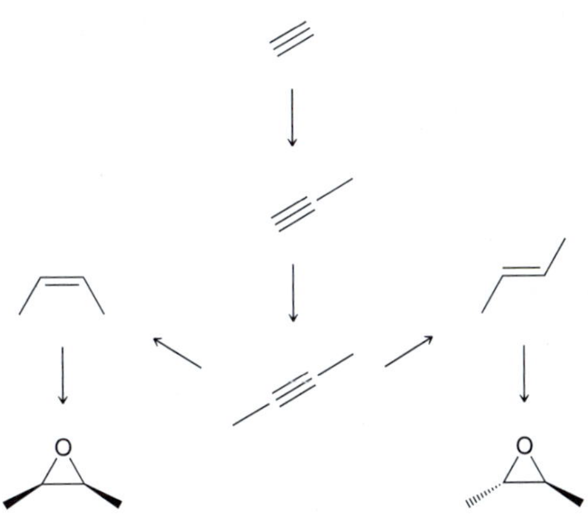


**14.40** When 5-bromo-2,2-dimethyl-1-pentanol is treated with sodium hydride, a compound with molecular formula $C_7H_{14}O$ is obtained. Identify the structure of this compound.

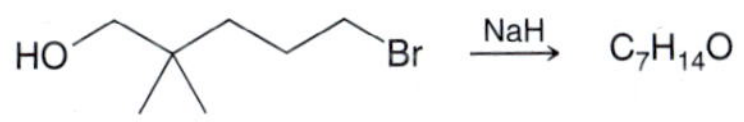

HO–…–Br $\xrightarrow{NaH}$ $C_7H_{14}O$

**14.41** Problem 14.39 outlines a general method for the preparation of *cis*- or *trans*-disubstituted epoxides. Using that method, identify what reagents you would use to prepare a racemic mixture of each of the following epoxides from acetylene:

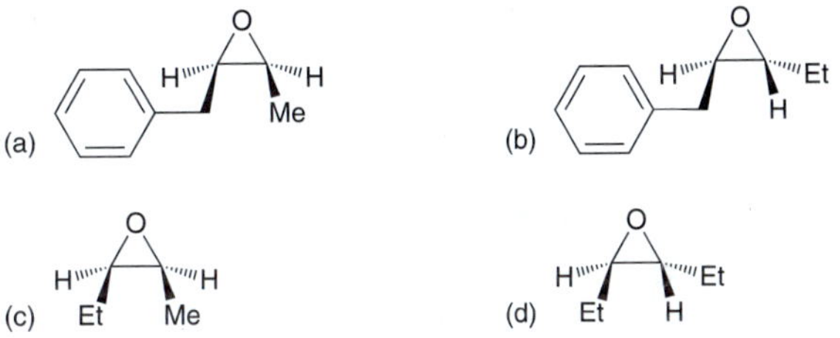

(a) (H, H, Me) (b) (H, Et, H) (c) (H, H, Et, Me) (d) (H, Et, Et, H)

**14.42** Predict the products for each of the following:

(a) 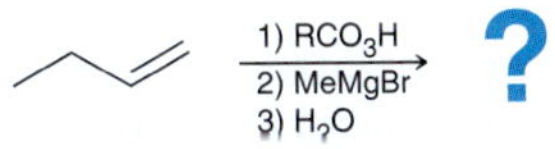 1) $RCO_3H$ 2) MeMgBr 3) $H_2O$ ?

(b) 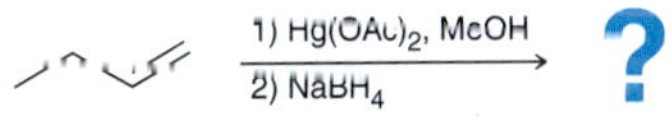 1) $Hg(OAc)_2$, MeOH 2) $NaBH_4$ ?

(c) 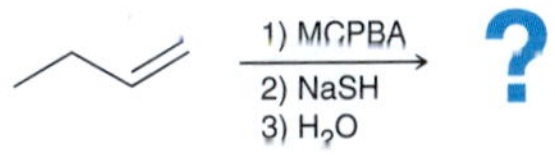 1) MCPBA 2) NaSH 3) $H_2O$ ?

(d) 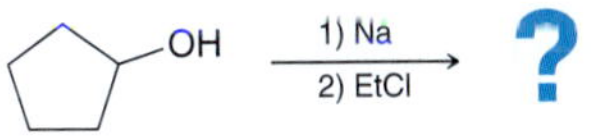 cyclopentanol (OH): 1) Na 2) EtCl ?

(e) 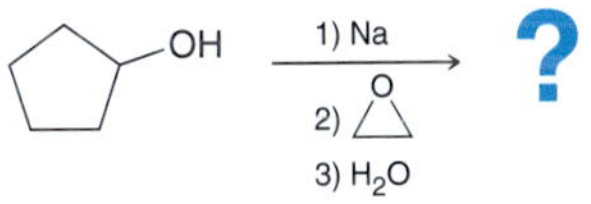 cyclopentanol (OH): 1) Na 2) ethylene oxide 3) $H_2O$ ?

(f) 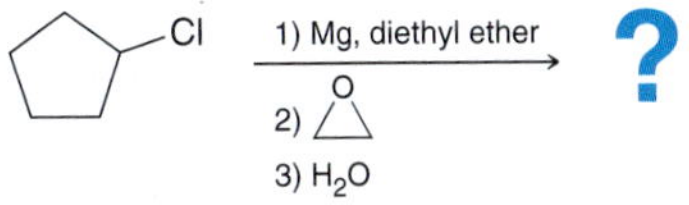 chlorocyclopentane (Cl): 1) Mg, diethyl ether 2) ethylene oxide 3) $H_2O$ ?

(g) 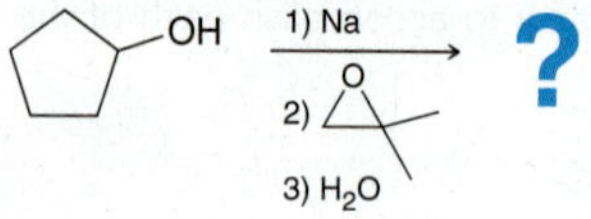 1) Na 2) [epoxide] 3) $H_2O$ ?

(h) 1) Mg, diethyl ether 2) [epoxide] 3) $H_2O$ ?

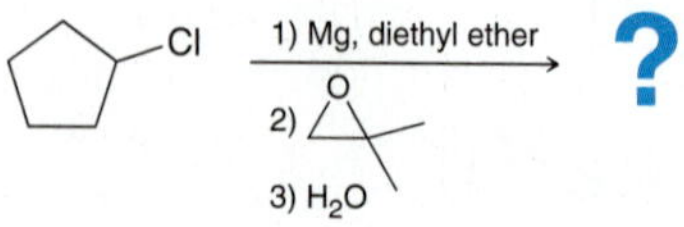

**14.43** Propose a plausible mechanism for each of the following transformations:

(a) 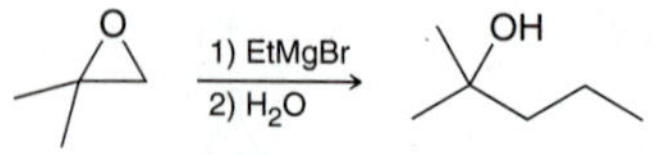 1) EtMgBr 2) $H_2O$ → OH

(b) 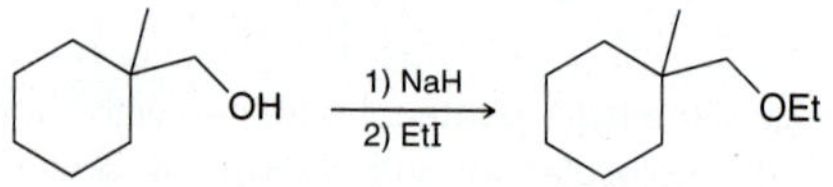 OH 1) NaH 2) EtI → OEt

(c) 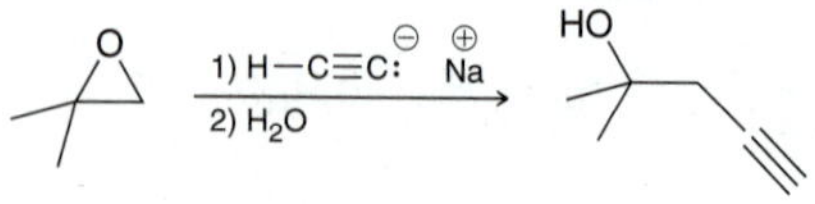 1) $H-C\equiv C:^{\ominus}$ $Na^{\oplus}$ 2) $H_2O$ → HO

(d) 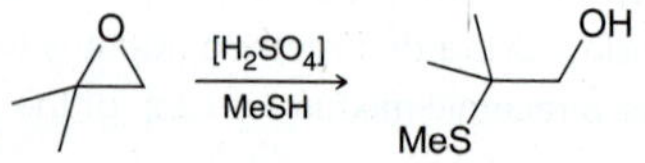 $[H_2SO_4]$ MeSH → OH MeS

(e) 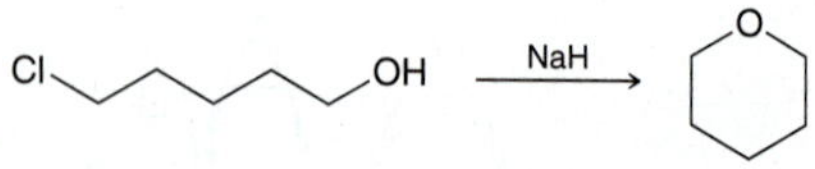 Cl OH NaH →

(f) 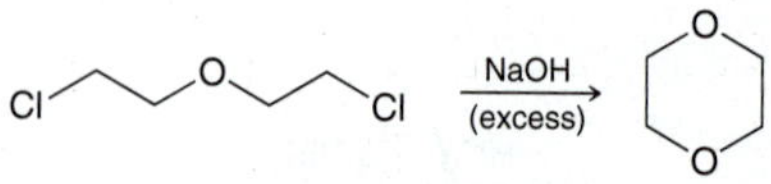 Cl O Cl NaOH (excess) →

**14.44** What product do you expect when tetrahydrofuran is heated in the presence of excess HBr?

**14.45** Compound **B** has molecular formula $C_6H_{10}O$ and does not possess any π bonds. When treated with concentrated HBr, *cis*-1,4-dibromocyclohexane is produced. Identify the structure of compound **B**.

**14.46** Propose a stepwise mechanism for the following transformation:

O Me Cl 1) Excess EtMgBr 2) $H_2O$ → Et Me OH Et

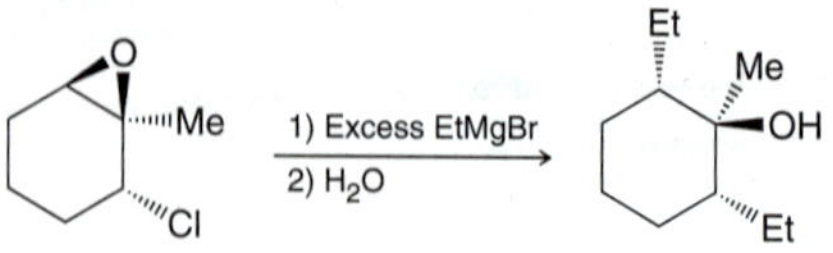

**14.47** Predict the products for each of the following:

(a) 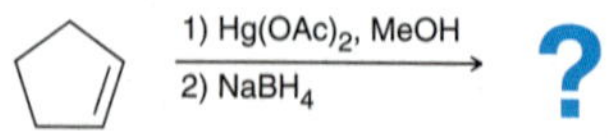 1) $Hg(OAc)_2$, MeOH 2) $NaBH_4$ ?

(b) 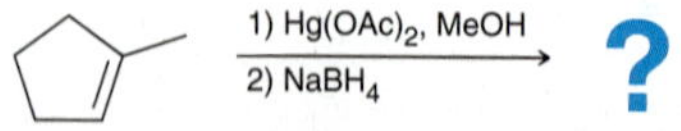 1) $Hg(OAc)_2$, MeOH 2) $NaBH_4$ ?

(c) 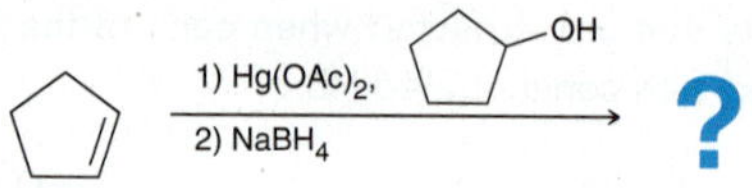 1) $Hg(OAc)_2$, [cyclopentanol] 2) $NaBH_4$ ?

(d) 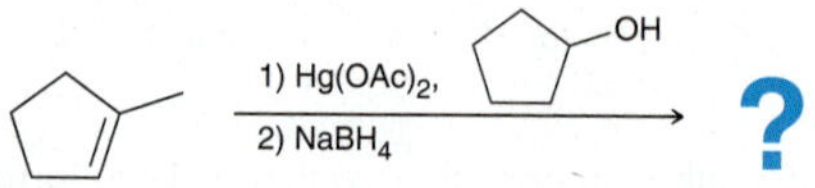 1) $Hg(OAc)_2$, [cyclopentanol] 2) $NaBH_4$ ?

**14.48** Using acetylene and ethylene oxide as your only sources of carbon atoms, propose a synthesis for each of the following compounds:

(a)  HO OH

(b) 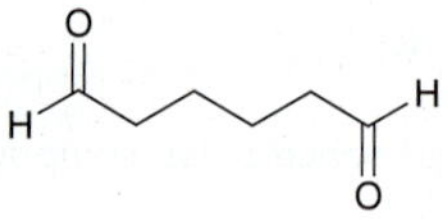 O H H O

**14.49** Fill in the missing reagents below.

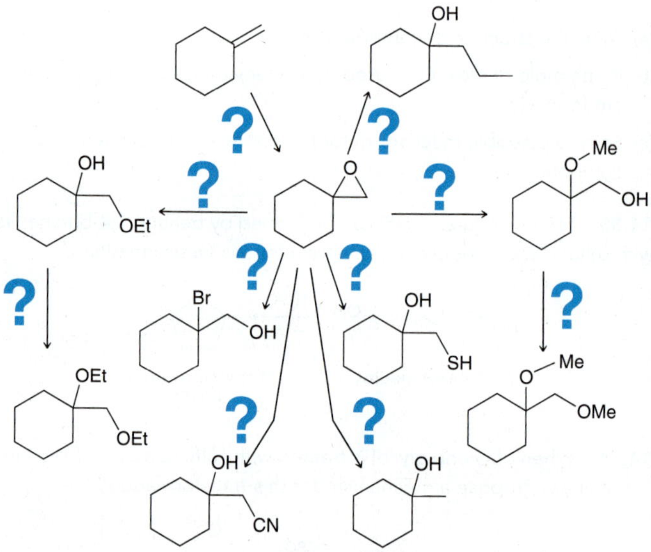

? ? ? ? ? ? ? ? ? ? ?

OH OEt; OEt OEt; Br OH; OH SH; O Me OH; O Me OMe; OH CN; OH

**14.50** Fill in the missing products below.

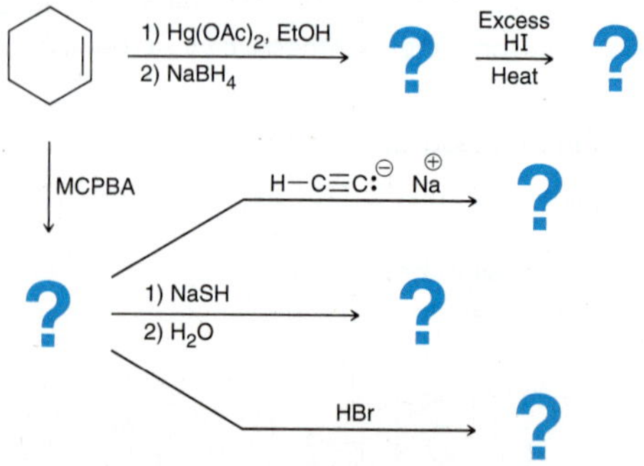

1) $Hg(OAc)_2$, EtOH 2) $NaBH_4$ ? Excess HI Heat ?

MCPBA ?

$H-C\equiv C:^{\ominus}$ $Na^{\oplus}$ ?

1) NaSH 2) $H_2O$ ?

HBr ?

# INTEGRATED PROBLEMS

**14.51** Propose an efficient synthesis for each transformation.

(a) 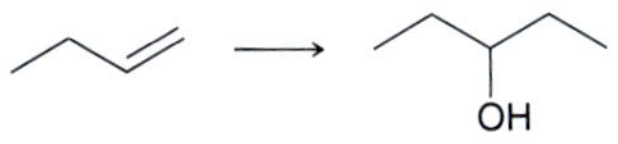 (b) 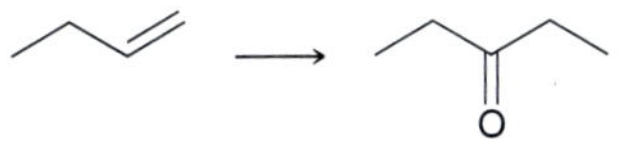 (c) 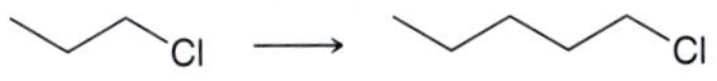 (d) 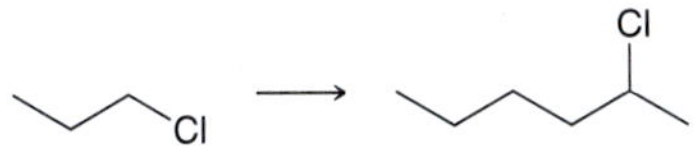 (e) 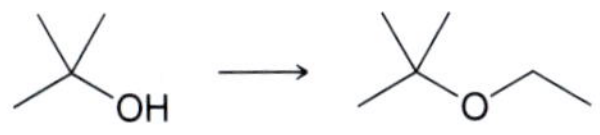 (f) 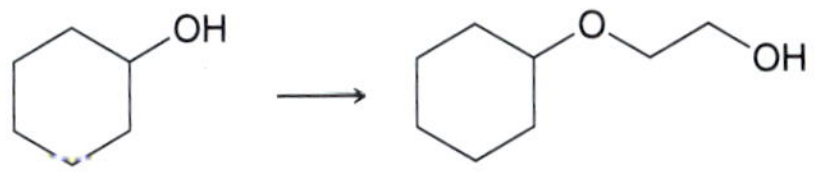 (g)  (h) 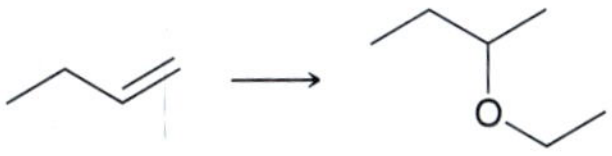 (i)  (j)  (k)  + En (l) 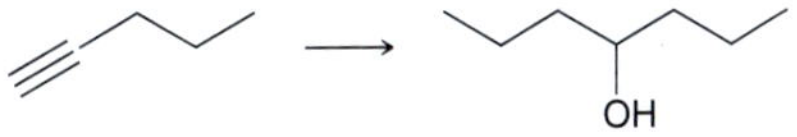 (m) 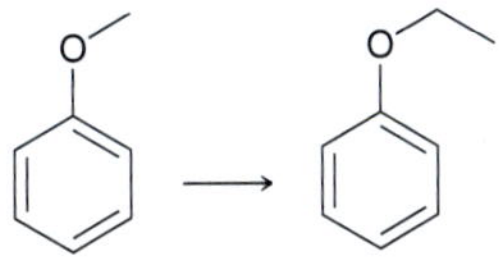 (n) 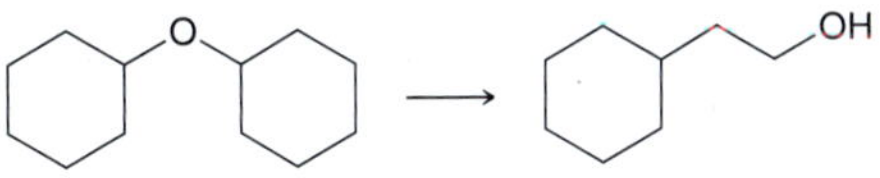 (o)  (p) 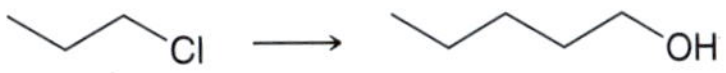 (q) 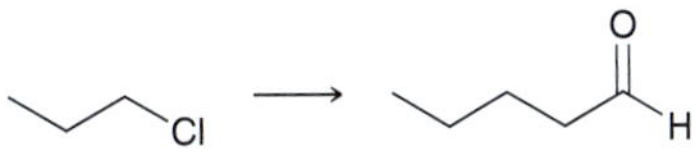 (r) 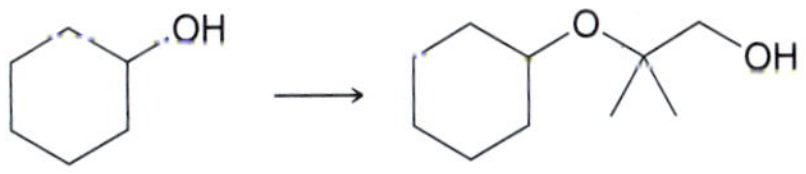 (s) 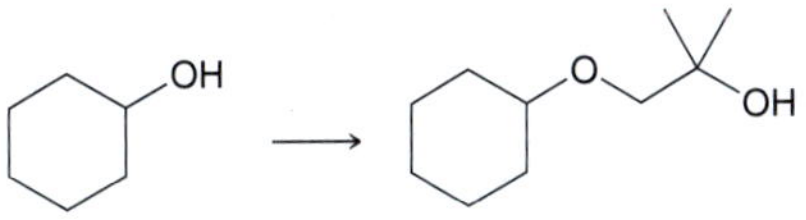 (t) 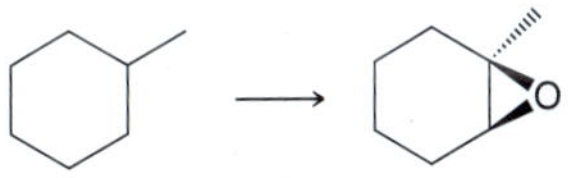 + En (u) 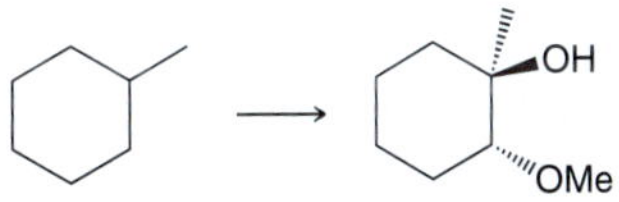 + En

**Problems 14.52–14.55 are intended for students who have already covered spectroscopy (Chapters 15 and 16).**

**14.52** Propose a structure for an ether with molecular formula $C_7H_8O$ that exhibits the following $^{13}C$ NMR spectrum:

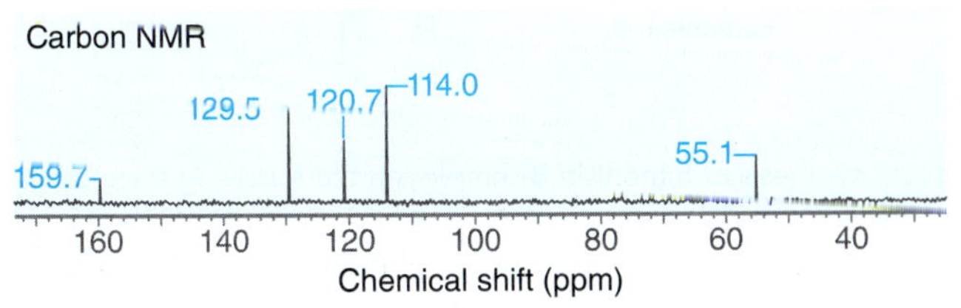


**14.53** Propose a structure for a compound with molecular formula $C_8H_{18}O$ that exhibits the following $^1H$ NMR and $^{13}C$ NMR spectra:

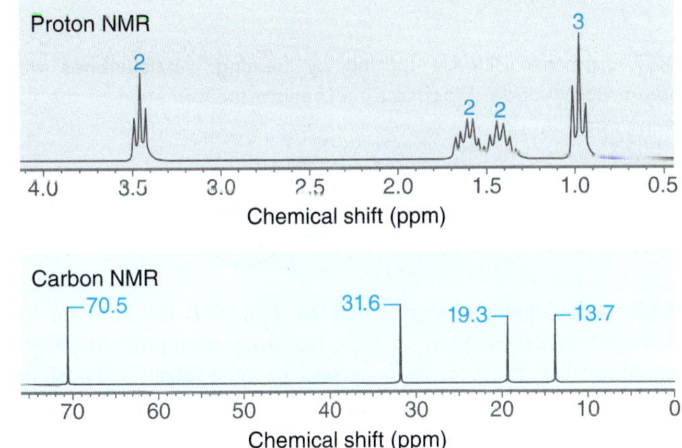

**14.54** Propose a structure for a compound with molecular formula $C_4H_8O$ that exhibits the following $^{13}C$ NMR and FTIR spectra:

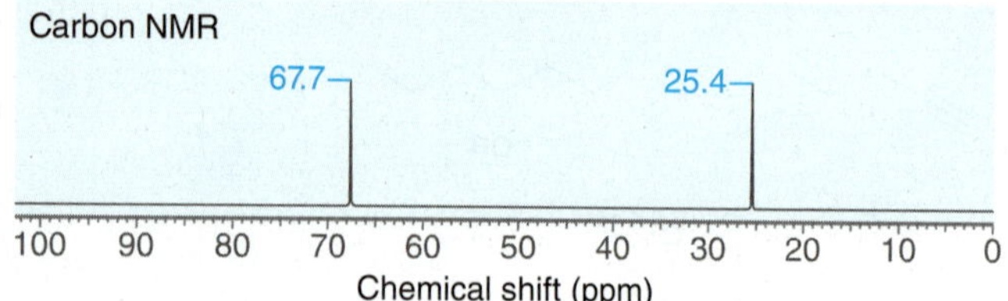


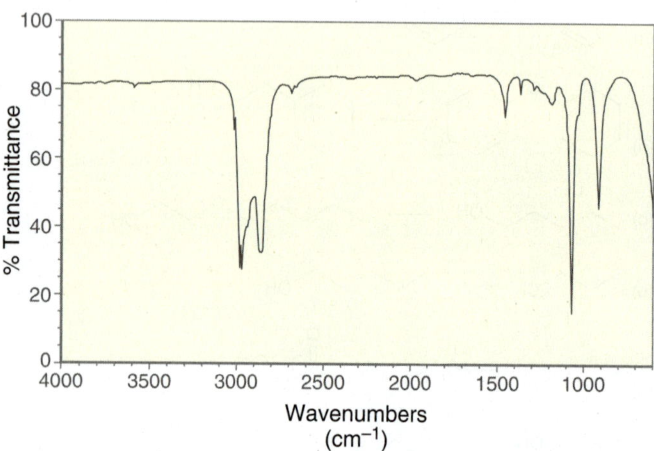


**14.55** Propose a structure for a compound with molecular formula $C_4H_{10}O$ that exhibits the following $^1H$ NMR spectrum:

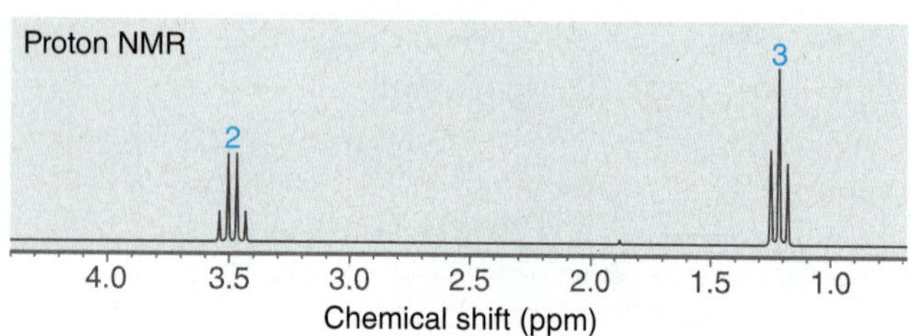


**14.56** Predict the product of the following reaction:

1) $LiAlD_4$
2) $H_2O$

**14.57** Epoxides can be formed by treating α-haloketones with sodium borohydride. Propose a mechanism for formation of the following epoxide:

$NaBH_4$

**14.58** When methyloxirane is treated with HBr, the bromide ion attacks the less substituted position. However, when phenyloxirane is treated with HBr, the bromide ion attacks the more substituted position.

Explain the difference in regiochemistry in terms of a competition between steric effects and electronic effects. (**Hint:** It may help to draw out the structure of the phenyl group.)

**14.59** Propose an efficient synthesis for the following transformation:

+ Enantiomer

**14.60** Using bromobenzene and ethylene oxide as your only sources of carbon, show how you could prepare *trans*-2,3-diphenyloxirane (a racemic mixture of enantiomers).

+ Enantiomer

**14.61** The $S_N2$ reaction between a Grignard reagent and an epoxide works reasonably well when the epoxide is ethylene oxide. However, when the epoxide is substituted with groups that provide steric hindrance, a competing reaction can dominate, in which an allylic alcohol is produced. Propose a mechanism for this transformation and use the principles discussed in Section 8.13 to justify why the allylic alcohol would be the major product in this case below.

1) EtMgBr
2) $H_2O$

Major product + Minor product

**14.62** The following sequence of reactions was employed during synthetic studies on reidispongiolide A, a cytotoxic marine natural product (*Tetrahedron Lett.* **2009,** *50,* **5012–5014**). Draw the structures of compounds **A**, **B**, and **C**.

1) $CH_2{=}CHMgBr$ 2) $H_2O$ → **A**; 1) NaH 2) $CH_3I$ → **B**; $OsO_4$, NMO → **C**

**14.63** Sphingolipids are a class of compounds that play an important role in signal transmission and cell recognition. Fumonisin $B_1$ is a potent sphingolipid biosynthesis inhibitor.

**Fumonisin $B_1$**

A recent synthesis of fumonisin $B_1$ employed the following transformation, involving the incorporation of carbons 7–9 via the formation of alkyne **2** from epoxide **1**, shown below (*Tetrahedron Lett.* **2012,** *53,* **3233–3236**). Starting with 3-bromo-1-propyne, show a synthesis of alkyne **2** from epoxide **1**.

**1** → **2**

# CHALLENGE PROBLEMS

**14.64** The following procedure (*J. Am. Chem. Soc.* **1982,** *104,* **3515–3516**) is part of a synthetic strategy for the enantioselective preparation of carbohydrates (Chapter 24):

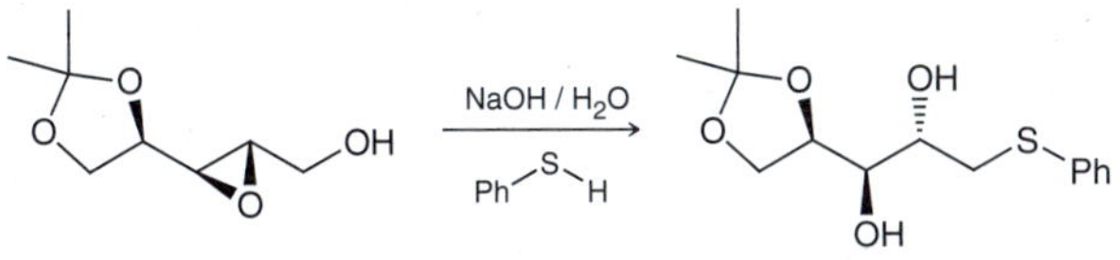

(a) Propose a plausible mechanism that explains the stereochemical outcome of this transformation.

(b) Predict the product of the following process:

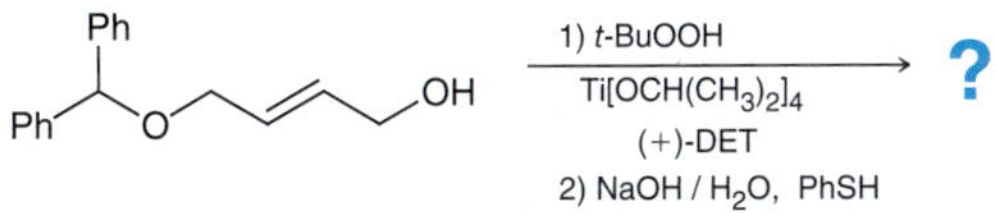

**14.65** In Section 14.11, we learned that **1** is a strong methyl transfer agent. However, when treated with a strong base (e.g., NaH), **1** is converted into a special type of carbanion, called a sulfur ylide, which can function as a methylene ($CH_2$) transfer agent. Specifically, treating the sulfur ylide with a ketone affords an epoxide, with dimethyl sulfide (DMS) being generated as a by-product (*Synth. Commun.* **2003,** *33,* **2135–2143**). Propose a plausible mechanism for this process, known as the Corey-Chaykovsky reaction.

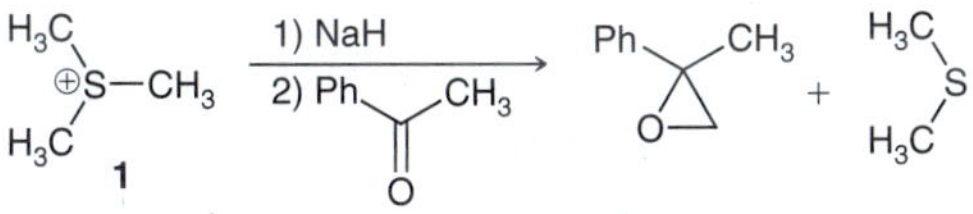

**14.66** Guggul is an herbal extract from the resin of the mukul myrrh tree, and it shows potential for treating high cholesterol. In a recent synthesis of (+)-myrrhanol A (a compound present in guggul), compound **1** was treated with MCPBA, followed by LAH to afford compound **2** as the only diastereomer detected (*J. Org. Chem.* **2009,** *74,* **6151–6156**). Draw a mechanism for the transformation of **1** to **2** and explain the stereochemical outcome with respect to the newly formed chirality center. Your explanation should include (i) a bond-line drawing that clearly shows the chair conformation adopted by each six-membered ring in the *trans* decalin system and (ii) a Newman projection that clearly shows the source of the observed stereochemical preference.

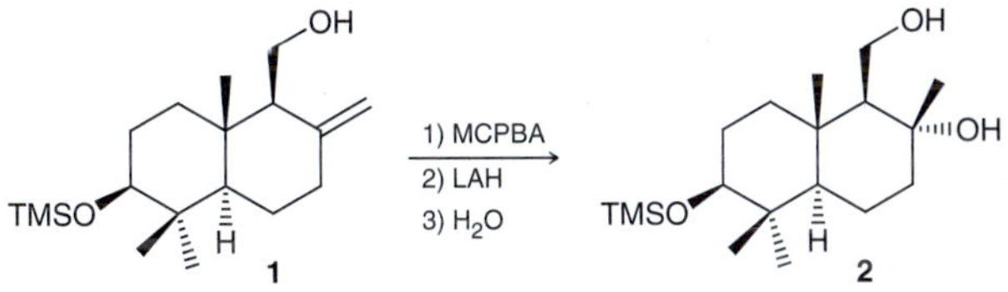

**14.67** Reboxetine mesylate is used in the treatment of depression and is currently marketed as the racemate. The (*S*,*S*)-enantiomer of reboxetine is being evaluated for the treatment of neuropathic pain. The following synthetic scheme was part of the synthesis of (*S*,*S*)-reboxetine completed by a research group at Pfizer (*Org. Proc. Res. & Devel.* **2007,** *11,* **354–358**):

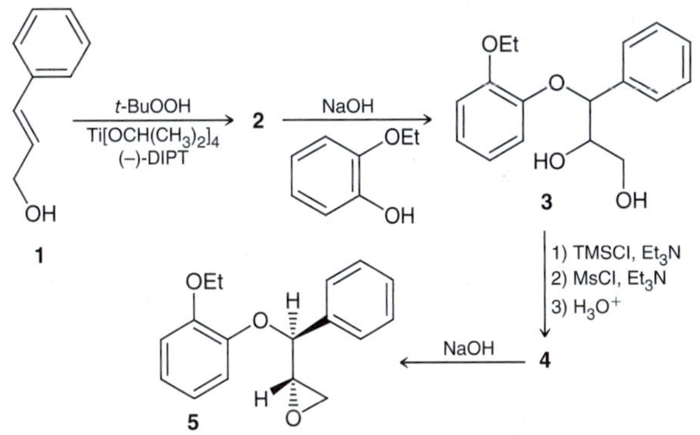

(a) In the conversion of **1** to **2**, (−)-DIPT serves the same function as (−)-DET. Provide the structure of **2**.

(b) Propose a plausible mechanism for the transformation of **2** to **3** and draw the expected configuration at each of the chirality centers in **3**.

(c) TMSCl is used to selectively protect the less hindered, primary OH group in **3**. MsCl functions very much like TsCl by converting an OH group into a good leaving group, called a mesylate (OMs) group (see Figure 7.28). Aqueous acid then removes the protecting group to give **4**. Draw the structure of **4**.

(d) Draw a plausible mechanism for the transformation of **4** to **5**.

**14.68** Laureatin is a natural product isolated from the marine algae *Laurencia nipponica* that exhibits potent insecticidal activity against mosquitos. During an investigation toward the synthesis of laureatin, an interesting skeletal rearrangement was observed. When compound **1** was treated with a source of electrophilic bromine (e.g., $Br_2$), the product isolated was epoxide **2** (*J. Org. Chem.* **2012,** *77,* **7883–7890**). Propose a plausible mechanism.

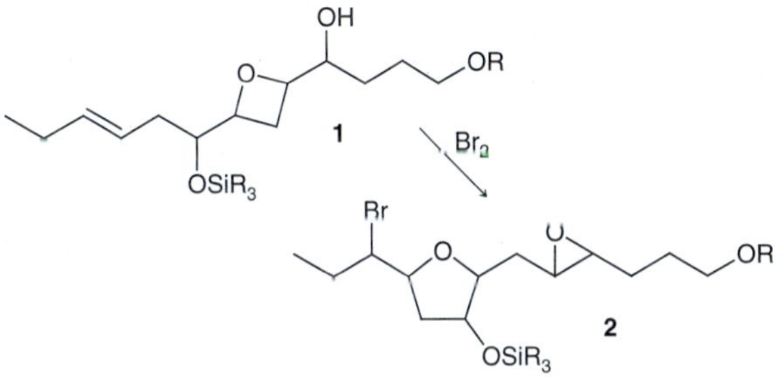

**14.69** Oxymercuration-demercuration of compound **1** affords the expected hydration product **2** in 96% yield. In contrast, oxymercuration-demercuration of compound **3** results in only a minor amount of the normal hydration product. Instead, compounds **7** and **8** are formed (*J. Org. Chem.* **1981,** *46,* **930–939**). It is believed that the initial product of oxymercuration (compound **4**) undergoes an intramolecular epoxide-opening step, giving two possible cyclic compounds, **5** and **6**. This step is likely catalyzed by the interaction between the epoxide group and $Hg(OAc)_2$, which functions as a Lewis acid.

(a) Draw the structure of **2**.

(b) Draw the structures of **4, 5,** and **6** and show a mechanism for the conversion of **4** → **5** as well as **4** → **6**.

(c) Explain why the product distribution is so different for epoxyalkenes **1** and **3**.

**14.70** The following electrophile-induced, ether transfer reaction was utilized in the synthesis of several structurally related natural products (*J. Org. Chem.* **2008,** *73,* **5592–5594**). Provide a plausible mechanism for this transformation.

**14.71** During a total synthesis of 2-methyl-D-erythritol (a sugar of importance to isoprenoid biosynthesis), epoxide **2** was required (*J. Org. Chem.* **2002,** *67,* **4856–4859**).

(a) Identify the reagents you would use to achieve a stereoselective synthesis of epoxide **2** from allylic alcohol **1**.

(b) Propose a plausible mechanism for the formation of orthoester **4** from epoxide **3** upon treatment with a catalytic amount of aqueous acid. (**Hint:** The O in the C=O bond of the ester group can function as a nucleophilic center.)

**14.72** Artemisinin (also known as Qinghaosu) is a peroxide-containing compound that has been used in traditional Chinese medicine as a treatment for malaria. Today, synthetic derivatives of artemisinin are standard treatments to fight against *Plasmodium falciparum* infections, the parasite which causes the most dangerous form of malaria. Epoxide **1** is a synthetic derivative of artemisinin; when treated with hexafluoro-2-propanol and catalytic acid, it undergoes a ring-opening reaction for which the regiochemical outcome and stereochemical outcome are both contrary to what you might have expected (*J. Org. Chem.* **2002,** *67,* **1253–1260**). Explain these observations with a plausible mechanism.

15

# Infrared Spectroscopy and Mass Spectrometry



## DID YOU EVER WONDER...

**how night vision goggles work?**

There are many different varieties of night vision goggles. One of these varieties depends on the principle that warm objects emit infrared radiation, which can be detected and then displayed as an image, called a thermogram (an example is shown below). We will soon see that a similar technology has been successfully implemented in the early screening of breast cancer. These technologies rely on the interactions between light and matter, the study of which is called spectroscopy. This chapter will serve as an introduction to spectroscopy and its relevance to the field of organic chemistry. After briefly reviewing how light and matter interact with each other, we will spend most of the chapter learning two very important spectroscopic techniques that are frequently used by organic chemists to verify the structures of compounds.

Less than one hundred years ago, structural determination was a difficult and time-consuming task. It was not uncommon for a chemist to spend years determining the structure of an unknown compound. The advent of modern spectroscopic techniques has completely transformed the field of chemistry, and structures can now be determined in several minutes. In this chapter, we will begin to appreciate how spectroscopy can be used to determine the structures of unknown compounds.

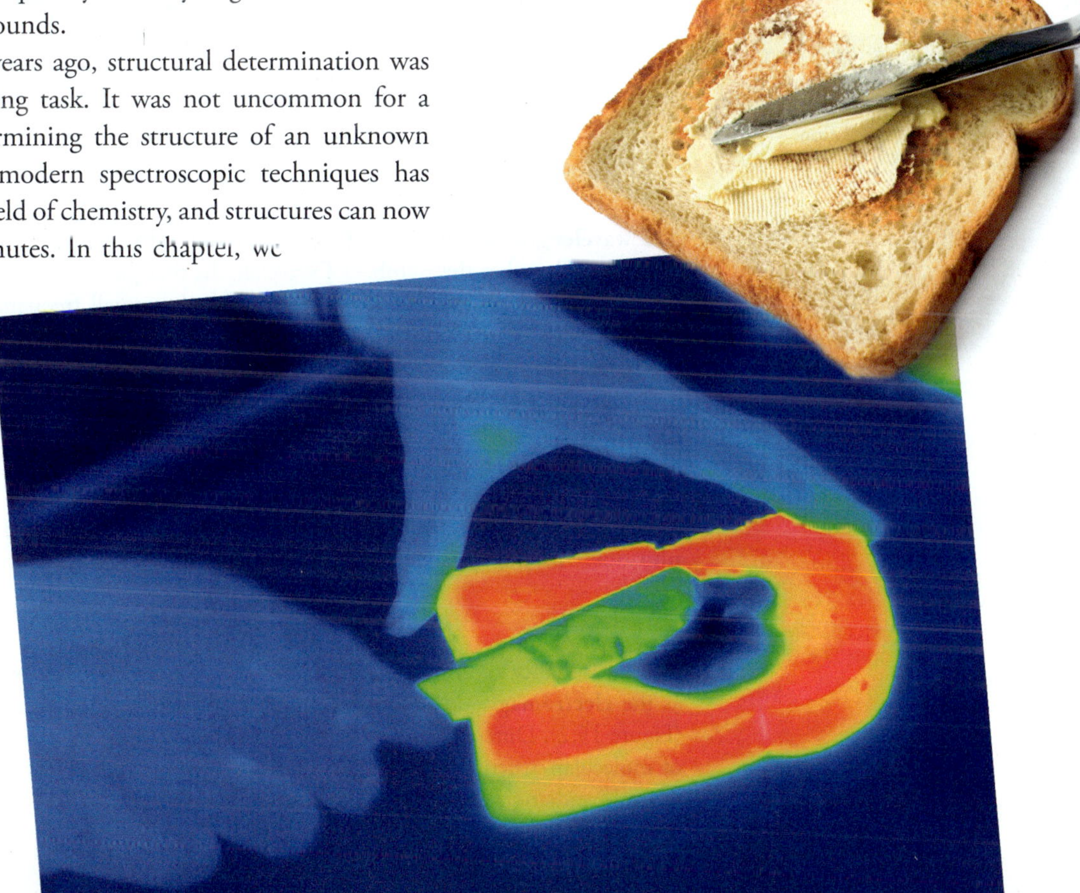

## DO YOU REMEMBER?

*Before you go on, be sure you understand the following topics.*
*If necessary, review the suggested sections to prepare for this chapter.*

- Quantum Mechanics (Section 1.6)
- Hydrogen Bonding (Section 1.12)
- Drawing Resonance Structures (Sections 2.7–2.10)
- Carbocation Stability (Section 6.11)

Take the DO YOU REMEMBER? QUIZ in **WileyPLUS** to check your understanding

## 15.1 Introduction to Spectroscopy

In order to understand how **spectroscopy** is used for structure determination, we must first review some of the basic features of light and matter.

### The Nature of Light

Electromagnetic radiation (light) exhibits both wave-like properties and particle-like properties. Consequently, electromagnetic radiation can be viewed as a wave or as a particle. When viewed as a wave, electromagnetic radiation consists of perpendicular oscillating, electric and magnetic fields (Figure 15.1).

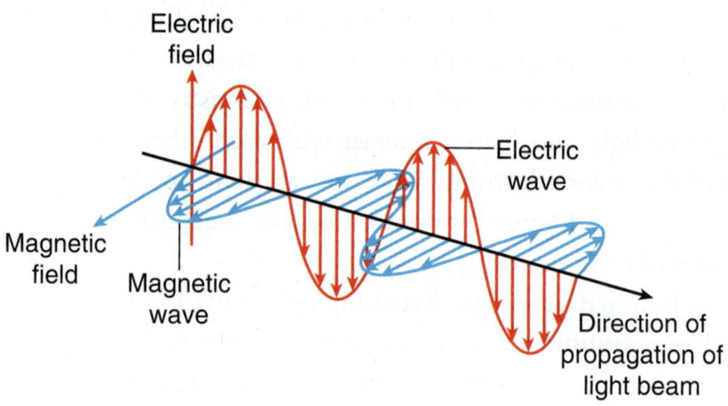


**FIGURE 15.1** The perpendicular, oscillating, electric and magnetic fields associated with electromagnetic radiation.

The **wavelength** describes the distance between adjacent peaks of an oscillating field, while the **frequency** describes the number of wavelengths that pass a particular point in space per unit time. Accordingly, a long wavelength corresponds with a small frequency, and a short wavelength corresponds with a large frequency. This inverse relationship is summarized in the following equation where frequency ($\nu$) and wavelength ($\lambda$) are inversely proportional. The constant of proportionality is the speed of light ($c$):

$$\nu = \frac{c}{\lambda}$$

When viewed as a particle, electromagnetic radiation consists of packets of energy, called **photons**. The energy of each photon is directly proportional to its frequency,

$$E = h\nu$$

where $h$ is Planck's constant ($h = 6.626 \times 10^{-34}$ J·s). The range of all possible frequencies is known as the **electromagnetic spectrum**, which is arbitrarily divided into several regions by wavelength (Figure 15.2). Different regions of the electromagnetic spectrum are used to probe different aspects of molecular structure (Table 15.1). In this chapter, we will learn about the information obtained when IR radiation interacts with a compound. NMR and UV-Vis spectroscopy will be discussed in subsequent chapters.

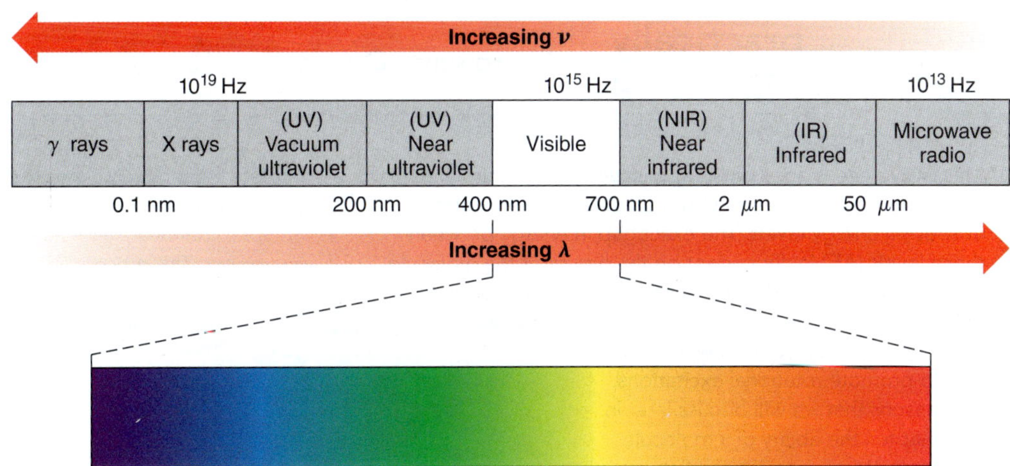


**FIGURE 15.2**
The electromagnetic spectrum.

**TABLE 15.1 SOME COMMON FORMS OF SPECTROSCOPY AND THEIR USES**

| TYPE OF SPECTROSCOPY | REGION OF ELECTROMAGNETIC SPECTRUM | INFORMATION OBTAINED |
|---|---|---|
| Nuclear magnetic resonance (NMR) spectroscopy | Radio waves | The specific arrangement of all carbon and hydrogen atoms in the compound |
| IR spectroscopy | Infrared | The functional groups present in the compound |
| UV-VIS spectroscopy | Visible and ultraviolet | Any conjugated π system present in the compound |

## The Nature of Matter

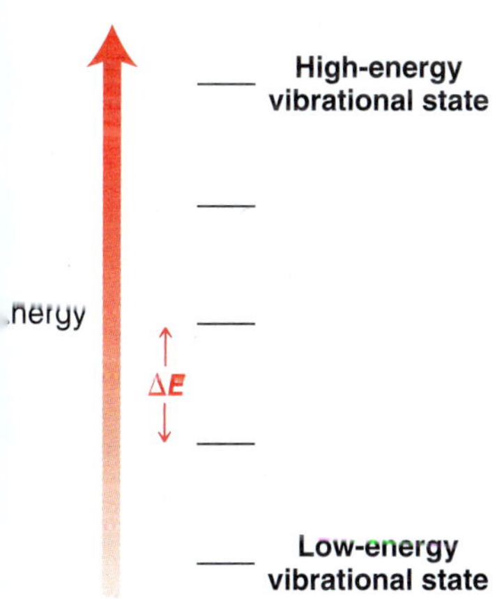


**FIGURE 15.3**
An energy diagram showing the energy gap between allowed vibrational states.

Matter, like electromagnetic radiation, also exhibits both wave-like properties and particle-like properties. In Section 1.6 we saw that quantum mechanics describes the wave-like properties of matter. According to the principles of quantum mechanics, the energy of a molecule is "quantized." To understand what this means, we will compare the rotation of a molecule with the rotation of an automobile tire connected directly to a motor. When the motor is turned on, the tire begins to rotate and a measuring device is used to determine the rate of rotation. Our ability to control the rate of rotation appears to be limited only by the precision of the motor and the sensitivity of the measuring device. It would be inconceivable to suggest that a measurable rotation rate would be unattainable by the laws of physics. If we want to spin the tire at exactly 60.06251 revolutions per second, there is nothing preventing us from doing so. In contrast, molecules behave differently. The rotation of a molecule appears to be restricted to specific energy levels. That is, a molecule can only rotate at specific rates, which are defined by the nature of the molecule. Other rotation rates are simply not allowed by the laws of physics. The rotational energy of a molecule is said to be quantized.

Molecules can store energy in a variety of ways. They rotate in space, their bonds vibrate like springs, their electrons can occupy a number of possible molecular orbitals, and so on. Each of these forms of energy is quantized. For example, a bond in a molecule can only vibrate at specific energy levels (Figure 15.3). The horizontal lines in the diagram represent allowed vibrational energy levels for a particular bond. The bond is restricted to these energy levels and cannot vibrate with an energy that is in between the allowed levels. The difference in energy ($\Delta E$) between allowed energy levels is determined by the nature of the bond.

## The Interaction between Light and Matter

In the previous section, we saw that the vibrational energy levels for a particular bond are separated from each other by an energy gap ($\Delta E$). If a photon of light possesses exactly this amount of energy, the bond can absorb the photon to promote a **vibrational excitation**. The energy of the photon is temporarily stored as vibrational energy until it is released back into the environment, usually in the form of heat. Each form of spectroscopy uses a different region of the electromagnetic spectrum

**practically speaking** | **Microwave Ovens**

Microwave ovens are a direct application of quantum mechanics and spectroscopy. We saw that molecules can only rotate at specific energy levels. The exact gap in energy between these levels ($\Delta E$) is dependent on the nature of the molecule but is generally equivalent to a photon in the microwave region of the electromagnetic spectrum. In other words, molecules will absorb microwave radiation to promote rotational excitations. For reasons that we will discuss later in this chapter, the ability of a molecule to absorb electromagnetic radiation depends very much on the presence of a permanent dipole moment. Water molecules, in particular, possess a strong dipole moment and are therefore extremely efficient at absorbing microwave radiation.

When frozen food is bombarded with microwave radiation, the water molecules in the food absorb the energy and begin to rotate more rapidly. As neighboring molecules collide with each other, the rotational energy is converted into translational energy (or heat), and the temperature of the food rises quickly. This process only works when water molecules are present. Plastic items (such as tupperware) generally do not get hot, because they are composed of long polymers that cannot freely rotate.

In addition to causing rotational excitation, microwave radiation can also cause electronic excitation in metals. That is, electrons can be promoted to higher energy orbitals. This type of electronic excitation accounts for the sparks that can be observed when metal objects are placed in a microwave oven.

and involves a different kind of excitation. In this chapter, we will focus mainly on the interaction between molecules and IR radiation, which promotes vibrational excitations of the bonds in a molecule.

## 15.2 IR Spectroscopy

### Vibrational Excitation

In the previous section, we saw that IR radiation causes vibrational excitation of the bonds in a molecule. There are many different kinds of vibrational excitation, because bonds store vibrational energy a number of ways. Bonds can *stretch*, very much the way a spring stretches, and bonds can *bend* in a number of ways.

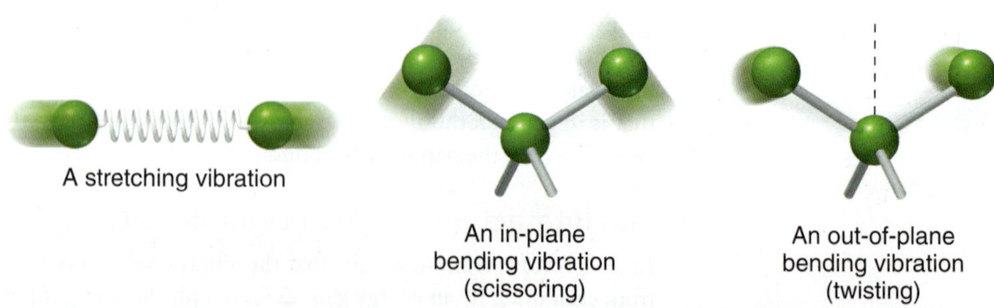


There are many other types of **bending** vibrations, but we will focus most of our attention in this chapter on **stretching** vibrations. Bending will only be mentioned briefly in this chapter.

## medically speaking

## IR Thermal Imaging for Cancer Detection

We have seen that vibrating bonds will absorb IR radiation to achieve a higher vibrational state, but it is worthwhile to note that the reverse process also occurs in nature. That is, vibrating bonds can be demoted to a *lower* vibrational state by *emitting* IR radiation. In this way, warm objects are able to release some of their energy in the form of IR radiation. This idea has found application in a wide variety of products. For example, some varieties of night vision goggles are able to produce an image in the total absence of visible light by detecting the IR radiation emitted by warm objects. Warmer areas emit more IR radiation, providing the contrast necessary to construct an image.

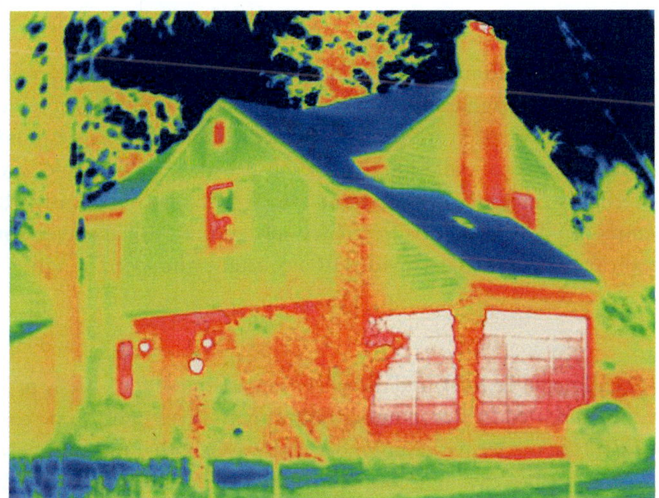

More recently, thermal imaging has found application in the detection of breast cancer. When a patient is exposed to cold air, the nervous system responds by reducing the flow of blood to the surface, thereby conserving heat. Cancerous cells and the surrounding tissue are less affected by this process and remain warm for a longer period of time. The contrast between warm and cool regions in the body can then be measured using IR sensors that detect the amount of IR radiation being emitted from different regions. This thermal imaging technique allows for the early detection of breast cancer. Various regions are color coded to indicate temperature, with the red areas being the warmest. The image shows the detection of a breast tumor.

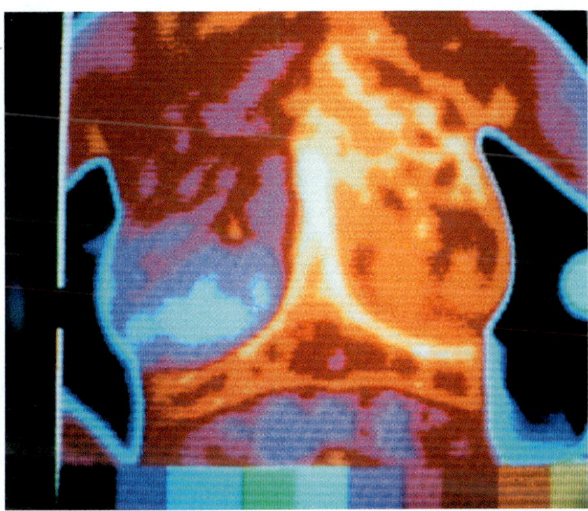

### Identification of Functional Groups with IR Spectroscopy

For each and every bond in a molecule, the energy gap between vibrational states is very much dependent on the nature of the bond. For example, the energy gap for a C—H bond is much larger than the energy gap for a C—O bond (Figure 15.4).

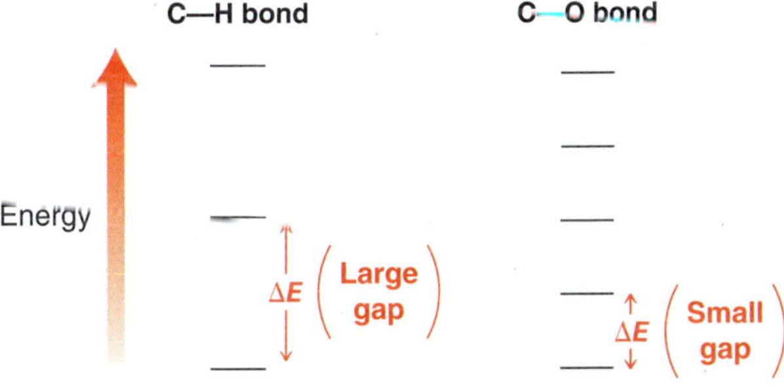


**FIGURE 15.4** The energy gap between vibrational energy levels is dependent on the nature of the bond.

Both bonds will absorb IR radiation but the C—H bond will absorb a higher energy photon. In fact, each type of bond will absorb a characteristic frequency, allowing us to determine which types of bonds are present in a compound. We simply irradiate the compound with all frequencies of IR radiation and then detect which frequencies were absorbed. For example, a compound containing an O—H bond will absorb a frequency of IR radiation characteristic of the O—H bond. In this way IR spectroscopy can be used to identify the presence of functional groups in a compound.

A salt plate used for IR spectroscopy. These plates are moisture sensitive and must be stored in a moisture-free environment.

## IR Spectrometer

In an IR spectrometer, a sample is irradiated with frequencies of IR radiation, and the frequencies that pass through (that are not absorbed by the sample) are detected. A plot is then constructed showing which frequencies were absorbed by the sample. The most commonly used type of spectrometer, called a Fourier transform (FT-IR) spectrometer, irradiates the sample with all frequencies simultaneously and then utilizes a mathematical operation called a Fourier transform to determine which frequencies passed through the sample.

Several techniques are used for preparing a sample for IR spectroscopy. The most common method involves the use of salt plates. These expensive plates are made from sodium chloride and are used because they are transparent to IR radiation. If the compound under investigation is a liquid at room temperature, a drop of the sample is sandwiched in between two salt plates and is called a *neat* sample. If the compound is a solid at room temperature, it can be dissolved in a suitable solvent and placed in between two salt plates. Alternatively, insoluble compounds can be mixed with powdered KBr and then pressed into a thin, transparent film, called a KBr pellet. All of these sampling techniques are commonly used for IR spectroscopy.

## The General Shape of an IR Absorbance Spectrum

An IR spectrometer measures the percent transmittance as a function of frequency. This plot is called an **absorption spectrum** (Figure 15.5). All the signals, called absorption bands, point down on an IR spectrum. The location of each signal on the spectrum can be specified either by the corresponding *wavelength* or by the corresponding *frequency* of radiation that was absorbed. Several decades ago, signals were reported by their wavelengths (measured in micrometers, or microns). Currently, the location of each signal is more often reported in terms of a frequency-related unit, called **wavenumber** ($\tilde{\nu}$). The wavenumber is simply the frequency of light divided by a constant (the speed of light, $c$):

$$\tilde{\nu} = \frac{\nu}{c}$$

The units of wavenumber are inverse centimeters ($cm^{-1}$), and the values range from 400 to 4000 $cm^{-1}$. All of the spectra in this chapter will be reported in wavenumber rather than wavelength, to be consistent with common practice. Don't confuse the terms wavenumber and wavelength. Wavenumber is proportional to frequency, and therefore, a larger wavenumber represents higher energy. Signals that appear on the left side of the spectrum correspond with higher energy radiation, while signals on the right side of the spectrum correspond with lower energy radiation.

Every signal in an IR spectrum has three characteristics: wavenumber, intensity, and shape. We will now explore each of these three characteristics, starting with wavenumber.

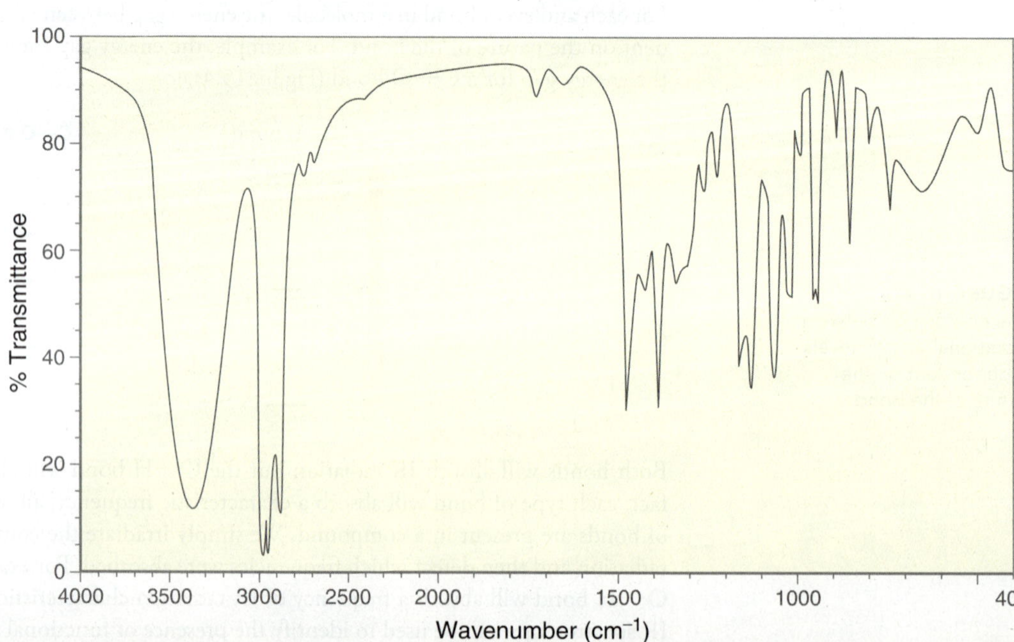


**FIGURE 15.5**
An example of an IR absorbance spectrum.

## 15.3 Signal Characteristics: Wavenumber

### Hooke's Law

For every bond, the wavenumber of absorption associated with bond stretching is dependent on two factors: (1) bond strength and (2) masses of the atoms sharing the bond. The impact of these two factors can be rationalized when we treat a bond as if it were a vibrating spring connecting two weights.

Using this analogy, we can construct the following equation, derived from Hooke's law, which enables us to approximate the frequency of vibration for a bond between two atoms of mass $m_1$ and $m_2$:

$$\tilde{\nu} = \left(\frac{1}{2\pi c}\right)\left(\frac{f}{m_{red}}\right)^{\frac{1}{2}}$$

force constant (bond strength)

$$\text{reduced mass} = \left(\frac{m_1 m_2}{m_1 + m_2}\right)$$

In this equation, $f$ is the force constant of the spring, which represents the bond strength of the bond, and $m_{red}$ is the *reduced mass* of the system. Use of the reduced mass in this equation allows us to treat the two atoms as one system. Notice that $m_{red}$ appears in the denominator. This means that smaller atoms give bonds that vibrate at higher frequencies, thereby corresponding to a higher wavenumber of absorption. For example, compare the following bonds. The C—H bond involves the smallest atom (H) and therefore appears at the highest wavenumber.

| C—H | C—D | C—O | C—Cl |
|---|---|---|---|
| ~3000 cm$^{-1}$ | ~2200 cm$^{-1}$ | ~1100 cm$^{-1}$ | ~700 cm$^{-1}$ |

While $m_{red}$ appears in the denominator of the equation, the force constant ($f$) appears in the numerator. This means that stronger bonds will vibrate at higher frequencies, thereby corresponding to a higher wavenumber of absorption. For example, compare the following bonds. The C≡N bond is the strongest of the three bonds and therefore appears at the highest wavenumber.

| C≡N | C=N | C—N |
|---|---|---|
| ~2200 cm$^{-1}$ | ~1600 cm$^{-1}$ | ~1100 cm$^{-1}$ |

Using the two trends shown, we can understand why different types of bonds will appear in different regions of an IR spectrum (Figure 15.6). Single bonds (except for X—H bonds) appear on the right side of the spectrum (below 1500 cm$^{-1}$) because single bonds are generally the weakest bonds. Double bonds appear at higher wavenumber (1600–1850 cm$^{-1}$) because they are stronger than single bonds, while triple bonds appear at even higher wavenumber (2100–2300 cm$^{-1}$) because they are even stronger than double bonds. And finally, the left side of the spectrum contains signals produced by X—H bonds (such as C—H, O—H, or N—H), all of which stretch at a high wavenumber because hydrogen has the smallest mass.

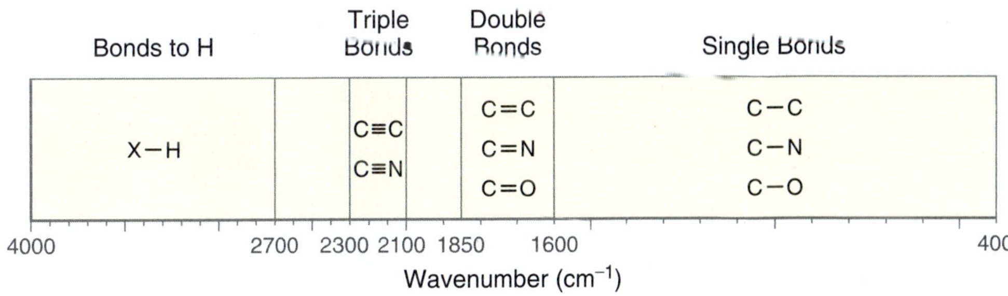


**FIGURE 15.6** An IR absorbance spectrum divided into regions based on bond strength and atomic mass.

IR spectra can be divided into two main regions (Figure 15.7). The **diagnostic region** generally has fewer peaks and provides the clearest information. This region contains all signals that arise from double bonds, triple bonds, and X—H bonds. The **fingerprint region** contains signals resulting from the vibrational excitation of most single bonds (stretching and bending). This

Diagnostic Region | Fingerprint Region

Bonds to H | Triple Bonds | Double Bonds | Single Bonds

4000 3500 3000 2500 2000 1500 1000 400

Wavenumber ($cm^{-1}$)

**FIGURE 15.7**
The diagnostic and fingerprint regions of an IR spectrum.

region generally contains many signals and is more difficult to analyze. What appears like a C—C stretch might in fact be another bond that is bending. This region is called the fingerprint region because each compound has a unique pattern of signals in this region, much the way each person has a unique fingerprint. To illustrate this point, compare the spectra for 2-butanol and 2-propanol (Figure 15.8). The diagnostic regions of these compounds are virtually indistinguishable (they both contain characteristic C—H and O—H signals), but the fingerprint regions of these compounds are very different.

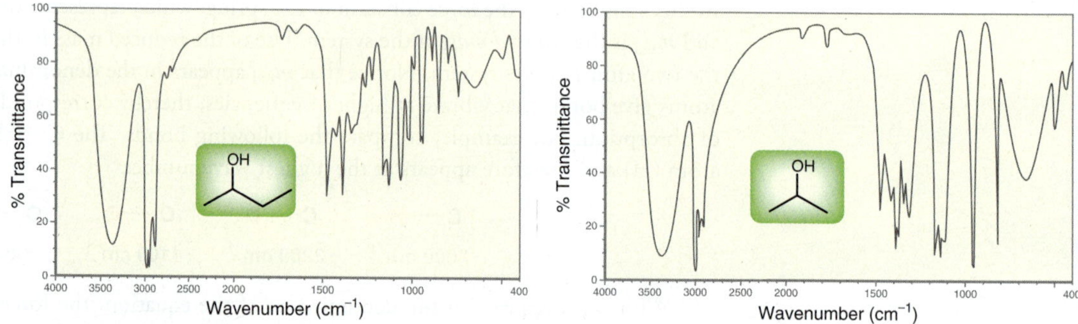


**FIGURE 15.8**
IR spectra for 2-butanol and 2-propanol.

## Effect of Hybridization States on Wavenumber of Absorption

In the previous section, we saw that bonds to hydrogen (such as C—H bonds) appear on the left side of an IR spectrum (high wavenumber). We will now compare various kinds of C—H bonds.

The wavenumber of absorption for a C—H bond is very much dependent on the hybridization state of the carbon atom. Compare the following three C—H bonds:

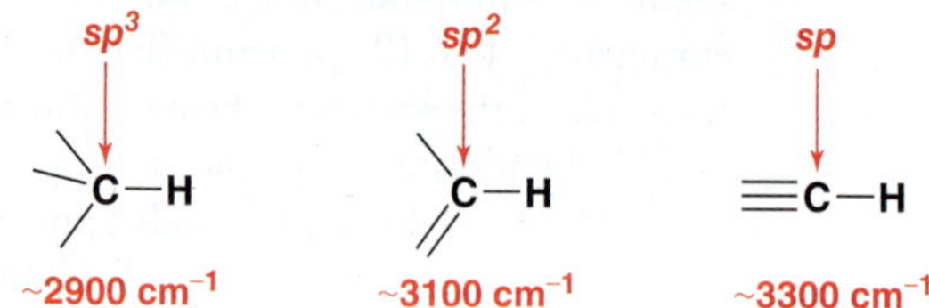


Of the three bonds shown, the $C_{sp}$—H bond produces the highest energy signal (~3300 $cm^{-1}$), while a $C_{sp^3}$—H bond produces the lowest energy signal (~2900 $cm^{-1}$). To understand this trend, we must revisit the shapes of the hybridized atomic orbitals (Figure 15.9). As illustrated,

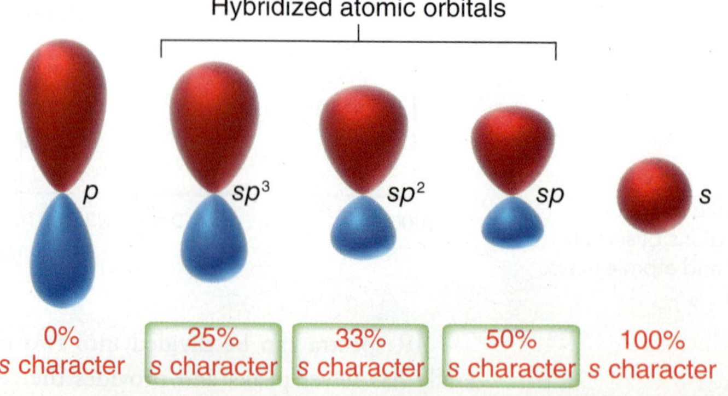


**FIGURE 15.9**
The shapes of atomic orbitals.

*sp* orbitals have more *s* character than the other hybridized atomic orbitals, and therefore *sp* orbitals more closely resemble *s* orbitals. Compare the shapes of the hybridized atomic orbitals and note that the electron density of an *sp* orbital is closest to the nucleus (much like an *s* orbital). As a result, a $C_{sp}$—H bond will be shorter than other C—H bonds. Compare the bond lengths of the following three C—H bonds:

**$sp^3$** C—H **109 pm**　　**$sp^2$** C—H **108 pm**　　**$sp$** ≡C—H **106 pm**

The $C_{sp}$—H bond has the shortest bond length and is therefore the strongest bond. The $C_{sp^3}$—H bond has the longest bond length and is therefore the weakest bond. Compare the spectra of an alkane, an alkene, and an alkyne (Figure 15.10). In each case, we draw a line at 3000 $cm^{-1}$.

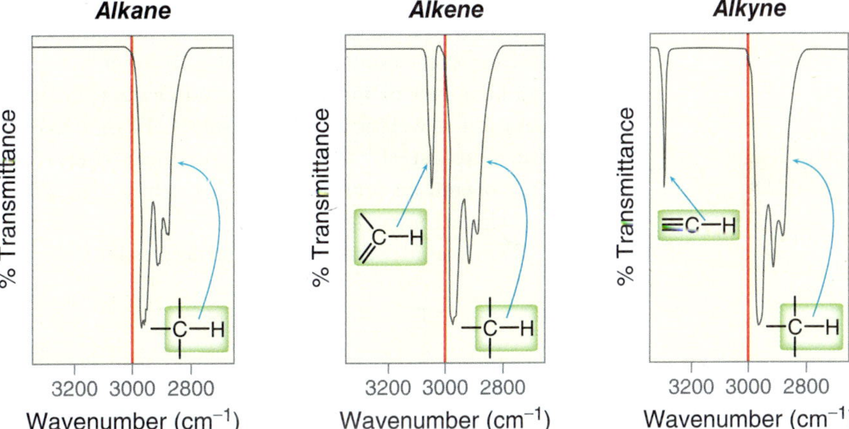


**FIGURE 15.10** The relevant portions of the IR spectra for an alkane, an alkene, and an alkyne.

All three spectra have signals to the right of the line, resulting from $C_{sp^3}$—H bonds. The key is to look for any signals to the left of the line. An alkane does not have a signal to the left of 3000 $cm^{-1}$. An alkene has a signal at 3100 $cm^{-1}$, and an alkyne has a signal at 3300 $cm^{-1}$. But be careful—the absence of a signal to the left of 3000 $cm^{-1}$ does *not* necessarily indicate the absence of a double bond or triple bond in the compound. Tetrasubstituted double bonds do not possess any $C_{sp^2}$—H bonds, and internal triple bonds also do not possess any $C_{sp}$—H bonds.

**No signal at 3100 $cm^{-1}$** (no $C_{sp^2}$—H)　　**No signal at 3300 $cm^{-1}$** (no $C_{sp}$—H)

## Effect of Resonance on Wavenumber of Absorption

We will now consider the effects of resonance on the wavenumber of absorption. As an illustration, compare the carbonyl groups (C═O bonds) in the following two compounds:

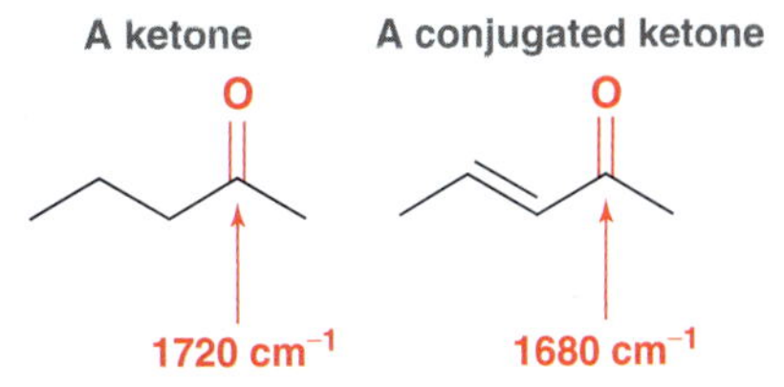

The second compound is called an unsaturated, conjugated ketone. It is *unsaturated* because of the presence of a C=C bond (we will see more on this terminology later in this chapter), and it is **conjugated** because the π bonds are separated from each other by exactly one sigma bond. We will explore conjugated π systems in more detail in Chapter 17. For now, we will just analyze the effect of conjugation on the IR absorption of the carbonyl group. As shown, the carbonyl group of an unsaturated, conjugated ketone produces a signal at lower wavenumber (1680 $cm^{-1}$) than the carbonyl group of a saturated ketone (1720 $cm^{-1}$). In order to understand why, we must draw resonance structures for each compound. Let's begin with the ketone.

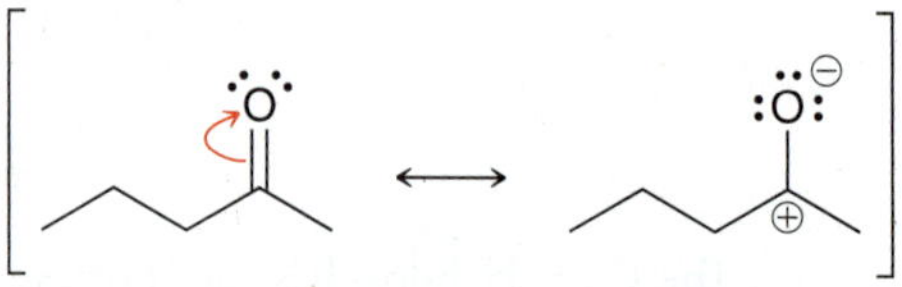

Ketones have two significant resonance structures. The carbonyl group is drawn as a double bond in the first resonance structure, and it is drawn as a single bond in the second resonance structure. This means that the carbonyl group has some double-bond character and some single-bond character. In order to determine the nature of this bond, we must consider the contribution from each resonance structure. In other words, does the carbonyl group have more double-bond character or more single-bond character? The second resonance structure exhibits charge separation as well as a carbon atom (C+) that has less than an octet of electrons. Both of these reasons explain why the second resonance structure contributes only a small amount of character to the overall resonance hybrid. Therefore, the carbonyl group of a ketone has mostly double-bond character.

Now consider the resonance structures for a conjugated, unsaturated ketone.

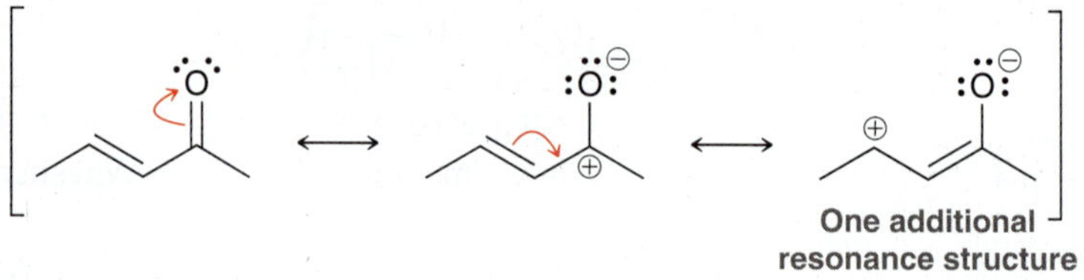

One additional resonance structure

Conjugated, unsaturated ketones have three resonance structures rather than two. In the third resonance structure, the carbonyl group is drawn as a single bond. Once again, this resonance structure exhibits charge separation as well as a carbon atom (C+) with less than an octet of electrons. As a result, this resonance structure also contributes only a small amount of character to the overall resonance hybrid. Nevertheless, this third resonance structure does contribute some character to the overall resonance hybrid, giving this carbonyl group slightly more single-bond character than the carbonyl group of a saturated ketone. With more single-bond character, it is a slightly weaker bond and therefore produces a signal at a lower wavenumber (1680 $cm^{-1}$ rather than 1720 $cm^{-1}$).

Esters exhibit a similar trend. An ester typically produces a signal at around 1740 $cm^{-1}$, but conjugated, unsaturated esters produce lower-energy signals, usually around 1710 $cm^{-1}$. Once again, the carbonyl group of a conjugated, unsaturated ester is a weaker bond, due to resonance.

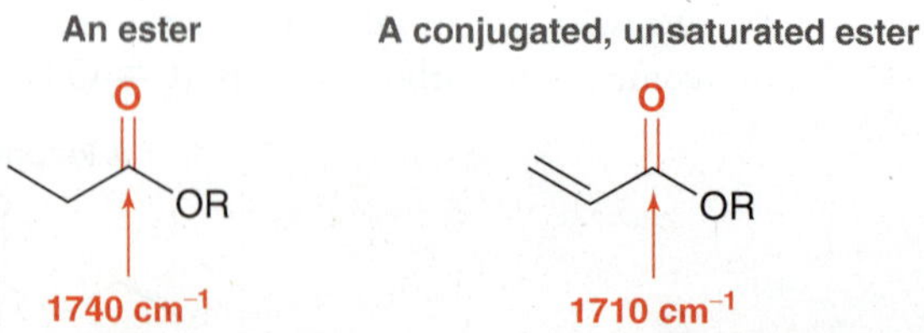

An ester

A conjugated, unsaturated ester

O

OR

1740 $cm^{-1}$

O

OR

1710 $cm^{-1}$

## CONCEPTUAL CHECKPOINT

**15.1** For each of the following compounds, rank the highlighted bonds in terms of increasing wavenumber:

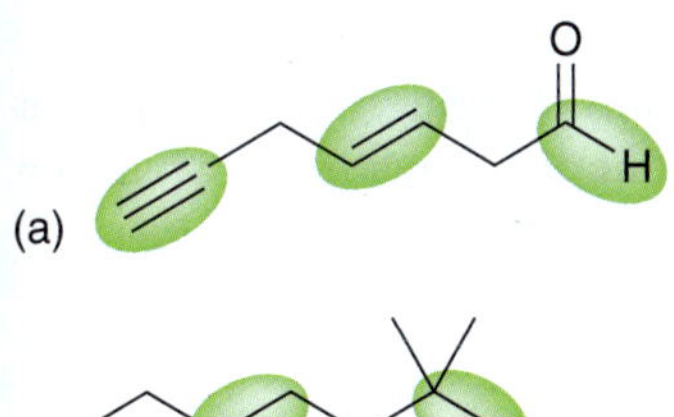


**15.2** For each of the following compounds, determine whether or not you would expect its IR spectrum to exhibit a signal to the left of 3000 $cm^{-1}$:

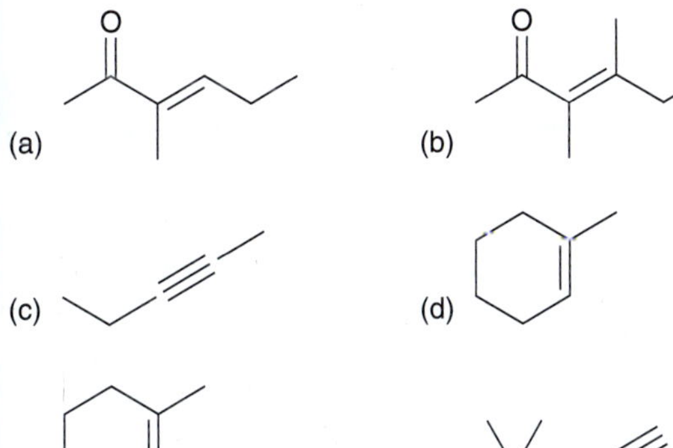


**15.3** Each of the following compounds contains two carbonyl groups. Identify which carbonyl group will exhibit a signal at lower wavenumber.

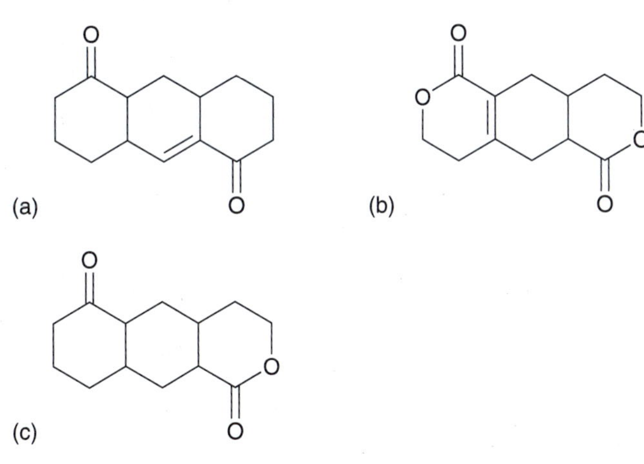


**15.4** Compare the wavenumber of absorption for the following two C=C bonds. Use resonance structures to explain why the C=C bond in the conjugated compound produces a signal at lower wavenumber.

O O

1650 $cm^{-1}$ 1600 $cm^{-1}$

## 15.4 Signal Characteristics: Intensity

In an IR spectrum, some signals will be very strong in comparison with other signals on the same spectrum (Figure 15.11). That is, some bonds absorb IR radiation very efficiently, while other bonds are less efficient at absorbing IR radiation. To understand the reason for this, we must

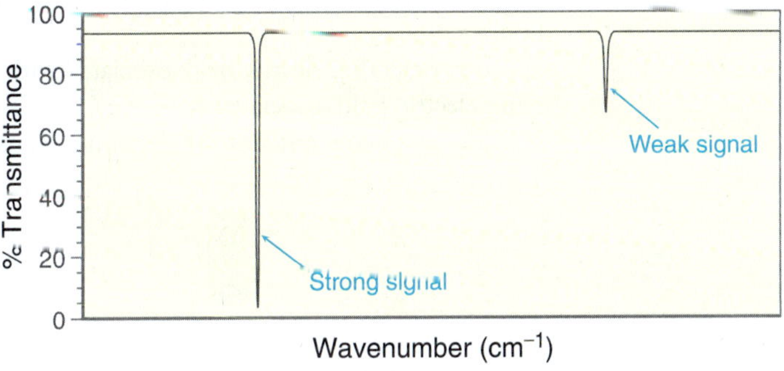


**FIGURE 15.11** A comparison of a strong signal and weak signal in an IR spectrum.

consider how the strength of a dipole moment changes when a bond vibrates. Recall that the dipole moment ($\mu$) of a bond is defined by the following equation:

$$\mu = e \times d$$

where $e$ is the strength of the partial charges (δ+ and δ−) and $d$ is the distance separating them. Now consider what happens to the strength of a dipole moment as a bond stretches.

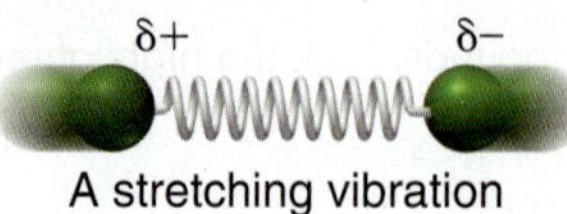


As a bond vibrates, the distance between the partial charges is constantly changing, which means that the strength of the dipole moment also changes with time. A plot of the dipole moment as a function of time shows an oscillating dipole moment (Figure 15.12).

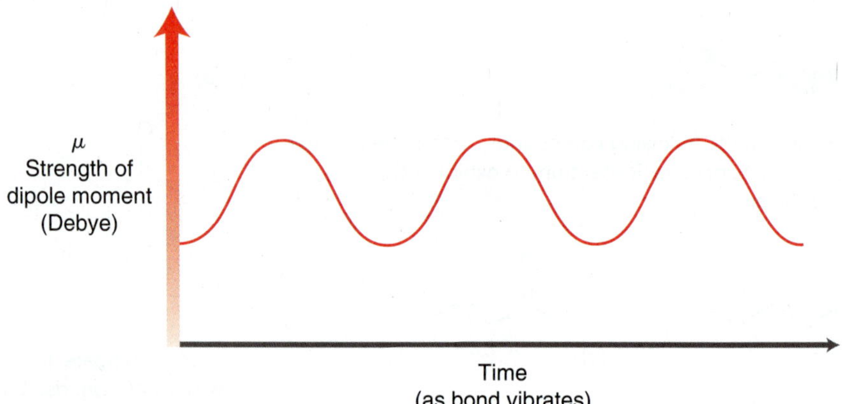


**FIGURE 15.12** For a vibrating bond, the strength of the dipole moment oscillates as a function of time.

The dipole moment is an electric field surrounding the bond. So as the dipole moment oscillates, the bond is essentially surrounded by an oscillating electric field, which serves as an antenna (so to speak) for absorbing IR radiation. Since electromagnetic radiation itself is comprised of an oscillating electric field, the bond can absorb a photon because the bond's oscillating electric field interacts with the oscillating electric field of the IR radiation.

The efficiency of a bond at absorbing IR radiation therefore depends on the strength of the dipole moment. For example, compare the following two highlighted bonds:

Each of these bonds has a measurable dipole moment, but they differ significantly in strength. Let's first analyze the carbonyl group (C=O bond). Due to resonance and induction, the carbon atom bears a large partial positive charge, and the oxygen atom bears a large partial negative charge. The carbonyl group therefore has a large dipole moment. Now let's analyze the C=C bond. One vinylic position is connected to alkyl groups, while the other vinylic position is connected to hydrogen atoms. As such, the two vinylic positions are not electronically identical, so there will be a small dipole moment.

The oscillating electric field associated with the carbonyl group is much stronger than the oscillating electric field associated with the C=C bond (Figure 15.13). Therefore, the carbonyl group functions as a better antenna for absorbing IR radiation. That is, the carbonyl group is more efficient

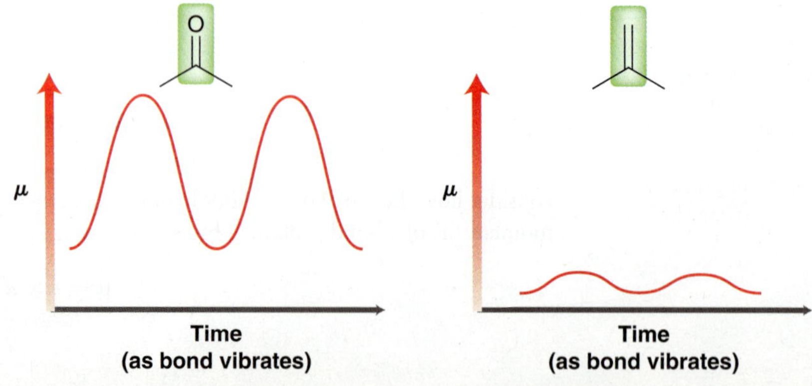


**FIGURE 15.13** A comparison of the oscillating electric fields for a C=O bond and a C=C bond.

at absorbing IR radiation, producing a stronger signal (Figure 15.14). Carbonyl groups often produce the strongest signals in an IR spectrum, while C═C bonds often produce fairly weak signals. In fact, some alkenes do not even produce any C═C signal at all. For example, consider the IR spectrum of 2,3-dimethyl-2-butene (Figure 15.15). This alkene is symmetrical. That is, both vinylic positions are electronically identical, and the bond has no dipole moment at all. As the C═C bond vibrates, there is no change in dipole moment, which means that the bond cannot function as an antenna for absorbing IR radiation. In other words, symmetrical C═C bonds are completely inefficient at absorbing IR radiation, and no signal is observed. The same is true for symmetrical C≡C bonds.

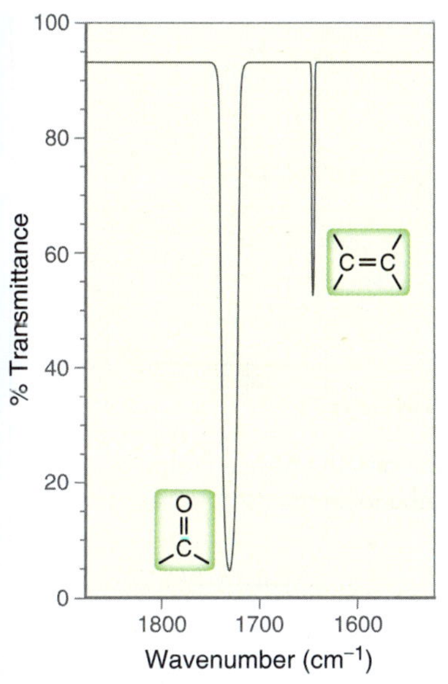


**FIGURE 15.14**
A comparison of the intensities of the signals arising from a C═O bond and a C═C bond.

**FIGURE 15.15**
An IR spectrum of 2,3-dimethyl-2-butene.

## LOOKING AHEAD

You might be wondering why there is more than one signal for these C—H bonds. This question will be answered in the next section.

There is one other factor that can contribute significantly to the intensity of signals in an IR spectrum. Consider the group of signals appearing just below 3000 cm$^{-1}$ in Figure 15.15. These signals are associated with the stretching of the C—H bonds in the compound. The intensity of these signals derives from the number of C—H bonds giving rise to the signals. In fact, the signals just below 3000 cm$^{-1}$ are typically among the strongest signals in an IR spectrum.

## CONCEPTUAL CHECKPOINT

**15.5** For each pair of compounds, determine which C═C bond will produce a stronger signal.

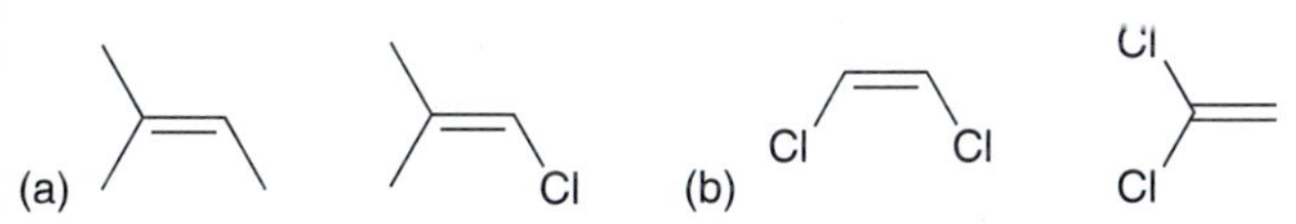


**15.6** The C═C bond in 2-cyclohexenone produces an unusually strong signal. Explain using resonance structures.

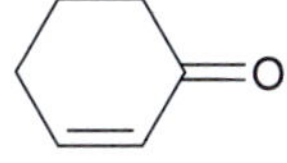


**15.7** *trans*-2-Butene does not exhibit a signal in the double-bond region of the spectrum (1600–1850 cm$^{-1}$); however, IR spectroscopy is still helpful in identifying the presence of the double bond. Identify the other signal that would indicate the presence of a C═C bond.

## IR Spectroscopy for Testing Blood Alcohol Levels

In the previous section, we saw that the intensity of a signal in an IR spectrum is dependent on how efficiently a bond absorbs IR radiation. The intensity of the signal is also dependent on other factors, such as the concentration of the sample being analyzed. This fact is used by law enforcement officials to measure blood alcohol levels accurately. One device for measuring blood alcohol levels is called the Intoxilyzer. It is essentially an IR spectrometer that is specifically tuned to measure the intensity of the signals for the C—H bonds in ethanol. For example, the Intoxilyzer 5000 measures the intensity of absorption at the five frequencies illustrated.

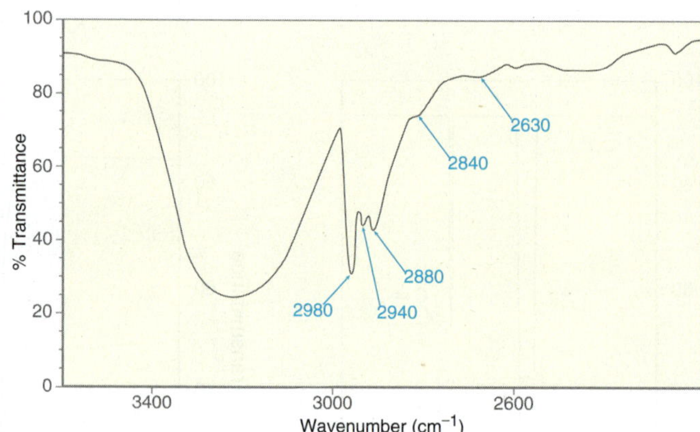


A breath sample is analyzed, and the device is able to calculate the concentration of ethanol in the sample based on the intensities of these signals.

## 15.5 Signal Characteristics: Shape

In this section, we will explore some of the factors that affect the shape of a signal. The shape of the signals in an IR spectrum varies—some are very broad while other signals are very narrow (Figure 15.16).

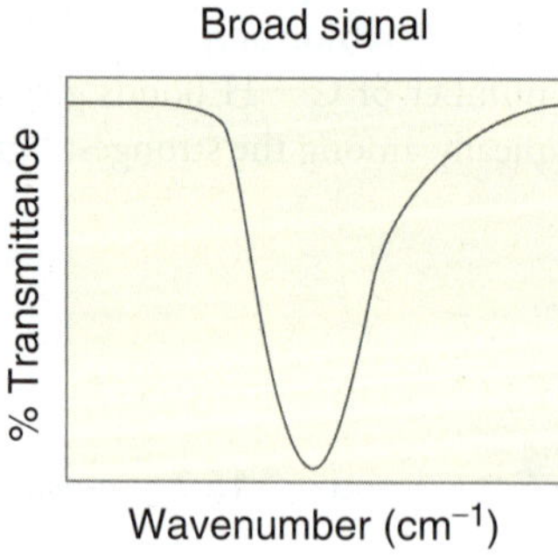


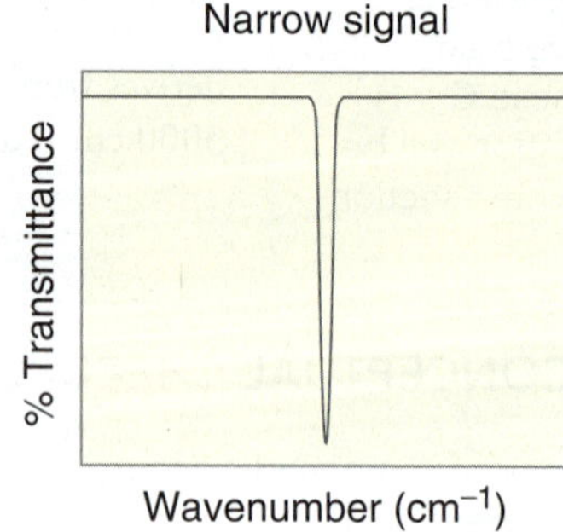


**FIGURE 15.16** A comparison of a broad signal and narrow signal in an IR spectrum.

### Effects of Hydrogen Bonding

Alcohols exhibit hydrogen bonding, as discussed in Section 13.1. One effect of H bonding is to weaken the existing O—H bond.

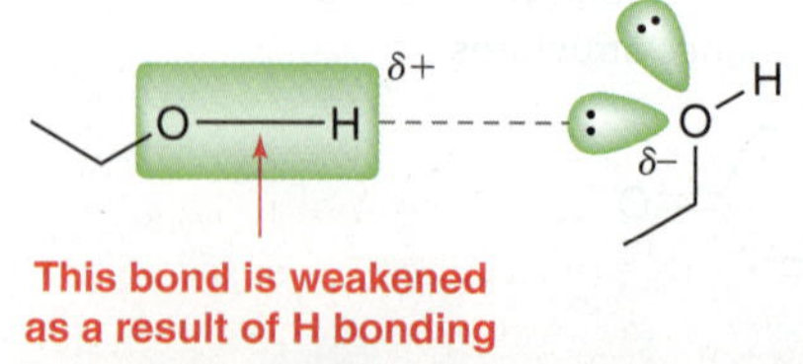

Concentrated alcohols give rise to broad signals because of this weakening effect. At any given moment in time, the O—H bond in each molecule is weakened to a different extent. As a result, the O—H bonds do not have a consistent bond strength, but rather, there is a *distribution* of bond strengths. That is, some molecules are barely participating in H bonding, while others are participating in H bonding to varying degrees. The result is a broad signal.

The shape of an OH signal is different when the alcohol is diluted in a solvent that cannot form hydrogen bonds with the alcohol. In such an environment, it is likely that the O—H bonds will not participate in an H-bonding interaction. The result is a narrow signal. When the solution is neither very concentrated nor very dilute, two signals are observed. The molecules that are not participating in H bonding will give rise to a narrow signal, while the molecules participating in H bonding will give rise to a broad signal. As an example, consider the spectrum of 2-butanol in which both signals can be observed (Figure 15.17). When O—H bonds do not participate in H bonding, they generally produce a signal at approximately 3600 $cm^{-1}$. That signal can be seen in the spectrum. When O—H bonds participate in H bonding, they generally produce a broad signal between 3200 and 3600 $cm^{-1}$. That signal can also be seen in the spectrum. Depending on the conditions, an alcohol will give a broad signal, or a narrow signal, or both.

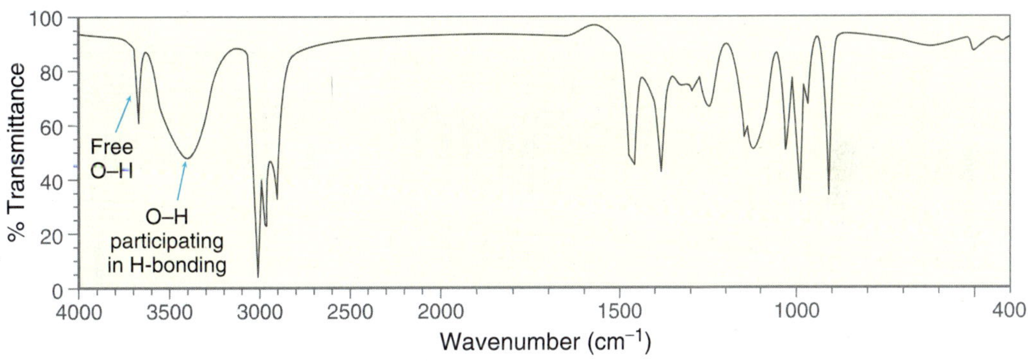


**FIGURE 15.17**
An IR spectrum of 2-butanol.

## CONCEPTUAL CHECKPOINT

**15.8** As explained previously, the concentration of an alcohol can be selected such that both a broad signal and a narrow signal appear simultaneously. In such cases, the broad signal is always to the *right* of the narrow signal, never to the left. Explain.

Carboxylic acids exhibit similar behavior, only more pronounced. For example, consider the spectrum of butyric acid (Figure 15.18). Notice the very broad signal on the left side of the spectrum, extending from 2200 to 3600 $cm^{-1}$. This signal is so broad that it extends over the usual C—H signals that appear around 2900 $cm^{-1}$. This very broad signal, characteristic of carboxylic acids, is a result of H bonding. The effect is more pronounced than alcohols, because molecules of the carboxylic acid can form two hydrogen-bonding interactions, resulting in a dimer.

The IR spectrum of a carboxylic acid is easy to recognize, because of the characteristic broad signal that covers nearly one third of the spectrum. This broad signal is also accompanied by a broad C=O signal just above 1700 $cm^{-1}$.

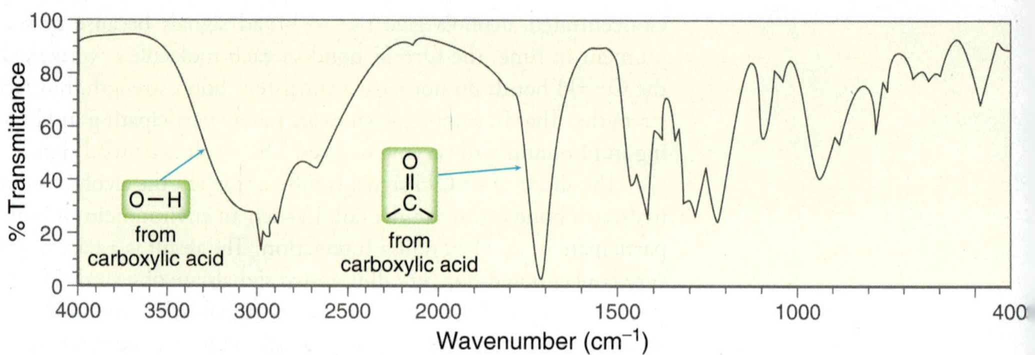


**FIGURE 15.18**
An IR spectrum of butyric acid.

## CONCEPTUAL CHECKPOINT

**15.9** For each of the following IR spectra, identify whether it is consistent with the structure of an alcohol, a carboxylic acid, or neither:

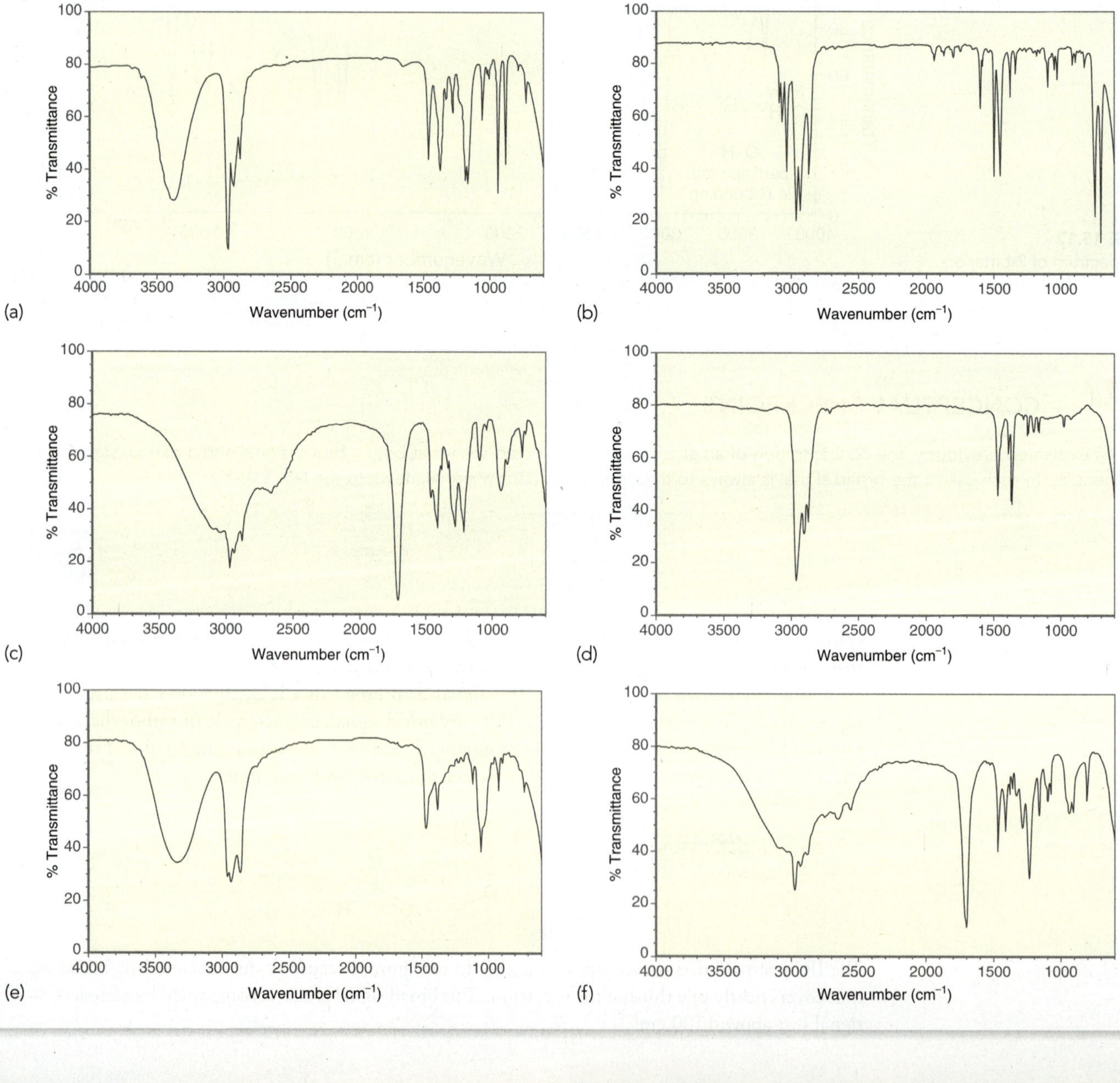

(a)

(b)

(c)

(d)

(e)

(f)

## Amines: Symmetrical vs. Asymmetrical

There is one other important factor in addition to H bonding that affects the shape of a signal. Consider the difference in shape of the N—H signals for primary and secondary amines (Figure 15.19).

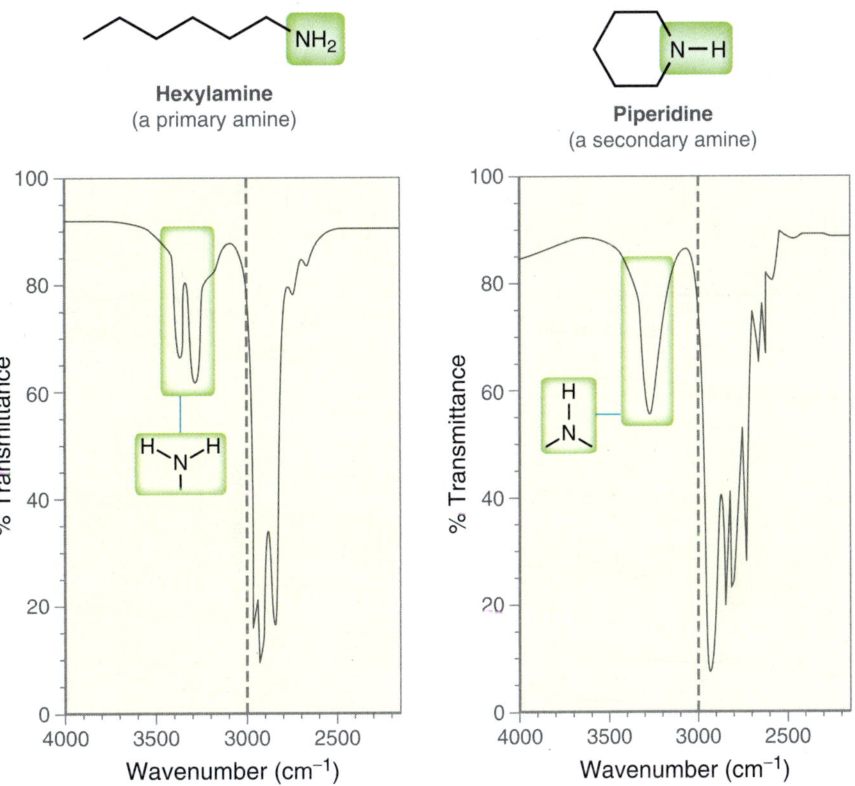


**FIGURE 15.19** The difference in shape of the signals for a primary amine and a secondary amine.

The primary amine exhibits two signals: one at 3350 $cm^{-1}$ and the other at 3450 $cm^{-1}$. In contrast, the secondary amine exhibits only one signal. It might be tempting to explain this by arguing that each N—H bond gives rise to a signal, and therefore a primary amine gives two signals because it has two N—H bonds. Unfortunately, the explanation is not quite that simple. In fact, both N—H bonds of a single molecule will together produce only one signal. The reason for the appearance of two signals is more accurately explained by considering the two possible ways in which the entire $NH_2$ group can vibrate. The N—H bonds can be stretching in phase with each other, called **symmetric stretching**, or they can be stretching out of phase with each other, called **asymmetric stretching**. At any given moment in time, approximately half of the molecules will be vibrating symmetrically, while the other half will be vibrating asymmetrically. The molecules vibrating symmetrically will absorb a particular frequency of IR radiation to promote a vibrational excitation, while the molecules vibrating asymmetrically will absorb a different frequency. In other words, one of the signals is produced by half of the molecules, and the other signal is produced by the other half of the molecules.

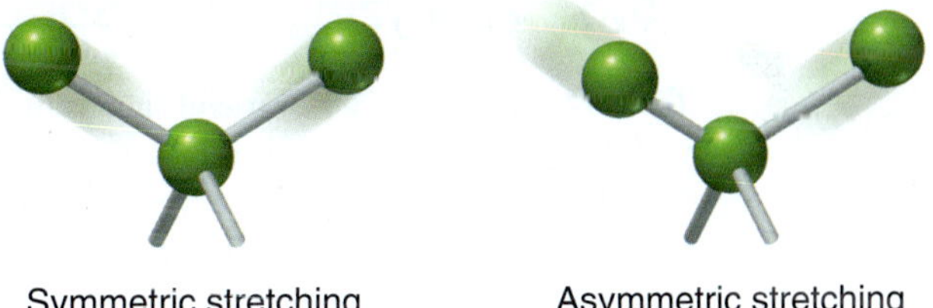


For a similar reason, the C—H bonds of a $CH_3$ group (appearing just below 3000 $cm^{-1}$ in an IR spectrum) generally give rise to a series of signals, rather than just one signal. These signals arise from the various ways in which a $CH_3$ group can be excited.

## CONCEPTUAL CHECKPOINT

**15.10** For each of the following IR spectra, determine whether it is consistent with the structure of a ketone, an alcohol, a carboxylic acid, a primary amine, or a secondary amine:

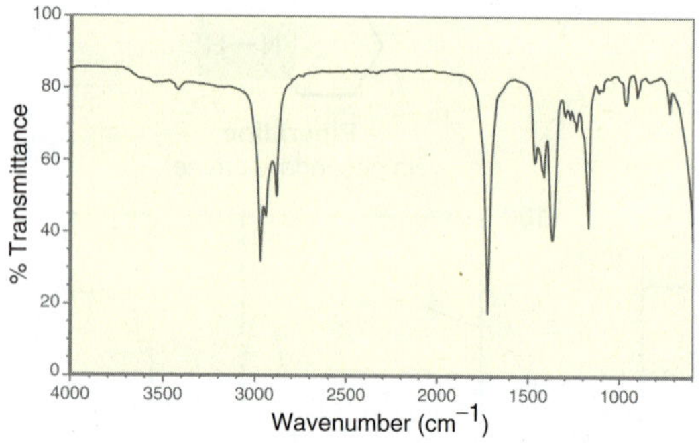

(a)

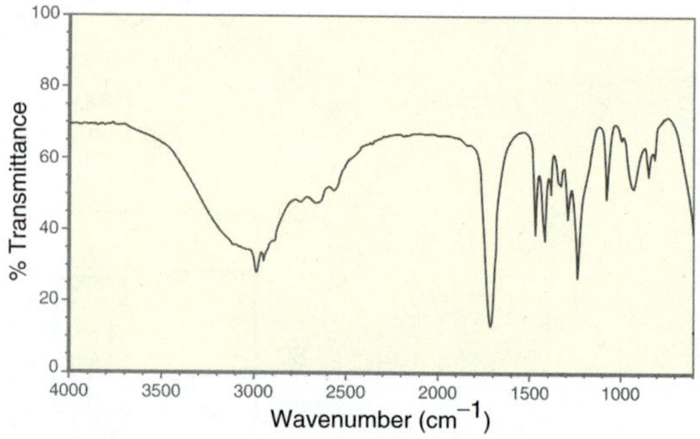

(b)

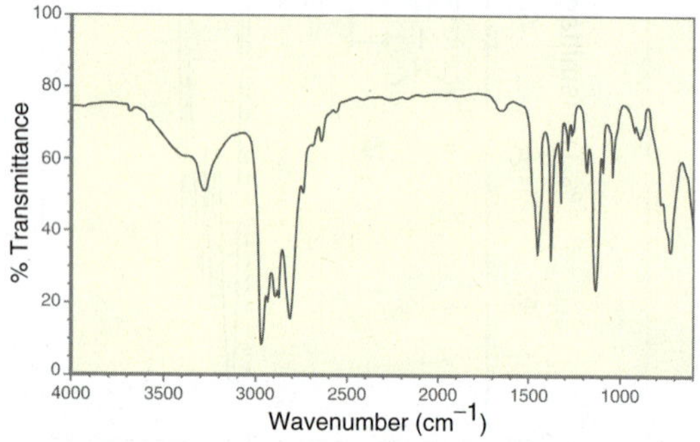

(c)

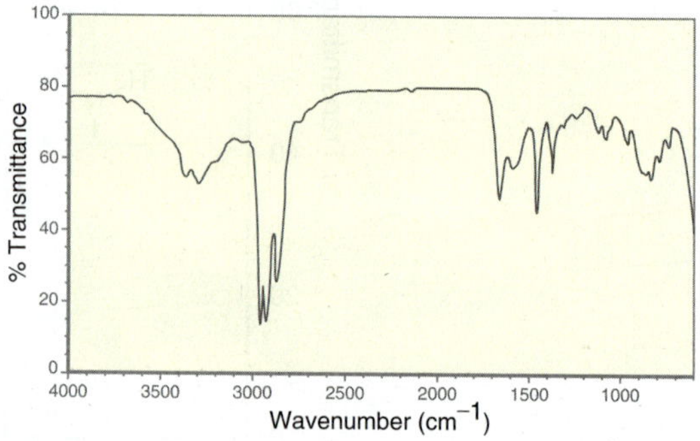

(d)

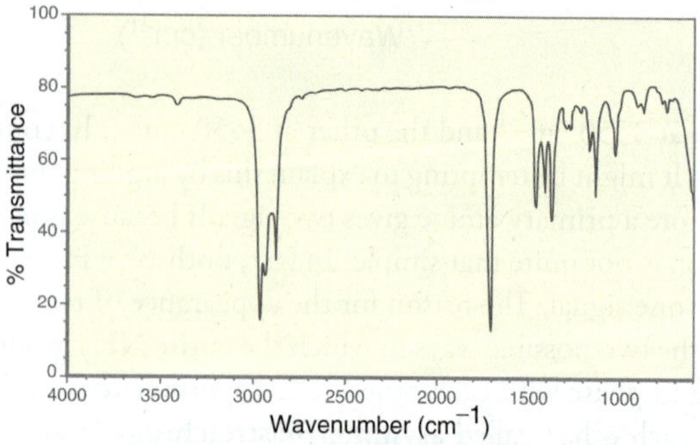

(e)

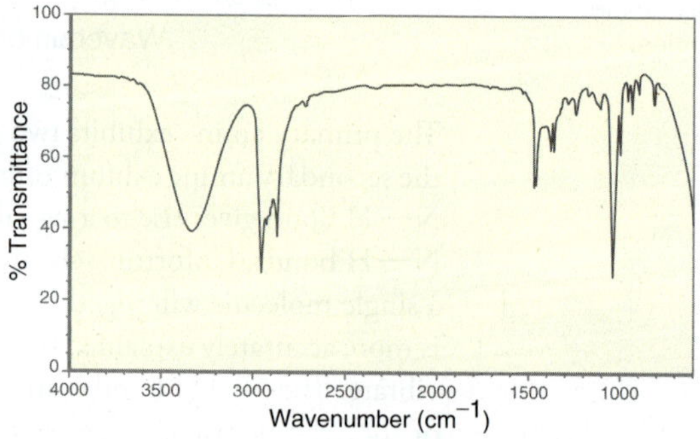

(f)

**15.11** Carefully consider the structure of 2,3-dimethyl-2-butene. There are twelve $C_{sp^3}$—H bonds, and they are all identical. Nevertheless, there is more than one signal just to the right of 3000 $cm^{-1}$ in the IR spectrum of this compound. Can you offer an explanation?

## 15.6 Analyzing an IR Spectrum

Table 15.2 is a summary of useful signals in the diagnostic region of an IR spectrum as well as some useful signals in the fingerprint region.

When analyzing an IR spectrum, the first step is to draw a line at 1500 $cm^{-1}$. Focus on any signals to the left of this line (the diagnostic region). In doing so, it will be extremely helpful if you can identify the following regions:

- Double bonds: 1600–1850 $cm^{-1}$
- Triple bonds: 2100–2300 $cm^{-1}$
- X—H bonds: 2700–4000 $cm^{-1}$

**TABLE 15.2 IMPORTANT SIGNALS IN IR SPECTROSCOPY**

**USEFUL SIGNALS IN THE DIAGNOSTIC REGION**

| STRUCTURAL UNIT | FREQUENCY ($cm^{-1}$) |
|---|---|
| **Single Bonds (X—H)** | |
| —O—H | 3200–3600 |
| O=C(—)—O—H | 2200–3600 |
| N—H | 3350–3500 |
| ≡C—H | ~3300 |
| =C—H | 3000–3100 |
| —C—H | 2850–3000 |
| O=C—H | 2750–2850 |
| **Triple Bonds** | |
| —C≡C— | 2100–2200 |
| —C≡N | 2200–2300 |

| STRUCTURAL UNIT | FREQUENCY ($cm^{-1}$) | |
|---|---|---|
| **Double Bonds** | | |
| Cl—C=O | 1750–1850 | Discussed in Chapter 21 |
| RO—C(R)=O | 1700–1750 | Discussed in Chapter 21 |
| HO—C(R)=O | 1700–1750 | Discussed in Chapter 21 |
| R—C(R)=O | 1680–1750 | Discussed in Chapter 21 |
| $H_2N$—C(R)=O | 1650–1700 | Discussed in Chapter 21 |
| C=C | 1600–1700 | |
| benzene ring | 1450–1600<br>1650–2000 | Discussed in Chapter 18 |

**USEFUL SIGNALS IN THE FINGERPRINT REGION**

| STRUCTURAL UNIT | FREQUENCY ($cm^{-1}$) |
|---|---|
| —C—Cl | 600–800 |
| —C—Br | 500–600 |
| —C—O— | 1000–1100 |
| O=C—O— | 1250–1350 |
| —C—N | 1000–1200 |

| STRUCTURAL UNIT | FREQUENCY ($cm^{-1}$) | |
|---|---|---|
| R(H)C=CH(H) | 900–920<br>980–1000 | *(bending)* |
| $R_2C=CH_2$ | 880–900 | *(bending)* |
| R—CH($CH_3$)$_2$ | 1370, 1380 | *(bending)* |
| R—C($CH_3$)$_3$ | 1370, 1390 | *(bending)* |

Remember that each signal appearing in the diagnostic region will have three characteristics (wavenumber, intensity, and shape). Make sure to analyze all three characteristics.

When looking for X—H bonds, draw a line at 3000 $cm^{-1}$ and look for signals that appear to the left of the line (Figure 15.20).

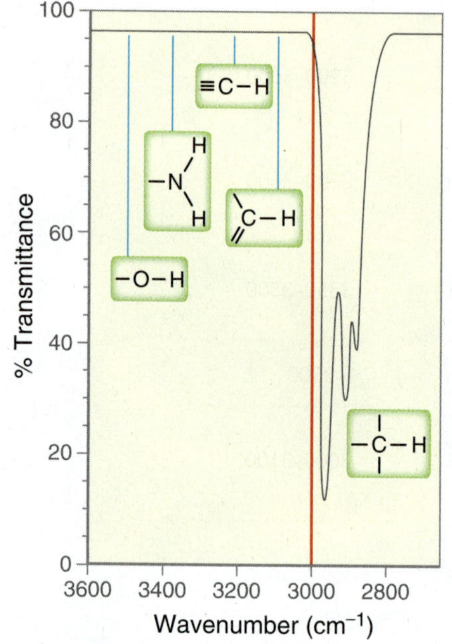


**FIGURE 15.20** The location of signals arising from several different X—H bonds.

## SKILLBUILDER

### 15.1 ANALYZING AN IR SPECTRUM

**LEARN** the skill

A compound with molecular formula $C_6H_{10}O$ gives the following IR spectrum:

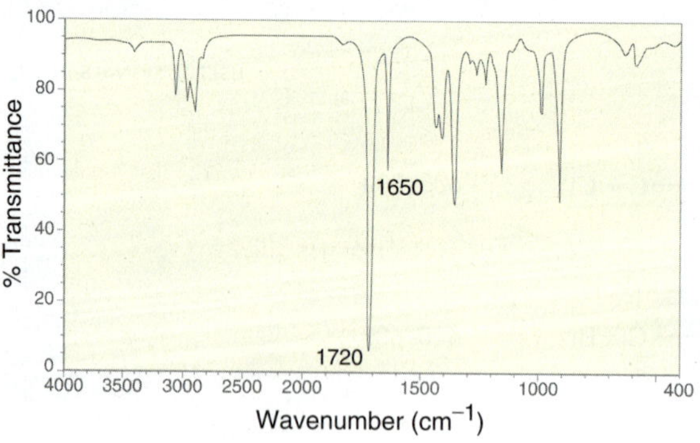


Identify the structure below that is most consistent with the spectrum:

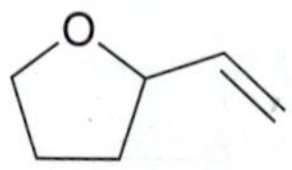

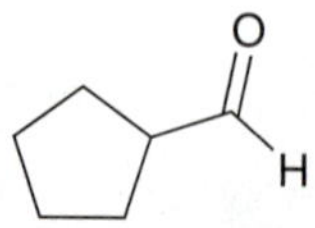

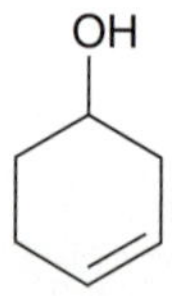

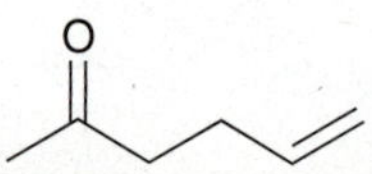

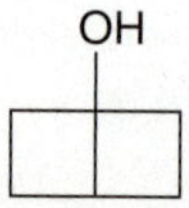

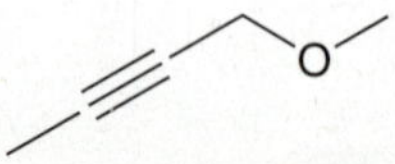

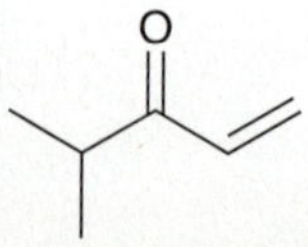

OH

O

H

O

O

OH

O

O

**STEPS 1 AND 2**
Look for double bonds ($1600–1850\ cm^{-1}$) and triple bonds ($2100–2300\ cm^{-1}$)

## SOLUTION

Draw a line at $1500\ cm^{-1}$ and focus on the diagnostic region (to the left of the line). Start by looking at the double-bond region and the triple-bond region.

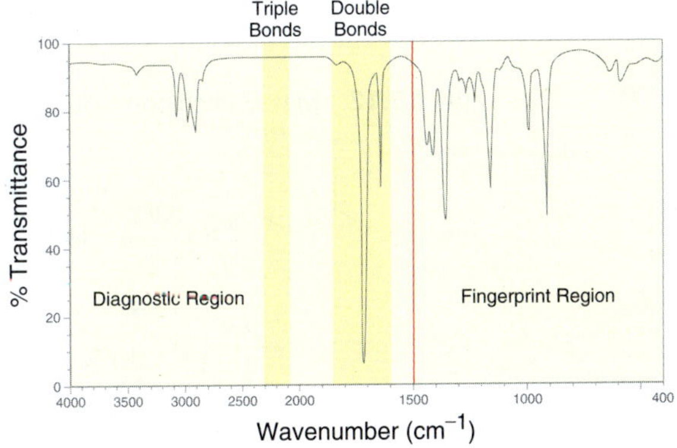


There are no signals in the triple-bond region, but there are two signals in the double-bond region. The signal at $1650\ cm^{-1}$ is narrow and weak, consistent with a C=C bond. The signal at $1720\ cm^{-1}$ is broad and strong, consistent with a C=O bond.

**STEP 3**
Look for X—H bonds. Draw a line at $3000\ cm^{-1}$.

Next, look for X—H bonds. Draw a line at $3000\ cm^{-1}$ and identify if there are any signals to the left of this line.

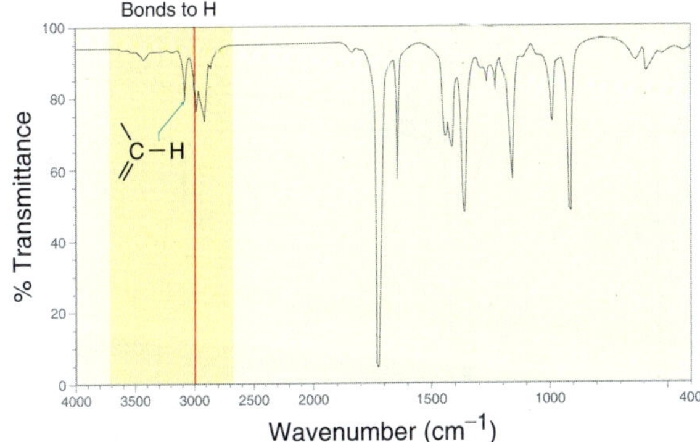


This spectrum exhibits one signal just above $3000\ cm^{-1}$, indicating a vinylic C—H bond.

The identification of a vinylic C—H bond is consistent with the observed C=C signal present in the double-bond region ($1650\ cm^{-1}$). There are no other signals above $3000\ cm^{-1}$, so the compound does not possess any OH or NH bonds.

The little bump between 3400 and $3500\ cm^{-1}$ is not strong enough to be considered a signal. These bumps are often observed in the spectra of compounds containing a C=O bond. The bump occurs at exactly twice the wavenumber of the C=O signal and is called an *overtone* of the C=O signal.

The diagnostic region provides the information necessary to solve this problem. Specifically, the compound must have the following bonds: C=C, C=O, and vinylic C—H. Among the possible choices, there are only two compounds that have these features.

To distinguish between these two possibilities, notice that the second compound is conjugated, while the first compound is *not* conjugated (the π bonds are separated by more than one sigma bond). Recall that ketones produce signals at approximately $1720\ cm^{-1}$, while conjugated ketones produce signals at approximately $1680\ cm^{-1}$.

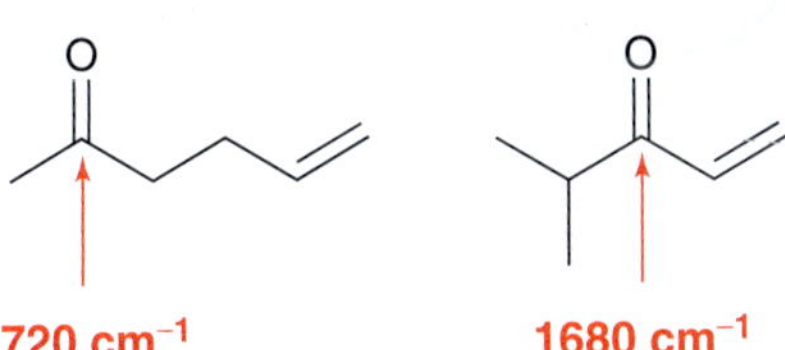

In the spectrum provided, the C=O signal appears at 1720 $cm^{-1}$, indicating that it is not conjugated. The spectrum is therefore consistent with the following compound:

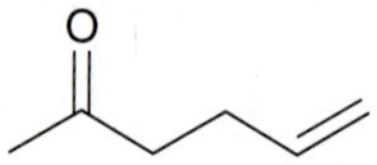

**PRACTICE the skill** **15.12** Match each compound with the appropriate IR spectrum:

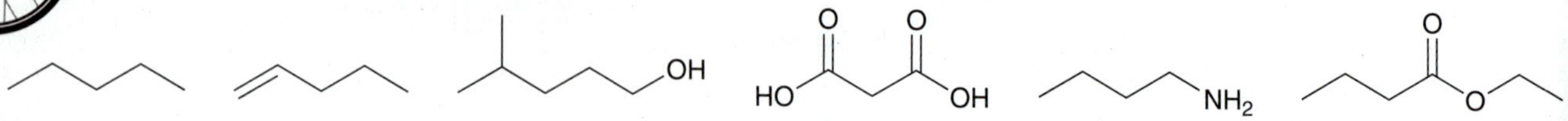

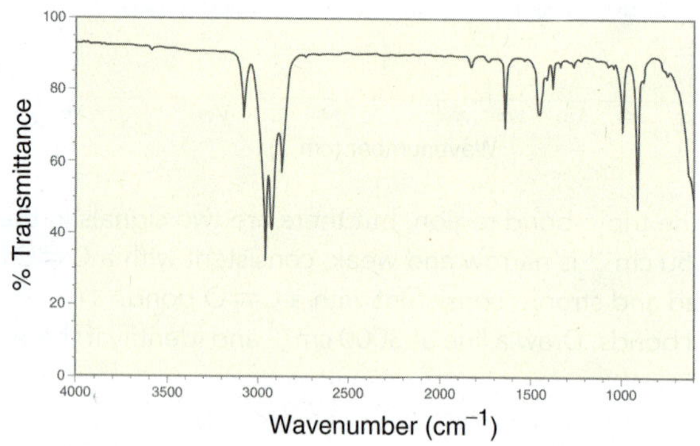

(a)

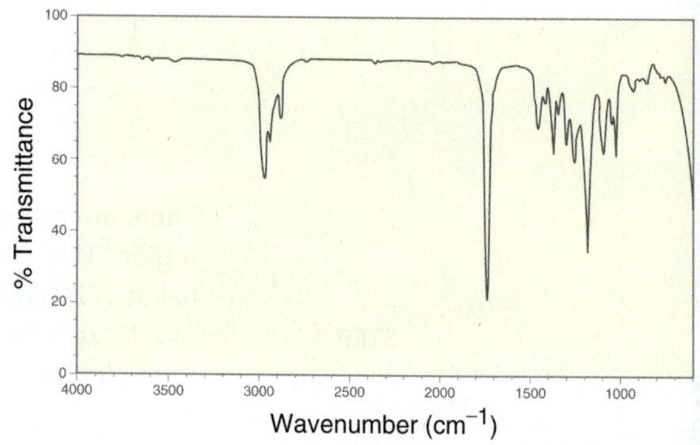

(b)

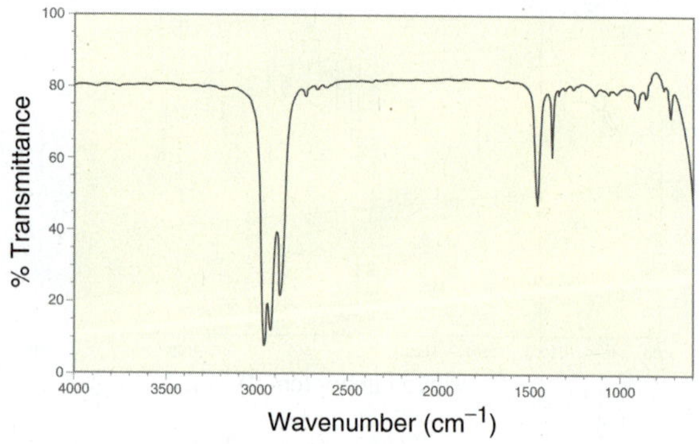

(c)

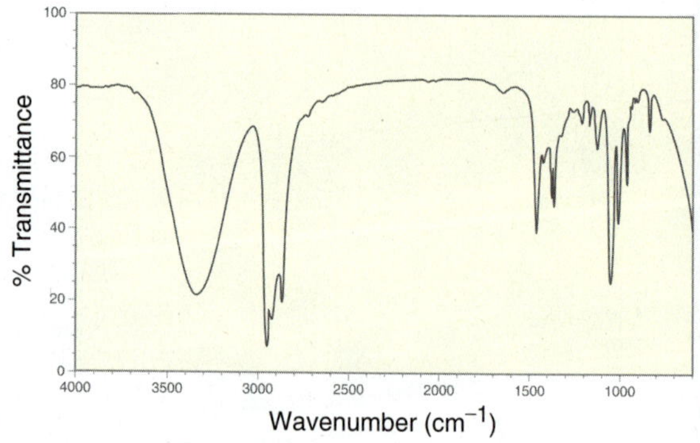

(d)

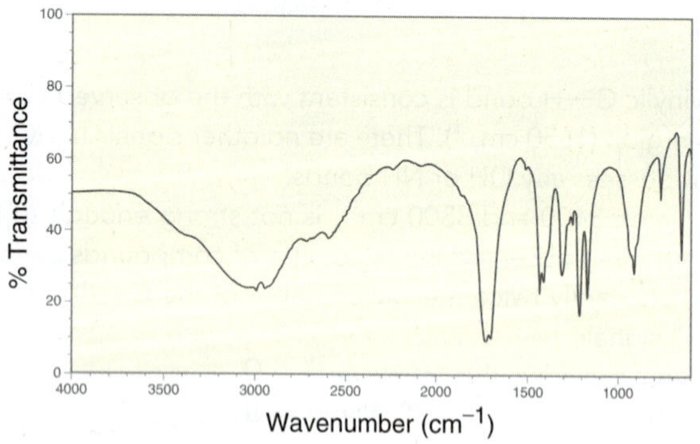

(e)

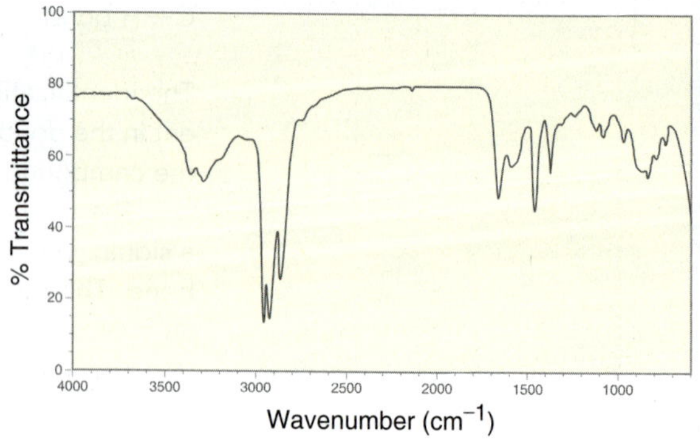

(f)

**APPLY the skill**

**15.13** Chrysanthemic acid is isolated from chrysanthemum flowers. The IR spectrum of chrysanthemic acid exhibits five signals above 1500 $cm^{-1}$. Identify the source for each of these signals.

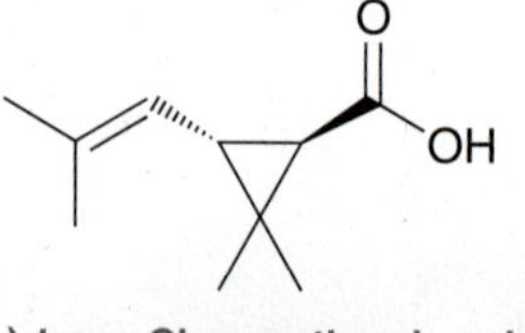


**(+)-*trans*-Chrysanthemic acid**

need more **PRACTICE?** **Try Problems 15.42, 15.57**

## 15.7 Using IR Spectroscopy to Distinguish between Two Compounds

IR spectroscopy can be used to differentiate between two compounds. This technique is often extremely helpful when performing a reaction in which a functional group is transformed into a different functional group. For example, consider the following reaction:

After running the reaction, it is possible to verify formation of the product by looking for the absence of an O—H signal and the presence of a C=O signal. In order to distinguish two compounds using IR spectroscopy, you must know what to look for in the spectrum.

### SKILLBUILDER

### 15.2 DISTINGUISHING BETWEEN TWO COMPOUNDS USING IR SPECTROSCOPY

**LEARN** the skill

Identify how IR spectroscopy might be used to monitor the progress of the following reaction.

#### SOLUTION

Carefully inspect both compounds (reactant and product) and think about what signals each compound would produce in its IR spectrum. It might be helpful to work your way through the various regions of the spectrum and determine if there would be a difference in each region.

Let's begin with the double-bond region (1600–1850 $cm^{-1}$). Each compound has a C=C bond, but consider the intensity of the expected signals. The reactant has an unsymmetrical C=C bond, while the product has a symmetrical C=C bond.

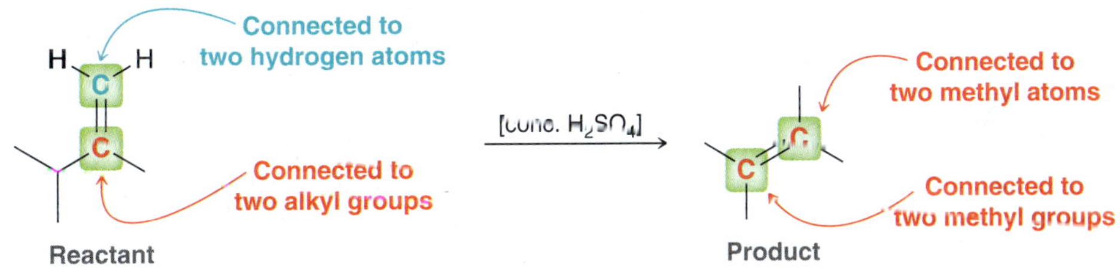


The reactant has a weak dipole moment because each vinylic carbon atom is in a slightly different electronic environment. A weak signal is therefore expected at 1650 $cm^{-1}$. In contrast, each vinylic position of the product is connected to two methyl groups, so there is no dipole moment associated with this bond. It is completely inefficient at absorbing IR radiation, and therefore, it will not produce a signal at 1650 $cm^{-1}$.

This difference can be used to monitor the progress of the reaction. A drop of the reaction mixture can be removed from the reaction flask at periodic intervals and analyzed in an IR spectrometer. As the product forms, the signal at 1650 $cm^{-1}$ should vanish.

Now let's consider the region containing signals from X—H bonds (2700–4000 $cm^{-1}$). Neither of the compounds has an O—H or N—H bond. Both compounds have $C_{sp^3}$—H bonds, but only the reactant has $C_{sp^2}$—H bonds.

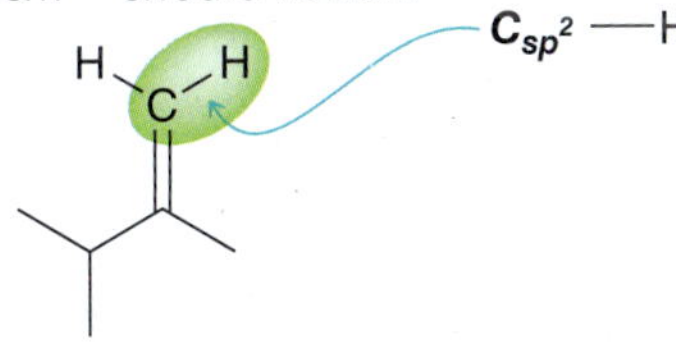

The product does not have this type of C—H bond. We therefore expect that the reactant will produce a signal at 3100 $cm^{-1}$ while the product will not. Once again, this difference can be used to monitor the progress of the reaction. As the product forms, the signal at 3100 $cm^{-1}$ should vanish.

Our analysis has produced two different ways to distinguish between these compounds. Following are the actual IR spectra of these two compounds with the relevant regions highlighted:

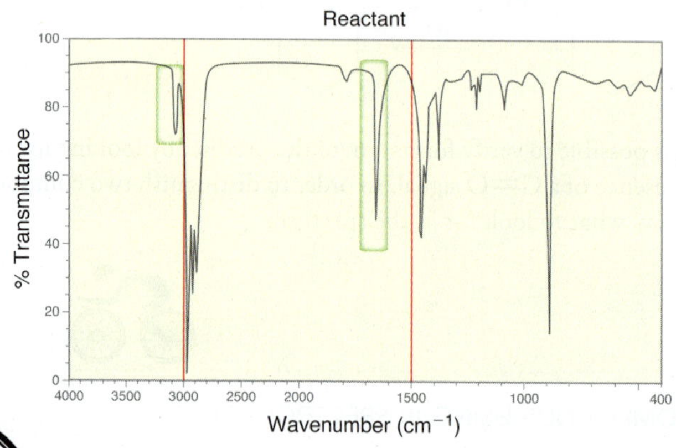


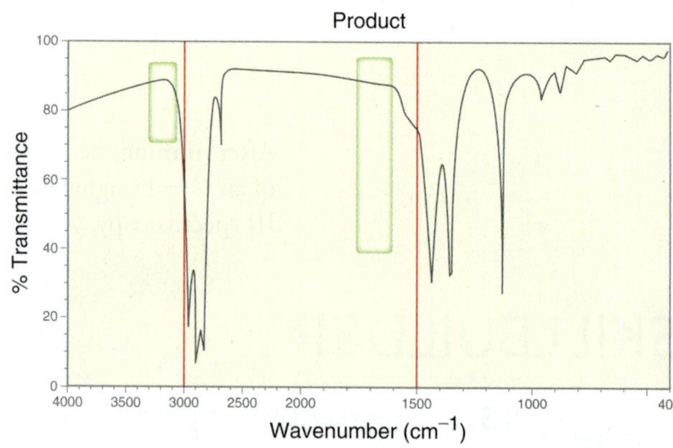


**PRACTICE the skill** **15.14** Describe how IR spectroscopy might be used to monitor the progress of each of the following reactions:

**(a)** OH $\xrightarrow[H_2SO_4,\ H_2O]{Na_2Cr_2O_7}$ OH

**(b)** $\xrightarrow{CH_3I}$

**(c)** $\xrightarrow[\text{Lindlar's catalyst}]{H_2}$

**(d)** $\xrightarrow[NH_3]{Na}$

**(e)**

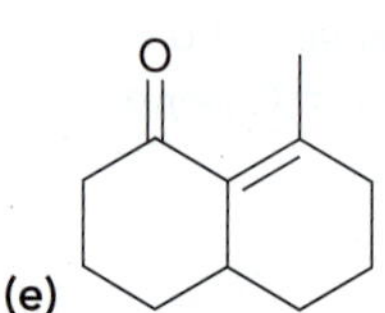

$\xrightarrow[Ni]{H_2}$

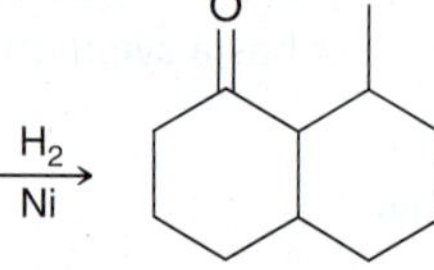

**APPLY the skill**

**15.15** In Chapter 21, we will explore how nitriles can be converted into carboxylic acids. How would you use IR spectroscopy to monitor the progress of this reaction?

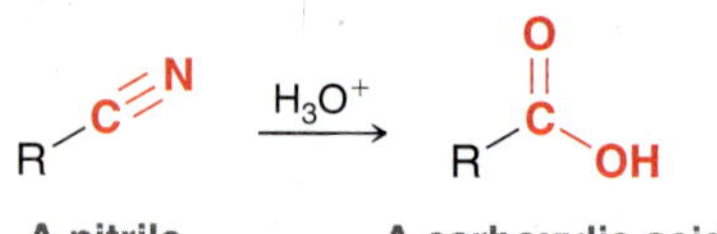


**15.16** 1-Butyne was treated with sodium amide followed by ethyl iodide. An IR spectrum of the product was acquired and then compared with the spectrum of the starting alkyne. A signal at 2200 $cm^{-1}$ was observed in the IR spectrum of the starting alkyne, but the same signal was not observed in the IR spectrum of the product. Explain this observation.

**15.17** Cyclopentanone was treated with lithium aluminum hydride followed by $H_3O^+$. Explain what you would look for in the IR spectrum of the product to verify that the expected reaction had occurred. Identify which signal should be present and which signal should be absent.

**15.18** When 1-chlorobutane is treated with sodium hydroxide, two products are formed. Identify the two products and explain how these products could be distinguished using IR spectroscopy.

need more **PRACTICE?** **Try Problems 15.38–15.41, 15.48d**

## 15.8 Introduction to Mass Spectrometry

In the beginning of this chapter, we defined spectroscopy as the study of the interaction between matter and electromagnetic radiation. In contrast, **mass spectrometry** is the study of the interaction between matter and an energy source other than electromagnetic radiation. Mass spectrometry is used primarily to determine the molecular weight and molecular formula of a compound.

In a **mass spectrometer**, a compound is first vaporized and converted into ions, which are then separated and detected. The most common ionization technique involves bombarding the compound with high-energy electrons. These electrons carry an extraordinary amount of energy, usually around 1600 kcal/mol, or 70 electron volts (eV). When a high-energy electron strikes the molecule, it causes one of the electrons in the molecule to be ejected. This technique, called **electron impact ionization** (**EI**), generates a high-energy intermediate that is both a radical and a cation.

$$\text{C–C} \xrightarrow{\text{Beam of electrons}} [\text{C} \cdot \text{C}]^{\oplus} + e^{\ominus}$$

A radical cation

It is a radical because it has an unpaired electron, and it is a cation because it bears a positive charge as a result of losing an electron. The mass of the ejected electron is negligible compared to the mass of the molecule, so the mass of the radical cation is essentially equivalent to the mass of the original molecule. This radical cation, symbolized by $(M)^{+\bullet}$, is called the **molecular ion**, or the **parent ion**. The molecular ion is often very unstable and is susceptible to **fragmentation**, which generates two distinct fragments. Most commonly, one fragment carries the unpaired electron, while the other fragment carries the charge.

$$[\text{C} \cdot \text{C}]^{\oplus} \longrightarrow \text{C}\cdot + \oplus\text{C}$$

Radical cation — Radical — Carbocation

In this way, the ionization process generates many different cations—the molecular ion as well as many different carbocation fragments. All of these ions are accelerated and then sent through a magnetic field, where they are deflected in curved paths (Figure 15.21).

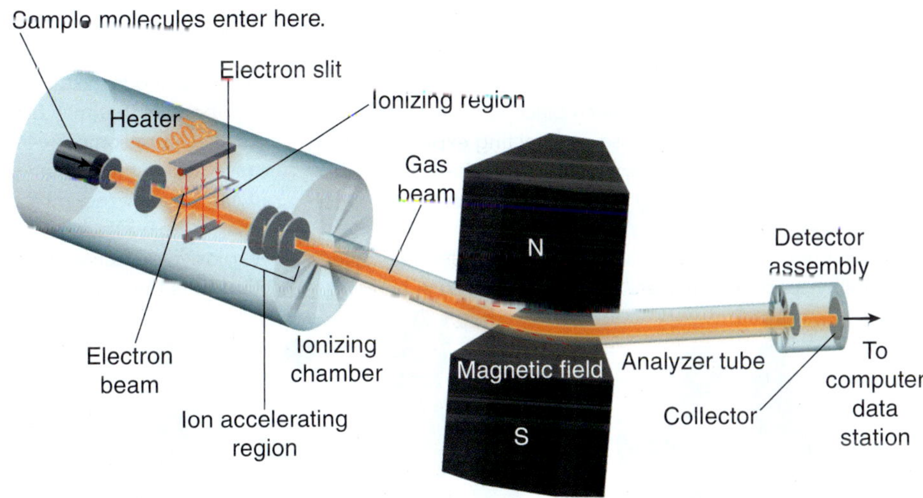


**FIGURE 15.21** A schematic of a mass spectrometer.

The uncharged radical fragments are not deflected by the magnetic field and are therefore not detected by the mass spectrometer. Only the molecular ion and the cationic fragments are deflected. Smaller ions are deflected more than larger ions, and ions with multiple charges are deflected more than ions

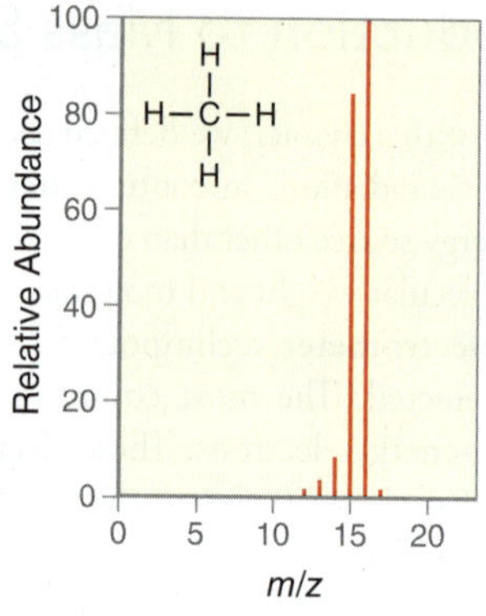


(a)

**MASS SPECTRUM DATA**

| m/z | RELATIVE HEIGHT (%) |
|---|---|
| 12 | 1.0 |
| 13 | 3.9 |
| 14 | 9.2 |
| 15 | 85.0 |
| 16 | 100 (base peak) |
| 17 | 1.1 |

(b)

**FIGURE 15.22**
A mass spectrum of methane.

with a charge of +1. In this way, the cations are separated by their **mass-to-charge ratio** (*m*/*z*). The charge (*z*) on most ions is +1, and therefore, *m*/*z* is effectively a measure of the mass (*m*) of each cation. A plot is then generated, called a **mass spectrum**, which shows the relative abundance of each cation that was detected. Figure 15.22 is a mass spectrum of methane. The tallest peak in the spectrum is assigned a relative value of 100% and is called the **base peak**. The height of every other peak is then described relative to the height of the base peak. In the case of methane, the molecular ion peak is the base peak, but this is not always the case for other compounds. In the mass spectra of larger compounds, it is very common for one of the fragments to produce the tallest peak. In such a case, the molecular ion peak is not the base peak. We will see examples of this in the upcoming sections.

## Mass Spectrometry for Detecting Explosives

Mass spectrometry is an incredibly important tool that has found a broad array of applications. The following is a summary of some major applications of mass spectrometry, categorized by the industry in which they are applied:

*Pharmaceutical:* drug discovery, drug metabolism, reaction monitoring
*Biotechnology:* amino acid sequencing, analysis of macromolecules
*Clinical:* neonatal screening, hemoglobin analysis
*Environmental:* drug testing, water quality testing, contamination level measurements in food
*Geological:* evaluation of oil compositions
*Forensic:* explosive detection

Significant advances have taken place in developing the field of mass spectrometry for use in detecting explosive materials at airports. In a post-9/11 world, there is a great demand for devices that can accurately and reliably detect the presence of explosive materials that might be present in a traveler's luggage or carry-on bags. Specialized mass spectrometers, called ion mobility spectrometers, are now being utilized at hundreds of major airports in U.S. cities. These devices collect chemicals from the surface of a traveler's luggage, subject these chemicals to a process that converts them into ions, and then measures the speed of these ions as they pass through an electric field. These spectrometers are designed to detect the presence of any ions that move at a speed consistent with known explosive materials. Several recent advances have made this possible, and new techniques are constantly being developed as this is an area of significant ongoing research. One such technique, developed by researchers at Purdue University, enables the acquisition of chemicals from the surface of luggage in just a few seconds. This method utilizes a technique called *desorption electrospray ionization (DESI)*, in which the surface of the suitcase is sprayed with a gaseous mixture that dislodges any residual explosive compounds that may be present on the surface of the suitcase as a result of someone loading the suitcase with explosives. The gaseous mixture is then sucked into a mass spectrometer where it can be analyzed in seconds. Many other advances in this field are emerging each year, and the role of mass spectrometry in explosive detection is likely to undergo further improvement in the years to come.

In the spectrum of methane, the peaks below 16 are formed by fragmentation of the molecular ion. Methane is a small molecule, and there are very few ways for the molecular ion to fragment. There are only C—H bonds, so fragmentation of methane simply involves the loss of hydrogen atoms.

Molecular ion
$m/z = 16$

Fragment
$m/z = 15$

The loss of a hydrogen atom produces a carbocation fragment with $m/z = 15$. The mass spectrum of methane (Figure 15.22) indicates that this fragment is almost as abundant as the molecular ion itself. This carbocation can then lose another hydrogen atom, resulting in a new fragment with $m/z = 14$, although this new fragment is not very abundant, as can be seen on the spectrum. This process can continue until all four hydrogen atoms have been lost, giving rise to a series of peaks with $m/z$ values ranging from 12 to 15.

We have explained all but one of the peaks in the mass spectrum of methane. Notice that there is a peak $m/z = 17$. This peak, called the $(M+1)^{+\bullet}$ peak, will be discussed in more detail in Section 15.10.

## 15.9 Analyzing the $(M)^{+\bullet}$ Peak

For some compounds, the $(M)^{+\bullet}$ is the base peak. Consider, for example, the mass spectrum of benzene (Figure 15.23). Clearly, the molecular ion ($m/z = 78$) is not very susceptible to fragmentation, as it is the most abundant ion to pass through the spectrometer. This is generally not the case. Most compounds will easily fragment, and the $(M)^{+\bullet}$ peak will not be the most abundant ion.

For example, consider the mass spectrum of pentane (Figure 15.24). In the mass spectrum of pentane, the $(M)^{+\bullet}$ peak (at $m/z = 72$) is very small indeed. In this case, the fragment at $m/z = 43$ is the most abundant ion, and that peak is therefore called the base peak and assigned a value of 100%. In some cases, it is possible for the $(M)^{+\bullet}$ peak to be entirely absent, if it is particularly susceptible to fragmentation. In such cases, there are more gentle methods of ionization (other than EI) that allow the parent ion to survive long enough to pass through the spectrometer. We will briefly review one of those methods later in this chapter.

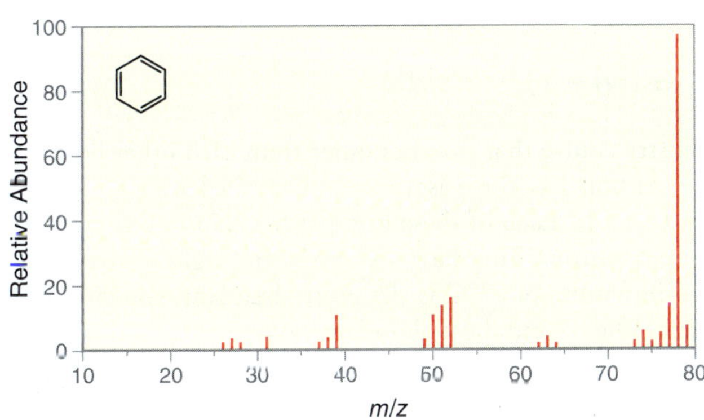


**FIGURE 15.23**
A mass spectrum of benzene.

Relative Abundance
m/z

**FIGURE 15.24**
A mass spectrum of pentane.

When analyzing a mass spectrum, the first step is to look for the $(M)^{+\bullet}$ peak, because it indicates the molecular weight of the molecule. This technique can be used to distinguish compounds. For example, compare the molecular weights of pentane and of 1-pentene.

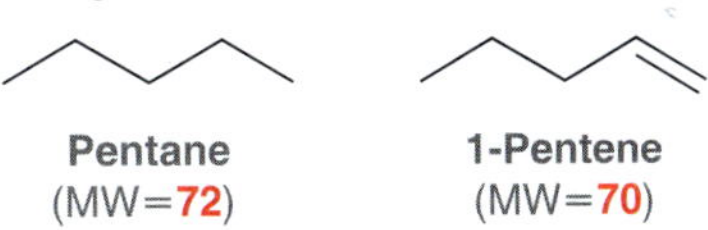

Pentane has 5 carbon atoms ($5 \times 12 = 60$) and 12 hydrogen atoms ($12 \times 1 = 12$) and therefore a molecular weight of 72. In contrast, 1-pentene only has 10 hydrogen atoms and therefore has a molecular weight of 70. As a result, we expect the mass spectrum of pentane to exhibit a molecular ion peak at 72, while the mass spectrum of 1-pentene should exhibit a molecular ion peak at 70.

Useful information can also be obtained by analyzing whether the molecular weight of the parent ion is odd or even. An odd molecular weight generally indicates an odd number of nitrogen atoms in the compound, while an even molecular weight indicates either the absence of nitrogen or an even number of nitrogen atoms. This is called the **nitrogen rule**, and it is illustrated in the following examples:

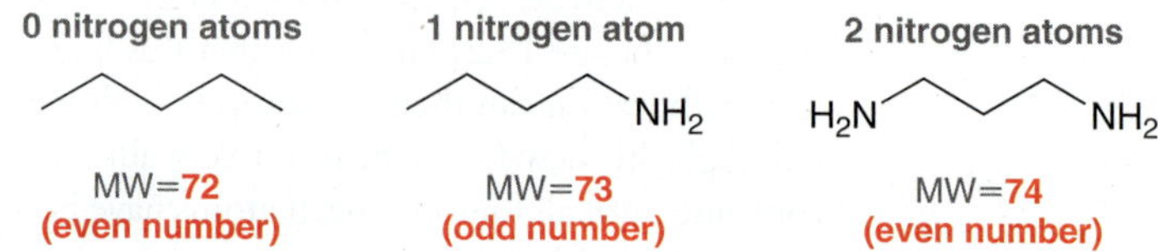

## CONCEPTUAL CHECKPOINT

**15.19** How would you distinguish between each pair of compounds using mass spectrometry?

(a) 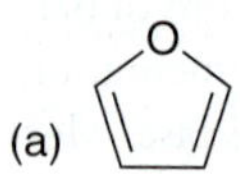 

(b) 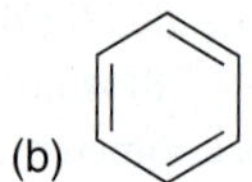 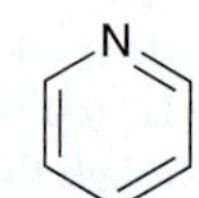

**15.20** For each of the following compounds, use the nitrogen rule to determine whether the molecular weight should be even or odd. Then calculate the expected *m/z* value for the molecular ion.

(a)

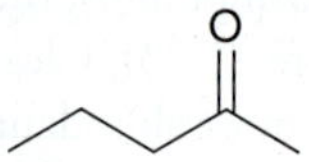

(b)

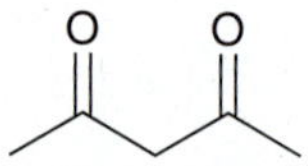

(c)

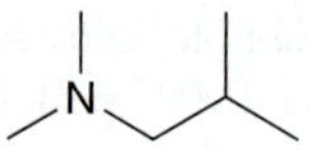

(d)

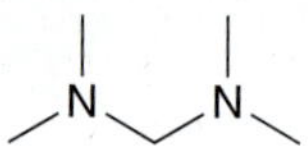

## 15.10 Analyzing the $(M+1)^{+\bullet}$ Peak

Recall from your general chemistry course that isotopes differ from each other only in the number of neutrons. For example, carbon has three isotopes: $^{12}C$ (called carbon 12), $^{13}C$ (called carbon 13), and $^{14}C$ (called carbon 14). Each of these isotopes has six protons and six electrons, but they differ in their number of neutrons. They have six, seven, and eight neutrons, respectively. All of these isotopes are found in nature, but $^{12}C$ is the most abundant, constituting 98.9% of all carbon atoms found on Earth. The second most abundant isotope of carbon is $^{13}C$, constituting approximately 1.1% of all carbon atoms. The amount of $^{14}C$ found in nature is very small (0.0000000001%).

In a mass spectrometer, each individual molecule is ionized and then passed through the magnetic field. When methane is analyzed, 98.9% of the molecular ions will contain a $^{12}C$ atom, while only 1.1% will contain a $^{13}C$ atom. The latter group of molecular ions are responsible for the observed peak at $(M+1)^{+\bullet}$. The relative height of this peak is approximately 1.1% as tall as the $(M)^{+\bullet}$ peak, just as expected. Larger compounds, containing more carbon atoms, will have a larger $(M+1)^{+\bullet}$ peak. For example, decane has 10 carbon atoms in its structure, so the chances that a molecule of decane will possess one $^{13}C$ atom are 10 times greater than the chances that a molecule of methane will possess a $^{13}C$ atom. Consequently, the $(M+1)^{+\bullet}$ peak in the mass spectrum of decane is 11% as tall as the molecular ion peak ($10 \times 1.1\%$). Similarly, the $(M+1)^{+\bullet}$ peak in the mass spectrum of icosane ($C_{20}H_{42}$) is 22% as tall as the molecular ion peak ($20 \times 1.1\%$) (Figure 15.25). Isotopes of other elements also contribute to the $(M+1)^{+\bullet}$ peak, but $^{13}C$ is the greatest contributor, and therefore, it is generally possible to determine the number of carbon atoms in an unknown compound by comparing the relative heights of the $(M)^{+\bullet}$ peak and the $(M+1)^{+\bullet}$ peak.

This piece of information can be very helpful in determining the molecular formula. The following SkillBuilder illustrates this technique.

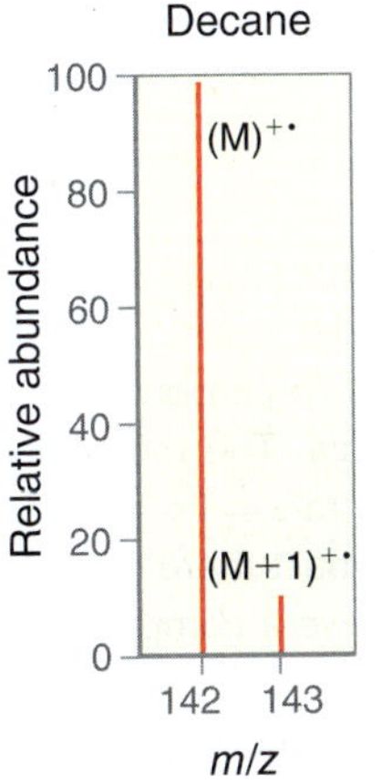


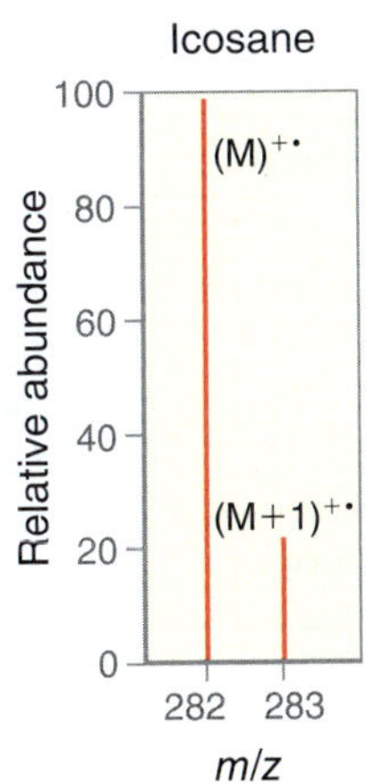


**FIGURE 15.25** The relative heights of the $(M)^{+\bullet}$ and $(M+1)^{+\bullet}$ peaks for decane and icosane.

## SKILLBUILDER

### 15.3 USING THE RELATIVE ABUNDANCE OF THE $(M+1)^{+\bullet}$ PEAK TO PROPOSE A MOLECULAR FORMULA

**LEARN the skill** Below is the mass spectrum as well as the tabulated mass spectrum data for an unknown compound. Propose a molecular formula for this compound.

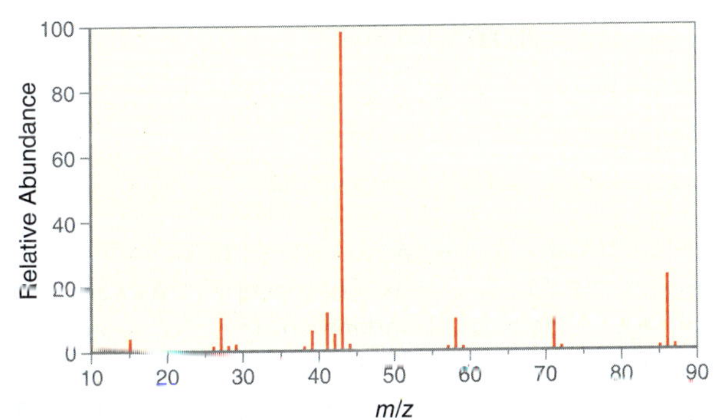


**MASS SPECTRUM DATA**

| *m/z* | RELATIVE HEIGHT (%) | *m/z* | RELATIVE HEIGHT (%) |
|---|---|---|---|
| 15 | 4.8 | 42 | 4.0 |
| 26 | 1.3 | 43 | 100 (base peak) |
| 27 | 10.5 | 44 | 2.3 |
| 28 | 1.3 | 58 | 10.3 |
| 29 | 1.9 | 71 | 11.0 |
| 38 | 1.2 | 86 | 20.9 ($M^{+\bullet}$) |
| 39 | 6.3 | 87 | 1.2 |
| 41 | 11.9 | | |

**SOLUTION**

To solve this problem, it is not necessary to visually inspect the spectrum. The data alone are sufficient. It is important to become accustomed to interpreting data even when the spectrum itself is not provided, much the way pilots learn to fly planes at night using instrument readings.

Let's begin with the molecular ion peak, which appears at $m/z = 86$. Now compare the relative abundance of this peak and the $(M+1)^{+\bullet}$ peak (which appears at $m/z = 87$). The relative height of the $(M+1)^{+\bullet}$ peak is 1.2%, but be careful here. In this case, the molecular ion is not the tallest peak. The tallest peak (base peak) appears at $m/z = 43$. The data indicate that the $(M+1)^{+\bullet}$ peak is 1.2% as tall as the base peak. But we need to know how the $(M+1)^{+\bullet}$ peak compares *to the molecular ion peak*. To do this, we take the relative height of the $(M+1)^{+\bullet}$ peak, divide by the relative height of the $(M)^{+\bullet}$ peak, and then multiply by 100%.

$$\frac{1.2\%}{20.9\%} \times 100\% = 5.7\%$$

**STEP 1**
Determine the number of carbon atoms in the compound by analyzing the relative abundance of the M+1 peak.

In other words, the $(M+1)^{+\bullet}$ peak is 5.7% as tall as the $(M)^{+\bullet}$ peak.

Recall that each carbon atom in the compound contributes 1.1% to the height of the $(M+1)^{+\bullet}$ peak, so we divide by 1.1% to determine the number of carbon atoms in the compound.

$$\text{Number of C} = \frac{5.7\%}{1.1\%} = 5.2$$

Obviously the compound cannot have a fractional number of carbon atoms, so the value must be rounded to the nearest whole number, or 5. This analysis suggests that the unknown compound has five carbon atoms. This information is incredibly useful in determining the molecular formula. The molecular weight is known to be 86, because the molecular ion peak appears at $m/z = 86$. Five carbon atoms equals $5 \times 12 = 60$, so the other elements in the compound must therefore give a total of $86 - 60 = 26$. The molecular formula cannot be $C_5H_{26}$, because a compound with five carbon atoms cannot have that many hydrogen atoms.

**STEP 2**
Analyze the mass of the molecular ion to determine if any heteroatoms are present.

Therefore, we conclude that there must be another element present. The two most common elements in organic chemistry (other than C and H) are nitrogen and oxygen. It cannot be a nitrogen atom, because that would give an odd molecular weight (remember the nitrogen rule). So, we try oxygen. This gives the following possible molecular formula:

$$C_5H_{10}O$$

**PRACTICE** the skill

**15.21** Propose a molecular formula for a compound that exhibits the following peaks in its mass spectrum:

**(a)** $(M)^{+\bullet}$ at $m/z = 72$, relative height = 38.3% of base peak
$(M+1)^{+\bullet}$ at $m/z = 73$, relative height = 1.7% of base peak

**(b)** $(M)^{+\bullet}$ at $m/z = 68$, relative height = 100% (base peak)
$(M+1)^{+\bullet}$ at $m/z = 69$, relative height = 4.3%

**(c)** $(M)^{+\bullet}$ at $m/z = 54$, relative height = 100% (base peak)
$(M+1)^{+\bullet}$ at $m/z = 55$, relative height = 4.6%

**(d)** $(M)^{+\bullet}$ at $m/z = 96$, relative height = 19.0% of base peak
$(M+1)^{+\bullet}$ at $m/z = 97$, relative height = 1.5% of base peak

**APPLY** the skill

**15.22** While $^{13}C$ is the main contributor to the $(M+1)^{+\bullet}$ peak, there are many other elements that can also contribute to the $(M+1)^{+\bullet}$ peak. For example, there are two naturally occurring isotopes of nitrogen. The most abundant isotope, $^{14}N$, represents 99.63% of all nitrogen atoms on Earth. The other isotope, $^{15}N$, represents 0.37% of all nitrogen atoms. In a compound with molecular formula $C_8H_{11}N_3$, if the molecular ion peak has a relative abundance of 24.5%, then what percent abundance do you expect for the $(M+1)^{+\bullet}$ peak?

need more **PRACTICE?** **Try Problems 15.45, 15.47**

## 15.11 Analyzing the $(M+2)^{+\bullet}$ Peak

Most elements have only one dominant isotope. For example, the dominant isotope of hydrogen is $^{1}H$, while $^{2}H$ (deuterium) and $^{3}H$ (tritium) represent only a small fraction of all hydrogen atoms. Similarly, the dominant isotope of carbon is $^{12}C$, while $^{13}C$ and $^{14}C$ represent only a small fraction of all carbon atoms. In contrast, chlorine has two major isotopes. One isotope of chlorine, $^{35}Cl$, represents 75.8% of all chlorine atoms; the other isotope of chlorine, $^{37}Cl$, represents 24.2% of all chlorine atoms. As a result, compounds that contain a chlorine atom will give a characteristically strong

$(M+2)^{+\bullet}$ peak. For example, consider the mass spectrum of chlorobenzene (Figure 15.26). The molecular ion appears at $m/z = 112$. The peak at (M+2) ($m/z = 114$) is approximately one-third the height of the parent peak. This pattern is characteristic of compounds containing a chlorine atom.

Compounds containing bromine also give a characteristic pattern. Bromine has two isotopes, $^{79}Br$ and $^{81}Br$, that are almost equally abundant in nature (50.7% and 49.3%, respectively). Compounds containing bromine will therefore have a characteristic peak at (M+2) that is approximately the same height as the molecular ion peak. For example, consider the mass spectrum of bromobenzene (Figure 15.27).

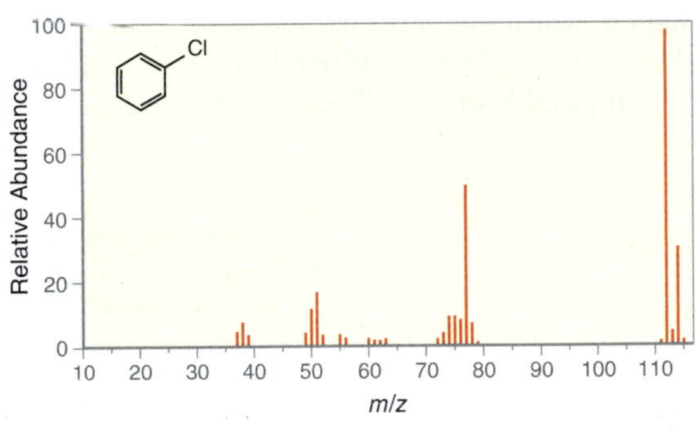


**FIGURE 15.26**
A mass spectrum of chlorobenzene.

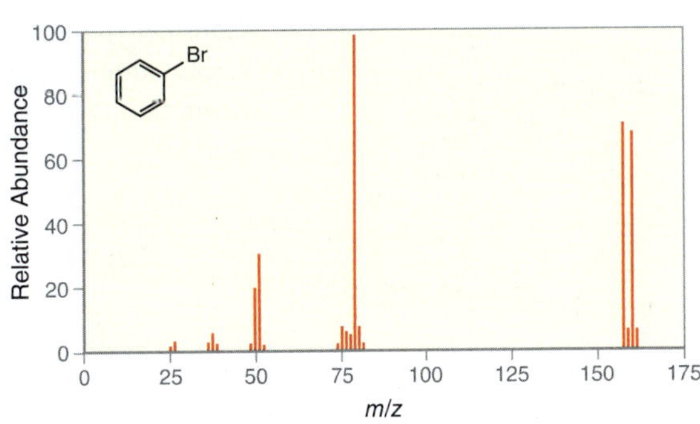


**FIGURE 15.27**
A mass spectrum of bromobenzene.

The presence of chlorine or bromine in a compound is readily identified by analyzing the height of the $(M+2)^{+\bullet}$ peak and comparing it with the height of the $(M)^{+\bullet}$ peak.

## CONCEPTUAL CHECKPOINT

**15.23** In the mass spectrum of bromobenzene (Figure 15.27), the base peak appears at $m/z = 77$.

(a) Does this fragment contain Br? Explain your reasoning.

(b) Draw the cationic fragment that represents the base peak.

**15.24** Below are mass spectra for four different compounds. Identify whether each of these compounds contains a bromine atom, a chlorine atom, or neither.

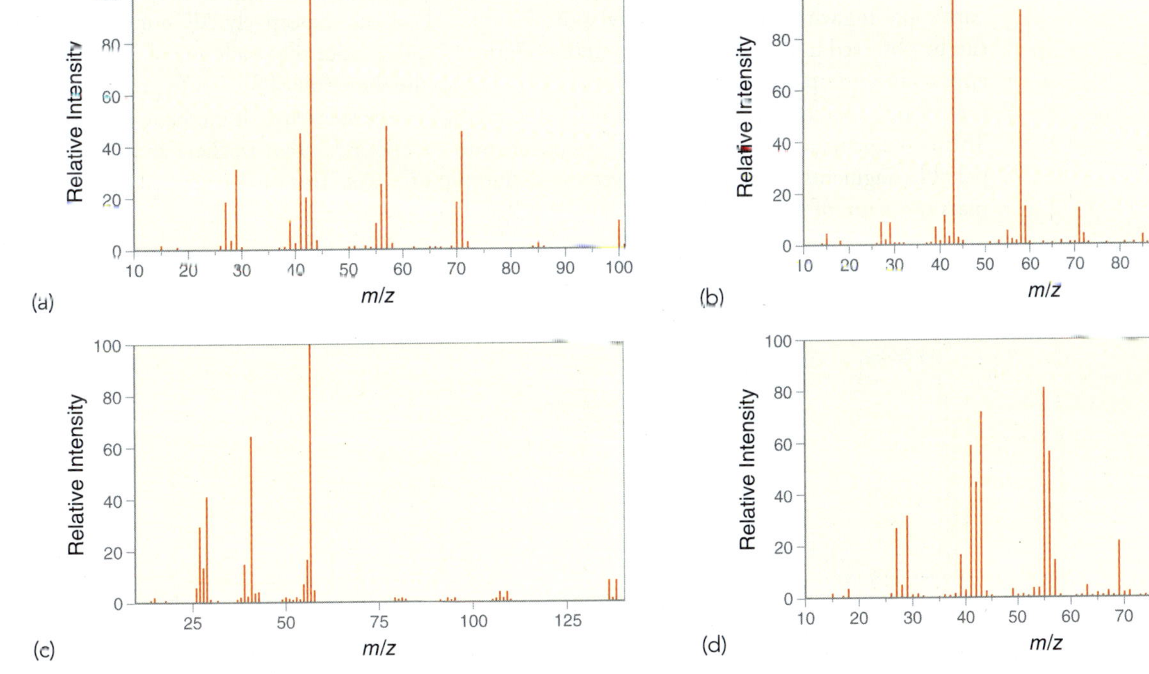

## 15.12 Analyzing the Fragments

The majority of the peaks in a mass spectrum are produced from fragmentation of the molecular ion. In this section, we will explore the characteristic fragmentation patterns for a number of compounds. These patterns are often helpful in identifying certain structural features of a compound, but it is generally not possible to use the fragmentation patterns to determine the entire structure of the compound.

### Fragmentation of Alkanes

Consider the different ways in which the molecular ion of pentane can fragment. Pentane has five carbon atoms connected by a series of four C—C bonds. Each of these bonds is susceptible to fragmentation, giving rise to four possible cations (Figure 15.28). Remember that a mass spectrometer

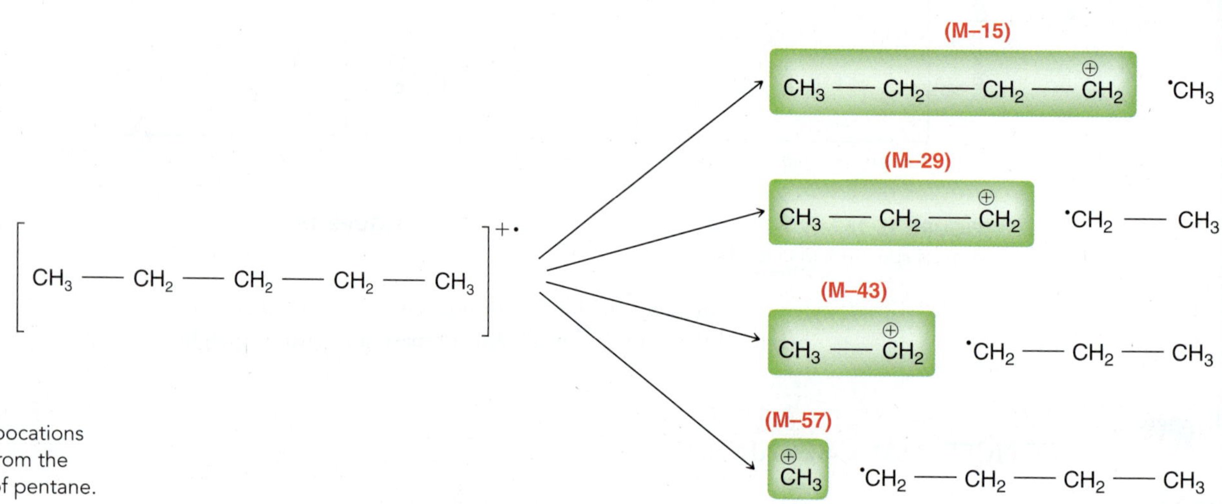


**FIGURE 15.28** The various carbocations that can result from the fragmentation of pentane.

does not detect the radical fragments; it only detects the ions. The first cation shown is formed from loss of a methyl radical. The methyl radical has a mass of 15, so the resulting cation appears as a peak at M–15. The second cation is formed from the loss of an ethyl radical (mass 29), so the resulting peak appears at M–29. In a similar way, the other two possible cations appear at M–43 and M–57, corresponding with the loss of a propyl radical and butyl radical, respectively. All four of these cations can be observed in the spectrum of pentane (Figure 15.29). Notice that each one of these four peaks appears in a group of smaller peaks. These smaller peaks are the result of further fragmentation of the carbocations via loss of hydrogen atoms, just as we saw in the spectrum of methane (Figure 15.22). This is the general trend in the mass spectra of most compounds. That is, there are many different possible fragments, each of which gives rise to a group of peaks. This can be very clearly seen in the mass spectrum of icosane, $C_{20}H_{42}$ (Figure 15.30).

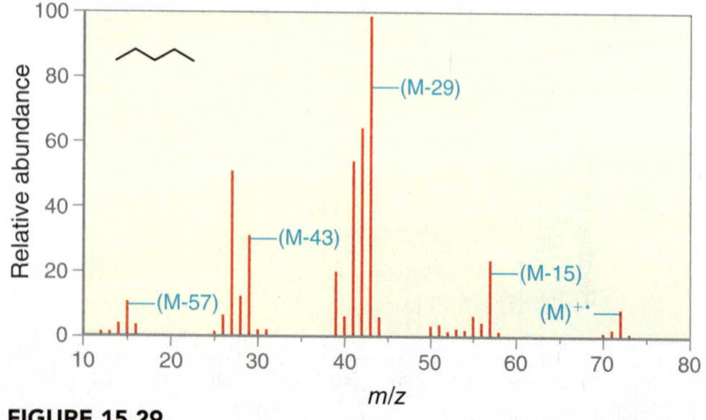


**FIGURE 15.29** A mass spectrum of pentane.

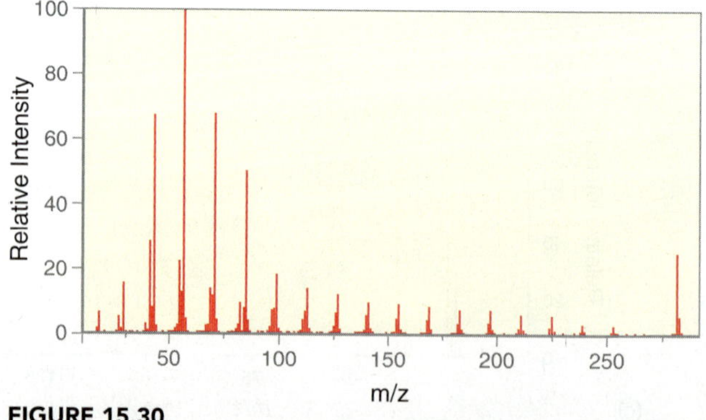


**FIGURE 15.30** A mass spectrum of icosane.

**LOOKING BACK**
For a review of carbocation stability, see Section 6.11.

Icosane is a straight-chain hydrocarbon consisting of 20 carbon atoms connected by 19 C—C bonds. Each of these bonds is susceptible to fragmentation to produce a cation that is detected by the mass spectrometer. Each one of these 19 possible carbocations appears in the spectrum as a group of peaks (except for the M–15 peak, which has a very small relative abundance).

When analyzing a fragment peak, the key is to look at its proximity to the molecular ion peak. For example, a signal at M–15 indicates the loss of a methyl group, and a signal at M–29 indicates the loss of an ethyl group. The likelihood of fragmentation increases with the stability of the carbocation formed as well as the stability of the radical that is ejected.

For example, look again at the various possible carbocations formed by fragmentation of pentane (Figure 15.28). The carbocation corresponding to M–57 is a methyl carbocation, which is less stable than the other possible carbocations (all of which are primary). That explains why the peak at M–57 is a fairly small peak in the spectrum. In general, fragmentation will occur in all possible locations but will typically favor the formation of the most stable carbocation.

As another example, consider the most likely fragmentation of the following molecular ion:

(M–43)
A tertiary carbocation
A primary radical

The most abundant peak in the spectrum of this compound is expected to be at M–43, corresponding with formation of a tertiary carbocation, via loss of a propyl radical. A tertiary carbocation can also be produced via loss of a methyl radical (from the left side of the molecular ion above); however, a methyl radical is less stable than a primary radical. Certainly, all possible fragmentations are observed under the high-energy conditions employed, but the most abundant peak will generally result from formation of the most stable carbocation via expulsion of the most stable possible radical. Therefore, it is generally possible to predict the location of the most abundant peak that is expected in the mass spectrum of a simple alkane or branched alkane.

## Fragmentation of Alcohols

Alcohols exhibit two common fragmentation patterns: alpha cleavage and dehydration. During alpha cleavage, a bond to the alpha position of the alcohol is cleaved to form a resonance-stabilized cation and a radical.

$\alpha$ Cleavage

Molecular ion

Resonance stabilized

Alternatively, alcohols can undergo dehydration via loss of a water molecule.

Molecular ion

(M–18)

This fragmentation results from an intramolecular elimination process, which can occur because of the high energy of the parent ion. This fragmentation does not eject a radical, but instead, it ejects a neutral molecule (water). Loss of a neutral molecule produces a new radical cation, which generates a peak at M–18 (because the molecular weight of water is 18). The water molecule itself does not appear on the spectrum, because the spectrometer only detects ions. A signal at M–18 is therefore characteristic of alcohols.

## Fragmentation of Amines

Much like alcohols, amines are also observed to undergo alpha cleavage to generate a resonance-stabilized cation and a radical.

$H_2N^{\oplus\bullet}$ … α Cleavage → [ $H_2N^{\oplus}$ ⟷ $H_2\ddot{N}$ ⊕ ] + $H_2\dot{C}-R$

Molecular ion — Resonance stabilized

## Fragmentation of Ketones and Aldehydes

Aldehydes and ketones containing a hydrogen atom at the gamma position generally undergo a characteristic fragmentation called the *McLafferty rearrangement*, which results in the loss of a neutral alkene fragment.

McLafferty rearrangement

The alkene fragment has an even mass. In contrast, most radical fragments encountered thus far have an odd mass (M–15 for loss of a methyl group, M–29 for loss of an ethyl group, M–43 for loss of a propyl group, etc.). Therefore, the mass spectrum of a ketone or aldehyde often contains a peak at M–*x*, where *x* is an even number.

Some of the most common fragments are listed in Table 15.3.

**TABLE 15.3 COMMON FRAGMENTS IN MASS SPECTROMETRY**

| Fragment | Description |
| --- | --- |
| M–15 | Loss of a methyl radical |
| M–29 | Loss of an ethyl radical |
| M–43 | Loss of a propyl radical |
| M–57 | Loss of a butyl radical |
| M–18 | Loss of water (from an alcohol) |
| M–x (where x = even number) | McLafferty rearrangement (ketone or aldehyde) |

### CONCEPTUAL CHECKPOINT

**15.25** Although 2,2-dimethylhexane has a molecular weight of 114, no peak is observed at $m/z = 114$. The base peak in the mass spectrum occurs at M–57.

(a) Draw the fragmentation responsible for formation of the M–57 ion.

(b) Explain why this cation is the most abundant ion to pass through the spectrometer.

(c) Explain why no molecular ions survive long enough to be detected.

(d) Can you offer an explanation as to why the M–15 peak is not the base peak?

**15.26** Identify two peaks that are expected to appear in the mass spectrum of 3-pentanol. For each peak, identify the fragment associated with the peak.

**15.27** Identify the expected base peak in the mass spectrum of 2,2,3-trimethylbutane. Draw the fragment associated with this peak and explain why the base peak results from this fragment.

**15.28** The following are mass spectra for the constitutional isomers ethylcyclohexane and 1,1-dimethylcyclohexane. Based on likely fragmentation patterns, match the compound with its spectrum.

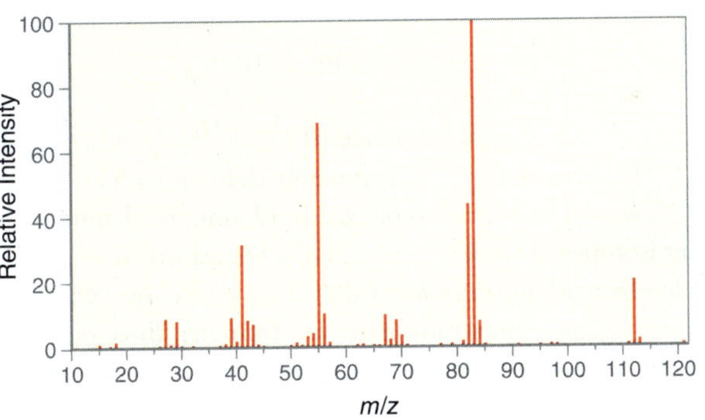


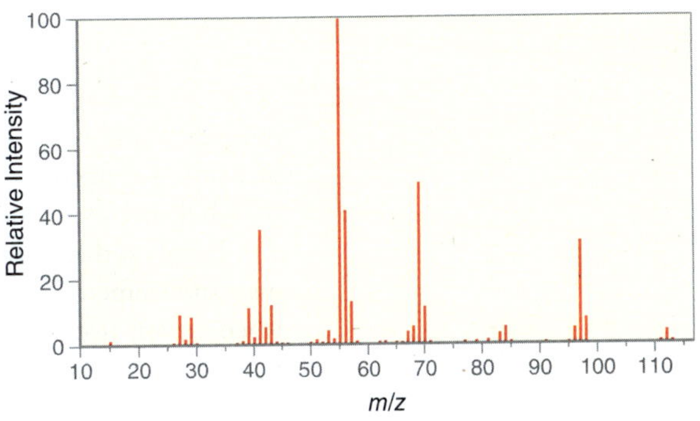


## 15.13 High-Resolution Mass Spectrometry

**High-resolution mass spectrometry** involves the use of a detector that can measure $m/z$ values to four decimal places. This technique allows for the determination of the molecular formula of an unknown compound. In order to analyze the data obtained from high-resolution mass spectrometry, we must first review some background information. Specifically, we must discuss why atomic masses are not whole numbers (despite the fact that we have treated them as such until now).

The mass of an atom is approximately equal to the sum of its protons and neutrons, because the mass of the electrons are negligible compared to the protons and neutrons. Originally, the mass of a proton was considered to be the same as the mass of a neutron, resulting in the *relative* atomic weights given in Table 15.4. With this simple model, the atomic weights appear as whole numbers.

**TABLE 15.4 RELATIVE ATOMIC WEIGHTS OF SEVERAL ELEMENTS**

| ELEMENT | NUMBER OF PROTONS | NUMBER OF NEUTRONS | RELATIVE ATOMIC WEIGHT |
|---|---|---|---|
| H | 1 | 0 | 1 |
| He | 2 | 2 | 4 |
| C | 6 | 6 | 12 |
| N | 7 | 7 | 14 |
| O | 8 | 8 | 16 |

These are the numbers we have used in this chapter so far, and they are a good approximation. But they are not accurate for two reasons:

1. Protons do not have exactly the same mass as neutrons:

$$\text{one proton} = 1.6726 \times 10^{-24}\text{ g}$$
$$\text{one neutron} = 1.6749 \times 10^{-24}\text{ g}$$

   As a result, a helium atom (which has two protons and two neutrons) is not exactly four times the weight of a hydrogen atom (which has one proton).

2. Protons repel each other but will nevertheless bind together in the nuclei of atoms under the influence of the strong nuclear force. These protons possess an enormous amount of potential energy, which is achieved at the expense of some mass. According to Einstein's famous equation ($E = mc^2$), matter and energy are interconvertible. When protons bind together to form the nucleus of an atom, some of their mass is converted into potential energy. As a result, two bound protons will have less mass than two individual protons. This explains why carbon, which has six protons and six neutrons, is less than six times the mass of a deuterium atom, which has one proton and one neutron.

For the two reasons given, atomic masses are not whole numbers. For reporting atomic mass, chemists use the **atomic mass unit (amu)**, which is equivalent to 1 g divided by Avogadro's number:

$$1 \text{ amu} = \frac{1 \text{ g}}{6.02214 \times 10^{23}} = 1.6605 \times 10^{-24} \text{g}$$

Avogadro's number is defined as the number of atoms in exactly 12 g of $^{12}C$. In other words, the value of 1 amu is defined with respect to $^{12}C$. Accordingly, $^{12}C$ is the only element with an atomic mass that is a whole number. One atom of $^{12}C$ has an atomic mass of exactly 12 amu, by definition.

Based on this definition, other isotopes of carbon can be precisely measured using high-resolution mass spectrometry, in which the detector can measure *m/z* values to four decimal places. For example, when carbon atoms are passed through a high-resolution mass spectrometer, three peaks are observed: one peak representing the $^{12}C$ atoms, one smaller peak representing the $^{13}C$ atoms, and one very small peak representing the $^{14}C$ atoms. By defining the $^{12}C$ peak as exactly 12.0000 amu, the other two peaks are measured at $m/z = 13.0034$ amu and $m/z = 14.0032$ amu. In this way, the isotopes of each element can be accurately "weighed," giving the values in Table 15.5.

**TABLE 15.5 RELATIVE ATOMIC MASS AND ABUNDANCE OF SEVERAL ELEMENTS**

| ISOTOPE | RELATIVE ATOMIC MASS (amu) | ABUNDANCE IN NATURE | ISOTOPE | RELATIVE ATOMIC MASS (amu) | ABUNDANCE IN NATURE |
|---|---|---|---|---|---|
| $^{1}H$ | 1.0078 | 99.99% | $^{16}O$ | 15.9949 | 99.76% |
| $^{2}H$ | 2.0141 | 0.01% | $^{17}O$ | 16.9991 | 0.04% |
| $^{3}H$ | 3.0161 | $<$0.01% | $^{18}O$ | 17.9992 | 0.20% |
| $^{12}C$ | 12.0000 | 98.93% | $^{35}Cl$ | 34.9689 | 75.78% |
| $^{13}C$ | 13.0034 | 1.07% | $^{37}Cl$ | 36.9659 | 24.22% |
| $^{14}C$ | 14.0032 | $<$0.01% | $^{79}Br$ | 78.9183 | 50.69% |
| $^{14}N$ | 14.0031 | 99.63% | $^{81}Br$ | 80.9163 | 49.31% |
| $^{15}N$ | 15.0001 | 0.37% | | | |

Note: Data obtained from the National Institute of Standards and Technology (NIST).

The values that typically appear on a periodic table are the weighted averages for each element, or **standard atomic weight**, which takes into account isotopic abundance. For example, the standard atomic weight of carbon is 12.011 amu (rather than 12.0000 amu), which is a weighted average of the various isotopes based on their abundance in nature. With high-resolution mass spectrometry, the values on the periodic table are irrelevant, because each molecular ion (or cationic fragment) passes through the spectrometer individually and strikes the detector in a particular location that is dependent on which isotopes are present in that specific molecule. The majority of the ions that strike the detector will be comprised of the most abundant isotopes ($^{1}H$, $^{12}C$, $^{14}N$, and $^{16}O$). Therefore, for purposes of interpreting the data from high-resolution mass spectrometry, the values in Table 15.5 must be used, rather than the values on the periodic table.

Now we are ready to see how high-resolution mass spectrometry can reveal the molecular formula of an unknown compound. As an example, compare the following two compounds:

$C_5H_8O$ (MW=84)

$C_6H_{12}$ (MW=84)

Both of these compounds have the same molecular weight when rounded to the nearest whole number. The low-resolution mass spectrum of each compound is therefore expected to give a molecular ion peak at $m/z = 84$. However, with high-resolution mass spectrometry, the molecular ion peaks of these compounds can be differentiated. The calculations for each compound are as follows:

$$C_5H_8O = (5 \times 12.0000) + (8 \times 1.0078) + (1 \times 15.9949) = \mathbf{84.0573} \text{ amu}$$

$$C_6H_{12} = (6 \times 12.0000) + (12 \times 1.0078) = \mathbf{84.0936} \text{ amu}$$

When measured to four decimal places, these compounds do not have the same mass, and the difference is detectable. In a high-resolution mass spectrometer, the molecular ion for the first compound will appear near $m/z = 84.0573$, while the molecular ion of the second compound will appear near $m/z = 84.0936$. In this way, the molecular formula of an unknown compound can be determined via high-resolution mass spectrometry. The mass of the molecular ion is measured accurately, and a simple computer program can then calculate the correct molecular formula. The computer program is not entirely necessary, because published data tables can be cross-referenced to find the matching molecular formula.

### CONCEPTUAL CHECKPOINT

**15.29** How would you distinguish between each pair of compounds using high-resolution mass spectrometry?

(a) (b)

**15.30** How would you distinguish between each pair of compounds in Problem 15.29 using IR spectroscopy?

## 15.14 Gas Chromatography–Mass Spectrometry

Mass spectrometry is ideally suited for analyzing pure compounds. However, when dealing with a mixture that contains several compounds, the compounds must first be separated from each other and then individually injected into the mass spectrometer to yield different spectra. This process has been greatly simplified by the advent of the **gas chromatograph–mass spectrometer** (Figure 15.31), often called "GC mass spec." This powerful diagnostic tool combines a gas chromatograph with a mass spectrometer. First, the mixture of compounds is separated in the gas chromatograph and then each compound is analyzed sequentially by the mass spectrometer. The gas chromatograph is comprised of a long, narrow tube containing a viscous, high-boiling liquid on a solid support, called the *stationary phase*. The tube is contained in an oven, allowing the temperature to be controlled. A syringe is used to inject the sample into the gas chromatograph where it is vaporized, mixed with an inert gas, and then carried through the tube. The various compounds in the mixture travel through the stationary phase at different rates based on their boiling points and their affinity for the stationary phase. Each compound in the mixture generally exhibits a unique **retention time**, which is the amount of time required for it to exit from the gas chromatograph. In this way, the compounds are separated from each other based on their different retention times. A plot, called a **chromatogram**, identifies the retention time of each compound in the mixture. The chromatogram in Figure 15.32 shows five different compounds exiting the gas chromatograph at different times. Each of these compounds is then passed through the mass spectrometer where its molecular weight can be measured.

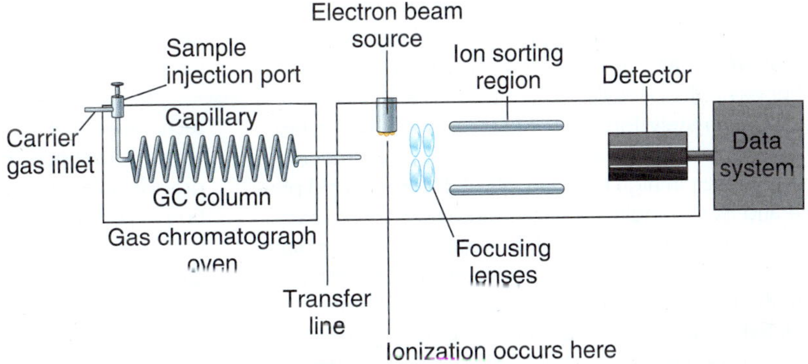


**FIGURE 15.31** A schematic of a gas chromatograph–mass spectrometer.

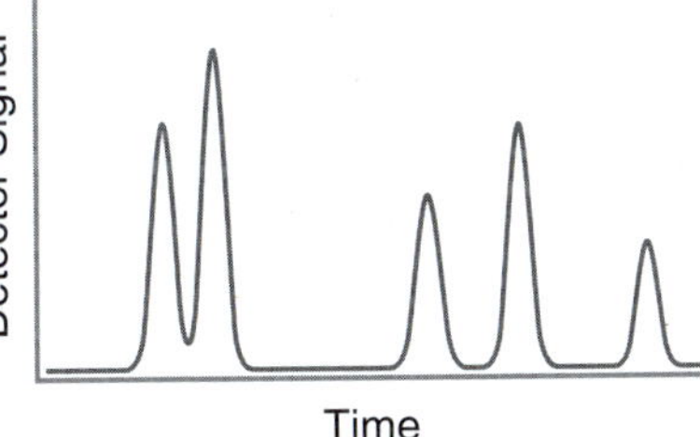


**FIGURE 15.32** An example of a chromatogram, showing five different compounds, each with a unique retention time.

GC-MS analysis is often used for drug screening, which can be performed on a urine sample. The organic compounds present in the urine are first extracted and then injected into the gas chromatograph. Each specific drug has a unique retention time, and its identity can be verified via mass spectrometry. This technique is used to test for marijuana, illegal steroids, and many other illegal drugs.

## 15.15 Mass Spectrometry of Large Biomolecules

Until about 30 years ago, mass spectrometry was limited to compounds with molecular weights under 1000 amu. Compounds with higher molecular weights could not be vaporized without undergoing decomposition. Over the last 30 years, a number of new techniques have emerged, expanding the scope of mass spectrometry. One such technique, called **electrospray ionization (ESI)**, is used most often for large molecules such as proteins and nucleic acids. In this technique, the compound is first dissolved in an ionic solvent and then sprayed via a high-voltage needle into a vacuum chamber. The tiny droplets of solution become charged by the needle, and subsequent evaporation forms gas-phase molecular ions that typically carry multiple charges. The resulting ions are then passed through a mass spectrometer and the *m/z* value for each ion is recorded. This technique has proven to be extremely effective for acquiring mass spectra of large biologically important compounds, particularly because the molecular ions formed generally do not undergo fragmentation. For the development of the ESI technique, Dr. John Fenn (Yale University) shared the 2002 Nobel Prize in Chemistry.

### Medical Applications of Mass Spectrometry

Mass spectrometry is currently being used in a variety of medical applications, several of which are described below.

#### Metabolic Defects in Newborns

A single drop of a newborn baby's blood can be analyzed via mass spectrometry to reveal approximately 30 genetic abnormalities of metabolism. Mass spectrometry is an effective tool in these cases, because it can be used to detect abnormal blood levels of amino acids (Chapter 25), fatty acids (Chapter 26), and sugars (Chapter 24), all in one analysis of a very small blood sample. For example, phenylketonuria (PKU) is a disease that occurs in approximately 1 in every 14,000 births in the United States. For a baby with this disease, a diet of typical amounts of protein will result in high blood levels of the amino acid phenylalanine and its metabolic products. Untreated, this disease causes severe mental retardation. If, however, the baby is maintained on a diet low in the amino acid phenylalanine promptly after birth, the child can develop with normal intelligence. Virtually all babies in the U.S. are now tested for high blood levels of phenylalanine shortly after birth. This was originally done with the Guthrie test, a laborious and time-consuming test, developed in 1963, utilizing bacterial growth to detect high levels of phenylalanine. The Guthrie test has largely been replaced with mass spectrometry for measuring blood levels of phenylalanine, as it provides rapid and reliable results.

#### Drug Analysis

Mass spectrometry can be used to assess drug purity and to characterize designer drugs whose structure is not known in advance. Mass spectrometry can also be used for measuring blood or urine levels of both legal and illegal drugs and their metabolites. For example, mass spectrometry can effectively measure blood levels of cotinine, the primary nicotine metabolite that results from tobacco use in saliva, and this application may see increased use as insurance companies base premiums on risk factors such as smoking.

#### Identification of Pathogenic Bacteria

For patients with bacterial infections, it is often critical to rapidly determine the identity of the bacterium. This is important because knowledge of the identity of a particular bacterial infection enables the administration of the most effective drugs against that particular bacterial strain. If the identity of the bacterial strain cannot be assessed in a timely fashion, the patient might be treated with less effective drugs, allowing the infection to progress in a life-threatening manner. Traditional identification techniques involve the use of cultures, which often require several days and do not always provide a definitive answer.

To overcome this problem, polymerase chain reaction (PCR) techniques have been developed to identify bacterial strains within a few hours of obtaining a sample. Mass spectrometry may provide an alternative technique for the rapid clinical identification of pathogenic bacterial strains, by using special techniques to identify bacterial proteins specific to various strains of pathogenic bacteria. Like PCR, mass spectrometry can often provide the identity of a bacterial strain within a few hours, instead of days. PCR techniques have reached the market first, but mass spectrometry has the potential for being a serious competitor.

#### Other Applications

At a medical research level, mass spectrometry is also being used to detect specific protein biomarkers for several types of cancer and can often detect those biomarkers earlier than any other techniques. Mass spectrometry has also been successfully used to separate variants of the lipoproteins involved in cholesterol transport (Chapter 26) which carry different risk of cardiovascular disease.

Although a mass spectrometer is an expensive piece of equipment, it requires only very small samples of blood and gives rapid results (often available in an hour or two, as opposed to days). A mass spectrometer dedicated to running clinical samples in a "high-throughput" fashion can provide definitive information cost effectively and much more rapidly than older techniques.

## 15.16 Hydrogen Deficiency Index: Degrees of Unsaturation

This chapter explained how mass spectrometry can be used to ascertain the molecular formula of an unknown compound. For example, it is possible to determine that the molecular formula of a compound is $C_6H_{12}O$, but that is still not enough information to draw the structure of the compound. There are many constitutional isomers that share the molecular formula $C_6H_{12}O$. IR spectroscopy can tell us whether or not a double bond is present and whether or not an O—H bond is present. But, once again, that is not enough information to draw the structure of the compound. In Chapter 16, we will learn about NMR spectroscopy, which provides even more information. But before we move on to NMR spectroscopy, there is still one piece of information that can be gleaned from the molecular formula. A careful analysis of the molecular formula can often provide us with a list of possible molecular structures. This skill will be important in the next chapter, because the molecular formula alone offers helpful clues about the structure of the compound. To see how this works, let's begin by analyzing the molecular formula of several alkanes.

Compare the structures of the following alkanes, paying special attention to the number of hydrogen atoms attached to each carbon atom:

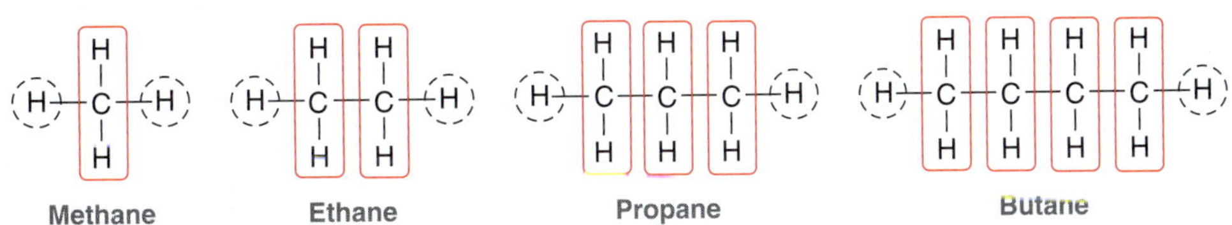


In each case there are two hydrogen atoms on the ends of the structures (circled), and there are two hydrogen atoms on every carbon atom. This can be summarized like this:

$$H—(CH_2)_n—H$$

where $n$ is the number of carbon atoms in the compound. Accordingly, the number of hydrogen atoms will be $2n+2$. In other words, all of the compounds shown have a molecular formula of $C_nH_{2n+2}$. This is true even for compounds that are branched rather than having a straight chain.

$C_5H_{12}$ $C_5H_{12}$ $C_5H_{12}$

These compounds are said to be **saturated** because they possess the maximum number of hydrogen atoms possible, relative to the number of carbon atoms present.

A compound with a π bond (a double or triple bond) will have fewer than the maximum number of hydrogen atoms. Such compounds are said to be **unsaturated**.

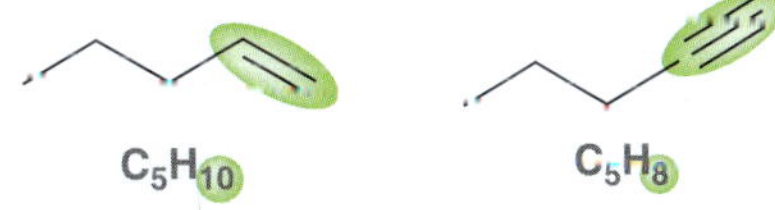


A compound containing a ring will also have fewer than the maximum number of hydrogen atoms, just like a compound with a double bond. For example, compare the structures of 1-hexene and cyclohexane.

1-Hexene ($C_6H_{12}$) Cyclohexane ($C_6H_{12}$)

Both compounds have molecular formula $C_6H_{12}$ because both are "missing" two hydrogen atoms [6 carbon atoms require $(2 \times 6) + 2 = 14$ hydrogen atoms]. Each of these compounds is said to have one **degree of unsaturation**. The **hydrogen deficiency index** (**HDI**) is a measure of the number of degrees of unsaturation. A compound is said to have one degree of unsaturation for every two hydrogen atoms that are missing. For example, a compound with molecular formula $C_4H_6$ is missing four hydrogen atoms (if saturated, it would be $C_4H_{10}$), so it has two degrees of unsaturation (HDI = 2).

There are several ways for a compound to possess two degrees of unsaturation: two double bonds, two rings, one double bond and one ring, or one triple bond. Let's explore all of these possibilities for $C_4H_6$ (Figure 15.33).

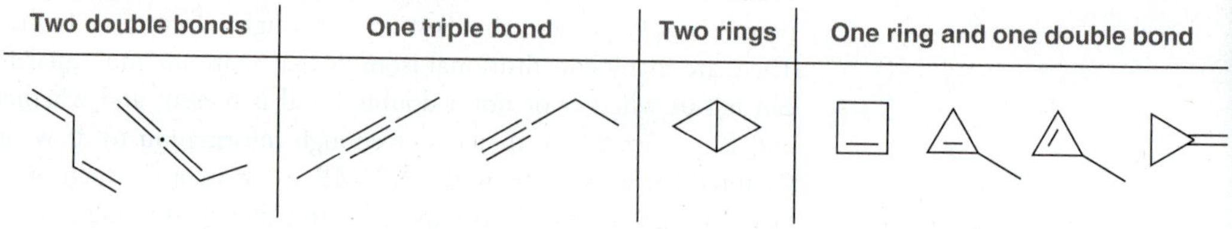

| Two double bonds | One triple bond | Two rings | One ring and one double bond |
|---|---|---|---|

**FIGURE 15.33** All possible constitutional isomers of $C_4H_6$.

With this in mind, let's expand our skills set. Let's explore how to calculate the HDI when other elements are present in the molecular formula.

1. ***Halogens:*** Compare the following two compounds. Notice that chlorine takes the place of a hydrogen atom. Therefore, for purposes of calculating the HDI, treat a halogen as if it were a hydrogen atom. In other words, $C_2H_5Cl$ should have the same HDI as $C_2H_6$.

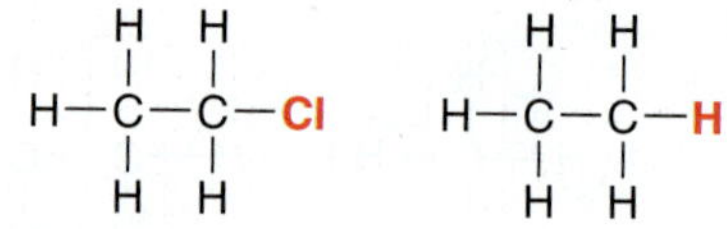

**Chloroethane** **Ethane**

2. ***Oxygen:*** Compare the following two compounds. Notice that the presence of the oxygen atom does not affect the expected number of hydrogen atoms. Therefore, whenever an oxygen atom appears in the molecular formula, it should be ignored for purposes of calculating the HDI. In other words, $C_2H_6O$ should have the same HDI as $C_2H_6$.

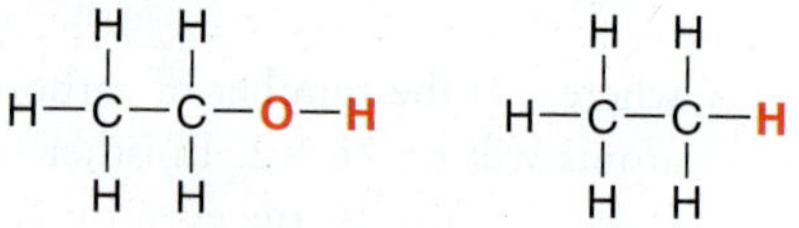

**Ethanol** **Ethane**

3. ***Nitrogen:*** Compare the following two compounds. Notice that the presence of a nitrogen atom changes the number of expected hydrogen atoms. It gives one more hydrogen atom than would be expected. Therefore, whenever a nitrogen atom appears in the molecular formula, one hydrogen atom must be subtracted from the molecular formula. In other words, $C_2H_7N$ should have the same HDI as $C_2H_6$.

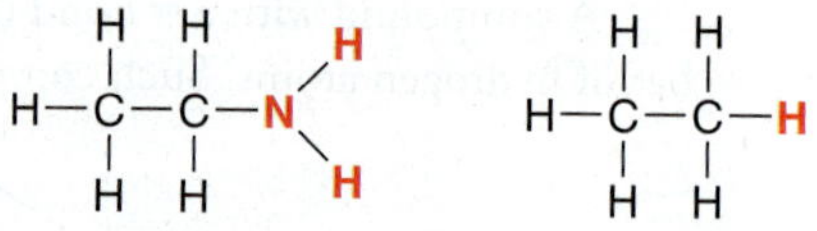

**Ethyl Amine** **Ethane**

In summary:

- Halogens: ***Add*** one H for each halogen.
- Oxygen: ***Ignore.***
- Nitrogen: ***Subtract*** one H for each N.

These rules will enable you to determine the HDI for most simple compounds. Alternatively, the following formula can be used:

$$\text{HDI} = \frac{1}{2}(2C + 2 + N - H - X)$$

where $C$ is the number of carbon atoms, $N$ is the number of nitrogen atoms, $H$ is the number of hydrogen atoms, and $X$ is the number of halogens. This formula will work for all compounds containing C, H, N, O, and X.

Calculating the HDI is particularly helpful, because it provides clues about the structural features of the compound. For example, an HDI of zero indicates that the compound cannot have any rings or π bonds. That is extremely useful information when trying to determine the structure of a compound, and it is information that is easily obtained by simply analyzing the molecular formula. Similarly, an HDI of 1 indicates that the compound must have either one double bond ***or*** one ring (but not both). If the HDI is 2, then there are a few possibilities: two rings, two double bonds, one ring and one double bond, or one triple bond. Analysis of the HDI for an unknown compound can often be a useful tool, but only when the molecular formula is known with certainty.

In Chapter 16, we will use this technique often. The following exercises are designed to develop the skill of calculating and interpreting the HDI of an unknown compound whose molecular formula is known.

## SKILLBUILDER

### 15.4 CALCULATING HDI

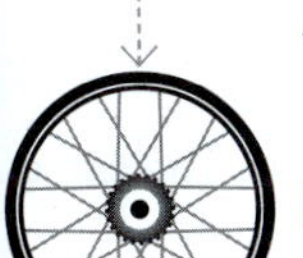

**LEARN the skill**

Calculate the HDI for a compound with molecular formula $C_4H_8ClNO_2$ and identify the structural information provided by the HDI.

**SOLUTION**

**STEP 1**
Identify the molecular formula of a hydrocarbon (a compound with only C and H) that will have the same HDI.

The calculation is:

| | |
|---|---|
| Number of H's: | 8 |
| Add 1 for each Cl: | +1 |
| Ignore each O: | 0 |
| Subtract 1 for each N: | −1 |
| Total: | 8 |

This compound will have the same HDI as a compound with molecular formula $C_4H_8$. To be fully saturated, four carbon atoms would require $(2 \times 4) + 2 = 10$ H's. According to our calculation, two hydrogen atoms are missing, and therefore, this compound has one degree of unsaturation: HDI = 1.

**STEP 2**
Determine how many hydrogen atoms are missing and assign an HDI.

Alternatively, the following formula can be used:

$$\text{HDI} = \frac{1}{2}(2C + 2 + N - H - X) = \frac{1}{2}(8 + 2 + 1 - 8 - 1) = \frac{2}{2} = 1$$

With one degree of unsaturation, the compound must contain either one ring or one double bond, but not both. The compound cannot have a triple bond, as this would require two degrees of unsaturation.

**PRACTICE the skill**

**15.31** Calculate the degree of unsaturation for each of the following molecular formulas:

**(a)** $C_6H_{10}$ **(b)** $C_5H_{10}O_2$ **(c)** $C_5H_9N$ **(d)** $C_3H_5ClO$ **(e)** $C_{10}H_{20}$
**(f)** $C_4H_6Br_2$ **(g)** $C_6H_6$ **(h)** $C_2Cl_6$ **(i)** $C_2H_4O_2$ **(j)** $C_{100}H_{200}Cl_2O_{16}$

**15.32** Identify which two compounds shown here have the same degree of unsaturation.

$C_3H_8O$ $C_3H_5ClO_2$ $C_3H_5NO_2$ $C_3H_6$

**APPLY the skill**

**15.33** Propose all possible structures for a compound with molecular formula $C_4H_8O$ that exhibits a signal at 1720 $cm^{-1}$ in its IR spectrum.

**15.34** Propose all possible structures for a compound with molecular formula $C_4H_8O$ that exhibits a broad signal between 3200 and 3600 $cm^{-1}$ in its IR spectrum and does not contain any signals between 1600 and 1850 $cm^{-1}$.

**15.35** Propose the only possible structure for a compound with molecular formula $C_4H_6$ that exhibits a signal at 2200 $cm^{-1}$ in its IR spectrum.

**15.36** What structural features are present in a compound with molecular formula $C_{10}H_{20}O$?

**15.37** Each pair of compounds below has the same number of carbon atoms. Without counting the hydrogen atoms, determine whether the pair of compounds has the same molecular formula. In each case, simply determine the HDI of each compound to make your decision. Then, count the number of hydrogen atoms to see if your analysis was correct.

**(a)** 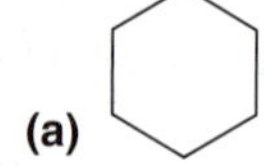 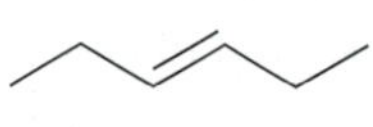 **(b)** 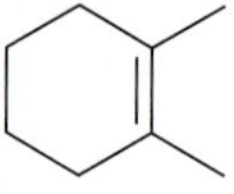 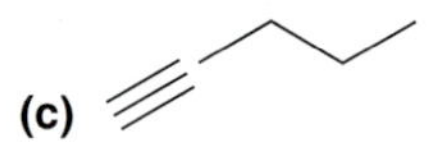 **(c)** 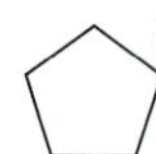

need more **PRACTICE?** **Try Problems 15.48b, 15.54–15.55**

# REVIEW OF CONCEPTS AND VOCABULARY

### SECTION 15.1

- **Spectroscopy** is the study of the interaction between light and matter.
- The **wavelength** of light describes the distance between adjacent peaks of an oscillating field, while the **frequency** describes the number of wavelengths that pass a particular point in space per unit time.
- The energy of each **photon** is determined by its frequency. The range of all possible frequencies is known as the **electromagnetic spectrum**.
- Molecules can store energy in a variety of ways. Each of these forms of energy is quantized.
- The difference in energy ($\Delta E$) between vibrational energy levels is determined by the nature of the bond. If a photon of light possesses exactly this amount of energy, the bond can absorb the photon to promote a **vibrational excitation**.

### SECTION 15.2

- There are many different kinds of vibrational excitation, including **stretching** and **bending**.
- IR spectroscopy can be used to identify which functional groups are present in a compound.
- An **absorption spectrum** is a plot that measures the percent transmittance as a function of frequency.
- The location of each signal in an IR spectrum is reported in terms of a frequency-related unit called **wavenumber**.

### SECTION 15.3

- The wavenumber of each signal is determined primarily by bond strength and the masses of the atoms sharing the bond.
- The **diagnostic region** contains all signals that arise from double bonds, triple bonds, and X—H bonds.
- The **fingerprint region** contains signals resulting from the vibrational excitation of most single bonds (stretching and bending).
- A **conjugated** C═O bond will produce a lower energy signal than an unconjugated C═O bond.

### SECTION 15.4

- The intensity of a signal is dependent on the dipole moment of the bond giving rise to the signal.
- C═O bonds produce strong signals in an IR spectrum, while C═C bonds often produce fairly weak signals.
- Symmetrical C═C bonds do not produce signals. The same is true for symmetrical triple bonds.

### SECTION 15.5

- Concentrated alcohols give rise to broad signals, while dilute alcohols give rise to narrow signals.
- Primary amines exhibit two signals resulting from **symmetric stretching** and **asymmetric stretching**.

### SECTIONS 15.6 AND 15.7

- When analyzing an IR spectrum, look for double bonds, triple bonds, and X—H bonds.
- Every signal has three characteristics—wavenumber, intensity, and shape. Analyze all three.
- When looking for X—H bonds, identify whether any signals appear to the left of a line drawn at 3000 $cm^{-1}$.

### SECTION 15.8

- **Mass spectrometry** is used to determine the molecular weight and molecular formula of a compound.
- In a **mass spectrometer**, a compound is converted into ions, which are then separated by a magnetic field.
- **Electron impact ionization (EI)** involves bombarding the compound with high-energy electrons, generating a radical cation that is symbolized by $(M)^{+\bullet}$ and is called the **molecular ion**, or the **parent ion**.
- The molecular ion is often very unstable and susceptible to **fragmentation**.
- Only the molecular ion and the cationic fragments are deflected, and they are then separated by their **mass-to-charge ratio** ($m/z$).
- A **mass spectrum** shows the relative abundance of each cation detected.
- The tallest peak in a mass spectrum is assigned a relative value of 100% and is called the **base peak**.

### SECTION 15.9

- The $(M)^{+\bullet}$ peak may be weak or entirely absent if it is particularly susceptible to fragmentation.
- The $(M)^{+\bullet}$ peak indicates the molecular weight of the molecule.
- According to the **nitrogen rule**, an odd molecular weight indicates an odd number of nitrogen atoms, while an even

molecular weight indicates either the absence of nitrogen or an even number of nitrogen atoms.

**SECTION 15.10**

- The relative heights of the $(M)^{+\bullet}$ peak and the $(M+1)^{+\bullet}$ peak indicate the number of carbon atoms.

**SECTION 15.11**

- Useful information can be obtained by comparing the relative heights of the $(M+2)^{+\bullet}$ peak and the $(M)^{+\bullet}$ peak. The structure likely contains a chlorine atom if the latter peak is approximately one-third as tall as the former peak. The presence of a bromine atom is indicated when these two peaks are similar in height.

**SECTION 15.12**

- A signal at M–15 indicates the loss of a methyl group; a signal at M–29 indicates the loss of an ethyl group.
- The likelihood of fragmentation increases with the stability of the carbocation formed.

**SECTION 15.13**

- The molecular formula of a compound can be determined with **high-resolution mass spectrometry**.
- The **atomic mass unit (amu)** is equivalent to 1 g divided by Avogadro's number.
- The **standard atomic weight** is a weighted average that takes into account relative isotopic abundance.

**SECTION 15.14**

- In a **gas chromatograph–mass spectrometer**, a mixture of compounds is first separated based on their boiling points and affinity for the *stationary phase*. Each compound is then analyzed individually.
- Each compound in the mixture generally exhibits a unique **retention time**, which is plotted on a **chromatogram**.

**SECTION 15.15**

- **Electrospray ionization (ESI)** is used most often for large molecules such as proteins and nucleic acids.

**SECTION 15.16**

- **Saturated** alkanes have a molecular formula of the form $C_nH_{2n+2}$.
- A compound possessing a π bond is **unsaturated**.
- Each double bond and each ring represents one **degree of unsaturation**.
- The **hydrogen deficiency index (HDI)** is a measure of the number of degrees of unsaturation.

# SKILLBUILDER REVIEW

## 15.1 ANALYZING AN IR SPECTRUM

**STEP 1** Look for double bonds between 1600 and 1850 $cm^{-1}$.

Guidelines:

- C=O bonds produce strong signals.
- C=C bonds generally produce weak signals. Symmetrical C=C bonds do not appear at all.
- The exact position of a signal indicates subtle factors that affect bond stiffness, such as resonance.

**STEP 2** Look for triple bonds between 2100 and 2300 $cm^{-1}$.

Guidelines:

- Symmetrical triple bonds do not produce signals.

**STEP 3** Look for X—H bonds between 2750 and 4000 $cm^{-1}$.

Guidelines:

- Draw a line at 3000 and look for vinylic or acetylenic C—H bonds to the left of the line.
- The shape of an O—H signal is affected by concentration (due to H bonding).
- Primary amines exhibit two N—H signals (symmetric and asymmetric stretching).

Try Problems **15.12, 15.13, 15.42, 15.57**

## 15.2 DISTINGUISHING BETWEEN TWO COMPOUNDS USING IR SPECTROSCOPY

**STEP 1** Work methodically through the expected IR spectrum of each compound.

**STEP 2** Determine if any signals will be present in one compound but absent in the other.

**STEP 3** For each expected signal, compare for any possible differences in wavenumber, intensity, or shape.

Try Problems **15.14–15.18, 15.38–15.41, 15.48d**

## 15.3 USING THE RELATIVE ABUNDANCE OF THE $(M + 1)^{+\bullet}$ PEAK TO PROPOSE A MOLECULAR FORMULA

**STEP 1** Determine the number of carbon atoms in the compound by analyzing the relative abundance of the $(M + 1)^{+\bullet}$ peak:

$$\frac{\left(\dfrac{\text{Abundance of } (M+1)^{+\bullet}\text{ peak}}{\text{Abundance of } (M)^{+\bullet}\text{ peak}}\right) \times 100\%}{1.1\ \%}$$

**STEP 2** Analyze the mass of the molecular ion to determine if any heteroatoms (such as oxygen or nitrogen) are present.

Try Problems **15.21, 15.22, 15.45, 15.47**

### 15.4 CALCULATING HDI

**STEP 1** Rewrite the molecular formula "as if" the compound had no elements other than C and H, using the following rules:

- Add one H for each halogen.
- Ignore all oxygen atoms.
- Subtract one H for each nitrogen.

**STEP 2** Determine whether any H's are missing. Every two H's represent one degree of unsaturation:

$C_4H_9Cl \longrightarrow C_4H_{10} \longrightarrow$ HDI=0

$C_4H_8O \longrightarrow C_4H_8 \longrightarrow$ HDI=1

$C_4H_9N \longrightarrow C_4H_8 \longrightarrow$ HDI=1

Try Problems **15.31–15.37, 15.48b, 15.54, 15.55**

## PRACTICE PROBLEMS

Note: Most of the Problems are available within **WileyPLUS**, an online teaching and learning solution.

**15.38** All of the following compounds absorb IR radiation in the range between 1600 and 1850 cm$^{-1}$. In each case, identify the specific bond(s) responsible for the absorption(s) and predict the approximate wavenumber of absorption for each of those bonds.

(a) 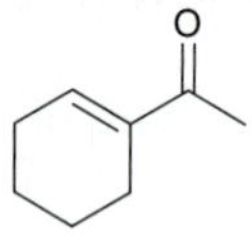 (b) 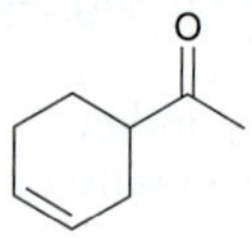 (c) 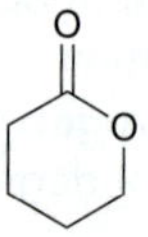 (d) 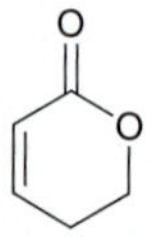 (e)

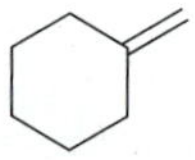

**15.39** Rank each of the bonds identified in order of increasing wavenumber.

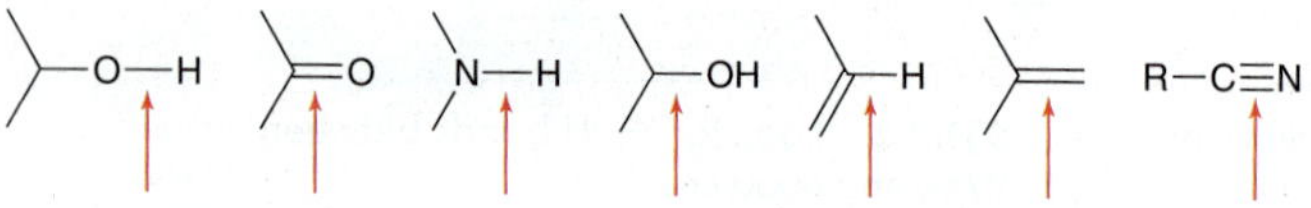

O—H C=O N—H OH H R—C≡N

**15.40** Identify the signals you would expect in the diagnostic region of the IR spectrum for each of the following compounds:

(a) 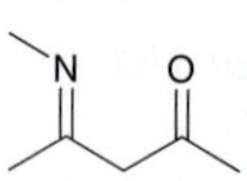 N O (b) 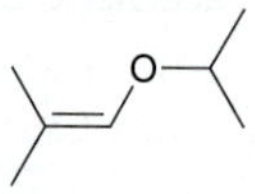 O (c) 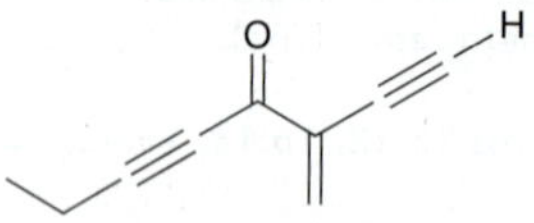 O H (d) 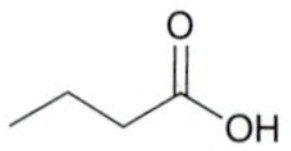 O OH

**15.41** Identify how IR spectroscopy might be used to monitor the progress of each of the following reactions:

(a) 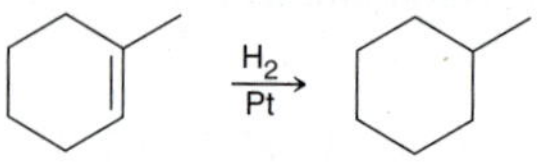 $H_2$, Pt (b) 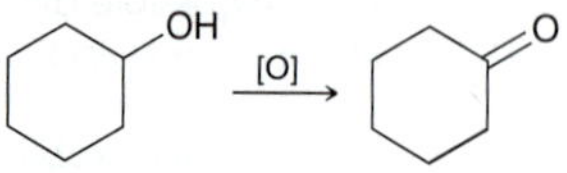 OH [O] O

(c) 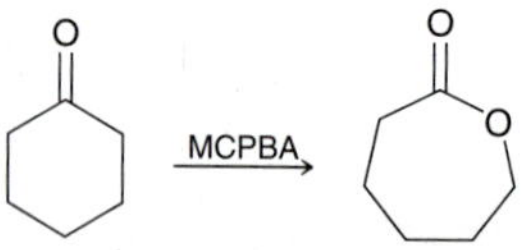 O MCPBA O O

(d) 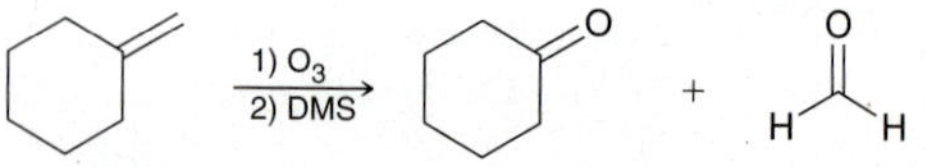 1) $O_3$ 2) DMS O + O H H

(e) 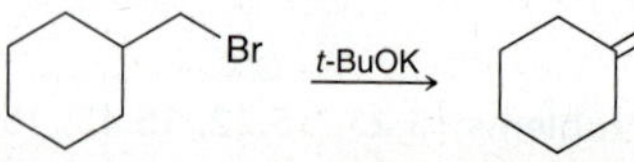 Br t-BuOK

**15.42** Identify the characteristic signals that you would expect in the diagnostic region of an IR spectrum of each of the following compounds:

(a) 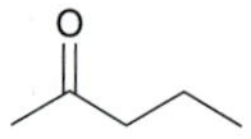 O (b) 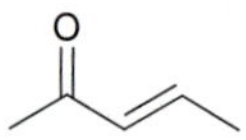 O (c) 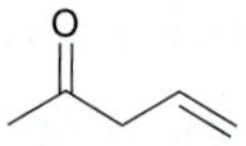 O

(d) 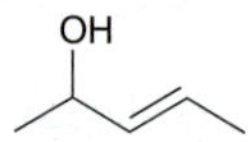 OH (e) 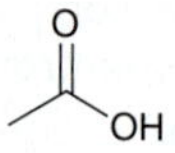 O OH (f) 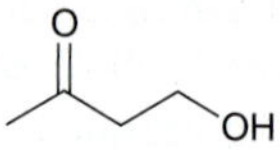 O OH

**15.43** Identify the molecular formula for each of the following compounds and then predict the mass of the expected molecular ion in the mass spectrum of each compound:

(a) 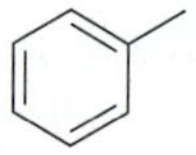 (b) 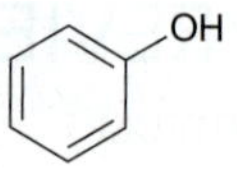 OH (c) 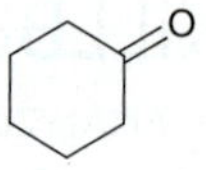 O

(d) 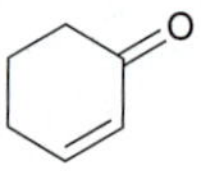 O (e) 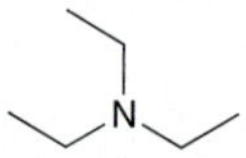 N

**15.44** Propose a molecular formula for a compound that has one degree of unsaturation and a mass spectrum that displays a molecular ion signal at $m/z = 86$.

**15.45** The mass spectrum of an unknown hydrocarbon exhibits an $(M+1)^{+\bullet}$ peak that is 10% as tall as the molecular ion peak. Identify the number of carbon atoms in the unknown compound.

**15.46** Match each compound with the appropriate spectrum.

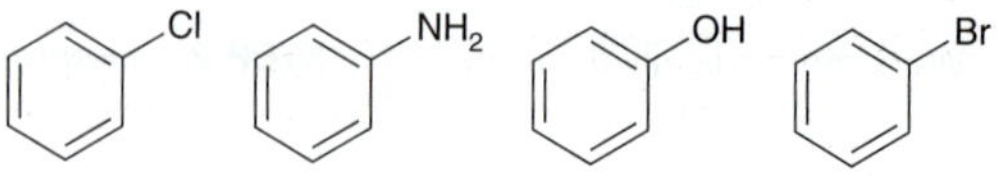

Cl NH$_2$ OH Br

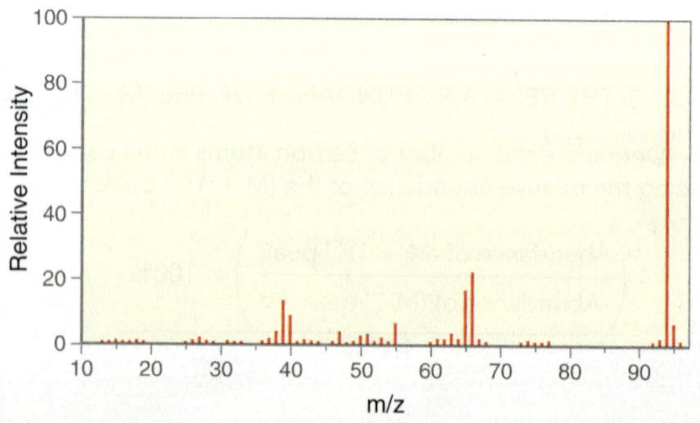

(a)

Relative Intensity

100 80 60 40 20 0

10 20 30 40 50 60 70 80 90

m/z

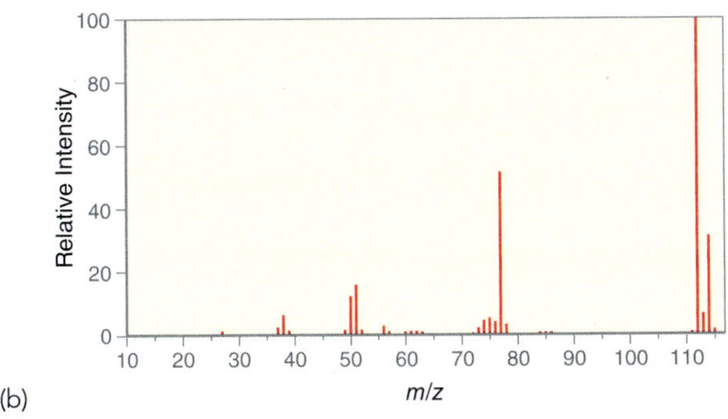


(b)

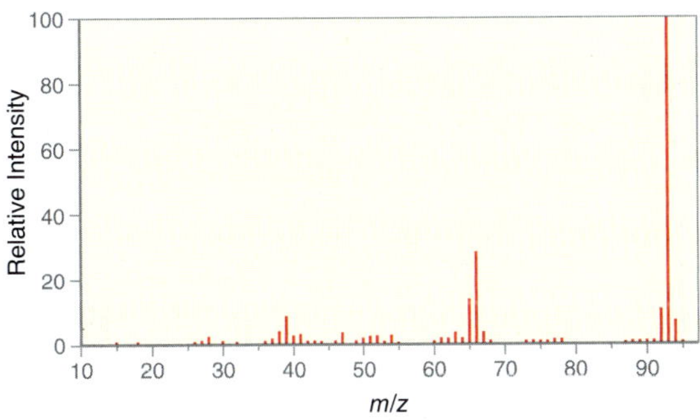


(c)

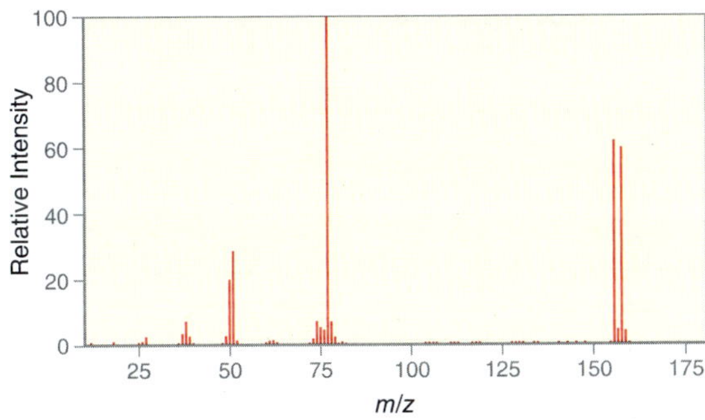


(d)

**15.47** The sex attractant of the codling moth gives an IR spectrum with a broad signal between 3200 and 3600 $cm^{-1}$ and two signals between 1600 and 1700 $cm^{-1}$. In the mass spectrum of this compound, the molecular ion peak appears at $m/z = 196$, and the relative abundances of the molecular ion and the $(M+1)^{+\bullet}$ peak are 27.2% and 3.9%, respectively.

(a) What functional groups are present in this compound?

(b) How many carbon atoms are present in the compound?

(c) Based on the information given, propose a molecular formula for the compound.

**15.48** Compare the structures of cyclohexane and 2-methyl-2-pentene.

(a) What is the molecular formula of each compound?

(b) What is the HDI of each compound?

(c) Can high-resolution mass spectrometry be used to distinguish between these compounds? Explain.

(d) How would you differentiate between these two compounds using IR spectroscopy?

**15.49** The mass spectrum of 1-ethyl-1-methylcyclohexane shows many fragments, with two in very large abundance. One appears at $m/z = 111$ and the other appears at $m/z = 97$. Identify the structure of each of these fragments.

**15.50** The mass spectrum of 2-bromopentane shows many fragments.

(a) One fragment appears at M–79. Would you expect a signal at M–77 that is equal in height to the M–79 peak? Explain.

(b) A fragment appears at M–15. Would you expect a signal at M–13 that is equal in height to the M–15 peak? Explain.

(c) One fragment appears at M–29. Would you expect a signal at M–27 that is equal in height to the M–29 peak? Explain.

**15.51** When treated with a strong base, 2-bromo-2,3-dimethylbutane will undergo an elimination reaction to produce two products. The choice of base (ethoxide vs. *tert*-butoxide) will determine which of the two products predominates. Draw both products and determine how you could distinguish between them using IR spectroscopy.

**15.52** Propose a molecular formula that fits the following data:

(a) A hydrocarbon ($C_xH_y$) with a molecular ion peak at $m/z = 66$

(b) A compound that absorbs IR radiation at 1720 $cm^{-1}$ and exhibits a molecular ion peak at $m/z = 70$

**15.53** The following is a mass spectrum of octane.

(a) Which peak represents the molecular ion?

(b) Which peak is the base peak?

(c) Draw the structure of the fragment that produces the base peak.

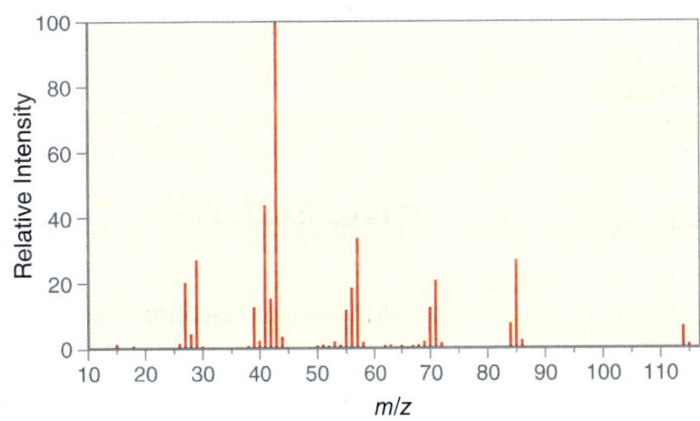


**15.54** Calculate the HDI for each molecular formula.

(a) $C_4H_6$ (b) $C_5H_8$ (c) $C_{40}H_{78}$ (d) $C_{72}H_{74}$
(e) $C_6H_6O_2$ (f) $C_7H_9NO_2$ (g) $C_8H_{10}N_2O$ (h) $C_5H_7Cl_3$
(i) $C_6H_5Br$ (j) $C_6H_{12}O_6$

**15.55** Propose two possible structures for a compound with molecular formula $C_5H_8$ that produces an IR signal at 3300 $cm^{-1}$.

**15.56** Limonene is a hydrocarbon found in the peels of lemons and contributes significantly to the smell of lemons. Limonene has a molecular ion peak at $m/z = 136$ in its mass spectrum, and it has two double bonds and one ring in its structure. What is the molecular formula of limonene?

**15.57** Explain how you would use IR spectroscopy to distinguish between 1-bromo-3-methyl-2-butene and 2-bromo-3-methyl-2-butene.

**15.58** A dilute solution of 1,3-pentanediol does not produce the characteristic IR signal for a dilute alcohol. Rather, it produces a signal that is characteristic of a concentrated alcohol. Explain.

**15.59** The following are IR and mass spectra of an unknown compound. Propose at least two possible structures for the unknown compound.

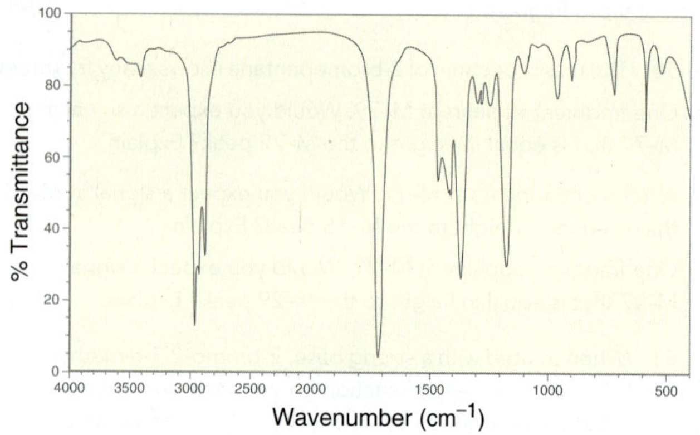


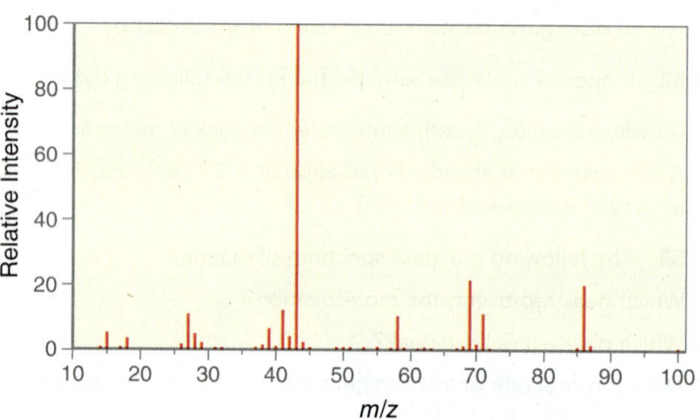


**15.60** The following are IR and mass spectra of an unknown compound. Propose at least two possible structures for the unknown compound.

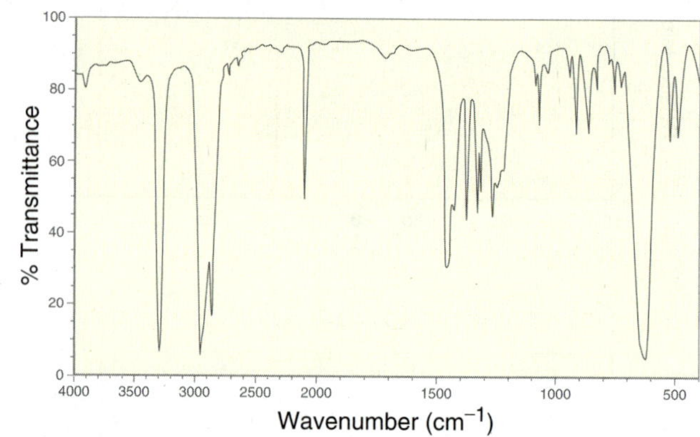


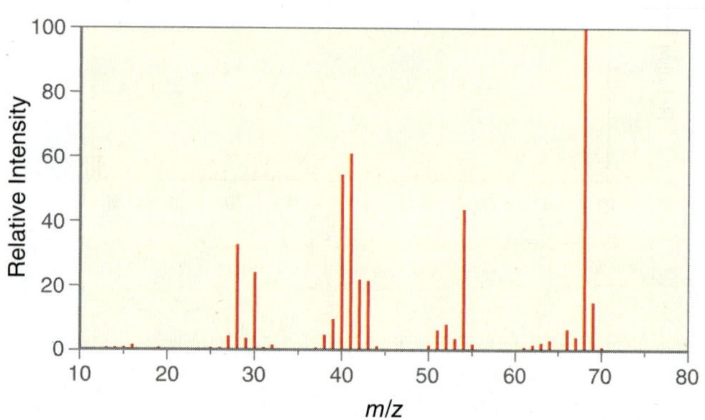


# INTEGRATED PROBLEMS

**15.61** Consider the following sequence of reactions:

(2-chloro-2-methylbutane) $\xrightarrow{\text{NaOEt}}$ **A** $\xrightarrow[\text{2) } H_2O_2\text{, NaOH}]{\text{1) } BH_3 \cdot \text{THF}}$ **B** $\xrightarrow[\text{py}]{\text{TsCl}}$ **C** $\xrightarrow{t\text{-BuOK}}$ **D** $\xrightarrow{\text{MCPBA}}$ **E** $\xrightarrow[\text{2) } H_2O]{\text{1) MeMgBr}}$ **F** $\xrightarrow[\text{2) MeI}]{\text{1) NaH}}$ **G**

(a) Explain how you could use IR spectroscopy to differentiate between compounds **F** and **G**.

(b) Explain how you could use IR spectroscopy to differentiate between compounds **D** and **E**.

(c) If you wanted to distinguish between compounds **B** and **F**, would it be more suitable to use IR spectroscopy or mass spectrometry? Explain.

(d) Would mass spectrometry be helpful for distinguishing between compounds **A** and **D**? Explain.

**15.62** There are five constitutional isomers with molecular formula $C_4H_8$. One of the isomers exhibits a particularly strong signal at M–15 in its mass spectrum. Identify this isomer and explain why the signal at M–15 is so strong.

**15.63** There are four isomers with molecular formula $C_4H_9Cl$. Only one of these isomers (compound **A**) has a chirality center. When compound **A** is treated with sodium ethoxide, three products are formed: compounds **B**, **C**, and **D**. Compounds **B** and **C** are diastereomers, with compound **B** being the less stable diastereomer. Do you expect compound **D** to exhibit a signal at approximately 1650 cm$^{-1}$ in its IR spectrum? Explain.

**15.64** Chloramphenicol is an antibiotic agent isolated from the *Streptomyces venezuelae* bacterium. Predict the expected isotope pattern in the mass spectrum of this compound (the relative heights of the molecular ion peak and surrounding peaks).

**Chloramphenicol**

**15.65** Ephedrine is a bronchodilator and decongestant obtained from the Chinese plant *Ephedra sinica*. A concentrated solution of ephedrine gives an IR spectrum with a broad signal between 3200 and 3600 $cm^{-1}$. An IR spectrum of a dilute solution of ephedrine is very similar to the IR spectrum of the concentrated solution. That is, the broad signal between 3200 and 3600 $cm^{-1}$ is not transformed into a narrow signal, as we might expect for a compound containing an OH group. Explain.

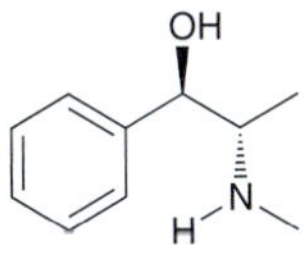

**Ephedrine**

**15.66** Predict the expected isotope pattern in the mass spectrum of a compound with molecular formula $C_{90}H_{180}Br_2$.

**15.67** Esters contain two C—O bonds and therefore will produce two separate stretching signals in the fingerprint region of an IR spectrum. One of these signals typically appears at approximately 1000 $cm^{-1}$, while the other appears at approximately 1300 $cm^{-1}$. Predict which of the two C—O bonds produces the higher energy signal. Using principles that we learned in this chapter (factors that affect the wavenumber of absorption), offer two separate explanations for your choice.

**15.68** Treating 1,2-cyclohexanediol with concentrated sulfuric acid yields a product with molecular formula $C_6H_{10}O$. An IR spectrum of the product exhibits a strong signal at 1720 $cm^{-1}$. Identify the structure of the product and show a mechanism for its formation.

**15.69** Draw the expected isotope pattern that would be observed in the mass spectrum of $CH_2BrCl$. In other words, predict the relative heights of the peaks at M, M+2, and M+4.

# CHALLENGE PROBLEMS

**15.70** The following is the proposed structure of a blue fabric dye, based on high-resolution mass spectrometry data (*Anal. Chem.* **2013,** *85,* **831–836**). The reported method employed a pulsed laser to desorb dye molecules directly from a sample of dyed fabric. These molecules then entered a high-resolution mass spectrometer in negative ion mode (where anions are detected instead of cations). The resulting spectrum included the following peaks and relative intensities: $m/z$ = 423.0647 (100%), 424.0681 (22.5%), and 425.0605 (4.21%).

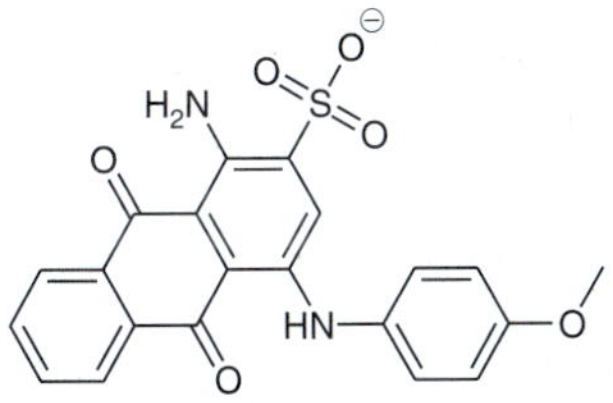

(a) Explain the origins and relative abundance of each of these peaks. Note that the two most abundant isotopes of sulfur are $^{32}S$ (31.9721 amu, 95.02%) and $^{34}S$ (33.9679 amu, 4.21%).

(b) The mass spectrum also includes a signal at $m/z$ = 425.0715. Explain the origin of this signal.

**15.71** The base peak of a low-resolution spectrum of cyclohexanone is at $m/z$ = 55. A high-resolution spectrum reveals that this peak actually consists of two peaks at 55.0183 and 55.0546 with relative intensities of 86.7 and 13.3, respectively (*Appl. Spectrosc.* **1960,** *14,* **95–97**). For each of these two peaks, propose a formula for the ion responsible for the peak.

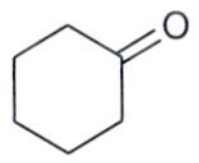

**Cyclohexanone**

**15.72** Compound **A** exists in equilibrium with its tautomeric form, compound **B**. An IR spectrum of a mixture of **A** and **B** exhibits four signals in the region 1600–1850 $cm^{-1}$. These signals appear at 1620, 1660, 1720, and 1740 $cm^{-1}$ (*J. Am. Chem. Soc.* **1952,** *74,* **4070–4073**).

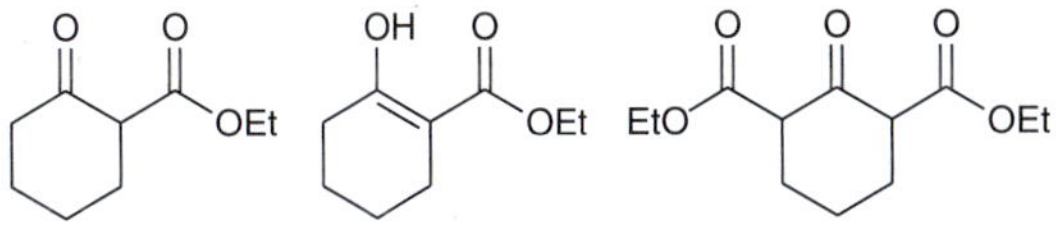

**A** **B** **C**

(a) Identify the source for each of these four signals and provide a justification for the location of the signal at 1660 $cm^{-1}$.

(b) For most simple ketones, the concentration of the enol tautomer is almost negligible at equilibrium. However, in this case, the concentration of **B** is significant. Provide a justification for this observation.

(c) The IR spectrum of compound **C** lacks a signal at 1720 $cm^{-1}$. Draw the tautomer of **C** and then determine which tautomeric form predominates at equilibrium. Explain your choice.

**15.73** As part of a study on cyclopropane derivatives of fatty acids, the following amide was subjected to mass spectrometric analysis (*J. Org. Chem.* **1977,** *42,* **126–129**). The most abundant peak in the spectrum was found at $m/z$ = 113. Suggest a reasonable structure and mechanism of formation for the ion consistent with the peak at $m/z$ = 113.

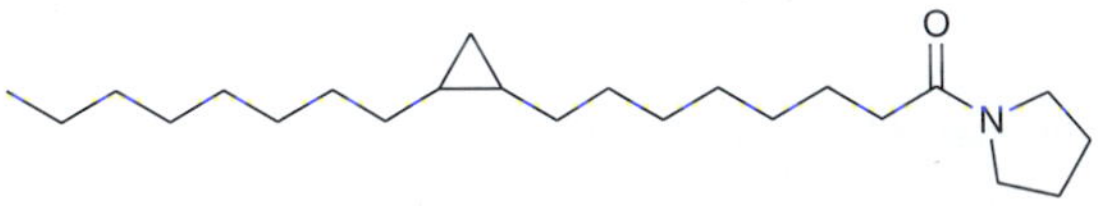

**15.74** Pinolenic acid ($C_{17}H_{29}CO_2H$) is an unbranched carboxylic acid with three *cis* alkene groups. It is found in pine nuts and is sometimes used in weight loss regimens as a hunger suppressant. Not every mass spectrometry technique requires ionization through ejection of single electrons; some instruments rely instead on milder ionization methods such as protonation or deprotonation to form charged species. Mass spectra of pinolenic acid were collected in positive ion mode (using conditions facilitating protonation) and in negative ion mode (using conditions facilitating deprotonation). A small amount of ozone was present in both experiments. Significant peaks in positive ion mode included $m/z$ = 279, 211, 171, and 117. In the absence of ozone, however, only the peak at 279 remains (*Anal. Chem.* **2011,** *83,* **4738–4744**).

(a) Draw the structure of pinolenic acid as well as the ionic structures consistent with each of the peaks listed.

(b) Predict *m/z* values for four significant peaks in negative ion mode, specifying the ion responsible for each peak.

**15.75** The following two isomers were each subjected to mass spectrometric analysis. Some of the significant peaks from each of the two spectra are presented below (*J. Chem. Ed.* **2008,** *85,* **832–833**). Match each isomer (**A**, **B**) to its corresponding mass spectrometry data (**X**, **Y**) and provide structural assignments for every peak.

Sample **X**, $m/z = 202, 200, 187, 185, 159, 157, 121$
Sample **Y**, $m/z = 202, 200, 171, 169, 121$

**A** **B**

**15.76** Myosmine can be isolated from tobacco, along with several other structurally similar compounds, including nicotine. The originally accepted structure for myosmine was disproven with IR spectroscopy (*Anal. Chem.* **1954,** *26,* **1428–1431**):

Originally accepted structure of myosmine

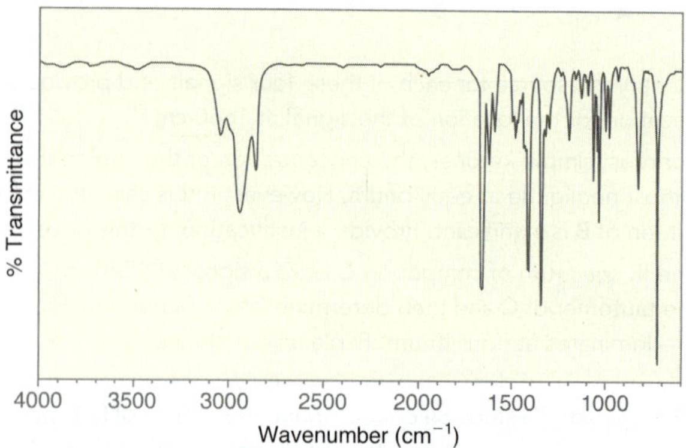


(a) Explain why the IR spectrum is not consistent with the originally proposed structure.

(b) The discord between the proposed structure and its IR spectrum was resolved by recognizing that myosmine exists primarily as a tautomer of the proposed structure. Draw this tautomer and explain why its structure is more consistent with the observed IR spectrum.

(c) Identify a likely source for the fairly intense signal at 1621 $cm^{-1}$.

**15.77** Compound **1** contains a tetrazole ring (a ring containing four nitrogen atoms), while its constitutional isomer, compound **2**, exhibits an azido group ($-N_3$). Compounds **1** and **2** rapidly interconvert, and IR spectroscopy provides evidence that the equilibrium greatly favors the tetrazole in this case (*J. Am. Chem. Soc.* **1959,** *81,* **4671–4673**). Specifically, the IR spectrum of compound **1** lacks a signal in the region 2120–2160 $cm^{-1}$. Explain why the absence of a signal in this region provides evidence that **1** is favored over **2**.

**1** **2**

**15.78** The IR spectrum of a dilute solution of compound **1** (in $CS_2$) exhibits a signal at 3617 $cm^{-1}$, while the IR spectrum of a dilute solution of compound **2** exhibits a signal at 3594 $cm^{-1}$ (*J. Am. Chem. Soc.* **1957,** *79,* **243–247**).

**1** **2**

(a) Explain why the latter signal appears at a lower wavenumber of absorption.

(b) Identify the signal that you expect to be broader. Explain your choice.

**15.79** Phospholipids are a class of compounds largely responsible for the bilayer structure of cell membranes of plants and animals (discussed in Chapter 26). The phospholipid shown below has two unbranched hydrocarbon chains, one of which contains an alkene group with a *cis* configuration. A mass spectrum of this phospholipid in negative ion mode (where anions are detected instead of cations) shows a peak for the molecular ion at $m/z = 673$. When ozone is present, a new peak appears at $m/z = 563$ (*J. Am. Chem. Soc.* **2006,** *128,* **58–59**).

(a) Draw the full structure of the phospholipid (indicating the position of the alkene group) consistent with these data.

(b) When methanol is used as a solvent in the experiment with ozone present, the appearance of the peak at *m/z* 563 is accompanied by a peak at $m/z = 611$. When the solvent is changed to ethanol, the peak at $m/z = 563$ remains, but the peak at $m/z = 611$ is shifted to 625. Propose an explanation consistent with these findings.

# Nuclear Magnetic Resonance Spectroscopy

16



## DID YOU EVER WONDER...

**whether it is scientifically possible for a human being to levitate?**

When any object (even plastic) is placed in a strong magnetic field, the electrons in the object begin to move in a way that produces a small magnetic field that opposes the external magnetic field. As a result, the object repels, and is repelled by, the external magnetic field. This weak effect, called diamagnetism, will cause objects to levitate in a strong magnetic field. Larger objects require a larger external magnetic field to induce levitation, but in theory, any object will levitate if it is placed in a strong enough magnetic field.

The strongest magnetic fields currently available are capable of causing small objects to levitate. In 1997, researchers at Radboud University (in the Netherlands) demonstrated that hazelnuts, strawberries, and even frogs will levitate in a magnetic field of 16 tesla. Unfortunately, our strongest magnets do not yet allow a human being to levitate. But in theory, stronger magnets would enable even people to enjoy weightlessness without having to travel to space.

In this chapter, we will see the role that diamagnetism plays in nuclear magnetic resonance (NMR) spectroscopy, which provides more structural information than any other form of spectroscopy. We will also learn how NMR spectroscopy is used as a powerful tool for structure determination.

## DO YOU REMEMBER?

*Before you go on, be sure you understand the following topics.*
*If necessary, review the suggested sections to prepare for this chapter:*

- Quantum Mechanics (Section 1.6)
- Symmetry and Chirality (Section 5.6)
- Hydrogen Deficiency Index (Section 15.16)
- Stereoisomeric Relationships: Enantiomers and Diastereomers (Section 5.5)
- Introduction to Spectroscopy (Section 15.1)

Take the DO YOU REMEMBER? QUIZ in **WileyPLUS** to check your understanding.

## 16.1 Introduction to NMR Spectroscopy

**Nuclear magnetic resonance (NMR)** spectroscopy is arguably the most powerful and broadly applicable technique for structure determination available to organic chemists. It provides the most information about molecular structure, and in some cases, the structure of a compound can be determined using only NMR spectroscopy. In practice, the structures of complicated molecules are determined through a combination of techniques that include NMR and IR spectroscopy and mass spectrometry.

NMR spectroscopy involves the study of the interaction between electromagnetic radiation and the nuclei of atoms. A wide variety of nuclei can be studied using NMR spectroscopy, including $^{1}$H, $^{13}$C, $^{15}$N, $^{19}$F, and $^{31}$P. In practice, $^{1}$H NMR spectroscopy and $^{13}$C NMR spectroscopy are used most often by organic chemists, because hydrogen and carbon are the primary constituents of organic compounds. Analysis of an NMR spectrum provides information about how the individual carbon and hydrogen atoms are connected to each other in a molecule. This information enables us to determine the carbon-hydrogen framework of a compound, much the way puzzle pieces can be assembled to form a picture.

A nucleus with an odd number of protons and/or an odd number of neutrons possesses a quantum mechanical property called *nuclear spin,* and it can be probed by an NMR spectrometer. Consider the nucleus of a hydrogen atom, which consists of just one proton and therefore has a spin. A spinning proton can be viewed as a rotating sphere of charge, which generates a magnetic field, called a **magnetic moment**. The magnetic moment of a spinning proton is similar to the magnetic field produced by a bar magnet (Figure 16.1). The nucleus of a $^{12}$C atom has an even number of

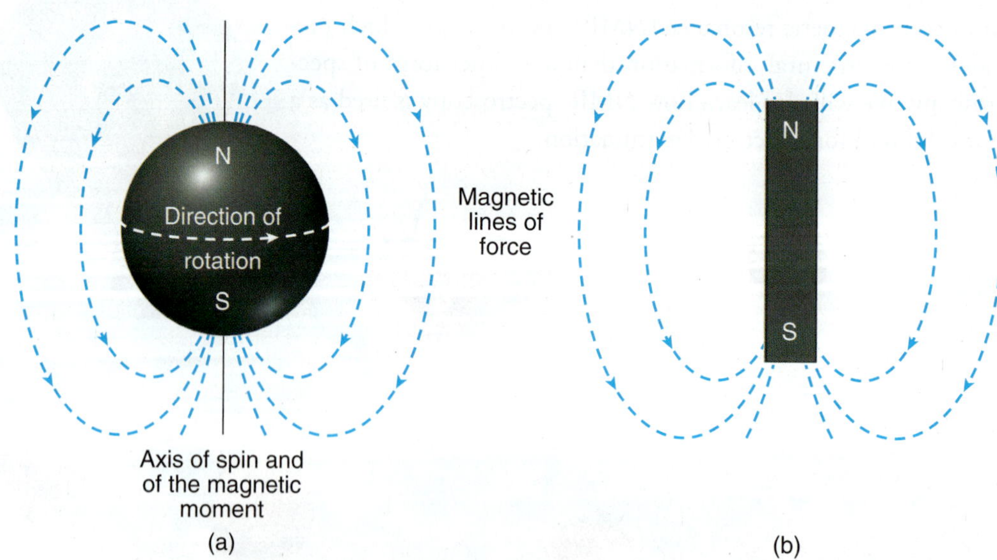


**FIGURE 16.1** (a) The magnetic moment of a spinning proton. (b) The magnetic field of a bar magnet.

protons and an even number of neutrons and therefore does not possess this property. In contrast, the nucleus of $^{13}$C has an odd number of neutrons and therefore exhibits spin. Our study of NMR spectroscopy will begin with $^{1}$H NMR spectroscopy and will conclude with $^{13}$C NMR spectroscopy.

When the nucleus of a hydrogen atom (a proton) is subjected to an external magnetic field, the interaction between the magnetic moment and the magnetic field is quantized, and the magnetic moment must align either with the field or against the field (Figure 16.2). A proton aligned with the field is said to occupy the alpha ($\alpha$) spin state, while a proton aligned against the field is said to

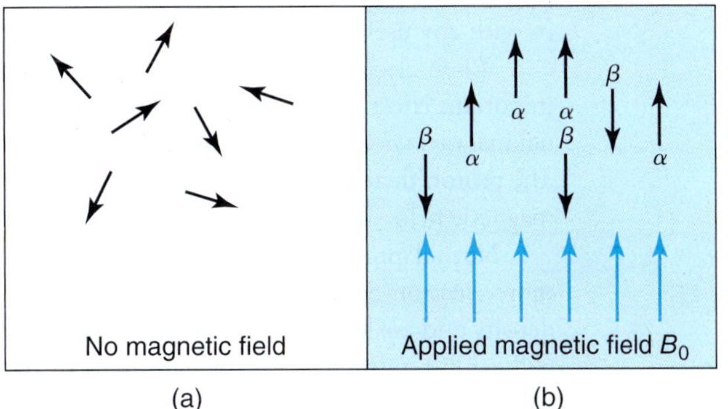


**FIGURE 16.2** The orientation of the magnetic moments of protons in (a) the absence of an external magnetic field or (b) the presence of an external magnetic field.

occupy the beta ($\beta$) spin state. The two spin states are not equivalent in energy, and there is a quantifiable difference in energy ($\Delta E$) between them (Figure 16.3).

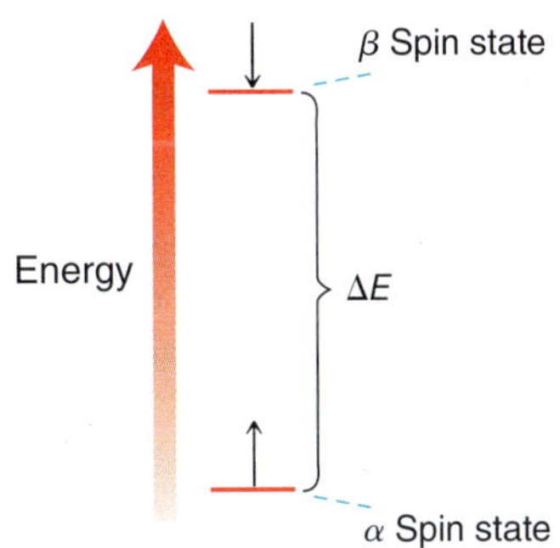


**FIGURE 16.3** The energy gap between the alpha and beta spin states as the result of an applied, external magnetic field.

**WATCH OUT**
Don't confuse the term "resonance" used here with resonance structures of compounds. These are two different concepts.

When a nucleus occupying the $\alpha$ spin state is subjected to electromagnetic radiation, an absorption can take place if the energy of the photon is equivalent to the energy gap between the spin states. The absorption causes the nucleus to *flip* to the $\beta$ spin state, and the nucleus is said to be in *resonance*, thus the term *nuclear magnetic resonance*. When a strong magnetic field is employed, the frequency of radiation typically required for nuclear resonance falls in the radio wave region of the electromagnetic spectrum [called radio frequency (rf) radiation].

At a particular magnetic field strength, we might expect all nuclei to absorb the same frequency of rf radiation. Luckily, this is not the case, as nuclei are surrounded by electrons. In the presence of an external magnetic field, the electron density circulates, which produces a local (induced) magnetic field that opposes the external magnetic field (Figure 16.4). This effect, called **diamagnetism**, was

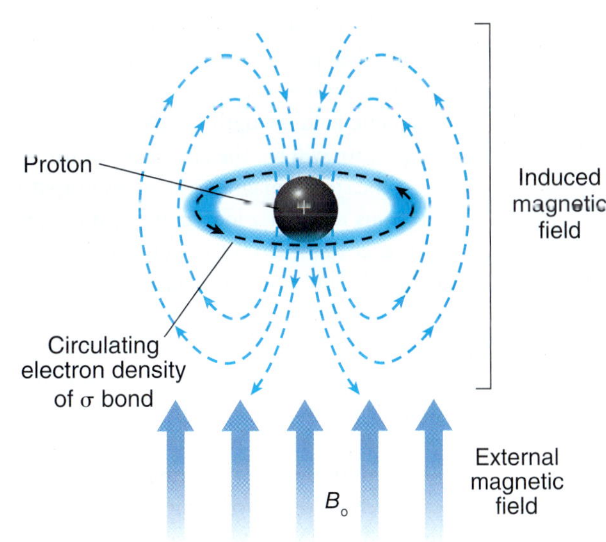


**FIGURE 16.4** The induced magnetic field that is generated as a result of the motion of electron density surrounding the proton.

discussed in the chapter opener. All materials possess diamagnetic properties, because all materials contain electrons. This effect is extremely important for NMR spectroscopy. Without this effect, all protons would absorb the same frequency of rf radiation, and NMR spectroscopy would not provide us with any useful information.

When electron density circulates around a proton, the induced magnetic field has a small but important effect on the proton. The proton is now subjected to two magnetic fields—the strong, external magnetic field and the weak induced magnetic field established by the circulating electron density. The proton therefore experiences a net magnetic field strength that is slightly smaller than the external magnetic field. The proton is said to be **shielded** by the electrons.

Not all protons occupy identical electronic environments. Some protons are surrounded by more electron density and are more shielded, while other protons are surrounded by less electron density and are less shielded, or **deshielded**. As a result, protons in different electronic environments will exhibit a different energy gap between the α and β spin states and will therefore absorb different frequencies of rf radiation. This allows us to probe the electronic environment of each hydrogen atom in a molecule.

## 16.2 Acquiring a $^1$H NMR Spectrum

### Magnetic Field Strength

We have seen that NMR spectroscopy requires a strong magnetic field as well as a source of rf radiation. The magnetic field establishes an energy gap ($\Delta E$) between spin states, which enables the nuclei to absorb rf radiation. The magnitude of this energy gap depends on the strength of the imposed external magnetic field (Figure 16.5). The energy gap increases with increasing magnetic field strength.

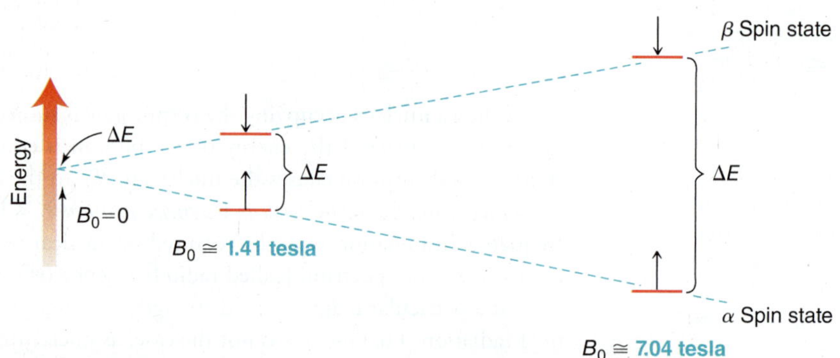


**FIGURE 16.5** The relationship between the strength of the magnetic field and the energy gap between the alpha and beta spin states.

With a magnetic field strength of 1.41 tesla, all of the protons in organic compounds will resonate over a narrow range of frequencies near 60 MHz (60,000,000 Hz). If, however, a magnetic field strength of 7.04 tesla is employed, all of the protons in organic compounds will resonate over a range of frequencies near 300 MHz (300,000,000 Hz). In other words, the strength of the magnetic field determines the range of frequencies that must be used. An NMR spectrometer that utilizes a magnetic field strength of 7.04 tesla is said to have an *operating frequency* of 300 MHz. Later in this chapter we will see the advantage of using an NMR spectrometer with a higher operating frequency.

**BY THE WAY**
Magnetic field strengths are often measured in gauss; 10,000 gauss = 1 tesla.

The strong magnetic fields employed in NMR spectroscopy are produced by passing a current of electrons through a loop composed of superconducting materials. These materials offer virtually zero resistance to the electric current, allowing for large magnetic fields to be produced. The superconducting materials only maintain their properties at extremely low temperatures (just a few degrees above absolute zero) and must therefore be kept in a very low temperature container. Most spectrometers employ three chambers to achieve this low temperature. The innermost chamber contains liquid helium, which has a boiling point of 4.3 K. Surrounding the liquid helium is a chamber containing liquid nitrogen (boiling point 77 K). The third and outermost chamber is a vacuum that minimizes heat transfer from the environment. Any heat developed in the system is transferred from the liquid helium to the liquid nitrogen. In this way, the innermost chamber can be maintained at 4 degrees above absolute zero. This extremely cold environment enables the use of superconductors that generate the large magnetic fields required in NMR spectroscopy.

## NMR Spectrometers

The previous generation of spectrometers, called **continuous-wave (CW) spectrometers**, held the magnetic field constant and slowly swept through a range of rf frequencies, monitoring which frequencies were absorbed. Alternatively, these devices could produce the same result by holding the frequency of rf radiation constant and slowly increasing the magnetic field strength, while monitoring which field strengths produced a signal. Both techniques work, but CW spectrometers are rarely used anymore. They have been replaced by pulsed **Fourier-transform NMR (FT-NMR)**.

In an FT-NMR spectrometer (Figure 16.6), the sample is irradiated with a short pulse that covers the entire range of relevant rf frequencies. All protons are excited simultaneously and then begin to return (or decay) to their original spin states. As each type of proton decays, it releases energy in a particular way, generating an electrical impulse in a receiver coil. The receiver coil records a complex signal, called a **free induction decay** (FID), which is a combination of all of the electrical impulses generated by each type of proton. The FID is then converted into a spectrum via a mathematical technique called a Fourier transform. Since each FID is acquired in 1–2 seconds, it is possible to acquire hundreds of FIDs in just a few minutes, and the FIDs can be averaged. Later in the chapter, we will see that signal averaging is the only practical way to produce a $^{13}$C NMR spectrum.

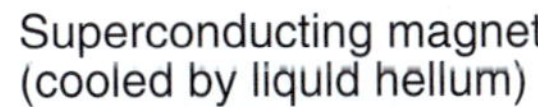


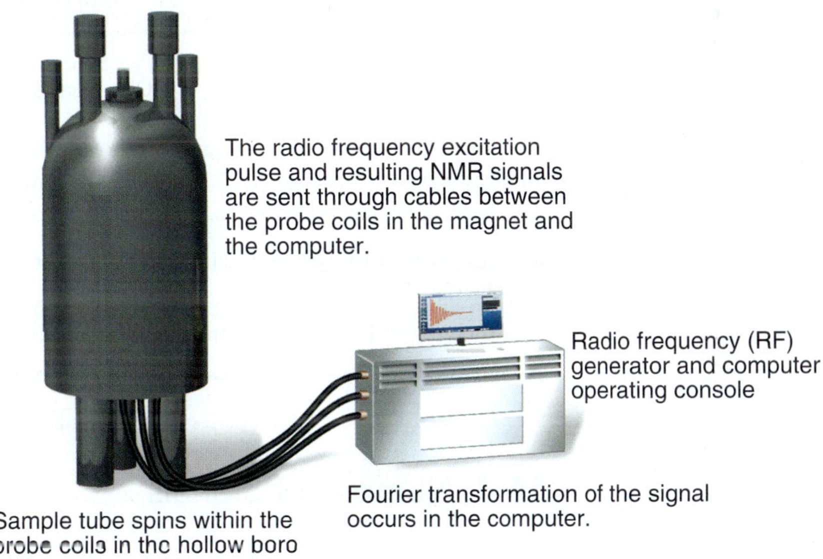


**FIGURE 16.6**
An FT-NMR spectrometer.

## Preparing the Sample

In order to acquire a $^1$H NMR (called a "proton NMR") spectrum of a compound, the compound is usually dissolved in a solvent and placed in a narrow glass tube, which is then inserted into the NMR spectrometer. If the solvent itself has protons, the spectrum will be overwhelmed with signals from the solvent, rendering it unreadable. As a result, solvents without protons must be used. Although there are several solvents that lack protons, such as $CCl_4$, these solvents do not dissolve all compounds. In practice, deuterated solvents are generally used.

Chloroform-d | Methylene chloride-$d_2$ | Acetonitrile-$d_3$ | Benzene-$d_6$ | Deuterium oxide

The nuclei of deuterium also exhibit nuclear spin, and therefore also resonate, but they absorb rf radiation over a very different range of frequencies than do protons. In an NMR spectrometer, a very narrow range of frequencies is used, covering just the frequencies absorbed by protons. For example, a 300-MHz spectrometer will use a pulse that consists only of frequencies between 300,000,000 and 300,005,000 Hz. The frequencies required for deuterium resonance do not fall in this range, so the deuterium atoms are invisible to the NMR spectrometer. All of the solvents shown above are routinely used, and many other deuterated solvents are also available commercially, although quite expensive.

## 16.3 Characteristics of a $^1$H NMR Spectrum

The spectrum produced by $^1$H NMR spectroscopy is generally rich with information that can be interpreted to determine a molecular structure. Consider the following $^1$H NMR spectrum:

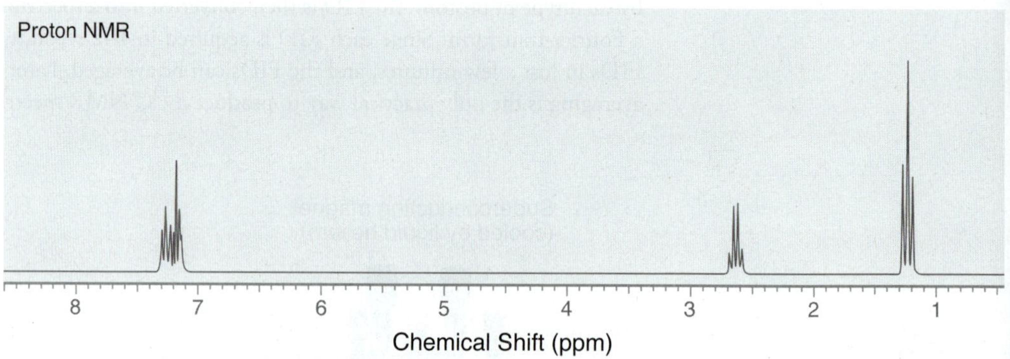


The first valuable piece of information is the number of signals. This spectrum appears to have three different signals. In Section 16.4, we will learn how to interpret this piece of information. In addition, each signal has three important characteristics:

1. The *location* of each signal indicates the electronic environment of the protons giving rise to the signal.
2. The *area* under each signal indicates the number of protons giving rise to the signal.
3. The *shape* of the signal indicates the number of neighboring protons.

We will discuss these characteristics in Sections 16.5–16.7.

## 16.4 Number of Signals

### Chemical Equivalence

**LOOKING BACK**
For a review of symmetry operations, see Section 5.6.

The number of signals in a $^1$H NMR spectrum indicates the number of different kinds of protons (protons in different electronic environments). Protons that occupy identical electronic environments are called **chemically equivalent**, and they will produce only one signal. Two protons are chemically equivalent if they can be interchanged via a symmetry operation—either rotation or reflection. The following examples will illustrate the relationship between symmetry and chemical equivalence. Let's begin by considering the two protons on the middle carbon of propane. Imagine that this molecule is rotated 180° about the following axis while your eyes are closed:

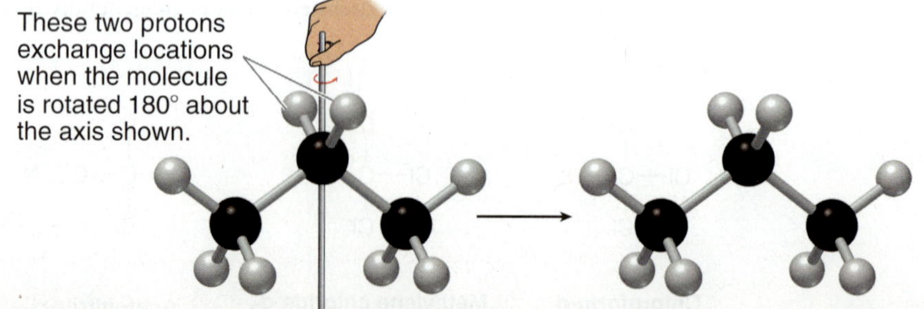

When you open your eyes, you cannot determine whether the molecule was rotated or not. From your point of view, the molecule appears exactly as it did before rotation, and it therefore has an axis of symmetry. The two protons on the middle carbon of propane are interchangeable by rotational symmetry and are therefore said to be **homotopic**. Homotopic protons are chemically equivalent. Below are other examples of homotopic protons.

In each of these examples, the two identified protons are homotopic, because they can be interchanged by rotational symmetry. (Can you identify the axis of rotation in each compound above?) If you are having trouble seeing axes of symmetry, there is a simple method, called the **replacement test**, that will allow you to verify whether or not two protons are homotopic. Draw the compound two times, each time replacing one of the protons with deuterium; for example:

Replacement test

and

Same compound

Then, determine the relationship between the two drawings. If they represent the same compound, then the protons are homotopic.

Now consider the two protons on the alpha carbon of ethanol and imagine that this molecule is rotated 180° while your eyes are closed.

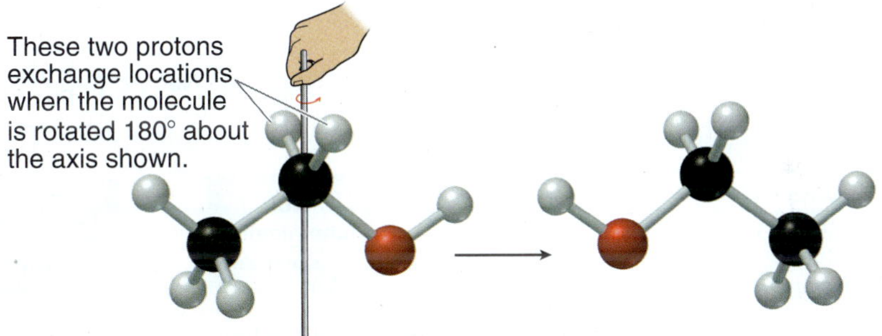


When you open your eyes, you *will* be able to determine that the molecule has been rotated. Specifically, the OH is now on the left side. That is, the two protons on the alpha carbon of ethanol are not interchangeable by rotational symmetry. These protons are therefore not homotopic. Nevertheless, they can be interchanged by reflectional symmetry. Imagine that the molecule is reflected about the plane of the page while your eyes are closed.

These two protons exchange locations when the molecule is reflected through the plane shown.

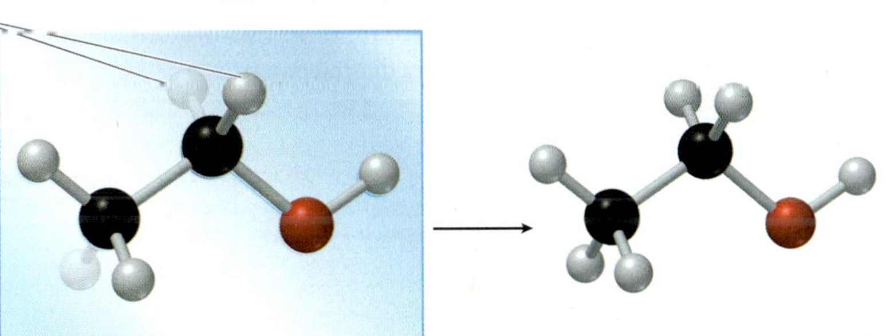

When you open your eyes, you cannot determine whether the molecule was reflected or not. The molecule appears exactly as it did before reflection. In this case, there is a plane of symmetry, and the protons are said to be **enantiotopic**. In the achiral environment of an NMR experiment,

### BY THE WAY

Enantiotopic protons are only chemically equivalent when the solvent is achiral, which is typically the case during routine NMR analysis. However, if a chiral solvent is used, then enantiotopic protons are no longer chemically equivalent and will indeed give different signals.

enantiotopic protons are chemically equivalent, because they are interchangeable by reflectional symmetry. Here are two other examples of enantiotopic protons:

In each of these examples the two highlighted protons are enantiotopic, because they are interchangeable by reflectional symmetry. (Can you see the plane of symmetry in each compound?) If you are having trouble seeing planes of symmetry, you can resort once again to the replacement test. Simply draw the compound twice, each time replacing one of the protons with deuterium. Then determine the relationship between the two drawings. If they are enantiomers, then the protons are enantiotopic.

When determining the relationship between two protons, always look for rotational symmetry first. Figure 16.7 indicates how to determine the relationship between two protons. First determine if there is an axis of symmetry that interchanges the protons. If there is, then the protons are homotopic, whether or not there is a plane of symmetry. If the protons cannot be interchanged by rotation, then look for reflectional symmetry. If there is a plane of symmetry, then the protons are enantiotopic.

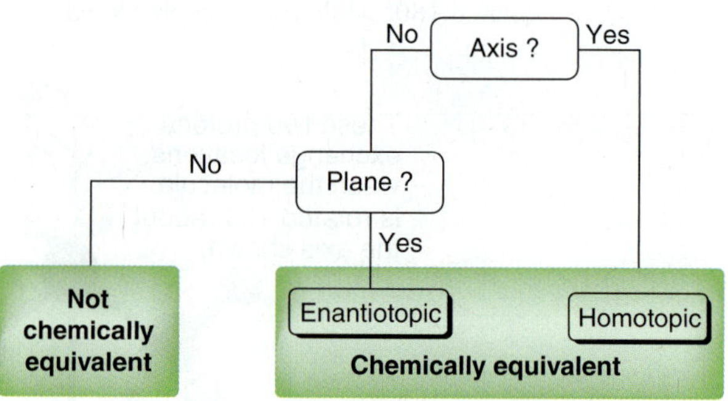


**FIGURE 16.7** A flow chart showing the process for determining whether two protons are chemically equivalent.

If two protons are neither homotopic nor enantiotopic, then they are not chemically equivalent. As an example, consider the protons on C3 of (*R*)-2-butanol.

These protons *cannot* be interchanged by either rotational symmetry or reflectional symmetry. Therefore, these two protons are not chemically equivalent. In this case, the replacement test produces diastereomers.

These protons are therefore said to be **diastereotopic**. They are not chemically equivalent, because they cannot be interchanged by symmetry.

# SKILLBUILDER

## 16.1 DETERMINING THE RELATIONSHIP BETWEEN TWO PROTONS IN A COMPOUND

**LEARN the skill**

Determine whether the two protons shown in red are homotopic, enantiotopic, diastereotopic, or simply not related at all.

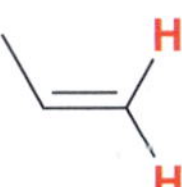

### SOLUTION

Begin by looking for an axis of symmetry. Do not look for any and all axes of symmetry, but rather the one that specifically interchanges the two protons of interest.

**STEP 1**
Determine if the protons are interchangeable by rotational symmetry.

If your eyes were closed during the operation, the position of the methyl group would indicate that the rotation had occurred. Since the molecule is not drawn exactly as it was before rotation, the protons are not interchangeable by rotation. They are not homotopic.

**STEP 2**
If the protons are not interchangeable by rotational symmetry, determine if they are interchangeable by reflectional symmetry.

Next, look for a plane of symmetry. Once again, do not look for any and all planes of symmetry, but rather the one that specifically interchanges the two protons of interest.

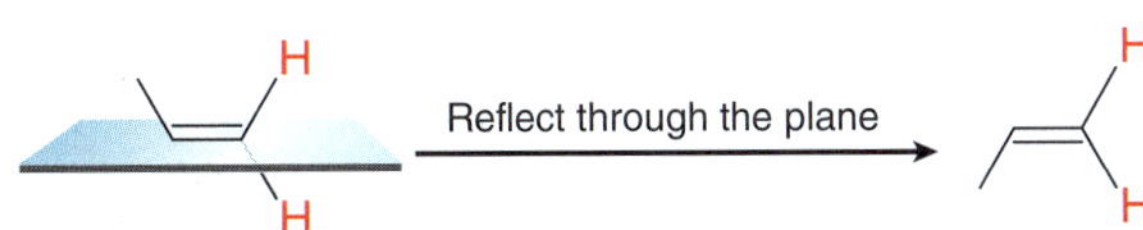


If your eyes were closed during the operation, the position of the methyl group would indicate that a reflection had occurred. Since the molecule is not drawn exactly as it was before reflection, the protons are not interchangeable by reflection, and therefore, they are *not* enantiotopic. In this problem, the protons are neither homotopic nor enantiotopic. Therefore, these protons are not chemically equivalent. They will each produce a different signal in an NMR spectrum. The two protons must either be diastereotopic or simply not related to each other at all. To determine if they are diastereotopic, do the replacement test.

**STEP 3**
If the protons cannot be interchanged via a symmetry operation, then they are not chemically equivalent.

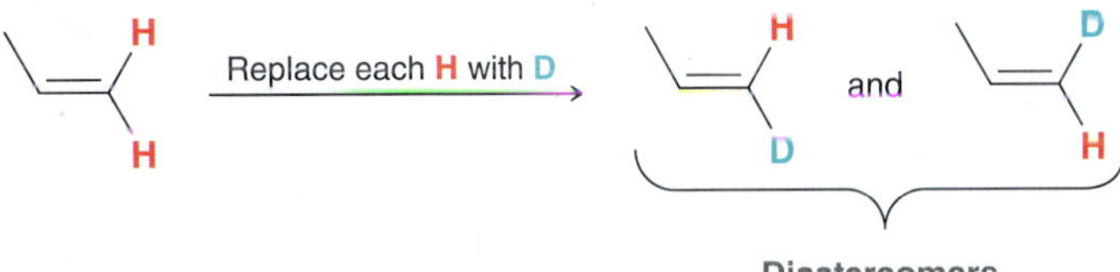


The two compounds are diastereomers, and therefore, the protons are diastereotopic.

**PRACTICE the skill**

**16.1** For each of the following compounds, determine whether the two protons shown in red are homotopic, enantiotopic, or diastereotopic:

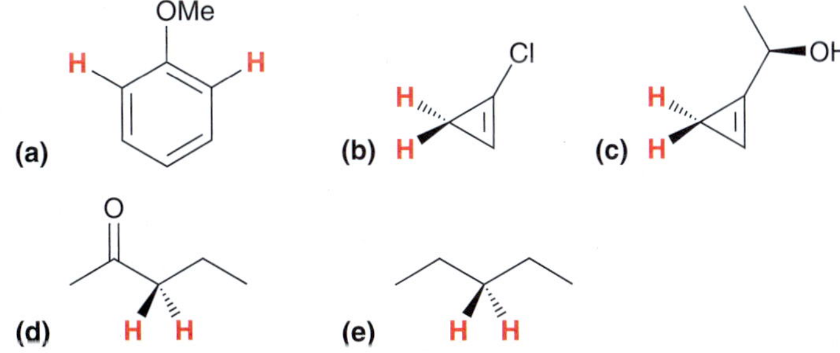

**APPLY the skill**

**16.2** Butane ($C_4H_{10}$) exhibits only two different kinds of protons, shown here in red and blue.

**(a)** Explain why all four protons shown in red are chemically equivalent.

**(b)** Explain why all six protons shown in blue are chemically equivalent.

**(c)** How many different kinds of protons are present in pentane?

**(d)** How many different kinds of protons are present in hexane?

**(e)** How many different kinds of protons are present in 1-chlorohexane?

**16.3** Identify the structure of a compound with molecular formula $C_5H_{12}$ that exhibits only one kind of proton. That is, all 12 protons are chemically equivalent.

need more **PRACTICE?** **Try Problem 16.40**

In order to predict the number of expected signals in the $^1$H NMR spectrum of a compound, it is not necessary to compare all of the protons and drive yourself crazy looking for axes and planes of symmetry. In general, it is possible to determine the number of expected signals for a compound using a few simple rules:

- The three protons of a $CH_3$ group are always chemically equivalent. Example:

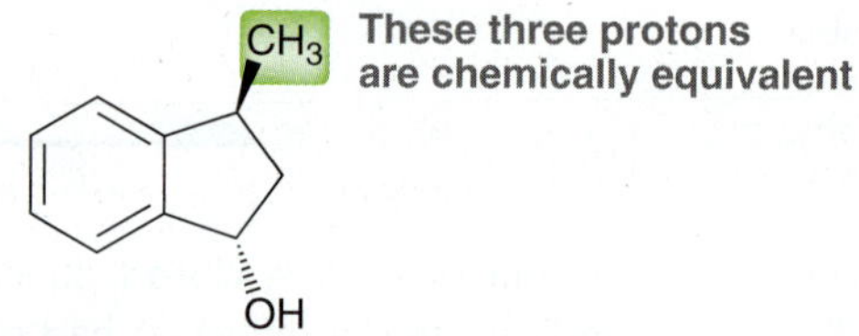


- The two protons of a $CH_2$ group will generally be chemically equivalent if the compound has no chirality centers. If the compound has a chirality center, then the protons of a $CH_2$ group will generally not be chemically equivalent. Examples:

These two protons <u>are</u> chemically equivalent

These two protons are <u>not</u> chemically equivalent

- Two $CH_2$ groups will be equivalent to each other (giving four equivalent protons) if the $CH_2$ groups can be interchanged by either rotation or reflection. Example:

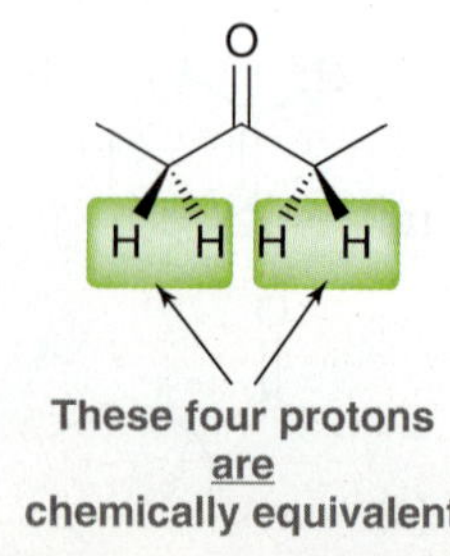

# SKILLBUILDER

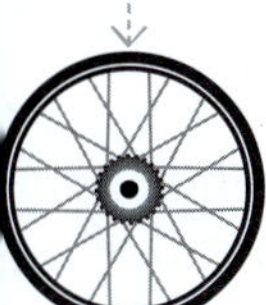

## 16.2 IDENTIFYING THE NUMBER OF EXPECTED SIGNALS IN A $^1$H NMR SPECTRUM

**LEARN the skill**

Identify the number of signals expected in the $^1$H NMR spectrum of the following compound:

OMe

**SOLUTION**

When looking for symmetry, don't be confused by the position of the double bonds in the aromatic ring. Recall that we can draw the following two resonance structures:

OMe ⟷ OMe

Neither resonance structure is more correct than the other. For purposes of looking for symmetry, it will be less confusing to draw the compound like this:

OMe

Begin with the methoxy group ($OCH_3$). These three protons are all connected to one carbon atom and are therefore equivalent. They will produce one signal.

$OCH_3$

Now look at the other remaining $CH_3$ groups in the compound. These two $CH_3$ groups are interchangeable by symmetry, which means that all six protons give rise to one signal.

OMe, $H_3C$, $CH_3$

Now look at each of the $CH_2$ groups. For each $CH_2$ group, the two protons are equivalent because the compound does not have any chirality centers. In addition, the two $CH_2$ groups are interchangeable by symmetry, so all four protons give rise to one signal.

H H OMe H H

Now look at the protons on the aromatic ring. Two of them are interchangeable by symmetry, giving rise to one signal.

OMe, H, H

The remaining aromatic proton gives one more signal, giving a total of five signals.

OMe, H

**PRACTICE the skill** **16.4** Identify the number of signals expected in the $^1H$ NMR spectrum of each of the following compounds:

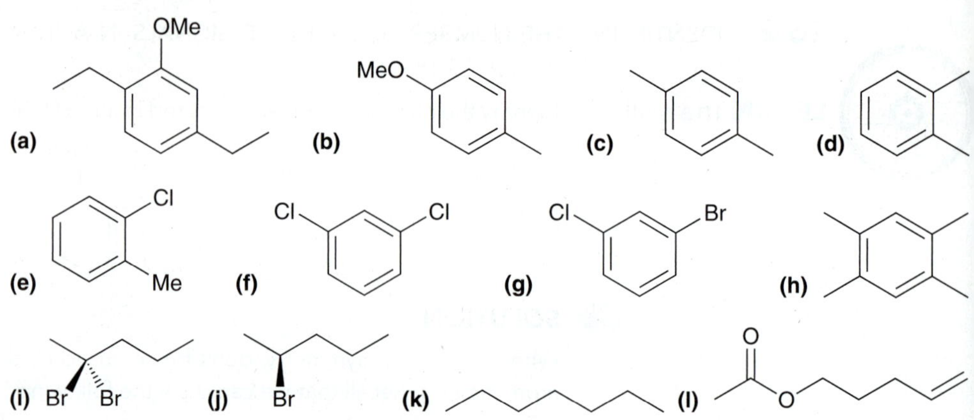

**APPLY the skill**

**16.5** We saw a general rule that the two protons of a $CH_2$ group will be chemically equivalent if there are no chirality centers in the compound. An example of an exception is 3-bromopentane. This compound does not possess a chirality center. Nevertheless, the two highlighted protons are not chemically equivalent. Explain.

**16.6** Identify the structure of a compound with molecular formula $C_9H_{20}$ that exhibits four $CH_2$ groups, all of which are chemically equivalent. How many total signals would you expect in the $^1H$ NMR spectrum of this compound?

need more **PRACTICE?** **Try Problems 16.34, 16.42a, 16.44, 16.48**

## Variable-Temperature NMR

In Chapter 4, we learned that the most stable conformation of cyclohexane is a chair conformation in which six of the protons occupy axial positions and six occupy equatorial positions.

H = axial

H = equatorial

Axial protons (shown in red) occupy a different electronic environment than equatorial protons (shown in blue). That is, axial and equatorial protons are not chemically equivalent, because they cannot be interchanged by a symmetry operation. Consequently, we might expect two signals in the $^1H$ NMR spectrum: one for the six axial protons and one for the six equatorial protons. Yet, the $^1H$ NMR spectrum of cyclohexane only exhibits one signal. Why?

The observation can be explained by considering the rapid rate at which ring flipping occurs at room temperature.

The NMR spectrometer is analogous to a camera with a slow shutter speed that produces a blurred picture when a fast-moving object is photographed. The NMR spectrometer is too slow to acquire a spectrum of a single chair conformation. While the spectrometer is acquiring the spectrum, the ring is flipping rapidly between the two chair conformations, producing a blurry picture. The spectrometer only "sees" the average electronic environment of the protons, and only one signal is observed. However, if the sample of cyclohexane is cooled inside the spectrometer, the ring-flipping process occurs at a slower rate. If the sample is cooled to −100°C, ring flipping occurs at a very slow rate, and separate signals are in fact observed for the axial and equatorial protons. By varying the temperature, it is possible to measure the rates and activation energies of many rapid processes.

## 16.5 Chemical Shift

### Chemical Shift Values

We will now begin exploring the three characteristics of every signal in an NMR spectrum. The first characteristic is the location of the signal, called its **chemical shift (δ)**, which is defined relative to the frequency of absorption of a reference compound, tetramethylsilane (TMS).

$CH_3$
$H_3C-Si-CH_3$
$CH_3$
**TMS**

In practice, deuterated solvents used for NMR spectroscopy typically contain a small amount of TMS, which produces a signal at a lower frequency than the signals produced by most organic compounds. The frequency of each signal is then described as the difference (in hertz) between the resonance frequency of the proton being observed and that of TMS divided by the operating frequency of the spectrometer:

$$\delta = \frac{\text{observed shift from TMS in hertz}}{\text{operating frequency of the instrument in hertz}}$$

For example, when benzene is analyzed using an NMR spectrometer operating at 300 MHz, the protons of benzene absorb a frequency of rf radiation that is 2181 Hz larger than the frequency of absorption of TMS. The chemical shift of these protons is then calculated in the following way:

$$\delta = \frac{2181 \text{ Hz}}{300 \times 10^6 \text{ Hz}} = 7.27 \times 10^{-6}$$

If a 60-MHz spectrometer is used instead, the protons of benzene absorb a frequency of rf radiation that is 436 Hz larger than the frequency of absorption of TMS. The chemical shift of these protons is then calculated in the following way:

$$\delta = \frac{436 \text{ Hz}}{60 \times 10^6 \text{ Hz}} = 7.27 \times 10^{-6}$$

Notice that the chemical shift of the protons is a constant, regardless of the operating frequency of the spectrometer. That is precisely why chemical shifts have been defined in relative terms, rather than absolute terms (hertz). If signals were reported in hertz (the precise frequency of rf radiation absorbed), then the frequency of absorption would be dependent on the strength of the magnetic field and would not be a constant.

In the previous two calculations, notice that the value obtained does not possess any dimensions (hertz divided by hertz gives a dimensionless number). The chemical shift for the protons of benzene is reported as 7.27 ppm (parts per million), which are dimensionless units indicating that signals are reported as a fraction of the operating frequency of the spectrometer. For most organic compounds, the signals produced will fall in a range between 0 and 12 ppm. In rare cases, it is possible to observe a signal occurring at a chemical shift below 0 ppm, which results from a proton that absorbs a lower frequency than TMS. Most protons in organic compounds absorb a higher frequency than TMS, so most chemical shifts that we encounter will be positive numbers.

The left side of an NMR spectrum is described as **downfield**, and the right side of the spectrum is described as **upfield**.

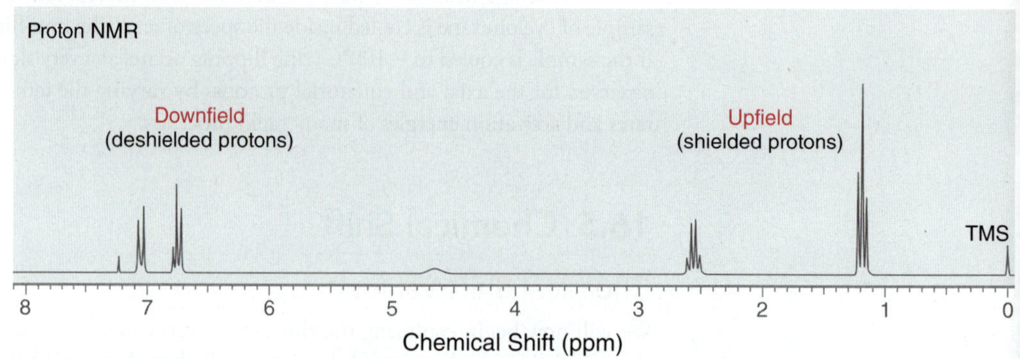


**LOOKING BACK**
Recall that a deshielded proton experiences a stronger magnetic field strength, thereby creating a larger difference in energy between the alpha and beta spin states (Figure 16.5). A shielded proton experiences a weaker magnetic field strength, thereby creating a smaller difference in energy between the alpha and beta spin states.

These terms are historical artifacts that reflect the way spectra were once acquired. As mentioned earlier in this chapter, continuous-wave spectrometers would hold the frequency of rf radiation constant and slowly increase the magnetic field strength while monitoring which field strengths produced a signal. Signals on the left side of the spectrum were produced at lower field strength (downfield), while signals on the right side of the spectrum were produced at higher field strength (upfield). With the advent of FT-NMR spectrometers, spectra are no longer acquired in this fashion. In modern spectrometers, the magnetic field strength is held constant while the sample is irradiated with a short pulse that covers the entire range of relevant rf frequencies. Accordingly, signals on the left side of the spectrum (downfield) are "high-frequency signals" because they result from deshielded protons that absorb higher frequencies of rf radiation. In contrast, signals on the right side of the spectrum (upfield) are "low-frequency signals" because they result from shielded protons that absorb lower frequencies of rf radiation. Despite the advent of modern spectrometers, the older terms "downfield" and "upfield" are still commonly used to describe the position of a signal in an NMR spectrum.

Protons of alkanes produce upfield signals (generally between 1 and 2 ppm). We will now explore some of the effects that can push a signal downfield.

## Inductive Effects

Recall from Section 1.11 that electronegative atoms, such as halogens, withdraw electron density from neighboring atoms (Figure 16.8). This inductive effect causes the protons of the methyl group to be deshielded (surrounded by less electron density), and as a result, the

$H_3C$—X
(X=F, Cl, Br, or I)

**FIGURE 16.8**
The inductive effect of an electronegative atom causes nearby protons to be deshielded.

signal produced by these protons is shifted downfield—that is, the signal appears at a higher chemical shift than the protons of an alkane. The strength of this effect depends on the electronegativity of the halogen. Compare the chemical shifts of the protons in the following compounds:

| $CH_4$ | $CH_3I$ | $CH_3Br$ | $CH_3Cl$ | $CH_3F$ |
|---|---|---|---|---|
| 1.0 ppm | 2.2 ppm | 2.7 ppm | 3.1 ppm | 4.3 ppm |

Fluorine is the most electronegative element and therefore produces the strongest effect. When multiple halogens are present, the effect is additive, as can be seen when comparing the following compounds:

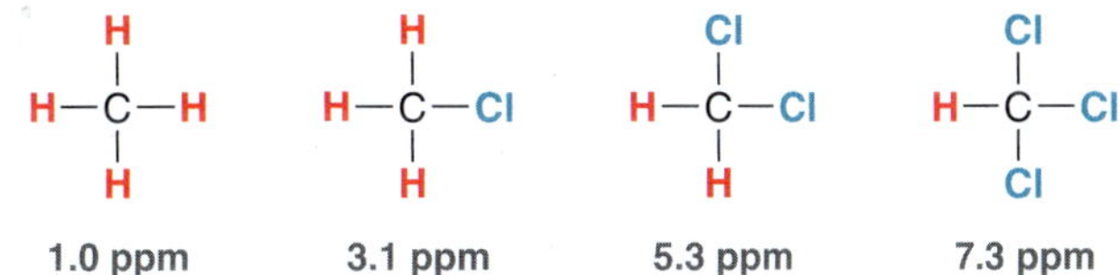


Each chlorine atom adds approximately 2 ppm to the chemical shift of the signal. The inductive effect tapers off drastically with distance, as can be seen by comparing the chemical shifts of the protons in 1-chloropropane.

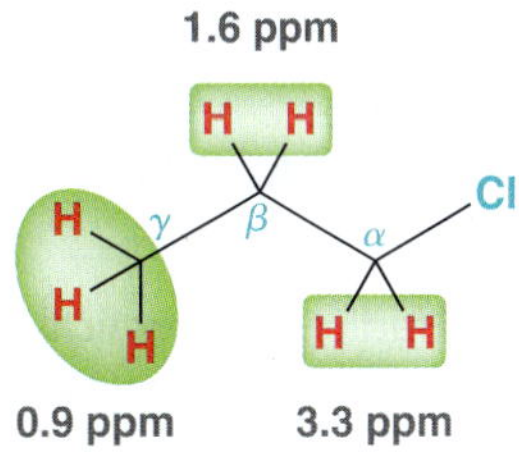


The effect is most significant for the protons at the alpha position. The protons at the beta position are only slightly affected, and the protons at the gamma position are virtually unaffected by the presence of the chlorine atom.

By committing a few numbers to memory, it is possible to predict the chemical shifts for the protons in a wide variety of compounds, including alcohols, ethers, ketones, esters, and carboxylic acids. The following numbers are used as benchmark values:

| Methyl | Methylene | Methine |
|---|---|---|
| ~ 0.9 ppm | ~ 1.2 ppm | ~ 1.7 ppm |

These are the expected chemical shifts for protons that lack neighboring electronegative atoms. In the absence of inductive effects, a methyl group ($CH_3$) will produce a signal near 0.9 ppm, a **methylene group** ($CH_2$) will produce a signal near 1.2 ppm, and a **methine group** (CH) will produce a signal near 1.7 ppm. These benchmark values are then modified by the presence of neighboring functional groups. Table 16.1 shows the effect of a few functional groups on the chemical shifts of alpha protons.

**TABLE 16.1 THE EFFECT OF NEIGHBORING FUNCTIONAL GROUPS ON CHEMICAL SHIFT**

| FUNCTIONAL GROUP | EFFECT ON ALPHA PROTONS | EXAMPLE |
|---|---|---|
| **Oxygen** of an alcohol or ether | +2.5 | Methylene group ($CH_2$) = 1.2 ppm<br>Next to oxygen = +2.5 ppm<br>3.7 ppm<br>Actual chemical shift = 3.7 ppm |
| **Oxygen** of an ester | +3 | Methylene group ($CH_2$) = 1.2 ppm<br>Next to oxygen = +3.0 ppm<br>4.2 ppm<br>Actual chemical shift = 4.1 ppm |
| **Carbonyl group** (C=O)<br>All carbonyl groups, including ketones, aldehydes, esters, etc. | +1 | Methylene group ($CH_2$) = 1.2 ppm<br>Next to carbonyl group = +1.0 ppm<br>2.2 ppm<br>Actual chemical shift = 2.4 ppm |

The effect on beta protons is generally about one-fifth of the effect on the alpha protons. For example, in an alcohol, the presence of an oxygen atom adds +2.5 ppm to the chemical shift of the alpha protons but adds only +0.5 ppm to the beta protons. Similarly, a carbonyl group adds +1 ppm to the chemical shift of the alpha protons but only +0.2 to the beta protons.

The three benchmark values, together with the three values shown in Table 16.1, enable us to predict the chemical shifts for the protons in a wide variety of compounds, as illustrated in the following exercise.

## SKILLBUILDER

### 16.3 PREDICTING CHEMICAL SHIFTS

**LEARN the skill**

Predict the chemical shifts for the signals in the $^1$H NMR spectrum of the following compound:

#### SOLUTION

First determine the total number of expected signals. In this compound, there are five different kinds of protons, giving rise to five distinct signals. For each type of signal, identify whether it represents methyl (0.9 ppm), methylene (1.2 ppm), or methine (1.7 ppm) groups.

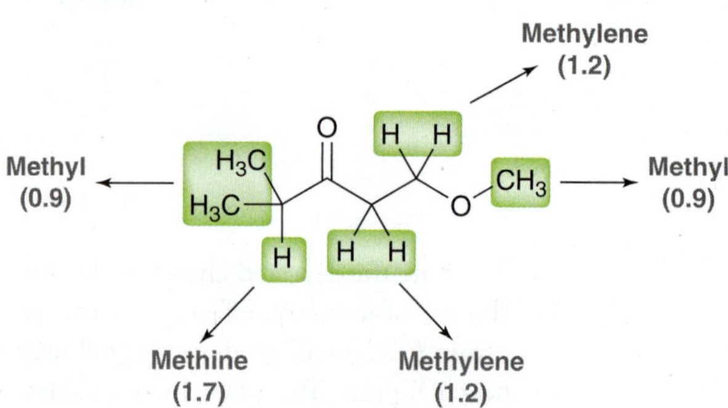


Finally, modify each of these numbers based on proximity to oxygen and the carbonyl group.

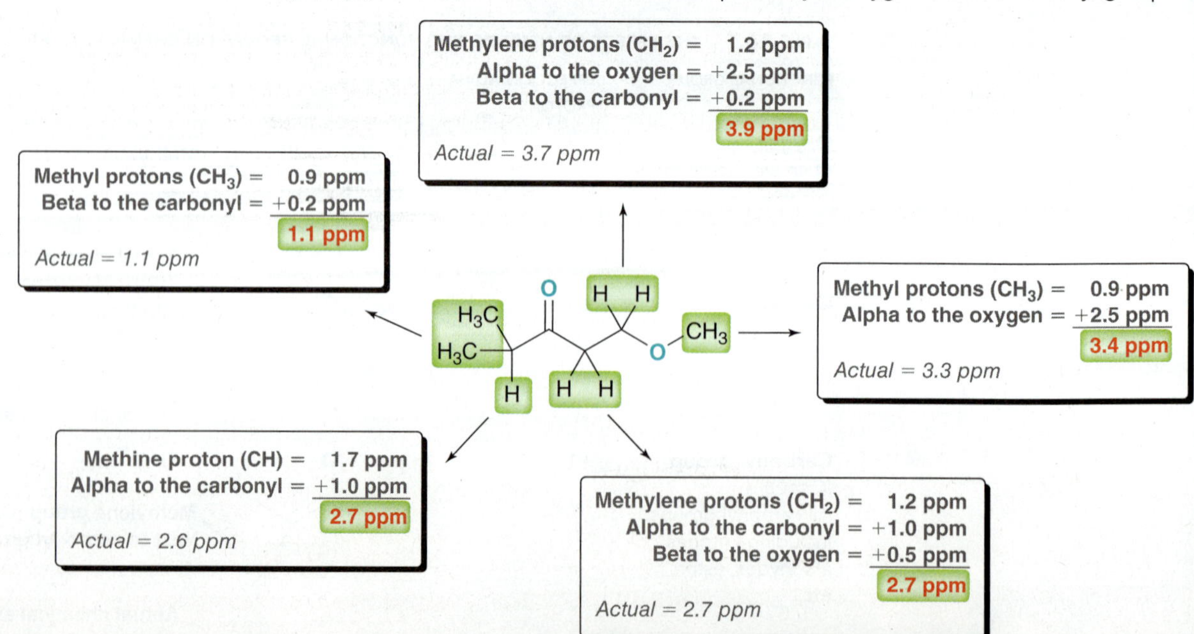

These values are only estimates, and the actual chemical shifts might differ slightly from the predicted values. The actual values are also shown, and they are indeed very close to the estimated values.

**PRACTICE the skill** **16.7** Predict the chemical shifts for the signals in the $^1$H NMR spectrum of each of the following compounds:

(a) 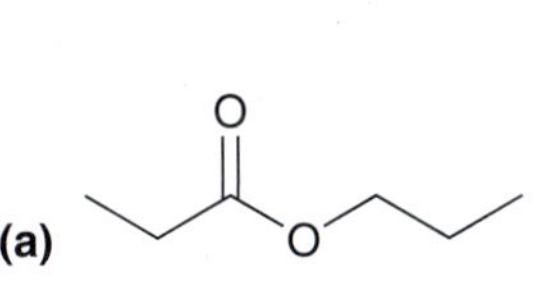 (b) 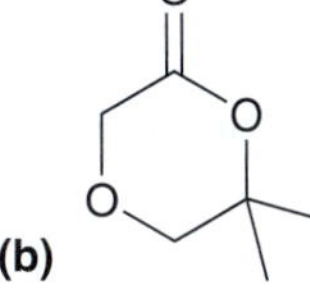 (c)

(d) (e) 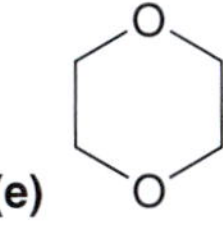

**APPLY the skill** **16.8** A $^1$H NMR spectrum was acquired for each of the following constitutional isomers. Comparison of the spectra reveals that only one of these spectra exhibits a signal between 6 and 7 ppm. Identify the structure that corresponds with this spectrum.

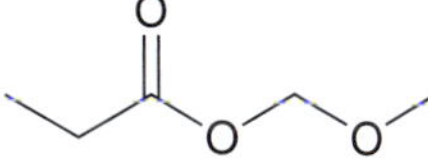 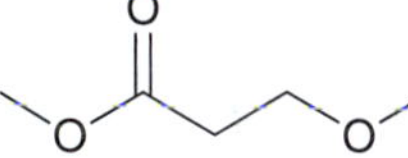

**16.9** A $^1$H NMR spectrum was acquired for each of the following two compounds. One spectrum exhibits two signals downfield of 2.0 ppm, while the other spectrum exhibits only one signal downfield of 2.0 ppm. Match each spectrum with its corresponding compound.

need more **PRACTICE?** **Try Problem 16.42b**

## Anisotropic Effects

The chemical shift of a proton is also sensitive to diamagnetic effects that result from the motion of nearby π electrons. As an example, consider what happens when benzene is placed in a strong magnetic field. The magnetic field causes the π electrons to circulate, and this flow of electrons creates an induced, local magnetic field (Figure 16.9). The result is **diamagnetic anisotropy**,

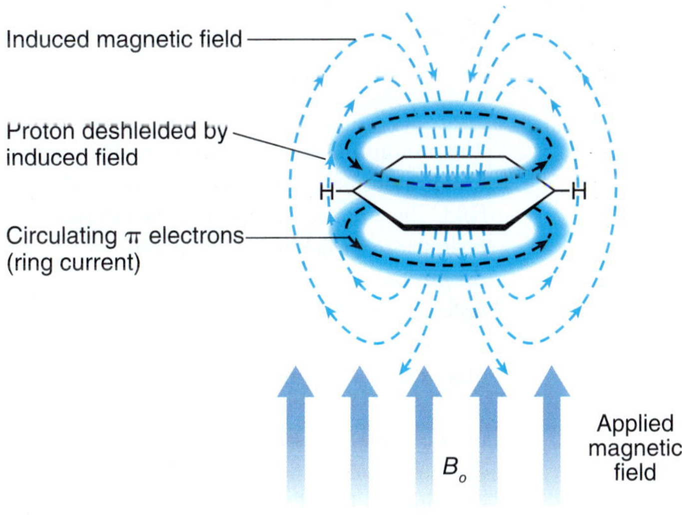


**FIGURE 16.9** The induced magnetic field that is generated as a result of the motion of the π electrons of an aromatic ring.

which means that different regions of space are characterized by different magnetic field strengths. Locations inside the ring are characterized by a local magnetic field that opposes the external field, while locations outside the ring are characterized by a local magnetic field that adds to the external field. The protons connected to the ring are permanently positioned outside of the ring, and as a result, they experience a stronger magnetic field. These protons experience the external magnetic field plus the local magnetic field. The effect is similar to a deshielding effect, and therefore, the protons are shifted downfield. Aromatic protons produce a signal in the neighborhood of 7 ppm (sometimes just above 7, sometimes just below 7) in an NMR spectrum. For example, consider the spectrum of ethylbenzene.

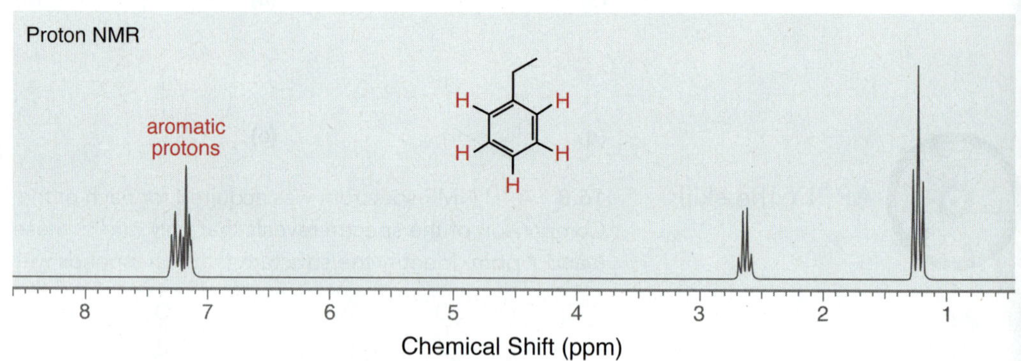


Ethylbenzene has three different kinds of aromatic protons (can you identify them in the structure?), producing a complex signal comprised of three overlapping signals just above 7 ppm. A complex signal around 7 ppm is characteristic of compounds with aromatic protons.

The methylene group ($CH_2$) in ethylbenzene produces a signal at 2.6 ppm, rather than the expected benchmark value of 1.2 ppm. These protons have been shifted downfield because of their position in the local magnetic field. They are not shifted as much as the aromatic protons themselves, because the methylene protons are farther away from the ring, where the induced, local magnetic field is weaker.

A similar effect is observed for [14] annulene.

**[14] Annulene**

The protons outside of the ring produce signals at approximately 8 ppm, but in this case, there are four protons positioned inside the ring, where the local magnetic field opposes the external magnetic field. These protons experience the external magnetic field minus the local magnetic field. The effect is similar to a shielding effect, because the protons experience a weaker magnetic field, and therefore, the protons are shifted upfield. This effect is quite strong, producing a signal at −1 ppm (even further upfield than TMS).

All π bonds exhibit a similar anisotropic effect. That is, π electrons circulate under the influence of an external magnetic field, generating a local magnetic field. For each type of π bond, the precise location of the nearby protons determines their chemical shift. For example, aldehydic protons produce characteristic signals at approximately 10 ppm. Table 16.2 summarizes important chemical shifts. It would be wise to become familiar with these numbers, as they will be required in order to interpret $^{1}H$ NMR spectra.

**TABLE 16.2 CHEMICAL SHIFTS FOR PROTONS IN DIFFERENT ELECTRONIC ENVIRONMENTS**

| TYPE OF PROTON | | CHEMICAL SHIFT (δ) | TYPE OF PROTON | | CHEMICAL SHIFT (δ) |
|---|---|---|---|---|---|
| Methyl | R—$CH_3$ | ~0.9 | Alkyl halide | R—C(H)(R)—X | 2–4 |
| Methylene | $>CH_2$ | ~1.2 | Alcohol | R—O—H | 2–5 |
| Methine | —CH | ~1.7 | Vinylic | | 4.5–6.5 |
| Allylic | | ~2 | Aryl | | 6.5–8 |
| Alkynyl | R—≡—H | ~2.5 | Aldehyde | | ~10 |
| Aromatic methyl | $—CH_3$ | ~2.5 | Carboxylic acid | | ~12 |

## CONCEPTUAL CHECKPOINT

**16.10** For each of the following compounds, identify the expected chemical shift for each type of proton:

(a) (b) (c) (d)

# 16.6 Integration

In the previous section, we learned about the first characteristic of every signal, chemical shift. In this section, we will explore the second characteristic, **integration**, or the area under each signal. This value indicates the number of protons giving rise to the signal. After acquiring a spectrum, the computer calculates the area under each signal and then displays this area as a numerical value placed either above or below the signal.

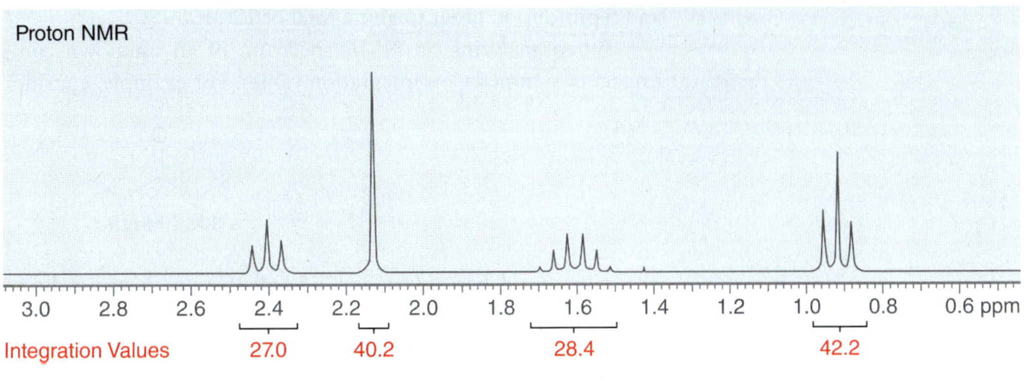

These numbers only have meaning when compared to each other. In order to convert these numbers into useful information, choose the smallest number (27.0 in this case) and then divide all integration values by this number:

$$\frac{27.0}{27.0} = 1 \qquad \frac{40.2}{27.0} = 1.49 \qquad \frac{28.4}{27.0} = 1.05 \qquad \frac{42.2}{27.0} = 1.56$$

These numbers provide the *relative number*, or ratio, of protons giving rise to each signal. This means that a signal with an integration of 1.5 involves one and a half times as many protons as a signal with an integration of 1. In order to arrive at whole numbers (there is no such thing as half a proton), multiply all the numbers by 2, giving the same ratio now expressed in whole numbers, 2:3:2:3. In other words, the signal at 2.4 ppm represents two equivalent protons, and the signal at 2.1 ppm represents three equivalent protons.

Integration is often represented by *step curves*:

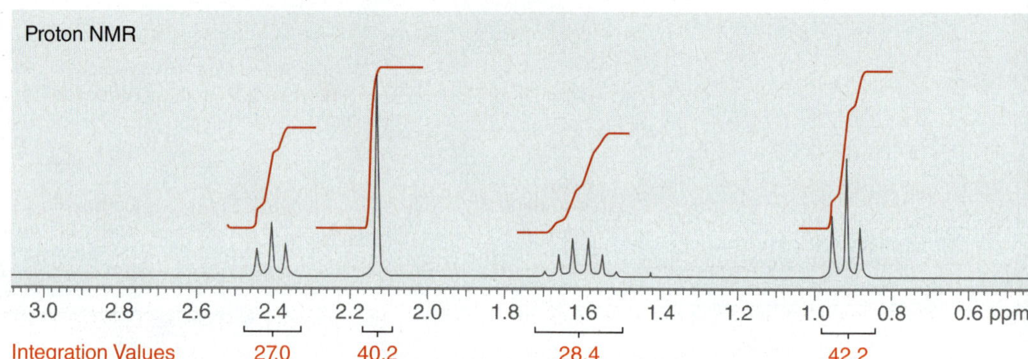


The height of each step curve represents the area under the signal. In this case, a comparison of the heights of the four step curves reveals a ratio of 2:3:2:3.

When interpreting integration values, don't forget that the numbers are only relative. To illustrate this point, consider the structure of *tert*-butyl methyl ether (MTBE).

**MTBE**

MTBE has two kinds of protons (the methyl group and the *tert*-butyl group) and will produce two signals in its [1]H NMR spectrum. The computer analyzes the area under each signal and gives numbers that allow us to calculate a ratio of 1:3. This ratio only indicates the relative number of protons giving rise to each signal, not the exact number of protons. In this case, the exact numbers are 3 (for the methyl group) and 9 (for the *tert*-butyl group). When analyzing the NMR spectrum of an unknown compound, the molecular formula provides extremely useful information because it enables us to determine the exact number of protons giving rise to each signal. If we were analyzing the spectrum of MTBE, the molecular formula ($C_5H_{12}O$) would indicate that the compound has a total of 12 protons. This information then allows us to determine that the ratio of 1:3 must correspond with 3 protons and 9 protons, in order to give a total of 12 protons.

When analyzing an NMR spectrum of an unknown compound, we must also consider the impact of symmetry on integration values. For example, consider the structure of 3-pentanone.

**3-Pentanone**

This compound has only two kinds of protons, because the methylene groups are equivalent to each other, and the methyl groups are equivalent to each other. The [1]H NMR spectrum is therefore expected to exhibit only two signals.

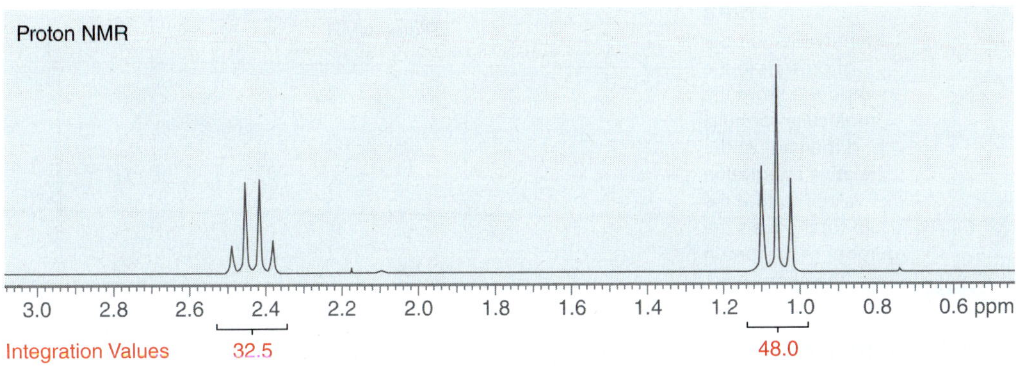


Compare the relative integration values: 32.5 and 48.0. These values give a ratio of 2:3, but again the values 2 and 3 are just relative numbers. They actually represent 4 protons and 6 protons. This can be determined by inspecting the molecular formula ($C_5H_{10}O$), which indicates a total of 10 protons in the compound. Since the ratio of protons is 2:3, this ratio must represent 4 and 6 protons, respectively, in order for the total number of protons to be 10. This analysis indicates that the molecule possesses symmetry.

## SKILLBUILDER

### 16.4 DETERMINING THE NUMBER OF PROTONS GIVING RISE TO A SIGNAL

**LEARN the skill**

A compound with molecular formula $C_5H_{10}O_2$ has the following $^1H$ NMR spectrum:

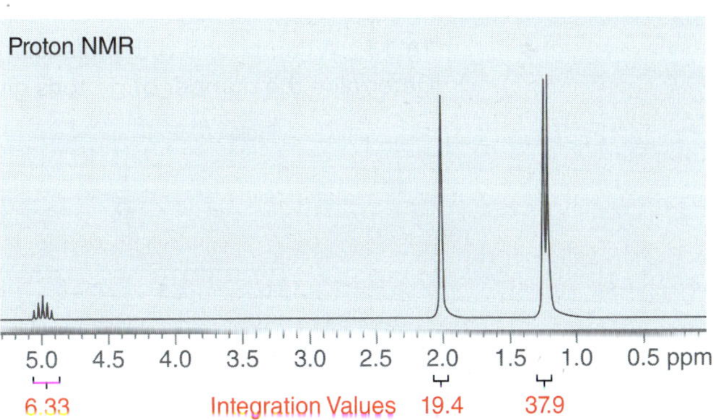


Determine the number of protons giving rise to each signal.

**STEP 1**
Compare the relative integration values and choose the lowest number.

**SOLUTION**

The spectrum exhibits three signals.

Begin by comparing the relative integration values: 6.33, 19.4, and 37.9. Divide each of these three numbers by the smallest number (6.33):

$$\frac{6.33}{6.33} = 1 \qquad \frac{1.94}{6.33} = 3.06 \qquad \frac{37.9}{6.33} = 5.99$$

**STEP 2**
Divide all the integration values by the number from step 1, which gives the ratio of protons.

This gives a ratio of 1:3:6, but these are just relative numbers. To determine the exact number of protons giving rise to each signal, look at the molecular formula, which indicates a total of 10 protons in the compound.

Therefore, the numbers 1:3:6 are not only relative values, but they are also the exact values. Exact integration values can be illustrated in the following way:

**STEP 3**
Identify the number of protons in the compound (from the molecular formula) and then adjust the relative integration values so that the sum total equals the number of protons in the compound.

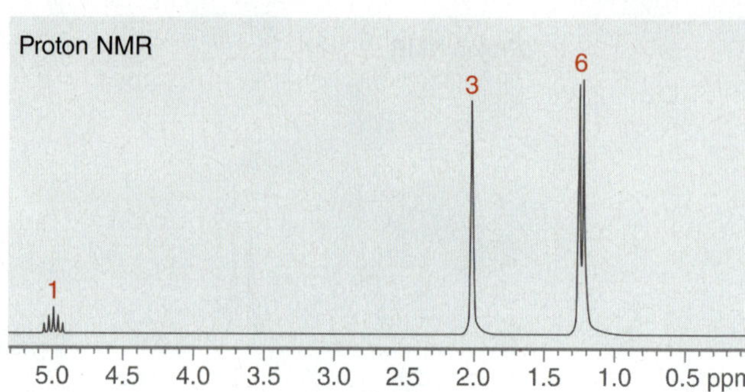


**PRACTICE the skill** **16.11** A compound with molecular formula $C_5H_{10}O_2$ has the following NMR spectrum. Determine the number of protons giving rise to each signal.

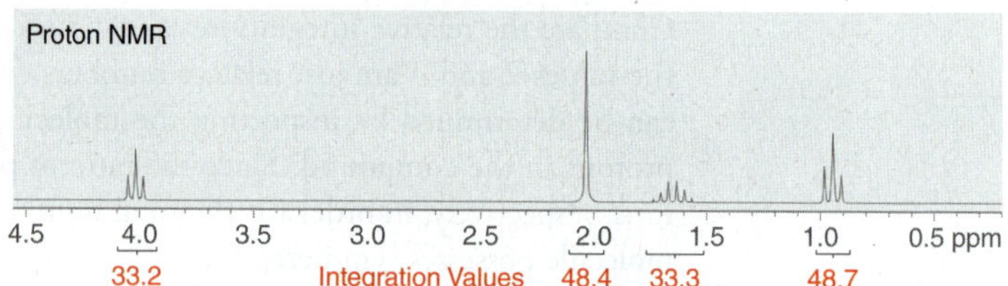


**16.12** A compound with molecular formula $C_{10}H_{10}O$ has the following NMR spectrum. Determine the number of protons giving rise to each signal.

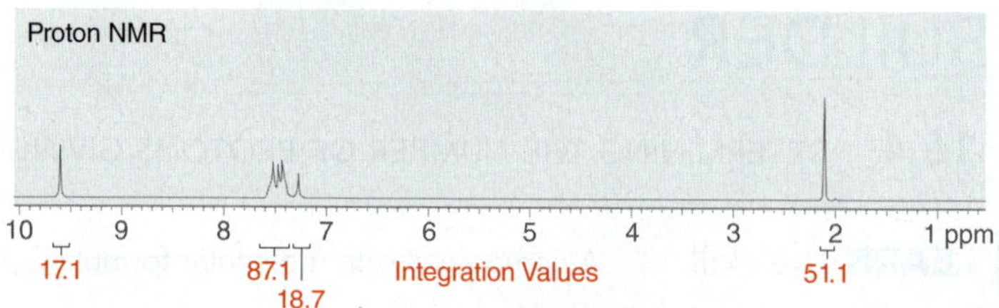


**16.13** A compound with molecular formula $C_4H_6O_2$ has the following NMR spectrum. Determine the number of protons giving rise to each signal.

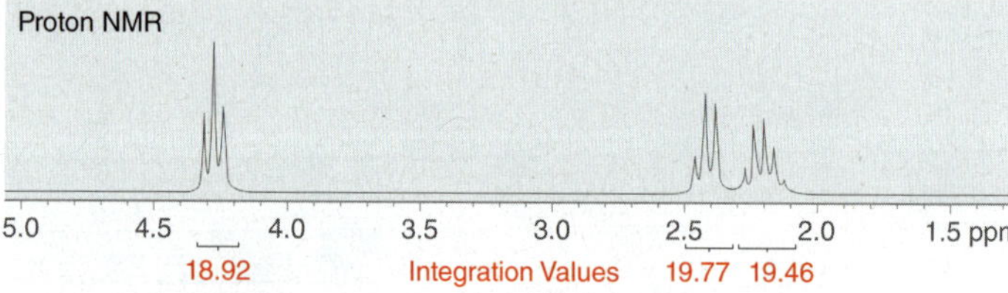

**APPLY the skill** **16.14** The [1]H NMR spectrum of a compound with molecular formula $C_7H_{15}Cl$ exhibits two signals with relative integration 2 : 3. Propose a structure for this compound.

need more **PRACTICE?** **Try Problem 16.54**

## 16.7 Multiplicity

### Coupling

The third, and final, characteristic of each signal is its **multiplicity**, which is defined by the number of peaks in the signal. A **singlet** has one peak, a **doublet** has two peaks, a **triplet** has three peaks, a **quartet** has four peaks, a **quintet** has five peaks, and so on.

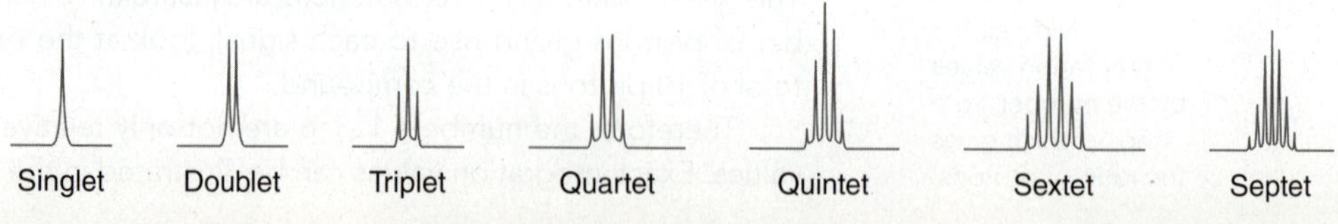

A signal's multiplicity is the result of the magnetic effects of neighboring protons and therefore indicates the number of neighboring protons. To illustrate this concept, consider the following example:

If $H_a$ and $H_b$ are not equivalent to each other, they will produce different signals. Let's focus on the signal produced by $H_a$. We have already seen a variety of factors that will affect the chemical shift of $H_a$, including inductive effects and diamagnetic anisotropy effects. All of these effects modify the magnetic field felt by $H_a$, thereby affecting the resonance frequency of $H_a$. The chemical shift of $H_a$ is also impacted by the presence of $H_b$, because $H_b$ has a magnetic moment that can either be aligned with or against the external magnetic field. $H_b$ is like a tiny magnet, and the chemical shift of $H_a$ is dependent on the alignment of this tiny magnet. In some molecules, $H_b$ will be aligned with the field, while in other molecules, $H_b$ will be aligned against the field. As a result, the chemical shift of $H_a$ in some molecules will be slightly different than the chemical shift of $H_a$ in other molecules, resulting in the appearance of two peaks. In other words, the presence of $H_b$ splits the signal for $H_a$ into a doublet (Figure 16.10).

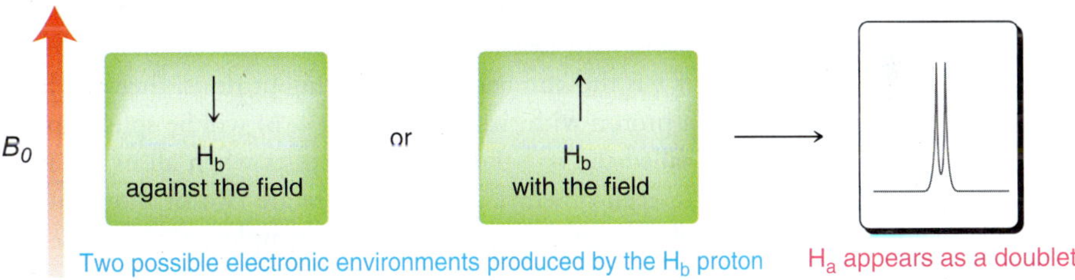


**FIGURE 16.10**
The source of a doublet.

$H_a$ has the same effect on the signal of $H_b$, splitting the signal for $H_b$ into a doublet. This phenomenon is called **spin-spin splitting**, or **coupling**.

Now consider a scenario in which $H_a$ has two neighboring protons.

The chemical shift of $H_a$ is impacted by the presence of both $H_b$ protons, each of which can be aligned either with or against the external field. Once again, each $H_b$ is like a tiny magnet and has an impact on the chemical shift of $H_a$. In each molecule, $H_a$ can find itself in one of three possible electronic environments, resulting in a triplet (Figure 16.11). If each peak of the triplet is separately integrated, a ratio of 1 : 2 : 1 is observed, consistent with statistical expectations.

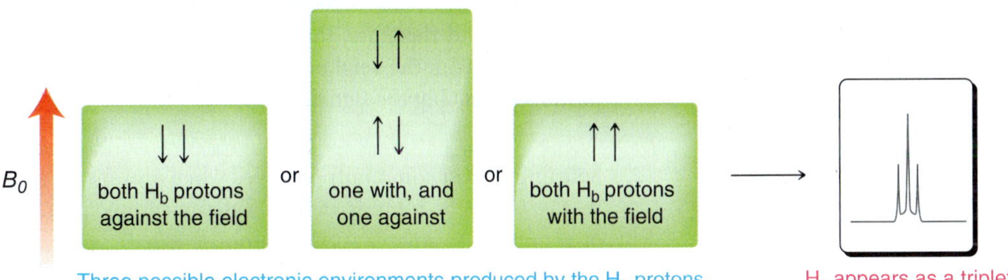


**FIGURE 16.11**
The source of a triplet.

Now consider a scenario in which $H_a$ has three neighbors.

The chemical shift of $H_a$ is impacted by the presence of all three $H_b$ protons, each of which can be aligned either with the field or against the field. Once again, each $H_b$ is like a tiny magnet and has an impact on the chemical shift of $H_a$. In each molecule, $H_a$ can find itself in one of four possible electronic environments, resulting in a quartet (Figure 16.12). If each peak of the quartet is integrated separately, a ratio of 1:3:3:1 is observed, consistent with statistical expectations.

**FIGURE 16.12**
The source of a quartet.

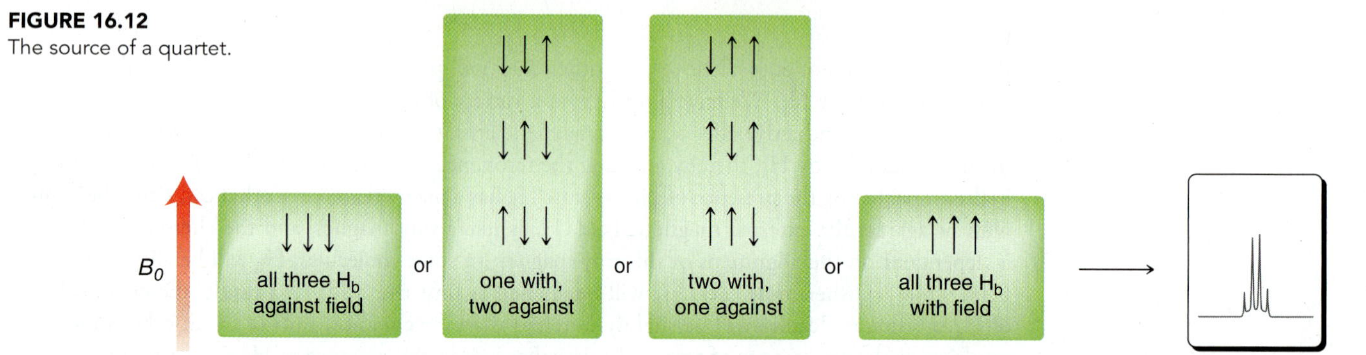


Table 16.3 summarizes the splitting patterns and peak intensities for signals that result from coupling with neighboring protons. When analyzing this information, a pattern emerges. Specifically, if $n$ is the number of neighboring protons, then the multiplicity will be $n + 1$. Extending this rule, a proton with six neighbors ($n = 6$) will be split into a septet (7 peaks, or $n+1$). This observation, called the **$n$ + 1 rule**, only applies when all of the neighboring protons are chemically equivalent to each other. If, however, there are two or more different kinds of neighboring protons, then the observed splitting will be more complex, as we will see at the end of this section.

**TABLE 16.3 MULTIPLICITY INDICATES THE NUMBER OF NEIGHBORING PROTONS**

| NUMBER OF NEIGHBORS | MULTIPLICITY | RELATIVE INTENSITIES OF INDIVIDUAL PEAKS |
|---|---|---|
| 1 | Doublet | 1:1 |
| 2 | Triplet | 1:2:1 |
| 3 | Quartet | 1:3:3:1 |
| 4 | Quintet | 1:4:6:4:1 |
| 5 | Sextet | 1:5:10:10:5:1 |
| 6 | Septet | 1:6:15:20:15:6:1 |

There are two major factors that determine whether or not splitting occurs:

1. Equivalent protons do not split each other. Consider the two methylene groups in 1,2-dichloroethane. All four protons are chemically equivalent, and therefore, they do not split each other. In order for splitting to occur, the neighboring protons must be different than the protons producing the signal.

Four equivalent protons
No splitting

2. Splitting is most commonly observed when protons are separated by either two or three σ bonds; that is, when the protons are either diastereotopic protons on the same carbon atom (geminal) or connected to adjacent carbon atoms (vicinal).

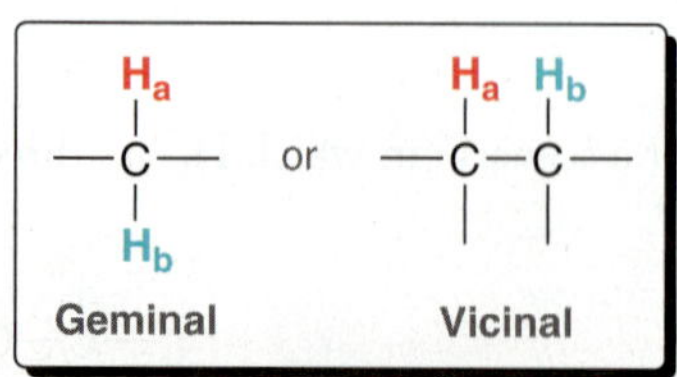


Splitting is observed

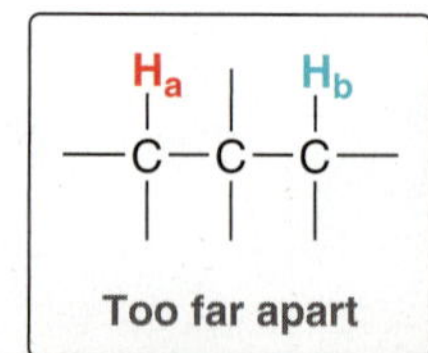


Splitting is generally not observed

When two protons are separated by more than three sigma bonds, splitting is generally not observed. Such long-range splitting is only observed in rigid molecules, such as bicyclic compounds, or in molecules that contain rigid structural groups, such as allylic systems. For purposes of this introductory treatment of NMR spectroscopy, we will avoid examples that exhibit long-range coupling.

# SKILLBUILDER

## 16.5 PREDICTING THE MULTIPLICITY OF A SIGNAL

**LEARN the skill**

Determine the multiplicity of each signal in the expected $^1H$ NMR spectrum of the following compound:

### SOLUTION

Begin by identifying the different kinds of protons. That is, determine the number of expected signals.

**STEP 1**
Identify all of the different kinds of protons.

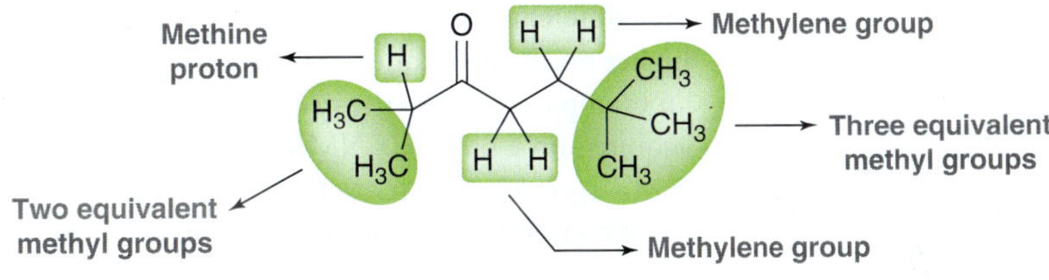


This compound is expected to produce five signals in its $^1H$ NMR spectrum. Now let's analyze each signal using the $n+1$ rule.

**STEP 2**
For each kind of proton, identify the number of neighbors ($n$). The multiplicity will follow the $n + 1$ rule.

Six neighbors
$n + 1 = 7$
Septet

Two neighbors
$n + 1 = 3$
Triplet

0 neighbors
$n + 1 = 1$
Singlet

One neighbor
$n + 1 = 2$
Doublet

Two neighbors
$n + 1 = 3$
Triplet

Notice that the *tert*-butyl group (on the right side of the molecule) appears as a singlet, because the following carbon atom has no protons:

No protons

This quaternary carbon atom is vicinal to each of the three methyl groups connected to it, and as a result, each of the three methyl groups has no neighboring protons. This is characteristic of *tert*-butyl groups.

**PRACTICE the skill** **16.15** For each of the following compounds, determine the multiplicity of each signal in the expected $^1$H NMR spectrum:

(a) (b) (c) (d)

**APPLY the skill** **16.16** Propose the structure for a compound that lacks a methyl group but nevertheless exhibits a quartet in its $^1$H NMR spectrum.

need more **PRACTICE?** **Try Problem 16.37**

## Coupling Constant

When signal splitting occurs, the distance between the individual peaks of a signal is called the **coupling constant**, or ***J* value**, and is measured in hertz. Neighboring protons always split each other with equivalent $J$ values. For example, consider the two kinds of protons in an ethyl group (Figure 16.13). The $H_a$ signal is split into a quartet under the influence of its three neighbors, while the $H_b$ signal is split into a triplet under the influence of its two neighbors. $H_a$ and $H_b$ are said to be coupled to each other. The coupling constant $J_{ab}$ is the same in both signals. $J$ values can range anywhere from 0 to 20 Hz, depending on the type of protons involved, and are independent of the operating frequency of the spectrometer. For example, if $J_{ab}$ is measured to be 7.3 Hz on one spectrometer, the value does not change when the spectrum is acquired on a different spectrometer that uses a stronger magnetic field. As a result, NMR spectrometers with higher operating frequencies produce spectra in which different signals are less likely to overlap. As an example, compare the two spectra of ethyl chloroacetate in Figure 16.14. The first spectrum was acquired on a 60-MHz NMR spectrometer, and the second spectrum was acquired on a 300-MHz NMR spectrometer. In each spectrum, the coupling constant ($J_{ab}$) is approximately 7 Hz. The coupling constant only appears to be larger in the 60-MHz $^1$H NMR spectrum, because each δ unit

X–C($H_a$)($H_a$)–C($H_b$)($H_b$)–$H_b$

$J_{ab}$ $J_{ab}$ $J_{ab}$ — Signal for $H_a$

$J_{ab}$ $J_{ab}$ — Signal for $H_b$

**FIGURE 16.13** The coupling constant ($J_{ab}$) is the same for both signals because these signals are splitting each other.

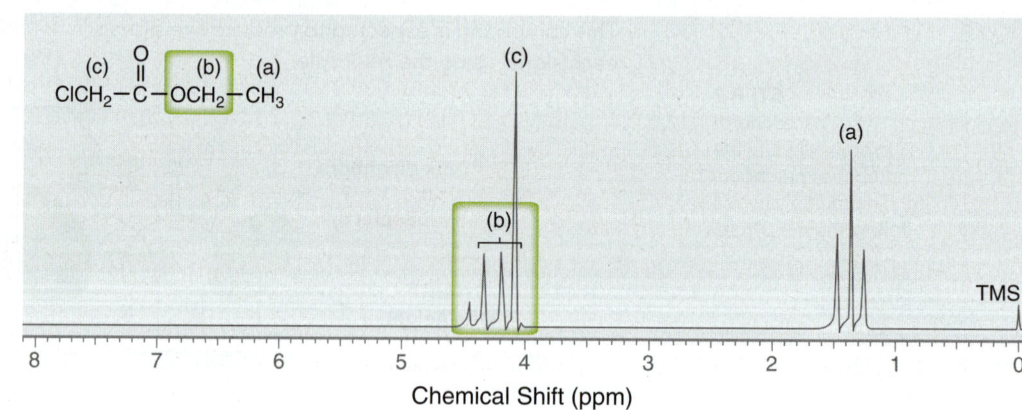


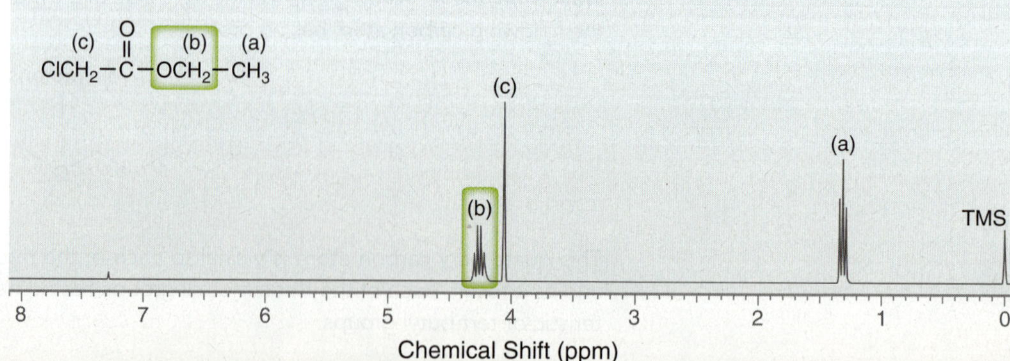


**FIGURE 16.14** The 60-MHz $^1$H NMR spectrum (top) of ethyl chloroacetate and the 300-MHz $^1$H NMR spectrum (bottom) of ethyl chloroacetate. The 300-MHz spectrum avoids overlapping signals.

corresponds with 60 Hz. The distance between each peak (7 Hz) is more than 10% of a δ unit. In contrast, the coupling constant appears much smaller in the 300-MHz $^1$H NMR spectrum, because each δ unit corresponds with 300 Hz, and as a result, the distance between each peak (7 Hz) is only 2% of a δ unit. This example illustrates why spectrometers with higher operating frequencies avoid overlapping signals. For this reason, 60-MHz NMR spectrometers are rarely used for routine research. They have been widely replaced with 300- and 500-MHz instruments.

## Pattern Recognition

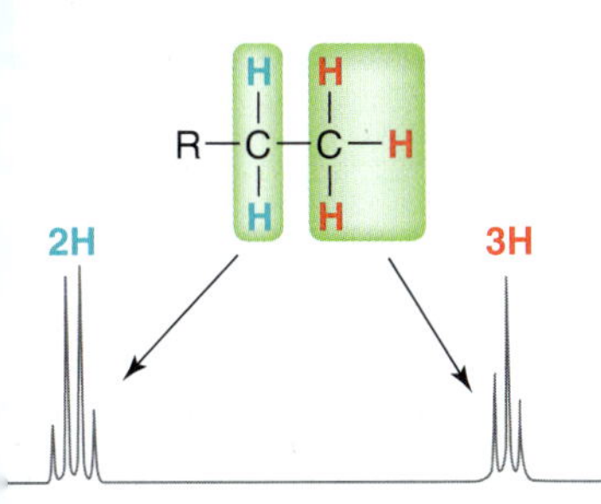


**FIGURE 16.15** The characteristic splitting pattern for an isolated ethyl group.

Specific splitting patterns are commonly seen in $^1$H NMR spectra, and recognizing these patterns allows for a more efficient analysis. Figure 16.15 shows the characteristic splitting pattern of an isolated ethyl group (*isolated* means that the $CH_2$ protons are only being split by the $CH_3$ protons). A compound containing an isolated ethyl group will display a triplet with an integration of 3, upfield from a quartet with an integration of 2 in its $^1$H NMR spectrum. The presence of these signals in a spectrum is strongly suggestive of the presence of an ethyl group in the structure of the compound. Since the two kinds of protons of an ethyl group are splitting each other, the *J* values for the triplet and quartet must be equivalent. If it is clear that the *J* values are different, then the two signals are not an ethyl group.

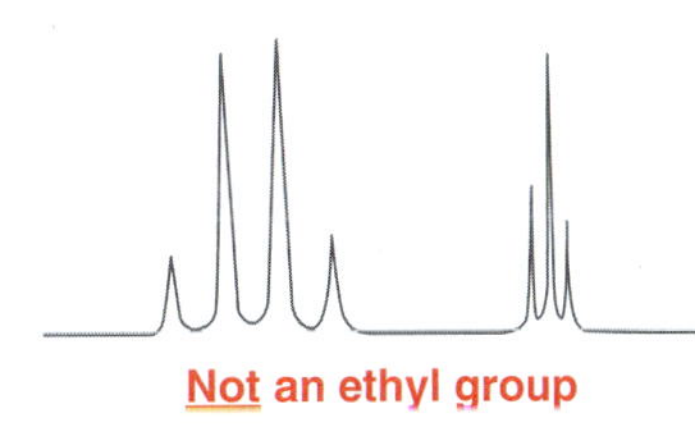


In this image, the quartet has a larger *J* value than the triplet, so these two signals are not splitting each other. In a situation like this, you would need to look at the rest of the NMR spectrum and find the signals that are splitting each other.

Another commonly observed splitting pattern is produced by isopropyl groups (Figure 16.16). A compound containing an isolated isopropyl group will display a doublet with an integration of 6, upfield from a septet (seven peaks) with an integration of 1. The presence of these signals in a $^1$H NMR spectrum is strongly suggestive of the presence of an isopropyl group in the structure of the compound. A septet is usually hard to see, since it is so small (integration of 1), so an enlarged reproduction (inset) of the signal is often displayed above the original signal (as shown in Figure 16.16).

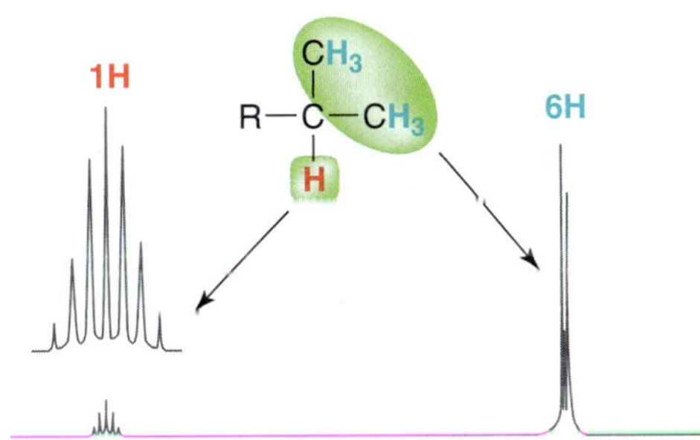


**FIGURE 16.16** The characteristic splitting pattern for an isolated isopropyl group.

Another commonly observed pattern is produced by *tert*-butyl groups (Figure 16.17). A compound containing a *tert*-butyl group will display a singlet with a relative integration of 9. The presence of this signal in a $^1$H NMR spectrum is strongly suggestive of the presence of a *tert*-butyl group in the structure of the compound.

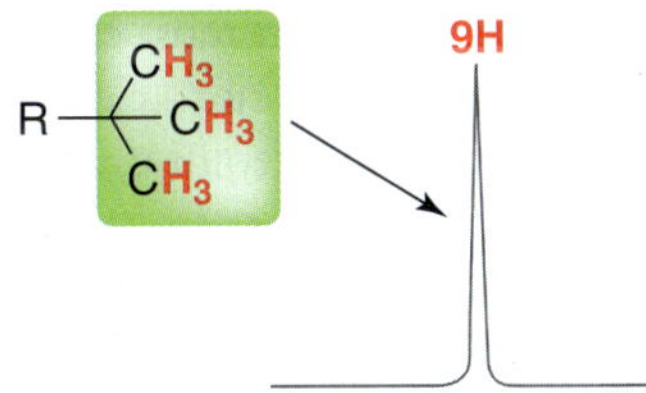


**FIGURE 16.17** The characteristic splitting pattern for a *tert*-butyl group.

Pattern recognition is an important tool when analyzing NMR spectra, so let's quickly review the patterns we have seen (Figure 16.18).

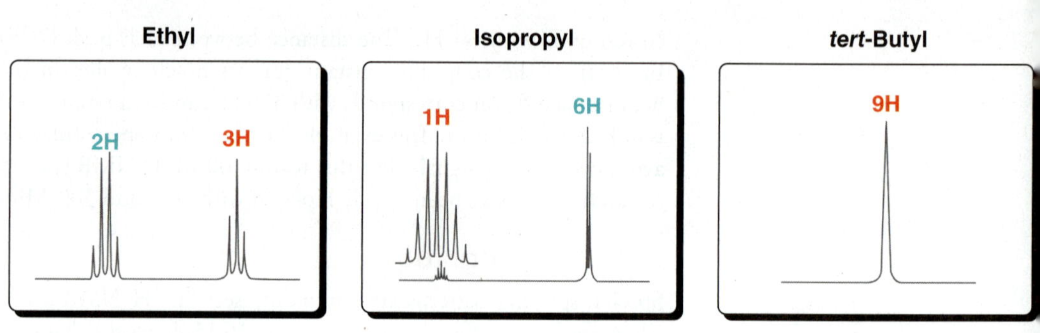


**FIGURE 16.18** The characteristic splitting patterns for ethyl, isopropyl, and *tert*-butyl groups.

## CONCEPTUAL CHECKPOINT

**16.17** Below are NMR spectra of several compounds. Identify whether these compounds are likely to contain ethyl, isopropyl, and/or *tert*-butyl groups:

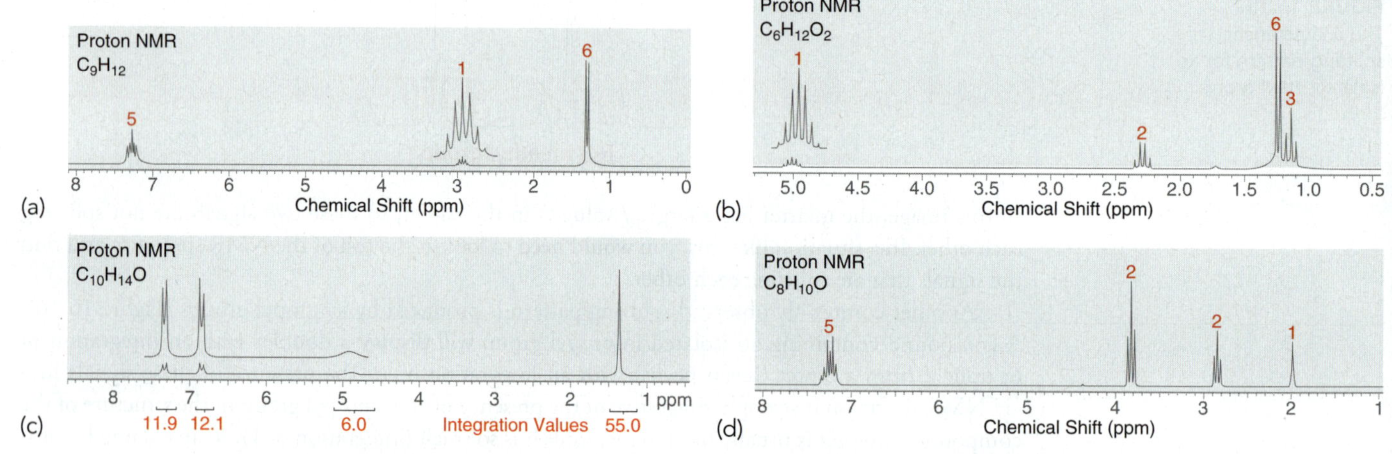


## Complex Splitting

Complex splitting occurs when a proton has two different kinds of neighboring protons. For example:

$$
\begin{array}{c}
\quad H_a \;\; H_b \;\; H_c \\
H_a\!-\!\overset{|}{\underset{|}{C}}\!-\!\overset{|}{\underset{|}{C}}\!-\!\overset{|}{\underset{|}{C}}\!-\!X \\
\quad H_a \;\; H_b \;\; H_c
\end{array}
$$

Consider the expected splitting pattern for $H_b$ in this example. The signal for $H_b$ is being split into a quartet because of the nearby $H_a$ protons, and it is being split into a triplet because of the nearby $H_c$ protons. The signal will therefore be comprised of 12 peaks ($4 \times 3$). The appearance of the signal will depend greatly on the $J$ values. If $J_{ab}$ is much greater than $J_{bc}$, then the signal will appear as a quartet of triplets. This is illustrated in the splitting tree shown in Figure 16.19.

If, however, $J_{bc}$ is much greater than $J_{ab}$, then the signal will appear as a triplet of quartets (Figure 16.20). In most cases, the $J$ values will be fairly similar, and we will observe neither a clean quartet of triplets nor a clean triplet of quartets. More often, several of the peaks will overlap, producing a signal that requires a more detailed analysis and is simply called a **multiplet**.

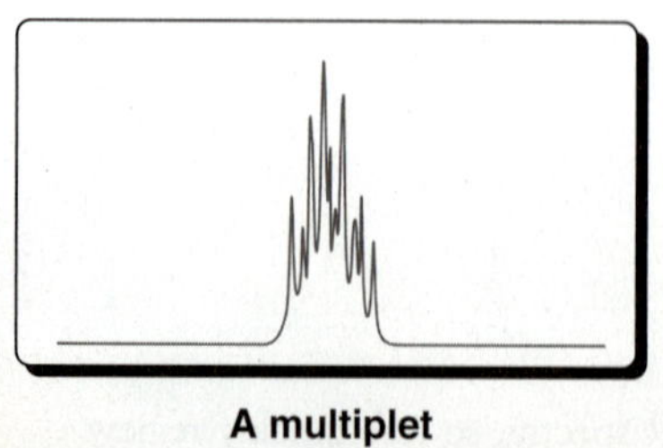

**A multiplet**

In some cases, $J_{ab}$ and $J_{ac}$ will be almost identical. For example, consider the $^1$H NMR spectrum of 1-nitropropane (Figure 16.21). Look carefully at the splitting pattern of the $H_b$ protons (at approximately 2 ppm). This signal looks like a sextet, because $J_{ab}$ and $J_{bc}$ are so close in value. In such a case, it appears "as if" there are five equivalent neighbors, even though all five protons are not equivalent.

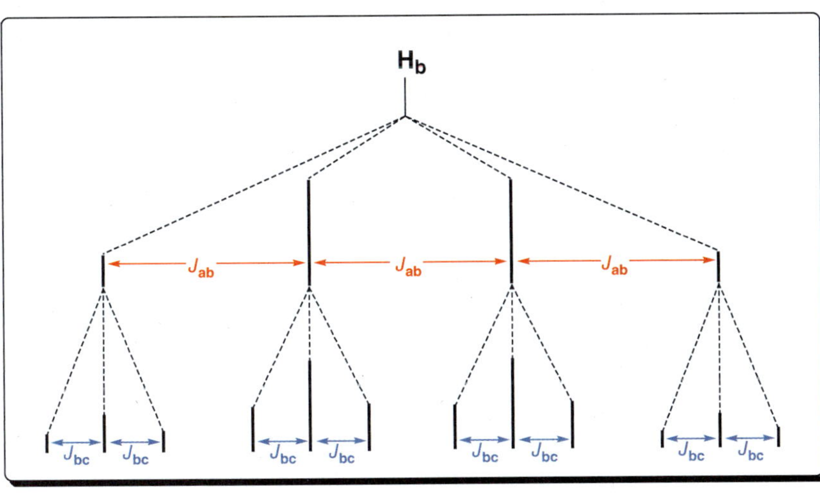


**FIGURE 16.19**
A splitting tree showing how a quartet of triplets is formed.

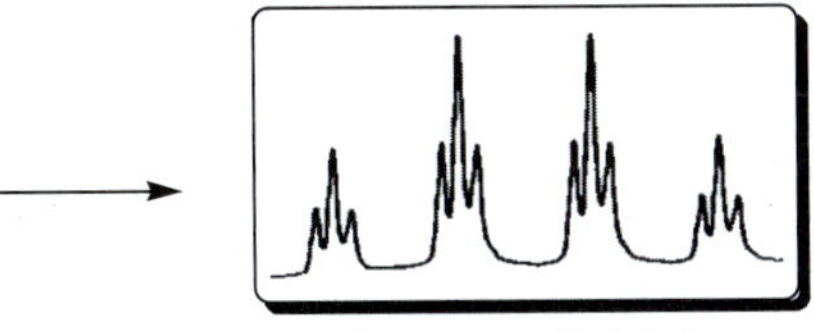

A quartet of triplets

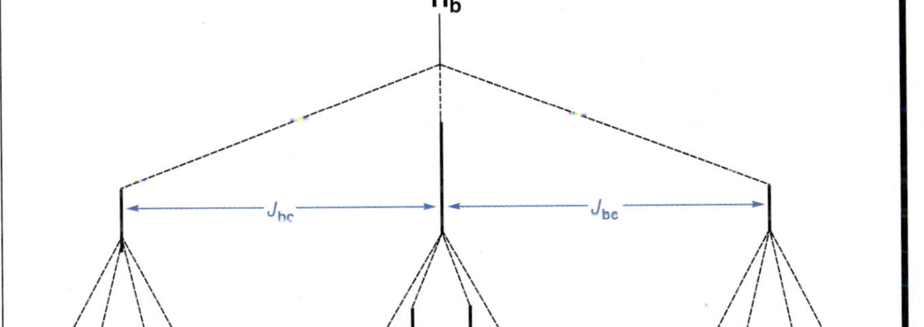


**FIGURE 16.20**
A splitting tree showing how a triplet of quartets is formed.

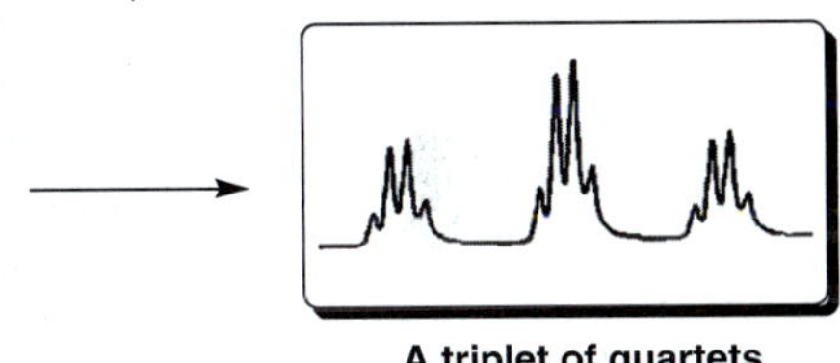

A triplet of quartets

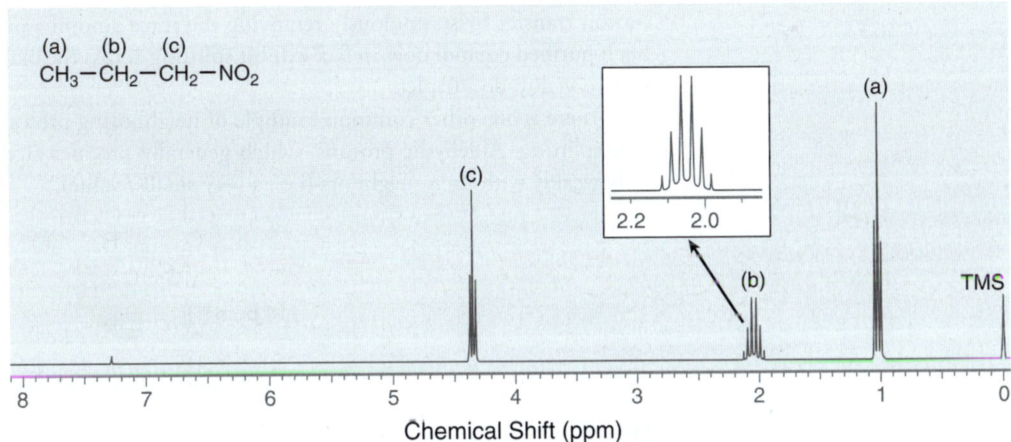


**FIGURE 16.21**
The [1]H NMR spectrum of 1-nitropropane, showing an "apparent" sextet for $H_b$.

## CONCEPTUAL CHECKPOINT

**16.18** Each of the three vinylic protons of styrene is split by the other two, and the $J$ values are found to be $J_{ab} = 11$ Hz, $J_{ac} = 17$ Hz, and $J_{bc} = 1$ Hz. Using this information, draw the expected splitting pattern for each of the three signals ($H_a$, $H_b$, and $H_c$).

Ha
Hb
Hc

Styrene

## Protons That Lack Observable Coupling Constants

We have seen that some protons can produce signals that exhibit complex splitting. In contrast, we will now explore cases in which a proton can produce a signal for which splitting is not observed despite the presence of neighboring protons. Consider the $^{1}$H NMR spectrum of ethanol:

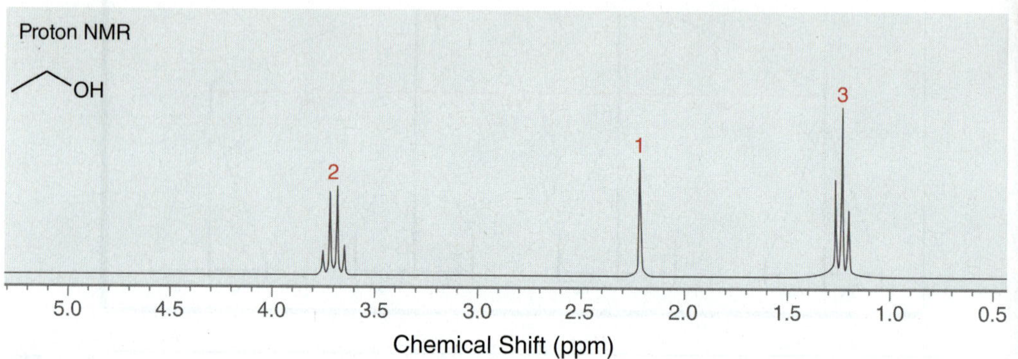


As expected, the spectrum exhibits the characteristic signals of an ethyl group. In addition, another signal is observed at 2.2 ppm, representing the hydroxyl proton (OH). Hydroxyl protons typically produce a signal between 2 and 5 ppm, and it is often difficult to predict exactly where that signal will appear. In this NMR spectrum notice that the hydroxyl proton is not split into a triplet from the neighboring methylene group. Generally, no splitting is observed across the oxygen atom of an alcohol, because proton exchange is a very rapid process that is catalyzed by trace amounts of acid or base.

Hydroxyl protons are said to be **labile**, because of the rapid rate at which they are exchanged. This proton transfer process occurs at a faster rate than the timescale of an NMR spectrometer, producing a blurring effect that averages out any possible splitting effect. It is possible to slow down the rate of proton transfer by scrupulously removing the trace amounts of acid and base dissolved in ethanol. Such purified ethanol does in fact exhibit splitting across the oxygen atom, and the signal at 2.2 ppm is observed to be a triplet.

There is one other common example of neighboring protons that often do not produce observable splitting. Aldehydic protons, which generally produce signals near 10 ppm, will often couple only weakly with their neighbors (i.e., a very small *J* value).

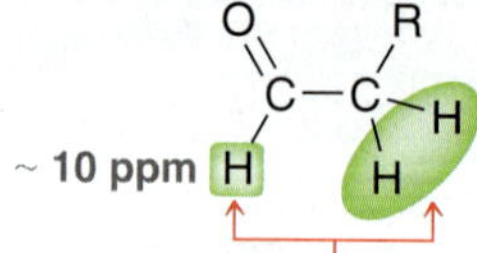


**This *J* value is often very small**

Depending on the size of the *J* value, splitting may or may not be noticeable. If the *J* value is too small, then the signal near 10 ppm will appear to be a singlet, despite the presence of neighboring protons.

# 16.8 Drawing the Expected $^{1}$H NMR Spectrum of a Compound

In the previous sections, we explored the three characteristics of every signal (chemical shift, multiplicity, and integration). In this section, we will apply the concepts and skills developed in the previous sections, and we will practice drawing the expected $^{1}$H NMR spectrum of a compound. The following exercise illustrates the procedure.

# SKILLBUILDER

## 16.6 DRAWING THE EXPECTED ¹H NMR SPECTRUM OF A COMPOUND

**LEARN the skill** Draw the expected ¹H NMR spectrum of isopropyl acetate.

**SOLUTION**

Begin by determining the number of signals:

**STEP 1**
Determine the number of signals.

**STEPS 2–4**
Predict the chemical shift, integration, and multiplicity of each signal.

**STEP 5**
Draw each signal.

This compound is expected to produce three signals in its ¹H NMR spectrum. For each signal, analyze all three characteristics methodically. Let's begin with the methyl group on the left side of the molecule. A methyl group is expected to produce a signal at 0.9 ppm, and the neighboring carbonyl group adds +1, so we expect the signal to appear at approximately 1.9 ppm. The integration should be 3, because there are three protons ($I = 3H$). The multiplicity should be a singlet, because there are no neighbors.

Now consider the signal of the methine proton on the right. The benchmark value for a methine proton is 1.7 ppm, and the neighboring oxygen atom adds +3, so we expect the signal to appear at 4.7 ppm. The integration should be 1, because there is one proton ($I = 1H$). The multiplicity should be a septet, because there are six neighbors.

The last signal is from the two methyl groups on the right. Methyl groups have a benchmark value of 0.9 ppm, and the distant oxygen atom adds +0.6, so we expect a signal at approximately 1.5 ppm. The integration should be 6 because there are six protons ($I = 6H$), and the multiplicity should be a doublet because there is only one neighbor. This information enables us to draw the expected ¹H NMR spectrum. On a separate piece of paper, try to draw the spectrum. Then compare your drawn spectrum to the actual spectrum below.

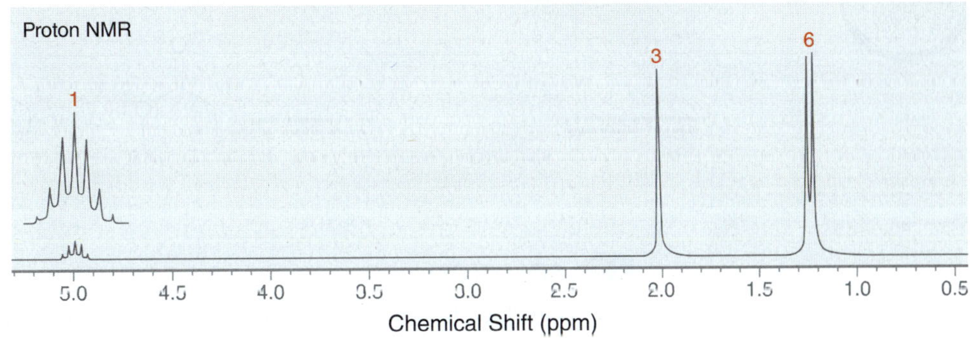


Note that our predicted chemical shifts are only estimates and should be treated as such. The actual chemical shifts will usually be fairly close to the predicted value.

**PRACTICE the skill** **16.19** Draw the expected ¹H NMR spectrum for each of the following compounds:

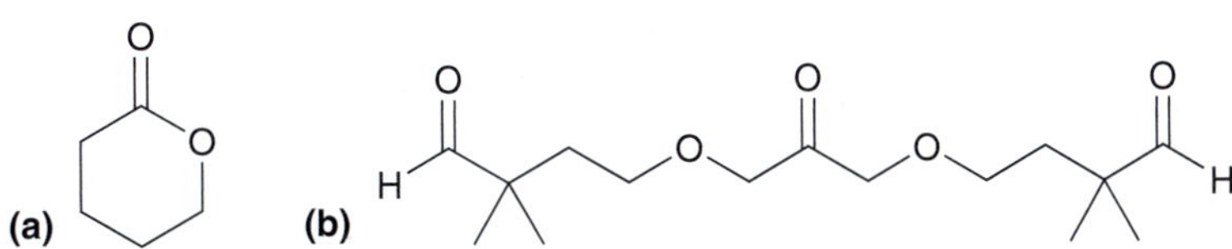

**APPLY the skill**

**16.20** The $^1H$ NMR spectrum of a compound with molecular formula $C_7H_{14}O_3$ exhibits only three signals, and all three signals appear downfield of 2 ppm on the spectrum. Propose a structure for this compound.

need more **PRACTICE?** **Try Problem 16.41**

## 16.9 Using $^1H$ NMR Spectroscopy to Distinguish between Compounds

NMR spectroscopy is a powerful tool for distinguishing compounds from each other. For example, consider the following three constitutional isomers, which produce different numbers of signals in their $^1H$ NMR spectra:

**Three signals** **Four signals** **Two signals**

These examples illustrate how similar compounds can have vastly different NMR spectra. Even compounds that produce the same number of signals can often be easily distinguished with NMR spectroscopy. You just need to identify one signal that would be different in the two spectra. The following exercise illustrates this point.

### 16.7 USING $^1H$ NMR SPECTROSCOPY TO DISTINGUISH BETWEEN COMPOUNDS

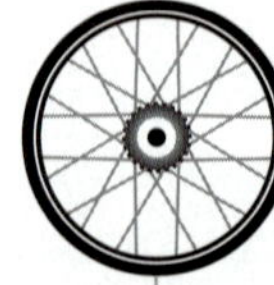

**LEARN the skill**

How would you use $^1H$ NMR spectroscopy to distinguish between the following compounds?

**SOLUTION**

First look to see if the two compounds will produce a different number of signals. For each of the structures, we expect four signals.

**STEP 1**
Identify the number of signals that each compound will produce.

Nevertheless, there are many other ways to distinguish these two compounds. In order to see all of the possible ways, let's methodically analyze all three characteristics for each signal:

**STEP 2**
Determine the expected chemical shift, integration, and multiplicity of each signal in both compounds.

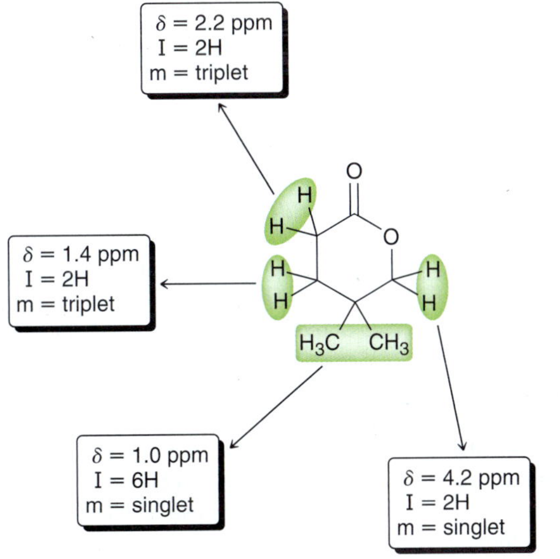


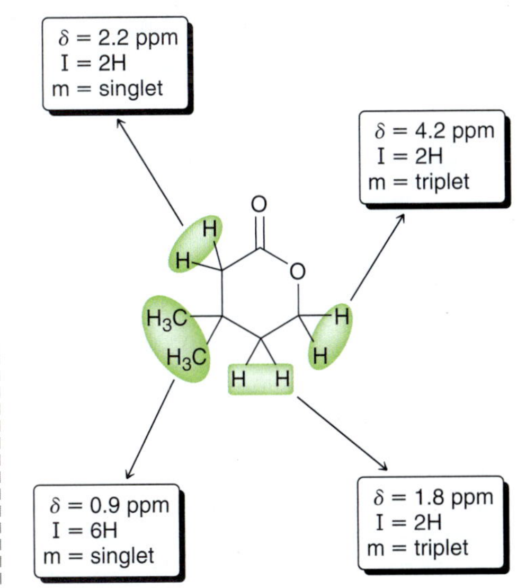


**STEP 3**
Look for differences in the chemical shifts, multiplicities, or integration values of the signals.

Now we can compare the predicted values above and look for differences. If we look at the chemical shifts for all four signals, we find that both compounds produce signals in similar locations. However, the nature of these signals is different. For example, both compounds will produce signals at 4.2 ppm, but the first compound produces a singlet, while the second compound produces a triplet. If we look at 2.2 ppm, we find that the first compound should have a triplet at that location, while the second compound should have a singlet.

Here is the bottom line: If you are able to predict all three characteristics of every signal, then you should be able to find several differences, even for compounds that are very similar in structure. At first, it might be helpful to actually write out all three characteristics of every signal and then compare the predicted values, as we did above. But after working several problems, you should find that you can identify differences without having to write out all of the values.

**PRACTICE the skill** **16.21** How would you use $^1$H NMR spectroscopy to distinguish between the following compounds?

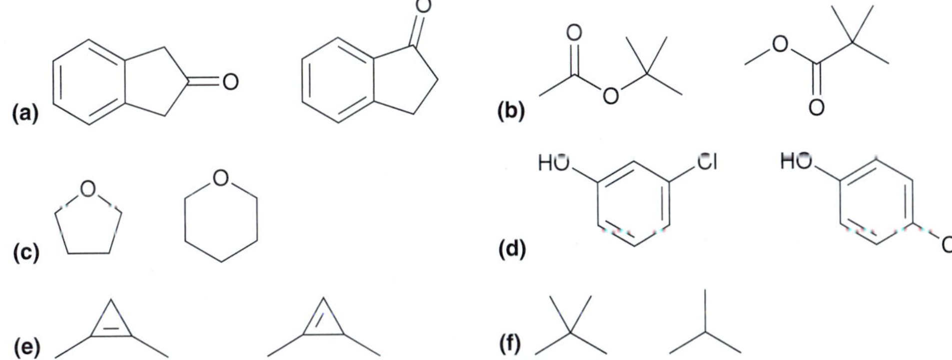

**APPLY the skill** **16.22** A chemist was attempting to achieve the following transformation:

HBr → Br

Mass spectrometry was then used to verify that the molecular formula of the major product was $C_7H_{15}Br$, just as expected. However, the $^1$H NMR spectrum of the product exhibited more than two signals. Identify what product was actually obtained. Can you suggest what might have caused formation of this product?

need more **PRACTICE?** **Try Problems 16.38, 16.47**

## medically speaking | Detection of Impurities in Heparin Sodium Using $^1$H NMR Spectroscopy

Heparin sodium is an anticoagulant that is derived from pig intestines. It has been used since the 1930s for the prevention of life-threatening blood clots. Because heparin is an animal-derived product, a number of steps are used in the manufacturing process to ensure that infectious agents and impurities are removed. In January 2008, the FDA announced a recall for certain lots of heparin sodium due to an increased number of reported adverse reactions (>350) in association with heparin administration. Some of the adverse events reported were serious in nature, including allergic or hypersensitivity-type reactions, with symptoms of oral swelling, shortness of breath, and severe hypotension (low blood pressure).

During the course of the FDA investigation, contaminants similar in structure to heparin were found in some of the recalled products. The contaminants were found in significant quantities (accounting for 5–20% of the total mass of each sample). In order to ensure a safe supply of this life-saving medication, the FDA released information on two analytical methods for detection of the contaminants.

One of these methods utilizes $^1$H NMR spectroscopy to differentiate products with and without the contaminants. In pure heparin sodium (no contaminants), a singlet peak appears upfield of 2 ppm without any additional features between 2.0 and 3.0 ppm. In heparin sodium products containing the contaminants, two additional peaks appear near 2 ppm. For future production of heparin, the FDA has recommended that manufacturers use $^1$H NMR spectroscopy in conjunction with other conventional methods to screen for impurities prior to the release of the final products into the market.

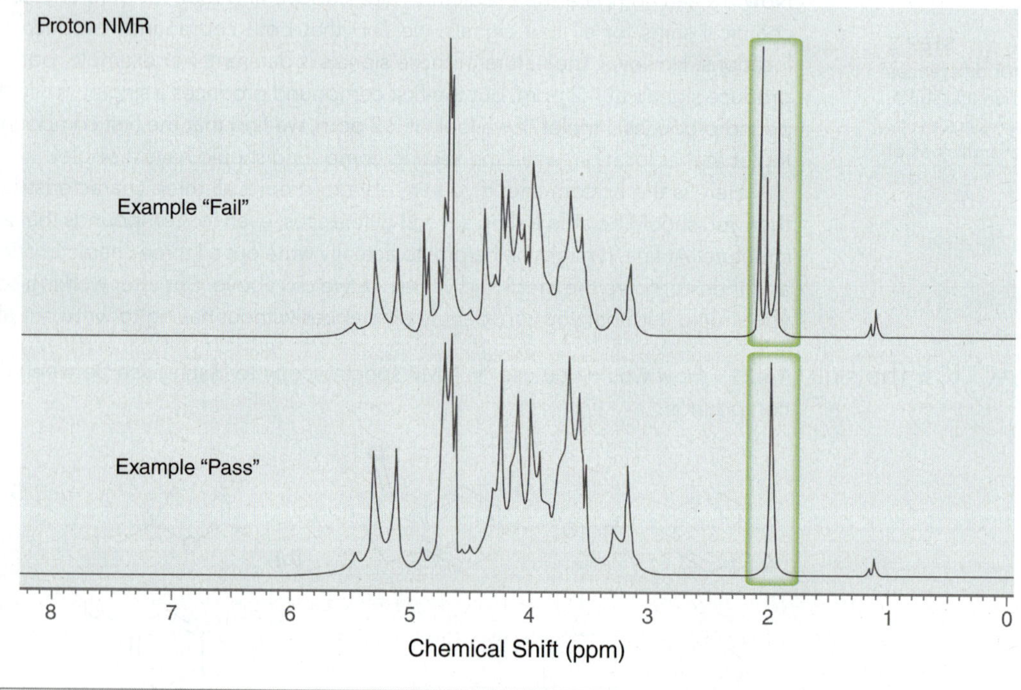


## 16.10 Analyzing a $^1$H NMR Spectrum

In this section, we will practice analyzing and interpreting NMR spectra, a process that involves four discrete steps:

**LOOKING BACK**
For a review of how to calculate and interpret the HDI of a compound, see Section 15.16.

1. Always begin by inspecting the molecular formula (if it is given), as it provides useful information. Specifically, calculating the hydrogen deficiency index (HDI) can provide important clues about the structure of the compound. An HDI of zero indicates that the compound does not possess any rings or π bonds. An HDI of 1 indicates that the compound has either one ring or one π bond. The larger the HDI, the less useful it is. However, an HDI of 4 or more should indicate the possible presence of an aromatic ring:

**Four degrees of unsaturation (HDI=4)**

2. Consider the number of signals and integration of each signal (gives clues about the symmetry of the compound).
3. Analyze each signal (chemical shift, integration, and multiplicity) and then draw fragments consistent with each signal. These fragments become our puzzle pieces that must be assembled to produce a molecular structure.
4. Assemble the fragments into a molecular structure.

The following exercise illustrates how this is done.

## SKILLBUILDER

### 16.8 ANALYZING A $^1$H NMR SPECTRUM AND PROPOSING THE STRUCTURE OF A COMPOUND

**LEARN the skill**

Identify the structure of a compound with molecular formula $C_9H_{10}O$ that exhibits the following $^1$H NMR spectrum:

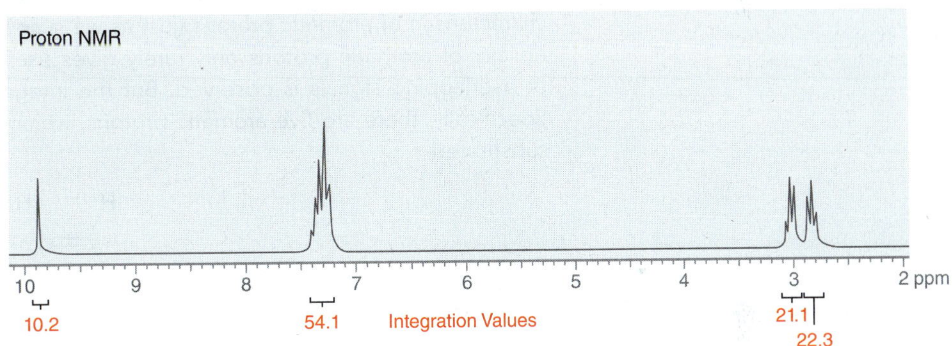


### SOLUTION

**STEP 1** Using the molecular formula, calculate and interpret the HDI.

Begin by calculating the HDI. The molecular formula indicates 9 carbon atoms, which would require 20 hydrogen atoms in order to be fully saturated. There are only 10 hydrogen atoms, which means that 10 hydrogen atoms are missing, and therefore, the HDI is 5. This is a large number, and it would not be efficient to think about all the possible ways to have five degrees of unsaturation. However, anytime we encounter an HDI of 4 or more, we should be on the lookout for an aromatic ring. We must keep this in mind when analyzing the spectrum. We should expect to see an aromatic ring (HDI = 4) plus one other degree of unsaturation (either a ring or a double bond).

**STEP 2** Consider the number of signals and integration values of each signal.

Next, consider the number of signals and the integration value for each signal. Be on the lookout for integration values that would suggest the presence of symmetry elements. For example, a signal with an integration of 4 would suggest two equivalent $CH_2$ groups.

In this spectrum, we see four signals. In order to analyze the integration of each signal, we must first divide by the lowest number (10.2):

$$\frac{10.2}{10.2} = 1 \qquad \frac{54.1}{10.2} = 5.30 \qquad \frac{21.1}{10.2} = 2.07 \qquad \frac{22.3}{10.2} = 2.19$$

The ratio is approximately 1 : 5 : 2 : 2. Now look at the molecular formula. There are 10 protons in the compound, so the relative integration values represent the actual number of protons giving rise to each signal.

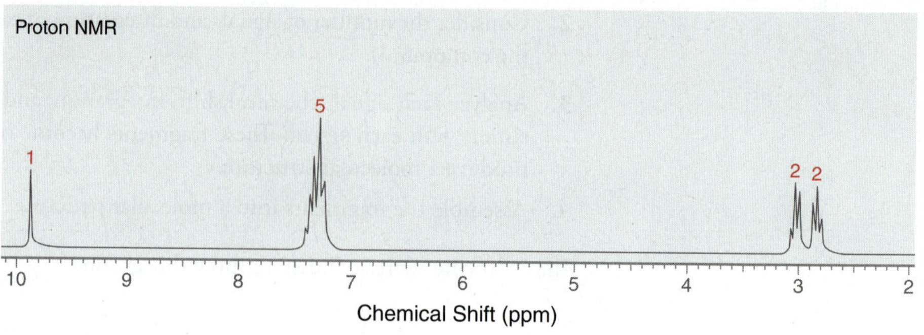


**STEP 3**
Analyze the chemical shift, multiplicity, and integration of each signal and then draw a fragment consistent with each signal.

Now analyze each signal. Starting upfield, there are two triplets, each with an integration of 2. This suggests that there are two adjacent methylene groups.

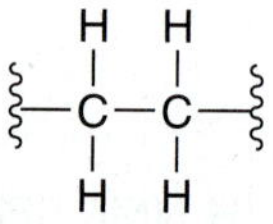

These signals do not appear at 1.2, where methylene groups are expected, so one or more factors are shifting these signals downfield. Our proposed structure must take this into account.

Moving downfield through the spectrum, the next signal appears just above 7 ppm, characteristic of aromatic protons (just as we suspected after analyzing the HDI). The multiplicity of aromatic protons only rarely gives useful information. More often, a multiplet of overlapping signals is observed. But the integration value gives valuable information. Specifically, there are five aromatic protons, which means that the aromatic ring is monosubstituted.

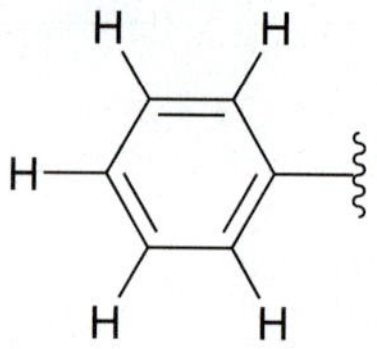

**Five aromatic protons**

**LOOKING BACK**
Recall from Section 16.7 that aldehydic protons often appear as singlets despite the presence of neighboring protons.

Next, move on to the last signal, which is a singlet at 10 ppm with an integration of 1. This is suggestive of an aldehydic proton. Our analysis has produced the following fragments:

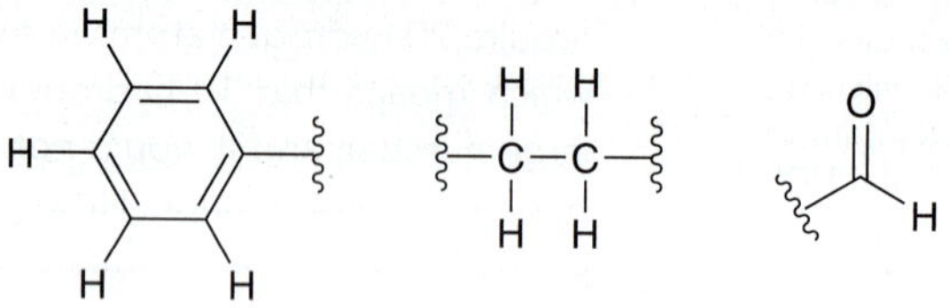

**STEP 4**
Assemble the fragments.

The final step is to assemble these fragments. Fortunately, there is only one way to assemble these three puzzle pieces.

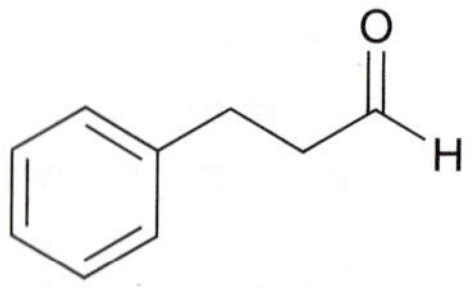

We mentioned before that each methylene group is being shifted downfield by one or more factors. Our proposed structure explains the observed chemical shifts. In particular, one methylene group is shifted significantly by the carbonyl group and slightly by the aromatic ring. The other methylene group is being shifted significantly by the aromatic ring and slightly by the carbonyl group.

**PRACTICE the skill** **16.23** Propose a structure that is consistent with each of the following $^1$H NMR spectra. In each case, the molecular formula is provided.

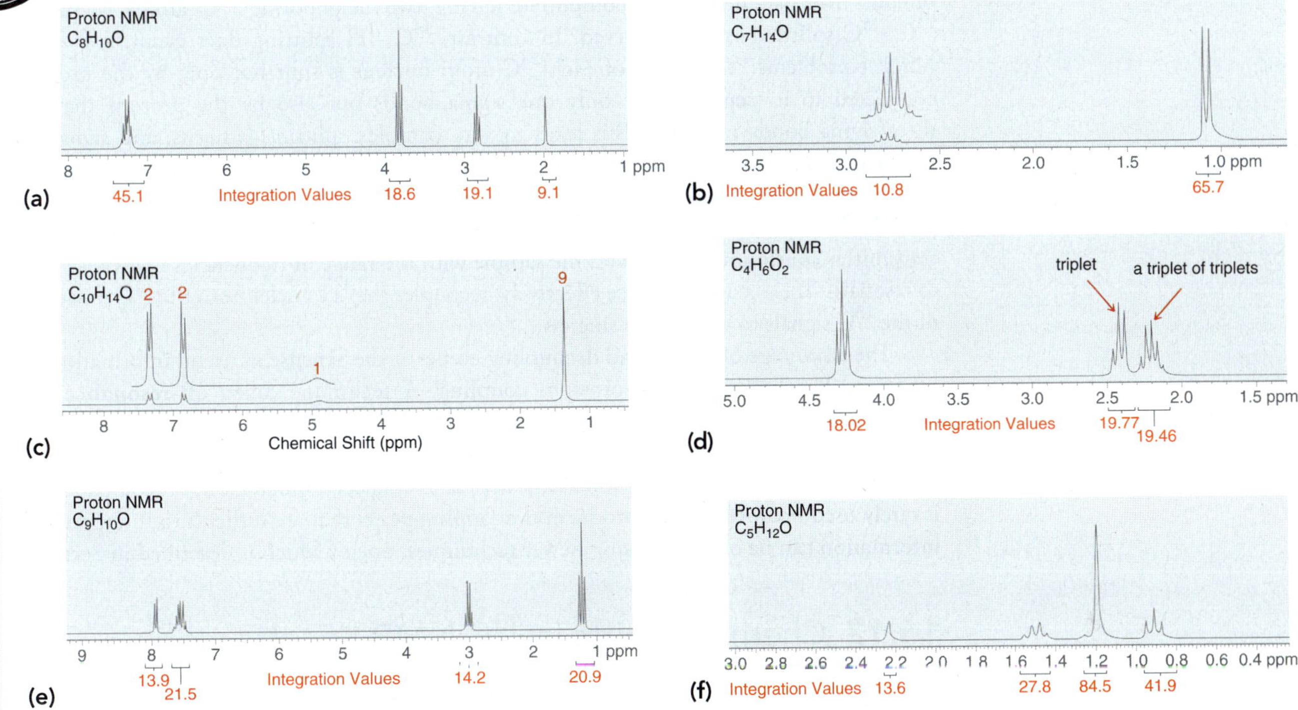

**APPLY the skill** **16.24** A compound with molecular formula $C_{10}H_{10}O_4$ produces a $^1$H NMR spectrum that exhibits only two signals, both singlets. One signal appears at 3.9 ppm with a relative integration value of 79. The other signal appears at 8.1 ppm with a relative integration value of 52. Identify the structure of this compound.

need more **PRACTICE?** **Try Problems 16.54, 16.57**

## 16.11 Acquiring a $^{13}$C NMR Spectrum

**LOOKING BACK**
Recall from Section 16.1 that $^{12}$C does not have a nuclear spin and therefore cannot be probed with NMR spectroscopy.

Many of the principles that apply to $^1$H NMR spectroscopy also apply to $^{13}$C NMR spectroscopy, but there are a few major differences, and we will focus on those. For example, $^1$H is the most abundant isotope of hydrogen, but $^{13}$C is only a minor isotope of carbon, representing about 1.1% of all carbon atoms found in nature. As a result, only one in every hundred carbon atoms will resonate, which demands the use of a sensitive receiver coil for $^{13}$C NMR spectroscopy.

In $^1$H NMR spectroscopy, we saw that each signal has three characteristics (chemical shift, integration, and multiplicity). In $^{13}$C NMR spectroscopy, only the chemical shift is generally reported. The integration and multiplicity of $^{13}$C signals are not reported, which greatly simplifies the interpretation of $^{13}$C NMR spectra. Integration values are not routinely calculated in $^{13}$C NMR spectroscopy because the pulse technique employed by FT NMR spectrometers has the undesired effect of distorting the integration values. Multiplicity is also not a common characteristic of $^{13}$C NMR spectra.

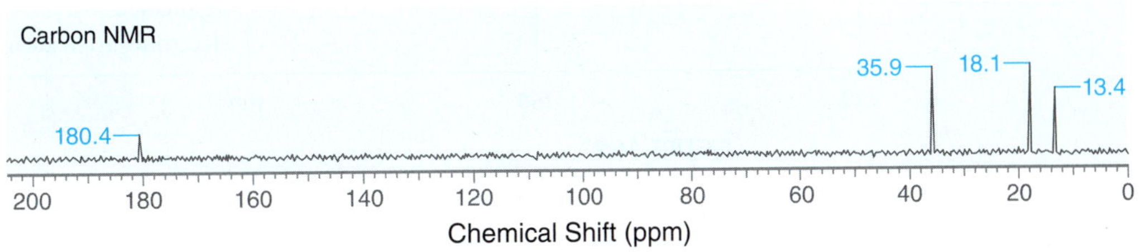

Notice that all of the signals are recorded as singlets. There are several good reasons for this. First, no splitting is observed between neighboring carbon atoms because of the low abundance of $^{13}$C. The likelihood of a compound having two neighboring $^{13}$C atoms is quite small, so $^{13}$C -$^{13}$C splitting is not observed. In contrast, $^{13}$C -$^{1}$H splitting does occur, and it creates significant problems. The signal of each $^{13}$C atom nucleus is split not only by the protons directly connected to it (separated by only one sigma bond) but also by the protons that are two or three sigma bonds removed. This leads to very complex splitting patterns, and signals overlap to produce an unreadable spectrum. To solve the problem, all $^{13}$C -$^{1}$H splitting is suppressed with a technique called **broadband decoupling**, which uses two rf transmitters. The first transmitter provides brief pulses in the range of frequencies that cause $^{13}$C nuclei to resonate, while the second transmitter continuously irradiates the sample with the range of frequencies that cause all $^{1}$H nuclei to resonate. This second rf source effectively decouples the $^{1}$H nuclei from the $^{13}$C nuclei, causing all of the $^{13}$C signals to collapse to singlets.

The advantage of broadband decoupling comes at the expense of useful information that would otherwise be obtained from spin-spin coupling. A technique called **off-resonance decoupling** allows us to retrieve some of this information. With this technique, only the one-bond couplings are observed, so $CH_3$ groups appear as quartets, $CH_2$ groups appear as triplets, CH groups appear as doublets, and quaternary carbon atoms appear as singlets. Nonetheless, off-resonance decoupling is rarely used because it often produces overlapping peaks that are difficult to interpret. The desired information can be obtained using newer techniques, one of which is described in Section 16.13.

## 16.12 Chemical Shifts in $^{13}$C NMR Spectroscopy

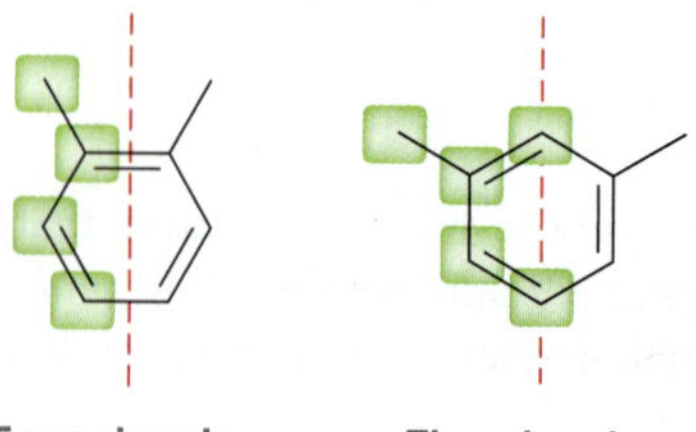


Three signals

The range of rf frequencies in $^{13}$C spectroscopy is different from that used in $^{1}$H NMR spectroscopy, because $^{13}$C atoms resonate over a different frequency range. Just as in $^{1}$H NMR spectroscopy, the position of each signal is defined relative to the frequency of absorption of a reference compound, TMS. With this definition, the chemical shift of each $^{13}$C atom is constant, regardless of the operating frequency of the spectrometer. In $^{13}$C NMR spectroscopy, chemical shift values typically range from 0 to 220 ppm.

The number of signals in a $^{13}$C NMR spectrum represents the number of carbon atoms in different electronic environments (not interchangeable by symmetry). Carbon atoms that are interchangeable by a symmetry operation (either rotation or reflection) will only produce one signal. To illustrate this point, consider the compounds in the margin. Each compound has eight carbon atoms but does not produce eight signals. The unique carbon atoms in each compound are highlighted. Each carbon atom that is not highlighted is equivalent to one of the highlighted carbon atoms.

The location of each signal is dependent on shielding and deshielding effects, just as we saw in $^{1}$H NMR spectroscopy. Figure 16.22 shows the chemical shifts of several important types of carbon atoms.

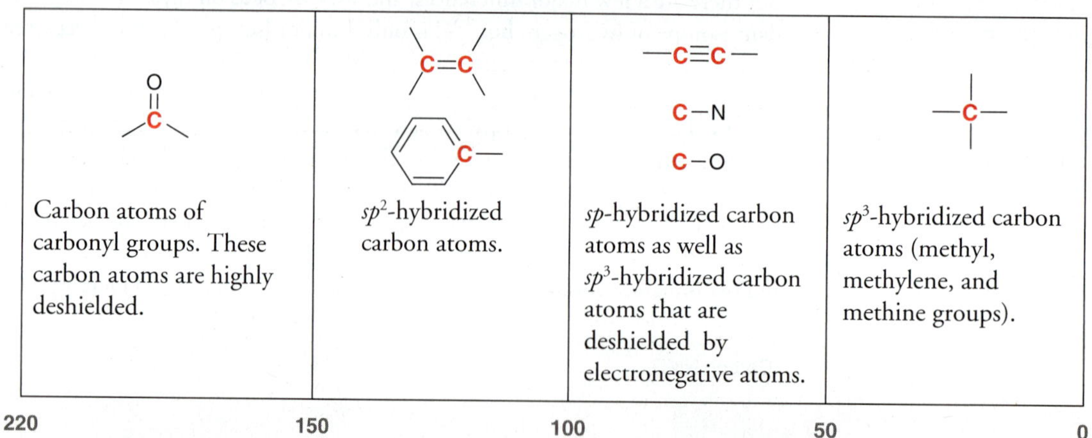


**FIGURE 16.22**
The general location of signals produced by different types of carbon atoms in $^{13}$C NMR spectroscopy.

Let's now use this information to analyze a $^{13}$C NMR spectrum.

# SKILLBUILDER

## 16.9 PREDICTING THE NUMBER OF SIGNALS AND APPROXIMATE LOCATION OF EACH SIGNAL IN A $^{13}$C NMR SPECTRUM

### LEARN the skill

Consider the following compound:

Predict the number of signals and the location of each signal in a $^{13}$C NMR spectrum of this compound.

### SOLUTION

**STEP 1**
Look to see if any of the carbon atoms are interchangeable by either rotational or reflectional symmetry and determine the number of expected signals.

Begin by determining the number of expected signals. The compound has nine carbon atoms, but we must look to see if any of these carbon atoms are interchangeable by a symmetry operation. In this case, there is both a plane and an axis of symmetry. As a result, we expect only five signals in the $^{13}$C NMR spectrum.

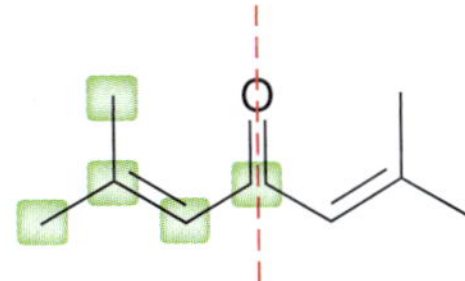



**STEP 2**
Predict the expected region in which each signal will appear, based on hybridization states and deshielding effects.

In general, when two methyl groups are attached to the same carbon atom, they will be equivalent. This is not the case in this example, because one methyl group is *cis* to the carbonyl group, while the other methyl group is *trans* to the carbonyl group.

The expected chemical shifts are shown below, categorized according to the region of the spectrum in which each signal is expected to appear:

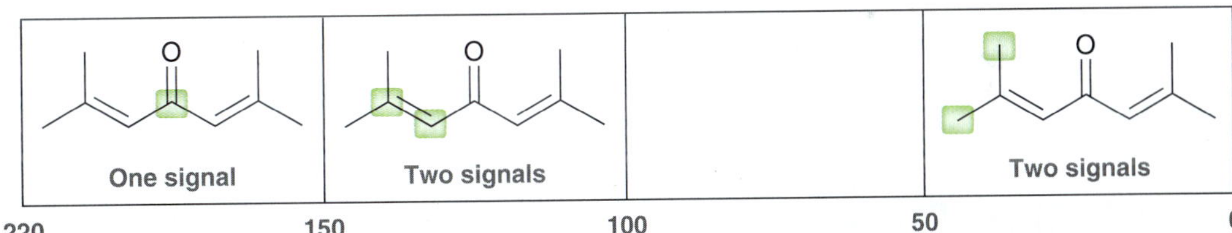

### PRACTICE the skill

**16.25** For each of the following compounds, predict the number of signals and location of each signal in a $^{13}$C NMR spectrum:

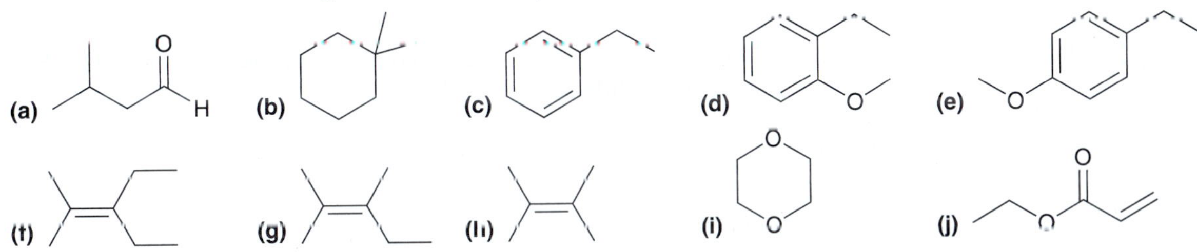

### APPLY the skill

**16.26** Compare the following two constitutional isomers. The $^{13}$C NMR spectrum of the first compound exhibits five signals, while the second compound exhibits six signals. Explain.

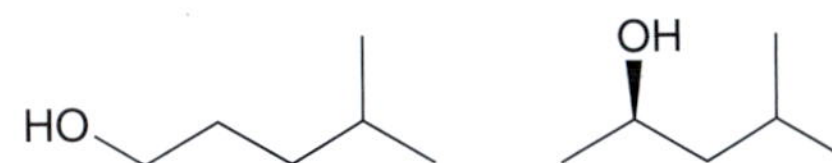



**16.27** Draw the structure of a compound with molecular formula $C_8H_{10}$ that exhibits five signals in its $^{13}$C NMR spectrum, four of which appear between 100 and 150 ppm.

need more **PRACTICE?** **Try Problems 16.35, 16.38, 16.46**

## 16.13 DEPT $^{13}C$ NMR Spectroscopy

As mentioned in Section 16.11, a broadband-decoupled $^{13}C$ spectrum does not provide information regarding the number of protons attached to each carbon atom in a compound. This information can be obtained through a variety of recently developed techniques, one of which is called distortionless enhancement by polarization transfer (DEPT). **DEPT $^{13}C$ NMR** spectroscopy utilizes two rf transmitters and relies on the fact that the intensity of each particular signal will respond to different pulse sequences in a predictable fashion, depending on the number of protons attached. This technique involves the acquisition of several spectra. First, a regular broadband-decoupled $^{13}C$ spectrum is acquired, indicating the chemical shifts associated with all carbon atoms in the compound. Then a special pulse sequence is utilized to produce a spectrum called a DEPT-90, in which only signals from CH groups appear. This spectrum does not show any signals resulting from $CH_3$ groups, $CH_2$ groups, or quaternary carbon atoms (C with no protons). Then, a different pulse sequence is employed to generate a spectrum, called a DEPT-135, in which $CH_3$ groups and CH groups appear as positive signals, $CH_2$ groups appear as negative signals (pointing down), and quaternary carbon atoms do not appear.

By comparing all of the spectra, it is possible to identify each signal in the broadband-decoupled spectrum as arising from either a $CH_3$ group, a $CH_2$ group, a CH group, or a quaternary carbon atom. This information is summarized in Table 16.4. Notice that each type of group exhibits a different absorption pattern when all three spectra are compared. For example, only CH groups give positive signals in all three spectra, while $CH_2$ groups are the only groups that give negative signals in the DEPT-135 spectrum. This technique therefore produces a series of spectra that collectively contain all of the information in an off-resonance decoupled spectrum, but without the disadvantage of overlapping signals. The following SkillBuilder illustrates how DEPT spectra can be interpreted.

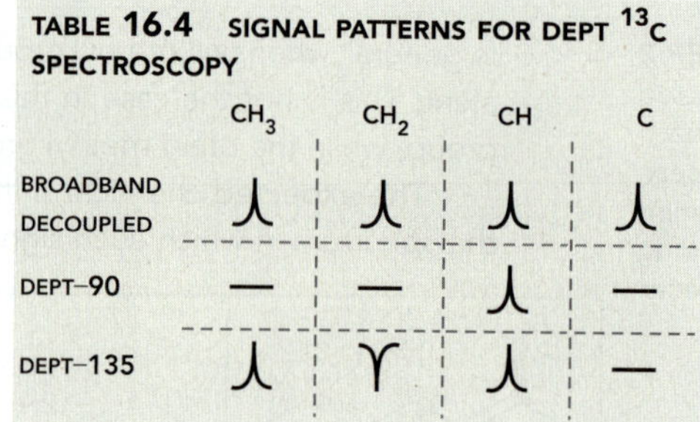

**TABLE 16.4 SIGNAL PATTERNS FOR DEPT $^{13}C$ SPECTROSCOPY**

| | $CH_3$ | $CH_2$ | CH | C |
|---|---|---|---|---|
| BROADBAND DECOUPLED | ⅄ | ⅄ | ⅄ | ⅄ |
| DEPT–90 | — | — | ⅄ | — |
| DEPT–135 | ⅄ | Y | ⅄ | — |

# SKILLBUILDER

### 16.10 DETERMINING MOLECULAR STRUCTURE USING DEPT $^{13}C$ NMR SPECTROSCOPY

**LEARN the skill**

Determine the structure of an alcohol with molecular formula $C_4H_{10}O$ that exhibits the following $^{13}C$ NMR spectra:

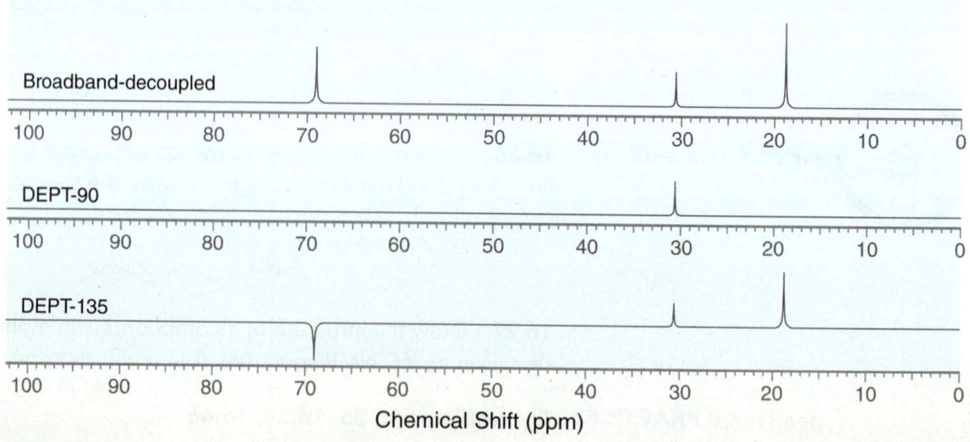

## SOLUTION

Whenever the molecular formula is known, the first step is always to calculate the HDI and determine the information that it provides. In this case, there are zero degrees of unsaturation, which means that the compound does not have any rings or $\pi$ bonds. Next, we focus on the number of signals in the broadband-decoupled spectrum. There are only three signals, but the molecular formula indicates that the compound has four carbon atoms. Therefore, we conclude that one of the signals must represent two chemically equivalent carbon atoms. Now we are ready to analyze each individual signal in the spectra. Using the information in Table 16.4, we can determine the following information:

- The signal at approximately 69 ppm is a $CH_2$ group (signal is negative in DEPT-135).
- The signal at approximately 30 ppm is a CH group (signal is positive in all spectra).
- The signal at approximately 19 ppm is a $CH_3$ group (signal is positive in the broadband-decoupled spectrum, absent in DEPT-90, and positive in DEPT-135).

We can now record this information on the broadband-decoupled spectrum to aid in our analysis.

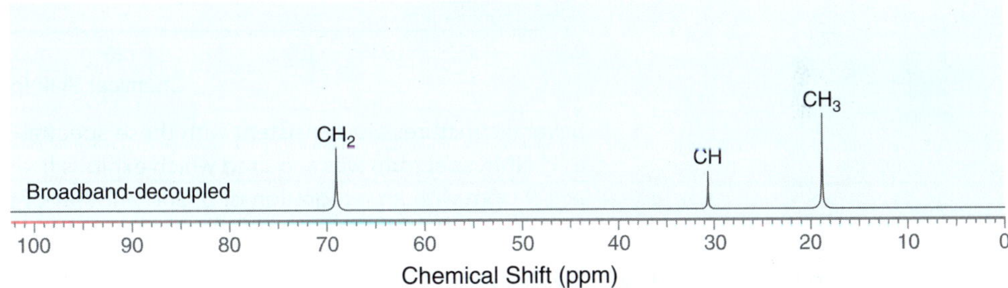


In order to determine which signal represents two carbon atoms, we notice that the molecular formula indicates that the structure must have 10 protons. So far, we have only accounted for 7 of the protons ($CH_2+CH+CH_3+OH=7$ protons). We need to account for 3 more protons. Therefore, we can conclude that the $CH_3$ signal must represent 2 equivalent carbon atoms.

We can now analyze the chemical shifts of each signal. The $CH_2$ signal is more downfield than the others (at 70 ppm), so this signal must represent the carbon atom attached directly to the oxygen atom. The signal at 30 ppm is also shifted slightly downfield (relative to the signal at 19 ppm), and therefore we expect this signal to represent the carbon atom that is beta to the OH group. Finally, the signal at 19 ppm represents two equivalent methyl groups, giving the following structure:

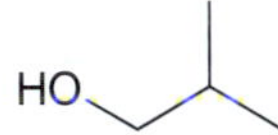

**16.28** Determine the structure of a compound with molecular formula $C_5H_{10}O$ that exhibits the following broadband-decoupled and DEPT-135 spectra. The DEPT-90 spectrum has no signals.

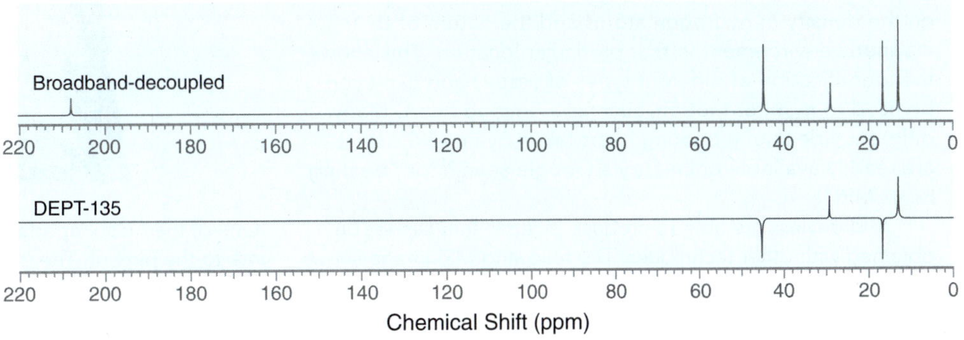

**16.29** Determine the structure of an alcohol with molecular formula $C_5H_{12}O$ that exhibits the following signals in its $^{13}C$ NMR spectra:

**(a)** Broadband decoupled: 73.8 δ, 29.1 δ, and 9.5 δ

**(b)** DEPT-90: 73.8 δ

**(c)** DEPT-135: positive signals at 73.8 δ and 9.5 δ; negative signal at 29.1 δ

**16.30** A compound with molecular formula $C_7H_{14}O$ exhibits the following $^{13}C$ NMR spectra:

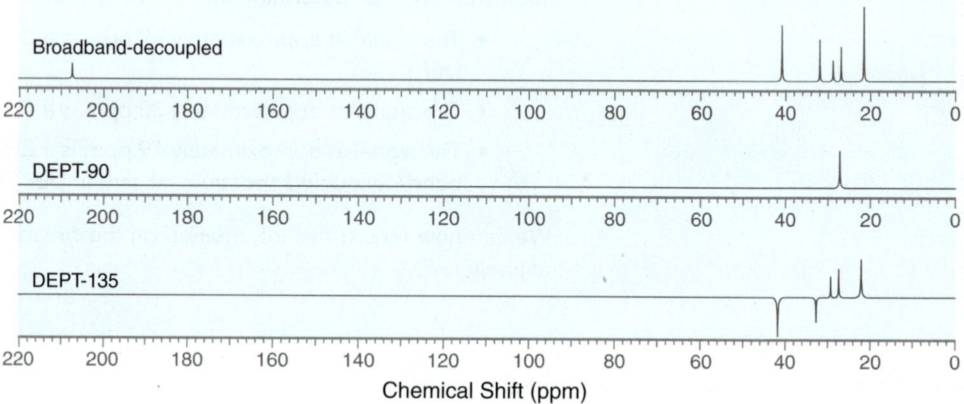


Several structures are consistent with these spectra. To determine which structure is correct, a $^{1}H$ NMR spectrum was acquired which exhibits five signals. One of those signals is a singlet at 1.9 ppm with an integration of 3, and another of the signals is a doublet at 0.9 ppm with an integration of 6. Using this information, identify the correct structure of the compound.

need more **PRACTICE?** **Try Problems 16.60, 16.61**

## Magnetic Resonance Imaging (MRI)

Magnetic resonance imaging has developed into an invaluable medical diagnostic tool for its ability to take pictures of internal organs. MRI devices are essentially NMR spectrometers that are large enough to accommodate a human being. In the presence of a strong magnetic field, the subject is irradiated with rf waves that cover the range of resonance frequencies for $^{1}H$ nuclei (protons). Protons are abundant in the human body where there is water or fat. When these protons are excited, the signals produced are monitored and recorded. Unlike NMR spectrometers, MRI devices use sophisticated techniques that analyze the three-dimensional location of each signal as well as its intensity. The intensity of each signal is dependent on the density of hydrogen atoms and the nature of their magnetic environment in that particular location. This allows the identification of different types of tissues and can even be used to monitor motion, such as a beating heart. Many different videos of a beating heart taken by an MRI device are readily available online (try a Google search for "beating heart MRI").

MRI devices are able to produce pictures that cannot be obtained with other techniques. The following MRI image shows a ruptured disc pressing on a patient's spinal cord.

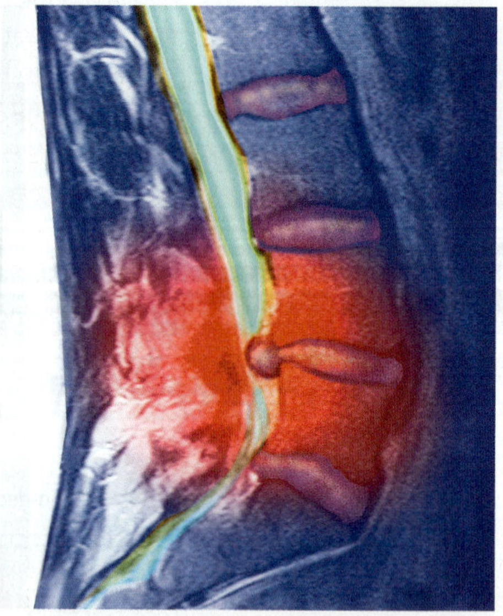

One of the most important features of MRI devices is the low risk to the patient. The magnetic field does not cause any known health concerns, and rf radiation is relatively harmless.

# REVIEW OF CONCEPTS AND VOCABULARY

### SECTION 16.1

- **Nuclear magnetic resonance** (NMR) spectroscopy provides information about how the individual carbon and hydrogen atoms in a molecule are connected to each other.
- A spinning proton generates a **magnetic moment**, which must align either with or against an imposed external magnetic field.
- When irradiated with rf radiation, the nucleus flips to a higher energy spin state and is said to be in *resonance*.
- All protons do not absorb the same frequency because of **diamagnetism**, a weak magnetic effect due to the motion of surrounding electrons that either **shield** or **deshield** the proton.

### SECTION 16.2

- The range of rf frequencies required to achieve resonance is dependent on the strength of the magnetic field, which determines the *operating frequency* of the spectrometer.
- **Continuous-wave spectrometers** have been replaced by **FT-NMR** spectrometers in which the sample is irradiated with a short pulse that covers the entire range of relevant rf frequencies, and a **free induction decay** (FID) is recorded and then converted into a spectrum.
- Deuterated solvents containing a small amount of TMS are generally used for acquiring NMR spectra.

### SECTION 16.3

- In a $^{1}H$ NMR spectrum, each signal has three important characteristics: location, area, and shape.

### SECTION 16.4

- **Chemically equivalent** protons occupy identical electronic environments and produce only one signal.
- When two protons are interchangeable by rotational symmetry, the protons are said to be **homotopic**.
- When two protons are interchangeable by reflectional symmetry, the protons are said to be **enantiotopic**.
- If two protons are neither homotopic nor enantiotopic, then they are not chemically equivalent.
- The **replacement test** is a simple technique for verifying the relationship between protons. If the replacement test produces diastereomers, then the protons are **diastereotopic**.

### SECTION 16.5

- **Chemical shift** (δ) is defined as the difference (in hertz) between the resonance frequency of the proton being observed and that of TMS divided by the operating frequency of the spectrometer.
- The left side of an NMR spectrum is described as **downfield**, and the right side is described as **upfield**.
- In the absence of inductive effects, a methyl group ($CH_3$) will produce a signal near 0.9 ppm, a **methylene group** ($CH_2$) will produce a signal near 1.2 ppm, and a **methine group** (CH) will produce a signal near 1.7 ppm. The presence of nearby groups increases these values somewhat predictably.
- **Diamagnetic anisotropy** causes the aromatic protons of benzene to be strongly deshielded.

### SECTION 16.6

- The **integration**, or area under each signal, indicates the number of protons giving rise to the signal. These numbers provide the relative number, or ratio, of protons giving rise to each signal.
- Integration is often represented by *step curves*, where the height of each step curve represents the area under the signal.

### SECTION 16.7

- **Multiplicity** represents the number of peaks in a signal. A **singlet** has one peak, a **doublet** has two, a **triplet** has three, a **quartet** has four, and a **quintet** has five.
- Multiplicity is the result of **spin-spin splitting**, also called **coupling**, which follows the ***n*+1 rule**.
- Equivalent protons do not split each other.
- In order for protons to split each other, they must be separated by no more than three sigma bonds.
- When signal splitting occurs, the distance between the individual peaks of a signal is called the **coupling constant**, or ***J* value**, and is measured in hertz.
- Generally, no splitting is observed across the oxygen atom of an alcohol, because hydroxyl protons are **labile**.
- Complex splitting occurs when a proton has two different kinds of neighbors, often producing a **multiplet**.

### SECTION 16.10

- Analyzing a $^{1}H$ NMR spectrum involves four discrete steps:
  - Calculate the HDI (if the molecular formula is provided).
  - Consider the number of signals and the integration of each signal.
  - Analyze each signal (δ, m, I), and then draw fragments consistent with each signal.
  - Assemble the fragments.

### SECTION 16.11

- $^{13}C$ is an isotope of carbon, representing 1.1% of all carbon atoms.
- In $^{13}C$ NMR spectroscopy, only the chemical shift is generally reported.
- All $^{13}C$-$^{1}H$ splitting is suppressed with a technique called **broadband decoupling**, causing all of the $^{13}C$ signals to collapse to singlets.
- **Off-resonance decoupling** allows the retrieval of some of the coupling information that was lost, but this technique is rarely used because it often produces overlapping peaks that are difficult to interpret.

### SECTION 16.12

- The position of each signal in a $^{13}C$ NMR spectrum is defined relative to the frequency of absorption of a reference compound, tetramethylsilane (TMS).
- In $^{13}C$ NMR spectroscopy, chemical shift values typically range from 0 to 220 ppm.
- The number of signals in a $^{13}C$ NMR spectrum represents the number of carbon atoms in different electronic environments (carbons that are not interchangeable by symmetry).

Carbon atoms that are interchangeable by a symmetry operation (either rotation or reflection) will only produce one signal.

- The location of a signal is dependent on a number of factors, including hybridization state and shielding effects.

**SECTION 16.13**

- **Distortionless enhancement by polarization transfer (DEPT)** involves the acquisition of several spectra and allows the determination of the number of protons attached to each carbon atom.

# SKILLBUILDER REVIEW

## 16.1 DETERMINING THE RELATIONSHIP BETWEEN TWO PROTONS IN A COMPOUND

### METHOD 1: LOOKING FOR SYMMETRY

**STEP 1** Determine if the protons are interchangeable by rotational symmetry.

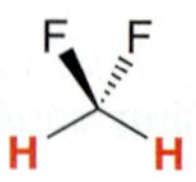


Homotopic
**Chemically equivalent**

**STEP 2** If there is no rotational symmetry, look for a plane of symmetry:

Enantiotopic
**Chemically equivalent**

**STEP 3** If the protons cannot be interchanged via a symmetry operation:

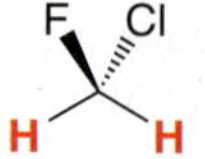


Diastereotopic
**Not chemically equivalent**

### METHOD 2: REPLACEMENT TEST

Replace each proton with a deuterium and compare the resulting structures.

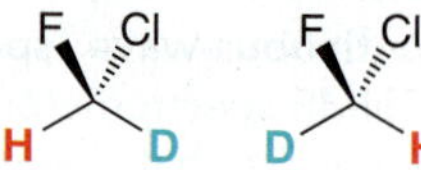


**Enantiomers**
Therefore, protons are enantiotopic.
**Chemically equivalent**

Try Problems **16.1–16.3, 16.40**

## 16.2 IDENTIFYING THE NUMBER OF EXPECTED SIGNALS IN A $^1$H NMR SPECTRUM

Count the number of different protons, using the following rules of thumb:

The three protons of a methyl group are always equivalent.

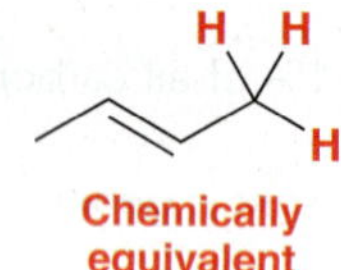


**Chemically equivalent**

The two protons of a methylene group ($CH_2$) are generally equivalent if the compound has no chirality centers.

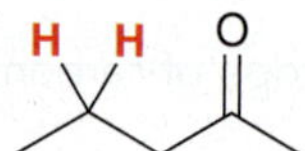


**Chemically equivalent**

The two protons of a methylene group ($CH_2$) are generally not equivalent if the compound has a chirality center.

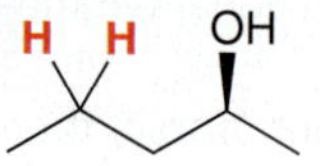


**Not chemically equivalent**

Two methylene groups will be equivalent if they can be interchanged by either rotation or reflection.

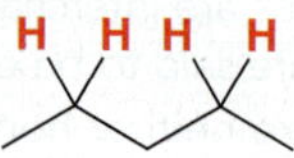


**Chemically equivalent**

Try Problems **16.4–16.6, 16.34, 16.42a, 16.44, 16.48**

## 16.3 PREDICTING CHEMICAL SHIFTS

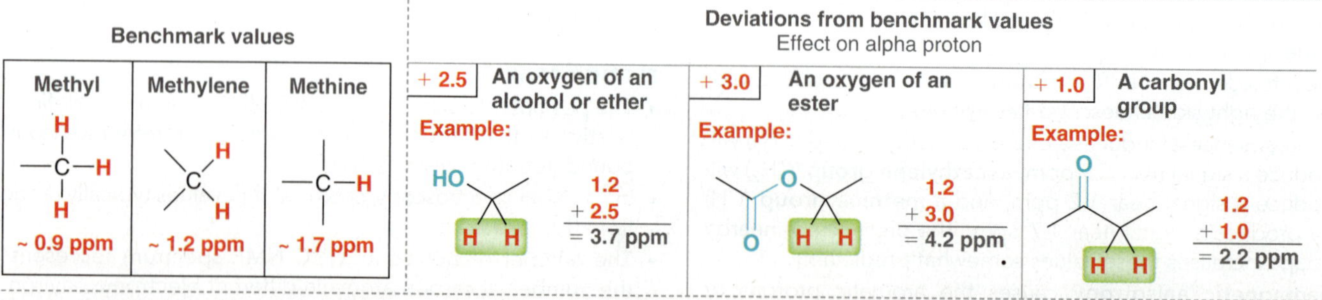


Try Problems **16.7–16.9, 16.42b**

## 16.4 DETERMINING THE NUMBER OF PROTONS GIVING RISE TO A SIGNAL

| | | |
|---|---|---|
| **STEP 1** Compare the relative integration values and choose the lowest number. | **STEP 2** Divide all integration values by the number from step 1, which gives the ratio of protons. | **STEP 3** Identify the number of protons in the compound (from the molecular formula) and then adjust the relative integration values so that the sum of the integrals equals the number of protons in the compound. |

Try Problems 16.11–16.14, 16.54

## 16.5 PREDICTING THE MULTIPLICITY OF A SIGNAL

**STEP 1** Identify all of the different kinds of protons.

Five kinds of protons

**STEP 2** For each kind of proton, identify the number of neighbors (N). The multiplicity will be $n + 1$.

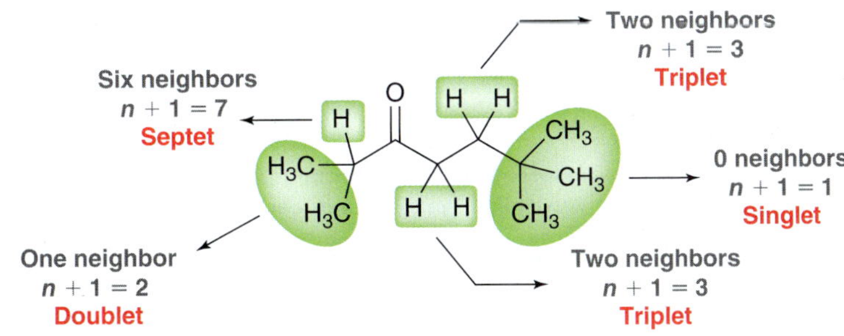


Try Problems 16.15, 16.16, 16.37

## 16.6 DRAWING THE EXPECTED $^{1}H$ NMR SPECTRUM OF A COMPOUND

| | | | | |
|---|---|---|---|---|
| **STEP 1** Identify the number of signals. | **STEP 2** Predict the chemical shift of each signal. | **STEP 3** Determine the integration of each signal by counting the number of protons giving rise to each signal. | **STEP 4** Predict the multiplicity of each signal. | **STEP 5** Draw each signal. |

Try Problems 16.19, 16.20, 16.41

## 16.7 USING $^{1}H$ NMR SPECTROSCOPY TO DISTINGUISH BETWEEN COMPOUNDS

| | | |
|---|---|---|
| **STEP 1** Identify the number of signals that each compound will produce. The simplest way to distinguish compounds is if they are expected to produce a different number of signals. | **STEP 2** If each compound is expected to produce the same number of signals, then determine the chemical shift, multiplicity, and integration of each signal in both compounds. | **STEP 3** Look for differences in the chemical shifts, multiplicities, or integration values of the expected signals. |

Try Problems 16.21, 16.22, 16.38, 16.47

## 16.8 ANALYZING A $^{1}H$ NMR SPECTRUM AND PROPOSING THE STRUCTURE OF A COMPOUND

| | | | |
|---|---|---|---|
| **STEP 1** Use the molecular formula to determine the HDI. An HDI of 4 or more indicates the possibility of an aromatic ring. | **STEP 2** Consider the number of signals and integration of each signal (gives clues about the symmetry of the compound). | **STEP 3** Analyze each signal (δ, m, I) and then draw fragments consistent with each signal. These fragments become our puzzle pieces that must be assembled to produce a molecular structure. | **STEP 4** Assemble the fragments. |

Try Problems 16.23, 16.24, 16.54, 16.57

## 16.9 PREDICTING THE NUMBER OF SIGNALS AND APPROXIMATE LOCATION OF EACH SIGNAL IN A $^{13}C$ NMR SPECTRUM

**STEP 1** Determine the number of expected signals. Look to see if any of the carbon atoms are interchangeable by either rotational or reflectional symmetry.

**STEP 2** Predict the expected region in which each signal will appear, based on hybridization states and deshielding effects.

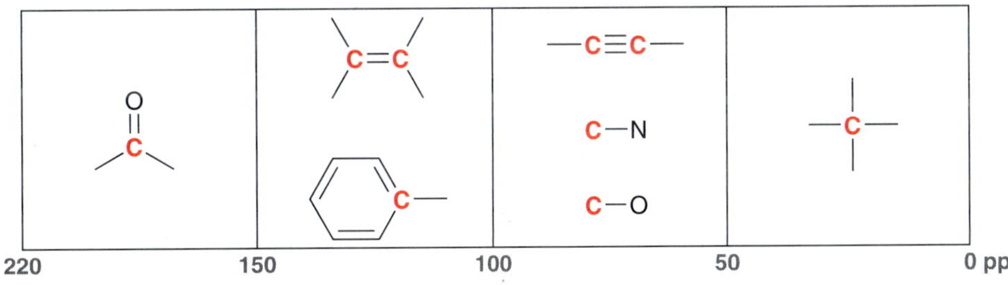


Try Problems 16.25–16.27, 16.35, 16.38, 16.46

### 16.10 DETERMINING MOLECULAR STRUCTURE USING DEPT $^{13}C$ NMR SPECTROSCOPY

| | $CH_3$ | $CH_2$ | CH | C |
|---|---|---|---|---|
| BROADBAND DECOUPLED | ⅄ | ⅄ | ⅄ | ⅄ |
| DEPT–90 | — | — | ⅄ | — |
| DEPT–135 | ⅄ | Y | ⅄ | — |

Try Problems 16.28–16.30, 16.60, 16.61

## PRACTICE PROBLEMS

Note: Most of the Problems are available within **WileyPLUS**, an online teaching and learning solution.

**16.31** Each of the following compounds exhibits a $^1H$ NMR spectrum with only one signal. Deduce the structure of each compound:

(a) $C_5H_{10}$ (b) $C_5H_8Cl_4$ (c) $C_{12}H_{18}$

**16.32** A compound with molecular formula $C_{12}H_{24}$ exhibits a $^1H$ NMR spectrum with only one signal and a $^{13}C$ NMR spectrum with two signals. Deduce the structure of this compound.

**16.33** A compound with molecular formula $C_{17}H_{36}$ exhibits a $^1H$ NMR spectrum with only one signal. How many signals would you expect in the $^{13}C$ NMR spectrum of this compound?

**16.34** How many signals would you expect in the $^1H$ NMR spectrum of each of the following compounds:

(a) 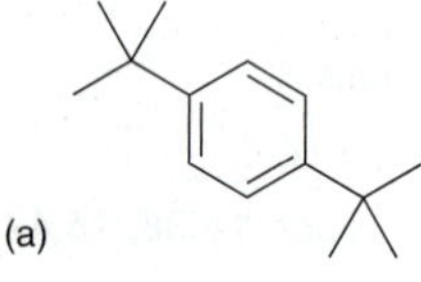

(b) 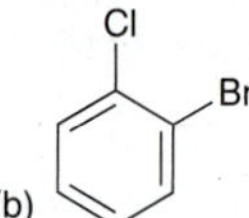


(c) 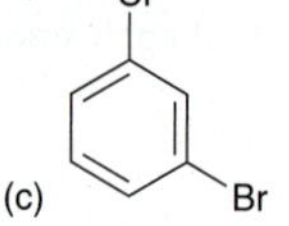


(d) 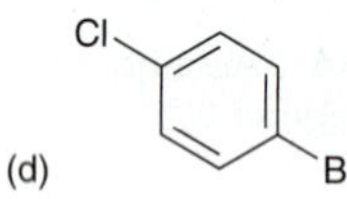


(e) 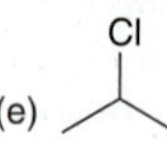


(f) Cl

**16.35** How many signals would you expect in the $^{13}C$ NMR spectrum of each of the compounds in Problem 16.34?

**16.36** How would you distinguish between the following compounds using $^{13}C$ NMR spectroscopy?

OH OH

**16.37** Predict the multiplicity of each signal in the $^1H$ NMR spectrum of the following compound:

O

**16.38** For each pair of compounds, identify how you would distinguish them using either $^1H$ NMR spectroscopy or $^{13}C$ NMR spectroscopy:

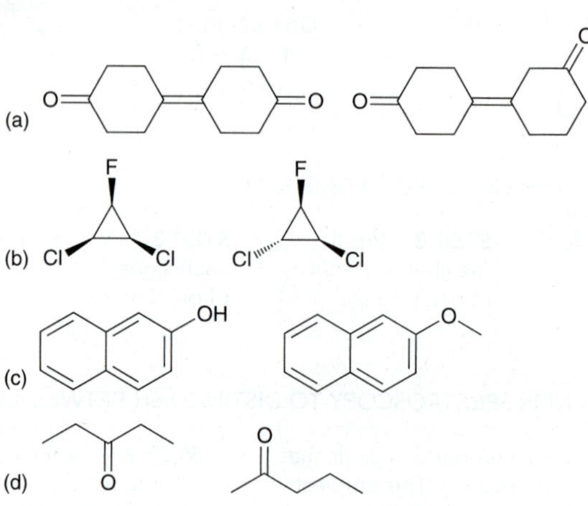


**16.39** A compound with molecular formula $C_8H_{18}$ exhibits a $^1H$ NMR spectrum with only one signal. How many signals would you expect in the $^{13}C$ NMR spectrum of this compound?

**16.40** For each of the following compounds, compare the two indicated protons and determine whether they are enantiotopic, homotopic, or diastereotopic:

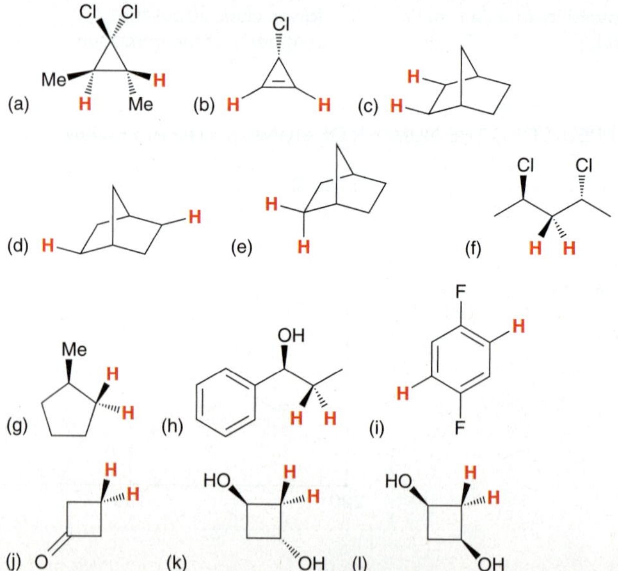

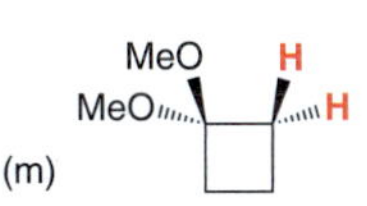


(m) (n) (o)

**16.41** Draw the expected $^1H$ NMR spectrum of the following compound:

**16.42** Consider the following compound:

(a) How many signals do you expect in the $^1H$ NMR spectrum of this compound?

(b) Rank the protons in terms of increasing chemical shift.

(c) How many signals do you expect in the $^{13}C$ NMR spectrum?

(d) Rank the carbon atoms in terms of increasing chemical shift.

**16.43** A compound with molecular formula $C_9H_{18}$ exhibits a $^1H$ NMR spectrum with only one signal and a $^{13}C$ NMR spectrum with two signals. Deduce the structure of this compound.

**16.44** How many signals do you expect in the $^1H$ NMR spectrum of each of the following compounds:

(a) **Geraniol** Isolated from roses and used in perfumes

(b) **Dopamine** A neurotransmitter that is deficient in Parkinson's disease

(c) **Isoprene** A precursor for natural rubber

**16.45** Rank the signals of the following compound in terms of increasing chemical shift. Identify the proton(s) giving rise to each signal:

**16.46** Predict the expected number of signals in the $^{13}C$ NMR spectrum of each of the following compounds. For each signal, identify where you expect it to appear in the $^{13}C$ NMR spectrum:

(a) (b) (c)

**16.47** When 1-methylcyclohexene is treated with HCl, a Markovnikov addition is observed. How would you use $^1H$ NMR spectroscopy to determine that the major product is indeed the Markovnikov product?

**16.48** How many signals are expected in the $^1H$ NMR spectrum of each of the following compounds?

(a) (b) (c) (d) (e) (f) (g) (h)

**16.49** Compare the structures of ethylene, acetylene, and benzene. Each of these compounds produces only one signal in its $^1H$ NMR spectrum. Arrange these signals in order of increasing chemical shift.

**16.50** Assuming a 300-MHz instrument is used, calculate the difference between the frequency of absorption (in hertz) of TMS and the frequency of absorption of a proton with a δ value of 1.2 ppm.

**16.51** A compound with molecular formula $C_{13}H_{28}$ exhibits a $^1H$ NMR spectrum with two signals: a septet with an integration of 1 and a doublet with an integration of 6. Deduce the structure of this compound.

**16.52** A compound with molecular formula $C_8H_{10}$ produces three signals in its $^{13}C$ NMR spectrum and only two signals in its $^1H$ NMR spectrum. Deduce the structure of the compound.

**16.53** A compound with molecular formula $C_3H_8O$ produces a broad signal between 3200 and 3600 $cm^{-1}$ in its IR spectrum and produces two signals in its $^{13}C$ NMR spectrum. Deduce the structure of the compound.

**16.54** A compound with molecular formula $C_4H_6O_4$ produces a broad signal between 2500 and 3600 $cm^{-1}$ in its IR spectrum and produces two signals in its $^1H$ NMR spectrum (a singlet at 12.1 ppm with a relative integration of 1 and a singlet at 2.4 ppm with a relative integration of 2). Deduce the structure of the compound.

**16.55** Propose the structure of a compound that exhibits the following $^1H$ NMR data:

(a) $C_5H_{10}O$
- 1.09 δ (6H, doublet)
- 2.12 δ (3H, singlet)
- 2.58 δ (1H, septet)

(b) $C_5H_{12}O$
- 0.91 δ (3H, triplet)
- 1.19 δ (6H, singlet)
- 1.50 δ (2H, quartet)
- 2.24 δ (1H, singlet)

(c) $C_4H_{10}O$
- 0.90 δ (6H, doublet)
- 1.76 δ (1H, multiplet)
- 3.38 δ (2H, doublet)
- 3.92 δ (1H, singlet)

(d) $C_4H_8O_2$
- 1.21 δ (6H, doublet)
- 2.59 δ (1H, septet)
- 11.38 δ (1H, singlet)

**16.56** A compound with molecular formula $C_8H_{10}O$ produces six signals in its $^{13}C$ NMR spectrum and exhibits the following $^1H$ NMR spectrum. Deduce the structure of the compound.

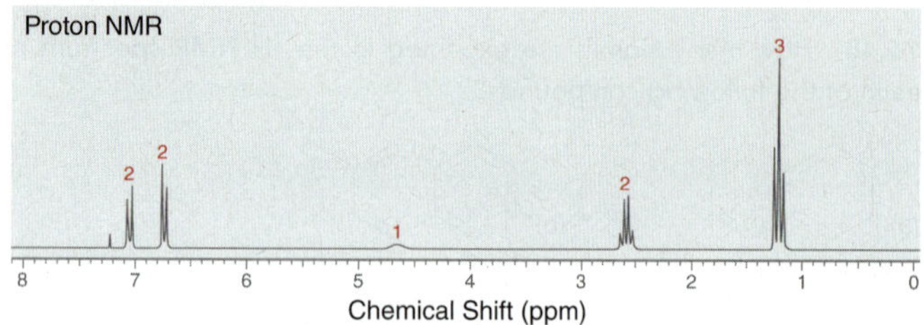


**16.57** Deduce the structure of a compound with molecular formula $C_9H_{12}$ that produces the following $^1H$ NMR spectrum:

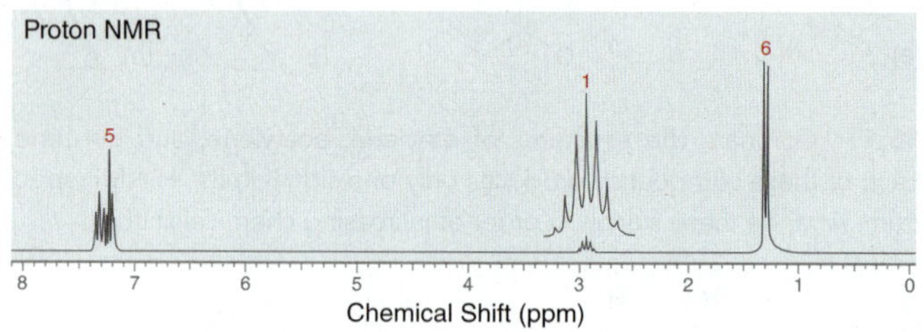


**16.58** Deduce the structure of a compound with molecular formula $C_9H_{10}O_2$ that produces the following $^1H$ NMR spectrum and $^{13}C$ NMR spectrum:

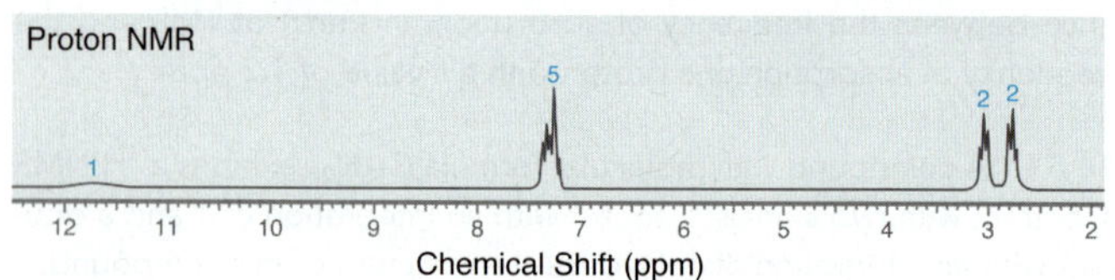


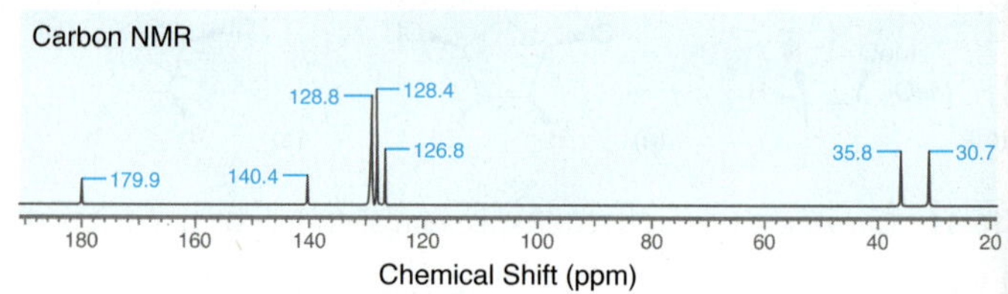


**16.59** Propose the structure of a compound consistent with the following data:

(a) $C_5H_{10}O$, broadband-decoupled $^{13}C$ NMR: 7.1, 34.6, 210.5 δ

(b) $C_6H_{10}O$, broadband-decoupled $^{13}C$ NMR: 70.8, 116.2, 134.8 δ

**16.60** Determine the structure of an alcohol with molecular formula $C_4H_{10}O$ that exhibits the following signals in its $^{13}C$ NMR spectra:

(a) Broadband decoupled: 69.3 δ, 32.1 δ, 22.8 δ, and 10.0 δ

(b) DEPT-90: 69.3 δ

(c) DEPT-135: positive signals at 69.3 δ, 22.8 δ, and 10.0 δ; negative signal at 32.1 δ

**16.61** Determine the structure of an alcohol with molecular formula $C_6H_{14}O$ that exhibits the following DEPT-135 spectrum:

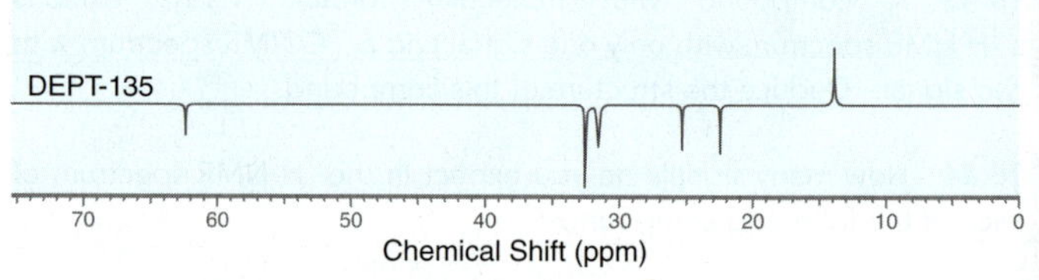


# INTEGRATED PROBLEMS

**16.62** Deduce the structure of a compound with molecular formula $C_6H_{14}O_2$ that exhibits the following IR, $^1H$ NMR, and $^{13}C$ NMR spectra:

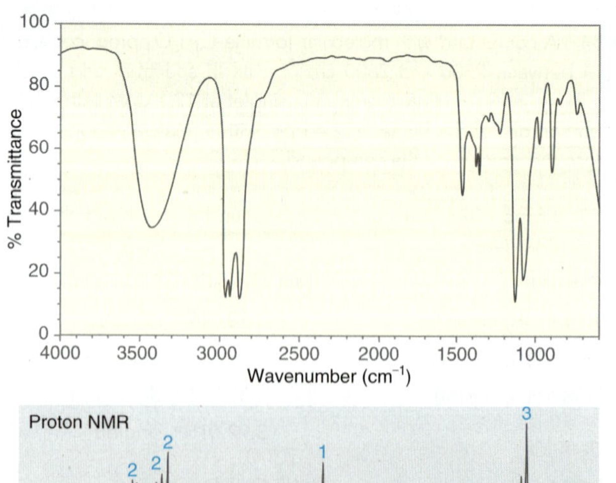


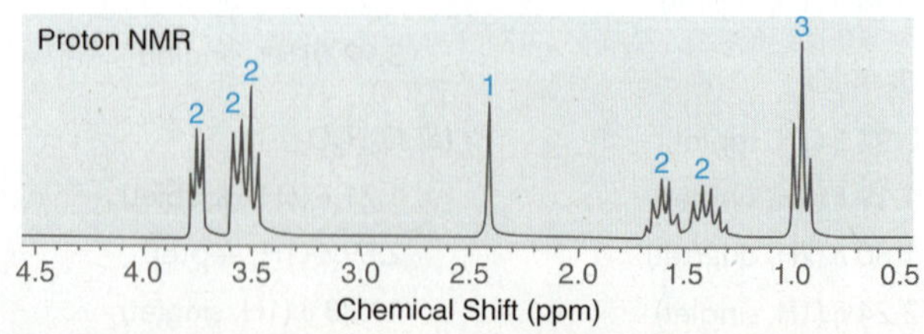


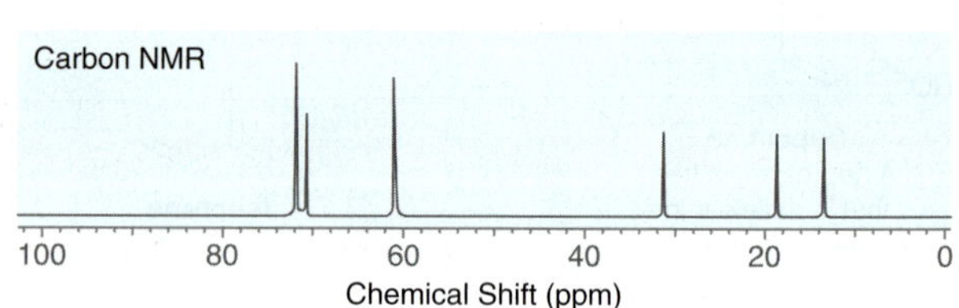


**16.63** Deduce the structure of a compound with molecular formula $C_8H_{10}O$ that exhibits the following IR, $^1H$ NMR, and $^{13}C$ NMR spectra:

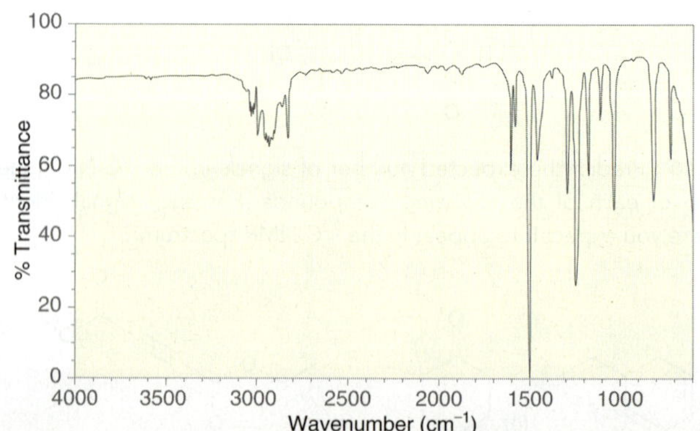

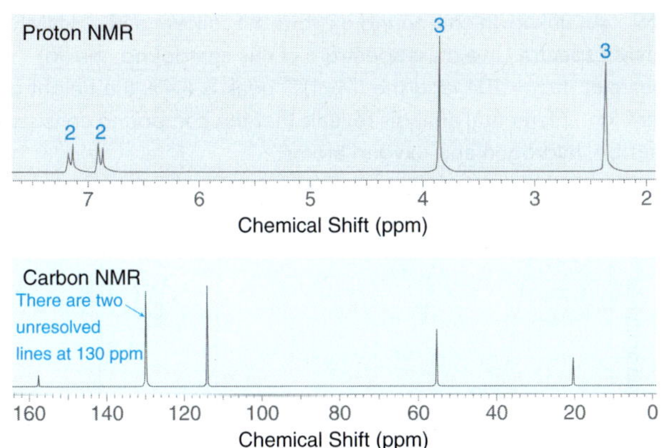


**16.64** Deduce the structure of a compound with molecular formula $C_5H_{10}O$ that exhibits the following IR, $^1H$ NMR, and $^{13}C$ NMR spectra. Data from the mass spectrum are also provided.

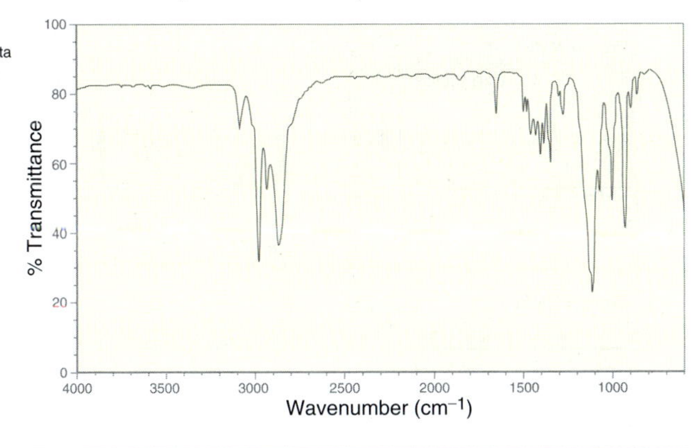


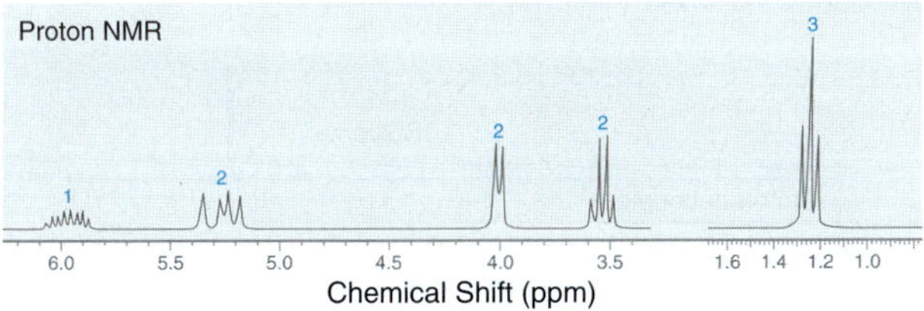


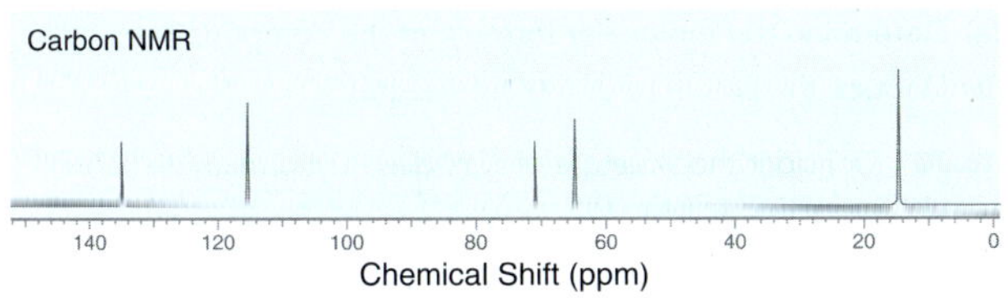


**16.65** Deduce the structure of a compound with molecular formula $C_8H_{14}O_3$ that exhibits the following IR, $^1H$ NMR, and $^{13}C$ NMR spectra:

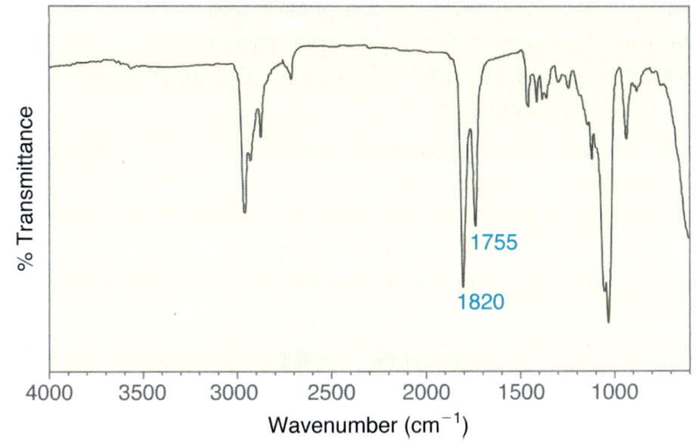


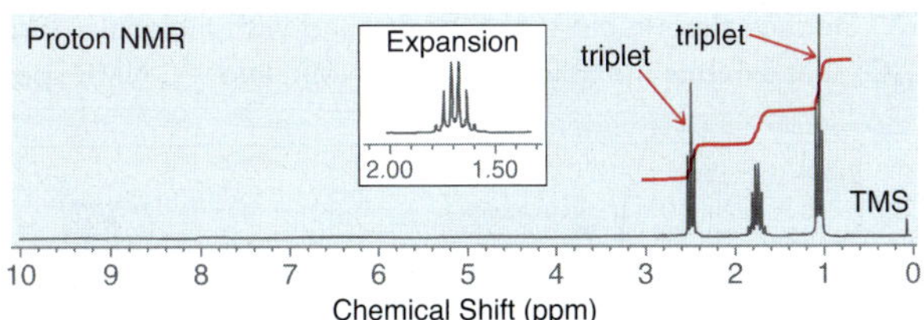


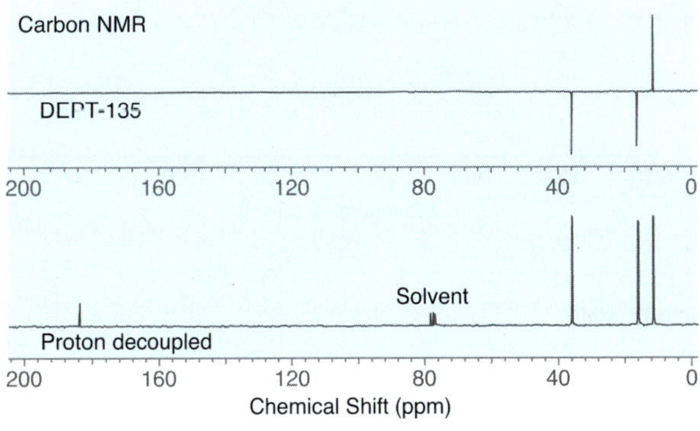


**16.66** Deduce the structure of a compound with molecular formula $C_6H_{10}O_4$ that exhibits the following IR, $^1H$ NMR, and $^{13}C$ NMR spectra (*Angew. Chem. Int. Ed.* **2011,** *50,* 8387–8390):

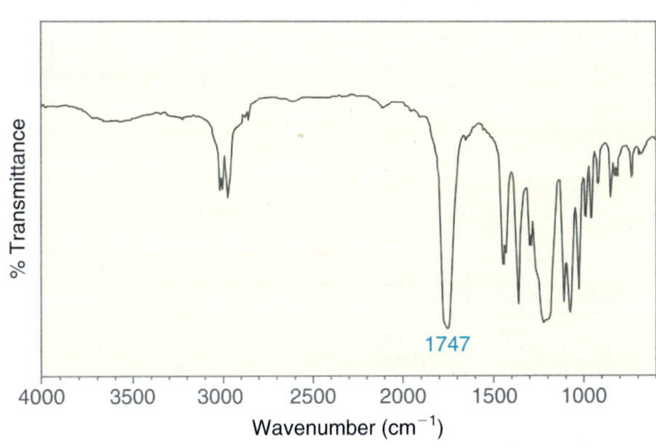


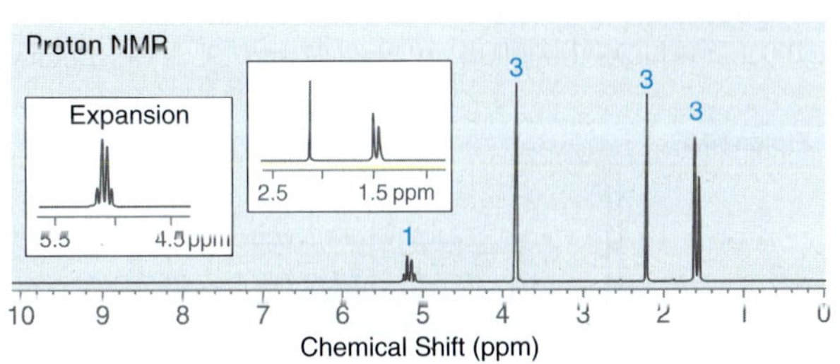


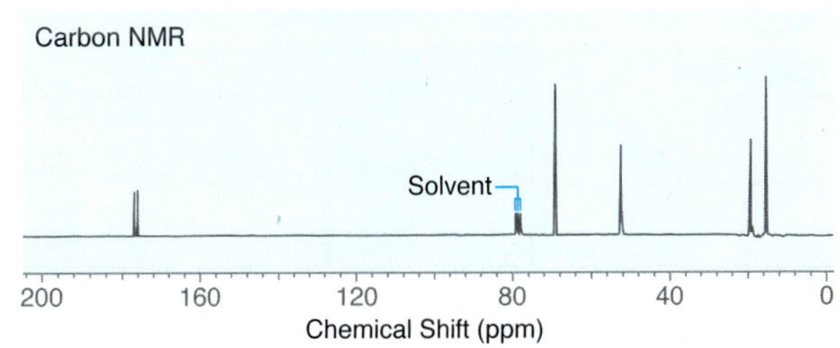

**16.67** Deduce the structure of a compound with molecular formula $C_8H_{14}O_4$ that exhibits the following IR, $^1H$ NMR, and $^{13}C$ NMR spectra:

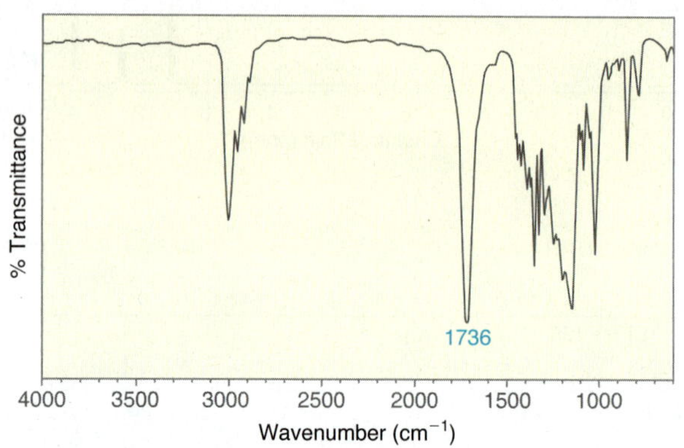


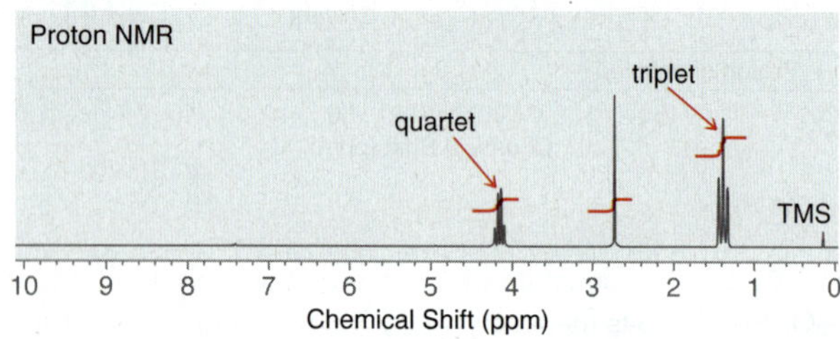


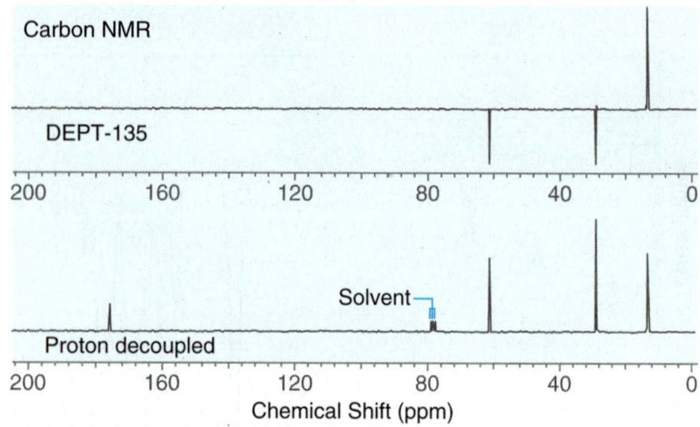


**16.68** Deduce the structure of a compound with molecular formula $C_{12}H_8Br_2$ that exhibits the following $^1H$ NMR and $^{13}C$ NMR spectra:

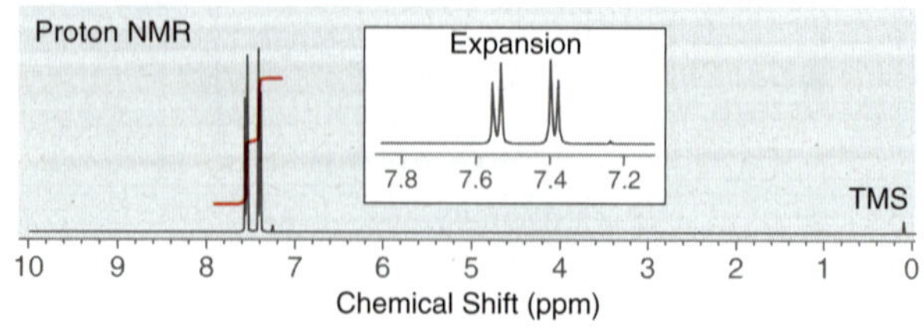


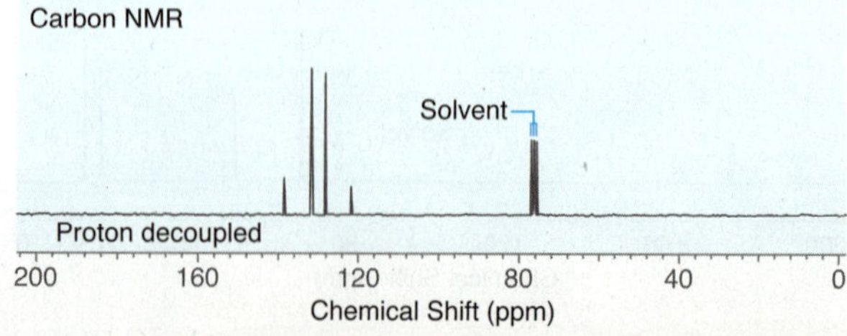


**16.69** An unknown compound exhibits the following IR, $^1H$ NMR, and $^{13}C$ NMR spectra. In a mass spectrum of this compound, the $(M)^{+\bullet}$ peak appears at $m/z = 104$, and the $(M+1)^{+\bullet}$ peak is 4.4% the height of the parent ion. Elemental analysis reveals that the compound consists only of carbon, hydrogen and oxygen atoms.

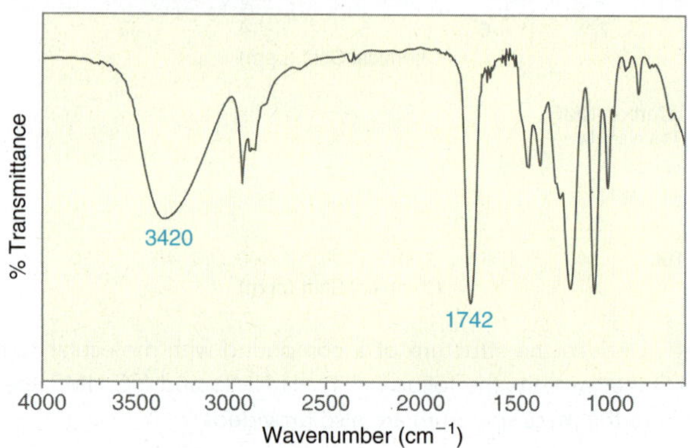


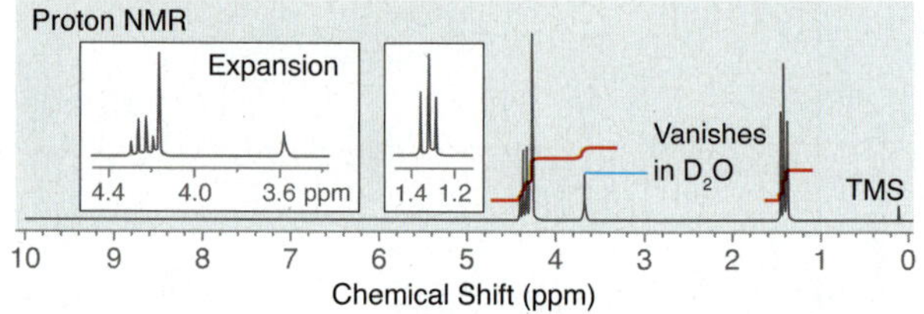


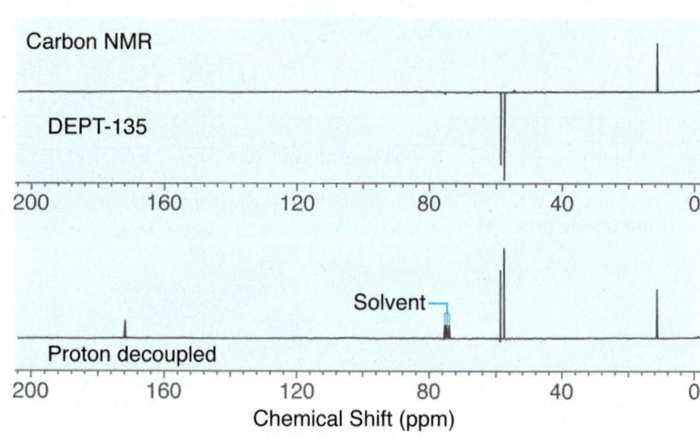


(a) Determine the molecular formula of the compound.

(b) Deduce the structure of the compound.

**16.70** Consider the structure of *N,N*-dimethylformamide (DMF):

We might expect the two methyl groups to be equivalent; however, both the proton and carbon NMR spectra of DMF show two separate signals for the methyl groups. Propose an explanation for the nonequivalence of the methyl groups. Would you expect the signals to collapse into one signal at high temperature?

(*Hint*: You may want to turn to Section 2.12 and consider the hybridization state of the nitrogen atom.)

**16.71** Consider the two methyl groups shown in the following compound. Explain why the methyl group on the right side appears at lower chemical shift.

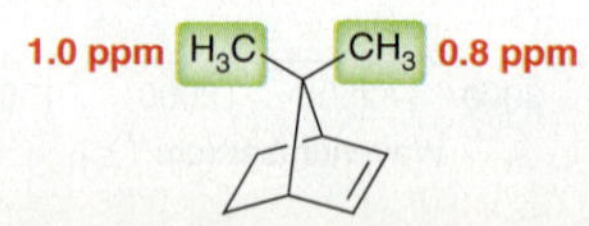

# CHALLENGE PROBLEMS

**16.72** Brevianamide S, a potent antitubercular natural product, was recently isolated from a marine sediment collected off the coast of China (*Org. Lett.* **2012,** *14,* 4770–4773). Predict the chemical shift and determine the multiplicity of each signal in the $^1$H NMR spectrum for brevianamide S (chemical shifts have been provided in situations where the relevant benchmark values were not covered in this chapter).

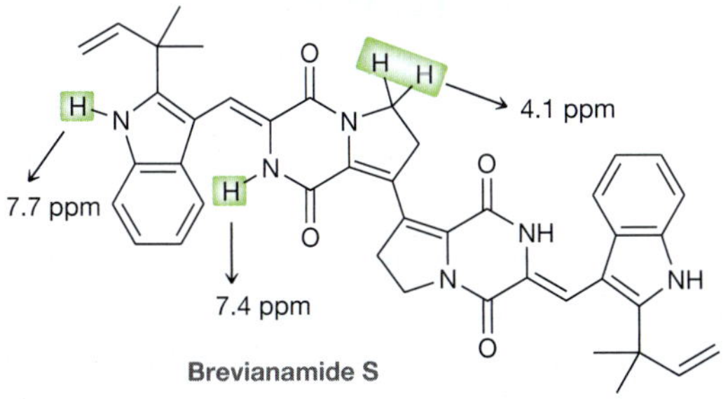

7.7 ppm 7.4 ppm 4.1 ppm

**Brevianamide S**

**16.73** The hydronaphthacene ring system consists of five fused six-membered rings, as shown below in compounds **3** and **4**. One method for construction of this ring system involves treatment of compound **1** with a Fischer carbene such as **2**. Two diastereomeric products are obtained, **3** and **4**, differing only in the configuration at one chirality center (*Tetrahedron Lett.* **2011,** *52,* 4182–4185). Explain how you could distinguish these two products using only $^{13}$C NMR.

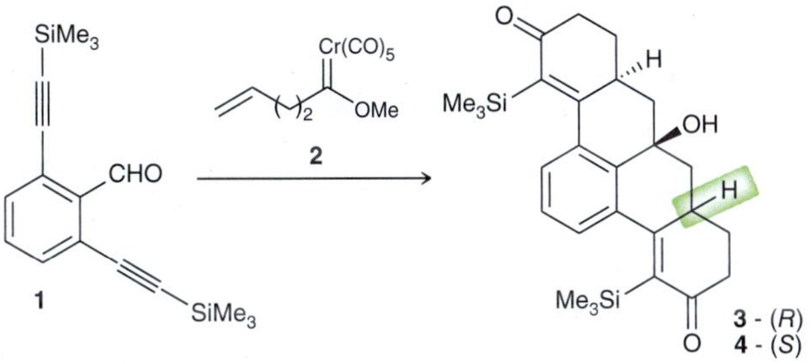

**3** - (*R*) **4** - (*S*)

**16.74** Treatment of 1,3,6-cyclononatriene (compound **1**), or its dimethyl derivative (compound **2**), with potassium amide ($KNH_2$) in liquid ammonia results in the formation of anion **1a** or **2a**, respectively (*J. Am. Chem. Soc.* **1973,** *95,* 3437–3438):

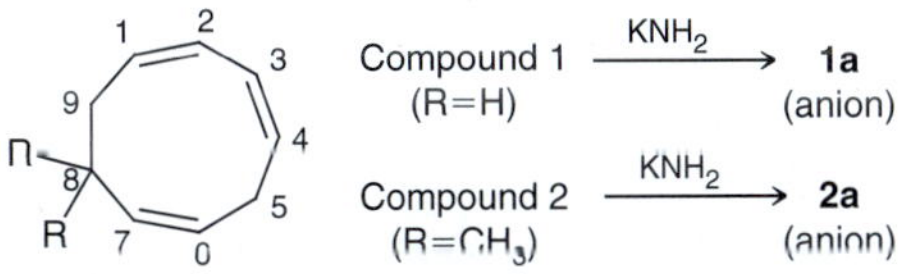

Compound 1 (R=H) $\xrightarrow{KNH_2}$ **1a** (anion)

Compound 2 (R=$CH_3$) $\xrightarrow{KNH_2}$ **2a** (anion)

(a) Draw all four resonance structures of **1a**.

(b) How many signals do you expect in the $^1$H NMR spectrum of **1a** and of **2a**?

(c) The $^1$H NMR spectrum of **1a** exhibits signals at 3.74 and 3.39 ppm, representing the protons at positions C1 and C3, respectively, while the protons at positions C2 and C4 produce signals at 5.63 and 5.52 ppm, respectively. Explain the two discrete groups of signals.

(d) In the $^1$H NMR spectrum of **2a**, how many signals do you expect to appear in the range 3–4 ppm, and how many signals in the range 5–6 ppm?

**16.75** Compound **1** can serve as a precursor in the synthesis of flutamide (**2**), a drug used in the treatment of prostate cancer (*J. Chem. Ed.* **2003,** *80,* 1439–1443):

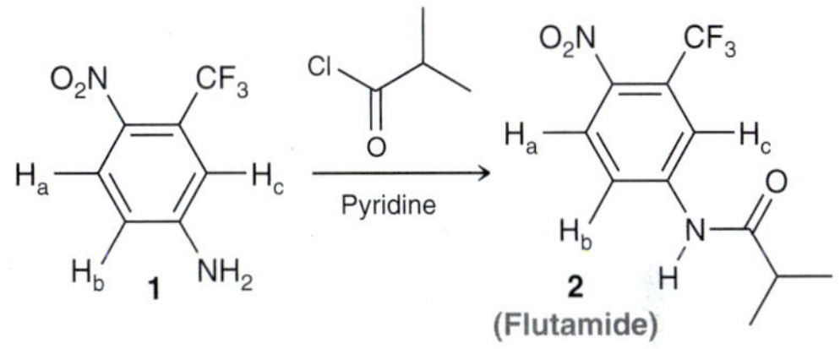

**(Flutamide)**

(a) Compound **1** has three distinct aromatic protons, labeled $H_a$, $H_b$, and $H_c$. Identify which aromatic proton gives rise to the signal most downfield. Justify your choice with an analysis of resonance effects and inductive effects.

(b) Consider the same three aromatic protons in compound **2**. Do you expect these three signals to be shifted upfield or downfield relative to the same three signals for compound **1**? Explain your reasoning and predict which of the protons are most affected by the transformation of **1** to **2**.

**16.76** Coumarin (**1**) and its derivatives exhibit a broad array of industrial applications, including, but not limited to, cosmetics, food preservatives, and fluorescent laser dyes. In the $^1$H NMR spectrum of compound **2** (a coumarin derivative), the two signals farthest downfield are at 7.38 and 8.42 ppm (*J. Chem. Ed.* **2006,** *83,* 287–289):

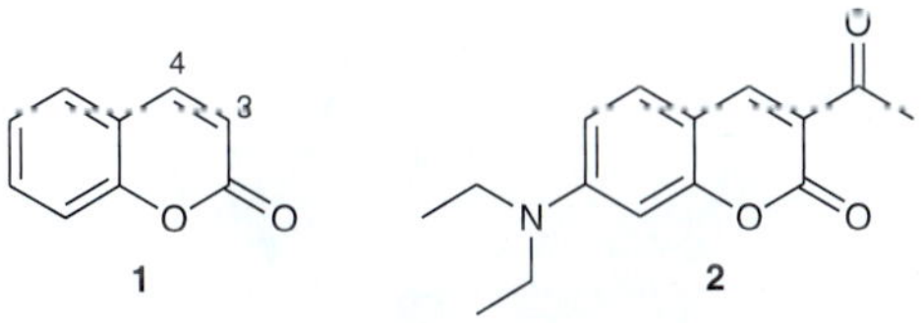

(a) Identify the two protons that most likely give rise to these signals and justify your decision using resonance structures.

(b) Do you expect the IR signal corresponding to the C3—C4 π bond to be a stronger signal in compound **1** or in compound **2**? Explain your reasoning.

17



# Conjugated Pi Systems and Pericyclic Reactions

## DID YOU EVER WONDER...

**how bleach removes stains that are not easily removed with regular detergents?**

Most bleaching agents do not actually remove stains, but instead, they react with the colored compounds in the stain to produce colorless compounds. The stain is still there, but it has been rendered "invisible." When illuminated with ultraviolet light, the invisible stain can be seen once again. In this chapter, we will learn about the structural features that cause organic compounds to absorb visible light. Specifically, we will see that color is a consequence of special conjugated π systems. In this chapter, we will learn about the structure, properties, and reactions of conjugated π systems.

## DO YOU REMEMBER?

*Before you go on, be sure you understand the following topics. If necessary, review the suggested sections to prepare for this chapter.*

- MO Theory and Pi Bonds (Sections 1.8, 1.9)
- Enthalpy, Entropy and Gibbs Free Energy (Sections 6.1–6.3)
- Conformations (Sections 4.7, 4.8)
- Equilibria, Kinetics, and Energy Diagrams (Sections 6.4–6.6)

Take the DO YOU REMEMBER? QUIZ in **WileyPLUS** to check your understanding

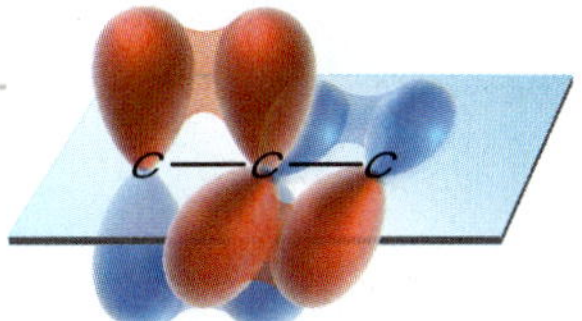


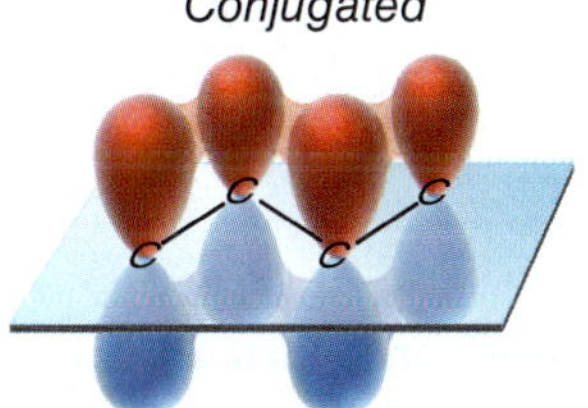


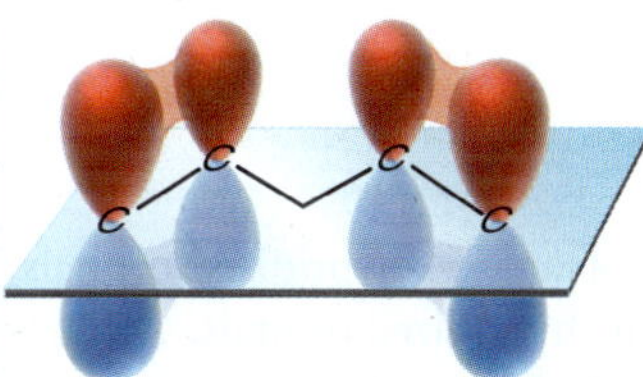


**FIGURE 17.1**
The relative spatial arrangement of the *p* orbitals in cumulated, conjugated, and isolated dienes

## 17.1 Classes of Dienes

**Dienes** are compounds that possess two C═C bonds. Depending on the proximity of the π bonds, they are classified as cumulated, conjugated, or isolated.

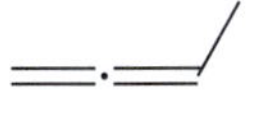


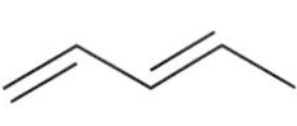


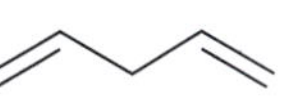


- In **cumulated dienes**, also called *allenes*, the π bonds are adjacent.
- In **conjugated dienes**, the π bonds are separated by exactly one σ bond.
- In **isolated dienes**, the π bonds are separated by two or more σ bonds.

The proximity of the π bonds is critical to understanding the structure and reactivity of a diene. In particular, consider the spatial arrangement of *p* orbitals in each of the dienes in Figure 17.1. Conjugated dienes represent a special category, because they contain one continuous system of overlapping *p* orbitals—that is, the two π bonds together comprise one functional group composed of four overlapping *p* orbitals. As such, conjugated dienes exhibit special properties and reactivity.

A C═C π bond can also be conjugated with other types of π bonds, such as carbonyl groups.

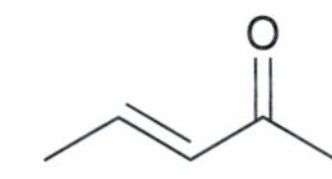


This chapter will focus on the properties and reactivity of conjugated π systems.

### CONCEPTUAL CHECKPOINT

**17.1** For each of the following compounds, identify whether each C═C π system is cumulated, conjugated, or isolated:

(a) 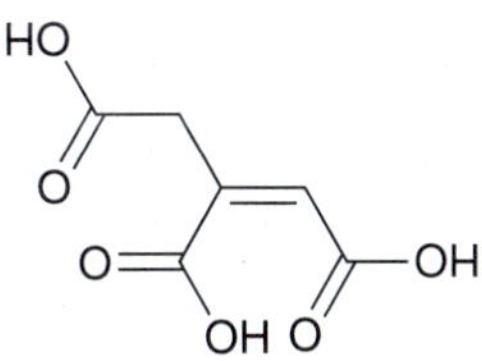


***cis*-Aconitic acid**
Plays a role in the citric acid cycle

(b) 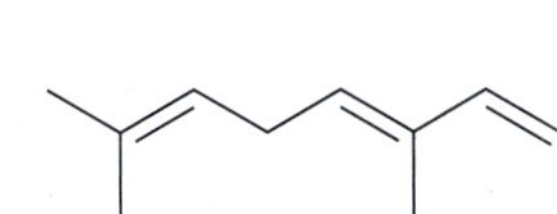

**Ocimene**
Present in the essential oils of many plants

(c) 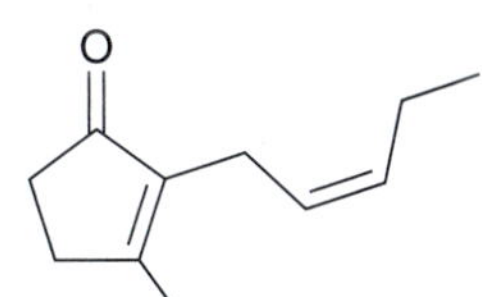


***cis*-Jasmone**
A constituent of many perfumes

(d) 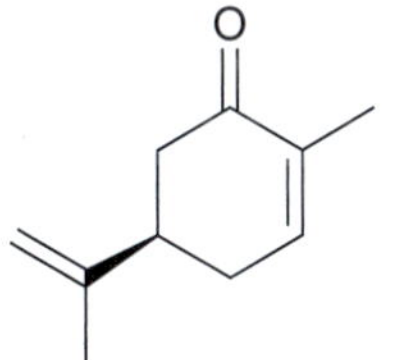


**(*R*)-Carvone**
Responsible for the flavor of spearmint

## 17.2 Conjugated Dienes

### Preparation

Conjugated dienes can be prepared from allylic halides via an elimination process.

Br

*t*-BuOK

A sterically hindered base, such as potassium *tert*-butoxide, is used to prevent the competing $S_N2$ reaction from occurring. Conjugated dienes can also be formed from dihalides via two successive elimination reactions.

Br

*t*-BuOK

Br

### Bond Lengths

Compare the bond lengths of the following single bonds:

$H_3C—CH_3$

148 pm

153 pm

The single bond in a conjugated diene is shorter than a typical C—C single bond. There are a couple of ways to explain this difference. The simpler approach derives from an analysis of the hybridization states involved:

$H_3C—CH_3$

$sp^2$ $sp^2$ $sp^3$ $sp^3$

The C—C bond of a conjugated diene is formed from the overlap of two $sp^2$-hybridized orbitals, while the C—C bond in ethane is formed from the overlap of two $sp^3$-hybridized orbitals. Orbitals that are $sp^2$-hybridized have more *s* character than $sp^3$-hybridized orbitals (Figure 17.2).

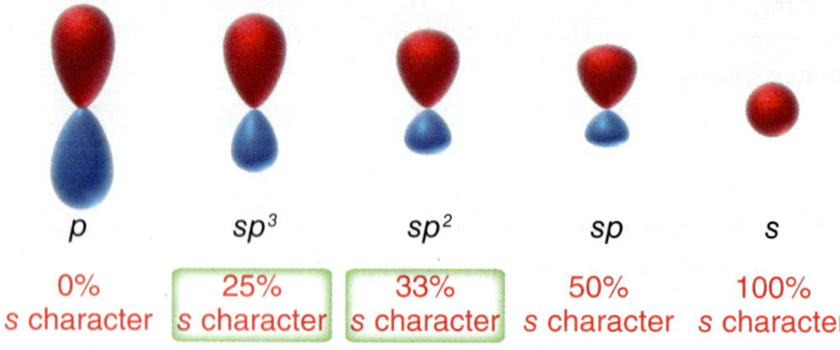


**FIGURE 17.2** The relative amount of "*s* character" for atomic orbitals.

Thus, the electron density of an $sp^2$-hybridized orbital will be closer to the nucleus than the electron density of an $sp^3$-hybridized orbital, and as a result, a bond between two $sp^2$-hybridized orbitals will be shorter than a bond between two $sp^3$-hybridized orbitals.

### Stability

We can compare the relative stability of conjugated and isolated dienes by comparing their heats of hydrogenation. For example, compare the heats of hydrogenation of 1-butene and 1,3-butadiene (Figure 17.3). We might expect that the heat of hydrogenation for 2 mol of 1-butene would be the same as the heat of hydrogenation for 1 mol of 1,3-butadiene. After all, both reactions produce butane as the sole product. Experiments reveal, however, that the heat

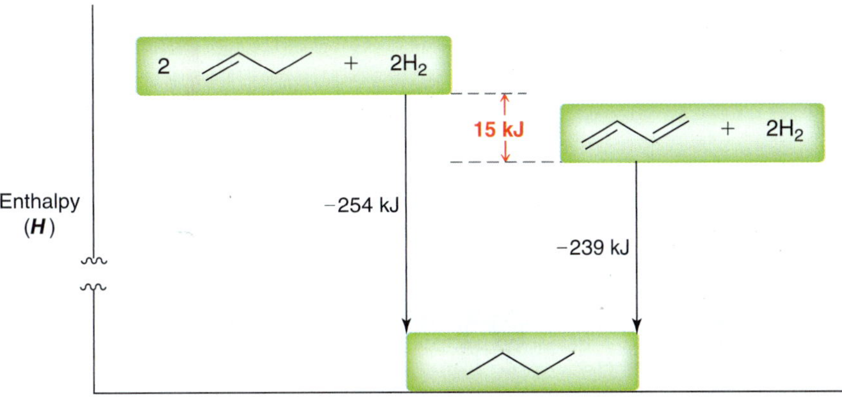


**FIGURE 17.3** A comparison of the heats of hydrogenation of isolated and conjugated π bonds.

of hydrogenation for the conjugated diene is less than expected. These results suggest that conjugated double bonds are more stable than isolated double bonds, and we can determine that the *stabilization energy* associated with a conjugated diene is approximately 15 kJ/mol. The source of this stabilization energy will be discussed in Section 17.3.

## CONCEPTUAL CHECKPOINT

**17.2** Identify reagents necessary to convert cyclohexane into 1,3-cyclohexadiene. Notice that the starting material has no leaving groups and cannot simply be treated with a strong base. You must first introduce functionality into the starting material (for help with introducing functional groups, see Section 12.2).

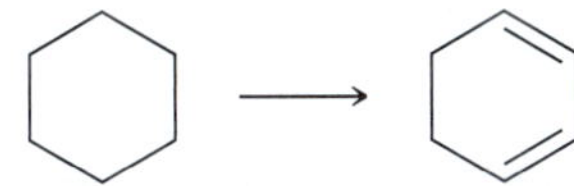

**17.3** Rank the three C—C bonds of 1-butene in terms of bond length (from shortest to longest).

**17.4** Compare the following three isomeric dienes:

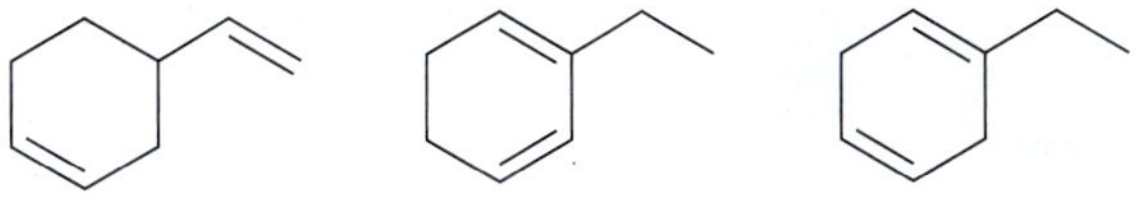

(a) Which compound will liberate the least heat upon hydrogenation with 2 mol of hydrogen gas? Why?

(b) Which compound will liberate the most heat upon hydrogenation with 2 mol of hydrogen gas? Why?
(For help, see Section 8.5.)

**17.5** Identify the most stable compound:

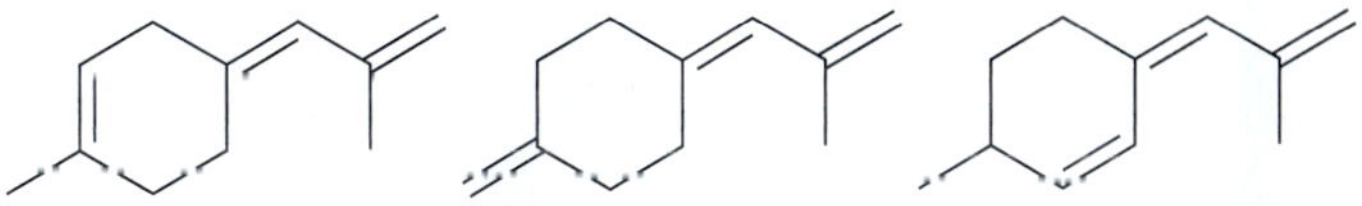

## Conformations of 1,3-Butadiene

At room temperature, 1,3-butadiene experiences free rotation about the C2—C3 bond, giving rise to two important conformers.

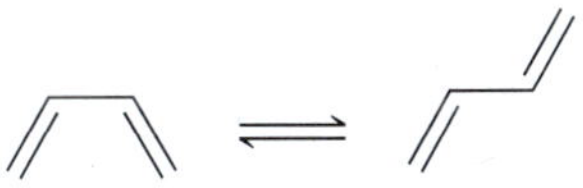

s-cis ⇌ s-trans

In the ***s-cis*** conformer, the disposition of the two π bonds with regard to the connecting single bond is *cis-like* (a dihedral angle of 0°). In the ***s-trans*** conformer, the disposition of the two π bonds with regard to the connecting single bond is *trans-like* (a dihedral angle of 180°). In each of these conformers, the *p* orbitals effectively overlap to give a continuous, conjugated π system (Figure 17.4).

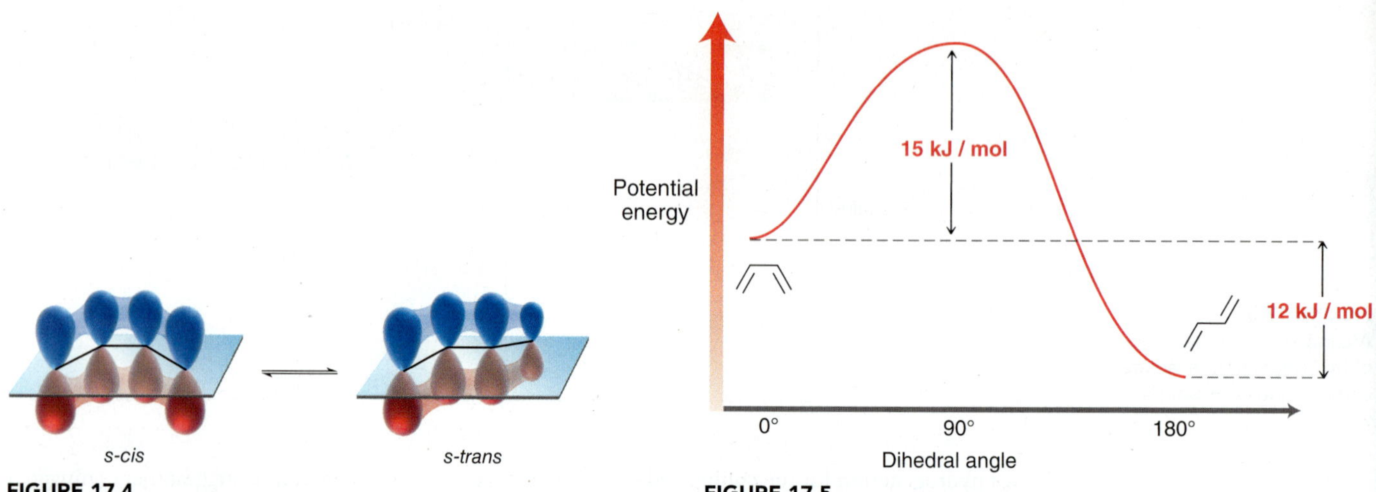


**FIGURE 17.4**
The *p* orbitals of 1,3-butadiene effectively overlap in both the *s-cis* and *s-trans* conformers.

**FIGURE 17.5**
An energy diagram that illustrates the equilibrium between the *s-cis* and *s-trans* conformations.

At room temperature, these two conformers rapidly interconvert, establishing an equilibrium (Figure 17.5). The activation energy for conversion from the *s-cis* to the *s-trans* conformer is approximately 15 kJ/mol, which is equivalent to the stabilization energy associated with conjugated double bonds. In other words, the stabilizing effect of conjugation is completely destroyed when the C2—C3 bond is rotated by 90°. In such a high-energy conformation, the two π bonds do not effectively overlap, so they are essentially equivalent to isolated double bonds (Figure 17.6).

Due to steric factors, the *s-trans* conformer is significantly lower in energy than the *s-cis* conformer, and as a result, the equilibrium favors the *s-trans* conformer. At any given time, approximately 98% of the molecules assume an *s-trans* conformation, while only 2% of the molecules assume an *s-cis* conformation.

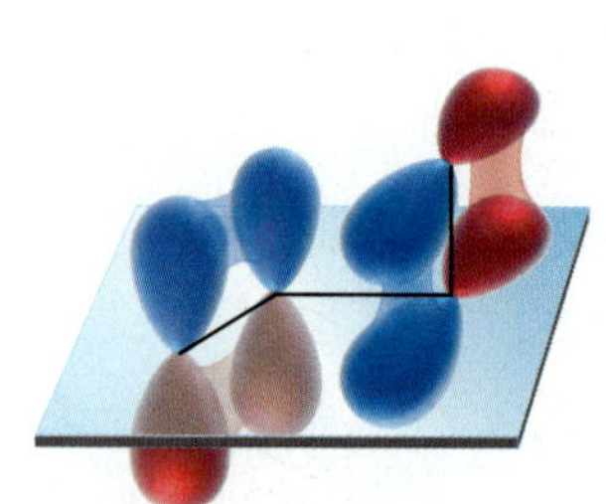

**FIGURE 17.6**
The *p* orbitals of 1,3-butadiene do not effectively overlap when the C2—C3 bond is rotated by 90°.

## 17.3 Molecular Orbital Theory

Throughout this chapter, we will invoke molecular orbital (MO) theory to explain the reactivity and properties of conjugated π systems. This section is designed to review and introduce some of the most basic features of MO theory.

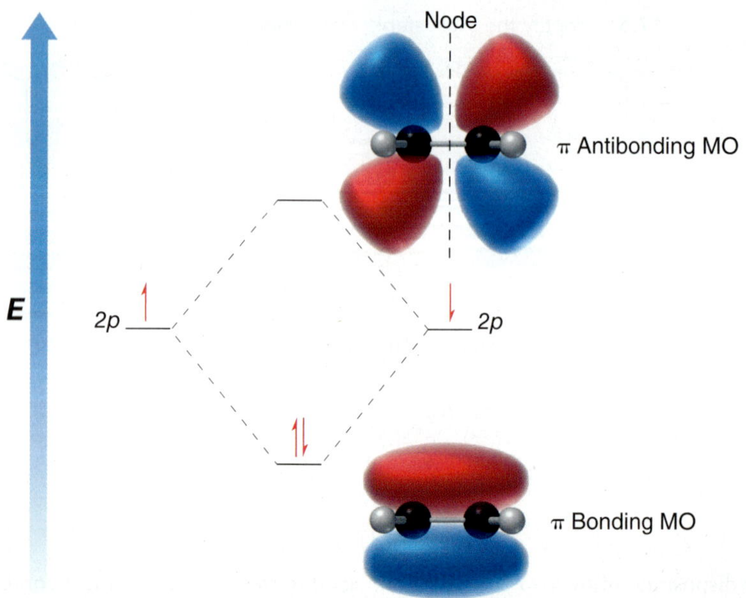


**FIGURE 17.7**
An energy diagram showing images of the bonding and antibonding MOs associated with the π bond in ethylene.

Recall that a π bond is constructed from overlapping *p* orbitals (Figure 1.29). According to MO theory, these two atomic *p* orbitals are mathematically combined to produce two new orbitals, called *molecular orbitals*. There is a profound difference between an atomic orbital and a molecular orbital. Specifically, electrons that occupy atomic orbitals are associated with an individual atom, while electrons that occupy a molecular orbital are associated with the entire molecule. In Section 1.9, we discussed the molecular orbitals associated with the π bond of ethylene (Figure 17.7). According to MO theory, the two atomic *p* orbitals are replaced by two molecular orbitals—a bonding MO and an antibonding MO. The antibonding MO exhibits a node that is absent in the bonding MO (shown in Figure 17.7). Accordingly, the antibonding MO is higher in energy than the bonding MO. The π bond results from the ability of the π electrons to achieve a lower energy state by occupying the bonding MO.

### Molecular Orbitals of Butadiene

A conjugated diene, such as 1,3-butadiene, is comprised of four overlapping *p* orbitals. According to MO theory, these four atomic *p* orbitals are mathematically combined to

produce four molecular orbitals (Figure 17.8). The lowest energy MO ($\psi_1$) has no vertical nodes. The next MO ($\psi_2$) has one vertical node, and each higher energy MO has one additional vertical node. These four MOs are often represented with the shorthand drawings in Figure 17.9. Although these drawings resemble *p* orbitals, they are actually used to represent *molecular* orbitals, rather than *atomic* orbitals. These drawings are just a quick, shorthand method for drawing the phases and nodes of the MOs from Figure 17.8. We will use this method several times in this chapter, so it is important to understand that these drawings are merely simplified representations of molecular orbitals.

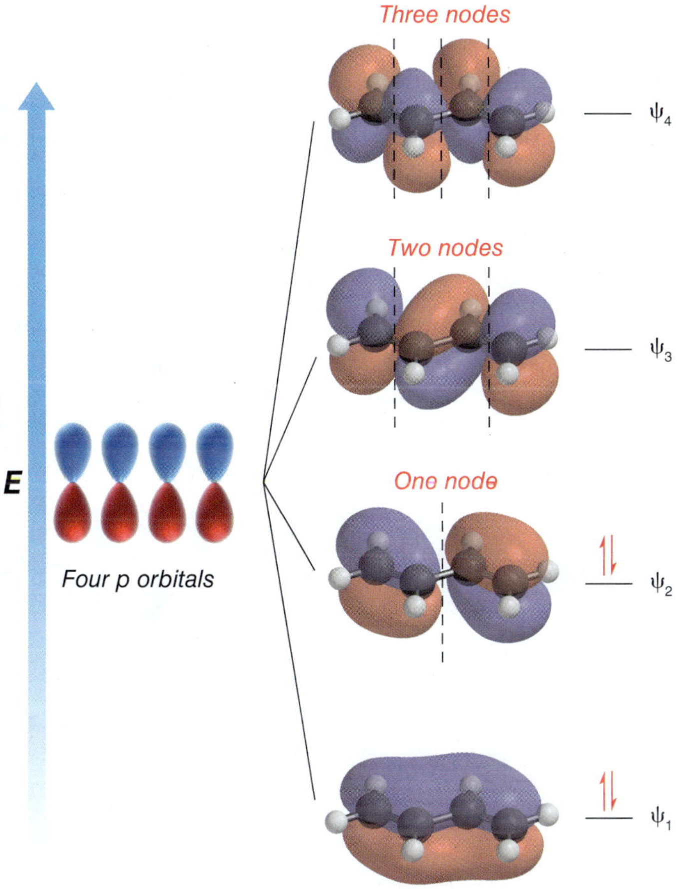

**FIGURE 17.8** The MOs of 1,3-butadiene.

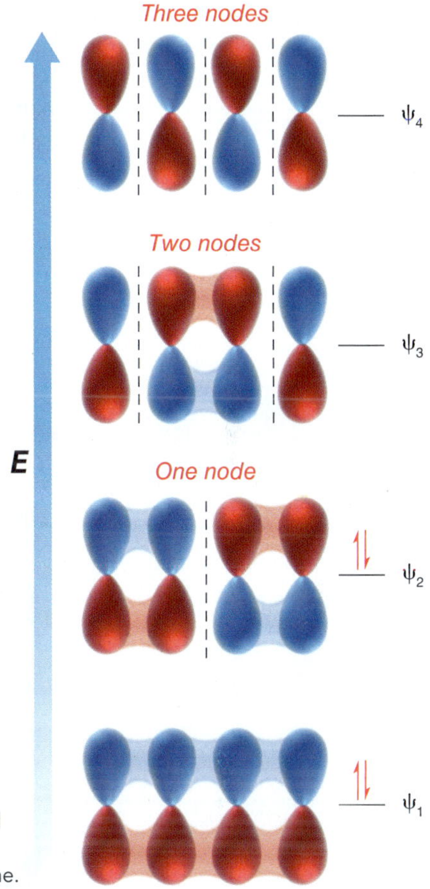

**FIGURE 17.9** A shorthand method for representing the MOs of 1,3-butadiene.

MO theory provides an explanation for the stability associated with conjugated dienes, as discussed in Section 17.2. As seen in Figure 17.9, the four $\pi$ electrons of a conjugated diene occupy the two lowest energy MOs (the bonding MOs). So let's focus our attention on those two MOs. The lowest energy MO ($\psi_1$) exhibits double-bond character at C2—C3, while the second MO ($\psi_2$) does not (Figure 17.10). All four $\pi$ electrons occupy these two MOs ($\psi_1$ and $\psi_2$).

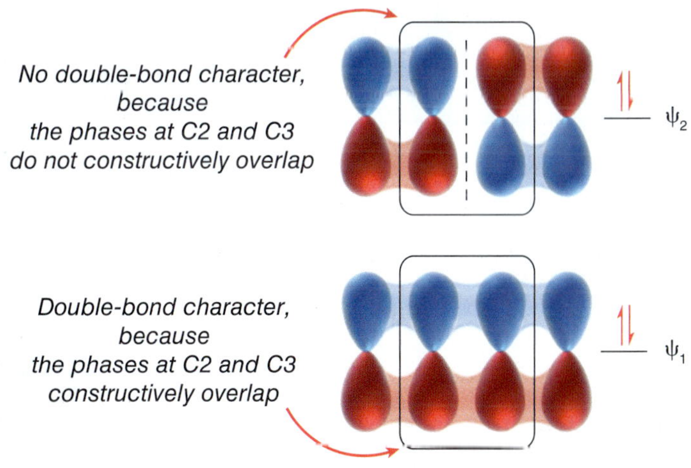


**FIGURE 17.10** The occupied MOs of 1,3-butadiene.

Therefore, the C2—C3 bond has some double-bond character and some single-bond character. MO theory therefore provides us with an alternate way of explaining the short bond distance of the C2—C3 bond. In addition, MO theory also explains the stabilization energy associated with a conjugated diene. Specifically, the two electrons that occupy $\psi_1$ are delocalized over four carbon atoms. This delocalization accounts for the observed stabilization energy discussed in the previous section.

## Molecular Orbitals of Hexatriene

A conjugated triene, such as 1,3,5-hexatriene, is comprised of six overlapping *p* orbitals. According to MO theory, these six atomic *p* orbitals are mathematically combined to produce six molecular orbitals (Figure 17.11). In this case, the six $\pi$ electrons occupy the three MOs that are lowest in energy (the bonding MOs). Of these three MOs, the highest in energy is called the

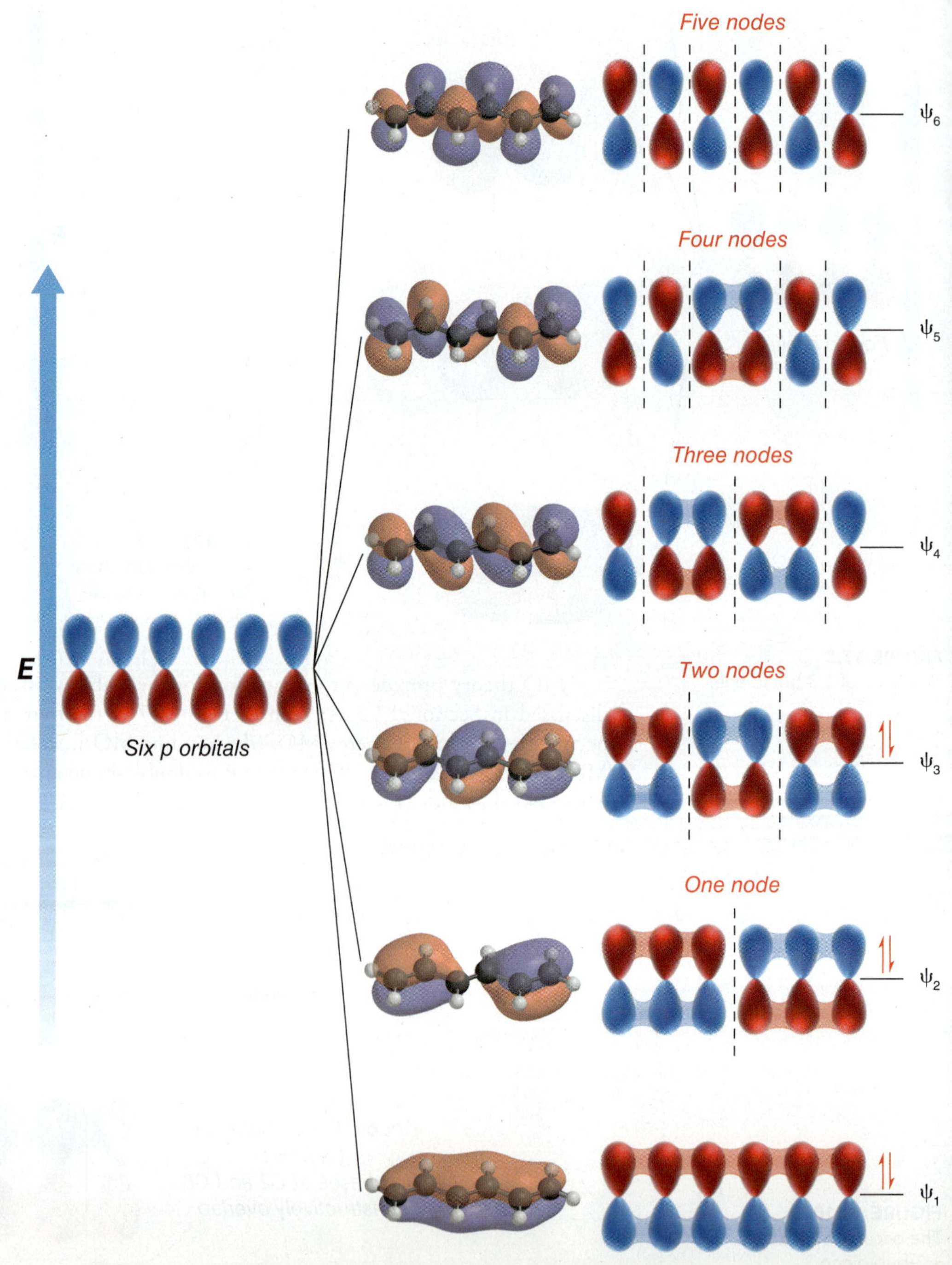


**FIGURE 17.11**
The MOs of 1,3,5-hexatriene.

*H*ighest *O*ccupied *M*olecular *O*rbital, or HOMO for short. Of the three unoccupied MOs (the antibonding MOs), the lowest in energy is called the *L*owest *U*noccupied *M*olecular *O*rbital, or LUMO for short.

For any conjugated polyene, the HOMO and the LUMO are the most important MOs to consider and are called the **frontier orbitals**. The HOMO contains the highest energy $\pi$ electrons, which are most readily available to participate in a reaction, and the LUMO is the lowest energy MO that is capable of accepting electron density. Throughout this chapter, we will explore the reactivity of conjugated polyenes by focusing on their frontier orbitals, the HOMO and LUMO. This approach, called **frontier orbital theory**, was first advanced in1954 by Kenichi Fukui (Kyoto University, Japan), a corecipient of the Nobel Prize in Chemistry in 1981.

Conjugated $\pi$ systems are capable of interacting with light, a phenomenon that will be discussed in greater detail at the end of this chapter. Under the right conditions, a $\pi$ electron in the HOMO can absorb a photon of light bearing the appropriate energy necessary to promote the electron to the LUMO. For example, consider the photochemical excitation of hexatriene (Figure 17.12). In the ground state of hexatriene (prior to excitation), the HOMO is $\psi_3$, but in the **excited state**, the HOMO is $\psi_4$. Excitation causes a change in the identities of the frontier orbitals. The ability of light to affect the frontier orbitals will be important in Section 17.9, when we will discuss reactions induced by light, called **photochemical reactions**.

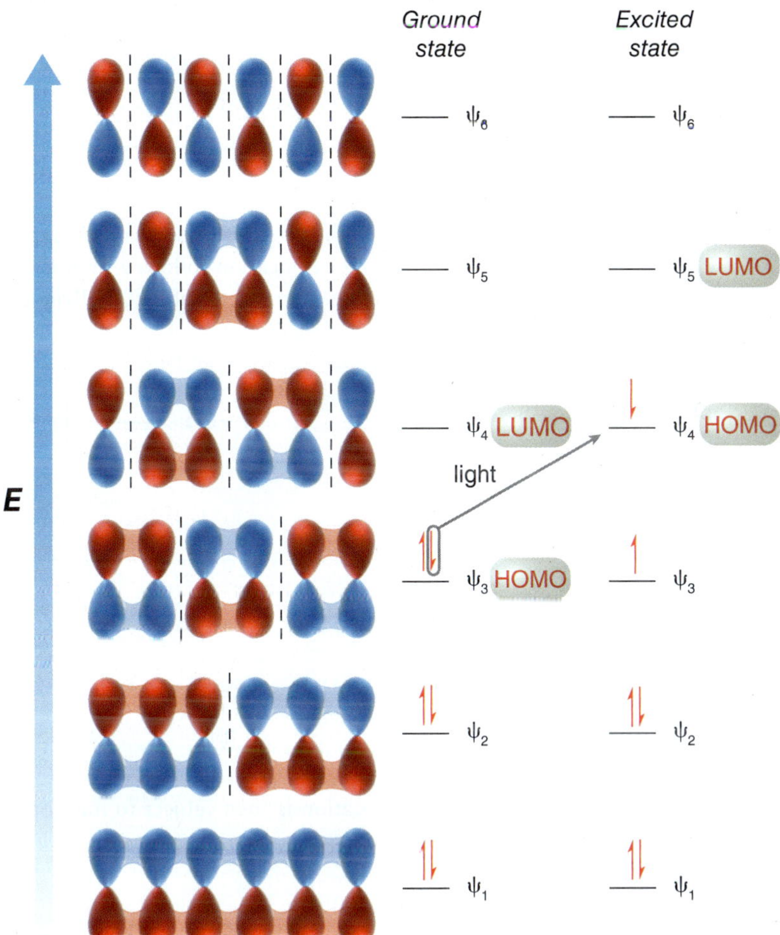


**FIGURE 17.12** An energy diagram showing the electron configuration of the ground state and excited state of 1,3,5-hexatriene.

## CONCEPTUAL CHECKPOINT

**17.6** Draw an energy diagram showing the relative energy levels of the MOs for 1,3,5,7-octatetraene and identify the HOMO and LUMO for both the ground state and the excited state.

## 17.4 Electrophilic Addition

### Addition of HX Across 1,3-Butadiene

In Section 9.3, we learned about the addition of HX across an alkene.

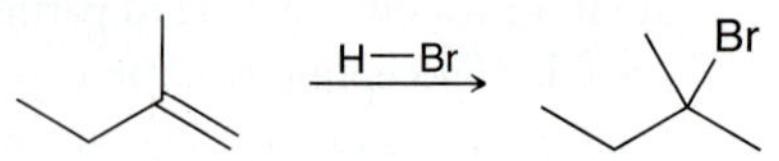

H—Br

Br

We saw that the reaction proceeds via Markovnikov addition, that is, the bromine atom is installed at the more substituted position. This regiochemical outcome was explained with the following proposed mechanism, comprised of two steps:

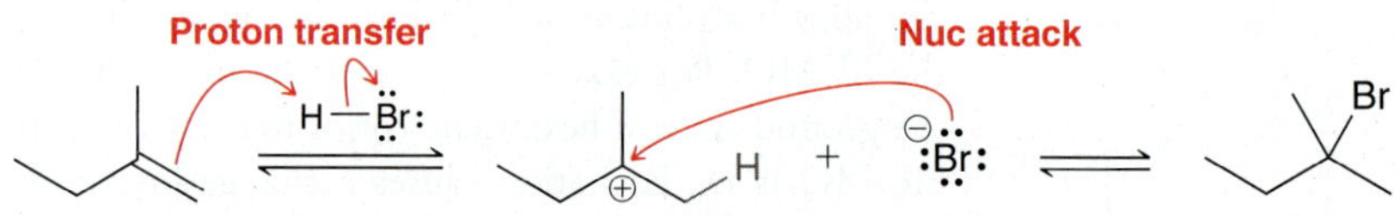

Proton transfer

Nuc attack

The first step of the mechanism controls the regiochemical outcome. Specifically, protonation gives the more stable, tertiary carbocation, rather than the less stable, primary carbocation.

When butadiene is treated with HBr, a similar process takes place, but two major products are observed.

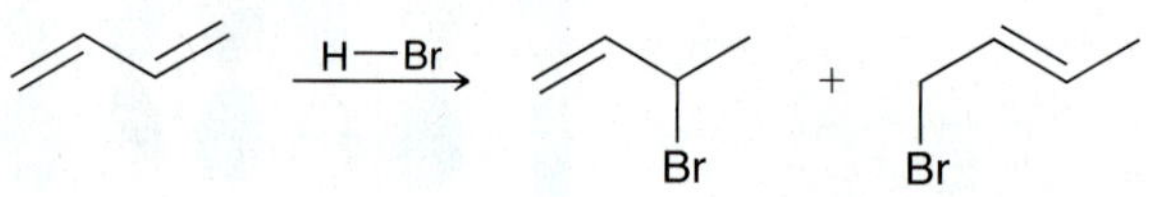

H—Br

Br + Br

The formation of these two products can be explained with a similar two-step mechanism: protonation to form a carbocation followed by nucleophilic attack. In the first step, protonation creates the more stable, resonance-stabilized, allylic carbocation, rather than an unstabilized primary carbocation.

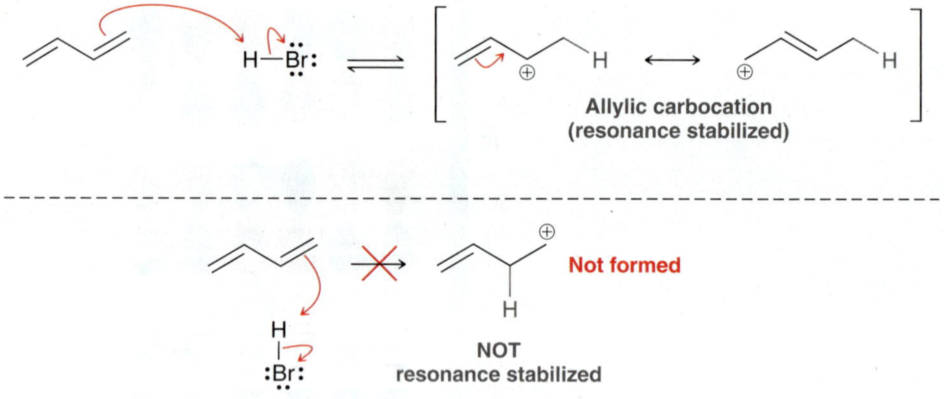

Allylic carbocation (resonance stabilized)

Not formed

NOT resonance stabilized

This allylic carbocation is then subject to nucleophilic attack in either of two positions, leading to two different products.

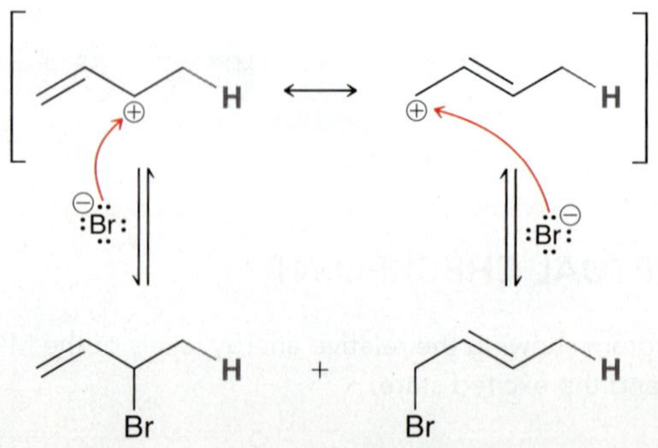

These compounds are said to be the products of **1,2-addition** and **1,4-addition**, respectively. This terminology derives from the fact that the starting diene contains a π system spread over four atoms, and the positions of H and Br are either at C1 and C2 or at C1 and C4. The products are called the **1,2-adduct** and the **1,4-adduct**, respectively.

1,2-Adduct 1,4-Adduct

The exact product distribution (the ratio of products) is temperature dependent. This temperature dependence will be explored in detail in the next section. For now, let's practice drawing both products and showing the mechanism of their formation.

# SKILLBUILDER

## 17.1 PROPOSING THE MECHANISM AND PREDICTING THE PRODUCTS OF ELECTROPHILIC ADDITION TO CONJUGATED DIENES

**LEARN the skill** Predict the products of the following reaction and propose a mechanism that explains the formation of each product:

### SOLUTION

This problem involves the addition of HBr across a conjugated π system. We therefore expect the mechanism to exhibit two steps: (1) proton transfer and (2) nucleophilic attack. In order to predict the products properly, we must explore the mechanism and identify the carbocation that is formed. In other words, we must carefully consider the regiochemistry of the protonation step. In this case, there are four distinct locations where the proton can be placed.

**STEP 1** Identify the possible locations where protonation can occur.

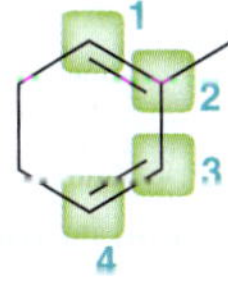


Protonation at either C2 or C3 will not produce an allylic carbocation.

| Protonation at C2 | Protonation at C3 |
| --- | --- |
| NOT resonance stabilized | NOT resonance stabilized |

In either case, protonation gives a secondary carbocation that is not resonance stabilized. Only protonation at C1 or C4 will produce an allylic carbocation:

**STEP 2**
Determine the locations where protonation will produce an allylic carbocation.

**Protonation at C1**

Resonance stabilized

**Protonation at C4**

Resonance stabilized

In this example, there are two different allylic carbocation intermediates that could be formed, each of which is resonance stabilized. In order to draw all possible products, consider the consequence of nucleophilic attack at each possible electrophilic position in each of the allylic carbocations:

**STEP 3**
Draw a nucleophilic attack occurring at each possible electrophilic position.

The last two drawings represent the same compound, giving a total of three expected constitutional isomers.

**PRACTICE the skill** **17.7** Predict the products for each of the following reactions and propose a mechanism that explains the formation of each product:

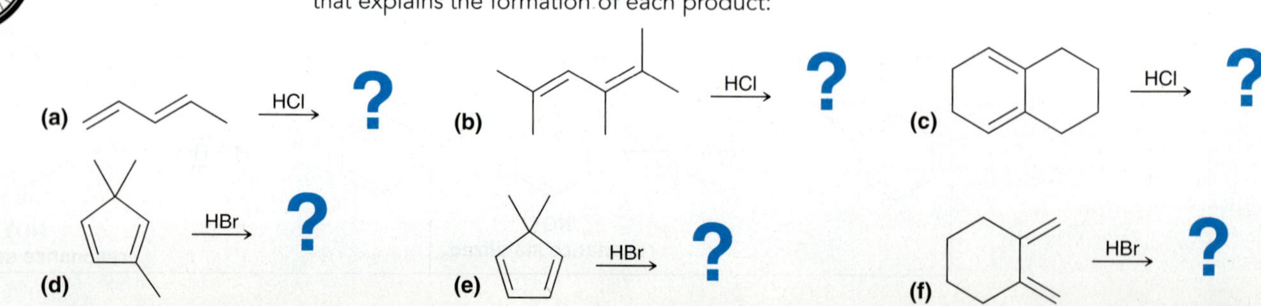

**APPLY the skill**

**17.8** Consider the following two dienes. When treated with HBr, one of these dienes yields four products, while the other diene yields only two products. Explain.

**17.9** Propose the structure of a conjugated diene for which 1,2-addition of HBr produces the same product as 1,4-addition of HBr.

need more **PRACTICE?** **Try Problems 17.35, 17.38**

### Addition of $Br_2$ Across 1,3-Butadiene

Many other electrophiles will also add across conjugated π systems to produce mixtures of 1,2- and 1,4-adducts. For example, bromine will add across 1,3-butadiene to produce the following mixture of products:

$Br_2$ → Br, Br (**1,2-Adduct**) + Br, Br (**1,4-Adduct**)

## 17.5 Thermodynamic Control vs. Kinetic Control

In the previous section, we saw that conjugated π systems will undergo 1,2-addition as well as 1,4-addition. The exact ratio of products is highly dependent on the temperature at which the reaction is performed.

H—Br → Br (**1,2-Adduct**) + Br (**1,4-Adduct**)

| | 1,2-Adduct | 1,4-Adduct |
|---|---|---|
| **At 0°C** | **71%** | **29%** |
| **At 40°C** | **15%** | **85%** |

When butadiene is treated with HBr at low temperature (0°C), the 1,2-adduct is favored. However, when the same reaction is performed at elevated temperature (40°C), the 1,4-adduct is favored. To understand the role temperature plays in this competition, we must look carefully at an energy diagram showing the formation of both products (Figure 17.13).

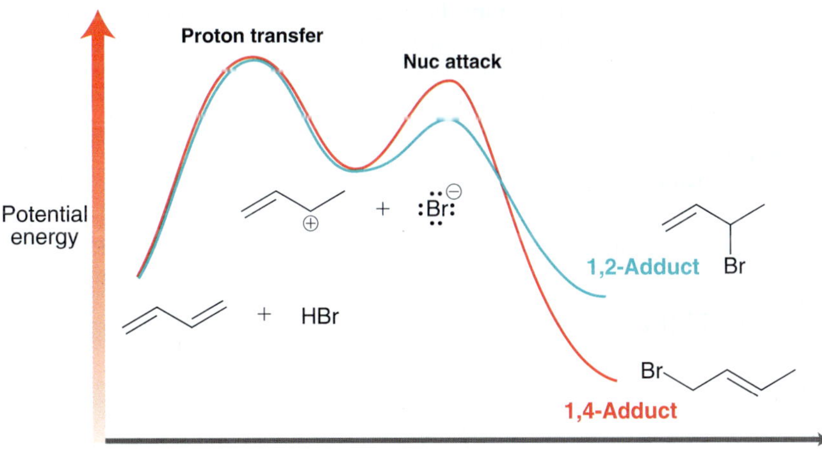


**FIGURE 17.13** An energy diagram illustrating the competing pathways for 1,2-addition and 1,4-addition.

The first step of the mechanism is identical for both 1,2-addition and 1,4-addition, namely, the conjugated diene is protonated to give a resonance-stabilized, allylic carbocation and a bromide ion. However, the second step of the mechanism can occur via either of two competing pathways (shown in blue and red). Comparing these pathways, we see that 1,4-addition leads to a more stable product (lower in energy), while 1,2-addition occurs more rapidly (lower energy of activation). The 1,4-adduct is lower in energy because it exhibits a more substituted double bond:

Less substituted alkene — Br — 1,2-Adduct

More substituted alkene — Br — 1,4-Adduct

**LOOKING BACK**
For a review of alkene stability, see Section 8.5.

The 1,2-adduct is believed to form more rapidly as a result of a *proximity effect*. Specifically, the carbocation and the bromide ion are initially very close to each other immediately after their formation in the first step of the mechanism. The bromide ion is simply closer in proximity to C2 than to C4, so attack at C2 occurs more rapidly.

Bromide ion is closer to C2 than to C4

The 1,2-adduct is favored as a result of the proximity effect, and at low temperatures, there is insufficient energy for it to be converted back into the allylic carbocation. Such a process would require loss of a leaving group (bromide), which is simply too slow at low temperatures. Under these conditions, the competing reaction pathways (1,2-addition and 1,4-addition) are both practically irreversible. The 1,2-addition pathway occurs more rapidly and therefore generates the 1,2-adduct as the major product. The reaction is said to be under **kinetic control**, which means that the product distribution is determined by the *relative rates* at which the products are formed.

At elevated temperatures, the two competing pathways are no longer irreversible. The products have sufficient energy to lose a leaving group, re-forming the allylic carbocation intermediate. Under such conditions, an equilibrium is established, and the product distribution will depend only on the relative energy levels of the two products. The product that is lower in energy (the 1,4-adduct) will predominate. The reaction is said to be under **thermodynamic control,** which means that the ratio of products is determined solely by the distribution of energy among the products. If such a reaction is monitored, it is found that the 1,2-adduct is initially formed at a faster rate (as a result of the proximity effect). However, equilibrium concentrations are quickly established, and the 1,4-adduct ultimately predominates.

## SKILLBUILDER

### 17.2 PREDICTING THE MAJOR PRODUCT OF AN ELECTROPHILIC ADDITION TO CONJUGATED DIENES

**LEARN** the skill

Predict the products of the following reaction and determine which product will predominate:

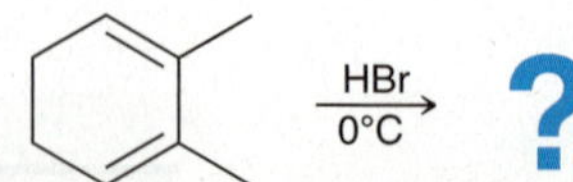

## SOLUTION

**STEP 1**
Identify the possible locations where protonation can occur.

First let's predict all of the products. This problem involves the addition of HBr across a conjugated π system. We therefore expect the mechanism to exhibit two steps: (1) proton transfer and (2) nucleophilic attack. In order to predict the products properly, we must explore the mechanism and identify the carbocation that is formed. Begin by carefully considering the regiochemistry of the protonation step. In this case, it might seem that there are four possibilities (highlighted).

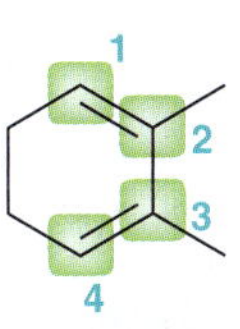


But this compound is symmetrical, with C1 being equivalent to C4, and C2 being equivalent to C3. Therefore, there are only two distinct possibilities for the protonation step: at C1 or at C2. The resonance-stabilized, allylic carbocation is only formed via protonation at C1:

**STEP 2**
Draw the structure of the allylic carbocation that is expected to form.

Protonation at C1

**STEP 3**
Attack each electrophilic position with a bromide ion and draw the possible products.

This allylic carbocation can then be attacked at either of two positions, leading to 1,2- or 1,4-addition. To predict which product predominates, we must determine which adduct is the kinetic product and which is the thermodynamic product. The 1,2-adduct is expected to be the kinetic product due to the proximity effect, while the 1,4-adduct exhibits a more highly substituted (tetrasubstituted) π bond and is therefore the thermodynamic product.

1,2-Adduct | 1,4-Adduct

**STEP 4**
Identify the kinetic and thermodynamic products and then choose which product predominates based on the temperature of the reaction.

Finally, look carefully at the temperature indicated. In this case, a low temperature is used, so the reaction is under kinetic control. Therefore, in this case, the 1,2-adduct is expected to be favored.

HBr, 0°C → Major + Minor

**PRACTICE the skill** **17.10** Predict the products for each of the following reactions, and in each case, determine which product will predominate:

(a) HBr, 40°C → ?

(b) 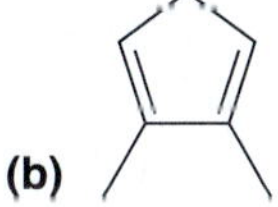 HCl, 0°C → ?

(c) 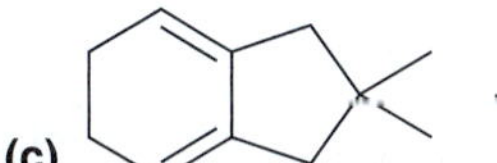 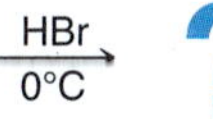


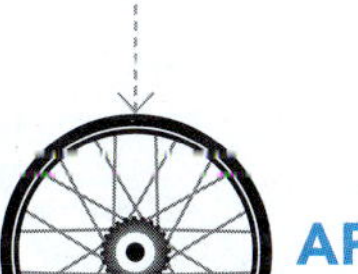

**APPLY the skill** **17.11** When 1,4-dimethylcyclohepta-1,3-diene is treated with HBr at elevated temperature, the 1,2-adduct predominates, rather than the 1,4-adduct. Explain this result.

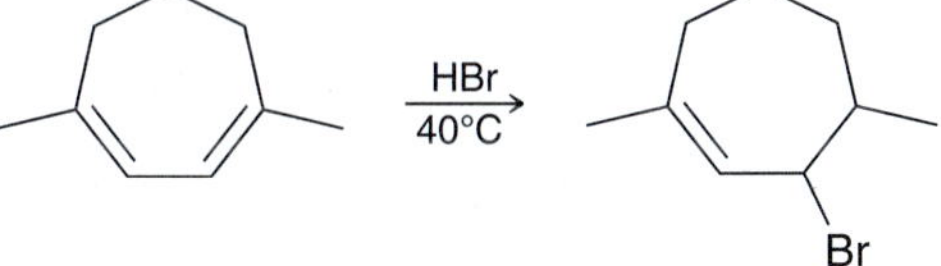


**17.12** When 1,3-cyclopentadiene is treated with HBr, the temperature does not have an influence on the identity of the major product. Explain.

need more **PRACTICE?** **Try Problems 17.36, 17.37, 17.39**

## practically speaking — Natural and Synthetic Rubbers

Natural rubber is produced from the polymerization of isoprene:

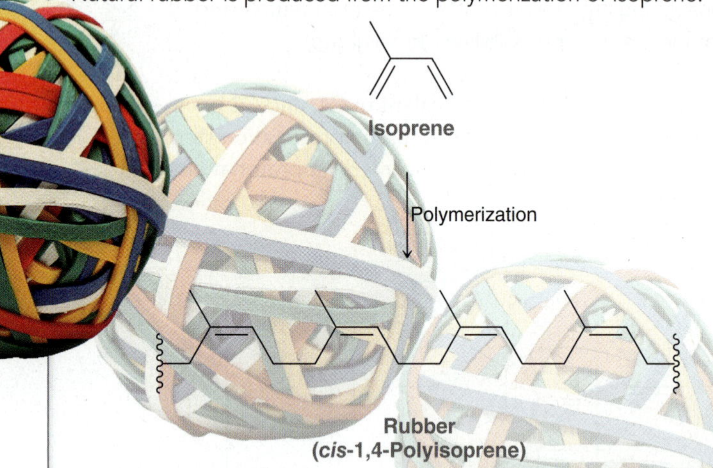


**Rubber (*cis*-1,4-Polyisoprene)**

Isoprene is used by plants in the biosynthesis of a wide variety of compounds. The importance of isoprene as a chemical precursor will be discussed in Section 26.7. During the polymerization of isoprene to form rubber, the monomers are joined by the head-to-tail linkage of C1 and C4 positions. The polymerization of isoprene is therefore a special case of 1,4-addition.

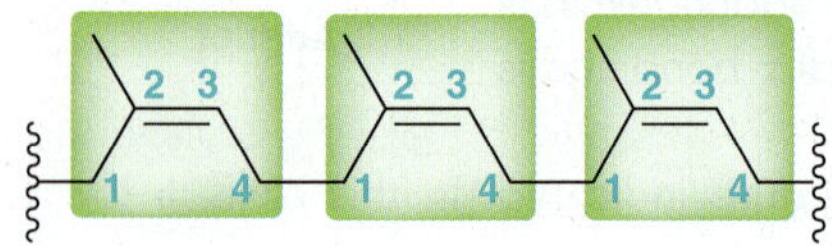


Rubber can also be made in the laboratory from polymerizing isoprene in the presence of suitable catalysts that favor formation of the *cis* polymer. Rubber produced in this way is virtually indistinguishable from natural rubber. Other substituted dienes can also be polymerized in the laboratory to produce a wide variety of rubberlike, synthetic polymers. In the early 1930s, chemists at the Du Pont chemical company produced a commercially important polymer via the free-radical polymerization of chloroprene. The resulting polymer is sold under the trade name Neoprene.

Cl

**Chloroprene**

Polymerization

Cl Cl Cl Cl

**Neoprene (*trans*-1,4-Polychloroprene)**

Neoprene is used in many applications, including insulation for electrical wiring, fan belts for automobiles, and scuba diving wetsuits.

When natural or synthetic rubber is stretched, neighboring polymer chains slide past one another and ultimately snap back into their original positions when the external force is removed. The ability of a polymer to return to its original shape after being stretched is called *elasticity*, and polymers that exhibit this quality are called *elastomers*. The elasticity of natural rubber is restricted to a specific temperature range, which limits its potential applications. In 1839, Charles Goodyear discovered that polymerization of isoprene in the presence of sulfur, a process called *vulcanization*, vastly improves the elasticity of the resulting polymer. Vulcanization is the result of the formation of disulfide linkages between neighboring polymeric chains:

S S S S

**Vulcanized rubber**

The sulfide linkages greatly improved the temperature range of rubber's elasticity. Vulcanized rubber maintains its elasticity even at high temperatures, because the disulfide linkages help snap the chains back into their original shape after the external force is removed. The vulcanized elastomer produced in greatest quantity is styrene-butadiene rubber (SBR). SBR is commercially prepared from styrene and butadiene via a free-radical polymerization process. It is called a copolymer, because it is made from two different monomers:

**Styrene** + **1,3-Butadiene**

**Styrene-butadiene rubber**

In the United States, approximately one billion pounds of vulcanized SBR is produced annually for the production of automobile tires.

CONCEPTUAL CHECKPOINT

**17.13** What monomer should be used to make each of the following polymers:

(a) (b)

## 17.6 An Introduction to Pericyclic Reactions

Most organic reactions, both in the laboratory and in living organisms, proceed via either ionic intermediates or radical intermediates. There is, however, a third category of organic reactions, called **pericyclic reactions**, which do not involve either ionic or radical intermediates. These reactions can be classified into three major groups: **cycloaddition reactions**, **electrocyclic reactions**, and **sigmatropic rearrangements** (Figure 17.14).

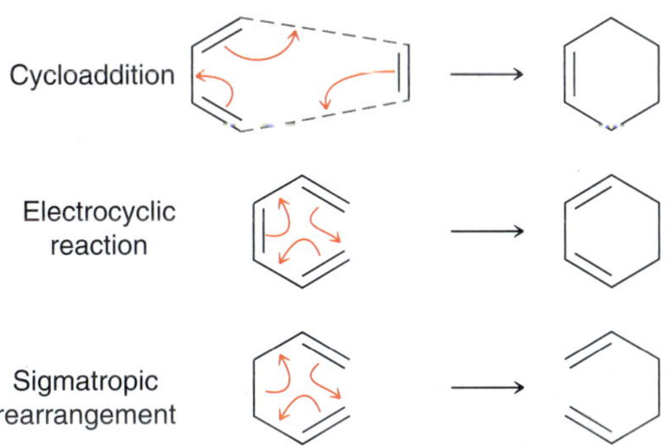


**FIGURE 17.14** Examples of the three major classes of pericyclic reactions.

Each of these reactions has the following features, characteristic of pericyclic reactions:

- The reaction proceeds via a concerted process, which means that all changes in bonding occur in a single step. As a result, the reaction mechanism has no intermediates.
- The reaction involves a ring of electrons moving around in a closed loop.
- The reaction occurs through a cyclic transition state.
- The polarity of the solvent generally does not have a large impact on the rate or yield of the reaction, suggesting that the transition state bears very little (if any) partial charge.

The three major types of pericyclic reactions differ in terms of the number and type of bonds that are broken and formed (Table 17.1). In a cycloaddition reaction, two $\pi$ bonds are converted into two $\sigma$ bonds, resulting in the addition of two reactants to form a ring. In an electrocyclic reaction, one $\pi$ bond is converted into a $\sigma$ bond, which effectively joins the ends of one reactant to form a ring. In a sigmatropic rearrangement, one $\sigma$ bond is formed at the expense of another, and the $\pi$ bonds change location. The following sections will explore each of the three categories of pericyclic reactions, beginning with cycloadditions.

**TABLE 17.1 A COMPARISON OF THE NUMBER OF BONDS BROKEN AND FORMED IN EACH OF THE THREE MAJOR TYPES OF PERICYCLIC REACTIONS**

| | CHANGE IN THE NUMBER OF $\sigma$ BONDS | CHANGE IN THE NUMBER OF $\pi$ BONDS |
|---|---|---|
| Cycloaddition | +2 | −2 |
| Electrocyclic | +1 | −1 |
| Sigmatropic | 0 | 0 |

## 17.7 Diels-Alder Reactions

**BY THE WAY**
Diels and Alder were awarded the 1950 Nobel Prize in Chemistry for their discovery of this reaction.

### The Mechanism of a Diels-Alder Reaction

The Diels-Alder reaction is an incredibly useful pericyclic reaction named after its discoverers, Otto Diels and Kurt Alder. In a Diels-Alder reaction, two C—C $\sigma$ bonds are formed simultaneously:

In the process, a ring is also formed, and the process is therefore a cycloaddition. Specifically, the Diels-Alder reaction is called a **[4+2] cycloaddition** because the reaction takes place between two different $\pi$ systems, one of which is associated with four atoms, while the other is associated with two atoms. The product of a Diels-Alder reaction is a substituted cyclohexene. As is the case for all pericyclic reactions, the Diels-Alder reaction is a concerted process.

The arrows can generally be drawn in either a clockwise or a counterclockwise fashion. Since the reaction takes place in just one step, the energy diagram has just one peak, where the top of the peak represents the transition state (Figure 17.15). The transition state is a six-membered ring in which three bonds are breaking and three bonds are forming simultaneously.

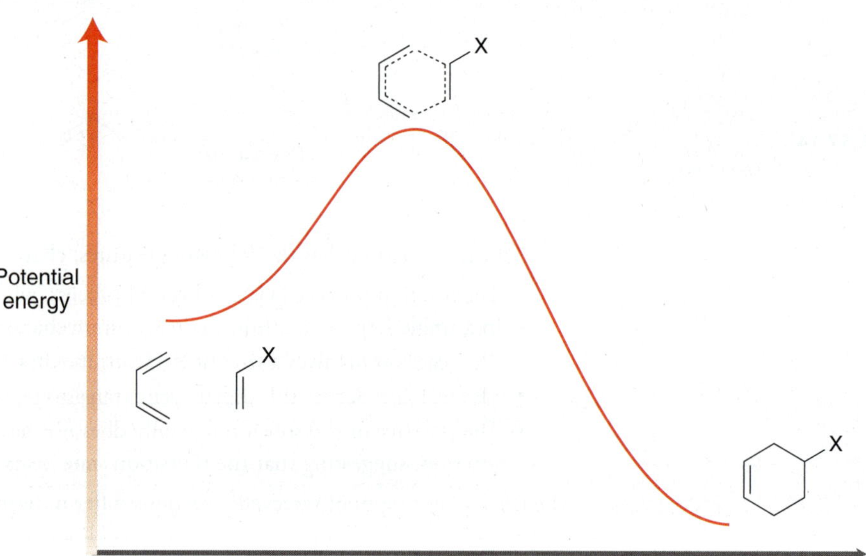


**FIGURE 17.15** An energy diagram of a Diels-Alder reaction.

### Thermodynamic Considerations

Moderate temperatures favor the formation of products in a Diels-Alder reaction, but very high temperatures (over 200°C) tend to disfavor product formation. In fact, in many cases, high temperatures can be used to achieve the reverse of a Diels-Alder reaction, called a **retro Diels-Alder reaction**.

Diels-Alder

Retro Diels-Alder (high temperature)

To understand why high temperatures favor ring opening rather than ring formation, recall from Section 6.3 that the sign of $\Delta G$ determines whether the equilibrium favors reactants or products. The sign of $\Delta G$ must be negative for the equilibrium to favor products. To determine the sign of $\Delta G$, there are two terms that must be considered:

$$\Delta G = \underbrace{(\Delta H)}_{\text{Enthalpy term}} + \underbrace{(-T\,\Delta S)}_{\text{Entropy term}}$$

Let's consider each term individually, beginning with the enthalpy term. There are many factors that contribute to the sign and magnitude of $\Delta H$, but the dominant factor is generally bond strength. We compare the bond strengths of the bonds broken and the bonds formed in a Diels-Alder reaction.

Bonds broken
3 $\pi$ bonds

Bonds formed
1 $\pi$ bond and 2 $\sigma$ bonds

Three $\pi$ bonds are broken and replaced by one $\pi$ bond and two $\sigma$ bonds. We saw in Chapter 1 that $\sigma$ bonds are stronger than $\pi$ bonds, and as a result, $\Delta H$ for this process has a *negative* value so the reaction is exothermic.

Now let's consider the second term, the entropy term $-T\,\Delta S$. This term will always be positive for a Diels-Alder reaction. Why? There are two main reasons: (1) two molecules are joining together to produce one molecule of product and (2) a ring is being formed. As described in Section 6.3, each of these factors represents a decrease in entropy, contributing to a negative value for $\Delta S$. Temperature (measured in kelvin) is always positive, and therefore, $-T\,\Delta S$ will be positive.

Now let's combine both terms (enthalpy and entropy). The sign of $\Delta G$ for a Diels-Alder reaction will be determined by the competition between these two terms:

$$\Delta G = \underbrace{(\Delta H)}_{\substack{\text{Enthalpy term} \\ \ominus}} + \underbrace{(-T\,\Delta S)}_{\substack{\text{Entropy term} \\ \oplus}}$$

In order for $\Delta G$ of a Diels-Alder reaction to be negative, the enthalpy term must be larger than the entropy term, which is temperature dependent. At moderate temperatures, the entropy term is small, and the enthalpy term dominates. As a result, $\Delta G$ will be negative, which means that products will be favored over reactants (the equilibrium constant $K$ will be greater than 1). In other words, Diels-Alder reactions are thermodynamically favorable at moderate temperatures.

However, at high temperatures, the entropy term will be large and will dominate the enthalpy term. As a result, $\Delta G$ will be positive, which means that reactants will be favored over products (the equilibrium constant $K$ will be less than 1). In other words, the reverse reaction (a retro Diels-Alder) will be thermodynamically favored at high temperature.

In summary, Diels-Alder reactions are generally performed at moderate temperatures, usually between room temperature and 200°C, depending on the specific case.

## The Dienophile

The starting materials for a Diels-Alder reaction are a diene and a compound that reacts with the diene, called the **dienophile**.

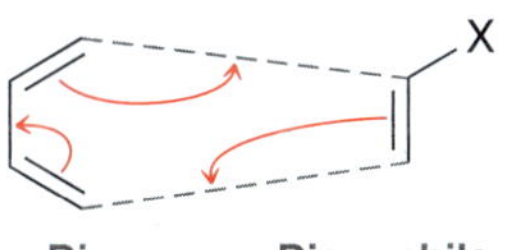


We will begin our discussion with the dienophile. When the dienophile does not contain any substituents, the reaction exhibits a large activation energy and proceeds slowly. If the temperature is raised to overcome the energy barrier, the starting materials are favored over the products, and the resulting yield is low:

200°C
20%

A Diels-Alder reaction will proceed more rapidly and with a much higher yield when the dienophile has an electron-withdrawing substituent such as a carbonyl group:

100%

The carbonyl group is electron withdrawing because of resonance. Can you draw the resonance structures? Other examples of dienophiles that possess electron-withdrawing substituents are shown here:

When the dienophile is a 1,2-disubstituted alkene, the reaction proceeds with stereospecificity. Specifically, a *cis* alkene produces a *cis* disubstituted ring, and a *trans* alkene produces a *trans* disubstituted ring.

A triple bond can also function as a dienophile, in which case the product is a ring with two double bonds (a 1,4-cyclohexadiene).

## SKILLBUILDER

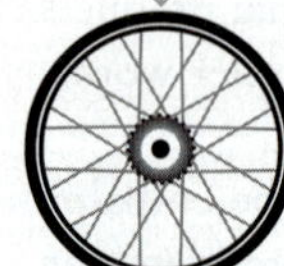

### 17.3 PREDICTING THE PRODUCT OF A DIELS-ALDER REACTION

**LEARN the skill**

Predict the product of the following reaction:

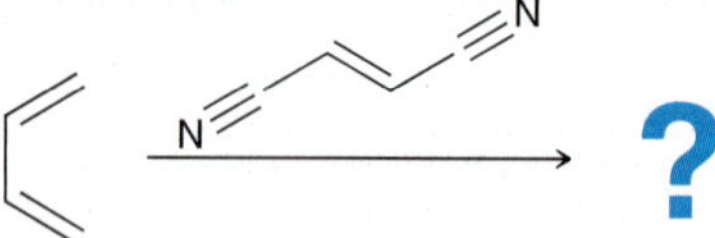


**STEP 1** Redraw and line up the ends of the diene with the dienophile.

**BY THE WAY** It is not necessary to draw the dotted lines between the diene and the dienophile, but you might find it helpful.

**SOLUTION**

First redraw the diene and the dienophile so that the ends of the diene are close to the dienophile:

**STEP 2** Draw three curved arrows, starting at the dienophile and going either clockwise or counterclockwise.

Next, draw three curved arrows that move around in a circle. Place the tail of the first curved arrow on the dienophile and then continue to draw all three arrows in a circle, either clockwise or counterclockwise:

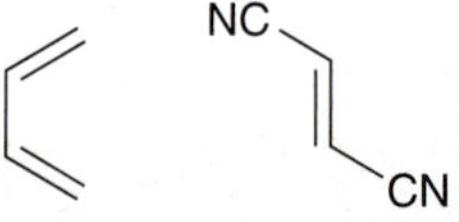

**STEP 3**
Draw the product with the correct stereochemical outcome.

Finally, draw the product with the correct stereochemical outcome. In this case, the starting dienophile is a *trans*-disubstituted alkene, so we expect the cyano groups to be *trans* to each other in the product:

CN CN + En

**PRACTICE the skill** **17.14** Predict the products for each of the following reactions:

(a) OMe OMe
(b) COOH COOH
(c)
(d)
(e)
(f) NC—≡—CN
(g)
(h) OH
(i)

**APPLY the skill** **17.15** Predict the product(s) obtained when benzoquinone is treated with excess butadiene:

(Excess)

**17.16** Propose a mechanism for the following transformation:

need more **PRACTICE?** **Try Problems 17.44, 17.46**

## The Diene

Recall that 1,3-butadiene exists as an equilibrium between the *s-cis* conformation and the *s-trans* conformation.

*s-cis* *s-trans*

The Diels-Alder reaction only occurs when the diene is in an *s-cis* conformation. When the compound is in an *s-trans* conformation, the ends of the diene are too far apart to react with the dienophile. Some dienes are incapable of adopting an *s-cis* conformation and are therefore unreactive toward a Diels-Alder reaction; for example:

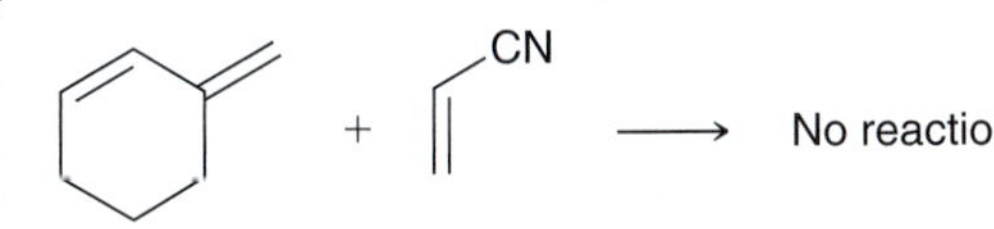

Other dienes, such as cyclopentadiene, are permanently locked in an *s-cis* conformation. Such dienes react extremely rapidly in Diels-Alder reactions.

Cyclopentadiene + X–CH=CH–X ⟶ (bicyclic product with X, H, H, X) + En

In fact, cyclopentadiene is so reactive toward Diels-Alder reactions that it even reacts with itself to form a dimer called dicyclopentadiene.

Cyclopentadiene + cyclopentadiene ⟶ Dicyclopentadiene + En

When cyclopentadiene is allowed to stand at room temperature, it is completely converted into the dimer in just a few hours. For this reason, cyclopentadiene cannot be stored at room temperature for long periods of time. When cyclopentadiene is to be used as a starting material in a Diels-Alder reaction, it must first be formed from dicyclopentadiene via a retro Diels-Alder reaction and then used immediately or stored at very low temperature.

Dicyclopentadiene —Heat→ cyclopentadiene + cyclopentadiene

## CONCEPTUAL CHECKPOINT

**17.17** Consider the following two isomers of 2,4-hexadiene. One isomer reacts rapidly as a diene in a Diels-Alder reaction, and the other does not. Identify which isomer is more reactive and explain your choice.

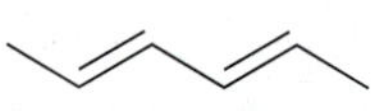

**(2*E*,4*E*)-Hexadiene**

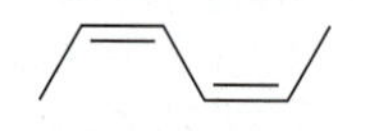

**(2*Z*,4*Z*)-Hexadiene**

**17.18** Rank the following dienes in terms of reactivity in Diels-Alder reactions (from least reactive to most reactive):

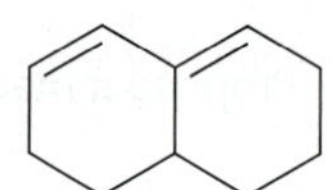

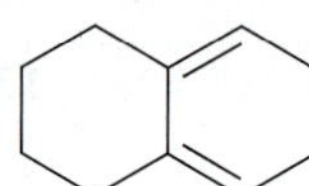

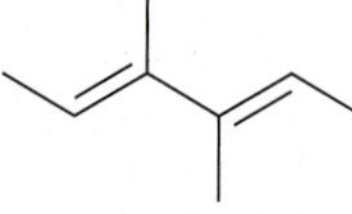

## *Endo* Preference

When cyclopentadiene is used as the starting diene, a bridged bicyclic compound is obtained as the product. In such a case, we might expect to obtain the following two products:

Cyclopentadiene + OHC–CH=CH–CHO ⟶ *Endo* (CHO, CHO) + *Exo* (CHO, CHO)

In one cycloadduct, the electron-withdrawing substituents occupy *endo* positions, and in the other cycloadduct, the substituents occupy *exo* positions. The ***endo*** positions are *syn* to the larger bridge, and the ***exo*** positions are *anti* to the larger bridge.

This bridge has one carbon atom

This bridge has two carbon atoms (larger bridge)

**Exo** (*anti* to larger bridge)

***Endo*** (*syn* to larger bridge)

In general, the *endo* cycloadduct is highly favored over the *exo* cycloadduct. In many cases, the *endo* product is the exclusive product. The explanation for this preference is based on an analysis of the transition states leading to the *endo* and *exo* products. During formation of the *endo* product, a favorable interaction exists between the electron-withdrawing substituents and the developing π bond.

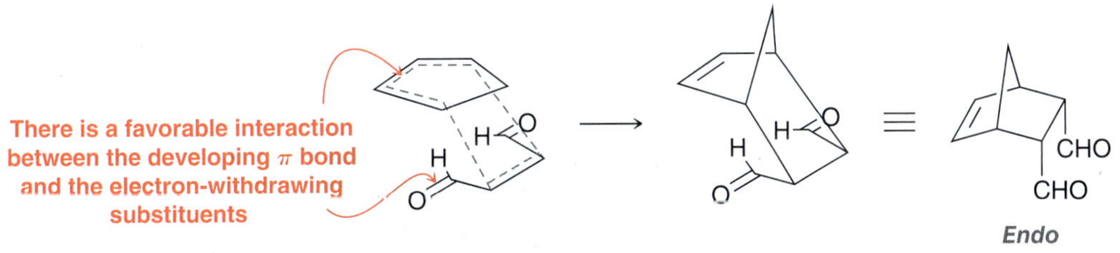

There is a favorable interaction between the developing π bond and the electron-withdrawing substituents

*Endo*

The transition state leading to the *exo* product does not exhibit this favorable interaction:

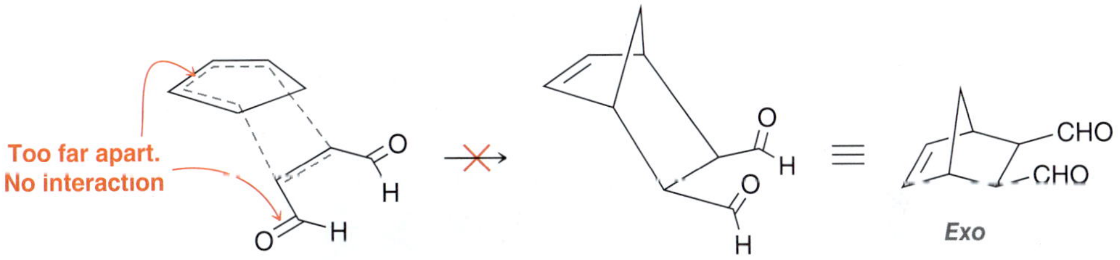

Too far apart. No interaction

*Exo*

The transition state leading to formation of the *endo* product is lower in energy than the transition state leading to formation of the *exo* product (Figure 17.16). As a result, the *endo* product is formed more rapidly.

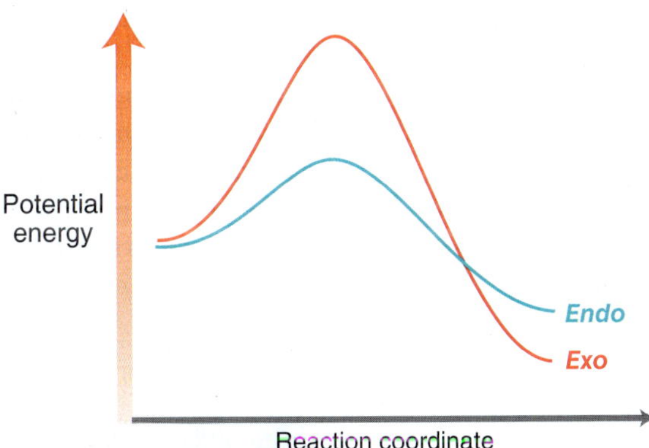


**FIGURE 17.16** An energy diagram illustrating the formation of *endo* products (blue) and *exo* products (red) in a Diels-Alder reaction.

## CONCEPTUAL CHECKPOINT

**17.19** Predict the products for each of the following reactions:

(a) 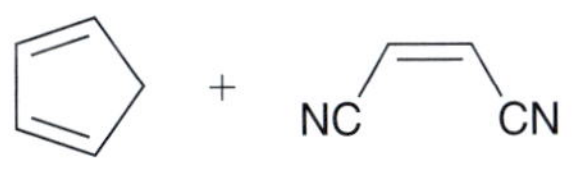 (b) 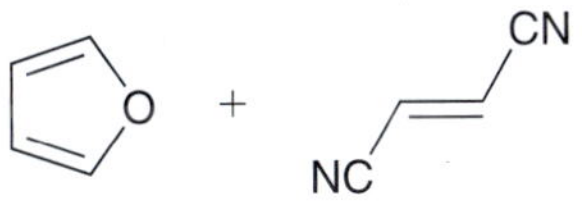 (c)

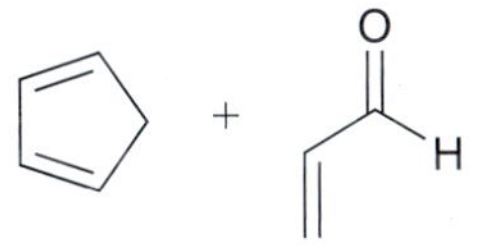

(d)  (e) 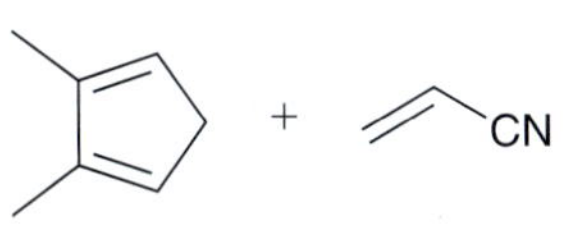 (f)

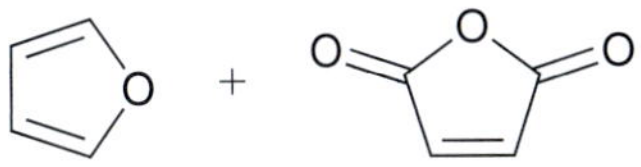

## 17.8 MO Description of Cycloadditions

We now turn our attention to the MOs involved in a Diels-Alder reaction. By studying the interactions between the relevant MOs we will gain insight into the reaction, and we will be able to explore the feasibility of reactions similar to Diels-Alder reactions.

Figure 17.17 illustrates the MOs for a conjugated diene (1,3-butadiene) and a simple dienophile (ethylene). The HOMO and LUMO are clearly labeled for both. According to frontier orbital

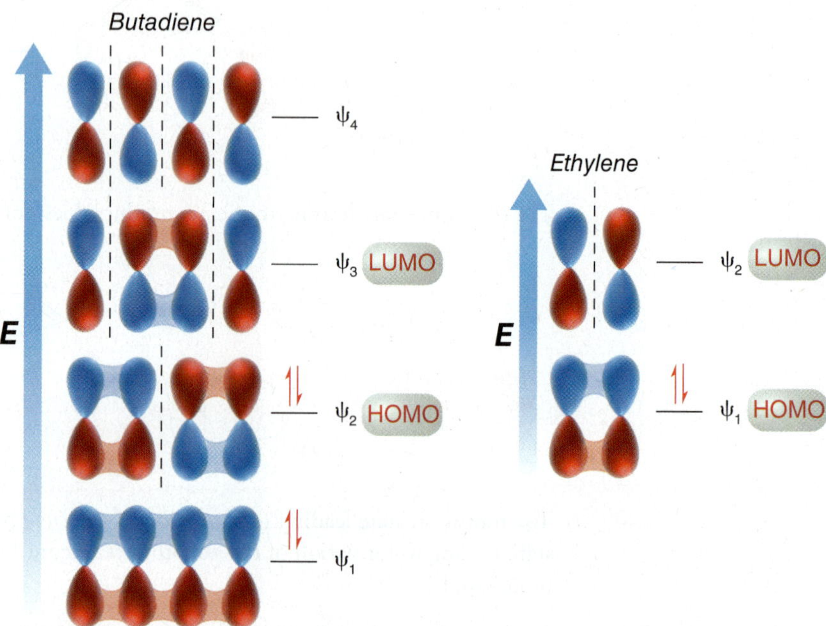


**FIGURE 17.17** The MOs of butadiene and ethylene.

theory, a Diels-Alder reaction is accomplished when the HOMO of one compound interacts with the LUMO of the other. That is, the electron density flows from a filled orbital of one compound (the HOMO) into an empty orbital of another compound (the LUMO). Since the dienophile in a Diels-Alder reaction generally has an electron-withdrawing substituent, we will treat the dienophile as the electron-poor species (the empty orbital) that accepts electron density. In other words, we will look at the LUMO of the dienophile and the HOMO of the diene. The reaction is therefore initiated when electron density is transferred from the HOMO of the diene to the LUMO of the dienophile (Figure 17.18).

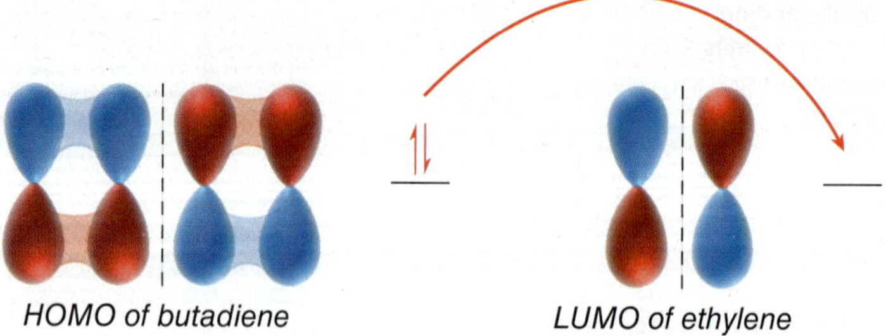


**FIGURE 17.18** In a Diels-Alder reaction, the electron density flows from the HOMO of the diene to the LUMO of the dienophile.

When we align these frontier orbitals, we find that the phases of the MOs overlap nicely (Figure 17.19). In the process, four of the carbon atoms rehybridize to give $sp^3$-hybridized orbitals

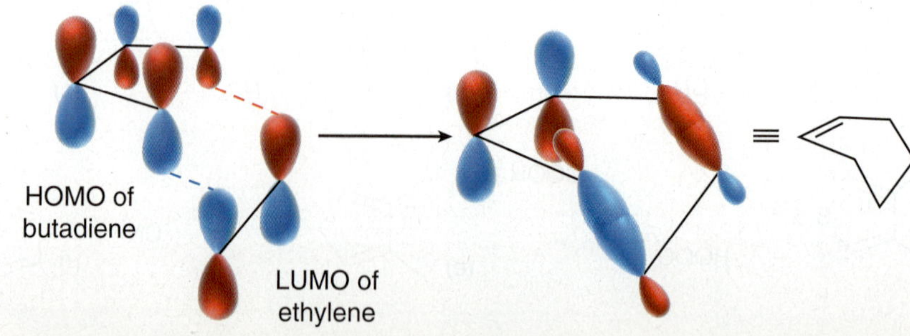


**FIGURE 17.19** The phases of the frontier orbitals align properly in a Diels-Alder reaction.

that form σ bonds. In order for this to occur, the phases of the MOs must overlap, that is, the phases must be symmetric. This requirement, called **conservation of orbital symmetry**, was first described by R. B. Woodward and Roald Hoffmann (both at Harvard University) in 1965. In the Diels-Alder reaction, because orbital symmetry is indeed conserved, and the reaction is therefore called a **symmetry-allowed** process.

Now let's use the same approach (analyzing the frontier orbitals) to determine whether the following reaction is possible:

Like the Diels-Alder reaction, this reaction is also a cycloaddition. Specifically, it is called a [2+2] cycloaddition, because it involves two different π systems, each of which is associated with two atoms. To determine if this reaction is feasible, we once again look at the frontier orbitals—the HOMO of one compound and the LUMO of the other (Figure 17.20). When we try to align these

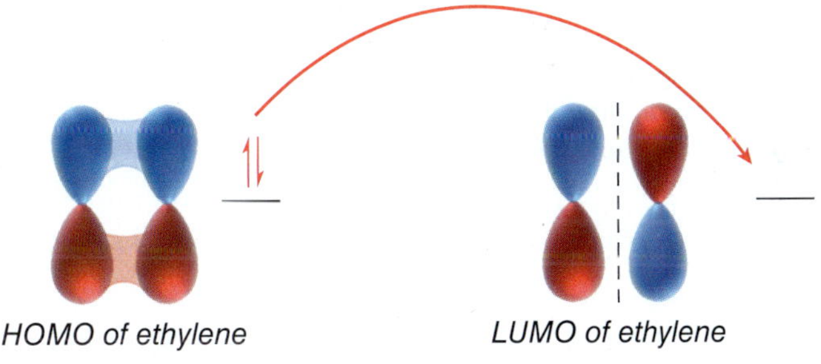


**FIGURE 17.20** In a [2+2] cycloaddition, the electron density must flow from the HOMO of one compound to the LUMO of the other compound.

frontier orbitals, we find that the phases of the MOs do not overlap (Figure 17.21). This reaction is therefore said to be **symmetry forbidden**, and the reaction does not occur. A [2+2] cycloaddition

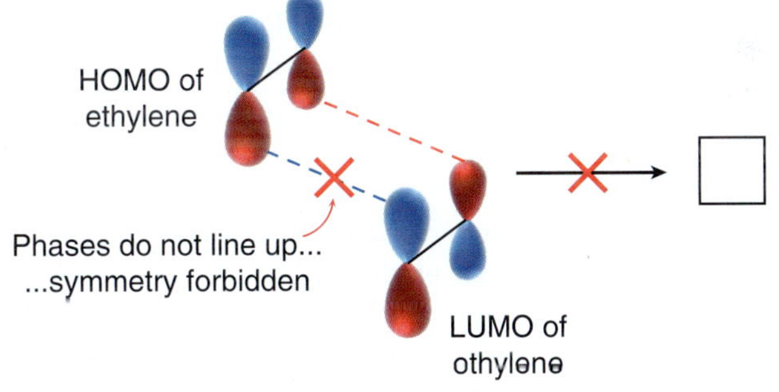


**FIGURE 17.21** The phases of the frontier orbitals do not align properly in a [2+2] cycloaddition.

can only be performed with photochemical excitation. When one of the compounds is subjected to UV light, it can absorb the light to promote a π electron to the next higher energy level (Figure 17.22). In this excited state, the HOMO is now considered to be $\psi_2$ rather than $\psi_1$. The HOMO of

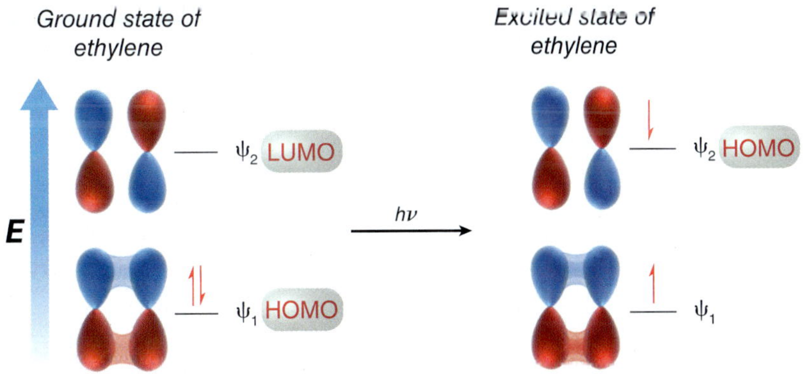


**FIGURE 17.22** The ground state of ethylene absorbs a photon of light to achieve the excited state, in which the HOMO is $\psi_2$.

the excited state can now interact with the LUMO of a ground-state molecule, and the reaction is symmetry allowed (Figure 17.23).

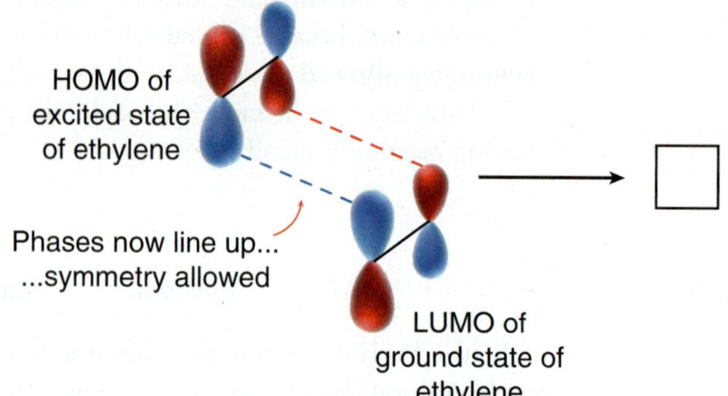


**FIGURE 17.23**
The phases of the frontier orbitals align properly in a photochemical [2+2] cycloaddition.

CONCEPTUAL CHECKPOINT

**17.20** Consider the following [4+4] cycloaddition process. Would you expect this process to occur through a thermal or photochemical pathway? Justify your answer with MO theory.

## 17.9 Electrocyclic Reactions

### An Introduction to Electrocyclic Reactions

An electrocyclic reaction is a pericyclic process in which a conjugated polyene undergoes cyclization. In the process, one π bond is converted into a σ bond, while the remaining π bonds all change their location. The newly formed σ bond joins the ends of the original π system, thereby creating a ring. Two examples are shown.

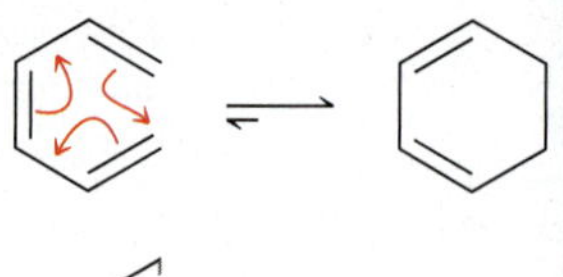

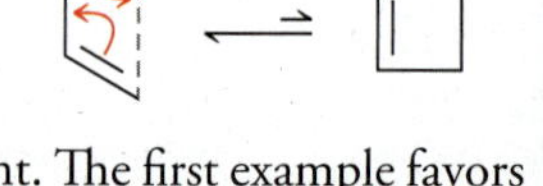

Both reactions are reversible, but the position of equilibrium is different. The first example favors the cyclic product, while the second example disfavors formation of the cyclic product as a result of the ring strain associated with a four-membered ring.

When substituents are present at the termini of the π system, the following stereochemical outcomes are observed:

Note that the configuration of the product is dependent not only on the configuration of the reactant but also on the conditions of ring closure. That is, a different outcome is observed when the reaction is performed under thermal conditions (using heat) or under photochemical conditions (using UV light).

The choice of whether to use thermal or photochemical conditions also has an impact on the stereochemical outcome of electrocyclic reactions with four $\pi$ electrons:

These stereochemical observations puzzled chemists for many years, until Woodward and Hoffmann developed their theory describing conservation of orbital symmetry. This single theory is capable of explaining all of the observations. We will first apply this theory to explain the stereochemical outcome of electrocyclic reactions taking place under thermal conditions, and then we will explore electrocyclic reactions taking place under photochemical conditions.

## The Stereochemistry of Thermal Electrocyclic Reactions

Despite their name, thermal electrocyclic reactions do not necessarily require the use of elevated temperature. They can, in fact, take place at or even below room temperature. In many cases, the heat available at room temperature is sufficient for a thermal electrocyclic process to occur.

Under thermal conditions, the configuration of the reactant determines the configuration of the product.

To explain these stereochemical outcomes, we focus on the symmetry of the HOMO for a $\pi$ system with three conjugated $\pi$ bonds. As seen in Section 17.3, the HOMO of a conjugated triene has two vertical nodes and can be drawn using our shorthand method (Figure 17.24). Focus, in particular, on the signs (illustrated with red and blue) of the outermost lobes, because these are the lobes that

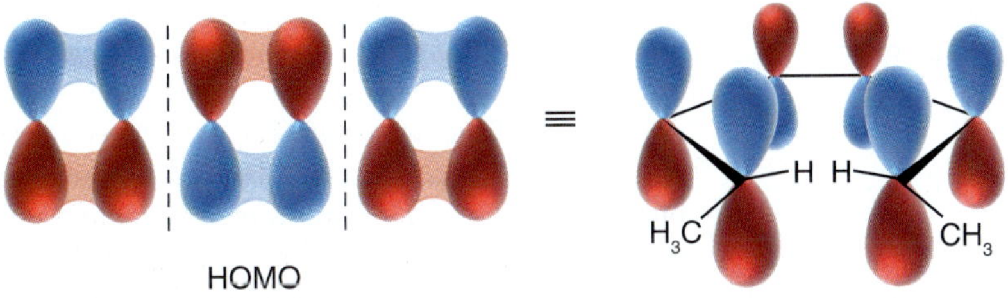


**FIGURE 17.24** A shorthand method for depicting the HOMO of a conjugated triene.

will participate in formation of the new σ bond. In order to form a bond, the lobes that interact with each other must exhibit the same sign. This requirement demands that the lobes must rotate in the manner illustrated in Figure 17.25.

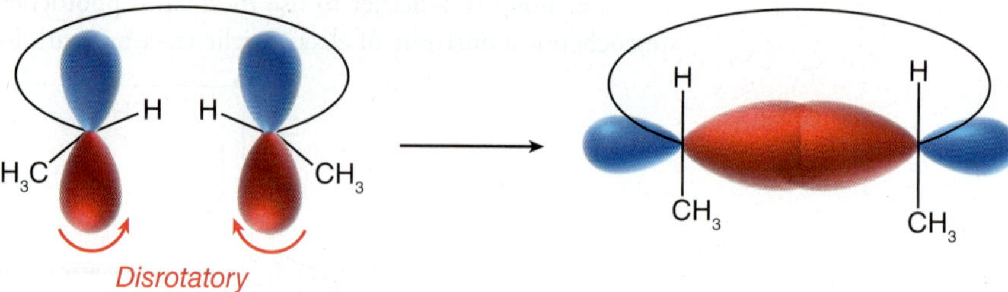


**FIGURE 17.25**
An illustration of disrotatory ring closure.

This type of rotation is called **disrotatory**, because one set of lobes rotates clockwise, while the other set rotates counterclockwise. The requirement for disrotatory ring closure determines the stereochemical outcome of the reaction. Specifically, in this example, the two methyl groups have a *cis* relationship in the product.

Now let's apply this approach to thermal electrocyclic reactions of π systems containing only four π electrons, rather than six π electrons. Once again, the configuration of the product is dependent on the configuration of the reactant.

$CH_3$ H H $CH_3$ ⇌ Heat $CH_3$ $CH_3$ + En H $CH_3$ H $CH_3$ ⇌ Heat $CH_3$ $CH_3$

To account for these results, we once again look at the outermost lobes of the HOMO. As seen in Section 17.3, the HOMO of a π system containing four π electrons has one node and can be drawn using our shorthand method (Figure 17.26). In order to form a bond, recall that the lobes that

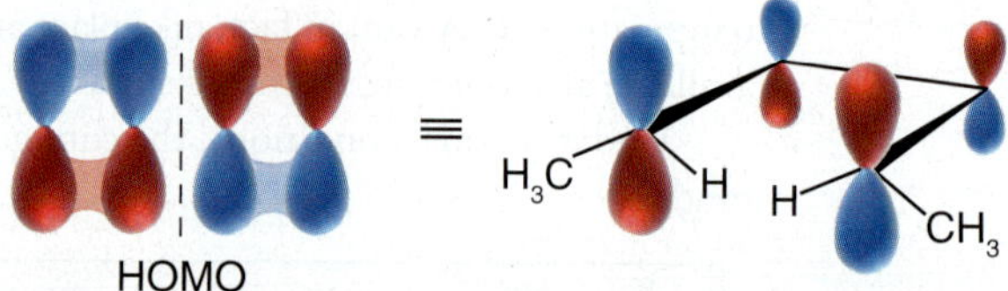


**FIGURE 17.26**
A shorthand method for depicting the HOMO of a conjugated diene.

interact must exhibit the same sign. In a case with only four π electrons, this requirement demands that the lobes must rotate in the manner illustrated in Figure 17.27. This type of rotation is called **conrotatory**, because both sets of lobes must rotate in the same way. The requirement for conrotatory ring closure determines the stereochemical outcome of the reaction. Specifically, in this example, the two methyl groups have a *trans* relationship in the product.

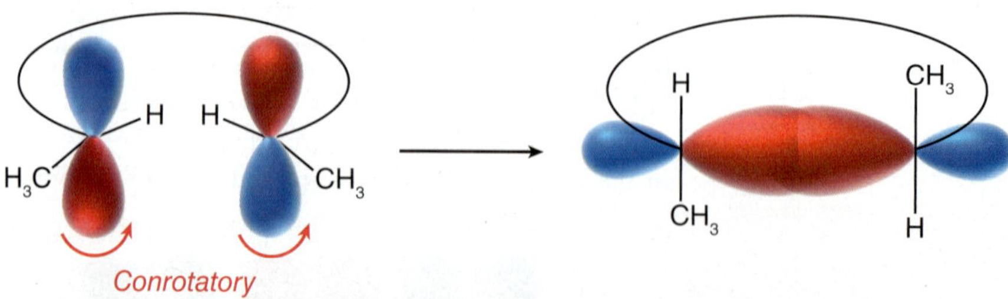


**FIGURE 17.27**
A conrotatory ring closure.

For butadiene, we mentioned that the equilibrium favors the open chain, so the conrotatory nature of this process can be observed by monitoring the stereochemical outcome for the ring-opening reaction of 3,4-disubstituted cyclobutene (Figure 17.28). In summary, conjugated systems with six π electrons undergo thermal electrocyclic reactions in a disrotatory fashion, while conjugated systems with four π electrons undergo thermal electrocyclic reactions in a conrotatory fashion.

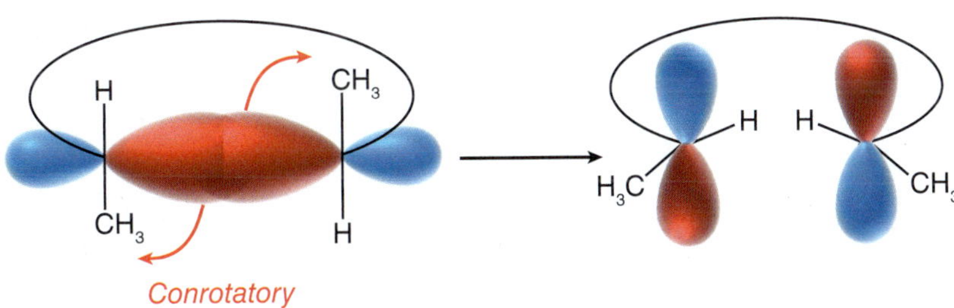


**FIGURE 17.28**
A conrotatory ring opening.

## CONCEPTUAL CHECKPOINT

**17.21** Predict the major product(s) for each of the following thermal electrocyclic reactions:

(a) Heat ? (b) Heat ? (c) Heat ?

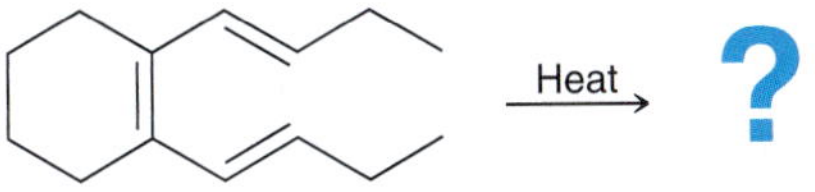

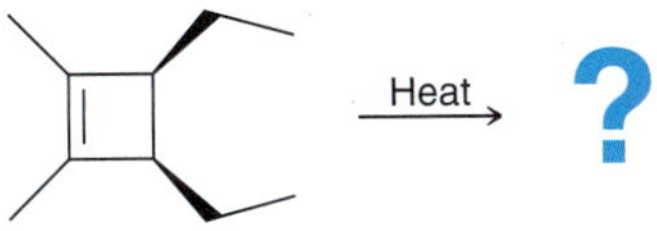

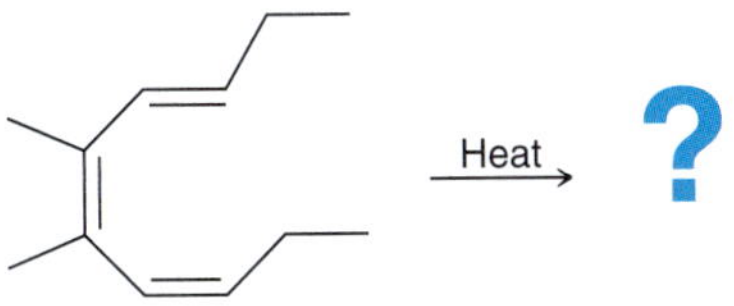

### Stereochemistry of Photochemical Electrocyclic Reactions

We have seen that the stereochemical outcome of an electrocyclic reaction depends on whether the reaction is performed under thermal conditions or under photochemical conditions.

$CH_3$ H H $CH_3$ — Heat → (cis-$CH_3$, $CH_3$); — $h\nu$ → (trans-$CH_3$, $CH_3$) + En

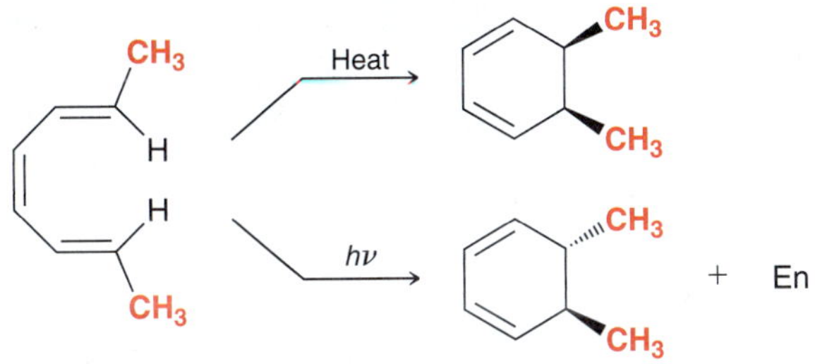

In this example, the configuration of the product depends on the conditions under which the reaction is performed. Once again, we can explain these observations with the theory of conservation of orbital symmetry. Specifically, we must focus on the symmetry of the HOMO involved in the reaction. Recall from Section 17.3 that excitation of an electron redefines the identity of the HOMO (Figure 17.29).

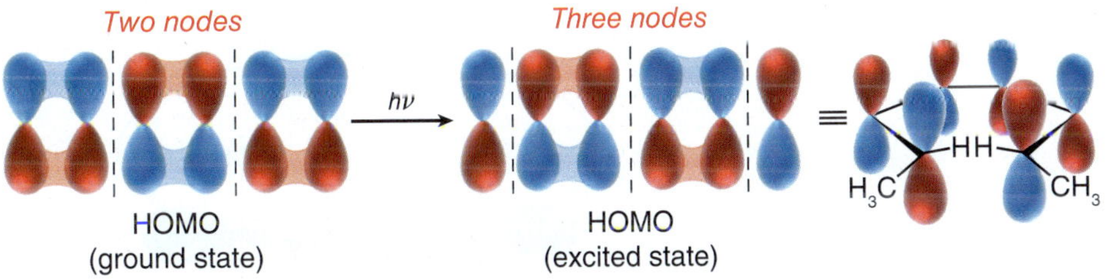

Two nodes — HOMO (ground state) — $h\nu$ → Three nodes — HOMO (excited state) ≡ H3C HH CH3

**FIGURE 17.29**
A shorthand method for depicting the HOMO of the excited state of a conjugated triene.

Focusing on the outer lobes, we find that photochemical ring closure of a system with six $\pi$ electrons must occur in a conrotatory fashion (Figure 17.30), as opposed to the disrotatory ring closure observed under thermal conditions. The requirement for conrotatory ring closure correctly explains the observed stereochemical outcome for this reaction.

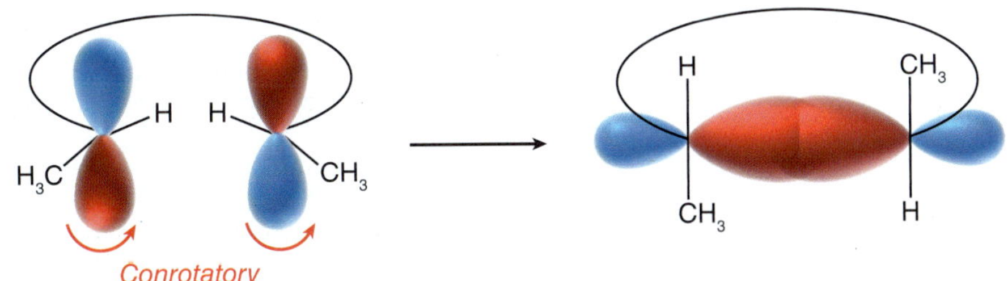


**FIGURE 17.30**
A system with six $\pi$ electrons will undergo conrotatory ring closure under photochemical conditions.

A similar type of explanation can be applied to photochemical electrocyclic reactions of systems with four $\pi$ electrons.

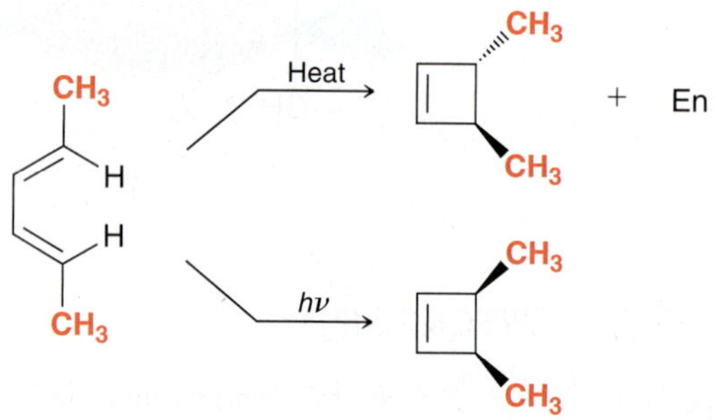

Once again, it is observed that the configuration of the product depends on the conditions under which the reaction is performed, and these observations can be explained based on conservation of orbital symmetry. Specifically, we must focus on the symmetry of the HOMO involved in the reaction. Recall from Section 17.3 that excitation of an electron redefines the identity of the HOMO (Figure 17.31). Focusing on the outer lobes, we find that photochemical ring closure of a system

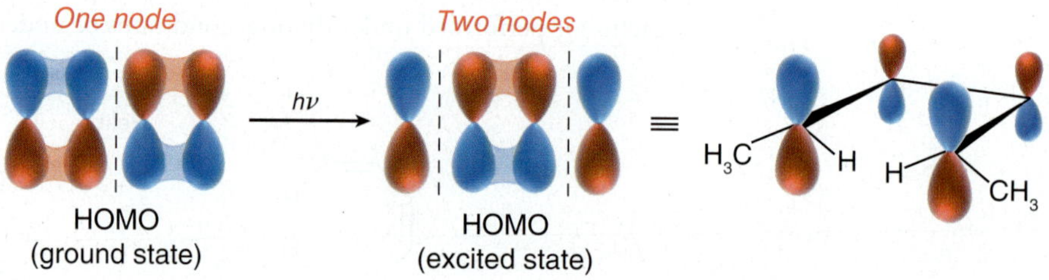

**FIGURE 17.31** A shorthand method for depicting the HOMO of the excited state of a conjugated diene.

with four $\pi$ electrons must occur in a disrotatory fashion, as opposed to the conrotatory ring closure observed under thermal conditions. For butadiene, the equilibrium favors the open chain, so the disrotatory nature of this process can be observed by monitoring the stereochemical outcome for the ring-opening reaction of 3,4-disubstituted cyclobutene under photochemical conditions (Figure 17.32). Table 17.2 summarizes the Woodward-Hoffmann rules for common electrocyclic reactions.

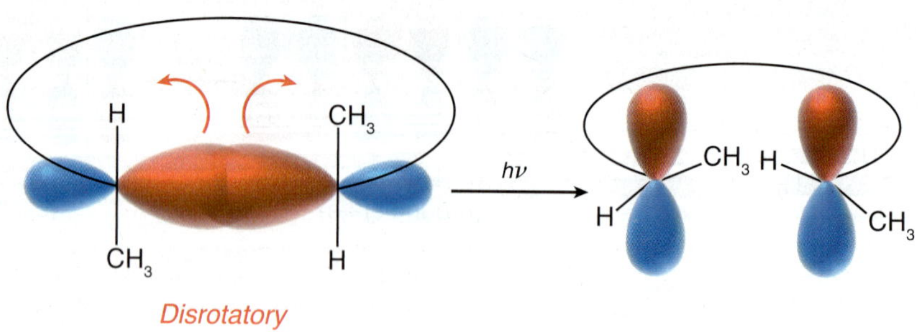


**FIGURE 17.32** A disrotatory ring opening.

**TABLE 17.2 WOODWARD-HOFFMANN RULES FOR THERMAL AND PHOTOCHEMICAL ELECTROCYCLIC REACTIONS**

| | THERMAL | PHOTOCHEMICAL |
|---|---|---|
| Four $\pi$ electrons | Conrotatory | Disrotatory |
| Six $\pi$ electrons | Disrotatory | Conrotatory |

# SKILLBUILDER

## 17.4 PREDICTING THE PRODUCT OF AN ELECTROCYCLIC REACTION

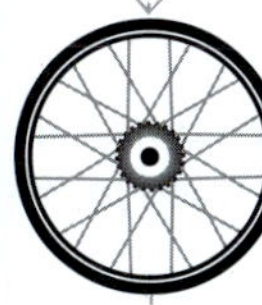

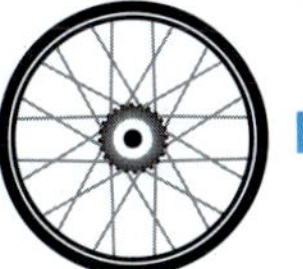

**LEARN the skill**

Predict the product(s) of the following electrocyclic reaction:

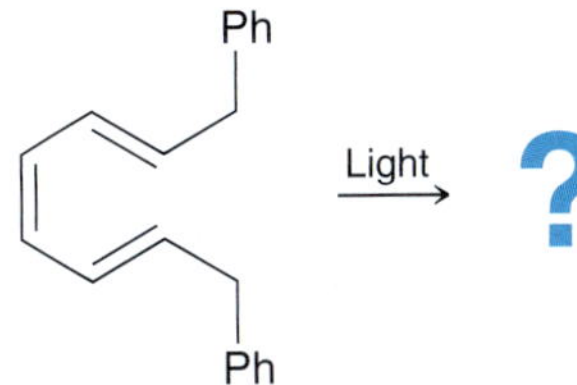


**STEP 1**
Count the number of $\pi$ electrons.

### SOLUTION

We begin by counting the number of $\pi$ electrons. In this case, there are six $\pi$ electrons (the $\pi$ electrons of the phenyl groups are not participating in the reaction).

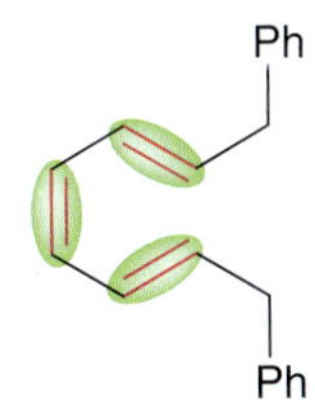


**STEP 2**
Determine whether the reaction is conrotatory or disrotatory.

Next, we must analyze the reaction conditions and determine whether the reaction occurs in a conrotatory or disrotatory fashion. As shown in Table 17.2, for a system with six $\pi$ electrons, photochemical conditions cause a conrotatory ring closure. This information is used to determine whether the substituents are *cis* or *trans* to each other in the product. In this case, the substituents are *trans* to each other in the product. A racemic mixture is expected because the conrotatory ring closure can occur either in a clockwise fashion (producing one enantiomer) or in a counterclockwise fashion (producing the other enantiomer).

**STEP 3**
Determine whether the substituents are *cis* or *trans* to each other in the product.

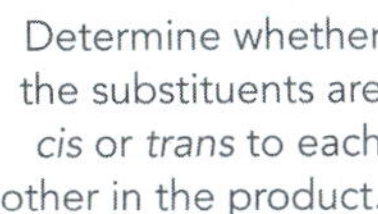

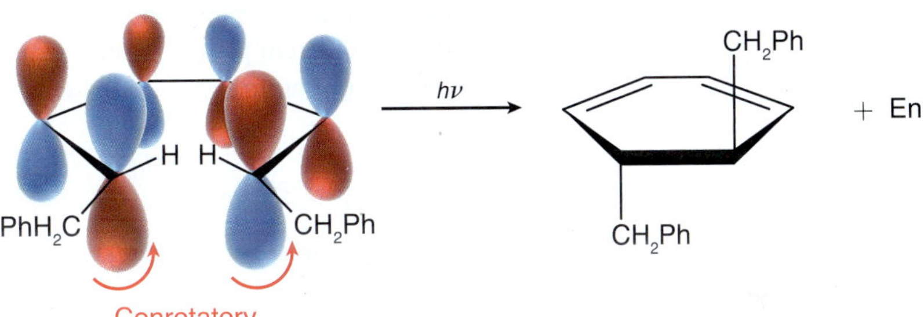

**PRACTICE the skill** **17.22** Predict the product(s) for each of the following reactions:

(a)

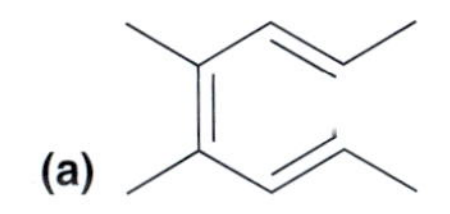

$h\nu$ ?

(b)

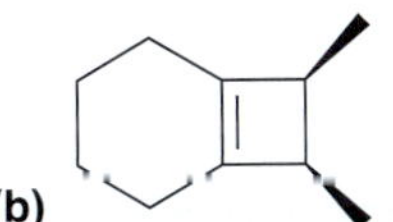

$h\nu$ ?

(c)

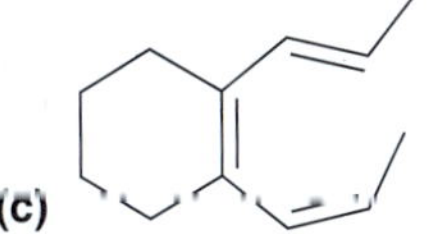

Heat ?

**APPLY the skill** **17.23** Draw the structure of *cis*-3,4-diethylcyclobutene.

**(a)** Conrotatory ring opening produces only one product. Draw the product and determine whether its formation is best achieved under thermal conditions or photochemical conditions.

**(b)** Disrotatory ring opening can produce two possible products, but only one of these products is obtained in practice. Identify the product that is formed and explain why the other possible product is not formed.

**17.24** Identify whether the product obtained from each of the following reactions is a *meso* compound or a pair of enantiomers:

**(a)** Irradiation of (2*E*,4*Z*,6*Z*)-4,5-dimethyl-2,4,6-octatriene with UV light

**(b)** Subjecting (2*E*,4*Z*,6*Z*)-4,5-dimethyl-2,4,6-octatriene to elevated temperature

**(c)** Irradiation of (2*E*,4*Z*,6*E*)-4,5-dimethyl-2,4,6-octatriene with UV light

need more **PRACTICE?** **Try Problems 17.55, 17.56**

## 17.10 Sigmatropic Rearrangements

### An Introduction to Sigmatropic Rearrangements

A sigmatropic rearrangement is a pericyclic reaction in which one σ bond is formed at the expense of another. In the process, the π bonds change their location. The term *sigmatropic* comes from the Greek word *tropos*, which means "change." A sigmatropic rearrangement is therefore a reaction in which a σ bond has undergone a change (in its location).

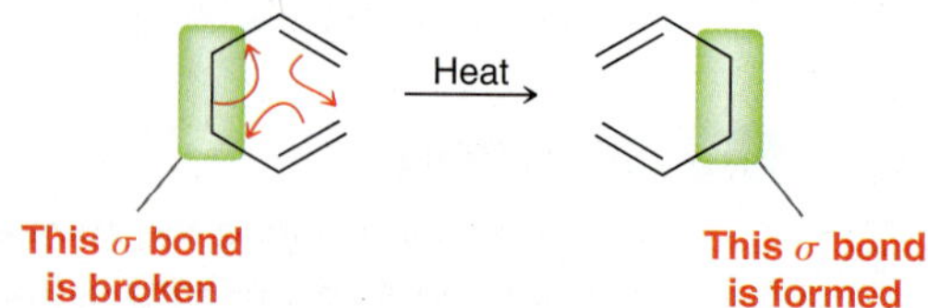


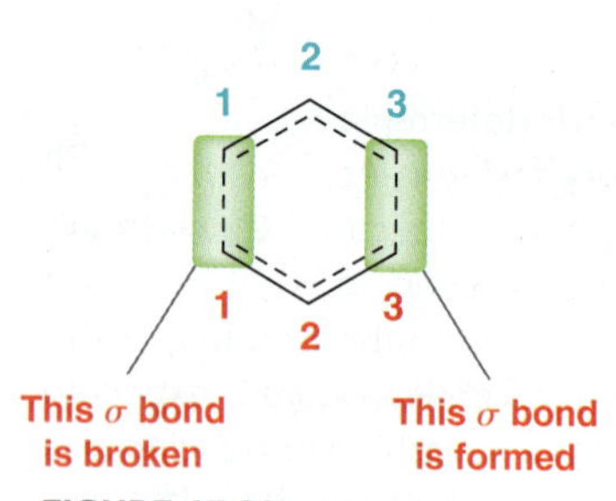


**FIGURE 17.33**
The transition state for a [3,3] sigmatropic rearrangement.

There are many different types of sigmatropic rearrangements. The previous example is called a [3,3] sigmatropic rearrangement. This notation is different than any other notation used thus far in this book. The two numbers in the brackets indicate the number of atoms separating the bond that is forming and the bond that is breaking in the transition state (Figure 17.33). Notice that the transition state is cyclic, which is a characteristic feature of all pericyclic reactions. In the transition state, the bond that is breaking and the bond that is forming are separated by two different pathways, each of which is comprised of three atoms (one path is labeled in red and the other in blue). Therefore, it is called a [3,3] sigmatropic rearrangement.

The following is an example of a [1,5] sigmatropic rearrangement, also sometimes referred to as a [1,5] hydrogen shift. In the transition state, the bond that is breaking and the bond that is forming are separated by two different pathways: one is comprised of five atoms (labeled in red), and the other is comprised of only one atom (labeled in blue).

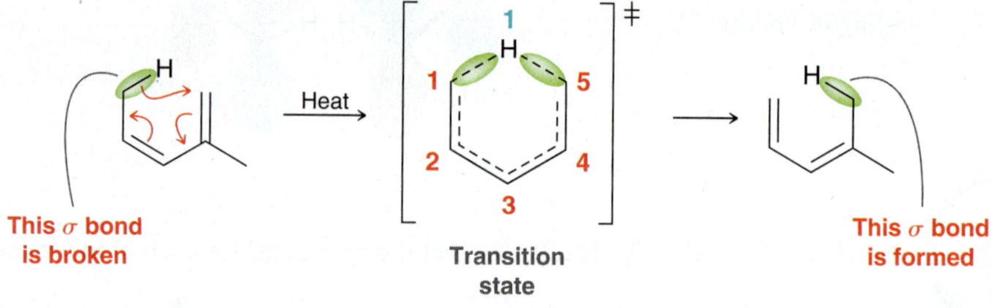


### The Cope Rearrangement

A [3,3] sigmatropic rearrangement is called a **Cope rearrangement** when all six atoms of the cyclic transition state are carbon atoms.

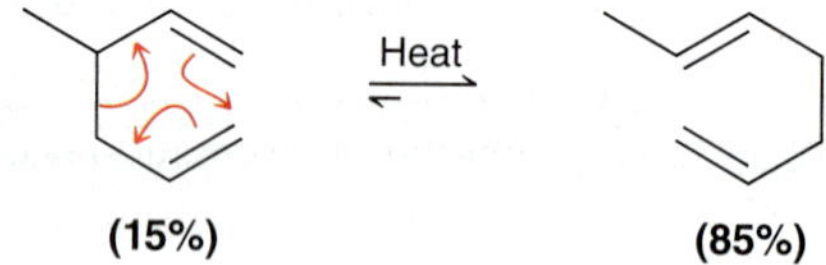


The equilibrium for a Cope rearrangement generally favors formation of the more substituted alkene. In the example above, both π bonds of the reactant are monosubstituted, but in the product, one of the π bonds is disubstituted. For this reason, the equilibrium for the reaction favors the product.

### The Claisen Rearrangement

The oxygen analogue of a Cope rearrangement is called a **Claisen rearrangement**.

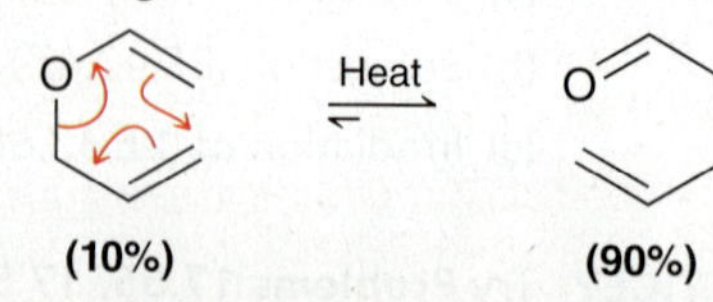

The Claisen rearrangement is a [3,3] sigmatropic rearrangement and is commonly observed for allylic vinylic ethers.

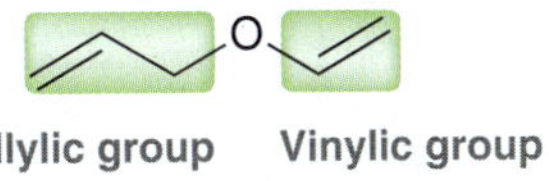

Allylic group Vinylic group

The equilibrium greatly favors the product because of formation of a C=O bond, which is thermodynamically more stable (lower in energy) than a C=C bond. The Claisen rearrangement is also observed for allylic aryl ethers.

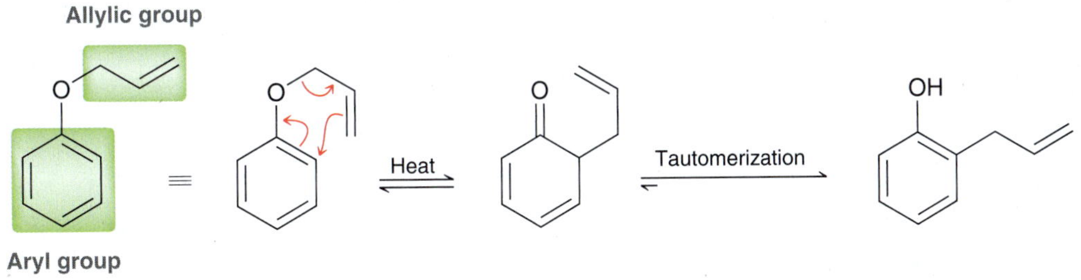

Allylic group

Aryl group

Heat

Tautomerization

Aryl groups possess an aromatic ring (a six-membered ring drawn with alternating double and single bonds), which is particularly stable, as we will see in the next chapter. In the case of allylic aryl ethers, the Claisen rearrangement initially destroys the aromatic ring but is rapidly followed by a spontaneous tautomerization process that regenerates the aromatic ring. In this tautomerization process, the conversion of a ketone into an enol is a testament to the stability of the aromatic ring system. Aromatic compounds, as well as their stability and reactivity, will be discussed in detail in the next two chapters.

## CONCEPTUAL CHECKPOINT

**17.25** For each of the following reactions, use brackets and two numbers to identify the type of sigmatropic rearrangement taking place:

(a) Heat

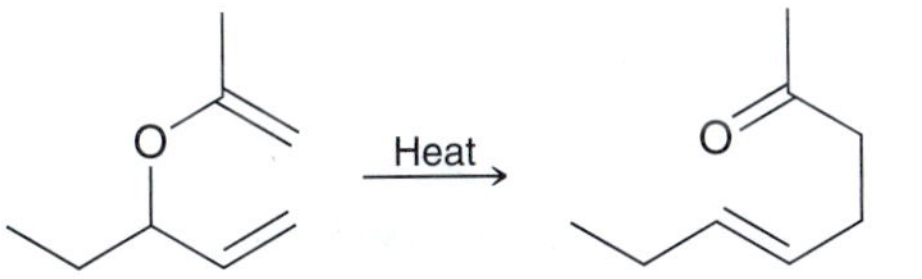

(b) Heat

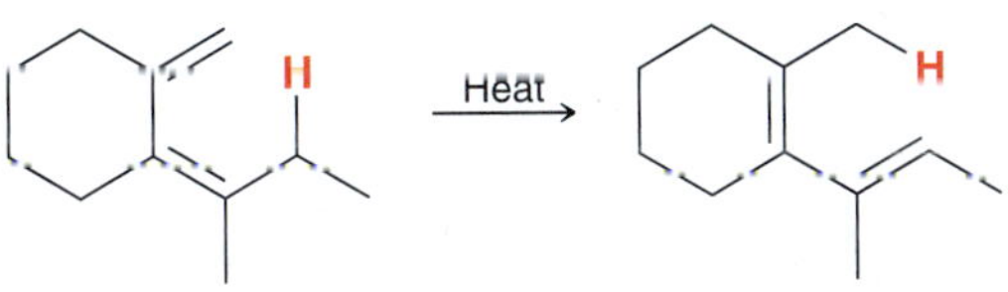

**17.26** Consider the structure of *cis*-1,2-divinylcyclopropane:

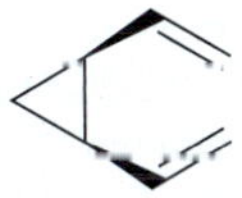

This compound is stable at low temperature but rearranges at room temperature to produce 1,4-cycloheptadiene.

(a) Draw a mechanism for this transformation.

(b) Using brackets and numbers, identify the type of sigmatropic rearrangement taking place in this case.

(c) This reaction proceeds to completion, which means that the concentration of the reactant is insignificant once the equilibrium has been established. Explain why this reaction proceeds to completion.

**17.27** Predict the product for each of the following reactions:

(a) Heat ?

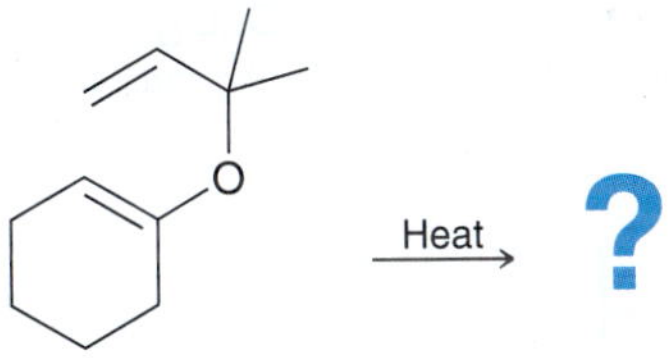

(b) Heat ?

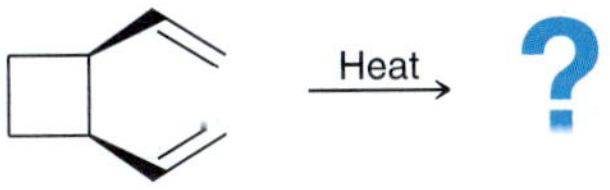

(c) Heat ?

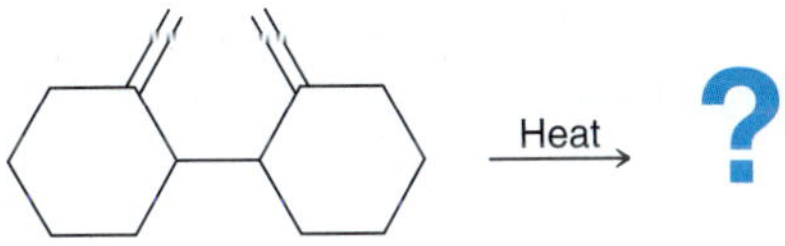

(d) Heat ?

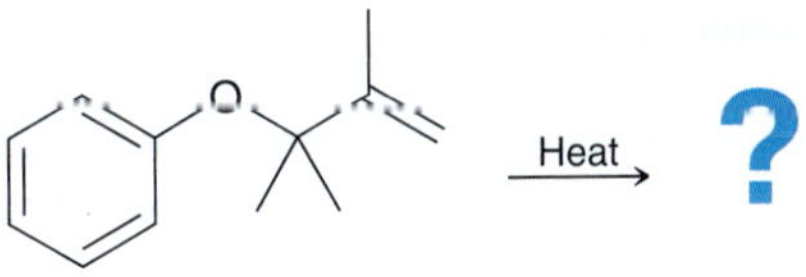

**17.28** Draw a plausible mechanism for the following transformation:

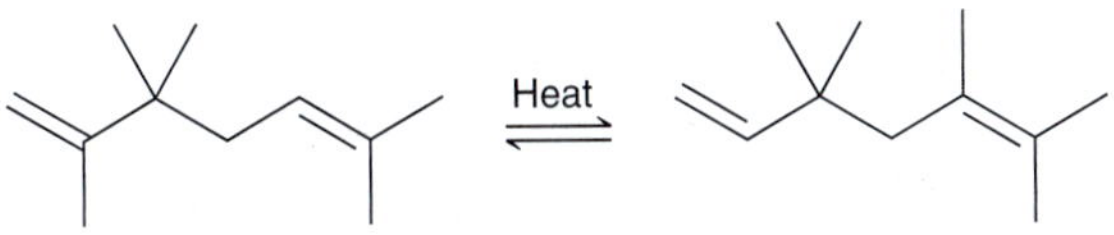

Heat

## The Photoinduced Biosynthesis of Vitamin D

Vitamin D is the general name for two related compounds called cholecalciferol (vitamin $D_3$) and ergocalciferol (vitamin $D_2$). These two compounds differ in the identity of the side chain, R:

Ergocalciferol (vitamin $D_2$) R =

Cholecalciferol (vitamin $D_3$) R =

Vitamin D serves an important function in that it increases the body's ability to absorb the calcium present in the foods we eat. Calcium is essential because it is used in the formation of bones and teeth. A deficiency of vitamin D can cause the childhood disease called rickets, which is characterized by abnormal bone growth.

Although most food does not contain substantial amounts of vitamin D, some food sources do contain the precursors that the body needs in order to synthesize vitamin D. Ergosterol is present in many vegetables, and 7-dehydrocholesterol is present in fish and dairy products.

Ergosterol R =

7-Dehydrocholesterol R =

When the skin is exposed to sunlight, these precursors, located just beneath the skin, are converted into ergocalciferol and cholecalciferol, respectively. For this reason, vitamin D is commonly called the "sunshine vitamin."

The precursors are converted into vitamin D via two successive pericyclic reactions.

$h\nu$ Electrocyclic reaction

[1,7] Sigmatropic rearrangement

The first step is an electrocyclic ring-opening reaction, and the second step is a [1,7] sigmatropic rearrangement. This sequence of steps not only occurs in our bodies but also is utilized commercially to enrich milk with vitamin $D_3$. All milk sold in the United States is irradiated with UV light, which converts the 7-dehydrocholesterol present in the milk into vitamin $D_3$.

# 17.11 UV-Vis Spectroscopy

## The Information in a UV-Vis Spectrum

As seen in the previous two chapters, spectroscopy allows us to probe molecular structure by studying the interaction between matter and electromagnetic radiation. Recall that the frequency of light determines the energy of a photon, and the range of all possible frequencies is known as the

*electromagnetic spectrum* (Figure 15.2). In this chapter, we will focus on the interaction between matter and the UV-Vis region of the electromagnetic spectrum. Specifically, organic compounds with conjugated π systems absorb UV or visible light to promote an electronic excitation. That is, a π electron absorbs the photon and is promoted to a higher energy level. As an example, consider what happens when butadiene is irradiated with UV light (Figure 17.34).

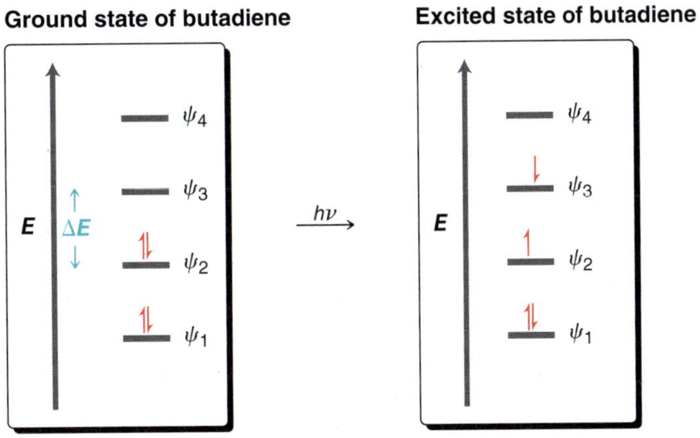

Ground state of butadiene

$E$ $\Delta E$ $\psi_4$ $\psi_3$ $\psi_2$ $\psi_1$

$\xrightarrow{h\nu}$

Excited state of butadiene

$E$ $\psi_4$ $\psi_3$ $\psi_2$ $\psi_1$

**FIGURE 17.34** The ground state of butadiene can absorb a photon of UV light to produce the excited state, in which an electron has been promoted to a higher energy MO.

The energy of the photon ($h\nu$) is absorbed by a π electron in the HOMO ($\psi_2$) and is consequently promoted to a higher energy MO ($\psi_3$). In order for this excitation to occur, the photon must possess the right amount of energy—the energy of the photon must be equivalent to the energy gap between the two MOs ($\psi_2$ and $\psi_3$). For most organic compounds containing conjugated π systems, the relevant energy gap corresponds with UV or visible light. This excitation is called a $\pi \rightarrow \pi^*$ transition (pronounced "pi to pi star"). The excited electron then returns to its original MO, releasing the energy in the form of heat or light. In this way, compounds with conjugated π systems will absorb UV or visible light.

A standard UV-Vis spectrophotometer irradiates a sample with wavelengths of light ranging from 200 to 800 nm. The light is first split into two beams. One beam passes through a cuvette (a small glass container) containing an organic compound dissolved in some solvent, while the other beam (the reference beam) is passed through a cuvette containing only the solvent. The spectrophotometer then compares the intensities of the beams at each particular wavelength, and the results are plotted on a chart that shows **absorbance** as a function of wavelength. Absorbance is defined as

$$A = \log \frac{I_0}{I}$$

where $I_0$ is the intensity of the reference beam and $I$ is the intensity of the sample beam. The plot that is generated is called a UV-Vis absorption spectrum. As an example, consider the absorption spectrum of butadiene (Figure 17.35). For our purposes, the most important feature of the absorption

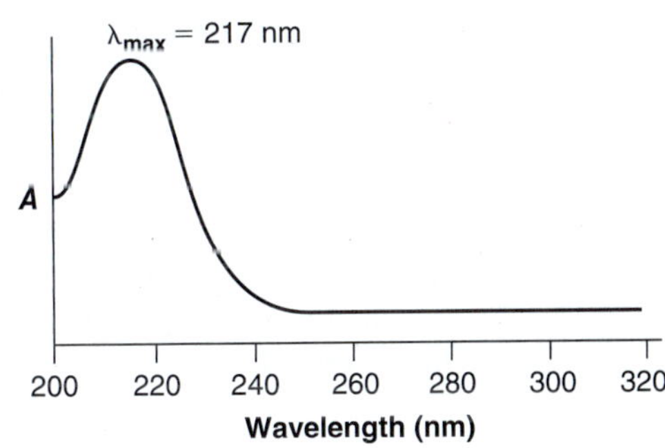


**FIGURE 17.35** An absorbance spectrum of butadiene.

spectrum is the $\lambda_{max}$ (pronounced **lambda max**), which indicates the wavelength of maximum absorption. For butadiene, $\lambda_{max}$ is 217 nm. The amount of UV light absorbed at the $\lambda_{max}$ for any compound is described by the **molar absorptivity** ($\varepsilon$) and is expressed by the following equation, called **Beer's law**:

$$\varepsilon = \frac{A}{C \times l}$$

where $A$ is the absorbance, $C$ is the concentration of the solution (mol/L), and $l$ is the sample path length (the width of the cuvette) measured in centimeters. The molar absorptivity ($\varepsilon$) is a physical characteristic of the particular compound being investigated and is typically in the range of 0–15,000.

The value of $\lambda_{max}$ for a particular compound is highly dependent on the extent of conjugation. To illustrate this point, compare the MOs of butadiene, hexatriene, and octatetraene (Figure 17.36).

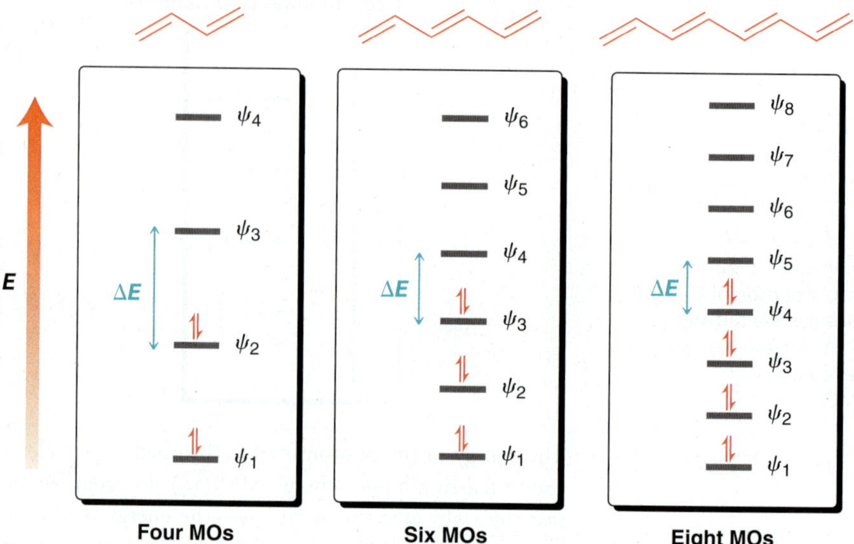


**FIGURE 17.36** An energy diagram showing the relative energy levels for MOs for various conjugated $\pi$ systems.

Compounds that are more highly conjugated will have more MOs and will exhibit smaller gaps between MOs. Compare the energy gaps ($\Delta E$) for each of the three compounds and notice the trend. Compounds with more highly conjugated systems require less energy to promote an electronic excitation. Recall that the energy of a photon is directly proportional to the frequency of the light but inversely proportional to the wavelength. As a result, a longer wavelength corresponds with less energy. We therefore find that compounds with a greater extent of conjugation will have a longer $\lambda_{max}$ (Table 17.3). The $\lambda_{max}$ of a compound therefore indicates the extent of conjugation present in the compound. Each additional conjugated double bond adds between 30 and 40 nm.

**TABLE 17.3 THE WAVELENGTH OF MAXIMUM ABSORPTION FOR A SIMPLE DIENE, TRIENE, AND TETRAENE**

| COMPOUND | $\lambda_{MAX}$ |
|---|---|
| | 217 nm |
| | 258 nm |
| | 290 nm |

## Woodward-Fieser Rules

In the previous section, we noted that a compound containing a conjugated $\pi$ system will absorb UV-Vis light. The region of the molecule responsible for the absorption (the conjugated $\pi$ system) is called the **chromophore**, while the groups attached to the chromophore are called **auxochromes**.

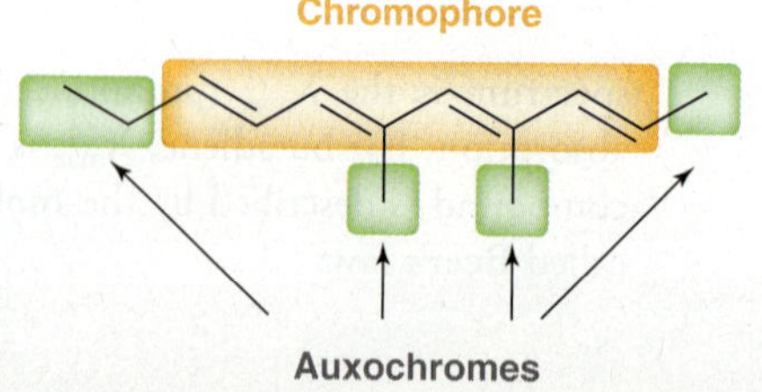

Auxochromes can also have a substantial effect on the value of $\lambda_{max}$. **Woodward-Fieser rules** can be used to make simple predictions (Table 17.4). The full list of Woodward-Fieser rules is quite long, but the few rules in Table 17.4 allow us to make simple predictions. Note that Woodward-Fieser rules provide us with an estimate for $\lambda_{max}$. In some cases, the estimate will be very close to the observed value, and in other cases, there will be a small discrepancy. The rules are only meant as a rough guide for making meaningful predictions, and they do not work well for compounds that contain more than six double bonds in conjugation.

**TABLE 17.4 WOODWARD–FIESER RULES FOR PREDICTING $\lambda_{MAX}$ FOR CONJUGATED π SYSTEMS**

| FEATURE TO LOOK FOR: | $\lambda_{MAX}$ | EXAMPLE | CALCULATION |
|---|---|---|---|
| Conjugated diene | Base value = 217 | | 217 nm |
| Each additional double bond | +30 | Two additional double bonds | Base: 217<br>+2 × 30<br>277 nm<br>(observed = 290 nm) |
| Each auxochromic alkyl group | +5 | Three alkyl groups connected to chromophore | Base: 217<br>+3 × 5<br>232 nm<br>(observed = 232 nm) |
| Each exocyclic double bond (a double bond where one vinylic position is part of a ring and the other vinylic position is outside the ring) | +5 | Exocyclic | Base: 217<br>+2 × 5 alkyl groups<br>+5 exocyclic double bonds<br>232 nm<br>(observed = 230 nm) |
| Homoannular diene—both double bonds are contained in one ring, so the diene is locked in an *s-cis* conformation | +39 | | Base: 217<br>+4 × 5 alkyl groups<br>+39 homoannular diene<br>276 nm<br>(observed = 269 nm) |

## SKILLBUILDER

### 17.5 USING WOODWARD-FIESER RULES TO ESTIMATE $\lambda_{MAX}$

**LEARN** the skill

Use Woodward-Fieser rules to estimate the expected $\lambda_{max}$ for the following compound:

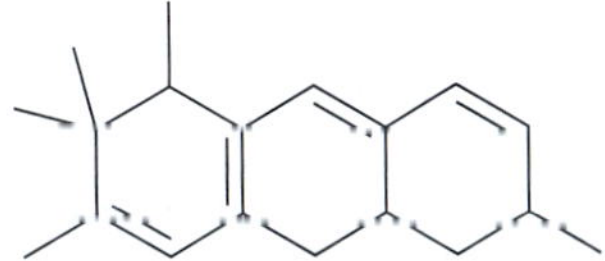

**SOLUTION**

**STEP 1**
Count the number of conjugated double bonds.

First count the number of conjugated double bonds. This compound has four double bonds in conjugation (highlighted in red). Two of them count toward the base value of 217 nm, and the other two double bonds will each add +30.

Next, look for any auxochromic alkyl groups. These are the carbon atoms connected directly to the chromophore. This compound has six auxochromic alkyl groups, each of which adds +5, giving another +30.

**STEP 2**
Count the number of auxochromic alkyl groups.

Next look for exocyclic double bonds. In this case, the double bond highlighted in green is exocyclic to the ring highlighted in red. This adds another +5.

**STEP 3**
Count the number of exocyclic double bonds.

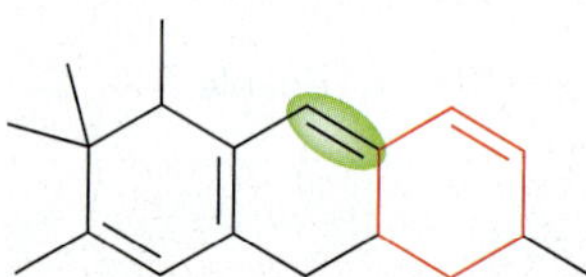

Finally, look for a homoannular diene. The first ring (on the left) exhibits two double bonds in the same ring, which adds +39. The second ring also exhibits two double bonds in the same ring, which adds another +39.

**STEP 4**
Look for homoannular dienes.

We predict this compound will have a $\lambda_{max}$ somewhere in the neighborhood of 390 nm:

**STEP 5**
Add up all factors.

Base = 217
Additional double bonds (2) = +60
Auxochromic alkyl groups (6) = +30
Exocyclic double bond (1) = +5
Homoannular dienes (2) = +78
Total = 390 nm

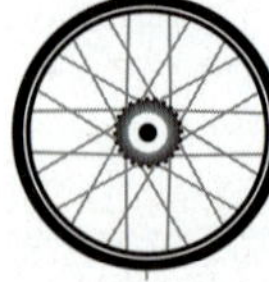

**PRACTICE the skill** **17.29** Use Woodward-Fieser rules to estimate the expected $\lambda_{max}$ for each of the following compounds:

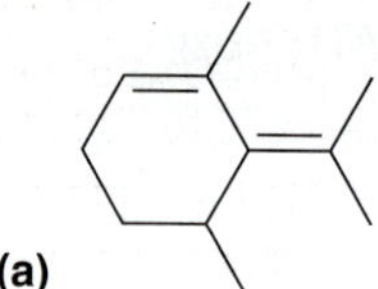

(a)

(b)

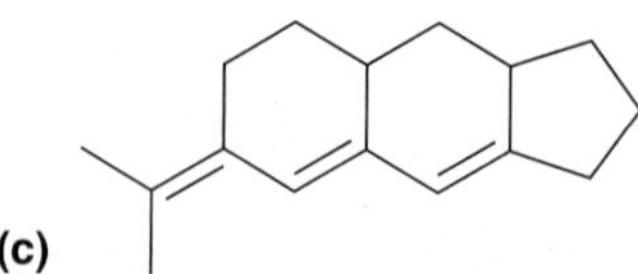

(c)

(d)

**APPLY the skill** **17.30** Identify which of the following compounds is expected to have a larger $\lambda_{max}$:

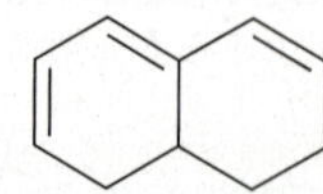

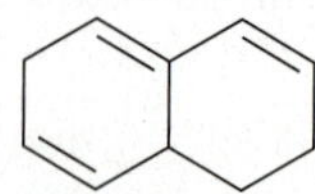

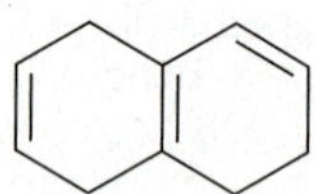

need more **PRACTICE?** **Try Problems 17.50–17.52, 17.60c**

## practically speaking | Sunscreens

In the United States, societal attitudes toward suntans have changed many times over the last two centuries. Before the industrial revolution, a suntan was considered a sign of lower class. Wealthy people stayed in the shade, while manual laborers worked in the sun. After the industrial revolution, manual laborers began to work indoors, and this distinction disappeared. In fact, societal attitudes took a 180° turn in the 1920s when sunlight was considered to be a cure-all for everything from acne to tuberculosis. By the 1940s, suntan lotions first appeared on the market, although the function of these products was not to block out the sun but to aid in the tanning process. In the 1960s, sun lamps were invented, so that people could tan in the winter.

In the last several decades, societal attitudes have shifted once again, with the discovery of the link between sunlight and skin cancer. Harmful UV radiation from the sun is mostly screened out by the ozone layer, although some harmful UV rays do make it to the earth's surface. There are two types of damaging UV light, called UV-A (315–400 nm) and UV-B (280–315 nm). UV-B is higher energy and more damaging, but UV-A is also dangerous, as it can penetrate the skin more deeply. It has been estimated that 1 in every 10 Americans will develop skin cancer, and it is believed that the situation will worsen over the coming decades, as a result of the slow destruction of the ozone layer by CFCs (Section 11.8). That means that there is a 10% chance that you will develop skin cancer at some point in your life unless you take precautions.

The mechanism by which UV light damages DNA has been extensively studied, and many pathways are responsible. One such pathway involves a thymine base of DNA absorbing UV light to give an excited state that is capable of undergoing a [2+2] cycloaddition with a neighboring thymine base.

In Section 17.8, we saw that [2+2] cycloadditions become symmetry allowed in the presence of UV light, and this is just a real-world example of that process. This reaction changes the structure of DNA and causes genetic code alterations that ultimately lead to the development of cancer cells.

Sunscreens are used to prevent UV light from causing damage to the structure of DNA. There are two basic types of sunscreens: inorganic and organic. Inorganic sunscreens (such as titanium dioxide or zinc oxide) reflect and scatter UV light. Organic sunscreens are conjugated π systems that absorb UV light and release the absorbed energy in the form of heat. The following compounds are examples of common organic sunscreens:

Octyl Methoxycinnamate

Oxybenzone

4-Methylbenzylidene camphor

Avobenzone

Homosalate

All of these compounds possess conjugated π systems. Both organic and inorganic sunscreens provide protection from UV-B radiation. Some sunscreens, although not all, also provide additional protection from UV-A radiation. Many commercially available sunscreens contain mixtures of both organic and inorganic compounds, thereby achieving high SPF (sun protection factor). However the SPF rating only indicates the strength of protection from UV-B, not UV-A. For products sold in the United States, there is no clear way to compare the abilities of sunscreens to filter UV-A radiation other than reading the list of ingredients. Among the organic sunscreens approved in the United States, it is believed that avobenzone offers the best protection against UV-A radiation.

Despite the established link between sunlight and skin cancer, many people continue to view suntans as sexy. This dangerous attitude will likely become less popular as the public becomes more aware of the dangers involved. As evidence of rising public awareness, annual sales of sunscreens rise each year in the United States and currently exceed $100 million.

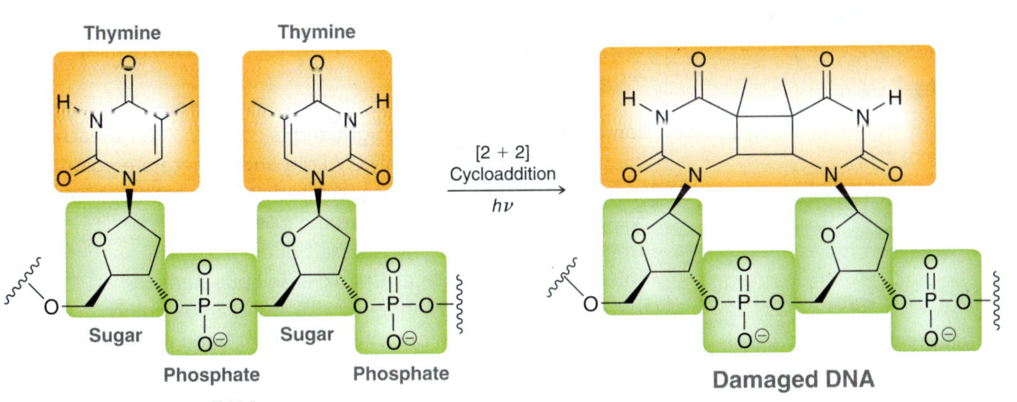

## 17.12 Color

**BY THE WAY**

What happened to indigo? You probably have learned at some point that the rainbow is composed of the colors ROYGBIV (red, orange, etc.). The "I" stands for indigo, but you might notice that indigo is absent from our color wheel. The truth is that indigo is not worth mentioning (it is intermediate between blue and violet). Indigo is mentioned as one of the colors of the rainbow, because Sir Isaac Newton, who described them, was a student of metaphysics and felt that the number 7 was more spiritually meaningful than 6. Thus, the "seven" colors of the rainbow.

The visible region of the electromagnetic spectrum has wavelengths of 400–700 nm. When a compound possesses a highly conjugated π system, it is possible for the $\lambda_{max}$ to be above 400 nm. Such a compound will absorb visible light, rather than UV light, and will therefore be colored. Consider the following two examples:

**Lycopene**

**β-Carotene**

Lycopene is responsible for the red color of tomatos; β-carotene is responsible for the orange color of carrots. Both compounds are colored because they possess highly conjugated π systems.

White light is composed of all visible wavelengths (400–700 nm) and can be divided into groups of complementary colors, which appear opposite from each other in the color wheel in Figure 17.37. Each pair of complementary colors (such as red and green) can be thought of as canceling each other (in a way) to produce white light. A compound will be colored if it absorbs a specific color more strongly than its complementary color. For example, β-carotene absorbs light of 455 nm (blue light) but reflects orange light. The compound therefore appears to be orange.

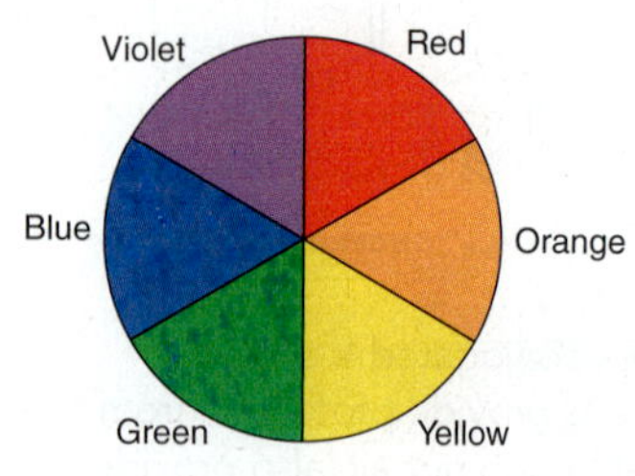


**FIGURE 17.37** The color wheel, showing pairs of complementary colors.

### Bleach

Now that we understand the origin of color in organic compounds, we are finally able to explore how bleaches function. Most bleaching agents react with conjugated π systems, disrupting the extended conjugation:

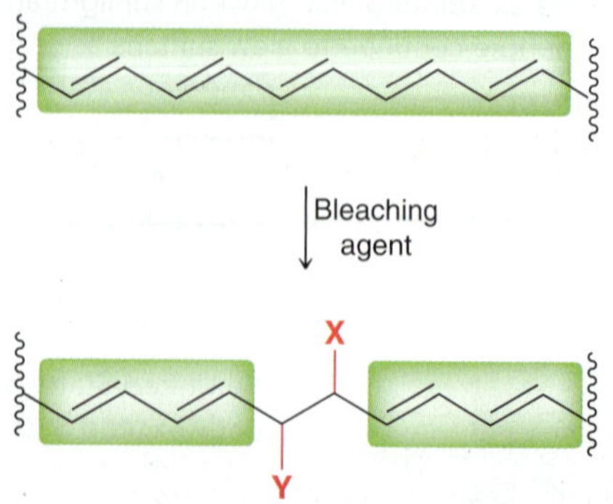


In this way, a conjugated π system that previously absorbed visible light is converted into smaller π systems that absorb UV light instead.

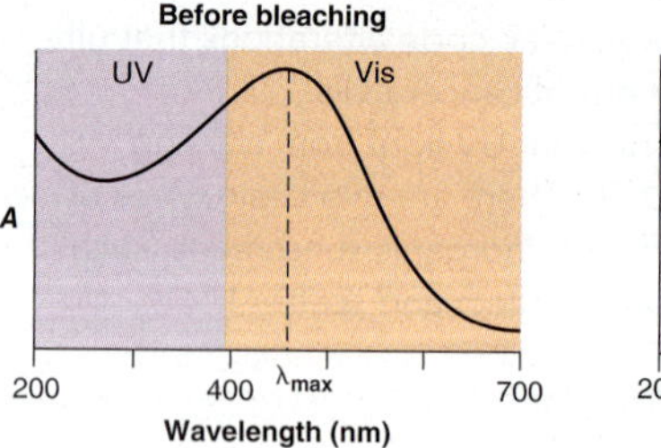


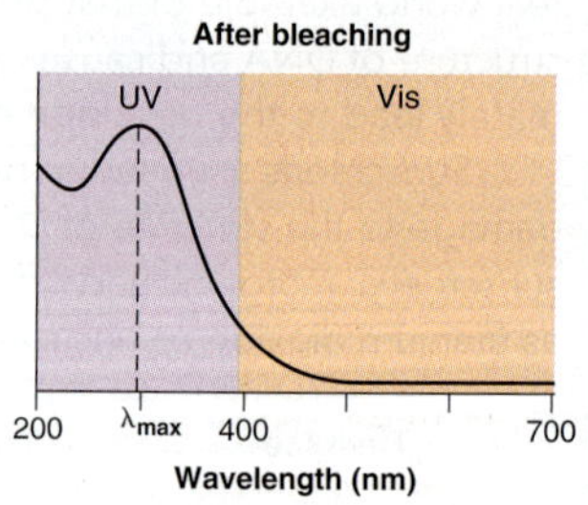


Notice the value of $\lambda_{max}$ in each case. Before bleaching, the compound absorbs visible light and is colored. After bleaching, the compound absorbs only UV light and is therefore colorless.

There are many different agents that will function as bleaches. Some act by oxidizing the double bonds, while others act by reducing the double bonds. Common household bleach (such as Clorox) is an aqueous solution of sodium hypochlorite (NaOCl) and is an oxidizing agent. When a stain is bleached, it has not been washed away. Rather, it has just been chemically altered so that it can no longer be seen. When placed under a UV lamp, the compound will glow, and its presence can be detected.

## CONCEPTUAL CHECKPOINT

**17.31** For each of the following tasks, use the color wheel in Figure 17.37.

(a) Identify the color of a compound that absorbs orange light.

(b) Identify the color of a compound that absorbs blue-green light.

(c) Identify the color of a compound that absorbs orange-yellow light.

## 17.13 Chemistry of Vision

The chemical reactions responsible for vision have been the subject of much investigation over the last 60 years.

Two types of light-sensitive cells function as photoreceptors: *rods* and *cones*. Rods do not detect color, and they function as the dominant light receptors in dim light. Cones contain the pigments necessary for color vision, but they only function as the dominant receptor cells in bright light. Humans possess both rods and cones, but some species possess only one of the two cell types. Pigeons, for example, have only cones. As a result, they see quite well in bright light, but they are blind at night. Owls, on the other hand, have only rods. Owls see well at night but are colorblind. In this section, we will focus solely on the chemistry of rods.

The light-sensitive compound in rods is called *rhodopsin*. In 1952, Nobel Laureate George Wald (Harvard University) and his co-workers showed that the chromophore in rhodopsin is the conjugated polyunsaturated system of 11-*cis*-retinal. Rhodopsin is produced by a chemical reaction between 11-*cis*-retinal and a protein called opsin.

$NH_2$ Opsin

H O

**11-*cis*-Retinal**

H N Opsin

**Rhodopsin**

The conjugated π system fits precisely into an internal cavity of opsin and absorbs light over a broad region of the visible spectrum (400–600 nm). Sources of 11-*cis*-retinal include vitamin A and β-carotene. A vitamin A deficiency causes night blindness, while a diet rich in β-carotene can improve vision.

OH

**Vitamin A**

**β-Carotene**

Wald demonstrated that rhodopsin made from 11-*cis*-retinal gave the same photoproduct as rhodopsin made from 9-*cis*-retinal. He concluded that the primary photochemical reaction must be a *cis-trans* isomerization.

*cis* N H Opsin

*hν*

*trans* N H Opsin

Further evidence supporting this photoisomerization step came from Koji Nakanishi's work at Columbia University. Nakanishi's group synthesized an analogue of 11-*cis*-retinal, in which the *cis* double bond was incorporated in a seven-membered ring, thereby preventing it from isomerizing to the *trans* configuration.

11-*cis*-Retinal

Nakanishi's analogue

Recall from Section 8.4 that rings of fewer than eight carbon atoms cannot accommodate a *trans* π bond at room temperature. When Nakanishi's analogue binds with opsin, the product is similar to rhodopsin but is unable to isomerize under the influence of light. When Nakanishi's analogue was administered to rats, their vision was severely impaired. Based on these and other convincing pieces of evidence, it is widely accepted that the first step in the chemistry of vision is a photoisomerization reaction.

When rhodopsin is excited to the high-energy all-*trans* isomer, the resulting change in rhodopsin's shape initiates a cascade of enzymatic reactions that causes the release of calcium ions. These ions block channels that normally allow the passage of billions of sodium ions per second. This regular flow of sodium ions is called the *dark current,* and it is reduced when calcium ions block the channels. This reduction in current culminates in a nerve impulse being sent to the brain. Human vision is extremely sensitive, because the absorption of a single photon can prevent the flow of millions of sodium ions. Our eyes are capable of sensing very dim light—even a few photons will trigger a nerve impulse. We can also adjust our eyes to a very bright environment within a minute. We have yet to create a photographic system that parallels the sensitivity and adaptability of the human eye.

# REVIEW OF REACTIONS

## Preparation of Dienes

### From Allylic Halides

Br

t-BuOK

### From Dihalides

Br

Br

t-BuOK

## Electrophilic Addition

### With HBr

H—Br

Br

Br

1,2-Adduct

1,4-Adduct

| | 1,2-Adduct | 1,4-Adduct |
|---|---|---|
| At 0°C | 71% | 29% |
| At 40°C | 15% | 85% |

### With Br2

$Br_2$

Br

Br

Br

Br

1,2-Adduct

1,4-Adduct

## Diels-Alder Reaction

Diels-Alder

Retro Diels-Alder (very high temperature)

+ En

Endo

## Electrocyclic Reactions

Heat

$h\nu$

+ En

## Sigmatropic Rearrangements

### Cope Rearrangement

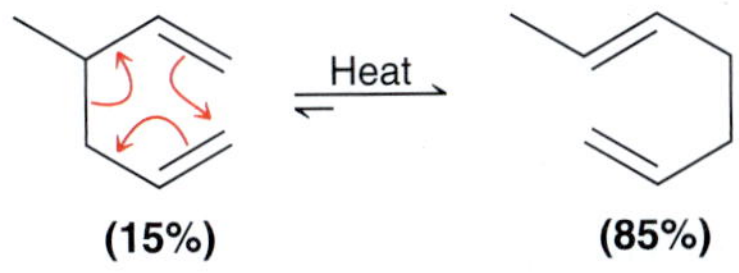


### Claisen Rearrangement

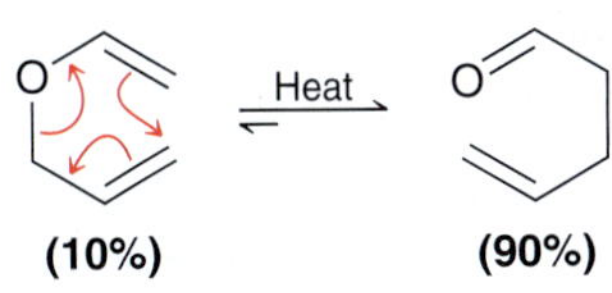

# REVIEW OF CONCEPTS AND VOCABULARY

### SECTION 17.1

- **Dienes** possess two C=C bonds and are classified as **cumulated**, **conjugated**, or **isolated**.
- Conjugated dienes contain one continuous system of overlapping *p* orbitals and therefore exhibit special properties and reactivity.

### SECTION 17.2

- Conjugated dienes can be prepared from allylic halides or from dihalides.
- Conjugated double bonds are more stable than isolated double bonds, with a *stabilization energy* of approximately 15 kJ/mol.
- Conjugated dienes experience free rotation about the C2—C3 bond, giving rise to two important conformations: ***s*-cis** and ***s*-trans**. The *s-trans* conformation is lower in energy.

### SECTION 17.3

- Electrons that occupy atomic orbitals are associated with an individual atom, while electrons that occupy a molecular orbital (MO) are associated with the entire molecule.
- The *H*ighest *O*ccupied *M*olecular *O*rbital, or HOMO, contains the $\pi$ electrons most readily available to participate in a reaction.
- The *L*owest *U*noccupied *M*olecular *O*rbital, or LUMO, is the lowest energy MO that is capable of accepting electron density.
- The HOMO and LUMO are referred to as **frontier orbitals**, and the reactivity of conjugated polyenes can be explained with **frontier orbital theory**.
- An **excited state** is produced when a $\pi$ electron in the HOMO absorbs a photon of light bearing the appropriate energy necessary to promote the electron to a higher energy orbital.
- Reactions induced by light are called **photochemical reactions**.

### SECTION 17.4

- When butadiene is treated with HBr, two major products are observed, resulting from **1,2-addition** and **1,4-addition**. These products are called the **1,2-adduct** and the **1,4-adduct**.
- Other electrophiles, such as $Br_2$, will also add across conjugated $\pi$ systems to produce 1,2- and 1,4-adducts.

### SECTION 17.5

- Conjugated dienes that undergo addition at low temperature are said to be under **kinetic control**.
- Conjugated dienes that undergo addition at elevated temperature are said to be under **thermodynamic control**.

### SECTION 17.6

- **Pericyclic reactions** proceed via a concerted process with a cyclic transition state, and they are classified as **cycloaddition reactions**, **electrocyclic reactions**, and **sigmatropic rearrangements**.

### SECTION 17.7

- The Diels-Alder reaction is a **[4+2] cycloaddition** in which two C—C bonds are formed simultaneously.
- The product of a Diels-Alder reaction is a substituted cyclohexene.
- Moderate temperatures favor the formation of products in a Diels-Alder reaction, but very high temperatures (over 200°C) tend to disfavor product formation. High temperatures can often be used to achieve the reverse of a Diels-Alder reaction, called a **retro Diels-Alder**.
- The starting materials for a Diels-Alder reaction are a diene and a **dienophile**.
- Diels-Alder reactions proceed rapidly and with a high yield when the dienophile has an electron-withdrawing substituent.
- When the dienophile is a 1,2-disubstituted alkene, the reaction proceeds with stereospecificity.
- A triple bond can also function as a dienophile.
- The Diels-Alder reaction only occurs when the diene adopts an *s-cis* conformation.
- When cyclopentadiene is used as the starting diene, a bridged bicyclic compound is obtained, and the ***endo*** cycloadduct is favored over the ***exo*** cycloadduct.

### SECTION 17.8

- In a Diels-Alder reaction, the electron density is transferred from the HOMO of the diene to the LUMO of the dienophile.
- **Conservation of orbital symmetry** requires that MOs have the same phases in order to effectively overlap during a reaction.
- Diels-Alder reactions are **symmetry allowed**.
- Thermal [2+2] cycloadditions are **symmetry forbidden**, because the phases of the frontier orbitals do not overlap.
- [2+2] Cycloadditions can only be performed with photochemical excitation in which the HOMO of an excited state interacts with the LUMO of a ground-state molecule.

### SECTION 17.9

- An electrocyclic reaction is a pericyclic process in which a conjugated polyene undergoes cyclization. In the process, one $\pi$ bond is converted into a $\sigma$ bond, while the remaining $\pi$ bonds all change their locations.
- The use of thermal conditions vs. photochemical conditions has a profound impact on the stereochemical outcome of the reaction.
- Conservation of orbital symmetry determines whether an electrocyclic reaction occurs in a **disrotatory** fashion or a **conrotatory** fashion.

### SECTION 17.10

- A sigmatropic rearrangement is a pericyclic reaction in which one $\sigma$ bond is formed at the expense of another.
- A [3,3] sigmatropic rearrangement is called a **Cope rearrangement** when all six atoms of the cyclic transition state are carbon atoms.
- The oxygen analogue of a Cope rearrangement is called a **Claisen rearrangement**.

### SECTION 17.11

- Compounds that possess a conjugated $\pi$ system will absorb UV or visible light to promote an electronic excitation called a $\pi \rightarrow \pi^*$ transition.
- A standard UV-Vis spectrophotometer will irradiate a sample with UV and visible light and will generate an absorption spectrum, which plots **absorbance** as a function of wavelength.

- The most important feature of the absorption spectrum is the $\lambda_{max}$ (pronounced **lambda max**), which indicates the wavelength of maximum absorption.
- The amount of UV light absorbed at the $\lambda_{max}$ for any compound is described by the **molar absorptivity** ($\varepsilon$) and is related to absorbance by an equation called **Beer's law**.
- Compounds with a greater extent of conjugation will have a longer $\lambda_{max}$.
- The region of the molecule responsible for the absorption (the conjugated $\pi$ system) is called the **chromophore**, while the groups attached to the chromophore are called **auxochromes**.
- The $\lambda_{max}$ for a simple compound can be predicted with **Woodward-Fieser rules**.

**SECTION 17.12**

- When a compound exhibits a $\lambda_{max}$ between 400 and 700 nm, the compound will absorb visible light, rather than UV light. Compounds that absorb light in this range will be colored.

**SECTION 17.13**

- The light-sensitive compound in rods is called rhodopsin.
- When rhodopsin absorbs light, a *cis-trans* isomerization occurs to give the all-*trans* isomer. The resulting change in rhodopsin's shape initiates a cascade of enzymatic reactions that culminate with a nerve impulse being sent to the brain.

# SKILLBUILDER REVIEW

## 17.1 PROPOSING THE MECHANISM AND PREDICTING THE PRODUCTS OF ELECTROPHILIC ADDITION TO CONJUGATED DIENES

**EXAMPLE** Predict the products and propose a mechanism.

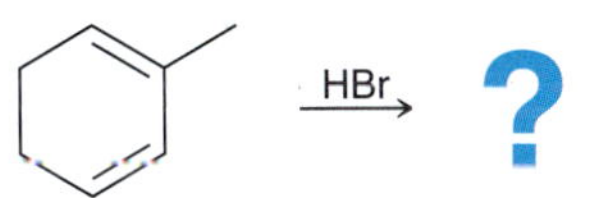


**STEP 1** Identify the possible locations where protonation can occur.

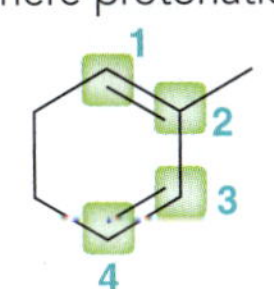


**STEP 2** Determine the location where protonation will produce an allylic carbocation.

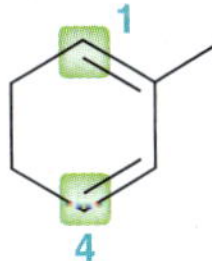


**STEP 3** Draw both resonance structures of each possible allylic carbocation and then draw a nucleophilic attack occurring at each possible electrophilic position.

Same compound

Try Problems **17.7–17.9, 17.35, 17.38**

## 17.2 PREDICTING THE MAJOR PRODUCT OF AN ELECTROPHILIC ADDITION TO CONJUGATED DIENES

**EXAMPLE**

HBr, 0°C

?

**STEP 1** Identify the possible locations for protonation.

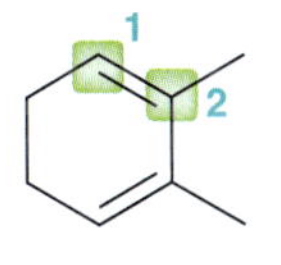


Only two unique positions

**STEP 2** Draw the resonance structures of the allylic carbocation that is expected to form.

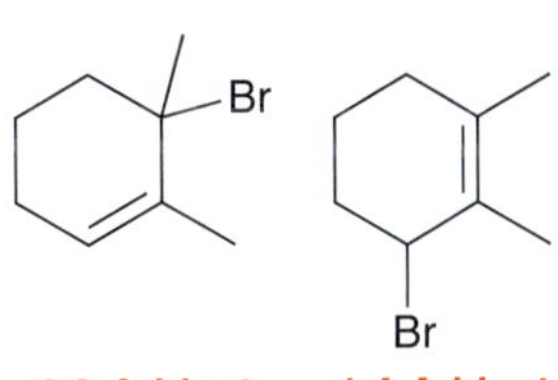

Allylic carbocation

**STEP 3** Attack each position with bromide and draw both possible products.

1,2-Adduct 1,4-Adduct

**STEP 4** Identify the kinetic and thermodynamic products and then choose which product predominates based on the temperature of the reaction.

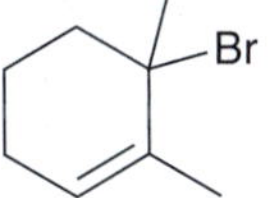


The 1,2-adduct is the kinetic product and is favored at low temperature

Try Problems **17.10–17.12, 17.36, 17.37, 17.39**

### 17.3 PREDICTING THE PRODUCT OF A DIELS-ALDER REACTION

**EXAMPLE** Predict the product:

**STEP 1** Redraw and line up the ends of the diene with the dienophile.

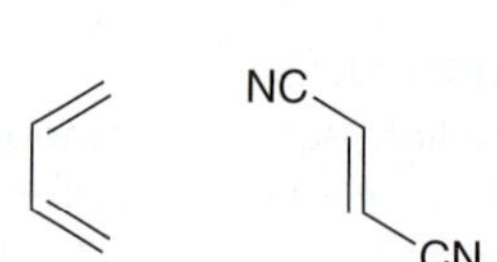


**STEP 2** Draw three curved arrows, starting at the dienophile and moving either clockwise or counterclockwise.

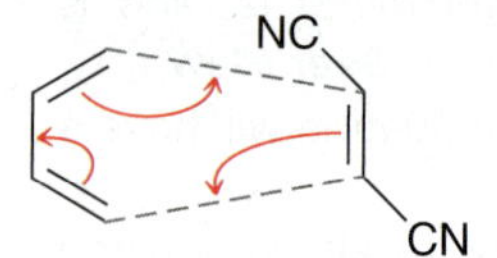


**STEP 3** Draw the product with the correct stereochemical outcome.

CN + En CN

Try Problems 17.14–17.16, 17.44, 17.46

### 17.4 PREDICTING THE PRODUCT OF AN ELECTROCYCLIC REACTION

**STEP 1** Count the number of π electrons.

Ph $h\nu$ Ph ?

6 π Electrons

**STEP 2** Determine whether the reaction is conrotatory or disrotatory.

| | Heat | $h\nu$ |
|---|---|---|
| 4π | Con | Dis |
| 6π | Dis | Con |

6π, $h\nu$ = Conrotatory

**STEP 3** Determine whether the substituents are *cis* or *trans* to each other in the product.

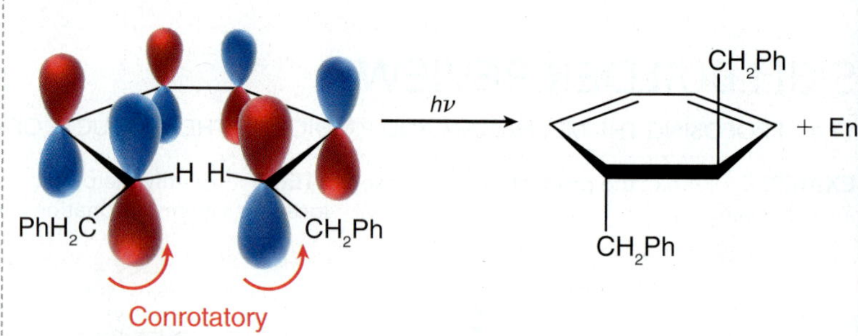


Try Problems 17.21–17.24, 17.55, 17.56

### 17.5 USING WOODWARD-FIESER RULES TO ESTIMATE $\lambda_{MAX}$

**STEP 1** Count the number of double bonds.

**STEP 2** Count the number of auxochromic alkyl groups.

**STEP 3** Count the number of exocyclic double bonds.

**STEP 4** Look for homoannular dienes.

**STEP 5** Add up all factors:

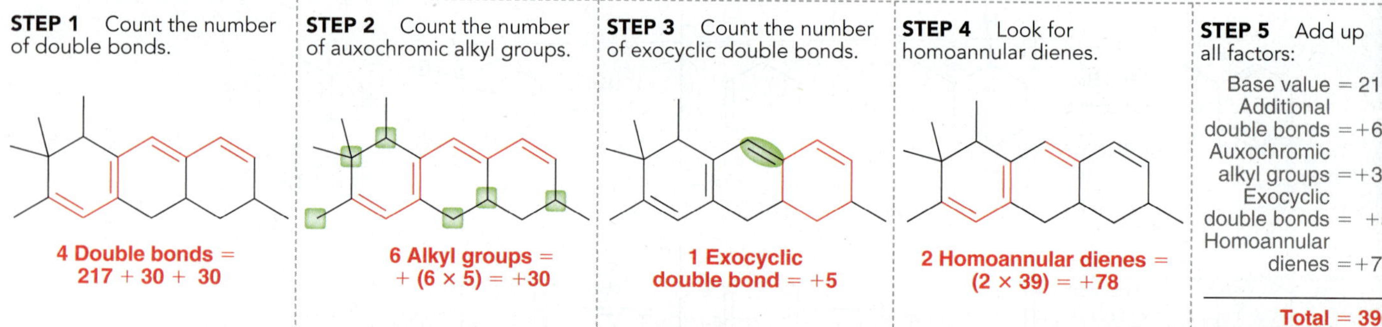


Base value = 217
Additional double bonds = +60
Auxochromic alkyl groups = +30
Exocyclic double bonds = +5
Homoannular dienes = +78

**Total = 390**

Try Problems 17.29, 17.30, 17.50–17.52, 17.60c

## PRACTICE PROBLEMS

Note: Most of the Problems are available within **WileyPLUS**, an online teaching and learning solution.

**17.32** Draw the structure of each of the following compounds:

(a) 1,4-Cyclohexadiene (b) 1,3-Cyclohexadiene

(c) (*Z*)-1,3-Pentadiene (d) (2*Z*,4*E*)-Hepta-2,4-diene

(e) 2,3-Dimethyl-1,3-butadiene

**17.33** Circle each compound that has a conjugated π system:

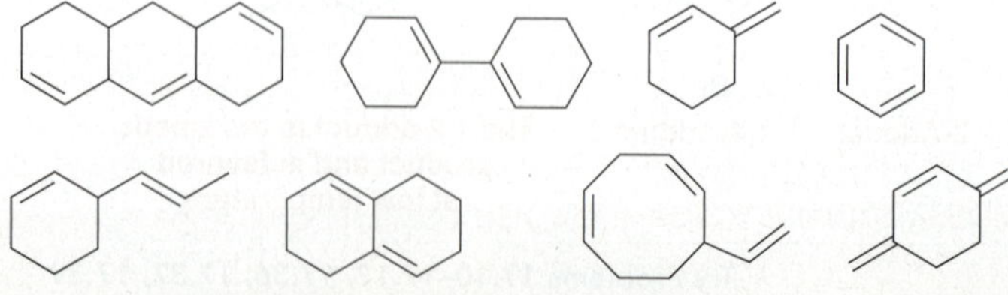

**17.34** For each of the following pairs of compounds determine whether the two compounds would be at equilibrium at room temperature or whether they could be isolated from one another:

(a)

(b)

(c)

**17.35** Identify the structure of the conjugated diene that will react with one equivalent of HBr to yield a racemic mixture of 3-bromocyclohexene.

**17.36** Draw the major product expected when 1,3-butadiene is treated with one equivalent of HBr at 0°C and show the mechanism of its formation.

**17.37** Draw the major product expected when 1,3-butadiene is treated with one equivalent of HBr at 40°C and show the mechanism of its formation.

**17.38** Draw all four products obtained when 2-ethyl-3-methyl-1,3-cyclohexadiene is treated with HBr at room temperature and show the mechanism of their formation.

**17.39** After performing the reaction from Problem 17.38, the reaction flask is heated to 40°C and two of the products become the major products. When the flask is then cooled to 0°C, no change occurs in the product distribution. Explain why an increase in temperature causes a change in the product distribution and explain why a subsequent reduction in temperature has no effect.

**17.40** Each of the following compounds does not participate as a diene in a Diels-Alder reaction. Explain why in each case.

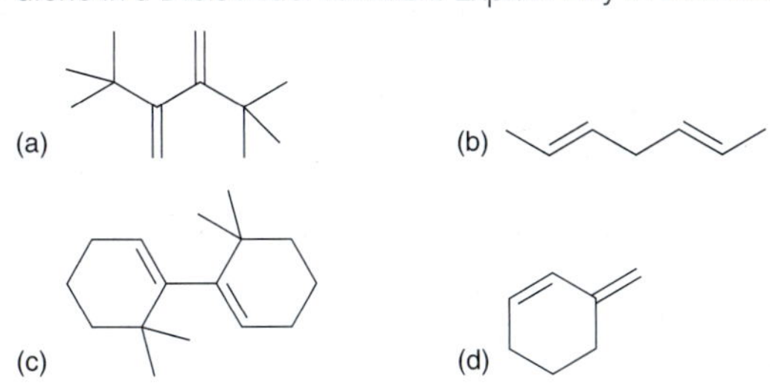


**17.41** Rank the following dienophiles (from least reactive to most reactive) in terms of reactivity in a Diels-Alder reaction:

**17.42** The absorbance spectrum of 1,3-butadiene displays an absorption in the UV region ($\lambda_{max}$ = 217 nm), while the absorption spectrum of 1,2-butadiene does not display a similar absorption. Explain.

**17.43** Predict the products for each of the following Diels-Alder reactions:

(a) + HOOC COOH (b) + HOOC CN

(c) + (d) +

(e) + NC CN

(f) MeO OMe +

**17.44** Identify the reagents you would use to prepare each of the following compounds via a Diels-Alder reaction:

(a) COOH COOH (b) CN CN CN CN (c) CHO CHO

(d) (e) (f)

(g) COOH COOH + En (h)

**17.45** Starting with 1,3-butadiene as your only source of carbon atoms and using any other reagents of your choice, design a synthesis of the following compound:

H CH3 + En

**17.46** The following triene reacts with excess maleic anhydride to produce a compound with molecular formula $C_{14}H_{12}O_6$. Draw the structure of this product (ignoring stereochemistry).

Maleic anhydride (excess) → $C_{14}H_{12}O_6$

**17.47** Chlordane is a powerful insecticide that was used in the United States during the second half of the twentieth century. Its use was discontinued in 1988 in recognition of its persistence and accumulation in the environment. Identify the reagents you would use to prepare chlordane via a Diels-Alder reaction.

**Chlordane**

**17.48** Cyclopentadiene reacts very rapidly in Diels-Alder reactions. In contrast, 1,3-cyclohexadiene reacts more slowly and 1,3-cycloheptadiene is practically unreactive. Can you offer an explanation for this trend?

**17.49** Compound **A** ($C_7H_{10}$) exhibits a $\lambda_{max}$ of 230 nm in its UV absorption spectrum. Upon hydrogenation with a metal catalyst, compound **A** will react with two equivalents of hydrogen gas. Ozonolysis of compound **A** yields the following two compounds. Identify two possible structures for compound **A**.

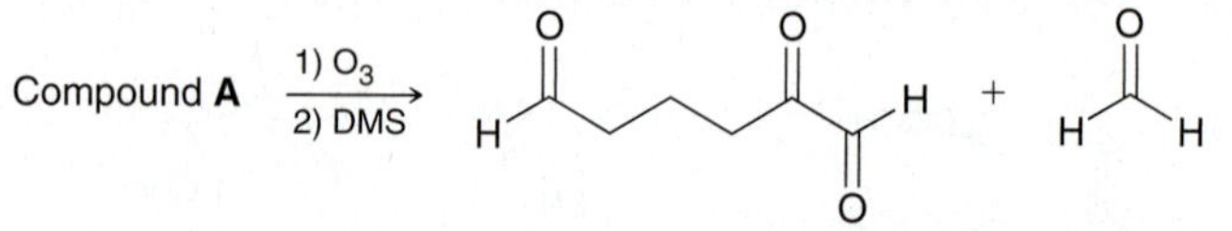

Compound **A** 1) $O_3$ 2) DMS

**17.50** Rank the following compounds in order of increasing $\lambda_{max}$:

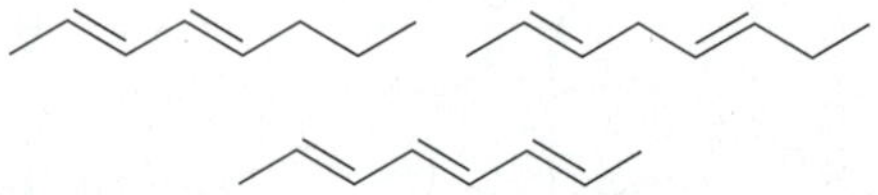

**17.51** Which of the following compounds do you expect to have a longer $\lambda_{max}$?

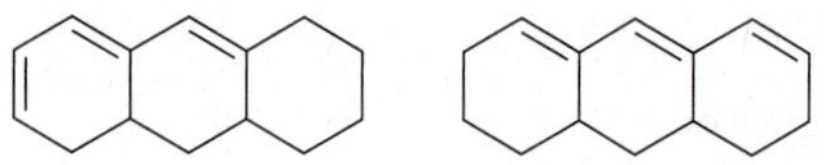

**17.52** Predict the expected $\lambda_{max}$ of the following compound:

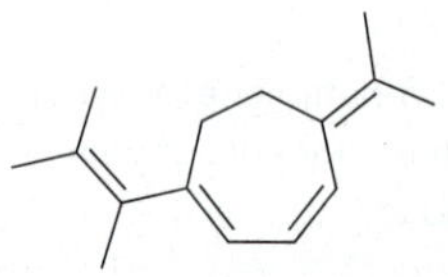

**17.53** When 5-deutero-5-methyl-1,3-cyclopentadiene is warmed to room temperature, it rapidly rearranges, giving an equilibrium mixture containing the original compound as well as two others. Propose a plausible mechanism for the formation of these other two compounds.

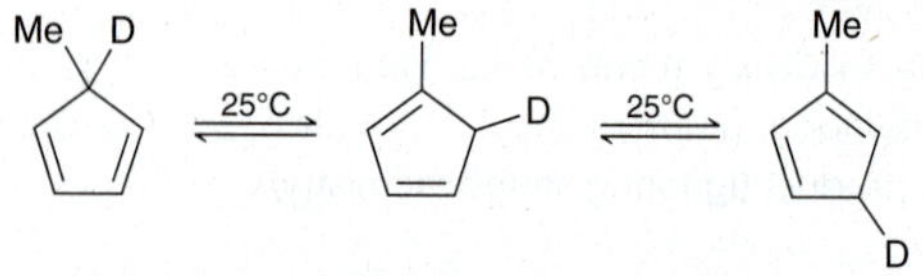

Me D 25°C Me D 25°C Me D

**17.54** Propose a plausible mechanism for the following transformation:

D D Heat D D

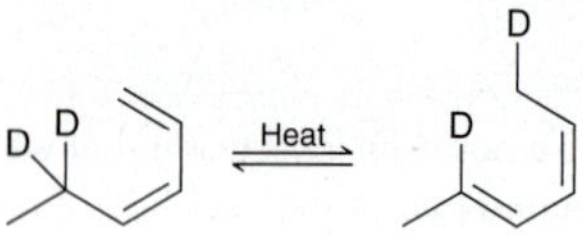

**17.55** Propose a method for achieving the following transformation:

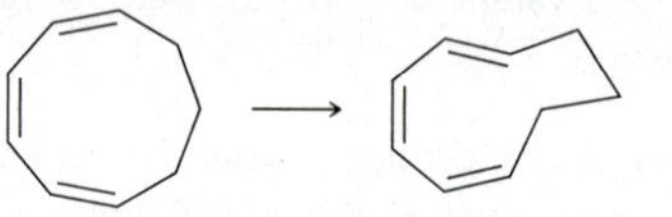

**17.56** Predict the product for each of the following electrocyclic reactions:

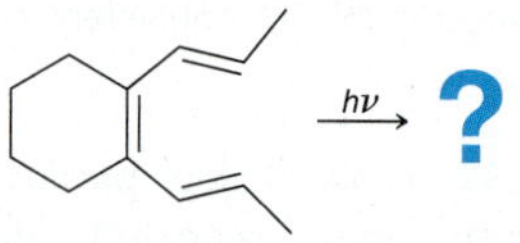

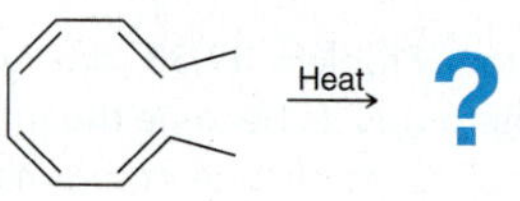

(a) Heat ? (b) $h\nu$ ? (c) Heat ? (d) $h\nu$ ?

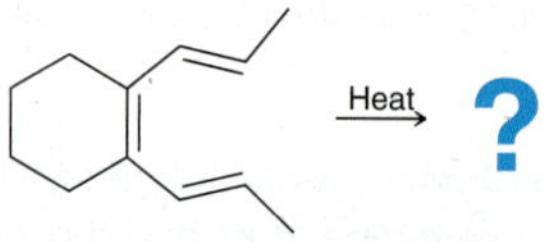

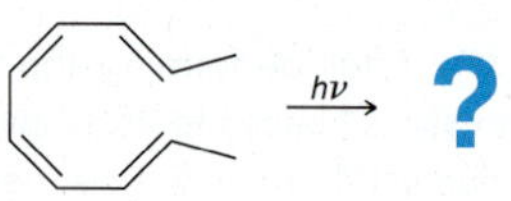

**17.57** Predict which side of the following equilibrium is favored and explain your choice.

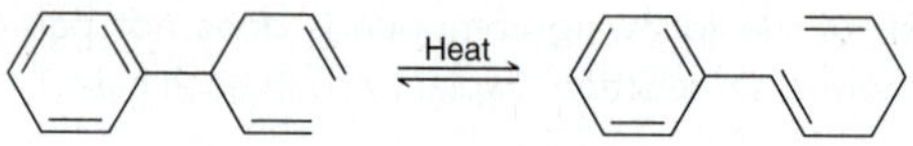

Heat

**17.58** When *trans*-3,4-dimethylcyclobutene is heated, conrotatory ring opening can produce two different products, yet only one is formed. Draw both products, identify which product is formed, and then explain why the other product is not formed.

**17.59** Predict the product for each of the following reactions:

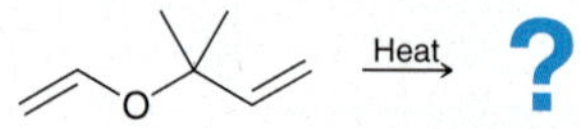

(a) Heat ? (b) Heat ? (c) Heat ?

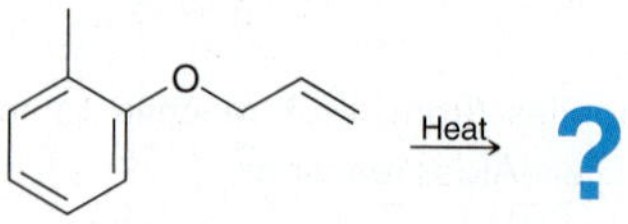

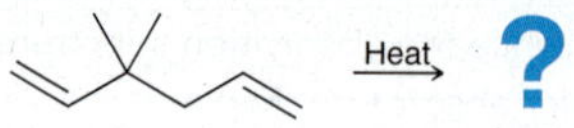

## INTEGRATED PROBLEMS

**17.60** α-Terpinene is a pleasant-smelling compound present in the essential oil of marjoram, a perennial herb. Upon hydrogenation with a metal catalyst, α-terpinene reacts with two equivalents of hydrogen gas to produce 1-isopropyl-4-methylcyclohexane. Ozonolysis of α-terpinene yields the following two compounds:

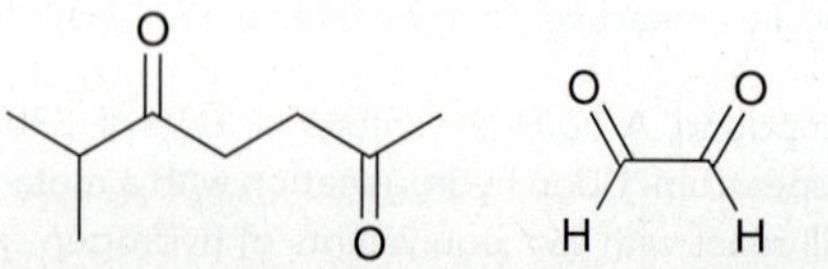

(a) How many double bonds are present in α-terpinene?

(b) Identify the structure of α-terpinene.

(c) Predict the wavelength of maximum absorption ($\lambda_{max}$) in the absorption spectrum of α-terpinene.

**17.61** Draw all possible conjugated dienes with molecular formula $C_6H_{10}$, taking special care not to draw the same compound twice.

**17.62** Treatment of 1,2-dibromocycloheptane with excess potassium *tert*-butoxide yields a product that absorbs UV light. Identify the product.

**17.63** In each of the following pairs of compounds identify the compound that liberates the most heat upon hydrogenation:

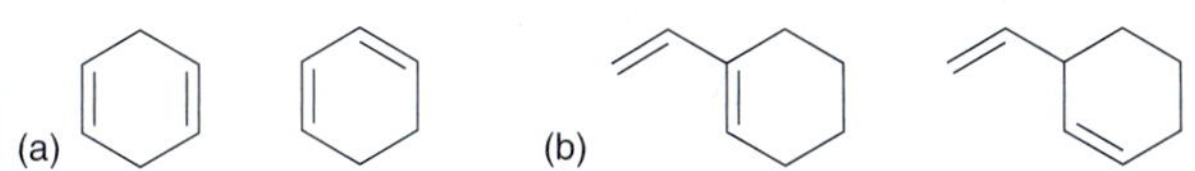


**17.64** Would you expect nitroethylene to be more or less reactive than ethylene in a Diels-Alder reaction? (*Hint*: Draw the resonance structures of nitroethylene.)

Ethylene Nitroethylene

**17.65** Propose a plausible mechanism for the following transformation. (*Hint:* Only two sequential pericyclic reactions are required.)

**17.66** When 2-methoxy-1,3-butadiene is treated with 3-buten-2-one, the resulting Diels-Alder reaction has two potential regiochemical outcomes:

Nevertheless, only one of these products is obtained. Draw resonance structures for the diene and the dienophile and use these resonance structures to make an educated guess regarding which product is formed.

**17.67** Propose a mechanism for the following transformation:

**17.68** Based on your answer to Problem 17.67, propose a mechanism for the following transformation:

**17.69** Compare the structures of 1,4-pentadiene and divinylamine:

1,4-Pentadiene Divinylamine

The first compound does not absorb UV light in the region between 200 and 400 nm. The second compound does absorb light above 200 nm. Using this information, identify the hybridization state of the nitrogen atom in divinylamine and justify your answer.

## CHALLENGE PROBLEMS

**17.70** The poison gelsemoxonine can be isolated from the leaves of a plant native to southeastern Asia (*Gelsemium elegans*). A key step in a synthesis of this natural product involves a thermally initiated (70°C) sigmatropic rearrangement of the compound shown below (R = protecting group) (*J. Am. Chem. Soc.* **2011,** *133,* **17634–17637**). Predict the product of this reaction and propose a mechanism for its formation.

**17.71** Upon irradiation, some organic compounds undergo a reversible change in molecular structure, such as *cis–trans* isomerization or cyclization of $\pi$-conjugated chromophores. This property, called *photochromism*, has shown strong potential application in the development of optical data storage materials. For example, upon treatment with UV light, compound **1** undergoes electrocyclic ring closure to give compound **2** (*J. Photochem. Photobiols. A.* **2006,** *184,* **177–183**). Further treatment of **2** with visible light causes a reversible ring-opening reaction, regenerating **1**. Draw the structure of **2**, determine whether it is chiral, and provide a mechanism for its formation.

**17.72** A hetero Diels-Alder reaction is a variation of the Diels-Alder reaction in which one or more of the carbon atoms of the diene and/or the dienophile are replaced by other atoms such as oxygen or nitrogen. With this in mind, propose a mechanism consistent with the following transformation in which a multifunctional acyclic reactant produces a macrocyclic (large ring) product (*Org. Lett.* **2001,** *3,* **723–726**):

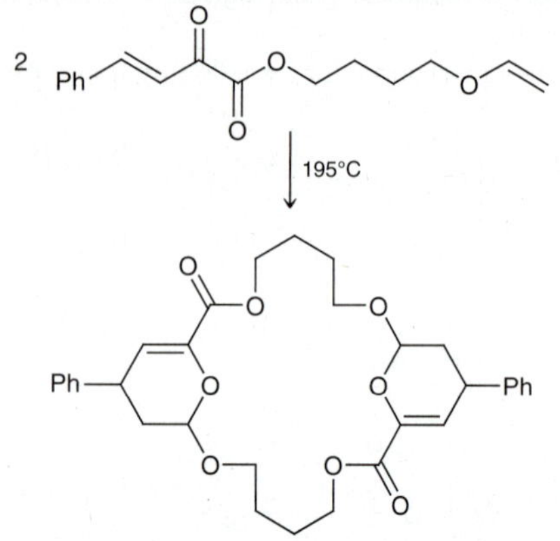

195°C

**17.73** During a recent investigation into the chemistry of oligofurans (polymers of the furan heterocycle, which show potential to be used in materials), the investigators observed an interesting reaction with dienophiles. When trifuran (below) is treated with maleimide, only the terminal furan participates in a [4+2] reaction (*Org. Lett.* **2012,** *14,* **502–505**). Provide a reasonable explanation for this observation.

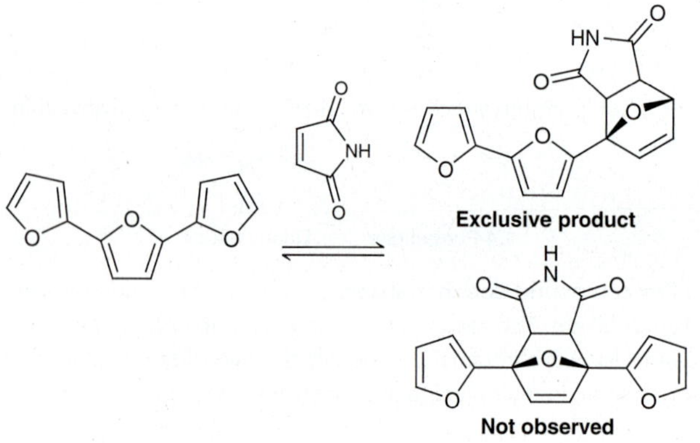

Exclusive product

Not observed

**17.74** During the first total synthesis of chelidonine, a cytotoxic natural product isolated from the root of *Chelidonium majus*, the investigators used only heat to achieve the following transformation (*J. Am. Chem. Soc.* **1971,** *93,* **3836–3837**). Propose a plausible mechanism for this thermal cyclization process.

120°C

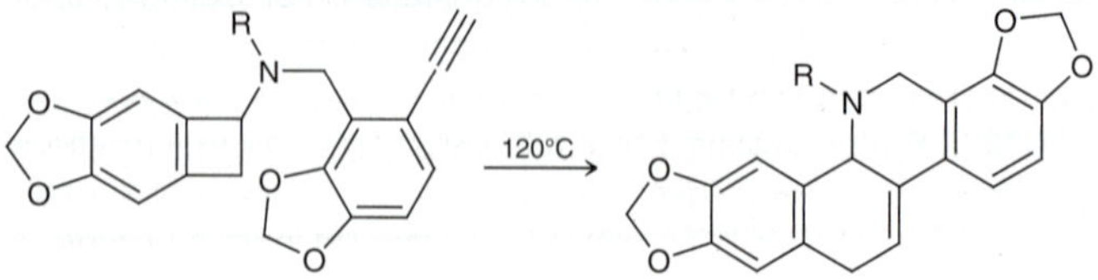

**17.75** During studies directed toward the synthesis of atropurpuran, a diterpene with interesting molecular architecture, the investigators used high temperature to convert an acyclic tetraene into a tricyclic compound, shown below (*Org. Lett.* **2010,** *12,* **1152–1155**). Propose a plausible mechanism for this thermal cyclization process.

135°C

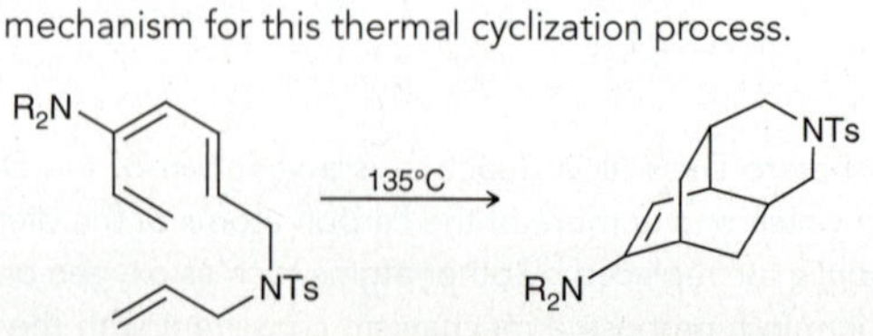

**17.76** When tricyclic cyclooctyne derivative **A** reacts with benzyl azide ($C_6H_5CH_2N_3$), a [3+2] cycloaddition occurs between the alkyne and the azide (called a *click reaction*) to install a new five-membered heterocycle (*J. Am. Chem. Soc.* **2012,** *134,* **9199–9208**):

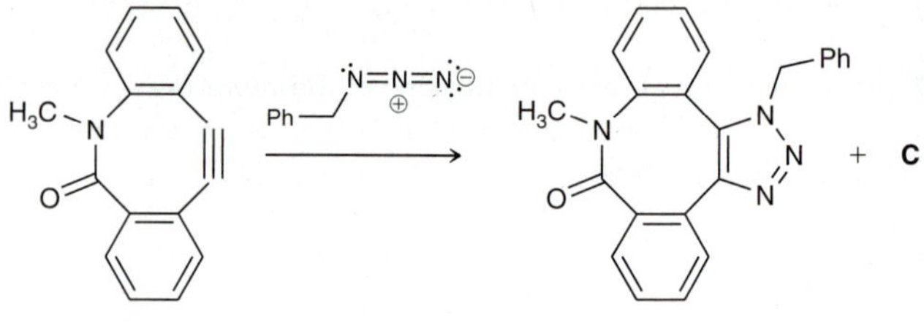

**A** → **B** + **C**

(a) Propose a plausible mechanism for the formation of **B**.

(b) Compounds **B** and **C** are constitutional isomers that are formed in roughly equal amounts. Draw the structure of compound **C**.

(c) Replacing alkyne **A** with each of the following alkynes results in a significantly slower reaction. Propose one or more reasons for this decrease in rate for each alkyne **D-F**.

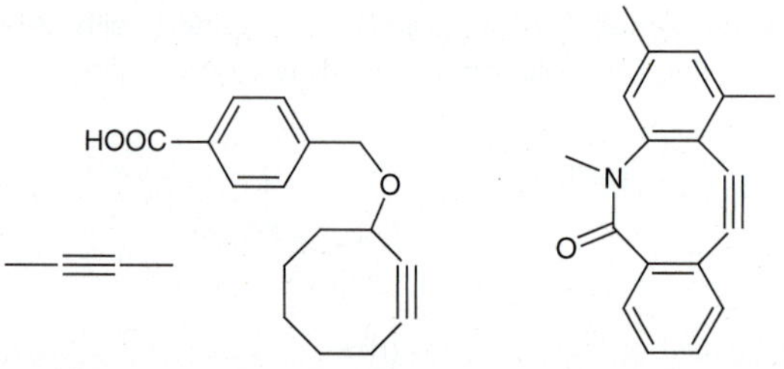

**D** Unreactive | **E** 375X Slower than A | **F** 1000X Slower than A

**17.77** Endiandric acids are a class of natural products isolated from the Australian plant *Endiandra introrsa*. Natural products containing chirality centers are generally found in nature as a single enantiomeric form (optically active), but the endiandric acids are an exception, as they are isolated as racemates. It has been suggested that they are formed via a series of pericyclic reactions from achiral starting materials, such as **1** (*Aust. J. Chem.* **1982,** *35,* **2247–2256**).

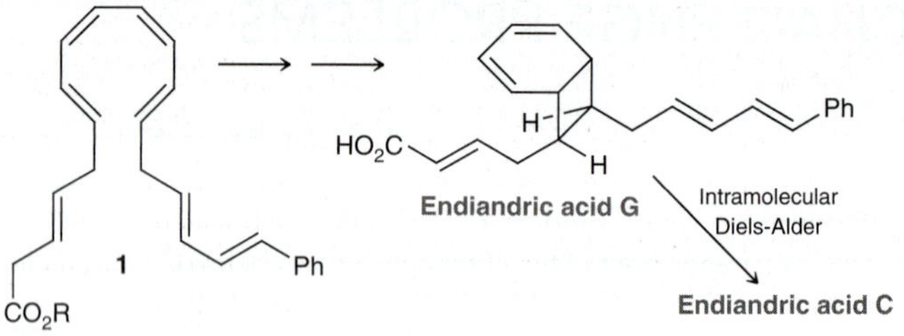

**1** → → **Endiandric acid G** → (Intramolecular Diels-Alder) **Endiandric acid C**

(a) The first step in the conversion of compound **1** to endiandric acids is believed to be a thermal electrocyclic reaction to form compound **2**, which possesses an eight-membered ring. Draw the structure of compound **2** and explain why it is formed as a racemic mixture.

(b) Compound **2** can undergo another electrocyclic reaction, thereby forming the skeletal structure of endiandric acid G. Draw a mechanism for this process.

(c) Endiandric acid C is formed from endiandric acid G via an intramolecular Diels-Alder reaction. Draw the structure of endiandric acid C and show a mechanism for its formation.

**17.78** Roquefortine C belongs to a family of natural products first isolated from cultures of the fungus *Penicillium roqueforti* in the mid-1970s. Since then, roquefortine C has also been found in other natural sources, including blue cheese. Curiously, isoroquefortine C (a diastereomer of roquefortine C) is not found in nature, despite studies revealing that it is more stable than roquefortine C (*J. Am. Chem. Soc.* **2008,** *130,* **6281–6287**). Provide a rational explanation for the difference in energy between these two compounds.

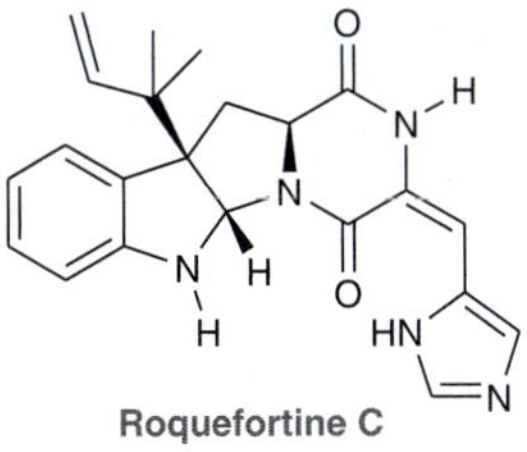

**Roquefortine C**

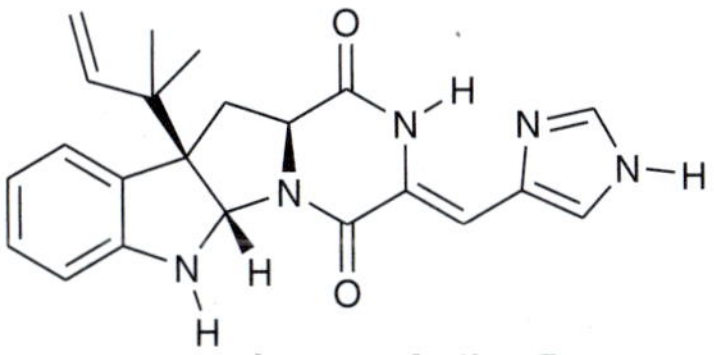

**Isoroquefortine C**

**17.79** As shown below, compound **1** can be transformed to compound **2**, which undergoes a thio-Claisen rearrangement (similar to a Claisen rearrangement, but with sulfur instead of oxygen) to give compound **3** (*J. Am. Chem. Soc.* **2000,** *122,* **190–191**).

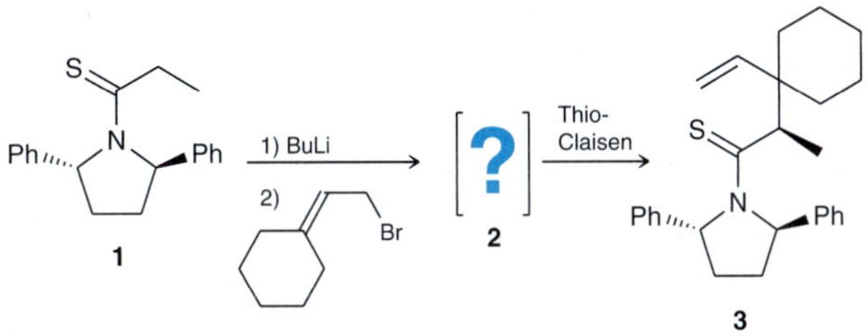

1) BuLi
2) Br
Thio-Claisen

(a) Identify the structure of compound **2**, which contains a carbon-carbon π bond in the *Z* configuration. (**Hint:** Work backward from compound **3**.)

(b) Propose a plausible mechanism for the transformation of **1** to **2**.

(c) Offer an explanation for the stereochemical outcome that is observed with respect to the newly formed chirality center. (**Hint:** You might find it helpful to use a molecular model set.)

# 18 Aromatic Compounds



## DID YOU EVER WONDER . . .
**what the difference is between antihistamines that cause drowsiness and those that don't?**

In this chapter, we expand our discussion of conjugated π systems and explore compounds that exhibit conjugated π bonds enclosed in a ring. For example, benzene and its derivatives belong to a class of compounds called *arenes* (aromatic hydrocarbons).

Benzene

Toluene

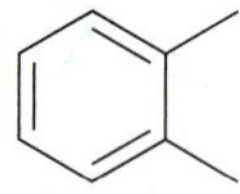
*ortho*-Xylene

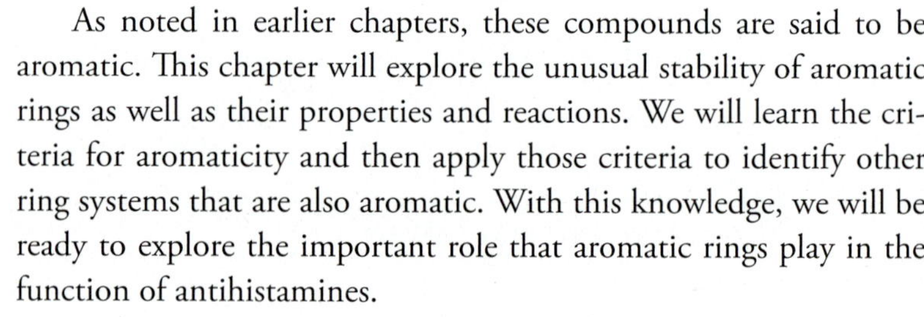

As noted in earlier chapters, these compounds are said to be aromatic. This chapter will explore the unusual stability of aromatic rings as well as their properties and reactions. We will learn the criteria for aromaticity and then apply those criteria to identify other ring systems that are also aromatic. With this knowledge, we will be ready to explore the important role that aromatic rings play in the function of antihistamines.

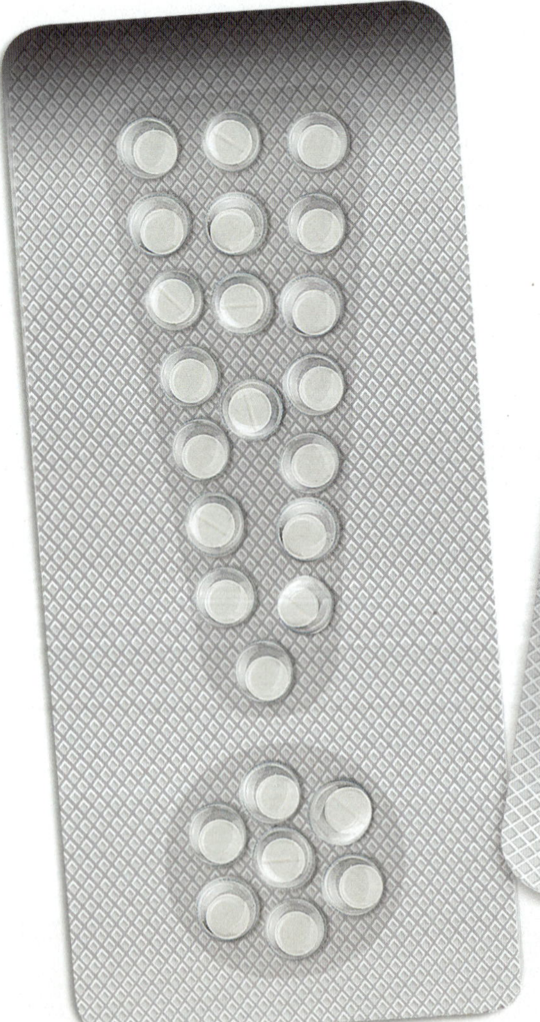

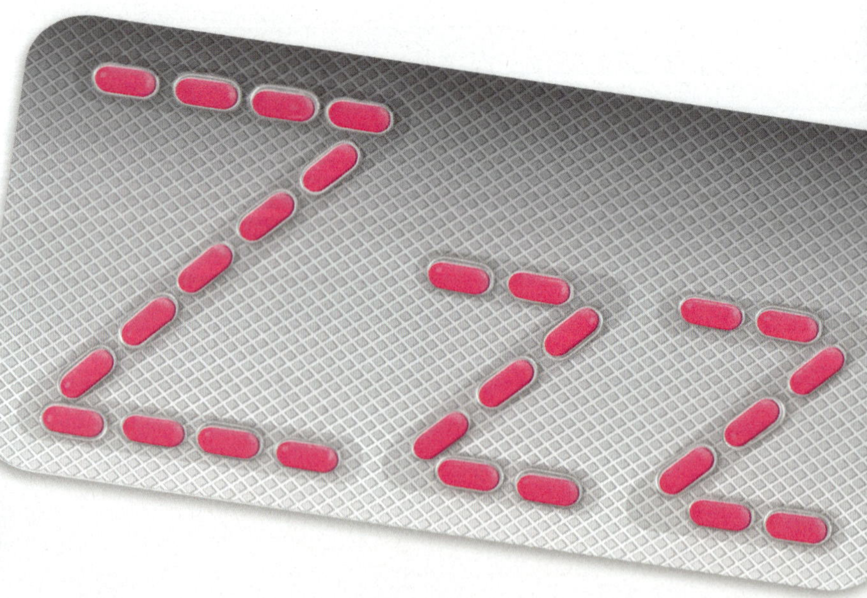

## DO YOU REMEMBER?

*Before you go on, be sure you understand the following topics.*
*If necessary, review the suggested sections to prepare for this chapter:*

- Molecular Orbital Theory (Section 1.8)
- Dissolving Metal Reduction (Section 10.5)
- Delocalized and Localized Lone Pairs (Section 2.12)
- MO Description of Conjugated Dienes (Section 17.3)

Take the DO YOU REMEMBER? QUIZ in **WileyPLUS** to check your understanding

## 18.1 Introduction to Aromatic Compounds

Many derivatives of benzene were originally isolated from the fragrant balsams obtained from trees and plants, and these compounds were therefore described as being **aromatic** in reference to their pleasant odors. Over time, chemists discovered that many derivatives of benzene are, in fact, odorless. Nevertheless, the term "aromatic" is still used to describe all derivatives of benzene, regardless of whether they are fragrant or odorless. The aromatic moiety—the benzene ring—is a particularly common structural feature in drugs. In fact, 8 of the 10 best-selling drugs in 2007 contained the benzene-like aromatic moiety (highlighted in red).

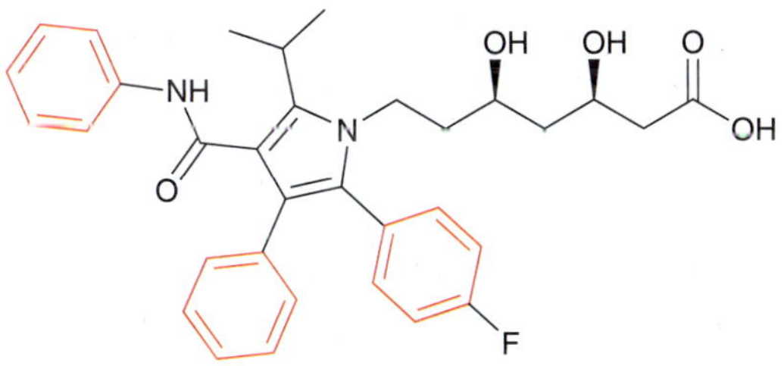

**Lipitor**
**(atorvastatin)**
Lowers cholesterol levels and reduces risk of heart attack and stroke

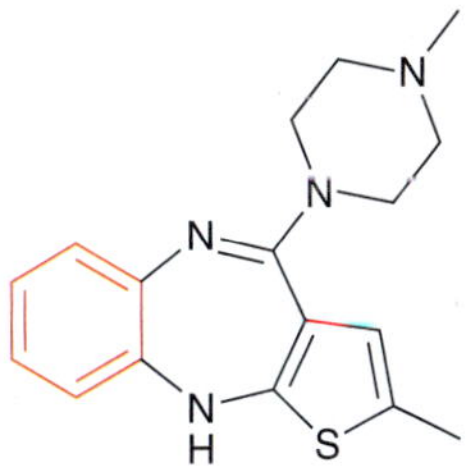

**Zyprexa**
**(olanzapine)**
An antipsychotic used in the treatment of schizophrenia and bipolar disorder

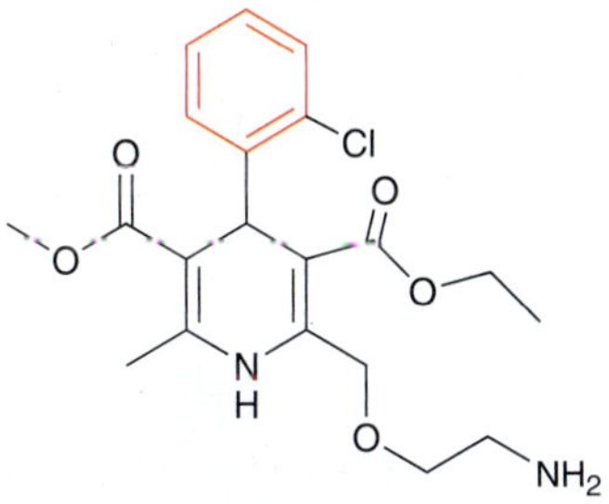

**Norvasc**
**(amlodipine)**
Used in the treatment of angina and high blood pressure

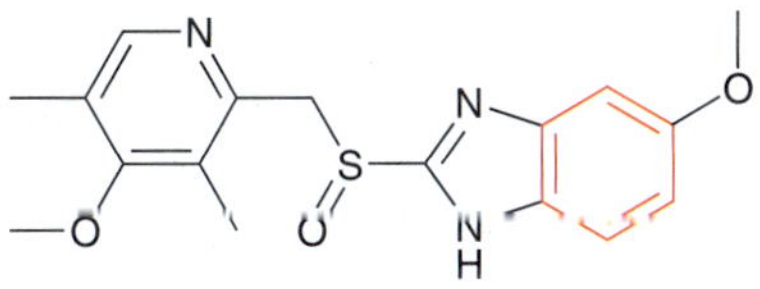

**Nexium**
**(omeprazole)**
A proton-pump inhibitor used in the treatment of ulcers and acid reflux

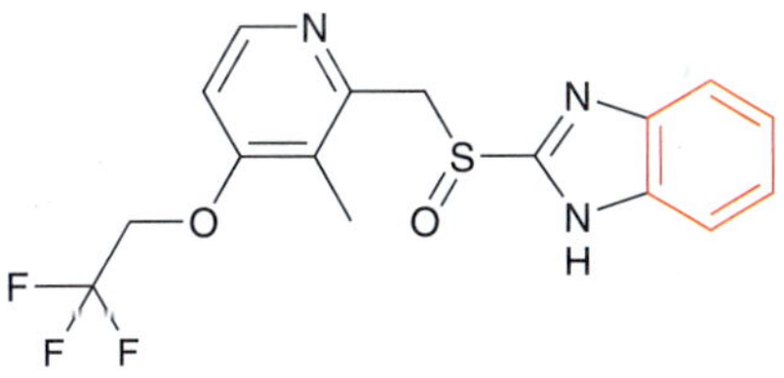

**Prevacid**
**(lansoprazole)**
A proton-pump inhibitor used in the treatment of ulcers and acid reflux

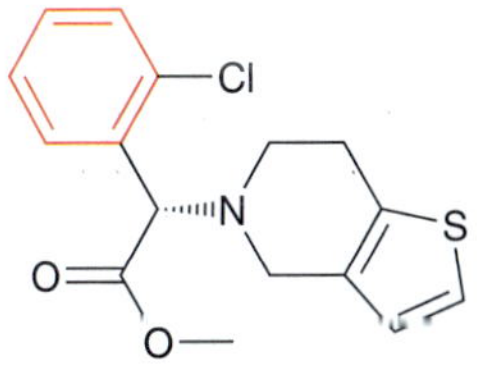

**Plavix**
**(clopidogrel)**
An antiplatelet agent (prevents formation of blood clots) used in the treatment of coronary artery disease

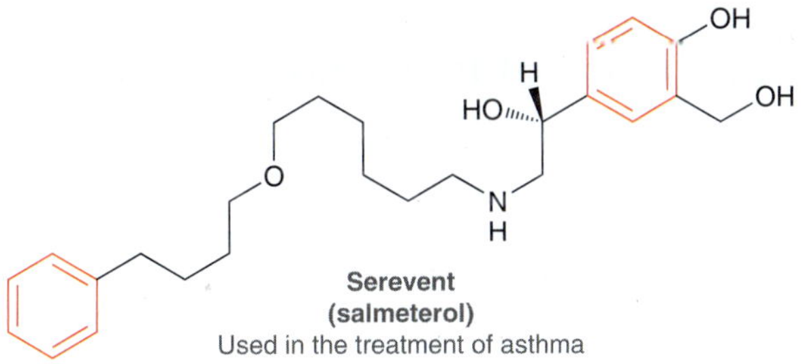

**Serevent**
**(salmeterol)**
Used in the treatment of asthma

**Zoloft**
**(sertraline)**
Used in the treatment of depression

In 2007, these eight drugs collectively generated over $38 billion in global sales. In this chapter, we will see that aromatic rings are particularly stable and significantly less reactive than originally expected.

## practically speaking

### What Is Coal?

It is estimated that 25% of the world's recoverable coal (coal that can be readily mined) is located in the United States. As such, it is believed that the United States has enough natural resources to meet all of its energy needs for at least 200 years. To provide some additional perspective, consider the fact that these natural resources contain more than double the energy contained in all of the combined oil reserves in the Middle East.

It is believed that the earth's coal deposits were formed as organic matter from prehistoric plant life was subjected to high pressures and temperatures over long periods of time. The molecular structure of coal involves haphazard arrangements of aromatic moieties (highlighted in orange boxes), connected to each other by nonaromatic links (shown in green boxes):

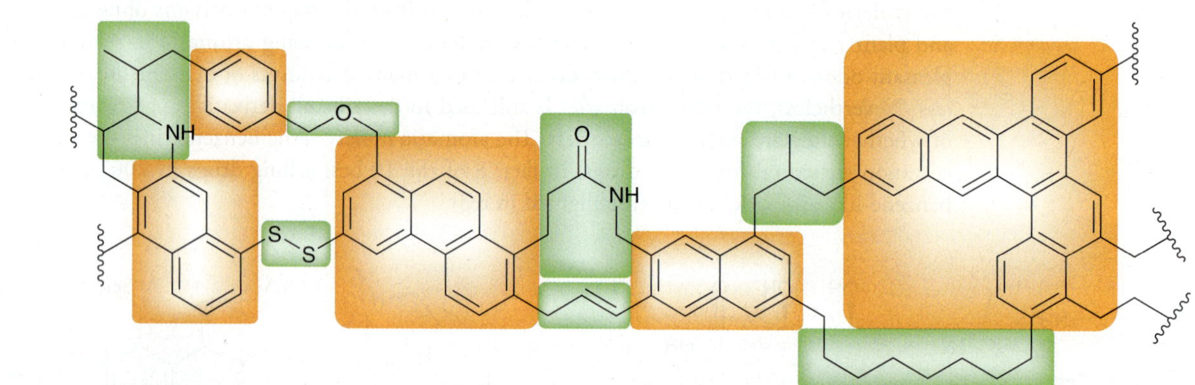


When heated in the presence of oxygen, a combustion process takes place, producing carbon dioxide, water, and many other products. During the combustion process, the C—C and C—H bonds are broken and replaced by lower energy C═O bonds, thereby releasing energy that can be used to generate electricity. In fact, more than half of the electricity consumed in the United States is produced from the combustion of coal.

When coal is heated to about 1000°C in the absence of oxygen, a mixture of compounds, called *coal tar*, is released. Distillation of coal tar yields many aromatic compounds. Although aromatic compounds can be obtained from coal, the primary source of aromatic compounds is from petroleum, using a process called *reforming* (Section 4.5).

**Coal** $\xrightarrow{\text{Heat}}$ Benzene (b.p. = 80°C) + Toluene (b.p. = 110°C) + Naphthalene (b.p. = 218°C) + **Many other aromatic compounds**

## 18.2 Nomenclature of Benzene Derivatives

### Monosubstituted Derivatives of Benzene

Monosubstituted derivatives of benzene are named systematically using benzene as the parent and listing the substituent as a prefix. Below are several examples.

The following are some monosubstituted aromatic compounds that have common names accepted by IUPAC. You must commit these names to memory, as they will be used extensively throughout the remaining chapters.

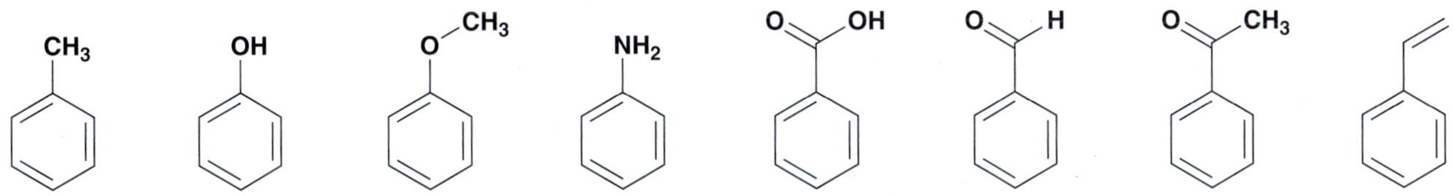

Toluene — Phenol — Anisole — Aniline — Benzoic acid — Benzaldehyde — Acetophenone — Styrene

If the substituent is larger than the benzene ring (i.e., if the substituent has more than six carbon atoms), then the benzene ring is treated as a substituent and is called a **phenyl group**.

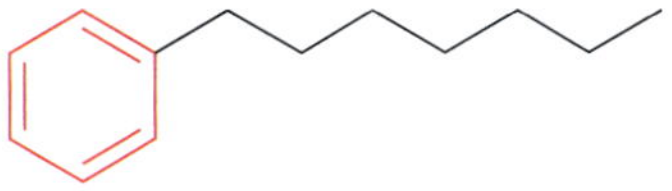

1-Phenylheptane

The presence of phenyl groups is often indicated with the letters Ph or with the Greek letter phi ($\phi$).

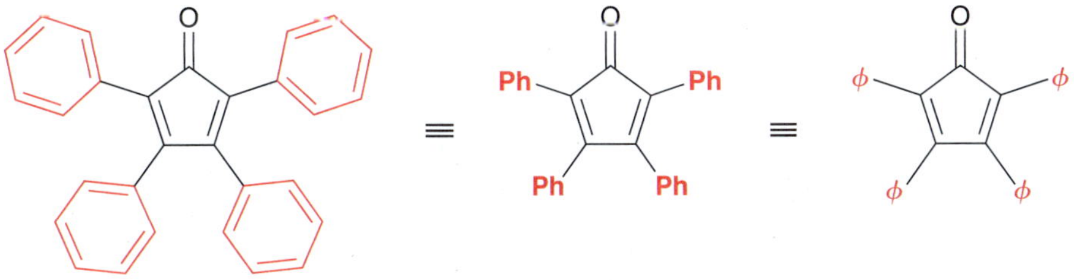

Tetraphenylcyclopentadienone

Phenyl groups bearing substituents are sometimes indicated with the letters Ar, indicating the presence of an aromatic ring.

## Disubstituted Derivatives of Benzene

Dimethyl derivatives of benzene are called xylene, and there are three constitutionally isomeric xylenes.

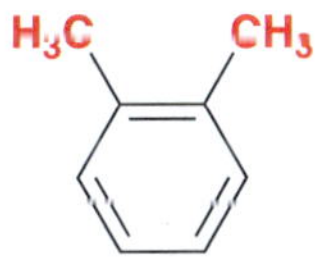

*ortho*-Xylene (1,2-dimethylbenzene) — *meta*-Xylene (1,3-dimethylbenzene) — *para*-Xylene (1,4-dimethylbenzene)

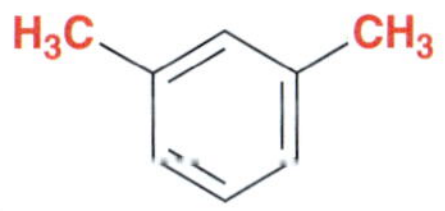

These isomers differ from each other in the relative positions of the methyl groups and can be named in two ways: (1) using the descriptors ***ortho***, ***meta***, and ***para*** or (2) using locants (i.e., 1,3 is the same as *meta*). Both methods can be used when the parent is a common name:

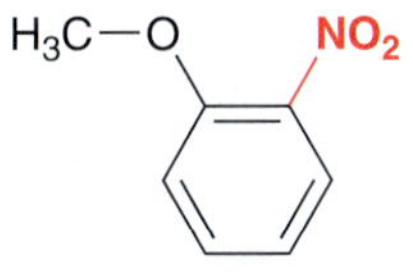

*ortho*-Nitroanisole (2-Nitroanisole) — *meta*-Bromotoluene (3-Bromotoluene) — *para*-Chlorobenzaldehyde (4-Chlorobenzaldehyde)

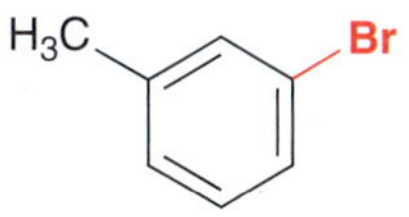

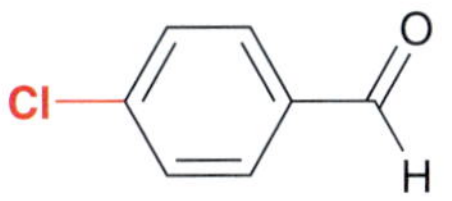

## Polysubstituted Derivatives of Benzene

The descriptors *ortho*, *meta*, and *para* cannot be used when naming an aromatic ring bearing three or more substituents. In such a case, locants are required. That is, each substituent is designated with a number to indicate its location on the ring.

When naming a polysubstituted benzene ring, we will follow the same four-step process used for naming alkanes, alkenes, alkynes, and alcohols.

1. Identify and name the parent.
2. Identify and name the substituents.
3. Assign a locant to each substituent.
4. Arrange the substituents alphabetically.

When identifying the parent, it is acceptable (and common practice) to choose a common name. Consider the following example:

OH, 1, 3, 5, Br, Br

**3,5-Dibromophenol**

This compound could certainly be named as a trisubstituted benzene ring. However, it is much more efficient to name the parent as phenol rather than benzene and to list the two bromine atoms as substituents. The choice to name this compound as a phenol dictates that the carbon atom connected to the OH group must receive the lowest locant (number 1). When a choice exists, place numbers so that the second substituent receives the lower possible number.

**Correct** — OH (1), Cl (2), 3, 4, Br (5)

**Incorrect** — OH (1), 2, Br (3), 4, 5, Cl (6)

When arranging the name, make sure to alphabetize the substituents. In the example above, *b*romo is listed before *c*hloro.

### 18.1 NAMING A POLYSUBSTITUTED BENZENE

**LEARN the skill** Provide a systematic name for TNT, a well-known explosive with the following molecular structure:

$CH_3$, $O_2N$, $NO_2$, $NO_2$

### SOLUTION

**STEP 1**
Identify and name the parent.

Apply the four-step method. First, identify and name the parent. In this case, we choose toluene as the parent (shown in red).

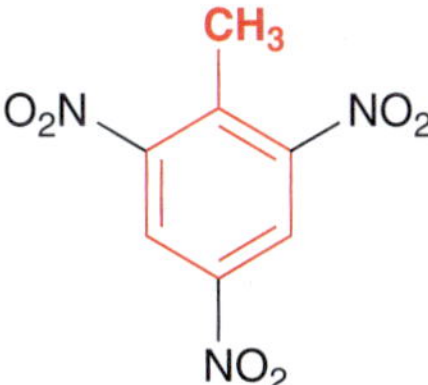


**STEPS 2 AND 3**
Identify the substituents and assign locants.

Next, identify the substituents and assign locants. The decision to name this compound as a trisubstituted toluene (rather than a tetrasubstituted benzene) dictates that the carbon atom bearing the methyl group be assigned the lowest locant (number 1). In this case, the remaining locants can be assigned either clockwise or counterclockwise because the result is the same.

**STEP 4**
Arrange the substituents alphabetically.

Finally, arrange the substituents alphabetically. In this case, there are three nitro groups. Make sure to indicate all three numbers in the name.

2,4,6-Trinitrotoluene

**PRACTICE the skill** **18.1** Provide a systematic name for each of the following compounds:

**(a)**
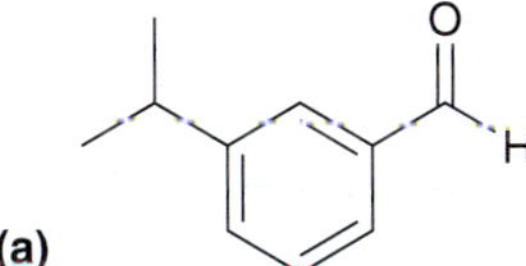


**(b)**
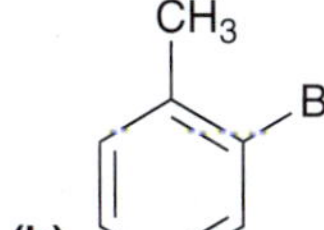


**(c)**
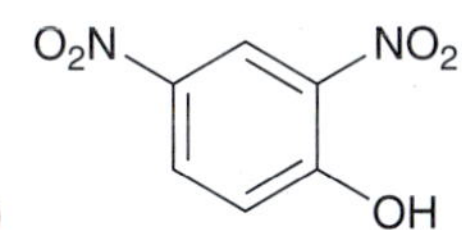


**(d)**

**(e)** OH, Br, Br, Cl

**APPLY the skill**

**18.2** Aromatic compounds often have multiple names that are all accepted by IUPAC. Provide three different systematic (IUPAC) names for the following compound:

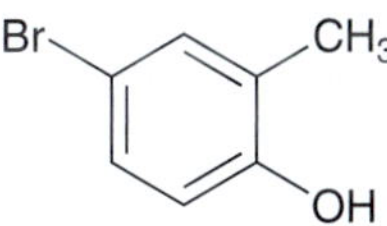


**18.3** For each of the following compounds, draw its structure:

**(a)** 2,6-Dibromo-4-chloroanisole

**(b)** *meta*-Nitrophenol

**18.4** Provide at least five different acceptable IUPAC names for the following compound:

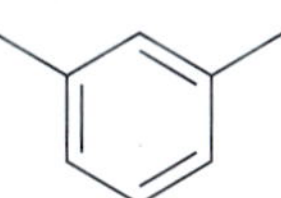

**18.5** In Chapter 9, we saw that *meta*-chloroperoxybenzoic acid (MCPBA) is a peroxy acid commonly used to convert alkenes into epoxides. Recall that peroxy acids have the following structure:

R–C(=O)–O–O–H

**(a)** Draw the structure of MCPBA.

**(b)** Provide a systematic name for the compound formed by replacing the chlorine atom in MCPBA with a methyl group.

need more **PRACTICE?** **Try Problems 18.28, 18.29, 18.33**

## 18.3 Structure of Benzene

In 1825, Michael Faraday isolated benzene from the oily residue left by illuminating gas in London street lamps. Further investigation showed that the molecular formula of this compound was $C_6H_6$: a hydrocarbon comprised of six carbon atoms and six hydrogen atoms.

In 1866, August Kekulé used his recently published structural theory of matter to propose a structure for benzene. Specifically, he proposed a ring comprised of alternating double and single bonds.

Kekulé described the exchange of double and single bonds to be an equilibrium process. Over time, this view was refined by the advent of resonance theory and molecular orbital concepts of delocalization. The two drawings above are now viewed as resonance structures, not as an equilibrium process.

Recall that resonance does not describe the motion of electrons, but rather, resonance is the way that chemists deal with the inadequacy of bond-line drawings. Specifically, each drawing alone is inadequate to describe the structure of benzene. The problem is that each C—C bond is neither a single bond nor a double bond, nor is it vibrating back and forth between these two states. Instead, each C—C bond has a bond order of 1.5, exactly midway between a single bond and a double bond. To avoid drawing resonance structures, benzene is often drawn like this:

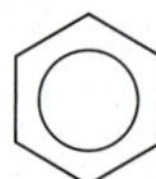

This type of drawing appears often in the literature and will likely appear on the chalkboard of your organic chemistry lecture hall. Nevertheless, these drawings should be avoided when proposing reaction mechanisms, which require scrupulous bookkeeping of electrons. Throughout the remainder of this textbook, we will draw benzene rings as alternating single and double bonds (Kekulé structures).

## 18.4 Stability of Benzene

### Evidence for Unusual Stability

There is much evidence that the aromatic moiety is particularly stable, much more so than expected. Below, we will explore two pieces of evidence for this stability.

Recall from Chapter 9 that alkenes readily undergo addition reactions, such as the addition of bromine to form a dihalide.

$Br_2$ + Enantiomer

We might therefore expect benzene to undergo a similar reaction, perhaps even three times. But in fact, no reaction is observed.

$Br_2$ **No reaction**

The aromatic moiety apparently exhibits a special stability that would be lost if an addition reaction were to occur. The stability of the aromatic moiety is also observed when comparing heats of hydrogenation for several similar compounds. Recall that hydrogenation of a π bond proceeds in the presence of a metal catalyst.

cyclohexene + $H_2$ $\xrightarrow{Pt}$ cyclohexane **(100%)** $\Delta H° = -120$ kJ/mol

The heat of hydrogenation ($\Delta H$) for this reaction is −120 kJ/mol. Benzene is generally stable to hydrogenation under standard conditions, but under forcing conditions (high pressure and high temperature), benzene also undergoes hydrogenation and reacts with three equivalents of molecular hydrogen to form cyclohexane. Therefore, we might expect $\Delta H = -360$ kJ/mol. But, in fact, the heat of hydrogenation for hydrogenation of benzene is only −208 kJ/mol.

benzene + 3 $H_2$ $\xrightarrow[\text{100 atm, 150°C}]{Ni}$ cyclohexane **(100%)**

**Expected:** $\Delta H° = -360$ kJ/mol

**Actual:** $\Delta H° = -208$ kJ/mol

In other words, benzene is much more stable than expected. This information is graphically represented on the energy diagram in Figure 18.1. This energy diagram shows the relative heats of hydrogenation for cyclohexene, 1,3-cyclohexadiene, and benzene. Notice that the heat of hydrogenation of 1,3-cyclohexadiene is not quite twice the value of the heat of hydrogenation of cyclohexene. This is attributed to conjugation. Extending this logic to an imaginary cyclohexatriene molecule, we would expect that the heat of hydrogenation would be just below −360 kJ/mol. In fact, the heat of hydrogenation for benzene is only −208 kJ/mol. The difference between the expected value (−360) and the observed value (−208) is 152 kJ/mol, which is called the *stabilization energy* of benzene. This value represents the amount of stabilization associated with aromaticity.

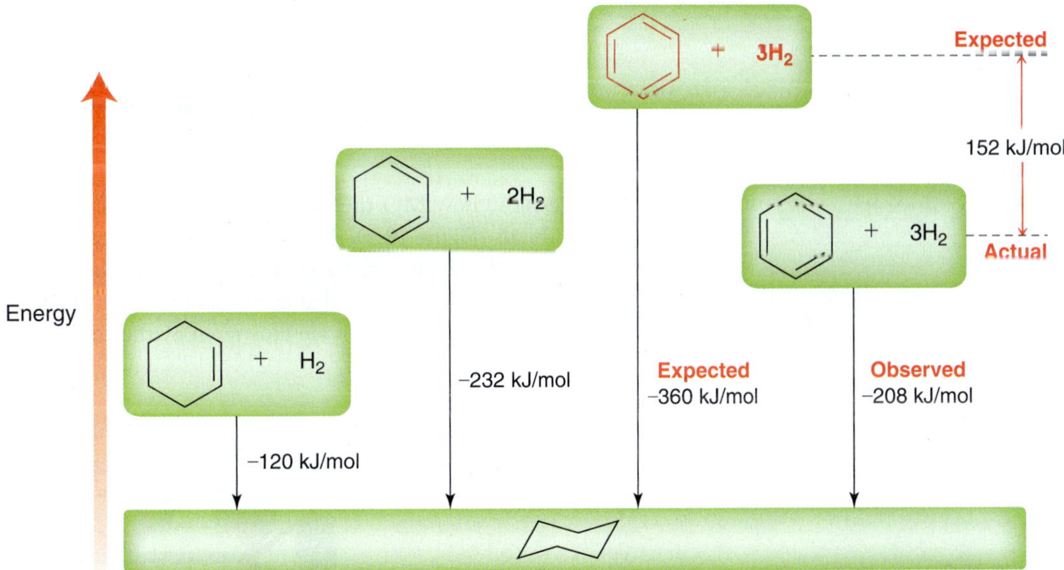


**FIGURE 18.1**
An energy diagram comparing the heats of hydrogenation for cyclohexene, cyclohexadiene, and benzene.

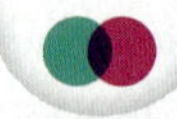

## CONCEPTUAL CHECKPOINT

**18.6** Compound **A** is an aromatic compound with molecular formula $C_8H_8$. When treated with excess $Br_2$, compound **A** is converted into compound **B**, with molecular formula $C_8H_8Br_2$. Identify the structures of compounds **A** and **B**.

**Compound A** ⟶ **Compound B**
($C_8H_8$) ($C_8H_8Br_2$)

**18.7** In some circumstances, dehydrogenation is observed. Dehydrogenation involves the loss of two hydrogen atoms (the reverse of hydrogenation). Analyze each of the following dehydrogenation reactions and then use the information in Figure 18.1 to predict whether each transformation will be downhill in energy (negative value for $\Delta H$) or uphill in energy (positive value for $\Delta H$).

(a) ⟶ + $H_2$

(b) ⟶ + $2\ H_2$

(c) ⟶ + $H_2$

## Source of Stability

In order to explain the stability of benzene, we will need to invoke MO theory. Recall that, according to MO theory, a π bond results when two atomic *p* orbitals overlap to form two new orbitals, called molecular orbitals (Figure 18.2). The lower energy MO is the bonding MO, while the higher energy MO is the antibonding MO. Both π electrons occupy the bonding MO and thereby achieve a lower energy state.

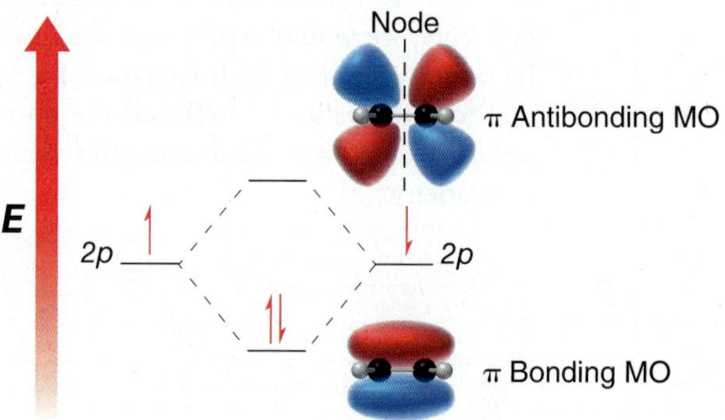


**FIGURE 18.2** The molecular orbitals of a π bond.

Now let's explore how MO theory describes the nature of benzene, which is comprised of six overlapping *p* orbitals (Figure 18.3). According to MO theory, these six atomic *p* orbitals are replaced with six molecular orbitals (Figure 18.4). The precise shape and energy level of each MO are determined by sophisticated mathematics that is beyond the scope of our current discussion. For now, let's focus on the important concepts that are illustrated by the energy diagram in Figure 18.4.

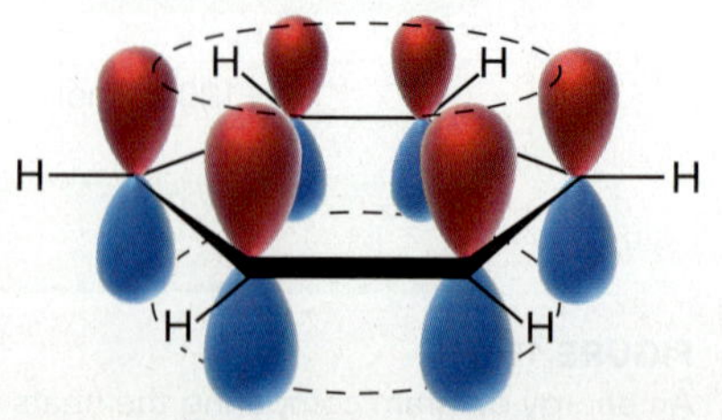


**FIGURE 18.3** The *p* orbitals of benzene overlap continuously around the ring.

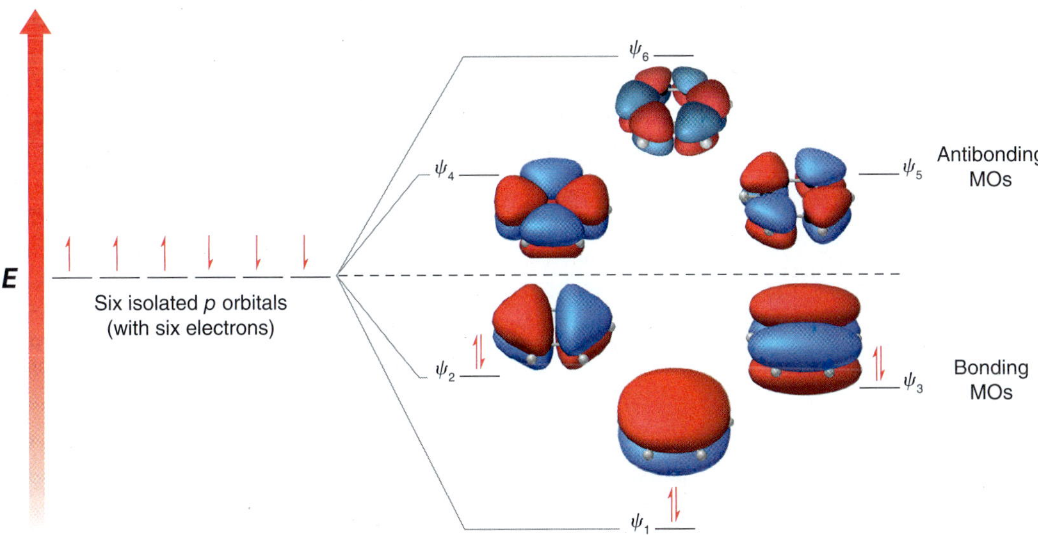


**FIGURE 18.4** The molecular orbitals of benzene.

- There are six molecular orbitals, each of which is associated with the entire molecule (rather than being associated with any specific bond).
- Three of the six MOs (those below the dashed line) are bonding MOs, while the other three MOs (those above the dashed line) are antibonding MOs.
- Since each MO can contain two electrons, the three bonding MOs can collectively accommodate up to six π electrons.
- By occupying the bonding MOs, all six electrons achieve a lower energy state and are said to be delocalized. This is the source of the stabilization energy associated with benzene.

## Hückel's Rule

We might expect the following two compounds to exhibit aromatic stabilization like benzene:

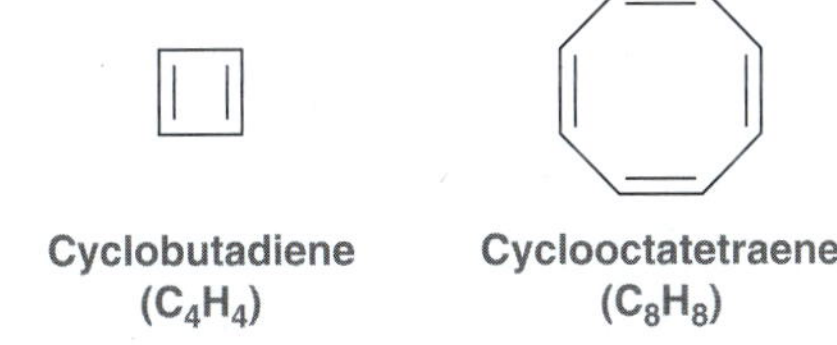


**BY THE WAY**
Willstätter was awarded the 1915 Nobel Prize in Chemistry for his research on elucidating the structure of chlorophyll.

After all, they are similar to benzene in that each compound is comprised of a ring of alternating single and double bonds. Cyclooctatetraene ($C_8H_8$) was first isolated by Richard Willstätter in 1911. The reactivity of this compound suggests that cyclooctatetraene does not exhibit the same stability exhibited by benzene. For example, the compound readily undergoes addition reactions, such as bromination.

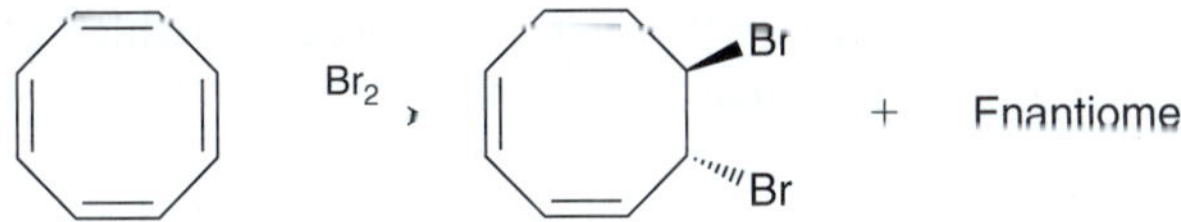


Cyclobutadiene ($C_4H_4$) also does not exhibit aromatic stability. In fact, quite the opposite. It is extremely unstable and resisted all attempts to prepare it until the second half of the twentieth century (see the Practically Speaking box that follows). Cyclobutadiene is so unstable that it reacts with itself at −78°C in a Diels-Alder reaction.

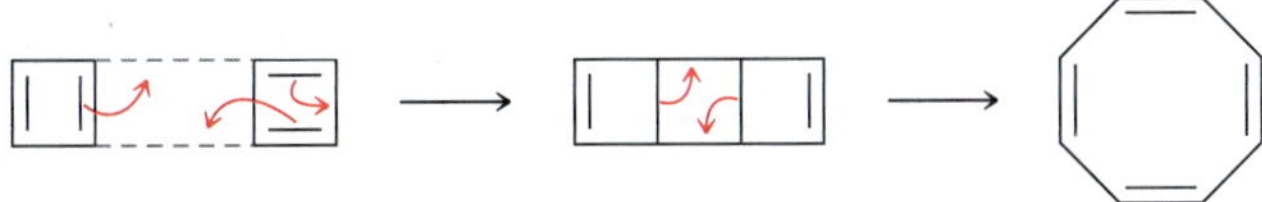

The Diels-Alder reaction generates an initial tricyclic product that is also unstable and rapidly rearranges to form cyclooctatetraene.

These observations indicate that the presence of a fully conjugated ring of π electrons is not the sole requirement for aromaticity; the number of π electrons in the ring is also important. Specifically, we have seen that an odd number of electron pairs is required for aromaticity.

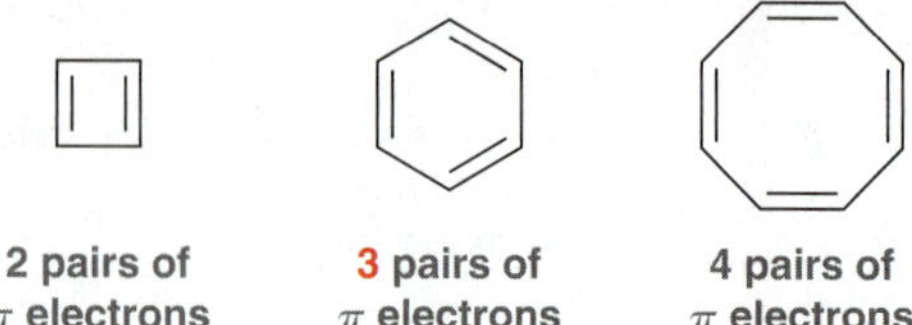


Benzene has a total of six π electrons, or *three* electron pairs. With an odd number of electron pairs, benzene is aromatic. In contrast, cyclobutadiene and cyclooctatetraene each have an even number of electron pairs (two pairs and four pairs, respectively) and do not exhibit aromatic stabilization.

The requirement for an odd number of electron pairs is called **Hückel's rule**. Specifically, a compound can only be aromatic if the number of π electrons in the ring is 2, 6, 10, 14, 18, and so on. This series of numbers can be expressed mathematically as $4n+2$, where $n$ is a whole number. In the upcoming section, we will use MO theory to explain why aromatic stabilization requires an odd number of electron pairs ($4n+2$ electrons).

## CONCEPTUAL CHECKPOINT

**18.8** Predict whether each of the following compounds should be aromatic:

(a) 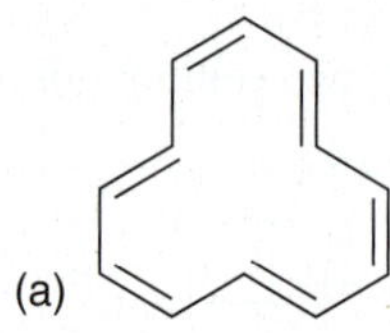

(b) 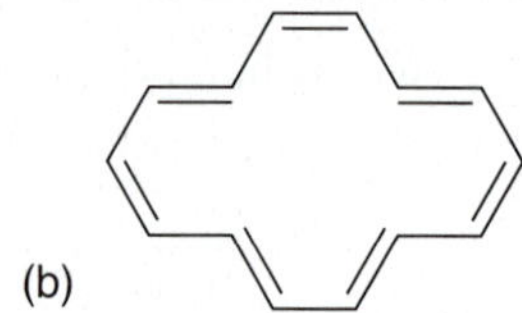

(c) 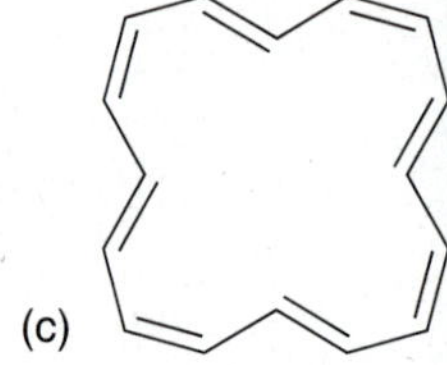

## MO Theory and Frost Circles

In this section, we will use MO theory to explain the source of Hückel's rule for planar, conjugated ring systems. We already analyzed the MOs of benzene (Figure 18.4) to explain its stability. Let's now turn our attention to the MOs of cyclobutadiene ($C_4H_4$) and cyclooctatetraene ($C_8H_8$) to see how MO theory can explain the observed instability of these compounds. We will start with square cyclobutadiene, which is a system constructed from the overlap of four atomic *p* orbitals (Figure 18.5).

**FIGURE 18.5** The *p* orbitals that are used in cyclobutadiene.

According to MO theory, these four atomic orbitals are replaced with four molecular orbitals, with the relative energy levels shown in Figure 18.6.

## Molecular Cages

Early attempts at preparing cyclobutadiene were met with failure again and again. In the second half of the twentieth century, several methods were developed. One such method is especially fascinating and will now be described.

The research of Donald Cram (UCLA) involved the investigation of cagelike molecules. Specifically, Cram prepared molecules shaped like hemispheres and then linked them together to form a cage:

Cram then used these cages to trap α-pyrone inside:

α-Pyrone

When irradiated with UV light, α-pyrone was known to liberate $CO_2$ to form cyclobutadiene:

The $CO_2$ molecules generated by this process were then able to escape through the spaces between the linkers holding the hemispheres together. However, cyclobutadiene is too large to escape through the holes in the cage. As a result, a single cyclobutadiene molecule would be trapped and unable to react with other molecules of cyclobutadiene. This extremely clever method for imprisoning cyclobutadiene worked beautifully and enabled its spectroscopic analysis. The many studies that followed provided strong evidence that cyclobutadiene rapidly equilibrates between two rectangular forms:

Cram's work with cagelike molecules, called *carcerands*, opened the doorway to a new field of chemistry. For his contributions, he shared a portion of the 1987 Nobel Prize in Chemistry.

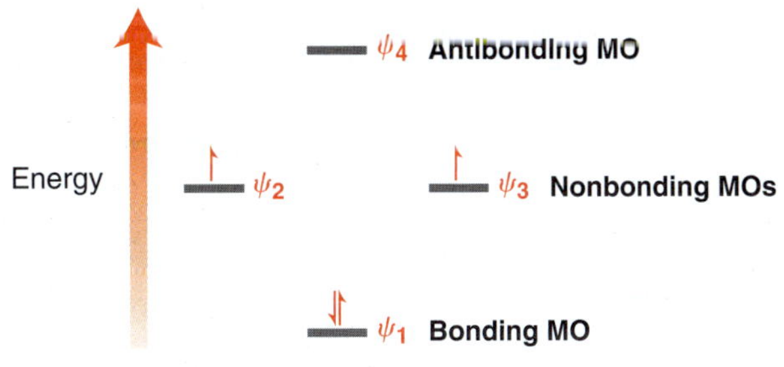


**FIGURE 18.6** The relative energy levels of the molecular orbitals of cyclobutadiene.

Of the four MOs, only one MO is a bonding MO, and it can accommodate no more than two π electrons. The remaining two π electrons must go into the degenerate nonbonding MOs. Each of these electrons occupies a different MO (according to Hund's rule), so each electron is unpaired. Unpaired electrons are very reactive, which explains why square cyclobutadiene is so unstable. Due to its unusual instability, this compound is said to be **antiaromatic**. Cyclobutadiene can alleviate some if its instability by adopting a rectangular shape (as described in the previous Practically Speaking box), in which the compound functions as two isolated π bonds, rather than a diradical. Nevertheless, cyclobutadiene is still extremely unstable and highly reactive.

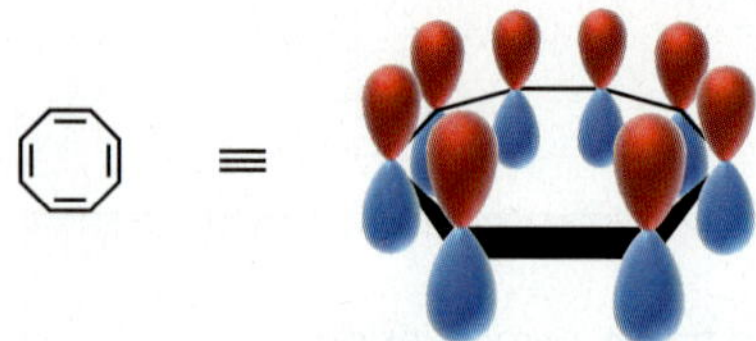

FIGURE 18.7
In a planar conformation, the $p$ orbitals of cyclooctatetraene overlap continuously around the ring.

Now consider how MO theory treats planar cyclooctatetraene, which is a system constructed from the overlap of eight atomic $p$ orbitals (Figure 18.7). According to MO theory, these eight atomic orbitals are replaced with eight molecular orbitals, with the relative energy levels shown in Figure 18.8. When we fill these MOs with eight π electrons, we encounter a situation similar to that which we encountered with square cyclobutadiene—specifically, two of the electrons are unpaired and occupy nonbonding orbitals. Once again, this electronic configuration leads to remarkable instability, and the compound should be antiaromatic. But the compound can avoid

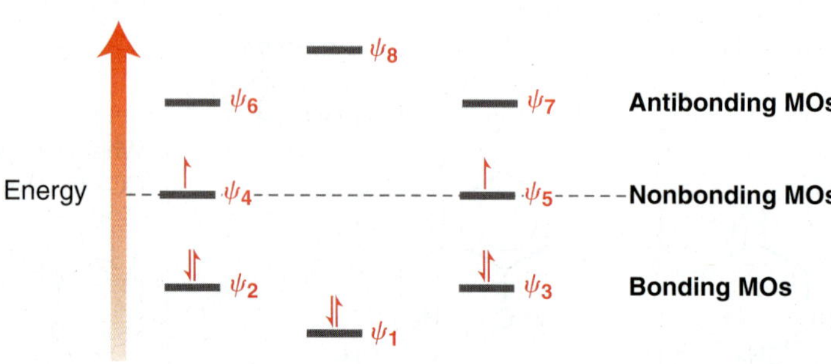


FIGURE 18.8
The relative energy levels of the molecular orbitals of cyclooctatetraene.

this unnecessary instability by assuming a tub shape (Figure 18.9). In this conformation, the eight $p$ orbitals no longer overlap as one continuous system, and the energy diagram in Figure 18.8 (of the eight MOs) simply does not apply. Rather, there are four isolated π bonds, each of which contains two electrons in a bonding MO (as shown in Figure 18.2).

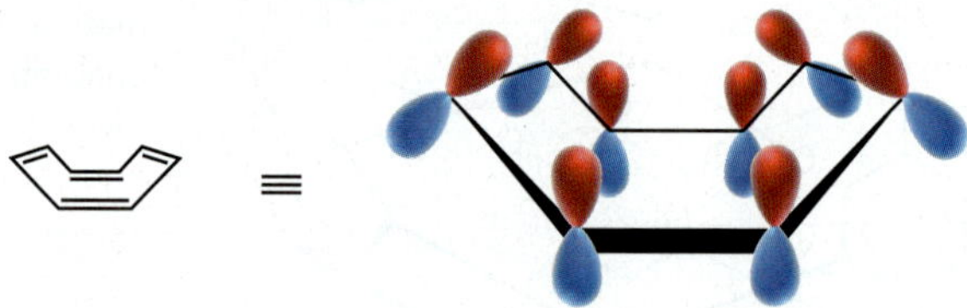

FIGURE 18.9
The tub-shaped structure of cyclooctatetraene.

The explanation above is predicated on our ability to draw the relative energy levels of the MOs, and as we already mentioned, that ability requires sophisticated mathematics that is beyond the scope of our course. Without getting heavily into the mathematics of MO theory, there is a simple method that will enable us to predict and draw the relative energy levels for any ring of overlapping p orbitals. This method is illustrated with Figure 18.10, in which we analyze a seven-membered ring comprised of continuous, overlapping $p$ orbitals.

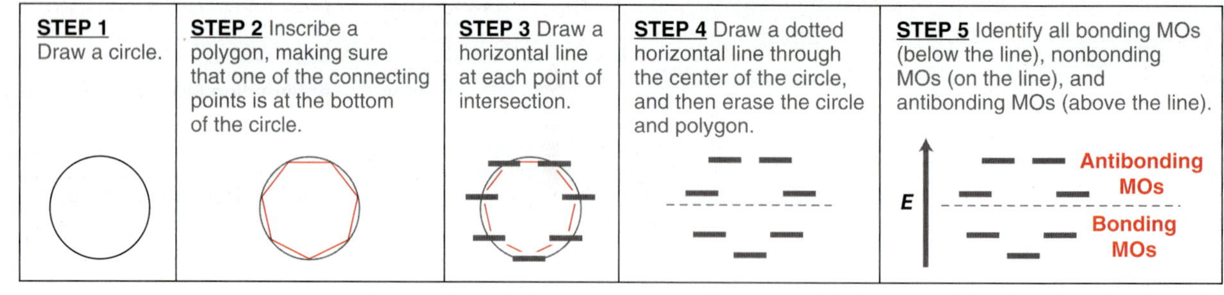


FIGURE 18.10
A method for determining the relative energy levels of the molecular orbitals for a ring system comprised of continuous, overlapping $p$ orbitals.

Begin by drawing a circle and inscribing a seven-membered polygon inside the circle, with one of the connecting points of the polygon at the bottom of the circle. Each location where the polygon touches the circle represents an energy level. All energy levels in the bottom half of the circle are the bonding MOs, and all energy levels in the top half of the circle are antibonding MOs. This method is extremely helpful, because the relative distances between the energy levels do in fact represent the relative difference in energy between the MOs. This method was first developed by Arthur Frost (Northwestern University), and it is referred to as a **Frost circle**.

If we accept that Frost circles accurately predict the relative energy levels of the MOs in a conjugated ring system, then we can use Frost circles to gain a better understanding of the $4n+2$

rule. Consider the Frost circles for ring systems containing anywhere from 4 to 10 carbon atoms (Figure 18.11). Look at the bonding MOs (highlighted in green boxes). Notice that the number of bonding MOs is always odd (1, 3, or 5). Recall that aromaticity is achieved when all of the π electrons are paired in bonding MOs. Depending on the size of the ring, the bonding MOs will accommodate 2, 6, or 10 π electrons. This explains the source of the $4n+2$ rule.

**FIGURE 18.11** Frost circles for different-size ring systems.

| Four-membered ring | Five-membered ring | Six-membered ring | Seven-membered ring | Eight-membered ring | Nine-membered ring | Ten-membered ring |
|---|---|---|---|---|---|---|
| | | | | | | |
| 1 Bonding MO | 3 Bonding MOs | 3 Bonding MOs | 3 Bonding MOs | 3 Bonding MOs | 5 Bonding MOs | 5 Bonding MOs |

## CONCEPTUAL CHECKPOINT

**18.9** The cyclopropenyl cation has a three-membered ring that contains a continuous system of overlapping *p* orbitals. This system contains a total of two π electrons. Using a Frost circle, draw an energy diagram showing the relative energy levels of all three MOs and then predict whether this cation is expected to exhibit aromatic stabilization.

# 18.5 Aromatic Compounds Other Than Benzene

## The Criteria for Aromaticity

Benzene is not the only compound that exhibits aromatic stabilization. A compound will be aromatic if it satisfies the following two criteria:

1. The compound must contain a ring comprised of continuously overlapping *p* orbitals.
2. The number of π electrons in the ring must be a Hückel number.

Compounds that fail the first criterion are called **nonaromatic**. Below are three examples, each of which fails the first criterion for a different reason.

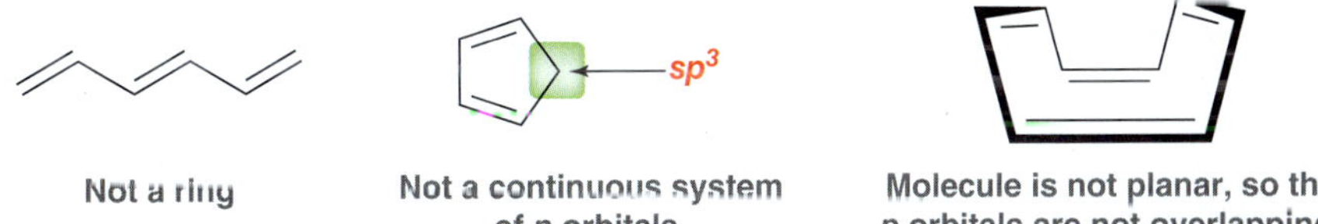


The first compound (1,3,5-hexatriene) is not aromatic, because it does not possess a ring. The second compound is not aromatic because the ring is not a continuous system of *p* orbitals (there is an intervening $sp^3$-hybridized carbon atom). The third compound is not planar, and the *p* orbitals do not effectively overlap.

Compounds that satisfy the first criterion but have $4n$ electrons (rather than $4n+2$) are antiaromatic. In practice, there are very few examples of antiaromatic compounds, because most compounds can change their geometry to avoid being antiaromatic, just as we saw in the case of cyclooctatetraene.

In the following sections, we will explore several examples of compounds that meet both criteria and are therefore aromatic.

## Annulenes

**Annulenes** are compounds consisting of a single ring containing a fully conjugated π system.

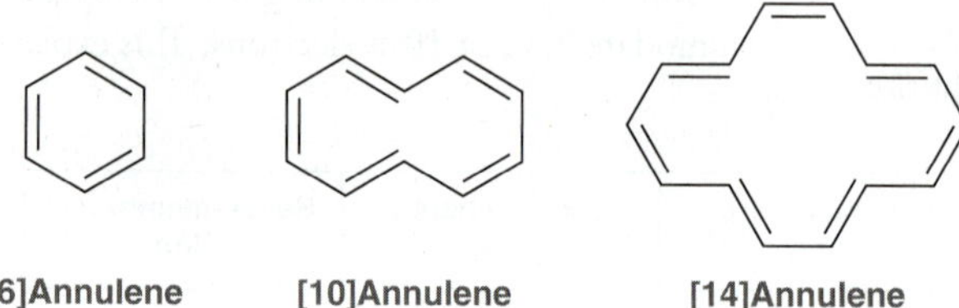


[6]Annulene is benzene, which is aromatic. We might also expect [10]annulene to be aromatic since it is a ring with a continuous system of *p* orbitals and it has a Hückel number of π electrons. However, the hydrogen atoms positioned inside the ring (shown in red) experience a steric interaction that forces the compound out of planarity:

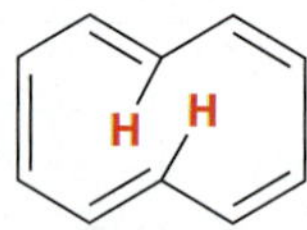


Since the molecule cannot adopt a planar conformation, the *p* orbitals cannot continuously overlap with each other to form one system, and as a result, [10]annulene does not meet the first criterion for aromaticity. It is nonaromatic. In a similar way, [14]annulene is also destabilized by a steric interaction between the hydrogen atoms positioned inside the ring, but to a lesser extent. Although [14]annulene is nonplanar, it does indeed exhibit aromatic stabilization, because the deviation from planarity is not too great. [18]Annulene has been found to be nearly planar, and is therefore aromatic, as we would expect.

### CONCEPTUAL CHECKPOINT

**18.10** Predict whether the following compound will be aromatic, nonaromatic, or antiaromatic. Explain your reasoning.

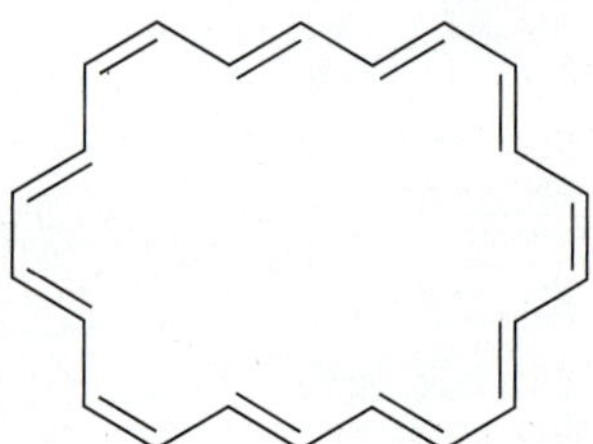

## Aromatic Ions

Previously, we used MO theory and Frost circles to explain the requirement for a Hückel number of π electrons. Let's now take a closer look at the Frost circle of a five-membered ring (Figure 18.12). A five-membered ring will be aromatic if it contains six π electrons (a Hückel number). In order

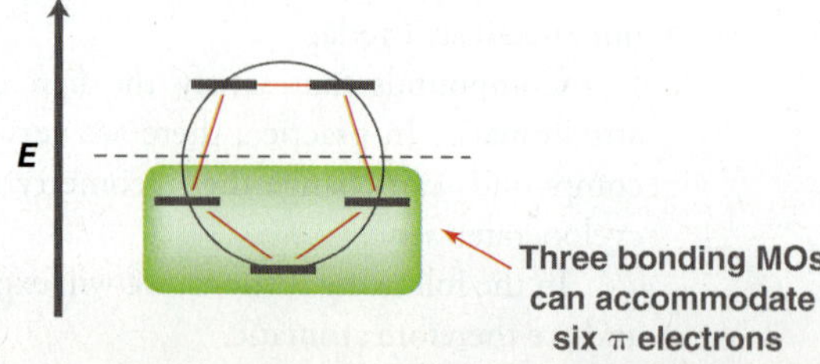


**FIGURE 18.12** The Frost circle for a five-membered ring system.

to have six π electrons, one of the carbon atoms must possess two electrons (a carbanion). The resulting anion, called the cyclopentadienyl anion, has five resonance structures.

The cyclopentadienyl anion is resonance stabilized, but that alone does not explain the observed stability. This anion is especially stable because it is aromatic. The lone pair occupies a *p* orbital (Section 2.12), and as a result, the cyclopentadienyl anion has a continuous system of overlapping *p* orbitals. With six π electrons, this anion fulfills both criteria for aromaticity. The stabilization energy of the cyclopentadienyl anion explains the remarkable acidity of cyclopentadiene.

$pK_a = 16$ $pK_a = 15.7$

Notice that the $pK_a$ of cyclopentadiene is very similar to the $pK_a$ of water (around 16), which is very unusual for a hydrocarbon (the $pK_a$ of cyclopentane is >50). The acidity of cyclopentadiene is attributed to the stability of its conjugate base, which is aromatic.

Now let's explore the Frost circle of a seven-membered ring (Figure 18.13). Once again, exactly six π electrons are necessary to achieve aromaticity.

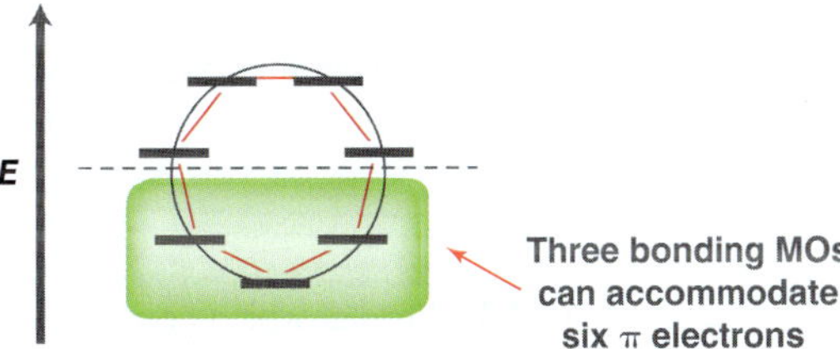


**FIGURE 18.13** The Frost circle for a seven-membered ring system.

In order to have six π electrons in a seven-membered ring, one of the carbon atoms must possess an empty *p* orbital (a carbocation). The resulting ion is called the tropylium cation, and it has seven resonance structures.

This cation is resonance stabilized, but that alone does not explain its observed stability. It is especially stable because it is aromatic—it exhibits a continuous system of overlapping *p* orbitals and has six π electrons, so both criteria for aromaticity are satisfied.

# SKILLBUILDER

## 18.2 DETERMINING WHETHER A STRUCTURE IS AROMATIC, NONAROMATIC, OR ANTIAROMATIC

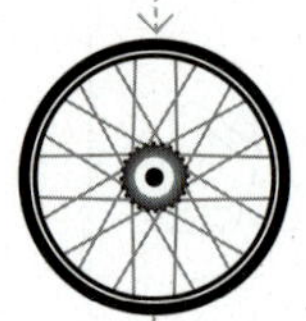

**LEARN the skill**

Determine whether the following anion is aromatic, nonaromatic, or antiaromatic:

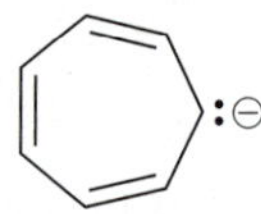

**SOLUTION**

To determine if this anion is aromatic, we must ask two questions:

1. Does the compound contain a ring comprised of continuously overlapping *p* orbitals?
2. Is there a Hückel number of $\pi$ electrons in the ring?

The answer to the first question appears to be yes, that is, the lone pair can occupy a *p* orbital, providing for continuous overlap of *p* orbitals around the ring. However, when we try to answer the second question, we discover that this anion would have eight $\pi$ electrons, which would render the anion antiaromatic. As such, the geometry of the anion will change to avoid the instability associated with antiaromaticity. Specifically, the lone pair can occupy an $sp^3$-hybridized orbital (rather than a *p* orbital), so the first criterion is no longer met. This renders the anion nonaromatic.

**PRACTICE the skill**

**18.11** Determine whether each of the following ions is aromatic, nonaromatic, or antiaromatic:

(a) 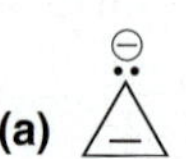 (b) 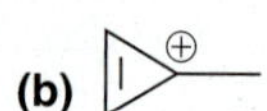 (c) 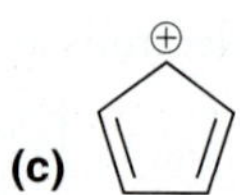 (d) 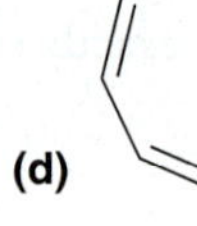

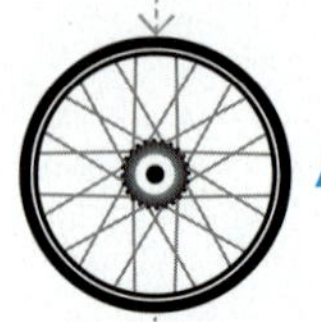

**APPLY the skill**

**18.12** Explain the vast difference in p$K_a$ values for the following two apparently similar compounds:

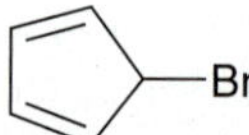

p$K_a$ = 16

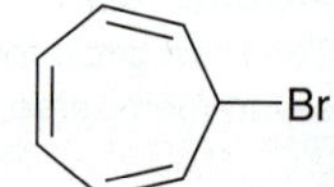

p$K_a$ = 36

**18.13** Predict which compound will react more readily in an $S_N1$ process and explain your choice:

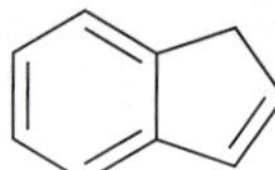


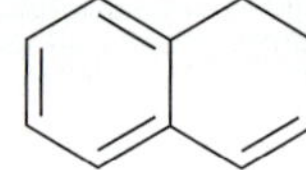


**18.14** Identify which of the following compounds is more acidic and explain your choice:

need more **PRACTICE?** **Try Problems 18.34a,c,e, 18.36e, 18.52**

## Aromatic Heterocycles

Cyclic compounds containing heteroatoms (such as S, N, or O) are called **heterocycles**. Below are two examples of nitrogen-containing heterocycles.

**Pyridine**

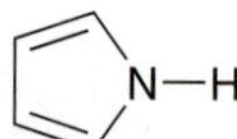


**Pyrrole**

Both of these compounds are aromatic, but for very different reasons. We will discuss them first separately and then compare them to each other.

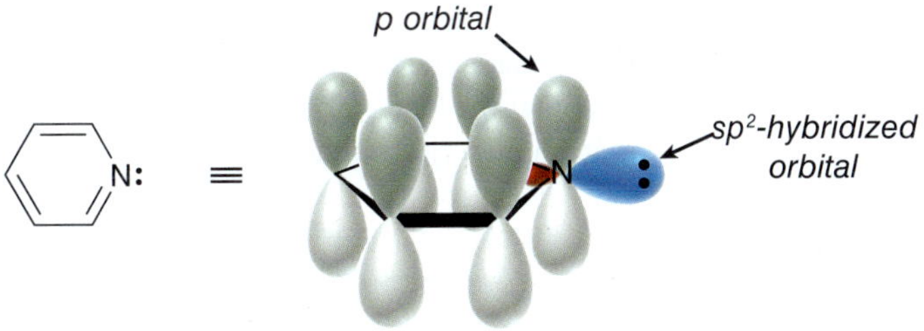

**FIGURE 18.14** The atomic *p* orbitals of pyridine that overlap to give an aromatic system. The lone pair on nitrogen occupies an $sp^2$-hybridized orbital.

Pyridine exhibits a continuous system of overlapping *p* orbitals (Figure 18.14) and therefore satisfies the first criterion for aromaticity. The nitrogen atom is $sp^2$ hybridized, and the lone pair on the nitrogen atom occupies an $sp^2$-hybridized orbital, which is pointing away from the ring. This lone pair is not part of the conjugated system and therefore is not included when we count the number of $\pi$ electrons. In this case, there are six $\pi$ electrons, so the compound is aromatic. Since the lone pair of pyridine does not participate in resonance or aromaticity, it is free to function as a base:

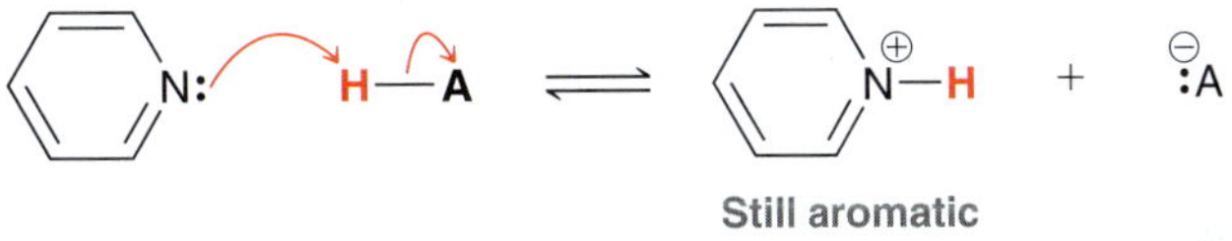

Still aromatic

Pyridine can function as a base because protonation of the nitrogen atom does not destroy aromaticity.

Now let's analyze the structure of pyrrole. Once again, the nitrogen atom is $sp^2$ hybridized, but in this case, the lone pair occupies a *p* orbital (Figure 18.15). In order for the pyrrole ring to achieve a continuous system of overlapping *p* orbitals, the lone pair of the nitrogen atom must occupy a *p* orbital. With six $\pi$ electrons (four from the $\pi$ bonds and two from the lone pair), this compound is aromatic. In this case, the lone pair is crucial in establishing aromaticity. The delocalized nature of the lone pair explains why pyrrole does not readily function as a base. Protonation of the lone pair on nitrogen effectively destroys the *p* orbital overlap, thereby destroying aromaticity.

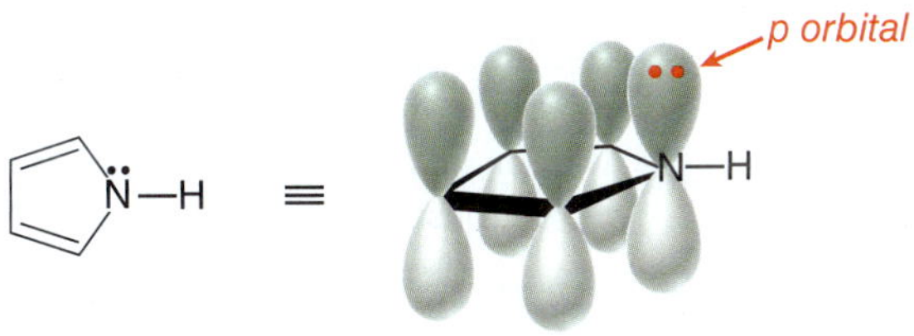

**FIGURE 18.15** The atomic *p* orbitals of pyrrole that overlap to give an aromatic system. The lone pair on nitrogen occupies one of the *p* orbitals.

**LOOKING BACK**

Recall from Section 2.12 that a lone pair occupies a *p* orbital if it participates in resonance.

The difference between pyridine and pyrrole can be seen in their electrostatic potential maps (Figure 18.16). In general, it is not safe to compare electrostatic potential maps throughout this book, because the color scale on the maps might differ. However, in this case, the same color scale was used to generate both electrostatic potential maps, which allows us to compare the location of electron density in the two compounds. The map of pyridine shows a high concentration of electron

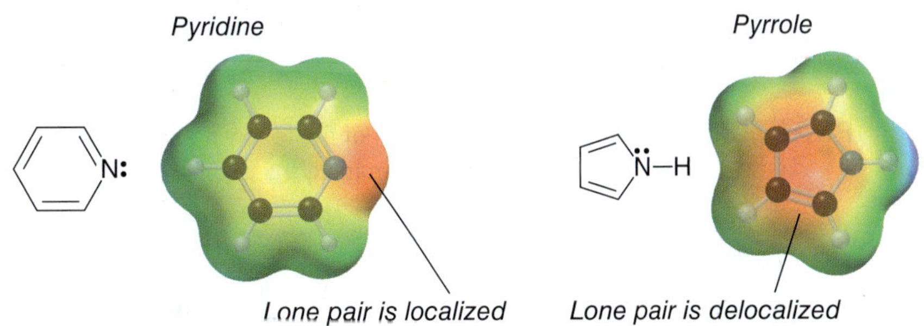


**FIGURE 18.16** Electrostatic potential maps of pyridine and pyrrole.

density on the nitrogen atom, which represents the localized lone pair. That lone pair is not part of the aromatic system and is therefore free to function as a base. In contrast, the map of pyrrole indicates that the lone pair has been delocalized into the ring, and the nitrogen atom does not represent a site of high electron density.

## SKILLBUILDER

### 18.3 DETERMINING WHETHER A LONE PAIR PARTICIPATES IN AROMATICITY

**LEARN the skill**

Histamine is responsible for many physiological responses and is known to mediate the onset of allergic reactions.

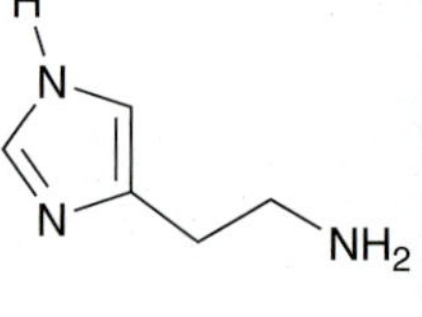


Histamine has three lone pairs. Determine which lone pair(s) participate in aromaticity.

**SOLUTION**

The lone pair on the side chain is certainly not participating in aromaticity, because it is not situated in a ring. Let's focus on the two nitrogen atoms in the ring, beginning with the nitrogen in the upper left corner.

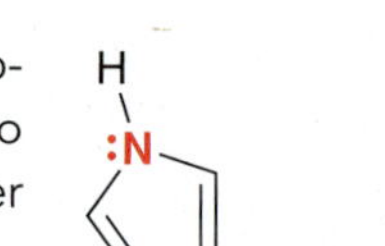


This lone pair must occupy a *p* orbital in order to have a continuous system of overlapping *p* orbitals. If this nitrogen were $sp^3$ hybridized, then the compound could not be aromatic. This lone pair is part of the aromatic system. Now look at the rest of the ring and count the $\pi$ electrons:

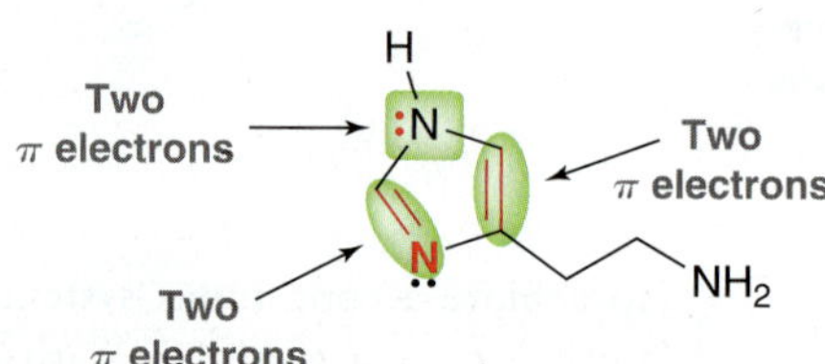


There are six $\pi$ electrons, without including the lone pair on the other nitrogen atom.

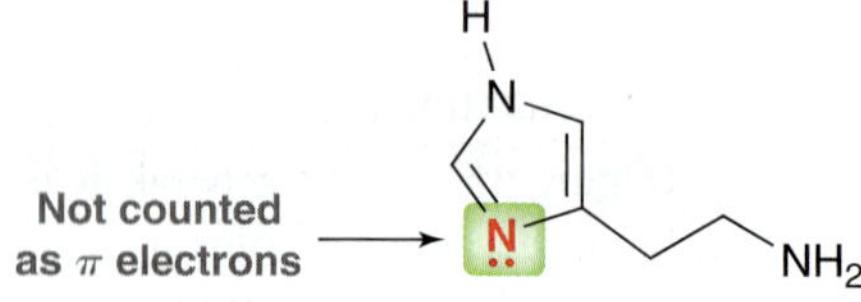


This lone pair cannot be included in the count, because each atom in the ring can only have one *p* orbital overlapping with the *p* orbitals of the ring. In this case, the nitrogen atom already has a *p* orbital from the double bond, and therefore, the lone pair cannot also occupy a *p* orbital.

**PRACTICE the skill**

**18.15** For each of the following compounds determine which (if any) lone pairs are participating in aromaticity:

(a) 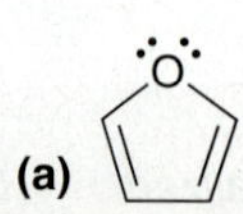


(b) 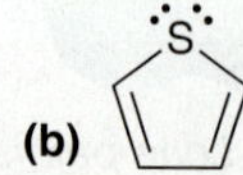


(c) 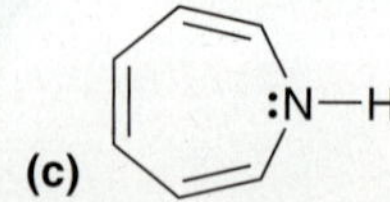


(d) 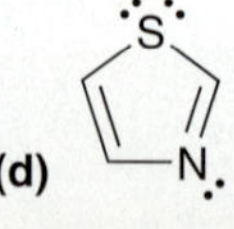


(e) 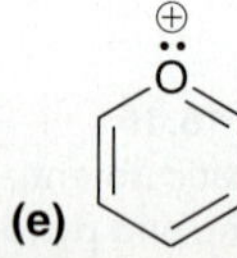

(f) 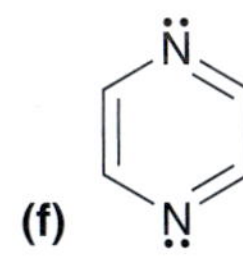 (g) 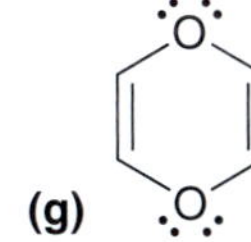 (h) 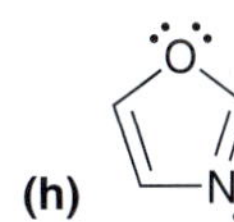

**APPLY** the skill

**18.16** Go to the beginning of Section 18.1 where eight best-selling drugs were shown. Review the structures of those compounds and identify all of the aromatic rings that are not already highlighted in red.

**18.17** Identify which compound is expected to have a lower p$K_a$. Justify your choice.

need more **PRACTICE?** **Try Problems 18.34b,d, 18.36c,d, 18.38, 18.41, 18.44, 18,62, 18.64**

## Polycyclic Aromatic Compounds

Hückel's rule ($4n+2$) can only be applied to compounds that exhibit a single ring of overlapping *p* orbitals (monocyclic compounds). Nevertheless, many **polycyclic aromatic hydrocarbons (PAHs)** are known to be stable and have been thoroughly studied:

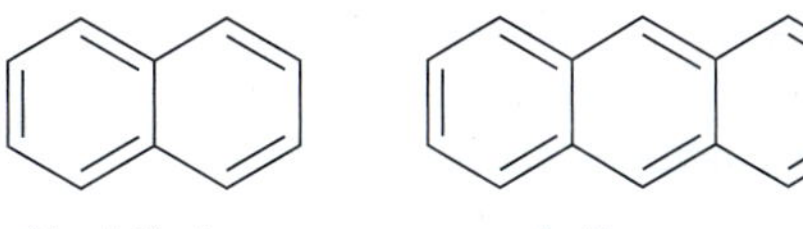

**Naphthalene** **Anthracene**

**Phenanthrene**

These compounds are comprised of fused benzene like rings, and they exhibit significant aromatic stabilization. For each compound, the stabilization energy can be measured by comparing heats of hydrogenation, just as we did with benzene. These values are summarized in Table 18.1. For polycyclic aromatic hydrocarbons, the stabilization energy *per ring* is dependent on the structure of the compound and is generally less than the stabilization energy of benzene, as seen in the last column of Table 18.1. In addition, the individual rings of a polycyclic aromatic hydrocarbon will often exhibit different levels of stabilization. For example, the middle ring of anthracene is known to be more reactive than either of the outer rings.

**TABLE 18.1 STABILIZATION ENERGY FOR A FEW POLYCYCLIC AROMATIC HYDROCARBONS**

| COMPOUND | STABILIZATION ENERGY (KJ/MOL) | AVERAGE STABILIZATION ENERGY PER RING (KJ/MOL) |
|---|---|---|
| Benzene | 152 | 152 |
| Naphthalene | 255 | 128 |
| Anthracene | 347 | 116 |
| Phenanthrene | 381 | 127 |

## medically speaking

## The Development of Nonsedating Antihistamines

Histamine (as seen in SkillBuilder 18.3) is an aromatic compound that plays many roles in the biological processes of mammals. It is involved in the mediation of allergic reactions and the regulation of gastric acid secretion in the stomach. Extensive pharmacological studies reveal the existence of at least three different types of histamine receptors, called $H_1$, $H_2$, and $H_3$ receptors. The binding of histamine to $H_1$ receptors is responsible for triggering allergic reactions. Compounds that compete with histamine to bind with $H_1$ receptors (without triggering the allergic reaction) are called $H_1$ antagonists and belong to a class of compounds called antihistamines. Examples of $H_1$ antagonists include Benadryl and Chlortrimeton.

**Benadryl (diphenhydramine)**

**Chlortrimeton (chlorpheniramine)**

Notice the structural similarities between these compounds. Both compounds exhibit two aromatic moieties, highlighted in red, as well as a tertiary amine (a nitrogen atom connected to three alkyl groups), highlighted in blue. Extensive research indicates that these structural features are necessary in order for a drug to exhibit $H_1$ antagonism. Studies also show that the two aromatic rings must be in close proximity (separated by either one or two carbon atoms). To understand this requirement, consider the structure of a simple compound that bears two phenyl groups separated by one carbon atom.

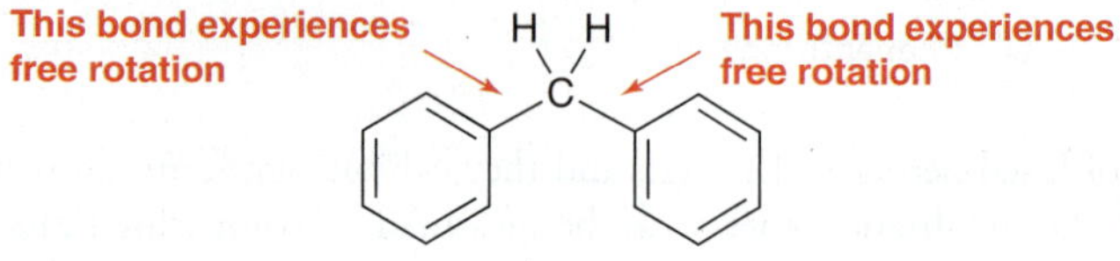


**Diphenylmethane**

The single bonds experience free rotation at room temperature, which enables the molecule to adopt different conformations. When the two rings are coplanar (when they lie in the same plane), there is a significant steric interaction between two of the aromatic hydrogen atoms.

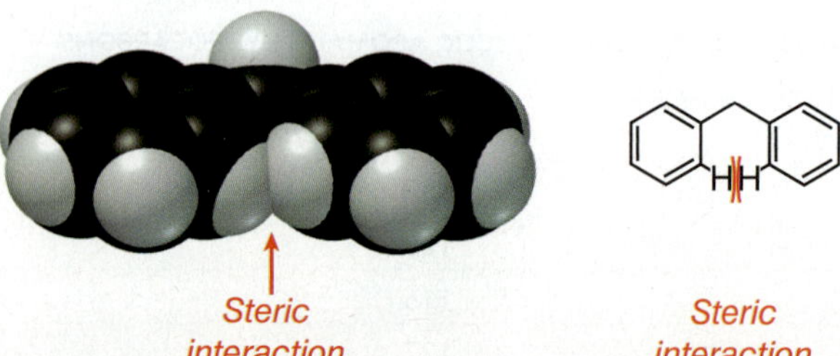


As a result, this conformation is very high in energy. The molecule can alleviate most of this strain by adopting a conformation in which the two rings are not coplanar. This noncoplanar conformation resembles a twisted waffle maker.

Rings are not coplanar — Resembles a waffle maker

This conformation is much lower in energy than the coplanar conformation. As a result, a molecule that contains two closely situated aromatic rings will spend most of its time in a noncoplanar conformation.

Antihistamines possess this "waffle maker" structural feature. For example, the structure of Benadryl exhibits two aromatic rings, and these rings adopt a noncoplanar conformation. This feature is important, because it enables the molecule to experience significant binding with the $H_1$ receptor.

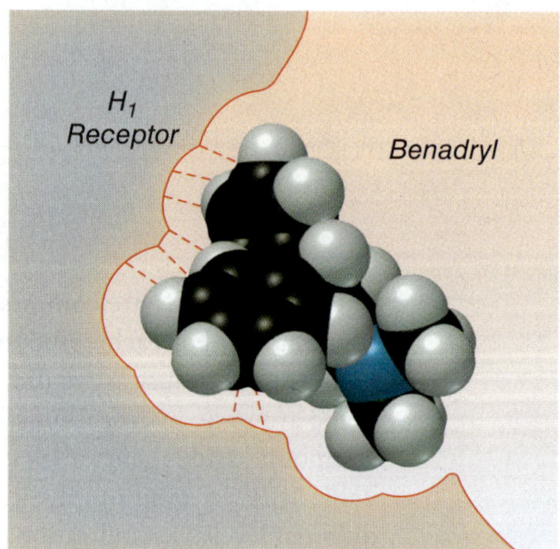


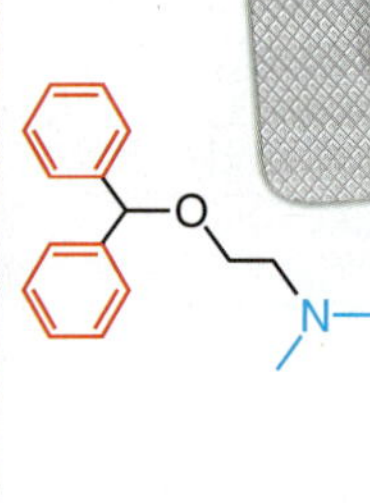


In a noncoplanar conformation, the two aromatic moieties achieve a combined binding force of approximately 40 kJ/mol (equivalent in strength to two hydrogen bonds). The nitrogen atom (shown in blue) also binds to the receptor. That interaction will be discussed in more detail in Chapter 23.

The table (next page) shows a sampling of antihistamines. Notice that each structure contains two aromatic moieties in close proximity (red) as well as the tertiary amine (blue).

In addition to their ability to bind with the $H_1$ receptors, these antihistamines also bind with receptors in the central nervous system, causing undesired side effects such as sedation (drowsiness).

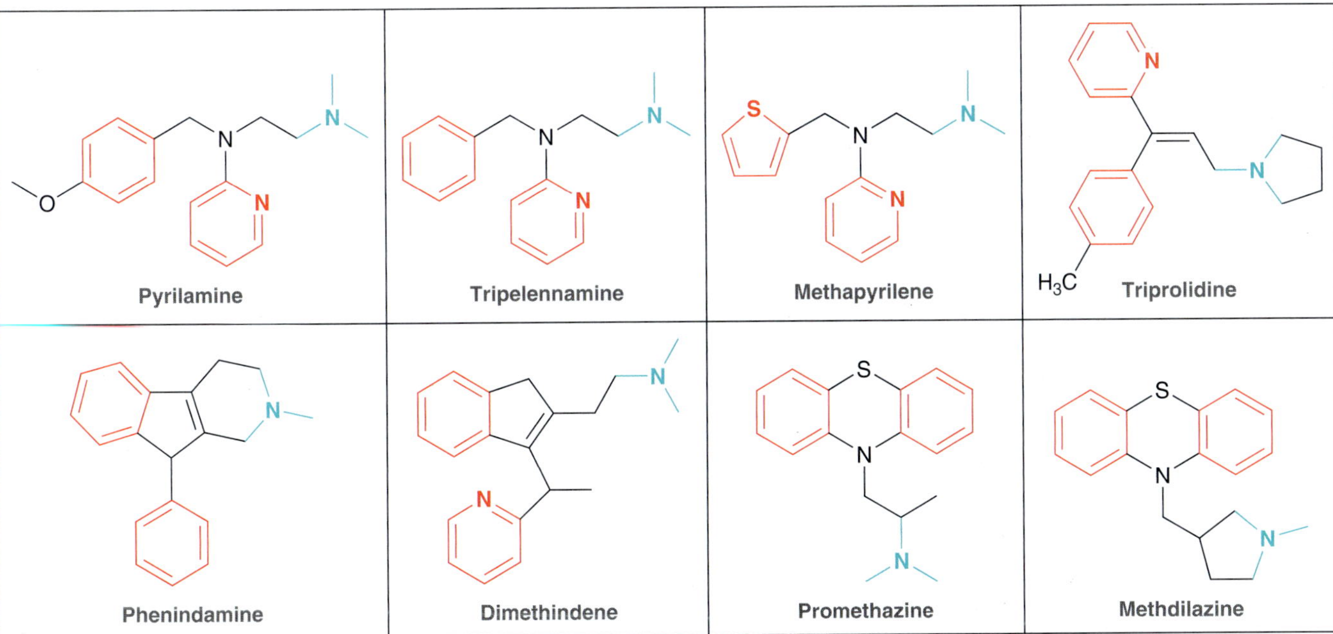


All of these compounds, called *first-generation antihistamines*, cause these undesired side effects. The sedative effect of first-generation antihistamines has been attributed to their ability to cross the blood-brain barrier, allowing them to interact with receptors in the central nervous system.

One of the goals of antihistamine research over the last two decades was the development of antihistamines that could bind with $H_1$ receptors but could not readily cross the blood-brain barrier and therefore could not reach the receptors that trigger sedation. Extensive research has led to the development of several new drugs, called *second-generation antihistamines*, which are nonsedating. As we might expect, these drugs contain the pharmacophore necessary for $H_1$ antagonism (two aromatic rings in close proximity and a tertiary amine). However, second-generation antihistamines also have polar functional groups that prevent the compound from crossing the nonpolar environment of the blood-brain barrier. One such example is fexofenadine.

Fexofenadine

This compound contains several polar groups (highlighted in green boxes). In addition, the COOH group is deprotonated under physiological pH to produce a carboxylate anion ($COO^-$). This ionic group, together with the polar hydroxyl groups, prevents the compound from crossing the blood-brain barrier and from reaching receptors in the central nervous system. As a result, fexofenadine does not produce the sedative effects that are typical of first-generation antihistamines. Many other nonsedating antihistamines have also been developed, such as loratidine, sold under the trade name Claritin. Apparently, the carbamate group (highlighted in green) is sufficiently polar to reduce the ability of this compound to cross the blood-brain barrier.

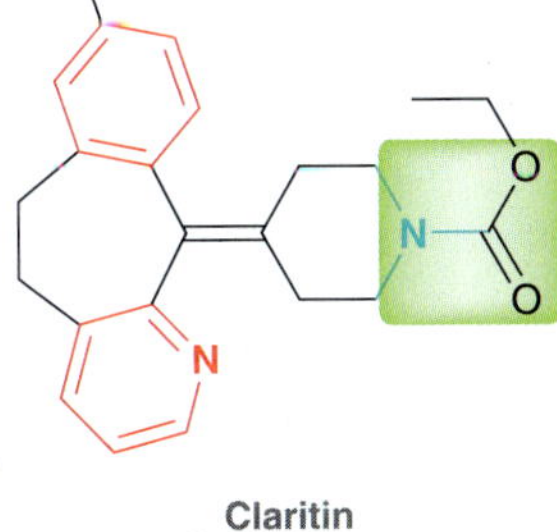

Claritin (loratadine)

## CONCEPTUAL CHECKPOINT

**18.18** Meclizine, shown below, is an antiemetic (prevents nausea and vomiting).

(a) Would you expect meclizine to be an antihistamine as well? Justify your answer.

(b) This drug is known to cause sedation. Describe the source of the sedative properties of meclizine.

(c) Can you suggest a structural modification that could possibly modify the sedative properties of meclizine?

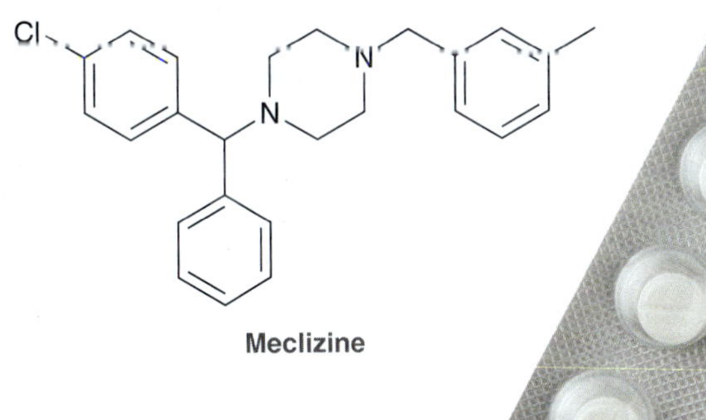

Meclizine

## 18.6 Reactions at the Benzylic Position

Any carbon atom attached directly to a benzene ring is called a **benzylic position**:

Benzylic positions

In the following sections, we will explore reactions that can occur at the benzylic position.

### Oxidation

Recall from Section 13.10 that chromic acid ($H_2CrO_4$) is a strong oxidizing agent used to oxidize primary or secondary alcohols. Chromic acid does not readily react with benzene or with alkanes:

$Na_2Cr_2O_7$ / $H_2SO_4$, $H_2O$ → No reaction

$Na_2Cr_2O_7$ / $H_2SO_4$, $H_2O$ → No reaction

Interestingly, however, alkylbenzenes are readily oxidized by chromic acid. The oxidation takes place selectively at the benzylic position:

$Na_2Cr_2O_7$ / $H_2SO_4$, $H_2O$ → O, OH

Although the aromatic moiety itself is stable in the presence of chromic acid, the benzylic position is particularly susceptible to oxidation. Notice that the alkyl group is entirely excised, leaving only the benzylic carbon atom behind. The product is benzoic acid, irrespective of the identity of the alkyl group. The only condition is that the benzylic position must have at least one proton. If the benzylic position lacks a proton, then oxidation does not occur.

$Na_2Cr_2O_7$ / $H_2SO_4$, $H_2O$ → No reaction

Oxidation of a proton-bearing benzylic position can also be accomplished with other reagents, including potassium permanganate ($KMnO_4$).

$KMnO_4$ / $H_2O$, heat → O, $O^{\ominus}$ $^{\oplus}K$ — $H_3O^+$ → O, OH

When an alkyl benzene is treated with potassium permanganate, a carboxylate salt is obtained, which must then be treated with a proton source in order to obtain benzoic acid.

CONCEPTUAL CHECKPOINT

**18.19** Draw the expected product when each of the following compounds is treated with chromic acid:

(a) (b) (c)

## Free-Radical Bromination

In Section 11.7 we explored the free-radical bromination of allylic positions. Similarly, free-radical bromination also occurs readily at benzylic positions.

NBS, heat

The reaction is highly regioselective (bromination occurs primarily at the benzylic position) due to resonance stabilization of the intermediate benzylic radical.

This reaction is extremely important, because it enables us to introduce a functional group at the benzylic position. Once introduced, that functional group can then be exchanged for a different group, as seen in the next section.

## Substitution Reactions of Benzylic Halides

As seen in Section 7.8, benzylic halides undergo $S_N1$ reactions very rapidly.

$H_2O$, $S_N1$ + HBr

The relative ease of the reaction is attributed to the stability of the carbocation intermediate. Specifically, a benzylic carbocation is resonance stabilized.

Benzylic halides also undergo $S_N2$ reactions very rapidly, provided that they are not sterically hindered.

Na⊕ ⊖:OH, $S_N2$ + NaBr

## Elimination Reactions of Benzylic Halides

Benzylic halides undergo elimination reactions very rapidly. For example, the following E1 reaction occurs readily:

conc. $H_2SO_4$, E1 + $H_2O$

The relative ease of the reaction is attributed to the stability of the carbocation intermediate (a benzylic carbocation), just as we saw with $S_N1$ reactions.

Benzylic halides also undergo E2 reactions very rapidly.

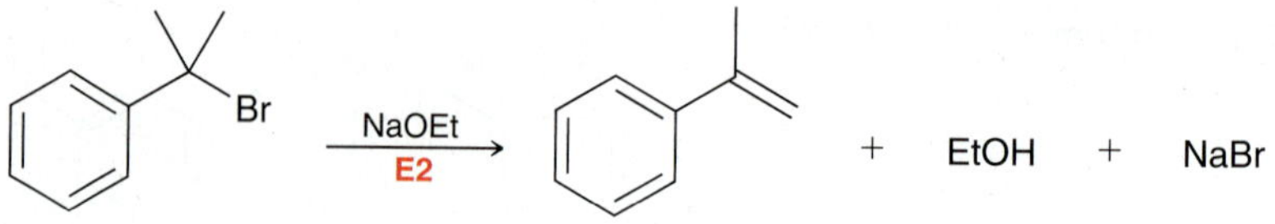

The relative ease of the reaction is attributed to the low energy of the transition state as a result of conjugation between the forming double bond and the aromatic ring.

## Summary of Reactions at the Benzylic Position

Figure 18.17 is a summary of reactions that can occur at the benzylic position. Note that there are two ways to introduce functionality at a benzylic position: (1) oxidation and (2) radical bromination. The first method produces benzoic acid, while the second produces a benzylic halide that can undergo substitution or elimination.

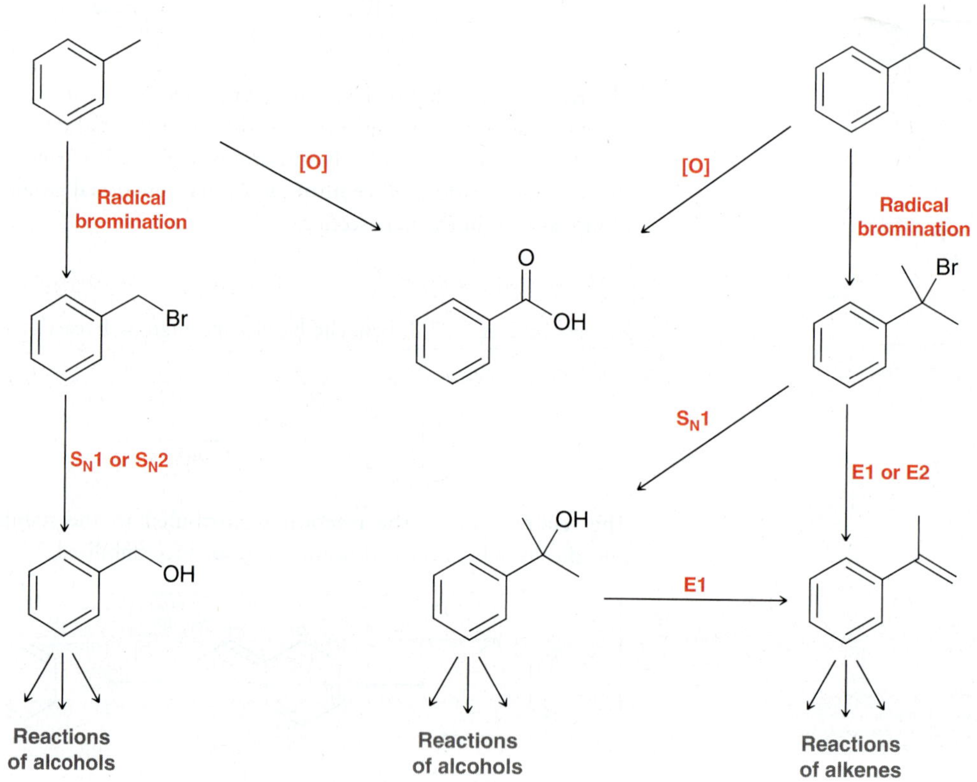

**FIGURE 18.17** Reactions that occur at the benzylic position.

# SKILLBUILDER

## 18.4 MANIPULATING THE SIDE CHAIN OF AN AROMATIC COMPOUND

**LEARN** the skill

Propose an efficient synthesis for the following transformation:

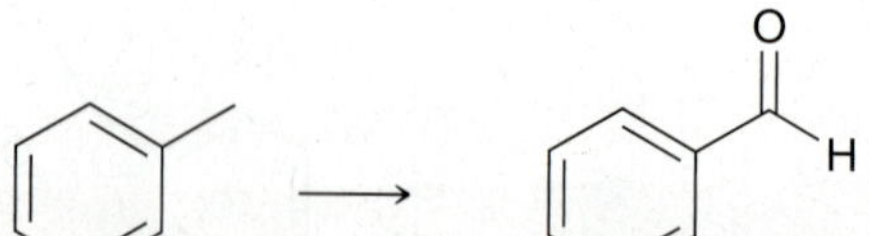

## SOLUTION

Recall from Chapter 12 the two questions to ask when approaching a synthesis problem:

1. Is there a change in the carbon skeleton?
2. Is there a change in the position or identity of the functional groups?

Always begin with the carbon skeleton. In this case, there is no change:

Now focus on the change in functional groups. In this example, the reactant does not have a functional group at the benzylic position, while the product does. Therefore, we must first introduce functionality at the benzylic position, and there are two ways to do that.

$Na_2Cr_2O_7$ / $H_2SO_4$, $H_2O$ → OH (carboxylic acid)

NBS / Heat → Br

The first method will be less helpful, because we do not yet know how to manipulate a carboxylic acid group (COOH). That topic will be covered extensively in Chapter 21. In the meantime, we will have to use radical bromination to introduce functionality. The resulting benzylic halide can then be manipulated. But can a Br be converted into an aldehyde? The bromine atom must be replaced with an oxygen, which requires a substitution reaction. Since the substrate is primary and benzylic, it is reasonable to choose an $S_N2$ process.

Br → NaOH → OH

Now the oxygen atom is attached, but the oxidation state is not correct. We must oxidize the alcohol to an aldehyde. This transformation can be accomplished with PCC (without further oxidizing the aldehyde to a carboxylic acid).

OH → PCC → aldehyde (O, H)

In summary, our synthesis has just a few steps:

1) NBS, heat
2) NaOH
3) PCC

**PRACTICE the skill** **18.20** Propose an efficient synthesis for each of the following transformations:

**(a)** → OH **(b)**

**(c)** → N **(d)** → O

**APPLY the skill**

**18.21** Propose an efficient synthesis for the following transformation:

**18.22** Starting with isopropyl benzene, propose a synthesis for acetophenone. (**Hint:** Make sure to count the number of carbon atoms in the starting material and product.)

O

**18.23** Propose an efficient synthesis for the following transformation:

OH

need more **PRACTICE?** **Try Problems 18.47, 18.56**

## 18.7 Reduction of the Aromatic Moiety

### Hydrogenation

Recall from earlier in this chapter that under forcing conditions benzene will react with three equivalents of molecular hydrogen to produce cyclohexane.

$$+ \quad 3\,H_2 \xrightarrow[\substack{100\text{ atm} \\ 150°\text{C}}]{\text{Ni}} \qquad \Delta H° = -208\text{ kJ/mol}$$

With some catalysts and under certain conditions, it is possible to selectively hydrogenate a vinyl group in the presence of an aromatic ring.

$$+ \quad H_2 \xrightarrow[\substack{2\text{ atm} \\ 25°\text{C}}]{\text{Pt}} \qquad \Delta H° = -117\text{ kJ/mol}$$

**(100%)**

The vinyl group is reduced, but the aromatic moiety is not. Notice that $\Delta H$ for this reaction is only slightly lower than what we expect for a double bond (−120 kJ/mol), which can be attributed to the fact that the double bond is conjugated in this case.

### Birch Reduction

Recall from Section 10.5 that alkynes can be reduced via the dissolving metal reduction. Benzene can also be reduced under similar conditions to give 1,4-cyclohexadiene.

$$\xrightarrow[\text{NH}_3]{\text{Na, CH}_3\text{OH}}$$

This reaction is called the **Birch reduction**, named after the Australian chemist Arthur Birch, who systematically explored the details of this reaction. The mechanism, which is believed to be very similar to the mechanism of alkyne reduction via a dissolving metal reduction, is comprised of four steps (Mechanism 18.1).

## MECHANISM 18.1 THE BIRCH REDUCTION

Nucleophilic attack | Proton transfer | Nucleophilic attack | Proton transfer

$\cdot$Na — A single electron is transferred from the sodium atom to the aromatic ring → **A radical anion**

Methanol donates a proton to the radical anion, generating a radical intermediate → **A radical**

$\cdot$Na — A single electron is transferred from the sodium atom to the radical intermediate, generating an anion → **An anion**

Methanol donates a proton to the anion, generating an isolated diene

In step 1, a single electron is transferred to the aromatic ring, giving a radical anion, which is then protonated in step 2. Steps 3 and 4 are similar to steps 1 and 2, with the transfer of an electron followed by protonation. In order to remember the mechanism of this reaction, it might be helpful to summarize the steps as (1) electron, (2) proton, (3) electron, and (4) proton.

In a Birch reduction, the ring is not completely reduced, since two double bonds remain. Specifically, only two of the carbon atoms in the ring are actually reduced. The other four carbon atoms remain $sp^2$ hybridized. Also notice that the two reduced carbon atoms are opposite each other:

Na, $CH_3OH$ / $NH_3$ → $sp^3$, $sp^3$

The product is a nonconjugated diene, rather than a conjugated diene.

When an alkyl benzene is treated with Birch conditions, the carbon atom connected to the alkyl group is not reduced:

R — Na, $CH_3OH$ / $NH_3$ → R — Not reduced ( R — Not observed )

Why not? Recall that alkyl groups are electron donating. This effect destabilizes the radical anion intermediate that is necessary to generate the product that is not observed.

When electron-withdrawing groups are used, a different regiochemical outcome is observed. For example, consider the structure of acetophenone, which has a carbonyl group (C=O bond) next to the aromatic ring:

Acetophenone $\xrightarrow[NH_3]{Na,\ CH_3OH}$ Reduced

Not observed

The carbonyl group is electron withdrawing via resonance. (Can you draw the resonance structures of acetophenone and explain why the carbonyl group is electron withdrawing?) This electron-withdrawing effect stabilizes the intermediate that is necessary to generate the observed product.

## SKILLBUILDER

### 18.5 PREDICTING THE PRODUCT OF A BIRCH REDUCTION

**LEARN the skill**

Predict the major product obtained when the following compound is treated with Birch conditions:

#### SOLUTION

Determine whether the groups attached to the aromatic ring are electron donating or electron withdrawing.

In this case, there are two groups. The carbonyl group is electron withdrawing, while the alkyl group is electron donating.

**STEP 1**
Determine whether each substituent is electron donating or electron withdrawing.

Carbonyl group is electron withdrawing

Alkyl group is electron donating

**STEP 2**
Identify which carbon atoms are reduced.

We therefore predict that the carbon atom next to the carbonyl group will be reduced, while the carbon atom next to the alkyl group will not be reduced.

Will be reduced

Will not be reduced

**STEP 3**
Draw the product, making sure that the two positions not reduced are 1,4 to each other.

Remember that only two positions are reduced in a Birch reaction, and these two positions must be on opposite sides of the ring (1,4 to each other). This requirement, together with the information above, dictates the following outcome:

Na
$CH_3OH$
$NH_3$

**PRACTICE the skill** **18.24** Predict the major product obtained when each of the following compounds is treated with Birch conditions:

(a) 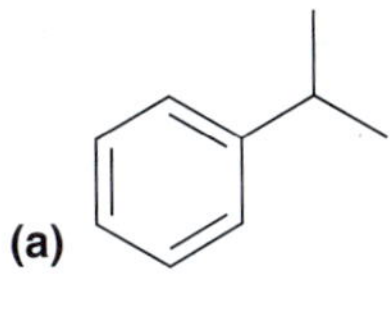

(b) 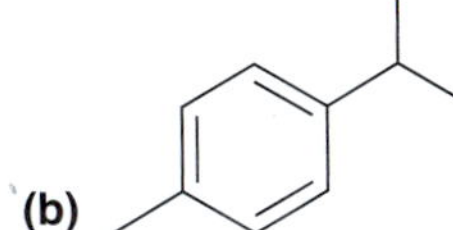

(c) 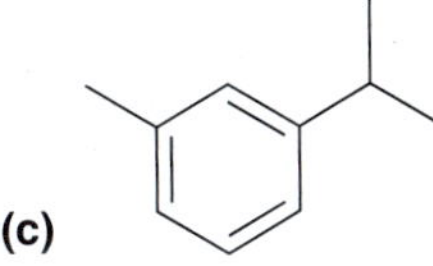

(d)

(e) 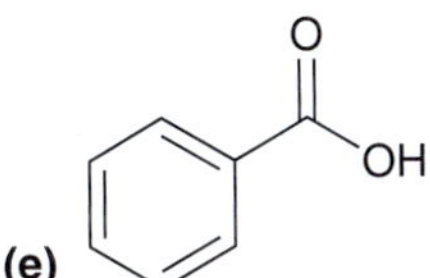


(f) 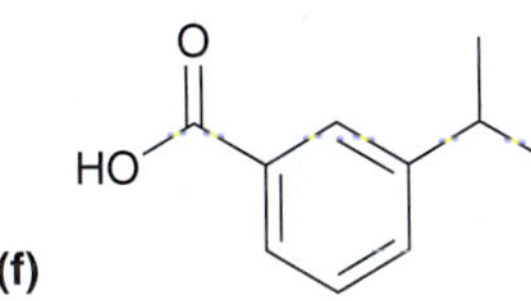

**APPLY the skill** **18.25** Consider the structure of anisole (also called methoxybenzene). In the next chapter, we will discuss whether a methoxy group is electron donating or electron withdrawing. We will see that there is a competition between two factors. Specifically, we will see that the methoxy group is electron withdrawing because of induction, but it is electron donating because of resonance.

**(a)** Draw the resonance structures of anisole.

**(b)** Whenever resonance and induction compete with each other, resonance is generally the dominant factor. Using this information, predict the regiochemical outcome of a Birch reduction of anisole.

need more **PRACTICE?** **Try Problems 18.49, 18.50**

## 18.8 Spectroscopy of Aromatic Compounds

### IR Spectroscopy

Benzene derivatives generally produce signals in five characteristic regions of an IR spectrum. These five regions and the vibrations associated with them are listed in Table 18.2 and can be seen in the IR spectrum of ethylbenzene (Figure 18.18). Notice that the signals just above 3000 $cm^{-1}$, corresponding with $C_{sp^2}$—H stretching, appear on the shoulder of the signals for all other C—H stretching (just below 3000 $cm^{-1}$). This is often the case, and these signal(s) can be identified by drawing a line at 3000 $cm^{-1}$ and looking for any signals to the left of that line (Section 15.3). Aromatic compounds also produce a series of signals between 1450 and 1600 $cm^{-1}$, resulting from stretching of the carbon-carbon bonds of the aromatic ring as well as ring vibrations. The pattern of signals in the other three characteristic regions (as seen in Table 18.2) can often be used to identify the specific substitution pattern of the aromatic ring (i.e., monosubstituted, *ortho* disubstituted, *meta* disubstituted, etc.), although this level of analysis will not be discussed in our current treatment of IR spectroscopy.

**TABLE 18.2 CHARACTERISTIC SIGNALS IN THE IR SPECTRA OF AROMATIC COMPOUNDS**

| ABSORPTION | FEATURE | COMMENTS |
|---|---|---|
| 3000–3100 $cm^{-1}$ | $C_{sp^2}$—H stretching | One or more signals just above 3000 $cm^{-1}$. Intensity is generally weak or medium. |
| 1700–2000 $cm^{-1}$ | Combination bands and overtones | A group of very weak signals |
| 1450–1650 $cm^{-1}$ | Stretching of carbon-carbon bonds as well as ring vibrations | Generally three signals (medium intensity) at around 1450, 1500, and 1600 $cm^{-1}$ |
| 1000–1275 $cm^{-1}$ | C—H bending (in plane) | Several signals of strong intensity |
| 690–900 $cm^{-1}$ | C—H bending (out of plane) | One or two strong signals |

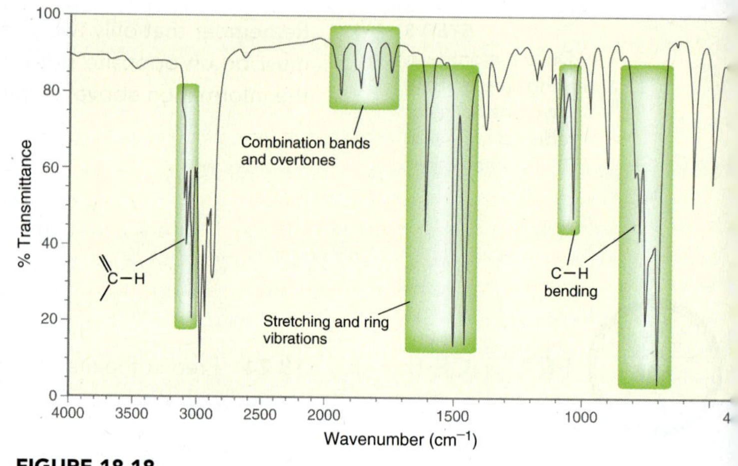


**FIGURE 18.18**
An IR spectrum of ethylbenzene indicating the five regions of absorption, characteristic of aromatic compounds.

**LOOKING BACK**
To see a picture of the shielding and deshielding regions established by an aromatic ring, see Figure 16.9.

## $^1$H NMR Spectroscopy

In Section 16.5, we first discussed the anisotropic effects of an aromatic ring. Specifically, the motion of the π electrons generates a local magnetic field that effectively deshields the protons connected directly to the ring.

The signals from these protons typically appear between 6.5 and 8 ppm. For example, consider the $^1$H NMR spectrum of ethylbenzene (Figure 18.19). The presence of a multiplet near 7 ppm is one of the best ways to verify the presence of an aromatic ring. Notice that the deshielding effects of the aromatic ring are felt most strongly for protons connected directly to the ring. Protons in benzylic positions are farther removed from the ring and are deshielded to a lesser extent. These protons generally produce signals between 2 and 3 ppm. Protons that are even farther removed from the aromatic ring exhibit almost no shielding effects.

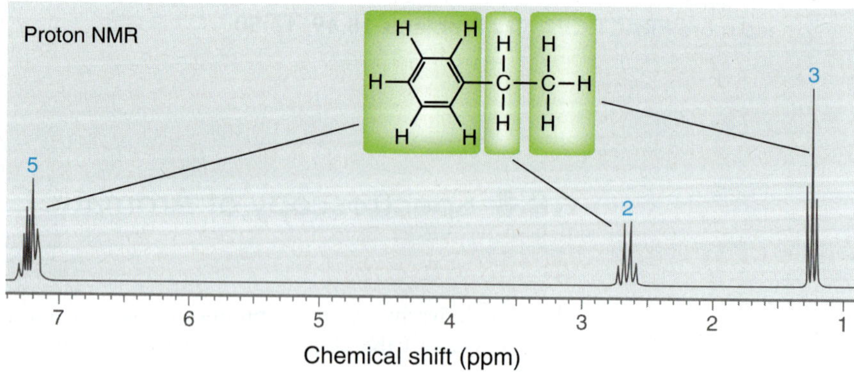


**FIGURE 18.19**
A $^1$H NMR spectrum of ethylbenzene showing that the deshielding effects of the aromatic ring are strongly dependent on proximity to the ring.

The integration value of the multiplet near 7 ppm is very useful information because it indicates the extent of substitution of the aromatic ring (monosubstituted, disubstituted, trisubstituted, etc.). An integration of 5 is indicative of a monosubstituted ring, an integration of 4 is indicative of a disubstituted ring, and so forth. The splitting pattern of this multiplet is generally too complex to analyze, because the rigid planar structure of benzene causes long-range coupling between all of the different aromatic protons (even non-neighbors). Therefore, the substitution pattern is usually complex, except in the following two *para*-disubstituted cases, both of which produce very distinctive and simple patterns:

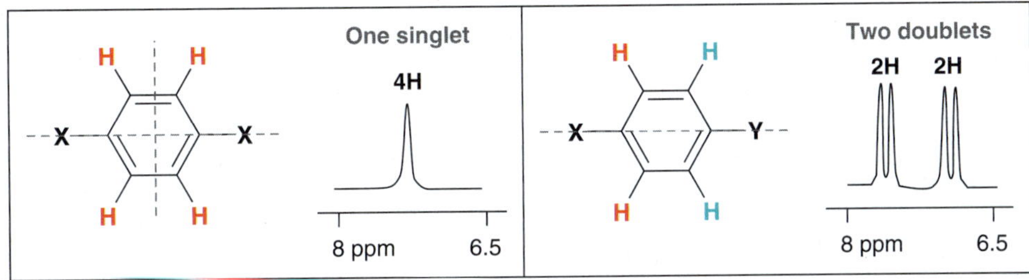


## $^{13}C$ NMR Spectroscopy

As first mentioned in Section 16.12, the carbon atoms of aromatic rings typically produce signals in the range of 100–150 ppm in a $^{13}C$ NMR spectrum. The number of signals is very helpful in determining the specific substitution pattern for substituted aromatic rings. Several common substitution patterns are shown below.

2 signals

3 signals

4 signals

6 signals

The number of signals in the region of 150–200 ppm can therefore provide valuable information.

### CONCEPTUAL CHECKPOINT

**18.26** A compound with molecular formula $C_8H_8O$ produces an IR spectrum with signals at 3063, 1686, and 1646 cm$^{-1}$. The $^1H$ NMR spectrum of this compound exhibits a singlet at 2.6 ppm (I = 3H) and a multiplet at 7.5 (I = 5H).

(a) Draw the structure of this compound.

(b) What is the common name of this compound?

(c) When this compound is treated with Na, $CH_3OH$, and $NH_3$, a reduction takes place, giving a new compound with molecular formula $C_8H_{10}O$. Draw the product of this reaction.

**18.27** A compound with molecular formula $C_8H_{10}$ produces an IR spectrum with many signals, including 3108, 3066, 3050, 3018, and 1608 cm$^{-1}$. The $^1H$ NMR spectrum of this compound exhibits a singlet at 2.2 ppm (I = 6H) and a multiplet at 7.1 ppm (I = 4H). The $^{13}C$ NMR spectrum of this compound exhibits signals at 19.7, 125.9, 129.6, and 136.4 ppm.

(a) Draw the structure of this compound.

(b) What is the common name of this compound?

(c) Treating this compound with chromic acid yields a product with molecular formula $C_8H_6O_4$. Draw the product of this reaction.

## Buckyballs and Nanotubes

Until the mid-1980s, only two forms of elemental carbon were known—diamond and graphite:

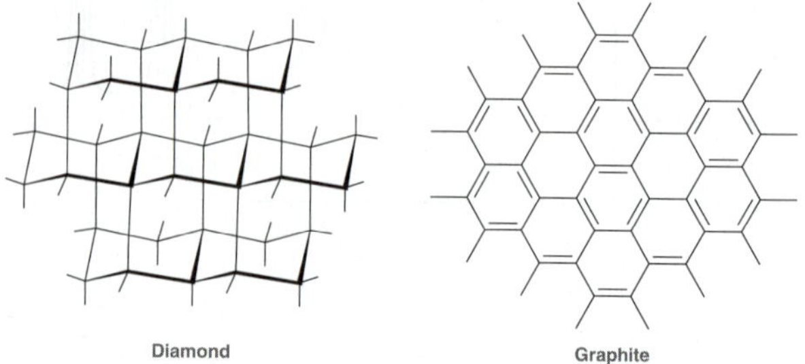


Diamond is a three-dimensional lattice of interlocking chair conformations. Graphite is comprised of flat sheets of interlocking benzene rings. These sheets adhere to one another due to van der Waals interaction, but they can readily slide past each other, which makes graphite an excellent lubricant.

In 1985, a new form of elemental carbon was discovered, very much by accident. Harold Kroto (University of Sussex) was investigating how certain organic compounds are formed in space. While visiting with Robert Curl and Richard Smalley (Rice University), they discussed a way to re-create the type of chemical transformations that might occur in stars that are carbon rich. Smalley had developed a method for laser-induced evaporation of metals, and the group of scientists agreed to apply this method to graphite in the hopes of creating polyacetylenic compounds:

The compounds generated by this procedure were then analyzed by mass spectrometry, and to everyone's surprise, certain conditions would reliably produce a compound with molecular formula $C_{60}$. They theorized that the structure of $C_{60}$ is based on alternating five- and six-membered rings. To illustrate this, imagine a five-membered ring completely surrounded by fused benzene rings:

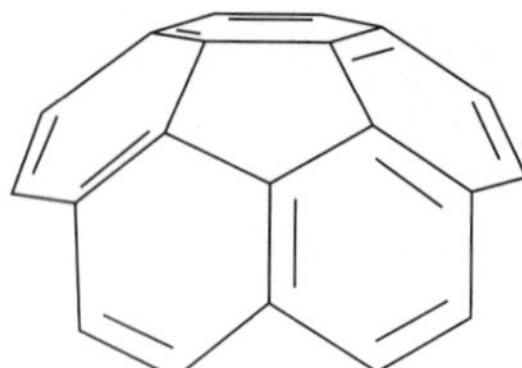

This group of atoms provides for a natural curvature. Extending this bonding pattern provides the possibility of a spherical molecule. Based on this reasoning, Kroto, Curl, and Smalley theorized that $C_{60}$ is comprised of fused rings (20 hexagons and 12 pentagons), resembling the pattern of the seams on a soccer ball.

They called this unusual compound buckminsterfullerene (or just fullerene, for short), named after American architect R. Buckminster Fuller, who was famous for building geodesic domes. This compound is also commonly referred to as a *buckyball*.

Shortly after the discovery of $C_{60}$, methods were developed for preparing $C_{60}$ in larger quantities, enabling its spectroscopic analysis as well as the exploration of its chemistry. A $^{13}C$ NMR spectrum of $C_{60}$ shows a single peak at 143 ppm, because all 60 carbon atoms are $sp^2$ hybridized and chemically equivalent.

Since buckyballs are comprised of interlocking aromatic rings, we might expect $C_{60}$ to function as if it were one large aromatic compound. However, this is not the case, because the curvature of the sphere prevents all of the *p* orbitals from overlapping with each other, so the first criterion for aromaticity is not met. This explains why $C_{60}$ does not exhibit the same stability as benzene. Specifically, $C_{60}$ readily undergoes addition reactions, much like alkenes.

Since 1990, chemists have been preparing and exploring larger fullerenes and their interesting chemistry. For example, $C_{60}$ can be prepared under conditions in which ions can be trapped inside the center of the sphere. Such compounds are superconductors at low temperature and offer the potential of many exciting applications. These compounds are also being explored as novel drug delivery systems. Tubular fullerenes have also been prepared:

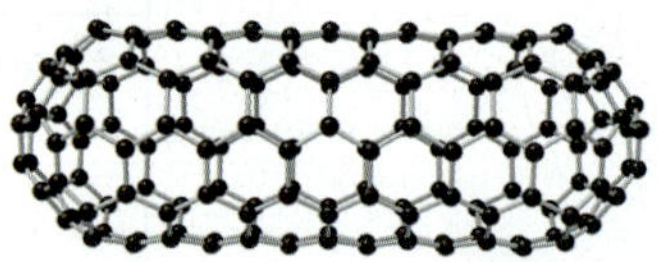

These compounds, which are called *nanotubes*, can be thought of as a rolled-up sheet of graphite capped on either end by half of a buckyball. Nanotubes have many potential applications. They can be spun into fibers that are stronger and lighter than steel, and they can also be made to carry electrical currents more efficiently than metals. The next several decades are likely to see many exciting applications of buckyballs and nanotubes. For their discovery of fullerenes, Kroto, Curl, and Smalley were awarded the 1996 Nobel Prize in Chemistry.

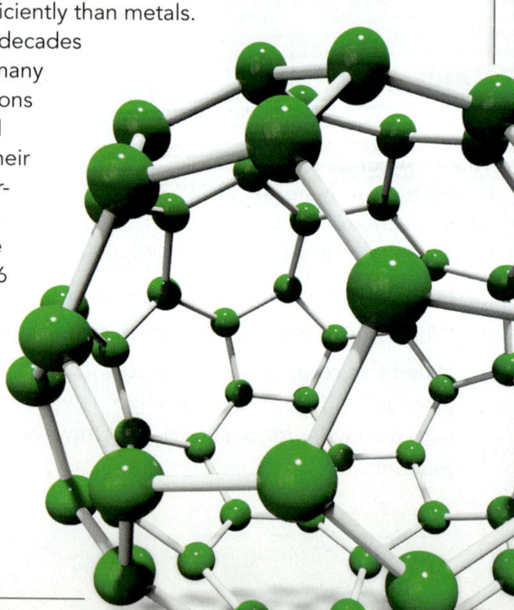

# REVIEW OF REACTIONS

## Reactions at the Benzylic Position

### Oxidation

$Na_2Cr_2O_7$ / $H_2SO_4$, $H_2O$ → OH (with O above)

1) $KMnO_4$, $H_2O$, heat
2) $H_3O^+$

### Free-Radical Bromination

NBS / Heat → Br

### Substitution Reactions

Br → $H_2O$ / $S_N1$ → OH + HBr

Br → NaOH / $S_N2$ → OH + NaBr

### Elimination Reactions

OH → Conc. $H_2SO_4$ / E1 → + $H_2O$

Br → NaOEt / E2 → + EtOH + NaBr

## Reduction

### Catalytic Hydrogenation

+ 3 $H_2$ → Ni / 100 atm / 150°C →

### Birch Reduction

R → Na, $CH_3OH$ / $NH_3$ → R

O → Na, $CH_3OH$ / $NH_3$ → O

# REVIEW OF CONCEPTS AND VOCABULARY

**SECTION 18.1**

- Derivatives of benzene are called **aromatic** compounds, regardless of whether they are fragrant or odorless.

**SECTION 18.2**

- Monosubstituted derivatives of benzene are named systematically using benzene as the parent and listing the substituent as a prefix.
- IUPAC also accepts many common names for monosubstituted benzenes.
- When a benzene ring is a substituent, it is called a **phenyl group**.
- Disubstituted derivatives of benzene can be differentiated by the use of the descriptors ***ortho***, ***meta***, and ***para*** or by the use of locants.
- Polysubstituted derivatives of benzene are named using locants. Common names can be used as parents.

### SECTION 18.3

- Benzene is comprised of a ring of six identical C—C bonds, each of which has a bond order of 1.5.
- No single Lewis structure adequately describes the structure of benzene. Resonance structures are required.

### SECTION 18.4

- Benzene exhibits unusual stability. It does not react with bromine in an addition reaction.
- The *stabilization energy* of benzene can be measured by comparing heats of hydrogenation.
- The stability of benzene can be explained with MO theory. The six $\pi$ electrons all occupy bonding MOs.
- The presence of a fully conjugated ring of $\pi$ electrons is not the sole requirement for aromaticity. The requirement for an odd number of electron pairs is called **Hückel's rule**.
- Cyclobutadiene is **antiaromatic**; cyclooctatetraene adopts a tub-shaped conformation and is nonaromatic.
- **Frost circles** accurately predict the relative energy levels of the MOs in a conjugated ring system.

### SECTION 18.5

- A compound is aromatic if it contains a ring comprised of continuously overlapping *p* orbitals and if it has a Hückel number of $\pi$ electrons in the ring.
- Compounds that fail the first criterion are called **nonaromatic**.
- Compounds that satisfy the first criterion but have $4n$ electrons (rather than $4n+2$) are antiaromatic.
- **Annulenes** are compounds consisting of a single ring containing a fully conjugated $\pi$ system. Due to steric hindrance, [10]annulene and [14]annulene do not meet the first criterion and are nonaromatic.
- The cyclopentadienyl anion exhibits aromatic stabilization, as does the tropylium cation.
- Cyclic compounds containing hetereoatoms, such as S, N, and O, are called **heterocycles**.
- The lone pair in pyridine is localized and does not participate in resonance, while the lone pair in pyrrole is delocalized and participates in aromaticity.
- Hückel's rule ($4n+2$) can only be applied to monocyclic compounds.
- Many **polycyclic aromatic hydrocarbons (PAHs)** are known to be stable.

### SECTION 18.6

- Any carbon atom attached directly to a benzene ring is called a **benzylic position**.
- Alkylbenzenes are oxidized at the benzylic position by chromic acid or potassium permanganate.
- Free-radical bromination occurs readily at benzylic positions.
- Benzylic halides readily undergo $S_N1$, $S_N2$, E1, and E2 reactions.

### SECTION 18.7

- Under certain conditions, a vinyl group can be selectively hydrogenated in the presence of an aromatic ring.
- In a **Birch reduction**, the aromatic moiety is reduced to give a nonconjugated diene. The carbon atom connected to an alkyl group is not reduced, while the carbon atom connected to an electron-withdrawing group is reduced.

### SECTION 18.8

- Aromatic compounds generally produce IR signals in five distinctive regions of the IR spectrum.
- Benzylic protons produce $^1H$ NMR signals between 2 and 3 ppm, while aromatic protons produce a characteristic signal (usually a multiplet) around 7 ppm.
- The *sp*$^2$-hybridized carbon atoms of an aromatic ring produce $^{13}C$ NMR signals between 100 and 150 ppm.

## SKILLBUILDER REVIEW

### 18.1 NAMING A POLYSUBSTITUTED BENZENE

**STEP 1** Identify and name the parent.

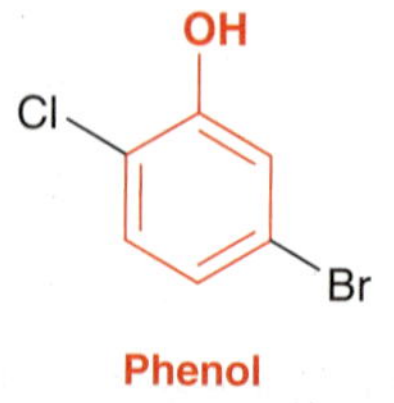

Cl, OH, Br — Phenol

**STEP 2** Identify the substituents.

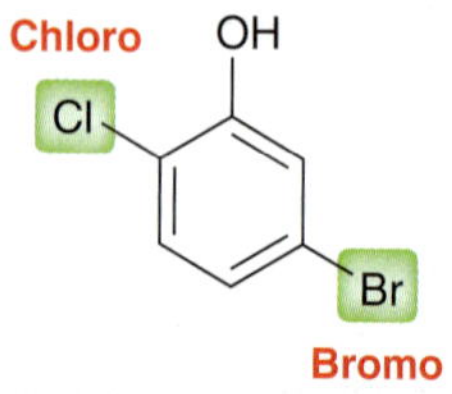

Chloro (Cl), OH, Bromo (Br)

**STEP 3** Assign locants.

Correct: OH 1, Cl 2, 3, 4, 5 Br

Incorrect: OH 1, 2, 3 Br, 4, 5, Cl 6

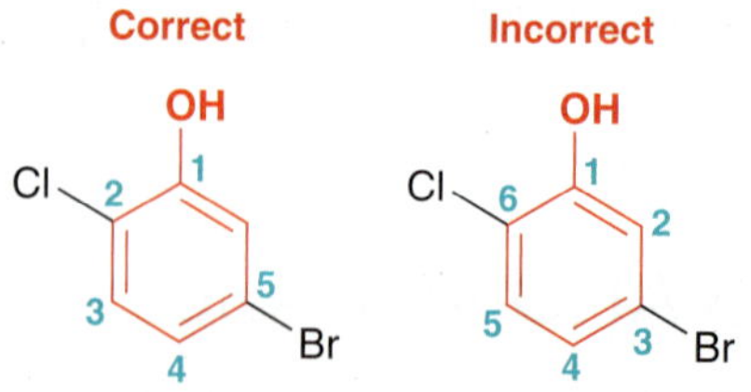

**STEP 4** Assemble the substituents alphabetically, with locants.

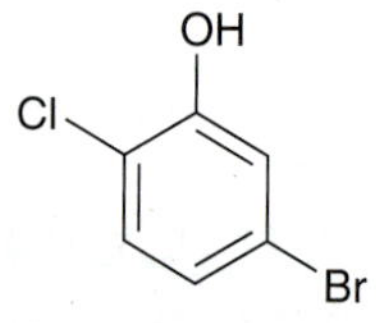

5-Bromo-2-chlorophenol

Try Problems 18.1–18.5, 18.28, 18.29, 18.33

**18.2** DETERMINING WHETHER A STRUCTURE IS AROMATIC, NONAROMATIC, OR ANTIAROMATIC

**STEP 1** Does the compound contain a ring comprised of continuously overlapping *p* orbitals?

**Examples that fail:**

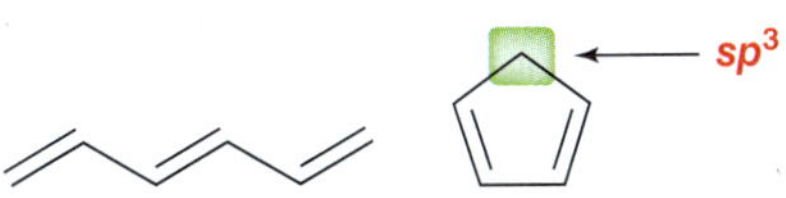


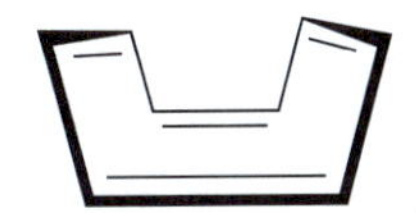


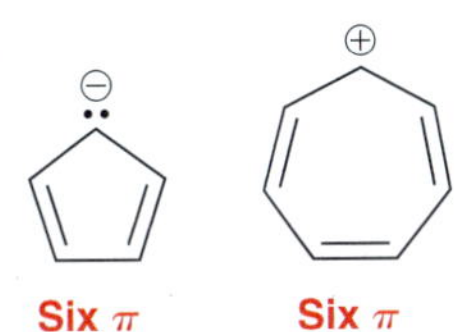


**STEP 2** Is there a Hückel number of $\pi$ electrons in the ring?

**Examples that pass:**

Six $\pi$ electrons

Six $\pi$ electrons

**STEP 3** Decision tree:

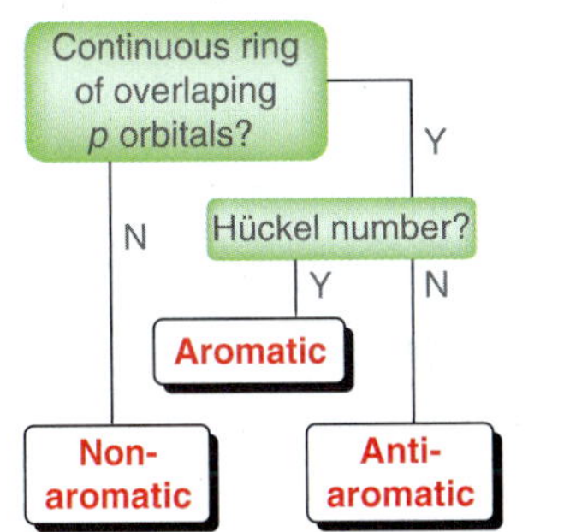


Try Problems 18.11–18.14, 18.34a,c,e, 18.36e, 18.52

**18.3** DETERMINING WHETHER A LONE PAIR PARTICIPATES IN AROMATICITY

**A lone pair can occupy the following atomic orbitals:**

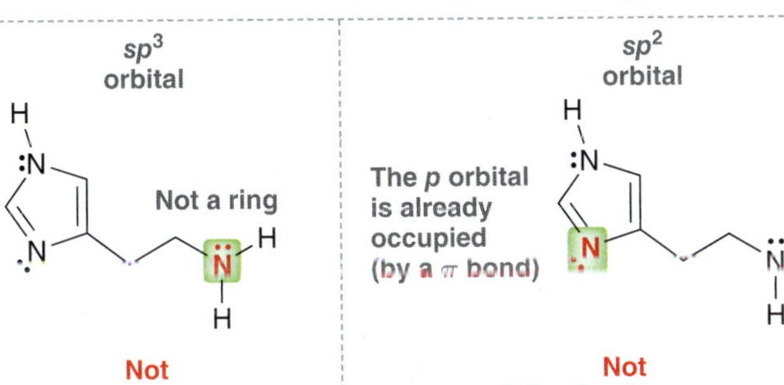


***p*** **orbital**

By occupying a *p* orbital, this lone pair establishes aromaticity (6 $\pi$ electrons)

**Participating in resonance**

Try Problems 18.15–18.17, 18.34b,d, 18.36c,d, 18.38, 18.41, 18.44, 18.62, 18.64

**18.4** MANIPULATING THE SIDE CHAIN OF AN AROMATIC COMPOUND

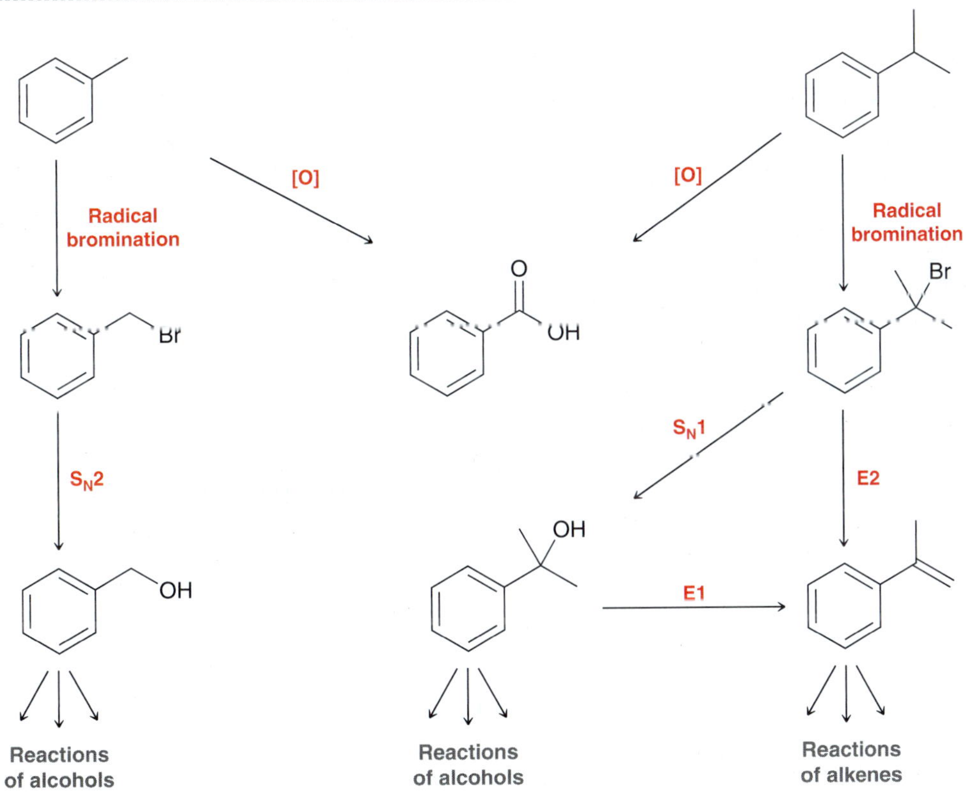


Try Problems 18.20–18.23, 18.47, 18.56

### 18.5 PREDICTING THE PRODUCT OF A BIRCH REDUCTION

**STEP 1** Identify whether each substituent is electron donating or electron withdrawing.

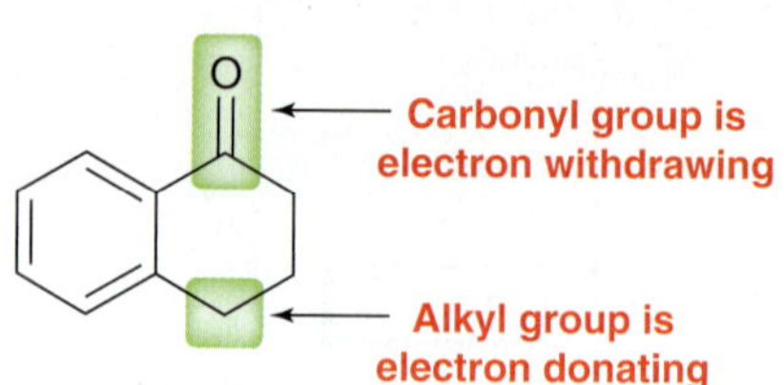


**STEP 2** Identify which carbon atoms are reduced.

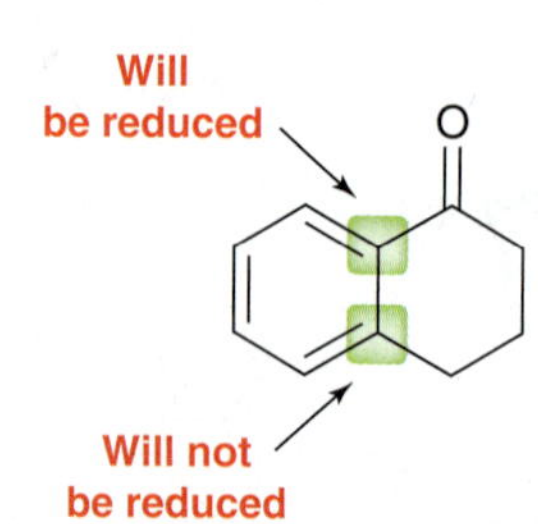


**STEP 3** Draw the product. Remember that the two positions not reduced must be 1,4 to each other.

O — $\xrightarrow[NH_3]{Na,\ CH_3OH}$ — O

Try Problems **18.24, 18.25, 18.49, 18.50**

## PRACTICE PROBLEMS

Note: Most of the Problems are available within **WileyPLUS**, an online teaching and learning solution.

**18.28** Provide a systematic name for each of the following compounds:

(a) (b) (c) (d) (e)

**18.29** Draw a structure for each of the following compounds:

(a) *ortho*-Dichlorobenzene (b) Anisole

(c) *meta*-Nitrotoluene (d) Aniline

(e) 2,4,6-Tribromophenol (f) *para*-Xylene

**18.30** Draw structures for the eight constitutional isomers with molecular formula $C_9H_{12}$ that contain a benzene ring.

**18.31** Draw structures for all constitutional isomers with molecular formula $C_8H_{10}$ that contain an aromatic ring.

**18.32** Draw all aromatic compounds that have molecular formula $C_8H_9Cl$.

**18.33** The systematic name of TNT, a well-known explosive, is 2,4,6-trinitrotoluene (as seen in SkillBuilder 18.1). There are only five constitutional isomers of TNT that contain an aromatic ring, a methyl group, and three nitro groups. Draw all five of these compounds and provide a systematic name for each.

**18.34** Identify the number of π electrons in each of the following compounds:

(a) 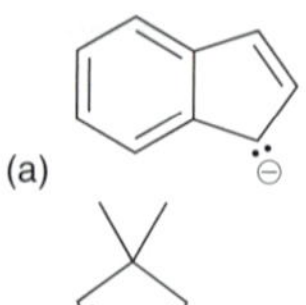

(b) 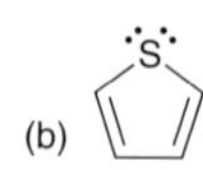


(c) 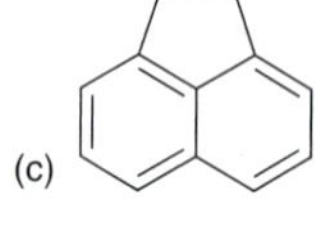

(d) 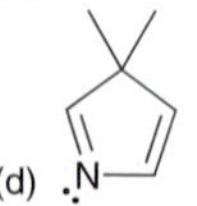


(e) 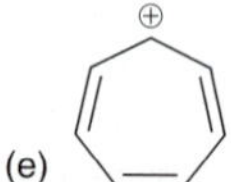

**18.35** Students often confuse cyclohexane and benzene.

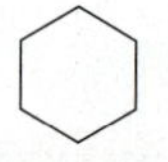

Cyclohexane

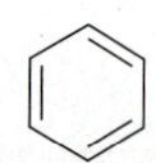

Benzene

In fact, these compounds have different properties, different geometry, and different reactivity. Each of these compounds also has a unique set of terminology. For each of the following terms, identify whether it is used in reference to benzene or to cyclohexane:

(a) *meta* (b) Frost circle

(c) $sp^2$ (d) Chair

(e) *ortho* (f) $sp^3$

(g) Resonance (h) π Electrons

(i) *para* (j) Ring flip

(k) Boat

**18.36** Identify which of the following compounds are aromatic:

(a)

(b) 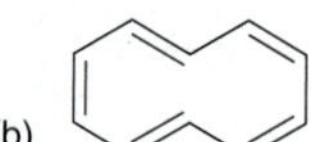

(c)

(d) 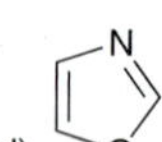


(e) 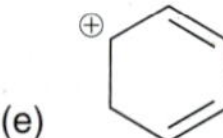

**18.37** Firefly luciferin is the compound that enables fireflies to glow.

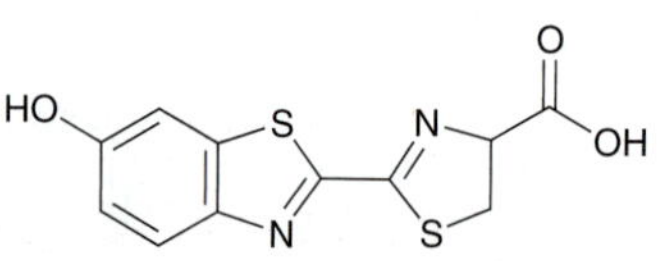


Firefly luciferin

(a) The structure exhibits three rings. Identify which of the rings are aromatic.

(b) Identify which lone pairs are participating in resonance.

**18.38** Identify each of the following compounds as aromatic, nonaromatic, or antiaromatic. Explain your choice in each case.

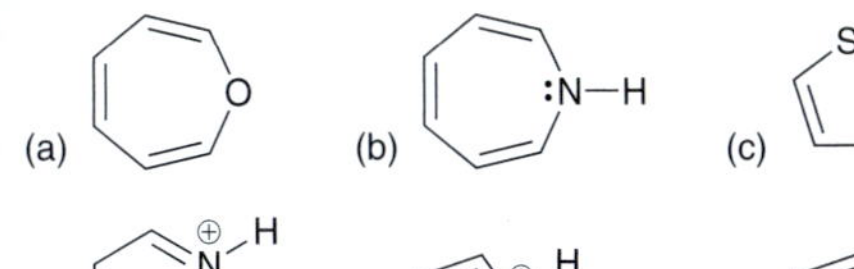

(a) (b) (c) (d)

(e) (f) (g) (h)

**18.39** Consider the structures of the following alkyl chlorides:

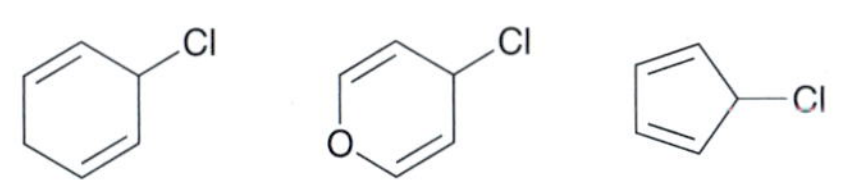


(a) Which compound would you expect to undergo an $S_N1$ process most readily? Justify your choice.

(b) Which compound would you expect to undergo an $S_N1$ process least readily? Justify your choice.

**18.40** Which of the following compounds would you expect to be most acidic? Justify your choice.

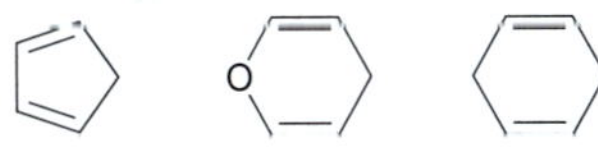


**18.41** Identify which of the following compounds is expected to be a stronger base. Justify your choice.

**18.42** Draw a Frost circle for the following cation and explain the source of instability of this cation:

**18.43** Do you expect the following dianion to exhibit aromatic stabilization? Explain.

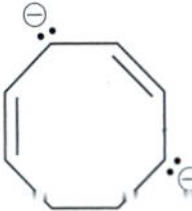

**18.44** Would you expect the following compound to be aromatic? Justify your answer.

N–O, N, OR

**18.45** Diphenylmethane exhibits two aromatic rings, which achieve coplanarity in the highest energy conformation. Explain.

**Diphenylmethane**

**18.46** The following two drawings are resonance structures of one compound:

But the following two drawings are not resonance structures:

**Not resonance structures**

They are, in fact, two different compounds. Explain.

**18.47** Predict the major product of the following reactions.

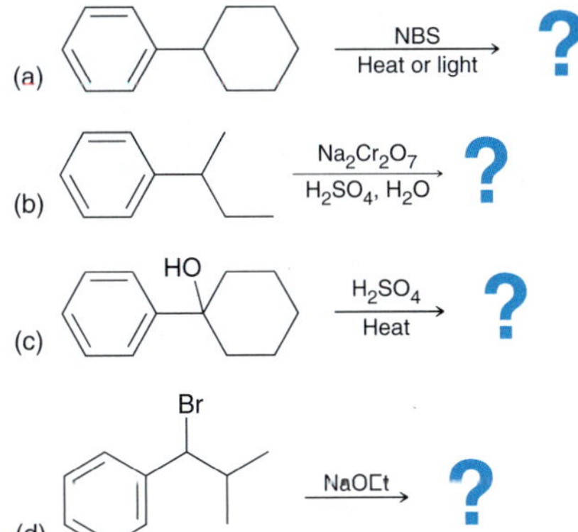


**18.48** How many signals do you expect in the $^{13}C$ NMR spectrum of each of the following compounds?

(a) (b) Br (c) (d)

**18.49** Predict the product of the following reaction and propose a mechanism for its formation:

Na, $CH_3OH$ / $NH_3$ ?

**18.50** One of the constitutional isomers of xylene was treated with sodium, methanol, and ammonia to yield a product that exhibited five signals in its $^{13}C$ NMR spectrum. Identify which constitutional isomer of xylene was used as the starting material.

**18.51** Consider the following two compounds:

How would you distinguish between them using:

(a) IR spectroscopy?

(b) $^1H$ NMR spectroscopy?

(c) $^{13}C$ NMR spectroscopy?

**18.52** Explain how the following two compounds can have the same conjugate base. Is this conjugate base aromatic?

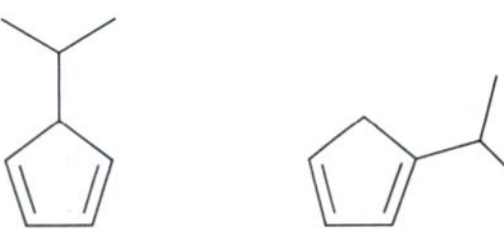

# INTEGRATED PROBLEMS

**18.53** Compare the electrostatic potential maps for cycloheptatrienone and cyclopentadienone.

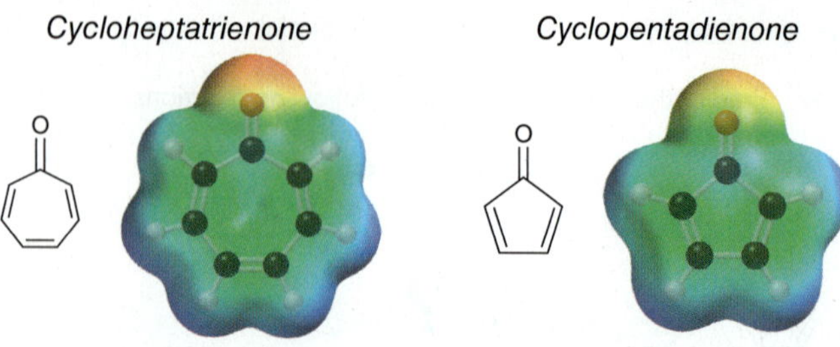


Both of these maps were created using the same color scale so they can be compared. Notice the difference between the oxygen atoms in these two compounds. There is more partial negative character on the oxygen in the first compound (cycloheptatrienone). Can you offer an explanation for this difference?

**18.54** Azulene exhibits an appreciable dipole moment, and an electrostatic potential map indicates that the five-membered ring is electron rich (at the expense of the seven-membered ring).

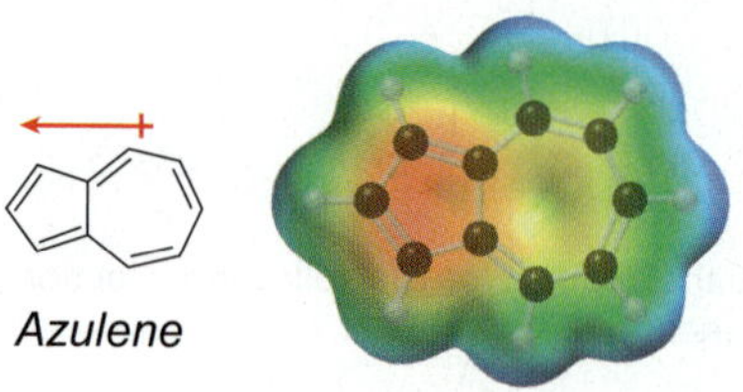


(a) In Chapter 2, we saw that a resonance structure will be insignificant if it has carbon atoms with opposite charges (C− and C+). Azulene represents an exception to this rule, because some resonance structures (with C− and C+) exhibit aromatic stabilization. With this in mind, draw resonance structures of azulene and use them to explain the observed dipole moment.

(b) Based on your explanation, determine which compound is expected to exhibit a larger dipole moment.

**18.55** Propose an efficient synthesis for the following transformation:

Br ⟶

**18.56** Propose a plausible mechanism for the following transformation:

OH $\xrightarrow{HCl}$ Cl

**18.57** Identify the structure of a compound with molecular formula $C_9H_{10}O_2$ that exhibits the following spectral data:

(a) IR: 3005 $cm^{-1}$, 1676 $cm^{-1}$, 1603 $cm^{-1}$

(b) $^1H$ NMR: 2.6 ppm (singlet, I = 3H), 3.9 ppm (singlet, I = 3H), 6.9 ppm (doublet, I = 2H), 7.9 ppm (doublet, I = 2H)

(c) $^{13}C$ NMR: 26.2, 55.4, 113.7, 130.3, 130.5, 163.5, 196.6 ppm

**18.58** Propose an efficient synthesis for each of the following transformations:

(a) ⟶ O, H

(b) ⟶ O

(c) OH ⟶ OH

(d) ⟶

**18.59** A compound with molecular formula $C_{11}H_{14}O_2$ exhibits the following spectra ($^1H$ NMR, $^{13}C$ NMR, and IR). Identify the structure of this compound.

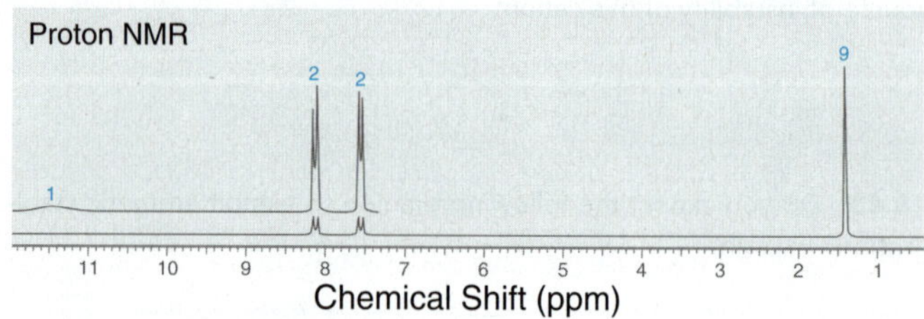


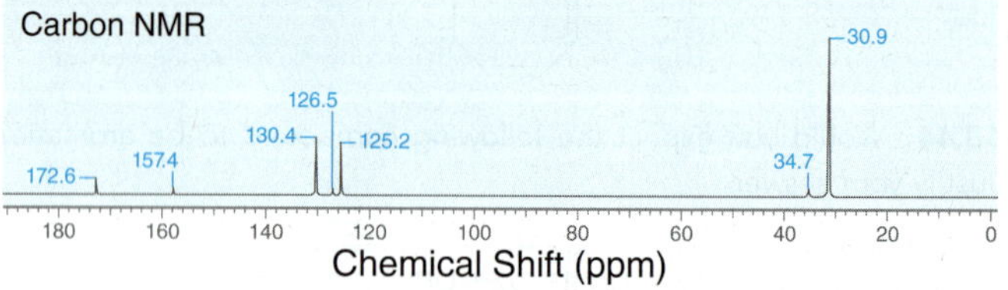


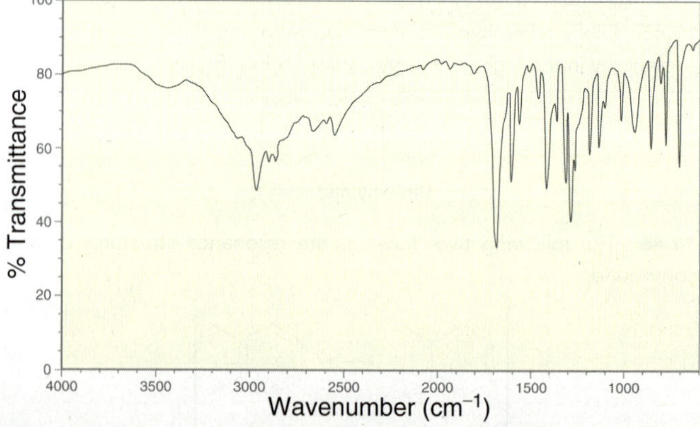

**18.60** A compound with molecular formula $C_9H_{10}O$ exhibits the following spectra ($^1H$ NMR, $^{13}C$ NMR, and IR). Identify the structure of this compound.

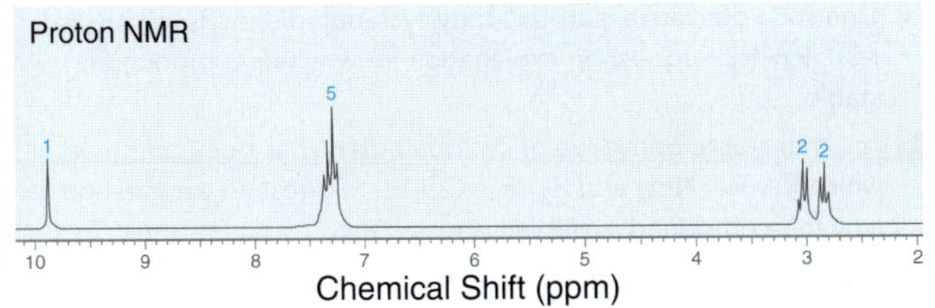


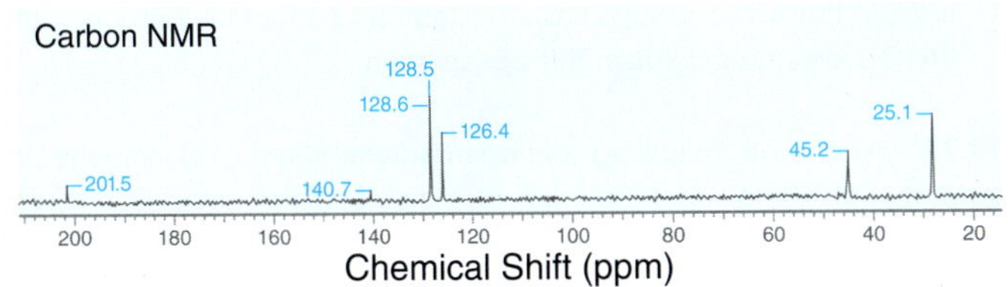


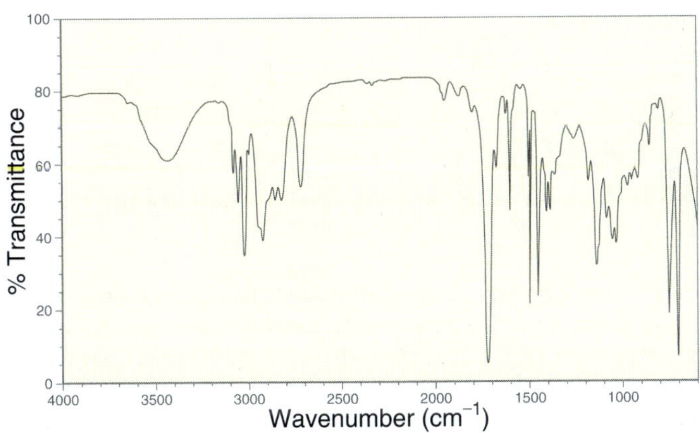


**18.61** Below are two hypothetical compounds.

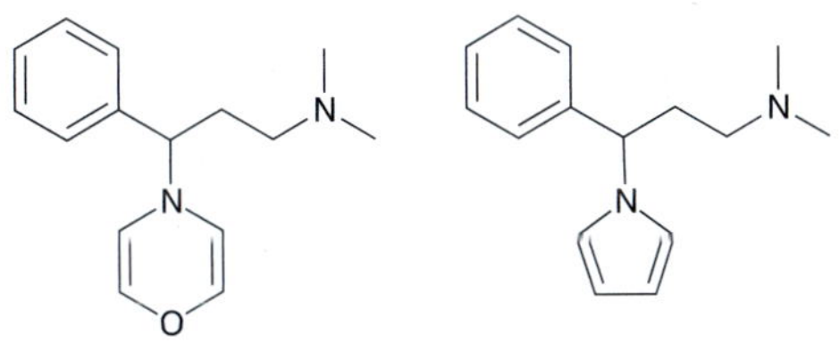

(a) Which compound would you expect to hold greater promise as a potential antihistamine? Explain your choice.

(b) Do you expect the compound you chose [in part (a)] to exhibit sedative properties? Explain your reasoning.

**18.62** Would you expect the following compound to be aromatic? Explain your answer.

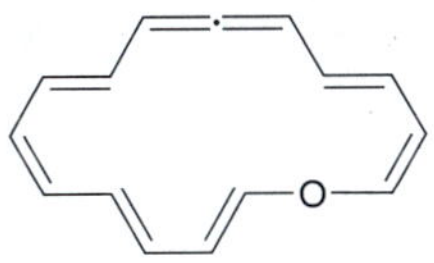

**18.63** Compounds **A**, **B**, **C**, and **D** are constitutionally isomeric, aromatic compounds with molecular formula $C_8H_{10}$. Deduce the structure of compound **D** using the following clues:

- The $^1H$ NMR spectrum of compound **A** exhibits two upfield signals as well as a multiplet near 7 ppm (with I = 5).
- The $^{13}C$ NMR spectrum of compound **B** exhibits four signals.
- The $^{13}C$ NMR spectrum of compound **C** exhibits only three signals.

**18.64** The following two compounds each exhibit two heteroatoms (one nitrogen atom and one oxygen atom):

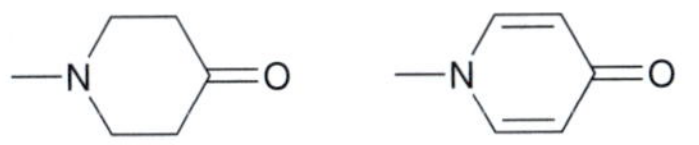

**Compound A** **Compound B**

In compound **A**, the lone pair on the nitrogen atom is more likely to function as a base. However, in compound **B**, a lone pair on the oxygen atom is more likely to function as a base. Explain this difference.

**18.65** Propose an efficient synthesis for the following transformation:

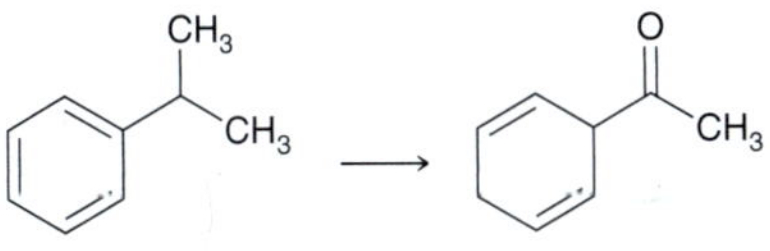

**18.66** Using toluene and acetylene as your only sources of carbon atoms, show how you would prepare the following compound:

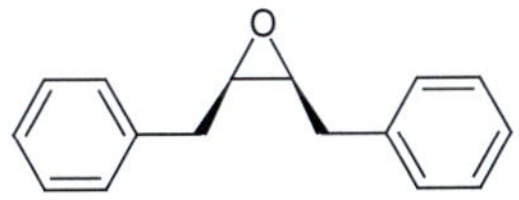

**18.67** Due to their potential application in organic electronic devices, such as television screens, a large number of compounds containing quinoidal pi systems (such as compound **A**) have been prepared. Reduction of **A** gives **B**, as shown (*Tetrahedron Lett.* **2012,** *53,* 5385–5388):

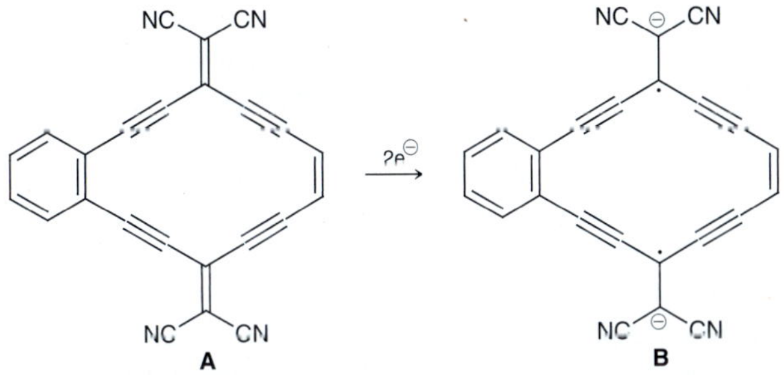

**A** **B**

(a) Determine whether the 14-membered ring in **B** is aromatic.

(b) Draw resonance structures showing how the negative charges in **B** are delocalized.

(c) Draw at least three resonance structures that demonstrate the delocalized nature of the unpaired electrons in **B**.

# CHALLENGE PROBLEMS

**18.68** Compound **1** contains a tetrazole ring (a five-membered ring containing four nitrogen atoms), while its constitutional isomer, compound **2**, exhibits an azido group ($—N_3$). There is evidence that compounds **1** and **2** rapidly interconvert, and exist in equilibrium with one another (*J. Am. Chem. Soc.* **1959,** *81,* **4671–4673**). Of the four nitrogen atoms in these compounds, only one of them remains incorporated in a ring in compound **2**. Determine whether this nitrogen atom undergoes a change in hybridization state as a result of the isomerization process. Explain.

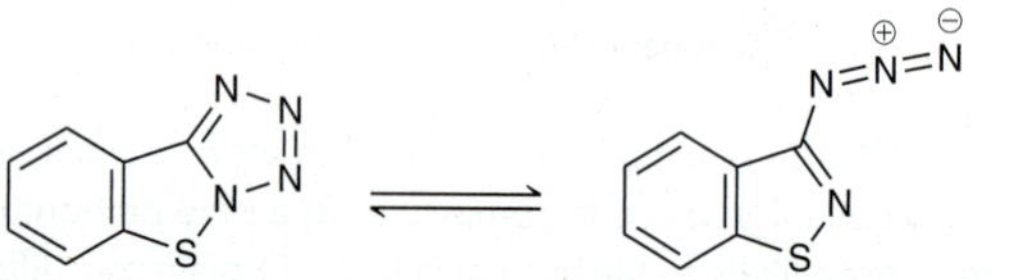

**1** ⇌ **2**

**18.69** Studies modeling chemical reactions in the atmosphere of Saturn's moon Titan suggest that the compound 1,2,5,6-tetracyanocyclooctatetraene can form through one of a series of gas-phase reactions mediated by monocationic magnesium ($Mg^{+\bullet}$). This organic compound forms a complex in which the lone pairs on all four nitrogen atoms interact with a single $Mg^{+\bullet}$ ion (*J. Am. Chem. Soc.* **2005,** *127,* **13070–13078**).

(a) Sketch the structure of this complex as described. Use dotted lines to show the interactions between the lone pairs and the magnesium ion.

(b) Explain why 1,2,4,5-tetracyanobenzene would not be able to form a comparable complex.

**18.70** Consider the structures for naphthalene and phenanthrene.

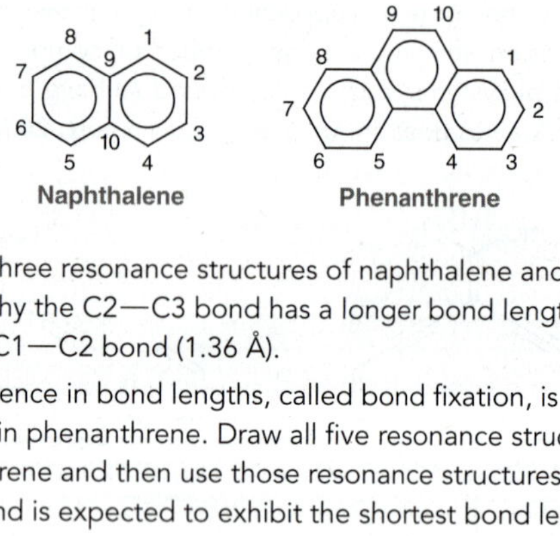

**Naphthalene** **Phenanthrene**

(a) Draw all three resonance structures of naphthalene and then explain why the C2—C3 bond has a longer bond length (1.42 Å) than the C1—C2 bond (1.36 Å).

(b) The difference in bond lengths, called bond fixation, is very pronounced in phenanthrene. Draw all five resonance structures of phenanthrene and then use those resonance structures to predict which bond is expected to exhibit the shortest bond length.

(c) Treatment of phenanthrene with $Br_2$ affords a compound with molecular formula $C_{14}H_{10}Br_2$ (*J. Am. Chem. Soc.* **1985,** *107,* **6678–6683**). Draw the structure of this product.

**18.71** Cyclobutadiene is not stable at room temperature. Upon formation, it rapidly dimerizes via a Diels-Alder reaction. However, several derivatives of cyclobutadiene are known to be stable at room temperature (*Angew Chem. Int. Ed.* **1988,** *27,* **1437–1455**):

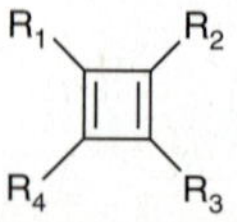

(a) One such derivative is tri-*tert*-butylcyclobutadiene ($R_1{=}R_2{=}R_3{=}$ *t*-Bu; $R_4{=}$H). Suggest an explanation for why this compound is stable.

(b) Another stable derivative of cyclobutadiene has been prepared in which $R_1{=}R_3{=}NEt_2$ and $R_2{=}R_4{=}CO_2Et$. Suggest an explanation for why this compound is stable.

(c) In part (b), we saw that cyclobutadiene is stabilized by the presence of neighboring $NEt_2$ and $CO_2Et$ groups. In contrast, the presence of these two groups has a destabilizing effect on systems with $4n+2$ π electrons. Explain this observation.

**18.72** When the following cyclopentadiene-fused [14]annulene is treated with KH, a green solution of a stable cyclopentadiene anion results. This transformation causes a dramatic change in part of the $^1$H NMR spectrum: the methyl groups shift from −3.9 ppm to −1.8 ppm (*J. Org. Chem.* **2004,** *69,* **549–554**). Provide an explanation for this observation.

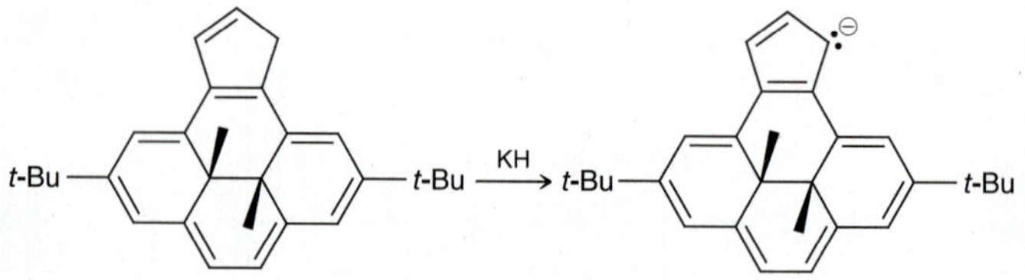

**18.73** The following compound, isolated from a New Zealand sea squirt, demonstrates activity against a malarial strain that is resistant to other treatments (*J. Nat. Prod.* **2011,** *74,* **1972–1979**). Consider the relative basicity of each nitrogen atom in this structure and draw the product expected when this compound is treated with two equivalents of a strong acid (e.g., HCl).

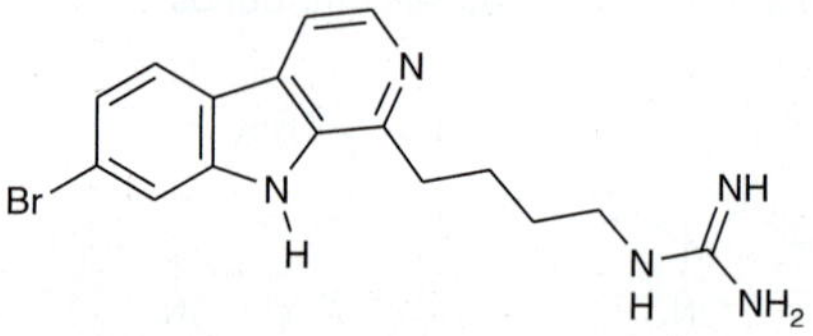

**18.74** In a mass spectrometric study of nitrogenous aromatic compounds, a peak at $m/z = 39$ was found for 17 of the 20 compounds investigated, including the 4 compounds in the first row below, with a relative intensity ranging from 5 to 84% of the base peak. This peak was absent in the mass spectrum of each of the three compounds in the bottom row (*J. Org. Chem.* **1964,** *29,* **2065–2066**). Propose a structure for the cationic fragment with $m/z = 39$ and an explanation consistent with the information provided.

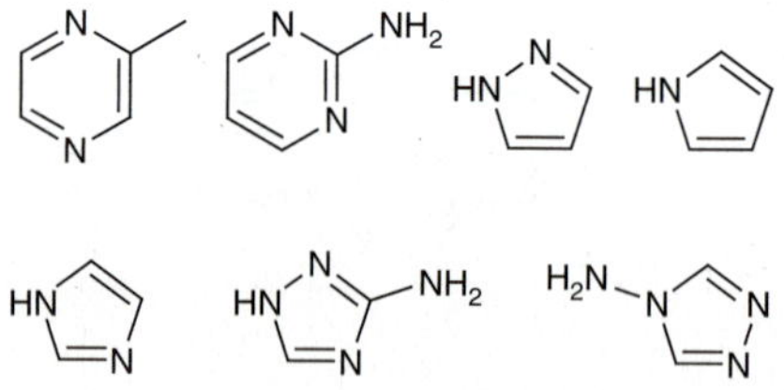

**18.75** Fluorenyldiene **1** is a highly conjugated, planar molecule that is readily oxidized to dication **2** (*J. Org. Chem.* **2002,** *67,* 7029–7036).

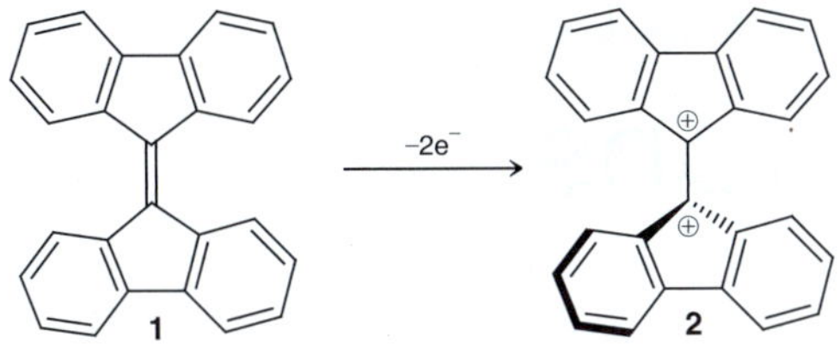

(a) Draw at least four other resonance structures of dication **2** and identify why the resonance structure shown here is the greatest contributor to the overall resonance hybrid, despite the presence of neighboring like charges (which is generally a destabilizing factor).

(b) Explain why the lowest energy conformation for **2** is not planar and instead exhibits a perpendicular arrangement between the two ring systems.

**18.76** Numerous herbicides and fungicides are known to contain an acetylenic group. For example, compound **A** is a pyrimidinone herbicide that functions by inhibiting the accumulation of both chlorophyll and β-carotene. During a synthesis of **A**, shown below, an inactive product was also formed (compound **B**) which was found to be a constitutional isomer of **A**, also containing an acetylenic group. The ratio of **A:B** was 1:4 (*Bioorg. Med. Chem.* **2009,** *17,* 4047–4063).

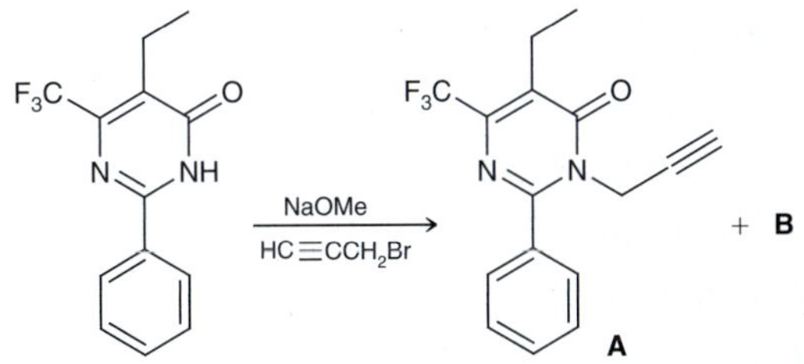

(a) Draw a plausible mechanism for the formation of compound **A**.

(b) Draw the likely structure of compound **B** and provide a mechanism for its formation.

(c) Explain why **B** is favored over **A** in this process.

# 19 Aromatic Substitution Reactions



## DID YOU EVER WONDER...

**what food coloring is? Look at the ingredients of Fruity Pebbles and you will find compounds such as Red #40 and Yellow #6. What are these substances that can be found in so many of the foods that we regularly eat?**

In this chapter, we will learn about the most common reactions of aromatic rings, with the main focus on electrophilic aromatic substitution reactions. During the course of our discussion, we will see that many common food colorings are aromatic compounds that are synthesized using this reaction type, and we will also see how extensive research of aromatic compounds in the early twentieth century made significant contributions to the field of medicine.

This chapter will paint Fruity Pebbles in a whole new light.

## DO YOU REMEMBER?

*Before you go on, be sure you understand the following topics.*
*If necessary, review the suggested sections to prepare for this chapter.*

- Resonance Structures (Sections 2.7–2.11)
- Lewis Acids (Section 3.9)
- Retrosynthetic Analysis (Section 12.5)
- Delocalized Lone Pairs (Section 2.12)
- Reading Energy Diagrams (Section 6.6)
- Aromaticity and Nomenclature of Aromatic Compounds (Sections 18.1–18.4)

Take the DO YOU REMEMBER? QUIZ in **WileyPLUS** to check your understanding

## 19.1 Introduction to Electrophilic Aromatic Substitution

In the previous chapter, we explored the remarkable stability of benzene. Specifically, we saw that while alkenes undergo an addition reaction when treated with bromine, benzene is inert under the same conditions.

+ $Br_2$ → Br, Br + Enantiomer

+ $Br_2$ ✗→ Br, Br + Enantiomer

Curiously, though, when Fe (iron) is introduced into the mixture, a reaction does in fact take place, although the product is not what we might have expected.

+ $Br_2$ —Fe→ Br (75%)

Rather than an addition reaction taking place, the observed reaction is an **electrophilic aromatic substitution** reaction in which one of the aromatic protons is replaced by an electrophile, and the aromatic moiety is preserved. In this chapter, we will see many other groups that can also be installed on an aromatic ring via an electrophilic aromatic substitution reaction.

Br Cl $NO_2$ $SO_3H$ R O R

## 19.2 Halogenation

Recall from Section 9.8 that during bromination of an alkene, $Br_2$ functions as an electrophile.

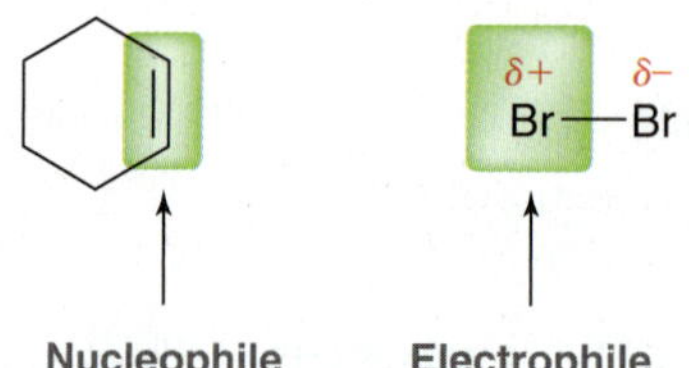


As it approaches the π electron cloud of the alkene, $Br_2$ becomes temporarily polarized, rendering one of the bromine atoms electrophilic (δ+). This bromine atom becomes sufficiently electrophilic to react with the alkene but is not sufficiently electrophilic to react with benzene. The presence of iron (Fe) in the reaction mixture enhances the electrophilicity of this bromine atom. To understand how iron accomplishes this task, we must recognize that iron itself is not the real catalyst. Rather, it first reacts with $Br_2$ to generate iron tribromide ($FeBr_3$).

$$2\ Fe + 3\ Br_2 \longrightarrow 2\ FeBr_3$$

Iron tribromide, a Lewis acid, is the real catalyst in the reaction between benzene and bromine. Specifically, $FeBr_3$ interacts with $Br_2$ to form a complex which reacts as if it were $Br^+$.

This complex serves as an electrophilic agent that achieves bromination of the aromatic ring via a two-step process (Mechanism 19.1).

### MECHANISM 19.1 BROMINATION OF BENZENE

Nucleophilic attack

Proton transfer

+ $FeBr_3$ + HBr

In the first step, the aromatic ring functions as a nucleophile, forming an intermediate sigma complex

In the second step, the sigma complex is deprotonated, restoring aromaticity

Sigma complex

In the first step, the aromatic moiety functions as a nucleophile and attacks the electrophilic agent, generating a positively charged intermediate called a **sigma complex**, or **arenium ion**, which is resonance stabilized. This step requires an input of energy because it involves the temporary loss of aromatic stabilization. The loss of stabilization occurs because the sigma complex is not aromatic—it does not possess a continuous system of overlapping *p* orbitals.

In the second step of the mechanism, the sigma complex is then deprotonated, thereby restoring aromaticity and regenerating the Lewis acid ($FeBr_3$). Notice that the Lewis acid is ultimately not consumed by the reaction and is, therefore, a catalyst. Aluminum tribromide ($AlBr_3$) is another common Lewis acid that can serve as a suitable alternative to $FeBr_3$.

$$+ \ Br_2 \xrightarrow{AlBr_3} \text{Br} \ + \ HBr$$

The observed substitution reaction is not accompanied by the formation of any addition products. Addition is not observed because it would involve a permanent loss of aromaticity, which is thermodynamically unfavorable (Figure 19.1). Notice that, overall, substitution is an exergonic process

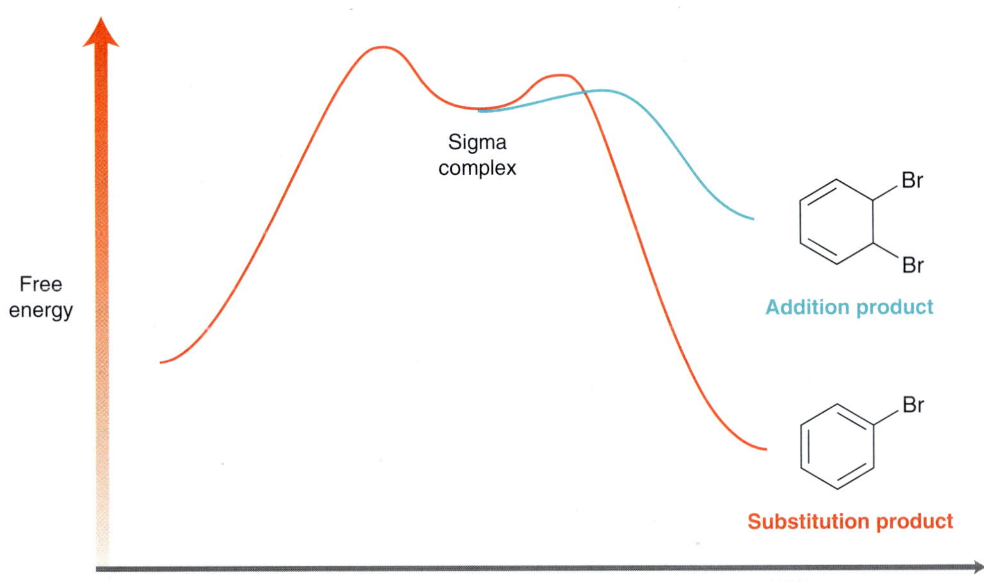


**FIGURE 19.1** An energy diagram comparing substitution and addition pathways for benzene.

(downhill in energy), while addition is an endergonic process (uphill in energy). For this reason, only substitution is observed.

A similar reaction occurs when chlorine is used rather than bromine. Chlorination of benzene is accomplished with a suitable Lewis acid, such as aluminum trichloride.

$$+ \ Cl_2 \xrightarrow{AlCl_3} \text{Cl} \ + \ HCl$$

**(85%)**

Chlorine reacts with $AlCl_3$ to form a complex, which reacts as if it were $Cl^+$.

$$Cl{-}Cl \ + \ AlCl_3 \rightleftharpoons Cl^{\oplus}{-}Cl{-}Al^{\ominus}Cl_3 \rightleftharpoons Cl^{\oplus} \ \ AlCl_4^{\ominus}$$

This complex is the electrophilic agent that achieves chlorination of the aromatic ring, as illustrated in Mechanism 19.2. This mechanism is directly analogous to the mechanism of bromination and involves the same two steps. In the first step, the aromatic moiety functions as a nucleophile and attacks the electrophilic agent, generating a sigma complex. Then, in the second step, the sigma complex is deprotonated, thereby restoring aromaticity and regenerating the Lewis acid ($AlCl_3$).

## MECHANISM 19.2 CHLORINATION OF BENZENE

$Cl{-}Cl \xrightarrow{AlCl_3} Cl{-}Cl^{\oplus}{-}AlCl_3^{\ominus}$

Nucleophilic attack

In the first step, the aromatic ring functions as a nucleophile, forming an intermediate sigma complex

Proton transfer

In the second step, the sigma complex is deprotonated, restoring aromaticity

+ $AlCl_3$

**Sigma complex**

Bromination and chlorination of aromatic rings are easily achieved, but fluorination and iodination are less common. Fluorination is a violent process and is difficult to control, while iodination is often slow, with poor yields in many cases. In Chapter 23, we will see more efficient methods for installing F or I on a benzene ring.

A variety of electrophiles ($E^+$) will react with a benzene ring, and we will explore many of them in the upcoming sections of this chapter. It will be helpful to realize that all of these reactions operate via the same general mechanism that has only two steps: (1) the aromatic ring functions as a nucleophile and attacks an electrophile to form a sigma complex followed by (2) deprotonation of the sigma complex to restore aromaticity (Mechanism 19.3).

## MECHANISM 19.3 A GENERAL MECHANISM FOR ELECTROPHILIC AROMATIC SUBSTITUTION

Nucleophilic attack

In the first step, the aromatic ring functions as a nucleophile and attacks $E^+$, forming an intermediate sigma complex

Proton transfer

:Base

In the second step, the sigma complex is deprotonated, restoring aromaticity

**Sigma complex**

## CONCEPTUAL CHECKPOINT

**19.1** When benzene is treated with $I_2$ in the presence of $CuCl_2$, iodination of the ring is achieved with modest yields. It is believed that $CuCl_2$ interacts with $I_2$ to generate $I^+$, which is an excellent electrophile. The aromatic ring then reacts with $I^+$ in an electrophilic aromatic substitution reaction. Draw a mechanism for the reaction between benzene and $I^+$. Make sure that your mechanism has two steps, and make sure to draw all of the resonance structures of the sigma complex.

$\xrightarrow[CuCl_2]{I_2}$

## medically speaking — Halogenation in Drug Design

Aromatic halogenation is a common technique used in drug design, as chemists attempt to modify the structure of a known drug to produce new drugs with enhanced properties. For example, consider the following three compounds:

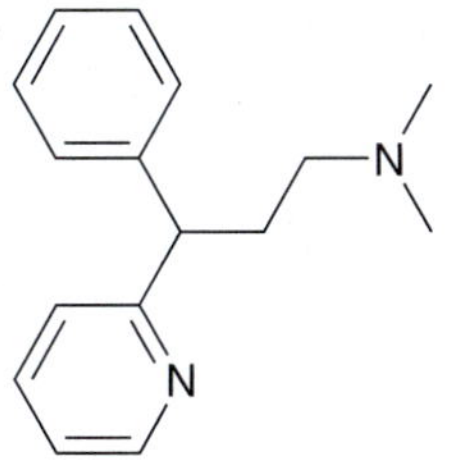

Pheniramine

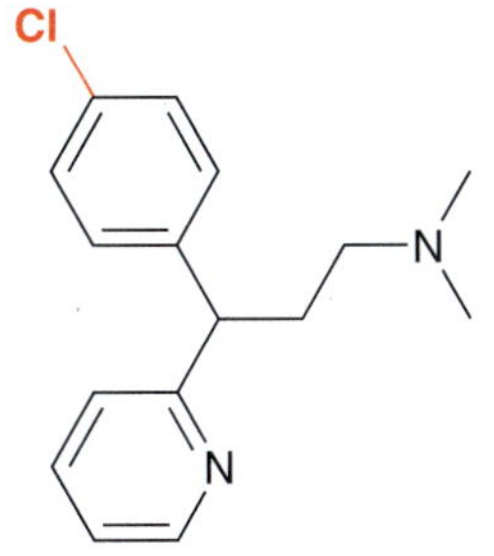

Chlorpheniramine

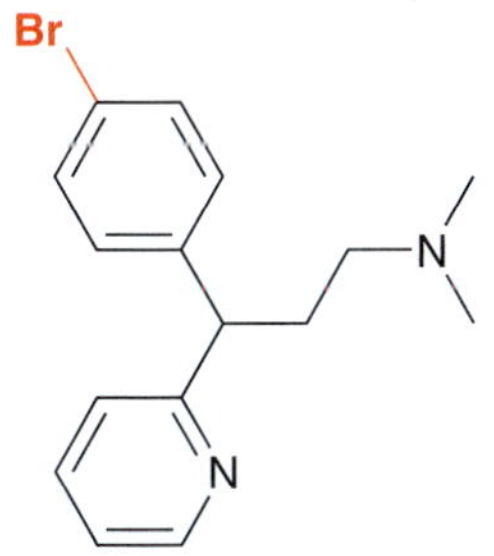

Brompheniramine

Pheniramine is an antihistamine (for a discussion on antihistamines, see Section 18.5). When a chlorine atom is installed in the *para* position of one of the rings, a new compound called chlorpheniramine is obtained. Chlorpheniramine is 10 times more potent than pheniramine and is marketed under the trade name Chlortrimeton. When a bromine atom is installed instead of chlorine, brompheniramine is obtained, which is marketed under the trade name Dimetane. This compound is one of the active ingredients in Dimetapp. It is similar in potency to Chlortrimeton, but its effects last almost twice as long.

Halogenation is a critical process in the design of many other types of drugs as well. For example, consider the antifungal agents chlotrimazole and econazole:

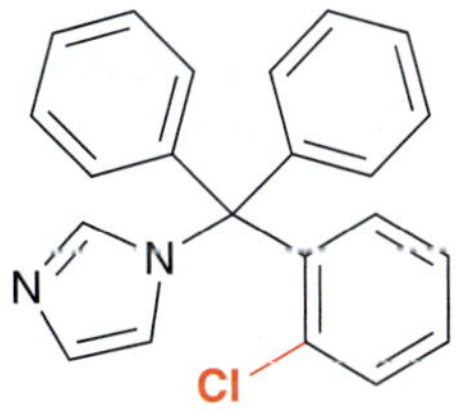

Chlotrimazole

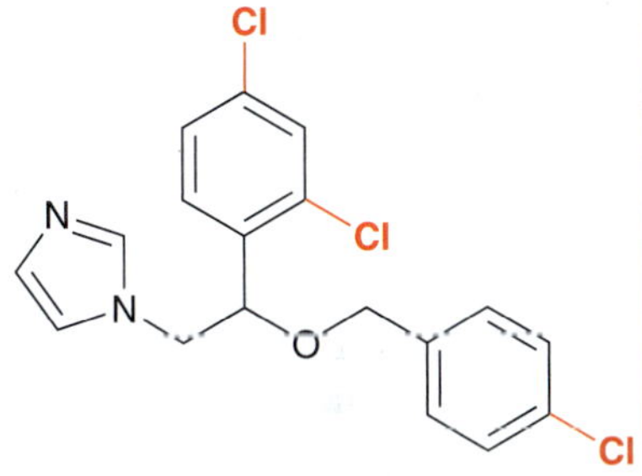

Econazole

Both of these compounds are examples of azole antifungal agents (azole is a five-membered aromatic ring containing two nitrogen atoms). Azole antifungal agents typically contain two or three additional aromatic rings, at least one of which is substituted with a halogen. Structure-activity studies have revealed that the presence of a halogen is critical for drug activity. Chlotrimazole is marketed under the trade name Lotrimin, and econazole is marketed under the trade name Spectazole. Notice that in both of these compounds the halogens are positioned in the *ortho* and *para* positions. In the later sections of this chapter, we will see why the *ortho* and *para* positions are more easily halogenated.

## 19.3 Sulfonation

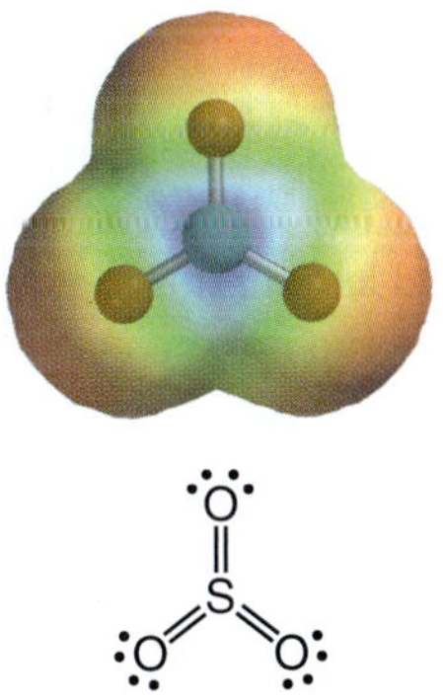

When benzene is treated with fuming sulfuric acid, a **sulfonation** reaction occurs and benzenesulfonic acid is obtained.

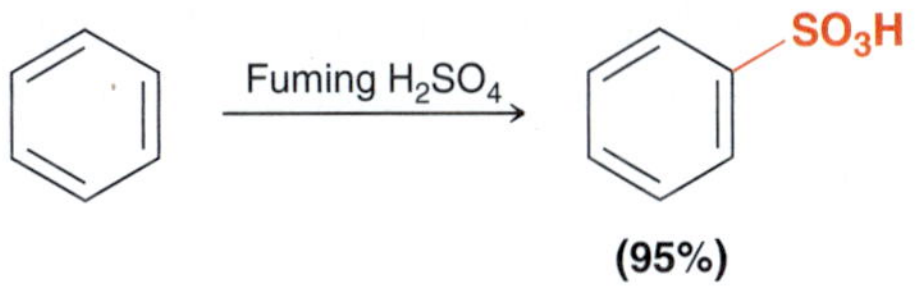

Fuming $H_2SO_4$ → $SO_3H$ (95%)

**FIGURE 19.2** An electrostatic potential map of sulfur trioxide.

Fuming sulfuric acid is a mixture of $H_2SO_4$ and $SO_3$ gas. Sulfur trioxide ($SO_3$) is a very powerful electrophile, as can be seen in the electrostatic potential map in Figure 19.2. This image illustrates that the sulfur atom is a site of low electron density (an electrophilic center). To understand the reason for this, we must explore the nature of S=O double bonds. Recall that a C=C double bond is formed from the overlap of *p* orbitals. The S=O double bond is also formed from the overlap of $p$ orbitals, but the overlap is less efficient because the $p$ orbitals are different sizes (Figure 19.3). The sulfur atom uses a $3p$ orbital (sulfur is in the third row of the periodic table), while the oxygen

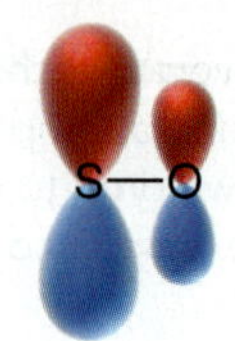


**FIGURE 19.3** The $p$ orbitals involved in an S=O bond.

atom uses a $2p$ orbital (oxygen is in the second row of the periodic table). The inefficient overlap of these orbitals suggests that we should consider the bond to be a single bond that exhibits charge separation ($S^+$ and $O^-$), rather than a double bond. Each of the S—O bonds in sulfur trioxide is highly polarized in this way, rendering the sulfur atom extremely electron poor and sufficiently electrophilic to react with benzene. The reaction involves the two steps that are characteristic of all electrophilic aromatic substitution reactions—a nucleophilic attack and a proton transfer (Mechanism 19.4). The product of these two steps exhibits a negative charge and is protonated in the presence of sulfuric acid.

## MECHANISM 19.4 SULFONATION OF BENZENE

Nucleophilic attack

In the first step, the aromatic ring functions as a nucleophile, forming an intermediate sigma complex

Proton transfer

In the second step, the sigma complex is deprotonated, restoring aromaticity

Proton transfer

The resulting anion is protonated

Sigma complex

The reaction between benzene and $SO_3$ is highly sensitive to the concentrations of the reagents and is, therefore, reversible. The reversibility of this process will be reexamined later in this chapter and will also be utilized heavily in the synthesis of polysubstituted aromatic compounds.

Concentrated fuming $H_2SO_4$

Dilute $H_2SO_4$

H → $SO_3H$

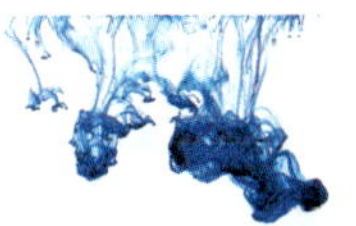

## practically speaking What Are Those Colors in Fruity Pebbles?

In the mid-nineteenth century, it was discovered that two aromatic moieties could be joined by an azo group (—N=N—) in a process called azo coupling:

(R = OH or NH$_2$)

Azo group

This process, which will be explored in more detail in Chapter 23, is believed to occur via an electrophilic aromatic substitution reaction:

Sigma complex (resonance stabilized)

$-H^+$

The resulting compound exhibits extended conjugation and is therefore colored (for more on the source of color, see Section 17.12).

By structurally modifying the starting materials (by placing substituents on the aromatic rings prior to azo coupling), a variety of products can be made, each exhibiting a unique color. Due to the variety of colors that can be prepared, a significant amount of research was aimed at designing compounds that could serve as fabric dyes. These compounds, called azo dyes, were produced in large quantities, and by the late nineteenth century, there was a very large market for them. While other types of dyes have been discovered since then, azo dyes still represent more than 50% of the synthetic dye market.

Among many other applications, azo dyes are currently used in paints, cosmetics, and food. Food colorings are regulated by the FDA (Food and Drug Administration) and include compounds such as Red #40 and Yellow #6, both of which are used in Fruity Pebbles:

**Red #40**

**Yellow #6**

Notice the presence of sulfonic acid groups (—$SO_3H$) in both of these compounds. These groups are necessary because they are easily deprotonated to give anions, rendering them water soluble. The sulfonic acid groups are introduced via the sulfonation process that we learned in this section.

Many azo dyes also contain nitro groups, such as Orange #1:

**Orange #1**

In the following section, we will learn how to install a nitro group on an aromatic ring.

## CONCEPTUAL CHECKPOINT

**19.2** Draw a mechanism for the following reaction. **Hint:** This reaction is the reverse of sulfonation, so you should read the sulfonation mechanism backward. Your mechanism should involve a sigma complex (positively charged).

$SO_3H$ — Dilute $H_2SO_4$ → H

**19.3** When benzene is treated with $D_2SO_4$, a deuterium atom replaces one of the hydrogen atoms. Propose a mechanism for this reaction. Once again, make sure that your mechanism involves a sigma complex.

$D_2SO_4$ → D

## 19.4 Nitration

When benzene is treated with a mixture of nitric acid and sulfuric acid, a **nitration** reaction occurs in which nitrobenzene is formed.

$HNO_3$ / $H_2SO_4$ → $NO_2$

**(95%)**

This reaction proceeds via an electrophilic aromatic substitution in which a **nitronium ion** ($NO_2^+$) is believed to be the electrophile. This strong electrophile is formed from the acid-base reaction that takes place between $HNO_3$ and $H_2SO_4$. Nitric acid functions as a base to accept a proton from sulfuric acid, followed by loss of water to produce a nitronium ion (Mechanism 19.5). It might seem strange that nitric acid functions as a base rather than an acid, but remember that acidity

## MECHANISM 19.5 NITRATION OF BENZENE

$-H_2O$

**Nitronium ion**

Nucleophilic attack

In the first step, the aromatic ring functions as a nucleophile, forming an intermediate sigma complex

**Sigma complex**

Proton transfer

In the second step, the sigma complex is deprotonated, restoring aromaticity is relative. Sulfuric acid is a much stronger acid than nitric acid, and it will protonate nitric acid when mixed together. The resulting nitronium ion then serves as an electrophile in an electrophilic aromatic substitution reaction.

This method can be used to install a nitro group on an aromatic ring. Once on the ring, the nitro group can be reduced to give an amino group ($NH_2$).

$NO_2$ — Fe or Zn / HCl → $NH_2$ **(95%)**

This provides us with a two-step method for installing an amino group on an aromatic ring: (1) nitration, followed by (2) reduction of the nitro group.

1) $HNO_3/H_2SO_4$ 2) Zn, HCl → $NH_2$

## CONCEPTUAL CHECKPOINT

**19.4** Draw a mechanism for the following reaction and make sure to draw all three resonance structures of the sigma complex:

$HNO_3$ / $H_2SO_4$ → $NO_2$

## 19.5 Friedel-Crafts Alkylation

In the previous sections, we have seen that a variety of electrophiles ($Br^+$, $Cl^+$, $SO_3$, and $NO_2^+$) will react with benzene in an electrophilic aromatic substitution reaction. In this section and the next, we will explore electrophiles in which the electrophilic center is a carbon atom.

The **Friedel-Crafts alkylation**, discovered by Charles Friedel and James Crafts in 1877, makes possible the installation of an alkyl group on an aromatic ring.

Cl / $AlCl_3$ → **(66%)**

Although an alkyl halide such as 2-chlorobutane is, by itself, electrophilic, it is not sufficiently electrophilic to react with benzene. However, in the presence of a Lewis acid, such as aluminum trichloride, the alkyl halide is converted into a carbocation.

**Carbocation**

## medically speaking — The Discovery of Prodrugs

We have seen that azo dyes are used in a variety of applications. One such application led to a discovery that had a profound impact on the field of medicine. Specifically, it was observed that certain bacteria absorbed azo dyes, making them more readily visible under a microscope. In an effort to find an azo dye that might be toxic to bacteria, Fritz Mietzsch and Joseph Klarer (at the German dye company, I.G. Farbenindustrie) began cataloguing azo dyes for possible antibacterial properties. A physician named Gerhard Domagk evaluated the dyes for potential activity, which led to the discovery of the potent antibacterial properties of prontosil.

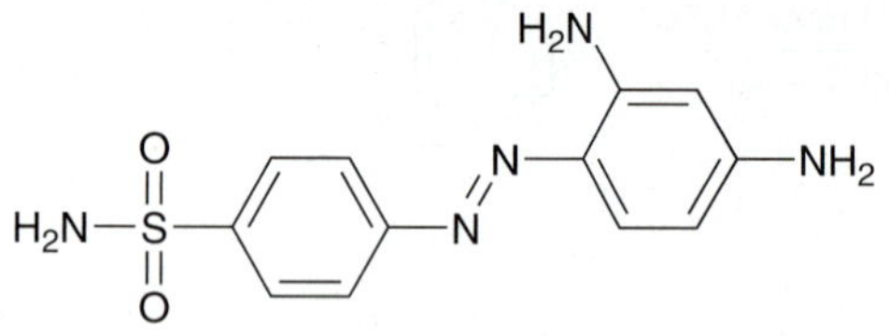

**Prontosil**

Domagk was able to demonstrate that prontosil cured streptococcal infections in mice. In 1933, physicians began using prontosil in human patients suffering from life-threatening bacterial infections. The success of this drug was extraordinary, and prontosil staked its claim as the first drug that was systematically used for the treatment of bacterial infections. The development of prontosil has been credited with saving thousands of lives. For his pioneering work that led to this discovery, Domagk was awarded the 1939 Nobel Prize in Physiology or Medicine.

Prontosil exhibited one very curious property that intrigued scientists. Specifically, it was found to be totally inactive against bacteria *in vitro* (literally "in glass," in bacterial cultures grown in glass dishes). Its antibacterial properties were only observed *in vivo* (literally "in life," when administered to living creatures, such as mice and humans). These observations inspired much research on the activity of prontosil, and in 1935, it was found that prontosil is metabolized in the body to produce a compound called sulfanilamide.

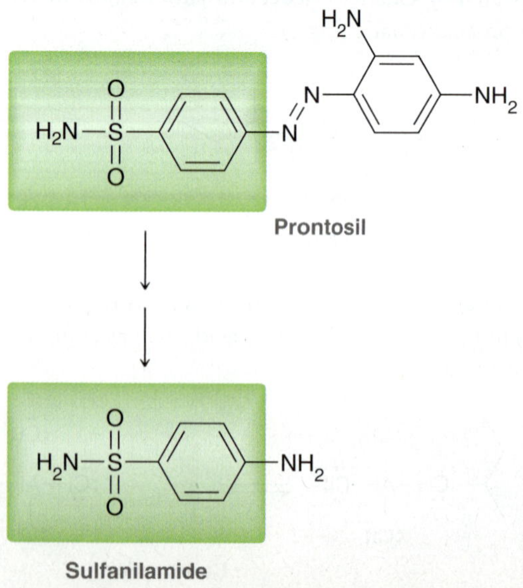

**Prontosil**

**Sulfanilamide**

Sulfanilamide was determined to be the active drug, as it interferes with bacterial cell growth. In a glass dish, prontosil is not converted into sulfanilamide, explaining why the antibacterial properties were only observed *in vivo*. This discovery ushered in the era of prodrugs. *Prodrugs* are pharmacologically inactive compounds that are converted by the body into active compounds. This discovery led scientists to direct their research in new directions. They began designing new potential drugs based on structural modifications to sulfanilamide rather than prontosil. Extensive research was directed at making these sulfanilamide analogues, called sulfonamides. By 1948, over 5000 sulfonamides were created, of which more than 20 were ultimately used in clinical practice.

The emergence of bacterial strains resistant to sulfanilamide, together with the advent of penicillins (discussed in Chapter 23), rendered most sulfonamides obsolete. Some sulfonamides are still used today to treat specific bacterial infection in patients with AIDS, as well as a few other applications. Despite their small role in current practice, sulfonamides occupy a unique role in history, because their development was based on the discovery of the first known prodrug.

There are currently a large number of prodrugs on the market. Prodrugs are often designed intentionally for a specific purpose. One such drug is used in the treatment of Parkinson's disease. The symptoms of Parkinson's disease are attributed to low levels of dopamine in a specific part of the brain. These symptoms can be treated by administering L-dopa to the patient. This prodrug is capable of reaching the desired destination (more effectively than dopamine), where it undergoes decarboxylation to produce the needed dopamine.

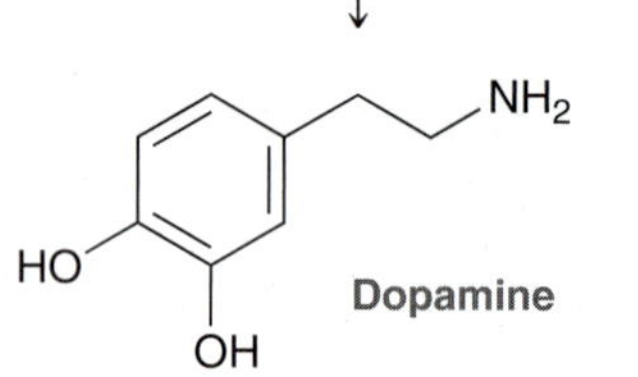

**L-dopa** → **In vivo** → **Dopamine**

There are many different varieties and classes of prodrugs, and a thorough treatment is beyond the scope of our discussion. The discovery of prodrugs was an extremely important achievement in the development of medicinal chemistry, and it all started with a careful analysis of azo dyes, just like the ones found in Fruity Pebbles.

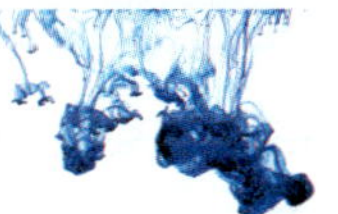



The catalyst functions exactly as expected (compare the role of $AlCl_3$ here to the role that it plays in Section 19.2). The result here is the formation of a carbocation, which is an excellent electrophile and is capable of reacting with benzene in an electrophilic aromatic substitution reaction (Mechanism 19.6).

## MECHANISM 19.6 FRIEDEL-CRAFTS ALKYLATION

Nucleophilic attack

Proton transfer

+ HCl + $AlCl_3$

the first step, aromatic ring s as a nucleophile, g an intermediate ma complex

In the second step, the sigma complex is deprotonated, restoring aromaticity

Sigma complex

After formation of the carbocation, two steps are involved in this electrophilic aromatic substitution. In the first step, the aromatic ring functions as a nucleophile and attacks the carbocation, forming a sigma complex. The sigma complex is then deprotonated to restore aromaticity.

Many different alkyl halides can be used in a Friedel-Crafts alkylation. Secondary and tertiary halides are readily converted into carbocations in the presence of $AlCl_3$. Primary alkyl halides are not converted into carbocations, since primary carbocations are extremely high in energy. Nevertheless, a Friedel-Crafts alkylation is indeed observed when benzene is treated with ethyl chloride in the presence of $AlCl_3$.

In this case, the electrophilic agent is presumed to be a complex between ethyl chloride and $AlCl_3$.

Electrophilic agent

This complex can be attacked by an aromatic ring, much like we saw during chlorination (Mechanism 19.2). Although a Friedel-Crafts alkylation is effective when ethyl chloride is used,

**LOOKING BACK**
For a review of carbocation rearrangements, see Section 6.11.

most other primary alkyl halides cannot be used effectively, because their complexes with $AlCl_3$ readily undergo rearrangement to form secondary or tertiary carbocations. For example, when 1-chlorobutane is treated with aluminum trichloride, a secondary carbocation is formed via a hydride shift.

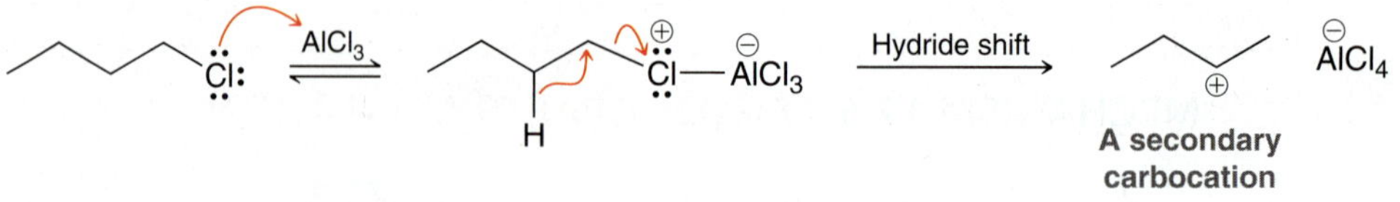

AlCl3 — Hydride shift — A secondary carbocation

In such a case, a mixture of products is obtained.

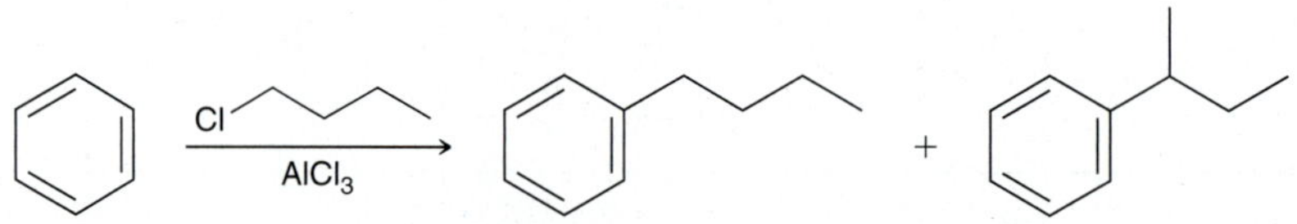

(35%) (65%)

The ratio of products depends on the conditions chosen (concentrations, temperature, etc.), but a mixture of products is unavoidable. Therefore, in practice, a Friedel-Crafts alkylation is only efficient when the substrate cannot undergo rearrangement.

There are several other limitations that must be noted.

1. When choosing an alkyl halide, the carbon atom connected to the halogen must be $sp^3$ hybridized. Vinyl carbocations and aryl carbocations are not sufficiently stable to be formed under Friedel-Crafts conditions.

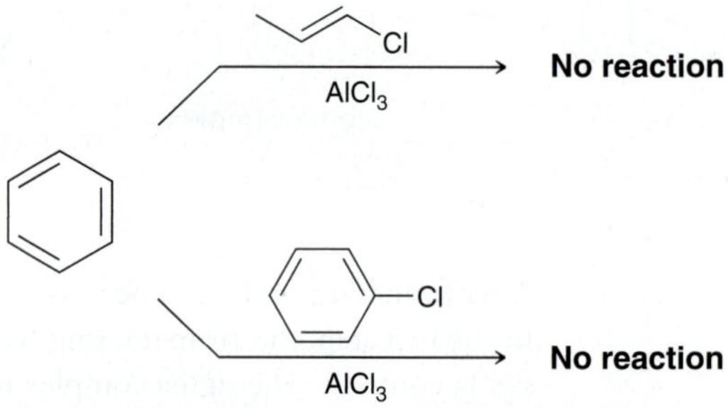

No reaction

No reaction

2. Installation of an alkyl group activates the ring toward further alkylation (for reasons that we will explore in the upcoming sections of this chapter). Therefore, polyalkylations often occur.

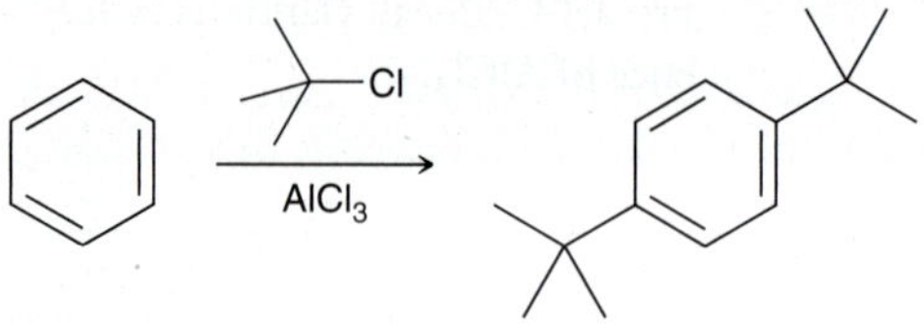

This problem can generally be avoided by choosing reaction conditions that favor monoalkylation. For the remainder of this chapter, assume that all Friedel-Crafts alkylations are performed under conditions that favor monoalkylation, unless otherwise specified.

3. There are certain groups, such as a nitro group, that are incompatible with a Friedel-Crafts reaction. In the upcoming sections of this chapter, we will explore the reason for this incompatibility.

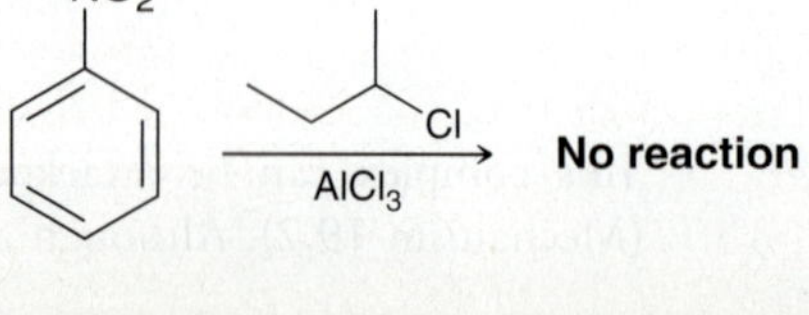

No reaction

## CONCEPTUAL CHECKPOINT

**19.5** Predict the expected product(s) when benzene is treated with each of the following alkyl halides in the presence of $AlCl_3$. In each case, assume conditions have been controlled to favor monoalkylation.

(a) (b) (c)

**19.6** Draw a mechanism for the following reaction, which involves two consecutive Friedel-Crafts alkylations. When drawing the mechanism, do not try to draw the two alkylations as occurring simultaneously (such a mechanism would have too many curved arrows and too many simultaneous charges). First draw the steps that install one alkyl group and then draw the steps that install the second alkyl group.

$AlCl_3$

**19.7** A Friedel-Crafts alkylation is an electrophilic aromatic substitution in which the electrophile ($E^+$) is a carbocation. In previous chapters, we have seen other methods of forming carbocations, such as protonation of an alkene using a strong acid. The resulting carbocation can also be attacked by a benzene ring, resulting in alkylation of the aromatic ring. With this in mind, draw a mechanism for the following transformation:

$H_2SO_4$

**(68%)**

## 19.6 Friedel-Crafts Acylation

In the previous section, we learned how to install an alkyl group on an aromatic ring. A similar method can be used to install an **acyl group**. The difference between an alkyl group and an acyl group is shown below.

Alkyl group

Acyl group

A reaction that installs an acyl group is called an *acylation*.

$AlCl_3$

This process, called a **Friedel-Crafts acylation**, is believed to proceed via a mechanism that is very similar to the mechanism presented for alkylation, in the previous section. An acyl chloride is treated with a Lewis acid to form a cationic species, called an **acylium ion**.

**Acylium ion**

Acylium ions are resonance stabilized and are therefore not susceptible to rearrangement.

**Resonance stabilized**

Rearrangement does not occur because a carbocation rearrangement would result in loss of resonance stabilization (an endergonic process). The acylium ion is an excellent electrophile, producing an electrophilic aromatic substitution reaction (Mechanism 19.7).

## MECHANISM 19.7 FRIEDEL-CRAFTS ACYLATION

Nucleophilic attack

In the first step, the aromatic ring functions as a nucleophile, forming an intermediate sigma complex

Sigma complex

Proton transfer

In the second step, the sigma complex is deprotonated, restoring aromaticity

The acylium ion is attacked by the benzene ring to produce an intermediate sigma complex, which is then deprotonated to restore aromaticity.

The product of a Friedel-Crafts acylation is an aryl ketone, which can be reduced using a **Clemmensen reduction**.

Zn(Hg)
HCl, heat

In the presence of HCl and amalgamated zinc (zinc that has been treated so that its surface is an alloy, or mixture, of zinc and mercury), the carbonyl group is completely reduced and replaced with two hydrogen atoms. When a Friedel-Crafts acylation is followed by a Clemmensen reduction, the net result is the installation of an alkyl group.

$AlCl_3$ ; Zn(Hg), HCl, heat — **(73%)**

This two-step process is a useful synthetic method for installing alkyl groups that cannot be efficiently installed with a direct alkylation process. If the product is made with a direct alkylation process, carbocation rearrangements would give a mixture of products. The advantage of the acylation process is that carbocation rearrangements are avoided because of the stability of the acylium ion.

Polyacylation is not observed, because introduction of an acyl group deactivates the ring toward further acylation. This will be explained in more detail in the upcoming sections.

CONCEPTUAL CHECKPOINT

**19.8** Identify whether each of the following compounds can be made using a direct Friedel-Crafts alkylation or whether it is necessary to perform an acylation followed by a Clemmensen reduction to avoid carbocation rearrangements:

(a) (b)

(c) (d)

**19.9** The following compound cannot be made with either a Friedel-Crafts alkylation or acylation. Explain.

**19.10** A Friedel-Crafts acylation is an electrophilic aromatic substitution in which the electrophile ($E^+$) is an acylium ion. There are other methods of forming acylium ions, such as treatment of an anhydride with a Lewis acid. The resulting acylium ion can also be attacked by a benzene ring, resulting in acylation of the aromatic ring. With this in mind, draw a mechanism for the following transformation:

## 19.7 Activating Groups

### Nitration of Toluene

Thus far, we have dealt only with reactions of benzene. We now expand our discussion to include reactions of aromatic compounds that already possess substituents. For example, consider the nitration of toluene:

In this nitration reaction, the presence of the methyl group raises two issues: (1) the effect of the methyl group on the rate of reaction and (2) the effect of the methyl group on the regiochemical outcome of the reaction. Let's begin with the rate of reaction.

Toluene undergoes nitration approximately 25 times faster than benzene. In other words, the methyl group is said to **activate** the aromatic ring. Why? Recall that alkyl groups are generally electron donating via hyperconjugation (see Section 6.8). As a result, a methyl group donates electron density to the ring, thereby stabilizing the positively charged sigma complex and lowering the energy of activation for its formation.

Now let's focus on the regiochemical outcome. The nitro group could be installed *ortho, meta*, or *para* to the methyl group, but the three possible products are not obtained in equal amounts. As seen in Figure 19.4, the *ortho* and *para* products predominate, while very little *meta* product is obtained.

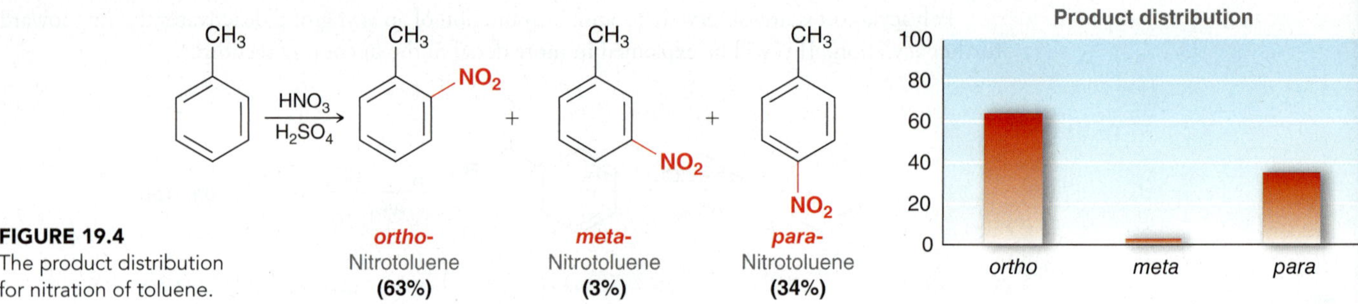


**FIGURE 19.4** The product distribution for nitration of toluene.

To explain this observation, we must compare the stability of the sigma complex formed for *ortho* attack, *meta* attack, and *para* attack (Figure 19.5). Notice that the sigma complex obtained from

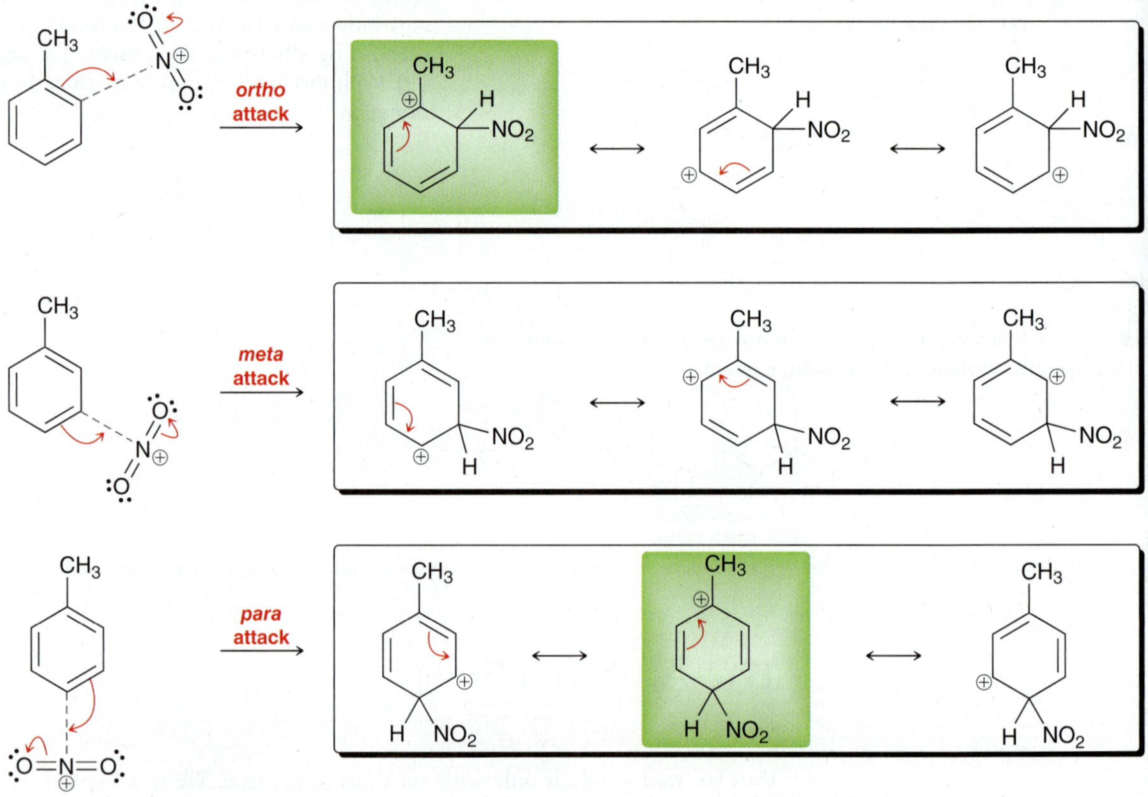

**FIGURE 19.5** A comparison of the sigma complexes formed for each of the possible regiochemical outcomes for the nitration of toluene.

*ortho* attack has additional stability because one of the resonance structures (highlighted in green) exhibits the positive charge directly adjacent to the electron-donating alkyl group. Similarly, the sigma complex obtained from *para* attack also exhibits this additional stability. The sigma complex obtained from *meta* attack does not exhibit this stability and is, therefore, higher in energy. The relative energy of each sigma complex is best visualized by comparing energy diagrams for *ortho* attack, *meta* attack, and *para* attack (Figure 19.6, as seen on the next page). Notice that the intermediate sigma complex formed from *meta* attack is the highest in energy and therefore requires the largest energy of activation ($E_a$). This explains why the *meta* product is only obtained in very small amounts. The products of the reaction are generated from *ortho* attack and *para* attack, both of which involve a lower $E_a$. If we compare $E_a$ for *ortho* attack and *para* attack, we see that *ortho* attack involves a slightly higher $E_a$ than *para* attack, because the methyl group and the nitro group are in close proximity and exhibit a small steric interaction. As a result, we might have expected the *para* product to be the major product. In fact, the *ortho* product predominates in this case for statistical reasons—there are two *ortho* positions and only one *para* position.

## BY THE WAY

The exact ratio of *ortho* and *para* products is often sensitive to the conditions employed, such as the choice of solvent. In some cases, nitration of toluene has been observed to favor the *para* product over the *ortho* product.

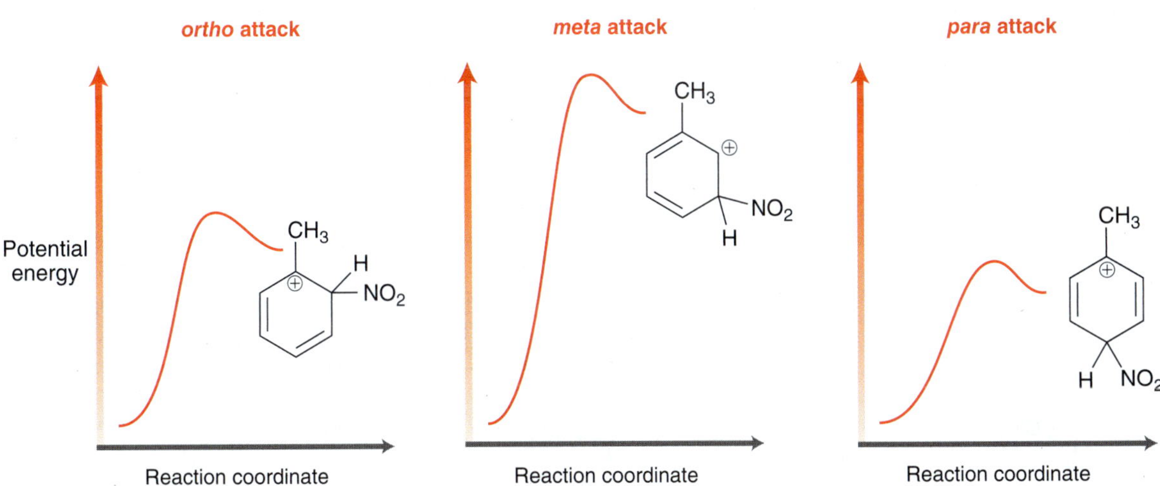


**FIGURE 19.6**
Energy diagrams comparing the relative energy levels of the possible sigma complexes that could be formed during nitration of toluene. The differences in energy between these three pathways have been slightly exaggerated for clarity of presentation.

Comparison of the energy diagrams in Figure 19.6 provides an explanation for the observation that the methyl group is an ***ortho-para*** **director**. In other words, the presence of the methyl group directs the incoming nitro group into the *ortho* and *para* positions.

## Nitration of Anisole

In the previous section, we saw that the presence of a methyl group activates the ring toward electrophilic aromatic substitution. There are many different groups that activate a ring, with some more activating than others. For example, a methoxy group is a more powerful activator than a methyl group, and methoxybenzene (anisole) undergoes nitration 400 times faster than toluene. To understand why a methoxy group is more activating than a methyl group, we must explore the electronic effects of a methoxy group connected to an aromatic ring.

A methoxy group is inductively electron withdrawing, because oxygen is more electronegative than carbon.

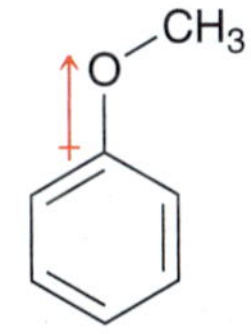

From this point of view, the methoxy group withdraws electron density from the aromatic ring. However, we see a different picture when we draw resonance structures of anisole.

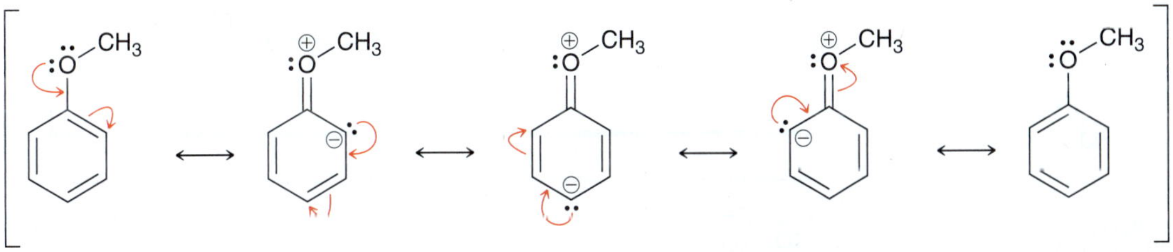

Three of the resonance structures exhibit a negative charge in the ring. This suggests that the methoxy group *donates* electron density to the ring. Clearly, there is a competition here between induction and resonance. Induction suggests that the methoxy group is electron withdrawing, while resonance suggests that the methoxy group is electron donating. Whenever resonance and induction compete with each other, resonance is generally the dominant factor and vastly overshadows any inductive effects. Therefore, the net effect of the methoxy group is to donate electron density to the ring. This effect stabilizes the positively charged sigma complex and lowers the energy of activation for its formation. In fact, the ring is so activated that treatment with excess bromine *without* a Lewis acid gives a trisubstituted product.

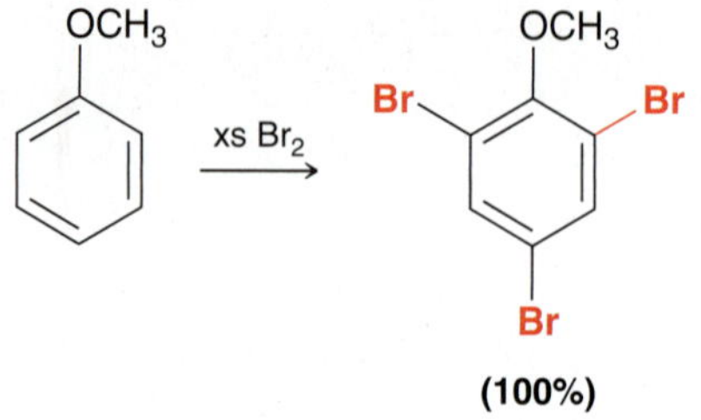

OCH3 → xs $Br_2$ → (2,4,6-tribromoanisole) (100%)

All three positions undergo bromination. Notice the preference, once again, for the reaction to occur at the *ortho* and *para* positions. This *ortho-para* directing effect is also observed when anisole undergoes nitration (Figure 19.7). Once again, explaining this observation requires that we compare the

$HNO_3$, $H_2SO_4$

*ortho*-Nitroanisole **(31%)** + *meta*-Nitroanisole **(2%)** + *para*-Nitroanisole **(67%)**

Product distribution: ortho, meta, para (0–100)

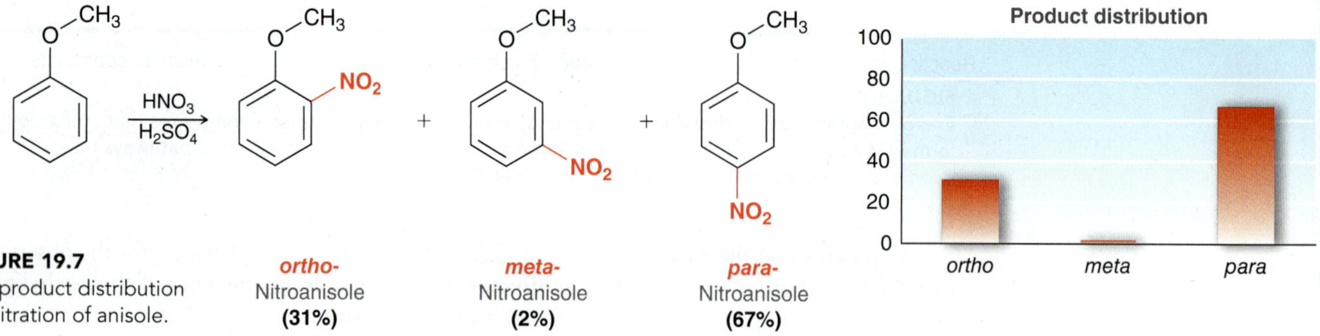

**FIGURE 19.7** The product distribution for nitration of anisole.

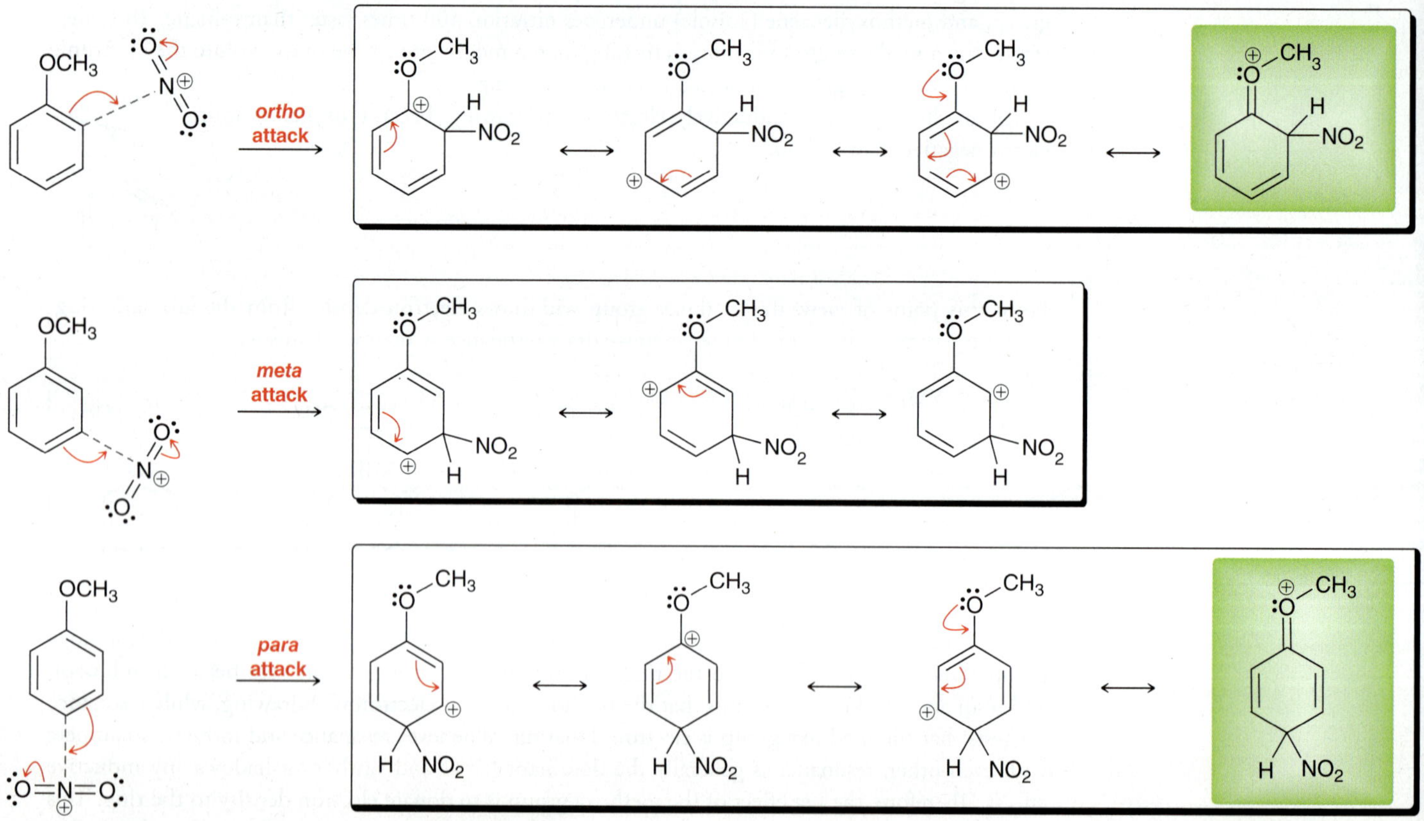

stability of the sigma complexes formed for *ortho*, *meta*, and *para* attack (Figure 19.8). Notice that the sigma complex obtained from *ortho* attack has one additional resonance structure (highlighted in green), which stabilizes the sigma complex. Similarly, the sigma complex obtained from *para* attack

*ortho* attack

*meta* attack

*para* attack

**FIGURE 19.8**
A comparison of the sigma complexes formed for the possible regiochemical outcomes for the nitration of anisole.

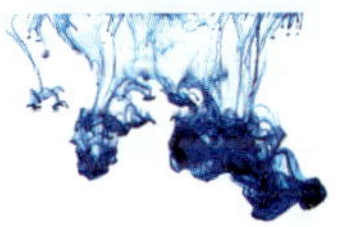


also exhibits this additional stability. The sigma complex obtained from *meta* attack does not exhibit this extra stability and is, therefore, higher in energy. Figure 19.9 shows a comparison of the energy

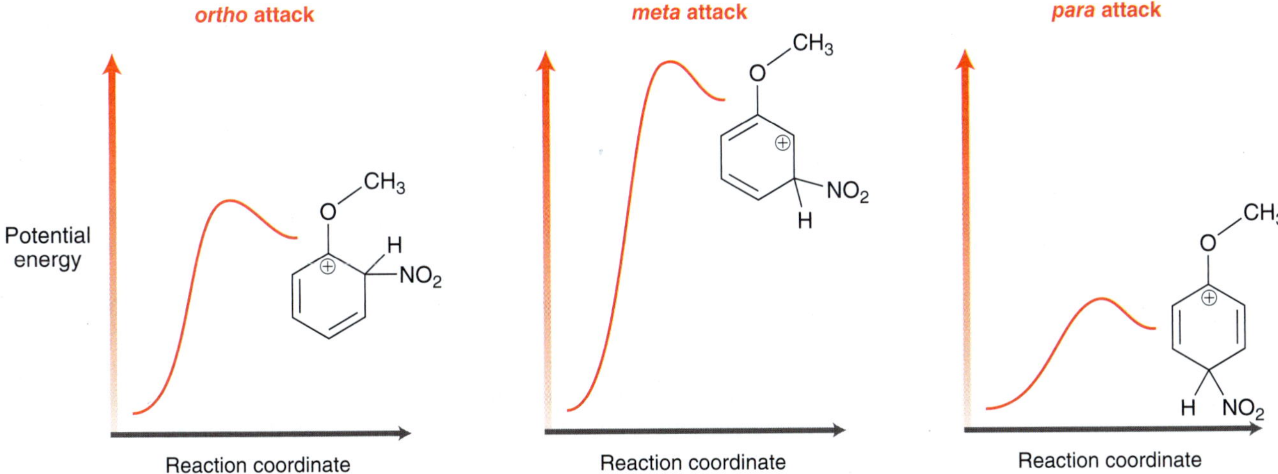


**FIGURE 19.9** Energy diagrams comparing the relative energy levels of the sigma complexes that could be formed during nitration of anisole. The differences in energy between these three pathways have been slightly exaggerated for clarity of presentation.

diagrams for *ortho*, *meta*, and *para* attack. Notice that the intermediate sigma complex formed from *meta* attack is the highest in energy and therefore requires the largest energy of activation ($E_a$). This explains why the *meta* product is only obtained in very small amounts. The products of the reaction are generated from *ortho* attack and *para* attack, both of which involve a lower $E_a$.

In the nitration of anisole, the *para* product is favored over the *ortho* product despite the fact that there are two *ortho* positions. Several factors contribute to this observation. One factor is most certainly a steric consideration. That is, the sigma complex resulting from *ortho* attack exhibits a greater steric interaction and is higher in energy than the sigma complex resulting from *para* attack.

In summary, we have seen that both a methyl group and a methoxy group activate the ring and are *ortho-para* directors. This is in fact a general rule that will be used extensively throughout the rest of this chapter: *All activators are ortho-para directors.*

## CONCEPTUAL CHECKPOINT

**19.11** Draw the two major products obtained when toluene undergoes monobromination.

**19.12** When ethoxybenzene is treated with a mixture of nitric acid and sulfuric acid, two products are obtained, each of which has the molecular formula $C_8H_9NO_3$.

(a) Draw the structure of each product.

(b) Propose a mechanism of formation for the major product.

## 19.8 Deactivating Groups

In the previous section, we saw that certain groups will activate the ring toward electrophilic aromatic substitution. In this section, we explore the effects of a nitro group, which is said to **deactivate** the ring toward electrophilic aromatic substitution. To understand why the nitro group deactivates the ring, we must explore the electronic effects of a nitro group connected to an aromatic ring.

A nitro group is inductively electron withdrawing, because a positively charged nitrogen atom is extremely electronegative.

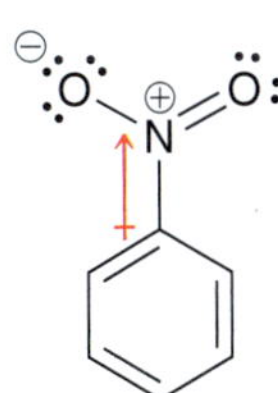

Now let's consider resonance. Many of the resonance structures exhibit a positive charge in the ring.

The positive charge indicates that the nitro group *withdraws* electron density from the ring. In this case, there is no competition between resonance and induction. Both factors suggest that the nitro group is a powerful electron-withdrawing group. By removing electron density from the ring, the nitro group destabilizes the positively charged sigma complex and raises the energy of activation for its formation. This effect is quite significant and can be observed by comparing rates of nitration. Specifically, nitrobenzene is 100,000 times less reactive than benzene toward nitration, and the reaction can only be accomplished at an elevated temperature. When the reaction is forced to proceed, the regiochemical outcome is different than what we have seen thus far (Figure 19.10). Notice that

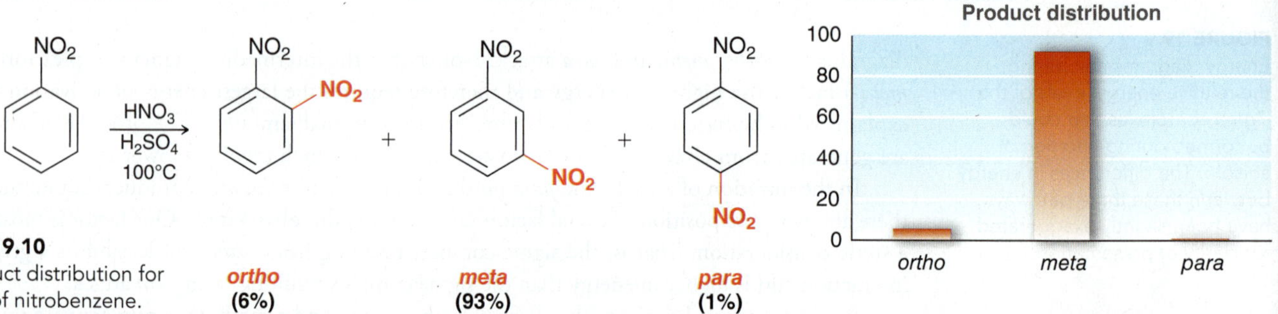


**FIGURE 19.10** The product distribution for nitration of nitrobenzene.

the *meta* product predominates, in stark contrast with the previous examples in which *ortho* and *para* products predominated. To explain this observation, we must compare the stability of the sigma complexes formed for *ortho*, *meta*, and *para* attack (Figure 19.11). Notice that the sigma complex

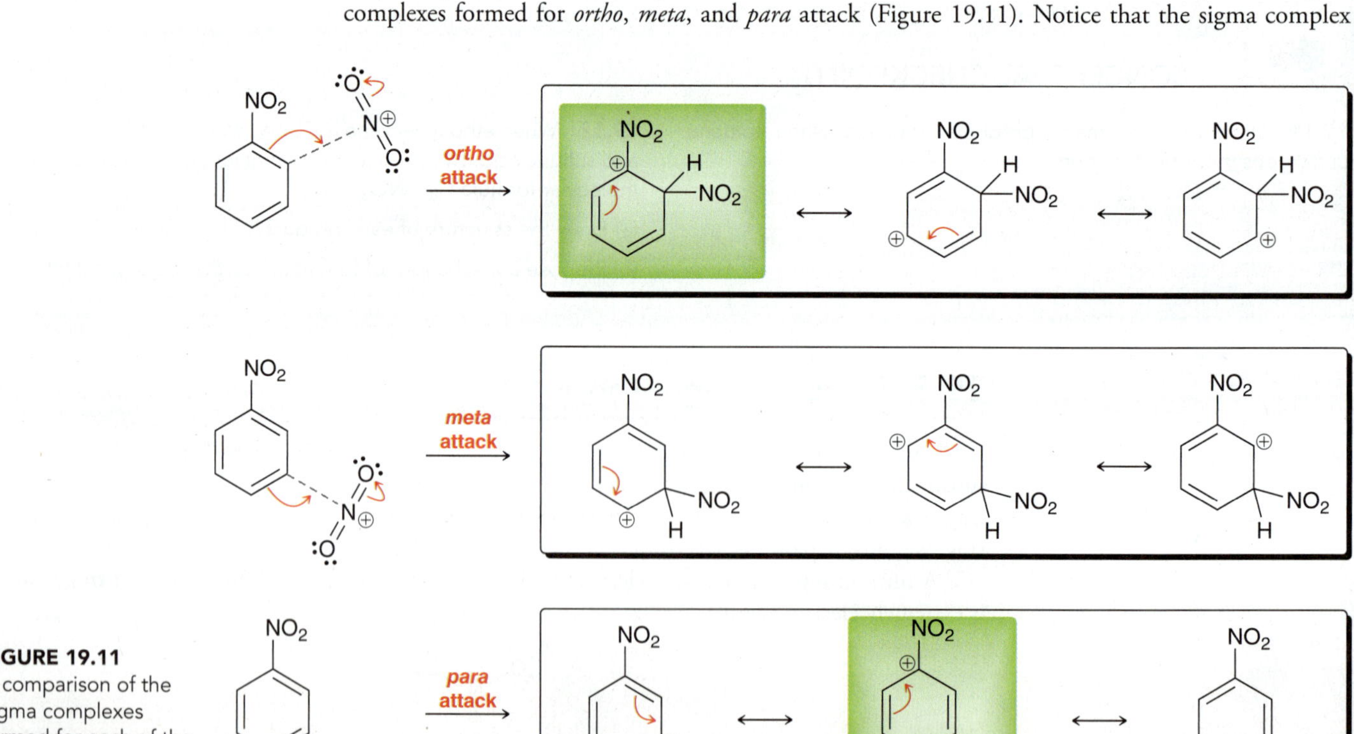


**FIGURE 19.11** A comparison of the sigma complexes formed for each of the possible regiochemical outcomes for the nitration of nitrobenzene.

obtained from *ortho* attack exhibits instability because one of the resonance structures (highlighted in green) has a positive charge directly adjacent to an electronegative, positively charged nitrogen atom. Similarly, the sigma complex obtained from *para* attack also exhibits this instability. The sigma complex obtained from *meta* attack does not exhibit this instability and is therefore lower in energy. Compare the energy diagrams for *ortho*, *meta*, and *para* attack (Figure 19.12). In summary, we have seen that a nitro group deactivates the ring and is a ***meta* director**. This is in fact a general rule that will be used extensively throughout the rest of this chapter: *Most deactivators are meta directors.*

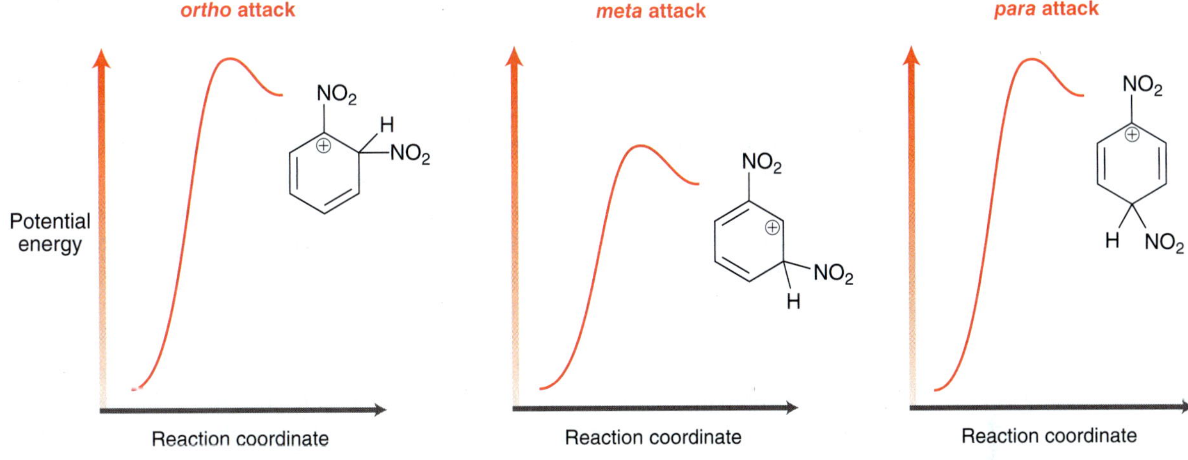


**FIGURE 19.12**
Energy diagrams comparing the relative energy levels of the sigma complexes that could be formed during nitration of nitrobenzene. The differences in energy between these three pathways have been slightly exaggerated for clarity of presentation.

## CONCEPTUAL CHECKPOINT

**19.13** When 1,3-dinitrobenzene is treated with nitric acid and sulfuric acid at elevated temperature, the product is 1,3,5-trinitrobenzene. Explain the regiochemical outcome of this reaction. In other words, explain why nitration takes place at the C5 position. Make sure to draw the sigma complex for each possible pathway and to compare the relative stability of each sigma complex.

## 19.9 Halogens: The Exception

In the previous sections, we have seen that activators are *ortho-para* directors and deactivators are *meta* directors.

Activator Deactivator

We will now explore one important exception to these general rules. Many of the halogens (including Cl, Br, and I) are *ortho-para* directors despite the fact that they are deactivators. To rationalize this curious exception, we must explore the electronic effects of a halogen connected to an aromatic ring. As we have seen several times, it is necessary to consider both inductive effects and resonance. Let's begin by exploring the inductive effects.

Cl

Halogens are fairly electronegative (more so than carbon) and are therefore inductively electron *withdrawing*. When we draw the resonance structures, a different picture emerges.

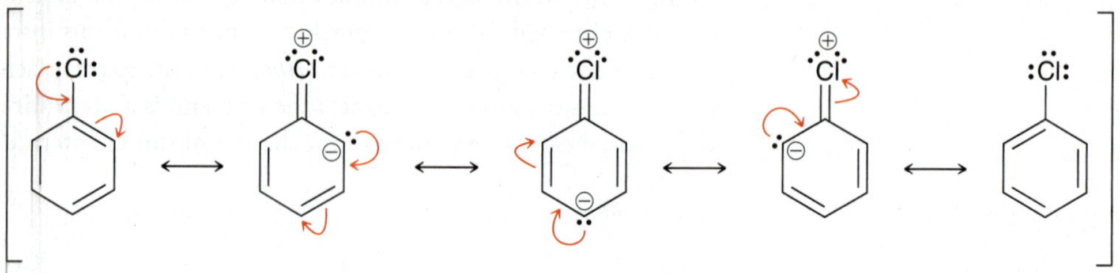

Three of the resonance structures exhibit a negative charge in the ring. This suggests that a halogen *donates* electron density to the ring. The competition between resonance and induction is very similar to the competition we saw when analyzing methoxybenzene. Induction suggests that a halogen is electron withdrawing, while resonance suggests that a halogen is electron donating. Although resonance is generally the dominant factor, this case is the exception. Induction is actually the dominant factor for halogens. As a result, halogens withdraw electron density from the ring, thereby destabilizing the positively charged sigma complex and raising the energy of activation for its formation.

To explain the fact that halogens are *ortho-para* directors even though they are deactivators, we must compare the stability of the sigma complexes formed for *ortho*, *meta*, and *para* attack (Figure 19.13). Notice that the sigma complex obtained from *ortho* attack has one additional resonance structure (highlighted in green), which stabilizes the sigma complex. Similarly, the sigma complex obtained from *para* attack also exhibits this additional stability, but the sigma complex from *meta* attack does not exhibit this extra stability, and therefore it is higher in energy. For this reason, halogens are *ortho-para* directors, despite the fact that they are deactivators.

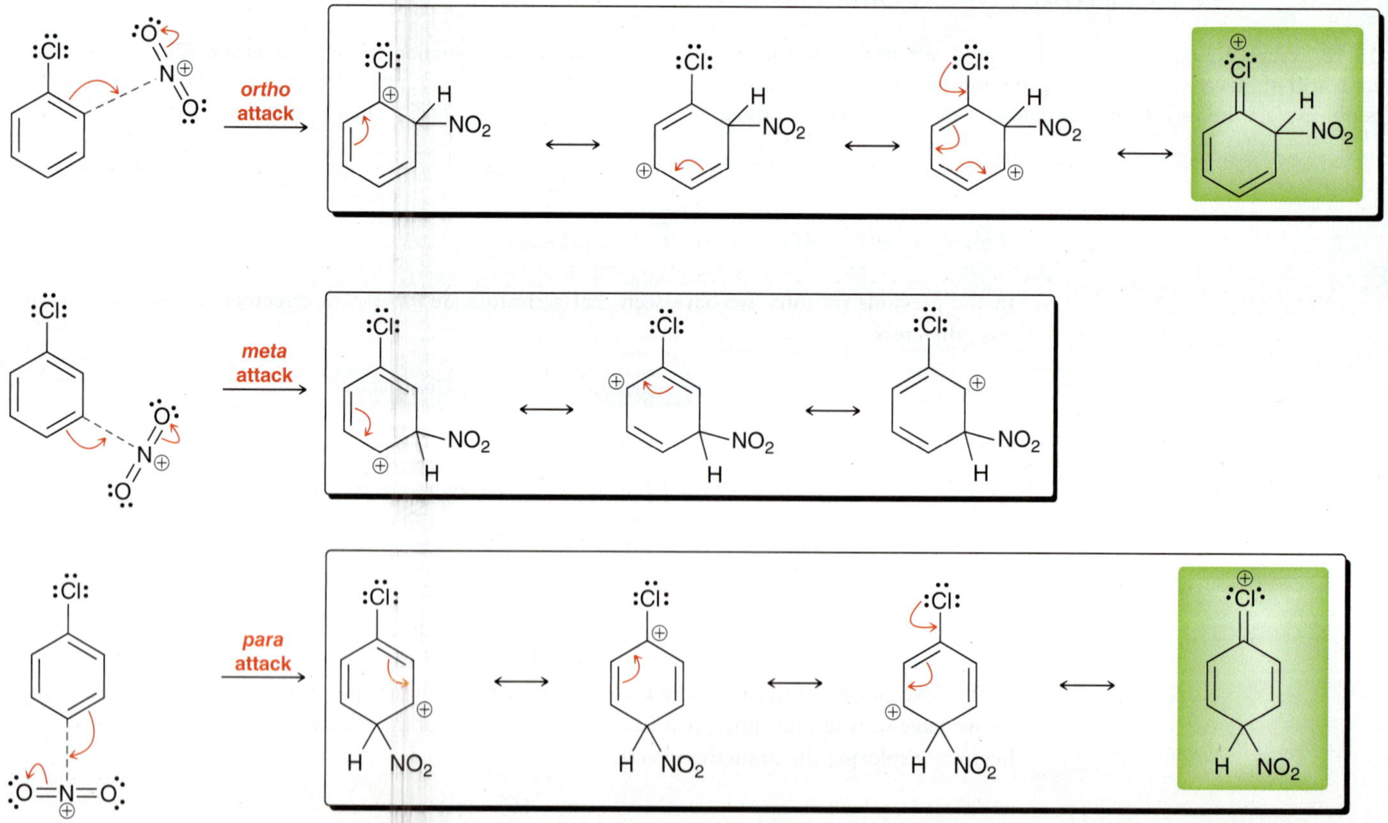

**FIGURE 19.13**
A comparison of the sigma complexes formed for each of the possible regiochemical outcomes for the nitration of chlorobenzene.

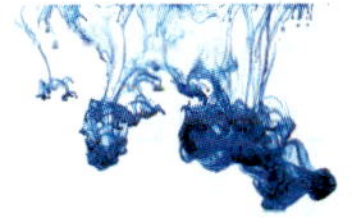

## CONCEPTUAL CHECKPOINT

**19.14** Does chlorination of chlorobenzene require the use of a Lewis acid? Explain why or why not?

**19.15** Predict and explain the regiochemical outcome for chlorination of bromobenzene.

## 19.10 Determining the Directing Effects of a Substituent

The previous sections focused on the directing effects of a few specific groups (methyl, methoxy, nitro, and halogens). In this section, we will learn how to predict the directing effects for any substituent. That skill will prove to be essential in subsequent sections that deal with synthesis.

Both activators and deactivators can be classified as strong, moderate, and weak. Each of these categories is described below, followed by a summary chart of all six categories. As discussed in the preceding sections, the activators are *ortho-para* directors, while the deactivators, except for the halogens, are *meta* directors.

### Activators

**Strong activators** are characterized by the presence of a lone pair immediately adjacent to the aromatic ring.

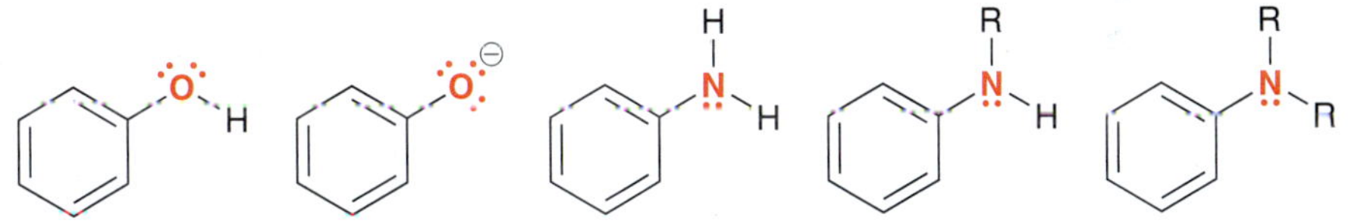

All of these groups exhibit a lone pair that is delocalized into the ring, as can be seen in their resonance structures. For example, phenol has the following resonance structures:

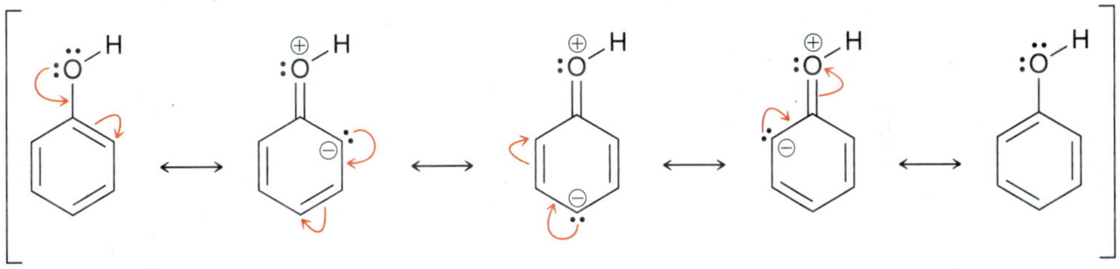

Many of these resonance structures have a negative charge in the ring, indicating that the OH group is donating electron density into the ring. This electron-donating effect strongly activates the ring.

**Moderate activators** exhibit a lone pair that is already delocalized outside of the ring.

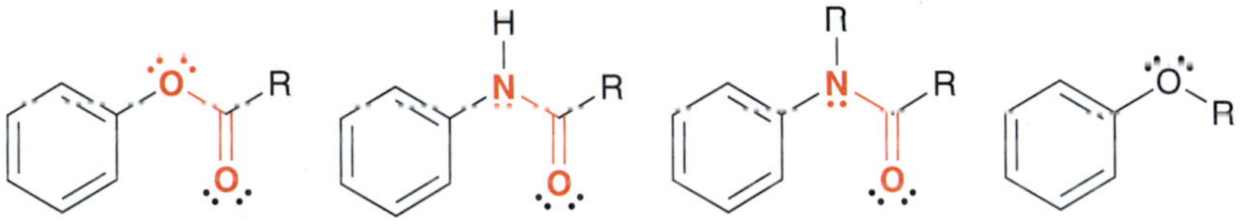

In the first three compounds, there is a lone pair next to the ring, but that lone pair is participating in resonance outside of the ring

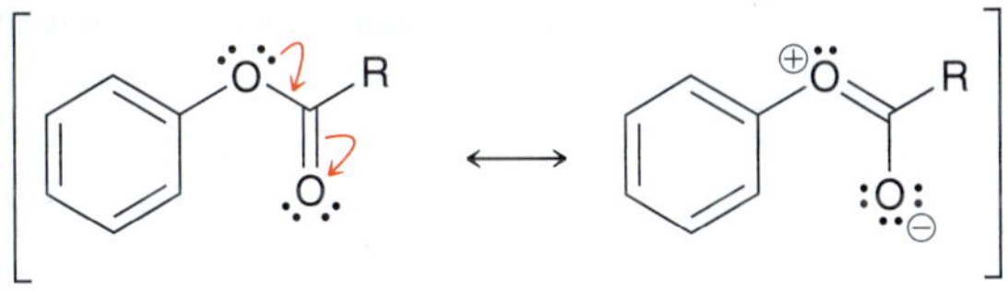

This effect diminishes the capability of the lone pair to donate electron density into the ring. These groups are activating, but they are moderate activators. The lone pair of an alkoxy group (OR) is not participating in resonance outside of the ring, and we might therefore expect that it would be a strong activator. Nevertheless, alkoxy groups belong to the class of moderate activators. Alkoxy groups are generally more activating than other moderate activators but less activating than strong activators (such as amino groups).

Alkyl groups are **weak activators**, because they donate electron density by the relatively weak effect of hyperconjugation (as described in Section 6.11).

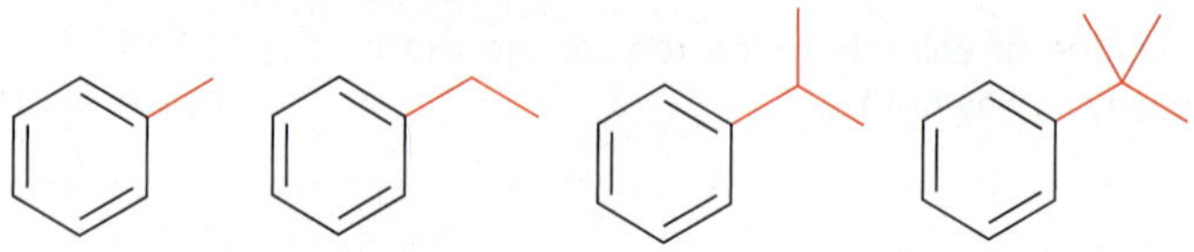

We will now turn our attention to deactivators, starting with weak deactivators and progressing to strong deactivators.

## Deactivators

As we have already seen, many of the halogens (Cl, Br, or I) are observed to deactivate a benzene ring:

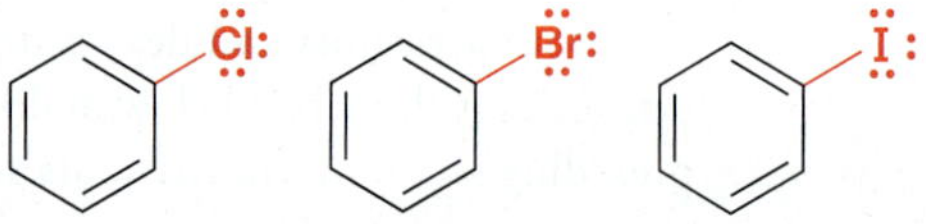

We have seen that the electronic effects of halogens are determined by the delicate competition between resonance and induction, with induction emerging as the dominant effect. As a result, halogens are **weak deactivators**.

**Moderate deactivators** are groups that exhibit a π bond to an electronegative atom, where the π bond is conjugated with the aromatic ring. Below are several examples.

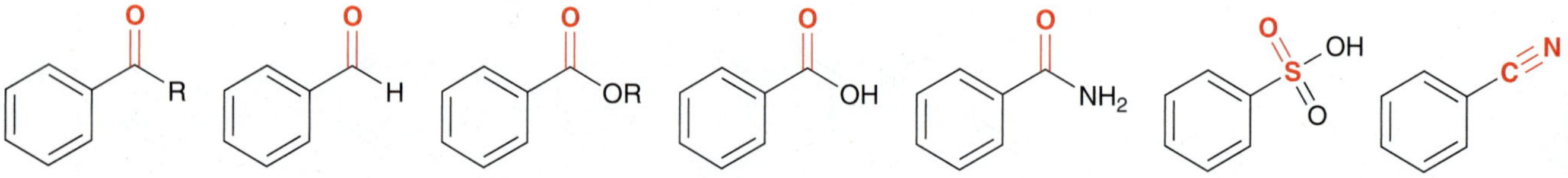

Each of these groups withdraws electron density from the ring via resonance. For example:

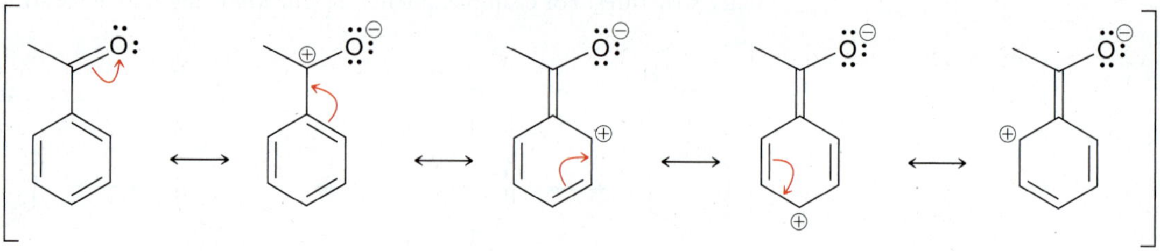

Three of the resonance structures have a positive charge in the ring, indicating that the group is withdrawing electron density from the ring. This electron-withdrawing effect moderately deactivates the ring.

There are only a few common substituents that are **strong deactivators**.

(X=Halogen)

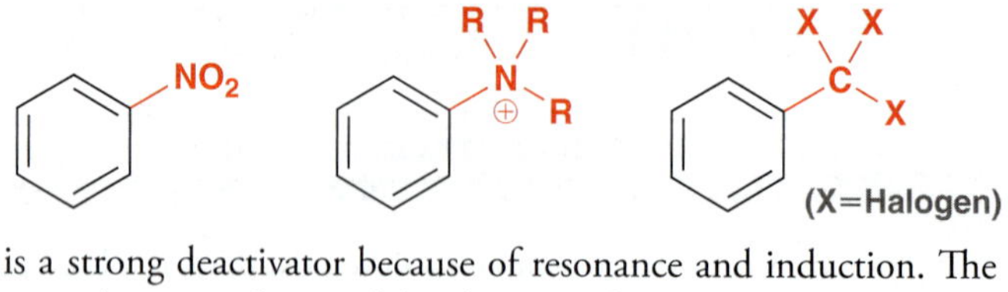

The nitro group is a strong deactivator because of resonance and induction. The other two groups are strong deactivators because of powerful inductive effects. A positively charged nitrogen atom is extremely electronegative, and $CX_3$ has three electron-withdrawing halogens. Do not confuse a $CX_3$ group with a halogen (X).

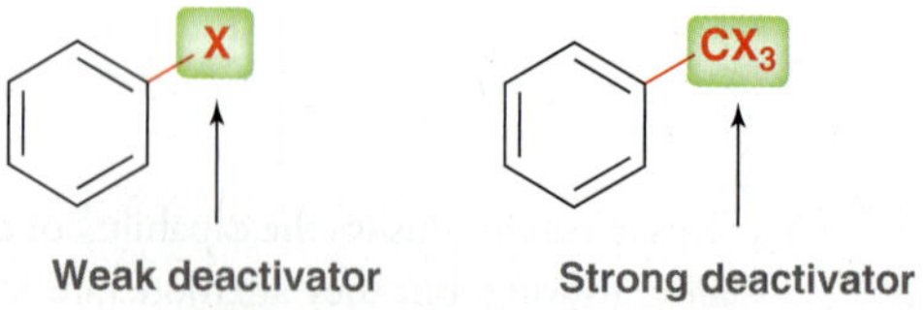

Weak deactivator

Strong deactivator

Table 19.1 summarizes the six categories of activators and deactivators. Notice the unique position of the halogens. In general, activators are *ortho-para* directors, while deactivators are *meta* directors, but halogens are the exception.

**TABLE 19.1 A LIST OF ACTIVATORS AND DEACTIVATORS BY CATEGORY**

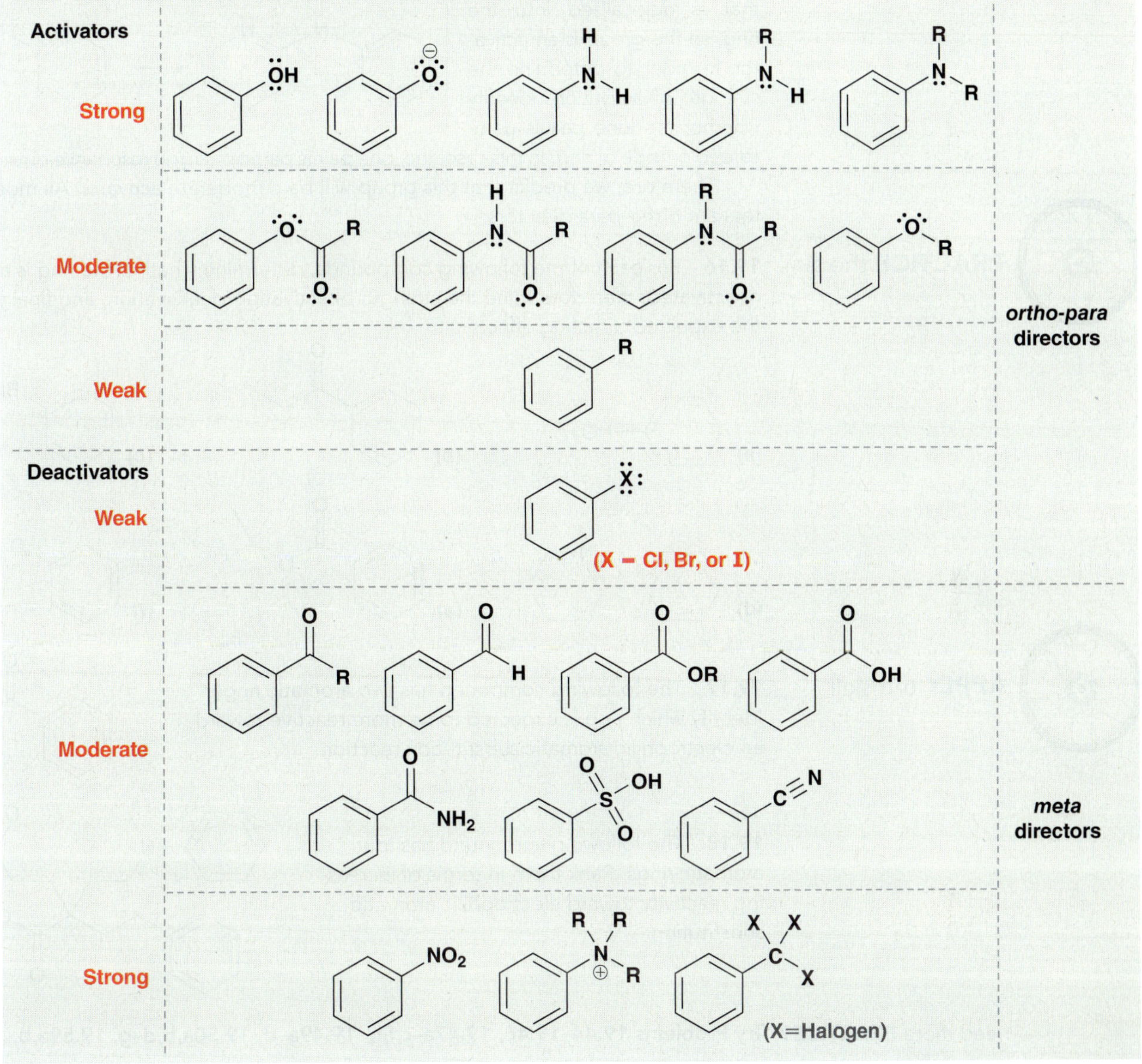

| |  |  |  |
|---|---|---|---|
| **Activators** | Strong | —OH; —$O^{\ominus}$; —NH(H); —N(R)H; —N(R)R | *ortho-para* directors |
| | Moderate | —O—C(=O)—R; —N(H)—C(=O)—R; —N(R)—C(=O)—R; —O—R | *ortho-para* directors |
| | Weak | —R | *ortho-para* directors |
| **Deactivators** | Weak | —X: (X = Cl, Br, or I) | *ortho-para* directors |
| | Moderate | —C(=O)R; —C(=O)H; —C(=O)OR; —C(=O)OH; —C(=O)$NH_2$; —$SO_3H$ (S(=O)(=O)OH); —C≡N | *meta* directors |
| | Strong | —$NO_2$; —$N^{\oplus}R_3$; —$CX_3$ (X=Halogen) | *meta* directors |

SKILLBUILDER

## 19.1 IDENTIFYING THE EFFECTS OF A SUBSTITUENT

**LEARN** the skill

Consider the following monosubstitued aromatic ring. Determine whether this aromatic ring is activated or deactivated, and determine the strength of activation/deactivation (i.e., is it strong, moderate, or weak). Finally, determine the directing effects of the group.

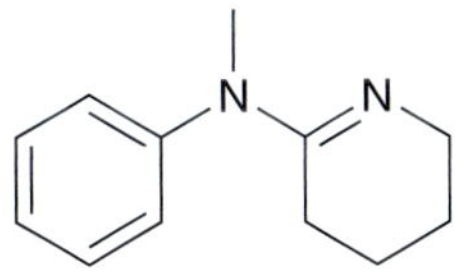

**SOLUTION**

First look for a lone pair immediately adjacent to the ring.

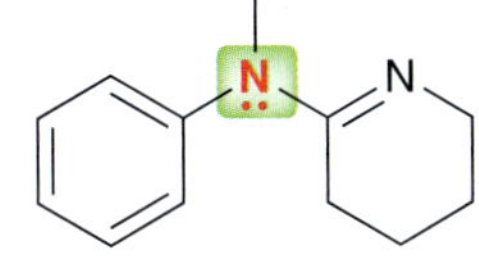

In this case, there is a lone pair that is delocalized into the ring, so the group is an activator. In order to determine the strength of activation, identify whether the lone pair is delocalized outside of ring. In this case, the lone pair is participating in resonance outside of the ring.

Therefore, we predict that this group will be a moderate activator. All moderate activators are *ortho-para* directors.

**PRACTICE** the skill **19.16** For each of the following compounds, determine whether the ring is activated or deactivated, then determine the strength of activation/deactivation, and finally determine the expected directing effects.

(a) $-NO_2$

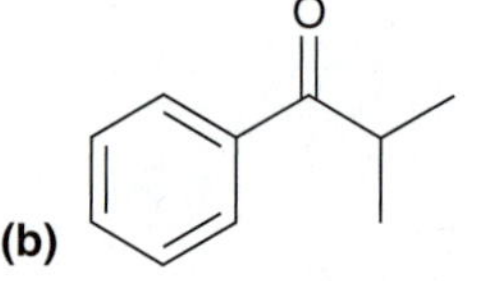

(b)

(c) Br

(d) N

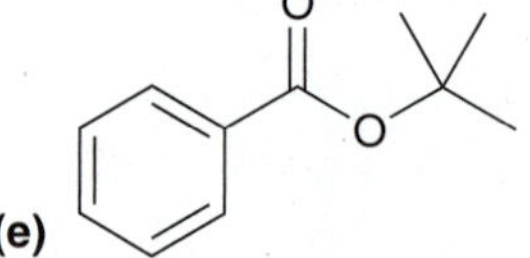

(e)

(f)

**APPLY** the skill **19.17** The following compound has two aromatic rings. Identify which ring is expected to be more reactive toward an electrophilic aromatic substitution reaction.

**19.18** The following compound has four aromatic rings. Rank them in terms of increasing reactivity toward electrophilic aromatic substitution.

need more **PRACTICE?** **Try Problems 19.44–19.46, 19.47a–c,f,h, 19.49a–d, 19.50a,b,d–g, 19.59a,b, 19.64, 19.66**

## 19.11 Multiple Substituents

### Directing Effects

We will now explore directing effects when multiple substituents are present on a ring. In some cases, the directing effects of all substituents reinforce each other; for example:

$CH_3$, $NO_2$ $\xrightarrow[FeBr_3]{Br_2}$ $CH_3$, Br, $NO_2$

In this case, the methyl group directs to the *ortho* positions (the *para* position is already occupied), and the nitro group directs to the positions that are *meta* to the nitro group. In this case, both the methyl group and the nitro group direct to the same two locations. Since the two locations are identical (because of symmetry), only one product is obtained.

In other cases, the directing effects of the various substituents may compete with each other. In such cases, the more powerful activating group dominates the directing effects.

$$\text{4-methylphenol} \xrightarrow[H_2SO_4]{HNO_3} \text{4-methyl-2-nitrophenol}$$

In this case, there are two substituents on the ring: an OH group (strong activator) and a methyl group (weak activator). The strong activator dominates, so the incoming nitro group is installed at a position that is *ortho* to the strong activator (the *para* position is already occupied).

## SKILLBUILDER

### 19.2 IDENTIFYING DIRECTING EFFECTS FOR DISUBSTITUTED AND POLYSUBSTITUTED BENZENE RINGS

**LEARN the skill**

In the following compound, identify the position that is most likely to undergo an electrophilic aromatic substitution reaction.

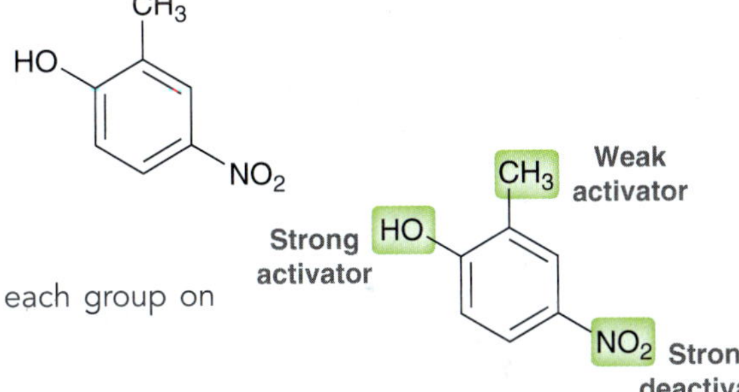

**SOLUTION**

**STEP 1** Identify the nature of each group.

Begin by identifying the effect of each group on the aromatic ring.

**STEP 2** Select the most powerful activator and identify the positions that are *ortho* or *para* to that group.

The most powerful activator will dominate the directing effects. In this case, the OH group is the strongest activator. Now consider the positions that are *ortho* and *para* to the OH group.

**STEP 3** Identify the unoccupied positions.

Two of these positions are already occupied. Only one position remains. We therefore predict that this position is most likely to undergo an electrophilic aromatic substitution reaction.

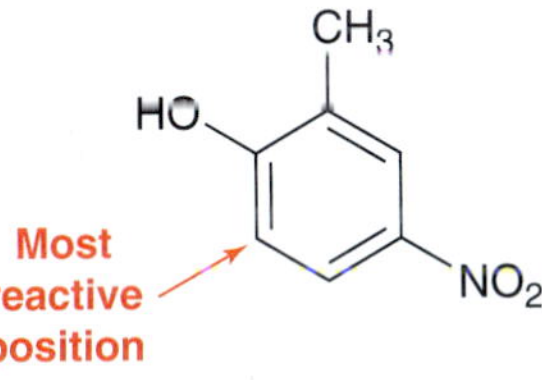


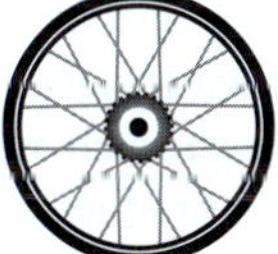

**PRACTICE the skill**

**19.19** For each compound below, identify which position(s) is/are most likely to undergo an electrophilic aromatic substitution reaction.

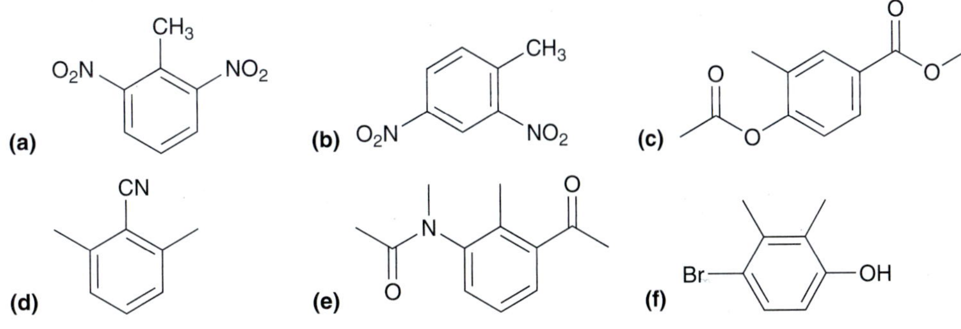

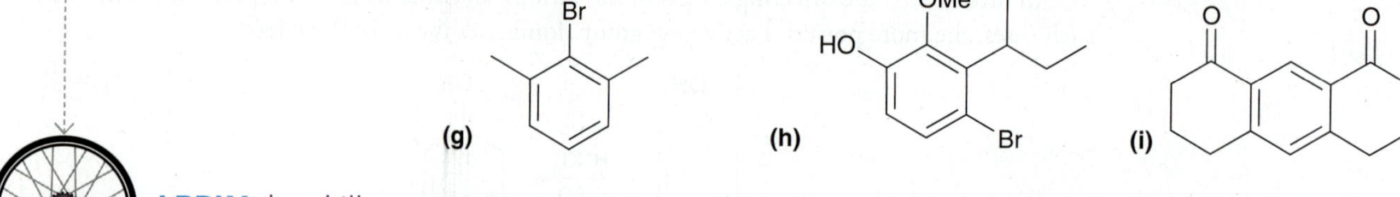


**APPLY the skill**

**19.20** Predict the product(s) for each of the following reactions:

(a) [structure] $\xrightarrow[H_2SO_4]{HNO_3}$ ?

(b) [structure] $\xrightarrow[FeBr_3]{Br_2}$ ?

(c) [structure] $\xrightarrow{\text{Fuming } H_2SO_4}$ ?

**19.21** When 2,4-dibromo-3-methyltoluene is treated with bromine in the presence of iron (Fe), a compound with molecular formula $C_8H_7Br_3$ is obtained. Identify the structure of this product.

need more **PRACTICE?** **Try Problems 19.47e, 19.50k, 19.56a,b, 19.59, 19.69**

## Steric effects

In many cases, steric effects can play an important role in determining product distribution. Let's begin with a simple case in which only one substituent is on the ring.

When an *ortho-para* director is present on the ring, it is difficult to predict the exact ratio of *ortho* and *para* products. Nevertheless, the following guidelines are helpful in most cases:

1. For most monosubstituted aromatic rings, the *para* product generally dominates over the *ortho* product as a result of steric considerations.

$\xrightarrow[H_2SO_4]{HNO_3}$ $NO_2$ (Major) + $NO_2$ (Minor)

Steric hindrance raises the energy of activation ($E_a$) for attack at the *ortho* position, and as a result, the *para* product is the major product. A notable exception is toluene (methylbenzene), for which the ratio of *ortho* and *para* products is sensitive to the conditions employed, such as the choice of solvent. In some cases, the *para* product is favored; in others, the *ortho* product is favored; Therefore, it is generally not wise to utilize the directing effects of a methyl group to favor a reaction at the *para* position over the *ortho* position.

2. For 1,4-disubstituted aromatic rings, steric effects again play a significant role. Consider the following case:

$\xrightarrow[H_2SO_4]{HNO_3}$ $NO_2$ (Major) + $NO_2$ (Minor)

The regiochemical outcome of this reaction is controlled by steric effects. Nitration is more likely to occur at the site that is less sterically hindered (*ortho* to the methyl group).

3. For 1,3-disubstituted aromatic rings, it is extremely unlikely that substitution will occur at the position between the two substituents. That position is the most sterically hindered position on the ring, and a reaction generally does not take place at that position.

$HNO_3$ / $H_2SO_4$ → Major ($NO_2$) + Minor ($O_2N$) (Not observed)

Using the previous three guidelines, let's get some practice predicting the product distribution in cases where steric effects control the outcome.

# SKILLBUILDER

## 19.3 IDENTIFYING STERIC EFFECTS FOR DISUBSTITUTED AND POLYSUBSTITUTED BENZENE RINGS

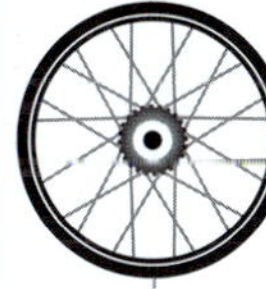

**LEARN the skill** In the following compound, determine the position that is most likely to be the site of an electrophilic aromatic substitution reaction.

### SOLUTION

**STEP 1** Identify the nature of each group.

Begin by identifying the effect of each group on the aromatic ring.

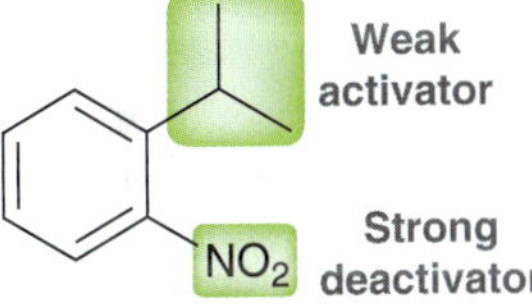


**STEP 2** Select the most powerful activator and identify the positions that are *ortho* or *para* to that group.

In this case, a weak activator is competing with a strong deactivator for directing effects. Recall that the more powerful *activator* controls the directing effects, so the isopropyl group determines the outcome in this case. The isopropyl group is *ortho-para* directing, so we must consider the positions that are *ortho* and *para* to the isopropyl group.

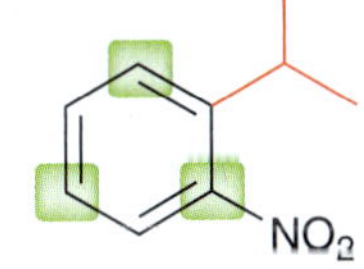


**STEP 3** Identify the unoccupied positions that are the least sterically hindered.

One of these locations is already occupied, leaving two choices. The *ortho* position is sterically hindered, while the *para* position is sterically unencumbered. In this case, we expect substitution to take place primarily at the *para* position and to a lesser extent at the *ortho* position.

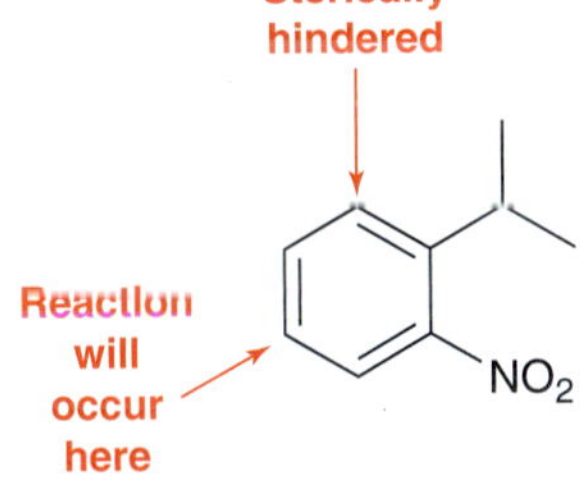

**PRACTICE the skill** **19.22** For each of the following compounds, determine the position that is most likely to be the site of an electrophilic aromatic substitution reaction:

(a) (b) (c)

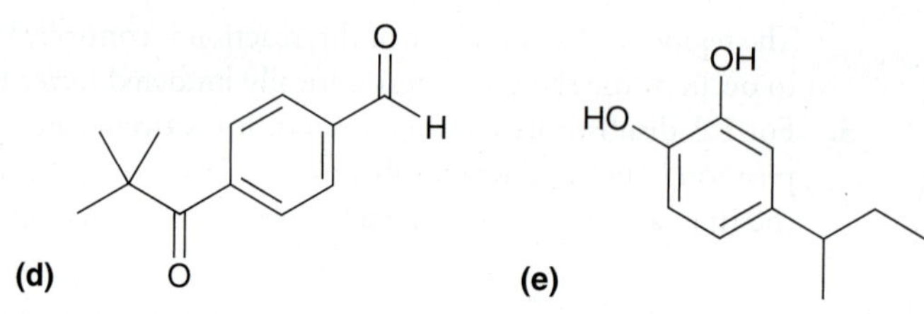

**19.23** The following compound is highly activated but nevertheless undergoes bromination very slowly. Explain.

**19.24** When the following compound is treated with $Br_2$ in the presence of a Lewis acid, one product predominates. Determine the structure of that product.

$Br_2$ / $FeBr_3$ → ?

need more **PRACTICE?** **Try Problems 19.59, 19.63, 19.69**

## Blocking Groups

Consider how the following transformation might be achieved:

? → Br

Direct bromination of *tert*-butylbenzene produces the *para* product as the major product, while the desired *ortho* product is the minor product. In such a situation, a **blocking group** can be used to direct the bromination toward the *ortho* position. In this case, the blocking group is first installed at the *para* position.

Install blocking group → Blocking group → Brominate → Br, Blocking group → Remove blocking group → Br

Once the *para* position is occupied, the desired reaction is forced to occur at the *ortho* position. Finally the blocking group is removed. In order for a group to function as a blocking group, it must

be easily removable after the desired reaction has been achieved. There are many different blocking groups that can be used. Sulfonation is commonly used for this purpose, because the sulfonation process is reversible.

Concentrated fuming $H_2SO_4$

Dilute $H_2SO_4$

Sulfonation provides a valuable blocking technique that enables us to achieve the desired transformation.

Fuming $H_2SO_4$ — $Br_2$, $FeBr_3$ — Dilute $H_2SO_4$

# SKILLBUILDER

## 19.4 USING BLOCKING GROUPS TO CONTROL THE REGIOCHEMICAL OUTCOME OF AN ELECTROPHILIC AROMATIC SUBSTITUTION REACTION

**LEARN** the skill

Identify whether a blocking group is necessary to accomplish the following transformation:

### SOLUTION

Analyze the starting material. The two substituents are the methoxy group, which is a moderate activator, and the acyl group, which is a moderate deactivator.

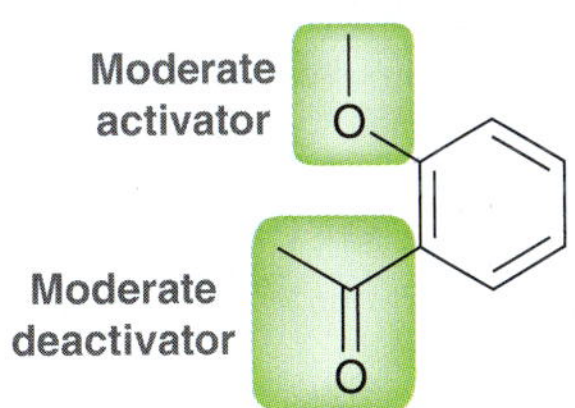


In this case, the methoxy group controls the directing effects, and therefore, the reactive centers are the unoccupied *ortho* and *para* positions.

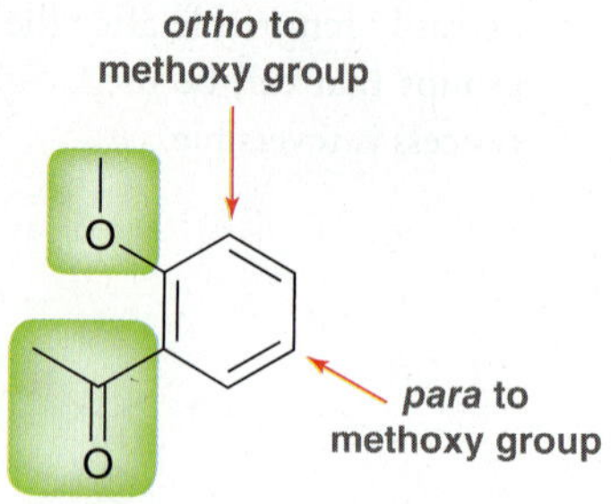

*ortho* to methoxy group

*para* to methoxy group

The *ortho* position is more sterically hindered, while the *para* position is not hindered. We therefore expect a substitution reaction to take place at the position that is *para* to the methoxy group. If we want to install a group in the *ortho* position, a blocking group would be required.

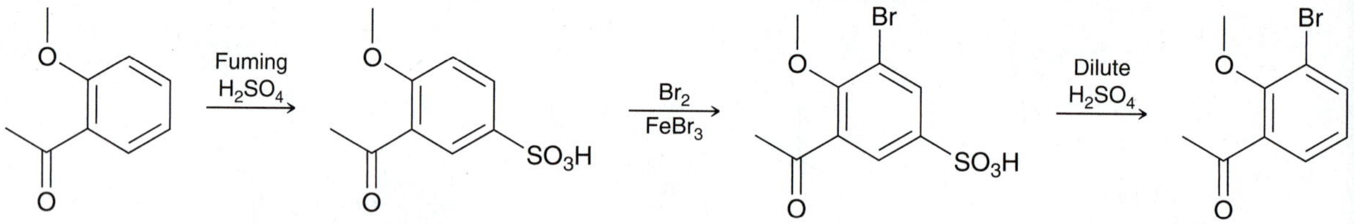

Fuming $H_2SO_4$ → $SO_3H$ ; $Br_2$, $FeBr_3$ → Br, $SO_3H$ ; Dilute $H_2SO_4$ → Br

**PRACTICE the skill** **19.25** Determine whether a blocking group is necessary to accomplish each of the following transformations:

(a) → $NO_2$ (b) → Br

(c) → Br (d) HO → HO, $NO_2$

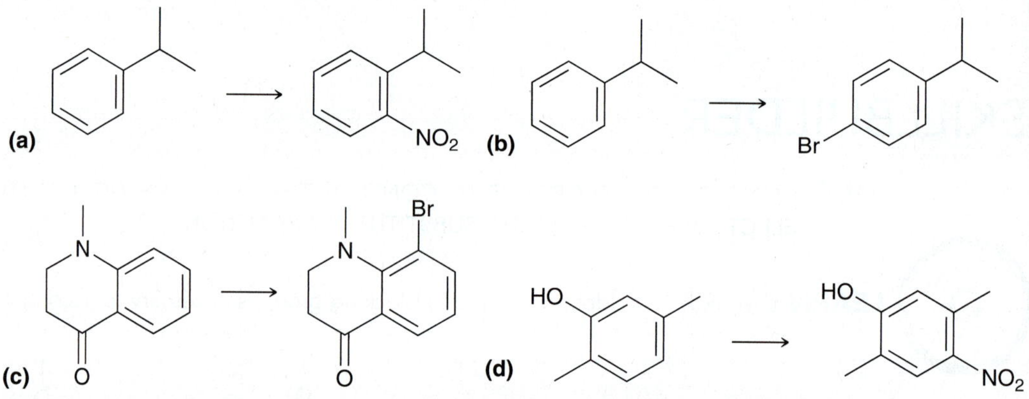

**APPLY the skill** **19.26** Predict the major product of the following reaction:

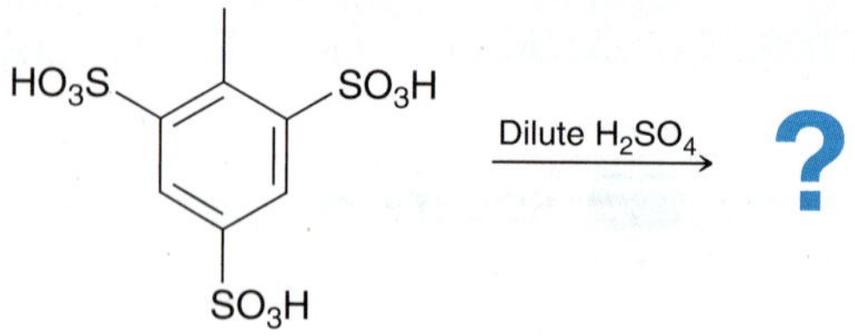

$HO_3S$, $SO_3H$, $SO_3H$ —Dilute $H_2SO_4$→ ?

**19.27** The following transformations cannot be accomplished, even with the help of blocking groups. In each case, explain why a blocking group will not help.

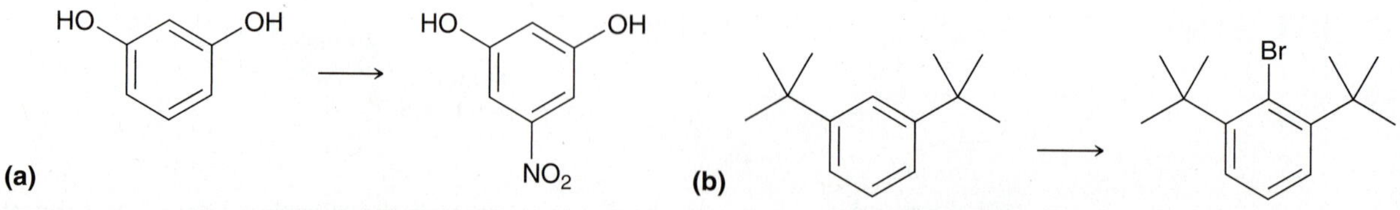

(a) HO, OH → HO, OH, $NO_2$ (b) → Br

need more **PRACTICE?** **Try Problems 19.58d, 19.68c**

## 19.12 Synthesis Strategies

### Monosubstituted Benzene Rings

The simplest kind of synthesis problem is one that requires formation of a monosubstituted benzene ring. Directing effects are irrelevant in such a case. You simply need to know what reagents are necessary to install the desired group. Figure 19.14 is a list of the reagents that we have seen thus far. This list should be committed to memory before moving on to more sophisticated synthesis problems. In total, we have seen 10 different groups that can be installed on an aromatic ring. Take special notice of the four groups shown in blue. Installation of these groups requires two steps.

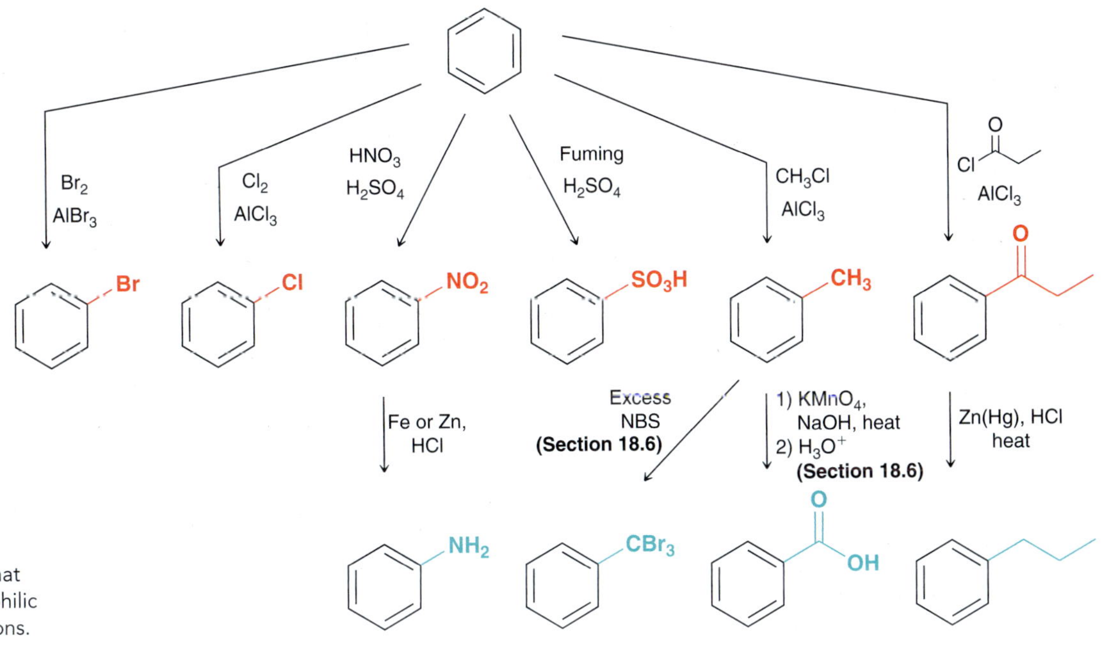

$Br_2$ / $AlBr_3$ → Br; $Cl_2$ / $AlCl_3$ → Cl; $HNO_3$ / $H_2SO_4$ → $NO_2$; Fuming $H_2SO_4$ → $SO_3H$; $CH_3Cl$ / $AlCl_3$ → $CH_3$; Cl(C=O)CH₂CH₃ / $AlCl_3$ → propanoyl

$NO_2$ → Fe or Zn, HCl → $NH_2$; $CH_3$ → Excess NBS (Section 18.6) → $CBr_3$; $CH_3$ → 1) $KMnO_4$, NaOH, heat 2) $H_3O^+$ (Section 18.6) → COOH; propanoyl → Zn(Hg), HCl, heat → propyl

**FIGURE 19.14** A list of functional groups that can be installed via electrophilic aromatic substitution reactions.

### CONCEPTUAL CHECKPOINT

**19.28** Identify the reagents necessary to convert benzene into each of the following compounds:

(a) Chlorobenzene (b) Nitrobenzene
(c) Bromobenzene (d) Ethylbenzene
(e) Propylbenzene (f) Isopropylbenzene
(g) Aniline (aminobenzene) (h) Benzoic acid
(i) Toluene

**19.29** Identify the product obtained when benzene is treated with each of the following reagents:

(a) Fuming sulfuric acid (b) $HNO_3/H_2SO_4$
(c) $Cl_2$, $AlCl_3$ (d) Ethyl chloride, $AlCl_3$
(e) $Br_2$, Fe
(f) $HNO_3/H_2SO_4$ followed by Zn, HCl

### Disubstituted Benzene Rings

Proposing a synthesis for a disubstituted benzene ring requires a careful analysis of directing effects to determine which group should be installed first. As an example, consider the following compound:

Br, $NO_2$ (meta-bromonitrobenzene structure)

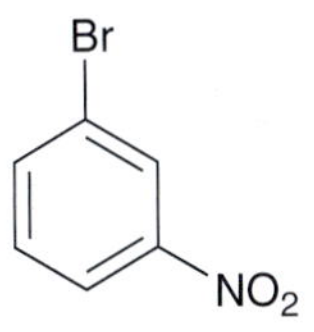

To make this compound from benzene requires two separate steps—bromination and nitration. Bromination followed by nitration will not produce the desired product, because a bromine substituent is *ortho-para* directing. In order to achieve the *meta* relationship between the two groups, nitration must be performed first. The nitro group is a *meta* director, which then directs the incoming bromine to the desired location.

The preceding example is fairly straightforward, because each group is installed with only one step. An extra consideration is necessary when installation of one of the groups requires two steps and involves a change in directing effects. These changes are summarized in Table 19.2.

**TABLE 19.2 FUNCTIONAL GROUP CONVERSIONS THAT CHANGE DIRECTING EFFECTS**

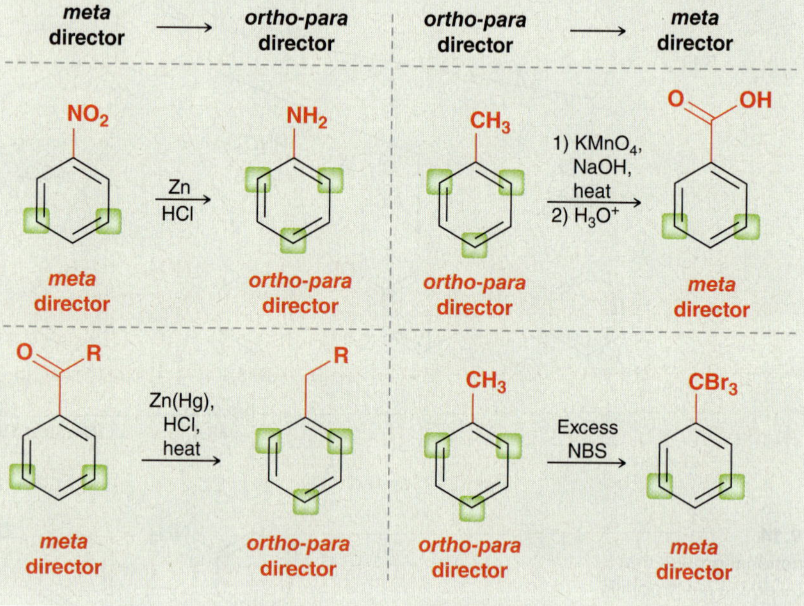

| *meta* director | → | *ortho-para* director | *ortho-para* director | → | *meta* director |
|---|---|---|---|---|---|
| $NO_2$ (*meta* director) | Zn, HCl | $NH_2$ (*ortho-para* director) | $CH_3$ (*ortho-para* director) | 1) $KMnO_4$, NaOH, heat 2) $H_3O^+$ | O, OH (*meta* director) |
| O, R (*meta* director) | Zn(Hg), HCl, heat | R (*ortho-para* director) | $CH_3$ (*ortho-para* director) | Excess NBS | $CBr_3$ (*meta* director) |

As an example, consider the installation of an amino group, which requires (1) nitration followed by (2) reduction. The reduction converts a *meta*-directing nitro group into an *ortho-para*-directing amino group. This change in directing effects must be considered when planning a synthesis that requires installation of an amino group. To illustrate this point, consider the following example:

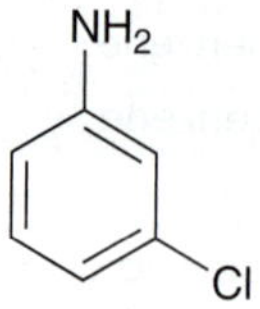

$NH_2$

Cl

This compound has two groups that are *meta* to each other, and we must decide which group to install first. The problem is that both groups are *ortho-para* directing, so neither group will direct the other group into the correct location. This problem can be solved if we recognize that installation of the amino group involves a change in directing effects.

$\xrightarrow[H_2SO_4]{HNO_3}$ $NO_2$ $\xrightarrow[HCl]{Zn}$ $NH_2$

*meta* director

*ortho-para* director

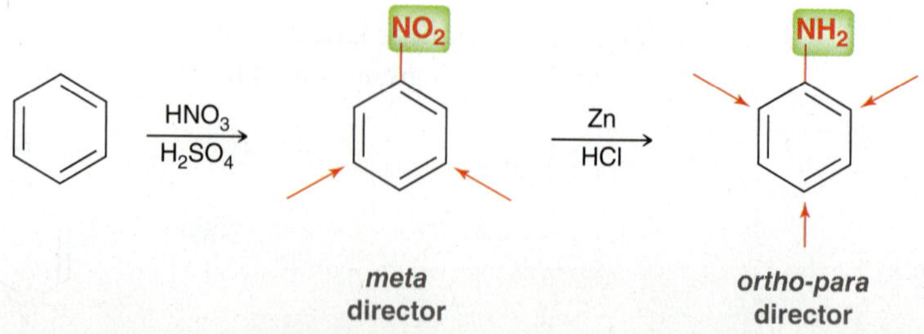

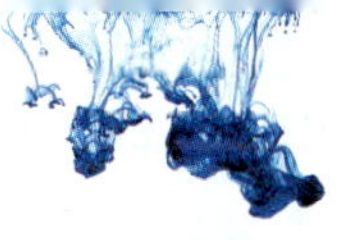

The nitro group is *meta* directing, while the amino group is *ortho-para* directing. The two steps above do not have to be consecutive, and we can exploit the *meta*-directing properties of the nitro group to install the chlorine substituent in the correct position. Specifically, the correct regiochemical outcome is achieved with the following order of events: (1) nitration, (2) chlorination, and (3) reduction.

1) $HNO_3/H_2SO_4$
2) $Cl_2$, $AlCl_3$
3) Zn, HCl

In addition to considering the order of events, the following limitations must also be considered when planning a synthesis.

1. Nitration cannot be performed on a ring that contains an amino group.

   The reagents for nitration (a mixture of $HNO_3$ and $H_2SO_4$) can oxidize the amino group, often leading to a mixture of undesirable products. Attempts to perform this reaction often produce a tarry substance.

2. A Friedel-Crafts reaction (either alkylation or acylation) cannot be accomplished on rings that are either moderately or strongly deactivated. The ring must be either activated or weakly deactivated in order for a Friedel-Crafts reaction to occur.

Moderate deactivator — $CH_3Cl$, $AlCl_3$ → No reaction

Strong deactivator — $CH_3Cl$, $AlCl_3$ → No reaction

# SKILLBUILDER

## 19.5 PROPOSING A SYNTHESIS FOR A DISUBSTITUTED BENZENE RING

**LEARN the skill** Starting with benzene and using any other necessary reagents of your choice, design a synthesis of the following compound:

### SOLUTION

**STEP 1** Identify the reagents necessary to install each substituent.

Installation of the amino group requires a two-step process—nitration followed by reduction. Installation of the propyl group also requires a two-step process—acylation followed by reduction (in order to avoid carbocation rearrangements that would occur during direct alkylation).

1) $HNO_3$, $H_2SO_4$
2) Zn, HCl

$NH_2$

1) $AlCl_3$, Cl
2) Zn(Hg), HCl, heat

**STEP 2**
Determine the order of events that achieves the desired regiochemical outcome.

Now let's consider the order of events. These two groups must be installed in an order that places them *meta* to each other. The amino group is *ortho-para* directing, so it cannot be installed first. However, the propyl group is also *ortho-para* directing, so it too cannot be installed first. In this case, we are forced to exploit the *meta*-directing effects of either the nitro group or the acyl group.

$HNO_3$ / $H_2SO_4$ → $NO_2$ ***meta* directing** → Zn / HCl → $NH_2$

Cl, $AlCl_3$ → O ***meta* directing** → Zn(Hg) / HCl, heat

By taking advantage of the *meta*-directing effects of either the nitro group or the acyl group, we have two possible routes to consider:

**Route 1**

Nitration → $NO_2$ → Acylation → $NO_2$, O → Reduction of both groups → $NH_2$

**The nitro group is *meta* directing**

**Route 2**

Acylation → O → Nitration → O, $NO_2$ → Reduction of both groups → $NH_2$

**The acyl group is *meta* directing**

When considering the viability of each route, we must make sure not to violate either of the following two limitations: (1) Nitration cannot be performed on a ring possessing an amino group and (2) Friedel-Crafts reactions cannot be performed on a moderately or strongly deactivated ring.

Neither of the proposed routes violates the first limitation, but one of the routes does, indeed, violate the second limitation. Specifically, the first route involves a Friedel-Crafts acylation with a strongly deactivated ring (nitrobenzene). This will not work. Therefore, only

the second route is viable. In the final step of this route, both groups are reduced under Clemmensen conditions.



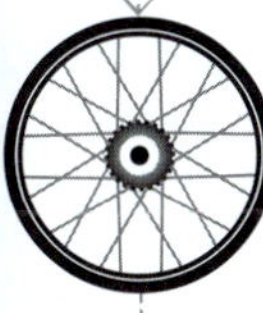

**PRACTICE the skill** **19.30** Starting with benzene and using any other necessary reagents of your choice, design a synthesis for each of the following compounds. Note: some of these problems have more than one plausible answer.

**APPLY the skill** **19.31** Using only reactions that we learned in this chapter, there are two different ways to prepare the following compound from benzene. Identify both ways and then determine which way is likely to produce a better yield of the desired product. Explain your choice.

**19.32** The following compounds cannot be made using only reactions that we learned in this chapter. For each compound, explain the issues that prevent its formation:



need more **PRACTICE?** **Try Problems 19.57, 19.58, 19.68, 19.75**

## Polysubstituted Benzene Rings

When designing a synthesis for a polysubstituted benzene ring, it is often most efficient to utilize a retrosynthetic analysis, as discussed in Section 12.5. The following example illustrates the process.

# SKILLBUILDER

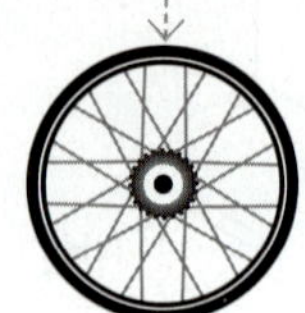

## 19.6 PROPOSING A SYNTHESIS FOR A POLYSUBSTITUTED BENZENE RING

**LEARN the skill**

Starting with benzene and using any other necessary reagents of your choice, design a synthesis for the following compound:

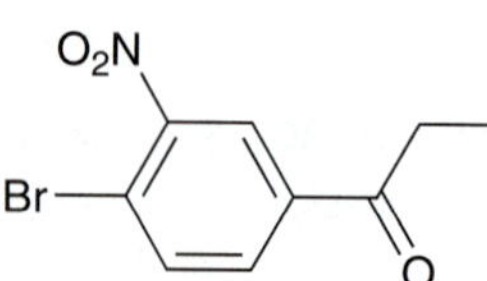

### SOLUTION

**STEP 1**
Determine the last step of the synthesis.

Approaching this problem from a retrosynthetic point of view, we begin by determining the last step of the synthesis. There are three possibilities: (1) the Br group is installed last, (2) the $NO_2$ group is installed last, or (3) the acyl group is installed last.
Let's begin by supposing that the Br group is installed last.

Recall that this arrow is a retrosynthetic arrow, and it means that the first compound might be made from the second compound. Our last step would therefore be a bromination reaction.

To consider the plausibility of this as our last step, we must first examine whether the desired regiochemical outcome will be achieved. In this case, we are trying to achieve the bromination of a disubstituted ring in which both groups (the nitro group and the acyl group) are *meta* directors. In such a case, the incoming Br group would be installed *meta* to both groups, which is not the desired location. Therefore, this step cannot be the last step of our synthesis.

Instead, let's consider installation of the acyl group as the last step. Once again, we must consider whether the desired regiochemical outcome would be achieved.

In this case, the desired regiochemical outcome would be achieved because both the Br group and the $NO_2$ group direct to the desired location. However, this transformation is a Friedel-Crafts acylation, so we must consider whether the proposed reaction violates any of the limitations for Friedel-Crafts acylation. In fact, there is a violation here because the ring is strongly deactivated by the presence of a nitro group, and the desired reaction simply cannot be achieved.

There is only one possibility left for the last step of our synthesis, which must be installation of the nitro group.

Both the Br group and the acyl group will direct the incoming nitro group to the desired location, and this reaction is plausible. This must be the last step.

Continuing to work backward, we must now consider the order in which the remaining two groups should be installed:

**STEP 2**
Determine the penultimate (second-to-last) step of the synthesis.

Once again, we must choose a sequence of events that achieves the desired regiochemical outcome. The Br and acyl groups are *para* to each other, so we must consider which group is a *para* director. Indeed, the Br group is an *ortho-para* director (with a preference for *para*), while the acyl group is a *meta* director. Therefore, we conclude that bromination be carried out first followed by acylation. Whenever performing an acylation step, it is necessary to consider the limitations of acylation. This case requires the acylation of bromobenzene. The Br group is only *weakly* deactivating, which does not interfere with the acylation process (acylation is only unattainable with moderately deactivated or strongly deactivated rings). In summary, our proposed synthesis has the following sequence of events:

**STEP 3**
Consider any limitations for a proposed step.

**STEP 4**
Redraw the synthesis from start to finish.

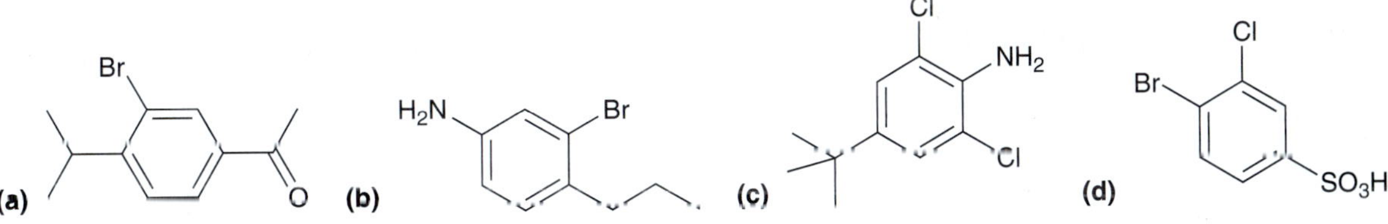

**PRACTICE the skill** **19.33** Starting with benzene and using any other necessary reagents of your choice, design a synthesis for each of the following compounds. In some cases, there may be more than one plausible answer.

**(a)** **(b)** **(c)** **(d)**

**APPLY the skill** **19.34** The compound below has a pentasubstituted benzene ring.

**(a)** Starting with benzene and using any other necessary reagents of your choice, design a synthesis for this compound.

**(b)** It is very difficult to install a sixth substituent. Explain.

**(c)** Is the ring activated or deactivated (relative to benzene)? Justify your answer.

need more **PRACTICE?** **Try Problems 19.68, 19.73**

## 19.13 Nucleophilic Aromatic Substitution

Thus far, we have only explored reactions in which the aromatic ring attacks an electrophile ($E^+$). Such reactions are called *electrophilic* aromatic substitution reactions. In this section, we consider reactions in which the ring is attacked by a nucleophile. Such reactions are called **nucleophilic aromatic substitution** reactions. In the following example, an aromatic compound is treated with a strong nucleophile (hydroxide), which displaces a leaving group (bromide):

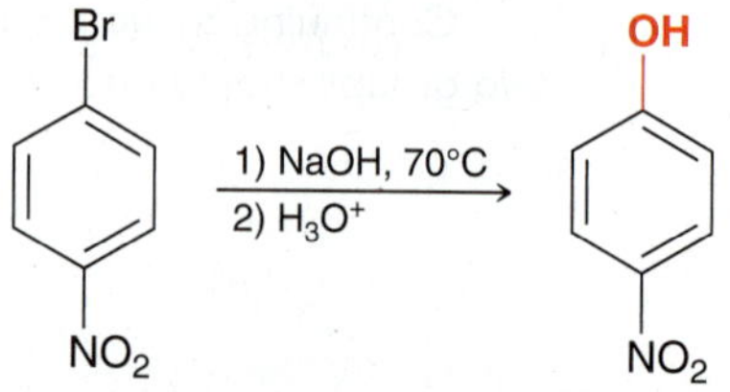

1) NaOH, 70°C
2) $H_3O^+$

In order for a reaction like this to occur, three criteria must be satisfied:

1. The ring must contain a powerful electron-withdrawing group (typically a nitro group).
2. The ring must contain a leaving group (usually a halide).
3. The leaving group must be either *ortho* or *para* to the electron-withdrawing group. If the leaving group is *meta* to the nitro group, the reaction is not observed.

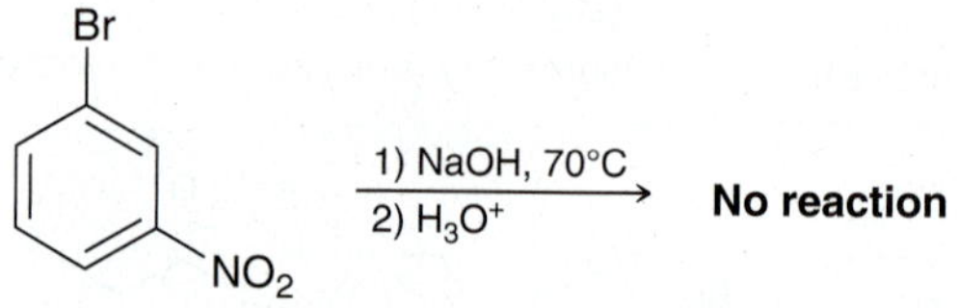

1) NaOH, 70°C
2) $H_3O^+$
**No reaction**

In this example, the first two criteria are met, but the last criterion is not met.

Any mechanism that we propose for nucleophilic aromatic substitution must successfully explain the three criteria. Mechanism 19.8 accomplishes this task and is called the $S_NAr$ mechanism.

### MECHANISM 19.8 NUCLEOPHILIC AROMATIC SUBSTITUTION ($S_NAr$)

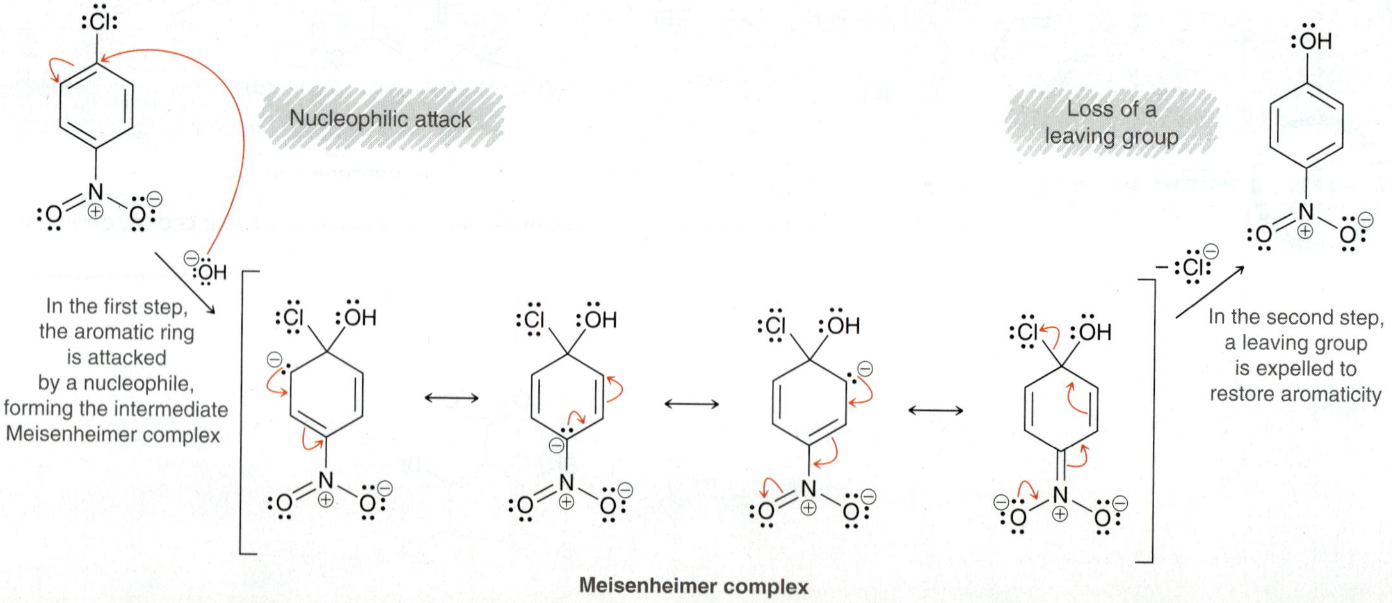

Nucleophilic attack

In the first step, the aromatic ring is attacked by a nucleophile, forming the intermediate Meisenheimer complex

Loss of a leaving group

In the second step, a leaving group is expelled to restore aromaticity

**Meisenheimer complex**

Much like the reactions we have seen thus far, this mechanism also involves two steps, but take special notice of the resonance-stabilized intermediate, called a **Meisenheimer complex**. This intermediate exhibits a negative charge that is resonance stabilized throughout the ring. This intermediate is very different from a sigma complex, which exhibits a positive charge that is resonance stabilized throughout the ring. The difference between these intermediates should make sense in that electrophilic aromatic substitution involves the ring attacking $E^+$, so the resulting intermediate will be positively charged; nucleophilic aromatic substitution involves the ring being attacked by a negatively charged nucleophile, so the resulting intermediate will be negatively charged. The second step of the $S_NAr$ mechanism involves loss of a leaving group to restore aromaticity.

In order to understand the role of the nitro group in this reaction, consider the last resonance structure of the Meisenheimer complex in Mechanism 19.8. In that resonance structure, the negative charge is removed from the ring and resides on an oxygen atom. This resonance structure stabilizes the Meisenheimer complex, and we can think of the nitro group as a temporary reservoir for electron density. That is, the nucleophile attacks the ring, dumping its electron density into the ring, where it is temporarily stored on the nitro group. Then, the nitro group releases the electron density to expel a leaving group. With this in mind, we can understand the requirement for the nitro group as well as the requirement for the leaving group. In addition, the $S_NAr$ mechanism also explains the requirement for the nitro group to be *ortho* or *para* to the leaving group. If the nitro group is *meta* to the leaving group, it cannot function as a reservoir. To convince yourself that this is the case, draw the structure of *meta*-chloronitrobenzene and then attack that structure with hydroxide at the position containing the chlorine atom. Try to draw the resonance structures of the intermediate that is generated and you will see that the negative charge cannot be placed on the nitro group.

When hydroxide is used as the attacking nucleophile, the resulting product is a substituted phenol, which will be deprotonated by hydroxide to give a phenolate ion. Therefore, acid is required in a separate step to protonate the phenolate ion and obtain a neutral product.

Br, NO2 — Excess NaOH, 70°C → :O:⁻, NO2 — $H_3O^+$ → :OH, NO2

## CONCEPTUAL CHECKPOINT

**19.35** Predict the product of the following reaction:

Br, Cl, NO2 — $NaOCH_3$, heat →

**19.36** Starting with benzene and using any other necessary reagents of your choice, design a synthesis for the following compound:

HO–C6H4–NH2

**19.37** The presence of additional nitro groups can have an impact on the temperature at which a nucleophilic aromatic substitution will readily occur. Consider the following example:

Cl, R, R, NO2 — 1) NaOH 2) $H_3O^+$ → OH, R, R, NO2

When both R groups are hydrogen atoms, the reaction readily occurs at 130°C. When one of the R groups is a nitro group, the reaction readily occurs at 100°C. When both R groups are nitro groups, the reaction readily occurs at 35°C.

(a) Provide an explanation that justifies the lower temperature requirement with additional nitro groups.

(b) If a fourth nitro group is placed on the ring, would you expect the temperature requirement to be significantly lowered further? Explain your answer.

## 19.14 Elimination-Addition

In the previous section, we explained why a nitro group is required in order for a nucleophilic aromatic substitution reaction to proceed. In the absence of a powerful electron-withdrawing substituent, the reaction simply does not occur.

Cl
NaOH / Heat → No reaction

However, if the temperature and pressure are raised significantly, a reaction is in fact observed.

Cl
1) NaOH, 350°C
2) $H_3O^+$
OH

This reaction was first discovered in 1928 by scientists at the Dow Chemical Company. The reaction can also be performed at lower temperatures using the amide ion ($H_2N^-$) as a nucleophile.

Cl
1) $NaNH_2$, $NH_3$ (*l*)
2) $H_3O^+$
$NH_2$

When other substituents are present on the ring, the regiochemical outcome is not what we might have expected.

Cl
1) $NaNH_2$, $NH_3$ (*l*)
2) $H_3O^+$
$NH_2$ + $NH_2$

In this case, two products are obtained. This regiochemical outcome initially baffled chemists, as it cannot be explained with a simple $S_NAr$ mechanism. Rather, a different mechanism is required to explain these results.

A clue to this puzzle comes from an isotopic labeling experiment. Chlorobenzene can be prepared such that the carbon bearing the chlorine atom is $^{14}C$, a radioactive isotope of carbon. The position of the isotopic label (indicated with an asterisk) can then be tracked before and after the reaction.

Cl
1) $NaNH_2$, $NH_3$ (*l*)
2) $H_3O^+$
$NH_2$ + $NH_2$
50% 50%

Notice the position of the isotopic label in the products. The proposed mechanism most consistent with these observations involves formation of a rather strange intermediate called **benzyne**.

Cl H :$NH_2^-$ ⇌ :Cl: → Benzyne

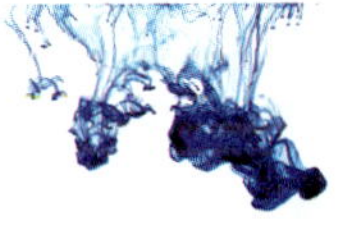



Elimination of H and Cl produces a very high energy intermediate called benzyne. This intermediate does not survive long because it is quickly attacked by the nucleophile, producing an addition reaction. Nucleophilic attack can take place at (a) the position of the isotopic label or (b) the other end of the triple bond.

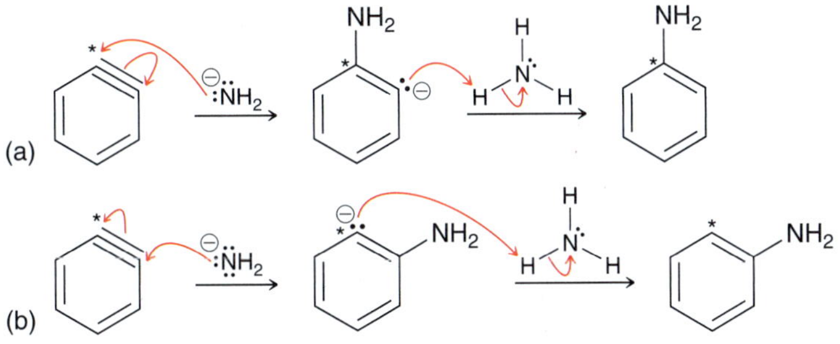

The attack can take place at either end of the triple bond with equal likelihood, explaining the observed results. This proposed mechanism is called **elimination-addition** (Mechanism 19.9).

## MECHANISM 19.9 ELIMINATION-ADDITION

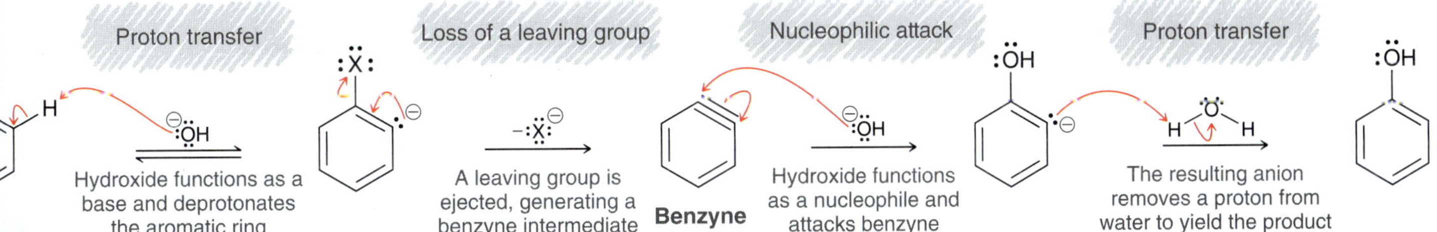


Evidence for this mechanism comes from a trapping experiment. When furan is added to the reaction mixture, a small amount of Diels-Alder cycloadduct is obtained.

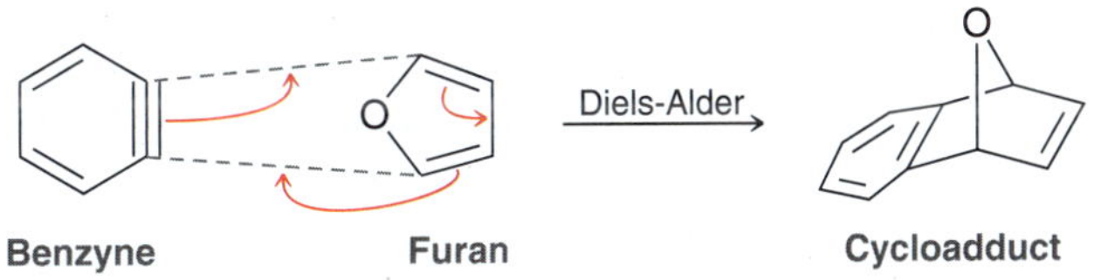

The presence of this cycloadduct can only be explained by invoking a benzyne intermediate, which is *trapped* by furan. The evidence requires that we explain how benzyne can exist, even for a brief moment. After all, a triple bond cannot be incorporated into a six-membered ring. This strange "triple bond" is best explained as resulting from the overlap of $sp^2$ orbitals rather than overlapping $p$ orbitals.

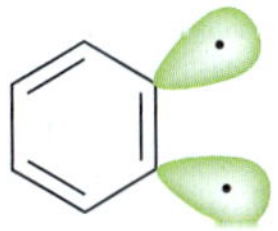

The overlap is very poor, and the intermediate more closely resembles a diradical than a triple bond. This explains why it is very unstable and so short-lived.

## CONCEPTUAL CHECKPOINT

**19.38** Draw both products that are obtained when 4-chloro-2-methyltoluene is treated with sodium amide followed by treatment with $H_3O^+$.

**19.39** Starting with benzene and using any other necessary reagents of your choice, design a synthesis for anisole (methoxybenzene).

## 19.15 Identifying the Mechanism of an Aromatic Substitution Reaction

We have seen three different mechanisms for aromatic substitution reactions (Figure 19.15).

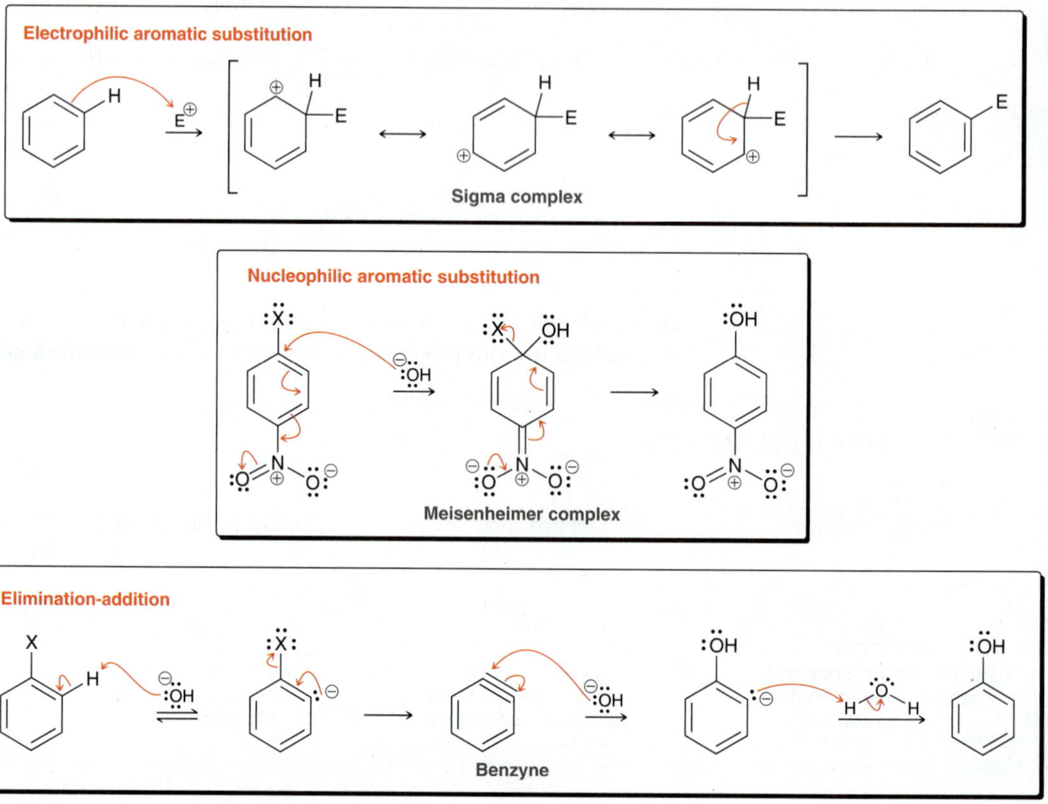

Electrophilic aromatic substitution

Sigma complex

Nucleophilic aromatic substitution

Meisenheimer complex

Elimination-addition

Benzyne

**FIGURE 19.15**
Three possible mechanisms for aromatic substitution.

All three mechanisms accomplish aromatic substitution, but there are a few key differences that warrant our attention:

1. *The intermediate:* Electrophilic aromatic substitution proceeds via a sigma complex, nucleophilic aromatic substitution proceeds via a Meisenheimer complex, and elimination-addition proceeds via a benzyne intermediate.
2. *The leaving group:* In electrophilic aromatic substitution, the incoming substituent replaces a proton. In the other two mechanisms, a negatively charged leaving group (such as a halide ion) is expelled.
3. *Substituent effects:* In electrophilic aromatic substitution, electron-withdrawing groups deactivate the ring toward attack, while in nucleophilic aromatic substitution, an electron-withdrawing group is required in order for the reaction to proceed.

Because of these fundamental differences, it is important to be able to determine which mechanism operates in any given situation. Figure 19.16 illustrates a decision tree for proposing a mechanism for an aromatic substitution.

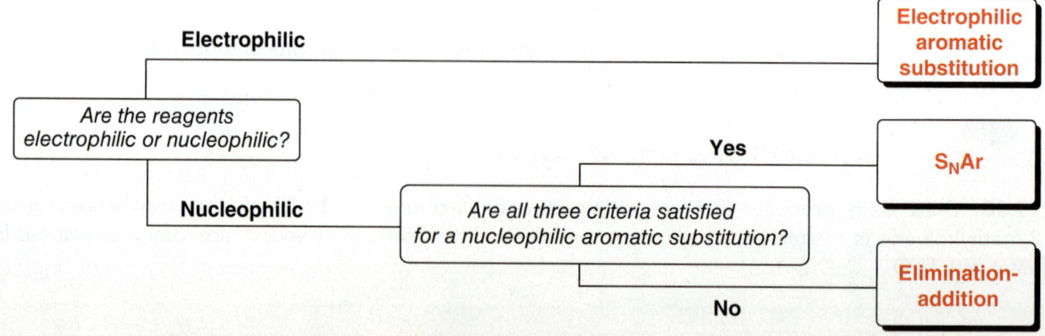


**FIGURE 19.16**
A decision tree for determining a mechanism for an aromatic substitution reaction.

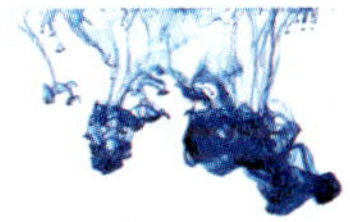

# SKILLBUILDER

## 19.7 DETERMINING THE MECHANISM OF AN AROMATIC SUBSTITUTION REACTION

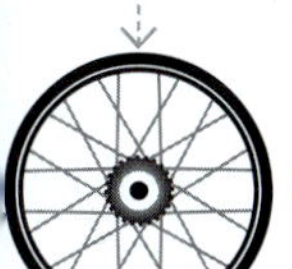

**LEARN the skill** Draw the most likely mechanism for the following transformation:

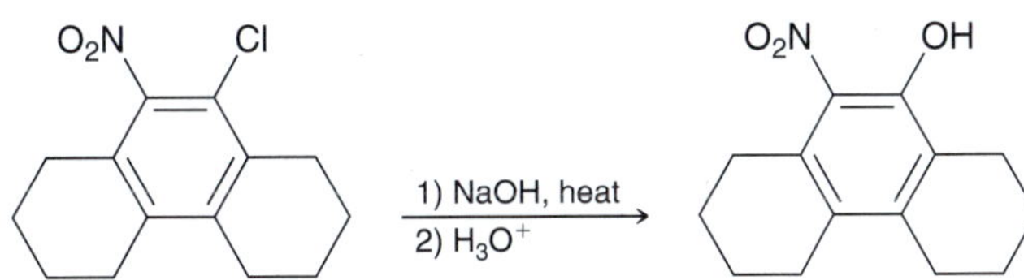

**SOLUTION**

**STEP 1**
Determine whether the reagents are electrophilic or nucleophilic.

First look at the reagents. NaOH is a common source of hydroxide ions ($Na^+$ is the counterion). Hydroxide is a nucleophile, not an electrophile, so we can rule out electrophilic aromatic substitution.

Next, look at the substrate to determine if all three criteria are present for a nucleophilic aromatic substitution: (1) there is a leaving group (chloride), (2) there is a nitro group, and (3) the nitro group is *ortho* to the leaving group. All three criteria are met, so the mechanism is likely to be $S_NAr$, which proceeds through a Meisenheimer complex.

**STEP 2**
If the reagent is nucleophilic, then determine if all three criteria are satisfied for a nucleophilic aromatic substitution.

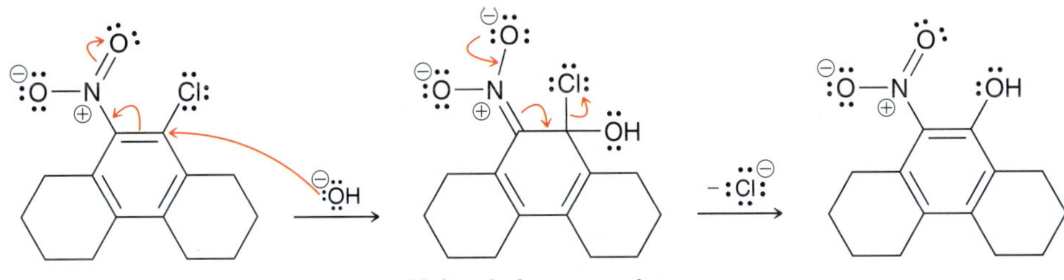


Meisenheimer complex

**PRACTICE the skill** **19.40** Draw the most likely mechanism for each of the following transformations:

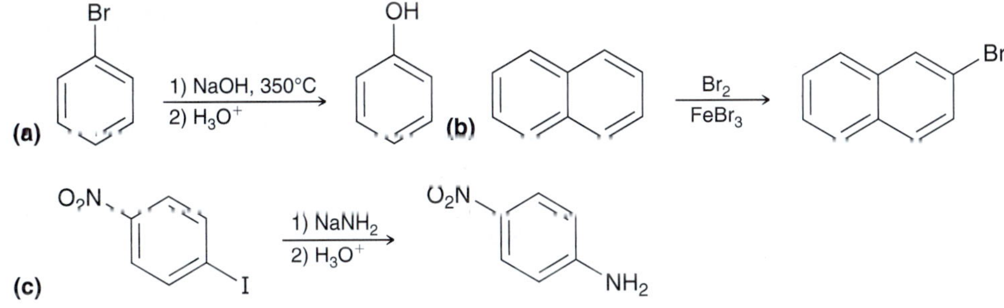


**19.41** When 2-ethyl-5-chlorotoluene was treated with sodium hydroxide at high temperature, followed by treatment with $H_3O^+$, three constitutional isomers with molecular formula $C_9H_{12}O$ were obtained. Draw all three products.

**APPLY the skill** **19.42** When *ortho*-bromonitrobenzene is treated with NaOH at elevated temperature, only one product is formed.

**(a)** Draw the product.

**(b)** Identify the intermediate formed en route to the product.

**(c)** Would the reaction occur if the starting compound were *meta*-bromonitrobenzene?

**(d)** Would the reaction occur if the starting compound were *para*-bromonitrobenzene?

need more **PRACTICE?** Try Problems 19.53–19.55, 19.60

# REVIEW OF REACTIONS

## Electrophilic Aromatic Substitution

Br2 AlBr3; Cl2 AlCl3; HNO3 H2SO4; Dilute H2SO4; Fuming H2SO4; CH3Cl AlCl3; AlCl3

Br 1; Cl 2; NO2 3; SO3H 4; CH3 5; O 6

Fe or Zn, HCl; Excess NBS (Section 18.6); 1) KMnO4, NaOH, heat 2) H3O+ (Section 18.6); Zn(Hg), HCl heat

NH2 7; CBr3 8; O OH 9; 10

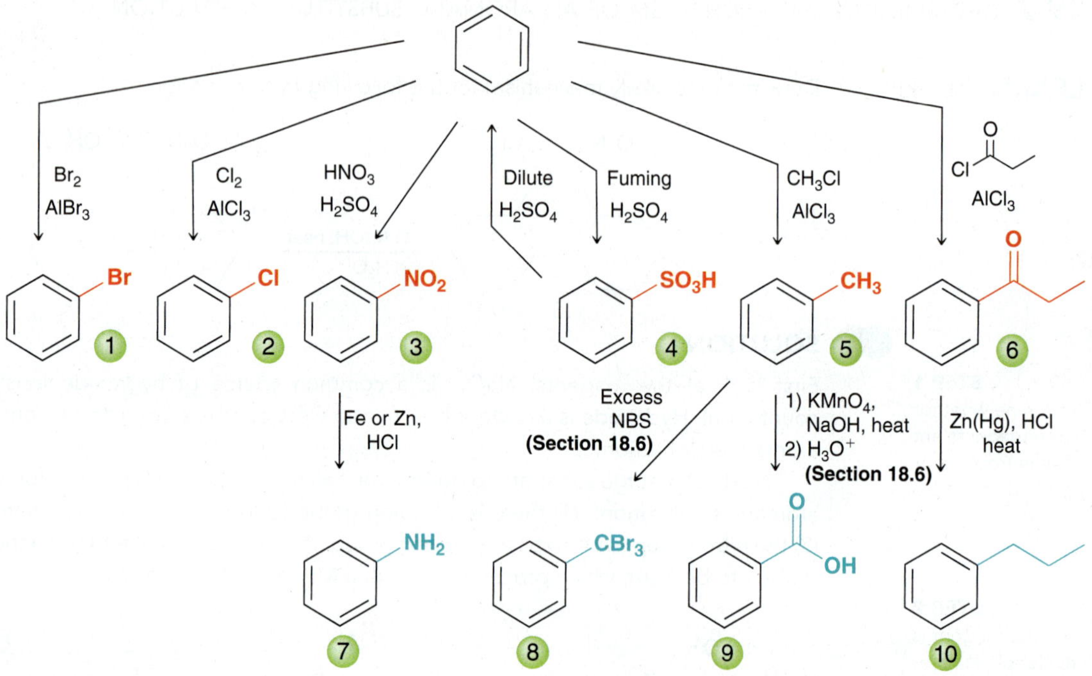

**1.** Bromination
**2.** Chlorination
**3.** Nitration
**4.** Sulfonation/desulfonation
**5.** Friedel-Crafts alkylation
**6.** Friedel-Crafts acylation
**7.** Reduction
**8.** Benzylic bromination
**9.** Oxidation
**10.** Clemmensen reduction

## Other Aromatic Substitution Reactions

### Nucleophilic Aromatic Substitution

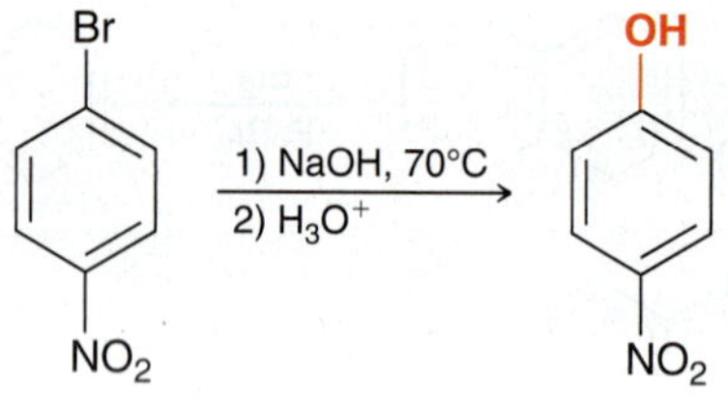

Br, NO2 → 1) NaOH, 70°C 2) $H_3O^+$ → OH, NO2

### Elimination-Addition

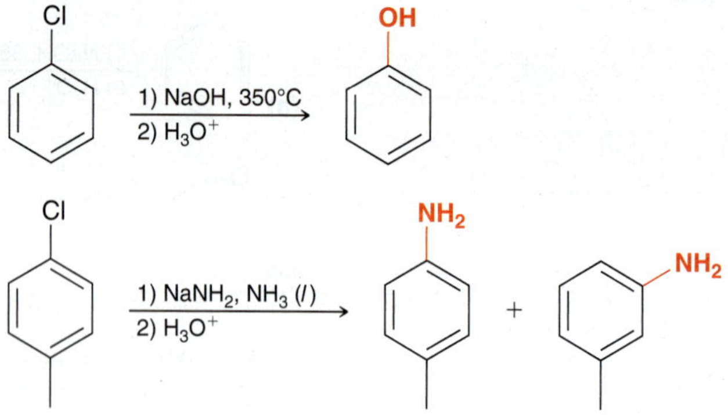

Cl → 1) NaOH, 350°C 2) $H_3O^+$ → OH

Cl → 1) $NaNH_2$, $NH_3$ (*l*) 2) $H_3O^+$ → NH2 + NH2

# REVIEW OF CONCEPTS AND VOCABULARY

### SECTION 19.1

- Alkenes undergo addition when treated with bromine, while benzene is inert under the same conditions.
- In the presence of iron, an **electrophilic aromatic substitution** reaction is observed between benzene and bromine.

### SECTION 19.2

- Iron tribromide is a Lewis acid that interacts with $Br_2$ and generates $Br^+$, which is sufficiently electrophilic to be attacked by benzene.

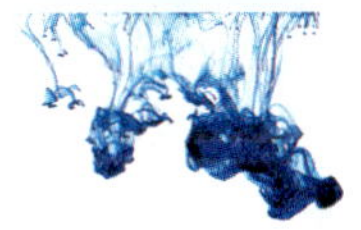



- Electrophilic aromatic substitution involves two steps:
  - Formation of the **sigma complex**, or arenium ion. This step is endergonic.
  - Deprotonation, which restores aromaticity.
- Aluminum tribromide ($AlBr_3$) is another common Lewis acid that can serve as a suitable alternative to $FeBr_3$.
- Chlorination of benzene is accomplished with a suitable Lewis acid, such as aluminum trichloride.

### SECTION 19.3

- Sulfur trioxide ($SO_3$) is a very powerful electrophile that is present in fuming sulfuric acid. Benzene reacts with $SO_3$ in a reversible process called **sulfonation**.

### SECTION 19.4

- A mixture of sulfuric acid and nitric acid produces a small amount of **nitronium ion** ($NO_2^+$). Benzene reacts with the nitronium ion in a process called **nitration**.
- A nitro group can be reduced to an amino group, providing a two-step method for installing an amino group.

### SECTION 19.5

- **Friedel-Crafts alkylation** enables the installation of an alkyl group on an aromatic ring.
- In the presence of a Lewis acid, an alkyl halide is converted into a carbocation, which can be attacked by benzene in an electrophilic aromatic substitution.
- A Friedel-Crafts alkylation is only efficient in cases where carbocation rearrangements cannot occur.
- When choosing an alkyl halide, the carbon atom connected to the halogen must be $sp^3$ hybridized.
- Polyalkylations are common and can generally be avoided by controlling the reaction conditions.

### SECTION 19.6

- **Friedel-Crafts acylation** enables the installation of an **acyl group** on an aromatic ring.
- When treated with a Lewis acid, an acyl chloride will generate an **acylium ion**, which is resonance stabilized and not susceptible to carbocation rearrangements.
- When a Friedel-Crafts acylation is followed by a **Clemmensen reduction**, the net result is the installation of an alkyl group. This two-step process is a useful synthetic method for installing alkyl groups that cannot be installed efficiently with a direct alkylation process.
- Polyacylation is not observed, because introduction of an acyl group deactivates the ring toward further acylation.

### SECTION 19.7

- A methyl group is said to **activate** an aromatic ring and is an ***ortho-para* director**.
- An aromatic ring is even more highly activated by a methoxy group, which is also an *ortho-para* director.
- All activators are *ortho-para* directors.

### SECTION 19.8

- A nitro group is said to **deactivate** an aromatic ring and is a *meta* director. Most deactivators are *meta* directors.

### SECTION 19.9

- Halogens (such as Cl, Br, or I) are an exception in that they are deactivators but are *ortho-para* directors.

### SECTION 19.10

- **Strong activators** are characterized by the presence of a lone pair immediately adjacent to the aromatic ring.
- **Moderate activators** exhibit a lone pair that is already delocalized outside of the ring. Alkoxy groups are an exception and are moderate activators.
- Alkyl groups are **weak activators**.
- Halogens (such as Cl, Br, or I) are **weak deactivators**.
- **Moderate deactivators** are groups that exhibit a $\pi$ bond to an electronegative atom, where the $\pi$ bond is conjugated with the aromatic ring.
- **Strong deactivators** are powerfully electron withdrawing, either by resonance or induction. There are three common groups that fall under this category.

### SECTION 19.11

- When multiple substituents are present, the more powerful activating group dominates the directing effects.
- Steric effects often play an important role in determining product distribution.
- A **blocking group** can be used to control the regiochemical outcome of an electrophilic aromatic substitution.

### SECTION 19.12

- Proposing a synthesis for a disubstituted benzene ring requires a careful analysis of directing effects to determine which group should be installed first.
- When designing a synthesis for a polysubstituted benzene ring, it is often most efficient to utilize a retrosynthetic analysis.

### SECTION 19.13

- In a **nucleophilic aromatic substitution** reaction, the aromatic ring is attacked by a nucleophile. This reaction has three requirements:
  - The ring must contain a powerful electron-withdrawing group (typically a nitro group).
  - The ring must contain a leaving group.
  - The leaving group must be either *ortho* or *para* to the electron-withdrawing group.
- Nucleophilic aromatic substitution involves two steps:
  - Formation of a **Meisenheimer complex**
  - Loss of a leaving group to restore aromaticity

### SECTION 19.14

- An **elimination-addition** reaction occurs via a **benzyne** intermediate. Evidence for this mechanism comes from isotopic labeling experiments as well as a trapping experiment.

### SECTION 19.15

- The three mechanisms for aromatic substitution differ in (1) the intermediate, (2) the leaving group, and (3) substituent effects.

# SKILLBUILDER REVIEW

## 19.1 IDENTIFYING THE EFFECTS OF A SUBSTITUENT

| ACTIVATORS | | | DEACTIVATORS | | |
|---|---|---|---|---|---|
| STRONG | MODERATE | WEAK | WEAK | MODERATE | STRONG |
| A lone pair immediately adjacent to the ring.<br>**Examples:** $-\ddot{N}H_2$ $-\ddot{O}H$<br>**Exception:** $-\ddot{O}R$ (Moderate activator) | A lone pair that is already participating in resonance outside of the ring.<br>**Example:** $-\ddot{O}-C(=O)-R$ | Alkyl groups:<br>$-R$ | Halogens:<br>$-Cl$<br>$-Br$<br>$-I$ | A $\pi$ bond to a heteroatom, where the $\pi$ bond is conjugated to the ring.<br>**Examples:** $-C(=O)-R$<br>$-C{\equiv}N$ | The following three groups:<br>$-NO_2$<br>$-NR_3^{\oplus}$<br>$-CX_3$ |
| ← *ORTHO-PARA* DIRECTORS → | | | | ← *META* DIRECTORS → | |

Try Problems 19.16–19.18, 19.44–19.46, 19.47a–c,f,h, 19.49a–d, 19.50a,b,d–g, 19.59a,b 19.64, 19.66

## 19.2 IDENTIFYING DIRECTING EFFECTS FOR DISUBSTITUTED AND POLYSUBSTITUTED BENZENE RINGS

**STEP 1** Identify the nature of each group.

**STEP 2** Select the most powerful activator and then identify the positions that are *ortho* and *para* to that group.

**STEP 3** Identify the unoccupied positions.

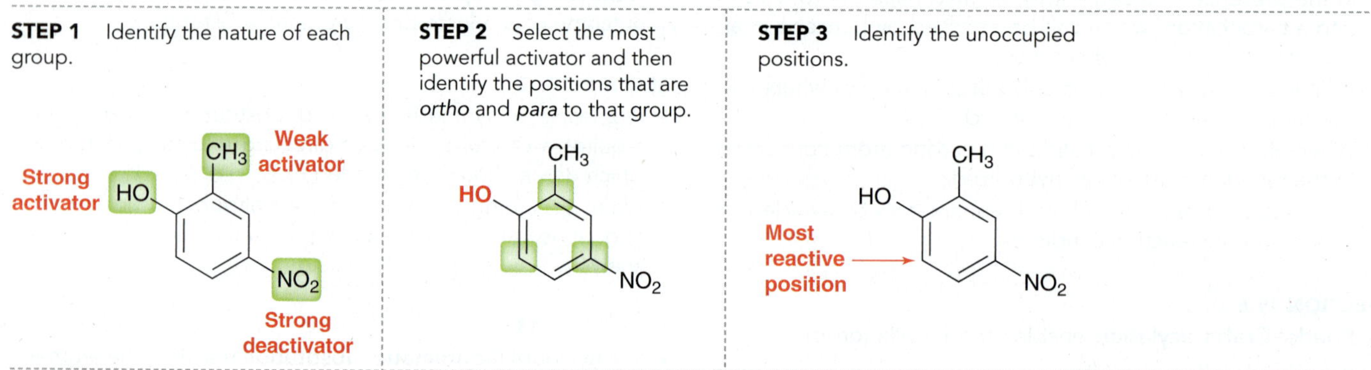


Try Problems 19.19–19.21, 19.47e, 19.50k, 19.56a,b, 19.59, 19.69

## 19.3 IDENTIFYING STERIC EFFECTS FOR DISUBSTITUTED AND POLYSUBSTITUTED AROMATIC BENZENE RINGS

**STEP 1** Identify the nature of each group.

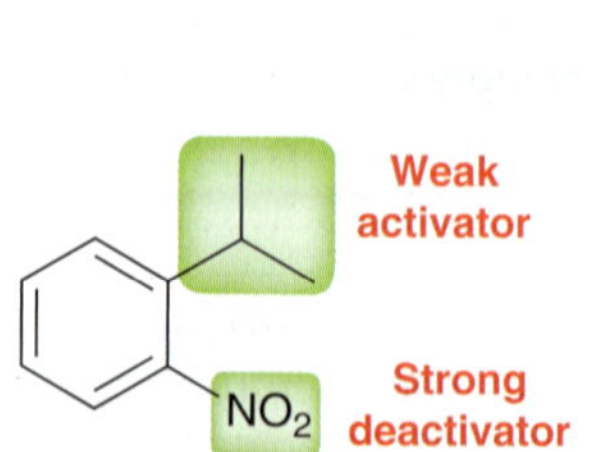


**STEP 2** Select the most powerful activator (shown in red below) and then identify the positions that are *ortho* and *para* to that group.

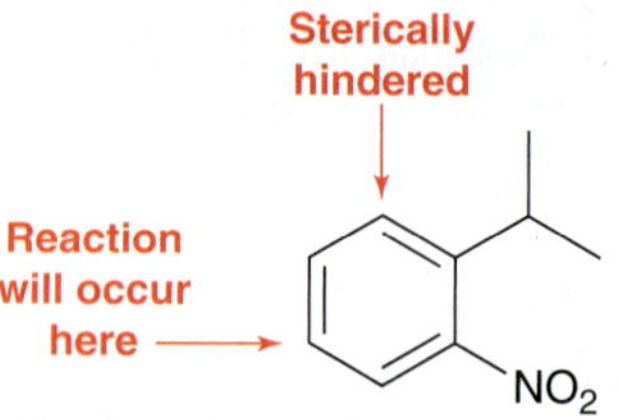


**STEP 3** Identify the unoccupied positions that are the least sterically hindered.

Sterically hindered

Reaction will occur here

$NO_2$

Try Problems 19.22–19.24, 19.59, 19.63, 19.69

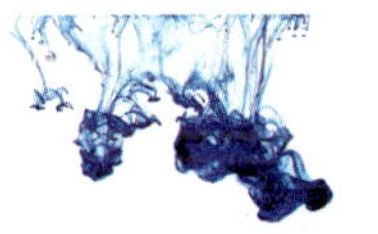



### 19.4 USING BLOCKING GROUPS TO CONTROL THE REGIOCHEMICAL OUTCOME OF AN ELECTROPHILIC AROMATIC SUBSTITUTION REACTION

**STEP 1** Determine whether or not a blocking group is necessary.

**STEP 2** Install the blocking group.

**STEP 3** Perform the desired reaction.

**STEP 4** Remove the blocking group.

Fuming $H_2SO_4$ → $SO_3H$ → $Br_2$, $FeBr_3$ → Br, $SO_3H$ → Dilute $H_2SO_4$ → Br

Try Problems 19.25–19.27, 19.58d, 19.68c

### 19.5 PROPOSING A SYNTHESIS FOR A DISUBSTITUTED BENZENE RING

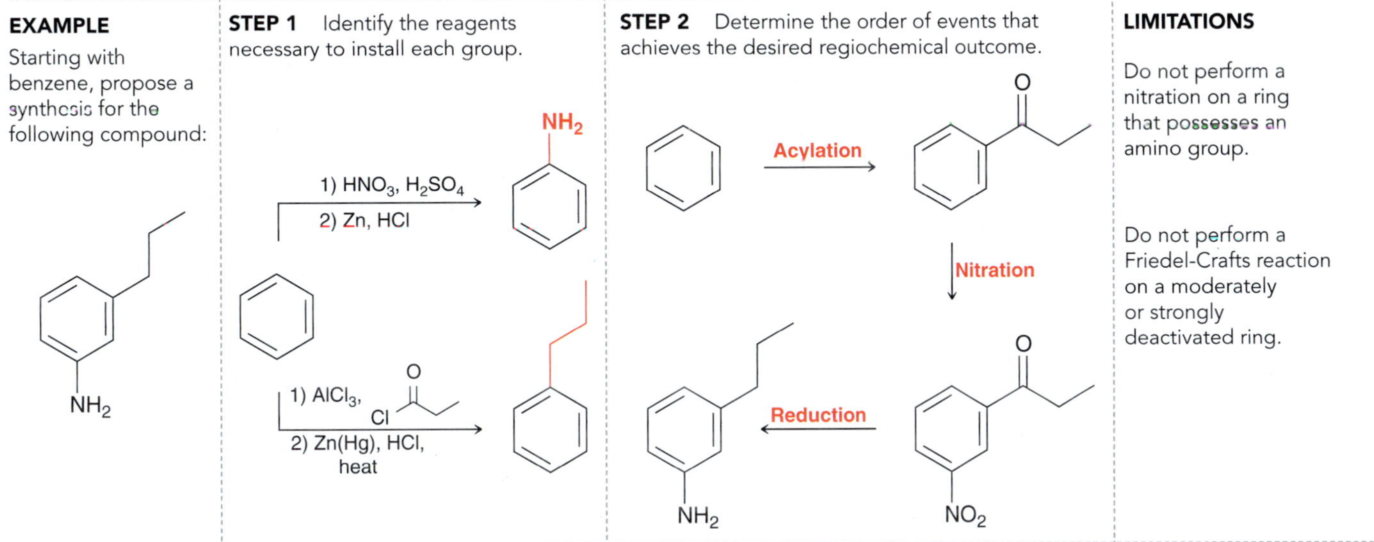


Try Problems 19.30–19.32, 19.57, 19.58, 19.68, 19.75

### 19.6 PROPOSING A SYNTHESIS FOR A POLYSUBSTITUTED BENZENE RING

**RETROSYNTHETIC ANALYSIS**

**Example:**

$O_2N$, Br, O ⟹ Br, O ⟹ Br ⟹ (benzene)

**FACTORS TO CONSIDER**

1. Electronic directing effects
2. Steric directing effects
3. Order of events
4. Limitations:
   - Cannot nitrate a ring that possesses an amino group
   - Cannot perform a Friedel-Crafts reaction with a moderately deactivated ring

Try Problems 19.33, 19.34, 19.68, 19.73

**19.7** DETERMINING THE MECHANISM OF AN AROMATIC SUBSTITUTION REACTION

**DECISION TREE**

Are the reagents nucleophilic or electrophilic?

**Electrophilic** → **Electrophilic aromatic substitution**

**Nucleophilic** → Are all three criteria satisfied for a nucleophilic aromatic substitution?

**Yes** → **$S_NAr$**

**No** → **Elimination-addition**

Try Problems 19.40–19.42, 19.53–19.55, 19.60

# PRACTICE PROBLEMS

Note: Most of the Problems are available within **WileyPLUS**, an online teaching and learning solution.

**19.43** Identify the reagents necessary to accomplish each of the following transformations:

**19.44** Rank the following compounds in order of increasing reactivity toward electrophilic aromatic substitution:

**19.45** Identify which of the following compounds is most activated toward electrophilic aromatic substitution. Which compound is least activated?

**19.46** Predict the product(s) obtained when each of the following compounds is treated with a mixture of nitric acid and sulfuric acid:

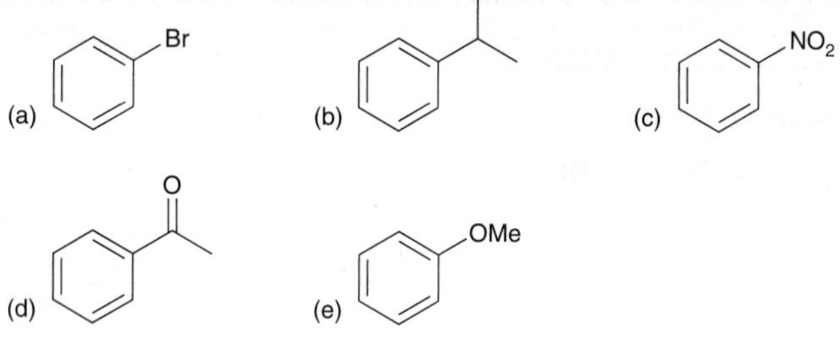


**19.47** Predict the major product obtained when each of the following compounds is treated with fuming sulfuric acid:

(a) Chlorobenzene
(b) Phenol
(c) Benzaldehyde
(d) *ortho*-Nitrophenol
(e) *para*-Bromotoluene
(f) Benzoic acid
(g) *para*-Ethyltoluene
(h) Benzene

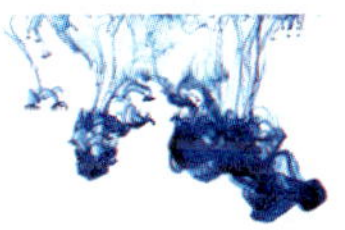

**19.48** For each of the following groups, identify whether it is an activator or a deactivator and determine its directing effects:

(a) –OMe (b) –O–C(=O)CH$_3$ (c) –$NH_2$ (d) –Cl (e) –$CCl_3$ (f) –$NO_2$ (g) –C(=O)OH (h) –C(=O)H (i) –Br (j) –$\overset{\oplus}{N}Me_3$

**19.49** Predict the product(s) obtained when each of the following compounds is treated with chloromethane and aluminum trichloride. Some of the compounds might be unreactive. For those that are reactive, assume that conditions are controlled to favor monoalkylation.

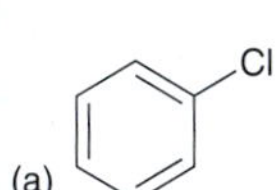


(a) (b)

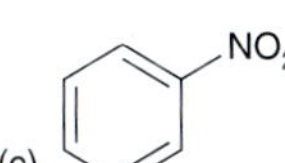


(c) (d)

(e) (f)

(g) (h)

**19.50** Predict the major product obtained when each of the following compounds is treated with bromine in the presence of iron tribromide:

(a) Bromobenzene (b) Nitrobenzene
(c) *ortho*-Xylene (d) *tert*-Butylbenzene
(e) Benzenesulfonic acid (f) Benzoic acid
(g) Benzaldehyde (h) *ortho*-Dibromobenzene
(i) *meta*-Nitrotoluene (j) *meta*-Dibromobenzene
(k) *para*-Dibromobenzene

**19.51** When the following compound is treated with a mixture of nitric and sulfuric acid at 50°C, nitration occurs to afford a compound with two nitro groups. Draw the structure of this product:

$\xrightarrow{HNO_3/H_2SO_4}$ ?

**19.52** When benzene is treated with 2-methylpropene and sulfuric acid, the product obtained is *tert*-butylbenzene. Propose a mechanism for this transformation.

**19.53** Draw a mechanism for each of the following transformations:

(a) $\xrightarrow[AlCl_3]{Cl_2}$

(b) $\xrightarrow[H_2SO_4]{HNO_3}$

(c) $\xrightarrow[H_2SO_4]{Fuming}$

(d) $\xrightarrow[AlCl_3]{CH_3Cl}$

(e) $\xrightarrow[Fe]{Br_2}$

**19.54** Propose a plausible mechanism for each of the following transformations:

(a) $\xrightarrow[AlCl_3]{I—Cl}$

(b) $\xrightarrow[AlCl_3]{CH_2Cl_2}$

**19.55** Propose a plausible mechanism for the following transformation:

$\xrightarrow{NaOMe}$ + NaCl

**19.56** Predict the product(s) of the following reactions:

(a) 1) $HNO_3$, $H_2SO_4$ 2) Zn, HCl ?

(b) 1) $AlCl_3$, Cl–C(=O)CH$_2$CH$_3$ 2) Zn(Hg), HCl, heat ?

(c) 1) $CH_3Cl$, $AlCl_3$ 2) $KMnO_4$, NaOH, heat 3) $H_3O^+$ ?

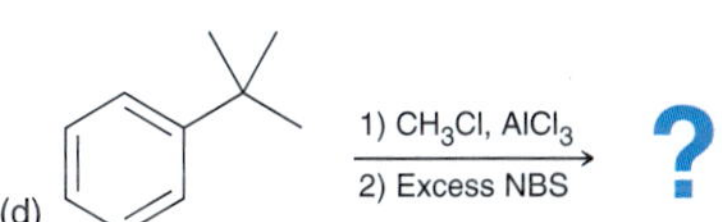


(d)

**19.57** Starting with benzene and using any other reagents of your choice, design a synthesis for each of the following compounds:

(a) (b) (c)

**19.58** Each of the following syntheses will not produce the desired product. In each case, identify the flaw in the synthesis.

(a) 1) $HNO_3$, $H_2SO_4$ 2) EtCl, $AlCl_3$ → $NO_2$ / Et

(b) 1) $Br_2$, $FeBr_3$ 2) Cl, $AlCl_3$ → Br

(c) 1) $Br_2$, $FeBr_3$ 2) O, Cl, $AlCl_3$ → Br / O

(d) 1) $AlCl_3$, Cl 2) $Br_2$, $FeBr_3$ → Br

**19.59** In each case, identify the most likely position at which monobromination would occur.

(a) O

(b) O, N

(c) O, O

**19.60** When *para*-bromotoluene is treated with sodium amide two products are obtained. Draw both products and propose a plausible mechanism for their formation.

**19.61** Picric acid is a military explosive formed via the nitration of phenol under conditions that install three nitro groups. Draw the structure and provide an IUPAC name for picric acid.

**19.62** Propose a plausible mechanism for the following transformation:

OH, OH, $H_2SO_4$

**19.63** Benzene was treated with isopropyl chloride in the presence of aluminum trichloride under conditions that favor dialkylation. Draw the major product expected from this reaction.

**19.64** Consider the structure of nitrosobenzene:

O, N

(a) Draw the resonance structures of the sigma complex formed when nitrosobenzene reacts with an electrophile ($E^+$) at the *ortho* position.

(b) Draw the resonance structures of the sigma complex formed when nitrosobenzene reacts with an electrophile ($E^+$) at the *meta* position.

(c) Draw the resonance structures of the sigma complex formed when nitrosobenzene reacts with an electrophile ($E^+$) at the *para* position.

(d) Compare the stability of the intermediates from parts a–c of this problem and then predict whether the nitroso group is *ortho-para* directing or *meta* directing.

(e) The nitroso group has a lone pair adjacent to the ring (suggesting that it could be an activator), yet it also has a π bond to a heteroatom in conjugation with the ring (suggesting that it could be a deactivator). Experiments reveal that this group is a deactivator. Based on this information, identify which of the following groups is expected to have properties most similar to the nitroso group and explain your choice:

—OH —$NO_2$ —$CH_3$ —Cl

**19.65** Draw all resonance structures of the sigma complex formed when toluene undergoes chlorination at the *para* position.

**19.66** For each of the following groups of compounds, identify which compound will react most rapidly with ethyl chloride in the presence of aluminum trichloride. Explain your choice in each case and then predict the expected products of that reaction.

(a) CN Cl $CH_3$

(b) Br OMe

**19.67** Compound **A** and compound **B** are both aromatic esters with molecular formula $C_8H_8O_2$. When treated with bromine in the presence of iron tribromide, compound **A** is converted into only one product, while compound **B** is converted into two different monobromination products. Identify the structure of compound **A** and compound **B**.

**19.68** Propose an efficient synthesis for each of the following transformations:

(a) $OCH_3$ → $OCH_3$, $NO_2$, Br

(b) $OCH_3$ ⟶ $OCH_3$, Br, $NO_2$

(c) ⟶ OH, O, $NO_2$

(d) ⟶ OH, O, $NO_2$

**19.69** Predict the product(s) for each of the following reactions:

(a) $\xrightarrow[AlCl_3]{Cl_2}$ ?

(b) $OCH_3$ $\xrightarrow[H_2SO_3]{HNO_3}$ ?

(c) $\xrightarrow[H_2SO_4]{Fuming}$ ?

(d) HO, Br, $NO_2$ $\xrightarrow[H_2SO_4]{HNO_3}$ ?

**19.70** Each of the following compounds can be made with a Friedel-Crafts acylation. Identify the acyl chloride and the aromatic compound you would use to produce each compound.

(a) $O_2N$, O, $OCH_3$

(b) $H_3CO$, O, $OCH_3$

**19.71** Aromatic heterocycles are also capable of undergoing electrophilic aromatic substitution. For example, when furan is treated with an electrophile, an electrophilic aromatic substitution reaction occurs in which the electrophile is installed exclusively at the C2 position. Explain why this reaction occurs at the C2 position, rather than the C3 position.

$\xrightarrow{E^{\oplus}}$ E

**19.72** Starting with benzene and using any other reagents of your choice, design a synthesis for each of the following compounds. In some cases, there may be more than one plausible answer.

(a) OMe, Br, Br, $NO_2$

(b) COOH, $NO_2$, COOH

(c) $NH_2$, Cl

(d) Cl, Br, $NO_2$

(e) O, $NO_2$, Cl

(f) COOH, Cl, $NO_2$, Br

(g) OEt, Cl, $NO_2$

(h) Cl, $NO_2$, O

(i) Br, $O_2N$

**19.73** When benzene is treated with methyl chloride and aluminum trichloride under conditions that favor trialkylation, one major product is obtained. Draw this product and provide an IUPAC name.

# INTEGRATED PROBLEMS

**19.74** Compound **A** has molecular formula $C_8H_8O$. An IR spectrum of compound **A** exhibits a signal at 1680 $cm^{-1}$. The $^1H$ NMR spectrum of compound **A** exhibits a group of signals between 7.5 and 8 ppm (with a combined integration of 5) and one upfield signal with an integration of 3. Compound **A** is converted into compound **B** under the following conditions:

**Compound A** —Zn(Hg), HCl, heat→ **Compound B**

When compound **B** is treated with $Br_2$ and $AlBr_3$, two different monobromination products are obtained. Identify both of these products and predict which one will be the major product. Explain your reasoning.

**19.75** Starting with benzene and using any other reagents of your choice, design a synthesis for each of the following compounds. Each compound has a Br and one other substituent that we did not learn how to install. In each case, you will need to choose one of the substituents that we learned in this chapter and then modify that substituent using reactions from previous chapters. Be careful to consider the order of events in each case.

(a) Br, OH (b) Br

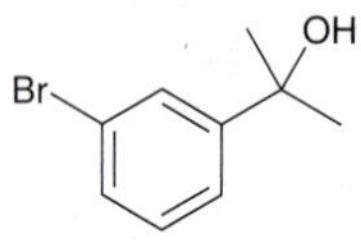

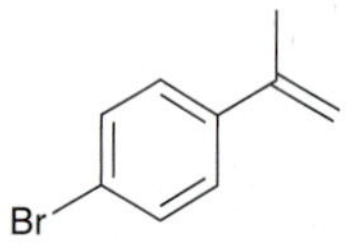

**19.76** Benzene was treated with (*R*)-2-chlorobutane in the presence of aluminum trichloride, and the resulting product mixture was found to be optically inactive.

(a) What products are expected, assuming that conditions are chosen to favor monoalkylation?

(b) Explain why the product mixture is optically inactive.

**19.77** The $^1H$ NMR spectrum of phenol exhibits three signals in the aromatic region of the spectrum. These signals appear at 6.7, 6.8, and 7.2 ppm. Use your understanding of shielding and deshielding effects (Chapter 16) to determine which signal corresponds with the *meta* protons. Explain your reasoning.

**19.78** When toluene is treated with a mixture of excess sulfuric acid and nitric acid at high temperature, a compound is obtained that exhibits only two signals in its $^1H$ NMR spectrum. One signal appears upfield and has an integration of 3. The other signal appears downfield and has an integration of 2. Identify the structure of this compound and assign an IUPAC name.

**19.79** Each of the following compounds is an aromatic compound bearing a substituent that we did not discuss in this chapter. Using the principles that we discussed in this chapter, predict the major product for each of the following reactions:

(a) —$HNO_3$, $H_2SO_4$→ ?

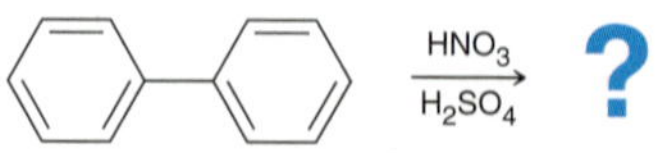

(b) O —$HNO_3$, $H_2SO_4$→ ?

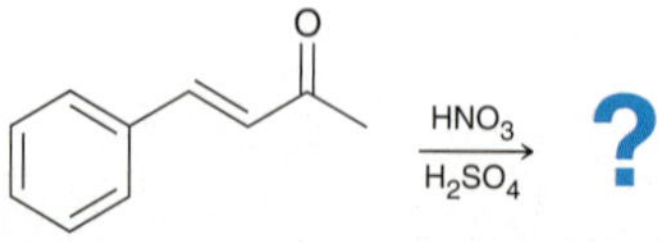

**19.80** Predict the major product of the following reaction:

+ Cl, Cl —$AlCl_3$→ ?

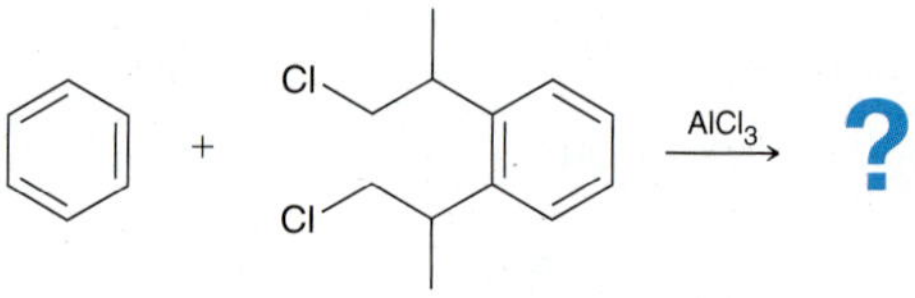

**19.81** Bakelite is one of the first known synthetic polymers and was used to make radio and telephone casings as well as automobile parts in the early twentieth century. Bakelite is formed by treating phenol with formaldehyde under acidic conditions. Draw a plausible mechanism for the formation of Bakelite.

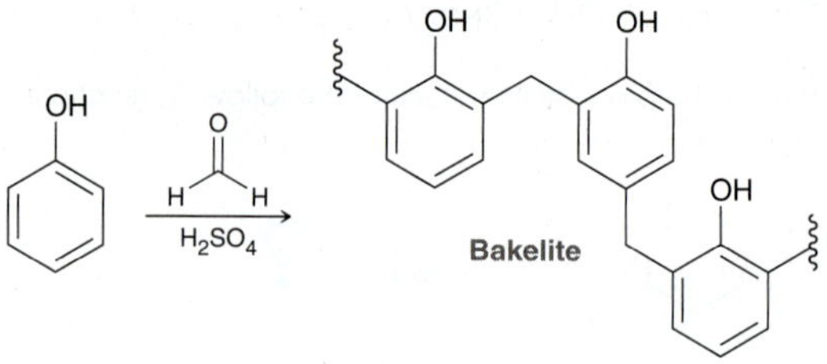

OH —H(C=O)H, $H_2SO_4$→ **Bakelite** (OH, OH, OH)

**19.82** Propose a plausible mechanism for the following transformation:

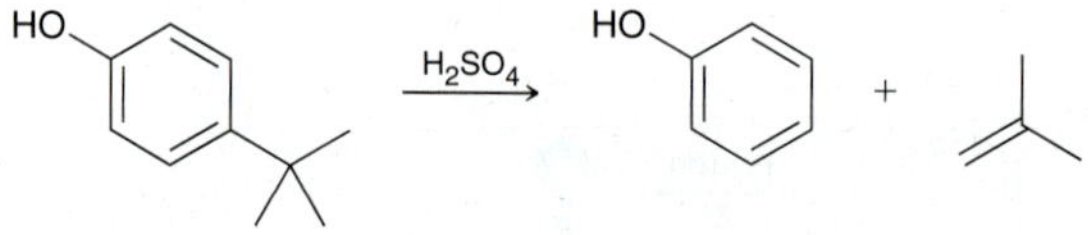

HO —$H_2SO_4$→ HO +

**19.83** When *N,N*-dimethylaniline is treated with bromine, *ortho* and *para* products are observed. Yet, when *N,N*-dimethylaniline is treated with a mixture of nitric and sulfuric acid, only the *meta* product is observed. Explain these curious results.

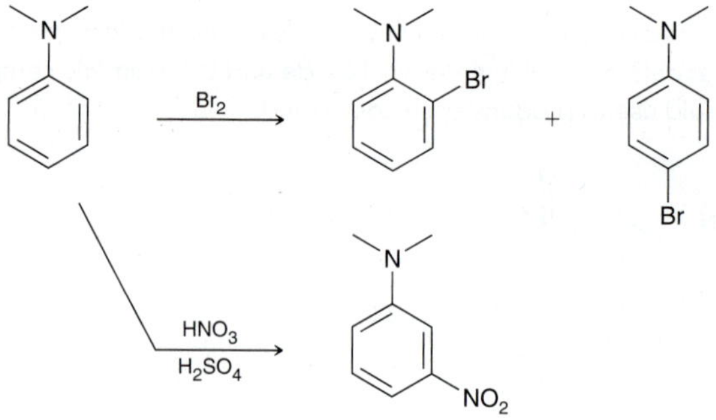

N —$Br_2$→ N, Br + N, Br

—$HNO_3$, $H_2SO_4$→ N, $NO_2$

# CHALLENGE PROBLEMS

**19.84** The rate constants for the bromination of several disubstituted stilbenes are given in the table below (*J. Org. Chem.* **1973,** *38,* **493–499**). Given that the double bond of stilbene acts as the nucleophile, provide a reasonable explanation for the trend observed among the rate constants.

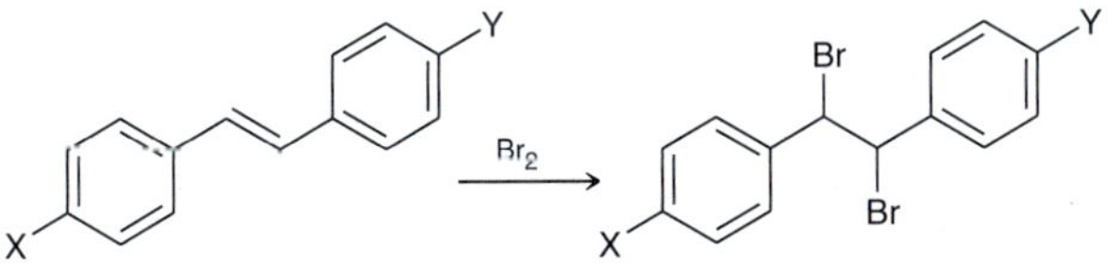

| X | Y | $k$ ($\times 10^3$ l/mol·min) |
|---|---|---|
| —OMe | —OMe | 220 |
| —OMe | —Me | 114 |
| —OMe | —Cl | 45 |
| —OMe | —$NO_2$ | 5.2 |

**19.85** Aminotetralins are a class of compounds currently being studied for their promise as antidepressant drugs. The following reaction was employed during a mechanistic study associated with a synthetic route for preparing aminotetralin drugs (*J. Org. Chem.* **2012,** *77,* **5503–5514**). Propose a plausible mechanism for this transformation.

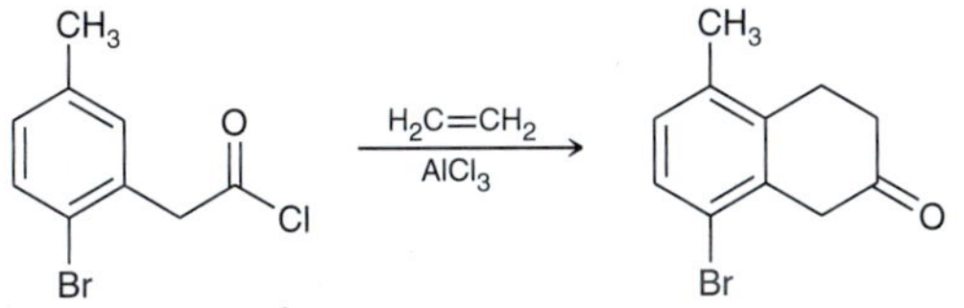

**19.86** Consider the following reaction, in which the product results from substitution of fluorine and not from substitution of chlorine (*Org. Lett.* **2007,** *9,* **2741–2743**):

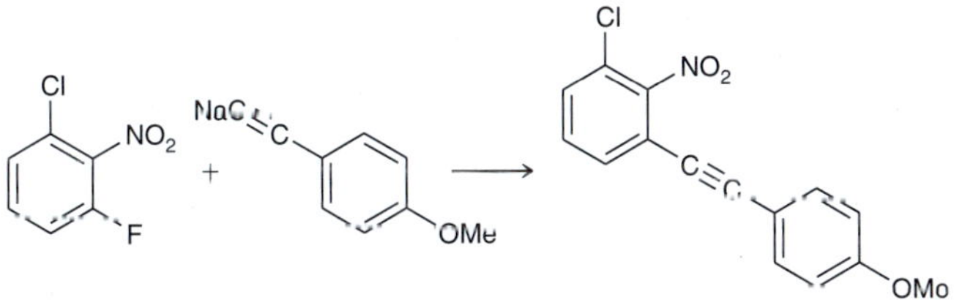

(a) Draw a mechanism for this reaction.

(b) Based on the observed regiochemical outcome, identify the step of the mechanism that is rate determining. Explain your answer.

**19.87** Compare the indicated bonds (***a*** and ***b***) in the following compound. Bond ***a*** has a bond length of 1.45 Å, while bond ***b*** has a bond length of 1.35 Å (*Prog. Stereochem.* **1958,** *2,* **125**). Suggest a reason for this difference in bond length for seemingly similar bonds.

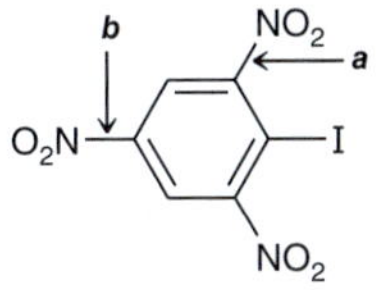

**19.88** Compound **1** and compound **2** both contain tritium (T), which is an isotope of hydrogen (tritium = $^3$H). Both compounds are stable upon treatment with aqueous base. However, upon prolonged treatment with aqueous acid, compounds **1** and **2** both lose tritium, to give 1,3-dimethoxybenzene and 1,3,5-trimethoxybenzene, respectively (*J. Am. Chem. Soc.* **1967,** *89,* **4418–4424**):

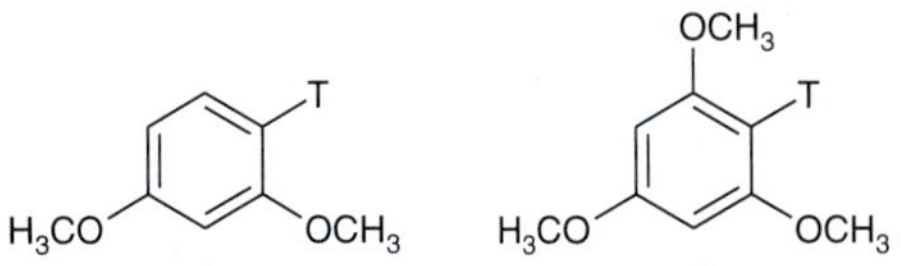

(a) Draw a mechanism showing how tritium is removed from compounds **1** and **2**.

(b) Use your mechanism to predict which compound (**1** or **2**) is expected to lose tritium at a faster rate. Justify your answer.

**19.89** During a synthesis of a potential anticancer agent, 7-hydroxynitidine, the investigators treated bromide **1** with two equivalents of a strong base to form compound **2** (*Org. Lett.* **1999,** *1,* **985–988**). Propose a plausible mechanism to account for the following ring-forming transformation:

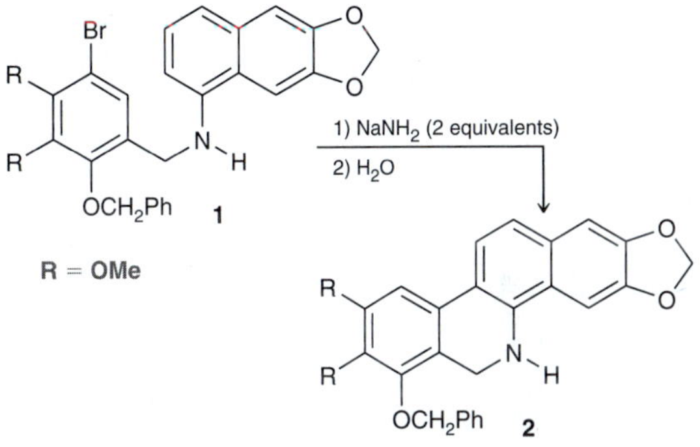

**19.90** In the following reaction, iodine monochloride (ICl) effectively serves as a source of an electrophilic iodonium species, $I^+$ (*J. Org. Chem.* **2005,** *70,* **3511–3517**):

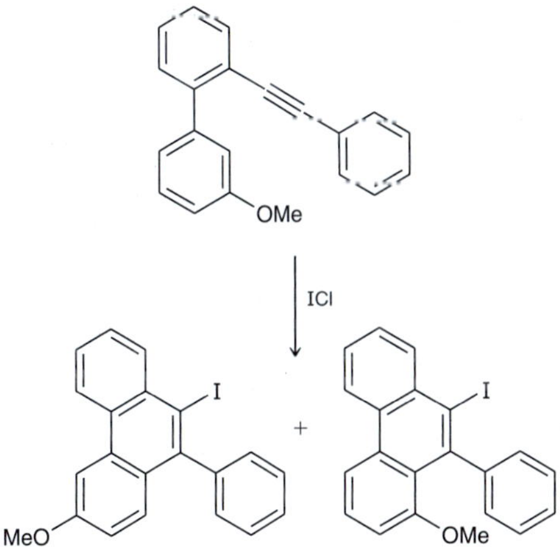

(a) Propose a mechanism for the formation of each of the two products, **A** and **B**.

(b) The ratio of **A**:**B** in the product mixture is found to be roughly 3:1. Explain why formation of **A** is favored over formation of **B**.

**19.91** Treatment of compound **1** with benzene in triflic acid ($CF_3SO_3H$) affords ammonium ion **3** (*J. Org. Chem.* **1999,** *64,* **6702–6705**). Triflic acid is an extremely strong acid ($pK_a = -14$), even more acidic than sulfuric acid; under these conditions, the transformation is believed to proceed via intermediate **2**, a highly electrophilic dication. Draw a complete mechanism for the conversion of **1** to **3**.

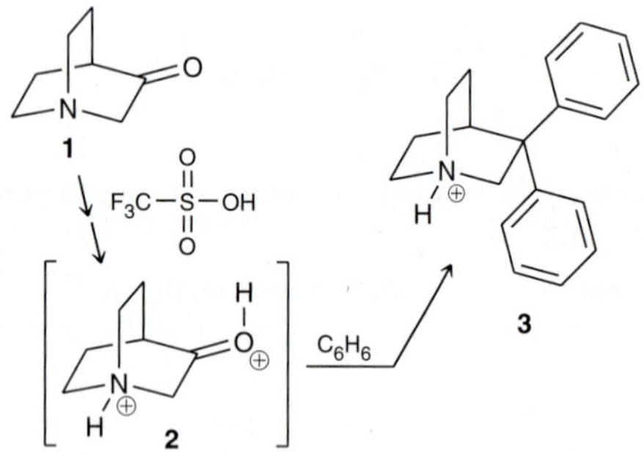

**19.92** One method for the synthesis of 9,9′-spirobifluorenes (rigid compounds with application to molecular electronics and light-emitting materials) involves protonation of ketone **1** with a catalytic amount of aqueous acid. Protonation of the C=O group generates an extremely powerful electrophile that facilitates a spontaneous cyclization (*Org. Lett.* **2004,** *6,* **2381–2383**). Propose a plausible mechanism for this transformation.

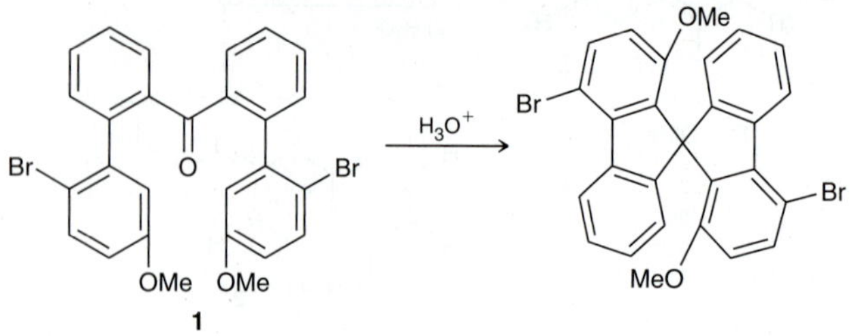

**19.93** The following synthesis was developed in an effort to prepare an analogue of a polycyclic aromatic hydrocarbon in which one of the C=C bonds was replaced with a B—N unit (*J. Am. Chem. Soc.* **2011,** *133,* **18614–18617**). Such derivatives are expected to display unique optical and electrical properties and thus may be useful in the preparation of novel organic electronic materials.

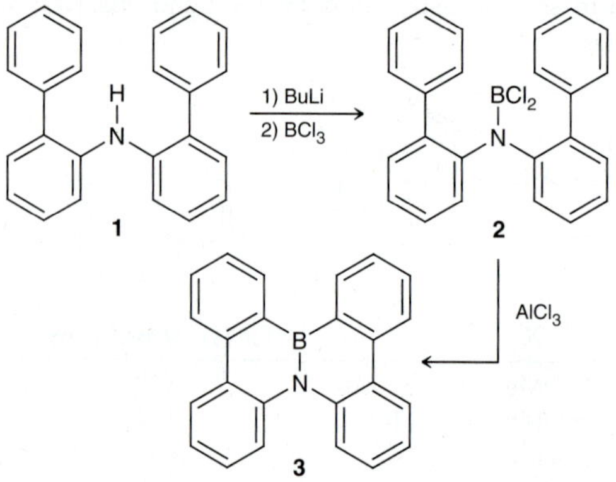

(a) Propose a mechanism for the conversion of compound **1** to compound **3**.

(b) In compound **3**, do you expect each of the six rings to be aromatic, giving one extended aromatic system for the compound? Explain your answer.

(c) Analysis of the actual structure of **3** (as well as its hydrocarbon analogue) reveals that the central two rings are slightly twisted (i.e., nonplanar). Provide a rationale for this finding.

# 20

# Aldehydes and Ketones



## DID YOU EVER WONDER...

**why beta-carotene, which makes carrots orange, is reportedly good for your eyes?**

This chapter will explore the reactivity of aldehydes and ketones. Specifically, we will see that a wide variety of nucleophiles will react with aldehydes and ketones. Many of these reactions are common in biological pathways, including the role that beta-carotene plays in promoting healthy vision. As we will see several times in this chapter, the reactions of aldehydes and ketones are also cleverly exploited in the design of drugs. The reactions and principles outlined in this chapter are central to the study of organic chemistry and will be used as guiding principles throughout the remaining chapters of this textbook.

## DO YOU REMEMBER?

*Before you go on, be sure you understand the following topics.*
*If necessary, review the suggested sections to prepare for this chapter.*

- Grignard Reagents (Section 13.6)
- Oxidation of Alcohols (Section 13.10)
- Retrosynthetic Analysis (Section 12.5)

Take the DO YOU REMEMBER? QUIZ in **WileyPLUS** to check your understanding

## 20.1 Introduction to Aldehydes and Ketones

Aldehydes (RCHO) and ketones ($R_2CO$) are similar in structure in that both classes of compounds possess a C=O bond, called a *carbonyl group*:

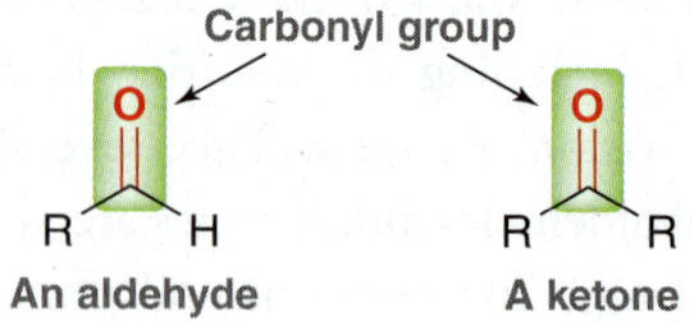

Carbonyl group

An aldehyde

A ketone

The carbonyl group of an aldehyde is flanked by a hydrogen atom, while the carbonyl group of a ketone is flanked by two carbon atoms.

Aldehydes and ketones are responsible for many flavors and odors that you will readily recognize:

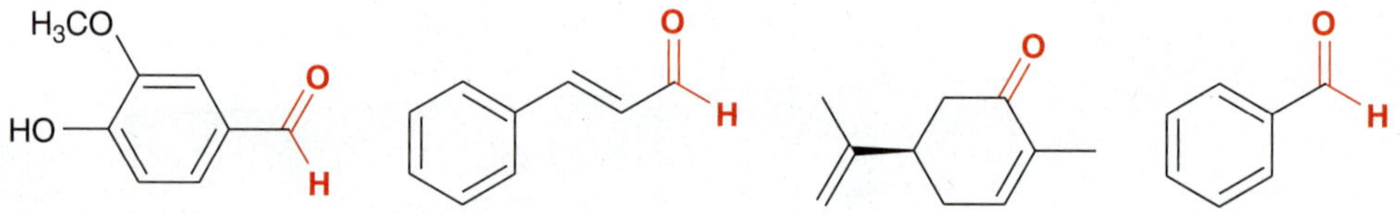

**Vanillin** (Vanilla flavor)

**Cinnamaldehyde** (Cinnamon flavor)

**(*R*)-Carvone** (Spearmint flavor)

**Benzaldehyde** (Almond flavor)

Many important biological compounds also exhibit the carbonyl group, including progesterone and testosterone, the female and male sex hormones.

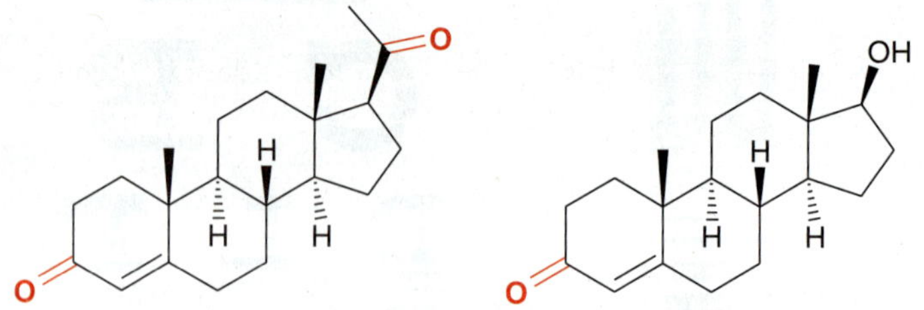

**Progesterone**

**Testosterone**

Simple aldehydes and ketones are industrially important; for example:

**Formaldehyde**

**Acetone**

Acetone is used as a solvent and is commonly found in nail polish remover, while formaldehyde is used as a preservative in some vaccine formulations. Aldehydes and ketones are also used as building blocks in the syntheses of commercially important compounds, including pharmaceuticals and polymers. Compounds containing a carbonyl group react with a large variety of nucleophiles, affording a wide range of possible products. Due to the versatile reactivity of the carbonyl group, aldehydes and ketones occupy a central role in organic chemistry.

## 20.2 Nomenclature

### Nomenclature of Aldehydes

Recall that four discrete steps are required to name most classes of organic compounds (as we saw with alkanes, alkenes, alkynes, and alcohols):

1. Identify and name the parent.
2. Identify and name the substituents.
3. Assign a locant to each substituent.
4. Assemble the substituents alphabetically.

Aldehydes are also named using the same four-step procedure. When applying this procedure for naming aldehydes, the following guidelines should be followed:

When naming the parent, the suffix "-al" indicates the presence of an aldehyde group:

Butane 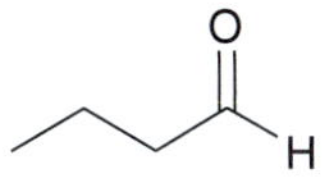    Butanal

When choosing the parent of an aldehyde, identify the longest chain *that includes the carbon atom of the aldehydic group*:

The parent must include this carbon atom

Parent = Octane 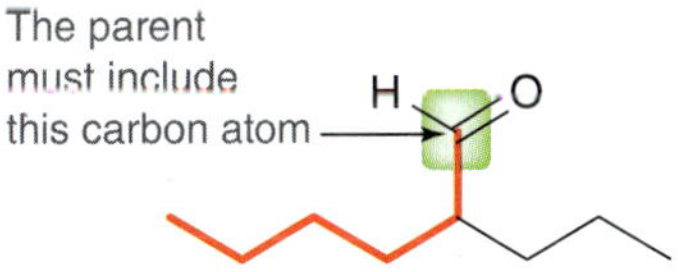    Parent = Hexanal

When numbering the parent chain of an aldehyde, the aldehydic carbon is assigned number 1, despite the presence of alkyl substituents, $\pi$ bonds, or hydroxyl groups:

Correct    Incorrect

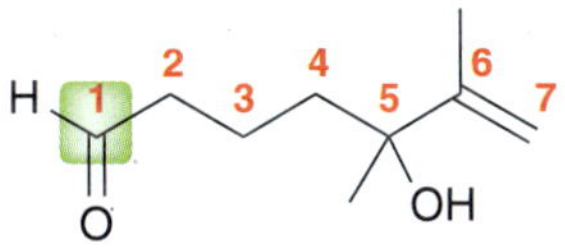

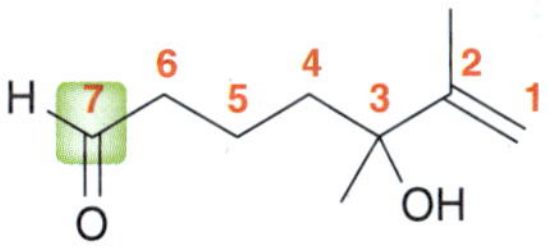

It is not necessary to include this locant in the name, because it is understood that the aldehydic carbon is the number 1 position.

As with all compounds, when a chirality center is present, the configuration is indicated at the beginning of the name; for example:

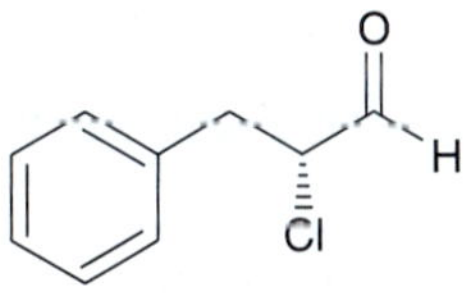

(*R*)-2-Chloro-3-phenylpropanal

A cyclic compound containing an aldehyde group immediately adjacent to the ring is named as a carbaldehyde:

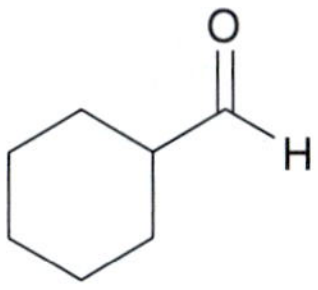

Cyclohexanecarbaldehyde

The International Union of Pure and Applied Chemistry (IUPAC) nomenclature also recognizes the common names of many simple aldehydes, including the three examples shown below:

**Formaldehyde** **Acetaldehyde** **Benzaldehyde**

## Nomenclature of Ketones

Ketones, like aldehydes, are named using the same four-step procedure. When naming the parent, the suffix "-one" indicates the presence of a ketone group:

**Butane** **Butanone**

The position of the ketone group is indicated using a locant. The IUPAC rules published in 1979 dictate that this locant be placed immediately before the parent, while the IUPAC recommendations released in 1993 and 2004 allow for the locant to be placed immediately before the suffix "-one":

**3-Heptanone** or **Heptan-3-one**

Both names above are acceptable IUPAC names. IUPAC nomenclature recognizes the common names of many simple ketones, including the three examples shown below:

**Acetone** **Acetophenone** **Benzophenone**

Although rarely used, IUPAC rules also allow simple ketones to be named as *alkyl alkyl ketones*. For example, 3-hexanone can also be called ethyl propyl ketone:

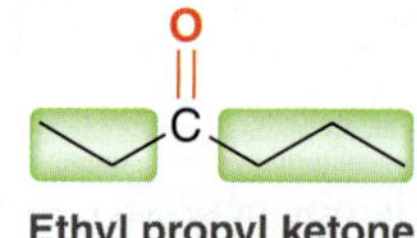


**Ethyl propyl ketone**

# SKILLBUILDER

## 20.1 NAMING ALDEHYDES AND KETONES

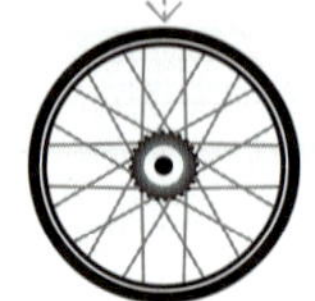

**LEARN the skill** Provide a systematic (IUPAC) name for the following compound:

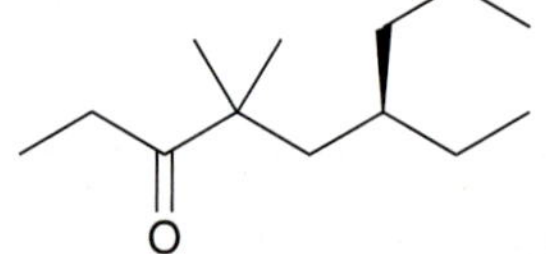


### SOLUTION

**STEP 1** Identify and name the parent.

The first step is to identify and name the parent. Choose the longest chain that includes the carbonyl group and then number the chain to give the carbonyl group the lowest number possible:

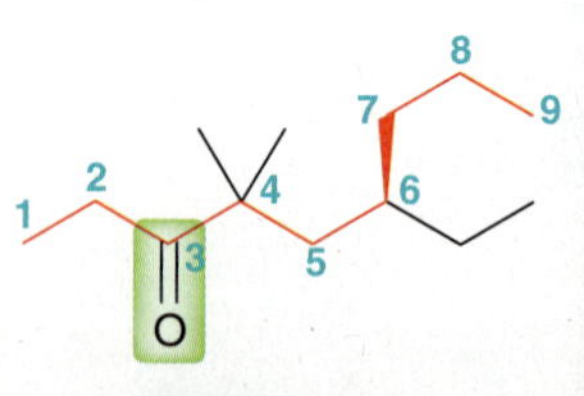


**3-Nonanone**

**STEP 2**
Identify and name the substituents.

**STEP 3**
Assign a locant to each substituent.

Next, identify the substituents and assign locants:

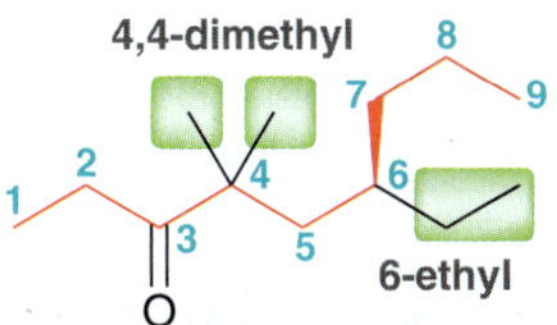


**STEP 4**
Assemble the substituents alphabetically.

Finally, assemble the substituents alphabetically: 6-ethyl-4,4-dimethyl-3-nonanone. Before concluding, we must always check to see if there are any chirality centers. This compound does exhibit one chirality center. Using the skills from Section 5.3, the *R* configuration is assigned to this chirality center:

**STEP 5**
Assign the configuration of any chirality centers.

Therefore, the complete name is (*R*)-6-ethyl-4,4-dimethyl-3-nonanone.

**PRACTICE the skill**

**20.1** Assign a systematic (IUPAC) name to each of the following compounds:

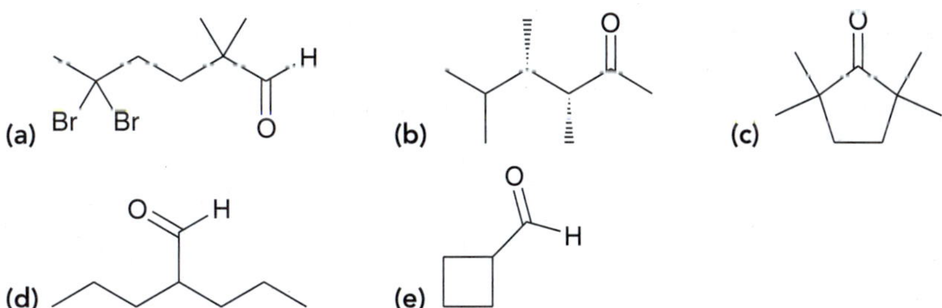

**APPLY the skill**

**20.2** Draw the structure of each of the following compounds:

**(a)** (*S*)-3,3-Dibromo-4-ethylcyclohexanone **(b)** 2,4-Dimethyl-3-pentanone

**(c)** (*R*)-3-Bromobutanal

**20.3** Provide a systematic (IUPAC) name for the compound below. Be careful: This compound has two chirality centers (can you find them?).

**20.4** Compounds with two ketone groups are named as alkane diones; for example:

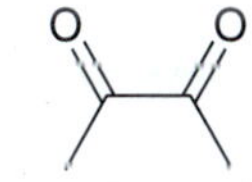


**2,3-Butanedione**

The compound above is an artificial flavor added to microwave popcorn and movie-theater popcorn to simulate the butter flavor. Interestingly, this very same compound is also known to contribute to body odor. Name the following compounds:

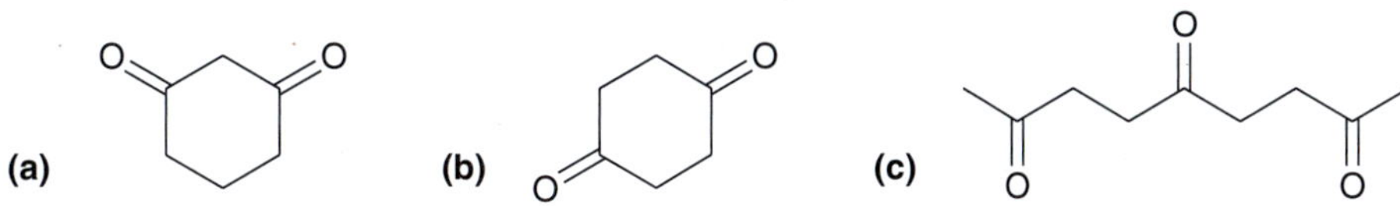


need more **PRACTICE?** **Try Problems 20.44–20.49**

## 20.3 Preparing Aldehydes and Ketones: A Review

In previous chapters, we have studied a variety of methods for preparing aldehydes and ketones, which are summarized in Tables 20.1 and 20.2, respectively.

**TABLE 20.1 A SUMMARY OF ALDEHYDE PREPARATION METHODS COVERED IN PREVIOUS CHAPTERS**

| REACTION | SECTION |
|---|---|
| Oxidation of Primary Alcohols<br>R–CH₂OH → (PCC, $CH_2Cl_2$) → RCHO<br>When treated with a strong oxidizing agent, primary alcohols are oxidized to carboxylic acids. Formation of an aldehyde requires an oxidizing agent, such as PCC, that will not further oxidize the resulting aldehyde. | 13.10 |
| Ozonolysis of Alkenes<br>1) $O_3$ 2) DMS<br>Ozonolysis will cleave a C=C double bond. If either carbon atom bears a hydrogen atom, an aldehyde will be formed. | 9.11 |
| Hydroboration-Oxidation of Terminal Alkynes<br>1) $R_2B–H$ 2) $H_2O_2$, NaOH<br>Hydroboration-oxidation results in an *anti*-Markovnikov addition of water across a π bond, followed by tautomerization of the resulting enol to form an aldehyde. | 10.7 |

**TABLE 20.2 A SUMMARY OF KETONE PREPARATION METHODS COVERED IN PREVIOUS CHAPTERS**

| REACTION | SECTION |
|---|---|
| Oxidation of Secondary Alcohols<br>$Na_2Cr_2O_7$, $H_2SO_4$, $H_2O$<br>A variety of strong or mild oxidizing agents can be used to oxidize secondary alcohols. The resulting ketone does not undergo further oxidation. | 13.10 |
| Ozonolysis of Alkenes<br>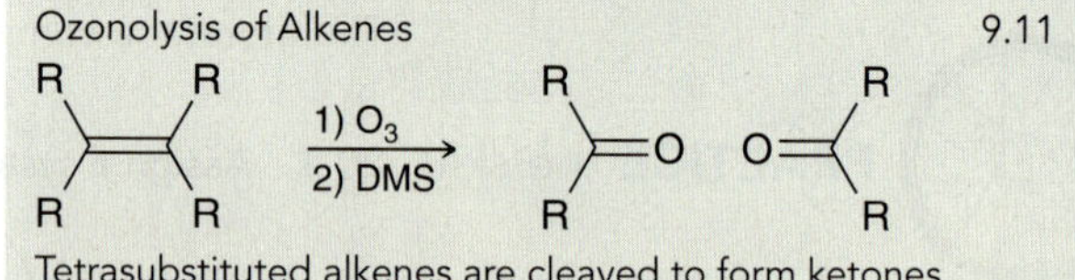<br>Tetrasubstituted alkenes are cleaved to form ketones. | 9.11 |
| Acid-Catalyzed Hydration of Terminal Alkynes<br>$H_2SO_4$, $H_2O$, $HgSO_4$<br>This procedure results in a Markovnikov addition of water across the π bond, followed by tautomerization to form a methyl ketone. | 10.7 |
| Friedel-Crafts Acylation<br>RCOCl, $AlCl_3$<br>Aromatic rings that are not too strongly deactivated will react with an acyl halide in the presence of a Lewis acid to produce an aryl ketone. | 19.6 |

### CONCEPTUAL CHECKPOINT

**20.5** Identify the reagents necessary to achieve each of the following transformations:

(a) (b) (c) (d) (e) (f)

## 20.4 Introduction to Nucleophilic Addition Reactions

The electrophilicity of a carbonyl group derives from resonance effects as well as inductive effects:

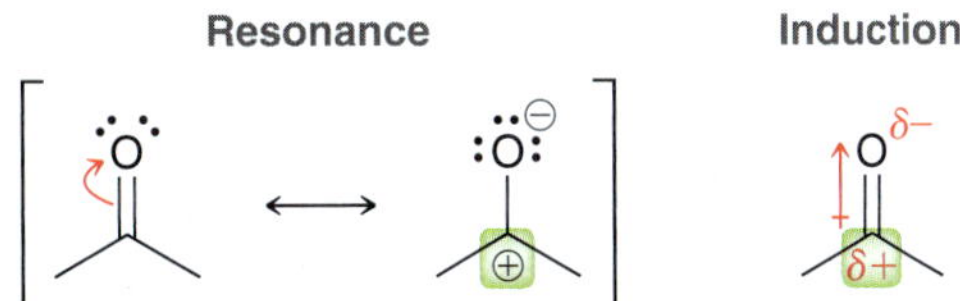


One of the resonance structures exhibits a positive charge on the carbon atom, indicating that the carbon atom is deficient in electron density (δ+). Inductive effects also render the carbon atom deficient in electron density. As a result, this carbon atom is particularly electrophilic and is susceptible to attack by a nucleophile. Molecular orbital calculations suggest that nucleophilic attack occurs at an angle of approximately 107° to the plane of the carbonyl group, and in the process, the hybridization state of the carbon atom changes (Figure 20.1).

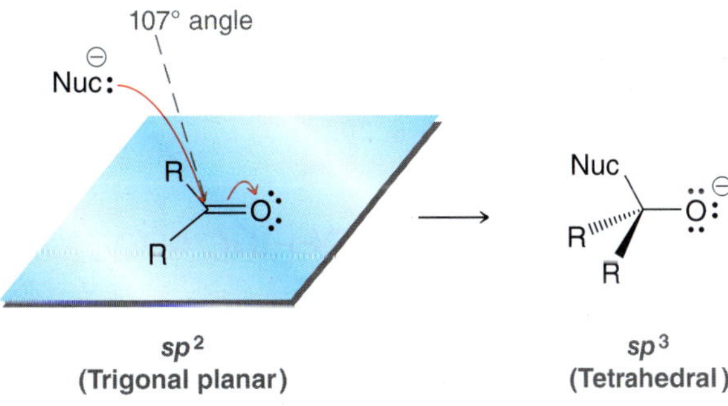


**FIGURE 20.1** When a carbonyl group is attacked by a nucleophile, the carbon atom undergoes a change in hybridization and geometry.

The carbon atom is originally $sp^2$ hybridized with a trigonal planar geometry. After the attack, the carbon atom is $sp^3$ hybridized with a tetrahedral geometry.

In general, aldehydes are more reactive than ketones toward nucleophilic attack. This observation can be explained in terms of both steric and electronic effects:

1. *Steric effects*. A ketone has two alkyl groups (one on either side of the carbonyl) that contribute to steric interactions in the transition state of a nucleophilic attack. In contrast, an aldehyde has only one alkyl group, so the transition state is less crowded and lower in energy.
2. *Electronic effects*. Recall that alkyl groups are electron donating. A ketone has two electron-donating alkyl groups that can stabilize the δ+ on the carbon atom of the carbonyl group. In contrast, aldehydes have only one electron-donating group:

**A ketone** has two electron-donating alkyl groups that stabilize the partial positive charge

**An aldehyde** has only one electron-donating alkyl group that stabilizes the partial positive charge

The δ+ charge of an aldehyde is less stabilized than a ketone. As a result, aldehydes are more electrophilic than ketones and therefore more reactive.

Aldehydes and ketones react with a wide variety of nucleophiles. As we will see in the coming sections of this chapter, some nucleophiles require basic conditions, while others require acidic conditions. For example, recall from Chapter 13 that Grignard reagents are very strong nucleophiles that will attack aldehydes and ketones to produce alcohols:

$$\text{RC(=O)R} \xrightarrow[\text{2) } H_2O]{\text{1) MeMgBr}} \text{R}_2\text{C(OH)Me}$$

The Grignard reagent itself provides for strongly basic conditions, because Grignard reagents are both strong nucleophiles and strong bases. This reaction cannot be achieved under acidic conditions, because, as explained in Section 13.6, Grignard reagents are destroyed in the presence of an acid. Mechanism 20.1, shown below, is a general mechanism for the reaction between a nucleophile and a carbonyl group under basic conditions. This general mechanism has two steps: (1) nucleophilic attack followed by (2) proton transfer.

## MECHANISM 20.1 NUCLEOPHILIC ADDITION UNDER BASIC CONDITIONS

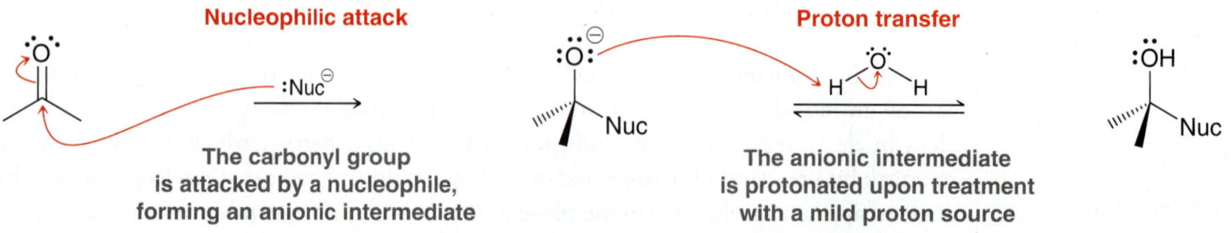


Aldehydes and ketones also react with a wide variety of other nucleophiles under acidic conditions. In acidic conditions, the same two mechanistic steps are observed, but in reverse order—that is, the carbonyl group is first protonated and then undergoes a nucleophilic attack (Mechanism 20.2).

## MECHANISM 20.2 NUCLEOPHILIC ADDITION UNDER ACIDIC CONDITIONS

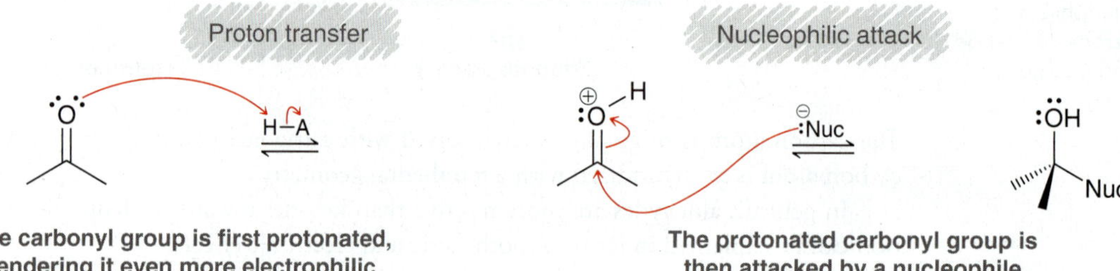


In acidic conditions, the first step plays an important role. Specifically, protonation of the carbonyl group generates a very powerful electrophile:

Very powerful electrophile

It is true that the carbonyl group is already a fairly strong electrophile; however, a protonated carbonyl group bears a full positive charge, rendering the carbon atom even more electrophilic. This is especially important when weak nucleophiles, such as $H_2O$ or ROH, are employed, as we will see in the upcoming sections.

When a nucleophile attacks a carbonyl group under either acidic or basic conditions, the position of equilibrium is highly dependent on the ability of the nucleophile to function as a leaving group. A Grignard reagent is a very strong nucleophile, but it does not function as a leaving group (a carbanion is too unstable to leave). As a result, the equilibrium so greatly favors products that the reaction effectively occurs in only one direction. With a sufficient amount of nucleophile present, the ketone is not observed in the product mixture. In contrast, halides are good nucleophiles, but they are also good leaving groups. Therefore, when a halide functions as the nucleophile, an equilibrium is established, with the starting ketone generally being favored:

$$R_2C{=}O + HCl \rightleftharpoons R_2C(OH)Cl$$

Once equilibrium has been achieved, the mixture consists primarily of the ketone, and only small quantities of the addition product.

In this chapter, we will explore a wide variety of nucleophiles, which will be classified according to the nature of the attacking atom. Specifically, we will see nucleophiles based on oxygen, sulfur, nitrogen, hydrogen, and carbon (Figure 20.2).

| Oxygen Nucleophiles | Sulfur Nucleophiles | Nitrogen Nucleophiles | Hydrogen Nucleophiles | Carbon Nucleophiles |
|---|---|---|---|---|
| $H_2O$ | $H_2S$ | $RNH_2$ | $AlH_4^-$ | RMgBr |
| ROH | HS–$CH_2CH_2$–SH | $R_2NH$ | $BH_4^-$ | $^-$:C≡N |
| HO–$CH_2CH_2$–OH | | | | $Ph_3P^+$–$CH_2^-$ |

**FIGURE 20.2** Various nucleophiles that can attack a carbonyl group.

The remainder of the chapter will be a methodical survey of the reactions that occur between the reagents in Figure 20.2 and ketones and aldehydes. We will begin our survey with oxygen nucleophiles.

### CONCEPTUAL CHECKPOINT

**20.6** Draw a mechanism for each of the following reactions:

(a) Cyclohexanone → 1) EtMgBr, 2) $H_2O$ → 1-ethylcyclohexanol

(b) Cyclopentanone + HCl ⇌ 1-chlorocyclopentanol

## 20.5 Oxygen Nucleophiles

### Hydrate Formation

When an aldehyde or ketone is treated with water, the carbonyl group can be converted into a **hydrate:**

$$R_2C{=}O + H_2O \rightleftharpoons R_2C(OH)_2$$

Hydrate

The position of equilibrium generally favors the carbonyl group rather than the hydrate, except in the case of very simple aldehydes, such as formaldehyde:

$$H_3C{-}CO{-}CH_3 + H_2O \rightleftharpoons (H_3C)_2C(OH)_2$$

99.9%

$$H{-}CO{-}H + H_2O \rightleftharpoons H_2C(OH)_2$$

> 99.9%

The rate of reaction is relatively slow under neutral conditions but is readily enhanced in the presence of either acid or base. That is, the reaction can be either acid catalyzed or base catalyzed, allowing the equilibrium to be achieved much more rapidly. Consider the base-catalyzed hydration of formaldehyde (Mechanism 20.3).

## MECHANISM 20.3 BASE-CATALYZED HYDRATION

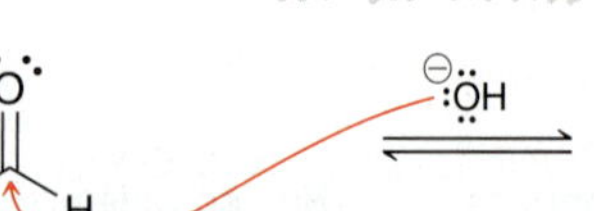


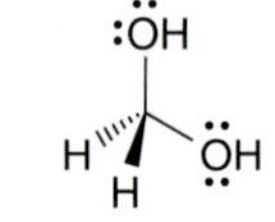

The carbonyl group is attacked by hydroxide, forming an anionic intermediate

The anionic intermediate is protonated by water to form the hydrate

In the first step, a hydroxide ion (rather than water) functions as a nucleophile. Then, in the second step, the intermediate is protonated with water, regenerating a hydroxide ion. In this way, hydroxide serves as a catalyst for the addition of water across the carbonyl group.

Now consider the acid-catalyzed hydration of formaldehyde (Mechanism 20.4).

## MECHANISM 20.4 ACID-CATALYZED HYDRATION



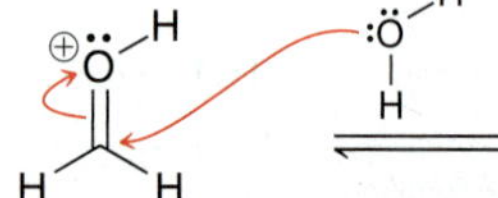


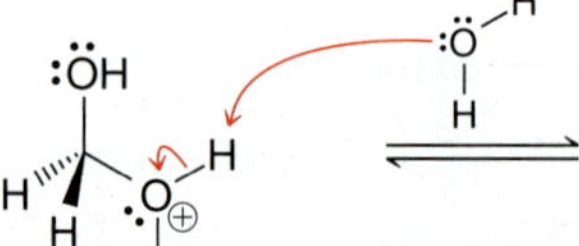


The carbonyl group is protonated, rendering it more electrophilic

The protonated carbonyl group is attacked by water, forming an oxonium intermediate

The oxonium intermediate is deprotonated by water to form the hydrate

Under acid-catalyzed conditions, the carbonyl group is first protonated, generating a positively charged intermediate that is extremely electrophilic (it bears a full positive charge). This intermediate is then attacked by water to form an oxonium ion (a cation in which the positive charge resides on an oxygen atom), which is deprotonated to give the product.

## CONCEPTUAL CHECKPOINT

**20.7** For most ketones, hydrate formation is unfavorable, because the equilibrium favors the ketone rather than the hydrate. However, the equilibrium for hydration of hexafluoroacetone favors formation of the hydrate: Provide a plausible explanation for this observation.

$F_3C$–CO–$CF_3$ + $H_2O$ ⇌ HO OH $F_3C$ $CF_3$

**> 99.99%**

## An Important Rule for Drawing Mechanisms

If we compare the mechanisms presented for base-catalyzed hydration (Mechanism 20.3) and acid-catalyzed hydration (Mechanism 20.4), an extremely important feature emerges. Under basic conditions, the mechanism employs a strong base (hydroxide) and a weak acid (water). Notice that a strong acid is not drawn (either as a reagent or as an intermediate) because it is unlikely to be present. In contrast, in acidic conditions, the mechanism employs a strong acid ($H_3O^+$) and a weak base (water). Notice that a strong base is not drawn (either as a reagent or as an intermediate) because it is unlikely to be present. This is an extremely important principle. When drawing mechanisms, it is important to consider the conditions being employed and to stay consistent with those conditions. That is, avoid drawing chemical entities that are unlikely to be present. This rule can be summarized as follows:

- *Under acidic conditions, a mechanism will only be reasonable if it avoids the use or formation of strong bases (only weak bases can be employed).*

- *Under basic conditions, a mechanism will only be reasonable if it avoids the use or formation of strong acids (only weak acids can be employed).*

It would be wise to internalize this rule, as we will see its application several times throughout this chapter and the upcoming chapters as well.

## Acetal Formation

The previous section discussed a reaction that can occur when water attacks an aldehyde or ketone. This section will explore a similar reaction, in which an alcohol attacks an aldehyde or ketone:

**BY THE WAY**
When the starting compound is a ketone, the product can also be called a "ketal." *Acetal* is a more general term, and it will be used exclusively for the remainder of this discussion.

O + 2 ROH $\underset{}{\overset{[H^+]}{\rightleftharpoons}}$ RO OR + $H_2O$

Acetal

In acidic conditions, an aldehyde or ketone will react with two molecules of alcohol to form an **acetal**. The brackets surrounding the $H^+$ indicate that the acid is a catalyst (brackets are not the standard convention for representing acid catalysis; nevertheless, we will use brackets to indicate acid catalysis throughout the remainder of this textbook). Common acids used for this purpose include *para*-toluenesulfonic acid (TsOH) and sulfuric acid ($H_2SO_4$):

**p-Toluenesulfonic acid** (TsOH) **Sulfuric acid**

As mentioned earlier, the acid catalyst serves an important role in this reaction. Specifically, in the presence of an acid, the carbonyl group is protonated, rendering the carbon atom even more electrophilic. This is necessary because the nucleophile (an alcohol) is weak; it reacts with the carbonyl group more rapidly if the carbonyl group is first protonated. A mechanism for acetal formation is shown in Mechanism 20.5. This mechanism has many steps, and it is best to divide it conceptually into two parts: (1) The first three steps produce an intermediate called a **hemiacetal** and (2) the last four steps convert the hemiacetal into an acetal:

## MECHANISM 20.5 ACETAL FORMATION

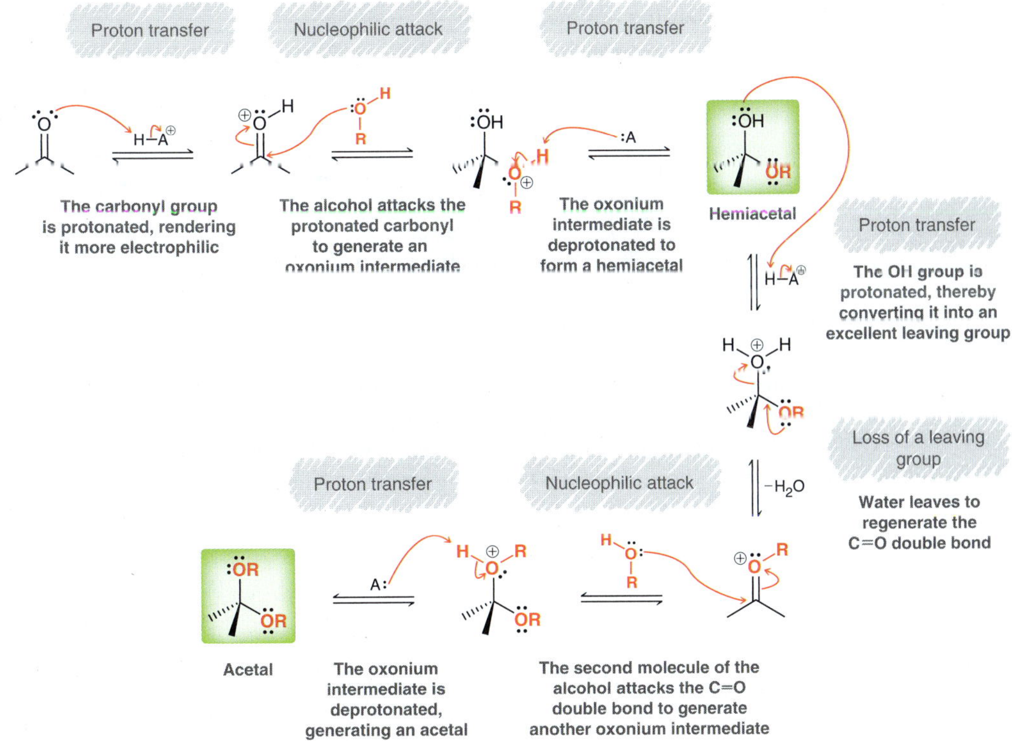

Let's begin our analysis of this mechanism by focusing on the first part: formation of the hemiacetal, which involves the three steps in Figure 20.3.

**FIGURE 20.3**
The sequence of steps involved in formation of a hemiacetal.

Proton transfer → Nucleophilic attack → Proton transfer

Notice that the sequence of steps begins and ends with a proton transfer. Let's focus on the details of these three steps:

1. The carbonyl group is protonated in the presence of an acid. The identity of the acid, $HA^+$, is most likely a protonated alcohol, which received its extra proton from the acid catalyst:

2. The protonated carbonyl group is a very powerful electrophile and is attacked by a molecule of alcohol (ROH) to form an oxonium ion.
3. The oxonium ion is deprotonated by a weak base (A), which is likely to be a molecule of alcohol present in solution.

Notice that the acid is not consumed in this process. A proton is used in step 1 and then returned in step 3, confirming the catalytic nature of the proton in the reaction.

Now let's focus on the second part of the mechanism, conversion of the hemiacetal into an acetal, which is accomplished with the four steps in Figure 20.4.

**FIGURE 20.4**
The sequence of steps that convert a hemiacetal into an acetal.

Notice, once again, that the sequence of steps begins and ends with a proton transfer. A proton is used in the first step and then returned in the last step, but this time there are two middle steps rather than just one. When drawing the mechanism of acetal formation, make sure to draw these two steps separately. Combining these two steps is incorrect and represents one of the most common student errors when drawing this mechanism:

Loss of a leaving group
$\overset{\oplus}{O}H_2$ ... OR + ROH ⟶✕
Nucleophilic attack
An $S_N2$ process cannot occur at this substrate

These two steps cannot occur simultaneously, because that would represent an $S_N2$ process occurring at a sterically hindered substrate. Such a process is disfavored and does not occur at an appreciable rate. Instead, the leaving group leaves first to form a resonance-stabilized intermediate, which is then attacked by the nucleophile in a separate step.

The equilibrium arrows in the full mechanism of acetal formation indicate that the process is governed by an equilibrium. For many simple aldehydes, the equilibrium favors formation of the acetal, so aldehydes are readily converted into acetals by treatment with two equivalents of alcohol in acidic conditions:

$$H_2C{=}O + 2\ EtOH \overset{[H^+]}{\rightleftharpoons} H_2C(OEt)_2 + H_2O$$

Products are favored at equilibrium

However, for most ketones, the equilibrium favors reactants rather than products:

$H_3C-C(=O)-CH_3$ + 2 EtOH $\underset{}{\overset{[H^+]}{\rightleftharpoons}}$ $H_3C-C(OEt)_2-CH_3$ + $H_2O$

**Reactants are favored at equilibrium**

**BY THE WAY**
This technique exploits Le Châtelier's principle, which was covered in your general chemistry course. According to Le Châtelier's principle, if a system at equilibrium is upset by some disturbance, the system will change in a way that restores equilibrium.

In such cases, formation of the acetal can be accomplished by removing one of the products (water) via a special distillation technique. By removing water as it is formed, the reaction can be forced to completion.

Notice that acetal formation requires two equivalents of the alcohol. That is, two molecules of ROH are required for every molecule of ketone. Alternatively, a compound containing two OH groups can be used, forming a cyclic acetal. This reaction proceeds via the regular seven-step mechanism for acetal formation: three steps for formation of the hemiacetal followed by four steps for formation of the cyclic acetal:

OH, OH, O —3 steps→ O, OH, OH —4 steps→ O, O + $H_2O$

**Hemiacetal** **Cyclic acetal**

The seven-step mechanism for acetal formation is very similar to other mechanisms that we will explore. It is therefore critical to master these seven steps. To help you draw the mechanism properly, remember to divide the entire mechanism into two parts, where each part begins and ends with a proton transfer step. Let's get some practice.

## SKILLBUILDER

### 20.2 DRAWING THE MECHANISM OF ACETAL FORMATION

**LEARN the skill**

Draw a plausible mechanism for the following transformation:

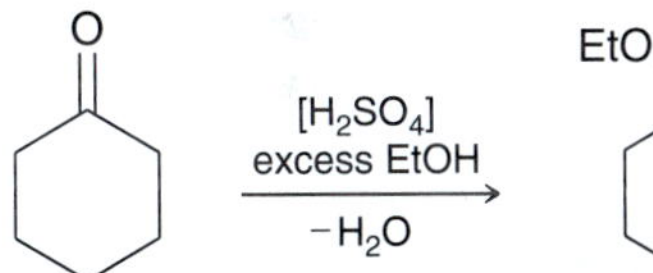

**SOLUTION**

The reaction above is an example of acid-catalyzed acetal formation, in which the product is favored by the removal of water. The mechanism can be divided into two parts: (1) formation of the hemiacetal and (2) formation of the acetal. Formation of the hemiacetal involves three mechanistic steps:

**STEP 1**
Draw the three steps necessary for hemiacetal formation.

Proton transfer — Nucleophilic attack — Proton transfer

When drawing these three steps, make sure to focus on proper arrow placement (as described in Chapter 6) and make sure to place all positive charges in their appropriate locations. Notice that every step requires two curved arrows.

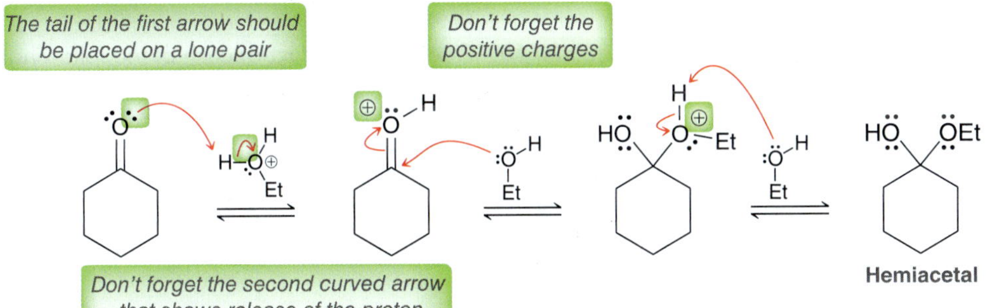

**STEP 2**
Draw the four steps necessary to convert the hemiacetal into an acetal.

Now let's focus on the last four steps of the mechanism, in which the hemiacetal is converted into an acetal:

Proton transfer → Loss of a leaving group → Nucleophilic attack → Proton transfer

Once again, this sequence of steps begins with a proton transfer and ends with a proton transfer. When drawing these four steps, make sure to draw the middle two steps separately, as discussed earlier. In addition, make sure to focus on proper arrow placement and make sure to place all positive charges in their appropriate locations:

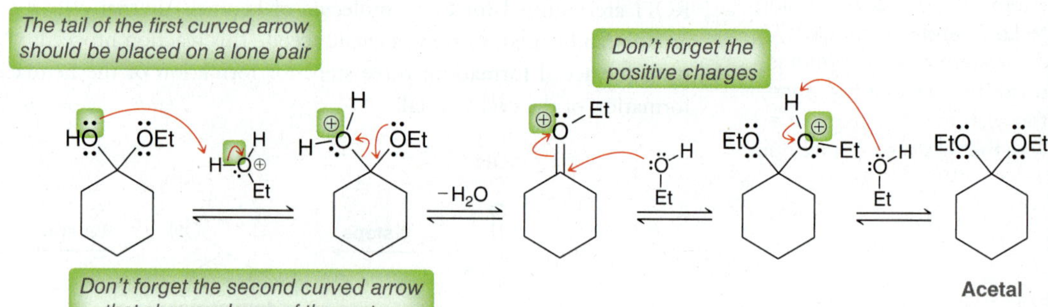

**PRACTICE the skill** **20.8** Draw a plausible mechanism for each of the following transformations:

(a) cyclopentanone $\xrightarrow[-H_2O]{[H_2SO_4],\ \text{excess MeOH}}$ 1,1-dimethoxycyclopentane (MeO, OMe)

(b) cyclobutanone $\xrightarrow[-H_2O]{[TsOH],\ \text{excess EtOH}}$ 1,1-diethoxycyclobutane (OEt, OEt)

(c) 3-methyl-2-butanone $\xrightarrow[-H_2O]{[H_2SO_4],\ \text{excess EtOH}}$ diethyl acetal (EtO, OEt)

(d) cyclohexyl methyl ketone $\xrightarrow[-H_2O]{[TsOH],\ \text{excess MeOH}}$ dimethyl acetal (MeO, OMe)

**APPLY the skill** **20.9** Draw a plausible mechanism for each of the following reactions:

(a) acetone $\xrightarrow[-H_2O]{HO\text{–}CH_2CH_2\text{–}OH,\ [H_2SO_4]}$ 2,2-dimethyl-1,3-dioxolane

(b) HO–(CH₂)₃–C(=O)–(CH₂)₃–OH $\xrightarrow[-H_2O]{[H_2SO_4]}$ spiroketal

**20.10** Predict the product of each of the following reactions:

(a) propiophenone $\xrightarrow[-H_2O]{[H_2SO_4],\ \text{excess MeOH}}$ **?**

(b) propiophenone $\xrightarrow[-H_2O]{HO\text{–}CH_2CH_2CH_2\text{–}OH,\ [H_2SO_4]}$ **?**

need more **PRACTICE?** **Try Problems 20.57, 20.62, 20.67**

## Acetals as Protecting Groups

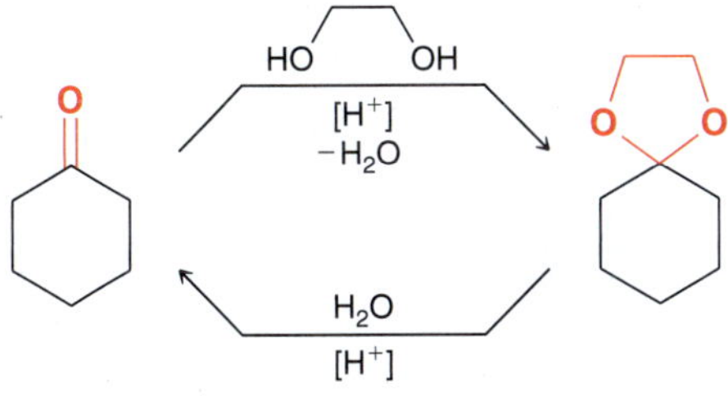


Acetal formation is a reversible process that can be controlled by carefully choosing reagents and conditions:

As mentioned in the previous section, acetal formation is favored by removal of water. To convert an acetal back into the corresponding aldehyde or ketone, it is simply treated with water in the presence of an acid catalyst. In this way, acetals can be used to protect ketones or aldehydes. For example, consider how the following transformation might be accomplished:

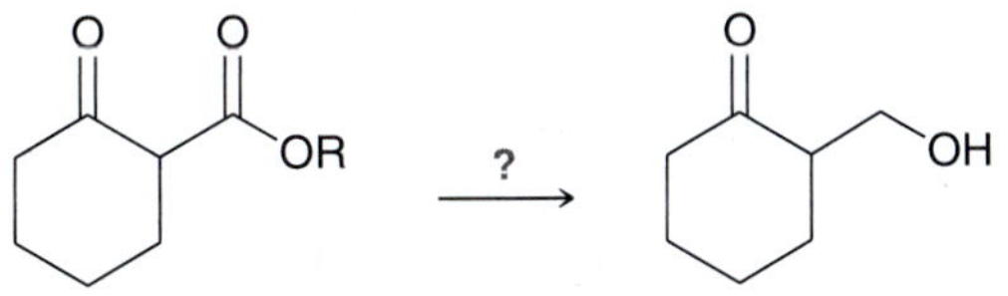

This transformation involves reduction of an ester to form an alcohol. Recall that lithium aluminum hydride (LAH) can be used to accomplish this type of reaction. However, under these conditions, the ketone group will also be reduced. The problem above requires reduction of the ester group without also reducing the ketone group. To accomplish this, a protecting group can be used. The first step is to convert the ketone into an acetal:

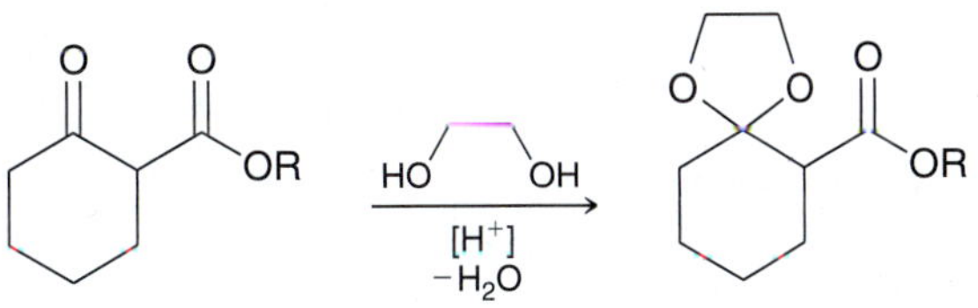

Notice that the ketone group is converted into an acetal, but the ester group is not. The resulting acetal group is stable under strongly basic conditions and will not react with LAH. This makes it possible to reduce only the ester, after which the acetal can be removed to regenerate the ketone. The three steps are summarized below:

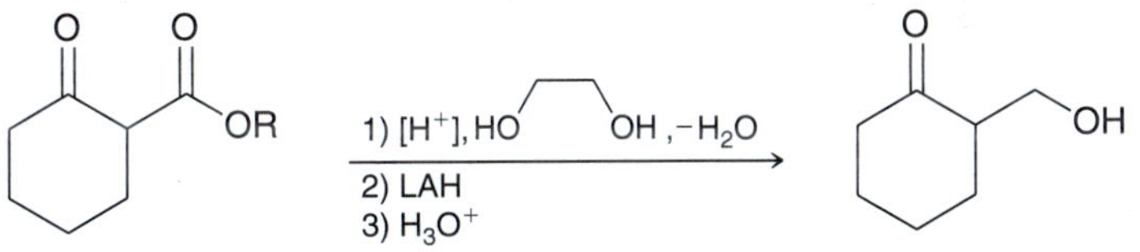

### CONCEPTUAL CHECKPOINT

**20.11** Propose an efficient synthesis for each of the following transformations:

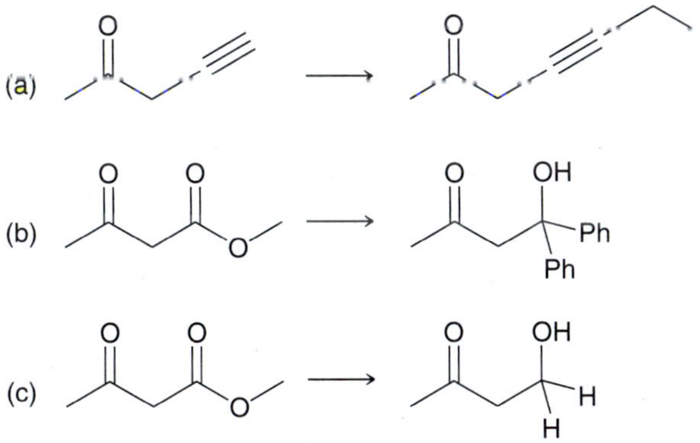

(a)

(b)

(c)

**20.12** Predict the product(s) for each reaction below:

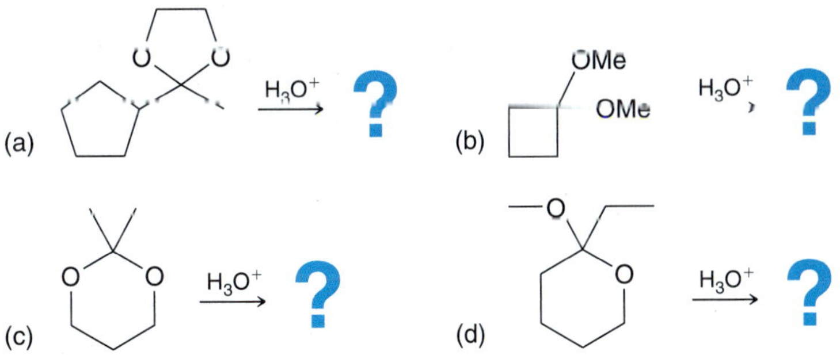

(a) $\xrightarrow{H_3O^+}$ ?

(b) $\xrightarrow{H_3O^+}$ ?

(c) $\xrightarrow{H_3O^+}$ ?

(d) $\xrightarrow{H_3O^+}$ ?

## medically speaking — Acetals as Prodrugs

In Chapter 19 we explored the concept of prodrugs—pharmacologically inactive compounds that are converted by the body into active compounds. Many strategies are used in the design of prodrugs. One such strategy involves an acetal group.

As an example, fluocinonide is a prodrug that contains an acetal group and is sold in a cream used for the topical treatment of eczema and other skin conditions.

Skin has several important functions, including preventing the absorption of foreign substances into the general circulation. This feature protects us from harmful substances, but it also prevents beneficial drugs from penetrating deep into the skin. This effect is most pronounced for drugs containing OH groups. Such drugs typically exhibit low dermal permeability (they are not readily absorbed by the skin). To circumvent this problem, two OH groups can be temporarily converted into an acetal. The acetal prodrug is capable of penetrating the skin more deeply, because it lacks the OH groups. Once the prodrug reaches its target, the acetal group is slowly hydrolyzed, thereby releasing the active drug:

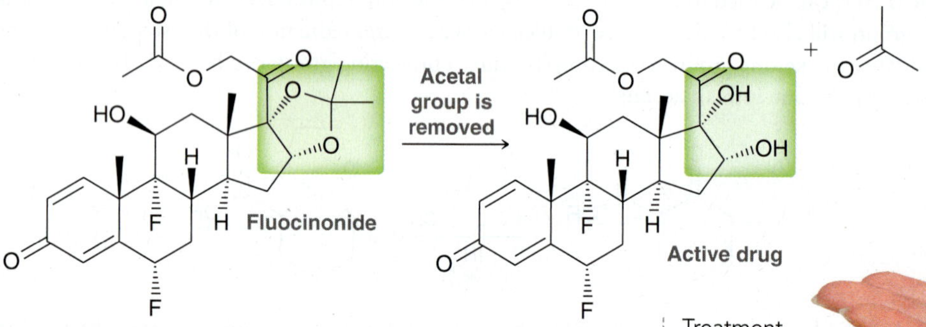

Fluocinonide → (Acetal group is removed) → Active drug + acetone

Treatment with fluocinonide is significantly more effective than direct treatment with the active drug, because the latter cannot reach all of the affected areas.

## Stable Hemiacetals

In the previous section, we saw how to convert an aldehyde or ketone into an acetal. In most cases, it is very difficult to isolate the intermediate hemiacetal:

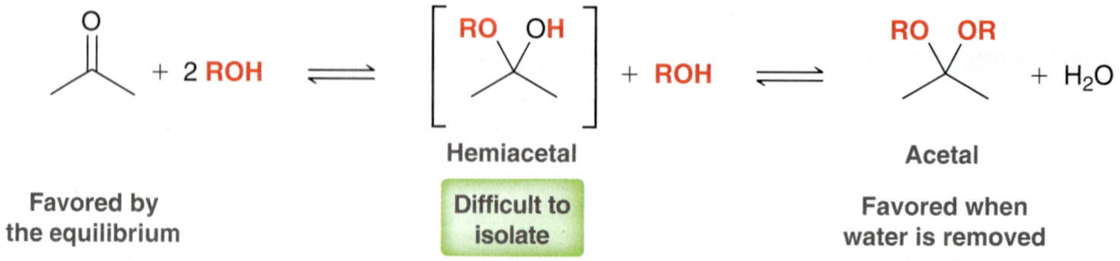

Ketone + 2 ROH ⇌ [Hemiacetal] + ROH ⇌ Acetal + $H_2O$

Favored by the equilibrium — Hemiacetal: Difficult to isolate — Acetal: Favored when water is removed

For ketones we saw that the equilibrium generally favors the reactants unless water is removed, which enables formation of the acetal. The hemiacetal is not favored under either set of conditions (with or without removal of water). However, when a compound contains both a carbonyl group and a hydroxyl group, the resulting cyclic hemiacetal can often be isolated; for example:

Cyclic hemiacetal

This will be important when we learn about carbohydrate chemistry in Chapter 24. Glucose, the major source of energy for the body, exists primarily as a cyclic hemiacetal:

Glucose (Open-chain)

Glucose (Cyclic hemiacetal)

## CONCEPTUAL CHECKPOINT

**20.13** Draw a plausible mechanism for the following transformation:

$[H_2SO_4]$

**20.14** Compound A has molecular formula $C_8H_{14}O_2$. Upon treatment with catalytic acid, compound A is converted into the cyclic hemiacetal shown below. Identify the structure of compound A.

**Compound A** $\overset{[H^+]}{\rightleftharpoons}$

## 20.6 Nitrogen Nucleophiles

### Primary Amines

In mildly acidic conditions, an aldehyde or ketone will react with a primary amine to form an **imine**:

$$\xrightarrow[\substack{CH_3NH_2 \\ -H_2O}]{[H^+]}$$

Imines are compounds that possess a C=N double bond and are common in biological pathways. Imines are also called Schiff bases, named after Hugo Schiff, a German chemist who first described their formation. A six-step mechanism for imine formation is shown in Mechanism 20.6. It is best to divide the mechanism conceptually into two parts (just as we did to conceptualize the mechanism of acetal formation): (1) The first three steps produce an intermediate called a **carbinolamine** and (2) the last three steps convert the carbinolamine into an imine:

## MECHANISM 20.6 IMINE FORMATION

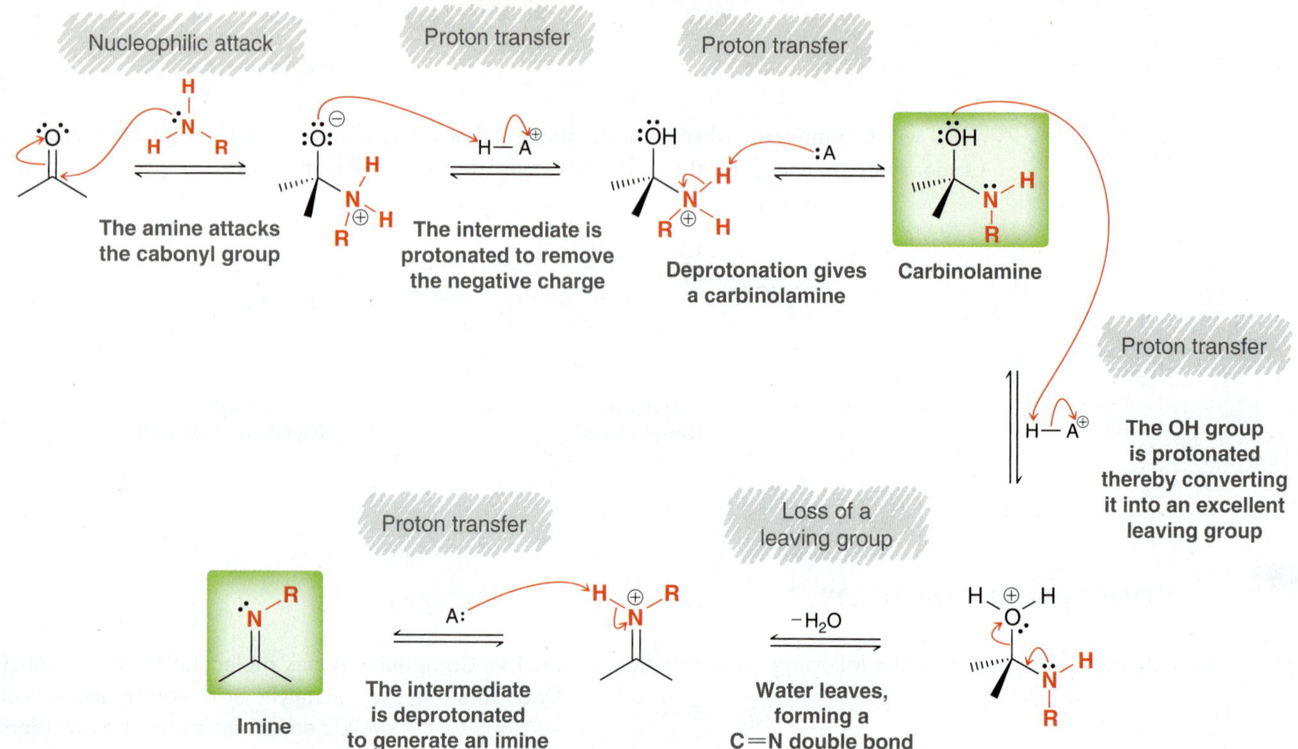


**FIGURE 20.5**
The sequence of steps involved in formation of a carbinolamine.

Nucleophilic attack → Proton transfer → Proton transfer

Let's begin our analysis of this mechanism by focusing on the first part: formation of the carbinolamine, which involves the three steps in Figure 20.5. Notice that these steps are similar to the first three steps of acetal formation (Mechanism 20.5), but the order of the steps has changed. Specifically, imine formation begins with a nucleophilic attack, while acetal formation begins with a proton transfer. To understand the difference, we must recognize that in the presence of an amine, any strong acid catalyst will transfer its proton to the amine, giving an ammonium ion:

$$RNH_2 + \underset{(pK_a = -7)}{H{-}Cl} \longrightarrow \underset{\substack{\textbf{Ammonium ion} \\ (pK_a = 10.5)}}{RNH_3^{\oplus}} + Cl^{\ominus}$$

This process is effectively irreversible as a result of the vast difference in $pK_a$ values. That is, the number of molecules of HCl present in solution is negligible, and instead, the acidic species will be ammonium ions. Under these conditions, it is very unlikely that a ketone will be protonated, because protonated ketones are highly acidic species ($pK_a \approx -7$). The concentration of protonated ketone is therefore negligible, so it is unlikely to serve as an intermediate in our mechanism.

The first step in Mechanism 20.6 is a nucleophilic attack in which a molecule of amine (that has not been protonated) functions as a nucleophile and attacks the carbonyl group. The resulting intermediate can then undergo two successive proton transfer steps, generating a carbinolamine. As explained a moment ago, the identity of the acid $HA^+$ is most likely an ammonium ion:

$$H{-}A^{\oplus} \equiv H{-}\overset{\oplus}{N}H_2{-}R$$

Once the carbinolamine has been formed, formation of the imine is accomplished with three steps (Figure 20.6).

**FIGURE 20.6** The sequence of steps that convert a carbinolamine into an imine.

The pH of the solution is an important consideration during imine formation, with the rate of reaction being greatest when the pH is around 4.5 (Figure 20.7). If the pH is too high (i.e., if no acid catalyst is used), the carbinolamine is not protonated (step 4 of the mechanism), so the reaction occurs more slowly. If the pH is too low (too much acid is used), most of the amine molecules will be protonated to give ammonium ions, which are not nucleophilic. Under these conditions, step 2 of the mechanism occurs too slowly. As a result, care must be taken to ensure optimal pH of the solution during imine formation.

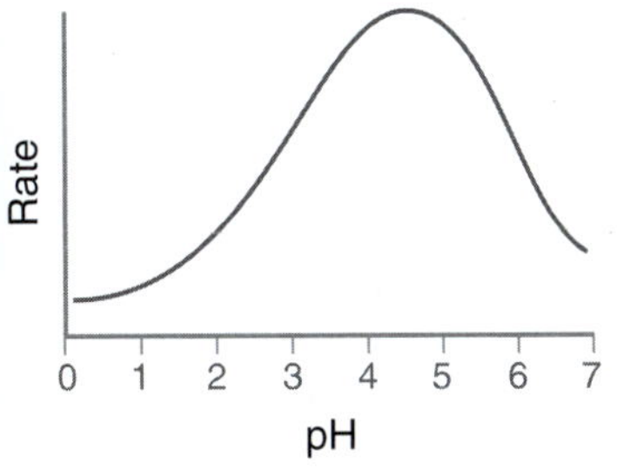


**FIGURE 20.7** The rate of imine formation as a function of pH.

## SKILLBUILDER

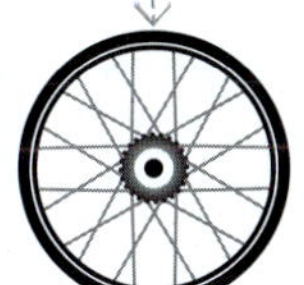

### 20.3 DRAWING THE MECHANISM OF IMINE FORMATION

**LEARN the skill** Draw a plausible mechanism for the following transformation:

Cyclohexanone $\xrightarrow[-H_2O]{[H_2SO_4],\ EtNH_2}$ N-ethyl imine of cyclohexanone (N–Et)

**SOLUTION**

The reaction above is an example of imine formation. The mechanism can be divided into two parts: (1) formation of the carbinolamine and (2) formation of the imine.

Formation of the carbinolamine involves three mechanistic steps:

Nucleophilic attack → Proton transfer → Proton transfer

When drawing these three steps, make sure to place the head and tail of every curved arrow in its precise location and make sure to place all positive charges in their appropriate locations. Notice that every step requires two curved arrows.

**STEP 1**
Draw the three steps necessary to form a carbinolamine.

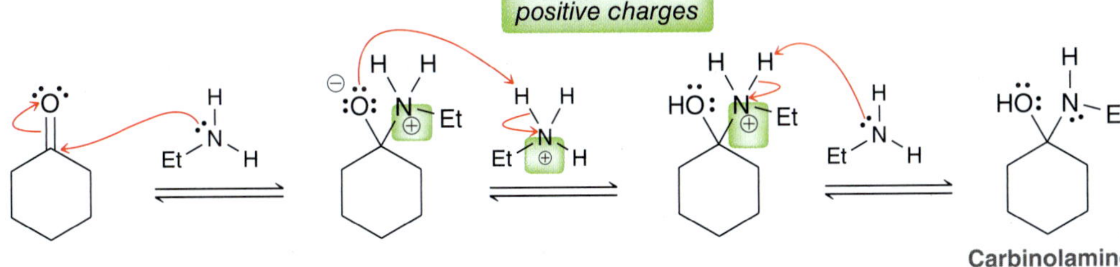

Now let's focus on the second part of the mechanism, in which the carbinolamine is converted into an imine. This requires three steps:

Proton transfer → Loss of a leaving group → Proton transfer

Make sure to place the head and tail of every curved arrow in its precise location and make sure to place all positive charges in their appropriate locations:

**STEP 2**
Draw the three steps necessary to convert the carbinolamine into an imine.

*The tail of the first curved arrow should be placed on a lone pair*

*Don't forget the positive charges*

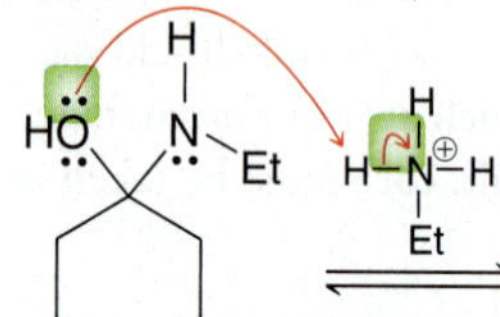


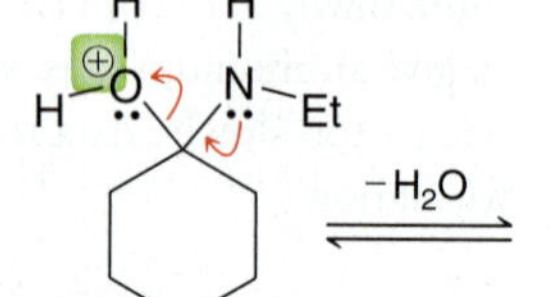


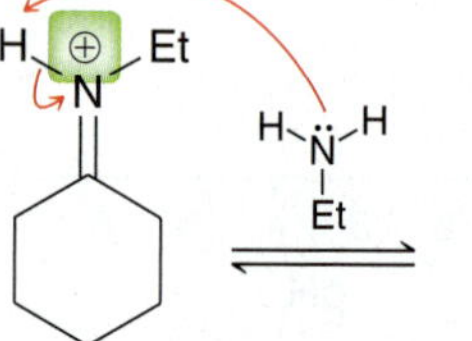


Imine

*Don't forget the second curved arrow that shows release of the proton*

**PRACTICE the skill** **20.15** Draw a plausible mechanism for each of the following transformations:

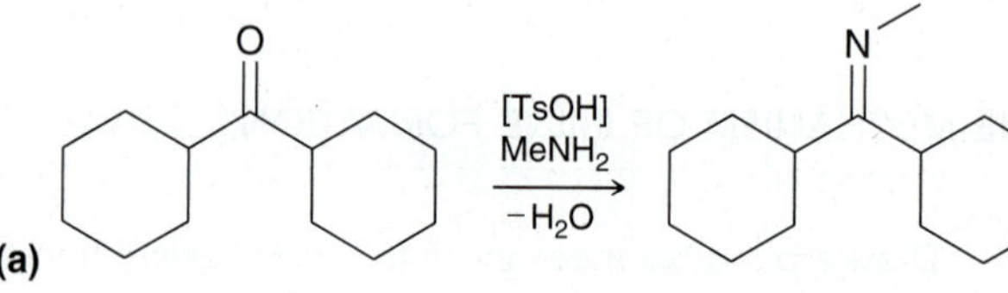

**APPLY the skill** **20.16** Predict the major product for each of the following reactions:

(a) [$H^+$], $NH_3$, −$H_2O$ → ? (b) $NH_2$, [$H^+$], −$H_2O$ → ?

**20.17** Predict the major product for each of the following intramolecular reactions:

(a) [$H^+$], −$H_2O$ → ? (b) [$H^+$], −$H_2O$ → ?

**20.18** Identify the reactants that you would use to make each of the following imines:

(a) (b) (c)

need more **PRACTICE?** **Try Problem 20.72**

Many different compounds of the form $RNH_2$ will react with aldehydes and ketones, including compounds in which R is not an alkyl group. In the following examples, the R group of the amine has been replaced with a group that has been highlighted in red:

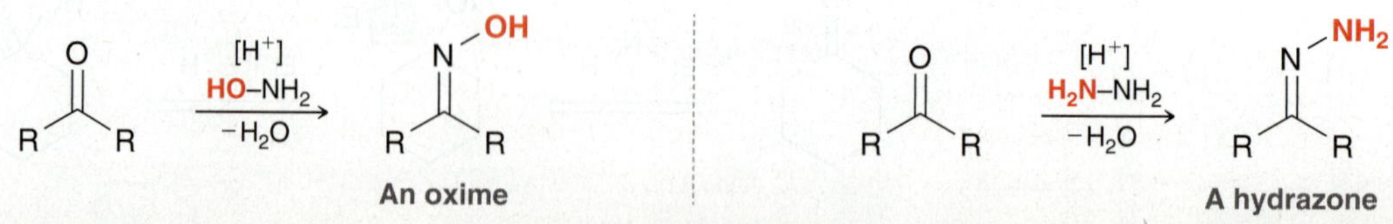

**LOOKING AHEAD**
Hydrazones are synthetically useful, as we will see in the discussion of the Wolff-Kishner reduction later in this chapter.

When hydroxylamine ($NH_2OH$) is used as a nucleophile, an **oxime** is formed. When hydrazine ($NH_2NH_2$) is used as a nucleophile, a **hydrazone** is formed. The mechanism for each of these reactions is directly analogous to the mechanism of imine formation.

## practically speaking — Beta-Carotene and Vision

Beta-carotene is a naturally occurring compound found in many orange-colored fruits and vegetables, including carrots, sweet potatoes, pumpkins, mangoes, cantaloupes, and apricots. As mentioned in the chapter opener, beta-carotene is known to be good for your eyes. To understand why, we must explore what happens to beta-carotene in your body. Imine formation plays an important role in the process.

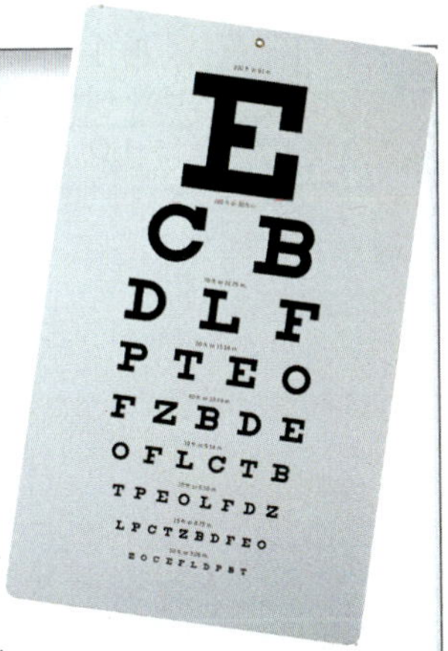

Beta-carotene is metabolized in the liver to produce vitamin A (also called retinol):

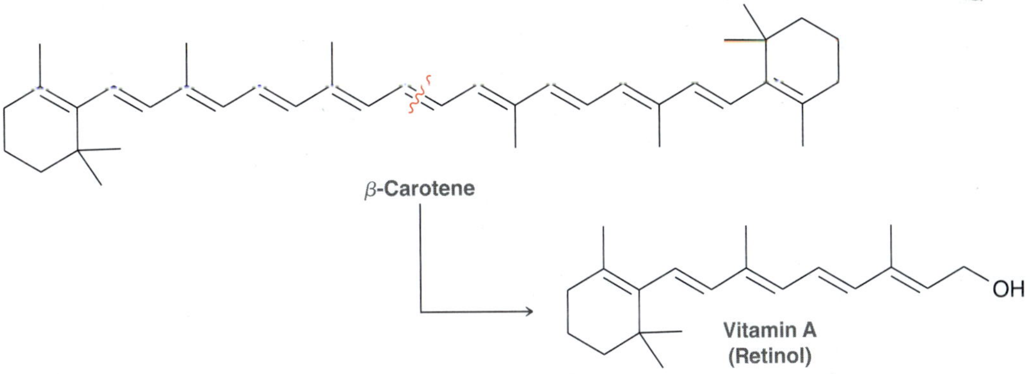

β-Carotene

Vitamin A (Retinol)

OH

Vitamin A is then oxidized, and one of the double bonds undergoes isomerization to produce 11-*cis*-retinal:

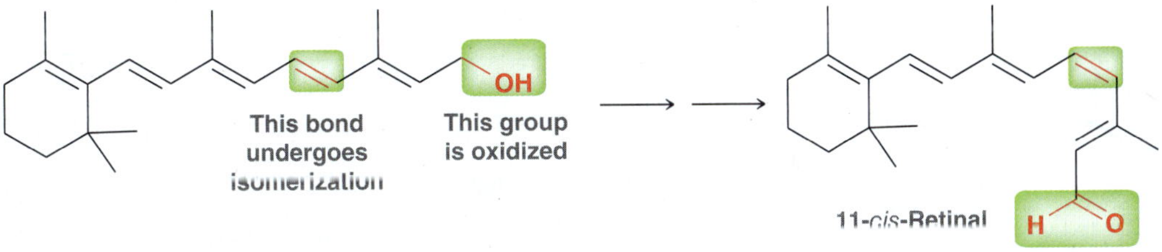

This bond undergoes isomerization

This group is oxidized

OH

11-*cis*-Retinal

H O

The resulting aldehyde then reacts with an amino group of a protein (called opsin) to produce rhodopsin, which possesses an imine group:

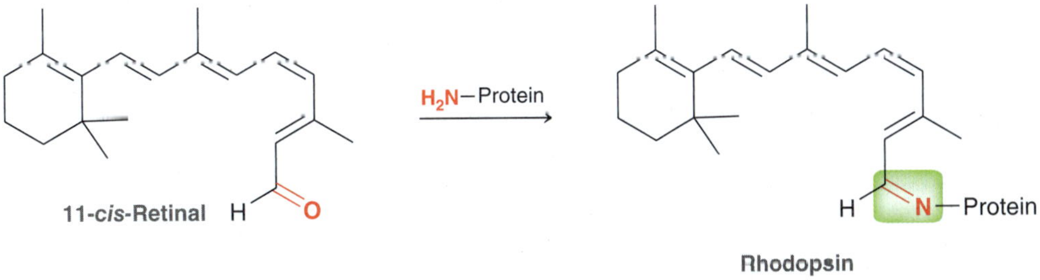

11-*cis*-Retinal H O

$H_2N$—Protein

H N—Protein

Rhodopsin

As described in Section 17.13, rhodopsin can absorb a photon of light, initiating a photoisomerization of the *cis* double bond to form a *trans* double bond. The resulting change in geometry triggers a signal that is ultimately detected by the brain and interpreted as vision.

A deficiency of vitamin A can lead to "night blindness," a condition that prevents the eyes from adjusting to dimly lit environments.

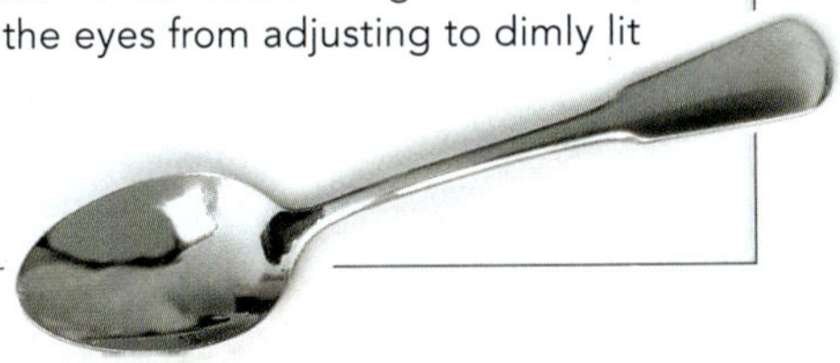

## CONCEPTUAL CHECKPOINT

**20.19** Predict the product of each of the following reactions:

(a) $[H^+]$, HO–$NH_2$, $-H_2O$ → ?

(b) $[H^+]$, $H_2N$–$NH_2$, $-H_2O$ → ?

**20.20** Identify the reactants that you would use to make each of the following compounds:

(a) N–OH

(b) N–$NH_2$

## Secondary Amines

In acidic conditions, an aldehyde or ketone will react with a secondary amine to form an **enamine:**

$[H^+]$, $R_2N$–H, $-H_2O$

An enamine

Enamines are compounds in which the nitrogen lone pair is delocalized by the presence of an adjacent C=C double bond. A mechanism for enamine formation is shown in Mechanism 20.7

## MECHANISM 20.7 ENAMINE FORMATION

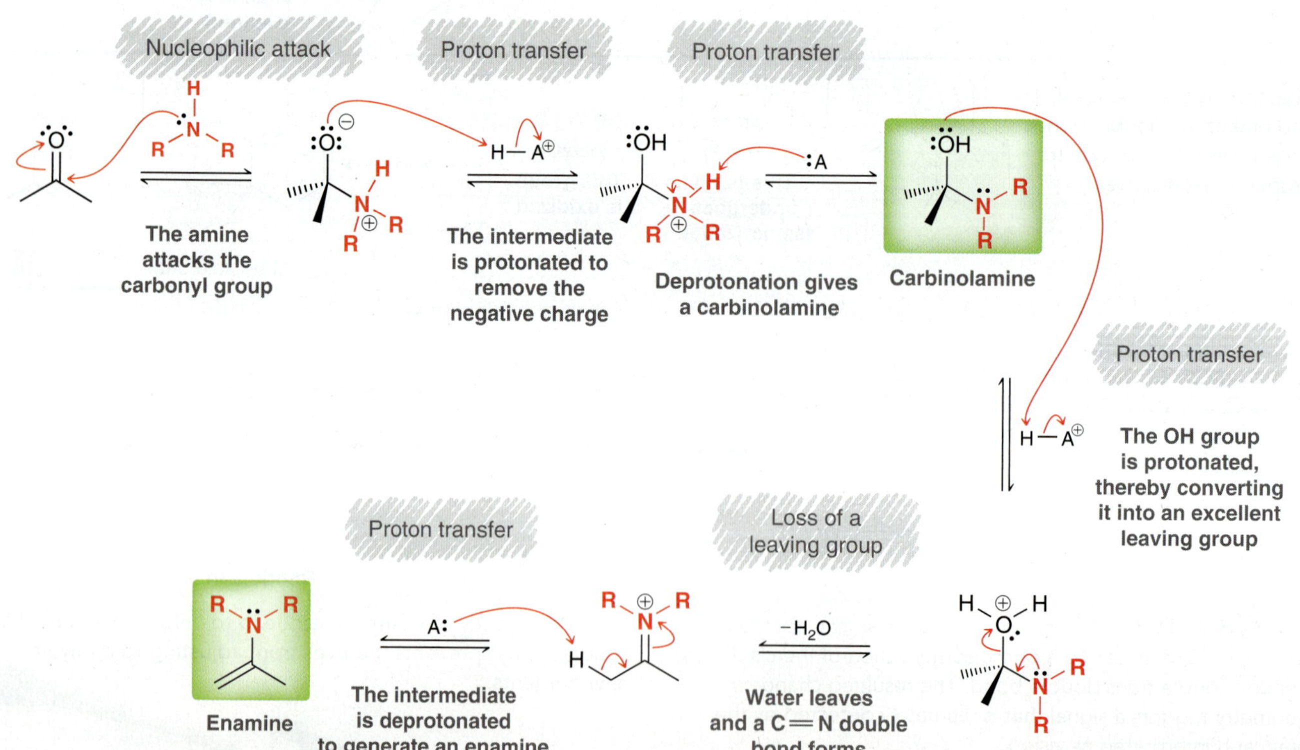

This mechanism of enamine formation is identical to the mechanism of imine formation except for the last step:

**An imine**

**An enamine**

The difference in the iminium ions explains the different outcomes for the two reactions. During imine formation, the nitrogen atom of the iminium ion possesses a proton that can be removed as the final step of the mechanism. In contrast, during enamine formation, the nitrogen atom of the iminium ion does not possess a proton. As a result, elimination from the adjacent carbon is necessary in order to yield a neutral species.

## 20.4 DRAWING THE MECHANISM OF ENAMINE FORMATION

**LEARN the skill**

Draw a plausible mechanism for the following reaction:

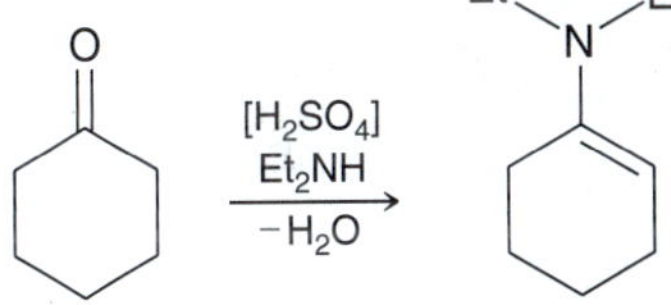

### SOLUTION

The reaction above is an example of enamine formation.

The mechanism can be divided into two parts: (1) formation of the carbinolamine and (2) formation of the enamine.

Formation of the carbinolamine involves three mechanistic steps:

Nucleophilic attack → Proton transfer → Proton transfer

**STEP 1**
Draw the three steps necessary to form a carbinolamine.

When drawing these three steps, make sure to place the head and tail of every curved arrow in its precise location and make sure to place all positive charges in their appropriate locations. Notice that every step requires two curved arrows.

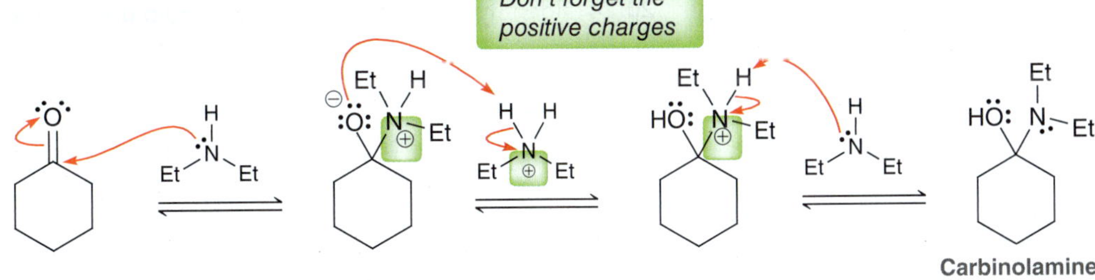

In the second part of the mechanism, the carbinolamine is converted into an enamine via a three-step process:

Proton transfer | Loss of a leaving group | Proton transfer

**STEP 2**
Draw the three steps necessary to convert the carbinolamine into an enamine.

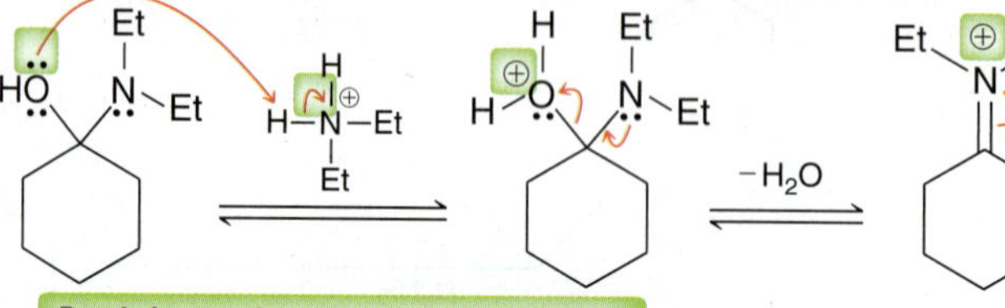
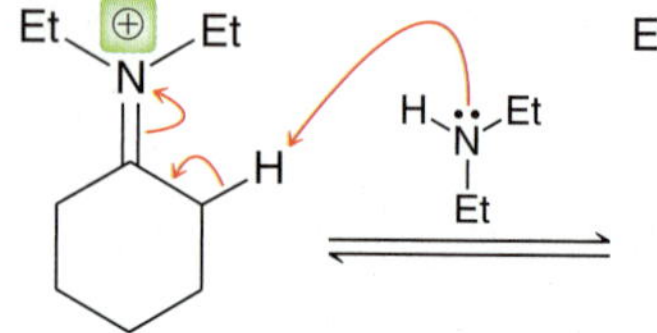
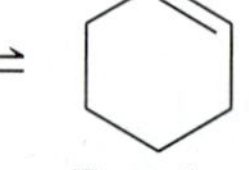

**PRACTICE the skill** **20.21** Draw a plausible mechanism for each of the following reactions:

(a) [$H_2SO_4$], $Et_2NH$, $-H_2O$

(b) [$H_2SO_4$], $Me_2NH$, $-H_2O$

**APPLY the skill** **20.22** Predict the major product for each of the following reactions:

(a) [$H^+$], $-H_2O$ ?

(b) [$H^+$], $-H_2O$ ?

**20.23** Predict the major product for each of the following intramolecular reactions:

(a) [$H^+$], $-H_2O$ ?

(b) [$H^+$], $-H_2O$ ?

**20.24** Identify the reactants that you would use to make each of the following enamines:

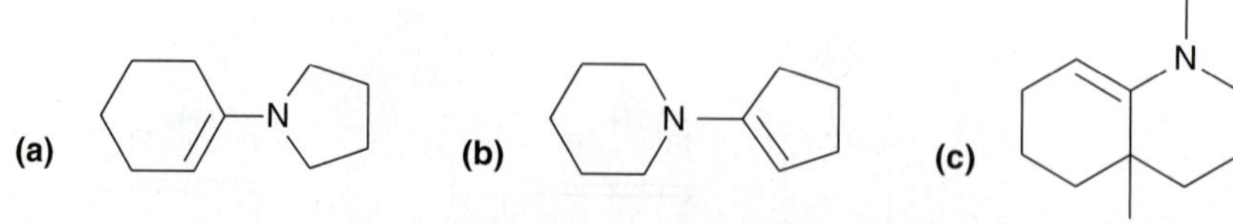


need more **PRACTICE?** Try Problems 20.64, 20.65a, 20.66e, 20.75g

## Wolff-Kishner Reduction

At the end of the previous section, we noted that ketones can be converted into hydrazones. This transformation has practical utility, because hydrazones are readily reduced under strongly basic conditions:

$NNH_2$ — $\xrightarrow[\text{Heat}]{KOH/H_2O}$ — H H

A hydrazone (82%)

This transformation is called the **Wolff-Kishner reduction,** named after the German chemist Ludwig Wolff and the Russian chemist N. M. Kishner. This provides a two-step procedure for reducing a ketone to an alkane:

O $\xrightarrow[-H_2O]{[H^+]\ H_2N\text{—}NH_2}$ N–$NH_2$ $\xrightarrow[\text{Heat}]{KOH/H_2O}$ H H + $N_2$

(80%)

The second part of the Wolff-Kishner reduction is believed to proceed via Mechanism 20.8.

### MECHANISM 20.8 THE WOLFF-KISHNER REDUCTION

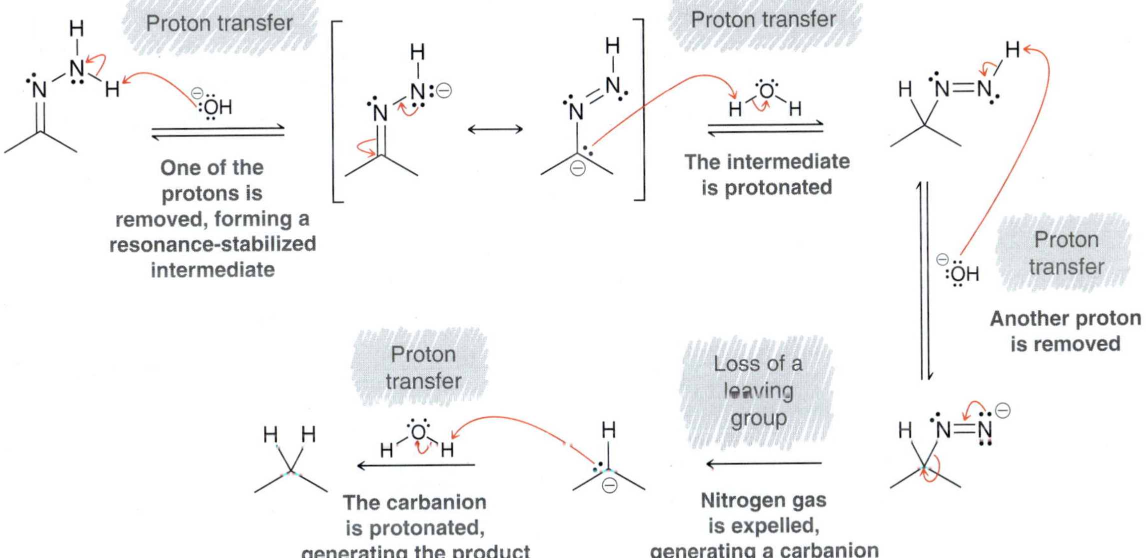


Notice that four of the five steps in the mechanism are proton transfers, the exception being the loss of $N_2$ gas to generate a carbanion. This step warrants special attention, because formation of a carbanion in a solution of aqueous hydroxide is thermodynamically unfavorable (*significantly* uphill in energy). Why, then, does this step occur? It is true that the equilibrium for this step greatly disfavors formation of the carbanion, and therefore, only a very small number of molecules will initially lose $N_2$ to form the carbanion. However, the resulting $N_2$ gas then bubbles out of the reaction mixture, and the equilibrium is adjusted to form more nitrogen gas, which again leaves the reaction mixture. The evolution of nitrogen gas ultimately renders this step irreversible and forces the reaction to completion. As a result, the yields for this process are generally very good.

## CONCEPTUAL CHECKPOINT

**20.25** Predict the product of the following two-step procedure and draw a mechanism for its formation:

1) $[H^+]$, $H_2N–NH_2$, $–H_2O$
2) KOH / $H_2O$, heat
?

## 20.7 Hydrolysis of Acetals, Imines, and Enamines

In our discussion of acetals as protecting groups (Section 20.5), we saw that treatment of an acetal with aqueous acid affords the corresponding aldehyde or ketone:

RO OR + $H_2O$ $\xrightarrow{[H^+]}$ O + 2 ROH

Acetal Ketone

This process is called a **hydrolysis** reaction, because bonds are cleaved (shown with red wavy lines) by treatment with water. Acetal hydrolysis generally requires acid catalysis. That is, acetals do not undergo hydrolysis under aqueous basic conditions:

RO OR $\xrightarrow[H_2O]{NaOH}$ **No reaction**

Acetal hydrolysis is believed to proceed via Mechanism 20.9. Since acidic conditions are employed, the mechanism does not involve any reagents or intermediates that are strong bases. For example, in the third step of the mechanism, water is used as a nucleophile, rather than hydroxide, because the latter is not present in substantial quantities. Similarly, $H_3O^+$ is used in all protonation steps (such as the first step), because the use of water as a proton source would generate a hydroxide ion, which is unlikely to form in acidic conditions. Always remember that a mechanism must be consistent with the conditions employed. In acidic conditions, strong bases should not be invoked as reagents or intermediates.

## MECHANISM 20.9 HYDROLYSIS OF ACETALS

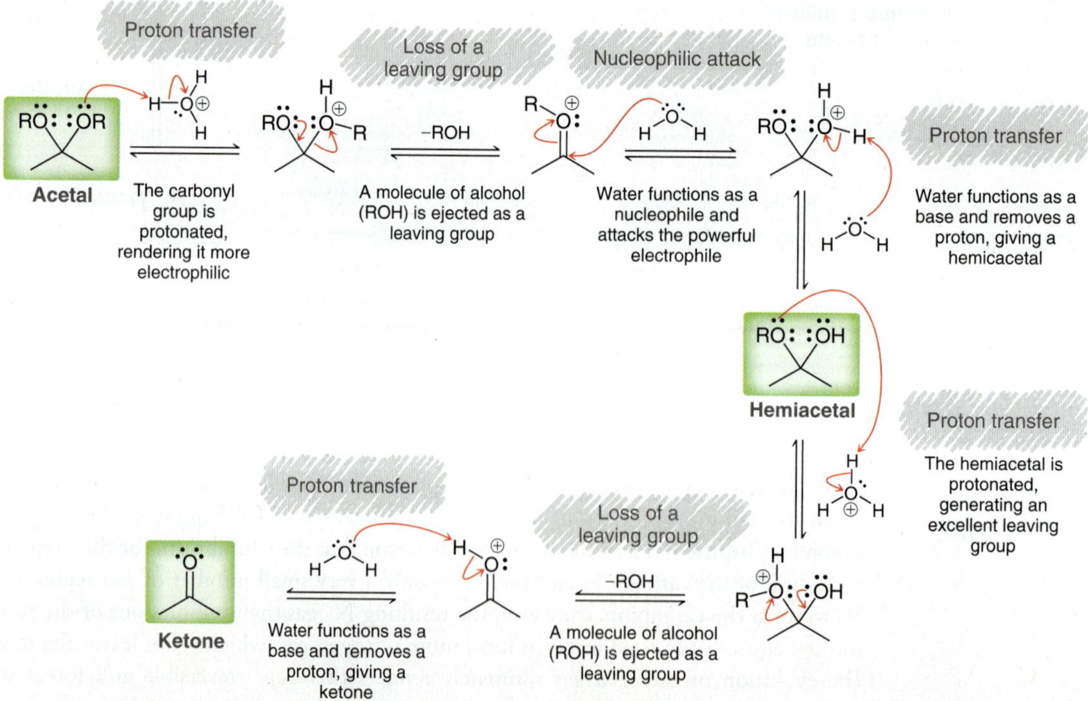

Notice that acetal hydrolysis proceeds via a hemiacetal intermediate, just as we saw with acetal formation. In fact, all of the intermediates involved in acetal hydrolysis are identical to the intermediates involved in acetal formation, but in reverse order. This is illustrated in the following scheme:

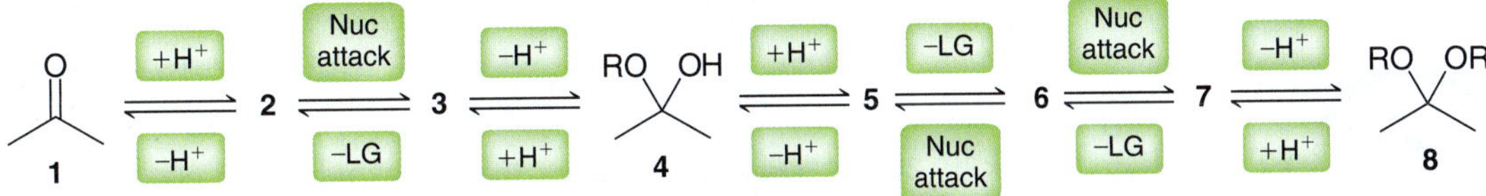


Conversion of a ketone (compound **1**) to an acetal (compound **8**) is achieved via intermediates **2**–**7**. The reverse process (conversion of **8** to **1**) is achieved via the same intermediates (**2**–**7**), but in reverse order (first **7**, then **6**, etc.).

Imines and enamines also undergo hydrolysis when treated with aqueous acid, and the red wavy lines (below) indicate the bonds that undergo cleavage:

$$\text{imine (N–R)} + H_2O \xrightarrow{[H^+]} \text{ketone} + RNH_2$$

$$\text{enamine (R}_2\text{N)} + H_2O \xrightarrow{[H^+]} \text{ketone} + R_2NH$$

Once again, the intermediates involved in imine hydrolysis are the same as the intermediates involved in imine formation, but in reverse order. Similarly, the intermediates involved in enamine hydrolysis are the same as the intermediates involved in enamine formation, but in reverse order.

Let's get some practice identifying the products of some hydrolysis reactions.

# SKILLBUILDER

## 20.5 DRAWING THE PRODUCTS OF A HYDROLYSIS REACTION

**LEARN the skill**

Draw the products that are expected from the following reaction:

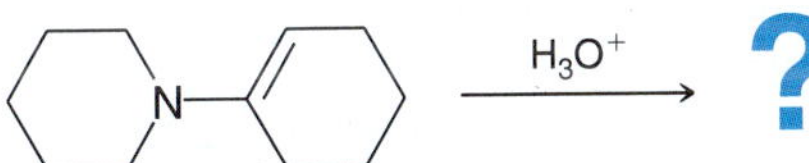

### SOLUTION

**STEP 1**
Identify which bond(s) undergo bond cleavage.

The starting compound is an enamine, and it is being treated with aqueous acid, so a hydrolysis reaction is expected. Begin by identifying the bond(s) that will undergo cleavage. When an enamine undergoes hydrolysis, cleavage occurs for the bond between the nitrogen atom and the $sp^2$-hybridized carbon atom to which it is attached:

$$\text{R}_2\text{N–C(=C) enamine} \longrightarrow \text{C=O}$$

In our case, this corresponds with the following bond:

[N–C bond between the piperidine nitrogen and the cyclohexene ring, marked with a wavy line]

**STEP 2**
Identify the carbon atom that is converted into a carbonyl group.

Next, identify the carbon atom that will ultimately become a carbonyl group in the product. For enamine hydrolysis, it is the $sp^2$-hybridized carbon atom connected to the nitrogen atom, highlighted below:

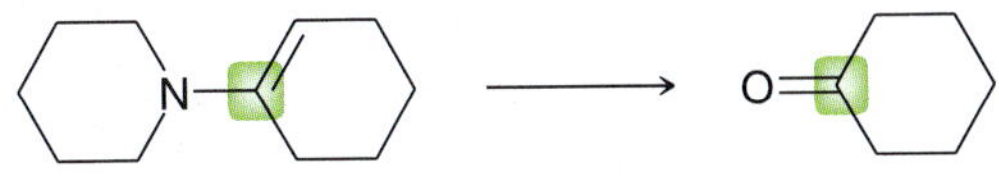

**STEP 3**
Draw the other fragment(s).

Finally, determine the identity of the other fragment(s). As a result of the C—N bond cleavage, the carbon atom becomes a carbonyl group, and the nitrogen atom will accept a proton to generate a secondary amine, as shown below:

**PRACTICE the skill** **20.26** Draw the products that are expected for each of the following reactions:

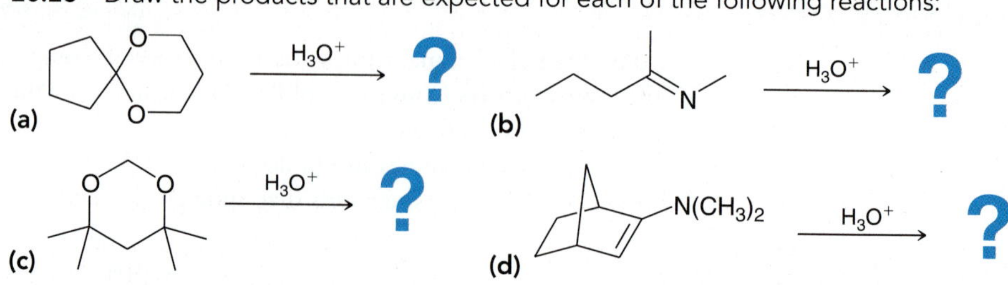


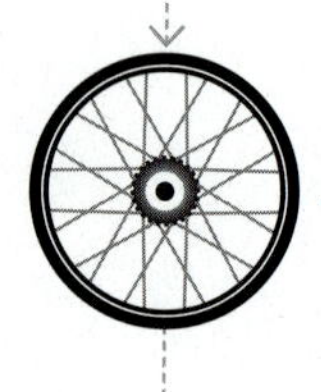

**APPLY the skill** **20.27** Draw the products that are expected for each of the following reactions:

(a) N— Excess $H_3O^+$ ? (b) O O $H_3O^+$ ?

need more **PRACTICE?** **Try Problems 20.63, 20.64**

The Medically Speaking box below provides two examples of hydrolysis reactions that are exploited in drug design.

**medically speaking** **Prodrugs**

### Methenamine as a Prodrug of Formaldehyde

Formaldehyde has antiseptic properties and can be employed in the treatment of urinary tract infections due to its ability to react with nucleophiles present in urine. However, formaldehyde can be toxic when exposed to other regions of the body. Therefore, the use of formaldehyde as an antiseptic agent requires a method for selective delivery to the urinary tract. This can be accomplished by using a prodrug called methenamine:

**Methenamine**

This compound is a nitrogen analogue of an acetal. That is, each carbon atom is connected to two nitrogen atoms, very much like an acetal in which a carbon atom is connected to two oxygen atoms. A carbon atom that is connected to two heteroatoms (O or N) can undergo acid-catalyzed hydrolysis:

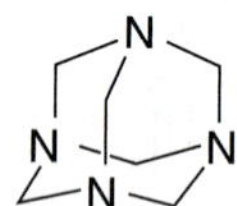


Z = O or N

Each of the carbon atoms in methenamine can be hydrolyzed, releasing formaldehyde:

$H_3O^+$ → 6 H(C=O)H + 4 $\overset{\oplus}{N}H_4$

**Formaldehyde**

Methenamine is placed in special tablets that do not dissolve as they travel through the acidic environment of the stomach but do dissolve once they reach the basic environment of the intestinal tract. Methenamine is thereby released in the intestinal tract, where it is stable under basic conditions. Once it reaches the acidic environment of the urinary tract, methenamine is hydrolyzed, releasing formaldehyde, as shown above. In this way, methenamine is used as a prodrug that enables delivery of formaldehyde specifically to the urinary tract. This method prevents the systemic release of formaldehyde in other organs of the body where it would be toxic.

### CONCEPTUAL CHECKPOINT

**20.28** As shown above, methenamine is hydrolyzed in aqueous acid to produce formaldehyde and ammonia. Draw a mechanism showing formation of one molecule of formaldehyde (the remaining five molecules of formaldehyde are each released via a similar sequence of steps). The release of each molecule of formaldehyde is directly analogous to the hydrolysis of an acetal. To get you started, the first two steps are provided below:

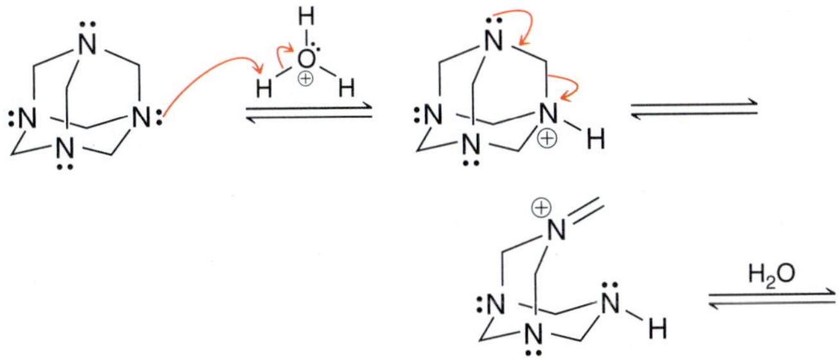

$H_2O$

### Imines as Prodrugs

The imine group is used in the design of many prodrugs. Here we will explore one such example.

The compound below, γ-aminobutyric acid, is an important neurotransmitter:

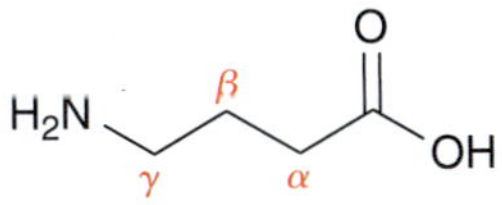

**γ-Aminobutyric acid**

A deficiency of this compound can cause convulsions. Administering γ-aminobutyric acid directly to a patient is not an effective treatment, because the compound does not readily cross the blood-brain barrier. Why not? At physiological pH, the amino group is protonated and the carboxylic acid group is deprotonated:

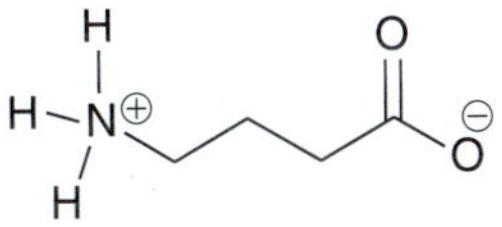

The compound exists primarily in this ionic form, which cannot cross the nonpolar environment of the blood-brain barrier. Progabide is a prodrug derivative used to treat patients who exhibit the symptoms of a deficiency of γ-aminobutyric acid:

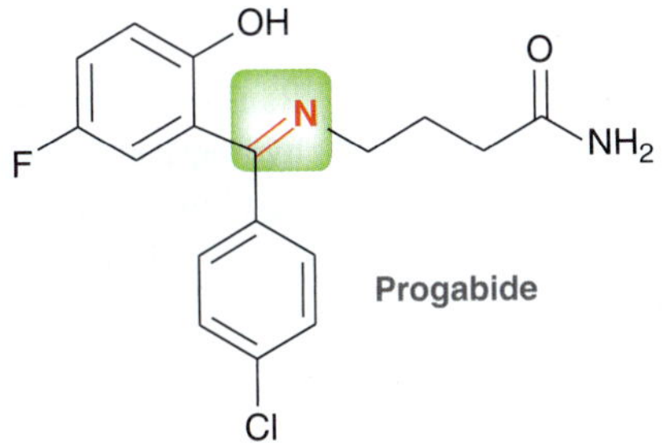

**Progabide**

The carboxylic acid has been converted to an amide, and the amino group has been converted into an imine (highlighted). At physiological pH, this compound exists primarily as a neutral compound (uncharged), and it can therefore cross the blood-brain barrier. Once in the brain, it is converted to γ-aminobutyric acid via hydrolysis of the imine and amide groups:

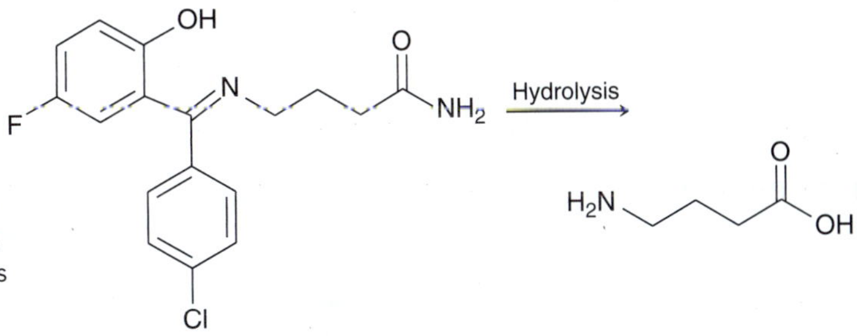

Hydrolysis

Progabide is just one example in which the imine group has been used in the design of a prodrug.

## 20.8 Sulfur Nucleophiles

In acidic conditions, an aldehyde or ketone will react with two equivalents of a thiol to form a **thioacetal:**

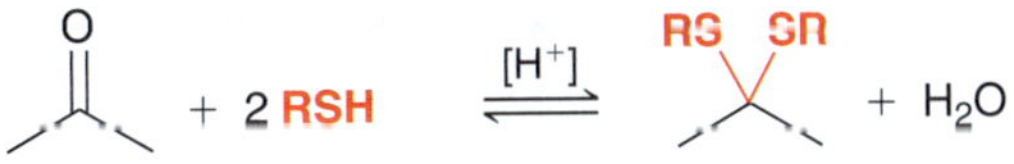

$$\text{ketone} + 2\,RSH \underset{}{\overset{[H^+]}{\rightleftharpoons}} \text{Thioacetal} + H_2O$$

**Thioacetal**

The mechanism of this transformation is directly analogous to acetal formation, with sulfur atoms taking the place of oxygen atoms. If a compound with two SH groups is used, a cyclic thioacetal is formed:

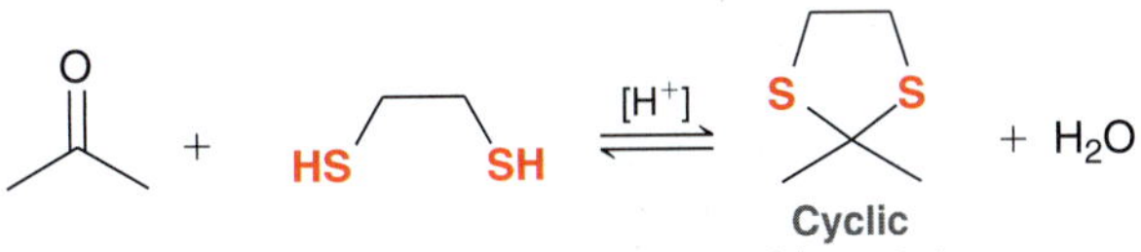

$$\text{ketone} + HS\text{–}CH_2CH_2\text{–}SH \underset{}{\overset{[H^+]}{\rightleftharpoons}} \text{Cyclic thioacetal} + H_2O$$

**Cyclic thioacetal**

When treated with Raney nickel, thioacetals undergo **desulfurization,** yielding an alkane:

Raney Ni

Raney Ni is a porous, Ni-Al alloy, in which the surface has adsorbed hydrogen atoms. It is these hydrogen atoms that ultimately replace the sulfur atoms, although a discussion of the mechanism for desulfurization is beyond the scope of this text.

The reactions above provide us with another two-step method for the reduction of a ketone:

1) [$H^+$], HS SH
2) Raney Ni

This method involves formation of the thioacetal followed by desulfurization with Raney nickel. It is the third method we have encountered for achieving this type of transformation. The other two methods are the Clemmensen reduction (Section 19.6) and the Wolff-Kishner reduction (Section 20.6).

## CONCEPTUAL CHECKPOINT

**20.29** Predict the major product for each of the following:

(a) 1) [$H^+$], HS SH 2) Raney Ni ?

(b) 1) [$H^+$], HS SH 2) Raney Ni ?

**20.30** Draw the structure of the cyclic compound that is produced when acetone is treated with 1,3-propanedithiol in the presence of an acid catalyst.

**Acetone** **1,3-Propanedithiol**

## 20.9 Hydrogen Nucleophiles

When treated with a hydride reducing agent, such as lithium aluminum hydride (LAH) or sodium borohydride ($NaBH_4$), aldehydes and ketones are reduced to alcohols:

1) LAH
2) $H_2O$

$NaBH_4$, MeOH

These reactions were discussed in Section 13.4, and we saw that LAH and $NaBH_4$ both function as delivery agents of hydride ($H^-$). The mechanism of action for these reagents has been heavily investigated and is somewhat complex. Nevertheless, the simplified version shown in Mechanism 20.10 will be sufficient for our purposes.

## MECHANISM 20.10 THE REDUCTION OF KETONES OR ALDEHYDES WITH HYDRIDE AGENTS

Nucleophilic attack

Lithium aluminum hydride (LAH) functions as a delivery agent of hydride ions ($H^-$)

Proton transfer

The resulting alkoxide intermediate is protonated to form an alcohol

In the first step of the mechanism, the reducing agent delivers a hydride ion, which attacks the carbonyl group, producing an alkoxide intermediate. This intermediate is then treated with a proton source to yield the product. This simplified mechanism does not take into account many important observations, such as the role of the lithium cation ($Li^+$). For example, when 12-crown-4 is added to the reaction mixture, the lithium ions are solvated (as described in Section 14.4), and reduction does not occur. Clearly, the lithium cation plays a pivotal role in the mechanism. However, a full treatment of the mechanism of hydride reducing agents is beyond the scope of this text, and the simplified version above will suffice.

The reduction of a carbonyl group with LAH or $NaBH_4$ is not a reversible process, because hydride does not function as a leaving group. Notice that the first step of the mechanism above employs an irreversible reaction arrow (rather than equilibrium arrows) to signify that the rate of the reverse process is insignificant.

**LOOKING BACK**

Hydride cannot function as a leaving group because it is too strongly basic. (See Section 7.8.)

## CONCEPTUAL CHECKPOINT

**20.31** Predict the major product for each of the following reactions:

(a) 1) LAH 2) $H_2O$ ?

(b) $NaBH_4$, MeOH ?

(c) 1) LAH 2) $H_2O$ ?

(d) $NaBH_4$, MeOH ?

**20.32** When 2 moles of benzaldehyde are treated with sodium hydroxide, a reaction occurs in which 1 mole of benzaldehyde is oxidized (giving benzoic acid) while the other mole of benzaldehyde is reduced (giving benzyl alcohol):

1) NaOH 2) $H_3O^+$

This reaction, called the Cannizzaro reaction, is believed to occur via the following mechanism: A hydroxide ion serves as a nucleophile to attack the carbonyl group of benzaldehyde. The resulting intermediate then functions as a hydride reducing agent by delivering a hydride ion to another molecule of benzaldehyde. In this way, one molecule is reduced while the other is oxidized.

(a) Draw a mechanism for the Cannizzaro reaction, consistent with the description above.

(b) What is the function of $H_3O^+$ in the second step?

(c) Water alone is not sufficient to accomplish the function of the second step. Explain.

## 20.10 Carbon Nucleophiles

### Grignard Reagents

When treated with a Grignard reagent, aldehydes and ketones are converted into alcohols, accompanied by the formation of a new C—C bond:

1) $CH_3MgBr$ 2) $H_2O$ — 1) $CH_3MgBr$ 2) $H_2O$

Grignard reactions were discussed in more detail in Section 13.6. The mechanism of action for these reagents has been heavily investigated and is fairly complex. The simplified version shown in Mechanism 20.11 will be sufficient for our purposes.

### MECHANISM 20.11 THE REACTION BETWEEN A GRIGNARD REAGENT AND A KETONE OR ALDEHYDE

Nucleophilic attack — The Grignard reagent functions as a nucleophile and attacks the carbonyl group

Proton transfer — The resulting alkoxide ion is protonated to form an alcohol

**LOOKING BACK**
Carbanions rarely function as leaving groups because they are generally strongly basic. (See Section 7.8.)

Grignard reactions are not reversible because carbanions generally do not function as leaving groups. Notice that the first step of the mechanism (nucleophilic attack) is shown with an irreversible reaction arrow to signify that the reverse process is insignificant.

### CONCEPTUAL CHECKPOINT

**20.33** Predict the major product for each of the following:

(a) 1) EtMgBr 2) $H_2O$ ?

(b) 1) PhMgBr 2) $H_2O$ ?

(c) 1) PhMgBr 2) $H_3O^+$ ?

**20.34** Identify the reagents necessary to accomplish each of the transformations below:

(a)

(b)

## Cyanohydrin Formation

When treated with hydrogen cyanide (HCN), aldehydes and ketones are converted into **cyanohydrins,** which are characterized by the presence of a cyano group and a hydroxyl group connected to the same carbon atom:

HCN

HO CN

A cyanohydrin

This reaction was studied extensively by Arthur Lapworth (University of Manchester) and was found to occur more rapidly in mildly basic conditions. In the presence of a catalytic amount of base, a small amount of hydrogen cyanide is deprotonated to give cyanide ions, which catalyze the reaction (Mechanism 20.12).

### MECHANISM 20.12 CYANOHYDRIN FORMATION

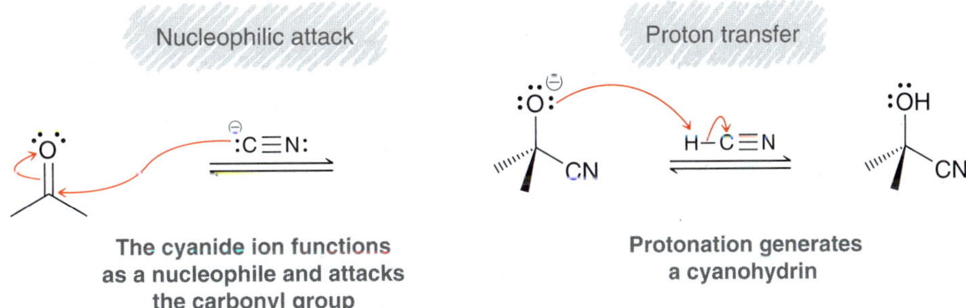


The cyanide ion functions as a nucleophile and attacks the carbonyl group

Protonation generates a cyanohydrin

In the first step, a cyanide ion attacks the carbonyl group. The resulting intermediate then abstracts a proton from HCN, regenerating a cyanide ion. In this way, cyanide functions as a catalyst for the addition of HCN to the carbonyl group.

Rather than using a catalytic amount of base to form cyanide ions, the reaction can simply be performed in a mixture of HCN and cyanide ions (from KCN). The process is reversible, and the yield of products is therefore determined by equilibrium concentrations. For most aldehydes and unhindered ketones, the equilibrium favors formation of the cyanohydrin:

$H_3C$ $CH_3$ — KCN, HCN → HO CN, $H_3C$ $CH_3$

78%

H — KCN, HCN → HO CN, H

88%

HCN is a liquid at room temperature and is extremely hazardous to handle because it is highly toxic and volatile (b.p. = 26°C). To avoid the dangers associated with handling HCN, cyanohydrins can also be prepared by treating a ketone or aldehyde with potassium cyanide and an alternate source of protons, such as HCl:

KCN, HCl

HO CN

Cyanohydrins are useful in syntheses, because the cyano group can be further treated to yield a range of products. Two examples are shown below:

1) LAH
2) $H_2O$

$H_3O^+$
Heat

In the first example, the cyano group is reduced to an amino group. In the second example, the cyano group is hydrolyzed to give a carboxylic acid. Both of these reactions and their mechanisms will be explored in more detail in the next chapter.

## CONCEPTUAL CHECKPOINT

**20.35** Predict the major product for each of the following reaction sequences:

(a) 1) KCN, HCN 2) LAH 3) $H_2O$ ?

(b) 1) KCN, HCl 2) $H_3O^+$, heat ?

**20.36** Identify the reagents necessary to accomplish each of the transformations below:

(a)

(b)

## practically speaking Cyanohydrin Derivatives in Nature

Amygdalin is a naturally occurring compound found in the pits of apricots, wild cherries, and peaches.

If ingested, this compound is metabolized to produce mandelonitrile, a cyanohydrin, which is converted by enzymes into benzaldehyde and HCN gas, a toxic compound:

Amygdalin → Mandelonitrile → Benzaldehyde + HCN (Toxic)

This last step (generation of HCN gas) is used as a defense mechanism by many species of millipedes. The millipedes manufacture and store mandelonitrile, and in a separate compartment, they store enzymes that are capable of catalyzing the conversion of mandelonitrile into benzaldehyde and HCN. To ward off predators, a millipede will mix the contents of the two compartments and secrete HCN gas.

## Wittig Reaction

Georg Wittig, a German chemist, was awarded the 1979 Nobel Prize in Chemistry for his work with phosphorus compounds and his discovery of a reaction with enormous synthetic utility. Below is an example of this reaction, called the **Wittig reaction** (pronounced Vittig):

This reaction can be used to convert a ketone into an alkene by forming a new C—C bond at the location of the carbonyl group. The phosphorus-containing reagent that accomplishes this transformation is called a phosphorus **ylide**. An ylide is a neutral molecule that contains a negatively charged atom ($C^-$ in this case) directly attached to a positively charged heteroatom ($P^+$ in this case). The phosphorus ylide shown above does, in fact, have a resonance structure that is free of any charges:

However, this resonance structure (with a C=P double bond) does not contribute much character to the overall resonance hybrid, because the *p* orbitals on C and P are vastly different in size and do not effectively overlap. A similar argument was used in describing S=O bonds in the previous chapter (Section 19.3). Despite this fact, the phosphorus ylide above, also called a **Wittig reagent,** is often drawn using either of the resonance structures shown above.

A mechanism for the Wittig reaction is shown in Mechanism 20.13. There is strong evidence that the first step involves a [2+2] cycloaddition process (Section 17.8), generating an **oxaphosphetane,** which then undergoes fragmentation to give the alkene product.

### MECHANISM 20.13 THE WITTIG REACTION

[2+2] Cycloaddition

The Wittig reagent reacts with the carbonyl group in a cycloaddition process

**An oxaphosphetane**

Fragmentation

The oxaphosphetane decomposes to produce an alkene and triphenylphosphine oxide

Wittig reagents are easily prepared by treating triphenylphosphine with an alkyl halide followed by a strong base:

1) $CH_3I$
2) BuLi

**Triphenylphosphine** **Wittig reagent**

The mechanism of formation for Wittig reagents involves an $S_N2$ reaction followed by deprotonation with a strong base:

$CH_3CH_2CH_2CH_2^{\ominus}\ Li^{\oplus}$

**Triphenylphosphine**

Since the first step is an $S_N2$ process, the regular restrictions of $S_N2$ processes apply. Specifically, primary alkyl halides will react more readily than secondary alkyl halides, and tertiary alkyl halides cannot be used. The Wittig reaction is useful for preparing mono-, di-, or trisubstituted alkenes. Tetrasubstituted alkenes are more difficult to prepare due to steric hindrance in the transition states.

The following exercise illustrates how to choose the reagents for a Wittig reaction.

## SKILLBUILDER

### 20.6 PLANNING AN ALKENE SYNTHESIS WITH A WITTIG REACTION

**LEARN the skill**

Identify the reagents necessary to prepare the compound below using a Wittig reaction:

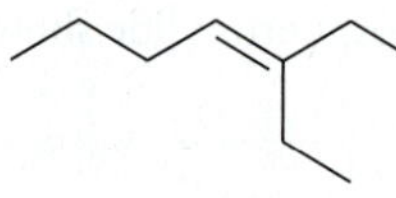

**SOLUTION**

Begin by focusing on the two carbon atoms of the double bond. One carbon atom must have been a carbonyl group, while the other must have been a Wittig reagent. This gives two potential routes to explore:

**STEP 1**
Using a retrosynthetic analysis, determine the two possible sets of reactants that could be used to form the C=C bond.

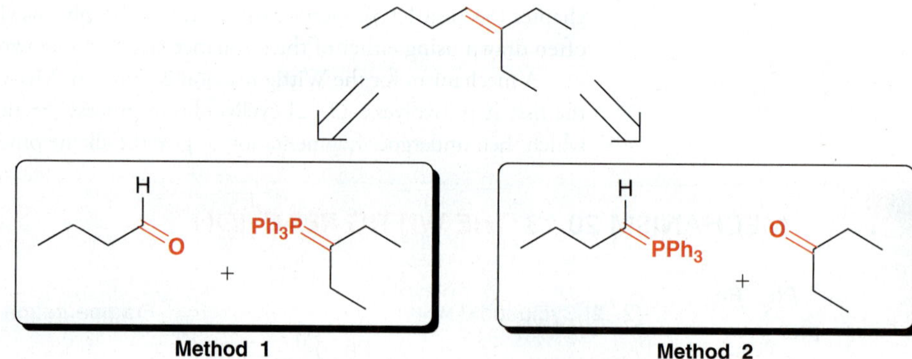


Let's compare these two methods by focusing on the Wittig reagent in each case. Recall that the Wittig reagent is prepared by an $S_N2$ process, and we therefore must consider steric factors during its preparation. Method 1 requires the use of a secondary alkyl halide:

X — 1) $PPh_3$, 2) BuLi → $Ph_3P$

2° Alkyl halide

but method 2 requires the use of a primary alkyl halide:

**STEP 2**
Consider how you would make each possible Wittig reagent and determine which method involves the less substituted alkyl halide.

X — 1) $PPh_3$, 2) BuLi → H, $PPh_3$

1° Alkyl halide

Method 2 is likely to be more efficient, because a primary alkyl halide will undergo $S_N2$ more rapidly than a secondary alkyl halide. Therefore, the following would be the preferred synthesis:

O + (⊖ ⊕ P, Ph, Ph, Ph) →

**PRACTICE the skill** **20.37** Identify the reagents necessary to prepare each of the following compounds using a Wittig reaction:

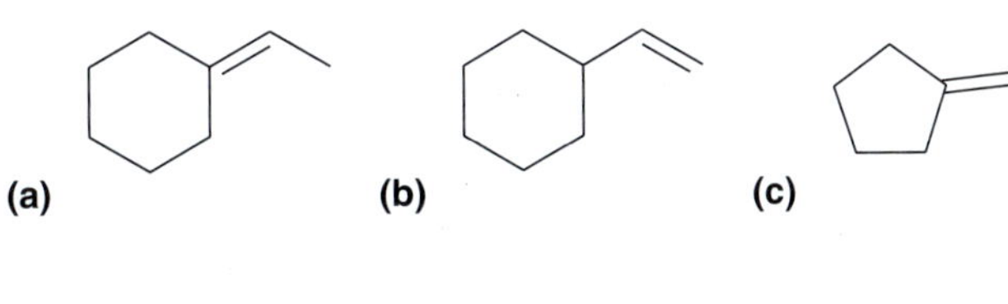

(a) (b) (c)

(d) (e)

**APPLY the skill** **20.38** Consider the structure of beta-carotene, mentioned earlier in this chapter:

**β-Carotene**

Design a synthesis of beta-carotene using the compound below as your only source of carbon atoms:

Br

**20.39** Identify the reagents necessary to accomplish each of the transformations below:

(a) OH → (b) →

need more **PRACTICE?** **Try Problems 20.51–20.53**

## 20.11 Baeyer-Villiger Oxidation of Aldehydes and Ketones

When treated with a peroxy acid, ketones can be converted into esters via the insertion of an oxygen atom:

$$\text{R(C=O)R} \xrightarrow{RCO_3H} \text{R(C=O)OR}$$

This reaction, discovered by Adolf von Baeyer and Victor Villiger in 1899, is called the **Baeyer-Villiger oxidation.** This process is believed to proceed via Mechanism 20.14.

## MECHANISM 20.14 THE BAEYER-VILLIGER OXIDATION

**Nucleophilic attack**

The peroxyacid functions as a nucleophile and attacks the carbonyl group

**Proton transfer**

A proton is transferred from one location to another. This step can occur intramolecularly, because it would involve a five-membered transition state

**Rearrangement**

The carbonyl group is reformed, with simultaneous migration of an alkyl group

The peroxy acid attacks the carbonyl group of the ketone, giving an intermediate that can undergo an intramolecular proton transfer step (or two successive intermolecular proton transfer steps). Finally, the C=O double bond is re-formed by migration of an R group. This rearrangement produces the ester.

In much the same way, treatment of a cyclic ketone with a peroxy acid yields a cyclic ester, or **lactone**.

$RCO_3H$

**A lactone**

When an unsymmetrical ketone is treated with a peroxy acid, formation of the ester is regioselective; for example:

$RCO_3H$

In this case, the oxygen atom is inserted on the left side of the carbonyl group, rather than the right side. This occurs because the isopropyl group migrates more rapidly than the methyl group during the rearrangement step of the mechanism. The migration rates of different groups, or **migratory aptitude,** can be summarized as follows:

$$\text{H} > 3° > 2°, \text{Ph} > 1° > \text{methyl}$$

A hydrogen atom will migrate more rapidly than a tertiary alkyl group, which will migrate more rapidly than a secondary alkyl group or phenyl group. Below is one more example that illustrates this concept:

$RCO_3H$

In this example, the oxygen atom is inserted on the right side of the carbonyl, because the hydrogen atom exhibits a greater migratory aptitude than the phenyl group.

CONCEPTUAL CHECKPOINT

**20.40** Predict the major product of each reaction below:

(a) $\xrightarrow{RCO_3H}$ ?

(b) $\xrightarrow{RCO_3H}$ ?

(c) $\xrightarrow{RCO_3H}$ ?

## 20.12 Synthesis Strategies

Recall from Chapter 12 that there are two main questions to ask when approaching a synthesis problem:

1. *Is there any change in the carbon skeleton?*
2. *Is there any change in the functional group?*

Let's focus on these issues separately, beginning with functional groups.

### Functional Group Interconversion

In previous chapters, we learned how to interconvert many different functional groups (Figure 20.8). The reactions in this chapter expand the playing field by opening up the frontier of aldehydes and ketones. You should be able to fill in the reagents for each transformation in Figure 20.8. If you are having trouble, refer to Figure 13.13 for help. Then, you should be able to make a list of the various products than can be made from aldehydes and ketones and identify the required reagents in each case.

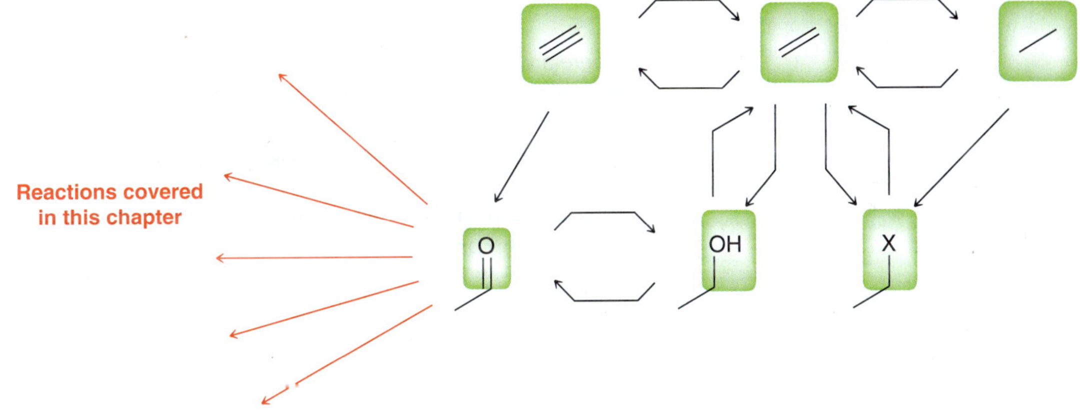


**FIGURE 20.8** Functional groups that can be interconverted using reactions that we have learned thus far.

### Reactions Involving a Change in Carbon Skeleton

In this chapter, we have seen three C—C bond-forming reactions: (1) a Grignard reaction, (2) cyanohydrin formation, and (3) a Wittig reaction:

1) $CH_3MgBr$
2) $H_2O$

KCN, HCl

$H_2\ddot{C}^{\ominus}-\overset{\oplus}{P}Ph_3$

We have only seen one C—C bond-breaking reaction: the Baeyer-Villiger oxidation:

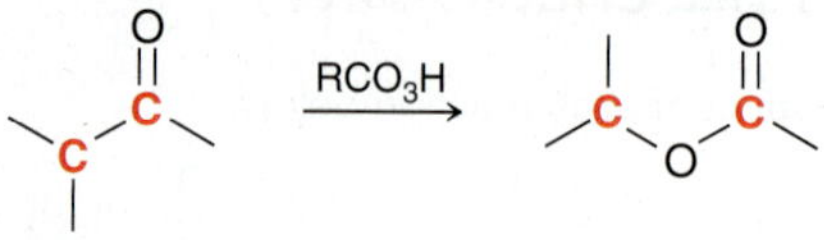

These four reactions should be added to your list of reactions that can change a carbon skeleton. Let's get some practice using these reactions.

## SKILLBUILDER

### 20.7 PROPOSING A SYNTHESIS

**LEARN the skill** Propose an efficient synthesis for the following transformation:

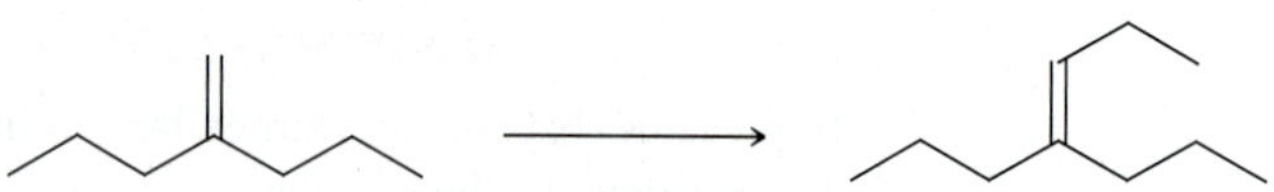

### SOLUTION

**STEP 1**
Inspect whether there is a change in the carbon skeleton and/or a change in the identity or location of the functional groups.

Always begin a synthesis problem by asking the following two questions:

1. *Is there any change in the carbon skeleton?* Yes. The product has two additional carbon atoms.
2. *Is there any change in the functional groups?* No. Both the starting material and the product have a double bond in the exact same location. If we destroy the double bond in the process of adding the two carbon atoms, we will need to make sure that we do so in such a way that we can restore the double bond.

Now let's consider how we might install the additional two-carbon atoms. The following C—C bond is the one that needs to be made:

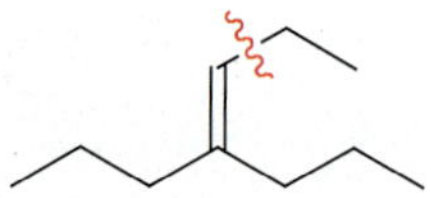

In this chapter, we have seen three C—C bond-forming reactions. Let's consider each one as a possibility.

**STEP 2**
When there is a change in the carbon skeleton, consider all of the C—C bond-forming reactions and all of the C—C bond-breaking reactions that you have learned so far.

We can immediately rule out cyanohydrin formation, as that process installs only one carbon atom, not two. So let's consider forming the C—C bond with either a Grignard reaction or a Wittig reaction.

A Grignard reagent won't attack a C=C double bond, so using a Grignard reaction would require first converting the C=C double bond into a functional group that can be attacked by a Grignard reagent, such as a carbonyl group:

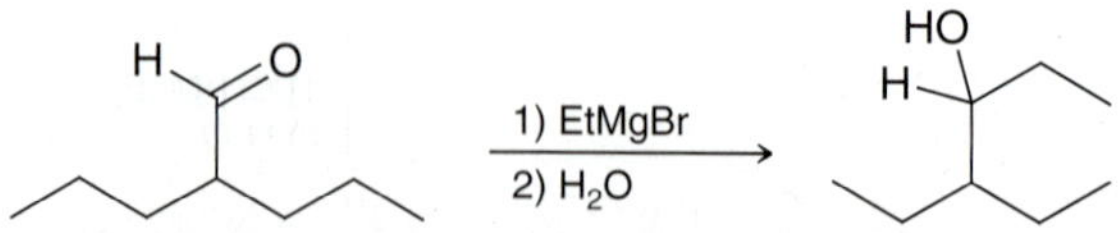

This reaction can indeed be used to form the crucial C—C bond. To use this method of C—C bond formation, we must first form the necessary aldehyde, then perform the Grignard reaction, and then finally restore the double bond in its proper location. This can be accomplished with the following reagents:

1) $BH_3 \cdot THF$
2) $H_2O_2$, NaOH

PCC

1) EtMgBr
2) $H_2O$

Conc. $H_2SO_4$
Heat

C—C Bond-forming reaction

This provides us with a four-step procedure, and this answer is certainly reasonable.

Let's now explore the possibility of proposing a synthesis with a Wittig reaction. Recall that a Wittig reaction can be used to form a C=C bond, so we focus on formation of this bond:

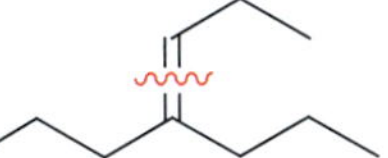

This bond can be formed if we start with a ketone and use the following Wittig reagent:

$Ph_3P$

To use this reaction, we must first form the necessary ketone from the starting alkene:

This can be accomplished with ozonolysis. This gives a two-step procedure for accomplishing the desired transformation: ozonolysis followed by a Wittig reaction. This approach is different than our first answer. In this approach, we are not attaching a two-carbon chain, but rather, we are first expelling a carbon atom and then attaching a three-carbon chain.

In summary, we have discovered two plausible methods. Both methods are acceptable, but the method employing the Wittig reaction is likely to be more efficient, because it requires fewer steps.

1) $BH_3 \cdot THF$
2) $H_2O_2$, NaOH
3) PCC
4) EtMgBr
5) $H_2O$
6) Conc. $H_2SO_4$, heat

1) $O_3$
2) DMS
3) $Ph_3P$

**PRACTICE the skill 20.41** Propose an efficient synthesis for each of the following transformations:

(a)

(b)

(c)

(d)

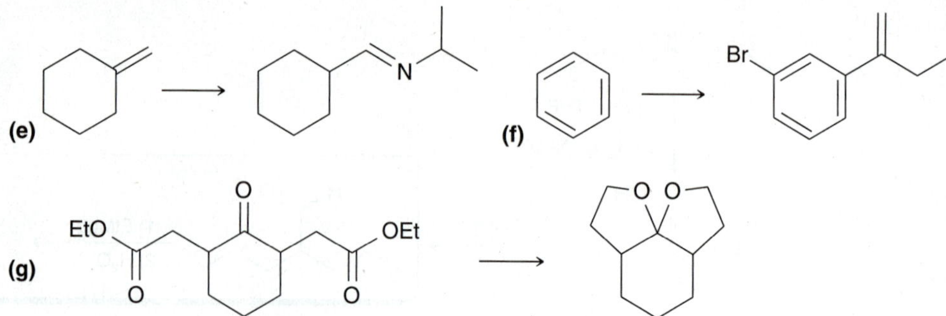

**20.42** Using any compounds of your choosing, identify a method for preparing each of the following compounds. *Your only limitation is that the compounds you use can have no more than two carbon atoms.* For purposes of counting carbon atoms, you may ignore the phenyl groups of a Wittig reagent. That is, you are permitted to use Wittig reagents.

NH (a) N (b) O O (c) O O (d)

HO N (e) (f) HO OH O (g) $H_2N$ OH (h)

need more **PRACTICE?** **Try Problems 20.55, 20.58, 20.67–20.69, 20.71, 20.75**

## 20.13 Spectroscopic Analysis of Aldehydes and Ketones

Aldehydes and ketones exhibit several characteristic signals in their infrared (IR) and nuclear magnetic resonance (NMR) spectra. We will now summarize these characteristic signals.

### IR Signals

The carbonyl group produces a strong signal in an IR spectrum, generally around 1715 or 1720 $cm^{-1}$. However, a conjugated carbonyl group will produce a signal at a lower wavenumber as a result of electron delocalization via resonance effects:

**LOOKING BACK**
For an explanation of this effect, see Section 15.3.

O O

1715 $cm^{-1}$ 1680 $cm^{-1}$

Ring strain has the opposite effect on a carbonyl group. That is, increasing ring strain tends to increase the wavenumber of absorption:

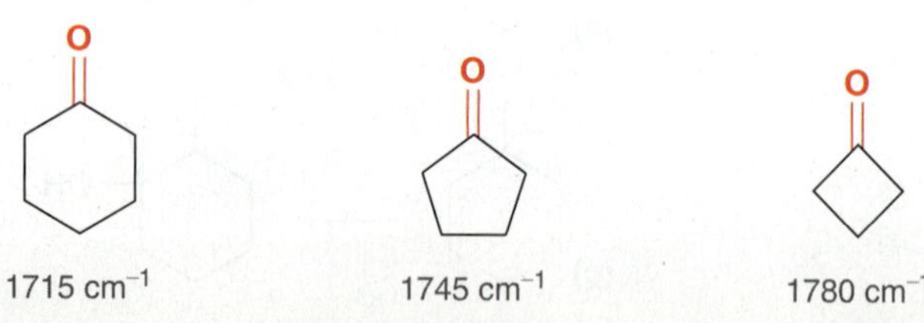

Aldehydes generally exhibit one or two signals (C—H stretching) between 2700 and 2850 $cm^{-1}$ (Figure 20.9) in addition to the C═O stretch.

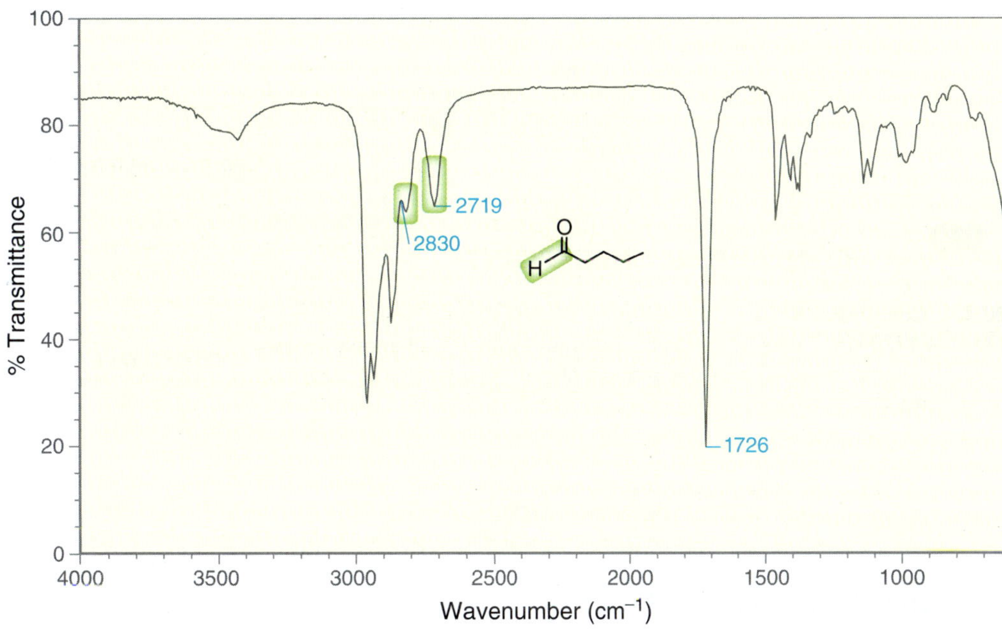


**FIGURE 20.9**
An IR spectrum of an aldehyde.

## $^1$H NMR Signals

In a $^1$H NMR spectrum, the carbonyl group itself does not produce a signal. However, it has a pronounced effect on the chemical shift of neighboring protons. We saw in Section 16.5 that a carbonyl group typically adds approximately +1 ppm to the chemical shift of its neighbors:

~1.2 ppm ~2.2 ppm

Aldehydic protons generally produce signals around 10 ppm. These signals can usually be identified with relative ease, because very few signals appear that far downfield in a $^1$H NMR spectrum (Figure 20.10).

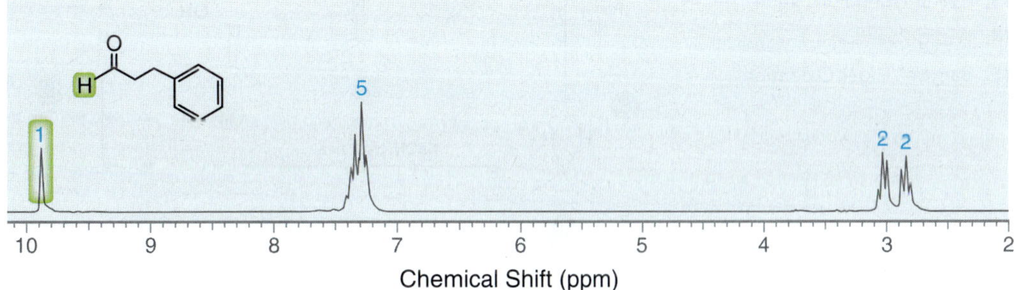


**FIGURE 20.10**
A $^1$H NMR spectrum of an aldehyde.

## $^{13}$C NMR Signals

In a $^{13}$C NMR spectrum, a carbonyl group of a ketone or aldehyde will generally produce a weak signal near 200 ppm. This signal can often be identified with relative ease, because very few signals appear that far downfield in a $^{13}$C NMR spectrum (Figure 20.11).

**FIGURE 20.11**
A $^{13}C$ NMR spectrum of a ketone.

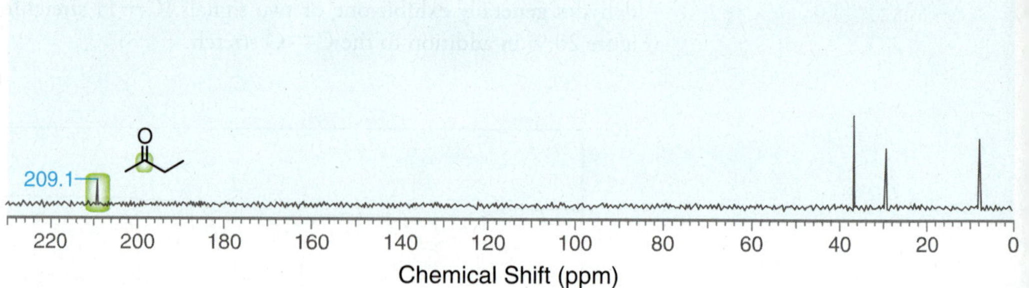

## CONCEPTUAL CHECKPOINT

**20.43** Compound A has molecular formula $C_{10}H_{10}O$ and exhibits a strong signal at 1720 $cm^{-1}$ in its IR spectrum. Treatment with 1,2-ethanedithiol followed by Raney nickel affords the product shown. Identify the structure of compound A.

**Compound A** 1) $[H^+]$, HS⁀SH 2) Raney Ni →

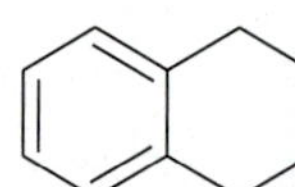

## REVIEW OF REACTIONS

SYNTHETICALLY USEFUL REACTIONS

1. Hydrate Formation
2. Acetal Formation
3. Cyclic Acetal Formation
4. Cyclic Thioacetal Formation
5. Desulfurization
6. Imine Formation
7. Enamine Formation
8. Oxime Formation
9. Hydrazone Formation
10. Wolff-Kishner Reduction
11. Reduction of a Ketone
12. Grignard Reaction
13. Cyanohydrin Formation
14. Wittig Reaction
15. Baeyer-Villiger Oxidation

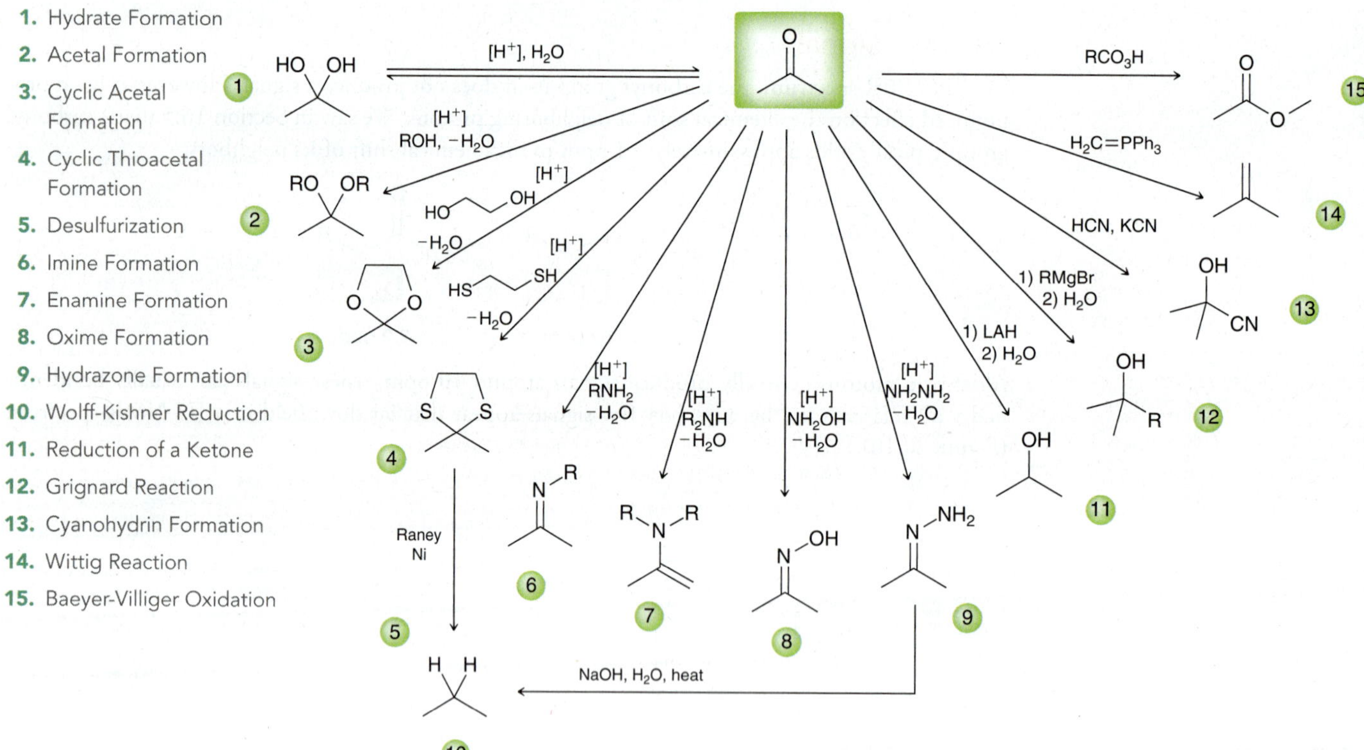


## REVIEW OF CONCEPTS AND VOCABULARY

### SECTION 20.1

- Both aldehydes and ketones contain a *carbonyl group*, and both are common in nature and industry, and both occupy a central role in organic chemistry.

### SECTION 20.2

- The suffix "-al" indicates an aldehydic group, and the suffix "-one" is used for ketones.

- In naming aldehydes and ketones, locants should be assigned so as to give the carbonyl group the lowest number possible.

### SECTION 20.3

- Aldehydes can be prepared via oxidation of primary alcohols, ozonolysis of alkenes, or hydroboration-oxidation of terminal alkynes.
- Ketones can be prepared via oxidation of secondary alcohols, ozonolysis of alkenes, acid-catalyzed hydration of terminal alkynes, or Friedel-Crafts acylation.

### SECTION 20.4

- The electrophilicity of a carbonyl group derives from resonance effects as well as inductive effects.
- Aldehydes are more reactive than ketones as a result of steric effects and electronic effects.
- A general mechanism for nucleophilic addition under basic conditions involves two steps:
  1. Nucleophilic attack
  2. Proton transfer
- The position of equilibrium is dependent on the ability of the nucleophile to function as a leaving group.

### SECTION 20.5

- When an aldehyde or ketone is treated with water, the carbonyl group can be converted into a **hydrate**. The equilibrium generally favors the carbonyl group, except in the case of very simple aldehydes, or ketones with strong electron-withdrawing substituents.
- Under acidic conditions, a mechanism will only be reasonable if it avoids the use or formation of strong bases (only weak bases can be employed).
- Under basic conditions, a mechanism will only be reasonable if it avoids the use or formation of strong acids (only weak acids can be employed).
- In acidic conditions, an aldehyde or ketone will react with two molecules of alcohol to form an **acetal**.
- In the presence of an acid, the carbonyl group is protonated to form a very powerful electrophile.
- The mechanism for acetal formation can be divided into two parts:
  1. The first three steps produce a **hemiacetal**.
  2. The last four steps convert the hemiacetal to an acetal.
- For many simple aldehydes, the equilibrium favors formation of the acetal; however, for most ketones, the equilibrium favors reactants rather than products.
- An aldehyde or ketone will react with one molecule of a diol to form a cyclic acetal.
- The reversibility of acetal formation enables acetals to function as protecting groups for ketones or aldehydes. Acetals are stable under strongly basic conditions.
- Hemiacetals are generally difficult to isolate unless they are cyclic.

### SECTION 20.6

- In acidic conditions, an aldehyde or ketone will react with a primary amine to form an **imine**.
- The first three steps in imine formation produce a **carbinolamine**, and the last three steps convert the carbinolamine into an imine.
- Many different compounds of the form $RNH_2$ will react with aldehydes and ketones; for example:
  1. When hydrazine is used as a nucleophile ($NH_2NH_2$), a **hydrazone** is formed.
  2. When hydroxylamine is used as a nucleophile ($NH_2OH$), an **oxime** is formed.
- In acidic conditions, an aldehyde or ketone will react with a secondary amine to form an **enamine**. The mechanism of enamine formation is identical to the mechanism of imine formation except for the last step.
- In the **Wolff-Kishner reduction**, a hydrazone is reduced to an alkane under strongly basic conditions.

### SECTION 20.7

- **Hydrolysis** of acetals, imines, and enamines under acidic conditions produces ketones or aldehydes.

### SECTION 20.8

- In acidic conditions, an aldehyde or ketone will react with two equivalents of a thiol to form a **thioacetal**. If a compound with two SH groups is used, a cyclic thioacetal is formed.
- When treated with Raney nickel, thioacetals undergo **desulfurization** to yield a methylene group.

### SECTION 20.9

- When treated with a hydride reducing agent, such as lithium aluminum hydride (LAH) or sodium borohydride ($NaBH_4$), aldehydes and ketones are reduced to alcohols.
- The reduction of a carbonyl group with LAH or $NaBH_4$ is not a reversible process, because hydride does not function as a leaving group.

### SECTION 20.10

- When treated with a Grignard reagent, aldehydes and ketones are converted into alcohols, accompanied by the formation of a new C—C bond.
- Grignard reactions are not reversible, because carbanions do not function as leaving groups.
- When treated with hydrogen cyanide (HCN), aldehydes and ketones are converted into **cyanohydrins**. For most aldehydes and unhindered ketones, the equilibrium favors formation of the cyanohydrin.
- The **Wittig reaction** can be used to convert a ketone to an alkene. The **Wittig reagent** that accomplishes this transformation is called a **phosphorus ylide**.
- The mechanism of a Wittig reaction involves initial formation of an **oxaphosphetane**, which rearranges to give the product.
- Preparation of Wittig reagents involves an $S_N2$ reaction, and the regular restrictions of $S_N2$ processes apply.

### SECTION 20.11

- A **Baeyer-Villiger oxidation** converts a ketone to an ester by inserting an oxygen atom next to the carbonyl group. Cyclic ketones produce cyclic esters, called **lactones**.
- When an unsymmetrical ketone is treated with a peroxy acid, formation of the ester is regioselective, and the product is determined by the **migratory aptitude** of each group next to the carbonyl.

**SECTION 20.12**

- This chapter explored three C—C bond-forming reactions: (1) a Grignard reaction, (2) cyanohydrin formation, and (3) a Wittig reaction.
- This chapter explored only one C—C bond-breaking reaction: the Baeyer-Villiger oxidation.

**SECTION 20.13**

- Carbonyl groups produce a strong IR signal around 1715 $cm^{-1}$. A conjugated carbonyl group produces a signal at a lower wavenumber, while ring strain increases the wavenumber of absorption.
- Aldehydic C—H bonds exhibit one or two signals between 2700 and 2850 $cm^{-1}$.
- In a $^{1}H$ NMR spectrum, a carbonyl group adds approximately +1 ppm to the chemical shift of its neighbors, and an aldehydic proton produces a signal around 10 ppm.
- In a $^{13}C$ NMR spectrum, a carbonyl group produces a weak signal near 200 ppm.

# SKILLBUILDER REVIEW

## 20.1 NAMING ALDEHYDES AND KETONES

**STEP 1** Choose the longest chain containing the carbonyl group and number the chain starting from the end closest to the carbonyl group.

**STEPS 2 AND 3** Identify the substituents and assign locants.

**STEP 4** Assemble the substituents alphabetically.

**STEP 5** Assign the configuration of any chirality center.

3-Nonanone

4,4-dimethyl

6-ethyl

6-ethyl-4,4-dimethyl

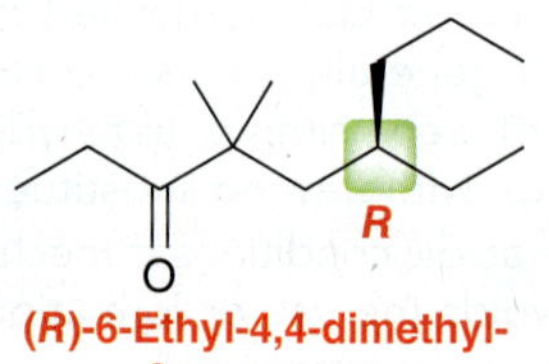


(*R*)-6-Ethyl-4,4-dimethyl-3-nonanone

Try problems 20.1–20.4, 20.44–20.49

## 20.2 DRAWING THE MECHANISM OF ACETAL FORMATION

**STEP 1** Draw the three steps necessary for hemiacetal formation.

Proton transfer — Nucleophilic attack — Proton transfer

**COMMENTS**

- Every step has two curved arrows. Draw them precisely.
- Do not forget the charges.
- Draw each step separately.

**STEP 2** Draw the four steps that convert the hemiacetal into an acetal.

Proton transfer — Loss of a leaving group ($-H_2O$) — Nucleophilic attack — Proton transfer

Try Problems 20.8–20.10, 20.57, 20.62, 20.67

### 20.3 DRAWING THE MECHANISM OF IMINE FORMATION

**STEP 1** Draw the three steps necessary for carbinolamine formation.

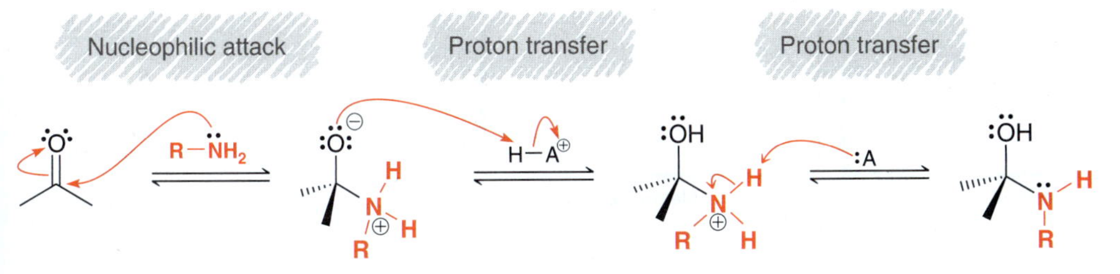


**STEP 2** Draw the three steps that convert the carbinolamine into an imine.

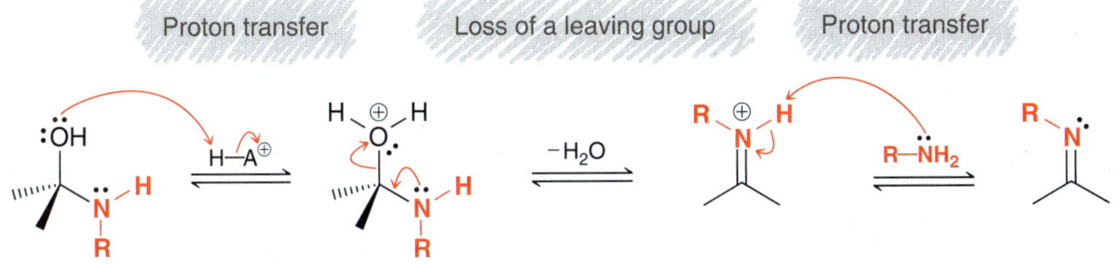


**COMMENTS**

- Every step has two curved arrows. Make sure to draw them precisely.
- Do not forget the positive charges.
- Draw each step separately, following the precise order of steps.

Try Problems **20.15–20.18, 20.72**

### 20.4 DRAWING THE MECHANISM OF ENAMINE FORMATION

**STEP 1** Draw the three steps necessary for carbinolamine formation.

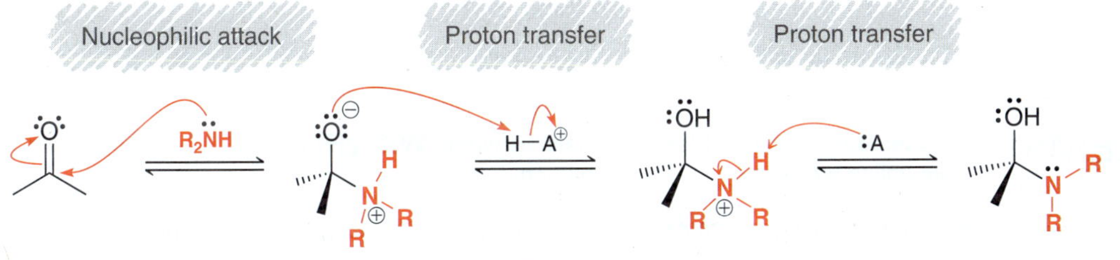


**STEP 2** Draw the three steps that convert the carbinolamine into an enamine.

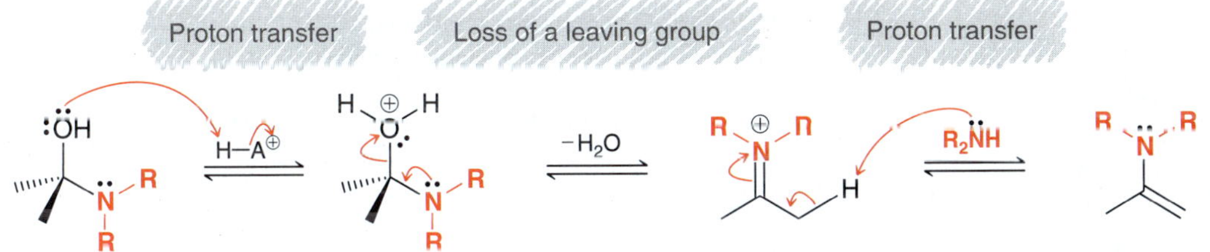


**COMMENTS**

- Every step has two curved arrows. Make sure to draw them precisely.
- Do not forget the positive charges.
- Draw each step separately following the precise order of steps.

Try Problems **20.21–20.24, 20.64, 20.65a, 20.66e, 20.75g**

### 20.5 DRAWING THE PRODUCTS OF A HYDROLYSIS REACTION

**EXAMPLE** Draw the products expected when the following compound is treated with aqueous acid:

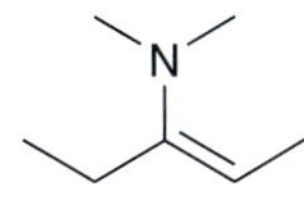


**STEP 1** Identify the bond(s) expected to undergo cleavage.

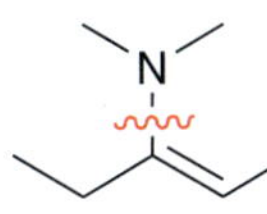


**STEP 2** Identify the carbon atom that will become a carbonyl group.

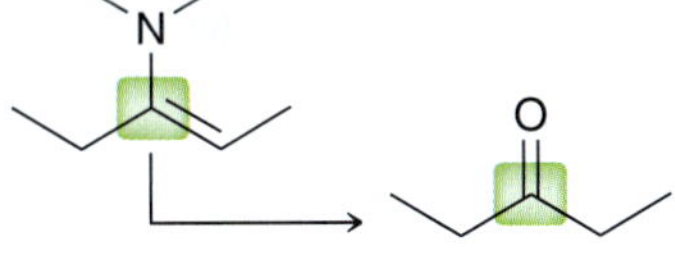


**STEP 3** Determine the identity of the other fragment(s).

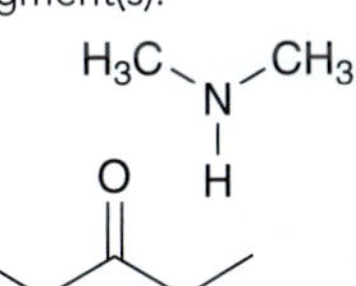


Try Problems **20.26, 20.27, 20.63, 20.64**

**20.6** PLANNING AN ALKENE SYNTHESIS WITH A WITTIG REACTION

**EXAMPLE** Identify the reactants you would use to prepare this compound via a Wittig reaction.

**STEP 1** Using a retrosynthetic analysis, determine the two possible sets of reactants that could be used to form the C=C bond.

Method 1

Method 2

**STEP 2** Consider how you would make each possible Wittig reagent and determine which method involves the less substituted alkyl halide.

2° Alkyl halide

1° Alkyl halide (Will undergo $S_N2$ more readily)

Try Problems **20.37–20.39, 20.51–20.53**

**20.7** PROPOSING A SYNTHESIS

**STEP 1** Begin by asking the following two questions:

1. Is there a change in the carbon skeleton?
2. Is there a change in the functional groups?

**STEP 2** If there is a change in the carbon skeleton, consider all of the C—C bond-forming reactions and all of the C—C bond-breaking reactions that you have learned so far

C—C bond-forming reactions in this chapter

- Grignard reaction
- Cyanohydrin formation
- Wittig reaction

C—C bond-breaking reactions in this chapter

- Baeyer-Villiger oxidation

**CONSIDERATIONS**

Remember that the desired product should be the major product of your proposed synthesis.

Make sure that the regiochemical outcome of each step is correct.

Always think backward (retrosynthetic analysis) as well as forward, and then try to bridge the gap.

Most synthesis problems will have multiple correct answers. Do not feel that you have to find the "one" correct answer.

Try Problems **20.41, 20.42, 20.55, 20.58, 20.67–20.69, 20.71, 20.75**

# PRACTICE PROBLEMS

**Note: Most of the Problems are available within WileyPLUS, an online teaching and learning solution.**

**20.44** Provide a systematic (IUPAC) name for each of the following compounds:

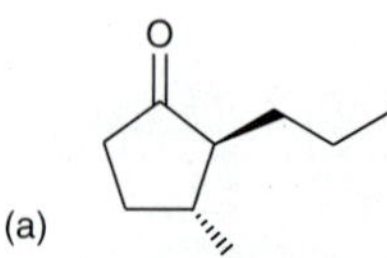

(a) (b)

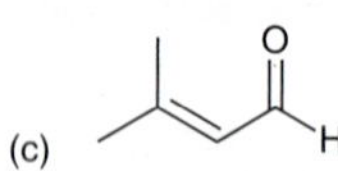

(c) (d)

**20.45** Draw the structure for each compound below:

(a) Propanedial

(b) 4-Phenylbutanal

(c) (*S*)-3-Phenylbutanal

(d) 3,3,5,5-Tetramethyl-4-heptanone

(e) (*R*)-3-Hydroxypentanal

(f) *meta*-Hydroxyacetophenone

(g) 2,4,6-Trinitrobenzaldehyde

(h) Tribromoacetaldehyde

(i) (3*R*,4*R*)-3,4-Dihydroxy-2-pentanone

**20.46** Draw all constitutionally isomeric aldehydes with molecular formula $C_4H_8O$ and provide a systematic (IUPAC) name for each isomer.

**20.47** Draw all constitutionally isomeric aldehydes with molecular formula $C_5H_{10}O$ and provide a systematic (IUPAC) name for each isomer. Which of these isomers possesses a chirality center?

**20.48** Draw all constitutionally isomeric ketones with molecular formula $C_6H_{12}O$ and provide a systematic (IUPAC) name for each isomer.

**20.49** Explain why the IUPAC name of a compound is unlikely to have the suffix "-1-one."

**20.50** For each pair of the following compounds, identify which compound would be expected to react more rapidly with a nucleophile:

(a)

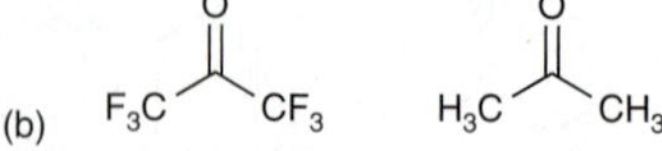


(b)

**20.51** Draw the products of each Wittig reaction below. If two stereoisomers are possible, draw both stereoisomers:

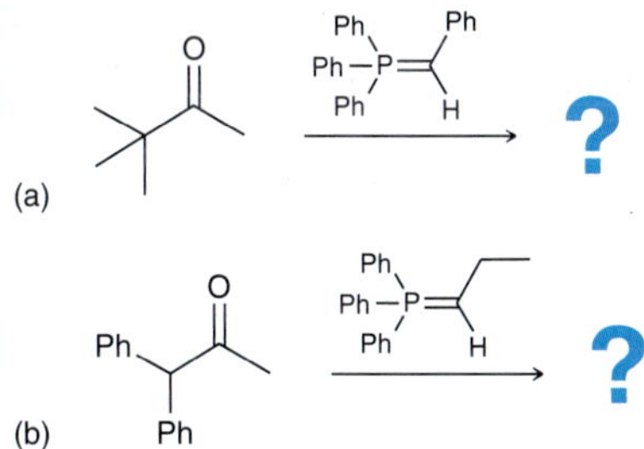


**20.52** Draw the structure of the alkyl halide needed to prepare each of the following Wittig reagents and then determine which Wittig reagent will be the more difficult to prepare. Explain your choice:

**20.53** Show how a Wittig reaction can be used to prepare each of the following compounds. In each case, also show how the Wittig reagent would be prepared:

(a) (b)

**20.54** Choose a Grignard reagent and a ketone that can be used to produce each of the following compounds:

(a) 3-Methyl-3-pentanol
(b) 1-Ethylcyclohexanol
(c) Triphenylmethanol
(d) 5-Phenyl-5-nonanol

**20.55** You are working in a laboratory, and you are given the task of converting cyclopentene into 1,5-pentanediol. Your first thought is simply to perform an ozonolysis followed by reduction with LAH, but your lab is not equipped for an ozonolysis reaction. Suggest an alternative method for converting cyclopentene into 1,5-pentanediol. For help, see Section 13.4 (reduction of esters to give alcohols).

**20.56** Predict the major product(s) from the treatment of acetone with the following:

(a) $[H^+]$, $NH_3$, $(-H_2O)$
(b) $[H^+]$, $CH_3NH_2$, $(-H_2O)$
(c) $[H^+]$, excess EtOH, $(-H_2O)$
(d) $[H^+]$, $(CH_3)_2NH$, $(-H_2O)$
(e) $[H^+]$, $NH_2NH_2$, $(-H_2O)$
(f) $[H^+]$, $NH_2OH$, $(-H_2O)$
(g) $NaBH_4$, MeOH
(h) MCPBA
(i) HCN, KCN
(j) EtMgBr followed by $H_2O$
(k) $(C_6H_5)_3P{=}CHCH_2CH_3$
(l) LAH followed by $H_2O$

**20.57** Propose a plausible mechanism for the following transformation:

HO ... H $\xrightarrow[\text{EtOH}]{[H^+]}$

**20.58** Devise an efficient synthesis for the following transformation (recall that aldehydes are more reactive than ketones):

**20.59** Treatment of catechol with formaldehyde in the presence of an acid catalyst produces a compound with molecular formula $C_7H_6O_2$. Draw the structure of this product.

Catechol

**20.60** Predict the major product(s) for each reaction below.

(a) 1) LAH 2) $H_2O$ ?

(b) 1) PhMgBr 2) $H_2O$ ?

(c) $(C_6H_5)_3P{=}CH_2$ ?

**20.61** Starting with cyclopentanone and using any other reagents of your choosing, identify how you would prepare each of the following compounds:

(a) (b) (c) (d)

**20.62** Glutaraldehyde is a germicidal agent that is sometimes used to sterilize medical equipment too sensitive to be heated in an autoclave. In mildly acidic conditions, glutaraldehyde exists in a cyclic form (below right). Draw a plausible mechanism for this transformation:

Glutaraldehyde $\underset{}{\overset{[H_3O^+]}{\rightleftharpoons}}$

**20.63** Predict the major product(s) obtained when each of the following compounds undergoes hydrolysis in the presence of $H_3O^+$:

(a) (b) (c) (d) (e)

**20.64** Identify all of the products formed when the compound below is treated with aqueous acid:

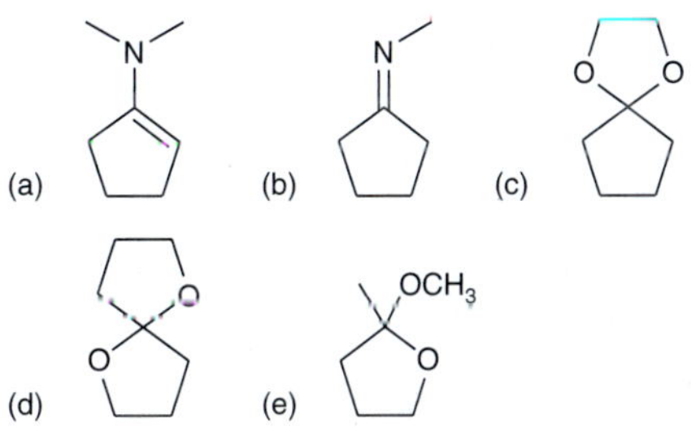

**20.65** Draw a plausible mechanism for each of the following transformations:

(a) $H_3O^+$ → H

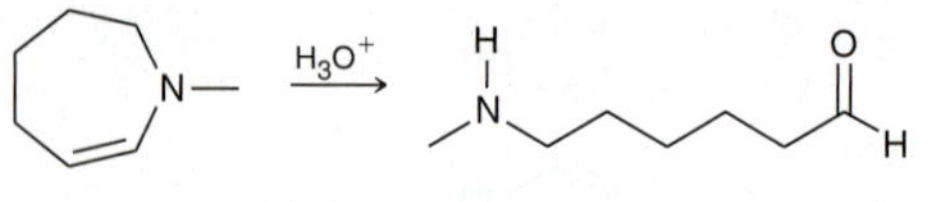

(b) $H_3O^+$ → $H_2N$ H

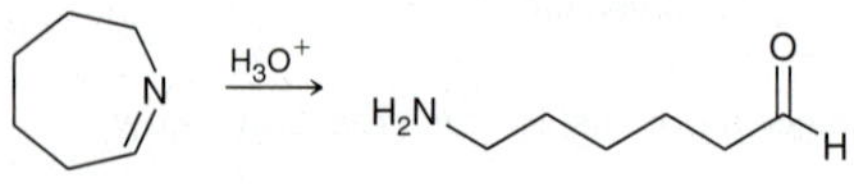

(c) [$H^+$] $H_2O$ → HO H OH

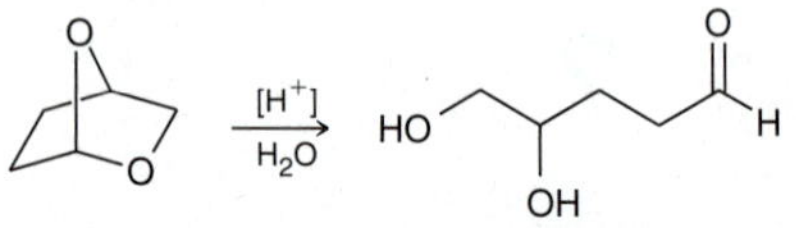

**20.66** Predict the major product(s) for each of the following reactions:

(a) $CH_3$ N $NH_2$ [$H^+$] (−$H_2O$) ?

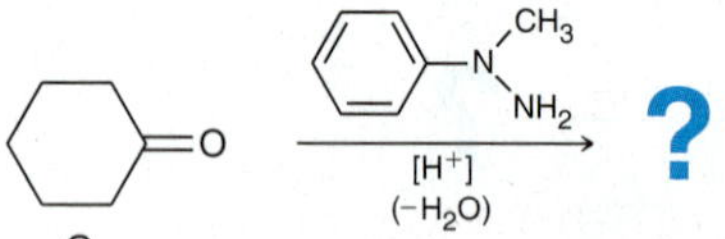

(b) 1) PhMgBr 2) $H_2O$ ?

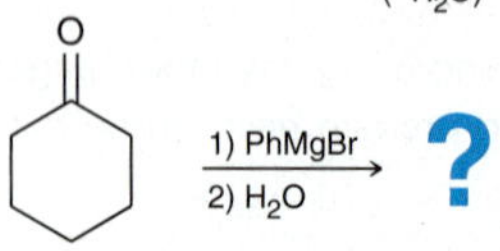

(c) $CH_3CO_3H$ ?

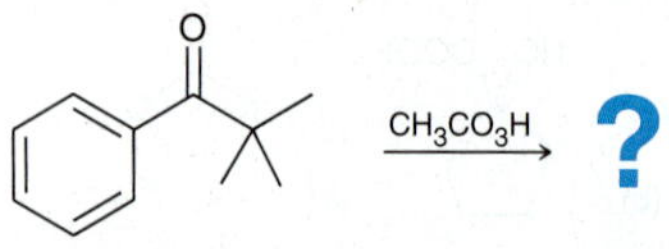

(d) $CH_3CO_3H$ ?

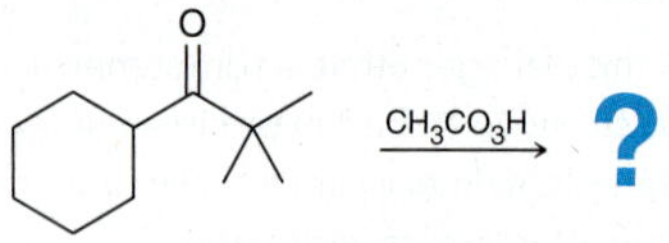

(e) [$H^+$] N−H (−$H_2O$) ?

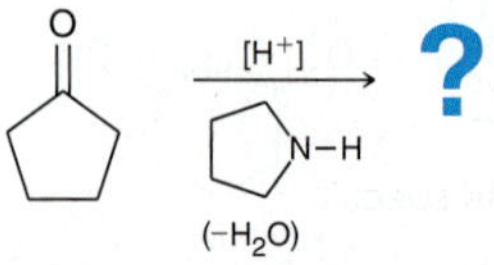

(f) [$H^+$] $NH_2$ (−$H_2O$) ?

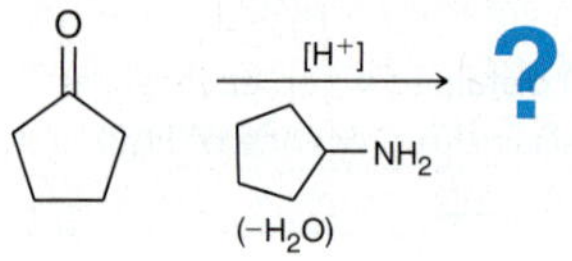

**20.67** Identify the starting materials needed to make each of the following acetals:

(a) (b) OEt (c)

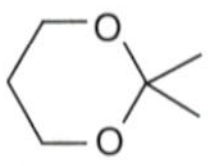

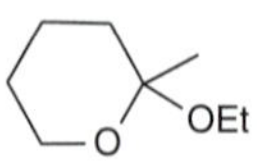

**20.68** Using ethanol as your only source of carbon atoms, design a synthesis for the following compound:

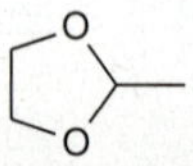

**20.69** Propose an efficient synthesis for each of the following transformations:

(a)

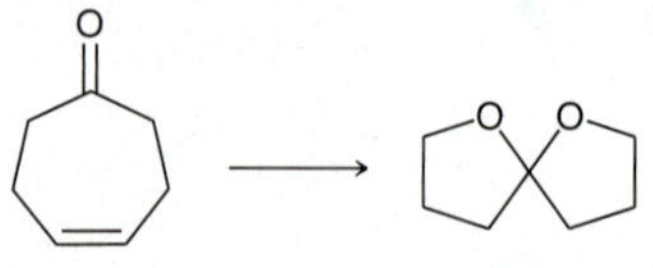

(b)

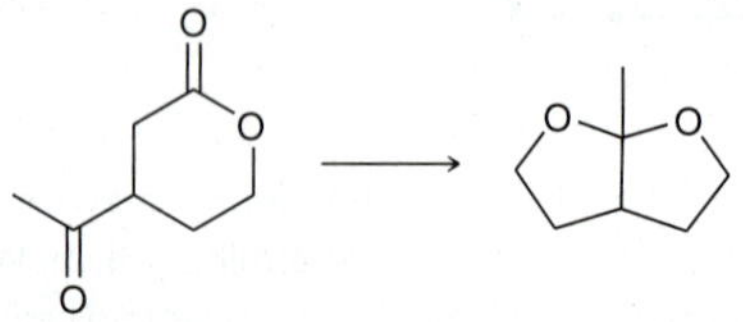

**20.70** The compound below is believed to be a wasp pheromone. Draw the major product formed when this compound is hydrolyzed in aqueous acid:

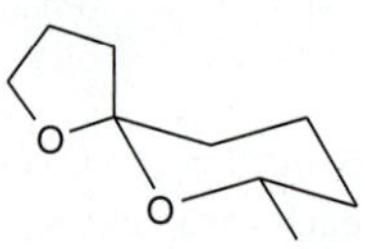

**20.71** Propose an efficient synthesis for each of the following transformations:

(a)

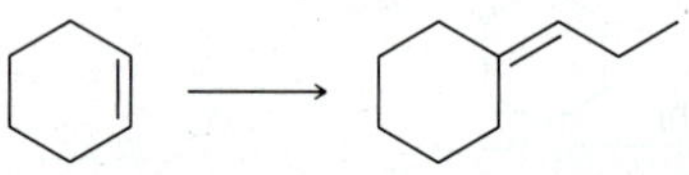

(b) Br Br → MeO OMe

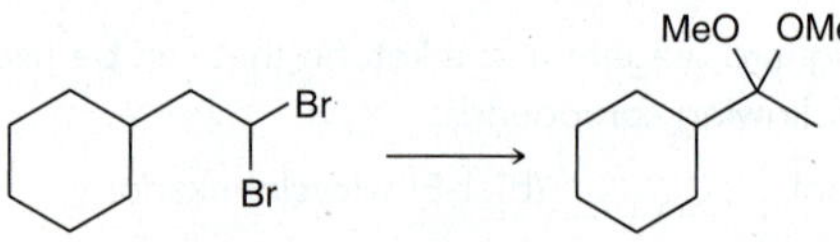

(c)

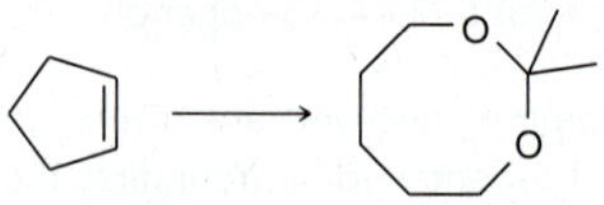

**20.72** Draw a plausible mechanism for the following transformation:

$NH_2$ $NH_2$ [$H_2SO_4$] (−$H_2O$) → N N

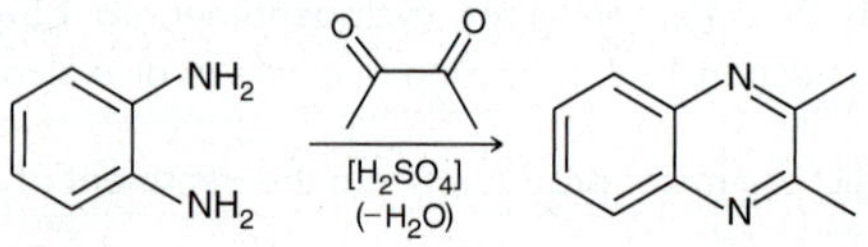

**20.73** When cyclohexanone is treated with $H_2O$, an equilibrium is established between cyclohexanone and its hydrate. This equilibrium greatly favors the ketone, and only trace amounts of the hydrate can be detected. In contrast, when cyclopropanone is treated with $H_2O$, the resulting hydrate predominates at equilibrium. Suggest an explanation for this curious observation.

**20.74** Consider the three constitutional isomers of dioxane ($C_4H_8O_2$):

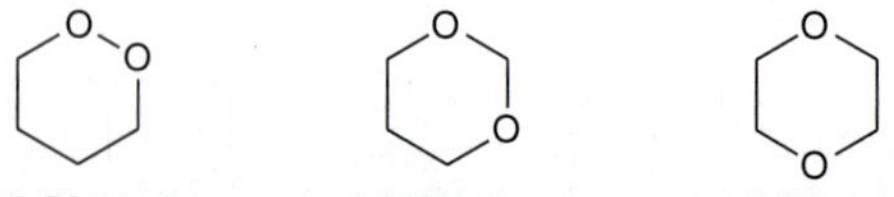

1,2-Dioxane 1,3-Dioxane 1,4-Dioxane

One of these constitutional isomers is stable under basic conditions as well as mildly acidic conditions and is therefore used as a common solvent. Another isomer is only stable under basic conditions but undergoes hydrolysis under mildly acidic conditions. The remaining isomer is extremely unstable and potentially explosive. Identify each isomer and explain the properties of each compound.

**20.75** Propose an efficient synthesis for each of the following transformations:

(a) 

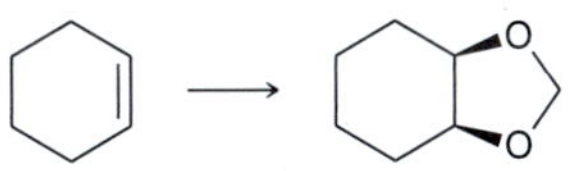

(b) 
Br

(c) 
O
O

(d) 
OH
CN

(e) 
NH

(f) 
Br
H
O

(g) 
N

(h) 
O
O

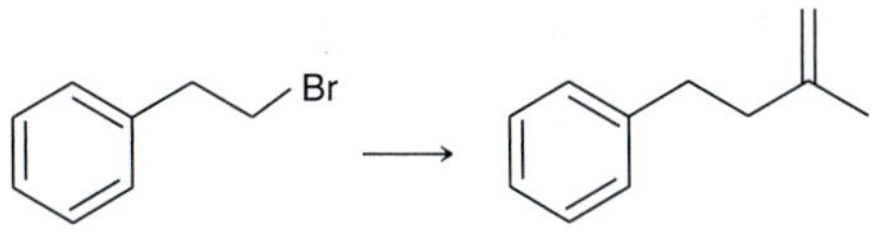

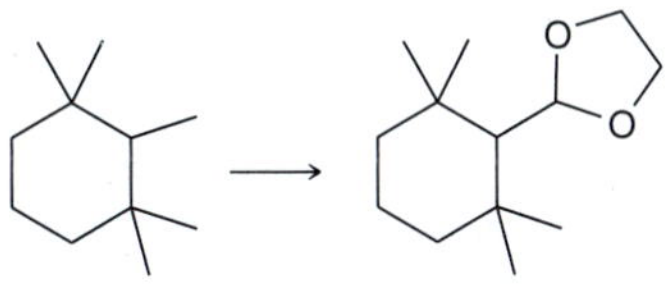

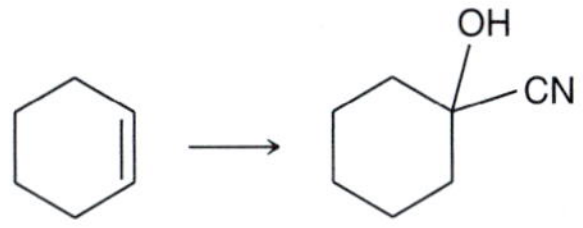

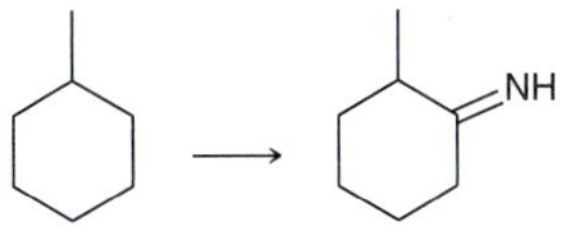

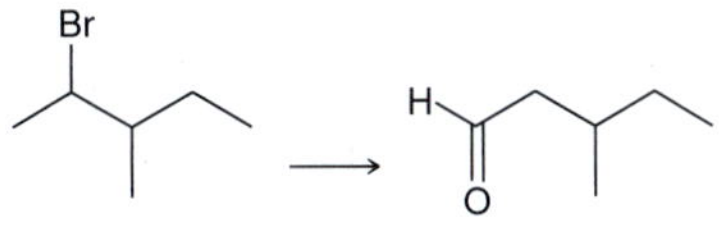

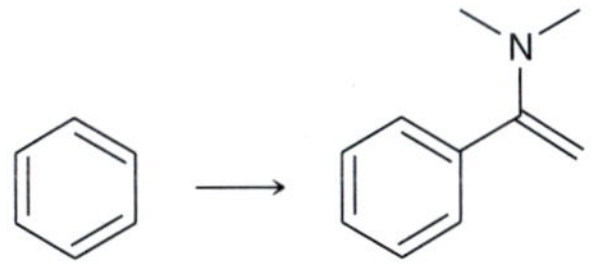

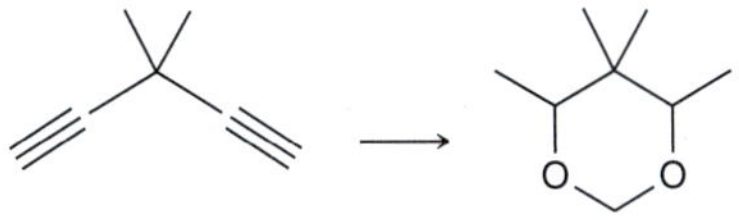

## INTEGRATED PROBLEMS

**20.76** Compound **A** has molecular formula $C_7H_{14}O$ and reacts with sodium borohydride in methanol to form an alcohol. The $^1H$ NMR spectrum of compound **A** exhibits only two signals: a doublet ($J = 12$) and a septet ($J = 2$). Treating compound **A** with 1,2-ethanedithiol ($HSCH_2CH_2SH$) under acidic conditions, followed by Raney nickel gives compound **B**.

(a) How many signals will appear in the $^1H$ NMR spectrum of compound **B**?

(b) How many signals will appear in the $^{13}C$ NMR spectrum of compound **B**?

(c) Describe how you could use IR spectroscopy to verify the conversion of compound **A** to compound **B**.

**20.77** Using the information provided below, deduce the structures of compounds **A**, **B**, **C**, and **D**:

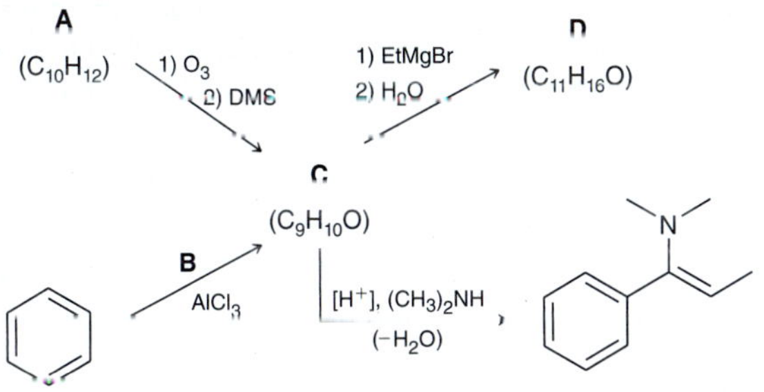

A
($C_{10}H_{12}$)
1) $O_3$
2) DMS
1) EtMgBr
2) $H_2O$
D
($C_{11}H_{16}O$)
C
($C_9H_{10}O$)
B
$AlCl_3$
[$H^+$], $(CH_3)_2NH$
($-H_2O$)
N

**20.78** Identify the structures of compounds **A** to **D** below and then identify the reagents that can be used to convert cyclohexene into compound **D** in just one step.

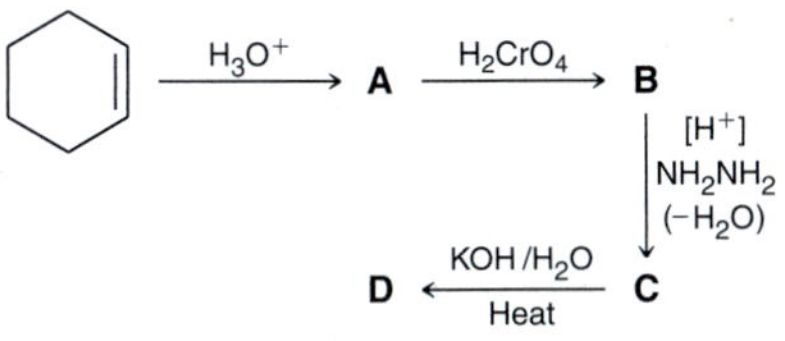

$H_3O^+$
A
$H_2CrO_4$
B
[$H^+$]
$NH_2NH_2$
($-H_2O$)
C
KOH/$H_2O$
Heat
D

**20.79** Identify the structures of compounds **A** to **E** below:

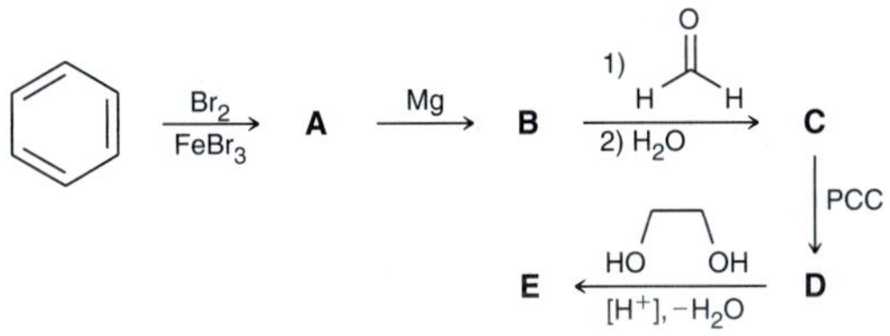

$Br_2$
$FeBr_3$
A
Mg
B
1) O, H, H
2) $H_2O$
C
PCC
D
HO OH
[$H^+$], $-H_2O$
E

**20.80** An aldehyde with molecular formula $C_4H_6O$ exhibits an IR signal at 1715 $cm^{-1}$.

(a) Propose two possible structures that are consistent with this information.

(b) Describe how you could use $^{13}C$ NMR spectroscopy to determine which of the two possible structures is correct.

**20.81** A compound with molecular formula $C_9H_{10}O$ exhibits a strong signal at 1687 $cm^{-1}$ in its IR spectrum. The $^1H$ and $^{13}C$ NMR spectra for this compound are shown below. Identify the structure of this compound.

Proton NMR
2
3
2
3
9
8
7
6
5
4
3
2
1
Chemical Shift (ppm)

Carbon 13 NMR
128.3
132.5
127.7
199.9
136.7
31.3
7.9
200
180
160
140
120
100
80
60
40
20
0
Chemical Shift (ppm)

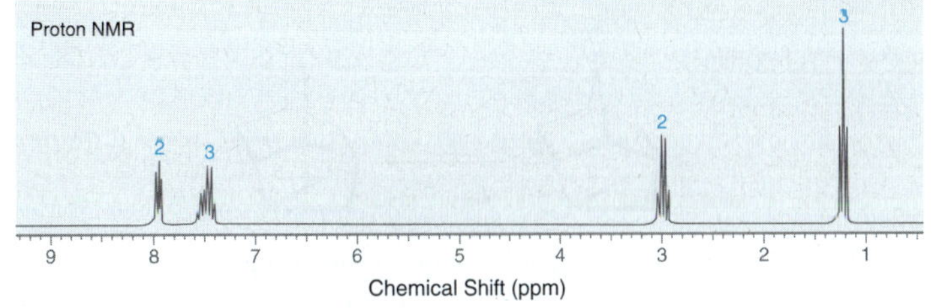

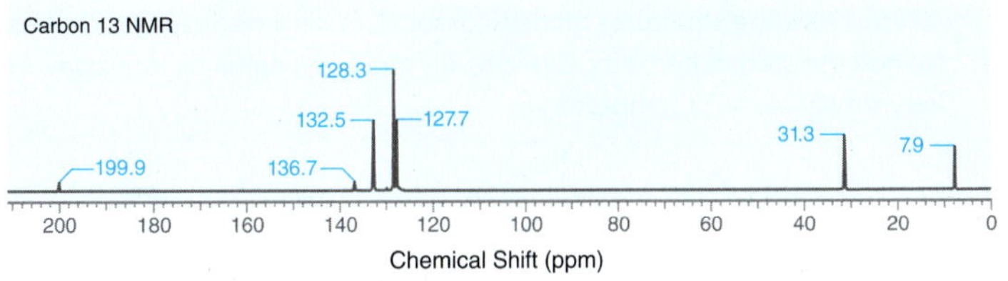

**20.82** A compound with molecular formula $C_{13}H_{10}O$ produces a strong signal at 1660 $cm^{-1}$ in its IR spectrum. The $^{13}C$ NMR spectrum for this compound is shown below. Identify the structure of this compound.

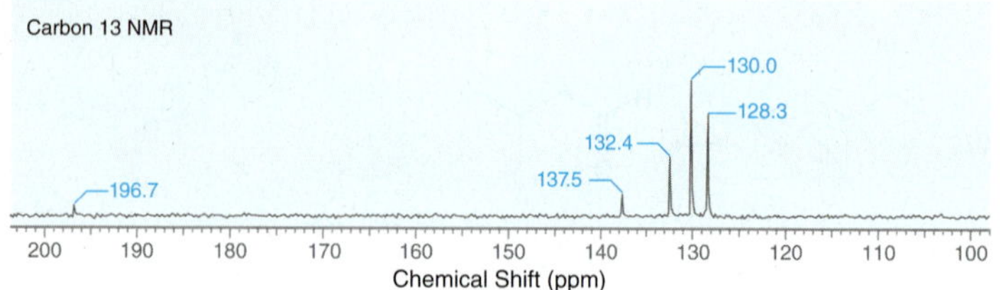


**20.83** A ketone with molecular formula $C_9H_{18}O$ exhibits only one signal in its $^1H$ NMR spectrum. Provide a systematic (IUPAC) name for this compound.

**20.84** Draw a plausible mechanism for each of the following transformations:

(a) (b) (c) (d) (e) (f)

**20.85** Under acid-catalyzed conditions, formaldehyde polymerizes to produce a number of compounds, including metaformaldehyde. Draw a plausible mechanism for this transformation:

Metaformaldehyde

**20.86** The following transformation was employed during synthetic studies toward the total synthesis of cyclodidemniserinol trisulfate, found to inhibit HIV-1 integrase (*Tetrahedron Lett.* **2009,** *50,* **4587–4591**). Propose a four-step synthesis to accomplish this transformation.

Four steps

## CHALLENGE PROBLEMS

**20.87** Propose an efficient synthesis for the conversion of the following ketone to the following ether. One of these compounds has nearly twice as many signals in its $^1H$ NMR spectrum than the other compound. Identify the compound that is expected to produce fewer signals, and explain your choice. (*J. Org. Chem.* **2001,** *66,* **2072–2077**).

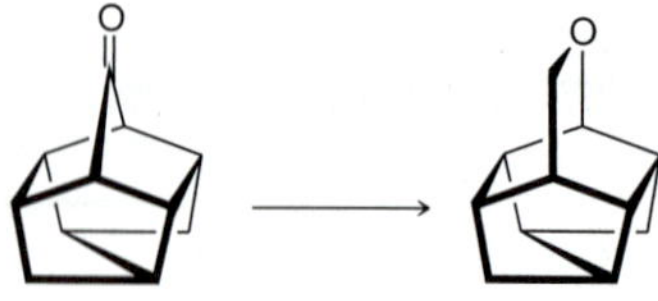

**20.88** The reaction of cyclohexanone with the given optically pure aminoester resulted in the formation of two enamines that were conformational isomers of each other (*Tetrahedron Lett.,* **1969,** *10,* **4233–4236**). Draw the structures of the two isomers using wedges and dashes to indicate stereochemistry. Explain why the two enamines interconvert very slowly at room temperature.

TsOH, $-H_2O$ ?

**20.89** Which of the following is the correct sequence of reactions needed to transform ketone **1** into aldehyde **2**, or will all three routes produce the desired product? (*J. Org. Chem.* **2003,** *68,* **6455–6458**). Explain your choice(s).

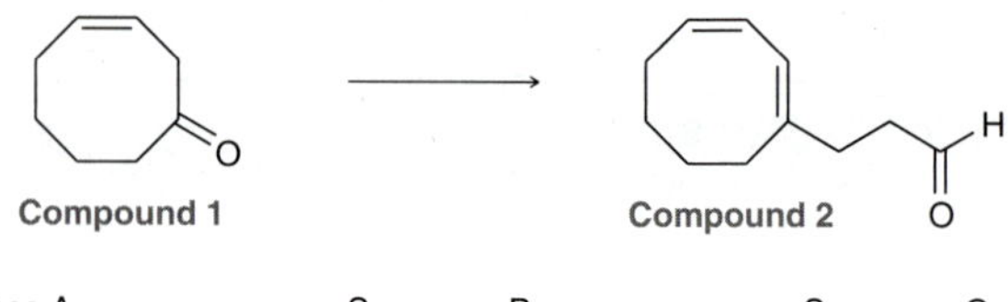


Sequence A
1) BrMg (dioxolane)
2) $H_2SO_4$, heat
3) AcOH, $H_2O$

Sequence B
1) BrMg (butenyl)
2) $H_2SO_4$, heat
3) $O_3$, DMS

Sequence C
1) BrMg (allyl)
2) $H_2SO_4$, heat
3) $BH_3$•THF
$H_2O_2$, NaOH
4) PCC

**20.90** Furo[2,3-*b*]indoles are compounds whose structural features are present in many natural products with biological activity. In a recently reported method for the synthesis of furo[2,3-*b*]indoles, the conversion of compound **1** to compound **2** was accomplished in just two steps (*Tetrahedron Lett.* **2010,** *51,* **4494–4496**). Propose a two-step synthesis for this transformation.

CHO NO2 1 → CHO NO2 2

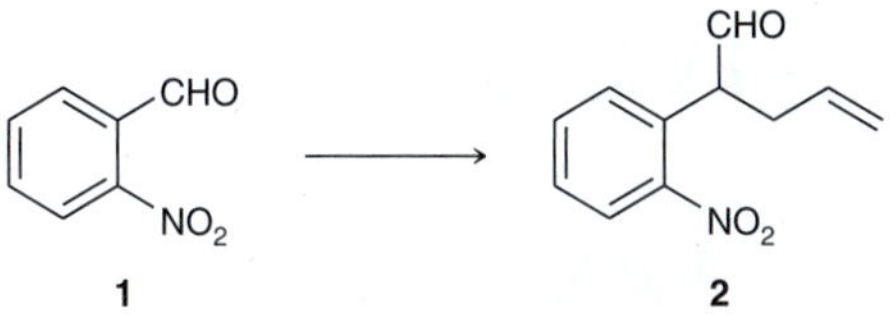

**20.91** In a Bamford-Stevens-Shapiro reaction, a sulfonylhydrazone (**1**) undergoes base-catalyzed decomposition to give a vinyl lithium species (**2**) which can then react with an electrophile, such as compound **3**. This reaction was recently employed in the total synthesis of frondosin B, a marine natural product with potential application in cancer chemotherapy (*Chem. Sci.* **2010,** *1,* **37–42**). Formation of **2** is believed to occur via a process that is mechanistically analogous to the Wolff-Kishner reduction. Draw a plausible mechanism for the conversion of **1** to **2**.

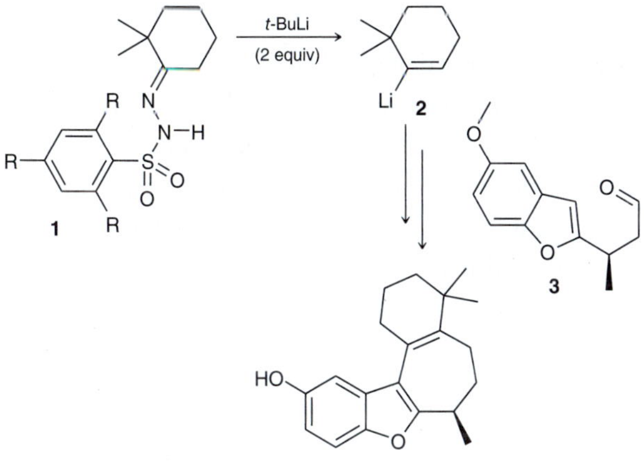

*t*-BuLi (2 equiv) 1 2 3 HO O **(+)-Frondosin B**

**20.92** A convenient method for achieving the transformation below involves treatment of the ketone with Wittig reagent **1** followed by acid-catalyzed hydration (*Chem. Ber.* **1962,** *95,* **2514–2525**):

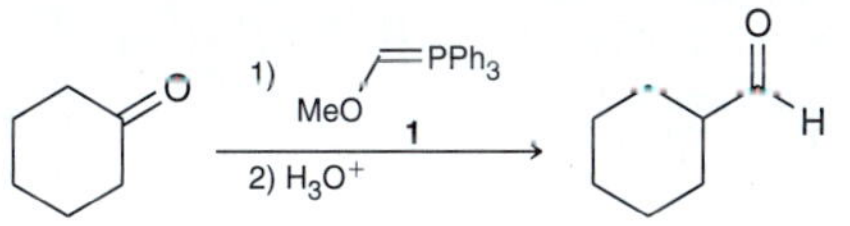

1) MeO =PPh3 1 2) $H_3O^+$ → H

(a) Predict the product of the Wittig reaction.

(b) Propose a plausible mechanism for the acid-catalyzed hydration to form the aldehyde.

**20.93** During a recent synthesis of hispidospermidin, a fungal isolate and an inhibitor of phospholipase C, the investigators employed a novel, aliphatic Friedel-Crafts acylation (*J. Am. Chem. Soc.* **1998,** *120,* **4039–4040**). The following acid chloride was treated with a Lewis acid, affording two products in a 3:2 ratio. Propose a plausible mechanism for the formation of compounds **1** and **2**.

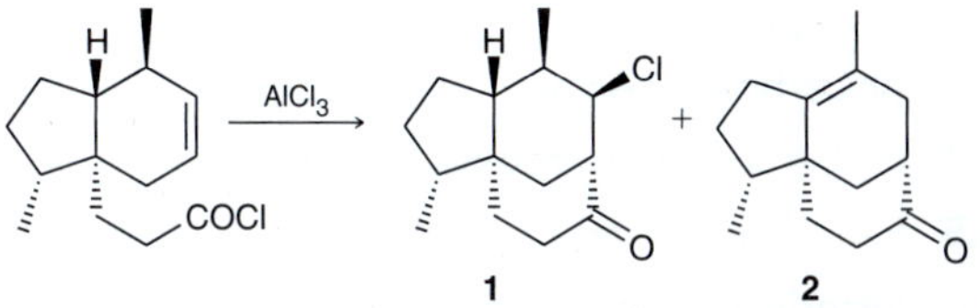

COCl $AlCl_3$ → Cl 1 + 2

**20.94** During a recent total synthesis of kibdelone C, a cytotoxic natural product isolated from soil bacteria, the investigators treated compound **1** with acetone and catalytic aqueous acid to afford acetal **2** (*J. Am. Chem. Soc.* **2011,** *133,* **9956–9959**). Propose a plausible mechanism for this transformation.

1 $H_3O^+$ Acetone → 2

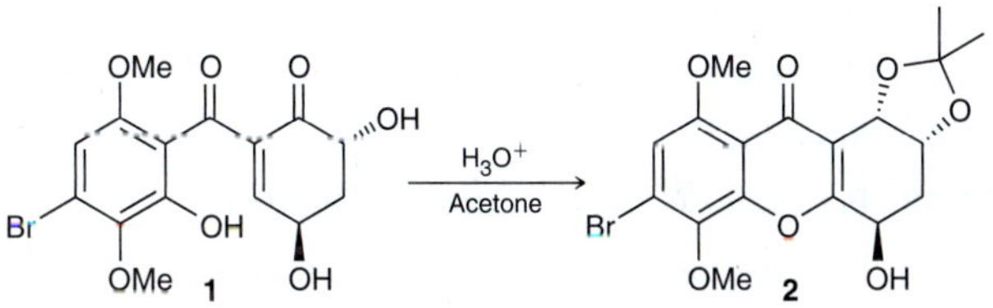

**20.95** Treatment of the following ketone with LAH affords two products, **A** and **B**. Compound **B** has a molecular formula of $C_8H_{14}O$ and exhibits strong signals at 3305 $cm^{-1}$ (broad) and 2117 $cm^{-1}$ in its IR spectrum (*J. Am. Chem. Soc.* **2006,** *128,* **6499–6507**):

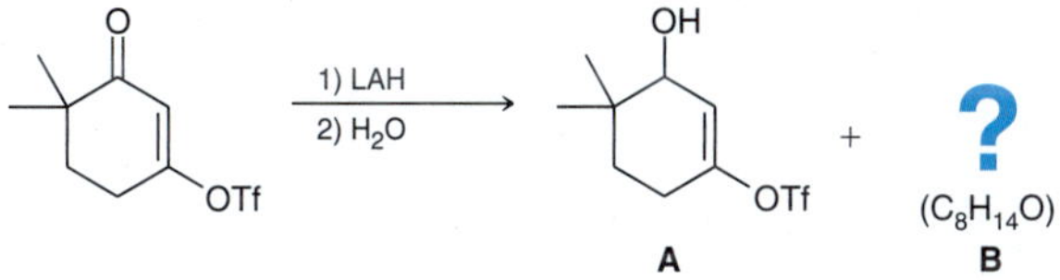

OTf 1) LAH 2) $H_2O$ → A + ? ($C_8H_{14}O$) B

(a) Using the following $^1$H NMR data, deduce the structure of compound **B**: 0.89 δ (6H, singlet), 1.49 δ (1H, broad singlet), 1.56 δ (2H, triplet), 1.95 δ (1H, singlet), 2.19 δ (2H, triplet), 3.35 δ (2H, singlet).

(b) Provide a plausible mechanism to account for the formation of compound **B**. (**Hint:** Tf = triflate, Section 7.8.)

# 21

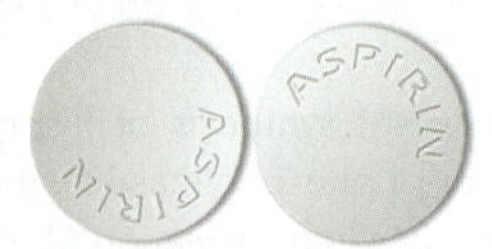




# Carboxylic Acids and Their Derivatives

## DID YOU EVER WONDER...

**how aspirin is able to reduce a fever?**

This chapter will explore the reactivity of carboxylic acids and their derivatives.

One reaction in particular will allow us to understand how aspirin works. We will study many similar reactions, which will also enable us to understand the life-saving properties of penicillin antibiotics.

This chapter is peppered with dozens of reactions, but don't be discouraged. By learning a few mechanistic principles, we will see that nearly all of these reactions are examples of a type of reaction called *nucleophilic acyl substitution*. By studying the mechanistic principles that guide this process, we will be able to unify the reactions presented in this chapter and greatly reduce the need for memorization.

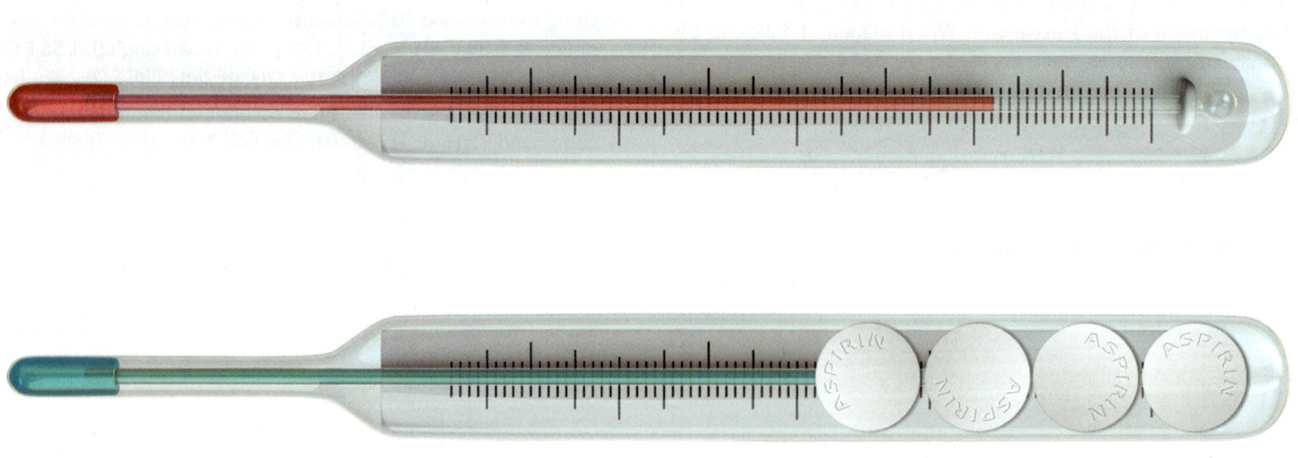

## DO YOU REMEMBER?

*Before you go on, be sure you understand the following topics.*
*If necessary, review the suggested sections to prepare for this chapter.*

- Mechanisms and Arrow Pushing (Section 6.8)
- Loss of a Leaving Group (Section 6.8)
- Drawing Curved Arrows (Section 6.10)
- Nucleophilic Attack (Section 6.8)
- Proton Transfers (Section 6.8)

Take the DO YOU REMEMBER? QUIZ in **WileyPLUS** to check your understanding

## 21.1 Introduction to Carboxylic Acids

Carboxylic acids, which were introduced in Section 3.4, are compounds with a —COOH group. These compounds are abundant in nature, where they are responsible for some familiar odors.

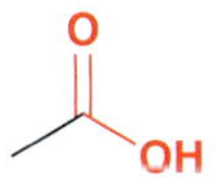

**Acetic acid** (Responsible for the pungent smell of vinegar)

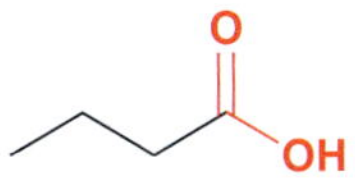

**Butanoic acid** (Responsible for the rancid odor of sour butter)

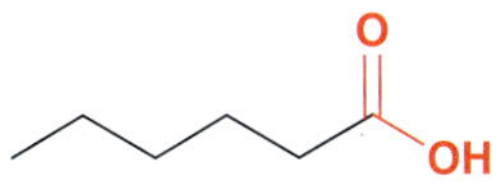

**Hexanoic acid** (Responsible for the odor of dirty socks)

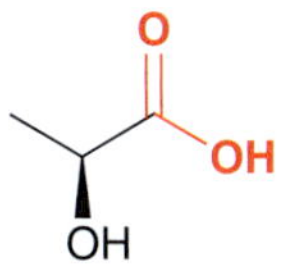

**Lactic acid** (Responsible for the taste of sour milk)

Carboxylic acids are also found in a wide range of pharmaceuticals that are used to treat a variety of conditions.

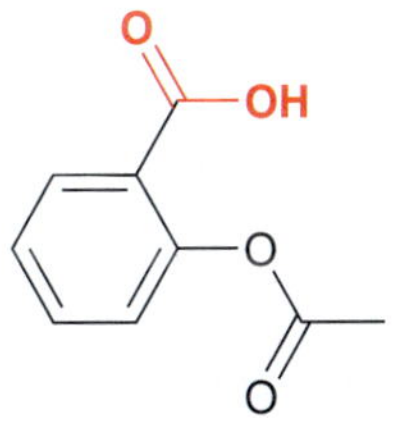

**Acetylsalicylic acid** (Aspirin, a widely used analgesic)

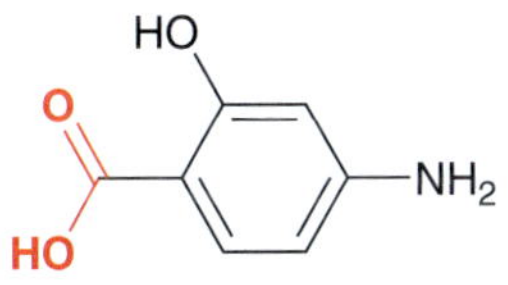

**4-Aminosalicylic acid** (Used in the treatment of tuberculosis)

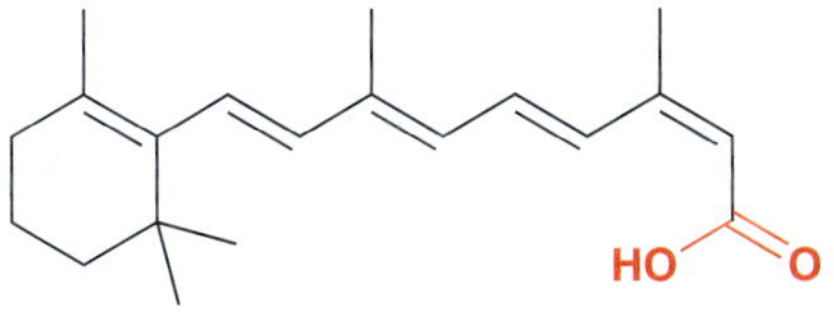

**Isotretinoin** (Used in the treatment of acne)

Each year the United States produces over 2.5 million tons of acetic acid from methanol and carbon monoxide. The primary use of acetic acid is in the synthesis of vinyl acetate, which is used in paints and adhesives.

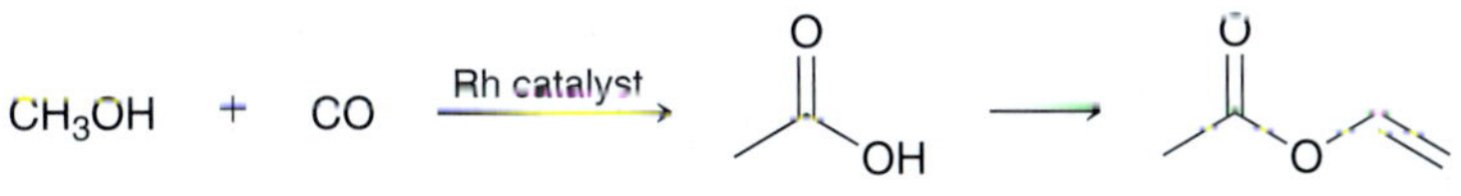

$$CH_3OH + CO \xrightarrow{\text{Rh catalyst}} \text{Acetic acid} \longrightarrow \text{Vinyl acetate}$$

Vinyl acetate is a derivative of acetic acid and is therefore said to be a *carboxylic acid derivative*. Carboxylic acids and their derivatives occupy a central role in organic chemistry, as we will see throughout this chapter.

## 21.2 Nomenclature of Carboxylic Acids

### Monocarboxylic Acids

Monocarboxylic acids, compounds containing one carboxylic acid group, are named with the suffix "oic acid":

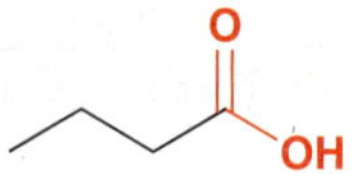

Butanoic acid

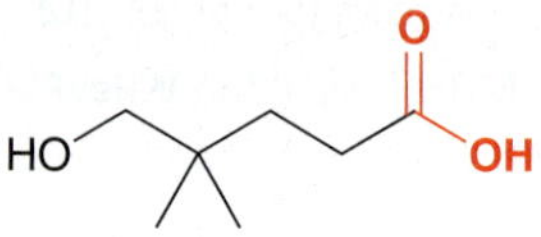

5-Hydroxy-4,4-dimethyl pentanoic acid

The parent is the longest chain that includes the carbon atom of the carboxylic acid group. That carbon atom is always assigned number 1 when numbering the parent.

When a carboxylic acid group is connected to a ring, the compound is named as an alkane carboxylic acid; for example:

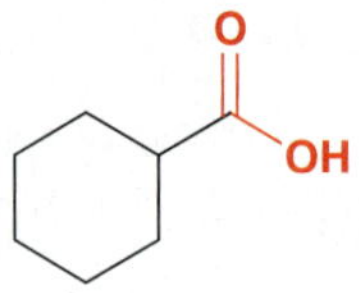

Cyclohexanecarboxylic acid

Many simple carboxylic acids have common names accepted by IUPAC. The examples shown here should be committed to memory, as they will appear frequently throughout the chapter.

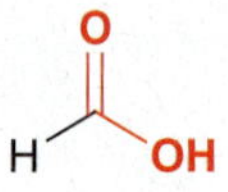

Formic acid 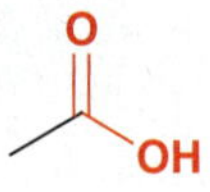    Acetic acid 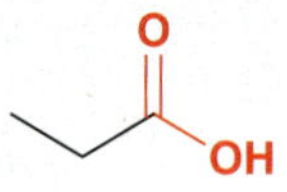    Propionic acid 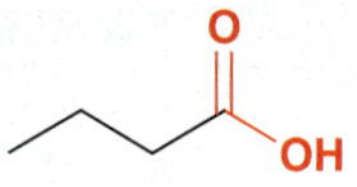    Butyric acid 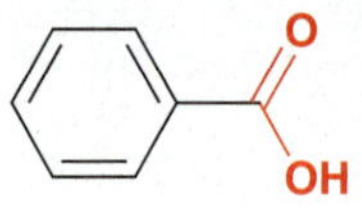    Benzoic acid

### Diacids

Diacids, compounds containing two carboxylic acid groups, are named with the suffix "dioic acid"; for example:

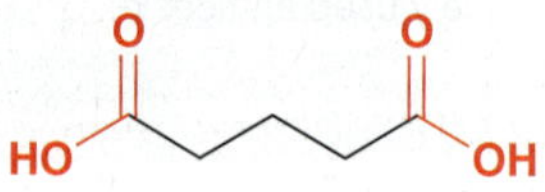

Pentanedioic acid

Many diacids have common names accepted by IUPAC.

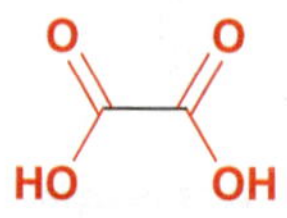

Oxalic acid 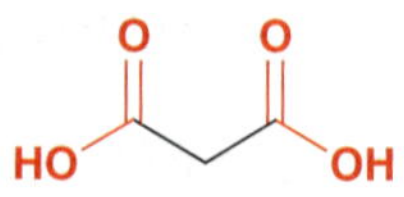    Malonic acid 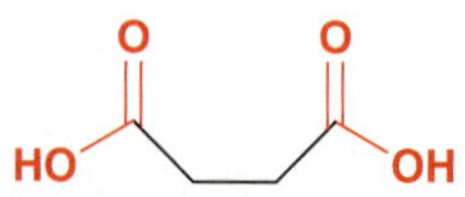    Succinic acid 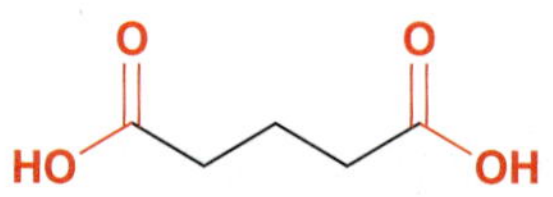    Glutaric acid

These compounds differ from each other only in the number of methylene ($CH_2$) groups separating the carboxylic acid groups. These names are used very often in the study of biochemical reactions and should therefore be committed to memory.

## CONCEPTUAL CHECKPOINT

**21.1** Provide both an IUPAC name and a common name for each of the following compounds:

(a) $HO_2C(CH_2)_3CO_2H$ (b) $CH_3(CH_2)_2CO_2H$

(c) $C_6H_5CO_2H$ (d) $HO_2C(CH_2)_2CO_2H$

(e) $CH_3COOH$ (f) $HCO_2H$

**21.2** Draw the structure of each of the following compounds:

(a) Cyclobutanecarboxylic acid (b) 3,3-Dichlorobutyric acid

(c) 3,3-Dimethylglutaric acid

**21.3** Provide an IUPAC name for each of the following compounds:

(a) (b) (c)

# 21.3 Structure and Properties of Carboxylic Acids

## Structure

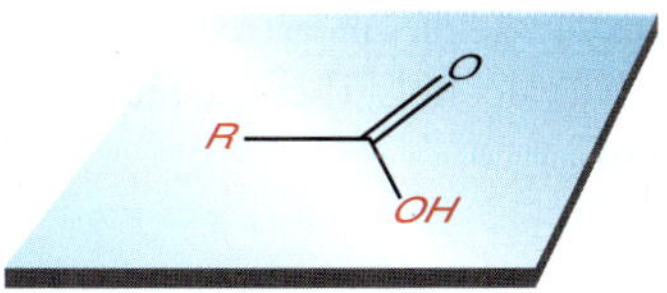


**FIGURE 21.1** The carbonyl group as well as the two atoms attached to the carbonyl carbon all reside in a plane.

The carbon atom of a carboxylic acid group is $sp^2$ hybridized and therefore exhibits trigonal planar geometry with bond angles that are nearly 120° (Figure 21.1). Carboxylic acids can form two hydrogen-bonding interactions, allowing molecules to associate with each other in pairs.

These hydrogen-bonding interactions explain the relatively high boiling points of carboxylic acids. For example, compare the boiling points of acetic acid and ethanol. Acetic acid has a higher boiling point as a result of stronger intermolecular forces.

**Acetic acid** b.p.=118°C

**Ethanol** b.p.=78°C

## Acidity of Carboxylic Acids

As their name implies, carboxylic acids exhibit mildly acidic protons. Treatment of a carboxylic acid with a strong base, such as sodium hydroxide, yields a carboxylate salt.

**A carboxylate salt**

**LOOKING BACK**
Carboxylate ions play an important role in how drugs are distributed throughout the body, as we saw in the Medically Speaking box at the end of Section 3.3 entitled "Drug Distribution and $pK_a$."

Carboxylate salts are ionic and are therefore more water-soluble than their corresponding carboxylic acids. Carboxylate ions are named by replacing the suffix "ic acid" with "ate"; for example:

**Benzoic acid** **Sodium benzoate**

You may recognize the name sodium benzoate, as it is commonly found in food products and beverages. It inhibits the growth of fungi and serves as a food preservative.

When dissolved in water, an equilibrium is established in which the carboxylic acid and the carboxylate ion are both present.

In most cases, the equilibrium significantly favors the carboxylic acid with a $K_a$ usually around $10^{-4}$ or $10^{-5}$. In other words, the p$K_a$ of most carboxylic acids is between 4 and 5.

p$K_a$=4.19 p$K_a$=4.76 p$K_a$=4.87

When compared to inorganic acids, such as HCl or $H_2SO_4$, carboxylic acids are extremely weak acids. But when compared to most classes of organic compounds, such as alcohols, they are relatively acidic. For example, compare the p$K_a$ values of acetic acid and ethanol.

p$K_a$=4.76 p$K_a$=16

Acetic acid is 11 orders of magnitude more acidic than ethanol (over a hundred billion times more acidic). As explained in Section 3.4, the acidity of carboxylic acids is primarily due to the stability of the conjugate base, which is resonance stabilized.

In the conjugate base of acetic acid, the negative charge is delocalized over two oxygen atoms, and it is therefore more stable than the conjugate base of ethanol. The delocalized nature of the charge can be seen in an electrostatic potential map of the acetate ion (Figure 21.2).

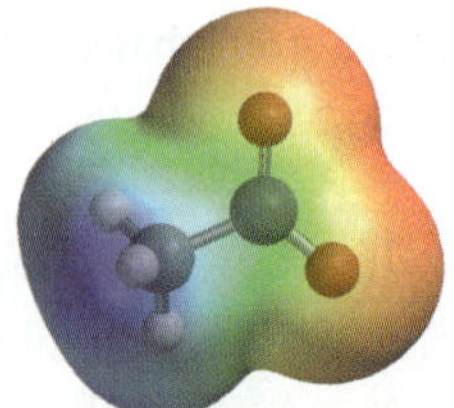

**FIGURE 21.2**
An electrostatic potential map of the acetate ion showing how the electron density is distributed over both oxygen atoms.

## CONCEPTUAL CHECKPOINT

**21.4** The following two compounds are constitutional isomers. Identify which of these is expected to be more acidic and explain your choice.

**21.5** Consider the structure of *para*-hydroxyacetophenone, which has a p$K_a$ value in the same range as a carboxylic acid, despite the fact that it lacks a COOH group. Offer an explanation for the acidity of *para*-hydroxyacetophenone.

***para*-Hydroxyacetophenone**
**p$K_a$=4.2**

**21.6** Based on your answer to the previous question, would you expect *meta*-hydroxyacetophenone to be more or less acidic than *para*-hydroxyacetophenone? Explain your answer.

***meta*-Hydroxyacetophenone** ***para*-Hydroxyacetophenone**

**21.7** When formic acid is treated with potassium hydroxide (KOH), an acid-base reaction occurs, forming a carboxylate ion. Draw the mechanism of this reaction and identify the name of the carboxylate salt.

## Carboxylic Acids at Physiological pH

Our blood is buffered to a pH of approximately 7.3, a value referred to as **physiological pH**. When dealing with buffered solutions, you may recall the **Henderson-Hasselbalch equation** from your general chemistry course.

$$\text{pH} = \text{p}K_a + \log \frac{[\text{conjugate base}]}{[\text{acid}]}$$

This equation is often employed to calculate the pH of buffered solutions, although for our purposes we will rearrange the equation in the following way:

$$\frac{[\text{conjugate base}]}{[\text{acid}]} = 10^{(\text{pH}-\text{p}K_a)}$$

This rearranged form of the Henderson-Hasselbalch equation provides a method for determining the extent to which an acid will dissociate to form its conjugate base in a buffered solution. When the p$K_a$ value of an acid is equivalent to the pH of a buffered solution into which it is dissolved, then

$$\frac{[\text{conjugate base}]}{[\text{acid}]} = 10^{(\text{pH}-\text{p}K_a)} = 10^{(0)} = 1$$

The ratio of the concentrations of conjugate base and acid will be 1. In other words, a carboxylic acid and its conjugate base will be present in approximately equal amounts when dissolved in a solution that is buffered such that pH = p$K_a$ of the acid.

Now let's apply this equation to carboxylic acids at physiological pH (7.3), so that we can determine which form predominates (the carboxylic acid or the carboxylate ion). Recall that carboxylic acids generally have a p$K_a$ value between 4 and 5. Therefore, at physiological pH:

$$\frac{[\text{conjugate base}]}{[\text{acid}]} = 10^{(\text{pH}-\text{p}K_a)} = 10^{(7.3-\text{p}K_a)} \approx 10^3$$

The ratio of the concentrations of the carboxylate ion and the carboxylic acid will be approximately 1000 : 1. That is, carboxylic acids will exist primarily as carboxylate salts at physiological pH. For example, pyruvic acid exists primarily as pyruvate ion at physiological pH.

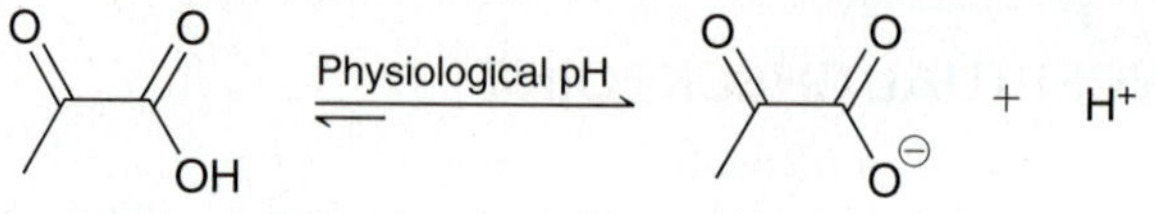

Physiological pH

Pyruvic acid ⇌ Pyruvate + $H^+$

Carboxylate ions play vital roles in many biological processes, as we will see in Chapter 25.

## Substituent Effects on Acidity

The presence of electron-withdrawing substituents can have a profound impact on the acidity of a carboxylic acid.

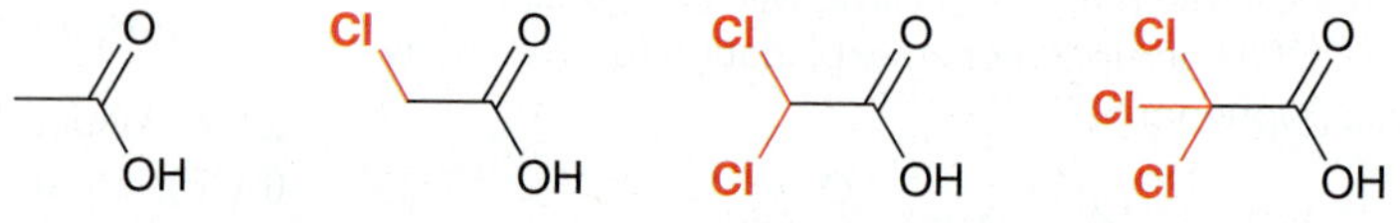

p$K_a$ = 4.8 p$K_a$ = 2.9 p$K_a$ = 1.3 p$K_a$ = 0.9

Notice that the p$K_a$ decreases with each additional chlorine substituent. This trend is explained in terms of the inductive effects of the chlorine atoms, which can stabilize the conjugate base (as explained in Section 3.4). The effect of an electron-withdrawing group depends on its proximity to the carboxylic acid group.

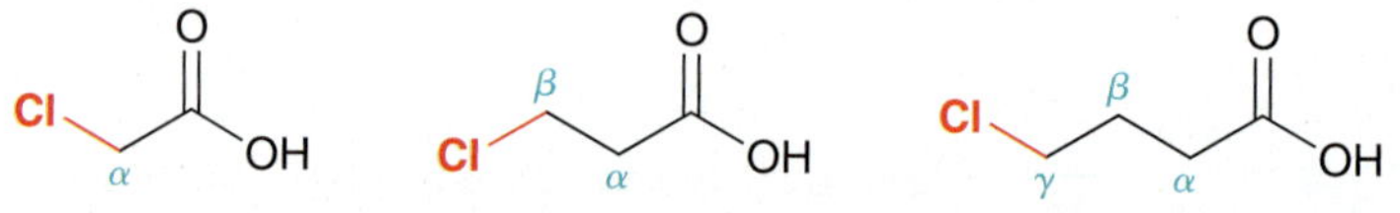

p$K_a$ = 2.9 p$K_a$ = 4.1 p$K_a$ = 4.5

The effect is most pronounced when the electron-withdrawing group is located at the α position. As the distance between the chlorine atom and the carboxylic acid group increases, the effect of the chlorine atom becomes less pronounced.

The effects of electron-withdrawing substituents are also observed for substituted benzoic acids (Figure 21.3). In Sections 19.7–19.10, we discussed the electronic effects of each of the substituents in Figure 21.3, and we saw that a nitro group is a powerful electron-withdrawing group. Consequently, the presence of the nitro group on the ring will stabilize the conjugate base, giving a low p$K_a$ value (relative to benzoic acid). In contrast, a hydroxy group is a powerful electron-donating group (Section 19.10), and therefore, the presence of the hydroxy group will destabilize the conjugate base, giving a high p$K_a$ value (relative to benzoic acid).

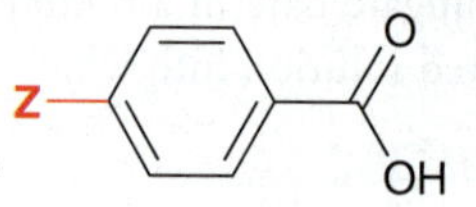

| Z | —$NO_2$ | —CHO | —Cl | —H | —$CH_3$ | —OH |
|---|---|---|---|---|---|---|
| p$K_a$ | 3.4 | 3.8 | 4.0 | 4.2 | 4.3 | 4.5 |

**FIGURE 21.3**
The p$K_a$ values of various *para*-substituted benzoic acids.

CONCEPTUAL CHECKPOINT

**21.8** Acetic acid was dissolved in a solution buffered to a pH of 5.76. Determine the ratio of the concentrations of acetate ion and acetic acid in this solution. Which species predominates under these conditions?

**21.9** Rank each set of compounds in order of increasing acidity:

(a) 2,4-Dichlorobutyric acid
2,3-Dichlorobutyric acid
3,4-Dimethylbutyric acid

(b) 3-Bromopropionic acid
2,2-Dibromopropionic acid
3,3-Dibromopropionic acid

## 21.4 Preparation of Carboxylic Acids

In previous chapters, we studied a variety of methods for preparing carboxylic acids (Table 21.1). In addition to the methods we have already seen, there are many other ways of preparing carboxylic acids. We will examine two of them.

**TABLE 21.1 A REVIEW OF METHODS FOR PREPARING CARBOXYLIC ACIDS**

| REACTION | SECTION NUMBER | COMMENTS |
|---|---|---|
| **Oxidative Cleavage of Alkynes** R—≡—R → 1) $O_3$ 2) $H_2O$ → RCOOH + RCOOH | 10.9 | Oxidative cleavage will break a C≡C triple bond forming two carboxylic acids. |
| **Oxidation of Primary Alcohols** RCH$_2$OH → $Na_2Cr_2O_7$, $H_2SO_4$, $H_2O$ → RCOOH | 13.10 | A variety of strong oxidizing agents can be used to oxidize primary alcohols and produce carboxylic acids. |
| **Oxidation of Alkylbenzenes** isopropylbenzene → $Na_2Cr_2O_7$, $H_2SO_4$, $H_2O$ → benzoic acid | 18.6 | Any alkyl group on an aromatic ring will be completely oxidized to give benzoic acid, provided that the benzylic position has at least one hydrogen atom. |

### Hydrolysis of Nitriles

When treated with aqueous acid, a *nitrile* (a compound with a cyano group) can be converted into a carboxylic acid.

$$R-C\equiv N \xrightarrow[\text{Heat}]{H_3O^+} RCOOH$$

This process is called *hydrolysis*, and the mechanism for nitrile hydrolysis will be discussed later in this chapter. This reaction provides us with a two-step process for converting an alkyl halide to a carboxylic acid.

$$PhCH_2Br \xrightarrow{:\overset{\ominus}{C}\equiv N} PhCH_2C\equiv N \xrightarrow[\text{Heat}]{H_3O^+} PhCH_2COOH$$

The first step is an $S_N2$ reaction in which cyanide acts as a nucleophile. The resulting nitrile is then hydrolyzed to yield a carboxylic acid that has one more carbon atom (shown in red) than the original alkyl halide. Since the first step is an $S_N2$ process, the reaction cannot occur with tertiary alkyl halides.

### Carboxylation of Grignard Reagents

Carboxylic acids can also be prepared by treating a Grignard reagent with carbon dioxide:

$$R-MgBr \xrightarrow[2)\ H_3O^+]{1)\ CO_2} RCOOH$$

A mechanism for this process is shown below:

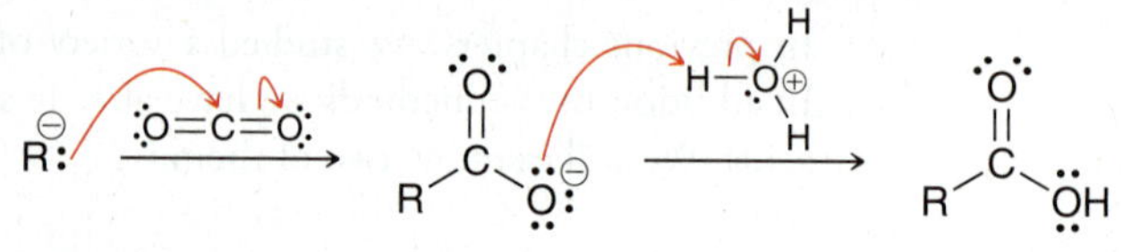

In the first step, the Grignard reagent attacks the electrophilic center of carbon dioxide, generating a carboxylate ion. Treating the carboxylate ion with a proton source affords the carboxylic acid. These two steps occur separately, as the proton source is not compatible with the Grignard reagent and can only be introduced after the Grignard reaction is complete. This reaction provides us with another two-step process for converting an alkyl (or vinyl or aryl) halide to a carboxylic acid.

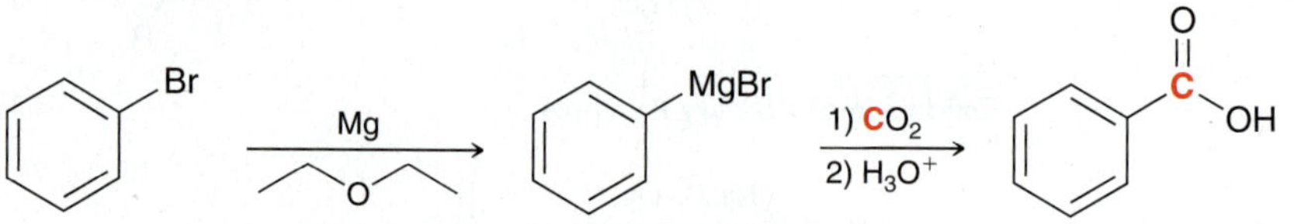

We have now seen two new methods for preparing carboxylic acids, both of which involve the introduction of one carbon atom.

## CONCEPTUAL CHECKPOINT

**21.10** Identify the reagents you would use to perform the following transformations:

(a) Ethanol → Acetic acid
(b) Toluene → Benzoic acid
(c) Benzene → Benzoic acid
(d) 1-Bromobutane → Pentanoic acid
(e) Ethylbenzene → Benzoic acid
(f) Bromocyclohexane → Cyclohexanecarboxylic acid

## 21.5 Reactions of Carboxylic Acids

Carboxylic acids are reduced to alcohols upon treatment with lithium aluminum hydride.

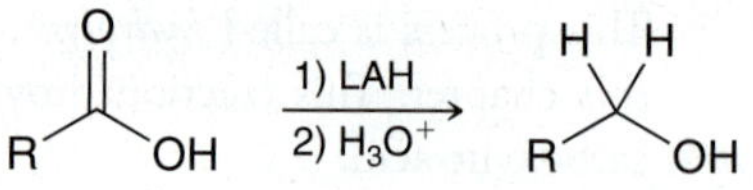

The first step of the mechanism is likely a proton transfer, because LAH is not only a powerful nucleophile, but it can also function as a strong base, forming a carboxylate ion.

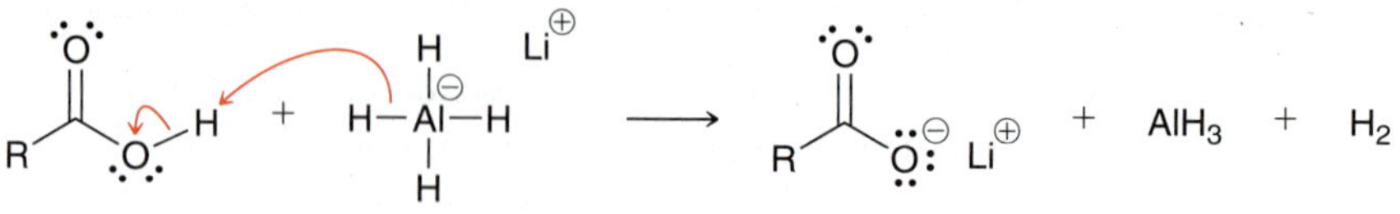

There are several possibilities for the rest of the mechanism. One possibility involves a reaction of the carboxylate ion with $AlH_3$ followed by elimination to form an aldehyde:

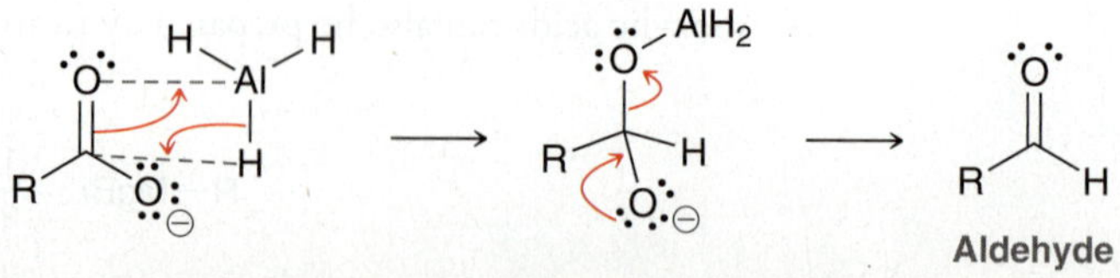

Aldehyde

Under these conditions, the aldehyde cannot be isolated. Instead, it is further attacked by LAH to form an alkoxide, which is then protonated when $H_3O^+$ is introduced into the reaction flask.

An alternative method for reducing carboxylic acids involves the use of borane ($BH_3$).

Reduction with borane is often preferred over reduction with LAH, because borane reacts selectively with a carboxylic acid group in the presence of another carbonyl group. As an example, if the following reaction were performed with LAH instead of borane, both carbonyl groups would be reduced:

(80%)

CONCEPTUAL CHECKPOINT

**21.11** Identify the reagents you would use to achieve each of the following transformations:

(a)

(b)

## 21.6 Introduction to Carboxylic Acid Derivatives

### Classes of Carboxylic Acid Derivatives

In the previous section, we learned about the reaction between a carboxylic acid and LAH. This reaction is a reduction, because the carbon atom of the carboxylic acid group is reduced in the process:

Reduction

Carboxylic acids also undergo many other reactions that do not involve a change in oxidation state:

No change in oxidation state

**LOOKING BACK**

For a review of oxidation states, see Section 13.4.

Replacement of the OH group with a different group (Z) does not involve a change in oxidation state if Z is a heteroatom (Cl, O, N, etc.). Compounds of this type are called **carboxylic acid derivatives**, and they will be the focus of the remainder of this chapter. The four most common types of carboxylic acid derivatives are shown below.

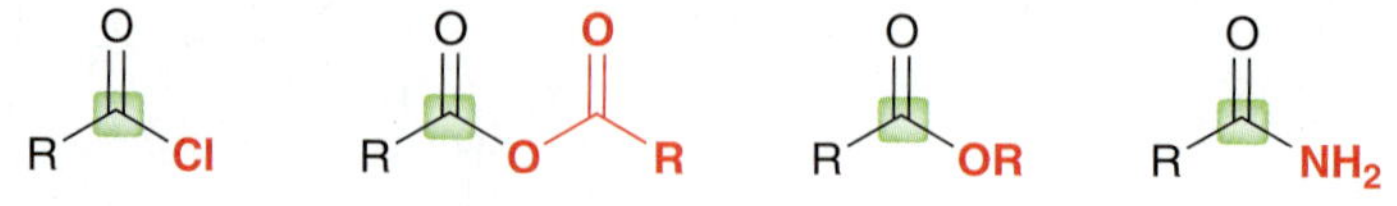

Acid halide　Acid anhydride　Ester　Amide

Notice that in each case there is a carbon atom (highlighted in green) with three bonds to heteroatoms. As a result, each of these carbon atoms has the same oxidation state as the carbon atom of a carboxylic acid. Although all of these derivatives exhibit a carbonyl group, the presence of a carbonyl group is not a necessary requirement to qualify as a carboxylic acid derivative. Any compound with a carbon atom that has three bonds to heteroatoms will be classified as a carboxylic acid derivative. For example, consider the structure of nitriles.

$$R{-}C{\equiv}N$$

A nitrile

Nitriles exhibit a carbon atom with three bonds to a heteroatom (nitrogen). As a result, the conversion of a nitrile into a carboxylic acid (or vice versa) is neither a reduction nor an oxidation. Nitriles are therefore considered to be carboxylic acid derivatives, and they will also be discussed in this chapter.

## Carboxylic Acid Derivatives in Nature

As we will soon see, acid halides and acid anhydrides are highly reactive and are therefore not very common in nature. In contrast, esters are more stable and are abundant in nature. Naturally occurring esters, such as the following three examples, often have pleasant odors and contribute to the aromas of fruits and flowers:

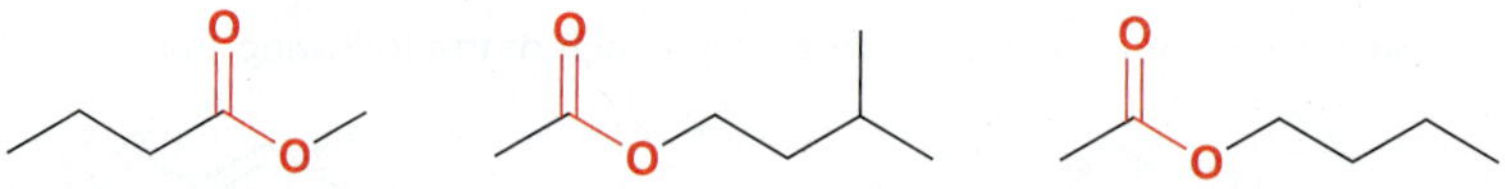

Methyl butanoate (pineapple)　Isopentyl acetate (banana)　Butyl acetate (pear)

Amides are abundant in living organisms. For example, proteins are comprised of repeating amide linkages.

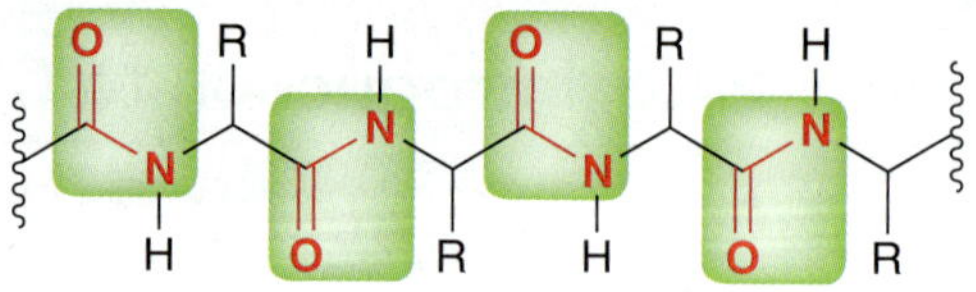

The structure of proteins

Chapter 26 will focus on the structure of proteins as well as the central role that proteins play in catalyzing most biochemical reactions.

## Naming Acid Halides

Acid halides are named as derivatives of carboxylic acids by replacing the suffix "ic acid" with "yl halide":

Acetic acid → Acetyl bromide

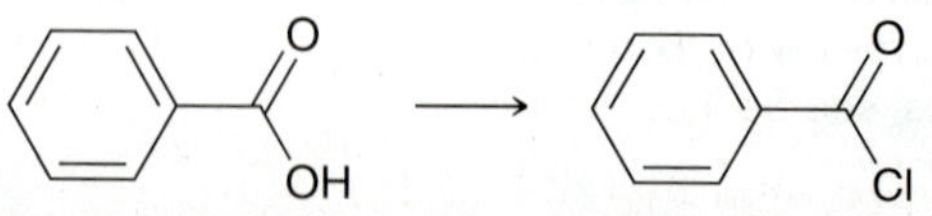

Benzoic acid → Benzoyl chloride

## medically speaking | Sedatives

Sedatives are compounds that reduce anxiety and induce sleep. Our bodies utilize many natural sedatives, including melatonin.

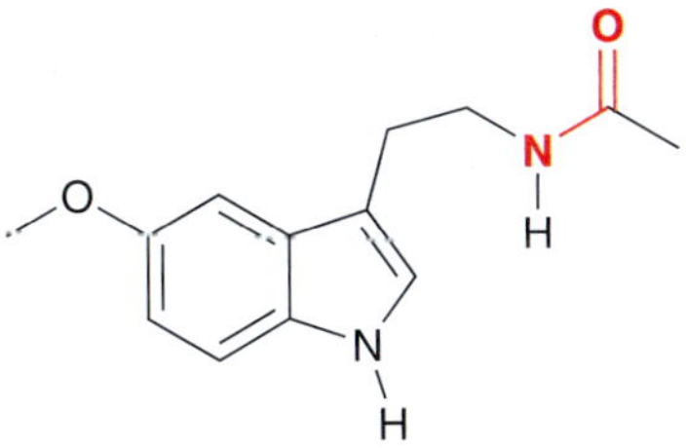

**Melatonin**

There is much evidence suggesting that melatonin plays an important role in regulating the body's natural sleep-wake cycle. For example, it has been observed that levels of melatonin for most people increase at night and then decrease in the morning. For this reason, many people take melatonin supplements to treat insomnia.

Notice the amide group in the structure of melatonin (shown in red). This group is a common feature in many drugs that are marketed as sedatives; for example:

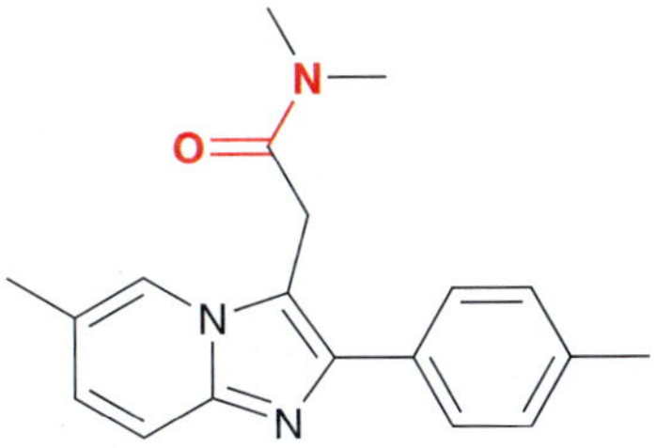

**Zolpidem (Ambien)** 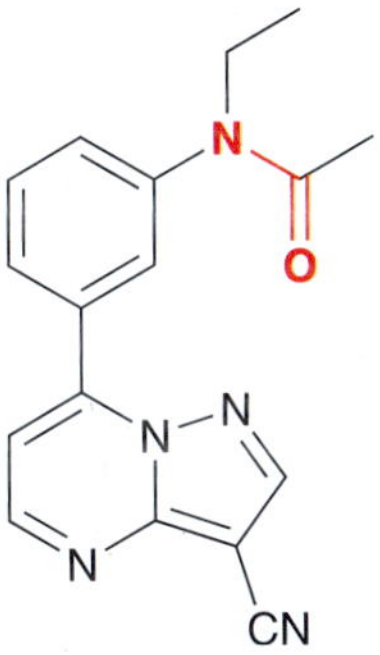 **Zaleplon (Sonata)**

These drugs, which are similar in structure to melatonin, are used to treat insomnia. Other sedatives are used primarily in the treatment of excessive anxiety:

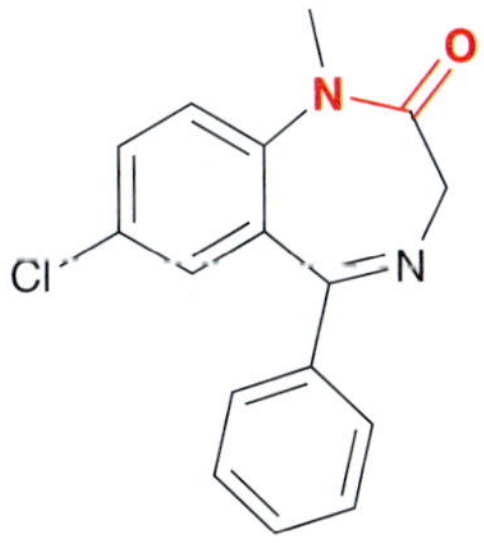

**Diazepam (Valium)** 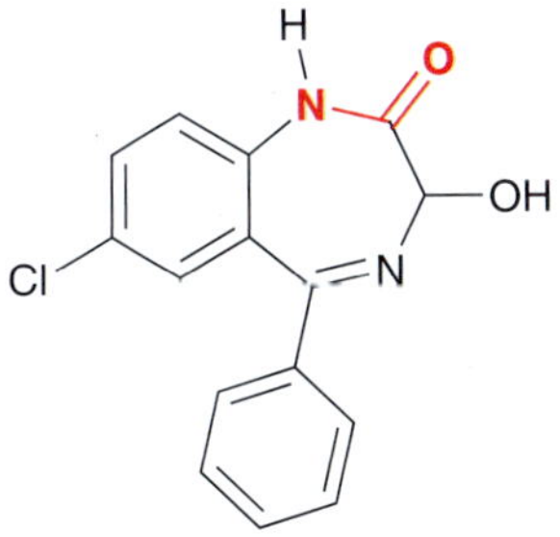 **Oxazepam (Serax)**

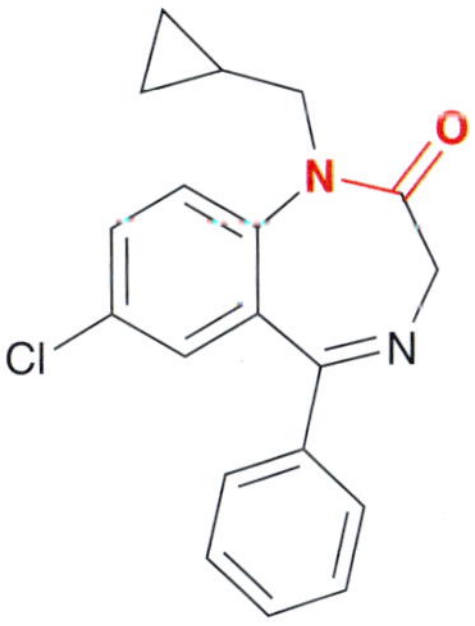

**Prazepam (Verstran)**

Drugs used in the treatment of anxiety are called anxiolytic agents. The three examples shown are all similar in structure and belong to a class of compounds called benzodiazepines. Extensive research has been undertaken to elucidate the relationship between the structure and activity of benzodiazepines. It was found that the amide group is not absolutely necessary, but its presence does increase the potency of these agents.

When an acid halide group is connected to a ring, the suffix "carboxylic acid" is replaced with "carbonyl halide"; for example:

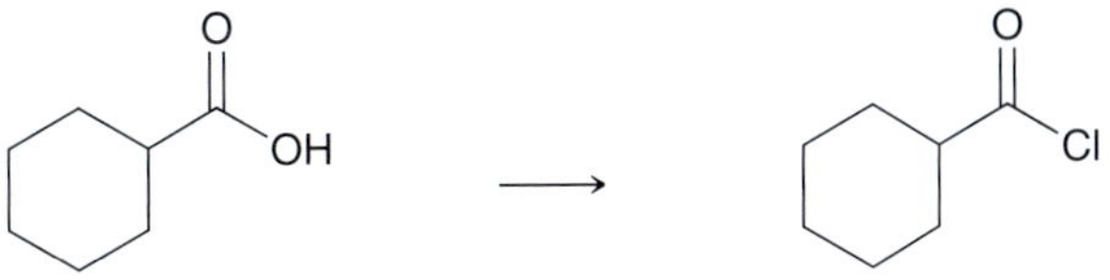

**Cyclohexanecarboxylic acid** → **Cyclohexanecarbonyl chloride**

## Naming Acid Anhydrides

Acid anhydrides are named as derivatives of carboxylic acids by replacing the suffix "acid" with "anhydride."

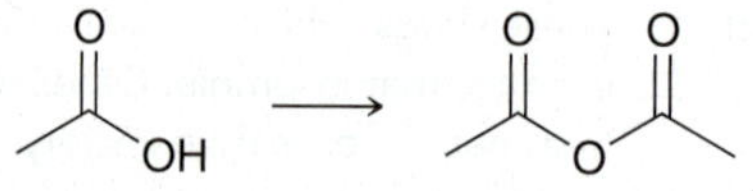

Acetic acid → Acetic anhydride

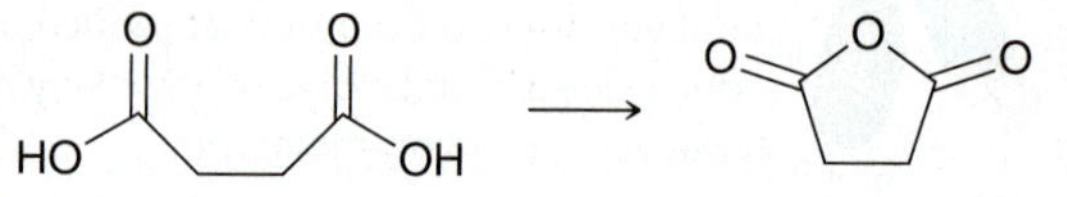

Succinic acid → Succinic anhydride

Unsymmetrical anhydrides are prepared from two different carboxylic acids and are named by indicating both acids alphabetically followed by the suffix "anhydride":

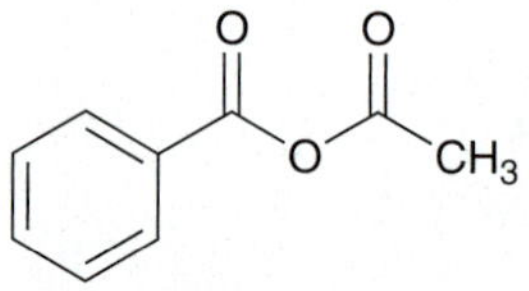

Acetic benzoic anhydride

## Naming Esters

Esters are named by first indicating the alkyl group attached to the oxygen atom followed by the carboxylic acid, for which the suffix "ic acid" is replaced with "ate."

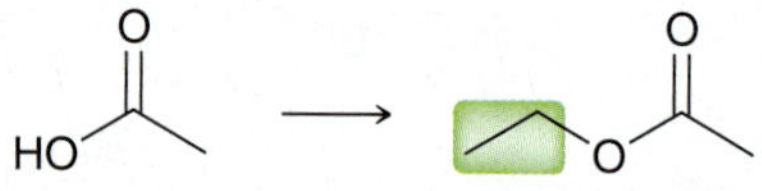

Acetic acid → Ethyl acetate

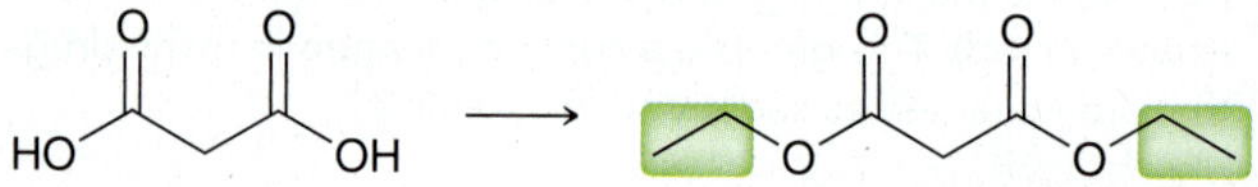

Malonic acid → Diethyl malonate

The same methodology is applied when the ester group is connected to a ring; for example:

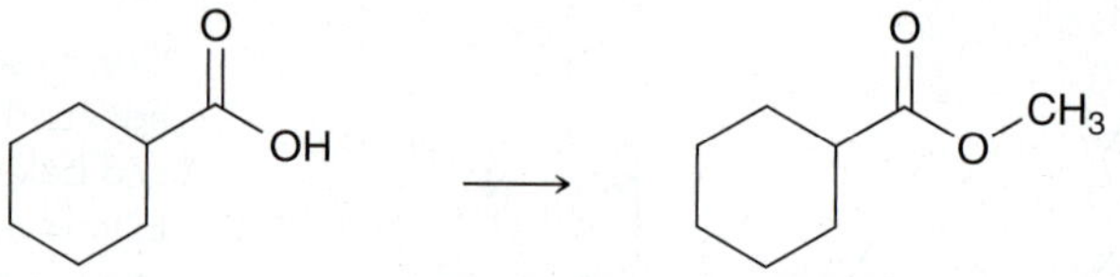

Cyclohexanecarboxylic acid → Methyl cyclohexanecarboxylate

## Naming Amides

Amides are named as derivatives of carboxylic acids by replacing the suffix "ic acid" or "oic acid" with "amide."

Acetic acid → Acetamide

Benzoic acid → Benzamide

When an amide group is connected to a ring, the suffix "carboxylic acid" is replaced with "carboxamide."

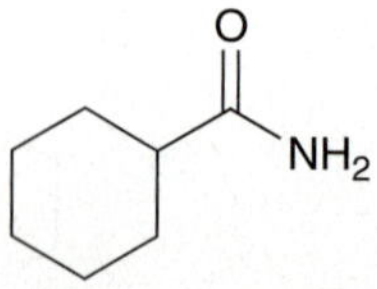

Cyclohexanecarboxamide

If the nitrogen atom bears alkyl groups, these groups are indicated at the beginning of the name, and the letter "*N*" is used as a locant to indicate that they are attached to the nitrogen.

*N*-Methylacetamide    *N*,*N*-Dimethylacetamide

## Naming Nitriles

Nitriles are named as derivatives of carboxylic acids by replacing the suffix "ic acid" or "oic acid" with "onitrile."

$H_3C-COOH \longrightarrow H_3C-C\equiv N$

Acetic acid    Acetonitrile

Benzoic acid    Benzonitrile

### CONCEPTUAL CHECKPOINT

**21.12** Provide a name for each of the following compounds.

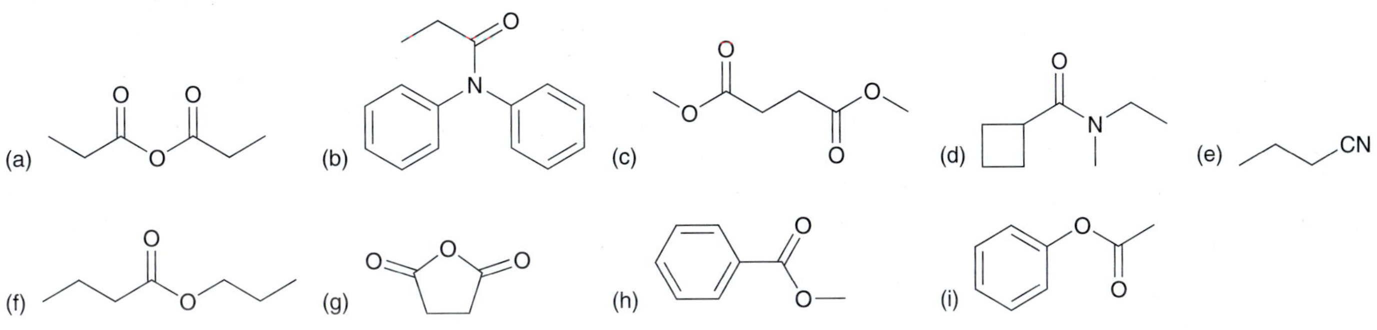


**21.13** Draw a structure for each of the following compounds:

(a) Dimethyl oxalate (b) Phenyl cyclopentanecarboxylate (c) *N*-Methylpropionamide (d) Propionyl chloride

## 21.7 Reactivity of Carboxylic Acid Derivatives

### Electrophilicity of Carboxylic Acid Derivatives

In the previous chapter, we saw that the carbon atom of a carbonyl group is electrophilic as a result of both inductive and resonance effects. The same is true of carboxylic acid derivatives, although there is a wide range of reactivity among the carboxylic acid derivatives, illustrated in Figure 21.4. Acid halides are the most reactive. To rationalize this, we must consider both inductive effects and resonance effects. Let's begin with induction. Chlorine is an electronegative atom and therefore withdraws electron density from the carbonyl group via induction.

$R-C(=O)-Cl$

This effect renders the carbonyl group even more electrophilic when compared with the carbonyl group of a ketone.

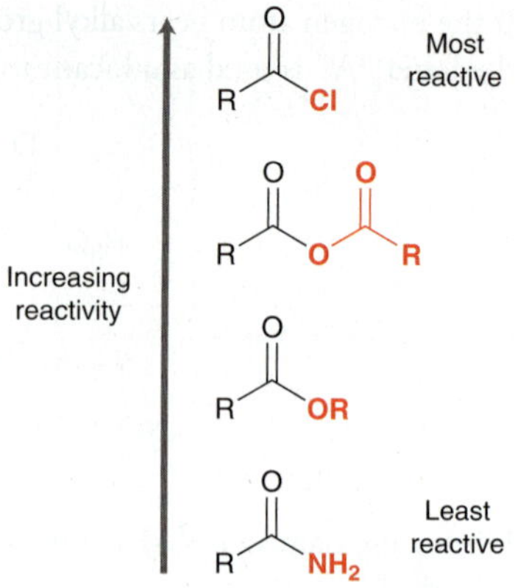

FIGURE 21.4
The relative order of reactivity of carboxylic acid derivatives.

Now let's consider resonance effects. An acid halide has three resonance structures.

Not a significant contributor

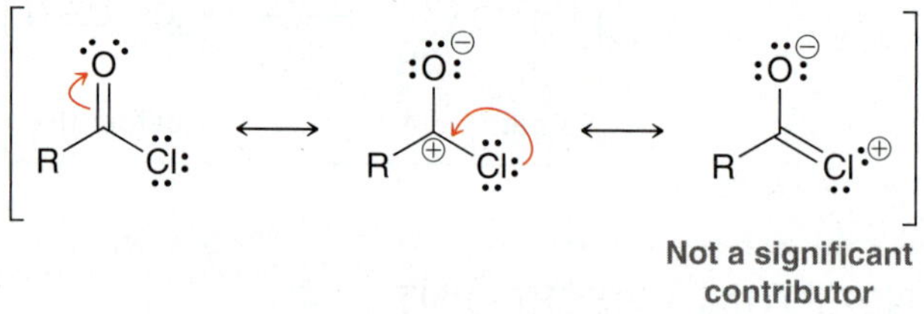

BY THE WAY
Note the similarity between this analysis and the analysis of the effect of a chlorine atom on the electronics of an aromatic ring (see Section 19.9).

The third resonance structure does not contribute much character to the overall resonance hybrid because the *p*-orbital overlap required for a C=Cl bond is not effective. This argument is similar to the argument provided in reference to the ineffective overlap of S=O bonds in Section 19.3. As a result, the chlorine atom does not donate much electron density to the carbonyl group via resonance. The net effect of the chlorine atom is to withdraw electron density, rendering the carbonyl group extremely electrophilic.

Amides are the least reactive of the carboxylic acid derivatives. To rationalize this observation, we must once again explore inductive effects and resonance effects. Let's begin with induction. Nitrogen is less electronegative than chlorine or oxygen and is not an effective electron-withdrawing group. The nitrogen atom does not withdraw much electron density from the carbonyl group, and inductive effects are not significant. However, resonance effects are substantial. Consider the three resonance structures of an amide.

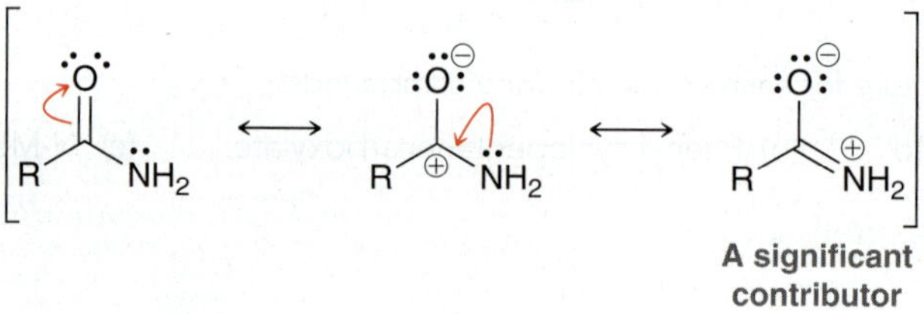

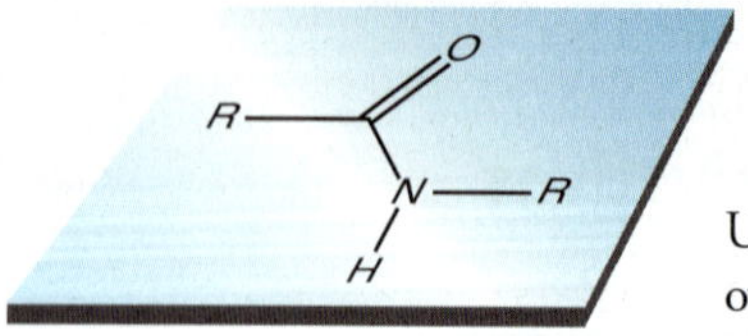

A significant contributor

FIGURE 21.5
An illustration of the planar geometry of amides.

Unlike an acid halide, the third resonance structure of an amide contributes significant character to the overall resonance hybrid. The *p* orbital on the carbon atom effectively overlaps with a *p* orbital on the nitrogen atom, and the nitrogen atom can easily accommodate the positive charge. The nitrogen atom is $sp^2$ hybridized, and the geometry of the nitrogen atom is trigonal planar. As a result, the entire amide group lies in a plane (Figure 21.5). The C—N bond of an amide has significant double-bond character, which can be verified by observing the relatively high barrier to rotation for the C—N bond.

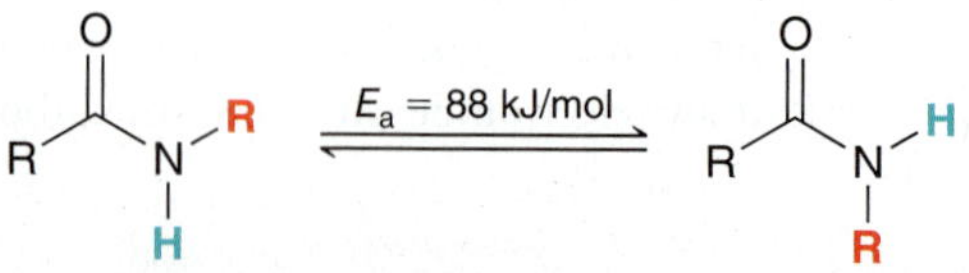

$E_a = 88$ kJ/mol

The restricted rotation of the C—N bond and the planar geometry of amide groups will be important when we discuss protein structure in Chapter 27.

## Nucleophilic Acyl Substitution

The reactivity of carboxylic acid derivatives is similar to the reactivity of aldehydes and ketones in a number of ways. In both cases, the carbonyl group is electrophilic and subject to attack by a nucleophile. In both cases, the same rules and principles govern the proton transfers that accompany the reactions, as we will soon see. Nevertheless, there is one critical difference between carboxylic acid derivatives and aldehydes/ketones. Specifically, carboxylic acid derivatives possess a heteroatom that can function as a leaving group, while aldehydes and ketones do not.

O R Z — **Can** function as a leaving group

O R R, O R H — **Cannot** function as a leaving group

When a nucleophile attacks a carboxylic acid derivative, a reaction can occur in which the nucleophile replaces the leaving group:

$$\mathrm{RCOZ} \xrightarrow{\mathrm{Nuc}^{\ominus}} \mathrm{RCONuc} + \mathrm{Z}^{\ominus}$$

This type of reaction is called a **nucleophilic acyl substitution**, and the rest of this chapter will be dominated by various examples of this type of reaction. The general mechanism has two core steps (Mechanism 21.1).

### MECHANISM 21.1 NUCLEOPHILIC ACYL SUBSTITUTION

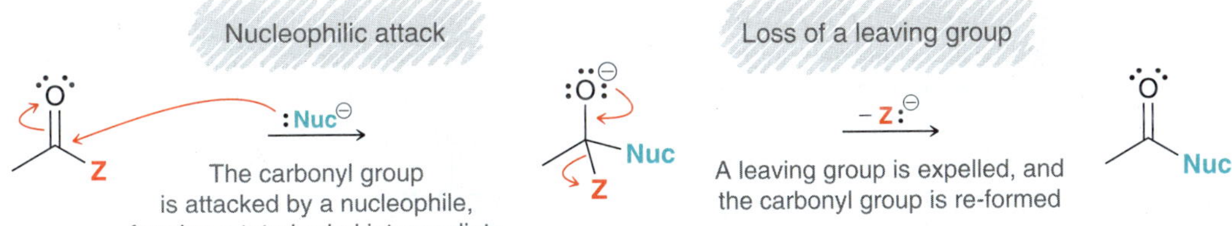


In the first step, a nucleophile attacks the carbonyl group, during which the hybridization state of the carbon atom changes. In both the starting material and product, the carbon atom is $sp^2$ hybridized with trigonal planar geometry, but the same atom in the intermediate is $sp^3$ hybridized with tetrahedral geometry. In recognition of this geometric change, the intermediate is often called a **tetrahedral intermediate**. In the second step, the carbonyl group is re-formed via loss of a leaving group. Re-formation of the C=O double bond is a powerful driving force, and even poor leaving groups (such as $RO^-$) can be expelled under certain conditions. Hydride ions ($H^-$) and carbanions ($C^-$) cannot function as leaving groups under any conditions, so this type of reaction is not observed for ketones or aldehydes. There are only a few rare exceptions when $H^-$ or $C^-$ do function as leaving groups, and we will specifically explain why those cases are exceptions when we discuss them in Chapter 22. For our purposes, the following rule will guide our discussion for the remainder of this chapter: *When a nucleophile attacks a carbonyl group to form a tetrahedral intermediate, the carbonyl group will always be re-formed, if possible, but* $H^-$ *and* $C^-$ *are generally not expelled as leaving groups.* There will be no exceptions to this rule in this chapter.

Let's explore a specific example of a nucleophilic acyl substitution, so that we can see how the rule applies. Consider the following transformation:

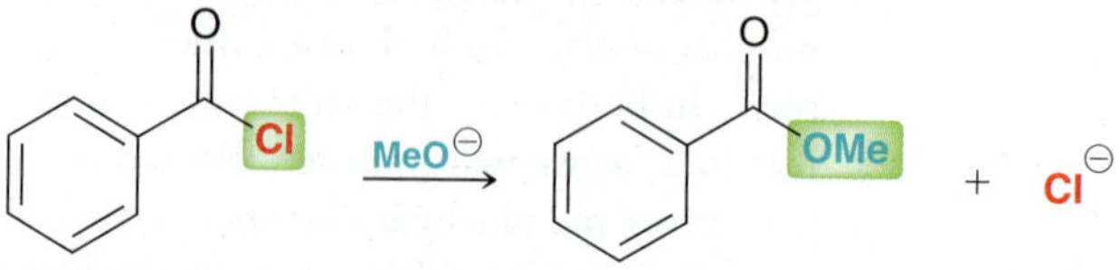

In this reaction, an acid chloride is converted into an ester. The mechanism of this transformation has two steps. In the first step, methoxide functions as a nucleophile and attacks the carbonyl group, forming a tetrahedral intermediate:

Nucleophilic attack

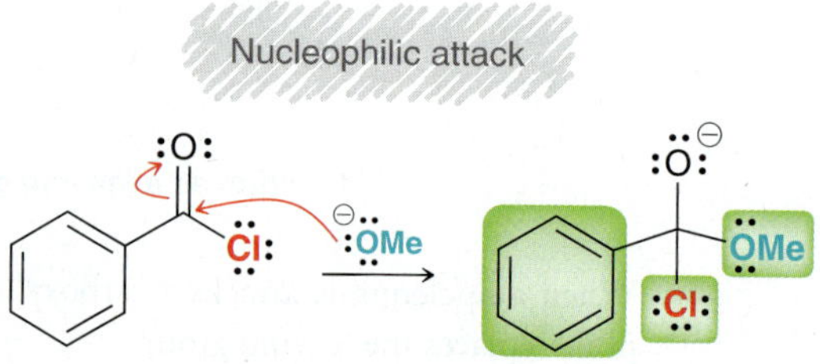

Now apply the rule: Re-form the carbonyl if possible, but avoid expelling $H^-$ or $C^-$. In order to re-form the carbonyl group in this case, one of the three highlighted groups must be expelled as a leaving group bearing a negative charge. The aromatic ring cannot leave, because that would involve expelling $C^-$. The remaining two choices (chloride or methoxide) are both viable options. Chloride is more stable than methoxide and is therefore a better leaving group, so the carbonyl will likely re-form to expel the chloride ion.

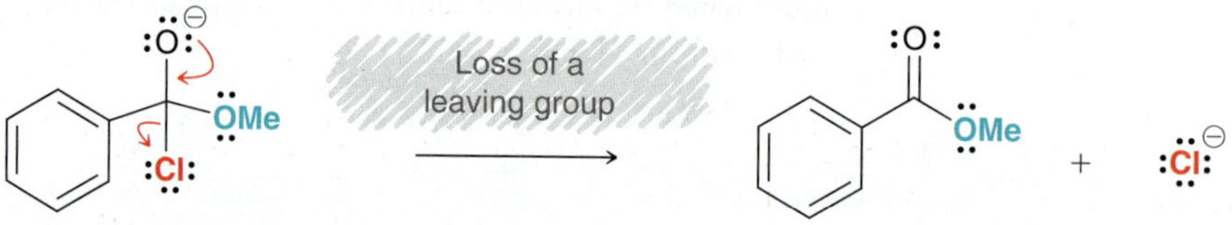

Loss of a leaving group

In summary, the mechanism for a nucleophilic acyl substitution reaction involves two core steps—nucleophilic attack and loss of a leaving group. Notice that these are the same two steps involved in an $S_N2$ process. However, there is one important difference. In an $S_N2$ process, the two steps occur in a concerted fashion (simultaneously), but in a nucleophilic acyl substitution reaction, the two steps occur separately. It is a common mistake to draw these two steps as occurring together.

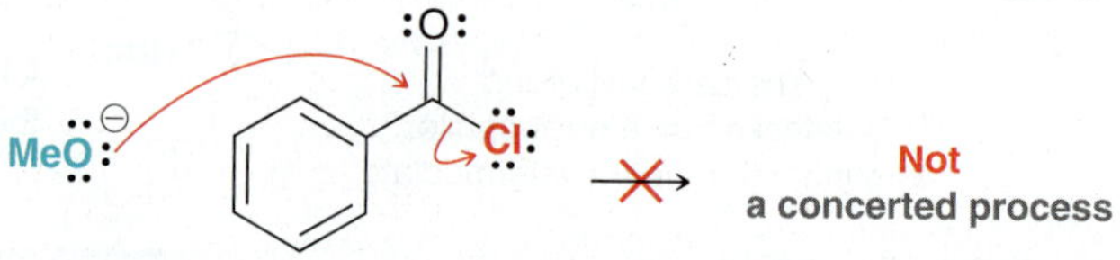

Not a concerted process

The reaction mechanism cannot be drawn like this because $S_N2$ reactions do not occur readily at $sp^2$-hybridized centers. When drawing a nucleophilic acyl substitution, make sure to draw the first step, which forms the tetrahedral intermediate, followed by the second step, which shows how the carbonyl group is re-formed.

The vast majority of the reactions in this chapter are nucleophilic acyl substitution reactions. All of these reactions will exhibit the two core steps of nucleophilic attack and loss of a leaving group to re-form the carbonyl group. But many of the reaction mechanisms will also exhibit proton transfers. In order to draw each mechanism properly, it is necessary to know why the proton transfers occur. The following rule will guide us in deciding whether or not to employ proton transfers in a particular mechanism: *In acidic conditions, avoid formation of a strong base. In basic conditions, such as hydroxide or methoxide, avoid formation of a strong acid.*

This rule dictates that all participants in a reaction (reactants, intermediates, and leaving groups) should be consistent with the conditions employed. As an example, consider the following reaction:

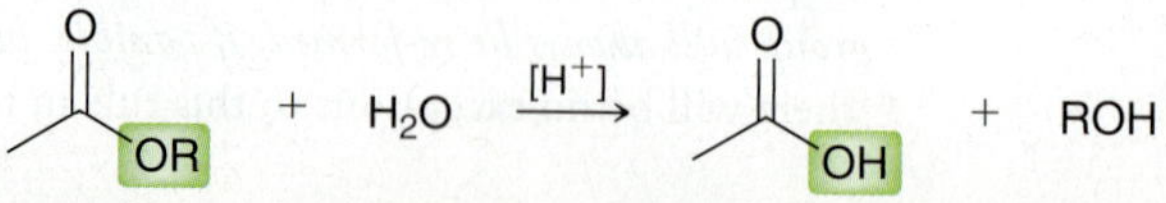

In this reaction, an ester is converted into a carboxylic acid under acid-catalyzed conditions. The nucleophile in this case is water ($H_2O$); however, the first step of the mechanism cannot simply be a nucleophilic attack (Figure 21.6). What is wrong with this step? The tetrahedral intermediate exhibits

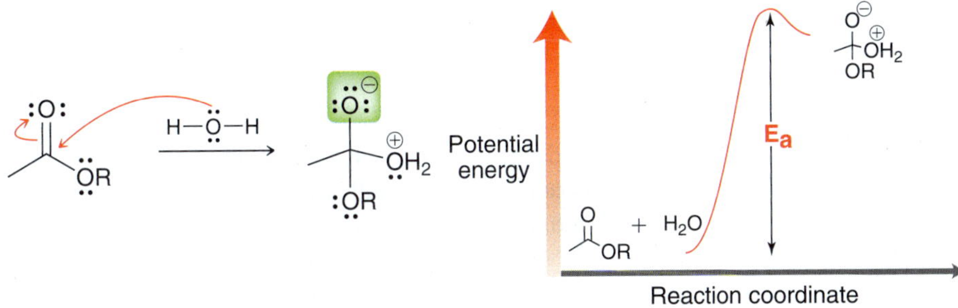


**FIGURE 21.6** An energy diagram showing the large energy of activation associated with water directly attacking an ester.

an oxygen atom with a localized negative charge, which is strongly basic and therefore inconsistent with acidic conditions. The energy diagram in Figure 21.7 illustrates a large energy of activation ($E_a$) for this step. To avoid formation of this intermediate, the acid catalyst is first employed to protonate the carbonyl group (just as we saw in the previous chapter with ketones and aldehydes).

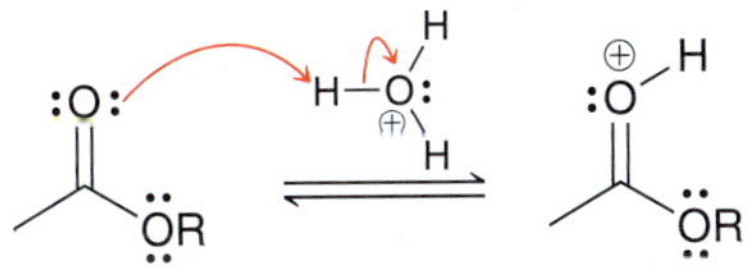

A protonated carbonyl group is significantly more electrophilic, and now when water attacks, no negative charge is formed. This step is now consistent with acidic conditions. As seen in the energy diagram in Figure 21.7, the energy of activation ($E_a$) is now much smaller, because the reactants are already high in energy and no strong base is being formed.

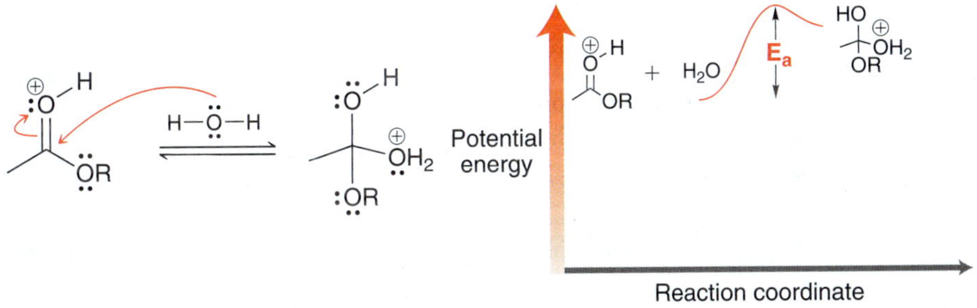

**FIGURE 21.7** An energy diagram showing the small energy of activation associated with water attacking a protonated ester.

Formation of a strong base must only be avoided in acidic conditions. The story is different under basic conditions. For example, consider what happens when an ester is treated with hydroxide ion. In that case, the carbonyl group is not protonated before the nucleophilic attack. In fact, protonation of the carbonyl group would involve formation of a strong acid, which is not consistent with basic conditions. Under basic conditions, hydroxide attacks the carbonyl group directly to give a tetrahedral intermediate. The energy diagram in Figure 21.8 shows that the energy of activation ($E_a$) is not very large, because a negative charge is present in both the reactants and the intermediate. In other words, a negative charge is not being formed but is merely being transferred from one location to another.

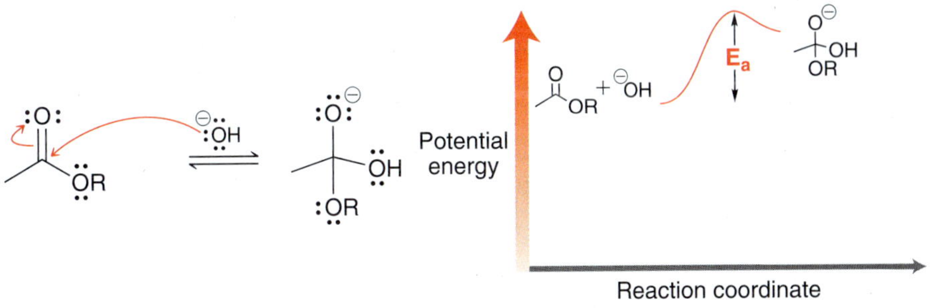

**FIGURE 21.8** An energy diagram showing the small energy of activation associated with hydroxide attacking an ester.

When using an amine as the nucleophile, it is acceptable to attack the carbonyl group directly (without first protonating the carbonyl group).

:O: :NH$_3$ ÖR :Ö: ⊖ ⊕ NH$_3$ :ÖR

This generates an intermediate with both a positive charge and a negative charge, which is OK in this case because amines are sufficiently nucleophilic to attack a carbonyl group directly. Just avoid drawing an intermediate with two of the same kind of charge (two positive charges or two negative charges).

To illustrate the rule further, consider what happens when a tetrahedral intermediate re-forms. In basic conditions, it is acceptable to eject a methoxide ion.

:Ö:⊖ ÖH :ÖMe :O: ÖH + MeÖ:⊖

This is not problematic, because the intermediate already exhibits a negative charge. In other words, a negative charge is not being formed but is merely being transferred from one location to another. However, in acidic conditions, the methoxy group must first be protonated in order to function as a leaving group (to avoid formation of a strong base).

:ÖH ÖH Me Ö: H H ⊕ O H :ÖH ÖH Me O ⊕ H :O ⊕ H ÖH + MeÖH

In summary, proton transfers are utilized in mechanisms in order to remain consistent with the conditions employed.

When drawing the mechanism of a nucleophilic acyl substitution reaction, there are three points where you must decide whether or not to consider a proton transfer.

**Proton transfer** ---- Nucleophilic attack ---- **Proton transfers** ---- Loss of a leaving group ---- **Proton transfer**

A proton transfer can occur (1) before the nucleophilic attack, (2) before loss of the leaving group, or (3) at the end of the mechanism. Some mechanisms will have proton transfers at all three points, while other mechanisms might only exhibit one proton transfer step at the end of the mechanism. As an example, consider the following reaction, which involves the conversion of an acid chloride into a carboxylic acid:

$$\mathrm{RCOCl} \xrightarrow{H_2O} \mathrm{RCOOH} + \mathrm{HCl}$$

In the following mechanism for this reaction, there is no proton transfer step before the nucleophilic attack (i.e., the carbonyl group is not first protonated), because the reagents are not acidic.

Nucleophilic attack    Loss of a leaving group    **Proton transfer**

:O: R :Cl: → H–Ö–H → :Ö:⊖ R :Cl: O⊕ H H → –:Cl:⊖ → :O: R ⊕O H H → H–Ö–H → :O: R ÖH

Similarly, there is no proton transfer step before loss of the leaving group, because the leaving group does not need to be protonated before it can leave. However, there is a proton transfer step at the end of the mechanism, in order to remove a proton and form the final product. The mechanism is comprised of the two core steps followed by a proton transfer. This pattern is typical for reactions of acid halides. We will see nearly a dozen reactions of acid halides, and all of their mechanisms will follow this pattern.

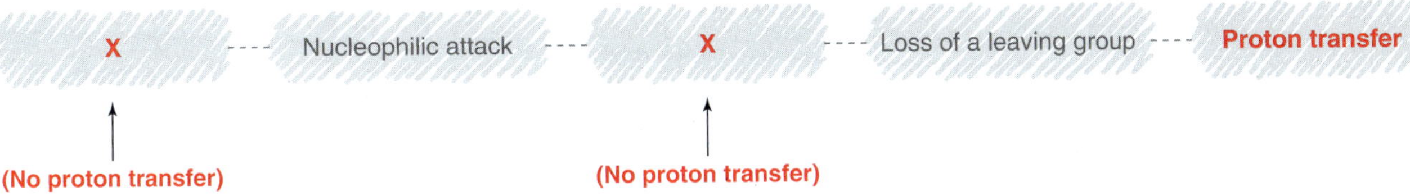


In contrast, reactions performed under acid-catalyzed conditions commonly have proton transfer steps at all three possible points. The following SkillBuilder illustrates such a case.

# SKILLBUILDER

## 21.1 DRAWING THE MECHANISM OF A NUCLEOPHILIC ACYL SUBSTITUTION REACTION

**LEARN** the skill

Propose a plausible mechanism for the following transformation:

(methyl propanoate) OMe $\xrightarrow[\text{EtOH}]{[H^+]}$ (ethyl propanoate) OEt + MeOH

### SOLUTION

In this reaction, a methoxy group is replaced by an ethoxy group:

OMe ⟶ OEt

As with all nucleophilic acyl substitution reactions, we expect the mechanism to have the following two core steps:

Nucleophilic attack ---- Loss of a leaving group

When drawing the mechanism, be sure to draw these steps separately. But we must also determine if any proton transfers are required. There are three points at which proton transfers might occur.

Proton transfer ---- Nucleophilic attack ---- Proton transfers ---- Loss of a leaving group ---- Proton transfer

Acidic conditions are employed in this case, so we must avoid formation of a strong base. This requirement dictates that we must perform proton transfer steps at all three possible points. The mechanism begins with a proton transfer in order to protonate the carbonyl group, rendering it a better electrophile:

:O: ... OMe + H–O⁺(H)–Et ⇌ ⁺:O–H ... OMe

Notice the proton source that we use to protonate the carbonyl group. We cannot use EtOH as the proton source, because transfer of a proton from ethanol would involve creation of an ethoxide ion, which should be avoided in acidic conditions.

The next step is a nucleophilic attack, in which ethanol functions as a nucleophile and attacks the protonated carbonyl group, forming a tetrahedral intermediate that does not bear a negative charge.

The tetrahedral intermediate formed in this step cannot immediately expel methoxide to re-form the carbonyl, because methoxide is a strong base, which should be avoided in acidic conditions. So, we must first protonate the methoxy group. However, protonating the methoxy group would involve the formation of two positive charges, which should also be avoided. As a result, two separate proton transfers are required.

First, a proton is removed to form a new tetrahedral intermediate without a charge, followed by protonation of the methoxy group. Be careful not to use ethoxide as a base in the first step (remember—no strong bases in acidic conditions).

The next step is loss of the leaving group to re-form the carbonyl:

Finally, a proton transfer is used to remove the positive charge and form the product.

In summary, the complete mechanism is shown here:

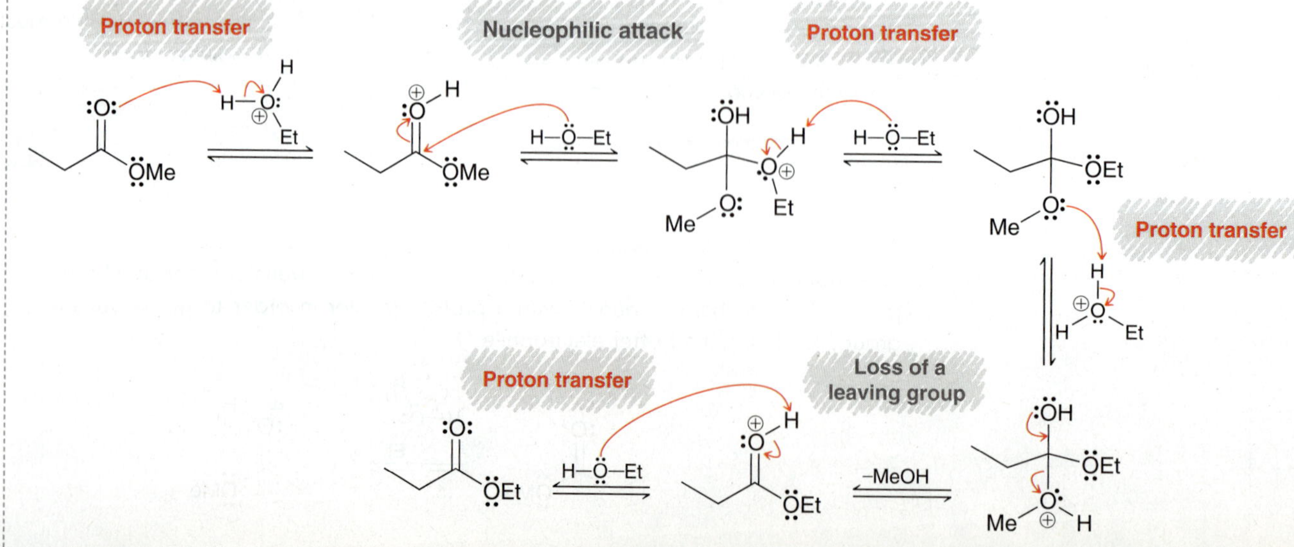

**PRACTICE the skill** **21.14** Propose a plausible mechanism for each of the following transformations. These reactions will all appear later in this chapter, so practicing their mechanisms now will serve as preparation for the rest of this chapter:

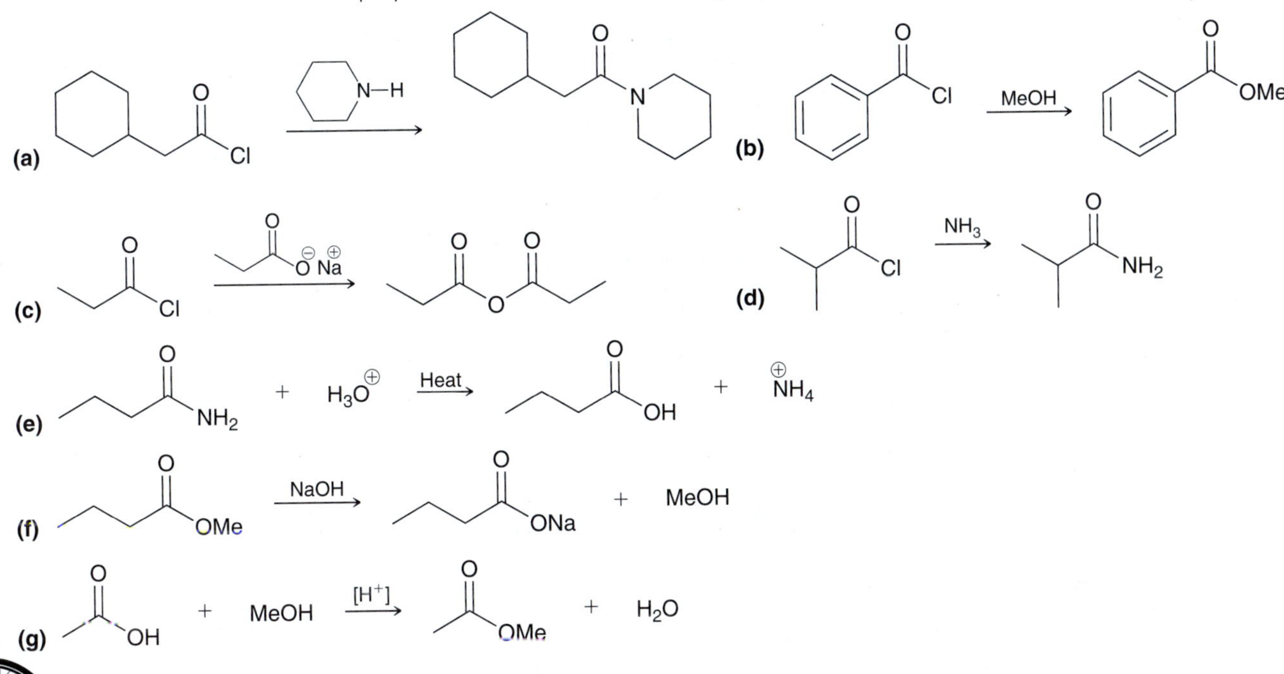


**APPLY the skill** **21.15** Propose a plausible mechanism for the following transformation:

**21.16** Propose a plausible mechanism for the following intramolecular transformation:

**21.17** Propose a plausible mechanism for the following transformation:

need more **PRACTICE?** **Try Problems 21.61, 21.72**

## 21.8 Preparation and Reactions of Acid Chlorides

### Preparation of Acid Chlorides

Acid chlorides can be formed by treating carboxylic acids with thionyl chloride ($SOCl_2$):

$$\text{RCOOH} \xrightarrow{SOCl_2} \text{RCOCl} + SO_2 + HCl$$

The mechanism for this transformation can be divided into two parts (Mechanism 21.2).

## MECHANISM 21.2 PREPARATION OF ACID CHLORIDES VIA THIONYL CHLORIDE

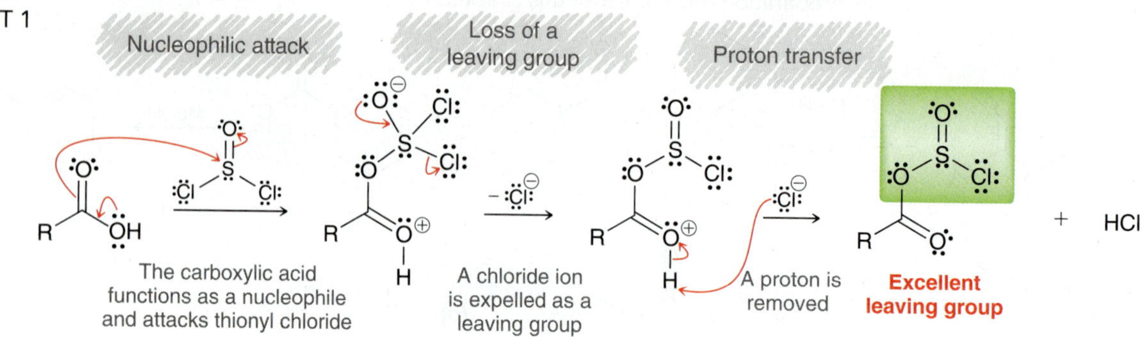


The first part of the mechanism converts the OH group into a better leaving group, which is accomplished in three steps. Each of these three steps should seem familiar if we focus on the chemistry of the S=O bond. In the three steps, the S=O bond behaves very much like a C=O bond of a carboxylic acid derivative (as described in the previous section). First, the S=O bond is attacked by a nucleophile, then it is re-formed to expel a leaving group, and finally a proton transfer is used to remove the charge. Part 2 of the mechanism is a typical nucleophilic acyl substitution, which is accomplished in two steps: nucleophilic attack followed by loss of a leaving group. In this case, the leaving group further degrades to form gaseous $SO_2$. Formation of a gas (which leaves the reaction mixture) forces the reaction to completion.

## Hydrolysis of Acid Chlorides

When treated with water, acid chlorides are hydrolyzed to give carboxylic acids.

$$\mathrm{RCOCl} \xrightarrow{H_2O} \mathrm{RCOOH} + \mathrm{HCl}$$

The mechanism of this transformation requires three steps (Mechanism 21.3).

## MECHANISM 21.3 HYDROLYSIS OF AN ACID CHLORIDE

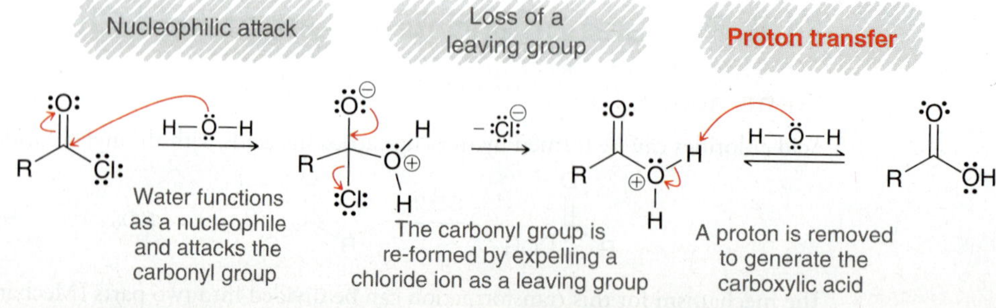

These are the same three steps used in part 1 of the previous mechanism. This reaction produces HCl as a by-product. The HCl can often produce undesired reactions with other functional groups that might be present in the compound, so pyridine is used to remove the HCl as it is produced.

Pyridine + H—Cl: ⟶ N⁺—H :Cl:⁻

Pyridine Pyridinium chloride

Pyridine is a base that reacts with HCl to form pyridinium chloride. This process effectively traps the HCl so that it is unavailable for any other side reactions.

## Alcoholysis of Acid Chlorides

When treated with an alcohol, acid chlorides are converted into esters.

ROH, Pyridine

The mechanism of this transformation is directly analogous to hydrolysis of an acid chloride (three mechanistic steps), and pyridine is used as a base to neutralize the HCl as it is produced. This reaction is viewed from the perspective of the acid chloride, but the same reaction can be written from the perspective of the alcohol.

Pyridine

When shown like this, the OH group is said to undergo acylation, because an acyl group has been transferred to the OH group to produce an ester. This process is sensitive to steric effects, which can be exploited to selectively acylate a primary alcohol in the presence of a secondary (more hindered) alcohol.

Pyridine

## Aminolysis of Acid Chlorides

When treated with ammonia, acid chlorides are converted into amides.

$NH_3$ (two equivalents)

Pyridine is not used in this reaction, because ammonia itself is a sufficiently strong base to neutralize the HCl as it is produced. For this reaction, two equivalents of ammonia are necessary: one for the nucleophilic attack and the other to neutralize the HCl. This reaction also occurs with primary and secondary amines to produce *N*-substituted amides.

$RNH_2$ (two equivalents)

$R_2NH$ (two equivalents)

The mechanism for each of these reactions is directly analogous to the hydrolysis of an acid chloride. There are three steps: (1) nucleophilic attack, (2) loss of a leaving group to re-form the carbonyl, and (3) proton transfer to remove the positive charge. Can you draw the mechanism?

## Reduction of Acid Chlorides

When treated with lithium aluminum hydride, acid chlorides are reduced to give alcohols:

Notice that two separate steps are required. First the acid chloride is treated with LAH, and then the proton source is added to the reaction flask. Water ($H_2O$) can serve as a proton source, although $H_3O^+$ can also be used as a proton source (Mechanism 21.4).

### MECHANISM 21.4 REDUCTION OF AN ACID CHLORIDE WITH LAH

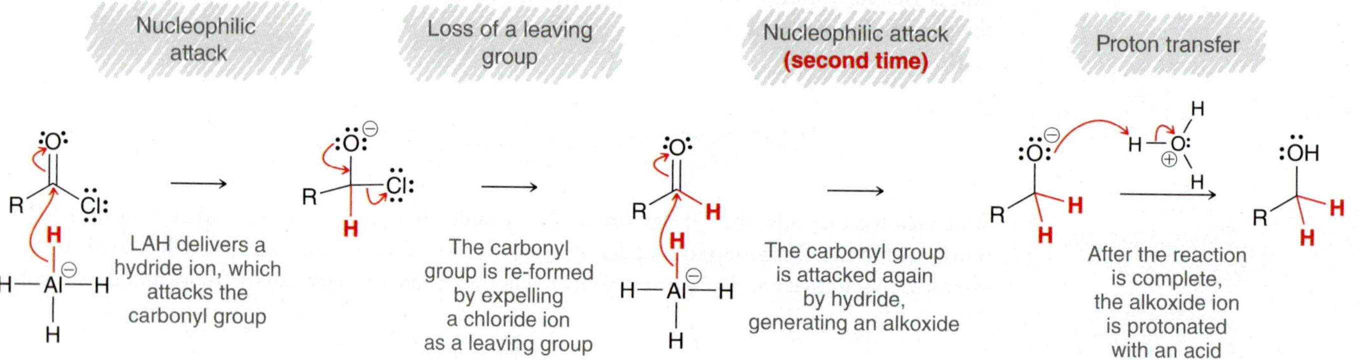


The first two steps of this mechanism are exactly what we might have expected: (1) nucleophilic attack followed by (2) loss of a leaving group to re-form the carbonyl. These two steps produce an aldehyde, which can be attacked again to give an alkoxide ion. Remember that the carbonyl group should always be re-formed, if possible, without expelling $H^-$ and $C^-$. The alkoxide ion produced after the second attack does not have any groups that can leave, and therefore, nothing else can occur until a proton source is provided to protonate the alkoxide ion. This reaction has little practical value, because acid chlorides are generally prepared from carboxylic acids, which can simply be treated directly with LAH to produce the alcohol.

The reaction between an acid chloride and LAH cannot be used to produce an aldehyde. Using one equivalent of LAH simply leads to a mess of products. Producing the aldehyde requires the use of a more selective hydride-reducing agent that will react with acid chlorides more rapidly than aldehydes. There are many such reagents, including lithium tri(*t*-butoxy) aluminum hydride.

**Lithium tri(*t*-butoxy) aluminum hydride**

Three of the four hydrogen atoms have been replaced with *tert*-butoxy groups, which modify the reactivity of the last remaining hydride group. This reducing agent will react with the acid chloride rapidly but will react with the aldehyde more slowly, allowing the aldehyde to be isolated. These conditions can be used to convert an acid chloride into an aldehyde.

$$\text{RCOCl} \xrightarrow[\text{2) } H_2O]{\text{1) } LiAl(OR)_3H} \text{RCHO}$$

## Reactions between Acid Chlorides and Organometallic Reagents

When treated with a Grignard reagent, acid chlorides are converted into alcohols, with the introduction of two alkyl groups.

$$\text{CH}_3\text{COCl} \xrightarrow[\text{2) } H_2O]{\text{1) Excess RMgBr}} \text{CH}_3\text{C(OH)R}_2$$

Just as with LAH, two separate steps are required. First the acid chloride is treated with the Grignard reagent, and then the proton source is added to the reaction flask. Water ($H_2O$) can serve as a proton source, although $H_3O^+$ can also be used (Mechanism 21.5).

### MECHANISM 21.5 THE REACTION BETWEEN AN ACID CHLORIDE AND A GRIGNARD REAGENT

Nucleophilic attack — A Grignard reagent functions as a nucleophile and attacks the carbonyl group

Loss of a leaving group — The carbonyl group is re-formed by expelling a chloride ion as a leaving group

Nucleophilic attack (second time) — The carbonyl group is attacked again by a Grignard reagent, generating an alkoxide

Proton transfer — After the reaction is complete, the alkoxide ion is protonated with an acid

The first two steps of the mechanism are exactly what we would expect: (1) nucleophilic attack followed by (2) loss of a leaving group to re-form the carbonyl. These two steps produce a ketone, which can be attacked again by another Grignard reagent to give an alkoxide ion. Remember that the carbonyl should always be re-formed, if possible, but $H^-$ and $C^-$ should not be expelled. The alkoxide ion produced after the second attack does not have any groups that can leave, so nothing else can occur until a proton source is provided to protonate the alkoxide ion.

The reaction between an acid chloride and a Grignard reagent cannot be used to produce a ketone. Using one equivalent of the Grignard reagent simply leads to a mess of products. Producing the ketone requires the use of a more selective carbon nucleophile that will react with acid chlorides but not with ketones. There are many such reagents. The most commonly used reagent for this purpose is a **lithium dialkyl cuprate**, also called a **Gilman reagent**.

$R_2CuLi$

**Lithium dialkyl cuprate**

The alkyl groups in this reagent are attached to copper rather than magnesium, and their carbanionic character is less pronounced (a C—Cu bond is less polarized than a C—Mg bond). This reagent can be used to convert acid chlorides into ketones with excellent yields. The resulting ketone is not further attacked under these conditions.

$$\mathrm{RC(O)Cl} \xrightarrow{\mathrm{Me_2CuLi}} \mathrm{RC(O)Me}$$

## Summary of Reactions of Acid Chlorides

Figure 21.9 summarizes the reactions of acid chlorides discussed in this section.

Reagents shown: ROH, Pyridine; xs $NH_3$; xs $RNH_2$; xs $R_2NH$; 1) xs LAH 2) $H_2O$; 1) $LiAl(OR)_3H$ 2) $H_2O$; 1) xs RMgBr 2) $H_2O$; $R_2CuLi$; $H_2O$, Pyridine; $SOCl_2$

**FIGURE 21.9**
Reactions of acid chlorides.

## CONCEPTUAL CHECKPOINT

**21.18** Predict the major product(s) for each of the following reactions:

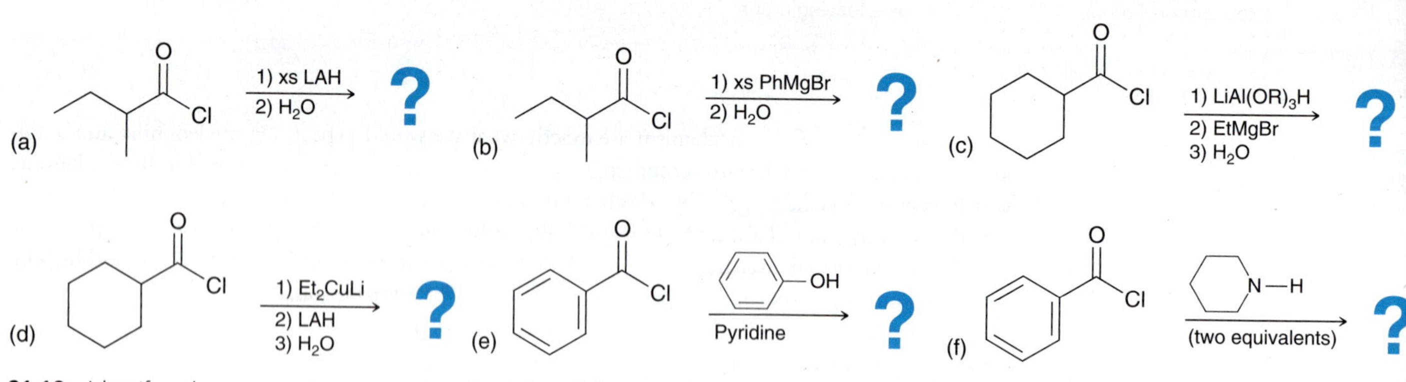


**21.19** Identify the reagents necessary for the following transformation:

**21.20** Propose a mechanism for the following transformation:

1) xs EtMgBr
2) conc. $H_2SO_4$, heat

## 21.9 Preparation and Reactions of Acid Anhydrides

### Preparation of Acid Anhydrides

Carboxylic acids can be converted into acid anhydrides with excessive heating.

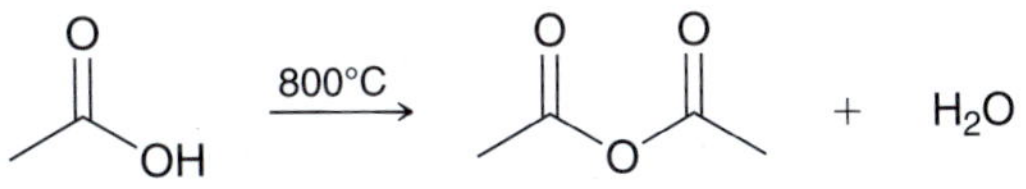

This method is only practical for acetic acid, as most other acids cannot survive the excessive heat. An alternative method for preparing acid anhydrides involves treating an acid chloride with a carboxylate ion, which functions as a nucleophile.

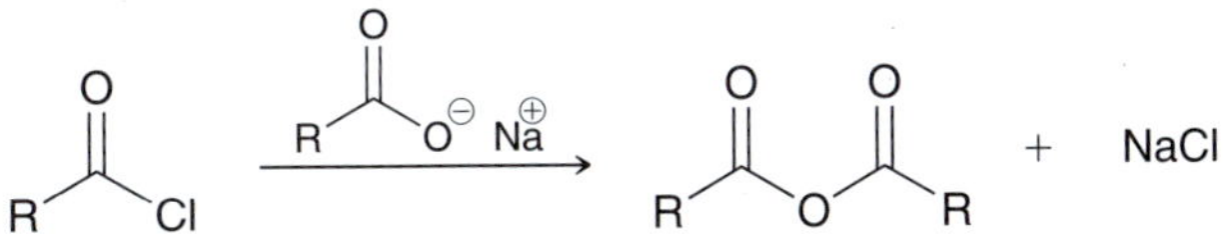

As we might expect, the mechanism of this transformation involves only two steps:

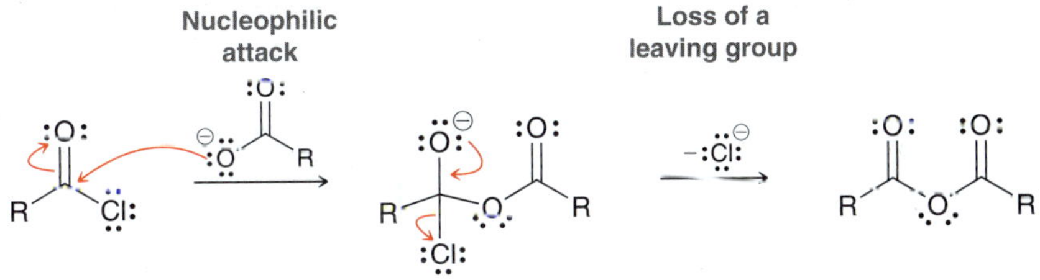

This method can be used to prepare symmetrical or unsymmetrical anhydrides.

### Reactions of Acid Anhydrides

The reactions of anhydrides are directly analogous to the reactions of acid chlorides. The only difference is in the identity of the leaving group.

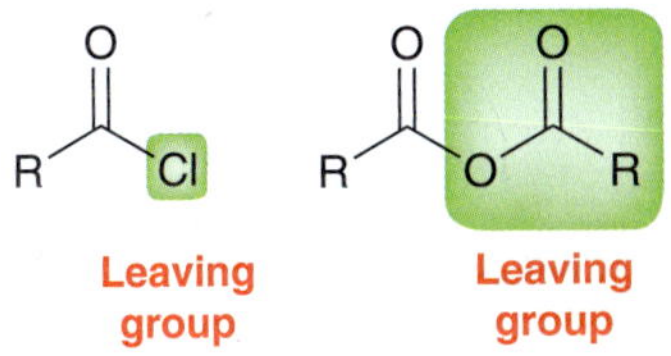

With an acid chloride, the leaving group is a chloride ion, and the by-product of the reaction is therefore HCl. With an acid anhydride, the leaving group is a carboxylate ion, and the by-product is therefore a carboxylic acid. As a result, it is not necessary to use pyridine in reactions with acid anhydrides, because HCl is not produced. Figure 21.10 summarizes the reactions of anhydrides.

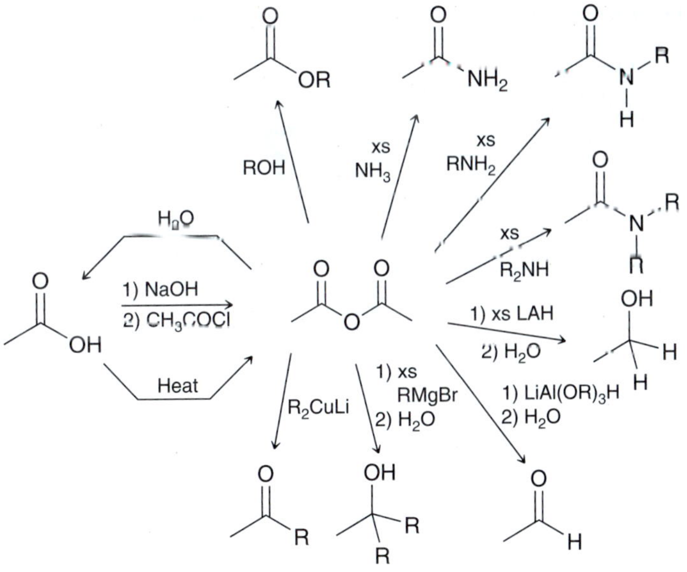

**FIGURE 21.10**
Reactions of acid anhydrides.

Each of these reactions produces acetic acid as a by-product. From a synthetic point of view, the use of anhydrides (rather than acid chlorides) involves the loss of half of the starting material, which is inefficient. For this reason, acid chlorides are more efficient as starting materials than acid anhydrides.

## Acetylation with Acetic Anhydride

Acetic anhydride is often used to acetylate an alcohol or an amine.

R—OH —Acetic anhydride→ R—O(C=O)CH₃

R—NH₂ —Acetic anhydride→ R—NH(C=O)CH₃

These reactions are utilized in the commercial preparation of aspirin and Tylenol.

Aspirin

Tylenol

## CONCEPTUAL CHECKPOINT

**21.21** Predict the major product(s) for each of the following reactions:

(a) ?

(b) (xs) ?

(c) ?

(d) (xs) ?

## medically speaking — How Does Aspirin Work?

Aspirin is prepared from salicylic acid, a compound found in the bark of the willow tree that has been used for its medicinal properties for thousands of years. Aspirin's mechanism of action remained unknown until the early 1970s, when John Vane, Bengt Samuelsson, and Sune Bergström elucidated its role in blocking the synthesis of prostaglandins, for which they were awarded the 1982 Nobel Prize in Physiology or Medicine. Prostaglandins, which are compounds containing five-membered rings, will be discussed in greater detail in Chapter 26. Prostaglandins have many important biological functions, including stimulating inflammation and inducing fever. They are produced in the body from arachidonic acid via a process that is catalyzed by an enzyme called cyclooxygenase:

Arachidonic acid

↓ Cyclooxygenase

PGG$_2$

Prostaglandins

An OH group of cyclooxygenase reacts with aspirin, resulting in the transfer of an acetyl group from aspirin to cyclooxygenase:

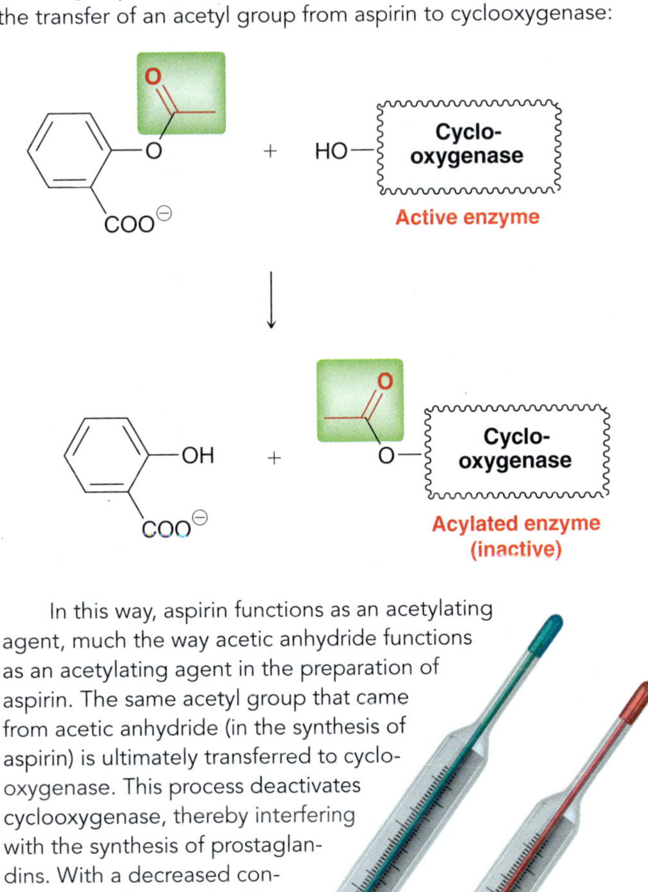


In this way, aspirin functions as an acetylating agent, much the way acetic anhydride functions as an acetylating agent in the preparation of aspirin. The same acetyl group that came from acetic anhydride (in the synthesis of aspirin) is ultimately transferred to cyclooxygenase. This process deactivates cyclooxygenase, thereby interfering with the synthesis of prostaglandins. With a decreased concentration of prostaglandins, the onset of inflammation is slowed and fevers are reduced.

## 21.10 Preparation of Esters

### Preparation of Esters via $S_N2$ Reactions

When treated with a strong base followed by an alkyl halide, carboxylic acids are converted into esters:

$$\mathrm{RCOOH} \xrightarrow[\text{2) } CH_3I]{\text{1) NaOH}} \mathrm{RCOOCH_3}$$

The carboxylic acid is first deprotonated to yield a carboxylate ion, which then functions as a nucleophile and attacks the alkyl halide in an $S_N2$ process. The expected limitations of $S_N2$ processes therefore apply. Specifically, tertiary alkyl halides cannot be used.

## Preparation of Esters via Fischer Esterification

Carboxylic acids are converted into esters when treated with an alcohol in the presence of an acid catalyst. This process is called the **Fischer esterification** (Mechanism 21.6).

$$\mathrm{RCOOH} + \mathrm{MeOH} \xrightleftharpoons{[H^+]} \mathrm{RCOOMe} + H_2O$$

## MECHANISM 21.6 THE FISCHER ESTERIFICATION PROCESS

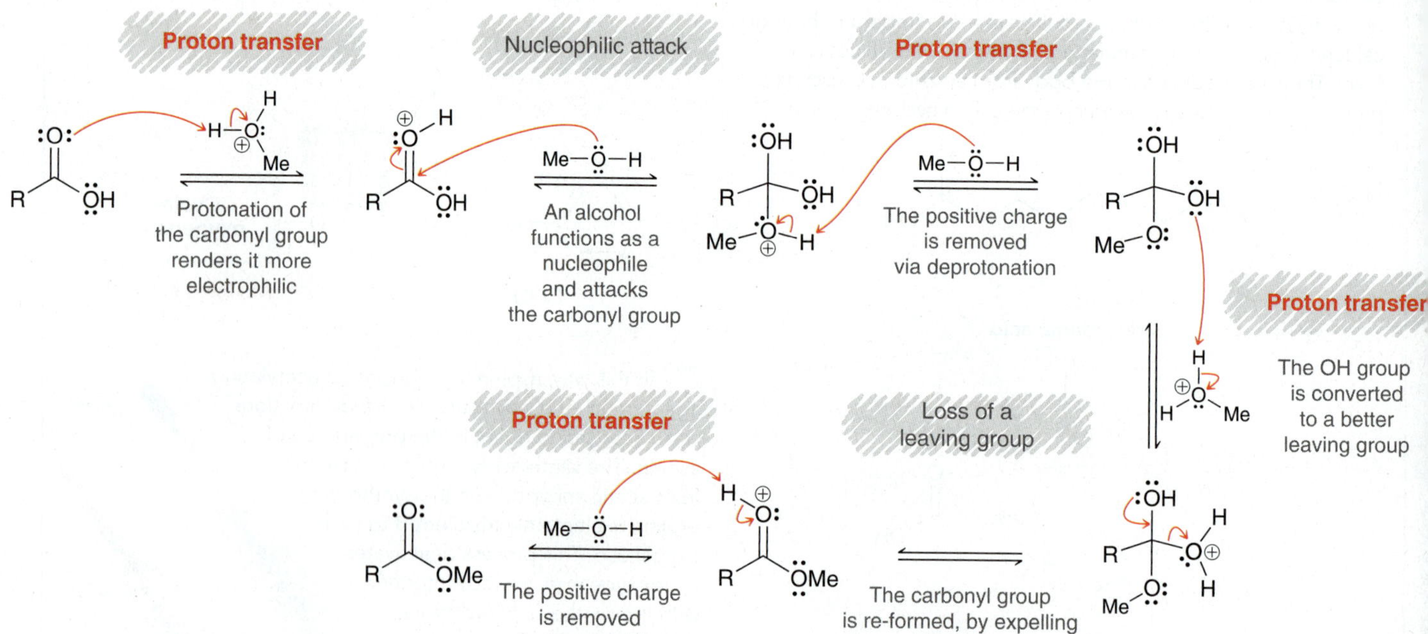


The accepted mechanism is exactly what we would expect for a nucleophilic acyl substitution that takes place under acidic conditions. Evidence for this mechanism comes from isotopic labeling experiments in which the oxygen atom of the alcohol is replaced with a heavier isotope of oxygen ($^{18}O$), and the location of this isotope is tracked throughout the reaction. The location of the isotope in the product (shown in red) supports Mechanism 21.6.

$$\mathrm{RCOOH} + \mathrm{Me\overset{*}{O}H} \xrightleftharpoons{[H^+]} \mathrm{RCO\overset{*}{O}Me} + H_2O \qquad (\overset{*}{O} = {}^{18}O)$$

**LOOKING BACK**
Le Châtelier's principle states that a system at equilibrium will adjust in order to minimize any stress placed on the system.

The Fischer esterification process is reversible and can be controlled by exploiting Le Châtelier's principle. That is, formation of the ester can be favored either by using an excess of the alcohol (i.e., using the alcohol as the solvent) or by removing water from the reaction mixture as it is formed.

$$\mathrm{RCOOH} \xrightleftharpoons[]{[H^+],\ \text{Excess MeOH}} \mathrm{RCOOMe} + H_2O$$

$H_2O$: Can be removed from the mixture

The reverse process, which is conversion of the ester into a carboxylic acid, can be achieved by using an excess of water, as we will see in Section 21.11.

## Preparation of Esters via Acid Chlorides

Esters can also be prepared by treating an acid chloride with an alcohol. We already explored this reaction in Section 21.8.

$$\text{RCOCl} \xrightarrow[\text{Pyridine}]{\text{ROH}} \text{RCOOR}$$

### CONCEPTUAL CHECKPOINT

**21.22** In this section, we have seen three ways to achieve the following transformation. Identify the reagents necessary for all three methods.

Benzoic acid (O, OH) → ? / ? / ? → ethyl benzoate (O, OEt)

**21.23** Identify reagents that can be used to accomplish each of the following transformations:

(a) Benzyl alcohol (OH) → ethyl benzoate (O, OEt)

(b) Styrene → ethyl benzoate (O, OEt)

## 21.11 Reactions of Esters

### Saponification

Esters can be converted into carboxylic acids by treatment with sodium hydroxide followed by an acid. This process is called **saponification** (Mechanism 21.7):

$$\text{RCOOR} \xrightarrow[\text{2) } H_3O^+]{\text{1) NaOH}} \text{RCOOH} + \text{ROH}$$

### MECHANISM 21.7 SAPONIFICATION OF ESTERS

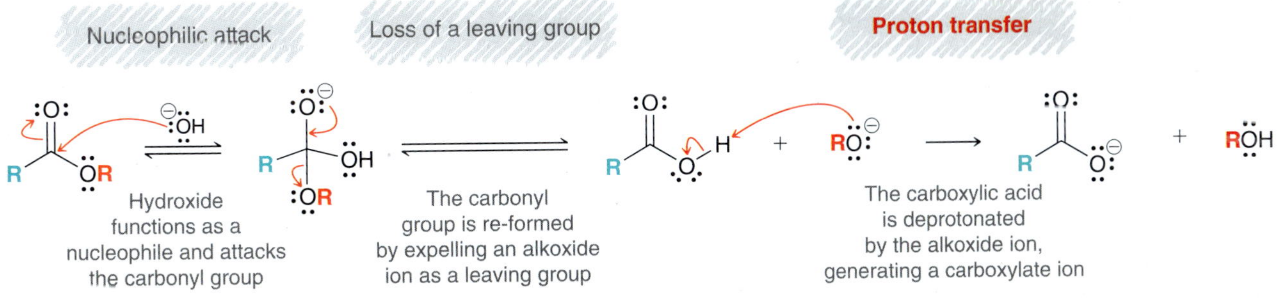

The first two steps of this mechanism are exactly what we would expect of a nucleophilic acyl substitution reaction occurring under basic conditions: (1) nucleophilic attack followed by (2) loss of a leaving group. In basic conditions, an alkoxide ion can function as a leaving group and is not protonated prior to its departure. Although alkoxide ions are not suitable leaving groups in $S_N2$ reactions, they can function as leaving groups in these circumstancess because the tetrahedral intermediate is sufficiently high in energy. The tetrahedral intermediate itself is an alkoxide ion, so expulsion of an alkoxide ion is not uphill in energy.

Under such strongly basic conditions, the carboxylic acid does not survive; it is deprotonated to produce a carboxylate salt. In fact, the formation of a stabilized carboxylate ion is a driving force that pushes the equilibrium to favor formation of products. After the reaction is complete, an acid is required to protonate the carboxylate ion to give the carboxylic acid.

Evidence for this mechanism comes from isotopic labeling experiments in which the oxygen atom of the alcohol is replaced with an oxygen isotope ($^{18}O$) that is tracked throughout the reaction. The location of the isotope in the alcohol by-product supports Mechanism 21.7.

$$\mathrm{R{-}C({=}O){-}\overset{*}{O}R \xrightarrow[2)\ H_3O^+]{1)\ NaOH} R{-}C({=}O){-}OH \;+\; R\overset{*}{O}H} \qquad \left(\overset{*}{O} = {}^{18}O\right)$$

## practically speaking — How Soap Is Made

Recall from Chapter 1 that soaps are compounds that contain a polar group on one end of the molecule and a nonpolar group on the other end:

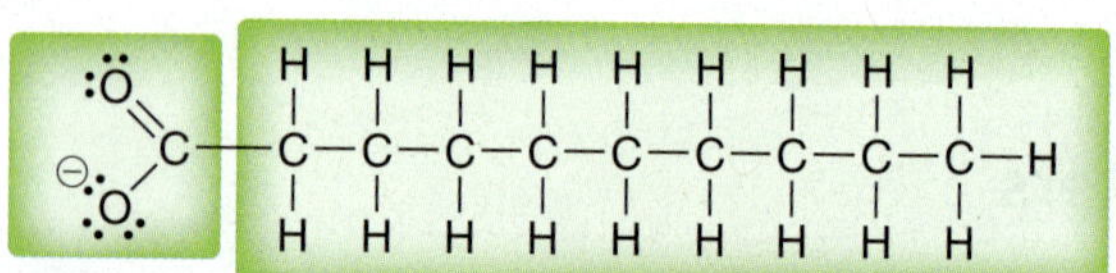

Polar group (hydrophilic) — Non-polar group (hydrophobic)

The hydrophobic tails of soap molecules surround oil molecules, forming a micelle, as described in Section 1.13.

Most soaps are produced from fats and oils, which contain three ester groups. Upon treatment with a strong base, such as sodium hydroxide, the ester groups are hydrolyzed, giving glycerol and three soap molecules:

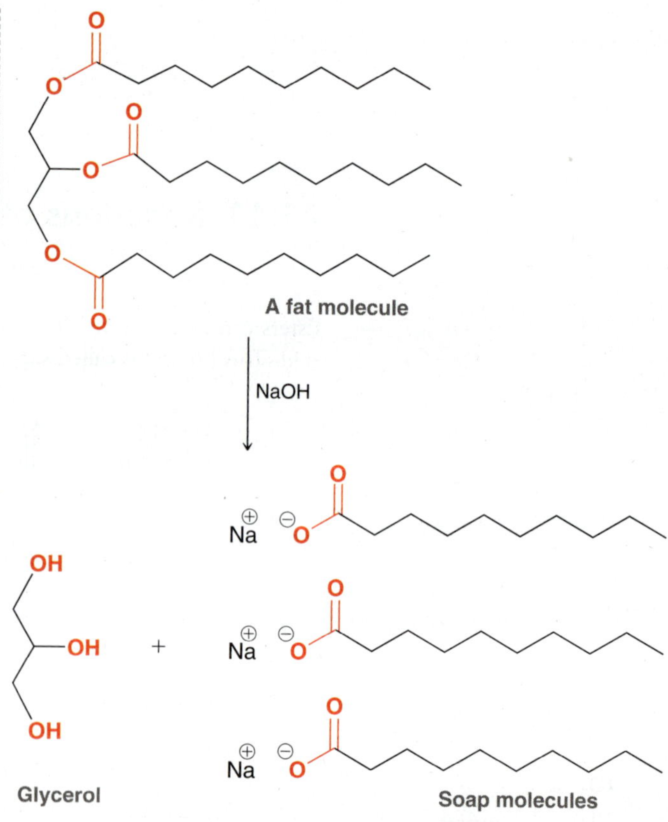


The identity of the alkyl chains can vary depending on the source of the fat or oil, but the concept is the same for all soaps. Specifically, the three ester groups are hydrolyzed under basic conditions to produce soap molecules. This process is called *saponification*, from the Latin word *sapo* (meaning soap).

## Acid-Catalyzed Hydrolysis of Esters

Esters can also be hydrolyzed under acidic conditions.

$$\mathrm{RCO_2Me} \xrightarrow{H_3O^+} \mathrm{RCO_2H} + \mathrm{MeOH}$$

This process (Mechanism 21.8) is the reverse of a Fischer esterification. The mechanism is exactly what we would expect for a nucleophilic acyl substitution that takes place under acidic conditions.

### MECHANISM 21.8 ACID-CATALYZED HYDROLYSIS OF ESTERS

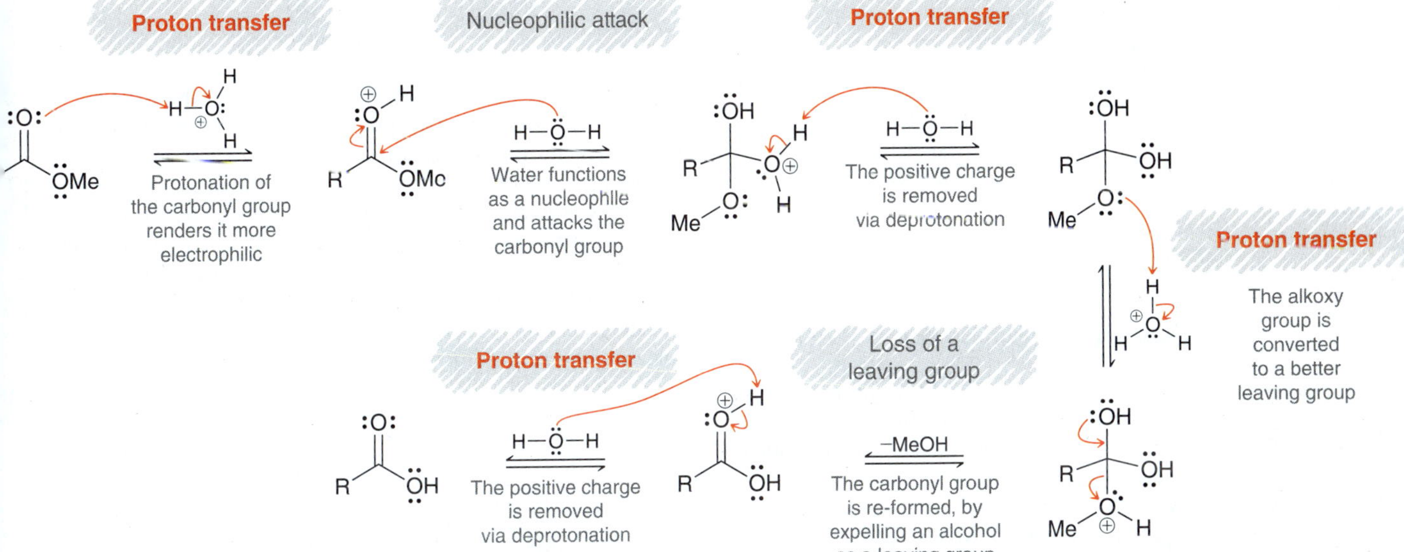


## Aminolysis of Esters

Esters react slowly with amines to yield amides.

$$\mathrm{RCO_2Me} \xrightarrow{NH_3} \mathrm{RCONH_2} + \mathrm{MeOH}$$

This process has little practical utility, because preparation of amides is achieved more efficiently from the reaction between acid chlorides and ammonia or primary or secondary amines.

## Reduction of Esters with Hydride-Reducing Agents

When treated with lithium aluminum hydride, esters are reduced to yield alcohols.

$$\mathrm{RCO_2Me} \xrightarrow[2)\ H_2O]{1)\ \text{Excess LAH}} \mathrm{RCH_2OH} + \mathrm{MeOH}$$

The mechanism for this process is somewhat complex, but the simplified version shown in Mechanism 21.9 will be sufficient for our purposes.

## Esters As Prodrugs

Drugs containing hydroxyl groups are often converted into prodrugs to achieve better absorption properties. For example, consider the structure of epinephrine, which is used in the treatment of glaucoma:

**Epinephrine hydrochloride**

A prodrug form of epinephrine has been developed in which the aromatic hydroxyl groups are acylated to form ester groups. This prodrug is called dipivefrin:

**Dipivefrin hydrochloride**

In this prodrug form, the hydrophobic *tert*-butyl groups enable the compound to cross the nonpolar membrane of the eye more readily. Once the drug reaches the other side of the membrane, it is hydrolyzed to release the active drug (epinephrine).

In fact, esters represent one of the most common types of prodrug, primarily because of the ease with which they are prepared and the ease with which they are hydrolyzed in the body. In addition, the precise ester group that is used in a particular situation can be chosen to control the rate and extent of hydrolysis. In the previous example, the *tert*-butyl groups were specifically chosen because of their steric bulk, which decreases the rate of hydrolysis, achieving the optimal delayed release of the drug.

Ester prodrugs are used for a variety of reasons, not just for increasing the rate of absorption for a drug. For example, consider the structure of the antibiotic agent chloramphenicol:

**Chloramphenicol**

This compound is able to dissolve in the mouth, where it can interact with taste receptors, producing a highly bitter taste. This property limits its usefulness in pediatric liquid suspensions, as children are less likely to drink something that tastes so bad. To avoid this problem, chloramphenicol has been converted to a palmitate ester. This prodrug form has a large, nonpolar tail that prevents it from dissolving in the mouth. As a result, it does not interact with taste receptors and is tasteless. Once it reaches the intestines, enzymes catalyze the hydrolysis of the ester group, releasing the active drug (chloramphenicol).

**Chloramphenicol palmitate**

## MECHANISM 21.9 REDUCTION OF AN ESTER WITH LAH

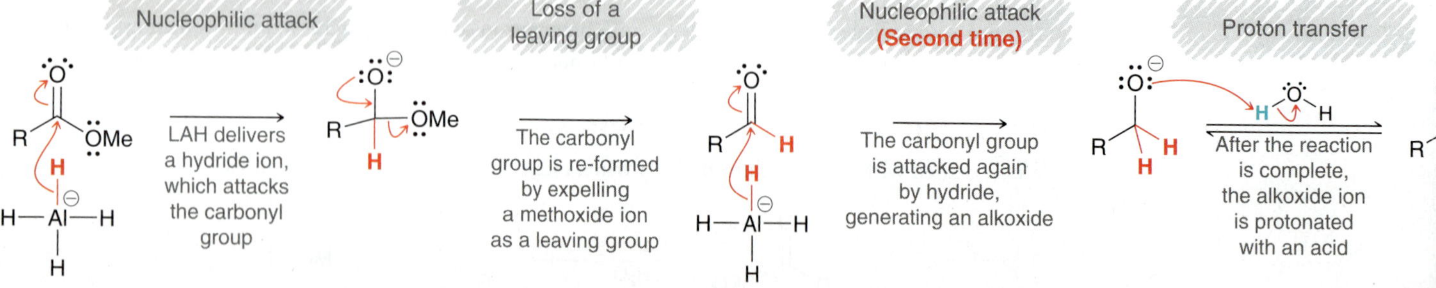

This mechanism is directly analogous to the mechanism for reduction of an acid chloride with LAH. The first equivalent of LAH reduces the ester to an aldehyde, and the second equivalent of LAH reduces the aldehyde to an alcohol. Treating an ester with only one equivalent of LAH is not an efficient method for preparing an aldehyde, because aldehydes are more reactive than esters and will react with LAH immediately after being formed. If the desired product is an aldehyde, then DIBAH can be used as a reducing agent instead of LAH. The reaction is performed at low temperature to prevent further reduction of the aldehyde.

RCOOMe —1) DIBAH, 2) $H_2O$→ RCHO + MeOH

DIBAH = diisobutylaluminum hydride (H–Al with two isobutyl groups)

## Reactions between Esters and Grignard Reagents

When treated with a Grignard reagent, esters are reduced to yield alcohols with the introduction of two alkyl groups.

$CH_3COOMe$ —1) Excess RMgBr, 2) $H_2O$→ $CH_3C(OH)R_2$

This process (Mechanism 21.10) is directly analogous to the reaction between a Grignard reagent and an acid chloride.

## MECHANISM 21.10 THE REACTION BETWEEN AN ESTER AND A GRIGNARD REAGENT

**Nucleophilic attack**

A Grignard reagent functions as a nucleophile and attacks the carbonyl group

**Loss of a leaving group**

The carbonyl group is re-formed by expelling an alkoxide ion as a leaving group

**Nucleophilic attack (Second time)**

The carbonyl group is attacked again by a Grignard reagent, generating an alkoxide

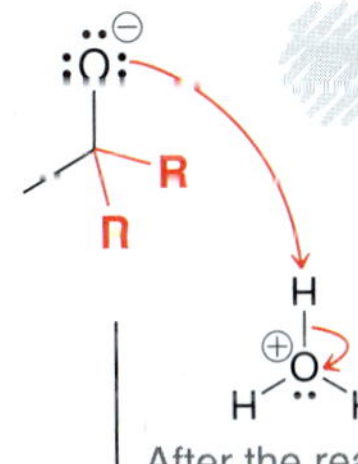

**Proton transfer**

After the reaction is complete, the alkoxide ion is protonated with an acid

### CONCEPTUAL CHECKPOINT

**21.24** Predict the major product(s) for each of the following reactions:

(a) 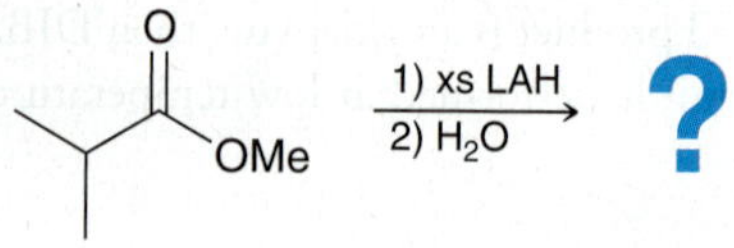 1) xs LAH 2) $H_2O$ ?

(b) 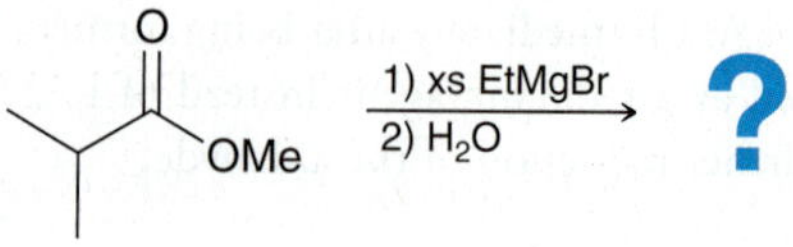 1) xs EtMgBr 2) $H_2O$ ?

(c) 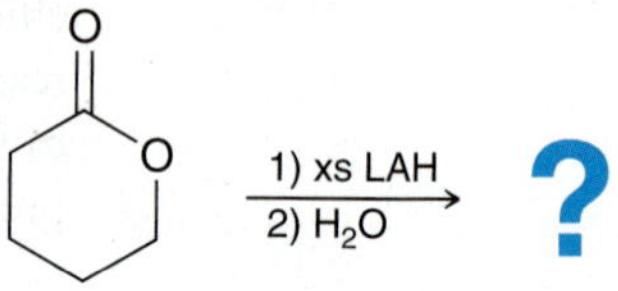 1) xs LAH 2) $H_2O$ ?

(d) 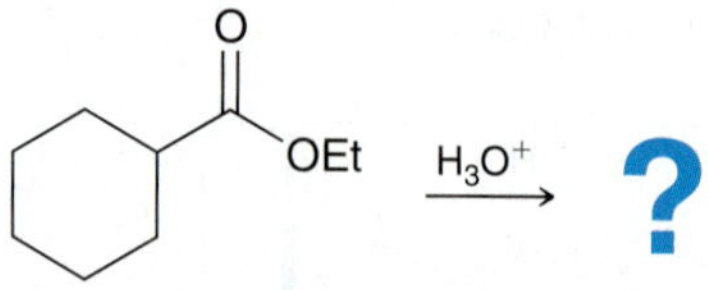 $H_3O^+$ ?

(e) 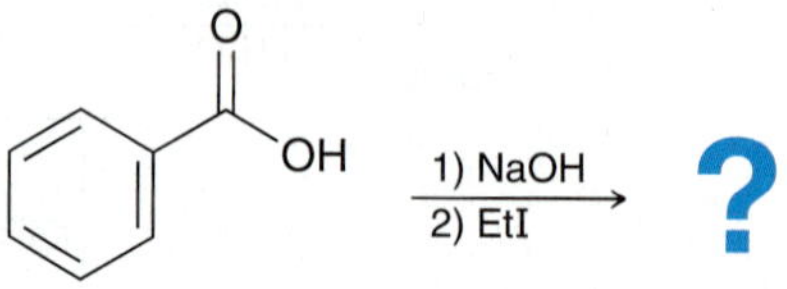 1) NaOH 2) EtI ?

(f) 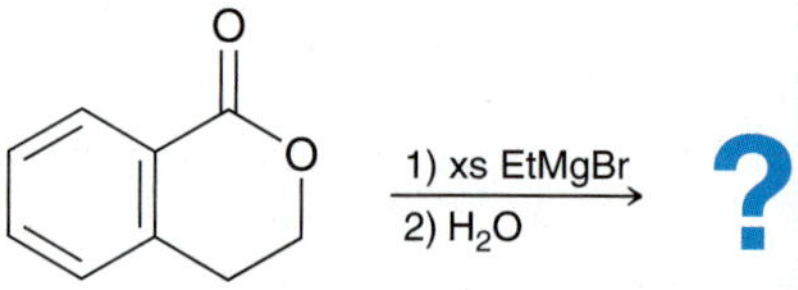 1) xs EtMgBr 2) $H_2O$ ?

**21.25** Propose a mechanism for the following transformation:

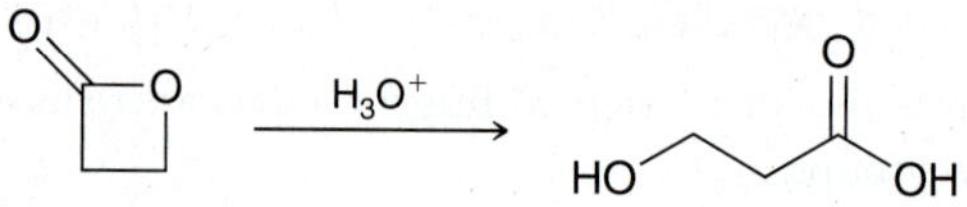

$H_3O^+$

## 21.12 Preparation and Reactions of Amides

### Preparation of Amides

Amides can be prepared from any of the carboxylic acid derivatives discussed earlier in this chapter.

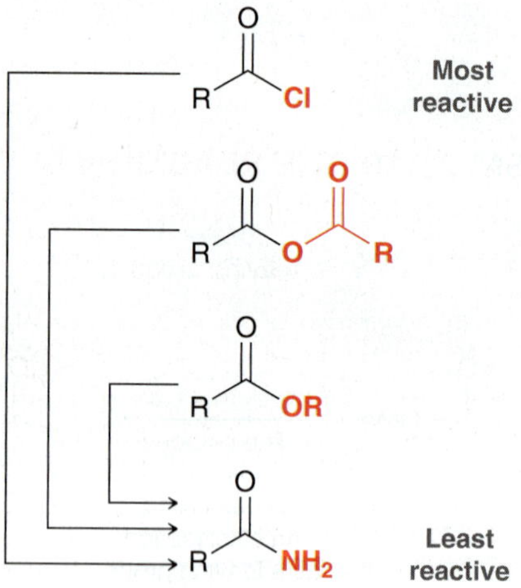

Most reactive

Least reactive

**LOOKING AHEAD** Amides can also be prepared efficiently from carboxylic acids using a reagent called DCC. This reagent and its action are discussed in Section 25.6.

Among these methods, amides are most efficiently prepared from acid chlorides.

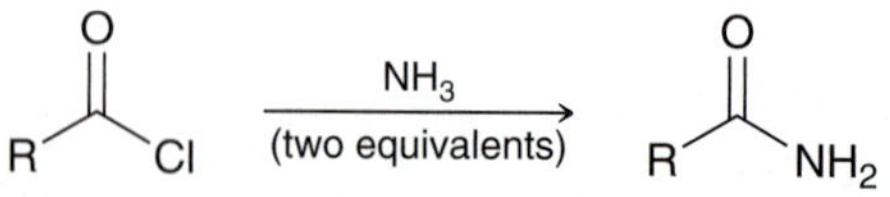

$NH_3$ (two equivalents)

Acid halides are the most reactive of the carboxylic acid derivatives, so the yields are best when an acid chloride is used as a starting material.

practically speaking

## Polyamides and Polyesters

Consider what happens when a diacid chloride and a diamine react together:

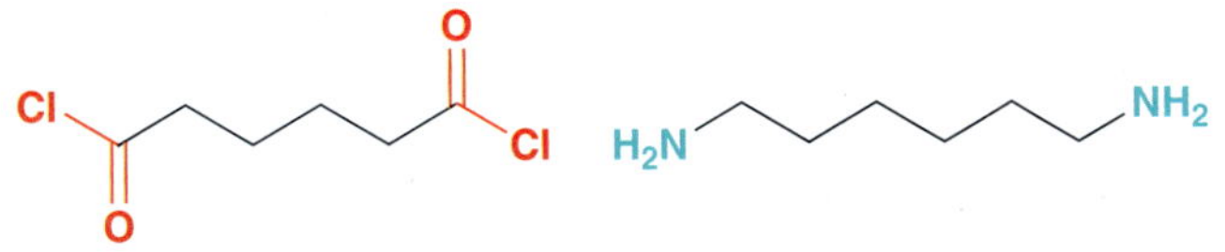

A diacid halide

A diamine

Each molecule has two reactive ends, allowing formation of a polymer:

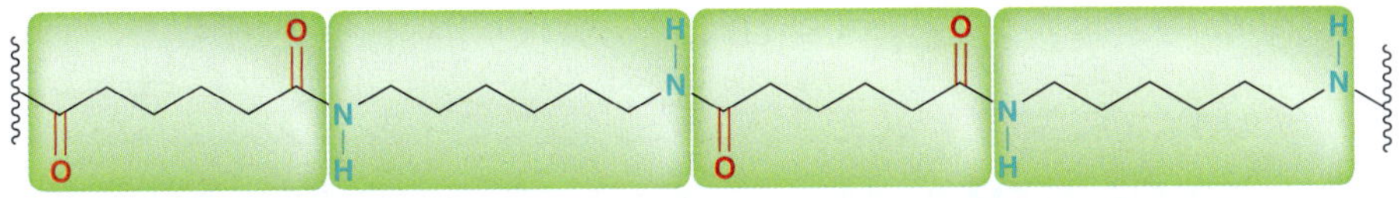

Nylon 6,6

The polymer exhibits multiple amide linkages and is therefore called a polyamide. This specific example is called Nylon 6,6 because it is created from two different compounds that both contain six carbon atoms. Nylon 6,6 was first used by the military to manufacture parachutes, but it quickly became popular as a replacement for silk in the manufacture of clothing.

Polyesters can be made in a similar way. Consider what happens when a diacid reacts with a diol:

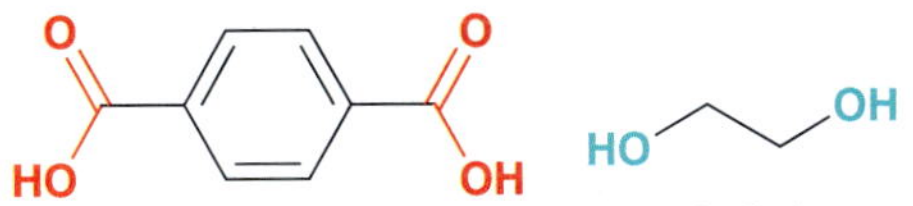

A diacid

A diol

Each molecule has two reactive ends, allowing formation of a polymer:

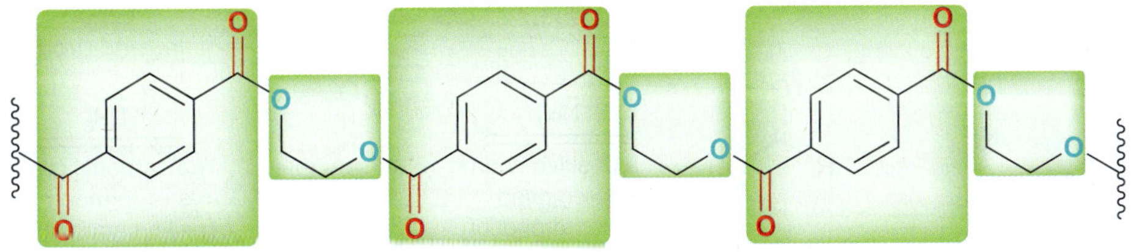

Polyethylene terephthalate (PET)

This polymer, polyethylene terephthalate (PET), exhibits multiple ester linkages and is therefore called a polyester. PET is sold under many trade names, including Dacron and Mylar. It is primarily used in the manufacture of clothing. There are many different kinds of polyamides and polyesters, serving a variety of purposes. Kevlar, for example, is stronger than steel and is used in bulletproof vests:

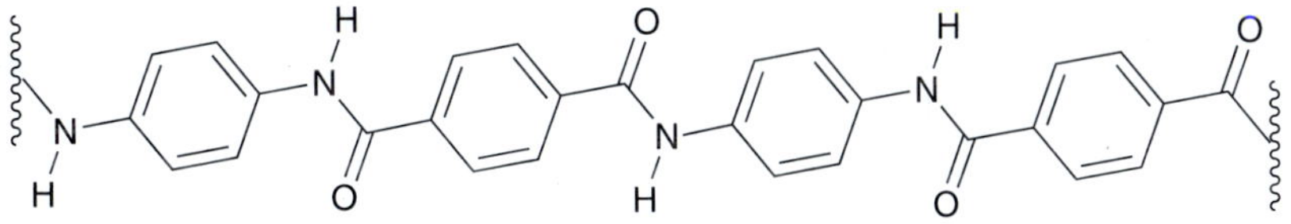

Kevlar

## Acid-Catalyzed Hydrolysis of Amides

Amides can be hydrolyzed to give carboxylic acids in the presence of aqueous acid, but the process is slow and requires heating to occur at an appreciable rate.

$$RC(=O)NH_2 + H_3O^{\oplus} \xrightarrow{\text{Heat}} RC(=O)OH + \overset{\oplus}{N}H_4$$

The mechanism for this transformation (Mechanism 21.11) is directly analogous to the acid-catalyzed hydrolysis of esters (Section 21.11).

In this reaction, notice that an ammonium ion ($NH_4^+$) is formed as a by-product. Since the ammonium ion ($pK_a = 9.2$) is a much weaker acid than $H_3O^+$ ($pK_a = -1.7$), the equilibrium greatly favors formation of products, rendering the process effectively irreversible.

### MECHANISM 21.11 ACID-CATALYZED HYDROLYSIS OF AN AMIDE

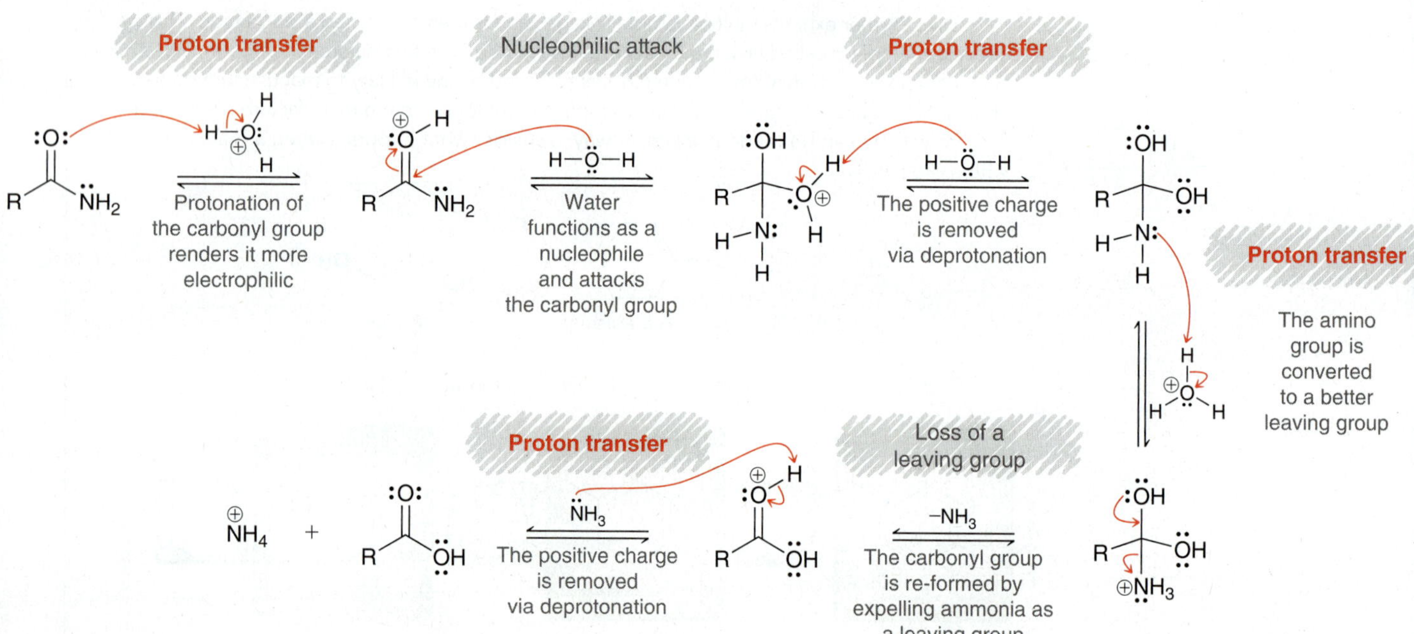


## Hydrolysis of Amides under Basic Conditions

Amides are also hydrolyzed when heated in basic aqueous solutions, although the process is very slow.

$$RC(=O)NH_2 \xrightarrow[\text{2) } H_3O^+]{\text{1) NaOH, heat}} RC(=O)OH$$

The mechanism for this process (Mechanism 21.12) is directly analogous to the saponification of esters (Section 21.11).

## MECHANISM 21.12 HYDROLYSIS OF AMIDES UNDER BASIC CONDITIONS

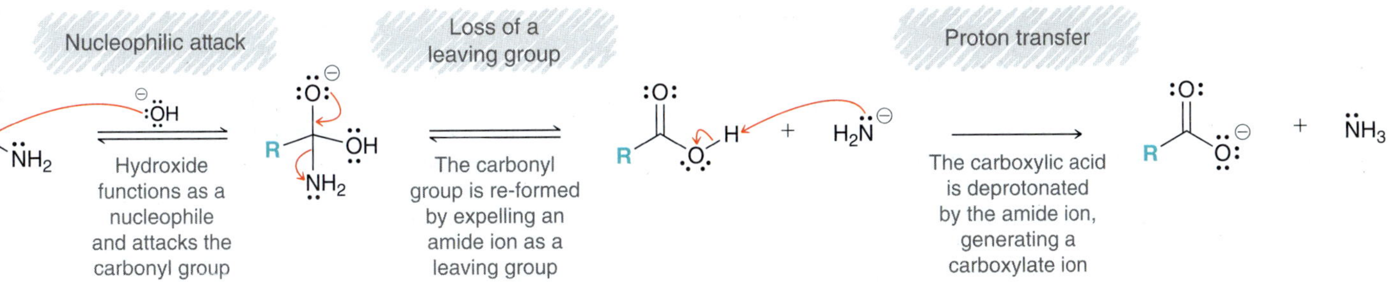


In the last step, formation of the carboxylate ion drives the reaction to completion and renders the process irreversible.

### Reduction of Amides

When treated with excess LAH, amides are converted into amines.

$$\text{RC(=O)NH}_2 \xrightarrow[\text{2) } H_2O]{\text{1) Excess LAH}} \text{RCH}_2\text{NH}_2$$

This is the first reaction we have seen that is somewhat different than the other reactions in this chapter. In this case, the carbonyl group is completely removed.

## CONCEPTUAL CHECKPOINT

**21.26** Predict the major product(s) for each of the following reactions:

(a) Amide $\xrightarrow[\text{2) } H_2O]{\text{1) Excess LAH}}$ ?

(b) Cyclohexanecarbonyl chloride $\xrightarrow{\text{Excess } NH_3}$ ?

(c) Cyclohexanecarboxamide $\xrightarrow[\text{Heat}]{H_3O^+}$ ?

**21.27** Identify the reagents necessary for the following transformation:

Benzoyl chloride → Benzylamine

**21.28** Propose a mechanism for each of the following transformations:

(a) β-Lactam $\xrightarrow[\text{Heat}]{H_3O^+}$ $H_3N^+CH_2CH_2COOH$

(b) β-Lactam $\xrightarrow[\text{Heat}]{\text{NaOH}}$ $H_2NCH_2CH_2COO^- \; Na^+$

## Beta-Lactam Antibiotics

In 1928, Alexander Fleming made a serendipitous discovery that had a profound impact on the field of medicine. He was growing a colony of *Staphylococcus* bacteria in a petri dish that was accidentally contaminated with spores of the *Penicillium notatum* mold. Fleming noticed that the spores of mold prevented the colony from growing, and he surmised that the mold was producing a compound with antibiotic properties, which he called penicillin. Penicillin was initially used to treat wounded soldiers as early as 1943 and shortly thereafter was used on the general population. It has been credited with saving millions of lives, and for his discovery, Fleming was a corecipient of the 1945 Nobel Prize in Physiology or Medicine.

Penicillin was initially believed to be one compound, as Fleming had suggested. However, in 1944, it became apparent that the *P. notatum* mold manufactures many structurally similar compounds, all of which exhibit antibacterial properties. This group of compounds is now said to belong to the penicillin family of drugs and can be represented with the following structural formula, where the R group can vary:

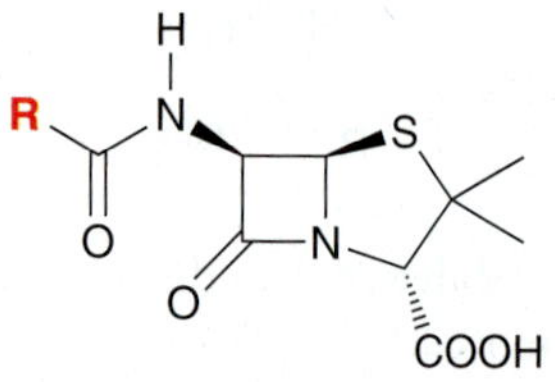

**Penicillin**

Over a dozen penicillin drugs are currently in clinical use, two of which are shown here:

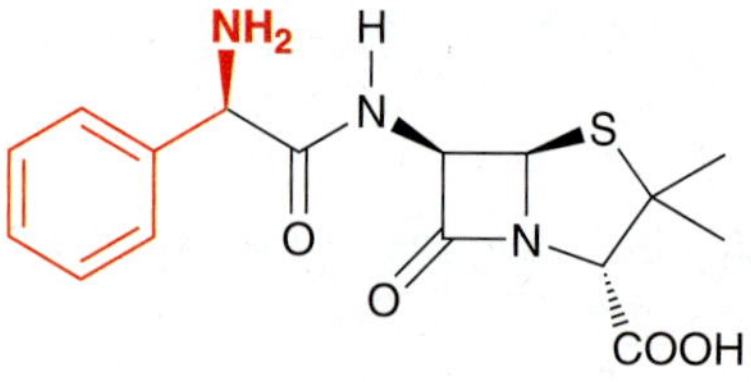

**Ampicillin**

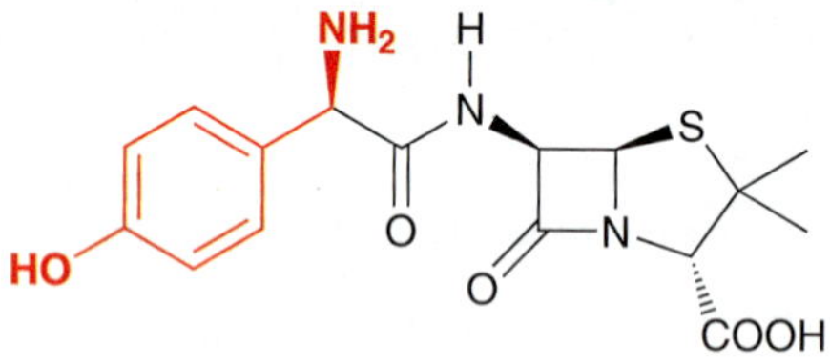

**Amoxicillin**

The key structural feature common to all penicillin drugs is an amide group contained in a four-membered ring, called a beta (β) lactam ring.

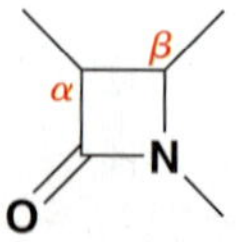

Amides are generally stable—they resist hydrolysis under most conditions. However β-lactams are particularly susceptible to hydrolysis because of the ring strain of the four-membered ring. Hydrolysis opens the ring and releases the ring strain. It is believed that the β-lactam ring reacts with *transpeptidase*, an enzyme that bacteria utilize in building their cell walls:

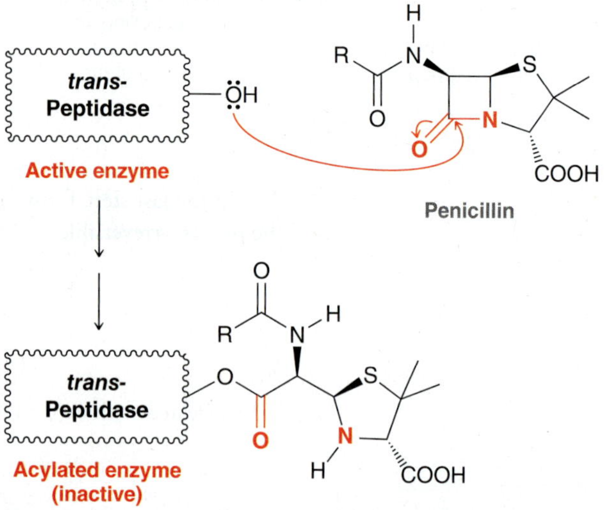

**Active enzyme** **Penicillin**

**Acylated enzyme (inactive)**

This reaction effectively acylates an OH group at the active site of the enzyme, and the acylated enzyme is inactive. By inactivating this enzyme, penicillin drugs are able to prevent bacteria from producing functional cell walls. Under these conditions, the bacteria are not able to reproduce, which allows the body's natural immune system to take control.

Some bacteria are resistant to penicillin drugs. These bacteria produce enzymes, called β-lactamases, that are capable of prematurely hydrolyzing the β-lactam ring before it has a chance to react with transpeptidase:

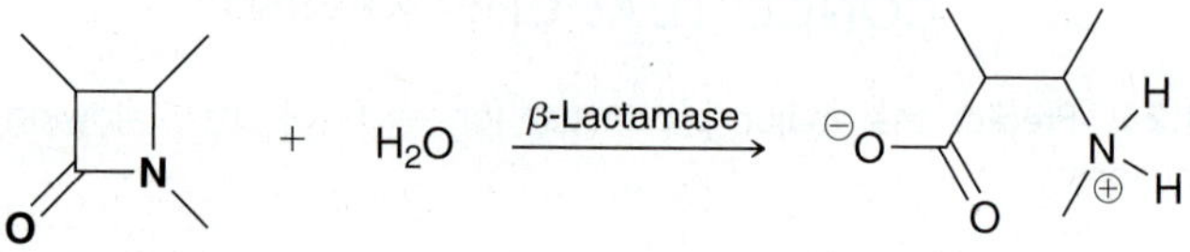

Once the β-lactam ring has been opened, the drug no longer has any antibiotic properties.

Cephalosporins are another family of antibiotics closely related in structure to penicillins:

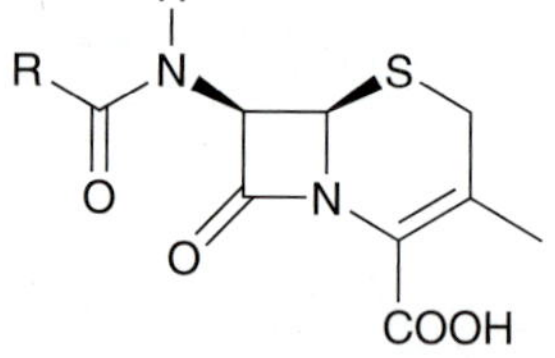

Cephalosporins also exhibit a β-lactam ring, but the lactam ring is fused to a six-membered ring. The cephalosporin family of antibiotics was first isolated from *Cephalosporium* fungi in 1945.

Extensive research is currently underway to identify other novel β-lactam antibiotics with superior properties. Over the decades to come, the list of known antibiotics is sure to grow.

## 21.13 Preparation and Reactions of Nitriles

### Preparation of Nitriles via $S_N2$ Reactions

Nitriles can be prepared by treating an alkyl halide with a cyanide ion.

$$RCH_2Br \xrightarrow{NaCN} RCH_2C{\equiv}N + NaBr$$

This process proceeds via an $S_N2$ mechanism, so tertiary alkyl halides cannot be used.

### Preparation of Nitriles from Amides

Nitriles can also be prepared via the dehydration of an amide. Many reagents can be used to accomplish the transformation. One such reagent is thionyl chloride ($SOCl_2$).

$$RC(=O)NH_2 \xrightarrow{SOCl_2} R{-}C{\equiv}N + SO_2 + 2\ HCl$$

This process (Mechanism 21.13) is useful for preparing tertiary nitriles, which cannot be prepared via an $S_N2$ process.

### MECHANISM 21.13 DEHYDRATION OF AMIDES

Nucleophilic attack: The amide functions as a nucleophile and attacks thionyl chloride

Loss of a leaving group: $-Cl^{\ominus}$ Chloride is expelled as a leaving group

Proton transfer: :Base The positive charge on the nitrogen atom is removed via deprotonation

Proton transfer: :Base Elimination of a proton and a leaving group affords the product

$R{-}C{\equiv}N:$ $\quad SO_2 \quad :\ddot{Cl}:^{\ominus}$

### Hydrolysis of Nitriles

In aqueous acidic conditions, nitriles are hydrolyzed to afford amides, which are then further hydrolyzed to yield carboxylic acids.

$$R{-}C{\equiv}N \xrightarrow[\text{heat}]{H_3O^+} RC(=O)NH_2 \xrightarrow[\text{heat}]{H_3O^+} RC(=O)OH + \overset{\oplus}{N}H_4$$

Formation of the amide occurs via Mechanism 21.14, and conversion of the amide into the carboxylic acid was discussed earlier (Mechanism 21.11).

Alternatively, nitriles can also be hydrolyzed in aqueous base.

$$R{-}C{\equiv}N \xrightarrow[\text{2) } H_3O^+]{\text{1) NaOH, } H_2O} RC(=O)OH$$

Once again, the nitrile is first converted to an amide (Mechanism 21.15), which is then converted to a carboxylic acid (see Mechanism 21.12).

## MECHANISM 21.14 ACID-CATALYZED HYDROLYSIS OF NITRILES

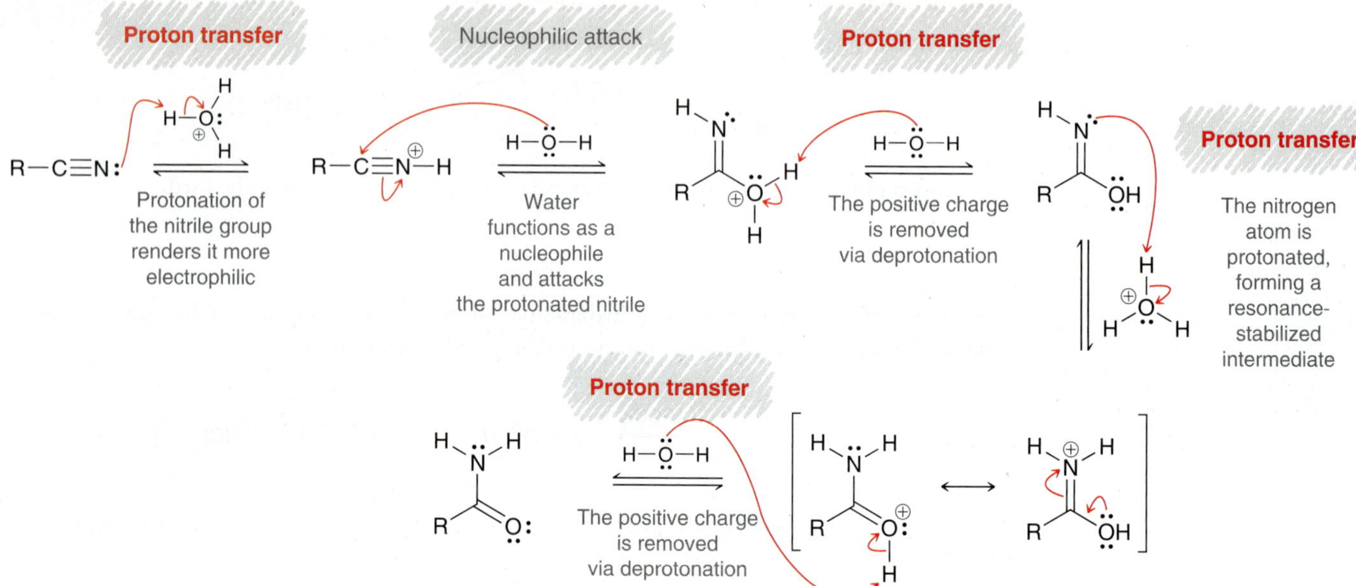

## MECHANISM 21.15 BASE-CATALYZED HYDROLYSIS OF NITRILES

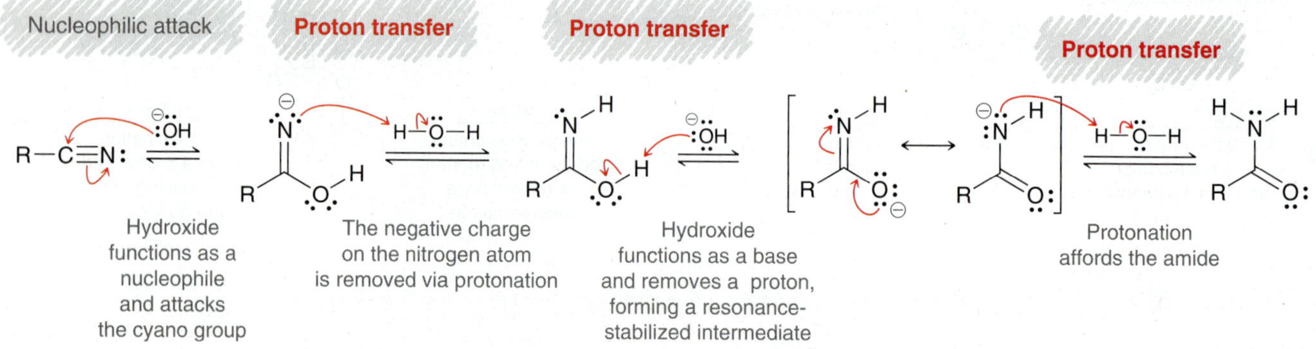


### Reactions between Nitriles and Grignard Reagents

A ketone is obtained when a nitrile is treated with a Grignard reagent, followed by aqueous acid.

$$\text{R—C}\equiv\text{N} \xrightarrow[\text{2) } H_3O^+]{\text{1) RMgBr}} \text{R—C(=O)—R}$$

The Grignard reagent attacks the nitrile, much like it attacks a carbonyl group.

$$\text{R—C}\equiv\text{N:} \;\; {}^{\ominus}\text{:R} \longrightarrow \text{R}_2\text{C=N:}^{\ominus}$$

The resulting anion is then treated with aqueous acid to give an imine, which is then hydrolyzed to a ketone under acidic conditions (see Section 20.6).

### Reduction of Nitriles

At the beginning of this chapter, we saw that carboxylic acids can be reduced to alcohols by treatment with LAH. Similarly, nitriles are converted to amines when treated with LAH.

$$R-C\equiv N \xrightarrow[\text{2) } H_2O]{\text{1) xs LAH}} RCH_2NH_2$$

## CONCEPTUAL CHECKPOINT

**21.29** Predict the major product(s) for each of the following reactions:

(a) 2-methylbutanenitrile: 1) xs LAH 2) $H_2O$ → ?

(b) Benzyl bromide: 1) NaCN 2) MeMgBr 3) $H_3O^+$ → ?

(c) Cyclohexanecarbonitrile: 1) EtMgBr 2) $H_3O^+$ 3) LAH 4) $H_2O$ → ?

(d) Cyclohexanecarbonitrile: $H_3O^+$, heat → ?

**21.30** Identify the reagents necessary for each of the following transformations:

(a) Benzoyl chloride → benzonitrile

(b) Benzyl bromide → 3-phenylpropanoic acid

**21.31** Propose a mechanism for the following transformation:

CN-substituted alkane $\xrightarrow[\text{Heat}]{H_3O^+}$ corresponding amide ($CONH_2$)

## 21.14 Synthesis Strategies

Recall from Chapter 12 that there are two primary considerations when approaching a synthesis problem: (1) a change in the carbon skeleton and (2) a change in the functional groups.

Let's focus on each of these issues separately, beginning with functional groups.

### Functional Group Interconversions

In this chapter, we have seen many different reactions that change the identity of a functional group without changing its location. Figure 21.11 summarizes how functional groups can be interconverted.

The figure shows many reactions, but there are a few key aspects of the diagram that are worth special attention:

- The figure is organized according to oxidation state. Carboxylic acids and their derivatives all have the same oxidation state and are shown at the top of the figure (for ease of presentation, carboxylic acids are shown above the derivatives but do have the same oxidation state). Aldehydes exhibit a lower oxidation state and are therefore shown below the carboxylic acid derivatives. Alcohols and amines are at the bottom of the figure, since they have the lowest oxidation state.

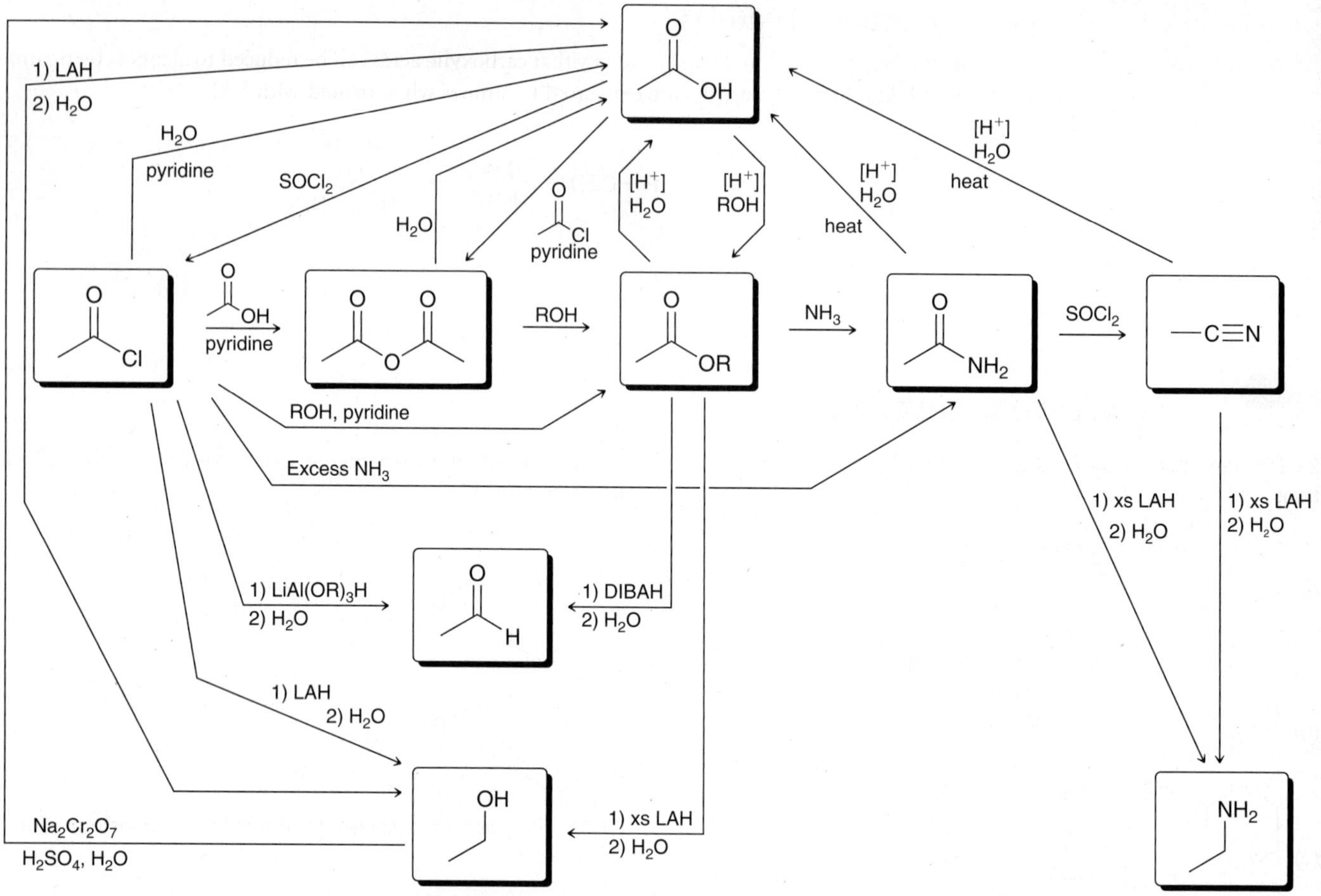

1) LAH
2) $H_2O$
$H_2O$
pyridine
$SOCl_2$
$H_2O$
pyridine
[$H^+$]
$H_2O$
[$H^+$]
ROH
[$H^+$]
$H_2O$
heat
[$H^+$]
$H_2O$
heat
OH
pyridine
ROH
$NH_3$
$SOCl_2$
ROH, pyridine
Excess $NH_3$
1) xs LAH
2) $H_2O$
1) xs LAH
2) $H_2O$
1) $LiAl(OR)_3H$
2) $H_2O$
1) DIBAH
2) $H_2O$
1) LAH
2) $H_2O$
$Na_2Cr_2O_7$
$H_2SO_4$, $H_2O$
1) xs LAH
2) $H_2O$

**FIGURE 21.11**
A summary of reactions from this chapter that enable the interconversion of functional groups.

- Among the carboxylic acid derivatives, the reactions are shown from left to right, from most reactive to least reactive. That is, acid chlorides (leftmost) are the most reactive and are most readily converted into the other derivatives:

Pyridine
ROH
$NH_3$
$SOCl_2$
ROH, pyridine
Excess $NH_3$

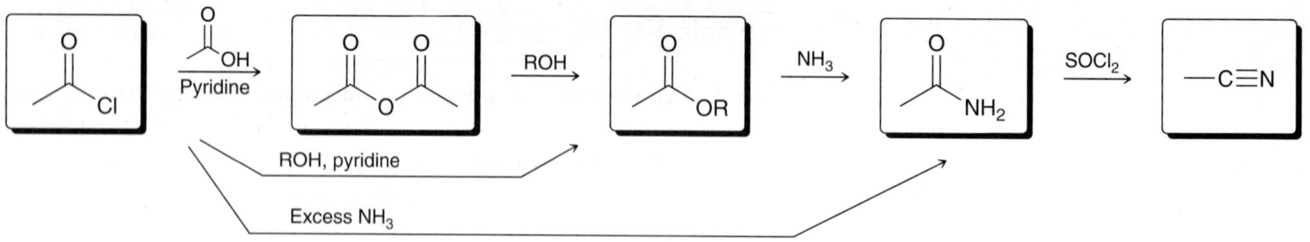

Going from right to left—for example, converting an ester into an acid chloride—requires two steps: first hydrolysis to form a carboxylic acid, then conversion of the acid into an acid chloride.

- We did not see a method for directly converting a carboxylic acid into either an amide or a nitrile. All other carboxylic acid derivatives (acid chlorides, acid anhydrides, and esters) can be made directly from a carboxylic acid.
- We have only seen two ways to make amines in this chapter (Chapter 24 will cover many other ways for making amines).

# SKILLBUILDER

## 21.2 INTERCONVERTING FUNCTIONAL GROUPS

**LEARN the skill** Propose an efficient synthesis for the following transformation:

### SOLUTION

This problem requires the transformation of an amide into an acid chloride. Amides are less reactive than acid chlorides, so this transformation requires a two-step process. The first step is hydrolysis of the amide to form a carboxylic acid:

$H_3O^+$, Heat

The second step is to convert the carboxylic acid into the desired acid chloride:

$SOCl_2$

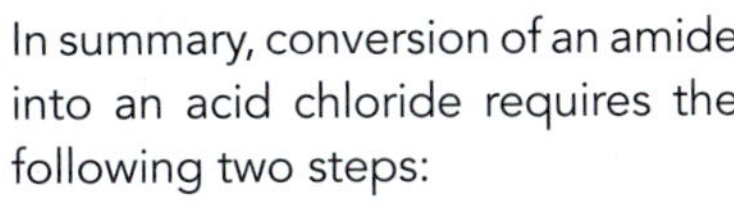

In summary, conversion of an amide into an acid chloride requires the following two steps:

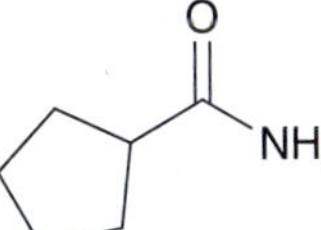

1) $H_3O^+$, heat
2) $SOCl_2$

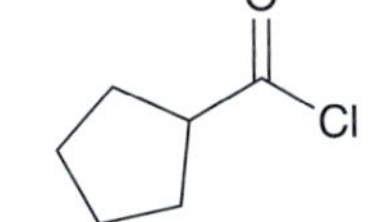

**PRACTICE the skill** **21.32** Identify reagents that will accomplish each of the following transformations:

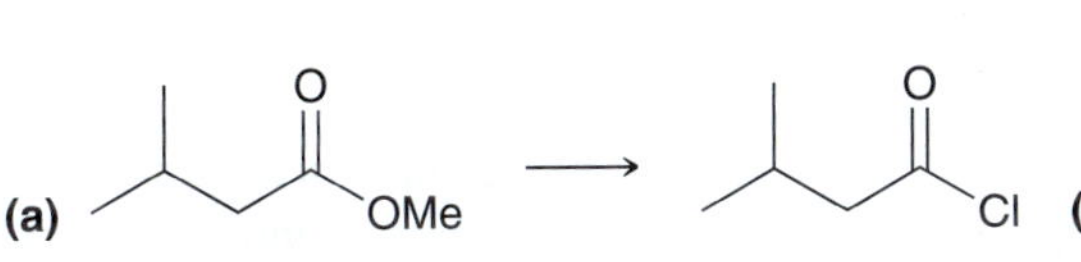

(a) (b) (c) (d) (e) (f) (g) (h) (i) (j)

**21.33** Using only reactions covered in this chapter (Figure 21.11), identify the minimum number of steps required to convert a primary alcohol into an amine (in Chapter 24, we will learn more efficient methods for achieving this type of transformation).

**21.34** Propose an efficient method for converting 1-hexene into hexanoyl chloride. You will need to use reactions from previous chapters at the beginning of your synthesis.

need more **PRACTICE?** **Try Problems 21.45a,b, 21.46a, 21.52, 21.53a,c,e, 21.57**

## C—C Bond-Forming Reactions

This chapter covered five new C—C bond-forming reactions, which can be placed into two categories (Table 21.2). In the first category (left), the location of the functional group remains unchanged. In each of the three reactions in this category, the product exhibits a functional group in the same location as the reactant.

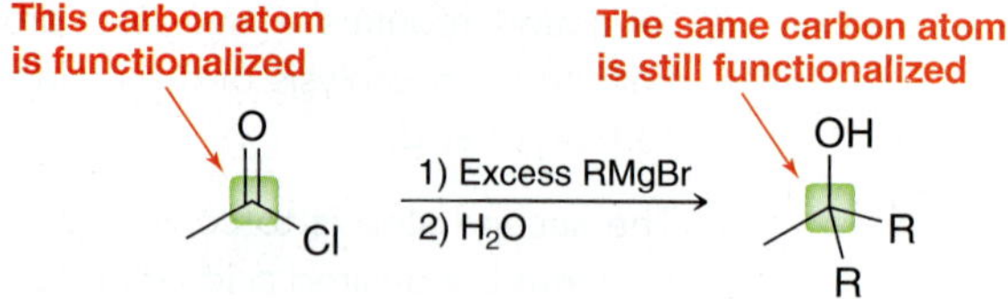


In the second category (right side of Table 21.2), the position of the functional group changes.

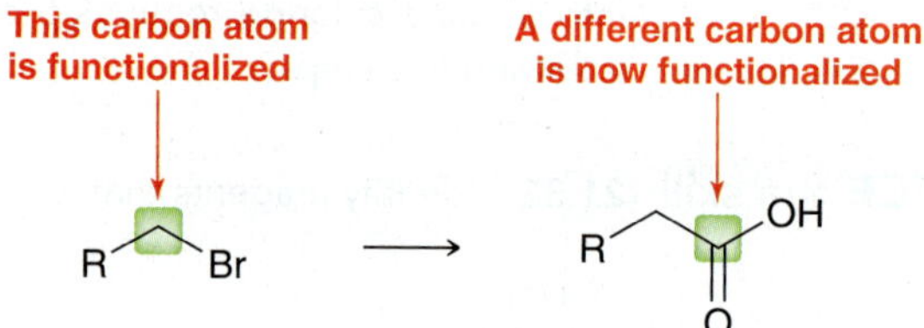


In this case, installation of the extra carbon atom is accompanied by a change in the location of the functional group. Specifically, the functional group moves to the newly installed carbon atom. In your mind, you should always categorize C—C bond-forming reactions in terms of the ultimate location of the functional group. It will be useful to keep this in mind when planning a synthesis, and it will be especially important in Chapter 22, which focuses on many C—C bond-forming reactions.

**TABLE 21.2 TWO KINDS OF C—C BOND-FORMING REACTIONS COVERED IN THIS CHAPTER**

| C—C Bond-Forming Reactions for Which the Functional Group Remains in the Same Location | C—C Bond-Forming Reactions Involving a Change in the Location of the Functional Group |
|---|---|
| Acyl derivative (Z) → 1) Excess RMgBr, 2) $H_2O$ → tertiary alcohol (OH, R, R) | Alkyl bromide (Br) → NaCN → nitrile (C≡N) → $H_3O^+$, heat → carboxylic acid |
| Acid chloride (Cl) → $R_2CuLi$ → ketone (R) | Alkyl bromide (Br) → 1) Mg, 2) $CO_2$, 3) $H_3O^+$ → carboxylic acid (OH, O) |
| Nitrile (—C≡N) → 1) RMgBr, 2) $H_3O^+$ → ketone (R) | |

In Section 14.12, we also saw a method for installing an alkyl chain while simultaneously moving the position of a functional group by two carbon atoms.

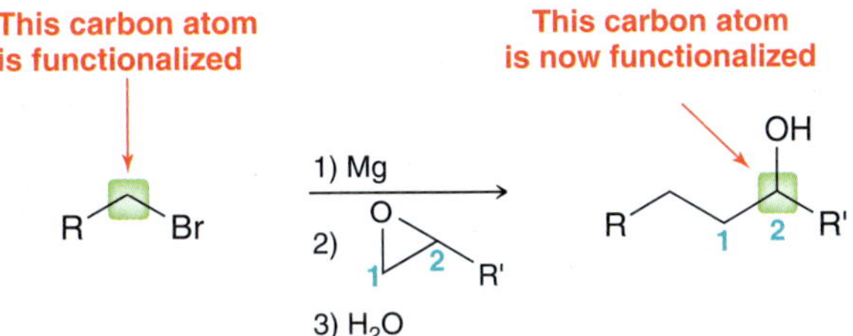


This is just another example of a C—C bond-forming reaction in which the ultimate location of the functional group is an important factor.

When planning a synthesis in which a C—C bond is formed, it is very important to consider the location of the functional group, as that will dictate which C—C bond-forming reaction to choose. Once the proper carbon atom has been functionalized, it is generally easy to change the identity of the functional group at that carbon atom (as we saw in the previous section). However, it is slightly more complicated to change the location of a functional group if it is installed in the wrong location. The following example illustrates this idea.

## SKILLBUILDER

### 21.3 CHOOSING THE MOST EFFICIENT C—C BOND-FORMING REACTION

**LEARN** the skill

Propose an efficient synthesis for the following transformation:

R–C(=O)OH ⟶ R–CH₂–C(=O)Cl

**SOLUTION**

Always begin a synthesis problem by asking the following two questions:

1. ***Is there any change in the carbon skeleton?*** Yes. The product has one additional carbon atom.
2. ***Is there any change in the identity or position of the functional group?*** Yes. The starting material is a carboxylic acid, and the product is an acid chloride. The position of the functional group has changed.

R–C(=O)OH ⟶ R–CH₂–C(=O)Cl

This carbon atom is functionalized | This carbon atom is now functionalized

If we focus on the first question without considering the answer to the second question, we might end up going down a long, inefficient pathway. For example, imagine that we first install a methyl group without considering the desired location of the functional group.

R–C(=O)OH $\xrightarrow{SOCl_2}$ R–C(=O)Cl $\xrightarrow{Me_2CuLi}$ R–C(=O)CH₃

Now the carbon skeleton is correct, but the position of the functional group is wrong.

Completing this synthesis involves moving the functional group with a multistep process.

This synthesis is unnecessarily inefficient. It is more efficient to consider both questions at the same time (the change in carbon skeleton and the location of the functional group), because it is possible to install the carbon atom in such a way that the functional group is placed in the desired location.

This approach forms the necessary C—C bond while simultaneously installing a functional group at the desired location. With this approach, it is possible to obtain the desired product using just one additional step. This provides for a shorter, more efficient synthesis without regiochemical complications.

**PRACTICE** the skill **21.35** Propose an efficient synthesis for each of the following transformations:

(a) (b) (c)

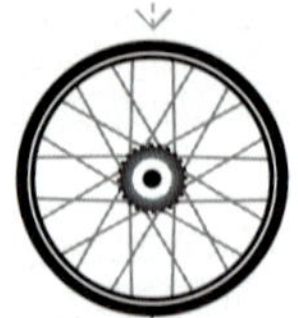

**APPLY** the skill

**21.36** Using reactions from this chapter and the previous chapter (ketones and aldehydes), propose an efficient synthesis for the following transformation:

**21.37** Using acetonitrile ($CH_3CN$) and $CO_2$ as your only sources of carbon atoms, identify how you could prepare each of the following compounds:

(a) (b) (c) (d)

need more **PRACTICE?** Try Problems 21.45b, 21.53b,d,f, 21.54, 21.55, 21.58, 21.74

# 21.15 Spectroscopy of Carboxylic Acids and Their Derivatives

## IR Spectroscopy

Recall that a carbonyl group produces a very strong signal between 1650 and 1850 $cm^{-1}$ in an IR spectrum. The precise location of the signal depends on the nature of the carbonyl group. Table 21.3 gives carbonyl stretching frequencies for each of the carboxylic acid derivatives.

**TABLE 21.3 IMPORTANT SIGNALS IN IR SPECTROSCOPY**

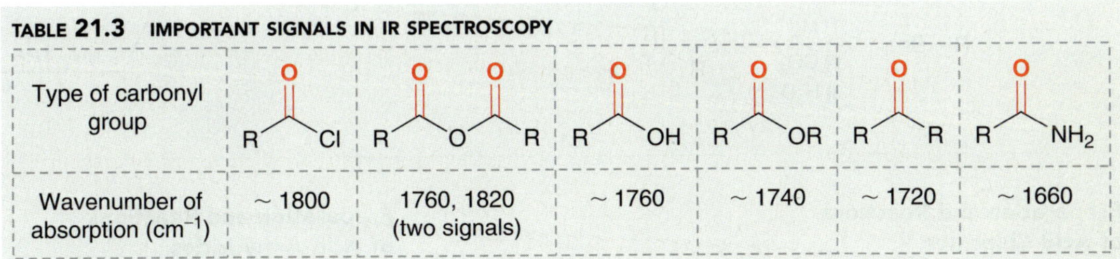

| Type of carbonyl group | R–C(=O)–Cl | R–C(=O)–O–C(=O)–R | R–C(=O)–OH | R–C(=O)–OR | R–C(=O)–R | R–C(=O)–$NH_2$ |
|---|---|---|---|---|---|---|
| Wavenumber of absorption ($cm^{-1}$) | ~ 1800 | 1760, 1820 (two signals) | ~ 1760 | ~ 1740 | ~ 1720 | ~ 1660 |

These numbers can be used to determine the type of carbonyl group in an unknown compound. When performing this type of analysis, recall that conjugated carbonyl groups will produce signals at lower frequencies (Section 15.33).

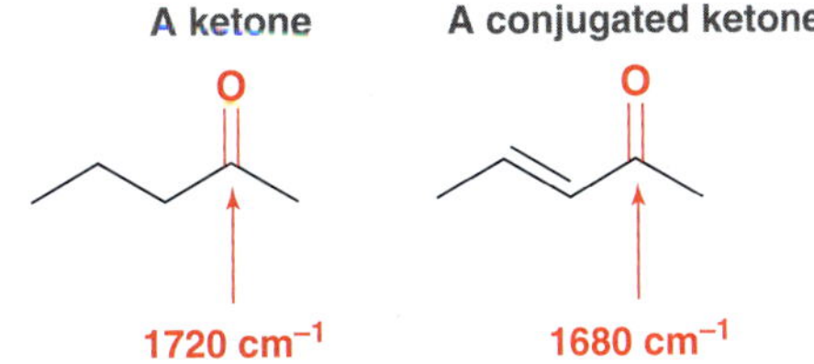


This shift to lower frequencies is observed for all conjugated carbonyl groups, including those present in carboxylic acid derivatives. For example, a conjugated ester will produce a signal below 1740 $cm^{-1}$, and this must be taken into account when analyzing a signal.

In addition, carboxylic acids show very broad O—H signals that span the distance between 2200 and 3600 $cm^{-1}$ (Section 15.5). The C—N triple bond of a nitrile appears in the triple-bond region of the spectrum, at approximately 2200 $cm^{-1}$ (Section 15.3).

## $^{13}C$ NMR Spectroscopy

The carbonyl group of a carboxylic acid derivative generally appears in the region between 160 and 185 ppm, and it is very difficult to use the precise location of a signal to determine the type of carbonyl group present in an unknown compound. The carbon atom of a nitrile typically produces a signal between 115 and 130 ppm in a $^{13}C$ NMR spectrum.

## $^{1}H$ NMR Spectroscopy

As discussed in Section 16.5, the proton of a carboxylic acid typically produces a signal at approximately 12 ppm in a $^{1}H$ NMR spectrum.

### CONCEPTUAL CHECKPOINT

**21.38** Compound **A** has molecular formula $C_9H_8O_2$ and exhibits a strong signal at 1740 $cm^{-1}$ in its IR spectrum. Treatment with two equivalents of LAH followed by water gives the following diol. Identify the structure of compound **A**.

**Compound A** → 1) LAH (2 eq.) 2) $H_2O$ → (diol: OH, OH)

# REVIEW OF REACTIONS

## Preparation of Carboxylic Acids

R–CH$_2$–Br → 1) NaCN 2) $H_3O^+$, heat → R–CH$_2$–COOH

R—Br → 1) Mg 2) $CO_2$ 3) $H_3O^+$ → RCOOH

## Reactions of Carboxylic Acids

RCOOH → 1) xs LAH 2) $H_3O^+$ → $RCH_2OH$

RCOOH → $BH_3$·THF → $RCH_2OH$

## Preparation and Reactions of Acid Chlorides

Carboxylic acid → $SOCl_2$ → acid chloride

Acid chloride → $H_2O$, Pyridine → carboxylic acid

Acid chloride → ROH, Pyridine → ester (OR)

Acid chloride → xs $NH_3$ → amide ($NH_2$)

Acid chloride → xs $RNH_2$ → amide (N–R, H)

Acid chloride → xs $R_2NH$ → amide (N–R, R)

Acid chloride → 1) xs LAH 2) $H_2O$ → primary alcohol (OH, H, H)

Acid chloride → $R_2CuLi$ → ketone (R)

Acid chloride → 1) xs RMgBr 2) $H_2O$ → tertiary alcohol (OH, R, R)

Acid chloride → 1) $LiAl(OR)_3H$ 2) $H_2O$ → aldehyde (H)

## Preparation and Reactions of Acid Anhydrides

Carboxylic acid → 1) NaOH 2) $CH_3COCl$ → anhydride

Carboxylic acid → Heat → anhydride

Anhydride → $H_2O$ → carboxylic acid

Anhydride → ROH → ester (OR)

Anhydride → xs $NH_3$ → amide ($NH_2$)

Anhydride → xs $RNH_2$ → amide (N–R, H)

Anhydride → xs $R_2NH$ → amide (N–R, R)

Anhydride → 1) xs LAH 2) $H_2O$ → primary alcohol (OH, H, H)

Anhydride → $R_2CuLi$ → ketone (R)

Anhydride → 1) xs RMgBr 2) $H_2O$ → tertiary alcohol (OH, R, R)

Anhydride → 1) $LiAl(OR)_3H$ 2) $H_2O$ → aldehyde (H)

## Preparation of Esters

RCOOH → 1) NaOH 2) $CH_3I$ → R–CO–O–$CH_3$

RCOOH → [$H^+$], MeOH → RCOOMe + $H_2O$

RCOCl → ROH, Pyridine → RCOOR

## Reactions of Esters

RCOOR → $H_3O^+$ → RCOOH + ROH

RCOOR → 1) NaOH, heat 2) $H_3O^+$ → RCOOH + ROH

RCOOR → $NH_3$ → $RCONH_2$

RCOOR → 1) xs LAH 2) $H_2O$ → $RCH_2OH$

RCOOR → 1) DIBAH 2) $H_2O$ → RCHO

RCOOR → 1) xs RMgBr 2) $H_2O$ → R–C(OH)(R)–R

### Preparation of Amides

$$RCOCl \xrightarrow[\text{(two equivalents)}]{NH_3} RCONH_2$$

### Reactions of Amides

$$RCONH_2 \xrightarrow{H_3O^+} RCOOH$$

$$RCONH_2 \xrightarrow[2)\ H_3O^+]{1)\ NaOH,\ heat} RCOOH$$

$$RCONH_2 \xrightarrow[2)\ H_2O]{1)\ xs\ LAH} RCH_2NH_2$$

### Preparation of Nitriles

$$RCH_2Br \xrightarrow{NaCN} RCH_2C{\equiv}N + NaBr$$

$$RCONH_2 \xrightarrow{SOCl_2} R{-}C{\equiv}N + SO_2 + 2\,HCl$$

### Reactions of Nitriles

$$R{-}C{\equiv}N \xrightarrow{H_3O^+} RCOOH$$

$$R{-}C{\equiv}N \xrightarrow[2)\ H_3O^+]{1)\ NaOH,\ heat} RCOOH$$

$$R{-}C{\equiv}N \xrightarrow[2)\ H_3O^+]{1)\ RMgBr} RCOR$$

$$R{-}C{\equiv}N \xrightarrow[2)\ H_3O^+]{1)\ xs\ LAH} RCH_2NH_2$$

## REVIEW OF CONCEPTS AND VOCABULARY

### SECTION 21.1

- Carboxylic acids are abundant in nature, and they are widely used in the pharmaceutical and other industries.
- For industrial purposes, acetic acid is converted into vinyl acetate, which is a **carboxylic acid derivative**.

### SECTION 21.2

- Compounds containing a carboxylic acid group are named with the suffix "oic acid."
- Compounds containing two carboxylic acid groups are named with the suffix "dioic acid."
- Many simple carboxylic acids and diacids have common names accepted by IUPAC.

### SECTION 21.3

- Carboxylic acids can form two hydrogen-bonding interactions.
- Treatment of a carboxylic acid with a strong base, such as sodium hydroxide, yields a carboxylate salt.
- Carboxylate ions are named by replacing the suffix "ic acid" with "ate."
- The $pK_a$ of most carboxylic acids is between 4 and 5.
- The acidity of carboxylic acids is due to the stability of the conjugate base, which is resonance stabilized.
- Using the **Henderson-Hasselbalch equation**, it can be shown that carboxylic acids exist primarily as carboxylate salts at **physiological pH**.
- Electron-withdrawing substituents can increase the acidity of a carboxylic acid; the strength of this effect depends on the distance between the electron-withdrawing substituent and the carboxylic acid group.

### SECTION 21.4

- When treated with aqueous acid, a nitrile will undergo *hydrolysis*, yielding a carboxylic acid.
- Carboxylic acids can also be prepared by treating a Grignard reagent with carbon dioxide.

### SECTION 21.5

- Carboxylic acids are reduced to alcohols upon treatment with lithium aluminum hydride or borane.

### SECTION 21.6

- Carboxylic acid derivatives exhibit the same oxidation state as carboxylic acids.
- Acid halides are named by replacing the suffix "ic acid" with "yl halide."
- Acid anhydrides are named by replacing the suffix "ic acid" with "anhydride."
- Esters are named by first indicating the alkyl group attached to the oxygen atom, followed by the carboxylic acid, for which the suffix "ic acid" is replaced with "ate."
- Amides are named by replacing the suffix "ic acid" or "oic acid" with "amide."
- Nitriles are named by replacing the suffix "ic acid" with "nitrile."

### SECTION 21.7

- Carboxylic acid derivatives differ in reactivity, with acid halides being the most reactive and amides the least reactive.
- The C—N bond of an amide has double-bond character and exhibits a relatively high barrier to rotation.
- When a nucleophile attacks a carboxylic acid derivative, a **nucleophilic acyl substitution** can occur in which the nucleophile replaces the leaving group. The mechanism of this reaction involves two core steps and often utilizes several proton transfer steps as well (especially in acidic conditions).
- When drawing a mechanism, avoid formation of a strong base in acidic conditions and avoid formation of a strong acid in basic conditions.
- When a nucleophile attacks a carbonyl group to form a **tetrahedral intermediate**, always re-form the carbonyl group if possible but avoid expelling $H^-$ or $C^-$.

### SECTION 21.8

- Acid chlorides can be formed by treating carboxylic acids with thionyl chloride.
- When treated with water, acid chlorides are hydrolyzed to give carboxylic acids.
- When treated with an alcohol, acid chlorides are converted into esters.
- When treated with ammonia, acid chlorides are converted into amides. Two equivalents of ammonia are required: one to serve as a nucleophile and the other to serve as a base.
- When treated with excess LAH, acid chlorides are reduced to give alcohols because two equivalents of hydride attack. Selective hydride-reducing agents, such as lithium tri(*t*-butoxy) aluminum hydride, can be used to prepare the aldehyde.
- When treated with a Grignard reagent, acid chlorides are converted into alcohols with the introduction of two alkyl groups. Two equivalents of Grignard reagent attack. Preparing a ketone requires the use of a more selective organometallic reagent, such as a **lithium dialkyl cuprate**, also called a **Gilman reagent**.

### SECTION 21.9

- Acetic acid can be converted into acetic anhydride with excessive heating.
- Acid anhydrides can be prepared by treating an acid chloride with a carboxylate ion.
- The reactions of anhydrides are the same as the reactions of acid chlorides except for the identity of the leaving group.

### SECTION 21.10

- When treated with a strong base followed by an alkyl halide, carboxylic acids are converted into esters.
- In a process called the **Fischer esterification**, carboxylic acids are converted into esters when treated with an alcohol in the presence of an acid catalyst. This process is reversible.
- Esters can also be prepared by treating an acid chloride with an alcohol in the presence of pyridine.

### SECTION 21.11

- Esters can be hydrolyzed to yield carboxylic acids upon treatment with either aqueous base or aqueous acid. Hydrolysis under basic conditions is also called **saponification**.
- When treated with lithium aluminum hydride, esters are reduced to yield alcohols. If the desired product is an aldehyde, then DIBAH is used as a reducing agent instead of LAH.
- When treated with a Grignard reagent, esters are reduced to yield alcohols, with the introduction of two alkyl groups.

### SECTION 21.12

- Amides can be efficiently prepared from acid chlorides.
- Amides are hydrolyzed to yield carboxylic acids by treatment with either aqueous base or aqueous acid.
- When treated with excess LAH, amides are converted into amines.

### SECTION 21.13

- Nitriles can be prepared by treating an alkyl halide with a cyanide ion or via the dehydration of an amide.
- Nitriles can be hydrolyzed to yield carboxylic acids by treatment with either aqueous base or aqueous acid.
- A ketone is obtained when a nitrile is treated with a Grignard reagent, followed by aqueous acid.
- Nitriles are converted to amines when treated with LAH.

### SECTION 21.14

- Carboxylic acids, their derivatives, aldehydes, alcohols, and amines can be readily interconverted using reactions covered in this chapter.
- When forming a C—C bond, always consider where you want the functional group to be located, as that will dictate which C—C bond-forming reaction to choose.

### SECTION 21.15

- In IR spectroscopy, the precise location of a carbonyl stretching signal, which appears between 1650 and 1850 $cm^{-1}$, can be used to determine the type of carbonyl group in an unknown compound.
- Conjugated carbonyl groups produce signals at lower frequencies.
- In a $^{13}C$ NMR spectrum, the carbonyl group of a carboxylic acid derivative will generally appear in the region between 160 and 185 ppm, and the carbon atom of a nitrile produces a signal between 115 and 130 ppm.
- In a $^{1}H$ NMR spectrum, the proton of a carboxylic acid produces a signal at approximately 12 ppm.

# SKILLBUILDER REVIEW

## 21.1 DRAWING THE MECHANISM OF A NUCLEOPHILIC ACYL SUBSTITUTION REACTION

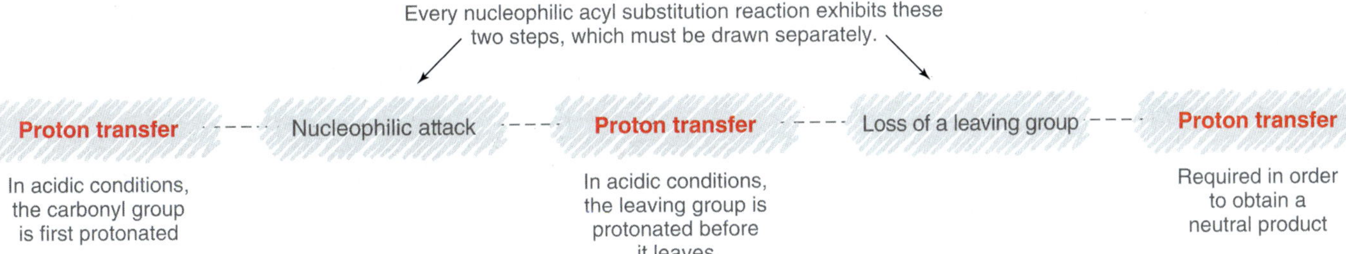


**Try Problems** 21.14–21.17, 21.61, 21.72

## 21.2 INTERCONVERTING FUNCTIONAL GROUPS

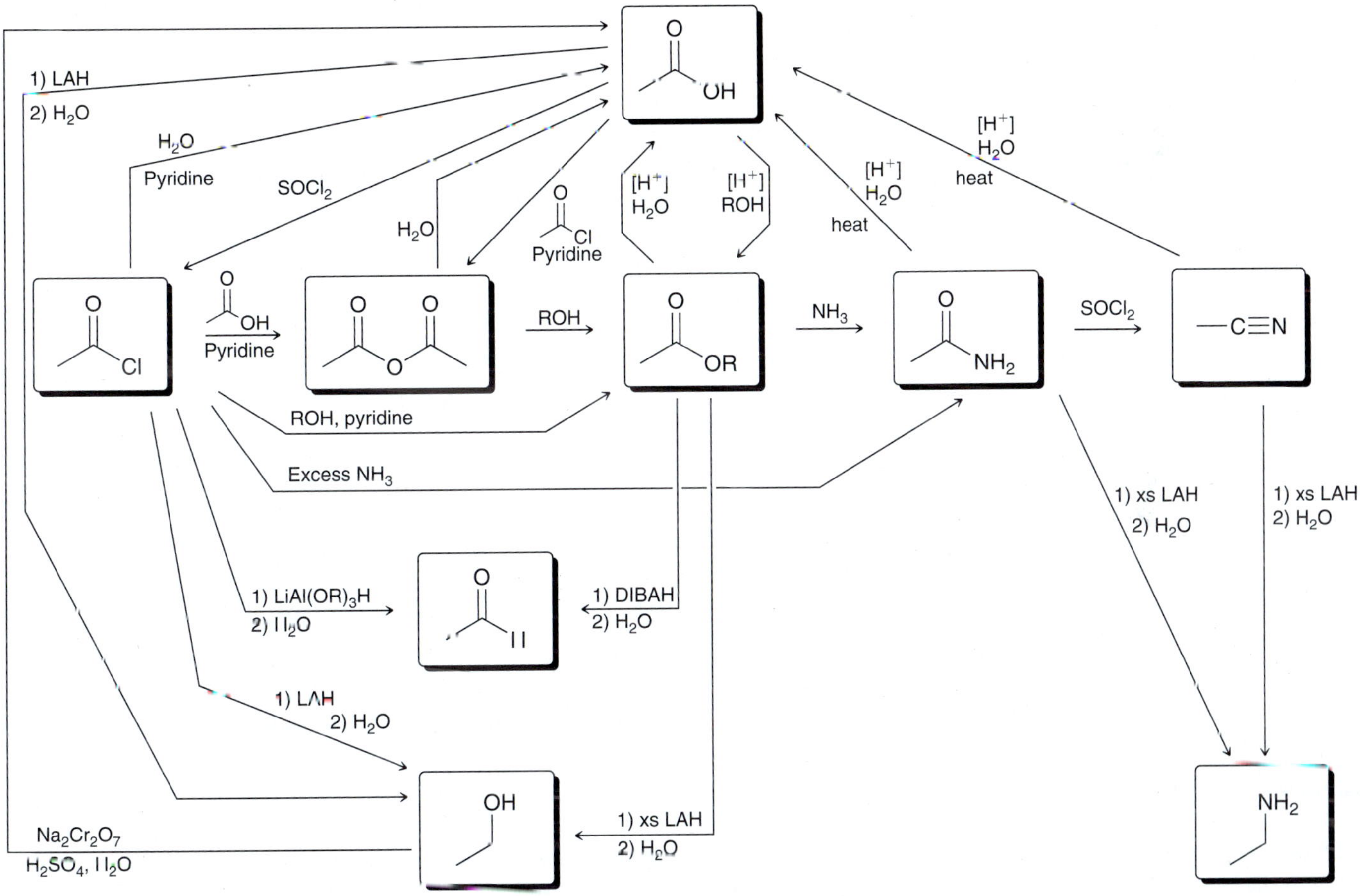

1) LAH
2) $H_2O$
$H_2O$
Pyridine
$SOCl_2$
$H_2O$
Pyridine
$[H^+]$ $H_2O$
$[H^+]$ ROH
$[H^+]$ $H_2O$ heat
$[H^+]$ $H_2O$ heat
Pyridine
ROH
$NH_3$
$SOCl_2$
ROH, pyridine
Excess $NH_3$
1) xs LAH 2) $H_2O$
1) xs LAH 2) $H_2O$
1) $LiAl(OR)_3H$ 2) $H_2O$
1) DIBAH 2) $H_2O$
1) LAH 2) $H_2O$
$Na_2Cr_2O_7$ $H_2SO_4$, $H_2O$
1) xs LAH 2) $H_2O$
OH
Cl
OR
$NH_2$
$—C{\equiv}N$
H
OH
$NH_2$

**Try Problems** 21.32–21.34, 21.45a,b, 21.46a, 21.52, 21.53a,c,e, 21.57

### 21.3 CHOOSING THE MOST EFFICIENT C—C BOND-FORMING REACTION

| C—C Bond-Forming Reactions for Which the Functional Group Remains in the Same Location | C—C Bond-Forming Reactions Involving a Change in the Location of the Functional Group |
|---|---|
| 1) Excess RMgBr, 2) $H_2O$; $R_2CuLi$; 1) RMgBr, 2) $H_3O^+$ | NaCN; $H_3O^+$, Heat; 1) Mg, 2) $CO_2$, 3) $H_3O^+$ |

**Try Problems** 21.35–21.37, 21.45b, 21.53b,d,f, 21.54, 21.55, 21.58, 21.74

## PRACTICE PROBLEMS

Note: Most of the Problems are available within **WileyPLUS**, an online teaching and learning solution.

**21.39** Rank each set of compounds in order of increasing acidity:

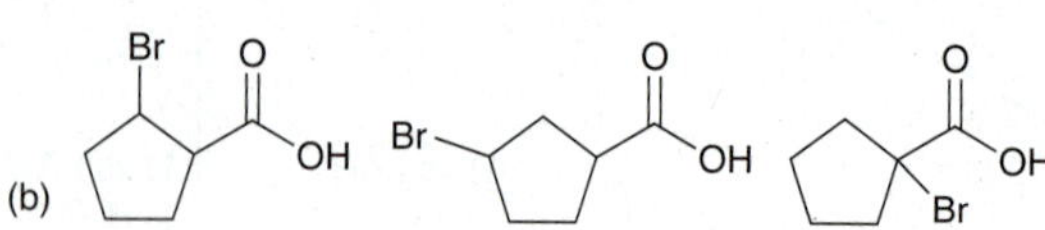


**21.40** Malonic acid has two acidic protons:

**Malonic acid**

The $pK_a$ of the first proton ($pK_1$) is measured to be 2.8, while the $pK_a$ of the second proton ($pK_2$) is measured to be 5.7.

(a) Explain why the first proton is more acidic than acetic acid ($pK_a = 4.76$).

(b) Explain why the second proton is less acidic than acetic acid.

(c) Draw the form of malonic acid that is expected to predominate at physiological pH.

(d) For succinic acid ($HO_2CCH_2CH_2CO_2H$), $pK_1 = 4.2$ (which is higher than $pK_1$ for malonic acid) and $pK_2 = 5.6$ (which is lower than $pK_2$ for malonic acid). In other words, the difference between $pK_1$ and $pK_2$ is not as large for succinic acid as it is for malonic acid. Explain this observation.

**21.41** Identify a systematic (IUPAC) name for each of the following compounds:

(a) (b)

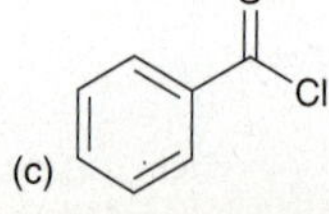


(c)

(d)

(e) $CH_3(CH_2)_4CO_2H$

(f) $CH_3(CH_2)_3COCl$

(g) $CH_3(CH_2)_4CONH_2$

**21.42** Identify the common name for each of the following compounds:

(a) (b) (c)

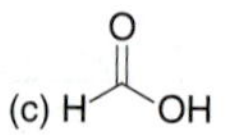


(d)

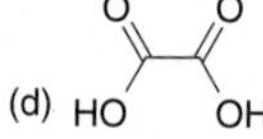


**21.43** Draw the structures of eight different carboxylic acids with molecular formula $C_6H_{12}O_2$. Then, provide a systematic name for each compound and identify which three isomers exhibit chirality centers.

**21.44** Draw and name all constitutionally isomeric acid chlorides with molecular formula $C_4H_7ClO$. Then provide a systematic name for each isomer.

**21.45** Identify the reagents you would use to convert pentanoic acid into each of the following compounds:

(a) 1-Pentanol (b) 1-Pentene (c) Hexanoic acid

**21.46** Identify the reagents you would use to convert each of the following compounds into pentanoic acid:

(a) 1-Pentene (b) 1-Bromobutane

**21.47** Careful measurements reveal that *para*-methoxybenzoic acid is less acidic than benzoic acid, while *meta*-methoxybenzoic acid is more acidic than benzoic acid. Explain these observations.

**21.48** Predict the major product(s) formed when hexanoyl chloride is treated with each of the following reagents:

(a) $CH_3CH_2NH_2$ (excess) (b) LAH (excess), followed by $H_2O$
(c) $CH_3CH_2OH$, pyridine (d) $H_2O$, pyridine
(e) $C_6H_5CO_2Na$ (f) $NH_3$ (excess)
(g) $Et_2CuLi$ (h) EtMgBr (excess), followed by $H_2O$

**21.49** Predict the major product(s) formed when cyclopentanecarboxylic acid is treated with each of the following reagents:

(a) $SOCl_2$ (b) LAH (excess), followed by $H_2O$
(c) NaOH (d) $[H^+]$, EtOH

**21.50** Predict the major product(s) for each of the following reactions:

(a) OH, O — 1) xs LAH 2) $H_2O$ → ?

(b) OH, O — 1) $SOCl_2$ 2) xs $(CH_3)_2NH$ → ?

(c) $NH_2$, O — $SOCl_2$ → ?

(d) OMe, O — 1) $H_3O^+$ 2) $CH_3COCl$, pyridine → ?

(e) OMe, O — 1) DIBAH 2) $H_2O$ → ?

(f) OH — acetic anhydride → ?

(g) H–N — acetyl chloride, Pyridine → ?

(h) O, O — 1) xs LAH 2) $H_2O$ → ?

(i) O, N — $H_3O^+$, Heat → ?

(j) O, O — $H_3O^+$ → ?

**21.51** Identify the carboxylic acid and the alcohol that are necessary in order to make each of the following compounds via a Fischer esterification:

(a) O, O (b) O, O

(c) $CH_3CH_2CO_2C(CH_3)_3$

**21.52** Determine the structures of compounds **A** through **F**:

OH (cyclopentylmethanol) — $Na_2Cr_2O_7$, $H_2SO_4$, $H_2O$ → **A**

**A** — $[H^+]$, EtOH → **D**

**A** — $SOCl_2$ → **B**

**D** — **E** → **F**

**B** — 1) $LiAl(OR)_3H$ 2) $H_2O$ → **F**

**B** — xs $NH_3$ → **C**

**21.53** Identify the reagents you would use to convert 1-bromopentane into each of the following compounds:

(a) Pentanoic acid (b) Hexanoic acid (c) Pentanoyl chloride
(d) Hexanamide (e) Pentanamide (f) Ethyl hexanoate

**21.54** Starting with benzene and using any other reagents of your choice, show how you would prepare each of the following compounds:

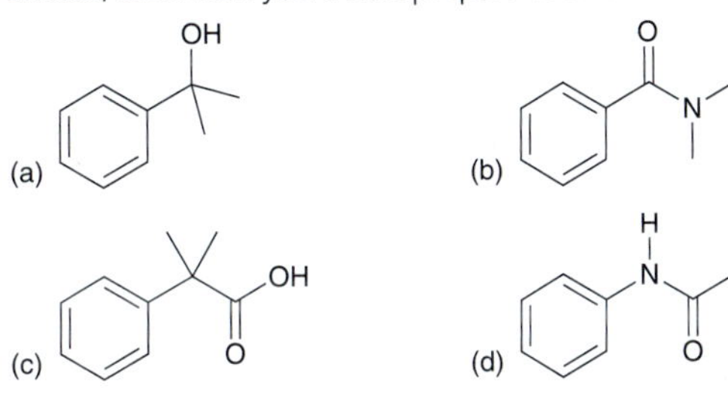


**21.55** Propose an efficient synthesis for each of the following transformations:

(a) Br → O, O

(b) OH → O

(c) Br → N, O

(d) Br → O

**21.56** When methyl benzoate bears a substituent at the *para* position, the rate of hydrolysis of the ester group depends on the nature of the substituent at the *para* position. Apparently, a methoxy substituent renders the ester less reactive, while a nitro substituent renders the ester more reactive. Explain this observation.

**21.57** Identify the reagents necessary to accomplish each of the following transformations:

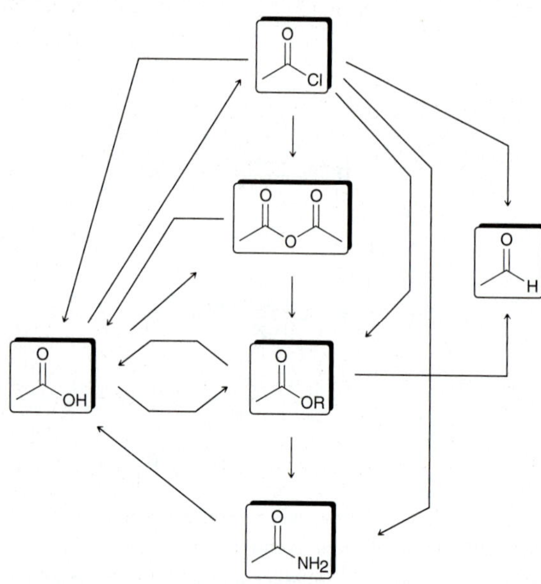


**21.58** DEET is the active ingredient in many insect repellants, such as OFF™. Starting with *meta*-bromotoluene and using any other reagents of your choice, devise an efficient synthesis for DEET.

**m-Bromotoluene** ? **N,N-Diethyl-m-toluamide (DEET)**

**21.59** Predict the products that are formed when diphenyl carbonate is treated with excess methyl magnesium bromide.

1) Excess MeMgBr
2) $H_3O^+$ ?

**21.60** When acetic acid is treated with isotopically labeled water ($^{18}O$, shown in red) in the presence of a catalytic amount of acid, it is observed that the isotopic label becomes incorporated at both possible positions of acetic acid. Draw a mechanism that accounts for this observation.

$[H^+]$, $H_2O$

**21.61** Phosgene is highly toxic and was used as a chemical weapon in World War I. It is also a synthetic precursor used in the production of many plastics.

**Phosgene**

(a) When vapors of phosgene are inhaled, the compound rapidly reacts with any nucleophilic sites present (OH groups, $NH_2$ groups, etc.), producing HCl gas. Draw a mechanism for this process.

(b) When phosgene is treated with ethylene glycol ($HOCH_2CH_2OH$), a compound with molecular formula $C_3H_4O_3$ is obtained. Draw the structure of this product.

(c) Predict the product that is expected when phosgene is treated with excess phenylmagnesium bromide, followed by water.

**21.62** Fluphenazine is an antipsychotic drug that is administered as an ester prodrug via intramuscular injection:

**Fluphenazine decanoate**

The hydrophobic tail of the ester is deliberately designed to enable a slow release of the prodrug into the bloodstream, where the prodrug is rapidly hydrolyzed to produce the active drug.

(a) Draw the structure of the active drug.

(b) Draw the structure of and assign a systematic name for the carboxylic acid that is produced as a by-product of the hydrolysis step.

**21.63** Benzyl acetate is a pleasant-smelling ester found in the essential oil of jasmine flowers and is used in many perfume formulations. Starting with benzene and using any other reagents of your choice, design an efficient synthesis for benzyl acetate.

**Benzyl acetate**

**21.64** Aspartame (below) is an artificial sweetener used in diet soft drinks and is marketed under many trade names, including Equal and Nutrasweet. In the body, aspartame is hydrolyzed to produce methanol, aspartic acid, and phenylalanine. The production of phenylalanine poses a health risk to infants born with a rare condition called phenylketonuria, which prevents phenylalanine from being digested properly. Draw the structures of aspartic acid and phenylalanine.

**Aspartame**

**21.65** Draw a plausible mechanism for each of the following transformations:

(a) Phenol, Pyridine

(b) 1) NaOH 2) $H_3O^+$

(c) 1) NaOH 2) $H_3O^+$

(d) $H_2N{-}NH_2$, Excess pyridine

(e) 1) Excess EtMgBr 2) $H_2O$

**21.66** Ethyl trichloroacetate is significantly more reactive toward hydrolysis than ethyl acetate. Explain this observation.

**21.67** Draw the structure of the diol that is produced when the following carbonate is heated under aqueous acidic conditions:

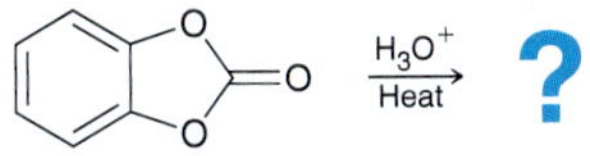

$H_3O^+$ Heat ?

**21.68** Pivampicillin is a penicillin prodrug:

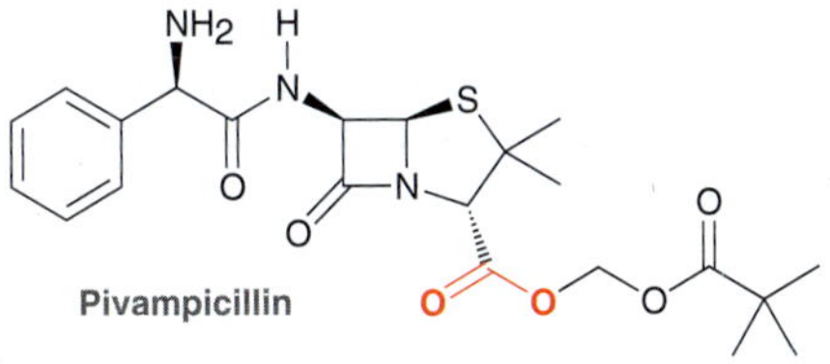

NH2 H N S O N O O O O O

**Pivampicillin**

The prodrug ester group (in red) enables a more rapid delivery of the prodrug to the bloodstream, where the ester group is subsequently hydrolyzed by enzymes, releasing the active drug.

(a) Draw the structure of the active drug.

(b) What is the name of the active drug (see the Medically Speaking box at the end of Section 21.12)?

**21.69** Dexon (below) is a polyester that is spun into fibers and used for surgical stitches that dissolve over time, eliminating the need for a follow-up procedure to remove the stitches. The ester groups are slowly hydrolyzed by enzymes present in the body, and in this way, the stitches are dissolved over a period of several months. Hydrolysis of the polymer produces glycolic acid, which is readily metabolized by the body. Draw the structure of glycolic acid. Identify a systematic name for glycolic acid.

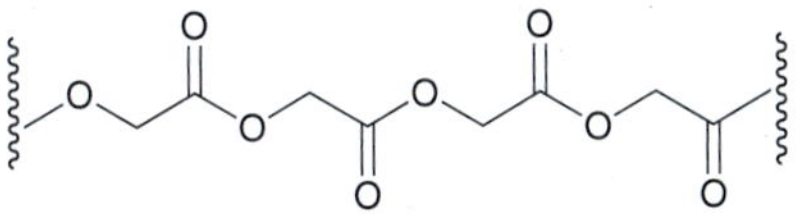

**Dexon**

**21.70** Draw the structure of the polymer produced when the following two monomers are allowed to react with each other:

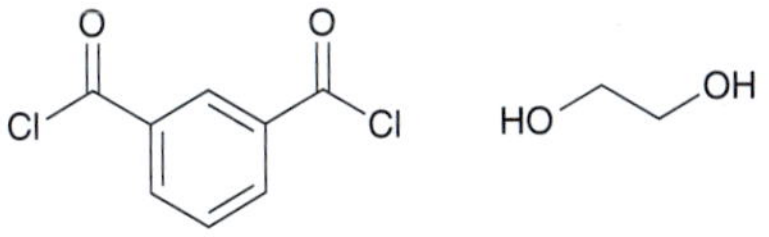

O O Cl Cl HO OH

**21.71** Identify what monomers you would use to produce the following polymer:

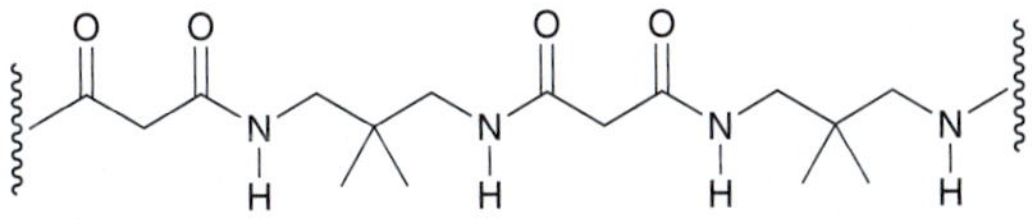

**21.72** *meta*-Hydroxybenzoyl chloride is not a stable compound, and it polymerizes upon preparation. Show a mechanism for the polymerization of this hypothetical compound.

# INTEGRATED PROBLEMS

**21.73** Propose an efficient synthesis for each of the following transformations:

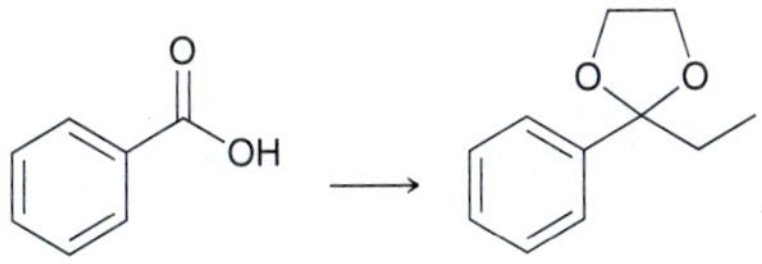

(a) OH →

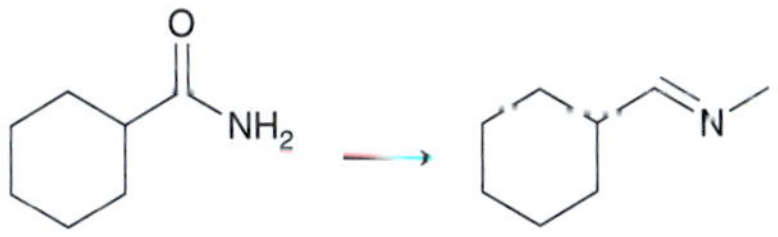

(b) NH2 → N

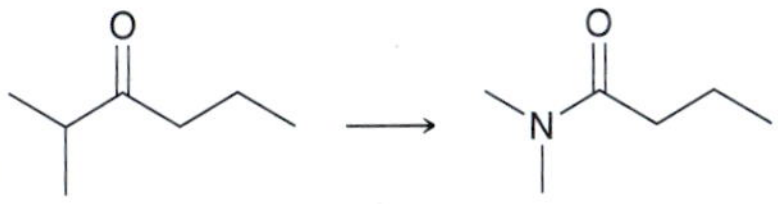

(c) → N

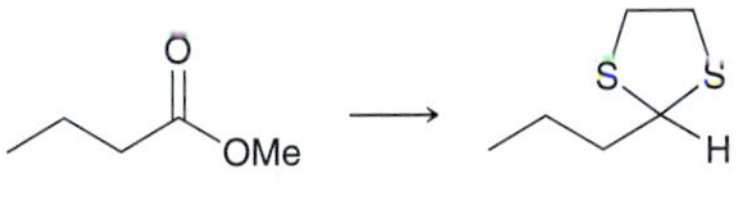

(d) OMe → S S H

(e) COOH →

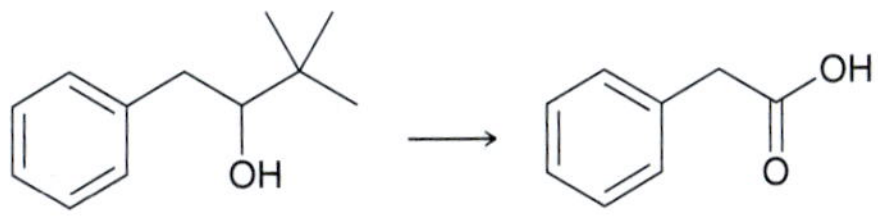

(f) OH → OH

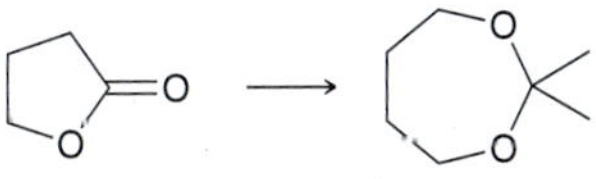

(g) O →

**21.74** Starting with benzene and using any other reagents of your choice, devise a synthesis for acetaminophen:

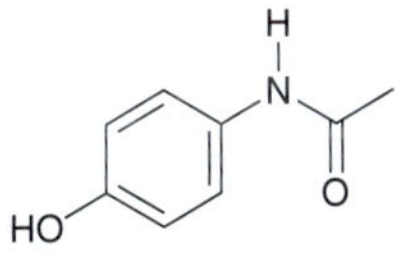

**Acetaminophen (Tylenol)**

**21.75** Draw a plausible mechanism for the following transformation.

$H_3O^+$

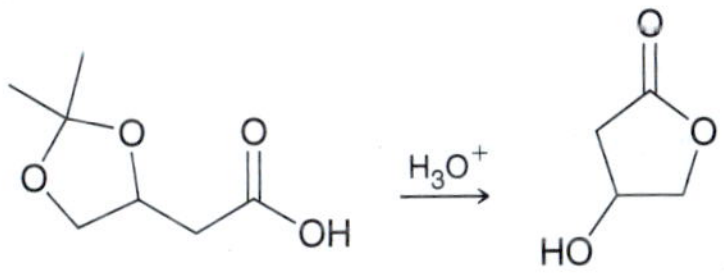

**21.76** A carboxylic acid with molecular formula $C_5H_{10}O_2$ is treated with thionyl chloride to give compound **A**. Compound **A** has only one signal in its $^1H$ NMR spectrum. Draw the structure of the product that is formed when compound **A** is treated with excess ammonia.

**21.77** A compound with molecular formula $C_{10}H_{10}O_4$ exhibits only two signals in its $^1H$ NMR spectrum: a singlet at 4.0 (I = 3H) and a singlet at 8.1 (I = 2H). Identify the structure of this compound.

**21.78** Describe how you could use IR spectroscopy to distinguish between ethyl acetate and butyric acid.

**21.79** Describe how you could use NMR spectroscopy to distinguish between benzoyl chloride and *para*-chlorobenzaldehyde.

**21.80** A compound with molecular formula $C_8H_8O_3$ exhibits the following IR, $^1H$ NMR, and $^{13}C$ NMR spectra. Deduce the structure of this compound.

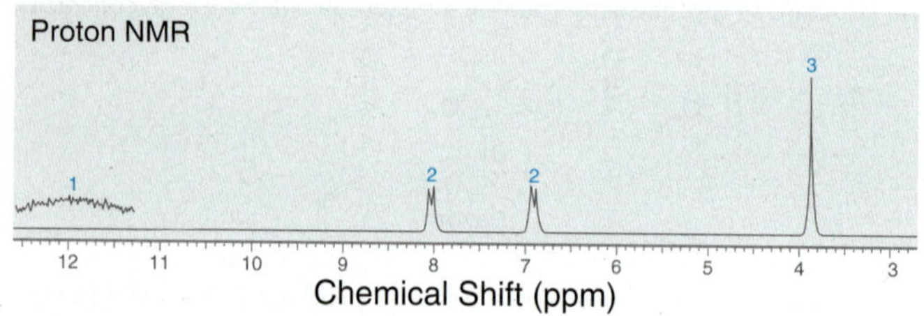


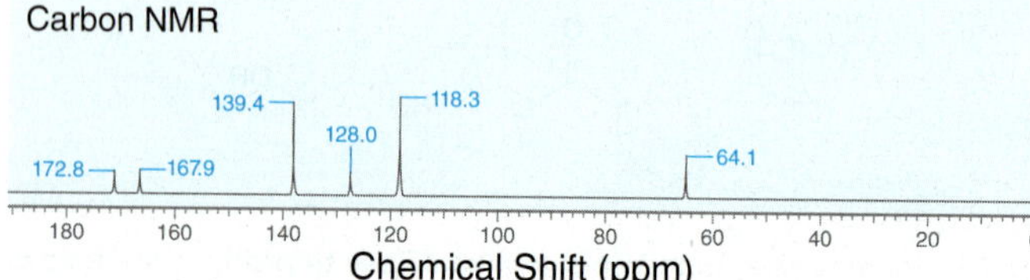


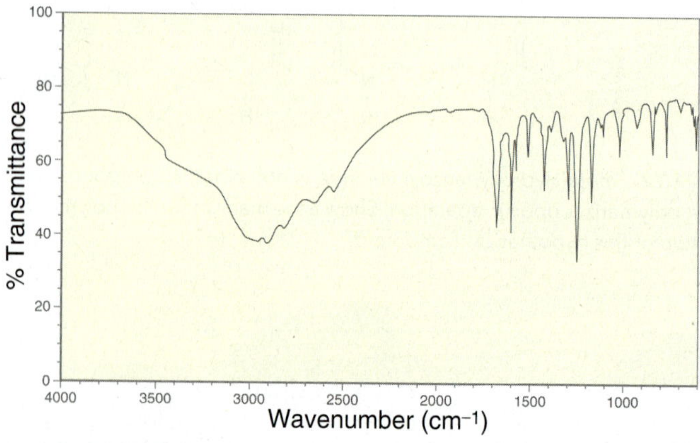


**21.81** Propose a mechanism for the following transformation, and explain how you could use an isotopic labeling experiment to verify your proposed mechanism:

[$H_2SO_4$]

**21.82** *N*-Acetylazoles undergo hydrolysis more readily than regular amides. Suggest a reason for the enhanced reactivity of *N*-acetylazoles toward nucleophilic acyl substitution:

**N-Acetylazole**

**21.83** Dimethylformamide (DMF) is a common solvent:

**Dimethylformamide**

(a) The $^1H$ NMR spectrum of DMF exhibits three signals. Upon treatment with excess LAH followed by water, DMF is converted into a new compound that exhibits only one signal in its $^1H$ NMR spectrum. Explain.

(b) Based on your answer to part a, how many signals do you expect in the $^{13}C$ NMR spectrum of DMF?

**21.84** Olefin metathesis (2005 Nobel Prize) using the Grubbs catalyst has emerged as one of the most important tools for the organic chemist (e.g., **2** to **3**). Metathesis substrate **2** can be prepared from **1** in three steps, shown below (*Org. Lett.* **2000**, *2*, 791–794). Show the product of each synthetic step.

1) NaH, then $BrCH(CH_3)CO_2Et$; 2) DIBAH; 3) $Ph_3P{=}CH_2$; Grubbs catalyst

**21.85** Aminotetralins are a class of compounds currently being studied for their promise as antidepressant drugs. The following conversion was employed during a mechanistic study associated with a synthetic route for preparing aminotetralin drugs (*J. Org. Chem.* **2012**, *77*, 5503–5514). Propose a three-step synthesis for this process.

Three steps

# CHALLENGE PROBLEMS

**21.86** Steroidal derivatives are important pharmacologically and biologically, and they continue to be a popular backbone in the search for new biologically active steroid hormone analogues. Propose a possible three-step synthesis for the conversion of steroid **1** to steroid **2**, an intermediate in the synthesis of pentacyclic steroids (*Tetrahedron Lett.* **2012**, *53*, 1859–1862).

Three steps

**21.87** Allylic bromide **2** was recently used as a key fragment in progress toward the total synthesis of an analogue of amphidinolide N, a potent cytotoxic agent isolated from the marine organism *Amphidinium* sp. (*Org. Biomol. Chem.* **2006**, *4*, 2119–2157). Propose an efficient synthesis of bromide **2** from carboxylic acid **1**.

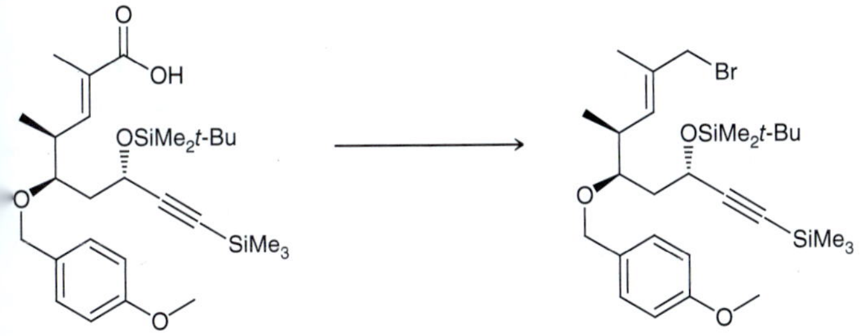

OH OSiMe$_2$t-Bu SiMe$_3$ Br OSiMe$_2$t-Bu SiMe$_3$

**Compound 1** **Compound 2**

**21.88** Compound **3** (below) was used as an intermediate in a recently reported synthesis of the γ-secretase inhibitor BMS-708163 (*Tetrahedron Lett.* **2010**, *51*, 6542–6544). Compound **3** was made from compound **1** via the two-step process shown below.

(a) Compound **1** cannot be directly converted into compound **3** upon treatment with aqueous sodium hydroxide, because undesired side products would occur. Identify the possible side products.

(b) Draw a complete mechanism for the two-step synthesis below.

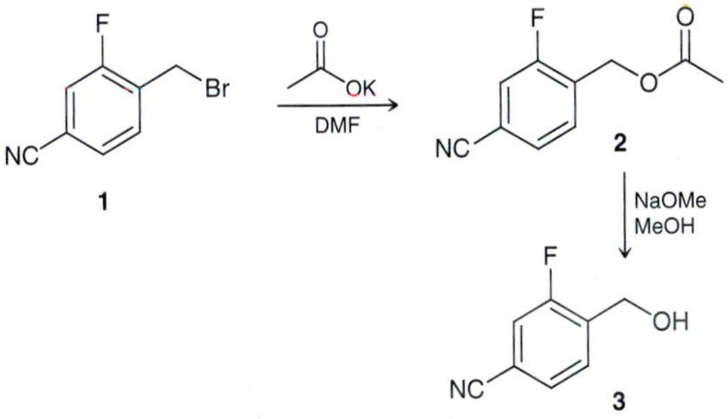

F Br NC **1** OK DMF F NC **2** NaOMe MeOH F OH NC **3**

**21.89** The *m*- and *p*-substituted methyl benzoates listed in the table below were treated with NaOH in dioxane and water. The rate constants of saponification, *k*, are also listed in the table (*J. Am. Chem. Soc.* **1961**, *83*, 4214–4216). Provide structural explanations for the trend observed in the rate constants by comparing the effects of the various substituents on the rate constant. Include all necessary resonance structures in your discussion.

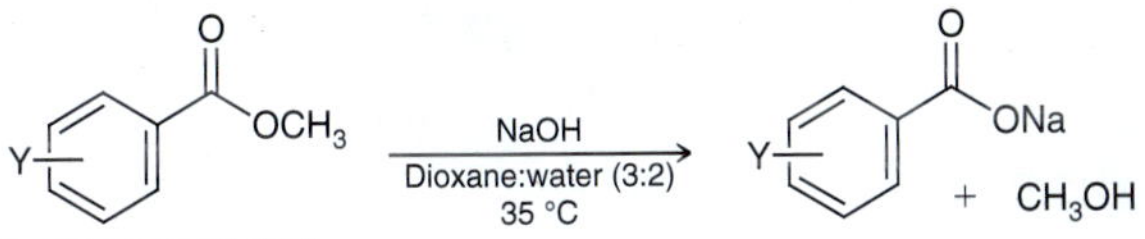

Y OCH$_3$ NaOH Dioxane:water (3:2) 35 °C Y ONa + $CH_3OH$

***m*- or *p*-substituted methyl benzoate**

| *p*- and *m*-Substituted Methyl Benzoates | *k* ($M^{-1}min^{-1}$) |
|---|---|
| Methyl *p*-nitrobenzoate | 102 |
| Methyl *m*-nitrobenzoate | 63 |
| Methyl *m*-chlorobenzoate | 9.1 |
| Methyl *m*-bromobenzoate | 8.6 |
| Methyl benzoate | 1.7 |
| Methyl *p*-methylbenzoate | 0.98 |
| Methyl *p*-methoxybenzoate | 0.42 |
| Methyl *p*-aminobenzoate | 0.06 |

**21.90** Prostaglandins (Section 26.7) are a group of structurally related compounds that function as biochemical regulators and exhibit a wide array of activity, including the regulation of blood pressure, blood clotting, gastric secretions, inflammation, kidney function, and reproductive systems. The following two-step transformation was employed by E. J. Corey during his classic prostaglandin synthesis (*J. Am. Chem. Soc.* **1970**, *92*, 397–398). Identify the reagents that you would use to convert compound **1** to compound **3** and draw the structure of compound **2**.

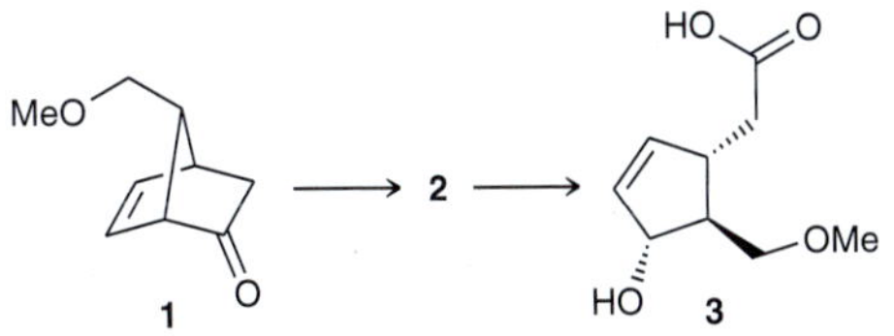

MeO **1** → **2** → HO O HO OMe **3**

**21.91** The following two-step rearrangement was the cornerstone of the first stereoselective total synthesis of quadrone, a biologically active natural product isolated from an *Aspergillus* fungus (*J. Org. Chem.* **1984**, *49*, 4094–4095). Propose a plausible mechanism for this transformation.

OMs $CO_2H$ 1) NaH 2) $H_2SO_2$ $H_2O$ O O

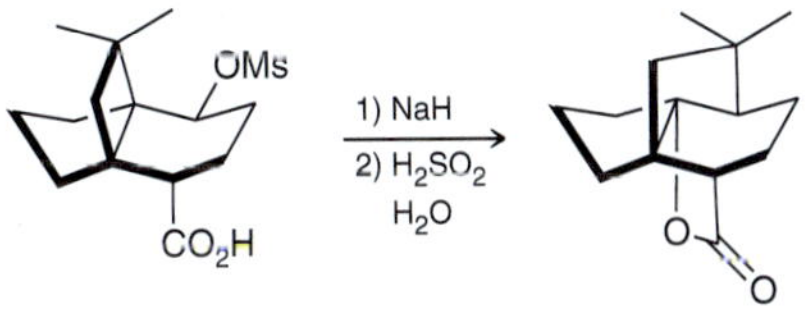

**21.92** During recent studies to determine the absolute stereochemistry of a bromohydrin, the investigators observed an unexpected skeletal rearrangement (*Tetrahedron Lett.* **2008**, *49*, 6853–6855). Provide a plausible mechanism for the formation of epoxide **2** from bromohydrin **1**.

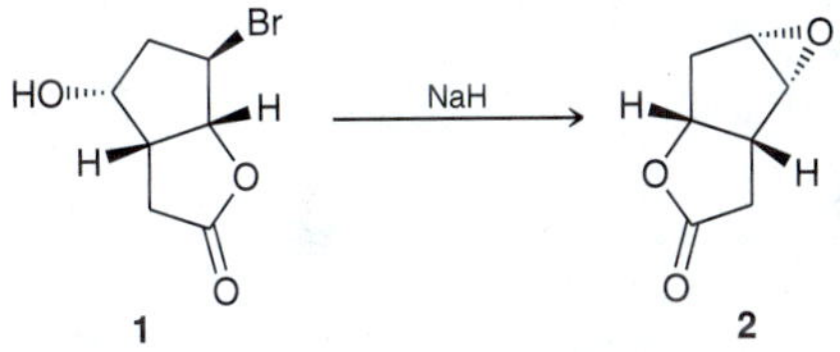

HO Br H H O O **1** NaH H O H O O **2**

**21.93** During a recent total synthesis of englerin A, a potent cytotoxic natural product isolated from the stem bark of a Tanzanian plant, the investigators observed the following (unusual) base-catalyzed ring contraction (*Angew. Chem. Int. Ed.* **2009**, *48*, 9105–9108). Propose a plausible mechanism for this transformation.

H O O H O NaOMe H O O H CHO

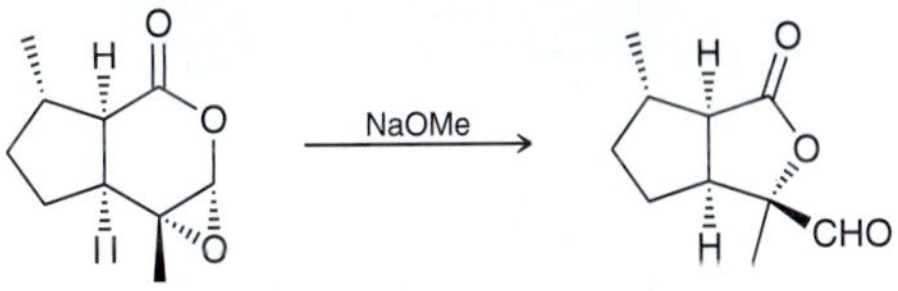

# 22



# Alpha Carbon Chemistry: Enols and Enolates

## DID YOU EVER WONDER...

**how our bodies convert food into energy that powers our muscles?**

In this chapter, we will explore a variety of C—C bond-forming reactions that are among the most versatile reactions available to synthetic organic chemists. Many of these reactions also occur in biochemical processes, including the metabolic pathways that produce energy for muscle contractions. The reactions in this chapter will greatly expand your ability to design syntheses for a wide variety of compounds.

## DO YOU REMEMBER?

*Before you go on, be sure you understand the following topics.*
*If necessary, review the suggested sections to prepare for this chapter:*

- Brønsted-Lowry Acidity (Section 3.4)
- Energy Diagrams: Thermodynamics vs. Kinetics (Section 6.6)
- Nucleophilic Acyl Substitution (Section 21.7)
- Position of Equilibrium and Choice of Reagents (Section 3.5)
- Keto-Enol Tautomerization (Section 10.7)
- Retrosynthetic Analysis (Section 12.5)

Take the DO YOU REMEMBER? QUIZ in **WileyPLUS** to check your understanding

## 22.1 Introduction to Alpha Carbon Chemistry: Enols and Enolates

### The Alpha Carbon

For compounds containing a carbonyl group, Greek letters are used to describe the proximity of each carbon atom to the carbonyl group.

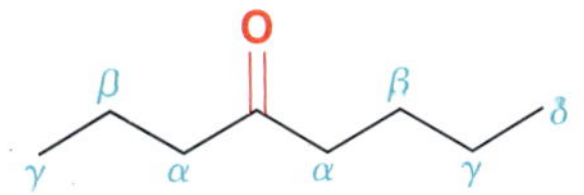

O β β δ γ α α γ

The carbonyl group itself does not receive a Greek letter. In this example there are two carbon atoms designated as alpha ($\alpha$) positions. Hydrogen atoms are designated with the Greek letter of the carbon to which they are attached; for example, the hydrogen atoms (protons) connected to the $\alpha$ carbon atoms are called $\alpha$ protons. This chapter will explore reactions occurring at the $\alpha$ position.

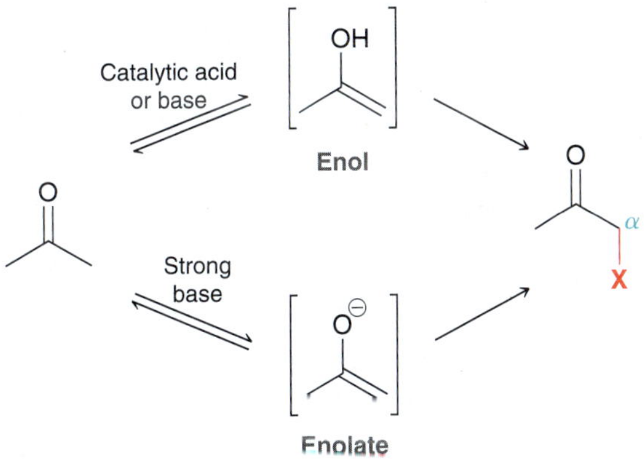

Catalytic acid or base
OH
Enol
O
Strong base
O⊖
Enolate
O
α
X

These reactions can occur via either an enol or an enolate intermediate. The vast majority of the reactions in this chapter will proceed via an enolate intermediate, but we will also explore some reactions that proceed via an enol intermediate.

### Enols

In the presence of catalytic acid or base, a ketone will exist in equilibrium with an enol.

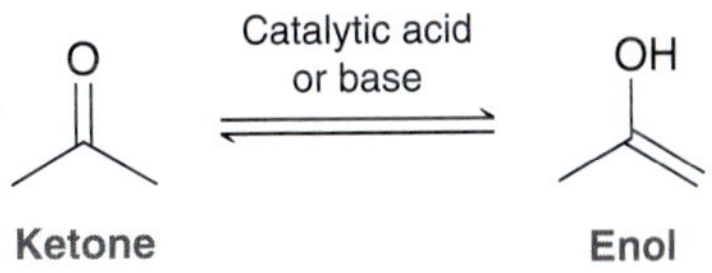

O
Catalytic acid or base
OH
Ketone
Enol

Recall that the ketone and enol shown are tautomers—rapidly interconverting constitutional isomers that differ from each other in the placement of a proton and the position of a double bond. Do not confuse tautomers with resonance structures. The two structures above are not resonance structures because they differ in the arrangement of their atoms. These structures represent two different compounds, both of which are present at equilibrium. In general, the position of equilibrium will significantly favor the ketone, as seen in the following example:

$>99.99\%$ $\quad$ $<0.01\%$

Cyclohexanone exists in equilibrium with its tautomeric enol form, which exhibits a very minor presence. This is the case for most ketones.

In some cases, the enol tautomer is stabilized and exhibits a more substantial presence at equilibrium. Consider, for example, the enol form of a beta-diketone, such as 2,4-pentanedione.

10–30% $\quad$ 70–90%

The equilibrium concentrations of the diketone and the enol depend on the solvent that is used, but the enol generally dominates. Two factors contribute to the remarkable stability of the enol in this case: (1) The enol has a conjugated $\pi$ system, which is a stabilizing factor (see Section 17.2), and (2) the enol can form an intramolecular H-bonding interaction between the hydroxyl proton and the nearby carbonyl group (shown with a gray dotted line above). Both of these factors serve to stabilize the enol.

Phenol is an extreme example in that the concentration of the ketone is practically negligible. In this case, the ketone lacks aromaticity, while the enol is aromatic and significantly more stable.

$<0.01\%$ $\quad$ $>99.99\%$

Tautomerization is catalyzed by trace amounts of either acid or base. The acid-catalyzed process is shown in Mechanism 22.1 (seen previously in Mechanism 10.2).

## MECHANISM 22.1 ACID-CATALYZED TAUTOMERIZATION

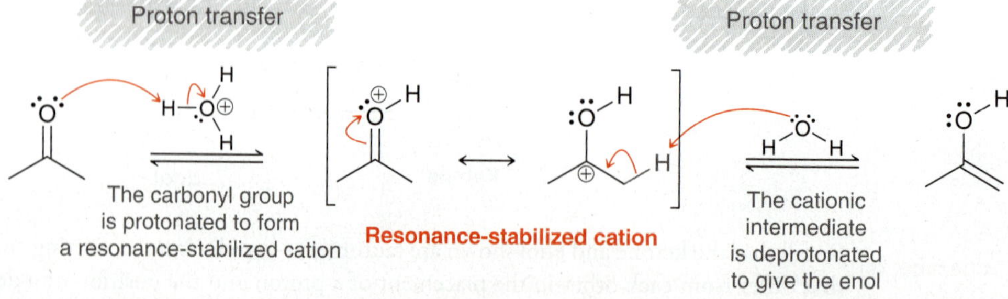

**LOOKING BACK**
This requirement was discussed in Section 21.7.

In the first step, the carbonyl group is protonated to form a resonance-stabilized cation, which is then deprotonated at the α position to give the enol. Notice that none of the reagents or intermediates are strong bases, which is consistent with acidic conditions.

The base-catalyzed process for tautomerization is shown in Mechanism 22.2.

## MECHANISM 22.2 BASE-CATALYZED TAUTOMERIZATION

The α position is deprotonated to form a resonance-stabilized anion

**Resonance-stabilized anion**

The anionic intermediate is protonated to give the enol

In the first step, the α position is deprotonated to form a resonance-stabilized anion, which is then protonated to give the enol. Notice that none of the reagents or intermediates are strong acids, in order to be consistent with basic conditions.

The mechanisms for acid-catalyzed and base-catalyzed tautomerization involve the same two steps (protonation at the carbonyl group and deprotonation of the α position). The difference between these mechanisms is the order of events. In acid conditions, the first step is protonation of the carbonyl group, giving a positively charged intermediate. In basic conditions, the first step is deprotonation of the α position, giving a negatively charged intermediate.

It is difficult to prevent tautomerization even if care is taken to remove all acids and bases from the solution. Tautomerization can still be catalyzed by the trace amounts of acid or base that are adsorbed to the surface of the glassware (even after washing the glassware scrupulously). Unless extremely rare conditions are employed, you should always assume that tautomerization will occur if possible, and an equilibrium will quickly be established favoring the more stable tautomer. The enol tautomer is generally present only in small amounts, but it is very reactive. Specifically, the α position is very nucleophilic due to resonance.

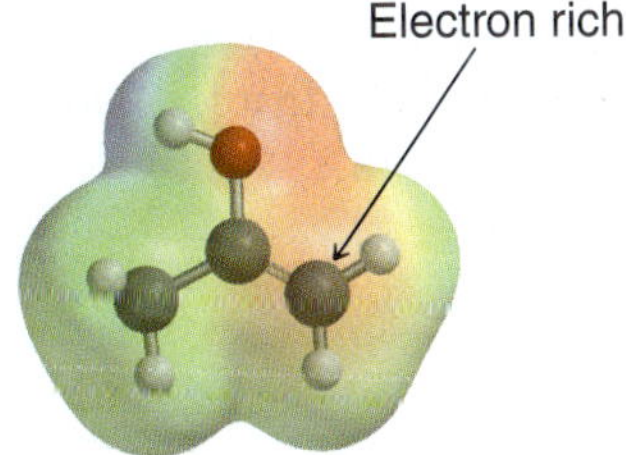


**FIGURE 22.1** An electrostatic potential map of the enol of acetone, showing the nucleophilic character of the α position.

In the second resonance structure, the α position exhibits a lone pair, rendering that position nucleophilic. The effect of an OH group in activating the α position is similar to its effect in activating an aromatic ring (as seen in Section 19.10). The electron-donating effect of the OH group can be visualized with an electrostatic potential map of a simple enol (Figure 22.1). Notice that the α position is somewhat red, representing an electron-rich site. In Section 22.2, we will see two reactions in which the α position of an enol functions as a nucleophile.

## CONCEPTUAL CHECKPOINT

**22.1** Draw a mechanism for the acid-catalyzed conversion of cyclohexanone into its tautomeric enol.

**22.2** Draw a mechanism for the reverse process of the previous problem. In other words, draw the acid-catalyzed conversion of 1-cyclohexenol to cyclohexanone.

**22.3** Draw the two possible enols that can be formed from 3-methyl-2-butanone and show a mechanism of formation of each under base-catalyzed conditions.

## Enolates

When treated with a strong base, the α position of a ketone is deprotonated to give a resonance-stabilized intermediate called an **enolate**.

Enolate (resonance-stabilized)

Enolates are ambident nucleophiles (*ambi* is Latin for "both" and *dent* is Latin for "teeth"), because they possess two nucleophilic sites, each of which can attack an electrophile. When the oxygen atom attacks an electrophile, it is called O-attack; and when the α carbon attacks an electrophile, it is called C-attack.

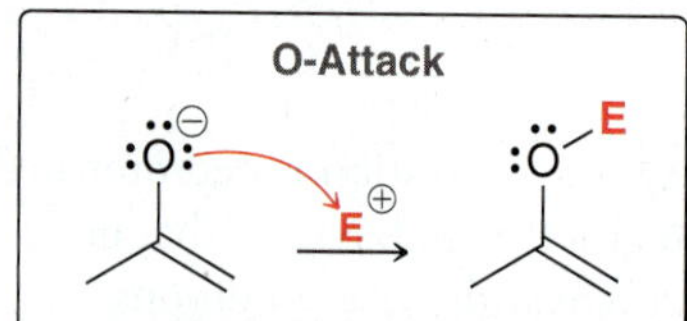


C-Attack

Although the oxygen atom of an enolate bears the majority of the negative charge, C-attack is nevertheless more common than O-attack. All of the reactions presented in this chapter will be examples of C-attack. When drawing the mechanism of an enolate undergoing C-attack, it is technically more appropriate to draw the resonance structure of the enolate in which the negative charge appears on the oxygen atom, because that drawing represents the more significant resonance contributor. Therefore, C-attack should be drawn like this:

Nevertheless, for simplicity, when drawing mechanisms, we will often draw the less significant resonance contributor of the enolate, in which the negative charge is on the α carbon. That is, C-attack will be drawn like this:

When drawn this way, fewer curved arrows are required, which will simplify many of the mechanisms in this chapter.

Enolates are more useful than enols because (1) enolates possess a full negative charge and are therefore more reactive than enols and (2) enolates can be isolated and stored for short periods of time, unlike enols, which cannot be isolated or stored. For these two reasons, the vast majority of reactions in this chapter proceed via enolate intermediates.

As we progress through the chapter, it is important to keep in mind that only the α protons of an aldehyde or ketone are acidic.

**LOOKING BACK**

For a review of the factors that affect the acidity of a proton, see Section 3.4.

Acidic protons

In the example shown, the beta (β) and gamma (γ) protons are not acidic, and neither is the aldehydic proton. Deprotonation at any of those positions does not lead to a resonance-stabilized anion.

## SKILLBUILDER

### 22.1 DRAWING ENOLATES

**LEARN the skill**

When the following ketone is treated with a strong base, an enolate ion is formed. Draw both resonance structures of the enolate.

**SOLUTION**

**STEP 1**
Identify all α positions.

Begin by identifying all α positions. In this case, there are two α positions, but only one of them bears a proton. The α position on the right side of the carbonyl group has no protons, and therefore, an enolate cannot form on that side of the carbonyl group. The other α carbon (on the left side) does have a proton, and an enolate can be formed at that location.

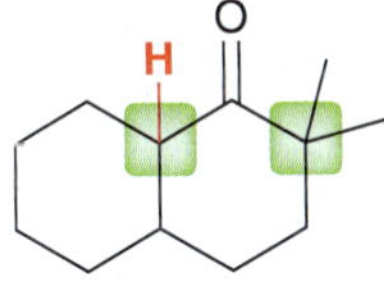

**STEP 2**
Remove the proton at the α position and draw the resulting anion.

Using two curved arrows, remove the proton, and then draw the enolate with a lone pair and a negative charge at the α position (in place of the proton).

**STEP 3**
Draw the resonance structure, showing the charge on the oxygen atom.

Finally, draw the other resonance structure using the skills developed in Section 2.10.

**PRACTICE the skill**

**22.4** Draw both resonance structures of the enolate formed when each of the following ketones is treated with a strong base:

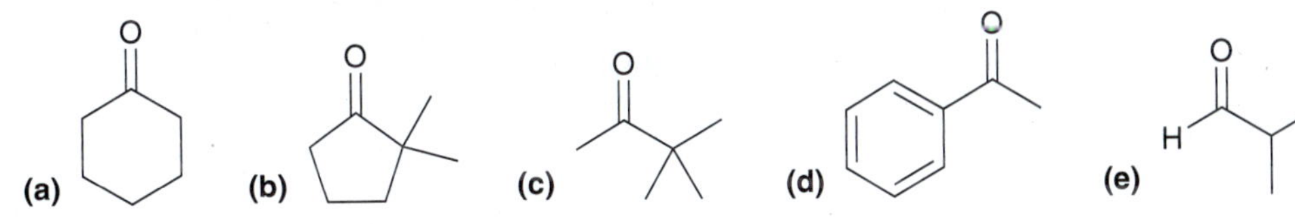

**APPLY the skill**

**22.5** When 2-methylcyclohexanone is treated with a strong base, two different enolates are formed. Draw both of them.

need more **PRACTICE?** **Try Problem 22.59**

## Choosing a Base for Enolate Formation

In this chapter, we will see many reactions involving enolates, and the choice of base will always be important. Some cases require a relatively mild base to form the enolate, while other cases demand a very strong base. In order to choose an appropriate base, we must take a careful look at p$K_a$ values. Aldehydes and ketones typically have p$K_a$ values in the range of 16–20, as seen in Table 22.1. The range of p$K_a$ values is similar to the range exhibited by alcohols (ethanol has a p$K_a$ of 16 and *tert*-butanol has a p$K_a$ of 18). As a result, when an alkoxide ion is used as the base, an equilibrium is established in which the alkoxide ion and the enolate ion are both present. The relative amount of alkoxide and enolate ions is determined by the relative p$K_a$ values, although there will usually be less of the enolate present at equilibrium. Consider the following example:

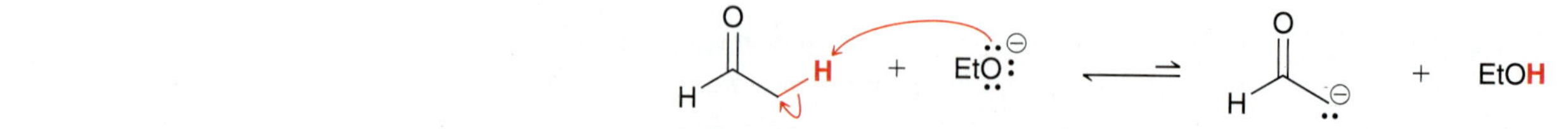

p$K_a$ = 16.7 + EtO⁻ ⇌ enolate + EtOH p$K_a$ = 15.9

In this example, ethoxide is used as a base to deprotonate acetaldehyde. Notice that the p$K_a$ values of acetaldehyde and ethanol are similar. The equilibrium slightly favors acetaldehyde, rather than its enolate, although both are present. The presence of both the aldehyde and its enolate is important because the enolate is a nucleophile and the aldehyde is an electrophile, and these two species will react with each other when they are both present, as we will see in Section 22.3.

In contrast, many other bases, such as sodium hydride, can irreversibly and completely convert the aldehyde into an enolate.

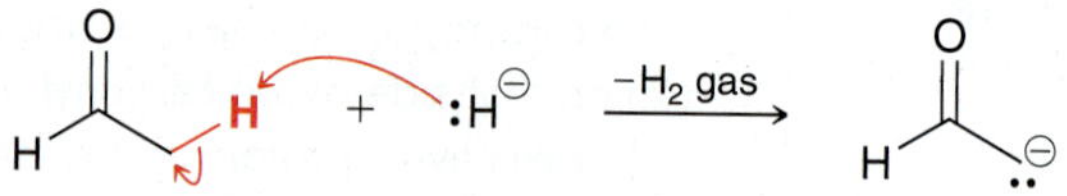

+ :H⁻ $\xrightarrow{-H_2\text{ gas}}$ enolate

When sodium hydride is used as the base, hydrogen gas is formed. It bubbles out of solution, as all of the aldehyde molecules are converted into enolate ions. Under these conditions, the aldehyde and the enolate are not both present. Only the enolate is present. Another base commonly used for irreversible enolate formation is lithium diisopropylamide (LDA), which is prepared by treating diisopropylamine with butyllithium.

**TABLE 22.1 p$K_a$ VALUES OF SOME COMMON KETONES AND ALDEHYDES**

| Compound | p$K_a$ | Enolate |
|---|---|---|
| 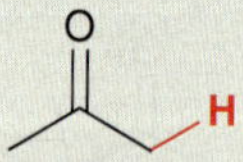 Acetone | 19.2 | 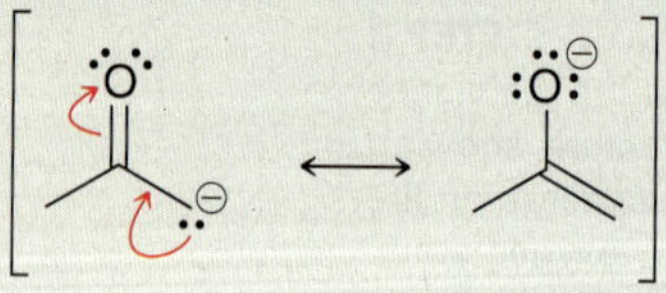 |
| 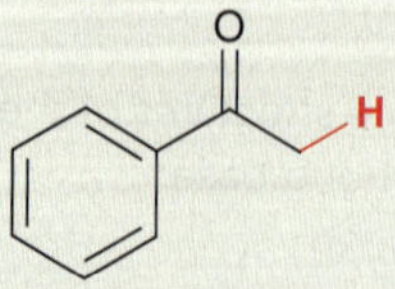 Acetophenone | 18.3 | 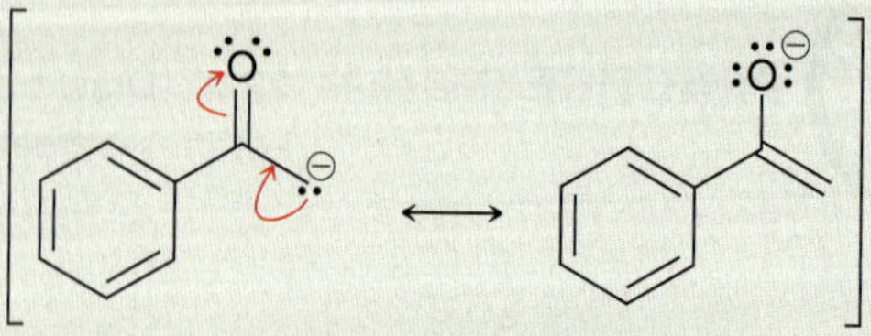 |
| 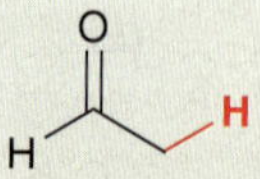 Acetaldehyde | 16.7 | 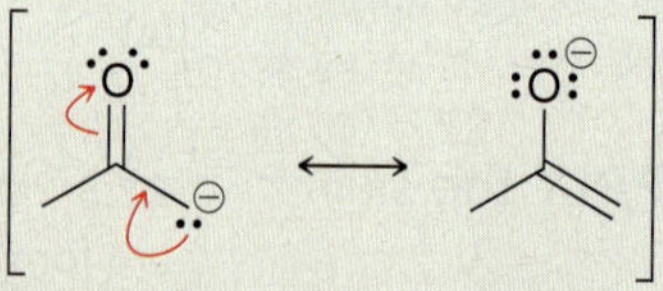 |

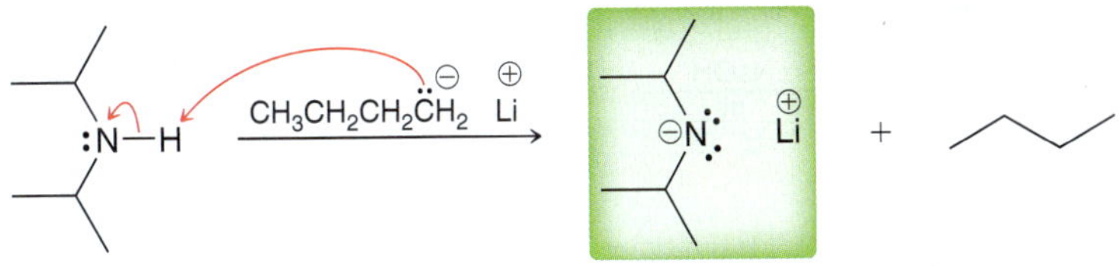

Diisopropylamine

Lithium diisopropylamide

The p$K_a$ of diisopropylamine is approximately 36, and therefore, LDA can be used to accomplish irreversible enolate formation.

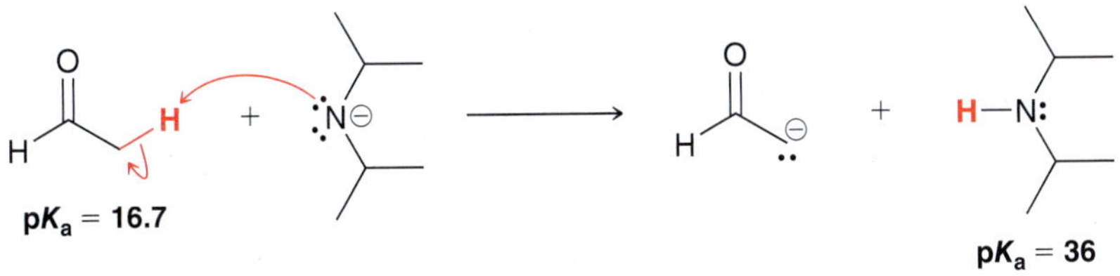

p$K_a$ = 16.7

p$K_a$ = 36

When LDA is used as a base, the amount of aldehyde present at equilibrium is negligible.

Now consider a compound with two carbonyl groups that are beta to each other:

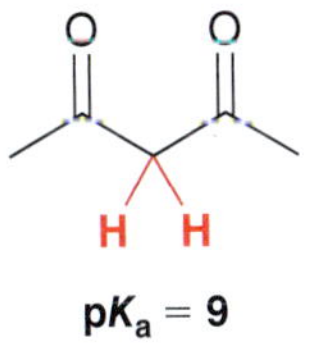

p$K_a$ = 9

In this case, the central $CH_2$ group is flanked by two carbonyl groups. The protons of the $CH_2$ group (shown in red) are therefore highly acidic. As opposed to most ketones (with a p$K_a$ somewhere in the range of 16–20), the p$K_a$ of this compound is approximately 9. The acidity of these protons can be attributed to the highly stabilized anion formed upon deprotonation.

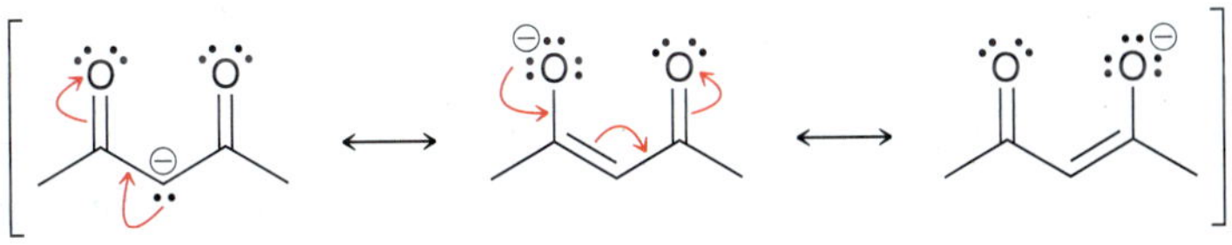

The anion is a doubly stabilized enolate ion with the negative charge being spread over two oxygen atoms and one carbon atom. Because of the relatively high acidity of beta diketones, LDA is not required to irreversibly deprotonate these compounds. Rather, treatment with hydroxide or an alkoxide ion is sufficient to ensure nearly complete enolate formation.

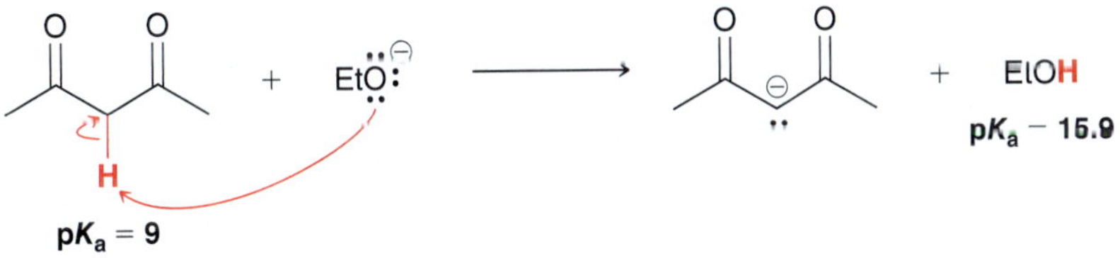

p$K_a$ = 9

p$K_a$ = 15.9

**LOOKING BACK**
To review how p$K_a$ values can be used to determine equilibrium concentrations for an acid-base reaction, see Section 3.3.

The difference in p$K_a$ values is about 7 (which represents seven orders of magnitude). In other words, when ethoxide is used as the base, 99.99999% of the diketone molecules are deprotonated to form enolates.

Figure 22.2 summarizes the relevant factors for choosing a base to form an enolate ion. The information summarized in this figure will be used several times in the upcoming sections of this chapter.

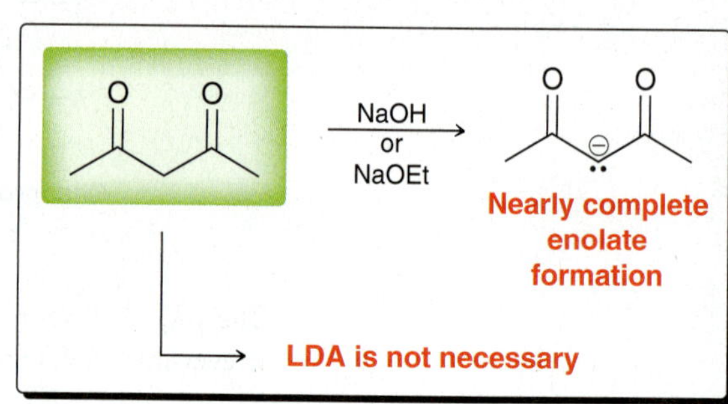


**FIGURE 22.2**
A summary of the outcomes observed when various bases are used to deprotonate ketones or 1,3-diketones.

## CONCEPTUAL CHECKPOINT

**22.6** Draw the enolate ion that is formed when each of the following compounds is treated with sodium ethoxide. In each case, draw all resonance structures of the enolate ion and predict whether a substantial amount of starting ketone will be present together with the enolate at equilibrium.

(a) (b) (c) (d)

**22.7** When the following compound is treated with sodium ethoxide, nearly all of it is converted into an enolate. Draw the resonance structures of the enolate that is formed and explain why enolate formation is nearly complete despite the use of ethoxide rather than LDA.

**22.8** For each pair of compounds, identify which compound is more acidic and explain your choice.

(a) 2,4-Dimethyl-3,5-heptanedione or 4,4-dimethyl-3,5-heptanedione

(b) 1,2-Cyclopentanedione or 1,3-cyclopentanedione

(c) Acetophenone or benzaldehyde

# 22.2 Alpha Halogenation of Enols and Enolates

## Alpha Halogenation in Acidic Conditions

Under acid-catalyzed conditions, ketones and aldehydes will undergo halogenation at the α position.

$[H_3O^+]$, $Br_2$ → (65%) + HBr

The reaction is observed for chlorine, bromine, and iodine, but not for fluorine. A variety of solvents can be used, including acetic acid, water, chloroform, and diethyl ether. The rate of halogenation is found to be independent of the concentration or identity of the halogen, indicating that the halogen does not participate in the rate-determining step. This information is consistent with Mechanism 22.3.

## MECHANISM 22.3 ACID-CATALYZED HALOGENATION OF KETONES

PART 1: ENOL FORMATION

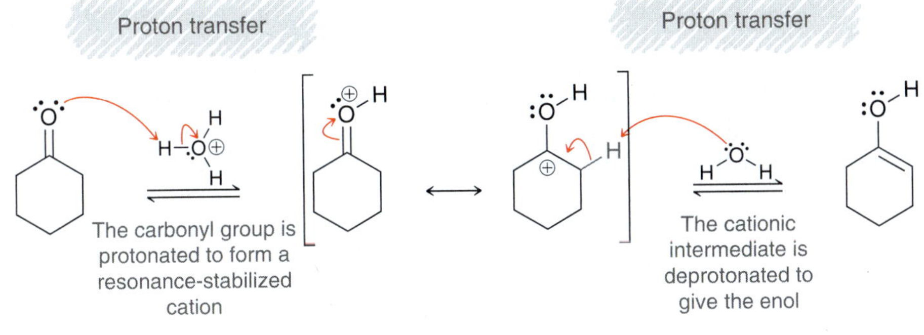


PART 2: HALOGENATION

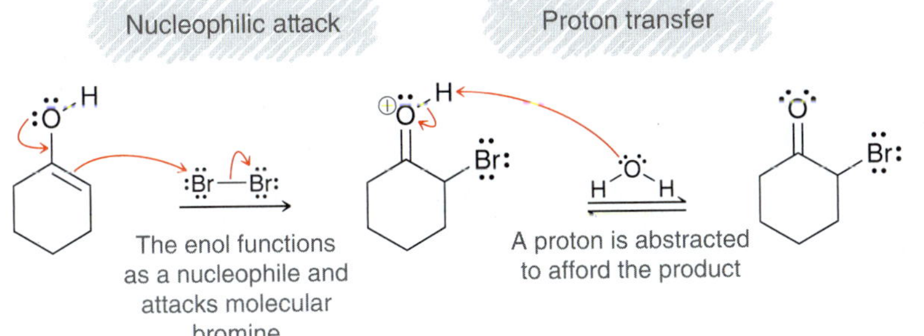


The mechanism has two parts. First the ketone undergoes tautomerization to produce an enol. Then, in the second part of the mechanism, the enol serves as a nucleophile and installs a halogen atom at the $\alpha$ position. The second part of the mechanism (halogenation) occurs more rapidly than the first part (enol formation), and therefore, enol formation represents the rate-determining step for the process. The halogen is not involved in enol formation, and therefore, the concentration of the halogen has no measurable impact on the rate of the overall process.

The by-product of $\alpha$ bromination is HBr, which is an acid and is capable of catalyzing the first part of the mechanism (enol formation). As a result, the reaction is said to be **autocatalytic**, that is, the reagent necessary to catalyze the reaction is produced by the reaction itself.

When an unsymmetrical ketone is used, bromination occurs primarily at the more substituted side of the ketone.

$[H_3O^+]$, $Br_2$

Major + Minor

In this example, a mixture of products is generally unavoidable. However, bromination occurs primarily at the side bearing the alkyl group, which can be attributed to the fact that the reaction proceeds more rapidly through the more substituted enol.

**More substituted (more stable)** **Less substituted (less stable)**

The halogenated product can undergo elimination when treated with a base.

Base

A variety of bases can be used, including pyridine, lithium carbonate ($Li_2CO_3$), or potassium *tert*-butoxide. This provides a two-step method for introducing α,β-unsaturation in a ketone. This procedure is only practical in some cases and yields are often low.

1) $[H_3O^+]$, $Br_2$
2) Pyridine

## CONCEPTUAL CHECKPOINT

**22.9** Predict the major product for each of the following transformations and propose a mechanism for its formation:

(a) 1) $[H_3O^+]$, $Br_2$ 2) Pyridine ?
(b) 1) $[H_3O^+]$, $Br_2$ 2) Pyridine ?
(c) 1) $[H_3O^+]$, $Br_2$ 2) Pyridine ?

**22.10** Identify the reagents that you would use to accomplish each of the following transformations:

(a) (b)

## Alpha Bromination of Carboxylic Acids: The Hell-Volhard-Zelinsky Reaction

Alpha halogenation, as described in the previous section, occurs readily with ketones and aldehydes, but not with carboxylic acids, esters, or amides. This is likely due to the fact that these functional groups are not readily converted to their corresponding enols. Nevertheless, carboxylic acids do undergo alpha halogenation when treated with bromine in the presence of $PBr_3$.

1) $Br_2$, $PBr_3$
2) $H_2O$

**(90%)**

This process, called the **Hell-Volhard-Zelinsky reaction**, is believed to occur via the following sequence of events:

$PBr_3$ → An acid halide ⇌ An acid halide enol → $Br_2$ → $H_2O$ →

The carboxylic acid first reacts with $PBr_3$ to form an acid halide, which exists in equilibrium with an enol. This enol then functions as a nucleophile and undergoes halogenation at the α position. Finally, hydrolysis regenerates the carboxylic acid.

## CONCEPTUAL CHECKPOINT

**22.11** Predict the major product for each of the following transformations:

(a) 1) $Br_2$, $PBr_3$ 2) $H_2O$ ?

(b) 1) $Br_2$, $PBr_3$ 2) $H_2O$ ?

**22.12** Identify the reagents that you would use to accomplish each of the following transformations (you will also need to use reactions from previous chapters).

(a) (b) (c)

### Alpha Halogenation in Basic Conditions: The Haloform Reaction

We have seen that ketones will undergo alpha halogenation in acid-catalyzed conditions. A similar result can also be achieved in basic conditions:

NaOH, $Br_2$

The base abstracts a proton to form the enolate, which then functions as a nucleophile and undergoes alpha halogenation.

With hydroxide as the base, the concentration of enolate is always low but is continuously maintained by the equilibrium as the reaction proceeds.

When more than one α proton is present, it is difficult to achieve monobromination in basic conditions, because the brominated product is more reactive and rapidly undergoes further bromination.

NaOH, $Br_2$

After the first halogenation reaction occurs, the presence of the halogen renders the α position more acidic, and the second halogenation step occurs more rapidly. As a result, it is often difficult to isolate the monobrominated product.

When a methyl ketone is treated with excess base and excess halogen, a reaction occurs in which a carboxylic acid is produced after acidic workup.

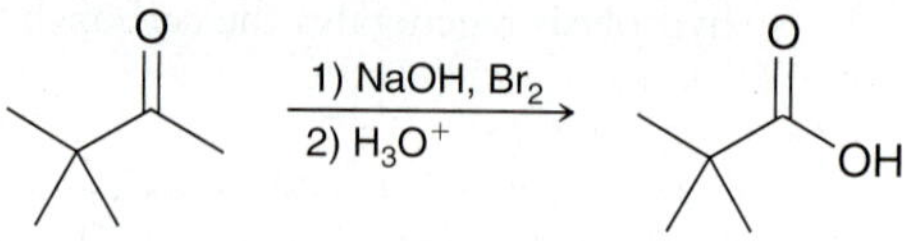

1) NaOH, $Br_2$
2) $H_3O^+$

OH

The mechanism is believed to involve several steps. First, the α protons are removed and replaced with bromine atoms, one at a time. Then, the tribromomethyl group can function as a leaving group, resulting in a nucleophilic acyl substitution reaction.

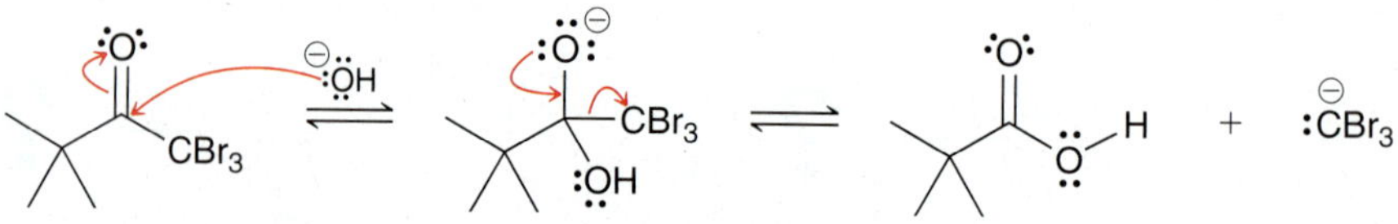

Notice that the leaving group is a carbanion, which violates the rule that we saw in Section 21.7 (re-form the carbonyl if possible but avoid expelling $C^-$). The discussion in Chapter 21 indicated that there would be some exceptions to the rule, and here is the first exception. The negative charge on carbon is stabilized in this case by the electron-withdrawing effects of the three bromine atoms, rendering $CBr_3^-$ a suitable leaving group. The resulting carboxylic acid is then deprotonated, producing a carboxylate ion and $CHBr_3$ (called bromoform). The formation of a carboxylate ion drives the reaction to completion.

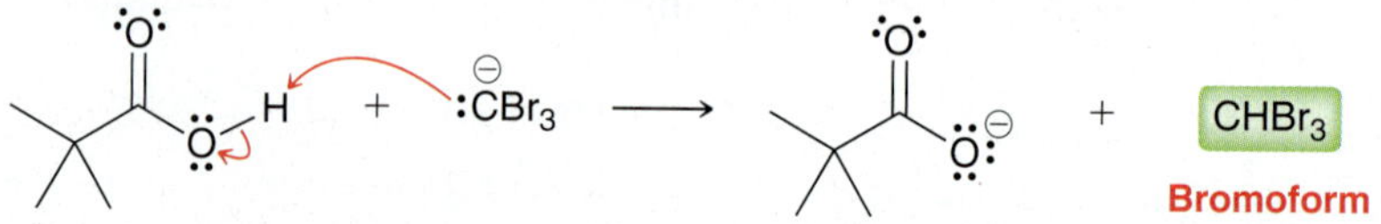

$CHBr_3$

Bromoform

The same process occurs with chlorine and iodine, and the by-products are chloroform and iodoform, respectively. This reaction is named after the by-product that is formed and is called the **haloform reaction**. The reaction must be followed by treatment with a proton source to protonate the carboxylate ion and form the carboxylic acid. This process is synthetically useful for converting methyl ketones into carboxylic acids.

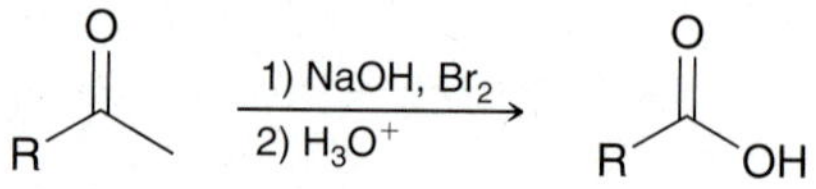

1) NaOH, $Br_2$
2) $H_3O^+$

The haloform reaction is most efficient when the other side of the ketone has no α protons.

## CONCEPTUAL CHECKPOINT

**22.13** Predict the major product obtained when each of the following compounds is treated with bromine ($Br_2$) together with sodium hydroxide (NaOH) followed by aqueous acid ($H_3O^+$):

(a) 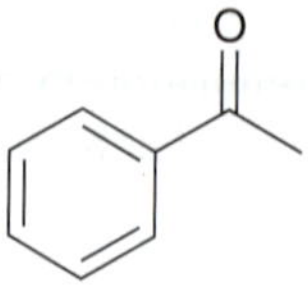 (b) 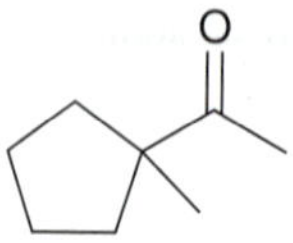 (c)

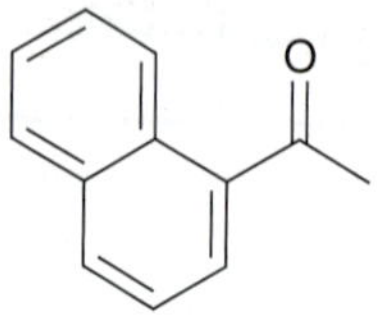

**22.14** Identify the reagents that you would use to accomplish each of the following transformations (you will need to use reactions from previous chapters):

(a) 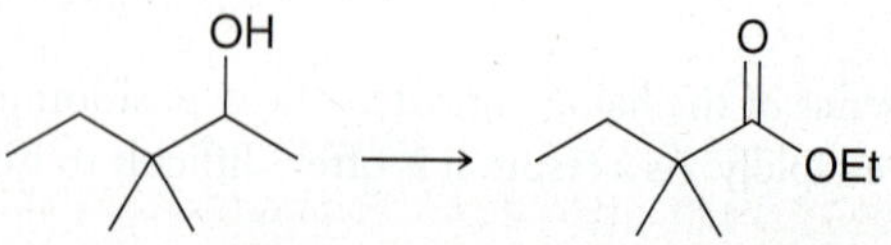 OH → OEt

(b) 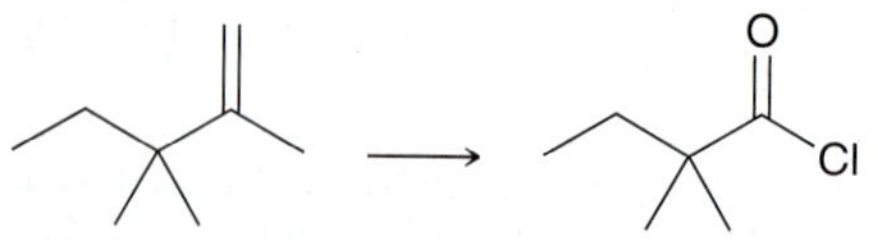 → Cl

(c) 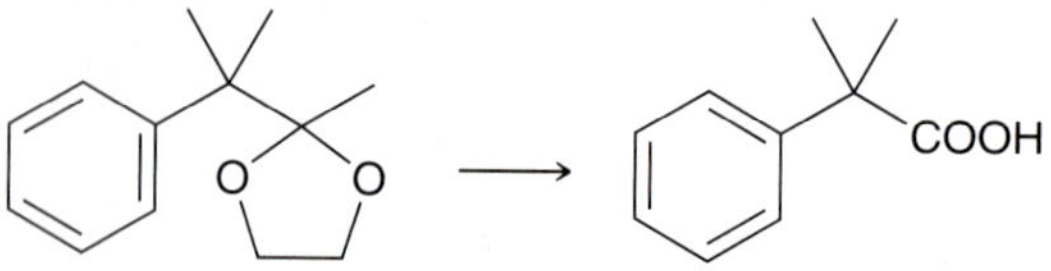 → COOH

(d) 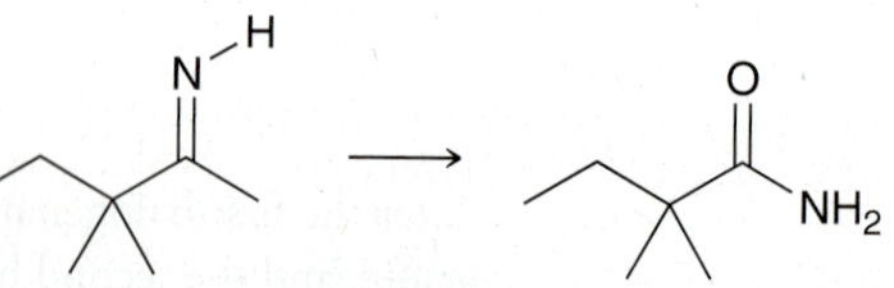 N–H → $NH_2$

## 22.3 Aldol Reactions

### Aldol Additions

Recall that when an aldehyde is treated with sodium hydroxide, both the aldehyde and the enolate will be present at equilibrium. Under such conditions, a reaction can occur between these two species. For example, treatment of acetaldehyde with sodium hydroxide gives 3-hydroxybutanal.

NaOH / $H_2O$ — α, β

The product exhibits both an aldehydic group and a hydroxyl group, and it is therefore called an aldol (*ald* for "aldehyde" and *ol* for "alcohol"). In recognition of the type of product formed, the reaction is called an **aldol addition reaction.** Notice that the hydroxyl group is located specifically at the β position relative to the carbonyl group. The product of an aldol addition reaction is always a β-hydroxy aldehyde or ketone (Mechanism 22.4).

### MECHANISM 22.4 ALDOL ADDITION

Proton transfer — The α position is deprotonated to form an enolate

Nucleophilic attack — The enolate serves as a nucleophile and attacks an aldehyde

Proton transfer — The resulting alkoxide ion is protonated to give the product

The mechanism for an aldol addition has three steps. In the first step, the aldehyde is deprotonated to form an enolate. Since hydroxide is used as the base, both the enolate and the aldehyde are present at equilibrium, and the enolate attacks the aldehyde. The resulting alkoxide ion is then protonated to yield the product. Notice the use of equilibrium arrows for every step of the mechanism. For most simple aldehydes, the position of equilibrium favors the aldol product.

NaOH, $H_2O$

(25%) (75%)

However, for most ketones, the aldol product is not favored, and poor yields are common.

NaOH, $H_2O$

(80%) (20%)

In this reaction, the reverse process is favored, that is, the β-hydroxy ketone is converted back into cyclohexanone more readily than the forward reaction. This reverse process, which is called a **retro-aldol reaction** (Mechanism 22.5), can be exploited in many situations (an example can be seen in the upcoming Practically Speaking box on muscle power).

## MECHANISM 22.5 RETRO-ALDOL REACTION

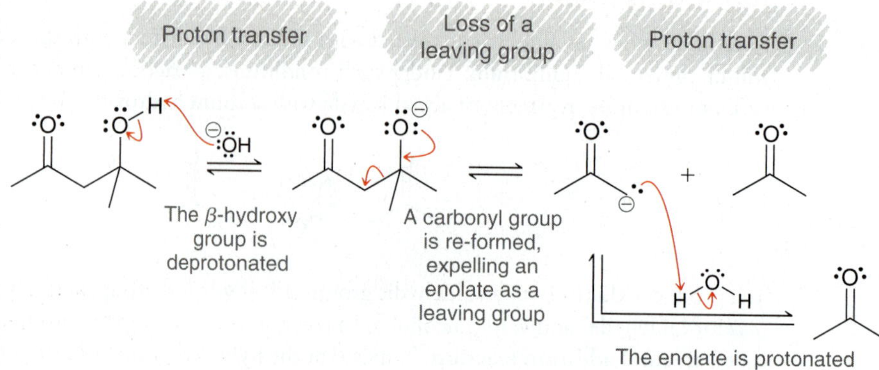


Once again, the mechanism has three steps. In fact, these three steps are simply the reverse of the three steps for an aldol addition. Notice that the second step in this mechanism involves re-formation of a carbonyl group to expel an enolate ion as a leaving group. This reaction therefore represents another exception to the rule about avoiding the expulsion of $C^-$. This exception is justified because an enolate is resonance stabilized, with the vast majority of the negative charge residing on the oxygen atom.

# SKILLBUILDER

## 22.2 PREDICTING THE PRODUCTS OF AN ALDOL ADDITION REACTION

**LEARN the skill**

Predict the product of the aldol addition reaction that occurs when the following aldehyde is treated with aqueous sodium hydroxide.

### SOLUTION

The best way to draw the product of an aldol addition reaction is to consider the mechanism. In the first step, the α position of the aldehyde is deprotonated to form an enolate.

**STEP 1**
Consider all three steps of the mechanism.

Remember that the aldehydic proton is not acidic. Only the α protons are acidic. In the second step of the mechanism, the enolate attacks an aldehyde, forming an alkoxide ion.

Finally, the alkoxide intermediate is protonated to give the product.

In total, there are three mechanistic steps: (1) deprotonation, (2) nucleophilic attack, and (3) protonation. Notice that the product is a β-hydroxy aldehyde. When drawing the product of an aldol addition reaction, always make sure that the hydroxyl group is located at the β position relative to the carbonyl group.

**STEP 2**
Double-check your answer to make sure that the product has an OH group at the β position.

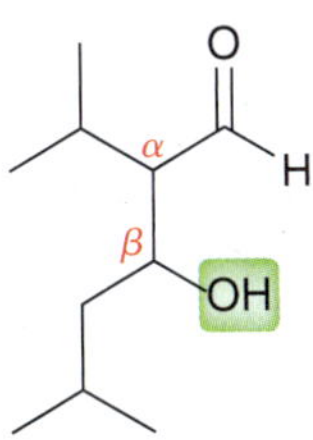

**PRACTICE the skill** **22.15** Predict the major product obtained when each of the following aldehydes is treated with aqueous sodium hydroxide:

(a) (b) (c) (d)

**22.16** When each of the following ketones is treated with aqueous sodium hydroxide, the aldol product is obtained in poor yields. In these cases, special distillation techniques are used to increase the yield of aldol product. In each case, predict the aldol addition product that is obtained and propose a mechanism for its formation:

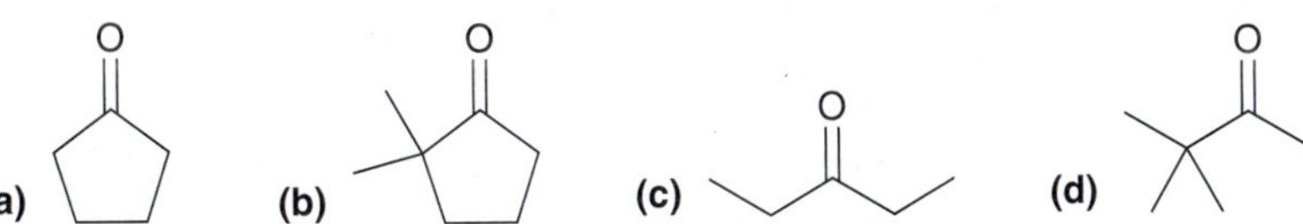

**22.17** When treated with aqueous sodium hydroxide, 2,2-dimethylbutanal does not undergo an aldol addition reaction. Explain this observation.

**22.18** An alcohol of molecular formula $C_4H_{10}O$ was treated with PCC to produce an aldehyde that exhibits exactly three signals in its $^1H$ NMR spectrum. Predict the aldol addition product that is obtained when this aldehyde is treated with aqueous sodium hydroxide.

**22.19** Using acetaldehyde as your only source of carbon, show how you would prepare 1,3-butanediol.

need more **PRACTICE?** **Try Problem 22.72**

## Aldol Condensations

When heated in acidic or basic conditions, the product of an aldol addition reaction will undergo elimination to produce unsaturation between the α and β positions:

$$\xrightarrow[\text{Heat}]{[H^+] \text{ or } [HO^-]} \quad + \; H_2O$$

## practically speaking — Muscle Power

Retro-aldol reactions play a vital role in many biochemical processes, including one of the processes by which energy is generated for our muscles. We first mentioned in Section 13.11 that energy in our bodies is stored in the form of ATP molecules. That is, energy from the food we eat is used to convert ADP into ATP, which is stored. When energy is needed, ATP is broken down to ADP, and the energy that is released can be used for various life processes, such as muscle contraction.

**Adenosine diphosphate (ADP)**

**Adenosine triphosphate (ATP)**

Notice that the structural difference between ATP and ADP is in the number of phosphate groups present (two in the case of ADP, three in the case of ATP). The ATP molecules in our muscles are used for any activity that requires a short burst of energy, such as a tennis serve, jumping, or throwing a ball. For activities that last longer than a second, such as sprinting, ATP molecules must be synthesized on the spot. This is initially achieved by a process called glycolysis, in which a molecule of glucose (obtained from metabolism of the carbohydrates we eat) is converted into two molecules of pyruvic acid.

**Glucose** → 2 **Pyruvic acid**

Glycolysis involves many steps and is accompanied by the conversion of two molecules of ADP into ATP. This metabolic process therefore generates the necessary ATP for muscle contraction. One of the steps involved in glycolysis is a retro-aldol reaction, which is achieved with the help of an enzyme called aldolase.

Aldolase

The end product of glycolysis is pyruvic acid, which is used as a starting material in a variety of biochemical processes.

Glycolysis provides ATP for activities lasting up to 1.5 min. Activities that exceed this time frame, such as long-distance running, require a different process for ATP generation, called the citric acid cycle. Unlike glycolysis, which can be achieved without oxygen (it is an anaerobic process), the citric acid cycle requires oxygen (it is an aerobic process). This explains why we breathe more rapidly during and after strenuous activity. The term "aerobic workout" is commonly used to refer to a workout that utilizes ATP that was generated by the citric acid cycle (an aerobic process) rather than glycolysis (an anaerobic process). A casual athlete can sense a shift from anaerobic ATP synthesis to aerobic ATP synthesis after about 1.5 min. Seasoned athletes will detect the shift occurring after about 2.5 min. If you watch the summer Olympics, you are likely aware of the distinction between sprinting and long-distance running. Athletes can run faster in a sprinting race, which relies mostly on glycolysis for energy production. Long-distance running requires the citric acid cycle for energy production.

Practically, this transformation is most readily achieved when an aldol addition is performed at elevated temperature. Under these basic conditions, the aldol addition reaction occurs, followed by dehydration to give an α,β-unsaturated product.

NaOH, $H_2O$
Heat
+ $H_2O$

This two-step process (aldol addition plus dehydration) is called an **aldol condensation**. The term *condensation* is used to refer to any reaction in which two molecules undergo addition accompanied by the loss of a small molecule such as water, carbon dioxide, or nitrogen gas. In the case of aldol condensations, water is the small molecule that is lost. Notice that the product of an aldol addition is a β-hydroxy aldehyde or ketone, while the product of an aldol condensation is an α,β-unsaturated aldehyde or ketone.

Aldol addition

β-Hydroxy aldehyde

α, β-Unsaturated aldehyde

+ $H_2O$

Aldol condensation

## MECHANISM 22.6 ALDOL CONDENSATION

### PART 1: ALDOL ADDITION

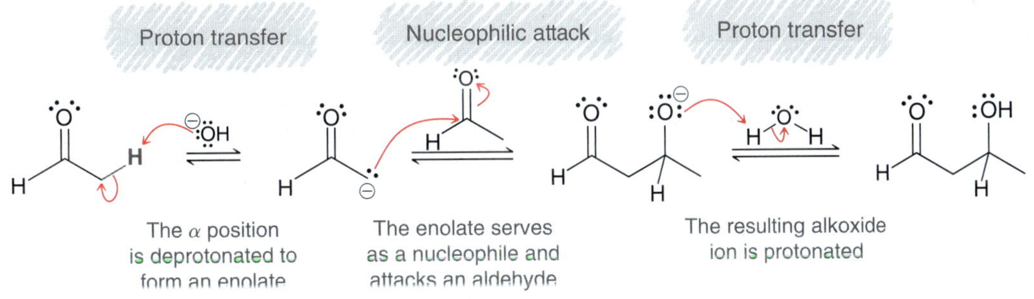


### PART 2: ELIMINATION OF $H_2O$

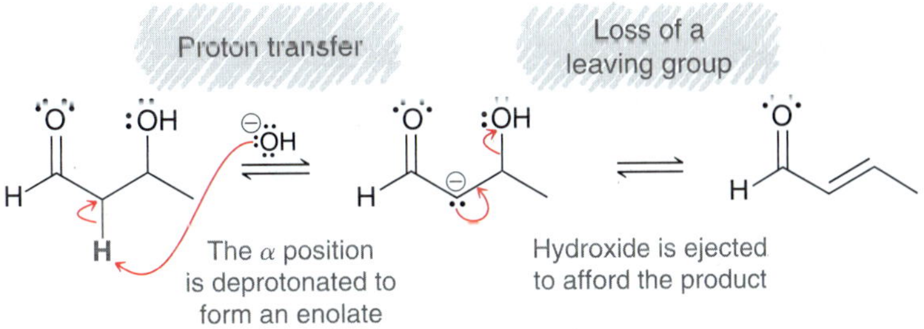

An aldol condensation (Mechanism 22.6) has two parts. The first part is just an aldol addition reaction, which has three mechanistic steps. The second part has two steps that accomplish the elimination of water. Normally, alcohols do not undergo dehydration in the presence of a strong base, but here, the presence of the carbonyl group enables the dehydration reaction to occur. The $\alpha$ position is first deprotonated to form an enolate ion, followed by expulsion of a hydroxide ion to produce $\alpha,\beta$ unsaturation. This two-step process, which is different from the elimination reactions we saw in Chapter 8, is called an **E1cb mechanism**. In an E1cb mechanism, the leaving group only leaves after deprotonation occurs.

In cases where two stereoisomeric $\pi$ bonds can be formed, the product with fewer steric interactions is generally the major product.

O H NaOH Heat O H **Major** + O H **Minor**

**LOOKING BACK**

The stability of conjugated $\pi$ systems was discussed in Section 17.2.

In this example, formation of the *trans* $\pi$ bond is favored over formation of the *cis* $\pi$ bond.

The driving force for an aldol condensation is formation of a conjugated system. The reaction conditions required for an aldol condensation are only slightly more vigorous than the conditions required for an aldol addition reaction. Usually, an aldol condensation can be achieved by simply performing the reaction at an elevated temperature. In fact, in some cases, it is not even possible to isolate the $\beta$-hydroxyketone. As an example, consider the following case:

O NaOH O OH **Not isolated** O

In this case, the aldol addition product cannot be isolated. Even at moderate temperatures, only the condensation product is obtained, because the condensation reaction involves formation of a highly conjugated $\pi$ system. Even in cases where the aldol addition product can be isolated (by performing the reaction at a low temperature), the yields for condensation reactions are often much greater than the yields for addition reaction. The following example illustrates this point:

O NaOH 10°C O OH **(20%)**

NaOH 100°C O **(90%)**

When the reaction is performed at low temperature, the aldol addition product is obtained, but the yield is very poor. As explained earlier, the starting material is a ketone, and the equilibrium does not favor formation of the aldol addition product. However, when the reaction is performed at an elevated temperature, the aldol condensation product is obtained in very good yield, because the equilibrium is driven by formation of a conjugated $\pi$ system.

# SKILLBUILDER

## 22.3 DRAWING THE PRODUCT OF AN ALDOL CONDENSATION

**LEARN the skill**

Predict the product of the aldol condensation that occurs when the following ketone is heated in the presence of aqueous sodium hydroxide:

### SOLUTION

One way to draw the product is to draw the entire mechanism, but for aldol condensations, there is a faster method for drawing the product. Begin by identifying the α protons:

**STEP 1**
Identify the α positions.

In order to achieve an aldol condensation, one of the α carbon atoms must bear at least two protons. In this case, one of the α positions has three protons, and the other α position has none.

Next draw two molecules of the ketone, oriented such that two α protons of one molecule are directly facing the carbonyl group of the other molecule:

**STEP 2**
Redraw two molecules of the ketone.

When drawn this way, it is easier to predict the product without having to draw the entire mechanism. Simply remove two α protons and the oxygen atom (shown in red) and replace them with a double bond:

**STEP 3**
Remove $H_2O$ and replace with a C=C bond.

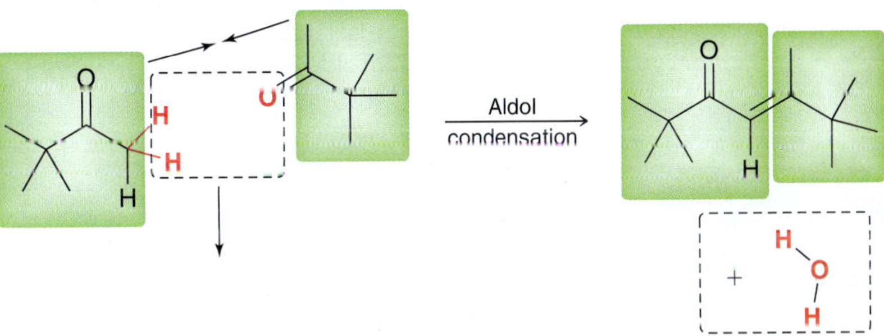


The product is an α,β-unsaturated ketone, and water is liberated as a by product. In this case, two stereoisomers are possible, so we draw the product with fewer steric interactions.

**STEP 4**
Draw the isomer with fewer steric interactions.

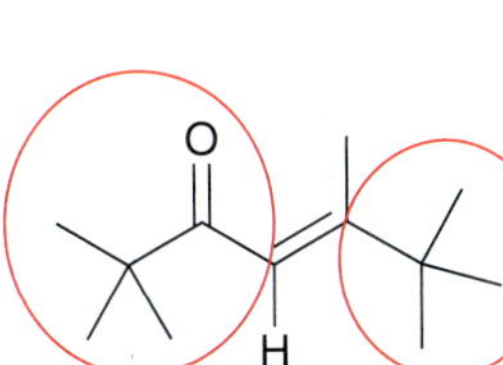

Major product

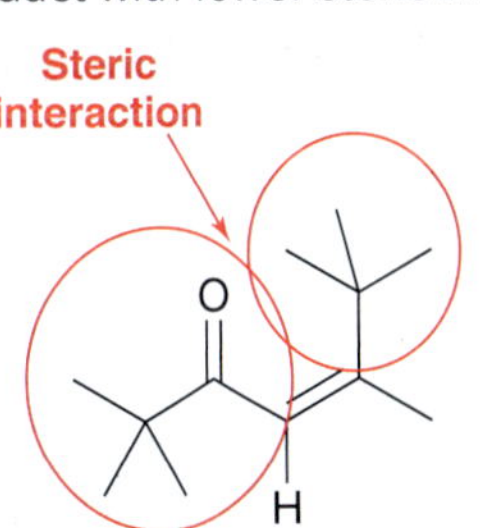


Not obtained

**PRACTICE the skill** **22.20** Draw the condensation product obtained when each of the following compounds is heated in the presence of aqueous sodium hydroxide:

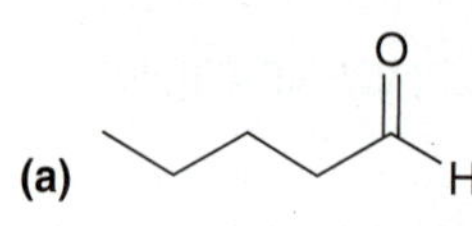


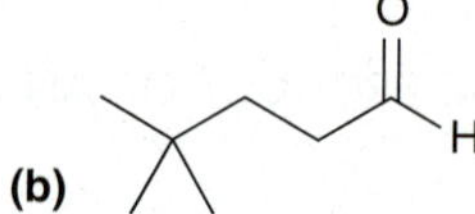


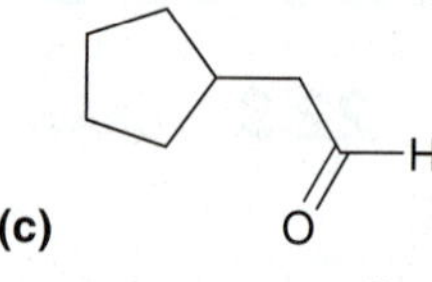


(d) (e) (f)

**APPLY the skill** **22.21** Identify the starting aldehyde or ketone needed to make each of the following compounds via an aldol condensation:

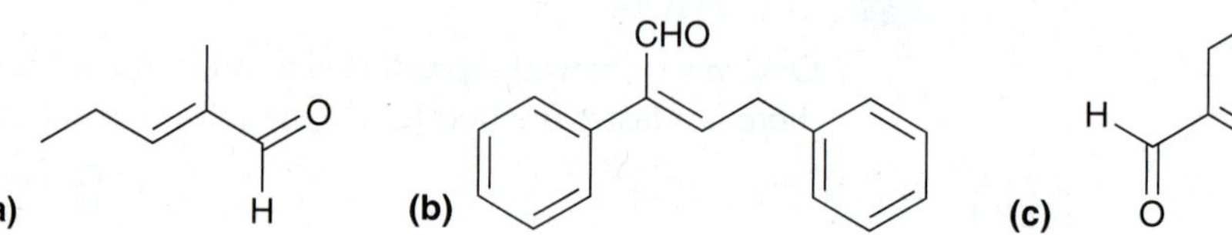


**22.22** When 2-butanone is heated in the presence of aqueous sodium hydroxide, four condensation products are obtained. Draw all four products.

need more **PRACTICE?** **Try Problems 22.71, 22.84c**

## practically speaking — Why Meat from Younger Animals Is Softer

As mentioned in Chapter 21, proteins are important biological compounds comprised of multiple amide groups. Proteins will be discussed in greater detail in Chapter 25. One of the most abundant proteins found in mammals is called collagen. It is found in skin, bones, and teeth and is used to make gelatin and gelatin-based desserts, such as Jell-O. Individual molecules of collagen can be isolated from young animals, but not from older animals, because with age, collagen molecules cross-link with each other via an aldol condensation reaction. First, amino groups located on side chains of collagen are converted to aldehyde groups via a process called oxidative deamination. Then, the resulting aldehyde groups can undergo an aldol condensation, as shown.

This process results in the cross-linking of two collagen molecules. As an animal ages, the number of cross-linked proteins increases at the expense of the individual collagen molecules. For this reason, meat obtained from older animals is usually tougher than meat obtained from younger animals.

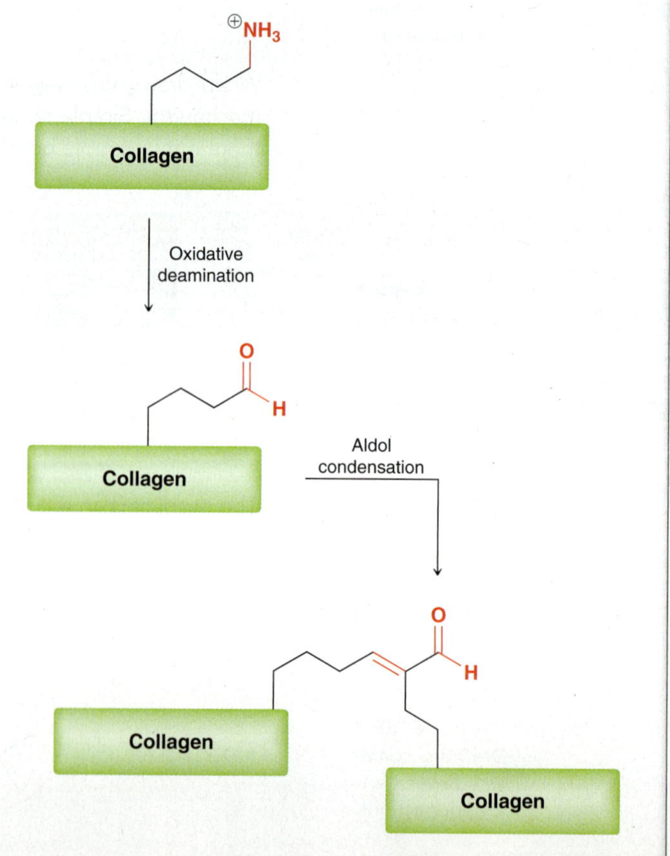

## Crossed Aldol Reactions

Until now, we have focused on symmetrical aldol reactions, that is, aldol reactions that occur between two identical partners. In this section, we explore **crossed aldol**, or **mixed aldol, reactions**, which are aldol reactions that can occur between different partners. As an example, consider what happens when a mixture of acetaldehyde and propionaldehyde is treated with a base. Under these circumstances, four possible aldol products can be formed (Figure 22.3). The first two products are formed from symmetrical aldol reactions, while the latter two products are formed from crossed aldol reactions. Reactions that form mixtures of products are of little use, and therefore, crossed aldol reactions are only efficient if they can be performed in a way that minimizes the number of possible products. This is best accomplished in either of the following ways:

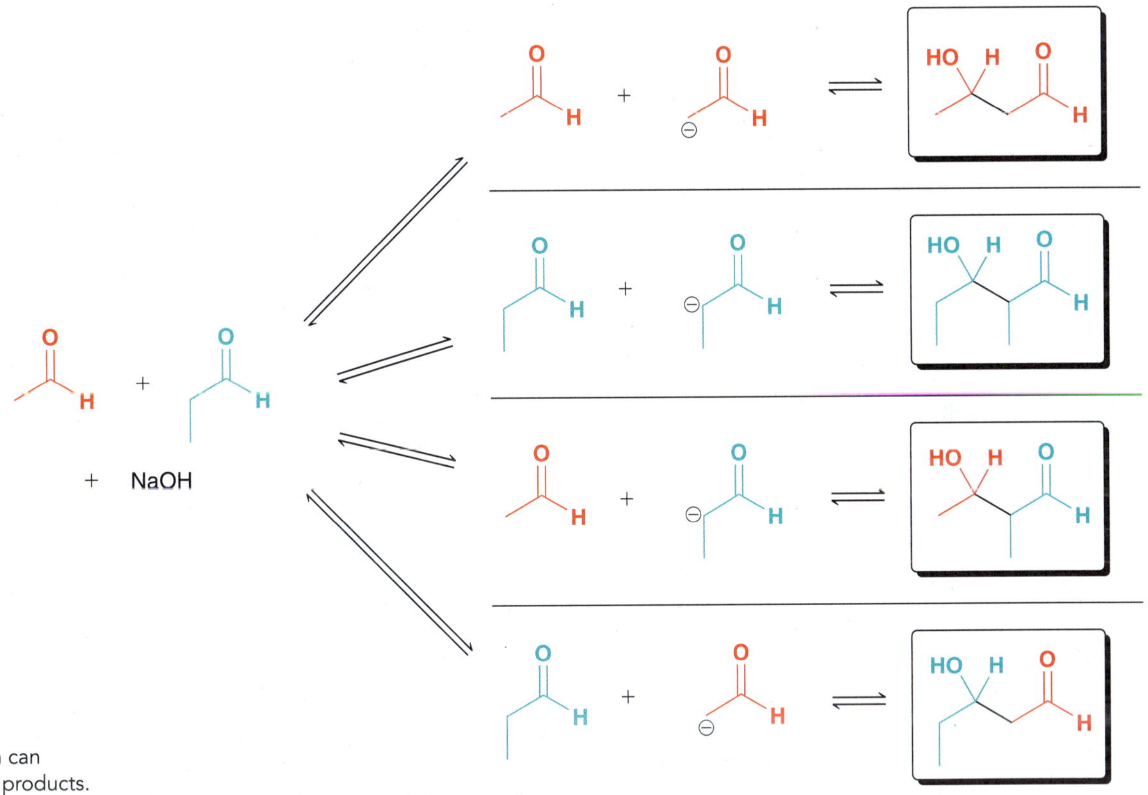

**FIGURE 22.3** A crossed aldol reaction can produce many different products.

1. If one of the aldehydes lacks α protons and possesses an unhindered carbonyl group, then a crossed aldol can be performed. As an example, consider what happens when a mixture of formaldehyde and propionaldehyde is treated with a base.

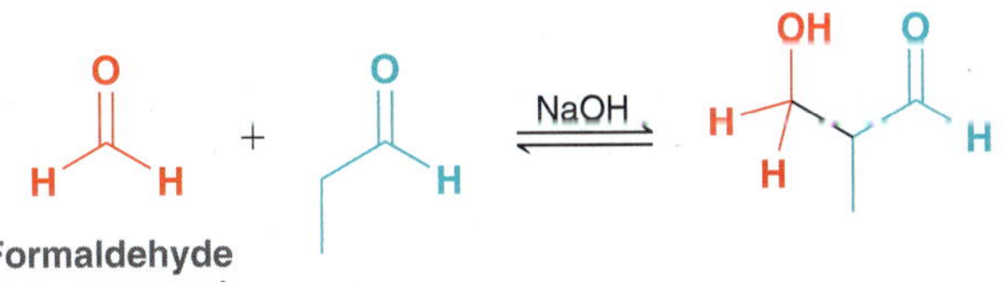

Formaldehyde (no $\alpha$ protons)

In this case, only one major aldol product is produced. Why? Formaldehyde has no α protons and therefore cannot form an enolate. As a result, only the enolate formed from propionaldehyde is present in solution. Under these conditions, there are only two possible products. The enolate can attack a molecule of propionaldehyde to produce a symmetrical aldol reaction, or the enolate can attack a molecule of formaldehyde to produce a crossed aldol reaction. The latter occurs more rapidly because the carbonyl group of formaldehyde is less hindered and more reactive than the carbonyl group of propionaldehyde. As a result, one product predominates.

Crossed aldol reactions can also be performed with benzaldehyde.

**Benzaldehyde (no α protons)**

When benzaldehyde is used, the dehydration step is spontaneous, and the equilibrium favors the condensation product rather than the addition product, because the condensation product is highly conjugated.

NaOH / Heat — Not isolated

Aldol reactions involving aromatic aldehydes generally produce condensation reactions.

2. Crossed aldol reactions can also be performed using LDA as a base.

LDA — $H_2O$

Recall that LDA causes irreversible enolate formation. If acetaldehyde is added dropwise to a solution of LDA, the result is a solution of enolate ions. Propionaldehyde can then be added dropwise to the mixture, resulting in a crossed aldol addition that produces one major product. This type of process is called a **directed aldol addition**, and its success is limited by the rate at which enolate ions can equilibrate. In other words, it is possible for an enolate ion to function as a base (rather than a nucleophile) and deprotonate a molecule of propionaldehyde. If this process occurs too rapidly, then a mixture of products will result.

## SKILLBUILDER

### 22.4 IDENTIFYING THE REAGENTS NECESSARY FOR A CROSSED ALDOL REACTION

**LEARN the skill** Identify the reagents necessary to produce the following compound via an aldol reaction:

**SOLUTION**

Begin by identifying the α and β positions.

**STEP 1**
Identify the α and β positions.

**STEP 2**
Use a retrosynthetic analysis that focuses on the α,β bond.

Now apply a retrosynthetic analysis. The bond between the α and β positions is the bond formed during an aldol reaction. To draw the necessary starting materials, break apart the bond between the α and β positions, drawing a carbonyl group in place of the OH group.

**STEP 3**
Identify an appropriate base.

These are the two starting carbonyl compounds. They are not identical, so a crossed aldol reaction is required.

The final step is to determine which base should be used. Both starting compounds have α protons, so hydroxide cannot be used because it would result in a mixture of products. In this case, a crossed aldol can only be accomplished if LDA is used to achieve a directed aldol reaction.

1) LDA
2) [aldehyde]
3) $H_2O$

In the first step, the symmetrical ketone is irreversibly and completely deprotonated by LDA to produce a solution of enolate ions. Then, the aldehyde is added dropwise to the solution to achieve a directed aldol addition.

**PRACTICE the skill** **22.23** Identify the reagents necessary to produce each of the following compounds via an aldol reaction.

(a) (b) (c) (d) (e)

**APPLY the skill** **22.24** Using formaldehyde and acetaldehyde as your only sources of carbon atoms, show how you could make each of the following compounds. You may find it helpful to review acetal formation (Section 20.5).

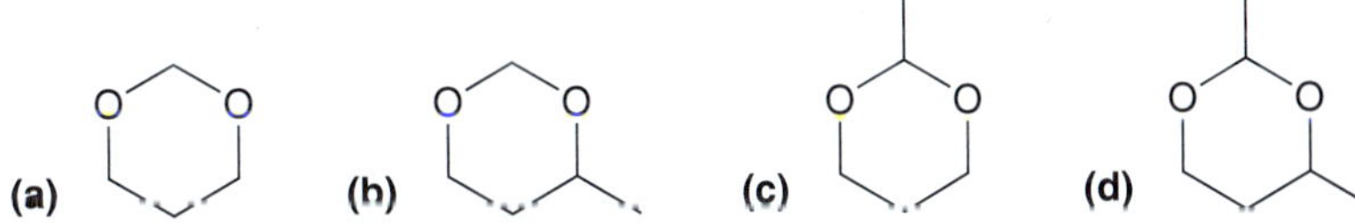


need more **PRACTICE?** **Try Problems 22.67, 22.68**

## Intramolecular Aldol Reactions

Compounds that possess two carbonyl groups can undergo intramolecular aldol reactions. Consider the reaction that occurs when 2,5-hexanedione is heated in the presence of aqueous sodium hydroxide.

NaOH, heat

In this case, a cyclic product is formed. The accepted mechanism for this process is nearly identical to the mechanism for any other aldol condensation but with one notable difference—the enolate and the carbonyl group are both present in the same structure, resulting in an intramolecular attack.

Nucleophilic site

Electrophilic site

Intramolecular aldol reactions show a preference for formation of five- and six-membered rings. Smaller rings are possible but are generally not observed.

Nucleophilic site

Electrophilic site

Not isolated (too strained)

It is conceivable that the three-membered ring might be formed initially, but recall that the products of aldol reactions are determined by equilibrium concentrations. Once the equilibrium has been achieved, the strained, three-membered ring is not present in substantial amounts because the equilibrium favors formation of a nearly strain-free product.

### CONCEPTUAL CHECKPOINT

**22.25** Draw a mechanism for the following transformation:

NaOH, heat

**22.26** The reaction in the previous problem is an equilibrium process. Draw a mechanism of the reverse process. That is, draw a mechanism showing conversion of the conjugated, cyclic enone into the acyclic dione.

**22.27** When 2,6-heptanedione is heated in the presence of aqueous sodium hydroxide, a condensation product with a six-membered ring is obtained. Draw the product and show a mechanism for its formation.

## 22.4 Claisen Condensations

### The Claisen Condensation

Like aldehydes and ketones, esters also exhibit reversible condensation reactions.

EtO — 1) NaOEt 2) $H_3O^+$ — EtO

**(75%)**

This type of reaction is called a **Claisen condensation**, and a mechanism for this process is shown in Mechanism 22.7.

## MECHANISM 22.7 CLAISEN CONDENSATION

Proton transfer — The α position is deprotonated to form an ester enolate

Nucleophilic attack — The enolate serves as a nucleophile and attacks an ester, forming a tetrahedral intermediate

Loss of a leaving group — The carbonyl group is re-formed by ejecting an alkoxide ion

Proton transfer — The α position is deprotonated to form a doubly stabilized enolate

The first two steps of this mechanism are much like an aldol addition. The ester is first deprotonated to form an enolate, which then functions as a nucleophile and attacks another molecule of the ester. The difference between an aldol reaction and a Claisen condensation is the fate of the tetrahedral intermediate. In a Claisen condensation, the tetrahedral intermediate can expel a leaving group to re-form a C=O bond. The Claisen condensation is simply a nucleophilic acyl substitution reaction in which the nucleophile is an ester enolate and the electrophile is an ester. The product of this reaction is a β-keto ester.

Notice that the last step in the mechanism is deprotonation of the β-keto ester to give a doubly stabilized enolate. This deprotonation step cannot be avoided, because the reaction occurs under basic conditions. Each molecule of base (alkoxide ion) is converted into the doubly stabilized enolate, which is a favorable transformation (downhill in energy). In fact, the deprotonation step at the end of the mechanism provides a driving force that causes the equilibrium to favor condensation. As a result, the base is not a catalyst but is actually consumed as the reaction proceeds. After the reaction is complete, it is necessary to use a mild acid in order to protonate the doubly stabilized enolate.

When performing a Claisen condensation, the starting ester must have two α protons. If it only has one α proton, then the driving force for condensation is absent (a doubly stabilized enolate cannot be formed).

Hydroxide cannot be used as the base for a Claisen condensation because it can cause hydrolysis of the starting ester.

If hydroxide is used as a base, the ester is hydrolyzed to form a carboxylic acid, which is then irreversibly deprotonated to form a carboxylate salt, thereby preventing the Claisen condensation from occurring. Instead, Claisen condensations are achieved by using an alkoxide ion as the base. Specifically, the alkoxide used must be the same OR group that is present in the starting ester in order to avoid *transesterification*.

For example, if an ethyl ester is treated with methoxide, transesterification can convert the ethyl ester into a methyl ester.

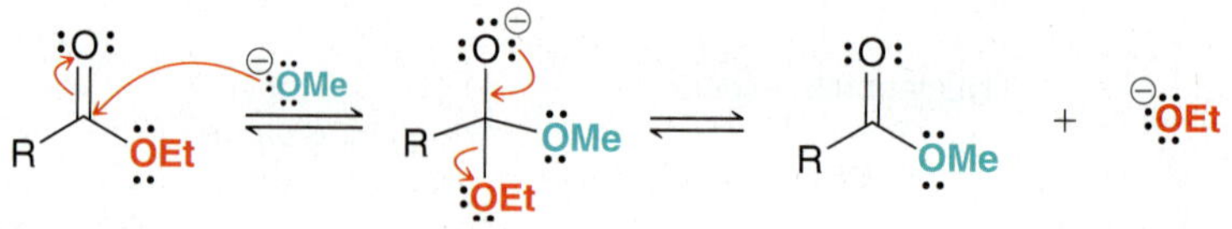

## Crossed Claisen Condensations

Claisen condensation reactions that occur between two different partners are called **crossed Claisen condensations**. Just as we saw with aldol reactions, crossed Claisen condensation reactions also produce a mixture of products and are only efficient if one of the two following criteria are met:

1. If one ester has no α protons and cannot form an enolate; for example:

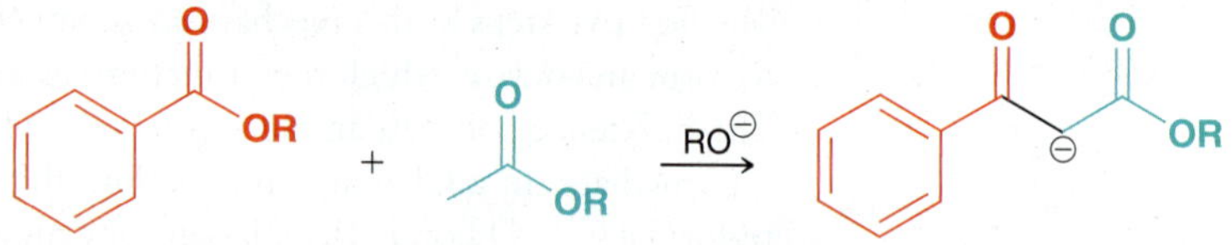

   In this reaction, the aryl ester lacks α protons and cannot form an enolate. Only the other ester is capable of forming an enolate, which reduces the number of possible products.

2. A directed Claisen condensation can be performed in which LDA is used as a base to irreversibly form an ester enolate, which is then treated with a different ester.

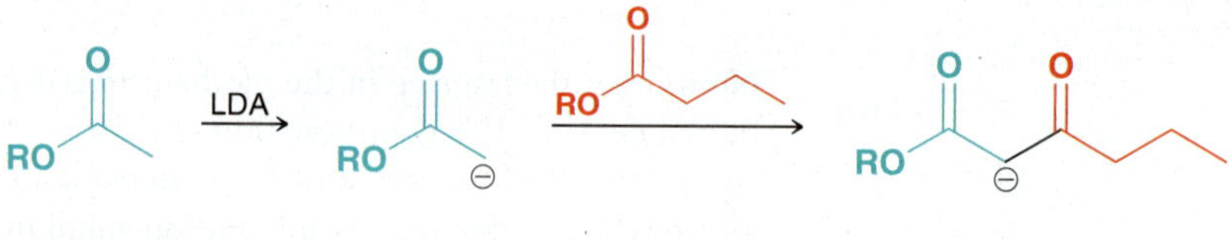

## Intramolecular Claisen Condensations: The Dieckmann Cyclization

Just as we saw with aldol reactions, Claisen condensations can also occur in an intramolecular fashion; for example:

**BY THE WAY**

When drawing the product of a Dieckmann cyclization, it is wise to use a numbering system to keep track of all carbon atoms, as shown here.

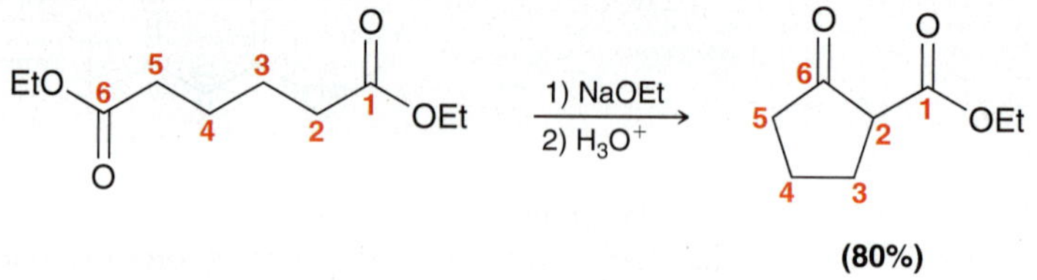

(80%)

This process is called a **Dieckmann cyclization**, and the product is a cyclic, β-keto ester. Notice that the ester enolate and the ester group are both present in the same molecule, resulting in an intramolecular attack.

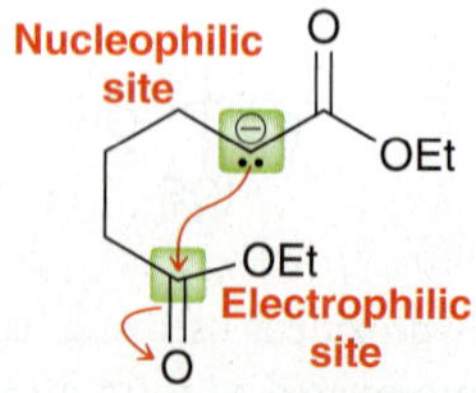

Intramolecular Claisen condensations show a preference for formation of five- and six-membered rings, just as we saw with intramolecular aldol reactions.

## CONCEPTUAL CHECKPOINT

**22.28** Identify the base you would use for each of the following transformations:

(a) O, OEt; 1) ? 2) $H_3O^+$ → O, O, OEt

(b) O, O; 1) ? 2) $H_3O^+$ → O, O, O

**22.29** Predict the major product obtained when each of the following compounds undergoes a Claisen condensation:

(a) OEt, O (b) OMe, O (c) O, OEt

**22.30** Identify the reagents that you would use to produce each of the following compounds using a Claisen condensation:

(a) O, O, OEt (b) O, O, OEt (c) COOMe, O (d) COOMe, O (e) O, O, O

**22.31** Predict the product of the Dieckmann cyclization that occurs when each of the following compounds is treated with sodium ethoxide, followed by acid work-up:

(a) O, OEt, OEt, O (b) EtO, OEt, O, O (c) O, O, EtO, OEt

**22.32** When the following compound is treated with sodium ethoxide, followed by acid work-up, two condensation products are obtained, both of which are produced via Dieckmann cyclizations. Draw both products.

O, O, EtO, OEt

## 22.5 Alkylation of the Alpha Position

### Alkylation via Enolate Ions

The α position of a ketone can be alkylated via a two-step process: (1) formation of an enolate followed by (2) treating the enolate with an alkyl halide.

O; 1) LDA 2) RX → O, R

In this process, the enolate ion functions as a nucleophile and attacks the alkyl halide in an $S_N2$ reaction.

O; R—X → O, R

The usual restrictions for $S_N2$ reactions apply, so that the alkyl halide should be either a methyl halide or a primary halide. With a secondary or tertiary alkyl halide, the enolate functions as a base instead of a nucleophile, and the alkyl halide undergoes elimination rather than substitution.

When forming the enolate for an alkylation process, the choice of base is important. Hydroxide or alkoxide ions cannot be used because under those conditions (1) both the ketone and its enolate are present at equilibrium and aldol reactions will compete with alkylation and (2) some of the base is not consumed (an equilibrium will be established) and the base can attack the alkyl halide directly, providing for competing $S_N2$ and E2 reactions.

Both of these problems are avoided by using a stronger base, such as LDA. With a stronger base, the ketone is irreversibly and quantitatively deprotonated to form enolate ions. Aldol reactions do not readily occur, because the ketone is no longer present. Also, the use of one equivalent of LDA ensures that none of the base survives after the enolate is formed.

With an unsymmetrical ketone, two possible enolates can be formed.

Base | Base

**Thermodynamic enolate (more substituted)** | **Kinetic enolate (less substituted)**

The more-substituted enolate is more stable and is called the *thermodynamic enolate*. The less-substituted enolate is less stable, but it is formed more rapidly and is therefore called the *kinetic enolate*. Figure 22.4 compares the pathways that lead to each of these enolates. Notice that formation of the kinetic enolate (red pathway) involves a lower energy barrier ($E_a$) and therefore occurs more rapidly, while the thermodynamic enolate (blue pathway) is the more stable enolate.

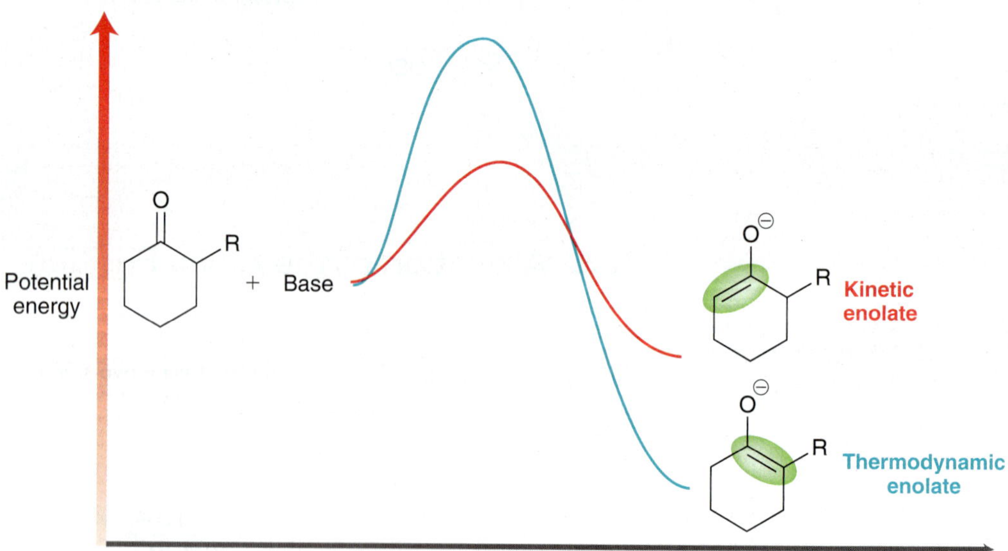


**FIGURE 22.4** An energy diagram showing two possible pathways for the deprotonation of an unsymmetrical ketone. One pathway leads to the kinetic enolate, and the other pathway leads to the thermodynamic enolate. The differences in energy between these two pathways has been exaggerated for clarity of presentation.

In general, it is possible to choose conditions that will favor formation of either enolate. When the kinetic enolate is desired, LDA is used at low temperature (−78°C). LDA is a sterically hindered base and can more readily deprotonate the less hindered α position, thereby forming the kinetic enolate. Low temperature is necessary to favor formation of the kinetic enolate and to prevent equilibration of the enolates via proton transfers. When the thermodynamic enolate is desired, a nonsterically hindered base (such as NaH) is used, and the reaction is performed at room temperature. Under these conditions, both enolates are formed and an equilibration process occurs that favors the more stable, thermodynamic enolate. In this way, alkylation can be achieved at either α position by carefully choosing the reagents and conditions.

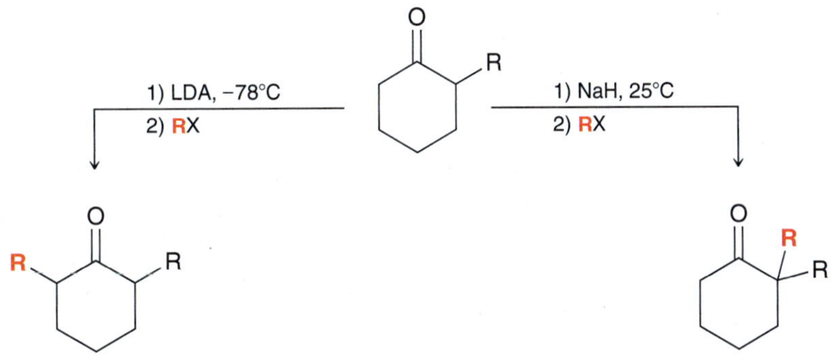

1) LDA, −78°C
2) RX

1) NaH, 25°C
2) RX

## CONCEPTUAL CHECKPOINT

**22.33** For each of the following reactions, predict the major product and propose a mechanism for its formation:

(a) 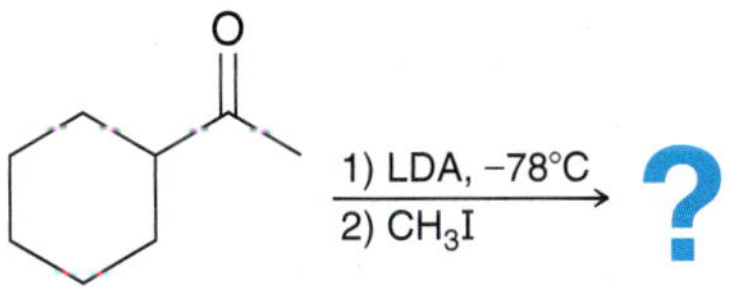 1) LDA, −78°C 2) $CH_3I$ ?

(b) 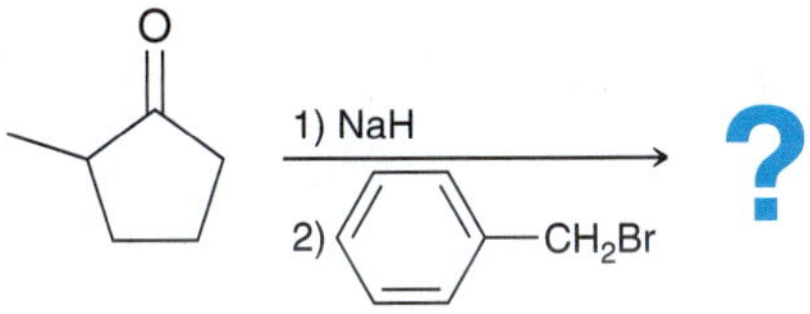 1) NaH 2) $C_6H_5CH_2Br$ ?

(c) 1) LDA 2) EtI 1) LDA, −78°C 2) $CH_3I$ ?

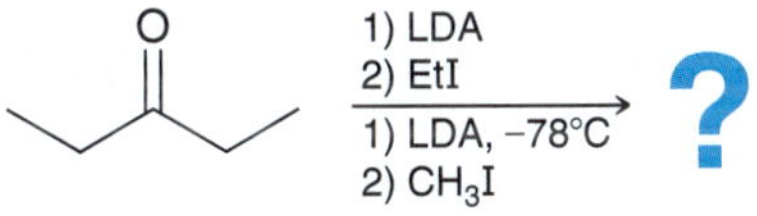

**22.34** Identify the reagents you would use to achieve the following transformation:

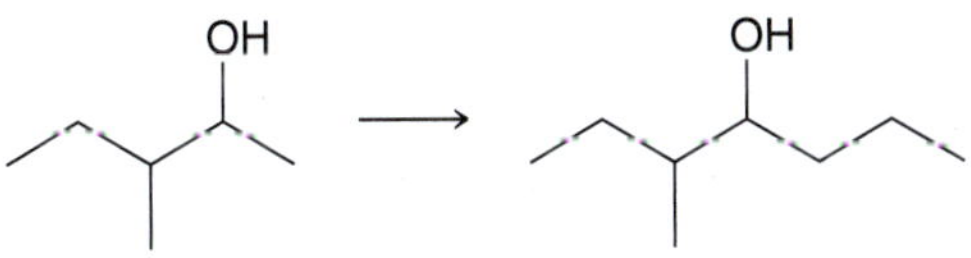

OH → OH

## The Malonic Ester Synthesis

The **malonic ester synthesis** is a technique that enables the transformation of a halide into a carboxylic acid with the introduction of two new carbon atoms.

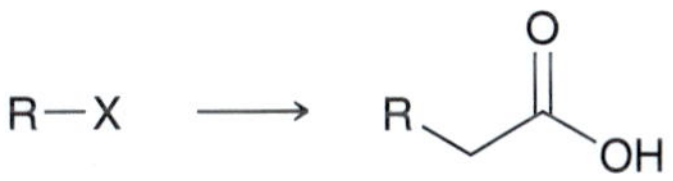

R—X ⟶ R–$CH_2$–COOH

One of the key reagents used to achieve this transformation is diethyl malonate.

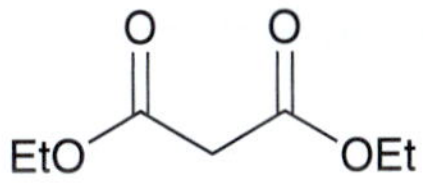

**Diethyl malonate**

This malonic ester is a fairly acidic compound. The first step of the malonic ester synthesis is the deprotonation of diethyl malonate, forming a doubly stabilized enolate.

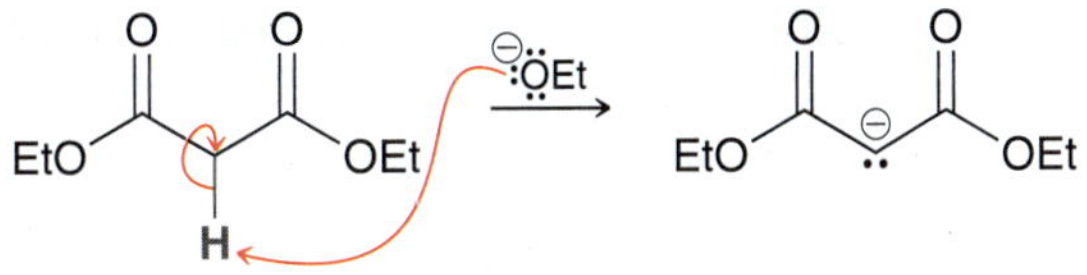

This enolate is then treated with an alkyl halide (RX), resulting in an alkylation step.

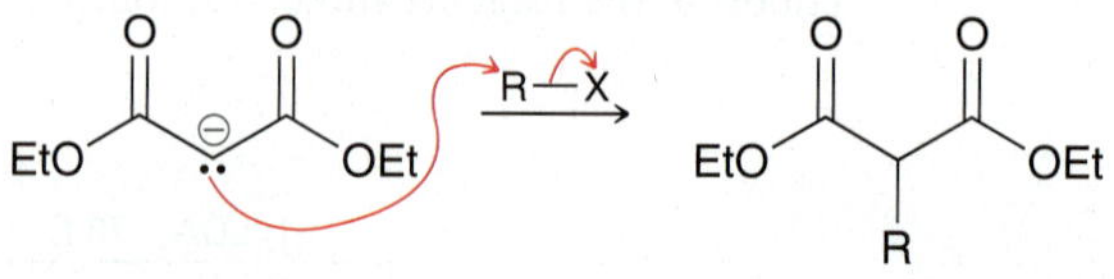

Treatment with aqueous acid then results in the hydrolysis of both ester groups:

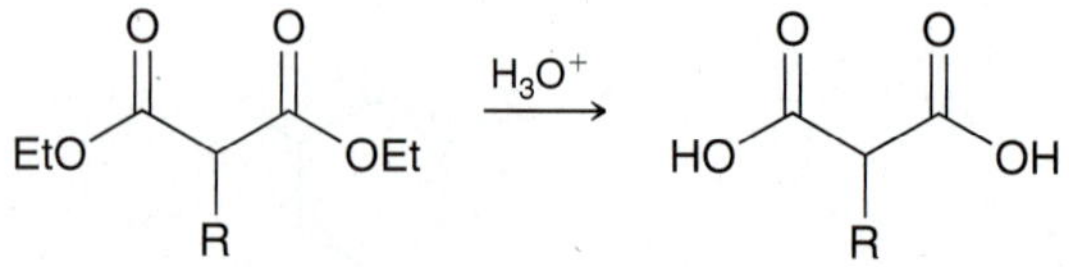

A mechanism for the acid-catalyzed hydrolysis of esters was discussed in Section 21.11. If hydrolysis is performed at elevated temperatures, the resulting 1,3-dicarboxylic acid will undergo **decarboxylation** to produce a monosubstituted acetic acid and carbon dioxide.

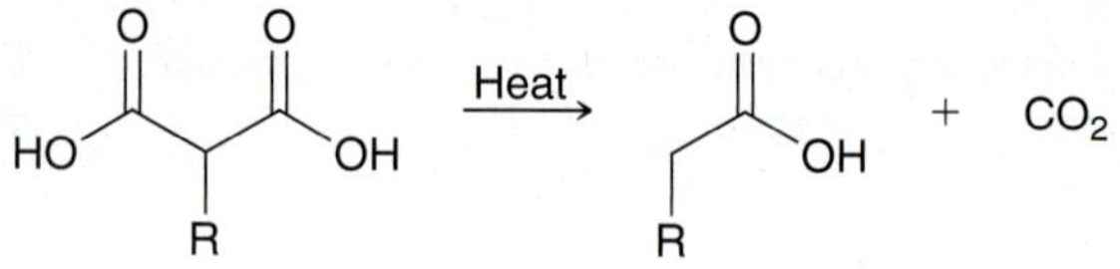

In this step, one of the carboxylic acid groups is expelled as shown in the following mechanism:

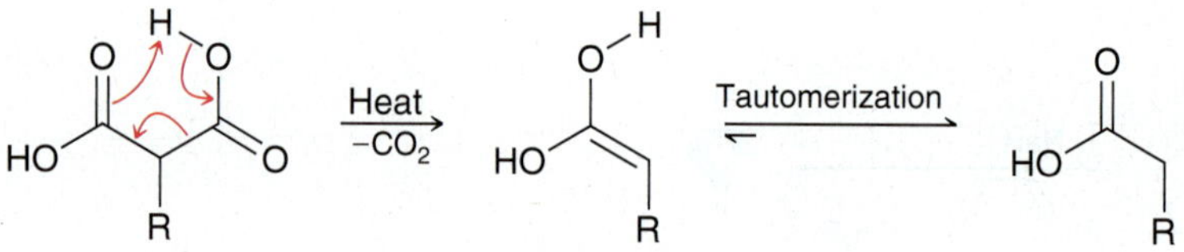

In the first step, a pericyclic reaction takes place, forming an enol, which then undergoes tautomerization. The process yields a monocarboxylic acid, which does not undergo further decarboxylation.

The following reaction is an example of a malonic ester synthesis:

**LOOKING BACK**
For a review of tautomerization, see Section 10.7.

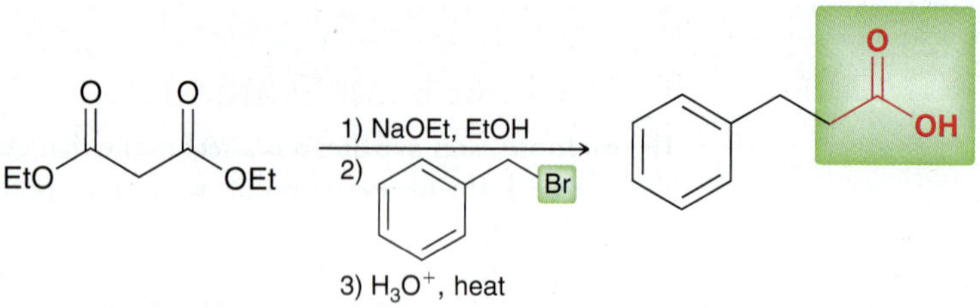

Diethyl malonate, the starting material, is first deprotonated, then treated with an alkyl halide, and then treated with aqueous acid at elevated temperature. Diethyl malonate can also be dialkylated, followed by treatment with aqueous acid, to produce carboxylic acids.

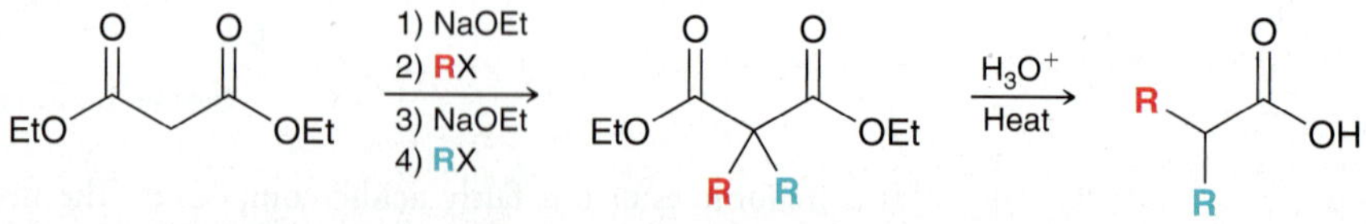

In this reaction sequence, diethyl malonate is alkylated twice, each time with a different alkyl group. Then, hydrolysis followed by decarboxylation produces a carboxylic acid with two alkyl groups at the $\alpha$ position.

# SKILLBUILDER

## 22.5 USING THE MALONIC ESTER SYNTHESIS

**LEARN** the skill

Show how you would use the malonic ester synthesis to prepare the following compound:

### SOLUTION

Begin by finding the carboxylic acid group and the α position.

Identify the alkyl groups connected to the α position.

**STEP 1**
Identify the alkyl groups connected to the α position.

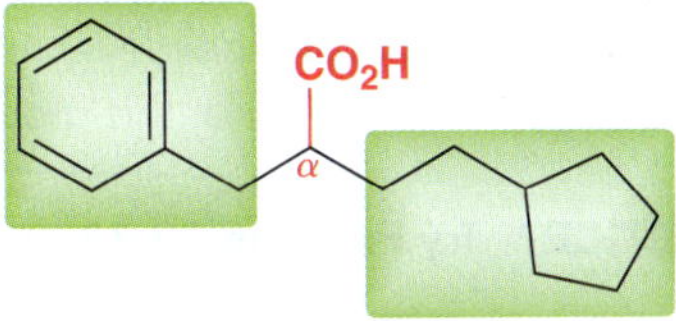


**STEP 2**
Identify the necessary alkyl halides and make sure they can serve as substrates in an $S_N2$ process.

In this case, there are two. Next, identify the alkyl halides that are necessary in order to install those two alkyl groups. Just redraw the alkyl groups connected to some halide (iodide and bromide are more commonly used as leaving groups, although chloride can be used). Below are alkyl bromides that can be used:

Analyze the structures of these alkyl halides and make sure that both will readily undergo an $S_N2$ process. If either substrate is tertiary, then a malonic ester synthesis will fail, because the alkylation step will not occur. In this case, both alkyl halides are primary, so the malonic ester synthesis can be used.

In order to perform the malonic ester synthesis, we will need diethyl malonate and the two alkyl halides shown above. It does not matter which alkyl group is installed first. The synthesis can be summarized as follows:

**STEP 3**
Identify all reagents, beginning with diethyl malonate.

1) NaOEt
2) [benzyl bromide]
3) NaOEt
4) [2-cyclopentylethyl bromide]
5) $H_3O^+$, heat

**PRACTICE the skill** **22.35** Propose an efficient synthesis for each of the following compounds using the malonic ester synthesis:

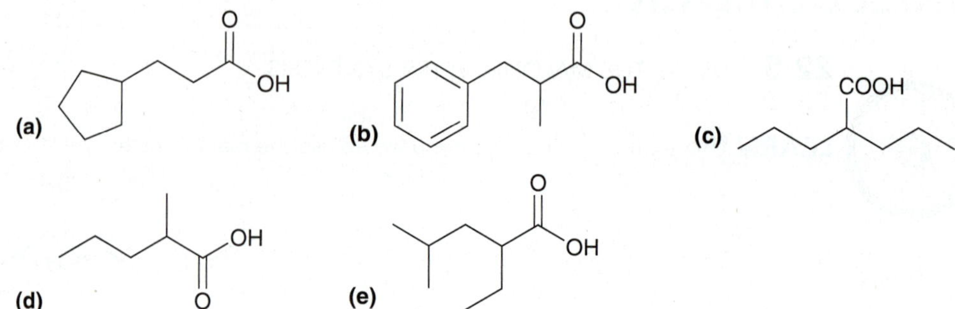

**APPLY the skill** **22.36** Starting with diethyl malonate and using any other reagents of your choice, propose an efficient synthesis for each of the following compounds:

(a) (b) OH (c) $NH_2$

**22.37** The malonic ester synthesis cannot be used to make 2,2-dimethylhexanoic acid. Explain why not.

**22.38** When a malonic ester synthesis is performed using excess base and 1,4-dibromobutane as the alkyl halide, an intramolecular reaction occurs, and the product contains a ring. Draw the product of this process.

need more **PRACTICE?** **Try Problem 22.78**

## The Acetoacetic Ester Synthesis

The **acetoacetic ester synthesis** is a useful technique that converts an alkyl halide into a methyl ketone with the introduction of three new carbon atoms.

R—X ⟶ R (methyl ketone)

This process is very similar to the malonic ester synthesis except that the key reagent is ethyl acetoacetate rather than diethyl malonate.

EtO ... **Ethyl acetoacetate** EtO ... OEt **Diethyl malonate**

Ethyl acetoacetate is very similar in structure to diethyl malonate, but one of the ester groups has been replaced with a ketone group.

The first step of the acetoacetic ester synthesis is analogous to the first step of the malonic ester synthesis. Ethyl acetoacetate is first deprotonated, forming a doubly stabilized enolate.

EtO, H, $^{\ominus}$:ÖEt ⟶ EtO (enolate)

This enolate is then treated with an alkyl halide (RX), resulting in an alkylation step.

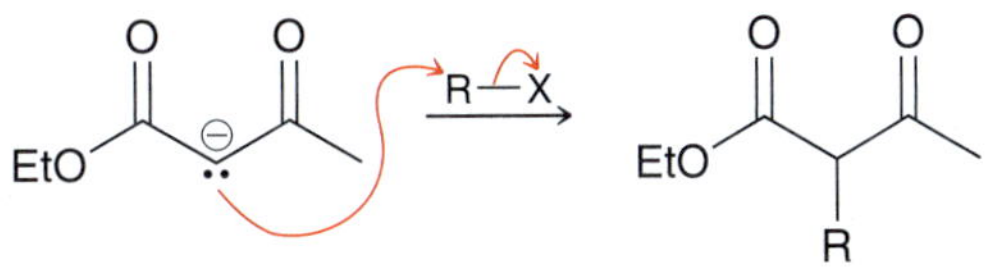

Treatment with aqueous acid then results in hydrolysis of the ester group.

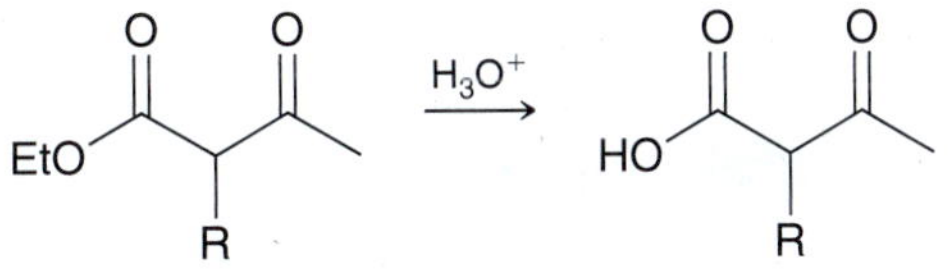

If hydrolysis is performed at an elevated temperature, the resulting β-keto acid will undergo decarboxylation to produce a ketone and carbon dioxide.

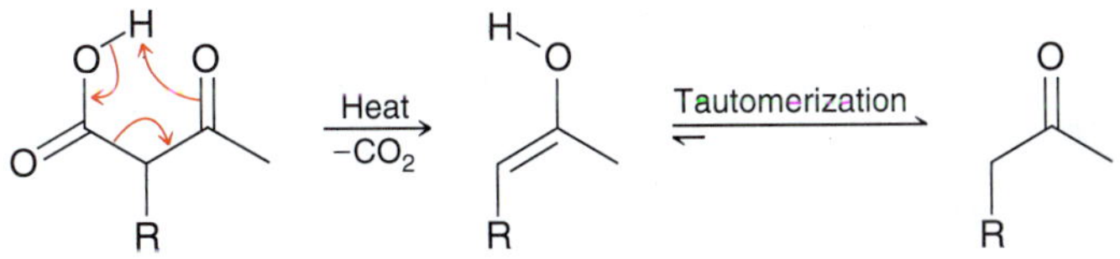

This process is very similar to the process we discussed for 1,3-dicarboxylic acids. In the first step, a pericyclic reaction forms an enol, which then undergoes tautomerization to form a ketone. Decarboxylation can occur because the carboxylic acid exhibits a carbonyl group in the β position, which enables the pericyclic reaction shown above. Below is an example of an acetoacetic ester synthesis.

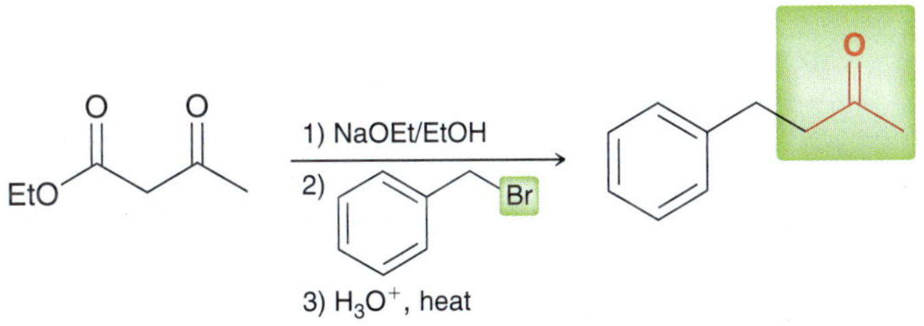

Notice the similarity to the malonic ester synthesis. Ethyl acetoacetate is the starting material. It is first deprotonated, then treated with an alkyl halide, and then treated with aqueous acid at elevated temperature. Ethyl acetoacetate can also be dialkylated, followed by treatment with aqueous acid, to produce ketones as follows:

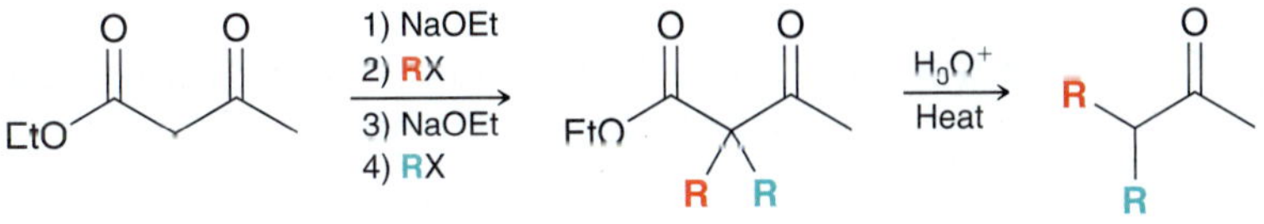

In this reaction sequence, ethyl acetoacetate is alkylated twice, each time with a different alkyl group. Then, hydrolysis followed by decarboxylation produces a derivative of acetone in which two alkyl groups are positioned at the α position. Why is it necessary to use an acetoacetic ester synthesis instead of simply treating the enolate of acetone with an alkyl halide? Wouldn't that be a more direct method for preparing substituted acetones? There are two answers to this question: (1) Direct alkylation of enolates is often difficult to achieve in good yields and (2) enolates are not only strong nucleophiles, but they are also strong bases, and as a result, enolates can react with alkyl halides to produce elimination products.

# SKILLBUILDER

## 22.6 USING THE ACETOACETIC ESTER SYNTHESIS

**LEARN** the skill

Show how you would use the acetoacetic ester synthesis to prepare the following compound:

**SOLUTION**

Find the $\alpha$ position connected to the methyl ketone group and identify the alkyl groups connected to the $\alpha$ position:

**STEP 1**
Identify the alkyl groups connected to the $\alpha$ position.

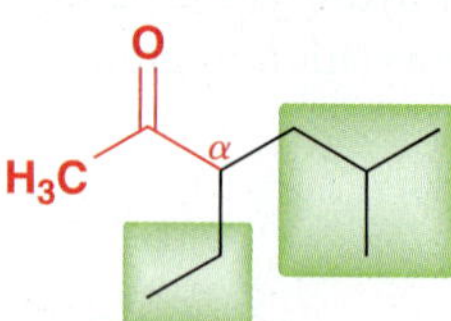


**STEP 2**
Identify the necessary alkyl halides and make sure they can serve as substrates in an $S_N2$ process.

In this case, there are two. Next, identify the alkyl halides needed to install those two alkyl groups. Just redraw those alkyl groups connected to some halide. Below are two alkyl iodides that can be used.

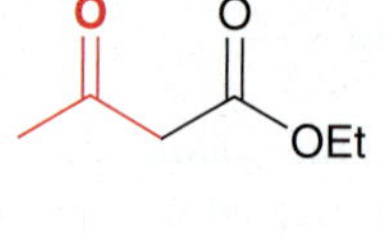


Analyze the structures of these alkyl halides and make sure that both will readily undergo $S_N2$ reactions. If either substrate is tertiary, then an acetoacetic ester synthesis will fail, because the alkylation step will not occur. In this case, both alkyl halides are primary, so the acetoacetic ester synthesis can be used. We have seen that the acetoacetic ester synthesis requires ethyl acetoacetate and the two alkyl halides shown. It does not matter which alkyl group is installed first. The synthesis can be summarized as follows:

**STEP 3**
Identify all reagents, beginning with ethyl acetoacetate.

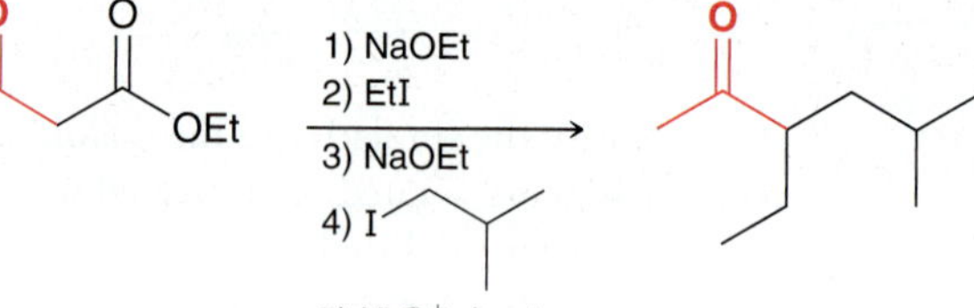

**PRACTICE** the skill

**22.39** Propose an efficient synthesis for each of the following compounds using the acetoacetic ester synthesis:

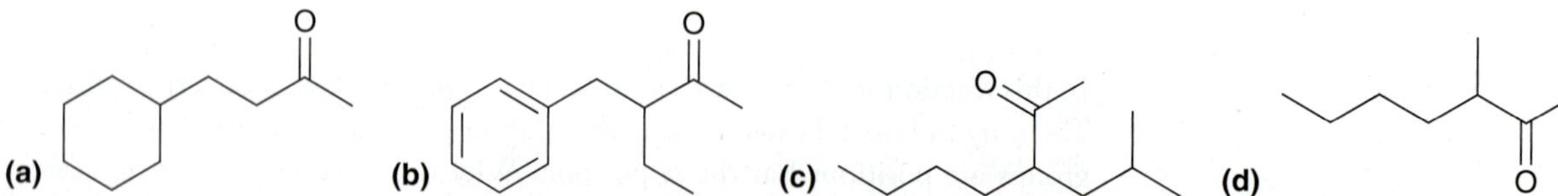


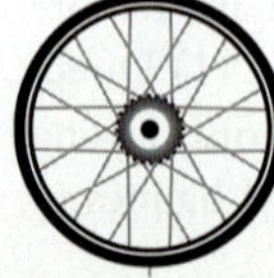

**APPLY** the skill

**22.40** In Problem 22.38, we saw an intramolecular example of a malonic ester synthesis using excess base and 1,4-dibromobutane. If this dibromide is used in an acetoacetic ester synthesis, an intramolecular process can also occur. Predict the product of that reaction.

**22.41** Starting with ethyl acetoacetate and using any other reagents of your choice, propose an efficient synthesis for each of the following compounds:

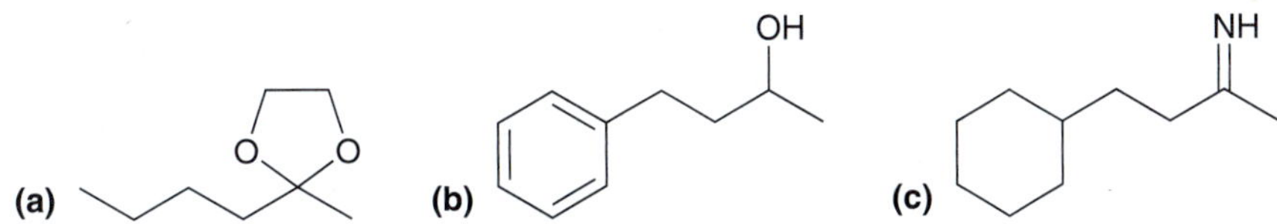


**22.42** The acetoacetic ester synthesis cannot be used to make 3,3-dimethyl-2-hexanone. Explain why not.

**22.43** The product of a Dieckmann cyclization can undergo alkylation, hydrolysis, and decarboxylation. This sequence represents an efficient method for preparing 2-substituted cyclopentanones (below) and cyclohexanones. Using this information, propose an efficient synthesis of 2-propylcyclohexanone using 1,7-heptanediol and propyl iodide.

EtO, OEt; 1) NaOEt 2) $H_3O^+$ → OEt; 1) NaOEt, EtOH 2) RX 3) $H_3O^+$, heat → R

need more **PRACTICE?** **Try Problem 22.79**

## 22.6 Conjugate Addition Reactions

### Michael Reactions

Recall that the product of an aldol condensation is an α,β-unsaturated aldehyde or ketone; for example:

H —Aldol condensation→ β α H

Aldehydes and ketones that possess α,β unsaturation, like the compound above, exhibit unique reactivity at the β position. In this section, we will explore reactions that can take place at the β position. To understand why the β position is reactive, we must draw resonance structures of the compound.

[ :O: H ⟷ :O:⊖ ⊕ H ⟷ ⊕ :O:⊖ H ]

Notice that two of the resonance contributors exhibit a positive charge. In other words, the compound will have two electrophilic positions: the carbon of the carbonyl group as well as the β position.

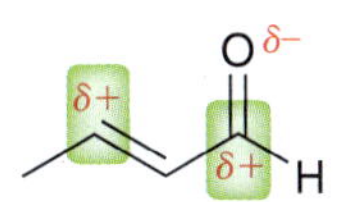


Two electrophilic positions

Both of these positions are therefore subject to attack by a nucleophile, and the nature of the nucleophile determines which position is attacked. For example, Grignard reagents tend to attack the carbonyl group rather than the β position.

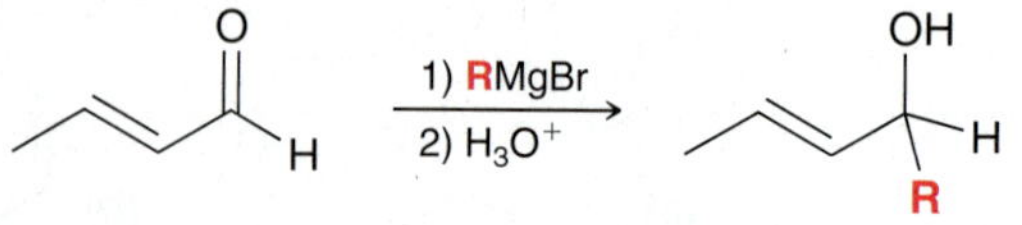

1) RMgBr
2) $H_3O^+$

In this reaction, the nucleophile attacks the carbonyl group, giving a tetrahedral intermediate, which is then treated with a proton source in a separate step. In this process, two groups (R and H) have added across the C=O π bond in a 1,2-addition (for a review of the difference between 1,2-additions and 1,4-additions, see Section 17.4).

A different outcome is observed when a lithium dialkyl cuprate ($R_2CuLi$) is employed.

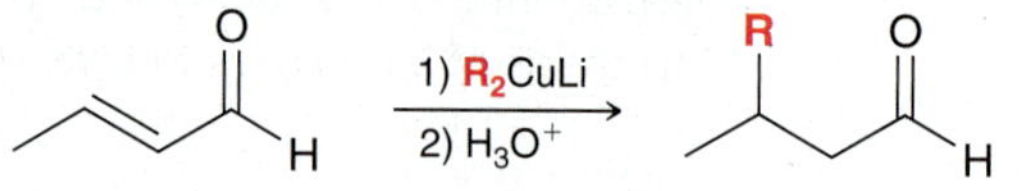

1) $R_2CuLi$
2) $H_3O^+$

In this case, the nucleophile is less reactive than a Grignard reagent, and the R group is ultimately positioned at the β position. A variety of nucleophiles can be used to attack the β position of an α,β-unsaturated aldehyde or ketone. The process involves attack at the β position followed by protonation to give an enol.

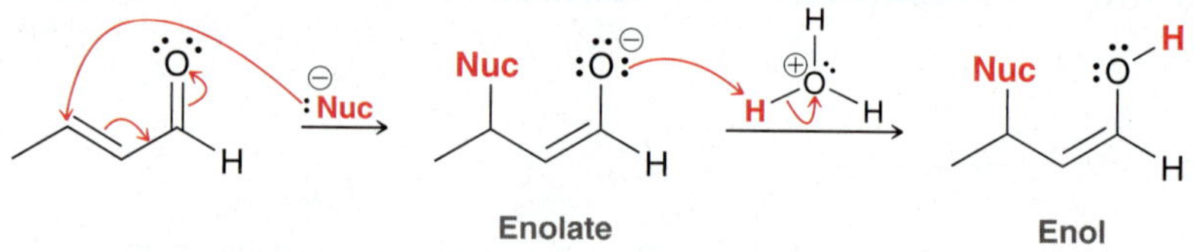

Enolate

Enol

This type of reaction is called a **conjugate addition,** or a *1,4-addition,* because the nucleophile and the proton have added across the ends of a conjugated π system. Conjugate addition reactions were first discussed in Section 17.4 when we explored the reactivity of conjugated dienes. A major difference between conjugate addition across a diene and conjugate addition across an enone is that the latter produces an enol as the product, and the enol rapidly tautomerizes to form a carbonyl group.

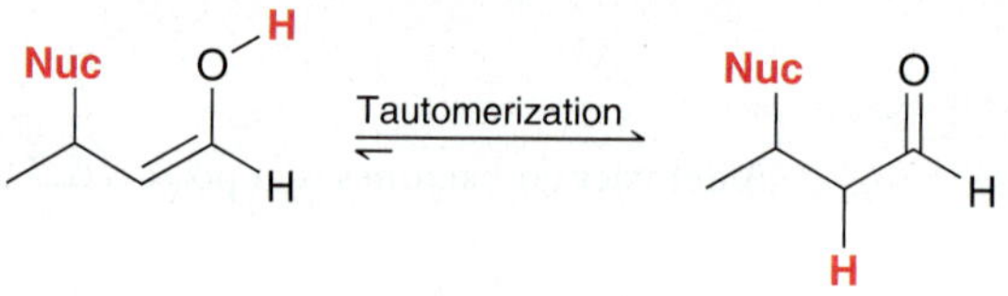

Tautomerization

After the reaction is complete, it might appear that the two groups (shown in red) have added across the α and β positions in a 1,2-addition. However, the actual mechanism likely involves a 1,4-addition followed by tautomerization to give the ultimate product shown.

With this in mind, let's now explore the outcome of a reaction in which an enolate ion is used as a nucleophile to attack an α,β-unsaturated aldehyde or ketone. In general, enolates are less reactive than Grignard reagents but more reactive than lithium dialkyl cuprates. As such, both 1,2-addition and 1,4-addition are observed, and a mixture of products is obtained. In contrast, doubly stabilized enolates are sufficiently stabilized to produce 1,4 conjugate addition exclusively.

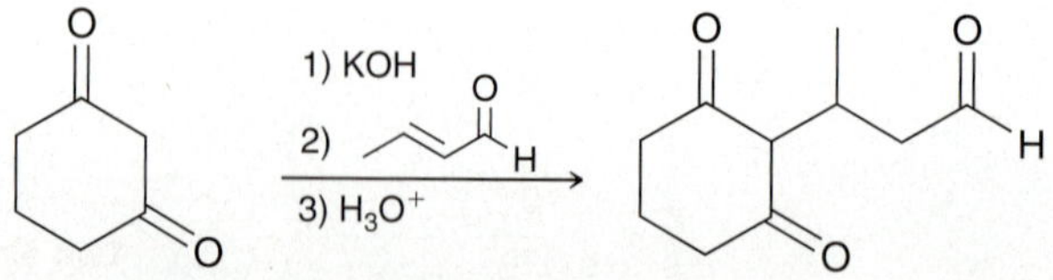

1) KOH
2)
3) $H_3O^+$

In this case, the starting diketone is deprotonated to form a doubly stabilized enolate ion, which then serves as a nucleophile in a 1,4-conjugate addition. This process is called a **Michael reaction.** The doubly stabilized enolate is called a **Michael donor**, while the α,β-unsaturated aldehyde is called a **Michael acceptor**.

Michael donor Michael acceptor

A variety of Michael donors and acceptors are observed to react with each other to produce a Michael reaction. Table 22.2 shows some common examples of Michael donors and acceptors. Any one of the Michael donors in this table will react with any one of the Michael acceptors to give a 1,4-conjugate addition reaction.

**TABLE 22.2 A LIST OF COMMON MICHAEL DONORS AND MICHAEL ACCEPTORS**

| MICHAEL DONORS | | MICHAEL ACCEPTORS | |
|---|---|---|---|
| (enolate of 2,4-pentanedione) | EtO, OEt (enolate of diethyl malonate) | δ+, H (α,β-unsaturated aldehyde) | δ+ (α,β-unsaturated ketone) |
| OEt (enolate of ethyl acetoacetate) | C≡N (enolate of 3-oxobutanenitrile) | δ+, OEt (α,β-unsaturated ester) | δ+, $NH_2$ (α,β-unsaturated amide) |
| R, $NO_2$ (nitronate anion) | $R_2CuLi$ | δ+, C≡N (α,β-unsaturated nitrile) | δ+, $NO_2$ (nitroalkene) |

## CONCEPTUAL CHECKPOINT

**22.44** Identify the major product formed when each of the following compounds is treated with $Et_2CuLi$ followed by mild acid:

(a) (b) CN (c) OEt

**22.45** Predict the major product of the three following steps and show a mechanism for its formation:

1) KOH
2) (cyclohexyl vinyl ketone)
3) $H_3O^+$
?

**22.46** In the previous section, we learned how to use diethyl malonate as a starting material in the preparation of substituted carboxylic acids (the malonic ester synthesis). That method employed a step in which the enolate of diethyl malonate attacked an alkyl halide to give an alkylation product. In this section, we saw that the enolate of diethyl malonate can attack many electrophilic reagents other than simple alkyl halides. Specifically, the enolate of diethyl malonate can attack any of the Michael acceptors in Table 22.2. Using diethyl malonate as your starting material and any other reagents of your choice, show how you would prepare each of the following compounds:

(a) HO (b) HO $NO_2$

## medically speaking | Glutathione Conjugation and Biochemical Michael Reactions

Glutathione is found in all mammals and plays many biological roles.

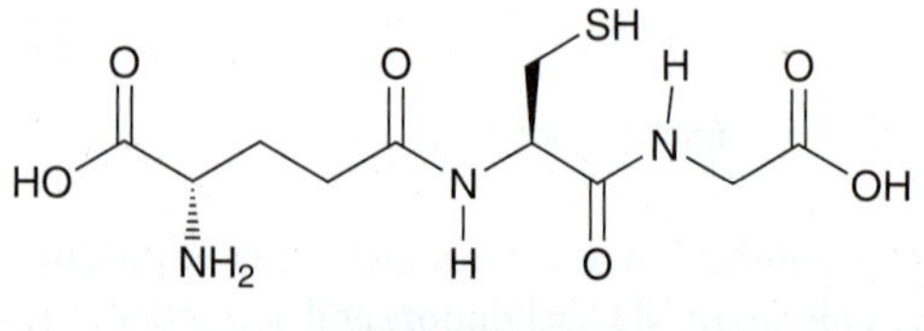

**Glutathione**

As we saw in the Medically Speaking box at the end of Section 11.9, glutathione functions as a radical scavenger by reacting with free radicals that are generated during metabolic processes. This compound has many other important biological functions. For example, the sulfur atom in glutathione is highly nucleophilic, which allows this compound to function as a scavenger for harmful *electrophiles* that are ingested or are produced by metabolic processes. Glutathione intercepts these electrophiles and reacts with them before they can react with the nucleophilic sites of vital biomolecules, such as DNA and proteins. Glutathione can react with many kinds of electrophiles through a variety of mechanisms, including $S_N2$, $S_NAr$, nucleophilic acyl substitution, and Michael addition.

A Michael addition occurs in the metabolism of morphine. One of the metabolites of morphine is generated via oxidation of the allylic alcohol group to produce an $\alpha,\beta$-unsaturated ketone:

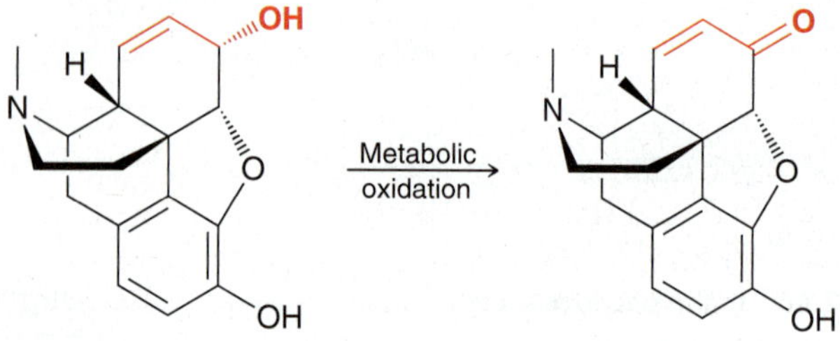

Metabolic oxidation

**Morphine** **A metabolite of morphine**

The $\beta$ position of this metabolite is now highly electrophilic and subject to attack by a variety of nucleophiles. The sulfur atom of glutathione attacks at the $\beta$ position, producing a Michael addition. This process is called glutathione conjugation, and the product is called a glutathione conjugate, which can be further metabolized to produce a mercapturic acid conjugate, a compound that is excreted in the urine.

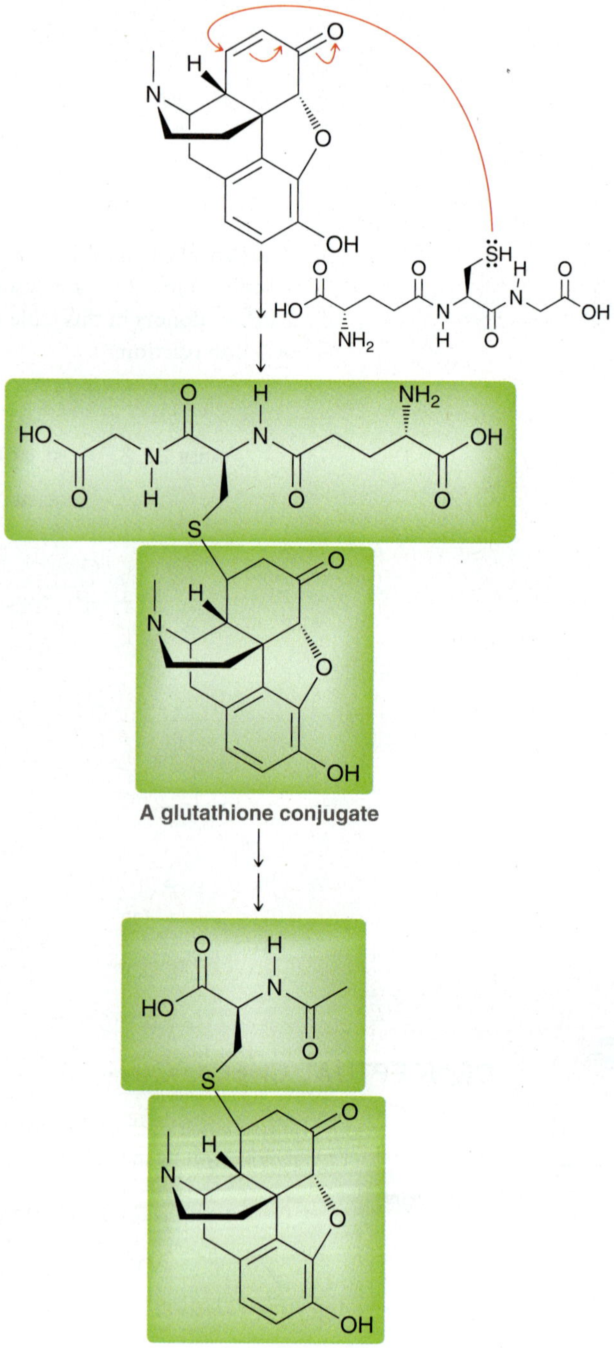

**A glutathione conjugate**

**A mercapturic acid conjugate**

Glutathione conjugation also occurs in the metabolism of acetaminophen (Tylenol). Metabolic oxidation produces an *N*-acetylimidoquinone intermediate, which is highly reactive.

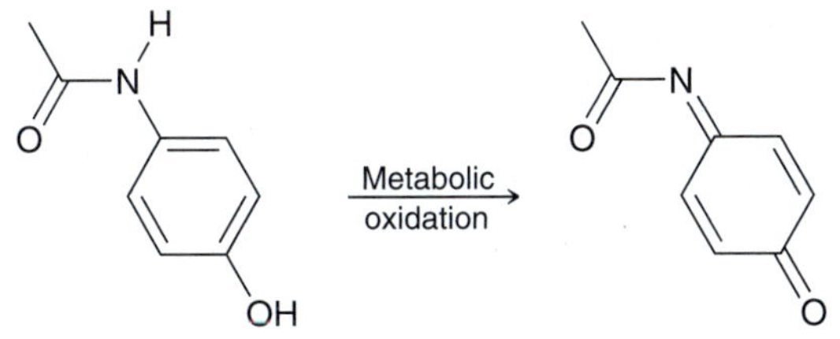

Metabolic oxidation

**Acetaminophen**

**A metabolite of acetaminophen (*N*-acetylimidoquinone)**

This metabolite undergoes glutathione conjugation followed by further metabolism that generates a water-soluble mercapturic acid derivative which is excreted in the urine.

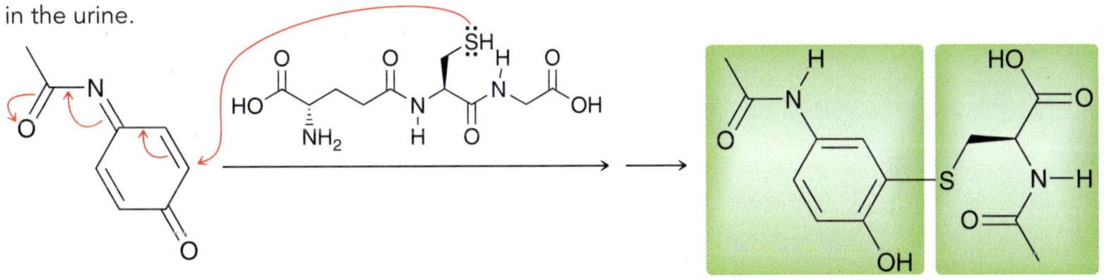

**A mercapturic acid conjugate of acetaminophen**

In our first discussion of glutathione and its role as a radical scavenger, we mentioned that an overdose of acetaminophen causes a temporary depletion of glutathione in the liver, during which time harmful radicals and electrophiles are free to run amok and cause permanent damage. Our current discussion of glutathione now explains why glutathione levels drop in the presence of too much acetaminophen. To review how an acetaminophen overdose is treated, see the Medically Speaking box at the end of Section 11.9.

## The Stork Enamine Synthesis

We saw in the previous section that doubly stabilized enolates can serve as Michael donors. It is important to point out that regular enolates do not serve as Michael donors.

**A Michael donor** **Not a Michael donor**

Therefore, the following synthesis will not be efficient:

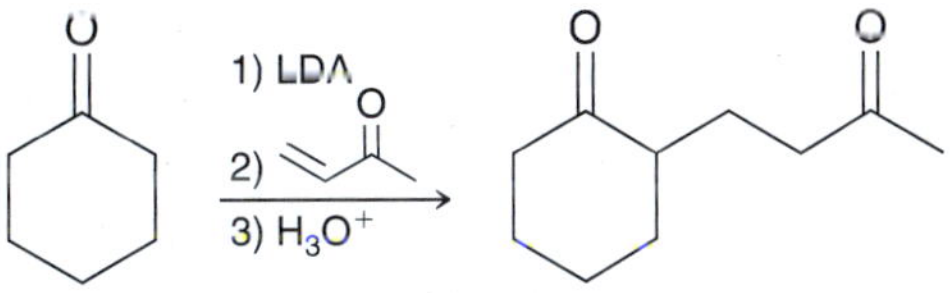

1) LDA
2)
3) $H_3O^+$

**Not obtained in good yields**

This synthetic route relies on an enolate serving as a Michael donor and attacking the β position of an α,β-unsaturated ketone. The process will not work because enolates are not stable enough to function as Michael donors. This transformation requires a stabilized enolate, or some species that will behave like a stabilized enolate. Gilbert Stork (Columbia University) developed a method for such a transformation in which the ketone is converted into an enamine by treatment with a secondary amine.

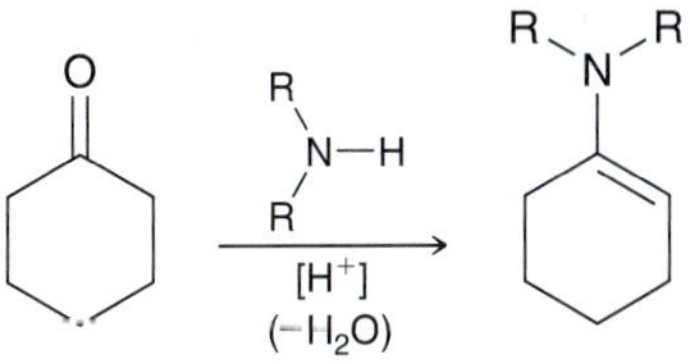

$R_2N-H$, $[H^+]$, $(-H_2O)$

The mechanism for conversion of a ketone into an enamine was discussed in Section 20.6. To see the similarity between an enolate and an enamine, compare the resonance structures of an enolate with the resonance structures of an enamine.

An enolate ion

An enamine

Much like enolates, enamines are also nucleophilic at the $\alpha$ position. However, enamines do not possess a net negative charge, as enolates do, and therefore, enamines are less reactive than enolates. As such, enamines are effective Michael donors and will participate in a Michael reaction with a suitable Michael acceptor.

Michael donor

Michael acceptor

This Michael reaction generates an intermediate that is both an iminium ion and an enolate ion, as shown below. When treated with aqueous acid, both groups are converted into carbonyl groups.

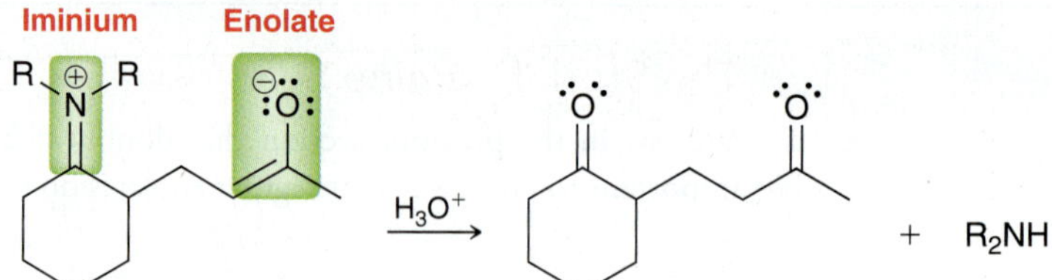


The iminium ion undergoes hydrolysis to form a carbonyl group, and the enolate ion is protonated to form an enol, which tautomerizes to form a carbonyl group.

The net result is a process for achieving the following type of transformation:

1) $R_2NH$, $[H^+]$, $(-H_2O)$
2)
3) $H_3O^+$

This process is called a **Stork enamine synthesis,** and it has three steps: (1) formation of an enamine, (2) a Michael addition, and (3) hydrolysis.

## SKILLBUILDER

### 22.7 DETERMINING WHEN TO USE A STORK ENAMINE SYNTHESIS

**LEARN** the skill

Using any reagents of your choosing, show how you might accomplish the following transformation:

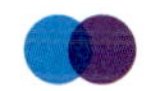

## SOLUTION

This transformation requires that we install the following (highlighted) group at the α position of the starting material:

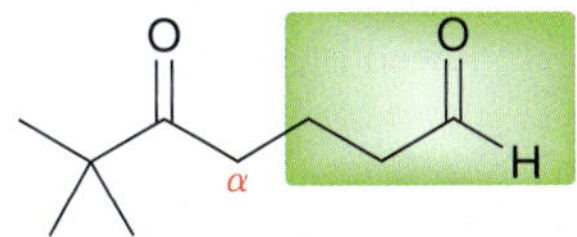


Using a retrosynthetic analysis, we determine that this transformation might be possible via a Michael addition:

Michael

However, a Michael reaction won't work because the enolate is not sufficiently stabilized to function as a Michael donor. This is a perfect example of a situation in which a Stork enamine synthesis might be used to achieve the desired Michael addition.

The starting ketone is first converted into an enamine and then used as a Michael donor followed by hydrolysis.

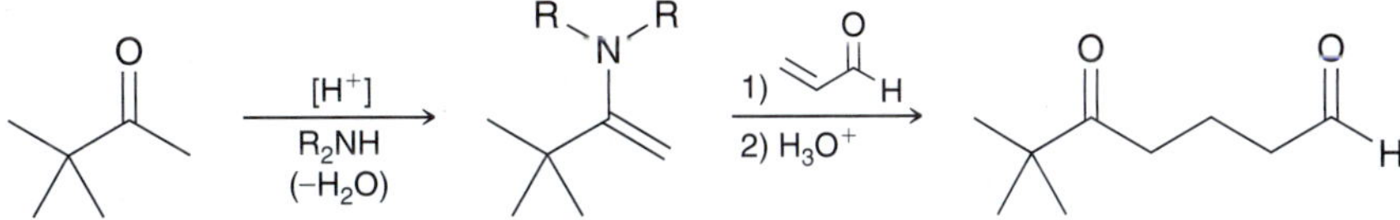

**PRACTICE the skill** **22.47** Using a Stork enamine synthesis, show how you might accomplish each of the following transformations:

(a)

(b)

(c)

**APPLY the skill** **22.48** Using acetophenone as your only source of carbon atoms, propose a synthesis for the following compound:

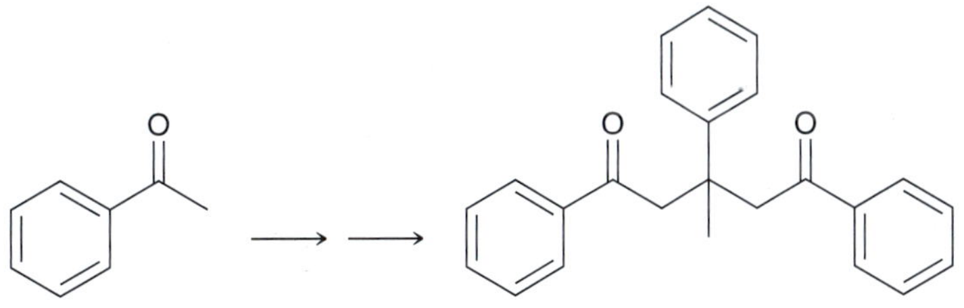


Acetophenone

need more **PRACTICE?** **Try Problem 22.87d**

### The Robinson Annulation Reaction

In this chapter, we have seen many reactions. When performed in combination, they can be very versatile. One such example is a two-step method for forming a ring, in which a Michael addition is followed by an intramolecular aldol condensation.

NaOH / Michael addition; NaOH, heat / Aldol condensation

This two-step method is called a **Robinson annulation**, named after Sir Robert Robinson (Oxford University), and is often used for the synthesis of polycyclic compounds. The term *annulation* is derived from the Latin word for ring (*annulus*).

CONCEPTUAL CHECKPOINT

**22.49** Draw a complete mechanism for the following transformation:

NaOH, heat

**22.50** Identify the reagents you would use to prepare the following compound via a Robinson annulation:

## 22.7 Synthesis Strategies

### Reactions That Yield Difunctionalized Compounds

In this chapter, we have seen many reactions that form C—C bonds. Three of those reactions are worth special attention, because of their ability to produce compounds with two functional groups. The aldol addition, the Claisen condensation, and the Stork enamine synthesis all produce difunctionalized compounds, yet they differ from each other in the ultimate positioning of the functional groups. The Stork enamine synthesis produces 1,5-difunctionalized compounds.

1) $R_2NH$, $[H^+]$, $(-H_2O)$ 2) methyl vinyl ketone 3) $H_2O$

In contrast, aldol addition reactions and Claisen condensation reactions both produce 1,3-difunctionalized compounds.

NaOH / $H_2O$

1) NaOEt 2) $H_3O^+$

Although the relative positioning of the functional groups is similar for aldol additions and Claisen condensations, the oxidation states are different. An aldol addition produces a carbonyl group and a hydroxyl group, while a Claisen condensation produces an ester and a carbonyl group.

These considerations can be very helpful when designing a synthesis. If the target compound has two functional groups, then you should look at their relative positions. If the target compound is 1,5-difunctionalized, then you should think of using a Stork enamine synthesis. If it is 1,3-difunctionalized, then you should think of using an aldol addition or a Claisen condensation. The choice (aldol vs. Claisen) will be influenced by the oxidation states of the functional groups. The following exercise illustrates this type of analysis.

## SKILLBUILDER

### 22.8 DETERMINING WHICH ADDITION OR CONDENSATION REACTION TO USE

**LEARN** the skill

Using 1-butanol as your only source of carbon, propose a synthesis for the following compound:

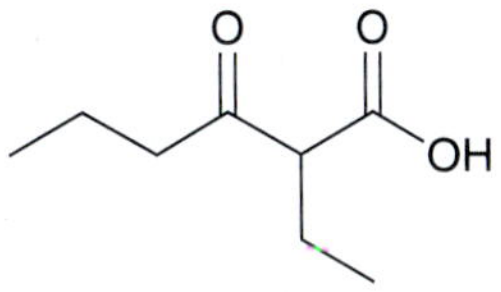

**SOLUTION**

Recall from Chapter 12 that there are always two things to analyze when approaching a synthesis problem: (1) any changes in the C—C framework and (2) the location and identity of the functional groups. Let's begin with the C—C framework. The target compound has a total of eight carbon atoms, and the starting material has only four carbon atoms. Therefore, our synthesis will require two molecules of butanol to construct the C—C framework of the target compound. In other words, we will need to use some kind of addition or condensation reaction. We have seen many such reactions. To determine which one will be most efficient in this case, we analyze the location and identity of the functional groups.

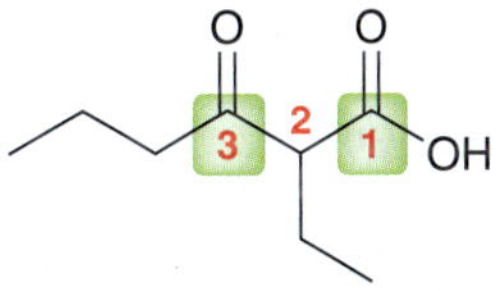

This compound is 1,3-difunctionalized, so we should focus our attention on either an aldol or a Claisen. To decide which reaction to use, look at the oxidation states of the two functional groups (a carbonyl group and a carboxylic acid group). In this case, a Claisen condensation seems like the likely candidate. The product of a Claisen condensation is always a β-keto ester, so if we use a Claisen condensation to form the carbon skeleton, then the last step of our synthesis will be hydrolysis of the ester to give a carboxylic acid.

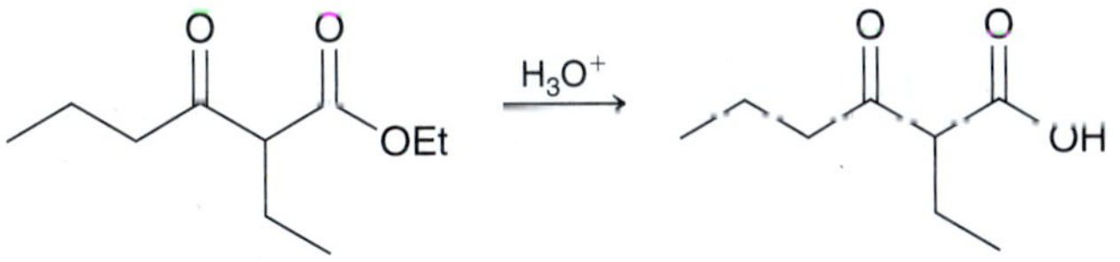

$\xrightarrow{H_3O^+}$

The penultimate step would be a Claisen condensation between two esters.

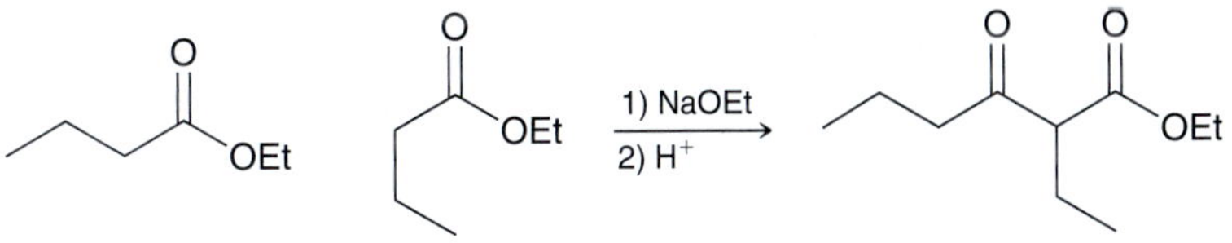

$\xrightarrow[2)\ H^+]{1)\ NaOEt}$

In this case, the two starting esters are identical, which simplifies the problem (it is not necessary to perform a mixed Claisen). This transformation is accomplished by simply treating the ester (ethyl butanoate) with sodium ethoxide.

To complete the synthesis, we need a method for making the necessary ester from 1-butanol. This can be accomplished by using reactions from previous chapters.

$\xrightarrow[H_2SO_4,\ H_2O]{Na_2Cr_2O_7}$ OH $\xrightarrow[EtOH]{[H^+]}$ OEt

Our proposed synthesis is summarized as follows:

OH → 1) $Na_2Cr_2O_7$, $H_2SO_4$, $H_2O$ 2) $[H^+]$, EtOH 3) NaOEt 4) $H_3O^+$ → OH

The last step has two functions: (1) to protonate the doubly stabilized enolate that is produced by the Claisen condensation and (2) to hydrolyze the resulting ester to give a carboxylic acid. Hydrolysis might require gentle heating, but excessive heating should be avoided as it would likely cause decarboxylation.

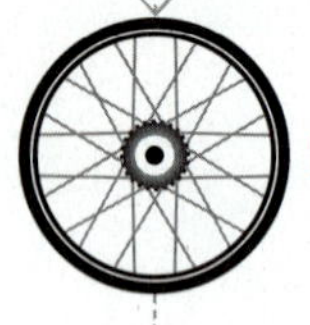

**PRACTICE the skill** **22.51** Using cyclopentanone as your starting material and using any other reagents of your choice, propose an efficient synthesis for each of the following compounds:

(a) (b) (c)

**22.52** Using 1-propanol as your only source of carbon, propose an efficient synthesis for each of the following compounds:

(a) (b) (c)

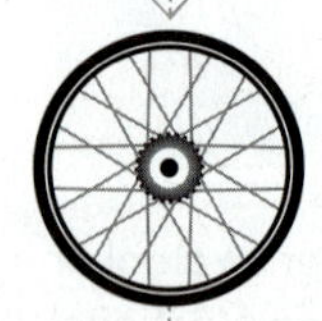

**APPLY the skill** **22.53** Using ethanol as your only source of carbon atoms, propose a synthesis for each of the following compounds:

(a) (b)

need more **PRACTICE?** **Try Problem 22.87d**

## Alkylation of the Alpha and Beta Positions

Recall that the initial product of a Michael addition is an enolate ion, which is then quenched with water to give the product.

$\xrightarrow{R_2CuLi}$ Enolate $\xrightarrow{H_2O}$

Instead of quenching with water, the enolate generated from a Michael addition can be treated with an alkyl halide, which results in alkylation of the α position.

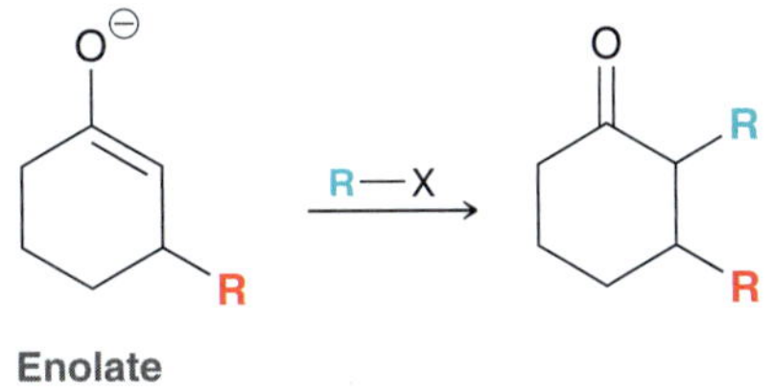

Enolate

This provides a method for alkylating both the α and β positions in one reaction flask. When using this method, the two alkyl groups need not be the same. Below is an example of such a reaction.

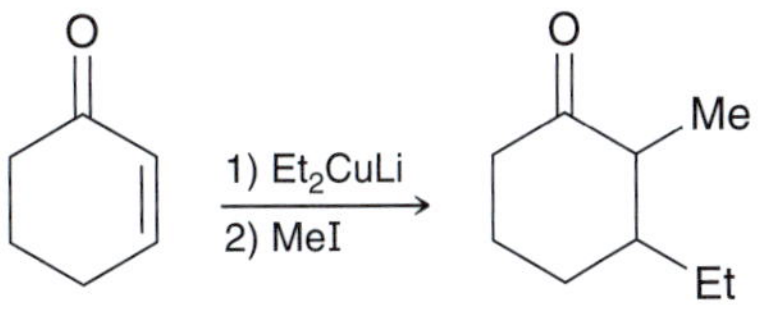

First, a carbon nucleophile is used to install an alkyl group at the β position, and then a carbon electrophile is used to install an alkyl group at the α position. The following exercise demonstrates the use of this technique.

## SKILLBUILDER

### 22.9 ALKYLATING THE ALPHA AND BETA POSITIONS

**LEARN the skill** Propose an efficient synthesis for the following transformation:

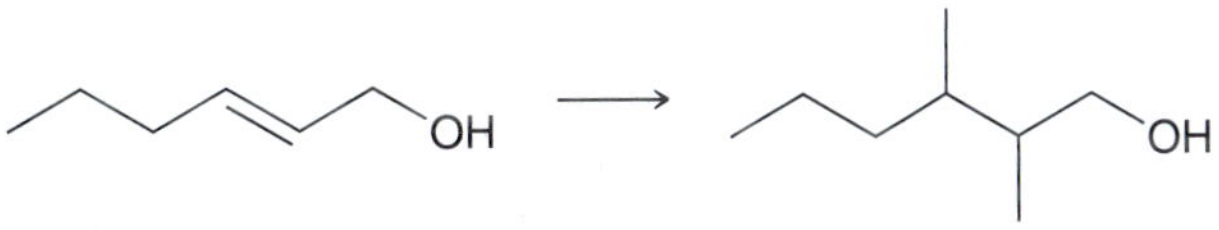

### SOLUTION

Recall that there are always two things to analyze when approaching a synthesis problem: (1) any changes in the C—C framework and (2) the location and identity of functional groups. In this case, let's begin with functional groups, because very little change seems to have taken place. The hydroxyl group remains in the same position, but the double bond is destroyed. Now let's focus on the C—C framework. It appears that two methyl groups have added across the double bond.

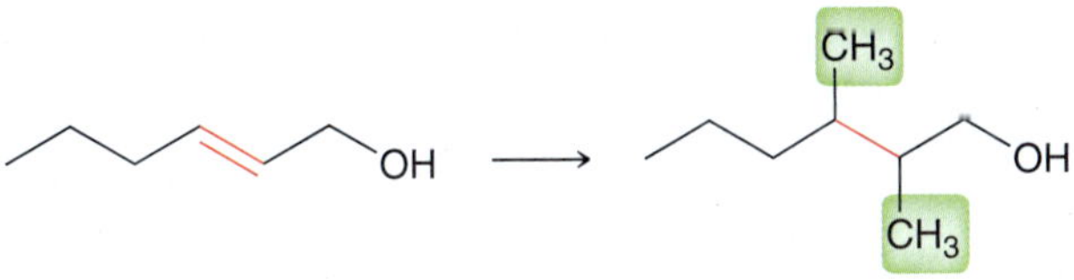

We did not see a way to add two alkyl groups across a double bond, but we did see a way to install two alkyl groups when the double bond is conjugated with a carbonyl group.

To use this method for installing the alkyl groups, we must first oxidize the alcohol to give a carbonyl group, which would then have to be reduced back to an alcohol at the end of the synthesis.

OH → PCC → O, H → 1) $Me_2CuLi$ 2) MeI → $CH_3$, O, H, $CH_3$ → 1) LAH 2) $H_2O$ → OH

Solving this problem required us to recognize that the hydroxyl group could be oxidized and then later reduced. The ability to interchange functional groups is an important skill, as this problem demonstrates.

**PRACTICE the skill** **22.54** Propose an efficient synthesis for each of the following transformations:

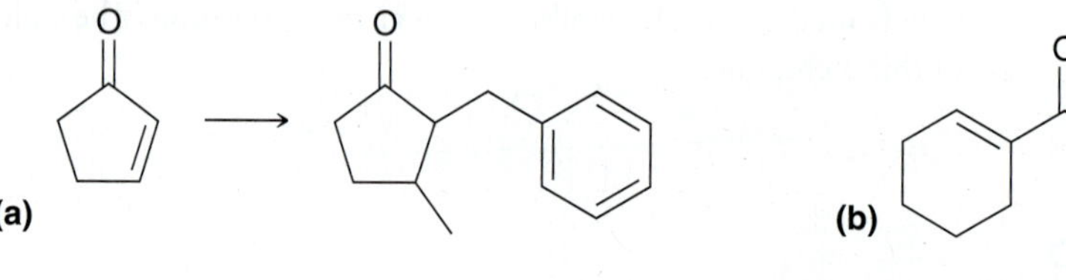


(b) O, H → OH

(c) O, H → O, Cl

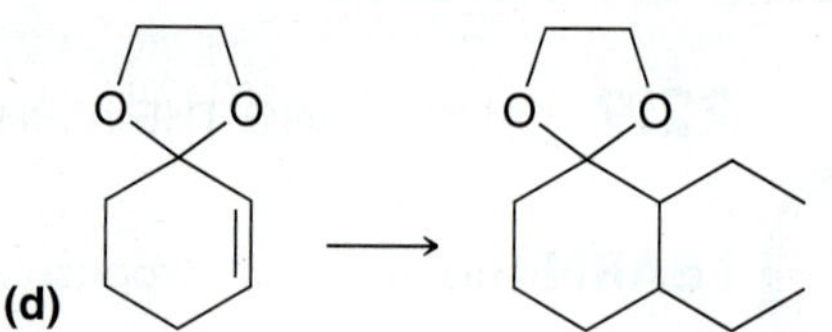


(e) OH → N

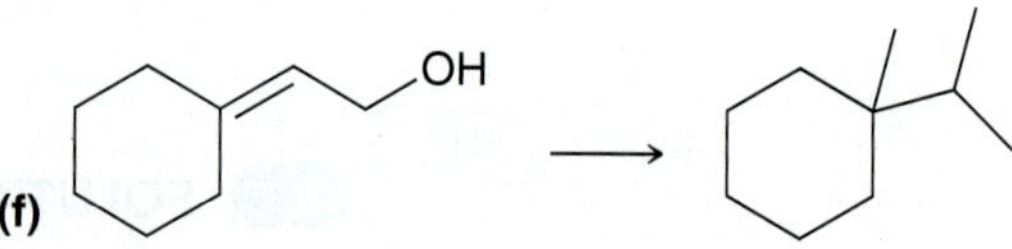

**APPLY the skill** **22.55** Propose an efficient synthesis for the following transformation:

O → O

**22.56** Propose an efficient synthesis for the following transformation. Take special notice of the fact that the starting material has a six-membered ring while the product has a five-membered ring.

→ O, H

need more **PRACTICE?** **Try Problems 22.82, 22.89**

# REVIEW OF REACTIONS

## Alpha Halogenation

### Of Ketones

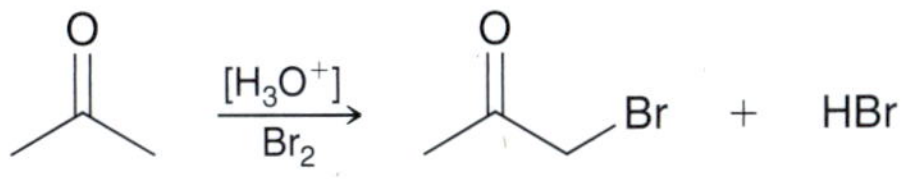

O

$[H_3O^+]$

$Br_2$

O

Br + HBr

### Haloform Reaction

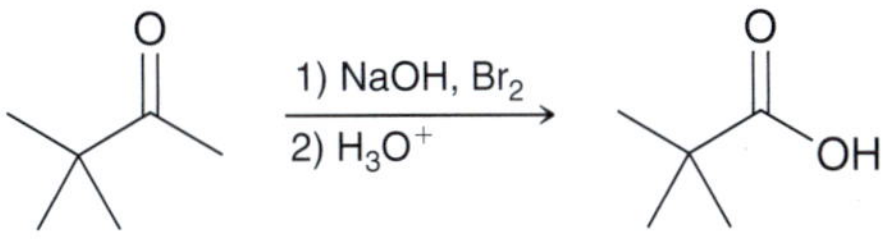

O

1) NaOH, $Br_2$

2) $H_3O^+$

O

OH

### Of Carboxylic Acids (Hell-Volhard-Zelinsky Reaction)

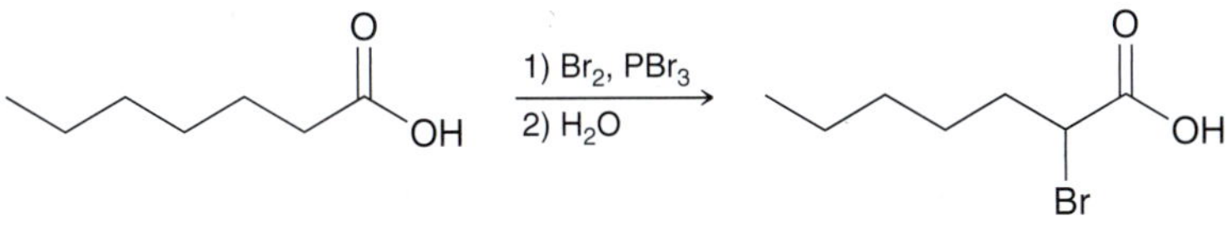

O

OH

1) $Br_2$, $PBr_3$

2) $H_2O$

O

OH

Br

## Aldol Reactions

### Aldol Addition and Condensation

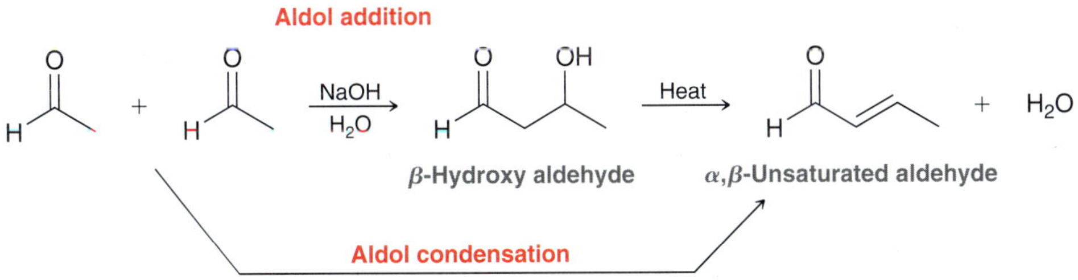

Aldol addition

O

H

+

O

H

NaOH

$H_2O$

O OH

H

Heat

O

H

+ $H_2O$

*β*-Hydroxy aldehyde

*α*,*β*-Unsaturated aldehyde

Aldol condensation

### Crossed Aldol Condensation

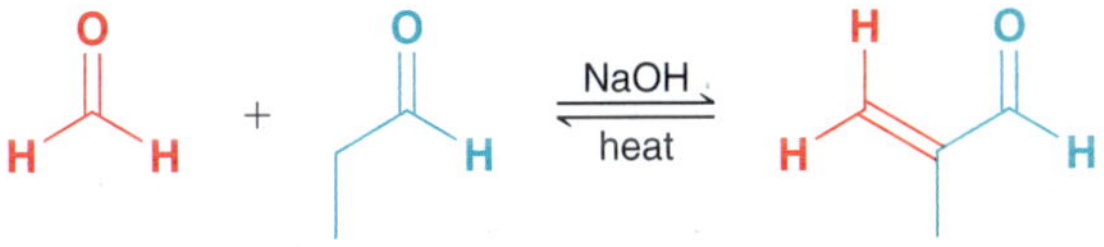

O

H H

+

O

H

NaOH

heat

H O

H H

### Intramolecular Aldol Condensation

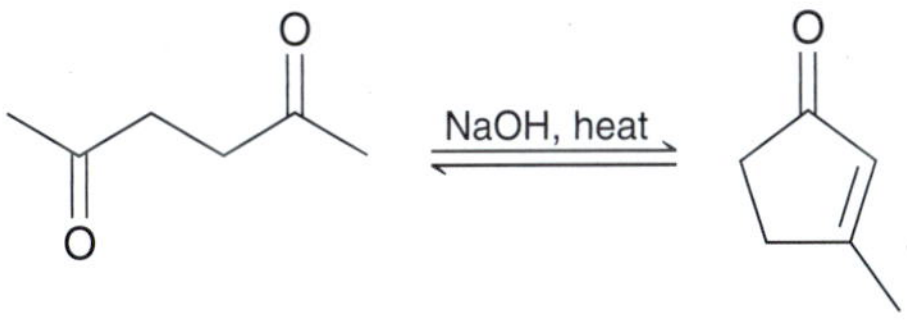

O

O

NaOH, heat

O

## Claisen Condensation

### Claisen Condensation

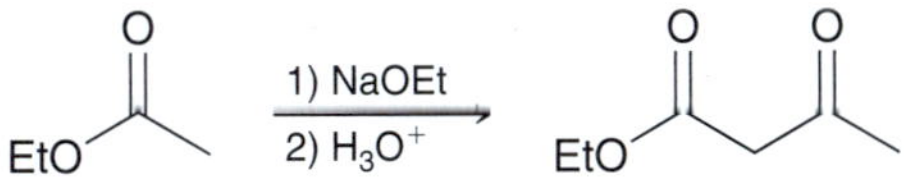

O

EtO

1) NaOEt

2) $H_3O^+$

O O

EtO

### Intramolecular Claisen Condensation (Dieckmann Cyclization)

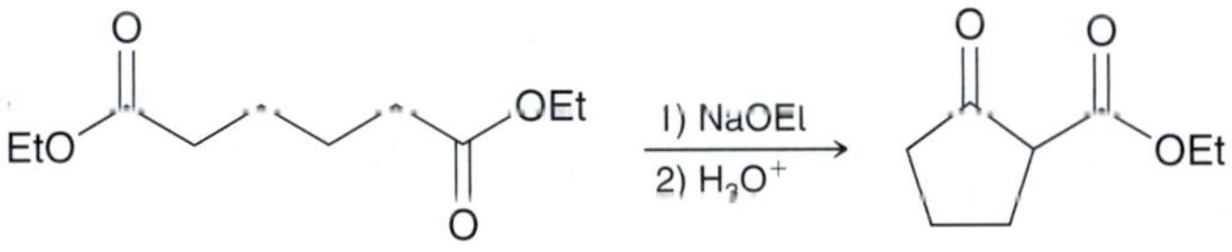

O

EtO

OEt

O

1) NaOEt

2) $H_3O^+$

O O

OEt

### Crossed Claisen Condensations

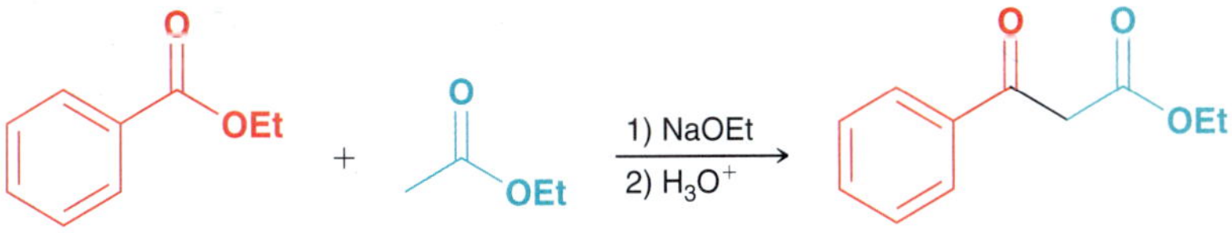

O

OEt

+

O

OEt

1) NaOEt

2) $H_3O^+$

O O

OEt

## Alkylation

**Via Enolate Ions**

1) LDA, −78°C
2) RX

1) NaH, 25°C
2) RX

**The Malonic Ester Synthesis**

EtO OEt O O
1) NaOEt, EtOH
2) RBr
3) $H_3O^+$, heat
R OH O

**The Acetoacetic Ester Synthesis**

EtO O O
1) NaOEt/EtOH
2) RBr
3) $H_3O^+$, heat
R O

## Michael Additions

**Stabilized Carbon Nucleophiles**

1) $R_2CuLi$
2) $H_3O^+$

1) KOH
2)
3) $H_3O^+$

**The Stork Enamine Synthesis**

1) $R_2NH$, $[H^+]$, $(-H_2O)$
2)
3) $H_3O^+$

**The Robinson Annulation**

NaOH, heat

# REVIEW OF CONCEPTS AND VOCABULARY

### SECTION 22.1

- Greek letters are used to describe the proximity of each carbon atom to the carbonyl group. Alpha (α) protons are the protons connected to the α carbon.
- In the presence of catalytic acid or base, a ketone will exist in equilibrium with an enol. In general, the equilibrium will significantly favor the ketone.
- The α position of an enol can function as a nucleophile.
- When treated with a strong base, the α position of a ketone is deprotonated to give an **enolate**.
- Sodium hydride or LDA will irreversibly and completely convert an aldehyde or ketone into an enolate.

### SECTION 22.2

- Ketones and aldehydes will undergo alpha halogenation in acidic or basic conditions.
- The acid-catalyzed process produces HBr and is therefore **autocatalytic**.
- In the **Hell-Volhard-Zelinsky reaction**, a carboxylic acid undergoes alpha halogenation when treated with bromine in the presence of $PBr_3$.
- In the **haloform reaction**, a methyl ketone is converted into a carboxylic acid upon treatment with excess base and excess halogen followed by acid workup.

### SECTION 22.3

- When an aldehyde is treated with sodium hydroxide, an **aldol addition reaction** occurs, and the product is a β-hydroxy aldehyde or ketone.
- For most simple aldehydes, the position of equilibrium favors the aldol product.
- For most ketones, the reverse process, called a **retro-aldol reaction**, is favored.
- When an aldehyde is heated in aqueous sodium hydroxide, an **aldol condensation reaction** occurs, and the product is an α,β-unsaturated aldehyde or ketone. Elimination of water occurs via an **E1cb mechanism**.

- **Crossed aldol,** or **mixed aldol, reactions** are aldol reactions that occur between different partners and are only efficient if one partner lacks α protons or if a **directed aldol addition** is performed.
- Intramolecular aldol reactions show a preference for formation of five- and six-membered rings.

SECTION 22.4

- When an ester is treated with an alkoxide base, a **Claisen condensation reaction** occurs, and the product is a β-keto ester.
- A Claisen condensation between two different partners is called a **crossed Claisen condensation**.
- An intramolecular Claisen condensation, called a **Dieckmann cyclization**, produces a cyclic, β-keto ester.

SECTION 22.5

- The α position of a ketone can be alkylated by forming an enolate and treating it with an alkyl halide.
- For unsymmetrical ketones, reactions with LDA at low temperature favor formation of the kinetic enolate, while reactions with NaH at room temperature favor the thermodynamic enolate.
- When LDA is used with an unsymmetrical ketone, alkylation occurs at the less hindered position.
- The **malonic ester synthesis** enables the conversion of an alkyl halide into a carboxylic acid with the introduction of two new carbon atoms. The **acetoacetic ester synthesis** enables the conversion of an alkyl halide into a methyl ketone with the introduction of three new carbon atoms.
- **Decarboxylation** occurs upon heating a carboxylic acid with a β-carbonyl group.

SECTION 22.6

- Aldehydes and ketones that possess α,β unsaturation are susceptible to nucleophilic attack at the β position. This reaction is called a **conjugate addition**, or *1,4-addition*, or a **Michael reaction**.
- The nucleophile is called a **Michael donor**, and the electrophile is called a **Michael acceptor**.
- Regular enolates do not serve as Michael donors, but the desired Michael reaction can be achieved with a **Stork enamine synthesis**.
- A **Robinson annulation** is a Michael addition followed by an intramolecular aldol and can be used to make cyclic compounds.

SECTION 22.7

- The Stork enamine synthesis produces 1,5-difunctionalized compounds.
- Aldol addition reactions and Claisen condensation reactions both produce 1,3-difunctionalized compounds.
- The initial product of a Michael addition is an enolate ion, which can be treated with an alkyl halide, thereby alkylating both the α and β positions in one reaction flask.

# SKILLBUILDER REVIEW

## 22.1 DRAWING ENOLATES

**STEP 1** Identify all α protons.

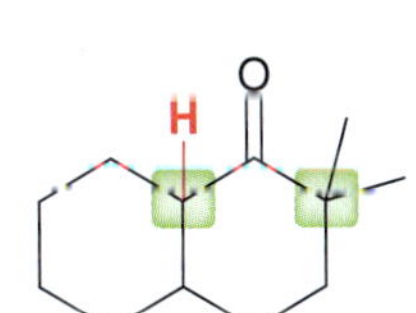


**STEP 2** Using two curved arrows, remove the proton, and then draw the enolate with a lone pair and a negative charge at the α position.

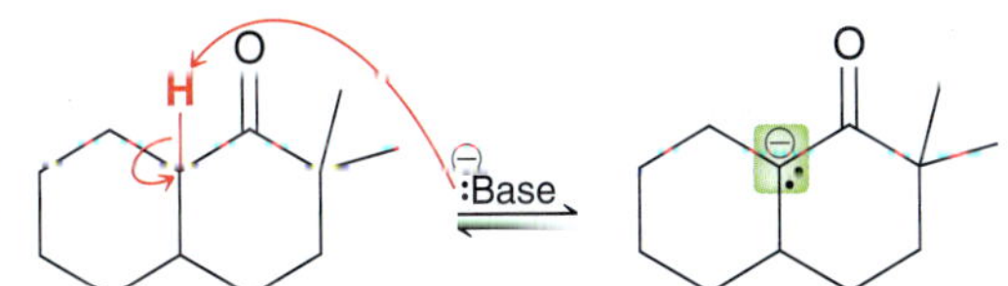


**STEP 3** Draw the other resonance structure of the enolate.

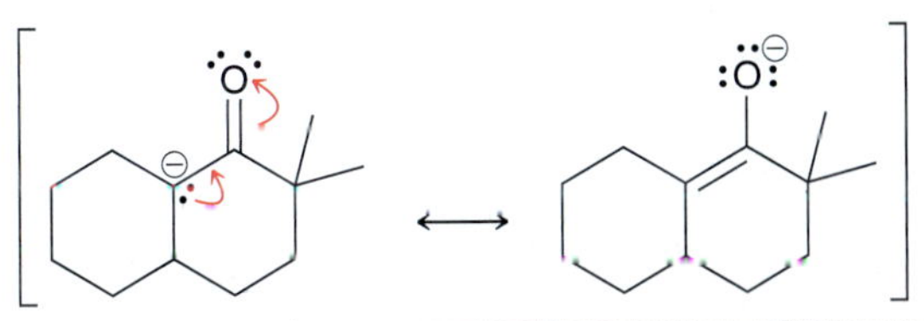

Try Problems **22.4, 22.5, 22.59**

## 22.2 PREDICTING THE PRODUCTS OF AN ALDOL ADDITION REACTION

**STEP 1** Draw all three steps of the mechanism as a guide for predicting the product.

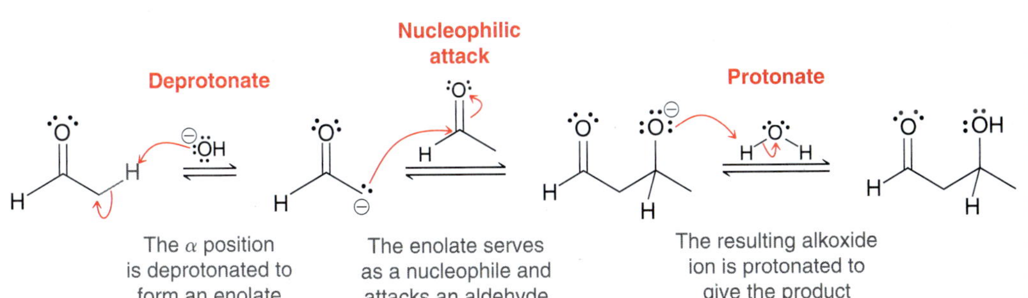


**STEP 2** Double-check your answer to ensure that the product has a hydroxyl group at the β position.

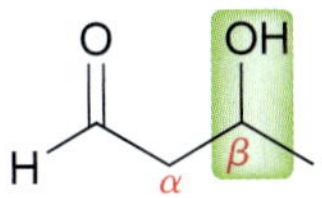


Try Problems **22.15–22.19, 22.72**

### 22.3 DRAWING THE PRODUCT OF AN ALDOL CONDENSATION

**STEP 1** Identify the $\alpha$ protons.

**STEP 2** Draw two molecules of the ketone, oriented such that two $\alpha$ protons of one molecule are directly facing the carbonyl group of the other molecule.

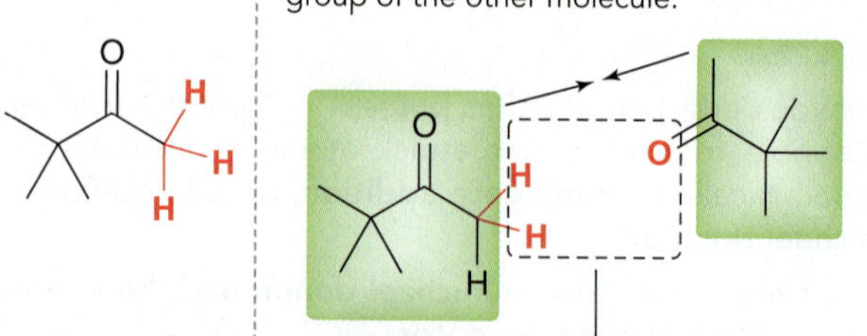


**STEP 3** Remove two $\alpha$ protons and the oxygen atom, forming a double bond in place of those groups.

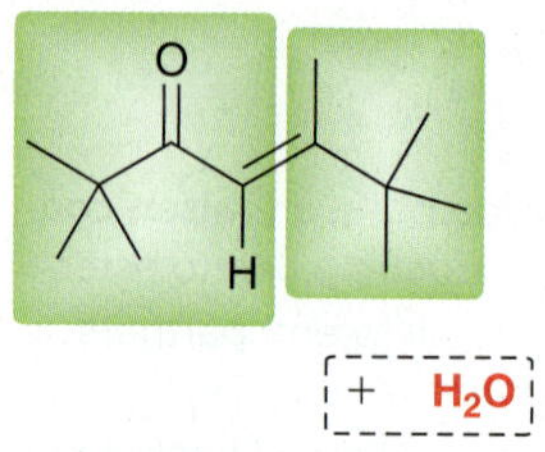


**STEP 4** Make sure to draw the product with fewer steric interactions.

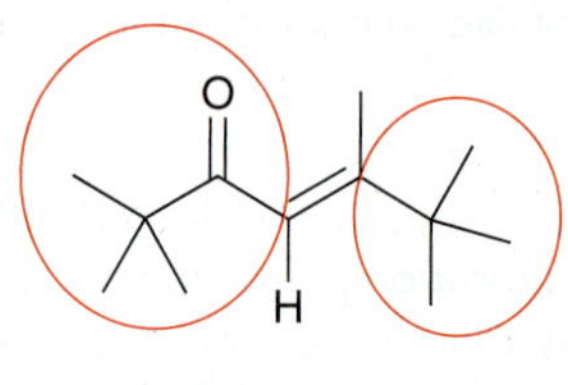


Try Problems 22.20–22.22, 22.71, 22.84c

### 22.4 IDENTIFYING THE REAGENTS NECESSARY FOR A CROSSED ALDOL REACTION

**STEP 1** Identify the $\alpha$ and $\beta$ positions.

**STEP 2** Using a retrosynthetic analysis, break apart the bond between the $\alpha$ and $\beta$ positions, placing a carbonyl group in place of the hydroxyl group.

**STEP 3** Determine which base should be used. A crossed aldol will require the use of LDA.

1) LDA
2) (butanal)
3) $H_2O$

Try Problems 22.23, 22.24, 22.67, 22.68

### 22.5 USING THE MALONIC ESTER SYNTHESIS

**STEP 1** Identify the alkyl groups that are connected to the $\alpha$ position of the carboxylic acid.

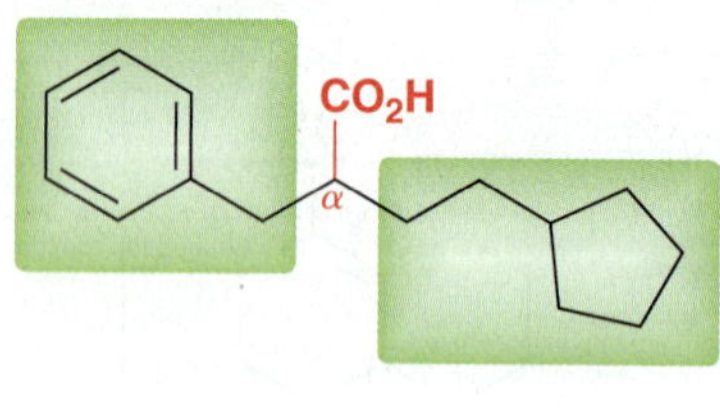


**STEP 2** Identify the alkyl halides necessary and ensure that both will readily undergo an $S_N2$ process.

**STEP 3** Identify the reagents. Begin with diethyl malonate as the starting material. Perform each alkylation and then heat with aqueous acid.

1) NaOEt
2) $PhCH_2Br$
3) NaOEt
4) Br–$CH_2CH_2$–cyclopentyl
5) $H_3O^+$, heat

Target compound

Try Problems 22.35–22.38, 22.78

### 22.6 USING THE ACETOACETIC ESTER SYNTHESIS

**STEP 1** Identify the alkyl groups that are connected to the $\alpha$ position of the methyl ketone.

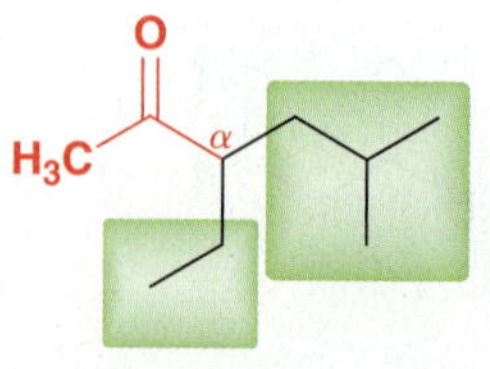


**STEP 2** Identify the alkyl halides necessary and ensure that both will readily undergo an $S_N2$ process.

**STEP 3** Identify the reagents. Begin with ethyl acetoacetate as the starting material. Perform each alkylation and then heat with aqueous acid.

1) NaOEt
2) EtI
3) NaOEt
4) I–$CH_2CH(CH_3)_2$
5) $H_3O^+$, heat

Try Problems 22.39–22.43, 22.79

## 22.7 DETERMINING WHEN TO USE A STORK ENAMINE SYNTHESIS

**STEP 1** Using a retrosynthetic analysis, identify whether it is possible to prepare the target compound with a Michael addition.

Michael

**STEP 2** If the Michael donor must be an enolate, then a Stork enamine synthesis is required.

1) $R_2NH$, $[H^+]$, $(-H_2O)$
2)
3) $H_3O^+$

Try Problems 22.47, 22.48, 22.87d

## 22.8 DETERMINING WHICH ADDITION OR CONDENSATION REACTION TO USE

For compounds with two functional groups, the relative positioning of the groups, as well as their oxidation states, will dictate which addition or condensation reaction to use.

### 1,5-DIFUNCTIONALIZED COMPOUNDS

**Stork enamine synthesis**

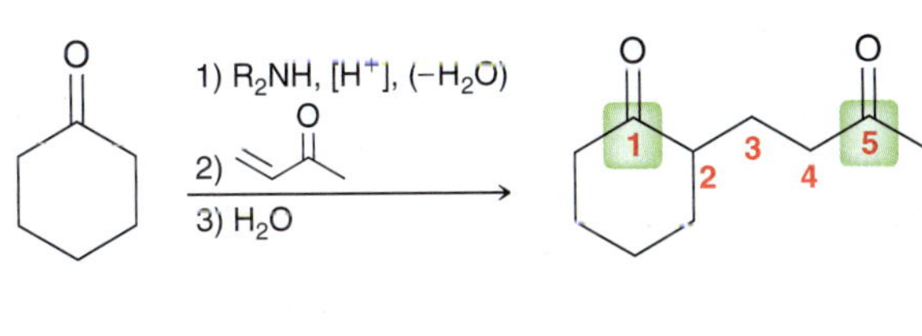


### 1,3-DIFUNCTIONALIZED COMPOUNDS

**Aldol addition**

NaOH, $H_2O$

**Claisen condensation**

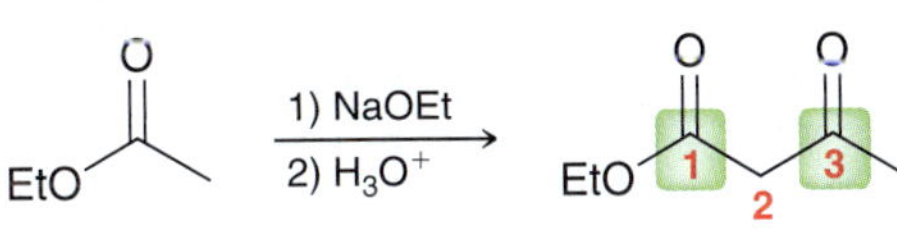


Try Problems 22.51–22.53, 22.87d

## 22.9 ALKYLATING THE α AND β POSITIONS

A strategy for installing two neighboring alkyl groups at the $\alpha$ and $\beta$ positions.

$R_2CuLi$ → Enolate → R—X

Try Problems 22.54–22.56, 22.82, 22.89

# PRACTICE PROBLEMS

Note: Most of the Problems are available within **WileyPLUS**, an online teaching and learning solution.

**22.57** Identify which of the following compounds are expected to have $pK_a < 20$. For each compound with $pK_a < 20$, identify the most acidic proton in the compound.

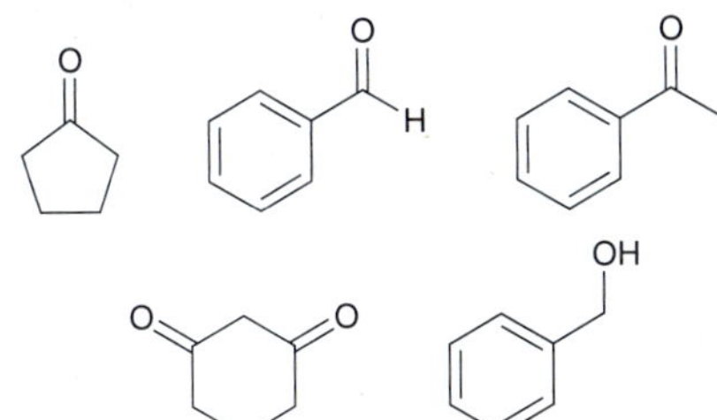

**22.58** One of the compounds from the previous problem has $pK_a < 10$. Identify that compound and explain why it is so much more acidic than all of the other compounds.

**22.59** Draw resonance structures for the conjugate base that is produced when each of the following compounds is treated with sodium ethoxide:

(a) (b) (c)

**22.60** Rank the following compounds in terms of increasing acidity:

**22.61** Draw the enol of each of the following compounds and identify whether the enol exhibits a significant presence at equilibrium. Explain.

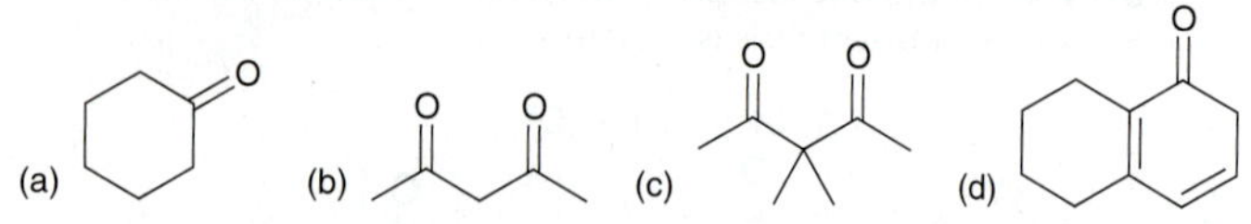


**22.62** Ethyl acetoacetate has three enol isomers. Draw all three.

**22.63** Draw the enolate that is formed when each of the following compounds is treated with LDA:

(a) (b) (c) (d)

**22.64** When 2-hepten-4-one is treated with LDA, a proton is removed from one of the gamma (γ) positions. Identify which γ position is deprotonated and explain why the γ proton is the most acidic proton in the compound.

**22.65** When optically active (*S*)-2-methylcyclopentanone is treated with aqueous base, the compound loses its optical activity. Explain this observation and draw a mechanism that shows how racemization occurs.

**22.66** The racemization process described in the previous problem also occurs in acidic conditions. Draw a mechanism for the racemization process in aqueous acid.

**22.67** Draw all four β-hydroxyaldehydes that are formed when a mixture of acetaldehyde and pentanal is treated with aqueous sodium hydroxide.

**22.68** Identify all of the different β-hydroxyaldehydes that are formed when a mixture of benzaldehyde and hexanal is treated with aqueous sodium hydroxide.

**22.69** Propose a plausible mechanism for the following isomerization and explain the driving force behind this reaction. In other words, explain why the equilibrium favors the product.

$H_3O^+$

**22.70** The isomerization in the previous problem can also occur in basic conditions. Draw a mechanism for the transformation in the presence of catalytic hydroxide.

**22.71** Draw the product obtained when each of the following compounds is heated in the presence of a base to give an aldol condensation:

(a) (b) (c)

**22.72** Trimethylacetaldehyde does not undergo an aldol reaction when treated with a base. Explain why not.

**22.73** Identify the reagents necessary to make each of the following compounds with an aldol condensation:

(a) (b) (c) (d)

**22.74** When acetaldehyde is treated with aqueous acid, an aldol reaction can occur. In other words, aldol reactions can also occur in acidic conditions, although the intermediate is different than the intermediate involved in the base-catalyzed reaction. Draw a plausible mechanism for the acid-catalyzed process.

$H_3O^+$

**22.75** Diethyl malonate (the starting material for the malonic ester synthesis) reacts with bromine in acid-catalyzed conditions to form a product with molecular formula $C_7H_{11}BrO_4$.

(a) Draw the structure of the product.

(b) Draw a mechanism of formation for the product.

(c) Would you expect this product to be more or less acidic than diethyl malonate?

**22.76** Cinnamaldehyde is one of the primary constituents of cinnamon oil and contributes significantly to the odor of cinnamon. Starting with benzaldehyde and using any other necessary reagents, show how you might prepare cinnamaldehyde.

**Cinnamaldehyde**

**22.77** Draw the condensation product that is expected when each of the following esters is treated with sodium ethoxide followed by acid workup:

(a) (b)

**22.78** Starting with diethyl malonate and using any other reagents of your choice, show how you would prepare each of the following compounds:

(a) (b) (c)

**22.79** Starting with ethyl acetoacetate and using any other reagents of your choice, show how you would prepare each of the following compounds:

(a) (b) (c)

**22.80** Propose a plausible mechanism for the following transformation:

$H_3O^+$

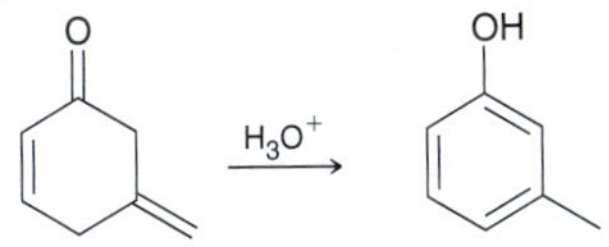

**22.81** Draw the condensation product obtained when the following compound is heated in the presence of aqueous sodium hydroxide:

$\xrightarrow[\text{Heat}]{NaOH,\ H_2O}$ $C_{12}H_{12}O$ + $H_2O$

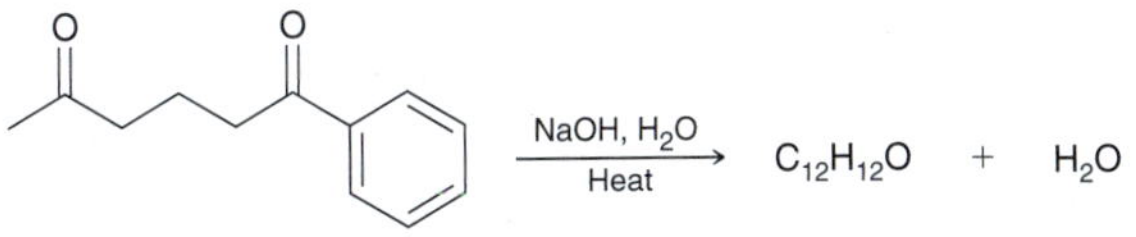

**22.82** Identify the reagents you would use to convert 3-pentanone into 3-hexanone.

**22.83** Identify the reagents necessary to achieve each of the following transformations:

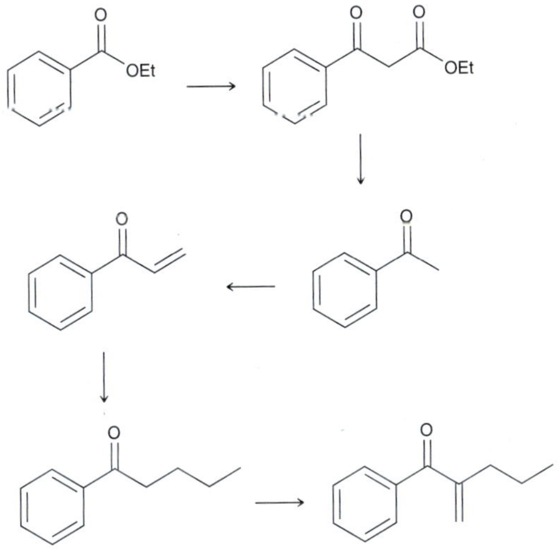

**22.84** Draw the structure of the product that is obtained when acetophenone is treated with each of following reagents:

(a) Sodium hydroxide and excess iodine followed by $H_3O^+$

(b) Bromine in acetic acid

(c) Aqueous sodium hydroxide at elevated temperature

**22.85** Draw a reasonable mechanism for the following transformation:

$\xrightarrow[\text{Heat}]{NaOH,\ H_2O}$ +

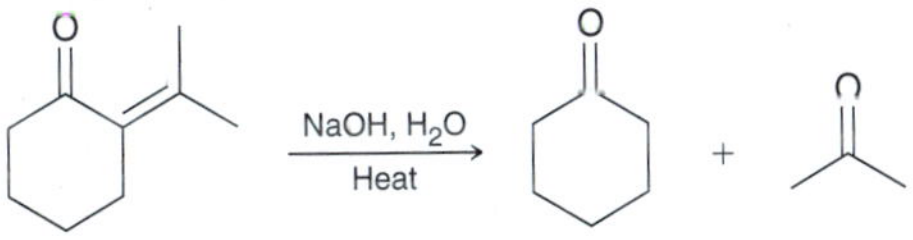

**22.86** Predict the major product for each of the following transformations:

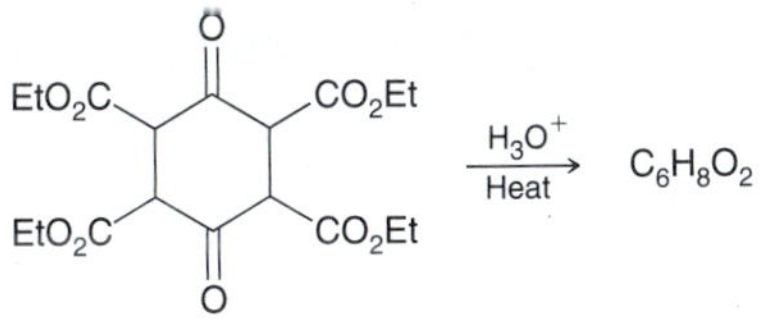

(a) $\xrightarrow[\text{Heat}]{H_3O^+}$ $C_6H_8O_2$

(b) 1) NaOEt 2) $H_3O^+$ → $C_{11}H_{16}O_3$ → ($H_3O^+$, Heat) → $C_9H_{12}O$

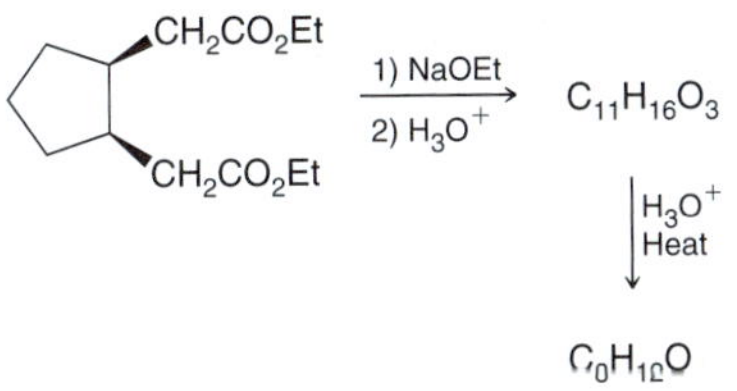

(c) $\xrightarrow{[H_3O^+],\ Br_2}$ $C_5H_9BrO$ → (Pyridine) → $C_5H_8O$ → (1) $Et_2CuLi$ 2) MeI) → $C_8H_{16}O$

(d) 1) LDA, –78°C 2) EtI → $C_9H_{16}O$

**22.87** Propose an efficient synthesis for each of the following transformations:

(a)

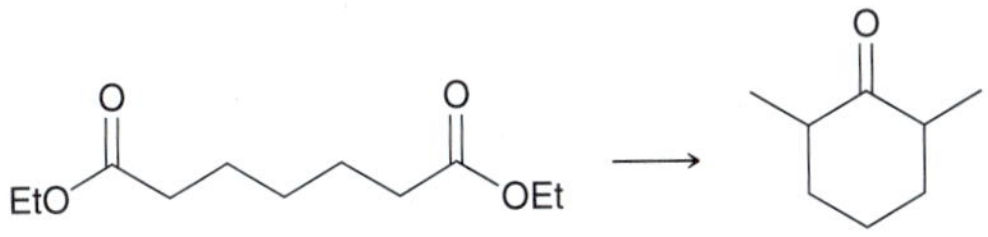

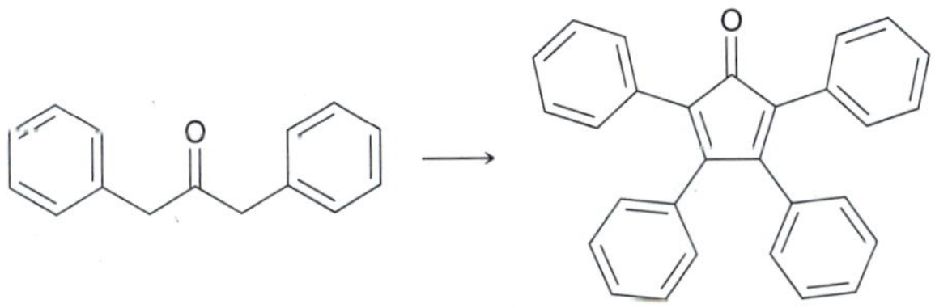

(b)

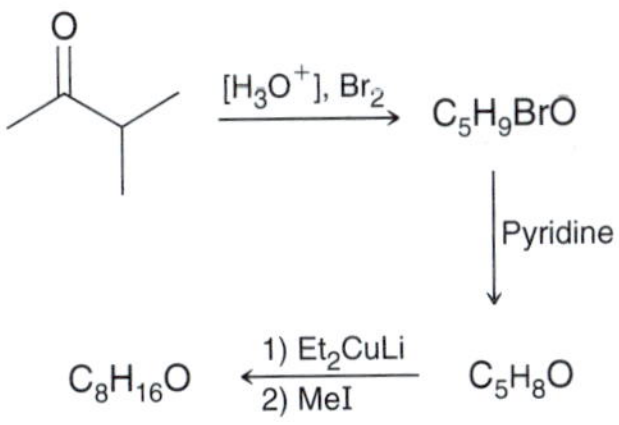

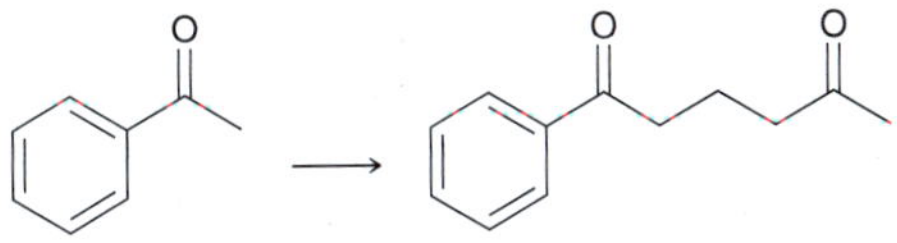

(c)

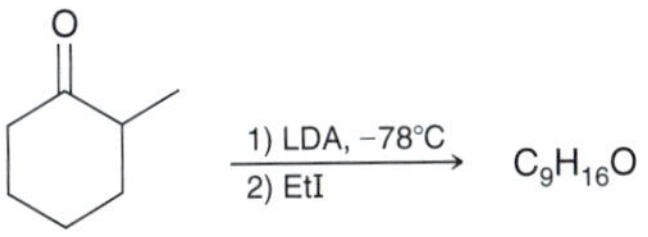

**22.88** The product of an aldol condensation is an $\alpha,\beta$-unsaturated ketone which is capable of undergoing hydrogenation to yield a saturated ketone. Using this technique, identify the reagents that you would need in order to prepare rheosmin via a crossed aldol reaction. Rheosmin is isolated from raspberries and is often used in perfume formulations for its pleasant odor. **Hint:** The presence of a phenolic proton will be problematic during an aldol reaction. (Can you explain why?) Consider using a protecting group (Section 13.7).

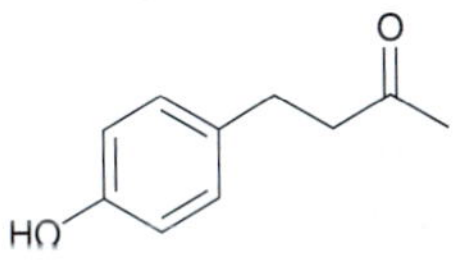

**Rheosmin**

**22.89** Identify the reagents you would use to convert cyclohexanone into each of the following compounds:

(a) 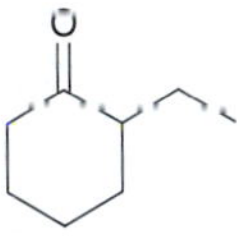 (b) 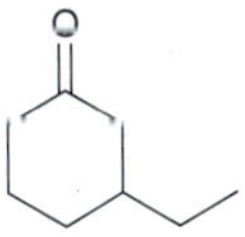 (c) 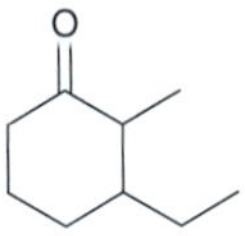 (d) 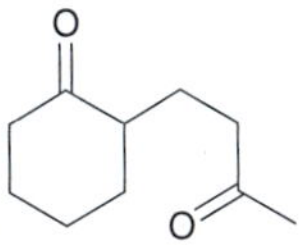 (e) 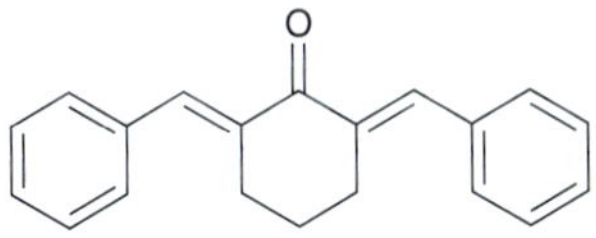 (f) 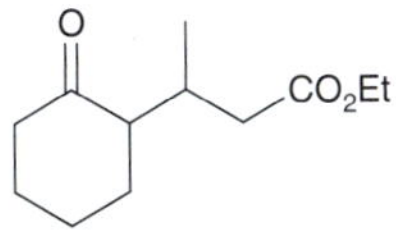 (g)

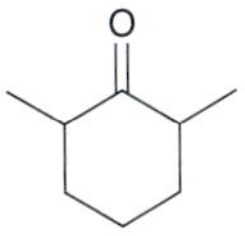

**22.90** The enolate of a ketone can be treated with an ester to give a diketone. Draw a mechanism for this Claisen-like reaction and explain why an acid source is required after the reaction is complete.

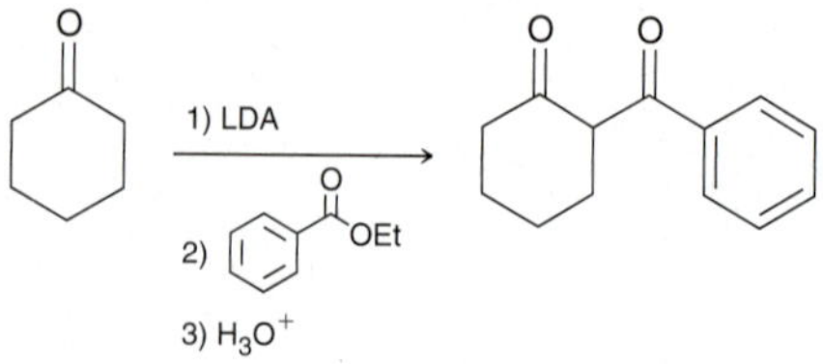

1) LDA
2) OEt
3) $H_3O^+$

**22.91** Beta-keto esters can be prepared by treating the enolate of a ketone with diethyl carbonate. Draw a plausible mechanism for this reaction.

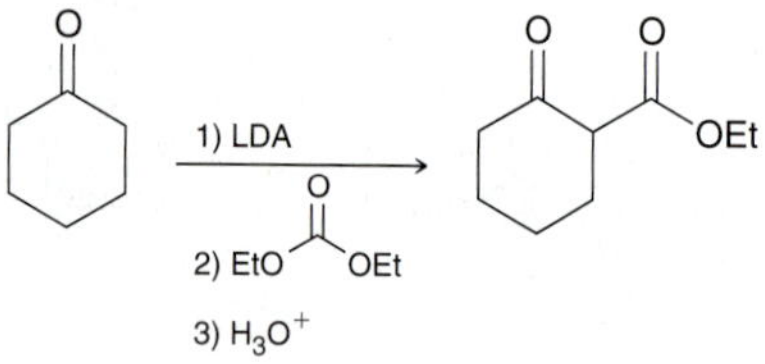

1) LDA
2) EtO OEt
3) $H_3O^+$

OEt

**22.92** The enolate of an ester can be treated with a ketone to give a β-hydroxy ester. Draw a mechanism for this aldol-like reaction.

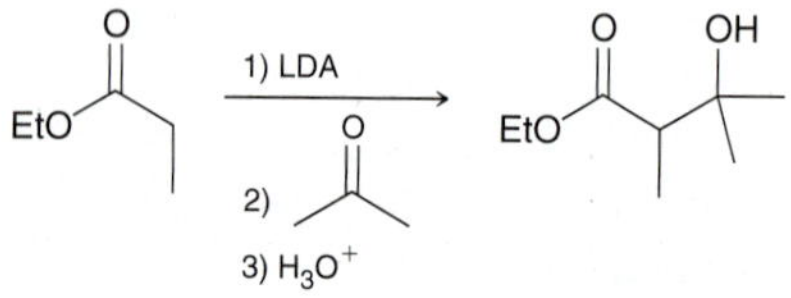

EtO
1) LDA
2)
3) $H_3O^+$
EtO OH

**22.93** Nitriles undergo alkylation at the α position much like ketones undergo alkylation at the α position.

CN 1) LDA 2) RX → CN R

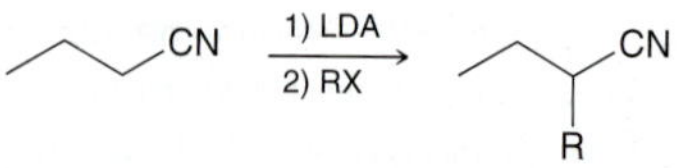

The α position of the nitrile is first deprotonated to give a resonance-stabilized anion (like an enolate), which then functions as a nucleophile to attack the alkyl halide.

(a) Draw the mechanism described above.

(b) Using this process, show the reagents you would use to achieve the following transformation:

Br → OH O

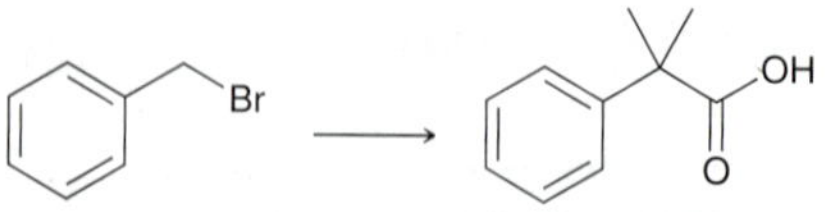

**22.94** Identify the Michael donor and Michael acceptor that could be used to prepare each of the following compounds via a Michael addition:

(a)

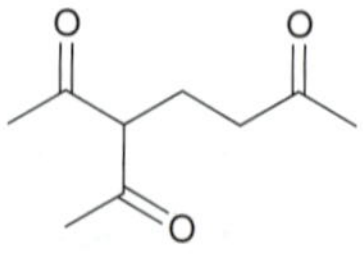

(b) 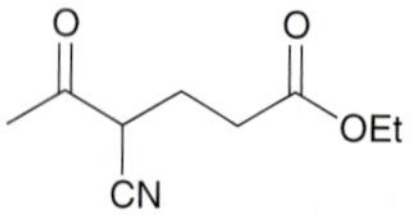 CN OEt

(c) 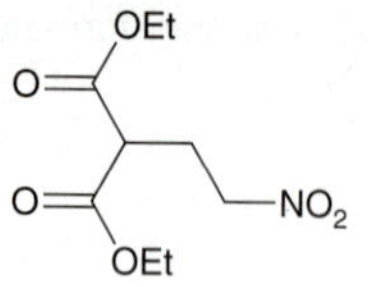 OEt NO2 OEt

(d) 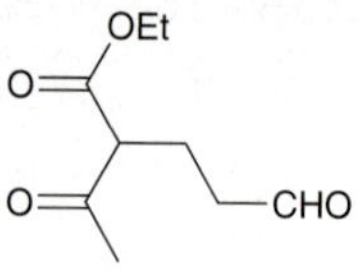 OEt CHO

(e)  $O_2N$ $NO_2$

**22.95** The conjugate base of diethyl malonate can serve as a nucleophile to attack a wide range of electrophiles. Identify the product that is formed when the conjugate base of diethyl malonate reacts with each of the following electrophiles followed by acid workup:

(a) 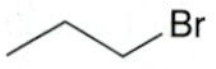 Br

(b)

(c) 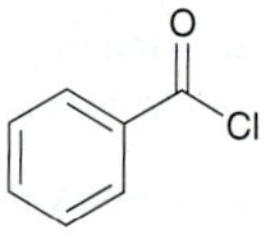 Cl

(d)

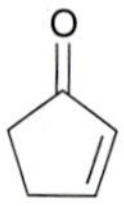

(e) 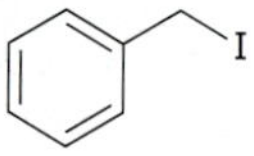 I

(f) 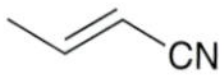 CN

(g)

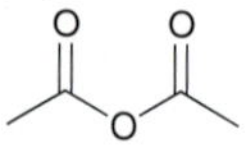

(h) 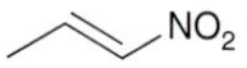 $NO_2$

**22.96** Draw the product of the Robinson annulation reaction that occurs when the following compounds are treated with aqueous sodium hydroxide:

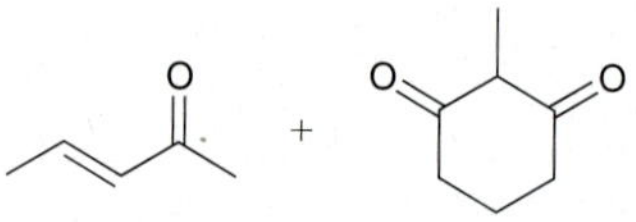

**22.97** Identify what reagents you would use to make the following compound with a Robinson annulation reaction:

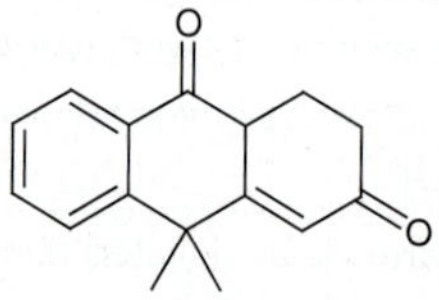

**22.98** Draw a plausible mechanism for the following transformation:

$NO_2$ NaOH, $H_2O$ Heat → $NO_2$

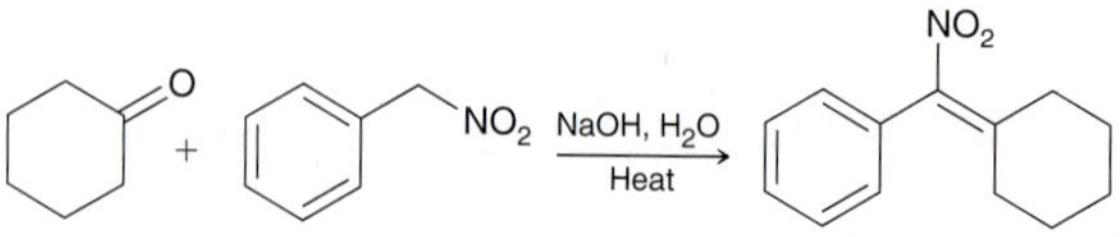

**22.99** Propose an efficient synthesis for the following transformation:

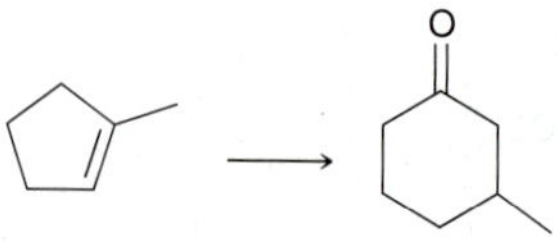

# INTEGRATED PROBLEMS

**22.100** For a pair of keto-enol tautomers, explain how IR spectroscopy might be used to identify whether the equilibrium favors the ketone or the enol.

**22.101** Acrolein is an α,β-unsaturated aldehyde that is used in the production of a variety of polymers. Acrolein can be prepared by treating glycerol with an acid catalyst. Propose a plausible mechanism for this transformation.

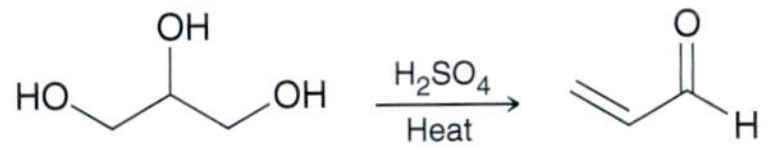

Glycerol $\xrightarrow[\text{Heat}]{H_2SO_4}$ Acrolein

**22.102** Draw the structure of the product with molecular formula $C_{10}H_{10}O$ that is obtained when the compound below is heated with aqueous acid.

$\xrightarrow[\text{Heat}]{H_3O^+}$ $C_{10}H_{10}O$

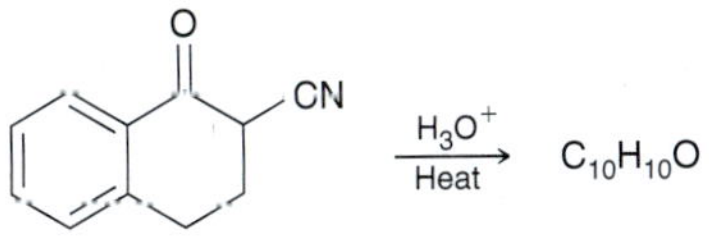

**22.103** Lactones can be prepared from diethyl malonate and epoxides. Diethyl malonate is treated with a base, followed by an epoxide, followed by heating in aqueous acid:

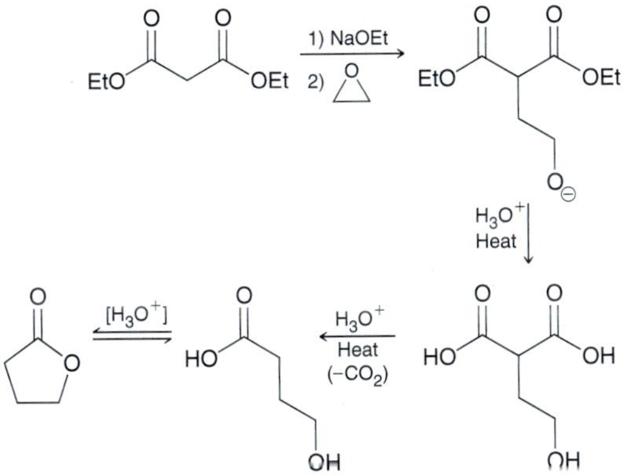

1) NaOEt 2) epoxide; $H_3O^+$, Heat; $H_3O^+$, Heat $(-CO_2)$; $[H_3O^+]$

Using this process, identify what reagents you would need to prepare the following compound:

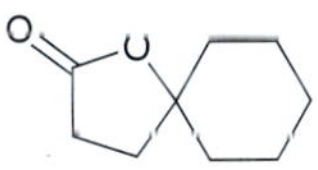

**22.104** Predict the major product of the following transformation.

$\xrightarrow[\text{Heat}]{H_3O^+}$ $C_{10}H_{10}O$

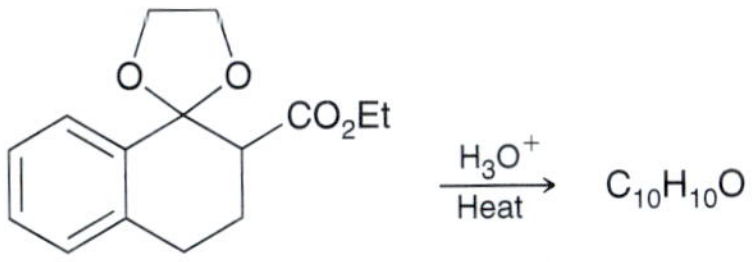

**22.105** Consider the structures of the constitutional isomers, compound **A** and compound **B**. When treated with aqueous acid, compound **A** undergoes isomerization to give a *cis* stereoisomer. In contrast, compound **B** does not undergo isomerization when treated with the same conditions. That is, compound **B** remains in the *trans* configuration. Explain the difference in reactivity between compound **A** and compound **B**.

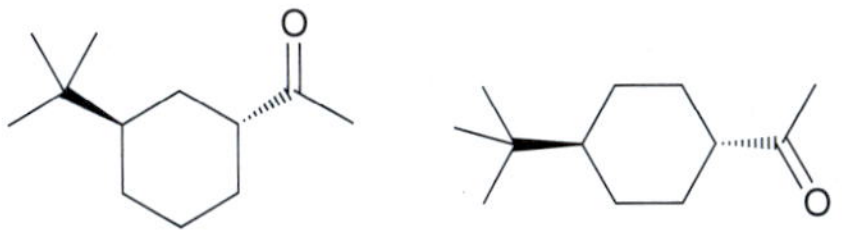

**Compound A** **Compound B**

**22.106** Propose a plausible mechanism for the following transformation:

$\xrightarrow{H_3O^+}$

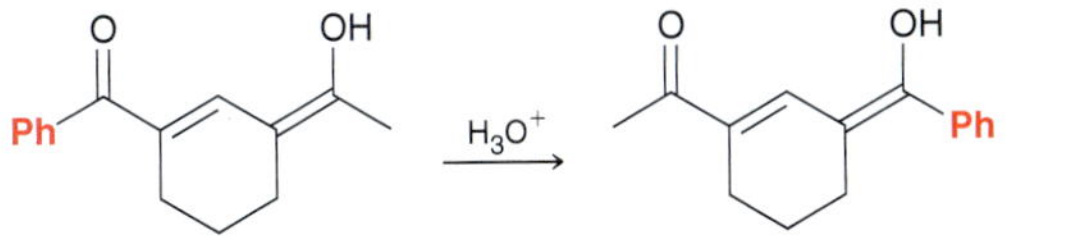

**22.107** Propose a plausible mechanism for the following transformation:

$\xrightarrow[\text{Heat}]{NaOH,\ H_2O}$

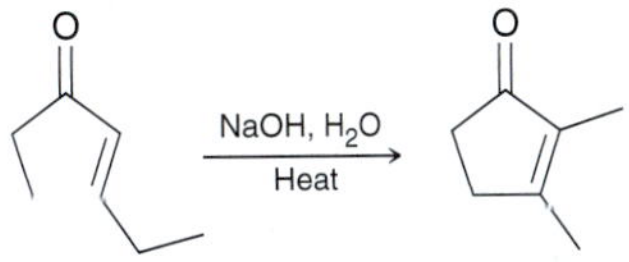

**22.108** This chapter covered many C—C bond-forming reactions, including aldol reactions, Claisen condensations, and Michael addition reactions. Two or more of these reactions are often performed sequentially, providing a great deal of versatility and complexity in the type of structures that can be prepared. Propose a plausible mechanism for each of the following transformations:

(a) $\xrightarrow{NaOH,\ EtOH}$

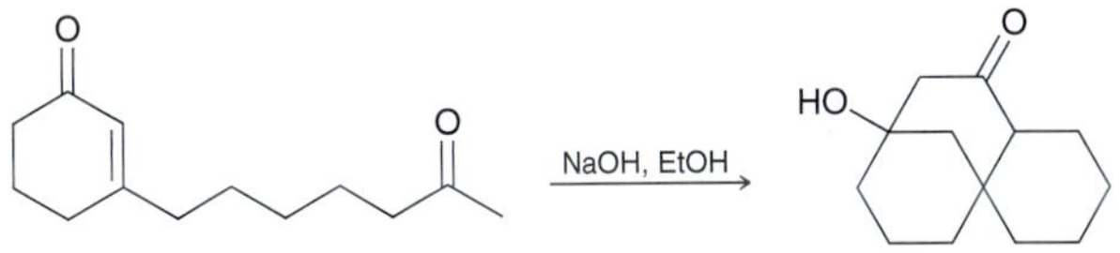

(b) $\xrightarrow[\text{EtOH}]{NaOEt}$

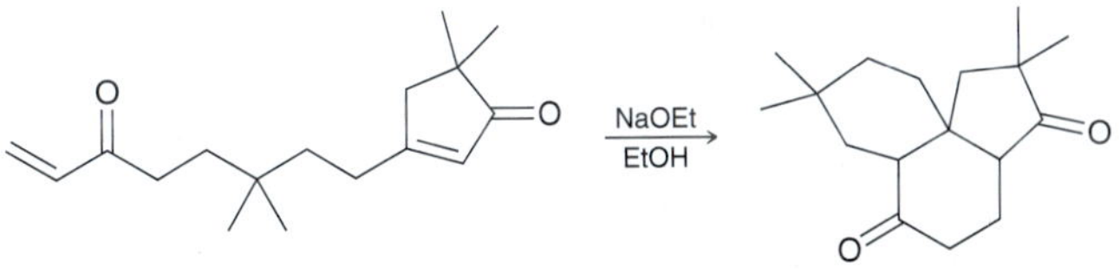

**22.109** The following transformation cannot be accomplished by direct alkylation of an enolate. Explain why not and then devise an alternate synthesis for this transformation.

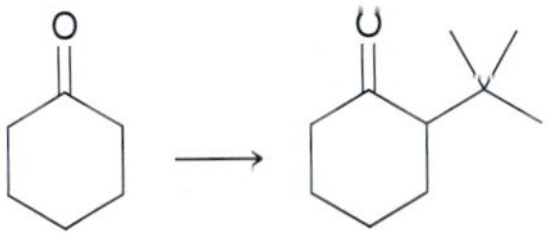

**22.110** We have seen that the alpha carbon atom of an enamine can function as a nucleophile in a Michael reaction, and in fact, enamines can function as nucleophiles in a wide variety of reactions. For example, an enamine will undergo alkylation when treated with an alkyl halide. Draw the structure of intermediate **A** and the alkylation product **B** in the following reaction scheme (*J. Am. Chem. Soc.* **1954**, *76*, 2029–2030):

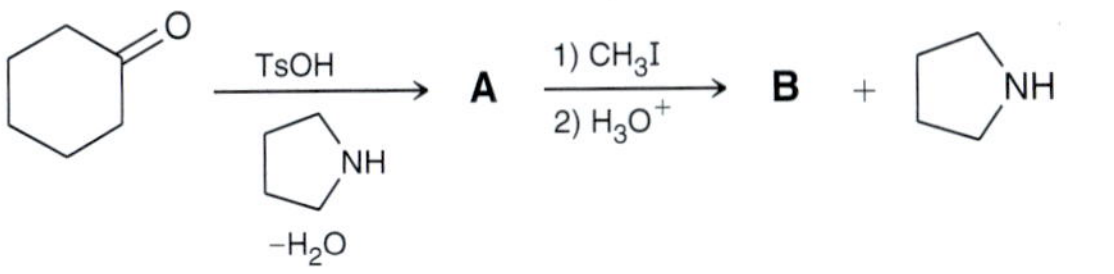

$\xrightarrow[-H_2O]{TsOH,\ \text{pyrrolidine (NH)}}$ **A** $\xrightarrow[2)\ H_3O^+]{1)\ CH_3I}$ **B** + NH (pyrrolidine)

**22.111** Consider the reaction between cyclohexanone and the optically pure amine shown below (*J. Org. Chem.*, **1977**, *42*, **1663–1664**). (a) Draw the structure of the resulting enamine and discuss how many isomer(s), if any, form in this reaction. (b) When the enamine(s) formed in part a is (are) treated with methyl iodide, followed by aqueous acid, 2-methylcyclohexanone with 83% *ee* is isolated. Explain the source of the observed enantiomeric excess (as compared with the racemic mixture obtained in Problem 22.110), and identify the enantiomer that predominates.

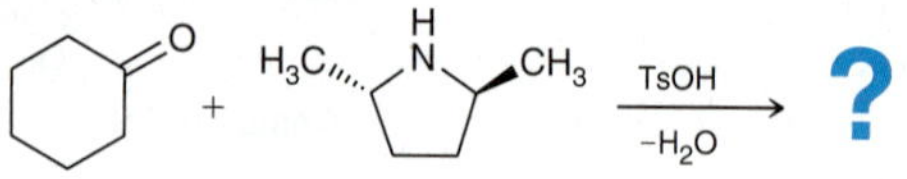

$H_3C$ ... $CH_3$; TsOH, $-H_2O$ → ?

**22.112** Predict the product of the following reaction sequence, which was recently used in a synthetic route toward a series of 1,3,6-substituted fulvenes (*J. Org. Chem.* **2012**, *77*, **6371–6376**):

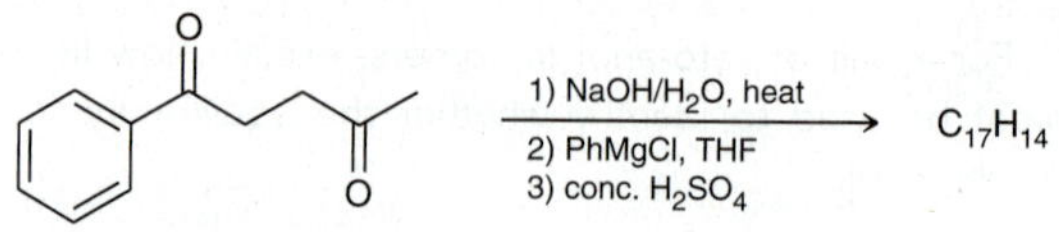

1) $NaOH/H_2O$, heat; 2) PhMgCl, THF; 3) conc. $H_2SO_4$ → $C_{17}H_{14}$

# CHALLENGE PROBLEMS

**22.113** The natural product (–)-*N*-methylwelwitindolinone C isothiocyanate is isolated from a blue-green algae and has been identified as a promising compound to treat drug-resistant tumors. Compound **X** (R = protecting group) was utilized as a key synthetic intermediate in the synthesis of the natural product (*J. Am. Chem. Soc.* **2011**, *133*, **15797–15799**). Compound **X** can be prepared in a ring-forming reaction from compound **Y** under strongly basic conditions (excess $NaNH_2$).

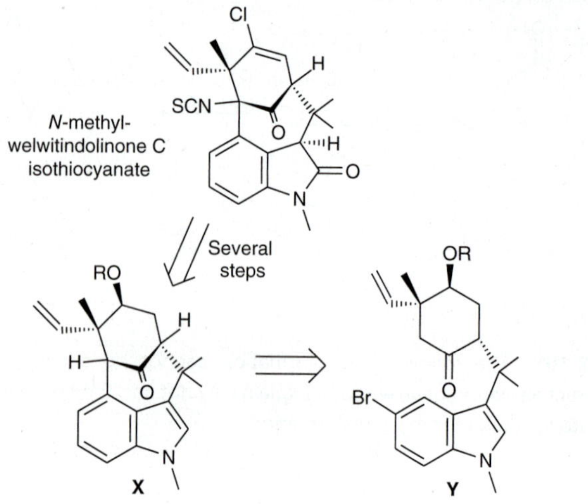

*N*-methyl-welwitindolinone C isothiocyanate; Several steps; **X**; **Y**

(a) Propose a plausible mechanism for conversion of **Y** into **X** under basic conditions.

(b) When **Y** is converted into **X**, a minor side product of the reaction is a constitutional isomer of **X** which lacks a ketone group. Propose a structure for this side product. (**Hint:** Recall that enolates are ambident nucleophiles.)

**22.114** When a mixture of conformationally isomeric enamines, shown below, is treated with methyl acrylate, followed by aqueous acid, the expected Stork enamine synthesis product **A** is isolated in 21% *ee* (*Tetrahedron Lett.*, **1969**, *10*, **4233–4236**). Assuming that the two enamine isomers are present in equal concentration, explain why such an enantiomeric mixture of **A** would result in this case by providing a mechanistic analysis for the Michael reaction possible via each enamine.

Based on your analysis, determine whether the *R* or *S* enantiomer of **A** predominates in the product mixture and explain your choice.

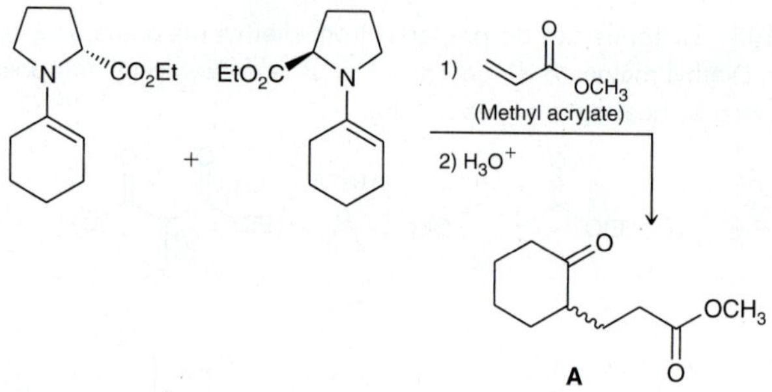

1) (Methyl acrylate); 2) $H_3O^+$ → **A**

**22.115** The following sequence, beginning with a cyclic hemiacetal (compound **A**), was part of a recently reported enantiospecific synthesis of a powerful sex pheromone (currently used in pest management) of the mealybug *Pseudococcus viburni* (*Tetrahedron Lett.* **2010**, *51*, **5291–5293**):

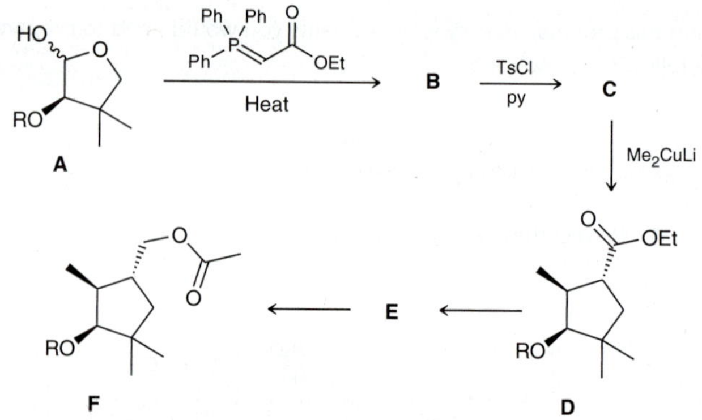

**A** → ($Ph_3P{=}CHCO_2Et$, Heat) → **B** → (TsCl, py) → **C** → ($Me_2CuLi$) → **D** → **E** → **F**

(a) Draw the structures of compounds **B** and **C**.

(b) Describe in words how the cyclopentyl ring is closed during the conversion of compound **C** into compound **D**.

(c) Identify the reagents you would use to convert compound **D** into compound **F** (in just two steps). Also identify the structure of compound **E**.

**22.116** The following synthetic step was utilized as part of a recent synthesis of the polycyclic natural product haouamine B (*J. Am. Chem. Soc.* **2012**, *134*, 9291–9295). In this reaction, the function of the first reagent (triflic anhydride) is to activate the Michael acceptor that is present in the starting compound (rendering it even more electrophilic), so that it can undergo an intramolecular Michael reaction. Draw a mechanism for this reaction and explain the observed stereochemistry of the newly formed chirality center.

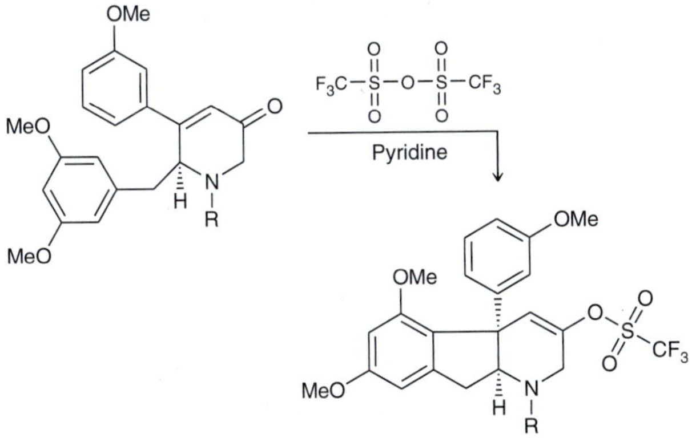

**22.117** In a recent effort to devise a new synthetic pathway for the preparation of useful building blocks for the synthesis of natural products, compound **2** was prepared by treating compound **1** with diethylmalonate in the presence of potassium carbonate (*Tetrahedron Lett.* **2010**, *51*, 6918–6920). Propose a plausible mechanism for this reaction.

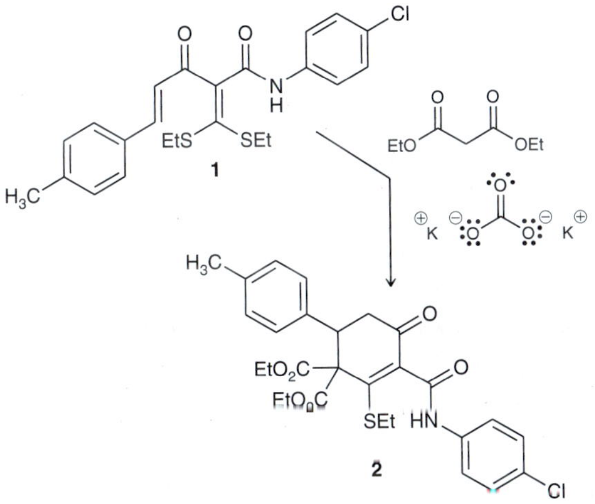

**22.118** The compound fusarisetin A (isolated from a soil fungus) displays significant anticancer activity without detectable cytotoxicity (toxicity to cells). A key step in a reported synthesis of the enantiomer of fusarisetin A involves a Dieckmann cyclization followed by the intramolecular formation of a hemiacetal under basic conditions (*J. Am. Chem. Soc.* **2012**, *134*, 920–923). Provide a mechanism consistent with this transformation.

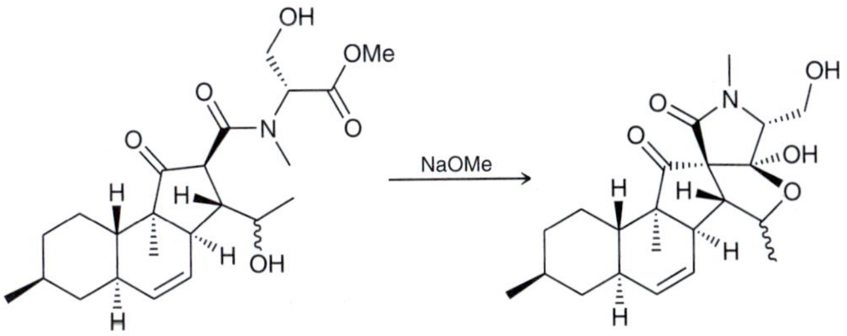

**22.119** Provide a reasonable mechanism to account for the formation of compound **2**, including a rationale to explain the observed stereochemistry of the product (*Org. Lett.* **2012**, *14*, 4738–4741). Note that NaHMDS is a strong base, comparable to LDA. (**Hint:** Consider both resonance structures of the intermediate formed in the first step of the mechanism.)

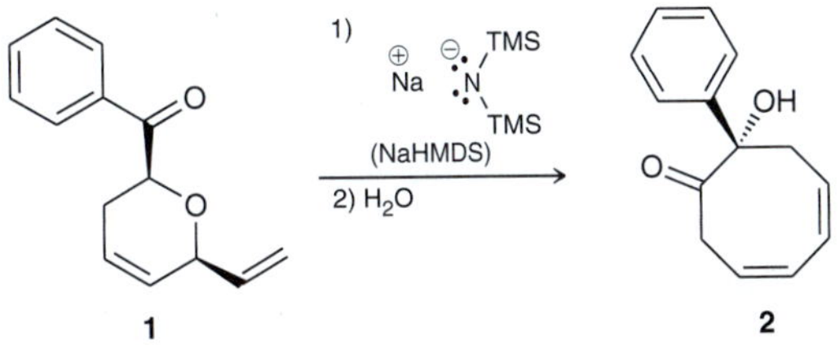

**22.120** Gibberellins (named $GA_1$ through $GA_n$ in order of discovery, over 120 are known) are plant hormones that regulate growth and influence various developmental processes, such as germination, flowering, etc. The modified gibberellin *exo*-16,17-dihydro-$GA_5$-13-acetate (**1**) has considerable promise in reducing the risk of lodging (harmful to a plant's growth) in cereal crops such as corn and wheat.

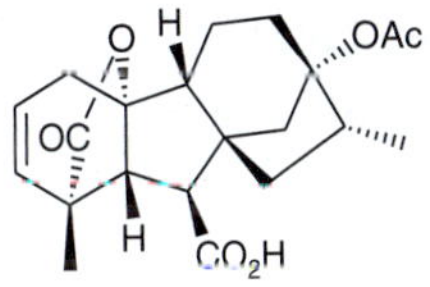

When developing a new drug for plant or human use, it is often important to introduce a *radiolabel*, or an isotope, into the structure so it can be traced as it does its job, and subsequently the metabolites (smaller fragments of the original compound formed as a molecule is metabolized) can be tested for toxicity to the host or environment. For the purpose of compound **1**, C3 will be labeled with $^{14}C$ (illustrated with an asterisk, *), and a key step in the synthesis of this labeled compound involves an intramolecular aldol-type reaction to form a ring (*Can. J. Chem.* **2004**, *82*, 293–300). Draw a detailed, stepwise mechanism for the transformation of aldehyde **2** to alcohol **3**.

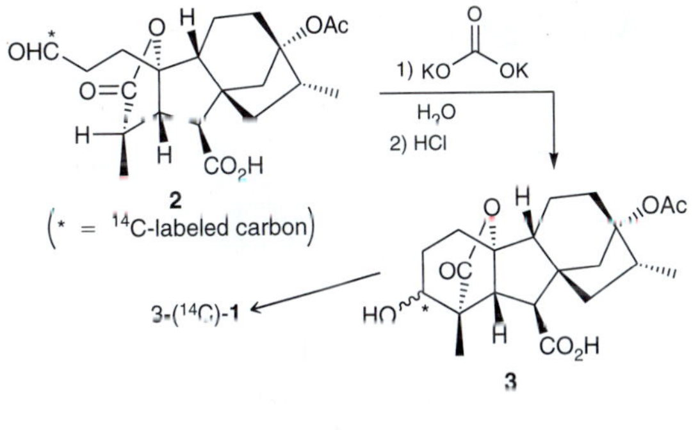

(* = $^{14}C$-labeled carbon)

# 23 Amines



## DID YOU EVER WONDER...

**how drugs like Tagamet, Zantac, and Pepcid are able to control stomach acid production and alleviate the symptoms of acid reflux disease?**

Tagamet, Zantac, and Pepcid are all compounds that contain several nitrogen atoms, which are important for the function of these drugs. In this chapter, we will explore the properties, reactions, and biological activity of many different nitrogen-containing compounds. The end of the chapter will revisit the structures and activity of the three drugs listed above, with a special emphasis on how medicinal chemists designed the first of these blockbuster drugs.

## DO YOU REMEMBER?

*Before you go on, be sure you understand the following topics.*
*If necessary, review the suggested sections to prepare for this chapter:*

- Delocalized and Localized Lone Pairs (Section 2.12)
- Aromatic Heterocycles (Section 18.5)
- Brønsted-Lowry Acidity: A Quantitative Perspective (Section 3.3)
- Activating Groups and Deactivating Groups (Sections 19.7 and 19.8)

Take the DO YOU REMEMBER? QUIZ in **WileyPLUS** to check your understanding

# 23.1 Introduction to Amines

## Classification of Amines

**Amines** are derivatives of ammonia in which one or more of the protons have been replaced with alkyl or aryl groups.

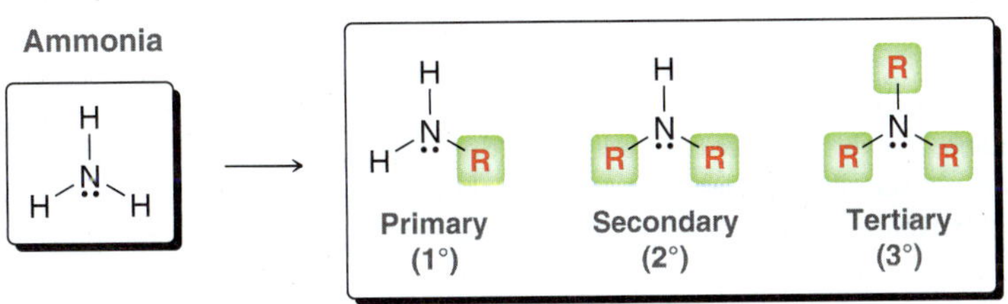


Amines are classified as primary, secondary, or tertiary, depending on the number of groups attached to the nitrogen atom. Note that these terms have a different meaning than when they were used in naming alcohols. A tertiary alcohol has three groups attached to the α carbon, while a tertiary amine has three groups attached to the nitrogen atom.

Amines are abundant in nature. Naturally occurring amines isolated from plants are called **alkaloids**. Below are examples of several alkaloids that have garnered public awareness as a result of their physiological activity:

**Morphine**
(A potent analgesic isolated from the unripe seeds of the poppy plant *Papaver somniferum*)

**Cocaine**
(A potent stimulant isolated from the leaves of the coca plant)

**Nicotine**
(An addictive and toxic compound found in tobacco)

Many amines also play vital roles in neurochemistry (chemistry taking place in the brain). Below are a few examples:

**Adrenaline**
(A "fight-or-flight" hormone, first discussed in Chapter 7)

**Noradrenaline**
(Regulates heart rate and dilates air passages)

**Dopamine**
(Regulates motor skills and emotions)

In fact, many pharmaceuticals are amines, as we will see throughout this chapter.

## medically speaking — Drug Metabolism Studies

Recall that many drugs are converted in our bodies to water-soluble compounds that can be excreted in the urine (see the Medically Speaking box in Section 13.9). This process, called drug metabolism, often involves the conversion of a drug into many different compounds called metabolites. Whenever a new drug is developed, medicinal chemists conduct drug metabolism studies in order to determine whether the observed activity is a property of the drug itself or whether the activity is derived from one of the drug's metabolites. Each of the metabolites is isolated, if possible, and tested for drug activity. In some cases, it is found that the drug and one of its metabolites both exhibit activity. In other cases, the drug itself has no activity, but one of its metabolites is found to be responsible for the observed activity. Drug metabolism studies have, in some cases, led to the development of safer drugs. One such example will now be described.

Terfenadine, marketed under the trade name Seldane, was one of the first nonsedating antihistamines. After being widely used, it was discovered to cause cardiac arrhythmia (abnormal heart rhythm) among patients who were concurrently using either the antifungal agent ketoconazole or the antibiotic agent erythromycin. These agents are believed to inhibit the enzyme responsible for metabolizing terfenadine so that the drug remains in the body for a longer time. After several doses of terfenadine, the drug accumulates, and the high concentrations ultimately lead to the observed arrhythmia.

Medicinal chemists were able to sidestep this problem by carefully analyzing and testing the metabolites of terfenandine. Specifically, one of the metabolites was isolated and was found to possess a carboxylic acid group. Further studies showed that this metabolite, named fexofenadine, is the active compound, and terfenadine simply functions as a prodrug. But unlike the prodrug, this metabolite is further metabolized in the presence of antifungal or antibiotic agents. As a result, terfenadine was removed from the market and replaced with fexofenadine, which is now marketed in the United States under the trade name Allegra. This example illustrates how drug metabolism studies can lead to the development of safer drugs.

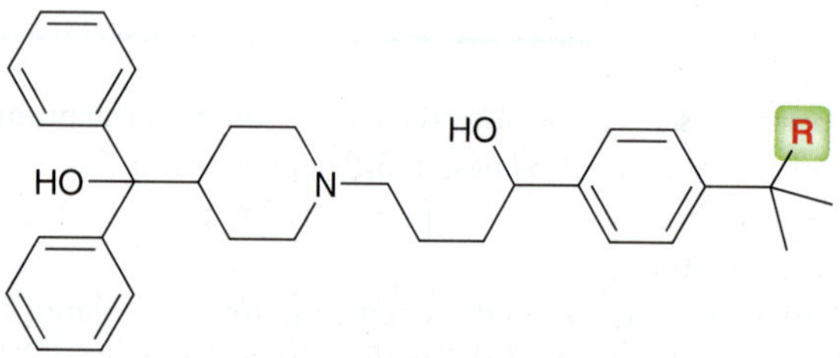

R = $CH_3$ — **Seldane** (terfenadine)

R = COOH — **Allegra** (fexofenadine)

## Reactivity of Amines

The nitrogen atom of an amine possesses a lone pair that represents a region of high electron density. This can be seen in an electrostatic potential map of trimethylamine (Figure 23.1). The presence of this lone pair is responsible for most of the reactions exhibited by amines. Specifically, the lone pair can function as a base or as a nucleophile:

**Functioning as a base**

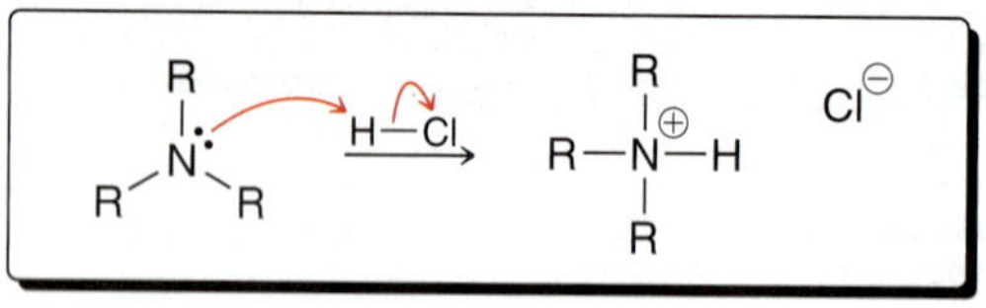

$$R_3N: + H{-}Cl \longrightarrow R_3N^{\oplus}{-}H \quad Cl^{\ominus}$$

**Functioning as a nucleophile**

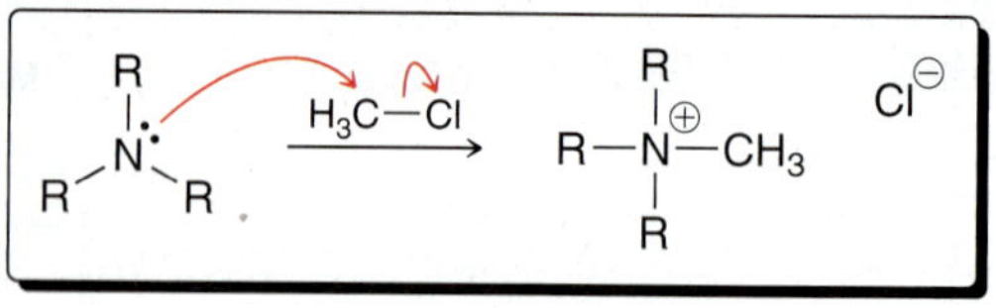

$$R_3N: \xrightarrow{H_3C{-}Cl} R_3N^{\oplus}{-}CH_3 \quad Cl^{\ominus}$$

Section 23.3 explores the basicity of amines, while Sections 23.8–23.11 explore reactions in which amines function as nucleophiles.

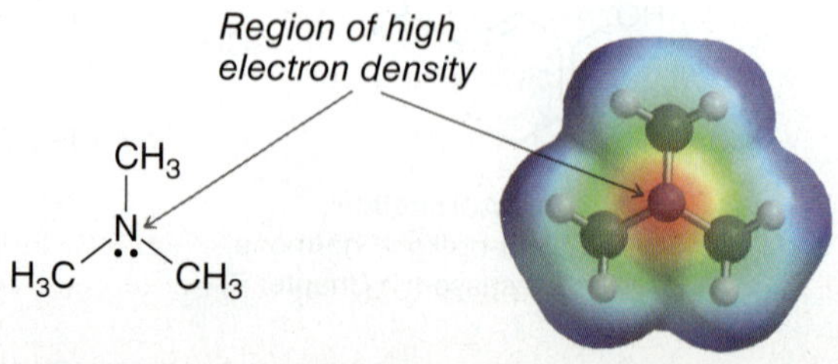

Region of high electron density

**FIGURE 23.1** An electrostatic potential map of trimethylamine. The red area indicates a region of high electron density.

## 23.2 Nomenclature of Amines

### Nomenclature of Primary Amines

A primary amine is a compound containing an $NH_2$ group connected to an alkyl group. IUPAC nomenclature allows two different ways to name primary amines. While either method can be used in most circumstances, the choice most often depends on the complexity of the alkyl group. If the alkyl group is rather simple, then the compound is generally named as an **alkyl amine**. With this approach, the alkyl substituent is identified followed by the suffix "amine." Below are several examples.

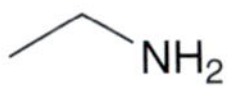

$NH_2$ $NH_2$ $NH_2$

Ethylamine 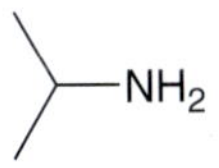 Isopropylamine 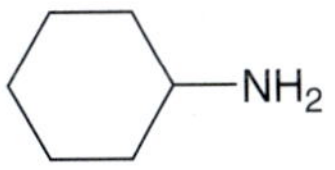 Cyclohexylamine

Primary amines containing more complex alkyl groups are generally named as **alkanamines**. With this approach, the amine is named much like an alcohol, where the suffix *-amine* is used in place of *-ol*.

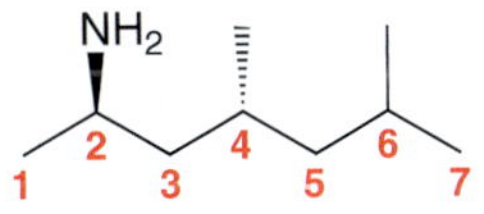

$NH_2$ 2 4 6 1 3 5 7 OH 2 4 6 1 3 5 7

(2*R*,4*R*)-4,6-Dimethyl-2-heptanamine 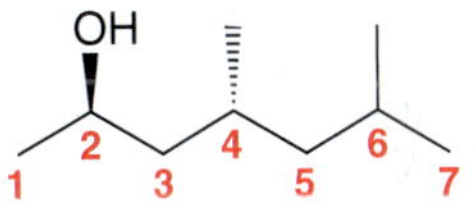 (2*R*,4*R*)-4,6-Dimethyl-2-heptanol

To use this approach, a parent is selected and the location of each substituent is identified with the appropriate locant. The parent chain is numbered starting with the side that is closer to the functional group. Any chirality centers are indicated at the beginning of the name. When another functional group is present in the compound, the amino group is generally listed as a substituent. The other functional group (with the exception of halogens) receives priority and its name becomes the suffix.

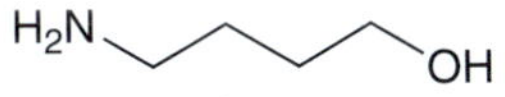

$H_2N$ OH O OH $H_2N$

4-Aminobutanol 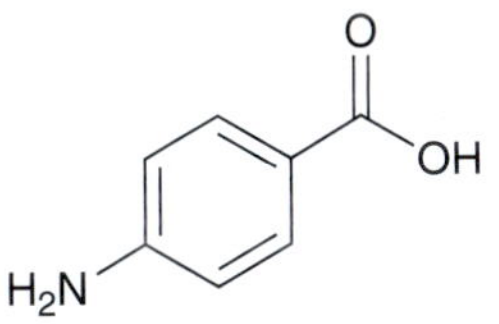 *para*-Aminobenzoic acid

Aromatic amines, also called **aryl amines**, are generally named as derivatives of aniline.

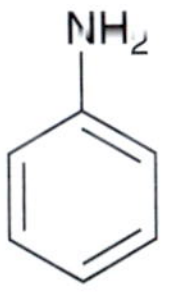

$NH_2$ $NH_2$ Cl F $NH_2$

Aniline 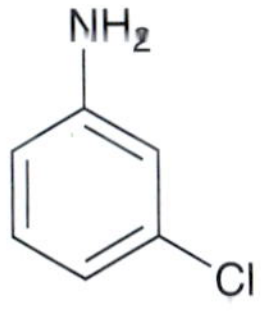 *meta*-Chloroaniline 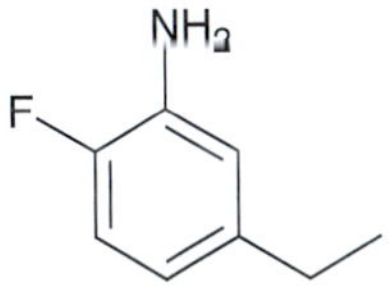 5-Ethyl-2-fluoroaniline

Notice in the last example that the numbering begins at the carbon attached to the amino group and continues in the direction that gives the lower number to the first point of difference (2-fluoro rather than 3-ethyl). When the name is assembled, all substituents are listed alphabetically, so ethyl is listed before fluoro.

### Nomenclature of Secondary and Tertiary Amines

Much like primary amines, secondary and tertiary amines can also be named as alkyl amines or as alkanamines. Once again, the complexity of the alkyl groups typically determines which system is chosen. If all alkyl groups are rather simple in structure, then the groups are listed in alphabetical order. The prefixes "di" and "tri" are used if the same alkyl group appears more than once.

Ethylmethylpropylamine    Diethylamine    Trimethylamine

If one of the alkyl groups is complex, then the compound is typically named as an alkanamine, with the most complex alkyl group treated as the parent and the simpler alkyl groups treated as substituents.

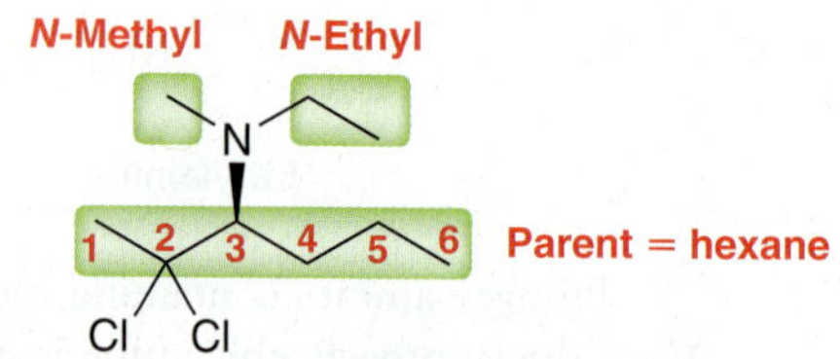


(*S*)-2,2-Dichloro-*N*-ethyl-*N*-methyl-3-hexanamine

In this example, the parent is a hexane chain and the suffix is *-amine*. The methyl and ethyl groups are listed as substituents using the locant "*N*" to identify that they are connected to the nitrogen atom.

## SKILLBUILDER

### 23.1 NAMING AN AMINE

**LEARN the skill** Assign a name for the following compound:

**SOLUTION**

**STEP 1**
Identify all of the alkyl groups connected to the nitrogen atom.

First identify all of the alkyl groups connected to the nitrogen atom: There are two alkyl groups, so this compound is a secondary amine.

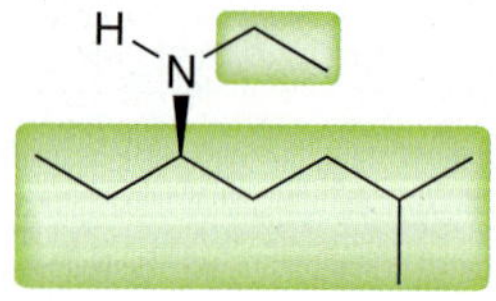

**STEP 2**
Determine what method to use.

**STEP 3**
If naming as an alkanamine, choose the more complex group to be the parent.

Next, determine whether to name the compound as an alkyl amine or as an alkanamine. If the alkyl groups are simple, then the compound can be named as an alkyl amine; otherwise, it should be named as an alkanamine. In this case, one of the alkyl groups is complex, so the compound will be named as an alkanamine. The complex alkyl group is chosen as the parent, and the other alkyl group is listed as a substituent, using the letter *N* as a locant. Notice that the parent is numbered starting from the side closer to the amine group.

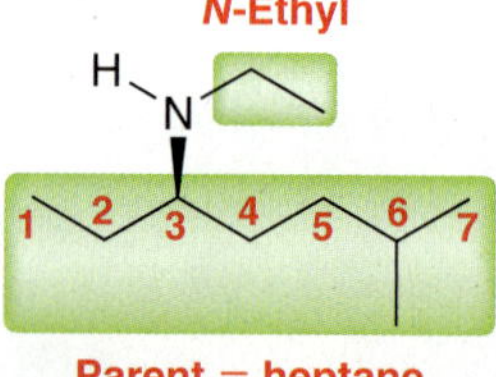


**STEP 4**
Assign locants, assemble the name, and assign configuration.

Finally, assign locants to each substituent, assemble the name, and assign the configuration of any chirality centers:

(*R*)-*N*-Ethyl-6-methyl-3-heptanamine

**PRACTICE the skill** **23.1** Assign a name for each of the following compounds:

(a) 

(b) 

(c) N

(d) 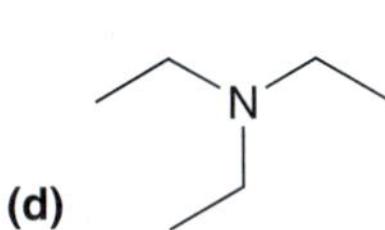


(e) $NH_2$

(f) HO, $NH_2$

**APPLY the skill**

**23.2** Draw the structure of each of the following compounds:

**(a)** Cyclohexylmethylamine **(b)** Tricyclobutylamine **(c)** 2,4-Diethylaniline
**(d)** (1*R*,2*S*)-2-Methylcyclohexanamine **(e)** *ortho*-Aminobenzaldehyde

**23.3** Draw all constitutional isomers with molecular formula $C_3H_9N$ and provide a name for each isomer.

need more **PRACTICE?** **Try Problems 23.42, 23.46, 23.47**

## 23.3 Properties of Amines

### Geometry

The nitrogen atom of an amine is typically $sp^3$ hybridized, with the lone pair occupying an $sp^3$-hybridized orbital. Consider trimethylamine as an example:

$sp^3$ Orbital

$H_3C$—N—$CH_3$, $CH_3$

The nitrogen atom exhibits trigonal pyramidal geometry (see Section 1.10), with bond angles of 108°. The C—N bond lengths are 147 pm, which is shorter than the average C—C bond of an alkane (153 pm) and longer than the average C—O bond of an alcohol (143 pm).

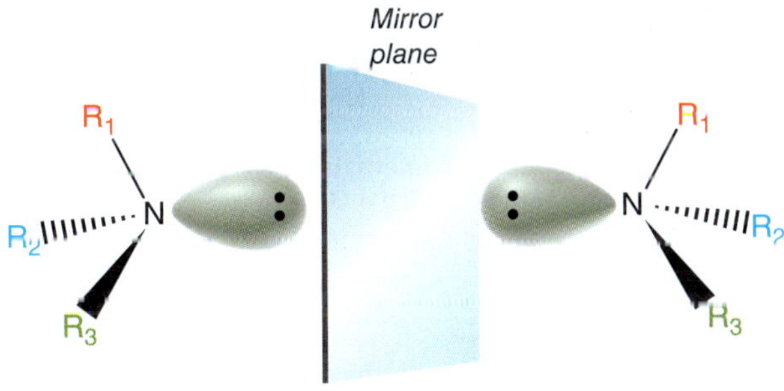


**FIGURE 23.2**
The mirror-image relationship of enantiomeric amines.

Amines containing three different alkyl groups are chiral compounds (Figure 23.2). The nitrogen atom has a total of four different groups (three alkyl groups and a lone pair) and is therefore a chirality center. Compounds of this type are generally not optically active at room temperature, because pyramidal inversion occurs quite rapidly, producing a racemic mixture of enantiomers (Figure 23.3). During pyramidal inversion, the compound passes through a transition state in which the nitrogen atom has $sp^2$ hybridization. This transition state is only about 25 kJ/mol (6 kcal/mol) higher in energy than the $sp^3$-hybridized geometry. This energy barrier is relatively small and is easily surmounted at room temperature. For this reason, most amines bearing three different alkyl groups cannot be resolved at room temperature.

### Colligative Properties

Amines exhibit solubility trends that are similar to the trends exhibited by alcohols. Specifically, amines with fewer than five carbon atoms per amino group will typically be water soluble, while amines with more than five carbon atoms per amino group will be only sparingly soluble. For example, ethylamine is soluble in water, while octylamine is not.

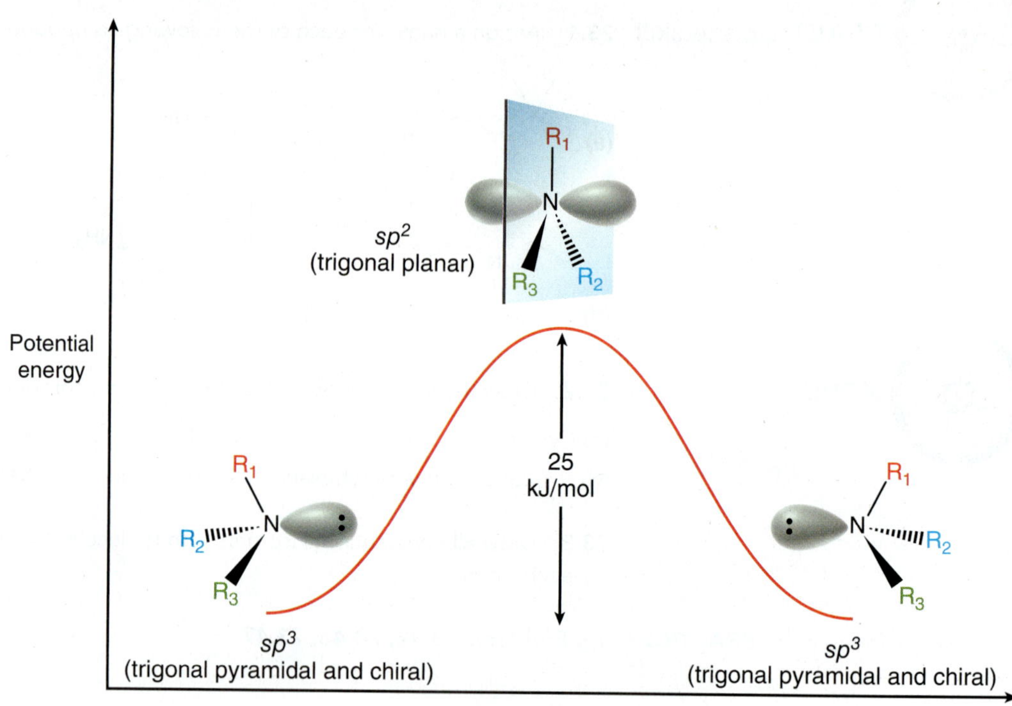


**FIGURE 23.3** Pyramidal inversion of an amine enables the enantiomers to rapidly interconvert at room temperature.

Primary and secondary amines can form intermolecular H bonds and typically have higher boiling points than analogous alkanes but lower boiling points than analogous alcohols.

| Propane | Ethylamine | Ethanol |
|---|---|---|
| bp = −42°C | bp = 17°C | bp = 78°C |

The boiling point of amines increases as a function of their capacity to form hydrogen bonds. As a result, primary amines typically have higher boiling points, while tertiary amines have lower boiling points. This trend can be observed by comparing the physical properties of the following three constitutional isomers:

| Propylamine | Ethylmethylamine | Trimethylamine |
|---|---|---|
| bp = 50°C | bp = 34°C | bp = 3°C |

## Other Distinctive Features of Amines

Amines with low molecular weights, such as trimethylamine, typically have a fishlike odor. In fact, the odor of fish is caused by amines that are produced when enzymes break down certain fish proteins. Putrescine and cadaverine are examples of compounds present in rotting fish. They are also present in urine, contributing to its characteristic odor.

**Putrescine (1,4-butanediamine)**

**Cadaverine (1,5-pentanediamine)**

## CONCEPTUAL CHECKPOINT

**23.4** Rank this group of compounds in order of increasing boiling point.

**23.5** Identify whether each of the following compounds is expected to be water soluble:

(a)

(b)

(c)

## medically speaking — Fortunate Side Effects

Most drugs produce more than one physiological response. In general, one response is the desired response, and the rest are considered to be undesirable side effects. However, in some cases, the side effects have turned out to be fortunate discoveries. These accidental discoveries have led to the development of novel treatments for a wide range of ailments. The following are some of the better-known historical examples of this phenomenon.

Dimenhydrinate, which was originally developed as an antihistamine, is a mixture of two compounds: diphenhydramine (Benadryl) and 8-chlorotheophylline:

Diphenhydramine + 8-Chlorotheophylline

Dimenhydrinate

The former is a first-generation (sedating) antihistamine, while the latter is a chlorinated derivative of caffeine (a stimulant). It was believed that the drowsiness of the former would be reduced by the effects of the latter. Unfortunately, the sedating effects of the antihistamine were too strong and overpowered the effects of the stimulant. However, an accidental discovery was made when this mixture was tested in 1947. During clinical studies at the Johns Hopkins University, one patient suffering from car sickness reported alleviation of his symptoms. Further studies were then conducted, and it was discovered that the mixture of compounds was effective in treating other forms of motion sickness, including air sickness and sea sickness. Marketed under the trade name Dramamine, this formulation quickly became one of the most widely used drugs for the treatment of motion sickness. It has since been replaced with a newer drug, called meclizine, which is sold under the trade name Dramamine II (less-drowsy formula).

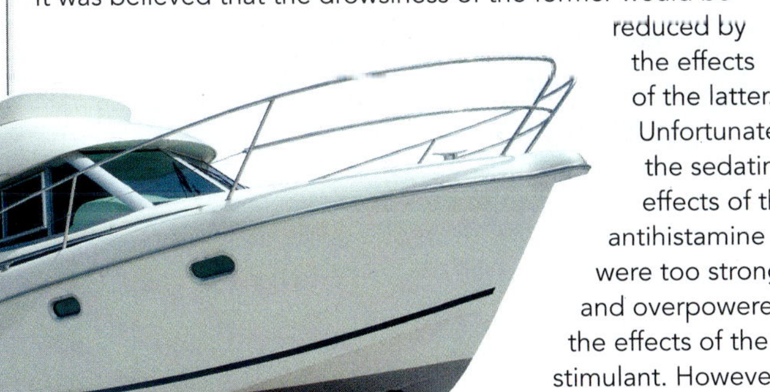

Bupropion and sildenafil are two other drugs that have useful side effects:

Zyban (bupropion)

Viagra (sildenafil)

Bupropion was developed as an antidepressant but was found to help patients quit smoking during clinical trials. It is now marketed as a smoking cessation aid under the trade name Zyban. Sildenafil was designed to treat angina (chest pain that occurs when the heart muscle receives insufficient oxygen). In clinical studies, many of the volunteers experienced increased erectile function. Since sildenafil was found to be ineffective in treating angina, its manufacturer (Pfizer) decided to market the drug instead as a treatment for erectile dysfunction, a condition that afflicts one in nine adult males. Sildenafil is now marketed and sold under the trade name Viagra, with annual sales in excess of $1 billion.

## Basicity of Amines

One of the most important properties of amines is their basicity. Amines are generally stronger bases than alcohols or ethers, and they can be effectively protonated even by weak acids.

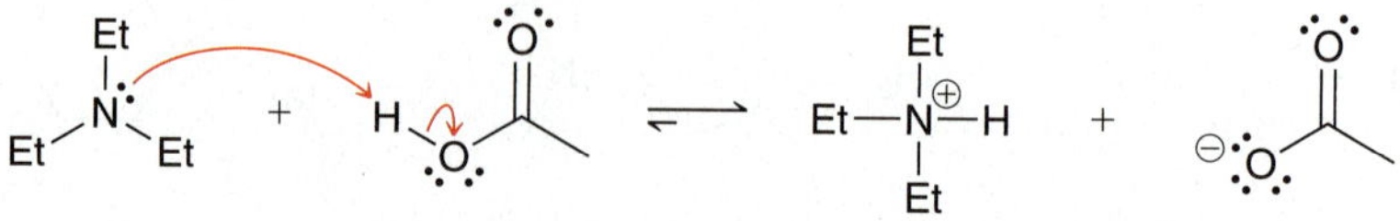

$$Et_3N: + CH_3COOH \rightleftharpoons Et_3NH^+ + CH_3COO^-$$

p$K_a$ = 4.76 (acetic acid) p$K_a$ = 10.76 (triethylammonium ion)

In this example, triethylamine is protonated using acetic acid. Compare the p$K_a$ values of acetic acid (4.76) and the ammonium ion (10.76). Recall that the equilibrium will favor the weaker acid. In this case, the ammonium ion is six orders of magnitude weaker than acetic acid, and therefore, the amine will exist almost completely in protonated form (one in every million molecules will be in the neutral form). This example illustrates how the basicity of an amine can be quantified by measuring the p$K_a$ of the corresponding ammonium ion. *A high p$K_a$ indicates that the amine is strongly basic, while a low p$K_a$ indicates that the amine is only weakly basic.* Table 23.1 shows p$K_a$ values for the ammonium ions of many amines.

**TABLE 23.1 THE p$K_a$ VALUES OF THE AMMONIUM IONS OF SEVERAL AMINES**

| Amine | p$K_a$ of Ammonium Ion | Amine | p$K_a$ of Ammonium Ion |
|---|---|---|---|
| *Alkyl amines* | | *Alkyl amines* | |
| Methylamine ($MeNH_2$) | $MeNH_3^+$ p$K_a$ = 10.6 | Dimethylamine ($Me_2NH$) | $Me_2NH_2^+$ p$K_a$ = 10.7 |
| Ethylamine ($EtNH_2$) | $EtNH_3^+$ p$K_a$ = 10.6 | Diethylamine ($Et_2NH$) | $Et_2NH_2^+$ p$K_a$ = 11.0 |
| Isopropylamine | p$K_a$ = 10.6 | Trimethylamine ($Me_3N$) | $Me_3NH^+$ p$K_a$ = 9.8 |
| 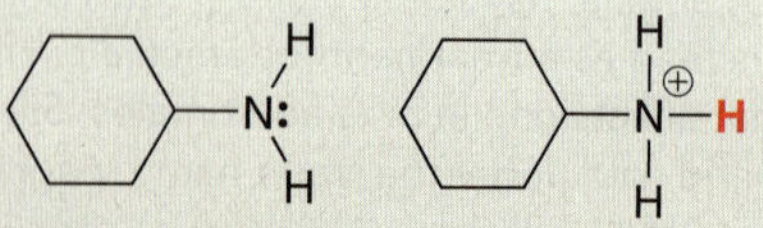 Cyclohexylamine | p$K_a$ = 10.7 | Triethylamine ($Et_3N$) | $Et_3NH^+$ p$K_a$ = 10.8 |

Continued

**TABLE 23.1 (CONTINUED)**

| AMINE | $pK_a$ OF AMMONIUM ION | AMINE | $pK_a$ OF AMMONIUM ION |
|---|---|---|---|
| *Aryl amines* | | *Heterocyclic amines* | |
| 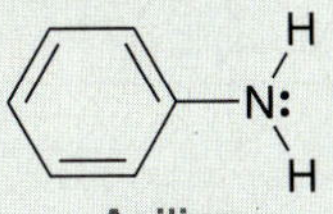 Aniline | 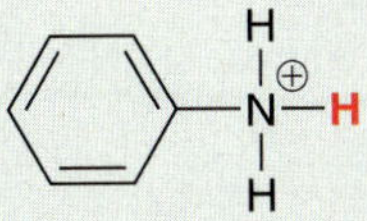 $pK_a = 4.6$ | 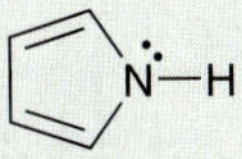 Pyrrole | 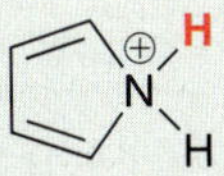 $pK_a = 0.4$ |
| 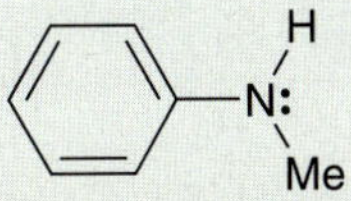 *N*-Methylaniline | 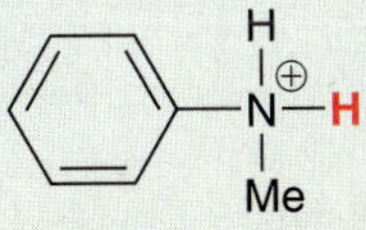 $pK_a = 4.8$ | 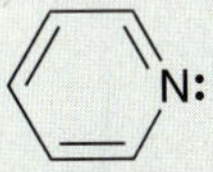 Pyridine | 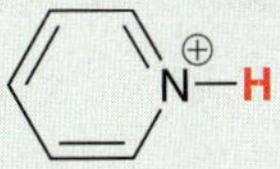 $pK_a = 5.3$ |
| 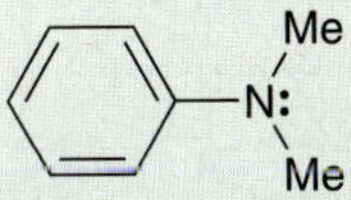 *N,N*-Dimethylaniline | 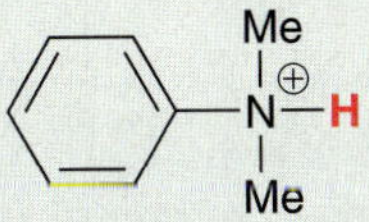 $pK_a = 5.1$ | | |

The ease with which amines are protonated can be used to remove them from mixtures of organic compounds. This process is called **solvent extraction**. A mixture of organic compounds is first dissolved in an organic solvent. A solution of aqueous acid is then added to the mixture, and the contents are shaken vigorously. Upon standing, the organic and aqueous layers separate from one another, much the way oil and water separate from each other in salad dressing bottles. Under these conditions, most of the amine molecules are protonated, forming ammonium ions. The ammonium ions bear charges and are therefore more soluble in the aqueous layer than the organic layer. Manual separation of the aqueous layer provides the ammonium ions which are then treated with base to regenerate the neutral amine. In this way, amines can be readily separated from a mixture of organic compounds. This process was used extensively in the nineteenth century to isolate amines from plant extracts. It was recognized early on that amines could be isolated from plants by exploiting their basic properties. For this reason, amines extracted from plants were called alkaloids (derived from the word *alkaline*, which means basic).

## Delocalization Effects

The ammonium ions of most alkyl amines are characterized by a $pK_a$ value between 10 and 11, but aryl amines are very different. Ammonium ions of aryl amines are more acidic (lower $pK_a$) than the ammonium ion of alkyl amines (Table 23.2). In other words, aryl amines are less basic. This can be rationalized by considering the delocalized nature of the lone pair of an aryl amine:

**LOOKING BACK**
For a review of delocalized lone pairs, see Section 2.12.

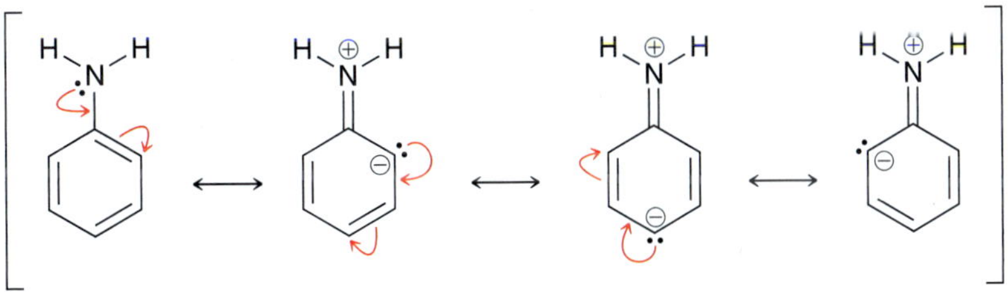

The lone pair occupies a *p* orbital and is delocalized by the aromatic system. This resonance stabilization is lost if the lone pair is protonated, and as a result, the nitrogen atom of an aryl amine is less basic than the nitrogen atom of an alkyl amine. If the aromatic ring bears a substituent, the basicity of the

**TABLE 23.2 THE p$K_a$ VALUES OF THE AMMONIUM IONS OF SEVERAL PARA-SUBSTITUTED ANILINES**

| *para*-Substituted aniline | Ammonium ion | p$K_a$ |
|---|---|---|
| $H_2N-C_6H_4-NH_2$ | $H_2N-C_6H_4-NH_3^+$ | p$K_a$ = 6.2 |
| $MeO-C_6H_4-NH_2$ | $MeO-C_6H_4-NH_3^+$ | p$K_a$ = 5.3 |
| $Me-C_6H_4-NH_2$ | $Me-C_6H_4-NH_3^+$ | p$K_a$ = 5.1 |
| $C_6H_5-NH_2$ | $C_6H_5-NH_3^+$ | p$K_a$ = 4.6 |
| $Cl-C_6H_4-NH_2$ | $Cl-C_6H_4-NH_3^+$ | p$K_a$ = 4.0 |
| $NC-C_6H_4-NH_2$ | $NC-C_6H_4-NH_3^+$ | p$K_a$ = 1.7 |
| $O_2N-C_6H_4-NH_2$ | $O_2N-C_6H_4-NH_3^+$ | p$K_a$ = 1.0 |

Increasing basicity (upward) — Increasing acidity (downward)

amino group will depend on the identity of the substituent (Table 23.2). Electron-donating groups, such as methoxy, slightly increase the basicity of aryl amines, while electron-withdrawing groups, such as nitro, can significantly decrease the basicity of aryl amines. This profound effect is attributed to the fact that the lone pair in *para*-nitroaniline is extensively delocalized:

Amides also represent an extreme case of lone-pair delocalization. The lone pair on the nitrogen atom of an amide is highly delocalized by resonance. In fact, the nitrogen atom exhibits very little electron density, as can be seen in an electrostatic potential map (Figure 23.4). For this reason, amides do not function as bases and are very poor nucleophiles.

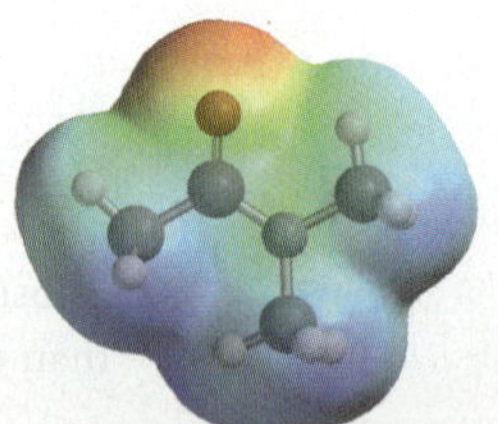

**FIGURE 23.4**
An electrostatic potential map of an amide. The nitrogen atom is not a region of high electron density.

**LOOKING BACK**
To visualize the orbitals occupied by the lone pairs in pyrrole and pyridine, see Figures 18.14 and 18.15.

Another example of electron delocalization can be seen in a comparison of the electrostatic potential maps of pyrrole and pyridine in Figure 18.16. The lone pair of pyrrole is delocalized due to its participation in establishing aromatic stabilization. As a result, pyrrole is an extremely weak base because protonation of pyrrole would result in the loss of aromaticity. In contrast, the lone pair in pyridine is localized and does not participate in establishing aromaticity. As a result, pyridine can function as a base without destroying its aromatic stabilization. For this reason, pyridine is a stronger base than pyrrole. In fact, inspection of the p$K_a$ values in Table 23.1 reveals that pyridine is five orders of magnitude (~100,000 times) more basic than pyrrole.

## Amines at Physiological pH

In Chapter 21, we used the Henderson-Hasselbalch equation to show that a carboxylic acid group (COOH) exists primarily as a carboxylate ion at physiological pH. A similar argument can be used to show that an amine group also exists primarily as a charged ammonium ion at physiological pH:

$$R_3N + H^+ \rightleftharpoons R_3NH^+$$

**The ammonium ion predominates at physiological pH**

This fact is important for understanding the activity of pharmaceuticals containing amine groups, a topic explored in the following Medically Speaking box.

## CONCEPTUAL CHECKPOINT

**23.6** For each of the following pairs of compounds, identify which compound is the stronger base:

(a) (b) (c) (d)

**23.7** Rank the following compounds in terms of increasing basicity.

**23.8** When (*E*)-4-amino-3-buten-2-one is treated with molecular hydrogen in the presence of platinum, the resulting amine is more basic than the reactant. Draw the reactant and the product and explain why the product is a stronger base than the reactant.

**23.9** For each of the following compounds, draw the form that predominates at physiological pH:

(a) **Sertraline (Zoloft)** An antidepressant

(b) **Amantadine** Used in the treatment of Parkinson's disease

(c) **Phenylpropanolamine** A nasal decongestant

## Amine Ionization and Drug Distribution

In Chapter 18 we saw that allergic reactions are triggered by the binding of histamine to $H_1$ receptors. Compounds that compete with histamine to bind with $H_1$ receptors without triggering the allergic reaction are called $H_1$ antagonists and belong to the class of compounds called antihistamines. For a compound to exhibit $H_1$ antagonism it must possess two aromatic rings in close proximity as well as a tertiary amine group.

The amine group of an antihistamine exists as an equilibrium between neutral and ionic forms. The equilibrium concentrations of the two forms depend on the $pK_a$ of the compound and the pH of the environment. Most antihistamines have an ammonium group with a $pK_a$ between 8 and 9. As a result, both the neutral and ionic forms are present in substantial amounts at physiological pH, although the ionic form slightly predominates. This is important because it is the ionic form that binds to the receptor. The positively charged ammonium ion binds with a negative charge in the receptor. The neutral form of the drug cannot bind with the receptor.

Although the ionic form is required to bind with the receptor, it is incapable of crossing the nonpolar environment of cell membranes. So how does the drug reach its target? The answer stems from the equilibrium between ionic and neutral forms of the drug. The neutral form crosses the membrane, and once on the other side, it reestablishes equilibrium, producing a large concentration of the ionic form, which then binds to the receptor. On the other side of the membrane, the ionic forms that did not cross the membrane also reestablish an equilibrium, generating more of the neutral form that is capable of crossing the membrane. In this way, the drug can cross the membrane in the neutral form and interact with the receptor in the ionic form. The rate at which this process takes place depends on the equilibrium concentrations. If the equilibrium concentration of the neutral form is too small, then the drug will not cross the cell membrane at an appreciable rate. Similarly, if the equilibrium concentration of the ionic form is too small, then the drug will not bind at an appreciable rate with the receptor. This delicate balance between drug distribution and receptor binding is therefore highly dependent on the equilibrium concentrations at physiological pH, which in turn is dependent on the $pK_a$ of the drug. Later in this chapter we will see an example of how $pK_a$ values must be taken into account when new drugs are designed.

## 23.4 Preparation of Amines: A Review

In previous chapters, we have seen reactions that can be used to make amines from alkyl halides, carboxylic acids, or benzene. This section will briefly summarize each of these three methods.

### Preparing an Amine from an Alkyl Halide

Amines can be prepared from alkyl halides in a two-step process in which the alkyl halide is converted to a nitrile, which is then reduced.

$$\text{Br} \xrightarrow[S_N2]{\text{NaCN}} \text{C}\equiv\text{N} \xrightarrow[\text{2) } H_2O]{\text{1) xs LAH}} \text{CH}_2\text{–NH}_2$$

The first step is an $S_N2$ process in which a cyanide ion functions as a nucleophile, displacing the halide leaving group. The limitations of $S_N2$ reactions apply in that tertiary alkyl halides and vinyl halides cannot be used. The second step involving reduction of the resulting nitrile can be accomplished with a strong reducing agent, such as LAH. Notice that installation of an amino group using this approach is accompanied by the introduction of one additional carbon atom (highlighted in red).

### Preparing an Amine from a Carboxylic Acid

Amines can be prepared from carboxylic acids using the following approach:

$$\text{OH} \xrightarrow[\text{2) xs } NH_3]{\text{1) } SOCl_2} \text{NH}_2 \xrightarrow[\text{2) } H_2O]{\text{1) xs LAH}} \text{NH}_2$$

The carboxylic acid is first converted into an amide, which is then reduced to give an amine (Section 21.14). This approach for installation of an amino group does not involve introduction of any additional carbon atoms; that is, the carbon skeleton is not changed.

## Preparing Aniline and Its Derivatives from Benzene

Aryl amines, such as aniline, can be prepared from benzene using the following approach:

Benzene $\xrightarrow[H_2SO_4]{HNO_3}$ nitrobenzene ($NO_2$) $\xrightarrow[Pt]{H_2}$ aniline ($NH_2$)

or nitrobenzene $\xrightarrow{\text{1) Fe, Zn, Sn, or } SnCl_2,\ H_3O^+ \quad \text{2) NaOH}}$ aniline

The first step involves nitration of the aromatic ring, and the second step involves reduction of the nitro group. Several reagents can be used to accomplish this reduction, including hydrogenation in the presence of a catalyst or reduction with iron, zinc, tin, or tin(II) chloride ($SnCl_2$) in the presence of aqueous acid. The latter method is a more mild approach that is used when other functional groups are present that would otherwise be susceptible to hydrogenation.

For example, a nitro group can be selectively reduced in the presence of a carbonyl group:

$O_2N$–C₆H₄–CHO $\xrightarrow{\text{1) } SnCl_2,\ H_3O^+ \quad \text{2) NaOH}}$ $H_2N$–C₆H₄–CHO

When reducing a nitro group in acidic conditions, the reaction must be followed up with a base, such as sodium hydroxide, because the resulting amino group will be protonated under acidic conditions (as we saw in Section 23.3).

### CONCEPTUAL CHECKPOINT

**23.10** Draw the structure of an alkyl halide or carboxylic acid that might serve as a precursor in the preparation of each of the following amines:

(a) butylamine ($NH_2$)

(b) 2-phenylethylamine ($NH_2$)

(c) cyclohexylmethylamine ($NH_2$)

**23.11** The following compound cannot be prepared from an alkyl halide or a carboxylic acid using the methods described in this section. Explain why each synthesis cannot be performed.

(2-cyclohexylpropan-2-amine, $NH_2$)

## 23.5 Preparation of Amines via Substitution Reactions

### Alkylation of Ammonia

Ammonia is a very good nucleophile and will readily undergo alkylation when treated with an alkyl halide.

Ammonia ($H–\ddot{N}H_2$) + $H_3C–I$ → $H_3N^{\oplus}–CH_3$ $I^{\ominus}$ $\xrightleftharpoons{:NH_3}$ $H–\ddot{N}(H)–CH_3$ (A primary amine)

This reaction proceeds via an $S_N2$ process followed by deprotonation to give a primary amine. As the primary amine is formed, it can undergo further alkylation to produce a secondary amine, which undergoes further alkylation to produce a tertiary amine. Finally, the tertiary amine undergoes alkylation one more time to produce a **quaternary ammonium salt**.

A primary amine

A secondary amine

A tertiary amine

A quaternary ammonium salt

If the quaternary ammonium salt is the desired product, then an excess of the alkyl halide is used, and ammonia is said to undergo *exhaustive alkylation*. However, monoalkylation is difficult to achieve because each successive alkylation renders the nitrogen atom more nucleophilic. If the primary amine is the desired product, then the process is generally not efficient because when 1 mol of ammonia is treated with 1 mol of the alkyl halide, a mixture of products is obtained. For this reason, the alkylation of ammonia is only useful when the starting alkyl halide is inexpensive and the desired product can be easily separated or when exhaustive alkylation is performed to yield the quaternary ammonium salt.

## The Azide Synthesis

The **azide synthesis** is a better method for preparing primary amines than alkylation of ammonia because it avoids the formation of secondary and tertiary amines. This method involves treating an alkyl halide with sodium azide followed by reduction (Mechanism 23.1).

$NaN_3$ ; $H_2$, Pt ; 1) LAH 2) $H_2O$

### MECHANISM 23.1 THE AZIDE SYNTHESIS

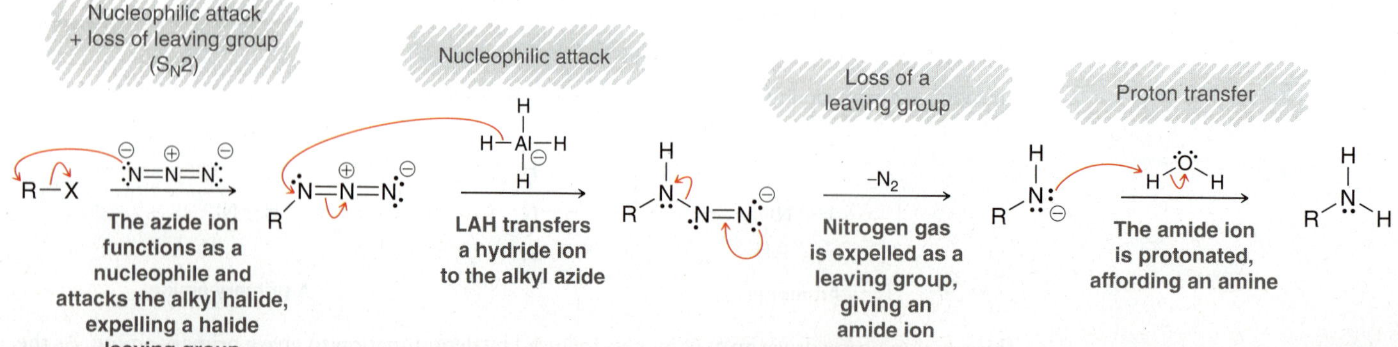

In the first step of this mechanism, the azide ion functions as a nucleophile and attacks the alkyl halide in an $S_N2$ process, producing an alkyl azide. The alkyl azide is then reduced using LAH. When performing this procedure, special precautions must be observed, because alkyl azides can be explosive, liberating nitrogen gas.

## The Gabriel Synthesis

The **Gabriel synthesis** is another method for preparing primary amines while avoiding formation of secondary and tertiary amines. The key reagent is potassium phthalimide, which is prepared by treating phthalimide with potassium hydroxide.

KOH

$K^{\oplus}$ + $H_2O$

Phthalimide

Potassium phthalimide

Hydroxide functions as a base and deprotonates phthalimide. The proton is relatively acidic ($pK_a = 8.3$), because it is flanked by two C=O groups (similar to a β-keto ester). Potassium phthalimide can function as a nucleophile and is readily alkylated to form a C—N bond (directly analogous to the acetoacetic ester synthesis, Section 22.5):

R—X

$S_N2$

Potassium phthalimide

This reaction proceeds via an $S_N2$ process, so it works best with primary alkyl halides. It can be performed with secondary alkyl halides in many cases, but tertiary alkyl halides cannot be used. Acid-catalyzed or base-catalyzed hydrolysis is then performed to release the amine. Acidic conditions are more common than basic conditions.

$H_3O^+$

Under acidic conditions, an ammonium ion is generated, which must be treated with a base to release the uncharged amine.

The mechanism of hydrolysis is directly analogous to the hydrolysis of amides, as seen in Section 21.12. The hydrolysis step is slow, and many alternative approaches have been developed. One such alternative employs hydrazine to release the amine and involves two successive nucleophilic acyl substitution reactions.

$H_2N—NH_2$

# SKILLBUILDER

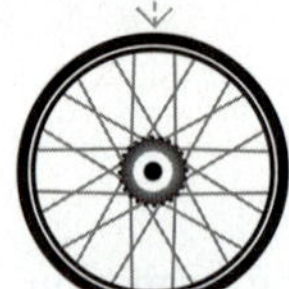

## 23.2 PREPARING A PRIMARY AMINE VIA THE GABRIEL REACTION

**LEARN the skill**

Amphetamine is a stimulant (stronger than caffeine) that increases alertness and decreases appetite. It was used extensively during World War II to prevent combat fatigue.

**Amphetamine**

Propose a method for preparing amphetamine that utilizes a Gabriel synthesis.

### SOLUTION

First identify an alkyl halide that can serve as a precursor in the preparation of amphetamine. To draw this precursor, simply replace the amino group with a halide:

**STEP 1** Identify the appropriate alkyl halide to use.

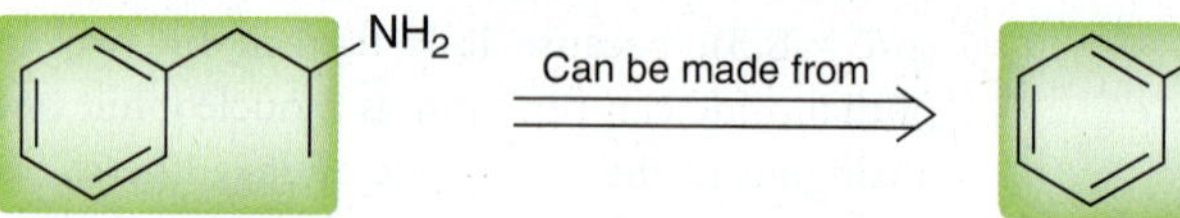


In the Gabriel synthesis, the nitrogen atom comes from phthalimide, which serves as the starting material. The process requires three steps: (1) deprotonation to form potassium phthalimide, (2) nucleophilic attack with an alkyl halide, and (3) releasing the amine via hydrolysis or via treatment with hydrazine. Draw the reagents for all three steps of the Gabriel synthesis:

**STEP 2** Draw all three steps, starting with phthalimide.

1) KOH
2) (alkyl bromide)
3) $H_2NNH_2$

*Note*: Planning a Gabriel synthesis requires that you identify the alkyl halide to use in the second step of the process. The first and third steps remain the same regardless of the identity of the desired product.

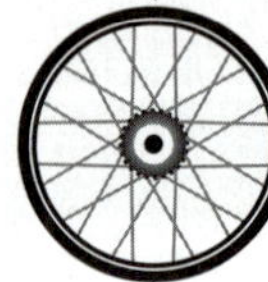

**PRACTICE the skill** **23.12** Using a Gabriel synthesis, show how you would make each of the following compounds:

(a) (b) (c) (d)

**APPLY the skill** **23.13** Using a Gabriel synthesis, propose an efficient synthesis for each of the following transformations. In each case, you will need to use reactions from previous chapters to convert the starting material into a suitable alkyl halide that can be used in a Gabriel synthesis.

(a) (b)

need more **PRACTICE?** **Try Problems 23.51a, 23.60, 23.61, 23.62b**

## 23.6 Preparation of Amines via Reductive Amination

Recall that ketones or aldehydes can be converted into imines when treated with ammonia in acidic conditions (Section 20.6). If the reaction is carried out in the presence of a reducing agent, such as $H_2$, then the imine is reduced as soon as it is formed, giving an amine.

A ketone $\xrightarrow[NH_3]{[H^+]}$ An imine $\xrightarrow[Pt]{H_2}$ An amine

This process, called **reductive amination**, can be accomplished in one reaction flask by hydrogenating a solution containing the ketone, ammonia, and a proton source. With this approach, the imine is reduced under the conditions of its formation. Many other reducing agents are also frequently employed instead of hydrogenation. The most common reducing agent for this purpose is **sodium cyanoborohydride** ($NaBH_3CN$), which is similar in structure to sodium borohydride ($NaBH_4$), but a cyano group has replaced one of the hydrogen atoms.

**Sodium borohydride ($NaBH_4$)** **Sodium cyanoborohydride ($NaBH_3CN$)**

Recall that a cyano group is electron withdrawing (Section 19.10), and its presence stabilizes the negative charge on the boron atom. As a result, $NaBH_3CN$ is a more selective hydride-reducing agent than $NaBH_4$. This selectivity can be exploited to achieve the reductive amination of a ketone or aldehyde:

$$\xrightarrow[NaBH_3CN]{[H^+], NH_3}$$

This process cannot be achieved with $NaBH_4$, which would simply reduce the ketone before it is converted into an imine. In contrast, $NaBH_3CN$ does not react with ketones but will reduce an iminium ion (a protonated imine). This process enables the conversion of a ketone or aldehyde into an amine in one reaction flask. The identity of the nucleophile ($NH_3$) can be changed, enabling the preparation of primary, secondary, or tertiary amines:

$NH_3$, $[H^+]$, $NaBH_3CN$ → 1° Amine

$R—NH_2$, $[H^+]$, $NaBH_3CN$ → 2° Amine

$R_2NH$, $[H^+]$, $NaBH_3CN$ → 3° Amine

# SKILLBUILDER

## 23.3 PREPARING AN AMINE VIA A REDUCTIVE AMINATION

**LEARN the skill**

Fluoxetine, sold under the trade name Prozac, is used in the treatment of depression and obsessive compulsive disorder. Show two different ways of preparing fluoxetine via reductive amination.

**Fluoxetine**

**STEP 1** Identify all C—N bonds.

**STEP 2** Apply a retrosynthetic analysis of each C—N bond.

### SOLUTION

Begin by identifying all C—N bonds. Then, using a retrosynthetic analysis, identify the starting amine and the carbonyl compound that could be used to make each of the C—N bonds:

**Method 1**

**Method 2**

**STEP 3** Draw all reagents necessary for each possible route.

Finally, show all of the reagents necessary for both of the synthetic routes shown. In each case, the desired compound can be produced in just one step, using $NaBH_3CN$:

**Method 1**

$[H^+]$, $H_2N-CH_3$ / $NaBH_3CN$

**Method 2**

$[H^+]$, $H_2C{=}O$ / $NaBH_3CN$

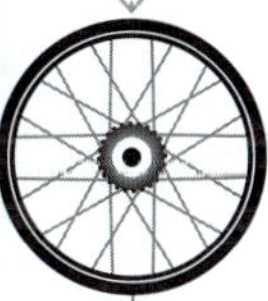

**PRACTICE the skill** **23.14** Show two different ways of preparing each of the following compounds via a reductive amination:

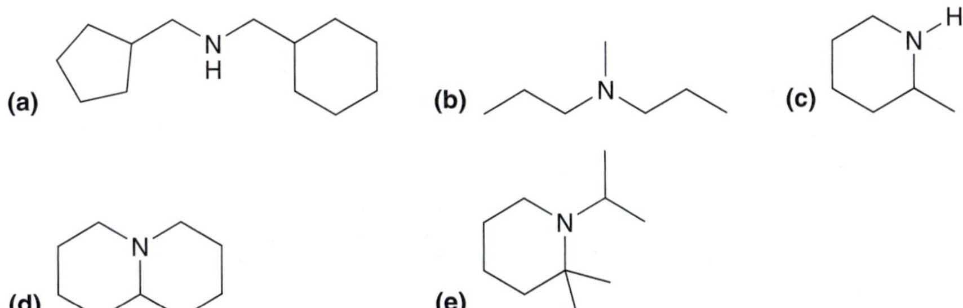

**APPLY the skill** **23.15** Methamphetamine is used in some formulations for the treatment of attention deficit disorder and can be prepared by reductive amination using phenylacetone and methylamine. Draw the structure of methamphetamine.

**23.16** Tri-*tert*-butylamine cannot be prepared via a reductive amination. Explain.

**23.17** Propose an efficient synthesis for the following transformation:

need more **PRACTICE?** **Try Problems 23.65, 23.66**

## 23.7 Synthesis Strategies

In the previous sections, we explored a variety of methods for preparing primary amines. Each of these methods utilizes a different source of nitrogen atoms.

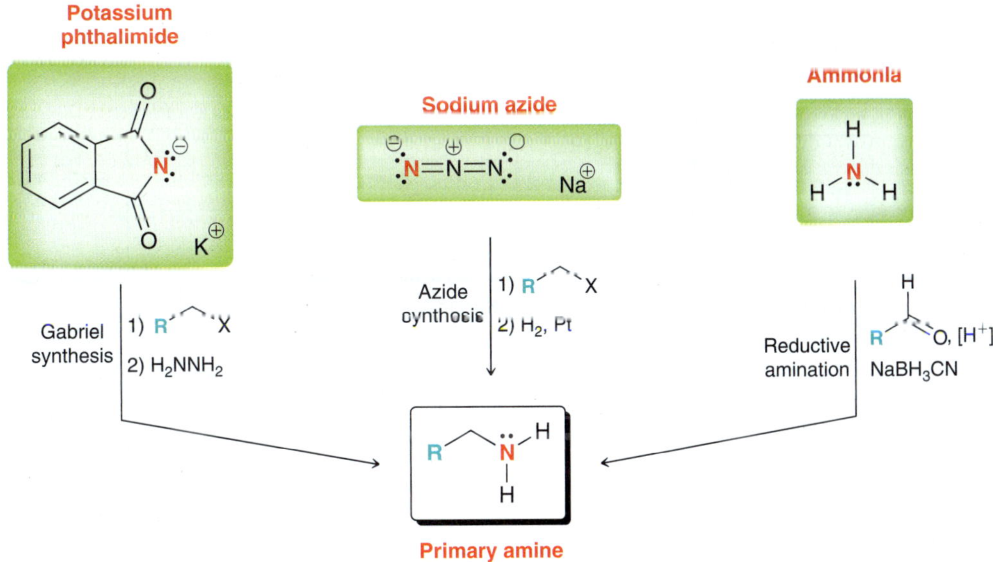


The Gabriel synthesis uses potassium phthalimide as the source of nitrogen, the azide synthesis uses sodium azide as the source of nitrogen, and reductive amination uses ammonia as the source of nitrogen. When preparing a primary amine, the available starting materials will dictate which of the three methods will be chosen.

Secondary amines are readily prepared from primary amines via reductive amination. Similarly, tertiary amines are readily prepared from secondary amines via reductive amination.

Primary amine —(Reductive amination; RCHO, $[H^+]$, $NaBH_3CN$)→ Secondary amine —(Reductive amination; RCHO, $[H^+]$, $NaBH_3CN$)→ Tertiary amine

Direct alkylation of the primary amine is not efficient because polyalkylation is generally unavoidable, as seen in Section 23.5. Therefore, it is more efficient to use reductive amination as an indirect method for alkylating the nitrogen atom of an amine.

Tertiary amines can be converted into quaternary ammonium salts via alkylation. This process is efficient because polyalkylation is not possible with tertiary amines.

Tertiary amine —(Alkylation; $RCH_2X$)→ Quaternary ammonium salt ($X^\ominus$)

Using just a few reactions it is possible to prepare a wide variety of amines from simple starting materials.

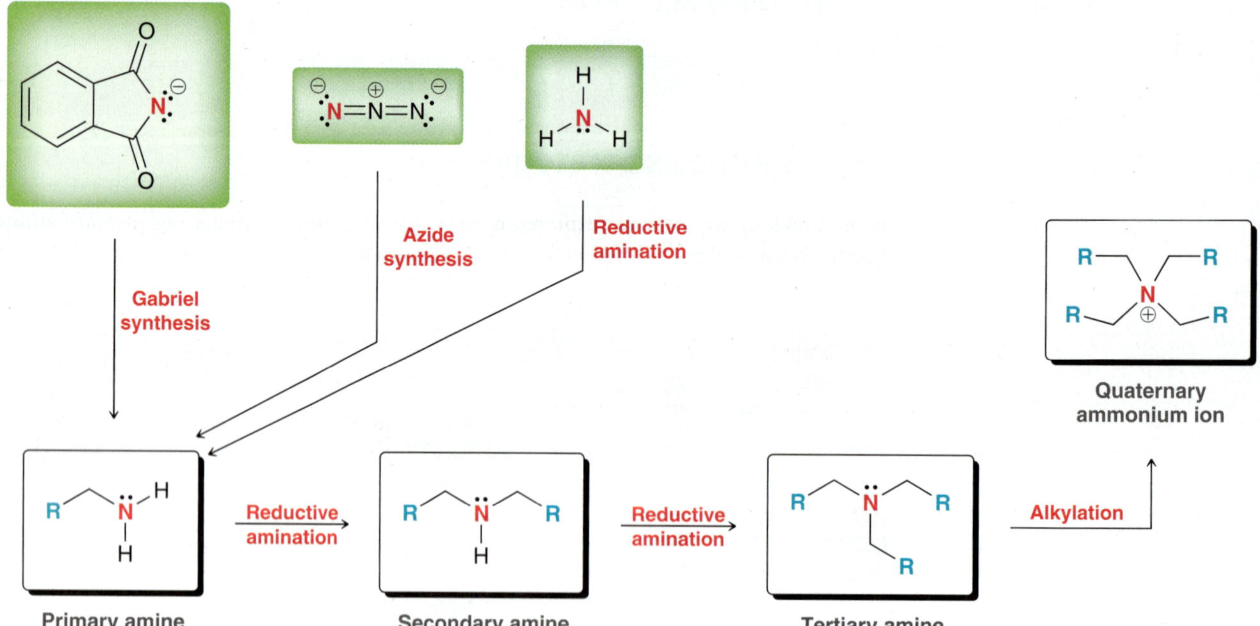


Using potassium phthalimide, sodium azide, or ammonia together with suitable alkyl halides, ketones, or aldehydes, it is possible to generate a large variety of primary, secondary, and tertiary amines as well as quaternary ammonium salts. Let's get some practice using these methods to prepare amines.

# SKILLBUILDER

## 23.4 PROPOSING A SYNTHESIS FOR AN AMINE

**LEARN** the skill

Using ammonia as your source of nitrogen, show what reagents you would use to prepare the following amine:

### SOLUTION

Begin by identifying the alkyl groups connected to the nitrogen atom and determine whether the amine is primary, secondary, or tertiary.

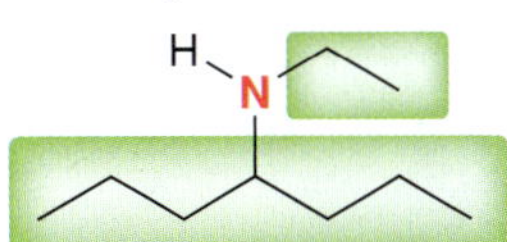

In this case, the amine is secondary because the nitrogen atom bears two alkyl groups and one hydrogen atom. Each of the highlighted groups must be installed separately. The source of nitrogen is ammonia, which dictates that both groups must be installed via reductive amination. The following retrosynthetic analysis reveals the necessary starting materials:

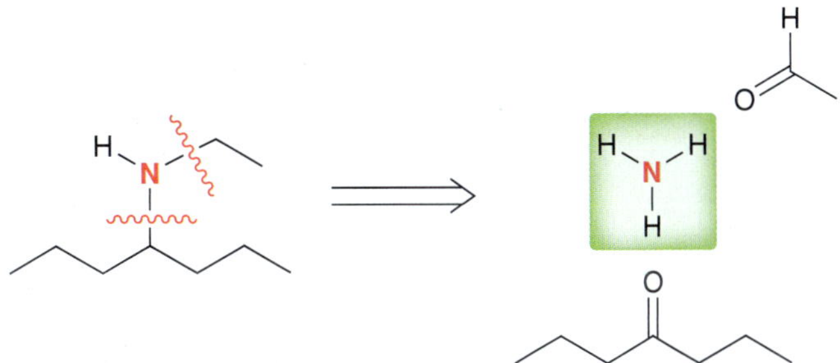

Each C—N bond in the product (left) can be formed via the reaction between ammonia and a ketone or aldehyde. This synthesis therefore requires two successive reductive aminations, which can be performed in either order. One reaction is used to install the first group,

$$\xrightarrow[\text{NaBH}_3\text{CN}]{\text{NH}_3,\ [\text{H}^+]}$$

and another reductive amination is used to install the second group,

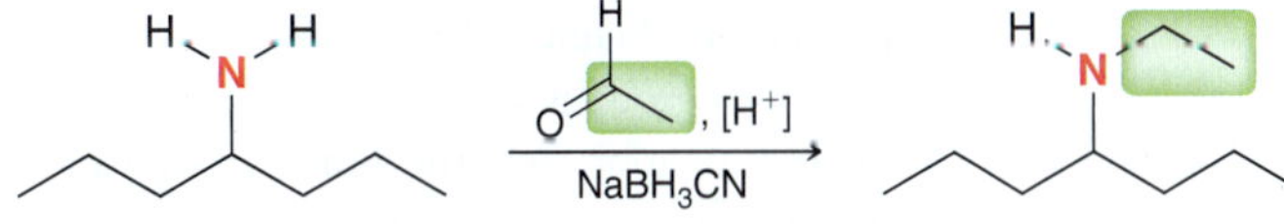

**PRACTICE** the skill

**23.18** Using ammonia as your source of nitrogen, show the reagents you would use to prepare each of the following amines:

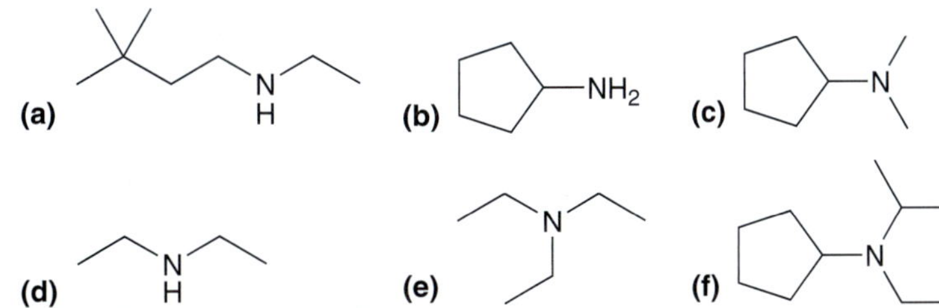

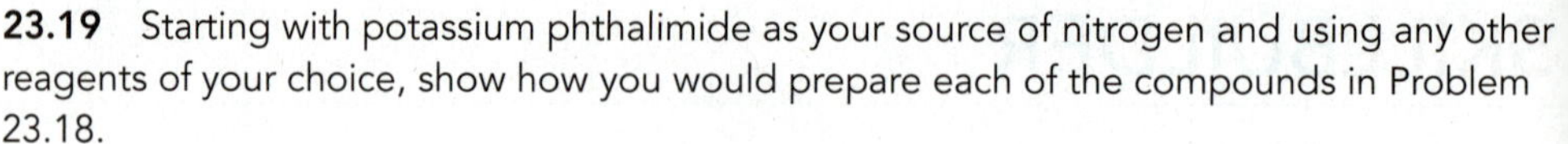

**23.19** Starting with potassium phthalimide as your source of nitrogen and using any other reagents of your choice, show how you would prepare each of the compounds in Problem 23.18.

**23.20** Starting with sodium azide as your source of nitrogen and using any other reagents of your choice, show how you would prepare each of the compounds in Problem 23.18.

**APPLY** the skill

**23.21** Show the reagents you would use to achieve the following transformation:

need more **PRACTICE?** **Try Problems 23.50, 23.51, 23.86**

## 23.8 Acylation of Amines

As discussed in Chapter 21, amines will react with acyl halides to yield amides.

The reaction takes place via a nucleophilic acyl substitution process, and HCl is a by-product of the reaction. As HCl is produced, the starting amine is protonated to form an ammonium ion, which will not attack the acyl halide. Therefore, two equivalents of the amine are required. Polyacylation does not occur, because the resulting amide is not nucleophilic.

Acylation of an amino group is an extremely useful technique when performing electrophilic aromatic substitution reactions, because it enables transformations that are otherwise difficult to achieve. For example, consider what happens when aniline is treated with bromine.

The amino group strongly activates the ring, and tribromination occurs readily. Monobromination of aniline is extremely difficult to achieve. Even when one equivalent of bromine is used, a mixture of products is obtained. This problem can be circumvented by first acylating the amino group. The resulting amide group is less activating than an amino group, and monobromination becomes possible. After the bromination step, the amide group can be removed by hydrolysis. This three-step route provides a method for the monobromination of aniline, which would otherwise be difficult to achieve.

As another example, consider what happens if we try to perform a Friedel-Crafts reaction (alkylation or acylation) using aniline as a starting material.

Ring is activated

Ring is deactivated

The amino group attacks the Lewis acid to form an acid-base complex in which the aromatic ring is now strongly deactivated. Recall that strongly deactivated rings are unreactive toward Friedel-Crafts reactions. As a result, it is not possible to achieve a Friedel-Crafts reaction with aniline as a starting material. Once again, this problem can be circumvented by first acylating the amino group. The nitrogen atom of the resulting amide is not nucleophilic. Under these conditions, a Friedel-Crafts reaction can be performed, and the amide group can then be removed via hydrolysis. This three-step route provides a method for alkylating aniline, something that cannot be accomplished directly.

## CONCEPTUAL CHECKPOINT

**23.22** When aniline is treated with a mixture of nitric acid and sulfuric acid, the expected nitration product (*para*-nitroaniline) is obtained in poor yield. Instead, the major product from nitration is *meta*-nitroaniline. Apparently, the amino group is protonated under these acidic conditions, and the resulting ammonium group is a *meta* director, rather than an *ortho para* director. Propose a plausible method for converting aniline into *para*-nitroaniline.

**23.23** Starting with nitrobenzene and using any other reagents of your choice, outline a synthesis of *para*-chloroaniline.

**23.24** Using acetic acid as your only source of carbon atoms, show how you could make *N*-ethyl acetamide.

## 23.9 Hofmann Elimination

Amines, like alcohols, can serve as precursors in the preparation of alkenes.

Recall that alcohols can only undergo an E2 process if the OH group is first converted into a better leaving group (such as a tosylate). Similarly, amines can also undergo an E2 reaction if the amino group is first converted into a better leaving group. This can be accomplished by treating the amine with excess methyl iodide.

Bad leaving group — Excess $CH_3I$ → Good leaving group

As we saw earlier in this chapter, amines will undergo exhaustive alkylation in the presence of excess methyl iodide, producing a quaternary ammonium salt. This process transforms the amino group into an excellent leaving group. Treating the quaternary ammonium salt with a strong base causes an E2 reaction that yields an alkene. The reagent most commonly used is aqueous silver oxide ($Ag_2O$):

1) Excess $CH_3I$
2) $Ag_2O$, $H_2O$, heat
(60%)

This process is called a **Hofmann elimination**. The function of silver oxide is to convert one ammonium salt into another ammonium salt by exchanging the iodide ion for a hydroxide ion.

Quaternary ammonium iodide — $Ag_2O$, $H_2O$, heat → Quaternary ammonium hydroxide + AgI

The newly formed hydroxide ion then serves as the base that triggers the E2 process:

E2 → + $H_2O$

Take special notice of the regiochemical outcome; it is not what we might have expected. Recall that E2 reactions generally proceed to form the more substituted alkene. But in this case, the major product is the less substituted alkene, and only trace amounts of the more substituted product are formed. This observation can be rationalized with a steric argument, because the leaving group is very bulky. To understand the effect of the leaving group in controlling the regiochemical outcome, recall that E2 reactions take place via a conformation in which the leaving group is anticoplanar with the proton being removed. Consider the Newman projection showing the anticoplanar conformation that leads to the more substituted alkene.

Look down the C2-C3 bond and draw a Newman projection

*gauche* Interaction

This conformation exhibits a *gauche* interaction, which raises the energy of the transition state. This *gauche* interaction is not present in the anticoplanar conformation that leads to the less substituted alkene.

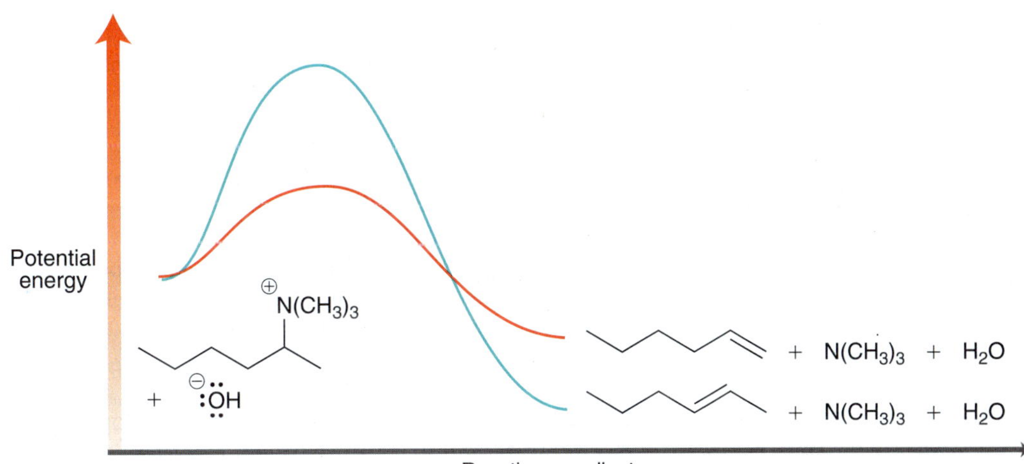


As a result, the transition state for formation of the less substituted alkene is lower in energy than the transition state for formation of the more substituted alkene (Figure 23.5). Formation of the less substituted alkene (shown in red) occurs more rapidly because it has a lower energy of activation, despite the fact that it is not the more stable product. This process is therefore said to be under kinetic control.



**FIGURE 23.5** An energy diagram showing the two possible pathways for a Hofmann elimination. The less substituted alkene is formed because the activation energy is lower for that pathway.

# SKILLBUILDER

## 23.5 PREDICTING THE PRODUCT OF A HOFMANN ELIMINATION

**LEARN the skill** Draw the major product obtained when 3-methyl-3-hexanamine is treated with excess methyl iodide followed by aqueous silver oxide and heat.

**SOLUTION**

Begin by drawing the starting amine and identifying all α and β positions. Then, consider forming a C—C double bond between each possible pair of α and β positions:

**STEP 1** Identify all α and β positions.

**STEP 2** Consider the regiochemical possibilities.

**STEP 3**
Identify the least substituted alkene.

Compare the products and identify the alkene that is least substituted:

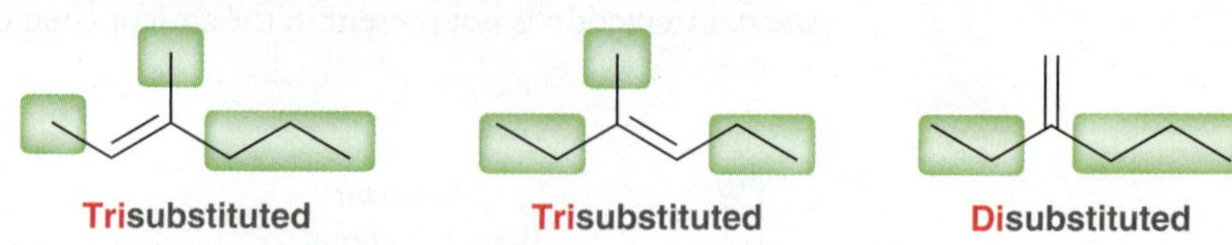


The disubstituted product is expected to be the major product:

$NH_2$ 1) Excess $CH_3I$ 2) $Ag_2O$, $H_2O$, heat

**PRACTICE the skill** **23.25** Draw the major product that is expected when each of the following compounds is treated with excess methyl iodide followed by aqueous silver oxide and heat:

**(a)** Cyclohexylamine **(b)** (*R*)-3-Methyl-2-butanamine **(c)** *N*,*N*-Dimethyl-1-phenylpropan-2-amine

**APPLY the skill** **23.26** Propose a synthesis for the following transformation (be sure to count the carbon atoms):

Br →

**23.27** Compound **A** is an amine that does not possess a chirality center. Compound **A** was treated with excess methyl iodide and then heated in the presence of aqueous silver oxide to produce an alkene. The alkene was further subjected to ozonolysis to produce butanal and pentanal. Draw the structure of compound **A**.

**23.28** Phencyclidine (PCP) was originally developed as an anesthetic for animals, but it has since become an illegal street drug because it is a powerful hallucinogen. Treatment of PCP with excess methyl iodide followed by aqueous silver oxide gives the following three principal products. Draw the structure of PCP.

N N

need more **PRACTICE?** **Try Problems 23.57, 23.69, 23.78**

## 23.10 Reactions of Amines with Nitrous Acid

This section will explore the reactions that occur when amines are treated with **nitrous acid**. Compare the structures of nitric acid ($HNO_3$) and nitrous acid ($HNO_2$):

O ⊕N HO O⊖ O N HO

**Nitric acid ($HNO_3$)** **Nitrous acid ($HNO_2$)**

Nitrous acid is unstable, and therefore, treating an amine with nitrous acid requires that the nitrous acid be prepared in the presence of the amine, called an *in situ* preparation. This is achieved by treating sodium nitrite ($NaNO_2$) with a strong acid, such as HCl or $H_2SO_4$. Under these con-

ditions nitrous acid is further protonated, followed by loss of water, to give a **nitrosonium ion** (Mechanism 23.2).

## MECHANISM 23.2 FORMATION OF NITROUS ACID AND NITROSONIUM IONS

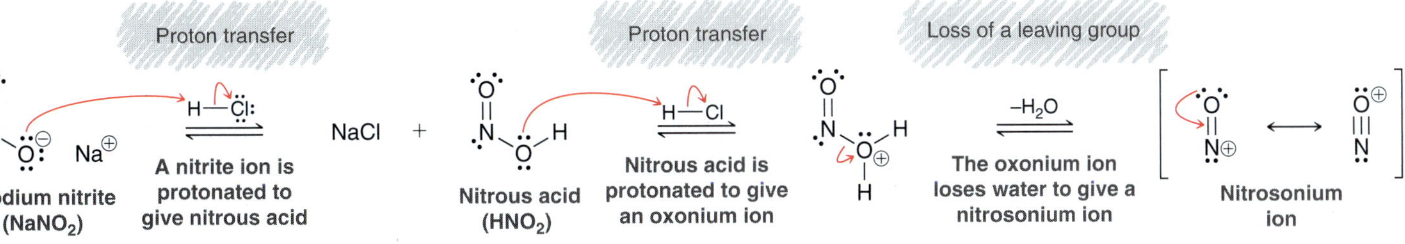


The nitrosonium ion is an extremely strong electrophile and is subject to attack by any amine that is present in solution. An equilibrium is established in which the concentration of nitrosonium ion is rather small. However, as the nitrosonium ion is formed, it can react with the amine that is present. The equilibrium then adjusts to produce more nitrosonium ion. The outcome of the reaction depends on whether the amine is primary or secondary. We will explore these possibilities separately, beginning with secondary amines.

### Secondary Amines and Nitrous Acid

When a secondary amine is treated with sodium nitrite and HCl, the reaction produces an ***N*-nitrosamine** (Mechanism 23.3):

$$R_2N{-}H \xrightarrow[HCl]{NaNO_2} R_2N{-}N{=}O$$

An *N*-nitrosamine

## MECHANISM 23.3 FORMATION OF N-NITROSAMINES

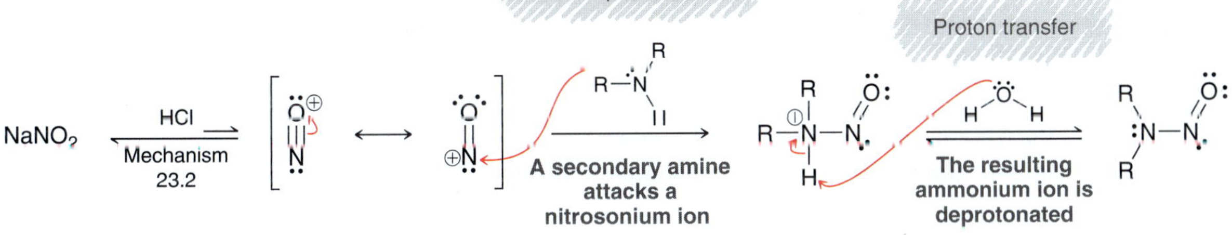


The amine functions as a nucleophile and attacks a nitrosonium ion that was generated *in situ* from sodium nitrite and HCl. The resulting ammonium ion is then deprotonated to give the product. Nitrosamines are known to be potent carcinogens. Several examples are shown below:

***N*-Nitrosodimethylamine**
(Found in many food products, including cured meat, fish, and beer)

***N*-Nitrosopyrrolidine**
(Found in fried bacon)

***N*-Nitrosonornicotine**
(Found in tobacco smoke)

## Primary Amines and Nitrous Acid

When a primary amine is treated with sodium nitrite and HCl, the reaction produces a **diazonium salt** (*azo* indicates a nitrogen atom, *diazo* indicates two nitrogen atoms, and *diazonium* indicates two nitrogen atoms with a positive charge).

$$R{-}NH_2 \xrightarrow[HCl]{NaNO_2} R{-}\overset{\oplus}{N}{\equiv}N \quad Cl^{\ominus}$$

A diazonium salt

This process is called **diazotization** and is believed to proceed via Mechanism 23.4. The amine functions as a nucleophile and attacks a nitrosonium ion which was generated *in situ* from sodium nitrite and HCl. Several proton transfers follow, and the last step involves loss of water to generate the diazonium salt.

### MECHANISM 23.4 DIAZOTIZATION

Nucleophilic attack

Proton transfer

$NaNO_2$ — HCl, Mechanism 23.2 → Nitrosonium ion

A primary amine attacks a nitrosonium ion

The resulting ammonium ion is deprotonated

Proton transfer

The oxygen atom is protonated

Proton transfer

A proton is removed, forming a double bond between the nitrogen atoms

Proton transfer

The oxygen atom is protonated, forming an excellent leaving group

Loss of a leaving group

$-H_2O$

Water leaves, producing a diazonium ion

$R{-}\overset{\oplus}{N}{\equiv}N{:}$

When the R group of the primary amine is an alkyl group as opposed to an aryl group, then the resulting diazonium salt is highly unstable and is too reactive to be isolated. It can spontaneously liberate nitrogen gas to form a carbocation, which then reacts in a variety of ways.

NaNO₂ / HCl → A diazonium salt ($Cl^{\ominus}$) → $-N_2$ → carbocation ($\oplus$)

For example, the carbocation can be captured by water to form an alcohol or it can lose a proton to form an alkene. The reaction generates a mixture of products and is therefore not useful. In addition, the process is also dangerous, because the expulsion of nitrogen gas can be an explosive process.

If, however, the primary amine is an aryl amine, then the resulting aryldiazonium salt is stable enough to be isolated. It does not liberate nitrogen gas, because that would involve formation of a high-energy aryl cation.

An aryldiazonium salt (stabilized)

A phenyl carbocation is high in energy

Aryldiazonium salts are extremely useful, because the diazonium group can be readily replaced with a number of other groups that are otherwise difficult to install on an aromatic ring. Reactions of aryldiazonium salts will be discussed in the next section.

## CONCEPTUAL CHECKPOINT

**23.29** Predict the major product obtained when each of the following amines is treated with a mixture of $NaNO_2$ and HCl:

(a) (b) (c) (d)

## 23.11 Reactions of Aryldiazonium Ions

In the previous section, we saw that aryl amines can be converted into aryldiazonium salts upon treatment with nitrous acid.

An aryldiazonium salt (stabilized)

As mentioned earlier, this reaction is extremely useful, because many different reagents will replace the diazo group, allowing for a simple procedure for installing a wide variety of groups on an aromatic ring:

In this section, we will explore some of the groups that can be installed on an aromatic ring using this procedure.

### Sandmeyer Reactions

The **Sandmeyer reactions** utilize copper salts (CuX) and enable the installation of a halogen or a cyano group on an aromatic ring:

Notice the installation of a cyano group. Recall that a cyano group can be hydrolyzed in aqueous acid or base, which provides a method for installing a carboxylic acid group on an aromatic ring.

1) $HNO_3$, $H_2SO_4$
2) Fe, $H_3O^+$
3) NaOH
4) $NaNO_2$, HCl

$N_2^{\oplus}$ $Cl^{\ominus}$

1) CuCN
2) $H_3O^+$, heat

COOH

## Fluorination

When treated with fluoroboric acid ($HBF_4$), an aryl diazonium salt is converted into a fluorobenzene. This reaction, called the **Schiemann reaction**, is useful for installing fluorine on an aromatic ring, which is not easy to accomplish with other methods.

$\oplus$N≡N

$HBF_4$

F

## Other Substitution Reactions of Aryl Diazonium Salts

When an aryl diazonium salt is heated in the presence of water, the diazo group is replaced with a hydroxyl group.

$\oplus$N≡N

$H_2O$
Heat

OH

This procedure is very useful, because there are not many other ways to install an OH group on an aromatic ring. An example of this process is shown below.

1) $HNO_3$, $H_2SO_4$
2) Fe, $H_3O^+$
3) NaOH
4) $NaNO_2$, HCl

$N_2^{\oplus}$ $Cl^{\ominus}$

$H_2O$, heat

OH

When treated with hypophosphorus acid ($H_3PO_2$), the diazo group of an aryldiazonium salt is replaced with a hydrogen atom:

$\oplus$N≡N

$H_3PO_2$

H

This reaction can be useful for manipulating the directing effects of a substituted aromatic ring. For example, consider the following synthesis of 1,3,5-tribromobenzene:

1) $HNO_3$, $H_2SO_4$
2) Fe, $H_3O^+$
3) NaOH

$NH_2$

$Br_2$

$NH_2$
Br Br
Br

1) $NaNO_2$, HCl
2) $H_3PO_2$

Br Br
Br

The amino group is first installed, its activating and directing effects are exploited, and then it is completely removed. The product of this sequence cannot be easily prepared from benzene via successive halogenation reactions because halogens are *ortho-para* directors.

The reactions of diazonium salts are believed to occur via radical intermediates, and their mechanisms are beyond the scope of our discussion.

## CONCEPTUAL CHECKPOINT

**23.30** Propose an efficient synthesis for each of the following transformations:

(a) (b) (c) (d) (e) (f)

NH2 → CN; NO2 → Br, Br; → OH; → OH, O; → Cl, F; Cl → Br

## Azo Coupling

Aryldiazonium salts are also known to react with activated aromatic rings.

Azo group

R (R = an activating group)

This process, called **azo coupling**, is believed to occur via an electrophilic aromatic substitution reaction.

HO N N Ph H Cl

Sigma complex
(resonance stabilized)

**LOOKING BACK**
For more on the source of color, see Section 17.12.

The reaction works best when the nucleophile is an activated aromatic ring. The resulting azo compound has extended conjugation and therefore exhibits color. By structurally modifying the starting materials (by placing substituents on the aromatic rings prior to azo coupling), a variety of products can be made, each of which will exhibit a unique color. These compounds, called **azo dyes**, were discussed in more detail in the Practically Speaking box in Section 19.3.

## SKILLBUILDER

### 23.6 DETERMINING THE REACTANTS FOR PREPARING AN AZO DYE

**LEARN the skill**

Identify the reactants you would use to prepare the following azo dye via an azo coupling reaction.

**SOLUTION**

Begin by identifying all substituents as activators or deactivators (for help, see Table 19.1).

**STEP 1**
Determine which ring is more strongly activated.

Moderate deactivator ($HO_3S$); Weak activator (Me); Moderate activator (MeO)

Based on this information, identify which ring is more strongly activated. In this case, it is the ring that bears the methoxy substituent. Using a retrosynthetic analysis, identify the starting nucleophile and electrophile needed to make this azo dye. The activated ring must be the nucleophile, and the other ring must function as the diazonium salt:

**STEP 2**
Identify the starting nucleophile and electrophile.

Nucleophile — Electrophile

Next, identify the starting aniline that is necessary in order to make the diazonium salt (the electrophile):

**STEP 3**
Identify the substituted aniline that is necessary to make the desired electrophile.

Electrophile

Finally, show all reagents. Draw the substituted aniline as the reactant and then show the two steps necessary to convert this compound into the desired product. The first step utilizes $NaNO_2$ and HCl to produce the diazonium salt, and the second step utilizes an azo coupling reaction:

**STEP 4**
Show all reagents.

1) $NaNO_2$, HCl
2) MeO–$C_6H_5$

**PRACTICE** the skill **23.31** Identify the reactants you would use to prepare each of the following azo dyes via an azo coupling reaction:

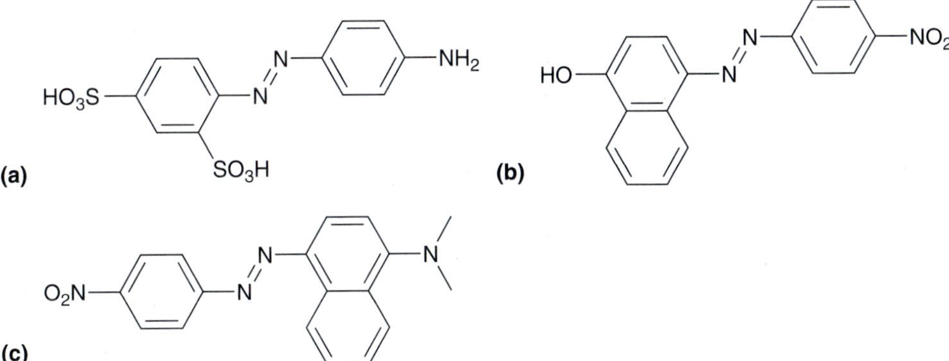

**APPLY** the skill **23.32** Starting with benzene and isopropyl chloride, show how you would prepare the following compound:

O2N–C6H4–N=N–C6H3(NH2)(isopropyl)

**23.33** Draw the product obtained when the diazonium salt formed from aniline is treated with each of the following compounds:

**(a)** Aniline **(b)** Phenol **(c)** Anisole (methoxybenzene)

need more **PRACTICE?** **Try Problems 23.68, 23.73d**

## 23.12 Nitrogen Heterocycles

A *heterocycle* is a ring that contains atoms of more than one element. Common organic heterocycles are comprised of carbon and either nitrogen, oxygen, or sulfur. Consider, for example, the structures of Viagra and Nexium. Both of these compounds contain nitrogen heterocycles, highlighted in red. Many different kinds of heterocycles are commonly found in the structures of biological molecules and pharmaceuticals. We will discuss just a few simple heterocycles in this section.

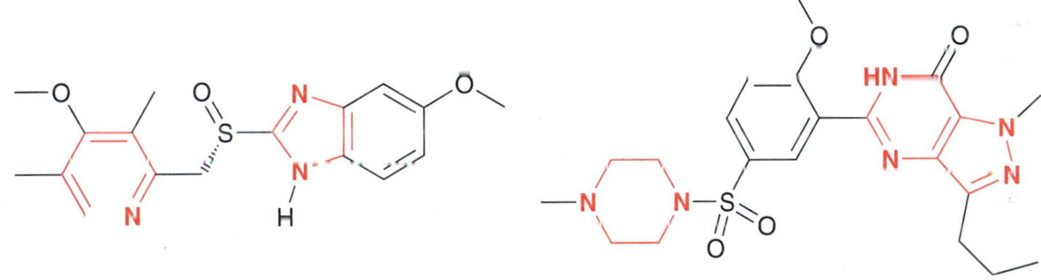

**Esomeprazole (Nexium)**
A proton pump inhibitor used in the treatment of ulcers and acid reflux disease

**Sildenafil (Viagra)**
Used in the treatment of erectile dysfunction and pulmonary arterial hypertension

### Pyrrole and Imidazole

Pyrrole, a five-membered aromatic ring containing one nitrogen atom, is numbered starting with the nitrogen atom. As seen in Chapter 18, the lone pair of this nitrogen atom participates in aromaticity (Figure 18.15) and is therefore significantly less basic and less nucleophilic than a typical amine. Pyrrole undergoes reactions that are

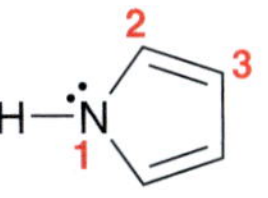


**Pyrrole**

expected for an aromatic system, such as electrophilic aromatic substitution. In fact, pyrrole is even more reactive than benzene, and low temperatures are often required to control the reaction.

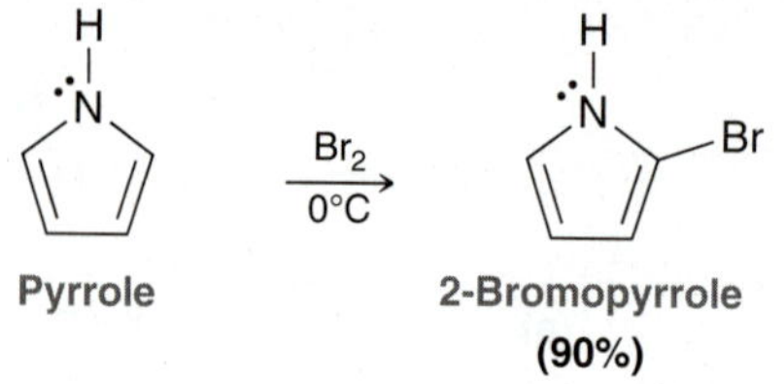

Pyrrole $\xrightarrow[0°C]{Br_2}$ 2-Bromopyrrole (90%)

Electrophilic aromatic substitution occurs primarily at C2, because the intermediate formed during attack at C2 is stabilized by resonance.

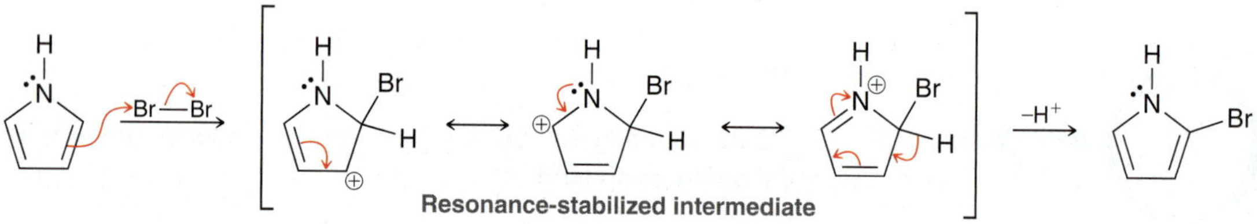

Resonance-stabilized intermediate

When attack takes place at C2, the intermediate has three resonance structures. In contrast, when attack takes place at C3, the intermediate has only two resonance structures.

An **imidazole** ring is like pyrrole but has one extra nitrogen atom at the 3 position. Histamine is an example of an important biological compound that contains an imidazole ring.

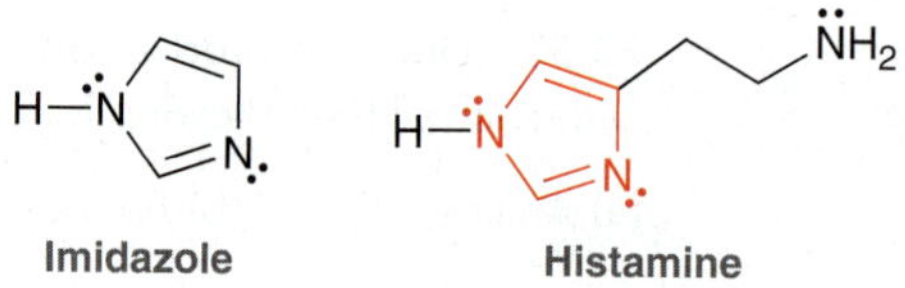

Imidazole Histamine

## medically speaking H₂-Receptor Antagonists and the Development of Cimetidine

Recall that histamine binds to the $H_1$ receptor, triggering allergic reactions (as seen in the Medically Speaking box in Section 18.5). It is now understood that histamine binds to other types of receptors as well, each of which elicits a different physiological response. The $H_2$ receptor has been associated with the stimulation of gastric (stomach) acid secretion. While histamine can bind with both $H_1$ and $H_2$ receptors, traditional antihistamines only bind with $H_1$ receptors. That is, they are $H_1$ antagonists, but they show no activity toward $H_2$ receptors. The discovery of the $H_2$ receptor naturally led to the search for new drugs that would selectively bind with the $H_2$ receptor and not with the $H_1$ receptor. The hope was that such drugs could theoretically be used to control the overproduction of gastric acid associated with acid reflux disease and ulcers. Years of research ultimately led to the development of cimetidine (Tagamet), the first selective $H_2$ antagonist.

The story of cimetidine begins with consideration of the features that might be required for a compound to exhibit selective $H_2$ antagonism. Medicinal chemists reasoned that the target compound must be sufficiently similar in structure to histamine to bind to the $H_2$ receptor, but it must be sufficiently different in order to be an antagonist, rather than an agonist. It was assumed that the imidazole ring was necessary for binding, so several hundred imidazole derivatives were prepared and tested. The first compound that showed any $H_2$ antagonistic activity was *N*-guanylhistamine.

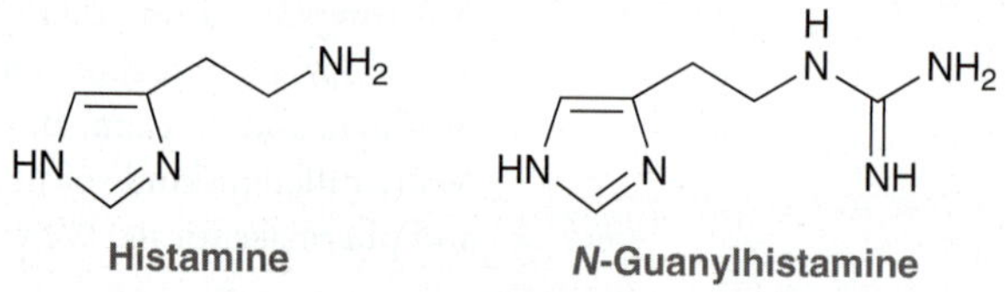

Histamine *N*-Guanylhistamine

Further studies revealed that *N*-guanylhistamine is not actually an $H_2$ antagonist, but rather, it is a partial $H_2$ agonist. It binds with the $H_2$ receptor and elicits the stimulation of gastric acid secretion, but to a smaller extent than histamine. This, of course, defeats the purpose. So the chemists continued their search, using *N*-guanylhistamine as a starting point. This compound was known to be protonated at physiological pH.

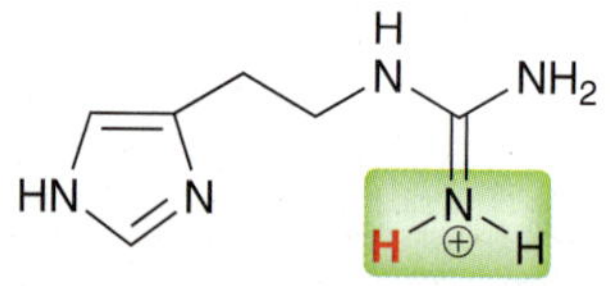

Medicinal chemists then tried to modify the guanyl group so that it would be neutral at physiological pH. After many different possibilities, a thiourea analogue was created that has one additional $CH_2$ group in the side chain, and the basic nitrogen atom of the guanyl group has been replaced with a sulfur atom that is not protonated at physiological pH.

This compound did, in fact, exhibit weak antagonistic activity. The search then focused on creating analogues that might exhibit greater potency. An additional $CH_2$ was inserted in the side chain, and the terminal nitrogen atom was methylated. The resulting compound, called burimamide, was shown to be effective in inhibiting gastric acid secretion in rats.

**Burimamide**

Burimamide was the first $H_2$ antagonist tested in humans. Unfortunately, it did not exhibit sufficient activity when taken orally, so the search continued, this time using the structure of burimamide as a starting point.

One approach was to compare the basicity of histamine and burimamide. In other words, we compare the p$K_a$ values of the protonated imidazole ring in histamine and burimamide:

p$K_a$ = 5.9 p$K_a$ = 7.3

**Histamine** **Burimamide**

The protonated imidazole ring of histamine has a p$K_a$ of 5.9, while the protonated imidazole ring of burimamide has a p$K_a$ of 7.3. Therefore, at physiological pH, the equilibrium concentrations of neutral and ionic forms are different for histamine and burimamide. Research efforts were then directed toward developing an analogue of burimamide that more closely resembles the basicity of histamine. It was reasoned that the presence of an electron-withdrawing group in the side chain would lower the p$K_a$ while an electron-donating group at the C4 position would increase the basicity of the nitrogen atom. This rationale led to the development of metiamide.

**Metiamide**

The structure of metiamide differs from that of burimamide in the presence of a methyl group at C4, and the replacement of one of the $CH_2$ groups in the side chain with an electron-withdrawing sulfur atom. This type of replacement is called an isosteric replacement, because the sulfur atom occupies roughly the same amount of space as a $CH_2$ group, but it has very different electronic properties. The net result is that the protonated form of metiamide has a p$K_a$ that is nearly identical with the protonated form of histamine. Clinical trials showed that metiamide exhibited $H_2$ antagonistic activity that was nine times greater than burimamide, and it was effective in relieving ulcer symptoms; however, a few patients developed granulocytopenia (lowering of the white blood cell count). This side effect was not acceptable, because it would temporarily impair the ability of the immune system to function properly. It was believed that the cases of granulocytopenia were caused by the presence of the thiourea group, so alternative groups were considered. Ultimately, the thiourea group was replaced with a cyanoguanidine group.

**Tagamet (cimetidine)**

Recall from earlier in our discussion that the presence of the guanyl group precluded antagonistic activity because the guanyl group is charged at physiological pH. However, in this derivative, the cyano group withdraws electron density from the neighboring nitrogen atom and delocalizes its lone pair. The resulting group is less basic and is not protonated at physiological pH. This compound, called cimetidine, was found to exhibit potent $H_2$ antagonistic properties and did not cause granulocytopenia in patients.

Cimetidine became available to the public in England in 1976 and in the United States in 1979, sold under the trade name Tagamet. Annual sales of this drug skyrocketed, ultimately reaching more than $1 billion in annual sales. Cimetidine earned its place in history as the first blockbuster drug. Since the development of cimetidine, other $H_2$ antagonists were developed and approved for use:

**Zantac (ranitidine)**

**Pepcid (famotidine)**

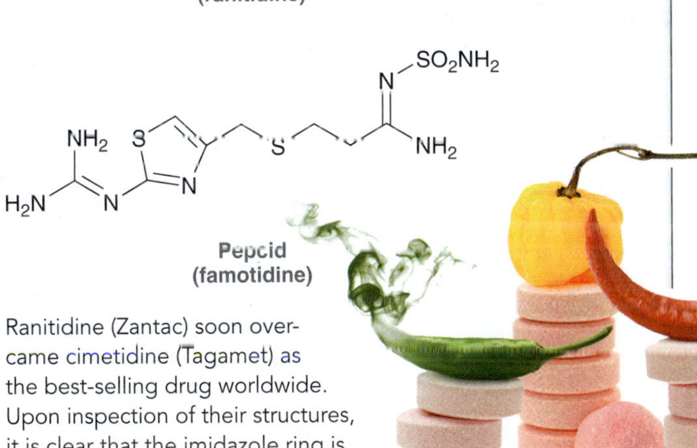

Ranitidine (Zantac) soon overcame cimetidine (Tagamet) as the best-selling drug worldwide. Upon inspection of their structures, it is clear that the imidazole ring is not absolutely essential for $H_2$ antagonistic activity, as analogous heterocyclic aromatic rings can replace the imidazole ring. However, all of these analogues appear to have some derivative of the guanyl group on the side chain.

## Pyridine and Pyrimidine

Pyridine is a six-membered aromatic ring containing one nitrogen atom and is numbered starting with the nitrogen atom:

Pyridine

The lone pair of the nitrogen atom is localized and occupies an $sp^2$-hybridized orbital, and as a result, pyridine is a stronger base than pyrrole. Nevertheless, pyridine is still a weaker base than alkyl amines, because the lone pair is housed in an $sp^2$-hybridized orbital, rather than an $sp^3$-hybridized orbital (as seen in Figure 18.14). By occupying an $sp^2$-hybridized orbital, the electrons of the lone pair have more *s* character and are therefore closer to the positively charged nucleus, rendering them less basic.

Since pyridine is an aromatic compound, we would expect it to exhibit reactions that are characteristic of aromatic systems. In fact, pyridine undergoes electrophilic aromatic substitution reactions; however, the yields are generally quite low, because the inductive effect of the nitrogen atom renders the ring electron poor. High temperatures are required:

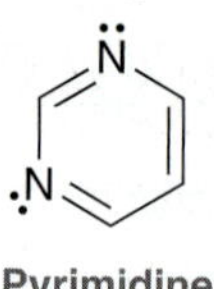


Pyridine → 3-Bromopyridine (30%)

**Pyrimidine** is similar in structure to pyridine but contains one extra nitrogen atom at the 3 position. Pyrimidine rings are very common in biological molecules. Pyrimidine is less basic than pyridine due to the inductive effect of the second nitrogen atom:

Pyrimidine

### CONCEPTUAL CHECKPOINT

**23.34** Pyridine undergoes electrophilic aromatic substitution at the C3 position. Justify this regiochemical outcome by drawing resonance structures of the intermediate produced from attack at C2, at C3, and at C4.

**23.35** Predict the product obtained when pyrrole is treated with a mixture of nitric acid and sulfuric acid at 0°C.

## 23.13 Spectroscopy of Amines

### IR Spectroscopy

In their IR spectra, primary and secondary amines exhibit signals between 3350 and 3500 cm$^{-1}$. These signals correspond with N—H stretching and are typically less intense than O—H signals. Primary amines give two peaks (symmetric and asymmetric stretching), while secondary amines give only one peak (Figure 15.19). This phenomenon was first explained in Section 15.5.

Tertiary amines lack an N—H bond and do not exhibit a signal in the region between 3350 and 3500 cm$^{-1}$. Tertiary amines can be detected with IR spectroscopy by treatment with HCl. The resulting N—H bond exhibits a characteristic signal between 2200 and 3000 cm$^{-1}$:

$R_3N: + H{-}Cl \longrightarrow R_3N^{+}{-}H \; Cl^{-}$

## NMR Spectroscopy

In the $^{1}H$ NMR spectrum of an amine, any proton attached directly to the nitrogen atom (for primary and secondary amines) will typically appear as a broad signal somewhere between 0.5 and 5.0 ppm. The precise location is sensitive to many factors, including the solvent, concentration, and temperature. Splitting is generally not observed for these protons, because they are labile and are exchanged at a rate that is faster than the timescale of the NMR spectrometer (as described in Section 16.7). The broad signal for these protons can generally be removed from the spectrum by dissolving the amine in $D_2O$, which results in a proton exchange that replaces these protons for deuterons.

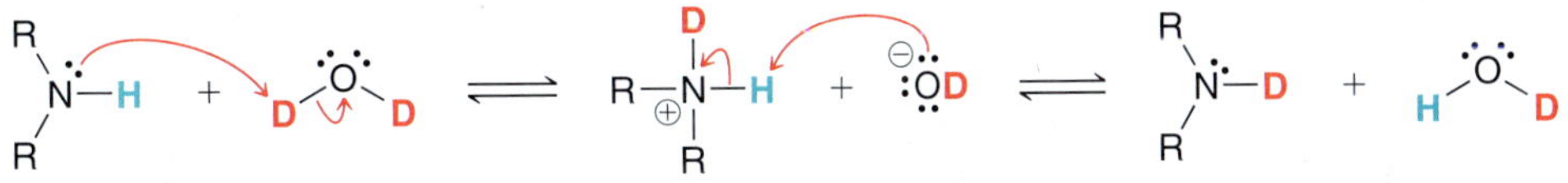

Protons connected to the α position typically appear between 2 and 3 ppm because of the deshielding effect of the nitrogen atom. As an example, consider the chemical shifts for the protons in propylamine:

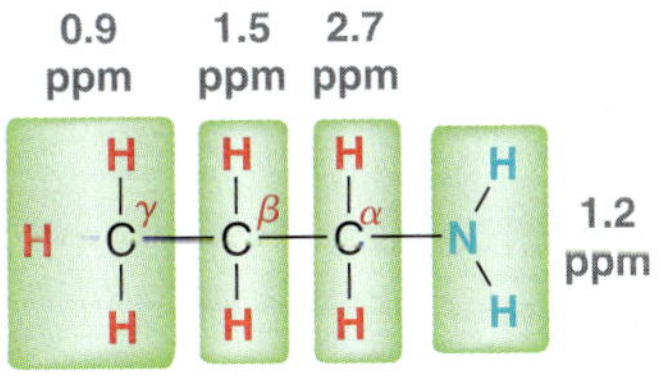

Notice that the deshielding effect of the nitrogen atom is greatest for the α protons (2.7 ppm) and tapers off with distance. The β protons are affected to a lesser extent (1.5 ppm), and the γ protons are not measurably affected (0.9 ppm).

In the $^{13}C$ NMR spectrum of an amine, the α carbon atoms typically appear between 30 and 50 ppm. That is, they are shifted about 20 ppm downfield due to the deshielding effect of the nitrogen atom.

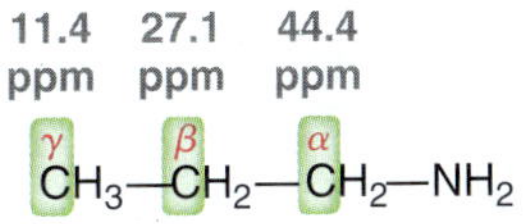

Once again, notice that the deshielding effect of the nitrogen atom is greatest for the α carbon (44.4 ppm) and tapers off with distance. The β carbon is affected to a lesser extent (27.1 ppm), and the γ carbon is not measurably affected (11.4 ppm).

## Mass Spectrometry

The mass spectrum of an amine is characterized by the presence of a parent ion with an odd molecular weight. This follows the nitrogen rule (see Section 15.9), which states that a compound with an odd number of nitrogen atoms will produce a parent ion with an odd molecular weight. In addition, amines generally exhibit a characteristic fragmentation pattern. They undergo α cleavage to generate a radical and a resonance-stabilized cation (see Section 15.12).

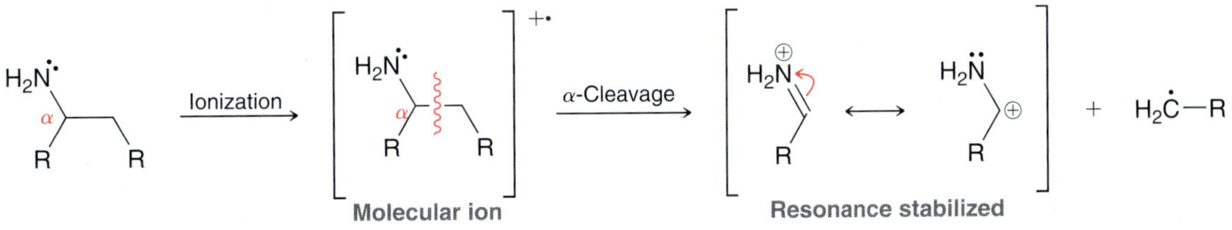

## CONCEPTUAL CHECKPOINT

**23.36** How would you use IR spectroscopy to distinguish between the following pairs of compounds?

(a) (b)

**23.37** How would you use NMR spectroscopy to distinguish between the following pairs of compounds?

(a)

(b)

# REVIEW OF REACTIONS

## Preparation of Amines

### From Alkyl Halides

Br $\xrightarrow[S_N2]{NaCN}$ C≡N $\xrightarrow[2)\ H_2O]{1)\ xs\ LAH}$ H H C N H H

### From Carboxylic Acids

O OH $\xrightarrow[2)\ xs\ NH_3]{1)\ SOCl_2}$ O $NH_2$ $\xrightarrow[2)\ H_2O]{1)\ xs\ LAH}$ $NH_2$

### From Benzene

$\xrightarrow[H_2SO_4]{HNO_3}$ $NO_2$ — $H_2$ / Pt, or 1) Fe, Zn, Sn, or $SnCl_2$, $H_3O^+$ 2) NaOH — $NH_2$

### The Azide Synthesis

X $\xrightarrow{NaN_3}$ $N_3$ — $H_2$ / Pt, or 1) LAH 2) $H_2O$ — $NH_2$

### The Gabriel Synthesis

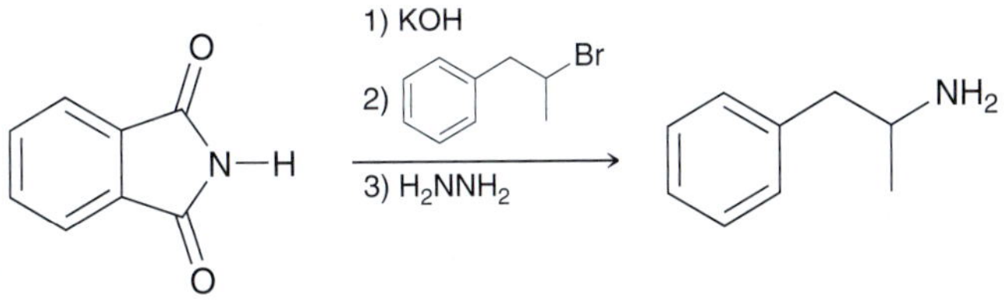

1) KOH
2) (1-phenyl-2-bromopropane)
3) $H_2NNH_2$

N–H → $NH_2$

### Via Reductive Amination

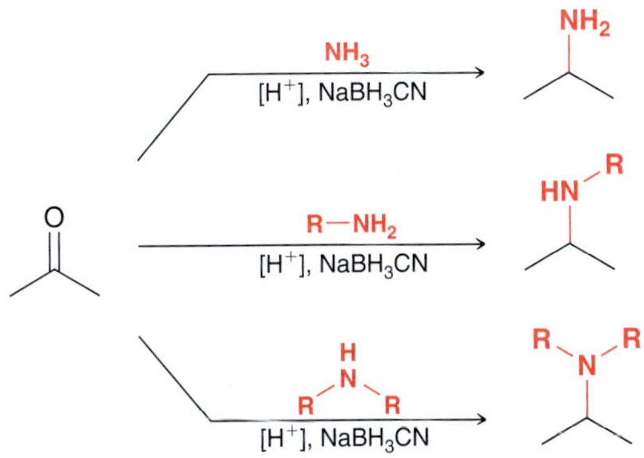

$NH_3$, [$H^+$], $NaBH_3CN$ → $NH_2$

R—$NH_2$, [$H^+$], $NaBH_3CN$ → HN–R

R–NH–R, [$H^+$], $NaBH_3CN$ → R–N–R

## Reactions of Amines

### Acylation

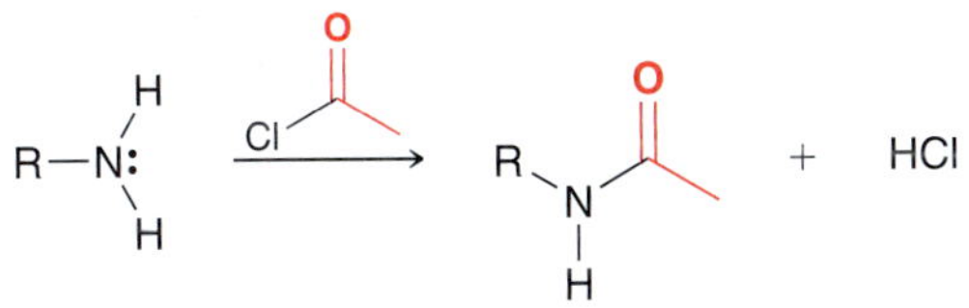

R—N: (H, H) + Cl–C(=O)– → R–N(H)–C(=O)– + HCl

### Hofmann Elimination

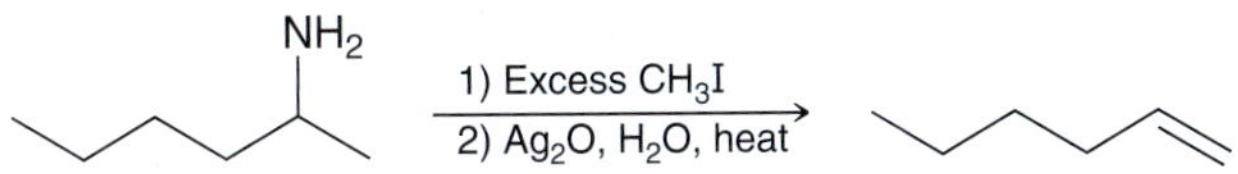

$NH_2$

1) Excess $CH_3I$
2) $Ag_2O$, $H_2O$, heat

### Reactions with Nitrous Acid

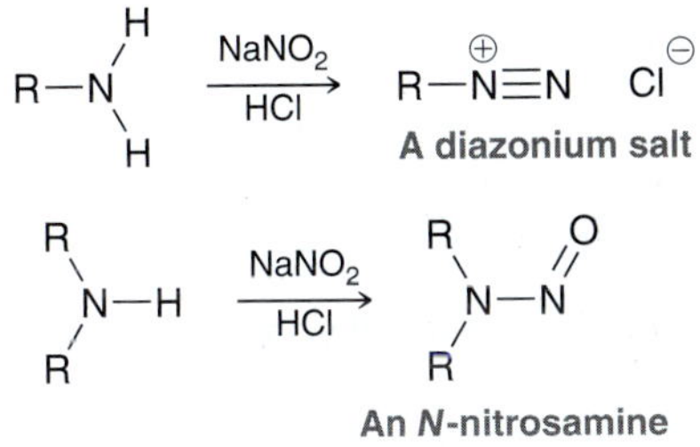

R—N(H)(H) → $NaNO_2$, HCl → R—$\overset{\oplus}{N}\equiv N$ $Cl^{\ominus}$

A diazonium salt

$R_2N$—H → $NaNO_2$, HCl → $R_2N$—N=O

An *N*-nitrosamine

## Reactions of Aryldiazonium Salts

### Sandmeyer Reactions

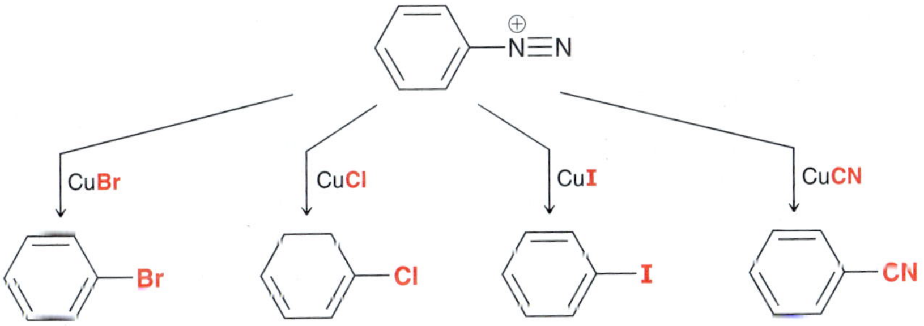

$\overset{\oplus}{N}\equiv N$

CuBr → Br

CuCl → Cl

CuI → I

CuCN → CN

### Fluorination (Schiemann Reaction)

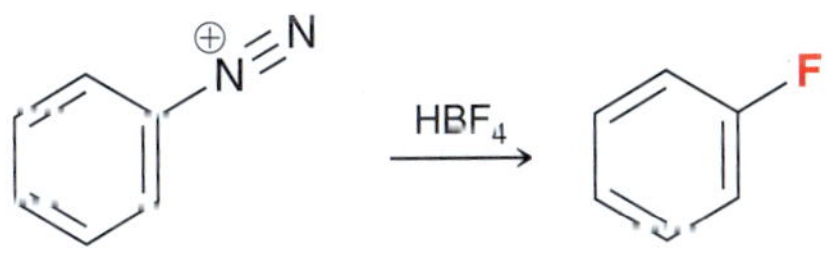

$\overset{\oplus}{N}\equiv N$ → $HBF_4$ → F

### Other Reactions of Aryldiazonium Salts

$\overset{\oplus}{N}\equiv N$ → $H_2O$, Heat → OH

$\overset{\oplus}{N}\equiv N$ → $H_3PO_2$ → H

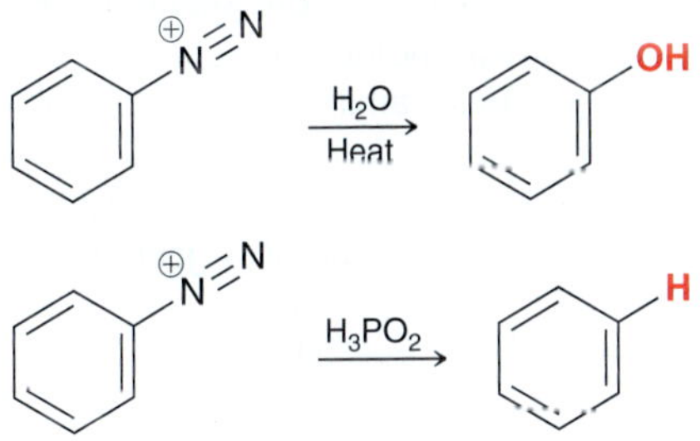

### Azo Coupling

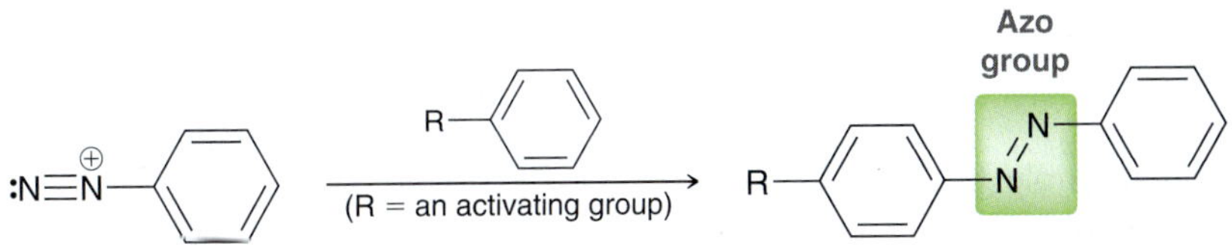

$:N\equiv\overset{\oplus}{N}$— + R— (R = an activating group) → R—C6H4—N=N—C6H5

Azo group

### Reactions of Nitrogen Heterocycles

Pyrrole $\xrightarrow[0°C]{Br_2}$ 2-Bromopyrrole

Pyridine $\xrightarrow[300°C]{Br_2}$ 3-Bromopyridine

## REVIEW OF CONCEPTS AND VOCABULARY

### SECTION 23.1

- **Amines** are derivatives of ammonia, in which one or more of the protons are replaced with alkyl or aryl groups.
- Amines are primary, secondary, or tertiary, depending on the number of groups attached to the nitrogen atom.
- Naturally occurring amines isolated from plants are called **alkaloids**.
- The lone pair on the nitrogen atom of an amine can function as a base or nucleophile.

### SECTION 23.2

- Amines can be named as **alkyl amines** or as **alkanamines**, depending on the complexity of the alkyl groups.
- When naming an amine as an alkanamine, the most complex alkyl group is chosen as the parent, and the other alkyl groups are listed as N substituents.
- Aromatic amines, also called **aryl amines**, are generally named as derivatives of aniline.

### SECTION 23.3

- The nitrogen atom of an amine is typically $sp^3$ hybridized; the lone pair occupies an $sp^3$-hybridized orbital.
- Amines containing three different alkyl groups are chiral, but they are generally not optically active at room temperature.
- Amines with fewer than five carbon atoms per functional group will typically be water soluble, while amines with more than five carbon atoms per functional group will be only sparingly soluble.
- The boiling point of an amine increases as a function of its capacity to form hydrogen bonds.
- Amines are effectively protonated even by weak acids.
- The basicity of an amine can be quantified by measuring the $pK_a$ of the corresponding ammonium ion. A large $pK_a$ indicates a strongly basic amine, while a low $pK_a$ indicates a weakly basic amine.
- The ease with which amines are protonated can be used to remove them from mixtures of organic compounds in a process called **solvent extraction**.
- Aryl amines are less basic than alkyl amines, because the lone pair is delocalized.
- Electron-donating groups slightly increase the basicity of aryl amines, while electron-withdrawing groups significantly decrease the basicity of aryl amines.
- Amides generally do not function as bases, and they are very poor nucleophiles.
- Pyridine is a stronger base than pyrrole, because the lone pair in pyrrole participates in aromaticity.
- An amine group exists primarily as a charged ammonium ion at physiological pH.

### SECTION 23.4

- Amines can be prepared from alkyl halides or from carboxylic acids.
- Aryl amines, such as aniline, can be prepared from benzene.

### SECTION 23.5

- Monoalkylation of ammonia is difficult to achieve because the resulting primary amine is even more nucleophilic than ammonia.
- If the **quaternary ammonium salt** is the desired product, then an excess of alkyl halide can be used, and ammonia undergoes *exhaustive alkylation*.
- The **azide synthesis** involves treating an alkyl halide with sodium azide followed by reduction.
- The **Gabriel synthesis** generates primary amines upon treatment of potassium phthalimide with an alkyl halide, followed by hydrolysis or reaction with $N_2H_4$.

### SECTION 23.6

- Amines can be prepared via **reductive amination**, in which a ketone or aldehyde is converted into an imine in the presence of a reducing agent, such as **sodium cyanoborohydride** ($NaBH_3CN$).

### SECTION 23.7

- Amines can be prepared by a variety of methods, each using a different source of nitrogen. The starting materials dictate which method will be used.
- Secondary amines are readily prepared from primary amines via reductive amination. Tertiary amines are readily prepared from secondary amines via reductive amination.
- Quaternary ammonium salts are prepared from tertiary amines via direct alkylation.

### SECTION 23.8

- Amines react with acyl halides to produce amides.
- Acylation of an amino group is a useful technique when performing electrophilic aromatic substitution reactions, because it enables transformations that are otherwise difficult to achieve.

### SECTION 23.9

- In the **Hofmann elimination,** an amino group is converted into a better leaving group which is expelled in an E2 process to form an alkene.
- The less substituted alkene is formed for steric reasons.

**SECTION 23.10**

- **Nitrous acid** is unstable and must be prepared *in situ* from sodium nitrite and an acid.
- In the presence of HCl, nitrous acid is protonated, followed by loss of water, to give a **nitrosonium ion**.
- Primary amines react with a nitrosonium ion to yield a **diazonium salt** in a process called **diazotization**.
- Alkyldiazonium salts are unstable, but aryldiazonium salts can be isolated.
- Secondary amines react with a nitrosonium ion to produce an ***N*-nitrosamine**.

**SECTION 23.11**

- Aryldiazonium salts are very useful, because many different reagents will replace the diazo group.
- **Sandmeyer reactions** utilize copper salts (CuX), enabling the installation of a halogen or a cyano group.
- In the **Schiemann reaction**, an aryl diazonium salt is converted into a fluorobenzene by treatment with fluoroboric acid ($HBF_4$).
- An aryldiazonium salt can be treated with water to install a hydroxyl group or with hypophosphorus acid ($H_3PO_2$) to replace the diazo group with a hydrogen atom.
- Aryldiazonium salts react with activated aromatic rings in a process called **azo coupling** to produce colored compounds called **azo dyes**.

**SECTION 23.12**

- A *heterocycle* is a ring that contains atoms of more than one element.
- Pyrrole undergoes electrophilic aromatic substitution reactions, which occur primarily at C2.
- An **imidazole** ring is like pyrrole but has one extra nitrogen atom at the 3 position.
- Pyridine will undergo electrophilic aromatic substitution; however, the yields are quite low.
- **Pyrimidine** is similar in structure to pyridine but contains one extra nitrogen atom at the 3 position.

**SECTION 23.13**

- In their IR spectra, primary and secondary amines exhibit signals between 3350 and 3500 $cm^{-1}$. Primary amines give two peaks (symmetric and asymmetric stretching), while secondary amines give only one peak.
- When treated with HCl, tertiary amines can be readily detected with IR spectroscopy.
- In the $^1H$ NMR spectrum of an amine, any proton attached directly to the nitrogen atom will typically appear as a broad signal somewhere between 0.5 and 5.0 ppm.
- The protons connected to the α position typically appear between 2 and 3 ppm.
- In the $^{13}C$ NMR spectrum of an amine, the α carbon atoms typically appear between 30 and 50 ppm.

# SKILLBUILDER REVIEW

## 23.1 NAMING AN AMINE

**STEP 1** Identify all of the alkyl groups connected to the nitrogen atom.

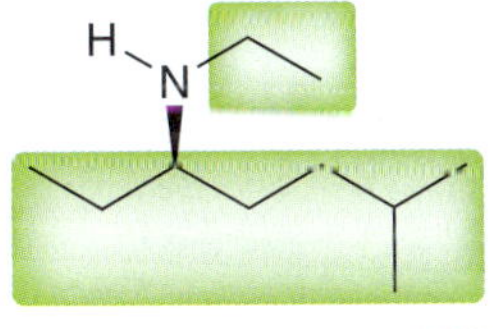


**STEP 2** Determine whether to name the compound as an alkyl amine or as an alkanamine:

**Alkyl amine** = if all alkyl groups are simple

**Alkanamine** = if one of the alkyl groups is complex

**STEP 3** If naming as an alkanamine, choose the complex alkyl group as the parent and list the other alkyl groups as substituents (using the letter "N" as a locant).

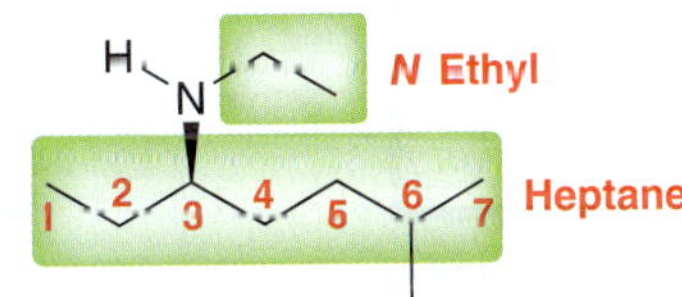


**STEP 4** Assign locants to each substituent, assemble the name, and assign the configuration of any chirality centers.

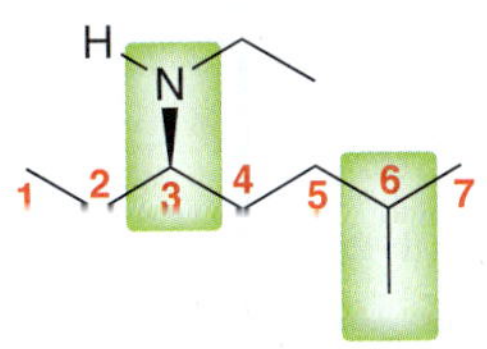


(*R*)-*N*-Ethyl-6-methyl-3-heptanamine

Try Problems 23.1–23.3, 23.42, 23.46, 23.47

## 23.2 PREPARING A PRIMARY AMINE VIA THE GABRIEL REACTION

**STEP 1** Identify an alkyl halide that can serve as a precursor for preparing the desired amine. Simply replace the amino group with a halide.

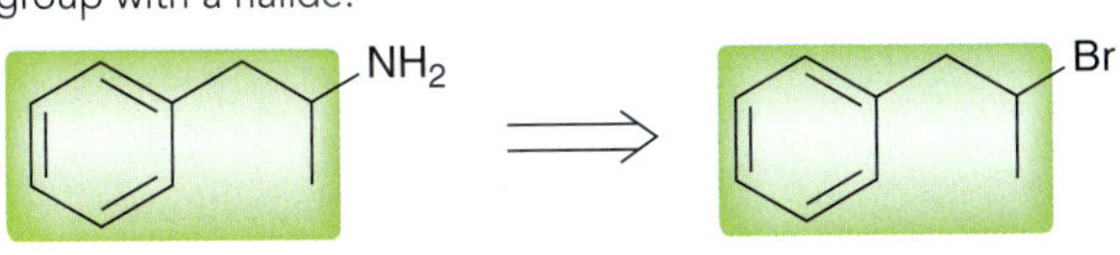


**STEP 2** Draw the reagents for all three steps, starting with phthalimide as the starting material.

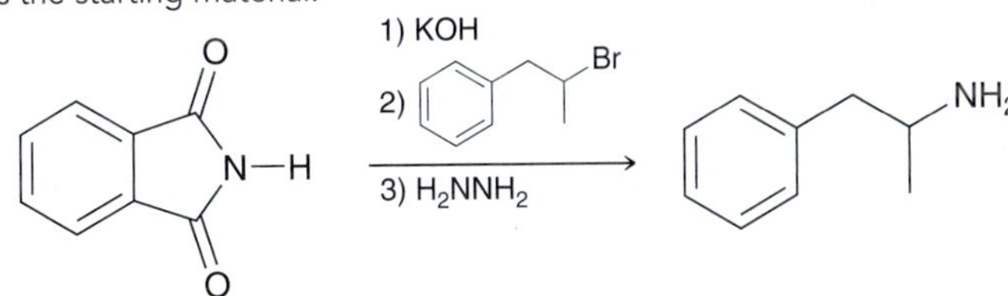


Try Problems 23.12, 23.13, 23.51a, 23.60, 23.61, 23.62b

**23.3** PREPARING AN AMINE VIA A REDUCTIVE AMINATION

**STEP 1** Identify all C—N bonds.

**STEP 2** Using a retrosynthetic analysis, identify the starting amine and the carbonyl group that could be used to make each of the C—N bonds.

**STEP 3** Show all reagents necessary for each synthetic route.

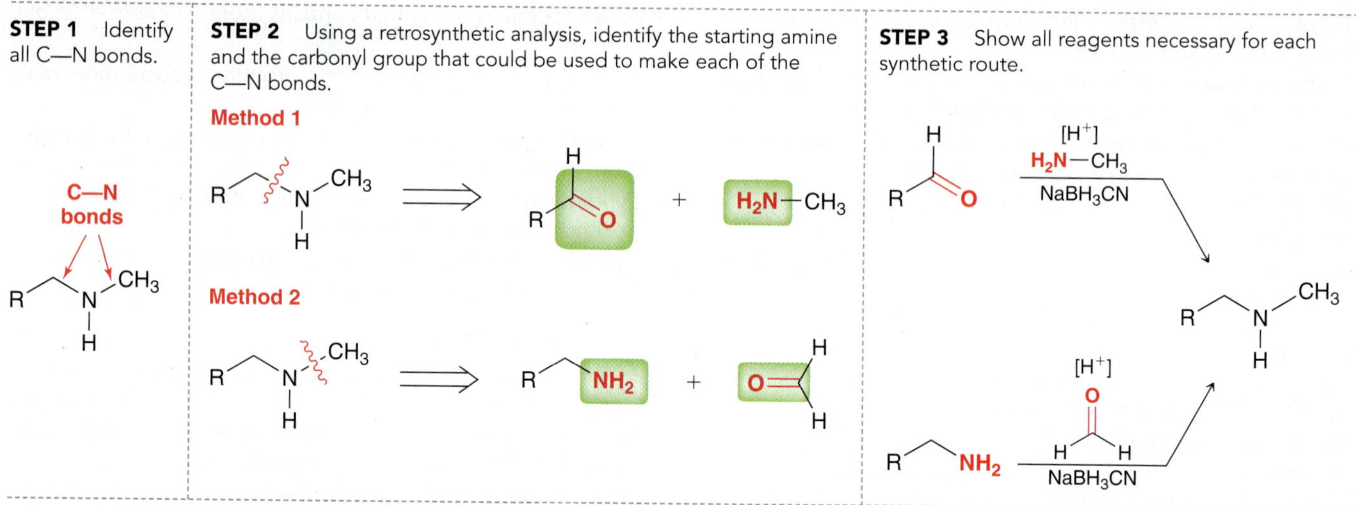


Try Problems 23.14–23.17, 23.65, 23.66

**23.4** SYNTHESIS STRATEGIES

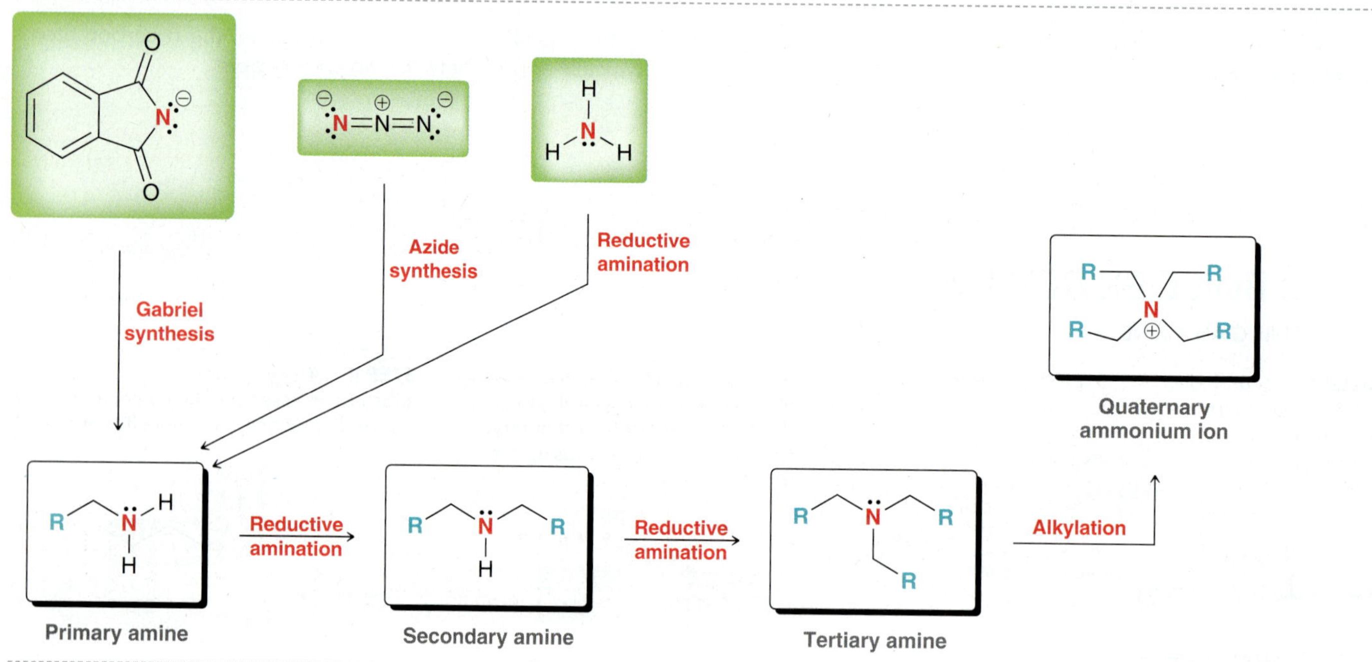


Try Problems 23.18–23.21, 23.50, 23.51, 23.86

**23.5** PREDICTING THE PRODUCT OF A HOFMANN ELIMINATION

**STEP 1** Identify all $\alpha$ and $\beta$ positions.

**STEP 2** Consider forming a C=C bond between each pair of $\alpha$ and $\beta$ positions.

**STEP 3** Compare all possible alkenes and identify which is the least substituted.

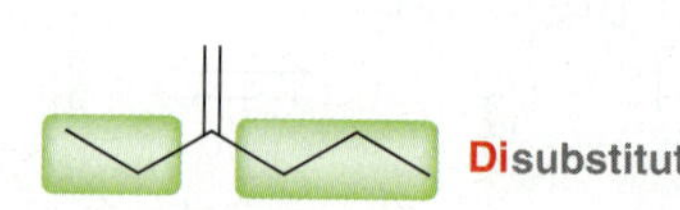


Try Problems 23.25–23.28, 23.57, 23.69, 23.78

### 23.6 DETERMINING THE REACTANTS FOR PREPARING AN AZO DYE

**STEP 1** Analyze all substituents and determine which ring is more activated.

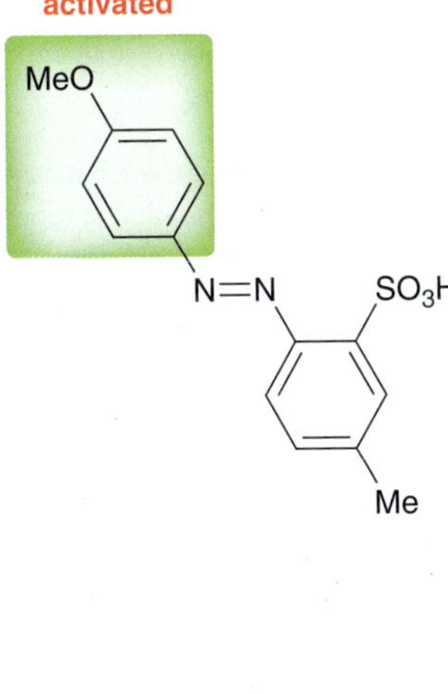


**STEP 2** Using a retrosynthetic analysis, identify the starting nucleophile and electrophile.

MeO

Nucleophile

N≡N⊕ SO3H

Me

Electrophile

**STEP 3** Identify the starting aniline that is necessary in order to make the diazonium salt.

N≡N⊕ SO3H

Me

⇓

H2N SO3H

Me

**STEP 4** Show all reagents.

H2N SO3H

Me

1) $NaNO_2$, HCl

2) MeO-benzene

MeO

N=N SO3H

Me

Try Problems 23.31–23.33, 23.68, 23.73d

## PRACTICE PROBLEMS

Note: Most of the Problems are available within **WileyPLUS**, an online teaching and learning solution.

**23.38** Spermine is a naturally occurring compound that contributes to the characteristic odor of semen. Classify each nitrogen atom in spermine as primary, secondary, or tertiary.

$H_2N$ N(H) N(H) $NH_2$

**Spermine**

**23.39** Clomipramine is marketed under the trade name Anafranil and is used in the treatment of obsessive compulsive disorder.

Cl N N

**Clomipramine**

(a) Identify which nitrogen atom in clomipramine is more basic and justify your choice.

(b) Draw the form of clomipramine that is expected to predominate at physiological pH.

**23.40** Cinchocaine is a long-acting local anesthetic used in spinal anesthesia. Identify the most basic nitrogen atom in cinchocaine.

O H N N N O

**Cinchocaine**

**23.41** For each pair of compounds, identify the stronger base.

(a) pyridine *vs.* trimethylamine

(b) pyridine *vs.* N,N-dimethylacetamide

(c) HN, H N *vs.* HN, NH

**23.42** Draw the structure of each of the following compounds:

(a) *N*-Ethyl-*N*-isopropylaniline

(b) *N*,*N*-Dimethylcyclopropylamine

(c) (2*R*,3*S*)-3-(*N*,*N*-Dimethylamino)-2-pentanamine

(d) Benzylamine

**23.43** Consider the structure of lysergic acid diethylamide (LSD), a potent hallucinogen containing three nitrogen atoms. One of these three nitrogen atoms is significantly more basic than the other two. Identify the most basic nitrogen atom in LSD and explain your choice.

**LSD**

**23.44** Identify the number of chirality centers in each of the following structures:

(a) (b) (c)

**23.45** Assign a name for each of the following compounds:

(a) (b) (c) (d) (e) (f)

**23.46** Draw all constitutional isomers with molecular formula $C_4H_{11}N$ and provide a name for each isomer.

**23.47** Draw all tertiary amines with molecular formula $C_5H_{13}N$ and provide a name for each isomer. Are any of these compounds chiral?

**23.48** Each pair of compounds below will undergo an acid-base reaction. In each case, identify the acid, identify the base, draw curved arrows that show the transfer of a proton, and draw the products.

(a) (b)

**23.49** Draw the structure of the major product obtained when aniline is treated with each of the following reagents:

(a) Excess $Br_2$

(b) $PhCH_2COCl$, py

(c) Excess methyl iodide

(d) $NaNO_2$ and HCl followed by $H_3PO_2$

(e) $NaNO_2$ and HCl followed by CuCN

**23.50** Identify how you would make each of the following compounds from 1-hexanol:

(a) Hexylamine (b) Heptylamine (c) Pentylamine

**23.51** Identify how you would make hexylamine from each of the following compounds:

(a) 1-Bromohexane (b) 1-Bromopentane

(c) Hexanoic acid (d) 1-Cyanopentane

**23.52** Tertiary amines with three different alkyl groups are chiral but cannot be resolved because pyramidal inversion causes racemization at room temperature. Nevertheless, chiral aziridines can be resolved and stored at room temperature. Aziridine is a three-membered heterocycle containing a nitrogen atom. The following is an example of a chiral aziridine. In this compound, the nitrogen atom is a chirality center. Suggest a reason why chiral aziridines do not undergo racemization at room temperature.

**23.53** Lidocaine is one of the most widely used local anesthetics. Draw the form of lidocaine that is expected to predominate at physiological pH.

**Lidocaine**

**23.54** Propose a mechanism for the following transformation:

$$H_2C{=}O + NH_3 \xrightarrow[NaBH_3CN]{[H^+]} CH_3NH_2$$

**23.55** When aniline is treated with fuming sulfuric acid, an electrophilic aromatic substitution reaction takes place at the *meta* position instead of the *para* position, despite the fact that the amino group is an *ortho-para* director. Explain this curious result.

Fuming $H_2SO_4$

**23.56** *para*-Nitroaniline is an order of magnitude less basic than *meta*-nitroaniline.

(a) Explain the observed difference in basicity.

(b) Would you expect the basicity of *ortho*-nitroaniline to be closer in value to *meta*-nitroaniline or to *para*-nitroaniline?

**23.57** Methadone is a powerful analgesic that is used to suppress withdrawal symptoms in the rehabilitation of heroin addicts. Identify the major product that is obtained when methadone is subjected to a Hofmann elimination.

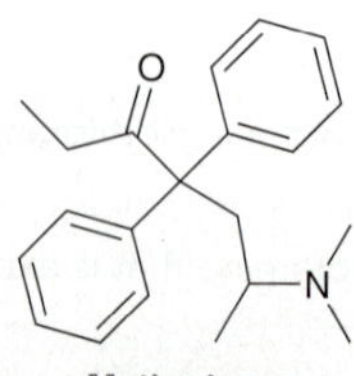

**Methadone**

**23.58** In general, nitrogen atoms are more basic than oxygen atoms. However, when an amide is treated with a strong acid, such as sulfuric acid, it is the oxygen atom of the amide that is protonated, rather than the nitrogen atom. Explain this observation.

**23.59** Propose a synthesis for each of the following transformations:

(a)

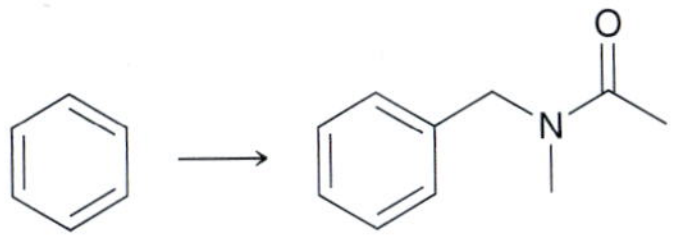

(b) COOH

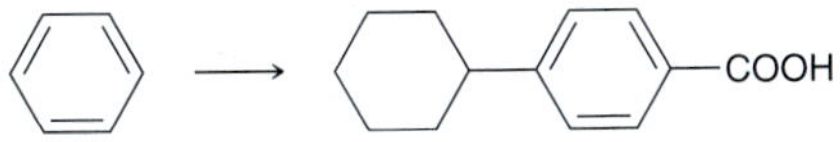

**23.60** Draw a mechanism for the last step of the Gabriel synthesis, performed under basic conditions.

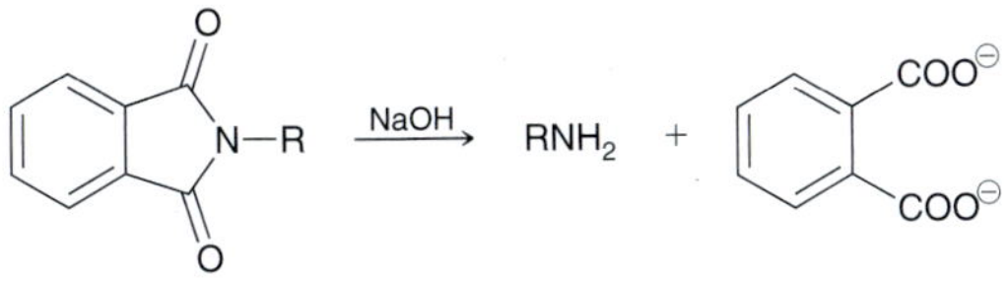

N—R $\xrightarrow{\text{NaOH}}$ $RNH_2$ + $COO^{\ominus}$ $COO^{\ominus}$

**23.61** One variation of the Gabriel synthesis employs hydrazine to free the amine in the final step of the synthesis. Draw the by-product obtained in this process.

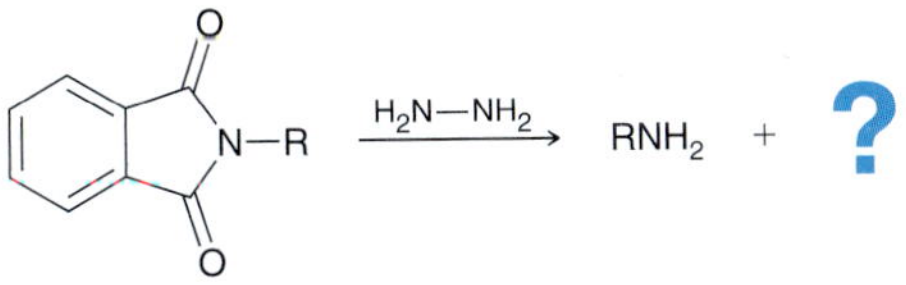

N—R $\xrightarrow{H_2N—NH_2}$ $RNH_2$ + ?

**23.62** Predict the major product for each of the following reactions:

(a) 1) Excess MeI 2) $Ag_2O$, $H_2O$, heat ?

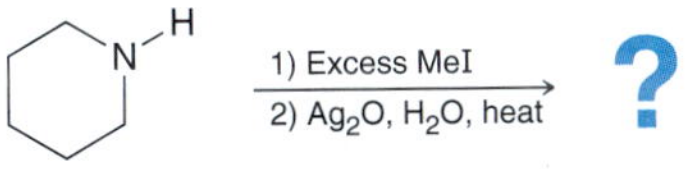

(b) 1) KOH 2) EtBr 3) $H_2NNH_2$ ?

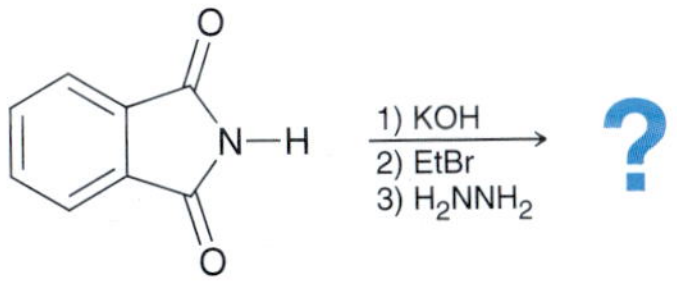

(c) Br 1) NaCN, DMSO 2) $H_3O^+$, heat 3) $SOCl_2$ 4) excess $NH_3$ ?

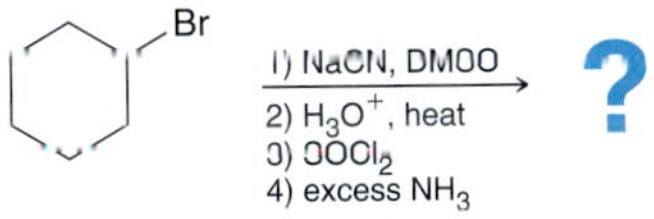

(d) 1) $HNO_3$, $H_2SO_4$ 2) Fe, $H_3O^+$ 3) $NaNO_2$, HCl 4) CuCN ?

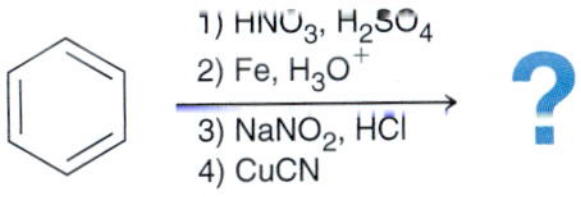

**23.63** Fill in the missing reagents:

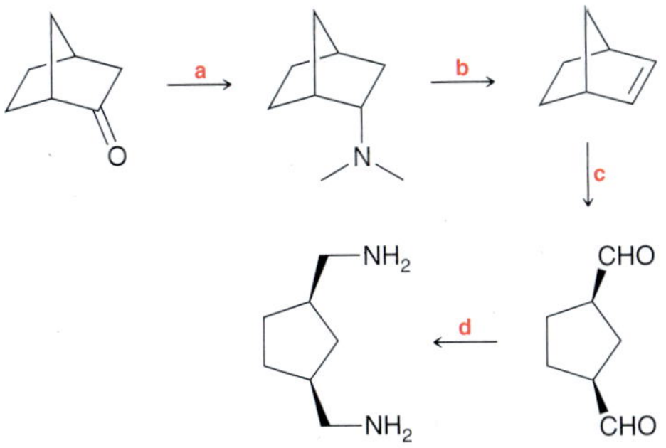

a, b, c, d; N; $NH_2$; $NH_2$; CHO; CHO

**23.64** In this chapter, we explained why pyrrole is such a weak base, but we did not discuss the acidity of pyrrole. In fact, pyrrole is 20 orders of magnitude more acidic than most simple amines. Draw the conjugate base of pyrrole and explain its relatively high acidity.

**23.65** Rimantadine is an antiviral drug used to treat people infected with life-threatening influenza viruses. Identify the starting ketone that would be necessary in order to prepare rimantadine via a reductive amination.

$NH_2$

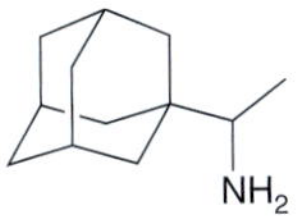

**Rimantadine**

**23.66** Benzphetamine is an appetite suppressant that is marketed under the trade name Didrex and used in the treatment of obesity. Identify at least two different ways to make benzphetamine via a reductive amination process.

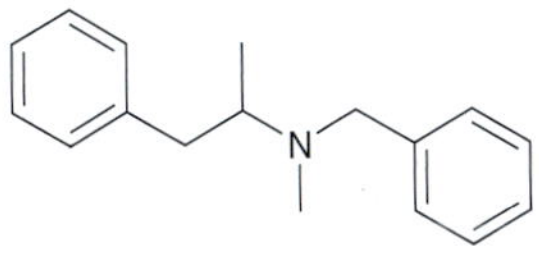

**Benzphetamine**

**23.67** Draw the product formed when each of the following compounds is treated with $NaNO_2$ and HCl:

(a) $NH_2$ (b) H N

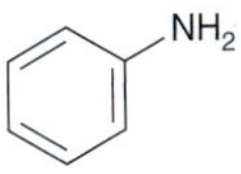

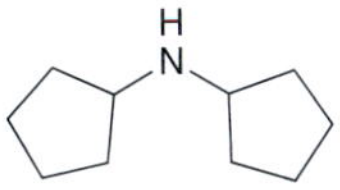

**23.68** Consider the structure of the azo dye called alizarine yellow R (below). Show the reagents you would use to prepare this compound via an azo coupling process.

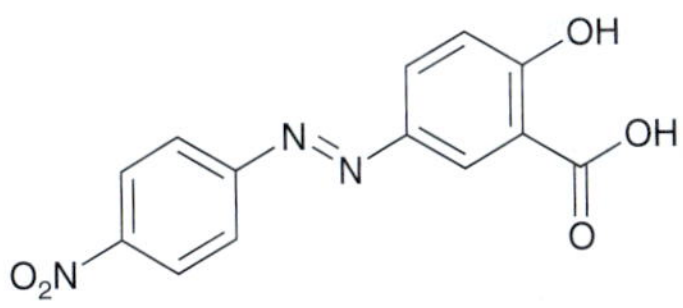

OH; OH; O; N=N; $O_2N$

**23.69** Draw the major product(s) that are expected when each of the following amines is treated with excess methyl iodide and then heated in the presence of aqueous silver oxide.

(a) $NH_2$ (b) $NH_2$

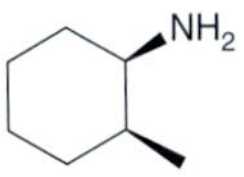

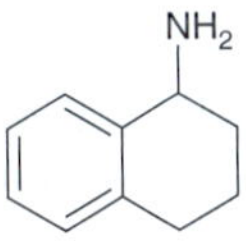

**23.70** Predict the major product(s) for each of the following reactions:

(a) $NO_2$ Br 1) Fe, $H_3O^+$ 2) NaOH ?

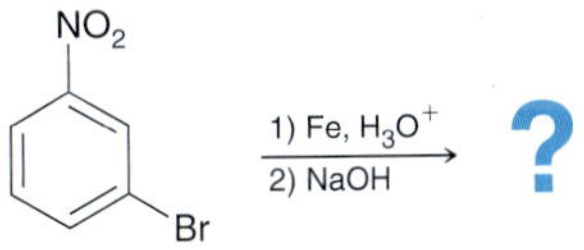

(b) $NH_2$ + O $\xrightarrow[NaBH_3CN]{[H^+]}$ ?

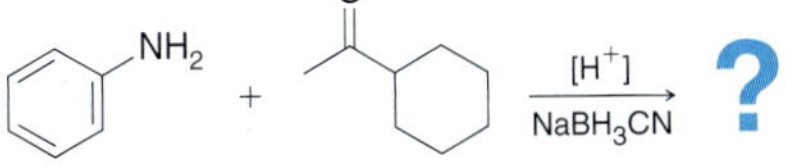

(c) 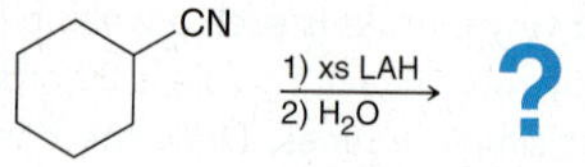 1) xs LAH 2) $H_2O$ ?

(d) 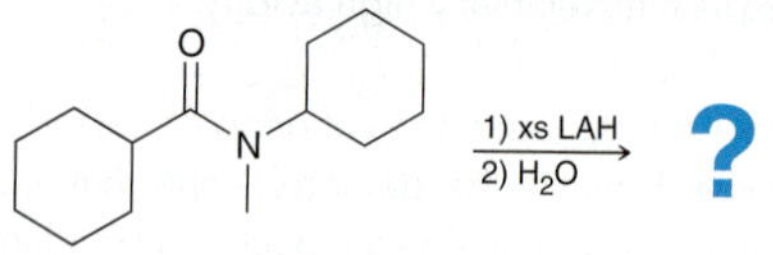 1) xs LAH 2) $H_2O$ ?

**23.71** *meta*-Bromoaniline was treated with $NaNO_2$ and HCl to yield a diazonium salt. Draw the product obtained when that diazonium salt is treated with each of the following reagents:

(a) $H_2O$ (b) $HBF_4$ (c) CuCN (d) $H_3PO_2$ (e) CuBr

**23.72** Draw the expected product of the following reductive amination:

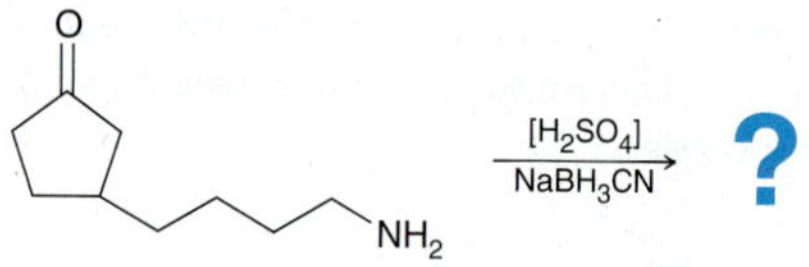

$[H_2SO_4]$ / $NaBH_3CN$ ?

**23.73** Starting with benzene and any reagents with three or fewer carbon atoms, show how you would prepare each of the following compounds:

(a) 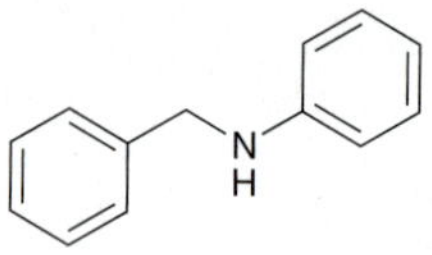 (b)

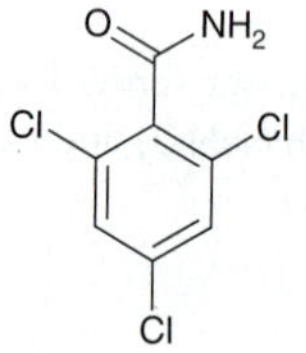

(c) 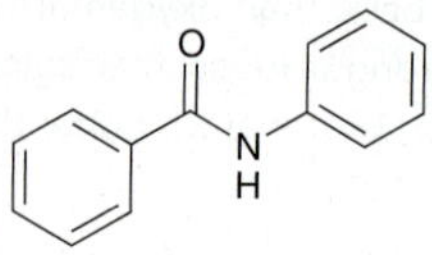 (d)

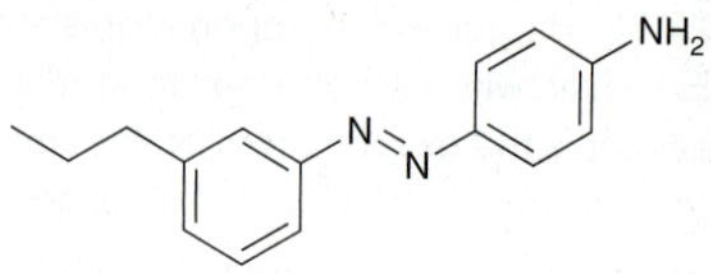

**23.74** Draw the structures of all isomeric amines with molecular formula $C_6H_{15}N$ that are not expected to produce any signal above 3000 $cm^{-1}$ in their IR spectra.

**23.75** When the following compound is treated with excess methyl iodide, a quaternary ammonium salt is obtained that bears only one positive charge. Draw the structure of the quaternary ammonium salt.

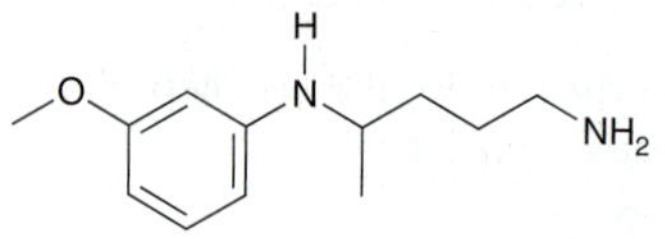

**23.76** Draw a mechanism for the following transformation:

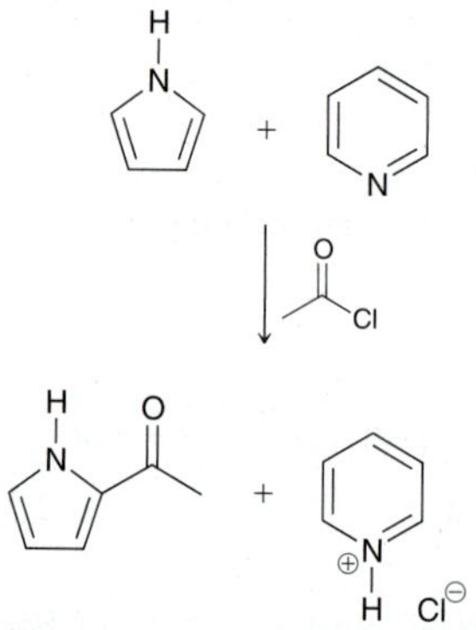

# INTEGRATED PROBLEMS

**23.77** A compound with molecular formula $C_5H_{13}N$ exhibits three signals in its proton NMR spectrum and no signals above 3000 $cm^{-1}$ in its IR spectrum. Draw two possible structures for this compound.

**23.78** Coniine has molecular formula $C_8H_{17}N$ and was present in the hemlock extract used to execute the Greek philosopher Socrates. Subjecting coniine to a Hofmann elimination produces (*S*)-*N*,*N*-dimethyloct-7-en-4-amine. Coniine exhibits one peak above 3000 $cm^{-1}$ in its IR spectrum. Draw the structure of coniine.

**23.79** Piperazine is an antihelminthic agent (a drug used in the treatment of intestinal worms) that has molecular formula $C_4H_{10}N_2$. The proton NMR spectrum of piperazine exhibits two signals. When dissolved in $D_2O$, one of these signals vanishes over time. Propose a structure for piperazine.

**23.80** Primary or secondary amines will attack epoxides in a ring-opening process:

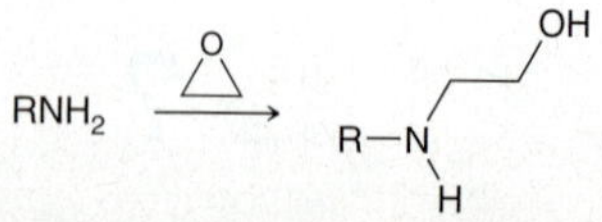

$RNH_2$ → R–NH–$CH_2CH_2$OH

For substituted epoxides, nucleophilic attack generally takes place at the less sterically hindered side of the epoxide. Using this type of reaction, show how you might prepare the following compound from benzene, ammonia, and any other reagents of your choice.

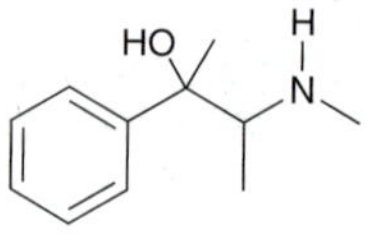

**23.81** Phenacetin was widely used as an analgesic before it was removed from the market in 1983 on suspicion of being a carcinogen. It was widely replaced with acetaminophen (Tylenol), which is very similar in structure but is not carcinogenic. Starting with benzene and using any other reagents of your choice, outline a synthesis of phenacetin.

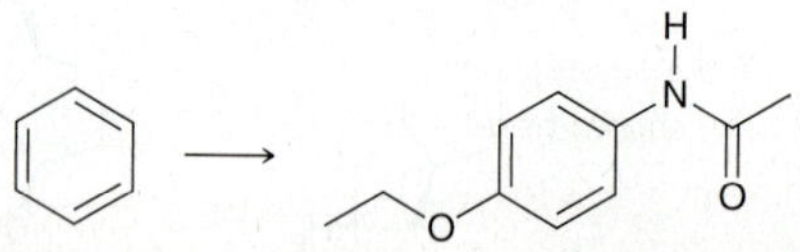

**Phenacetin**

**23.82** Propose a mechanism for the following process:

Heat

$+ N_2 + CO_2$

**23.83** Draw the structure of the compound with molecular formula $C_6H_{15}N$ that exhibits the following $^1H$ NMR and $^{13}C$ NMR spectra:

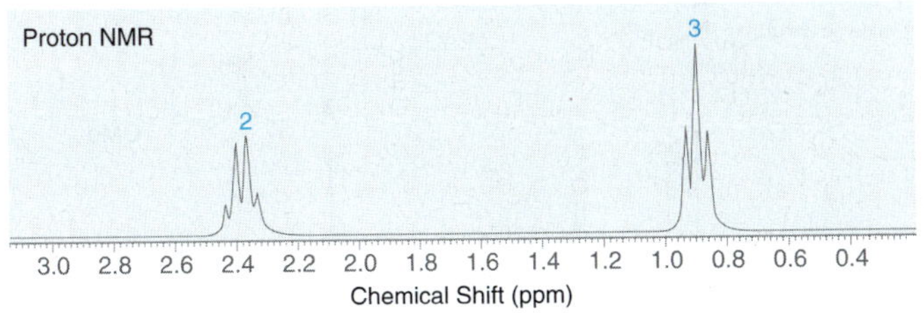


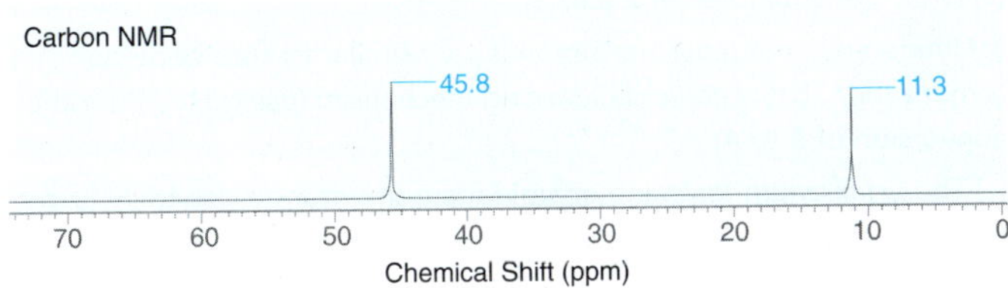


**23.84** Draw the structure of the compound with molecular formula $C_8H_{11}N$ that exhibits the following $^1H$ NMR and $^{13}C$ NMR spectra:

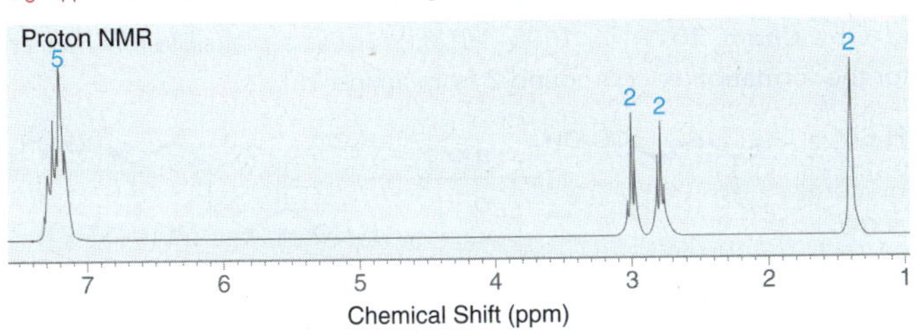


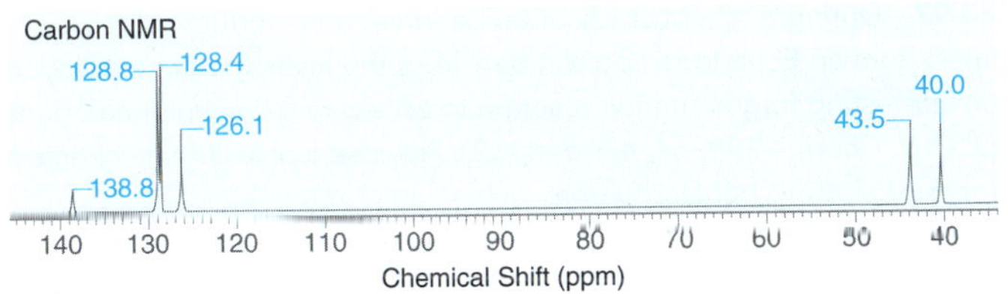


**23.85** Using benzene as your only source of carbon atoms and ammonia as your only source of nitrogen atoms, propose an efficient synthesis for the following compound:

**23.86** Propose an efficient synthesis for the following transformation:

**23.87** When 3-methyl-3-phenyl-1-butanamine is treated with sodium nitrite and HCl, a mixture of products is obtained. The following compound was found to be present in the reaction mixture. Account for its formation with a complete mechanism (make sure to show the mechanism of formation for a nitrosonium ion).

**23.88** Guanidine lacks a negative charge but is an extremely powerful base. In fact, it is almost as strong a base as a hydroxide ion. Identify which nitrogen atom in guanidine is so basic and explain why guanidine is a much stronger base than most other amines.

**Guanidine**

**23.89** In the first asymmetric synthesis of (−)-(*S*,*S*)-homaline, a symmetrical alkaloid isolated in the early 1970s, a key intermediate was compound **2** (*Tetradhedron Lett.* **2012,** *53,* **1119–1121**). Provide reagents for converting compound **1** into compound **2**.

Two steps

**1** **2**

**23.90** Nicotine is well known for its addictive characteristic in cigarettes. Interestingly, it has also been suggested to have potential therapeutic potential in central nervous system disorders such as Alzheimer's disease, Parkinson's disease, and depression. However, due to the toxic side effects of nicotine, derivatives are being developed, such as compound **2**, which shows much lower toxicity and exhibits analgesic (pain-inhibiting) properties (*Bioorg. Med. Chem. Lett.* **2006,** *16,* **2013–2016**). Suggest an efficient synthesis of compound **2** from compound **1**.

**1** **2**

# CHALLENGE PROBLEMS

**23.91** Halosaline is a naturally occurring alkaloid isolated from the herb *Lobelia inflata*. Its synthesis was recently completed by an azide reduction followed by an amide reduction (*Org. Lett.* **2007,** *9,* **2673–2676**). The first of these steps was a ring contraction that converted a lactone to a lactam, shown here:

$H_2$, Pd

1) xs LAH
2) $H_2O$

**Halosaline**

(a) Provide a structure for the alkaloid halosaline.

(b) Why is it not possible to achieve the transformation above in just one step (rather than two), using LAH to reduce both the azide and the amide, in one reaction flask?

(c) Draw a mechanism to account for the lactone-to-lactam ring contraction.

**23.92** The formal synthesis of quinine in 1944 by Woodward and Doering was a landmark achievement (*J. Am. Chem. Soc.* **1945**, *67*, 860–874). During their synthesis, the following compound was treated with excess methyl iodide, followed by a strong alkali solution of NaOH. Under these conditions, the initial Hofmann elimination product undergoes further conversion to afford a product with molecular formula $C_{10}H_{17}NO_2$. Identify the structure of the product.

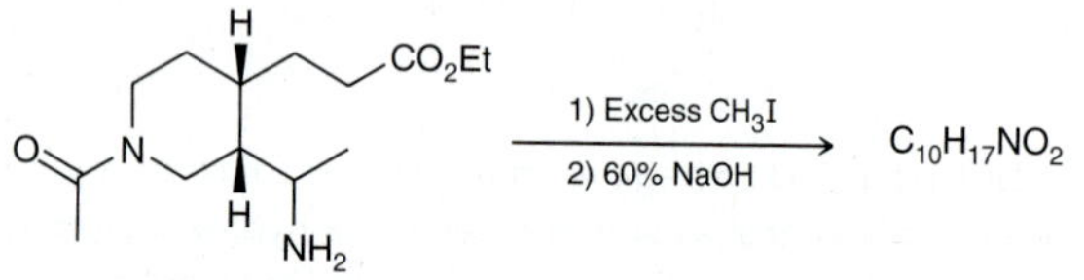

1) Excess $CH_3I$
2) 60% NaOH
→ $C_{10}H_{17}NO_2$

**23.93** Piperazinone derivatives, such as compound **4**, are being investigated for their potential use in the treatment of migraines as well as a variety of other diseases, including hypertension and sepsis. The following two-step synthetic procedure was developed as a general method for producing a variety of piperazinone derivatives (*Tetrahedron Lett.* **2009**, *50*, 3817–3819). Compound **1** is a readily available starting material that can be converted to compound **2** via an $S_N2$ process. Reductive amination of compound **2** affords compound **3** which is short-lived and rapidly undergoes acid-catalyzed amide formation to the cyclic product **4**, a piperazinone derivative.

$NaHCO_3$, DMF; $PhCH_2NH_2$, $NaBH_3CN$, AcOH, heat

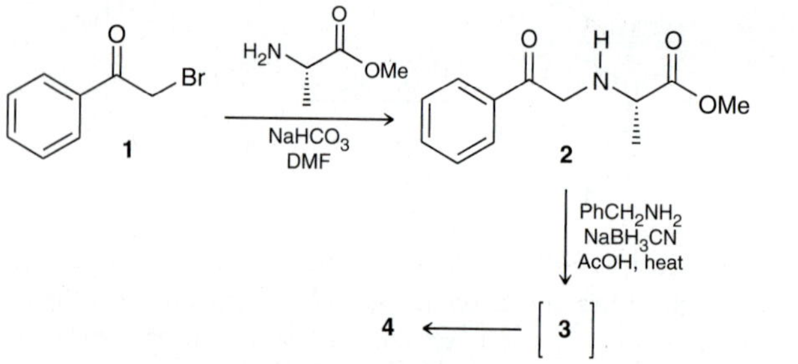

(a) Draw the structure of **3** and explain why it is produced as a diastereomeric mixture.

(b) Draw the structure of **4** and determine whether it will be produced as a mixture of diastereomers. Explain your reasoning.

**23.94** Alkylating agents are compounds capable of transferring an alkyl group to a suitable nucleophile, and they represent an important class of anticancer agents. These agents can transfer an alkyl group to DNA (in a cancer cell), causing structural changes that ultimately lead to cell death. The product of the following transformation, which occurs via two successive substitution reactions, was designed to act as an anticancer alkylating agent (*Tetrahedron Lett.* **2004**, *45*, 5807–5810). Draw a structure for this product.

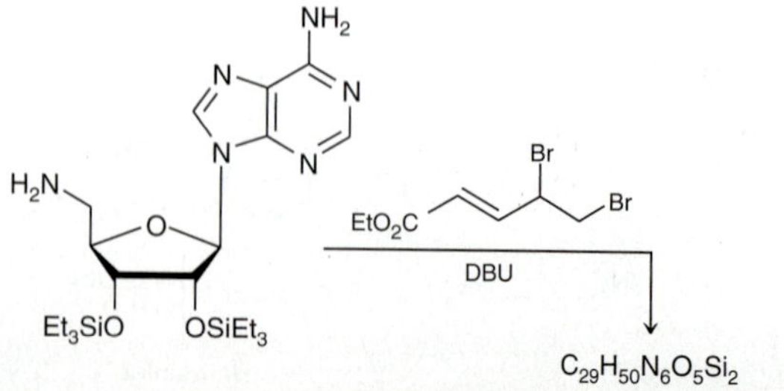

DBU → $C_{29}H_{50}N_6O_5Si_2$

**23.95** Enaminoesters (also called vinylogous carbamates), such as compound **4** below, can serve as building blocks in the synthesis of nitrogen heterocycles. Compound **4** was prepared from compound **1** in a one-pot method (the entire transformation took place in one reaction vessel). Compound **1** was first converted to compound **2**, but in the presence of methyl acetoacetate, compound **2** was further converted into imine **3**, even without the introduction of an acid catalyst. Upon its formation, compound **3** then tautomerized to give enamine **4** (*Tetrahedron Lett.* **2005**, *46*, 979–982).

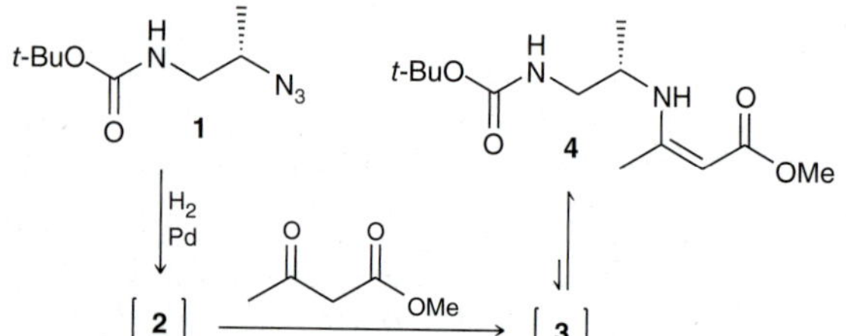

$H_2$, Pd

(a) Draw the structures of **2** and **3**.

(b) Imine-enamine tautomerization is very similar to keto-enol tautomerization. Draw an acid-catalyzed mechanism (using $H_3O^+$) for the conversion of **3** to **4**.

(c) The equilibrium for imine-enamine tautomerization generally favors the imine, but in this case, the enamine is favored. Give at least two reasons why the equilibrium favors enamine **4** over imine **3**.

**23.96** The following reaction was attempted during the total synthesis of crispine A, a cytotoxic alkaloid isolated from a Mongolian thistle (*J. Org. Chem.* **2011**, *76*, 1605–1613). Propose a plausible mechanism for the formation of compound **2** from amine **1**.

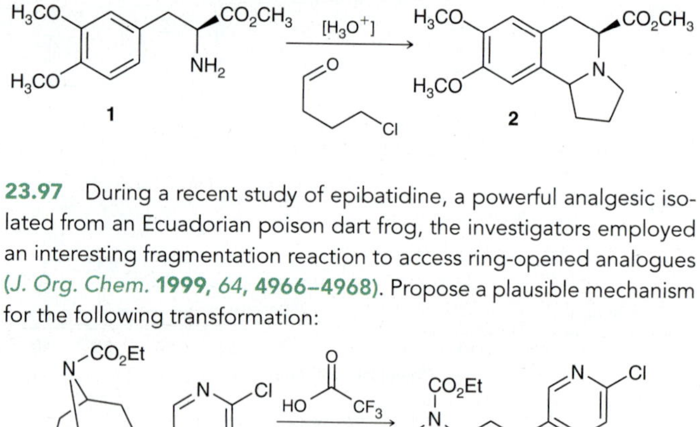

$[H_3O^+]$

**23.97** During a recent study of epibatidine, a powerful analgesic isolated from an Ecuadorian poison dart frog, the investigators employed an interesting fragmentation reaction to access ring-opened analogues (*J. Org. Chem.* **1999**, *64*, 4966–4968). Propose a plausible mechanism for the following transformation:

$NaBH_3CN$

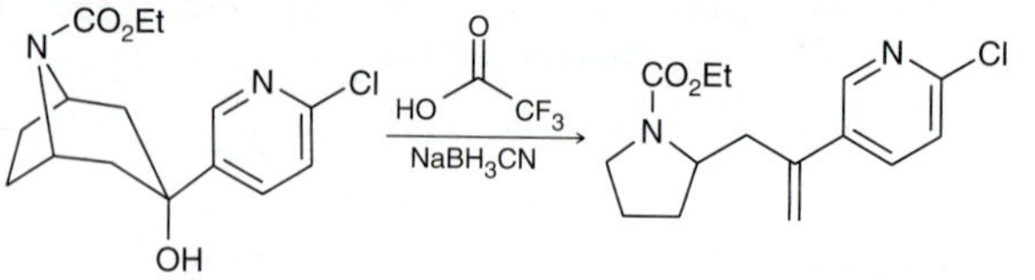

**23.98** Propose a plausible mechanism for the following transformation and justify the stereochemical outcome (*J. Org. Chem.* **1999**, *64*, 4617–4626):

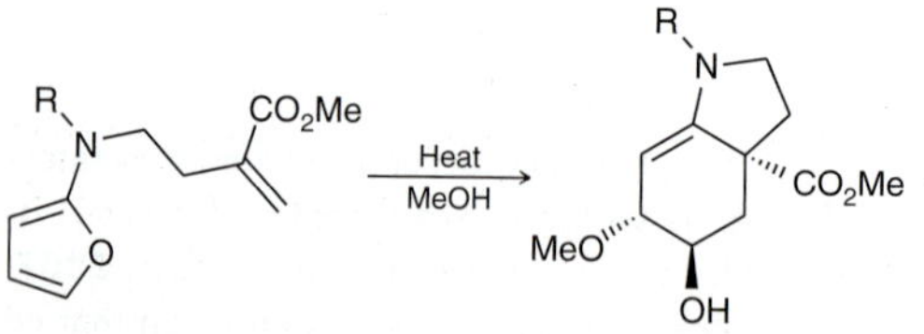

Heat, MeOH

# 24 Carbohydrates



## DID YOU EVER WONDER...
### what Neosporin is and how it works?

Neosporin is an antibiotic cream that contains three active ingredients. One of them, called neomycin, is a carbohydrate derivative. In this chapter, we will investigate carbohydrates and the various roles they play in nature. After an introduction to the structure and reactivity of carbohydrates, we will return to analyze the structure of neomycin and related antibiotics in more detail.

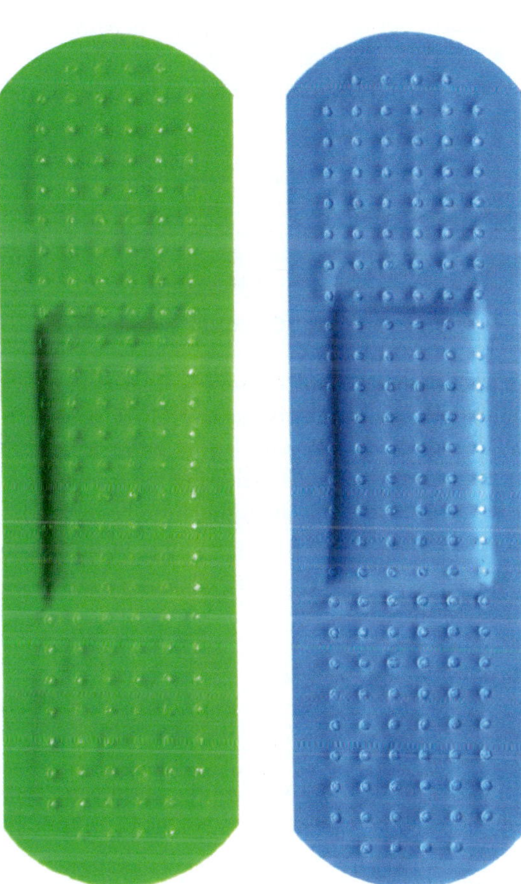

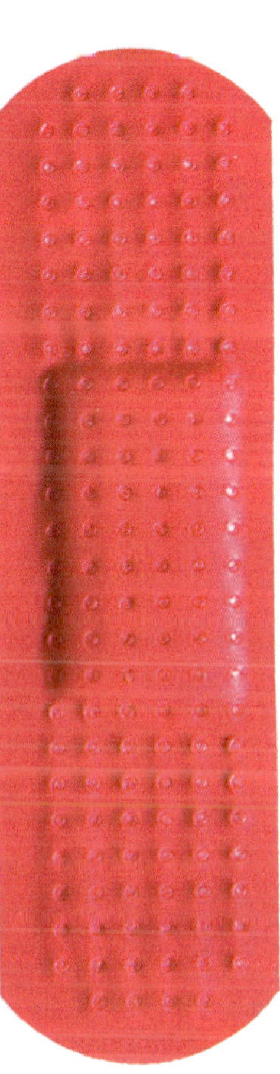

## DO YOU REMEMBER?

*Before you go on, be sure you understand the following topics.*
*If necessary, review the suggested sections to prepare for this chapter:*

- Drawing Chair Conformations (Section 4.11)
- Fischer Projections (Section 5.7)
- Reduction of Aldehydes and Ketones (Section 20.9)
- Haworth Projections (Section 4.14)
- Hemiacetals and Acetals (Section 20.5)

Take the DO YOU REMEMBER? QUIZ in **WileyPLUS** to check your understanding

## 24.1 Introduction to Carbohydrates

**Carbohydrates**, commonly referred to as sugars, are abundant in nature. They represent a significant portion of the foods that we eat, providing us with the energy necessary to drive most of the biochemical processes in our bodies. Carbohydrates are the building blocks used to provide structural rigidity for living organisms, including the wood found in trees and the shells of lobsters. Even our DNA is assembled from derivatives of carbohydrates, as we will see at the end of this chapter.

The earliest carbohydrates to be isolated and purified were originally considered to be hydrates of carbon. For example, glucose was known to have molecular formula $C_6H_{12}O_6$, which could be rearranged as $C_6(H_2O)_6$ to indicate six carbon atoms and six water molecules. As the structure of glucose was elucidated, this view was discarded, but the term *carbohydrate* persisted. Carbohydrates are now understood to be polyhydroxy aldehydes or ketones. For example, consider the structure of naturally occurring glucose. Trees and plants convert carbon dioxide and water into glucose during photosynthesis.

$$6\,CO_2 + 6\,H_2O \xrightarrow{\text{Sunlight}} \text{Glucose} + 6\,O_2$$

Glucose

Energy from the sun is absorbed by vegetation and is used to convert $CO_2$ molecules into larger organic compounds. These organic compounds, such as glucose, have C—C and C—H bonds, which are higher in energy than the C=O bonds in $CO_2$. Our bodies utilize a series of chemical reactions to convert these compounds back into $CO_2$, thereby releasing the stored solar energy. In the process, we return carbon dioxide and water back to the environment to be recycled. In essence, our bodies are powered, in large part, by solar energy that has been stored in the form of glucose molecules.

## 24.2 Classification of Monosaccharides

### Aldoses vs. Ketoses

**LOOKING BACK**
For a review of Fischer projections and the skills necessary to interpret these drawings refer to Section 5.7.

Simple sugars are called **monosaccharides**, a term that derives from the Latin word for sugar, *saccharum*. Complex sugars, such as disaccharides and polysaccharides, are made by joining monosaccharides together, and they will be discussed later in this chapter.

Monosaccharides generally contain multiple chirality centers, and Fischer projections are used to indicate the configuration at each chirality center. Glucose and fructose are two examples of simple sugars.

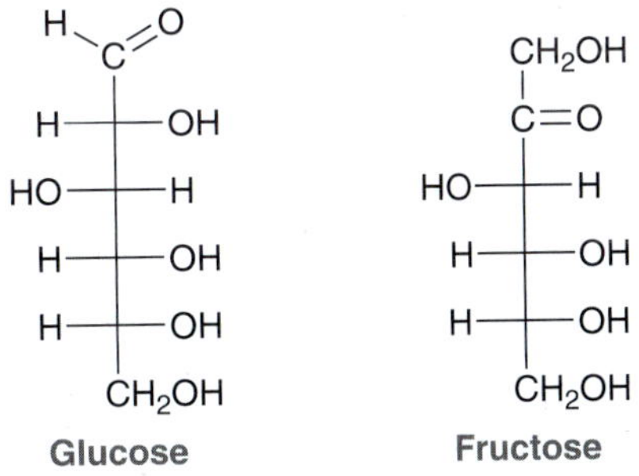

Glucose

Fructose

The suffix *-ose* is used to signify a carbohydrate. Hundreds of different monosaccharides are known, each of which can generally be classified as either an *aldose* or a *ketose*. **Aldoses** contain an aldehyde group, while **ketoses** contain a ketone group. According to this classification scheme, glucose is an aldose and fructose is a ketose.

Aldoses and ketoses can be further classified based on the number of carbon atoms they contain. This is accomplished by inserting a term (tri-, tetr-, pent-, hex-, or hept-) immediately before the suffix *-ose*.

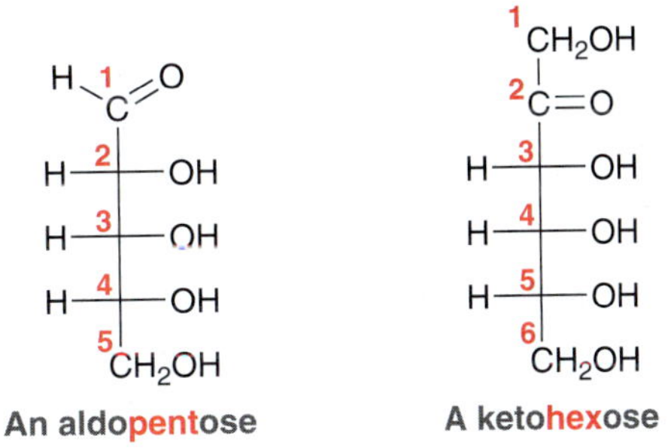

An aldopentose

A ketohexose

The first compound is an aldose with five carbon atoms, and it is therefore called an *aldopentose*. The second compound is a ketose with six carbon atoms, and it is therefore called a *ketohexose*. In this way, carbohydrates are classified using three descriptors:

1. *aldo* or *keto*, indicating whether the compound is an aldehyde or a ketone
2. *tri-*, *tetr-*, *pent-*, *hex-*, or *hept-*, indicating the number of carbon atoms
3. *-ose*, indicating a carbohydrate

## D and L Sugars

Glyceraldehyde is one of the smallest compounds considered to be a carbohydrate. It has only one chirality center and therefore can exist as a pair of enantiomers.

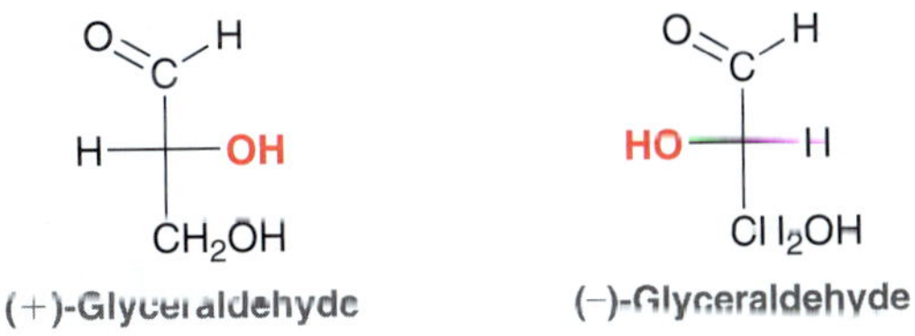

(+)-Glyceraldehyde

(−)-Glyceraldehyde

As discussed in Section 5.4, enantiomers rotate plane-polarized light in opposite directions. One enantiomer rotates plane-polarized light in a clockwise fashion (dextrorotatory) and is designated as (+); the other enantiomer rotates plane-polarized light in a counterclockwise direction (levorotatory) and is designated as (−). Only (+) or dextrorotatory glyceraldehyde is abundant in nature, so glyceraldehyde obtained from natural sources is generally referred to as D-glyceraldehyde. Levorotatory or L-glyceraldehyde can be made in the laboratory, but it is generally not observed in nature.

Early studies with carbohydrates revealed that most naturally occurring carbohydrates can be degraded (broken down) to produce D-glyceraldehyde. For example, degradation of naturally occurring glucose yields D-glyceraldehyde (Figure 24.1). During the degradation process, the carbon

H—C=O
H—OH
HO—H
H—OH
H—OH
$CH_2OH$
Glucose
(naturally occurring)

→ → →

H—C=O
H—OH
$CH_2OH$
D-Glyceraldehyde

**FIGURE 24.1** Degradation of most naturally occurring sugars produces D-glyceraldehyde.

atoms are removed one at a time (from the top of the Fischer projection). The loss of three carbon atoms from naturally occurring glucose yields D-glyceraldehyde. The same observation is made for other naturally occurring carbohydrates.

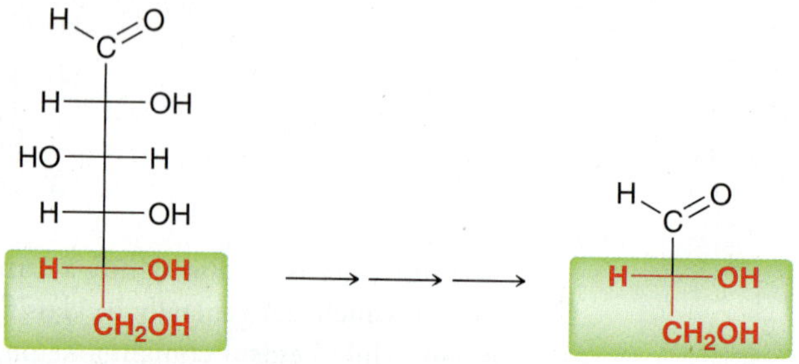

In contrast, when synthetic sugars (those prepared in the laboratory) are degraded, they produce a mixture of D- and L-glyceraldehyde. In response to these observations, chemists began using the Fischer-Rosanoff convention in which the letter D designates any sugar that degrades to (+)-glyceraldehyde. In accord with this convention, almost all naturally occurring carbohydrates are **D sugars**—that is, the chirality center farthest from the carbonyl group will have an OH group pointing to the right in the Fischer projection, as in the following examples:

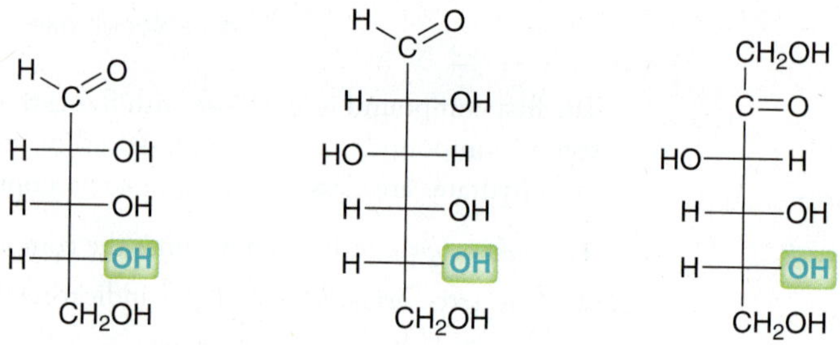

H—C=O
H—OH
H—OH
H—OH
$CH_2OH$
D-Ribose

H—C=O
H—OH
HO—H
H—OH
H—OH
$CH_2OH$
D-Glucose

$CH_2OH$
C=O
HO—H
H—OH
H—OH
$CH_2OH$
D-Fructose

While D-glyceraldehyde is dextrorotatory (by definition), other D sugars are not necessarily dextrorotatory. For example, D-erythrose and D-threose are actually levorotatory.

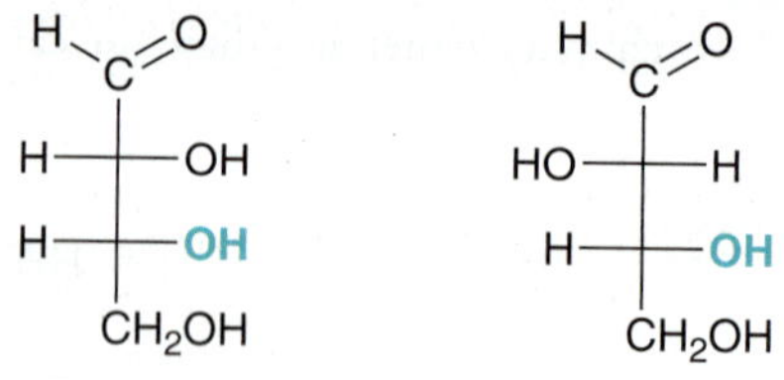

H—C=O
H—OH
H—OH
$CH_2OH$
D-Erythrose
$[\alpha]_D^{27} = -32.7°$

H—C=O
HO—H
H—OH
$CH_2OH$
D-Threose
$[\alpha]_D^{20} = -12.2°$

In this context, the D no longer refers to the direction in which plane-polarized light is rotated. Rather, it means that the lowest chirality center (the one farthest from the carbonyl group) has the *R* configuration just as it does in (+)-glyceraldehyde. Similarly, **L sugars** are not necessarily levorotatory, but rather an L sugar is simply the enantiomer of the corresponding D sugar.

## CONCEPTUAL CHECKPOINT

**24.1** Classify each of the following carbohydrates as an aldose or ketose and then insert the appropriate term to indicate the number of carbon atoms present (e.g., an aldopentose):

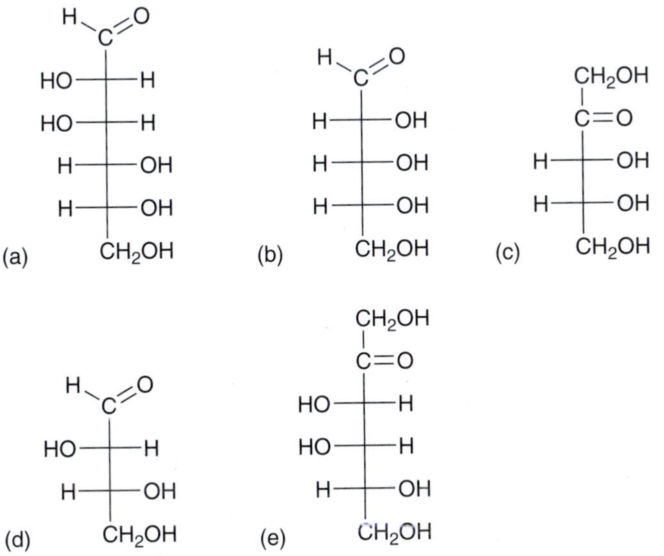

(a) H–C=O; HO–H; HO–H; H–OH; H–OH; $CH_2OH$

(b) H–C=O; H–OH; H–OH; H–OH; $CH_2OH$

(c) $CH_2OH$; C=O; H–OH; H–OH; $CH_2OH$

(d) H–C=O; HO–H; H–OH; $CH_2OH$

(e) $CH_2OH$; C=O; HO–H; HO–H; H–OH; $CH_2OH$

**24.2** Would you expect an aldohexose and a ketohexose to be constitutionally isomeric? Explain why or why not.

**24.3** Determine whether each of the following carbohydrates is a D sugar or an L sugar and assign a configuration for each chirality center. After assigning the configuration for all of the chirality centers, do you notice any trend that would enable you to assign the configuration of a chirality center in a carbohydrate more quickly?

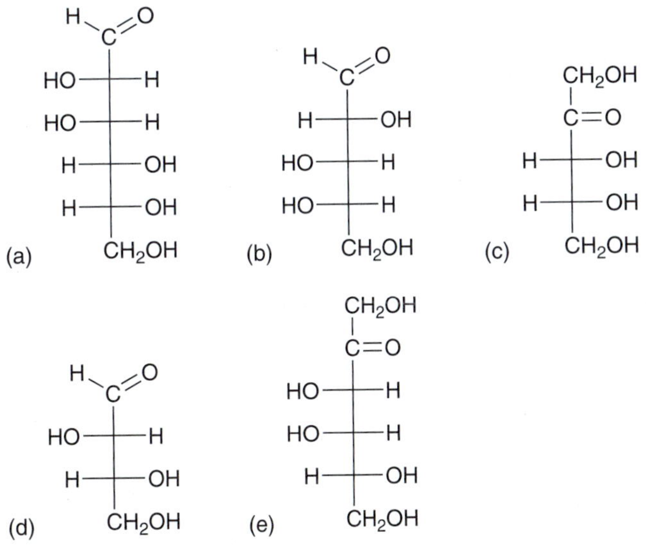

(a) H–C=O; HO–H; HO–H; H–OH; H–OH; $CH_2OH$

(b) H–C=O; H–OH; HO–H; HO–H; $CH_2OH$

(c) $CH_2OH$; C=O; H–OH; H–OH; $CH_2OH$

(d) H–C=O; HO–H; H–OH; $CH_2OH$

(e) $CH_2OH$; C=O; HO–H; HO–H; H–OH; $CH_2OH$

**24.4** D-Allose is an aldohexose in which all four chirality centers have the *R* configuration. Draw a Fischer projection of each of the following compounds:

(a) D-Allose (b) L-Allose

**24.5** There are only two stereoisomeric ketotetroses.

(a) Draw both of them.

(b) Identify their stereoisomeric relationship.

(c) Identify which is a D sugar and which is an L sugar.

**24.6** There are four stereoisomeric aldotetroses.

(a) Draw all four and arrange them in pairs of enantiomers.

(b) Identify which stereoisomers are D sugars and which are L sugars.

## 24.3 Configuration of Aldoses

### Aldotetroses

Aldotetroses have two chirality centers, and there are only four possible aldotetroses (two pairs of enantiomers). Two of the possible aldotetroses are D sugars, while the other two are L sugars. The D sugars are called D-erythrose and D-threose (Figure 24.2). The L sugars, called L-erythrose and L-threose, are the enantiomers of the D aldotetroses.

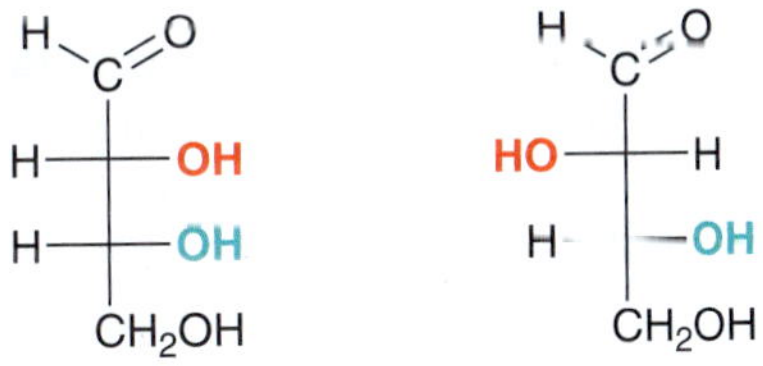

D-Erythrose: H–C=O; H–OH; H–OH; $CH_2OH$

D-Threose: H–C=O; HO–H; H–OH; $CH_2OH$

**FIGURE 24.2** The structures of the D aldotetroses.

## Aldopentoses

**LOOKING BACK**
For a review of the $2^n$ rule, see Section 5.5.

Aldopentoses have three chirality centers, which means that there are eight ($2^3$) possible aldopentoses (four pairs of enantiomers). Four of the possible aldopentoses are D sugars, while the other four are L sugars. Figure 24.3 shows the D sugars. D-Ribose is a key building block of RNA, as we will see at the end of this chapter. D-Arabinose is produced by most plants, and D-xylose is found in wood.

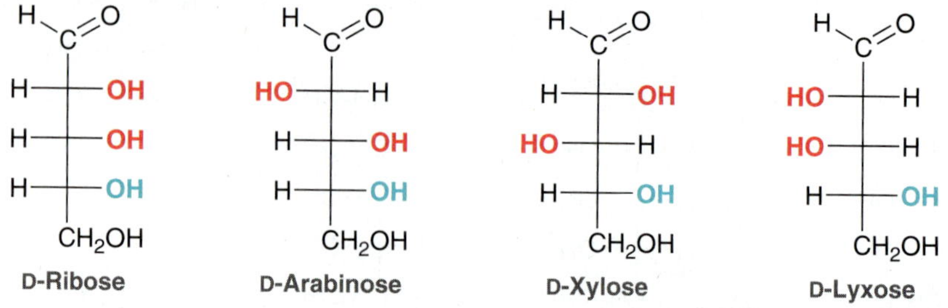

**D-Ribose** **D-Arabinose** **D-Xylose** **D-Lyxose**

**FIGURE 24.3** The structures of the D-aldopentoses.

## Aldohexoses

Aldohexoses have four chirality centers, giving rise to $2^4$ (or 16) possible stereoisomers. These 16 stereoisomers can be grouped into 8 pairs of enantiomers, where each pair consists of one D sugar and one L sugar. Since D sugars are observed in nature, let's focus on the D sugars. There are eight of them (Figure 24.4).

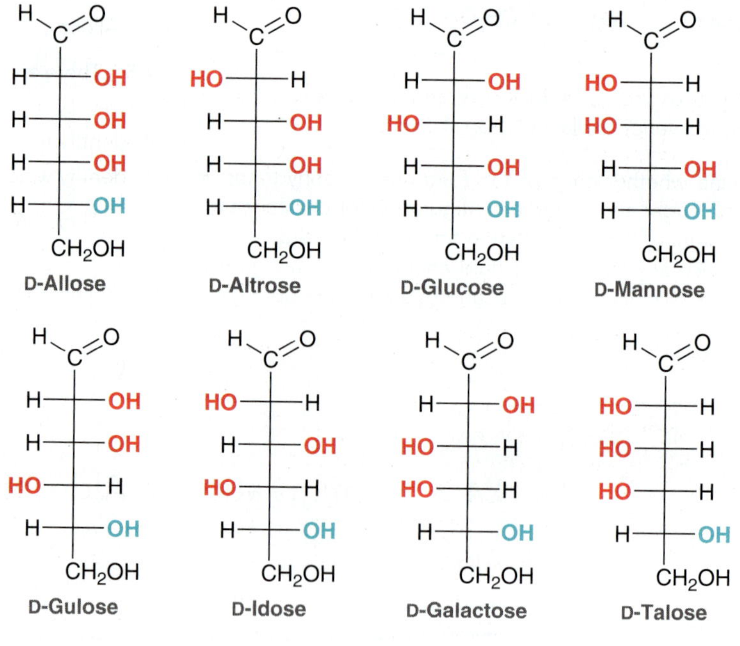

**D-Allose** **D-Altrose** **D-Glucose** **D-Mannose**

**D-Gulose** **D-Idose** **D-Galactose** **D-Talose**

**FIGURE 24.4** The structures of the D-aldohexoses.

Of the eight D aldohexoses, D-glucose is the most common and most important, and the rest of this chapter will focus primarily on D-glucose. Its structure should be at your fingertips.

Figure 24.5 summarizes the family of D aldoses discussed in this section. The family tree is constructed in the following way: Starting with D-glyceraldehyde, a new chirality center is inserted just below the carbonyl group, generating two possible aldotetroses. A new chirality center is then inserted just below the carbonyl group of each aldotetrose, generating four possible aldopentoses. In a similar way, each of the aldopentoses leads to two aldohexoses, giving a total of eight aldohexoses. Each compound in Figure 24.5 has a corresponding L enantiomer that is not shown.

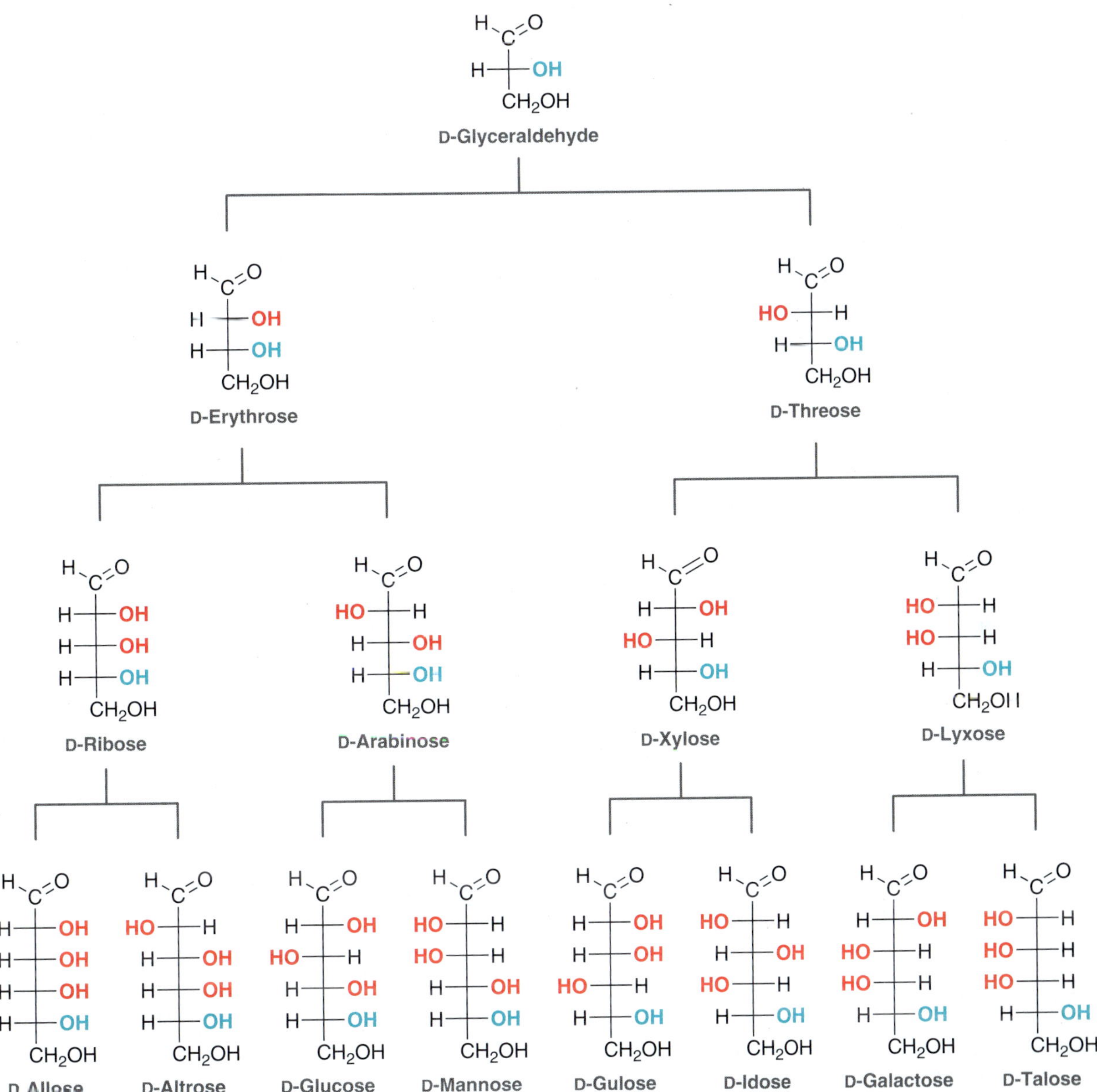


**FIGURE 24.5**
A family tree of the D-aldoses that have between three and six carbon atoms.

## 24.4 Configuration of Ketoses

Figure 24.6 summarizes the family of D ketoses that have between three and six carbon atoms. This family tree is constructed in much the same way as the family tree of aldoses, beginning with dihydroxyacetone as the parent. However, because the carbonyl group is at C2 rather than C1, ketoses have one less chirality center than aldoses of the same molecular formula. As a result, there are only four D ketohexoses, rather than eight. The most common naturally occurring ketose is D-fructose. Each compound in Figure 24.6 has a corresponding L enantiomer that is not shown.

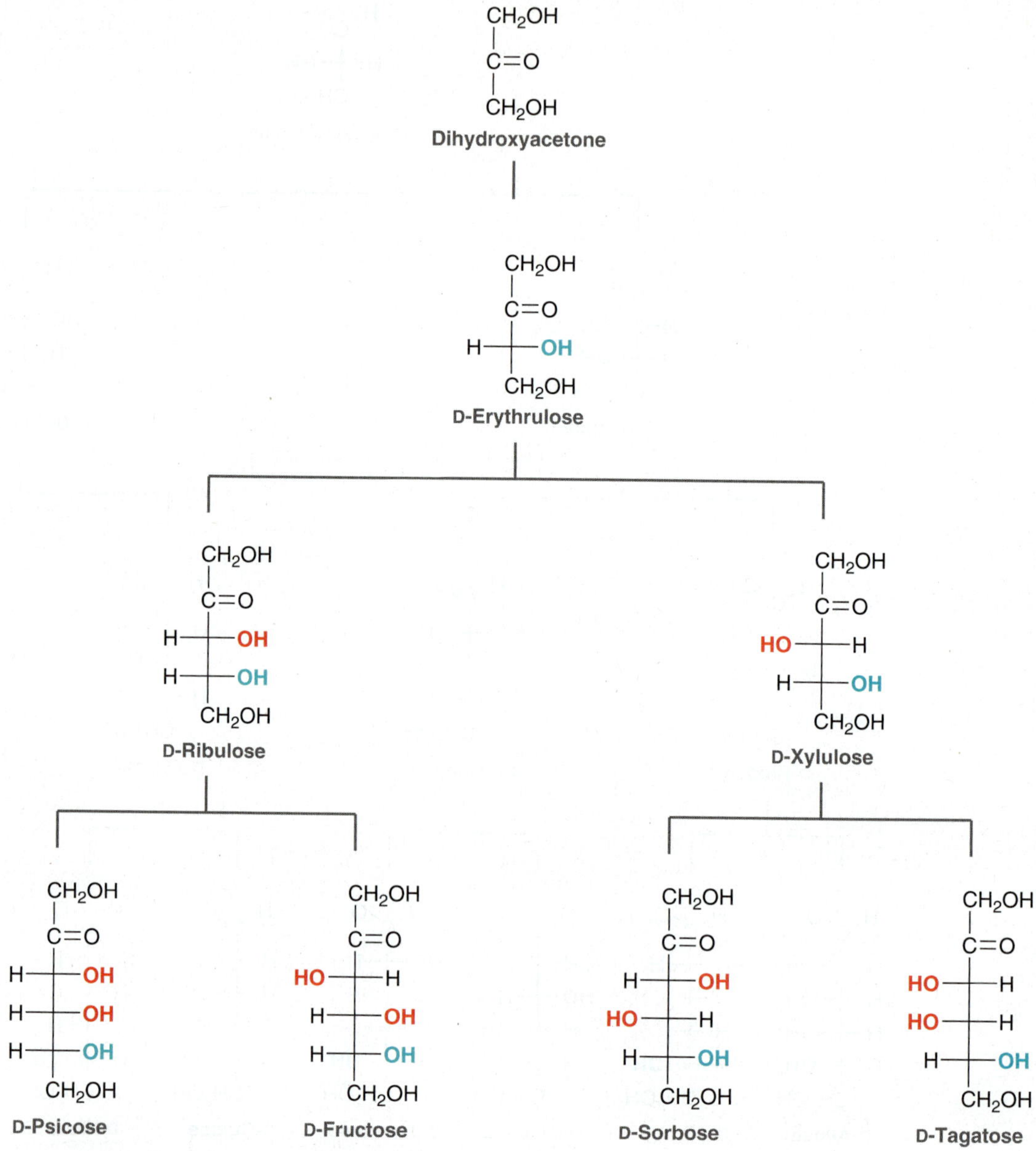

$CH_2OH$–C=O–$CH_2OH$

**Dihydroxyacetone**

$CH_2OH$, C=O, H–C–**OH**, $CH_2OH$

**D-Erythrulose**

$CH_2OH$, C=O, H–C–**OH**, H–C–**OH**, $CH_2OH$

**D-Ribulose**

$CH_2OH$, C=O, **HO**–C–H, H–C–**OH**, $CH_2OH$

**D-Xylulose**

$CH_2OH$, C=O, H–C–**OH**, H–C–**OH**, H–C–**OH**, $CH_2OH$

**D-Psicose**

$CH_2OH$, C=O, **HO**–C–H, H–C–**OH**, H–C–**OH**, $CH_2OH$

**D-Fructose**

$CH_2OH$, C=O, H–C–**OH**, **HO**–C–H, H–C–**OH**, $CH_2OH$

**D-Sorbose**

$CH_2OH$, C=O, **HO**–C–H, **HO**–C–H, H–C–**OH**, $CH_2OH$

**D-Tagatose**

**FIGURE 24.6**
A family tree of the D ketoses that have between three and six carbon atoms.

## CONCEPTUAL CHECKPOINT

**24.7** Draw and name the enantiomer of D-fructose.

**24.8** Which of the following terms best describes the relationship between D-fructose and D-glucose? Explain your choice.

(a) Enantiomers

(b) Diastereomers

(c) Constitutional isomers

# 24.5 Cyclization of Monosaccharides

## Cyclization of Hydroxyaldehydes

Recall from Section 20.5 (Mechanism 20.5) that an aldehyde can react with an alcohol in the presence of an acid catalyst to produce a hemiacetal.

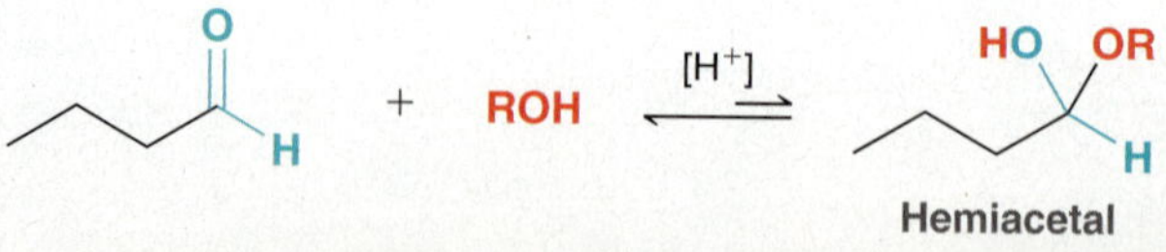

aldehyde (O, H) + ROH $\underset{}{\overset{[H^+]}{\rightleftharpoons}}$ HO, OR, H

**Hemiacetal**

We saw that the equilibrium does not favor formation of the hemiacetal. However, when the aldehyde group and the hydroxyl group are contained in the same molecule, an intramolecular process can occur to form a cyclic hemiacetal with a more favorable equilibrium constant.

$[H^+]$

Cyclic hemiacetal

Six-membered rings are relatively strain free, and the equilibrium favors formation of the cyclic hemiacetal. This type of reaction is characteristic of bifunctional compounds containing both a hydroxyl group and a carbonyl group (an aldehyde or ketone group). When drawing the hemiacetal of a hydroxyaldehyde or hydroxyketone, it is crucial to keep track of all carbon atoms. It is a common mistake to draw the hemiacetal with either too many or too few carbon atoms. The following exercises are designed to help you avoid those mistakes.

# SKILLBUILDER

## 24.1 DRAWING THE CYCLIC HEMIACETAL OF A HYDROXYALDEHYDE

**LEARN the skill**

Draw the cyclic hemiacetal that is formed when the following hydroxyaldehyde undergoes cyclization under acidic conditions:

**SOLUTION**

This compound contains both a hydroxyl group and an aldehyde group, so it is expected to form a hemiacetal. We must first determine the size of the ring that is formed. In order to account for all carbon atoms and to avoid drawing an incorrect ring, it is best to assign numbers to the carbon atoms so we can keep track of them. Begin with the carbon atom of the aldehyde group and continue until reaching the carbon atom connected to the OH group.

**STEP 1** Using a numbering system, determine the size of the ring that is formed.

Begin counting here

For now, we will not focus on the methyl groups as we will account for them after drawing the proper size ring. In this case, there are four carbon atoms that will be incorporated into the ring. The ring is formed when the oxygen atom of the OH group attacks the carbonyl group , and as a result, the oxygen atom of the OH group must also be incorporated into the ring—that is, the ring will be comprised of four carbon atoms and one oxygen atom, generating a five-membered ring:

**STEP 2** Draw the ring, making sure to incorporate an oxygen atom in the ring.

Notice that the first carbon atom (the C1 position) is connected to the oxygen atom of the ring as well as to the newly formed OH group. In other words, the hemiacetal group is located at C1. This numbering system not only helps with drawing the appropriate ring, but it also facilitates the proper placement of the substituents. In this case, there are two methyl groups, both attached to the C3 position.

**STEP 3** Place the substituents.

**PRACTICE the skill** **24.9** Draw the cyclic hemiacetal that is formed when each of the following bifunctional compounds is treated with aqueous acid:

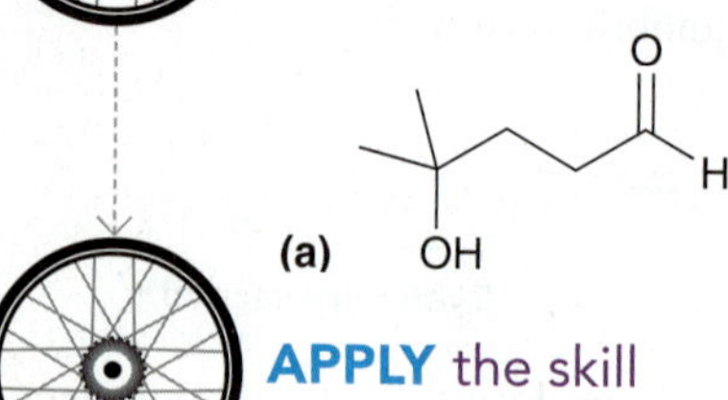

**(a)** **(b)** **(c)** **(d)**

**APPLY the skill** **24.10** Identify the hydroxyaldehyde that will cyclize under acidic conditions to give the following hemiacetal:

**24.11** The following compound has one aldehyde group and two OH groups. Under acidic conditions, either one of the OH groups can function as a nucleophile and attack the carbonyl group, giving rise to two possible ring sizes.

**(a)** Ignoring stereochemistry (for now), draw both possible rings.

**(b)** In Section 4.9, we discussed the ring strain associated with different size rings. Based on those concepts, predict which cyclic hemiacetal will be favored in this case.

need more **PRACTICE?** **Try Problems 24.46, 24.47**

## Pyranose Forms of Monosaccharides

**LOOKING BACK**
For an introduction to Haworth projections, see Section 2.6.

We have seen that a compound containing both an OH group and an aldehyde group will undergo an intramolecular process to form a cyclic hemiacetal, with a favorable equilibrium constant. A similar equilibrium between open and closed forms is observed for carbohydrates. For example, D-glucose can exist in both open and closed forms. The following Haworth projections indicate the configuration at each chirality center:

Open form ⇌ $[H^+]$ Closed form

In the cyclic form, the ring is called a **pyranose ring**, named after pyran, a simple compound that possesses a six-membered ring with an oxygen atom incorporated in the ring. This equilibrium greatly favors formation of the cyclic hemiacetal. That is, D-glucose exists almost exclusively in the cyclic hemiacetal form called D-glucopyranose, and only a trace amount of the open-chain form is present at equilibrium.

When D-glucose is closed into a cyclic hemiacetal, two different stereoisomers of D-glucopyranose can be formed (Figure 24.7). The aldehydic carbon (C1) becomes a chirality center, and both configurations for that new chirality center are possible. The C1 position is called the **anomeric carbon**, and the stereoisomers are called **anomers**. The two anomers are designated as α-D-glucopyranose and β-D-glucopyranose. In the **α anomer**, the hydroxyl group at the anomeric position is *trans* to the $CH_2OH$ group, while in the **β anomer**, the hydroxyl group is *cis* to the $CH_2OH$ group (Figure 24.8). An equilibrium is established between α-D-glucopyranose and β-D-glucopyranose with the equilibrium favoring the β anomer (63% β and 37% α). The open-chain form exhibits only a minimal presence at equilibrium (<0.01%), but it is important nonetheless, because it serves as the intermediate for equilibration between the two anomers.

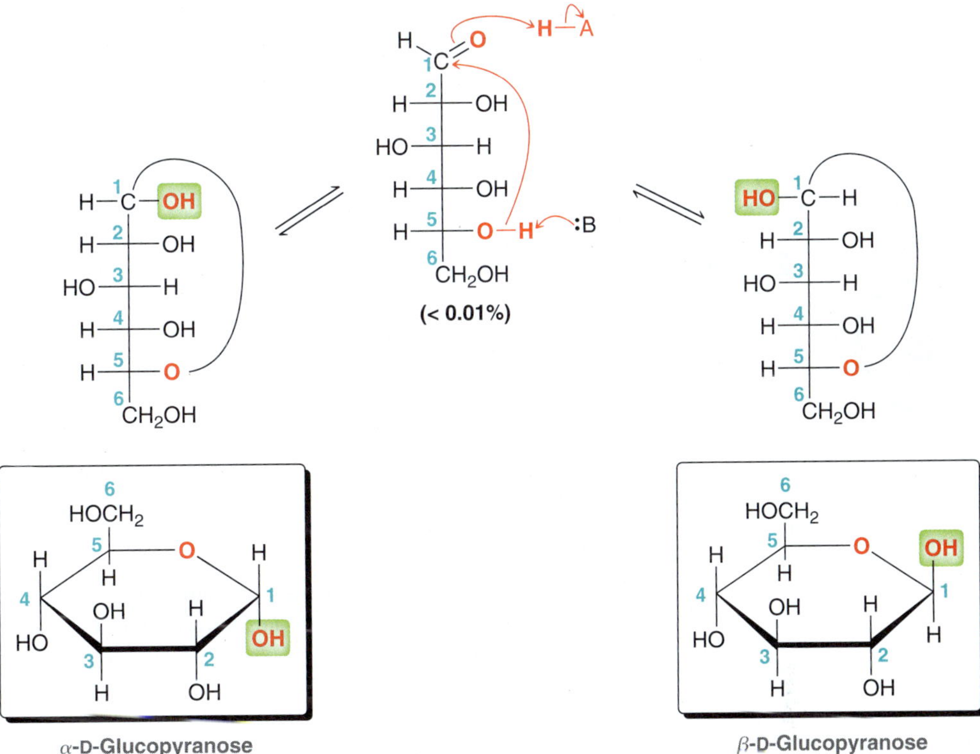


**FIGURE 24.7** Formation of the two possible cyclic hemiacetals of D-glucose.

As we might expect, α-D-glucopyranose and β-D-glucopyranose have different physical properties because their relationship is diastereomeric. For example, compare the specific rotation for these two compounds:

α-D-Glucopyranose
$[\alpha]_D = +112.2°$

β-D-Glucopyranose
$[\alpha]_D = +18.7°$

When the pure α anomer is dissolved in water, it begins to equilibrate with the β anomer, resulting in a mixture that ultimately achieves the expected equilibrium concentrations. At first, the pure anomer exhibits a specific rotation of +112°, but as it equilibrates with the β anomer, the specific rotation changes until the equilibrium is established, and the specific rotation is measured to be +52.6°. A similar result is achieved when the pure β anomer is dissolved in water. At first, the pure anomer has a specific rotation of +18.7°, but as the mixture equilibrates, the specific rotation changes and ultimately is measured to be +52.6°. This phenomenon is called **mutarotation**, a term commonly used to describe the fact that α and β anomers can equilibrate via the open-chain form. Mutarotation occurs more rapidly in the presence of catalytic acid or base.

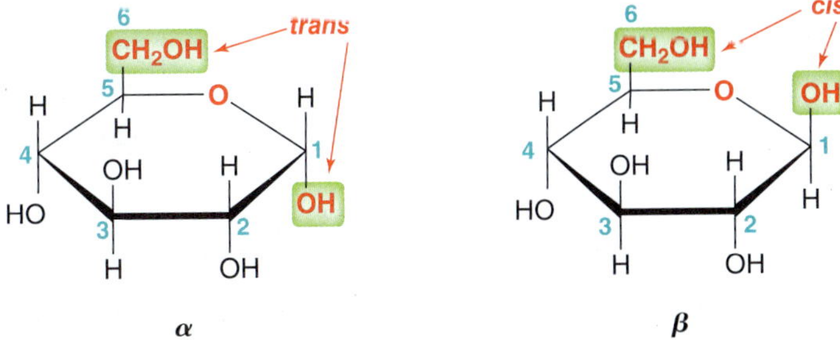


**FIGURE 24.8** The α and β anomers are distinguished based on the relative position of the OH group at the anomeric position and the $CH_2OH$ group at C5.

It is not always possible to determine by inspection which pyranose form (α or β) will predominate. For D-glucose, the β form predominates once equilibrium has been achieved, while for D-mannose, the α form dominates at equilibrium.

# SKILLBUILDER

## 24.2 DRAWING A HAWORTH PROJECTION OF AN ALDOHEXOSE

**LEARN the skill**

Draw the structure of α-D-galactopyranose.

**SOLUTION**

**STEP 1** Draw the skeleton of a Haworth projection.

Analyze all parts of the name. This compound is D-galactose that has undergone cyclization to form an α-pyranose ring. Let's begin by drawing the skeleton of a pyranose ring:

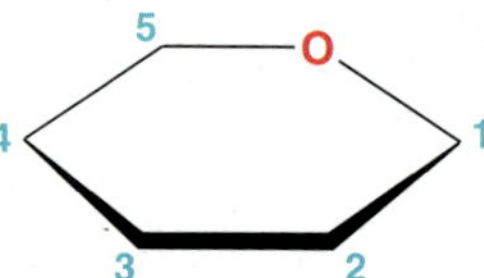


When drawing a pyranose ring, the accepted convention is to place the oxygen atom in the back right position. Now we must account for all carbon atoms. A pyranose ring contains only five carbon atoms, but D-galactose has six carbon atoms:

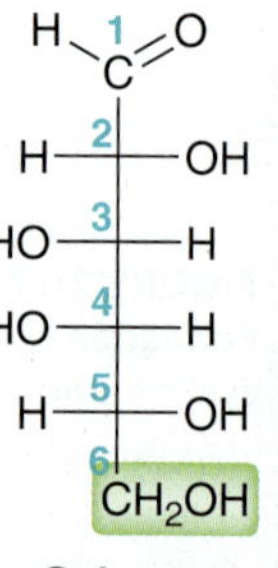


**D-Galactose**

**STEP 2** Draw the $CH_2OH$ group.

The sixth carbon atom (C6) in D-galactose is a $CH_2OH$ group, and it is connected to the C5 position. Therefore, we draw a $CH_2OH$ group connected to the C5 position in the pyranose ring. The "D" indicates that the $CH_2OH$ group must be in the up position in the pyranose ring:

**STEP 3** Draw the OH group at the anomeric position.

That takes care of the configuration at one of the positions of the pyranose ring. Now let's focus on the remaining positions. The configuration at C1 (the anomeric carbon) is conveyed in the name of the pyranose ring form. The term *alpha* means that the OH group at the C1 position must be down (*trans* to the $CH_2OH$ group):

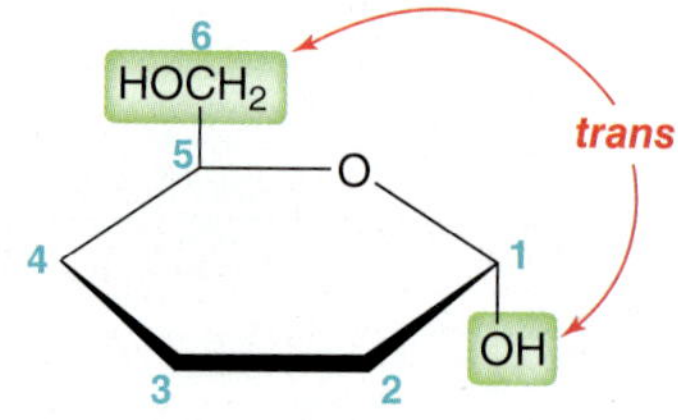


**STEP 4** Draw the remaining groups.

The remaining chirality centers (C2, C3, and C4) are drawn using the following rule: Any OH groups on the right side of the Fischer projection of the open-chain form will be pointing down in the Haworth projection of the cyclic form, while any OH groups on the left side of the Fischer projection will be pointing up in the Haworth projection:

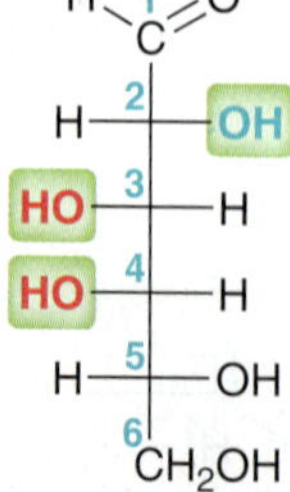


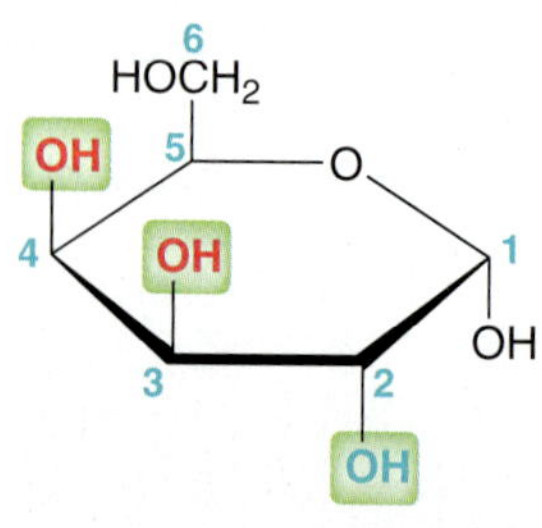

**PRACTICE the skill** **24.12** Draw a Haworth projection for each of the following compounds:

**(a)** β-D-Galactopyranose **(b)** α-D-Mannopyranose **(c)** α-D-Allopyranose

**(d)** β-D-Mannopyranose **(e)** β-D-Glucopyranose **(f)** α-D-Glucopyranose

**24.13** Provide a complete name for the following compound:

**APPLY** the skill

**24.14** Mutarotation causes the conversion of β-D-mannopyranose to α-D-mannopyranose. Using Haworth projections, draw the equilibrium between the two pyranose forms and the open-chain form of D-mannose.

**24.15** When D-talose is dissolved in water, an equilibrium is established in which two pyranose forms are present. Draw both pyranose forms and name them.

need more **PRACTICE?** **Try Problems 24.48a, 24.53b,c,d**

## Three-Dimensional Representations of Pyranose Rings

We have seen that Haworth projections are often used for drawing the cyclic forms of D-glucose. While Haworth projections are extremely useful for showing configurations, they are not effective for communicating the conformation of a compound. A cyclic form of D-glucose will spend the majority of its time in a chair conformation.

**LOOKING BACK**
For a review of chair conformations and axial/equatorial positions, see Section 4.11.

When drawing either a Haworth projection or a chair conformation, the convention is to place the oxygen atom in the upper, rear-right position, as highlighted in red. Notice that in a cyclic form of D-glucose all substituents can occupy equatorial positions, which renders the compound particularly stable and explains why D-glucose is the most common naturally occurring monosaccharide.

## 24.3 DRAWING THE MORE STABLE CHAIR CONFORMATION OF A PYRANOSE RING

**LEARN** the skill

Draw the more stable chair conformation of α-D-galactopyranose.

**SOLUTION**

The first step is to draw the Haworth projection of the compound, which was done in the previous SkillBuilder:

**STEP 1**
Draw the Haworth projection.

**STEP 2**
Draw the skeleton of a chair conformation with an oxygen atom in the upper, rear-right position.

This compound can adopt two different chair conformations, and we must determine which one is more stable. We begin by drawing one of the chair conformations. The oxygen atom occupies the upper, rear-right position:

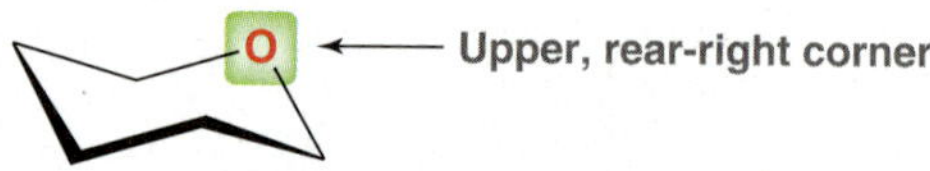


The following alternatives are not conventional for the reasons indicated:

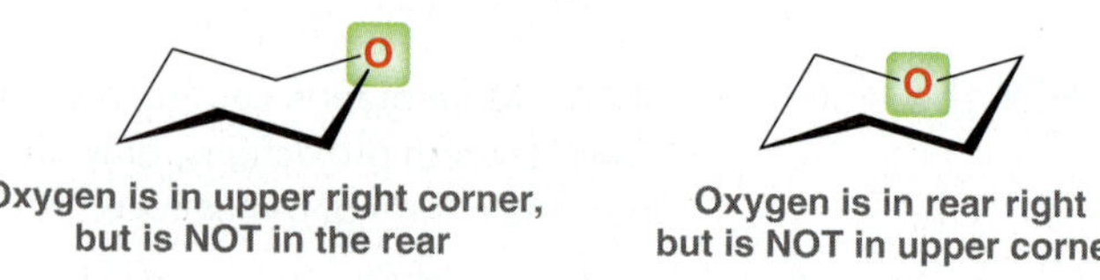


**Oxygen is in upper right corner, but is NOT in the rear**  **Oxygen is in rear right but is NOT in upper corner**

Drawing the appropriate skeleton is critical for drawing the chair conformation properly. The next step is to draw all substituents on the chair using the skills we learned in SkillBuilders 4.12 and 4.13. Each substituent is labeled as UP or DOWN and then placed in the appropriate position on the chair conformation:

**STEP 3**
Label each substituent as UP or DOWN and draw each substituent on the chair.

This is one of the two chair conformations that the compound can adopt. Before drawing the ring flip to give the other chair conformation, let's first analyze this one. Specifically, we look at each substituent and determine whether it occupies an axial or equatorial position. In this case, two of the OH groups occupy axial positions and two occupy equatorial positions; the $CH_2OH$ group occupies an equatorial position.

**STEP 4**
Verify that this chair conformation is the more stable chair conformation.

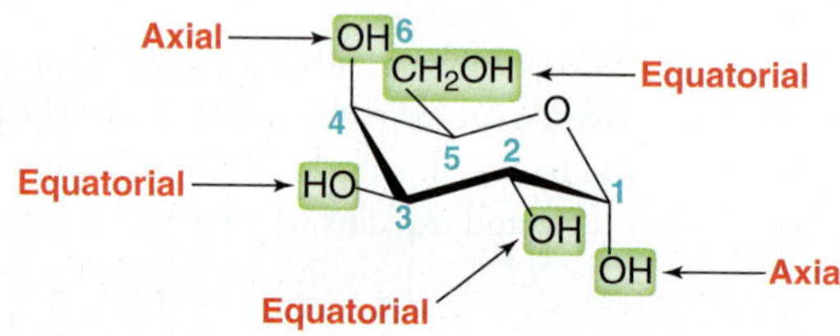


Recall that a ring flip converts all axial positions into equatorial positions and all equatorial positions into axial positions (see Section 4.10). Also recall that the more stable chair conformation will be the one in which the largest groups occupy equatorial positions. In this case, the $CH_2OH$ is the largest group, so the most stable chair conformation will be the one in which this group occupies an equatorial position, as shown. The other chair conformation will be less stable, so it is not necessary to draw the ring flip. This is generally the case for D-aldohexoses (for an exception, see Problem 24.85).

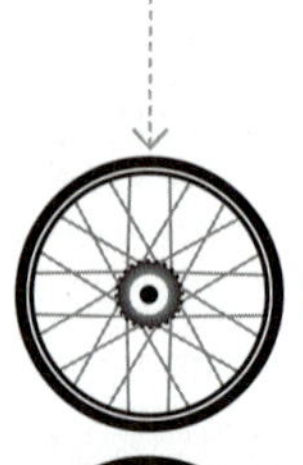

**PRACTICE the skill** **24.16** Draw the more stable chair conformation for each of the following compounds:

**(a)** β-D-Galactopyranose **(b)** α-D-Glucopyranose **(c)** β-D-Glucopyranose

**APPLY the skill** **24.17** Draw the open-chain form of the following cyclic monosaccharide:

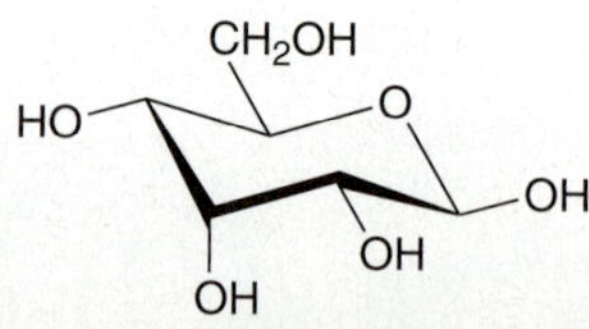

**24.18** There are two chair conformations for β-D-glucopyranose. Draw the less stable chair conformation.

need more **PRACTICE?** **Try Problem 24.60**

## Furanose Forms of Monosaccharides

As seen in SkillBuilder 24.1, hydroxyaldehydes can form five-membered cyclic hemiacetals.

**Cyclic hemiacetal**

In much the same way, many carbohydrates can also form five-membered rings, called **furanose** rings, named after furan, a simple compound that possesses a five-membered ring with an oxygen atom incorporated in the ring. As an example, D-fructose can form a furanose ring that results from the reaction between the carbonyl group and the hydroxyl group connected to C5.

**D-Fructose** **A furanose ring**

D-Fructose can also cyclize to give pyranose forms that result from the reaction involving the hydroxyl group at C6.

**D-Fructose** **A pyranose ring**

Therefore, D-fructose exists as an equilibrium between an open-chain form, two pyranose forms (α and β), and two furanose forms (α and β). When D-fructose is dissolved in water, the following equilibrium concentrations are observed: 70% β-pyranose, 2% α-pyranose, 23% β-furanose, 5% α-furanose, and 0.7% open chain. Despite the fact that the β-pyranose ring predominates at equilibrium as observed in the laboratory, it is the β-furanose form of D-fructose that participates in most biochemical pathways.

### CONCEPTUAL CHECKPOINT

**24.19** Consider the structures of the following two D-aldotetroses:

Each of these compounds exists as a furanose ring, which is formed when the OH at C4 attacks the aldehyde group. Draw each of the following furanose rings:

(a) α-D-Erythrofuranose (b) β-D-Erythrofuranose

(c) α-D-Threofuranose (d) β-D-Threofuranose

**D-Erythrose** **D-Threose**

**24.20** Draw a mechanism for the acid-catalyzed cyclization of L-threose to give β-L-threofuranose. (*Hint*: You may want to first review the mechanism for acid-catalyzed hemiacetal formation, Mechanism 20.5.)

**24.21** Draw a mechanism for the acid-catalyzed cyclization of D-fructose to give β-D-fructofuranose.

**24.22** Draw the open-chain form of the carbohydrate that can undergo acid-catalyzed cyclization to produce α-D-fructopyranose.

## 24.6 Reactions of Monosaccharides

### Ester and Ether Formation

Monosaccharides are highly soluble in water (due to the presence of many hydroxyl groups) and are generally insoluble in most organic solvents. This property makes them difficult to purify by conventional methods. However, the ester derivatives of monosaccharides are soluble in most organic solvents and are easily purified. Monosaccharides are converted into their ester derivatives when treated with an excess of acid chloride or acid anhydride in the presence of a base, such as pyridine. Under these conditions, all five hydroxyl groups of β-D-glucopyranose are converted into ester groups.

β-D-Glucopyranose —(Excess $Ac_2O$, Py)→ Penta-O-acetyl-β-D-glucopyranose (Ac = $H_3C$–C(=O)–)

Monosaccharides can also be converted into their ether derivatives via the Williamson ether synthesis. As discussed in Section 14.5, this process generally involves treating an alcohol with a strong base to form an alkoxide ion followed by treatment with an alkyl halide (an $S_N2$ process). When dealing with monosaccharides, a strong base cannot be used, for reasons that we will discuss later, and instead, a mild base such as silver oxide is used. Under these conditions, all five hydroxyl groups of β-D-glucopyranose are converted into ether groups.

β-D-Glucopyranose —(Excess $CH_3I$, $Ag_2O$)→ β-D-Glucopyranose pentamethyl ether

CONCEPTUAL CHECKPOINT

**24.23** Draw the product obtained when each of the following compounds is treated with acetic anhydride in the presence of pyridine:

(a) α-D-Galactopyranose (b) α-D-Glucopyranose

(c) β-D-Galactopyranose

**24.24** Draw the product obtained when each of the compounds from the previous problem is treated with methyl iodide in the presence of silver oxide ($Ag_2O$).

### Glycoside Formation

Recall from Section 20.5 that a hemiacetal will react with an alcohol in the presence of an acid catalyst to produce an acetal.

Hemiacetal + ROH ⇌ ($[H^+]$) Acetal + $H_2O$

As mentioned earlier in this chapter, monosaccharides exist primarily as cyclic hemiacetals; they are therefore converted into acetals when treated with an alcohol under acid-catalyzed conditions. The acetal products obtained are called **glycosides**.

$\beta$-D-Glucopyranose (a cyclic hemiacetal) $\xrightleftharpoons[HCl]{CH_3OH}$ Methyl $\alpha$-D-glucopyranoside (66%) + Methyl $\beta$-D-glucopyranoside (33%)

Glycosides are named by placing the alkyl group as a prefix and the term *-oside* as a suffix. During glycoside formation, only the anomeric hydroxyl group is replaced. This is best explained by exploring the mechanism of glycoside formation, which is directly analogous to the mechanism of acetal formation described in Section 20.5. The anomeric hydroxyl group is protonated, followed by loss of water to generate a carbocation.

$-H_2O$

Resonance-stabilized carbocation

Notice that the carbocation intermediate is resonance stabilized. This stabilization only occurs when the reaction takes place at the anomeric position. The carbocation intermediate is then attacked by the alcohol, followed by loss of a proton to generate the product.

This mechanism shows formation of the $\beta$ anomer. The $\alpha$ anomer is also observed, because the alcohol can attack the carbocation from the other face of the planar intermediate.

The product distribution is independent of the identity of the starting anomer—that is, $\alpha$-D-glucopyranose and $\beta$-D-glucopyranose each yield the same ratio of anomeric glycosides when treated with an alcohol in acid-catalyzed conditions. Once isolated, the glycoside products are stable under neutral or basic conditions. However, upon treatment with aqueous acid, they are readily converted back to a hemiacetal.

## CONCEPTUAL CHECKPOINT

**24.25** When α-D-galactopyranose is treated with ethanol in the presence of an acid catalyst, such as HCl, two products are formed. Draw both products and account for their formation with a mechanism.

**24.26** Methyl α-D-glucopyranoside is a stable compound that does not undergo mutarotation under neutral or basic conditions. However, when subjected to acidic conditions, an equilibrium is established consisting of both methyl α-D-glucopyranoside and methyl β-D-glucopyranoside. Draw a mechanism that accounts for this observation.

## Epimerization

When D-glucose is exposed to strongly basic conditions, it is converted into a mixture containing both D-glucose and D-mannose via the following process:

D-Glucose ⇌ ($NaOH$, $H_2O$) [An enediol] ⇌ ($NaOH$, $H_2O$) D-Glucose + D-Mannose

D-Glucose first undergoes base-catalyzed tautomerization to form an enediol. This intermediate can undergo tautomerization once again to revert back to the aldose, but in the process, the configuration at C2 is lost, leading to a mixture of D-glucose and D-mannose. D-Glucose and D-mannose are said to be **epimers** because they are diastereomers that differ from each other in the configuration of only one chirality center. When either pure D-glucose or pure D-mannose is treated with a strong base, epimerization occurs, giving a mixture containing both D-glucose and D-mannose. For this reason, chemists generally avoid exposing carbohydrates to strongly basic conditions.

## CONCEPTUAL CHECKPOINT

**24.27** Draw and name the structure of the aldohexose that is epimeric with D-glucose at each of the following positions:

(a) C2 (b) C3 (c) C4

## Reduction of Monosaccharides

The carbonyl group of an aldose or ketose can be reduced upon treatment with sodium borohydride to yield a product called an **alditol**. Consider, for example, the reduction of D-glucose.

β-D-Glucopyranose ⇌ D-Glucose (open form) → ($NaBH_4$, $H_2O$) D-Glucitol (an alditol)

The starting monosaccharide exists primarily as a hemiacetal, which does not react with sodium borohydride because it lacks a carbonyl group. However, a small amount of the open-chain form is present at equilibrium, and it is that form that reacts with sodium borohydride to produce an alditol. As the open-chain form of glucose is converted into the alditol, the equilibrium is perturbed. According to Le Châtelier's principle, this causes more of the glucose molecules to adopt an open-chain form, which then also undergo reduction. This process continues until nearly all of the glucose molecules have been reduced to D-glucitol. D-Glucitol is found in many fruits and berries. It is commonly referred to as D-sorbitol, or just sorbitol, and is often used as a sugar substitute in processed foods.

## CONCEPTUAL CHECKPOINT

**24.28** The same product is obtained when either D-altrose or D-talose is treated with sodium borohydride in the presence of water. Explain this observation.

**24.29** The same product is obtained when either D-allose or L-allose is treated with sodium borohydride in the presence of water. Explain this observation.

**24.30** Of the eight D-aldohexoses, only two of them form optically inactive alditols when treated with sodium borohydride in the presence of water. Identify these two aldohexoses and explain why their alditols are optically inactive.

## Oxidation of Monosaccharides

When treated with a suitable oxidizing agent, the aldehyde group of an aldose can be oxidized to yield a compound called an **aldonic acid**.

An aldose + Oxidizing agent ⟶ An aldonic acid + Reduced form of oxidizing agent

**An aldose** **An aldonic acid**

This reaction is observed for a wide variety of oxidizing agents. When a high yield is desired, a useful oxidizing agent is an aqueous solution of bromine buffered to a pH of 6. For example:

β-D-Glucopyranose ⇌ D-Glucose $\xrightarrow[\text{pH} = 6]{Br_2,\ H_2O}$ D-Gluconic acid (an aldonic acid)

**β-D-Glucopyranose** **D-Glucose** **D-Gluconic acid (an aldonic acid)**

The starting monosaccharide exists primarily as a hemiacetal, and the OH groups are not oxidized by this mild oxidizing agent. However, a small amount of the open-chain form is present at equilibrium, and it is the aldehyde group of the open-chain form that reacts with the oxidizing agent to produce an aldonic acid. As the open-chain form of glucose is converted into the aldonic acid, the equilibrium is perturbed. This causes more of the glucose molecules to adopt an open-chain form, which then undergo oxidation, a process that continues until nearly all of the glucose molecules have been oxidized.

The oxidizing agent above will oxidize aldoses but does not react with ketoses. A ketose lacks the aldehydic hydrogen that is characteristic of an aldose.

**An aldose** **A ketose**

It is true that ketoses slowly equilibrate with aldoses via a process that is similar to epimerization, and we might therefore expect that a small concentration of the ketose would be converted into an aldose, which would then undergo oxidation, providing a pathway for the ketose to be oxidized as well. However, the mildly acidic conditions employed (a buffered solution of pH = 6) prevent isomerization from occurring. Under these conditions, a ketose does not undergo isomerization to give an aldose at an appreciable rate, and ketoses are therefore not oxidized by a buffered solution of aqueous bromine. This, in fact, provides a method for distinguishing between aldoses and ketoses. In contrast, there are other oxidizing agents that employ conditions under which isomerization can occur. These alternative reagents can oxidize ketoses as well as aldoses. Examples of such oxidizing agents include the Tollens' reagent ($Ag^+$ in aqueous ammonia), Fehling's reagent ($Cu^{2+}$ in aqueous sodium tartrate), and Benedict's reagent ($Cu^{2+}$ in aqueous sodium citrate). All three can be used as chemical tests for the presence of an aldose or ketose.

The Tollens' reagent indicates the presence of an aldose or ketose when the surface of the reaction flask is coated with metallic silver (looking like a mirror), while the other two tests indicate the presence of an aldose or ketose with the formation of a reddish precipitate ($Cu_2O$). These three reactions are used only as tests to obtain structural information about unknown carbohydrates but are not efficient as preparative methods when sufficient quantities of the aldonic acid are desired. A carbohydrate that tests positively for any of these three tests is said to be a **reducing sugar** because the carbohydrate can reduce the oxidizing agent.

These chemical tests are useful for distinguishing between aldoses or ketoses and their glycosides. While aldoses and ketoses are reducing sugars, their corresponding glycosides are not. Glycosides do not exhibit ring-opening reactions under neutral or basic conditions and therefore do not produce even small quantities of an open-chain form containing an aldehyde group. As such they are not oxidized by any of the oxidizing agents discussed.

In the presence of a stronger oxidizing agent, such as $HNO_3$, the primary hydroxyl group is also oxidized, giving a dicarboxylic acid, called an **aldaric acid**. This reaction involves the oxidation of both of the highlighted functional groups.

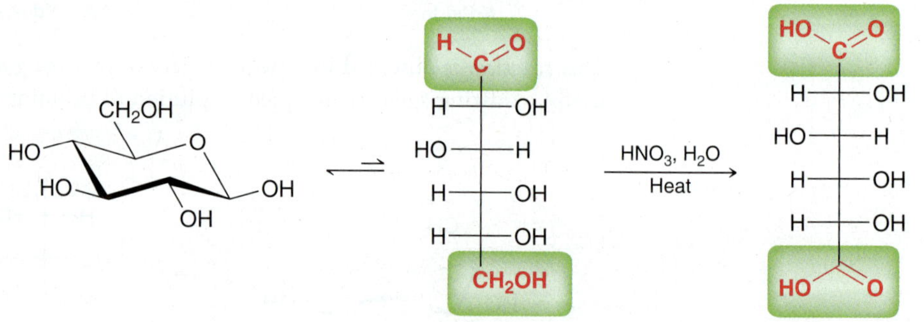

$\beta$-D-Glucopyranose ⇌ D-Glucose $\xrightarrow{HNO_3,\ H_2O,\ \text{Heat}}$ D-Glucaric acid (an aldaric acid)

SKILLBUILDER

**24.4** IDENTIFYING A REDUCING SUGAR

**LEARN** the skill

Identify whether the following compound is a reducing sugar:

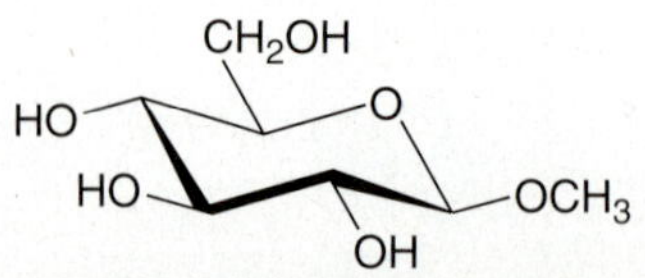

## SOLUTION

Begin by identifying the anomeric position:

**STEP 1**
Identify the anomeric position.

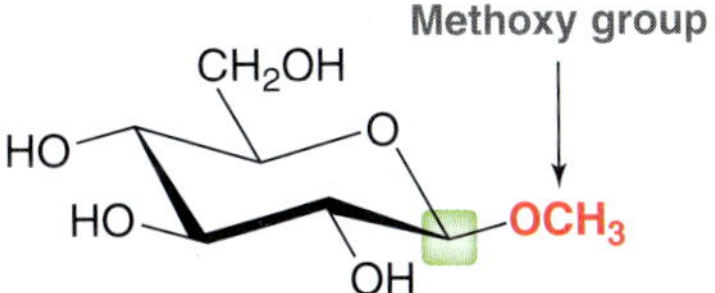


Now determine whether this compound is a hemiacetal or an acetal. Specifically, look to see if the group attached to the anomeric position is a hydroxy group (OH) or an alkoxy group (OR). If it is a hydroxy group, then the compound is a hemiacetal and will be a reducing sugar. If it is an alkoxy group, then the compound is an acetal and will not be a reducing sugar. In this case, the group attached to the anomeric position is a methoxy group. Therefore, this compound is an acetal and is not a reducing sugar.

**STEP 2**
Determine if the group at the anomeric position is a hydroxy group or alkoxy group.



**PRACTICE the skill** **24.31** Determine whether each of the following compounds is a reducing sugar:

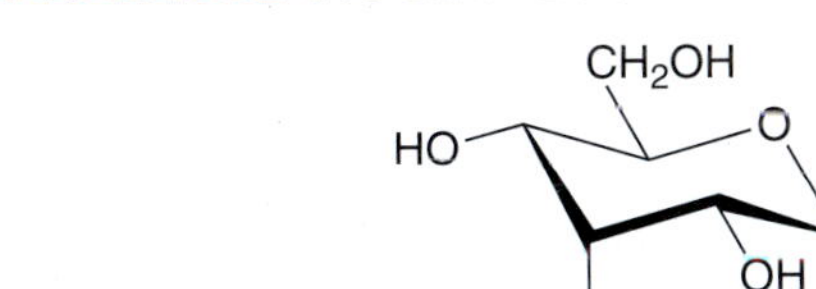


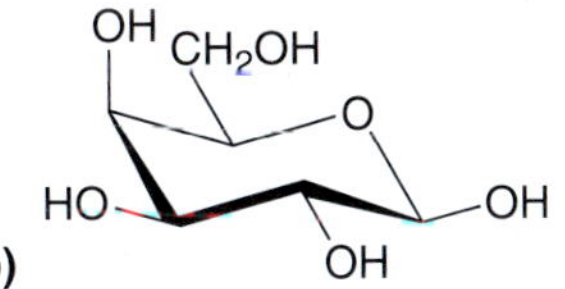


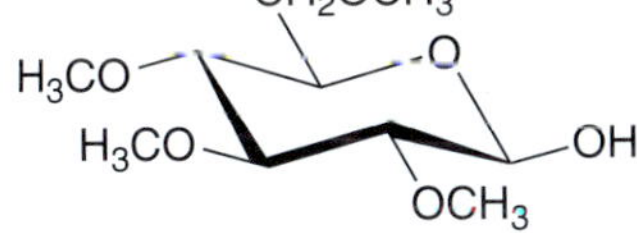


**APPLY the skill** **24.32** Draw and name the product obtained when each of the following compounds is treated with aqueous bromine (at pH = 6):

**(a)** α-D-Galactopyranose **(b)** β-D-Galactopyranose

**(c)** α-D-Glucopyranose **(d)** β-D-Glucopyranose

**24.33** Do you expect β-D-glucopyranose pentamethyl ether to be a reducing sugar? Explain your reasoning.

need more **PRACTICE?** **Try Problems 24.69, 24.76a**

## Chain Lengthening: The Kiliani-Fischer Synthesis

In 1886, Heinrich Kiliani (University of Freiburg, Germany) observed that an aldose will react with HCN to form a pair of stereoisomeric cyanohydrins.

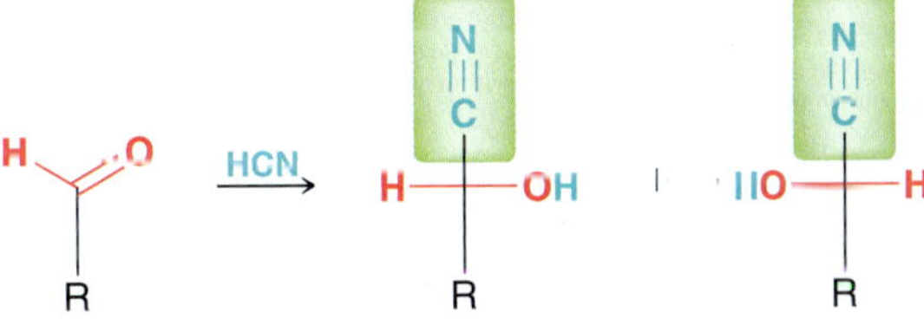


This process occurs via a nucleophilic acyl addition. A mechanism for cyanohydrin formation was discussed in Section 20.10. Building on Kiliani's observation, Emil Fischer (the same Fischer who gave us Fischer projections) then devised a multistep method for converting a cyano group to an aldehyde.

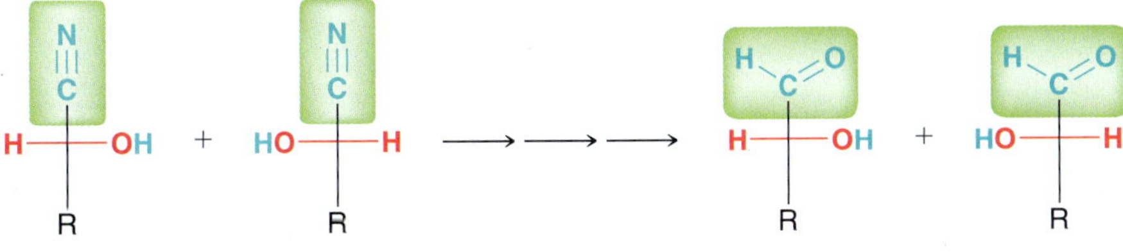

Overall, the net result is the ability to lengthen the chain of a carbohydrate by one carbon atom. For example, an aldopentose can be converted into an aldohexose using this process, called the **Kiliani-Fischer synthesis**. A more modern version of the Kiliani-Fischer synthesis enables the conversion of a cyano group into an aldehyde in just one step (rather than several steps). This is accomplished via hydrogenation of the cyano group using a poisoned catalyst in an aqueous solution. Under these conditions, the cyano group is reduced to give an imine, which then undergoes hydrolysis to give an aldehyde.

$H_2$, Pd/$BaSO_4$; $H_2O$

As an example of this process, consider the outcome achieved when the modern version of the Kiliani-Fischer synthesis is performed with D-arabinose as the starting material:

1) HCN
2) $H_2$, Pd/$BaSO_4$, $H_2O$

**D-Arabinose** → **D-Glucose** + **D-Mannose**

D-Arabinose has five carbon atoms (an aldopentose), and the products each have six carbon atoms (aldohexoses). In the process, C1 of the starting material is transformed into C2 in the products. Notice that two products are formed because the first step of the process is not stereoselective; that is, the new chirality center (C2) can have either configuration, giving D-glucose and D-mannose, which are C2 epimers.

## CONCEPTUAL CHECKPOINT

**24.34** Draw and name the pair of epimers formed when the following aldopentoses undergo a Kiliani-Fischer chain-lengthening process:

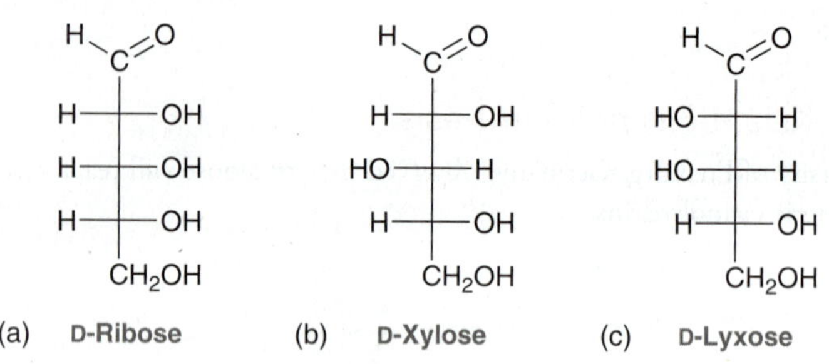


**24.35** Identify the reagents you would use to convert D-erythrose (Problem 24.22) into D-ribose. What other product is also formed in this process?

## Chain Shortening: The Wohl Degradation

The **Wohl degradation** is the reverse of the Kiliani-Fischer synthesis and involves the removal of a carbon atom from an aldose. The aldehyde group is first converted to a cyanohydrin, followed by loss of HCN in the presence of a base.

Base (–HCN)

Overall, the net result is the ability to shorten a carbohydrate chain by one carbon atom. Conversion of the aldehyde into a cyano group is accomplished via oxime formation (as seen in Section 20.6) followed by dehydration. The resulting cyanohydrin then loses HCN when treated with a strong base to afford the new carbohydrate that has one less carbon atom than the starting carbohydrate:

$NH_2OH$ $Ac_2O$ $\ddot{O}Me^{\ominus}$

D-Glucose An oxime A cyanohydrin D-Arabinose

This chain reduction process generates only one product, in contrast with chain lengthening (Kiliani-Fischer), which gives two products. The yields of the Wohl degradation are generally not very high, but this process and others like it were incredibly useful during early investigations aimed at elucidating the structures of monosaccharides.

## CONCEPTUAL CHECKPOINT

**24.36** Draw and name the two aldohexoses that can be converted into D-ribose (Problem 24.34a) using a Wohl degradation.

**24.37** Identify the reagents you would use to convert D-ribose into D-erythrose (Problem 24.22).

**24.38** When D-glucose undergoes a Wohl degradation followed by a Kiliani-Fischer chain-lengthening process, a mixture of two epimeric products are obtained. Identify both epimers.

## 24.7 Disaccharides

### Maltose

Maltose
(a 1→4 α-glycoside)

**Disaccharides** are carbohydrates comprised of two monosaccharide units joined via a glycosidic linkage between the anomeric carbon of one monosaccharide and a hydroxyl group of the other monosaccharide. For example, consider the structure of maltose.

Maltose is obtained from the hydrolysis of starch. It is comprised of two α-D-glucopyranose units joined between the anomeric carbon (C1) of one glucopyranose unit and the OH at C4 of the other glucopyranose unit. This type of linkage is called a 1→4 link. Two monosaccharides can link together in a variety of ways, but the 1→4 link is especially common.

Maltose undergoes mutarotation, because one of the rings (bottom right) is a hemiacetal. As a result, the anomeric position of that ring is capable of opening and closing, allowing for equilibration between the two possible anomers.

$H_2O$

These anomers are interchanged as a result of mutarotation, which occurs more rapidly in the presence of catalytic acid or base.

Maltose is also a reducing sugar because one of the rings is a hemiacetal and exists in equilibrium with the open-chain form.

In the presence of an oxidizing agent, the aldehyde group of the open-chain form is oxidized to give an aldonic acid.

# SKILLBUILDER

## 24.5 DETERMINING WHETHER A DISACCHARIDE IS A REDUCING SUGAR

**LEARN the skill**

Determine whether lactose is a reducing sugar:

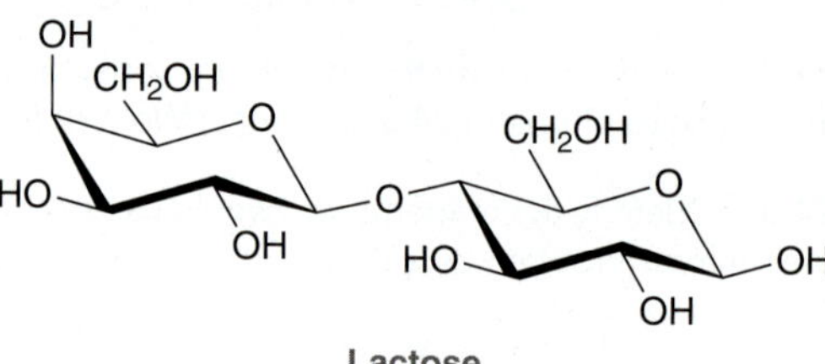


**Lactose**

### SOLUTION

Begin by identifying the anomeric carbon on each ring. Remember the anomeric position of a ring is the position that is connected to two oxygen atoms.

**STEP 1**
Identify the anomeric position on each ring.

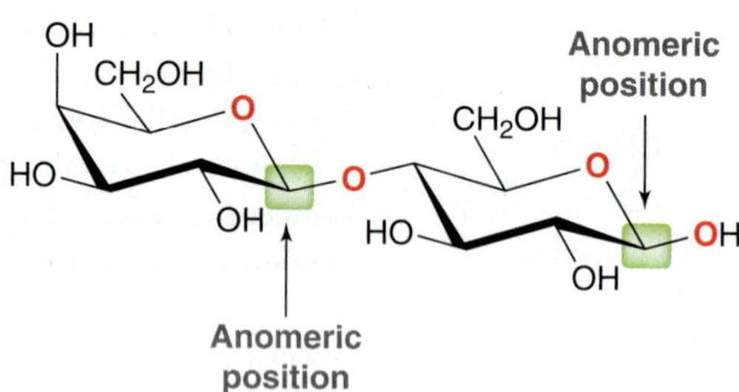


**STEP 2**
Determine if the group at each anomeric position is a hydroxy group or alkoxy group.

Now determine whether either of these positions is a hemiacetal. Specifically, identify whether the group attached to each anomeric position is a hydroxy group (OH) or an alkoxy group (OR). If even one of the anomeric positions has an OH group, then the compound has a hemiacetal moiety and will be a reducing sugar. If neither of the anomeric positions has an OH group, then the compound lacks a hemiacetal moiety and will not be a reducing sugar. In this case, one of the anomeric positions has an OH group. Therefore, this compound is a reducing sugar.

**An OH group at an anomeric position signifies a reducing sugar**

**PRACTICE the skill** **24.39** Determine whether each of the following disaccharides is a reducing sugar:

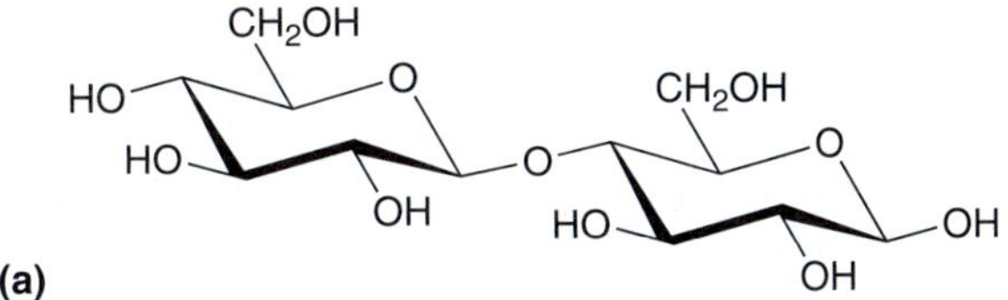


(a)

(b)

(c) Sucrose

**APPLY the skill**

**24.40** Draw the structure of the product obtained when the following disaccharide is treated with $NaBH_4$ in methanol:

need more **PRACTICE?** **Try Problem 24.73**

## Cellobiose

Another example of a disaccharide is cellobiose, which is obtained from the hydrolysis of cellulose. It is comprised of two β-D-glucopyranose units joined by a 1→4 link.

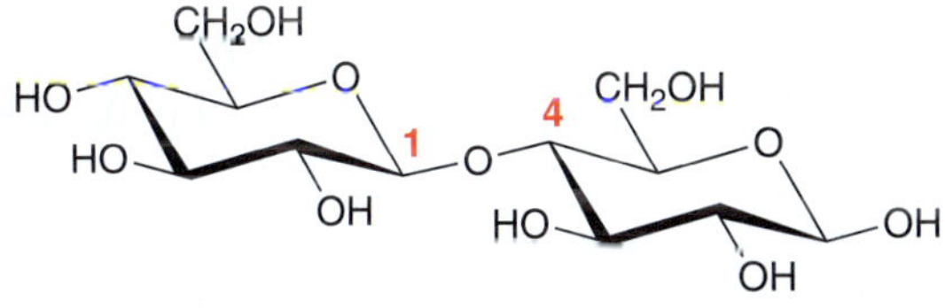


**Cellobiose (a 1→4 β-glycoside)**

This compound is very similar in structure to maltose, only it is comprised of β-D-glucopyranose units rather than α-D-glucopyranose units. Much like maltose, cellobiose also exhibits mutarotation and is a reducing sugar, because the ring on the right is a hemiacetal and is therefore capable of opening and closing.

## CONCEPTUAL CHECKPOINT

**24.41** Predict the product that is obtained when cellobiose is treated with each of the following reagents:

(a) $NaBH_4$, $H_2O$ (b) $Br_2$, $H_2O$ (pH = 6) (c) $CH_3OH$, HCl (d) $Ac_2O$, pyridine

## Lactose

Lactose, commonly called milk sugar, is a naturally occurring disaccharide found in milk. Unlike maltose or cellobiose, lactose is comprised of two different monosaccharides—galactose and glucose.

The two monosaccharide units are joined by a 1→4 link, between C1 of galactose and C4 of glucose. Like the other disaccharides we have seen so far, lactose also exhibits mutarotation and is a reducing sugar, because the ring on the bottom right is a hemiacetal and is therefore capable of opening and closing.

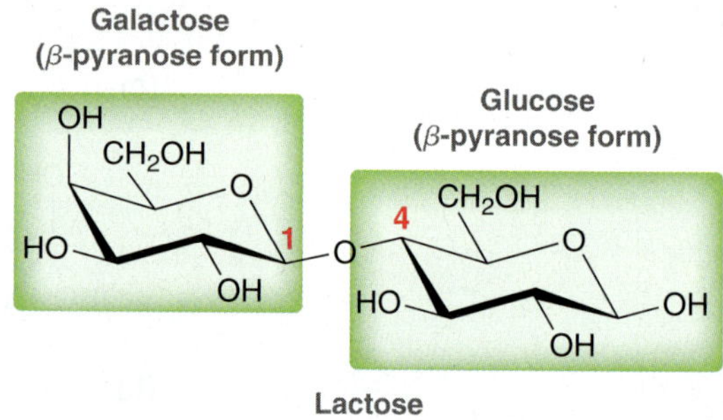


**Lactose**
**(a 1→ 4 β-glycoside)**

## medically speaking — Lactose Intolerance

In our bodies, an enzyme called lactase (β-galactosidase) catalyzes the hydrolysis of lactose to form glucose and galactose.

**Lactose**

↓ β-Galactosidase

**D-Galactose** + **D-Glucose**

The galactose produced in this process is subsequently converted by the liver into additional glucose, which is then further metabolized to produce energy. Many people do not produce a sufficient amount of lactase and are incapable of hydrolyzing large quantities of lactose. Instead, lactose accumulates and is ultimately broken down into $CO_2$ and $H_2$ by bacteria present in the intestines. Bacterial degradation of lactose produces several by-products, including lactic acid.

**Lactic acid**

The buildup of lactic acid and other acidic by-products causes cramps, nausea, and diarrhea. This condition is called lactose intolerance, and it is estimated that between 30 and 50 million Americans are lactose intolerant. Different ethnic and racial groups are affected to different extents, with the highest incidence of lactose intolerance occurring among Asians and the lowest incidence occurring among Europeans.

Lactose intolerance develops with time. Lactase production begins to decline for most children at about age two, although many people do not experience symptoms of lactose intolerance until later in life. It is estimated that 75% of the adult population worldwide will develop lactose intolerance. This condition also affects other species, including dogs and cats, both of which are particularly susceptible to developing lactose intolerance.

Lactose intolerance is easily treated through a controlled diet that minimizes the intake of food products containing lactose. Several dairy products are produced via processes that remove the lactose, and these products are marketed as being "lactose free." In addition, the enzyme lactase is available in tablet form without a prescription and can be taken prior to eating any products containing lactose.

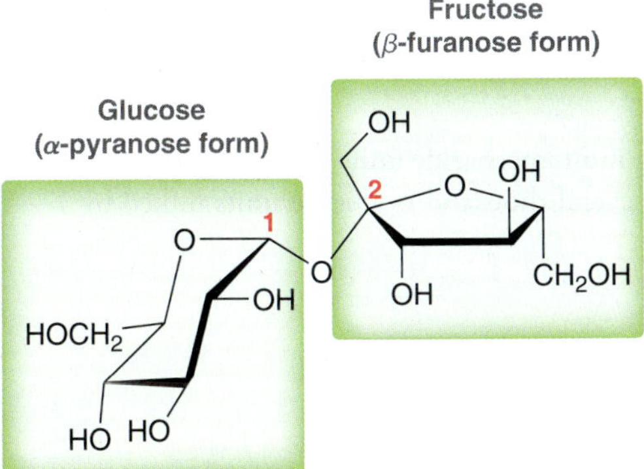


Sucrose
(a 1→2 glycoside)

## Sucrose

Sucrose, commonly referred to as table sugar, is a disaccharide comprised of glucose and fructose linked at C1 of glucose and C2 of fructose. Hydrolysis of sucrose yields glucose and fructose. Honeybees have enzymes that catalyze this hydrolysis, allowing them to convert sucrose into honey, which is primarily a mixture of sucrose, glucose, and fructose. Honey is sweeter than table sugar, because fructose is sweeter than sucrose.

Unlike the other disaccharides we have seen thus far, sucrose is not a reducing sugar and does not undergo mutarotation. This can be explained by noting that sucrose is comprised of two units that are linked to each other via their anomeric positions. As such, neither unit has a hemiacetal group and neither unit is capable of adopting an open-chain form.

## practically speaking — Artificial Sweeteners

A variety of health problems have been associated with excessive consumption of sucrose, including diabetes and tooth decay. These issues, together with the desire of many people to reduce their caloric intake, have fueled the development of many artificial sweeteners, such as the following compounds.

Saccharin

Aspartame

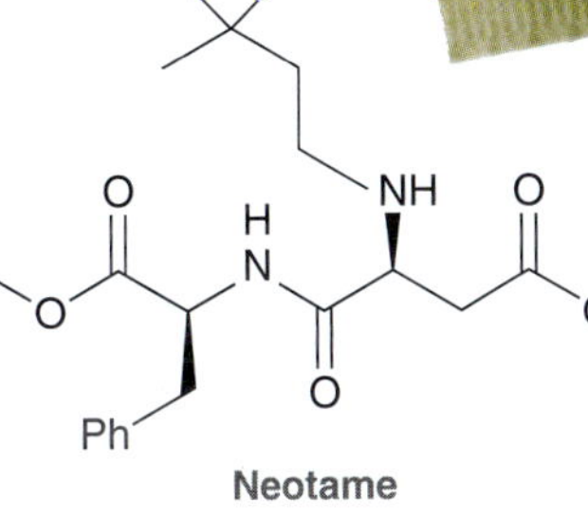


Neotame

Acesulfame K

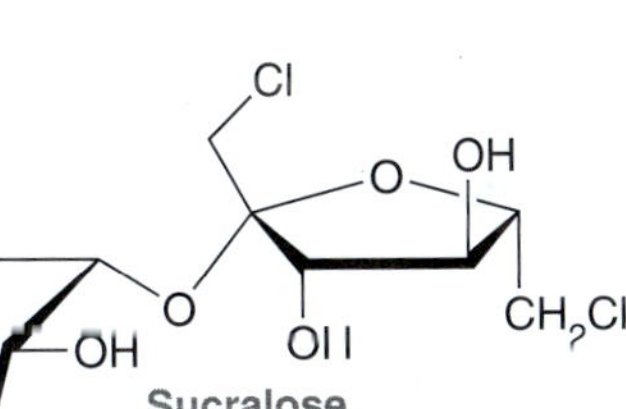


Sucralose

Saccharin is the oldest known artificial sweetener. It was discovered by accident in 1879 and was originally used by diabetics as an alternative to sucrose. In the 1970s some studies suggested that it might be carcinogenic. Extensive research since then has indicated that it is safe for consumption. Nevertheless, due to its metallic aftertaste, it has been largely replaced by many of the other artificial sweeteners on the market.

Aspartame (sold under the trade name NutraSweet) was approved by the FDA in 1981 and has been widely used in soft drinks ever since. It is about 200 times sweeter than sucrose. People suffering from a condition called *phenylketonuria* are incapable of fully metabolizing aspartame and must avoid consuming this compound, as discussed in Problem 21.61. A derivative of aspartame called Neotame was approved by the FDA in 2001 and is about 10,000 times sweeter than sucrose. Neotame must also be avoided by people with phenylketonuria.

Acesulfame potassium, also called acesulfame K, was approved by the FDA in 1998 for use in beverages and is now commonly blended with aspartame in soft drinks. The mixture of aspartame and acesulfame K is reputed to have less of a bitter aftertaste than either compound alone.

Among the common artificial sweeteners, sucralose (600 times sweeter than sucrose) is the only compound that resembles the structure of a carbohydrate. It is prepared from sucrose, whereby three of the hydroxyl groups are replaced with chlorine atoms. The presence of the chlorine atoms prevents the body from effectively metabolizing the compound and releasing its energy content. Sucralose is especially popular for use in baked goods, because it does not decompose upon heating, as is the case for many of the other artificial sweeteners.

Many of the artificial sweeteners on the market were discovered by accident. Nevertheless, extensive ongoing research is aimed at finding new sweeteners with even better properties (none of the artificial sweeteners taste exactly like sucrose, to the dismay of many consumers). Over the decades to come, we are likely to see many other sugar substitutes enter the market.

## 24.8 Polysaccharides

### Cellulose

**Polysaccharides** are polymers made up of repeating monosaccharide units linked together by glycoside bonds. For example, cellulose is comprised of several thousand D-glucose units linked by 1→4 β-glycoside bonds.

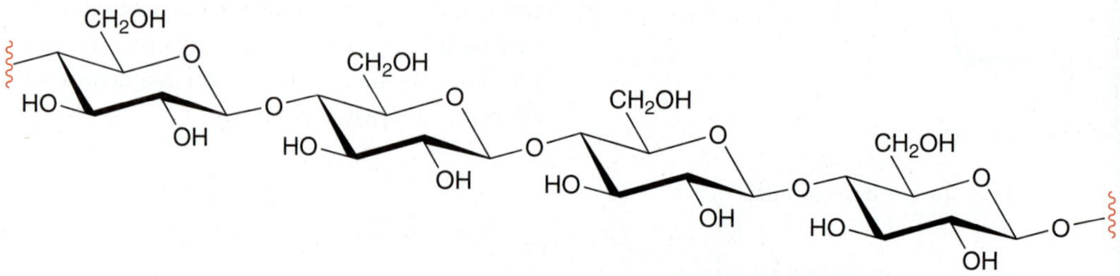

Cellulose
[a 1 → 4 *O*-(β-glucopyranoside) polymer]

The average cellulose strand is comprised of approximately 7,000 glucose units but can be as large as 12,000 units. The strands of the polymer interact with each other via hydrogen bonding, providing structural rigidity for trees and plants. Wood is comprised of 30–40% cellulose, and cotton is comprised of 90% cellulose.

### Starch

Starch is the major component of many of the foods that we eat, including potatoes, corn, and cereal grain. Starch can be separated into two components: amylose, which is insoluble in cold water, and amylopectin, which is soluble in cold water. Amylose is comprised of glucose units linked by 1→4 α-glycoside bonds.

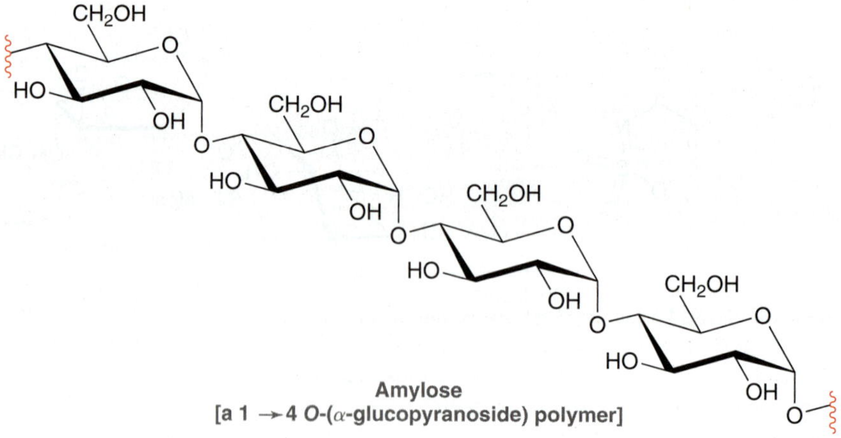

Amylose
[a 1 → 4 *O*-(α-glucopyranoside) polymer]

A strand of amylose is linear, like cellulose, but is comprised of repeating α-D-glucopyranose units rather than β-D-glucopyranose units. Amylose accounts for approximately 20% of starch. The remaining 80% of starch is amylopectin, which is similar to amylose but also contains 1→6 α-glycoside branches approximately every 25 glucose units:

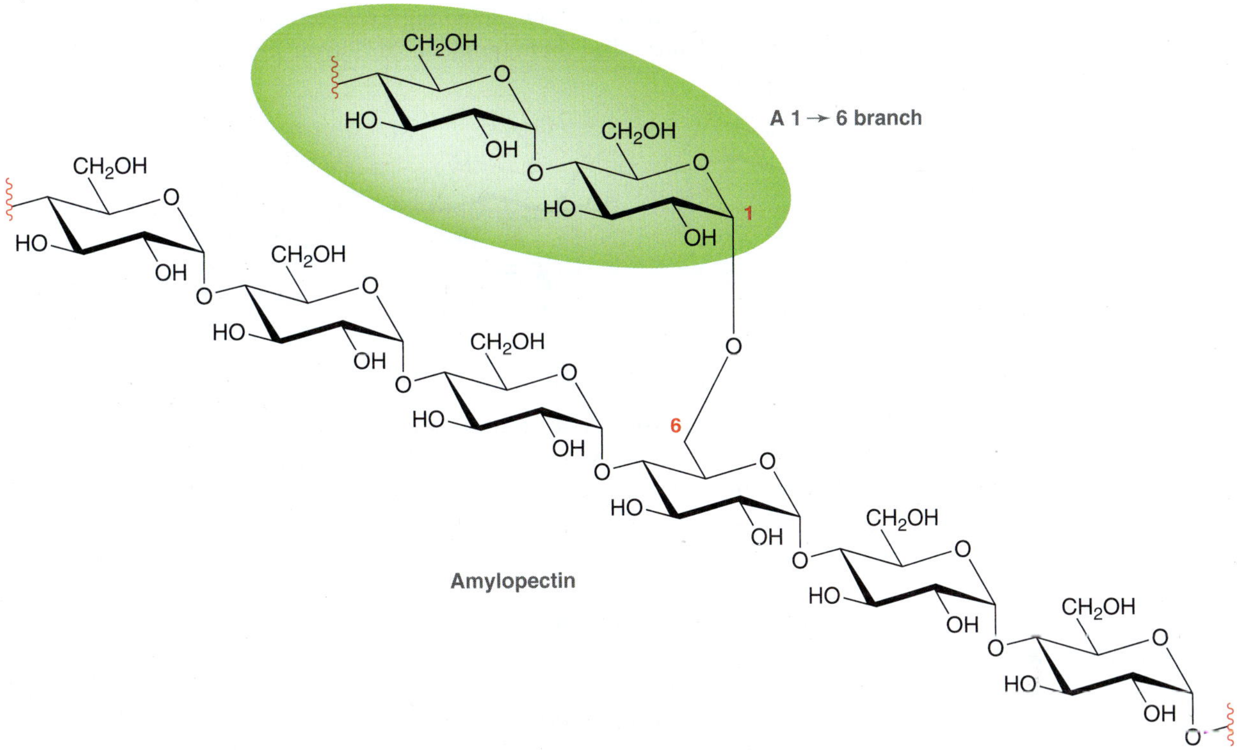

A 1 → 6 branch

Amylopectin

Starch is digested by glycosidase enzymes that catalyze its hydrolysis, releasing individual molecules of glucose. These enzymes are highly selective in their activity and will not catalyze the hydrolysis of cellulose. There are organisms that have enzymes that can catalyze the hydrolysis of cellulose, but human beings lack those enzymes, which is why we can eat corn and potatoes but we cannot eat grass. Cows are also incapable of synthesizing the enzymes necessary to digest cellulose; however, the stomachs of cows contain microorganisms that do produce enzymes necessary to hydrolyze cellulose. As a result, cows are able to digest grass.

### Glycogen

Glucose serves as a fuel to provide the energy needs for both plants and animals. Glucose molecules not required for immediate energy needs are stored in the form of a polymer. Plants store excess glucose in the form of starch, while animals store excess glucose in the form of glycogen. Glycogen is similar in structure to amylopectin (the major component of starch), but its polymer chain exhibits more regular branching. While amylopectin has branches approximately every 25 glucose units, glycogen has branches approximately every 10 glucose units. Individual glycogen molecules can contain up to 100,000 glucose units.

## 24.9 Amino Sugars

**Amino sugars** are carbohydrate derivatives in which an OH group has been replaced with an amino group. Amino sugars are quite common in nature and serve as important building blocks for biological polymers. One such example is β-D-glucosamine, which is biosynthesized from D-glucose:

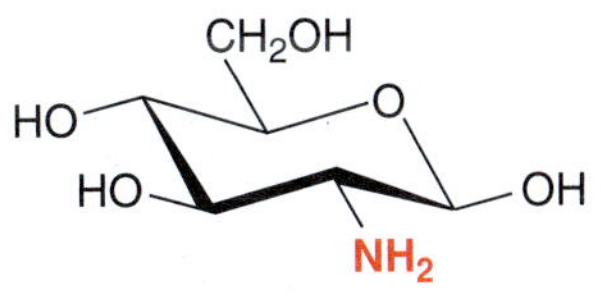

**β-D-Glucosamine**
**(an amino sugar)**

The *N*-acetyl derivative of this amino sugar serves as the repeating monosaccharide unit in the important biopolymer called chitin (pronounced "kite-in"):

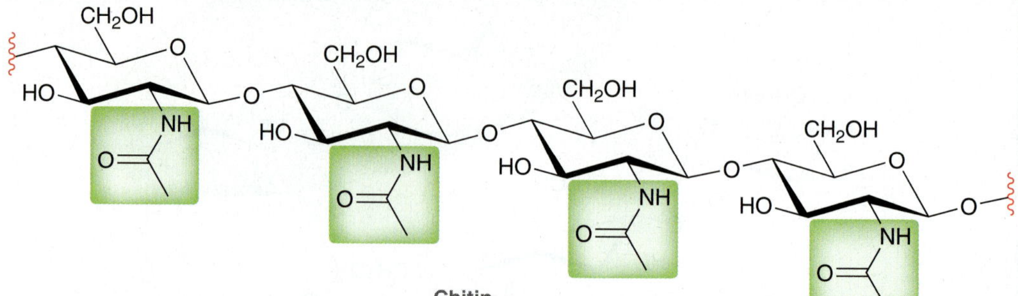

Chitin

Chitin is similar in structure to cellulose, but the amide groups allow for more significant hydrogen bonding between neighboring strands, which renders the polymer even stronger than cellulose (wood). Chitin is the material used in the exoskeletons of arthropods and insects, and over a trillion pounds of this polymer are produced by living organisms each year.

## 24.10 *N*-Glycosides

When treated with an amine in the presence of an acid catalyst, monosaccharides are converted into their corresponding ***N*-glycosides**:

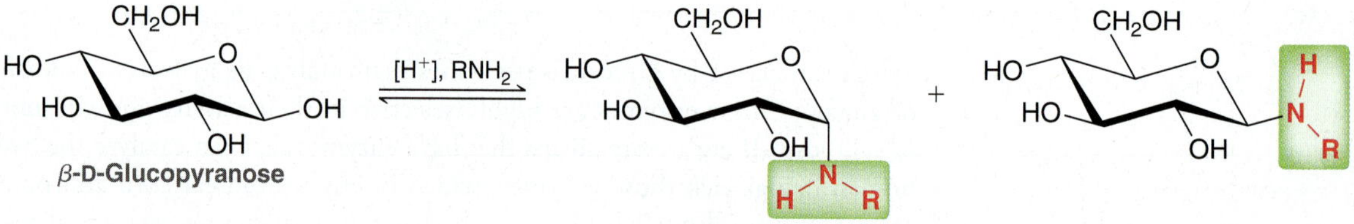

$\beta$-D-Glucopyranose $\xrightleftharpoons{[H^+],\ RNH_2}$ An $\alpha$-*N*-glycoside + A $\beta$-*N*-glycoside

This process likely occurs via a mechanism that is analogous to the mechanism that was presented for glycoside formation. Two carbohydrates, in particular, D-ribose and 2-deoxy-D-ribose, form especially important *N*-glycosides.

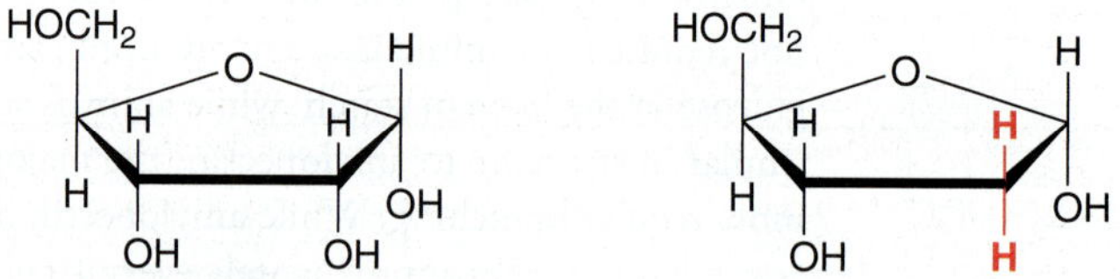

D-ribose ($\alpha$-furanose form)

2-Deoxy-D-ribose ($\alpha$-furanose form)

These two carbohydrates serve as the building blocks for RNA and DNA, respectively. These compounds are biologically coupled with certain nitrogen heterocycles (called bases) to yield special *N*-glycosides, called **nucleosides**. In each case, the β anomers are formed exclusively.

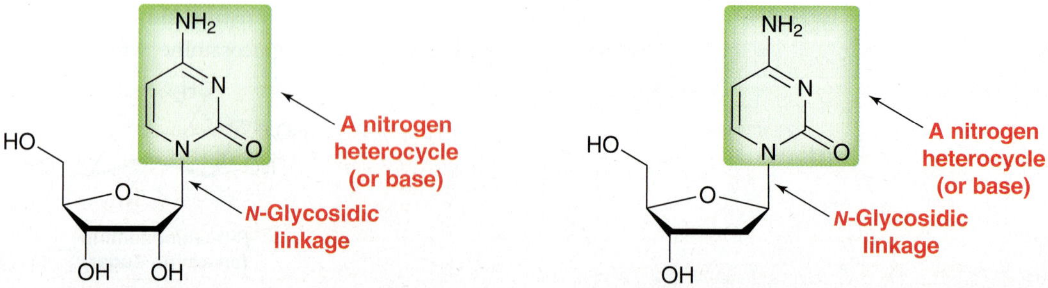

A nitrogen heterocycle (or base)

*N*-Glycosidic linkage

A ribonucleoside

A nitrogen heterocycle (or base)

*N*-Glycosidic linkage

A deoxyribonucleoside

## medically speaking

## Aminoglycoside Antibiotics

Aminoglycoside antibiotics are antibiotic compounds that contain both an amino sugar and a glycosidic linkage. The first known example, called streptomycin, was isolated in 1944 from the genus *Streptomyces*:

**Streptomycin**

A glycosidic bond

An amino hexose

Notice that the structure of streptomycin contains an amino sugar (highlighted). Interestingly, this amino hexose is a derivative of L-glucosamine, rather than D-glucosamine, which indicates that *Streptomyces* has developed a pathway for synthesizing L-glucose; this is somewhat unusual, although there are other known examples of L sugars in nature.

Many other antibiotic compounds have also been isolated from the genus *Streptomyces*, all of which are closely related in structure to streptomycin. Six of them are currently in use in the United States: kanamycin, neomycin, paromomycin, gentamicin, tobramycin, and netilmicin. All of these compounds exhibit at least one aminohexose in their structure, and extensive research on the relationship between structure and activity indicates that at least one amino sugar is necessary for antibiotic activity.

As shown, kanamycin and neomycin both contain two amino sugars. As mentioned in the chapter opener, neomycin is one of the three active ingredients in Neosporin.

Many studies have been undertaken to elucidate the mechanism of antibiotic action of aminoglycosides. It is believed that all of these compounds act by inhibiting protein synthesis in bacteria. In other words, these compounds prevent bacteria from synthesizing the enzymes that are necessary for proper functioning. The development of strains of antibiotic-resistant bacteria is a very serious problem in the field of medicine. The aminoglycosides are no exception to this problem. As clinical use of aminoglycoside antibiotics became more popular, some strains of bacteria have become resistant to them. Specifically, many strains of bacteria have developed the ability to produce enzymes that catalyze the modification of hydroxyl and amino groups of these antibiotics. For example, bacteria have emerged that are capable of transforming six different functional groups in kanamycin B. Some of these functional groups undergo enzyme-catalyzed acetylation, while others are phosphorylated. If any one of these groups is acetylated or phosphorylated, the modified antibiotic is no longer capable of binding to the bacterial RNA. Much of the current research on aminoglycoside antibiotics is directed toward designing structures that are less susceptible to inactivation by bacterial enzymes.

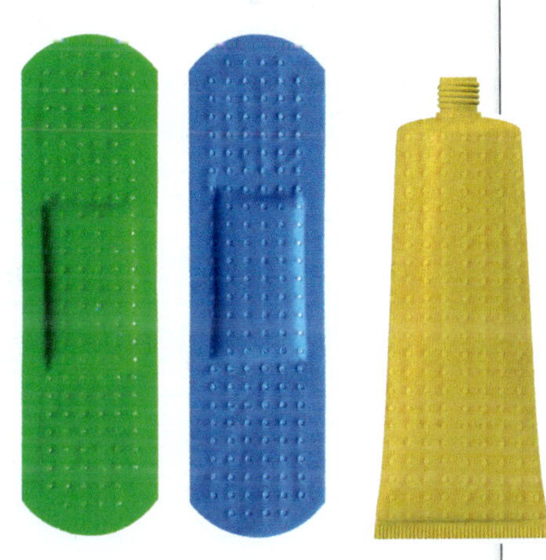

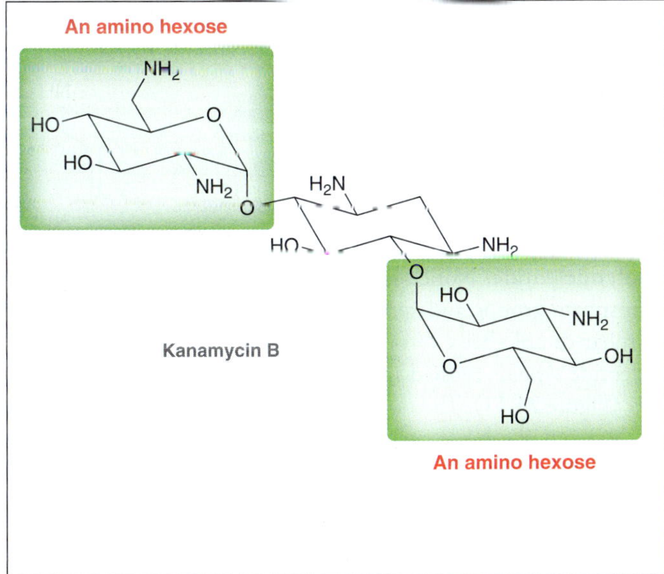


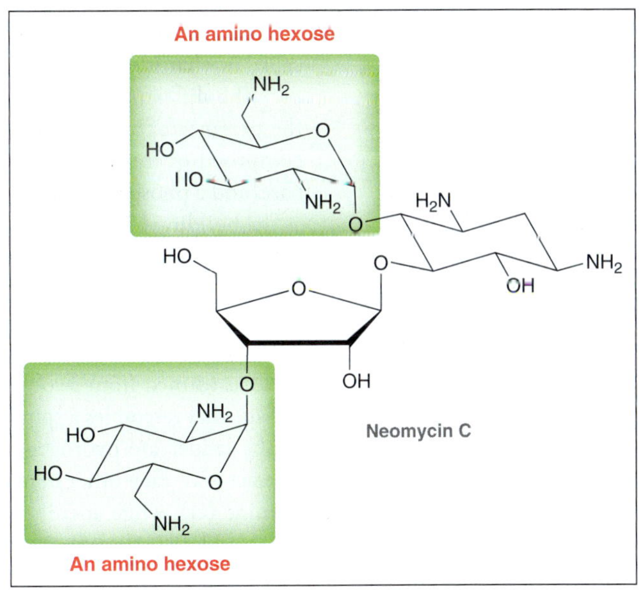

## DNA

Four kinds of heterocyclic amines are found as bases in DNA: cytosine (C), thymine (T), adenine (A), and guanine (G):

Cytosine C | Thymine T | Adenine A | Guanine G

Each of these four bases can couple to 2-deoxyribose, giving rise to four possible deoxyribonucleosides (Figure 24.9).

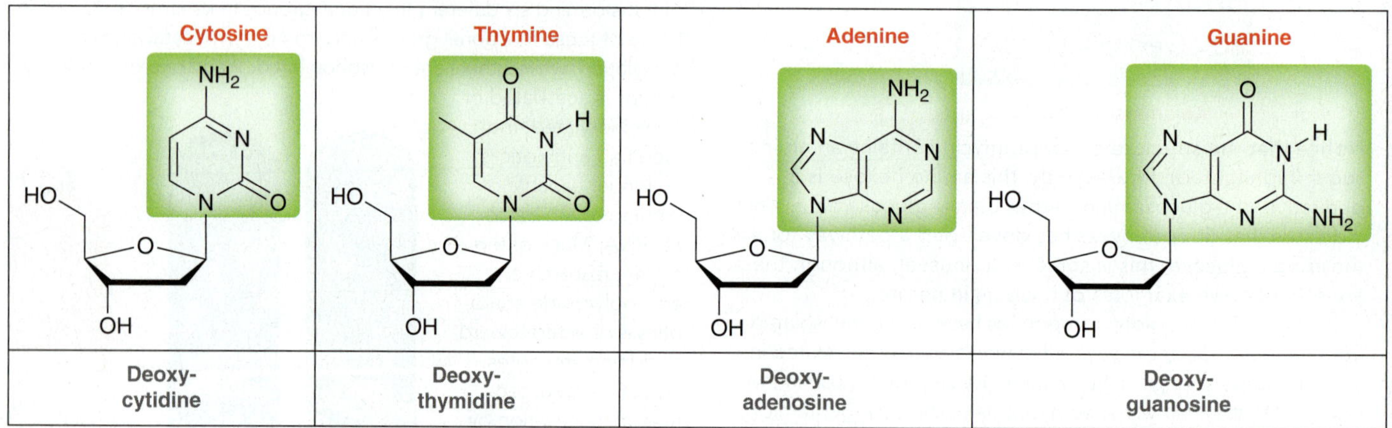


**FIGURE 24.9**
Structures of the four naturally occurring deoxyribonucleosides present in DNA.

Each of these four nucleosides can then be coupled to a phosphate group, giving compounds called **nucleotides** (rather than nucleosides). One of the four possible deoxyribonucleotides is shown:

The difference between a nucleoside and a nucleotide is the presence of a phosphate group. Deoxyribonucleotides are comprised of three parts: deoxyribose, a nitrogen-containing base, and a phosphate group. When linked together, they serve as the building blocks of DNA (Figure 24.10).

**A deoxyribonucleotide (deoxycytidine monophosphate)**

The segment of DNA in Figure 24.10 shows nucleotides linked together in a polymer, or **polynucleotide**. Notice that each sugar is connected to two phosphate groups and serves as the backbone for DNA. The familiar double-helix form of DNA is formed from two polynucleotide strands twisted into a shape that resembles a spiral ladder (Figure 24.11). The rungs of the ladder are hydrogen bonds between the bases, which interact with each other in pairs. As shown in Figure 24.12, cytosine (C) forms hydrogen bonds with guanine (G), while adenine (A) forms hydrogen bonds with thymine (T). The two strands of the spiral ladder are therefore complementary strands. They can be separated from each other much the way a zipper can be opened. DNA encodes all of our genetic information and serves as a template for assembling RNA, described in the following section.

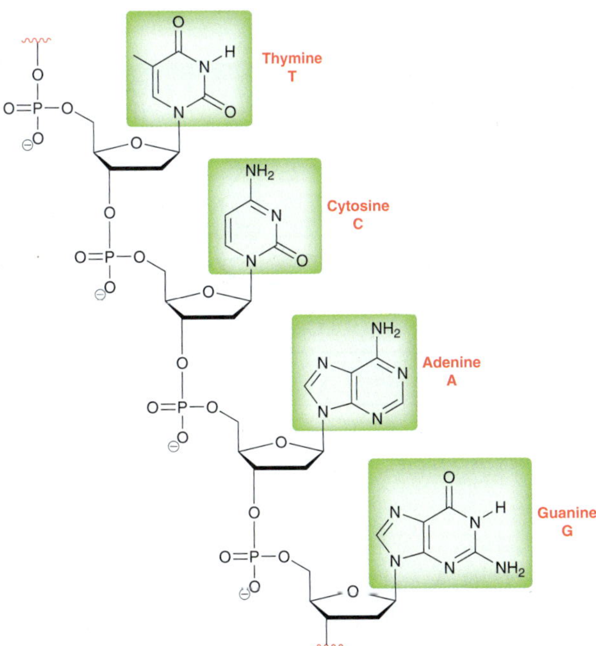


**FIGURE 24.10**
A segment of DNA, consisting of deoxynucleotides linked together in a polymer.

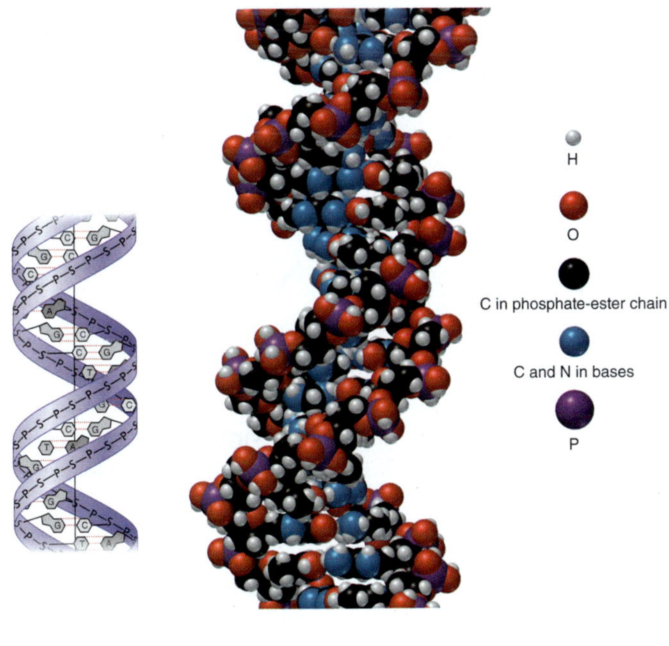


**FIGURE 24.11**
Two illustrations of the double-helix structure of DNA. Left, a structure showing the identity of the individual bases. Right, a color-coded space-filling model of DNA.

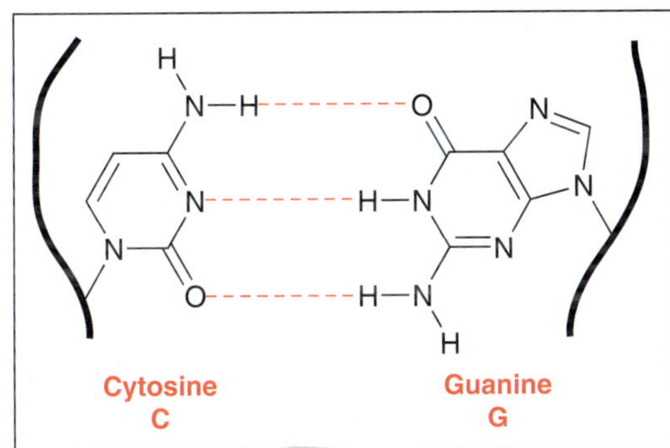


Adenine
A
Thymine
T

**FIGURE 24.12**
The hydrogen-bonding interactions that occur between complementary base pairs in DNA.

## RNA

A strand of RNA is very similar in structure to a single strand of DNA. Both are comprised of repeating nucleotide units where each unit consists of a sugar, a phosphate, and a base. There are two important differences between DNA and RNA: (1) the sugar in DNA is D-deoxyribose while the sugar in RNA is D-ribose and (2) in place of the thymine found in DNA, RNA contains a base called uracil (U). The difference between thymine and uracil is that thymine contains a methyl group that is lacking in uracil. Like thymine, uracil forms hydrogen bonds with adenine (A).

Thymine
T

Uracil
U

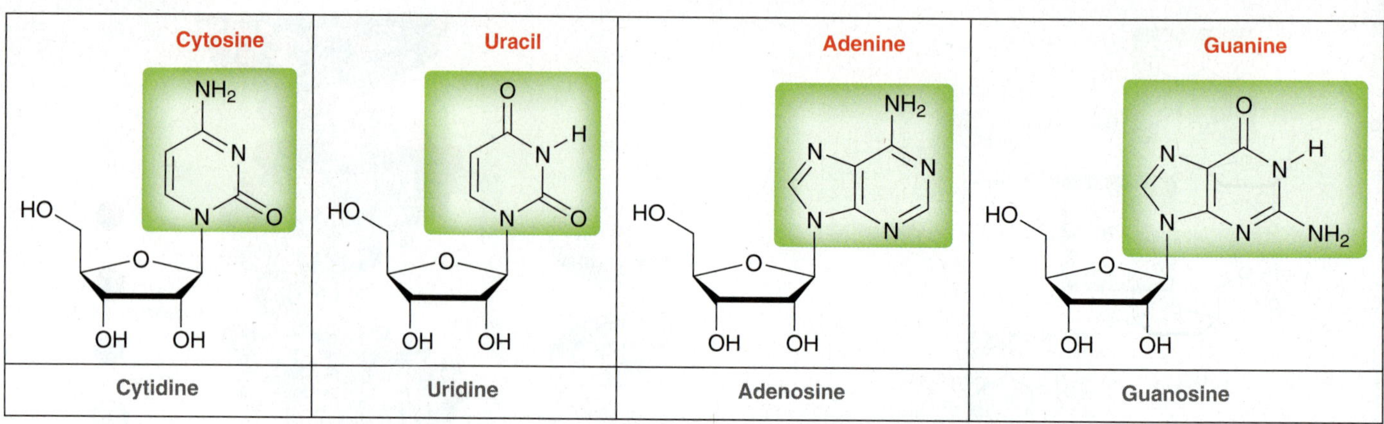


**FIGURE 24.13**
Structures of the four naturally occurring ribonucleosides present in RNA.

The four bases from which RNA is comprised are C, G, A, and U, giving rise to four possible nucleosides (Figure 24.13). Each of these nucleosides is coupled with a phosphate group, giving rise to four possible nucleotides which are the building blocks of RNA. A segment of RNA is shown in Figure 24.14. RNA directs the assembly of the proteins and enzymes that are used to catalyze the chemical reactions occurring in cells. Proteins and enzymes are the subjects of Chapter 26.

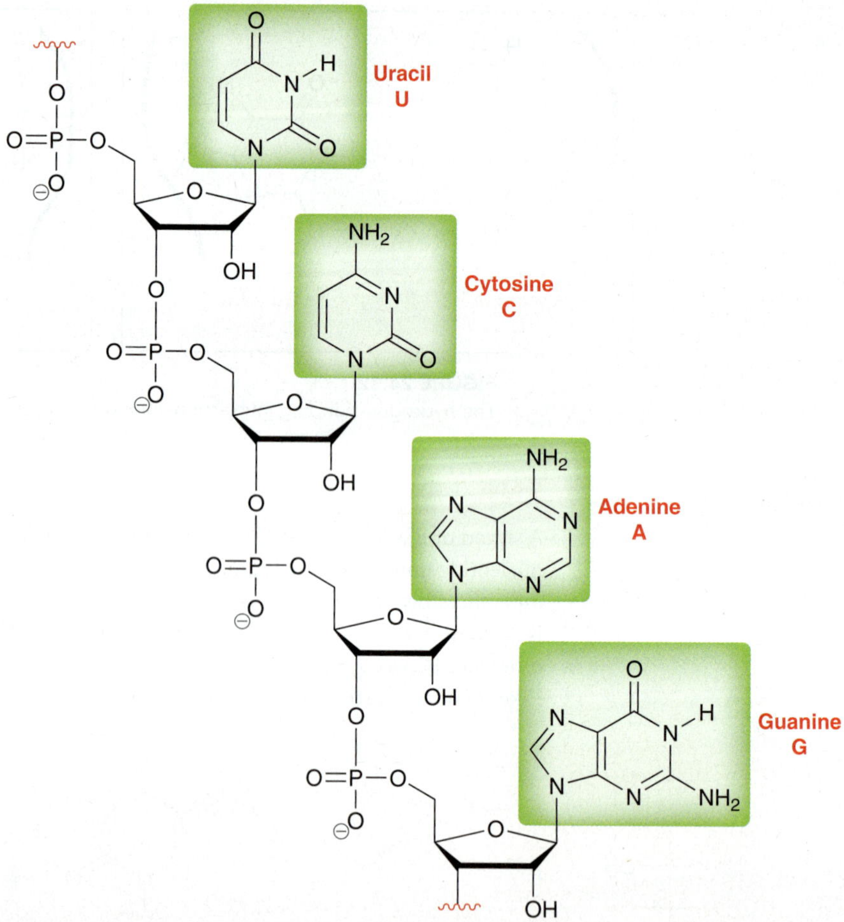

**FIGURE 24.14**
A segment of RNA, consisting of nucleotides linked together in a polymer.

## medically speaking — Erythromycin Biosynthesis

The genetic code (DNA) of all living things, including bacteria and fungi, is a blueprint for the biosynthesis of a wide variety of structurally complex compounds with a wide variety of functions. Some of these compounds can have powerful medicinal applications, as demonstrated by Alexander Fleming's discovery of penicillin in 1928. That discovery fueled an intensive search for other antibiotic compounds produced by microorganisms. Many such compounds have been discovered, including the broad-spectrum antibiotic erythromycin, identified in the 1950s.

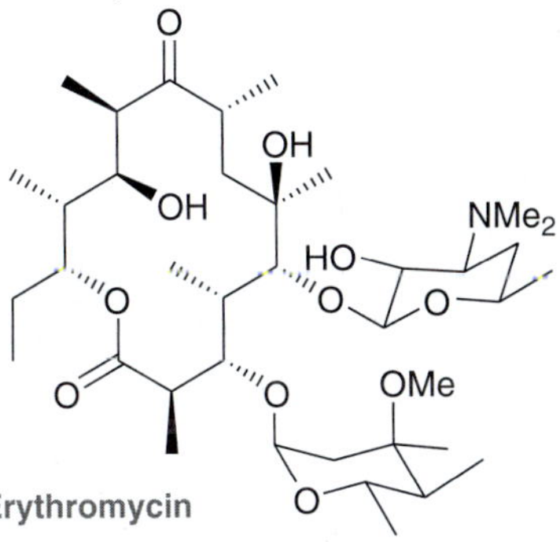

Erythromycin

Erythromycin is effective against certain kinds of bacteria, including *Staphylococcus aureus* and *Streptococcus pneumonia*, two pathogenic bacteria responsible for numerous infections in humans. Erythromycin is produced by fermentation cultures of the bacterium *Saccharopolyspora erythraea*, and strains of this organism are still used today for the large-scale industrial production of erythromycin.

Erythromycin is one member of a class of compounds called *macrolide* antibiotics, so named because of the presence of a large ring that forms the central part of their structures. Attached to the central macrolide structure of erythromycin are two carbohydrate rings, which are believed to be essential for its antibiotic activity. Erythromycin functions by inhibiting the formation of proteins essential for the survival of the bacteria.

Much work has been done to understand how erythromycin is produced by *S. erythraea*. One of the most exciting developments has been the complete sequencing of the genes in *S. erythraea* that are responsible for erythromycin biosynthesis. By introducing specific modifications to those genes, scientists have created modified bacteria that produce analogues of erythromycin that might also have potent antibiotic properties. Examples include the introduction of additional carbon atoms (analogue **1**) or an aromatic ring (analogue **2**) or even a halogen (analogue **3**). These derivatives, and many others produced in a similar manner, are being investigated in the hopes of finding new antibiotic compounds that may be more potent than erythromycin. In other words, scientists have learned how to use bacteria as miniature production facilities for compounds that would otherwise be very difficult to prepare in the laboratory. Modified bacteria can be used to achieve the biosynthesis of complex structures that would take synthetic organic chemists years to make through conventional synthetic techniques.

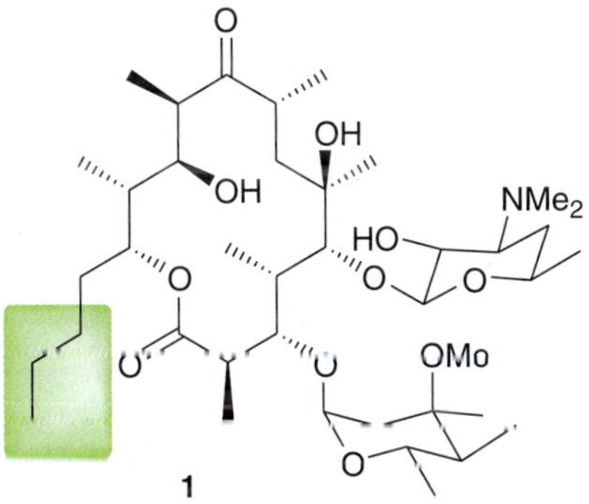

**1** 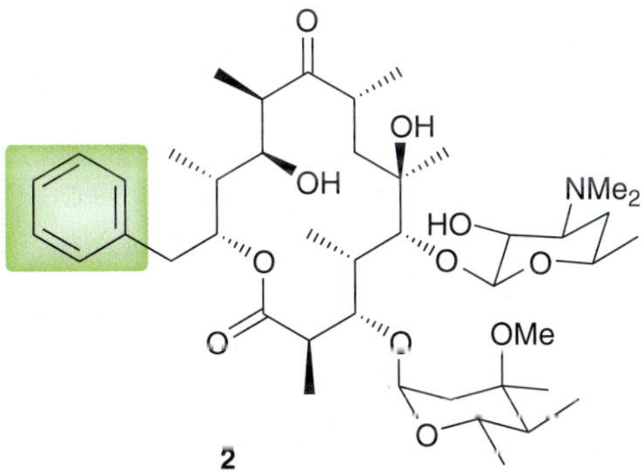 **2** 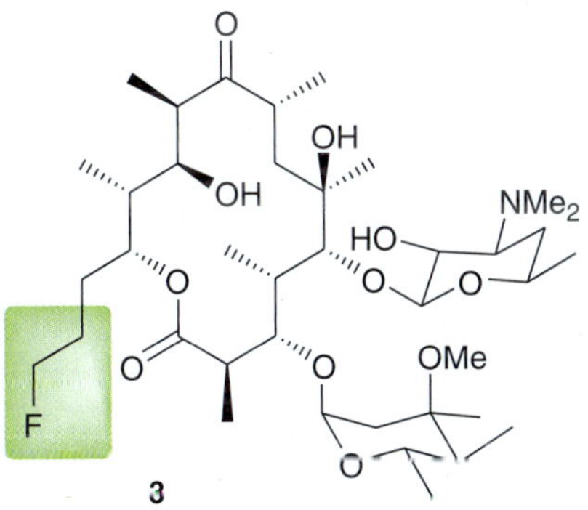 **3**

# REVIEW OF REACTIONS

### Hemiacetal Formation

**Pyranose Rings**

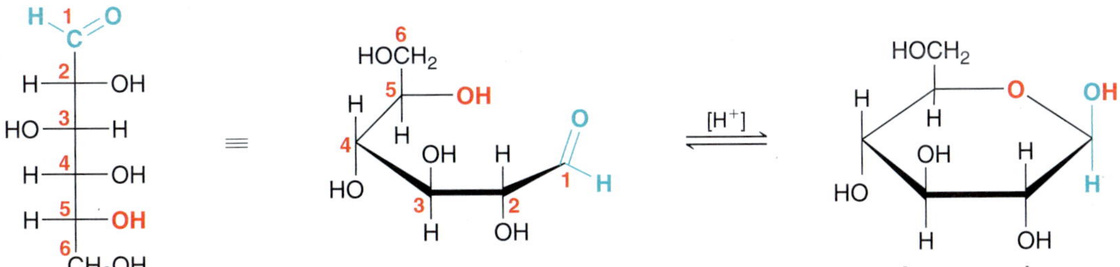

D-Glucose

$[H^+]$

A pyranose ring

### Furanose Rings

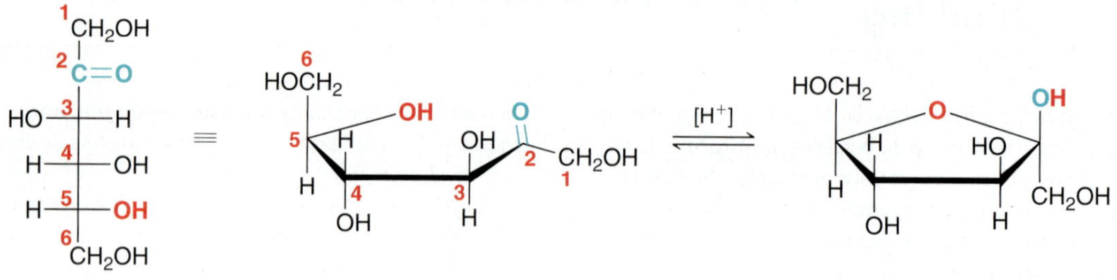

1 $CH_2OH$
2 C=O
HO 3 H
H 4 OH
H 5 OH
6 $CH_2OH$

≡

6 $HOCH_2$
5 H OH
H 4 OH
OH 3 H
O 2 $CH_2OH$ 1

$[H^+]$

$HOCH_2$ O OH
H HO
H $CH_2OH$
OH H

D-Fructose

A furanose ring

### Reactions of Monosaccharides

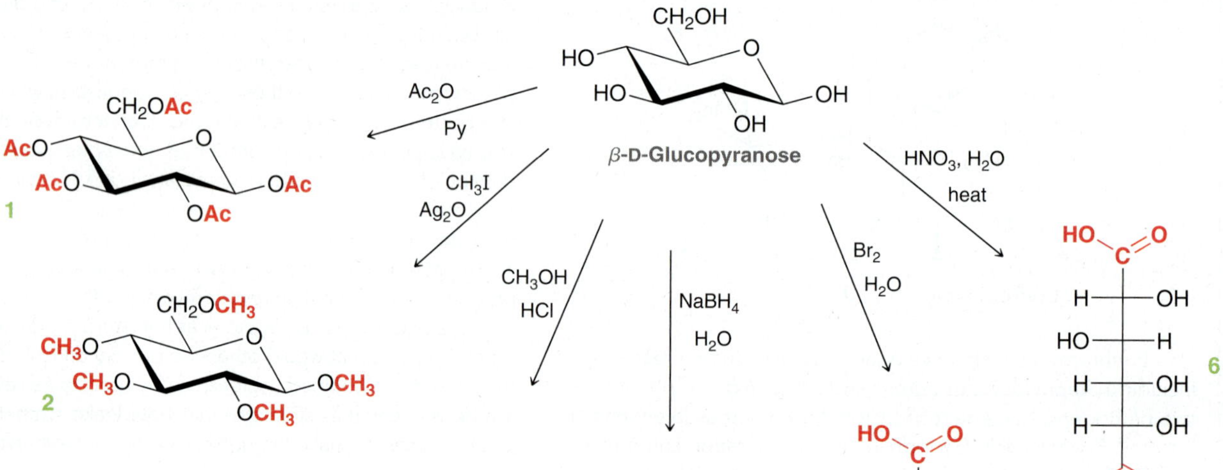

$CH_2OH$
HO O
HO OH
OH

*β*-D-Glucopyranose

$Ac_2O$
Py

$CH_2OAc$
AcO O
AcO OAc
OAc
1

$CH_3I$
$Ag_2O$

$CH_2OCH_3$
$CH_3O$ O
$CH_3O$ $OCH_3$
$OCH_3$
2

$CH_3OH$
HCl

$CH_2OH$
HO O
HO $OCH_3$
OH
3

$NaBH_4$
$H_2O$

$CH_2OH$
H OH
HO H
H OH
H OH
$CH_2OH$
4

$Br_2$
$H_2O$

HO C=O
H OH
HO H
H OH
H OH
$CH_2OH$
5

$HNO_3$, $H_2O$
heat

HO C=O
H OH
HO H
H OH
H OH
HO C=O
6

1. Acetylation
2. Alkylation
3. Glycoside formation
4. Reduction
5. Oxidation to an aldonic acid
6. Oxidation to an aldaric acid

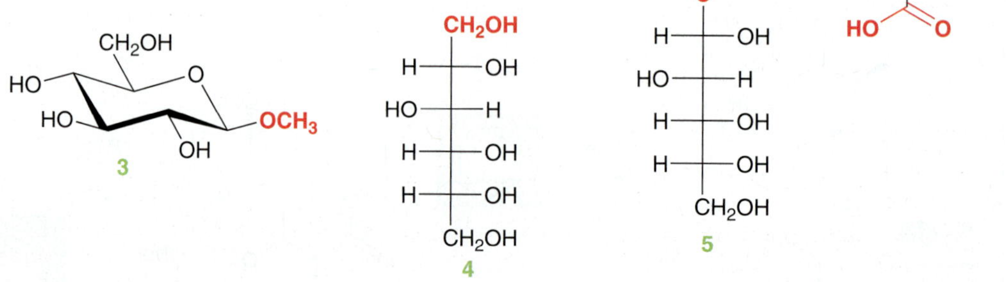

### Chain Lengthening and Chain Shortening

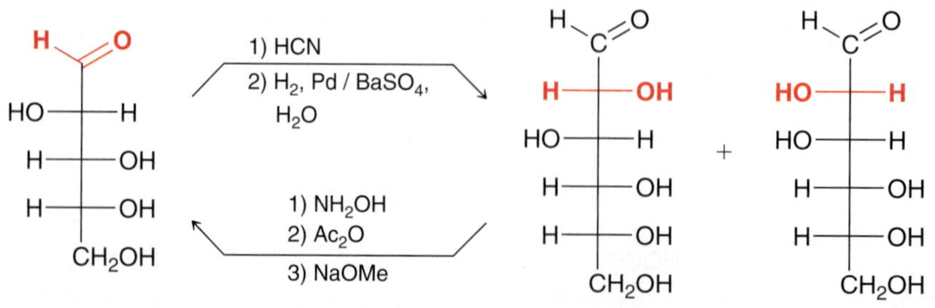

H C=O
HO H
H OH
H OH
$CH_2OH$

D-Arabinose

1) HCN
2) $H_2$, Pd / $BaSO_4$, $H_2O$

1) $NH_2OH$
2) $Ac_2O$
3) NaOMe

H C=O
H OH
HO H
H OH
H OH
$CH_2OH$

D-Glucose

+

H C=O
HO H
HO H
H OH
H OH
$CH_2OH$

D-Mannose

# REVIEW OF CONCEPTS AND VOCABULARY

### SECTION 24.1

- **Carbohydrates** are polyhydroxy aldehydes or ketones.
- In nature, carbohydrates are used for energy storage and structural rigidity.

### SECTION 24.2

- Simple sugars are called **monosaccharides** and are generally classified as **aldoses** and **ketoses.**
- Aldoses and ketoses are further classified according to the number of carbon atoms they contain.
- (+)-Glyceraldehyde, called D-glyceraldehyde, is abundant in nature, but its enantiomer is not.
- For all **D sugars,** the chirality center farthest from the carbonyl group has the *R* configuration.
- An **L sugar** is the enantiomer of the corresponding D sugar and is not necessarily levorotatory.

### SECTION 24.3

- There are two D aldotetroses called D-erythrose and D-threose.
- There are four D aldopentoses called D-ribose, D-arabinose, D-xylose, and D-lyxose.
- There are eight D aldohexoses, of which D-glucose is the most abundant in nature.

### SECTION 24.4

- There is one D ketotetrose called D-erythrulose.
- There are two D ketopentoses called D-ribulose and D-xylulose.
- There are four D ketohexoses called D-psicose, D-fructose, D-sorbose, and D-tagatose.

### SECTION 24.5

- Aldohexoses can form cyclic hemiacetals that exhibit a **pyranose** ring.
- Cyclization produces two stereoisomeric hemiacetals, called **anomers.** The newly created chirality center is called the **anomeric carbon.**
- In the $\alpha$ **anomer,** the hydroxyl group at the anomeric position is *trans* to the $CH_2OH$ group, while in the $\beta$ **anomer,** the hydroxyl group is *cis* to the $CH_2OH$ group.
- The open-chain form exhibits only a minimal presence at equilibrium.
- Anomers equilibrate by a process called **mutarotation,** which is catalyzed by either acid or base.
- Some carbohydrates, such as D-fructose, can also form five-membered rings, called **furanose** rings.

### SECTION 24.6

- Monosaccharides are converted into their ester derivatives when treated with excess acid chloride or anhydride.
- Monosaccharides are converted into their ether derivatives when treated with excess alkyl halide and silver oxide.
- When treated with an alcohol under acid-catalyzed conditions, monosaccharides are converted into acetals, called **glycosides.** Both anomers are formed.
- Upon treatment with sodium borohydride, an aldose or ketose can be reduced to yield an **alditol.**
- When treated with a suitable oxidizing agent, an aldose can be oxidized to yield an **aldonic acid.**
- Aldoses and ketoses are **reducing sugars.**
- When treated with $HNO_3$, an aldose is oxidized to give a dicarboxylic acid called an **aldaric acid.**
- D-Glucose and D-mannose are **epimers** and are interconverted under strongly basic conditions.
- The **Kiliani-Fischer synthesis** can be used to lengthen the chain of an aldose.
- The **Wohl degradation** can be used to shorten the chain of an aldose.

### SECTION 24.7

- **Disaccharides** are comprised of two monosaccharide units joined together via a glycosidic linkage.
- Examples of disaccharides are maltose, cellobiose, lactose, and sucrose.

### SECTION 24.8

- **Polysaccharides** are polymers consisting of repeating monosaccharide units linked by glycoside bonds.
- Examples of polysaccharides are starch, cellulose, and glycogen.

### SECTION 24.9

- **Amino sugars** are carbohydrate derivatives in which one or more OH groups have been replaced with amino groups.
- The *N*-acetyl derivative of $\beta$-D-glucosamine serves as the repeating monosaccharide unit in chitin.

### SECTION 24.10

- When treated with an amine in the presence of an acid catalyst, monosaccharides are converted into their corresponding ***N*-glycosides.**
- D-Ribose and D-2-deoxyribose are carbohydrates that form especially important *N*-glycosides called **nucleosides.**
- There are four naturally occurring nucleosides of 2-deoxyribose, each of which can be coupled to a phosphate group to form a **nucleotide.**
- DNA is comprised of nucleotides linked together in a polymer, or **polynucleotide.**
- RNA differs from DNA in that it contains D-ribose rather than deoxyribose, it contains uracil in place of thymine, and it is single stranded rather than double stranded like DNA.

# SKILLBUILDER REVIEW

## 24.1 DRAWING THE CYCLIC HEMIACETAL OF A HYDROXYALDEHYDE

**STEP 1** Assign numbers to the carbon atoms, beginning with the carbon of the aldehyde group, and continue until reaching the carbon connected to the OH group.

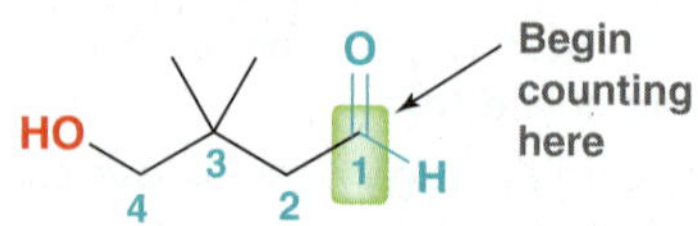


**STEP 2** Using the numbering system from step 1, draw a ring that incorporates an oxygen atom in the ring.

**STEP 3** Using the numbering system from step 1, place the substituents in the correct locations.

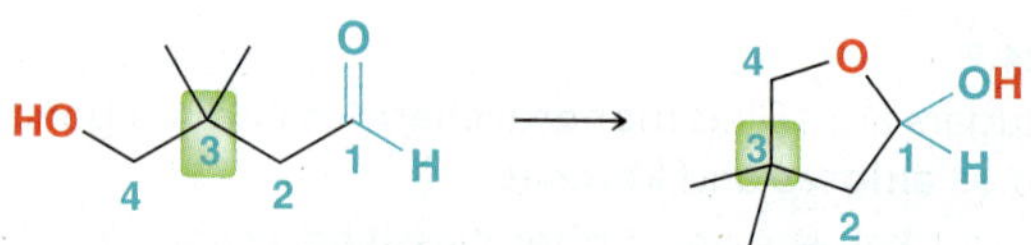


Try Problems 24.9–24.11, 24.46, 24.47

## 24.2 DRAWING A HAWORTH PROJECTION OF AN ALDOHEXOSE

**STEP 1** Draw the skeleton of a Haworth projection, with the ring oxygen in the back right corner. Number each position.

**STEP 2** For D sugars, place the $CH_2OH$ group pointing up at C5.

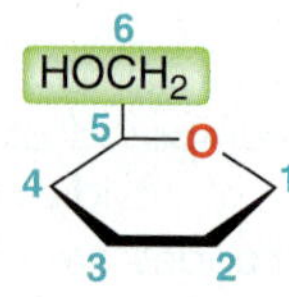


**STEP 3** Place the OH group at the anomeric position. For the $\alpha$ anomer, the OH group should be *trans* to the $CH_2OH$ group.

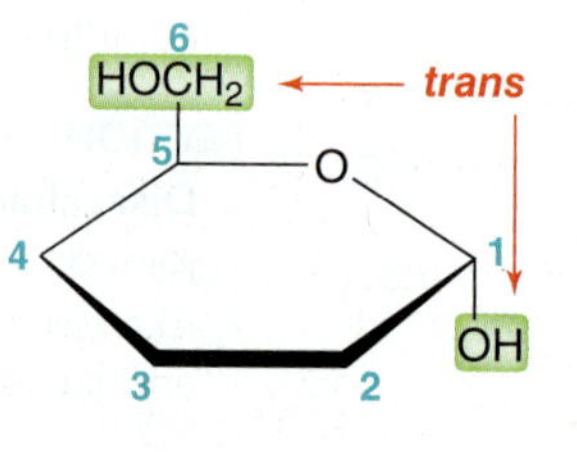


**STEP 4** Place the remaining OH groups at C2, C3, and C4. Groups on the left side of the Fischer projection should point up. Groups on the right side of the Fischer projection should point down.

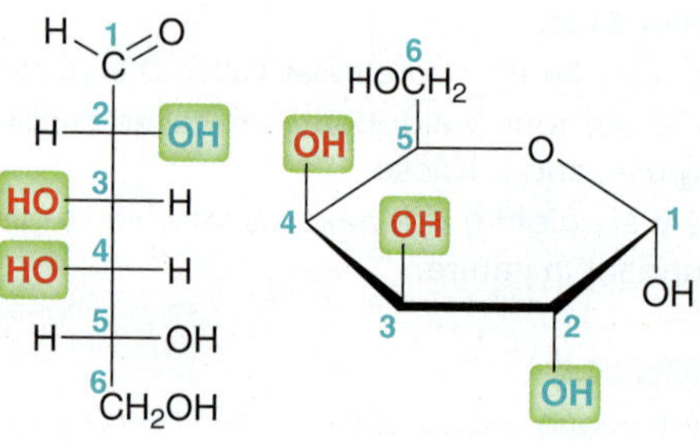


Try Problems 24.12–24.15, 24.48a, 24.53b,c,d

## 24.3 DRAWING THE MORE STABLE CHAIR CONFORMATION OF A PYRANOSE RING

**STEP 1** Draw the Haworth projection.

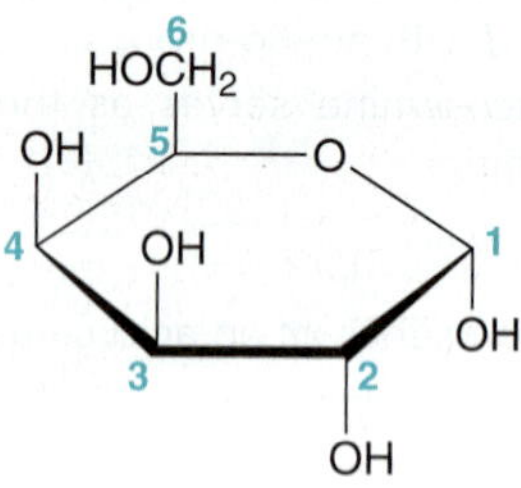


**STEP 2** Draw the skeleton of a chair conformation, with the oxygen atom in the upper, rear-right corner.

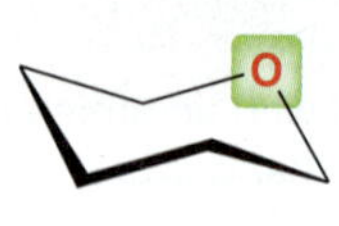


**STEP 3** Label each substituent as either "up" or "down" and then place each substituent on the chair accordingly.

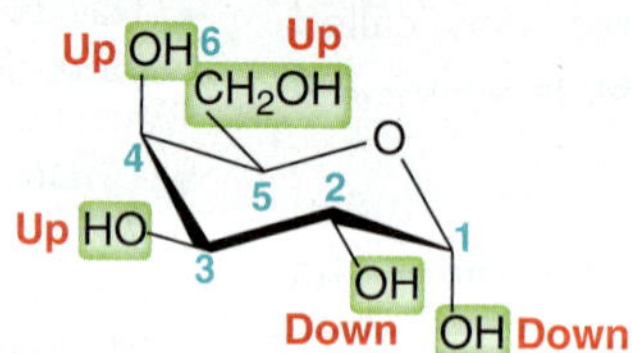


**STEP 4** Verify that this chair conformation is more stable. Look for the largest group and make sure it is equatorial.

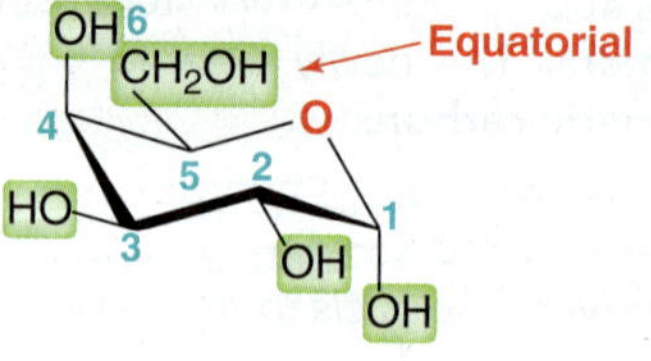


Try Problems 24.16–24.18, 24.60

## 24.4 IDENTIFYING A REDUCING SUGAR

**STEP 1** Identify the anomeric position.

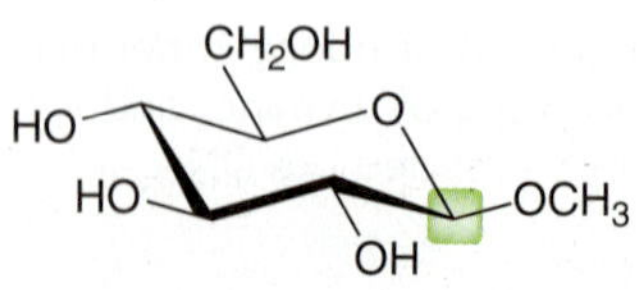


**STEP 2** Determine if the group attached to the anomeric position is a hydroxy group or alkoxy group:

- If hydroxy group, then the compound is a reducing sugar.
- If alkoxy group, then the compound is not a reducing sugar.

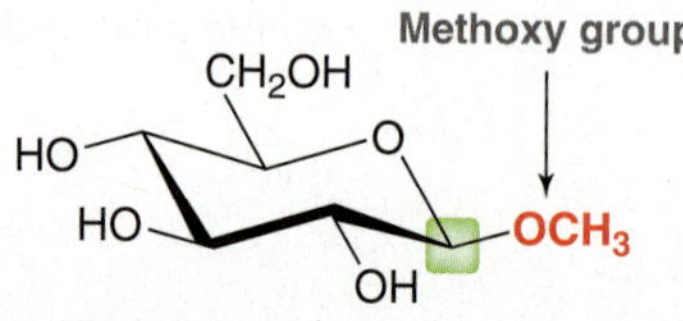


Try Problems 24.31–24.33, 24.69, 24.76a

**24.5 DETERMINING WHETHER A DISACCHARIDE IS A REDUCING SUGAR**

**STEP 1** Identify the anomeric positions.

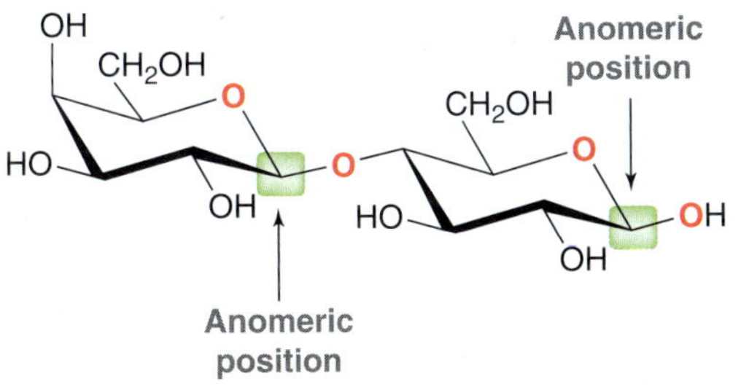


**STEP 2** Determine if the groups attached to the anomeric positions are hydroxy groups or alkoxy groups:

- If at least one is a hydroxy group, then the compound is a reducing sugar.
- If neither are hydroxy groups, then the compound is not a reducing sugar.

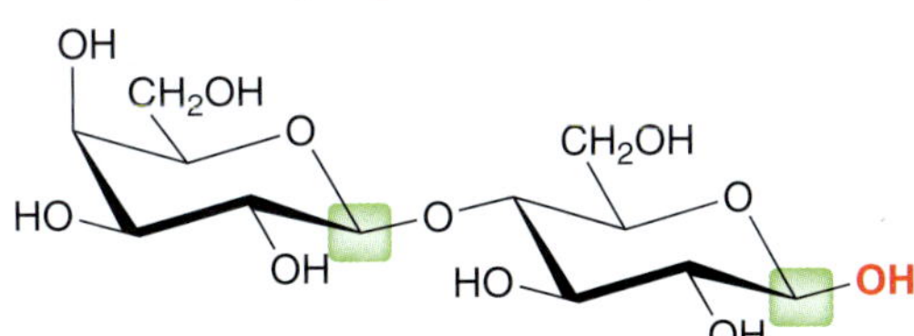


Try Problems **24.39, 24.40, 24.73**

## PRACTICE PROBLEMS

Note: Most of the Problems are available within **WileyPLUS**, an online teaching and learning solution.

**24.42** Classify each of the following monosaccharides as either D or L, as either an aldo or a keto sugar, and as a tetrose, pentose, or hexose:

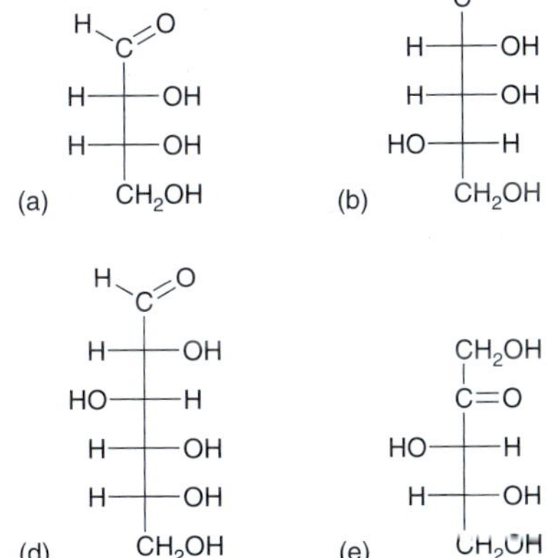


(b) (c) (e)

**24.43** Identify each of the following structures as either D- or L-glyceraldehyde:

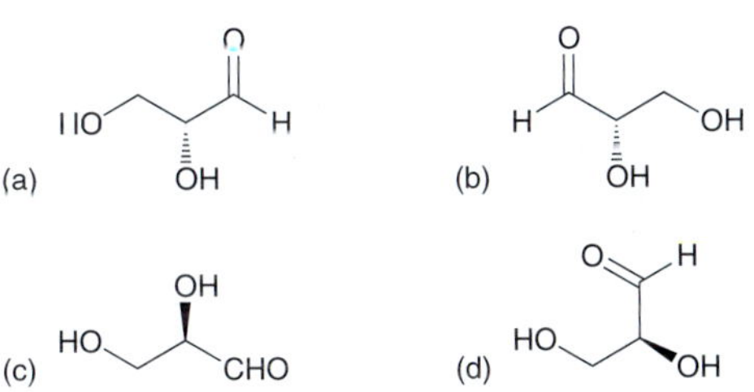


(b) (d)

**24.44** Name each of the following aldohexoses:

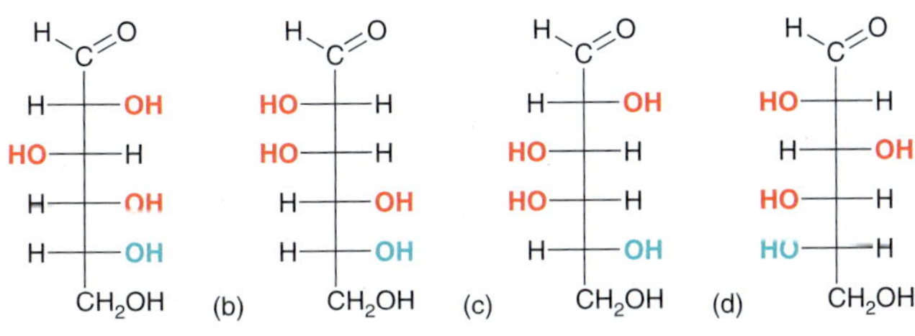


**24.45** Consider the structures of the D aldopentoses:

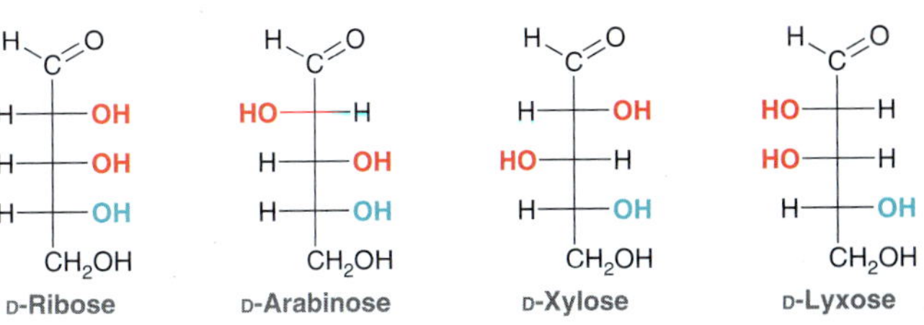


(a) Identify the aldopentose that is epimeric with D-arabinose at C2.

(b) Identify the aldopentose that is epimeric with D-lyxose at C3.

(c) Draw the enantiomer of D-ribose.

(d) Identify the relationship between the enantiomer of D-arabinose and the C2 epimer of L-ribose.

(e) Identify the relationship between D-ribose and D-lyxose.

**24.46** Draw the cyclic hemiacetal that is formed when each of the following bifunctional compounds is treated with aqueous acid.

(a) (b) (c)

**24.47** Identify the hydroxyaldehyde that will cyclize under acidic conditions to give the following hemiacetal:

**24.48** D-Ribose can adopt two pyranose forms and two furanose forms.

(a) Draw both pyranose forms of D-ribose and identify each as α or β.

(b) Draw both furanose forms of D-ribose and identify each as α or β.

**24.49** For each of the following pairs of compounds, determine whether they are enantiomers, epimers, diastereomers that are not epimers, or identical compounds:

(a)

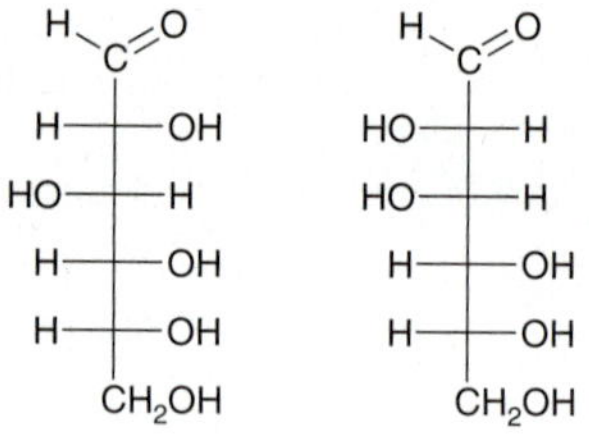

(b)

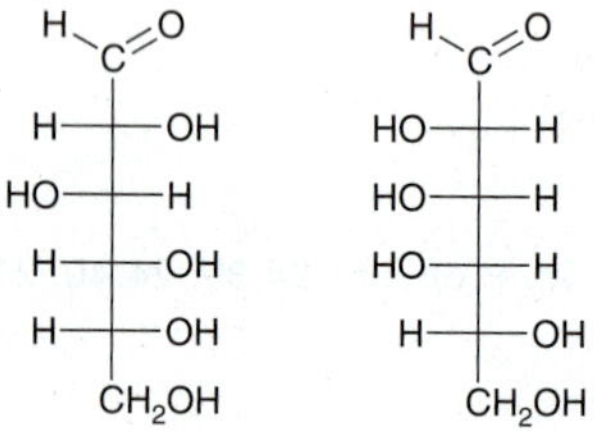

(c)

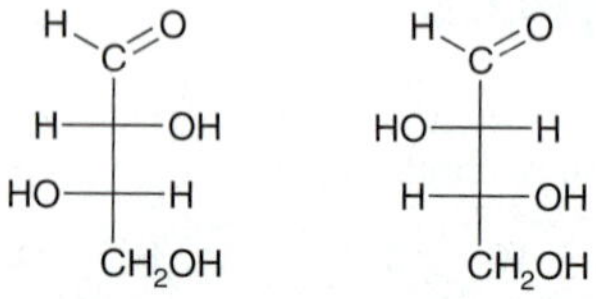

(d)

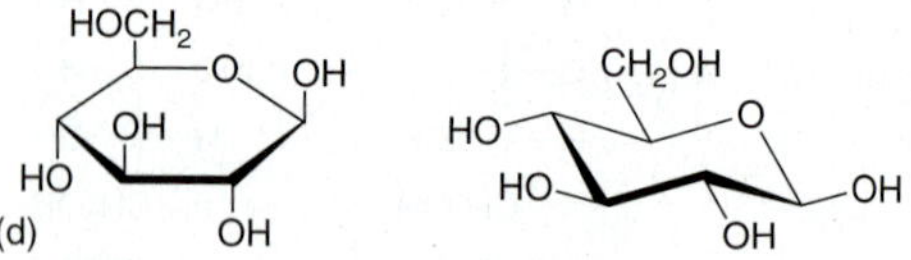

**24.50** Draw the open-chain form of the compound formed when methyl β-D-glucopyranoside is treated with aqueous acid.

**24.51** Assign the configuration of each chirality center in the following compounds:

(a) (b) 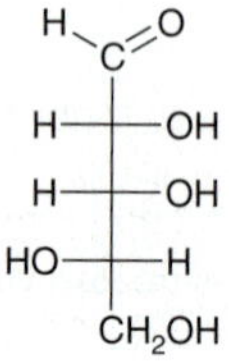 (c)

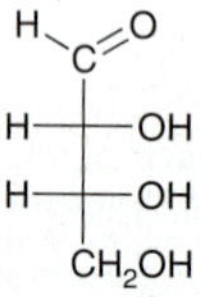

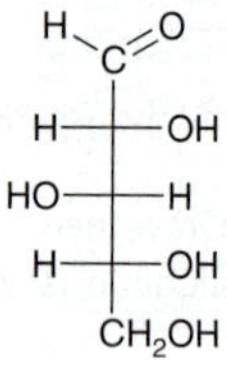

(d) 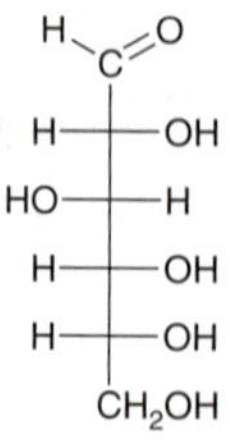 (e)

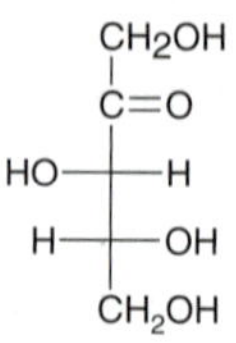

**24.52** Draw a Fischer projection for each of the following compounds:

(a) D-Glucose

(b) D-Galactose

(c) D-Mannose

(d) D-Allose

**24.53** Draw a Haworth projection for each of the following compounds:

(a) β-D-Fructofuranose

(b) β-D-Galactopyranose

(c) β-D-Glucopyranose

(d) β-D-Mannopyranose

**24.54** Draw a Haworth projection showing the α-pyranose form of the D-aldohexose that is epimeric with D-glucose at C3.

**24.55** Provide a complete name for each of the following compounds:

(a) (b)

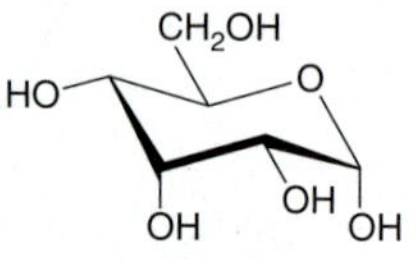

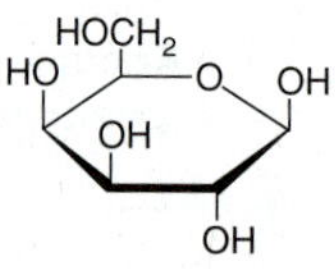

(c)

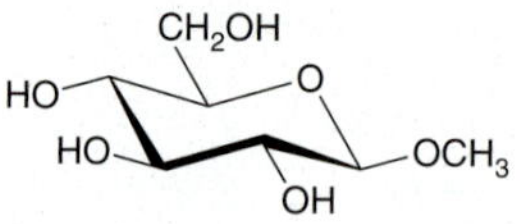

**24.56** Draw the open-chain form of each of the compounds in the previous problem.

**24.57** Draw the products that are expected when β-D-allopyranose is treated with each of the following reagents:

(a) Excess $CH_3I$, $Ag_2O$

(b) Excess acetic anhydride, pyridine

(c) $CH_3OH$, HCl

**24.58** When D-galactose is heated in the presence of nitric acid, an optically inactive compound is obtained. Draw the structure of the product and explain why it is optically inactive.

**24.59** In addition to D-galactose, one other D-aldohexose also forms an optically inactive aldaric acid when treated with nitric acid. Draw the structure of this aldohexose.

**24.60** Draw the more stable chair conformation of α-D-altropyranose and label all substituents as axial or equatorial.

**24.61** Draw the products that are expected when α-D-galactopyranose is treated with excess methyl iodide in the presence of silver oxide, followed by aqueous acid.

**24.62** For each of the following pairs of compounds, determine whether they are enantiomers, epimers, diastereomers that are not epimers, or identical compounds:

(a) D-Glucose and D-gulose

(b) 2-Deoxy-D-ribose and 2-deoxy-D-arabinose

**24.63** Draw all possible 2-ketohexoses that are D sugars.

**24.64** Identify the two aldohexoses that will undergo a Wohl degradation to yield D-ribose. Draw a Fischer projection of the open-chain form for each of these two aldohexoses.

**24.65** Identify the two aldohexoses that are obtained when D-arabinose undergoes a Kiliani-Fischer synthesis.

**24.66** Identify the two products obtained when D-glyceraldehyde is treated with HCN and determine the relationship between these two products.

**24.67** When treated with sodium borohydride, D-glucose is converted into an alditol.

(a) Draw the structure of the alditol.

(b) Which L-aldohexose gives the same alditol when treated with sodium borohydride?

**24.68** Which of the D-aldohexoses are converted into optically inactive alditols upon treatment with sodium borohydride?

**24.69** Determine whether each of the following compounds is a reducing sugar:

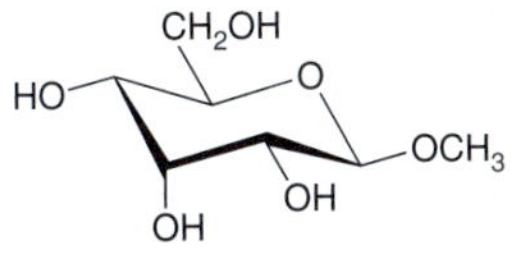

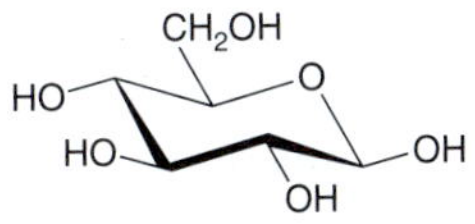

**24.70** Identify the reagents that you would use to convert β-D-glucopyranose into each of the following compounds:

(a) (b) 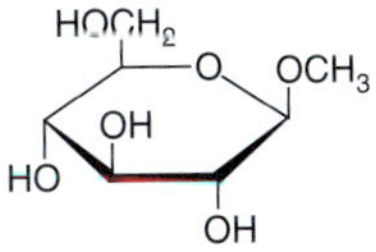 (c) 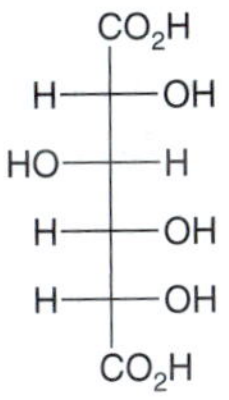 (d)

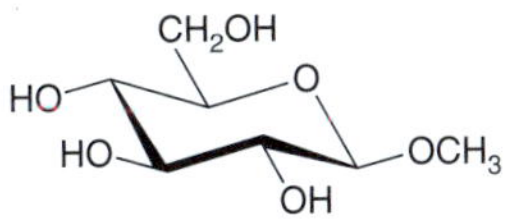

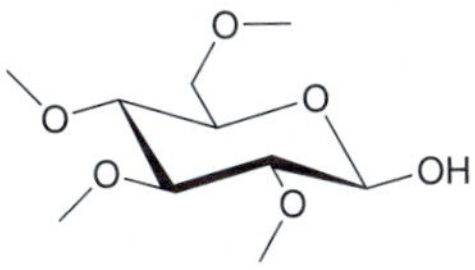

**24.71** Identify the product(s) that would be formed when each of the following compounds is treated with aqueous acid:

(a) Methyl α-D-glucopyranoside

(b) Ethyl β-D-galactopyranoside

**24.72** Consider the structures of the four D-aldopentoses (Figure 24.3).

(a) Which D-aldopentose produces the same aldaric acid as D-lyxose?

(b) Which D-aldopentoses yield optically inactive alditols when treated with sodium borohydride?

(c) Which D-aldopentose yields the same alditol as L-xylose?

(d) Which D-aldopentose can close into a β-pyranose form in which all substituents are equatorial?

**24.73** Trehalose is a naturally occurring disaccharide found in bacteria, insects, and many plants. It protects cells from dry conditions because of its ability to retain water, thereby preventing cellular damage from dehydration. This property of trehalose has also been exploited in the preparation of food and cosmetics. Trehalose is not a reducing sugar, it is hydrolyzed to yield two equivalents of D-glucose, and it does not have any β-glycoside linkages. Draw the structure of trehalose.

**24.74** Xylitol is found in many kinds of berries. It is approximately as sweet as sucrose but with fewer calories. It is often used in sugarless chewing gum. Xylitol is obtained upon reduction of D-xylose. Draw a structure of xylitol.

**24.75** Isomaltose is similar in structure to maltose, except that it is a 1→6 α-glycoside, rather than a 1→4 α-glycoside. Draw the structure of isomaltose.

**24.76** Salicin is a natural analgesic present in the bark of willow trees, and it has been used for thousands of years to treat pain and reduce fevers.

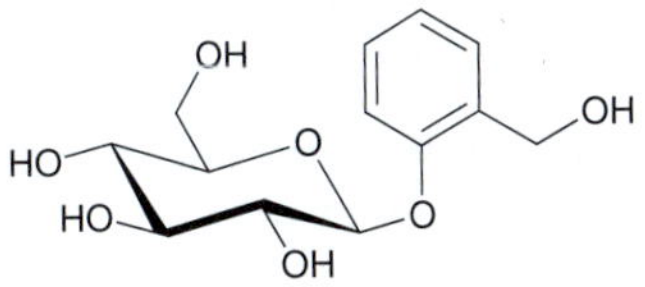

**Salicin**

(a) Is salicin a reducing sugar?

(b) Identify the products obtained when salicin is hydrolyzed in the presence of an acid.

(c) Is salicin an α-glycoside or a β-glycoside?

(d) Draw the major product expected when salicin is treated with excess acetic anhydride in the presence of pyridine.

(e) Would you expect salicin to exhibit mutarotation when dissolved in neutral water?

**24.77** Draw a mechanism for the following transformation:

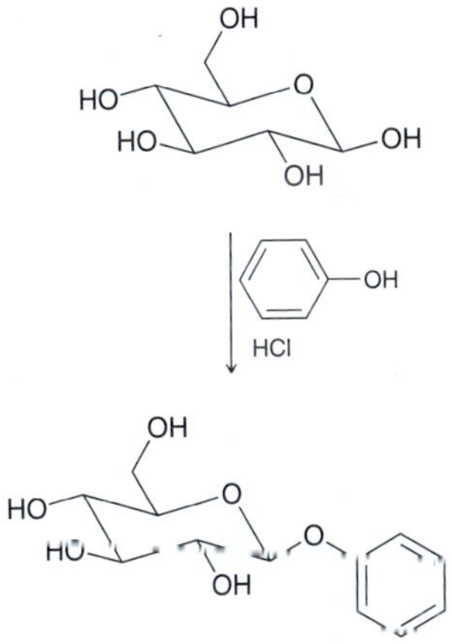

HCl

**24.78** Draw the α-*N*-glycoside and the β-*N*-glycoside formed when D-glucose is treated with aniline ($C_6H_5NH_2$).

**24.79** Draw the nucleoside formed from each of the following pairs of compounds and name the nucleoside.

(a) 2-Deoxy-D-ribose and adenine

(b) D-Ribose and guanine

**24.80** Compound **A** is a D-aldopentose that is converted into an optically active alditol upon treatment with sodium borohydride. Draw two possible structures for compound **A**.

# INTEGRATED PROBLEMS

**24.81** When D-glucose is treated with an aqueous bromine solution (buffered to a pH of 6), an aldonic acid is formed called D-gluconic acid. Treatment of D-gluconic acid with an acid catalyst produces a lactone (cyclic ester) with a six-membered ring.

(a) Draw the structure of D-gluconic acid.

(b) Draw the structure of the lactone formed from D-gluconic acid, showing the configuration at each chirality center.

(c) Would you expect this lactone to be optically active?

(d) Explain how you would distinguish between the D-gluconic acid and the lactone using IR spectroscopy.

**24.82** When each of the D-aldohexoses assumes an α-pyranose form, the $CH_2OH$ group occupies an equatorial position in the more stable chair conformation. The one exception is D-idose, for which the $CH_2OH$ group occupies an axial position in the more stable chair conformation. Explain this observation and then draw the more stable chair conformation of L-idose in its α-pyranose form.

**24.83** Draw the product that is expected when the β-pyranose form of compound **A** is treated with excess ethyl iodide in the presence of silver oxide. The following information can be used to determine the identity of compound **A**:

1. The molecular formula of compound **A** is $C_6H_{12}O_6$.
2. Compound **A** is a reducing sugar.
3. When compound **A** is subjected to a Wohl degradation two times sequentially, D-erythrose is obtained.
4. Compound **A** is epimeric with D-glucose at C3.
5. The configuration at C2 is *R*.

**24.84** Explain why glucose is the most common monosaccharide observed in nature.

# CHALLENGE PROBLEMS

**24.85** When D-glucose is treated with aqueous sodium hydroxide, a complex mixture of carbohydrates is formed, including D-mannose and D-fructose. Over time, almost all aldohexoses will be present in the mixture. Even L-glucose can be detected, albeit in very small concentrations. Using the fewest number of mechanistic steps possible, draw a reasonable mechanism showing the formation of L-glucose from D-glucose:

H–C=O
H–C–OH
HO–C–H
H–C–OH
H–C–OH
$CH_2OH$
**D-Glucose**

$\xrightarrow[H_2O]{NaOH}$

H–C=O
HO–C–H
H–C–OH
HO–C–H
HO–C–H
$CH_2OH$
**L-Glucose**

**24.86** Compound **X** is a D-aldohexose that can adopt a β-pyranose form with only one axial substituent. Compound **X** undergoes a Wohl degradation to produce an aldopentose, which is converted into an optically active alditol when treated with sodium borohydride. From the information presented, there are only two possible structures for compound **X**. Identify the two possibilities and then propose a chemical test that would allow you to distinguish between the two possibilities and thereby determine the structure of compound **X**.

**24.87** Compound **A** is a D-aldopentose. When treated with sodium borohydride, compound **A** is converted into an alditol that exhibits three signals in its $^{13}C$ NMR spectrum. Compound **A** undergoes a Kiliani-Fischer synthesis to produce two aldohexoses, compounds **B** and **C**. Upon treatment with nitric acid, compound **B** yields compound **D**, while compound **C** yields compound **E**. Both **D** and **E** are optically active aldaric acids.

(a) Draw the structure of compound **A**.

(b) Draw the structures of compounds **D** and **E** and describe how you might be able to distinguish between these two compounds using $^{13}C$ NMR spectroscopy.

# 25

# Amino Acids, Peptides, and Proteins



## DID YOU EVER WONDER...

**how crime-scene investigators are able to find invisible fingerprints and render them visible?**

A fingerprint can often be the most important piece of evidence left behind at the scene of a crime. Police investigators use a variety of methods to visualize fingerprints. One such method involves the use of a chemical agent called ninhydrin, which reacts with the amino acids present in the fingerprint to produce colored compounds that can be seen. But what are amino acids, why are they present in our fingerprints, and what function do amino acids serve?

In this chapter, we will explore the structure and properties of amino acids, and we will see how they function as the building blocks that nature employs to assemble important biological compounds called peptides and proteins. These compounds serve a wide array of functions, as we will see later in this chapter. This chapter focuses on the structure, properties, function, and synthesis of amino acids, peptides, and proteins.

## DO YOU REMEMBER?

*Before you go on, be sure you understand the following topics.*
*If necessary, review the suggested sections to prepare for this chapter:*

- Designating Configuration Using the Cahn-Ingold-Prelog System (Section 5.3)
- Structure and Properties of Carboxylic Acids (Section 21.3)
- Nucleophilic Acyl Substitution (Section 21.7)
- Properties of Amines (Section 23.3)

Take the DO YOU REMEMBER? QUIZ in **WileyPLUS** to check your understanding

## 25.1 Introduction to Amino Acids, Peptides, and Proteins

Throughout this book, particularly in the Medically Speaking boxes, we have explored the relationship between the structure of a compound and its medicinal activity. The relationship between structure and activity is perhaps most striking for biological molecules called proteins. **Proteins** are polymers that are assembled from amino acid monomers that have been linked together, much like jigsaw puzzle pieces (Figure 25.1). Each **amino acid** contains an amino group and a carboxylic acid group. It is the presence of these two functional groups that enables amino acids to link together.

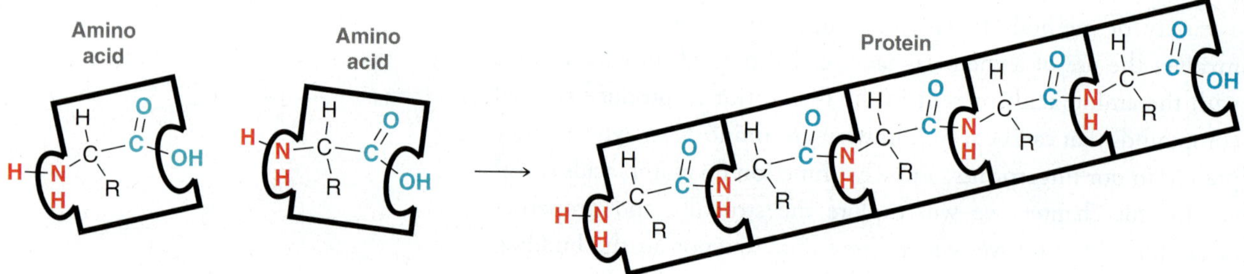


**FIGURE 25.1**
An illustration showing how amino acids serve as building blocks for proteins.

An amino acid can have any number of carbon atoms separating the two functional groups, but of particular interest are the **alpha (α) amino acids** in which the two functional groups are separated by exactly one carbon atom.

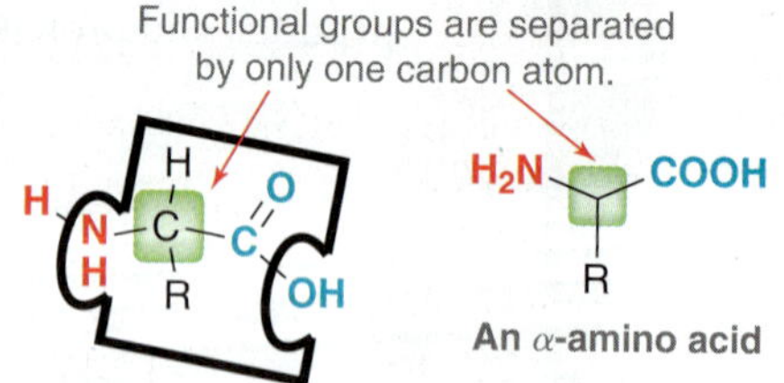


Amino acids of this type are called α-amino acids because the amino group is connected to the carbon atom that is alpha (α) to the carboxylic acid group. Notice that this α carbon is a chirality center, provided that the R group is not simply a hydrogen atom. The configuration of the α position will be discussed in the coming sections.

Amino acids are coupled together by amide linkages, also called **peptide bonds:**

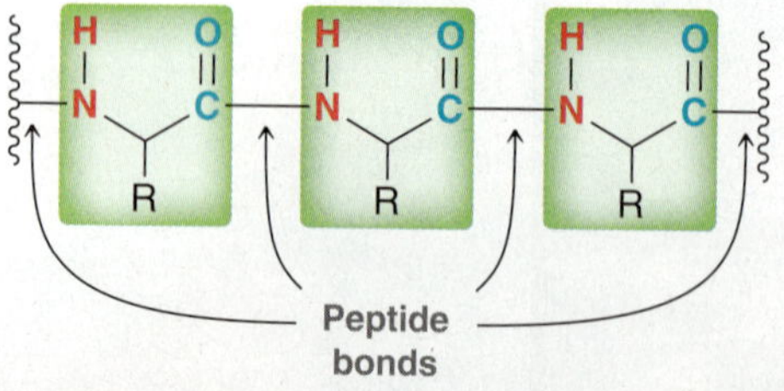

Relatively short amino acid chains are called **peptides**. A dipeptide is formed when two amino acids are coupled together, a tripeptide from three amino acids, a tetrapeptide from four, and so on. Chains comprised of fewer than 40 or 50 amino acids are often called polypeptides, while still larger chains are called proteins. Proteins serve a wide array of important biological functions, as we will discuss in the final section of this chapter. Certain proteins, called enzymes, serve as catalysts for most of the reactions that occur in living cells, and it is estimated that more than 50,000 different enzymes are needed for our bodies to function properly.

In order to understand the structure and function of proteins, we must first explore the structure and properties of the most basic building blocks, amino acids.

## 25.2 Structure and Properties of Amino Acids

### Naturally Occurring Amino Acids

Hundreds of different amino acids are observed in nature, but only 20 amino acids are abundantly found in proteins. These twenty α-amino acids differ from each other only in the identity of the side chain (the R group, highlighted).

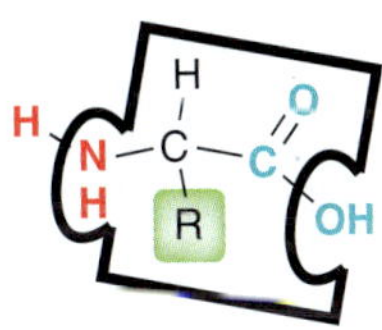


The structures of all 20 amino acids are shown in Table 25.1, together with the accepted three-letter abbreviation and one-letter abbreviation for each amino acid. Except for glycine (R = H), all of these amino acids are chiral, and nature typically employs only one enantiomer of each. The amino acids primarily observed in nature are called **L amino acids**, because their Fischer projections resemble the Fischer projections of L sugars.

HO, C=O; $H_2N$—C—H; $CH_3$ **L-Alanine**

HO, C=O; $H_2N$—C—H; $CH_2OH$ **L-Serine**

H, C=O; HO—C—H; $CH_2OH$ **L-Glyceraldehyde**

There are some examples of D amino acids found in nature, but for the most part, the peptides and proteins found in humans and other mammals are constructed almost exclusively from L amino acids.

### CONCEPTUAL CHECKPOINT

**25.1** Although most naturally occurring proteins are made up only of L amino acids, proteins isolated from bacteria will sometimes contain D amino acids. Draw Fischer projections for D-alanine and D-valine. In each case, assign the configuration (*R* or *S*) of the chirality center.

**25.2** Draw a bond-line structure for each of the following amino acids:

(a) L-Leucine (b) L-Tryptophan

(c) L-Methionine (d) L-Valine

**25.3** Of the 20 naturally occurring amino acids shown in Table 25.1, identify any amino acids that exhibit the following:

(a) A cyclic structure (b) An aromatic side chain

(c) A side chain with a basic group

(d) A sulfur atom

(e) A side chain with an acidic group

(f) A side chain containing a proton that will likely participate in hydrogen bonding

**TABLE 25.1 THE STRUCTURES OF THE TWENTY NATURALLY OCCURRING AMINO ACIDS THAT ARE FOUND IN PROTEINS**

| Name | Structure | Abbreviation |
|---|---|---|
| **Amino acids with nonpolar side chains** | | |
| Glycine | 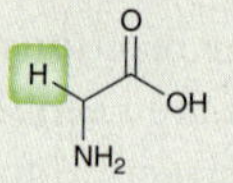 | Gly G |
| Alanine | 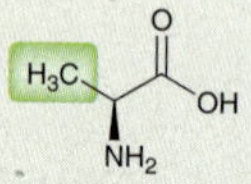 | Ala A |
| Valine | 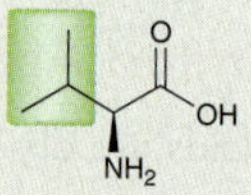 | Val V |
| Leucine | 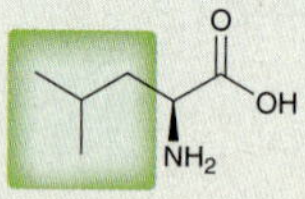 | Leu L |
| Isoleucine | 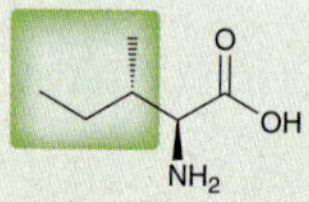 | Ile I |
| Methionine | 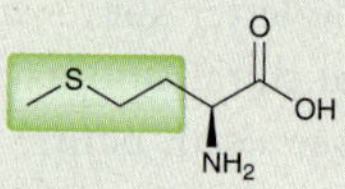 | Met M |
| Proline | 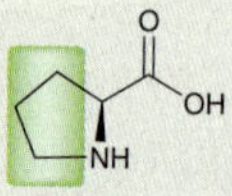 | Pro P |
| Phenylalanine | 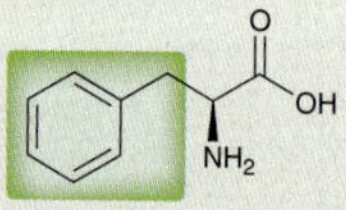 | Phe F |
| Tryptophan | 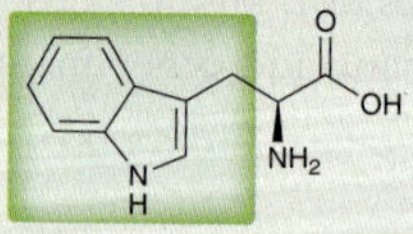 | Trp W |
| **Amino acids with polar side chains** | | |
| Asparagine | 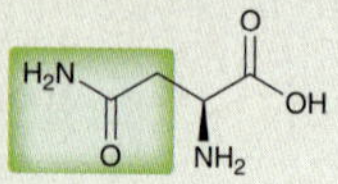 | Asn N |
| Glutamine | 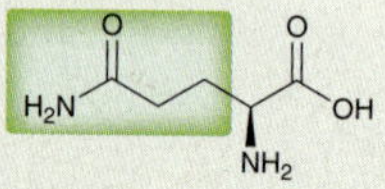 | Gln Q |
| Serine | 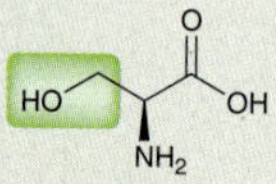 | Ser S |
| Threonine | 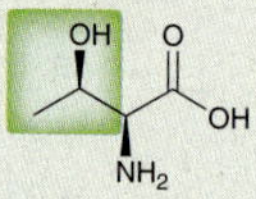 | Thr T |
| Tyrosine | 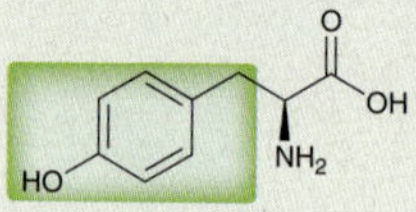 | Tyr Y |
| Cysteine | 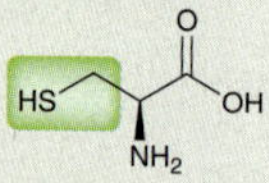 | Cys C |
| **Amino acids with acidic side chains** | | |
| Aspartic acid | 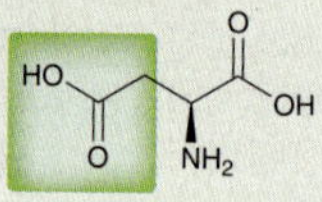 | Asp D |
| Glutamic acid | 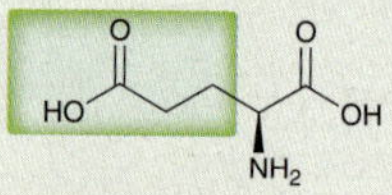 | Glu E |
| **Amino acids with basic side chains** | | |
| Arginine | 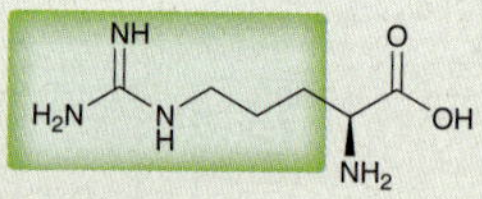 | Arg R |
| Histidine | 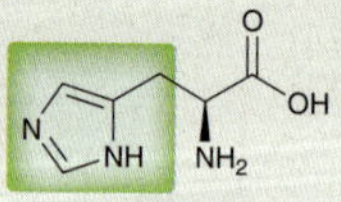 | His H |
| Lysine | 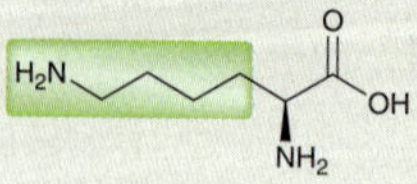 | Lys K |

## Other Naturally Occurring Amino Acids

In addition to the 20 amino acids found in proteins (Table 25.1), there are other amino acids used by organisms for a variety of functions. For example, consider the structures of γ-aminobutyric acid (GABA) and thyroxine.

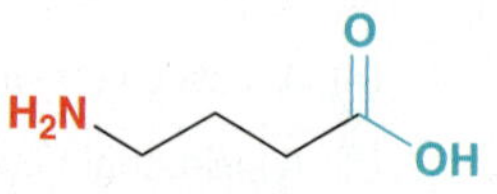

**γ-Aminobutyric acid (GABA)**

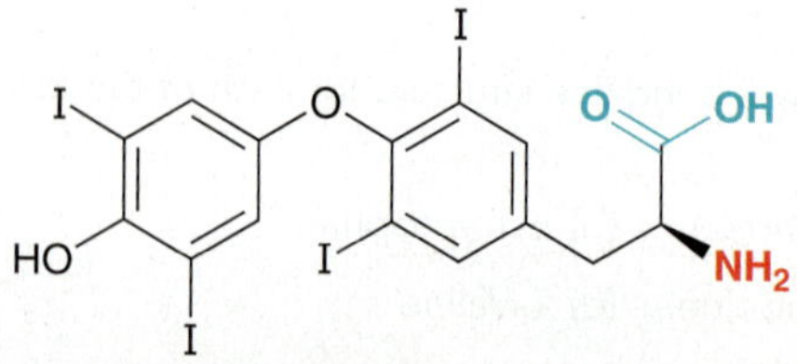

**Thyroxine**

## practically speaking — Nutrition and Sources of Amino Acids

In Section 25.1, we saw that 20 different L amino acids serve as the building blocks of proteins. Our bodies can synthesize 10 of these amino acids in sufficient quantity, but the other 10, called *essential amino acids*, must be obtained from our diet. The essential amino acids are isoleucine, leucine, methionine, phenylalanine, threonine, tryptophan, valine, arginine, histidine, and lysine. These amino acids are obtained from digesting foods containing proteins.

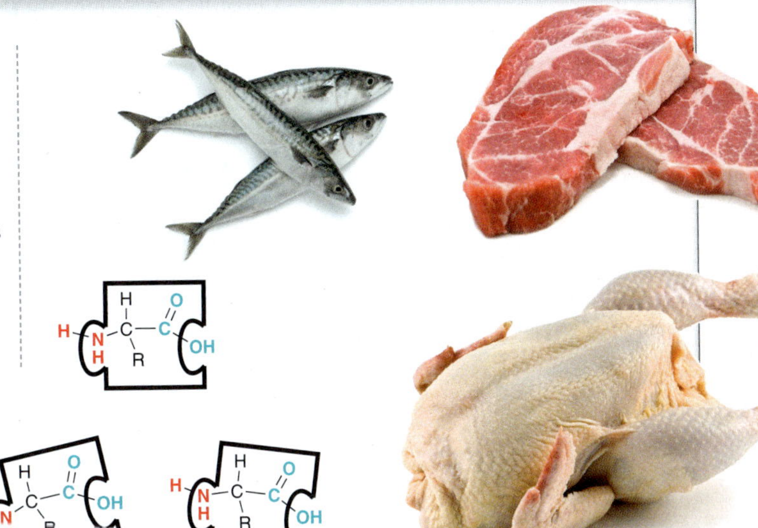

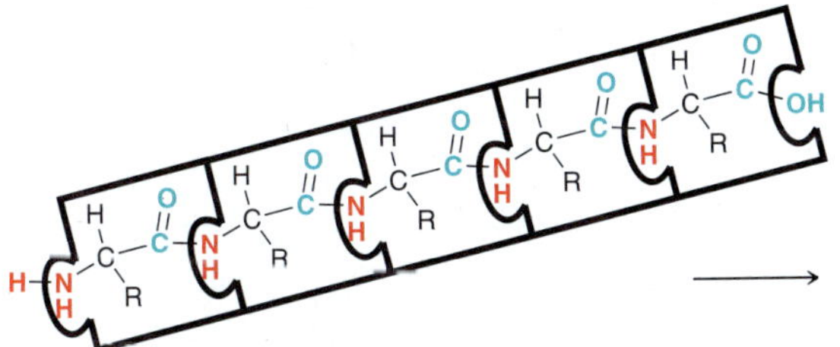

Proteins that contain all 10 essential amino acids are called *complete proteins*, as they provide us with all of the building blocks that we need. Examples of complete proteins include meat, fish, milk, and eggs. Proteins that are deficient in one or more of the essential amino acids are called *incomplete proteins*. Examples include rice (deficient in lysine and threonine), corn (deficient in lysine and tryptophan), and beans and peas (deficient in methionine).

Inadequate intake of the essential amino acids can lead to a host of diseases, which can be avoided with proper nutrition. Meat eaters receive all of the necessary amino acids from a piece of meat, while vegetarians must eat a variety of plant foods that complement each other. Alternatively, vegetarians can supplement their diet with a source of complete proteins, such as milk or eggs.

Both of these compounds are amino acids, but they are not found in proteins. GABA is found in the brain and acts as a neurotransmitter, while thyroxine is found in the thyroid gland and acts as a hormone. There are many other examples as well, but this chapter will focus almost exclusively on the 20 amino acids found in proteins.

### Acid-Base Properties

When an amino acid is dissolved in a solution at a pH of 1, both functional groups exist primarily in their protonated forms.

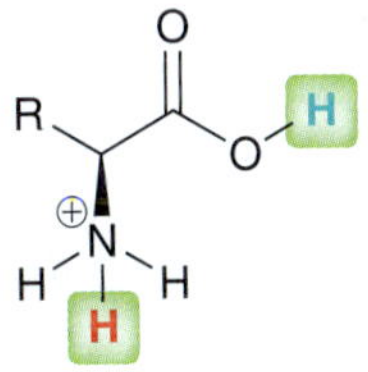

Each of the highlighted protons has its own unique p$K_a$ value, often called p$K_{a1}$ and p$K_{a2}$.

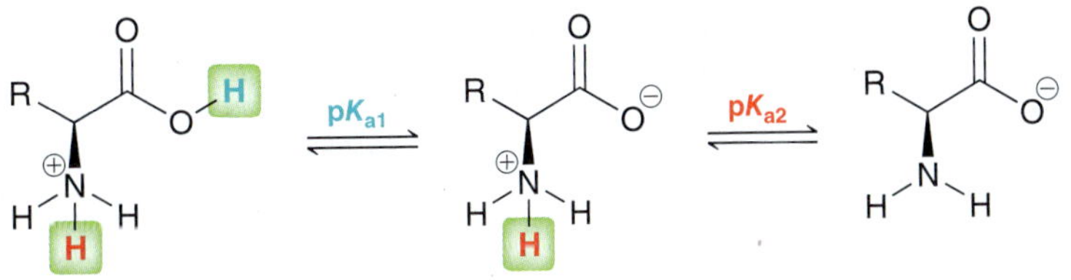

Notice that the carboxylic acid group is deprotonated first. That is, the first p$K_a$ value refers to the acidity of the carboxylic acid group, while the second p$K_a$ value refers to the acidity of the ammonium group. Some of the amino acids have side chains containing either basic or acidic groups. These amino acids will have a third p$K_a$ value associated with the side chain, as can be seen in Table 25.2.

**TABLE 25.2 THE P$K_A$ VALUES FOR TWENTY NATURALLY OCCURRING AMINO ACIDS**

| AMINO ACID | α-COOH | α-$NH_3^+$ | SIDE CHAIN |
|---|---|---|---|
| Alanine | 2.34 | 9.69 | — |
| Arginine | 2.17 | 9.04 | 12.48 |
| Asparagine | 2.02 | 8.80 | — |
| Aspartic acid | 1.88 | 9.60 | 3.65 |
| Cysteine | 1.96 | 10.28 | 8.18 |
| Glutamic acid | 2.19 | 9.67 | 4.25 |
| Glutamine | 2.17 | 9.13 | — |
| Glycine | 2.34 | 9.60 | — |
| Histidine | 1.82 | 9.17 | 6.00 |
| Isoleucine | 2.36 | 9.60 | — |
| Leucine | 2.36 | 9.60 | — |
| Lysine | 2.18 | 8.95 | 10.53 |
| Methionine | 2.28 | 9.21 | — |
| Phenylalanine | 1.83 | 9.13 | — |
| Proline | 1.99 | 10.60 | — |
| Serine | 2.21 | 9.15 | — |
| Threonine | 2.09 | 9.10 | — |
| Tryptophan | 2.83 | 9.39 | — |
| Tyrosine | 2.20 | 9.11 | 10.07 |
| Valine | 2.32 | 9.62 | — |

Notice that the p$K_a$ values for the carboxylic acid groups are in the range of 2–3 for nearly all of the amino acids shown. For example, alanine has a p$K_{a1}$ of 2.34. This value indicates that the uncharged form (COOH) and the anionic form ($COO^-$) will be present in equal amounts at a pH of 2.34. At any pH below 2.34 (highly acidic conditions), the uncharged form will predominate. At any pH above 2.34, the carboxylate anion will predominate. Indeed, this anionic form predominates at physiological pH.

O OH Physiological pH O O⊖ + $H^+$

Carboxylate ion

**LOOKING BACK**
Recall that the pH of blood is approximately 7.4, which is called physiological pH (see Section 21.3).

Now let's focus on the p$K_a$ values for the ammonium groups of amino acids. Table 25.2 indicates that these values are in the range of 9–10 for nearly all of the amino acids. For example, alanine has a p$K_{a2}$ of 9.69. This value indicates that the uncharged and cationic forms of the ammonium group will be present in equal amounts at a pH of 9.69. At any pH above 9.69, the uncharged form will predominate. At any pH below 9.69, the cationic ammonium group will predominate. Indeed, the cationic form predominates at physiological pH.

$-NH_2$ + $H^+$ Physiological pH ⊕ $-NH_3$

Ammonium ion

In summary, consider the structure of an amino acid at physiological pH. The amino group is protonated, while the carboxylic acid group is deprotonated.

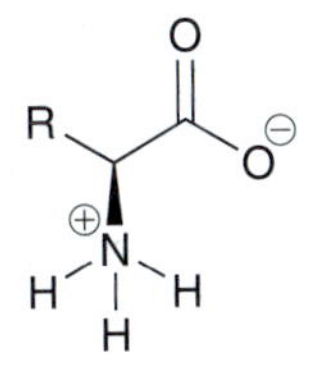

In this form, the amino acid is said to be a **zwitterion**, which is a net neutral compound that exhibits charge separation. An amino acid can be thought of as an internal salt, and as such, it will exhibit many of the physical properties of salts. For example, amino acids are highly soluble in water and have very high melting points. Amino acids are also said to be **amphoteric**, because they will react with either acids or bases. When treated with a base, an amino acid will function as an acid by giving up a proton (the ammonium group is deprotonated).

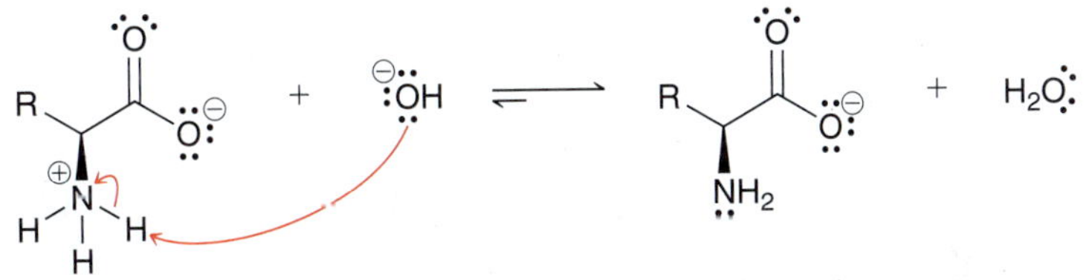

However, when treated with an acid, an amino acid will serve as a base by abstracting a proton (the carboxylate group is protonated).

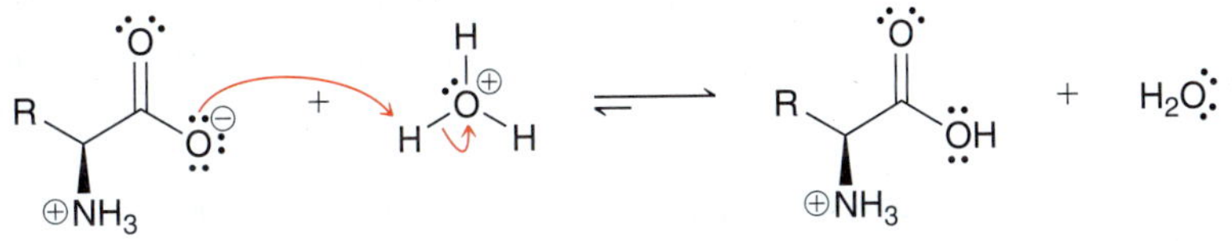

The two reactions show that, in its zwitterionic form, an amino acid can function either as an acid or as a base.

## 25.1 DETERMINING THE PREDOMINANT FORM OF AN AMINO ACID AT A SPECIFIC pH

**LEARN** the skill

Draw the form of lysine that predominates at a pH of 9.5.

**SOLUTION**

The identity of the side chain of lysine can be found in Table 25.1. This amino acid has a basic side chain, which means that the compound has three locations that we must consider.

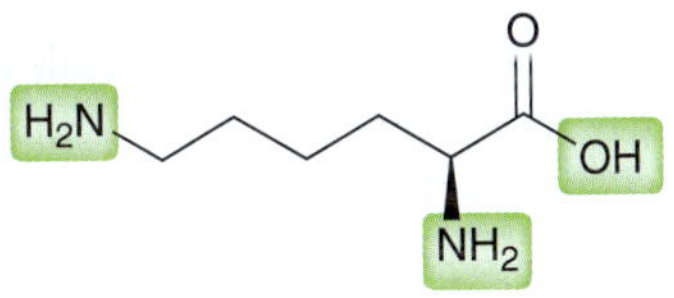

**STEP 1**
Identify the p$K_a$ of the carboxylic acid group and determine which form predominates.

We must consider whether or not each amino group is protonated and whether or not the carboxylic acid group is deprotonated at the stated pH. Let's begin with the carboxylic acid group. Table 25.2 indicates that p$K_{a1}$ for lysine is 2.18. At a pH below 2.18 (highly acidic conditions), we expect the COOH group to have its proton. But at higher pH, we expect the carboxylate anion to predominate. Accordingly, at a pH of 9.5, we expect the deprotonated form (the carboxylate) to predominate.

**STEP 2**
Identify the $pK_a$ of the α-amino group and determine which form predominates.

Next consider the α-amino group. Table 25.2 indicates that $pK_{a2}$ for lysine is 8.95. At a pH below 8.95 (lower pH = more acidic), we expect the amino group to be protonated, but at a higher pH, we expect the amino group to predominate in its uncharged form. Accordingly, at a pH of 9.5, we expect the uncharged form of the α-amino group to predominate.

Finally, consider the side chain, which has a $pK_a$ of 10.53. At a pH below 10.53 (lower pH = more acidic), we expect the amino group to be protonated, but at a higher pH, we expect the amino group to predominate in its uncharged form. Accordingly, at a pH of 9.5, we expect the protonated form of the side-chain amino group to predominate. Therefore, the following form of lysine is expected to predominate at a pH of 9.5:

**STEP 3**
Identify the $pK_a$ of the side chain, if necessary, and determine which form predominates.

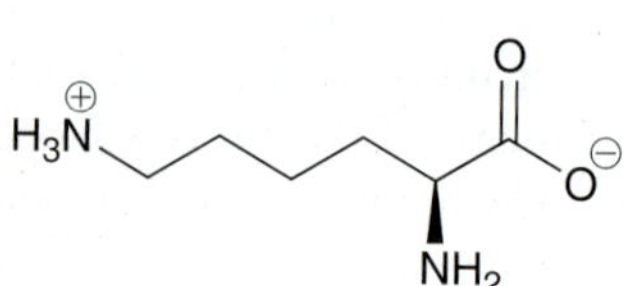

**PRACTICE the skill** **25.4** Draw the form of the amino acid that is expected to predominate at the stated pH.

**(a)** Alanine at a pH of 10
**(b)** Proline at a pH of 10
**(c)** Tyrosine at a pH of 9
**(d)** Asparagine at physiological pH
**(e)** Histidine at physiological pH
**(f)** Glutamic acid at a pH of 3

**APPLY the skill** **25.5** At a pH of 11, arginine is a more effective proton donor than asparagine. Explain.

**25.6** The OH group on the side chain of serine is not deprotonated at a pH of 12. However, the OH group on the side chain of tyrosine is deprotonated at a pH of 12. This can be verified by inspecting the $pK_a$ values in Table 25.2. Suggest an explanation for the difference in acid-base properties between these two OH groups.

need more **PRACTICE?** **Try Problems 25.40, 25.47, 25.48**

## Isoelectric Point

For each amino acid, there is a specific pH at which the concentration of the zwitterionic form reaches its maximum value. This pH is called the **isoelectric point (pI)**, and each amino acid has its own unique pI. For amino acids that lack an acidic or basic side chain, the pI is simply the average of the two $pK_a$ values. The following example shows the calculation for the pI of alanine.

R–COOH ← $pK_a = 2.34$

$pK_a = 9.69$ → $\oplus NH_3$

$$\text{pI} = \frac{2.34 + 9.69}{2} = 6.02$$

For amino acids with acidic or basic side chains, the pI is the average of the two $pK_a$ values that correspond with the similar groups. For example, the pI of lysine is determined by the two amino groups, while the pI of glutamic acid is determined by the two carboxylic acid groups.

**Lysine**

$H_3N^{\oplus}$ ($pK_a = 10.53$) … COOH; $\oplus NH_3$ ($pK_a = 8.95$)

$$\text{pI} = \frac{10.53+8.95}{2} = 9.74$$

**Glutamic acid**

HOOC ($pK_a = 4.25$) … COOH ($pK_a = 2.19$); $\oplus NH_3$

$$\text{pI} = \frac{4.25+2.19}{2} = 3.22$$

## Separation of Amino Acids via Electrophoresis

Amino acids can be separated from each other by a variety of techniques. One method, called **electrophoresis**, relies on a difference in pI values and can be used to determine the number of different amino acids present in a mixture. In practice, a few drops of the mixture are applied to filter paper or gel that is placed in a buffered solution between two electrodes. When an electric field is applied, the amino acids separate based on their different pI values. If the pI of an amino acid is greater than the pH of the solution, the amino acid will exist predominantly in a form that bears a positive charge and will migrate toward the cathode. The greater the difference between pI and pH, the faster it will migrate. An amino acid with a pI that is lower than the pH of the solution will exist predominantly in a form that bears a negative charge and will migrate toward the anode. The greater the difference between pI and pH, the faster it will migrate (Figure 25.2). If two amino acids have very similar pI values (such as glycine and leucine), the amino acid with the larger molecular weight will move more slowly, because the charge has to carry a greater mass.

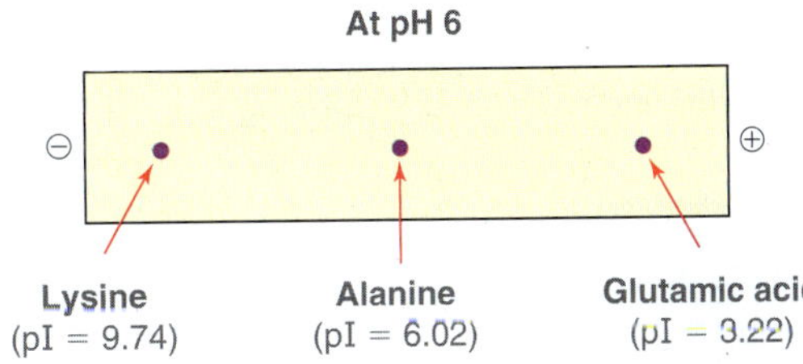

At pH 6

⊖ ⊕

Lysine (pI = 9.74) Alanine (pI = 6.02) Glutamic acid (pI = 3.22)

**FIGURE 25.2** Separation of amino acids via electrophoresis.

Amino acids are colorless, so a detection technique is necessary in order to visualize the location of the various spots. The most common method involves treating the filter paper or gel with a solution containing ninhydrin followed by heating in an oven. Ninhydrin reacts with amino acids to produce a purple product.

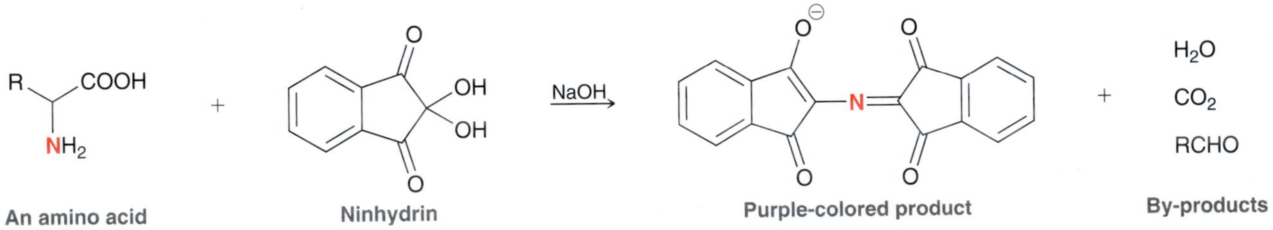

An amino acid + Ninhydrin $\xrightarrow{NaOH}$ Purple-colored product + By-products ($H_2O$, $CO_2$, RCHO)

The nitrogen atom of the amino acid is ultimately incorporated into the purple product, and the rest of the amino acid is degraded into a few by-products (water, carbon dioxide, and an aldehyde). The purple compound is obtained regardless of the identity of the amino acid, provided that the amino acid is primary (i.e., not proline). The number of purple spots indicates the number of different kinds of amino acids present.

Electrophoresis cannot be used to separate large quantities of amino acids. It is used just as an analytical method for determining the number of amino acids in a mixture. In order to actually separate an entire mixture of amino acids, other laboratory techniques are used, such as column chromatography.

### CONCEPTUAL CHECKPOINT

**25.7** Using the data in Table 25.2, calculate the pI of the following amino acids:

(a) Aspartic acid (b) Leucine

(c) Lysine (d) Proline

**25.8** For each group of amino acids, identify the amino acid with the lowest pI (try to solve this problem by inspecting their structures, rather than performing calculations).

(a) Alanine, aspartic acid, or lysine

(b) Methionine, glutamic acid, or histidine

**25.9** Identify which 2 of the 20 naturally occurring amino acids are expected to have the same pI.

**25.10** A mixture containing phenylalanine, tryptophan, and leucine was subjected to electrophoresis. Determine which of the amino acids moved the farthest distance assuming that the experiment was performed at the pH indicated:

(a) pH 6.0 (b) pH 5.0

**25.11** Draw the aldehyde that is obtained as a by-product when L-leucine is treated with ninhydrin.

## practically speaking — Forensic Chemistry and Fingerprint Detection

There are many popular TV shows that portray the work of crime-scene investigators and the methods they employ to analyze evidence. The use of chemicals to visualize latent (invisible) fingerprints is a particularly common practice, and many compounds can be used for this purpose, including ninhydrin.

Latent fingerprints are created by the residue of sweat found on the surface of the skin. Sweat is comprised primarily of water (99%), but it also contains a variety of organic compounds, including amino acids. These amino acids are present only in very small concentrations, but they are relatively stable over long periods of time. When treated with ninhydrin, a reaction occurs between the amino acids and the ninhydrin, forming a fluorescent purple image.

To develop a fingerprint, a solution of ninhydrin is applied as a spray, and then mild heating is used to accelerate the reaction. The process has several undesirable features, including background staining (which reduces the contrast of the image) as well as light-induced fading of the image. For best results, the ninhydrin solution should be applied in the dark and at room temperature. Under such conditions, the development process can take up to two weeks, which is not feasible for most cases in which a rapid analysis is required.

Over the past 50 years, many attempts have been made to design ninhydrin analogues with better properties. The following are a few examples.

Some of these reagents do have slightly improved properties, but the improvements are not significant enough to justify the added expense associated with the replacement of ninhydrin as a standard method used by forensic chemists to develop latent fingerprints.

**Ninhydrin**

**Ninhydrin analogues**

## 25.3 Amino Acid Synthesis

Over the last century, many methods have been used to prepare amino acids in the laboratory. In this section, we will explore a few of those methods.

### Amino Acid Synthesis via α-Haloacids

One of the oldest methods for preparing racemic mixtures of α-amino acids involves the use of the Hell-Volhard-Zelinsky reaction (Section 22.2) to functionalize the α position of a carboxylic acid.

1) $Br_2$, $PBr_3$
2) $H_2O$

**(Racemic)**

The halogen is then replaced with an amino group in an $S_N2$ reaction.

R–CH(Br)–COOH —[Excess $NH_3$, $S_N2$]→ R–CH($NH_2$)–COOH (Racemic)

In Section 23.5, we saw that polyalkylation is often unavoidable when ammonia is treated with an alkyl halide. However, in this case, polyalkylation is not a problem because the alkyl halide is fairly large and steric hindrance prevents subsequent alkylations.

## CONCEPTUAL CHECKPOINT

**25.12** Identify the reagents necessary to make each of the following amino acids using a Hell-Volhard-Zelinsky reaction:

(a) Leucine (b) Alanine (c) Valine

**25.13** Each of the following carboxylic acids was treated with bromine and $PBr_3$ followed by water, and the resulting α-haloacid was then treated with excess ammonia. In each case, draw and name the amino acid that is produced.

(a) (isopentyl)–COOH (b) 3-methylbutanoic acid (OH) (c) Ph–CH₂CH₂–COOH (d) Acetic acid

## Amino Acid Synthesis via the Amidomalonate Synthesis

Recall that the malonic ester synthesis (Section 22.5) can be used to prepare carboxylic acids, starting with diethyl malonate.

Diethyl malonate —[1) NaOEt 2) RX]→ EtO₂C–CH(R)–CO₂Et —[$H_3O^+$, Heat]→ R–CH₂–COOH

A clever adaptation of this process, called the **amidomalonate synthesis**, employs diethyl acetamidomalonate as the starting material, enabling the preparation of racemic α-amino acids.

Diethyl acetamidomalonate —[1) NaOEt 2) RX]→ —[$H_3O^+$, Heat]→ R–CH($NH_3^+$)–COOH (Racemic)

The process involves the same three steps used in the malonic ester synthesis: (1) deprotonation, (2) alkylation, and (3) hydrolysis and decarboxylation. During the hydrolysis step, the amide is hydrolyzed as well. As an example, consider the synthesis of phenylalanine via the amidomalonate synthesis.

Diethyl acetamidomalonate —[1) NaOEt 2) benzyl bromide (Br)]→ —[$H_3O^+$, Heat]→ phenylalanine ($NH_3^+$, OH) (Racemic)

The identity of the amino acid obtained is determined by the choice of the alkyl halide in the second step.

# SKILLBUILDER

## 25.2 USING THE AMIDOMALONATE SYNTHESIS

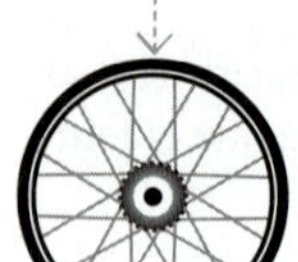

**LEARN the skill**

Show how you would use the amidomalonate synthesis to prepare a racemic mixture of tryptophan.

**SOLUTION**

Begin by identifying the side chain attached to the α position of tryptophan (see Table 25.1).

**STEP 1**
Identify the side chain connected to the α position.

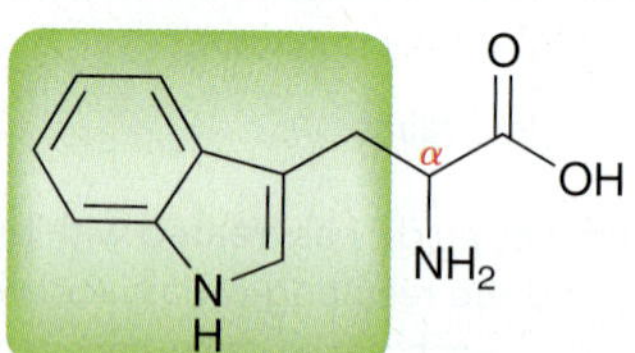


Next, identify the alkyl halide needed to install this group. Just redraw the alkyl group connected to a leaving group. Iodide and bromide are commonly used as leaving groups.

**STEP 2**
Identify the necessary alkyl halide and ensure that it will readily undergo $S_N2$.

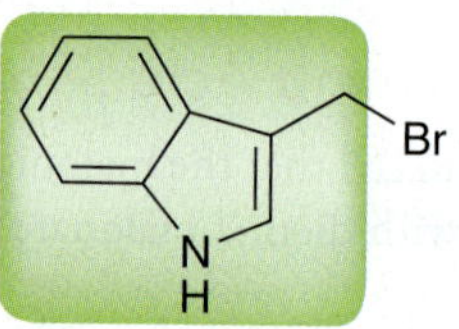


Analyze the structure of this alkyl halide and make sure that it will readily participate in an $S_N2$ reaction. If the substrate is tertiary, then an amidomalonate synthesis will fail, because the alkylation step will not occur. In this case, the alkyl halide is primary and would be expected to undergo an $S_N2$ process quite rapidly.

**STEP 3**
Identify the reagents beginning with acetamidomalonate.

In order to perform the amidomalonate synthesis, we must first deprotonate diethyl acetamidomalonate with a strong base and then treat the resulting anion with the alkyl halide. The resulting product is then heated in the presence of aqueous acid to achieve hydrolysis and decarboxylation.

1) NaOEt
2) (3-(bromomethyl)indole)
$H_3O^+$, Heat

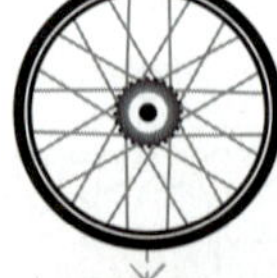

**PRACTICE the skill** **25.14** Identify the reagents necessary to make each of the following amino acids via the amidomalonate synthesis:

**(a)** Isoleucine **(b)** Alanine **(c)** Valine

**APPLY the skill** **25.15** An amidomalonate synthesis was performed using each of the following alkyl halides. In each case, draw and name the amino acid that was produced.

**(a)** Methyl chloride **(b)** Isopropyl chloride **(c)** 2-Methyl-1-chloropropane

**25.16** Both leucine and isoleucine can be prepared via the amidomalonate synthesis, although one of these amino acids can be produced in higher yields. Identify the higher yield process and explain your choice.

need more **PRACTICE?** **Try Problems 25.57, 25.58, 25.59c, 25.60b, 25.84**

## Amino Acid Synthesis via the Strecker Synthesis

Racemic α-amino acids can also be prepared from aldehydes via a two-step process called the **Strecker synthesis**.

$$\text{RCHO} \xrightarrow[\text{NaCN}]{\text{NH}_4\text{Cl}} \underset{\textbf{(Racemic)}}{\text{R–CH(NH}_2\text{)–C}\equiv\text{N}} \xrightarrow{\text{H}_3\text{O}^+} \underset{\textbf{(Racemic)}}{\text{R–CH(NH}_3^{\oplus}\text{)–COOH}}$$

The first step involves conversion of the aldehyde into an α-amino nitrile, and the second step involves hydrolysis of the cyano group to yield a carboxylic acid group.

Formation of the α-amino nitrile likely proceeds via Mechanism 25.1. The resulting α-amino nitrile then undergoes hydrolysis via a mechanism found in Section 21.13.

### MECHANISM 25.1 FORMATION OF AN α-AMINO NITRILE

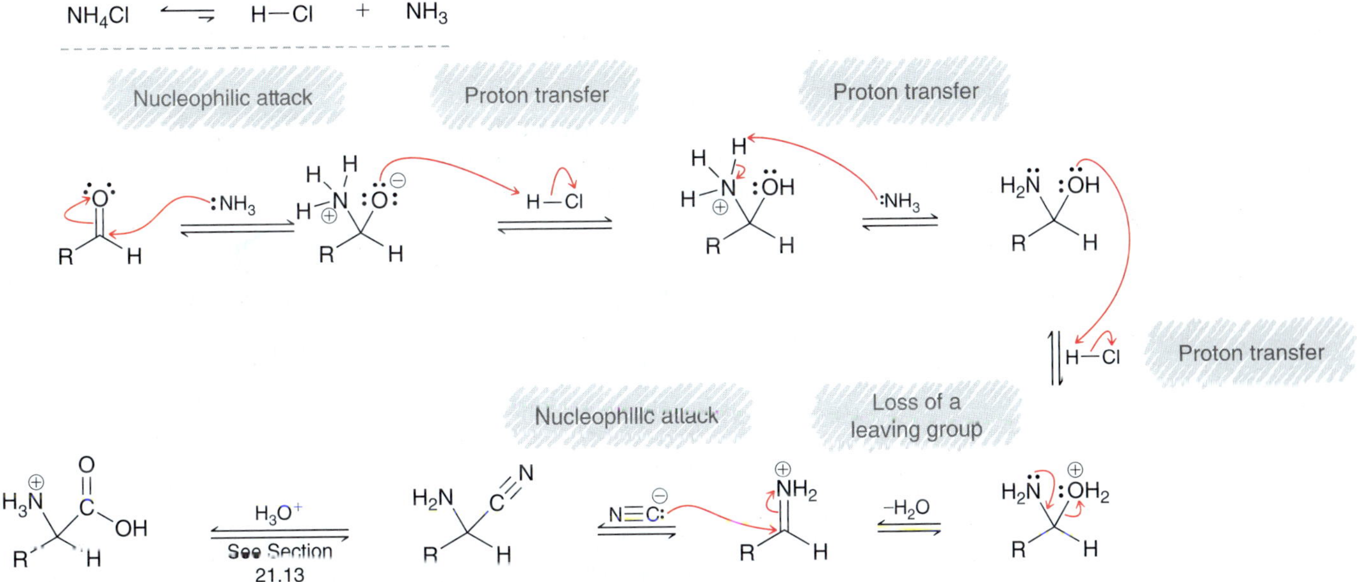


As an example, consider the synthesis of alanine via the Strecker synthesis:

$$\text{H}_3\text{C–CHO} \xrightarrow[\text{2) H}_3\text{O}^+]{\text{1) NH}_4\text{Cl, NaCN}} \text{H}_3\text{C–CH(NH}_3^{\oplus}\text{)–COOH}$$

**Alanine (racemic)**

The identity of the amino acid obtained is determined by the choice of the starting aldehyde.

## CONCEPTUAL CHECKPOINT

**25.17** Identify the reagents necessary to make each of the following amino acids using a Strecker synthesis:

(a) Methionine (b) Histidine

(c) Phenylalanine (d) Leucine

**25.18** Each of the following aldehydes was converted into an α-amino nitrile followed by hydrolysis to yield an amino acid. In each case, draw and name the amino acid that was produced.

(a) Acetaldehyde (b) 3-Methylbutanal

(c) 2-Methylpropanal

## Enantioselective Synthesis of L Amino Acids

Racemic mixtures of amino acids are produced from each of the three methods described thus far. Obtaining optically active amino acids requires either resolution of the racemic mixture or enantioselective synthesis. Resolution is less efficient and more costly because half of the starting material is wasted in the process. Resolution is not necessary if an enantioselective synthesis is performed. In Section 9.7 we saw how Knowles developed a procedure for achieving asymmetric hydrogenation and then used that procedure for the enantioselective synthesis of L-dopa.

**Asymmetric hydrogenation of this π bond**

COOH HN O O O → $H_2$ Chiral catalyst → (S) COOH HN → → (S) COOH $NH_2$ HO OH

**(S)-3,4-Dihydroxyphenylalanine (L-dopa)**

This technique employs a chiral catalyst and can be used to prepare amino acids with very high enantiomeric excess (% *ee*). Many of the chiral catalysts that have been developed contain a ruthenium atom (Ru) complexed to a chiral ligand, such as BINAP.

$PPh_2$ $PPh_2$ → Ph Ph P Cl Ru Cl P Ph Ph

**(R)-(+)-BINAP**
(R)-2,2'-Bis(diphenylphosphino)-1,1'-binaphthyl

**(R)-(+)-Ru(BINAP)$Cl_2$**
**(chiral catalyst)**

Only small amounts of a chiral catalyst are required to prepare amino acids with very high enantioselectivity. For example, preparation of D-phenylalanine proceeds with 99% *ee* when this chiral catalyst is employed.

O OH NHAc → $H_2$, (R)-(+)-Ru(BINAP)$Cl_2$ → O OH NHAc **99% ee** → 1) NaOH, $H_2O$ 2) $H_3O^+$ → O OH $^{\oplus}NH_3$ **D-Phenylalanine**

**(R)-(+)-Ru(BINAP)$Cl_2$**

## CONCEPTUAL CHECKPOINT

**25.19** Identify the starting alkene necessary to make each of the following amino acids using an asymmetric catalytic hydrogenation:

(a) L-Alanine (b) L-Valine (c) L-Leucine (d) L-Tyrosine

**25.20** Explain why it is inappropriate to use a chiral catalyst in the preparation of glycine.

# 25.4 Structure of Peptides

## The Assembly of Peptides from Amino Acids

As mentioned earlier in this chapter, amino acids are joined together to form peptides. For example, glutathione is a tripeptide assembled from glutamic acid, cysteine, and glycine (Figure 25.3).

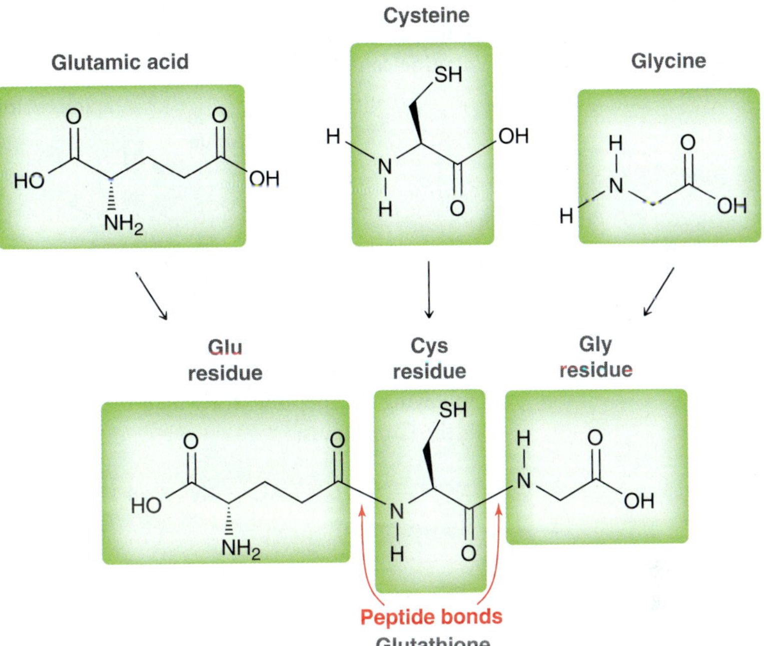


**FIGURE 25.3** The tripeptide glutathione is assembled from three amino acids.

Glutathione is found in all mammals, and it serves an important role as a scavenger of harmful radicals and electrophiles. Glutathione exhibits two peptide bonds and is comprised of three **amino acid residues**. When amino acids join together to form a peptide, the order in which they are connected is important. For example, consider a simple dipeptide made by joining alanine and glycine. The peptide bond can be formed between the COOH group of alanine and the $NH_2$ group of glycine, or from the COOH group of glycine and the $NH_2$ group of alanine.

Alanine Glycine Glycine Alanine

Ala-Gly Gly-Ala

These two dipeptides are not the same compound. They are, in fact, constitutional isomers.

Peptide chains always have an amino group on one end, called the **N terminus,** and a COOH group on the other end, called the **C terminus** (Figure 25.4). By convention, peptides are always drawn with the N terminus on the left side.

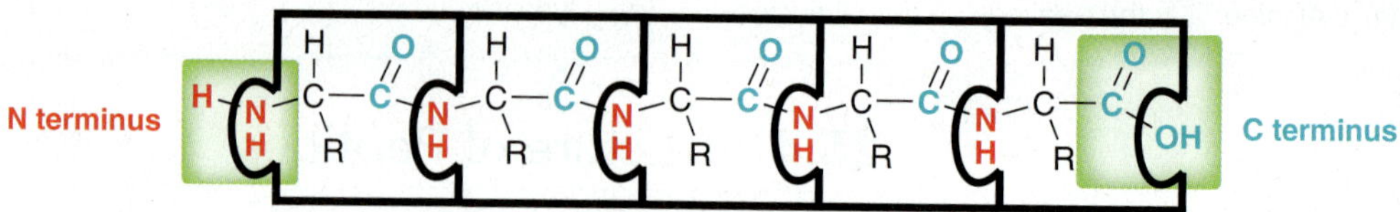


**FIGURE 25.4**
The N terminus and C terminus of a peptide chain.

The sequence of amino acid residues in a peptide can be abbreviated with one- or three-letter abbreviations, starting with the N terminus. For example, a dipeptide of glycine and alanine can be written as follows:

Glycine residue

Alanine residue

$H_2N$ ... OH ... $CH_3$ ≡ Gly — Ala

N terminus C terminus

Simple peptide chains will have one N terminus and one C terminus. For example, consider the following decapeptide, for which the alanine residue is the N terminus and the leucine residue is the C terminus:

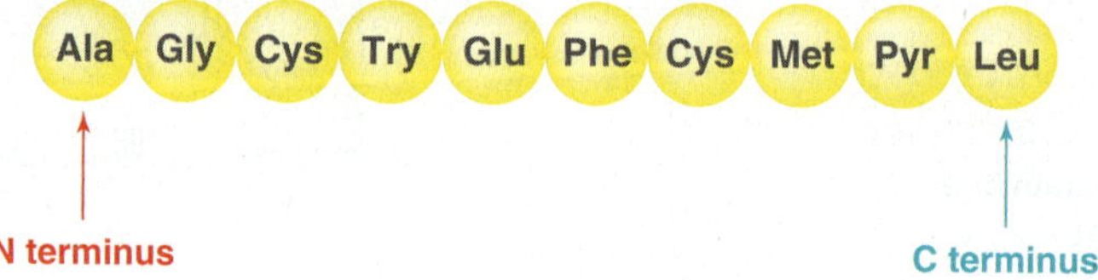

## SKILLBUILDER

### 25.3 DRAWING A PEPTIDE

**LEARN the skill**

Draw a bond-line structure showing the tripeptide Phe-Val-Trp (assume that all three residues are L amino acids).

**SOLUTION**

Begin by drawing a peptide comprised of three residues with the N terminus on the left and the C terminus on the right.

**STEP 1**
Draw a peptide with the correct number of residues.

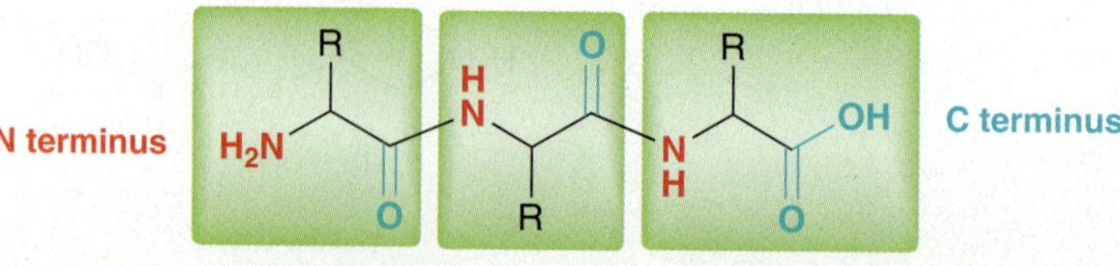

Next, identify the side chain (R group) associated with each residue. The side chains associated with Phe, Trp, and Val are highlighted.

**STEP 2**
Identify the side chain associated with each residue.

Finally, assign the proper configuration for each side chain.

**STEP 3**
Assign the proper configuration for each α position.

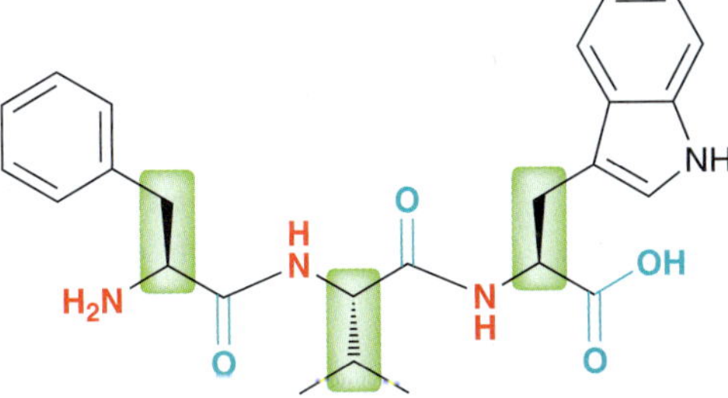


Recall that all naturally occurring amino acids will have the *S* configuration, except for cysteine, which has the *R* configuration. With this information, it is possible to draw each chirality center using the Cahn-Ingold-Prelog convention (Section 5.3). Alternatively, the following stereochemical shortcut can be applied to save time: When the peptide is drawn conventionally with the N terminus on the left and the C terminus on the right, all side chains at the top of the drawing will be on wedges, and all side chains at the bottom of the drawing will be on dashes.

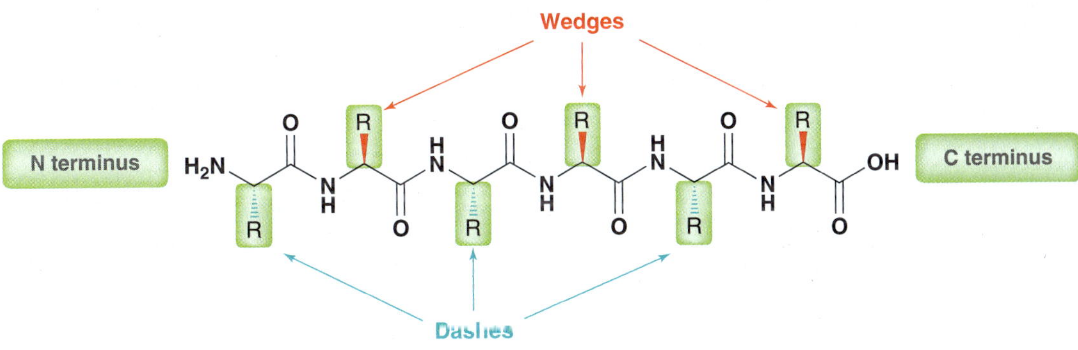


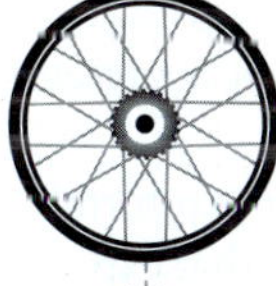

**PRACTICE the skill** **25.21** Draw the structure of each of the following peptides:

**(a)** Leu-Ala-Gly **(b)** Cys-Asp-Ala-Gly **(c)** Met-Lys-His-Tyr-Ser-Phe-Val

**APPLY the skill** **25.22** Using three- and one-letter abbreviations, show the sequence of amino acid residues in the following pentapeptide:

**25.23** Determine which of the following peptides will have a higher molecular weight (**Hint:** It is not necessary to actually calculate the molecular weight of each peptide, but rather, just compare the side chains.)

Cys-Tyr-Leu or Cys-Phe-Ile

**25.24** Compare the following tripeptides and determine whether they are constitutional isomers or the same compound:

Ala-Gly-Leu and Leu-Gly-Ala

need more **PRACTICE?** **Try Problems 25.65, 25.66**

## The Geometry of Peptide Bonds

In order to understand the three-dimensional geometry of peptides, we must first explore the geometry of peptide bonds. Recall that peptide bonds are amide linkages and that amides have a significant resonance structure that endows the C—N bond with some double-bond character.

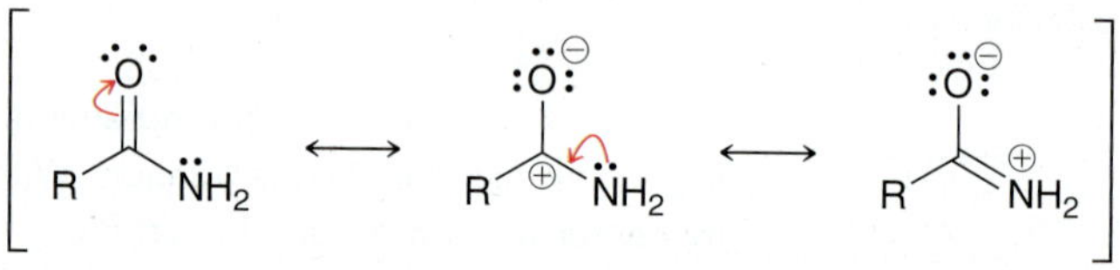

**A significant contributor**

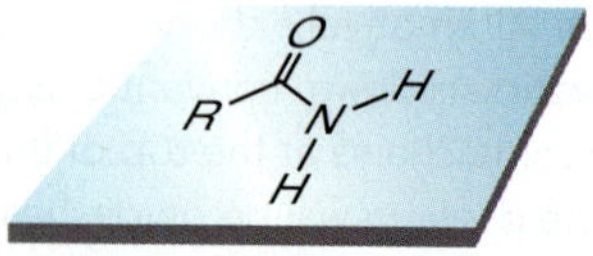

**FIGURE 25.5** The planar structure of an amide.

As a result, the nitrogen atom is $sp^2$ hybridized and planar (Figure 25.5). Peptide bonds are simply amides and therefore exhibit double-bond character just like regular amides. As such, a peptide bond experiences restricted rotation, giving rise to two possible conformations called *s-trans* and *s-cis*.

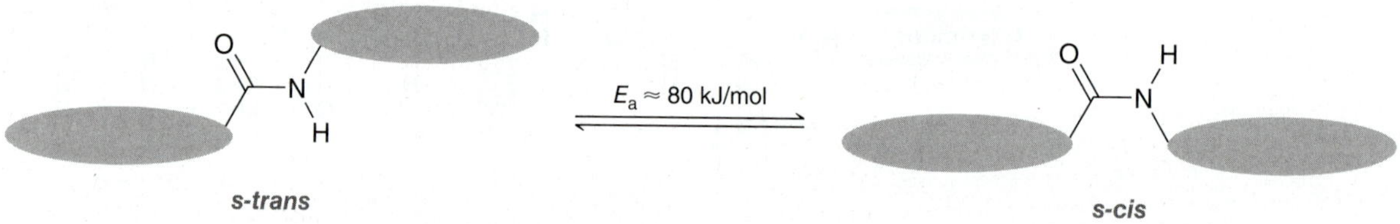

$E_a \approx 80$ kJ/mol

***s-trans*** ***s-cis***

The *s-trans* conformation is usually more stable because it lacks the steric interaction that is present in the *s-cis* conformation.

In a polypeptide, each peptide bond will show a preference for adopting an *s-trans* conformation. Nevertheless, polypeptides are not entirely planar, because they still possess σ bonds that experience free rotation. Only the peptide bonds experience restricted rotation.

**Only these bonds experience restricted rotation**

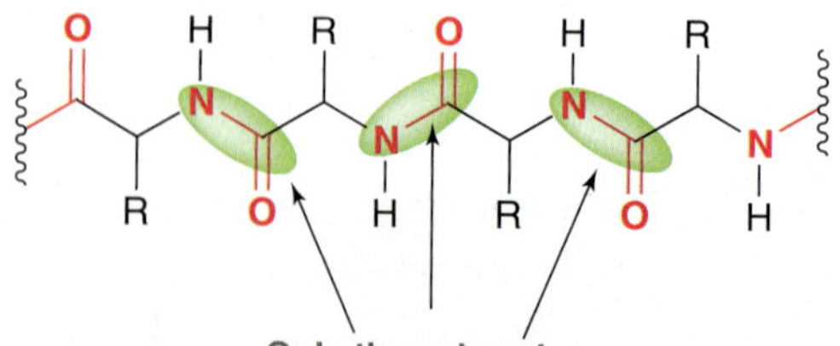

As a result, polypeptides can assume a variety of conformations, as will be discussed in upcoming sections of this chapter.

## Disulfide Bridges

In Section 14.11 we saw that two thiols can be joined via an oxidation process to form a disulfide.

Oxidation

Reduction

**A disulfide**

Of the 20 amino acids found in proteins, only cysteine contains a thiol group. As such, cysteine residues are uniquely capable of being joined to one another via **disulfide bridges**.

Oxidation

**A disulfide bridge**

Disulfide bridges are commonly observed between cysteine residues in the same strand or between cysteine residues in different strands.

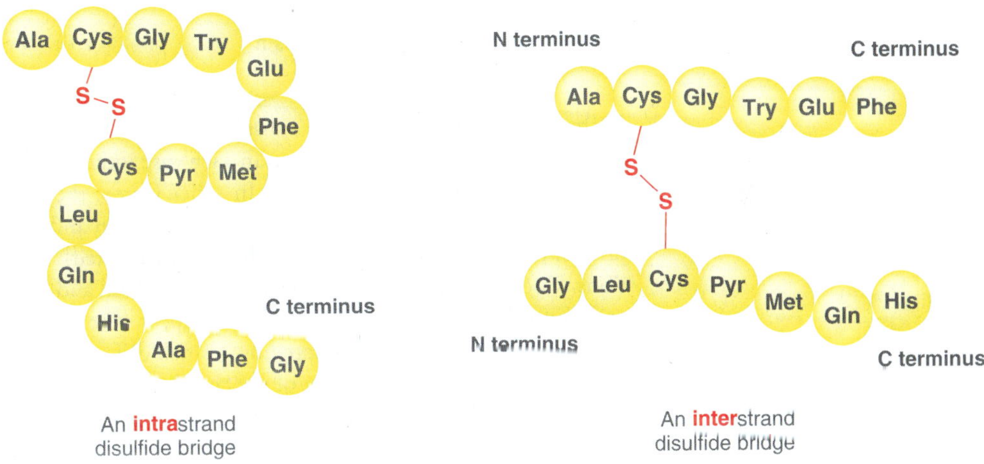


An **intra**strand disulfide bridge

An **inter**strand disulfide bridge

These disulfide bridges greatly affect the three-dimensional structure and properties of peptides and proteins, as will be discussed later in this chapter.

## Some Interesting Peptides

Peptides play a variety of important biological roles. For example, consider the structures of the following two pentapeptides:

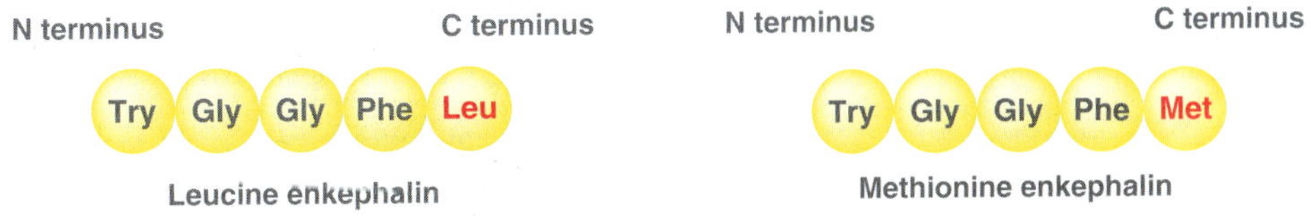


**Leucine enkephalin**

**Methionine enkephalin**

Leucine enkephalin and methionine enkephalin are both found in the brain, where they function in the control of pain. It is believed that they interact with the same receptor as morphine. Notice that the sequences of these compounds are the same except for the last residue (Leu vs. Met).

Bradykinin is another biologically important peptide. It functions as a hormone that dilates blood vessels and acts as an anti-inflammatory agent. Bradykinin is a nonapeptide (nine residues) with the following sequence:

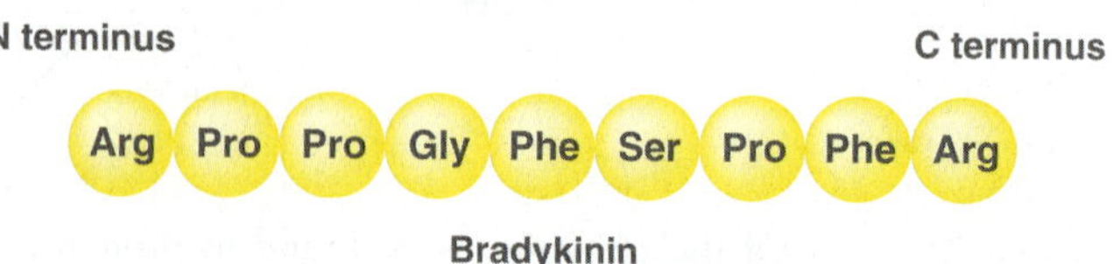


Vasopressin and oxytocin are very similar in structure and represent two more biologically important peptides.

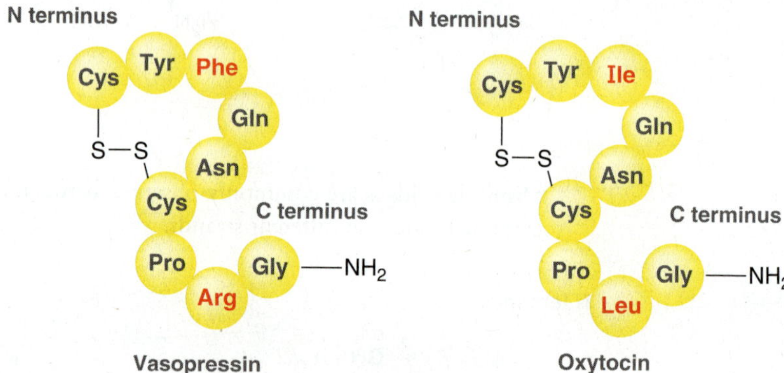


Each of these peptides contains a disulfide bridge. Also notice that the C-terminal end in each case has been modified. The $NH_2$ group in each case indicates that the OH group of the C terminus has been replaced with an $NH_2$ group. In other words, the COOH group has been converted into a $CONH_2$ group (an amide group). While vasopressin and oxytocin are similar in structure (differing only in the identity of two residues, highlighted in red), they nevertheless have very different functions. Vasopressin is released from the pituitary gland and causes readsorption of water by the kidneys and controls blood pressure. Oxytocin induces labor and stimulates milk production in nursing mothers. These examples illustrate how a small difference in structure can produce a large difference in function. Throughout the rest of the chapter, we will see several other examples of the relationship between structure and function.

## CONCEPTUAL CHECKPOINT

**25.25** Draw the *s-trans* conformation of the dipeptide Phe-Leu and identify both the N terminus and the C terminus.

**25.26** Draw the *s-cis* conformation of the dipeptide Phe-Phe and identify the source of the large steric interaction associated with that conformation.

**25.27** Using a bond-line structure, show the tetrapeptide obtained when two molecules of Cys-Phe are joined by a disulfide bridge.

**25.28** Aspartame is an artificial sweetener sold under the trade name NutraSweet. Aspartame is the methyl ester of a dipeptide formed from L-aspartic acid and L-phenylalanine and can be summarized as Asp-Phe-$OCH_3$. Surprisingly, the analogous ethyl ester (Asp-Phe-$OCH_2CH_3$) is not sweet. If either L residue is replaced with its enantiomeric D residue, the resulting compound is bitter instead of sweet.

(a) Draw a bond-line structure of aspartame.

(b) Draw a bond-line structure for each of the three bitter stereoisomers of aspartame.

## medically speaking | Polypeptide Antibiotics

Some naturally occurring polypeptides are among the most powerful bactericidal antibiotics. Most have cyclic structures, and they often contain D amino acids not typically found in plants and animals. In addition, many polypeptide antibiotics contain regions comprised of carbohydrates or heterocycles instead of amino acids.

In the previous chapter, we explored the structure of neomycin, one of the principal components of Neosporin. The other two ingredients in Neosporin are polypeptide antibiotics, called bacitracin and polymyxin.

Bacitracin is a complex mixture of polypeptides that was first isolated in 1945 from a strain of *Bacillus subtilis*. The commercial product called bacitracin is a mixture comprised primarily of bacitracin A.

This compound is believed to bind to an enzyme that bacteria use for synthesis of their cell walls, thereby inhibiting cell wall synthesis. Under these conditions, bacteria cannot sustain the high internal osmotic pressure (4–20 atm), which causes the bacteria to burst. It has been found that the effects of bacitracin are enhanced by the presence of zinc, and therefore, bacitracin zinc is used commonly in topical preparations used to treat local infections. It is not absorbed by the gastrointestinal tract, so oral administration is generally ineffective.

Polymyxin B was first isolated from the bacterium *Bacillus polymyxa* in 1947.

Polymyxin B is also commonly used in topical preparations to treat local infections. Like bacitracin A, it is also not absorbed by the gastrointestinal tract, so oral administration is ineffective except in treatment of intestinal infections. Bacitracin A and polymyxin B target different bacteria and are therefore commonly mixed together in topical preparations, such as Neosporin.

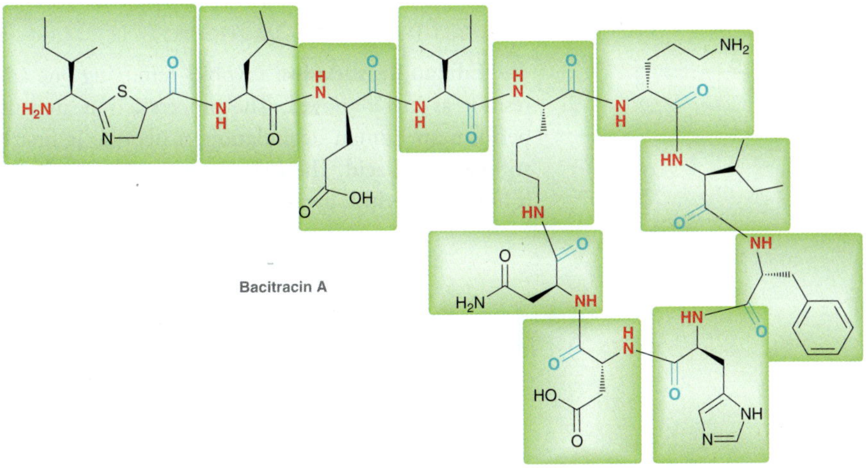

Bacitracin A

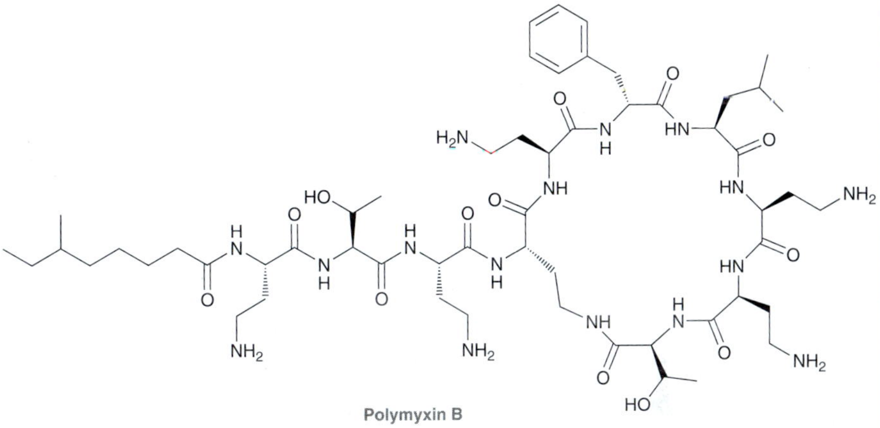

Polymyxin B

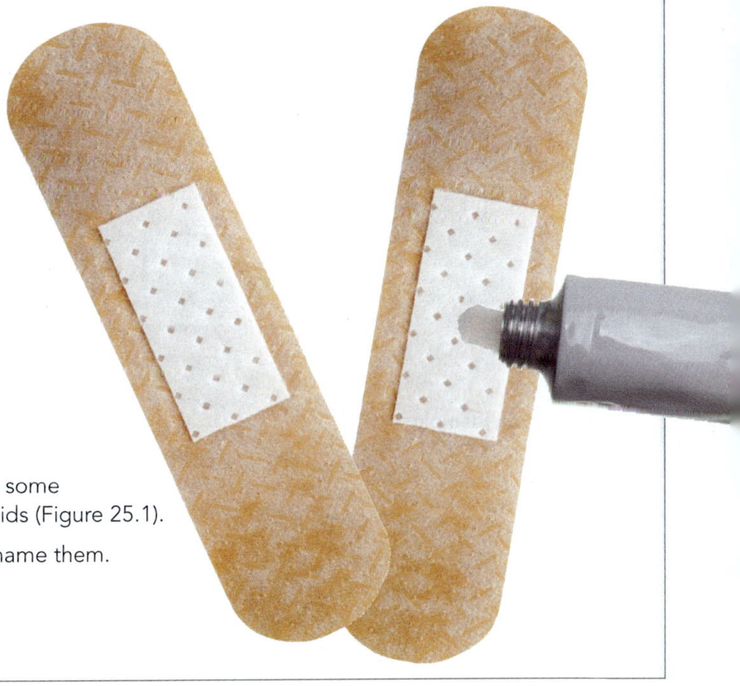

### CONCEPTUAL CHECKPOINT

**25.29** Bacitracin A is produced by bacteria and therefore contains some residues that are not from the list of 20 naturally occurring amino acids (Figure 25.1).

(a) Identify which amino acid residues are found in Table 25.1 and name them.

(b) Identify which of these amino acids are D rather than L.

(c) Identify all residues that are not listed in Table 25.1.

## 25.5 Sequencing a Peptide

The sequence of amino acids in a peptide can be analyzed using a variety of techniques. One such method is called the Edman degradation.

### The Edman Degradation

The **Edman degradation** involves removing one amino acid residue at a time and identifying each residue as it is removed. The process is repeated until the entire peptide has been sequenced. To perform an Edman degradation, a peptide is first treated with phenyl isothiocyanate followed by trifluoroacetic acid. These two reagents remove the amino acid residue at the N terminus and convert it into a phenylthiohydantoin (PTH) derivative.

PEPTIDE; 1) Ph—N=C=S; 2) $CF_3CO_2H$; This residue is removed; PTH derivative

The PTH derivative can then be analyzed by a variety of different techniques to determine the identity of the amino acid residue that was removed. The transformation is believed to occur via Mechanism 25.2.

### MECHANISM 25.2 THE EDMAN DEGRADATION

Nucleophilic attack

The free amino group at the end of the peptide chain functions as a nucleophile and attacks phenyl isothiocyanate

Nucleophilic attack

An intramolecular nucleophilic attack forms a ring

Proton transfers

Loss of a leaving group

$-H^+$

$CF_3COO—H$

Rearrangement

$H_3O^+$

Phenylthiohydantoin (PTH) derivative

The process can be repeated on the chain-shortened peptide to remove another amino acid residue. In this way, the amino acid residues are removed one at a time until the entire sequence has been identified. This entire process has been fully automated, and automated peptide sequencers are capable of sequencing a peptide chain with as many as 50 amino acid residues.

## CONCEPTUAL CHECKPOINT

**25.30** Draw the structure of the initial PTH derivative formed when the tripeptide Ala-Phe-Val undergoes an Edman degradation.

## Enzymatic Cleavage

For peptides containing more than 50 residues, the removal of each residue one at a time is not practical because unwanted side products accumulate and interfere with the results. Larger peptides are analyzed by first cleaving them into smaller fragments and then sequencing the fragments. A variety of enzymes, called **peptidases**, selectively hydrolyze specific peptide bonds. For example, trypsin is a digestive enzyme that catalyzes the hydrolysis of the peptide bond at the carboxyl side of the basic amino acids arginine and lysine.

Ala-Phe-**Lys** ⁝ Pro-Met-Tyr-Gly-**Arg** ⁝ Ser-Trp-Leu-His —Trypsin→ Ala-Phe-**Lys**, Pro-Met-Tyr-Gly-**Arg**, Ser-Trp-Leu-His

**Trypsin cleaves the peptide at these locations**

Chymotrypsin, another digestive enzyme, selectively hydrolyzes the carboxyl end of amino acids containing aromatic side chains, which includes phenylalanine, tyrosine, and tryptophan.

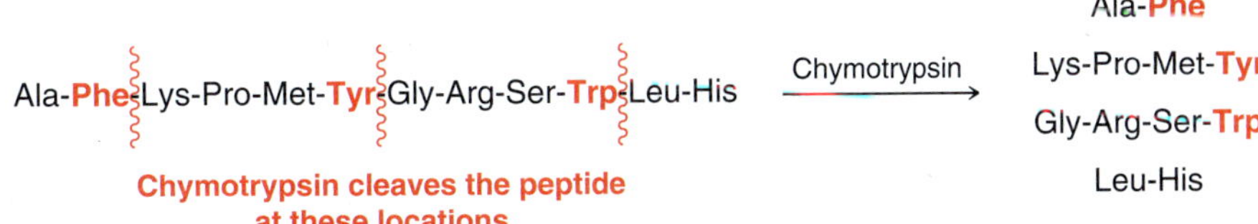


Many other digestive enzymes are known, and their ability to cleave peptide bonds selectively has been exploited. The fragments obtained are then individually sequenced and compared. In this way, even large peptides can be fully sequenced.

## SKILLBUILDER

### 25.4 PEPTIDE SEQUENCING VIA ENZYMATIC CLEAVAGE

**LEARN the skill**

A peptide with 16 amino acid residues is treated with trypsin to give three fragments, while treatment with chymotrypsin gives four fragments (shown). Identify the sequence of the 16 amino acid residues in the starting peptide.

| TRYPSIN FRAGMENTS | CHYMOTRYPSIN FRAGMENTS |
|---|---|
| Ala-Ser-Ala-Gly-Phe-Lys | Ile-Trp |
| Pro-Cys | Lys-Pro-Cys |
| Ile-Trp-Met-His-Phe-Met-Cys-Arg | Met-His-Phe |
| | Met-Cys-Arg-Ala-Ser-Ala-Gly-Phe |

**SOLUTION**

First consider the fragments generated upon cleavage with trypsin. Recall that trypsin will hydrolyze a peptide bond at the carboxyl side of arginine and lysine. We therefore expect that all fragments will have a C terminus that is either arginine or lysine, except for the fragment that contains the C terminus of the original peptide. This allows us to identify which fragment was last in the peptide sequence.

Pro—Cys

**This residue must have been the C terminus of the original peptide**

There are only two other fragments, so there are only two possibilities for the sequence of the original peptide. The correct sequence can be determined by analyzing the chymotrypsin fragments. Specifically, we look for the fragment that ends with Pro-Cys, and we find that this fragment has a lysine residue immediately before the proline residue.

Lys—Pro—Cys

This information makes it possible to determine the order of the trypsin fragments.

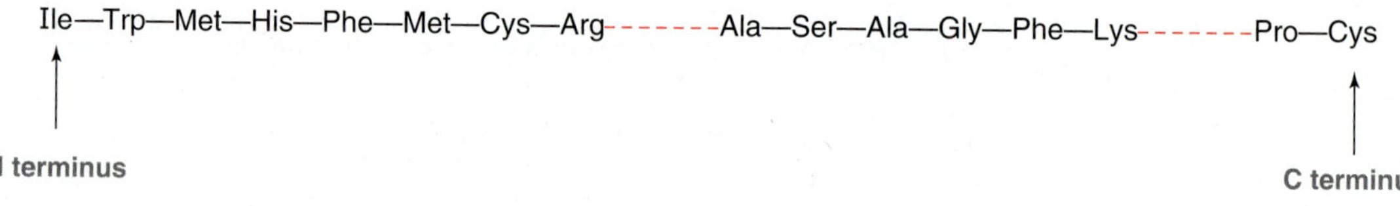

**PRACTICE the skill** **25.31** A peptide with 22 amino acid residues is treated with trypsin to give four fragments, while treatment with chymotrypsin yields six fragments. Identify the sequence of the 22 amino acid residues in the starting peptide.

| TRYPSIN FRAGMENTS | | CHYMOTRYPSIN FRAGMENTS | |
|---|---|---|---|
| Trp-His-Phe-Met-Cys-Arg | Pro-Val-Ile-Leu-Arg | Lys-Pro-Val-Ile-Leu-Arg-Trp | His-Phe |
| Met-Phe-Val-Ala-Tyr-Lys | Gly-Pro-Phe-Ala-Val | Val-Ala-Tyr | Ala-Val |
| | | Met-Cys-Arg-Gly-Pro-Phe | Met-Phe |

**APPLY the skill**

**25.32** The tetrapeptide Val-Lys-Ala-Phe is cleaved into two fragments upon treatment with trypsin. Identify the sequence of a tetrapeptide that will produce the same two fragments when treated with chymotrypsin.

**25.33** Consider the structure of the following cyclic octapeptide. Would cleavage of this peptide with trypsin produce different fragments than cleavage with chymotrypsin? Explain.

Phe, Arg, Phe, Arg, Phe, Arg, Phe, Arg

need more **PRACTICE?** **Try Problems 25.71, 25.74**

## 25.6 Peptide Synthesis

### Peptide Bond Formation

Forming a peptide bond requires the coupling of a carboxylic acid group and an amino group.

**A peptide bond**

Several reagents are available to achieve this type of transformation. One common reagent is dicyclohexylcarbodiimide (DCC).

Formation of a peptide bond via DCC is believed to proceed via Mechanism 25.3.

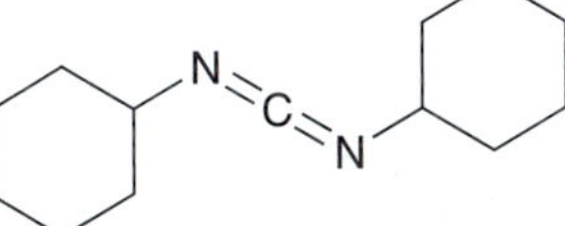

**Dicyclohexylcarbodiimide (DCC)**

## MECHANISM 25.3 PEPTIDE BOND FORMATION VIA DCC

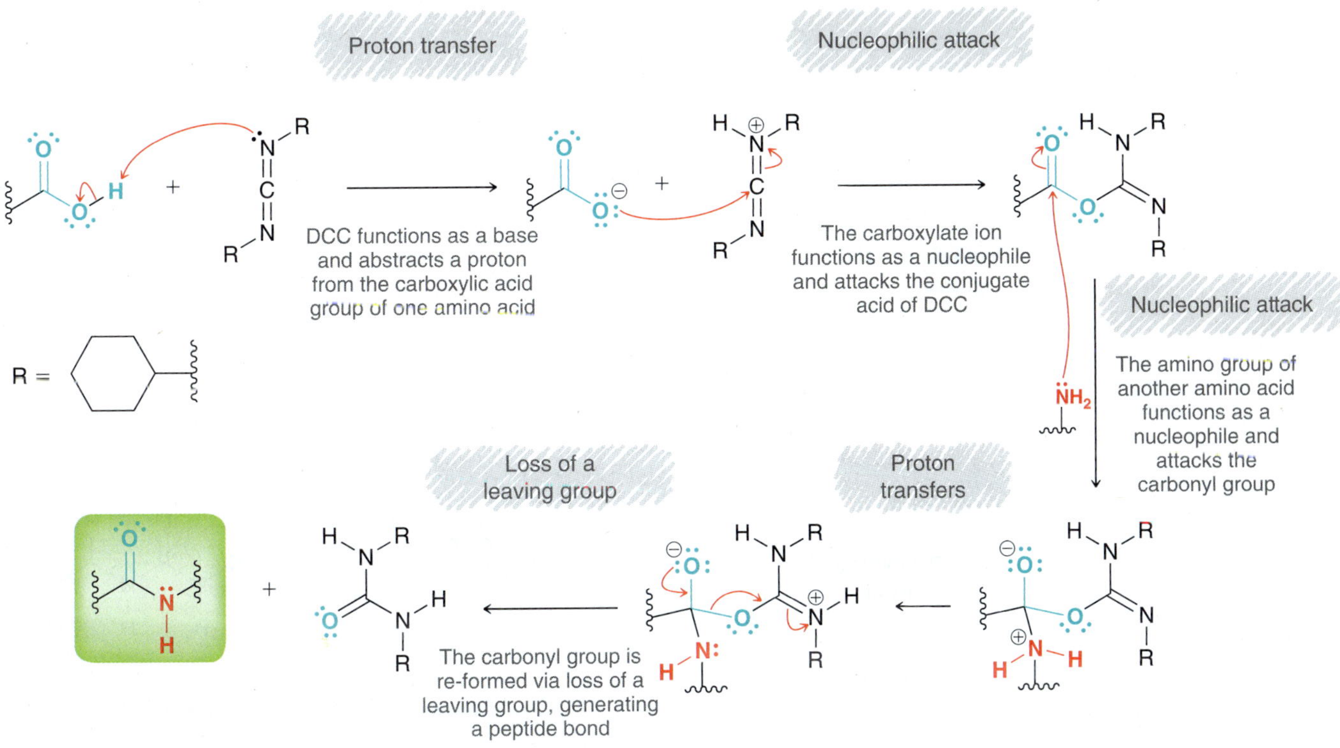


A lack of regioselectivity becomes problematic when DCC is used to couple two different amino acids. Specifically, four different dipeptides are possible:

This problem can be circumvented by first protecting the $NH_2$ group of one amino acid and the COOH group of the other amino acid (Figure 25.6): The protected amino acids can be coupled regioselectively followed by removal of the protecting groups. The following sections describe how amino groups and carboxylic acid groups can be protected.

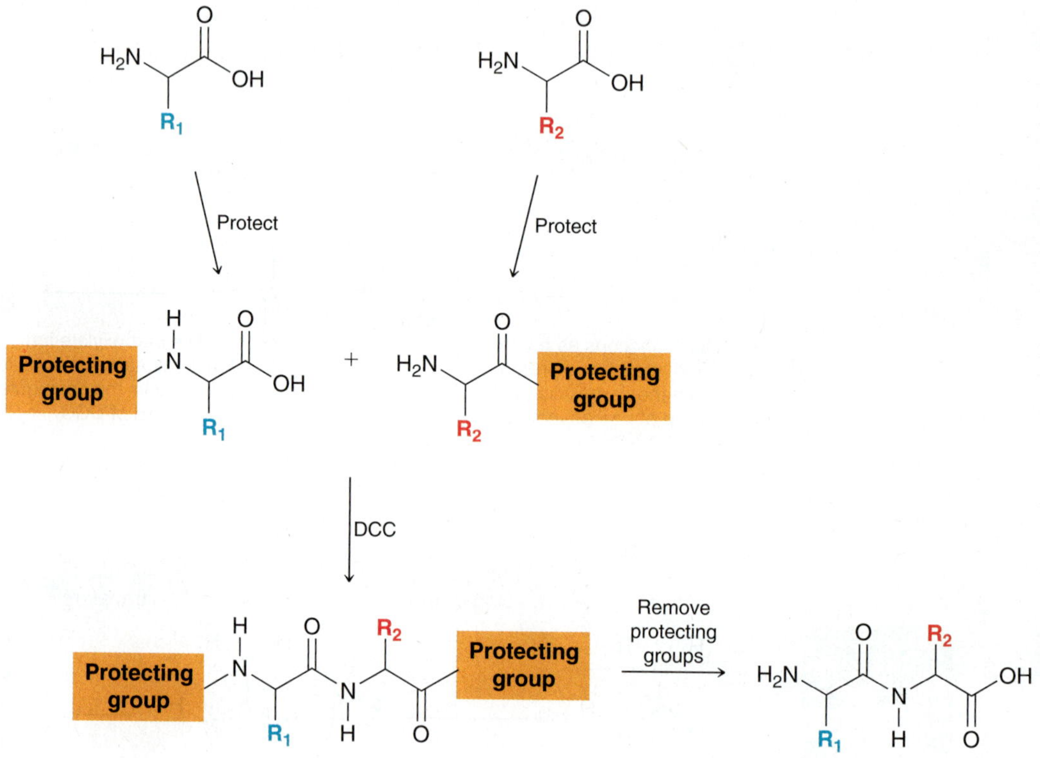

Protect

Protect

Protecting group

Protecting group

DCC

Protecting group

Protecting group

Remove protecting groups

**FIGURE 25.6**
The overall strategy for synthesizing dipeptides involves protection, coupling, and then deprotection.

## Protecting the Amino Group of an Amino Acid

An amino group can be protected by converting it into a carbamate.

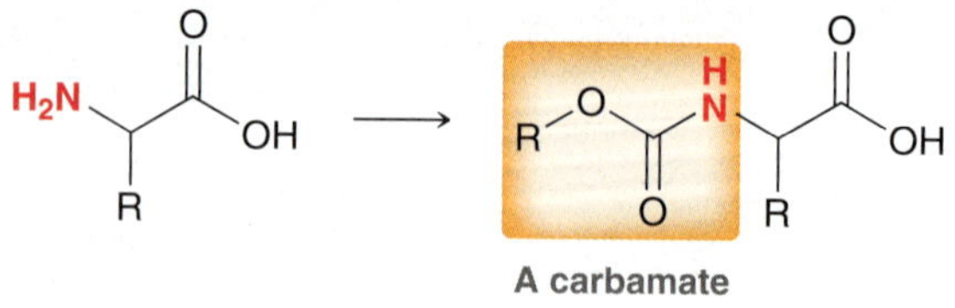

**A carbamate**

The nitrogen atom of a carbamate is less nucleophilic because its lone pair is delocalized by the neighboring carboxyl group. One of the most common carbamate protecting groups is the *tert*-butoxycarbonyl group (Boc), formed via treatment with di-*tert*-butyl dicarbonate.

**Di-*tert*-butyl dicarbonate**
**$(Boc)_2O$**

**Boc protecting group**

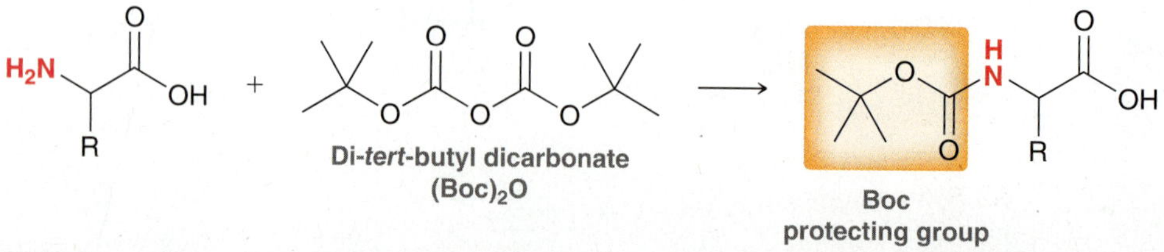

A likely mechanism for this process involves the amino group functioning as a nucleophile and performing a nucleophilic acyl substitution reaction (Mechanism 25.4).

## MECHANISM 25.4 INSTALLATION OF A BOC PROTECTING GROUP

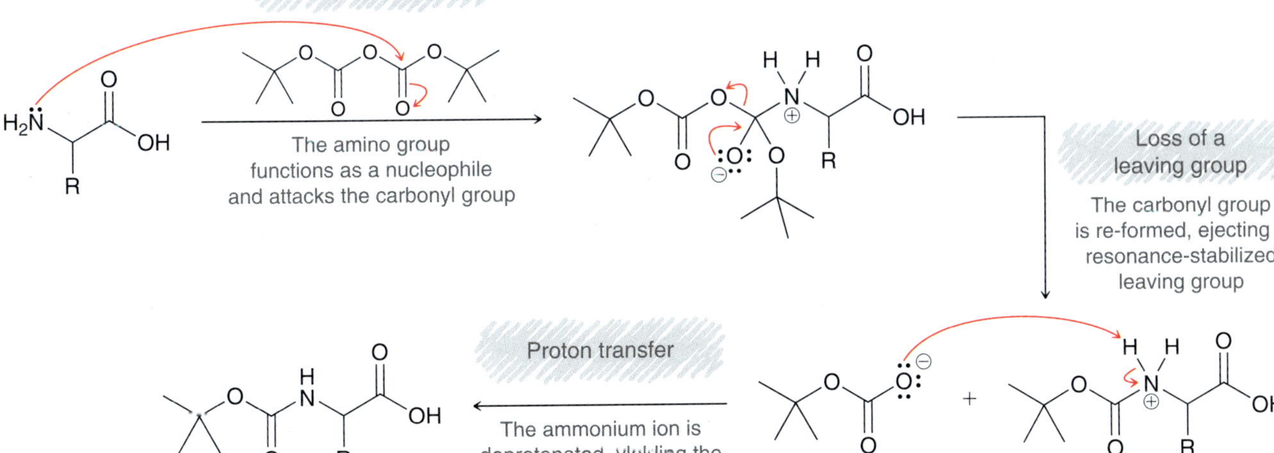

## MECHANISM 25.5 REMOVAL OF A BOC PROTECTING GROUP

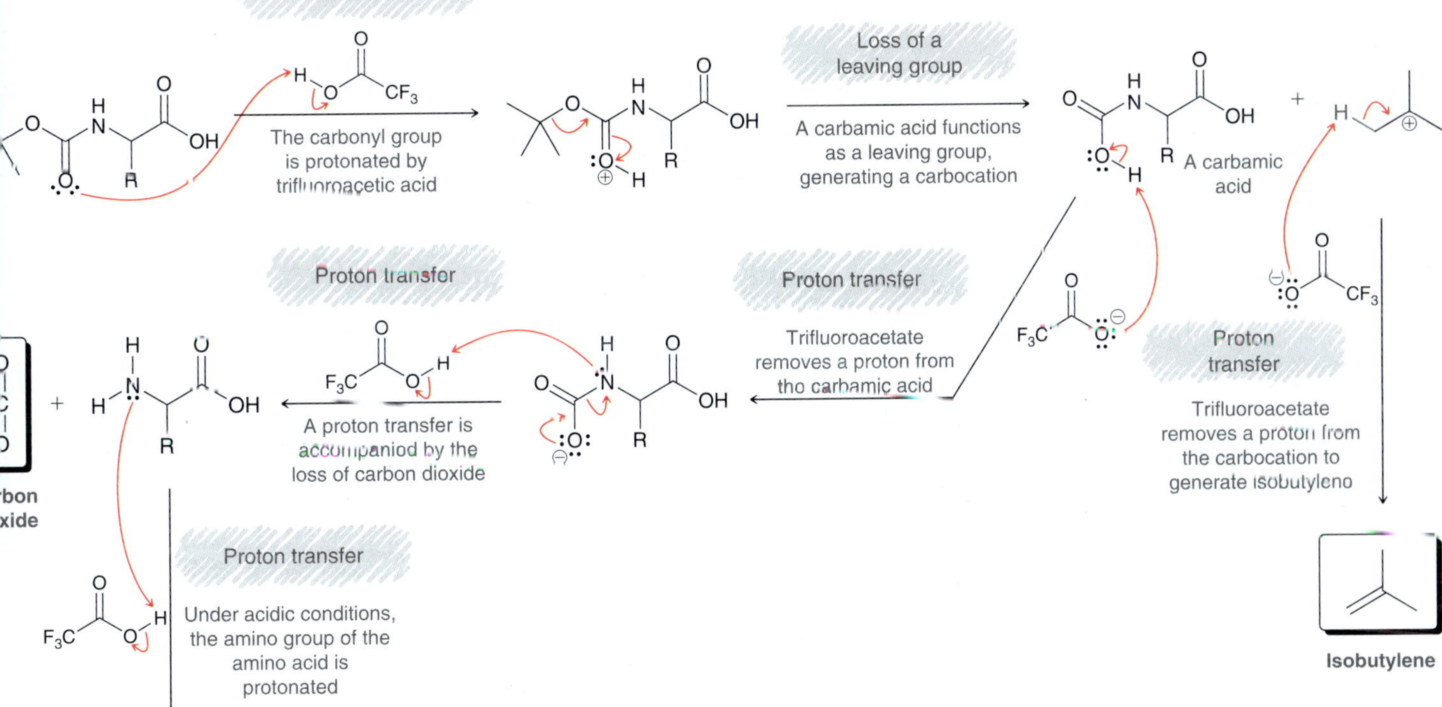

The Boc group is one of the most commonly used protecting groups because of the ease with which it can be removed. Several reagents can be used to remove a Boc group, including trifluoroacetic acid.

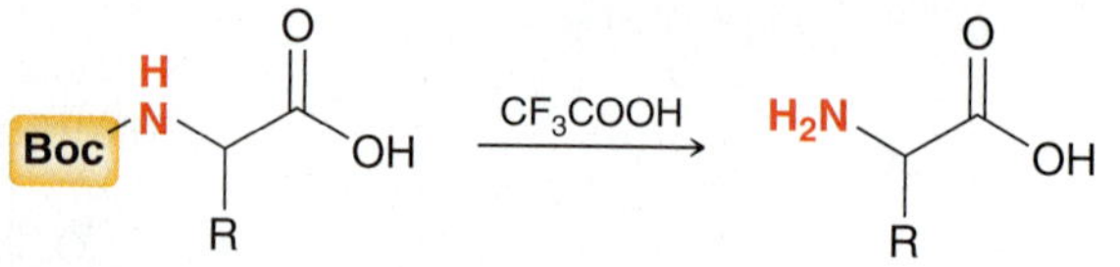

Boc–NH–CH(R)–C(=O)OH $\xrightarrow{CF_3COOH}$ $H_2N$–CH(R)–C(=O)OH

The mechanism is believed to involve protonation of the carbonyl group followed by formation of a carbocation and loss of carbon dioxide (Mechanism 25.5). This process generates isobutylene and carbon dioxide, both of which are gases. The evolution of gases effectively drives the reaction to completion, and the yield for removal of a Boc group is often 100%.

An alternative protecting group is known as Fmoc; it is formed by treating an amino acid with 9-fluorenylmethyl chloroformate.

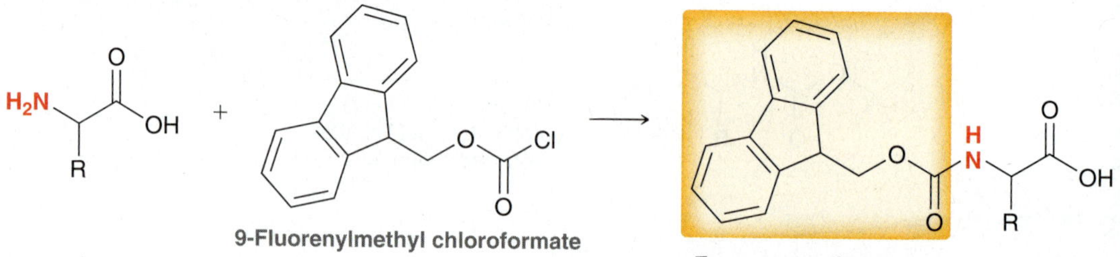

9-Fluorenylmethyl chloroformate

Fmoc protecting group

This process likely occurs via a mechanism that is similar to Mechanism 25.4; that is, the amino group functions as a nucleophile and performs a nucleophilic acyl substitution reaction (where chloride is the leaving group that is replaced). The Fmoc protecting group can be removed with a base, such as piperidine, which initiates the process by deprotonating the acidic benzylic position (Mechanism 25.6). Deprotonation at this position generates an anion that is similar to the anion of cyclopentadiene. The anion is aromatic and is therefore highly stabilized. Loss of carbon dioxide then drives the deprotection process to completion.

## MECHANISM 25.6 REMOVAL OF A FMOC PROTECTING GROUP

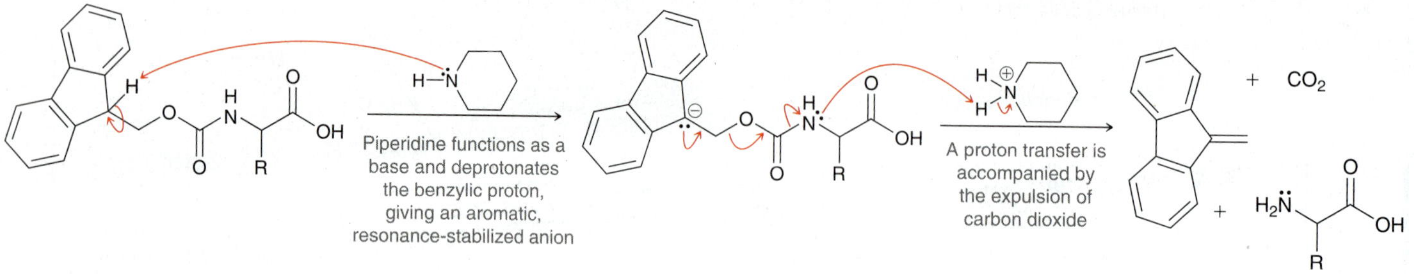

Piperidine functions as a base and deprotonates the benzylic proton, giving an aromatic, resonance-stabilized anion

A proton transfer is accompanied by the expulsion of carbon dioxide

+ $CO_2$

## Protecting the Carboxylic Acid Group of an Amino Acid

A carboxylic acid group can be protected by converting it into an ester, which is achieved upon treatment with an alcohol under acidic conditions.

**LOOKING BACK**

For a review of this process, see Section 21.10.

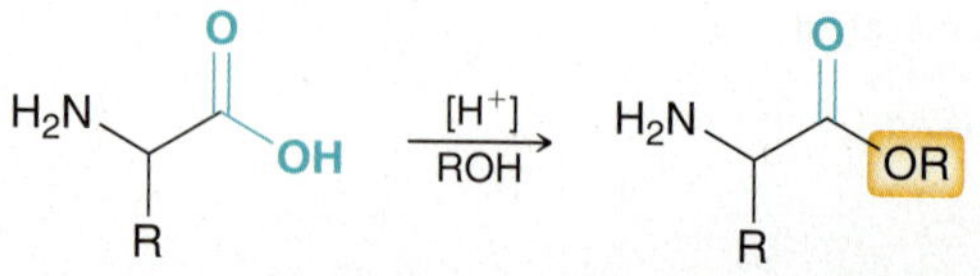

$H_2N$–CH(R)–C(=O)OH $\xrightarrow[ROH]{[H^+]}$ $H_2N$–CH(R)–C(=O)OR

An ester

Carboxylic acid groups are most commonly converted to methyl esters by treatment with MeOH or to benzyl esters by treatment with $PhCH_2OH$.

The ester protecting group can be removed with aqueous base ($H_2O$, NaOH).

$$H_2N\text{-}CH(R)\text{-}C(=O)\text{-}OR \xrightarrow[H_2O]{NaOH} H_2N\text{-}CH(R)\text{-}C(=O)\text{-}O^{\ominus}$$

Benzyl esters can also be removed with hydrogenolysis or with HBr in acetic acid.

$$H_2N\text{-}CH(R)\text{-}C(=O)\text{-}OCH_2C_6H_5 \xrightarrow[\text{or HBr, Acetic acid}]{H_2,\ Pt} H_2N\text{-}CH(R)\text{-}C(=O)\text{-}OH$$

**Benzyl ester**

## Preparing a Dipeptide

The overall synthesis of a dipeptide requires several steps: (1) protect the COOH group of one amino acid and the $NH_2$ group of the other amino acid, (2) couple the protected amino acids with DCC, and (3) remove both protecting groups. This process is illustrated in the synthesis of Ala-Gly.

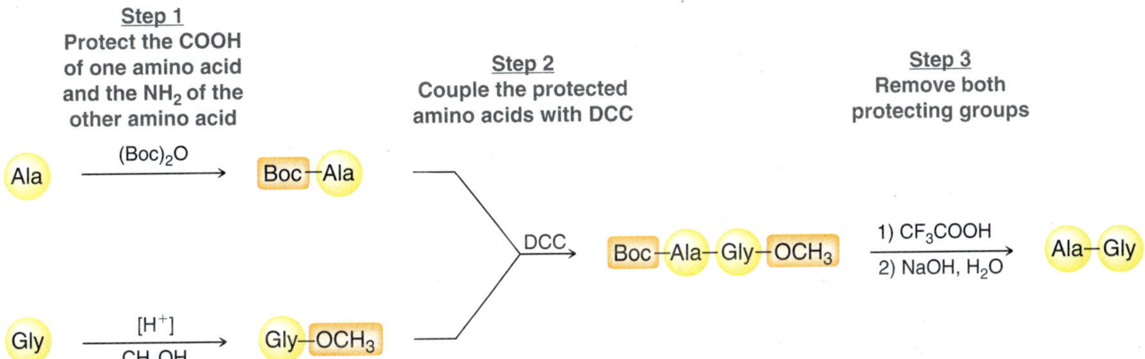


A similar method can be used to prepare tripeptides and tetrapeptides.

# SKILLBUILDER

## 25.5 PLANNING THE SYNTHESIS OF A DIPEPTIDE

**LEARN the skill** Propose a synthesis for the following dipeptide: Phe-Val.

**SOLUTION**

Draw the two amino acids necessary and identify the functional groups that must be joined to form a peptide bond.

**Phenylalanine (Phe)** **Valine (Val)** → **Phe-Val**

Identify the groups that must be protected (the groups that are not being coupled). The amino group is protected by installing a Boc protecting group.

**STEP 1**
Install the appropriate protecting groups.

The carboxylic acid group is protected by converting it to an ester, such as a benzyl ester.

**STEP 2**
Couple the protected amino acids using DCC.

With the protecting groups in place, the next step is to form the desired peptide linkage using DCC.

The final step is to remove the protecting groups to yield the desired product.

**STEP 3**
Remove the protecting groups.

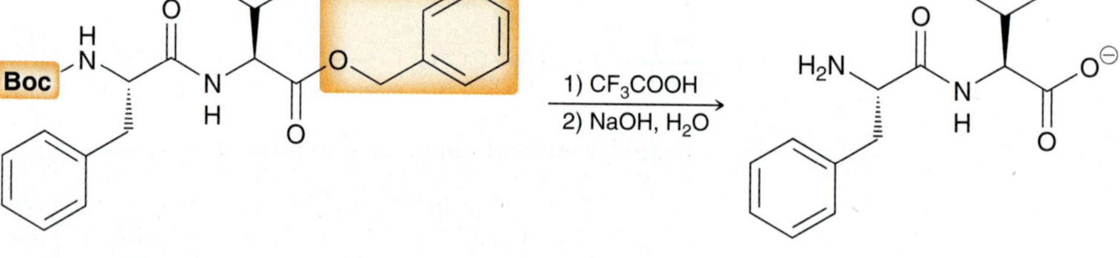

**PRACTICE the skill** **25.34** Draw all of the steps and reagents necessary to prepare each of the following dipeptides from their corresponding amino acids:

**(a)** Trp-Met **(b)** Ala-Ile **(c)** Leu-Val

**APPLY the skill** **25.35** Draw all of the steps and reagents necessary to prepare a tripeptide with the sequence Ile-Phe-Gly.

**25.36** Draw all of the steps and reagents necessary to prepare a pentapeptide with the sequence Leu-Val-Phe-Ile-Ala.

need more **PRACTICE?** **Try Problems 25.77–25.79**

## The Merrifield Synthesis

The method outlined in the previous section works well for very small peptides, but it is not feasible to use it for larger peptides, because each step requires isolation and purification due to the accumulation of unwanted side products. Larger peptides can be prepared with the **Merrifield synthesis**, developed by R. Bruce Merrifield (Rockefeller University). A protected amino acid is first tethered to beads of an insoluble polymer. One such polymer is a polystyrene derivative with $CH_2Cl$ groups on some of the aromatic rings.

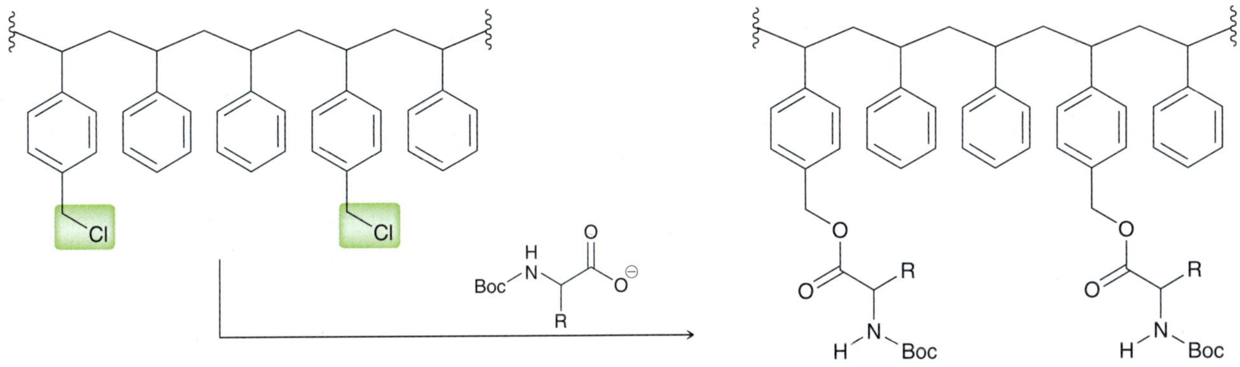

An amino acid protected with a Boc group is first attached to the polymer via an $S_N2$ reaction. In this way, the polymer serves as a handle that holds the amino acid in place. The Boc protecting group is then removed, and DCC is used to couple the next Boc-protected amino acid:

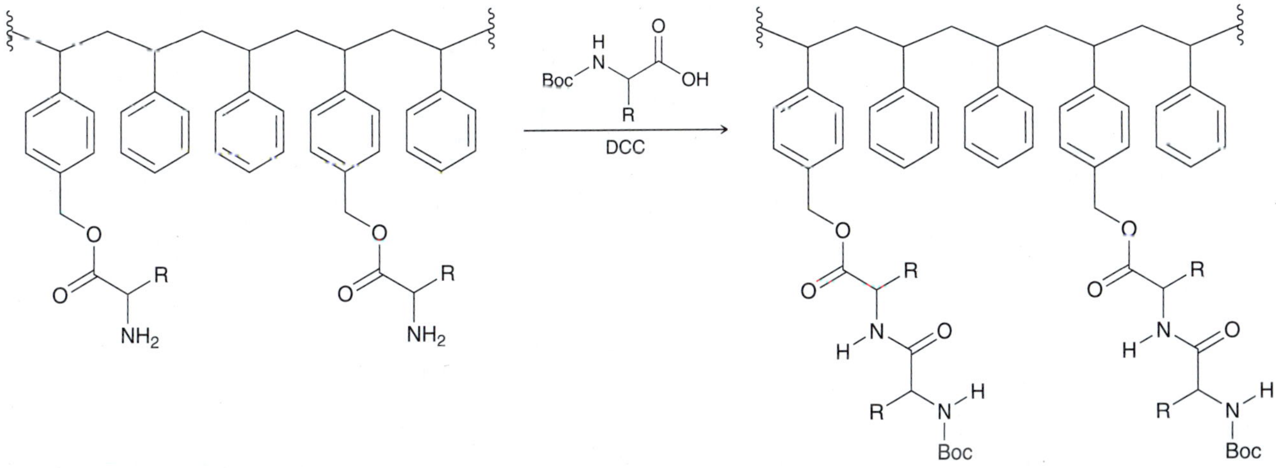

After the coupling step, impurities and by-products are simply washed away, while the peptide chain remains tethered to the insoluble polymer. The process can then be repeated many times, each time lengthening the chain of the peptide by one amino acid residue. When the desired polypeptide has been created, it is removed from the polymer using hydrofluoric acid (HF).

In 1969, Merrifield used this technique in the preparation of a protein called ribonuclease, which is comprised of 128 residues. The procedure required 369 separate reactions, which were performed in just six weeks. The overall yield of the process was 17%, implying that each individual step had a yield greater than 99%. Merrifield's ingenious method set the stage for an entirely new way of performing chemical reactions. For his groundbreaking work, he received the 1984 Nobel Prize in Chemistry. Merrifield's solid-phase method has now been completely automated and is accomplished by machines called peptide synthesizers.

## SKILLBUILDER

### 25.6 PREPARING A PEPTIDE USING THE MERRIFIELD SYNTHESIS

**LEARN the skill**

Identify the steps you would use to prepare the following peptide via the Merrifield synthesis:

Ile-Gly-Leu-Ala-Phe

**SOLUTION**

Begin with the C terminus. In this case, the C terminus must be a phenylalanine residue, so the first step is to attach a Boc-protected phenylalanine to the polymer via an $S_N2$ process.

**STEP 1**
Attach the appropriate Boc-protected residue to the polymer.

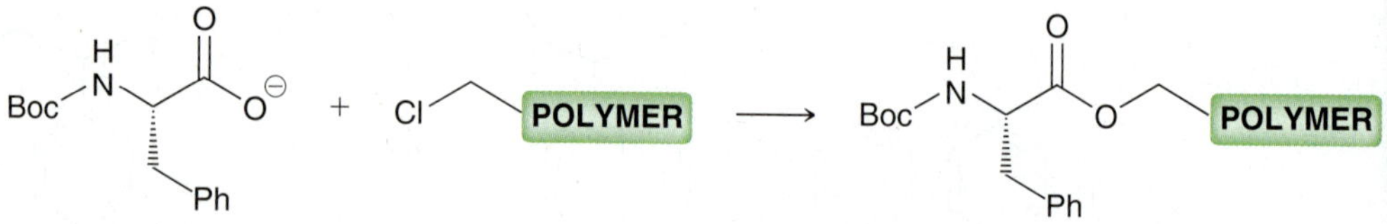

The second step is to remove the protecting group, exposing the N terminus of the phenylalanine residue.

**STEP 2**
Remove the Boc protecting group.

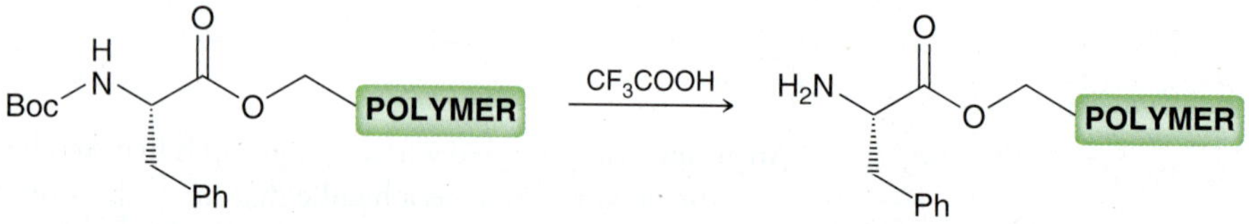

**STEP 3**
Using DCC, create a new peptide bond with a Boc-protected amino acid.

In the third step, the next residue (Ala) is installed as a Boc-protected residue using DCC to form the peptide bond.

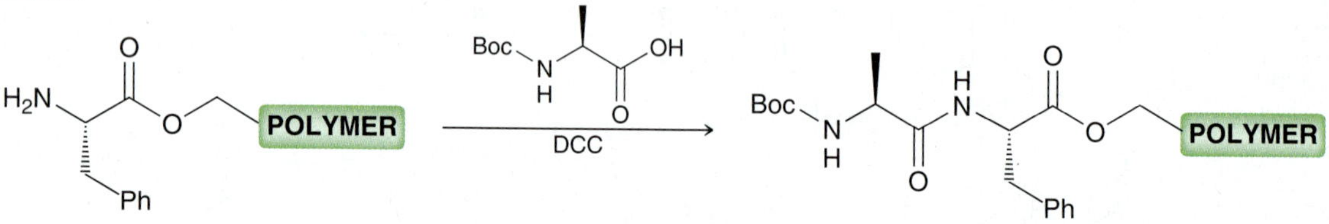

Steps 2 and 3 are then repeated for the installation of each additional residue until the desired peptide chain has been assembled.

**STEP 4**
Repeat steps 2 and 3 for each residue that is to be added to the growing peptide chain.

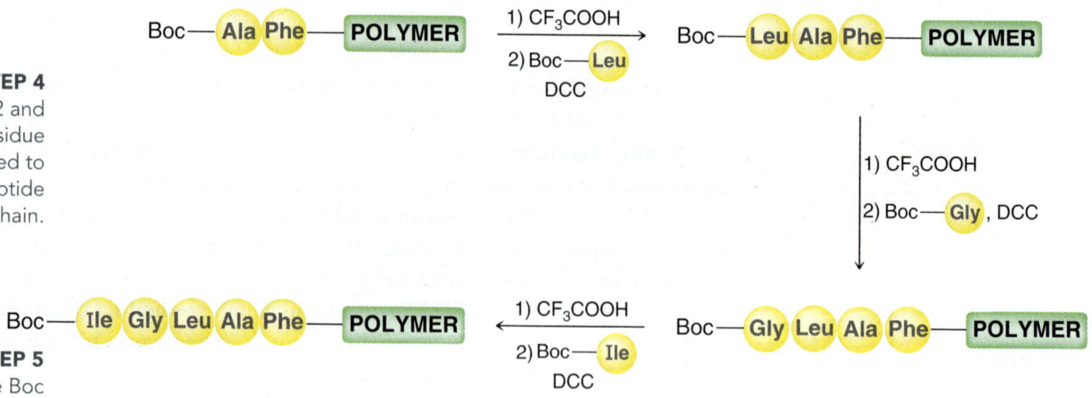

**STEP 5**
Remove the Boc protecting group and detach the peptide from the polymer.

Finally, the protecting group is removed, and the peptide is released from the polymer.

**PRACTICE** the skill

**25.37** Identify all of the steps necessary to prepare each of the following peptides with a Merrifield synthesis:

**(a)** Phe-Leu-Val-Phe **(b)** Ala-Val-Leu-Ile

**APPLY** the skill

**25.38** Identify the sequence of the tripeptide that would be formed from the following order of reagents. Clearly label the C terminus and N terminus of the tripeptide.

Cl–CH₂–POLYMER

1) Boc–NH–CH(CH₂Ph)–C(=O)O⁻
2) $CF_3COOH$
3) Boc–NH–CH($CH_3$)–C(=O)OH, DCC
4) $CF_3COOH$
5) Boc–NH–CH(CH($CH_3$)₂)–C(=O)OH, DCC
6) $CF_3COOH$
7) HF

→ **?**

need more **PRACTICE?** **Try Problems 25.80, 25.81**

## 25.7 Protein Structure

Proteins are fairly large organic compounds, and as such, the structure of a protein is more complex than the structures of simple organic compounds. In fact, proteins are generally described in terms of four levels of structure: primary, secondary, tertiary, and quaternary structure.

### Primary Structure

The **primary structure** of a protein is the sequence of amino acid residues. Figure 25.7 shows the primary structure of human insulin. Insulin is comprised of two polypeptide chains, called chains A and B, which are linked together by disulfide bridges.

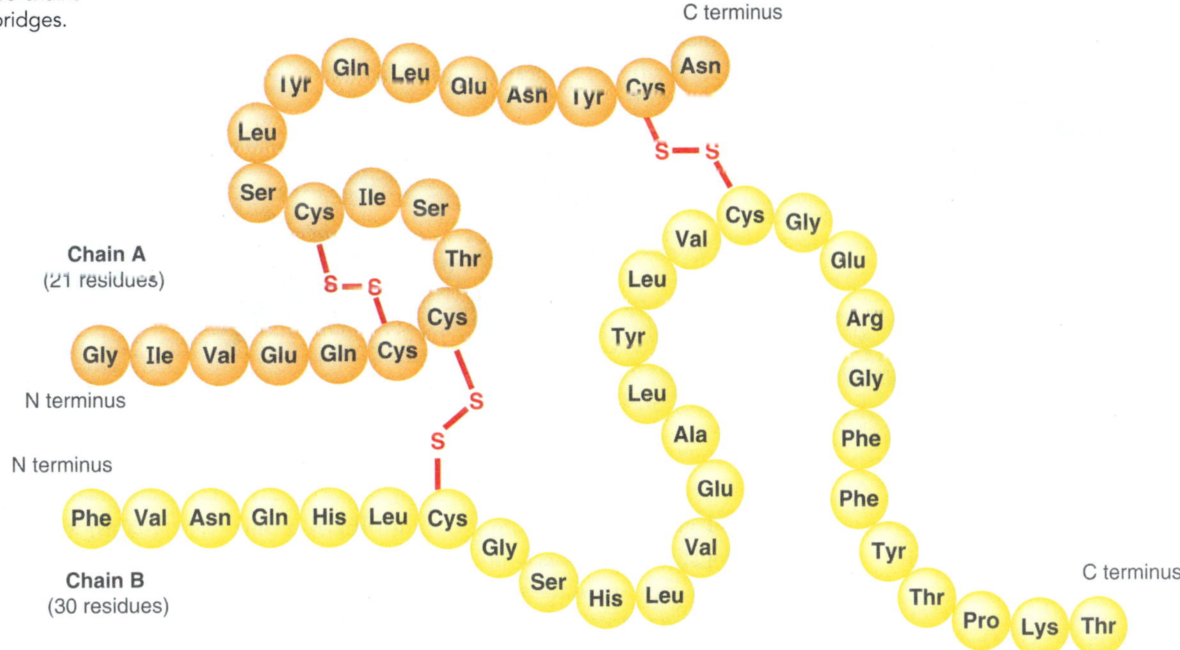


**FIGURE 25.7** The primary structure of human insulin consists of two chains linked by disulfide bridges.

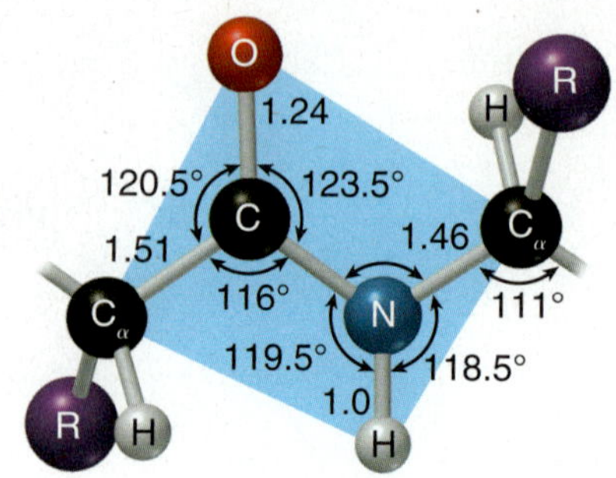


**FIGURE 25.8**
The planar geometry of a peptide bond.

## Secondary Structure

The **secondary structure** of a protein refers to the three-dimensional conformations of localized regions of the protein. Recall that each peptide bond exhibits restricted rotation and planar geometry (Figure 25.8). This is the case for each amino acid residue in a protein (Figure 25.9). Although the individual planes are free to rotate with respect to each other, their presence limits the conformations available to the protein. Depending on the sequence of amino acid residues, localized regions of peptides often adopt particular shapes. Two particularly stable arrangements are the **α helix** and **β pleated sheet**. An α helix forms when a portion of the protein twists into a clockwise spiral (Figure 25.10). Each turn has approximately four amino acid residues, and each C═O group experiences hydrogen bonding with an N—H group that is four residues farther along on the chain. In an α helix, the R groups (side chains) extend outward, away from the helix. A proline residue cannot be part of an α helix, because it lacks an N—H proton and does not participate in hydrogen bonding. Many proteins contain α helices. For example, α-keratin, the major component of hair, is comprised almost entirely of α helices. The composition of hair will be further discussed in Section 25.8.

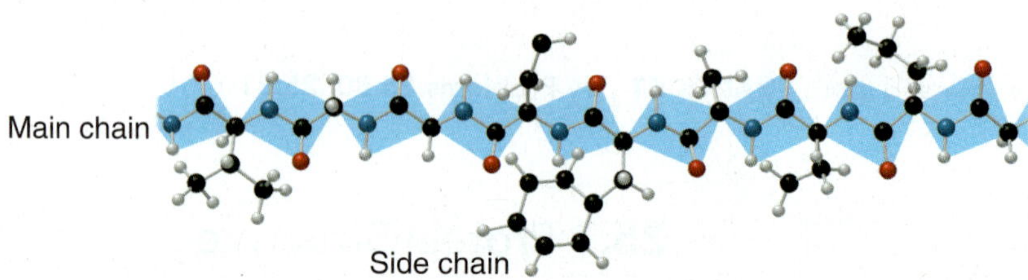


**FIGURE 25.9**
Proteins are comprised of repeating planar units, each of which can freely rotate with respect to the others.

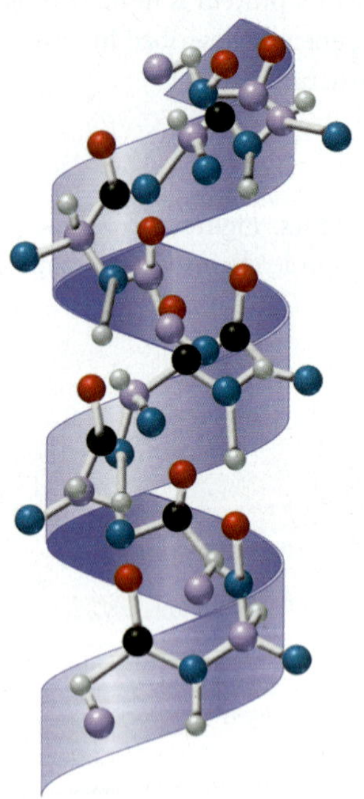

**FIGURE 25.10**
An α helix. (Illustration, Irving Geis. Image from Irving Geis Collection/Howard Hughes Medical Institute. Rights owned by ***HHMI***. Not to be reproduced without permission.)

Beta pleated sheets are formed when two or more protein chains line up side by side (Figure 25.11). In a β pleated sheet, hydrogen bonding occurs between the C═O group and N—H group of neighboring strands. The R groups (side chains) are positioned above and below the plane of the sheet, in an alternating pattern. Beta sheets are commonly formed from segments that bear residues with small R groups, such as Ala and Gly. With larger side chains, steric hindrance prevents the strands from getting close enough to participate in hydrogen bonding. Spider webs are comprised primarily of fibroin, a protein containing a pleated sheet arrangement that endows fibroin with structural strength. Fibroin is also the major component of silk.

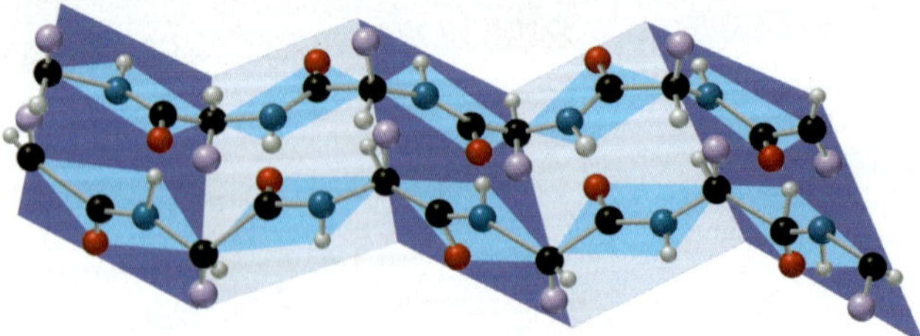

**FIGURE 25.11**
A β pleated sheet. (Illustration, Irving Geis. Image from Irving Geis Collection/Howard Hughes Medical Institute. Rights owned by ***HHMI***. Not to be reproduced without permission.)

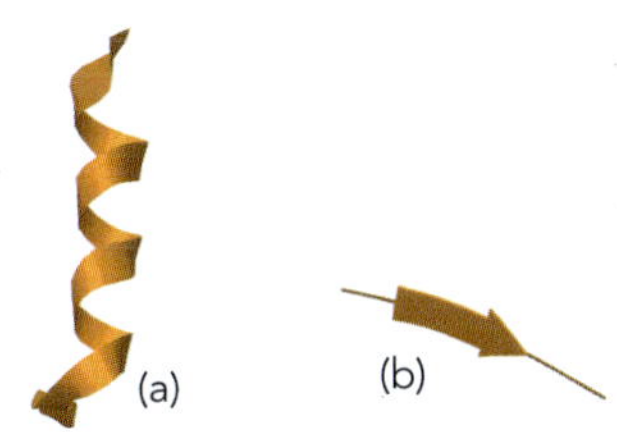


**FIGURE 25.12** Symbols used when illustrating the secondary structure of proteins: (a) an α helix and (b) a β pleated sheet.

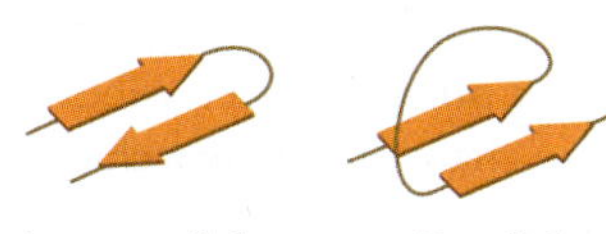


**FIGURE 25.13** Parallel and antiparallel β sheets.

Proteins are generally comprised of many structural domains, including α helices as well as β pleated sheets. In order to draw a protein in a way that illustrates its secondary structure, the shorthand symbols given in Figure 25.12 are frequently used. The flat helical ribbon represents an α helix, while the flat wide arrow represents a β sheet. Beta sheets can be either parallel or antiparallel, depending on the direction of neighboring segments (Figure 25.13). Both chains of human insulin exhibit α helices (Figure 25.14).

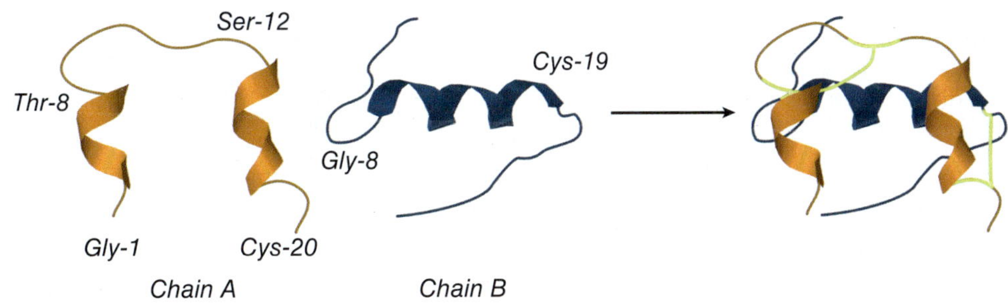


**FIGURE 25.14** Alpha helices in the two peptide chains of human insulin.

## CONCEPTUAL CHECKPOINT

**25.39** Below is the primary structure for a peptide. Identify the regions that are most likely to form β pleated sheets.

Trp-His-Pro-Ala-Gly-Gly-Ala-Val-His-Cys-Asp-Ser-Arg-Arg-Ala-Gly-Ala-Phe

## Tertiary Structure

The **tertiary structure** of a protein refers to its three-dimensional shape. Proteins typically adopt conformations that maximize stability. In aqueous solution, a protein will adopt a conformation in which the polar groups are pointing toward the exterior of the protein and the nonpolar groups are located in the interior. Consider the structure of myoglobin, a protein used to store molecular oxygen (Figure 25.15). The regions marked in red represent polar side groups, and they are positioned on the surface of the three-dimensional structure so that they can interact with water, thereby maximizing the stability of the protein.

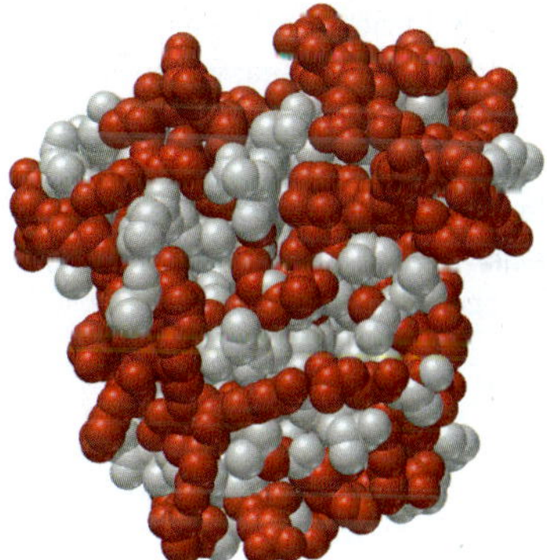

**FIGURE 25.15** A space-filling model of myoglobin with hydrophilic groups highlighted in red.

The intramolecular forces holding a typical protein in its folded shape are relatively weak, and under conditions of mild heating, a protein can unfold, in a process called **denaturation** (Figure 25.16). A common example is the cooking of an egg, during which the protein albumin is denatured and changes from a clear solution to a white solid. The secondary and tertiary structures are lost, but the primary structure remains intact. That is, the sequence of amino acid residues does not change. Loss

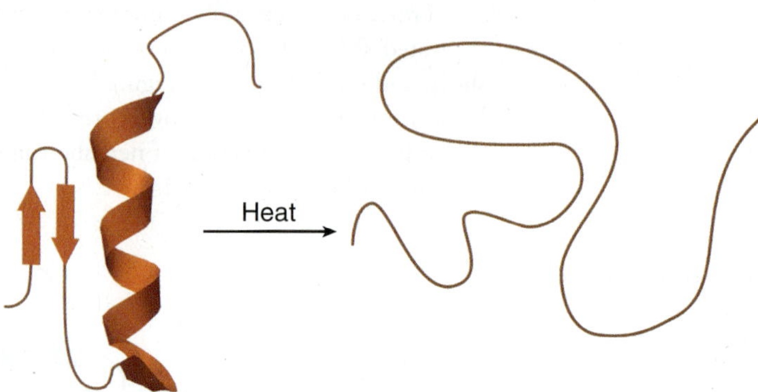


**FIGURE 25.16**
A protein loses its secondary and tertiary structures when it is denatured.

of tertiary structure is typically accompanied by a loss of function, because the function of a protein is very much dependent on its three-dimensional shape. The denaturation process is irreversible in most cases. Factors other than heat can also cause denaturation, including a change in pH or a change in solvent.

## medically speaking | Diseases Caused by Misfolded Proteins

Until recently, it was widely believed that all contagious diseases were caused by living pathogens, such as viruses, bacteria, or fungi. For a small number of contagious diseases, scientists were baffled by their inability to identify the pathogen responsible. Examples include *Creutzfeldt-Jakob disease* in humans and *scrapie* in sheep. Both diseases are similar, causing loss of mental function and death. Autopsies of victims revealed plaques of amyloid protein surrounded by spongelike tissue in the brain. These diseases came to be known as spongiform encephalopathies. The prevailing belief was that some virus or other organism would soon be isolated and identified as the causitive pathogen.

In the 1980s Stanley Prusiner, a neurologist at the Univeristy of California at San Francisco, systematically and scrupulously separated all of the components of scrapie-infected sheep brains. He found the infectious agent to be a particular protein. He suggested that scrapie and other related diseases are caused by *infectious proteins*, which he called *prions*. This conclusion was highly significant, because it represented the first example of a nonliving pathogen capable of causing deadly infections without employing DNA or RNA. For his groundbreaking work, Prusiner was awarded the 1997 Nobel Prize in Physiology or Medicine.

Prion diseases attracted significant public awareness in the 1990s when cows in England, Canada, and the United States became infected with a disease called bovine spongiform encephalopathy, which caused them to move around erratically and eventually die. This disease, commonly referred to as mad cow disease, was presumably caused by using the remains of scrapie-infected sheep to feed cows. Mad cow disease represents a significant risk to humans because eating infected meat could trigger a form of Creutzfeldt-Jakob disease in humans, which is a fatal disease with no cure.

Prions are believed to be normal proteins that have become misfolded. In other words, the primary structure of the protein remains unaltered, but the tertiary structure is different. The normal protein unfolds and then refolds into a new shape (a prion). When animals eat the remains of scrapie-infected sheep, the prion is not digested and is not detected by the immune system as a harmful substance. The infectious form of the prion then causes other normal proteins to misfold as well. As the amount of misfolded proteins increases, they begin to aggregate in the brain, causing the plaques and spongelike tissues associated with spongiform encephalopathies.

Over the last decade, progress in finding a cure for prion diseases has been slow but steady. Current research efforts are focused primarily on the design of drugs that can potentially inhibit misfolding. It is now understood that the body utilizes a variety of proteins called chemical *chaperones* to guide the folding of other proteins. By studying how these chaperones function, scientists believe that they will be able to synthesize chaperones that will prevent formation of the misfolded proteins characteristic of prion diseases. Prion diseases are an extreme example of the relationship between structure and function.

## Quaternary Structure

A **quaternary structure** arises when a protein consists of two or more folded polypeptide chains, called *subunits*, that aggregate to form one protein complex. For example, consider the quaternary structure of hemoglobin, which is comprised of four different subunits (Figure 25.17). Each subunit in hemoglobin is displayed in a different color. The subunits have many nonpolar amino acid residues, and as such, there are many nonpolar groups on their surfaces. These nonpolar groups experience van der Waals interactions, which hold the quaternary structure together.

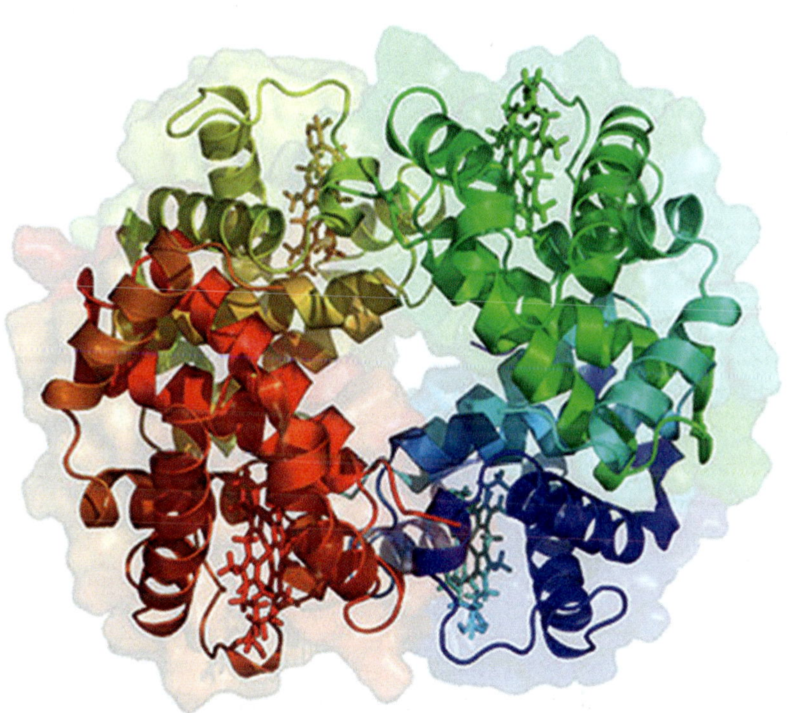

**FIGURE 25.17**
The quaternary structure of hemoglobin. For clarity, each subunit is presented in a different color.

# 25.8 Protein Function

Based on their shape, proteins are often classified as fibrous or globular. **Fibrous proteins** consist of linear chains that are bundled together. **Globular proteins** are chains that are coiled into compact shapes.

Proteins are also classified according to their function. A few different classes of proteins are described in the following sections.

## Structural Proteins

**Structural proteins** are fibrous proteins that are used for their structural rigidity. For example, α-keratins are structural proteins found in hair, nails, skin, feathers, and wool. Hair is comprised of many polypeptide chains of α-keratin, each coiled into a right-handed α helix, as seen in Figure 25.18. Four of these right-handed helices are then wrapped into left-handed superhelices, similar to the way that rope is made from twisting individual threads. These superhelices are then clustered together to form individual hairs. In general, α-keratins contain many cysteine residues, which allows them to form disulfide bridges. The number of disulfide bridges determines the strength of the protein. For example, the α-keratins in fingernails have more disulfide bridges than the α-keratins in hair.

**FIGURE 25.18**
Polypeptide chains of α-keratin, coiled into a right-handed α helix.

## Enzymes

**Enzymes** are among the most important biological molecules because they catalyze virtually all cellular processes. For example, glycosidase enzymes speed up the hydrolysis of polysaccharides by a factor of $10^{17}$. In other words, a reaction that would normally take millions of years is accomplished in milliseconds in the presence of an enzyme. Enzymes speed up reactions through their ability to bind to the transition state of a particular reaction, thereby lowering the energy of the transition state (Figure 25.19).

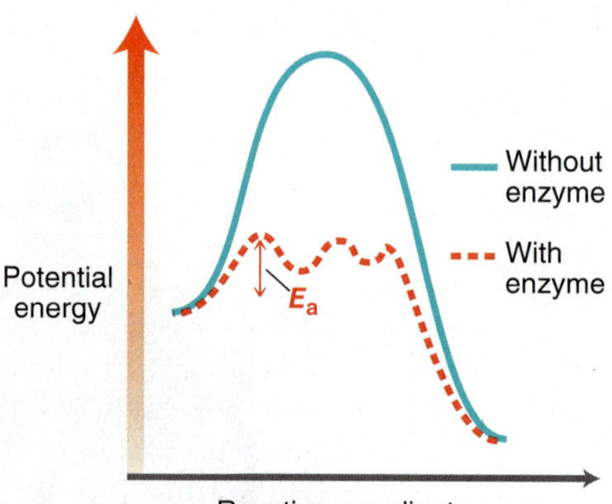


**FIGURE 25.19**
An energy diagram showing that an enzyme speeds up a reaction by lowering the energy of the transition state.

By lowering the energy of the transition state, the energy of activation is lowered, and the reaction occurs more rapidly. Even the simplest bacteria require thousands of enzymes to carry out their life processes, and humans require over fifty thousand.

## Transport Proteins

**Transport proteins** are used to transport molecules or ions from one location to another. Hemoglobin is a classic example of a transport protein, used to transport molecular oxygen from the lungs to all the tissues of the body. Hemoglobin consists of a protein unit bound to a nonprotein unit called a **prosthetic group**. The protein portion of hemoglobin was shown in Figure 25.17. The prosthetic group of hemoglobin binds oxygen.

Heme

A small change in primary structure can result in a large change in function. Hemoglobin is a good example of this phenomenon. Hundreds of variations of hemoglobin have been discovered. Most humans have the common type, called HbA, but some people have genetic variations in which one amino acid residue is different. In many of these cases, the genetic variation has no effect on the structure or activity of hemoglobin. However, some variations can be deadly. In one such variation, called HbS, the sixth residue from the N terminus of the β chain (consisting of 146 residues) is valine instead of glutamic acid. This variation greatly affects the structure and activity of hemoglobin, causing red blood cells to be distorted, or sickle shaped (Figure 25.20).

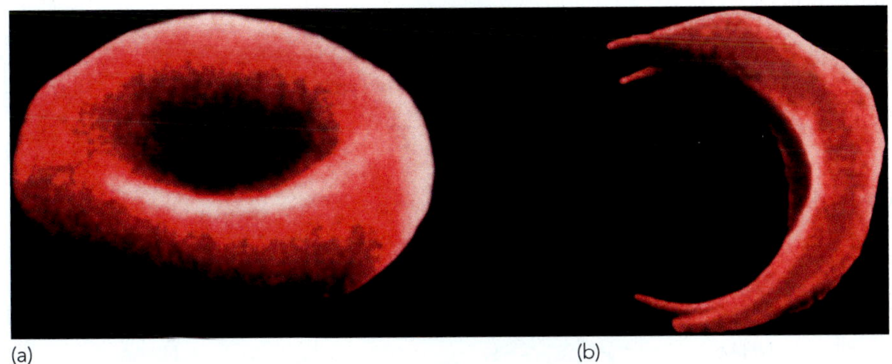

**FIGURE 25.20**
(a) Healthy red blood cells and
(b) sickle-shaped red blood cells.

(a) (b)

These distorted cells are highly susceptible to rupturing, and they interfere with the regular flow of blood. This condition is called sickle-cell anemia, and it can be fatal. A person with sickle-cell anemia has inherited two copies of the gene for abnormal hemoglobin, one from each parent. A person who carries only one copy of the sickle-cell gene is said to have the sickle-cell trait, which is a less severe form of the disease, but it can still cause serious problems under conditions of stress. Interestingly, the sickle-cell trait offers an advantage in that it seems to be accompanied by a resistance to malaria. This example illustrates how a change in a single amino acid residue can greatly affect the structure and function of a protein.

# REVIEW OF REACTIONS

## Analysis of Amino Acids

**Reaction with Ninhydrin**

An amino acid + Ninhydrin $\xrightarrow{NaOH}$ Purple-colored product + $H_2O$, $CO_2$, RCHO (By-products)

## Synthesis of Amino Acids

**Via an α-Haloacid**

$RCH_2COOH \xrightarrow[2)\ H_2O]{1)\ Br_2,\ PBr_3} RCH(\mathbf{Br})COOH \xrightarrow[S_N2]{\text{Excess } NH_3} RCH(\mathbf{NH_2})COOH$

**Via the Amidomalonate Synthesis**

Diethyl acetamidomalonate (EtO, OEt, N–H) $\xrightarrow[2)\ \mathbf{R}X]{1)\ NaOEt}$ alkylated amidomalonate (EtO, OEt, **R**, N–H) $\xrightarrow[\text{Heat}]{H_3O^+}$ **R**–CH($\oplus NH_3$)–COOH

**Via the Strecker Synthesis**

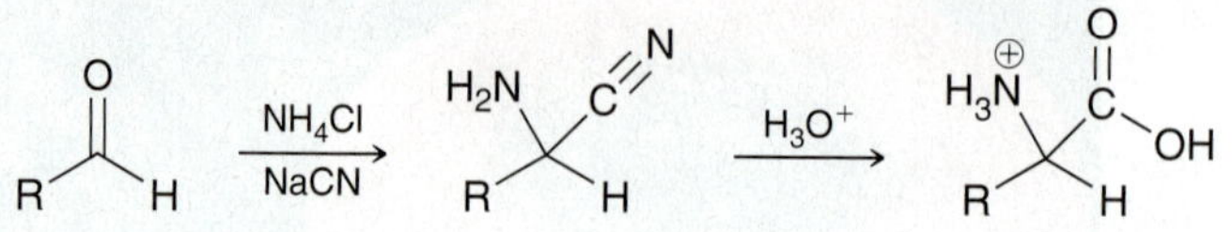

R H O
$NH_4Cl$
NaCN
$H_2N$ C≡N R H
$H_3O^+$
$H_3\overset{\oplus}{N}$ C(=O) OH R H

**Enantioselective Synthesis**

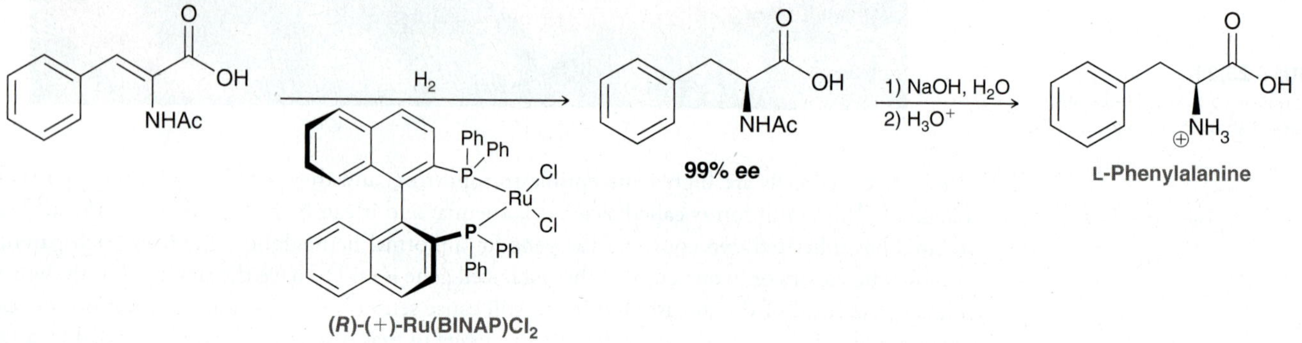

O OH NHAc
$H_2$
Ph Ph P Cl Ru Cl P Ph Ph
**(*R*)-(+)-Ru(BINAP)$Cl_2$**
O OH NHAc
**99% ee**
1) NaOH, $H_2O$
2) $H_3O^+$
O OH $\oplus NH_3$
**L-Phenylalanine**

## Analysis of Amino Acids

**Edman Degradation**

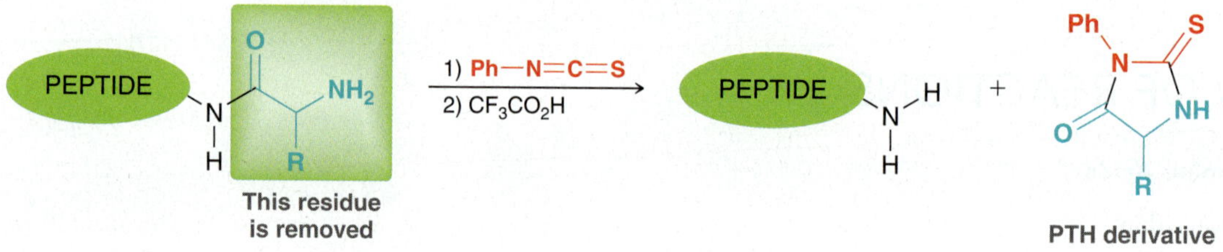

PEPTIDE N H O $NH_2$ R
This residue is removed
1) Ph—N=C=S
2) $CF_3CO_2H$
PEPTIDE N H H
+
Ph N S NH O R
**PTH derivative**

## Synthesis of Peptides

**Peptide Bond Formation**

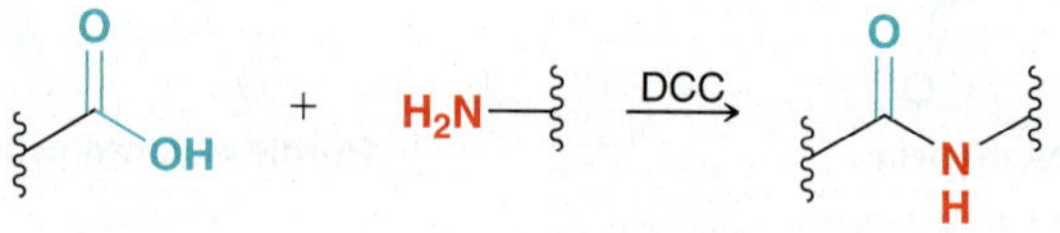

O OH + $H_2N$—
DCC
O N H

**Protection and Deprotection of the N terminus**

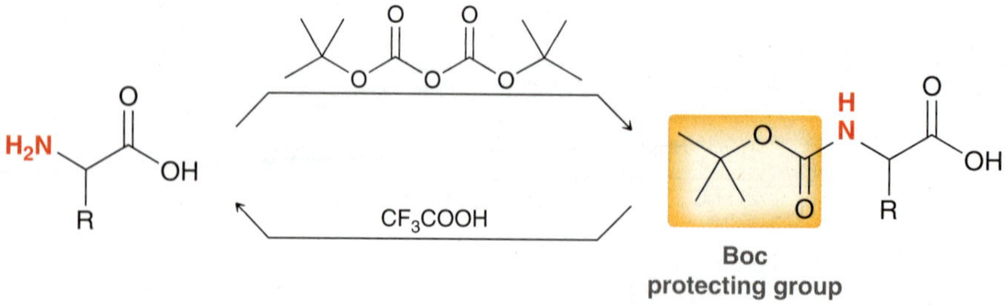

$H_2N$ O OH R
O O O O O
$CF_3COOH$
O O H N O OH R
**Boc protecting group**

**Protection and Deprotection of the C terminus**

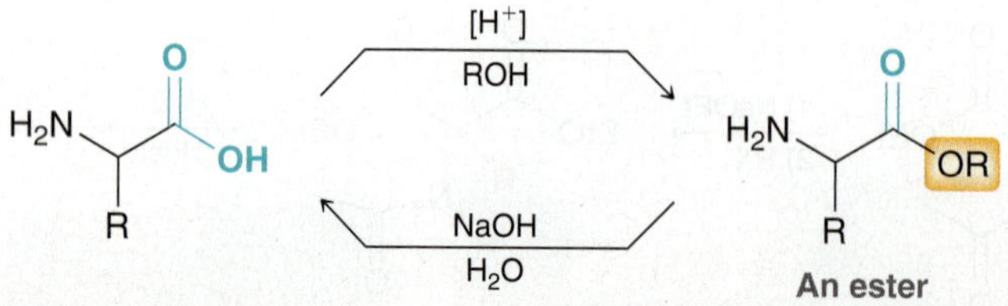

$H_2N$ O OH R
$[H^+]$
ROH
NaOH
$H_2O$
$H_2N$ O OR R
**An ester**

# REVIEW OF CONCEPTS AND VOCABULARY

### SECTION 25.1

- **Proteins** are large compounds formed from **amino acids** linked together.
- Each amino acid contains one amino group and one carboxylic acid group. Amino acids in which the two functional groups are separated by exactly one carbon atom are called **alpha amino acids**.
- Amino acids are coupled together by amide linkages called **peptide bonds.**
- Relatively short chains of amino acids are called **peptides**.

### SECTION 25.2

- Amino acids are said to be **amphoteric**, because they can function either as acids or as bases.
- Only 20 amino acids are abundantly found in proteins, all of which are **L amino acids**, except for glycine, which lacks a chirality center.
- Amino acids have two p$K_a$ values, one for the carboxylic acid group and one for the ammonium group.
- Amino acids with basic or acidic side chains will exhibit a third p$K_a$ value.
- Amino acids exist primarily as **zwitterions** at physiological pH.
- The **isoelectric point (pI)** of an amino acid is the pH at which the concentration of the zwitterionic form reaches its maximum value. During **electrophoresis**, amino acids are separated based on pI values.

### SECTION 25.3

- Racemic mixtures of amino acids can be prepared in the laboratory via α-haloacids, via the **amidomalonate synthesis**, or via the **Strecker synthesis**. Optically active amino acids are obtained either via resolution of a racemic mixture or via enantioselective synthesis.

### SECTION 25.4

- Peptides are comprised of **amino acid residues** joined by peptide bonds.
- Peptide chains have an amino group called the **N terminus** and a COOH group called the **C terminus**.
- Peptide bonds experience restricted rotation, giving rise to two possible conformations, called *s-trans* and *s-cis*. The *s-trans* conformation is generally more stable.
- Cysteine residues are uniquely capable of being joined to one another via **disulfide bridges.**

### SECTION 25.5

- The sequence of amino acids in a peptide can be analyzed using an **Edman degradation.**
- Large peptides are first cleaved into smaller fragments using **peptidases**, such as trypsin or chymotrypsin.

### SECTION 25.6

- DCC is commonly used to form peptide bonds.
- Protecting groups are used to control regioselectivity.
- Amino groups can be protected by conversion to carbamates, while carboxylic acid groups can be protected by conversion to esters.
- In the **Merrifield synthesis**, a peptide chain is assembled while tethered to an insoluble polymer.

### SECTION 25.7

- The **primary structure** of a protein is the sequence of amino acid residues.
- The **secondary structure** of a protein refers to the three-dimensional conformations of localized regions of the protein. Two particularly stable arrangements are the **α helix** and **β pleated sheet**.
- The **tertiary structure** of a protein refers to its three-dimensional shape.
- Under conditions of mild heating, a protein can unfold, a process called **denaturation**.
- **Quaternary structure** arises when a protein consists of two or more folded polypeptide chains, called *subunits*, that aggregate to form one protein complex.

### SECTION 25.8

- **Fibrous proteins** consist of linear chains that are bundled together.
- **Globular proteins** are chains that are coiled into compact shapes.
- **Structural proteins**, such as α-keratin, are fibrous proteins that provide structural rigidity.
- **Enzymes** catalyze virtually all cellular processes.
- **Transport proteins**, such as hemoglobin, are used to transport molecules or ions from one location to another. Hemoglobin consists of a protein unit bound to a nonprotein unit called a **prosthetic group**.
- A small change in primary structure can result in a large change in function.

# SKILLBUILDER REVIEW

## 25.1 DETERMINING THE PREDOMINANT FORM OF AN AMINO ACID AT A SPECIFIC pH

**STEP 1** Identify the p$K_a$ associated with the carboxylic acid group and determine whether this group will predominate as the uncharged form or as the carboxylate anion at the specified pH.

p$K_{a1}$ = 2.18

pH < 2.18 pH > 2.18

**STEP 2** Identify the p$K_a$ associated with the α-amino group and determine whether this group will predominate as the uncharged form or as the cationic form at the specified pH.

p$K_{a2}$ = 8.95

pH < 8.95 pH > 8.95

**STEP 3** If the side chain has an ionizable group, identify the p$K_a$ associated with the side chain and determine the form of the side chain that will predominate at the specified pH.

p$K_a$ = 10.53

pH < 10.53 pH > 10.53

Try Problems 25.4–25.6, 25.40, 25.47, 25.48

## 25.2 USING THE AMIDOMALONATE SYNTHESIS

**STEP 1** Identify the side chain that is connected to the α position.

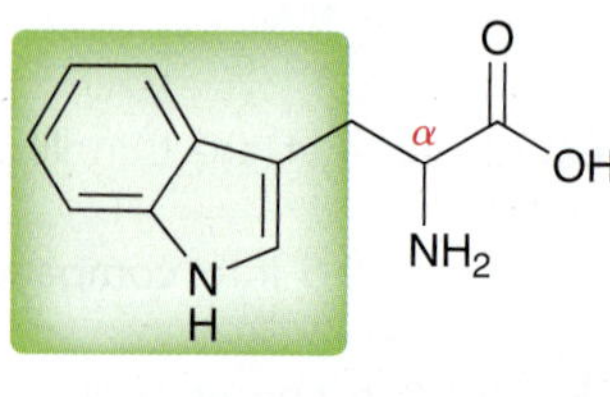


**STEP 2** Identify the alkyl halide necessary and ensure it will readily undergo an $S_N2$ process.

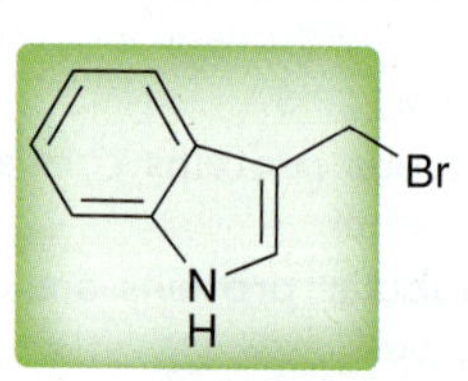


**STEP 3** Identify the reagents. Begin with diethyl acetamidomalonate as the starting material. Perform the alkylation and then heat with aqueous acid.

1) NaOEt
2) [indol-3-ylmethyl bromide]
3) $H_3O^+$, heat
→ **Target compound**

Try Problems 25.14–25.16, 25.57, 25.58, 25.59c, 25.60b, 25.84

## 25.3 DRAWING A PEPTIDE

**STEP 1** Draw a peptide with the correct number of residues.

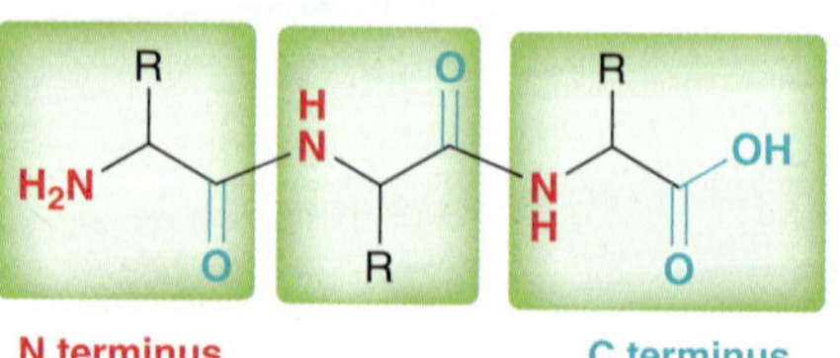


**STEP 2** Identify the side chain associated with each residue.

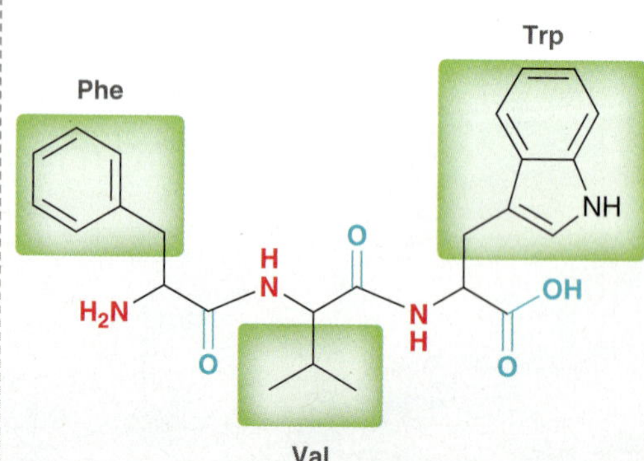


**STEP 3** Assign the proper configuration for each α position.

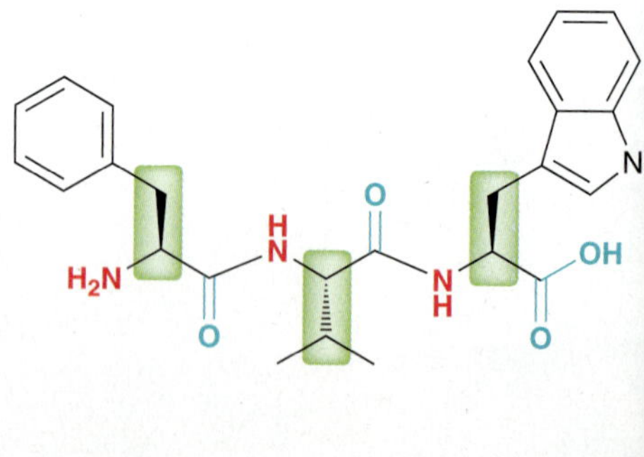


Try Problems 25.21–25.24, 25.65, 25.66

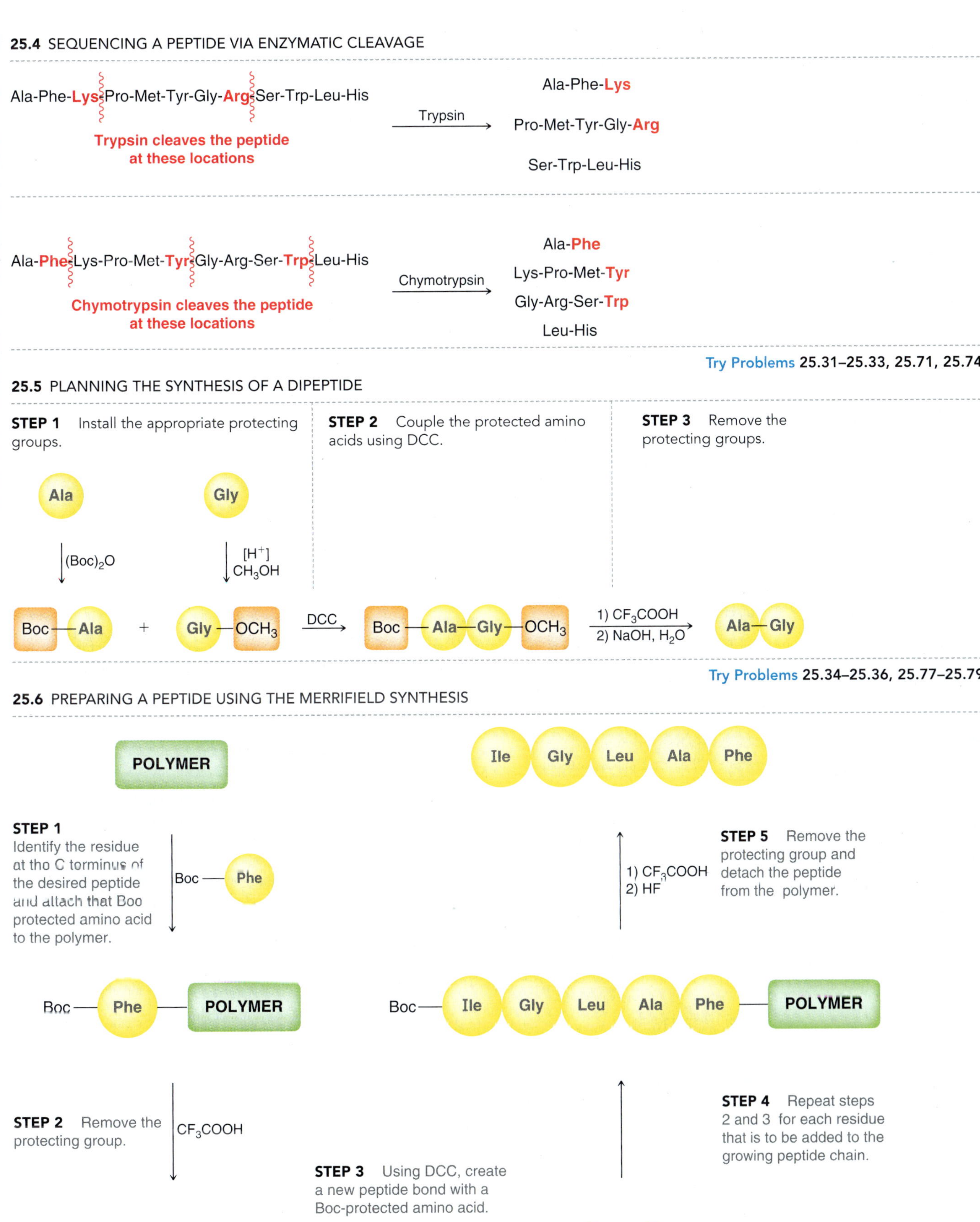


Try Problems 25.37, 25.38, 25.80, 25.81

# PRACTICE PROBLEMS:



**25.40** Draw a bond-line structure showing the zwitterionic form of each of the following amino acids:

(a) L-Valine (b) L-Tryptophan

(c) L-Glutamine (d) L-Proline

**25.41** The 20 naturally occurring amino acids (Table 25.1) are all L amino acids, and they all have the *S* configuration, with the exception of glycine (which lacks a chirality center) and cysteine. Naturally occurring cysteine is an L amino acid, but it has the *R* configuration. Explain.

**25.42** Draw a Fischer projection for each of the following amino acids:

(a) L-Threonine (b) L-Serine

(c) L-Phenylalanine (d) L-Asparagine

**25.43** Seventeen of the 20 naturally occurring amino acids (Table 25.1) exhibit exactly one chirality center. Of the remaining three amino acids, glycine has no chirality center, and the other two amino acids each have two chirality centers.

(a) Identify the amino acids with two chirality centers.

(b) Assign the configuration of each chirality center in these two amino acids.

**25.44** Draw all stereoisomers of L-isoleucine. In each stereoisomer, assign the configuration (*R* or *S*) of all chirality centers.

**25.45** Arginine is the most basic of the 20 naturally occurring amino acids. At physiological pH, the side chain of arginine is protonated. Identify which nitrogen atom in the side chain is protonated. (**Hint:** Consider all three possibilities and draw all resonance structures.)

**25.46** Histidine possesses a basic side chain which is protonated at physiological pH. Identify which nitrogen atom in the side chain is protonated.

**25.47** Draw the form of L-glutamic acid that predominates at each pH:

(a) 1.9 (b) 2.4 (c) 5.8 (d) 10.4

**25.48** For each of the following amino acids, draw the form that is expected to predominate at physiological pH:

(a) L-Isoleucine (b) L-Tryptophan

(c) L-Glutamine (d) L-Glutamic acid

**25.49** Using the data in Table 25.2, calculate the p*I* of the following amino acids:

(a) L-Alanine (b) L-Asparagine

(c) L-Histidine (d) L-Glutamic acid

**25.50** Just as each amino acid has a unique p*I* value, proteins also have an overall observable p*I*. For example, lysozyme (present in tears and saliva) has a p*I* of 11.0 while pepsin (used in our stomachs to digest other proteins) has a p*I* of 1.0. What information does this give you about the types of amino acid residues that predominantly make up each of these proteins?

**25.51** For each amino acid, draw the structure that predominates at the isoelectric point:

(a) L-Glutamine (b) L-Phenylalanine

(c) L-Proline (d) L-Threonine

**25.52** Optically active amino acids undergo racemization at the α position when treated with strongly basic conditions. Provide a mechanism that supports this observation.

**25.53** A mixture containing glycine, L-glutamine, and L-asparagine was subjected to electrophoresis. Identify which of the amino acids moved the farthest distance assuming that the experiment was performed at the pH indicated:

(a) 6.0 (b) 5.0

**25.54** Draw the products that are expected when each of the following amino acids is treated with ninhydrin:

(a) L-Aspartic acid (b) L-Leucine

(c) L-Phenylalanine (d) L-Proline

**25.55** A mixture of amino acids was treated with ninhydrin, and the following aldehydes were all observed in the product mixture:

(a) Identify the structure and name all three amino acids in the starting mixture.

(b) In addition to the aldehydes, a purple-colored product is obtained. Draw the structure of the compound responsible for the purple color.

(c) Describe the origin of the purple color.

**25.56** Show how you would use a Strecker synthesis to make valine.

**25.57** Under similar conditions, alanine and valine were each prepared with an amidomalonate synthesis, and alanine was obtained in higher yields than valine. Explain the difference in yields.

**25.58** The amidomalonate synthesis can be used to prepare amino acids from alkyl halides. When the amidomalonate synthesis is used to make glycine, no alkyl halide is required. Explain.

**25.59** Predict the major product(s) in each of the following cases:

(a) 1) $Br_2$, $PBr_3$ 2) $H_2O$ 3) Excess $NH_3$ ?

(b)

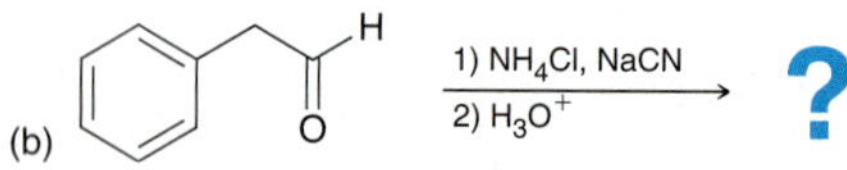


(c)

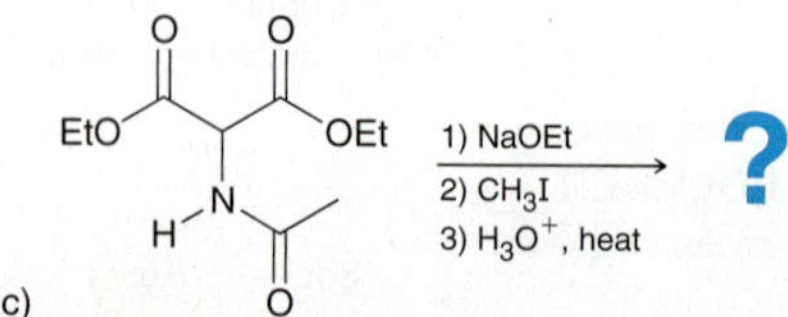

**25.60** Show how racemic valine can be prepared by each of the following methods:

(a) The Hell-Volhard-Zelinsky reaction

(b) The amidomalonate synthesis

(c) The Strecker synthesis

**25.61** How many different pentapeptides can be constructed from the 20 naturally occurring amino acids in Table 25.1?

**25.62** Using an asymmetric catalytic hydrogenation, identify the starting alkene that you would use to make L-histidine.

**25.63** Using three-letter abbreviations, identify all possible acyclic tripeptides containing L-leucine, L-methionine, and L-valine.

**25.64** Draw the predominant form of Asp-Lys-Phe at physiological pH.

**25.65** Draw a bond-line structure of the peptide that corresponds with the following sequence of amino acid residues and identify the N terminus and C terminus:

Trp-Val-Ser-Met-Gly-Glu

**25.66** Methionine enkephalin is a pentapeptide that is produced by the body to control pain. From the sequence of its amino acid residues, draw a bond-line structure of methionine enkephalin.

N terminus C terminus

Tyr-Gly-Gly-Phe-Met

**Methionine enkephalin**

**25.67** From its amino acid sequence, draw the form of aspartame that is expected to predominate at physiological pH:

**Asp-Phe**-$OCH_3$

**25.68** It is believed that penicillin antibiotics are biosynthesized from amino acid precursors. Identify the two amino acids that are most likely utilized during the biosynthesis of penicillin antibiotics:

H N S N O COOH

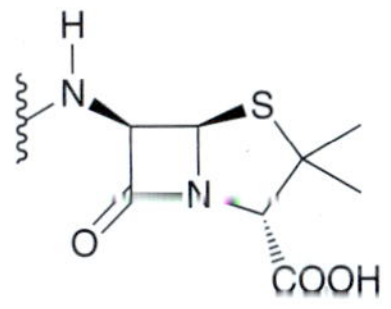

**25.69** Green fluorescent protein (GFP), first isolated from bioluminescent jellyfish, is a protein containing 238 amino acid residues. The discovery of GFP has revolutionized the field of fluorescence microscopy, which enables biochemists to monitor the biosynthesis of proteins. The 2008 Nobel Prize in Chemistry was awarded to Martin Chalfie, Osamu Shimomura, and Roger Tsien for the discovery and development of GFP. The structural subunit of GFP responsible for fluorescence, called the fluorophore, results when three amino acid residues undergo cyclization. Identify the three amino acids that are necessary for the biosynthesis of this fluorophore:

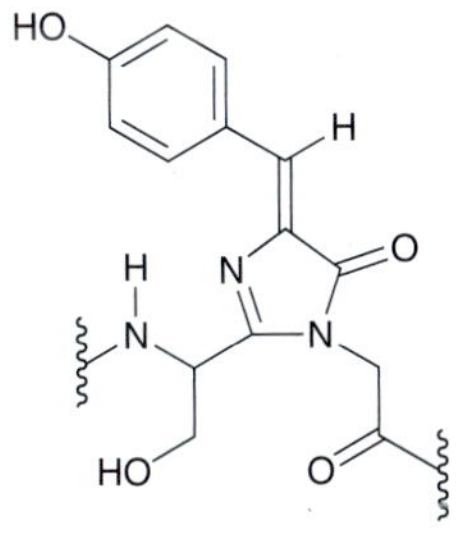

HO H H N N O N HO O

**25.70** Propose two structures for a tripeptide that contains glycine, L-alanine, and L-phenylalanine but does not react with phenyl isothiocyanate.

**25.71** Bradykinin has the following sequence:

(*N terminus*) ***Arg-Pro-Pro-Gly-Phe-Ser-Pro-Phe-Arg*** (*C terminus*)

Identify all fragments that will be produced when bradykinin is treated with:

(a) Trypsin (b) Chymotrypsin

**25.72** Identify the N-terminal residue of a peptide that yields the following PTH derivative upon Edman degradation:

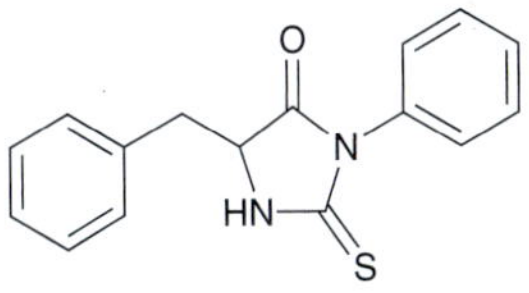

O N HN S

**25.73** Treatment of a tripeptide with phenyl isothiocyanate yields compound **A** and a dipeptide. Treatment of the dipeptide with phenyl isothiocyanate yields compound **B** and glycine. Identify the structure of the starting tripeptide.

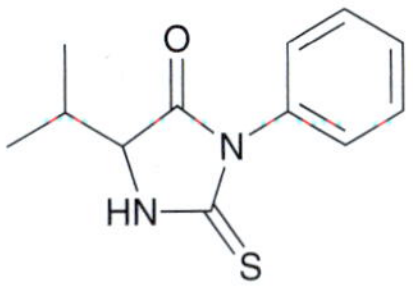

O N HN S **Compound A**

O N HN S **Compound B**

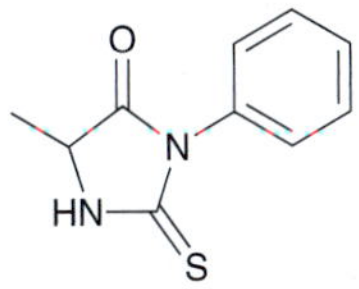

**25.74** Glucagon is a peptide hormone produced by the pancreas that, with insulin, regulates blood glucose levels. Glucagon is comprised of 29 amino acid residues. Treatment with trypsin yields four fragments, while treatment with chymotrypsin yields six fragments. Identify the sequence of amino acid residues for glucagon and determine whether any disulfide bridges are present.

***Trypsin fragments***

His-Ser-Gln-Gly-Thr-Phe-Thr-Ser-Asp-Tyr-Ser-Lys
Ala-Gln-Asp-Phe-Val-Gln-Trp-Leu-Met-Asn-Thr
Tyr-Leu-Asp-Ser-Arg
Arg

***Chymotrypsin fragments***

His-Ser-Gln-Gly-Thr-Phe
Thr-Ser-Asp-Tyr
Leu-Met-Asn-Thr
Ser-Lys-Tyr
Leu-Asp-Ser-Arg-Arg-Ala-Gln-Asp-Phe
Val-Gln-Trp

**25.75** When the N terminus of a peptide is acetylated, the peptide derivative that is formed is unreactive toward phenyl isothiocyanate. Explain.

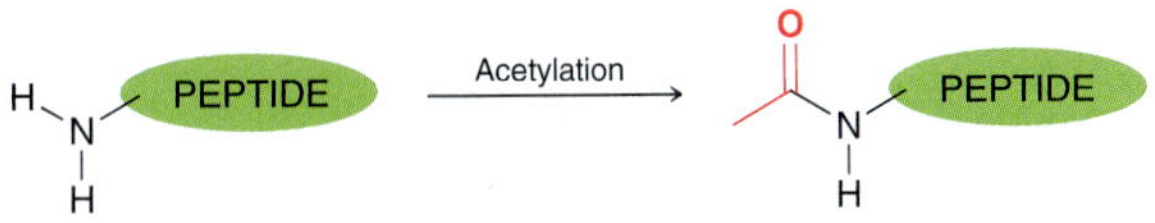

H N H PEPTIDE —Acetylation→ O N H PEPTIDE

**25.76** Predict the major product(s) of the reaction between L-valine and:

(a) MeOH, $H^+$ (b) Di-*tert*-butyl-dicarbonate

(c) NaOH, $H_2O$ (d) HCl

**25.77** Show all steps necessary to make the dipeptide Phe-Ala from L-phenylalanine and L-alanine.

**25.78** Draw all four possible dipeptides that are obtained when a mixture of L-phenylalanine and L-alanine is treated with DCC.

**25.79** Identify the steps you would use to combine Val-Leu and Phe-Ile to form the tetrapeptide Phe-Ile-Val-Leu.

**25.80** Identify all of the steps necessary to prepare the tripeptide Leu-Val-Ala with a Merrifield synthesis.

**25.81** Draw the structure of the protected amino acid that must be anchored to the solid support in order to use a Merrifield synthesis to prepare leucine enkephalin.

(*N terminus*) ***Try-Gly-Gly-Phe-Leu*** (*C terminus*)

**25.82** A proline residue will often appear at the end of an α helix but will rarely appear in the middle. Explain why proline generally cannot be incorporated into an α helix.

**25.83** Draw a mechanism for the following reaction:

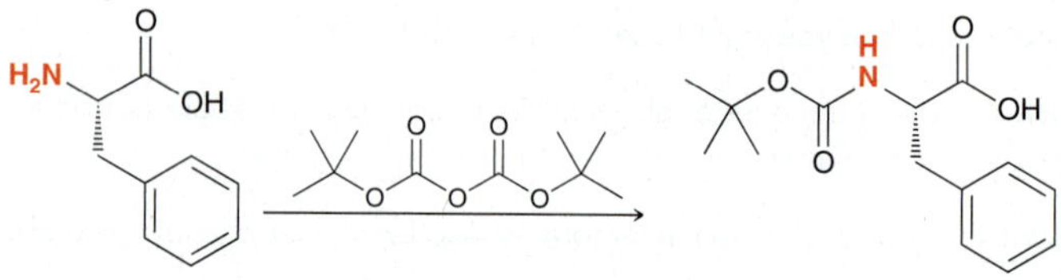

## INTEGRATED PROBLEMS

**25.84** Draw the alkyl halide that would be necessary to make the amino acid tyrosine using an amidomalonate synthesis. This alkyl halide is highly susceptible to polymerization. Draw the structure of the expected polymeric material.

**25.85** When leucine is prepared with an amidomalonate synthesis, isobutylene (also called 2-methylpropene) is a gaseous by-product. Draw a mechanism for the formation of this by-product.

**25.86** The side chain of tryptophan is not considered to be basic, despite the fact that it possesses a nitrogen atom with a lone pair. Explain.

**25.87** Consider a process that attempts to prepare tyrosine using a Hell-Volhard-Zelinsky reaction:

(a) Identify the necessary starting carboxylic acid.

(b) When treated with $Br_2$, the starting carboxylic acid can react with two equivalents to produce a compound with molecular formula $C_9H_8Br_2O_3$, in which neither of the bromine atoms is located at the α position. Identify the likely structure of this product.

## CHALLENGE PROBLEMS

**25.88** Proton NMR spectroscopy provides evidence for the restricted rotation of a peptide bond. For example, *N,N*-dimethylformamide exhibits three signals in its proton NMR spectrum at room temperature. Two of those signals are observed upfield, at 2.9 and 3.0 ppm. As the temperature is raised, the two signals begin to merge together. Over 180°C, these two signals combine into one signal. Explain how these results are evidence for the restricted rotation of peptide bonds.

**25.89** We saw in Section 25.6 that DCC can be used to form a peptide bond. We explored the mechanism, and we saw that DCC activates the COOH group so that it readily undergoes nucleophilic acyl substitution. An alternative method for activating a COOH group involves converting it into an activated ester, such as a *para*-nitrophenyl ester:

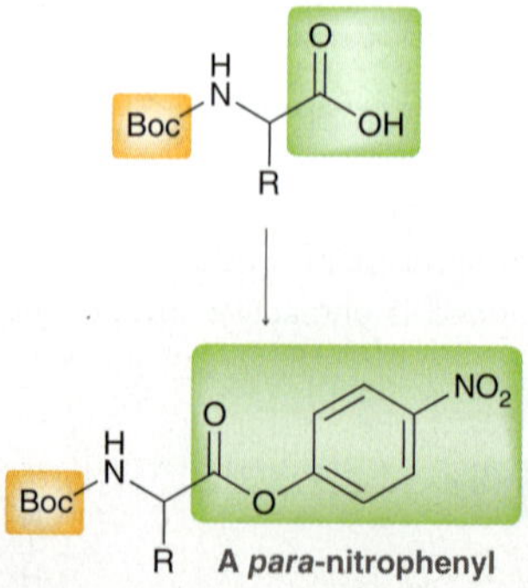

**A *para*-nitrophenyl ester**

The activated ester is readily attacked by a suitably protected amino acid to form a peptide bond:

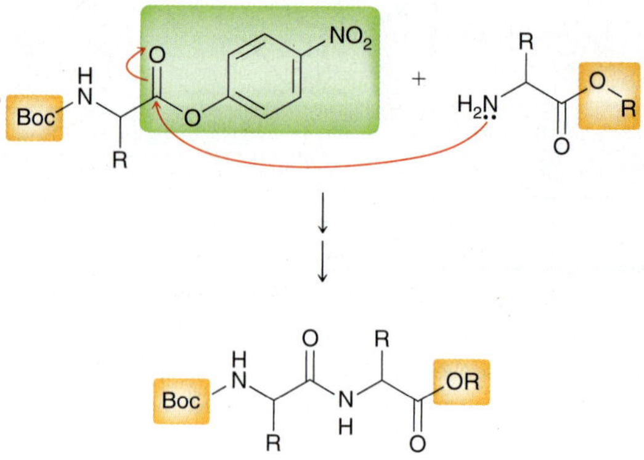

(a) Explain how the *p*-nitrophenyl ester activates the carbonyl group toward nucleophilic acyl substitution.

(b) What is the function of the nitro group?

(c) A *meta*-nitrophenyl ester is less activating than a *para*-nitrophenyl ester. Explain.

**25.90** When the following compound is treated with concentrated HCl at 100°C for several hours, hydrolysis occurs, producing one of the 20 naturally occurring amino acids. Identify which one.

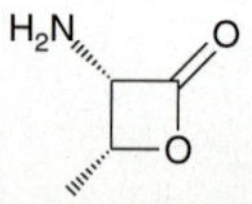

# 26 Lipids

## DID YOU EVER WONDER...

**why high cholesterol levels increase the risk of a heart attack?**

Cholesterol is a type of steroid that belongs to a larger class of naturally occurring compounds called lipids. Cholesterol serves many important functions, as we will see in this chapter, and maintaining healthy cholesterol levels can truly be a matter of life and death. In this chapter, we will explore the properties and functions of steroids as well as the properties and functions of many other classes of lipids.

## DO YOU REMEMBER?

*Before you go on, be sure you understand the following topics.*
*If necessary, review the suggested sections to prepare for this chapter:*

- Partially Hydrogenated Fats (Section 9.7)
- Autooxidation and Antioxidants (Section 11.9)
- Base-Catalyzed Hydrolysis of Esters: Saponification (Section 21.11)

Take the DO YOU REMEMBER? QUIZ in **WileyPLUS** to check your understanding

## 26.1 Introduction to Lipids

Throughout the chapters of this book, we have regularly classified organic compounds based on their functional groups (e.g., alkenes, alkynes, alcohols). **Lipids** represent a somewhat unique category because they are not defined by the presence or absence of a particular functional group. Instead, they are defined by a physical property—solubility. Specifically, lipids are naturally occurring compounds that can be extracted from cells using nonpolar organic solvents. An extremely large number of biological compounds are considered to be lipids, so it is helpful to classify and categorize them. **Complex lipids** are lipids that readily undergo hydrolysis in aqueous acid or base to produce smaller fragments, while **simple lipids** do not readily undergo hydrolysis. The terms "simple" and "complex" can be somewhat misleading in this context, since many complex lipids have rather simple structures, while many simple lipids have more complex structures. Complex lipids are so named because they contain one or more ester groups, which can be hydrolyzed to produce a carboxylic acid and an alcohol (see Section 21.11). In this chapter, we will explore three classes of complex lipids and three classes of simple lipids, as seen in Figure 26.1.

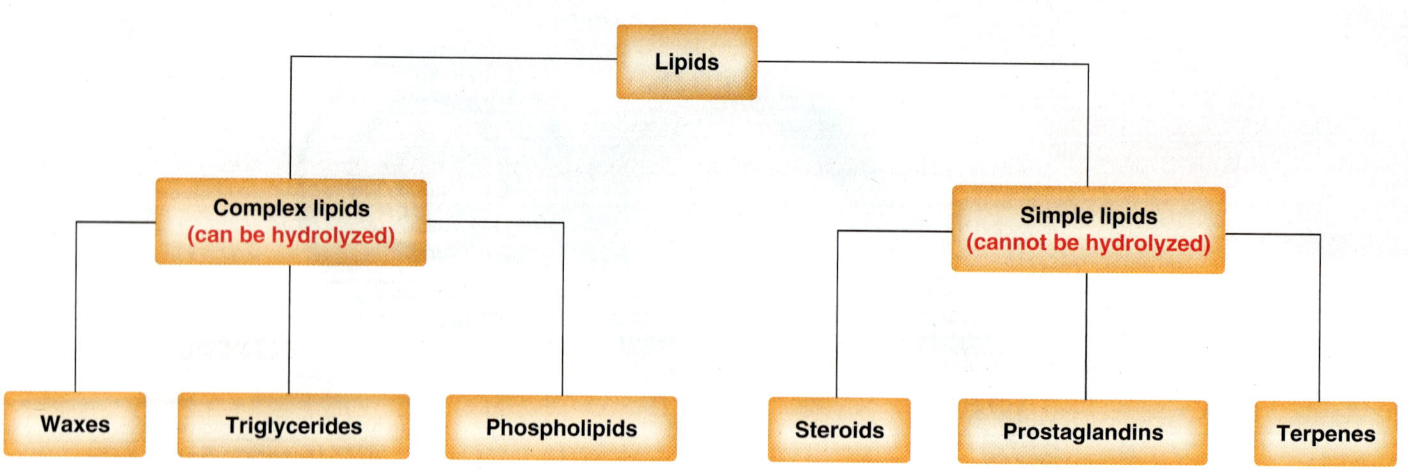


**FIGURE 26.1**
The six classes of lipids that will be discussed in this chapter.

The three major classes of complex lipids are waxes, triglycerides, and phospholipids. One example of each class is shown. Notice that all three classes contain ester groups, rendering them easily hydrolyzed, and they also contain long, hydrocarbon chains, rendering them soluble in organic solvents.

**TABLE 26.1 MELTING POINTS OF COMMON SATURATED AND UNSATURATED FATTY ACIDS**

| STRUCTURE AND NAME | NUMBER OF CARBON ATOMS | NUMBER OF CARBON-CARBON DOUBLE BONDS | MELTING POINT (°C) |
|---|---|---|---|
| ***SATURATED*** | | | |
| 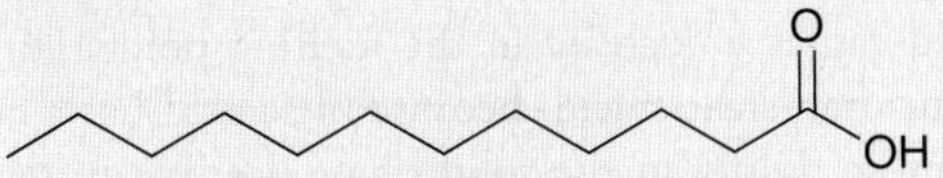 Lauric acid | 12 | 0 | 43 |
| 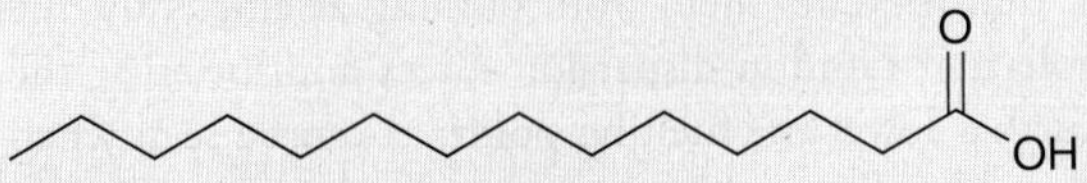 Myristic acid | 14 | 0 | 54 |
| 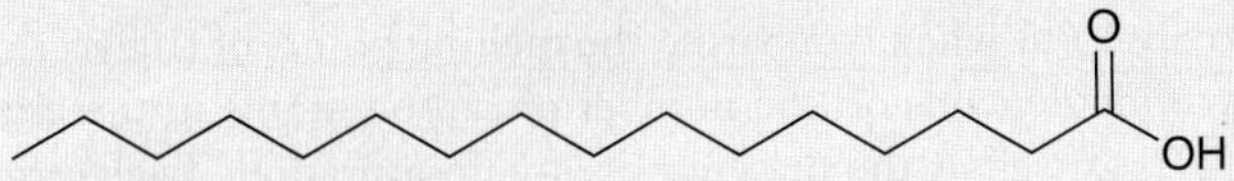 Palmitic acid | 16 | 0 | 63 |
| 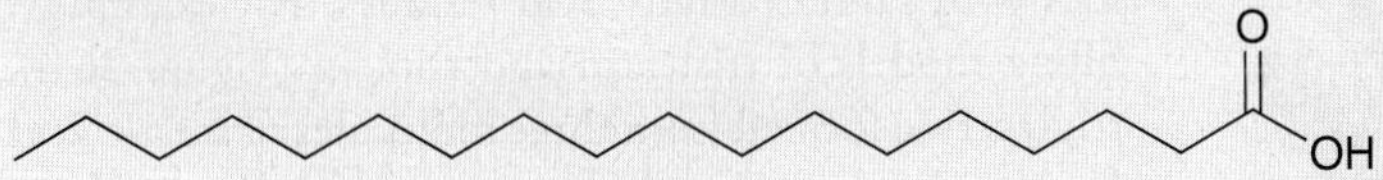 Stearic acid | 18 | 0 | 69 |
| 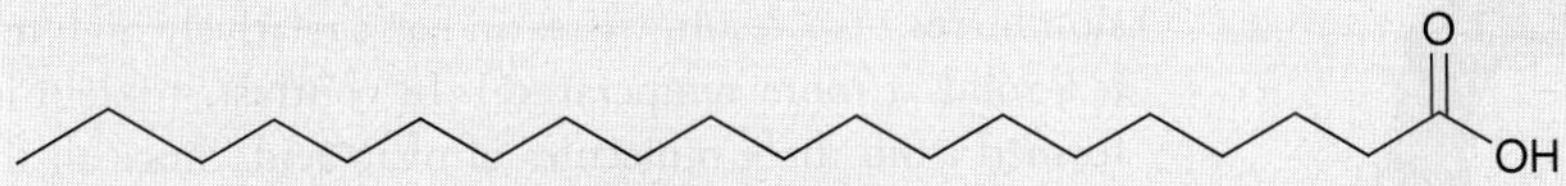 Arachidic acid | 20 | 0 | 77 |
| ***UNSATURATED*** | | | |
| 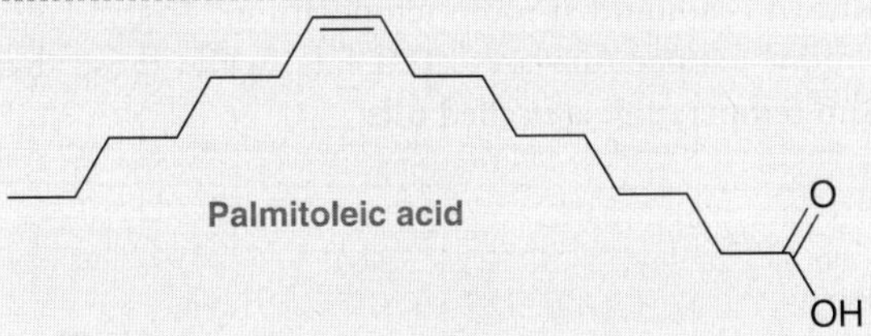 Palmitoleic acid | 16 | 1 | 0 |
| 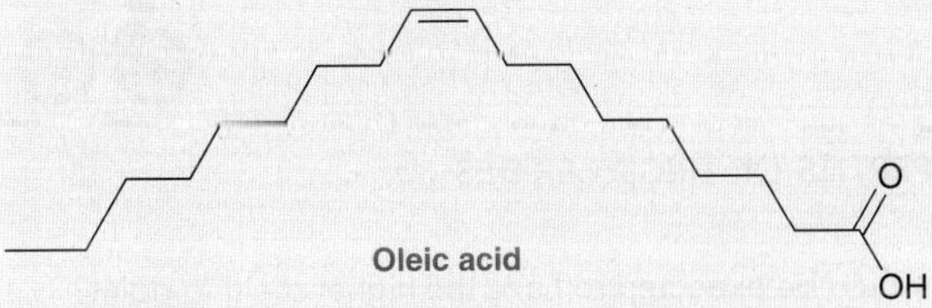 Oleic acid | 18 | 1 | 13 |
| 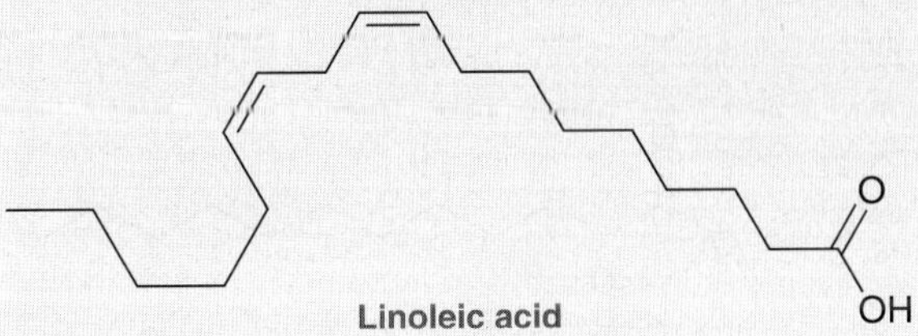 Linoleic acid | 18 | 2 | −5 |
| 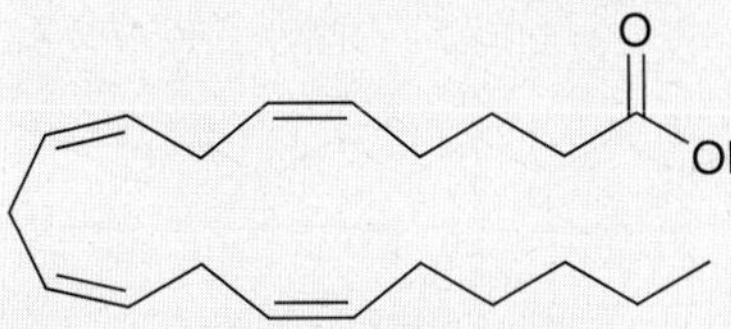 Arachidonic acid | 20 | 4 | −50 |

Two important trends emerge from the data in Table 26.1:

1. For saturated fatty acids, the melting point increases with increasing molecular weight. For example, myristic acid exhibits a higher melting point than lauric acid, because myristic acid has more carbon atoms.
2. The presence of a *cis* double bond causes a decrease in the melting point. For example, oleic acid exhibits a lower melting point than stearic acid, because oleic acid has a *cis* double bond. Stearic acid molecules can pack more tightly in the solid phase, leading to stronger intermolecular London dispersion forces and, consequently, a higher melting point. In contrast, oleic acid molecules have a kink that prevents them from packing as efficiently, thereby decreasing the strength of the intermolecular London dispersion forces and lowering the melting point. Additional double bonds further lower the melting point, as can be seen when comparing the melting points of oleic acid and linoleic acid.

These two trends are also observed when comparing the melting points of triglycerides. That is, the melting point of a triglyceride depends on the number of carbon atoms in the fatty acid residues as well as the presence of any unsaturation. Triglycerides with unsaturated fatty acid residues have lower melting points than triglycerides with saturated fatty acid residues. For example, compare space-filling models of tristearin and triolein (Figure 26.4). Tristearin is a triglyceride formed from three molecules of stearic acid. Since all three fatty acids are saturated, the resulting triglyceride can pack efficiently in the solid state, allowing for strong intermolecular London dispersion forces. As a result, tristearin has a relatively high melting point (it is a solid at room temperature). In contrast, triolein is a triglyceride formed from three molecules of oleic acid. Since all three fatty acids are unsaturated, the resulting triglyceride cannot pack as efficiently in the solid state. As a result, triolein experiences weaker intermolecular London dispersion forces and therefore has a relatively low melting point (it is a liquid at room temperature). Triglycerides that are solids at room temperature are called **fats**, while those that are liquids at room temperature are called **oils**.

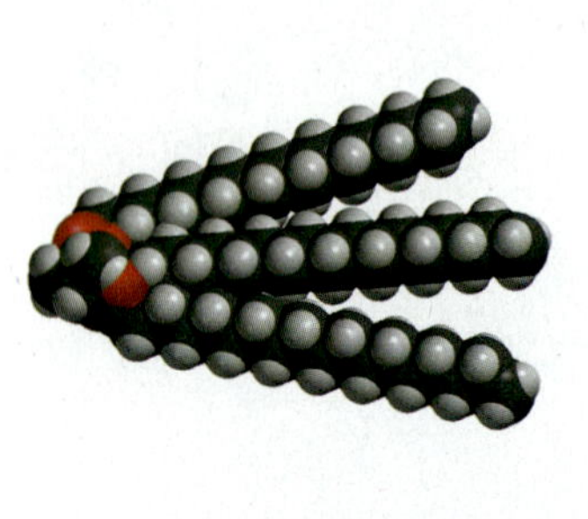


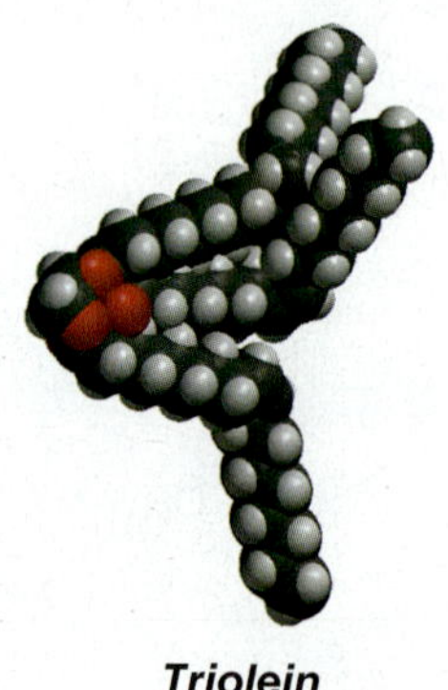


**FIGURE 26.4**
Space-filling models of tristearin and triolein.

## SKILLBUILDER

### 26.1 COMPARING MOLECULAR PROPERTIES OF TRIGLYCERIDES

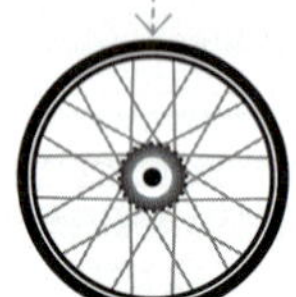

**LEARN** the skill Which triglyceride would be expected to have the higher melting point?

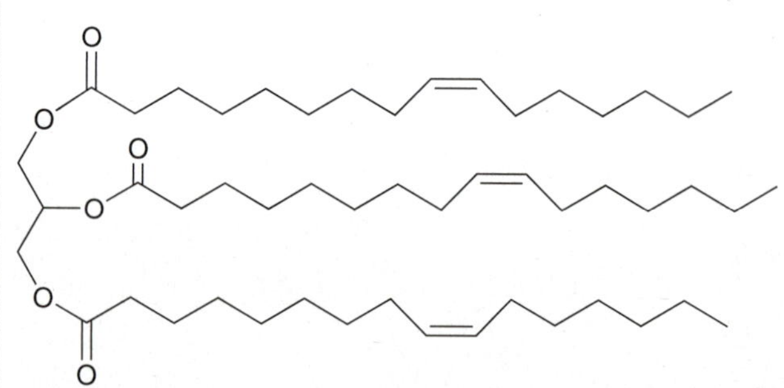


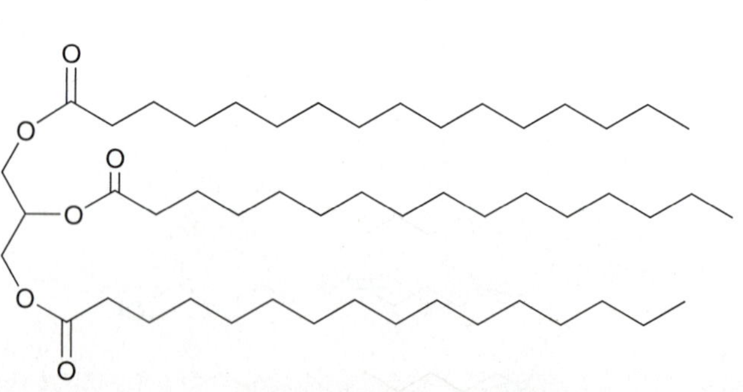

## SOLUTION

First identify the fatty acid residues in each triglyceride:

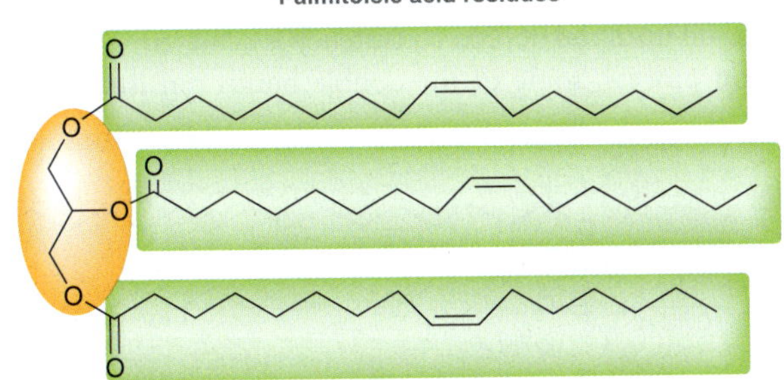


Palmitic acid residues

Tripalmitin

**STEP 1**
Identify the fatty acid residues.

As the names imply, tripalmitolein contains three palmitoleic acid residues, and tripalmitin contains three palmitic acid residues. We must compare these fatty acid residues, keeping in mind that two features contribute to a higher melting point:

**STEP 2**
Compare the length and saturation of the residues.

1. Length of fatty acid residues (longer chain = higher melting point)
2. Absence of C═C unsaturation (no unsaturation = higher melting point)

The first feature cannot be used to distinguish these triglycerides, because palmitoleic acid and palmitic acid each have 16 carbon atoms. However, the second feature does distinguish these triglycerides. Specifically, palmitoleic acid is an unsaturated carboxylic acid, while palmitic acid is a saturated carboxylic acid. We have seen that unsaturation causes inefficient packing, thereby lowering the melting point. Therefore, we expect tripalmitolein will have the lower melting point and tripalmitin the higher melting point.

**PRACTICE** the skill

**26.3** For each pair of triglycerides, identify the one that is expected to have the higher melting point. Consult Table 26.1 to determine which fatty acid residues are present in each triglyceride.

**(a)** Trilaurin and trimyristin **(b)** Triarachidin and trilinolein

**(c)** Triolein and trilinolein **(d)** Trimyristin and tristearin

**APPLY** the skill

**26.4** Arrange the following three triglycerides in order of increasing melting point:

Tristearin, tripalmitin, and tripalmitolein

**26.5** Tristearin has a melting point of 72 °C. Based on this information, would you expect triarachadin to be classified as a fat or as an oil?

**26.6** Identify each of the following compounds as a fat or an oil. Explain your answers.

**(a)** A triglyceride containing one palmitic acid residue and two stearic acid residues

**(b)** A triglyceride containing one oleic acid residue and two linoleic acid residues

need more **PRACTICE?** **Try Problems 26.40a,d,e, 26.41a,d,e, 26.43**

## 26.4 Reactions of Triglycerides

### Hydrogenation of Triglycerides

Triglycerides containing unsaturated fatty acid residues will undergo hydrogenation (see Section 9.7).

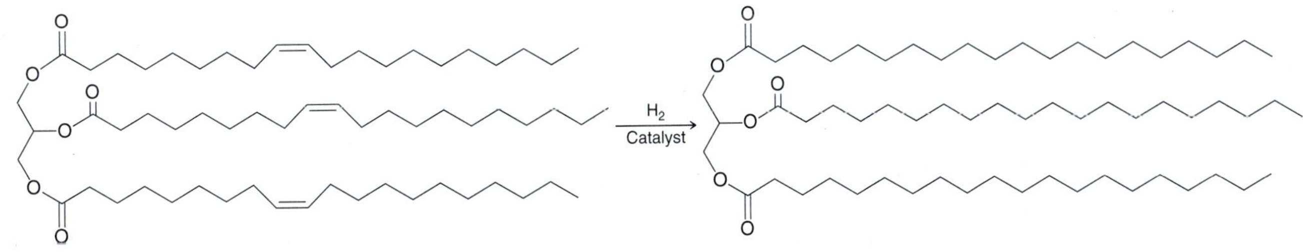

This type of transformation is generally achieved using high temperatures and a catalyst such as nickel (Ni). In the previous example, all of the carbon-carbon π bonds are hydrogenated, but it is also possible to control the conditions so that only some of the carbon-carbon π bonds are hydrogenated to give partially hydrogenated vegetable oils. For example, margarine is produced by hydrogenating soybean, peanut, or cottonseed oil until the desired consistency is achieved.

During the hydrogenation process, some of the double bonds can isomerize to give *trans* π bonds.

$H_2$, Catalyst

*trans*

The hydrogenation process typically gives a mixture containing 10–15% *trans* unsaturated fatty acids. These "trans fats" have been implicated as culprits in raising cholesterol levels, thereby increasing the risk of heart attacks (as described later in this chapter). The FDA now requires food product labels to indicate the amount of trans fats present in food.

**Nutrition Facts**

Serving Size 1 ounce Servings in bag 4

| Amount Per Serving | |
|---|---|
| **Calories 155** Calories from Fat 93 | |
| | **% Daily Value*** |
| **Fat 13g** | 20% |
| Saturated Fat 3g + Trans Fat 2g | 25% |
| **Cholesterol 0mg** | 0% |
| **Sodium 148mg** | 6% |
| **Total Carbohydrate 14g** | 5% |
| Dietary Fiber 1g | 5% |
| Sugars 1g | |
| **Protein 2g** | |
| Vitamin A 0% • Vitamin C | 9% |
| Calcium 1% • Iron | 3% |

*Percent Daily Values are based on a 2,000 calorie diet. Your daily values may be higher or lower depending on your calorie needs.

## CONCEPTUAL CHECKPOINT

**26.7** Triolein was treated with molecular hydrogen at high temperature in the presence of nickel. At completion, the reaction had consumed three equivalents of molecular hydrogen.

(a) Draw the structure of the product.

(b) Identify the name of the product.

(c) Determine whether the melting point of the product is higher or lower than triolein.

(d) When the product is treated with aqueous base, three equivalents of a fatty acid are produced. Identify this fatty acid.

**26.8** Partial hydrogenation of triolein produces several different *trans* fats. Draw all possible *trans* fats that might be obtained in the process.

## Autooxidation of Triglycerides

In the presence of molecular oxygen, triglycerides that contain unsaturated fatty acid residues are particularly susceptible to autooxidation at the allylic position to yield hydroperoxides (see Section 11.9).

OOH

**A hydroperoxide**

This process occurs via a radical mechanism initiated by a hydrogen abstraction at the allylic position to give a resonance-stabilized allylic radical (Mechanism 11.2).

As with all radical mechanisms, the net reaction is the sum of the propagation steps:

**Coupling** R· + ·O–O· ⟶ R–O–O·

**Hydrogen abstraction** R–O–O· + H–R ⟶ R–O–OH + R·

**Net Reaction** R–H + $O_2$ ⟶ R–O–O–H

The resulting hydroperoxide is responsible for the rancid smell that develops over time in foods containing unsaturated oils. In addition, hydroperoxides are also toxic. Therefore, food products containing unsaturated oils have a short shelf-life unless radical inhibitors are used to slow the formation of hydroperoxides. The role of radical inhibitors in food chemistry was discussed in Section 11.9.

## Hydrolysis of Triglycerides

When treated with aqueous base, triglycerides undergo hydrolysis, yielding glycerol and three carboxylate ions.

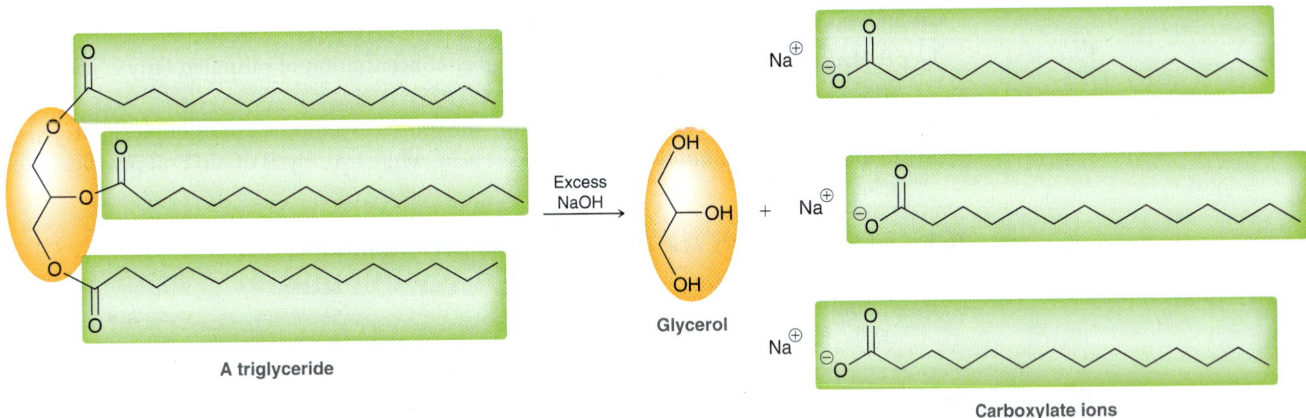


Each of the three ester groups of the triglyceride is hydrolyzed via a nucleophilic acyl substitution reaction (Mechanism 26.1).

## MECHANISM 26.1 HYDROLYSIS OF AN ESTER UNDER BASIC CONDITIONS

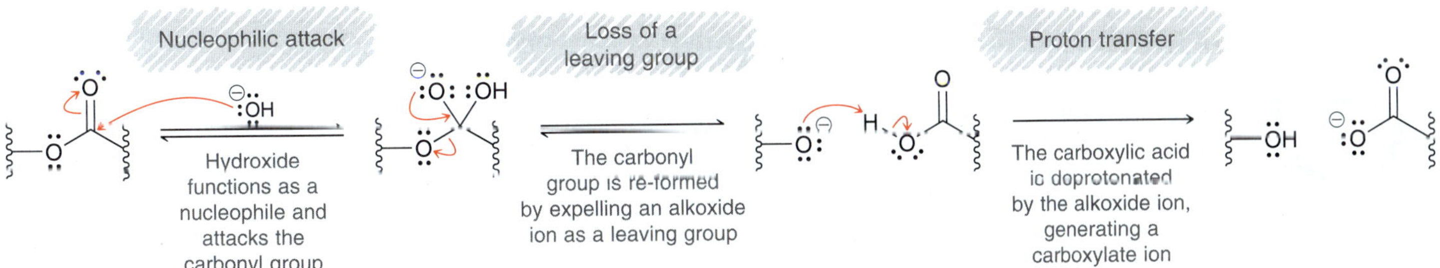


In the first step, a hydroxide ion functions as a nucleophile and attacks the carbonyl group of the ester, generating a tetrahedral intermediate. This tetrahedral intermediate then re-forms the carbonyl group by expelling an alkoxide ion as a leaving group. Alkoxide ions are generally poor leaving groups, but in this case, the tetrahedral intermediate itself is an alkoxide ion, so expulsion of an alkoxide ion is not uphill in energy. In the final step, the alkoxide ion functions as a base and deprotonates the carboxylic acid, generating a carboxylate ion. For a triglyceride, this three-step process is repeated until all three fatty acid residues have been released.

Naturally occurring fats and oils are usually complex mixtures of many different triglycerides. In fact, the three fatty acid residues within a single triglyceride are often different, as in the following example:

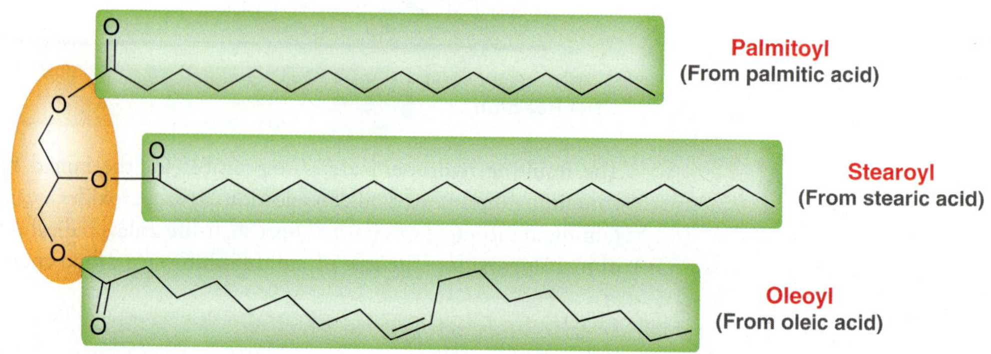


When heated with aqueous base, fats and oils can be completely hydrolyzed to provide a mixture of fatty acid carboxylates. Table 26.2 gives the approximate fatty acid composition of several common fats and oils. Two important observations emerge from the data in Table 26.2:

1. Triglycerides found in animals have a higher concentration of saturated fatty acids than triglycerides from vegetable origins.
2. The exact ratio of fatty acids differs from one source to another. For example, hydrolysis of corn oil produces more linoleic acid than oleic acid, while hydrolysis of olive oil produces more oleic acid.

**TABLE 26.2 APPROXIMATE FATTY ACID COMPOSITIONS FOR SEVERAL FATS AND OILS**

| SOURCE | PERCENT SATURATED FATTY ACIDS | PERCENT OLEIC ACID | PERCENT LINOLEIC ACID | SOURCE | PERCENT SATURATED FATTY ACIDS | PERCENT OLEIC ACID | PERCENT LINOLEIC ACID |
|---|---|---|---|---|---|---|---|
| **Animal Fat** | | | | **Vegetable Oil** | | | |
| Beef fat | 55 | 40 | 3 | Corn oil | 14 | 34 | 48 |
| Milk fat | 37 | 33 | 3 | Olive oil | 11 | 82 | 5 |
| Lard | 41 | 50 | 6 | Canola oil | 9 | 54 | 30 |
| Human fat | 37 | 46 | 10 | Peanut oil | 12 | 60 | 20 |

## SKILLBUILDER

### 26.2 IDENTIFYING THE PRODUCTS OF TRIGLYCERIDE HYDROLYSIS

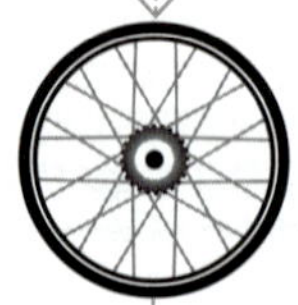

**LEARN** the skill

Identify the products expected when the following triglyceride is hydrolyzed with aqueous base:

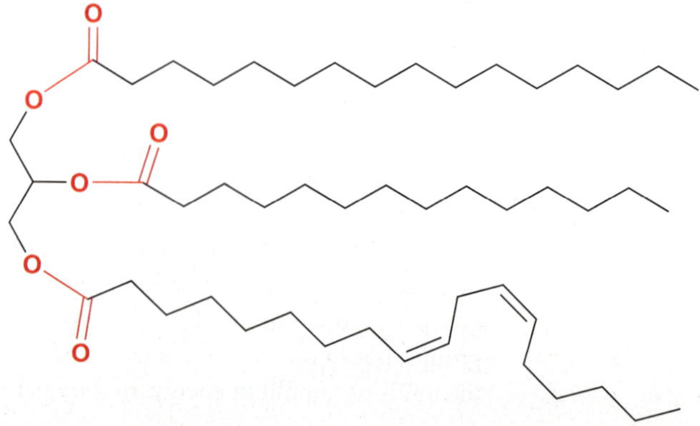

## SOLUTION

**STEP 1**
Identify the ester groups.

First identify the three ester groups (highlighted in red). Each ester group is hydrolyzed to give an alcohol and a carboxylate ion, according to the following mechanism:

Alcohol

Carboxylate ion

**STEP 2**
Draw glycerol and the three appropriate carboxylate ions.

Therefore, we expect the following products:

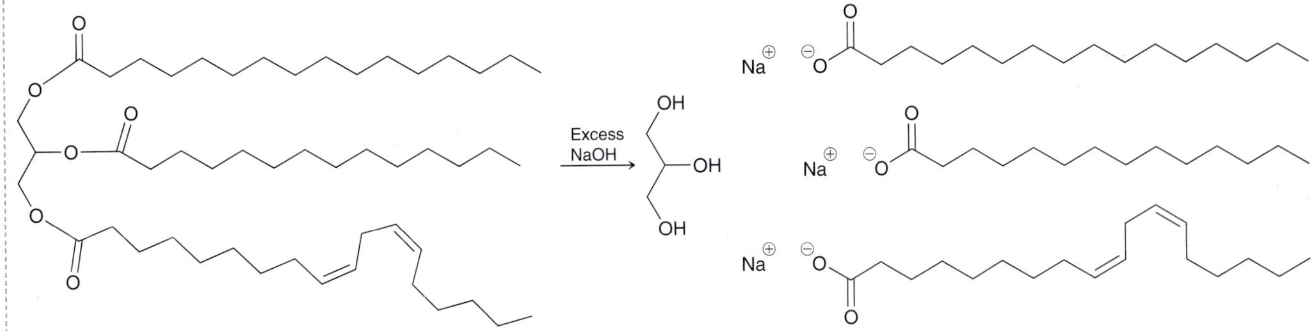


The products are glycerol and three carboxylate ions. Using Table 26.1, we can identify these carboxylate ions as the conjugate bases of palmitic acid, myristic acid, and linoleic acid.

**PRACTICE the skill** **26.9** Identify the products that are expected when the following triglyceride is hydrolyzed with aqueous sodium hydroxide:

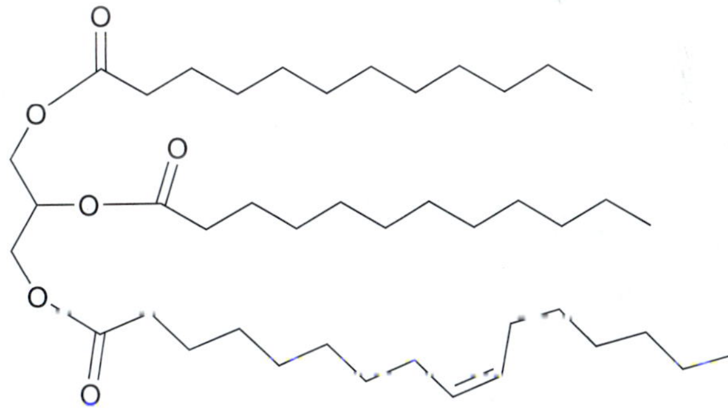

**APPLY the skill**

**26.10** A triglyceride was treated with sodium hydroxide to yield glycerol and three equivalents of sodium laurate (the conjugate base of lauric acid). Draw the structure of the triglyceride.

**26.11** An optically inactive triglyceride was hydrolyzed to yield one equivalent of palmitic acid and two equivalents of lauric acid. Draw the structure of the triglyceride.

need more **PRACTICE?** **Try Problems 26.30b, 26.40c, 26.41c**

Hydrolysis of triglycerides is commercially valuable, because soap is prepared via this process. Accordingly, the hydrolysis of triglycerides in the presence of aqueous base is also called saponification, derived from the Latin word for soap, *saponis*. Saponification of triglycerides produces carboxylate ions, which are the primary ingredients of soap. Recall from Chapter 1 that carboxylate ions function as soap because they have both a polar, hydrophilic group and a nonpolar, hydrophobic group.

Polar group (hydrophilic)

Nonpolar group (hydrophobic)

When dissolved in water, compounds of this type assemble around nonpolar substances to form spheres called *micelles* (Figure 1.54). The nonpolar substance is located at the center of the micelle, where it interacts with the nonpolar ends of the soap molecules via intermolecular London dispersion forces. The surface of the micelle is comprised of polar groups, which interact with the polar solvent. That is, the micelle acts as a unit that is solvated by the polar solvent. In this way, soap molecules can solvate nonpolar substances, such as grease, in polar solvents, such as water.

Most soaps are produced by boiling either animal fat or vegetable oil together with a strong alkaline solution, such as aqueous sodium hydroxide. The identity of the alkyl chains can vary, depending on the source of the fat or oil, but the concept is the same for all soaps.

## practically speaking — Soaps Versus Synthetic Detergents

Soap has been used for over two millennia, as people discovered long ago that soap could be made by heating animal fat together with wood ashes, which contain alkaline substances. Nevertheless, the usefulness of soap is diminished in the presence of water that contains high concentrations of calcium ions ($Ca^{2+}$) or magnesium ions ($Mg^{2+}$). When soap is used with such water, called hard water, a precipitate is formed as a result of the following ion exchange reaction:

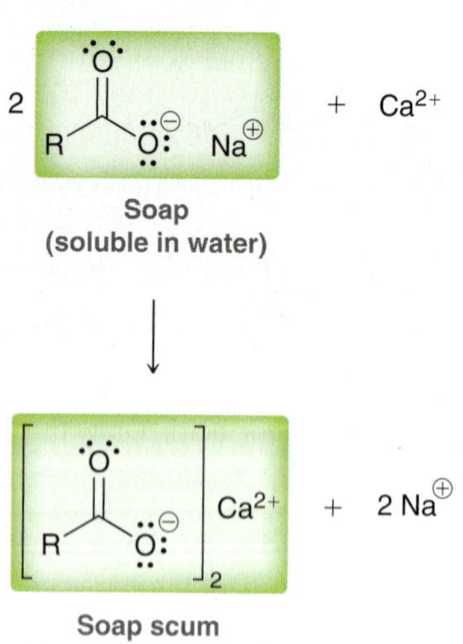


The generation of a precipitate, often called soap scum, limits the usefulness of soap. To circumvent this problem, chemists have developed synthetic detergents that do not form precipitates when used with hard water. Like soap, synthetic detergents also contain both hydrophobic and hydrophilic regions, but the identity of the hydrophilic region has been modified.

Rather than using a carboxylate group, synthetic detergents use a different group. For example, consider the structure of sodium lauryl sulfate.

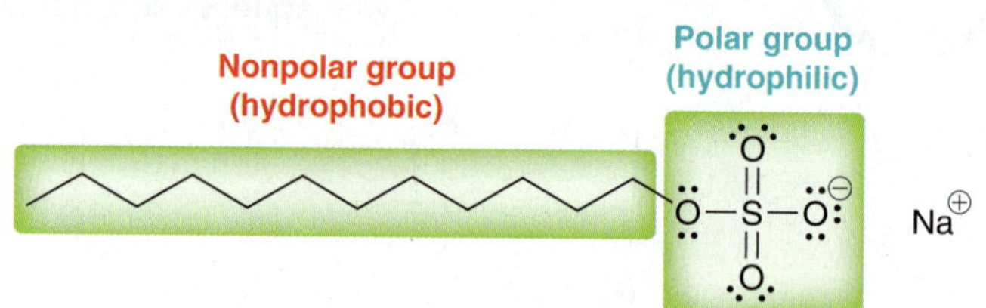


Sodium lauryl sulfate

Like soap molecules, this compound also has a hydrophobic group and a hydrophilic group. However, in this case, an ion exchange reaction does not generate a precipitate, because the calcium salt is water soluble, hence, no soap scum. Sodium lauryl sulfate is in fact a common ingredient found in many shampoo formulations.

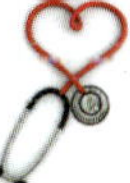


## Transesterification of Triglycerides

We have seen that triglycerides are extremely efficient at storing energy. For this reason, our bodies use triglycerides as a source of fuel. It should therefore come as no surprise that diesel engines can be modified to use cooking oil as a fuel. In fact, coconut oil was used extensively as fuel for vehicles in World War I and World War II, when the supply of gasoline was scarce. This technique cannot be used in colder climates, because many oils will solidify at low temperatures. An alternative to vegetable oil is biodiesel, which is formed from the transesterification of vegetable oils to produce a mixture of fatty acid methyl esters.

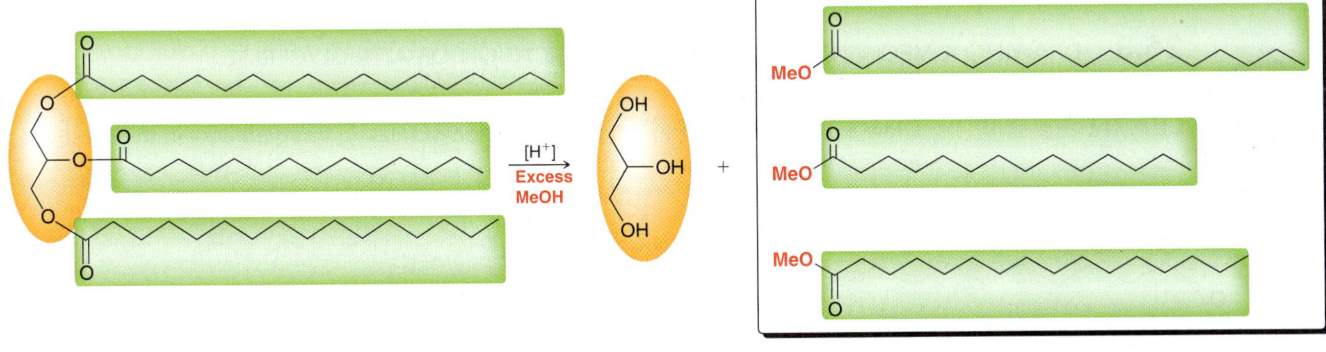


**Biodiesel**
**(a mixture of fatty acid methyl esters)**

This transformation can be achieved with either acid catalysis or base catalysis. Mechanism 26.2 shows the acid-catalyzed mechanism.

### MECHANISM 26.2 ACID-CATALYZED TRANSESTERIFICATION

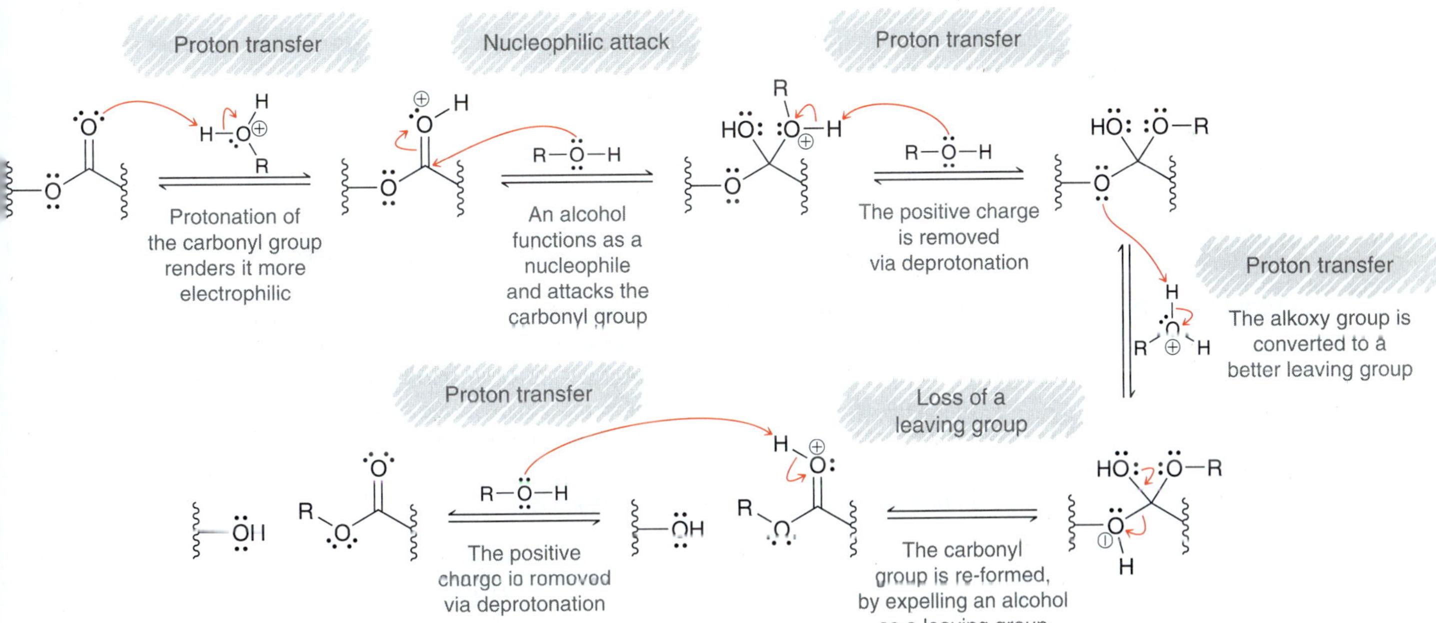


**LOOKING BACK**
To review the function of these four proton transfers, see Section 21.7.

In the first step, the carbonyl group of the ester is protonated, which renders the carbonyl group more electrophilic. An alcohol, such as methanol, then functions as a nucleophile and attacks the protonated carbonyl group, giving a tetrahedral intermediate. After two proton transfer steps, the carbonyl group can re-form, followed by a final proton transfer step. This mechanism is identical to Mechanism 21.8 (Section 21.11). The mechanism consists of two core steps (nucleophilic attack and loss of a leaving group) and four proton transfers.

Transesterification makes it possible to convert triglycerides into biodiesel. Since biodiesel can be derived from plants (vegetable oil), it serves as a potential alternative to petroleum-based gasoline as a renewable source of energy. Unfortunately, the current cost associated with producing biodiesel outweighs the cost of producing an equivalent amount of gasoline, so it is unlikely that biodiesel will completely replace petroleum-based fuels in the near future. However, as the price of crude oil increases, biodiesel will become an attractive alternative.

## SKILLBUILDER

### 26.3 DRAWING A MECHANISM FOR TRANSESTERIFICATION OF A TRIGLYCERIDE

**LEARN** the skill

Draw a mechanism for the transesterification of trilaurin using ethanol in the presence of an acid catalyst.

$[H^+]$, Excess EtOH

### SOLUTION

Each ester group undergoes transesterification via a mechanism that has two core steps and four proton transfers.

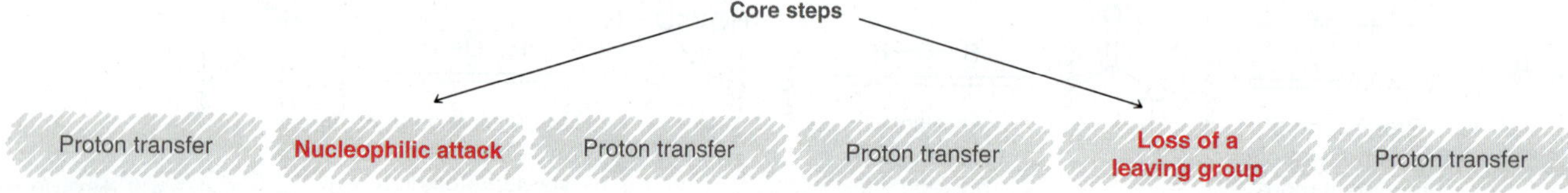


**LOOKING BACK**
For a review of how this rule is applied when drawing mechanisms, see Section 21.7.

Each of the four proton transfers serves a function, as we first saw in Chapter 21 when discussing the reactions of carboxylic acid derivatives. Recall that there is one guiding rule that determines when proton transfers are used: *In acidic conditions, a mechanism will only be reasonable if it avoids the use or formation of strong bases.* The first step is to protonate one of the ester groups. This renders the carbonyl group more electrophilic, and it avoids the formation of a strong base, which would result if the carbonyl group were attacked by the nucleophile without first being protonated.

Proton transfer

The second step is a nucleophilic attack. Ethanol functions as a nucleophile and attacks the protonated carbonyl group.

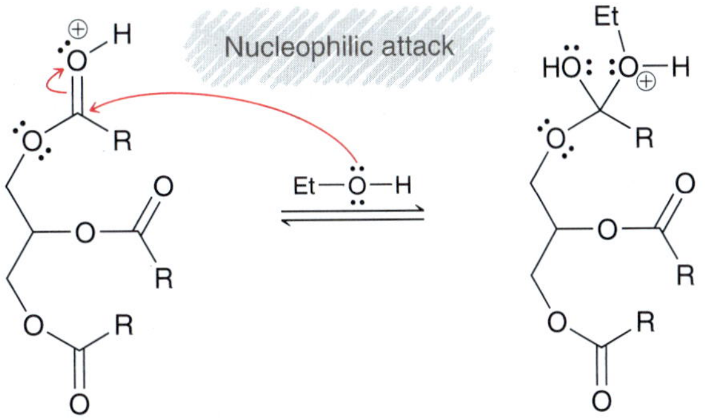

The next two steps are both proton transfers.

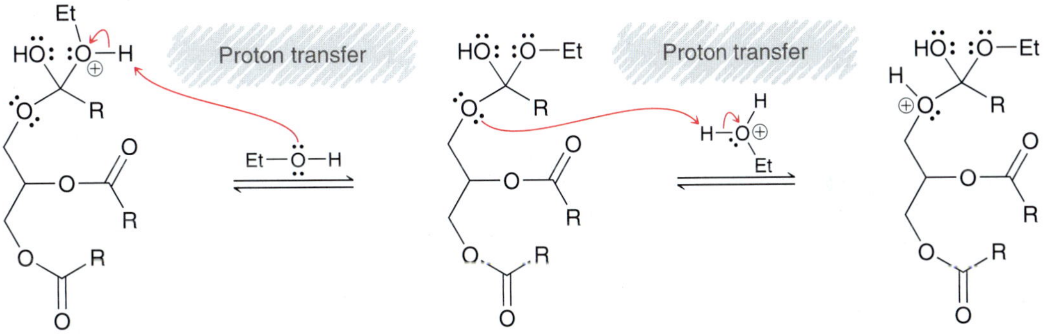

The first proton transfer removes the positive charge so as to avoid the formation of two positive charges that would result from the second proton transfer. The second proton transfer occurs to protonate the leaving group. This avoids the formation of a strong base when the leaving group leaves in the next step.

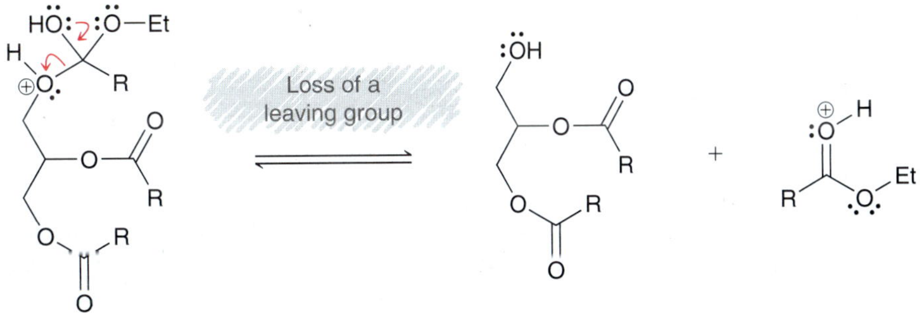

The final proton transfer removes the positive charge.

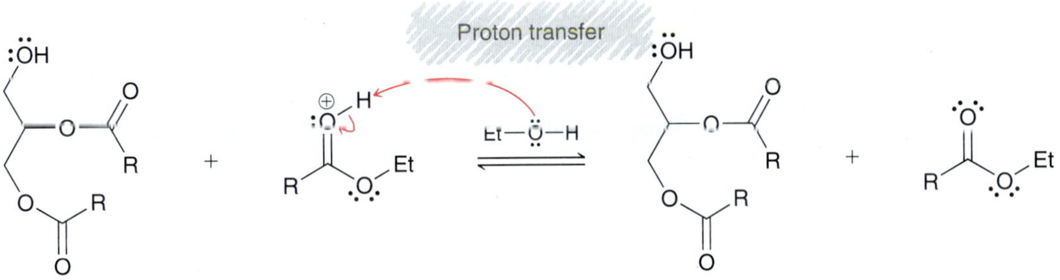

These six steps are then repeated for each of the other remaining fatty acid residues.

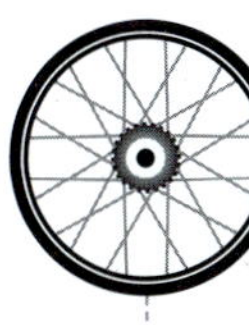

**PRACTICE the skill** **26.12** Draw a mechanism for the transesterification of tristearin using methanol in the presence of catalytic acid.

**26.13** Draw the products obtained when triolein undergoes transesterification using isopropyl alcohol in the presence of catalytic sulfuric acid.

**26.14** The conversion of triglycerides into biodiesel can be achieved in the presence of either catalytic acid or catalytic base. We have seen a mechanism for transesterification with catalytic acid. In contrast, the mechanism for base-catalyzed transesterification has fewer steps. The base, such as hydroxide, functions as a catalyst by establishing an equilibrium in which some alkoxide ions are present.

$$HO^{-} + H{-}OEt \rightleftharpoons H_2O + {}^{-}OEt$$

Ethoxide

This equilibrium favors the hydroxide ions. Nevertheless, some ethoxide ions are present at equilibrium. These ethoxide ions are strong nucleophiles that can attack each ester group of the triglyceride according to the following mechanism:

Alcohol Ester

In the final step, water is deprotonated, regenerating the catalyst.

**(a)** Draw a mechanism for the following process:

[NaOH]
Excess EtOH

**(b)** When sodium hydroxide is used as a catalyst for transesterification, it is essential that only a small amount of the catalyst is present. Explain what would happen in the presence of too much sodium hydroxide.

need more **PRACTICE?** **Try Problems 26.44, 26.45**

## 26.5 Phospholipids

**Phospholipids** are esterlike derivatives of phosphoric acid.

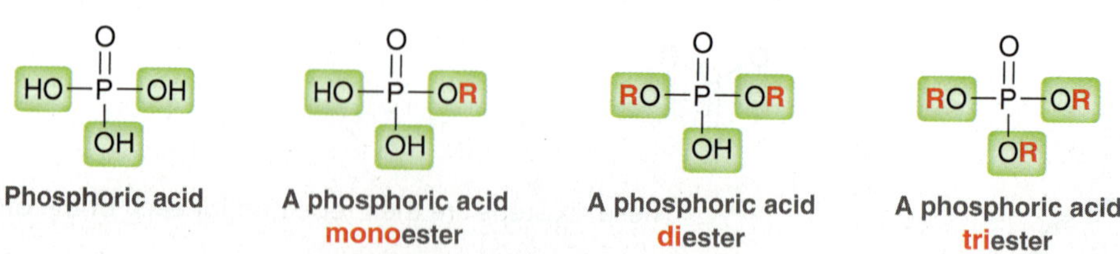


The most common phospholipids are phosphoglycerides.

## Phosphoglycerides

**Phosphoglycerides** are very similar in structure to triglycerides, with the main difference being that in phosphoglycerides one of the three fatty acid residues is replaced by a phosphoester group. The simplest kind of phosphoglyceride is a phosphoric monoester called a **phosphatidic acid**.

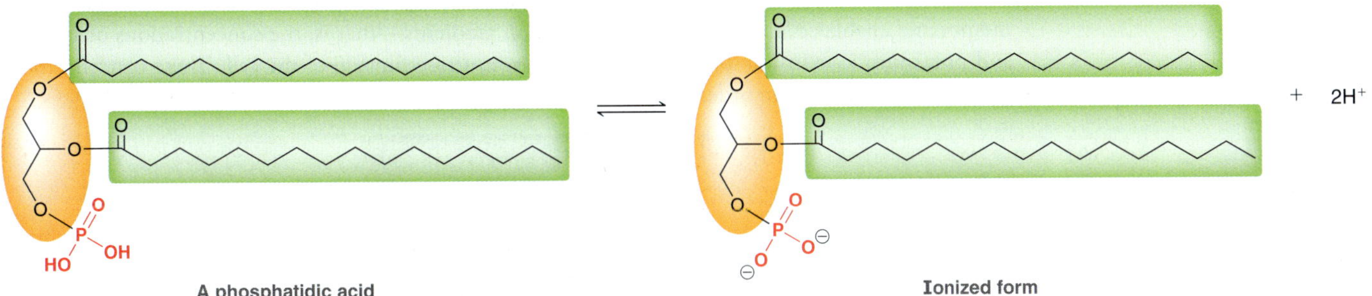


At physiological pH, the ionized form of a phosphatidic acid predominates.

The most abundant phosphoglycerides are phosphoric acid diesters:

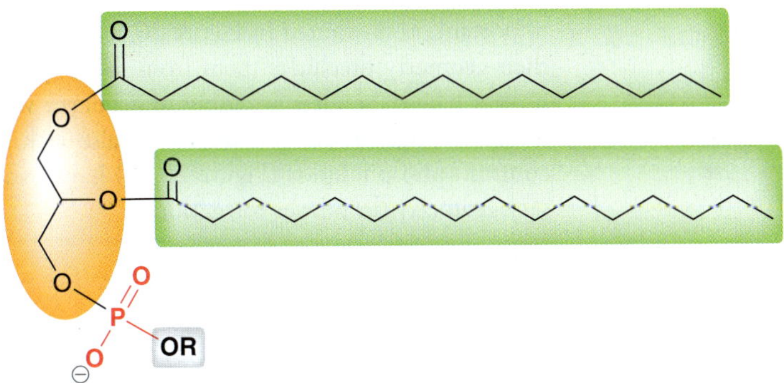


The identity of the alkoxy group can vary. Phosphoglycerides derived from ethanolamine and from choline are particularly abundant in the cells of plants and animals.

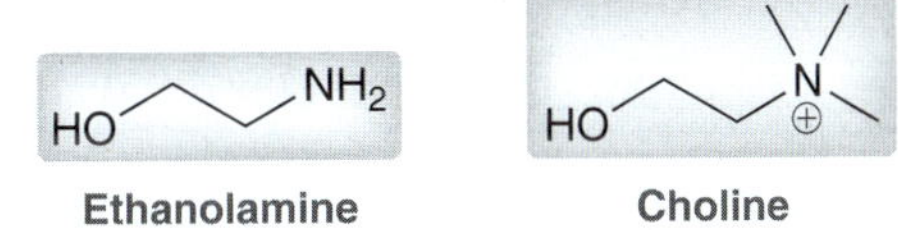


Phosphoglycerides that contain ethanolamine are called **cephalins**, while those that contain choline are called **lecithins**.

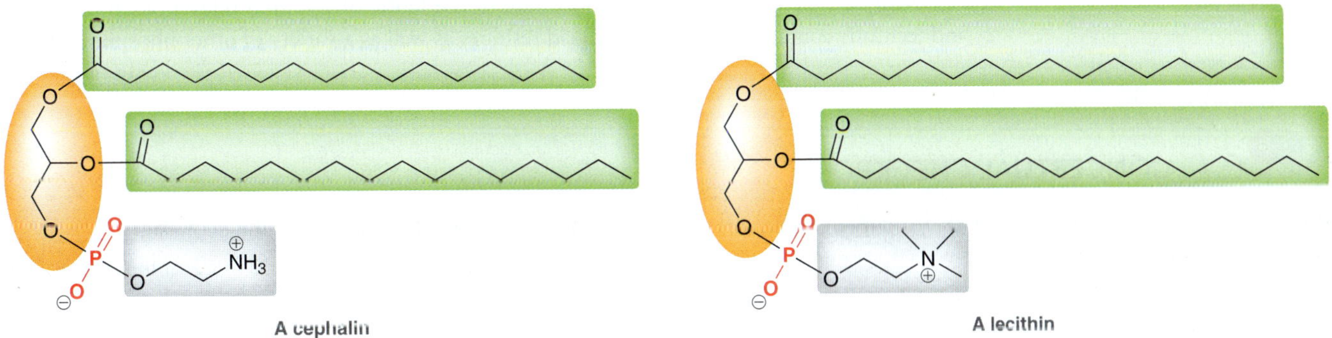


These compounds contain a chirality center (C2 of the glycerol unit) and generally exhibit the *R* configuration. Cephalins and lecithins have two nonpolar, hydrophobic tails and one polar head group. The polar head consists of the glycerol backbone as well as the phosphodiester, while the nonpolar tails are hydrocarbon chains. These features determine their function in cells, as will be described in the upcoming section.

## Lipid Bilayers

In water, phosphoglycerides self-assemble into a **lipid bilayer** (Figure 26.5). In this way, the hydrophobic tails avoid contact with water and interact with each other via London dispersion forces. The surface of the bilayer is polar and therefore water soluble. Lipid bilayers constitute the main fabric of cell membranes, where they function as barriers that restrict the flow of water and ions. Cell membranes enable cells to maintain concentration gradients—that is, the concentrations of sodium and potassium ions inside the cell are different from those outside of the cell. These concentration gradients are necessary in order for a cell to function properly.

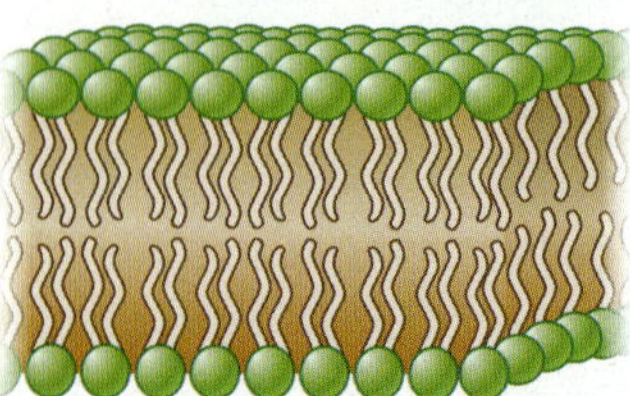

**FIGURE 26.5** A graphic illustration of the three-dimensional structure of a lipid bilayer.

In order for compounds to self-assemble and form lipid bilayers, the compounds must have both a polar head and nonpolar tails. In addition, the three-dimensional shape of the compounds are also important, as illustrated in Figure 26.6. Fatty acids have a polar head group and a nonpolar tail, but their geometry precludes them from forming a bilayer. As seen in Figure 26.6a, fatty acids cannot form a bilayer because assembly of a bilayer would leave empty space in between each fatty acid. Similarly, triglycerides (Figure 26.6c) also lack the appropriate geometry for bilayer formation. In contrast, phospholipids (Figure 26.6b) have two hydrophobic tails and therefore exhibit the necessary geometry for bilayer formation.

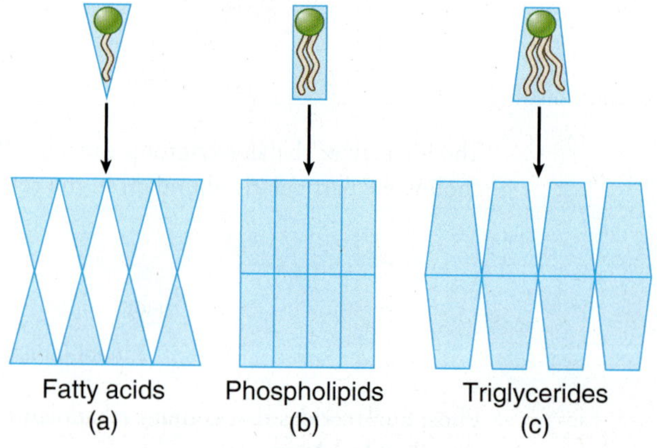


**FIGURE 26.6** (a) The assembly of fatty acids in a lipid bilayer would leave empty space between the molecules, thereby destabilizing the potential bilayer. (b) Phospholipids have just the right geometry to form lipid bilayers, with no empty space between the molecules. (c) The assembly of triglycerides in a lipid bilayer would leave empty space between the molecules, thereby destabilizing the potential bilayer.

Many different phospholipids contribute to the fabric of the lipid bilayer, including cephalins and lecithins containing various fatty acid residues as well as a variety of other phospholipids.

### CONCEPTUAL CHECKPOINT

**26.15** A lecithin was hydrolyzed to yield two equivalents of myristic acid.

(a) Draw the structure of the lecithin.

(b) This compound is chiral, but only one enantiomer predominates in nature. Draw the enantiomer that is found in nature.

(c) The phosphodiester is generally located at C3 of the glycerol unit. If the phosphodiester was located at C2, would the compound be chiral?

**26.16** A cephalin was hydrolyzed to yield one equivalent of palmitic acid and one equivalent of oleic acid.

(a) Draw two possible structures of the cephalin.

(b) If the phosphodiester was located at C2 of the glycerol unit, would the compound be chiral?

**26.17** Draw the resonance structures of a fully deprotonated phosphatidic acid.

**26.18** Octanol is more efficient than hexanol at crossing a cell membrane and entering a cell. Explain.

**26.19** Would you expect glycerol to readily cross a cell membrane?

## medically speaking

## Selectivity of Antifungal Agents

Many different fungi are known to cause infections in human skin and nails. The conditions caused by these fungi are called *tinea*.

| Type | Location |
|---|---|
| Tinea manuum | Hand |
| Tinea cruris | Groin |
| Tinea sycosis | Beard |
| Tinea capitis | Scalp |
| Tinea unguium | Nails |

Treatment of these conditions can often be difficult because the cell membranes of fungi are nearly identical to the cell membranes of human cells, and many of the biochemical processes taking place at the cell membrane are also similar for both organisms. Therefore, any agents that interfere with the membrane integrity of fungi will generally also interfere with the membrane integrity of human cells, and agents that are toxic to fungi will also be toxic to humans. There are, however, small differences between the cell membranes of fungal and human cells. These differences have enabled medicinal chemists to develop agents that are toxic to fungal cells and less toxic to human cells. To understand how these agents work, we must focus on the nature of the hydrophobic region of the lipid bilayer.

The center of the lipid bilayer (the hydrophobic region) is highly fluid and resembles a liquid hydrocarbon.

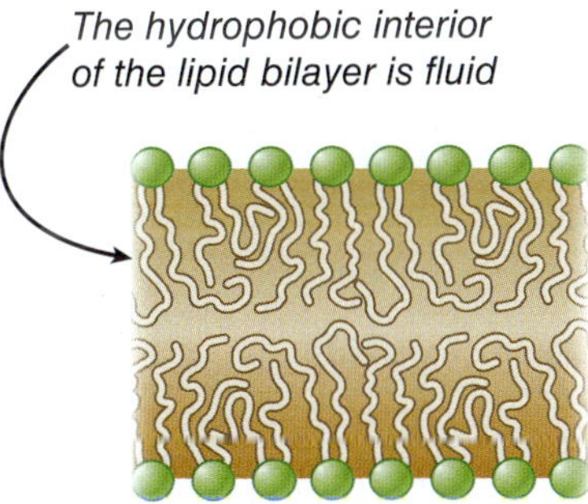


The hydrophobic tails are in rapid motion. This is an important feature, because it imparts flexibility to the lipid bilayer. However, if the bilayer is too fluid, then it becomes unstable. It would be incapable of holding its shape and carrying out its vital functions. In order to prevent the bilayer from becoming too fluid, it contains embedded stiffening agents. These agents are typically lipids with rigid geometry, and their effect is to limit the movement of the hydrocarbon tails.

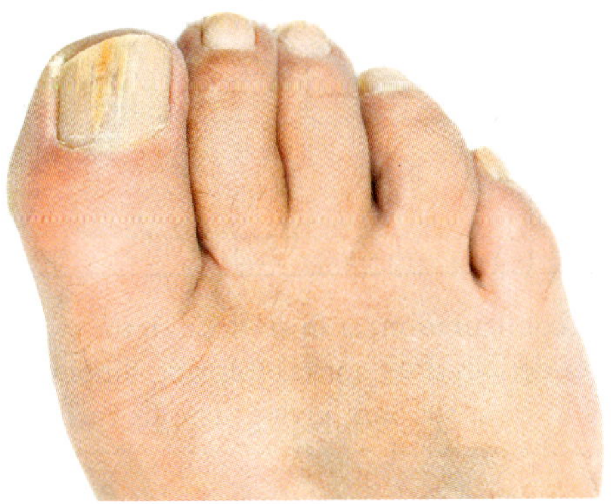

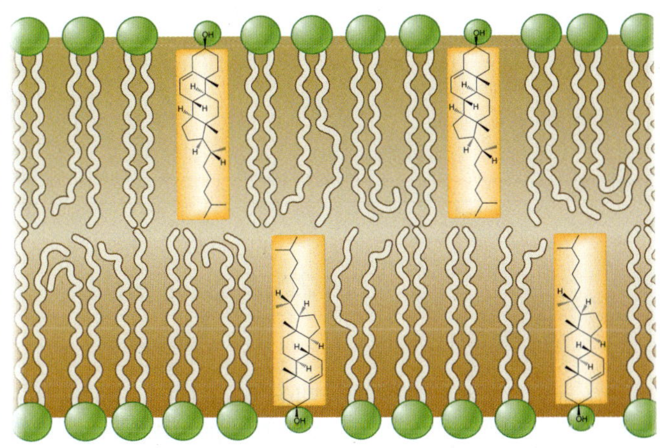

The hydrophobic region of the bilayer is still fluid, but motion is more limited, rendering the cell membrane more stiff. In human cells, the predominant stiffening agent is cholesterol, while in fungi, the predominant stiffening agent is ergosterol. Each of these compounds contains a small polar head and a large, rigid, nonpolar tail.

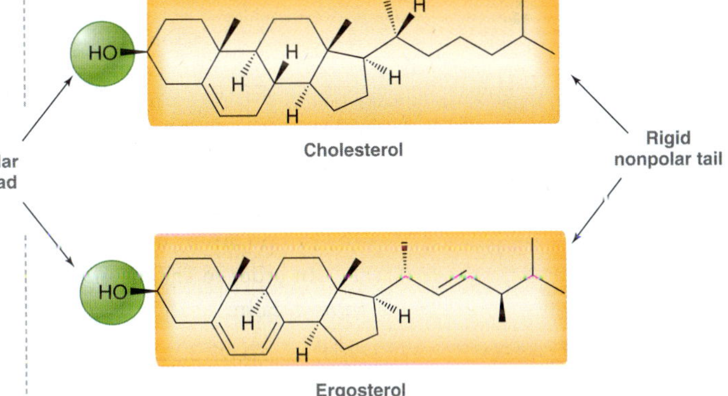


These stiffening agents are naturally occurring steroids, discussed in the upcoming section. Each of these compounds has an OH group that serves as a very small polar head group, while the rest of the compound is hydrophobic. Although they are very similar in structure, there are subtle differences. Specifically, ergosterol has two $\pi$ bonds that are not present in cholesterol. These small differences provide a source of selectivity in treating fungal infections. Most antifungal drug research has focused on the development of agents that are capable of binding more effectively with ergosterol than cholesterol, thereby interfering with the function of ergosterol as a stiffening agent more so than they interfere with the function of cholesterol as a stiffening agent. For example, amphotericin B is a potent antifungal agent that is capable of penetrating the fungal cell membrane and binding closely with ergosterol, thereby disrupting the integrity of the fungal cell membranes.

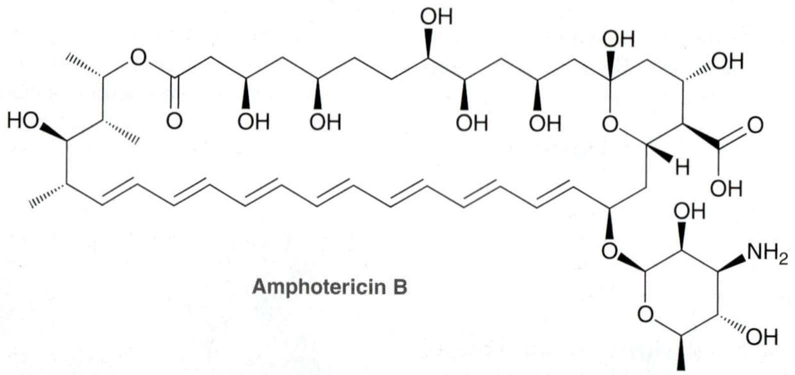

Amphotericin B

## 26.6 Steroids

### Introduction to Steroids

Most steroids function as chemical messengers, or hormones, that are secreted by endocrine glands and transported through the bloodstream to their target organs. Steroids and their derivatives are also among the most widely used therapeutic agents. They are used in birth control and hormone replacement therapy and in the treatment of inflammatory conditions and cancer.

The structures of **steroids** are based on a tetracyclic ring system involving three six-membered rings and one five-membered ring.

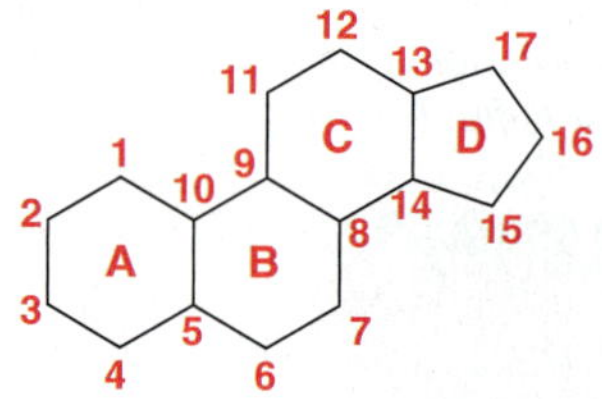

The tetracyclic skeleton of steroids

These rings are labeled using the letters A, B, C, and D, where the D ring is the five-membered ring. The carbon atoms in this system are numbered as shown.

In order to analyze the configuration of each ring fusion, recall the structures of *cis*-decalin and *trans*-decalin.

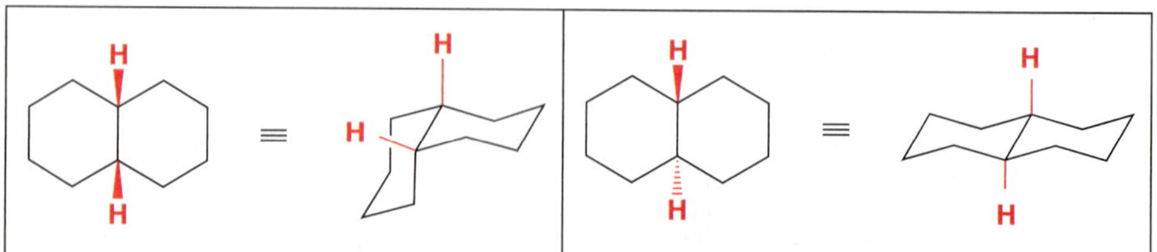

*cis*-Decalin

*trans*-Decalin

When six-membered rings are fused with a *cis* configuration, as in *cis*-decalin, the two rings are both free to exhibit ring flipping. In contrast, *trans*-decalin does not exhibit ring flipping and is a more rigid structure. The ring fusions are all *trans* in most steroids, giving steroids their rigid geometry.

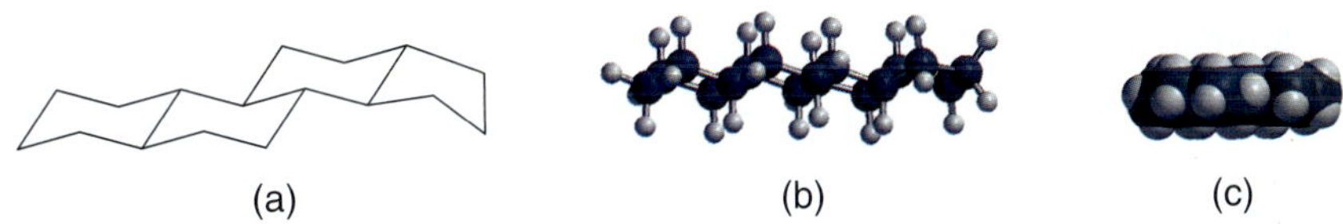


**FIGURE 26.7**
(a) The carbon skeleton of most steroids, (b) a ball-and-stick model of the carbon skeleton of a steroid, and (c) a space-filling model of the carbon skeleton of a steroid.

For some steroids, the A-B ring fusion is *cis*, although the B-C and C-D fusions are almost always *trans* in naturally occurring steroids. As a result, the carbon skeleton of most steroids provides for a fairly flat compound (Figure 26.7). The rigid geometry of cholesterol serves as an example.

**Cholesterol**
(a steroid)

In the Medically Speaking box on antifungal drugs, we saw that the rigid structure of cholesterol enables the compound to function as a stiffening agent for lipid bilayers. Cholesterol contains two methyl groups (one attached to C10 and the other attached to C13) as well as a side chain attached to C17. This substitution pattern is common among many steroids. The two methyl groups occupy axial positions and are therefore perpendicular to the nearly planar skeleton, while the side chain occupies an equatorial position. This can be seen more clearly with three-dimensional representations of cholesterol (Figure 26.8). Cholesterol has eight chirality centers, giving rise to $2^8$, or 256, possible stereoisomers. Nevertheless, only the one stereoisomer shown exists in nature.

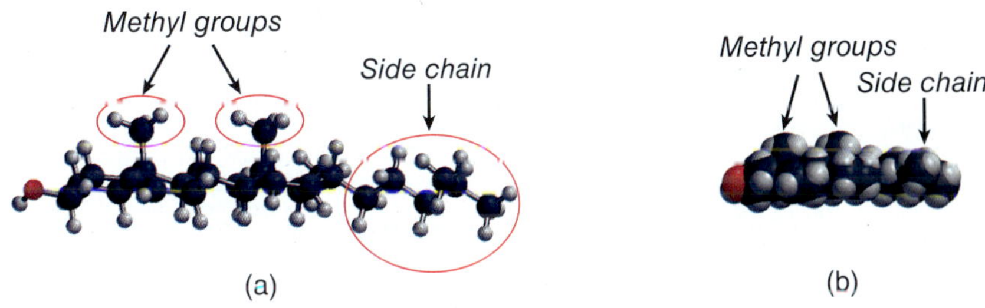


**FIGURE 26.8**
(a) A ball-and-stick representation of cholesterol which clearly shows that the two methyl groups occupy axial positions while the side chain occupies an equatorial position.
(b) A space-filling model of cholesterol which also shows the locations of the two methyl groups and the side chain.

## Biosynthesis of Cholesterol

The biosynthesis of cholesterol involves many steps, some of which are shown in Mechanism 26.3. The starting material is squalene, which undergoes asymmetric epoxidation via enzymatic catalysis. The epoxide then opens to generate a carbocation, which undergoes a series of intramolecular cyclization reactions. All of these steps involve a $\pi$ bond functioning as a nucleophile and attacking a carbocation in an intramolecular process. Each of these steps occurs within the cavity of an enzyme, where the configuration of each new chirality center is carefully controlled. Then, a series of rearrangements occurs that includes both methyl shifts and hydride shifts. Finally, deprotonation yields lanosterol, which is a precursor for cholesterol and all other steroids.

**LOOKING BACK**
For a discussion of the role of chiral catalysts in asymmetric epoxidation, see Section 14.9.

## MECHANISM 26.3 BIOSYNTHESIS OF CHOLESTEROL

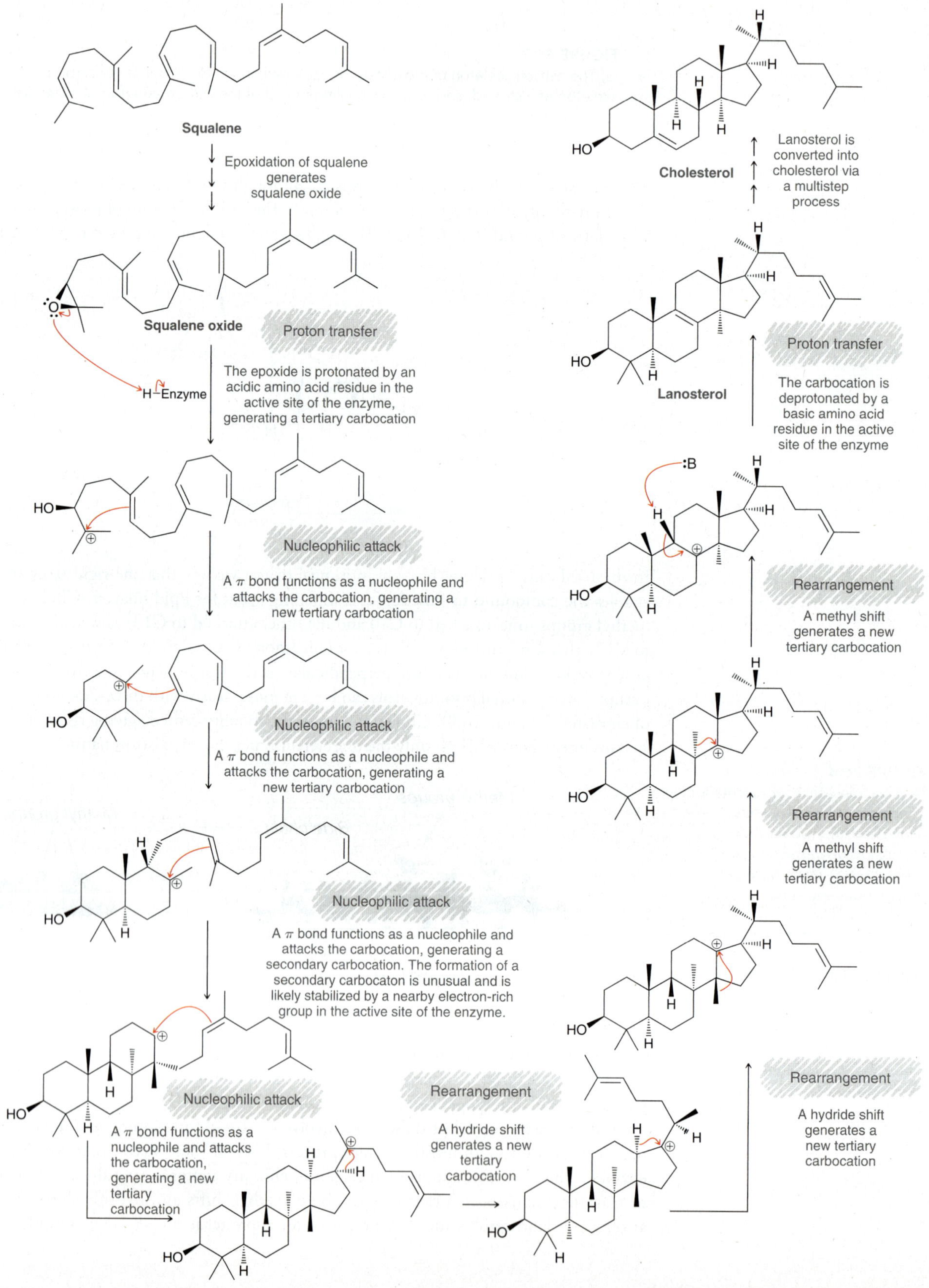

Squalene

Epoxidation of squalene generates squalene oxide

Squalene oxide

H–Enzyme

**Proton transfer**

The epoxide is protonated by an acidic amino acid residue in the active site of the enzyme, generating a tertiary carbocation

HO

**Nucleophilic attack**

A $\pi$ bond functions as a nucleophile and attacks the carbocation, generating a new tertiary carbocation

HO

**Nucleophilic attack**

A $\pi$ bond functions as a nucleophile and attacks the carbocation, generating a new tertiary carbocation

H

HO

H

**Nucleophilic attack**

A $\pi$ bond functions as a nucleophile and attacks the carbocation, generating a secondary carbocation. The formation of a secondary carbocaton is unusual and is likely stabilized by a nearby electron-rich group in the active site of the enzyme.

H

HO

H

**Nucleophilic attack**

A $\pi$ bond functions as a nucleophile and attacks the carbocation, generating a new tertiary carbocation

H

H

HO

H

**Rearrangement**

A hydride shift generates a new tertiary carbocation

H

H

H

HO

H

**Rearrangement**

A hydride shift generates a new tertiary carbocation

H

H

H

HO

H

**Rearrangement**

A methyl shift generates a new tertiary carbocation

H

H

H

HO

H

**Rearrangement**

A methyl shift generates a new tertiary carbocation

:B

H

H

H

HO

H

**Proton transfer**

The carbocation is deprotonated by a basic amino acid residue in the active site of the enzyme

H

H

HO

H

Lanosterol

Lanosterol is converted into cholesterol via a multistep process

H

H

H

H

H

HO

Cholesterol

## medically speaking — Cholesterol and Heart Disease

As described in the previous Medically Speaking box, cholesterol plays a vital role in maintaining the integrity of cell membranes. Cholesterol also has many other important biological functions. For instance, it is a precursor in the biosynthesis of most other steroids, including sex hormones. Our bodies can produce all the cholesterol we need, but we also obtain cholesterol in the diet. If dietary intake of cholesterol is high, our bodies produce less of it, in an attempt to compensate. If intake is excessively large, then cholesterol levels can rise, increasing the risk of heart attack and stroke. In order to understand the reason for this, we must consider how cholesterol is transported throughout the body.

Cholesterol is a lipid, which means that it is not water soluble. As a result, cholesterol cannot, by itself, be dissolved in the aqueous medium of the blood. Cholesterol is produced in the liver and then transported in the blood by large particles called lipoproteins. Lipoproteins consist of many proteins and lipids bound together via intermolecular interactions. Cholesterol and ester derivatives of cholesterol are contained in the center (the lipophilic region). Lipoproteins are soluble in the aqueous environment of the blood and are used by the body to transport lipids, such as cholesterol, throughout the body.

Lipoproteins are categorized into several different categories based on their density. High-density lipoproteins, or HDLs, have a higher proportion of proteins, while low-density lipoproteins, or LDLs, have a higher proportion of lipids. It is now understood that HDLs and LDLs have different functions. LDLs transport cholesterol from the liver to cells throughout the body, while HDLs transport cholesterol for the reverse journey, back to the liver, where they are used as precursors for the synthesis of other steroids. It is important that the concentration of HDLs is greater than the concentration of LDLs. If the LDL concentration is too high, then some of the LDLs will be unable to unload their cargo, and instead, they accumulate and form deposits in the arteries. These deposits restrict blood flow, which can lead to a heart attack (caused by an inadequate flow of blood to the heart) or a stroke (caused by an inadequate flow of blood to the brain).

There are many drugs available for the treatment of high levels of LDL cholesterol. Most notably, the *statins* are a family of compounds that were all developed based on the chemical structure of a single natural product, called lovastatin.

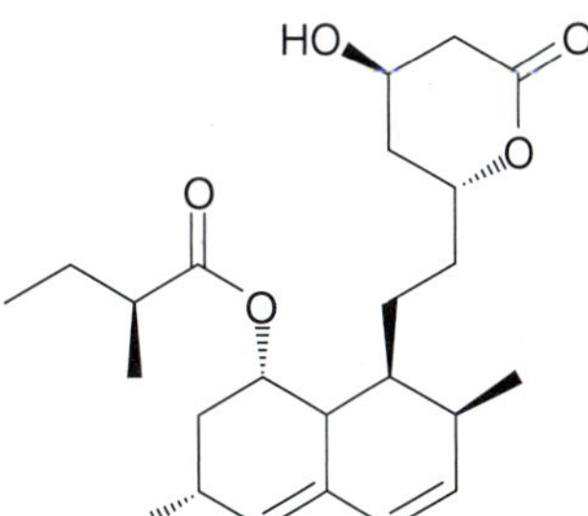

**Lovastatin (Mevacor)**

Scientists at Merck Research laboratories in the United States discovered lovastatin in the late 1970s in fermentation cultures of the fungus *Aspergillus terreus*. In clinical trials, lovastatin was found to be very effective at lowering LDL cholesterol levels, by inhibiting an enzyme (HMG-CoA reductase) that catalyzes a critical step in the biosynthesis of cholesterol in humans. Lovastatin, sold under the trade name Mevacor, once enjoyed annual sales in excess of $1 billion. Due to the success of lovastatin, a search was undertaken for analogues that could also be used for treating high cholesterol.

A molecule very similar to lovastatin, called compactin (missing only a methyl group), had been isolated from fermentation cultures of the fungus *Penicillium citrinum* by scientists in Japan a few years before lovastatin was discovered. Although compactin failed during clinical testing, it was discovered that biotransformation with a bacteria would introduce an alcohol group into the ring, and this simple modification eliminated the problems observed with compactin. The modified structure, called pravastatin, became a widely prescribed cholesterol-lowering drug, sold under the trade name Pravacol.

**Compactin**

**Pravastatin (Pravacol)**

This drug is produced by a two-stage fermentation process, where the compactin isolated from a fungal fermentation culture is then fed to a bacterial culture to produce pravastatin.

In recent years, synthetic chemistry has been used to generate a series of analogues of lovastatin. A "semisynthetic" analogue of lovastatin is the molecule simvastatin (Zocor), which is made by introducing a methyl group, in a single synthetic step, into the side chain of lovastatin. Other analogues of lovastatin include fluvastatin (Lescol) and atorvastatin (Lipitor), both of which are made entirely by chemical synthesis. Atrovastatin is currently one of the most widely prescribed drugs in the world and has proven extremely successful in lowering cholesterol.

The statins serve as an example of how a single natural product can act as inspiration for an entirely new class of drugs.

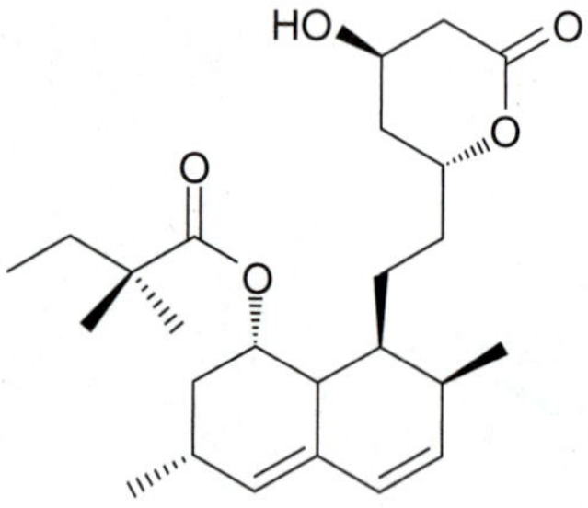

Simvastatin (Zocor)

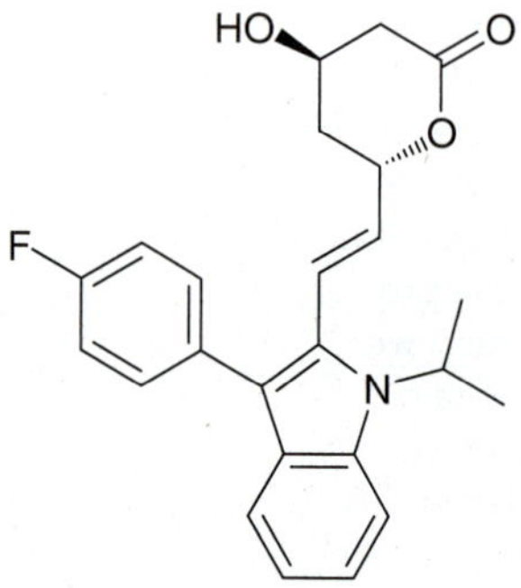

Fluvastatin (Lescol)

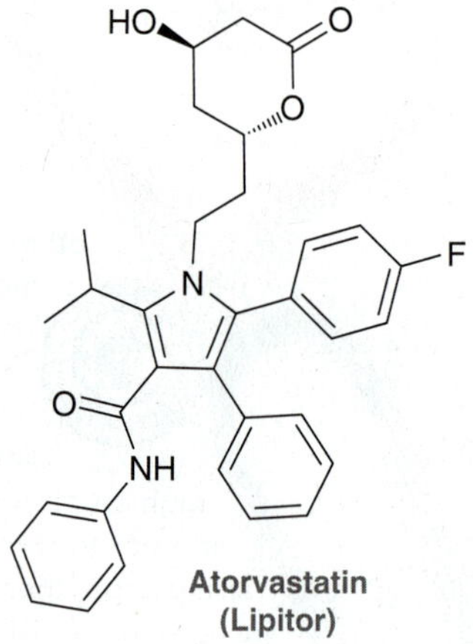

Atorvastatin (Lipitor)

## Sex Hormones

Human sex hormones are steroids that regulate tissue growth and reproductive processes. Male sex hormones are called **androgens**, and there are two types of female sex hormones called **estrogens** and **progestins**. Table 26.3 shows the most important male and female sex hormones. All five of these sex hormones are present in both males and females. Estrogens and progestins are produced

**TABLE 26.3 THE MOST IMPORTANT MALE AND FEMALE SEX HORMONES**

**Male sex hormones**

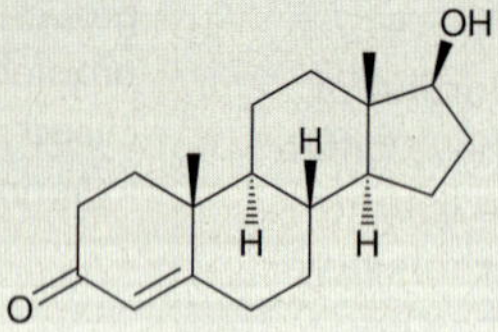

Testosterone

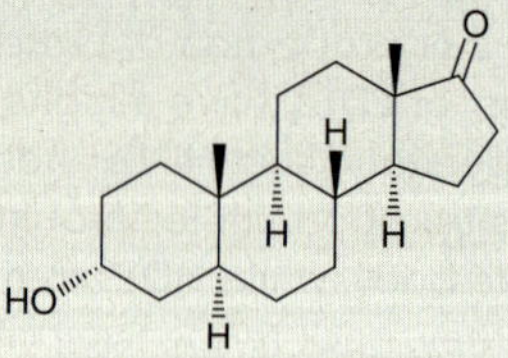

Androsterone

Androgens

**Female sex hormones**

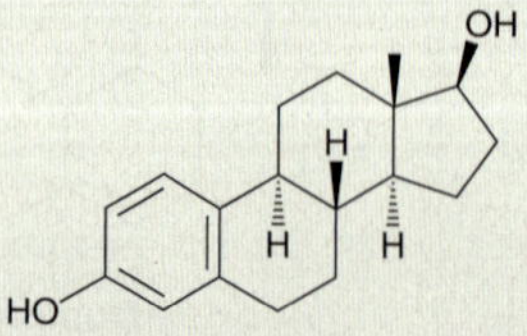

Estradiol

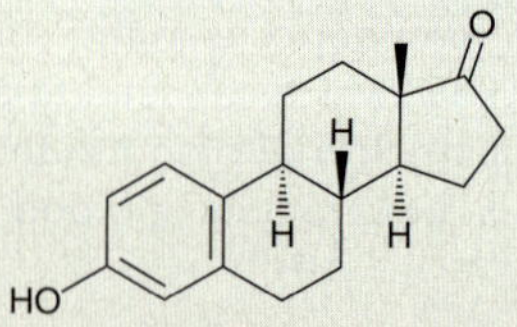

Estrone

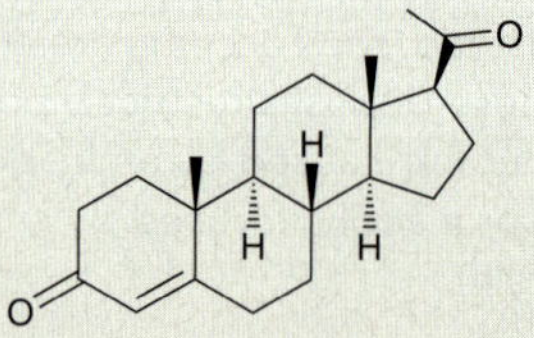

Progesterone

Estrogens

A progestin

in greater concentrations in women, while androgens are produced in greater concentrations in men. Testosterone and androsterone are among the more potent androgens, and they control the development of secondary sex characteristics in males. Estradiol and estrone are estrogens, which are characterized by an aromatic A ring and the absence of a methyl group at C10. These hormones are produced in the ovaries from testosterone and play important roles in regulating a woman's menstrual cycle and controlling the development of secondary sex characteristics. Progesterone is a progestin that prepares the uterus for nurturing a fertilized egg during pregnancy.

During pregnancy, ovulation is inhibited by the release of estrogens and progestins from the placenta and ovaries. This process is mimicked by most birth control formulations, which generally contain a mixture of a synthetic estrogen (such as ethynyl estradiol) and a synthetic progestin (such as norethindrone).

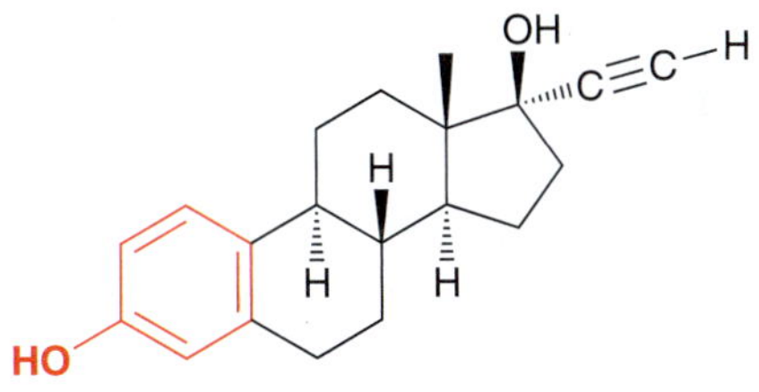

Ethynyl estradiol

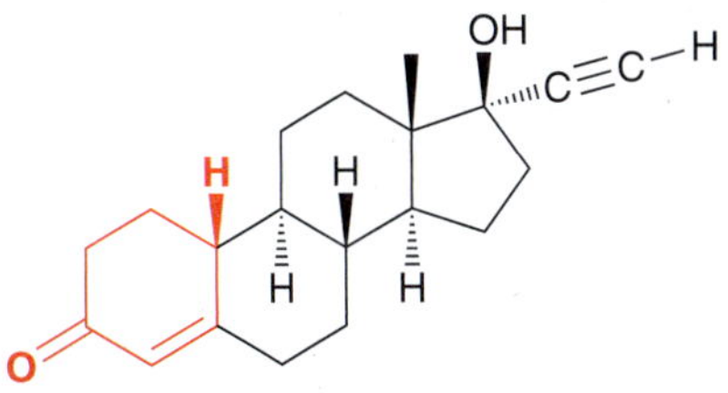

Norethindrone

This mixture of compounds inhibits ovulation in much the same way that the body naturally inhibits ovulation during pregnancy.

## Adrenocortical Hormones

**Adrenocortical hormones** are thus named because they are secreted by the cortex (the outer layer) of the adrenal glands. Adrenocortical hormones are typically characterized by a carbonyl group or hydroxyl group at C11. Examples include cortisone and cortisol.

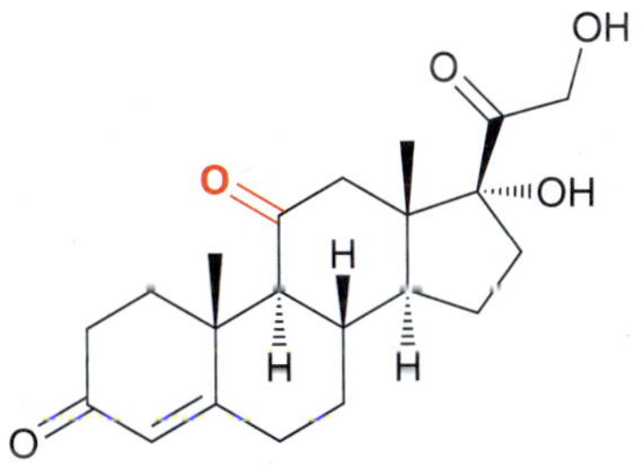

Cortisone

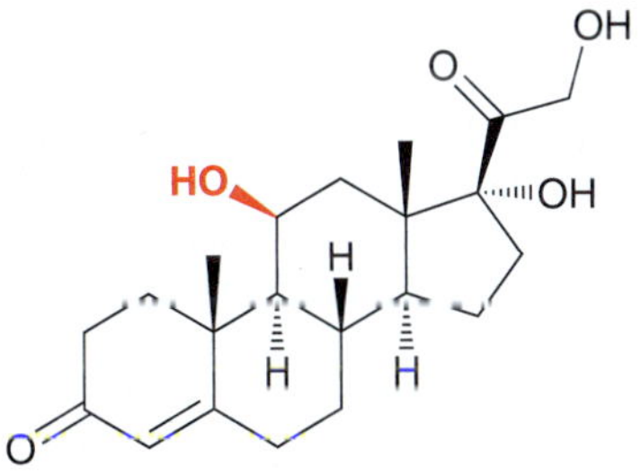

Cortisol

Cortisone and cortisol differ only in the identity of one functional group (highlighted in red). Cortisone has a ketone group at C11 (indicated by the "-one" suffix), while cortisol has a hydroxy group at C11 (indicated by the "-ol" suffix). Cortisol is more abundant in nature, but cortisone is better known because of its therapeutic use. Both agents are used to treat the effects of inflammatory diseases, including psoriasis (inflammation of the skin), arthritis (inflammation of the joints), and asthma (inflammation of the lungs).

Many synthetic corticoids have been developed that are even more potent than their natural analogues. Synthetic corticoids are used in the treatment of rashes caused by poison oak, poison ivy, and eczema as well as inflammatory diseases such as psoriasis, arthritis, and asthma.

## Anabolic Steroids and Competitive Sports

Anabolic steroids are compounds that promote muscle growth by mimicking the tissue-building effect of testosterone. Many synthetic androgen analogues have been created that are even more potent than testosterone, including stanozolol, nandrolone, and methandrostenolone.

**Stanozolol**

**Nandrolone**

**Methandrostenolone**

Synthetic anabolic steroids were originally developed in the 1930s for use in treating illnesses and injuries that involved muscle deterioration. Unfortunately, the development of these drugs led to rampant abuse by athletes and bodybuilders, even as early as the 1940s. The problem received wide public attention during the 1988 Olympics when Ben Johnson (Canada) was disqualified as the gold medal winner of the 100-meter dash because his urine tested positive for traces of stanozolol.

Steroid abuse is still a major problem, particularly among bodybuilders and professional baseball and football players. Many studies have been conducted to determine whether there is any correlation between steroid use and increased athletic performance. Some studies have shown a small connection, while other studies show no connection whatsoever. It appears that the health risks associated with steroid use outweigh the uncertain benefits. The health risks include increased risk of heart disease, stroke, liver cancer, sterility, as well as adverse behavioral changes caused by increased aggressive tendencies (called "steroid rage").

### CONCEPTUAL CHECKPOINT

**26.20** The following compounds are steroids. One is an anabolic steroid called oxymetholone and the other, called norgestrel, is used in oral contraceptive formulations. Identify these compounds based on their structural features.

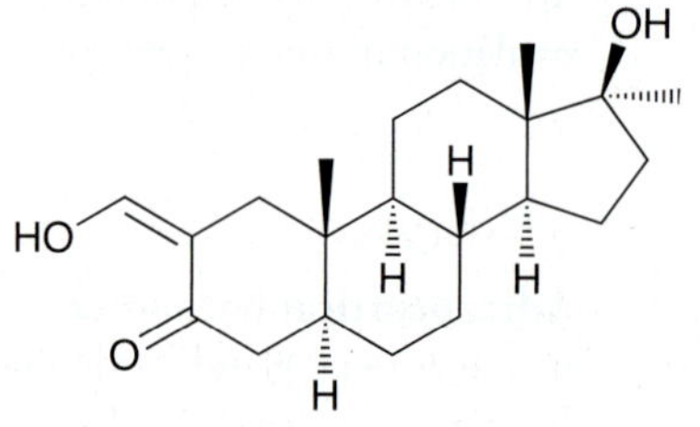


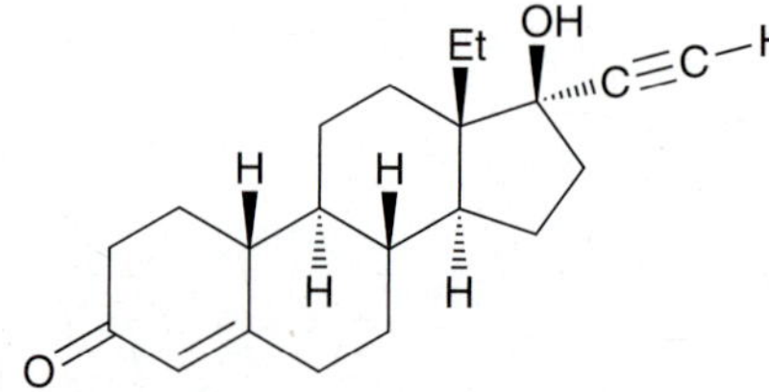

### CONCEPTUAL CHECKPOINT

**26.21** Draw the hypothetical ring-flip of *trans*-decalin and explain why it does not occur. Use this analysis to explain why cholesterol has a fairly rigid three-dimensional geometry.

**26.22** Draw chair conformations for each of the following compounds and then identify whether each substituent is axial or equatorial:

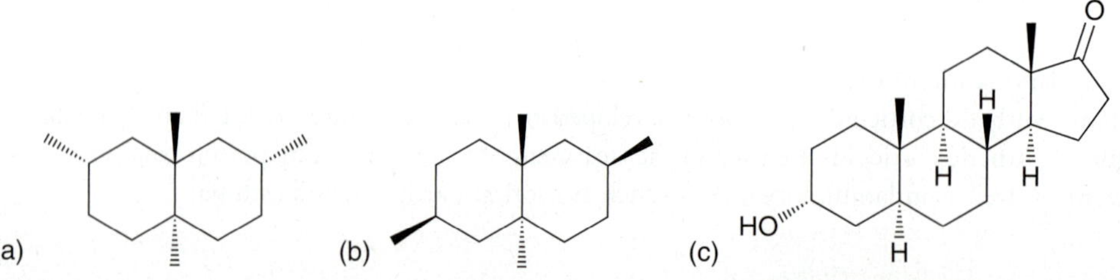


**26.23** Prednisolone acetate is an anti-inflammatory agent in clinical use. It is similar in structure to cortisol, with the following two differences: (1) Prednisolone acetate exhibits a double bond between C1 and C2 of the A ring and (2) the primary hydroxyl group has been acetylated. Using this information, draw the structure of prednisolone acetate.

## 26.7 Prostaglandins

In the early 1930s, it was observed that human seminal fluid could induce muscle contractions in uterine tissue. It was believed that this phenomenon was caused by an acidic substance produced in the prostate gland. This unknown substance was called **prostaglandin** by Ulf von Euler (Karolina Institute in Sweden). In the 1950s, the acidic extract from sheep prostate glands was found to contain not one, but many structurally related prostaglandin substances. These prostaglandin compounds were separated, purified, and characterized and are now known to be present in all body tissues and fluid in very small concentrations. Despite our current understanding of the ubiquitous nature of these compounds, the original name is still used.

Prostaglandins contain 20 carbon atoms and are characterized by a five-membered ring with two side chains. Many substitution patterns are observed in nature, the most common of which are shown in Figure 26.9. In each case, the letters PG indicate that the compound is a prostaglandin

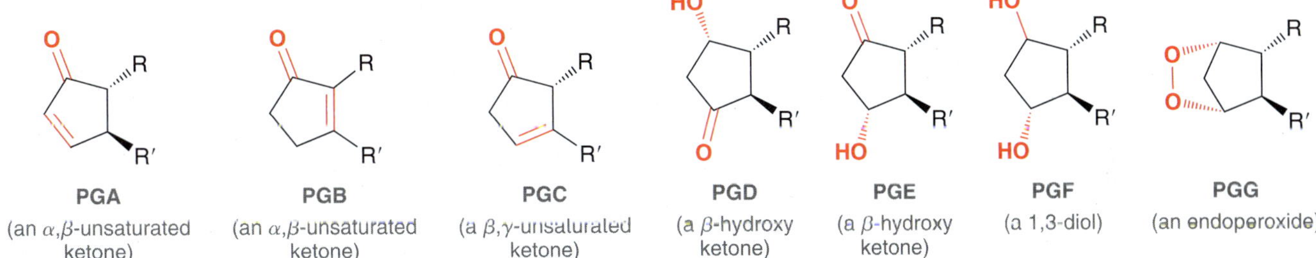


**FIGURE 26.9** Common substitution patterns for prostaglandins.

and the third letter indicates the substitution pattern. PGAs, PGBs, and PGCs all exhibit a carbonyl group and a carbon-carbon π bond in the five-membered ring. These three substitution patterns differ from each other only in the location of the carbon-carbon π bond. PGDs and PGEs are β-hydroxyketones, PGFs are 1,3-diols, and PGGs are endoperoxides. The number of carbon-carbon π bonds in the side chains is indicated with a subscript after the letter for the substitution pattern. For example, $PGE_2$ (dinoprostone) has the E substitution pattern and contains two double bonds in the side chains.

$PGE_2$

This particular prostaglandin regulates muscle contractions during labor and can be administered in larger doses to terminate pregnancies.

For the PGF substitution pattern, an additional descriptor is added to the name to indicate the configuration of the OH groups. A *cis* diol is designated as "α," while a *trans* diol is designated as "β."

$PGF_{2\alpha}$ $PGF_{2\beta}$

## CONCEPTUAL CHECKPOINT

**26.24** Classify each prostaglandin according to the instructions provided in Section 26.7.

(a) O, COOH, HO, OH

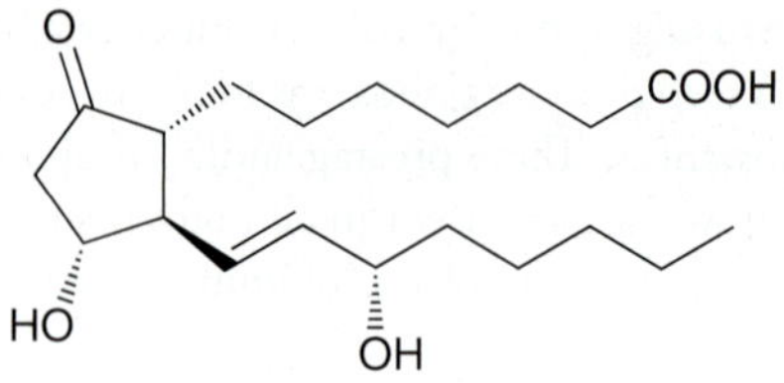

(b) HO, COOH, HO, OH

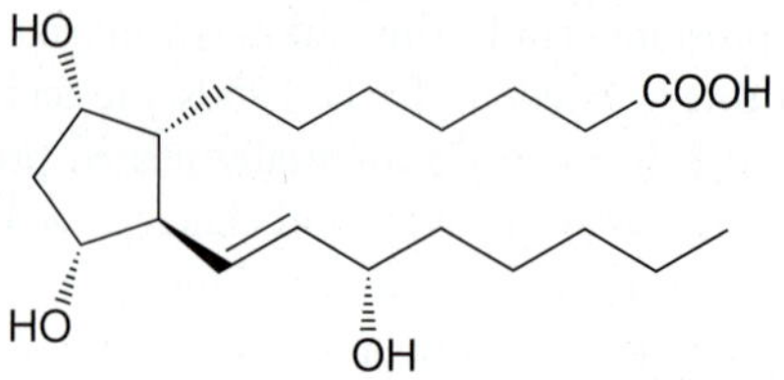

Prostaglandins are biochemical regulators that are even more powerful than steroids. Unlike hormones, which are produced in one location and then transported to another location in the body, prostaglandins are called local mediators because they perform their function where they are synthesized. Prostaglandins exhibit a wide array of biological activity, including the regulation of blood pressure, blood clotting, gastric secretions, inflammation, kidney function, and reproductive systems. Prostaglandins are biosynthesized from arachidonic acid with the help of enzymes called cyclooxygenases. A few key steps of this process are outlined in Mechanism 26.4.

## MECHANISM 26.4 BIOSYNTHESIS OF PROSTAGLANDINS FROM ARACHIDONIC ACID

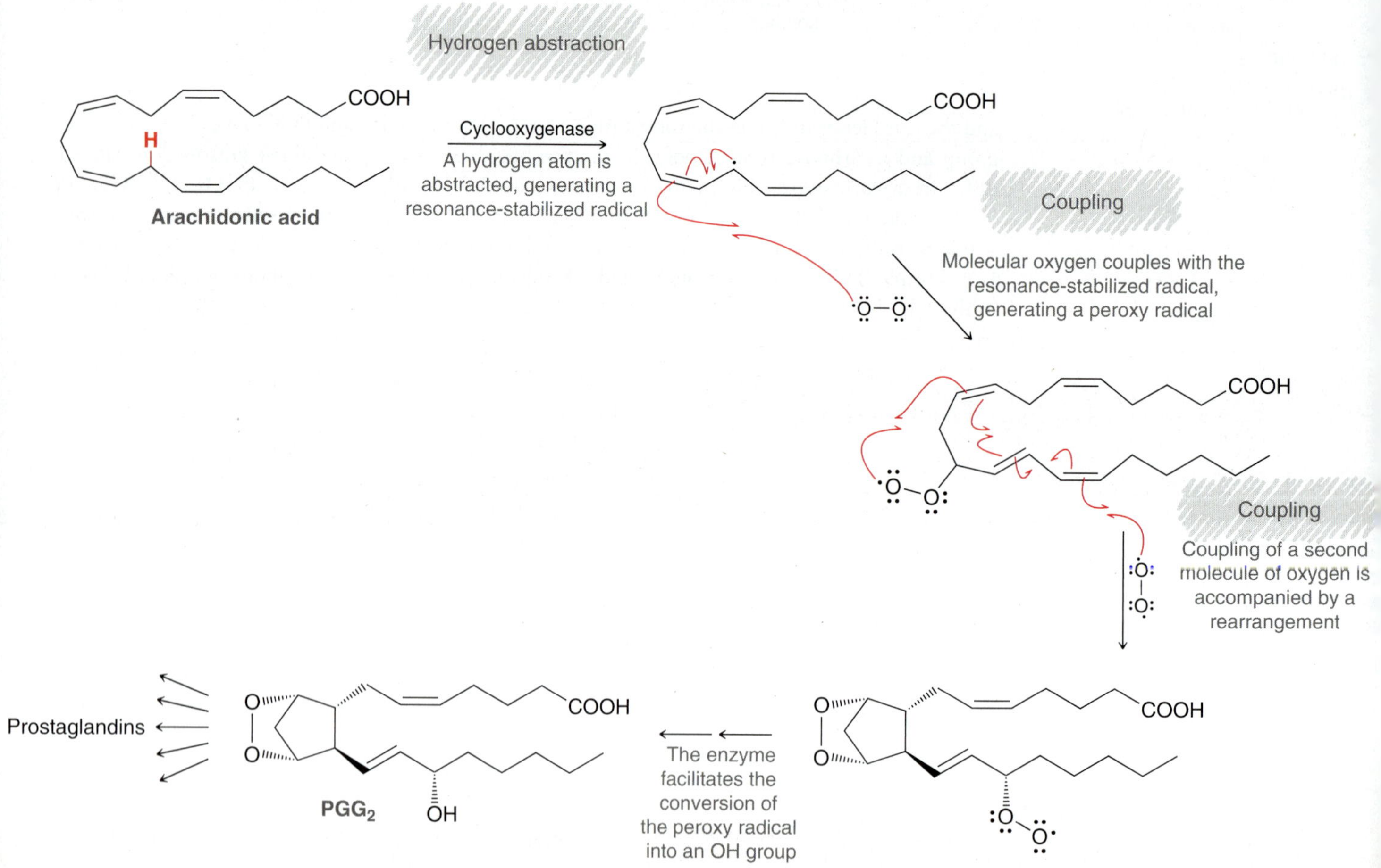

Hydrogen abstraction

COOH
H
**Arachidonic acid**

Cyclooxygenase →
A hydrogen atom is abstracted, generating a resonance-stabilized radical

COOH

Coupling

Molecular oxygen couples with the resonance-stabilized radical, generating a peroxy radical

·Ö–Ö·

COOH

Coupling

Coupling of a second molecule of oxygen is accompanied by a rearrangement

:Ö–Ö:

COOH

The enzyme facilitates the conversion of the peroxy radical into an OH group

COOH
**$PGG_2$**
OH

Prostaglandins

The release of arachidonic acid is stimulated in response to trauma (tissue damage). It is believed that the anti-inflammatory effects of adrenocortical steroids derive from their ability to suppress the enzymes that cause the release of arachidonic acid, thereby preventing the biosynthesis of prostaglandins.

## medically speaking

## NSAIDs and COX-2 Inhibitors

As previously mentioned, the action of anti-inflammatory steroids stems from their ability to prevent the release of arachidonic acid. There is another class of therapeutic agents that also exhibit anti-inflammatory properties, but their mode of action is entirely different. These drugs are called nonsteroidal anti-inflamma-tory drugs, or NSAIDs, and the most common examples are aspirin, ibuprofen, and naproxen.

Aspirin

Ibuprofen

Naproxen

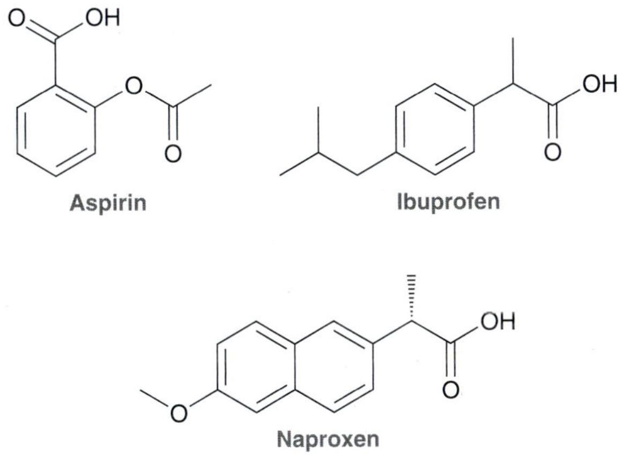

NSAIDs do not inhibit the enzymes that cause the release of arachidonic acid, but rather, they inhibit the cyclooxygenase enzymes that catalyze the conversion of arachidonic acid into prostaglandins. Ibuprofen and naproxen deactivate the cyclo-oxygenase enzymes by binding to them (thereby preventing the binding of arachidonic acid), while aspirin deactivates the cyclooxygenase enzymes by transferring an acetyl group to a serine residue within the active site of the enzymes.

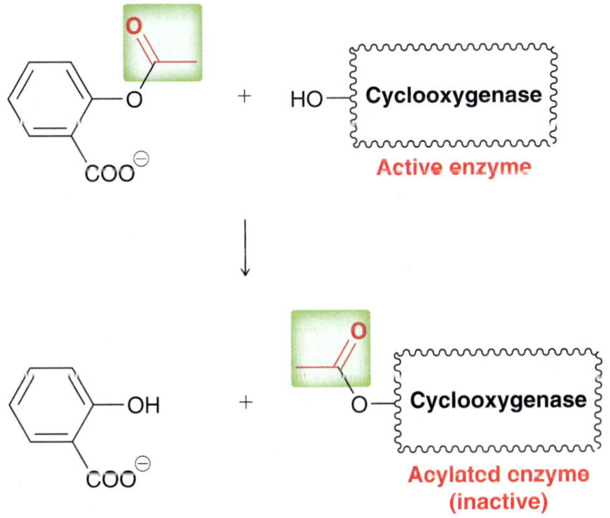

HO— Cyclooxygenase

Active enzyme

Cyclooxygenase

Acylated enzyme (inactive)

In this way, aspirin functions as an acetylating agent, which effectively deactivates the enzymes, thereby inhibiting the production of prostaglandins. With a decreased concentration of prostaglandins, the onset of inflammation is slowed, and fevers are reduced.

More recently, it has been discovered that there are two different kinds of cyclooxygenase enzymes, called COX-1 and COX-2. The primary function of COX-2 is to catalyze the synthesis of prostaglandins that cause inflammation and pain, while the primary function of COX-1 is to catalyze the synthe-sis of prostaglandins that help protect the stomach. NSAIDs inhibit the action of both COX-1 and COX-2 enzymes, and it has been discovered that the inhibition of COX-1 can result in gastric irritation. This understanding instigated an intensive search for therapeutic agents capable of selectively inhibit-ing COX-2 without also inhibiting COX-1. Extensive research efforts culminated in the release of several COX-2 inhibitors on the market in the late 1990s.

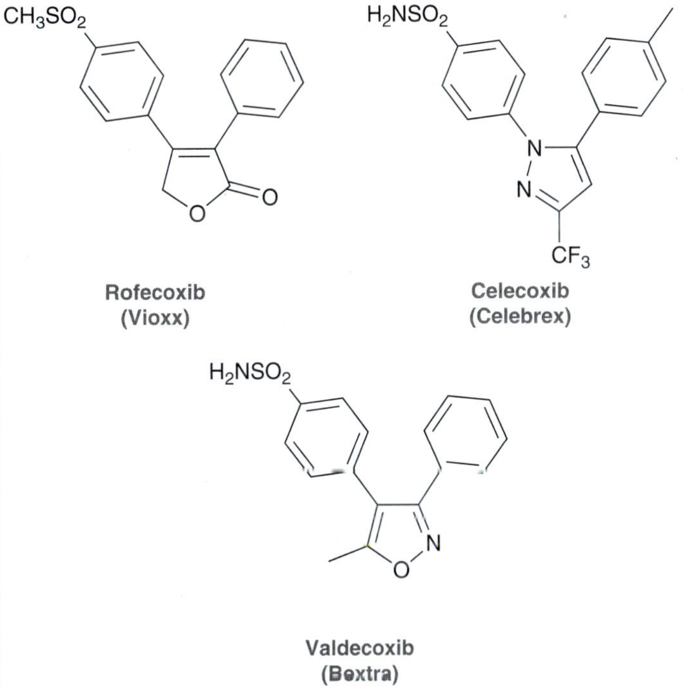

Rofecoxib (Vioxx)

Celecoxib (Celebrex)

Valdecoxib (Bextra)

Unfortunately, it was later discovered that many of these drugs caused an increased risk of heart attacks and strokes, especially in elderly patients. As a result, Vioxx and Bextra were removed from the market in 2004 and 2005, respectively.

Prostaglandins belong to a larger class of compounds called **eicosanoids**, which include leukotrienes, prostaglandins, thromboxanes, and prostacyclins, all of which are biosynthesized from arachidonic acid.

O
OH
Arachidonic Acid
$PGG_2$
Leukotrienes
Prostaglandins
Thromboxanes
Prostacyclins

These four classes of compounds exhibit a wide array of biological activity. In some cases, their biological functions oppose each other. For example, thromboxanes are generally vasoconstrictors that trigger blood clotting, while prostacyclins are generally vasodilators that inhibit blood clotting. Our bodies are dependent on the proper balance between the effects of these compounds.

## 26.8 Terpenes

**Terpenes** are a diverse class of naturally occurring compounds that share one feature in common. According to the isoprene rule, all terpenes can be thought of as being assembled from **isoprene** units, each of which contains five carbon atoms.

Isoprene

Consequently, the number of carbon atoms present in terpenes will be a multiple of five. The following examples have either 10 or 15 carbons:

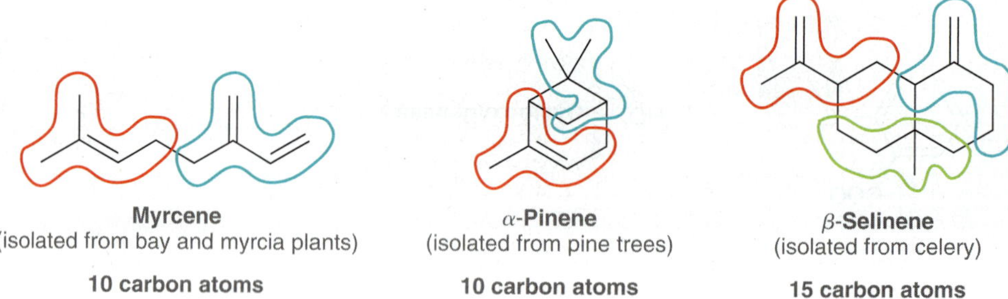

**Myrcene**
(isolated from bay and myrcia plants)
**10 carbon atoms**

**α-Pinene**
(isolated from pine trees)
**10 carbon atoms**

**β-Selinene**
(isolated from celery)
**15 carbon atoms**

A wide variety of terpenes are isolated from the essential oils of plants. Terpenes generally have a strong fragrance and are often used as flavorants and odorants in a wide variety of applications, including food products and cosmetics.

During the biosynthesis of terpenes, isoprene units are generally connected head to tail.

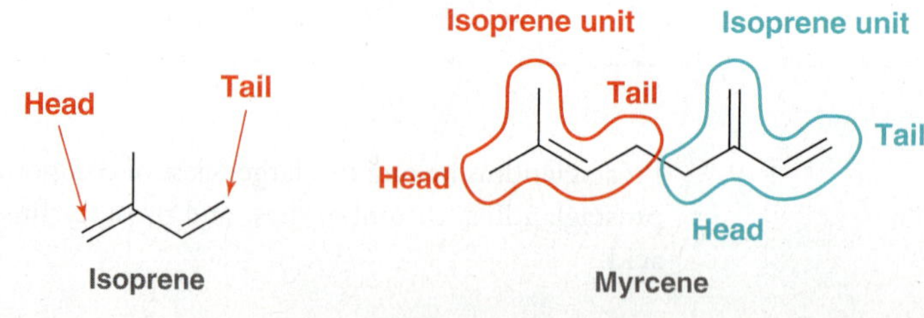

Many terpenes also contain functional groups:

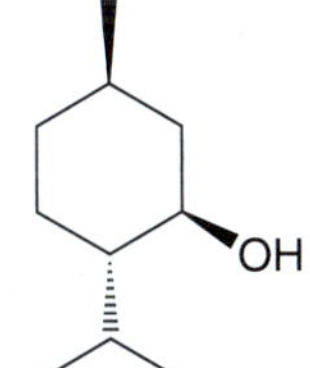


**Menthol**
**(isolated from peppermint oil)**

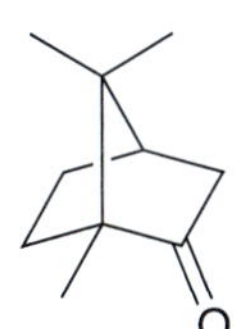


**Camphor**
**(isolated from evergreen trees)**

***R*-Carvone**
**(flavor of spearmint)**

## Classification of Terpenes

**TABLE 26.4 CLASSIFICATION OF TERPENES**

| CLASS | NO. OF CARBON ATOMS |
|---|---|
| Monoterpene | 10 |
| Sesquiterpene | 15 |
| Diterpene | 20 |
| Triterpene | 30 |
| Tetraterpene | 40 |

Terpenes are classified based on units of 10 carbon atoms (two isoprene units). For example, a terpene with 10 carbon atoms is called a monoterpene, while a terpene with 20 carbon atoms is called a diterpene. As seen in Table 26.4, a compound with 15 carbon atoms is called a sesquiterpene. An example of a sesquiterpene is α-farnesene, found in the waxy coating on apple skins.

**α-Farnesene**

Beta-carotene and lycopene each contain 40 carbon atoms and are therefore classified as tetraterpenes.

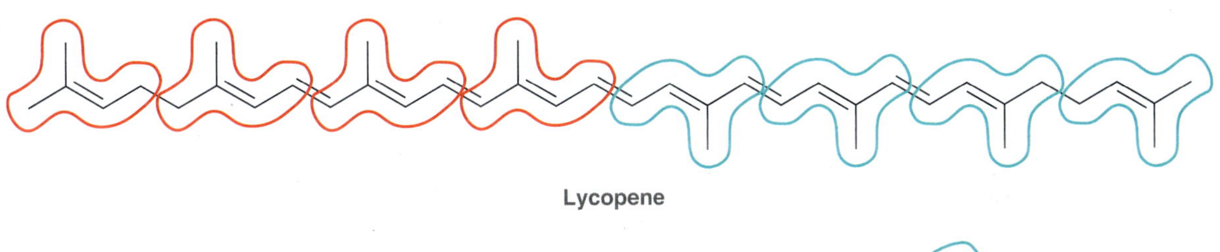

**Lycopene**

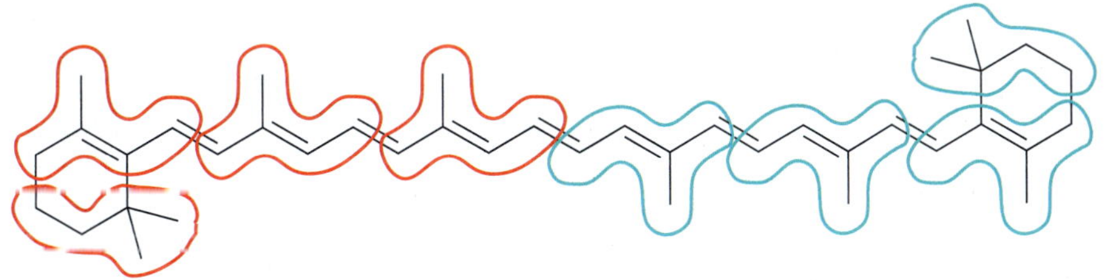

**β-Carotene**

Each of these compounds is assembled from two diterpenes.

# SKILLBUILDER

## 26.4 IDENTIFYING ISOPRENE UNITS IN A TERPENE

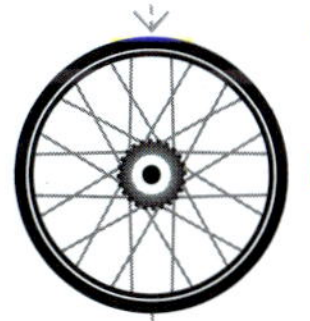

**LEARN the skill**

Identify the isoprene units in camphor:

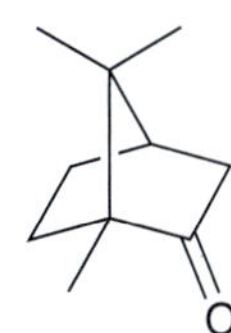

## SOLUTION

**STEP 1**
Count the carbon atoms and identify the number of isoprene units.

First count the number of carbon atoms. This compound is a monoterpene, because it has 10 carbon atoms. Therefore, we are looking for two isoprene units. Each isoprene unit may or may not have double bonds, but there must be four carbon atoms in a straight chain with exactly one branch.

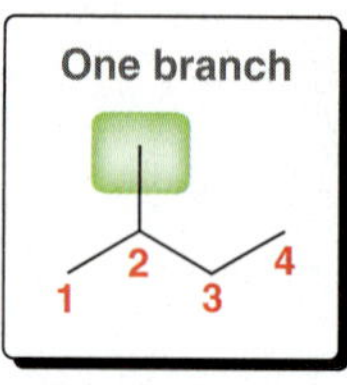


Two branches — NOT an isoprene unit

No branches — NOT an isoprene unit

**STEP 2**
Look for methyl groups.

To identify the isoprene units, it is best to focus on any methyl groups as well as the carbon atoms to which they are attached. For example, the following carbon atoms must be grouped together:

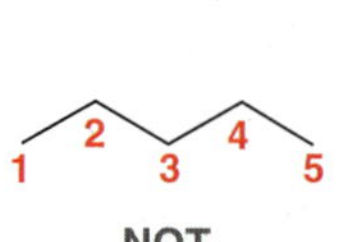


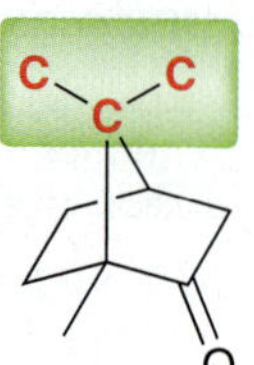


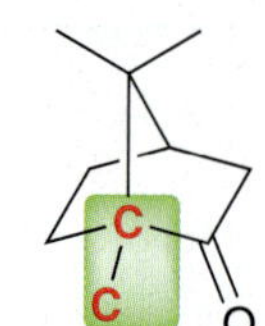


We now consider whether it is possible for the group of five carbon atoms to represent one isoprene unit. Are these five carbon atoms connected in a way that gives four carbon atoms in a straight chain with one branch?

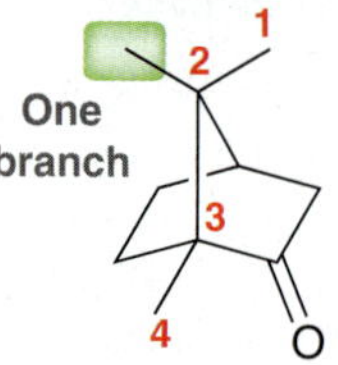


Indeed, these five carbon atoms could be an isoprene unit because they have the correct branching pattern. However, when we analyze the remaining five carbon atoms, they do not have the correct branching pattern.

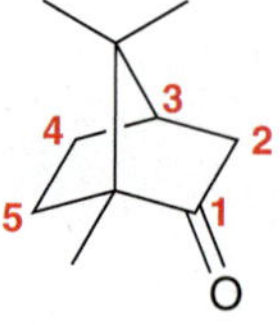


**STEP 3**
Use trial and error to find the isoprene units.

The remaining five carbon atoms cannot represent an isoprene unit because they do not exhibit a branch. Therefore, the original five carbon atoms that we first identified must in fact be part of two separate isoprene units.

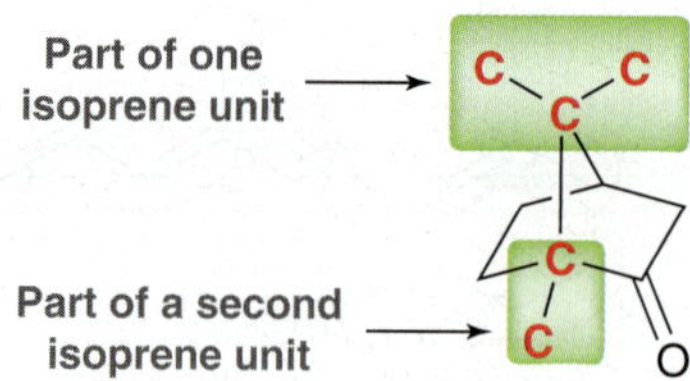


Finding each isoprene unit sometimes requires a little bit of trial and error. In this case, there are two acceptable solutions.

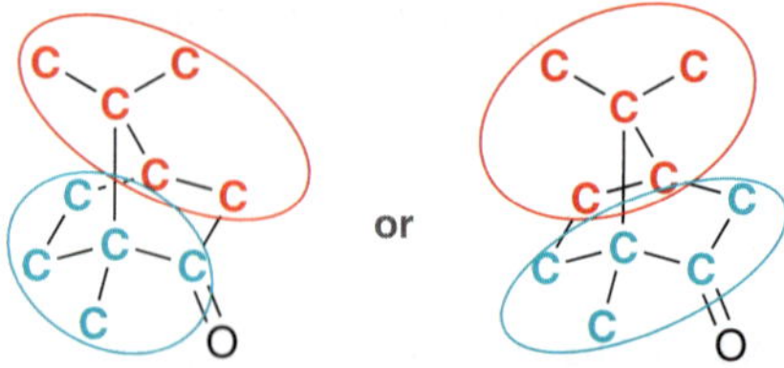

**PRACTICE the skill** **26.25** Circle the isoprene units in each of the following compounds:

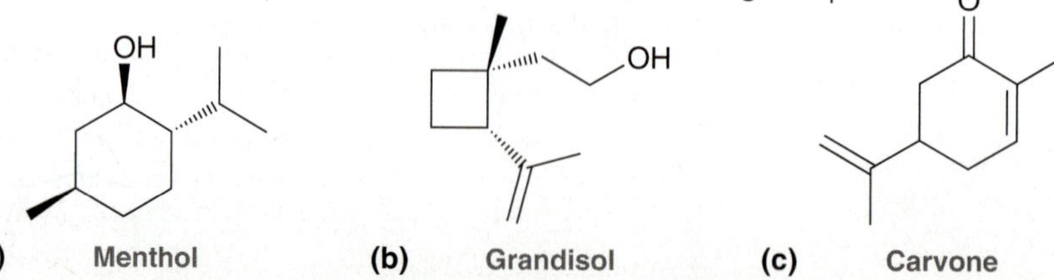

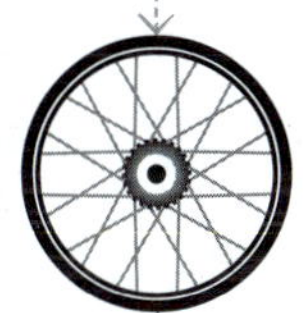

**APPLY the skill**

**26.26** Determine whether each of the following compounds is a terpene:

(a) OH (b) OH

(c) OH (d) OH

need more **PRACTICE?** Try Problem 26.55

## Biosynthesis of Terpenes

Although the isoprene rule considers terpenes to be constructed from isoprene units, the actual building blocks are dimethylallyl pyrophosphate and isopentenyl pyrophosphate.

OPP OPP

**Dimethylallyl pyrophosphate** **Isopentenyl pyrophosphate**

All terpenes are biosynthesized from these two starting materials. In each of these compounds, the OPP group represents a biological leaving group, called pyrophosphate. OPP is a good leaving group because it is a weak base.

**LOOKING BACK**
For a review of the connection between leaving groups and basicity, see Section 7.8.

R—OPP ≡ R—O—P(=O)(O⁻)—O—P(=O)(O⁻)—O⁻ ⟶ $R^{\oplus}$ + ⁻O—P(=O)(O⁻)—O—P(=O)(O⁻)—O⁻

**Pyrophosphate (good leaving group)** **Weak base**

The reaction between dimethylallyl pyrophosphate and isopentenyl pyrophosphate yields a monoterpene called geranyl pyrophosphate, which is the starting material for all other monoterpenes (Mechanism 26.5).

## MECHANISM 26.5 BIOSYNTHESIS OF GERANYL PYROPHOSPHATE

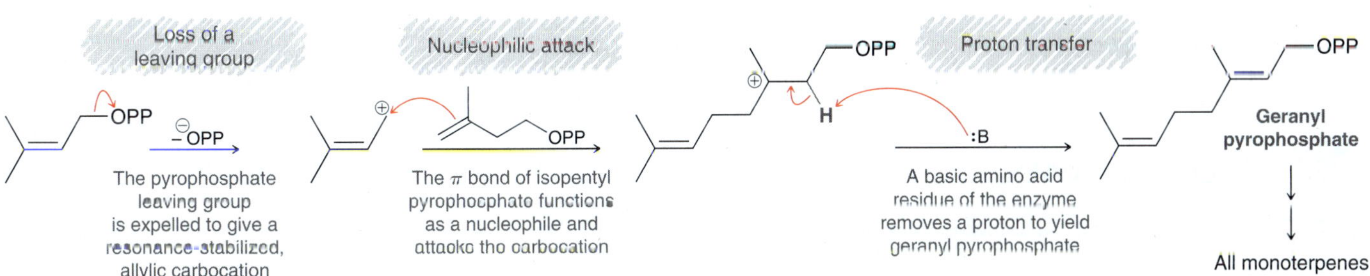


The biosynthesis of geranyl pyrophosphate is achieved in just three steps. The first step involves loss of a leaving group (pyrophosphate) to generate a resonance-stabilized carbocation. The second step is a nucleophilic attack in which the carbocation is attacked by the π bond of isopentyl pyrophosphate to generate a new carbocation. Finally, a proton transfer yields geranyl pyrophosphate.

The same three mechanistic steps can be repeated to add another isoprene unit to geranyl pyrophosphate, yielding the sesquiterpene farnesyl pyrophosphate. Once again, loss of a leaving group is followed by nucleophilic attack and then a proton transfer (Mechanism 26.6). Farnesyl pyrophosphate is the starting material for all other sesquiterpenes and diterpenes.

## MECHANISM 26.6 BIOSYNTHESIS OF FARNESYL PYROPHOSPHATE

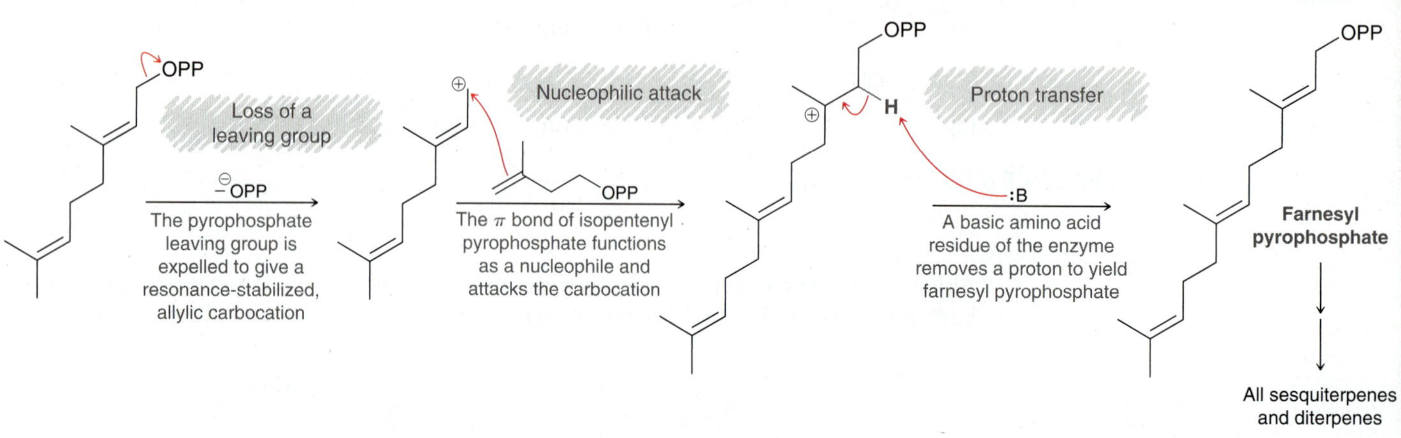


Squalene, the biological precursor for all steroids (as seen in Mechanism 26.3), is biosynthesized from the coupling of two molecules of farnesyl pyrophosphate.

OPP + PPO

**Farnesyl pyrophosphate** **Farnesyl pyrophosphate**

**Squalene**

See Mechanism 26.3

**All steroids**

## CONCEPTUAL CHECKPOINT

**26.27** Draw a mechanism for the following transformation:

OPP → (OPP) → OPP

**26.28** Draw a mechanism for the biosynthesis of α-farnesene starting with dimethylallyl pyrophosphate and isopentenyl pyrophosphate.

**α-Farnesene**

# REVIEW OF REACTIONS SYNTHETICALLY USEFUL REACTIONS

## Reactions of Triglycerides

### Hydrogenation (production of margarine)

$\xrightarrow[\text{Catalyst}]{H_2}$

### Saponification (production of soap)

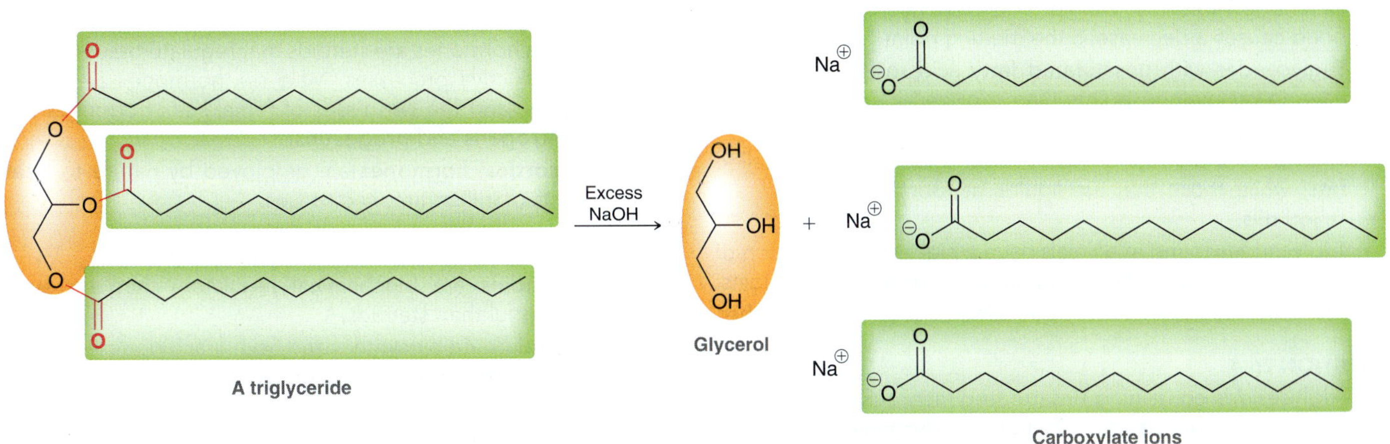


### Transesterification (production of biodiesel)

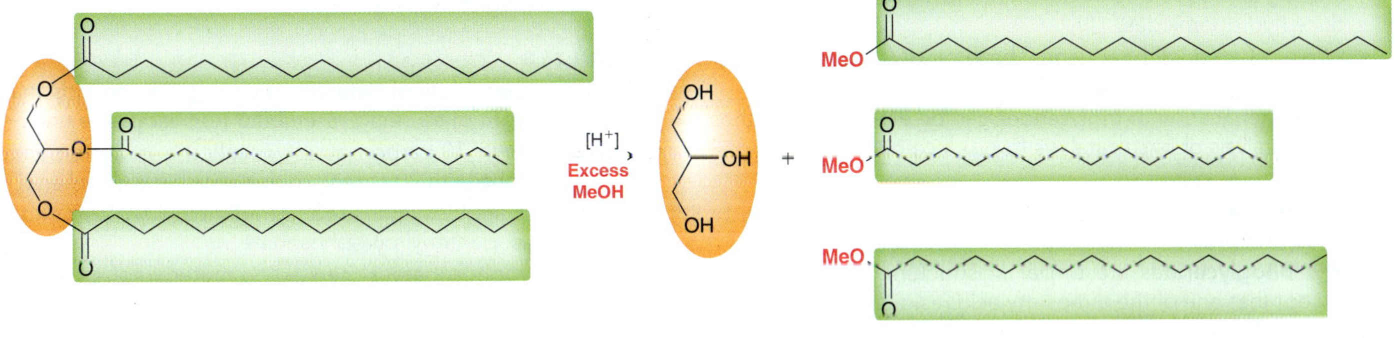

## REVIEW OF CONCEPTS AND VOCABULARY

### SECTION 26.1

- **Lipids** are naturally occurring compounds that are extracted from cells using nonpolar solvents.
- **Complex lipids** readily undergo hydrolysis, while **simple lipids** do not undergo hydrolysis.

### SECTION 26.2

- **Waxes** are high-molecular-weight esters that are constructed from carboxylic acids and alcohols.

### SECTION 26.3

- **Triglycerides** are the triesters formed from glycerol and three long-chain carboxylic acids, called **fatty acids**. The resulting triglyceride is said to contain three *fatty acid residues*.
- For saturated fatty acids, the melting point increases with increasing molecular weight. The presence of a *cis* double bond causes a decrease in the melting point.
- Triglycerides with unsaturated fatty acid residues generally have lower melting points than triglycerides with saturated fatty acid residues.
- Triglycerides that are solids at room temperature are called **fats**, while those that are liquids at room temperature are called **oils**.
- Triglycerides found in animals have a higher concentration of saturated fatty acids than triglycerides from vegetable origins.
- The exact ratio of fatty acids differs from one source to another.

### SECTION 26.4

- Triglycerides containing unsaturated fatty acid residues will undergo hydrogenation. During the hydrogenation process, some of the double bonds can isomerize to give *trans* $\pi$ bonds.
- In the presence of molecular oxygen, triglycerides are particularly susceptible to oxidation at the allylic position to produce hydroperoxides.
- When treated with aqueous base, triglycerides undergo hydrolysis, also called saponification.
- Transesterification of triglycerides can be achieved via either acid catalysis or base catalysis to produce biodiesel.

### SECTION 26.5

- **Phospholipids** are esterlike derivatives of phosphoric acid.
- **Phosphoglycerides** are similar in structure to triglycerides except that one of the three fatty acid residues is replaced by a phosphoester group.
- The simplest kind of phosphoglyceride is a phosphoric monoester, called a **phosphatidic acid**.
- Phosphoglycerides that contain ethanolamine are called **cephalins,** while phosphoglycerides that contain choline are called **lecithins**.
- In water, phosphoglycerides will self-assemble to form a **lipid bilayer**.

### SECTION 26.6

- The structures of **steroids** are based on a tetracyclic ring system, involving three six-membered rings and one five-membered ring.
- The ring fusions are all *trans* in most steroids, giving steroids their rigid geometry.
- All steroids, including cholesterol, are biosynthesized from squalene.
- Human sex hormones are steroids that regulate tissue growth and reproductive processes. Male sex hormones are called **androgens**, and the two types of female sex hormones are called **estrogens** and **progestins**.
- **Adrenocortical hormones** are employed by nature to treat the effects of inflammatory diseases.

### SECTION 26.7

- **Prostaglandins** are biochemical regulators that are even more powerful than steroids.
- Prostaglandins are biosynthesized from arachidonic acid with the help of enzymes called cyclooxygenases.
- Prostaglandins contain 20 carbon atoms and are characterized by a five-membered ring with two side chains.
- Prostaglandins belong to a larger class of compounds called **eicosanoids**, which include leukotrienes, prostaglandins, thromboxanes, and prostacyclins, all of which exhibit a wide array of biological activity.

### SECTION 26.8

- **Terpenes** are a class of naturally occurring compounds that can be thought of as being assembled from **isoprene** units.
- A terpene with 10 carbon atoms is called a monoterpene, while a terpene with 20 carbon atoms is called a diterpene.
- All terpenes are biosynthesized from dimethylallyl pyrophosphate and isopentenyl pyrophosphate.
- Geranyl pyrophosphate is the starting material for all monoterpenes, and farnesyl pyrophosphate is the starting material for all sesquiterpenes.

# SKILLBUILDER REVIEW

## 26.1 COMPARING MOLECULAR PROPERTIES OF TRIGLYCERIDES

**STEP 1** Identify the fatty acid residues in each triglyceride.

Palmitoleic acid residues

Tripalmitolein

Palmitic acid residues

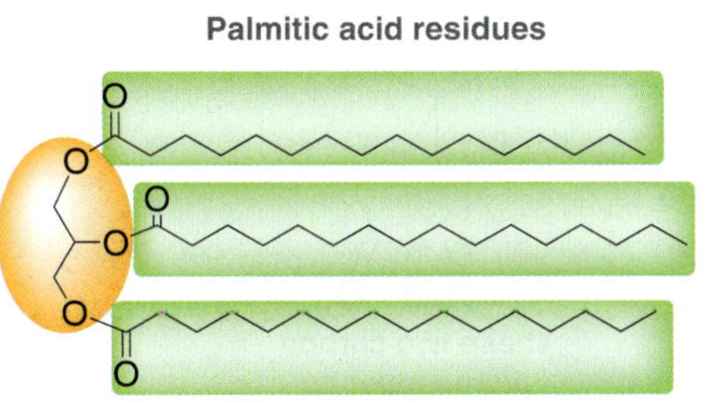

Tripalmitin

**STEP 2** Compare the residues, keeping in mind that the melting point is affected by:

1) Length of the chain (longer chain = higher melting point).

2) Saturated residues (no C—C $\pi$ bonds) have higher melting points.

Try Problems 26.3–26.6, 26.40a,d,e, 26.41a,d,e, 26.43

## 26.2 IDENTIFYING THE PRODUCTS OF TRIGLYCERIDE HYDROLYSIS

**STEP 1** Identify the three ester groups.

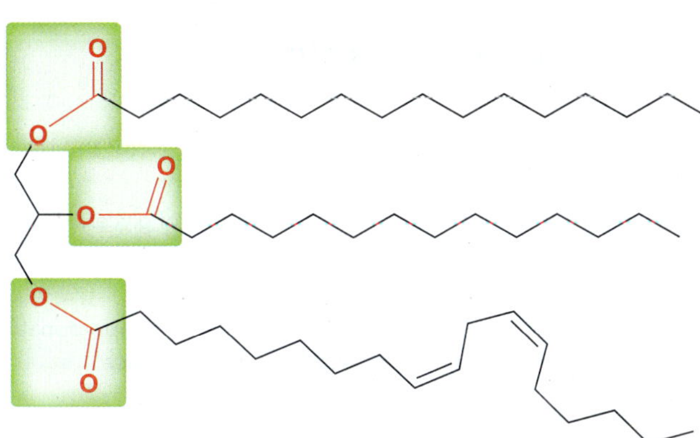

**STEP 2** Each ester group is hydrolyzed, generating glycerol and three carboxylate ions.

Try Problems 26.9–26.11, 26.30b, 26.40c, 26.41c

## 26.3 DRAWING A MECHANISM FOR TRANSESTERIFICATION OF A TRIGLYCERIDE

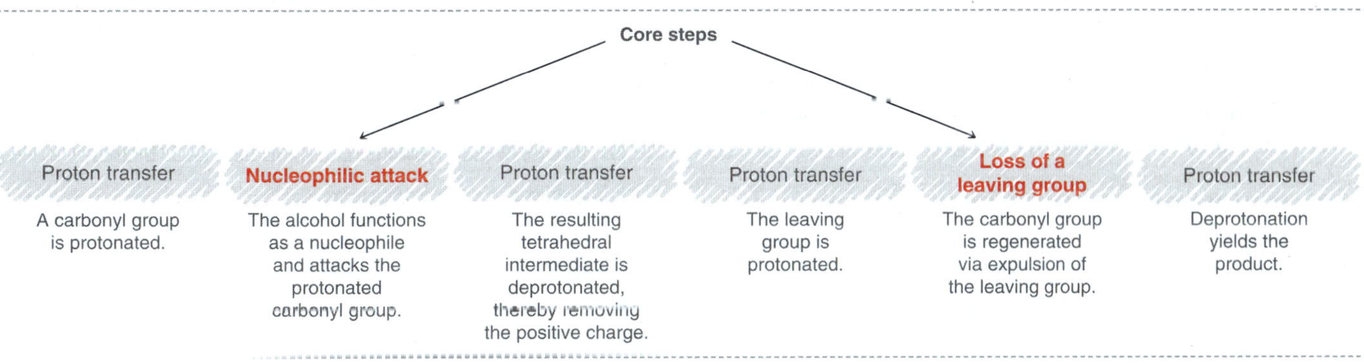


Try Problems 26.12–26.14, 26.44, 26.45

**26.4** IDENTIFYING ISOPRENE UNITS IN A TERPENE

**STEP 1** Count the number of carbon atoms in order to identify the number of isoprene units.

10 Carbon atoms = 2 Isoprene units

**STEP 2** Look for any methyl groups and the carbon atoms to which they are attached.

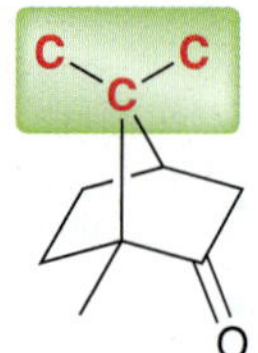


These three carbon atoms must be grouped together

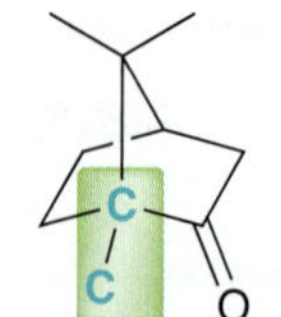


These two carbon atoms must be grouped together

**STEP 3** Using trial and error, identify the isoprene units that have the correct branching structure.

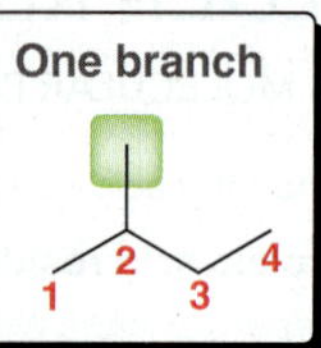


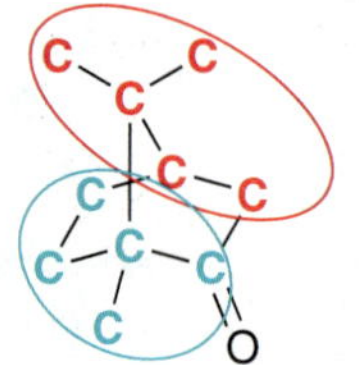

or

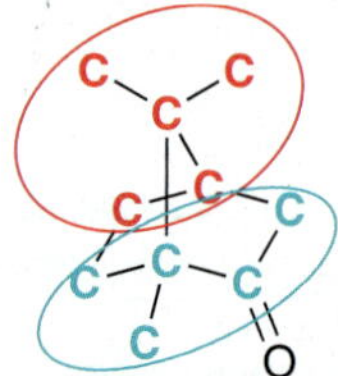

Try Problems **26.25, 26.26, 26.55**

## PRACTICE PROBLEMS

Note: Most of the Problems are available within **WileyPLUS**, an online teaching and learning solution.

**26.29** Identify each of the following compounds as a wax, triglyceride, phospholipid, steroid, prostaglandin, or terpene:

(a) Stanozolol
(b) Lycopene
(c) Tristearin
(d) Distearoyl lecithin
(e) $PGF_2$
(f) Pentadecyl octadecanoate

**26.30** Predict the product(s) formed when tripalmitolein is treated with each of the following reagents:

(a) Excess $H_2$, Ni

(b) Excess NaOH, $H_2O$

**26.31** Draw two different cephalins that contain one lauric acid residue and one myristic acid residue. Are both of these compounds chiral?

**26.32** Draw an optically active triglyceride that contains one palmitic acid residue and two myristic acid residues. Will this compound react with molecular hydrogen in the presence of a catalyst?

**26.33** Which of the following compounds are lipids?

(a) L-Threonine
(b) 1-Octanol
(c) Lycopene
(d) Trimyristin
(e) Palmitic acid
(f) D-Glucose
(g) Testosterone
(h) D-Mannose

**26.34** Draw the structure of *trans*-oleic acid.

**26.35** Tristearin is less susceptible to becoming rancid than triolein. Explain.

**26.36** Arrange the following compounds in order of increasing water solubility:

(a) A triglyceride constructed from one equivalent of glycerol and three equivalents of myristic acid

(b) A diglyceride constructed from one equivalent of glycerol and two equivalents of myristic acid

(c) A monoglyceride constructed from one equivalent of glycerol and one equivalent of myristic acid

**26.37** Identify whether hexane or water would be more appropriate for extracting terpenes from plant tissues. Explain your choice.

**26.38** Identify each of the following fatty acids as saturated or unsaturated:

(a) Palmitic acid
(b) Myristic acid
(c) Oleic acid
(d) Lauric acid
(e) Linoleic acid
(f) Arachidonic acid

**26.39** Which of the fatty acids in the previous problem has four carbon-carbon double bonds?

**26.40** Which of the following statements applies to triolein?

(a) It is a solid at room temperature.
(b) It is unreactive toward molecular hydrogen in the presence of Ni.
(c) It undergoes hydrolysis to produce unsaturated fatty acids.
(d) It is a complex lipid.
(e) It is a wax.
(f) It has a phosphate group.

**26.41** Identify which of the following statements applies to tristearin:

(a) It is a solid at room temperature.
(b) It is unreactive toward molecular hydrogen in the presence of Ni.
(c) It undergoes hydrolysis to produce unsaturated fatty acids.
(d) It is a complex lipid.
(e) It is a wax.
(f) It has a phosphate group.

**26.42** One of the compounds present in carnauba wax was isolated, purified, and then treated with aqueous sodium hydroxide to yield an alcohol with 30 carbon atoms and a carboxylate ion with 20 carbon atoms. Draw the likely structure of the compound.

**26.43** Draw the structures of trimyristin and tripalmitin and determine which is expected to have the lower melting point. Explain your choice.

**26.44** Draw a mechanism for the transesterification of trimyristin using excess isopropanol in the presence of an acid catalyst.

**26.45** Draw a mechanism for the base-catalyzed transesterification of trimyristin using ethanol in the presence of sodium hydroxide.

**26.46** Draw the structure of an optically inactive triglyceride that contains two oleic acid residues and one palmitic acid residue.

**26.47** Draw the structure of an optically active triglyceride that contains two oleic acid residues and one palmitic acid residue.

**26.48** Draw the enantiomer of cholesterol.

**26.49** Circle the isoprene units in each of the following compounds:

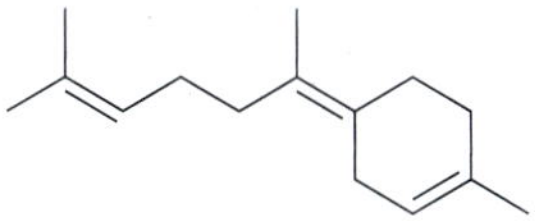

(a) **Bisabolene** 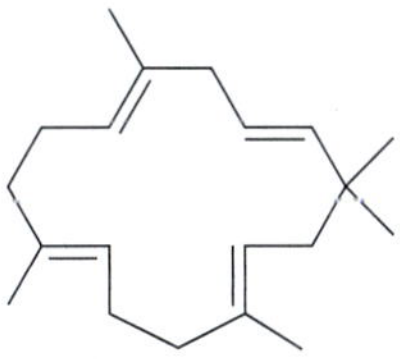 (b) **Flexibilene**

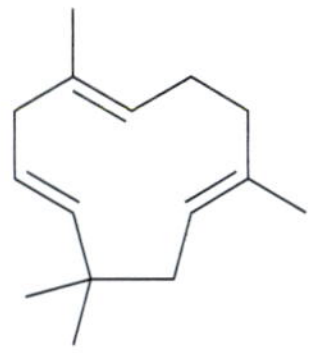

(c) **Humulene** 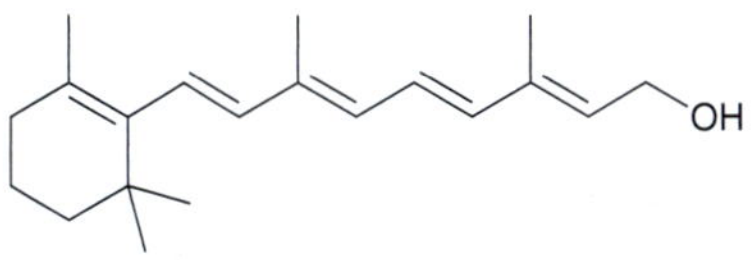 (d) **Vitamin A**

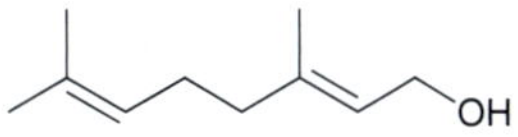

(e) **Geraniol** 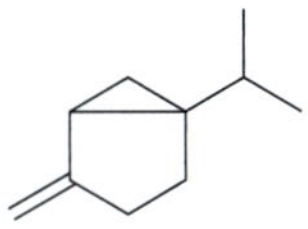 (f) **Sabinene**

**26.50** Sphingomyelins are lipids with the following general structure:

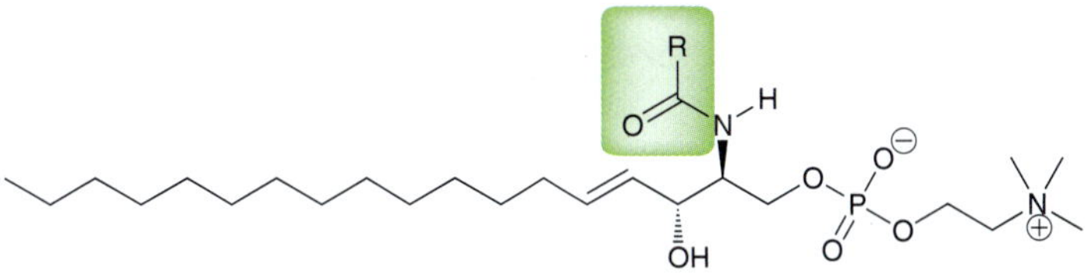

**Fatty acid residue**

**Sphingomyelins**

(a) Identify the polar head and all hydrophobic tails in sphingomyelins.
(b) Do sphingomyelins have the appropriate structural features and three-dimensional geometry necessary to be major constituents of lipid bilayers?

# INTEGRATED PROBLEMS

**26.51** Treatment of cholesterol with MCPBA could potentially produce two diastereomeric epoxides.

(a) Draw both diastereomeric epoxides.
(b) Only one of these epoxides is formed. Predict which one and explain why the other is not formed.

**26.52** Identify the products expected when estradiol is treated with each of the following reagents:

(a) Excess $Br_2$
(b) PCC
(c) A strong base followed by excess ethyl iodide
(d) Excess acetyl chloride in the presence of pyridine

**26.53** Olestra is a noncaloric oil substitute that is produced by esterifying sucrose with eight equivalents of fatty acids obtained from the hydrolysis of vegetable oils. The eight fatty acid residues give Olestra the consistency and flavor of cooking oil, but the steric bulk of the compound prevents digestive enzymes from hydrolyzing the ester groups. As a result, Olestra passes through the digestive tract unaltered. Draw the structure of a molecule of Olestra that contains eight lauric acid residues. Is this compound chiral?

**26.54** Identify the reagents you would use to convert oleic acid into each of the following compounds:

(a) Stearic acid
(b) Ethyl stearate
(c) 1-Octadecanol
(d) Nonanedioic acid
(e) 2-Bromostearic acid

**26.55** Limonene is an optically active compound isolated from the peels of lemons and oranges.

(a) Is limonene a monoterpene or a diterpene?
(b) Treatment of limonene with excess HBr yields a compound with molecular formula $C_{10}H_{18}Br_2$. Identify the structure of this compound and determine whether it is chiral or achiral.
(c) Draw the products obtained when limonene is treated with $O_3$ followed by DMS.

**Limonene**

# CHALLENGE PROBLEMS

**26.56** Starting with the following compound and using any other reagents of your choice, outline a synthesis for trimyristin:

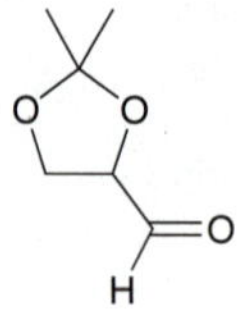

**26.57** The following compound was isolated from nerve cells:

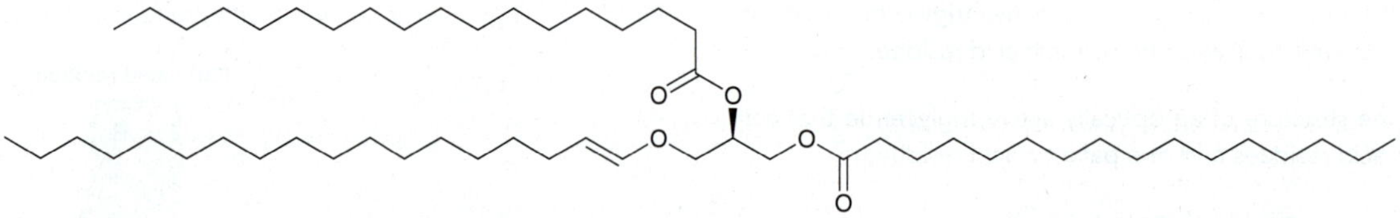

(a) Describe how this compound differs in structure from fats and oils.

(b) Three products are obtained when this compound is hydrolyzed with aqueous sodium hydroxide. Draw the structures of all three products.

(c) Four products are obtained when this compound is hydrolyzed with aqueous acid. Draw the structures of all four products.

# 27

# Synthetic Polymers



## DID YOU EVER WONDER...

### how bulletproof glass works?

Bulletproof glass, such as that used in armored vehicles, consists of several sheets of glass sandwiched between transparent polymer sheets. The resulting material is comprised of many layers, and the outer layers are capable of absorbing the impact of the bullet, thereby preventing it from penetrating the inner layers. The ability of the glass to resist bullets depends on many factors, including the type of polymer used, the thickness of the glass, the number of layers sandwiched together, the type of bullet, and the range from which the bullet is fired. For this reason, the glass is generally described as "bullet resistant" rather than "bulletproof."

This chapter will focus on the preparation, classification, and properties of a variety of synthetic polymers, including those found in bullet-resistant glass as well as a large variety of other applications.

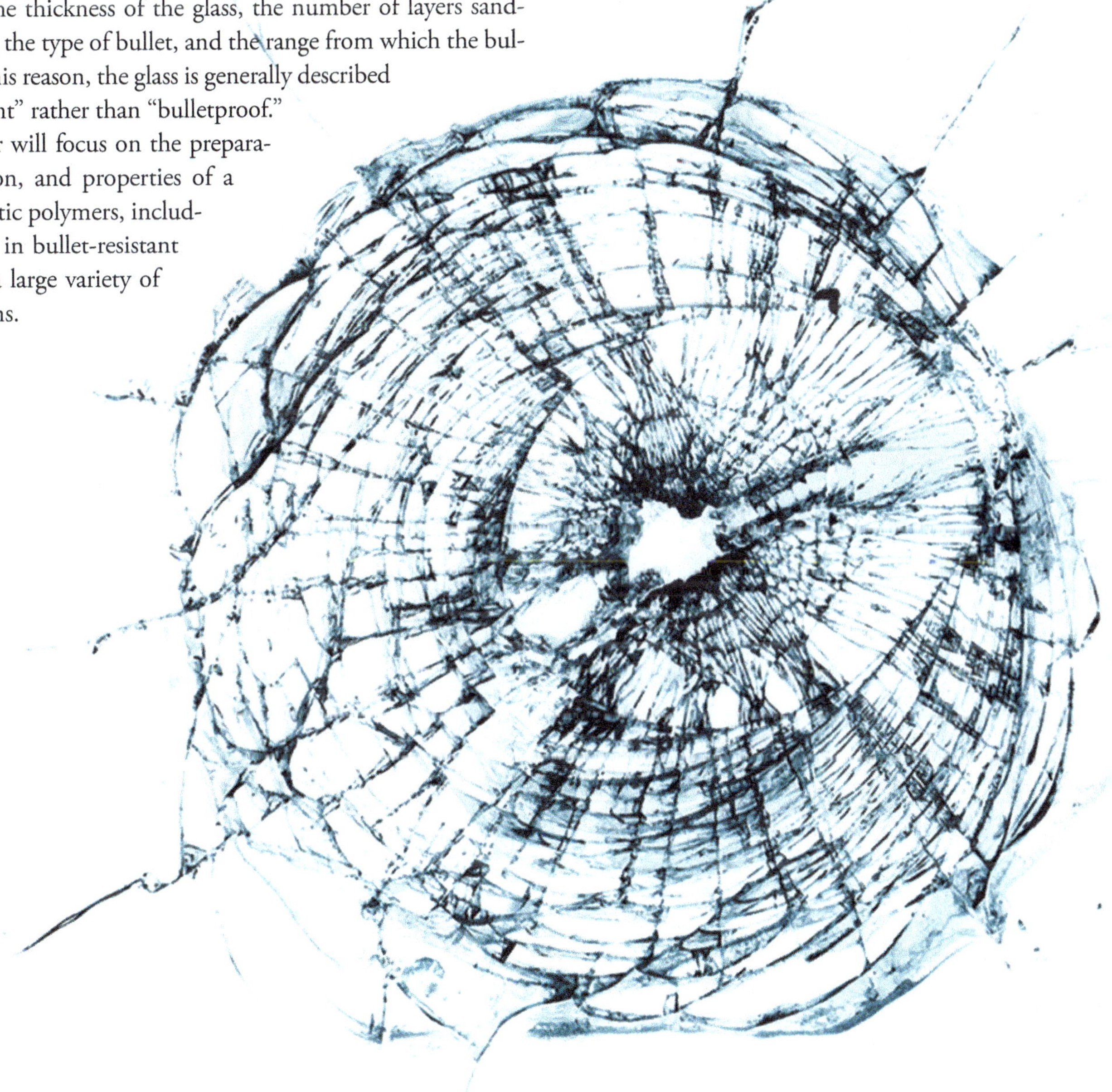

## DO YOU REMEMBER?

*Before you go on, be sure you understand the following topics.*
*If necessary, review the suggested sections to prepare for this chapter:*

- Sources and Uses of Alkanes: An Introduction to Polymers (Section 4.5)
- Cationic Polymerization and Polystyrene (Section 9.3)
- Radical Polymerization (Section 11.11)

Take the DO YOU REMEMBER? QUIZ in **WileyPLUS** to check your understanding

## 27.1 Introduction to Synthetic Polymers

Over the past century, our society has become heavily reliant on polymers. Coffee cups, soft-drink bottles, synthetic fabrics for clothing, CDs and DVDs, trash bags, artificial heart valves, and automobile parts are all manufactured from polymers. Polymers represent a multibillion-dollar industry, with more than 50 trillion pounds of synthetic polymers being manufactured each year in the United States alone.

Recall that polymers are comprised of repeating units that are constructed by joining monomers together.

Repeating units

Monomers

Polymer

Polymers are drawn most efficiently by placing brackets around the repeating unit, where the subscript "*n*" indicates that the polymer is constructed from a large number of repeating units. Polymers have been discussed many times throughout this textbook, as indicated in Table 27.1.

**TABLE 27.1 REFERENCES TO PRIOR DISCUSSIONS OF POLYMERS IN THIS TEXTBOOK**

| TOPIC | SECTION |
|---|---|
| Introduction to polymers | 4.5 |
| Cationic polymerization | 9.3 |
| Conducting organic polymers | 10.1 |
| Radical polymerization | 11.11 |
| Polymers from dienes | 17.5 |
| Polyesters and polyamides | 21.12 |
| Polysaccharides | 24.8 |
| *N*-Glycosides (DNA and RNA) | 24.10 |
| Proteins | 25.1 |

Polymers can be divided into two major categories: synthetic polymers and biopolymers. The former are prepared by scientists in the laboratory or in factories, while the latter are produced by living organisms. Biopolymers include polysaccharides, DNA, RNA, and proteins, all of which were discussed in the previous chapters. This chapter will focus exclusively on synthetic polymers.

## 27.2 Nomenclature of Synthetic Polymers

IUPAC has established rules for assigning systematic names to polymers, but these names are rarely used by scientists. An alternative and more common system, also recognized by IUPAC, involves naming the polymer based on the monomers from which it was derived, as shown in the following examples:

| Propylene | → | Polypropylene | Styrene | → | Polystyrene |
|---|---|---|---|---|---|
| $CH_2=CH(Me)$ | → | $[CH_2-CH(Me)]_n$ | $CH_2=CH(Ph)$ | → | $[CH_2-CH(Ph)]_n$ |

When writing the name of a polymer whose monomer contains two words, parentheses are used around the name of the monomer. Thus, polymerization of vinyl chloride yields poly(vinyl chloride).

Polymers are also often called by their trade name, such as Teflon and Kevlar. Table 27.2 provides a list of many common polymers and their uses.

**TABLE 27.2 SEVERAL COMMON POLYMERS AND THEIR USES**

| NAME OF POLYMER | MONOMER STRUCTURE | POLYMER STRUCTURE | USES |
|---|---|---|---|
| Polyethylene | $H_2C=CH_2$ | $[CH_2-CH_2]_n$ | Bottles and trash bags |
| Polypropylene | $H_2C=CH(CH_3)$ | $[CH_2-CH(CH_3)]_n$ | Carpet fibers, appliances, car tires |
| Polyisobutylene | $H_2C=C(CH_3)_2$ | $[CH_2-C(CH_3)_2]_n$ | Caulks and sealants, bicycle inner tubes, basketballs |
| Polystyrene | $H_2C=CH(Ph)$ | $[CH_2-CH(Ph)]_n$ | Foam insulation, televisions, radios |
| Poly(vinyl chloride) | $H_2C=CH(Cl)$ | $[CH_2-CH(Cl)]_n$ | Water pipes, vinyl plastics |
| Poly(methyl-α-cyanoacrylate) | $H_2C=C(CO_2CH_3)(CN)$ | $[CH_2-C(C(=O)OMe)(CN)]_n$ | Superglue |
| Polytetrafluoroethylene | $F_2C=CF_2$ | $[CF_2-CF_2]_n$ | Nonstick coating for frying pans (Teflon) |

## CONCEPTUAL CHECKPOINT

**27.1** Draw and name the polymer that results when each of the following monomers undergoes polymerization:

(a) **Vinyl acetate** (b) **Vinyl bromide**

(c) **$\alpha$-Butylene**

**27.2** Sodium polyacrylate is a synthetic polymer used in diapers, because of its ability to absorb several hundred times its own mass of water. This extraordinary polymer is made from another polymer, poly(methyl acrylate), via hydrolysis of the ester groups. Draw and name the monomer used to make poly(methyl acrylate).

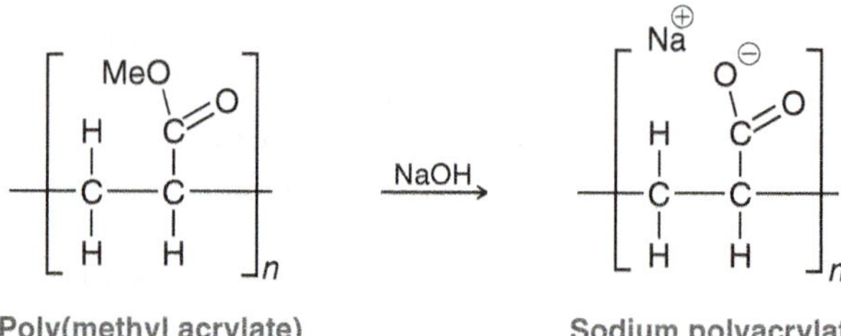


**Poly(methyl acrylate)** **Sodium polyacrylate**

## 27.3 Copolymers

The polymers discussed thus far are all called homopolymers. A **homopolymer** is a polymer constructed from a single type of monomer. In contrast, polymers constructed from two or more different types of monomers are called **copolymers**. A common example of a copolymer is Saran, which is made from vinyl chloride and vinylidene chloride.

**Vinyl chloride** + **Vinylidene chloride** ⟶ **Saran**

Copolymers often have different properties than either of their corresponding homopolymers. A list of some common copolymers and their applications is presented in Table 27.3. Copolymers are often classified based on the order in which the monomers are joined together. **Alternating copolymers** contain an alternating distribution of repeating units, while **random copolymers** contain a random distribution of repeating units.

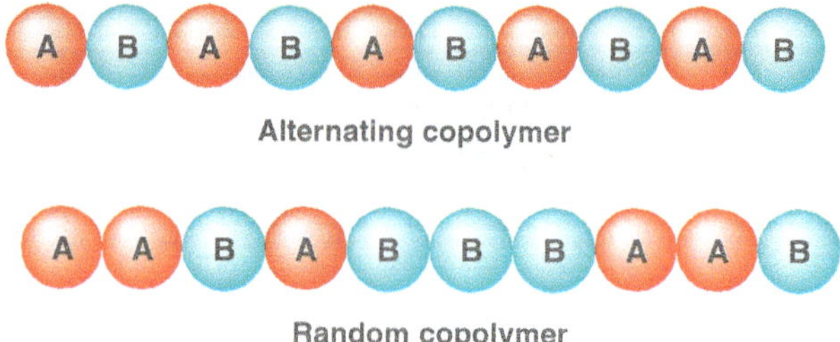


**Alternating copolymer**

**Random copolymer**

The precise distribution of any copolymer is dependent on the conditions of its formation. In practice, however, even alternating copolymers will exhibit regions in which the repeating units are distributed randomly.

**TABLE 27.3 SOME COMMON COPOLYMERS AND THEIR USES**

| MONOMERS | COPOLYMER NAME | USES |
|---|---|---|
| $H_2C{=}CHCl$ + $H_2C{=}CCl_2$<br>Vinyl chloride + Vinylidene chloride | Saran | Food packaging |
| $H_2C{=}CHPh$ + $H_2C{=}CHCN$<br>Styrene + Acrylonitrile | SAN | Dishwasher-safe kitchenware and battery cases |
| $H_2C{=}CHPh$ + $H_2C{=}CHCN$ + $H_2C{=}CH{-}CH{=}CH_2$<br>Styrene + Acrylonitrile + 1,3-Butadiene | ABS | Crash helmets, luggage, and car bumpers |
| $H_2C{=}C(CH_3)_2$ + $H_2C{=}C(CH_3){-}CH{=}CH_2$<br>Isobutylene + Isoprene | Butyl rubber | Inner tubes, balls, sporting goods |
| $F_2C{=}CF(CF_3)$ + $H_2C{=}CF_2$<br>Hexafluoropropylene + Vinylidene fluoride | Viton | Gaskets, seals, and automotive fuel lines |

Copolymers comprised of homopolymer subunits are classified as either block copolymers or graft copolymers.

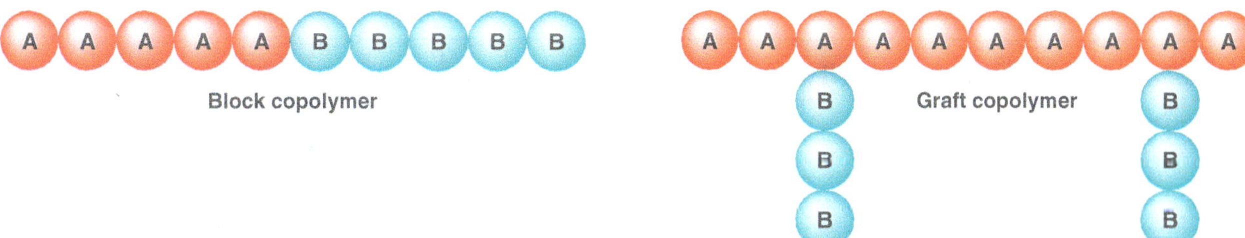


In a **block copolymer**, the different homopolymer subunits are connected together in one chain. In contrast, a **graft copolymer** contains sections of one homopolymer that have been grafted onto a chain of the other homopolymer. Block copolymers and graft copolymers are formed by controlling the conditions under which polymerization occurs.

## CONCEPTUAL CHECKPOINT

**27.3** Draw a region of an alternating copolymer constructed from styrene and ethylene.

**27.4** Draw a region of a block copolymer constructed from propylene and vinyl chloride.

**27.5** Identify the monomers required to make the following alternating copolymer:

$$\left[ -CH_2-C(CH_3)_2-CH_2-CH(Ph)- \right]_n$$

## 27.4 Polymer Classification by Reaction Type

Polymers can be classified in a variety of ways. Common classification schemes focus on either the type of reaction used to make the polymer, the mode of assembly, the structure, or the properties of the polymer. We will explore each of these classification schemes, beginning with the type of reaction used to make the polymer.

**LOOKING BACK**
Cationic addition was discussed in Chapter 9, free-radical addition was discussed in Chapter 11, and anionic addition was discussed in Chapter 22.

### Addition Polymers

As seen in previous chapters, addition reactions involving π bonds can occur through a variety of mechanisms, including cationic addition, anionic addition, and free-radical addition. In much the same way, monomers can join together to form polymers via cationic addition, anionic addition, or free-radical addition. Polymers formed via any one of these processes are called **addition polymers**. A mechanism for radical polymerization was first discussed in Section 11.11 and is reviewed in Mechanism 27.1.

MECHANISM 27.1 RADICAL POLYMERIZATION

Initiation

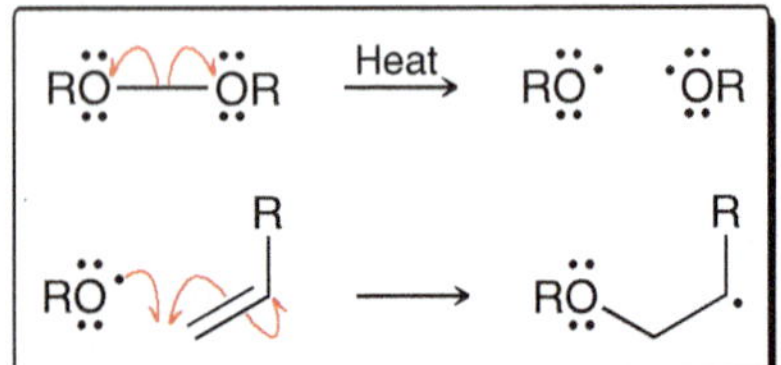


A radical initiator is formed, which reacts with a monomer to give a carbon radical

Propagation

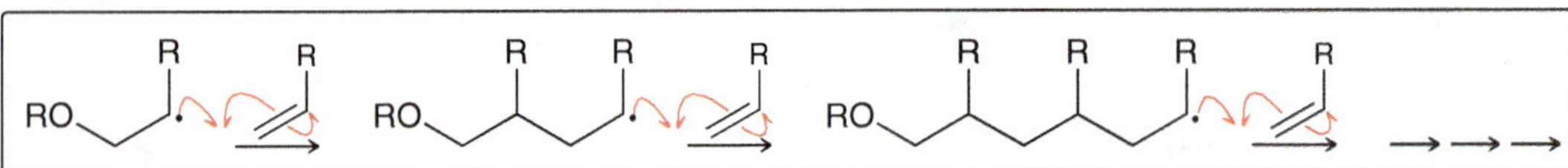


The radical addition process is repeated, building up the polymer chain

Termination

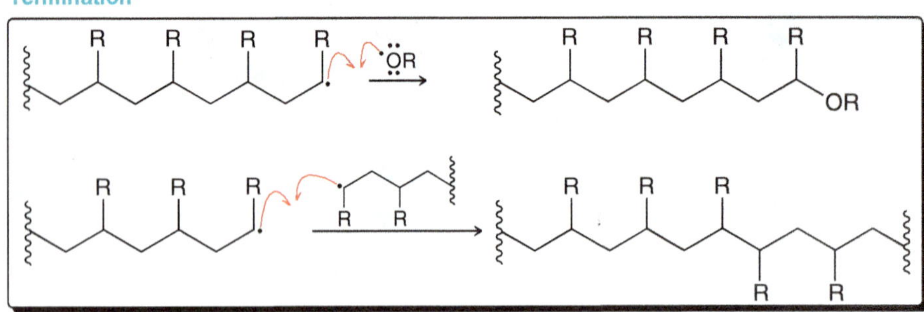


Each of these coupling reactions destroys radicals and is therefore an example of a termination step

In the initiation steps, a radical initiator is formed and then couples to one of the monomers, forming a carbon radical. This highly reactive intermediate then undergoes a propagation step in which it couples with another monomer. This process repeats itself, causing the polymer chain to grow. The process ends with a termination step in which two radicals couple together. Since the concentration of radicals is quite low at all times, the probability of two radicals coupling is rather small. As a result, thousands of monomers can be strung together in a single polymer chain before a termination step occurs. In this way, ethylene can be converted into polyethylene in the presence of a radical initiator. In fact, most derivatives of ethylene will also undergo radical polymerization under suitable conditions.

In contrast, cationic addition is only efficient with ethylene derivatives that contain an electron-donating group, such as the following examples:

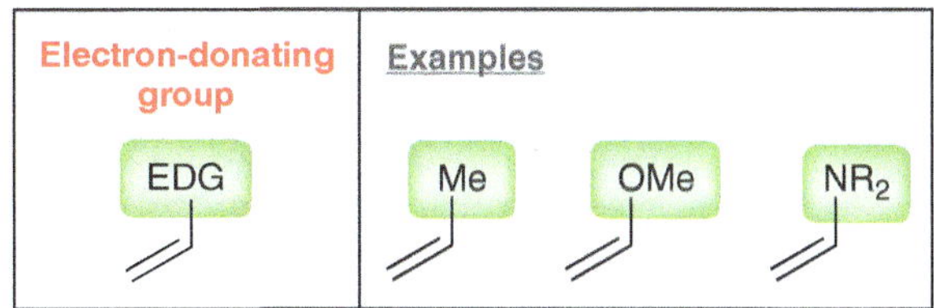


A mechanism for cationic polymerization was first discussed in the Practically Speaking box in Section 9.3 and is reviewed here in Mechanism 27.2.

## MECHANISM 27.2 CATIONIC POLYMERIZATION

### Initiation

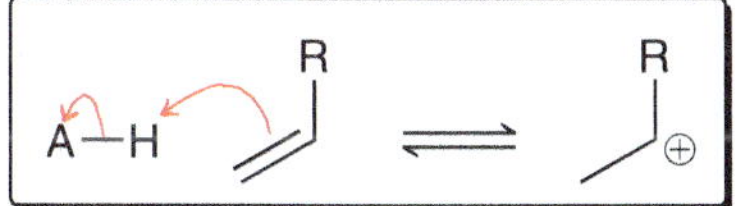


The π bond of a monomer is protonated, generating a carbocation intermediate

### Propagation

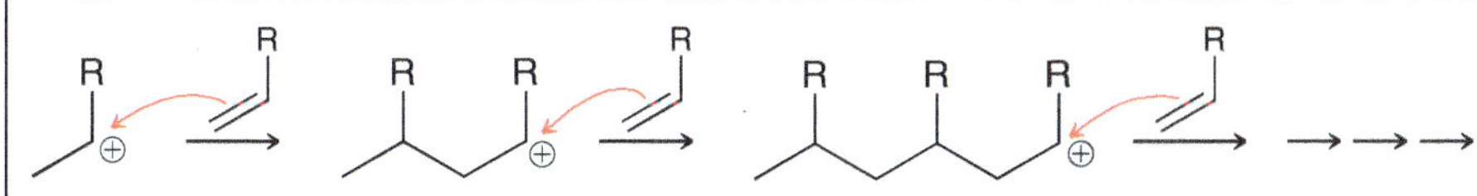


The π bond of a monomer acts as a nucleophile and attacks the carbocation. This process is repeated, building up the polymer chain

### Termination

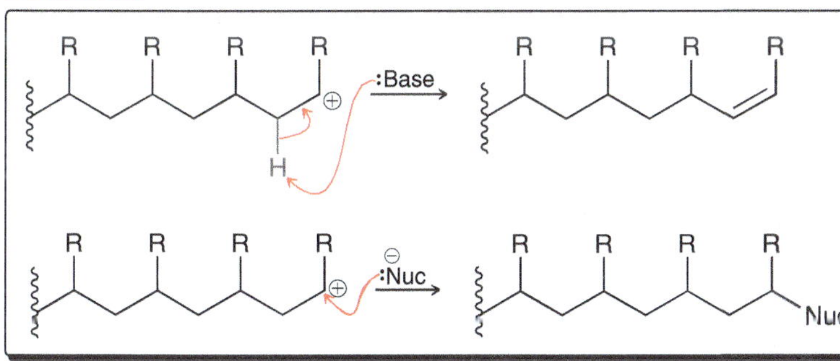


The polymerization process is terminated when either a base deprotonates the carbocation or a nucleophile attacks the carbocation

The cationic polymerization process is initiated in the presence of an acid, which transfers a proton to the π bond of a monomer, thereby generating a carbocation. The acid catalyst most often used is formed by treating $BF_3$ with water.

Acid catalyst

The carbocation generated during the initiation stage is then attacked by another monomer in a propagation step, and the process repeats itself, enabling the polymer chain to grow. The function of the electron-donating group is to stabilize the carbocation intermediate that is formed after the addition of each monomer to the growing polymer chain. For example, isobutylene readily undergoes cationic polymerization because a tertiary carbocation intermediate is formed during each propagation step.

Isobutylene 3° Carbocation 3° Carbocation 3° Carbocation

In contrast, ethylene does not readily undergo cationic polymerization, because the process would involve formation of a primary carbocation, which is not sufficiently stable to form at an appreciable rate.

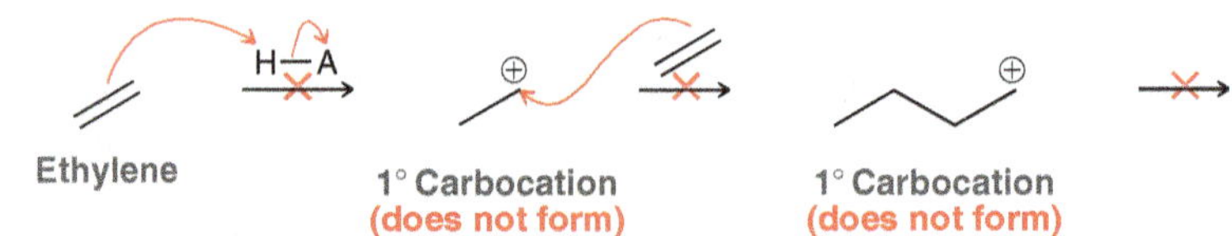


For monomers that undergo cationic polymerization, the process is terminated when the carbocation intermediate is deprotonated by a base or attacked by a nucleophile, as seen in Mechanism 27.2.

Unlike cationic addition, anionic addition is only efficient with ethylene derivatives that contain an electron-withdrawing group, such as the following examples:

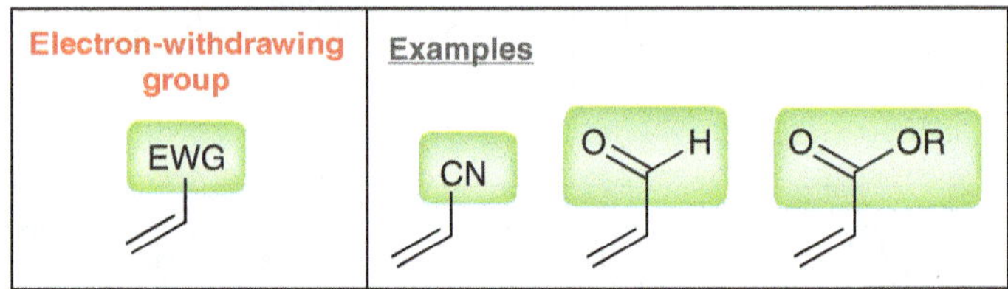


Electron-withdrawing groups were first discussed in Chapter 19, and a list can be found in Section 19.10. Anionic polymerization is believed to proceed via Mechanism 27.3.

## MECHANISM 27.3 ANIONIC POLYMERIZATION

**Initiation**

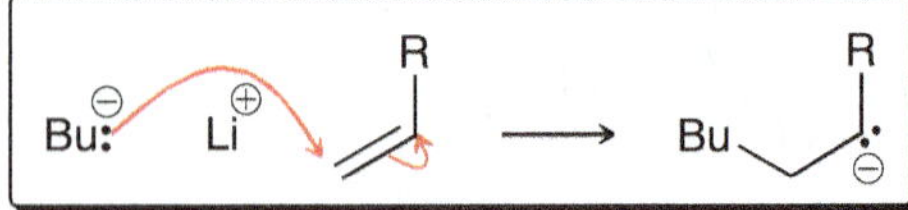


A carbon nucleophile initiates the process by attacking the $\pi$ bond, generating a carbanion

**Propagation**

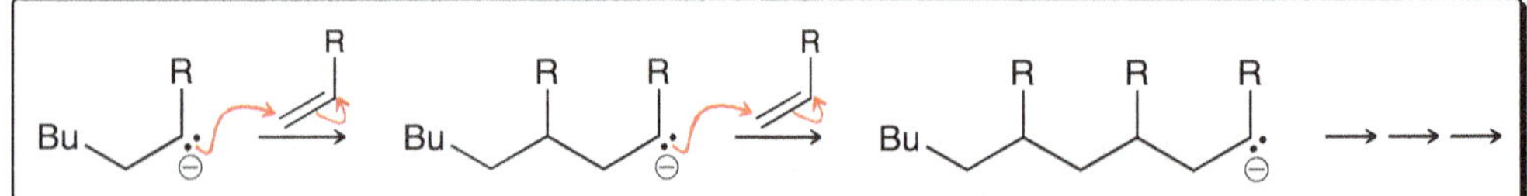


The carbanion acts as a nucleophile and attacks a monomer, creating a new carbanion. This process is repeated, building up the polymer chain

**Termination**

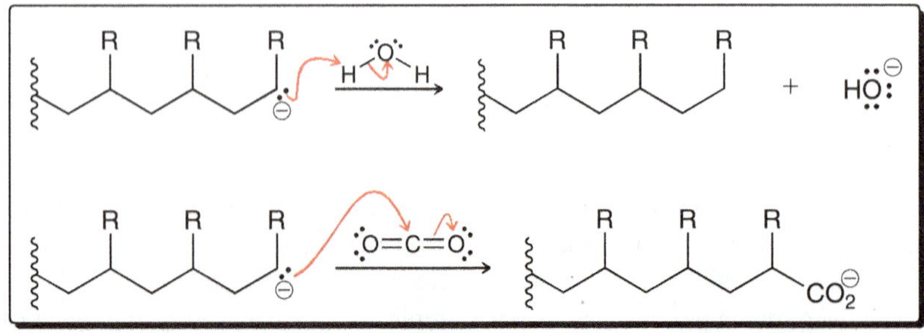


The polymerization process is terminated when the growing polymer is treated with a weak acid (such as $H_2O$) or with an electrophile (such as $CO_2$)

Anionic polymerization is initiated in the presence of a highly reactive anion, such as butyl lithium. The anion functions as a nucleophile and attacks the π bond of a monomer, thereby generating a new carbanion. The carbanion generated during the initiation stage then functions as a Michael donor and attacks another monomer, which functions as a Michael acceptor:

**LOOKING BACK**
For a review of Michael donors and Michael acceptors, see Section 22.6.

EWG EWG → EWG EWG

Michael donor | Michael acceptor

This propagation step is repeated, enabling the polymer chain to grow. The function of the electron-withdrawing group is to stabilize the carbanion intermediate that is formed after the addition of each monomer to the growing polymer chain.

EWG EWG → EWG EWG EWG → EWG EWG EWG

Stabilized by the EWG | Stabilized by the EWG | Stabilized by the EWG

For monomers that undergo anionic polymerization, the process continues until all of the monomers have been consumed. But even after all of the monomers have been exhausted, the process is not actually terminated until a suitable acid (such as water) or electrophile (such as $CO_2$) is added to the reaction mixture. In the absence of a suitable acid or electrophile, the end of each polymer chain will possess a stabilized carbanion site, and the polymerization process can continue if more monomers are added to the reaction mixture. For this reason polymers generated through this process are often called **living polymers**.

Superglue is a common example of a monomer that readily polymerizes via an anionic addition process. Superglue is a pure solution of methyl-α-cyanoacrylate, which is a monomer containing two electron-withdrawing groups.

$CO_2Me$, CN → [ $MeO_2C$ CN ]$_n$

Methyl α-cyanoacrylate (superglue)

With two electron-withdrawing groups, the compound is so reactive toward anionic polymerization that even a weak nucleophile, such as water, will initiate the polymerization process. When super glue is applied to a metal surface, the moisture on the surface of the metal is sufficient to catalyze polymerization, which then occurs very rapidly. In fact, the water and other nucleophiles present in your skin will initiate the polymerization process, explaining why superglue bonds to skin so tightly. In some cases, doctors utilize compounds like superglue to close wounds, instead of using stitches. These compounds are structurally very similar to superglue, except that the methyl group of the ester is replaced with a slightly larger alkyl group. For example, Dermabond, a cyanoacrylate ester with a 2-octyl group, was developed to replace stitches in certain situations.

O, O, CN → [ O, O, CN ]$_n$

2-Octyl α-cyanoacrylate (Dermabond)

# SKILLBUILDER

## 27.1 DETERMINING THE MORE EFFICIENT POLYMERIZATION TECHNIQUE

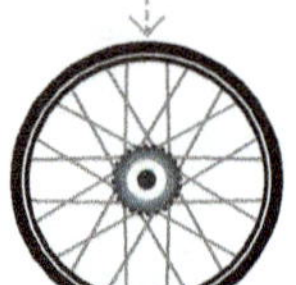

**LEARN the skill**

Determine whether preparation of the following polymer would be best achieved via cationic addition or anionic addition:

### SOLUTION

First identify the repeating units in the polymer structure.

**STEP 1**
Identify the repeating units.

Repeating units

Next identify the monomer necessary to prepare a polymer with those repeating units.

**STEP 2**
Identify the required monomer.

**STEP 3**
Determine the nature of the vinylic group.

Now determine whether the group attached to the vinyl position is an electron-withdrawing group or an electron-donating group, as first described in Section 19.10. In this case, the necessary monomer has a carbonyl group, which is an electron-withdrawing group because it can stabilize a negative charge via resonance. We therefore expect this monomer to polymerize most efficiently under basic conditions.

**STEP 4**
Use anionic conditions if the vinylic group is electron withdrawing and cationic conditions for an electron-donating group.

Resonance-stabilized

**PRACTICE the skill**

**27.6** Determine whether preparation of each of the following polymers would best be achieved via cationic addition or anionic addition:

(a) (polymer with CN substituents)
(b) (polymer with OMe substituents)
(c)
(d) (polymer with OAc substituents)
(e) (polymer with $NO_2$ substituents)
(f) (polymer with $CCl_3$ substituents)

**27.7** Arrange the following monomers in order of reactivity toward cationic polymerization:

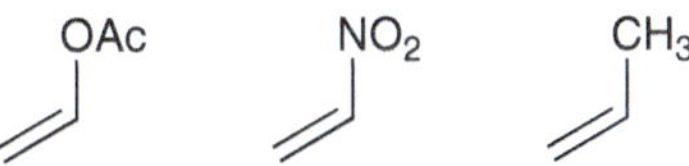


**27.8** Arrange the following monomers in order of reactivity toward anionic polymerization:

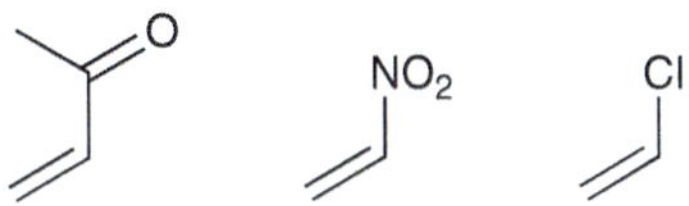


**27.9** Many monomers that readily undergo cationic polymerization, such as isobutylene, will not readily undergo anionic polymerization. Styrene, however, can be effectively polymerized via cationic, anionic, or radical addition. Explain.

**27.10** A tube of superglue will harden if it remains open too long in a humid environment. Draw a mechanism that shows how superglue polymerizes in the presence of atmospheric moisture. When drawing your mechanism, make sure to draw the initiation step, at least two propagation steps, and a termination step involving water as the acid.

need more **PRACTICE?** **Try Problems 27.27, 27.28, 27.34, 27.38–27.40**

## Condensation Polymers

As seen in previous chapters, the term “condensation” is used to characterize any reaction in which two molecules undergo addition accompanied by the loss of a small molecule such as water, carbon dioxide, or nitrogen gas. Throughout this book, we have seen many condensation reactions. One such example is the Fischer esterification process, as seen in Section 21.10.

Fischer esterification $[H^+]$ + $H_2O$

In this reaction, a carboxylic acid is treated with an alcohol in the presence of an acid catalyst, forming an ester and a water molecule. This process is considered to be a condensation reaction because the reactants undergo addition accompanied by the loss of a water molecule.

Now consider the condensation reactions that can occur when a diacid is treated with a diol. Each compound is capable of reacting twice, enabling formation of a polymer. As an example, consider the formation of poly(ethylene terephthalate), also known simply as PET, which is used to make soft-drink bottles.

Terephthalic acid + Ethylene glycol $\xrightarrow{[H^+]}$ Poly(ethylene terephthalate) (PET) + $H_2O$

As shown, PET is prepared by successive Fischer esterification reactions. Since the polymer is generated via condensation reactions, it is called a **condensation polymer**. Because PET has repeating ester groups, the polymer is also classified as a polyester. Like PET, nylon 6,6 is also prepared via condensation reactions and is also a condensation polymer. Nylon 6,6 is a polyamide, which is prepared from adipic acid and 1,6-hexanediamine.

Adipic acid + 1,6-Hexanediamine

[$H^+$], heat

Nylon 6,6 + $H_2O$

**LOOKING BACK**
For a review of polyesters and polyamides, see the Practically Speaking box in Section 21.12.

## CONCEPTUAL CHECKPOINT

**27.11** Draw the mechanism of formation of PET in acidic conditions. It might be helpful to first review the mechanism for the Fischer esterification process.

**27.12** Draw the polymer that would be generated from the acid-catalyzed reaction between oxalic acid and resorcinol.

Oxalic acid + Resorcinol [$H^+$] →

**Polycarbonates** are similar in structure to polyesters, but with repeating carbonate groups instead of repeating ester groups:

Carbonate group

Ester group

Carbonates can be formed from the reaction between phosgene and an alcohol.

Phosgene + 2 H—OR (An alcohol) ⟶ A carbonate + 2 HCl

Polycarbonates are formed in a similar way, by treating phosgene with a diol. As an example, consider the reaction between phosgene and bisphenol A.

Phosgene + Bisphenol A → Lexan + HCl

The resulting polymer is a condensation polymer sold under the trade name Lexan. It is a lightweight, transparent polymer with high impact strength and is used to make bicycle safety helmets, bullet-resistant glass, and traffic lights. Lexan is also used in the production of safety goggles, because it is both lightweight and shatterproof. In recent years, polycarbonates have also become quite popular for the production of CDs and DVDs.

## SKILLBUILDER

### 27.2 IDENTIFYING MONOMERS FOR A CONDENSATION POLYMER

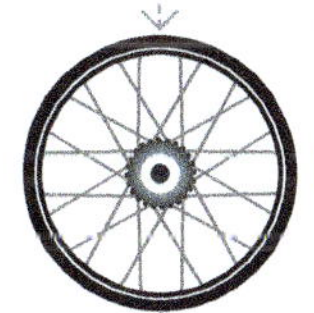

**LEARN the skill**

Determine which monomers you would use to prepare the following condensation polymer:

**SOLUTION**

First identify the type of functional group that repeats itself:

**STEP 1** Identify the repeating functional group.

A polyester

In this case, the repeating functional group is an ester group, so this condensation polymer is a polyester. Recall that ester groups can be prepared from the reaction between a carboxylic acid and an alcohol via a Fischer esterification process.

**STEP 2** Identify the type of reaction that will produce the desired functional group.

$RCOOH$ + H—O—R $\xrightarrow[\text{[H}^+]]{\text{Fischer esterification}}$ $RCOOR$ + $H_2O$

Preparing a polymer that exhibits repeating ester groups requires reactants that are difunctional; specifically, a suitable diacid and diol must be selected. To determine the starting diol and diacid, it is best to work backward from the structure of the polyester; that is, imagine hydrolyzing each ester group (the reverse of a Fischer esterification).

**STEP 3**
For a condensation polymer, identify the two difunctional monomers that are necessary.

Inspection of the individual monomers reveals that two different monomers are required. The following diacid and diol can be used to prepare the desired polymer.

HO, O, O, OH + HO, OH

**PRACTICE the skill** **27.13** Kevlar is a condensation polymer used in the manufacture of bulletproof vests. Identify the monomers required for the preparation of Kevlar.

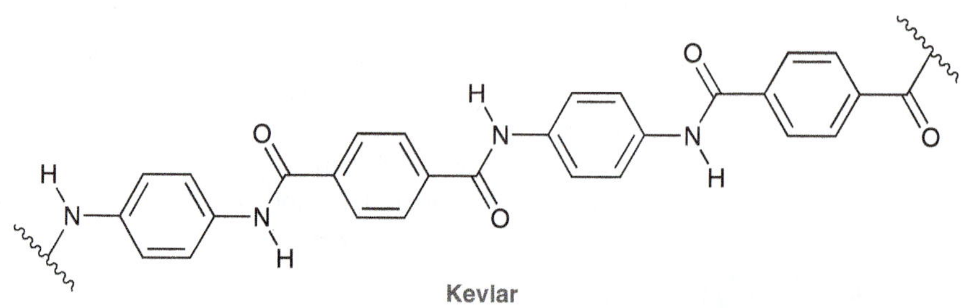


**27.14** Identify the monomers required to make each of the following condensation polymers:

(a)

(b)

**APPLY the skill** **27.15** Identify the condensation polymer that would be produced when phosgene is treated with 1,4-butanediol.

**27.16** Nylon 6 is a polyamide used in the manufacture of ropes. It can be prepared via hydrolysis of ε-caprolactam to form ε-aminocaproic acid followed by acid-catalyzed polymerization.

ε-Caprolactam —Hydrolysis→ ε-Aminocaproic acid —Polymerization→ Nylon 6

**(a)** Draw the structure of Nylon 6.

**(b)** Compare the structure of Nylon 6 with the structure of Nylon 6,6 (shown just before Conceptual Checkpoint 27.11). Which polymer exhibits a smaller repeating unit?

need more **PRACTICE?** **Try Problems 27.23, 27.30a,b,d, 27.31, 27.32**

## 27.5 Polymer Classification by Mode of Assembly

### Chain-Growth Polymers

Polymers are also often classified as either chain-growth or step-growth polymers. These terms are used to signify the way in which the polymers are assembled. **Chain-growth polymers** are formed under conditions in which the monomers do not react directly with each other, but rather, each monomer is added to the growing chain one at a time. The growing polymer chain generally has only one reactive site, called a *growth point*, and the monomers attach to the chain at the growth point.

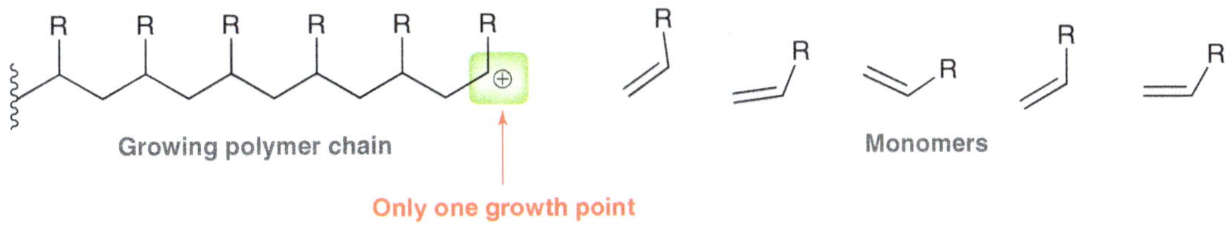

Growing polymer chain

Monomers

Only one growth point

All of the addition reactions seen earlier in this chapter (radical, cationic, and anionic polymerization) are examples of chain-growth processes. Polymers created by any of these addition processes are chain-growth polymers. In contrast, most condensation polymers are not chain-growth polymers, but instead belong to a class called step-growth polymers.

## Step-Growth Polymers

**Step-growth polymers** are formed under conditions in which the individual monomers react with each other to form **oligomers** (compounds constructed from just a few monomers), which are then joined together to form polymers.

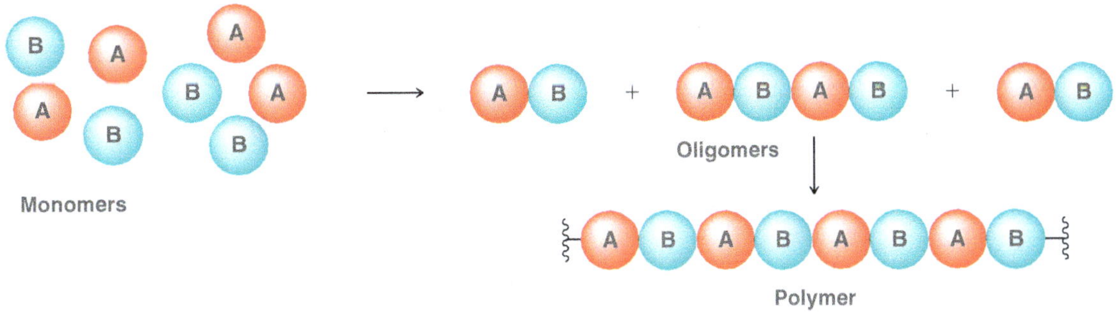

Monomers

Oligomers

Polymer

This mode of assembly occurs when difunctional monomers are used, just as we saw with condensation polymers in the previous section. In such a case, all monomers and oligomers have two growth points instead of just one. As an example, consider the formation of PET.

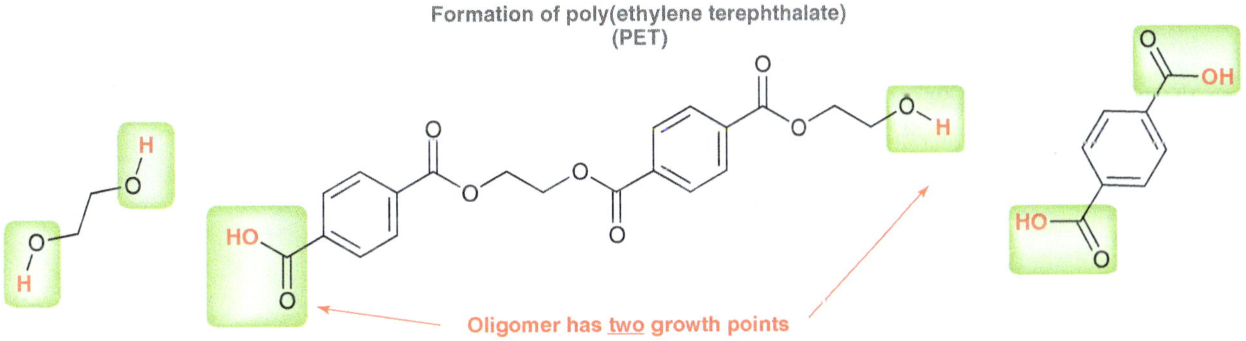

Formation of poly(ethylene terephthalate) (PET)

Oligomer has two growth points

In general, step-growth polymers are usually condensation polymers, while chain-growth polymers are usually addition polymers, but there are exceptions, such as polyurethanes. **Polyurethanes** are made up of repeating urethane groups, also sometimes called carbamate groups.

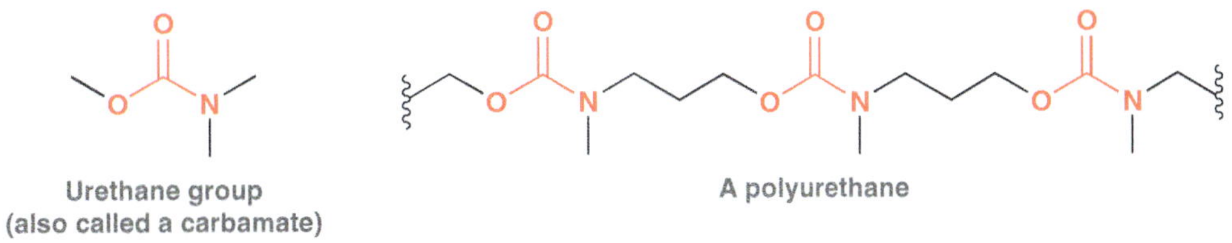

Urethane group (also called a carbamate)

A polyurethane

Urethanes can be prepared by treating an isocyanate with an alcohol.

An isocyanate + An alcohol ⟶ A urethane

Polyurethanes can be prepared by treating a diisocyanate with a diol. In preparing polyurethanes, the diols most commonly used are small polymers that bear hydroxyl ends.

A diisocyanate + A diol ⟶ A polyurethane

This process is not a condensation, because it does not involve the loss of a small molecule. Polyurethanes are technically addition polymers, but nevertheless, they are step-growth polymers because the growing polymer chain has two growth points rather than one. This example illustrates that step-growth polymers cannot always be classified as condensation polymers. Polyurethanes are used for insulation in the construction of homes and portable coolers.

### CONCEPTUAL CHECKPOINT

**27.17** Would these pairs of monomers form chain-growth or step-growth polymers?

(a)

(b)

**27.18** Should this polymer be classified as a chain-growth polymer or a step-growth polymer?

## 27.6 Polymer Classification by Structure

### Branched Polymers

Polymers are also often classified by their structure. **Branched polymers** contain a large number of branches connected to the main chain of the polymer.

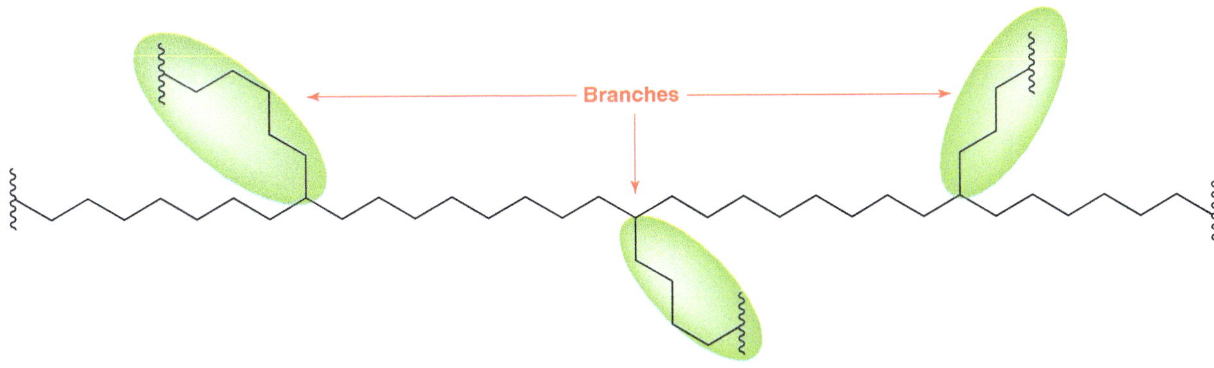


Branching occurs when the location of a growth point is moved during the polymerization process. For example, during radical addition, a hydrogen abstraction step can produce a growth point in the middle of an existing chain:

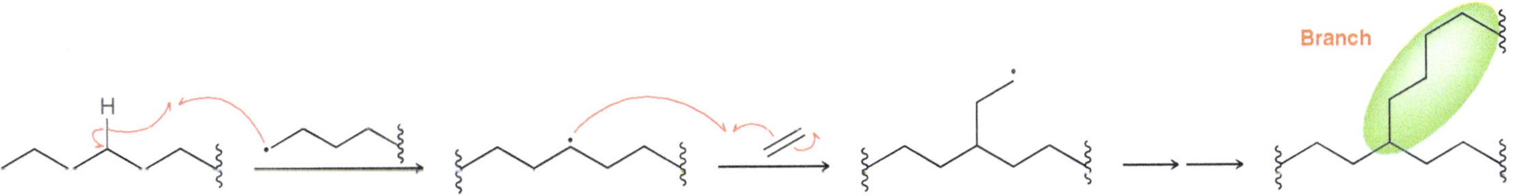


Branching is very common and is observed during many polymerization processes. Polymers can be differentiated from each other based on the extent of branching. Some polymers are highly branched, while other polymers have only a minimal amount of branching or no branching at all. The former are called branched polymers, while the latter are called **linear polymers**. The synthesis of linear polymers often requires the use of special catalysts, described in the following section.

## Linear Polymers

The field of polymer chemistry was revolutionized in 1953 with the development of Ziegler-Natta catalysts. These catalysts are organometallic complexes that can be prepared by treating an alkyl aluminum compound, such as $Et_3Al$, with $TiCl_4$. The development of Ziegler-Natta catalysts enabled scientists to control the polymerization process in the following ways:

1. *Extent of branching.* Free radicals are not involved as intermediates when Ziegler-Natta catalysts are employed, and therefore very little branching occurs. Polymers produced with these catalysts are generally linear polymers, which have very different properties than branched polymers. Without a Ziegler-Natta catalyst, ethylene polymerizes to form a polymer that exhibits about 20 branches per thousand carbon atoms. The resulting polymer chains cannot effectively pack, giving rise to a material with a relatively low density ($\sim$0.92 g/cm$^3$) called low-density polyethylene, or LDPE. However, in the presence of a Ziegler-Natta catalyst, ethylene polymerizes to form a polymer that exhibits about five branches per thousand carbon atoms. The resulting polymer chains can pack more efficiently, giving rise to a material with a relatively high density, $\sim$0.96 g/cm$^3$, called high-density polyethylene, or HDPE. LDPE is used to make trash bags, while HDPE has greater strength and is used to make plastic squeeze bottles and Tupperware.

2. *Stereochemical control.* When polymerization is performed using monosubstituted ethylenes as monomers, the resulting polymer has a large number of chirality centers. The relative configurations of these chirality centers are classified as isotactic, syndiotactic, or atactic.

Isotactic
(same configuration)

Syndiotactic
(alternating configuration)

Atactic
(random)

The term **isotactic** is used when all of the chirality centers have the same configuration, **syndiotactic** is used when the chirality centers have alternating configuration, and **atactic** is used when the chirality centers are not arranged in a pattern (they have random configurations). Isotactic, syndiotactic, and atactic polymers exhibit different properties, and all three can be made with the appropriate Ziegler-Natta catalyst.

## CONCEPTUAL CHECKPOINT

**27.19** Polyisobutylene cannot be described as isotactic, syndiotactic, or atactic. Explain.

**27.20** Polyethylene is used to make Ziploc bags and folding tables. Identify which of these applications is most likely to be made from HDPE and which is most likely to be made from LDPE.

## Cross-Linked Polymers

In the Practically Speaking box at the end of Section 17.5, we discussed natural and synthetic rubbers. Recall from that discussion that the vulcanization process introduces disulfide bridges between neighboring chains within the polymer.

Vulcanized rubber

The resulting polymer is said to be a **crossed-linked polymer**, which markedly affects the properties of the polymer. Disulfide bridges are not the only way for chains to cross-link. For example, chains can also be cross-linked with branches.

The conditions of polymer formation can be controlled so as to favor either a greater or lesser amount of cross-linking. This difference can have a profound impact on the properties of the resulting polymer, as we will see in the next section.

## Crystalline vs. Amorphous Polymers

Many polymers contain regions, called **crystallites**, in which the chains are linearly extended and close in proximity to one another, resulting in van der Waals forces that hold the chains close together.

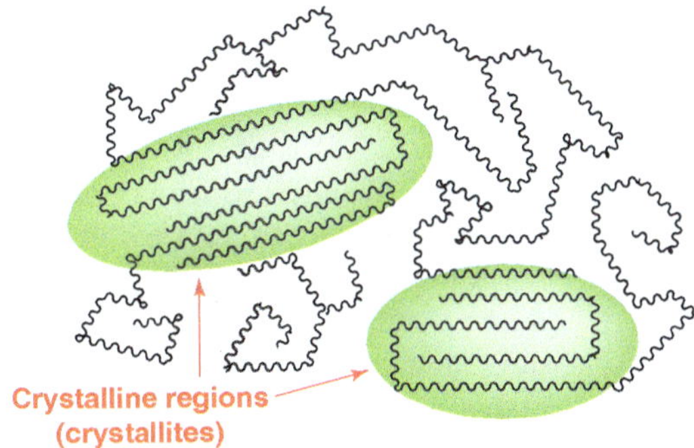


The regions that are not crystalline are called **amorphous** regions. Crystalline regions render a polymer hard and durable, while amorphous regions render a polymer flexible. The degree of crystallinity of a polymer, and therefore its physical properties, greatly depends on the steric requirements of the substituent(s) present in the repeating unit of the polymer. For example, compare the structures of polyethylene and polyisobutylene.

$$\left[ -\mathrm{CH_2}-\mathrm{CH_2}- \right]_n \qquad \left[ -\mathrm{CH_2}-\mathrm{C(CH_3)_2}- \right]_n$$

**Polyethylene** **Polyisobutylene**

Linear polyethylene (HDPE) exhibits a high degree of crystallinity because there are no substituents to prevent the chains from closely packing. In contrast, polyisobutylene exhibits a low degree of crystallinity, because there are two methyl groups that provide steric bulk, preventing the chains from closely packing.

When highly crystalline polymers like HDPE are heated, the crystalline regions become amorphous at a specific temperature called the **melt transition temperature** ($T_m$). When noncrystalline polymers like polyisobutylene are heated, they become very soft. The temperature at which this transition takes place is called the **glass transition temperature** ($T_g$).

# 27.7 Polymer Classification by Properties

Polymers are also sometimes classified by their properties. The four most common categories are thermoplastics, elastomers, fibers, and thermosetting resins.

## Thermoplastics

**Thermoplastics** are polymers that are hard at room temperature but soft when heated; that is, they exhibit a high $T_g$. Polymers in this category are very useful because they can be easily molded and are commonly used to make toys and storage containers. For example, PET is a thermoplastic used in the manufacture of soft-drink bottles. Other examples of thermoplastics include polystyrene, polyvinyl chloride (PVC), and low-density polyethylene.

Many thermoplastics become brittle at room temperature, which severely limits their utility. This is true of PVC. Pure PVC is highly susceptible to cracking at room temperature and is therefore useless for most applications. To avoid this problem, the polymer can be prepared in the presence of small molecules called **plasticizers**. These molecules become trapped between the polymer chains where they function as lubricants. Common plasticizers are dialkyl phthalates, such as di-2-ethylhexyl phthalate used in vinyl upholstery, raincoats, shower curtains, inflatable boats, and garden hoses. Some plasticizers evaporate slowly with time, and the polymer ultimately returns to a brittle state in which it can be easily cracked. Intravenous (IV) drip bags used in hospitals are typically made from PVC with plasticizers.

## Elastomers

**Elastomers** are polymers that return to their original shape after being stretched. Elastomers are typically amorphous polymers that have a small degree of cross-linking. Natural rubber (described in Section 17.5) is one of the most common examples of an elastomer. A more contemporary example is Spandex, which is a polyurethane with a small degree of cross-linking. Spandex fibers are used to make bathing suits and athletic gear, among other applications.

Chewing gum is a complex mixture of polymers, many of which are elastomers. The polymer mixture, called the gum base, serves as a delivery vehicle for the flavors that are present in the mixture. Most gum manufacturers do not reveal the identity of the polymers present in their gum base, as this information is regarded as a trade secret. One thing is certain though: When you chew gum, you are chewing on a glob of synthetic polymers.

## Fibers

**Fibers** are generated when certain polymers are heated, forced though small holes, and then cooled. The resulting fibers exhibit crystalline regions that are oriented along the axis of the fiber, which endow the fibers with significant tensile strength. Examples include Nylon, Dacron, and polyethylene, all of which have the appropriate degree of crystallinity to form fibers.

## Thermosetting Resins

**Thermosetting resins** are highly cross-linked polymers that are generally very hard and insoluble. One such example is Bakelite, first produced in 1907 from the reaction between phenol and formaldehyde.

OH
O
H H
$[H^+]$
OH OH OH OH HO
Bakelite

Phenol functions as a nucleophile, and formaldehyde (activated by protonation) functions as an electrophile in an electrophilic aromatic substitution reaction. The resulting alcohol then functions as an electrophile (again activated by protonation) and reacts with phenol in another electrophilic aromatic substitution reaction.

**LOOKING BACK**
For a justification of *ortho-para* directing effects, see Section 19.7.

OH
O
H H
$[H^+]$
EAS
OH OH
OH
$[H^+]$
EAS
OH OH

This process continues and can occur at the *ortho* and *para* positions of each ring,

Thermosetting resins can generally withstand high temperatures and are often used in high-temperature applications, such as missile nose cones.

### Safety Glass and Car Windshields

Automobile windshields are specifically designed so that they do not create free shards of glass when broken. This is accomplished in very much the same way that bullet-resistant glass is manufactured, as described in the chapter opener. Two sheets of glass are pressed together with a thin layer of polymer between them. The polymer most often used is poly(vinyl butyral), often called PVB.

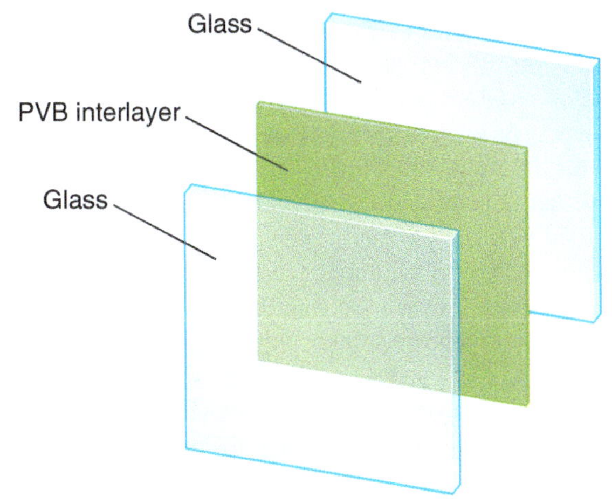


The PVB layer is generally prepared by treating polyvinyl alcohol with butyraldehyde.

OH OH OH OH

Poly(vinyl alcohol)

$[H^+]$

O

H

O O O O

Poly(vinyl butyral)
(PVB)

The carbonyl group of butyraldehyde reacts with two hydroxyl groups to form a cyclic acetal. The resulting polyacetal adheres to both sheets of glass. In the event of a collision, the glass breaks, but the individual pieces remain stuck to the polymer interlayer, thereby preventing the pieces of glass from causing injury. Bullet-resistant glass is very similar, but multiple sheets of glass are used, rather than just two. Alternatively, bullet-resistant glass can also be made using thicker polycarbonate interlayers such as Lexan.

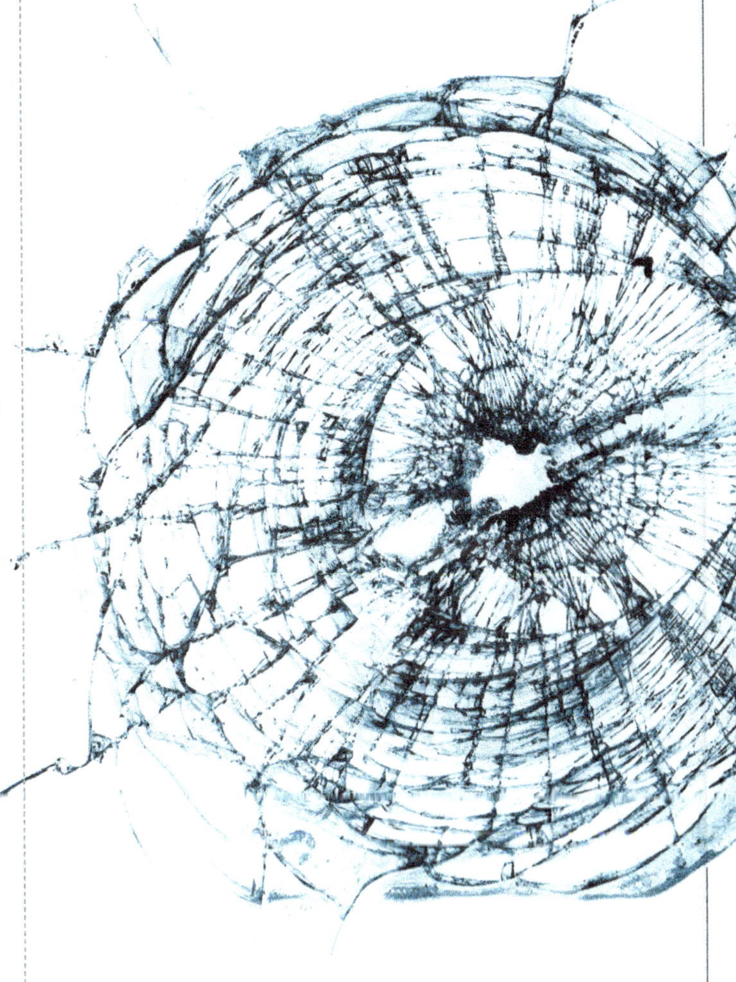

## 27.8 Polymer Recycling

Polymers have certainly increased our quality of life in countless ways, but their preparation and use have caused serious environmental concerns. Specifically, most synthetic polymers are not biodegradable, which means that they persist and accumulate in the environment. To address this growing problem, many polymers can be recycled. Some polymers are recycled with greater ease than others, depending on the identity of the polymer. For example, in the recycling of soft-drink bottles (made from PET) the bottles are chopped into small chips that are washed to remove the labels and adhesives, then treated with aqueous acid. A reaction occurs in which the polymer chips are broken down into the corresponding monomers. This process is essentially the reverse of the polymerization process that formed PET in the first place. The ester groups undergo hydrolysis to produce terephthalic acid and ethylene glycol.

Poly(ethylene terephthalate) (PET) $\xrightarrow[H_2O]{[H^+]}$ Terephthalic acid + Ethylene glycol

These monomers are purified and then used as feedstock for the production of new PET. PET is somewhat unique in its ease of recycling, because the polymer can be converted back into monomers. As of yet, there is no way to break down addition polymers, such as polyethylene or polystyrene, into their corresponding monomers. These polymers can be melted down and remolded into a new shape or they can be broken down into smaller fragments in a process that is similar to the cracking of petroleum.

There are many logistical problems that limit the effectiveness of polymer recycling, most significantly the collection and sorting of used polymer products. Different kinds of polymers must be recycled in different ways. When recycling PET, a small amount of a different polymer present in the batch will interfere with the recycling process. As such, polymer recycling requires that polymer products be sorted by hand. To facilitate the sorting process, most polymer products are labeled with recycling codes that indicate their composition. These codes (1–7) indicate the type of polymer used and are arranged in order of ease with which the polymer can be recycled (1 being the easiest and 7 being the most difficult). Table 27.4 indicates the seven recycling codes, the polymers that correspond

**TABLE 27.4 RECYCLING CODES AND USES OF RECYCLED PRODUCTS**

| RECYCLING CODE | POLYMER | STRUCTURE | RECYCLED PRODUCT |
|---|---|---|---|
| 01 PET | Poly(ethylene terephthalate) | | Clothing, carpet fibers, detergent bottles, audio and video tapes |
| 02 PE-HD | High-density polyethylene | | Tyvek insulation, clothing, picnic tables |
| 03 PVC | Poly(vinyl chloride) | Cl | Floor mats, garden hoses, shower curtains, plumbing pipes |
| 04 PE-LD | Low-density polyethylene | | Trash bags |
| 05 PP | Polypropylene | | Rope, fishing nets, carpets |
| 06 PS | Polystyrene | Ph | Styrofoam packaging materials, coat hangers, trash cans |
| 07 O | All other polymers | Polycarbonates, polyurethanes, polyamides, etc. | Auto parts |

with each code, and several uses for the recycled products. In many cases, the recycled polymer can be contaminated with adhesives and other materials that may have survived the washing stage. Therefore, recycled polymers cannot be used for food packaging.

## CONCEPTUAL CHECKPOINT

**27.21** Propose a mechanism for the acid-catalyzed hydrolysis of PET to regenerate the monomers terephthalic acid and ethylene glycol.

Another, more promising, method of reducing the accumulation of polymers in the environment is to develop polymers that are biodegradable and can be recycled by natural processes. **Biodegradable polymers** are polymers that can be broken down by enzymes produced by soil microorganisms. Much research is directed at the development of such polymers, and many have already been developed. Most biodegradable polymers exhibit ester or amide groups, which can be hydrolyzed by enzymes. Examples include a class of compounds called polyhydroxyalkanoates (PHAs), which are polymers of β-hydroxy carboxylic acids.

A β-hydroxy carboxylic acid → A polyhydroxyalkanoate (PHA)

There are many different PHAs, depending on the identity of the R group. The salient feature in all PHAs is the presence of the repeating ester groups, which are readily hydrolyzed. These ester groups serve as the "weak link" in the polymer. Much research has been devoted to creating polymers with weak links that render the polymers biodegradable. The decades to come are likely to see the emergence of new biodegradable polymers with properties that will rival the properties of traditional, nonbiodegradable polymers.

# REVIEW OF REACTIONS

## Reactions for Formation of Chain-Growth Polymers

### Radical Addition

ROOR; Heat or light

### Anionic Addition

1) BuLi; 2) $H_2O$ or $CO_2$

### Cationic Addition

$BF_3$, $H_2O$

## Reactions for Formation of Step-Growth Polymers

### Fischer Esterification

\+ H—O—R $\xrightarrow{[H^+]}$ ... + $H_2O$

### Carbonate Formation

Phosgene + 2 H—OR (An alcohol) → A carbonate + 2 HCl

**Amide Formation**

A carboxylic acid (RCOOH) + H—NHR (An amine) $\xrightarrow[\text{heat}]{[H^+]}$ RC(O)NHR (An amide) + $H_2O$

**Urethane Formation**

R—N=C=O (An isocyanate) + ROH (An alcohol) ⟶ RN(H)C(O)OR (A urethane)

# REVIEW OF CONCEPTS AND VOCABULARY

### SECTION 27.1

- Polymers are comprised of repeating units that are constructed by joining monomers together.
- Synthetic polymers are prepared by scientists in the laboratory or in factories, while biopolymers are produced by living organisms.

### SECTION 27.2

- The names of polymers are most commonly based on the monomers from which they are derived.

### SECTION 27.3

- A **homopolymer** is a polymer made up of a single type of monomer. Polymers made from two or more different types of monomers are called **copolymers**.
- **Alternating copolymers** contain an alternating distribution of repeating units, while **random copolymers** contain a random distribution of repeating units.
- In a **block copolymer**, different homopolymer subunits are connected together in one chain. In a **graft copolymer**, sections of one homopolymer have been grafted onto a chain of another homopolymer.

### SECTION 27.4

- Monomers can join together to form **addition polymers** by cationic, anionic, or free-radical addition.
- Most derivatives of ethylene will undergo radical polymerization under suitable conditions.
- Cationic addition is only efficient with derivatives of ethylene that contain an electron-donating group.
- Anionic addition is only efficient with derivatives of ethylene that contain an electron-withdrawing group, and polymers generated through this process are often called **living polymers**.
- Polymers generated via condensation reactions are called **condensation polymers**.
- **Polycarbonates** are similar in structure to polyesters, but with repeating carbonate groups instead of repeating ester groups.

### SECTION 27.5

- **Chain-growth polymers** are formed under conditions in which each monomer is added to the growing chain one at a time. The monomers do not react directly with each other.
- **Step-growth polymers** are formed under conditions in which the individual monomers react with each other to form **oligomers**, which are then joined together to form polymers.
- In general, step-growth polymers are usually condensation polymers, while chain-growth polymers are usually addition polymers, but there are exceptions, such as **polyurethanes**.

### SECTION 27.6

- **Branched polymers** contain a large number of branches connected to the main chain of the polymer, while **linear polymers** have only a minimal amount of branching.
- Polymers produced with Ziegler-Natta catalysts are generally linear polymers.
- When polymerization is performed using monosubstituted ethylenes as monomers, **isotactic**, **syndiotactic**, or **atactic** polymers can be formed, depending on the catalyst used.
- **Crossed-linked polymers** contain disulfide bridges or branches that connect neighboring chains.
- Most polymers contain some degree of cross-linking, and the conditions of polymer formation can be controlled so as to favor either more or less cross-linking.
- Many polymers contain regions called **crystallites** in which the chains are linearly extended and close in proximity to one another. The regions that are not crystalline are said to be **amorphous**.
- Crystalline regions become amorphous at the **melt transition temperature** ($T_m$).
- Noncrystalline polymers become soft at the **glass transition temperature** ($T_g$).

### SECTION 27.7

- **Thermoplastics** are polymers that are hard at room temperature but soft when heated. They are often prepared in the presence of **plasticizers** to prevent the polymer from being brittle.
- **Elastomers** are polymers that return to their original shape after being stretched.
- **Fibers** are generated when certain polymers are heated, forced though small holes, and then cooled.
- **Thermosetting resins** are highly cross-linked polymers that are generally very hard and insoluble.

### SECTION 27.8

- Most synthetic polymers persist and accumulate in the environment.
- For recycling purposes, polymer products are labeled with recycling codes that indicate their composition.
- **Biodegradable polymers** can be broken down by enzymes produced by microorganisms in the soil.

# SKILLBUILDER REVIEW

### 27.1 DETERMINING WHICH POLYMERIZATION TECHNIQUE IS MORE EFFICIENT

**STEP 1** Identify the repeating units.

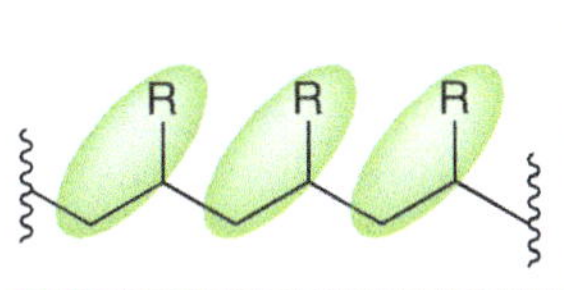


**STEP 2** Identify the monomer necessary to make this polymer.

R

**STEP 3** Determine whether the monomer bears an electron-withdrawing group or an electron-donating group:

EDGs

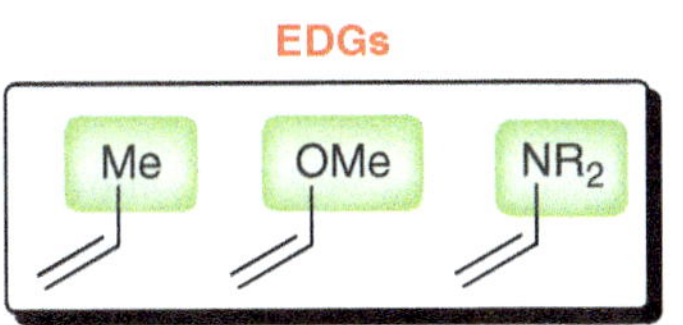


EWGs

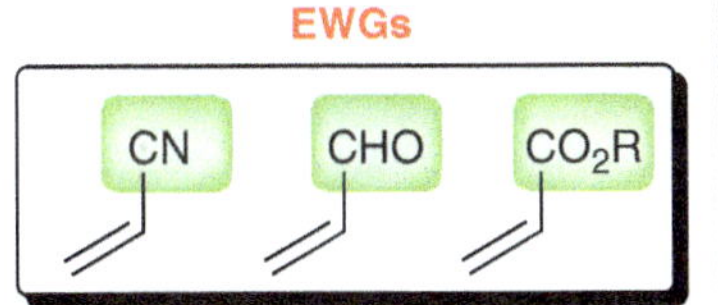


**STEP 4** If EWG, then use anionic addition.

If EDG, then use cationic addition.

Try Problems **27.6–27.10, 27.27, 27.28, 27.34, 27.38–27.40**

### 27.2 IDENTIFYING THE MONOMERS REQUIRED TO PRODUCE A DESIRED CONDENSATION POLYMER

**STEP 1** Identify the functional group that repeats itself.

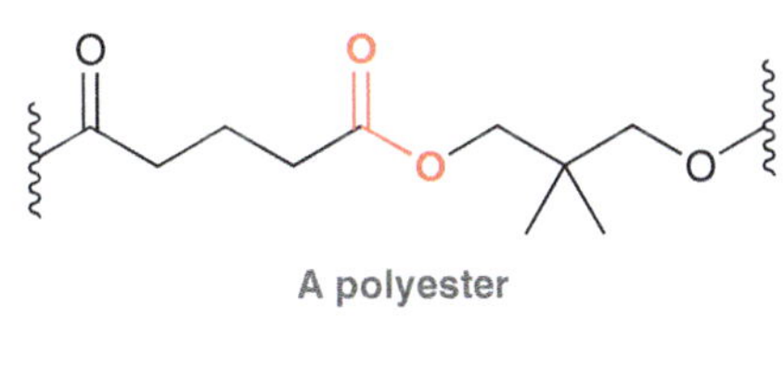


**A polyester**

**STEP 2** Identify the type of reaction that will produce the desired functional group.

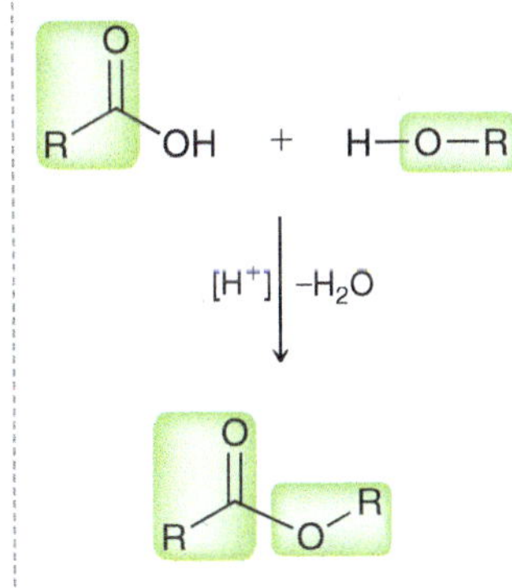


**STEP 3** Identify the two difunctional monomers necessary to form the desired condensation polymer.

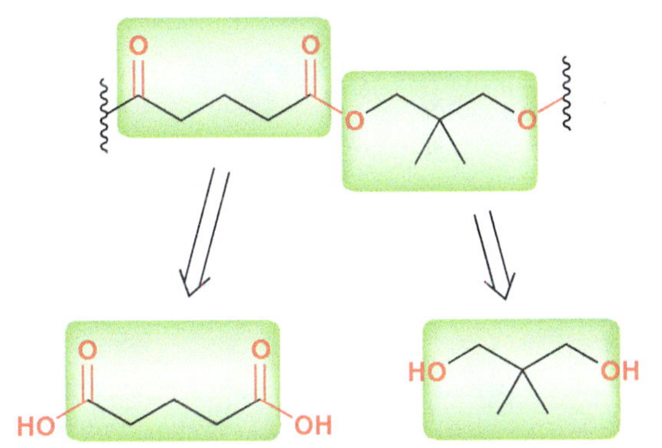


Try Problems **27.13–27.16, 27.23, 27.30a,b,d, 27.31, 27.32**

# PRACTICE PROBLEMS

Note: Most of the Problems are available within **WileyPLUS**, an online teaching and learning solution.

**27.22** Draw and name the polymer that results when each of the following monomers undergoes polymerization:

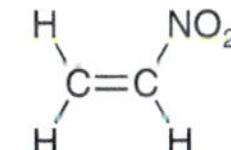


(a) **Nitroethylene**

(b) **Acrylonitrile**

(c) **Vinylidene fluoride**

**27.23** Kodel is a synthetic polyester with the following structure:

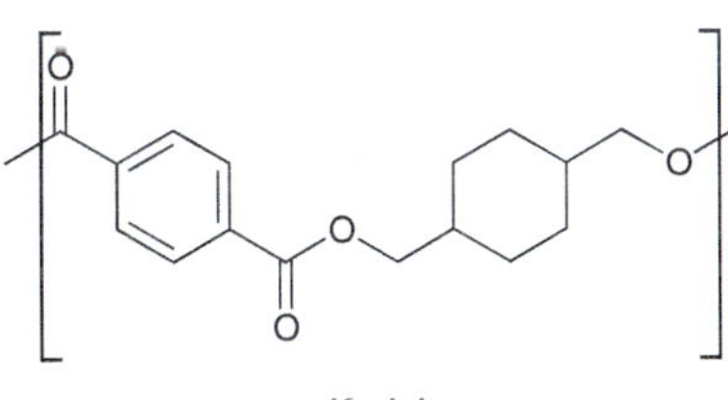


**Kodel**

(a) Identify what monomers you would use to make Kodel.

(b) Would you use acidic conditions or basic conditions for this polymerization process?

**27.24** Draw the structures of the monomers required to make the following alternating copolymer:

$$\left[ -\mathrm{CH_2}-\mathrm{CCl_2}-\mathrm{CH_2}-\mathrm{CPh_2}- \right]_n$$

**27.25** Draw a region of a block copolymer constructed from isobutylene and styrene.

**27.26** Draw a region of an alternating copolymer constructed from vinyl chloride and ethylene.

**27.27** Identify which of the following monomers would be most reactive toward cationic polymerization:

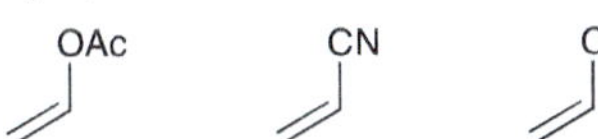


**27.28** Identify which of the following monomers would be most reactive toward anionic polymerization:

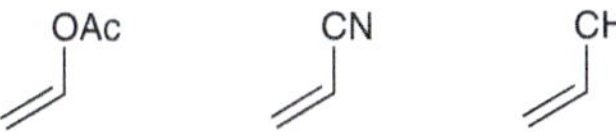

**27.29** Identify the repeating unit of the polymer formed from each of the following reactions and then determine whether the polymer is a chain-growth or a step-growth polymer.

(a) $CO_2H$, $CO_2H$ + $NH_2$, $NH_2$ $\xrightarrow{[H^+]}$ ?

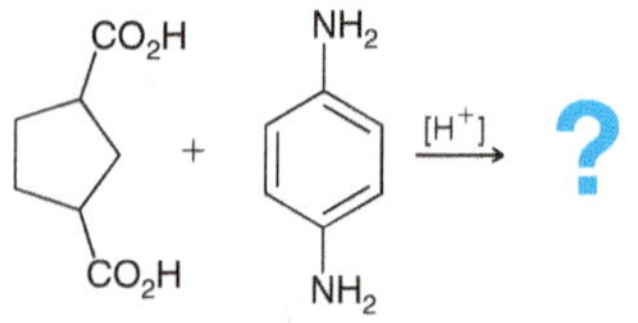

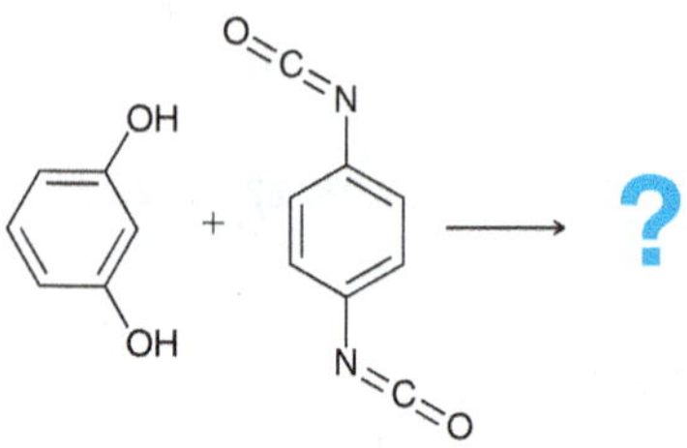

(b) OH, OH + O=C=N, N=C=O ⟶ ?

(c) HO, HO + Cl, O, Cl ⟶ ?

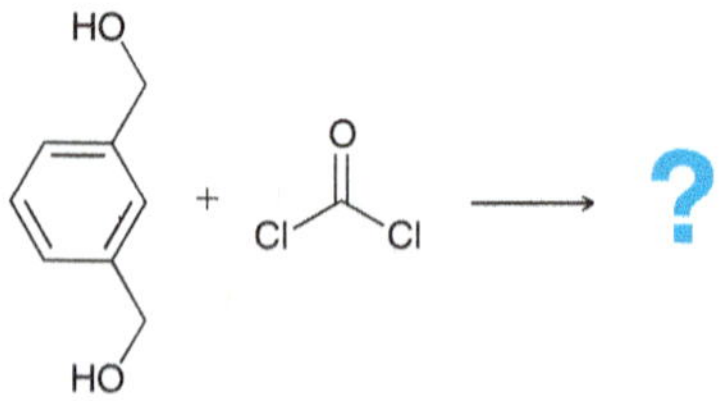

**27.30** Quiana is a synthetic polymer that can be used to make fabric that mimics the texture of silk. It can be prepared from the following monomers:

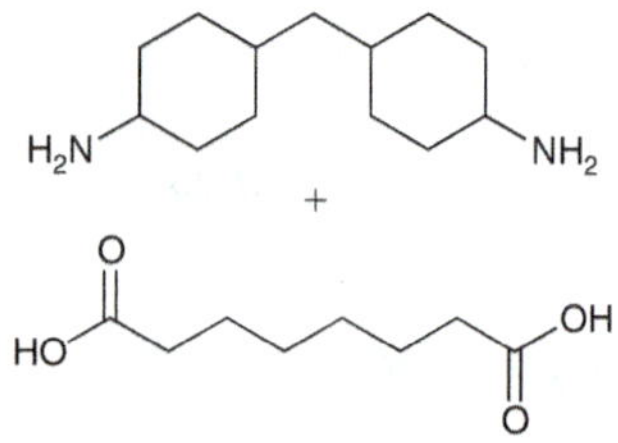

$H_2N$, $NH_2$ + HO, O, OH, O

(a) Draw the structure of Quiana.

(b) Is Quiana a polyester, a polyamide, a polycarbonate, or a polyurethane?

(c) Is Quiana a step-growth polymer or a chain-growth polymer?

(d) Is Quiana an addition polymer or a condensation polymer?

**27.31** Draw the monomer(s) required to make each of the following condensation polymers:

(a) H, N, O, N, H, H, N, O, O

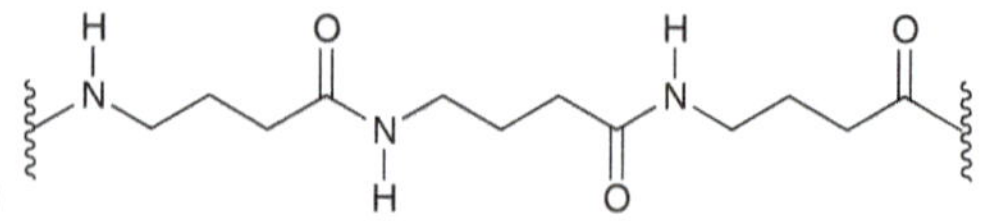

(b) O, O, O, O, O, O

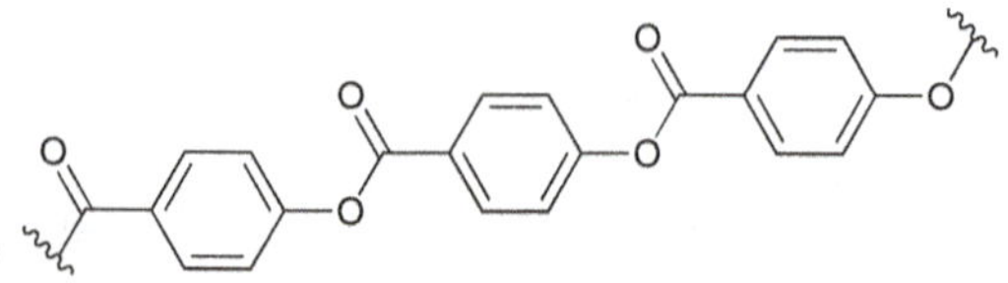

**27.32** Draw the condensation polymer produced when 1,4-cyclohexanediol is treated with phosgene.

**27.33** Determine whether the following pairs of monomers would form chain-growth polymers or step-growth polymers.

(a) HO, O, O, OH + HO—◇—OH

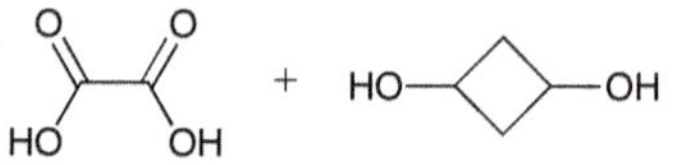

(b) $NO_2$ + Ph

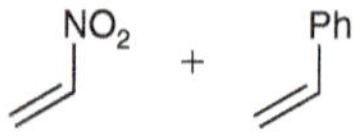

**27.34** Nitroethylene undergoes anionic polymerization so rapidly that it is difficult to isolate nitroethylene without it polymerizing. Explain.

**27.35** Explain why vinyl shower curtains develop cracks over time.

**27.36** Polyformaldehyde, sold under the trade name Delrin, is a strong polymer used in the manufacture of many guitar picks. It is prepared via the acid-catalyzed polymerization of formaldehyde.

(a) Draw the structure of polyformaldehyde.

(b) Would you describe polyformaldehyde as a polyether or a polyester?

(c) Is polyformaldehyde a step-growth polymer or a chain-growth polymer?

(d) Is polyformaldehyde an addition polymer or a condensation polymer?

**27.37** The following monomer can be polymerized under either acidic or basic conditions. Explain.

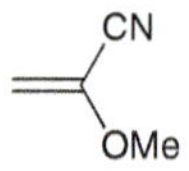

CN, OMe

**27.38** Anionic polymerization of *para*-nitrostyrene occurs much more rapidly than anionic polymerization of styrene. Explain this difference in rate.

**27.39** Cationic polymerization of *para*-methoxystyrene occurs much more rapidly than cationic polymerization of styrene. Explain this difference in rate.

**27.40** Cationic polymerization of *para*-methoxystyrene occurs much more rapidly than cationic polymerization of *meta*-methoxystyrene. Explain this difference in rate.

**27.41** Draw segments of the following polymers, indicating the stereochemistry with wedges and/or dashes.

(a) Syndiotactic poly(vinyl chloride)

(b) Isotactic polystyrene

**27.42** Consider the structure of the following polymer:

O, N, H, N, H, O, O, O, n

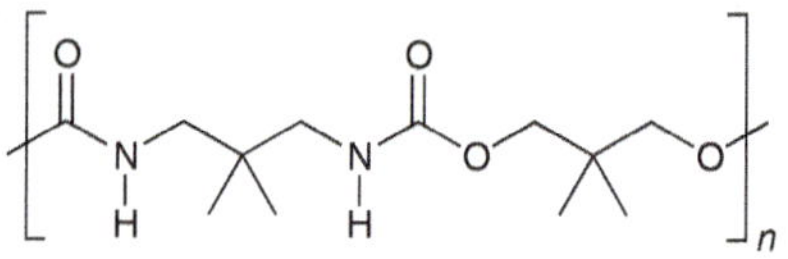

(a) Draw the monomers you would use to prepare this polymer.

(b) Determine whether this polymer is a step-growth polymer or a chain-growth polymer.

(c) Determine whether this polymer is an addition polymer or a condensation polymer.

## INTEGRATED PROBLEMS

**27.43** Draw the polymer that is expected when the following monomers react under acidic conditions. You may find it helpful to review Section 20.6.

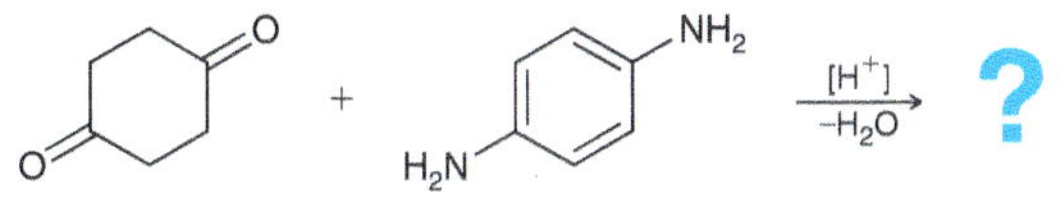

O, O, + $NH_2$, $H_2N$, $[H^+]$, $-H_2O$, ?

**27.44** As described in the Practically Speaking box in Section 27.7, poly(vinyl alcohol) is used as a precursor to make PVB for use in automobile windshields. Poly(vinyl alcohol) can be prepared by polymerizing vinyl acetate and then removing the acetate groups via hydrolysis.

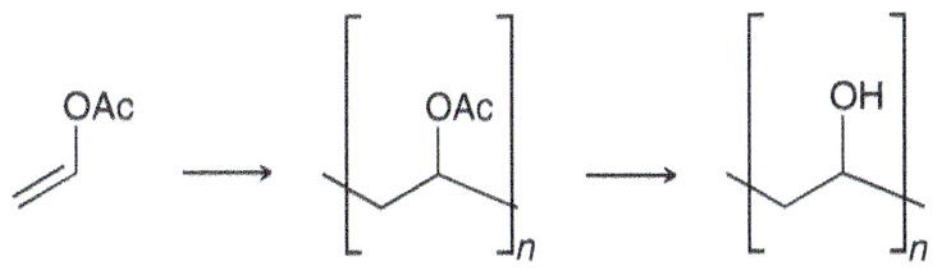

OAc, OAc, OH, n, n

Vinyl acetate Poly(vinyl acetate) Poly(vinyl alcohol)

This two-step process is required because poly(vinyl alcohol) cannot be made from its corresponding monomer. Explain why vinyl alcohol will not directly undergo polymerization.

**27.45** When a solution of aqueous sodium hydroxide is spilled on polyester clothing, a hole develops in the fabric. Describe how the polyester is destroyed.

## CHALLENGE PROBLEMS

**27.46** When 3-methyl-1-butene is treated with a catalytic amount of $BF_3$ and $H_2O$, a cationic polymerization process occurs, but the expected homopolymer is not formed. Instead, a random copolymer is obtained.

(a) Explain why there are two different repeating units.

(b) Draw a segment of the random copolymer that is formed, clearly showing the two different repeating units.

(c) Would you expect cationic polymerization of 3,3-dimethyl-1-butene to produce a random copolymer as well? Justify your answer.

**27.47** When ethylene oxide is treated with a strong nucleophile, the epoxide ring is opened to form an alkoxide ion that can function as a nucleophile to attack another molecule of ethylene oxide. This process repeats itself, and a polymer is formed.

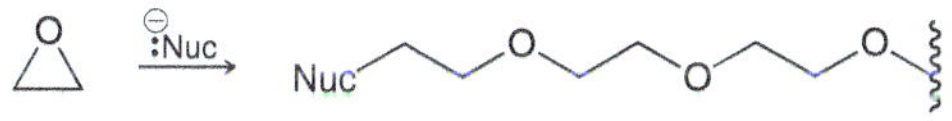

O, :Nuc⁻, Nuc, O, O, O

Ethylene oxide Poly(ethylene oxide)

The resulting polymer is called poly(ethylene oxide) or poly(ethylene glycol). It is sold under the trade name Carbowax and is used as an adhesive and a thickening agent.

(a) Draw a mechanism showing the formation of a segment of poly(ethylene oxide).

(b) Ethylene oxide can also be polymerized when treated with an acid. Draw the mechanism of formation of a segment of poly(ethylene oxide) under acidic conditions.

(c) Identify the monomer you would use to prepare the following polymer:

O, O, O, O

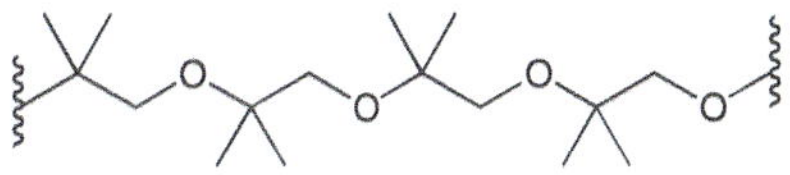

(d) Determine whether you would use basic conditions or acidic conditions to prepare the following polymer. Explain your choice.

O, O, O, O

**27.48** Using vinyl acetate as your only source of carbon atoms, design a synthesis for the following polymer:

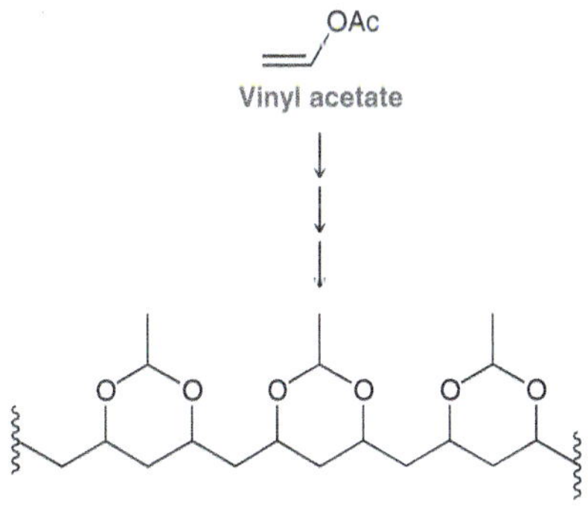

OAc

Vinyl acetate

O, O, O, O, O, O

28

# Introduction to Organometallic Compounds



## DID YOU EVER WONDER...

**how long it takes organic chemists to synthesize new drugs?**

Common sense dictates that it should depend on the complexity of the desired structure (a true assertion), although popular TV might have you believe that even complex structures can be made overnight. This is simply not true! Organic synthesis takes time, often a lot of time. Why? Each step of the synthesis requires detailed planning, assembling the glassware apparatus, running the reaction, isolating the desired product from the product mixture, purifying the product, and then confirming its structure via spectroscopic techniques, such as NMR and IR. A synthetic chemist must first complete all of these activities successfully before moving on to the next step of the synthesis. As a result, each step of the synthesis can take days—and that's in a best-case scenario, where everything works according to plan. Often, a planned step will fail, requiring the chemist to devise a way to circumvent the problematic step. This process can take weeks, months, or even years. In a multistep synthesis (with dozens of planned steps), the entire synthesis can often span years, if not decades.

The boundaries of organic chemistry are constantly evolving; the nonstop emergence and discovery of new reactions, methods, and techniques have enabled chemists to prepare increasingly complex structures in ever shorter time frames. In this chapter, we will explore a group of reactions, all discovered in recent decades, that have allowed for the construction of relatively sophisticated structures in relatively few steps. Many of these reactions employ special catalysts that belong to a class of compounds, called *organometallic compounds*. This chapter is therefore arranged to serve as a brief survey of organometallic compounds and their unique chemistry.

## DO YOU REMEMBER?

*Before you go on, be sure you understand the following topics.*
*If necessary, review the suggested sections to prepare for this chapter:*

- Induction and Polar Covalent Bonds (Section 1.5)
- Stereoisomerism in Alkenes – *E* and *Z* Stereoisomers (Section 8.4)
- Preparation of Alcohols via Grignard Reagents (Section 13.6)
- Carbon Nucleophiles (Section 20.10)

Take the DO YOU REMEMBER? QUIZ in **WileyPLUS** to check your understanding

## 28.1 General Properties of Organometallic Compounds

Compounds containing a bond between a carbon atom and a metal (C—M) are called **organometallic compounds**, and we have already encountered many such compounds in previous chapters, including the following:

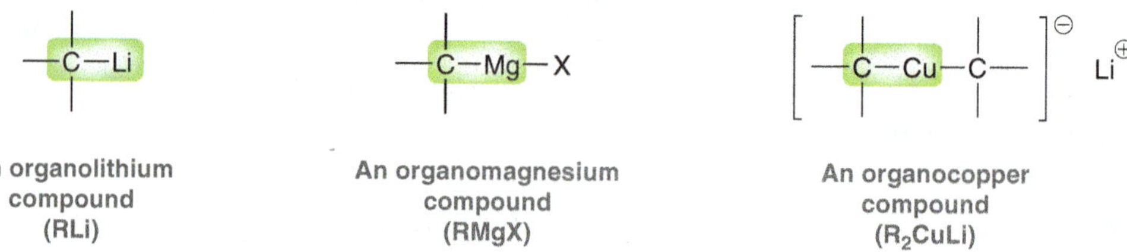


In each case, the highlighted carbon atom is more electronegative than the metal to which it is attached, and as a result, each of the highlighted carbon atoms is electron rich. This can be visualized if we compare electrostatic potential maps for methyl chloride and methyllithium:

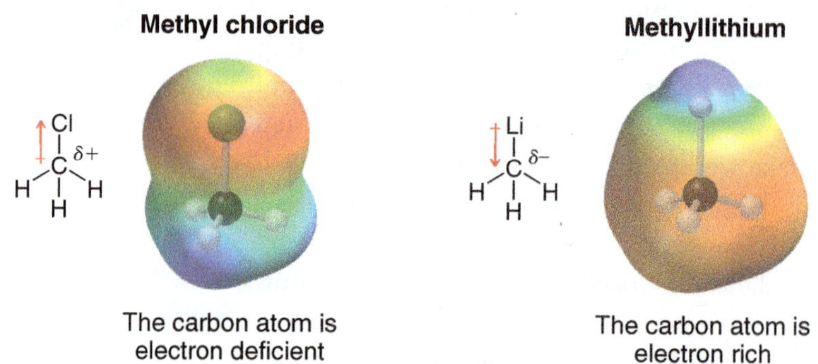


In methyl chloride, the chlorine atom withdraws electron density away from the carbon atom via induction, rendering the carbon atom electron deficient (δ+). In contrast, the carbon atom in methyllithium is electron rich (δ−), because the inductive effect occurs in the opposite direction (carbon withdraws electron density from the metal). Indeed, methyllithium is both a very strong base and a very strong nucleophile. This property makes organometallic reagents particularly useful in synthesis as a source of nucleophilic carbon.

For any C—M bond, the magnitude of the partial charge (δ−) on the carbon atom is dependent on the identity of the metal. Consider the electronegativity values of elements that we will encounter in this chapter:

Increasing electronegativity →

| Li | Mg | Zn | Mo | Cu | B | Ru | Pd | C |
|---|---|---|---|---|---|---|---|---|
| 1.0 | 1.2 | 1.6 | 1.8 | 1.9 | 2.0 | 2.2 | 2.2 | 2.5 |
| | | | | | | | Pt | |
| | | | | | | | 2.2 | |

The difference in electronegativity between C (2.5) and Mg (1.2) is much greater than the difference in electronegativity between C (2.5) and Cu (1.9), for example. As such, organomagnesium compounds (RMgX) are much stronger nucleophiles and bases than organocopper reagents ($R_2CuLi$). Indeed, in Chapter 21, we encountered a difference in the reactivity of these reagents. Recall that acid halides react with two equivalents of an organomagnesium (Grignard) reagent, but acid halides react with only one equivalent of an organocopper reagent (giving a ketone):

1) MeMgBr (2 eq.)
2) $H_2O$

$Me_2CuLi$

We have encountered yet another difference in reactivity between these reagents. In Section 22.6, we saw that organocopper reagents are effective Michael donors, while organomagnesium reagents are not. For example, the following transformation (a Michael reaction) cannot be efficiently achieved with an organomagnesium reagent (see Section 22.6):

1) $Me_2CuLi$
2) $H_2O$

### CONCEPTUAL CHECKPOINT

**28.1** Identify which of the following reagents is expected to be a stronger nucleophile and explain your choice:

PhMgBr PhZnBr

The use of transition metals (as described in the upcoming sections of this chapter) has revolutionized the field of synthetic organic chemistry, and the number of advancements in this field has grown exponentially in recent decades. One chapter is surely insufficient to cover the enormous field of organometallic chemistry, so this chapter is meant to serve only as an introduction.

## 28.2 Organolithium and Organomagnesium Compounds

### The Ionic Nature of RLi and RMgX

**LOOKING BACK**
Recall from Section 1.5 that a C—Li bond can be drawn either as polar covalent or as ionic (both drawings are acceptable). The same is true of a C—Mg bond; it can be drawn either as polar covalent or as ionic, as shown here.

Both organolithium (RLi) and organomagnesium (RMgX) compounds exhibit a C—M bond with a high degree of ionic character. As such, organolithium and organomagnesium compounds behave very much like carbanions:

$H_3C{-}Li \equiv H_3C{:}^{\ominus}\ Li^{\oplus}$ $\quad H_3C{-}MgX \equiv H_3C{:}^{\ominus}\ {}^{\oplus}MgX$

Carbanions are strong bases and strong nucleophiles. In fact, RLi and RMgX are among the strongest bases and strongest nucleophiles. Victor Grignard was awarded the 1912 Nobel Prize in Chemistry for his pioneering work on the preparation and use of organomagnesium compounds, now commonly called Grignard reagents. Since the reactivity of Grignard reagents is similar to the reactivity of organolithium reagents (RLi), we will cover both in the same section, beginning with their preparation.

## Preparation of RLi

Organolithium reagents can be prepared by treating an organohalide (RX) with two equivalents of Li:

$$\text{R—X} \xrightarrow{\text{Li (2 eq.)}} \text{R—Li} + \text{LiX}$$

The R group of the organohalide can be alkyl, aryl, or vinyl, as seen in the following examples:

$$\text{CH}_3\text{CH}_2\text{CH}_2\text{CH}_2\text{Cl} \xrightarrow[\text{(2 eq.)}]{\text{Li}} \text{CH}_3\text{CH}_2\text{CH}_2\text{CH}_2\text{Li} + \text{LiCl}$$

$$\text{C}_6\text{H}_5\text{—Br} \xrightarrow[\text{(2 eq.)}]{\text{Li}} \text{C}_6\text{H}_5\text{—Li} + \text{LiBr}$$

$$\text{CH}_2\text{=CH—I} \xrightarrow[\text{(2 eq.)}]{\text{Li}} \text{CH}_2\text{=CH—Li} + \text{LiI}$$

In each case, the transformation is believed to occur via radical intermediates, and a detailed mechanism is beyond the scope of our current discussion. A nonpolar solvent, such as hexane or diethyl ether, is typically used.

## Preparation of RMgX

Grignard reagents can be prepared by treating an organohalide (RX) with Mg in an ether solvent, such as diethyl ether, which results in the insertion of magnesium into the C—X bond:

$$\text{R—X} \xrightarrow[\text{Et}_2\text{O}]{\text{Mg}} \text{R—Mg—X}$$

Much as we saw in the formation of organolithium compounds, the formation of organomagnesium compounds is also believed to occur via radical intermediates. The R group can be alkyl, aryl, or vinyl, as seen in the following examples:

$$\text{CH}_3\text{CH}_2\text{CH}_2\text{CH}_2\text{Cl} \xrightarrow[\text{Et}_2\text{O}]{\text{Mg}} \text{CH}_3\text{CH}_2\text{CH}_2\text{CH}_2\text{MgCl}$$

$$\text{C}_6\text{H}_5\text{—Br} \xrightarrow[\text{Et}_2\text{O}]{\text{Mg}} \text{C}_6\text{H}_5\text{—MgBr}$$

$$\text{CH}_2\text{=CH—I} \xrightarrow[\text{Et}_2\text{O}]{\text{Mg}} \text{CH}_2\text{=CH—MgI}$$

Iodides are the most reactive (I > Br > Cl), although bromides are used more commonly, because bromides are less expensive and less toxic than iodides (and bromides get the job done!).

The preparation of a Grignard reagent requires the use of a solvent, such as diethyl ether, that can function as a Lewis base and stabilize the Grignard reagent as it is formed:

R—X $\xrightarrow[\text{Et}_2\text{O}]{\text{Mg}}$ R—Mg—X (δ+)

For less reactive halides, such as vinyl chlorides, tetrahydrofuran (THF) can be used in place of diethyl ether. THF has a higher boiling point, so the process can be performed at a higher temperature:

Cl → (Mg, THF) → MgCl

(THF)

The solvent, whether it is diethyl ether or THF, must be anhydrous (free of water). That is, care must be taken to remove even the trace amounts of water that are normally dissolved in ether, because the presence of even trace amounts of water will consume the Grignard reagent upon its formation. Similarly, the preparation of organolithium reagents must also be performed with anhydrous solvents, for the same reason. The reaction between Grignard reagents (or organolithium reagents) and water is shown below.

## RLi and RMgX Are Strong Bases

Grignard reagents (RMgX) and organolithium reagents (RLi) are both very strong bases, and each of them reacts rapidly with water to give a hydrocarbon (R—H), as shown:

R: MgX + H–O–H ⟶ R—H + HOMgX

($\mathrm{p}K_a = 15.7$) ($\mathrm{p}K_a \sim 50$)

Note the enormous difference in $\mathrm{p}K_a$ values between water (15.7) and the hydrocarbon (~50), rendering the process irreversible. A similar irreversible reaction occurs when RMgX or RLi is treated with $D_2O$. This reaction can be used to install a deuterium atom into a compound, in a process called *deuteration*:

Br $\xrightarrow[\text{Et}_2\text{O}]{\text{Mg}}$ MgBr $\xrightarrow{D_2O}$ D

Grignard reagents (and organolithium reagents) will not only react with $H_2O$ and $D_2O$, but they will also react with any functional group that exhibits a weakly acidic proton (OH, NH, SH, C≡CH, or COOH). Therefore, none of these functional groups can be present in the starting RX. For example, a Grignard reagent cannot be prepared if an OH group is present in the compound:

OH ... Br $\xrightarrow{\text{Mg}}$ (✗) OH ... MgBr

Not compatible

Cannot form this Grignard reagent

## CONCEPTUAL CHECKPOINT

**28.2** Draw the structures of compounds **A–E** in the following reaction scheme:

(4-iodotoluene) —Li (2 eq.)→ **A** —$H_2O$→ **B**

(benzyl iodide) —Mg, $Et_2O$→ **C** —$D_2O$→ **E**

**C** —**D**→ **B**

**28.3** Draw a mechanism for the conversion of compound **C** to compound **E** in the previous problem.

## RLi and RMgX Are Strong Nucleophiles

In previous chapters, we have encountered the use of Grignard reagents in many different reactions, which are summarized in Table 28.1. Organolithium reagents behave very much like Grignard reagents and will also undergo similar reactions.

**TABLE 28.1 PREVIOUSLY SEEN REACTIONS INVOLVING GRIGNARD REAGENTS**

| REACTION | SECTION |
|---|---|
| *Reaction with ketones and aldehydes*<br>ketone/aldehyde —1) RMgBr 2) $H_2O$→ alcohol (HO, R) | 13.6, 21.10 |
| *Reaction with epoxides*<br>RMgBr —1) epoxide 2) $H_2O$→ R–CH₂CH₂–OH | 14.10 |
| *Reaction with $CO_2$*<br>RMgBr —1) $CO_2$ 2) $H_3O^+$→ RCOOH | 21.4 |
| *Reaction with esters*<br>ester (OMe) —1) RMgBr (2 eq.) 2) $H_2O$→ alcohol (HO, R, R) | 21.11 |
| *Reaction with nitriles*<br>R'–C≡N —1) RMgBr 2) $H_3O^+$→ ketone (R) | 21.13 |

CONCEPTUAL CHECKPOINT

**28.4** Draw the structures of compounds **A–E** in the following reaction scheme:

1) MeMgBr 2) $H_3O^+$ → **A**; ethyl acetate, 1) **D** (2 eq.) 2) $H_2O$ → **B**; **A**, 1) PhMgBr 2) $H_2O$ → **B**; **A**, 1) EtMgBr 2) $H_2O$ → **C**; 1) **E** 2) $H_2O$ → **C**

**28.5** Draw the structures of compounds **A–C** in the following reaction scheme:

1) EtMgBr 2) $H_2O$ → **A**; 1) **B** 2) $H_2O$ → **A**; 1) **C** 2) $H_2O$ → **A**

**28.6** Draw the structure of the product that is expected when the following lactone (cyclic ester) is treated with two equivalents of MeMgBr followed by aqueous workup:

1) MeMgBr (2 eq.) 2) $H_2O$ → ?

## 28.3 Lithium Dialkyl Cuprates (Gilman Reagents)

### Preparation of Gilman Reagents

Lithium dialkyl cuprates ($R_2CuLi$) were previously encountered as carbon nucleophiles (Chapters 21 and 22). Specifically, we saw that these nucleophiles react with certain carbon electrophiles in a selective manner, as seen in Table 28.2.

Lithium dialkyl cuprates are often called Gilman reagents, in tribute to Henry Gilman (Iowa State University), who first prepared and studied them. These reagents can be prepared by treating an organohalide with lithium to give an organolithium, followed by treatment of the organolithium with CuX:

$$\underset{\text{An organohalide}}{R-X} \xrightarrow[(-\text{LiX})]{\text{Li (2 eq.)}} \underset{\text{An organolithium}}{R-Li} \xrightarrow[(-\text{LiX})]{\text{CuX (0.5 eq.)}} \underset{\text{Gilman reagent}}{[R-Cu-R]^{\ominus}\,Li^{\oplus}}$$

**TABLE 28.2 PREVIOUSLY SEEN REACTIONS INVOLVING LITHIUM DIALKYL CUPRATES**

| REACTION | CHAPTER | SELECTIVITY |
|---|---|---|
| Acid chloride $\xrightarrow{R_2CuLi}$ ketone (R) | Section 21.8 | An acid halide is converted into a ketone upon treatment with a lithium dialkyl cuprate. In contrast, treating an acid halide with a Grignard reagent initially gives a ketone that cannot survive under the conditions of its formation; the ketone is further attacked by a Grignard reagent, giving an alcohol. |
| α,β-Unsaturated ketone $\xrightarrow[\text{2) } H_2O]{\text{1) } R_2CuLi}$ β-R ketone | Section 22.6 | If a Grignard reagent had been used, it would have attacked the α,β-unsaturated ketone predominantly via 1,2-addition (attacking the carbonyl group). In contrast, lithium dialkyl cuprates are more selective and will attack the β position exclusively, in a 1,4-addition. |

The R group can be alkyl, aryl, or vinyl, as seen in the following examples:

1) Li (2 eq.)
2) CuI (0.5 eq.)
Cu
Li

Br
1) Li (2 eq.)
2) CuBr (0.5 eq.)
Cu
Li

I
1) Li (2 eq.)
2) CuI (0.5 eq.)
Cu
Li

Gilman reagents are nucleophiles, and as such, they will attack a variety of electrophiles. Of notable importance is the reaction that occurs when a Gilman reagent reacts with an organohalide, as we will now explore.

## Reaction with R′X (Coupling)

Gilman reagents ($R_2CuLi$) react with organohalides (R′X) to give a product in which the R and R′ groups have been coupled together:

$$R_2CuLi + R'{-}X \longrightarrow R{-}R' + RCu + LiX$$

Coupling product — By-products

This *coupling* reaction, called the Corey-Posner/Whitesides-House reaction, can be performed in diethyl ether or THF (as solvent) at low temperature (due to the thermal instability of Gilman reagents). The halogen can be I, Br, or Cl, with iodides being the most reactive and chlorides being the least reactive (I > Br > Cl). The R′ group of the organohalide (R′X) can be methyl, primary, vinyl, or aryl, as seen in the following general examples:

$R_2CuLi$ + I ⟶ R

$R_2CuLi$ + I ⟶ R

$R_2CuLi$ + I ⟶ R

R′X cannot be a tertiary alkyl halide, as elimination products predominate. Even secondary alkyl halides are generally ineffective, also because of competing elimination reactions. Nevertheless, some secondary alkyl halides, such as cyclic halides, are observed to couple effectively with Gilman reagents:

$R_2CuLi$ + I ⟶ R

In each of the previous four general reactions, R (of $R_2CuLi$) can be alkyl, aryl, or vinyl, giving rise to a wide variety of bonds that can be made with this process. During the process, only one of the two R groups (from $R_2CuLi$) is coupled to R′ of the organohalide (R′X).

The coupling process is stereospecific because it is observed to proceed with retention of configuration at each of the C=C units (in the Gilman reagent and in the organohalide), as seen in the following example:

(E) —Cu— $Li^{\oplus}$ + I (Z) ⟶ (E) (Z)

The mechanism of this process is not well understood and is likely dependent on the identity of R′ as well as the reaction conditions. The coupling process can occur in the presence of a variety of functional groups, including keto, ester, or amide groups. For example, consider the following transformation, in which the presence of the ester group does not interfere with the reaction:

$Ph_2CuLi$ + I ... OMe ⟶ Ph ... OMe

## Retrosynthesis

From a retrosynthetic perspective, coupling allows us to make R—R′ from two organohalides, RX and R′X, highlighted in the following retrosynthetic scheme:

R—R′ ⟹ $R_2CuLi$ + R′X
⇓
RX

The forward reaction sequence can be shown in the following way:

R—X (Organohalide) —1) Li (2 eq.) 2) CuI (0.5 eq.)→ $R_2CuLi$ —R′—X (Organohalide)→ R—R′ (Coupling product)

When R ≠ R′, the process is called a *cross-coupling* reaction. When these coupling reactions were first discovered, Gilman reagents were commonly used by chemists in the synthesis of a wide range of target compounds. But in the past few decades, more versatile reagents have been developed for achieving coupling reactions, as will be discussed in Sections 28.5–28.8.

SKILLBUILDER

## 28.1 IDENTIFYING THE PARTNERS FOR A COREY-POSNER/WHITESIDES-HOUSE COUPLING REACTION

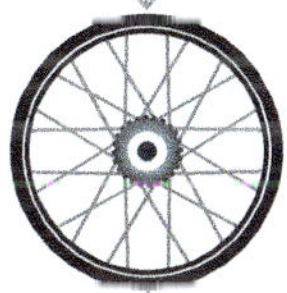

**LEARN the skill**

Show how you would prepare 1-butylcyclopentene from 1-bromobutane and any other organohalide of your choice:

1-Butylcyclopentene ⟹ Br (1-Bromobutane) + RX

### SOLUTION

**STEP 1**
Analyze the structure of the product and determine which bond must be made via a coupling process.

In order to determine which bond (in the product) will be made via a coupling process, we compare the structure of the product (1-butylcyclopentene) and the structure of the given starting material (1-bromobutane):

Product (1, 2, 3, 4) ⟹ Br (1, 2, 3, 4) Starting material

Note that the desired product contains a butyl group, so this butyl group must come from 1-bromobutane. Therefore, we draw a wavy line in between the ring and the butyl group to indicate the bond that must be made via coupling.

**STEP 2**
Identify the two organohalides that are necessary.

Once we have identified which bond will be made via coupling, we can draw the two organohalides that are necessary for the coupling process. One of the organohalides has already been identified in the problem statement (1-bromobutane). The other organohalide (of our choice) must be a vinyl halide, so we draw an iodide (because iodides are more reactive than bromides or chlorides):

I + Br

**STEP 3**
Draw the forward scheme, showing both possible synthetic routes (either organohalide can be converted into a Gilman reagent).

Now that we have identified the two starting organohalides, one of them must be converted into a Gilman reagent and then treated with the other. This leads to two possible synthetic routes, both of which are viable. In the first possibility, the vinyl iodide is converted into a Gilman reagent and then treated with 1-bromobutane, as shown:

**Possibility 1**

I — 1) Li (2 eq.) 2) CuI (0.5 eq.) → [ —Cu— ]⊖ Li⊕ — Br → 

In the second possibility, 1-bromobutane is converted into a Gilman reagent and then treated with a vinyl iodide, as shown:

**Possibility 2**

Br — 1) Li (2 eq.) 2) CuI (0.5 eq.) → [ —Cu— ]⊖ Li⊕ — I → 

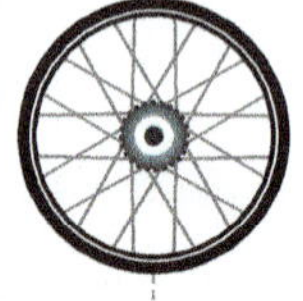

**PRACTICE the skill** **28.7** For each of the following cases, show how you would make the desired product from the organohalide shown and any other organohalide (RX) of your choice:

(a) ⟹ I + RX

(b) ⟹ I + RX

(c) ⟹ Br + RX

(d) ⟹ I + RX

**28.8** Using any two organohalides of your choice (where each organohalide can have no more than six carbon atoms), show how you would prepare each of the following compounds:

(a) (b) (c)

**APPLY the skill**

**28.9** Using styrene as your only source of carbon atoms, show how you would prepare 1,4-diphenylbutane:

**1,4-Diphenylbutane** $\Longrightarrow$ **Styrene**

need more **PRACTICE?** **Try Problems 28.40a, 28.43, 28.48**

## 28.4 The Simmons-Smith Reaction and Carbenoids

Compounds with a C—Zn bond are called organozinc compounds, and they can be prepared in much the same way that Grignard reagents are prepared. Specifically, an alkyl halide is treated with the metal (Zn in this case) which inserts itself into the C—X bond:

$$\text{R—X} \xrightarrow[\text{Et}_2\text{O}]{\text{Zn}} \text{R—Zn—X}$$

An ether, such as diethyl ether ($Et_2O$), is used to stabilize the resulting organozinc compound. Carbon is more electronegative than zinc, so induction gives the carbon atom a partial negative charge ($\delta-$). This was also the case with Grignard reagents, although Grignard reagents have much greater ionic character than organozinc reagents. A C—Zn bond should be viewed as polar covalent rather than ionic.

One very useful organozinc reagent, called (iodomethyl)zinc iodide ($ICH_2ZnI$), was developed in 1959 by Howard E. Simmons and Ronald D. Smith, working for DuPont. This reagent can be prepared by treating diiodomethane ($CH_2I_2$) with a zinc copper couple (Zn–Cu), which is formed by treating zinc dust with CuCl or $CuSO_4$:

$$\text{I—CH}_2\text{—I} \xrightarrow[\text{Et}_2\text{O}]{\text{Zn–Cu}} \text{I—CH}_2\text{—Zn—I}$$

**(Iodomethyl)zinc iodide**

The structure of $ICH_2ZnI$ is actually more complex than shown and is beyond the scope of our current discussion. Simmons and Smith demonstrated that alkenes react with $ICH_2ZnI$ to form a three-membered ring. This process is called *cyclopropanation*:

$$\text{R}_2\text{C=CR}_2 \xrightarrow[\text{Et}_2\text{O}]{\text{CH}_2\text{I}_2,\ \text{Zn–Cu}} \text{cyclopropane (R, R, R, R)}$$

Notice that one methylene ($CH_2$) group is added across the $\pi$ bond, thereby converting the alkene into a cyclopropane derivative. The following are several examples of Simmons-Smith cyclopropanation:

$CH_2I_2$, Zn–Cu / $Et_2O$

$CH_2I_2$, Zn–Cu / $Et_2O$

$CH_2I_2$, Zn–Cu / $Et_2O$

The cyclopropanation process is observed to be stereospecific, with the configuration of the alkene being preserved in the cyclopropane product, as seen in the following examples:

(trans) $CH_2I_2$, Zn–Cu / $Et_2O$ (trans) + En

(cis) $CH_2I_2$, Zn–Cu / $Et_2O$ (cis)

In all of the previous examples, $ICH_2ZnI$ functions as a source of a methylene ($CH_2$) group. In this way, the reactivity of $ICH_2ZnI$ resembles the reactivity of carbenes, which can also be used to achieve cyclopropanation. Carbenes are generally short-lived intermediates in which a carbon atom has two bonds and two unshared electrons:

**A carbene**

This carbon atom lacks an octet of electrons, accounting for the electrophilicity of carbene intermediates. One useful carbene, called dichlorocarbene, can be prepared by treating chloroform ($CHCl_3$) with a strong base, such as potassium *tert*-butoxide. Under these conditions, deprotonation is followed by loss of a leaving group, giving dichlorocarbene:

**$CHCl_3$** **Dichlorocarbene**

This process is called $\alpha$-elimination (as opposed to $\beta$-elimination, seen in Chapter 8), because the proton and the leaving group are removed from the same carbon atom, rather than from adjacent carbon atoms. When dichlorocarbene is prepared in the presence of an alkene, the carbene reacts with the alkene to give a cyclopropane ring, as shown:

Notice that the carbene intermediate functions both as a nucleophile (it has a lone pair) and as an electrophile (it lacks an octet of electrons). This ring-forming process, called cyclopropanation, is observed to be stereospecific, with the configuration of the alkene being preserved in the cyclopropane product, as seen in the following examples:

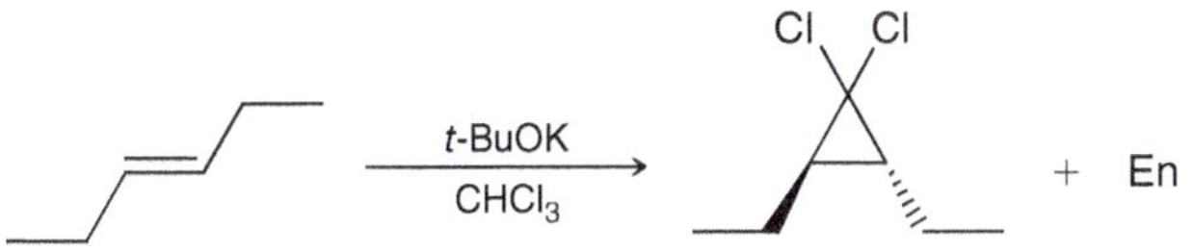

(*trans*) $\xrightarrow[CHCl_3]{t\text{-BuOK}}$ (*trans*) + En

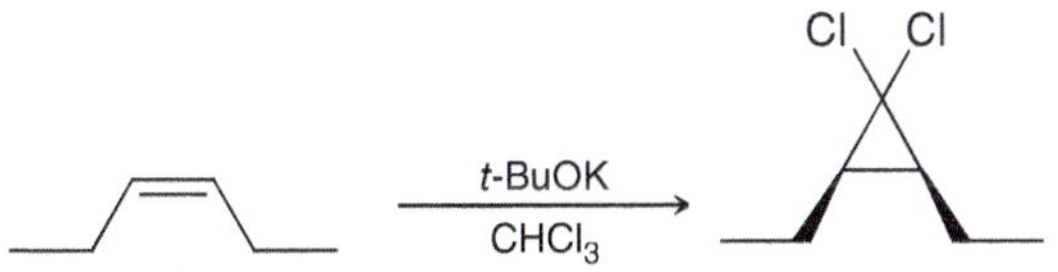

(*cis*) $\xrightarrow[CHCl_3]{t\text{-BuOK}}$ (*cis*)

Direct cyclopropanation of alkenes (without the chlorine atoms present) is more difficult to achieve with carbenes, but it can be achieved readily with the procedure developed by Simmons and Smith. That is, the Simmons-Smith reaction resembles the reactions between carbenes and alkenes, although a different mechanism operates. As seen in Mechanism 28.1, free carbenes are not involved under Simmons-Smith conditions.

## MECHANISM 28.1 SIMMONS-SMITH CYCLOPROPANATION

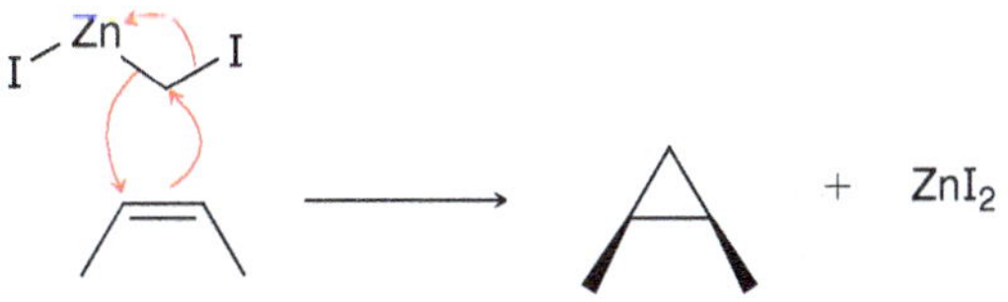

$\longrightarrow$ + $ZnI_2$

The reaction is believed to occur in a concerted manner, thus explaining why the configuration of the starting alkene is preserved throughout the process. The Simmons-Smith procedure can be used to achieve cyclopropanation in good yields, while avoiding the formation of undesired by-products that often accompany reactions involving reactive carbene intermediates. Since the action of $ICH_2ZnI$ resembles the action of carbenes, it is commonly referred to as a *carbenoid* reagent.

## CONCEPTUAL CHECKPOINT

**28.10** Draw the product that is expected when each of the following alkenes is treated with $CH_2I_2$ and Zn-Cu:

(a) 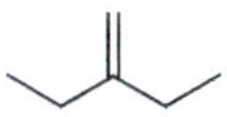 (b) 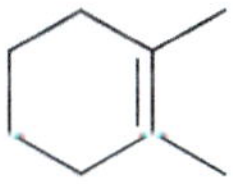 (c)

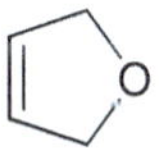

(d) 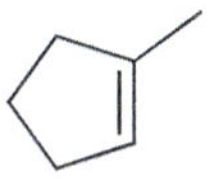 (e)

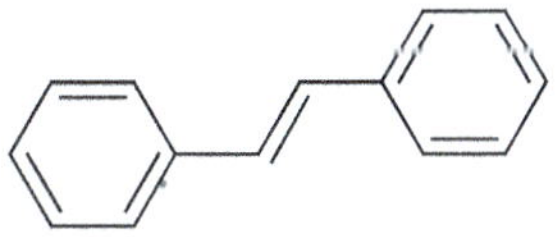

**28.11** When treated with molecular hydrogen ($H_2$) in the presence of $PtO_2$, cyclopropanes readily undergo carbon-carbon bond cleavage as a means to relieve ring strain, in a process known as *hydrogenolysis*:

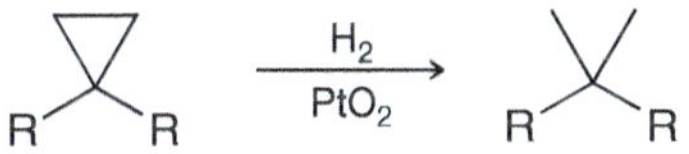

$\xrightarrow[PtO_2]{H_2}$

This hydrogenolysis procedure was successfully employed during the total synthesis of spongian-16-one, a unique sponge metabolite that exhibits potential anticancer properties (*Tet. Lett.*, **1996**, *37*, 4191–4194). Show how compound **1** can be converted into spongian-16-one:

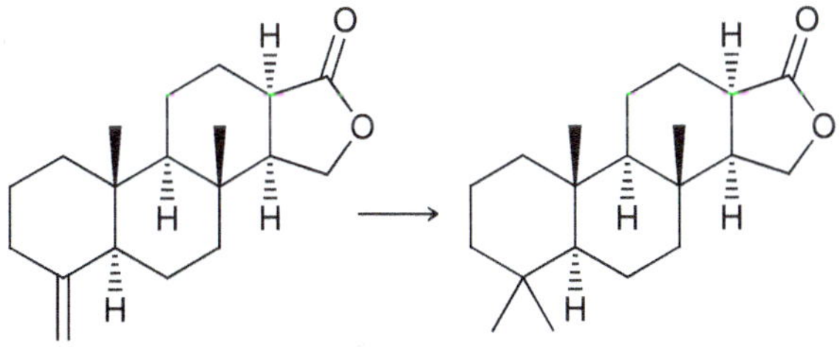

**1** $\longrightarrow$ **Spongian-16-one**

## 28.5 Stille Coupling

### Introduction to Palladium Compounds

In the periodic table shown in Figure 28.1, the orange region represents *transition metal elements*, which are characterized by partly filled *d* (or *f*) orbitals.

Transition metal elements can react with molecules or ions, giving rise to complexes that exhibit useful properties and behavior. A very large number of transition metal complexes have been developed for use as catalysts in performing synthetic transformations that would otherwise be difficult or even impossible to achieve. Over the past several decades, the advent of these incredibly useful catalysts has transformed the field of synthetic organic chemistry. Indeed, as we will soon see, several Nobel prizes in chemistry have been awarded for the development of synthetic techniques that employ transition metal catalysts. In particular, palladium (Pd) catalysts have been among the most important catalysts available to practicing synthetic organic chemists, and as such, most of the remainder of this chapter will focus on the chemistry of palladium catalysts.

As is characteristic of transition metals, Pd reacts with a variety of molecules or ions, called *ligands*, to give transition metal complexes. For example, in the reaction below, the first structure has two ligands, while the second structure has four ligands:

Cl—Pd—Cl (Two ligands) → :P(Ph)₃ (2 eq.) → Cl—Pd—Cl with PPh$_3$ above and below (Four ligands)

The second structure illustrates the accepted convention to omit formal charges on the metal and on the ligands (for example, each of the phosphorus atoms has a positive formal charge that is not indicated). Ligands donate electron density to vacant orbitals, giving rise to bonds (Pd—L) that are generally polar covalent. These ligands are weakly associated with the palladium center (they are labile). As a result, the number of ligands that are bonded to palladium can change, depending on the

**FIGURE 28.1**
A periodic table of the elements, in which the orange region represents transition metals.

PERIODIC TABLE OF THE ELEMENTS

Atomic number : 6
Symbol : C
Name (IUPAC) : Carbon
Atomic mass : 12.011

IUPAC recommendations:
Chemical Abstracts Service group notation :

| 1 IA | 2 IIA | 3 IIIB | 4 IVB | 5 VB | 6 VIB | 7 VIIB | 8 VIIIB | 9 VIIIB | 10 VIIIB | 11 IB | 12 IIB | 13 IIIA | 14 IVA | 15 VA | 16 VIA | 17 VIIA | 18 VIIIA |
|---|---|---|---|---|---|---|---|---|---|---|---|---|---|---|---|---|---|
| 1 H Hydrogen 1.0079 | | | | | | | | | | | | | | | | | 2 He Helium 4.0026 |
| 3 Li Lithium 6.941 | 4 Be Berylium 9.0122 | | | | | | | | | | | 5 B Boron 10.811 | 6 C Carbon 12.011 | 7 N Nitrogen 14.007 | 8 O Oxygen 15.999 | 9 F Fluorine 18.998 | 10 Ne Neon 20.180 |
| 11 Na Sodium 22.990 | 12 Mg Magnesium 24.305 | | | | | | | | | | | 13 Al Aluminum 26.982 | 14 Si Silicon 28.086 | 15 P Phosphorus 30.974 | 16 S Sulfur 32.065 | 17 Cl Chlorine 35.453 | 18 Ar Argon 39.948 |
| 19 K Potassium 39.098 | 20 Ca Calcium 40.078 | 21 Sc Scandium 44.956 | 22 Ti Titanium 47.867 | 23 V Vanadium 50.942 | 24 Cr Chromium 51.996 | 25 Mn Manganese 54.938 | 26 Fe Iron 55.845 | 27 Co Cobalt 58.933 | 28 Ni Nickel 58.693 | 29 Cu Copper 63.546 | 30 Zn Zinc 65.409 | 31 Ga Gallium 69.723 | 32 Ge Germanium 72.64 | 33 As Arsenic 74.922 | 34 Se Selenium 78.96 | 35 Br Bromine 79.904 | 36 Kr Krypton 83.798 |
| 37 Rb Rubidium 85.468 | 38 Sr Strontium 87.62 | 39 Y Yttrium 88.906 | 40 Zr Zirconium 91.224 | 41 Nb Niobium 92.906 | 42 Mo Molybdenum 95.94 | 43 Tc Technetium (98) | 44 Ru Ruthenium 101.07 | 45 Rh Rhodium 102.91 | 46 Pd Palladium 106.42 | 47 Ag Silver 107.87 | 48 Cd Cadmium 112.41 | 49 In Indium 114.82 | 50 Sn Tin 118.71 | 51 Sb Antimony 121.76 | 52 Te Tellurium 127.60 | 53 I Iodine 126.90 | 54 Xe Xeno 131.29 |
| 55 Cs Caesium 132.91 | 56 Ba Barium 137.33 | 57 *La Lanthanum 138.91 | 72 Hf Hafnium 178.49 | 73 Ta Tantalum 180.95 | 74 W Tungsten 183.84 | 75 Re Rhenium 186.21 | 76 Os Osmium 190.23 | 77 Ir Iridium 192.22 | 78 Pt Platinum 195.08 | 79 Au Gold 196.97 | 80 Hg Mercury 200.59 | 81 Tl Thallium 204.38 | 82 Pb Lead 207.2 | 83 Bi Bismuth 208.98 | 84 Po Polonium (209) | 85 At Astatine (210) | 86 Rn Radon (222) |

conditions, although transition metals are most stable when they achieve the electronic configuration of a noble gas. We have seen similar behavior with second-row elements, although second-row elements can only accommodate 8 valence electrons (the valence shell has only one *s* orbital and three *p* orbitals, each of which can accommodate 2 electrons). In contrast, the valence shell of transition metal elements can accommodate 18 electrons (the valence shell has one *s* orbital, three *p* orbitals, and five *d* orbitals). Thus, the *18-electron rule* for transition metals parallels the octet rule for second-row elements.

A transition metal is said to be *coordinatively saturated* if it has 18 electrons. For example, $Pd(PPh_3)_4$ features a palladium atom that is coordinatively saturated because palladium has 10 valence electrons (when the oxidation state is 0, as it is in this case), and each of the ligands provides 2 electrons, for a total of 18 electrons.

$PPh_3$ / $Ph_3P—Pd—PPh_3$ / $PPh_3$

Coordinatively saturated
(18 $e^-$)

Complexes in which the palladium atom has *fewer* than 18 electrons are said to be *coordinatively unsaturated*, and such complexes will often function as reactive intermediates in the reactions that we will soon encounter. Throughout the remainder of this chapter, we will see many useful synthetic transformations that utilize Pd complexes as catalysts.

## Stille Coupling

Stille coupling, pioneered by John Stille at Colorado State University, has emerged as a powerful and versatile transformation used frequently by practicing synthetic organic chemists. Stille coupling, which occurs in the presence of a suitable palladium catalyst, is the reaction between an organostannane (a compound with a C—Sn bond, also called an organotin compound) and an organic electrophile to form a new C—C σ bond:

$$R—SnBu_3 + R'—X \xrightarrow{\text{Pd catalyst}} R—R' + X—SnBu_3$$

This coupling reaction between R and R′ can be achieved with a catalyst such as $Pd(PPh_3)_4$, in which the palladium atom has an oxidation state of 0. Alternatively, Pd(II) complexes can be used (where the oxidation state of Pd is +2), such as $Pd(OAc)_2$ or $PdCl_2(PPh_3)_2$. When a Pd(II) complex is used, it is converted into a Pd(0) complex, which is believed to function as the active catalyst.

In the Stille coupling process, each of the coupling partners (R and R′) can be aryl or vinyl groups, and X can be a halogen (Cl, Br, or I) or a triflate (OTf).

**LOOKING BACK**
Triflate (OTf) groups were first introduced in Figure 7.28.

When both partners are aryl groups, the product is a substituted biaryl system, as seen in the following example:

Ph—$SnBu_3$ + I—$C_6H_4$—OMe $\xrightarrow{Pd(OAc)_2}$ Ph—$C_6H_4$—OMe + I—$SnBu_3$

New σ bond

In this way, Stille coupling provides a convenient method for joining two aromatic systems into biaryl structures (a common structural feature in many pharmaceutical agents).

When one coupling partner is aryl and the other coupling partner is vinyl, the product is a styrene derivative, as seen in the following example:

$CH_2$=CH—$SnBu_3$ + Br—$C_6H_4$—$NO_2$ $\xrightarrow{Pd(OAc)_2}$ $CH_2$=CH—$C_6H_4$—$NO_2$ + Br—$SnBu_3$

New σ bond

If both coupling partners are vinyl, then the product will have a conjugated diene group, as seen in the following example:

SnBu$_3$ + TfO– (cyclohexenyl) —Pd(PPh$_3$)$_4$→ (product) + TfO—SnBu$_3$

New σ bond

(TfO = –O–S(=O)(=O)–CF$_3$)

Trifluoromethane sulfonate (triflate)

As such, Stille coupling is a useful way to prepare conjugated dienes, and the process is observed to be stereospecific. That is, the process generally proceeds with retention of configuration at each of the C=C units, as seen in the following example:

SnBu$_3$ + I —Pd(PPh$_3$)$_4$→ + I—SnBu$_3$

E E E E

Notice that in all of the previous examples of Stille coupling, the new σ bond is formed between two $sp^2$-hybridized centers, enabling the assembly of the following types of bonds:

$C_{sp^2}$—$C_{sp^2}$ $C_{sp^2}$—$C_{sp^2}$ $C_{sp^2}$—$C_{sp^2}$

Stille coupling can be extended to include formation of a σ bond between $sp^3$- and $sp^2$-hybridized centers when an allylic or benzylic stannane is used, as seen in the following examples:

SnBu$_3$ + I —Pd(PPh$_3$)$_4$→ + I—SnBu$_3$

An allylic stannane

New σ bond ($C_{sp^3}$—$C_{sp^2}$)

SnBu$_3$ + I —Pd(PPh$_3$)$_4$→ + I—SnBu$_3$

A benzylic stannane

New σ bond ($C_{sp^3}$—$C_{sp^2}$)

These reactions further extend the utility of the Stille coupling process. Indeed, the Stille coupling process has been extended in many other ways, providing great versatility to organic chemists, and a full treatment is beyond the scope of our discussion.

## Preparation and Stability of Organostannanes

Many organostannanes are commercially available, and they can be stored for periods of time, as they are not sensitive to oxygen or moisture. They can also be readily prepared from Grignard reagents, upon treatment with tributyltin chloride ($Bu_3SnCl$), as shown:

$$R{-}X \xrightarrow[Et_2O]{Mg} RMgX \xrightarrow{Bu_3SnCl} R{-}SnBu_3$$

Organostannanes are so useful because they are extremely selective in their reactivity. In the presence of a palladium catalyst, they react with aryl and vinyl halides or triflates, even when other functional groups are present, including carboxylic acids, amides, esters, nitro groups, ethers, amines, hydroxyl groups, ketones, and aldehydes. Therefore, the process can be performed even when each of the coupling partners bears other functional groups, as seen in following example:

SnBu3 + I CF3 → Pd(PPh3)4 → CF3 + I—SnBu3

## A Mechanism for Stille Coupling

The detailed mechanism of Stille coupling is still debated, but it is accepted to be comprised of three stages: oxidative addition, transmetallation, and reductive elimination. We will now briefly explore each of these stages:

1. **Oxidative Addition** The electrophile (R′X) reacts with an active palladium complex giving a new complex in which both R′ and X are each bonded directly to palladium:

$$R'{-}X \;+\; PdL_2 \xrightarrow{\text{Oxidative addition}} L{-}Pd(R')(L){-}X$$

During this addition process, the oxidation state of the palladium atom increases, and the process is therefore called an oxidative addition. The ligands (L) are not identified, because their identity (and, indeed, even the number of ligands) can change during the process. So the convention is to show only the crucial reacting groups.

2. **Transmetallation** The R group of the organostannane is transferred to the palladium complex and takes the place of the X group:

$$R{-}SnBu_3 \;+\; L{-}Pd(R')(L){-}X \xrightleftharpoons{\text{Transmetallation}} X{-}SnBu_3 \;+\; L{-}Pd(R')(L){-}R$$

The term transmetallation indicates that the R group was transferred from one metal (Sn) to another (Pd). This step is generally the rate-determining step of the overall coupling process, and the rate of transmetallation is dependent on the identity of the R group. The following relative rates have been observed:

vinyl > aryl > allyl ~ benzyl >>>> alkyl

Vinyl groups migrate most rapidly, while alkyl groups migrate most slowly. Indeed, the low rate of migration of alkyl groups is exploited during Stille coupling, which typically employs organostannanes with three alkyl groups (either three butyl groups or three methyl groups):

$R-SnBu_3$ $\qquad$ $R-SnMe_3$

A vinyl, aryl, allyl, or benzyl R group will migrate much more rapidly than the other three alkyl groups (Bu or Me), enabling the selective migration of the R group during the transmetallation step.

3. **Reductive Elimination** Both R and R′ are released from the palladium complex, with formation of a new σ bond between the coupling partners:

$$L_2Pd(R')(R) \xrightarrow{\text{Reductive elimination}} R-R' + PdL_2$$

During this process, the oxidation state of palladium is reduced, thereby regenerating the active Pd(0) catalyst, $PdL_2$.

Mechanism 28.2 illustrates that all three stages (oxidative addition, transmetallation, and reductive elimination) comprise a catalytic cycle that consumes the starting materials (an organostannane and an organic electrophile), produces the coupling product (R—R′), and regenerates the catalyst. Notice that curved arrows do not appear in Mechanism 28.2, because the mechanism for each stage of the catalytic cycle is likely dependent on conditions. Throughout the rest of this chapter, we will explore many other coupling reactions, and in each case, the catalytic cycle will be shown without curved arrows.

## MECHANISM 28.2 STILLE COUPLING

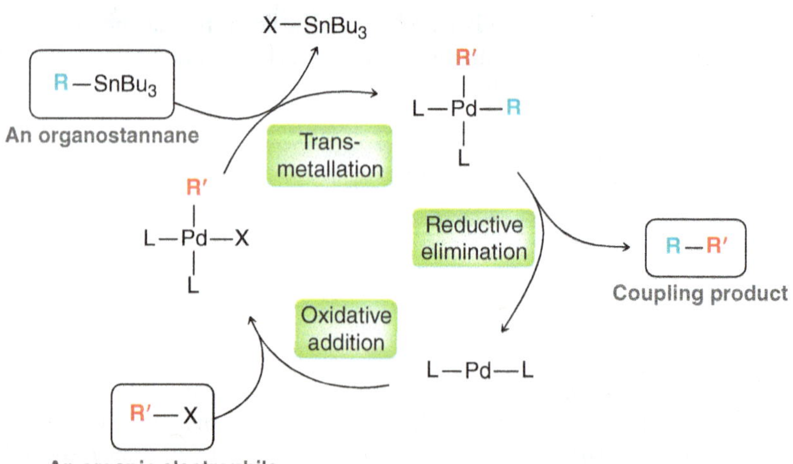

# SKILLBUILDER

## 28.2 PREDICTING THE PRODUCT OF A STILLE COUPLING REACTION

**LEARN the skill**

Draw the coupling product that is expected when the following aryl iodide is treated with the following organostannane in the presence of $Pd(PPh_3)_4$:

### SOLUTION

**STEP 1** Identify the carbon atoms that will be joined via a new σ bond.

Begin by identifying the carbon atoms that will be joined. The carbon atom connected directly to iodide (in the aryl iodide) will be joined with the carbon atom connected directly to tin (Sn) in the organostannane, highlighted here:

**STEP 2** Rotate the coupling partners.

In order to draw the coupling product more easily, rotate each of the coupling partners, so that they are properly aligned to form a bond between them, as shown:

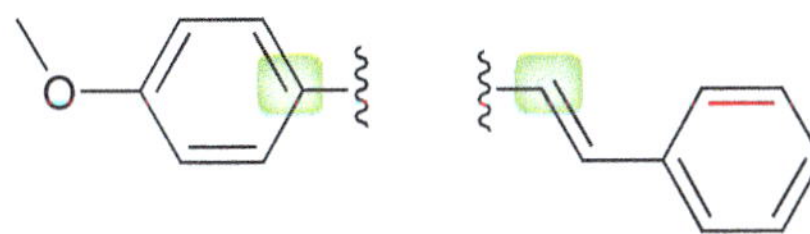

**STEP 3** Draw the product, making sure to retain the configuration of any C=C units.

Finally, when drawing the product, make sure that the configuration of each C=C unit is retained. In this case, notice that the *E* configuration of the organostannane is preserved in the coupling product.

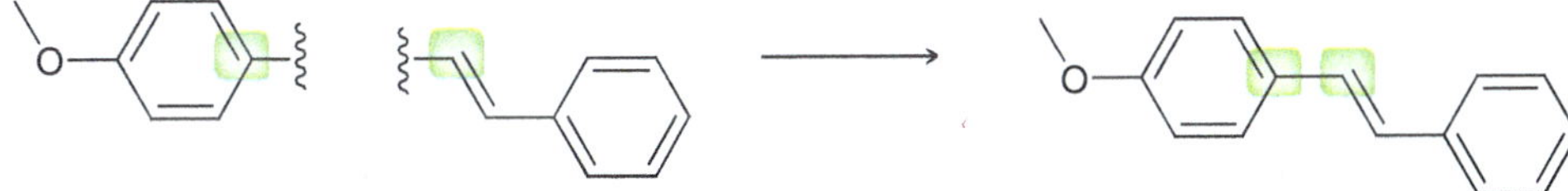

**PRACTICE the skill**

**28.12** For each of the following cases, draw the coupling product that is expected when the organic electrophile is treated with the organostannane in the presence of catalytic $Pd(PPh_3)_4$:

**(a)** 

**(b)** 

**(c)** 

**(d)** 

**(e)** 

**(f)** 

**APPLY the skill**

**28.13** A compound possessing a vinyl triflate group as well as a trimethylstannane group, such as compound **1**, can undergo an intramolecular Stille coupling reaction (*Tetrahedron*, **1991**, *47*, 4555–4570). This annulation (ring-making) procedure provides an efficient pathway for the construction of a variety of bicyclic diene systems. Draw the structure of the coupling product (compound **2**) that is formed when **1** is treated with $Pd(PPh_3)_4$:

OTf
$SnMe_3$
$CO_2Me$
**1**
$Pd(PPh_3)_4$
**2**
$C_{13}H_{18}O_2$

**28.14** (*S*)-Zearalenone, a natural product isolated from the fungus *Gibberella zeae*, exhibits useful biological activity, including antibiotic properties. In a total synthesis of (*S*)-zearalenone (*J. Org. Chem.*, **1991**, *56*, 2883-2894), an intramolecular Stille coupling process was employed as the penultimate procedure. The resulting coupling product was then treated with HCl to cleave the OR groups, thereby converting them into OH groups (as seen in Section 14.6), giving (*S*)-zearalenone. Draw the structure of (*S*)-zearalenone:

OR O O
$SnBu_3$
O
RO I
1) $Pd(PPh_3)_4$
2) HCl
(*S*)-Zearalenone
$(C_{18}H_{22}O_5)$

**28.15** (+)-Jatrophone, a natural product isolated from the flowering plant *Jatropha gossypiifolia*, is known to function as a tumor inhibitor and has been used to treat cancerous growths. As part of the first total synthesis of (+)-jatrophone, compound **1** was prepared and then treated with compound **2** in the presence of $Pd(PPh_3)_4$ to give compound **3** (*J. Am. Chem. Soc.*, **1992**, *114*, 7692–7697). Draw the structure of **3**.

O
O
OTf
**1**
$Me_3Sn$ OTMS
**2**
$Pd(PPh_3)_4$
**3**
O
O
(+)-Jatrophone O

need more **PRACTICE?** **Try Problems 28.33, 28.53, 28.65**

## 28.6 Suzuki Coupling

### Reagents and Conditions for Suzuki Coupling

Suzuki coupling, named after Professor Akira Suzuki (Hokkaido University), is very similar to Stille coupling, but an organoboron compound is used instead of an organotin compound:

R—B
**An organoboron compound**

R—Sn—
**An organotin compound**

Much like Stille coupling, a palladium catalyst such as $Pd(PPh_3)_4$ is employed, but Suzuki coupling also requires basic conditions (we will soon explore the function of the base). For a Suzuki coupling, the coupling partners are an organoboron compound ($RBY_2$) and an organic electrophile (R′X), giving R—R′ as the coupling product:

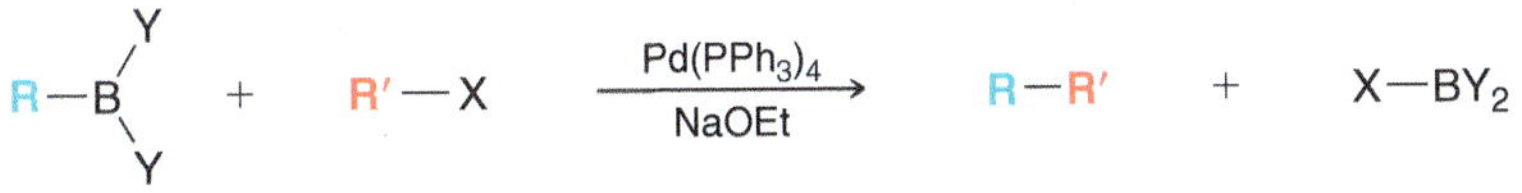

$$R-BY_2 + R'-X \xrightarrow[\text{NaOEt}]{Pd(PPh_3)_4} R-R' + X-BY_2$$

In the organoboron compound, the Y groups can be alkyl groups, alkoxy groups, or hydroxyl groups, depending on the method of preparation. These compounds are called organoboranes, boronic esters, and boronic acids, respectively:

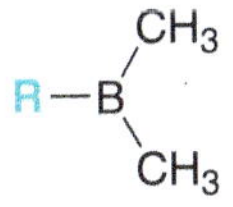

$R-B(CH_3)_2$ — An organoborane

$R-B(OCH_3)_2$ — A boronic ester

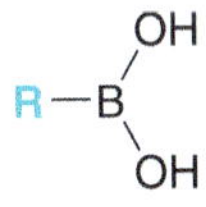

$R-B(OH)_2$ — A boronic acid

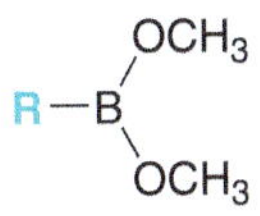

Just as we saw for Stille coupling, the R group (of the organoboron compound) can be vinyl, aryl, or allyl. The following are all suitable organoboron compounds for Suzuki coupling:

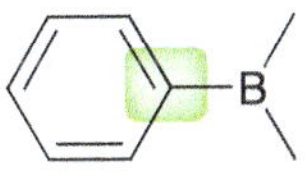

An aryl borane

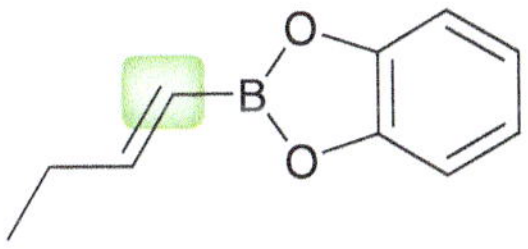

A vinyl boronic ester

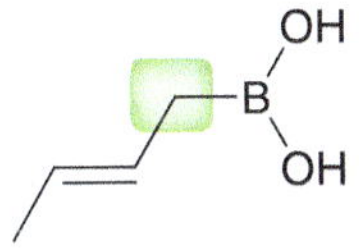

An allyl boronic acid

Now let's explore the structure of the organic electrophile (R′X) in Suzuki coupling. Again, just as we saw with Stille coupling, R′ can be aryl or vinyl, and X can be a halide or a triflate group (although chlorides are generally much less reactive). The following are all suitable organic electrophiles (R′X) for Suzuki coupling:

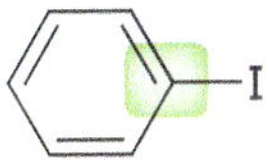

An aryl iodide

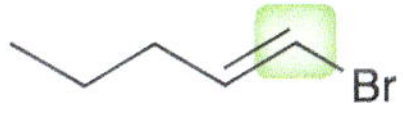

A vinyl bromide

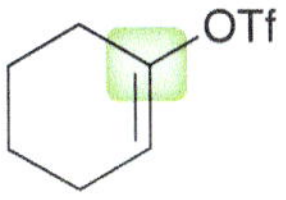

A vinyl triflate (OTf)

Two examples of Suzuki coupling are shown below:

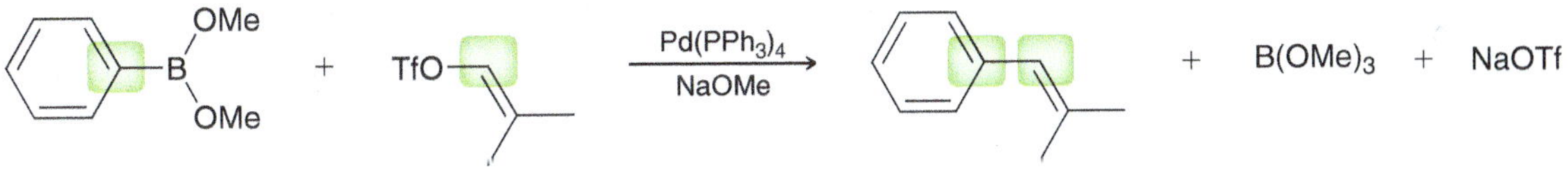

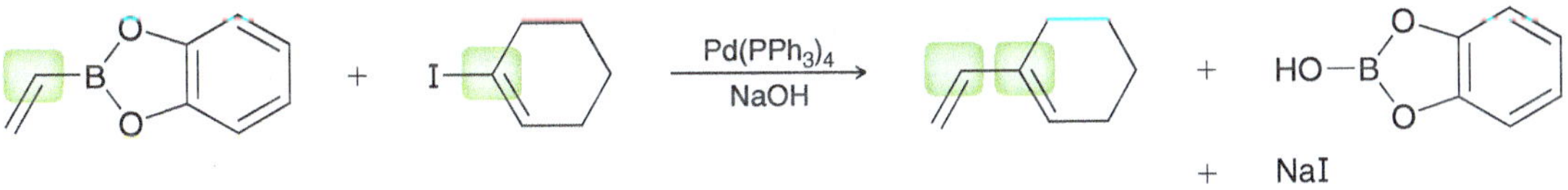

$$PhB(OMe)_2 + TfO\text{–vinyl} \xrightarrow[\text{NaOMe}]{Pd(PPh_3)_4} \text{product} + B(OMe)_3 + NaOTf$$

$$\text{vinyl boronic ester} + I\text{–cyclohexenyl} \xrightarrow[\text{NaOH}]{Pd(PPh_3)_4} \text{product} + HO-B(\text{catechol}) + NaI$$

Suzuki coupling offers several advantages over Stille coupling. One significant advantage of Suzuki coupling is the ability to use an alkyl borane in the coupling process, as seen in the following example:

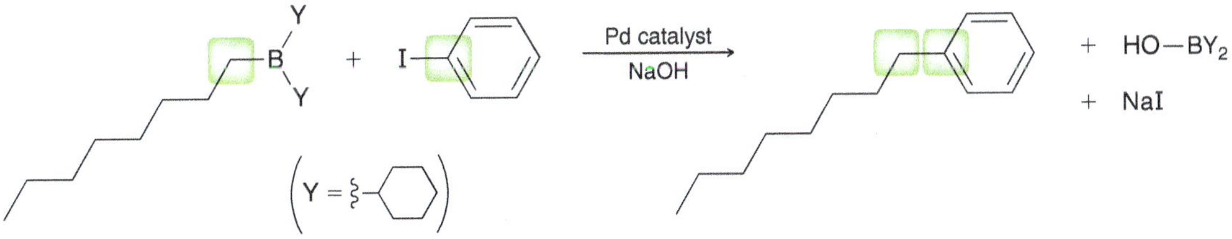

$$R-BY_2 + I-Ph \xrightarrow[\text{NaOH}]{\text{Pd catalyst}} R-Ph + HO-BY_2 + NaI$$

(Y = cyclohexyl)

This provides a general method for making a bond between $sp^3$- and $sp^2$-hybridized centers. Suzuki coupling therefore provides greater versatility than Stille coupling in the types of bonds that can be made, as summarized in Table 28.3. Other advantages of Suzuki coupling include the ease with

**TABLE 28.3 BONDS THAT CAN BE MADE VIA STILLE COUPLING AND/OR VIA SUZUKI COUPLING**

| Stille or Suzuki Coupling | | | Suzuki Coupling |
|---|---|---|---|
| C—C ($sp^2$, $sp^2$) | C—C ($sp^2$, $sp^2$) | C—C ($sp^2$, $sp^2$) | C—C ($sp^3$, $sp^2$) |
| C—C ($sp^3$, $sp^2$) | C—C ($sp^3$, $sp^2$) | | C—C ($sp^3$, $sp^2$) |

which inorganic by-products are removed as well as the lower cost and lower toxicity of organoboron compounds relative to organotin compounds. One disadvantage of the Suzuki coupling reaction (relative to Stille coupling) is the requirement for basic conditions. Under these conditions, if either coupling partner has a keto or aldehyde group that bears α protons, aldol-type reactions can occur to give undesired side products. Both Stille coupling and Suzuki coupling are frequently employed by practicing synthetic organic chemists, and the choice (of which method to use) will generally be determined by the details of the particular transformation that is desired.

Suzuki coupling is particularly useful for making conjugated dienes, and the process is observed to be stereospecific. That is, the process generally proceeds with retention of configuration at each of the C=C units, as seen in the following example:

(*E*) + Br— (*Z*) $\xrightarrow[\text{NaOEt}]{\text{Pd(PPh}_3)_4}$ (*E*) (*Z*) + EtO—B + NaBr

## Preparation of Organoboron Compounds

Organoboron compounds can be readily prepared from a variety of starting materials. For example, vinyl boronic esters and vinyl boranes can be prepared via hydroboration of alkynes using catechol borane or 9-BBN, respectively:

H—B (catecholborane)

H—B (9-BBN)

Similarly, alkyl boronic esters and alkyl boranes can be prepared via hydroboration of alkenes using catechol borane or 9-BBN, respectively:

(catecholborane)

(9-BBN)

And aryl boronic esters can be made by treating the corresponding aryllithium compound with trimethylborate, $B(OMe)_3$:

$B(OMe)_3$ + LiOMe

## A Mechanism for Suzuki Coupling

Suzuki coupling is believed to occur via a mechanism that is similar to the mechanism shown for Stille coupling in the previous section. That is, the catalytic cycle begins with oxidative addition and ends with transmetallation and reductive elimination (thereby regenerating the active palladium catalyst). But with Suzuki coupling, which occurs in basic conditions, there is an additional step that occurs as part of the catalytic cycle (see Mechanism 28.2). In this additional step, called displacement, the X group (halide or triflate) is displaced by an alkoxide ion, thereby rendering the palladium complex more reactive toward transmetallation.

### MECHANISM 28.3 SUZUKI COUPLING

An organoboron compound

R—$BY_2$

EtO—$BY_2$

Transmetallation

Displacement

NaX

NaOEt

Reductive elimination

Oxidative addition

R—R′

Coupling product

R′—X

An organic electrophile

During Suzuki coupling, the base also plays another critical role. Not only does it activate the catalyst for the transmetallation step, but it also activates the organoboron compound by converting it into an organoborate anion, which undergoes transmetallation more rapidly:

R—B(Y)(Y) + $Na^{\oplus}$ $^{\ominus}$:ÖEt → R—$B^{\ominus}$(Y)(Y)—OEt $Na^{\oplus}$

An organoborate anion

Commonly used bases include NaOH and NaOEt, although many other bases are often used.

## Suzuki Coupling in Synthesis

Suzuki coupling tolerates the presence of a wide range of functional groups, although as mentioned earlier, ketones and aldehydes can give aldol-type reactions in basic conditions. This issue can be mitigated by using milder conditions. Indeed, a variety of modified reagents and conditions have been developed to further enhance the versatility of Suzuki coupling. A full treatment of Suzuki coupling is beyond the scope of our discussion.

Suzuki coupling is generally inefficient when the organic electrophile (R′X) is an alkyl halide. Under the basic conditions required for Suzuki coupling, an alkyl halide can undergo undesired elimination (E2) reactions. Nevertheless, recent advances have begun to circumvent this problem as well. Specifically, conditions and reagents have been developed that allow for the coupling between two $sp^3$-hybridized centers under certain circumstances. These developments, and many others, continue to improve the significant versatility of Suzuki coupling.

For his pioneering work on palladium-catalyzed cross-coupling reactions, Suzuki shared in the 2010 Nobel Prize in Chemistry.

# SKILLBUILDER

## 28.3 PREDICTING THE PRODUCT OF A SUZUKI COUPLING REACTION

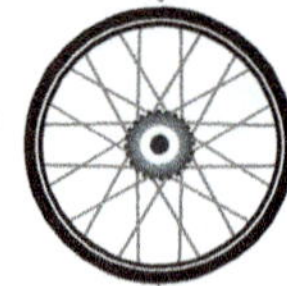

**LEARN** the skill

Bombykol is a sex pheromone used by silkworms (females release it to attract the males), and it can be prepared in the laboratory via a Suzuki coupling reaction between the two partners shown below:

HO–(chain)–CH=CH–B(OH)(OH) + vinyl bromide (Br)

The synthetic production of bombykol provides farmers with powerful methods for preventing outbreaks that can destroy crops. Specifically, bombykol can be released to confuse the males (thereby interfering with the mating process) or it can be used as a lure to trap the males. These methods reduce the need for powerful insecticides. Draw the structure of bombykol.

### SOLUTION

**STEP 1**
Identify the carbon atoms that will be joined via a new σ bond.

Begin by identifying the carbon atoms that will be joined. The carbon atom connected directly to boron in the organoboron compound will be joined with the carbon atom connected directly to bromide (in the vinyl bromide), highlighted here:

HO–(chain)–CH=CH–B(OH)(OH) + vinyl bromide (Br)

**STEP 2**
Rotate the coupling partners.

In order to draw the coupling product more easily, rotate the coupling partners, so that they are properly aligned to form a bond between them, as shown:

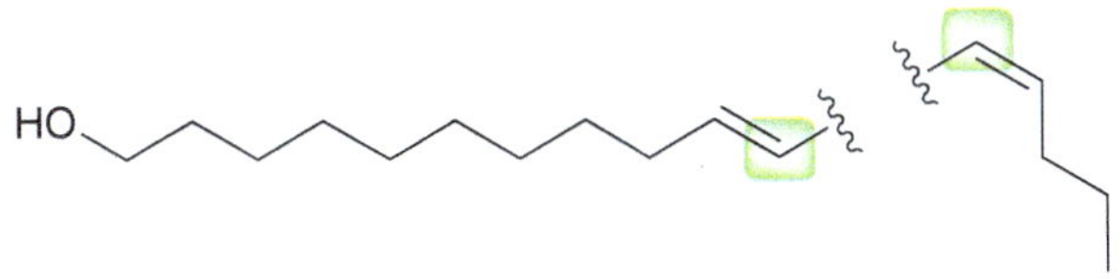

HO

**STEP 3**
Draw the product, making sure to retain the configuration of any C=C units.

Finally, when drawing the product, make sure that the configuration of each C=C unit is retained. In this case, notice that the *E* configuration of the organoboron compound and the *Z* configuration of the vinyl bromide are both preserved in the coupling product:

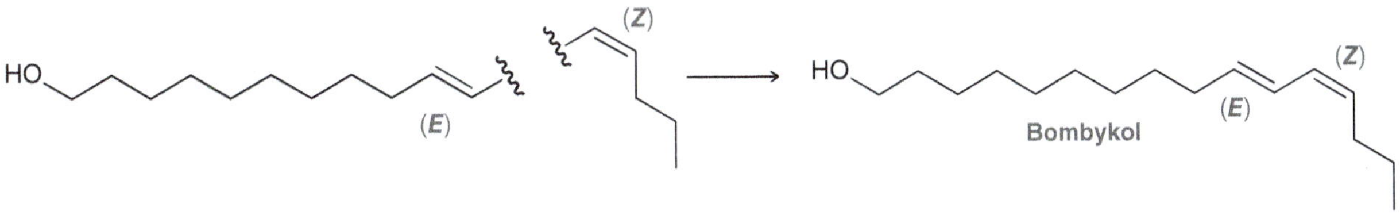

HO (*E*) (*Z*) → HO (*E*) (*Z*) **Bombykol**

**PRACTICE the skill** **28.16** For each of the following cases, draw the coupling product that is expected when the organic electrophile is treated with the organoboron compound in the presence of a base and catalytic $Pd(PPh_3)_4$:

**(a)** OTf + O–B–O OMe

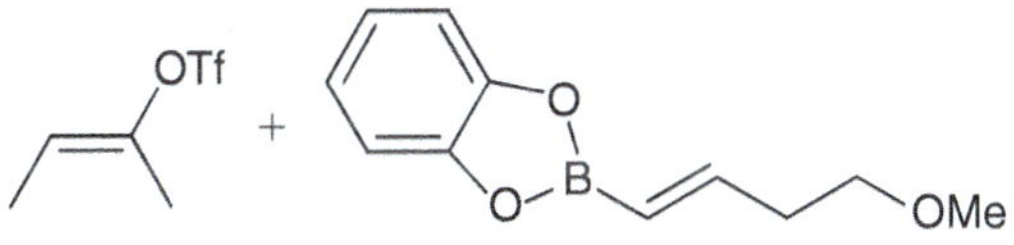

**(b)** Br + $B(OH)_2$

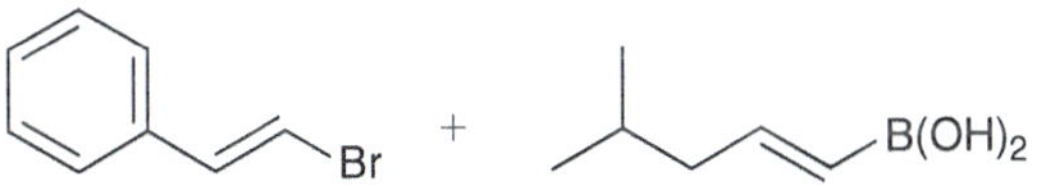

**(c)** I + HO–B–OH O

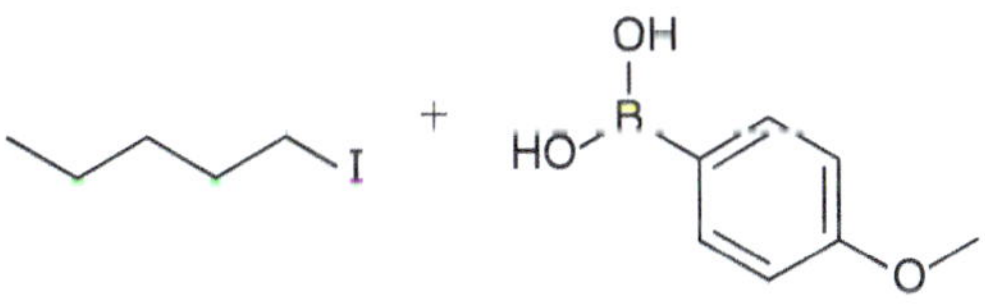

**(d)** MeO O I + MeO–B–OMe

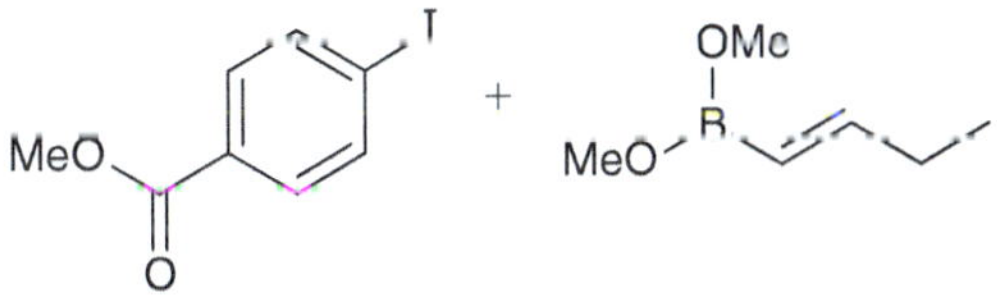

**28.17** Draw the structures of boronic ester A and coupling product B in the following reaction sequence:

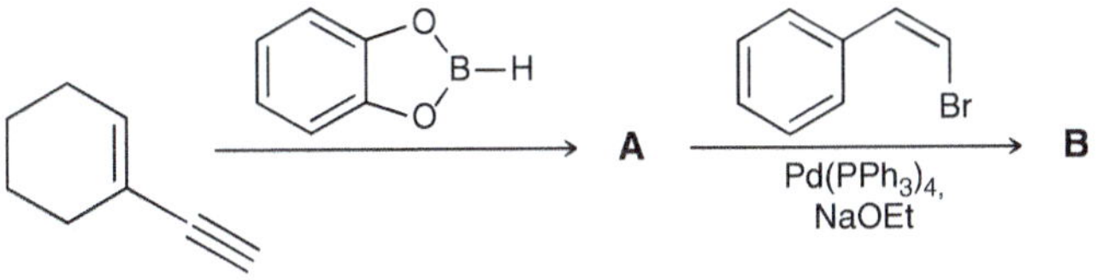

O–B–H–O → **A** Br, $Pd(PPh_3)_4$, NaOEt → **B**

**APPLY** the skill

**28.18** Taxol, used in the treatment of breast cancer, is produced by fungi in the bark of the Pacific Yew tree. Taxol is part of a class of structurally related natural compounds called taxanes that represent particularly challenging synthetic targets for organic chemists due to the complexity of their structure. In an effort to prepare a key eight-membered ring of taxol, compound **1** was prepared and treated with 9-BBN to give compound **2**, followed by addition of a palladium catalyst under basic conditions, giving compound **3** (*Org. Lett.*, **2004**, *6*, **4491–4494**). Draw structures for compounds **2** and **3**.

**1** $\xrightarrow{\text{9-BBN}}$ **2** $\xrightarrow{\text{Pd(PPh}_3)_4\text{, NaOH}}$ **3**

**28.19** By virtue of noncovalent π–π interactions, compound **1** represents a novel blue-transparent frequency doubler that has proven useful in the field of nonlinear optics. Compound **1** can be prepared via sequential Suzuki cross-coupling reactions (*Angew. Chem. Int. Ed.*, **1995**, *34*, **1485–1488**). First, compound **A** is treated with an aryl iodide to give compound **B**, which is then treated with an aryl boronic acid to produce **1**.

**1** ⟹ **B** ⟹ **A**

**(a)** Draw structures for the requisite aryl iodide, boronic acid, and compound **B**.

**(b)** Explain why **A** cross couples with the aryl iodide, rather than two molecules of **A** coupling together.

need more **PRACTICE?** **Try Problems 28.40b–c, 28.44, 28.45, 28.68, 28.71**

## 28.7 Negishi Coupling

### Reagents and Scope for Negishi Coupling

The Negishi coupling is another versatile reaction that utilizes an organometallic species containing zinc, aluminum, or zirconium:

R—Zn— **An organozinc**  R—Al **An organoaluminum**  R—Zr— **An organozirconium**

The organometallic reagent is coupled to an organic electrophile in the presence of a transition metal catalyst (palladium or nickel). With a few exceptions, this section will focus primarily on a major subset of this reaction: palladium-catalyzed couplings of organozinc reagents, as shown below.

$$\text{R—ZnX} + \text{R}'\text{—X} \xrightarrow[\text{(cat)}]{\text{Pd(PPh}_3)_4} \text{R—R}' + \text{ZnX}_2$$

This class of reactions is named for Professor Ei-ichi Negishi (Purdue University), who shared the 2010 Nobel Prize in Chemistry with Akira Suzuki (Section 28.6) and Richard Heck (Section 28.8) for their work on palladium-catalyzed cross couplings.

Organozinc compounds are generally more reactive than the corresponding organostannanes and organoboranes, in part due to the low electronegativity of zinc (1.6) relative to tin (1.8) and boron (2.0). The greater difference in electronegativity between the metal and carbon gives rise to a more polar carbon-metal bond and thus a more reactive species. This increase in reactivity in some cases leads to efficient Negishi coupling reactions where the analogous Stille and Suzuki reactions fail or are sluggish. A disadvantage of the increased reactivity is that many organozinc reagents are air sensitive and water sensitive, necessitating the use of equipment to ensure that the reaction is run in an inert atmosphere. Recent advances, including the preparation of salt-stabilized organozinc reagents, have somewhat diminished this problem.

Successful Negishi coupling reactions have been reported using a wide variety of organozinc reagents (RZnX), including those where R = vinyl, aryl, alkynyl, benzyl, and alkyl.

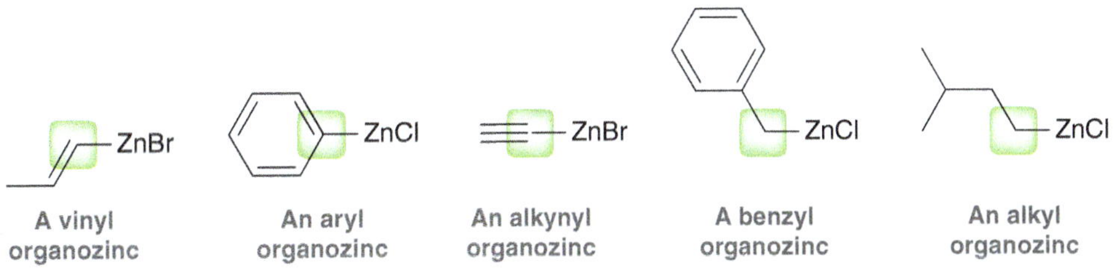

ZnBr ZnCl ZnBr ZnCl ZnCl

A vinyl organozinc | An aryl organozinc | An alkynyl organozinc | A benzyl organozinc | An alkyl organozinc

Likewise, a broad range of organic electrophiles (R′X) are known to be successful coupling partners such as those where R′ = allyl, benzyl, vinyl, aryl, and alkyl.

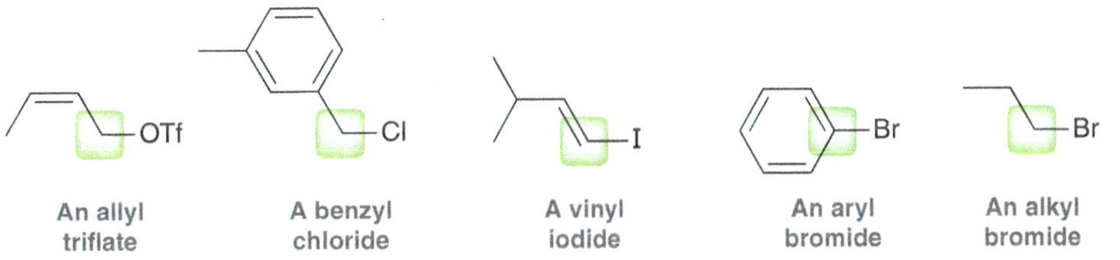

OTf Cl I Br Br

An allyl triflate | A benzyl chloride | A vinyl iodide | An aryl bromide | An alkyl bromide

The general order of reactivity of carbon electrophiles is as follows: allyl > benzyl > vinyl > aryl > alkyl. In the following Negishi coupling, the vinyl iodide is reported to be ≥100 times as reactive as the primary alkyl iodide, resulting in the formation of the product shown in high yield (96%). Note that the reaction proceeds with retention of configuration of the C=C unit (>98% *E*).

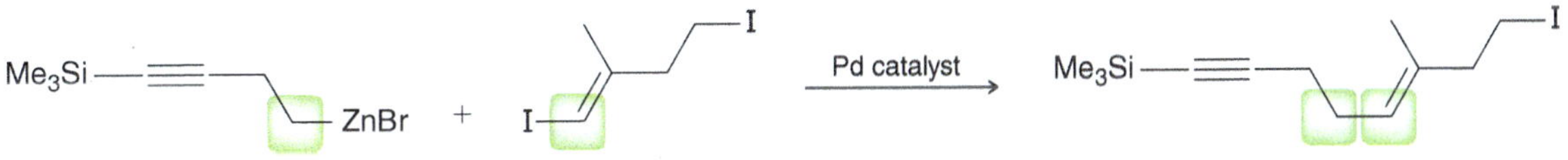

$Me_3Si$ ZnBr + I I —Pd catalyst→ $Me_3Si$ I

The stereoselectivity of Negishi coupling is often dependent on the metal catalyst. For example, consider the coupling reaction between an *E* vinyl organoaluminum and a *Z* vinyliodide, shown here:

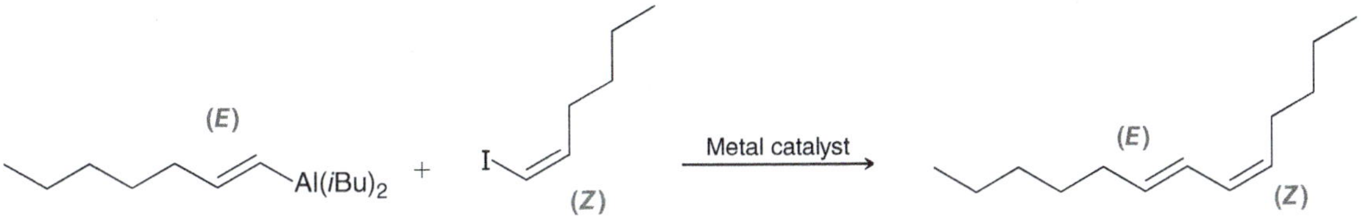

(*E*) $Al(iBu)_2$ + I (*Z*) —Metal catalyst→ (*E*) (*Z*)

Use of a palladium catalyst results in a highly stereoselective process, where the configuration of each C=C unit is retained in the product (>99% *E,Z*). When a nickel transition metal catalyst is used instead, the stereospecificity is diminished (90% *E,Z*). Note that in the organoaluminum reactant, the aluminum has one vinyl group and two alkyl groups (*i*Bu). As in Stille coupling, the vinyl group migrates more rapidly from the aluminum to the palladium in the transmetallation step, thus resulting in selective coupling of the vinyl group over the alkyl groups.

The scope of effective Negishi reactions, which includes combinations of the coupling partners listed above among others, is too broad to fully cover here. While not every combination of organozinc reagent with an organic electrophile produces the desired cross-coupling product in high yield, and some pairings require special catalysts or additives, many of them are successful. For example, alkynylzincs are known to successfully couple with a variety of organic electrophiles in the presence of a transition metal catalyst, as seen in the following reactions:

$R^1$—≡—ZnX → (X–Ph) → $R^1$—≡—Ph

$R^1$—≡—ZnX → (X—≡—$R^2$) → $R^1$—≡—≡—$R^2$

$R^1$—≡—ZnX → (X–C($R^2$)=C($R^3$)($R^4$)) → $R^1$—≡—C($R^2$)=C($R^3$)($R^4$)

As with Suzuki coupling, Negishi coupling can even be used to form a new bond between two $sp^3$-hybridized carbons in certain circumstances, often with the use of unconventional catalysts and/or additives.

## Preparation of Organozinc Reagents

Organozinc reagents are generally prepared in one of two ways. One common method is to treat a carbanion (organolithium compound or Grignard reagent) with $ZnX_2$, where X = Cl, Br, or I:

$$\text{MeO–C}_6\text{H}_4\text{–Li} \xrightarrow{ZnBr_2} \text{MeO–C}_6\text{H}_4\text{–ZnBr} + \text{LiBr}$$

$$\text{CH}_3\text{CH}_2\text{CH=CH–MgBr} \xrightarrow{ZnBr_2} \text{CH}_3\text{CH}_2\text{CH=CH–ZnBr} + \text{MgBr}_2$$

Alternatively, organozinc reagents can be prepared by the addition of zinc metal to an ether solution of the aryl, vinyl, or alkyl halide.

$$\text{vinyl–Br} \xrightarrow[Et_2O]{Zn} \text{vinyl–ZnBr}$$

$$\text{CH}_2\text{=CH–CH}_2\text{CH}_2\text{CH}_2\text{–I} \xrightarrow[Et_2O]{Zn} \text{CH}_2\text{=CH–CH}_2\text{CH}_2\text{CH}_2\text{–ZnI}$$

Palladium-catalyzed cross-coupling reactions using organometallic reagents containing Al, Zr, or Cu can often be greatly accelerated by the addition of $ZnCl_2$ or $ZnBr_2$. The addition of a $ZnX_2$ catalyst can increase the reaction rate several thousand times or in some cases can produce a successful cross coupling that would otherwise fail in the absence of the additive. It has been proposed that the organometallic reagents undergo *in situ* reversible transmetallation with $ZnX_2$ to produce the more reactive organozinc reagents. Recently, $InCl_3$ (which generates organoindium reagents *in situ*) has been demonstrated as an even more effective additive than $ZnBr_2$ or $ZnCl_2$ for the acceleration of similar cross-coupling reactions.

## A Mechanism for Negishi Coupling

The generally accepted mechanism for Negishi coupling is a catalytic cycle analogous to that of Stille coupling. Oxidative addition of an organic electrophile to the active palladium species results in a new (oxidized) palladium complex. In the transmetallation step, the organic group from the organozinc reagent is transferred to the palladium complex and $ZnX_2$ is expelled. Finally, the reductive elimination step produces the coupled product (R—R′) and regenerates the active palladium complex.

### MECHANISM 28.4 NEGISHI COUPLING

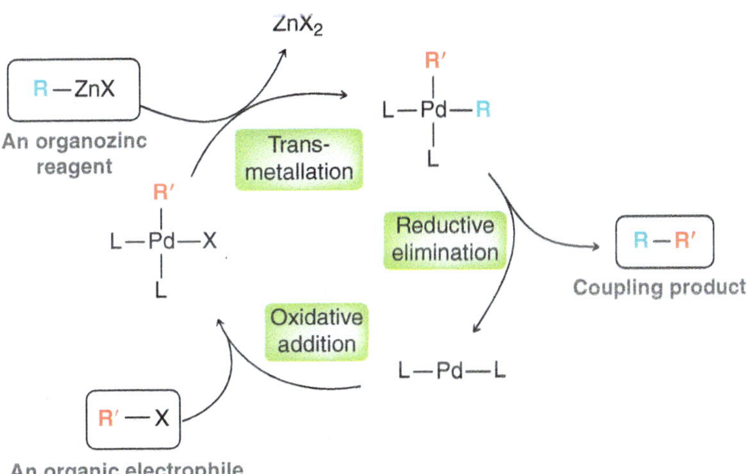


## Negishi Coupling in Synthesis

The carbon-carbon bond formation of the Negishi coupling provides access to a diversity of molecular structures and has been used extensively in the preparation of natural products and pharmaceuticals. In 2001, a total synthesis of several carotenoids, including vitamin A, γ-carotene, and β-carotene (shown below), was reported. Each of these natural products was prepared by the use of a series of organometallic reactions. In fact, each of the indicated C—C bonds in β-carotene was produced by a Pd- and Zn-catalyzed Negishi cross coupling.

β-Carotene

The final step in the synthesis of β-carotene is show here:

$AlMe_2$ (2 eq.)

(*E*) I–CH=CH–Br, $ZnCl_2$, Pd catalyst, phosphine ligand

(*E*)

β-Carotene

Two equivalents of the vinylaluminum reagent were combined with (*E*)-1-bromo-2-iodoethene in this tandem palladium-catalyzed cross coupling. The addition of $ZnCl_2$ accelerates the reaction by undergoing a transmetallation with the vinylaluminum as described above. The (*E*)-1-bromo-2-iodoethene allows for two sequential cross couplings: first with the more reactive vinyl iodide followed by the vinyl bromide to produce the symmetric β-carotene in 68% yield. Note that the *E* configuration of the bromoiodoethylene is retained in the product.

## 28.4 PREDICTING THE PRODUCT OF A NEGISHI COUPLING REACTION

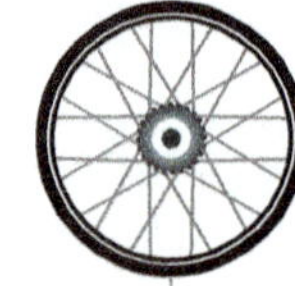

**LEARN** the skill

Coenzyme $Q_{10}$ (also called ubiquinone) is a natural product that is found throughout the human body. It resides in the nonpolar regions of biological membranes and is involved in a number of essential biochemical processes, including energy generation. It can be prepared using a Negishi cross-coupling reaction between the following two partners:

$Me_2Al$ ( )$_9$ + MeO, MeO, O, O, Cl

Coenzyme $Q_{10}$ is sold as a dietary supplement that is purported to produce a number of benefits, including an energy boost after exercise and an immune system aid. Draw the structure of coenzyme $Q_{10}$.

### SOLUTION

**STEP 1** Identify the carbon atoms that will be joined via a new σ bond.

Begin by identifying the carbon atoms that will be joined. The carbon atom connected directly to aluminum will be joined with the carbon atom connected to the chlorine atom, highlighted here:

$Me_2Al$ ( )$_9$ + MeO, MeO, O, O, Cl

**STEP 2**
Redraw the coupling partners, so that they are oriented in a way that more clearly shows where the bond will be formed.

Redraw the coupling partners so that the two highlighted carbon atoms are adjacent and in an orientation that will make it easy to draw the product.

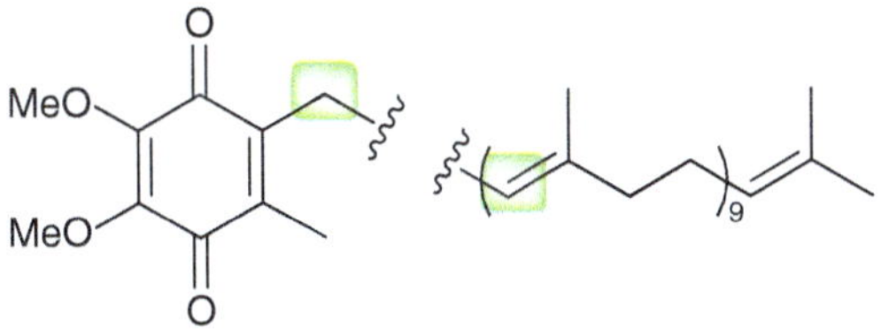

**STEP 3**
Draw the product, making sure to retain the configuration of any C=C units.

Finally, when drawing the product, make sure that the configuration of each C=C unit is retained. In this case, notice that the *E* configuration of the organoaluminum is preserved in the coupling product.

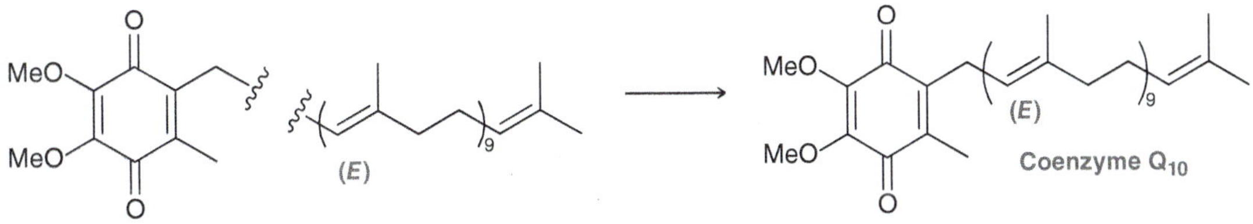

Coenzyme $Q_{10}$

**PRACTICE the skill** **28.20** For each of the following cases, draw the coupling product that is expected when the organic electrophile is treated with the organozinc in the presence of a base and catalytic $Pd(PPh_3)_4$:

(a)

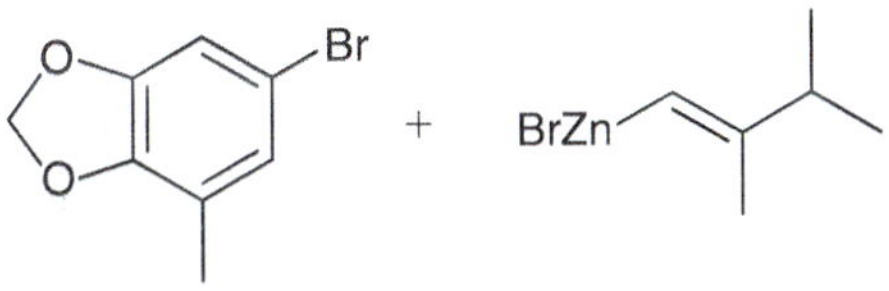

(b)

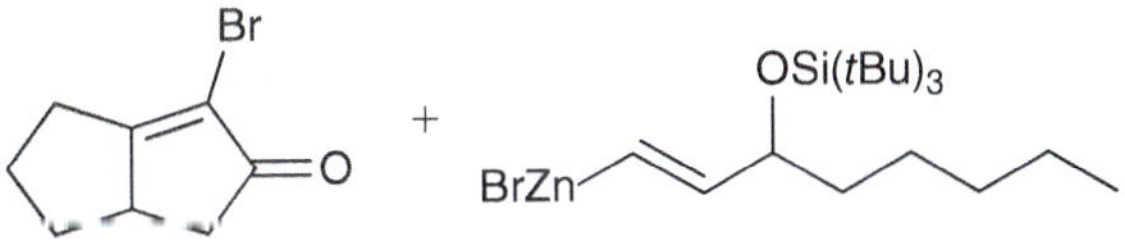

(c)

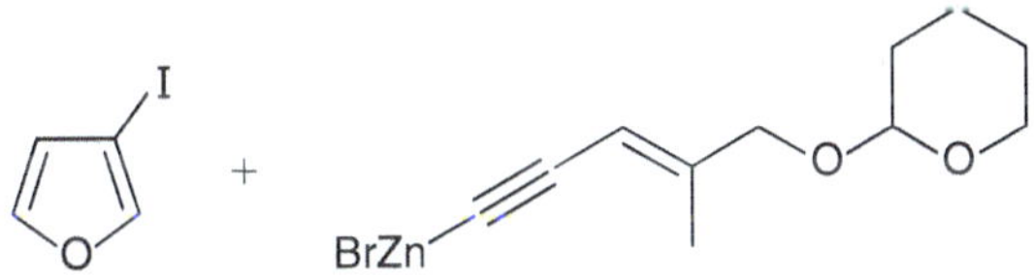

(d)

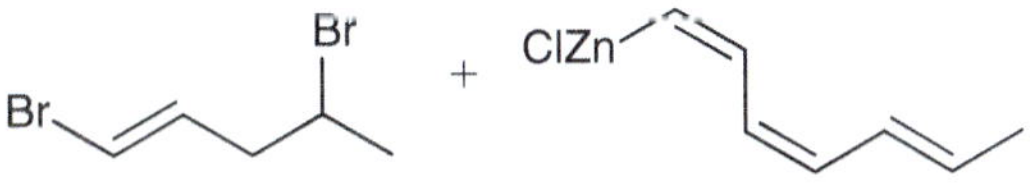

**28.21** Keeping in mind the relative reactivity of vinyl iodides and vinyl bromides, draw the structures of coupling products **A** and **B** in the following reaction sequence:

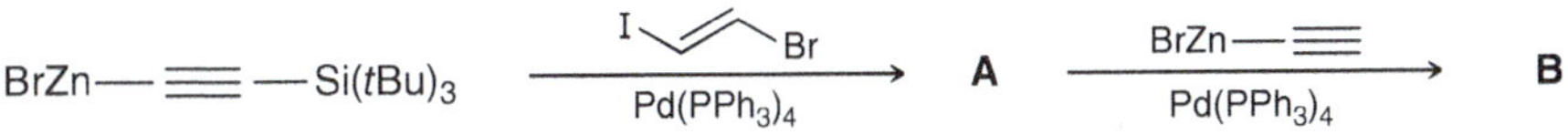

**APPLY** the skill

**28.22** The antitumor agent FR901464 is a natural product isolated from a bacterium found in a Japanese soil sample. In a reported total synthesis of this molecule, compound **1** was treated with compound **2**, in the presence of $ZnCl_2$ and $Pd(PPh_3)_4$, to produce compound **3**, a synthetic intermediate en route to the natural product (*J. Am. Chem. Soc.*, **2000**, *122*, 10482–10483). Draw the structure of **3**.

$ClR_2Zr$ ... I, $Et_3SiO$, O — **1** + **2** ($N_3$, O, I)

$ZnCl_2$, $Pd(PPh_3)_4$ → **3**

Additional steps

**FR901464**

**28.23** Savinin and gadain are two naturally occurring compounds with structures that differ only in the configuration of a C═C unit. Organozinc compound **1** was treated with vinyl bromide **2** in the presence of $Pd(PPh_3)_4$ to produce **3**, which was subsequently converted to savinin (*Synthesis*, **1997**, *9*, 1061–1066). **(a)** Draw the structure of **2**. **(b)** Draw the structure of gadain and the corresponding vinyl bromide that was used in its preparation.

ZnCl — **1**

**2**, $Pd(PPh_3)_4$ → **3** ($CO_2Me$)

Additional steps

**Savinin**

need more **PRACTICE?** Try Problems 28.55–28.59

## 28.8 The Heck Reaction

### The General Reaction

In the late 1960s, Richard Heck (working at the Hercules chemical manufacturing company and then later at the University of Delaware) discovered and developed the coupling reaction between an aryl, vinyl, or benzyl halide (RX) and an alkene in the presence of an appropriate Pd catalyst:

R—X + alkene $\xrightarrow[\text{Base}]{PdL_2}$ R-substituted alkene + HX

(R = aryl, vinyl, or benzyl; X = I, Br, or OTf)

This process replaces a vinylic hydrogen atom of the alkene with an R group. The Heck reaction is particularly useful as there are no other general ways for achieving this type of transformation directly. The reaction is most efficient when the organohalide (RX) is an iodide, although bromides and triflates are also very effective (I > Br ~ OTf >> Cl). The following are examples of Heck reactions:

An aryl iodide + ethylene $\xrightarrow[\text{Base}]{PdL_2}$ product (New C—C bond)

A vinyl triflate + alkene $\xrightarrow[\text{Base}]{PdL_2}$ product (New C—C bond)

Notice that, in each case, the newly formed C—C bond is between two $sp^2$-hybridized centers, which is a characteristic feature of the Heck reaction. We have seen other reactions that can also create a C—C bond between two $sp^2$-hybridized centers (such as Stille or Suzuki coupling), but the Heck reaction has the advantage that it does not require the preparation of an organotin compound or an organoboron compound. Indeed, the Heck reaction is one of the most widely used methods for C—C bond formation. For his pioneering work on Pd-catalyzed coupling reactions, Richard Heck shared the 2010 Nobel Prize in Chemistry with Akira Suzuki (see Section 28.6) and Ei-ichi Negishi (Section 28.7).

### The Catalyst

Like Stille coupling and Suzuki coupling, the Heck reaction can be achieved with a catalyst such as $Pd(PPh_3)_4$, in which the palladium atom has an oxidation state of 0. Alternatively, Pd(II) complexes can be used, such as $Pd(OAc)_2$. When a Pd(II) complex is used, it is converted into a Pd(0) complex, which is believed to function as the active catalyst. As we will see in Mechanism 28.5, the function of the base is to regenerate the active catalyst ($PdL_2$), thereby restarting the catalytic cycle. A commonly used base is triethylamine ($Et_3N$), although other bases are often used, including NaOAc, KOAc, and $NaHCO_3$.

### The Alkene

In a Heck reaction, the reactivity of the alkene decreases with increasing substitution, and the highest yields are obtained for ethylene and monosubstituted ethylenes. In particular, the most reactive monosubstituted ethylenes are those in which the substituent is electron withdrawing, as in the following examples:

Z ⟸ Electron-withdrawing group

OR R $NH_2$ C≡N $N^{\oplus}$ O $O^{\ominus}$

When a monosubstituted alkene is used in a Heck reaction, the process is generally observed to be highly regioselective, with the R group (of RX) being installed at the less substituted vinylic position of the alkene:

Pd(OAc)$_2$, PPh$_3$
Et$_3$N
(*E*)

**LOOKING BACK**
See Figure 8.17 to review the difference between the terms "stereoselective" and "stereospecific"

Notice also that this reaction is stereoselective with respect to the alkene, with the *E* isomer being formed (rather than the *Z* isomer), often exclusively. The reaction is also stereospecific with respect to RX (when RX is a vinyl halide). That is, the configuration of the C=C unit in the vinyl halide is retained, as seen in the following examples:

(*E*) Pd(OAc)$_2$, PPh$_3$ NaOAc (*E*)

(*Z*) Pd(OAc)$_2$, PPh$_3$ NaOAc (*Z*)

## A Mechanism for the Heck Reaction

The Heck reaction occurs via a catalytic cycle that is believed to be comprised of several stages, shown in Mechanism 28.5. For each stage of the catalytic cycle, the mechanism may vary, depending on the conditions, so curved arrows are not shown. In the first stage of the catalytic cycle, the organohalide (RX) reacts with the active catalyst via oxidative addition (as we have seen in previous mechanisms involving Pd catalysts):

R—X + PdL$_2$ → (Oxidative addition) L—Pd(R)(L)—X

The resulting complex then reacts with the starting alkene, adding R and Pd across the $\pi$ bond in a *syn* addition, with the R group being installed at the less substituted position:

Installed via *syn*-addition
*syn* addition

The resulting adduct then undergoes a conformational change (via C—C bond rotation), followed by a *syn* reductive elimination that gives the product:

Removed via *syn* elimination
C—C bond rotation
Reductive elimination

The catalytic cycle is completed when the base removes HX, via reductive elimination, to regenerate the active catalyst:

$$\mathrm{H{-}PdL_2{-}X} \xrightarrow[\text{-HX}]{\text{Base}} \mathrm{L{-}Pd{-}L}$$

Active catalyst

The entire catalytic cycle is shown in Mechanism 28.5.

## MECHANISM 28.5 THE HECK REACTION

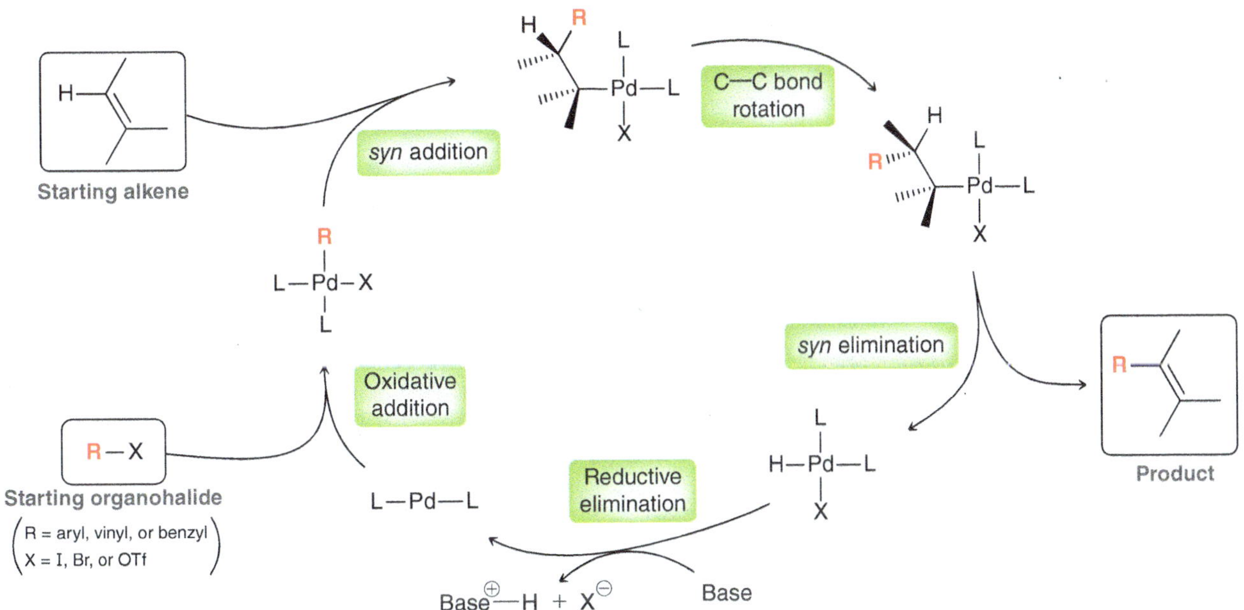


In some cases, the regiochemical outcome of a Heck reaction can be affected by the *syn* requirement of stage 2 (*syn* addition). For example, consider a Heck reaction in which cyclohexene is used as the alkene:

$$\text{R—X} + \text{cyclohexene} \xrightarrow[\text{Et}_3\text{N}]{\text{Pd(OAc)}_2,\ \text{PPh}_3} \text{R—cyclohexene}$$

Notice that the π bond is not located in the position that we might expect. This regiochemical outcome can be rationalized if we analyze the structure of the intermediate (shown in brackets below) that undergoes a *syn* elimination to give the product:

$$\text{cyclohexene} \xrightarrow{\text{R—X}} \xrightarrow{\text{PdL}_2} [\text{intermediate: R, H, } \text{L}_2\text{XPd, H}] \xrightarrow{\textit{syn}\text{ elimination}} \text{R—cyclohexene}$$

Notice that the intermediate has only one hydrogen atom (highlighted) that is properly positioned to undergo a *syn* elimination with the Pd group. The other neighboring hydrogen atoms are not *syn* to the Pd group, and as a result, there is only one possible regiochemical outcome that can accommodate *syn* elimination. This explains the ultimate position of the π bond in the product.

Much as we saw with Stille, Negishi, and Suzuki coupling, the Heck reaction also tolerates a wide variety of functional groups, including alcohols, ethers, aldehydes, ketones, and esters. This feature contributes to the enormous synthetic utility of the Heck reaction.

# SKILLBUILDER

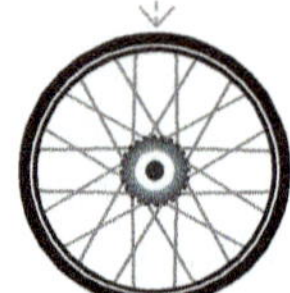

## 28.5 IDENTIFYING THE PARTNERS FOR A HECK REACTION

**LEARN the skill**

Identify the coupling partners that can be used to prepare the following compound via a Heck reaction:

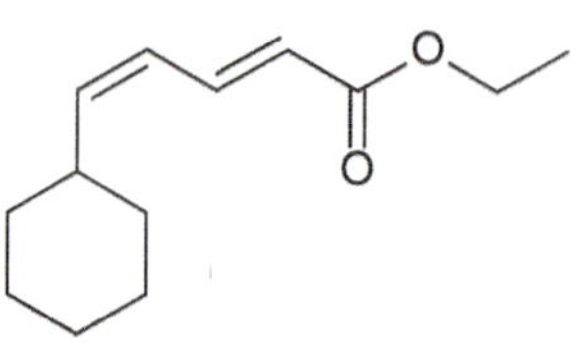

### SOLUTION

**STEP 1** Analyze the structure of the product and determine which bond(s) can be made via a Heck reaction.

Recall that a Heck reaction forms a bond between a vinyl position and either an aryl, vinyl, or benzyl position. In the desired compound, there is only one bond that fits this criterion:

**STEP 2** Identify the organohalide and alkene necessary.

This bond, indicated with a wavy line, is between two vinyl positions, so we will explore the reagents necessary to make this C—C bond via a Heck reaction. Note that in many cases there will be more than one bond that can be made via a Heck reaction. In such cases, multiple synthetic pathways may be viable.

In order to make the desired C—C bond via a Heck reaction, we must draw an alkene and an organohalide. There are two possibilities, shown here:

**Possibility 1**

**Possibility 2**

The first possibility is not viable because Heck reactions are stereoselective with respect to the starting alkene, and the *E* isomer (rather than the *Z* isomer) would be expected:

$Pd(OAc)_2$, $PPh_3$ / $Et_3N$ → (*E*)

This is not the desired stereochemical outcome, so we turn our attention to the second possibility. This possibility works, and it has the added advantage that the starting monosubstituted alkene has an electron-withdrawing substituent (so we expect the reaction to be very effective):

(*Z*) + → $Pd(OAc)_2$, $PPh_3$ / $Et_3N$ → (*Z*) (*E*)

**PRACTICE the skill** **28.24** Identify the coupling partners that can be used to prepare each of the following compounds via a Heck reaction:

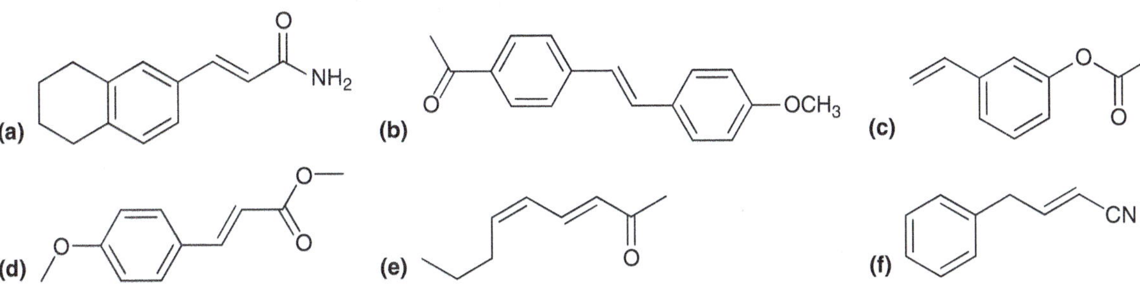

**APPLY the skill** **28.25** Predict the product of the following intramolecular Heck reaction:

$Pd(OAc)_2$, $PPh_3$
$Et_3N$
?

**28.26** Strychnine is a highly toxic, heptacyclic alkaloid isolated from seeds of the *Strychnos* genus of plant which are distributed throughout Asia, Africa, and the Americas. In addition to being a human toxin, it has also been used for centuries as a vertebrate pesticide. Since the 1950s, several total syntheses have been reported, and in one landmark synthesis, an intramolecular Heck reaction was utilized to produce compound **2**—a key intermediate. Provide a rationale for the observed stereochemical outcome as well as the observed regiochemical outcome for this process. (*J. Org. Chem.*, **1994**, *59*, 2685–2686).

**1** **2**

need more **PRACTICE?** Try Problems **28.37, 28.40d, 28.42, 28.46, 28.47, 28.50, 28.51, 28.72**

## 28.9 Alkene Metathesis

**BY THE WAY**
The term "olefin" refers to unsaturated compounds, such as alkenes and alkynes. The term is used in this context because metathesis is observed for both alkenes and alkynes, although we will only explore alkene metathesis in this section.

**Alkene metathesis**, also called **olefin metathesis**, has recently emerged as one of the most powerful synthetic techniques available to practicing synthetic organic chemists.

The term *metathesis* is derived from the Greek words *meta* and *thesis* (which can be translated as "change" and "position"). Alkene metathesis is characterized by the redistribution (changing of position) of carbon-carbon double bonds, as seen in the following general reaction:

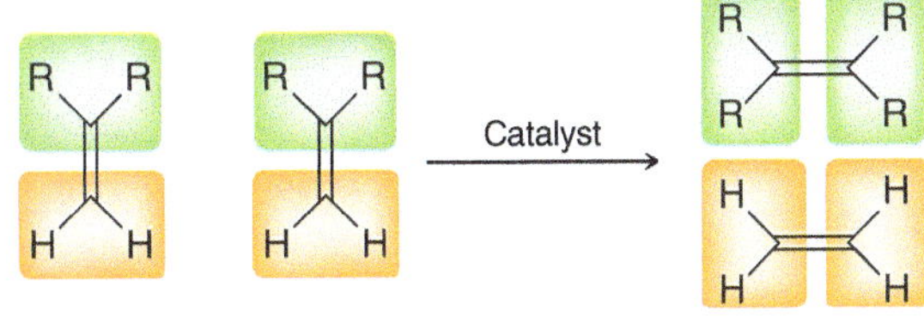

We will soon see that this process provides extraordinary synthetic utility, and indeed, the 2005 Nobel Prize in Chemistry was awarded to Robert Grubbs (Caltech), Richard Schrock (MIT), and Yves Chauvin (French Institute of Petroleum) for their contributions to developing alkene metathesis into a versatile synthetic technique.

## The Catalyst

Many catalysts have been developed for achieving alkene metathesis. All of these catalysts contain a transition metal element with a double bond to a carbon atom (C═M), and are therefore called *metal alkylidenes*. Schrock catalysts are based on molybdenum (Mo), while Grubbs catalysts are based on ruthenium (Ru). In each case, a variety of ligands have been used, with the most commonly utilized catalysts shown below:

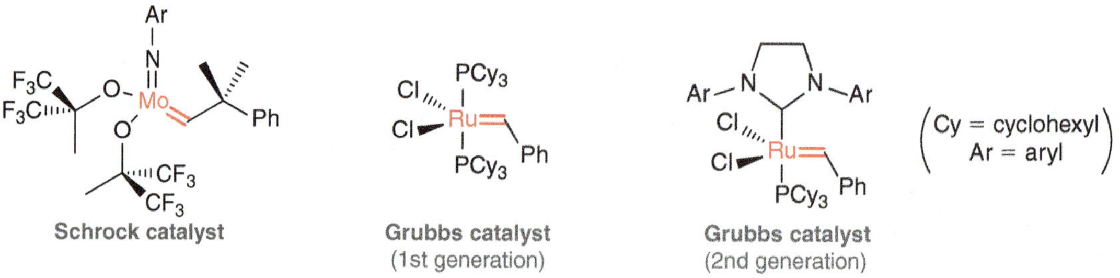

**Schrock catalyst**

**Grubbs catalyst** (1st generation)

**Grubbs catalyst** (2nd generation)

(Cy = cyclohexyl
Ar = aryl)

The C═M bond has been identified (in red) in each case. As we have seen in earlier sections, it is the convention to omit formal charges on the metal and ligands of organometallic compounds. While all three of these catalysts can participate in alkene metathesis reactions, each has a specialized use that distinguishes it from the others. In many cases, the most reactive of these catalysts toward metathesis is the Schrock catalyst. However, the Schrock catalyst is not easily handled in an oxygen-rich environment, therefore hindering its ability to be employed in routine synthesis. As a result, the Grubbs catalysts, which are both air and moisture stable, have risen to prominence among modern synthetic organic chemists.

## The Products

When an alkene is treated with a metal alkylidene catalyst (Schrock or Grubbs), the two vinyl positions of the alkene are separated from each other and the fragments can then be recombined:

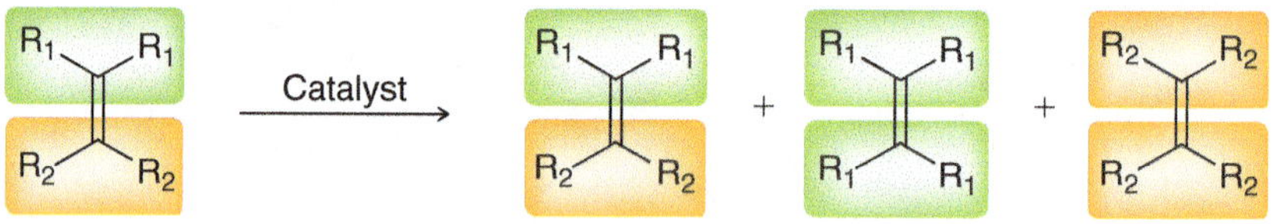

Catalyst

This process is particularly effective in (although not limited to) cases that involve terminal alkenes (monosubstituted alkenes or 2,2-disubstituted alkenes):

**Monosubstituted alkene** **2,2-Disubstituted alkene**

In such cases, one of the products is ethylene gas ($H_2C{=}CH_2$), which bubbles out of solution and is effectively removed from the product mixture. The evolution of ethylene gas drives the reaction toward the formation of one alkene with excellent yields.

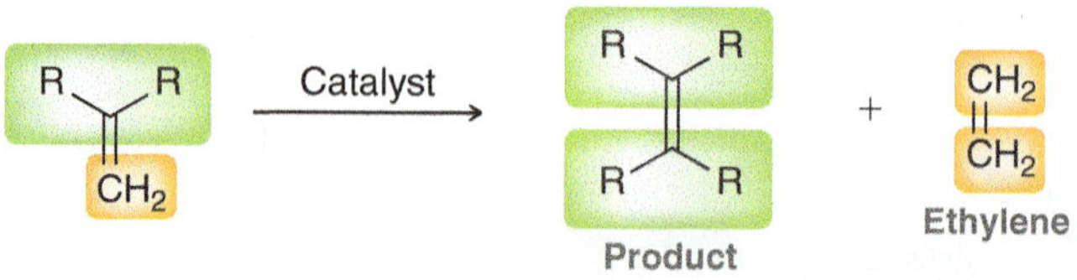

Catalyst

**Product** **Ethylene**

Below are specific examples of alkene metathesis. Notice that mixtures of *E* and *Z* isomers are obtained:

2 [alkene] —Catalyst→ (*E* + *Z* isomers) + $CH_2=CH_2$

2 [alkene] —Catalyst→ (*E* + *Z* isomers) + $CH_2=CH_2$

## CONCEPTUAL CHECKPOINT

**28.27** Draw the products that are expected when each of the following compounds is treated with a Grubbs catalyst:

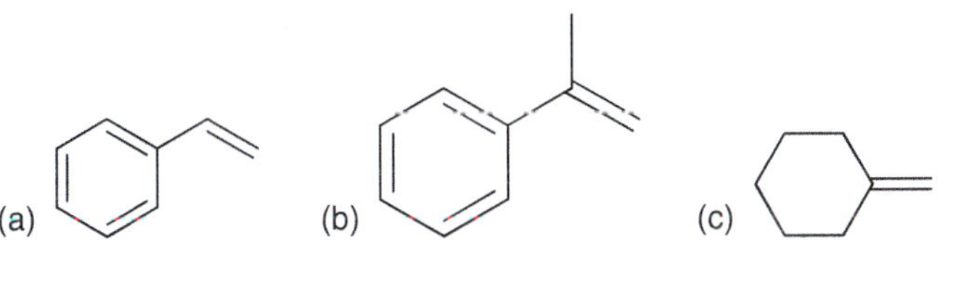


**28.28** Draw all of the products that are expected when *trans*-2-pentene is treated with a Grubbs catalyst.

**28.29** Draw the alkene that you would use to prepare the following compound via an alkene metathesis reaction:

## A Mechanism for Alkene Metathesis

Alkene metathesis can be divided into two stages. In the first stage, the catalyst reacts with the starting material (in one of two possible ways) to give two different metal alkylidene complexes (intermediates **A** and **B**, highlighted below):

Intermediate A + By-product

Intermediate B + By-product

(Ⓜ = metal together with any ligands attached to it)

Each intermediate is formed via a [2+2] cycloaddition, followed by a fragmentation process. Notice that by-products are formed along with **A** and **B**. Only small amounts of these by-products are formed because the metal complex is only present in a catalytic amount. Once **A** and **B** are formed, they function as catalysts for stage 2 of the process, in which the starting material is converted into the product. Both **A** and **B** can react with the starting alkene, so let's first explore the reaction between intermediate **A** and the starting alkene. This reaction gives the product as well as intermediate **B**:

Intermediate A

Starting alkene

Intermediate B

Product

Notice that the cycloaddition step has been drawn to show the metal being installed at the less substituted carbon of the starting alkene. If, instead, the metal had been installed at the more substituted carbon of the starting alkene, the result would be the production of the starting alkene and intermediate **A**. Although this can certainly occur, it does not involve a net change (starting alkene + **A** → starting alkene + **A**), so it can be ignored.

Now let's explore the reaction between intermediate **B** and the starting alkene, which regenerates intermediate **A** and produces ethylene gas:

Intermediate B

Starting alkene

Intermediate A

Ethylene

Notice that the cycloaddition step has been drawn to show the metal being installed at the more substituted carbon of the starting alkene. If, instead, the metal had been installed at the less substituted carbon of the starting alkene, the result would be the production of the starting alkene and intermediate **B**. Once again, this can certainly occur, but it does not involve a net change (starting alkene + **B** → starting alkene + **B**), so it can be ignored.

Stage 2 of the process can thus be summarized as follows: **A** is converted into **B** (with conversion of starting alkene into product), and then **B** is converted back to **A** (with conversion of starting alkene into ethylene). In this way, the starting alkene is continuously converted into the product and ethylene. This catalytic cycle is summarized here:

$CH_2=CH_2$

**A**

Starting alkene

Starting alkene

**B**

Product

The entire mechanism, as described in this section, is summarized in Mechanism 28.6.

## MECHANISM 28.6 ALKENE METATHESIS

### Stage 1 (Formation of Intermediates A and B)

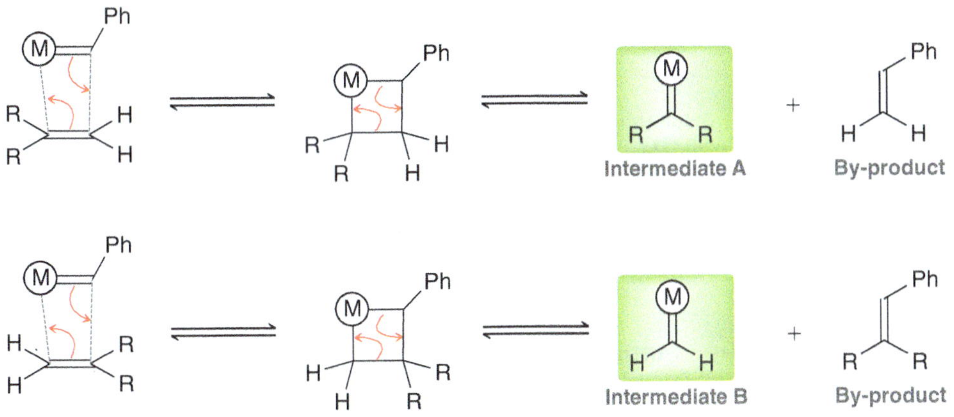

Intermediate A

By-product

Intermediate B

By-product

### Stage 2 (Catalytic Cycle)

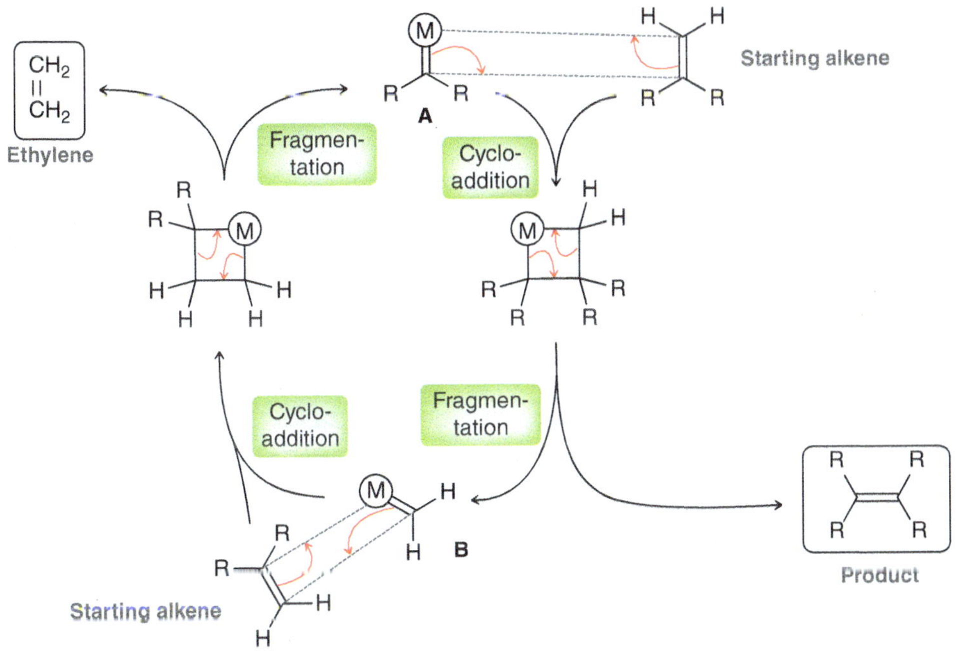

Ethylene

Fragmentation

Cycloaddition

Starting alkene

Cycloaddition

Fragmentation

Product

Starting alkene

## Ring-Closing Metathesis and Ring-Opening Metathesis

When the starting material is a diene and the reaction is conducted in dilute solutions (thereby favoring intramolecular processes over intermolecular processes), alkene metathesis can serve as a method for ring formation. This process is called **ring-closing metathesis**:

Grubbs catalyst

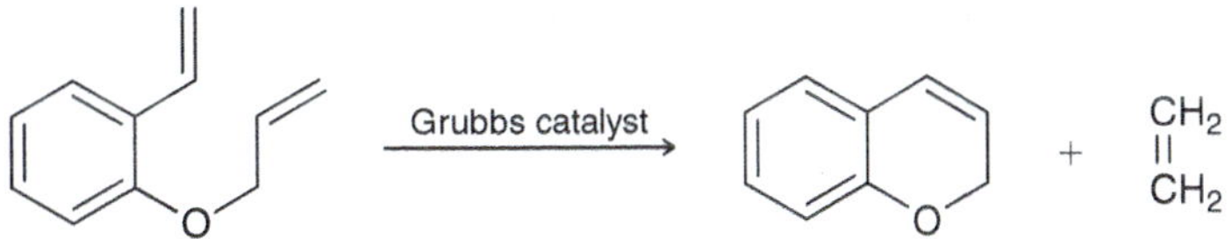

This procedure can be used to make a ring of virtually any size (rings of more than 20 carbon atoms have been prepared), and indeed, this procedure has been used creatively and frequently in the construction of a wide variety of synthetic targets.

Similarly, **ring-opening metathesis** can be achieved in the presence of ethylene. Under these conditions, a cycloalkene can be opened to give a diene, with the introduction of two methylene ($CH_2$) groups; for example:

Alkene metathesis can be achieved in the presence of a wide variety of functional groups, including ketones, aldehydes, esters, amines, amides, sulfonate esters, ethers, and even alcohols. This feature contributes to the enormous synthetic utility of alkene metathesis.

## SKILLBUILDER

### 28.6 IDENTIFYING THE STARTING MATERIAL FOR A RING-CLOSING METATHESIS

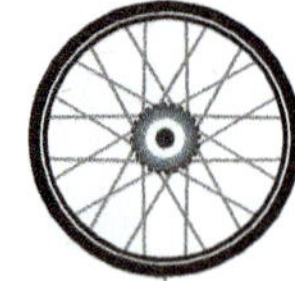

**LEARN the skill** Draw the diene that you would use to prepare the following compound via a ring-closing metathesis reaction.

**SOLUTION**

**STEP 1** Identify any C=C unit contained in a ring.

Identify any C=C unit that is incorporated into a ring. In this case, there is a π bond incorporated into an eight-membered ring:

**STEP 2** Break the C=C unit apart and attach a methylene ($CH_2$) group to each of the two vinyl positions of the broken C=C unit.

In order to draw the starting diene, we erase the highlighted C=C bond of the ring and then connect a methylene ($CH_2$) group to each of the two vinyl positions, like this:

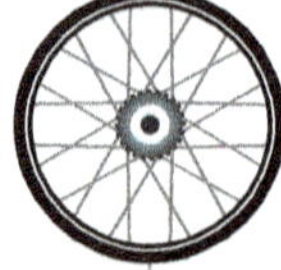

**PRACTICE the skill** **28.30** Draw the diene that you would use to prepare each of the following compounds via a ring-closing metathesis reaction:

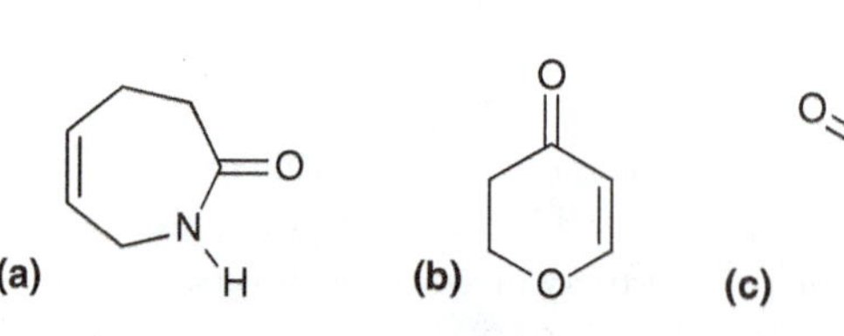


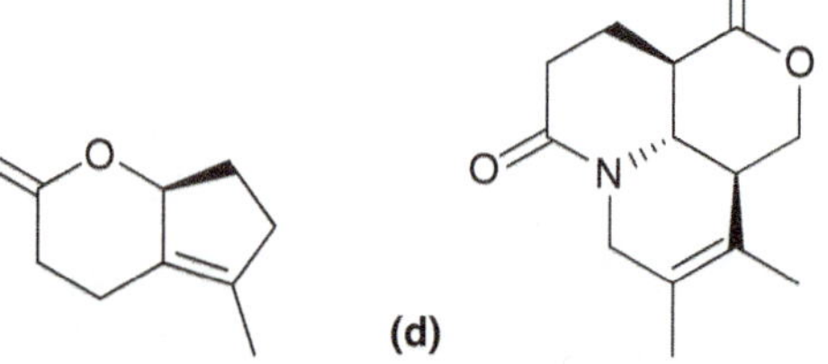

**28.31** Draw the product that is expected when each of the following compounds is treated with a Grubbs catalyst:

**(a)** 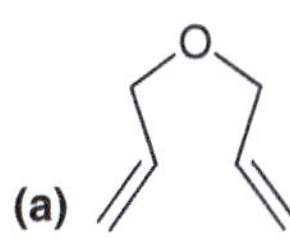 **(b)** 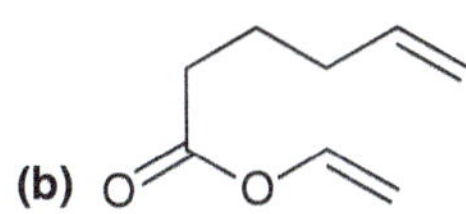 **(c)** 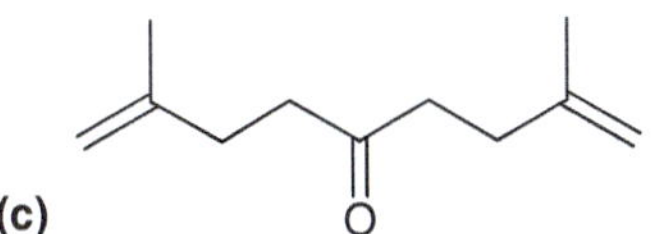 **(d)** 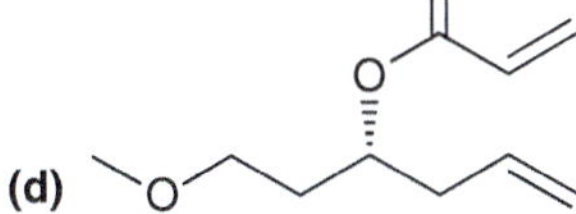

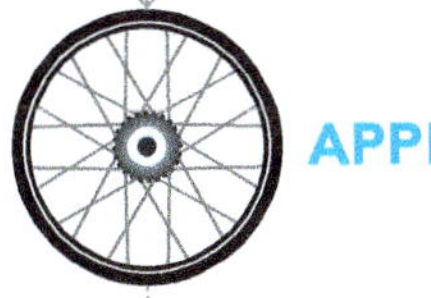

**APPLY the skill**

**28.32** In a study exploring the utility of olefin metathesis reactions, each of the following trienes was prepared and subjected to a Grubbs catalyst (*J. Am. Chem. Soc.*, **1996**, *118*, 6634–6640). Each molecule underwent a tandem ring-opening/ring-closing reaction resulting in the cleavage of the C=C unit within the ring, the formation of two new cyclic C=C groups, and the expulsion of ethylene gas. Determine the structure of the product in each case.

**(a)** **(b)**

need more **PRACTICE?** **Try Problems 28.36, 28.38, 28.39, 28.52, 28.54, 28.70, 28.74**

## practically speaking — Improving Biodiesel via Alkene Metathesis

Alkene metathesis has certainly become an invaluable tool for practicing synthetic chemists, but it has also found a wide variety of industrial applications. For example, alkene metathesis can be used to improve the properties of biodiesel, a renewable energy source. Recall from Section 26.4 that biodiesel can be made from triacylglycerides, which are comprised of a glycerol backbone and three fatty acid residues:

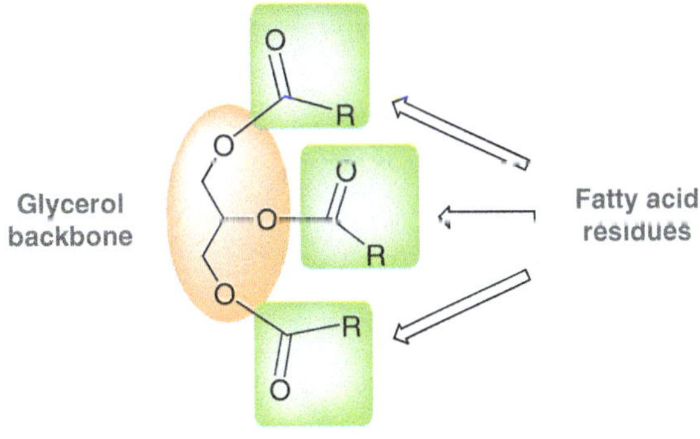


**A triacylglyceride**

Triacylglycerides can be obtained from a wide variety of natural sources, including oilseed crops, such as canola (*Brassica napus*). Treatment of a triacylglyceride with methanol, under acidic conditions, gives biodiesel. As seen is Section 26.4, this process is called transesterification and it results in the release of the fatty acid residues from the glycerol backbone, giving methyl esters (biodiesel), shown here:

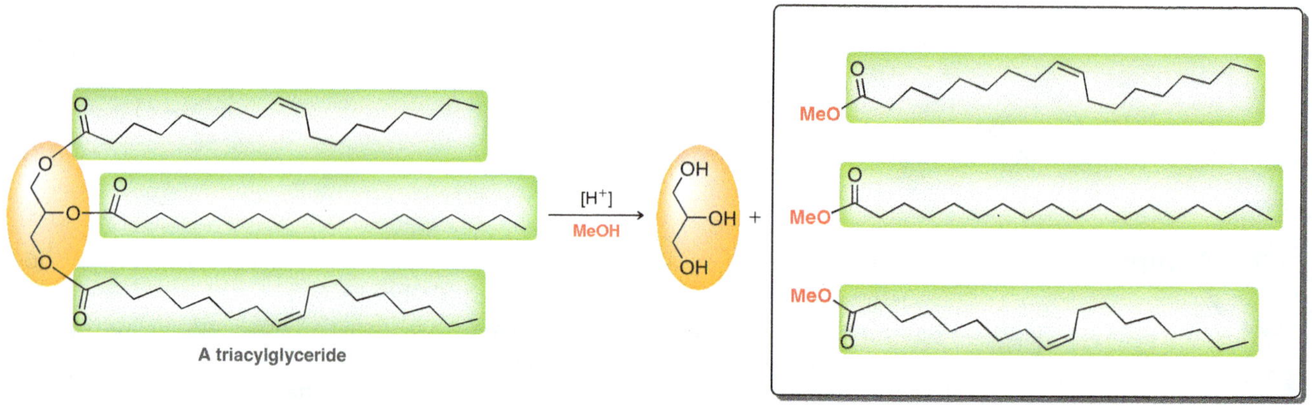

A triacylglyceride

$[H^+]$
MeOH

Biodiesel
(a mixture of fatty acid methyl esters)

In canola oil, one of the most abundant fatty acid residues is oleic acid, so biodiesel made from canola oil will have a large quantity of oleic acid methyl ester, shown below. This methyl ester can be used directly in specially modified internal combustion engines; however, it has some properties, especially in cold weather, that make it less than optimal as a fuel. These undesirable properties are the result of the presence of the ester group. One of the applications of metathesis catalysts is to remove the ester group, giving a hydrocarbon with improved combustion properties. For example, the reaction of two molecules of oleic acid methyl ester, in the presence of a metathesis catalyst, produces an alkene and a diester:

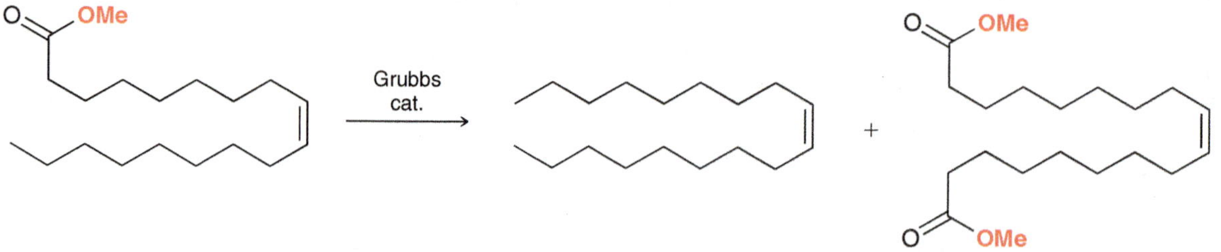

Grubbs cat.

Oleic acid methyl ester (2 eq.) → An alkene + A diester

Distillation of the reaction mixture allows for the alkene to be separated from the diester (which also has useful industrial applications). The alkene can then be used directly in an unmodified diesel engine. Indeed, this alkene exhibits significantly improved cold-weather performance, compared to oleic acid methyl ester. This example illustrates just one of the many exciting ways in which alkene metathesis can be used to solve global challenges.

# REVIEW OF REACTIONS

## Organolithium and Organomagnesium Compounds

**Preparation**

$$R{-}X \xrightarrow{\text{Li (2 eq.)}} R{-}Li \; + \; LiX$$

$$R{-}X \xrightarrow[\text{Et}_2\text{O}]{\text{Mg}} R{-}Mg{-}X$$

## Organocuprates

### Preparation of Gilman Reagents

$$R{-}X \xrightarrow[(-LiX)]{Li\ (2\ eq.)} R{-}Li \xrightarrow[(-LiX)]{CuX\ (0.5\ eq.)} R_2CuLi$$

R = alkyl, aryl, or vinyl
X = I, Br, or Cl

### Coupling Reaction of a Gilman Reagent with an Organohalide

$$R_2CuLi + R'{-}X \longrightarrow R{-}R' + \underbrace{RCu + LiX}_{\text{By-products}}$$

R = alkyl, aryl, or vinyl

R′ = methyl, primary, aryl, or vinyl
X = I, Br, or Cl

Coupling product

By-products

(E) Cu Li (Z) I (E) (Z)

## Simmons-Smith Reaction

$CH_2I_2$, Zn-Cu, $Et_2O$

R R R R → R R R R

(trans) $CH_2I_2$, Zn-Cu / $Et_2O$ → (trans) + En

(cis) $CH_2I_2$, Zn-Cu / $Et_2O$ → (cis)

## Stille Coupling

### Preparation of Organostannanes

$$R{-}X \xrightarrow[Et_2O]{Mg} RMgX \xrightarrow{Bu_3SnCl} R{-}SnBu_3$$

### Coupling Reaction of an Organostannane with an Organohalide

$$R{-}SnBu_3 + R'{-}X \xrightarrow[Pd(OAc)_2]{Pd(PPh_3)_4\ \text{or}} R{-}R' + X{-}SnBu_3$$

R = aryl, vinyl, allyl, or benzyl

R′ = aryl or vinyl
X = I, Br, Cl, or OTf

Coupling product

By-product

$SnBu_3$ + I → $Pd(PPh_3)_4$ → + $I{-}SnBu_3$

E E E E

## Suzuki Coupling

### Preparation of Organoboron Compounds

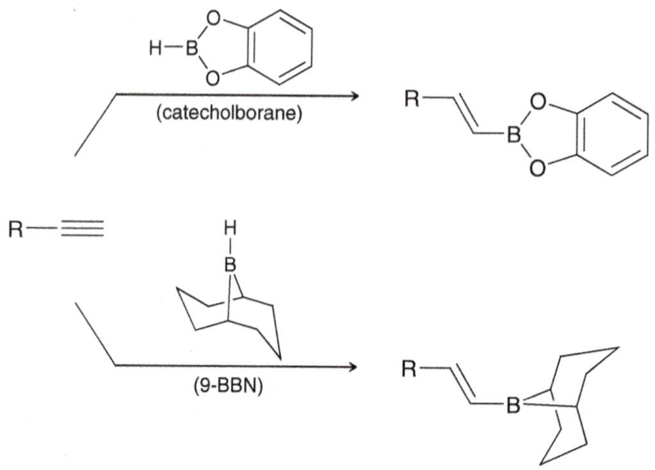

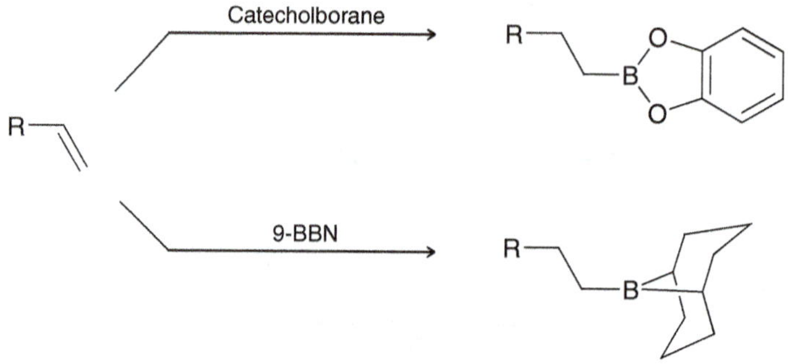

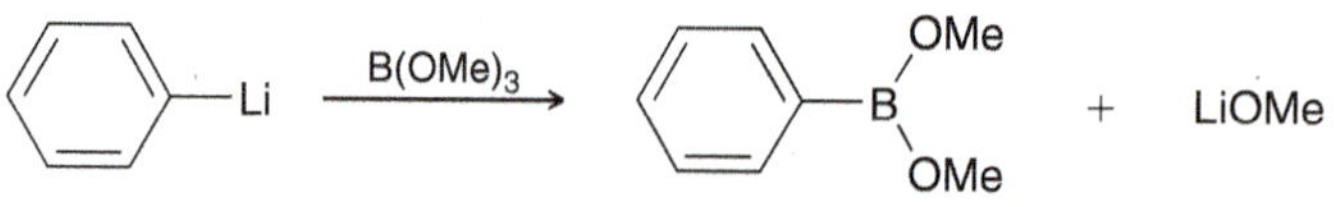

H—B (catecholborane)

R—≡

(9-BBN)

R—B

Catecholborane

9-BBN

Li $\xrightarrow{B(OMe)_3}$ B(OMe)(OMe) + LiOMe

### Coupling Reaction of an Organoboron Compound with an Organohalide

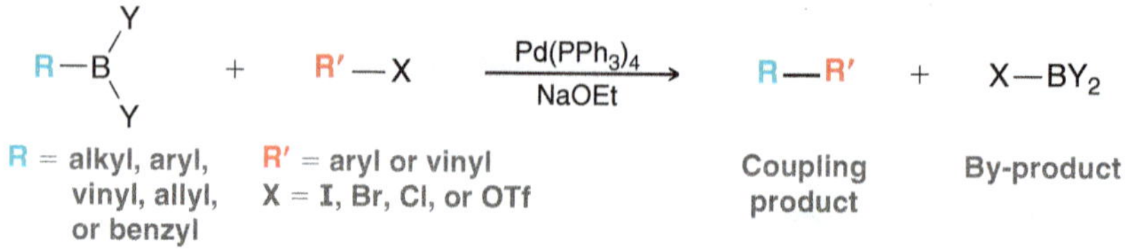

R—BY$_2$ + R′—X $\xrightarrow[NaOEt]{Pd(PPh_3)_4}$ R—R′ + X—BY$_2$

R = alkyl, aryl, vinyl, allyl, or benzyl

R′ = aryl or vinyl
X = I, Br, Cl, or OTf

Coupling product

By-product

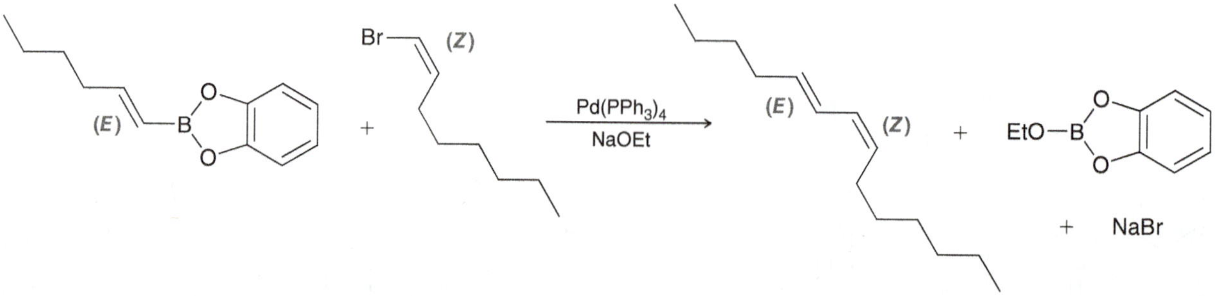

(E) + Br (Z) $\xrightarrow[NaOEt]{Pd(PPh_3)_4}$ (E) (Z) + EtO—B + NaBr

## Negishi Coupling

### Preparation of Organozinc Compounds

4-methoxyphenyl–Li $\xrightarrow{ZnBr_2}$ 4-methoxyphenyl–ZnBr + LiBr

MgBr $\xrightarrow{ZnBr_2}$ ZnBr + $MgBr_2$

Br $\xrightarrow[Et_2O]{Zn}$ ZnBr

I $\xrightarrow[Et_2O]{Zn}$ ZnI

### Coupling Reaction of an Organozinc Compound with an Organohalide

R—ZnX + R′—X $\xrightarrow[(cat)]{Pd(PPh_3)_4}$ R—R′ + $ZnX_2$

Ph–C≡C–ZnX $\xrightarrow{I–Ph}$ Ph–C≡C–Ph

### Coupling Reaction of an Organoaluminum Compound with an Organohalide

(E) Al(iBu)$_2$ + I (Z) $\xrightarrow{\text{Metal catalyst}}$ (E) (Z)

## The Heck Reaction

R—X + alkene $\xrightarrow[Base]{Pd(PPh_3)_4 \text{ or } Pd(OAc)_2}$ R–alkene + HX

R = aryl, vinyl, or benzyl
X = I, Br, or OTf

Ph–I + O $\xrightarrow[Et_3N]{Pd(OAc)_2,\ PPh_3}$ (E) O

(E) I + ethylene $\xrightarrow[NaOAc]{Pd(OAc)_2,\ PPh_3}$ (E)

I (Z) + ethylene $\xrightarrow[NaOAc]{Pd(OAc)_2,\ PPh_3}$ (Z)

R—X + cyclohexene $\xrightarrow[Et_3N]{Pd(OAc)_2,\ PPh_3}$ R–cyclohexene

### Alkene Metathesis

2 [alkene] —Catalyst→ (E + Z isomers) + $CH_2{=}CH_2$

2 [alkene] —Catalyst→ (E + Z isomers) + $CH_2{=}CH_2$

### Ring-Closing Metathesis

—Grubbs catalyst→ + $CH_2{=}CH_2$

### Ring-Opening Metathesis

—Grubbs catalyst, $H_2C{=}CH_2$→

# REVIEW OF CONCEPTS AND VOCABULARY

### SECTION 28.1

- Compounds containing a bond between a carbon atom and a metal (C—M) are called **organometallic compounds**.
- For any C—M bond, the magnitude of the partial charge ($\delta-$) on the carbon atom is dependent on the identity of the metal.

### SECTION 28.2

- Organolithium (RLi) and organomagnesium (RMgX) compounds are strong bases and strong nucleophiles.
- As strong bases, organolithium and organomagnesium compounds react rapidly with water to give hydrocarbons.
- Grignard reagents (and organolithium reagents) react rapidly with $H_2O$ or $D_2O$ or with any functional group that exhibits a weakly acidic proton (OH, NH, SH, C≡CH, or COOH).

### SECTION 28.3

- Gilman reagents can be prepared by treating an organohalide with lithium to give an organolithium, followed by treatment of the organolithium with CuX.
- Gilman reagents ($R_2CuLi$) react with organohalides (R′X) to give a product in which the R and R′ groups have been coupled together.
- The coupling process is stereospecific. It is observed to proceed with retention of configuration at each of the C=C units.
- When R ≠ R′, the process is called a *cross-coupling* reaction.

### SECTION 28.4

- Alkenes react with $ICH_2ZnI$ to form a three-membered ring in a process called *cyclopropanation*.
- The cyclopropanation process is observed to be stereospecific, with the configuration of the alkene being preserved in the cyclopropane product.
- Since the action of $ICH_2ZnI$ resembles the action of carbenes, it is commonly referred to as a *carbenoid* reagent.

### SECTION 28.5

- Stille coupling, which occurs in the presence of a suitable palladium catalyst, is the reaction between an organostannane (a compound with a C—Sn bond, also called an organotin compound) and an organic electrophile to form a new C—C σ bond.
- The coupling process is stereospecific. It is observed to proceed with retention of configuration at each of the C=C units.
- Organostannanes react with aryl and vinyl halides or triflates, even when other functional groups are present, including carboxylic acids, amides, esters, nitro groups, ethers, amines, hydroxyl groups, ketones, and aldehydes.

### SECTION 28.6

- Suzuki coupling, which occurs in the presence of a suitable palladium catalyst, is the reaction between an organoboron compound ($RBY_2$) and an organic electrophile (R′X) to form a new C—C σ bond.

- The coupling process is stereospecific. It is observed to proceed with retention of configuration at each of the C=C units.

SECTION 28.7

- Negishi coupling, which occurs in the presence of a suitable palladium catalyst, is the reaction between an organic electrophile (R′X) and an organometallic species containing zinc, aluminum, or zirconium, to form a new C—C σ bond.
- The coupling process is stereospecific. It is observed to proceed with retention of configuration at each of the C=C units.

SECTION 28.8

- The Heck reaction is a coupling reaction that occurs between an aryl, vinyl, or benzyl halide (RX) and an alkene in the presence of an appropriate Pd catalyst.
- When a monosubstituted alkene is used in a Heck reaction, the process is observed to be highly regioselective, with the R group (of RX) being installed at the less substituted vinylic position of the alkene.
- When a monosubstituted alkene is used in a Heck reaction, the process is observed to be highly stereoselective with respect to the alkene, with the *E* isomer being formed (rather than the *Z* isomer), often exclusively.
- The Heck reaction is also stereospecific with respect to RX (when RX is a vinyl halide). That is, the configuration of the C=C unit in the vinyl halide is retained.

SECTION 28.9

- **Alkene metathesis**, also called **olefin metathesis**, is characterized by the redistribution (changing of position) of carbon-carbon double bonds.
- When terminal alkenes are used, the evolution of ethylene gas drives the reaction toward the formation of one alkene with excellent yields.
- When the starting material is a diene and the reaction is conducted in dilute solutions (thereby favoring an intramolecular process over intermolecular processes), alkene metathesis can serve as a method for ring formation. This process is called **ring-closing metathesis**. Similarly, **ring-opening metathesis** can be achieved in the presence of ethylene.
- Alkene metathesis can be achieved in the presence of a wide variety of functional groups, including ketones, aldehydes, esters, amines, amides, sulfonate esters, ethers, and even alcohols.

## SKILLBUILDER REVIEW

### 28.1 IDENTIFYING THE PARTNERS FOR A COREY-POSNER/WHITESIDES-HOUSE COUPLING REACTION

**EXAMPLE** Show how you would prepare 1-butylcyclopentene from 1-bromobutane and any other organohalide of your choice.

1-Butyl-cyclopentene

1-Bromobutane

+ RX

**STEP 1** Analyze the structure of the product and determine which bond must be made via a coupling process.

Product

Starting material

**STEP 2** Identify the two organohalides that are necessary.

**STEP 3** Draw the forward scheme, showing both possible synthetic routes (either organohalide can be converted into a Gilman reagent).

Possibility 1: 1) Li (2 eq.) 2) CuI (0.5 eq.) 3) 1-bromobutane

Possibility 2: 1) Li (2 eq.) 2) CuI (0.5 eq.) 3) 1-iodocyclopentene

Try Problems 28.7–28.9, 28.40a, 28.43, 28.48

### 28.2 PREDICTING THE PRODUCT OF A STILLE COUPLING REACTION

**STEP 1** Identify the carbon atoms that will be joined via a new σ bond.

$SnBu_3$

**STEPS 2 and 3** Rotate the coupling partners and draw the product, making sure to retain the configuration of any C=C units.

(*E*)

(*E*)

Try Problems 28.12–28.15, 28.33, 28.53, 28.65

**28.3** PREDICTING THE PRODUCT OF A SUZUKI COUPLING REACTION

**STEP 1** Identify the carbon atoms that will be joined via a new σ bond.

**STEPS 2 and 3** Rotate the coupling partners and draw the product, making sure to retain the configuration of any C═C units.

Try Problems **28.16–28.19, 28.40b–c, 28.44, 28.45, 28.68, 28.71**

**28.4** PREDICTING THE PRODUCT OF A NEGISHI COUPLING REACTION

**STEP 1** Identify the carbon atoms that will be joined via a new σ bond.

**STEPS 2 and 3** Rotate the coupling partners and draw the product, making sure to retain the configuration of any C═C units.

Try Problems **28.20–28.23, 28.55–28.59**

**28.5** IDENTIFYING THE PARTNERS FOR A HECK REACTION

**STEP 1** Analyze the structure of the product and determine which bond(s) can be made via a Heck reaction.

**STEP 2** Identify the organohalide and alkene necessary. Keep in mind that Heck reactions are stereoselective with respect to the starting alkene and stereospecific with respect to the organohalide (if it is a vinyl halide, as in this case).

Try Problems **28.24–28.26, 28.37, 28.40d, 28.42, 28.46, 28.47, 28.50, 28.51, 28.72**

**28.6** IDENTIFYING THE STARTING MATERIAL FOR A RING-CLOSING METATHESIS

**STEP 1** Identify any C═C unit contained in a ring.

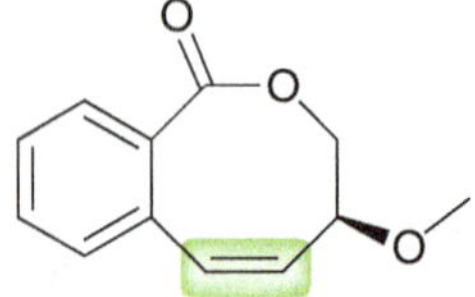

**STEP 2** Break the C═C unit apart and attach a methylene ($CH_2$) group to each of the two vinyl positions of the broken C═C unit.

Try Problems **28.30–28.32, 28.36, 28.38, 28.39, 28.52, 28.54, 28.70, 28.74**

# PRACTICE PROBLEMS

**28.33** Stille coupling has been extended to include the coupling between *sp*- and *sp*$^2$-hybridized carbon atoms, when an alkynyl stannane (R—≡—$SnBu_3$) is used. Draw the coupling product that is expected when *para*-nitrobromobenzene is treated with the following alkynyl stannane in the presence of $Pd(PPh_3)_4$:

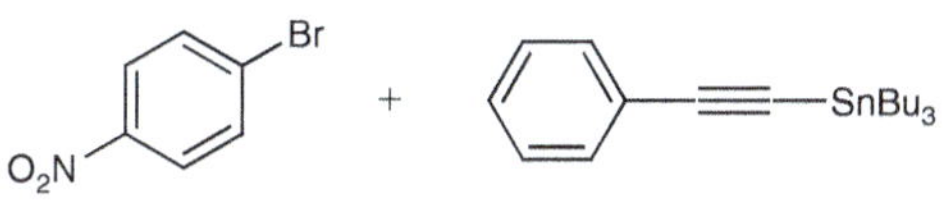

Br, $O_2N$ + $SnBu_3$

**28.34** Draw the structures of compounds **A**, **B**, and **C** in the following reaction sequence:

O; Grubbs cat. → **A**; $Pd(PPh_3)_4$, I → **B**; Grubbs cat., H H C=C H H → **C**

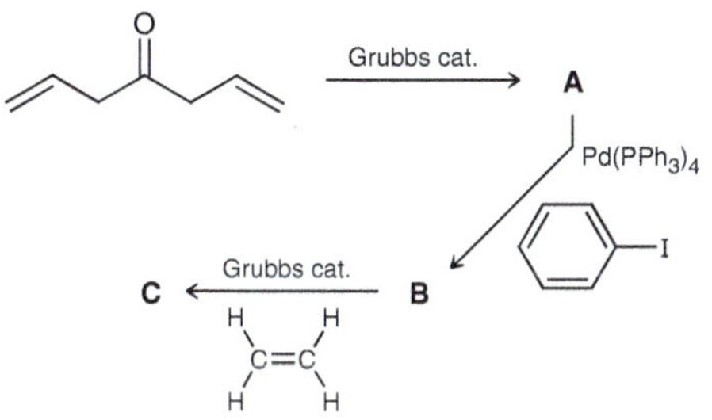

**28.35** Draw the structures of compounds **A**, **B**, and **C** in the following reaction sequence:

Br; 1) Li (2 eq.) 2) $B(OMe)_3$ → **A**; $Pd(PPh_3)_4$, I, O, O → **B**; Grubbs cat. → **C**

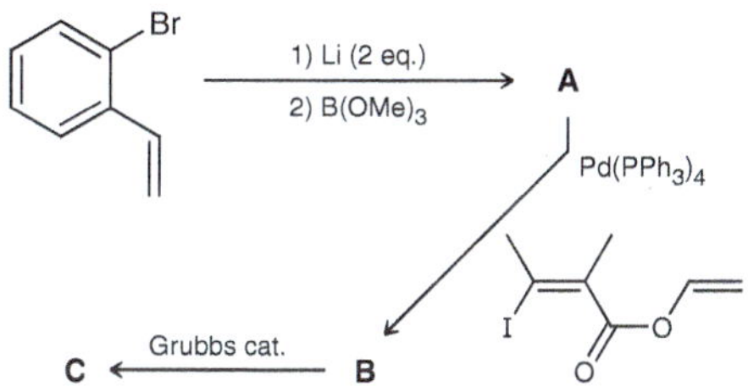

**28.36** Exaltolide is a lactone (cyclic ester) used in perfume formulations, and it can be prepared as shown below. Draw the structure of exaltolide:

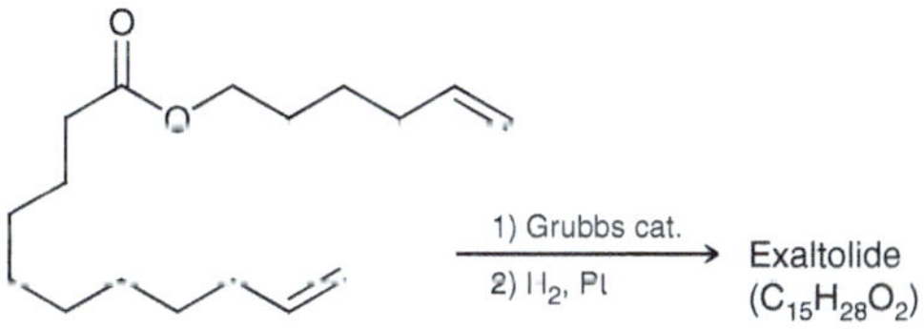

O, O; 1) Grubbs cat. 2) $H_2$, Pt → Exaltolide ($C_{15}H_{28}O_2$)

**28.37** Draw the expected product for each of the following coupling reactions.

(a) I, I + 2; $Pd(OAc)_2$, $PPh_3$, $Et_3N$ → ?

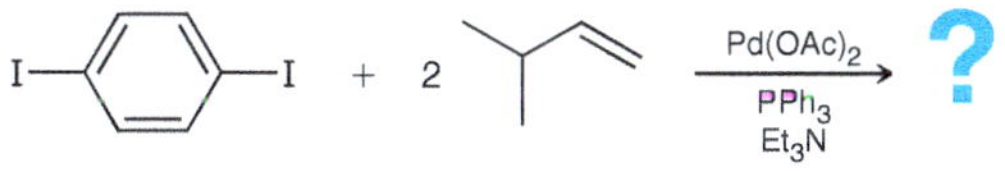

(b) O, O + I, N, O; $Pd(OAc)_2$, $PPh_3$, $Et_3N$ → ?

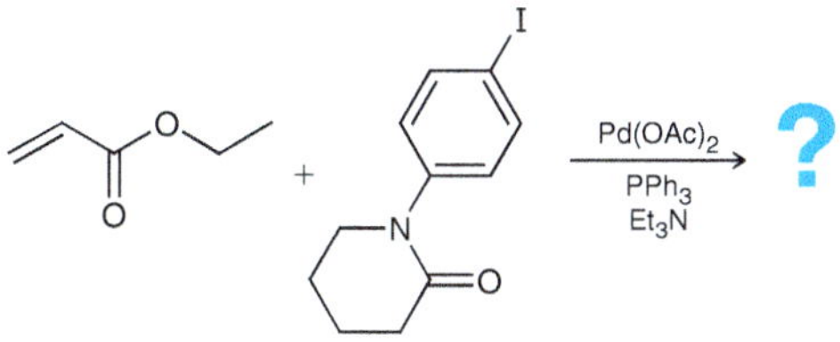

(c) + OTf; $Pd(OAc)_2$, $PPh_3$, $Et_3N$ → ?

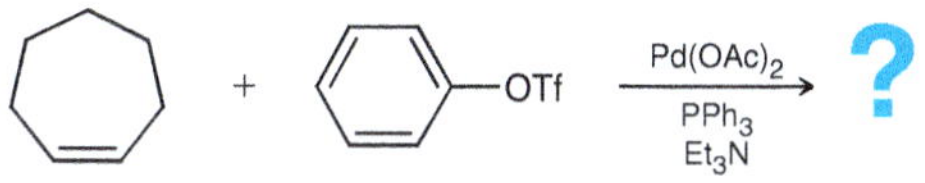

(d) O + O, I; $Pd(OAc)_2$, $PPh_3$, $Et_3N$ → ?

**28.38** Draw the product that is expected when each of the following undergoes a ring-opening metathesis (ROM) in the presence of ethylene gas:

(a) (b)  (c) H, H

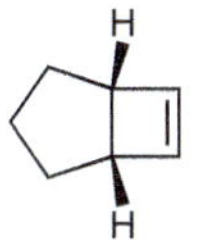

(d) O, H (e) O, O (f) O, O, O

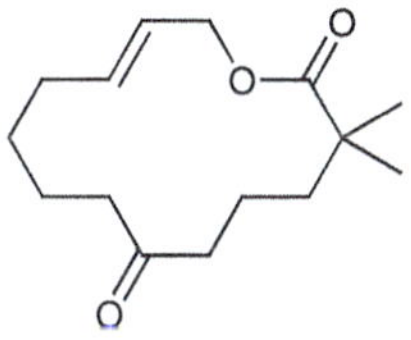

**28.39** Draw the product that is expected when each of the following undergoes a ring-closing metathesis (RCM):

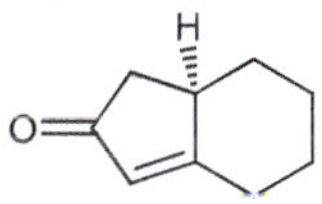

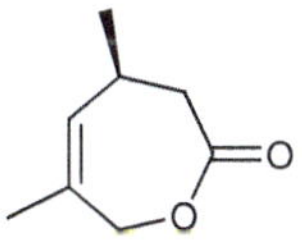

(a) H—N, O; Grubbs cat. → ?

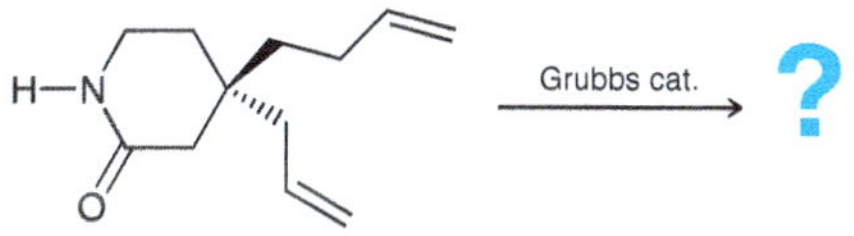

(b) OH, OH; Grubbs cat. → ?

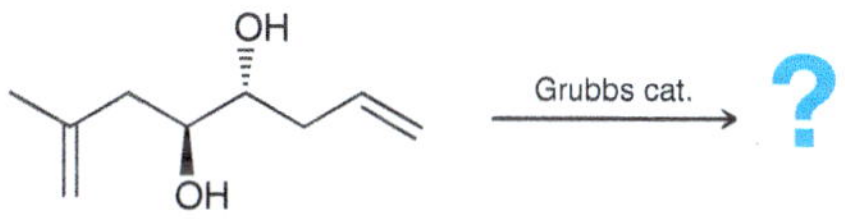

(c) OMe, O, H; Grubbs cat. → ?

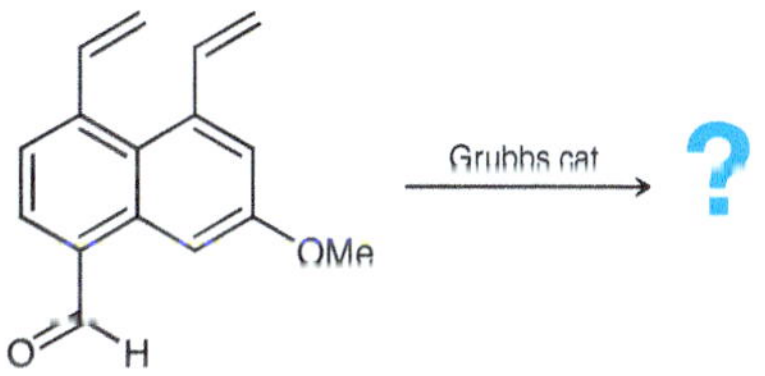

(d) O, O; Grubbs cat. → ?

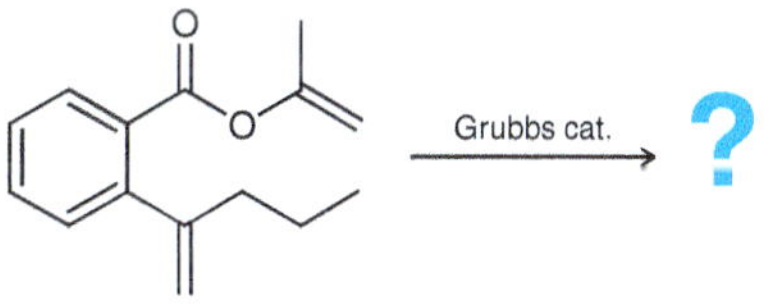

**28.40** Draw the expected product for each of the following coupling reactions:

(a) I; $(Ph)_2CuLi$ → ?

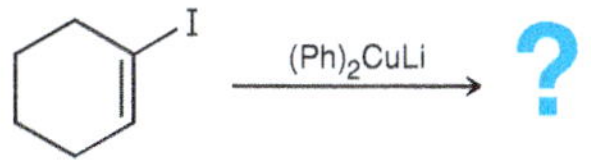

(b) 1) Catechol borane 2) $C_6H_5Br$, $Pd(PPh_3)_4$, NaOH → ?

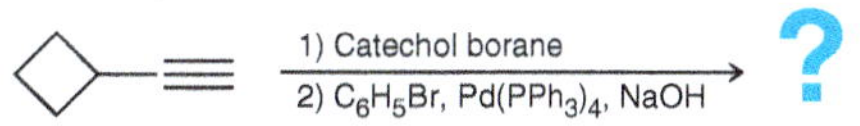

(c) 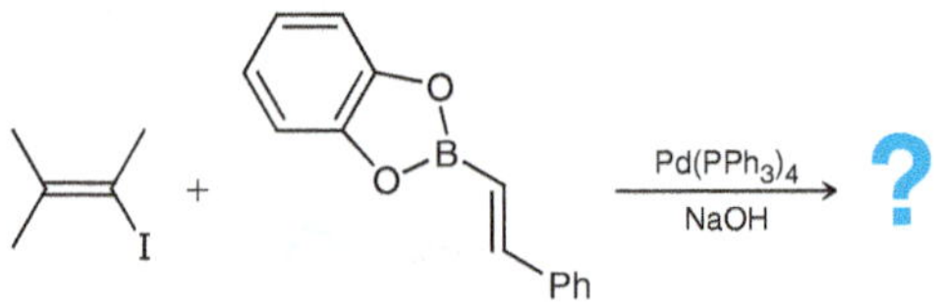 + Ph $\xrightarrow[\text{NaOH}]{Pd(PPh_3)_4}$ ?

(d) 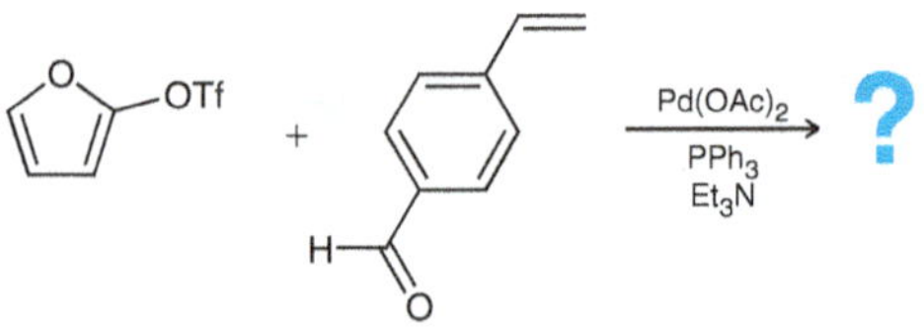 OTf + H O $\xrightarrow[\text{PPh}_3,\ Et_3N]{Pd(OAc)_2}$ ?

**28.41** When 5-*tert*-butylcycloheptene undergoes a Simmons-Smith reaction, two products are formed, **A** and **B**, each of which has the molecular formula $C_{12}H_{22}$. **(a)** Draw structures for **A** and **B**. **(b)** Identify the relationship between **A** and **B**. **(c)** Taking steric considerations into account, which product would you expect to predominate? Explain your choice.

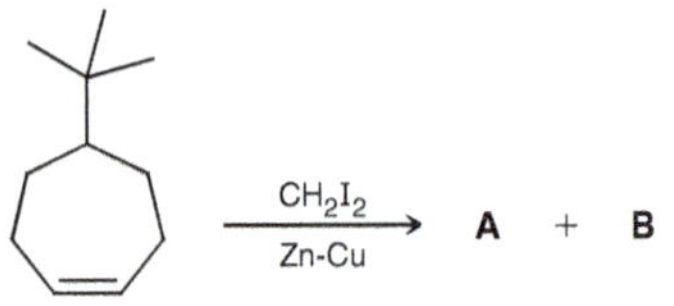

$\xrightarrow[\text{Zn-Cu}]{CH_2I_2}$ **A** + **B**

**28.42** Compounds **A** and **B** will each undergo an intramolecular Heck reaction, although each compound generates a different product. Draw the product of each reaction and explain why different products are obtained.

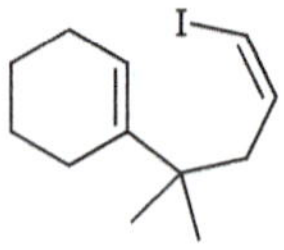

**A** 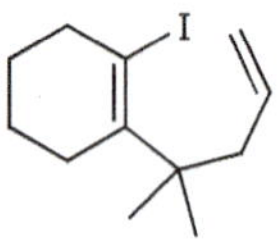 **B**

**28.43** Using lithium diphenylcuprate ($Ph_2CuLi$) and any other reagents of your choice, show how you would prepare each of the following compounds:

(a) 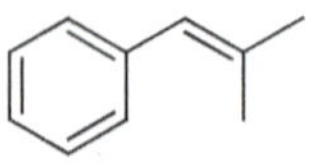 (b) 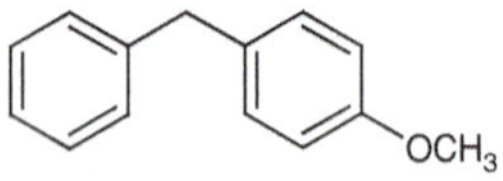 $OCH_3$ (c) 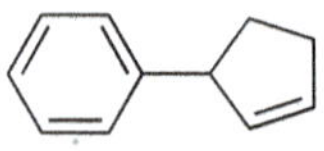 (d)

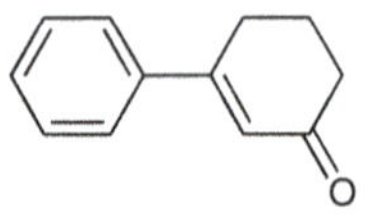

**28.44** Starting with 1-butyne and using any other reagents of your choice, show how you would use a Suzuki coupling reaction to make each of the following compounds:

(a) 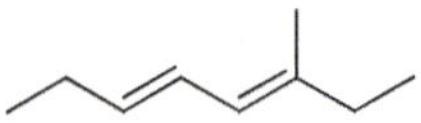 (b) 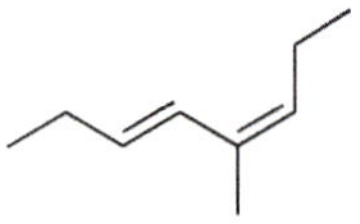 (c)

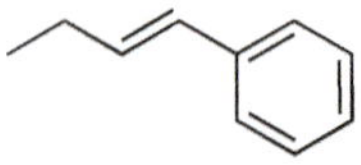

**28.45** Explain why the following compound cannot easily be made from 1-butyne via a Suzuki coupling reaction:

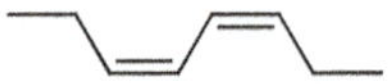

**28.46** Starting with 4-nitrostyrene and using any other reagents of your choice, show how you would make each of the following:

(a) 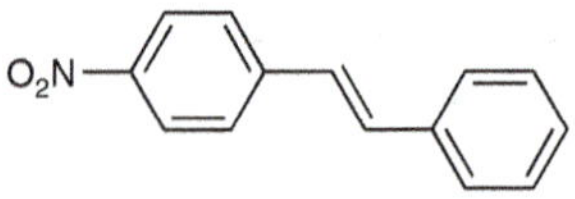 $O_2N$ (b) 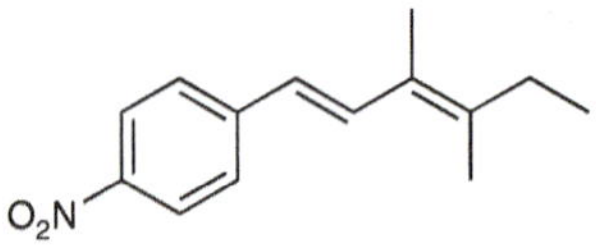 $O_2N$ (c) 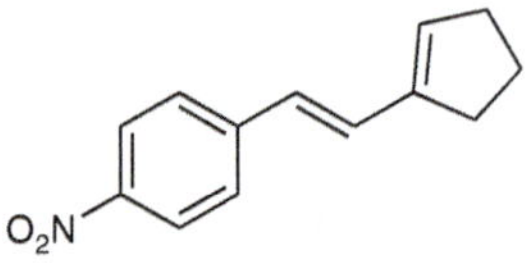 $O_2N$

**28.47** Propose a two-step synthesis that will achieve the following transformation:

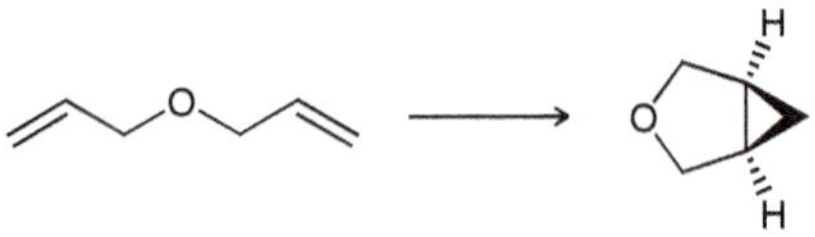

**28.48** Using 1-pentene as your only source of carbon atoms, show how you would use a Corey-Posner/Whitesides-House reaction to prepare decane.

**28.49** Starting with benzene and any other reagents of your choice, propose a synthesis for each of the following compounds:

(a) 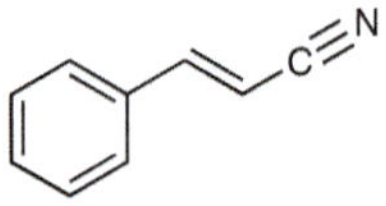 (b) 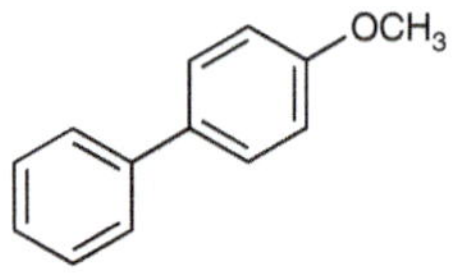 $OCH_3$ (c) 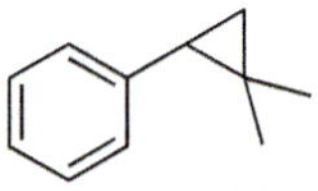 (racemic) (d) 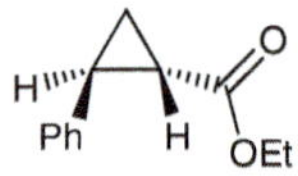 + En

**28.50** The following compound will undergo an intramolecular Heck reaction to give a single product with two chirality centers. This process is observed to occur diastereoselectively (a new chirality center is formed, but only one stereoisomer is obtained). Draw the product and rationalize the source of the observed diastereoselectivity:

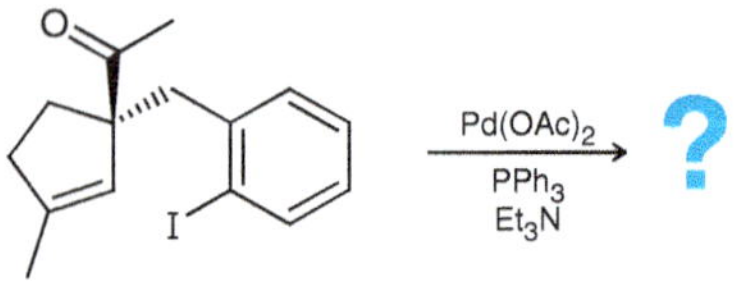

$\xrightarrow[\text{PPh}_3,\ Et_3N]{Pd(OAc)_2}$ ?

**28.51** Starting with cyclopentene and using any other reagents of your choice, show how you would make each of the following compounds:

(a) 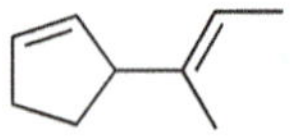 (b) 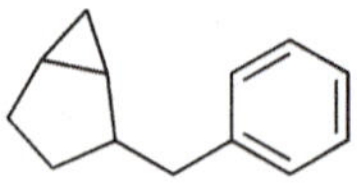 (c)  HO OH

**28.52** When treated with a Grubbs catalyst, compound **A** (under conditions of high dilution) is converted into one product (compound **1**) with the molecular formula $C_6H_8O_2$. Under similar conditions, compound **B** is converted into two products, compounds **2** and **3**, each of which has the molecular formula $C_8H_{12}O_2$. Draw the structures of compounds **1**, **2**, and **3** and explain why compound **A** gives only one metathesis product while compound **B** gives two metathesis products.

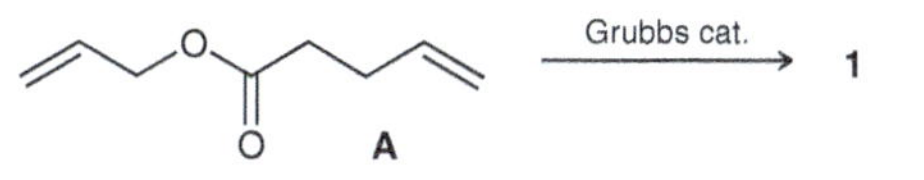

Grubbs cat. → 1

A

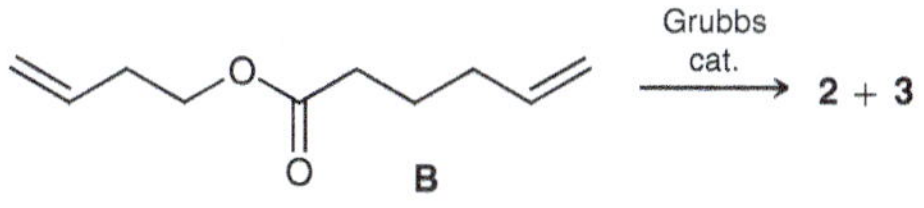

Grubbs cat. → 2 + 3

B

**28.53** When a Stille coupling reaction is performed in the presence of carbon monoxide (CO), a ketone is obtained, as shown here:

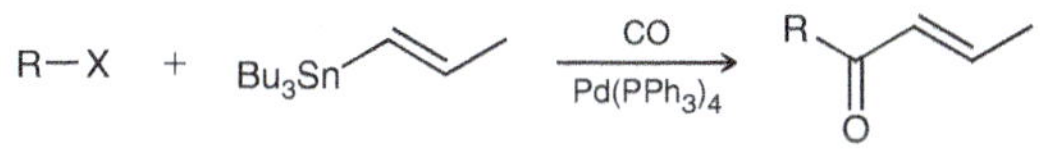

R—X + $Bu_3Sn$ —CO, $Pd(PPh_3)_4$→ R

Using this procedure, show how you would make each of the following compounds from bromobenzene:

(a) (b) $OCH_3$

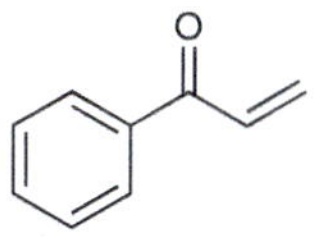

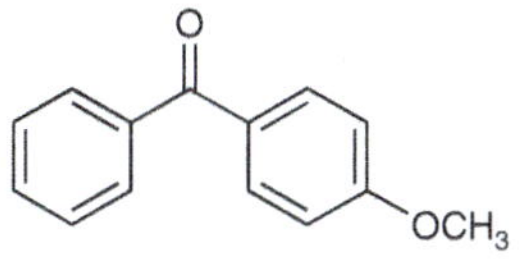

(c)

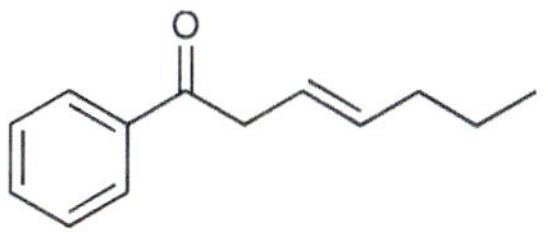

(d) $OCH_3$

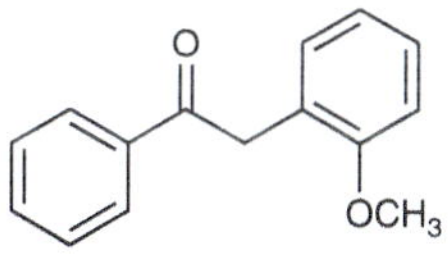

**28.54** When treated with a Grubbs catalyst, 1-pentene is converted into two products, **A** and **B**, each of which has the molecular formula $C_8H_{16}$. Compound **A** undergoes a Simmons-Smith reaction to give **1**, while **B** undergoes a Simmons-Smith reaction to give compounds **2** and **3**.

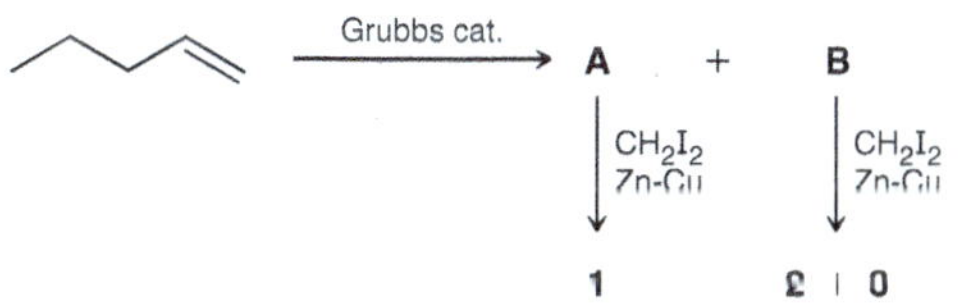

Grubbs cat. → A + B

$CH_2I_2$, Zn-Cu → 1; $CH_2I_2$, Zn-Cu → 2 + 3

**(a)** Draw the structures of compounds **1**, **2**, and **3** and explain why compound **A** gives only one Simmons-Smith product while compound **B** gives two Simmons-Smith products. **(b)** Identify the relationship between compounds **2** and **3**.

**28.55** The following synthesis was developed as a method to prepare 4-aryl piperidine derivatives, a group of compounds containing a common structural unit found in a variety of active pharmaceutical agents (*J. Org. Chem.*, **2004**, *69*, 5120). Draw the structures of compounds **A** and **B**.

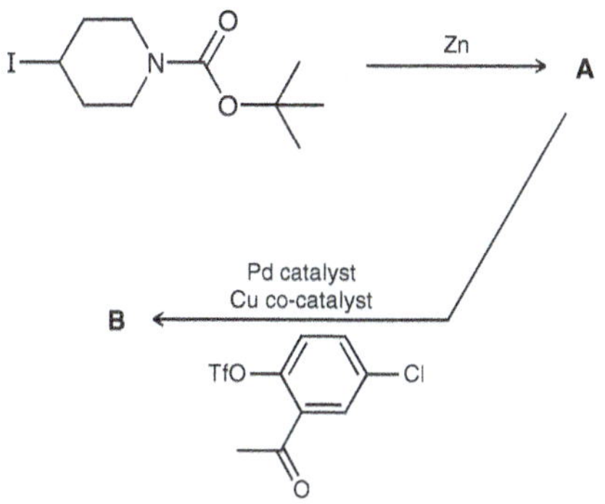

I— —N— O — Zn → A

A — Pd catalyst, Cu co-catalyst; TfO— —Cl → B

**28.56** A calix[4]arene is a cone-shaped macrocycle (large ring) composed of four arenes connected by intervening $CH_2$ groups. The calix[4]arene below was subjected to the following reaction sequence: (1) excess *t*-BuLi; (2) excess $ZnCl_2$; (3) excess 4-iodotoluene, $Pd(PPh_3)_4$ (*J. Org. Chem.*, **1997**, *62*, 4171). In the first step of this sequence, *t*-BuLi was used (in place of excess lithium) to convert the C—Br bonds into C—Li bonds. Draw the structure of the product of this reaction sequence.

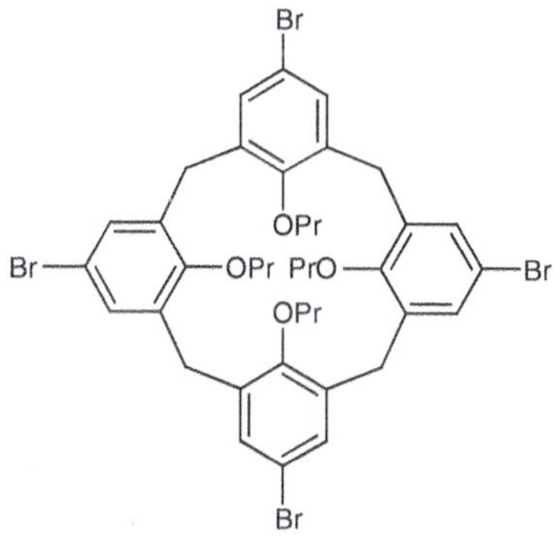

Br, Br, Br, Br; OPr, OPr, PrO, OPr

**28.57** Compounds **1–3**, called alkenyl phosphates, were investigated as electrophiles in Negishi coupling reactions with organozinc compounds **4** and **5** (*J. Org. Chem.*, **2007**, *72*, 6464). The phosphate group, —OP(=O)(OPh)$_2$, serves as the leaving group in the reaction. Draw the structures of the products from each of the following combinations of reactants in the presence of a Pd catalyst:

**(a)** **1** (1 equivalent) + **4** (1 equivalent)
**(b)** **1** (1 equivalent) + **5** (1 equivalent)
**(c)** **2** (1 equivalent) + **4** (1 equivalent)
**(d)** **3** (1 equivalent) + **5** (2 equivalents)

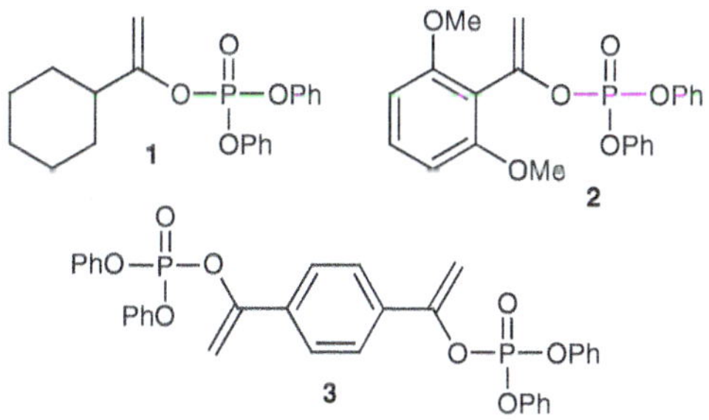

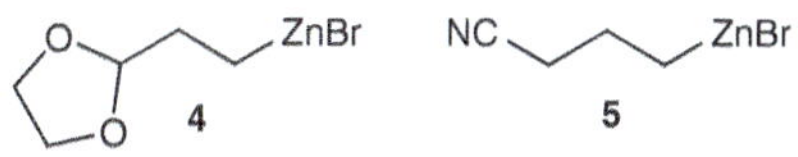

O—P(=O)—OPh, OPh — **1**

OMe, OMe, O—P(=O)—OPh, OPh — **2**

PhO—P(=O)—O, PhO; O—P(=O)—OPh, OPh — **3**

ZnBr — **4**; NC ZnBr — **5**

**28.58** Bipyridine compounds are used as bidentate ligands in a broad range of metal complexes. Asymmetric bipyridines, in which the two pyridine units are not identical, can be produced by Negishi coupling between an arylzinc and an aryl triflate as shown (*J. Org. Chem.*, **1998**, *63*, 10048).

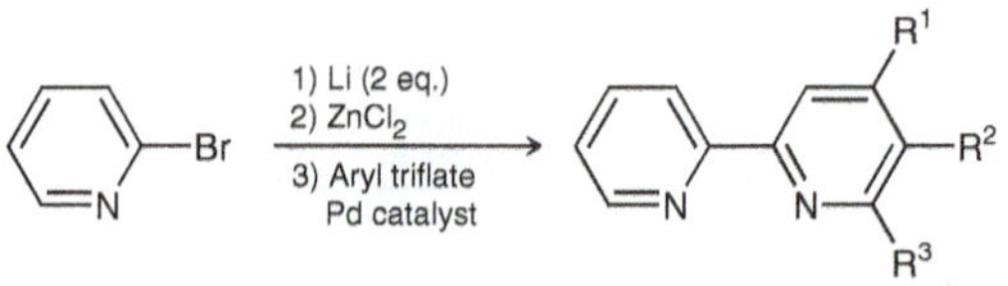

Provide the structure of the aryl triflate used to produce each of the following bipyridines:

**(a)** $R_1 = Me$, $R_2 = H$, $R_3 = H$
**(b)** $R_1 = H$, $R_2 = Me$, $R_3 = H$
**(c)** $R_1 = H$, $R_2 = H$, $R_3 = Me$

**28.59** K-13 is a naturally occurring molecule with a cyclic structure that constrains its tripeptide backbone. A key step in a synthesis of K-13 involves an intramolecular Pd-catalyzed Negishi coupling to produce compound **1**, which is subsequently converted to the natural product (*J. Org. Chem.*, **2009**, *74*, 8280). Draw the structure of compound **1**.

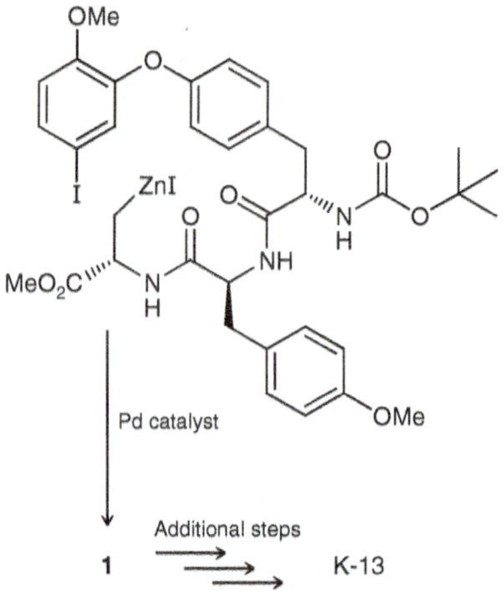

# INTEGRATED PROBLEMS

**28.60** The reaction sequence below allows for the preparation of novel liquid crystalline materials using a series of organometallic reactions (*J. Org. Chem.*, **2008**, *73*, 830). The aryllithium shown is converted to organozinc **A**. This compound then undergoes Negishi coupling with an organic electrophile containing a tributyltin group to produce compound **B**. The tin group survives the mild conditions (room temperature) of this reaction. A subsequent Stille coupling at elevated temperature produces compound **C**. Draw the structures of compounds **A, B,** and **C**.

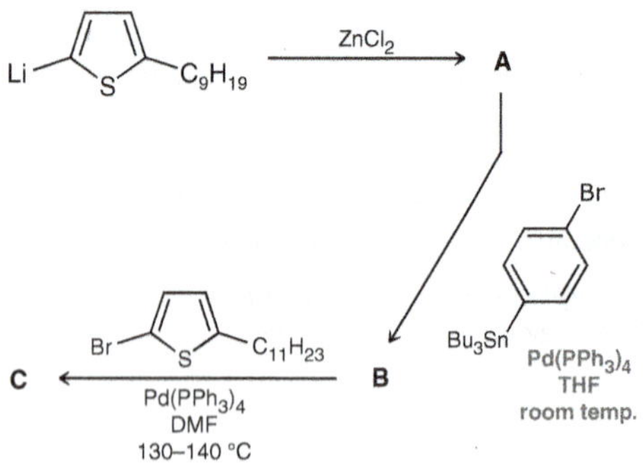

**28.61** Saudin is a naturally occurring compound that has been found to induce hypoglycemia (low blood sugar) in mice and may thus serve as a potential lead structure for the development of new drugs to help control diabetes.

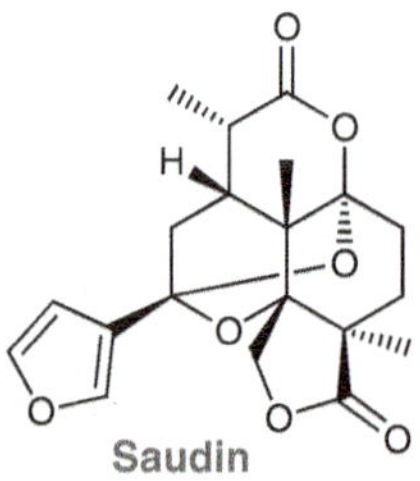

Saudin

A section of the fused polycyclic structure of saudin has been synthesized using the Stille reaction shown below (*Org. Lett.*, **2005**, *7*, 2413–2416). The vinyl iodide and vinyl organotin reactants were combined in the presence of a palladium catalyst to produce the Stille product, which spontaneously undergoes a rearrangement to produce the product shown below. Draw the initial product of the Stille reaction and then draw a mechanism consistent with the formation of the product shown. Also draw the transition state of the rearrangement.

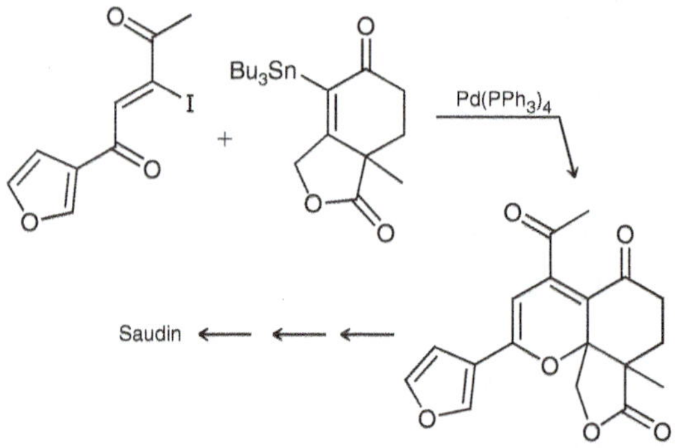

**28.62** The compound below was exposed to a palladium catalyst to produce a macrocycle (large ring). The macrocycle subsequently underwent a rearrangement to produce a fused hexacyclic (six-ring) structure (*Org. Lett.*, **2002**, *4*, 3391–3393). The ring system of the final product is one that is found in a number of polycyclic natural products. Draw the initial product of the palladium-catalyzed reaction, the final product, and a mechanism consistent with the rearrangement step.

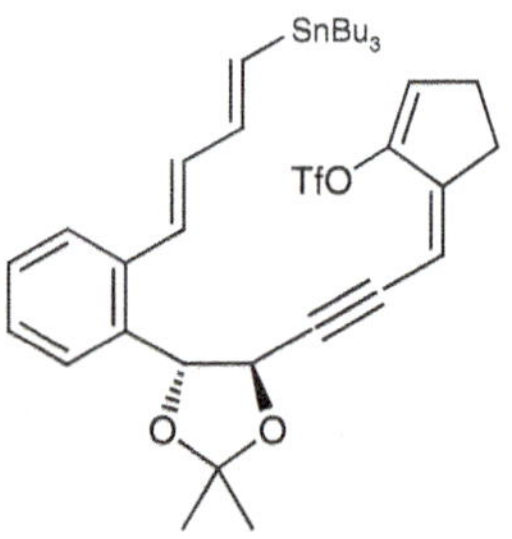

**28.63** Disorazoles are a family of structurally related natural products first isolated in 1994 from the fermentation broth of the bacterium *Sorangium cellulosum*. These natural products exhibit anticancer properties, thereby fueling the search for more potent analogs of these compounds. Recently, a cyclopropane-containing analog, called $CP_2$-disorazole $C_1$, was synthesized and its biological activity assessed (*Org. Lett.*, **2011**, *13*, 4088–4091). This new compound was found to have significant activity against human colon cancer cell lines. During the synthesis of $CP_2$-disorazole $C_1$, compound **1** was prepared from alkenes **2** and **3**. Provide a reasonable synthesis of compound **1** from alkenes **2** and **3**.

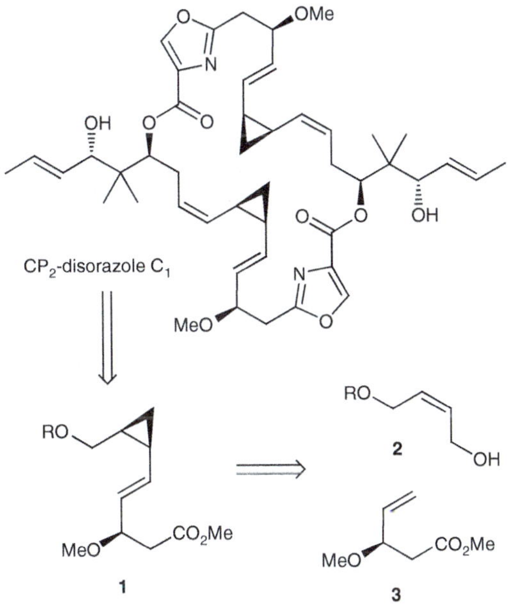

$CP_2$-disorazole $C_1$

**28.64** During a recent total synthesis of asteriscanolide, a sesquiterpene lactone with unprecedented molecular architecture, compound **A** was heated with the Grubbs 2nd generation catalyst under an atmosphere of ethylene gas to form compound **B**. Compound **B** is not isolated, because at elevated temperature (80°C) it undergoes a rearrangement to give compound **C** (*J. Am. Chem. Soc.*, **2000**, *122*, 8071–8072).

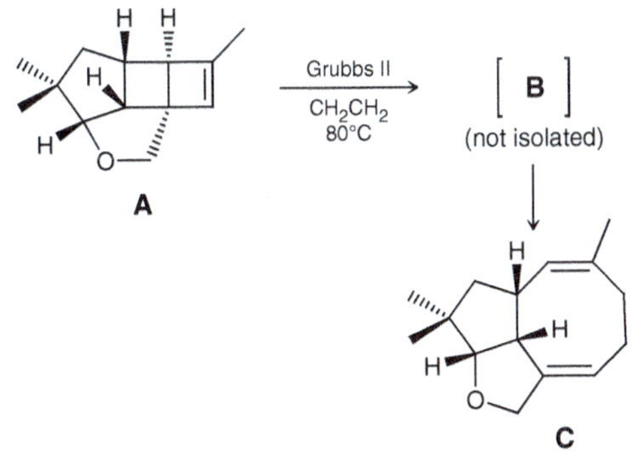

**(a)** Draw the structure of **B**.

**(b)** Propose a plausible mechanism for the conversion of **B** into **C**.

**(c)** Identify the driving force for the conversion of **A** into **B**.

**(d)** Identify the driving force for the conversion of **B** into **C**.

**(e)** Propose an alternate route to synthesize **C** from **A** that does not involve ring-opening metathesis.

## CHALLENGE PROBLEMS

**28.65** (–)-Rapamycin, a powerful immunosuppressive and antibiotic agent, is produced by the bacterium *Streptomyces hygroscopicus*, found in the soil native to Easter Island (in the South Pacific). K. C. Nicolaou was the first of several investigators to report a total synthesis of (–)-rapamycin (*J. Am. Chem. Soc.*, **1993**, *115*, 4419). The final maneuver involved a novel cyclization procedure in which compound **1** was treated with a distannane (compound **2**) in the presence of a suitable palladium catalyst to afford (–)-rapamycin. Under these conditions, an initial intermolecular Stille coupling process occurs followed by an intramolecular Stille coupling process. This tandem inter/intramolecular Stille coupling effectively "stitches up" compound **1** into a macrolide (a large ring). Draw the structure of (–)-rapamycin.

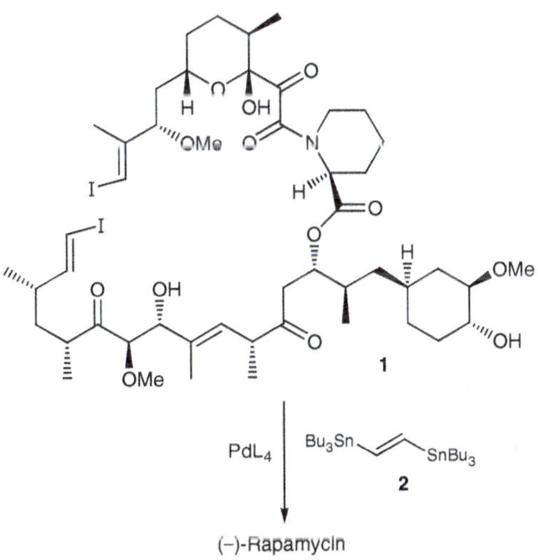

**28.66** The bicyclic compound *cis*-sabinene hydrate is a natural product that is one of the main molecules responsible for the flavor of the herb marjoram. It can be prepared as a racemic mixture via the one-pot synthesis shown below (*Org. Lett.*, **2001**, *3*, 2121–2123). Step 2 is diastereoselective, giving rise to the product shown and its enantiomer.

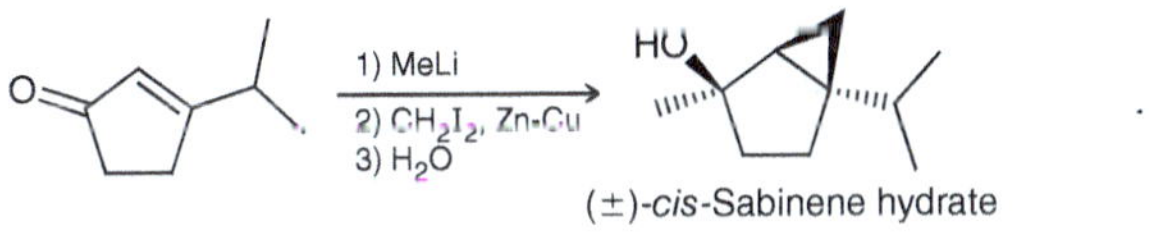

(±)-*cis*-Sabinene hydrate

**(a)** Draw a mechanism for each step in the reaction sequence.

**(b)** Suggest a source for the observed diastereoselectivity.

**(c)** If steps 2 and 3 are reversed, the reaction sequence proves unsuccessful. That is, if the starting material is treated with methyllithium followed by water, the product resulting from those two steps would be very unstable and would decompose quickly in the presence of a small trace of acid or under the conditions of the final step. Propose an explanation for this observation.

**28.67** Compound **1** undergoes an intramolecular Simmons-Smith type reaction to afford a fused bicyclic *meso* product (compound **3**), which can be used as a synthetic intermediate for subsequent ring-opening reactions (*J. Am. Chem. Soc.*, **2010**, *132*, 1895–1902). The cyclopropanation process is initiated by replacement of one of the iodine atoms with a ZnEt group to give the carbenoid intermediate shown.

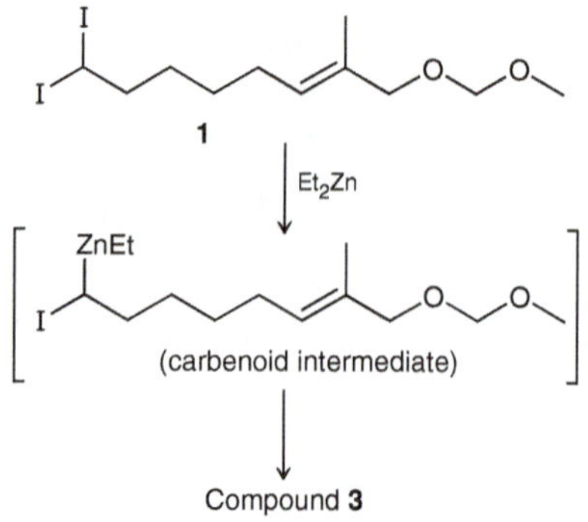

I I 1 $Et_2Zn$ ZnEt I (carbenoid intermediate) Compound **3**

Unlike compound **1**, compound **2** produces only a small amount of the cyclopropanation product:

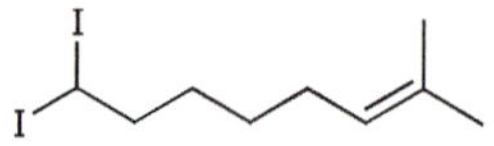

I I 2

Draw the structure of the cyclopropanation product derived from **1** and propose an explanation for the difference in reactivity of compounds **1** and **2**.

**28.68** *N*-Acetylcolchinol is a highly potent anticancer drug, and its characteristic 6,7,6 fused ring system can also be seen in the structure of dibenzo[*a*,*c*]cyclohepten-5-one.

HO H N O MeO MeO MeO O

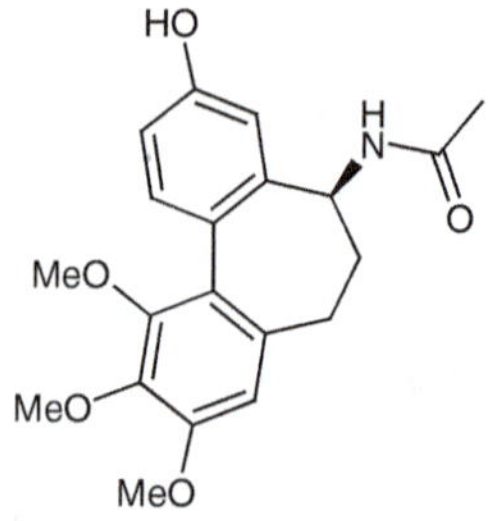

***N*-acetylcolchinol** 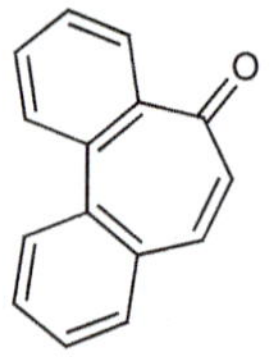 **Dibenzo[*a*,*c*]cyclohepten-5-one**

This ketone (and derivatives of it) may thus serve as a useful synthetic intermediate toward analogs of the cancer drug (*J. Org. Chem.*, **2009**, *74*, 3948–3951). Using starting materials with eight or fewer carbon atoms, propose a multistep synthesis of dibenzo[*a*,*c*]cyclohepten-5-one that includes a Suzuki coupling.

**28.69** Diene **1** is useful in that it can be utilized in sequential Stille and Suzuki coupling reactions.

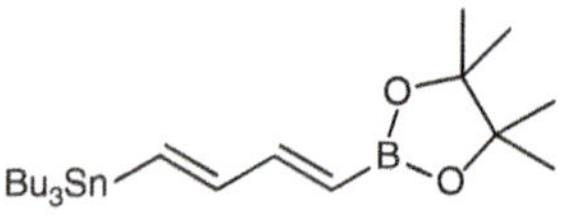

$Bu_3Sn$ O B O 1

The researchers who developed this reagent recognized that a Stille coupling could be selectively accomplished on the tin-bearing end of the molecule because of the requirement of basic conditions for transmetallation at the boron-bearing end of the structure (*Org. Lett.*, **2005**, *7*, 2289–2291). **(a)** Determine the product of the reaction of **1** with 4-iodoanisole and a palladium catalyst, followed by the addition of CsF (which serves as a base), and 4-*tert*-butyliodobenzene. **(b)** Propose a synthesis of **2** (which contains the side chain of the natural product and antitumor agent lucilactaene) using a similar protocol.

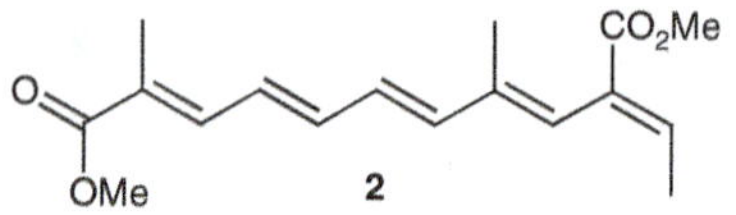

O OMe 2 $CO_2Me$

**28.70** In a study exploring the utility of olefin metathesis reactions, the following triene was prepared and subjected to a Grubbs catalyst (*J. Am. Chem. Soc.*, **1996**, *118*, 6634–6640). The triene underwent a tandem ring-opening/ring-closing reaction resulting in the cleavage of the C=C unit within the ring, the formation of two new cyclic C=C groups, and the expulsion of ethylene gas. Draw the structure of the product, showing all chirality centers:

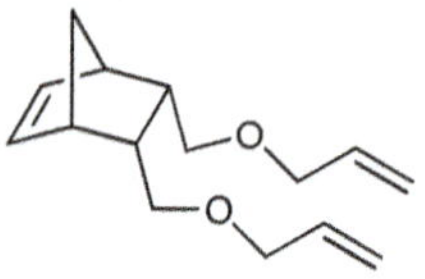

O O

**28.71** Like alcohols, boronic acids [$RB(OH)_2$)] can be protected using *N*-methyliminodiacetic acid (MIDA). The MIDA boronates that result are unreactive to normal Suzuki coupling conditions. When treated with NaOH, the boronic acid can be revealed and utilized in a Suzuki reaction.

R—B(OH)(OH) MIDA MeN R—B O O O O ⇌ ⊕ MeN ⊖ R—B O O O O NaOH R—B(OH)(OH)

**MIDA boronate**
**(unreactive toward Suzuki coupling)**

This technique enables synthetic organic chemists to perform iterative Suzuki coupling reactions. For example, consider a starting material that contains both a MIDA boronate group as well as a halide group. This type of bifunctional compound can be used to achieve successive (iterative) Suzuki coupling reactions, as shown:

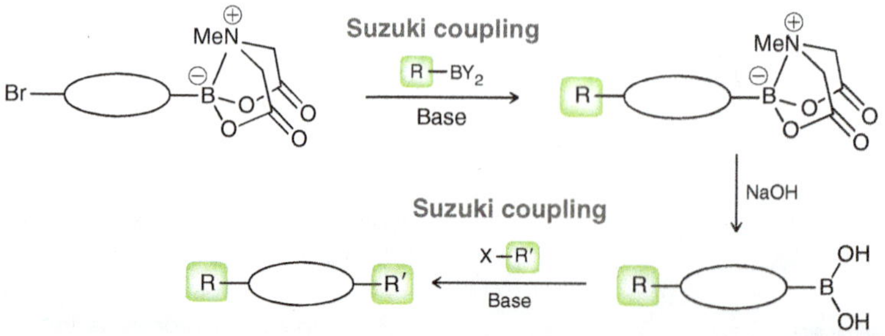

Br ⊕ MeN ⊖ B O O O O **Suzuki coupling** R—$BY_2$ Base R ⊕ MeN ⊖ B O O O O NaOH

**Suzuki coupling** R R′ X—R′ Base R B(OH)(OH)

This strategy (iterative Suzuki coupling) was recently employed in a synthesis of ratanhine, the most complex member of a family of compounds isolated from the medicinal plant *Ratanhiae radix*. Propose a reasonable synthesis of ratanhine using the four building blocks shown (*J. Am. Chem. Soc.*, **2007**, *129*, 6716–6717). Note: R is a protecting group that can be cleaved upon treatment with HCl to give an OH group.

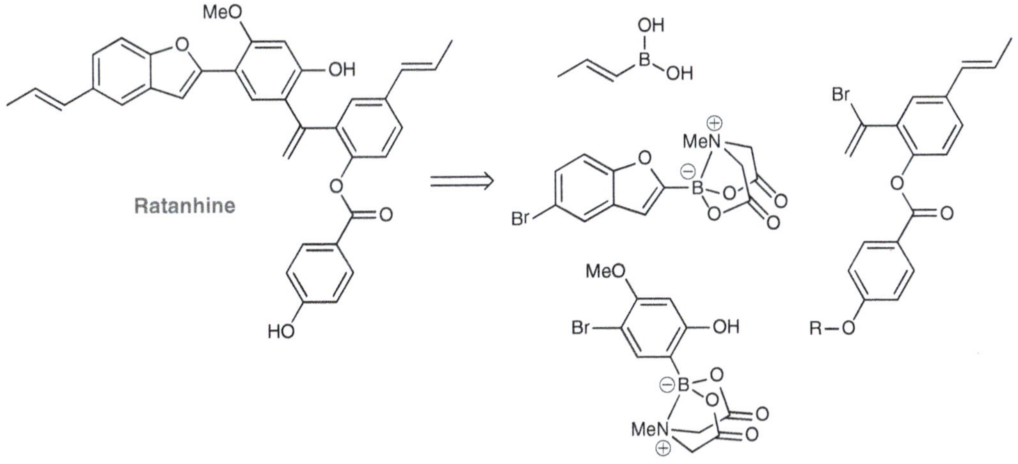

Ratanhine

**28.72** First isolated in 1969, ecteinascidin 743 belongs to a family of cytotoxic natural products isolated from a marine sponge of genus *Ecteinascidia*. These natural products were not fully identified until 1990, when exhaustive NMR studies provided synthetic chemists with the overall structures of these compounds. During a total synthesis of ecteinascidin 743, compound **1** was treated with a palladium catalyst under Heck reaction conditions to form compound **A** (*J. Am. Chem. Soc.*, **2002**, *124*, 6552–6554). Draw the structure of compound **A** and provide a rationale for the observed stereochemical outcome as well as the observed regiochemical outcome for this process.

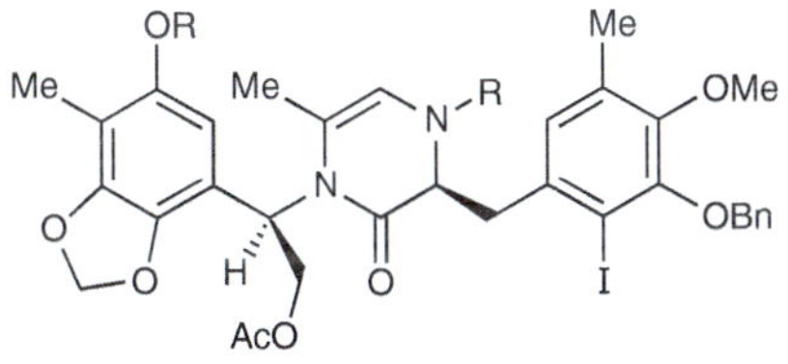

**28.73** Compound **1**, called 5-*epi*-hydroxycornexistin, is a diastereoisomer of the herbicidal natural product hydroxycornexistin. The nine-membered carbocyclic core of compound **1** was recently synthesized using two sequential organometallic operations. Provide a reasonable synthesis of compound **2** from compound **3** and any other reagents of your choice (*Org. Biomol. Chem.*, **2008**, *6*, 4012–4025.)

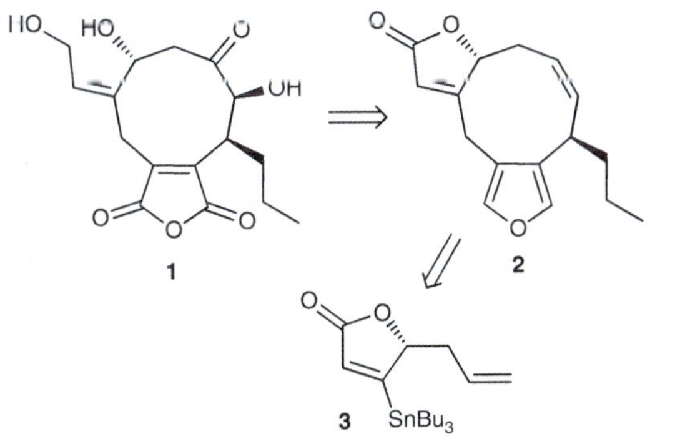

**28.74** Treatment of compound **1** with a Grubbs catalyst affords a mixture of two cyclic ethers, **A** and **B**.

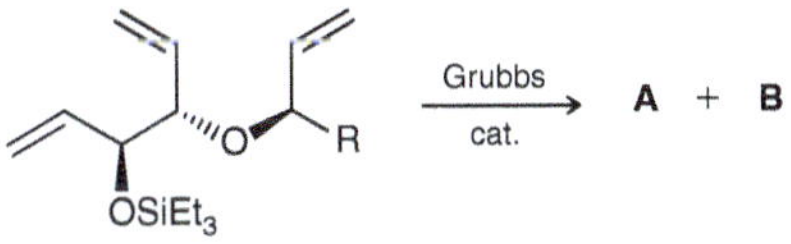

**(a)** Draw the structures of the products **A** and **B**. Note: A four-membered ring will not form if the option of a five-membered ring (or larger ring) is possible.

**(b)** Compound **A** was recently employed as a key intermediate in the synthesis of mucocin, a compound with noted inhibitory activity against a number of cancer cell lines (*Org. Lett.*, **2006**, *8*, 2369–2372). In order to prepare **A** selectively (without the concurrent formation of **B**), compound **2** was used in place of compound **1**. Compound **2** has three monosubstituted C=C units, but one of them (the allyl ether group) is significantly less hindered than the others. The Grubbs catalyst will preferentially metathesize that C=C unit first (giving intermediate **3**) and then "relay" to the internal alkene via a ring-closing metathesis event, giving intermediate **4** and cyclic compound **5**. This process, called *relay ring-closing metathesis*, enables the selective formation of cyclic ether **A**. Identify which of the two cyclic ethers (from part a of this problem) is compound **A** and draw the structure of compound **5**.

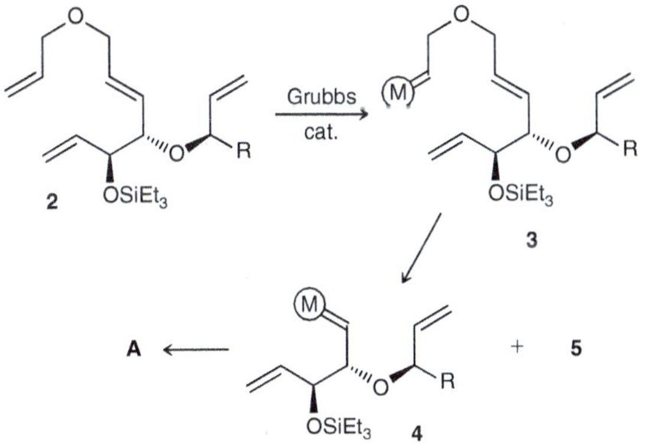

# Nomenclature of Polyfunctional Compounds

Rules of nomenclature were first introduced for naming alkanes in Chapter 4. Those rules were then further developed in subsequent chapters to include the naming of a variety of functional groups. Table A.1 contains a list of sections in which nomenclature rules were discussed.

**TABLE A.1 RULES OF NOMENCLATURE COVERED IN PREVIOUS CHAPTERS**

| GROUP | SECTION |
|---|---|
| Alkanes | 4.2 |
| Alkyl halides | 7.2 |
| Alkenes | 8.3 |
| Alkynes | 10.2 |
| Alcohols and phenols | 13.1 |
| Ethers | 14.2 |
| Epoxides | 14.7 |
| Thiols and sulfides | 14.11 |
| Benzene derivatives | 18.2 |
| Aldehydes and ketones | 20.2 |
| Carboxylic acids | 21.2 |
| Carboxylic acid derivatives | 21.6 |
| Amines | 23.2 |

As you progress through the topics in Table A.1, you might notice some patterns that emerge. For example, the systematic name of an organic compound will generally have at least three parts, as seen in each of the following examples:

| $CH_4$ | ethanol | cyclohexene | $H_3C-C\equiv C-H$ | bicyclic ketone |
|---|---|---|---|---|
| meth an e | eth an ol | cyclohex en e | prop yn e | bicyclo[3.1.0]hex an -3-one |

Each of these compounds has three parts (highlighted) in its name.

parent unsaturation suffix

First is the *parent* (meth, eth, etc.) which indicates the number of carbon atoms in the parent structure (see Section 4.2). The next part of the name, called *unsaturation*, indicates the presence or absence of C=C or C≡C bonds (where "an" indicates the absence of any C=C or C≡C bonds; "en" indicates the presence of a C=C bond; and "yn" indicates the presence of a C≡C bond). The final part of the name, or *suffix*, indicates the presence of a functional group (for example, "-ol" indicates the presence of an OH group, while "-one" indicates a ketone group). If no functional group is present (other than C=C or C≡C bonds), then "e" is used as the suffix, such as in methan**e** or propyn**e**.

If substituents are present, then the systematic name will have at least four parts, rather than just three, for example:

5-chloro-4,4,6-trimethyl hept -5-en -2-one

Substituents Parent Unsaturation Suffix

Notice that all of the substituents appear before the parent. Locants are used to indicate the locations of all substituents, as well as the locations of any unsaturation and the location of the group named in the suffix. Notice also that locants are separated from letters with hyphens (5-chloro), but separated from each other with commas (4,4,6). When multiple substituents are present, they are organized in alphabetical order (see Section 4.2 for a review of the non-intuitive alphabetization rules; for example, "isopropyl" is alphabetized by "i" rather than "p," while *tert*-butyl is alphabetized by "b" rather than "t").

Compounds with chirality centers or stereoisomeric C=C bonds must include stereodescriptors to identify their configuration. These stereodescriptors (*R* or *S*, *E* or *Z*, *cis* or *trans*) must be placed at the beginning of the name, and therefore constitute a fifth part to consider when assigning a compound's name. For example, each of the following compounds has five parts to its name, where the first part indicates configuration:

(2*R*,3*S*) -2,3-dibromo pent an e

*E* or *trans* -4-methyl pent -2-en e

The more complex a compound is, the longer its name will be. But even for very long names that are a paragraph in length, it should still be possible to dissect the name into its five parts:

stereoisomerism substituents parent unsaturation suffix

Table A.2 contains a list of sections in which the rules for each of these five parts were introduced.

**TABLE A.2 THE PREVIOUS SECTIONS IN WHICH EACH OF THE FIVE PARTS OF A COMPOUND'S NAME WERE FIRST INTRODUCED**

| PART OF NAME | SECTIONS |
|---|---|
| Stereoisomerism | 5.3, 8.4 |
| Substituents | 4.2 |
| Parent | 4.2 |
| Unsaturation | 8.3, 10.2 |
| Suffix | 13.1, 14.2, 14.7, 14.11, 20.2, 21.2, 21.6, and 23.2 |

When a compound contains multiple functional groups, we must choose which functional group has the highest priority and is assigned the suffix of the name. The other functional groups must be listed as substituents. For example, consider the following compound

5-oxo hexan al

Substituent Suffix

Notice that two functional groups are present, but the suffix of a compound's name can only indicate one functional group. In this case, the aldehyde group gets the priority (suffix = "al"), and the ketone group must be named as a substituent (oxo). Table A.3 shows the terminology that is used for several commonly encountered functional groups, in order of priority, with carboxylic acids having the highest priority.

**TABLE A.3 THE NAMING OF COMMON FUNCTIONAL GROUPS, EITHER AS THE SUFFIX OR AS SUBSTITUENTS. GROUPS ARE ARRANGED IN ORDER OF PRIORITY, WITH CARBOXYLIC ACIDS HAVING THE HIGHEST PRIORITY AND ETHERS HAVING THE LOWEST PRIORITY.**

| FUNCTIONAL GROUP | NAME AS SUFFIX | NAME AS SUBSTITUENT |
|---|---|---|
| Carboxylic acids | -oic acid | carboxy |
| Esters | -oate | alkoxycarbonyl |
| Acid halides | -oyl halide | halocarbonyl |
| Amides | -amide | carbamoyl |
| Nitriles | -nitrile | cyano |
| Aldehydes | -al | oxo |
| Ketones | -one | oxo |
| Alcohols | -ol | hydroxy |
| Amines | -amine | amino |
| Ethers | ether | alkoxy |

When assigning locants, the functional group in the suffix should be assigned the lowest possible number, for example:

Correct

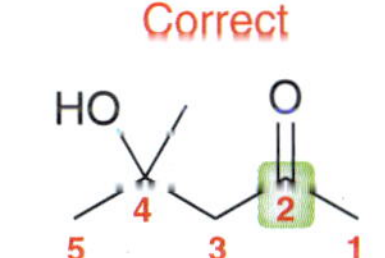


4-hydroxy-4-methylpentan-2-one

Incorrect

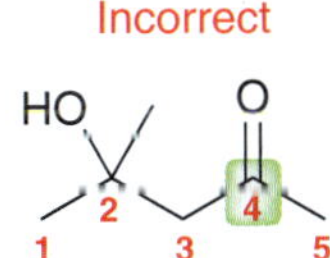


2-hydroxy-2-methylpentan-4-one

In this example, there are two functional groups. The ketone group takes priority, so the suffix will be "-one," and the OH group gets named as a substituent (hydroxy). Notice that the locant for the ketone group is 2, rather than 4.

When there is a choice, the parent chain should include the functional group that did not receive the priority, for example:

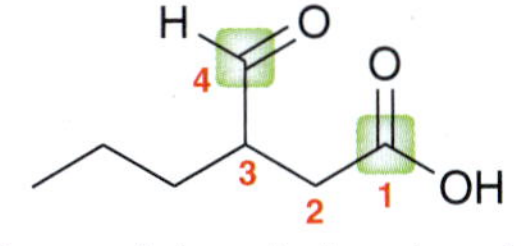


3-propyl-4-oxobutanoic acid

In this case, there are two functional groups – a carboxylic acid group and an aldehyde group. The former will command the suffix, while the latter will be indicated as a substituent. It is clear that the carbon atom of the carboxylic acid group must represent the beginning of the parent chain, but notice that the carbon atom of the aldehyde group should also be contained in the parent chain, even though doing so results in a smaller chain size (a parent of four carbon atoms, rather than a parent of six, in this case).

When assigning locants in bicyclic systems, keep in mind that locants must begin at a bridgehead atom and then continue along the longest bridge, followed by the second longest bridge, and ending with the shortest bridge; even if it means that the locant assigned to the suffix will be a high number, as seen in the following case:

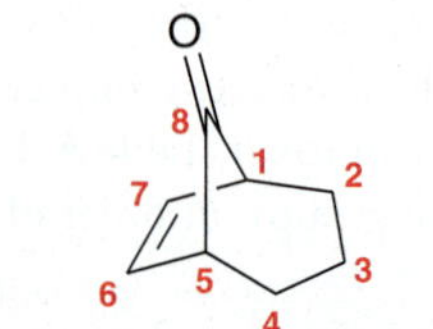

bicyclo[3.2.1]oct-6-en-8-one

This example exhibits a ketone group which will command the suffix. Nevertheless, this functional group is at C8 (the highest possible locant in this case), because it is located on the shortest bridge. If there was a hydroxyl group at C3 of this compound, we would name that group as a substituent (3-hydroxy-), even while the ketone group would remain at C8. This is often a feature of rigid systems with defined numbering systems (such as bicyclic compounds).

# Credits

**All chapters** SkillBuilders: *bicycle rider, wheel*: Created by DeMarinis Designs, LLC, for John Wiley & Sons, Inc. Medically Speaking, *pills*, Norph/Shutterstock.

**Chapter 1** Opener: *kite string*, Dave King/Dorling Kindersley/Getty Images; *bolt of lightning striking Empire State Building*, Chang W. Lee/The New York Times/Redux Pictures; *brass skeleton key hanging from string*, Gary S Chapman/Photographer's Choice/Getty Images; *lightning striking the Empire State Building during a storm*, Keystone/Hulton Archive/Getty Images; *polaroid frame*, Cole Vineyard/iStockphoto. Practically Speaking, p. 36: *Tokay gecko*, Eric Isselée/iStockphoto. Medically Speaking, p. 42: *bottle of Propofol*, Frederick M. Brown/Getty Images. Figures 1.6, 1.13, 1.23–1.25, 1.27–1.29, 1.32, 1.33, 1.36–1.40, 1.42, 1.44, 1.47, 1.48, Table 1.1, and Table 1.2 are Reprinted with permission of John Wiley & Sons, Inc. from Solomons, G., *Organic Chemistry, 10e*, © 2011. Running head: *lightning strikes*, jhorrocks/iStockphoto; *brass key*, J. R. Bale/Stockphotopro, Inc.

**Chapter 2** Opener: *poppy flower*, Neil Fletcher/Getty Images; *hand drawing molecule*, susoy/iStockphoto; *capsule, tablet, gel tab*, Norph/Shutterstock. Medically Speaking, p. 61: *corals*, © Bob Shanley/Palm Beach Post/ZUMAPRESS.com. Medically Speaking, p. 68: *poppy flower*, Neil Fletcher/Getty Images. Running head: *capsule, tablet, gel tab*, Norph/Shutterstock.

**Chapter 3** Opener: *spoons of assorted raising agents*, Ken Karp; *batch of baking powder biscuits*, Ken Karp; *lemon*, Shane White/iStockphoto; *whole meal flour*, Masterfile. Medically Speaking, p. 100: *pink antacid*, © Lawrence Manning/Corbis/Glow Images; *spoonful of medicine (pink antacid)*, © Radius Images/Glow Images. Practically Speaking, p. 128: *spoons of assorted raising agents*, Ken Karp; *batch of baking powder biscuits*, Ken Karp. Running head: *lemon slice*, Shane White/iStockphoto.

**Chapter 4** Opener: *loose ribbon*, macida/Getty Images; *overlapped ribbon*, JamesBrey/iStockphoto; *AIDS ribbon*, JamesBrey/iStockphoto. Practically Speaking, p. 145: *moth*, Tanya Back/iStockphoto. Practically Speaking, p. 158: *Empty blue plastic bag*, t300/iStockphoto; *plastic grocery bags*, NoDerog/iStockphoto. Medically Speaking, p. 166: *loose ribbon*, macida/Getty Images; *overlapped ribbon*, JamesBrey/iStockphoto; *AIDS ribbon*, JamesBrey/iStockphoto Figure 4.19 Reprinted with permission of John Wiley & Sons, Inc. from Solomons, G., *Organic Chemistry, 10e*, © 2011. Running head: JamesBrey/iStockphoto.

**Chapter 5** Opener: *people in circle*, diez artwork/Shutterstock, *green capsule*, Dimitris66/iStockphoto. Practically Speaking, p. 202: *spice mix*, Tatyana Nyshko/iStockphoto; *spices*, Adam Gryko/iStockphoto. Medically Speaking, p. 205: *people in circle*, diez artwork/Shutterstock, *green capsule*, Dimitris66/iStockphoto. Figures 5.2, 5.5, 5.6, 5.10, and the art in the Medically Speaking box on p. 208 Reprinted with permission of John Wiley & Sons, Inc. from Solomons, G., *Organic Chemistry, 10e*, © 2011. Figure 5.9 Reprinted with permission of John Wiley & Sons, Inc. from Holum, J.R., *Organic Chemistry: A Brief Course*, p.316, © 1975. Running head: *capsule*, Dimitris66/iStockphoto.

**Chapter 6** Opener: *dynamite bomb*, O.V.D./Shutterstock; *floating dollars*, Lukiyanova Natalia/frenta/Shutterstock. Practically Speaking, p. 244: *round bomb with fuse*, Andrzej Tokarski/iStockphoto. Practically Speaking, p. 245: *growing plant*, Loren Evans/iStockphoto. Medically Speaking, p. 251: *dynamite*, O.V.D./Shutterstock. Practically Speaking, p. 252: *wheat overfilled glass of beer*, Dinamir Predov/iStockphoto; *wheat berries with ears*, Kenan Savas/iStockphoto. Figure 6.27 Reprinted with permission of John Wiley & Sons, Inc. from Solomons, G., *Organic Chemistry, 10e*, © 2011. Running head: *dynamite*, O.V.D./Shutterstock.

**Chapter 7** Opener: *green syringe*, Stockcam/Getty Images, *red blood cells*, Andril Muzyka/Shutterstock, *ink*, mkurtbas/iStockphoto, *single cancer cell*, Lightspring/Shutterstock. Medically Speaking, p. 320: *PET scan*, Courtesy Brookhaven National Laboratory. Running head: *green syringe*, Stockcam/Getty Images.

**Chapter 8** Opener: *light bulb*, Evgeny Terentev/Getty Images, *baby bracelet*, © kroach/Age Fotostock America, Inc. Practically Speaking, p.343: *codling moth*, © Nigel Cattlin/Alamy; *apple on a branch*, hans slegers/iStockphoto; *branch of apple tree blossoms*, BORTEL Pavel/Shutterstock; *codling moth larvae*, N A Callow/Photoshot. Medically Speaking, p. 353: *jaundiced baby*, DAVID JONES/PA Photos/Landov LLC, *baby under blue light*, TommyIX/Vetta/Getty Images. Figure 8.4 Reprinted with permission of John Wiley & Sons, Inc. from Solomons, G., *Organic Chemistry, 10e*, © 2011. Running head: © kroach/Age Fotostock America, Inc.

**Chapter 9** Opener: *Styrofoam peanuts*, CrackerClips/Shutterstock. Practically Speaking, p. 415: *Styrofoam peanuts*, CrackerClips/Shutterstock. Practically Speaking, p. 420: *fuel pump with last drop of gasoline*, Grafissimo/iStockphoto. Practically Speaking, p. 434: *Kentucky Fried Chicken bucket and McDonald's french fries*, © AP/Wide World Photos. Running head: *Styrofoam peanuts*, CrackerClips/Shutterstock.

**Chapter 10** Opener: *headache*, Sebastian Kaulitzki/Shutterstock; *neuron*, Imrich Farkas/Shutterstock. Margin, p. 465: *welding torch with flying sparks*, Fertnig/iStockphoto. Medically Speaking, p. 466: *headache*, Sebastian Kaulitzki/Shutterstock; *neuron*, Imrich Farkas/Shutterstock. Practically Speaking, p. 467: *digital camera*, Marek Mnich/iStockphoto; *LEDs in red, green and blue*, © Media Bakery. Running head: *neuron*, Imrich Farkas/Shutterstock.

**Chapter 11** Opener: *hand holding fire extinguisher*, Sabine Scheckel/Photodisc/Getty Images; *burn hole in white paper*, Robyn Mackenzie/Shutterstock; *fire flame*, Valeev/Shutterstock; *molecules*, David Klein. Practically Speaking, p. 526: *hand holding fire extinguisher*, Sabine Scheckel/Photodisc/Getty Images; *fire flame*, Valeev/Shutterstock. Margin, p. 531: *anti-aging nourishing cream*, Jakub Pavlinec/iStockphoto. Running head: *fire flame*, Valeev/Shutterstock.

**Chapter 12** Opener: *pills*, Matej Pribelsky/iStockphoto, Josiah Lewis/iStockphoto, paul kline/iStockphoto, btolgak/iStockphoto, Photolink/Shutterstock. Medically Speaking, p. 570: *yew needles*, Dennis Flaherty/Science Source. Running head: *pills*, Matej Pribelsky/iStockphoto, Josiah Lewis/iStockphoto, paul kline/iStockphoto, btolgak/iStockphoto, Photolink/Shutterstock.

**Chapter 13** Opener: *glass, blue umbrella,* Michael Gray/iStockphoto; *yellow umbrella,* Doug Cannell/iStockphoto; *icebag*, Richard Cano/iStockphoto. Practically Speaking, p. 596: *overheated car*, © Media Bakery. Medically Speaking, p. 603: *legs*, Jens Handt/iStockphoto. Practically Speaking, p. 610: *breathalyzer*, Bridget McGill/iStockphoto. Practically Speaking, p. 614: *glass/umbrella*, Michael Gray/iStockphoto; *icebag*, Richard Cano/iStockphoto. Spectra in Problems 13.53-13.56 © Dr. Richard A. Tomasi. Running head: *glass, blue umbrella,* Michael Gray/iStockphoto *ice bag*, Richard Cano/iStockphoto.

**Chapter 14** Opener: *cigarette butt*, Peter vd Rol/Shutterstock; *smoke trail*, stavklem/Shutterstock; *cigarette*, Oliver Hoffmann/Shutterstock. Practically Speaking, p. 657: *instruments/tray*, Ruslan Kerlmov/iStockphoto. Spectra in Problems 14.52-14.55 © Dr. Richard A. Tomasi. Running head: *cigarette butt*, Peter vd Rol/Shutterstock.

**Chapter 15** Opener: *toast thermogram*, Tony McConnell/Science Source, *bread and butter*, slobo/iStockphoto. Practically Speaking, p. 686: *microwave oven*, 2happy/Shutterstock; *kernel popping*, Bill Grove/iStockphoto; *popcorn in bag*, Joao Virissimo/Shutterstock; *unpopped kernel*, Charles B. Ming Onn/Shutterstock. Medically Speaking, p. 687: *house*, Ted Kinsman/Science Source, Inc.; *torso*, SPL/Science Source. Margin, p. 688: *salt plate*, Martyn F. Chillmaid/Science Source. Practically Speaking, p. 696: *Intoxilizer 5000*, © AP/Wide World Photos. p. 708: *people*, © Media Bakery. Figures 15.1, 15.2, 15.9, 15.11, 15.18, and 15.31 reprinted with permission of John Wiley & Sons, Inc. from Solomons, G., *Organic Chemistry 10e*, © 2011. Spectra in Problems 15.9, 15.10, and 15.12 © Dr. Richard A. Tomasi. Figure 15.21 Reprint with the permission of John Wiley & Sons, Inc. from Holum, J.R., *Organic Chemistry: A Brief Course*, © 1975. Spectra in Problems 15.24, 15.46, 15.53, 15.59, 15.60, 15.76, and Figure 15.30 Reprinted with permission of the SDBS, National Institute of Advanced Industrial Science and Technology. Running head: John Wiley & Sons, Inc.

**Chapter 16** Opener: *magnet*, Sideways Design/Shutterstock; *blue sky/clouds*, Andresr/Shutterstock; *strawberry*, ranplett/iStockphoto; *strawberry with stem,* Stephen Rees/iStockphoto; *frog*, Marcus Jones/iStockphoto; *hazelnuts*, fotosav/iStockphoto; *hot-air balloon*, Mike Sonnenberg/iStockphoto. Medically Speaking, p. 772: *MRI*, Chris Bjornberg/Science Source. Figures 16.1, 16.2, 16.4, 16.5, 16.6, 16.9, 16.13, 16.14, and 16.21 Reprinted with permission of John Wiley & Sons, Inc. from Solomons, G., *Organic Chemistry 10e*, © 2011. Spectra on pages 731, 747, 748, and 757 and in Problems 16.11-16.13, 16.17, 16.23, 16.30, 16.56-16.58. 16.61, 16.62-16.64 © Dr. Richard A. Tomasi. Spectra in Problems 16.65, 16.66, 16.67, 16.68, and 16.69 reprinted with permission of John Wiley & Sons, Inc. from Field, L.D., Sternhell, S., and Kalman, J.R., *Organic Structures from Spectra*, 5e, © 2013. Running head: *magnet*, Sideways Design/Shutterstock.

**Chapter 17** Opener: *white t-shirt*, Kayros Studio Be Happy/Veer, *coffee stains*, R-studio/Shutterstock, *pouring cup of coffee*, ifong/Shutterstock. Practically Speaking, p. 796: *rubberband ball*, Sam Cornwell/Shutterstock. Practically Speaking, p. 820: *white t-shirt*, Kayros Studio Be happy/Veer, *coffee stains*, R-studio/Shutterstock. Figure 17.7 Reprinted with permission of John Wiley & Sons, Inc. from Solomons, G., *Organic Chemistry 10e*, © 2011. Running head: *white t-shirt*, Kayros Studio Be happy/Veer, *coffee stains*, R-studio/Shutterstock.

**Chapter 18** Opener: *pink pills*, frenc/iStockphoto, *white pills in blister pack*, topseller/Shutterstock, *v-shaped blister pack*, A-R-T/Shutterstock. Practically Speaking, p.834: *pack of pills*, Roman Sigaev/Shutterstock; *coal*, Marek Kosmal/iStockphoto; *charcoal/flame*, Don Nichols/iStockphoto. Medically Speaking, p. 852: *pink pills*, frenc/iStockphoto, *white pills in blister pack*, topseller/Shutterstock, *v-shaped blister pack*, A-R-T/Shutterstock. p. 853: Roman Sigaev/Shutterstock. Practically Speaking, p. 864; *green ball model*, Fabrizio Denna/iStockphoto; line art reprinted with permission from Diederic, F., and Whette, R.L. Accounts of Chemical Research, Vol 25, pp. 119-125, 1991. Copyright 1992 by American Chemical Society. Figures 18.2-18.4 Reprinted with permission of John Wiley & Sons, Inc. from

Solomons, G., *Organic Chemistry 10e*, © 2011. Figure 18.19 and spectra in Problems 18.59 and 18.60 © Dr. Richard A. Tomasi. Running head: A-R-T/Shutterstock, Roman Sigaev/Shutterstock, franc/iStockphoto.

**Chapter 19** Opener: *cereal*, Andy Waashnik; *blue spurt*, Auke Holwerda/iStockphoto; *blue liquid*, Pulse/Corbis/Jupiter Images Corp.; *green liquid*, Pulse/Corbis/Jupiter Images Corp.; *yellow liquid*, Pulse/Corbis/Jupiter Images Corp.; *red liquid*, Stuart Burford/iStockphoto; *bottles*, John A. Meents/iStockphoto. Practically Speaking, p. 881: *cereal*, Andy Waashnik; *blue liquid*, Pulse/Corbis/Jupiter Images Corp.; *green liquid*, Pulse/Corbis/Jupiter Images Corp.; *yellow liquid*, Pulse/Corbis/Jupiter Images Corp.; *red liquid*, Stuart Burford/iStockphoto; *bottles*, John A. Meents/iStockphoto. Running head: *blue spurt*, Auke Holwerda/iStockphoto.

**Chapter 20** Opener: *eye chart*, © Media Bakery; *large whole carrot*, Ints Vikmanis/iStockphoto; *smaller carrot*, Ints Vikmanis/Shutterstock; *flatware*, Lane V. Erickson/Shutterstock; *carrot slice*, ZTS/Shutterstock. Medically Speaking, p. 946: *hands/lotion*, LoaDao/Shutterstock. Practically Speaking, p. 951: *eyechart*, © Media Bakery; *spoon*, Lane V. Erickson/Shutterstock; *carrot slice*, Ints Vikmanis/Shutterstock. Practically Speaking, p. 964: *worm*, © Media Bakery. Figures 20.10-20.12 and spectra in Problems 20.81 and 20.82 © Dr. Richard A. Tomasi. Running head: *eye chart*, © Media Bakery; *carrot*, EddWestmacott/iStockphoto.

**Chapter 21** Opener: *thermometers*, EdwardHarkonen/iStockphoto; *aspirin*, Bruce Roiff/Shutterstock. Medically Speaking, p.1013, *thermometers*, EdwardHarkonen/iStockphoto; *aspirin*, Bruce Roiff/Shutterstock. Practically Speaking, p. 1016, *soap*, Gabor Izso/iStockphoto. Spectra in Problem 21.80 © Dr. Richard A. Tomasi. Running head: Bruce Roiff/Shutterstock.

**Chapter 22** Opener: *pineapple barbell*, Daniel Loiselle/iStockphoto; *pineapple and watermelon barbells*, Daniel Loiselle/iStockphoto. Practically Speaking, p. 1060, *pineapple and watermelon barbells*, Daniel Loiselle/iStockphoto. Practically Speaking, p. 1064, *jello*, Lauri Patterson/iStockphoto. Running head: *pineapple and watermelon barbells*, Daniel Loiselle/iStockphoto.

**Chapter 23** Opener: *pair red peppers*, Doug Cannell/iStockphoto; *single red chili*, Katja Zgonc/iStockphoto; *pink tablets short stack*, Bryan Sikora/Shutterstock; *pink tablets tall stack*, Bryan Sikora/Shutterstock; *green pepper*, Reinhold Leitner/Shutterstock; *yellow pepper*, J.D. Rogers/Shutterstock. Medically Speaking, p. 1109, *boat*, Carlos Arranz/iStockphoto. Medically Speaking box, p. 1137, *single red chili*, Katja Zgonc/iStockphoto; *pink tablets short stack*, Bryan Sikora/Shutterstock; *pink tablets tall stack*, Bryan Sikora/Shutterstock; *green pepper*, Reinhold Leitner/Shutterstock; *yellow pepper*, J.D. Rogers/Shutterstock. Spectra in Problems 23.83 and 23.84 © Dr. Richard A. Tomasi. Running head: *single red chili*, Katja Zgonc/iStockphoto; *pink tablets short stack*, Bryan Sikora/Shutterstock.

**Chapter 24** Opener: *tube*, ILYA AKINSHIN/123RF.com, *red medical box*, negative productions/Shutterstock, *bandages*, Carlos Caetano/Shutterstock. Medically Speaking, p. 1176, *ice cream*, Donald Erickson/iStockphoto. Practically Speaking, p. 1177: *Sweetplus*, Martin Bond/Science Source; *Equal*, Rachel Epstein/PhotoEdit; *Splenda*, KRT Photograph by Ken Lambert/Seattle Times/NewsCom. Medically Speaking, p. 1181, *tube*, ILYA AKINSHIN/123RF.com, *bandages*, Carlos Caetano/Shutterstock. Figure 21.11 (left) Reprinted with permission of John Wiley & Sons, Inc. from Solomons, G., *Organic Chemistry 10e*, © 2011, (right) Reprinted with permission of The McGraw-Hill Companies from Neal, L., *Chemistry and Biochemistry: A COmprehensive Introduction*, © 1971.; Running head: negative productions/Shutterstock.

**Chapter 25** Opener: *magnifying glass*, dem10/iStockphoto; *fingerprints*, blaneyphoto/iStockphoto; *hand*, TommL/iStockphoto. Practically Speaking, p. 1197: *chicken*, Ivan Kmit/iStockphoto; *steak*, Ivan Kmit/iStockphoto; *fish*, PicturePartners/iStockphoto. Practically Speaking, p. 1202: *fingerprints*, blaneyphoto/iStockphoto; *hand*, TommL/iStockphoto. Medically Speaking, p. 1213: *bandage*, Ales Veluscek/iStockphoto; *tube*, motorolka/Shutterstock. Medically Speaking, p. 1228: *cow*, Supertrooper/Shutterstock. Figure 25.8 Reprinted with permission of John Wiley & Sons, Inc. from Voet, D. and Voet, J.G., *Biochemistry*, Second edition. Copyright 1995 Voet, D. and Voet, J.G. Figure 25.9 Reprinted with permission of John Wiley & Sons, Inc. from Solomons, G., *Organic Chemistry 10e*, © 2011. Figures 25.10 and 25.11: Illustration, Irving Geis. Image from Irving Geis Collection/Howard Hughes Medical Institute. Rights owned by HHMI. Not to be reproduced without permission. Figure 25.15 Reprinted with permission of John Wiley & Sons, Inc. from Pratt, C. and Cornely, K., *Essential Biochemistry, 1e*, © 2004. Figure 25.18 Reprinted with permission of John Wiley & Sons, Inc., from Black, Jacquelyn G., *Microbiology, 7e*, © 2008. Figure 25.20, Mary Martin/Science Source. Running head: *magnifying glass*, dem10/iStockphoto.

**Chapter 26** Opener: *bacon*, Craig Veltri/iStockphoto; *stethoscope*, Mustafa Deliormanili/iStockphoto; *egg*, Debbi Smirnoff/iStockphoto; *shrimp*, Alex Staroseitsev/Shutterstock. Practically Speaking, p. 1250: *baby*, Tim Kimberley/iStockphoto. Medically Speaking, p. 1257: *foot*, Ziga Lisjak/iStockphoto. Medically Speaking, p. 1261: *stethoscope*, Mustafa Deliormanili/iStockphoto; *egg*, Debbi Smirnoff/iStockphoto. Medically Speaking, p. 1264: *arm*, Damir Spanic/iStockphoto. Figures 26.5 and 26.6 Reprinted with permission of John Wiley & Sons, Inc. from Pratt, C. and Cornely, K., *Essential Biochemistry, 1e*, © 2004. Running head: *stethoscope*, Mustafa Deliormanili/iStockphoto.

**Chapter 27** Opener: *glass*, Irina Tischenko/123.RF.com; *bullet*, Vladimir Kirlenko/Shutterstock. Practically Speaking, p. 1299, *glass*, Irina Tischenko/123.RF.com. Running head: *bullet*, Vladimir Kirlenko/Shutterstock.

# Glossary

**[4+2]-cycloaddition** (Sect. 17.7): A pericyclic reaction, also called a Diels-Alder reaction, that takes place between two different π systems, one of which is associated with four atoms while the other is associated with two atoms.

**1,2-addition** (Sect. 17.4): A reaction involving the addition of two groups to a conjugated π system in which one group is installed at the C1 position and the other group is installed at the C2 position.

**1,2-adduct** (Sect. 17.4): The product obtained from 1,2-addition across a conjugated π system.

**1,2-elimination** (Sect. 8.1): An elimination reaction in which a proton from the beta (β) position is removed together with the leaving group, forming a double bond.

**1,3-diaxial interaction** (Sect. 4.12): Steric interactions that occur between axial substituents in a chair conformation.

**1,4-addition** (Sect. 17.4): A reaction involving the addition of two groups to a conjugated π system in which one group is installed at the C1 position and the other group is installed at the C4 position.

**1,4-adduct** (Sect. 17.4): The product obtained from 1,4-addition across a conjugated π system.

## A

**absorbance** (Sect. 17.11): In UV-Vis spectroscopy, the value log $(I_0/I)$ where $I_0$ is the intensity of the reference beam and $I$ is the intensity of the sample beam.

**absorption spectrum** (Sect. 15.2): In IR spectroscopy as well as UV-VIS spectroscopy, a plot that measures the percent transmittance or absorption as a function of frequency.

**acetal** (Sect. 20.5): A functional group characterized by two alkoxy (OR) groups connected to the same carbon atom.Acetals can be used as protecting groups for aldehydes or ketones.

**acetoacetic ester synthesis** (Sect. 22.5): A three-step process that converts an alkyl halide into a methyl ketone with the introduction of three new carbon atoms.

**acetylide ion** (Sect. 10.3): The conjugate base of acetylene or any terminal alkyne.

**acid-catalyzed hydration** (Sect. 9.4): A reaction that achieves the addition of water across a double bond in the presence of an acid catalyst.

**acidic cleavage** (Sect. 14.6): A reaction in which bonds are broken in the presence of an acid. For example, in the presence of a strong acid, an ether is converted into two alkyl halides.

**activate** (Sect. 19.7): For a substituted aromatic ring, the effect of an electron-donating substituent that increases the rate of electrophilic aromatic substitution.

**acyl group** (Sect. 19.6): The term describing a carbonyl group (C=O bond) connected to an alkyl group or aryl group.

**acyl peroxide** (Sect. 11.3): A peroxide for which each oxygen atom is connected to an acyl group. Acyl peroxides are often used as radical initiators, because the O—O bond is especially weak.

**acylium ion** (Sect. 19.6): The resonance-stabilized, cationic intermediate of a Friedel-Crafts acylation, formed by treating an acyl halide with aluminum trichloride.

**addition polymers** (Sect. 27.4): Polymers that are formed via cationic addition, anionic addition, or free-radical addition.

**addition reactions** (Sect. 9.1): Reactions that are characterized by the addition of two groups across a double bond. In the process, the pi (π) bond is broken.

**addition to π bond** (Sect. 11.2): One of the six kinds of arrow-pushing patterns used in drawing mechanisms for radical reactions. A radical adds to a π bond, destroying the π bond and generating a new radical.

**adrenocortical hormones** (Sect. 26.6): Hormones that are secreted by the cortex (the outer layer) of the adrenal glands. Adrenocortical hormones are typically characterized by a carbonyl group or hydroxyl group at C11 of the steroid skeleton.

**alcohol** (Sect. 13.1): A compound that possesses a hydroxyl group (OH).

**aldaric acid** (Sect. 24.6): A dicarboxylic acid that is produced when an aldose or ketose is treated with a strong oxidizing agent, such as $HNO_3$.

**alditol** (Sect. 24.6): The product obtained when the aldehyde group of an aldose is reduced.

**aldol addition reaction** (Sect. 22.3): A reaction that occurs when an aldehyde or ketone is attacked by an enolate ion. The product of an aldol addition reaction is always a β-hydroxy aldehyde or ketone.

**aldol condensation** (Sect. 22.3): An aldol addition followed by dehydration to give an α,β-unsaturated ketone or aldehyde.

**aldonic acid** (Sect. 24.6): The product obtained when the aldehyde group of an aldose is oxidized.

**aldose** (Sect. 24.2): A carbohydrate that contains an aldehyde group.

**alkaloids** (Sect. 23.1): Naturally occurring amines isolated from plants.

**alkanamine** (Sect. 23.2): A format for naming primary amines containing a complex alkyl group.

**alkane** (Sect. 4.1): A hydrocarbon that lacks π bonds.

**alkene** (Sect. 8.1): A compound that possesses a carbon-carbon double bond.

**alkoxide** (Sect. 13.2): The conjugate base of an alcohol.

**alkoxy substituent** (Sect. 14.2): An OR group.

**alkoxymercuration-demercuration** (Sect. 14.5): A two-step process that achieves Markovnikov addition of an alcohol (H and OR) across an alkene. The product of this process is an ether.

**alkyl amines** (Sect. 23.2): A format for naming amines containing simple alkyl groups.

**alkyl group** (Sect. 4.2): A substituent lacking π bonds and comprised of only carbon and hydrogen atoms.

**alkyl halide** (Sect. 7.2): An organic compound containing at least one halogen.

**alkylation** (Sect. 10.10): A reaction that achieves the installation of an alkyl group. For example, an $S_N2$ reaction in which an alkyl group is connected to an attacking nucleophile.

**alkylthio group** (Sect. 14.11): An SR group.

**alkynes** (Sect. 10.1): Compounds containing a carbon-carbon triple bond.

**alkynide ion** (Sect. 10.3): The conjugate base of a terminal alkyne.

**allylic** (Sect. 2.10): The positions that are adjacent to the vinylic positions of a carbon-carbon double bond.

**allylic bromination** (Sect. 11.7): A radical reaction that achieves installation of a bromine atom at an allylic position.

**allylic carbocation** (Sect. 6.11): A carbocation in which the positive charge is adjacent to a carbon-carbon double bond.

**alpha (α) amino acid** (Sect. 25.1): A compound containing a carboxylic acid group (COOH) as well as an amino group ($NH_2$), both of which are attached to the same carbon atom.

**alpha (α) anomer** (Sect. 24.5): The cyclic hemiacetal of an aldose in which the hydroxyl group at the anomeric position is *trans* to the $CH_2OH$ group.

**alpha (α) helix** (Sect. 25.7): For proteins, a feature of secondary structure that forms when a portion of the protein twists into a spiral.

**alpha (α) position** (Sect. 7.2): The position immediately adjacent to a functional group.

**alternating copolymers** (Sect. 27.3): A copolymer that contains an alternating distribution of repeating units.

**amidomalonate synthesis** (Sect. 25.3): A synthetic method that employs diethyl acetamidomalonate as the starting material and enables the preparation of racemic α-amino acids.

**amine** (Sect. 23.1): Compounds containing a nitrogen atom that is connected to one, two, or three alkyl or aryl groups.

**amino acid** (Sect. 25.1): A compound containing a carboxylic acid group (COOH) as well as an amino group ($NH_2$).

**amino acid residue** (Sect. 25.4): The individual repeating units in a polypeptide chain or protein.

**amino sugars** (Sect. 24.9): Carbohydrate derivatives in which an OH group has been replaced with an amino group.

**amorphous** (Sect. 27.6): A region of a polymer in which nearby chains are not linearly extended and are not parallel to one another.

**amphoteric** (Sect. 25.2): Compounds that will react with either acids or bases. Amino acids are amphoteric.

**androgens** (Sect. 26.6): Male sex hormones.

**angle strain** (Sect. 4.9): The increase in energy associated with a bond angle that has deviated from the preferred angle of 109.5°.

**annulenes** (Sect. 18.5): Compounds consisting of a single ring containing a fully conjugated π system. Benzene is [6]annulene.

**anomeric carbon** (Sect. 24.5): The C1 position of the cyclic hemiacetal of an aldose or the C2 position of the cyclic hemiacetal of a ketose.

**anomers** (Sect. 24.5): Stereoisomeric cyclic hemiacetals of an aldose or ketose that differ from each other in their configuration at the anomeric carbon.

***anti* addition** (Sect. 9.8): An addition reaction in which two groups are installed on opposite sides of a π bond.

**anti conformation** (Sect. 4.8): A conformation in which the dihedral angle between two groups is 180°.

**antiaromatic** (Sect. 18.4): Instability that arises when a planar ring of continuously overlapping *p* orbitals contains $4n$ π electrons.

**antibonding MO** (Sect. 1.8): A high-energy molecular orbital resulting from the destructive interference between atomic orbitals.

***anti*-coplanar** (Sect. 8.7): A conformation in which a hydrogen atom and a leaving group are separated by a dihedral angle of exactly 180°.

***anti*-Markovnikov addition** (Sect. 9.3): An addition reaction in which a hydrogen atom is installed at the more substituted vinylic position and another group (such as a halogen) is installed at the less substituted vinylic position.

**antioxidants** (Sect. 11.9): Radical scavengers that prevent autooxidation by preventing radical chain reactions from beginning.

***anti*-periplanar** (Sect. 8.7): A conformation in which a hydrogen atom and a leaving group are separated by a dihedral angle of approximately 180°.

**arenium ion** (Sect. 19.2): The positively charged, resonance-stabilized, intermediate of an electrophilic aromatic substitution reaction. Also called a sigma complex.

**aromatic** (Sect. 18.1): A compound containing a planar ring of continuously overlapping *p* orbitals with $4n+2$ π electrons.

**aryl amine** (Sect. 23.2): An amine in which the nitrogen atom is connected directly to an aromatic ring.

**asymmetric hydrogenation** (Sect. 9.7): The addition of $H_2$ across only one face of a π bond.

**asymmetric stretching** (Sect. 15.5): In IR spectroscopy, when two bonds are stretching out of phase with each other.

**atactic** (Sect. 27.6): A polymer in which the repeating units contain chirality centers which are not arranged in a pattern (they have random configurations).

**atomic mass unit (amu):** (Sect. 15.13): A unit of measure equivalent to 1 g divided by Avogadro's number.

**atomic orbital** (Sect. 1.6): A three-dimensional plot of $\psi^2$ of a wavefunction. It is a region of space that can accommodate electron density.

**Aufbau principle** (Sect. 1.6): A rule that determines the order in which orbitals are filled by electrons. Specifically, the lowest energy orbital is filled first.

**autocatalytic** (Sect. 22.2): A reaction for which the reagent necessary to catalyze the reaction is produced by the reaction itself.

**autooxidation** (Sect. 11.9): The slow oxidation of organic compounds that occurs in the presence of atmospheric oxygen.

**auxochrome** (Sect. 17.11): When applying Woodward-Fieser rules, the groups attached to the chromophore.

**axial position** (Sect. 4.11): For chair conformations of substituted cyclohexanes, a position that is parallel to a vertical axis passing through the center of the ring.

**axis of symmetry** (Sect. 5.6): An axis about which a compound possesses rotational symmetry.

**azide synthesis** (Sect. 23.5): A method for preparing primary amines that avoids the formation of secondary and tertiary amines.

**azo coupling** (Sect. 23.11): An electrophilic aromatic substitution reaction in which an aryldiazonium salt reacts with an activated aromatic ring.

**azo dyes** (Sect. 23.11): A class of colored compounds that are formed via azo coupling.

## B

**back-side attack** (Sect. 7.4): In $S_N2$ reactions, the side opposite the leaving group, which is where the nucleophile attacks.

**Baeyer-Villiger oxidation** (Sect. 20.11): A reaction in which a ketone is treated with a peroxy acid and is converted into an ester via the insertion of an oxygen atom.

**base peak** (Sect. 15.8): In mass spectrometry, the tallest peak in the spectrum, which is assigned a relative value of 100%.

**Beer's law** (Sect. 17.11): In UV-Vis spectroscopy, an equation describing the relationship between molar absorptivity (ε), absorbance (*A*), concentration (*C*), and path length (*l*):

$$\varepsilon = \frac{A}{(C \times l)}$$

**bending** (Sect. 15.2): In IR spectroscopy, a type of vibration that generally produces a signal in the fingerprint region of an IR spectrum.

**bent** (Sect. 1.10): A type of geometry resulting from an $sp^3$-hybridized atom that has two lone pairs. For example, the oxygen atom in $H_2O$.

**benzylic position** (Sect. 18.6): A carbon atom that is immediately adjacent to a benzene ring.

**benzyne** (Sect. 19.14): A high-energy intermediate formed during the elimination-addition reaction that occurs between chlorobenzene and either NaOH (at high temperature) or $NaNH_2$.

**beta (β) anomer** (Sect. 24.5): The cyclic hemiacetal of an aldose, in which the hydroxyl group at the anomeric position is *cis* to the $CH_2OH$ group.

**beta (β) pleated sheet** (Sect. 25.7): For proteins, a feature of secondary structure that forms when two or more protein chains line up side-by-side.

**beta (β) position** (Sect. 7.2): The position immediately adjacent to an alpha (α) position.

**beta elimination** (Sect. 8.1): An elimination reaction in which a proton from the beta (β) position is removed together with the leaving group, forming a double bond.

**bicyclic** (Sect. 4.2): A structure containing two rings that are fused together.

**bimolecular** (Sect. 7.4): For mechanisms, a step that involves two chemical entities.

**biodegradable polymers** (Sect. 27.8): Polymers that can be broken down by enzymes produced by soil microorganisms.

**Birch reduction** (Sect. 18.7): A reaction in which benzene is reduced to give 1,4-cyclohexadiene.

**block copolymer** (Sect. 27.3): A copolymer in which the different homopolymer subunits are connected together in one chain.

**blocking group** (Sect. 19.11): A group that can be readily installed and uninstalled. Used for regiochemical control during synthesis.

**boat conformation** (Sect. 4.10): A conformation of cyclohexane in which all bond angles are fairly close to 109.5° and many hydrogen atoms are eclipsing each other.

**bond cleavage** (Sect. 12.3): The breaking of a bond, either homolytically or heterolytically.

**bond dissociation energy** (Sect. 6.1): The energy required to achieve homolytic bond cleavage (generating radicals).

**bonding MO** (Sect. 1.8): A low-energy molecular orbital resulting from the constructive interference between atomic orbitals.

**bond-line structures** (Sect. 2.2): The most common drawing style employed by organic chemists. All carbon atoms and most hydrogen atoms are implied but not explicitly drawn in a bond-line structure.

**branched polymer** (Sect. 27.6): A polymer that contains a large number of branches connected to the main chain of the polymer.

**Bredt's rule** (Sect. 8.4): A rule that states that it is not possible for a bridgehead carbon of a bicyclic system to possess a carbon-carbon double bond if it involves a *trans* π bond being incorporated in a ring comprised of fewer than eight atoms.

**bridgeheads** (Sect. 4.2): In a bicyclic system, the carbon atoms where the rings are fused together.

**broadband decoupling** (Sect. 16.11): In $^{13}C$ NMR spectroscopy, a technique in which all

$^{13}C—^1H$ splitting is suppressed with the use of two rf transmitters.

**bromohydrin** (Sect. 9.8): A compound containing a Br group and a hydroxyl group (OH) on adjacent carbon atoms.

**bromonium ion** (Sect. 9.8): A positively charged, bridged intermediate formed during the addition reaction that occurs when an alkene is treated with molecular bromine ($Br_2$).

**Brønsted-Lowry acid** (Sect. 3.1): A compound that can serve as a proton donor.

**Brønsted-Lowry base** (Sect. 3.1): A compound that can serve as a proton acceptor.

## C

**C terminus** (Sect. 25.4): For a peptide chain, the end that contains the COOH group.

**carbinolamine** (Sect. 20.6): A compound containing a hydroxyl group (OH) and a nitrogen atom, both of which are connceted to the same carbon atom.

**carbocation** (Sect. 6.7): An intermediate containing a positively charged carbon atom.

**carbohydrates** (Sect. 24.1): Polyhydroxy aldehydes or ketones with molecular formula $C_xH_{2x}O_x$.

**carbonyl group** (Sect. 13.4): A C=O bond.

**carboxylic acid derivative** (Sect. 21.6): A compound that is similar in structure to a carboxylic acid (RCOOH) but the OH group of the carboxylic acid has been replaced with a different group, Z, where Z is a heteroatom such as Cl, O, N, etc. Nitriles (R—C≡N) are also considered to be carboxylic acid derivatives because they have the same oxidation state as carboxylic acids.

**catalyst** (Sect. 6.5): A compound that can speed up the rate of a reaction without itself being consumed by the reaction.

**catalytic hydrogenation** (Sect. 9.7): A reaction that involves the addition of molecular hydrogen ($H_2$) across a double bond in the presence of a metal catalyst.

**cation** (Sect. 3.8): A structure that bears a positive charge.

**cellular respiration** (Sect. 13.12): A process by which molecular oxygen is used to convert food into $CO_2$, water, and energy.

**cephalins** (Sect. 26.5): Phosphoglycerides that contain ethanolamine.

**chain branching** (Sect. 11.11): During polymerization, the growth of a branch connected to the main chain.

**chain reaction** (Sect. 11.3): A reaction (generally involving radicals) in which one chemical entity can ultimately cause a chemical transformation for thousands of molecules.

**chain-growth polymer** (Sect. 27.5): A polymer that is formed under conditions in which the monomers do not react directly with each other, but rather, each monomer is added to the growing chain, one at a time.

**chair conformation** (Sect. 4.10): The lowest energy conformation for cyclohexane, in which all bond angles are fairly close to 109.5° and all hydrogen atoms are staggered.

**chemical shift** (δ) (Sect. 16.5): In an NMR spectrum, the location of a signal, defined relative to the frequency of absorption of a reference compound, tetramethylsilane (TMS).

**chemically equivalent** (Sect. 16.4): In NMR spectroscopy, protons (or carbon atoms) that occupy identical electronic environments and produce only one signal.

**chiral** (Sect. 5.2): An object that is not superimposable on its mirror image.

**chirality center** (Sect. 5.2): A tetrahedral carbon atom bearing four different groups.

**chlorofluorocarbons (CFCs)** (Sect. 11.8): Compound containing only carbon, chlorine, and fluorine.

**chlorohydrin** (Sect. 9.8): A compound containing a Cl group and a hydroxyl group (OH) on adjacent carbon atoms.

**chromatogram** (Sect. 15.14): In gas chromatography, a plot that identifies the retention time of each compound in the mixture.

**chromophore** (Sect. 17.11): In UV-Vis spectroscopy, the region of the molecule responsible for the absorption (the conjugated π system).

**Claisen condensation** (Sect. 22.4): A nucleophilic acyl substitution reaction in which the nucleophile is an ester enolate and the electrophile is an ester.

**Claisen rearrangement** (Sect. 17.10): A [3,3] sigmatropic rearrangement that is observed for allylic vinylic ethers.

**Clemmensen reduction** (Sect. 19.6): A reaction in which a carbonyl group is completely reduced and replaced with two hydrogen atoms.

**column chromatography** (Sect. 5.9): A technique by which compounds are separated from each other based on a difference in the way they interact with the medium (the adsorbent) through which they are passed.

**complex lipid** (Sect. 26.1): A lipid that readily undergoes hydrolysis in aqueous acid or base to produce smaller fragments.

**condensation polymer** (Sect. 27.4): A polymer that is formed via a condensation reaction.

**condensed structure** (Sect. 2.1): A drawing style in which none of the bonds are drawn. Groups of atoms are clustered together when possible. For example, isopropanol has two $CH_3$ groups, both of which are connected to the central carbon atom, shown like this: $(CH_3)_2CHOH$.

**configuration** (Sect. 5.3): The 3D spatial orientation of the groups connected to a chirality center (*R* or *S*) or of the groups in a stereoisiomeric alkene (E or Z).

**conformation** (Sect. 4.6): A three-dimensional shape that can be adopted by a compound as a result of rotation about single bonds.

**conjugate acid** (Sect. 3.1): In an acid-base reaction, the product that results when a base is protonated.

**conjugate addition** (Sect. 22.6): An addition reaction in which a nucleophile and a proton are added across the two ends of a conjugated π system.

**conjugate base** (Sect. 3.1): In an acid-base reaction, the product that results when an acid is deprotonated.

**conjugated** (Sect. 2.10, 15.3): A compound in which two π bonds are separated from each other by exactly one σ bond.

**conjugated diene** (Sect. 17.1): A compound in which two carbon-carbon π bonds are separated from each other by exactly one σ bond.

**conrotatory** (Sect. 17.9): In electrocyclic reactions, a type of rotation in which the orbitals being used to form the new σ bond must rotate in the same way.

**conservation of orbital symmetry** (Sect. 17.8): During a reaction, the requirement that the phases of the frontier MOs must be aligned.

**constitutional isomers** (Sect. 1.2): Compounds that have the same molecular formula but differ in the way the atoms are connected.

**constructive interference** (Sect. 1.7): When two waves interact with each other in a way that produces a wave with a larger amplitude.

**continuous-wave (CW) spectrometer** (Sect. 16.2): An NMR spectrometer that holds the magnetic field constant and slowly sweeps through a range of rf frequencies, monitoring which frequencies are absorbed.

**Cope rearrangement** (Sect. 17.10): A [3,3] sigmatropic rearrangement in which all six atoms of the cyclic transition state are carbon atoms.

**coplanar** (Sect. 8.7): Atoms that lie in the same plane.

**copolymer** (Sect. 27.3): A polymer that is constructed from more than one repeating unit.

**coupling** (of protons) (Sect. 16.7): A phenomenon observed most commonly for nonequivalent protons connected to adjacent carbon atoms in which the multiplicity of each signal is affected by the other.

**coupling** (of radicals) (Sect. 11.2): A radical process in which two radicals join together and form a bond.

**coupling constant** (Sect. 16.7): When signal splitting occurs in NMR spectroscopy, the distance between the individual peaks of a signal.

**covalent bond** (Sect. 1.3): A bond that results when two atoms share a pair of electrons.

**crossed aldol reaction** (Sect. 22.3): An aldol reaction that occurs between different partners.

**crossed Claisen condensation** (Sect. 22.4): A Claisen condensation reaction that occurs between different partners.

**crossed-linked polymer** (Sect. 27.6): A polymer in which neighboring chains are linked together, for example, by disulfide bonds.

**crown ether** (Sect. 14.4): Cyclic polyethers whose molecular models resemble crowns.

**crystallite** (Sect. 27.6): A region of a polymer in which the chains are linearly extended and close in proximity to one another, resulting in van der Waals forces that hold the chains close together.

**cumulated diene** (Sect. 17.1): A compound containing two adjacent π bonds.

**curved arrows** (Sect. 2.8): Tools that are used for drawing resonance structures and for showing the flow of electron density during each step of a reaction mechanism.

**cyanohydrin** (Sect. 20.10): A compound containing a cyano group and a hydroxyl group connected to the same carbon atom.

**cycloaddition reactions** (Sect. 17.6): Reactions in which two π systems are joined together in a way that forms a ring. In the process, two π bonds are converted into two σ bonds.

**cycloalkane** (Sect. 4.2): An alkane whose structure contains a ring.

## D

**D sugar** (Sect. 24.2): A carbohydrate for which the chirality center farthest from the carbonyl group will have an OH group pointing to the right in the Fischer projection.

**dash** (Sect. 2.6): In bond-line structures, a group going behind the page.

**deactivate** (Sect. 19.8): For a substituted aromatic ring, the effect of an electron-withdrawing substituent that decreases the rate of electrophilic aromatic substitution.

**debye (D)** (Sect. 1.11): A unit of measure for dipole moments, where 1 debye $= 10^{-18}$ esu · cm.

**decarboxylation** (Sect. 22.5): A reaction involving loss of $CO_2$, characteristic of compounds containing a carbonyl group that is beta to a COOH group.

**degenerate** (Sect. 4.7): Having the same energy.

**degenerate orbitals** (Sect. 1.6): Orbitals that have the same energy.

**degree of substitution** (Sect. 8.3): For alkenes, a classification method that refers to the number of alkyl groups connected to the double bond.

**degree of unsaturation** (Sect. 15.16): The absence of two hydrogen atoms associated with a ring or a π bond.

**dehydration** (Sect. 8.1): An elimination reaction involving the loss of H and OH.

**dehydrohalogenation** (Sect. 8.1): An elimination reaction involving the loss of H and a halogen (such as Cl, Br, or I).

**delocalization** (Sect. 2.7): The spreading of a charge or lone pair as described by resonance theory.

**delocalized** (Sect. 2.12): A lone pair or charge that is participating in resonance.

**denaturation** (Sect. 25.7): A process during which a protein unfolds under conditions of mild heating.

**DEPT $^{13}$C NMR** (Sect. 16.13): In $^{13}$C NMR spectroscopy, a technique that utilizes two rf radiation emitters and provides information regarding the number of protons attached to each carbon atom in a compound.

**deshielded** (Sect. 16.1): In NMR spectroscopy, protons or carbon atoms whose surrounding electron density is poor.

**desulfurization** (Sect. 20.8): The conversion of a thioacetal into an alkane in the presence of Raney nickel.

**dextrorotatory** (Sect. 5.4): A compound that rotates plane-polarized light in a clockwise direction (+).

**diagnostic region** (Sect. 15.3): The region of an IR spectrum that contains signals that arise from double bonds, triple bonds, and X—H bonds.

**diamagnetic anisotropy** (Sect. 16. 5): An effect that causes different regions of space to be characterized by different magnetic field strengths.

**diamagnetism** (Sect. 16.1): The circulation of electron density in the presence of an external magnetic field, which produces a local (induced) magnetic field that opposes the external magnetic field.

**diastereomers** (Sect. 5.5): Stereoisomers that are not mirror images of one another.

**diastereotopic** (Sect. 16.4): Nonequivalent protons for which the replacement test produces diastereomers.

**diazonium salt** (Sect. 23.10): An ionic compound that is formed upon treatment of a primary amine with $NaNO_2$ and HCl.

**diazotization** (Sect. 23.10): The process of forming a diazonium salt by treating a primary amine with $NaNO_2$ and HCl.

**diborane** (Sect. 9.6): $B_2H_6$. A dimeric structure formed when one borane molecule reacts with another.

**Dieckmann cyclization** (Sect. 22.4): An intramolecular Claisen condensation.

**diene** (Sect. 17.1): A compound containing two carbon-carbon π bonds.

**dienophile** (Sect. 17.7): A compound that reacts with a diene in a Diels-Alder reaction.

**dihedral angle** (Sect. 4.7): The angle by which two groups are separated in a Newman projection.

**dihydroxylation** (Sect. 9.9): A reaction characterized by the addition of two hydroxyl groups (OH) across an alkene.

**diol** (Sect. 13.5): A compound containing two hydroxyl groups (OH).

**dipole moment (μ)** (Sect. 1.11): The amount of partial charge (*d*) on either end of a dipole multiplied by the distance of separation (δ):

$$\mu = d \cdot \delta$$

**dipole-dipole interactions** (Sect. 1.12): The resulting net attraction between two dipoles.

**directed aldol addition** (Sect. 22.3): A technique for performing a crossed aldol addition that produces one major product.

**disaccharide** (Sect. 24.7): Carbohydrates comprised of two monosaccharide units joined via a glycosidic linkage between the anomeric carbon of one monosaccharide and a hydroxyl group of the other monosaccharide.

**disrotatory** (Sect. 17.9): In electrocyclic reactions, a type of rotation in which the orbitals being used to form the new σ bond must rotate in opposite directions (one rotates clockwise while the other rotates counterclockwise).

**dissolving metal reduction** (Sect. 10.5): A reaction in which an alkyne is converted into a trans alkene.

**disulfide** (Sect. 14.11): A compound with the structure R—S—S—R.

**disulfide bridge** (Sect. 25.4): The group that is formed when two cysteine residues of a polypeptide or protein are joined together.

**divalent** (Sect. 1.2): An element that forms two bonds, such as oxygen.

**doublet** (Sect. 16.7): In NMR spectroscopy, a signal that is comprised of two peaks.

**downfield** (Sect. 16.5): The left side of an NMR spectrum.

## E

***E*** (Sect. 8.4): For alkenes, a stereodescriptor that indicates that the two priority groups are on opposite sides of the π bond.

**E1** (Sect. 8.9): A unimolecular elimination reaction.

**E1cb mechanism** (Sect. 22.3): An elimination reaction in which the leaving group only leaves after deprotonation occurs. This process occurs at the end of an aldol condensation.

**E2** (Sect. 8.7): A bimolecular elimination reaction.

**eclipsed conformation** (Sect. 4.7): A conformation in which groups are eclipsing each other in a Newman projection.

**Edman degradation** (Sect. 25.5): A method for analyzing the sequence of amino acids in a peptide by removing one amino acid residue at a time and identifying each residue as it is removed.

**eicosanoids** (Sect. 26.7): A class of lipids which includes leukotrienes, prostaglandins, thromboxanes, and prostacyclins.

**elastomers** (Sect. 27.7): Polymers that return to their original shape after being stretched.

**electrocyclic reaction** (Sect. 17.6): A pericyclic process in which a conjugated polyene undergoes cyclization. In the process, one π bond is converted into a σ bond, while the remaining π bonds all change their location. The newly formed σ bond joins the ends of the original π system, thereby creating a ring.

**electromagnetic spectrum** (Sect. 15.1): The range of all frequencies of electromagnetic radiation, which is arbitrarily divided into several regions, most commonly by wavelength.

**electron density** (Sect. 1.6): A term associated with the probability of finding an electron in a particular region of space.

**electron impact ionization (EI)** (Sect. 15.8): In mass spectrometry, an ionization technique that involves bombarding a compound with high-energy electrons.

**electrophile** (Sect. 6.7): A compound containing an electron-deficient atom that is capable of accepting a pair of electrons.

**electrophilic aromatic substitution** (Sect. 19.1): A substitution reaction in which an aromatic proton is replaced by an electrophile and the aromatic moiety is preserved.

**electrophoresis** (Sect. 25.2): A technique for separating amino acids from each other based on a difference in pI values.

**electrospray ionization (ESI):** (Sect. 15.15): In mass spectrometry, an ionization technique in which the compound is first dissolved in a solvent and then sprayed via a high-voltage needle into a vacuum chamber. The tiny droplets of solution become charged by the needle, and subsequent evaporation forms gas-phase molecular ions that typically carry one or more charges.

**electrostatic potential maps** (Sect. 1.5): A three-dimensional, rainbowlike image used to visualize partial charges in a compound.

**elimination** (Sect. 7.1): A reaction involving the loss of a leaving group and formation of a $\pi$ bond.

**elimination (of radicals)** (Sect. 11.2): In radical reaction mechanisms, a step in which a bond forms between the alpha ($\alpha$) and beta ($\beta$) positions. As a result, a single bond at the $\beta$ position is cleaved, causing the compound to fragment into two pieces.

**elimination-addition** (Sect. 19.14): A reaction that occurs between chlorobenzene and either NaOH (at high temperature) or $NaNH_2$.

**enamine** (Sect. 20.6): A compound containing a nitrogen atom directly connected to a carbon-carbon $\pi$ bond.

**enantiomer** (Sect. 5.2): A nonsuperimposable mirror image.

**enantiomeric excess** (Sect. 5.4): For a mixture containing two enantiomers, the difference between the percent concentration of the major enantiomer and the percent concentration of its mirror image.

**enantiomerically pure** (Sect. 5.4): A substance that consists of a single enantiomer, and not its mirror image.

**enantiotopic** (Sect. 16.4): Protons that are not interchangeable by rotational symmetry but are interchangeable by reflectional symmetry.

**endergonic** (Sect. 6.3): Any process with a positive $\Delta G$.

***endo*** (Sect. 17.7): In Diels-Alder reactions that produce bicyclic structures, the positions that are *syn* to the larger bridge.

**endothermic** (Sect. 6.1): Any process with a positive $\Delta H$ (the system receives energy from the surroundings).

**energy of activation** (Sect. 6.5): In an energy diagram, the height of the energy barrier (the hump) between the reactants and the products.

**enol** (Sect. 10.7): A compound containing a hydroxyl group (OH) connected directly to a carbon-carbon double bond.

**enolate** (Sect. 22.1): The resonance-stabilized conjugate base of a ketone, aldehyde, or ester.

**enthalpy** (Sect. 6.1): A measure of the exchange of energy between the system and its surroundings during any process.

**entropy** (Sect. 6.2): The measure of disorder associated with a system.

**enzymes** (Sect. 25.8): Important biological molecules that catalyze virtually all cellular processes.

**epimer** (Sect. 24.6): Diastereomers that differ from each other in the configuration of only one chirality center.

**epoxide** (Sect. 9.9): A cyclic ether containing a three-membered ring system. Also called an oxirane (also see Sect. 14.7).

**equatorial position** (Sect. 4.11): For chair conformations of substituted cyclohexanes, a position that is approximately along the equator of the ring.

**equilibrium** (Sect. 3.3): For a reaction, a state in which there is no longer an observable change in the concentrations of reactants and products.

**estrogens** (Sect. 26.6): Female sex hormones.

**ether** (Sect. 14.1): A compound with the structure R—O—R.

**excited state** (Sect. 17.3): A state that is achieved when a compound absorbs energy.

**exergonic** (Sect. 6.3): Any process with a negative $\Delta G$.

***exo*** (Sect. 17.7): In Diels-Alder reactions that produce bicyclic structures, the positions that are anti to the larger bridge.

**exothermic** (Sect. 6.1): Any process with a negative $\Delta H$ (the system gives energy to the surroundings).

## F

**fats** (Sect. 26.3): Triglycerides that are solids at room temperature.

**fatty acids** (Sect. 26.3): Long-chain carboxylic acids.

**fibers** (Sect. 27.7): Strands of a polymer that are generated when the polymer is heated, forced through small holes, and then cooled.

**fibrous proteins** (Sect. 25.8): Proteins that consist of linear chains that are bundled together.

**fingerprint region** (Sect. 15.3): The region of an IR spectrum that contains signals resulting from the vibrational excitation of most single bonds (stretching and bending).

**first order** (Sect. 6.5, 7.5): A reaction that has a rate equation in which the sum of all exponents is one.

**Fischer esterification** (Sect. 21.10): A process in which a carboxylic acid is converted into an ester when treated with an alcohol in the presence of an acid catalyst.

**Fischer projections** (Sect. 2.6): A drawing style that is often used when dealing with compounds bearing multiple chirality centers, especially for carbohydrates. (See also Sect. 5.7.)

**fishhook arrow** (Sect. 10.5): A curved arrow with only one barb, indicating the motion of just one electron (also see Sect. 11.1).

**flagpole interactions** (Sect. 4.10): For cyclohexane, the steric interactions that occur between the flagpole hydrogen atoms in a boat conformation.

**formal charge** (Sect. 1.4): A charge associated with any atom that does not exhibit the appropriate number of valence electrons.

**Fourier-transform NMR (FT-NMR)** (Sect. 16.2): In nuclear magnetic resonance (NMR) spectroscopy, a technique in which the sample is irradiated with a short pulse that covers the entire range of relevant rf frequencies.

**fragmentation** (Sect. 15.8): In mass spectrometry, when the molecular ion breaks apart into fragments.

**free induction decay** (Sect. 16.2): In NMR spectroscopy, a complex signal which is a combination of all of the electrical impulses generated by each type of proton.

**Freons** (Sect. 11.8): CFCs that were heavily used for a wide variety of commercial applications, including as refrigerants, as propellants, in the production of foam insulation, as fire-fighting materials, and many other useful applications.

**frequency** (Sect. 15.1): For electromagnetic radiation, the number of wavelengths that pass a particular point in space per unit time.

**Friedel-Crafts acylation** (Sect. 19.6): An electrophilic aromatic substitution reaction that installs an acyl group on an aromatic ring.

**Friedel-Crafts alkylation** (Sect. 19.5): An electrophilic aromatic substitution reaction that installs an alkyl group on an aromatic ring.

**frontier orbital theory** (Sect. 17.3): The analysis of a reaction using MO theory, where only the frontier orbitals (HOMO and LUMO) are considered.

**frontier orbitals** (Sect. 17.3): The highest occupied molecular orbital (HOMO) and lowest unoccupied molecular orbital (LUMO) that participate in a reaction.

**Frost circles** (Sect. 18.4): A simple method for drawing the relative energy levels of the MOs for a ring assembled from continuously overlapping *p* orbitals.

**functional group** (Sect. 2.3): A characteristic group of atoms/bonds that possess a predictable chemical behavior.

**furanose** (Sect. 24.5): A five-membered cyclic hemiacetal form of a carbohydrate.

## G

**Gabriel synthesis** (Sect. 23.5): A method for preparing primary amines that avoids formation of secondary and tertiary amines.

**gas chromatograph – mass spectrometer** (Sect. 15.14): A device used for the analysis of a mixture that contains several compounds.

***gauche* conformation** (Sect. 4.8): A conformation that exhibits a *gauche* interaction.

***gauche* interaction** (Sect. 4.8): The steric interaction that results when two groups in a Newman projection are separated by a dihedral angle of 60°.

**geminal** (Sect. 10.4): Two groups connected to the same carbon atom. For example, a geminal dihalide is a compound with two halogens connected to the same carbon atom.

**Gibbs free energy** (*G*) (Sect. 6.3): The ultimate arbiter of the spontaneity of a reaction, where $\Delta G = \Delta H - T\Delta S$.

**Gilman reagent** (Sect. 21.8): A lithium dialkyl cuprate ($R_2CuLi$).

**glass transition temperature** ($T_g$) (Sect. 27.6): The temperature at which noncrystalline polymers become very soft.

**globular proteins** (Sect. 25.8): Proteins that consist of chains that are coiled into compact shapes.

**glycoside** (Sect. 24.6): An acetal that is obtained by treating the cyclic hemiacetal form of a monosaccharide with an alcohol under acid-catalyzed conditions.

**graft copolymer** (Sect. 27.3): A polymer that contains sections of one homopolymer that have been grafted onto a chain of the other homopolymer.

**Grignard reagent** (Sect. 13.6): A carbanion with the structure RMgX.

## H

**haloalkane** (Sect. 7.2): An organic compound containing at least one halogen.

**haloform reaction** (Sect. 22.2): A reaction in which a methyl ketone is converted into a carboxylic acid upon treatment with excess base and excess halogen, followed by aqueous acid.

**halogen abstraction** (Sect. 11.2): In radical reactions, a type of arrow-pushing pattern in which a halogen atom is abstracted by a radical, generating a new radical.

**halogenation** (Sect. 9.8): A reaction that involves the addition of $X_2$ (either $Br_2$ or $Cl_2$) across an alkene.

**halohydrin formation** (Sect. 9.8): A reaction which involves the addition of a halogen and a hydroxyl group (OH) across an alkene.

**Hammond postulate** (Sect. 6.6): In an exothermic process the transition state is closer in energy to the reactants than to the products, and therefore the structure of the transition state more closely resembles the reactants. In contrast, the transition state in an endothermic process is closer in energy to the products, and therefore the transition state more closely resembles the products.

**Haworth projection** (Sect. 2.6): For substituted cycloalkanes, a drawing style used to clearly identify which groups are above the ring and which groups are below the ring. (See also Sect. 4.14.)

**heat of combustion** (Sect. 4.4): The heat given off during a reaction in which an alkane reacts with oxygen to produce CO2 and water.

**heat of reaction** (Sect. 6.1): The heat given off during a reaction.

**Hell-Volhard-Zelinsky reaction** (Sect. 22.2): A reaction in which a carboxylic acid undergoes α-halogenation when treated with bromine in the presence of $PBr_3$.

**hemiacetal** (Sect. 20.5): A compound containing a hydroxyl group (OH) and an alkoxy group (OR) connected to the same carbon atom.

**Henderson-Hasselbalch equation** (Sect. 21.3): An equation that is often employed to calculate the pH of buffered solutions:

$$\text{pH} = \text{p}K_a + \log \frac{[\text{conjugated base}]}{[\text{acid}]}$$

**heterocycle** (Sect. 18.5): A cyclic compound containing at least one heteroatom (such as S, N, or O) in the ring.

**heterogeneous catalyst** (Sect. 9.7): A catalyst that does not dissolve in the reaction medium.

**heterolytic bond cleavage** (Sect. 6.1): Bond breaking that results in the formation of ions.

**high-resolution mass spectrometry** (Sect. 15.13): A technique that involves the use of a detector that can measure the *m/z* values to four decimal places.This technique allows for the determination of the molecular formula of an unknown compound.

**Hofmann elimination** (Sect. 23.9): A reaction in which an amino group is treated with excess methyl iodide, thereby converting it into an excellent leaving group, followed by treatment with a strong base to give an E2 reaction that yields an alkene.

**Hofmann product** (Sect. 8.7): The less substituted product (alkene) of an elimination reaction.

**HOMO** (Sect. 1.8): The highest occupied molecular orbital.

**homogeneous catalysts** (Sect. 9.7): A catalyst that dissolves in the reaction medium.

**homolitic bond cleavage** (Sects. 6.1, 11.2): Bond breaking that results in the formation of unchanged species called radicals.

**homopolymer** (Sect. 27.3): A polymer constructed from a single type of monomer.

**homotopic** (Sect. 16.4): Protons that are interchangeable by rotational symmetry.

**Hückel's rule** (Sect. 18.4): The requirement for an odd number of π electron pairs in order for a compound to be aromatic.

**Hund's rule** (Sect. 1.6): When considering electrons in atomic orbitals, a rule that states that one electron is placed in each degenerate orbital first, before electrons are paired up.

**hydrate** (Sect. 20.5): A compound containing two hydroxyl groups (OH) connected to the same carbon atom.

**hydration** (Sect. 9.4): A reaction in which a proton and a hydroxyl group (OH) are added across a π bond.

**hydrazone** (Sect. 20.6): A compound with the structure $R_2C{=}N{-}NH_2$.

**hydride shift** (Sect. 6.8): A type of carbocation rearrangement that involves the migration of a hydride ion ($H^-$).

**hydroboration-oxidation** (Sect. 9.6): A two-step process that achieves an *anti*-Markovnikov addition of a proton and a hydroxyl group (OH) across an alkene.

**hydrochlorofluorocarbons, (HCFCs)** (Sect. 11.8): Compounds that are similar in structure to CFCs but also possess at least one C—H bond.

**hydrocracking** (Sect. 11.12): A process performed in the presence of hydrogen gas by which large alkanes in petroleum are converted into smaller alkanes that are more suitable for use as gasoline.

**hydrofluorocarbons (HFCs)** (Sect. 11.8): Compounds that contain only carbon, fluorine, and hydrogen (no chlorine).

**hydrogen abstraction** (Sect. 11.2): In radical reactions, a type of arrow-pushing pattern in which a hydrogen atom is abstracted by a radical, generating a new radical.

**hydrogen bonding** (Sect. 1.12): A special type of dipole-dipole interaction that occurs between an electronegative atom and a hydrogen atom that is connected to another electronegative atom.

**hydrogen deficiency index (HDI)** (Sect. 15.16): A measure of the number of degrees of unsaturation in a compound.

**hydrohalogenation** (Sect. 9.3): A reaction that involves the addition of H and X (either Br or Cl) across an alkene.

**hydrolysis** (Sect. 20.7): A reaction in which bonds are cleaved by treatment with water.

**hydroperoxide** (Sect. 11.9): A compound with the structure R—O—O—H.

**hydrophilic** (Sect. 1.13): A polar group that has favorable interactions with water.

**hydrophobic** (Sect. 1.13): A nonpolar group that does not have favorable interactions with water.

**hydroxyl group** (Sect. 13.1): An OH group.

**hyperconjugation** (Sect. 6.8): An effect that explains why alkyl groups stabilize a carbocation.

## I

**imidazole** (Sect. 23.12): A compound containing a five-membered ring that is similar to pyrrole but has one extra nitrogen atom at the 3 position.

**imine** (Sect. 20.6): A compound containing a C=N bond.

**induction** (Sect. 1.5): The withdrawal of electron density that occurs when a bond is shared by two atoms of differing electronegativity.

**initiation** (Sect. 11.2): In radical reaction mechanisms, a step in which radicals are created.

**integration** (Sect. 16.6): In $^1H$ NMR spectroscopy, the area under a signal indicates the number of protons giving rise to the signal.

**intermediate** (Sect. 6.6): A structure corresponding to a local minimum (valley) in an energy diagram.

**intermolecular forces** (Sect. 1.12): The attractive forces between molecules.

**internal alkyne** (Sect. 10.2): A compound with the structure R—C≡C—R, where each R group is not a hydrogen atom.

**inversion of configuration** (Sect. 7.4): During a reaction, when the configuration of a chirality center is changed.

**ionic bond** (Sect. 1.5): A bond that results from the force of attraction between two oppositely charged ions.

**ionic reaction** (Sect. 6.7): A reaction that involves the participation of ions as reactants, intermediates, or products.

**isoelectric point (pI)** (Sect. 25.2): For an amino acid, the specific pH at which the concentration of the zwitterionic form reaches its maximum value.

**isolated diene** (Sect. 17.1): A compound containing two carbon-carbon π bonds that are separated by two or more σ bonds.

**isoprene** (Sect. 26.8): 2-Methyl-1,3-butadiene.

**isotactic** (Sect. 27.6): A polymer in which the repeating units contain chirality centers which all have the same configuration.

**IUPAC** (Sect. 4.2): The International Union of Pure and Applied Chemistry.

## J

***J* value** (Sect. 16.7): When signal splitting occurs in $^1$H NMR spectroscopy, the distance (in hertz) between the individual peaks of a signal.

## K

**$K_a$** (Sect. 3.3): A measure of the strength of an acid:

$$K_a = K_{eq}[H_2O] = \frac{[H_3O^+][A^-]}{[HA]}$$

**$K_{eq}$** (Sect. 3.3): A term that describes the position of equilibrium for a reaction:

$$K_{eq} = \frac{[H_3O^+][A^-]}{[HA][H_2O]}$$

**keto-enol tautomerization** (Sect. 10.7): The equilibrium that is established between an enol and a ketone in either acid-catalyzed or base-catalyzed conditions.

**ketose** (Sect. 24.2): A carbohydrate that contains a ketone group.

**Kiliani-Fischer synthesis** (Sect. 24.6): A process by which the chain of a carbohydrate is lengthened by one carbon atom.

**kinetic control** (Sect. 17.5): A reaction for which the product distribution is determined by the relative rates at which the products are formed.

**kinetics** (Sect. 6.5): A term that refers to the rate of a reaction.

## L

**L amino acid** (Sect. 25.2): Amino acids with Fischer projections that resemble the Fischer projections of L sugars.

**L sugar** (Sect. 24.2): A carbohydrate for which the chirality center farthest from the carbonyl group will have an OH group pointing to the left in the Fischer projection.

**labile** (Sect. 16.7): Protons that are exchanged at a rapid rate.

**lactone** (Sect. 20.11): A cyclic ester.

**lambda max ($\lambda_{max}$)** (Sect. 17.11): In UV-Vis spectroscopy, the wavelength of maximum absorption.

**leaving group** (Sect. 7.1): A group capable of separating from a compound.

**lecithins** (Sect. 26.5): Phosphoglycerides that contain choline.

**leveling effect** (Sect. 3.6): An effect that prevents the use of bases stronger than hydroxide when the solvent is water.

**levorotatory** (Sect. 5.4): A compound that rotates plane-polarized light in a counterclockwise direction (−).

**Lewis acid** (Sect. 3.9): A compound capable of functioning as an electron pair acceptor.

**Lewis base** (Sect. 3.9): A compound capable of functioning as an electron pair donor.

**Lewis structures** (Sect. 1.3): A drawing style in which the electrons take center stage.

**linear polymer** (Sect. 27.6): A polymer that has only a minimal amount of branching or no branching at all.

**lipid** (Sect. 26.1): Naturally occurring compounds that can be extracted from cells using nonpolar organic solvents.

**lipid bilayer** (Sect. 26.5): The main fabric of cell membranes, assembled primarily from phosphoglycerides.

**lithium dialkyl cuprate** (Sect. 21.8): A nucleophilic compound with the general structure $R_2CuLi$.

**living polymer** (Sect. 27.4): A polymer that is formed via anionic polymerization.

**localized lone pair** (Sect. 2.12): A lone pair that is not participating in resonance.

**locant** (Sect. 4.2): In nomenclature, a number used to identify the location of a substituent.

**London dispersion forces** (Sect. 1.12): Attractive forces between transient dipole moments, observed in alkanes.

**lone pair** (Sect. 1.3): A pair of unshared, or nonbonding, electrons.

**loss of a leaving group** (Sect. 6.8): One of the four arrow-pushing patterns for ionic reactions.

**LUMO** (Sect. 1.8): The lowest unoccupied molecular orbital.

## M

**magnetic moment** (Sect. 16.1): A magnetic field generated by a spinning proton.

**malonic ester synthesis** (Sect. 22.5): A synthetic technique that enables the transformation of a halide into a carboxylic acid with the introduction of two new carbon atoms.

**Markovnikov addition** (Sect. 9.3): In addition reactions, the observation that the hydrogen atom is generally placed at the vinylic position already bearing the larger number of hydrogen atoms.

**mass spectrometer** (Sect. 15.8): A device in which a compound is first vaporized and converted into ions, which are then separated and detected.

**mass spectrometry** (Sect. 15.8): The study of the interaction between matter and an energy source other than electromagnetic radiation. Mass spectrometry is used primarily to determine the molecular weight and molecular formula of a compound.

**mass spectrum** (Sect. 15.8): In mass spectrometry, a plot that shows the relative abundance of each cation that was detected.

**mass-to-charge ratio(*m/z*)** (Sect. 15.8): The determining factor by which ions are separated from each other in mass spectrometry.

**Meisenheimer complex** (Sect. 19.13): The resonance-stabilized intermediate of a nucleophilic aromatic substitution reaction.

**melt transition temperature (*Tm*)** (Sect. 27.6): The temperature at which the crystalline regions of a polymer become amorphous.

**mercapto group** (Sect. 14.11): An SH group.

**mercurinium ion** (Sect. 9.5): The intermediate formed during oxymercuration.

**Merrifield synthesis** (Sect. 25.6): A method for building a peptide from protected building blocks.

***meso* compound** (Sect. 5.6): A compound that possesses chirality centers and an internal plane of symmetry.

***meta*** (Sect. 18.2): On an aromatic ring, the C3 position.

***meta* director** (Sect. 19.8): An electron-withdrawing group that directs the regiochemistry of an electrophilic aromatic substitution reaction such that the incoming electrophile is installed at the *meta* position.

**methine group** (Sect. 16.5): A CH group.

**methyl shift** (Sect. 6.8): A type of carbocation rearrangement in which a methyl group migrates.

**methylene group** (Sect. 16.5): A $CH_2$ group.

**micelle** (Sect. 1.13): A group of molecules arranged in a sphere such that the surface of the sphere is comprised of polar groups, rendering the micelle water soluble.

**Michael acceptor** (Sect. 22.6): The electrophile in a Michael reaction.

**Michael donor** (Sect. 22.6): The nucleophile in a Michael reaction.

**Michael reaction** (Sect. 22.6): A reaction in which a nucleophile attacks a conjugated p system, resulting in a 1,4-addition.

**migratory aptitude** (Sect. 20.11): In a Baeyer-Villiger oxidation, the migration rates of different groups, which determine the regiochemical outcome of the reaction.

**miscible** (Sect. 13.1): Two liquids that can be mixed with each other in any proportion.

**mixed aldol reaction** (Sect. 22.3): An aldol reaction that occurs between different partners.

**moderate activators** (Sect. 19.10): Groups that moderately activate an aromatic ring toward an electrophilic aromatic substitution reaction.

**moderate deactivators** (Sect. 19.10): Groups that moderately deactivate an aromatic ring toward an electrophilic aromatic substitution reaction.

**molar absorptivity** (Sect. 17.11): Determines the amount of UV light absorbed at the $\lambda_{max}$ of a compound, as described by Beer's law.

**molecular dipole moment** (Sect. 1.11): The vector sum of the individual dipole moments in a compound.

**molecular ion** (Sect. 15.8): In mass spectrometry, the ion that is generated when the compound is ionized.

**molecular orbital (MO) theory** (Sect. 1.8): A description of bonding in terms of molecular orbitals, which are orbitals associated with an entire molecule rather than an individual atom.

**molecular orbitals** (Sect. 1.8): Orbitals associated A measure of the strength of with an entire molecule rather than an individual atom.

**monosaccharides** (Sect. 24.2): Simple sugars that generally contain multiple chirality centers.

**monovalent** (Sect. 1.2): An element that is capable of forming one bond (such as hydrogen).

**multiplet** (Sect. 16.7): In $^{1}$H NMR spectroscopy, a signal whose multiplicity requires a more detailed analysis.

**multiplicity** (Sect. 16.7): In $^{1}$H NMR spectroscopy, the number of peaks in a signal.

**mutarotation** (Sect. 24.5): A term used to describe the fact that $\alpha$ and $\beta$ anomers of carbohydrates can equilibrate via the open-chain form.

## N

**N terminus** (Sect. 25.4): For a peptide chain, the end that contains the amino group.

***n*+1 rule** (Sect. 16.7): In NMR spectroscopy, if $n$ is the number of neighboring protons, then the multiplicity will be $n+1$.

***N*-bromosuccinimide** (Sect. 11.7): A reagent used for allylic bromination to avoid a competing reaction in which bromine adds across the $\pi$ bond.

**Newman projection** (Sect. 4.6): A drawing style that is designed to show the conformation of a molecule.

***N*-glycoside** (Sect. 24.10): The product obtained when a monosaccharide is treated with an amine in the presence of an acid catalyst.

**nitration** (Sect. 19.4): An electrophilic aromatic substitution reaction that involves the installation of a nitro group ($NO_2$) on an aromatic ring.

**nitrogen rule** (Sect. 15.9): In mass spectrometry, an odd molecular weight indicates an odd number of nitrogen atoms in the compound, while an even molecular weight indicates either an even number of nitrogen atoms or the absence of nitrogen.

**nitronium ion** (Sect. 19.4): The $NO_2^+$ ion, which is present in a mixture of nitric acid and sulfuric acid.

**nitrosonium ion** (Sect. 23.10): The $NO^+$ ion, which is formed when $NaNO_2$ is treated with HCl.

**nitrous acid** (Sect. 23.10): A compound with molecular formula HONO.

***N*-nitrosamine** (Sect. 23.10): A compound with the structure $R_2N{-}N{=}O$.

**node** (Sect. 1.6): In atomic and molecular orbitals, a location where the value of $\psi$ is zero.

**nomenclature** (Sect. 4.1): A system for naming organic compounds.

**nonaromatic** (Sect. 18.5): A compound that lacks a ring with a continuous system of overlapping *p* orbitals.

**norbornane** (Sect. 4.15): The common name for bicyclo[2.2.1]heptane.

**nuclear magnetic resonance (NMR)** (Sect. 16.1): A form of spectroscopy that involves the study of the interaction between electromagnetic radiation and the nuclei of atoms.

**nucleophile** (Sect. 6.7): A compound containing an electron-rich atom that is capable of donating a pair of electrons.

**nucleophilic acyl substitution** (Sect. 21.7): A reaction in which a nucleophile attacks a carboxylic acid derivative.

**nucleophilic aromatic substitution** (Sect. 19.13): A substitution reaction in which an aromatic ring is attacked by a nucleophile, which replaces a leaving group.

**nucleophilic attack** (Sect. 6.8): One of the four arrow-pushing patterns for ionic reactions.

**nucleosides** (Sect. 24.10): The product formed when either D-ribose or 2-deoxy-D-ribose is coupled with certain nitrogen heterocycles (called bases).

**nucleotides** (Sect. 24.10): The product formed when a nucleoside is coupled to a phosphate group.

## O

**observed rotation** (Sect. 5.4): The extent to which plane-polarized light is rotated by a solution of a chiral compound.

**octet rule** (Sect. 1.3): The observation that second-row elements (C, N, O, and F) will form the necessary number of bonds so as to achieve a full valence shell (eight electrons).

**off-resonance decoupling** (Sect. 16.11): In NMR spectroscopy, a technique in which only the one-bond couplings are observed. $CH_3$ groups appear as quartets, $CH_2$ groups appear as triplets, CH groups appear as doublets, and quaternary carbon atoms appear as singlets.

**oils** (Sect. 26.3): Triglycerides that are liquids at room temperature.

**oligomers** (Sect. 27.5): During the polymerization process, compounds constructed from just a few monomers.

**optically active** (Sect. 5.4): A compound that rotates plane-polarized light.

**optically inactive** (Sect. 5.4): A compound that does not rotate plane-polarized light.

**optically pure** (Sect. 5.4): A solution containing just one enantiomer, but not its mirror image.

**organohalide** (Sect. 7.2): An organic compound containing at least one halogen.

***ortho*** (Sect. 18.2): On an aromatic ring, the C2 position.

***ortho-para* director** (Sect. 19.7): A group that directs the regiochemistry of an electrophilic aromatic substitution reaction such that the incoming electrophile is installed at the *ortho* or *para* positions.

**oxaphosphetane** (Sect. 20.10): An intermediate that is believed to be formed during Wittig reactions.

**oxidation** (Sect. 13.10): A reaction in which one compound undergoes an increase in oxidation state.

**oxidation state** (Sect. 13.4): A method of electron book-keeping in which all bonds are treated as if they were purely ionic.

**oxime** (Sect. 20.6): A compound with the structure $R_2C{=}N{-}OH$.

**oxirane** (Sect. 14.7): A cyclic ether containing a three-membered ring system. Also called an epoxide.

**oxonium ion** (Sect. 9.4): An intermediate with a positively charged oxygen atom.

**oxymercuration-demercuration** (Sect. 9.5): A two-step process for the Markovnikov addition of water across an alkene. With this process, carbocation rearrangements do not occur.

**ozonolysis** (Sect. 9.11): A reaction in which the C=C bond of an alkene is cleaved to form two C=O bonds.

## P

***para*** (Sect. 18.2): On an aromatic ring, the C4 position.

**parent ion** **(Sect. 15.8):** In mass spectrometry, the ion that is generated when the compound is ionized.

**partially condensed structures** (Sect. 2.1): A drawing style in which the CH bonds are not drawn explicitly, but all other bonds are drawn.

**Pauli exclusion principle** (Sect. 1.6): The rule that states that an atomic orbital or molecular orbital can accommodate a maximum of two electrons with opposite spin.

**peptidases** (Sect. 25.5): A variety of enzymes that selectively hydrolyze specific peptide bonds.

**peptide** (Sect. 25.1): A chain comprised of a small number of amino acid residues.

**peptide bond** (Sect. 25.1): The amide linkage by which two amino acids are coupled together to form peptides.

**pericylic reactions** (Sect. 17.6): Reactions that occur via a concerted process and do not involve either ionic or radical intermediates.

**periplanar** (Sect. 8.7): A conformation in which a hydrogen atom and a leaving group are approximately coplanar.

**peroxides** (Sect. 11.3): Compounds with the general structure R—O—O—R.

**phenolate** (Sect. 13.2): The conjugate base of phenol or a substituted phenol.

**phenoxide** (Sect. 13.2): The conjugate base of phenol or a substituted phenol.

**phenyl group** (Sect. 18.2): A C6H5 group.

**phosphatidic acid** (Sect. 26.5): A phosphoric monoester, which is the simplest kind of phosphoglyceride.

**phosphoglycerides** (Sect. 26.5): Compounds that are very similar in structure to triglycerides, with the main difference being that one of the three fatty acid residues is replaced by a phosphoester group.

**phospholipids** (Sect. 26.5): Esterlike derivatives of phosphoric acid.

**photochemical reaction** (Sect. 17.3): A reaction that is performed with photochemical excitation (usually UV light).

**photon** (Sect. 15.1): When electromagnetic radiation is viewed as a particle, an individual packet of energy.

**physiological pH** (Sect. 21.3): The pH of blood (approximately 7.3).

**pi (π) bond** (Sect. 1.9): A bond formed from adjacent, overlapping *p* orbitals.

**plane of symmetry** (Sect. 5.6): A plane that bisects a compound into two halves that are mirror images of each other.

**plane-polarized light** (Sect. 5.4): Light for which all photons have the same polarization, generally formed by passing light through a polarizing filter.

**plasticizers** (Sect. 27.7): Small molecules that are trapped between polymer chains where they function as lubricants, preventing the polymer from being brittle.

**polar aprotic solvent** (Sect. 7.8): A solvent that lacks hydrogen atoms connected directly to an electronegative atom.

**polar covalent bond** (Sect. 1.5): A bond in which the difference in electronegative values of the two atoms is between 0.5 and 1.7.

**polar reaction** (Sect. 6.7): A reaction that involves the participation of ions as reactants, intermediates, or products.

**polarimeter** (Sect. 5.4): A device that measures the rotation of plane-polarized light caused by optically active compounds.

**polarizability** (Sect. 6.7): The ability of an atom or molecule to distribute its electron density unevenly in response to external influences.

**polarization** (Sect. 5.4): For light, the orientation of the electric field.

**polycarbonates** (Sect. 27.4): Polymers that are similar in structure to polyesters but with repeating carbonate groups (—O—$CO_2$—) instead of repeating ester groups (—$CO_2$—).

**polycyclic aromatic hydrocarbons (PAHs)** (Sect. 18.5): Compounds containing multiple aromatic rings fused together.

**polyether** (Sect. 14.4): A compound containing several ether groups.

**polynucleotide** (Sect. 24.10): A polymer constructed from nucleotides linked together.

**polysaccharides** (Sect. 24.8): Polymers made up of repeating monosaccharide units linked together by glycoside bonds.

**polyurethanes** (Sect. 27.5): Polymers made up of repeating urethane groups, also sometimes called carbamate groups (—N—$CO_2$—).

**polyvinyl chloride, (PVC)** (Sect. 11.11): A polymer formed from the polymerization of vinyl chloride ($H_2C{=}CHCl$).

**primary** (Sect. 6.8): A term used to indicate that exactly one alkyl group is attached directly to a particular position. For example, a primary carbocation has one alkyl group (not more) attached directly to the electrophilic carbon atom ($C_+$).

**primary alkyl halide** (Sect. 7.2): An organohalide in which the alpha (α) position is connected to only one alkyl group.

**primary structure** (Sect. 25.7): For proteins, the sequence of amino acid residues.

**progestins** (Sect. 26.6): Female sex hormones.

**propagation** (Sect. 11.2): For radical reactions, the steps whose sum gives the net chemical reaction.

**prostaglandins** (Sect. 26.7): Lipids that contain 20 carbon atoms and are characterized by a five-membered ring with two side chains.

**prosthetic group** (Sect. 25.8): A nonprotein unit attached to a protein, such as heme in hemoglobin.

**protecting group** (Sect. 13.7): A group that is used during synthesis to protect a functional group from the reaction conditions.

**proteins** (Sect. 25.1): Polypeptide chains comprised of more than 40 or 50 amino acids.

**protic solvent** (Sect. 7.8): A solvent that contains at least one hydrogen atom connected directly to an electronegative atom.

**proton transfer** (Sect. 6.8): One of the four arrow-pushing patterns for ionic reactions.

**pyranose ring** (Sect. 24.5): A six-membered cyclic hemiacetal form of a carbohydrate.

**pyrimidine** (Sect. 23.12): A compound that is similar in structure to pyridine but contains one extra nitrogen atom at the 3 position.

## Q

**quantum mechanics** (Sect. 1.6): A mathematical description of an electron that incorporates its wavelike properties.

**quartet** (Sect. 16.7): In NMR spectroscopy, a signal that is comprised of four peaks.

**quaternary ammonium salt** (Sect. 23.5): An ionic compound containing a positively charged nitrogen atom connected to four alkyl groups.

**quaternary structure** (Sect. 25.7): The structure that arises when a protein consists of two or more folded polypeptide chains that aggregate to form one protein complex.

**quintet** (Sect. 16.7): In NMR spectroscopy, a signal that is comprised of five peaks.

## R

***R*** (Sect. 5.3): A term used to designate the configuration of a chirality center, determined in the following way: Each of the four groups is assigned a priority, and the molecule is then rotated (if necessary) so that the #4 group is directed behind the page (on a dash). A clockwise sequence for 1-2-3 is designated as R.

**racemic mixture** (Sect. 5.4): A solution containing equal amounts of both enantiomers.

**radical** (Sect. 6.1): A chemical entity with an unpaired electron.

**radical anion** (Sect. 10.5): An intermediate that has both a negative charge and an unpaired electron.

**radical inhibitor** (Sect. 11.3): A compound that prevents a radical chain process from either getting started or continuing.

**radical initiator** (Sect. 11.3): A compound with a weak bond that undergoes homolytic bond cleavage with great ease, producing radicals that can initiate a radical chain process.

**random copolymer** (Sect. 27.3): A polymer, comprised of more than one kind of repeating unit, in which there is a random distribution of repeating units.

**rate equation** (Sect. 6.5): An equation that describes the relationship between the rate of a reaction and the concentration of reactants.

**rate-determining step** (Sect. 7.5): The slowest step in a multistep reaction which determines the rate of the reaction.

**reaction mechanism** (Sect. 3.2): A series of intermediates and curved arrows that show how the reaction occurs in terms of the motion of electrons.

**rearrangement** (Sect. 6.8): One of the four arrow-pushing patterns for ionic reactions.

**reducing agent** (Sect. 13.4): A compound that reduces another compound and in the process is itself oxidized. Sodium borohydride and lithium aluminum hydride are reducing agents.

**reducing sugar** (Sect. 24.6): A carbohydrate that is oxidized upon treatment with Tollens' reagent, Fehling's reagent, or Benedict's reagent.

**reduction** (Sect. 13.4): A reaction in which a compound undergoes a decrease in oxidation state.

**reductive amination** (Sect. 23.6): The conversion of a ketone or aldehyde into an imine under conditions in which the imine is reduced as soon as it is formed, giving an amine.

**regiochemistry** (Sect. 8.7): A term describing a consideration that must be taken into account for a reaction in which two or more constitutional isomers can be formed.

**regioselective** (Sect. 8.7): A reaction that can produce two or more constitutional isomers but nevertheless produces one as the major product.

**replacement test** (Sect. 16.4): A test for determining the relationship between two protons. The compound is drawn two times, each time replacing one of the protons with deuterium. If the two compounds are identical, the protons are homotopic. If the two compounds are enantiomers, the protons are enantiotopic. If the two compounds are diastereomers, the protons are diastereotopic.

**resolution** (Sect. 5.9): The separation of enantiomers from a mixture containing both enantiomers.

**resolving agents** (Sect. 5.9): A compound that can be used to achieve the resolution of enantiomers.

**resonance** (Sect. 2.7): A method that chemists use to deal with the inadequacy of bond-line drawings.

**resonance hybrid** (Sect. 2.7): A term used to describe the character of a chemical entity (molecule, ion, or radical) exhibiting more than one significant resonance structure.

**resonance stabilization** (Sect. 2.7): The stabilization associated with the delocalization of electrons via resonance.

**resonance structures** (Sect. 2.7): A series of structures that are melded together (conceptually) to circumvent the inadequacies of bond-line drawings.

**retention of configuration** (Sect. 7.5): During a reaction, when the configuration of a chirality center remains unchanged.

**retention time** (Sect. 15.14): The amount of time required for a compound to exit from a gas chromatograph.

**retro-aldol reaction** (Sect. 22.3): The reverse of an aldol reaction. A β-hydroxyketone or aldehyde is converted into two ketones or aldehydes.

**retro Diels-Alder reaction** (Sect. 17.7): The reverse of a Diels-Alder reaction, achieved at high temperature. A cyclohexene derivative is converted into a diene and a dienophile.

**retrosynthetic analysis** (Sect. 12.5): A systematic set of principles that enable the design of a synthetic route by working backward from the desired product.

**ring flip** (Sect. 4.12): A conformational change in which one chair conformation is converted into the other.

**Robinson annulation** (Sect. 22.6): The combination of a Michael addition followed by an aldol condensation to form a ring.

## S

***S*** (Sect. 5.3): A term used to designate the configuration of a chirality center, determined in the following way: Each of the four groups is assigned a priority, and the molecule is then rotated (if necessary) so that the #4 group is directed behind the page (on a dash). A counterclockwise sequence for 1-2-3 is designated as S.

**Sandmeyer reactions** (Sect. 23.11): Reactions that utilize copper salts (CuX) and enable the installation of a halogen or a cyano group on an aromatic ring.

**saponification** (Sect. 21.11): The base-catalyzed hydrolysis of an ester. This method is used to make soap.

**saturated** (Sect. 15.16): A compound that contains no π bonds.

**saturated hydrocarbon** (Sect. 4.1): A hydrocarbon that contains no π bonds.

**Schiemann reaction** (Sect. 23.11): The conversion of an aryl diazonium salt into fluorobenzene upon treatment with fluoroboric acid ($HBF_4$).

***s-cis*** (Sect. 17.2): A conformation of a conjugated diene in which the disposition of the two π bonds with regard to the connecting single bond is *cis-like* (a dihedral angle of 0°).

**second order** (Sect. 6.5, 7.4): A reaction that has a rate equation in which the sum of all exponents is two.

**secondary** (Sect. 6.8): A term used to indicate that exactly two alkyl groups are attached directly to a particular position. For example, a secondary carbocation has two alkyl groups attached directly to the electrophilic carbon atom ($C^+$).

**secondary alkyl halide** (Sect. 7.2): An organohalide in which the alpha (α) position is connected to exactly two alkyl groups.

**secondary structure** (Sect. 25.7): The three-dimensional conformations of localized regions of a protein, including helices and β-pleated sheets.

**Sharpless asymmetric epoxidation** (Sect. 14.9): A reaction that converts an alkene into an epoxide via a stereospecific pathway.

**shielded** (Sect. 16.1): In NMR spectroscopy, protons or carbon atoms whose surrounding electron density is rich.

**sigma (σ) bond** (Sect. 1.7): A bond that is characterized by circular symmetry with respect to the bond axis.

**sigma complex** (Sect. 19.2): The positively charged intermediate of an electrophilic aromatic substitution reaction.

**sigmatropic rearrangements** (Sect. 17.6): A pericyclic reaction in which one σ bond is formed at the expense of another.

**simple lipid** (Sect. 26.1): A lipid that does not undergo hydrolysis in aqueous acid or base to produce smaller fragments.

**singlet** (Sect. 16.7): In NMR spectroscopy, a signal that is comprised of only one peak.

**$S_N1$** (Sect. 7.5): A unimolecular nucleophilic substitution reaction.

**$S_N2$** (Sect. 7.4): A bimolecular nucleophilic substitution reaction.

**sodium cyanoborohydride** (Sect. 23.6): A selective reducing agent ($NaBH_3CN$) that can be used for reductive amination.

**soluble** (Sect. 13.1): A term used to indicate that a certain volume of a compound will dissolve in a specified amount of a liquid at room temperature.

**solvent extraction** (Sect. 23.3): A process by which one or more compounds are removed from a mixture of organic compounds, based on a difference in solubility and/or acid-base properties.

**solvolysis** (Sect. 7.6): A substitution reaction in which the solvent functions as the nucleophile.

***$sp^2$*-hybridized orbitals** (Sect. 1.9): Atomic orbitals that are achieved by mathematically averaging one *s* orbital with two *p* orbitals to form three hybridized atomic orbitals.

***$sp^3$*-hybridized orbitals** (Sect. 1.9): Atomic orbitals that are achieved by mathematically averaging one *s* orbital with three *p* orbitals to form four hybridized atomic orbitals.

**specific rotation** (Sect. 5.4): For a chiral compound that is subjected to plane-polarized light, the observed rotation when a standard concentration (1 g/mL) and a standard path length (1 dm) are used.

**spectroscopy** (Sect. 15.1): The study of the interaction between matter and electromagnetic radiation.

***sp*-hybridized** (Sect. 1.9): Atomic orbitals that are achieved by mathematically averaging one *s* orbital with only one *p* orbital to form two hybridized atomic orbitals.

**spin-spin splitting** (Sect. 16.7): A phenomenon observed most commonly for nonequivalent protons connected to adjacent carbon atoms, in which the multiplicity of each signal is affected by the other.

**spontaneous** (Sect. 6.2): A reaction with a negative $\Delta G$, which means that products are favored at equilibrium.

**staggered conformation** (Sect. 4.7): A conformation in which nearby groups in a Newman projection have a dihedral angle of 60°.

**standard atomic weight** (Sect. 15.13): The weighted averages for each element, which takes into account isotopic abundance.

**step-growth polymers** (Sect. 27.5): Polymers that are formed under conditions in which the individual monomers react with each other to form oligomers, which are then joined together to form polymers.

**stereoisomers** (Sect. 4.14): Compounds that have the same constitution but differ in the 3D arrangement of atoms.

**stereoselective** (Sect. 8.7): A reaction in which one substrate produces two stereoisomers in unequal amounts.

**stereospecific** (Sect. 7.4): A reaction in which the configuration of the product is dependent on the configuration of the starting material.

**steric number** (Sect. 1.10): The total of (single bonds + lone pairs) for an atom in a compound.

**sterically hindered** (Sect. 3.7): A compound or region of a compound that is very bulky.

**steroids** (Sect. 26.6): Lipids that are based on a tetracyclic ring system involving three six-membered rings and one five-membered ring. Cholesterol is an example.

**Stork enamine synthesis** (Sect. 22.6): A Michael reaction in which an enamine functions as a nucleophile.

***s-trans*** (Sect. 17.2): A conformation of a conjugated diene in which the disposition of the two π bonds with regard to the connecting single bond is *trans-like* (a dihedral angle of 180°).

**Strecker synthesis** (Sect. 25.3): A synthetic technique for preparing racemic α-amino acids from aldehydes.

**stretching** (Sect. 15.2): In IR spectroscopy, a type of vibration that generally produces a signal in the diagnostic region of an IR spectrum.

**strong activators** (Sect. 19.10): Groups that strongly activate an aromatic ring toward electrophilic aromatic substitution, thereby significantly enhancing the rate of the reaction.

**strong deactivators** (Sect. 19.10): Groups that strongy deactivate an aromatic ring toward electrophilic aromatic substitution, thereby significantly decreasing the rate of the reaction.

**structural proteins** (Sect. 25.8): Fibrous proteins that are used for their structural rigidity. Examples include α-keratins found in hair, nails, skin, feathers, and wool.

**substituents** (Sect. 4.2): In nomenclature, the groups connected to the parent chain.

**substitution reactions** (Sect. 7.1): Reactions in which one group is replaced by another group.

**substrate** (Sect. 7.1): The starting alkyl halide in a substitution or elimination reaction.

**sulfide** (Sect. 14.11): A compound that is similar in structure to an ether, but the oxygen atom has been replaced with a sulfur atom. Also called a thioether.

**sulfonate ions** (Sect. 7.8): Common leaving groups. Examples include tosylate, mesylate, and triflate ions.

**sulfonation** (Sect. 19.3): An electrophilic aromatic substitution reaction in which an $SO_3H$ group is installed on an aromatic ring.

**sulfone** (Sect. 14.11): A compound that contains a sulfur atom that has double bonds with two oxygen atoms and is flanked on both sides by R groups.

**sulfoxide** (Sect. 14.11): A compound containing an S=O bond that is flanked on both sides by R groups.

**superimposable** (Sect. 5.2): Two objects that are identical.

**symmetric stretching** (Sect. 15.5): In IR spectroscopy, when two bonds are stretching in phase with each other.

**symmetrical ether** (Sect. 14.2): An ether (R—O—R) where both R groups are identical.

**symmetry allowed** (Sect. 17.8): A reaction that obeys conservation of orbital symmetry.

**symmetry forbidden** (Sect. 17.8): A reaction that disobeys conservation of orbital symmetry.

***syn* addition** (Sect. 9.6): An addition reaction in which two groups are added to the same face of a π bond.

***syn*-coplanar** (Sect. 8.7): A conformation in which a hydrogen atom and a leaving group are separated by a dihedral angle of exactly 0°.

**syndiotactic** (Sect. 27.6): A polymer in which the repeating units contain chirality centers which have alternating configuration.

**systematic name** (Sect. 4.2): A name that is assigned using the rules of IUPAC nomenclature.

## T

**tautomers** (Sect. 10.7): Constitutional isomers that rapidly interconvert via the migration of a proton.

**terminal alkynes** (Sect. 10.2): Compounds with the following structure: R—C≡C—H

**termination** (Sect. 11.2): In radical reactions, a step in which two radicals are joined to give a compound with no unshared electrons.

**termolecular** (Sect. 10.6): For mechanisms, a step that involves three chemical entities.

**terpenes** (Sect. 26.8): A diverse class of naturally occurring compounds that can be thought of as being assembled from isoprene units, each of which contains five carbon atoms.

**tertiary** (Sect. 6.8): A term used to indicate that exactly three alkyl groups are attached directly to a particular position. For example, a tertiary carbocation has three alkyl groups attached directly to the electrophilic carbon atom ($C_+$).

**tertiary alkyl halide** (Sect. 7.2): An organohalide in which the alpha (α) position is connected to three alkyl groups.

**tertiary structure** (Sect. 25.7): The three-dimensional shape of a protein.

**tetrahedral** (Sect. 1.10): The geometry of an atom with four bonds separated from each other by 109.5°.

**tetrahedral intermediate** (Sect. 21.7): An intermediate with tetrahedral geometry. This type of intermediate is formed when a nucleophile attacks the carbonyl group of a carboxylic acid derivative.

**tetravalent** (Sect. 1.2): An element, such as carbon, that forms four bonds.

**thermodynamic control** (Sect. 17.5): A reaction for which the ratio of products is determined solely by the distribution of energy among the products.

**thermodynamics** (Sect. 6.4): The study of how energy is distributed under the influence of entropy. For chemists, the thermodynamics of a reaction specifically refers to the study of the relative energy levels of reactants and products.

**thermoplastics** (Sect. 27.7): Polymers that are hard at room temperature but soft when heated.

**thermosetting resins** (Sect. 27.7): Highly cross-linked polymers that are generally very hard and insoluble.

**thioacetal** (Sect. 20.8): A compound that contains two SR groups, both of which are connected to the same carbon atom.

**thiolate** (Sect. 14.11): The conjugate base of a thiol.

**thiols** (Sect. 14.11): Compounds containing a mercapto group (SH).

**third order** (Sect. 6.5): A reaction that has a rate equation in which the sum of all exponents is three

**three-center, two-electron bonds** (Sect. 9.6): A bond in which two electrons are associated with three atoms, such as in diborane ($B_2H_6$).

**torsional angle** (Sect. 4.7): The angle between two groups in a Newman projection, also called the dihedral angle.

**torsional strain** (Sect. 4.7): The difference in energy between staggered and eclipsed conformations (for example, in ethane).

**tosylate** (Sect. 7.8): An excellent leaving group (OTs).

**transition state** (Sect. 6.6): A state through which a reaction passes. On an energy diagram, a transition state corresponds with a local maximum.

**transport protein** (Sect. 25.8): A protein used to transport molecules or ions from one location to another. Hemoglobin is a classic example of a transport protein, used to transport molecular oxygen from the lungs to all the tissues of the body.

**triglyceride** (Sect. 26.3): A triester formed from glycerol and three long-chain carboxylic acids.

**trigonal planar** (Sect. 1.10): A geometry adopted by an atom with a steric number of 3. All three groups lie in one plane and are separated by 120°.

**trigonal pyramidal** (Sect. 1.10): A geometry adopted by an atom that has one lone pair and a steric number of 4.

**triplet** (Sect. 16.7): In NMR spectroscopy, a signal that is comprised of three peaks.

**trivalent** (Sect. 1.2): An element, such as nitrogen, that forms three bonds.

**twist boat** (Sect. 4.10): A conformation of cyclohexane that is lower in energy than a boat conformation but higher in energy than a chair conformation.

## U

**unimolecular** (Sect. 7.5): For mechanisms, a step that involves only one chemical entity.

**unsaturated** (Sect. 15.16): A compound containing one or more π bonds.

**unsymmetrical ether** (Sect. 14.2): An ether (R—O—R) where the two R groups are not identical.

**upfield** (Sect. 16.5): The right side of an NMR spectrum.

## V

**valence bond theory** (Sect. 1.7): A theory that treats a bond as the sharing of electrons that are associated with individual atoms, rather than being associated with the entire molecule.

**vibrational excitation** (Sect. 15.1): In IR spectroscopy, the energy of a photon is absorbed and temporarily stored as vibrational energy.

**vicinal** (Sect. 10.4): A term used to describe two identical groups attached to adjacent carbon atoms.

**vinylic** (Sect. 2.10): The carbon atoms of a carbon-carbon double bond.

**vinylic carbocation** (Sect. 10.6): A carbocation in which the positive charge resides on a vinylic carbon atom. This type of carbocation is very unstable and will not readily form in most cases.

**VSEPR theory** (Sect. 1.10): Valence Shell Electron Pair Repulsion theory, which can be used to predict the geometry around an atom.

## W

**wavelength** (Sect. 15.1): The distance between adjacent peaks of an oscillating magnetic or electric field.

**wavenumber** (Sect. 15.2): In IR spectroscopy, the location of each signal is reported in terms of this frequency-related unit.

**waxes** (Sect. 26.2): High-molecular-weight esters that are constructed from carboxylic acids and alcohols.

**weak activators** (Sect. 19.10): Groups that weakly activate an aromatic ring toward electrophilic aromatic substitution, thereby enhancing the rate of the reaction.

**weak deactivators** (Sect. 19.10): Groups that weakly deactivate an aromatic ring toward electrophilic aromatic substitution, thereby decreasing the rate of the reaction.

**wedge** (Sect. 2.6): In bond-line structures, a group in front of the page.

**Williamson ether synthesis** (Sect. 14.5): A method for preparing an ether from an alkoxide ion and an alkyl halide (via an $S_N2$ process).

**Wittig reaction** (Sect. 20.10): A reaction that converts an aldehyde or ketone into an alkene, with the introduction of one or more carbon atoms.

**Wittig reagent** (Sect. 20.10): A reagent used to perform a Wittig reaction.

**Wohl degradation** (Sect. 24.6): A process that involves the removal of a carbon atom from an aldose. The aldehyde group is first converted to a cyanohydrin, followed by loss of HCN in the presence of a base.

**Wolff-Kishner reduction** (Sect. 20.6): A method for converting a carbonyl group into a methylene group (CH2) under basic conditions.

**Woodward-Fieser rules** (Sect. 17.11): Rules for predicting the wavelength of maximum absorption for a compound with extended conjugation.

## Y

**ylide** (Sect. 20.10): A compound with two oppositely charged atoms adjacent to each other.

## Z

***Z*** (Sect. 8.4): For alkenes, a stereodescriptor that indicates that the two priority groups are on the same side of the $\pi$ bond.

**Zaitsev product** (Sect. 8.7): The more substituted product (alkene) of an elimination reaction.

**zwitterion** (Sect. 25.2): A net neutral compound that exhibits charge separation. Amino acids exist as zwitterions at physiological pH.

# Index

Note: In this index, *f* and *t* following the page number indicate figures and tables, respectively.

## A

## B

## C

## F

## G



## H

## J



## K



## L

## T

# PERIODIC TABLE OF THE ELEMENTS

Atomic number → 6
Symbol → C
Name (IUPAC) → Carbon
Atomic mass → 12.011

IUPAC recommendations →
Chemical Abstracts Service group notation →

| 1<br>IA | 2<br>IIA | 3<br>IIIB | 4<br>IVB | 5<br>VB | 6<br>VIB | 7<br>VIIB | 8<br>VIIIB | 9<br>VIIIB | 10<br>VIIIB | 11<br>IB | 12<br>IIB | 13<br>IIIA | 14<br>IVA | 15<br>VA | 16<br>VIA | 17<br>VIIA | 18<br>VIIIA |
|---|---|---|---|---|---|---|---|---|---|---|---|---|---|---|---|---|---|
| 1<br>**H**<br>Hydrogen<br>1.0079 | | | | | | | | | | | | | | | | | 2<br>**He**<br>Helium<br>4.0026 |
| 3<br>**Li**<br>Lithium<br>6.941 | 4<br>**Be**<br>Berylium<br>9.0122 | | | | | | | | | | | 5<br>**B**<br>Boron<br>10.811 | 6<br>**C**<br>Carbon<br>12.011 | 7<br>**N**<br>Nitrogen<br>14.007 | 8<br>**O**<br>Oxygen<br>15.999 | 9<br>**F**<br>Fluorine<br>18.998 | 10<br>**Ne**<br>Neon<br>20.180 |
| 11<br>**Na**<br>Sodium<br>22.990 | 12<br>**Mg**<br>Magnesium<br>24.305 | | | | | | | | | | | 13<br>**Al**<br>Aluminum<br>26.982 | 14<br>**Si**<br>Silicon<br>28.086 | 15<br>**P**<br>Phosphorus<br>30.974 | 16<br>**S**<br>Sulfur<br>32.065 | 17<br>**Cl**<br>Chlorine<br>35.453 | 18<br>**Ar**<br>Argon<br>39.948 |
| 19<br>**K**<br>Potassium<br>39.098 | 20<br>**Ca**<br>Calcium<br>40.078 | 21<br>**Sc**<br>Scandium<br>44.956 | 22<br>**Ti**<br>Titanium<br>47.867 | 23<br>**V**<br>Vanadium<br>50.942 | 24<br>**Cr**<br>Chromium<br>51.996 | 25<br>**Mn**<br>Manganese<br>54.938 | 26<br>**Fe**<br>Iron<br>55.845 | 27<br>**Co**<br>Cobalt<br>58.933 | 28<br>**Ni**<br>Nickel<br>58.693 | 29<br>**Cu**<br>Copper<br>63.546 | 30<br>**Zn**<br>Zinc<br>65.409 | 31<br>**Ga**<br>Gallium<br>69.723 | 32<br>**Ge**<br>Germanium<br>72.64 | 33<br>**As**<br>Arsenic<br>74.922 | 34<br>**Se**<br>Selenium<br>78.96 | 35<br>**Br**<br>Bromine<br>79.904 | 36<br>**Kr**<br>Krypton<br>83.798 |
| 37<br>**Rb**<br>Rubidium<br>85.468 | 38<br>**Sr**<br>Strontium<br>87.62 | 39<br>**Y**<br>Yttrium<br>88.906 | 40<br>**Zr**<br>Zirconium<br>91.224 | 41<br>**Nb**<br>Niobium<br>92.906 | 42<br>**Mo**<br>Molybdenum<br>95.94 | 43<br>**Tc**<br>Technetium<br>(98) | 44<br>**Ru**<br>Ruthenium<br>101.07 | 45<br>**Rh**<br>Rhodium<br>102.91 | 46<br>**Pd**<br>Palladium<br>106.42 | 47<br>**Ag**<br>Silver<br>107.87 | 48<br>**Cd**<br>Cadmium<br>112.41 | 49<br>**In**<br>Indium<br>114.82 | 50<br>**Sn**<br>Tin<br>118.71 | 51<br>**Sb**<br>Antimony<br>121.76 | 52<br>**Te**<br>Tellurium<br>127.60 | 53<br>**I**<br>Iodine<br>126.90 | 54<br>**Xe**<br>Xeno<br>131.29 |
| 55<br>**Cs**<br>Caesium<br>132.91 | 56<br>**Ba**<br>Barium<br>137.33 | 57<br>***La**<br>Lanthanum<br>138.91 | 72<br>**Hf**<br>Hafnium<br>178.49 | 73<br>**Ta**<br>Tantalum<br>180.95 | 74<br>**W**<br>Tungsten<br>183.84 | 75<br>**Re**<br>Rhenium<br>186.21 | 76<br>**Os**<br>Osmium<br>190.23 | 77<br>**Ir**<br>Iridium<br>192.22 | 78<br>**Pt**<br>Platinum<br>195.08 | 79<br>**Au**<br>Gold<br>196.97 | 80<br>**Hg**<br>Mercury<br>200.59 | 81<br>**Tl**<br>Thallium<br>204.38 | 82<br>**Pb**<br>Lead<br>207.2 | 83<br>**Bi**<br>Bismuth<br>208.98 | 84<br>**Po**<br>Polonium<br>(209) | 85<br>**At**<br>Astatine<br>(210) | 86<br>**Rn**<br>Radon<br>(222) |
| 87<br>**Fr**<br>Francium<br>(223) | 88<br>**Ra**<br>Radium<br>(226) | 89<br>**#Ac**<br>Actinium<br>(227) | 104<br>**Rf**<br>Rutherfordium<br>(261) | 105<br>**Db**<br>Dubnium<br>(262) | 106<br>**Sg**<br>Seaborgium<br>(266) | 107<br>**Bh**<br>Bohrium<br>(264) | 108<br>**Hs**<br>Hassium<br>(277) | 109<br>**Mt**<br>Meitnerium<br>(268) | 110<br>**Ds**<br>Darmstadtium<br>(281) | 111<br>**Rg**<br>Roentgenium<br>(272) | | | | | | | |

| | | | | | | | | | | | | | | |
|---|---|---|---|---|---|---|---|---|---|---|---|---|---|---|
| *Lanthanide Series | 58<br>**Ce**<br>Cerium<br>140.12 | 59<br>**Pr**<br>Praseodymium<br>140.91 | 60<br>**Nd**<br>Neodymium<br>144.24 | 61<br>**Pm**<br>Promethium<br>(145) | 62<br>**Sm**<br>Samarium<br>150.36 | 63<br>**Eu**<br>Europium<br>151.96 | 64<br>**Gd**<br>Gadolinium<br>157.25 | 65<br>**Tb**<br>Terbium<br>158.93 | 66<br>**Dy**<br>Dysprosium<br>162.50 | 67<br>**Ho**<br>Holmium<br>164.93 | 68<br>**Er**<br>Erbium<br>167.26 | 69<br>**Tm**<br>Thulium<br>168.93 | 70<br>**Yb**<br>Ytterbium<br>173.04 | 71<br>**Lu**<br>Lutetium<br>174.97 |
| # Actinide Series | 90<br>**Th**<br>Thorium<br>232.04 | 91<br>**Pa**<br>Protactinium<br>231.04 | 92<br>**U**<br>Uranium<br>238.03 | 93<br>**Np**<br>Neptunium<br>(237) | 94<br>**Pu**<br>Plutonium<br>(244) | 95<br>**Am**<br>Americium<br>(243) | 96<br>**Cm**<br>Curium<br>(247) | 97<br>**Bk**<br>Berkelium<br>(247) | 98<br>**Cf**<br>Californium<br>(251) | 99<br>**Es**<br>Einsteinium<br>(252) | 100<br>**Fm**<br>Fermium<br>(257) | 101<br>**Md**<br>Mendelevium<br>(258) | 102<br>**No**<br>Nobelium<br>(259) | 103<br>**Lr**<br>Lawrencium<br>(262) |

Wiley Custom Learning Solutions

*9781119303329*
TP
APEX